SIXTH EDITION

Thermal Radiation Heat Transfer

SIXTH EDITION

Thermal Radiation Heat Transfer

JOHN R. HOWELL • M. PINAR MENGÜÇ
ROBERT SIEGEL

CRC Press is an imprint of the
Taylor & Francis Group, an **informa** business

CRC Press
Taylor & Francis Group
6000 Broken Sound Parkway NW, Suite 300
Boca Raton, FL 33487-2742

Printed on acid-free paper
Version Date: 20150623

International Standard Book Number-13: 978-1-4665-9326-8 (Hardback)

Visit the Taylor & Francis Web site at
http://www.taylorandfrancis.com

and the CRC Press Web site at
http://www.crcpress.com

To my wife, Susan, and my children and grandchildren, and to Bob Siegel, who made possible this continuing enterprise.

John R. Howell

To my children, Yiğit and Tuğçe, who gave me the pleasure of life all along, and to all of my mentors, students, and colleagues who helped me to gather each and every piece of knowledge and wisdom reflected here.

M. Pinar Mengüç

Contents

Preface

This sixth edition of *Thermal Radiation Heat Transfer* extends the major changes made in the fifth edition. These included redrafting of many figures (some dating from the first edition), changing the nomenclature and symbols to agree with the major heat transfer journals, and adding four new chapters with updated and contemporary topics.

The fundamental aspects of engineering radiation transfer remain the core of the book and are updated throughout the text. Changes have been made specifically on properties of surfaces and of absorbing/emitting/scattering materials, radiative transfer among surfaces, and radiative transfer in participating media. The chapters added to the fifth edition (Inverse Methods; Electromagnetic Theory; Scattering and Absorption by Particles; and Near-Field Radiative Transfer) are retained, updated, and expanded. In particular, the chapter on near-field effects, which is very important in treating nanoscale effects for various applications, is augmented. New applications are addressed, including enhanced solar cell performance and self-regulating surfaces for thermal control. Updated references, useful for historical context and contemporary research, are to make the book a comprehensive reference for engineers and researchers.

The overall subject arrangement of the fifth edition has been retained. The text consists of three main sections. The first section (Chapters 1 through 3) presents blackbody radiative properties and the radiative properties of opaque materials as predicted by the electromagnetic theory and obtained by measurements. The second section (Chapters 4 through 8) covers radiative exchange in enclosures with no participating media; that is, there is no radiating medium between the surfaces; yet heat conduction and convection at and through boundaries are included. The third section (Chapters 9 through 17) deals with the radiative properties of gases and particles and discusses the radiative energy exchange when gases or other participating materials are present, such as in furnaces, spacecraft atmospheric reentry, and solar absorption within solar photovoltaic cells. Within this basic framework, the fifth edition material is streamlined, condensed, and/or consolidated. A considerable amount of older material has been removed from the main flow of the text and incorporated as appendices to aid in readability.

Chapter 1 on fundamentals of radiation and the blackbody is an introduction to the propagation of radiation in participating media and the radiative transfer equation. Chapters 2 and 3 provide information on the definition and characteristics of radiative surface properties and on comparisons with the behavior of real properties as required for applying the methods for analysis that are developed. This foundation on properties must be established because choosing the best procedure for analysis depends on the nature of the radiative properties involved. Property prediction based on electrical properties is in this section, but derivations of the electromagnetic theory property predictions are in Chapter 14 to aid in the flow of discussion.

The analysis of radiative exchange among multiple surfaces separated by transparent media is developed in Chapters 4 through 8. This discussion includes surfaces with black, gray, spectral, diffuse, directional, and specular properties. For enclosures with diffuse surfaces, the geometric configuration factors are required. A short appendix (Appendix C) provides convenient analytical expressions for a number of configuration factors that are useful for homework problems and basic geometries. A greatly extended catalog of factors is posted in the open-source online catalog of configuration factors at http://www.thermalradiation.net. The catalog now allows built-in numerical computation of values for many factors. Sections on numerical techniques are also updated in the text, with new material on the emerging use of graphical program units (GPUs) for parallel processing in radiative transfer solutions. Chapter 8 describes inverse problems, how they appear in radiative transfer and the design of nonparticipating radiating systems, and various solution techniques available for inverse problems. This chapter has been augmented and includes many recent advances.

Chapter 9 introduces the radiative properties of gases, liquids, and nonopaque solids, including references and descriptions of contemporary methods for calculating the spectral and total properties of gases. An updated section on the k-distribution method is included, along with an introduction to problems where nonlocal thermodynamic equilibrium is present. The development of radiative transfer in participating media is contained in updated Chapters 10 through 12 with an emphasis on gaseous media. This material is reorganized, with a new section that emphasizes the close relationship among the many differential methods (diffusion, two-flux, P_N, SP_N, M_N, and others). Chapter 13 deals with multimode problems where radiation in participating media is important. Chapters 14 and 15 deal with the electromagnetic theory in radiative transfer and with scattering. Absorption and scattering by particles and agglomerates of particles are discussed extensively and the new developments in the field are included. Chapter 16 is on the important developing area of near-field radiation effects within matter. Applications to radiation interactions with nanopatterned surfaces are included and significantly enhanced since the publication of the fifth edition. Chapter 17 considers participating media that have a refractive index larger than unity such as glass, where reflections and other important effects at interfaces need to be included. This material is extended to discuss the near-field radiative interactions in multiple nanolayers.

Some material discussed in the previous editions of the text still remains useful, but of less direct importance due to the increasing availability of computer-based mathematical programs. Material on the exponential wideband properties of gases, geometric-mean beam lengths, and certain approximations to solving the radiative transfer equation (exponential kernel approximation, Curtis–Godson approximation) is now less used in practice. Because software packages such as MATLAB® and Mathcad have built-in routines for numerical integration, material on numerical integration has also been moved to the online appendix. Similarly, material on applications that are more limited in scope but still important such as radiation in porous media and radiative cooling technology has also been placed in the web appendix. To shorten the written text, this material has been moved to the dedicated site http://www.thermalradiation.net, where it is available for download.

Teachers using this book as a text should be aware that a full solution manual is available from the publisher. In addition, continually updated errata, the appendix material moved from earlier editions, and the full catalog of configuration factors are available at http://www.thermalradiation.net.

The authors hope that this sixth edition continues the tradition of providing both a comprehensive textbook for those interested in the study of thermal radiative transfer and being a source of important references to the literature for researchers.

The authors also wish to recognize Prof. Hakan Ertürk and his students at Boğaziçi University, Istanbul, for coordinating the editing and revision of the homework solution manual for this edition. We further acknowledge the contributions of Mathieu Francoeur to the original version of Chapter 16 in the fifth edition of the book. Since then, several of our students (Azadeh Didari in particular) have contributed to the description of near-field effects.

Additional appendix material is available on the authors' website: http://www.thermalradiation.net. Relevant material will be continually posted.

John R. Howell
M. Pinar Mengüç
Robert Siegel

The MathWorks, Inc.
3 Apple Hill Drive
Natick, MA 01760-2098 USA
Tel: 508-647-7000
Fax: 508-647-7001
E-mail: info@mathworks.com
Web: www.mathworks.com

Authors

John R. Howell received all of his academic degrees from Case Western Reserve University (then Case Institute of Technology), Clevland, Ohio. He began his engineering career as a researcher at NASA Lewis (now Glenn) Research Center (1961–1968) and then took academic positions at the University of Houston (1978–1988) and the University of Texas at Austin, where he remained until retirement in 2012. He is presently Ernest Cockrell, Jr., Memorial Chair emeritus at The University of Texas.

He pioneered the use of the Monte Carlo method for the analysis of radiative heat transfer in complex systems that contain absorbing, emitting, and scattering media. Jack has concentrated his research on computational techniques for radiative transfer and combined-mode problems for over 65 years. Recently, he has adapted inverse solution techniques to combined-mode problems and to radiation at the nanoscale. Along with many awards in heat transfer, he was elected a member of the U.S. National Academy of Engineering and a foreign member of the Russian Academy of Sciences. In 2013, he received the Elsevier Poynting Award from the *Journal of Quantitative Spectroscopy and Radiative Transfer* for his pioneering work in radiation heat transfer; the same year he also became an honorary member of the American Society of Mechanical Engineering (ASME). He has continued an active research program since retirement from teaching.

M. Pinar Mengüç completed his BSc and MS in mechanical engineering from the Middle East Technical University (METU) in Ankara, Turkey. He earned his PhD in mechanical engineering from Purdue University in 1985. He joined the faculty at the University of Kentucky the same year and was promoted to the ranks of associate and full professor in 1988 and 1993, respectively. In 2008, he became an Engineering Alumni Association professor. Later that year he joined Özyegin University in Istanbul as the founding head of the mechanical engineering department and the founding director of the Center for Energy, Environment, and Economy (CEEE). He still carries on both titles.

Pinar Mengüç is a fellow of the American Society of Mechanical Engineering (ASME) and the International Centre for Heat and Mass Transfer (ICHMT). He is currently one of the three editors in chief of the *Journal of Quantitative Spectroscopy and Radiative Transfer*. He has organized several conferences, including five international symposia on radiation transfer as the chair and co-chair. His research expertise includes radiation transfer in multidimensional geometries, light scattering–based particle characterization and diagnostic systems, nanoscale transport phenomena, near-field radiation transfer, applied optics, and sustainable energy and its applications to buildings and cities. He has guided more than 50 MS and PhD students and postdoctoral fellows both in the United States and in Turkey and has four assigned and two pending patents. He is the co-owner of a start-up company on particle characterization. He served as the director of the Nanoscale Engineering Certificate Program at the University of Kentucky. He is the Turkish delegate to the European Framework programs (FP-7 and Horizon 2020) on energy-related topics. He carries on an active research program in Istanbul.

Robert Siegel received his ScD in mechanical engineering from Massachusetts Institute of Technology in 1953. For two years, he worked at General Electric Co. in the Heat Transfer Consulting Office and on analyzing the heat transfer characteristics of the Seawolf submarine nuclear reactor. He joined NASA in 1955 and was a senior research scientist at the Lewis/Glenn Research Center for more than 50 years until he retired in 1999. He has worked in heat transfer research for more than 50 years. He has written 150 technical papers and has given graduate heat transfer courses as an adjunct professor at three universities. He was an associate editor for the *Journal of Heat Transfer* and the *Journal of Thermophysics and Heat Transfer.* He is a fellow of the American Society of Mechanical Engineering (ASME) and the American Institute of Aeronautics and Astronautics (AIAA). He received the ASME Heat Transfer Memorial Award in 1970, the NASA Exceptional Scientific Achievement Award in 1986, the Space Act Award in 1993, the AIAA Thermophysics Award in 1993, and the ASME/AIChE Max Jacob Memorial Award in 1996.

List of Symbols

This is a consolidated list of symbols for the entire text. Some symbols that are used in only a local development are defined where they are used and are not included here. The symbols used in radiative transfer have evolved from many different disciplines where radiation is important. This has led to the same quantity being defined by a variety of symbols and to multiple quantities designated with the same symbol. The symbols listed here are typical of those used for engineering heat transfer and follow, where possible, those adopted formally by the major heat transfer journals (Howell 1999). The study of radiative transfer combined with conduction and convection involves many types of applications and hence requires definitions for a large number of different quantities and parameters. There are an insufficient number of convenient symbols that can be used, so some symbols must be used for multiple quantities. Attention has been devoted to making the particular definition clear from the context of its use. Some typical units have been indicated. Some care must be observed as quantities may have multiple units, such as a spectral bandwidth that can be in terms of wavelength, wave number, or frequency; some of these are designated by (mu), meaning *multiple units*. A length could, for example, be in m, cm, μm, nm, or other units, so that only a typical unit is shown. Some quantities are nondimensional; these are designated by (nd).

Symbol	Definition
a	Quantity in reflectivity relations (nd); spacing between surfaces, m; thickness, m; coefficient in phase velocity of electromagnetic wave (nd)
a_0	Autocorrelation distance of surface roughness, m
a_{kj}	Matrix elements (mu or nd)
$\mathbf{a}$	Matrix of elements a_{kj}
$\mathbf{a}^{-1}$	Inverse matrix (mu or nd)
A	Surface area, m^2; absorptance of a translucent plane layer (nd)
A, B, C, D	Field amplitude coefficients
A_R	Aspect ratio of rectangle (nd)
A_l^m	Coefficients in spherical harmonics expansion
$\bar{A}_{ij}$	Equivalent spectral line width (mu, wavelength, wave number, frequency)
$\bar{A}_l, \bar{A}$	Equivalent spectral bandwidth (mu)
b	Spacing, m; width of a base, m; a dimension, m; coefficient in phase velocity of electromagnetic wave (nd); quantity in reflectivity relations (nd)
B	Pressure broadening parameter (nd); length dimension, m
$\mathbf{B}$	Magnetic induction vector, Wb/m^2
c	Speed of electromagnetic radiation propagation in medium other than a vacuum, m/s
c_0	Speed of electromagnetic radiation propagation in vacuum, m/s
c, c_p, c_v	Specific heat, J/(kg·K)
C	A coefficient or constant (mu or nd); clearance between particles, m; particle volume fraction (nd)
C_1	Constant in Planck's spectral energy distribution (Table A.4), W · μm^4/(m^2·sr)
C_2	Constant in Planck's spectral energy distribution (Table A.4), μm · K
C_3	Constant in Wien's displacement law (Table A.4), μm · K
C_i	Concentrations of component i in a mixture (nd)
C_{kj}	Elements of matrix $\mathbf{C}$ (mu or nd)
C_{CO_2}, C_{H_2O}	Pressure-correction coefficients (nd)
d	Number of diffuse surfaces (nd); a dimension, m

dA^*	Differential element on the same surface area as dA, m^2
D	Thickness of a layer or plate, m; a dimension, m; diameter of tube or hole, m; diameter of atom or molecule, m; number of dimensions (nd)
D_f	Fractal dimension (nd)
D_p	Particle diameter, m
$\mathbf{D}$	Electric displacement, C/m^2
$\mathbf{e}$	Energy level of a quantized state or photon, J
E	Emissive power (usually with a subscript), W/m^2; amplitude of electric intensity wave, N/C; overall emittance of a translucent layer (nd); the quantity $(1-\varepsilon)/\varepsilon$, (nd)
En	Exponential integral (Appendix D) (nd)
Ew	Weighted error (nd)
Eff	Absorption efficiency for a grooved directional absorber (nd)
$\mathbf{E}$	Electric field vector, V/m
$f(\xi)$	Frequency distribution of events occurring at ξ (nd)
F	Configuration factor (nd); objective function in optimization (mu); separation variable in Equation 11.52
$F_{0\to\lambda}$	Fraction of total blackbody intensity or emissive power in spectral region 0 to λ (nd)
$\boldsymbol{F}$	Transfer factors in enclosure (nd)
g	Gravitational acceleration, m/s^2
$g(k_\eta)$	Cumulative distribution function in k-distribution method, Equation 9.47
$\overline{gg}, \overline{gs}$	Gas–gas and gas–surface direct exchange areas, m^2
$\overline{g}$	Weyl component of the dyadic Green's function, m
G	Incident radiative flux onto a surface, W/m^2; Green's function
$\overline{\overline{\mathbf{G}}}$	Dyadic Green's function, 1/m
h	Planck's constant, J·s (Table A.1); height dimension, m; convective heat transfer coefficient, W/(m^2·K); enthalpy, J/kg
h_v	Volumetric heat transfer coefficient, W/(m^3·K)
H	Wave amplitude of magnetic intensity, C/(m·s); convection-radiation parameter (nd)
$\mathbf{H}$	Magnetic field vector, C/(m·s)
I_λ	Spectral radiation intensity, W/(m^2·μm·sr)
I	Radiation intensity, W/(m^2·sr)
$\mathbf{i}, \mathbf{j}, \mathbf{k}$	Unit vectors in x, y, z coordinate directions (nd)
i	Number of increments (nd)
$\hat{I}$	Source function, W/m^2
Im	Imaginary part
J	Radiosity; outgoing radiative flux from a surface; auxiliary variational function (mu or nd); number of increments (nd)
$\mathbf{J}$	Current density vector, A/m^2
$\mathbf{J}^r$	Random current density vector, A/m^2
k	Thermal conductivity, W/(m·K); extinction index for electromagnetic radiation, m^{-1}; wave vector $(= k' + ik'')$ rad/m
k_B	Boltzmann constant (Table A.1), J/K
K	Dielectric constant $(= K' + iK'')$ (nd); kernel of integral equation (mu or nd)
K_{ij}	Finite element function defined in Equation 12.175
l	A length, m, or a dimensionless length (nd)
l, m, n	Direction cosines for normal direction used in contour integration method (nd)
l_1, l_2, l_3	Direction cosines for rectangular coordinates designated as x_1, x_2, x_3 (nd)
l_m	Mean penetration distance, m
L	Length dimension, m
L_e	Mean beam length of gas volume, m

$L_{e,0}$	Mean beam length for the limit of very small absorption, m
L	Ladenberg–Reiche function (nd)
M	Mass of a molecule or atom; molecular weight
M_{kj}	Minor of matrix element a_{kj} (mu or nd)
n	Index of refraction of a lossless material c_0/c (nd); ratio n_2/n_1 in a few equations (nd); index in summation (nd); sample index (nd); number of a surface (nd); ordinate directions in S_n approximation (nd); normal direction (nd)
$\bar{n}$	Complex refractive index $n-ik$ (nd)
$\mathbf{n}$	Unit normal vector (nd)
N	Number of surfaces in an enclosure (nd); number of sample bundles per unit time, s^{-1}; number of particles per unit volume, m^{-3}; density of electromagnetic states, $s/(m^3 \cdot rad)$
N_{CR}	Conduction-radiation parameter (nd)
N_x, N_y	Number of x and y grid points (nd)
Nu	Nusselt number hD/k (nd)
P	Partial pressure of gas in mixture, atm
$\hat{\mathbf{p}}$	TM-polarized unit vector
P	Perimeter, m; probability density function (nd); total pressure of gas, atm
P, Q, R	Functions used in contour integration, m^{-1}
P_0	Pressure of 1 atm
P_e	Effective broadening pressure (nd)
P_l^m	Associated Legendre polynomials (nd)
Pr	Prandtl number $c_p\mu_f/k$ (nd)
q	Energy flux, energy per unit area and per unit time, W/m^2
$\dot{q}$	Internal energy generation per unit volume, W/m^3
q_c	Energy per unit area per unit time resulting from heat conduction, W/m^2
q_l	Radiative flux in a spectral band (mu)
q_r	Net radiant energy per unit area per unit time leaving a surface element, W/m^2
$\mathbf{q}_r$	Radiative flux vector, W/m^2
Q	Energy per unit time, W; ray origin point
Q_a	Absorption efficiency factor (nd)
Q_s	Scattering efficiency factor (nd)
r	Radial coordinate, m; radius, m
r_e	Electrical resistivity, $N \cdot m^2 \cdot s/C^2 = \Omega \cdot m$
r_{ij}	Fresnel's reflection coefficient at interface i–j
$\mathbf{r}$	Position vector (mu or nd)
R	Radius, m; overall reflectance of a translucent plane layer or a group of multiple layers (nd); random number in range 0 to 1 (nd)
Re	Reynolds number $Du_m\rho_f/\mu_f$ (nd); real part
s_λ	Scattering cross section, m^2
$\mathbf{s}$	Unit vector in S direction (nd)
$\hat{\mathbf{s}}$	TE-polarized unit vector
$\overline{s_j\gamma_\gamma}$	Surface–gas direct exchange area in zonal method
$\overline{s_js_k}$	Surface–surface direct exchange area in zonal method
S	Coordinate along the path of radiation, m; distance between two locations or areas, m; surface, m^2; number of sample energy bundles per unit time, s^{-1}
$\mathbf{S}$	Poynting vector, W/m^2; energy per unit area and time, W/m^2
$\dot{S}$	Dimensionless internal energy source (nd)
S_c	Collisional line intensity (mu)
S_{dv}	Number of energy bundles absorbed per unit time in volume dV (nd)
S_{ij}	Spectral line intensity (mu)

$\bar{S}_{k-j}$	Geometric-mean beam length from A_k to A_j, m
S_n	Two-dimensional radiation integral functions (Appendix D) (nd); singular values from matrix decomposition
S_r	Dimensionless radiative heat source (nd)
St	Stanton number, Nu/(Re·Pr) (nd)
t	Time, s
t_{ij}	Fresnel's transmission coefficient at interface i–j
t	Dimensionless time (nd)
$t(S)$	Transmittance of a medium (nd)
$\bar{t}_{j-k}$	Geometric-mean transmittance (nd)
T	Absolute temperature, K; overall transmittance of a plane layer or a group of multiple layers (nd)
$\bar{T}_l$	Mean transmission in a spectral band (nd)
T_{w1}, T_{w2}	Temperatures of walls 1 and 2, K
u	Fluid velocity, m/s; the variable $X\alpha/\omega$ (nd); energy density, J/m^3
u_k	Spectral band parameter $hc\eta_k/kT$ (nd)
$\bar{u}, u_m$	Mean fluid velocity, m/s
u, v	Velocity, m/s
$\mathbf{U}$, $\mathbf{V}$	Orthogonal matrices resulting from singular value decomposition
U	Total number of unknowns for an enclosure (nd); radiant energy density, J/m^3
$U(x, y)$	Approximate solution in finite-element method (mu)
U_v	Spectral radiant energy density, J/(m^3·um)
V	Volume, m^3; voltage signal, $V = N/C$
V_γ	Volume of element γ in zoning method, m^3
w	Width, m; energy carried by the sample Monte Carlo bundle, J; weighting factors (nd)
W	Weighting function in finite-element method (nd); width dimension, m
x, y, z	Coordinates in Cartesian system, m
X	Coordinate, m, or dimensionless coordinate (nd); mass path length, g/m^2
X, Y, Z	Optical or dimensionless coordinates (nd)
Y_l^m	Normalized spherical harmonics (nd)
Z	Height of surface roughness, m

Greek Symbols

α	Absorptivity (nd); thermal diffusivity, m^2/s; coefficient in soot scattering correlations
α(S)	Absorptance of a medium (nd)
α, α_0	Regularization parameter in Tikhonov regularization
α, β, γ	Direction cosines (nd)
α, δ, γ	Angles measured from normal direction in contour integration method, rad
$\bar{\alpha}_{j-k}$	Geometric-mean absorptance, m^{-1}
β	Extinction or attenuation coefficient $\kappa + \sigma_S$, m^{-1}; angle in x-y plane, rad; coefficient of volume expansion, K^{-1}; the parameter $\pi\gamma_c/\delta$ (nd)
β_i	Coefficients in shape function in the finite-element method (nd)
β_R	Rosseland mean attenuation coefficient, m^{-1}
γ	Electrical permittivity, C^2/(N · m^2); polynomial coefficients (nd); half-width of a spectral line (mu)
γ^2	Variance in a statistical solution (nd)
Γ	Number of gas elements (nd)
Γ	Factors in Gebhart's method (fraction of energy leaving one surface that is absorbed by another) (nd); separation variable in Equation 11.53; function in integral equation (Equation 11.62), W/m^2

δ	Propagation angle in medium, rad; boundary layer thickness, m; average spacing between lines in absorption band (mu); penetration distance of evanescent waves, m
δ_{kj}	Kronecker delta; = 1 when $j = k$; = 0 when $j \neq k$
$\delta(r'-r'')$	Dirac delta function
Δ	Distance above a radiating body
$\Delta\varepsilon$	Correction for spectral overlap (nd)
$\Delta\varphi$	Intermediate function in alternating direction implicit method (mu)
ε	Emissivity of a surface (nd)
$\varepsilon\ (S)$	Emittance of a medium (nd)
ε_{d}	Deformatiion parameter in T-Matrix method
ε_h	Eddy diffusivity for turbulent flow, m^2/s
ε_ρ	Porosity (nd)
$\bar{\varepsilon}$	Electrical permittivity
ζ	Arbitrary direction (nd); the quantity $C_2/\lambda T$ (nd)
η	Fin efficiency (nd); Blasius similarity variable (nd); wave number, $1/\lambda$, m^{-1}
θ	Polar or cone angle measured from the surface normal, rad
θ_o	Scattering angle, rad
Θ	Separation variable in Equations 11.52 and 11.53; mean energy of a Planck oscillator, J
ϑ	Dimensionless temperature T/T_{ref} (nd) or $T(\sigma/q_{\max})^{1/4}$
κ	Absorption coefficient, m^{-1}
κ_e	Effective mean absorption coefficient, m^{-1}
κ_i	Incident mean absorption coefficient; absorption coefficients in weighted-sum-of-gray-gases emittance model, m^{-1}
κ_P	Planck mean absorption coefficient, m^{-1}
κ_R	Rosseland mean absorption coefficient
κ_λ	Spectral absorption coefficient, m^{-1}
λ	Wavelength, m
λ_m	Wavelength in a *medium* other than vacuum, m
μ	Magnetic permeability, N/A^2; dimensionless fin conduction parameter (nd); the quantity $\cos\theta$ (nd); the quantity SS_c/δ (mu)
μ_f	Fluid viscosity, kg/m·s
ν	Frequency, $c_0/\lambda_0 = c_0/\lambda = c/\lambda_m$, s^{-1} = Hz
ξ	Length coordinate, m; parameter $\pi D/\lambda$ for scattering (nd); parameter $SS_{ij}/2\pi\gamma_c$ for equivalent line width
ξ, η	Dimensionless coordinates (nd)
ξ_C	Clearance parameter for particle separation criteria, $\pi C/\lambda$
ρ	Reflectivity (nd); gas density, kg/m^3
ρ_e	Electric charge density, C/m^3
ρ_{ij}	Reflectivity at interface i–j
ρ_f	Density of a fluid, kg/m^3
ρ_M	Density of a material, kg/m^3
ρ_S	Specular reflectivity, (nd)
ρ^*, ρ_0	Distances between points, m or (nd)
σ	Stefan–Boltzmann constant, Equation 1.31 and Table A.4, W/(m^2·K^4)
σ_s	Scattering coefficient, m^{-1}
σ_0	Root-mean-square height of surface roughness, m
τ	Roughness correlation length, m; optical thickness (nd); transmittance (Chapter 17) (nd)
τ_D	Optical thickness for path length D (nd)
ϕ	Circumferential or azimuthal angle, rad; dimensionless function, Equation 11.70 (nd)

Φ	Scattering phase function (nd); shape function in finite-element method (nd); function in integral equation (mu or nd)
Φ_d	Viscous dissipation function, J/(kg·m^2)
χ	Angle of refraction, rad
ψ	Dimensionless heat flux (nd); stream function
ψ_1	Temperature jump coefficient (nd)
$\psi^{(3)}$	Pentagamma function (nd)
ψ_b	Dimensionless energy flux for black walls (nd)
ω	Albedo for scattering (nd); angular frequency, rad/s; width of spectral band (mu); band width parameter, cm^{-1}
Ω	Solid angle, sr
Ω_i	Incident solid angle, sr
$\mathscr{F}$	Transfer factor, Equation 5.35

Subscripts

α	Absorbed; absorption; absorber; apparent value
$\alpha_0, \alpha_1, \ldots$	Coefficients
abs	Absorbed
A	Property of surface A
b	On a base surface; at the base of a fin; bottom
b, black	Blackbody condition
$bi\text{-}d$	Bidirectional
c	Evaluated at cutoff wavelength; corrected values; collision broadening; at a collector (absorber) plate; cylinder; cross section
cond	Conduction
c	Coating
CO_2	Carbon dioxide
D	Disk
d, dif	Diffuse
d_1, d_2	Evaluated at differential elements d_1, d_2
$d\text{-}h$	Directional hemispherical
D	Doppler broadening
e	Emitted or emitting; entering; environment; element of area; energy input; electrical; effective value
eq	At thermal equilibrium
evan	Evanescent wave
E	Electric
f	Fluid
fc	Free convection
fd	Fully developed
$\boldsymbol{F}$	Final
g	Gas
h	Hemispherical
$\boldsymbol{H}$	Magnetic
H_2O	Water vapor
i	Incident; inner; incoming
i,j	Energy states
$\boldsymbol{I}$	Initial
j, k	Property of surface A_j or A_k
l	Spectral band; layer index
L	Long wavelength region

LO	Longitudinal optical
m	Mean value; in a medium; maximum value; metal
m, m'	Outgoing and incoming angular directions
m, n	Number of identical semitransparent plates in a system
mc	Metal on cold side
mh	Metal on hot side
$m\boldsymbol{P}$	Evaluated at midpoint
max	Corresponding to maximum energy; maximum value; maximum refraction angle
min	Minimum
$\boldsymbol{M}$	Maximum value; material
n	Normal direction; natural broadening
nd	Nondiffuse
N, S, E, W	Directions in Figures 12.13 and 12.14
o	Outer; outgoing; evaluated in vacuum
p	Projected; particle
prop	Propagating wave
$\boldsymbol{P}$	Planck mean value; perimeter; point in discrete ordinates method
r	Reflected; reduced temperature; reservoir; radiative
rad	Radiation
ref	Reference value
$\boldsymbol{R}$	Radiator; radiating source; Rosseland mean value; radiative
s	Surface of a sphere, sun, scattering; surroundings; source; solid; specular
sol	Solar
sub	Substrate
S	Short-wavelength region
t	Transmitted; top
TE	Transverse electric
TM	Transverse magnetic
TO	Transverse optical
u	Uniform conditions
w	Wall; window
x, y, z	Components in x, y, z directions
η	Wave number dependent
λ	Wavelength (spectrally) dependent
$\Delta\lambda$	For a wavelength band $\Delta\lambda$
$\lambda_1 \rightarrow \lambda_2$	In wavelength region from λ_1 to λ_2
λT	Evaluated at λT
ν	Frequency dependent
ω	Angular frequency dependent
0	In vacuum
1, 2	Surface or medium 1 or 2
$\perp$	Perpendicular component
$\parallel$	Parallel component
$\cap$	Hemisphere of solid angles

Superscripts

i	Inside of an interface
n	nth time interval
o	Inlet value; outside of an interface
s	Specular exchange factor
(0), (1), (2)	Zeroth-, first-, or second-order term; designation for moments

+	Along directions having positive cos θ
−	Along directions having negative cos θ
−	(Overbar) averaged over all incident or outgoing directions; mean value; complex value
~	Dimensionless quantity
*	Complex conjugate

1 Introduction to Radiative Transfer

Energy radiates from one object to another one under all conditions and all times. The source of the emitted radiation is a combination of electronic and molecular oscillations and transitions in the emitting material, as well as lattice vibrations. Once the energy is radiated, it propagates as an electromagnetic (EM) wave, regardless of whether there is vacuum or matter along its path. After these waves reach another object, they partially lose their energy and give rise to electronic or molecular transitions and/or lattice vibrations in the receiving body. This is called absorption, which causes the energy level of the second object to rise. Both the emitted and absorbed radiation are functions of the physical and chemical properties of the material as well as its energy level (as quantified by its temperature). The interaction between emitting and absorbing bodies via EM waves is the essence of radiative energy transfer.

If the objects are at different temperatures, there will be net energy transfer between them. An obvious example is radiation emitted by the sun, which enters the Earth's atmosphere where it is partially absorbed and scattered. A significant part of the sun's radiation reaches the Earth's surface, where again, some is absorbed and some is reflected (Figure 1.1). In these processes, three different physical mechanisms play roles: emission, absorption, and scattering, which are all spectral in nature; that is, they all depend on the wavelength or frequency of the EM wave. Emission and absorption can be described by an elegant, yet complicated formulation using quantum electrodynamics (QED). Yet, for most thermal radiation transfer problems, there is no need to use QED. Instead, the Planck blackbody radiation distribution can be used to correlate the temperature of an object with its emission spectrum and energy. Scattering is the third leg of the trilogy of radiative transfer. It describes the propagation of EM waves from one object to another, as well as any redirection of radiative energy that is not converted to thermal energy. Reflection, refraction, diffraction, and rescattering of energy by any particle, structure, or inhomogeneity along the path of the wave are all considered under the term of "scattering."

Thermal radiative energy transfer occurs in a small part of the EM wave spectrum. Yet, radiative transfer is the reason for everything we see and sense in terms of light and heat. On Earth, the reason for all these interactions originates with the sun. The sun's energy is spectral in nature and peaks around 550 nm in wavelength, just in the middle of the spectral range where our eyes are most sensitive. Interestingly, this wavelength is where liquid water absorbs the least. This is critical for all living organisms on Earth whose eyes are within some kind of water-filled biological system (Figure 1.2).

Note that the short-wavelength (blue) part of the visible spectrum is much more weakly absorbed than the longer-wavelength (red) part. This gives the observed blue shade common to underwater scenes.

The propagation and scattering of EM waves, including all effects of reflection, refraction, transmission, polarization, and coherence, are governed by the Maxwell equations describing a propagating EM wave, as we discuss later in the book. These physical phenomena need to be understood to explain how the solar radiation is selectively scattered in the atmosphere by gases and particles and reflected by the Earth. This energy is also absorbed to a degree by everything it shines onto, heating them up and triggering other modes of energy transport, including conduction and

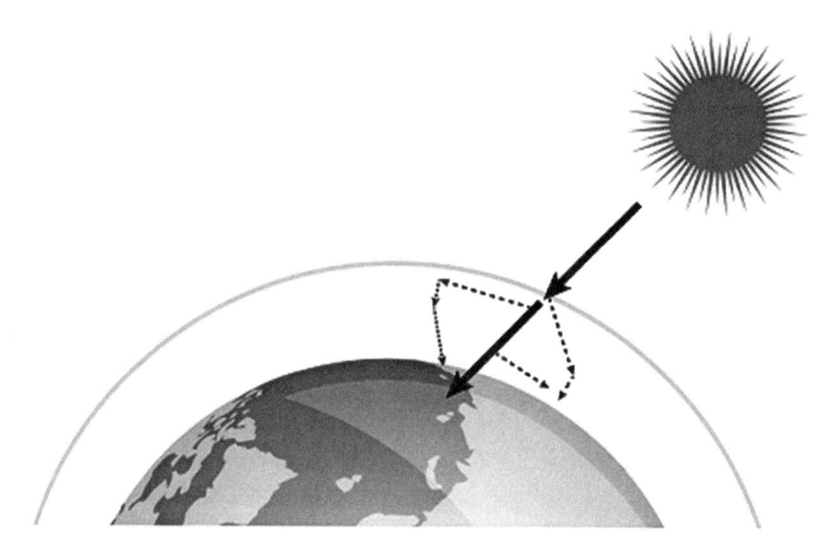

FIGURE 1.1 Atmospheric effects on radiation from the sun.

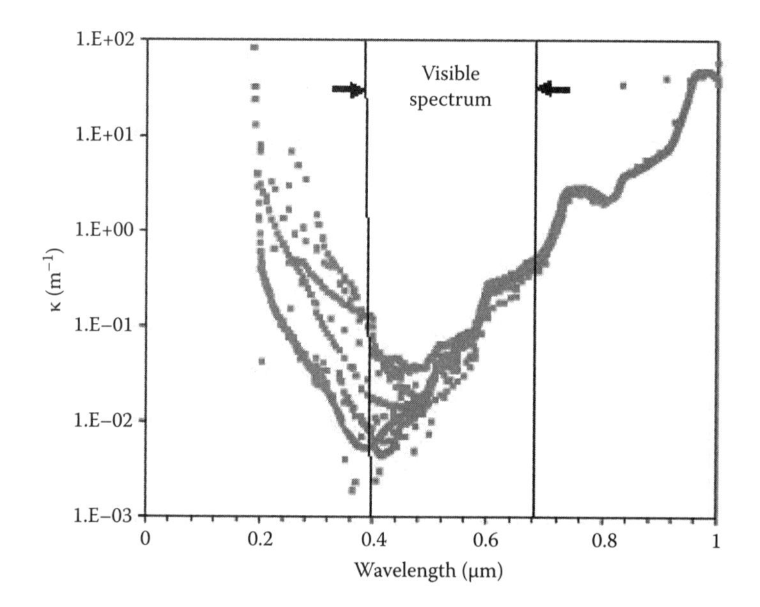

FIGURE 1.2 Representative data for the absorption coefficient of pure water collected from various sources. (Redrawn from Irvine, W.M. and Pollack, J.B., *Icarus,* 8, 324, 1968.)

convection. The change in the resulting energy balance yields circulations in the atmosphere and oceans, fires, chemical reactions, photosynthesis, and of course the life itself.

The strength of the emitted energy is correlated with the internal energy state of the emitter. If the size of the object is relatively large compared to the size of the atoms and molecules, the object's internal energy can be described starting from the principles of statistical thermodynamics. Statistical averaging allows the definition of average thermodynamic and thermophysical properties, including internal energy and temperature. Yet, as we see later, we can use relatively simple laws, such as Planck blackbody law and Kirchhoff's law to explain these averaged phenomena. Each emitting object exchanges radiant energy with the others that it can "see." A fraction of the incident energy that it receives from other bodies is absorbed and converted into its internal energy. The exchange of this energy is again explained by Planck's law if the objects are far from each other.

For systems and separation distances much larger than the wavelength of radiation, radiation energy can be expressed in terms of light quanta, or photons. Then, radiative transfer can be described using geometric optics approximations and ray-tracing methods. The concept of photons is often used to explain the complex interactions; yet, physically, it is questionable. We discuss this further.

If the distance between the objects is less than approximately the wavelength of radiation, then the tunneling of the EM waves into the objects significantly enhances the radiative transfer. At that scale, Planck law loses its applicability. These near-field effects can also be accounted for using Maxwell equations; however, as we see in Chapter 16, more detailed analysis needs to be followed.

The Maxwell equations do not predict thermal emission. The energy emitted by an object is described by Planck function. However, when the object size approaches the nanoscale regime, fluctuational electrodynamics is used to include an emission term into the governing equations. We discuss these effects in Chapters 15 and 16.

Understanding the emission, absorption, and scattering of radiative energy and its interaction with the matter is essential for engineering of advanced devices and processes. Radiative transfer is important in solar thermal collectors, photovoltaic energy generation systems, industrial furnaces, internal combustion engines, materials processing, laser–tissue interactions, and manufacturing processes and for all kinds of diagnostics tools. This text aims to provide the necessary theories, concepts, tools, and properties to provide the necessary understanding and skill levels to address analysis and design of such systems and devices.

As we discussed earlier, the quantitative calculation of the exchange of EM energy between objects at different energy levels constitutes radiative transfer—the focus of this book. We start our discussion with a section on the importance and fundamentals of radiation transfer, and then provide a heuristic view, tying the traditional radiation exchange formulation in enclosures with the radiative transfer equation (RTE) in participating media. In this context, the word "participating" refers to a medium such as the Earth's atmosphere (Figure 1.1) that alters the radiative energy propagating through it because of absorption, scattering, or emission mechanisms. We introduce the fundamentals of blackbody radiation, define radiation intensity, and outline the phenomena of emission, absorption, and propagation, including reflection, refraction, and scattering. Throughout the book, both the fundamentals and applications of radiative transfer are emphasized and several practical engineering problems are discussed, ranging from combustion chambers to solar radiators to nanoscale devices.

1.1 IMPORTANCE OF THERMAL RADIATION IN ENGINEERING

Thermal radiation is important in many engineering applications, particularly at high temperatures. For conduction and convection, energy transfer between two objects depends approximately on the temperature difference between them. For natural (free) convection, or when variable property effects are included, the energy transferred can be a function of the difference between $(T_1)^n$ and $(T_2)^n$, where the power n may be slightly larger than one, and can reach two. Thermal radiation transfer that takes place between two distant bodies depends on the difference between the fourth power of their absolute temperatures ($n = 4$). If the temperature-dependent material properties are accounted for in the calculations, the radiative flux can be proportional to an even higher power n of absolute temperatures, and consequently the importance of radiation transfer is significantly enhanced at high temperatures. Because of this, radiation contributes substantially to energy transfer in furnaces, combustion chambers, fires, rocket plumes, spacecraft atmospheric entry, high-temperature heat exchangers, and during explosions of chemicals. Conversely, if conduction and convection are suppressed, then radiation can be the important heat transfer mode even at low temperatures. This is the case in spacecraft thermal control and thermos bottles. A thorough consideration of radiation transfer would improve the design and the operation of such devices. To this end, radiative transfer calculations need to be made with rigor, which requires the use of accurate radiative properties and appropriate radiation models.

Radiative transfer governs the temperature distribution of the sun and the spectral and directional nature of solar emission. Understanding this spectral and directional nature of radiative energy propagation is important for solar energy utilization. Given that we need cleaner and more sustainable energy production for the growing population of Earth, implementation of efficient designs and optimum radiation transfer principles for solar energy storage and utilization is expected to have a significant impact on our surroundings and the environment. In addition, understanding the directional and spectral principles of radiation transfer allows us to design and operate most fossil-fuel burning combustion systems more efficiently, which means less energy use and minimum effects on the environment.

Another distinguishing feature of radiation transfer is that no intervening medium is required between two objects for radiation exchange to occur. Radiative energy can propagate through a vacuum. This is in contrast to conduction and convection, where a physical medium must be present to carry energy between two objects. When no medium is present, radiation becomes the only mode of energy transfer, as in the case of heat leakage through the evacuated space in the walls of a Dewar flask or thermos bottle. It is also the most important transport mechanism for devices operating in outer space or in the Earth's orbit, where waste energy must ultimately be rejected by radiation. For large-space power plants, such as a space station or an extraterrestrial colony, the heat rejection radiators are major components, requiring careful design for optimized performance. The temperature control of satellites also depends on radiative transfer. The equilibrium temperature of each surface in these systems is obtained by making an energy balance of absorbed solar energy, emitted and reflected energy from the Earth for near-Earth orbits, radiation from satellite surfaces, and sources of internal energy such as onboard electronics. Without detailed spatial, angular, and spectral calculations of radiative flux, reliable operation of satellites under intense solar radiation would be jeopardized.

Radiation transfer can be important even at typical room and air temperatures. In a letter to a Cleveland newspaper, a florist "noted the recurrence of a phenomenon he has observed for two seasons since using plastic coverings over (flower) flats. Water collecting in the plastic, formed ice, a quarter-inch thick at night when the official temperature reading was well about freezing. "I would like an answer to that; I supposed you could not get ice without freezing temperatures."" The oversight was in considering only convection to the air and omitting nighttime radiation loss between the water-covered surface and the cold heat sink of the night sky. A similar example is the discomfort a person experiences in a room where cold surfaces are present. Cold window surfaces without a shade or drape can have a chilling effect as the body radiates heat to them without receiving compensating radiated energy from them. Indeed, radiative cooling via chilled walls is quite effective and ecologically appealing for warmer climates. Another example is the warming effect of sunshine felt by a person outside in cold weather. In northern climates, the "wind-chill" factor refers to wind-augmented convective heat losses from the human body. In addition, the "heat-index" includes humidity, and the "solar index" accounts for sunburn. With the same idea, a compensating "sun-warmth" factor can be introduced for evaluating solar comfort.

Understanding the spectral nature of radiation is a key issue in design and construction of sustainable buildings. Windows are often a major part of the shell of residential and commercial buildings, which consume 35%–40% of the total world energy. The air-conditioning (AC) load is significant for most houses in moderate and warmer climates during the summer. Once the solar radiation enters through the windows, AC is needed to cool a house, which impacts the use of electricity. The key requirement is to allow sufficient light but little heat through windows. This means that the engineering, architecture, and material selection for windows need to be done hand in hand, and the spectral properties of windows and coatings are used as part of overall design.

Since radiation acts from a distance, its absorption alters the temperature profile on a surface and inside a volume. Consequently, it influences other modes of energy transfer, that is, by conduction and by free or forced convection. Radiative energy transfer can heat the walls of an enclosure, producing free convection where it would not ordinarily occur. In boundary-layer flows of fluids that absorb radiation, the presence of external radiative sources can alter the convective heat transfer. Radiation can also penetrate into porous media and act as a volumetric heat generation source; further, its impact is quite well recognized in the design of fiberglass insulation systems, where even at room temperatures, it can be as significant as the energy flow by conduction.

1.2 THERMAL ENERGY TRANSFER

Analysis of energy transfer requires the application of conservation of energy principles. Most forms of the energy conservation laws are expressed in terms of temperature. Temperature is a yardstick to measure the internal energy of matter. This requires the assumption of local thermodynamic equilibrium (LTE), defined over a volume element and specific time duration. However, LTE cannot be assumed at atomic scales or for very fast processes, such as femto- or atto-second laser beam heating, or in regions with extremely large gradients in properties such as near-shock layers. In addition, it is not possible to assume LTE when an object is so small that the number of atoms or molecules that make it up does not allow a statistically meaningful averaging. For most practical engineering systems of interest in this text, however, the LTE assumption is valid.

The energy transferred into and from an infinitesimal volume of solid or fluid by conduction or convection depends on the temperature gradients and physical properties in the immediate vicinity of the volume.

Radiation transfer is one of the three modes of transferring thermal energy. Conduction and convection energy transfer are significantly different from radiation transfer at macroscales, where dimensions are much larger than those for atoms and molecules. At atomic levels, conduction and radiation transfer have similar equations and follow the rules of statistical thermodynamics.

For heat conduction in a material with temperature distribution $T(x,y,z)$ and constant thermal conductivity k, the energy equation is derived by applying the macroscopic empirical Fourier's conduction law:

$$q_{\text{cond}}(S) = -k\frac{dT}{dS} \tag{1.1}$$

where $q_{\text{cond}}(S)$ is in the direction S and has units of W/m^2.

For an elemental cube within a solid as in Figure 1.3a, the time-dependent energy change in cubical element dV due to a volumetric heat source term $\dot{q}$ is

$$\rho c\frac{\partial T}{\partial t}dV + \nabla(k\nabla T)dV = \dot{q}dV \tag{1.2}$$

where ∂ represents partial differentiation. For a homogenous steady-state system, only the heat conduction in and out of all the faces of the cubical element needs to be considered, which yields Laplace equation governing the heat conduction within the material:

$$\frac{\partial^2 T}{\partial x^2} + \frac{\partial^2 T}{\partial y^2} + \frac{\partial^2 T}{\partial z^2} = \frac{\dot{q}}{k} \tag{1.3}$$

Equation 1.3 is written for a medium with uniform thermal conductivity. Here, $\dot{q}$ is a volumetric heat source or sink and may correspond to radiative energy either deposited in (absorbed, positive) or rejected by (emitted, negative) the elemental cube. As we discuss later in this book, $\dot{q}$ is obtained from the solution of the RTE, as it is equal to the divergence of radiative flux. The terms on the left-hand side of the energy balance equation (Equation 1.3) depend only on local temperature derivatives in the material. A similar, although more complex, analysis can be made for convection,

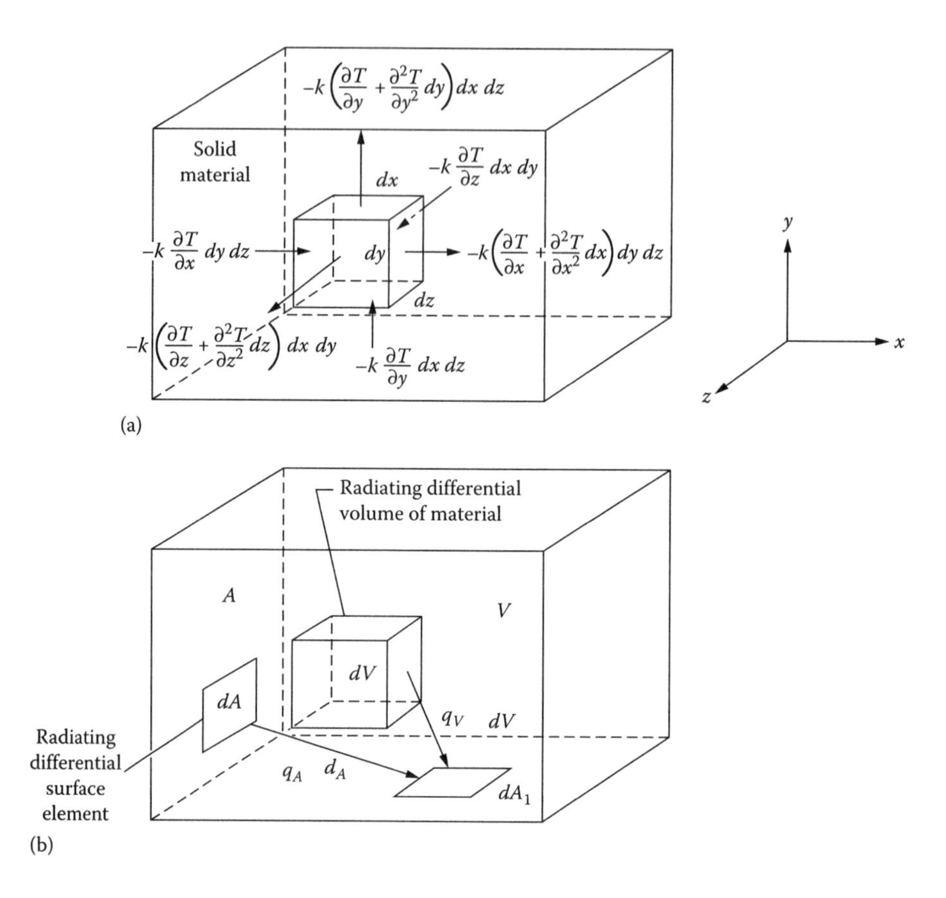

FIGURE 1.3 Terms for conduction and radiation energy balances: (a) energy conduction terms for volume element in solid and (b) radiation terms for enclosure filled with radiating material.

again demonstrating that the energy balance depends only on the conditions in the immediate vicinity of the location being considered.

Radiative energy is transmitted between the distant elements without requiring a medium between them. As an example, let us consider a heated enclosure with surface A and volume V, filled with radiating material such as hot gas, a cloud of hot particles, or hot glass as shown in Figure 1.3b. If $q_A dA$ is defined as the rate of radiant energy arriving at dA_1 from a surface element dA and the energy rate $q_V dV$ arrives at dA_1 from a volume element dV, then the rate of radiative energy from the entire enclosure arriving at dA_1 is

$$Q_{\text{rad}}(dA_1) = \int_A q_A\, dA + \int_V q_V\, dV \tag{1.4}$$

Calculation of the radiative energy rate requires integration of energy incident on a surface coming from all directions. Equation 1.4 suggests that integration is to be carried over all surfaces and volumes. Because of the directional nature of radiation, combined conduction, convection, and radiation transfer problems are quite complicated, and their solution requires extensive computational resources. In addition, radiation transfer calculations have to be performed at each wavelength or frequency. The *total* radiative energy is determined by integrating over all wavelengths or frequencies of the EM spectrum.

Radiative properties of surfaces, particles, and gases usually vary substantially over the wavelength (or frequency) spectrum. To account for these variations, the solution is first obtained for each wavelength, and then the spectral intensity and flux profiles are integrated over the entire spectrum to determine the total energy quantities. The boundary conditions may be specified in terms of total energies, but detailed spectral energy flows at the boundaries are usually unknown. This means that solutions must be obtained iteratively to satisfy the total energy flow conditions at the boundaries.

In addition to mathematical complexities, the lack of detailed physical property databases also hampers the accuracy of radiative transfer solutions, and consequently the energy transfer predictions. This shortcoming is usually a major bottleneck and underscores the difficulty of engineering practice. Radiative properties for solids and surfaces depend on many variables such as surface roughness and degree of polish, purity of the material, and thickness of any thin film (e.g., oxide) or paint coating on the surface. The coating thickness, its material properties, and temperature may affect the radiation absorbed by the material or radiation leaving the surface. Unfortunately, most available experimental data for surface radiative properties do not include the necessary information on the coatings, making their use guesswork at best. For radiating gases, radiative properties display even more complicated and spectrally irregular behavior. These properties are functions not only of the constituent molecules but also of gas pressure and temperature. In addition, the atmospheric or combustion gases that attenuate radiation are usually laden with various particles, including soot, dust, pollen, fly ash, or char particles. The particles absorb and scatter radiation depending on their relative size with respect to incoming radiation and show different characteristics according to their size and shape distributions and material properties. Even though some of these physical characteristics can be approximated for engineering applications, their relative importance needs to be established carefully for each problem to be solved.

Even if all properties are precisely known, the solution of governing integrodifferential equations of radiation transfer in a spectrally absorbing, emitting, and scattering medium is not trivial and requires a systematic study. In this text, we try to shed some light on these complex problems.

1.3 THERMAL RADIATIVE TRANSFER

Radiative energy propagation can be considered from two viewpoints: quantum mechanics and the classical EM wave theory. Quantum mechanics, in its strict sense, is almost impossible to apply to complicated engineering problems. Yet, with its guidance, the radiation energy carriers can be described as quanta of radiation or photons. Their propagation in a medium and interaction with matter are approximated using statistical mechanics. Even though the photon was first described by Einstein as "light quantum" (*Lichtquanten*), the word "photon" was coined by G. N. Lewis. The photon concept introduces a simplification that allows the treatment of radiation in simple terms. Although this simplification makes practical engineering problems tractable, it fails to explain some of the fundamental physics predicted from full-fledged quantum mechanics that includes polarization, coherence, and tunneling. A number of researchers question the use of the photon to describe radiation (e.g., Kidd et al. 1989, Lamb 1995, Mishchenko et al. 2006).

The classical EM wave view of radiation, in most cases, provides conservation equations similar to those obtained from quantum mechanics. Therefore, thermal radiation can be described by EM waves; even then, the computations required for radiative transfer calculations can be extensive. In addition, important effects such as the spectral distribution of the energy emitted from a body and the spectral properties of gases cannot be fully described by the EM theory. These effects can be explained and the corresponding equations can be derived only on the basis of quantum effects.

Further, EM wave phenomena have been explored extensively by many researchers since Maxwell in the late nineteenth century, and it is still an active research area. Within the framework of wave theory, EM radiation follows the laws for waves oscillating perpendicular to their direction of propagation. Energy emitted by an object is effectively a transverse wave comprised of interwoven electric and magnetic fields. These two fields are always perpendicular to each other, as well as to the direction of propagation. However, Maxwell equations describe this wave propagation in a beautifully symmetric way. Light is considered as a wave, and so is any radiative energy.

Moreover, EM waves propagate with the speed of light in vacuum; indeed, light itself is an EM wave within the narrow visible range of the spectrum. The light speed c_0 in vacuum (as designated by the subscript 0) is 2.9979×10^8 m/s ($\approx$ 186,000 miles/s). The speed in a vacuum is related to the speed of light in matter through the refractive index n, where $n = c_0/c$. The refractive index, which is the real part of the complex index of refraction, $\bar{n} = n - ik$ is greater than 1 for almost all natural materials within the visible and the infrared spectra. For example, for glass, it is about 1.5 in the visible spectrum, for water about 1.33, and for most gases about 1.0. The imaginary part of the index is known as the *absorption index*, and is a measure of how an EM wave is attenuated during propagation through an absorbing medium. The real part of the index of refraction can sometimes be less than 1 within a narrow wavelength interval, which is the indication of strong absorption. This does not mean that speed of the propagating waves would be greater than c_0. The propagation velocity is determined using a group velocity, which we discuss in Chapter 14.

Any property used for radiative transfer calculations must be spectral in nature, specified at a given wavelength or frequency. Also, EM radiation can be classified according to its wavelength λ_0 in vacuum, or its frequency ν, where $\nu = c_0/\lambda_0$. Wave number is also considered to represent the spectral effects, especially in quantitative spectroscopy applications, and is defined as $\eta_0 = 1/\lambda_0$. As the EM radiation propagates from one medium to another, its wavelength changes as a function of the index of refraction, as $n = \lambda_0/\lambda = (c_0\nu)/(c/\nu)$. The frequency of the oscillations, ν, however, remains the same in each medium; for that reason, the use of frequency to describe the spectral behavior of matter is common. The wavelength is usually given in the units of micrometers (μm) or nanometers (nm), where 1 μm is 10^{-6} m or 10^{-4} cm and 1 nm is 10^{-9} m. Sometimes, the units are expressed in angstroms (Å), where 1 μm = 10^4 Å or 1 nm = 10 Å. Wave number is given in units of cm^{-1}; the wavelength spectrum can be converted to wave number by using the multiplier $10{,}000/\lambda$ where λ in μm yields η in cm^{-1}. Spectral units of electron volts (eV) are sometimes used, where eV $= h\nu = hc/\lambda$ and h is Planck constant (Table A.1 in Appendix A). The conversion hence yields 1 eV = 1.602×10^{-19} J. Additional conversion factors and units for radiative transfer applications are provided in Tables A.2 and A.3 in Appendix A.

The range of EM waves is shown in Figure 1.4. Radio and television wavelengths may be as long as thousands of meters. At the other end of the spectrum, the wavelengths of cosmic rays can be as

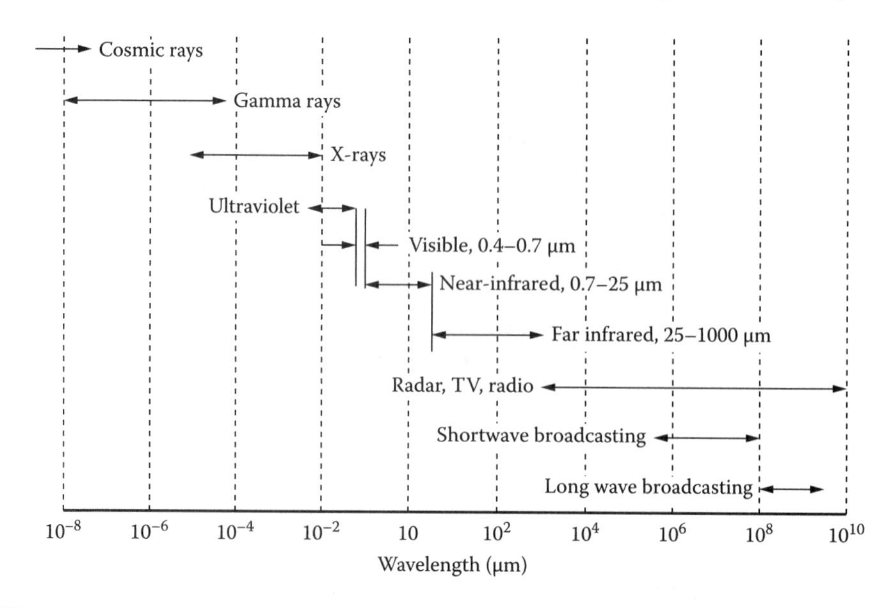

FIGURE 1.4 Spectrum of EM radiation (wavelength in vacuum).

small as 10^{-14} m. Within this vast range spanning more than 18 orders of magnitude, visible light is confined to 400–740 nm (i.e., 0.4–0.7 μm). What we call "red light" corresponds to the narrow wavelength range of roughly 620–740 nm; the green is roughly for λ of 510–530 nm, and violet is for 400–430 nm. For most practical purposes, thermal radiation is concerned with a wavelength range of 100 nm to 100 μm, including the ultraviolet and visible (λ = 0.1–0.7 μm), near-infrared (λ = 0.7–10 μm), and far-infrared spectra (λ = 10–100 μm).

Different physical mechanisms produce EM waves at different wavelengths. Some of these mechanisms include atoms or molecules undergoing transitions from one energy state to another lower energy state. These transitions may occur spontaneously or may be induced by the presence of an external radiation field. The details of these discussions can be found in the physics literature; they are outside the scope of present text.

1.4 RADIATIVE ENERGY EXCHANGE AND RADIATIVE INTENSITY

Before further discussing the fundamentals of radiative heat transfer, it is instructive to provide an overview of a physical problem where radiative exchange is important. The overall objective in radiative transfer calculations is to determine the amount of energy leaving one surface and reaching another after traveling through an intervening space, which may be comprised of particles, gases, some other material, or a vacuum. If the medium is altering the amount of radiation reaching the second surface, then it is called "participating." For example, gases such as water vapor and carbon dioxide may interact with radiative energy as they selectively absorb radiation in certain wavelength bands. In addition, the same combustion gases, including water vapor or carbon dioxide, emit at infrared wavelengths, enhancing the energy reaching to the second surface. Particles in the medium also absorb and emit radiation. In addition, they change the direction of radiative energy due to scattering, which plays a role in decreasing or increasing the amount of radiative energy propagating along a given direction. A complete radiative transfer prediction requires the analyses of all these interactions, which are highlighted in Figure 1.5. These interactions will be used as the basis of our discussion throughout this chapter.

Emitted radiation that leaves a surface is always spectral and directional in nature, meaning that it may change both as a function of wavelength and direction (as in the case of surface 1, in Figure 1.5). If the medium is participating, the radiative energy will change along the path from surface 1 to 2. The small cylindrical volume element shown is going to help us in our discussion

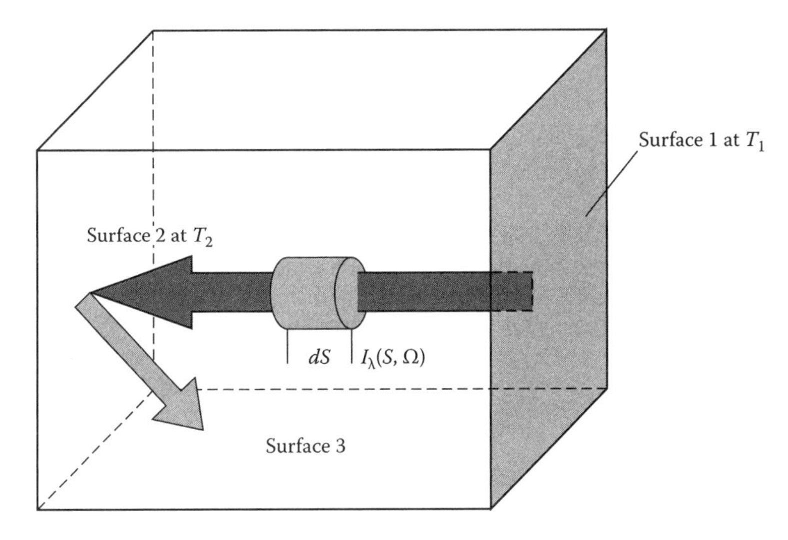

FIGURE 1.5 General schematic for radiative transfer discussions.

to define the conservation of radiative energy, which is expressed only in a specific direction Ω as indicated. Once the radiative energy is incident on the second surface, it will be either absorbed (hence, the energy will be converted to thermal energy in the body) or reflected to travel in another direction, for example, toward surface 3.

The schematic shown in Figure 1.5 can be used to discuss the radiative exchange in several practical cases. For example, it represents the propagation of a laser beam through a semitransparent medium, if the emission from surface 1 is replaced by a laser beam. Alternatively, we can consider the emitting body as the sun; then, the problem is the same as the propagation of solar rays through the Earth's atmosphere, which includes absorbing and scattering gases and particles. In this case, the temperature of the air is relatively low so that the radiative energy emitted by the atmosphere is small compared to the incident solar radiation or that reflected by the Earth. As a more practical engineering problem, we can visualize a three-dimensional (3D) configuration with several surfaces. This can simply be a kitchen oven where the air inside does not reduce the strength of radiation propagating through it. The problem, however, is still complicated as the exchange of energy from one surface element to another needs to be carefully accounted for in order to determine the energy balance everywhere. For a combustion chamber, the flame or hot gases or particles inside will absorb, emit, and scatter the energy incident on them. In this case, energy may be emanating from any point in the medium or on any surface; therefore, radiation propagating in any direction carries the radiative fingerprints of all particles, gas molecules, and surfaces in the enclosure. Radiative transfer needs to be determined after accounting for all directional effects, making the governing equation a complicated one.

It is obvious that to be able to quantitatively analyze any of these problems, we need to properly define and quantify the radiative properties, including the absorption, emission, and scattering coefficients. For this, we first introduce the concepts of solid angle and radiation intensity. After that, we consider emission from a single surface and discuss the idealized emission from a surface, including the maximum possible value. Such a maximum value is emitted from an idealized "blackbody." Following that, absorption and scattering coefficients and the scattering phase function are introduced, and the RTE is outlined. The RTE, which is the expression for the conservation of radiative energy along a line of sight, is the backbone of the discussion provided throughout this book.

1.4.1 Solid Angle

To develop the understanding of radiative energy propagating from one surface to another, consider the spectral radiative energy propagating along a direction S and incident on a small control volume dV at $S(x,y,z)$. This energy is confined to a small conical region, which is called the *solid angle*, Ω.

The solid angle is defined in 3D space, but is analogous to a planar angle. The planar angle is the ratio of the arc length to the distance from the apex to the base, while the solid angle is the ratio of a projected area to the square of the chord length from the apex to the area. For the definition of solid angle, consider an elemental surface area dA surrounded by a hemisphere of radius R as in Figure 1.6. A hemisphere has a surface area $2\pi R^2$, and it subtends a solid angle of 2π steradians (sr) about the center of its base. Hence, for a hemisphere of unit radius of $R = 1$, the solid angle around the center of the base equals the area on the unit hemisphere. Direction is measured by the zenith and azimuthal angles θ and ϕ, respectively, where θ is measured from the direction *normal* to dA. The angular position for $\phi = 0$ is arbitrary, but is usually measured from the x-axis. By definition, a solid angle anywhere above dA is equal to the intercepted area on the unit hemisphere. As shown in Figure 1.6, an element of this hemispherical area is given by $\sin\theta d\phi d\theta$ so that

$$d\Omega = \sin\theta d\theta d\phi \tag{1.5}$$

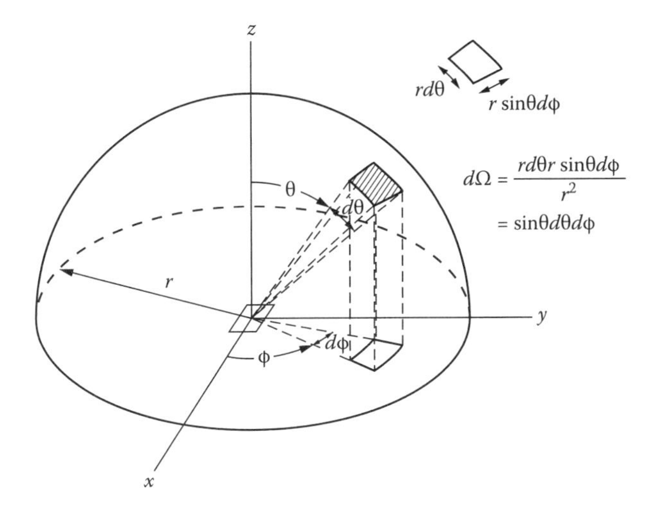

FIGURE 1.6 Hemisphere showing solid angle relations. Schematic for the definition of solid angle.

The integral of $d\Omega$ over all angles of a surrounding hemisphere yields 4π steradians (sr), units for solid angle in analogy to radians for planar angles:

$$\int_{4\pi} d\Omega = \int_{\theta=0}^{\pi/2} \int_{\phi=0}^{2\pi} \sin\theta d\theta d\phi = 4\pi \tag{1.6}$$

1.4.2 Spectral Radiative Intensity

Now we introduce the concept of radiative intensity, starting from radiative energy $dQ_\lambda(S,\Omega,t)$ propagating in direction Ω along path S.

The spectral radiative intensity I_λ is defined as the spectral radiative energy dQ_λ per unit projected area dA along a path S, per the solid angle $d\Omega$ around Ω, per unit time interval dt at time t, and a small wavelength interval $d\lambda$ around λ.

$$I_\lambda(S,\Omega,t) = \lim \frac{dQ_\lambda(S,\Omega,t)}{dA\,d\lambda\, d\Omega dt} \tag{1.7}$$

Here, we use "limit" to indicate that intensity is indeed a mathematical quantity. The spectral radiative intensity is the fundamental quantity for all radiation calculations. Its value is spectrally dependent.

The units of the spectral radiative intensity $I_\lambda(S,\Omega,t)$ are W/(m^2·μm·sr). The measureable quantity is usually the radiative energy, $Q(S,\Omega,t)$, which requires an integration of the radiative intensity over a finite area element, solid angle, wavelength interval, and time duration, yielding units of *Joule* (J).

The spectral intensity is a function of seven independent parameters, including three space coordinates (x,y,z) along S and two angular coordinates (θ, ϕ) for direction Ω, the wavelength (λ) (or frequency, ν), and time (t). It is always specified at a wavelength λ per unit wavelength interval around that λ. The temperature dependence of intensity is implicit in these expressions, as temperature is specified with the spatial coordinates within the medium. Specifically, the spectral intensity refers to radiation in an interval $d\lambda$ around a single wavelength, while the total intensity refers to combined radiation including *all* wavelengths.

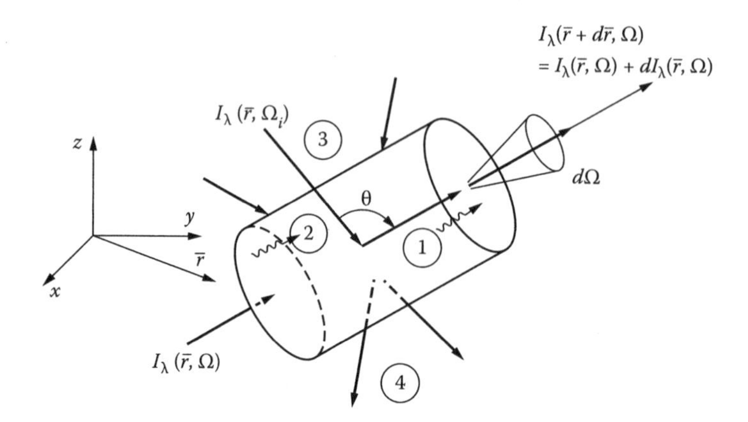

FIGURE 1.7 Radiation intensity propagating through a volume element.

As it propagates along the direction Ω through a path interval dS, radiative energy loses some of its strength due to scattering and absorption. On the other hand, it gains additional strength because of in-scattering of radiation and emission into the direction Ω. Along this direction and path, we can tally the change in radiative energy, which provides a conservation equation of radiative energy within a small directional cone along a given direction, for a small wavelength interval around a fixed wavelength, and within an infinitesimal time duration at a given time (see Figure 1.7) This formulation leads us to the radiative transfer equation (RTE), expressed in terms of I_λ.

Before discussing the RTE further, we need to introduce the definitions of absorption, scattering, and emission coefficients. We first discuss the fundamentals of emission, the Planck blackbody radiation, and the EM spectrum. We then present an analysis in which we trace the radiative energy propagating in a medium comprised of emitting, absorbing, and scattering gases and particles. This approach allows us to conceptualize the attenuation and enhancement of the radiative intensity due to respective mechanisms. Once we obtain the conservation of radiative energy along a line of sight, it can be written over a finite length of the medium, from one surface to another. After these fundamental expressions are outlined, we are poised to model and solve the expressions for engineering calculations.

1.5 CHARACTERISTICS OF EMISSION

Every object at a finite temperature emits radiative energy. In principle, emission from a given body is a function of the material properties, the temperature, and direction:

$$E_\lambda = f(T, \lambda, \theta, \phi) \tag{1.8}$$

where

$T = T(x,y,z)$

θ and ϕ are the zenith and azimuthal angles, respectively

An ideal body would emit the maximum amount of energy uniformly in all directions and at each wavelength. The concept of such an ideal body, which is the so-called *blackbody*, is basic to the study of radiative energy transfer. An ideal blackbody absorbs all radiation incidents on it. As an ideal absorber, it serves as a standard with which real materials can be compared, as they never absorb all energy at all wavelengths.

The blackbody derives its name from observing that good absorbers of incident visible light appear black to the eye. However, except for the visible region, the eye is not a good indicator of absorption capability over the entire wavelength range corresponding thermal radiation. For example, a surface coated with white oil-base paint is a very good absorber for infrared radiation, although it is a poor absorber for the shorter-wavelength region characteristic of visible light, as evidenced by its white appearance. Only a few materials, such as carbon black, carborundum, platinum black, gold black, some specialty formulated black paints on absorbing substrates, and some nanostructured surfaces, approach the blackbody in their ability to absorb radiant energy. The quantitative radiative properties of the ideal blackbody have been well established by quantum mechanics and also verified by experiments. Aside from being a perfect absorber for all incident radiation, the blackbody has other important features pertinent to our understanding of radiative transfer. These are discussed later.

The interaction of radiant energy at the surface of a body depends not only on properties of the surface itself but also on the material beneath the surface. When a beam is incident on a homogeneous body, some of the radiation is reflected and the remainder penetrates into the body. The radiation may then be absorbed and scattered as it travels through the medium as discussed earlier. If the material thickness is smaller than that required to absorb all the radiation, then a significant amount of the radiation passes through the body. If, on the other hand, the material is a strong absorber, no radiation may escape the medium; in this case, the radiation not reflected at the surface would be converted into internal energy of the medium. This conversion varies as a function of the distance from the surface depending on the spectral properties of the medium. This absorption mechanism increases the temperature of the body. A body is called *opaque* if the entire incident radiation that passes through its surface is internally absorbed. A material can be opaque if the material thickness is quite large, but can be semitransparent if its thickness is a mere few wavelengths long. One of the best examples of this is gold; because of the spectrally selective transmissivity of thin gold films, they are used in many practical applications, including protecting astronauts from excessive solar energy input through their helmets (see Figure 1.8). Yet we all know that bulk gold is nontransparent; its yellowish color is due to high reflectivity of that narrow wavelength range within visual spectrum.

Metallic objects strongly internally absorb incident radiation. That strong absorption is contrary to the strong reflection of metals, and only a small fraction of the energy penetrates into a metallic object. This fraction is eventually completely absorbed. The rest is reflected at the surface. Nonmetallic objects may exhibit the opposite tendency. They may allow a substantial portion of the incident radiation to pass into the material. This means that the surface reflection would be small,

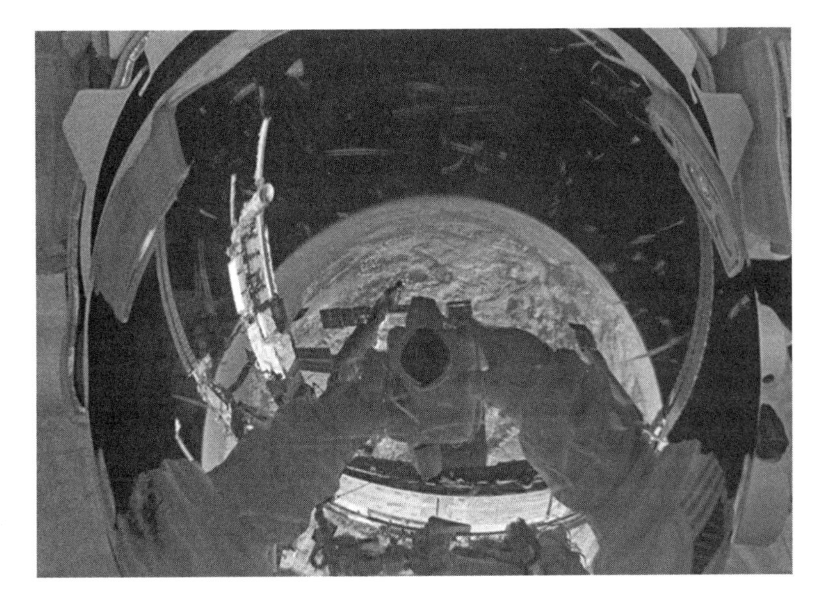

FIGURE 1.8 Astronaut helmet showing reflection of the Earth due to thin gold film coating. (Courtesy of NASA Photo Department-Expedition 15 Astronaut Clayton Anderson visiting the International Space Station.) An early selfie.

and the absorption requires longer depth into the material. A glass window readily allows visible radiation to pass through its surface. However, it is a poor absorber at visible wavelengths, and most of the incident visible radiation is transmitted.

To be a good absorber for incident energy, a body must have a low surface reflectivity and high internal absorption to prevent radiation from leaving it. As an example, if very fine metallic particles are deposited on a surface, the resulting textured surface has low reflectivity. This effect, combined with the high internal absorption of metal particles, causes this type of coating to be a good absorber. This is the basis for formation of metallic "blacks" such as platinum or gold black. These materials can have less than 1% reflection for the solar spectrum and provide complete internal absorption within a few micrometers of thickness.

The tailoring of radiative characteristics can be considered as an engineering challenge and can be achieved by altering the surface properties of an object. Particularly, with the development of manufacturing and patterning techniques at the nanoscale, new materials and structures are being designed to achieve novel specific requirements of absorption, reflection, transmission, and emission. For example, a composite of vertically aligned carbon nanotubes was reported by researchers at Rice University in 2008 to reflect only 0.045% of incident radiation in the visible region of the spectrum (absorptivity of 99.955% of visible radiation), very closely approaching blackbody characteristics. Such a perfect black material can be used extensively for spacecraft thermal management applications, among others. For radiative transfer applications, we also consider other types of surfaces and media that are opaque or partially transmitting, including particle and bubble clouds and absorbing gases. Examples of such applications are discussed throughout the text.

1.5.1 Perfect Emitter

Consider a blackbody at a uniform temperature within a perfectly insulated evacuated enclosure of arbitrary shape whose walls are also blackbodies at uniform but different temperatures (Figure 1.9). After a period of time, the blackbody and the enclosure attain a common uniform equilibrium temperature. In equilibrium, the blackbody must emit as much energy as it absorbs. To prove this,

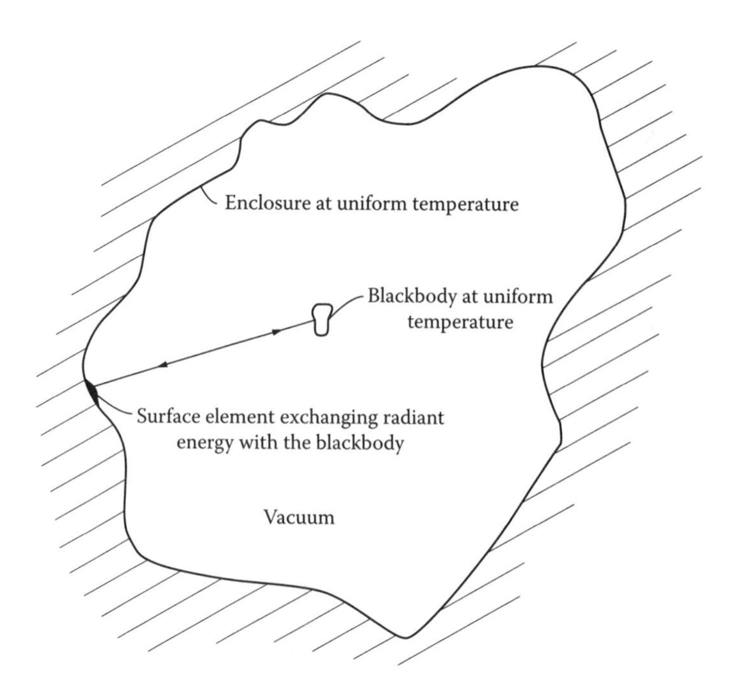

FIGURE 1.9 Enclosure geometry for derivation of blackbody properties.

consider what would happen if the incoming and outgoing amounts of radiation were not equal with the system at a uniform temperature. Then, the enclosed blackbody would either increase or decrease in temperature. This would involve a net amount of energy transfer between two bodies at the same temperature. This will be impossible according to the second law of thermodynamics unless there is external work input to the system. It follows that, because the blackbody is, by definition, absorbing the maximum possible radiation incident from the enclosure at each wavelength and from each direction, it must also be emitting the maximum total amount of radiation. This deduction becomes clear by considering any less-than-perfect absorber, which must consequently emit less energy than the blackbody to remain in equilibrium. The fact that a body emits radiation even when in thermal equilibrium with its surroundings is called *Prevost's law*.

1.5.2 Radiation Isotropy in a Black Enclosure

Now, consider the isothermal enclosure with black walls and arbitrary shape in Figure 1.9; this time, we move the blackbody to another position and rotate it to another orientation. The blackbody must still be at the same temperature because the whole enclosure remains isothermal. Consequently, the blackbody must be emitting the same amount of radiation as before. To be in equilibrium, the body must still be receiving the same amount of radiation from the enclosure walls. Thus, the total radiation received by the blackbody is independent of the orientation or position throughout the enclosure. Hence, the radiation traveling through any point within the enclosure is independent of the position or direction. This means that the radiation filling the black enclosure is *isotropic*.

1.5.3 Perfect Emitter in Each Direction and Wavelength

Consider an area element on the surface of the black isothermal enclosure and an elemental blackbody within the enclosure. Some of the radiation emitted by the surface element strikes the elemental body at a random direction. All this radiation, by definition, is absorbed by the blackbody. To maintain thermal equilibrium and isotropic radiation throughout the enclosure, the radiation emitted back into the incident direction must be equal to that received. Since the body is absorbing the maximum radiation from any direction, it must be emitting the maximum in any direction. Because the radiation filling the black enclosure is isotropic, the radiation received and hence emitted in *any* direction by the enclosed black surface, per unit projected area normal to that direction, must be the same as that in any other direction.

By considering the detailed balance of absorbed and emitted energy in each small wavelength interval, we also reach the conclusion that the blackbody is a perfect emitter at each wavelength.

1.5.4 Total Radiation into Vacuum

If the enclosure temperature is altered, the enclosed blackbody temperature also changes and becomes equal to the new temperature of the enclosure. The absorbed and emitted energy by the blackbody are again equal to each other, although the new equilibrium temperature differs from the original temperature of the enclosure. Since by definition a blackbody absorbs, and hence emits, the maximum amounts of radiation corresponding to its temperature, the characteristics of the surroundings do not affect its emissive behavior. Hence, the *total radiant energy emitted by a blackbody in vacuum is a function only of its temperature.*

As mentioned earlier, the second law of thermodynamics forbids net energy transfer from a cooler to a hotter surface without any external work being done on the system. If the radiant energy emitted by a blackbody increased with decreasing temperature, we could build a device to violate this law. Without considering the proof in detail, it is found that radiant energy emitted by a blackbody must *increase* with temperature. The total radiant energy emitted by a blackbody is therefore proportional to a monotonically increasing function of temperature.

1.5.5 Blackbody Intensity and Its Directional Independence

As we defined previously, the intensity is the fundamental quantity to describe radiation propagating in any direction. The spectral intensity of a blackbody is denoted as $I_{\lambda b}$, where the subscripts correspond, respectively, that a spectral quantity is being considered and that the properties are for a blackbody. The emitted spectral intensity from a surface is defined as the radiative energy emitted per unit time per unit small wavelength interval around the wavelength λ, per unit elemental projected surface area normal to (θ, ϕ) direction and into the elemental solid angle centered on the direction (θ, ϕ).

The directional independence of $I_{\lambda b}$ and I_b can be shown by considering a spherical isothermal blackbody enclosure of radius R with a blackbody element dA at its center (Figure 1.10a). The enclosure and the central element are in thermal equilibrium, and therefore, all radiation throughout the enclosure is isotropic. Consider radiation in a wavelength interval $d\lambda$ about λ that is emitted by an element dA_s on the enclosure surface and travels toward the central element dA (Figure 1.10b). The emitted energy in this direction per unit solid angle and time is written $I_{\lambda b,n} dA_s d\lambda$.

The normal spectral intensity of a blackbody is used because the energy is emitted normal to the black wall element dA_s of the spherical enclosure. The amount of energy per unit time that strikes dA depends on the solid angle that dA occupies when viewed from dA_s. This solid angle is the projected area of dA normal to the (θ, ϕ) direction, $dA_p = dA \cos\theta$, divided by R^2. Then, the energy absorbed by dA is $I_{\lambda b,n}(\theta, \phi) dA_s (dA\cos\theta/R^2) d\lambda$. The energy emitted by dA in the (θ, ϕ) direction and incident on dA_s in solid angle dA_s/R^2 (Figure 1.10c) must equal that absorbed from dA_s, or equilibrium would be disturbed; hence,

$$I_{\lambda b}(\theta, \phi) dA_p \left(\frac{dA_s}{R^2} \right) d\lambda = I_{\lambda b}(\theta, \phi) dA \cos\theta \left(\frac{dA_s}{R^2} \right) d\lambda$$

$$= I_{\lambda b,n}(\theta, \phi) dA_s \left(\frac{dA \cos\theta}{R^2} \right) d\lambda \tag{1.9}$$

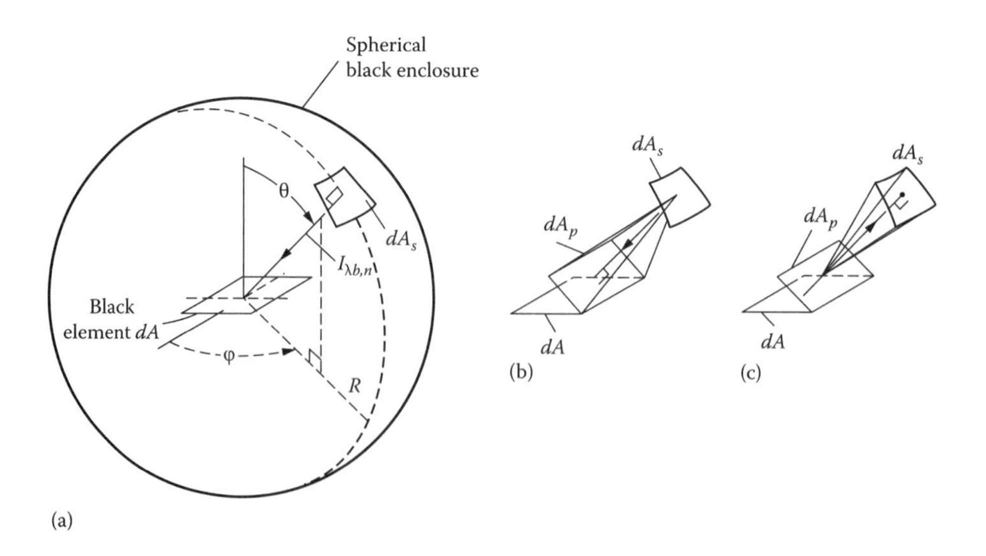

FIGURE 1.10 Energy exchange between element of enclosure surface and element within enclosure. (a) Black element dA within black spherical enclosure, (b) energy transfer from dA to dA_p, and (c) energy transfer from dA_p to dA_s.

This shows that

$$I_{\lambda b}(\theta,\phi) = I_{\lambda b,n} = I_{\lambda b} \tag{1.10}$$

The intensity of radiation from a blackbody, $I_{\lambda b}$, as defined on the basis of its projected area, is independent of the direction of emission. Neither the subscript n nor the (θ, ϕ) notation is needed for the intensity emanating from a blackbody. Since the blackbody is always a perfect absorber and emitter, its properties are independent of its surroundings. Hence, these results are independent of both assumptions used in the derivation of a spherical enclosure and thermodynamic equilibrium with the surroundings.

Some exceptions exist for most of the blackbody "laws"; however, they are of minor importance in almost any practical engineering situation but need to be considered when extremely rapid transients or extreme temperature gradients (e.g., near shocks) are present in a radiative transfer process. For example, if the transient period is of the order of the timescale of the process that governs the emission of radiation from the body in question, then the emission properties of the body may lag behind the absorption properties. In such a case, the concepts of temperature used in the derivation of the blackbody laws no longer hold rigorously. The treatment of such non-local thermodynamic equilibrium (non-LTE) problems is generally outside the scope of this work. (Some general comments are in Sections 9.2.2 and 10.6.)

1.5.6 Blackbody Emissive Power: Cosine-Law Dependence

As discussed earlier, the radiative intensity is defined on the basis of the normal area. This means that energy leaving an emitting surface is a maximum if the surface is normal to a receiving surface or a detector. Since the intensity of blackbody emission is uniform in all directions, we can write the angular profile of blackbody energy per unit area as a function of zenith and azimuthal angles. A uniformly emitting blackbody does not display any ϕ dependence; therefore, $E_{\lambda b}(\theta, \phi)$ is reduced to $E_{\lambda b}(\theta)$, which is the *directional spectral emissive power* for a black surface:

$$E_{\lambda b}(\theta,\phi) = I_{\lambda b}\cos\theta = E_{\lambda b}(\theta) \tag{1.11}$$

In Figure 1.11, both the blackbody intensity and blackbody energy are depicted as a function of zenith angle θ, although $I_{\lambda b}$ is independent of θ. In the case of nonblack surfaces, emitted energy may depend on both the zenith and azimuthal angles and can be expressed more properly as $E_{\lambda}(\theta, \phi)$.

Equation 1.11 is known as *Lambert's cosine law.* Surfaces having a directional emissive power that follows this relation are called *Lambertian, diffuse*, or *cosine-law* surfaces. A blackbody, because it is always diffuse, serves as a standard for comparison with the directional properties of real surfaces that do not follow the cosine law.

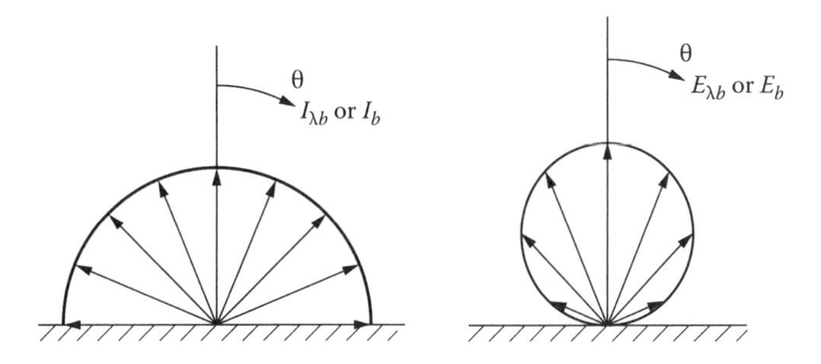

FIGURE 1.11 Cosine law: The angular distribution of blackbody intensity (independent of θ) and blackbody directional emissive power ($\cos\theta$ function).

1.5.7 Hemispherical Spectral Emissive Power of a Blackbody

In calculations of radiative energy loss by a surface, the angular profile of emission needs to be determined through a hemispherical envelope placed over a black surface. The *hemispherical spectral emissive power,* $E_{\lambda b}$, is the energy emitted by a black surface per unit time, per unit area, and per unit wavelength interval around λ. Figure 1.10 shows the area element dA at the center of the base of a unit hemisphere.

Hence, the spectral emission from dA per unit time and unit surface area passing through the element on the hemisphere is given by use of Equation 1.10 as

$$I_{\lambda b}\cos\theta d\Omega d\lambda = I_{\lambda b}\cos\theta\sin\theta d\theta d\phi d\lambda \tag{1.12}$$

To obtain the blackbody spectral emission passing though the entire hemisphere in terms of the blackbody intensity, Equation 1.12 is integrated over all solid angles of the hemisphere to give

$$E_{\lambda b}d\lambda = I_{\lambda b}d\lambda \int_{\phi=0}^{2\pi}\int_{\theta=0}^{\pi/2}\cos\theta\sin\theta\, d\theta d\phi = \pi I_{\lambda b}d\lambda \tag{1.13}$$

which is a remarkably simple relation: *the blackbody hemispherical emissive power is* π *times the blackbody intensity.* This relation for a blackbody is useful in relating directional and hemispherical quantities in the chapters that follow.

The emission may be desired through only part of the hemispherical solid angle enclosing an area element. The emission within the solid angle extending from θ_1 to θ_2 and ϕ_1 to ϕ_2 is found by modifying the integration limits in Equation 1.13 and integrating to obtain

$$\begin{aligned} E_{\lambda b}(\theta_1 \text{ to } \theta_2, \phi_1 \text{ to } \phi_2)d\lambda &= I_{\lambda b}d\lambda \int_{\phi=\phi_1}^{\phi_2}\int_{\theta=\theta_1}^{\theta_2}\cos\theta\sin\theta\, d\theta d\phi \\ &= (\phi_2-\phi_1)\left[\frac{\sin^2\theta_2-\sin^2\theta_1}{2}\right]I_{\lambda b}d\lambda \end{aligned} \tag{1.14}$$

1.5.8 Planck Law: Spectral Distribution of Emissive Power

So far, we have discussed the blackbody characteristics based on thermodynamic arguments. However, the spectral nature of blackbody emission cannot be obtained from purely thermodynamic arguments. Indeed, this was Planck's impetus in exploring a more exact theoretical expression, which eventually became the foundation of quantum theory. In Section 1.9, we outline the history of the blackbody formulation by Planck (1901), following Boltzmann's (1884) statistical thermodynamics. Here, we adopt the result of Planck blackbody formulation to express the spectral distribution of hemispherical emissive power and radiant intensity in vacuum. This expression, which was verified experimentally, is based on wavelength and the absolute temperature of the blackbody:

$$E_{\lambda b}(T) = \pi I_{\lambda b}(T) = \frac{2\pi h c_o^2}{n^2\lambda^5\left[\exp\left(\dfrac{hc_o}{nk_B\lambda T}\right)-1\right]} = \frac{2\pi C_1}{n^2\lambda^5\left[\exp\left(\dfrac{C_2}{n\lambda T}\right)-1\right]} \tag{1.15}$$

Here, λ is the wavelength in the medium of refractive index n. This is Planck spectral distribution of emissive power. For most engineering applications, radiative emission is into air or other gases, for which the index of refraction, $n = c_0/c$, is unity. Note that Planck blackbody function, which provides quantitative values for the emission, is expressed in terms of two universal constants: Planck constant, $h = 6.62606957 \times 10^{-34}$ J·s, and Boltzmann constant, $k_B = 1.3806488 \times 10^{-23}$ J/K. Two auxiliary radiation constants are defined as $C_1 = hc_0^2$ and $C_2 = hc_0/k_B$. The values of these constants are given in Table A.4 for two common unit systems.

Example 1.1

A black surface at temperature 1000°C is radiating into vacuum. At a wavelength of 6 μm, what are the emitted intensity, the directional spectral emissive power of the blackbody at an angle of 60° from the surface normal, and the hemispherical spectral emissive power?

From Equation 1.15,

$$I_{\lambda b}(6\ \mu\text{m}) = \frac{2 \times 0.59552 \times 10^8\ \text{W} \cdot \mu\text{m}^4/\text{m}^2}{6^5\ \mu\text{m}^5 \left(e^{14{,}388/[6\times 1273]} - 1\right)\ \text{sr}} = 2746\ \text{W}/(\text{m}^2 \cdot \mu\text{m} \cdot \text{sr})$$

From Equation 1.11, the directional spectral emissive power is

$$E_{\lambda b}(6\ \mu\text{m}, 60°) = 2746 \cos 60° = 1373\ \text{W}/\left(\text{m}^2 \cdot \mu\text{m} \cdot \text{sr}\right)$$

The hemispherical spectral emissive power is $E_{\lambda b}(6\ \mu\text{m}) = \pi I_{\lambda b}(6\ \mu\text{m}) = 8627\ \text{W}/(\text{m}^2 \cdot \mu\text{m})$.

Example 1.2

To a good approximation, the sun can be considered to emit as a blackbody at 5780 K. Determine the sun's intensity at the center of the visible spectrum.

From Figure 1.4, the wavelength of interest is 0.55 μm. Then, from Equation 1.15,

$$I_{\lambda b}(0.55\ \mu\text{m}) = \frac{2 \times 0.59552 \times 10^8\ \left(\text{W} \cdot \mu\text{m}^4/\text{m}^2\right)}{0.55^5\ \left(\mu\text{m}^5\right)\left[e^{(14.388/0.55\times 5780)} - 1\right]\ (\text{sr})}$$
$$= 0.256 \times 10^8\ \left[\text{W}/(\text{m}^2 \cdot \mu\text{m} \cdot \text{sr})\right]$$

Alternative forms of Equation 1.15 are employed where frequency or wave number is used rather than wavelength. Using frequency has an advantage when spectral radiation travels from one medium into another medium with different refractive index. Frequency remains constant while wavelength changes because of the change in propagation speed. We can rewrite Equation 1.15 in terms of frequency after noting that the wavelength in the medium is related to the wavelength in vacuum and the frequency through $n\lambda = \lambda_o = c_o/\nu$, and hence $d\lambda_o = -(c_o/\nu^2)d\nu$. Then, the emissive power in the wavelength interval $d\lambda_o$ becomes

$$E_{\lambda b}(T)d\lambda_o = \frac{2\pi n^2 C_1}{\lambda_0^5\left[\exp\left(\dfrac{C_2}{\lambda_o T}\right) - 1\right]} d\lambda_o = \frac{-2\pi n^2 C_1 \nu^3}{c_o^4\left[\exp\left(\dfrac{C_2 \nu}{c_o T}\right) - 1\right]} d\nu = -E_{\nu b}(T)d\nu \qquad (1.16)$$

The quantity $E_{\nu b}(\nu)$ is the emissive power per unit frequency interval about ν. The intensity is $I_{\nu b}(\nu) = E_{\nu b}(\nu)/\pi$, so that

$$I_{\nu b}(T) = \frac{2C_1 n^2 \nu^3}{c_o^4\left[\exp\left(\frac{C_2 \nu}{c_o T}\right) - 1\right]} \tag{1.17}$$

The wave number $\eta = 1/\lambda$. Then,

$$d\lambda = -\left(\frac{1}{\eta^2}\right) d\eta \tag{1.18}$$

The quantity $E_{\lambda b}(\eta)$ is the emissive power per unit wave number interval about η. The intensity is then

$$I_{\eta b} = \frac{E_{\eta b}(\eta)}{\pi} \tag{1.19}$$

The spectral nature of Planck law is observed in Figure 1.12, where the hemispherical spectral emissive power (Equation 1.15) is shown as a function of wavelength for several temperatures. We have already shown that total radiated energy, which is the spectral energy integrated over all wavelengths, increases with temperature. Figure 1.12 shows this for the energy emitted at different wavelengths. Also, the maximum spectral emissive power shifts toward smaller wavelengths as temperature increases. The energy emitted at shorter wavelengths increases more rapidly with temperature than the energy at long wavelengths.

For a blackbody at 555 K (1000°R), only a small amount of energy is in the visible region; however, that energy is very difficult to detect by the eye. The curves at lower temperatures slope downward toward the violet end of the spectrum. Red light first becomes visible from a heated object in darkened surroundings at the *Draper point* of 798 K (1436°R) (Draper 1847). At higher temperatures, additional wavelengths in the visible light spectral range appear. At a sufficiently high temperature, the visible

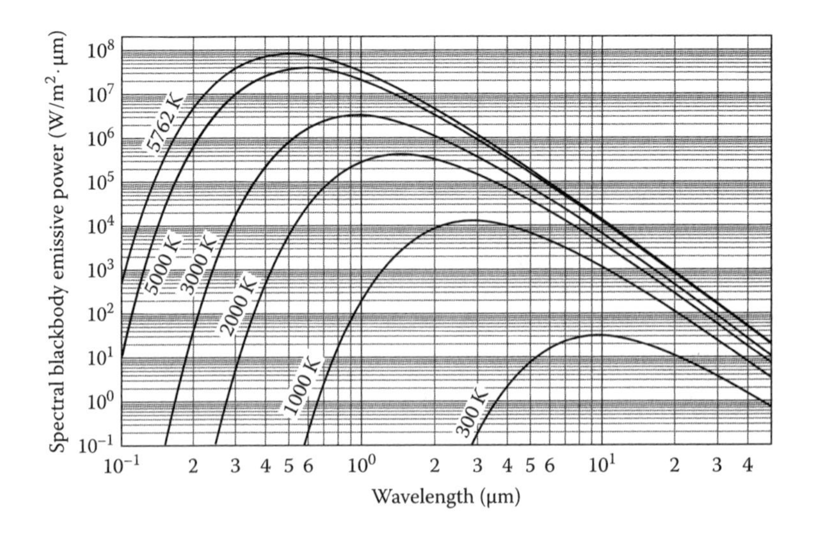

FIGURE 1.12 Hemispherical spectral emissive power, $E_{\lambda b}$, of a blackbody at several temperatures.

emitted light becomes white as it is composed of a mixture of all the visible wavelengths. The visible brightness of an object also increases substantially with its increased temperature.

It is also desirable to mention some tangible examples from our everyday experiences. For example, for the filament of an incandescent lamp to operate efficiently, the temperature must be high; otherwise, too much of the electrical energy is dissipated as infrared radiation rather than in the visible range. Most tungsten filament lamps operate at about 3000 K (5400°R) and thus do emit a large fraction of their energy in the infrared, but their filament vaporization rate limits the temperature to near this value. The sun emits a spectrum quite similar to that of a blackbody at 5780 K (10,400°R), and an appreciable amount of the solar energy is within what we call the visible range.

Interestingly, water absorbs poorly within the visible spectrum (Figure 1.2). Most of the biological systems on Earth have their vision systems (eyes) immersed in water. The coincidence of these two spectral responses may have been the reason why most mammalian eyes are responsive to the visible light spectrum. If the eyes were sensitive in other regions such as the infrared, so that we could see thermal images in the "dark" then our definition of "visible region of the spectrum" would change. Although human vision is well tuned to the solar spectrum, this is not true of all life on Earth. Turtles have eyes sensitive to infrared but not to blue; bees have eyes sensitive to ultraviolet but not red. If we find life in other solar systems having a sun with effective temperature different from ours, it will be interesting to discover what wavelength range encompasses the "visible spectrum" if the beings there posses sight.

It is important to have easy-to-use expressions to calculate the energy available in different wavelength intervals. For this, first we would like to obtain a single universal expression. Equation 1.15 can be placed in a convenient form that eliminates the need for a separate curve or table of values for each T. This is done by dividing by T^5 to obtain

$$\frac{E_{\lambda b}(T)}{T^5} = \frac{\pi I_{\lambda b}(T)}{T^5} = \frac{2\pi C_1}{n^2(\lambda T)^5\left[\exp\left(\frac{C_2}{n\lambda T}\right)-1\right]} \tag{1.20}$$

This gives $E_{\lambda b}(T)/T^5$ in terms of the single variable λT. The plot in Figure 1.13 replaces the multiple curves in Figure 1.12. Illustrative numerical values are shown in Table A.5 for $n = 1$.

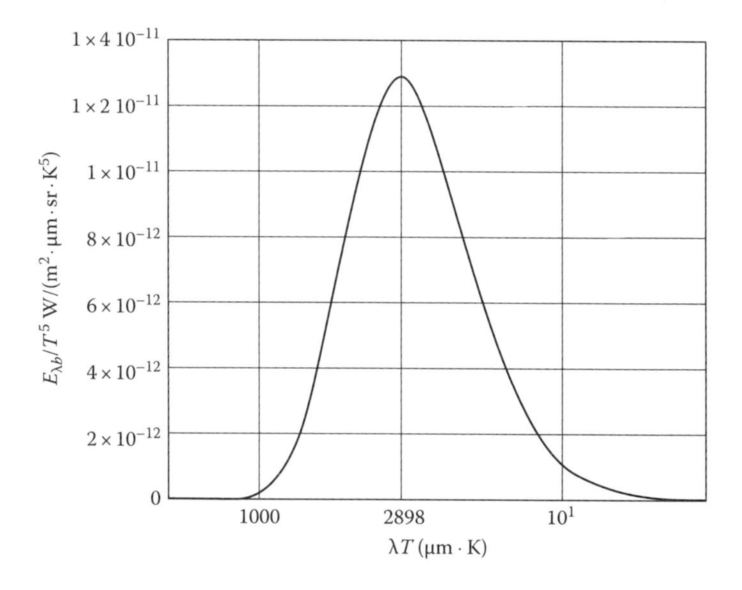

FIGURE 1.13 Spectral distribution of blackbody hemispherical emissive power as a function of λT.

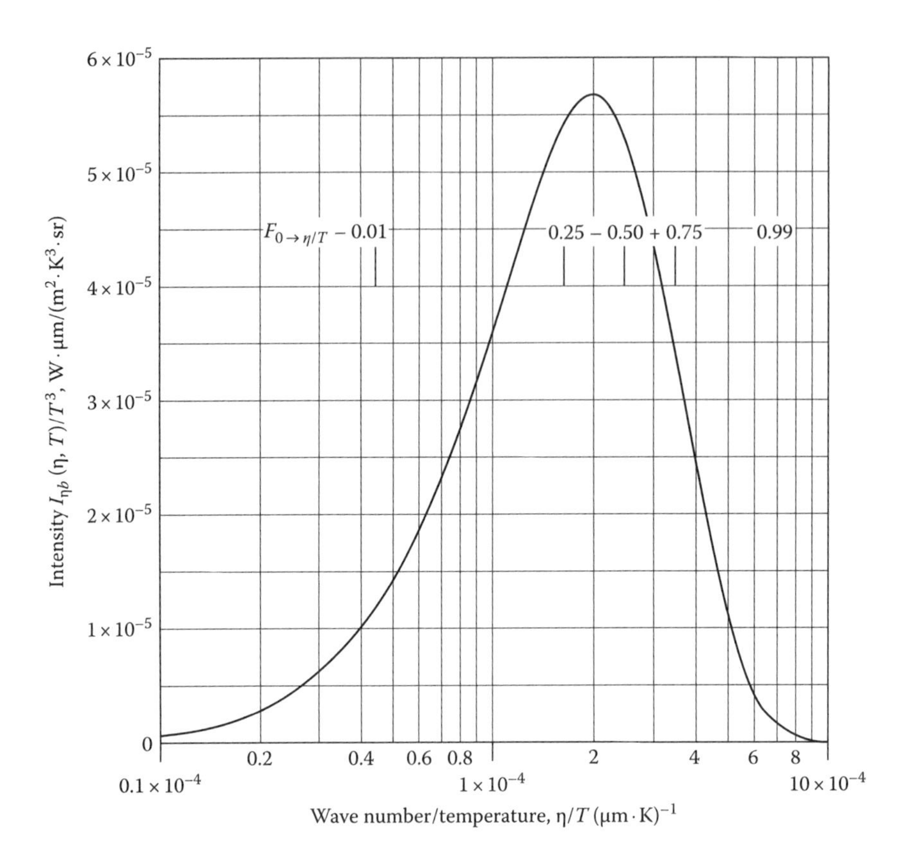

FIGURE 1.14 Spectral distribution of blackbody intensity as a function of η/T.

Equation 1.19 can also be placed in a more universal form in terms of the variable η/T ($= 1/\lambda T$), where η is in the medium:

$$\frac{I_{\eta b}(T)}{T^3} = \frac{2C_1\left(\dfrac{\eta}{T}\right)^3}{n^2\left[\exp\left(\dfrac{C_2\eta}{nT}\right)-1\right]} \tag{1.21}$$

This function is shown in Figure 1.14. Note again that Planck spectral distribution gives the maximum (blackbody) radiant intensity that a body can emit in vacuum at a given wavelength and temperature. This expression serves as a yardstick with which real surface performance can be compared.

1.5.9 Approximations for Blackbody Spectral Distribution

Some approximate forms of Planck distribution are occasionally useful in derivations because of their simplicity. Care must be taken to use them only in range where their accuracy is acceptable.

1.5.9.1 Wien's Formula

If the term $\exp(C_2/\lambda T)$ is much larger than 1, Equation 1.20 reduces to

$$\frac{I_{\lambda b}(T)}{T^5} = \frac{2C_1}{n^2(\lambda T)^5 \exp\left(\dfrac{C_2}{n\lambda T}\right)} \tag{1.22}$$

which is known as *Wien's formula*. It is accurate to within 1% for λT less than 3000 μm K (5400 μm·°R).

1.5.9.2 Rayleigh–Jeans Formula

Another approximation is found by expanding the exponential in the denominator of Equation 1.20 in a series (for $n = 1$) to obtain

$$e^{C_2/\lambda T} - 1 = 1 + \frac{C_2}{\lambda T} + \frac{1}{2!}\left(\frac{C_2}{\lambda T}\right)^2 + \frac{1}{3!}\left(\frac{C_2}{\lambda T}\right)^3 + \cdots - 1 \tag{1.23}$$

For λT much larger than C_2, the term $(e^{C_2/\lambda T} - 1)$ can be approximated by the single term $C_2/\lambda T$, and Equation 1.20 becomes

$$\frac{I_{\lambda b}(T)}{T^5} = \frac{2C_1}{C_2 \cdot (\lambda T)^4} \tag{1.24}$$

This is the *Rayleigh–Jeans formula* and is accurate to within 1% for λT greater than 7.78×10^5 μm·K (=14×10^5 μm·°R). This is well outside the range generally encountered in thermal radiation problems, since a blackbody emits over 99.9% of its energy at λT values below this.

1.5.10 Wien's Displacement Law

A useful quantity is the wavelength λ_{max} at which the blackbody intensity $I_{\lambda b}(T)$ is a maximum for a given temperature. This maximum shifts toward shorter wavelengths as the temperature is increased (Figure 1.12). The value of $\lambda_{max}T$ is at the peak of the distribution curve in Figure 1.13. It is obtained by differentiating Planck distribution in Equation 1.20 with respect to λT and setting the left side equal to zero. This gives the transcendental equation, if the index of refraction $n = 1$:

$$\lambda_{max}T = \frac{C_2}{5}\frac{1}{1 - e^{-C_2/(\lambda_{max}T)}} \tag{1.25}$$

The solution to this equation for $\lambda_{max}T$ is a constant

$$\lambda_{max}T = C_3 = 2897.7721\ \mu\text{m} \cdot \text{K} \tag{1.26}$$

which is one form of *Wien's displacement law*. Note that this value is roughly 3000 μm·K, a simple number to remember to have the back-of-the-envelope calculations to determine the peak emission wavelength or temperature in designing spectrally selective systems. Values of the constant C_3 in other units are listed in Table A.4. Equation 1.26 indicates that with increased temperature, maximum emissive power and the intensity shift to a shorter wavelength in inverse proportion to T.

The intensity at a given wavelength is found from Planck spectral distribution. Substituting Wien's displacement law Equation 1.26 into Equation 1.20 gives the maximum intensity as

$$I_{\lambda_{\max},b} = \frac{2C_1}{C_3^5\left(e^{C_2/C_3}-1\right)}T^5 = 4.09570\times10^{-12}\left(\frac{W}{m^2\cdot\mu m\cdot K^5}\right)T^5 = C_4T^5 \tag{1.27}$$

where

$C_4 = 4.09570 \times 10^{-12}$ W/m²·μm·K

$n = 1$

In Table A.4, C_4 is also listed in other units. Equation 1.27 shows that the maximum blackbody intensity, $I_{\lambda_{\max},b}$, increases as temperature to the *fifth power.*

Example 1.3

What temperature is needed for a blackbody to radiate its maximum intensity $I_{\lambda b}$ at the center of the visible spectrum?

Figure 1.4 shows that the visible spectrum spans the range λ = 0.4–0.7 μm with its center at 0.55 μm. From Equation 1.26,

$$T = \frac{C_3}{\lambda_{\max}} = \frac{2897.8\ \mu m\cdot K}{0.55\ \mu m} = 5269\ K\ (9484°R)$$

This is near the observed effective radiating surface temperature of the sun, which is 5780 K (10.400°R).

Example 1.4

At what wavelength do you observe the maximum emission from a blackbody at room temperature?

Using a room temperature of 21°C (69.8°F, 294 K), Wien's displacement law gives

$$\lambda_{\max} = \frac{C_3}{T} = \frac{2897.8\ \mu m\cdot K}{294\ K} = 9.86\ \mu m$$

which is in the middle of the near-infrared region. About 25% of radiative emission at room temperature is at wavelengths smaller than this value.

1.5.11 Total Blackbody Intensity and Emissive Power

The previous discussion provided the energy per unit wavelength interval that a blackbody radiates into vacuum at each wavelength. Now the *total intensity* is determined, where a "total" quantity includes radiation for all wavelengths.

The intensity emitted in the wavelength interval $d\lambda$ is $I_{\lambda b}(T)d\lambda$. Integrating over all wavelengths gives the *total blackbody intensity*:

$$I_b(T) = \int_0^\infty I_{\lambda b}(T)d\lambda \tag{1.28}$$

This is evaluated by substituting Planck distribution from Equation 1.15 with $n = 1$ and transforming variables using $\zeta = C_2/\lambda T$. Equation 1.28 then becomes

$$I_b = \int_0^\infty \frac{2C_1}{\lambda^5\left(e^{\frac{C_2}{\lambda T}} - 1\right)} d\lambda = \frac{2C_1T^4}{C_2^4}\int_0^\infty \frac{\zeta^3}{e^\zeta - 1} d\zeta \tag{1.29}$$

From a table of definite integrals, this is evaluated as

$$I_b = \frac{2C_1T^4}{C_2^4}\frac{\pi^4}{15} = \frac{\sigma}{\pi}T^4 \tag{1.30}$$

where the constant σ is

$$\sigma \equiv \frac{2C_1\pi^5}{15C_2^4} = 5.670373\times10^{-8}\ \mathrm{W/(m^2\cdot K^4)}$$

$$= 0.17123\times10^{-8}\ (\mathrm{Btu/h\cdot ft^2\cdot R^4}) \tag{1.31}$$

The *hemispherical total emissive power* of a blackbody radiating into vacuum is then

$$E_b = \int_0^\infty E_{\lambda b}d\lambda = \int_0^\infty \pi I_{\lambda b}d\lambda = \pi I_b = \sigma T^4 \tag{1.32}$$

which is the emitted blackbody energy flux (W/m²). Equation 1.32 is *Stefan–Boltzmann law,* and σ is the *Stefan–Boltzmann constant.*

Example 1.5

Energy emitted in the normal direction from a blackbody surface is found to have a total radiation per unit solid angle and per unit surface area of 10,000 W/(m²·sr). What is the surface temperature?

The hemispherical total emissive power is related to the total intensity in the normal direction by $E_b = \pi I_b$. Hence, from Equation 1.32, $T = (\pi I_b/\sigma)^{1/4} = (10{,}000\pi/5.67040\times10^{-8})^{1/4} = 862.7$ K

Example 1.6

A black surface is radiating with a hemispherical total emissive power of 20 kW/m². What is its surface temperature? At what wavelength is its maximum spectral intensity?

From Stefan–Boltzmann law, the temperature of the blackbody is $T = (E_b/\sigma)^{1/4} = (20{,}000/5.67040) \times 10^{-8})^{1/4} = 770.6$ K. Then, from Wien's displacement law, $\lambda_{max} = C_3/T = 2898/770.6 = 3.76\ \mu$m.

Example 1.7

An electric flat-plate heater that is square with a 0.1 m edge length is radiating 100 W from its face. If the heater can be considered black, what is its temperature?

Using the Stefan–Boltzmann law,

$$T = \left(\frac{Q}{A\sigma}\right)^{1/4} = \left[\frac{100\,\text{W}}{(0.1\text{m})^2 \times 5.67040 \times 10^{-8}\,\text{W/m}^2\cdot\text{K}^4}\right]^{1/4} = 648\,\text{K}$$

The spectral intensity of a black surface $I_{\lambda b}$ was shown in Equation 1.10 to be independent of the angle of emission. Integrating over all wavelengths did not, of course, change this angular independence. The intensity of a surface is what the eye interprets as "brightness" since it is the energy per unit of projected area that is the apparent viewed area. A black surface will exhibit the same brightness when viewed from any angle.

1.5.12 Blackbody Radiation within a Spectral Band

The hemispherical total emissive power of a blackbody radiating into vacuum is given by Stefan–Boltzmann law (Equation 1.32). In calculating radiative exchange between two objects, it is often necessary to determine the fraction of the total emissive power emitted in a wavelength band, as illustrated in Figure 1.15. This fraction is designated by $F_{\lambda_1 T \to \lambda_2 T}$ and is given by the ratio (the arrow in the subscript means "to")

$$F_{\lambda_1 T \to \lambda_2 T} = \frac{\int_{\lambda=\lambda_1}^{\lambda_2} E_{\lambda b}(T)\,d\lambda}{\int_{\lambda=0}^{\infty} E_{\lambda b}(T)\,d\lambda} = \frac{\int_{\lambda=\lambda_1}^{\lambda_2} E_{\lambda b}(T)\,d\lambda}{\sigma T^4} = \frac{1}{\sigma}\int_{\lambda T=(\lambda T)_1}^{(\lambda T)_2} \frac{2\pi C_1}{(\lambda T)^5\left[\exp(C_2/\lambda T)-1\right]}\,d(\lambda T) \quad (1.33)$$

The wavelength band integral in Equation 1.33 can be expressed by two integrals, each with a lower limit $\lambda = 0$, so that

$$F_{\lambda_1 T \to \lambda_2 T} = \frac{1}{\sigma T^4}\left[\int_0^{\lambda_2} E_{\lambda b}\,d\lambda - \int_0^{\lambda_1} E_{\lambda b}\,d\lambda\right] = F_{0\to\lambda_2 T} - F_{0\to\lambda_1 T} \quad (1.34)$$

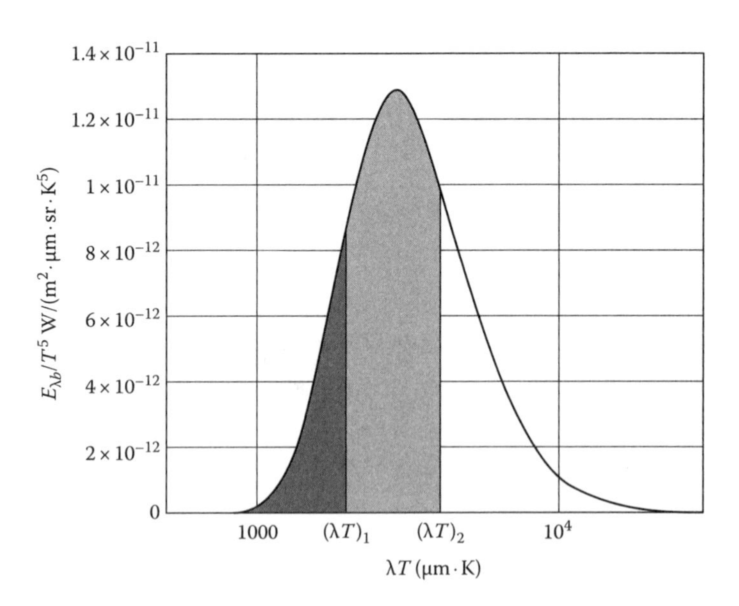

FIGURE 1.15 Emitted energy in wavelength band.

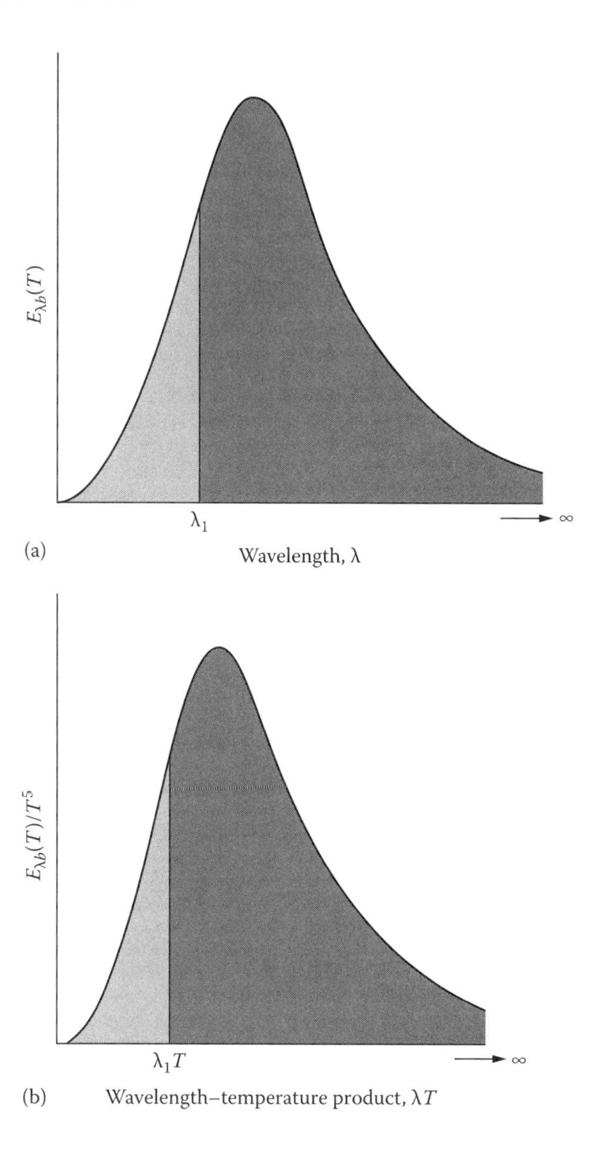

FIGURE 1.16 Physical representation of F factor, where $F_{0\to\lambda_1}$ or $F_{0\to\lambda_1 T}$ is the ratio of crosshatched to total shaded area, (a) in terms of curve for specific temperature (entire area under curve is σT^4) and (b) in terms of universal curve. Entire area under universal curve is σ.

The fraction of the emissive power for any wavelength band can therefore be found by having $F_{0\to\lambda_1 T}$ as a function of λ. The $F_{0\to\lambda_1 T}$ function is illustrated by Figure 1.16a, where it equals the crosshatched area divided by the total area (shaded) under the curve.

For a blackbody, the hemispherical emissive power is related to intensity (Equation 1.13) and the $F_{\lambda_1 T\to\lambda_2 T}$ function gives the fraction of the total intensity within the wavelength interval λ_1 to λ_2. Expression of the F function in terms of a single variable λT simplifies the calculations significantly (Figure 1.16b). This is done by rewriting Equation 1.34 as

$$F_{\lambda_1 T\to\lambda_2 T} = \frac{1}{\sigma}\left[\int_0^{\lambda_2 T} \frac{E_{\lambda b}}{T^5}\, d(\lambda T) - \int_0^{\lambda_1 T} \frac{E_{\lambda b}}{T^5}\, d(\lambda T)\right] = F_{0\to\lambda_2 T} - F_{0\to\lambda_1 T} \tag{1.35}$$

As shown by Equation 1.20, $E_{\lambda b}/T^5$ is a function only of λT, so the integrands in Equations 1.34 and 1.35 depend only on the λT variable. A convenient series form for $F_{0\to\lambda T}$ is found by using the substitution $\zeta = C_2/\lambda T$ to obtain, from Equation 1.29,

$$F_{0\to\lambda T} = \frac{2\pi C_1}{\sigma T^4}\int_0^{\lambda} \frac{d\lambda}{\lambda^5\left(e^{\frac{C_2}{\lambda T}}-1\right)} = \frac{2\pi C_1}{\sigma C_2^4}\int_{\zeta}^{\infty} \frac{\zeta^3}{e^{\zeta}-1}d\zeta$$

Using the definition of σ and the value of the definite integrals in Equations 1.29, this reduces to

$$F_{0\to\lambda T} = 1-\frac{15}{\pi^4}\int_0^{\zeta} \frac{\zeta^3}{e^{\zeta}-1}d\zeta \tag{1.36}$$

By using the series expansion $(e^{\zeta}-1)^{-1} = e^{-\zeta}+e^{-2\zeta}+e^{-3\zeta}+\cdots$ and then integrating by parts, the $F_{0\to\lambda T}$ becomes (Chang and Rhee 1984)

$$F_{0\to\lambda T} = \frac{15}{\pi^4}\sum_{m=1}^{\infty}\left[\frac{e^{-m\zeta}}{m}\left(\zeta^3+\frac{3\zeta^2}{m}+\frac{6\zeta}{m^2}+\frac{6}{m^3}\right)\right] \tag{1.37}$$

where $\zeta = C_2/\lambda T$. The series in Equation 1.37 converges very rapidly, and using the first three terms of the summation, m = 1–3, gives good results over most of the range of $F_{0\to\lambda T}$. As λT becomes large (small ζ), a larger number of terms is required. The series is useful in computer solutions, and using available computational software packages is usually more convenient than using tabulated results for $F_{0\to\lambda T}$. For computer solutions, direct numerical integration can also be used if a limited number of $F_{0\to\lambda T}$ values is needed. Illustrative $F_{0\to\lambda T}$ values from Equation 1.37 are in Table A.5. A plot of $F_{0\to\lambda T}$ as function of λT is in Figure 1.17.

When working with wave number or frequency, the fraction $F_{0\to\eta/T} = F_{0\to\nu/c_0T} = F_{0\to 1/\lambda T}$ is needed. The inverse relation between wave number and wavelength gives $F_{0\to\eta/T} = 1 - F_{0\to\lambda T}$. The use of the $F_{0\to\lambda T}$ function is illustrated in the examples that follow.

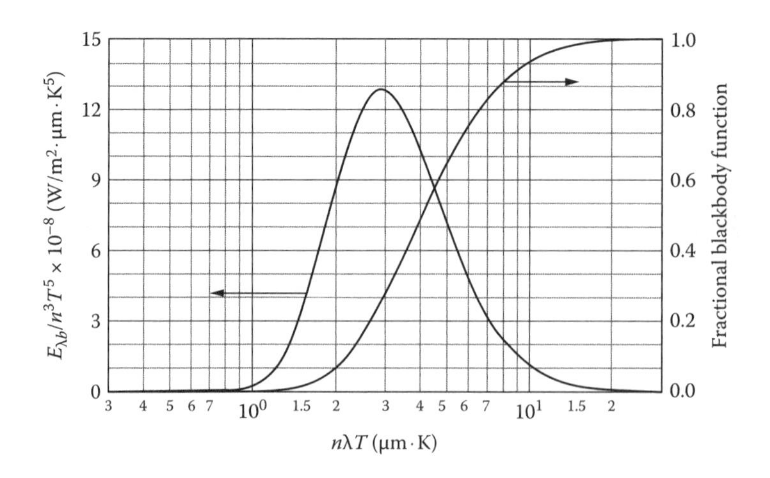

FIGURE 1.17 Fractional blackbody emissive power in the range 0 to λT.

Example 1.8

A blackbody is radiating at a temperature of 3000 K. An experimenter wants to measure total radiant emission by using a radiation detector. The detector absorbs all radiation in the range of $\lambda = 0.8$–5 μm but detects no energy outside that range. What percentage correction does the experimenter need to apply to the energy measurement? If the sensitivity of the detector could be extended in the range of λ by 0.5 μm at only one end of its sensitivity range, which end should be extended?

Using $\lambda_1 T = 0.8 \times 3000 = 2400$ μm·K and $\lambda_2 T = 5 \times 3000 = 15{,}000$ μm·K in Equation 1.37 gives the fraction of energy outside the sensitivity range as $F_{0\to\lambda_1 T} + F_{\lambda_2 T\to\infty} = F_{0\to\lambda_1 T} + \left(1 - F_{0\to\lambda_2 T}\right) = \left(0.14026 + 1 - 0.96893\right) = 0.17133$, or a correction of 17.1% of the total incident energy. Extending the sensitive range to the longer-wavelength side of the measurement interval adds little accuracy because of the small slope of the curve of F against λT in that region, so extending to shorter wavelengths would provide the greatest increase in the detected energy. The quantitative results to demonstrate this are found from the factors $F_{0\to 0.3\times 3000} = 0.00009$ and $F_{0\to 5.5\times 3000} = 0.97581$. The correction factors after extending the short- and long-wavelength boundaries of the interval are then 0.00009 + (1 – 0.96893) = 0.03116 and 0.14026 + (1 – 0.97581) = 0.16445.

Example 1.9

The experimenter of the previous example has designed a radiant energy detector that can be made sensitive over only a 1 μm range of wavelength. He or she wants to measure the total emissive power of two blackbodies, one at 2500 K and the other at 5000 K, and plans to adjust the 1 μm interval to give a 0.5 μm sensitive band on each side of the peak blackbody emissive power. For which blackbody would you expect to detect the greatest percentage of the total emissive power? What will the percentage be in each case?

Wien's displacement law tells us that the peak emissive power exists at $\lambda_{max} = 2897.8/T$ μm in each case. For the higher temperature, a wavelength interval of 1 μm will give a wider spread of λT values around the peak of $\lambda_{max}T$ on the normalized blackbody curve (Figure 1.13), so the measurement should be more accurate for the 5000 K case. For the 5000 K blackbody, $\lambda_{max} = 0.5796$ μm, and $\lambda_1 T = (0.5796 - 0.5000) \times 5000 = 398$ μm·K. Similarly, $\lambda_2 T = 1.0796 \times 5000 = 5398$ μm·K. The percentage of detected emissive power is then $100(F_{0\to 5398} - F_{0\to 398}) = 100 \times (0.6801 - 0) = 68.0\%$

A similar calculation for the 2500 K blackbody shows that 48.6% of the emissive power is detected.

Example 1.10

A light-bulb filament is at 3600 K. Assuming that the filament radiates as a blackbody, what fraction of its emission is in the visible region?

The visible region is within $\lambda = 0.4 \to 0.7$ μm. The desired fraction is then

$$F_{0\to 2520\mu m\cdot K} - F_{0\to 1440\mu m\cdot K} = 0.1657 - 0.0096 = 0.1561$$

This very low efficiency of converting energy into visible light is a compelling reason for replacing incandescent bulbs with more efficient lighting.

Some commonly used values of the $F_{0\to\lambda T}$ and $F_{0\to\eta/T}$ are listed in Table 1.1 for quick reference. It is interesting to note that very nearly one-fourth of the total emissive power lies in the wavelength range below the peak of Planck spectral distribution at any temperature.

TABLE 1.1
Fraction of Blackbody Emission Contained in the Region 0 to λT or 0 to η/T

λT				
μm·°R	μm·K	$F_{0\to\lambda T}$	η/T, (μm·K)$^{-1}$	$F_{0\to\eta/T}$
2,606	1,448	0.01	0.4369×10^{4}	0.01
$5{,}216 = \lambda_{max} T$	$2{,}898 = \lambda_{max} T$	0.25	1.627×10^{-4}	0.25
7,393	4,107	0.50	2.435×10^{-4}	0.50
11,067	6,148	0.75	3.451×10^{-4}	0.75
41,200	22,890	0.99	6.906×10^{-4}	0.99

1.5.13 Summary of Blackbody Properties

It has been shown that the blackbody posses certain fundamental properties that make it a standard for comparing real radiating bodies. These properties, listed here for convenience, are the following:

1. The blackbody is the maximum absorber and emitter of radiant energy at any wavelength and in any direction.
2. The total (including all wavelengths) radiant intensity and hemispherical total emissive power of a blackbody into a medium with constant index of refraction n are given by *Stefan–Boltzmann law*:

$$\pi I_b = E_b = n^2 \sigma T^4$$

3. The blackbody spectral and total intensities are independent of the direction so that emission of energy into a direction θ away from the surface normal direction is proportional to the projected area of the emitting element, $dA\cos\theta$. This is known as *Lambert's cosine law*.
4. The blackbody spectral distribution of intensity for emission into medium with refractive index n is given by *Planck distribution*:

$$\frac{E_{\lambda b}(T)}{\pi} = I_{\lambda b}(T) = \frac{2C_1}{n^2\lambda^5\left[\exp\left(\dfrac{C_2}{n\lambda T}\right) - 1\right]}$$

5. The wavelength at which the maximum spectral intensity of radiation into a medium with refractive index n for a blackbody occurs is given by *Wien's displacement law:*

$$\lambda_{max,n} = \frac{\lambda_{max,o}}{n} = \frac{C_3}{nT}$$

For vacuum or air, n will equal 1, so $\lambda_{max,vacuum} = \lambda_{max,o}$.

The definitions for blackbody properties introduced in this chapter are summarized in Table 1.2. The formulas for the quantities are given in terms of either the spectral intensity $I_{\lambda b}$, which is computed from Planck's law, or the surface temperature T.

TABLE 1.2
Blackbody Radiation Quantities (n = 1)

Symbol	Name	Definition	Geometry	Formula
$I_{\lambda b}(T)$	Spectral intensity	Emission in any direction per unit of projected area normal to that direction, and per unit time, wavelength interval about λ, and solid angle	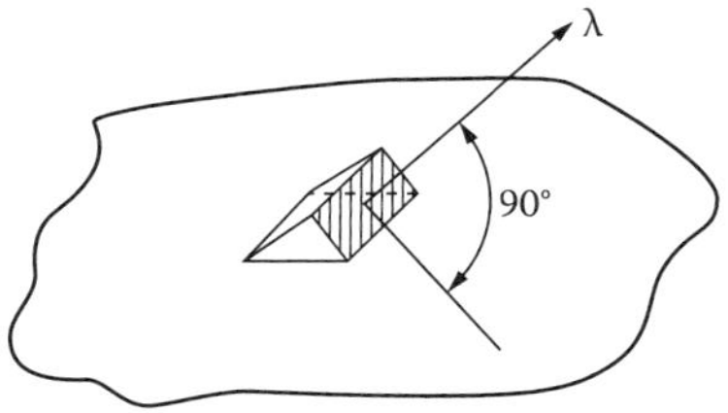	$2C_1/[\lambda^5(e^{C_2/\lambda T}-1)]$
$I_b(T)$	Total intensity	Emission, including all wavelengths, in any direction per unit of projected area normal to that direction, and per unit time and solid angle	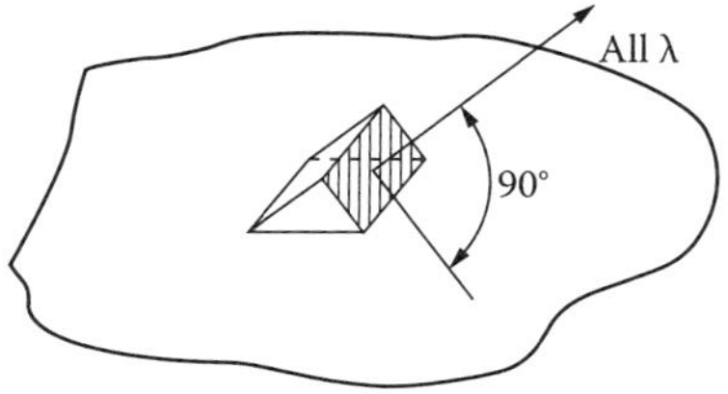	$\sigma T^4/\pi$
$E_{\lambda b}(\theta, T)$	Directional spectral emissive power	Emission per unit solid angle in direction θ per unit surface area, wavelength interval, and time	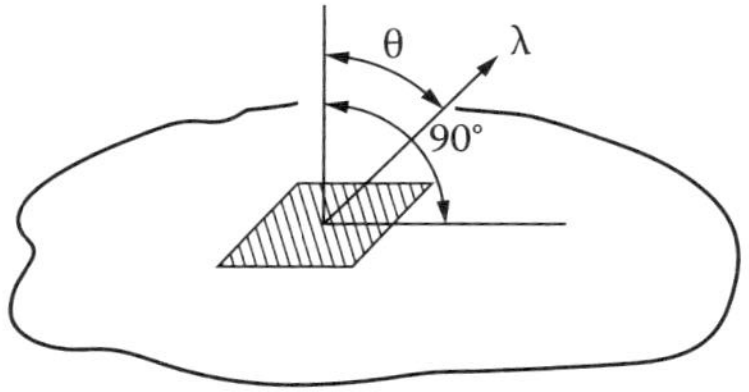	$I_{\lambda b}(T)\cos\theta$
$E_b(\theta, T)$	Directional total emissive power	Emission, including all wavelengths, in direction θ per unit surface area, solid angle, and time	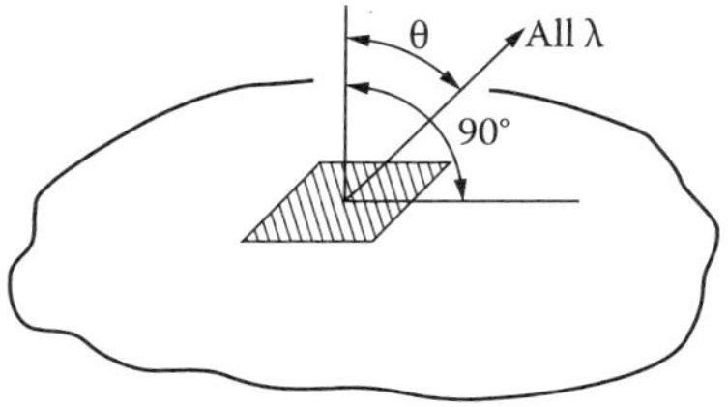	$(\sigma T^4/\pi)\cos\theta$

(Continued)

TABLE 1.2 (*Continued*)
Blackbody Radiation Quantities (n = 1)

Symbol	Name	Definition	Geometry	Formula
$E_{\lambda b}(\theta_1 \to \theta_2, \phi_1 \to \phi_2, T)$	Finite solid-angle spectral emissive power	Emission in solid angle $\theta_1 \le \theta \le \theta_2$, $\phi_1 \le \phi \le \phi_2$ per unit surface area, wavelength interval, and time	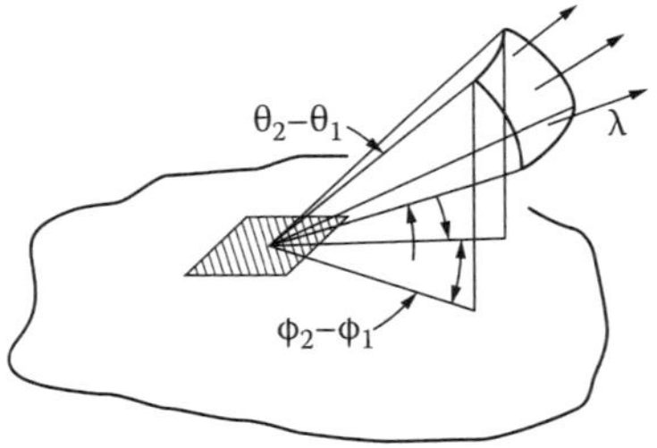 $\theta_2-\theta_1$, λ, $\phi_2-\phi_1$	$I_{\lambda b}(T)[(\sin^2\theta_2 - \sin^2\theta_1)/2] \times (\phi_2 - \phi_1)$
$E_b(\theta_1 \to \theta_2, \phi_1 \to \phi_2, T)$	Finite solid-angle total emissive power	Emission, including all wave lengths, in solid angle $\theta_1 \le \theta \le \theta_2$, $\phi_1 \le \phi \le \phi_2$ per unit surface area and time.	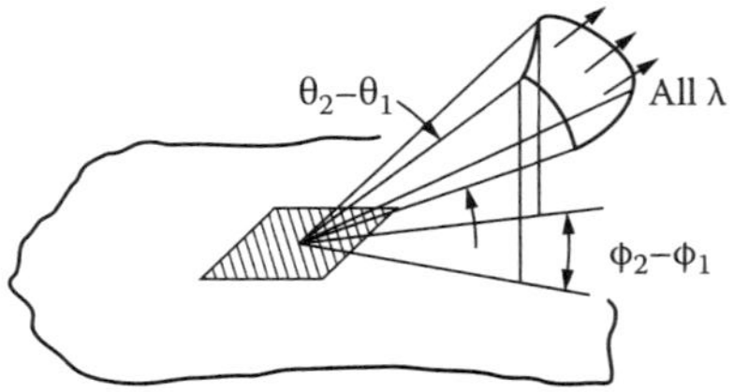 $\theta_2-\theta_1$, All λ, $\phi_2-\phi_1$	$\dfrac{\sigma T^4}{\pi}\dfrac{\sin^2\theta_2 - \sin^2\theta_1}{2}(\phi_2 - \phi_1)$
$E_{\lambda b}(\lambda_1 \to \lambda_2, \theta_1 \to \theta_2, \phi_1 \to \phi_2, T)$	Finite solid-angle band emissive power	Emission in solid angle $\theta_1 \le \theta \le \theta_2$, $\phi_l \le \phi \le \phi_2$ and wavelength band $\lambda_1 \to \lambda_2$ per unit surface area and time	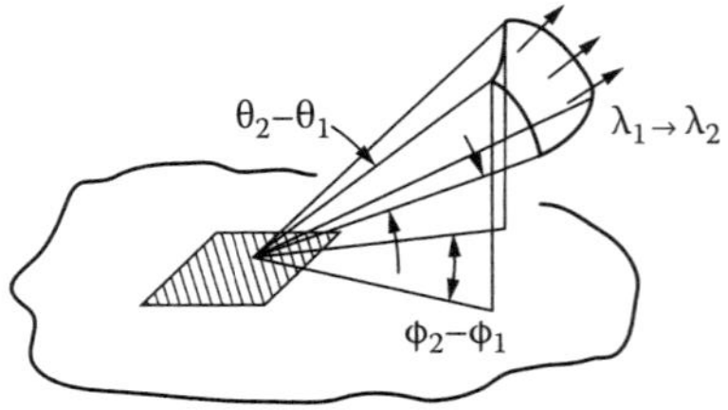 $\theta_2-\theta_1$, $\lambda_1 \to \lambda_2$, $\phi_2-\phi_1$	$\dfrac{\sigma T^4}{\pi}\dfrac{\sin^2\theta_2 - \sin^2\theta_1}{2}(\phi_2 - \phi_1) \times (F_{0\to\lambda_2} - F_{0\to\lambda_1})$

(*Continued*)

TABLE 1.2 (*Continued*)
Blackbody Radiation Quantities ($n = 1$)

Symbol	Name	Definition	Geometry	Formula
$E_{\lambda b}(T)$	Hemispherical spectral emissive power	Emission into hemispherical solid angle per unit surface area, wavelength interval, and time	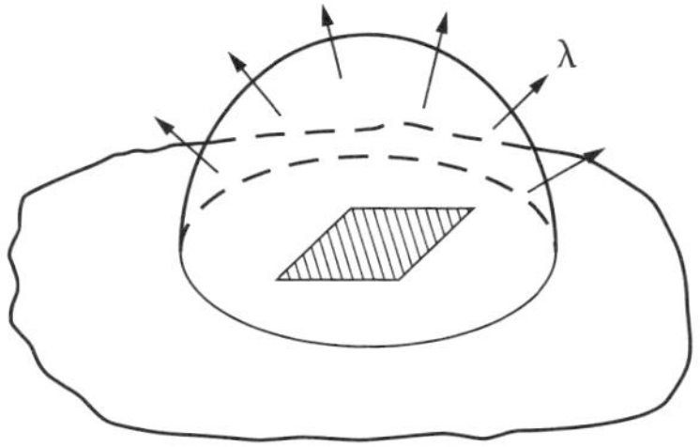 λ	$\pi I_{\lambda b}(T)$
$E_{\lambda b}(\lambda_1 \rightarrow \lambda_2, T)$	Hemispherical band emissive power	Emission in wavelength band $\lambda_1 \rightarrow \lambda_2$ into hemispherical solid angle per unit surface area and time	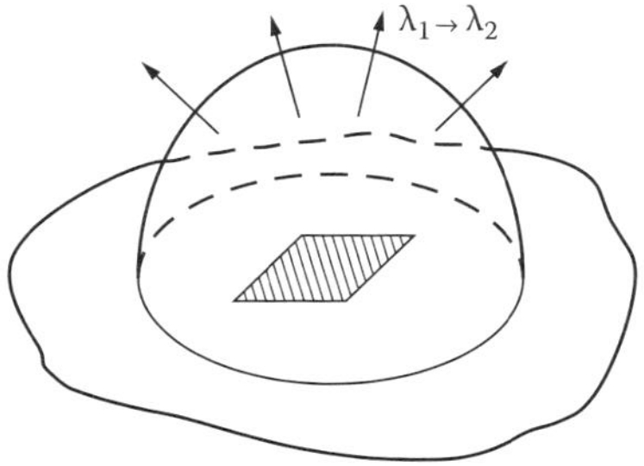 $\lambda_1 \rightarrow \lambda_2$	$(F_{0\rightarrow\lambda_2} - F_{0\rightarrow\lambda_1})\sigma T^4$
$E_b(T)$	Hemispherical total emissive power	Emission, including all wavelengths, into hemispherical solid angle per unit surface area and time	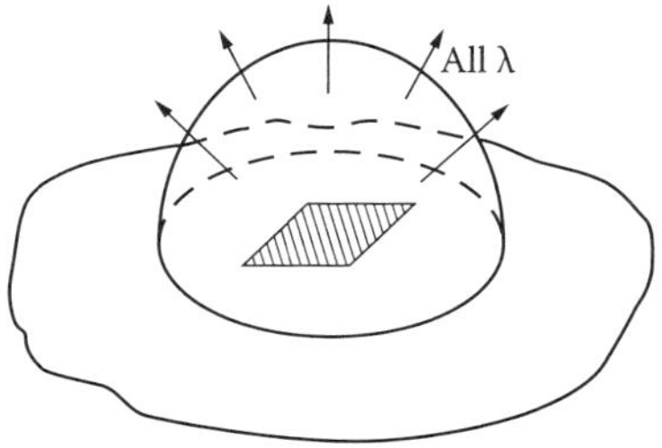 All λ	σT^4

1.6 RADIATIVE ENERGY ALONG A LINE-OF-SIGHT

When we introduced the discussion about the RTE, we mentioned that the radiation intensity can be attenuated and augmented as it propagates along a path. Now, let us return to our discussion of radiative transfer as depicted in Figure 1.5. Here, we discuss the fate of radiative transfer in a medium as it propagates after emission from a surface.

1.6.1 Radiative Energy Loss due to Absorption and Scattering

Consider spectral radiation of intensity I_λ incident on a volume element of length dS along the direction of propagation S. As radiation passes through dS, its intensity is reduced by absorption and scattering. This reduction is also proportional to the magnitude of the local intensity, as can be demonstrated by experiments. The decrease in radiation intensity can be expressed after introducing a medium property β_λ:

$$dI_\lambda(S,\Omega) = -\beta_\lambda(S)I_\lambda(S,\Omega)dS \tag{1.38}$$

where β_λ is called the spectral *extinction* or *attenuation coefficient* of the medium and includes losses due to both absorption and scattering. β_λ is a physical property, has units of reciprocal length, and depends on the local properties of the medium, including temperature T, pressure P, composition of the material (indicated here in terms of the concentrations C_i of the i components), and wavelength of the incident radiation. Therefore, it depends on local parameters, as indicated by S dependency, and in explicit form, it is expressed as $\beta_\lambda(S) = \beta_\lambda(T,P,C_i)$. If β_λ in a given medium is constant, it can be interpreted as the inverse of the mean penetration distance (often called the mean free path) of radiation, as will be shown later.

The extinction coefficient consists of two parts, the *absorption* coefficient κ_λ and the *scattering* coefficient $\sigma_{s,\lambda}$:

$$\beta_\lambda = \kappa_\lambda + \sigma_{s,\lambda} \tag{1.39}$$

The subscript s on the scattering coefficient is included to avoid confusion with Stefan–Boltzmann constant. All these parameters have units of reciprocal length and are called *linear,* or *volumetric, coefficients.* Absorption describes how the radiative energy is converted to the internal energy of the matter. It is one of two mechanisms coupling the radiative energy propagation with the thermodynamic state of the matter. The other is radiative emission, which is proportional to the internal energy of matter and to its temperature. Scattering, on the other hand, causes a change in the direction of radiation propagating along S. During a scattering event, no radiative energy is converted to thermal energy (i.e., scattering in radiation is considered to be perfectly elastic except in some very special cases). However, scattering changes the balance of the energy propagating in a given direction.

In a nonparticipating medium, β_λ is zero, and there would be no change in intensity along a path. In a translucent, or participating, medium, however, the initial radiative intensity changes as it propagates along a line of sight. In order to determine how the energy decreases, we can integrate Equation 1.38 over a finite path length S to obtain

$$\int_{I_\lambda=I_\lambda(0)}^{I_\lambda(S)} \frac{dI_\lambda}{I_\lambda} = -\int_{S^*=0}^{S} \beta_\lambda(S^*)dS^* \tag{1.40}$$

where

$I_\lambda(0)$ is the intensity incident on the control volume at $S = 0$

S^* is a dummy variable of integration

Carrying out the integration yields the intensity at location S for a given direction Ω

$$I_\lambda(S) = I_\lambda(0)\exp\left[-\int_0^S \beta_\lambda(S^*)dS^*\right] \tag{1.41}$$

Equation 1.41 is known as *Bouguer's, Beer's,* or *Lambert–Bouguer's* law.* This law shows that the intensity of spectral radiation is attenuated exponentially as it passes through an absorbing–scattering medium. It has extensive use in many fields, from astrophysics to spectroscopy. Note that Bouguer's law does not account for local emission or scattering into direction S.

If the medium is only absorbing, then the intensity will be converted to internal energy of the matter along the line of sight of the beam without being redirected to other directions due to scattering. Neglecting scattering in Equation 1.41, we obtain

$$\frac{dI_\lambda}{dS} = -\kappa_\lambda(S)I_\lambda \tag{1.42}$$

which again is in the direction Ω. This expression can be integrated to determine the intensity leaving the boundary at $S = 0$ after it travels along the path S:

$$I_\lambda(S) = I_\lambda(0)\exp\left[-\int_0^S \kappa_\lambda(S^*)dS^*\right] \tag{1.43}$$

It can be further simplified if the absorption coefficient is uniform in the medium:

$$I_\lambda(S) = I_\lambda(0)\exp(-\kappa_\lambda S) \tag{1.44}$$

We can obtain a similar expression if the medium is scattering only. Then, κ_λ is zero and $\beta_\lambda = \sigma_{s,\lambda}$. The resulting expressions are the same as Equations 1.42 through 1.44 where κ_λ is replaced with $\sigma_{s,\lambda}$:

$$\frac{dI_\lambda}{dS} = -\sigma_{s,\lambda}(S)I_\lambda \tag{1.45}$$

which can be integrated to determine the intensity leaving the medium after it travels the distance S along Ω.

$$I_\lambda(S) = I_\lambda(0)\exp\left[-\int_0^S \sigma_{s,\lambda}(S^*)dS^*\right] \tag{1.46}$$

The S dependence of $\sigma_{s\lambda}(S)$ indicates that medium properties are changing from location to location.

* Named after Pierre Bouguer (bōo'gâr) (1698–1758) who first showed on a quantitative basis how light intensities could be compared. Equation 1.41 is sometimes called Lambert's law or Bouguer–Lambert law. To avoid confusion with Lambert's cosine law, Equation 1.41 is referred to here as Bouguer's law.

For a *homogenous scattering medium,* $\sigma_{s,\lambda}$ is spatially constant, and Equation 1.44 becomes

$$I_\lambda(S) = I_\lambda(0)\exp(-\sigma_{s,\lambda}S) \tag{1.47}$$

Equations 1.44 and 1.47 are the same as Bouguer's law discussed earlier, where either the absorption coefficient κ_λ or the scattering coefficient $\sigma_{s,\lambda}$ is substituted for the extinction coefficient β_λ. Note that, in writing these expressions, we implicitly assume that there are no multiple scattering effects. Otherwise, some energy scattered out of the beam path along S can find its way back to the same direction after multiple scattering events. This requires a more thorough analysis of the conservation of radiative energy and yields the integrodifferential RTE, as discussed later.

1.6.2 Mean Penetration Distance

We can use the previous analysis to find out how much radiation would penetrate into a medium. Starting from Equation 1.38, the fraction of the radiation attenuated in the small volume element from S to $S + dS$ is expressed as

$$-\frac{dI_\lambda(S)}{I_\lambda(0)} = \beta_\lambda(S)\frac{I_\lambda(S)}{I_\lambda(0)}dS = \beta_\lambda(S)\exp\left[-\int_0^S \beta_\lambda(S^*)dS^*\right]dS \tag{1.48}$$

A mean penetration distance of the radiation, l_m, is obtained by multiplying the fraction absorbed at S by the distance S and integrating over all path lengths from $S = 0$ to ∞:

$$l_m = \int_{S=0}^{\infty} S\beta_\lambda(S)\exp\left[-\int_0^S \beta_\lambda(S^*)dS^*\right]dS \tag{1.49}$$

If β_λ is uniform along the path, carrying out the integration yields

$$l_m = \beta_\lambda \int_{S=0}^{\infty} S\exp(-\beta_\lambda S)dS = \frac{1}{\beta_\lambda} \tag{1.50}$$

demonstrating that the average penetration distance is the reciprocal of β_λ. This is a very useful expression for the interpretation of the experimental measurements in many applications. Similar penetration depths can be obtained for absorbing-only or scattering-only media by replacing the extinction coefficient β_λ with either the absorption coefficient κ_λ or the scattering coefficient $\sigma_{s,\lambda}$.

In many fields, for example, in spectroscopy, Bouguer's law can be written using *mass extinction, absorption,* and *scattering coefficients.* In this case,

$$\beta_{\lambda,m} = \kappa_{\lambda,m} + \sigma_{s,\lambda,m} = \frac{\beta_\lambda}{\rho} = \frac{\kappa_\lambda}{\rho} + \frac{\sigma_{s,\lambda}}{\rho} \tag{1.51}$$

where ρ is the local density of the absorbing and scattering medium. Mass coefficients have units of area per unit mass and are analogous to the concept of cross sections in molecular physics. Because the extinction coefficient β_λ increases as the density of the absorbing or scattering species is changed, the $\beta_{\lambda,m} = \beta_\lambda/\rho$ is more uniform with density changes than β_λ.

1.6.3 Optical Thickness

The exponential factor in Equation 1.41 is often written by defining a useful dimensionless quantity, $\tau_\lambda(S)$, called the *optical thickness* or *opacity* along the path length S:

$$\tau_\lambda = \int_0^S \beta_\lambda(S^*)dS^* \tag{1.52}$$

For a medium with uniform properties, Equation 1.52 is reduced to

$$\tau_\lambda(S) = \beta_\lambda S \tag{1.53}$$

Then, Equation 1.41 becomes

$$I_\lambda(S) = I_\lambda(0)\exp\left[-\tau_\lambda(S)\right] \tag{1.54}$$

which, if $\tau_\lambda(S)$ is given by Equation 1.52, is an integral function accounting for all of the values of β_λ between 0 and S. Because β_λ is a function of the local parameters P, T, and C_i, the optical thickness is also a function of them along the path between 0 and S. The optical thickness indicates how strongly a medium attenuates radiation at a given wavelength. If $\tau_\lambda(S) \ll 1$, the medium is called *optically thin*, whereas if $\tau_\lambda \gg 1$, the medium is *optically thick*. With decreasing optical thickness, the participating medium approaches a nonparticipating one; on the other hand, with increasing τ_λ, less and less energy is transferred and the medium eventually becomes *opaque*.

The optical thickness can also be given as the number of mean penetration distances within the path length, l_m:

$$\tau_\lambda(S) = \frac{S}{l_m} \tag{1.55}$$

The notation for the optical thickness τ_λ should not be confused with the surface transmissivity defined in Chapter 2, which uses the same symbol τ.

1.6.4 Radiative Energy Gain due to Emission

As radiative energy propagates in a medium along a given direction, its strength increases due to volumetric emission from every point on its path. If we denote the strength of the emission at a given location along S as $j_\lambda(S,\theta,\phi)$, then the incremental increase in radiative energy due to emission is

$$dI_\lambda(\theta,\phi) = j_\lambda(S,\theta,\phi)dS \tag{1.56}$$

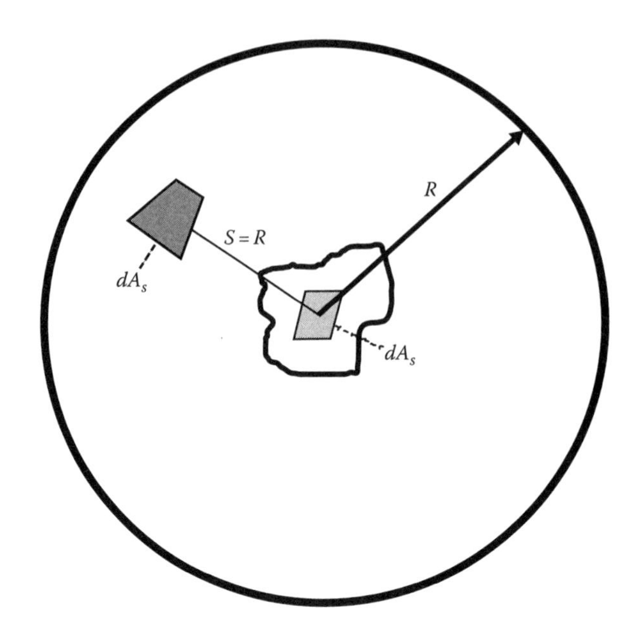

FIGURE 1.18 Emission from a volume element.

Here $j_\lambda(S,\theta,\phi)$ is any function representing the physics of the emission mechanism and has units of W/m³·μm·sr. Under LTE, the thermal radiation strength is directly related to how much energy is absorbed by the medium and is correlated with Planck blackbody function:

$$j_\lambda\left(S,\theta,\phi\right) = \kappa_\lambda I_{\lambda b}(S) \tag{1.57}$$

In some radiation literature, j_λ is called the emission coefficient.

For most practical engineering problems, the radiative emission can be considered isotropic, and does not have (θ,ϕ) dependence. With the advances in micro- and nanomanufacturing techniques, however, novel structures are currently built to obtain directional emission (Greffet et al. 2002). We discuss such developments in Chapters 16 and 17.

For a detailed analysis of the emission mechanism, consider an elemental volume dV with a spectral absorption coefficient κ_λ, which can be a function of local temperature, pressure, and species concentration (T,P,C_i). Let dV be at the center of a large black hollow sphere of radius R at uniform temperature T (Figure 1.18). Assume the space between dV and the sphere boundary is filled with nonparticipating (transparent) material. Then, the spectral intensity incident at the location of dA_s on dV, from element dA on the enclosure surface (at $S = 0$), is

$$I_\lambda\left(0\right) = I_\lambda\left(S=0\right) = I_{\lambda b}\left(S=0\right) = I_{\lambda b}(0) \tag{1.58}$$

The change of this intensity in dV as a result of absorption is

$$-I_\lambda(0)\kappa_\lambda dS = -I_{\lambda b}(0)\kappa_\lambda dS \tag{1.59}$$

The energy absorbed by the differential subvolume $dV = dSdA_s$ is $I_{\lambda b}(S = 0)\kappa_\lambda dSdA_s d\lambda d\Omega$, where $d\Omega = dA/R^2$ is the solid angle subtended by dA when viewed from dV and dA_s and dA is the projected area normal to $I_\lambda(0)$. The energy emitted by dA and absorbed by all of dV is found by integrating over dV (over all dA_s *and* dA elements) to obtain $\kappa_\lambda I_{\lambda b} dV d\lambda d\Omega$. To account for all energy incident upon dV from the entire spherical enclosure, integration over all solid angles gives $4\pi\kappa_\lambda I_{\lambda b} dV d\lambda$.

To maintain equilibrium in the enclosure, dV at location S must emit energy equal to that absorbed. Hence, the *spectral emission* by an isothermal volume element is

$$4\pi\kappa_\lambda I_{\lambda b}(S)dVd\lambda = 4\kappa_\lambda E_{\lambda b}(S)dVd\lambda \tag{1.60}$$

This is a very important result for emission within a volume. The shape of dV is arbitrary, but small enough that energy emitted within dV escapes before any reabsorption occurs within dV. The emitting material must be in thermodynamic equilibrium with respect to its internal energy; we always assume that the system is in LTE.

For all conditions discussed here, the spontaneous thermal emission is uniform over all directions (isotropic), so the spectral intensity emitted by a volume element into any direction, $dI_{\lambda,e}$, is obtained by dividing by $4\pi d\lambda$ and the cross-sectional area to give

$$dI_{\lambda,e}(S) = \kappa_\lambda(S)I_{\lambda b}(S)dS = j_\lambda(S,\theta,\varphi)dS \tag{1.61}$$

If the refractive index of the medium is $n \neq 1$, the spectral volumetric emission is modified by a factor of n_λ^2:

$$dI_{\lambda,e}(S) = n_\lambda^2(S)\kappa_\lambda(S)I_{\lambda b}(S)dS \tag{1.62}$$

1.6.5 Radiative Energy Density and Radiation Pressure

Radiation traversing a volume element dV produces a radiative energy density within the element. The radiative energy per unit volume at a particular frequency, the *spectral radiant energy density*, U_ν, is given by integrating the intensity passing through an element at location $\boldsymbol{r}$ to give

$$U_\nu(\nu,\mathbf{r}) = \frac{1}{c}\int_{\Omega=0}^{4\pi} I_\nu(\mathbf{r},S)d\Omega = \frac{4\pi}{c}\bar{I}_\nu(\mathbf{r}) \tag{1.63}$$

where $\bar{I}_\nu(\mathbf{r})$ is the local mean spectral intensity.

When the energy is emitted from a surface, an opposite reaction force provides a pressure on a surface due to the emission from it. The pressure on a body from a surrounding isotropic blackbody radiation field is expressed using the photon concept (Feynman 1963):

$$P = \frac{U}{3} = \frac{4\pi}{3c}I_b = \frac{4}{3c}E_b \tag{1.64}$$

The normal pressure exerted because of emission from a surface with emissivity ε is (Adhya 2005):

$$P = \frac{2\varepsilon\pi}{3c}I_b = \frac{2\varepsilon}{3c}E_b = \frac{2\varepsilon\sigma T^4}{3c} \tag{1.65}$$

Because of the speed of light c in the denominator of Equations 1.64 and 1.65, radiative pressures are usually small in comparison with pressures caused by molecular interactions.

1.6.6 Radiative Energy Gain due to In-Scattering

If radiative energy is scattered away from the original direction of radiative energy, it constitutes a "loss" that is expressed by an "out-scattering" term in the RTE. If the scatterers (particles, molecules, impurities, voids, or any inhomogeneities) in the small volume element direct the energy coming from other angles to the direction of the original beam, then it is a "gain" for the radiative energy propagating in the direction *S*; this is considered as the "in-scattering" term.

Consider the radiation within solid angle $d\Omega_i$, that is incident on *dA*, as shown in Figure 1.7. The portion of the incident intensity scattered away within the increment *dS* is $dI_\lambda(S) = -\sigma_{s,\lambda} I_\lambda dS$ (Equation 1.45). More specifically, the term $dI_\lambda(S)$ is the spectral energy scattered within path length *dS* per unit incident solid angle and area normal to the incident direction. This scattered energy is a function of the zenith and azimuthal angles (θ, ϕ) measured relative to the direction (θ_i, ϕ_i) of the incident intensity. The scattered intensity in direction θ, ϕ is defined as the energy scattered in that direction per unit solid angle of the scattered direction and per unit normal area and solid angle of the incident radiation:

$$dI_\lambda(S) = dI_\lambda(\theta,\phi) = \frac{\text{Spectral energy scattered in direction } (\theta,\phi)}{d\Omega_s dA d\Omega_i d\lambda} \tag{1.66}$$

Now we introduce the concept of the scattering phase function, which is a measure of the amount of radiative energy propagating in incident direction $\Omega_i(\theta_i, \phi_i)$ that is redirected into $\Omega(\theta, \phi)$ and can be expressed as

$$\Phi\left[\Omega_i(\theta_i,\phi_i),\Omega(\theta,\phi)\right] = \Phi(\Omega_i,\Omega) = \Phi\left[(\theta_i,\phi_i),(\theta,\phi)\right] \tag{1.67}$$

The angular profile of $\Phi\left[(\theta_i,\phi_i),(\theta,\phi)\right]$ is directly related to the size, shape, and material properties of the scatterer and varies with the wavelength of the incident radiation. In Chapter 15, we provide a more rigorous derivation of the scattering phase function starting from Maxwell equations.

Using the definition of the phase function, the directional magnitude of $dI_\lambda(S) = dI_\lambda(\theta, \phi)$ is related to the entire intensity $I_\lambda(S)$ scattered away from the incident radiation as

$$dI_{\lambda,s}(\theta,\phi) = \frac{1}{4\pi}\Phi_\lambda(\theta_i,\phi_i,\theta,\phi)\, dI_\lambda(S) = \frac{\sigma_{s,\lambda}}{4\pi}\Phi_\lambda(\theta_i,\phi_i,\theta,\phi)\, I_\lambda(S)\, dS \tag{1.68}$$

The spectral energy scattered into all $d\Omega$, per unit $d\lambda$, $d\Omega_i$, and *dA* is

$$dI_{\lambda,s} = \int_{\Omega_s=0}^{4\pi} dI_{\lambda,s}(\theta,\phi)\, d\Omega_s \tag{1.69}$$

Using Equation 1.66 to eliminate $dI_{\lambda,s}$ in Equation 1.68 yields the expression for the phase function as

$$\Phi(\theta,\phi) = \frac{4\pi\, dI_{\lambda,s}(\theta,\phi)}{dI_{\lambda,s}} = \frac{dI_{\lambda,s}(\theta,\phi)}{(1/4\pi)\int_{\Omega_s=0}^{4\pi} dI_{\lambda,s}(\theta,\phi)\, d\Omega_s} \tag{1.70}$$

Thus $\Phi(\theta,\phi)$ is the scattered intensity in a direction (θ,ϕ), divided by the mean intensity scattered in all directions. Integrating Equation 1.68 over all possible directions allows the normalization of $\Phi(\theta,\phi)$ such that

$$\frac{1}{4\pi}\int_{\Omega_s=0}^{4\pi}\Phi_\lambda(\theta,\phi)\,d\Omega_s=\frac{1}{4\pi}\int_0^{2\pi}\int_0^{\pi}\Phi_\lambda(\theta,\phi)\sin\theta\,d\theta\,d\phi=1 \tag{1.71}$$

If the scattered intensity is *isotropic*, that is, the same for all scattered directions, $\Phi(\theta,\phi)$ is equal to unity. Isotropic scattering is a convenient mathematical approximation, which significantly simplifies the radiative transfer calculations. However, it should be used with care as there are no isotropic single scatterers in nature; on the other hand, multiple scattering systems can often be adequately approximated by isotropic scattering. Examples of multiple scattering systems include sand, flour, granulated sugar, and some types of clouds.

1.7 RADIATIVE TRANSFER EQUATION

So far we have discussed the change in radiative energy in absorbing and scattering media. We have introduced the concepts of radiation intensity, absorption, scattering and extinction coefficients, and the scattering phase function. We have also provided a formulation for the attenuation of the radiative energy of an energy beam (like a laser beam) along the direction of its propagation. This theoretical background is sufficient for description of many practical problems and the working principles of several diagnostic tools where a single beam is incident on a medium. Equipped with this formulation, we can now determine the amount of radiative energy reaching a surface element or a detector at the observation location.

If the medium is scattering or if there are several sources of radiative energy incident on it (like several laser beams arriving from different directions), then the recorded signal at the observation point has the combined effect of all of these energy sources. Similarly, if the radiant energy is emitted by the walls of the enclosure, then each emitting surface contributes to the detected signal, and each volume element dV redirects energy incident toward the detector, regardless of the original direction of radiative energy. In addition, each beam arrives at other surface elements after being scattered by the absorbing and emitting particles and gases in the medium or after reflection by other surfaces. The combined effect of these multiple beams needs to be tallied for calculating the net radiative exchange. In addition, the emission process effectively "cools" each surface element or volume element in the medium. These complex mechanisms are interrelated and take place practically simultaneously, as the radiative energy propagates with the speed of light.

To be able to express this series of events, we write the conservation of radiative energy along a line of sight for a given wavelength. This conservation equation constitutes the radiative transfer equation (RTE), which is a fundamental relation of radiation heat transfer. In deriving the RTE, we consider a small volume element dV with macroscopic parameters such as absorption, scattering, and extinction coefficients, κ_λ, $\sigma_{s,\lambda}$, and β_λ, and the scattering phase function $\Phi(\theta_i, \phi_i, \theta, \phi)$. Also, we consider emission by the matter within each small volume element. Now, we can write conservation of radiative energy within a small volume element dV for a small path increment dS along S within a small solid angle $d\Omega$ around the direction (θ,ϕ) for a small time interval dt at time t and within a small wavelength interval $d\lambda$ around a given wavelength λ as

Change in radiative energy = gain due to emission − loss due to absorption
− loss due to out-scattering + gain due to in-scattering

This conservation relation is expressed mathematically as

$$
\begin{aligned}
&I_\lambda(S+dS,\Omega,t+dt)-I_\lambda(S,\Omega,t)\\
&=\kappa_\lambda I_{\lambda b}(S,t)dS-\kappa_\lambda I_\lambda(S,\Omega,t)dS\\
&-\sigma_{s,\lambda}I_\lambda(S,\Omega,t)dS+\frac{1}{4\pi}\int_{\Omega_i=4\pi}\sigma_{s,\lambda}I_\lambda(S,\Omega_i,t)\Phi_\lambda(\Omega_i,\Omega)d\Omega_i dS
\end{aligned}
\tag{1.72}
$$

We can recast this expression into a differential form:

$$
\begin{aligned}
&\frac{\partial I_\lambda(S,\Omega,t)}{c\partial t}+\frac{\partial I_\lambda(S,\Omega,t)}{\partial S}\\
&=\kappa_\lambda I_{\lambda b}(S,t)-\kappa_\lambda I_\lambda(S,\Omega,t)-\sigma_{s,\lambda}I_\lambda(S,\Omega,t)+\frac{1}{4\pi}\int_{\Omega_i=4\pi}\sigma_{s,\lambda}I_\lambda(S,\Omega_i,t)\Phi_\lambda(\Omega_i,\Omega)d\Omega_i
\end{aligned}
\tag{1.73}
$$

where ∂ represents the partial differential with respect to time or space. This energy balance is called the RTE. Note that in Equation 1.73, the first term is important for ultrafast phenomena; however, for many practical applications, it is neglected due to the large value of the speed of light c appearing in the denominator.

1.8 RADIATIVE TRANSFER IN ENCLOSURES WITH NONPARTICIPATING MEDIA

The RTE expresses the general conservation of radiative energy along a line of sight in an absorbing, emitting, and scattering medium. The solution of this integrodifferential equation is by no means trivial, especially in 3D geometries. It is the stepping stone to determining the contribution of radiation to an overall energy conservation relation. However, it is not necessary to have its full-fledged solution for every engineering problem. It is indeed possible to introduce a number of physical approximations for a given application, which reduces this equation to much simpler forms. Among all complications, it is the scattering by particles and inhomogeneities in a given medium that makes the directional dependency of the RTE more complicated. Scattering couples radiative energy propagating in a given direction with those in other directions, resulting in the integral nature of the RTE. It should be noted that scattering is a natural phenomena, and physically it is always present unless there is pure vacuum between the surfaces exchanging radiation. All gases scatter light, and the blue color of the atmosphere is due to scattering by air molecules (O_2, NO_2, and others). However, gas scattering is orders of magnitude smaller than that by dust and other particles and can be neglected in many cases without introducing any significant error to calculations. If there are no scattering particles, the RTE is reduced to a partial differential equation. Applications abound for nonscattering media, ranging from kitchen ovens to satellite energy management systems, and therefore constitute an important class of problems.

Now, let us return to Figure 1.7 and simplify the governing equations we developed in Section 1.7 in a heuristic way. First, we consider an absorbing, but cold and nonscattering, medium. Here, cold refers to the medium itself, rather than the surfaces that exchange radiative energy with each other. A medium at typical room or atmospheric conditions can be assumed cold, as the energy it emits would not be significant compared to solar radiation or a hot surface of an enclosure, such as an oven. The energy transfer in this simplified case is limited to that between the surfaces, each of which emits, mostly uniformly, in all directions. In addition, each surface reflects the energy received from other surfaces. Emission characteristics of surfaces and discussion of reflection (whether it is diffuse, i.e., with uniform intensity in all directions, or specular, i.e., mirrorlike, or a combination) are covered in Chapter 3.

We start the analysis with the radiative intensity $I_\lambda(\theta,\phi)$ leaving surface element dA_1. To analyze the radiative exchange between two finite surfaces, we need to carry out integration over the entire area of each surface. For this, consider radiative energy leaving a small area element dA_1 and traveling in a nonparticipating (nonabsorbing, nonemitting, nonscattering) medium. Assume that this energy is incident on a second small area element dA_2, on finite area A_2, at distance S_{12} from dA_1. The projected areas are formed by taking the area that the energy is passing through and projecting it normal to the direction of the radiation; therefore, $dA_1\cos\theta_1$ and $dA_2\cos\theta_2$ are the normal components of the infinitesimal areas along direction S_{12} as shown in Figure 1.19a. The elemental solid angle is centered about the direction of the radiant path and has its origin at dA. Using the definition of spectral intensity $I_{\lambda,1}$ as the rate of energy passing through dA_l per unit projected area per unit solid angle and per unit wavelength interval, the energy $dQ_{\lambda,1}$ from dA passing through dA_l in the direction of S_{12} is written as

$$dQ_{\lambda,1} = I_{\lambda,1}dA_1\cos\theta_1 d\Omega_{21}d\lambda = I_{\lambda,1}dA_1\cos\theta_1\left(\frac{dA\cos\theta}{S_{12}^2}\right)d\lambda \tag{1.74}$$

If dA_1 is placed at a farther distance along the same direction, the rate of energy passing through would be smaller as in Figure 1.19b. Let's assume another small area element $dA_2 = dA_l$ is located at a distance S_2. The energy rate at this new position is $I_{\lambda,2}dA_1\cos\theta_1 d\Omega_2 d\lambda = I_{\lambda,2}dA_1\cos\theta_1\left(dA_2\cos\theta_2/S_2^2\right)d\lambda$. If all areas are perpendicular to the direction of propagation, then all $\cos\theta$ terms would be equal to unity, and the ratio of energy rates for distances S_1 and S_2 is given as $I_{\lambda,1}S_1^2/I_{\lambda,2}S_2^2$.

Now, consider a differential source emitting energy equally into all directions. Assume we draw two concentric spheres around this source as in Figure 1.19c. If $dQ_{\lambda,s}d\lambda$ is the entire spectral energy leaving the differential source, the energy flux crossing the inner sphere is $dQ_{\lambda,s}d\lambda/4\pi S_1^2$, and that crossing the outer sphere is $dQ_{\lambda,s}d\lambda/4\pi S_2^2$. The ratio of the energies passing through the two elements dA_1 at S_1 and $dA_2 = dA_1$ at S_2 is then $[(dQ_{\lambda,s}d\lambda/4\pi S_1^2)dA_2]/[(dQ_{\lambda,s}d\lambda/4\pi S_2^2)dA_2] = S_2^2/S_1^2$. This is indeed expected as the radiative energy density decreases with the square of the distance from the origin (the inverse square law). However, in the previous paragraph, this energy ratio was derived as $I_{\lambda,1}S_1^2/I_{\lambda,2}S_2^2$. For these two expressions to be equal, the intensities must remain constant along the line of sight:

$$I_{\lambda,1} = I_{\lambda,2} \tag{1.75}$$

Thus, the intensity in a given direction in a nonattenuating and nonemitting (nonparticipating) medium is independent of position along that direction, including that at the starting point, that is, $I_{\lambda,s}$ originating at the surface dA_s. This demonstrates again the invariance of intensity with position along a path in a nonparticipating medium. The invariance of intensity provides a convenient way to specify the magnitude of attenuation or emission when they are present in a medium, as their effects are directly shown by the change of intensity with distance along a path. If the medium is absorbing, scattering, and/or emitting, the intensity will differ at any two locations along a line of sight.

Next, consider a cold *absorbing* medium, which is more complicated, but allows us to start our discussion straight from the RTE (Equation 1.73). Since the medium is not emitting or scattering, the equation will not have any gain terms due to emission or in-scattering, and the only loss is due to absorption along the path S. The governing equation is a first-order differential equation, as given by Equation 1.42, and solution is relatively simple. It can be obtained by integrating the

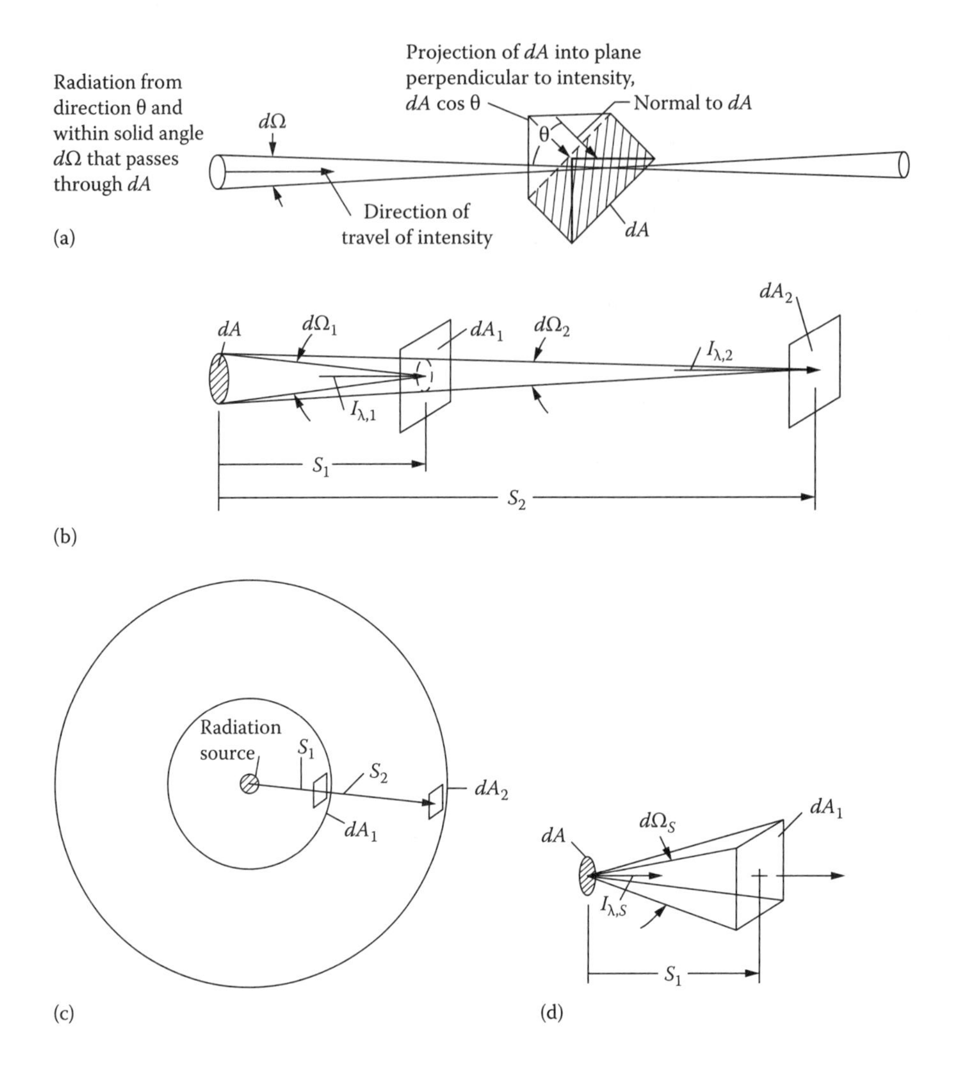

FIGURE 1.19 Demonstration that intensity is invariant along a path in a transparent medium. (a) Black element dA within black spherical enclosure, (b) energy transfer from dA to dA_1, (c) energy transfer from dA_1 to dA, and (d) solid angle subtended by dA_1 on sphere surface.

equation from $S = S_1 = 0$ on dA_1 to $S = S_2$ on dA_2. Using the boundary condition at $S = S_1 = 0$, which is $I_\lambda\left(S=0,\theta,\phi\right) = I_{\lambda,dA_1}\left(\theta,\phi\right) = I_{\lambda,\text{emission},dA_1}\left(\theta,\phi\right) + I_{\lambda,\text{reflection},dA_1}\left(\theta,\phi\right)$, we obtain

$$I_{\lambda,dA_2}\left(S_2,\theta,\phi\right) = I_{\lambda,dA_1}\left(S_1,\theta,\phi\right)\exp\left[-\int_{S=0}^{S_2}\kappa_\lambda(S)dS\right] \tag{1.76}$$

Equation 1.76 is further simplified to $I_{\lambda,dA_2}\left(S_2,\theta,\phi\right) = I_{\lambda,dA_1}\left(S_1,\theta,\phi\right)\exp\left[-\kappa_\lambda(S_2 - S_1)\right]$ for a medium with spatially uniform absorption coefficient. This is a restatement of Equation 1.43.

In these equations, S_1 and S_2 correspond to the central locations of infinitesimal area elements dA_1 and dA_2 on the directly opposed surfaces A_1 and A_2, respectively. Equation 1.76 is valid only along a single line connecting S_1 to S_2. If we want to determine the radiative energy leaving from a finite area and reaching another one, then we need to integrate Equation 1.76 over each of these areas. These area integrations are equivalent to those carried over solid angles subtended by each area. First, consider infinitesimally small area elements; spectral radiative energy leaving dA_1 toward dA_2 is given as

$$dQ_{\lambda,1} = I_{\lambda,1}\cos\theta_1 dA_1 d\lambda d\Omega_2 = I_{\lambda,1}\cos\theta_1 dA_1 d\lambda \frac{\cos\theta_2 dA_2}{(S_2 - S_1)^2}$$

where

$$\Omega_2 = \frac{\cos\theta_2 dA_2}{\left(S_2 - S_1\right)^2}$$

The radiative energy arriving to dA_2 is then found from

$$\begin{aligned} dQ_{\lambda,1} &= I_{\lambda,1}\cos\theta_1 dA_1 \exp\left[-\int_{S=0}^{S_2} \kappa_\lambda(S)dS\right] d\lambda d\Omega_2 \\ &= I_{\lambda,1}\cos\theta_1 dA_1 \frac{\cos\theta_2 dA_2}{(S_2 - S_1)^2} \exp\left[-\int_{S=0}^{S_2} \kappa_\lambda(S)dS\right] d\lambda \end{aligned} \tag{1.77}$$

To write a similar expression for finite surfaces, Equation 1.77 is integrated over both surfaces A_1 and A_2.

$$Q_{\lambda,1\to2} = \iint_{A_1,A_2} dQ_1 = \iint_{A_1,A_2} I_{\lambda,1}\cos\theta_1 \frac{\cos\theta_2}{(S_2 - S_1)^2} \exp\left[-\int_{S=0}^{S_2} \kappa_\lambda(S)dS\right] d\lambda dA_1 dA_2 \tag{1.78}$$

If the medium absorption approaches zero, that is, if the medium becomes optically thin and eventually completely nonabsorbing, the exponential term drops out. If radiation leaving surface A_1 is directionally uniform, then the intensity term can be taken outside the integral sign, yielding

$$Q_{\lambda,1\to2} = I_{\lambda,1} d\lambda \iint_{A_1,A_2} \cos\theta_1 \frac{\cos\theta_2}{\left(S_2 - S_1\right)^2} dA_1 dA_2 = E_{\lambda,1} d\lambda \iint_{A_1,A_2} \frac{\cos\theta_1 \cos\theta_2}{\pi\left(S_2 - S_1\right)^2} dA_1 dA_2 \tag{1.79}$$

The double integral in this equation is a purely geometrical factor and can be evaluated independently of the spectral surface properties:

$$Q_{\lambda 1\to2} = E_{\lambda,1} F_{1-2} A_1 d\lambda \tag{1.80}$$

where

$$A_1 F_{1-2} = \iint_{A_1,A_2} \frac{\cos\theta_1 \cos\theta_2}{\pi\left(S_2 - S_1\right)^2} dA_1 dA_2 \tag{1.81}$$

Here, F_{1-2} is called a *shape, view*, or *configuration factor* and is treated extensively in later chapters.

This formulation can be expanded to absorbing and scattering media. If the medium properties are uniform, then the attenuation of radiation along a path connecting one surface to another one would be included in the configuration factor analysis with little difficulty. We return to these discussions later.

1.9 CONCLUDING REMARKS AND HISTORICAL NOTES

In this chapter, we discussed the importance, definitions, and equations of thermal radiative transfer. We derived the fundamental RTE and discussed its simplified forms for different medium properties. We also presented an intuitive relationship between the RTE formulations and the configuration factor analyses, which is used extensively for nonparticipating media calculations.

The historical development of the blackbody relations in this chapter differed from the sequence in which they are presented here. The derivation of the approximate spectral distributions of Wien and of Rayleigh and Jeans, Stefan–Boltzmann law, and Wien's displacement law are all presented as logical consequences of the fundamental spectral distribution of intensity derived by Max Planck. However, it is interesting to note that all these relations were formulated *prior* to publication of Planck's work (1901) and were originally derived through fairly complex thermodynamic arguments.

Joseph Stefan (1879) proposed, after study of some experimental results, that emissive power was related to the fourth power of the absolute temperature of a radiating body. His student, Ludwig Eduard Boltzmann (1884), was able to derive the same relation by analyzing a Carnot cycle in which radiation pressure was assumed to act as the pressure of the working fluid.

Wilhelm Carl Werner Otto Fritz Franz (Willy) Wien (1894) derived the displacement law by consideration of a piston moving within a mirrored cylinder. He found that the spectral energy density in an isothermal enclosure and the spectral emissive power of a blackbody are both directly proportional to the fifth power of the absolute temperature when "corresponding wavelengths" are chosen.

Wien (1896) also derived his spectral distribution of intensity through thermodynamic argument plus assumptions concerning the absorption and emission processes. Lord Rayleigh (1900) and Sir James Jeans (1905) based their spectral distribution on the assumption that the classical idea of equipartition of energy was valid.

The fact that measurements and some theoretical considerations indicated Wien's expression for the spectral distribution was invalid at high temperatures and/or large wavelengths led Planck to an investigation of harmonic oscillators that were assumed to be the emitters and absorbers of radiant energy. Various further assumptions as to the average energy of the oscillators led Planck to derive both the Wien and Rayleigh–Jeans distributions. Planck finally found an empirical equation that fit the measured energy distributions over the entire spectrum. In determining what modifications to the theory would allow derivation of this empirical equation, he was led to the assumptions that form the basis of quantum theory. His equation leads directly to all the results derived previously by Wien, Stefan, Boltzmann, Rayleigh, and Jeans.

Gustav Mie used the EM theory to predict the scattering coefficient and phase function for radiation interacting with small spherical particles, and a historical overview of his contributions is in Horvath (2009a).

For an interesting and informative comprehensive review of the history of the field of thermal radiation, the article by Barr (1960) is recommended. Lewis (1973) and Crepeau (2009) discuss the derivation of Planck law.

HOMEWORK

1.1 What are the wave number range in vacuum and the frequency range for the visible spectrum (0.4–0.7 μm)? What are the wave number and frequency values at the spectral boundaries between the near- and the far-infrared regions?

Answer: 2.5×10^6 to 1.4286×10^6 m^{-1}; 4.2827×10^{14} to 7.4948×10^{14} s^{-1}; 4×10^4 m^{-1}; 1.1992×10^{13} s^{-1}

1.2 Radiant energy at a wavelength of 2.0 μm is traveling through a vacuum. It then enters a medium with a refractive index of 1.24.

(a) Find the following quantities for the radiation in the vacuum: speed, frequency, and wave number.

(b) Find the following quantities for the radiation in the medium: speed, frequency, wave number, and wavelength.

Answer: (a) 2.9979×10^8 m/s; 1.4990×10^{14} s^{-1}; 5×10^5 m^{-1}. (b) 2.4177×10^8 m/s; 1.4989×10^{14} s^{-1}; 6.20×10^5 m^{-1}; 1.6129×10^{-6} m

1.3 Radiation propagating within a medium is found to have a wavelength within the medium of 1.570 μm and a speed of 2.500×10^8 m/s.

(a) What is the refractive index of the medium?

(b) What is the wavelength of this radiation if it propagates into a vacuum?

Answer: (a) 1.199; (b) 1.882 μm

1.4 What range of radiation wavelengths are present within a glass sheet that has a wavelength-independent refractive index of 1.33 when the sheet is exposed in vacuum to incident radiation in the visible range $\lambda_0 = 0.4$–0.7 μm?

Answer: 0.301–0.526 μm

1.5 A material has an index of refraction $n(x)$ that varies with position x within its thickness. Obtain an expression in terms of c_o and $n(x)$ for the transit time for radiation to pass through a thickness L. If $n(x) = n_i(1 + kx)$, where n_i and k are constants, what is the relation for transit time? How does wave number (relative to that in a vacuum) vary with position within the medium?

Answer: $t = \frac{n_i}{c_o}\left(L + \frac{kL^2}{2}\right)$; $\quad n_i\left(1 + kx\right)$

1.6 Derive Equation 1.26 by analytically finding the maximum of the $E_{\lambda b}/T^5$ versus λT relation (Equation 1.20).

1.7 A blackbody is at a temperature of 1250 K and is in air.

(a) What is the spectral intensity emitted in a direction normal to the black surface at $\lambda =$ 3.75 μm?

(b) What is the spectral intensity emitted at $\theta = 45°$ with respect to the normal of the black surface at $\lambda = 3.75$ μm?

(c) What is the directional spectral emissive power from the black surface at $\theta = 45°$ and $\lambda = 3.75$ μm?

(d) At what λ is the maximum spectral intensity emitted from this blackbody, and what is the value of this intensity?

(e) What is the hemispherical total emissive power of the blackbody?

Answers: (a) 7823 W/(m^2·μm·sr); (b) 7823 W/(m^2·μm·sr); (c) 5532 W/(m^2·μm·sr); (d) 2.3182 μm, 12,499 W/(m^2·μm·sr); (e) 138,437 W/m^2

1.8 Plot the hemispherical spectral emissive power $E_{\lambda b}$ for a blackbody in air [W/(m^2·μm)] as a function of wavelength (μm) for surface temperatures of 2000 and 6250 K.

1.9 For a blackbody at 2250 K that is in air, find

(a) The maximum emitted spectral intensity (kW/m^2·μm·sr)

(b) The hemispherical total emissive power (kW/m^2)

(c) The emissive power in the spectral range between $\lambda_o = 2$ and 8 μm

(d) The ratio of spectral intensity at $\lambda_o = 2$ μm to that at $\lambda_o = 8$ μm

Answers: (a) 236.2 kW/m^2·μm·sr; (b) 1453 kW/m^2; (c) 600.7 kW/m^2; (d) 53.4

1.10 Determine the fractions of blackbody energy that lie below and above the peak of the blackbody curve.

Answer: 25% and 75%

1.11 A blackbody at 1250 K is radiating in the vacuum of outer space.

(a) What is the ratio of the spectral intensity of the blackbody at $\lambda = 2.0$ μm to the spectral intensity at $\lambda = 5$ μm?

(b) What fraction of the blackbody emissive power lies between the wavelengths of $\lambda = 2.0$ μm and $\lambda = 5$ μm?

(c) At what wavelength does the peak energy in the radiated spectrum occur for this blackbody?

(d) How much energy is emitted by the blackbody in the range $2.0 \le \lambda \le 5$ μm?

Answer: (a) 2.790; (b) 0.0.597; (c) 2.138 μm; (d) 82,587 W/m^2

1.12 Solar radiation is emitted by a fairly thin layer of hot plasma near the sun's surface. This layer is cool compared with the interior of the sun, where nuclear reactions are occurring. Various methods can be used to estimate the resulting effective radiating temperature of the sun, such as determining the best fit of a blackbody spectrum to the observed solar spectrum. Use two other methods (as follows), and compare the results to the oft-quoted value of $T_{solar} = 5780$ K.

(a) Using Wien's law and taking the peak of the solar spectrum as 0.50 μm, estimate the solar radiating temperature.

(b) Given the measured solar constant in the Earth's orbit of 1368 W/m^2 and using the "inverse square law" for the drop in heat flux with distance, estimate the solar temperature. The mean radius of the Earth's orbit around the sun is 149×10^6 km and the diameter of the sun is 1.392×10^6 km.

Answer: (a) 5796 K; (b) 5766 K

1.13 The surface of the sun has an effective blackbody radiating temperature of 5780 K.

(a) What percentage of the solar radiant emission lies in the visible range $\lambda = 0.4$–0.7 μm?

(b) What percentage is in the ultraviolet?

(c) At what wavelength and frequency is the maximum energy per unit wavelength emitted?

(d) What is the maximum value of the solar hemispherical spectral emissive power?

Answer: (a) 36.7%; (b) 12.2%; (c) 0.5013 μm, 5.98×10^{14} Hz; (d) 8.301×10^7 W/(m^2·μm)

1.14 A blackbody radiates such that the wavelength at its maximum emissive power is 2.00 μm. What fraction of the total emissive power from this blackbody is in the range $\lambda = 0.7$ to $\lambda = 5$ μm?

Answer: 0.821

1.15 A blackbody has a hemispherical spectral emissive power of 0.03500 W/(m^2·μm) at a wavelength of 90 μm. What is the wavelength for the maximum emissive power of this blackbody?

Answer: 18.73 μm

1.16 A radiometer is sensitive to radiation only in the interval $3.6 \le \lambda \le 8.5$ μm. The radiometer is used to calibrate a blackbody source at 1000 K. The radiometer records that the emitted energy is 4600 W/m^2. What percentage of the blackbody radiated energy in the prescribed wavelength range is the source actually emitting?

Answer: 17.2%

1.17 What temperature must a blackbody have for 25% of its emitted energy to be in the visible wavelength region?

Answer: 4,343 K, 12,460 K (Note: two solutions are possible!)

1.18 Show that the blackbody spectral intensity $I_{\lambda b}$ increases with T at any fixed value of λ

1.19 Blackbody radiation is leaving a small hole in a furnace at 1200 K (see the figure.) What fraction of the radiation is intercepted by the annular disk? What fraction passes through the hole in the disk?

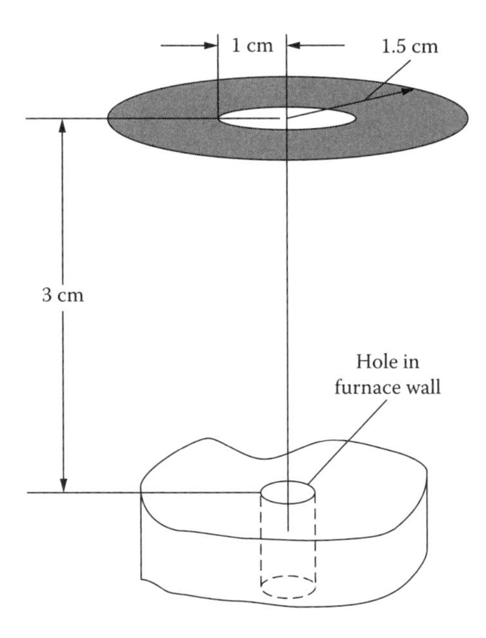

Answer: 0.1000; 0.1000

1.20 A sheet of silica glass transmits 85% of the radiation that is incident in the wavelength range between 0.38 and 2.7 μm and is essentially opaque to radiation having longer and shorter wavelengths. Estimate the percent of solar radiation that the glass will transmit. (Consider the sun as a blackbody at 5780 K.)

If the garden in a greenhouse radiates as a black surface and is at 40°C, what percent of this radiation will be transmitted through the glass?

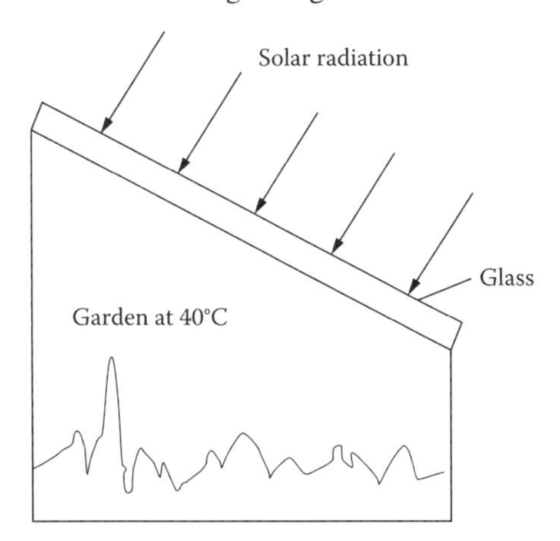

Answer: 73.9%; 0.003%

1.21 Derive Wien's displacement law in terms of wave number by differentiation of Planckspectral distribution in terms of wave number, and show that $T/\eta_{max} = 5099.4$ μm·K.

1.22 A student notes that the peak emission of the sun according to Wien's displacement law is at a wavelength of about $\lambda_{max} = C_3/5780$ K = 2897.8/ 5780 = 0.501 μm. Using $\eta_{max} = 1/0.501$ μm, the student solves again for the solar temperature using the result derived in Homework Problem 1.21. Does this computed temperature agree with the solar temperature? Why? (This is not trivial—put some thought into *why*.)

1.23 Derive the relation between the wave number and the wavelength at the peak of the blackbody emission spectrum. (You may use the result of Homework Problem 1.22.)
Answer: η_{max} (cm^{-1}) = 5682.6/λ_{max} (μm)

1.24 A solid copper sphere 3.0 cm in diameter has on it a thin black coating. Initially, the sphere is at 850 K, and it is then placed in a vacuum with very cold surroundings. How long will it take for the sphere to cool to 200 K? Because of the high thermal conductivity of copper, it is assumed that the temperature within the sphere is uniform at any instant during the cooling process. (Properties of copper: density, ρ = 8950 kg/m^3; specific heat, c = 383 J/kg·K.)
Answer: 3.45 h

1.25 A 6 by 10 cm black rectangular sheet of metal is heated uniformly with 2350 W by passing an electric current through it. One face of the rectangle is well insulated. The other face is exposed to vacuum and very cold surroundings. At thermal equilibrium, what fraction of the emitted energy is in the wave number range from 0.33 to 2 μm^{-1}?
Answer: 0.6225

1.26 Spectral radiation at λ = 2.445 μm and with intensity 7.35 kW/(m^2·μm·sr) enters a gas and travels through the gas along a path length of 21.5 cm. The gas is at uniform temperature 1000 K and has an absorption coefficient $\kappa_{2.445\,\mu m}$ = 0.557 m^{-1}. What is the intensity of the radiation at the end of the path? Neglect scattering, but include emission by the gas.
Answer: 6.950 kW/(m^2·μm·sr)

1.27 Radiation from a blackbody source at 2,000 K is passing through a layer of air at 12,000 K and 1 atm. Considering only the transmitted radiation (i.e., not accounting for emission by the air), what path length is required to attenuate by 35% the energy at the wavelength corresponding to the maximum emission by the blackbody source? At this λ, take $\kappa_\lambda = 1.2 \times 10^{-1}$ cm^{-1} for air at 12,000 K and 1 atm.
Answer: 3.59 cm

1.28 A gas layer at constant pressure P has a linearly decreasing temperature across the layer and a constant mass absorption coefficient κ_m (no scattering). For radiation passing in a normal direction through the layer, what is the ratio I_2/I_1 as a function of T_1, T_2, and L? The temperature range T_2 to T_1 is low enough that emission from the gas can be neglected. The gas constant is R.

Answer: $\dfrac{I_2}{I_1} = e^{-\frac{\kappa_m PL}{R(T_1-T_2)}\ln(T_1/T_2)}$

1.29 Hawking (1974) predicts that black holes should emit a perfect blackbody radiation spectrum in proportion to the surface gravity of the black hole. Further predictions are that the surface gravity of a black hole is inversely proportional to its mass. Thus, very small black holes should emit very large blackbody radiation. If a black hole with 1 Earth mass has a predicted blackbody temperature of 10^{-7} K, what fraction of an Earth mass black hole will emit at T = 2.7 K (equal to the background temperature of space) and thus possibly be detectable?
Answer: $3.7 \times 10^{-8}\, m_{Earth}$

1.30 Observers at NASA noticed an unexpected small deceleration of the early deep space probes Pioneer 10 and 11 as they moved away from the Earth. This was known as "the Pioneer Anomaly." After many speculations and false starts, researchers determined that thermal radiation pressure exerted by radiation from thermoelectric generators heated by small nuclear power sources plus radiated waste heat from the electronics was the culprit. Assuming a projected radiating area of 0.2 m^2 facing along the spacecraft trajectory, a spacecraft mass of 258 kg, and a mean radiating temperature on a the outward facing surface (effective emissivity = 0.3) of 375 K, what deceleration was exerted on the spacecraft? (Much more detailed analysis can be found in Turyshev et al. 2012. The detailed thermal model predicts a deceleration of 7 to 10×10^{-10} m/s^2.)
Answer: 8.7×10^{-10} m/s

2 Radiative Properties at Interfaces

2.1 INTRODUCTION

The radiative behavior of a blackbody was presented in Chapter 1. The ideal blackbody is a standard with which the performance of real radiating bodies can be compared. The radiative behavior of a real body depends on many factors such as composition, surface finish, temperature, radiation wavelength, angle at which radiation is either emitted or intercepted, and spectral distribution of the incident radiation. We need to define spectral, directional, or averaged emissive, absorptive, and reflective properties to describe the radiative behavior of real materials relative to blackbody behavior.

The definitions of the surface radiative properties of materials are introduced in this chapter. A complete discussion of radiative transfer requires a large number of parameters to be defined clearly; however, the introduction of the many surface properties and their dependence on wavelength, angle, and temperature makes absorbing them all at once quite burdensome. Given this, the reader is encouraged to scan the information first, and then use the details as necessary for a given application. This reference chapter is the foundation for the property definitions used for radiation/surface interactions discussed in the rest of the text.

To understand the radiative property definitions, the most general definitions are presented first. Consider Figure 2.1, where an infinitesimal area element dA is part of an object at temperature T, and is exchanging radiation with other objects. The location of dA can be specified in any coordinate system; for convenience, consider a Cartesian system so the temperature of dA can be specified as $T(x, y, z)$. The emitted energy from dA is thus an implicit function of location through its dependence on $T(x, y, z)$ and may also be a function of wavelength and direction of emission.

The ability of the surface to absorb and reflect radiation depends not only on the temperature of the surface, $T(x, y, z)$, but on the direction and spectral characteristics of the source of incident radiation. This intriguing effect will become obvious through the spectral calculations. The source characteristics will be denoted by indicating the dependence of the absorption properties on T_i, the temperature and spectral characteristics of the incident source. A blackbody source at 6000 K will have peak radiation near 0.5 μm (Wien's law, Equation 1.26) while a source at 300 K will have a spectral peak near 10 μm. The amount of incident radiation absorbed by dA will depend on whether its absorption ability is highest near 0.5 or 10 μm.

Directional and spectral reflection from and transmission through dA also depend on the spectral characteristics of the source in addition to the spectral, directional, and temperature-dependent properties of dA. In general, then, reflected and transmitted radiation are functions of the source spectral characteristics, the direction of the source, the direction of the reflected or transmitted radiation, and the particular wavelength considered. Time dependence may be important in ultrafast processes such as femto- or picosecond laser applications, where the incident energy may change the temperature and/or properties of the surface.

Generally, emission properties will have dependence on $(\lambda, \theta, \phi, T, t)$, absorption will depend on $(\lambda, \theta_i, \phi_i, T, t)$, reflection on $(\lambda, \theta_i, \phi_i, \theta_r, \phi_r, T, t)$, and transmission on $(\lambda, \theta_i, \phi_i, \theta_t, \phi_t, T, t)$.

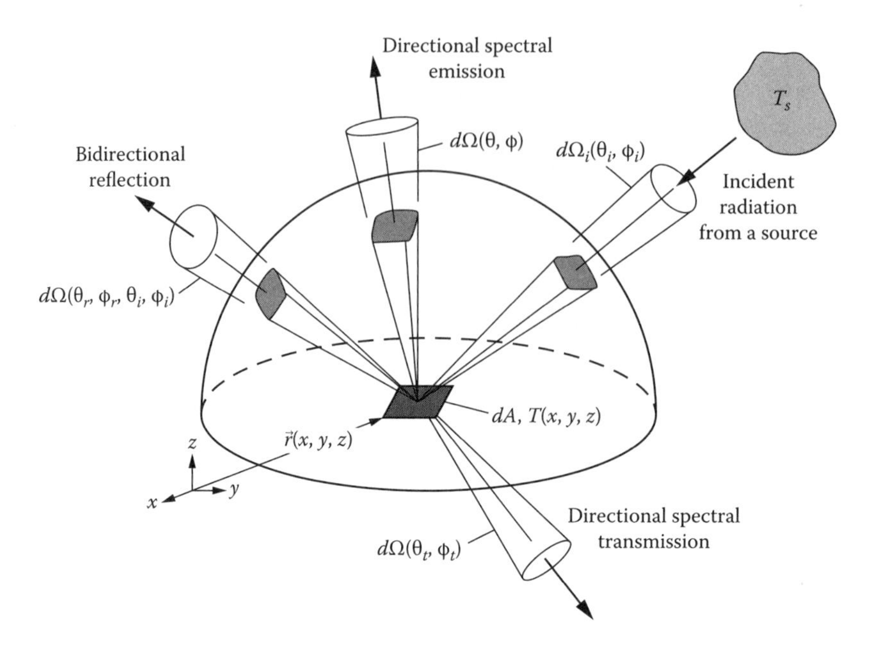

FIGURE 2.1 Description of radiation interactions at a surface.

To simplify the notation, the time dependence is dropped, and the wavelength dependence will be shown through the presence or absence of the subscript λ.

The detailed examination of radiative property definitions arises from the need to interpret and use available property data in energy transfer computations. Experimental data in the literature that provide both detailed directional and spectral data are scarce. Because of the complexities in making these measurements, (for example, see Shafey et al. 1982, Makino 2002, Makino and Wakabayashi 2003) most tabulated property values are averaged over all directions, all wavelengths, or both. The definitions reveal relations among averaged properties that enable maximum use of available property information; for example, absorptivity data can be obtained from emissivity data if certain restrictions are observed.

Table 2.1 lists each property, its symbolic notation, the equation number of its definition, and the figure showing a physical interpretation.

Emissivity ε and absorptivity α are considered to be surface properties. For an opaque material, the portion of incident radiation that is not reflected is transmitted through the surface and then absorbed within material that extends below the surface. This layer can be quite thin (tens of nanometers) for a material with high internal absorptance, such as a metal, or can be a few millimeters or more for a less strongly internally absorbing dielectric material. Emission comes from within a material. The amount of energy leaving the surface depends on how much of the energy emitted by each internal volume element can reach and then be transmitted through the surface. As discussed in Section 17.5, some of the energy reaching the surface can be internally reflected from the surface back into the body, depending on the refractive index of the body relative to that of the external medium. Thus, the emissive ability of a body depends on the refractive index of its surrounding medium. With rare exceptions, emissivity measurements are made for emission into air or vacuum, so the tabulated values of spectral or total emissivity ε_λ or ε are for an external medium with $n \approx 1$. If the medium adjacent to the surface has $n > 1$, the ε_λ or α_λ, and hence ε and α, can be different from the tabulated values.

In this chapter, the properties are discussed for the surface of a medium that has sufficient thickness to attenuate and absorb all radiation that passes through the surface; thus, the medium

TABLE 2.1
Summary of Surface Radiative Property Definitions

Quantity	Symbol	Defining Equation	Descriptive Figure
Emissivity			
Directional spectral	$\varepsilon_\lambda(\theta, \phi)$	(2.3)	2.2a
Directional total	$\varepsilon(\theta, \phi)$	(2.4)	2.2a
Hemispherical spectral	ε_λ	(2.7)	2.2b
Hemispherical total	ε	(2.8), (2.9), (2.10)	2.2b
Absorptivity			
Directional spectral	$\alpha_\lambda(\theta, \phi)$	(2.14)	2.2c
Directional total	$\alpha(\theta, \phi)$	(2.20), (2.21)	2.2c
Hemispherical spectral	α_λ	(2.24), (2.25)	2.2d
Hemispherical total	α	(2.28) through (2.31)	2.2d
Reflectivity			
Bidirectional spectral	$\rho_\lambda(\theta_r, \phi_r, \theta_i, \phi_i)$	(2.33)	2.2e
Directional-hemispherical spectral	$\rho_\lambda(\theta_i, \phi_i)$	(2.39)	2.2f
Hemispherical-directional spectral	$\rho_\lambda(\theta_r, \phi_r)$	(2.41)	2.2g
Hemispherical spectral	ρ_λ	(2.44)	2.2h
Bidirectional total	$\rho(\theta_r, \phi_r, \theta_i, \phi_i)$	(2.52), (2.53)	2.2e
Directional-hemispherical total	$\rho(\theta_i, \phi_i)$	(2.56)	2.2f
Hemispherical-directional total	$\rho(\theta_r, \phi_r)$	(2.57)	2.2g
Hemispherical total	ρ	(2.59)	2.2h
Transmissivity			
Bidirectional spectral	$\tau_\lambda(\theta_t, \phi_t, \theta_i, \phi_i)$	(2.61)	2.2i
Directional-hemispherical spectral	$\tau_\lambda(\theta_i, \phi_i)$	(2.62, 2.63)	2.2j
Hemispherical-directional spectral	$\tau_\lambda(\theta_i, \phi_i)$	(2.64, 2.65)	2.2k
Hemispherical spectral	τ_λ	(2.68, 2.69)	2.2l
Bidirectional total	$\tau(\theta_t, \phi_t, \theta_i, \phi_i)$	(2.70)	2.2i
Directional-hemispherical total	$\tau(\theta_i, \phi_i)$	(2.71, 2.72)	2.2j
Hemispherical-directional total	$\tau(\theta_t, \phi_t)$	(2.73, 2.74)	2.2k
Hemispherical total	τ	(2.77, 2.78)	2.2l

is opaque. The radiant energy transmitted through the surface of the medium is then equal to the radiant energy absorbed by the medium. Although the ability of a surface to transmit radiation is not important for radiative transfer among opaque surfaces, the definitions describing transmission across a surface or interface are provided for reference when nonopaque materials are treated. Layers of material that are not opaque, and therefore transmit a portion of the incident radiation, have an emissive ability that depends on their thickness.

It is common practice in most fields of science to assign the *-ivity* ending to intensive properties such as electrical resistivity, thermal conductivity, or thermal diffusivity. The *-ivity* ending is used throughout this book for radiative properties of opaque materials, whether for ideal surfaces or for properties with a given surface condition. The *-ance* ending is for an extensive property such as the emittance of a partially transmitting isothermal layer of water or glass, where the emittance depends on layer thickness. It should be noted that the *-ance* ending is often found in literature dealing with the experimental determination of surface properties. The term emittance is also used in some references to describe what we have called emissive power. Over the years, suggestions have been made to standardize radiation nomenclature. The U.S. National Institute of Standards and Technology (NIST) has published a nomenclature useful in the fields of illumination and measurement (Nicodemus et al. 1977). That nomenclature is close to the set used in this text, here, and we follow the notation recommended by the editors of the major heat transfer journals (Howell 1999).

Because of the many independent variables that must be specified for radiative properties, a concise but accurate notation is required. The notation used here is an extension of that in Chapter 1. A functional notation is used to give explicitly the variables upon which a quantity depends. For example, $\varepsilon_\lambda(\theta, \phi, T)$ shows that the spectral emissivity is directional and can depend on four variables. The λ subscript specifies that the quantity is *spectral* and is convenient when the functional notation is omitted. A *total* quantity does not have a λ subscript. Certain quantities depend on *two* directions and have *four angles* in their functional notation. A hemispherical-directional or directional-hemispherical quantity, for example, has only two angles. Bidirectional properties require four explicit angles. In later chapters, the functional notation may be abbreviated or omitted to simplify the equation format when the functional dependencies are evident.

Additional notation is needed for Q, the energy rate for a *finite area*, to keep consistent mathematical forms for energy balances. The $Q_\lambda d\lambda$ is used to designate spectral energy in the interval $d\lambda$; this is consistent with blackbody spectral energy where terms such as $E_{\lambda b} d\lambda$ are used in Chapter 1. In $dQ_\lambda(\theta, \phi)d\lambda$ the Q has a derivative to indicate that the spectral energy is also of differential order in solid angle. The total energy $dQ(\theta, \phi)$ is a differential quantity with respect to solid angle. If the energy is for a *differential area*, the order of the derivative is correspondingly increased.

The notation $\int_\cap d\Omega$ signifies integration over the hemispherical solid angle where $d\Omega = \sin\theta \, d\theta \, d\phi$. Hence, for any function $f(\theta, \phi)$,

$$\int_{\phi=0}^{2\pi} \int_{\theta=0}^{\pi/2} f(\theta,\phi) \sin\theta \, d\theta d\phi \equiv \int_\cap f(\theta,\phi) d\Omega$$

Each of the four main sections of this chapter deals with a different property: emissivity, absorptivity, reflectivity, and transmissivity. In each section, the most basic property is presented first; for example, the first section begins with the directional spectral emissivity. Then, averaged quantities are obtained by integration. The section on absorptivity also contains various forms of Kirchhoff's law that relate absorptivity to emissivity. The sections on reflectivity and transmissivity include reciprocity relations.

2.2 EMISSIVITY

Emissivity specifies radiation from a real surface as compared with emission from a blackbody at the same temperature. It can be spectral, and in general depends on surface temperature and direction. In detailed radiative exchange calculations, data for spectral-directional emissivity at the correct surface temperature is needed. Such detailed data for particular materials is scarce. Most available data are for a limited spectral range and are mostly for the normal direction. Because of this, averaged emissivity values are often used. Averages may be over direction, wavelength, or both.

Values averaged with respect to all wavelengths are termed *total* quantities; averages with respect to all directions are termed *hemispherical* quantities. In this section, the basic definition of the directional spectral emissivity is given. This emissivity is then averaged with respect to wavelength, direction, or both wavelength and direction simultaneously. In this section, the basic definition of the directional spectral emissivity is given. This emissivity is then averaged with respect to wavelength, direction, and wavelength and direction simultaneously.

2.2.1 Directional Spectral Emissivity $\varepsilon_\lambda(\theta, \phi, T)$

Consider the geometry for emitted radiation in Figure 2.2a. As discussed in Chapter 1, the radiation intensity is the energy per unit time emitted in direction (θ, ϕ) per unit of *projected area dA_p normal* to this direction, per unit solid angle, and per unit wavelength interval. Basing the intensity on the projected area has the advantage that for a black surface the intensity has the same value in all directions. Unlike the intensity from a blackbody, the intensity emitted from a real body *does*

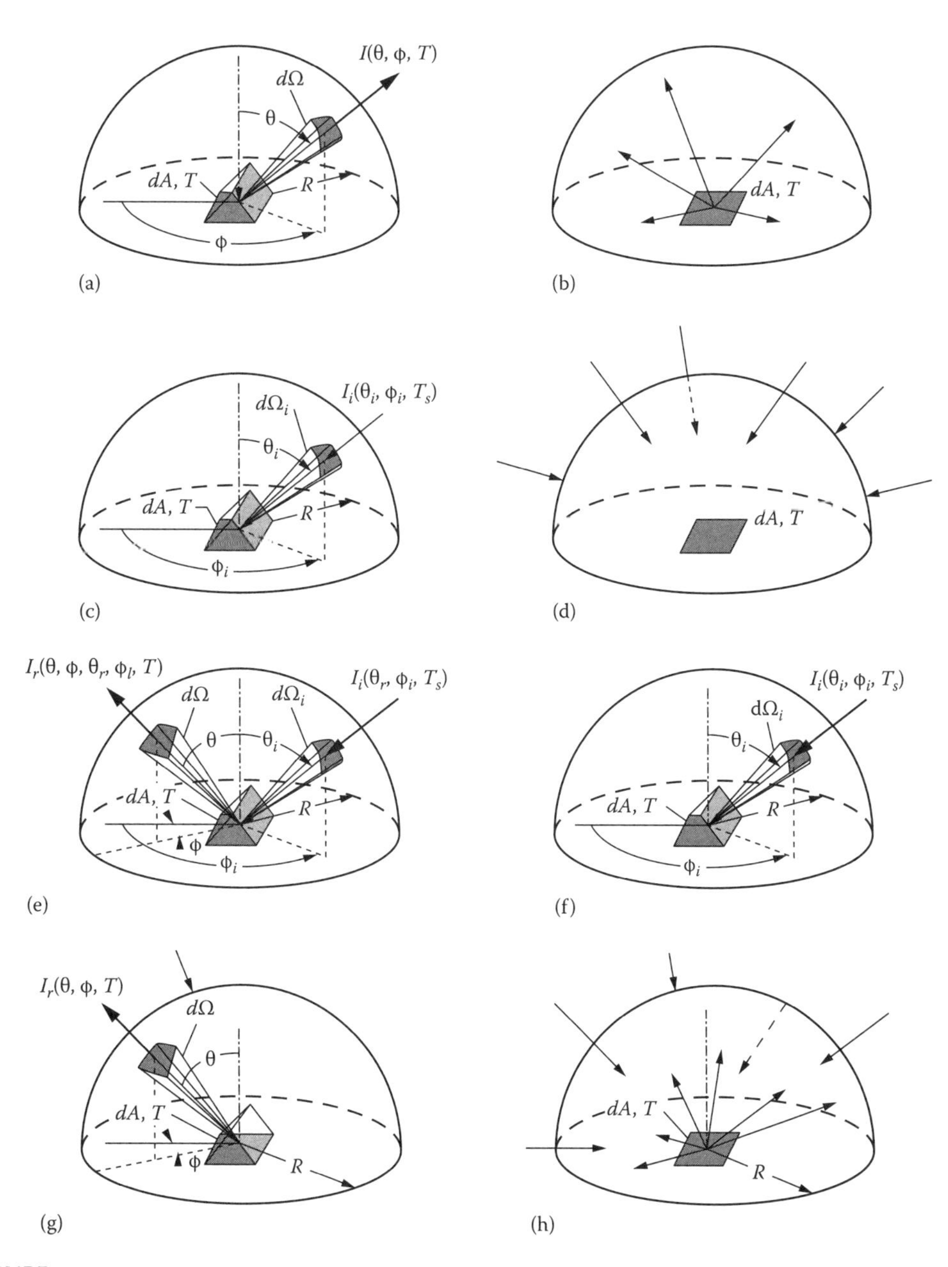

FIGURE 2.2 Pictorial descriptions of directional and hemispherical radiation properties. For spectral properties, the subscript λ is added the property definitions: (a) directional emissivity $\varepsilon(\theta, \phi, T)$; (b) hemispherical emissivity $\varepsilon(T)$; (c) directional absorptivity $\alpha(\theta_i, \phi_i, T)$; (d) hemispherical absorptivity $\alpha(T)$; (e) bi-directional reflectivity $\rho(\theta, \phi, \theta_i, \phi_i, T)$; (f) directional-hemispherical reflectivity $\rho(\theta_i, \phi_i, T)$; (g) hemispherical-directional reflectivity $\rho(\theta, \phi, T)$; (h) hemispherical reflectivity $\rho(T)$. *(Continued)*

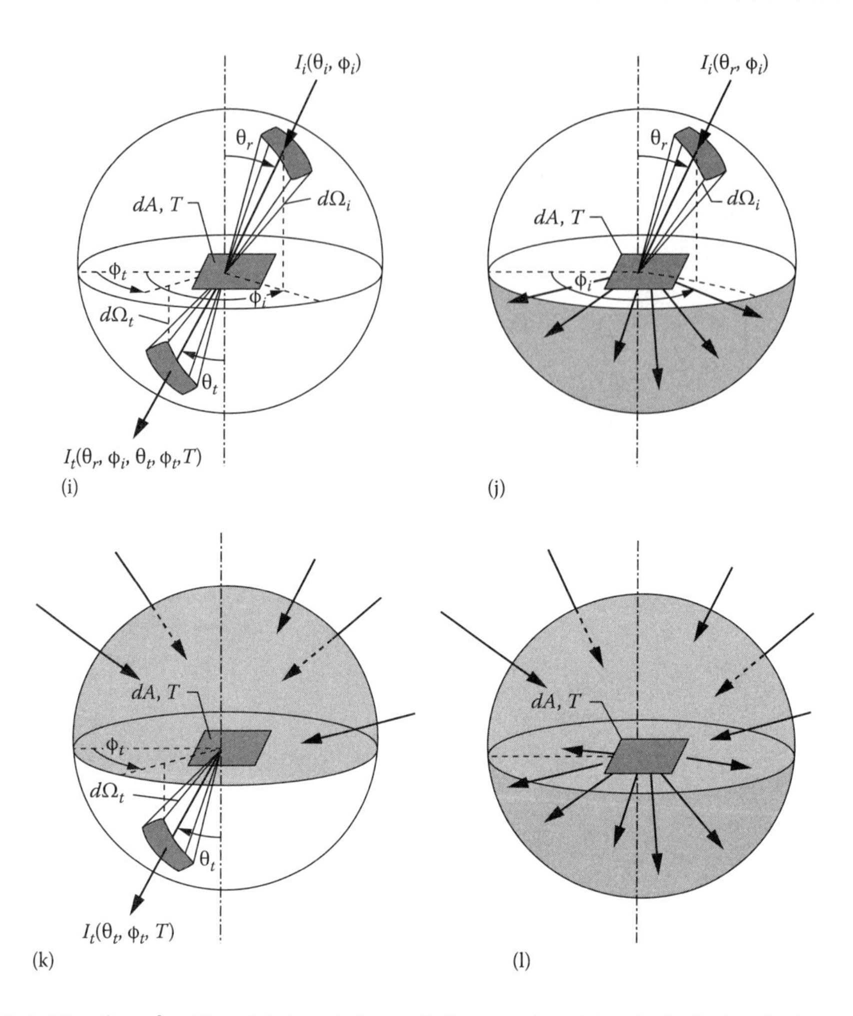

FIGURE 2.2 (*Continued*) Pictorial descriptions of directional and hemispherical radiation properties. For spectral properties, the subscript λ is added the property definitions: (i) bidirectional transmissivity $\tau(\theta_i, \phi_i, \theta_t, \phi_t, T)$; (j) directional-hemispherical transmissivity $\tau(\theta_i, \phi_i, T)$; (k) hemispherical-directional transmissivity $\tau(\theta_t, \phi_t, T)$; and (l) hemispherical transmissivity $\tau(T)$.

depend on direction. The energy leaving a real surface dA of temperature T per unit time in the wavelength interval $d\lambda$ and within the solid angle $d\Omega = \sin\theta d\theta d\phi$ is then given by

$$d^2Q_\lambda(\theta,\phi,T)d\lambda = I_\lambda\left(\theta, \phi, T\right)dA\cos\theta\, d\Omega d\lambda \tag{2.1}$$

For a blackbody the intensity $I_{\lambda b}(T)$ is independent of direction. The T notation is introduced here to clarify when quantities are temperature dependent, so the blackbody intensity is $I_{\lambda b}(T)$. It is important to note that the T-dependence specifies the spatial variation (x, y, z in Cartesian coordinates) when the temperature of the emitting surface is defined. The energy leaving a black area element per unit time within $d\lambda$ and $d\Omega$ is

$$d^2Q_{\lambda b}(\theta,\phi,T)d\lambda = I_{\lambda b}\left(T\right)dA\cos\theta\, d\Omega d\lambda \tag{2.2}$$

The emissivity is then defined as the ratio of the emissive ability of the real surface to that of a blackbody:

$$\textit{Directional spectral emissivity} \equiv \varepsilon_\lambda(\theta,\phi,T) = \frac{d^2Q_\lambda(\theta,\phi,T)\,d\lambda}{d^2Q_{\lambda b}(\theta,\phi,T)\,d\lambda} = \frac{I_\lambda(\theta,\phi,T)}{I_{\lambda b}(T)} \tag{2.3}$$

This is the most fundamental emissivity expression because it includes dependencies on wavelength, direction, and surface temperature.

Example 2.1

A surface heated to 1000 K in vacuum has a directional spectral emissivity of 0.70 at a wavelength of 5 μm. The emissivity is independent of azimuthal angle ϕ. What is the emitted spectral intensity in this direction?

From Equation 2.3, I_λ(5 μm, 60°, 1000 K) = ε_λ(5 μm, 60°, 1000 K) $I_{\lambda b}$(5 μm, 1000 K). Using Equation 1.15 for $I_{\lambda b}$(5 μm, 1000 K),

$$I_\lambda(5\ \mu\text{m}, 60°, 1000\ \text{K}) = 0.70 \times \frac{2C_1}{\lambda^5\left(e^{C_2/\lambda T}-1\right)} = 0.70 \times \frac{2 \times 0.59552 \times 10^8}{5^5\left(e^{14387.75/5000}-1\right)}$$

$$= 0.70 \times 2272.61 = 1591\ \text{W}/\left(\text{m}^2 \cdot \mu\text{m} \cdot \text{sr}\right)$$

From the directional spectral emissivity in Equation 2.3, an averaged emissivity can be derived by two approaches: averaging over all wavelengths or averaging over all directions.

2.2.2 Directional Total Emissivity $\varepsilon(\theta, \phi, T)$

The radiation emitted at all wavelengths into direction (θ, ϕ) is found by integrating the directional spectral intensity to obtain the *directional total intensity* (the term *total* denotes that radiation at all wavelengths is included):

$$I(\theta,\phi,T) = \int_{\lambda=0}^{\infty} I_\lambda(\theta,\phi,T)\,d\lambda$$

Using the expression from Equation 1.30, the total intensity is

$$I_b(T) = \int_{\lambda=0}^{\infty} I_{\lambda b}(T)\,d\lambda = \frac{\sigma T^4}{\pi}$$

The directional total emissivity is the ratio of $I(\theta, \phi, T)$ for the real surface to $I_b(T)$ emitted by a blackbody at the same temperature; that is

$$\textit{Directional total emissivity} \equiv \varepsilon(\theta,\phi,T) = \frac{I(\theta,\phi,T)}{I_b(T)} = \frac{\pi\int_{\lambda=0}^{\infty} I_\lambda(\theta,\phi,T)\,d\lambda}{\sigma T^4} \tag{2.4}$$

The $I_\lambda(\theta, \phi, T)$ in the numerator can be replaced in terms of $\varepsilon_\lambda(\theta, \phi, T)$ by using Equation 2.3 to give

Directional total emissivity (in terms of directional spectral emissivity)

$$\equiv \varepsilon(\theta,\phi,T) = \frac{\pi\int_{\lambda=0}^{\infty} \varepsilon_\lambda(\theta,\phi,T) I_{\lambda b}(T)\,d\lambda}{\sigma T^4} = \frac{\int_{\lambda=0}^{\infty} \varepsilon_\lambda(\theta,\phi,T) E_{\lambda b}(T)\,d\lambda}{\sigma T^4} \tag{2.5}$$

Thus, if the wavelength dependence of $\varepsilon_\lambda(\theta, \phi, T)$ is known, the $\varepsilon(\theta, \phi, T)$ is obtained from Equation 2.5. The total directional emissivity can be calculated accurately only if the spectral-directional emissivity is available. Total emissivity is an integrated average weighted by the blackbody intensity $I_{\lambda b}(T)$. Therefore, it strongly correlates with the ε_λ values within the spectral range where $I_{\lambda b}(T)$ peaks, which varies with the temperature of the emitting surface, T.

Example 2.2

For a surface at 700 K, the $\varepsilon_\lambda(\theta, \phi, T)$ can be approximated by 0.8 for $\lambda = 0$–$5\ \mu m$, and $\varepsilon_\lambda(\theta, \phi, T) = 0.4$ for $\lambda > 5\ \mu m$. What is the value of $\varepsilon(\theta, \phi, T)$?

From Equation 2.5, and with $E_{\lambda b} = \pi I_{\lambda b}$,

$$\varepsilon(\theta,\phi,T) = \int_0^{5T} 0.8\frac{E_{\lambda b}(T)}{T^5}d(\lambda T) + \int_{5T}^{\infty} 0.4\frac{E_{\lambda b}(T)}{T^5}d(\lambda T)$$

From Equation 1.35 and by use of Equation 1.34,

$$\varepsilon(\theta,\phi,T) = 0.8F_{0-3500} + 0.4F_{3500-\infty} = 0.8\times 0.38291 + 0.4\times 0.61709 = 0.553$$

Because 61.7% of the emitted blackbody radiation at 700 K is in the region of 5 μm, the result is heavily weighted toward the emissivity value of 0.4.

2.2.3 Hemispherical Spectral Emissivity $\varepsilon_\lambda(T)$

Now return to Equation 2.3 and consider the average obtained by integrating the directional spectral quantities over all directions of a hemispherical envelope centered over dA (Figure 2.2b). The spectral radiation emitted by a unit surface area into all directions of the hemisphere is the *hemispherical spectral emissive power* found by integrating the spectral energy per unit solid angle over all solid angles. This is analogous to Equation 1.13 for a blackbody and is given by (see integration notation in Section 2.1.2)

$$E_\lambda(T) = \int_{\phi=0}^{2\pi}\int_{\theta=0}^{\pi/2} I_\lambda(\theta,\phi,T)\cos\theta\sin\theta\,d\theta\,d\phi = \int_{\cap} I_\lambda(\theta,\phi,T)\cos\theta\,d\Omega$$

Using Equation 2.3, this becomes

$$E_\lambda(T) = I_{\lambda b}(T)\int_{\cap} \varepsilon_\lambda(\theta,\phi,T)\cos\theta\,d\Omega \tag{2.6}$$

For a blackbody, the hemispherical spectral emissive power is $E_{\lambda b}(T) = \pi I_{\lambda b}(T)$. The ratio of actual to blackbody emission from the surface is then

$$\textit{Hemispherical spectral emissivity}\left(\text{in terms of directional spectral emissivity}\right)$$
$$\equiv \varepsilon_\lambda(T) = \frac{E_\lambda(T)}{E_{\lambda b}(T)} = \frac{1}{\pi}\int_{\cap} \varepsilon_\lambda(\theta,\phi,T)\cos\theta d\Omega \tag{2.7}$$

2.2.4 Hemispherical Total Emissivity $\varepsilon(T)$

Hemispherical total emissivity represents the average of the spectral directional emissivity over both wavelength and direction. To derive the hemispherical total emissivity, consider that from a unit area, the spectral emissive power in any direction is derived from Equation 2.3 as $\varepsilon_\lambda(\theta, \phi, T)I_{\lambda b}(T)\cos\theta$. This is integrated over all λ and all directions to give the hemispherical total emissive power. Dividing by σT^4, the hemispherical emissive power of a blackbody, results in the emissivity:

$$\textit{Hemispherical total emissivity}\left(\text{in terms of directional spectral emissivity}\right)$$
$$\equiv \varepsilon(T) = \frac{E(T)}{E_b(T)} = \frac{\int_{\cap}\int_{\lambda=0}^{\infty}[\varepsilon_\lambda(\theta,\phi,T)I_{\lambda b}(T)d\lambda]\cos\theta d\Omega}{\sigma T^4} \tag{2.8}$$

Using Equation 2.5, this can be placed in a second form:

$$\textit{Hemispherical total emissivity}\left(\text{in terms of directional total emissivity}\right)$$
$$\equiv \varepsilon(T) = \frac{1}{\pi}\int_{\cap}\varepsilon(\theta,\phi,T)\cos\theta d\Omega \tag{2.9}$$

If the order of integrations in Equation 2.8 is interchanged and Equation 2.7 is then used, a third form is obtained:

$$\textit{Hemispherical total emissivity}\left(\text{in terms of hemispherical spectral emissivity}\right)$$
$$\equiv \varepsilon(T) = \frac{\pi\int_{\lambda=0}^{\infty}\varepsilon_\lambda(T)I_{\lambda b}(T)d\lambda}{\sigma T^4} = \frac{\int_{\lambda=0}^{\infty}\varepsilon_\lambda(T)E_{\lambda b}(T)d\lambda}{\sigma T^4} \tag{2.10}$$

To interpret Equation 2.10 physically, consider Figure 2.3. The spectral profile of $\varepsilon_\lambda(T)$ is provided at surface temperature $T = 1000$ K in Figure 2.3a. The solid curve in Figure 2.3b is the hemispherical spectral emissive power for a blackbody at T. The area under the solid curve is σT^4, which is the denominator in Equation 2.10 and equals the radiation emitted per unit area by a blackbody including all wavelengths and directions. The dashed curve in Figure 2.3b is the product $\varepsilon_\lambda(T) E_{\lambda b}(T)$, and the area under this curve is the numerator of Equation 2.10, which is the emission from the real surface. Hence, $\varepsilon(T)$ is the ratio of the area under the dashed curve to that under the solid curve. At each λ, the $\varepsilon_\lambda(T)$ is the ordinate of the dashed curve divided by the ordinate of the solid curve. In Figure 2.3b, the hemispherical spectral emissivity at 7.5 μm is ε_λ ($\lambda = 7.5$ μm, $T = 1000$ K) $= b/a$.

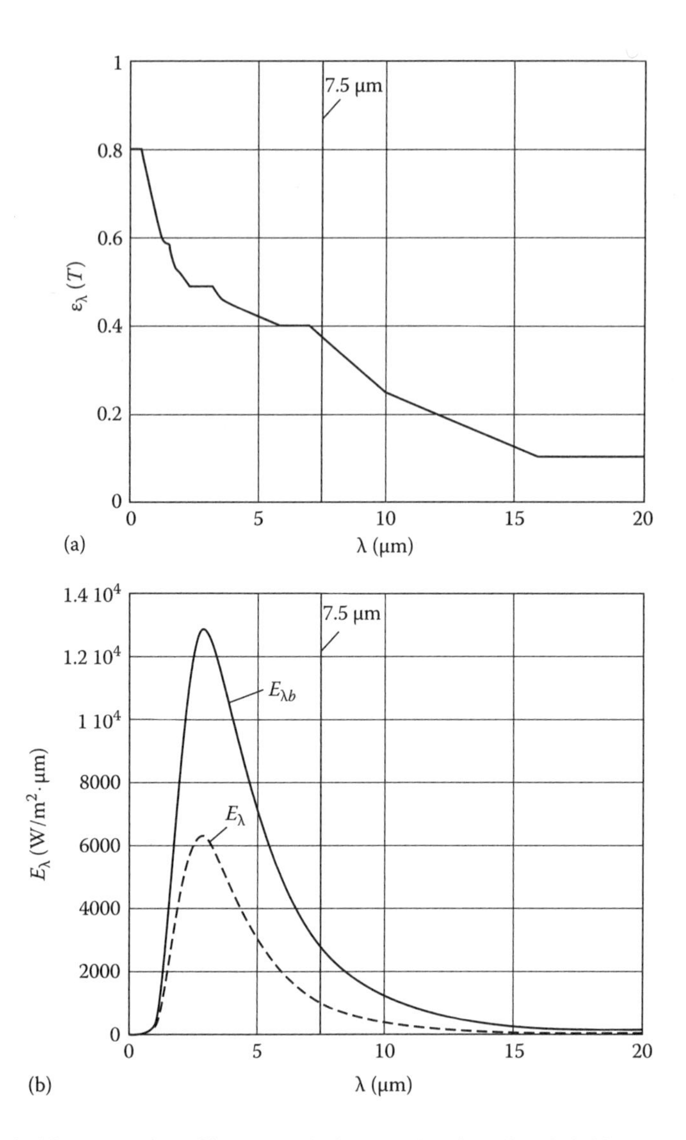

FIGURE 2.3 Physical interpretation of hemispherical spectral and total emissivities. (a) Measured emissivity values at T = 1000 K; (b) interpretation of emissivity as ratio of actual emissive power (dashed curve) to blackbody emissive power (solid curve).

Example 2.3

A surface at 1000 K has an $\varepsilon(\theta, \phi, T_A)$ that is independent of ϕ, but depends on θ as in Figure 2.4. What are the hemispherical total emissivity and the hemispherical total emissive power?

The $\varepsilon(\theta, 1000\text{ K})$ is approximated by the function $0.85\cos\theta$. Then, from Equation 2.9, the total hemispherical emissivity is

$$\varepsilon(T) = \int_{\cap} 0.85\cos^2\theta d\Omega = \frac{1}{\pi}\int_{\phi=0}^{2\pi}\int_{\theta=0}^{\pi/2} 0.85\cos^2\theta\sin\theta d\theta d\phi$$

$$= -1.70\frac{\cos^3\theta}{3}\bigg|_0^{\pi/2} = 0.567$$

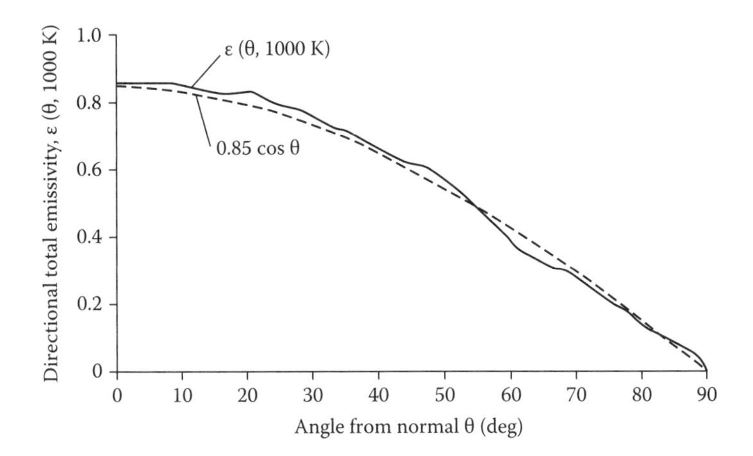

FIGURE 2.4 Directional total emissivity at 1000 K for Example 2.3.

The hemispherical total emissive power is then

$$E(T) = \varepsilon(T)\sigma T^4 = 0.567 \times 5.6704 \times 10^{-8} \times 1{,}000^4 = 32{,}150 \text{ W/m}^2$$

Generally, $\varepsilon(\theta, T)$ cannot be approximated very well by a convenient analytical function, and numerical integration of Equation 2.9 is necessary.

Example 2.4

The $\varepsilon_\lambda(\lambda, T_A)$ for a surface at $T_A = 1000$ K is approximated by the solid line as shown in Figure 2.5. What are the hemispherical total emissivity and the hemispherical total emissive power of the surface?

From Equation 2.10,

$$\varepsilon(T) = \frac{1}{\sigma T^4} \int_{\lambda=0}^{\infty} \varepsilon_\lambda(T) E_{\lambda b}(T) d\lambda = \frac{1}{\sigma} \int_{\lambda_T=0}^{2000} 0.1 \frac{E_{\lambda b}(T)}{T^5} d(\lambda T)$$

$$+ \frac{1}{\sigma} \int_{\lambda_T=2000}^{6000} 0.4 \frac{E_{\lambda b}(T)}{T^5} d(\lambda T) + \frac{1}{\sigma} \int_{\lambda_T=6000}^{\infty} 0.2 \frac{E_{\lambda b}(T)}{T^5} d(\lambda T)$$

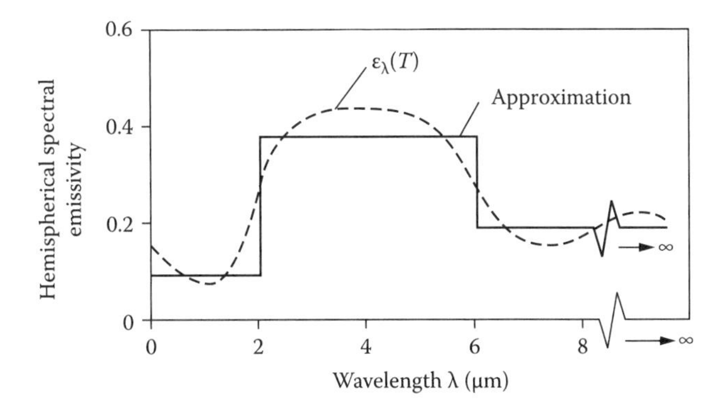

FIGURE 2.5 Hemispherical spectral emissivity for Example 2.4. Surface temperature $T = 1000$ K.

From Equations 1.35 and 1.37, this is evaluated as

$$\varepsilon(T_A) = 0.1F_{0\to2000} + 0.4(F_{0\to6000} - F_{0\to2000}) + 0.2(1 - F_{0\to6000}) = 0.3275$$

It is often easier and/or more accurate to numerically integrate Equation 2.10 using the actual $\varepsilon_\lambda(T)$. The hemispherical total emissive power is

$$E(T) = \varepsilon(T)\sigma T^4 = 0.3275 \times 5.67040 \times 10^{-8} \times 1000^4 = 18.57\ \text{kW/m}^2$$

Example 2.5

The $\varepsilon_\lambda(T)$ for a surface at T = 650 K is approximated as shown in Figure 2.6. What is the hemispherical total emissivity?

Numerical integration is used for the portion where $3.5 \le \lambda \le 9.5$ μm. This part of the spectral emissivity is given by $\varepsilon_\lambda(T) = 1.27917 - 0.10833\ \lambda$. The λT bounds are 3.5 × 650 = 2275 μm·K and 9.5 × 650 = 6175 μm·K. Then, as in Example 2.4, and by use of Equations 1.29, 1.31, 1.35, and 1.37,

$$\varepsilon(650\ \text{K}) = 0.90F_{0\to2275} + \frac{1}{\sigma 650^4}\int_{3.5}^{9.5}(1.27917 - 0.10833\lambda)\frac{2\pi C_1}{\lambda^5\left(e^{C_2/650\lambda} - 1\right)}d\lambda + 0.25(1 - F_{0\to6175})$$

Evaluating the integral numerically yields

$$\varepsilon(650\ \text{K}) = 0.90 \times 0.11514 + 0.40141 + 0.25(1 - 0.75214) = 0.5670$$

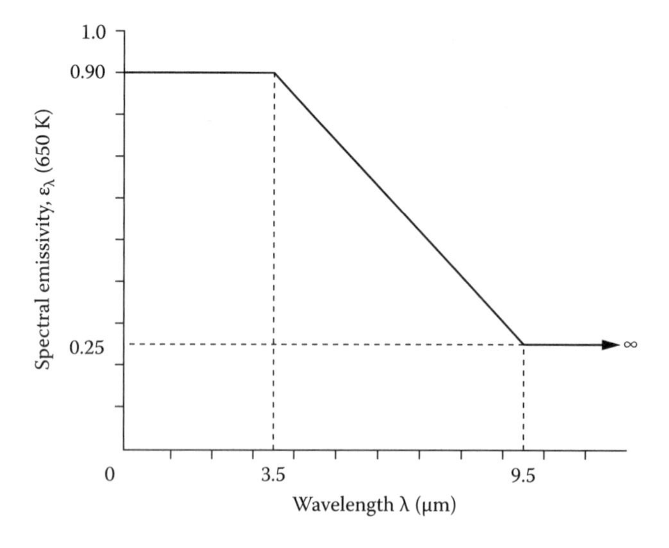

FIGURE 2.6 Spectral emissivity for Example 2.5.

2.3 ABSORPTIVITY

The *absorptivity* is defined as the fraction of the energy incident on a body that is absorbed by the body. The incident radiation depends on the radiative conditions (spectral intensity) at the *source* of the incident energy. The spectral distribution of the incident radiation is independent of the temperature or physical nature of the absorbing surface unless radiation emitted from the surface is partially reflected from the source or surroundings back to the surface. Compared with emissivity, the absorptivity has additional complexities because directional and spectral characteristics of the incident radiation must be included along with the absorbing surface temperature. It is desirable to have relations between emissivity and absorptivity so that measured values of one will allow the other to be calculated. Such relations are developed in this section, along with the absorptivity definitions.

2.3.1 Directional Spectral Absorptivity $\alpha_\lambda(\theta_i, \phi_i, T)$

Figure 2.2c illustrates the energy incident on a surface element dA from the (θ_i, ϕ_i) direction. The line from dA in the direction (θ_i, ϕ_i) passes normally through an area element dA_i on the surface of a hemisphere of radius R centered over dA. The incident spectral intensity passing through dA_i is $I_{\lambda,i}(\theta_i, \phi_i)$. This is the energy per unit area of the hemisphere, per unit solid angle $d\Omega_i$, per unit time, and per unit wavelength interval. The energy within the incident solid angle $d\Omega_i$ strikes the area dA of the absorbing surface. Hence, the energy per unit time incident from the direction (θ_i, ϕ_i) in the wavelength interval $d\lambda$ is

$$d^2Q_{\lambda,i}(\theta_i,\phi_i)\,d\lambda = I_{\lambda,i}(\theta_i,\phi_i)\,dA_i\,d\Omega\,d\lambda = I_{\lambda,i}(\theta_i,\phi_i)\,dA_i\frac{dA\cos\theta_i}{R^2}\,d\lambda \tag{2.11}$$

where $dA \cos \theta/R^2$ is the solid angle $d\Omega$ subtended by dA when viewed from dA_i.

Equation 2.11 can also be expressed in terms of the solid angle $d\Omega_i$ in Figure 2.7b. This is the solid angle subtended by dA_i when viewed from dA. The $d\Omega_i$ has its vertex at dA and hence is the convenient solid angle to use when integrating to obtain energy incident on dA from more than one direction. For a *nonabsorbing medium* in the region above the surface, as is being considered here, the incident intensity *does not change along the path* from dA_i to dA (see Section 1.8). For these reasons, in the figures that follow, the energy incident on dA from dA_i will be pictured as that arriving in $d\Omega_i$ as shown in Figure 2.7b rather than as the energy leaving dA in $d\Omega$ as in Figure 2.7a. To place Equation 2.11 in terms of $d\Omega_i$, note that

$$dA_i\frac{dA\cos\theta_i}{R^2} = \frac{dA_i}{R^2}dA\cos\theta_i = d\Omega_i dA\cos\theta_i \tag{2.12}$$

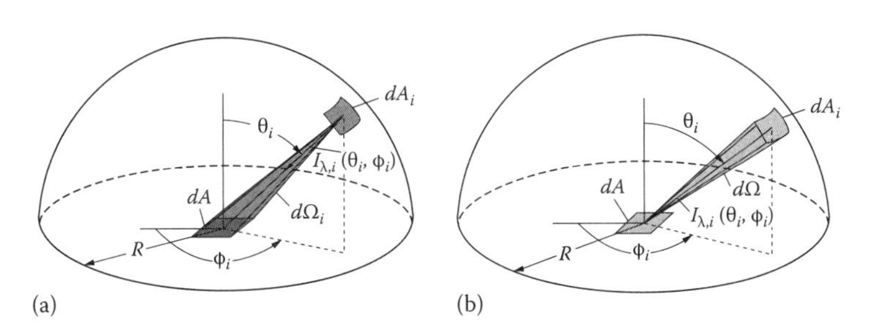

FIGURE 2.7 Equivalent ways of showing energy from dA_i that is incident upon dA. (a) Incidence within solid angle $d\Omega_i$ having origin at dA_i; (b) incidence within solid angle $d\Omega$ having origin at dA.

Equation 2.11 is then

$$d^2Q_{\lambda,i}\left(\theta_i,\phi_i\right)d\lambda = I_{\lambda,i}\left(\theta_i,\phi_i\right)d\Omega_i dA\cos\theta_i d\lambda \tag{2.13}$$

The fraction of the incident energy $d^2Q_{\lambda,i}(\theta_i, \phi_i, T)d\lambda$ that is absorbed is defined as the *directional spectral absorptivity* $\alpha_\lambda(\theta_i, \phi_i, T)$. In addition to depending on the wavelength and direction of the incident radiation, the spectral absorptivity is a function of the absorbing surface temperature. The amount of the incident energy that is absorbed is $d^2Q_{\lambda,a}(\theta_i, \phi_i, T)d\lambda$. Then, the ratio is formed

$$\textit{Directional spectral absorptivity} \equiv \alpha_\lambda(\theta_i,\phi_i,T) = \frac{d^2Q_{\lambda,a}(\theta_i,\phi_i,T)d\lambda}{I_{\lambda,i}(\theta_i,\phi_i)dA\cos\theta_i d\Omega_i d\lambda} \tag{2.14}$$

If the incident energy is from *black surroundings* at uniform temperature T_b, then there is the special case

$$\alpha_\lambda(\theta_i,\phi_i,T) = \frac{d^2Q_{\lambda,a}(\theta_i,\phi_i,T)d\lambda}{I_{\lambda b}(T_b)dA\cos\theta_i d\Omega_i d\lambda} \tag{2.15}$$

2.3.2 Kirchhoff's Law

Kirchhoff's law relates the emitting and absorbing abilities of a body. The law has necessary conditions imposed on it, depending on whether spectral, total, directional, or hemispherical quantities are being considered. From Equations 2.1 and 2.3, the energy emitted per unit time by an element dA in a wavelength interval $d\lambda$ and solid angle $d\Omega$ is

$$d^2Q_{\lambda,e}d\lambda = \varepsilon_\lambda\left(\theta,\phi,T\right)I_{\lambda b}\left(T\right)dA\cos\theta d\Omega d\lambda \tag{2.16}$$

If the element dA at temperature T is placed in an isothermal black enclosure also at temperature T, then the intensity of the energy incident on dA from the direction (θ_i, ϕ_i) (recalling the isotropy of intensity in a black enclosure) is $I_{\lambda b}(T)$. To maintain radiation isotropy within the black enclosure (i.e., to maintain an energy balance with the element of the black enclosure that is emitting the intensity that is incident on dA at λ, θ_i, ϕ_i), the absorbed and emitted energies in Equations 2.15 and 2.16 must be equal. Equating gives

$$\varepsilon_\lambda\left(\theta,\phi,T\right) = \alpha_\lambda\left(\theta,\phi,T\right) \tag{2.17}$$

This relation between the properties of the material holds without restriction. This is the most *general form of Kirchhoff's law.*

As discussed in Chapters 3 and 14, radiation is polarized, having two wave components vibrating at right angles to each other and the propagation direction. For the special case of blackbody radiation, the two components are equal. To be strictly accurate, Equation 2.17 holds only for each component of polarization, and for Equation 2.17 to be valid as written for all incident energy, the incident radiation must have equal components of polarization.

2.3.3 Directional Total Absorptivity $\alpha(\theta_i, \phi_i, T)$

The directional total absorptivity is the energy including all wavelengths that is absorbed from a given direction, divided by the energy incident from that direction. The total energy incident from the given direction is obtained by integrating the spectral incident energy (Equation 2.13) over all wavelengths to obtain

$$d^2Q_i(\theta_i,\phi_i) = dA\cos\theta_i d\Omega_i \int_{\lambda=0}^{\infty} I_{\lambda,i}(\theta_i,\phi_i)d\lambda \tag{2.18}$$

The radiation absorbed is determined from Equation 2.14 by integrating over all λ,

$$d^2Q_a(\theta_i,\phi_i,T_i) = dA\cos\theta_i d\Omega_i \int_{\lambda=0}^{\infty} \alpha_\lambda(\theta_i,\phi_i,T)I_{\lambda,i}(\theta_i,\phi_i)d\lambda \tag{2.19}$$

The ratio is then formed:

$$\text{Directional total absorptivity} \equiv \alpha(\theta_i,\phi_i,T) = \frac{d^2Q_a(\theta_i,\phi_i,T)}{d^2Q_i(\theta_i,\phi_i)}$$

$$= \frac{\int_{\lambda=0}^{\infty} \alpha_\lambda(\theta_i,\phi_i,T)I_{\lambda,i}(\theta_i,\phi_i)d\lambda}{\int_{\lambda=0}^{\infty} I_{\lambda,i}(\theta_i,\phi_i)d\lambda} \tag{2.20}$$

Using Kirchhoff's law, Equation 2.17, an alternative form is

$$\alpha(\theta_i,\phi_i,T_i) = \frac{\int_{\lambda=0}^{\infty} \varepsilon_\lambda(\theta_i,\phi_i,T)I_{\lambda,i}(\theta_i,\phi_i)d\lambda}{\int_{\lambda=0}^{\infty} I_{\lambda,i}(\theta_i,\phi_i)d\lambda} \tag{2.21}$$

Kirchhoff's law is derived for thermodynamic equilibrium in an isothermal enclosure and is strictly true only when there is no net heat transfer at a surface. When heat transfer is present, Equation 2.17 is an approximation. Experimental evidence and detailed analysis indicates that in most practical applications, $\alpha_\lambda(\theta_i, \phi_i, T)$ and $\varepsilon_\lambda(\theta, \phi, T)$ are not significantly influenced by the surrounding radiation field. A solid or liquid is able to maintain itself in a local thermodynamic equilibrium in which the populations of energy states that govern the emission and absorption processes are given to a very close approximation by their equilibrium distributions at the local temperature.

2.3.4 Kirchhoff's Law for Directional Total Properties

The general form of Kirchhoff's law (Equation 2.17) shows that the directional-spectral properties $\varepsilon_\lambda(\theta, \phi, T)$ and $\alpha_\lambda(\theta_i, \phi_i, T)$ are equal. It is now of interest to examine this equality for the *directional total* quantities. This can be accomplished by comparing a special case of Equation 2.21 with

Equation 2.5. If in Equation 2.21 the incident radiation has a spectral distribution proportional to that of a blackbody at T, then $I_{\lambda,i}(\theta_i, \phi_i) = C(\theta_i, \phi_i)I_{\lambda b}(T)$. Equation 2.21 becomes

$$\alpha(\theta_i, \phi_i, T) = \frac{\int_{\lambda=0}^{\infty} \varepsilon_\lambda(\theta = \theta_i, \phi = \phi_i, T) I_{\lambda b}(T) d\lambda}{\int_{\lambda=0}^{\infty} I_{\lambda b}(T) d\lambda \left(= \sigma T^4/\pi\right)} = \varepsilon(\theta = \theta_i, \phi = \phi_i, T)$$

Hence, when $\varepsilon_\lambda(\theta_i, \phi_i, T)$ and $\alpha_\lambda(\theta_i, \phi_i, T)$ are dependent on wavelength, there is the equality $\alpha(\theta_i, \phi_i, T) = \varepsilon(\theta_i, \phi_i, T)$ only when the incident radiation meets the restriction $I_{\lambda,i}(\theta_i, \phi_i) = C(\theta_i, \phi_i)I_{\lambda b}(T)$, where C is independent of wavelength.

There is another important case when the equality $\alpha(\theta_i, \phi_i, T) = \varepsilon(\theta_i, \phi_i, T)$ is valid. If the directional emission from a surface has the same wavelength dependence as the emission from a blackbody, $I_{\lambda,i}(\theta_i, \phi_i) = C(\theta_i, \phi_i)I_{\lambda b}(T)$, then $\varepsilon_\lambda(\theta_i, \phi_i, T)$ is independent of λ. From Equations 2.5 and 2.21, if $\varepsilon_\lambda(\theta_i, \phi_i, T)$ and hence $\alpha_\lambda(\theta_i, \phi_i, T)$ do not depend on λ, then, *for the direction* (θ_i, ϕ_i), $\varepsilon_\lambda(\theta_i, \phi_i, T)$, $\alpha_\lambda(\theta_i, \phi_i, T)$, $\varepsilon(\theta_i, \phi_i, T)$ and $\alpha(\theta_i, \phi_i, T)$ are all equal. A surface exhibiting such behavior is termed a *directional gray* surface.

2.3.5 Hemispherical Spectral Absorptivity $\alpha_\lambda(T)$

Now, consider energy in a wavelength interval $d\lambda$. The hemispherical spectral absorptivity is the fraction of the spectral energy that is absorbed from the spectral energy that is incident from all directions of a surrounding hemisphere (Figure 2.2d). The spectral energy from an element dA_i on the hemisphere that is intercepted by a surface element dA is given by Equation 2.13. The incident energy on dA from all directions of the hemisphere is then given by the integral

$$dQ_{\lambda,i} d\lambda = dA d\lambda \int_{\cap} I_{\lambda,i}(\theta_i, \phi_i) \cos\theta_i d\Omega_i \tag{2.22}$$

The amount absorbed is given by integrating Equation 2.14 over the hemisphere:

$$dQ_{\lambda,a}(T) d\lambda = dA d\lambda \int_{\cap} \alpha_\lambda(\theta_i, \phi_i, T) I_{\lambda,i}(\theta_i, \phi_i) \cos\theta_i d\Omega_i \tag{2.23}$$

The ratio of these quantities gives

$$\textit{Hemispherical spectral absorptivity} \equiv \alpha_\lambda(T) = \frac{dQ_{\lambda,a}(T) d\lambda}{dQ_{\lambda,i} d\lambda}$$

$$= \frac{\int_{\cap} \alpha_\lambda(\theta_i, \phi_i, T) I_{\lambda,i}(\theta_i, \phi_i) \cos\theta_i d\Omega_i}{\int_{\cap} I_{\lambda,i}(\theta_i, \phi_i) \cos\theta_i d\Omega_i} \tag{2.24}$$

or, in terms of emissivity by using Kirchhoff's law (Equation 2.17),

$$\alpha_\lambda(T) = \frac{\int_{\cap} \varepsilon_\lambda(\theta = \theta_i, \phi = \phi_i, T) I_{\lambda,i}(\theta_i, \phi_i) \cos\theta_i d\Omega_i}{\int_{\cap} I_{\lambda,i}(\theta_i, \phi_i) \cos\theta_i d\Omega_i} \tag{2.25}$$

The hemispherical spectral absorptivity and emissivity can now be compared looking at Equations 2.25 and 2.7. For the general case where α_λ and ε_λ are functions of (λ, θ, ϕ) and T, $\alpha_\lambda(T) = \varepsilon_\lambda(T)$ only if $I_{\lambda,i}(\theta_i, \phi_i)$ is independent of θ_i and ϕ_i, that is, if the incident spectral intensity is uniform from all directions. If this is so, the $I_{\lambda,i}$ can be canceled in Equation 2.25 and the denominator becomes π, and Equation 2.25 then compares with Equation 2.7.

For the case where $\alpha_\lambda(\theta, \phi, T) = \varepsilon_\lambda(\theta, \phi, T) = \alpha_\lambda(T) = \varepsilon_\lambda(T)$, that is, the directional spectral properties are independent of angle, the hemispherical spectral properties are related by $\alpha_\lambda(T) = \varepsilon_\lambda(T)$ for any angular variation of incident spectral intensity. Such a surface is termed a *diffuse spectral surface.*

2.3.6 Hemispherical Total Absorptivity $\alpha(T)$

The hemispherical total absorptivity represents the fraction of energy absorbed that is incident from all directions of the enclosing hemisphere and for all wavelengths, as shown in Figure 2.2d. The total incident energy that is intercepted by a surface element dA is determined by integrating Equation 2.13 over all λ and all (θ_i, ϕ_i) of the hemisphere, which results in

$$dQ_i = dA \int_\cap \left[\int_{\lambda=0}^{\infty} I_{\lambda,i}(\theta_i, \phi_i) d\lambda \right] \cos\theta_i d\Omega_i \tag{2.26}$$

Similarly, by integrating Equation 2.14, the total amount of energy absorbed is

$$dQ_a(T) = dA \int_\cap \left[\int_{\lambda=0}^{\infty} \alpha_\lambda(\theta_i, \phi_i, T) I_{\lambda,i}(\theta_i, \phi_i) d\lambda \right] \cos\theta_i d\Omega_i \tag{2.27}$$

The ratio of absorbed to incident energy provides the definition

Hemispherical total absorptivity (in terms of directional spectral absorptivity or emissivity)

$$\equiv \alpha(T) = \frac{dQ_a(T)}{dQ_i} = \frac{\int_\cap \left[\int_{\lambda=0}^{\infty} \alpha_\lambda(\theta_i, \phi_i, T) I_{\lambda,i}(\theta_i, \phi_i) d\lambda \right] \cos\theta_i d\Omega_i}{\int_\cap \left[\int_{\lambda=0}^{\infty} I_{\lambda,i}(\theta_i, \phi_i) d\lambda \right] \cos\theta_i d\Omega_i} \tag{2.28}$$

or, in terms of emissivity by using Kirchhoff's law (Equation 2.17),

$$\alpha(T) = \frac{\int_\cap \left[\int_{\lambda=0}^{\infty} \varepsilon_\lambda\left(\theta = \theta_i, \phi = \phi_i, T\right) I_{\lambda,i}\left(\theta_i, \phi_i\right) d\lambda \right] \cos\theta_i d\Omega_i}{\int_\cap \left[\int_{\lambda=0}^{\infty} I_{\lambda,i}\left(\theta_i, \phi_i\right) d\lambda \right] \cos\theta_i d\Omega_i} \tag{2.29}$$

Equation 2.29 can be compared with Equation 2.8 to determine the conditions when the hemispherical total absorptivity and emissivity are equal. From Equation 2.8, $\int_\cap \left[\int_{\lambda=0}^{\infty} I_{\lambda b}(T) d\lambda \right] \cos\theta d\Omega = \sigma T^4$. The comparison reveals that, for the general case when $\varepsilon_\lambda(\theta, \phi, T)$ and $\alpha_\lambda(\theta_i, \phi_i, T)$ vary with both wavelength and angle, $\alpha(T) = \varepsilon(T)$ *only when the*

TABLE 2.2
Summary of Kirchhoff's Law Relations between Absorptivity and Emissivity

Type of Quantity	Equality	Restrictions
Directional spectral	$\alpha_\lambda(\theta, \phi, T) = \varepsilon_\lambda(\theta, \phi, T)$	None
Directional total	$\alpha(\theta, \phi, T) = \varepsilon(\theta, \phi, T)$	Incident radiation must have a spectral distribution proportional to that of a blackbody at T, $I_{\lambda,i}(\theta, \phi) = C(\theta, \phi)\, I_{\lambda,b}(T)$ or $\alpha_\lambda(\theta, \phi, T) = \varepsilon_\lambda(\theta, \phi, T)$ are independent of wavelength (directional-gray surface)
Hemispherical spectral	$\alpha_\lambda(T) = \varepsilon_\lambda(T)$	Incident radiation must be independent of angle, $I_{\lambda,i} = C(\lambda)$, or $\alpha_\lambda(T) = \varepsilon_\lambda(T)$ do not depend on angle (diffuse-spectral surface)
Hemispherical total	$\alpha(T) = \varepsilon(T)$	Incident radiation must be independent of angle and have a spectral distribution proportional to that of a blackbody at T, $I_{\lambda,i} = CI_{\lambda b}(T)$, or incident radiation independent of angle and $\alpha_\lambda(\theta, \phi, T) = \varepsilon_\lambda(\theta, \phi, T)$ are independent of λ (directional-gray surface), or incident radiation from each direction has spectral distribution proportional to that of a blackbody at T_A and $\alpha_\lambda(T) = \varepsilon_\lambda(T)$ are independent of angle (diffuse-spectral surface), or $\alpha_\lambda(T) = \varepsilon_\lambda(T)$ are independent of wavelength and angle (diffuse-gray surface)

incident intensity is independent of the incident angle and has the same spectral form as that emitted by a blackbody with temperature equal to the surface temperature T, that is, only when $I_{\lambda,i}(\theta_i, \phi_i) = CI_{\lambda b}(T)$, where C is a positive constant. Some more restrictive cases are in Table 2.2.

Substituting Equation 2.20 into Equation 2.28 gives the alternative form:

Hemispherical total absorptivity (in terms of directional total absorptivity)

$$\equiv \alpha(T) = \frac{\int_{\cap} \alpha(\theta_i, \phi_i, T) I_i(\theta_i, \phi_i) \cos\theta_i d\Omega_i}{\int_{\cap} I_i(\theta_i, \phi_i) \cos\theta_i d\Omega_i} \tag{2.30}$$

where $I_i(\theta_i, \phi_i) = \int_{\lambda=0}^{\infty} I_{\lambda,i}(\theta_i, \phi_i) d\lambda$ is the total incident intensity from direction (θ_i, ϕ_i). Changing the order of integration in Equation 2.28 and substituting into Equation 2.24 gives

Hemispherical total absorptivity (in terms of hemispherical spectral absorptivity)

$$\equiv \alpha(T) = \frac{\int_{\lambda=0}^{\infty} \alpha_\lambda(T) dQ_{\lambda,i} d\lambda}{\int_{\lambda=0}^{\infty} dQ_{\lambda,i} d\lambda} \tag{2.31}$$

where $dQ_{\lambda,i} d\lambda = dA d\lambda \int_{\cap} I_{\lambda,i}(\theta_i, \phi_i) \cos\theta_i d\Omega_i$ is the spectral radiation incident from all directions that is intercepted by the surface element dA.

A special case of hemispherical total absorptivity is asked as an assignment in Homework Problem 2.1 at the end of this chapter. If there is a uniform incident intensity from a gray source at T_i, and if $\varepsilon_\lambda(T)$ is independent of T, then the hemispherical total absorptivity for the incident radiation is equal to the hemispherical total emissivity of the material evaluated at the source temperature T_i, $\alpha(T) = \varepsilon(T_i)$.

Example 2.6

The hemispherical spectral emissivity of a surface at 300 K is approximated by a step function: $\varepsilon_\lambda(\lambda, T) = 0.8$ for $0 \le \lambda \le 3$ μm, and 0.2 for $\lambda > 3$ μm. What is the hemispherical total absorptivity for diffuse incident radiation from a black source at $T_i = 1000$ K? What is it for diffuse incident solar radiation?

For diffuse incident radiation, $\alpha_\lambda(T) = \varepsilon_\lambda(T)$, and for a black source at T_i, Equation 2.29 becomes

$$\alpha(T) = \frac{\int_{\lambda=0}^{\infty} \varepsilon_\lambda(T) I_{\lambda b,i}(T_i)\, d\lambda}{\int_{\lambda=0}^{\infty} I_{\lambda b,i}(T_i)\, d\lambda} = \frac{\int_{\lambda=0}^{\infty} \varepsilon_\lambda(T) E_{\lambda b,i}(T_i)\, d\lambda}{\sigma T_i^4}$$

Expressing $\alpha(T)$ in terms of the two wavelength regions over which $\varepsilon_\lambda(T)$ is constant gives $\alpha(T) = 0.8F_{0\to 3T_i} + 0.2(1-F_{0\to 3T_i})$. From Equation 1.37 $F_{0\to 3\ \mu m \cdot 1000\ K} = 0.27323$ so that

$$\alpha(T) = 0.8F_{0\to 3000} + 0.2(1 - F_{0\to 3000}) = 0.364$$

For incident solar radiation, $T_i = 5780$ K, $F_{0\to 3\ \mu m \cdot 5780\ K} = 0.97880$ so that

$$\alpha(T) = 0.8F_{0\to 17.340} + 0.2(1 - F_{0\to 17,340}) = 0.787$$

Hence, for a surface with this type of spectral emissivity variation, there is a considerable increase in $\alpha(T)$ by the shift of the incident-energy spectrum toward shorter wavelengths as the source temperature is raised.

2.3.7 Diffuse-Gray Surface

As discussed in Chapters 4 and 5, a common assumption in enclosure calculations is that surfaces are *diffuse-gray. Diffuse* signifies that the directional emissivity and directional absorptivity do not depend on direction. Hence, for emission, the emitted intensity is uniform over all directions as for a blackbody. The term *gray* signifies that the spectral emissivity and absorptivity do not depend on wavelength. They can, however, depend on temperature. Thus at each surface temperature, for any wavelength the emitted spectral radiation is a fixed fraction of blackbody spectral radiation.

The diffuse-gray surface therefore *absorbs a fixed fraction of incident radiation from any direction and at any wavelength.* It emits radiation that is a *fixed fraction of blackbody radiation for all directions and all wavelengths* (this is the motivation for the term "gray"). The directional-spectral absorptivity and emissivity are then independent of λ, θ, ϕ so that $\alpha_\lambda(\theta_i, \phi_i, T) = \alpha(T)$ and $\varepsilon_\lambda(\theta, \phi, T) = \varepsilon(T)$, and from Kirchhoff's law, Equation 2.17, $\alpha(T) = \varepsilon(T)$. In Equations 2.8, 2.28, and 2.29, since the $\alpha_\lambda(\theta_i, \phi_i, T)$ and $\varepsilon_\lambda(\theta, \phi, T)$ are not functions of either direction or wavelength, they can be taken out of the integrals. Then, for a diffuse-gray surface, the directional-spectral and hemispherical-total values of absorptivity and emissivity are *all equal*, and the hemispherical total absorptivity is *independent* of the nature of the incident radiation. This provides a great simplification for analysis at the expense of accuracy.

For diffuse-gray and other surface characteristics, the restrictions for applying Kirchhoff's law are summarized in Table 2.2.

Example 2.7

The surface in Example 2.5 at $T = 650$ K is subject to incident radiation from a diffuse-gray source at $T_i = 925$ K. What is the hemispherical-total absorptivity?

For a diffuse-gray source at T_i, $I_{\lambda,i}(\theta_i, \phi_i) = CI_{\lambda b}(T_i)$ where C is a constant. From Table 2.2, since there is no angular dependence, $\alpha_\lambda(T) = \varepsilon_\lambda(T)$. Then from Equation 2.30,

$$\alpha(T) = \frac{\int_\cap \left[\int_{\lambda=0}^{\infty} \varepsilon_\lambda(T) CI_{\lambda b}(T_i)\, d\lambda\right] \cos\theta_i d\Omega_i}{\int_\cap \left[\int_{\lambda=0}^{\infty} CI_{\lambda b}(T_i)\, d\lambda\right] \cos\theta_i d\Omega_i} = \frac{1}{\sigma T_i^4}\int_{\lambda=0}^{\infty} \varepsilon_\lambda(T) E_{\lambda b}(T_i)\, d\lambda$$

As in Example 2.5, using λT_i values of $3.5 \times 925 = 3238$ μm·K and $9.5 \times 925 = 8788$ μm·K,

$$\alpha = 0.90 F_{0\to 3238} + \frac{1}{\sigma 925^4}\int_{3.5}^{9.5} (1.27917 - 0.10833\lambda)\frac{2\pi C_1}{\lambda^5 (e^{C_2/925\lambda} - 1)} d\lambda + 0.25(1 - F_{0\to 8788})$$

Numerical integration yields

$$\alpha = 0.90 \times 0.32650 + 0.37872 + 0.25(1 - 0.88377) = 0.7016$$

2.4 REFLECTIVITY

The reflective properties of a surface are more complicated to specify than the emissivity or absorptivity. This is because reflected energy depends not only on the angle at which the incident energy impinges on the surfaces but also on the direction being considered for the reflected energy. Reflectivities are functions of the reflective surface temperature T. We omit explicitly showing this temperature-dependence for clarity, but it should be understood as for emissivity and absorptivity. The important reflectivity quantities are now defined.

2.4.1 Spectral Reflectivities

2.4.1.1 Bidirectional Spectral Reflectivity $\rho_\lambda(\theta_r, \phi_r, \theta_i, \phi_i)$

Consider spectral radiation incident on a surface from direction (θ_i, ϕ_i) as in Figure 2.2e. Part of this energy is reflected into the (θ_r, ϕ_r) direction and provides part of the reflected intensity in the (θ_r, ϕ_r) direction. The subscript r denotes quantities evaluated at the reflected angle. The entire $I_{\lambda,r}(\theta_r, \phi_r)$ is the result of summing the reflected intensities produced by the incident intensities $I_{\lambda,i}(\theta_i, \phi_i)$ from all incident directions (θ_i, ϕ_i) of the hemisphere surrounding the surface element. The contribution to $I_{\lambda,r}(\theta_r, \phi_r)$ produced by the incident energy from only one direction (θ_i, ϕ_i) is designated as $I_{\lambda,r}(\theta_r, \phi_r, \theta_i, \phi_i)$.

The energy from direction (θ_i, ϕ_i) intercepted by dA per unit area and wavelength is, from Equation 2.13,

$$\frac{d^2 Q_{\lambda,i}(\theta_i, \phi_i) d\lambda}{dA\, d\lambda} = I_{\lambda,i}(\theta_i, \phi_i) \cos\theta_i d\Omega_i \tag{2.32}$$

The *bidirectional spectral reflectivity* is a ratio expressing the contribution $I_{\lambda,i}(\theta_i, \phi_i)\cos\theta_i d\Omega_i$ makes to the reflected spectral intensity in the (θ_r, ϕ_r) direction:

$$\textit{Bidirectional spectral reflectivity} \equiv \rho_\lambda(\theta_r, \phi_r, \theta_i, \phi_i) = \frac{I_{\lambda,r}(\theta_r, \phi_r, \theta_i, \phi_i)}{I_{\lambda,i}(\theta_i, \phi_i)\cos\theta_i d\Omega_i} \tag{2.33}$$

The reflectivity also depends on surface temperature, but the T notation is omitted for simplicity. The ratio in Equation 2.33 is a reflected intensity divided by the intercepted energy arriving within solid angle $d\Omega_i$. Having $\cos\theta_i\, d\Omega_i$ in the denominator means that when $\rho_\lambda(\theta_r, \phi_r, \theta_i, \phi_i)I_{\lambda,i}(\theta_i, \phi_i)\cos\theta_i d\Omega_i$ is integrated over all incidence angles (θ_i, ϕ_i) to provide the reflected intensity $I_{\lambda,r}(\theta_r, \phi_r)$, the reflected intensity is properly weighted by the amount of energy intercepted from each direction. Since $I_{\lambda,r}(\theta_r, \phi_r, \theta_i, \phi_i)$ is generally one differential order smaller than $I_{\lambda,i}(\theta_i, \phi_i)$, the $d\Omega_i$ in the denominator prevents $\rho_\lambda(\theta_r, \phi_r, \theta_i, \phi_i)$ from being a differential quantity. The form of Equation 2.33 also enables some convenient reciprocity relations to be obtained. For a *diffuse reflection* the incident energy from each (θ_i, ϕ_i) contributes equally to the reflected intensity for all (θ_r, ϕ_r). A mirror-like (specular) reflection is a special case and is discussed separately. The measurement and application of the bidirectional reflectivity is discussed by Zaworski et al. (1996a,b).

2.4.1.2 Reciprocity for Bidirectional Spectral Reflectivity

The $\rho_\lambda(\theta_r, \phi_r, \theta_i, \phi_i)$ is symmetric with regard to reflection and incidence angles; the reflectivity for energy incident at (θ_i, ϕ_i) and reflected into (θ_r, ϕ_r) is equal to that for energy incident at (θ_r, ϕ_r) and reflected into (θ_i, ϕ_i). This is demonstrated by considering a nonblack element dA_2 in an isothermal black enclosure as in Figure 2.8. For the isothermal condition, the net energy exchange between black elements dA_1 and dA_3 must be zero. The energy exchange is by two paths. The direct exchange along the dashed line is between black elements at the same temperature and hence is zero. If the net exchange along this path is zero and net exchange including all paths between dA_1 and dA_3 is zero, then net exchange along the path having reflection from dA_2 must be zero. For the energy traveling along the reflected path,

$$d^3Q_{\lambda,1-2-3}d\lambda = d^3Q_{\lambda,3-2-1}d\lambda \tag{2.34}$$

The energy reflected from dA_2 that reaches dA_3 is

$$d^3Q_{\lambda,1-2-3}d\lambda = I_{\lambda,r}(\theta_r,\phi_r,\theta_i,\phi_i)\cos\theta_r dA_2(dA_3\cos\theta_3/S_2^2)d\lambda,$$

or, using Equation 2.33,

$$d^3Q_{\lambda,1-2-3}d\lambda = \rho_\lambda(\theta_r,\phi_r,\theta_i,\phi_i)I_{\lambda,1}(T)\cos\theta_i\left(\frac{dA_1\cos\theta_1}{S_1^2}\right)\cos\theta_r dA_2\left(\frac{dA_3\cos\theta_3}{S_2^2}\right)d\lambda \tag{2.35}$$

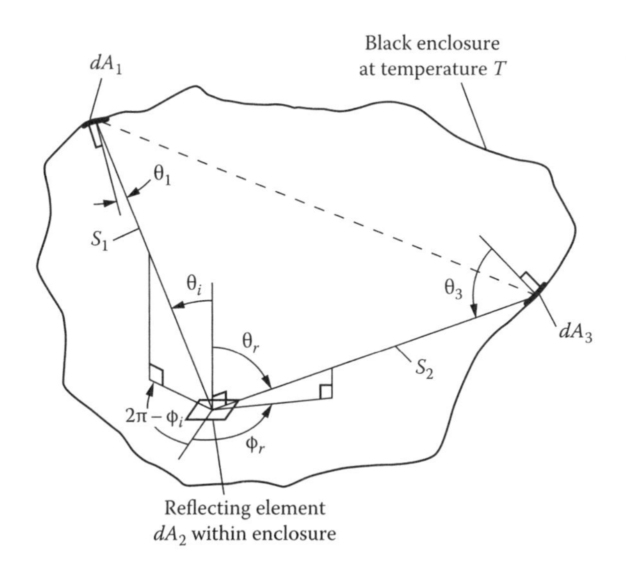

FIGURE 2.8 Enclosure used to examine reciprocity of bidirectional spectral reflectivity.

Similarly,

$$d^3Q_{\lambda,3-2-1}d\lambda = \rho_\lambda(\theta_i,\phi_i,\theta_r,\phi_r)I_{\lambda,3}(T)\cos\theta_r\left(\frac{dA_3\cos\theta_3}{S_2^2}\right)\cos\theta_i dA_2\left(\frac{dA_1\cos\theta_1}{S_1^2}\right)d\lambda \tag{2.36}$$

Then from Equation 2.34 and because $I_{\lambda,1}(T) = I_{\lambda,3}(T) = I_{\lambda b}(T)$, there is the reciprocity relation

$$\rho_\lambda(\theta_r,\phi_r,\theta_i,\phi_i) = \rho_\lambda(\theta_i,\phi_i,\theta_r,\phi_r) \tag{2.37}$$

The bidirectional reflectivity is important for interpreting signals in remote sensing of the Earth's characteristics (Walthall et al. 1985, Kimes et al. 1986). This and other experimental uses for the bidirectional reflectivity has initiated further investigations into the reciprocity relations, and whether the use of thermodynamic "detailed balancing" provides an adequate proof of bidirectional reciprocity (Clarke and Parry 1985, Snyder et al. 1998). It is presently believed that reciprocity is valid for conventional reflecting materials, but there may be exceptions for materials with very unusual optical characteristics, or in the presence of a strong magnetic field. An analysis of the bidirectional reflectivity distribution was made by Greffet and Nieto-Vesperinas (1998) using electromagnetic theory (basic electromagnetic theory relations are derived in Chapter 14 and applied in Chapter 3). The results developed from these basic principles rather than from thermodynamic arguments also indicate that reciprocity and Kirchhoff's law are indeed valid. The analysis accounts for penetration of electromagnetic waves into the medium beyond the reflecting surface, the scattering of these waves within the medium, and their re-emergence through the surface as part of the reflected component.

2.4.1.3 Directional Spectral Reflectivities

Multiplying $I_{\lambda,r}(\theta_r, \phi_r, \theta_i, \phi_i)$ by $d\lambda\cos\theta_r\, dA\, d\Omega_r$ and integrating over all θ_r and ϕ_r gives the energy per unit time reflected into the entire hemisphere from an incident intensity from a single direction (θ_i, ϕ_i):

$$d^2Q_{\lambda,r}(\theta_i,\phi_i)d\lambda = d\lambda dA\int_{\cap} I_{\lambda,r}(\theta_r,\phi_r,\theta_i,\phi_i)\cos\theta_r d\Omega_r$$

From Equation 2.33, this equals

$$d^2Q_{\lambda,r}(\theta_i,\phi_i)d\lambda = d\lambda dA I_{\lambda,i}(\theta_i,\phi_i)\cos\theta_i d\Omega_i\int_{\cap}\rho_\lambda(\theta_r,\phi_r,\theta_i,\phi_i)\cos\theta_r d\Omega_r \tag{2.38}$$

The *directional-hemispherical spectral reflectivity* is defined as the energy reflected into all solid angles divided by the energy incident from a particular direction (θ_i, ϕ_i) (Figure 2.2f). This gives Equation 2.38 divided by the incident energy from Equation 2.32:

Directional-hemispherical spectral reflectivity (in terms of bidirectional spectral reflectivity)

$$\equiv \rho_\lambda(\theta_i,\phi_i) = \frac{d^2Q_{\lambda,r}(\theta_i,\phi_i)d\lambda}{d^2Q_{\lambda,i}(\theta_i,\phi_i)d\lambda} = \int_{\cap}\rho_\lambda(\theta_r,\phi_r,\theta_i,\phi_i)\cos\theta_r d\Omega_r \tag{2.39}$$

By integrating Equation 2.33 over all incident directions(θ_i, ϕ_i):

$$I_{\lambda,r}(\theta_r,\phi_r) = \int_{\cap} \rho_\lambda(\theta_r,\phi_r,\theta_i,\phi_i) I_{\lambda,i}(\theta_i,\phi_i)\cos\theta_i d\Omega_i \tag{2.40}$$

The *hemispherical-directional spectral reflectivity* is then defined as the reflected intensity into the (θ_r, ϕ_r) direction divided by the mean incident intensity:

Hemispherical-directional spectral reflectivity (in terms of bidirectional spectral reflectivity)

$$\equiv \rho_\lambda(\theta_r,\phi_r) = \frac{\int_{\cap} \rho_\lambda(\theta_r,\phi_r,\theta_i,\phi_i) I_{\lambda,i}(\theta_i,\phi_i)\cos\theta_i d\Omega_i}{(1/\pi)\int_{\cap} I_{\lambda,i}(\theta_i,\phi_i)\cos\theta_i d\Omega_i} \tag{2.41}$$

2.4.1.4 Reciprocity for Directional Spectral Reflectivity

When the incident intensity is uniform over all incident directions (θ_i, ϕ_i), Equation 2.41 reduces to

Hemispherical-directional spectral reflectivity (for uniform incident intensity)

$$= \rho_\lambda(\theta_r,\phi_r) = \int_{\cap} \rho_\lambda(\theta_r,\phi_r,\theta_i,\phi_i)\cos\theta_i d\Omega_i \tag{2.42}$$

Comparing Equations 2.39 and 2.42 and noting Equation 2.37, the reciprocal relation for ρ_λ is obtained (restricted to uniform incident intensity):

$$\rho_\lambda(\theta_i,\phi_i) = \rho_\lambda(\theta_r,\phi_r) \tag{2.43}$$

where (θ_r, ϕ_r) and (θ_i, ϕ_i) are the same angles, that is, $\theta_i = \theta_r$ and $\phi_i = \phi_r$. This means that the reflectivity of a material irradiated at a given angle of incidence (θ_i, ϕ_i), as measured by the energy collected over the entire hemisphere of reflection, is equal to the reflectivity for *uniform* irradiation from the hemisphere as measured by collecting the reflected energy at a single angle of reflection (θ_r, ϕ_r) when (θ_r, ϕ_r) are the *same* angles as (θ_i, ϕ_i). This relation is used for the design of hemispherical reflectometers for measuring radiative properties (Brandenberg 1963).

2.4.1.5 Hemispherical Spectral Reflectivity $\rho_\lambda(\lambda)$

If the incident spectral radiation arrives from all angles of the hemisphere (Figure 2.2h), then all the radiation intercepted by the area element dA of the surface is given by Equation 2.22 as

$$dQ_{\lambda,i} d\lambda = dA d\lambda \int_{\cap} I_{\lambda,i}(\theta_i,\phi_i)\cos\theta_i d\Omega_i$$

The amount of $dQ_i d\lambda$ that is reflected is, by integration using Equation 2.39 to include all incident directions

$$dQ_{\lambda,r} d\lambda = dA d\lambda \int_{\cap} \rho_\lambda(\theta_i,\phi_i) I_{\lambda,i}(\theta_i,\phi_i)\cos\theta_i d\Omega_i$$

The fraction of $dQ_{\lambda,i}d\lambda$ that is reflected provides the definition.

Hemispherical spectral reflectivity (in terms of directional-hemispherical spectral reflectivity)

$$\equiv \rho_\lambda = \frac{dQ_{\lambda,r}d\lambda}{dQ_{\lambda,i}d\lambda} = \frac{\int_\cap \rho_\lambda(\theta_i,\phi_i)I_{\lambda,i}(\theta_i,\phi_i)\cos\theta_i d\Omega_i}{\int_\cap I_{\lambda,i}(\theta_i,\phi_i)\cos\theta_i d\Omega_i} \tag{2.44}$$

2.4.1.6 Limiting Cases for Spectral Surfaces

In the following, we use a subscript notation for clarification: *d-h* denotes directional-hemispherical, and *bi-d* is for bidirectional.

2.4.1.6.1 Diffuse Surfaces

For a diffuse surface, the incident energy from the direction (θ_i, ϕ_i) that is reflected produces a reflected intensity that is uniform over all (θ_r, ϕ_r) directions. When an irradiated diffuse surface is viewed, the surface is equally bright from all viewing directions. The bidirectional spectral reflectivity is independent of (θ_r, ϕ_r), and Equation 2.39 simplifies to $\rho_{\lambda,d\text{-}h} = \rho_{\lambda,bi\text{-}d}\int_\cap \cos\theta_r d\Omega_r = \pi\rho_{\lambda,bi\text{-}d}(\theta_i,\phi_i)$. However, note that for an opaque surface, $\rho_{\lambda,d-h}(\theta_i, \phi_i) = 1 - \alpha_\lambda$, where for a diffuse surface α_λ is not a function of incidence angle (θ_i, ϕ_i). Hence, the directional-hemispherical reflectivity must be independent of angle of incidence, so that for a diffuse surface,

$$\rho_{\lambda,d\text{-}h} = \pi\rho_{\lambda,bi\text{-}d} \tag{2.45}$$

for any incidence angle. The π arises because $\rho_{\lambda,d-h}$ accounts for energy reflected into all (θ_r, ϕ_r) directions, while $\rho_{\lambda,bi\text{-}d}$ accounts for the reflected intensity into only one direction. This is analogous to the relation between blackbody hemispherical emissive power and intensity, $E_{\lambda b} = \pi I_{\lambda b}$.

Equation 2.40 provides the intensity in the (θ_r, ϕ_r) direction when the incident radiation is distributed over (θ_i, ϕ_i) values. If the surface is *diffuse, and if the incident intensity is uniform for all incident angles*, Equation 2.40 reduces to

$$I_{\lambda,r} = \rho_{\lambda,bi\text{-}d}I_{\lambda,i}\int_\cap \cos\theta_i d\Omega_i = \pi\rho_{\lambda,bi\text{-}d}I_{\lambda,i} \tag{2.46}$$

But, using Equation 2.45,

$$I_{\lambda,r} = \rho_{\lambda,d\text{-}h}I_{\lambda,i} = (1-\alpha_\lambda)I_{\lambda,i} \tag{2.47}$$

so that the reflected intensity in any direction for a diffuse surface is the uniform incident intensity times either the directional-hemispherical reflectivity or the hemispherical reflectivity. For angularly nonuniform irradiation on a diffuse surface, the incident intensity can be replaced by $dQ_{\lambda,i}/\pi$, where $dQ_{\lambda,i}$ is the spectral energy per unit wavelength and time intercepted by the surface element dA from all angular directions of the hemisphere.

2.4.1.6.2 Specular Surfaces

Mirror-like, or *specular*, surfaces obey well-known Fresnel laws of reflection. For incident radiation from a single direction, a specular reflector, by definition, provides reflected radiation at the same magnitude of the angle from the surface normal as for the incident intensity, and in the same plane as the incident intensity and the surface normal. Hence

$$\theta_r = \theta_i; \quad \phi_r = \phi_i + \pi \tag{2.48}$$

At all other angles, the bidirectional reflectivity of a specular surface is zero; then

$$\rho_\lambda(\lambda,\theta_r,\phi_r,\theta,\phi)_{\text{specular}} = \rho_\lambda(\theta_r = \theta_i,\phi_r = \phi_i + \pi,\theta_i,\phi_i) \equiv \rho_{\lambda,bi-d}(\theta_i,\phi_i) \tag{2.49}$$

and the bidirectional reflectivity is a function of only the incident direction, as this direction also provides the reflected direction.

For the intensity of radiation reflected from a specular surface into the solid angle around (θ_r, ϕ_r), Equation 2.40 becomes, for an *arbitrary* directional distribution of incident intensity, $I_{\lambda,r}(\theta_r,\phi_r) = \int_\cap \rho_{\lambda,bi\text{-}d}(\theta_i,\phi_i) I_{\lambda,i}(\theta_i,\phi_i)\cos\theta_i d\Omega_i$. The integrand has a nonzero value only in the small solid angle around (θ_i, ϕ_i) because of the properties of $\rho_{\lambda,bi-d}(\theta_i, \phi_i)$ for a specular surface. When examining the radiation reflected from a specular surface in a given direction, only that radiation at the incident direction (θ_i, ϕ_i) defined by Equation 2.48 need be considered as contributing to the reflected intensity. Hence

$$I_{\lambda,r}(\theta_r = \theta_i,\phi_r = \phi_i + \pi) = \rho_{\lambda,bi\text{-}d}(\theta_i,\phi_i) I_{\lambda,i}(\theta_i,\phi_i)\cos\theta_i d\Omega_i \tag{2.50}$$

If a surface tends to be specular, the use of the bidirectional reflectivity can become less practical than for a situation with a more diffuse reflection behavior. For a specular reflection the reflected intensity can be of the same order as the incident intensity; hence, the bidirectional reflectivity becomes very large because of the $d\Omega_i$ on the right side of Equation 2.50. A reflectivity that equals the well-behaved quantity $\rho_{\lambda,bi-d}(\theta_i, \phi_i)\cos\theta_i d\Omega_i$ in Equation 2.50 is more useful and is now defined as $\rho_{\lambda,s}(\theta_i, \phi_i)$. From Equation 2.50, it is this quantity that equals the ratio of reflected to incident intensities. Using reciprocity, we can then define a convenient *specular* reflectivity as

$$\frac{I_{\lambda,r}(\theta_r = \theta_i,\phi_r = \phi_i + \pi)}{I_{\lambda,i}(\theta_i,\phi_i)} = \rho_{\lambda,s}(\theta_i,\phi_i) = \rho_{\lambda,s}(\theta_r,\phi_r) \tag{2.51}$$

The angular relations of Equation 2.48 apply so that all the reflected radiation is into one direction.

2.4.2 Total Reflectivities

The previous expressions for spectral radiation can be readily generalized to include all wavelengths.

2.4.2.1 Bidirectional Total Reflectivity $\rho(\theta_r, \phi_r, \theta_i, \phi_i)$

This gives the contribution made by the total *energy* incident from direction (θ_i, ϕ_i) to the reflected total *intensity* into the direction (θ_r, ϕ_r). By analogy with Equation 2.33,

$$\textit{Bidirectional total reflectivity} \equiv \rho(\theta_r,\phi_r,\theta_i,\phi_i) = \frac{\int_{\lambda=0}^{\infty} I_{\lambda,r}(\theta_r,\phi_r,\theta_i,\phi_i)\,d\lambda}{\cos\theta_i d\Omega_i \int_{\lambda=0}^{\infty} I_{\lambda,i}(\theta_i,\phi_i)\,d\lambda}$$

$$= \frac{I_r(\theta_r,\phi_r,\theta_i,\phi_i)}{I_i(\theta_i,\phi_i)\cos\theta_i d\Omega_i} \tag{2.52}$$

Using Equation 2.33, another form is

Bidirectional total reflectivity (in terms of bidirectional spectral reflectivity)

$$\equiv \rho(\theta_r,\phi_r,\theta_i,\phi_i) = \frac{\int_{\lambda=0}^{\infty} \rho_\lambda(\theta_r,\phi_r,\theta_i,\phi_i) I_{\lambda,i}(\theta_i,\phi_i)\,d\lambda}{\int_{\lambda=0}^{\infty} I_{\lambda,i}(\theta_i,\phi_i)\,d\lambda} \tag{2.53}$$

2.4.2.2 Reciprocity for Bidirectional Total Reflectivity

Rewriting Equation 2.53 for energy incident from direction (θ_r, ϕ_r) and reflected into direction (θ_i, ϕ_i) gives

$$\rho(\theta_i,\phi_i,\theta_r,\phi_r) = \frac{\int_{\lambda=0}^{\infty} \rho_\lambda(\theta_i,\phi_i,\theta_r,\phi_r) I_{\lambda,i}(\theta_r,\phi_r)\,d\lambda}{\int_{\lambda=0}^{\infty} I_{\lambda,i}(\theta_r,\phi_r)\,d\lambda} \tag{2.54}$$

Comparing Equations 2.53 and 2.54 shows that

$$\rho(\theta_i,\phi_i,\theta_r,\phi_r) = \rho(\theta_r,\phi_r,\theta_i,\phi_i) \tag{2.55}$$

if the spectral distribution of incident intensity is the same for all directions or, less restrictively, if $I_{\lambda,i}(\theta_i, \phi_i) = CI_{\lambda,i}(\theta_r, \phi_r)$.

2.4.2.3 Directional Total Reflectivity $\rho(\theta_i, \phi_i) = \rho(\theta_r, \phi_r)$

The *directional-hemispherical total reflectivity* is the fraction of the total energy incident from a single direction that is reflected into all angular directions. The spectral energy from a given direction (θ_i, ϕ_i) that is intercepted by the surface is $I_{\lambda,i}(\theta_i, \phi_i)\cos\theta_i d\Omega_i d\lambda dA$. The reflected portion is $\rho_\lambda(\theta_i, \phi_i)I_{\lambda,i}(\theta_i, \phi_i)\cos\theta_i d\Omega_i d\lambda dA$. Integrating over all wavelengths and forming a ratio gives

Directional-hemispherical total reflectivity (in terms of directional-hemispherical spectral reflectivity)

$$\equiv \rho(\theta_i,\phi_i) = \frac{\int_{\lambda=0}^{\infty} \rho_\lambda(\theta_i,\phi_i) I_{\lambda,i}(\theta_i,\phi_i)\,d\lambda}{\int_{\lambda=0}^{\infty} I_{\lambda,i}(\theta_i,\phi_i)\,d\lambda} \tag{2.56}$$

Another directional total reflectivity specifies the fraction of radiation reflected into the single (θ_r, ϕ_r) direction when there is diffuse irradiation. The total radiation intensity reflected into the (θ_r, ϕ_r) direction when the incident intensity is uniform over all incident directions is

$$I_r(\theta_r,\phi_r) = \int_{\lambda=0}^{\infty} \rho_\lambda(\theta_r,\phi_r) I_{\lambda,i}\,d\lambda$$

where $\rho_\lambda(\theta_r, \phi_r)$ is in Equation 2.42. The total reflectivity is defined as the reflected intensity divided by the uniform incident intensity:

Hemispherical-directional total reflectivity (for diffuse irradiation)

$$\equiv \frac{I_r(\theta_r,\phi_r)}{I_i} = \frac{\int_{\lambda=0}^{\infty} \rho_\lambda(\theta_r,\phi_r) I_{\lambda,i} d\lambda}{\int_{\lambda=0}^{\infty} I_{\lambda,i} d\lambda} \tag{2.57}$$

2.4.2.4 Reciprocity for Directional Total Reflectivity

Equations 2.56 and 2.57 are now compared, bearing in mind that the latter is restricted to uniform incident intensity. With this restriction, from Equation 2.43 it is found that

$$\rho(\theta_r,\phi_r) = \rho(\theta_i,\phi_i) \tag{2.58}$$

where (θ_r, ϕ_r) and (θ_i, ϕ_i) are the same angles, that is, $\theta_i = \theta_r$ and $\phi_i = \phi_r$. In addition, there is a fixed spectral distribution of the incident intensity such that $I_{\lambda i}$ in Equation 2.56 is related to that in Equation 2.57 by $I_{\lambda,i}(\theta_i, \phi_i) = CI_{\lambda,i}$.

2.4.2.5 Hemispherical Total Reflectivity, ρ

If the incident total radiation arrives from all angles of the hemisphere, the total radiation intercepted by a unit area at the surface is given by Equation 2.26. The amount of this radiation that is reflected is $dQ_r = dA\int_{\cap} \rho(\theta_i,\phi_i) I_i(\theta_i,\phi_i)\cos\theta_i d\Omega_i$. The fraction of all the incident energy that is reflected, including all directions of reflection, is then

Hemispherical total reflectivity (in terms of directional-hemispherical total reflectivity)

$$\equiv \rho = \frac{dQ_r}{dQ_i} = \frac{dA}{dQ_i}\int_{\cap} \rho(\theta_i,\phi_i) I_i(\theta_i,\phi_i)\cos\theta_i d\Omega_i \tag{2.59}$$

Another form is found by using the incident hemispherical spectral energy. The amount reflected is $\rho_\lambda dQ_{\lambda,i} d\lambda$ where ρ_λ is the hemispherical spectral reflectivity from Equation 2.44. Integrating hence yields

Hemispherical total reflectivity (in terms of hemispherical spectral reflectivity)

$$\equiv \rho = \frac{\int_0^{\infty} \rho_\lambda dQ_{\lambda,i} d\lambda}{dQ_i = \int_0^{\infty} dQ_{\lambda,i} d\lambda} \tag{2.60}$$

2.4.3 Summary of Restrictions on Reflectivity Reciprocity Relations

Table 2.3 summarizes the restrictions necessary for applying the reflectivity reciprocity relations.

TABLE 2.3
Summary of Reciprocity Relations between Reflectivities

Type of Quantity	Equality	Restrictions
Bidirectional spectral (Equation 2.37)	$\rho_\lambda(\theta_i, \phi_i, \theta_r, \phi_r) = \rho_\lambda(\theta_r, \phi_r, \theta_i, \phi_i)$	None
Directional spectral (Equation 2.43)	$\rho_\lambda(\theta_i, \phi_i) = \rho_\lambda(\theta_r, \phi_r)$, where $\theta_i = \theta_r, \phi_i = \phi_r$	$\rho_\lambda(\theta_r, \phi_r)$ is for uniform incident intensity, or ρ_λ is independent of θ_i, ϕ_i, θ_r, and ϕ_r
Bidirectional total (Equation 2.55)	$\rho(\theta_i, \phi_i, \theta_r, \phi_r) = \rho_\lambda(\theta_r, \phi_r, \theta_i, \phi_i)$	$I_{\lambda,i}(\theta_i, \phi_i) = CI_{\lambda,i}(\theta_r, \phi_r)$ or $\rho_\lambda(\theta_i, \phi_i, \theta_r, \phi_r)$ independent of wavelength
Directional total (Equation 2.58)	$\rho(\theta_i, \phi_i) = \rho(\theta_r, \phi_r)$	For uniform incident intensity, or ρ_λ is independent of incident angle; independent of wavelength

2.5 TRANSMISSIVITY AT AN INTERFACE

As for reflectivity, the transmissivity of radiation across a single interface depends on two sets of angles (the angle of incidence on the interface and the direction at which the radiation is transmitted after crossing the surface) and the wavelength of the radiation. The properties defined in this section pertain only to the *fraction of incident radiation crossing the single interface*; for many real configurations, radiation that is transmitted across the interface may return to the interface by reflection from another nearby surface as is the case in a pane of glass, or by scattering from material in the region of the interface. These situations are considered in later chapters. For opaque surfaces, the transmissivity across the surface of the material is not needed in radiative transfer calculations, but the definitions are included here for later reference.

Given that the transmissivity is defined here for a single interface, it follows that the radiation incident on a surface is either reflected from the interface or transmitted through the surface, so that the transmissivity is directly related to the reflectivity. Because the definitions in this section are derived in parallel to those for reflectivity, the derivations are given more tersely.

2.5.1 Spectral Transmissivities

2.5.1.1 Bidirectional Spectral Transmissivity $\tau_\lambda(\theta_i, \phi_i, \theta_t, \phi_t)$

The ratio of intensity that is transmitted across an interface into the direction (θ_t, ϕ_t) to the incident spectral energy on the interface from a particular incident direction (θ_i, ϕ_i) (Figure 2.2i) is thus defined by the relation

$$\textit{Bidirectional spectral transmissivity} \equiv \tau_\lambda(\theta_t, \phi_t, \theta_i, \phi_i) = \frac{I_{\lambda,t}(\theta_t, \phi_t, \theta_i, \phi_i)}{I_{\lambda,i}(\theta_i, \phi_i)\cos\theta_i d\Omega_i} \tag{2.61}$$

which, as for the other properties, depends on the surface temperature T, and that notation is again omitted for simplicity.

2.5.1.2 Directional Spectral Transmissivities, $\tau_\lambda(\theta_i, \phi_i)$

The directional hemispherical spectral transmissivity $\tau_\lambda(\theta_i, \phi_i)$ is the fraction of the incident spectral energy that is transmitted across the interface into all directions (Figure 2.2j). By energy conservation, it is related to the directional-hemispherical spectral reflectivity, so that

Directional-hemispherical spectral transmissivity

$$\equiv \tau_\lambda(\theta_i, \phi_i) = \frac{\int_\cap I_{\lambda,t}(\theta_t, \phi_t, \theta_i, \phi_i)\cos\theta_t d\Omega_t}{I_{\lambda,i}(\lambda, \theta_i, \phi_i)\cos\theta_i d\Omega_i} = 1 - \rho_\lambda(\theta_i, \phi_i) \tag{2.62}$$

Using Equation 2.61, this becomes

$$\tau_\lambda(\theta_i,\phi_i) = \frac{I_{\lambda,i}(\theta_i,\phi_i)\cos\theta_i d\Omega_i \int_{\cap} \tau_\lambda(\theta_t,\phi_t,\theta_i,\phi_i)\cos\theta_t d\Omega_t}{I_{\lambda,i}(\theta_i,\phi_i)\cos\theta_i d\Omega_i}$$

$$= \int_{\cap} \tau_\lambda(\theta_t,\phi_t,\theta_i,\phi_i)\cos\theta_t d\Omega_t \tag{2.63}$$

Following Equations 2.40 and 2.41, the hemispherical-directional spectral transmissivity $\tau_\lambda(\theta_t, \phi_t)$ is the spectral intensity transmitted into the (θ_t, ϕ_t) direction over the integrated average spectral intensity incident on the surface (Figure 2.2k), or

Hemispherical-directional spectral transmissivity

$$\equiv \tau_\lambda(\theta_t,\phi_t) = \frac{I_{\lambda,t}(\theta_t,\phi_t)}{\left(\frac{1}{\pi}\right)\int_{\cap} I_{\lambda,i}(\theta_i,\phi_i)\cos\theta_i d\Omega_i} \tag{2.64}$$

or, using Equation 2.61,

Hemispherical-directional spectral transmissivity (in terms of the bidirectional spectral transmissivity)

$$\equiv \tau_\lambda(\theta_t,\phi_t) = \frac{\int_{\cap} \tau_\lambda(\theta_t,\phi_t,\theta_i,\phi_i) I_{\lambda,i}(\theta_i,\phi_i)\cos\theta_i d\Omega_i}{\left(\frac{1}{\pi}\right)\int_{\cap} I_{\lambda,i}(\theta_i,\phi_i)\cos\theta_i d\Omega_i} \tag{2.65}$$

For uniform incident spectral intensity from all directions, Equation 2.64 reduces to

Hemispherical-directional spectral transmissivity (for uniform incident intensity)

$$\equiv \tau_\lambda(\theta_t,\phi_t) = \int_{\cap} \tau_\lambda(\theta_t,\phi_t,\theta_i,\phi_i)\cos\theta_i d\Omega_i \tag{2.66}$$

Comparing Equation 2.66 with Equation 2.63 shows that, for uniform incident radiation,

$$\tau_\lambda(\theta_t,\phi_t) = \tau_\lambda(\theta_i,\phi_i) = 1 - \rho_\lambda(\theta,\phi) \tag{2.67}$$

2.5.1.3 Hemispherical Spectral Transmissivity τ_λ

The fraction of spectral energy incident on the surface from all directions that is transmitted across the surface into all directions (Figure 2.2l) is the hemispherical spectral transmissivity. It is related through energy conservation to the hemispherical spectral reflectivity:

$$\textit{Hemispherical spectral transmissivity} \equiv \tau_\lambda = \frac{\int_{\cap} I_{\lambda,t}(\theta_t,\phi_t)\cos\theta_t d\Omega_t}{\int_{\cap} I_{\lambda,i}(\theta_i,\phi_i)\cos\theta_i d\Omega_i} = 1 - \rho_\lambda \tag{2.68}$$

or, using Equation 2.62,

$$\textit{Hemispherical-spectral transmissivity}\left(\text{in terms of directional-hemispherical transmissivity}\right)$$

$$\equiv \tau_\lambda = \frac{\int_{\cap} \tau_\lambda(\theta_i,\phi_i) I_{\lambda,i}(\theta_i,\phi_i)\cos\theta_i d\Omega_i}{\int_{\cap} I_{\lambda,i}(\lambda,\theta_i,\phi_i)\cos\theta_i d\Omega_i} \tag{2.69}$$

2.5.2 Total Transmissivities

Total transmissivities are obtained by direct analogy with the spectral property definitions.

2.5.2.1 Bidirectional Total Transmissivity $\tau(\theta_i, \phi_i, \theta_t, \phi_t)$

The ratio of total intensity that is transmitted across an interface into the direction (θ_t, ϕ_t) to the incident total energy on the interface from a particular direction (θ_i, ϕ_i) (Figure 2.2i) is defined by

$$\textit{Bidirectional spectral transmissivity} \equiv \tau(\theta_t,\phi_t,\theta_i,\phi_i) = \frac{I_t(\theta_t,\phi_t,\theta_i,\phi_i)}{I_i(\theta_i,\phi_i)\cos\theta_i d\Omega_i} \tag{2.70}$$

2.5.2.2 Directional Total Transmissivities, $\tau(\theta_i, \phi_i)$

The directional-hemispherical total transmissivity is the fraction of the incident total energy that is transmitted across the interface into all directions (Figure 2.2j). By energy conservation, it is related to the directional-hemispherical total reflectivity, so that

$$\textit{Directional-hemispherical total transmissivity}$$

$$\equiv \tau(\theta_i,\phi_i) = \frac{\int_{\cap} I_t(\theta_t,\phi_t,\theta_i,\phi_i)\cos\theta_t d\Omega_t}{I_i(\theta_i,\phi_i)\cos\theta_i d\Omega_i} = 1-\rho(\theta_i,\phi_i) \tag{2.71}$$

Using Equation 2.70, this can be expressed as

$$\textit{Directional-hemispherical total transmissivity}\left(\text{in terms of bidirectional transmissivity}\right)$$

$$\equiv \tau(\theta_i,\phi_i) = \frac{I_i(\theta_i,\phi_i)\cos\theta_i d\Omega_i \int_{\cap} \tau(\theta_t,\phi_t,\theta_i,\phi_i)\cos\theta_t d\Omega_t}{I_i(\theta_i,\phi_i)\cos\theta_i d\Omega_i} = \int_{\cap} \tau(\theta_t,\phi_t,\theta_i,\phi_i)\cos\theta_t d\Omega_t \tag{2.72}$$

2.5.2.3 Hemispherical-Directional Total Transmissivity $\tau(\theta_t, \phi_t)$

The hemispherical-directional total transmissivity $\tau(\theta_t, \phi_t)$ is the total intensity transmitted into the (θ_t, ϕ_t) direction over the integrated average total intensity incident on the surface (Figure 2.2k), or

$$\textit{Hemispherical-directional total transmissivity}$$

$$\equiv \tau(\theta_t,\phi_t) = \frac{I_t(\theta_t,\phi_t)}{\left(\dfrac{1}{\pi}\right)\int I_i(\theta_i,\phi_i)\cos\theta_i d\Omega_i} \tag{2.73}$$

or, using Equation 2.70,

Hemispherical-directional total transmissivity (in terms of the bidirectional total transmissivity)

$$\equiv \tau(\theta_t,\phi_t) = \frac{\int_{\cap} \tau(\theta_t,\phi_t,\theta_i,\phi_i) I_i(\theta_i,\phi_i)\cos\theta_i d\Omega_i}{\left(\frac{1}{\pi}\right)\int_{\cap} I_i(\theta_i,\phi_i)\cos\theta_i d\Omega_i} \tag{2.74}$$

For uniform incident total intensity from all directions, Equation 2.74 reduces to

Hemispherical-directional total transmissivity (for uniform incident intensity)

$$\equiv \tau(\theta_t,\phi_t) = \int_{\cap} \tau(\theta_t,\phi_t,\theta_i,\phi_i)\cos\theta_i d\Omega_i \tag{2.75}$$

Comparing Equation 2.75 with Equation 2.72 shows that, for uniform incident radiation and because of the reciprocity of $\tau(\theta_t, \phi_t, \theta_i, \phi_i)$,

$$\tau(\theta_t,\phi_t) = \tau(\theta_i,\phi_i) \tag{2.76}$$

2.5.2.4 Hemispherical Total Transmissivity, τ

The fraction of total energy incident on the surface from all directions that is transmitted across the surface into all directions (Figure 2.21) is the hemispherical total transmissivity:

$$\textit{Hemispherical total transmissivity} \equiv \tau = \frac{\int_{\cap} I_t(\theta_t,\phi_t)\cos\theta_t d\Omega_t}{\int_{\cap} I_i(\theta_i,\phi_i)\cos\theta_i d\Omega_i} \tag{2.77}$$

or, using Equation 2.74,

Hemispherical total transmissivity (in terms of directional-hemispherical transmissivity)

$$\equiv \tau = \frac{\int_{\cap} \tau(\theta_i,\phi_i) I_i(\theta_i,\phi_i)\cos\theta_i d\Omega_i}{\int_{\cap} I_i(\theta_i,\phi_i)\cos\theta_i d\Omega_i} \tag{2.78}$$

For uniform incident intensity, this reduces to

Hemispherical total transmissivity (for uniform incident intensity)

$$\equiv \tau = \frac{1}{\pi}\int_{\cap} \tau(\theta_i,\phi_i)\cos\theta_i d\Omega_i \tag{2.79}$$

If the material beneath the interface has sufficient thickness to completely absorb the transmitted radiation (i.e., it is an opaque layer), then the transmissivity of the surface is equal to the absorptivity of the surface without restriction, giving the relations in Table 2.4.

TABLE 2.4
Summary of Relations among Reflectivity, Transmissivity, and Absorptivity

Property Type	Equality	Equation
Directional-hemispherical spectral	$\tau_\lambda(\theta_i, \phi_i) = \alpha_\lambda(\theta_i, \phi_i) = 1 - \rho_\lambda(\theta_i, \phi_i)$	(2.81)
Hemispherical spectral	$\tau_\lambda = \alpha_\lambda = 1 - \rho_\lambda$	(2.87)
Directional-hemispherical total	$\tau(\theta_i, \phi_i) = \alpha(\theta_i, \phi_i) = 1 - \rho(\theta_i, \phi_i)$	(2.84)
Hemispherical total	$\tau = \alpha = 1 - \rho$	(2.90)

2.6 RELATIONS AMONG REFLECTIVITY, ABSORPTIVITY, EMISSIVITY, AND TRANSMISSIVITY

From the definitions of absorptivity, reflectivity, and transmissivity as fractions of incident energy absorbed, reflected, or transmitted, it is evident that some relations exist among these properties. For an opaque material, all energy that is transmitted across a surface is absorbed by the medium beneath the surface, so it follows that transmissivity and absorptivity are equal. This will not be true when translucent materials are considered. With the use of Kirchhoff's law and its restrictions (Table 2.2), further relations between emissivity and reflectivity can be obtained.

The spectral energy $d^2Q_{\lambda,i}(\theta_i, \phi_i)d\lambda$ incident per unit time on an element dA of an opaque body from within a solid angle $d\Omega_i$ at (θ_i, ϕ_i) is either absorbed or reflected, so that

$$\frac{d^2Q_{\lambda,a}(\theta_i,\phi_i)d\lambda}{d^2Q_{\lambda,i}(\theta_i,\phi_i)d\lambda}+\frac{d^2Q_{\lambda,r}(\theta_i,\phi_i)d\lambda}{d^2Q_{\lambda,i}(\theta_i,\phi_i)d\lambda}=1 \tag{2.80}$$

Because the radiation is incident from the angles (θ_i, ϕ_i), the two energy ratios in Equation 2.80 are the directional spectral absorptivity (Equation 2.14) (or the directional spectral transmissivity [Equation 2.62]) and the directional hemispherical spectral reflectivity (Equation 2.39). Substituting gives

$$\alpha_\lambda(\theta_i,\phi_i)+\rho_\lambda(\theta_i,\phi_i)=\tau_\lambda(\theta_i,\phi_i)+\rho_\lambda(\theta_i,\phi_i)=1 \tag{2.81}$$

Kirchhoff's law (Equation 2.17) can be applied without restriction to give

$$\varepsilon_\lambda(\theta_i,\phi_i)+\rho_\lambda(\theta_i,\phi_i)=\tau_\lambda(\theta_i,\phi_i)+\rho_\lambda(\theta_i,\phi_i)=1 \tag{2.82}$$

When the total energy is considered arriving at dA in $d\Omega_i$ from the given direction (θ_i, ϕ_i), Equation 2.80 becomes

$$\frac{d^2Q_a(\theta_i,\phi_i)}{d^2Q_i(\theta_i,\phi_i)}+\frac{d^2Q_r(\theta_i,\phi_i)}{d^2Q_i(\theta_i,\phi_i)}=1 \tag{2.83}$$

Substituting Equation 2.20 (or Equation 2.71) and Equation 2.56 results in

$$\alpha(\theta_i,\phi_i)+\rho(\theta_i,\phi_i)=\tau(\theta_i,\phi_i)+\rho(\theta_i,\phi_i)=1 \tag{2.84}$$

The absorptivity (and transmissivity) are directional-total values, and the reflectivity is directional-hemispherical total. Kirchhoff's law for directional total properties (Section 2.3.4) is applied to give

$$\varepsilon(\theta_i,\phi_i)+\rho(\theta_i,\phi_i)=\tau(\theta_i,\phi_i)+\rho(\theta_i,\phi_i)=1 \tag{2.85}$$

under the restriction that the incident radiation obeys the relation $I_{\lambda,i}(\theta_i, \phi_i) = C(\theta_i, \phi_i)I_{\lambda b}(T)$ or the surface is directional-gray.

If the incident spectral intensity is arriving at dA from all directions over the hemisphere, then Equation 2.80 gives

$$\frac{d^2Q_{\lambda,a}d\lambda}{d^2Q_{\lambda,i}d\lambda}+\frac{d^2Q_{\lambda,r}d\lambda}{d^2Q_{\lambda,i}d\lambda}=1 \tag{2.86}$$

Equation 2.86, using the property definitions, is

$$\alpha_\lambda+\rho_\lambda=\tau_\lambda+\rho_\lambda=1 \tag{2.87}$$

where the properties are hemispherical spectral values from Equations 2.25, 2.44, and 2.68. Substitution of the hemispherical spectral emissivity ε_λ for α_λ in this relation is only valid if the incident intensity is independent of incident angle (diffuse irradiation), or if the ε_λ and α_λ do not depend on angle (see Table 2.2). Under these restrictions, Equation 2.87 becomes

$$\varepsilon_\lambda+\rho_\lambda=\tau_\lambda+\rho_\lambda=1 \tag{2.88}$$

If the energy incident on dA is integrated over all wavelengths and directions, then Equation 2.80 becomes

$$\frac{dQ_a}{dQ_i}+\frac{dQ_r}{dQ_i}=1 \tag{2.89}$$

or

$$\alpha+\rho=\tau+\rho=1 \tag{2.90}$$

and ε may be substituted for α under the restrictions of Table 2.2.

Table 2.4 may be used to develop further the relationship of surface transmissivity to the other radiative properties of opaque surfaces (see Table 2.3).

Example 2.8

Radiation from the sun is incident on a surface in orbit above the Earth's atmosphere. The surface is at 1000 K, and the directional total emissivity is given in Figure 2.4. If the solar energy is incident at an angle of 25° from the normal to the surface, what is the reflected energy flux?

From Figure 2.4, ε(25°, 1000 K) = 0.8. Table 2.2 shows that for directional total properties α(25°, 1000 K) = ε(25°, 1000 K) only when the incident spectrum is proportional to that emitted by a blackbody at 1000 K. This is not the case here, since the solar spectrum is like that of a blackbody at 5780 K. Hence α(25°, 1000 K) ≠ 0.8, and without α(25°, 1000 K) we cannot determine ρ(25°, 1000 K). The emissivity data given are insufficient to solve the problem; for, spectral values are needed.

Example 2.9

A surface at T = 500 K has a spectral emissivity in the normal direction that is approximated as shown in Figure 2.9. The surface is maintained at 500 K by cooling water and is enclosed by a black hemisphere heated to T_i = 1500 K. What is the reflected radiant intensity in the direction normal to the surface?

From Equation 2.82, $\rho_\lambda(\theta_i = 0°) = 1 - \varepsilon_\lambda(\theta = 0°)$, which is the reflectivity into the hemisphere for radiation arriving from the normal direction. From reciprocity, for uniform incident intensity over the hemisphere, $\rho_\lambda(\theta_r = 0°) = \rho_\lambda(\theta = 0°)$. Hence, the reflectivity into the normal direction resulting from the incident radiation from the hemisphere is (by use of Figure 2.9):

$$\rho_\lambda(0 \le \lambda < 2, \theta_r = 0°, T) = 0.7; \quad \rho_\lambda(2 \le \lambda \le 5, \theta_r = 0°, T) = 0.2; \quad \rho_\lambda(5 \le \lambda \le \infty, \theta_r = 0°, T) = 0.5$$

The incident intensity is $I_{\lambda,i}(T_i) = I_{\lambda b}(1500\text{ K})$. From the relation preceding Equation 2.57, the reflected intensity is

$$I_r(\theta_r = 0°)\int_0^\infty I_{\lambda b}(T_i)\rho_\lambda(\theta_r = 0°, T)d\lambda$$

From Equations 1.35 and 1.37, this becomes

$$\begin{aligned} I_r(\theta_r = 0°) &= \frac{\sigma T_i^4}{\pi}(0.7F_{0\to 2T_i} + 0.2F_{2T_i\to 5T_i} + 0.5F_{5T_i\to\infty}) \\ &= \frac{5.67040\times 10^{-8}}{\pi}(1500)^4[0.7(0.27323) \\ &\quad + 0.2(0.83437 - 0.27323) + 0.5(1 - 0.83437)] \\ &= 35.3\text{ kW/(m}^2\cdot\text{sr)} \end{aligned}$$

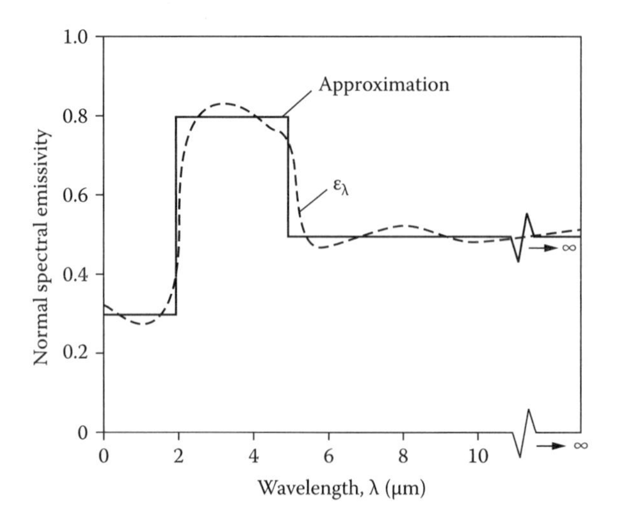

FIGURE 2.9 Directional spectral emissivity in normal direction for Example 2.9.

Example 2.10

A large flat plate at $T = 925$ K has a hemispherical spectral emissivity that can be approximated by a straight line decreasing from 0.850 to 0 as λ increases from 0 to 10.5 μm. The flat plate faces a second large plate that has a reflectivity of 0.35 for $0 < \lambda < 4.5$ μm and 0.820 for $\lambda > 4.5$ μm. Both plates are diffuse. Consider the emitted energy from the first plate that arrives at the second plate and is then reflected back to the first plate. What is the energy flux being reabsorbed by the first plate as a result of this single reflection?

The spectral energy emitted is $\varepsilon_\lambda(T)E_{\lambda b}(T)d\lambda$. For the geometry of large plates all of this energy reaches the second plate and the fractions 0.350 and 0.820 are reflected back for the respective ranges $\lambda < 4.5$ μm and $\lambda > 4.5$ μm. From Kirchhoff's law, the spectral absorptivity is equal to $\varepsilon_\lambda(T)$. Hence, the reabsorbed energy is

$$q_{abs} = 0.350\int_0^{4.5} \varepsilon_\lambda^2(T)E_{\lambda b}(T)d\lambda + 0.820\int_{4.5}^{\infty} \varepsilon_\lambda^2(T)E_{\lambda b}(T)d\lambda$$

The $\varepsilon_\lambda(T) = 0.850[1-(\lambda/10.5)]$ for $\lambda \le 10.5$ μm and is zero for $\lambda \le 10.5$ μm. Then

$$q_{abs} = 0.350\times 0.850^2\int_0^{4.5} (1-\lambda/10.5)^2 E_{\lambda b}(T)d\lambda$$

$$+0.820\times 0.850^2\int_{4.5}^{10.5} (1-\lambda/10.5)^2 E_{\lambda b}(T)d\lambda$$

where $E_{\lambda b}(T) = 2\pi C_1/\lambda^5(e^{C2/\lambda T} - 1)$. Numerical integration yields

$$q_{abs} = 0.350\times 0.850^2\times 10{,}541 + 0.820\times 0.850^2\times 2{,}745 = 4{,}292 \text{ W/m}^2$$

HOMEWORK

2.1 A material has a hemispherical spectral emissivity that varies considerably with wavelength but is fairly independent of surface temperature (see, e.g., the behavior of tungsten in Figures 3.31 and 3.32). Radiation from a gray source at T_i is incident on the surface uniformly from all directions. Show that the total absorptivity for the incident radiation is equal to the total emissivity of the material evaluated at the source temperature T_i.

2.2 Using Figure 3.31, estimate the hemispherical total emissivity of tungsten at 2600 K. *Answer*: 0.284.

2.3 Suppose that ε_λ is independent of λ (gray-body radiation). Show that $F_{0\to\lambda T}$ represents the fraction of the total radiant emission of the gray body in the range from 0 to λT.

2.4 For a surface with hemispherical spectral emissivity ε_λ, does the maximum of the E_λ distribution occur at the same λ as the maximum of the $E_{\lambda b}$ distribution at the same temperature? (*Hint*: examine the behavior of $dE_\lambda/d\lambda$.) Plot the distributions of E_λ as a function of λ for the data of Figure 2.9 at 600 K and for the property data at 700 K. At what λ is the maximum of E_λ? How does this compare with the maximum of $E_{\lambda b}$?

2.5 Find the emissivity at 400 K and the solar absorptivity of the diffuse material with the measured spectral emissivity shown in the figure.

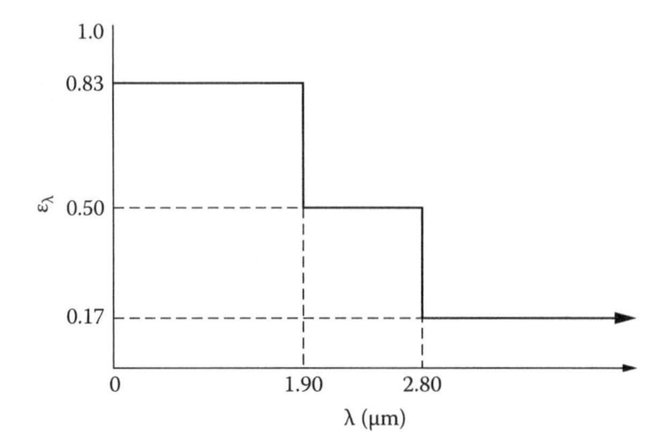

Answer: $\varepsilon(T = 400\text{ K}) = 0.17$; $\alpha_s(T = 5780\text{ K}) = 0.799$.

2.6 The surface temperature-independent hemispherical spectral absorptivity of a surface is measured when it is exposed to isotropic incident spectral intensity, and the results are approximated as shown in the following text. What is the total hemispherical emissivity of this surface when it is at a temperature of 1000 K?

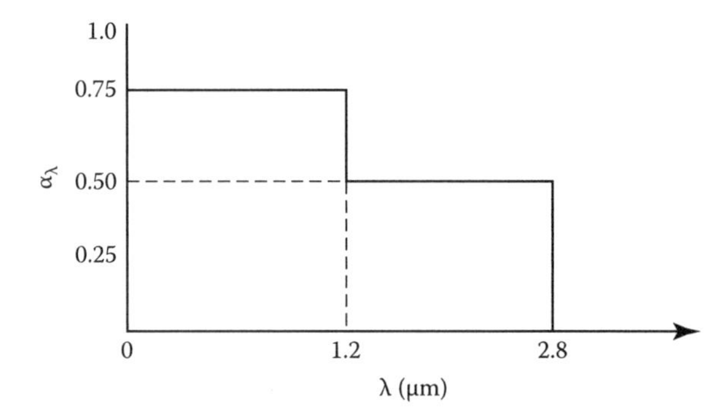

Answer: 0.034.

2.7 (a) Obtain the total absorptivity of a diffuse surface with properties given in the figure for incident radiation from a blackbody with a temperature of 6200 K.

(b) What is the total emissivity of the diffuse surface with properties given in the figure if the surface temperature is 500 K?

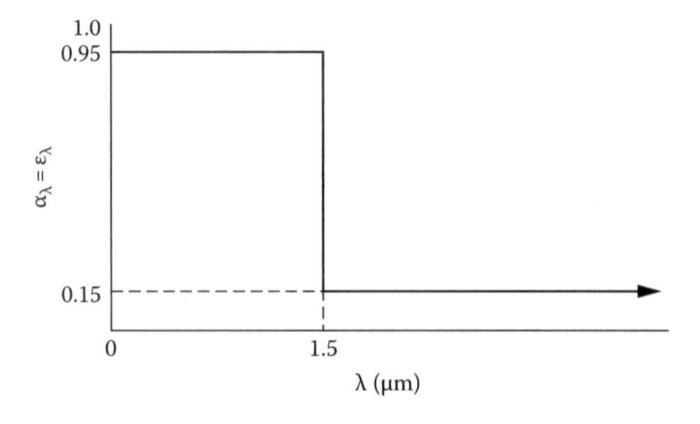

Answer: (a) 0.8680; (b) 0.15.

2.8 For the spectral properties given in the figure for a diffuse surface:

(a) What is the solar absorptivity of the surface (assume the solar temperature is 5800 K)?

(b) What is the total hemispherical emissivity of the surface if the surface temperature is 700 K?

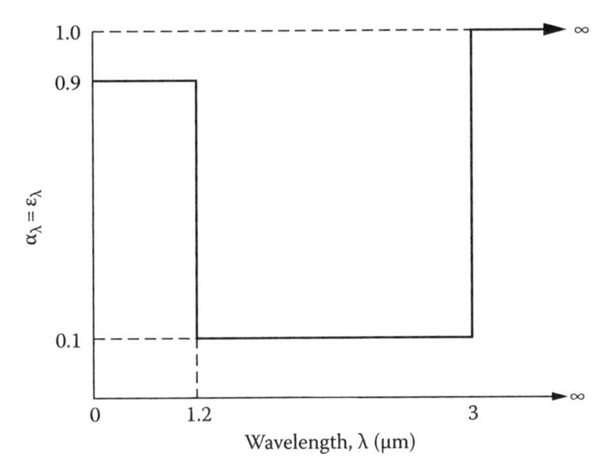

Answer: (a) 0.766; (b) 0.925.

2.9 A white ceramic surface has a hemispherical spectral emissivity distribution at 1600 K as shown. What is the hemispherical total emissivity of the surface at this surface temperature?

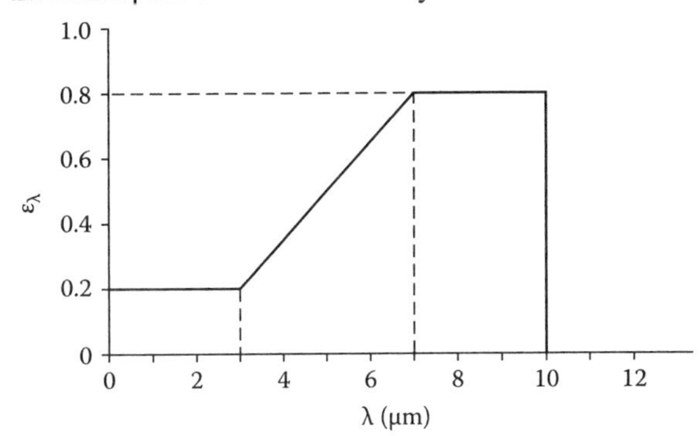

Answer: 0.281.

2.10 A surface has the following values of hemispherical spectral emissivity at a temperature of 800 K.

λ (μm)	**ε_λ(800 K)**
<1	0
1	0
1.5	0.2
2	0.4
2.5	0.6
3	0.8
3.5	0.8
4	0.8
4.5	0.7
5	0.6
6	0.4
7	0.2
8	0
>8	0

(a) What is the hemispherical total emissivity of the surface at 800 K?
(b) What is the hemispherical total absorptivity of the surface at 800 K if the incident radiation is from a gray source at 1800 K that has an emissivity of 0.815? The incident radiation is uniform over all incident angles.

Answer: (a) 0.438; (b) 0.427.

2.11 Find the emissivity at 950 K and the solar absorptivity of the diffuse material with the measured spectral emissivity shown in the figure. This will require numerical integration.

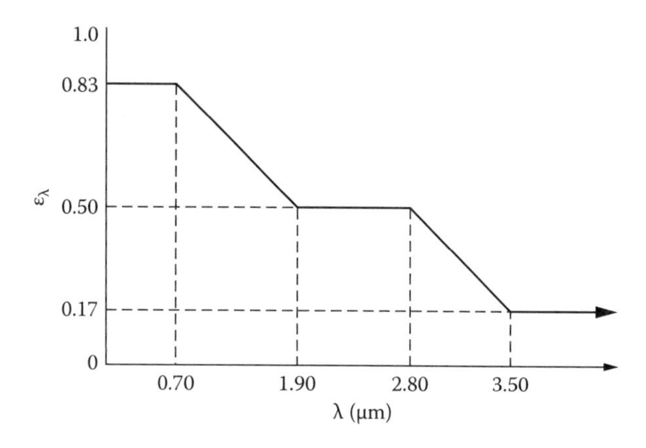

Answer: $\varepsilon(950\text{ K}) = 0.3248$; $\varepsilon(5780\text{ K}) = \alpha_s(5780\text{ K}) = 0.7564$.

2.12 A diffuse surface at 1000 K has a hemispherical spectral emissivity that can be approximated by the solid line shown.
(a) What is the hemispherical-total emissive power of the surface? What is the total intensity emitted in a direction 60° from the normal to the surface?
(b) What percentage of the total emitted energy is in the wavelength range $5 < \lambda < 10$ μm? How does this compare with the percentage emitted in this wavelength range by a gray body at 1000 K with an emissivity $\varepsilon = 0.611$?

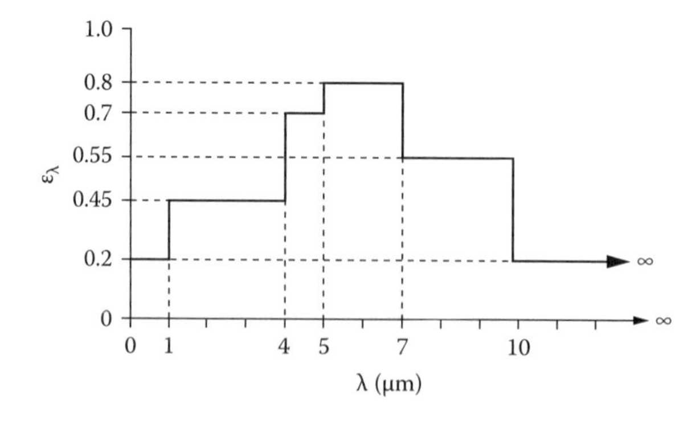

Answer: (a) 30, 507 W/m^2; 9711 W/m$^2\cdot$sr (b) 36.7%; 28.0%.

2.13 The ε_λ for a metal at 1000 K is approximated as shown, and it does not vary significantly with the metal temperature. The surface is diffuse.

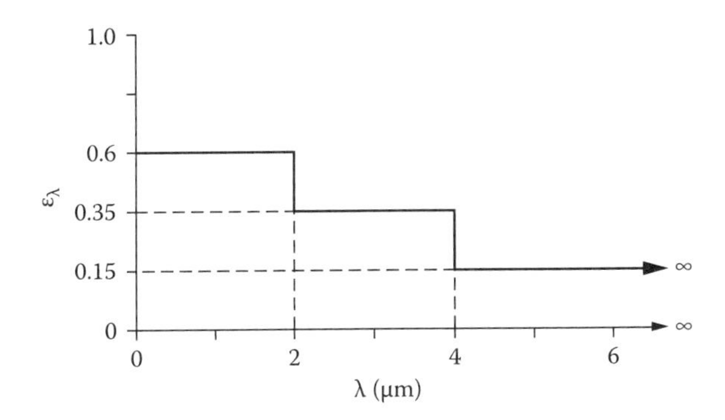

(a) What is α for incident radiation from a gray source at 1200 K with $\varepsilon_{source} = 0.822$?
(b) What is α for incident radiation from a source at 1200 K made from the same metal as the receiving plate?

Answer: (a) 0.3066; (b) 0.3802.

2.14 The directional total absorptivity of a gray surface is given by the expression $\alpha(\theta) = 0.450 \cos^2\theta$ where θ is the angle relative to the normal to the surface.
(a) What is the hemispherical total emissivity of the surface?
(b) What is the hemispherical-hemispherical total reflectivity of this surface for diffuse incident radiation (uniform incident intensity)?
(c) What is the hemispherical-directional total reflectivity for diffuse incident radiation reflected into a direction 75° from the normal?

Answer: (a) 0.225; (b) 0.775; (c) 0.970.

2.15 Using Figure 3.24, estimate the total absorptivity of typewriter paper for normally incident radiation from a blackbody source at 1200 K.

Answer: 0.892.

2.16 A gray surface has a directional emissivity as shown. The properties are isotropic with respect to circumferential angle ϕ.
(a) What is the hemispherical emissivity of this surface?
(b) If the energy from a blackbody source at 650 K is incident uniformly from all directions, what fraction of the incident energy is absorbed by this surface?
(c) If the surface is placed in a very cold environment, at what rate must energy be added per unit area to maintain the surface temperature at 1000 K?

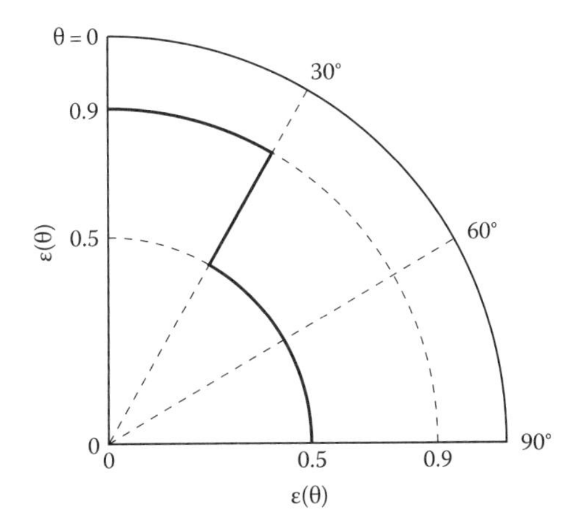

Answer: (a) 0.600; (b) 0.600; (c) 34,022 W/m^2.

2.17 Using Figure 3.44, estimate the ratio of normal total solar absorptivity to hemispherical total emissivity for aluminum at a surface temperature of 650 K with a coating of 0.1 μm dendritic lead sulfide crystals. Assume the surface is diffuse. (The solar temperature can be taken as 5780 K.)

Answer: 3.67.

2.18 A gray surface has a directional total emissivity that depends on angle of incidence as $\varepsilon(\theta) = 0.788 \cos \theta$. Uniform radiant energy from a single direction normal to the cylinder axis is incident on a long cylinder of radius R. What fraction of energy striking the cylinder is reflected? What is the result if the body is a sphere rather than a cylinder?

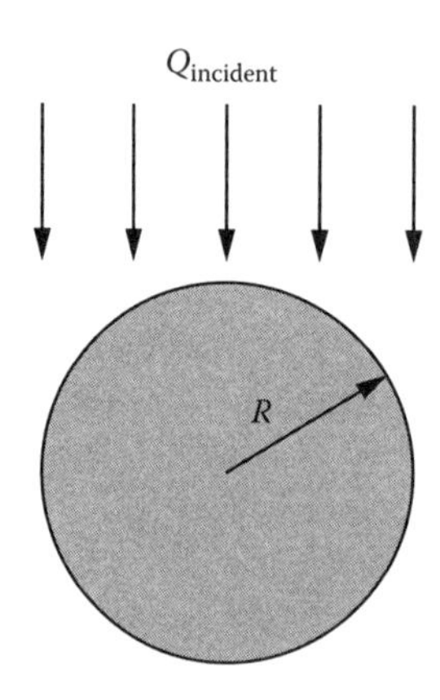

Answer: 0.381; 0.475.

2.19 A flat metal plate 0.1 m wide by 1.0 m long has a temperature that varies only along the long direction. The temperature is 900 K at one end, and decreases linearly over the one meter length to 350 K. The hemispherical spectral emissivity of the plate does not change significantly with temperature but is a function of wavelength. The wavelength dependence is approximated by a linear function decreasing from $\varepsilon_\lambda = 0.85$ at $\lambda = 0$ to $\varepsilon_\lambda = 0.02$ at $\lambda = 10$ μm. What is the rate of radiative energy loss from one side of the plate? The surroundings are at a very low temperature.

Answer: 417 W.

2.20 A thin ceramic plate, insulated on one side, is radiating energy from its exposed side into a vacuum at very low temperature. The plate is initially at 1200 K, and is to cool to 300 K. At any instant, the plate is assumed to be at uniform temperature across its thickness and over its exposed area. The plate is 0.25 cm thick, and the surface hemispherical-spectral emissivity is as shown and is independent of temperature. What is the cooling time? The density of the ceramic is 3200 kg/m^3, and its specific heat is 710 J/(kg · K).

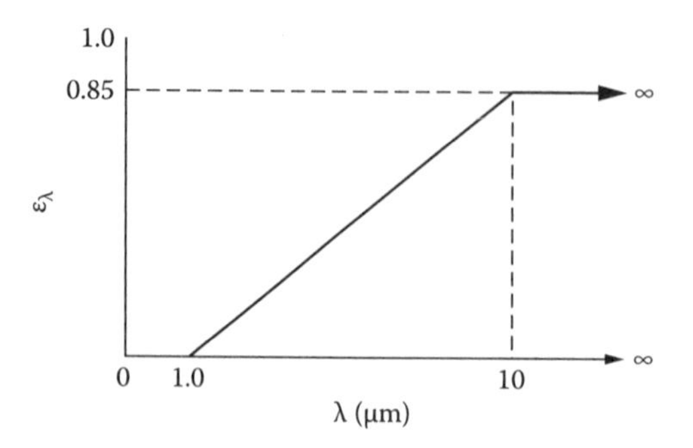

Answer: $t = 29.9$ min.

3 Radiative Properties of Opaque Materials

3.1 INTRODUCTION

For detailed radiation transfer predictions, we need to have spectral or total radiative properties of surfaces and the media between them. These properties are usually considered as the bottleneck in predictions as they are not readily available for many materials. It is possible to obtain some radiative properties from experiments, although the experimental conditions may limit the usefulness of such data. On the other hand, it is possible to derive these properties for homogeneous, optically smooth media using the electromagnetic (EM) wave theory. As we discuss in Chapter 14, the EM theory yields useful trends and provides a unifying basis to help explain radiation phenomena. However, the analyses are inadequate to predict the properties of surfaces that are not perfectly clean or have varying amounts of contaminants or oxides on them. Surfaces in practical applications almost always have varying degrees of surface roughness, which may not be modeled properly. In this chapter, we focus on the properties of real materials and report radiative properties of several opaque materials as a function of different parameters. In addition, we also demonstrate how radiative properties can be selected for radiative exchange calculations.

We begin this chapter with a brief introduction to dielectric and electrically conducting materials. Examples of measured properties of real opaque materials are presented along with comparisons against the idealized surfaces obtained from the EM theory predictions. The discussion in this chapter is to show the sensitivity of radiative transfer predictions to radiative properties of surfaces and surface conditions. We also present predictions of reflectivity and absorptivity of opaque materials, where roughness is much smaller than the incident wavelength (optically smooth materials). A general discussion of the EM theory is left to Chapter 14, to avoid breaking the continuity of the underlying discussion. In some cases, surface roughness and nanoscale structures can influence the radiative properties, particularly the directional-spectral emissivity and reflectivity, but these discussions are left to Chapter 16. Most uses of surface property data are for the case of incident or emitted radiation for a surface in air or vacuum (both of which have refractive indices very close to unity); consequently, the majority of the discussion is limited to the cases of refractive index $n_1 = 1$ adjacent to the surface.

We will not make any attempt to provide comprehensive radiative property data in this text. However, a selective property table is given in Appendix B. Extensive tabulations and graphs of radiative properties are available in the literature, including Wood et al. (1964), Svet (1965), Touloukian and Ho (1970), Touloukian and DeWitt (1972a), Touloukian et al. (1972b), Henninger (1984), Palik (1998), Fox (2001), and Lide (2008), as well as several recent journal articles for particular materials.

3.2 ELECTROMAGNETIC WAVE THEORY PREDICTIONS

James Clark Maxwell (1864) provided a crowning achievement of classical physics and explained the relation between electric and magnetic fields. He also showed that EM waves propagate with the speed of light, indicating that light itself is in the form of an EM wave. Further, Maxwell EM wave equations provide the required framework for obtaining the radiative properties of opaque materials and surfaces. For "optically smooth" surfaces, spectral-directional reflectivity, transmissivity, absorptivity, and emissivity can be calculated by entering the optical and electrical properties to these equations. In this context, "optically smooth" means that surface imperfections, roughness, or texture are much smaller than the wavelength of the incident radiation to the extent that they do not affect the far-field wave fronts. As we discuss in Chapter 16, for near-field radiation transfer applications where the optically smooth criterion is violated, surface roughness or nanoscale textures become important in determining the corresponding radiative properties. Nanoscale architectures on surfaces also allow us to custom-build surfaces with specific spectral-directional emissivities and reflectivities. Such surfaces can also be analyzed and constructed starting from Maxwell equations, by accounting for the near-field effects, including polarization and interference of the incident and reflected waves.

The wave–surface interactions for ideal surface conditions can be obtained using Maxwell equations. Although it is possible to perform more accurate computations for nonideal surfaces, such specific conditions are better suited for specific cases, as they may not be generalized to all possible real surfaces. The departures of real materials from ideal conditions assumed in the theory are often responsible for large variations of measured property values from theoretical predictions. These departures are caused by factors such as surface roughness, surface contamination, impurities, and crystal-structure modifications on surfaces.

Although there may be large effects of surface conditions on the radiative properties, the EM theory for ideal surfaces does serve a number of useful purposes. It provides an understanding of why there are basic differences in the radiative properties of insulators and electrical conductors and reveals general trends that help unify presentation of experimental data. These trends are also useful as a guide to extrapolate limited experimental data into another range. As discussed in Chapter 14, the EM wave theory helps to understand the directional behavior of the directional reflectivity, absorptivity, and emissivity. Since the theory applies to pure substances with ideally smooth surfaces, it provides a means for computing a limit of attainable properties, such as maximum reflectivity or minimum emissivity of a metallic surface.

3.2.1 Dielectric Materials

Maxwell equations provide the basic guidelines for reflection and refraction of EM waves at the interface between two media. To carry out these calculations, the complex index of refraction for each medium is needed. This index, given as $\bar{n}_\lambda = n_\lambda - ik_\lambda$, is always wavelength dependent. The real part, the so-called refractive index n_λ, and the complex part, the so-called extinction or attenuation index k_λ, are related to the complex dielectric constant $K_\lambda = K'_\lambda - iK''_\lambda$ and the magnetic permeability $\mu_\lambda = \mu'_\lambda - i\mu''_\lambda$. These relationships are discussed in Chapter 14. First, we discuss surface radiative properties without referring to the fundamental Maxwell equations. Later, we provide more general expressions.

3.2.1.1 Reflection and Refraction at the Interface between Two Perfect Dielectrics (No Wave Attenuation, $k \to 0$)

A medium is called a dielectric if it does not absorb incident radiation. Then, the complex index of refraction of a dielectric medium is expressed only with its real component: $\bar{n}_\lambda = n_\lambda$ as k_λ is zero. Such cases are quite common within a narrow spectrum of wavelengths when reflection of light at an interface of two dielectric surfaces is considered. Typical examples are the air–water or air–glass interfaces for the visible spectrum, but only in the portion of spectrum where $k_\lambda = 0$.

Consider a beam of radiation, an EM wave traveling through a dielectric medium with refractive index n_1, which encounters the smooth planar interface of a second medium with a refractive index n_2. If the angle of the incidence onto the surface of medium 2 is θ_i, then the wave is refracted

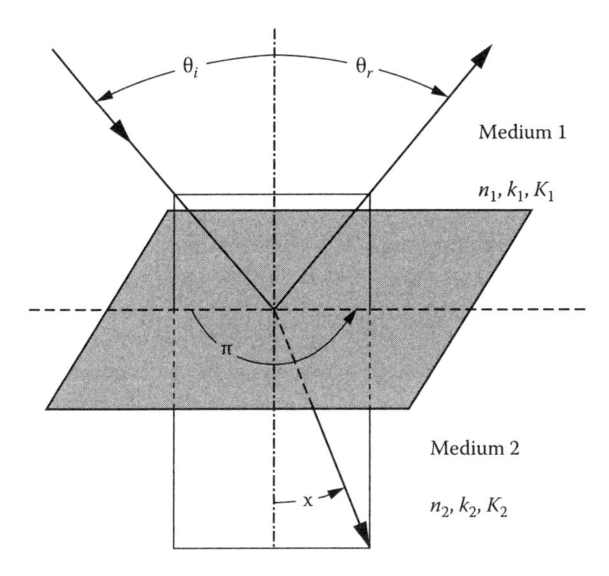

FIGURE 3.1 Interaction of the EM wave at an interface between media 1 and 2.

into medium 2 at angle χ. The angle of incidence of θ_i, reflection θ_r, and refraction χ, which are all in the same plane (Figure 3.1), are related through *Snell's* law:

$$n_1 \sin\theta_i = n_1 \sin\theta_r = n_2 \sin\chi \tag{3.1}$$

which results in

$$\theta_i = \theta_r \tag{3.2}$$

The angle of reflection of an EM wave from an ideal interface is thus equal to its angle of incidence rotated about the normal to the interface through a circumferential angle of $\theta = \pi$. These are the relations that define mirrorlike or *specular* (not to be confused with spectral- or wavelength-dependent) reflections.

Equation 3.1 yields the following relation between θ_i and χ:

$$\frac{\sin\chi}{\sin\theta_i} = \frac{n_1}{n_2} \tag{3.3}$$

The energy carried by a wave is proportional to the square of the wave amplitude E. Squaring the amplitude ratio of reflected to incident amplitude gives the ratio of energy reflected from a surface to energy incident from a given direction.

The refractive index of a medium is a function of wavelength, and thus the calculated radiative properties are wavelength dependent. If, however, the refractive index is calculated from the electric permittivity $\bar{\varepsilon}$ or the *dielectric constant* K (where $K = \bar{\varepsilon}/\bar{\varepsilon}_0$), which are not generally given as functions of wavelength, the spectral dependence is lost. For simplicity, the notation for signifying spectral λ dependence is omitted, but this dependence can be examined if the spectral dependence of the optical or EM properties is known. Also, here we introduce the orthogonal fields, which correspond to polarization components of the electric field associated with the incident radiation. The total intensity is the sum of the two polarized components. Incident light can be polarized in any plane; however, a traditional nomenclature for parallel and perpendicularly polarized components is adopted here. These components correspond to oscillating planes of the electric field, and both are perpendicular to the direction of the propagation.

3.2.1.2 Reflectivity

The specular reflectivity of a wave incident on a surface at angle θ and polarized parallel or perpendicular to the plane of incidence is obtained from the following equations:

$$\rho_{\parallel}(\theta_i) = \left[\frac{\tan(\theta_i - \chi)}{\tan(\theta_i + \chi)}\right]^2 \tag{3.4}$$

$$\rho_{\perp}(\theta_i) = \left[\frac{\sin(\theta_i - \chi)}{\sin(\theta_i + \chi)}\right]^2 \tag{3.5}$$

where χ as shown in Figure 3.1 is the angle of refraction in the medium on which the wave impinges. For a given θ_i, the χ can be determined from Equation 3.3 as

$$\frac{\sin\chi}{\sin\theta_i} = \frac{n_1}{n_2} = \frac{\sqrt{\bar{\varepsilon}_1}}{\sqrt{\bar{\varepsilon}_2}} = \frac{\sqrt{K_1}}{\sqrt{K_2}} \tag{3.6}$$

The n, $\bar{\varepsilon}$, and K are assumed not to have any angular dependence. Useful forms containing only θ_i are obtained by eliminating χ in Equations 3.4 and 3.5 by using Equation 3.6:

$$\rho_{\parallel}(\theta_i) = \left\{\frac{(n_2/n_1)^2\cos\theta_i - [(n_2/n_1)^2 - \sin^2\theta_i]^{1/2}}{(n_2/n_1)^2\cos\theta_i + [(n_2/n_1)^2 - \sin^2\theta_i]^{1/2}}\right\}^2 \tag{3.7}$$

$$\rho_{\perp}(\theta_i) = \left\{\frac{[(n_2/n_1)^2 - \sin^2\theta_i]^{1/2} - \cos\theta_i}{[(n_2/n_1)^2 - \sin^2\theta_i]^{1/2} + \cos\theta_i}\right\}^2 \tag{3.8}$$

The $\rho_{\parallel}(\theta_i) = 0$ when $\theta_i = \tan^{-1}(n_2/n_1)$; this θ_i is called *Brewster's* angle. Radiation reflected from energy incident at this angle is all perpendicularly polarized.

The reflectivity for unpolarized incident radiation is given by Fresnel equation (see Chapter 14):

$$\rho(\theta_i) = \frac{1}{2}\frac{\sin^2(\theta_i - \chi)}{\sin^2(\theta_i + \chi)}\left[1 + \frac{\cos^2(\theta_i + \chi)}{\cos^2(\theta_i - \chi)}\right] \tag{3.9}$$

The average of Equations 3.7 and 3.8 is often used to obtain the angular reflectivity profile $\rho(\theta_i)$ in terms of θ_i only. For normal incidence, $\theta_i = 0$, reflectivity becomes

$$\rho_n = \rho(\theta_i = 0) = \left(\frac{n_2 - n_1}{n_2 + n_1}\right)^2 = \left[\frac{(n_2/n_1) - 1}{(n_2/n_1) + 1}\right]^2 \tag{3.10}$$

The reflectivities are always spectral as n_1 and n_2 are function of λ.

Note that for most cases of interest we can take $n_1 = 1$ (e.g., for incidence in air or a vacuum). Also, these relations are for an interface on an opaque material, so that there is no reflection from a second nearby internal surface in medium 2. The cases of nonopaque layers and materials are considered in Chapter 17.

TABLE 3.1
Optical Property Values and Normal Spectral Reflectivity of Various Dielectric Materials at T = 300 K and λ = 0.589 μm

Material	n (Refractive Index)	K (Dielectric Constant)	$\rho_{\lambda,n}$ (Equation 3.10)
SiO_2 (glass)	1.458	4.42	0.035
SiO_2 (fused quartz)	1.544	3.75	0.046
NaCl	1.5441	5.90	0.046
KCl	1.4902	4.86	0.039
H_2O (liquid)	1.332	77.78	0.020
H_2O (ice, 0°C)	1.309	91.60	0.018
Vacuum	1.000	1.00	0.000

Source: Data from Lide, D.R. (ed.), *Handbook of Chemistry and Physics*, 88th edn., CRC Press, Boca Raton, FL, 2008.

The EM theory has a number of limitations for calculations of radiative properties of surfaces used in many engineering applications. In general, the theory is invalid for frequencies of the order of molecular vibrational frequencies due to unavailable dielectric constants. The theory also neglects effects of surface conditions. This is a significant limitation, since perfectly clean, optically smooth interfaces between dielectrics are present only in certain applications such as glass lenses, windows, and other optical components. Predictions of the normal reflectivity from the EM theory for various dielectrics are listed in Table 3.1.

Example 3.1

Unpolarized radiation is incident on a dielectric surface (medium 2) in air (medium 1) at angle $\theta = 30°$ from the normal. The dielectric has refractive index $n_2 = 3.0$. Find the reflectivity of each polarized component and of the unpolarized incident radiation.

Because the incident intensity is in air, $n_1 \approx 1$, and from Equation 3.6, $n_1/n_2 = 1/3.0 = \sin\chi/\sin 30°$; therefore, $\chi = 9.6°$. The reflectivity of the parallel component is, from Equation 3.4, $\rho_{\|}(\theta_i = 30°) = (\tan 20.4°/\tan 39.6°)^2 = 0.202$. and that of the perpendicular component is, from Equation 3.5, $\rho_{\perp}(\theta_i = 30°) = (\sin 20.4°/\sin 39.6)^2 = 0.301$. The reflectivity for the unpolarized incident intensity, obtained from Equation 3.9 or from the average of the components, is $\rho(\theta_i = 30°) = (0.202 + 0.301)/2 = 0.252$.

Example 3.2

What fraction of light is reflected for normal incidence in air on a glass surface or on the surface of water?

For glass in the visible spectrum, $n \approx 1.55$, and for water $n \approx 1.33$. Then, from Equation 3.10,

$$\rho_n(\text{glass}) = \left(\frac{n-1}{n+1}\right)^2 = \left(\frac{0.55}{2.55}\right)^2 = 0.047\ (<5\%)$$

$$\rho_n(\text{water}) = \left(\frac{0.33}{2.33}\right)^2 = 0.020\ (<2\%)$$

Note that these results are only for the given portion of the spectrum.

3.2.1.3 Emissivity

One can obtain the directional emissivity starting from the directional reflectivity data and by using the radiative transfer equation along a given direction and for a given wavelength. For a nontransparent (opaque) medium, we can write $\varepsilon(\theta) = 1 - \rho(\theta)$ where spectral dependence is omitted for convenience. As for directional reflectivity, directional emissivity is also a function of the index of refraction of the emitting body and the medium surrounding it. In Figure 3.2, different curves are shown for various ratios of (n_2/n_1), where $n_2 > n_1$. When $n_2 < n_1$, there is a limiting angle such that radiation at larger incidence angles is totally reflected; this is discussed in Section 17.5.2. When $\rho(\theta)$ is computed for incident radiation in air ($n_1 \approx 1$), the ratio n_2/n_1 reduces to the refractive index of the material on which the radiation is incident. Figure 3.2 gives the emissivity of a dielectric into air when n_2/n_1 in the figure is set equal to the simple refractive index n of the emitting dielectric. For $(n_2/n_1) = 1$, the emissivity is one (blackbody case), and the curve in Figure 3.2 is circular with a radius of one.

As (n_2/n_1) increases, the curves remain circular up to about $\theta = 70°$ and then decrease rapidly to zero at $\theta = 90°$. Thus, dielectric materials emit poorly at large angles from the normal direction. For angles less than 70°, the emissivities are quite high, so that, in a hemispherical sense, dielectrics are good emitters. The assumptions used in applying Maxwell equations restrict these findings to wavelengths longer than the visible spectrum, as verified by comparisons with experimental measurements.

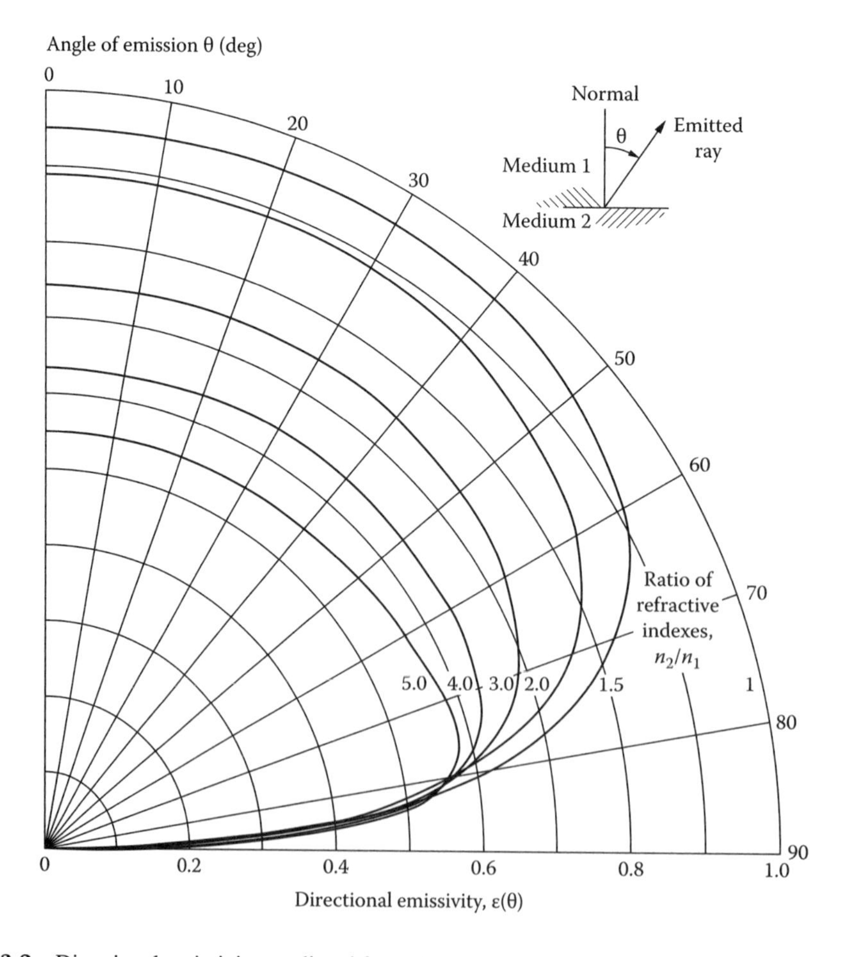

FIGURE 3.2 Directional emissivity predicted from the EM theory.

From the directional-spectral emissivity, the hemispherical-spectral emissivity can be computed from Equation 2.7 as $\varepsilon_\lambda(T)=(1/\pi)\int_\cap \varepsilon_\lambda(T,\theta,\phi)\cos\theta d\Omega$. Then an integration can be performed over all wavelengths to obtain the hemispherical total emissivity as in Equation 2.8.

The integration of $\varepsilon_\lambda(\theta)$ over all directions to evaluate the hemispherical ε_λ is complicated, but the integration has been carried out to yield ε in terms of n_2/n_1:

$$\varepsilon = \frac{1}{2} - \frac{(3n+1)(n-1)}{6(n+1)^2} - \frac{n^2(n^2-1)^2}{(n^2+1)^3}\ln\left(\frac{n-1}{n+1}\right) + \frac{2n^3(n^2+2n-1)}{(n^2+1)(n^4-1)} - \frac{8n^4(n^4+1)}{(n^2+1)(n^4-1)^2}\ln n \quad \left[n = \left(\frac{n_2}{n_1}\right) > 1\right] \tag{3.11}$$

The normal emissivity is a convenient value to which the hemispherical value may be compared:

$$\varepsilon_n = 1-\rho_n = 1-\left(\frac{n-1}{n+1}\right)^2 = \frac{4n}{(n+1)^2} \quad \left(n = \frac{n_2}{n_1} > 1\right) \tag{3.12}$$

The ε_n is shown as a function of n_2/n_1 in Figure 3.3a. Note that normal emissivities less than about 0.50 correspond to $n_2/n_1 > 6$. Such large n_2/n_1 values are uncommon for dielectrics, so the

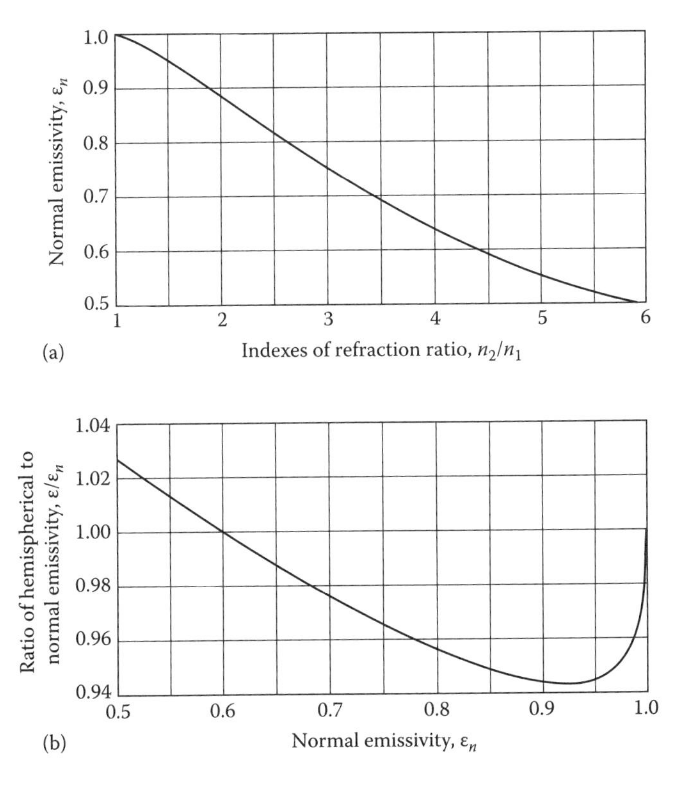

FIGURE 3.3 Predicted emissivities of dielectric materials for emission into medium with refractive index n_1 from medium with refractive index n_2 where $n_2 \geq n_1$. (a) Normal emissivity as a function of n_2/n_1 and (b) comparison of hemispherical and normal emissivity.

curve is not extended to smaller ε_n. The hemispherical emissivity for dielectrics is compared to the normal value in Figure 3.3b.

For large n_2/n_1, the ε_n values are relatively low, and with increasing n_2/n_1, the curves in Figure 3.2 depart more from circular form. Figure 3.3b reveals that the flattening of the curves of Figure 3.2 in the region near the normal causes the hemispherical emissivity to exceed the normal value at large n_2/n_1 ($\varepsilon_n \approx$ 0.5–0.6). For n_2/n_1 near 1 (ε_n near 1), the hemispherical ε is lower than ε_n because of poor emission at large θ.

Example 3.3

A dielectric medium has $n = 1.41$. What is its hemispherical emissivity into air at the wavelength at which the n was measured?

From Equation 3.11, for $n_1/n_2 = n = 1.41$, the $\varepsilon = 0.92$. Alternatively, from Equation 3.12, the hemispherical normal $\varepsilon_n = 1-(0.41/2.41)^2 = 0.97$, and from Figure 3.2b, $\varepsilon/\varepsilon_n = 0.95$; then, the hemispherical emissivity is $\varepsilon = 0.97 \times 0.95 = 0.92$.

3.2.2 Radiative Properties of Metals

3.2.2.1 Electromagnetic Relations for Incidence on an Absorbing Medium

In the preceding section, we have dealt with perfect dielectrics where there is no attenuation ($k = 0$) of radiation as it travels within the material. In other words, we assumed that the material is perfectly transparent except for interface reflections. For real dielectrics, there is attenuation, so k is nonzero, although small.

The propagation of an EM wave in an infinite medium that attenuates the EM wave is governed by the same relations as in a nonattenuating medium if the refractive index n is replaced by the complex index, $\bar{n}$, although this leads to some complexities in interpretation. For example, Snell's law (Equation 3.3) becomes

$$\frac{\sin\chi}{\sin\theta_i} = \frac{\bar{n}_1}{\bar{n}_2} = \frac{n_1 - ik_1}{n_2 - ik_2} \tag{3.13}$$

Because this relation is complex, sin χ is complex, and the angle χ can no longer be interpreted physically as a simple angle of refraction for propagation into the material. In addition, except for the special case of normal incidence, n is no longer directly related to the propagation velocity.

Now consider radiation incident in air or vacuum on a material with $\bar{n} = n - ik$. The two polarized components of reflectivity are

$$\rho_{\lambda,\parallel}(\theta_i) = \frac{a^2 + b^2 - 2a\sin\theta_i\tan\theta_i + \sin^2\theta_i\tan^2\theta_i}{a^2 + b^2 + 2a\sin\theta_i\tan\theta_i + \sin^2\theta_i\tan^2\theta_i}\rho_{\lambda,\perp}(\theta_i) \tag{3.14}$$

$$\rho_{\lambda,\perp}(\theta_i) = \frac{a^2 + b^2 - 2a\cos\theta_i + \cos^2\theta_i}{a^2 + b^2 + 2a\cos\theta_i + \cos^2\theta_i} \tag{3.15}$$

where

$$a^2 = \frac{1}{2}\left\{\left[(n^2 - k^2 - \sin^2\theta_i)^2 + 4n^2k^2\right]^{1/2} + n^2 - k^2 - \sin^2\theta_i\right\} \tag{3.16}$$

$$b^2 = \frac{1}{2}\left\{\left[(n^2 - k^2 - \sin^2\theta_i)^2 + 4n^2k^2\right]^{1/2} - (n^2 - k^2 - \sin^2\theta_i)\right\} \tag{3.17}$$

If the incident beam has no specific polarization, the reflectivity is an average of the parallel and perpendicular components. Some results are given by Harpole (1980) for the more complex case of oblique incidence from an adjacent absorbing material rather than from vacuum or a material with $n \approx 1$ such as air.

The reflectivity for normal incidence reduces to

$$\rho_{\lambda,n} = \left[\frac{n_2 - ik_2 - (n_1 - ik_1)}{n_2 - ik_2 + n_1 - ik_1}\right]\left[\frac{n_2 + ik_2 - (n_1 + ik_1)}{n_2 + ik_2 + n_1 + ik_1}\right]$$

$$= \frac{(n_2 - n_1)^2 + (k_2 - k_1)^2}{(n_2 + n_1)^2 + (k_2 + k_1)^2} \tag{3.18}$$

For an incident ray in air, $n_1 \approx 1$ and $k_1 \approx 0$. When the material is also nonattenuating (perfectly transparent) $k_2 \to 0$, and Equation 3.18 reduces to Equation 3.10.

Metals are usually highly absorbing; therefore, the complex index of refraction $\bar{n} = n - ik$ must be used in the theoretical relations. With k included, Equations 3.14 and 3.15 can be used with a and b from Equations 3.16 and 3.17. Hering and Smith (1968) give the following alternative form *for incidence from air or vacuum* ($n_1 = 1$, $k_1 = 0$, $n = n_2$, $k = k_2$):

$$\rho_\perp(\theta_i) = \frac{(n\beta - \cos\theta_i)^2 + (n^2 + k^2)\alpha - n^2\beta^2}{(n\beta + \cos\theta_i)^2 + (n^2 + k^2)\alpha - n^2\beta^2} \tag{3.19}$$

$$\rho_\parallel(\theta_i) = \frac{(n\gamma - \alpha/\cos\theta_i)^2 + (n^2 + k^2)\alpha - n^2\gamma^2}{(n\gamma + \alpha/\cos\theta_i)^2 + (n^2 + k^2)\alpha - n^2\gamma^2} \tag{3.20}$$

wherein

$$\alpha^2 = \left(1 + \frac{\sin^2\theta}{n^2 + k^2}\right) - \frac{4n^2}{n^2 + k^2}\left(\frac{\sin^2\theta}{n^2 + k^2}\right) \tag{3.21}$$

$$\beta^2 = \frac{n^2 + k^2}{2n^2}\left(\frac{n^2 - k^2}{n^2 + k^2} - \frac{\sin^2\theta}{n^2 + k^2} + \alpha\right) \tag{3.22}$$

$$\gamma = \frac{n^2 - k^2}{n^2 + k^2}\beta + \frac{2nk}{n^2 + k^2}\left(\frac{n^2 + k^2}{n^2}\alpha - \beta\right)^{1/2} \tag{3.23}$$

For unpolarized incident radiation, $\varepsilon(\theta) = 1 - [\rho_\perp(\theta) + \rho_\parallel(\theta)]/2$ was evaluated and then integrated over the hemisphere to obtain the hemispherical emissivity, $\varepsilon = \int_0^1 \varepsilon(\theta)\, d(\sin^2\theta)$, as a function of n and k. Equations 3.19 and 3.20 were evaluated and presented graphically in detail in Hering and Smith (1968). One set of results, for $n \geq 1$, is in Figure 3.4 as a function of n and k/n. This shows how the hemispherical emittance decreases as both n and k increase.

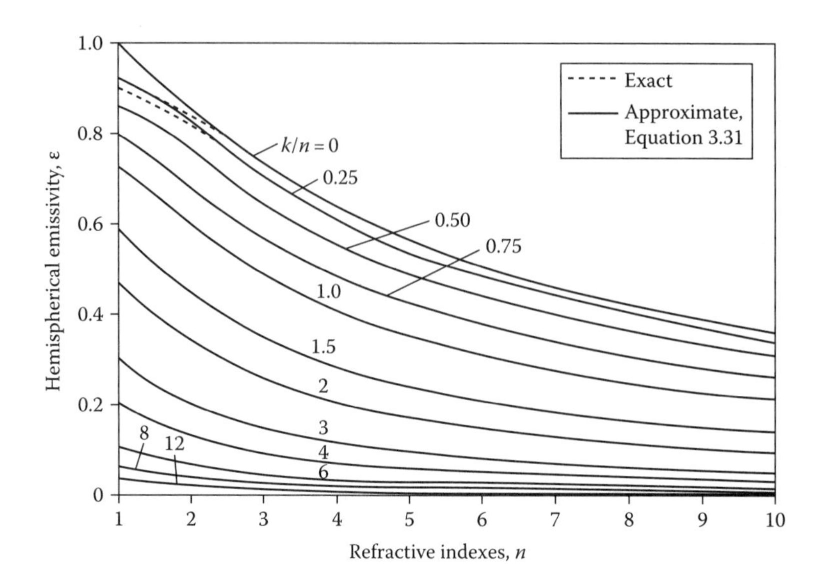

FIGURE 3.4 Exact and approximate hemispherical emissivity results ($n \geq 1.0$). (From Hering, R.G. and Smith, T.F., *IJHMT*, 11(10), 1567, 1968.)

3.2.2.2 Reflectivity and Emissivity Relations for Metals (Large *k*)

Usually, metals are internally highly absorbing; this means that the extinction coefficient k is large, which allows simplifying assumptions that lead to more convenient equations than the preceding general results. Therefore, the following expressions should not be used for absorbing dielectrics where k is not large enough. This elimination is not always straightforward, as the optical properties are difficult to obtain and they can be inaccurate because of experimental measurement problems. For large k, the $\sin^2\theta$ terms in Equations 3.21 and 3.22 can be neglected relative to $n^2 + k^2$. Then $\alpha = \beta = \gamma = 1$ and Equations 3.19 and 3.20 yield, for incidence from air or vacuum onto a medium with $\bar{n} = n - ik$,

$$\rho_{\parallel}(\theta_i) = \frac{(n\cos\theta_i - 1)^2 + (k\cos\theta_i)^2}{(n\cos\theta_i + 1)^2 + (k\cos\theta_i)^2} \tag{3.24}$$

$$\rho_{\perp}(\theta) = \frac{(n - \cos\theta_i)^2 + k^2}{(n + \cos\theta_i)^2 + k^2} \tag{3.25}$$

For unpolarized incident radiation,

$$\rho(\theta_i) = \frac{\rho_{\parallel}(\theta_i) + \rho_{\perp}(\theta_i)}{2} \tag{3.26}$$

For the normal direction ($\theta_i = 0$), $\rho_n = [(n - 1)^2 + k^2]/[(n + 1)^2 + k^2]$, which is the same as the exact relation Equation 3.24.

The corresponding emissivity values are obtained using $\varepsilon(\theta) = 1 - \rho(\theta)$, which gives

$$\varepsilon_{\parallel}(\theta) = \frac{4n\cos\theta}{(n^2 + k^2)\cos^2\theta + 2n\cos\theta + 1} \tag{3.27}$$

$$\varepsilon_{\perp}(\theta)=\frac{4n\cos\theta}{\cos^2\theta+2n\ \cos\theta+n^2+k^2} \tag{3.28}$$

Emission is assumed to be unpolarized, so the emissivity is given by

$$\varepsilon(\theta)=\frac{\varepsilon_{\parallel}(\theta)+\varepsilon_{\perp}(\theta)}{2} \tag{3.29}$$

In the normal direction ($\theta = 0$), this becomes

$$\varepsilon_n=\frac{4n}{(n+1)^2+k^2} \tag{3.30}$$

These emissivity relations are demonstrated in Figure 3.5 for a pure smooth platinum surface at a wavelength of 2 μm. (The data for n and k for platinum, taken from Lide (2008), are n = 5.29 and k = 6.71 at λ = 2 μm, and the data and prediction are in quite good agreement). It is evident by comparison with this and other experimental data from Harrison et al. (1961), Brandenberg (1963), Brandenberg and Clausen (1965), Price (1947), Palik (1998), and Lide (2008) that the general shape of the curve predicted by Equation 3.29 is correct.

The complex refractive index can be defined in other ways than $\tilde{n}=n-ik$ as used here. It is also common to define $\tilde{n}=n(1-ink)$. Occasionally this definition has a positive sign in front of the extinction factor. When consulting references, care should be taken in determining which definition is used.

For metals, as illustrated in Figure 3.5, the emissivity is essentially constant for about 50° away from the normal and then increases to a maximum within a few degrees of the tangent to the surface.

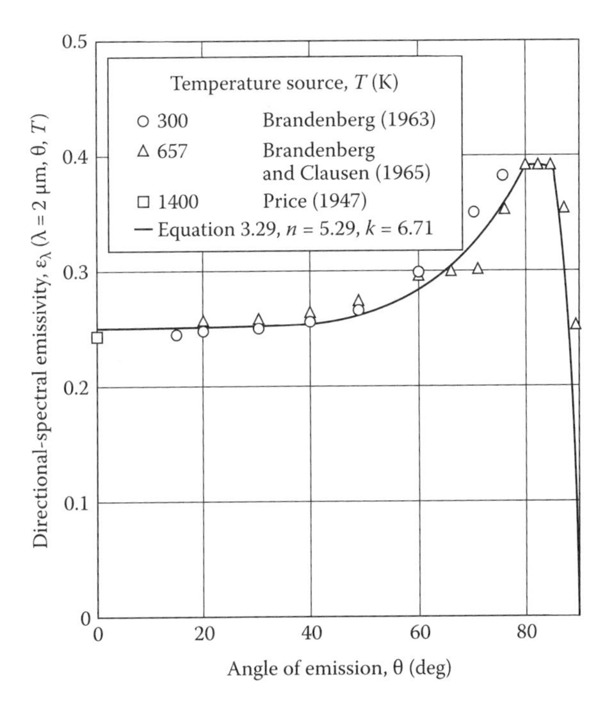

FIGURE 3.5 Directional-spectral emissivity of platinum at wavelength λ = 2 μm.

TABLE 3.2
Optical Property Values and Normal Spectral Reflectivity of Various Metals at T = 300 K and λ = 0.589 and 10 μm

Material	Refractive Index[a], n	Extinction Index[a], k	Electrical Resistivity[a], r_e (Ohm·cm)	Normal Spectral Reflectivity, $\rho_{\lambda,n}$ Equation 3.30	Equation 3.35	Measured[b]
Aluminum (0.589 μm)	1.15	7.147	2.73×10^{-6}	0.917	0.923	0.91
(10 μm)	25.4	90.0	"	0.988	0.981	0.988
Chromium (0.589 μm)	3.33	4.39	12.7×10^{-6}	0.650	0.841	0.57
(10 μm)	13.9	27.7	"	0.944	0.960	0.946
Copper (0.589 μm)	0.47	2.81	1.72×10^{-6}	0.813	0.939	0.76
(10 μm)	28	68	"	0.980	0.985	0.985
Gold (0.589 μm)	0.18	2.84	2.27×10^{-6}	0.924	0.930	0.85
(10 μm)	6.7	73	"	0.995	0.983	0.994
Iron (0.589 μm)	2.80	3.34	9.98×10^{-6}	0.562	0.858	0.517
(10 μm)	6.34	28.2	"	0.970	0.964	0.953
Nickel (0.589 μm)	1.85	3.48	7.20×10^{-6}	0.634	0.878	0.655
(10 μm)	7.59	38.5	"	0.980	0.969	0.965
Platinum (0.589 μm)	2.23	3.92	10.8×10^{-6}	0.654	0.852	0.673
(10 μm)	10.8	38.2	"	0.973	0.963	0.970
Silver (0.589 μm)	0.26	3.96	1.63×10^{-6}	0.940	0.941	0.95
(10 μm)	8.22	79	"	0.995	0.985	0.995
Titanium (0.589 μm)	2.01	2.77	39×10^{-6}	0.520	0.734	0.490
(10 μm)	4.1	19.7	"	0.960	0.930	0.970
Tungsten (0.589 μm)	3.54	2.84	5.44×10^{-6}	0.506	0.893	0.51
(10 μm)	11.6	48.4	"	0.981	0.973	0.97

[a] Lide, D.R. (ed.), *Handbook of Chemistry and Physics*, 88th edn., CRC Press, Boca Raton, FL, 2008.
[b] Measured data are the largest reflectivity value from multiple references cited by Touloukian (1970).

This angular dependence for emission by metals is in contrast to the behavior for dielectrics, for which emission decreases substantially as the angle from the normal becomes larger than about 70°.

In Table 3.2, normal spectral emissivities calculated using Equation 3.30 are compared with measured values. A wavelength of λ = 0.589 μm is used for some of the comparisons because data are available from experiments conducted using a sodium-vapor lamp, which emits at this wavelength. Since this wavelength is in the visible range, it is in the borderline short-wavelength region where the EM theory becomes inaccurate.

The values in Table 3.2 show agreement between the values predicted by Equation 3.30 and measured $\varepsilon_{\lambda,n}$ to be good at λ = 10 μm, but in greater error at λ = 0.589 μm. For the cases of poor agreement, it is difficult to ascribe the error specifically to the optical constants, to the measured emissivity, or to the theory itself, although the theory is expected to be less accurate at shorter wavelengths. Most probably the optical constants are somewhat in error, and the experimental samples do not meet the standards of perfection in surface preparation demanded by the theory.

Palik (1998) provides extensive data for n and k for metals such as copper, gold, silver, nickel, aluminum, and tungsten. This database also includes semiconductors such as germanium, indium arsenide, silicon, and lead telluride, as well as some insulating materials such as lithium fluoride, glass, and salt (sodium chloride). The information on k is also useful for obtaining the spectral absorption coefficient κ used to describe exponential attenuation by absorption in a medium according to the relation $\exp(-\kappa x)$ (Equations 1.43 and 1.44). The relation between k and κ is $\kappa = 4\pi k/\lambda_0$, where both k and κ are functions of wavelength. Some additional data about these properties are provided by Brewster (1992).

Within the approximation of neglecting $\sin^2\theta$ relative to $n^2 + k^2$, the hemispherical emissivity for a metal in air or vacuum is found by substituting Equation 3.29 into Equation 2.9. After integration, this yields

$$\varepsilon = 4n - 4n^2 \ln\frac{1+2n+n^2+k^2}{n^2+k^2} + \frac{4n(n^2-k^2)}{k}\tan^{-1}\frac{k}{n+n^2+k^2} + \frac{4n}{n^2+k^2}$$
$$- \frac{4n^2}{(n^2+k^2)^2}\ln\left(1+2n+n^2+k^2\right) - \frac{4n(k^2-n^2)}{k(n^2+k^2)}\tan^{-1}\frac{k}{1+n} \quad (3.31)$$

Without neglecting $\sin^2\theta$, the hemispherical ε was calculated by numerical integration by Hering and Smith (1968) as shown in Figure 3.4, and the results were compared with those from Equation 3.31 for various n and k values. They found that for the accuracy of Equation 3.31 to be within 1%, 2%, 5%, or 10% of the numerically integrated values, $n^2 + k^2$ should be larger than 40, 3.25, 1.75, and 1.25, respectively. Based on the optical constants given in Table 3.2, Equation 3.31 should usually be accurate for most metals, within a few percent.

From Equation 3.30, the normal emissivity from both metals and attenuating dielectrics into air can be computed as a function of n and k. This is shown in Figure 3.6, and more complete results are provided by Hering and Smith (1968). The hemispherical emissivity from Equation 3.31 is very important as it provides the radiation emission rate into all directions. For polished metals, when ε_n is less than about 0.5, the hemispherical emissivity is usually larger than the normal value because of the increase in emissivity in the direction near tangency to the surface, as was pointed out in Figure 3.5. Hence, in a table listing emissivity values for polished metals, if ε_n is given, it should be multiplied by a factor larger than one such as is obtained by comparing Equations 3.30 and 3.31. In the visible wavelength region, the $\varepsilon_\lambda/\varepsilon_{\lambda,n}$ value is ≈ 1, but in the infrared (IR), some values approach 1.2. Real surfaces that have roughness or may be slightly oxidized often tend to have a directional emissivity that is more nearly diffuse than for polished specimens.

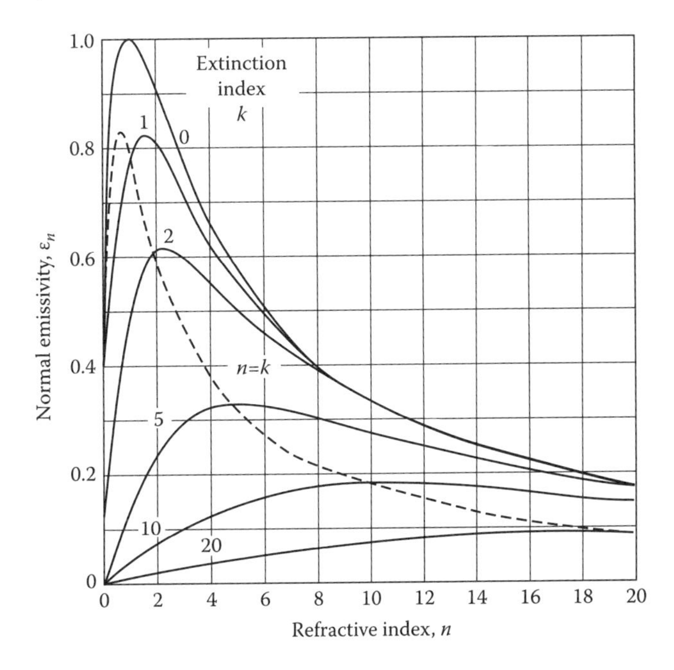

FIGURE 3.6 Emissivity of attenuating medium emitting into air.

3.2.2.3 Relations between Radiative Emission and Electrical Properties

The wave solutions to Maxwell equations provide a means for determining n and k from the electric and magnetic properties of a material. For metals where electrical resistivity, r_e, is small, and for relatively long wavelengths, $\lambda_0 > \sim 5$ μm (μ_0 is the magnetic permeability):

$$n = k = \sqrt{\frac{\lambda_0 \mu_0 c_o}{4\pi r_e}} = \sqrt{\frac{0.003\lambda_0}{r_e}} \quad (\lambda_0 \text{ in } \mu\text{m}, r_e \text{ in Ohm} \cdot \text{cm}) \tag{3.32}$$

This is *Hagen–Rubens equation* (Hagen and Rubens 1900). Predictions of n and k from this equation can be greatly in error, particularly at short wavelengths, as shown in Table 3.2. Nevertheless, some useful results are obtained if care is exercised.

With the conditions in Equation 3.32 that $n = k$, Equation 3.30 reduces to Equation 3.33 for a material with refractive index n radiating in the normal direction into air or vacuum:

$$\varepsilon_{\lambda,n} = 1 - \rho_{\lambda,n} = \frac{4n}{2n^2 + 2n + 1} \tag{3.33}$$

Substituting Equation 3.32 into Equation 3.33 and expanding into a series gives *Hagen–Rubens emissivity relation*:

$$\varepsilon_{\lambda,n} = \frac{2}{\sqrt{0.003}}\left(\frac{r_e}{\lambda_0}\right)^{1/2} - \frac{2}{0.003}\frac{r_e}{\lambda_0} + \cdots \tag{3.34}$$

Because the index of refraction of metals as predicted from Equation 3.32 is generally large at the longer wavelengths so that (r_e/λ_o) is small, that is, for $\lambda > \sim 5$ μm (see Table 3.2), one or two terms of the series are usually adequate. A two-term approximation is made by adjusting the coefficients of the second term for the remaining terms in the series to yield

$$\varepsilon_{\lambda,n} = 36.5\left(\frac{r_e}{\lambda_0}\right)^{1/2} - 464\frac{r_e}{\lambda_0} \tag{3.35}$$

Data for polished nickel are shown in Figure 3.7, and the extrapolation to long wavelengths by Equations 3.34 or 3.35 appears reasonable. The predictions of normal spectral emissivity at long wavelengths, as presented in Table 3.2, are much better than the predictions of the optical constants.

The angular behavior of the spectral emissivity can be obtained by substituting Equation 3.32 into Equations 3.27 and 3.28 to yield for the two components of polarization:

$$\varepsilon_{\lambda\|}(\theta) = \frac{4(0.003\lambda_0/r_e)^{1/2}\cos\theta}{0.006(\lambda_0/r_e)\cos^2\theta + 2(0.006\lambda_0/r_e)^{1/2}\cos\theta + 1} \tag{3.36}$$

$$\varepsilon_{\lambda\perp}(\theta) = \frac{4(0.003\lambda_0/r_e)^{1/2}\cos\theta}{\cos^2\theta + 2(0.003\lambda_0/r_e)^{1/2}\cos\theta + 0.006\lambda_0/r_e} \tag{3.37}$$

The normal spectral emissivity in Equation 3.34 can be integrated with respect to wavelength to yield a normal total emissivity, $\varepsilon_n(T) = \pi\int_0^\infty \varepsilon_{\lambda,n}(T) I_{\lambda b}(T)\, d\lambda/\sigma T^4$. Note that Equation 3.34 is

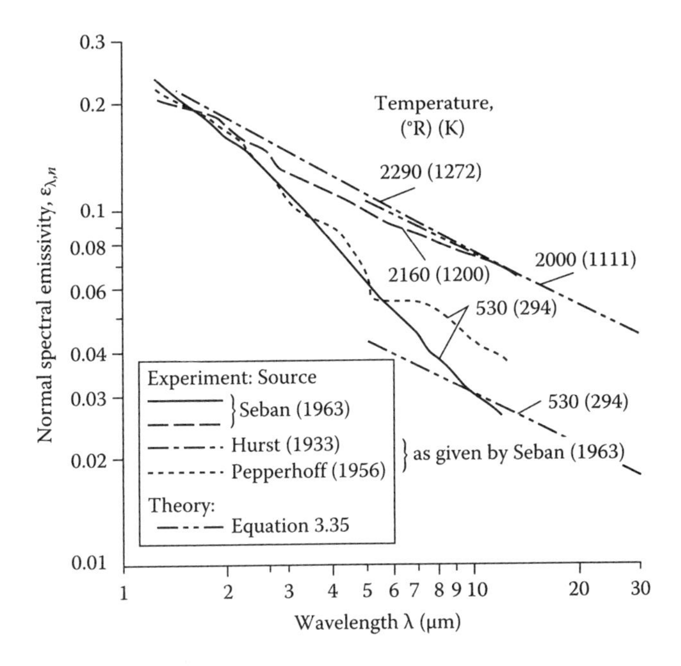

FIGURE 3.7 Comparison of measured values with theoretical predictions for normal spectral emissivity of polished nickel.

accurate only for $\lambda_0 > \sim 5$ μm, so in performing the integration starting from $\lambda = 0$, the energy radiated from $\lambda_0 = 0$–5 μm is assumed to be small compared with that at wavelengths longer than 5 μm. Then, substituting into the integral, the first term of Equations 3.34 and 1.15 for $I_{\lambda b}$ provides

$$\varepsilon_n(T) \approx \pi \frac{\int_0^\infty 2\left(\frac{r_e}{(0.003\lambda_0)^{\frac{1}{2}}}\right) 2C_1 \Big/ \left[\lambda_0^5(e^{C_2/\lambda_0 T}-1)\right] d\lambda_0}{\sigma T^4}$$

$$= \frac{4\pi C_1 (Tr_e)^{1/2}}{(0.003)^{1/2}\sigma C_2^{4.5}} \int_0^\infty \frac{\zeta^{3.5}}{e^{\zeta}-1} d\zeta \tag{3.38}$$

where $\zeta = C_2/\lambda_0 T$ as was used in conjunction with Equation 1.29. The integration is carried out by use of gamma functions to yield

$$\varepsilon_n(T) \approx \frac{4\pi C_1 (Tr_e)^{1/2}}{(0.003)^{1/2}\sigma C_2^{4.5}}(12.27) = 0.575(r_e T)^{1/2} \tag{3.39}$$

If additional terms in the series Equation 3.34 are retained, the recommended three-term approximation for total normal emissivity of a metal from Parker and Abbott (1964) is

$$\varepsilon_n(T) = 0.578\,(r_e T)^{1/2} - 0.178\,r_e T + 0.584(r_e T)^{3/2} \tag{3.40}$$

where

T is in Kelvins

r_e in Ohm-cm

For pure metals, r_e near room temperature is approximately described by

$$r_e \approx r_{e,273} \frac{T}{273} \tag{3.41}$$

where $r_{e,273}$ is the electrical resistivity in Ohm-cm at 273 K. Substituting Equation 3.41 into 3.39 gives the approximate result

$$\varepsilon_n(T) \approx 0.0348T\sqrt{r_{e,273}} \tag{3.42}$$

This indicates that the total emissivity of pure metals increases directly with absolute temperature. This expression was originally derived by Aschkinass (1905). In some cases, it holds to unexpectedly high temperatures where considerable emitted radiation is in the short-wavelength region (for platinum, to near 1800 K), but often it applies only below about 550 K. This is illustrated in Figure 3.8 for platinum and tungsten (data from Lide 2008).

In Figure 3.9, normal emissivity for various polished surfaces of pure metals is compared with predictions based on Equation 3.42. The experimental data are obtained from three standard compilations (Hottel 1954, Eckert and Drake 1959, Lide 2008). Overall, agreement between the theory and the experimental data is satisfactory.

Using the angular dependence of emissivity from Equations 3.36 and 3.37 combined with Equation 3.39, integration over all directions provides hemispherical quantities. The following approximate equations for the hemispherical total emissivity fit the results in two ranges:

$$\varepsilon_n(T) = 0.751\sqrt{r_eT} - 0.396r_eT \quad 0 < r_eT < 0.2 \tag{3.43}$$

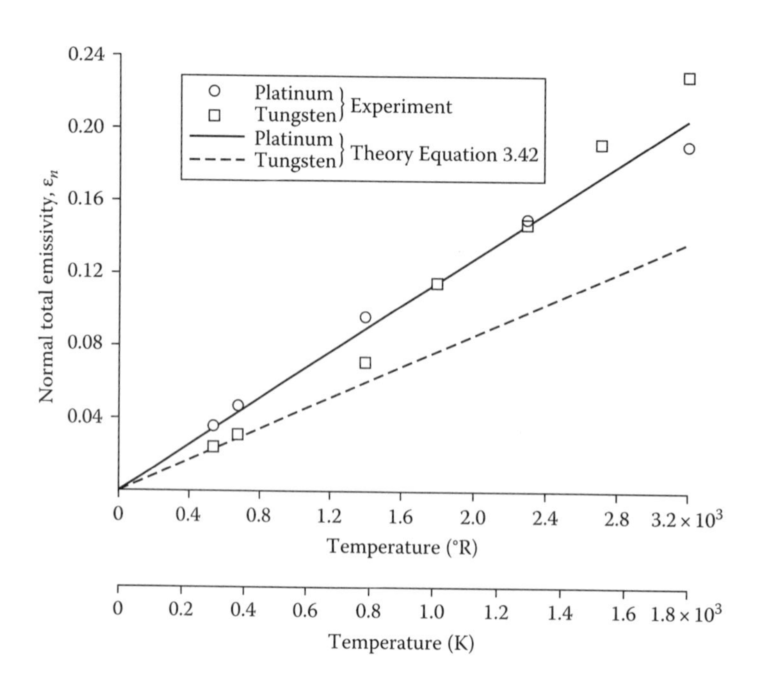

FIGURE 3.8 Temperature dependence of normal total emissivity of polished metals.

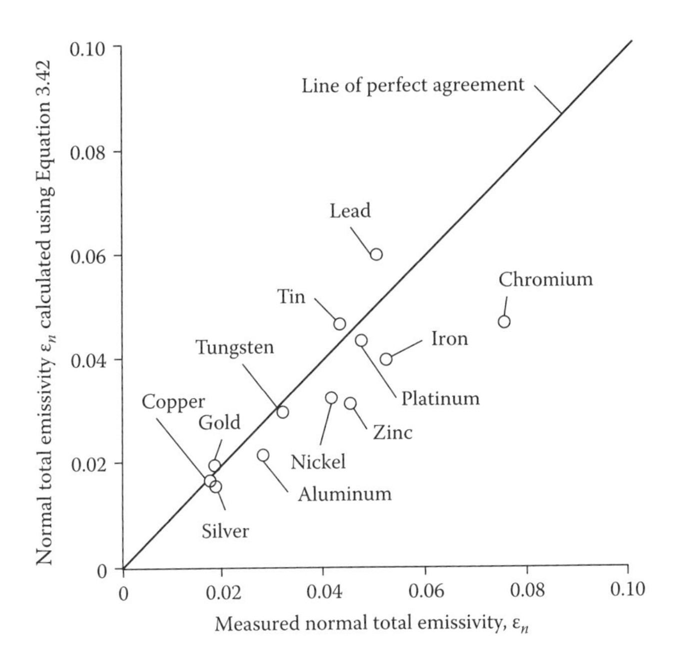

FIGURE 3.9 Comparison of data with calculated normal total emissivity for polished metals at 373 K.

and

$$\varepsilon_n(T) = 0.698\sqrt{r_e T} - 0.266 r_e T \quad 0.2 < r_e T < 0.5 \tag{3.44}$$

where the numerical factors and those used in specifying the ranges of validity apply for T in K and r_e in Ohm-cm. The resistivity r_e depends approximately on T to the first power so that the first term of each of these equations provides a T proportionality, indicating that the temperature dependence for energy emission by metals is higher than the blackbody dependence of T^4. A different hemispherical total emissivity expression suggested by Parker and Abbott (1964) is

$$\varepsilon_n(T) = 0.766\sqrt{r_e T} - \left[0.309 - 0.0889\ln(r_e T)\right] r_e T - 0.0175(r_e T)^{3/2} \tag{3.45}$$

Comparison of Equations 3.40 and 3.45 with data is in Figures 3.10 and 3.11.

Example 3.4

A polished platinum surface is maintained at $T = 250$ K. Energy is incident upon the surface from a black enclosure at $T_i = 500$ K that surrounds the surface. What is the hemispherical-directional total reflectivity into the direction normal to the surface?

Equation 2.67 shows that the directional-hemispherical total reflectivity for the normal direction can be found from

$$\rho_n(T = 250\ \text{K}) = 1 - \alpha_n(T = 250\ \text{K})$$

where $\alpha_n(T = 250\ \text{K})$ is the normal total absorptivity of a surface at 250 K for incident black radiation at 500 K.

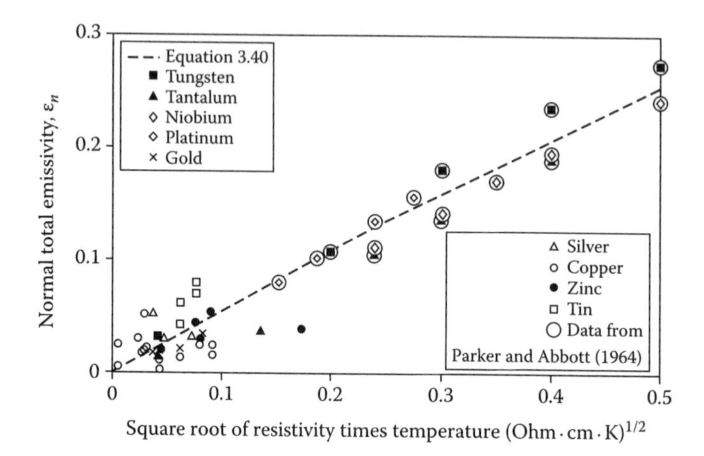

FIGURE 3.10 Normal total emissivity of various metals compared with theory. (From Parker, W.J. and Abbott, G.L., Theoretical and experimental studies of the total emittance of metals, *Symposium on Thermal Radiation Solids*, NASA SP-55, pp. 11–28, 1964.)

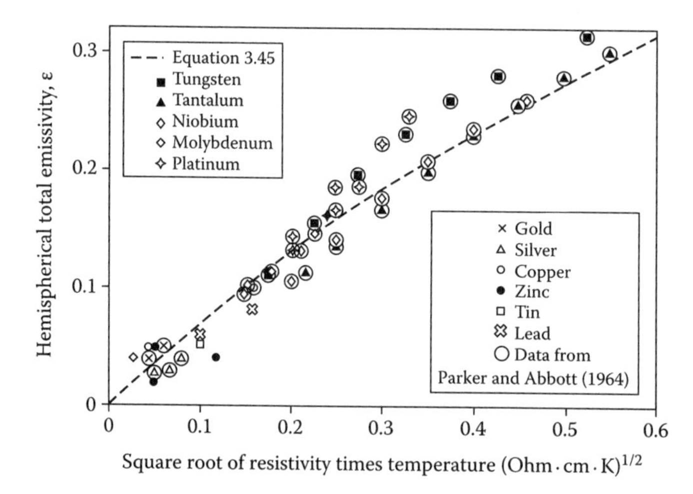

FIGURE 3.11 Hemispherical total emissivity of various metals compared with theory. (From Parker, W.J. and Abbott, G.L., Theoretical and experimental studies of the total emittance of metals, *Symposium on Thermal Radiation Solids*, NASA SP-55, pp. 11–28, 1964.)

$$\alpha_n(T = 250\text{ K}) = \frac{\int_0^\infty \alpha_{\lambda,n}(T = 250\text{ K}) I_{\lambda,b}(500\text{ K})\, d\lambda}{\int_0^\infty I_{\lambda,b}(500\text{ K})\, d\lambda}$$

For spectral quantities, $\alpha_n(T = 250\text{ K}) = \varepsilon_{\lambda,n}(T = 250\text{ K})$. From Equation 3.34, the near-linear variation of r_e with temperature provides the approximate emissivity variation $\varepsilon_{\lambda,n}(T) \propto T^{1/2}$. Then $\varepsilon_{\lambda,n}(T = 250\text{ K}) = \varepsilon_{\lambda,n}(T = 500\text{ K})(250/500)^{1/2}$, and we obtain

$$\alpha_n(T = 250\text{ K}) = \frac{\sqrt{1/2}\int_0^\infty \varepsilon_{\lambda,n}(T = 250\text{ K}) I_{\lambda,b}(500\text{ K})\, d\lambda}{\int_0^\infty I_{\lambda,b}(500\text{ K})\, d\lambda}$$

$$= \frac{\varepsilon_n(T = 500\text{ K})}{\sqrt{2}}$$

where the last equality is obtained by examining the emissivity definition, Equation 2.5. The normal total emissivity of platinum at 500 K is given by Equation 3.42, as plotted in Figure 3.8, as

$$\varepsilon_n(T = 500\text{ K}) = 0.348\sqrt{r_{e,273}T} = 0.348\sqrt{r_{e,293}\sqrt{\frac{273}{293}}T}$$

$$= 0.348\sqrt{10\times10^{-6}\sqrt{\frac{273}{293}}\times 500} = 0.053$$

Note that Equation 3.42 is to be used only when temperatures are such that most of the energy involved is at wavelengths greater than 5 μm. From the blackbody functions, for T = 500 K about 10% of the energy is below λ = 5 μm, so a small amount of error is introduced. The reciprocity relation of Equation 2.44 for uniform incident intensity can now be employed to give the desired result for the hemispherical-directional total reflectivity:

$$\rho_n(T = 250\text{ K}) = 1-\alpha_n(T = 250\text{ K}) \approx 1-\frac{1}{\sqrt{2}}\varepsilon_n(T = 500\text{ K})$$

$$= 1-\frac{0.053}{\sqrt{2}} = 0.963$$

3.3 EXTENSIONS OF THE THEORY FOR RADIATIVE PROPERTIES

The theory for radiative properties of materials has been improved significantly over the years based on using both the classical wave approach and quantum mechanics. The contributions of Davisson and Weeks (1924), Foote (1915), Schmidt and Eckert (1935), and Parker and Abbot (1964) extended the emissivity relations for metals to shorter wavelengths and higher temperatures, and Mott and Zener (1934) derived predictions for metal emissivity at very short wavelengths on the basis of quantum-mechanical relations. Kunitomo (1984) provides accurate predictions of high- and low-temperature properties of metals and alloys by including the effect of bound electrons on optical properties. Edwards (1969) and Kunitomo (1984) review advances in predicting surface properties, and Sievers (1978) and Kunitomo give additional theoretical expressions and results.

Chen and Ge (2000) introduced a damping frequency term into the Drude model of free-electron oscillations, giving a more accurate prediction of the complex dielectric constant. This modifies the relations between the complex refractive index and the complex dielectric constant. Chan and Ge found much better agreement with experimental total directional emissivity data using their method than for the predictions of Equations 3.36 and 3.37. They present comparisons for aluminum in the range $0 \le \theta \le 85.8°$ at 423 K integrated over $3.27 \le \lambda \le 50.23$ μm (Schmidt and Eckert 1935); the normal spectral emissivity at 4 μm, 293 K (Tables 3.1 and 3.2); and the normal total emissivity of copper integrated over $4.95 \le \lambda \le 75.94$ μm at 293 K and for the normal total emissivity of indium tin oxide films at room temperature (data from Chan and Ge 2000).

None of the treatments discussed so far accounts for surface conditions. Because of the difficulty of specifying surface conditions and controlling surface preparation, comparisons of theories with experiments are not always adequate for even the refined theories. Makino and Kaga (1991) modeled the radiative properties of real surfaces by combining the EM theory for pure substances with models for surface microgeometry and roughness and for surface films. Encouraging agreement was reported by Makino et al. (1991) of theoretical predictions with measured properties showing the effect of oxidation layer growth on metals (Makino et al. 1988). Further, EM scattering theory was used by Cohn et al. (1997) to predict reflection from a microcontoured surface with roughness scale elements in the same size range as the radiation wavelengths of interest in the heat transfer process. The inverse situation of using microcontouring to obtain desired selective radiative properties is investigated by Sentenac and Greffet (1994). For example, a contouring is proposed to increase the emissivity of platinum for IR radiation.

A subject of increased importance in relation to semiconductor and photonic devices is the prediction of properties for very thin films of metals, semiconductors, and dielectrics with thicknesses comparable to the radiation wavelengths significant in heat transfer processes (Chen 1995). Thin films introduce the complexity of interference between incident and reflected waves within the film material; however, this depends on the film thickness relative to the wavelength, as considered by Chen and Tien (1992). The behavior of thin films is discussed in Section 17.4.2 in connection with the radiative behavior of windows and other single- or multi-layered structures.

Discussion of these effects requires a more detailed study of the wave approach, to account for the interference, tunneling, and polarization effects for films and textures much smaller than the wavelength of radiation. We return to these problems and to discussion of near-field radiative transfer in Chapter 16.

3.4 MEASURED PROPERTIES OF REAL DIELECTRIC MATERIALS

The EM theory provides guidance for the radiative property behavior of dielectrics and metals. For clean, pure, dielectric materials with optically smooth surfaces, the EM theory predicts reflectivity, which can be used to infer a material's emissivity following Kirchhoff's law as $\varepsilon_\lambda = 1 - \rho_\lambda$. The results obtained in this way show a decrease in emissivity with increasing angle from the normal. On the other hand, the wavelength dependence of emissivity is generally weak, and the temperature dependence is small. This behavior is not surprising as the reflectivity, and consequently emissivity, depends on index of refraction, which is a weak function of wavelength and temperature for perfect dielectrics.

The difficulty with these generalizations is that most nonmetals cannot be made smooth enough for their surfaces to be considered ideal, although important exceptions exist, such as glass, large crystals of various types, gemstones, and some plastics (some of these are not opaque bodies as are being discussed here). Because of nonideal surface finishes, many nonmetals have behavior that deviates radically from EM theory predictions. Available property measurements for nonmetals are less detailed than for metals, and specifications such as surface composition and texture are often lacking.

A complication in the interpretation of measured properties of nonmetals is that radiation passing into such a material may penetrate quite far before being absorbed (this is evident for visible wavelengths in glass). To be opaque, a specimen must have sufficient thickness to absorb essentially all the radiation that enters through adjacent boundaries; otherwise, radiation transmitted through and reflected from adjacent boundaries must be accounted for. Transmission is not considered in the present discussion. Often, samples of nonmetals such as paints are sprayed onto a metallic or other opaque base (substrate), and the properties of the composite are measured. If it is desired to have the surface behave completely as the coating material, then the coating thickness must be large enough that no significant radiation is transmitted through it. Otherwise, in reflectivity measurements, some of the incident radiation can be reflected from the substrate and transmitted back through the coating to reappear as measured reflected energy. The data will then depend on both the coating material and thickness and the substrate. Liebert (1965) examined the spectral emissivity of zinc oxide coatings on various substrates for one such case. In Figure 3.12, the effect of coating thickness is shown on the spectral emissivity of a composite of zinc oxide on a substrate that has an approximately constant normal spectral emissivity. The effect of increasing coating thickness becomes small in the thickness range of 0.2–0.4 mm, indicating that the emissivity approaches that of zinc oxide alone.

Another example is in Figure 3.13, where the measured effects of different substrates are shown. A white paint film was placed on a dense thick white paint substrate, on a black substrate, or on specularly reflecting aluminum (Shafey et al. 1982). The paint film was 14.4 μm thick and the wavelength of the incident radiation varied from 0.5 to 10 μm. The spectral reflectivity for normal incidence is plotted showing that the substrate has a considerable effect on reflectivity. For the black substrate, as the wavelength decreases and becomes small relative to the film thickness, the reflectivity increases and becomes more like that of the white paint film and less like that of the black substrate.

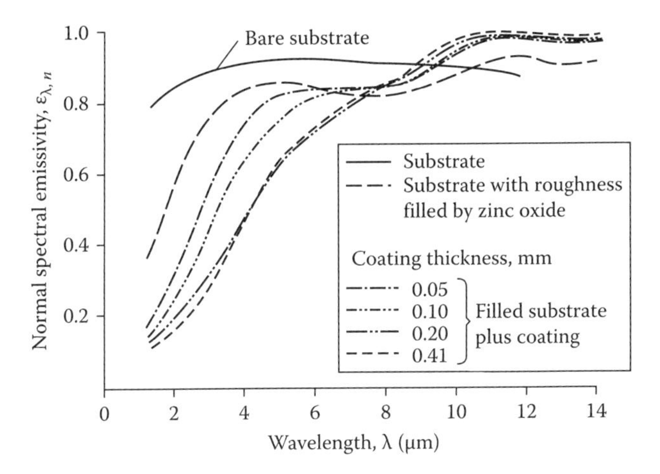

FIGURE 3.12 Spectral emissivity of zinc oxide coatings on oxidized stainless steel substrate. Surface temperature, 880 ± 8 K. (From Liebert, C.H., Spectral emittance of aluminum oxide and zinc oxide on opaque substrates, NASA TN D-3115, 1965.)

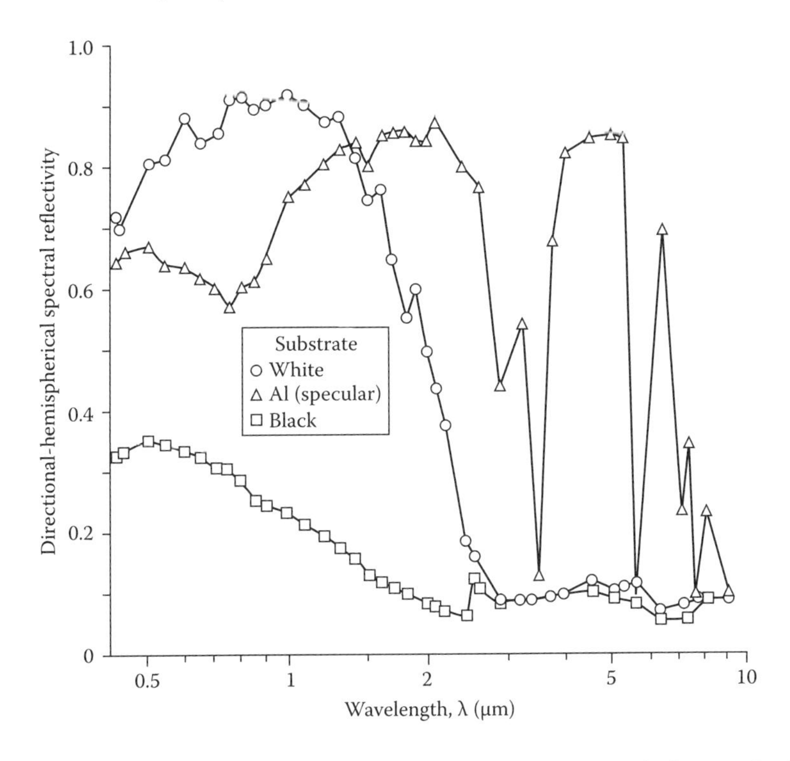

FIGURE 3.13 Effect of substrate reflectivity characteristics on the hemispherical-spectral reflectivity of a TiO_2 paint film for normal incidence. Film thickness, 14.4 μm; volume concentration of pigment, 0.017. (From Shafey, H.M. et al., *AIAA J.*, 20(12), 1747, 1982.)

Figure 3.14 shows the hemispherical normal spectral reflectivity of three paint coatings on steel (Ohlsen and Etamad 1957). Kirchhoff's law and the reflectivity reciprocity relations dictate that the normal spectral emissivity can be simply calculated as $1-\rho_\lambda$. Note that the three paints exhibit different characteristics. White paint has a high reflectivity (low emissivity) at short wavelengths, and the reflectivity decreases at longer wavelengths. Black paint, however, has a relatively low reflectivity over the entire wavelength region. By using aluminum powder in a silicone base as paint, the

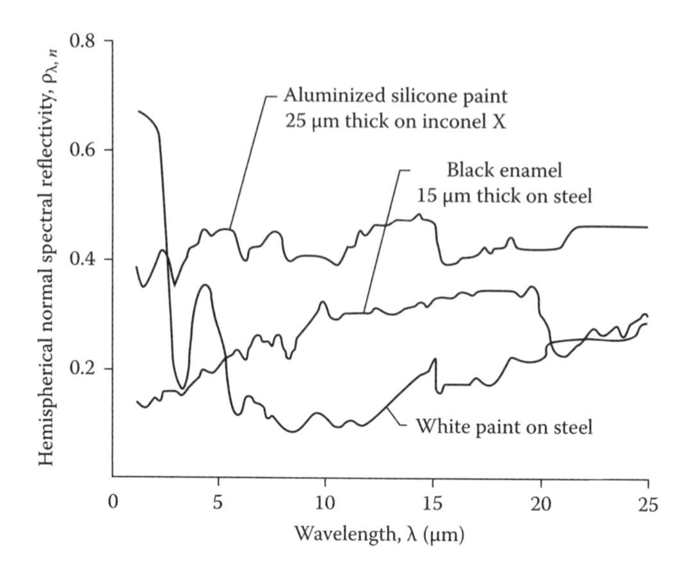

FIGURE 3.14 Spectral reflectivity of paint coatings. Specimens at room temperature. (From Ohlsen, P.E. and Etamad, G.A., Spectral and total radiation data of various aircraft materials, Report NA57-330, North American Aviation, July 23, 1957.)

reflectivity is increased, as expected for a more metallic coating. This specimen of aluminized paint acts approximately as a "gray" surface as its properties are reasonably independent of wavelength. Because of the large variation in spectral emissivity at short wavelengths, the gray approximation is poor for the white paint unless a small fraction of the participating radiation is at short wavelengths (25% below 0.5 μm).

Figure 3.15 illustrates that the reflectivity for some nonmetals decreases substantially at the short wavelengths in the visible range. This is important for considering the suitability of a specific nonmetallic coating for reflecting or absorbing radiation from a high-temperature source such as the sun, where much of the energy is at short wavelengths (25% below 0.5 μm).

The spectral behavior of the normal emissivities of several nonmetals (Touloukian and Ho 1970, 1972a, Postlethwait et al. 1994) is shown in Figure 3.16. The $\varepsilon_{\lambda,n}$ values for nitrides vary mostly over

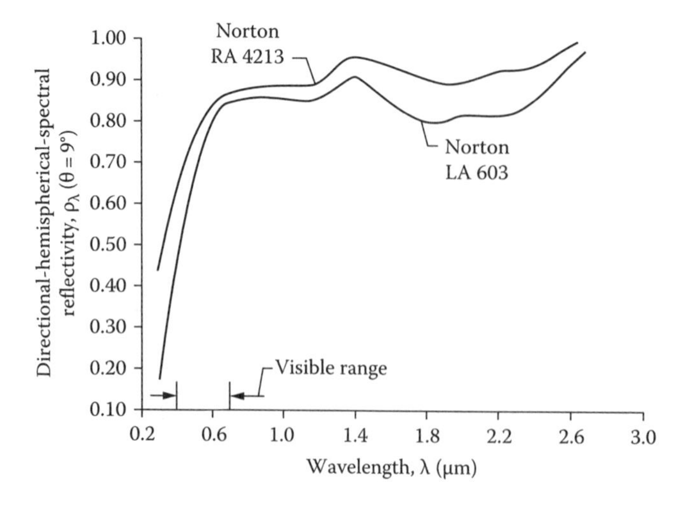

FIGURE 3.15 Directional-hemispherical-spectral reflectivity of aluminum oxide. Incident angle, 9°; specimens at room temperature. (From Wood, W.D. et al., *Thermal Radiative Properties*, Plenum Press, New York, 1964.)

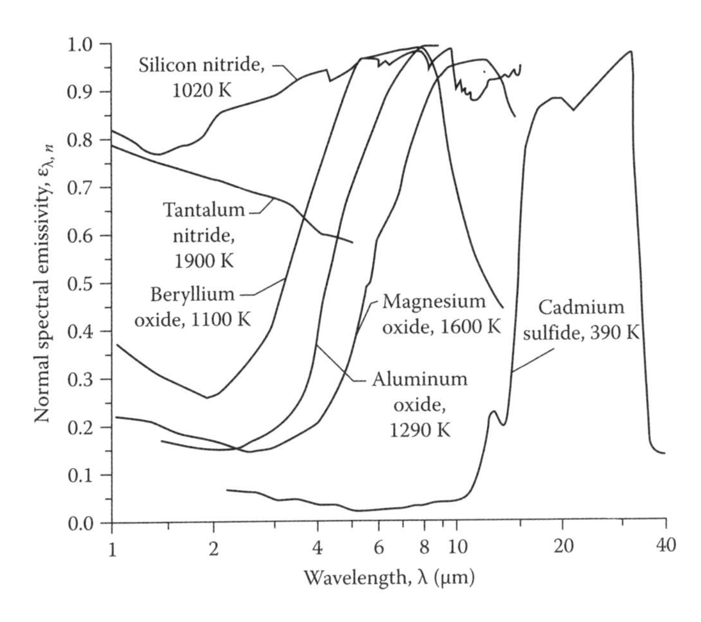

FIGURE 3.16 Normal spectral emissivity of some nonmetal samples at the temperatures shown. (From Touloukian, Y.S. and Ho, C.Y., eds., Thermophysical properties of matter, TRPC data services, in Y.S. Touloukian and D.P. DeWitt, eds., *Thermal Radiative Properties: Metallic Elements and Alloys*, vol. 7, 1970; Y.S. Touloukian and D.P. DeWitt, eds., *Thermal Radiative Properties: Nonmetallic Solids*, vol. 8, 1972a; Postlethwait, M.A. et al., *JTHT*, 8(3), 412, 1994.)

a range of about 0.2, but the other three materials have strong spectral variations with λ. They each exhibit large variations in $\varepsilon_{\lambda,n}$ in some wavelength regions and have a high emitting and absorbing region in the IR.

Experimental data for the emissive properties of copper oxide were given by Jones et al. (1996) as a function of wavelength, temperature, and direction. Metallic copper was oxidized until the properties no longer depended on the substrate. Figure 3.17 shows the directional behavior of the spectral emittance at two wavelengths and at four temperatures. This behavior can be compared with that predicted for a dielectric shown in Figure 3.2. At $\lambda = 9.5$ μm, the measured emissivity does not exhibit the decreasing trend predicted by the theory as the angle from the normal exceeds about 65°, but this trend is indicated for $\lambda = 3.5$ μm. The hemispherical-spectral emissivity increases with both temperature and wavelength, as shown in Figure 3.18.

3.4.1 Variation of Total Properties with Surface Temperature

The effect of surface temperature on the total emissivity of several nonmetallic materials is shown in Figures 3.19 through 3.21. Both increasing and decreasing trends with temperature are observed. Some of these effects may be caused by the dielectric coating being rather thin; hence, the properties are influenced by the temperature and spectral characteristics of the underlying material (substrate). For example, in Figure 3.19, the emissivity of aluminum oxide refractory decreases significantly with increasing temperature. For a silicon carbide coating on graphite, however, the emissivity increases with temperature; this may be partly caused by the emissive behavior of the graphite substrate that increases with temperature (see Figure 3.32).

White and black paints have high emissivities for the temperature up to 500 K, as is typical for ordinary oil-based paint. Aluminized paint has a much lower emissivity as it behaves partly like a metal. The emissivity for aluminized paint in Figure 3.19 is about one-half that in Figure 3.14. This illustrates the variation in properties that can be found for samples having the same general

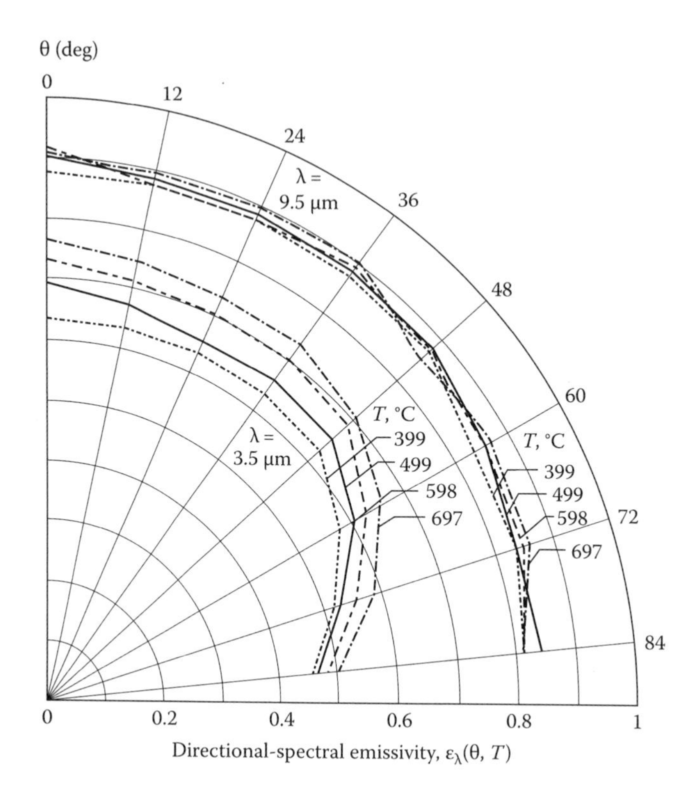

FIGURE 3.17 Directional-spectral emissivity of fully oxidized copper as a function of direction and temperature for $\lambda = 3.5$ μm and $\lambda = 9.5$ μm. (From Jones, P.D. et al., *JTHT*, 10(2), 343, 1996.)

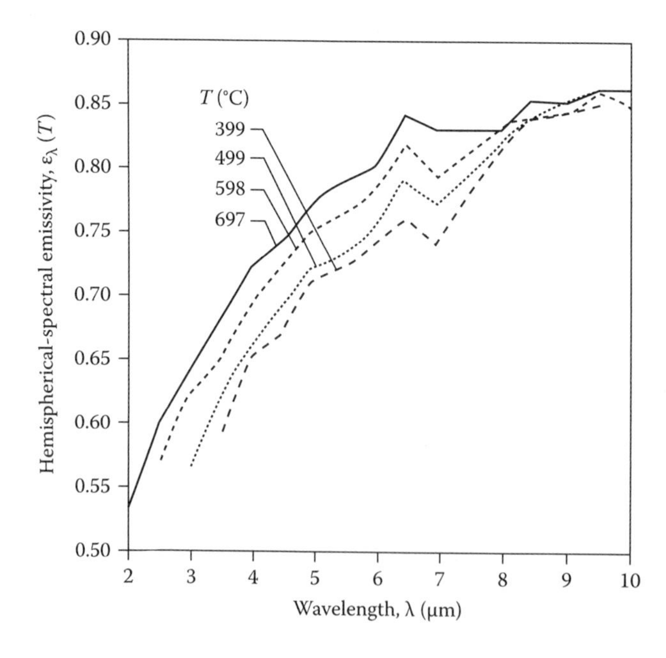

FIGURE 3.18 Hemispherical-spectral emissivity of fully oxidized copper as a function of wavelength. (From Jones, P.D. et al., *JTHT*, 10(2), 343, 1996.)

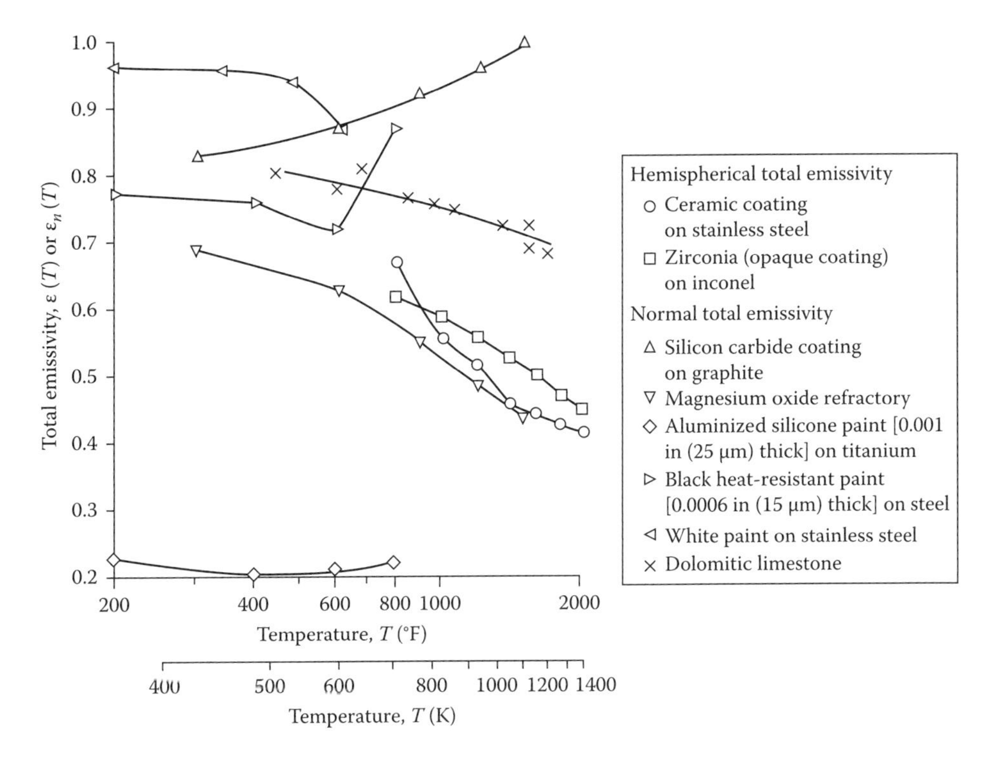

FIGURE 3.19 Effect of surface temperature on total emissivity of dielectrics. (From Touloukian, Y.S. and Ho, C.Y., Thermophysical properties of matter, in Y.S. Touloukian and D.P. DeWitt (eds.), *Thermal Radiative Properties: Metallic Elements and Alloys*, 7, 1970; and *Thermal Radiative Properties: Coatings*, 9, 1972b; Grewal, N.S. and Kestoras, M., *IJHMT*, 31(1), 207, 1988.)

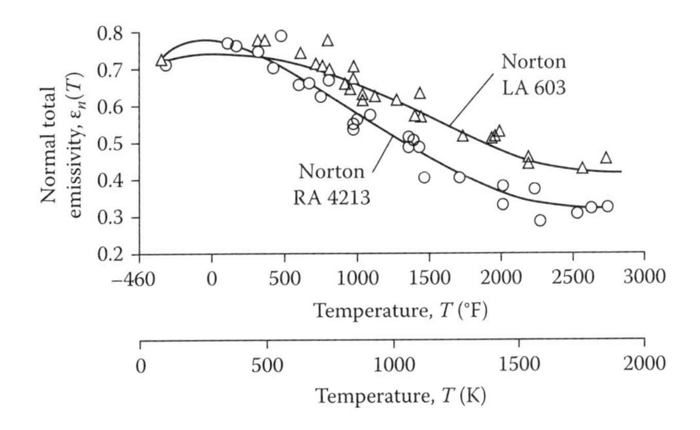

FIGURE 3.20 Effect of surface temperature on normal total emissivity of aluminum oxide. (From Wood, W.D. et al., *Thermal Radiative Properties*, Plenum Press, New York, 1964.)

description. For applications where property values are critical, it is advisable to make radiation measurements for the specific materials being used. For example, to study fluidized-bed combustion of high-sulfur coal, the emissivity of dolomitic limestone was needed at high temperatures. The normal total emissivity measured by Grewal and Kestoras (1988) was found to decrease linearly as T increased from 539 to 1223 K.

The spectral properties of a dielectric often vary slowly with temperature since the index of refraction is not a strong function of temperature. This is illustrated by the normal spectral

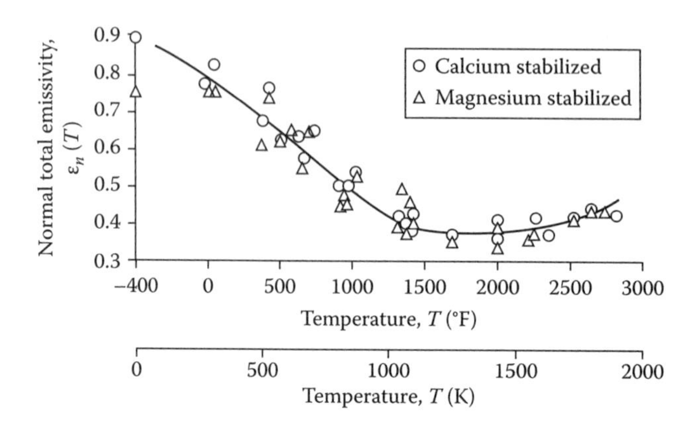

FIGURE 3.21 Effect of surface temperature on normal total emissivity of zirconium oxide. (From Wood, W.D. et al., *Thermal Radiative Properties*, Plenum Press, New York, 1964.)

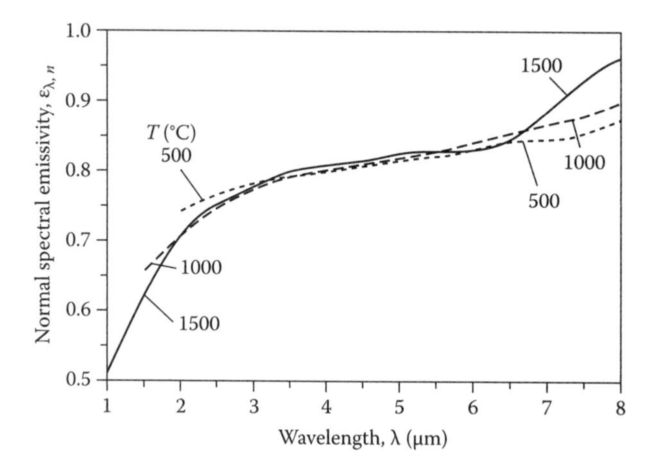

FIGURE 3.22 Normal spectral emissivity for various temperatures of a silicon carbide material, Hexoloy ST (as received). (From Postlethwait, M.A. et al., *JTHT*, 8(3), 412, 1994.)

emissivity of silicon carbide in Figure 3.22 (from Postlethwait et al. 1994). Then Equation 2.10 for the total emissivity is $\varepsilon(T) = \int_0^\infty \varepsilon_\lambda(T) E_{\lambda b}(T) d\lambda / \sigma T^4$, where the spectral variation ε_λ is approximately independent of temperature. This demonstrates that the most significant factor in the variation of ε with T is the wavelength shift in the blackbody function $E_{\lambda b}$ as T changes. This provides the very useful result that spectral data for ε measured at one temperature can be used in the integral to calculate accurate $\varepsilon(T)$ values over a range of nearby temperatures.

The normal total absorptivity is depicted in Figure 3.23 for a few materials for blackbody radiation incident from sources at various temperatures. White paper is a good absorber for the spectrum emitted at low temperatures, yet it is a poor absorber of radiation emitted at higher temperatures of a few thousand degrees K. It is thus a reasonably good reflector for solar energy. Asphalt pavement or a gray slate roof, on the other hand, absorb solar energy very well. Absorptivity expressions for solar radiation are summarized in Appendix B as needed for solar collector, satellite, and space vehicle design.

3.4.1.1 Effect of Surface Roughness

An important parameter in characterizing surface roughness effects is the *optical roughness*. This is the ratio σ_0/λ of a characteristic roughness height, usually the root-mean-square (rms) roughness σ_0, to the radiation wavelength. If the surface imperfections are much smaller than the

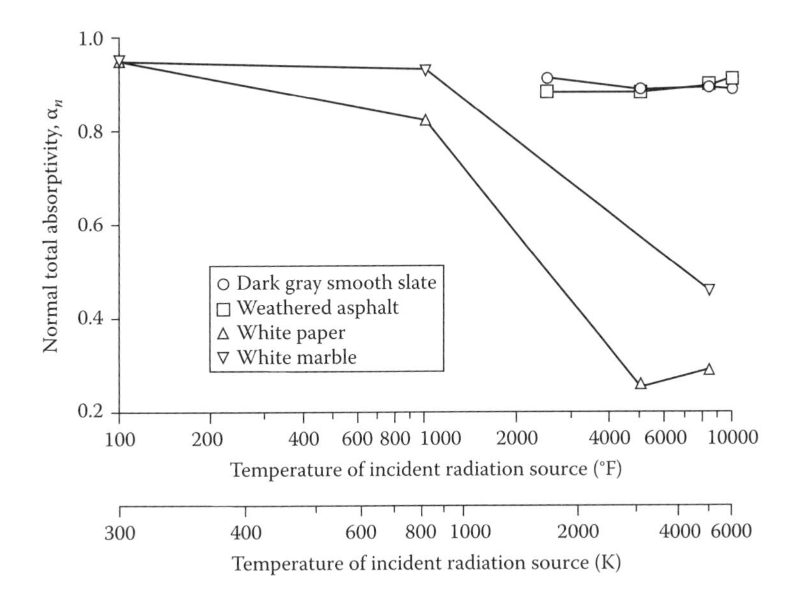

FIGURE 3.23 Normal total absorptivity of nonmetals at room temperature for incident black radiation from the sources at temperatures indicated. (From Touloukian, Y.S. and Ho, C.Y., Thermophysical Properties of Matter, TRPC Data Services, Volume 7, *Thermal Radiative Properties: Metallic Elements and Alloys*, Touloukian, Y.S. and DeWitt, D.P. (1970); Volume 8, *Thermal Radiative Properties: Nonmetallic Solids*, Touloukian, Y.S. and DeWitt, D.P. (1972a).)

radiation wavelength, the material is optically smooth. A surface that is optically smooth for long wavelengths may be optically quite rough at short wavelengths. The influence of additional geometric scales that characterize surface roughness is discussed by Yang and Buckius (1995).

In Figure 3.24, the bidirectional total reflectivity of typewriter paper is shown for three angles of incidence (Munch 1955). For an ideal (polished, smooth) surface, a specular (mirrorlike) peak would be expected with the angles of reflection and incidence symmetric about the normal; obviously, the surface finish of typewriter paper is not ideal, since the reflected intensity occupies a

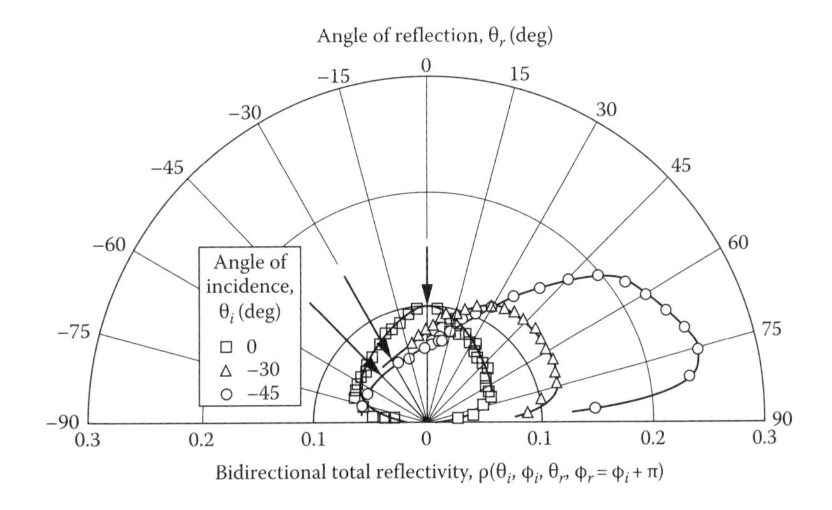

FIGURE 3.24 Bidirectional total reflectivity of typewriter paper in plane of incidence. Source temperature, 1178 K (2120°R). (From Munch, B., Directional distribution in the reflection of heat radiation and its effect in heat transfer, PhD thesis, Swiss Technical College of Zurich, Zurich, Switzerland, 1955.)

rather large angular envelope around the direction of specular reflection. Detailed bidirectional reflectivity measurements for magnesium oxide ceramic with optical roughness σ_0/λ varying from 0.46 to 11.6 are listed by Torrance and Sparrow (1966). As the roughness and incidence angle of the incoming radiation are increased, off-specular peaks are obtained, as discussed in connection with Figure 3.36. Bidirectional-spectral reflectivity measurements are reported by DeSilva and Jones (1987) for six materials used for solar energy absorption. Peaks in the specular direction were found.

Dielectrics generally show a slight increase in emissivity with roughness. However, Cox (1965) showed that for materials with $\sigma_0\beta_\lambda$, a ratio of roughness to radiation mean penetration distance in the dielectric of about 0.05 (see Section 1.6.2 for definition of mean penetration distance, $l = 1/\beta_\lambda$), the emissivity may be less than for a smooth surface. This result was predicted by analysis and confirmed by experiment. Because of the porosity of many dielectrics, it is difficult to study roughness size parameters below a certain limiting smoothness.

The shape of curves in Figure 3.24 suggests that characterization of reflected energy as a combination of a diffuse plus a specular component is possible. This approximation has merit in some cases and results in a simplification of radiant interchange calculations compared with using exact directional properties (Sparrow and Lin 1965, Sarofim and Hottel 1966); in other cases, however, the approximation would fail. An example is in Figure 3.25, which shows the observed bidirectional total reflectivity for visible light reflected from the moon (Orlova 1956). These particular curves are for mountainous regions of the moon, but similar curves have been obtained for other areas. The interesting feature is that the peak reflected radiation is back toward the radiation source in opposition to the direction of the incident radiation. The peak reflection is located at a circumferential angle ϕ_r of 180° away from where a specular peak would occur.

Observations confirm that curves of this type characterize lunar reflectivity. At full moon, when the sun, Earth, and moon are almost in a straight line (Figure 3.26), the moon appears equally bright across its face. For this to be true, an observer on Earth must see equal intensities from all locations on the moon. However, the solar energy incident upon a unit area of the lunar surface varies as the cosine of the angle θ_i between the sun and the normal to the lunar surface. The angle θ_i varies from 0° to 90°, as the position of the incident energy varies from the center to the edge of the lunar disk. To reflect a constant intensity to an observer on Earth from all observable points on the lunar surface requires that the product $\rho(\theta_i,\theta_r)\cos\theta_i$ be constant. Consequently, the bidirectional reflectivity in the direction of incidence must increase approximately in proportion to $1/\cos\theta_i$ (shown by the dashed

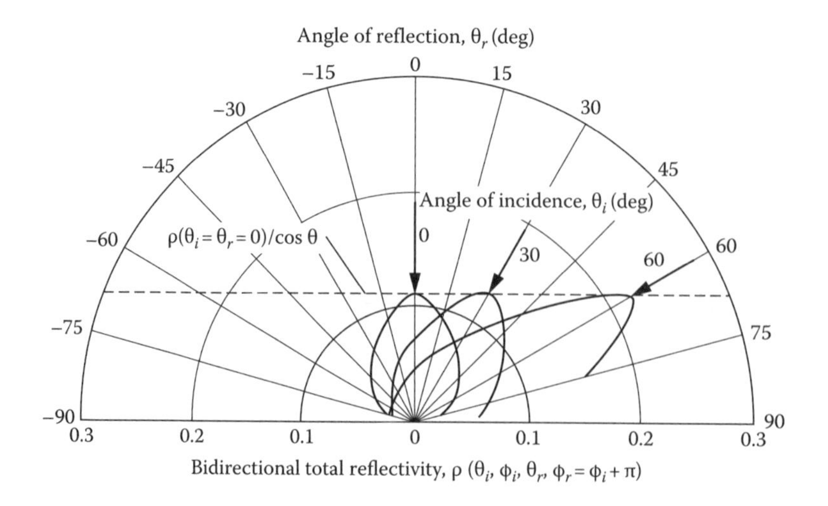

FIGURE 3.25 Bidirectional total reflectivity in plane of incidence ($\phi_r = \phi_i + \pi$) for mountainous regions of the lunar surface. (After Orlova, N.S., *Astron. Z.*, 33(1), 93, 1956.)

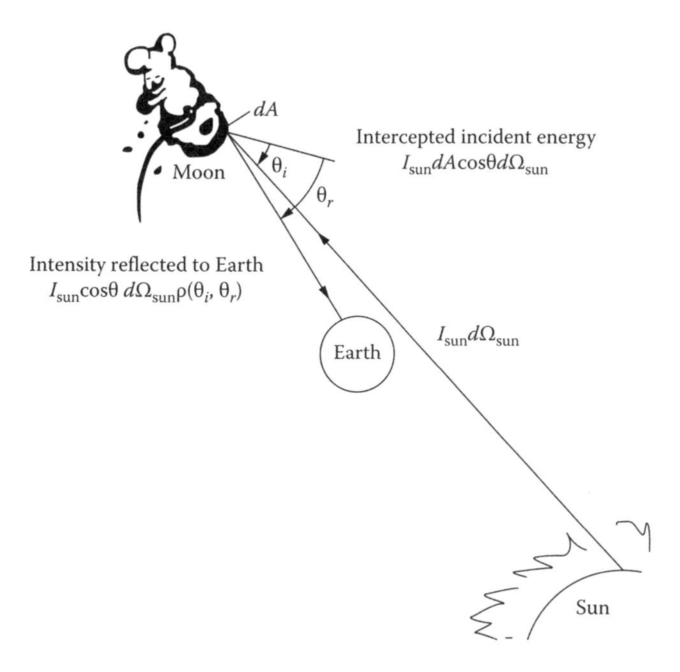

FIGURE 3.26 Reflected energy at full moon (not to scale).

line in Figure 3.25) as the angle of incidence increases. This compensates for the reduced energy incident per unit area on the moon at the large angles. The reflectivity behavior is confirmed by the curves in Figure 3.25. Hence, a uniformly bright moon does not imply that the moon is a diffuse reflector. A diffuse moon would appear bright at the center and dark at the edges. The strong backscattering of the lunar surface causes the moon's brightness to peak at full moon when the sun is almost directly behind the observer. Further discussion of lunar IR behavior is given by Saari and Shorthill (1967), Harrison (1969), and Birkebak (1974).

Drolen (1992) provides bidirectional reflectivity measurements for 12 materials commonly used for spacecraft thermal control, such as white paint, black Kapton, and aluminized Kapton. The "specularity" given is the fraction of the directional-hemispherical reflectivity contained within the specular solid angle. Specularity values can be used in surface property models that assume a combination of diffuse and specular reflectivities (Section 6.4). Most measurements reported by Drolen are at a wavelength of $\lambda = 0.488$ μm, but some white paints were measured at four discrete wavelengths covering the range $0.488 \le \lambda \le 10.63$ μm. Detailed bidirectional reflectance data are also reported by Zaworski et al. (1996a). An example for white paint is shown in Figure 3.27, where the paint is seen to be quite diffuse until incidence angles of 60° or greater are reached. Detailed data of this type were used to predict radiative transfer through a rectangular gap (Zaworski et al. 1996b), illustrating the usefulness of more detailed data that can be included as computational capabilities increase.

3.4.2 Properties of Semiconductors and Superconductors

Semiconductors are arbitrarily discussed here along with nonmetals, but they behave partly as metals. Liebert and Thomas (1967) have shown that the radiative properties can be determined from the EM theory by treating semiconductors as metals with high resistivity. Figure 3.28 shows the normal spectral emissivity of a silicon semiconductor. Predictions based on Hagen–Rubens relation, shown for comparison, are based on the dc resistivity measured for the same sample, one of the few cases where such comparable emissive and electrical data are available. Agreement between the experiments and Hagen–Rubens curves is not good until wavelengths are reached that are much greater than those

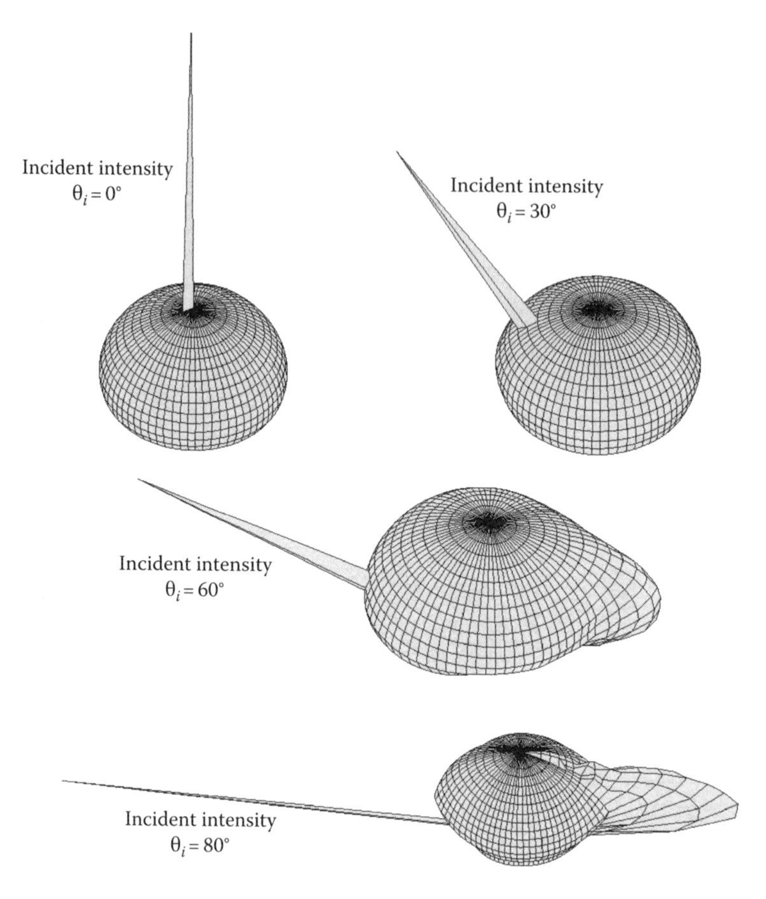

FIGURE 3.27 Bidirectional reflectivity of Krylon white paint. Plots are of $\ln\left[\rho_\lambda\left(\lambda,\theta_r,\phi_r,\theta_i,\phi_i\right)\right]$ at $\lambda = 0.6328$ μm. (From Zaworski, J.R. et al., *IJHMT*, 39(6), 1149, 1996a.)

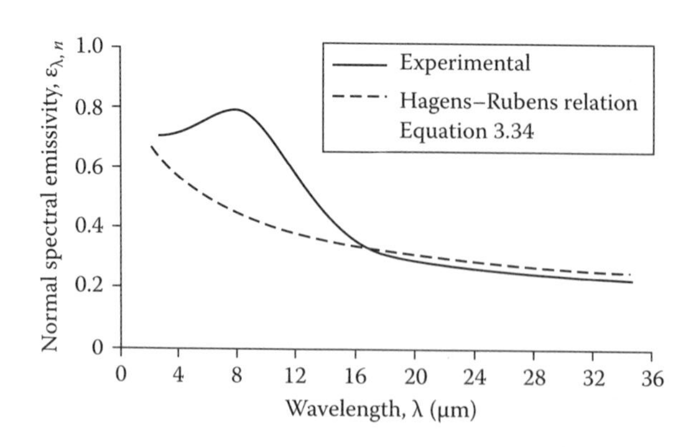

FIGURE 3.28 Normal spectral emissivity of a highly doped silicon semiconductor at room temperature. (From Liebert, C.H. and Thomas, R.D., Spectral emissivity of highly doped silicon, in G.B. Heller (ed.), *Thermophysics of Spacecraft and Planetary Bodies*, vol. 20, *AIAA Progress in Astronautics and Aeronautics*, Academic Press, New York, pp. 17–40, 1967. Also, NASA TN D-4303, April, 1968.)

giving agreement for metals. This difference in range of agreement can be traced to the assumption used in deriving Hagen–Rubens equation; $(\lambda_0/2\pi c_o r_e \gamma)^2 \gg 1$ where c_o and λ_0 are in vacuum. For semiconductors, where the resistivity is larger than for metals, this inequality cannot hold until larger wavelengths than for metals are reached. The shape of the curve measured for silicon (Figure 3.28) resembles what would be expected for a polished metal (e.g., the tungsten data in Figure 3.31). The emissivity increases with decreasing wavelength over much of the spectrum, with a peak occurring at shorter wavelengths. However, most of the features of the semiconductor curve occur at longer wavelengths than for a metal; the peak emissivity, for example, is well outside the visible region. Liebert and Thomas (1968) were also able to show excellent agreement between the measured emissivity and predictions from the EM theory that included the effects of free electrons and were more sophisticated than those in Section 3.2.2. In their study, the theoretical equations were evaluated by using physical properties measured from the specific samples used for the emissivity measurements.

The theory for spectral reflectivity of high-temperature superconductors is examined by Phelan et al. (1992) and compared with experimental data. The material Y–Ba–Cu–O is a ceramic in the form of a thin film at liquid nitrogen temperature. This material has almost perfect reflection in the far-IR region, and the reflectivity decreases to 0.85 at a wavelength of about 6 μm. The very high reflectivity in the far IR can be useful in space applications as a radiation shield for storing liquid nitrogen. A quantum-mechanical theory (Mattis and Bardeen 1958) provides predictions within 12% of experimental results, with better agreement as wavelength increases. Relations given by Phelan et al. (1992) provide excellent predictions for engineering use.

3.5 MEASURED PROPERTIES OF METALS

Predictions of the radiative property behavior of metals with optically smooth surfaces through the EM theory indicate that emissivity increases with the increasing angle from the normal; the emissivity decreases with the increasing wavelength, and total emissivity increases with absolute temperature. In this section, comparisons of measured data with the EM theory show the applicability and limitations of the theoretical predictions.

Pure metals with smooth surfaces often have low emissivity and absorptivity and therefore high reflectivity. Figure 3.9 demonstrates that the emissivity normal to the surface is quite low for various polished metals. However, low emissivity values are not an absolute rule for metals; in some of the examples that will be given below, the spectral emissivity rises to 0.5 or larger at short wavelengths, and the total emissivity becomes large at high surface temperatures (for examples, see Figures 3.29, 3.34, 3.38, and 3.39).

3.5.1 Directional and Spectral Variations

Behavior typical of polished metals is that the directional emissivity tends to increase with increasing angle θ from the surface normal, except near $\theta = 90°$. This is predicted by the EM theory, as shown for platinum in Figure 3.5. At wavelengths shorter than the range for which the simple EM theory of Section 3.2 applies, deviation from this behavior might be expected, as illustrated by the directional-spectral emissivity of polished titanium in Figure 3.29. At wavelengths greater than about 1 μm, the directional-spectral emissivity of titanium does tend to increase with increasing θ over most of the θ range. The increase with θ becomes smaller as wavelength decreases; finally, at wavelengths less than about 1 μm, the directional-spectral emissivity decreases with increasing θ over the entire range of θ. For polished metals, the typical behavior of increased emission for directions nearly tangent to the surface may not occur at short wavelengths.

In the IR region, it was shown in Section 3.2 that the spectral emissivity of metals tends to increase as wavelength decreases. This trend is illustrated for several metals in Figure 3.30 for the spectral emissivity in the normal direction (Seban 1963). For other directions, the same effect is

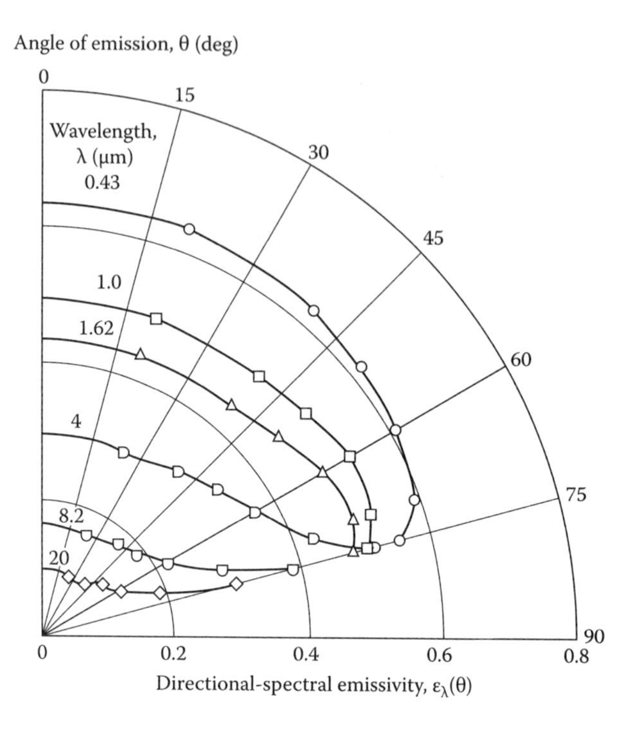

FIGURE 3.29 Effect of a wavelength on directional-spectral emissivity of pure titanium. Surface ground to 0.4 μm rms. (From Edwards, D.K. and Catton, I., Radiation characteristics of rough and oxidized metals, in S. Gratch (ed.), *Advances in Thermophysical Properties at Extreme Temperatures and Pressures*, ASME, New York, pp. 189–199, 1964.)

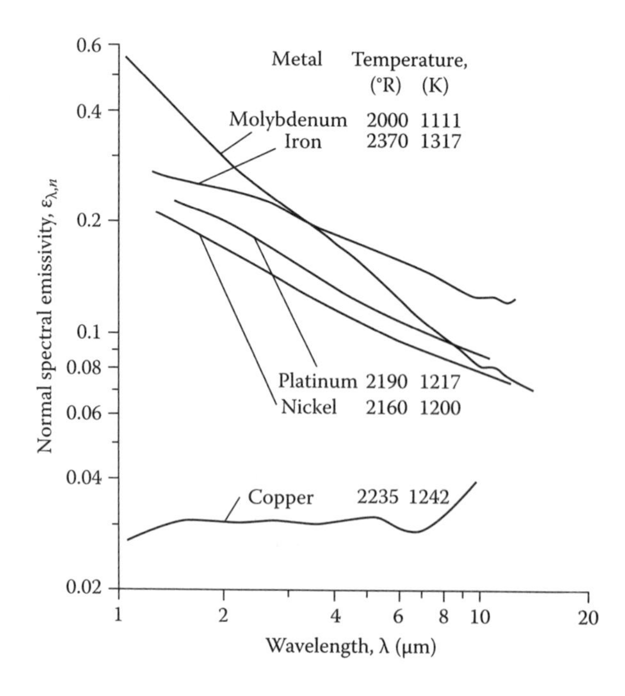

FIGURE 3.30 Variation with wavelength of normal spectral emissivity for polished metals. (From Seban, R.A., *Thermal Radiation Properties of Materials*, pt. III, WADD-TR-60-370, University of California, Berkeley, CA, August, 1963.)

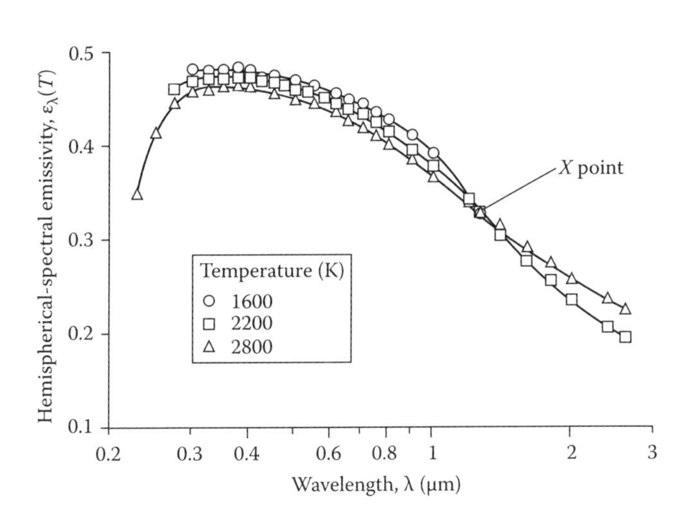

FIGURE 3.31 Effect of wavelength and surface temperature on hemispherical-spectral emissivity of tungsten. (From De Vos, J.C., *Physica*, 20, 690, 1954.)

illustrated in Figure 3.29 except at large angles from the normal, where various spectral profiles may cross. The curve for the copper sample in Figure 3.30 illustrates an exception as the emissivity remains relatively constant with wavelength. At very short wavelengths, the assumptions for the simplified EM theory of Section 3.2 become invalid. Most metals exhibit a peak emissivity somewhere near the visible region, and the emissivity then decreases rapidly with further decrease in wavelength. This is illustrated by the behavior of tungsten in Figure 3.31.

3.5.2 Effect of Surface Temperature

Hagen–Rubens relation (Equation 3.34) shows that for wavelengths that are not too short, $\lambda > \sim 5$ μm, the spectral emissivity of a metal tends to be proportional to the metal resistivity to the one-half power. Hence, we can expect the spectral emissivity of pure metals to increase with temperature as does its resistivity, and this is usually found to be the behavior. Figure 3.31 is an example for the hemispherical-spectral emissivity of tungsten (De Vos 1954). The expected trend is observed for $\lambda > 1.27$ μm.

Figure 3.31 also illustrates a characteristic of many metals, as discussed by Sadykov (1965). At short wavelengths (in the case of tungsten, $\lambda > 1.27$ μm), the temperature effect is reversed and the spectral emissivity decreases with increased temperature. The emissivity curves all cross at the same point, which has been called the "X point." Other X points for different metals are iron, 1.0 μm; nickel, 1.5 μm; copper, 1.7 μm; and platinum, 0.7 μm. At very high temperatures in thermionic energy conversion devices, alloys are used of tungsten and rhenium compounded with thoria and hafnium carbide. Their normal spectral emittances were measured by Tsao et al. (1992), and results are given for temperatures from 1500 to 2500 K. Most of the results are at $\lambda = 0.535$ μm, and the normal spectral emittances were found to decrease linearly with increasing temperature. This is consistent with the results for tungsten shown in Figure 3.31 as the wavelength is smaller than at the X point and hence is in the spectral region where spectral emittance decreases as temperature increases.

The increase of spectral emissivity with decreasing wavelength for metals in the IR region (wavelengths longer than in the visible region), as discussed in Section 3.2.2, accounts for the increase in total emissivity with temperature. With increased temperature, the peak of the blackbody radiation distribution (Figure 1.12) moves toward shorter wavelengths. Consequently, as the surface temperature increases, proportionately more radiation is emitted in the region of higher spectral emissivity,

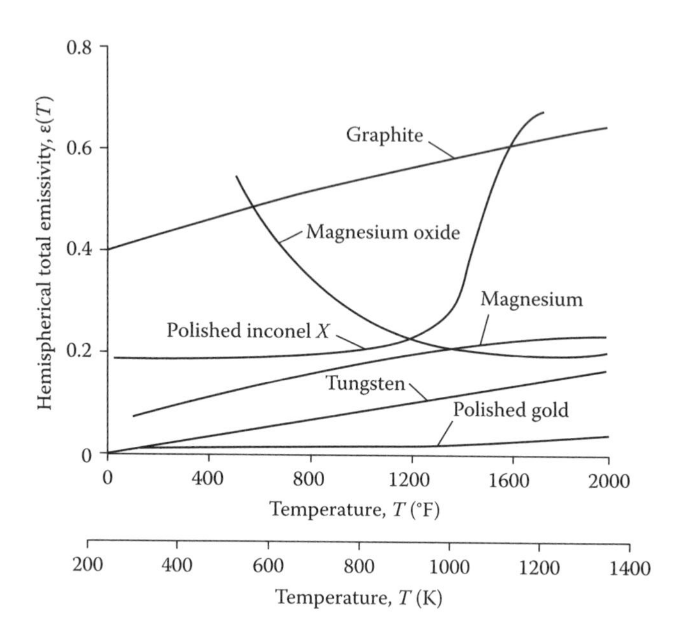

FIGURE 3.32 Effect of temperature on hemispherical total emissivity of several metals and one dielectric. (From Touloukian, Y.S. and Ho, C.Y. (eds.), Thermophysical Properties of Matter, TRPC Data Services; Volume 7, *Thermal Radiative Properties: Metallic Elements and Alloys*, Touloukian, Y.S. and DeWitt, D.P. (1970); Volume 8, *Thermal Radiative Properties: Nonmetallic Solids*, Touloukian, Y.S. and DeWitt, D.P. (1972a).)

which increases the total emissivity. Some examples are in Figure 3.32. Here the behavior of metals is compared against that of a dielectric, magnesium oxide, for which the emissivity decreases with increasing temperature.

In insulation systems consisting of a series of radiation shields in vacuum, radiation is the significant heat loss mechanism. At the low temperatures involved in the insulation of cryogenic systems, there is a lack of experimental property data, especially at the long wavelengths characteristic of emission at these temperatures. Hagen–Rubens result indicates that the normal spectral emissivity is proportional to $(r_e/\lambda)^{1/2}$, and the normal total emissivity is then proportional to $(r_eT)^{1/2}$. If electrical resistivity is directly proportional to temperature, then $\varepsilon_{\lambda,n}(T) \propto (T/\lambda)^{1/2}$ and $\varepsilon_n(T) \propto T$. This indicates that emissivities should become quite small at low T and large λ. Experimental measurements summarized by Toscano and Cravalho (1976) for copper, silver, and gold indicate that ε does not decrease to such small values. For refinements to the theory, the Drude free-electron theory and anomalous skin-effect theory, described by Toscano and Cravalho, include additional electron and quantum-mechanical interactions. Drude theory, which reduces to Hagen–Rubens relation at long wavelengths, predicts ε values that decrease much more rapidly with temperature than has been observed. The anomalous skin-effect model with diffuse electron reflections was found to predict the emissivity most accurately. Figure 3.33 shows results of the two models. The limited data at low temperatures lie somewhat above the anomalous skin-effect model. Similar results for gold films are reported by Tien and Cunnington (1973).

3.5.3 Effect of Surface Roughness

The radiative properties of optically smooth materials can be predicted within the limitations of the EM theory as discussed in Section 3.2. When the surface optical roughness σ_0/λ is greater than about 1, multiple reflections occur in the cavities between roughness elements. This increases the trapping of incident radiation, thereby increasing the observed surface absorptivity and consequently

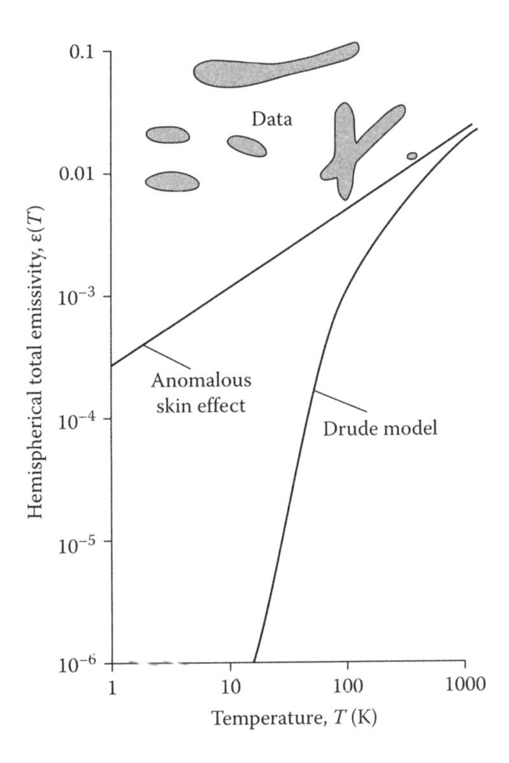

FIGURE 3.33 Effect of low temperatures on hemispherical total emissivity of copper. (Results from Toscano, W.M. and Cravalho, E.G., *JHT*, 98(3), 438, 1976.)

the emissivity. When roughness is large, it has a considerable effect on directional emission and reflection. For $\sigma_0/\lambda \gg 1$, geometric optics can be used to trace the radiation paths reflected within cavities formed by the roughness elements. For σ_0/λ near 1, the geometric optics approximation is compared with EM theory predictions of Tang and Buckius (1998). If the roughness geometry is completely specified, it is possible in certain cases to predict the directional behavior. An example is the directional emissivity of a parallel-grooved surface in Section 3.2. Ordinarily, the roughness is very irregular, and a statistical model must be assumed. For example, roughness can be represented as randomly oriented facets that each reflect in a specular manner. A roughness model composed of spherical cap indentations was also suggested by Demont et al. (1982).

When the optical roughness is small ($\sigma_0/\lambda < 1$), effects of multiple reflections in roughness cavities are usually small, and the hemispherical properties approach those for optically smooth surfaces. However, as a result of diffraction effects, the directional properties (especially the bidirectional reflectivity) can be significantly influenced by the roughness.

Various synthetic surfaces have been numerically and experimentally tested by different researchers with the intent of predicting the effects of surface roughness on radiative properties, and they have provided a basis for understanding many of the effects observed. A significant difficulty in the prediction of radiative properties is in the precise definition of the surface characteristics for use in the analytical relations. A common way to characterize surface roughness is by the method of preparation (lapping, grinding, etching, etc.) plus a specification of rms roughness. The latter is usually obtained with a profilometer that traverses a sharp stylus over the surface and reads out the vertical perturbations in terms of an rms value. It does not account for the horizontal spacing of the roughness and does not give the *distribution* of the roughness size around the rms value, or information on the average slope of the sides of the roughness peaks that influences the behavior of the cavities. At present, there is no generally accepted method of accurately and adequately specifying surface characteristics other than the use of microscale approaches.

Some analytical approaches are now discussed and comparisons made with experimental data. Davies (1954) used diffraction theory to examine the reflecting properties of a surface with roughness assumed to be distributed according to a Gaussian (normal) probability distribution, specified as a probability $p(z)$ of having a roughness of height z given by $p(z)=\left(\frac{1}{\sigma_0\sqrt{2\pi}}\right)\exp(-z^2/2\sigma_0^{\;2})$. Individual surface irregularities were assumed to be of sufficiently small slope that shadowing could be neglected, and σ_0 was assumed very much smaller than the wavelength of incident radiation. The material was assumed to be a perfect electrical conductor so that, from the EM theory, the extinction coefficient is infinite. This provides perfect reflection, and consequently, the theory gives the directional distribution of the energy that is reflected rather than the amount reflected. The reflected distribution was found to consist of a specular component and a distribution about the specular peak.

A similar derivation, with σ_0 being assumed much larger than λ, again yielded a distribution of reflected intensity about the specular peak, this time of larger angular spread than for $\sigma_0 \ll \lambda$. This would be expected since the surface should behave less like an ideal specular reflector as the roughness increases compared with the wavelength of the incident radiation. However, Davies treatment is inaccurate at near-grazing angles because shadowing by the roughness elements is neglected.

Porteus (1963) extended Davies approach by removing the restrictions on the relation between σ_0 and λ and including more parameters for specifying the surface roughness characteristics. Some success was obtained in predicting the roughness characteristics of prepared samples from measured reflectivity data. Measurements were mainly at normal incidence, and the neglect of shadowing makes the results doubtful at near-grazing angles. An improved treatment by Beckmann and Spizzichino (1963) includes the autocorrelation distance of the roughness. This is a measure of the spacing of the characteristic roughness peaks on the surface and hence is related to the rms slope of the roughness elements. The method provides better data correlation than the earlier analyses. A critical evaluation of Davies and Beckmann analyses is given by Houchens and Hering (1967).

Some observed effects of surface roughness for small σ_0/λ are displayed in Figures 3.34 and 3.35. The former shows the directional emissivity of titanium (Edwards and Catton 1964) at a wavelength of 2 μm for three roughnesses. The maximum roughness is 0.4 μm, so relative to the 2 μm wavelength, the specimens are all smooth, and the emissivity changes only a small amount with the roughness. There is very little effect on the directional variation of the emissivity. Edwards and Catton also give data for sandblasted surfaces that produced larger emissivity increases.

Figure 3.35 provides the reflectivity of nickel for energy reflected into the specular direction from radiation incident at 10° from the normal (Birkebak and Eckert 1965). The reflectivities are expressed as a ratio to the reflectivity of a polished surface to show the effect of roughness on the directional characteristics rather than on the magnitude of the reflectivity. The polished surface for comparison had about 10 times less roughness than any of the rough specimens. A high value of the ordinate means that the specimen behaves more like a polished surface. Data are shown for ground nickel with four roughnesses, each with $\sigma_0/\lambda < 1$. The reflectivity rises as wavelength increases because, for a given roughness, the surface appears smoother with decreasing σ_0/λ ratio. As expected, for a fixed wavelength, the reflectivity for the specular direction decreases as the roughness appears smoother with the decreasing σ_0/λ ratio increased. Data for aluminum have been well correlated in Smith and Hering (1970) by use of Beckmann theory.

For an optical roughness $\sigma_0/\lambda > 1$, some detailed experimental measurements of the bidirectional reflectivity, along with an analysis using geometric optics, are in Torrance and Sparrow (1966, 1967). The analysis substantiated the important trends of the data. Typical results for the bidirectional reflectivity of aluminum are in Figure 3.36 for an optical roughness of $\sigma_0/\lambda = 2.6$.

The bidirectional reflectivity is given as a function of reflection angle θ_r in the plane formed by the incident direction and the normal to the surface. The reflectivity is divided by the value in the specular direction. For a diffuse surface, the reflected intensity is independent of θ_r so the ratio is 1,

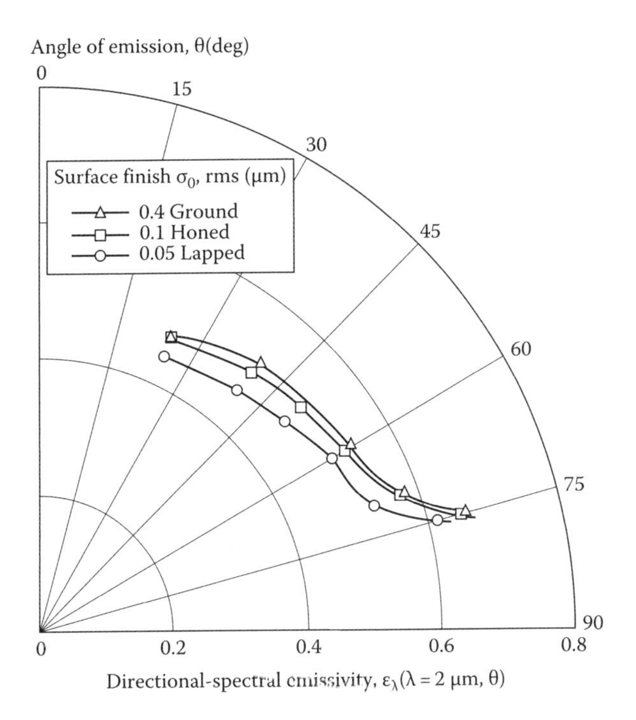

FIGURE 3.34 Roughness effects for small optical roughness, $\sigma_0/\lambda < 1$; effect of surface finish on directional-spectral emissivity of pure titanium. Wavelength, 2 μm. (From Edwards, D.K. and Catton, I., Radiation characteristics of rough and oxidized metals, in S. Gratch (ed.), *Advances in Thermophysical Properties at Extreme Temperatures and Pressures*, ASME, New York, pp. 189–199, 1964.)

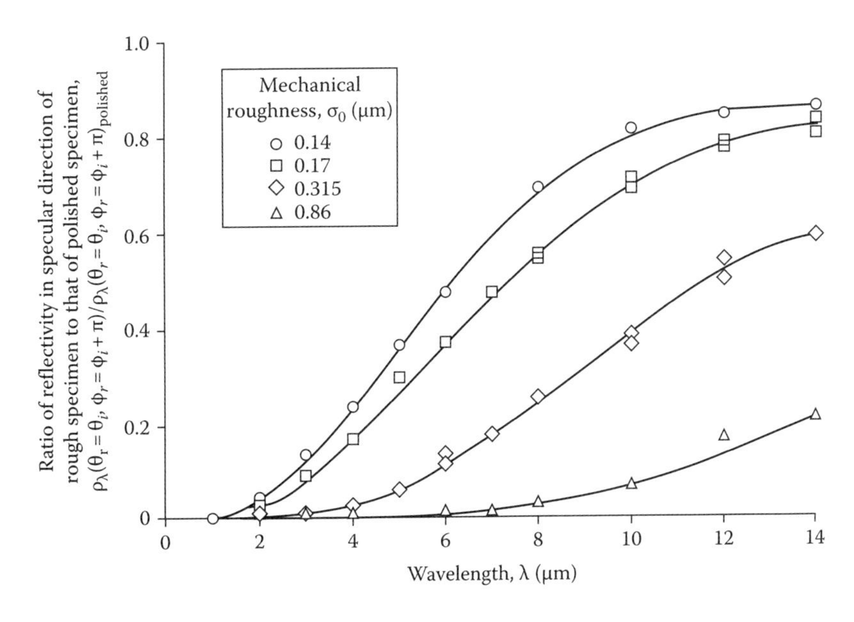

FIGURE 3.35 Effect of roughness on bidirectional reflectivity in specular direction for ground nickel specimens. Mechanical roughness for polished specimen, 0.015 μm. (From Birkebak, R.C. and Eckert, E.R.G., *JHT*, 87(1), 85, 1965.)

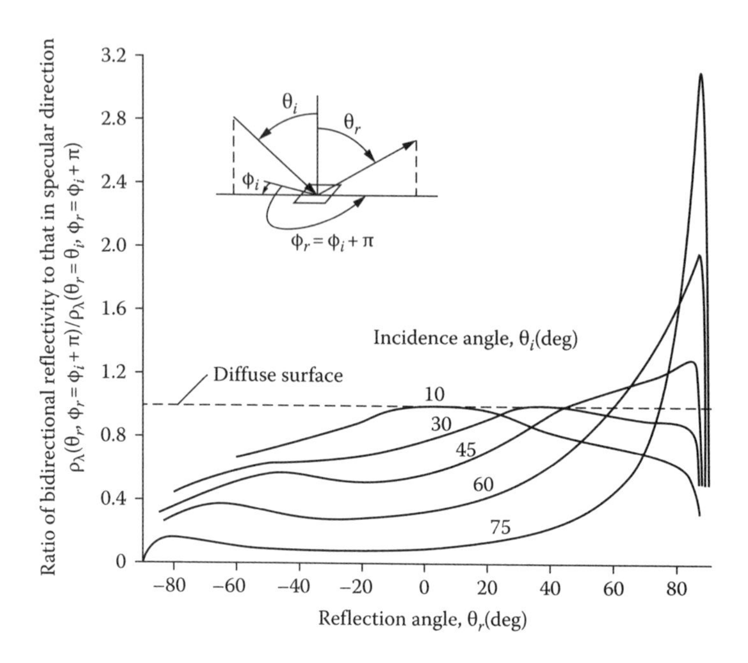

FIGURE 3.36 Bidirectional reflectivity in plane of incidence for various incidence angles; material, aluminum (2024-T4), aluminum coated; rms roughness, $\sigma_0 = 1.3$ μm; wavelength of incident radiation, $\lambda = 0.5$ μm. (From Torrance, K.E. and Sparrow, E.M., *JOSA*, 57(9), 1105, 1967.)

as shown by the dashed line. A specular reflection would appear as a sharp high peak at $\theta_r = \theta_i$. For incidence at 30°, the reflected intensity peaks in the specular direction ($\theta_r = 30°$). However, at larger incidence angles, the maximum shifts to a θ_r that is larger than the specular angle; for example, when $\theta_i = 60°$, the reflectivity peak is at $\theta_r = 85°$. This is in contrast to a smooth surface, where the peak would be at $\theta_r = \theta_i$. The theory indicates that this off-specular reflection, which occurs for large optical roughness and large incidence angles, is the result of shadowing by the roughness elements.

Ody Sacadura (1972) and Kanayama (1972) analyze the angular distribution of the emissivity from a rough surface by considering emission from V-grooves in a parallel or circular pattern. A more realistic model is the randomly roughened metal surface in Abdulkadir and Birkebak (1978) and Wolff and Kurlander (1990) where multiple reflections between surface elements were neglected. Calculations for gold and chromium indicate that for θ_i less than about 60°, the emissivity increases with roughness. For larger θ_i, the directional emissivity is less than for an ideal smooth surface; this is a result of both roughness and the behavior of ideal smooth metallic surfaces where the emissivity becomes large for directions nearly tangent to the surface. The roughness models are also used with ray tracing methods to simulate realistic visual reflection from objects (Wolff and Kurlander 1990).

Tang and Buckius (1999) review literature on surface reflection (surface scattering) models and experiments. The influence of incidence angle, radiation wavelength, surface materials, and surface geometry on scattering from a surface is discussed. Both deterministic and probabilistic surface models are explained. It is noted that polarized energy incident on tailored 2D rough surfaces can provide extremely strong spikes in reflectivity, particularly when the wavelength of the incident radiation is such that resonance occurs within the surface cavities. Comparisons between predictions and measurements are in good agreement, adding confidence in the use of the EM theory for these predictions.

A regime map is given by Dimenna and Buckius (1994a) to show the region of validity of the various approaches for predicting surface reflectivity of a highly conducting material with roughness that varies only in one direction. The exact solution of Maxwell equations is compared with a

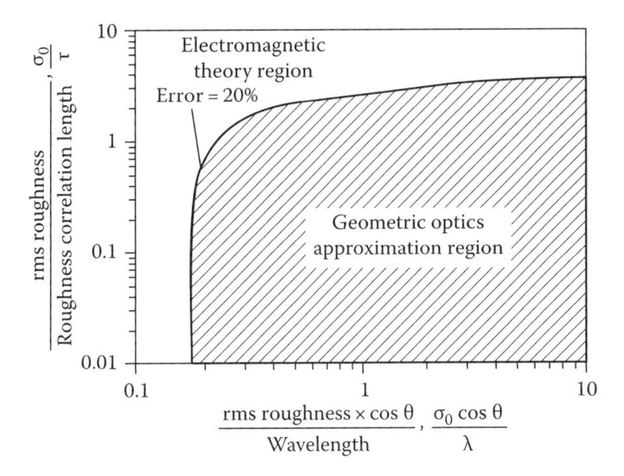

FIGURE 3.37 Region of validity (within 20% error) for using geometric optics analysis of reflectivity for rough surfaces. (From Tang, K. et al., *IJHMT*, 40, 49, 1977.)

geometric optics solution using Fresnel reflection equations applied to the tangent plane of the surface roughness at each point of incidence (*Kirchhoff's approximation*) and with the approximation that the reflection from each point of incidence is purely specular (*Fresnel approximation*).

Another regime map, shown in Figure 3.37, depicts when geometric optics provides a specified acceptable level of accuracy for one-dimensional (1D) roughness on a perfectly conducting surface (Tang et al. 1977). The axes on the plot are the surface rms roughness σ_0 over the roughness correlation length τ, versus the ratio of roughness times the cosine of the incidence angle divided by the wavelength, $\sigma_0\cos\theta_i/\lambda$. The correlation length τ is the width of a Gaussian correlation function over which the roughness correlation decreases by a factor of e. The shaded region is delineated by comparing the results of EM theory computations with geometric optics results for over 40 surfaces. The boundary is drawn where the error is less than 20%; outside of the shaded region, detailed analysis using the EM theory is required. This map was derived for perfectly conducting surfaces; similar results for dielectric surfaces indicate that the region increases where geometric optics provides good agreement. For a given roughness, if the wavelength is small relative to the roughness, the geometric optics approximation applies. This is also true if the correlation length is large so that the roughness has a less random irregularity. A more recent regime map is provided by Fu and Hsu (2008).

3.5.4 Effect of Surface Impurities

The most common surface impurities are thin layers deposited either by absorption, such as water vapor, or by chemical reaction, such as a thin oxide layer. Because dielectrics generally have high emissivities, an oxide layer, or other nonmetallic contaminant, usually increases the emissivity of an otherwise ideal metallic surface.

Figure 3.38 shows the directional-spectral emissivity of titanium at $\theta = 25°$ (Edwards and Catton 1964, Edwards and Bayard de Volo 1965). The data points are for unoxidized metal, and the solid line is the ideal emissivity from the EM theory. The dashed curve is the observed emissivity when oxide only 0.06 μm thick is present. The emissivity is increased by a factor of almost 2 from the pure titanium over much of the wavelength range. Figure 3.39 shows a similar increase in the normal spectral emissivity of Inconel X for an oxidized surface as compared with the polished metal.

Figures 3.40 and 3.41a illustrate the effect of an oxide coating on the normal total emissivity of stainless steel and on the hemispherical total emissivity of copper. The details of the oxide coatings are not specified; however, the large effect of surface oxidation is apparent. More precise indications of oxide effects are in Figures 3.41b and 3.42 for the normal total emissivity of copper and the

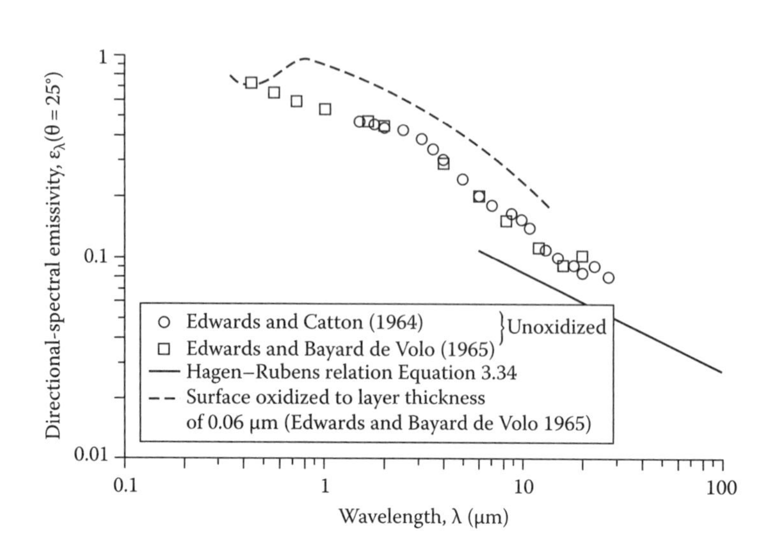

FIGURE 3.38 Effect of oxide layer on directional-spectral emissivity of titanium. Emission angle, $\theta = 25°$; surface lapped to 0.05 μm rms; temperature, 294 K.

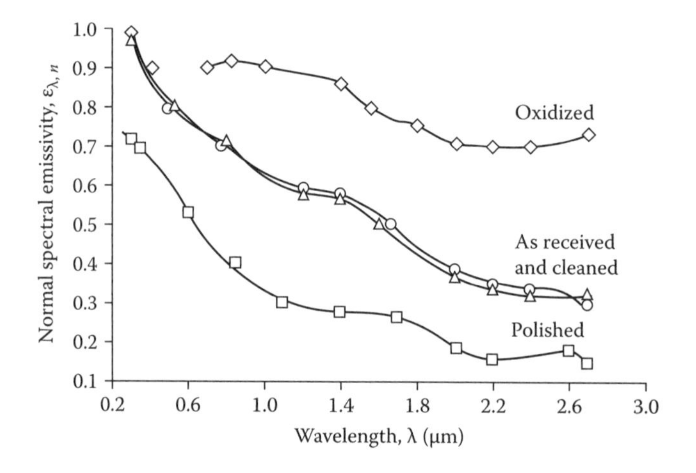

FIGURE 3.39 Effect of oxidation on normal spectral emissivity of Inconel X. (Data from Wood, W.D. et al., *Thermal Radiative Properties*, Plenum Press, New York, 1964.)

hemispherical total emissivity of aluminum. It is obvious that oxidation, even a few micrometers thick, produces a very substantial emissivity increase. For oxidized aluminum, some results similar to Figure 3.42 are in Brannon and Goldstein (1970).

Makino et al. (1988) report the measurements of the spectral near-normal bidirectional reflectivity of a polished chromium surface as a function of time during the growth of a surface oxide layer. These measurements are for wavelengths from 0.35 to 10 μm. For a surface at 1000 K oxidizing in air over a period of 1.5 h, the reflectivity follows definite trends that are explained through diffraction/interference effects within the oxide film and a model of the surface microgeometry.

Figure 3.43 illustrates the directional total absorptivity of anodized aluminum for radiation incident from various θ directions and originating from sources at various temperatures. The $\rho_s(\theta_i)$ is the fraction of the incident energy that is reflected into the specular direction; hence, $1-\rho_s(\theta_i)$ is the fraction of incident energy that is absorbed plus the fraction of incident energy reflected into

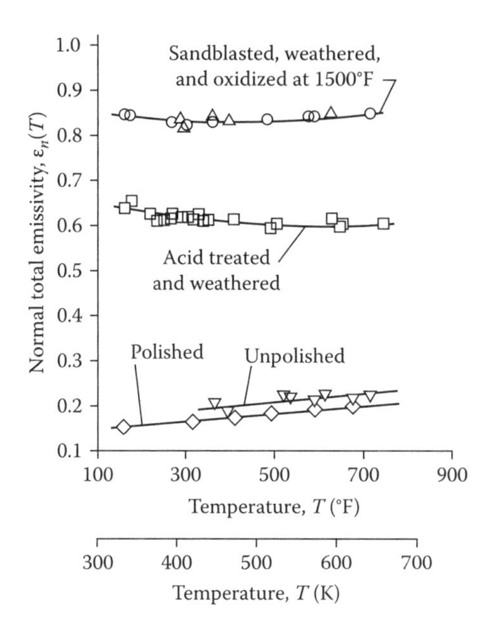

FIGURE 3.40 Effect of surface condition and oxidation on normal total emissivity of stainless steel type 18-8. (From Wood, W.D. et al., *Thermal Radiative Properties*, Plenum Press, New York, 1964.)

directions other than the specular direction. For the specimens tested, only a small percentage of the energy was reflected into directions other than the specular direction. Thus, in Figure 3.43, the $1-\rho_s(\theta_i)$ can be regarded as the directional total absorptivity. The curves are normalized to correspond to 1 at $\theta_i = 0$. Rather than the absolute values, we want to explain the angular behavior of directional absorption. At low source temperatures, the incident radiation is predominantly at long wavelengths. This incident radiation is barely influenced by the thin oxide film on the anodized surface; consequently, the specimen acts like bare metal and has large absorption at large θ_i. At high source temperatures where the incident radiation is predominantly at shorter wavelengths, the thin oxide has a significant effect, and the surface behaves as a nonmetal where absorptivity decreases with increasing θ_i. The structure of the surface coating can also have a substantial effect on the radiative behavior. Figure 3.44 shows the hemispherical-spectral reflectivity of aluminum coated with lead sulfide (Williams et al. 1963). The mass of the coating per unit area of surface is the same for both sets of data. The difference in crystal size and structure causes the reflectivity of the coated specimens to differ by a factor of 2 at wavelengths longer than 3 μm.

3.5.5 Molten Metals

The normal spectral and total emissivities of liquid sodium are reported by Hattori et al. (1984). The liquid used in this study was blanketed in argon to avoid oxidation. Unlike the behavior of many solid metals, the normal total emissivity shows only a small temperature dependence. For temperatures increasing from 450 to 715 K, the ε_n increased from 0.045 to 0.057 and then decreased to 0.044 as the temperature further increased to 795 K. The spectral emissivity in the range from 3 to 14 μm does not vary with λ for temperatures below 700 K. Above 700 K the $\varepsilon_{\lambda,n}$ has an increasing trend with λ.

The emissivity of molten uranium is reported by Havstad et al. (1993), and details are given for the variations with wavelength, angle from the surface normal, and temperature. The melting temperature of uranium is 1406 K, and results for normal spectral emissivity are in Figure 3.45 for

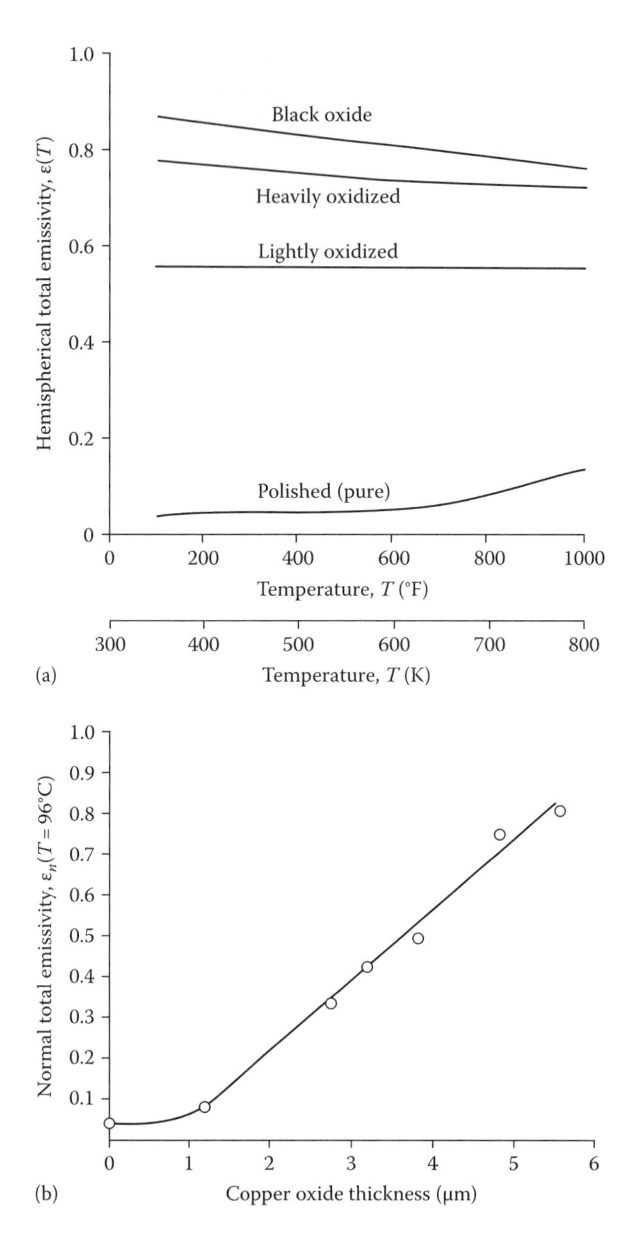

FIGURE 3.41 Effect of oxide coating on emissive properties of copper. (a) Effect of oxide coating on hemispherical total emissivity of copper. (From Touloukian, Y.S. and Ho, C.Y. (eds.), Thermophysical Properties of Matter, TRPC Data Services; Volume 7, *Thermal Radiative Properties: Metallic Elements and Alloys*, Touloukian, Y.S. and DeWitt, D.P. (1970); Volume 9, *Thermal Radiative Properties: Coatings*, Touloukian, Y.S., DeWitt, D.P., and Hernicz, R.S. (1972b), Plenum Press, New York.) (b) Effect of oxide thickness on normal total emissivity of copper at 369 K. (From Brannon, R.R. Jr. and Goldstein, R.J., *JHT*, 92(2), 257, 1970.)

temperatures from 1410 to 1630 K and for λ from 0.4 to 10 μm. There is no significant variation with temperature, and the emissivity decreases with wavelength similar to a material like iron in Figure 3.30; this is the expected trend for metals discussed in Section 3.2.2.

Measurements reported by Moscowitz et al. (1972) for three rare Earth liquid metals show a similar lack of temperature dependence, as given in the first three lines of Table 3.3. These are normal spectral emissivity values at $\lambda = 0.645$ μm. Normal spectral emissivities at this λ were measured

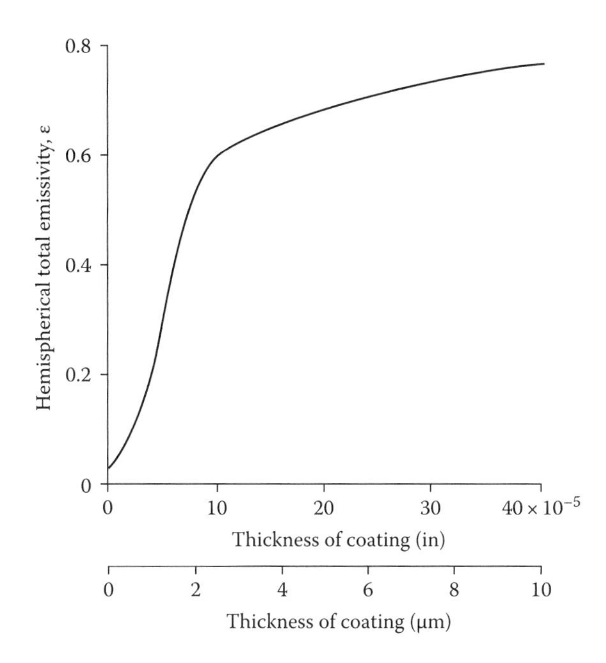

FIGURE 3.42 Typical curve illustrating effect of electrolytically produced oxide thickness on hemispherical total emissivity of aluminum. Temperature, 311 K. (From Touloukian, Y.S. and Ho, C.Y. (eds.), Thermophysical Properties of Matter, TRPC Data Services, Volume 9, *Thermal Radiative Properties: Coatings*, Touloukian, Y.S., DeWitt, D.P., and Hernicz, R.S. (1972b).)

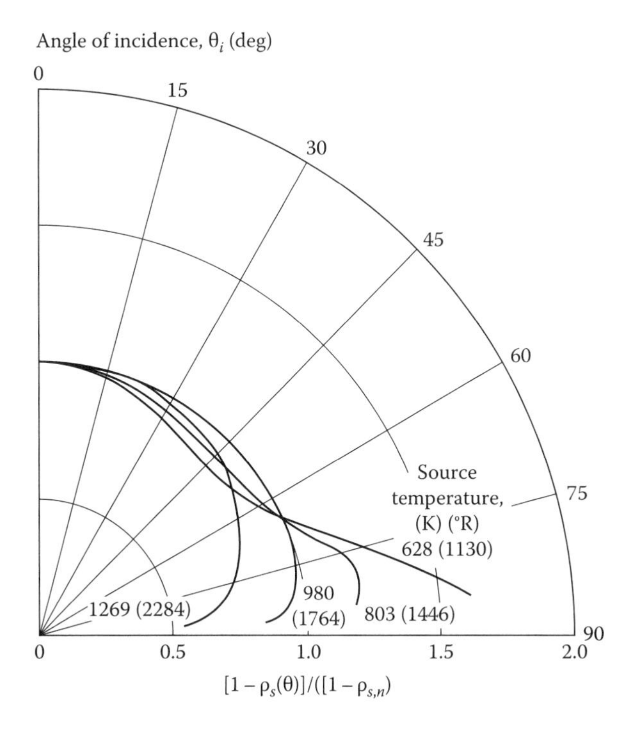

FIGURE 3.43 Approximate directional total absorptivity of anodized aluminum at room temperature relative to value for normal incidence. (Redrawn data from Munch, B., Directional distribution in the reflection of heat radiation and its effect in heat Ttransfer, PhD thesis, Swiss Technical College of Zurich, Zurich, Switzerland, 1955.)

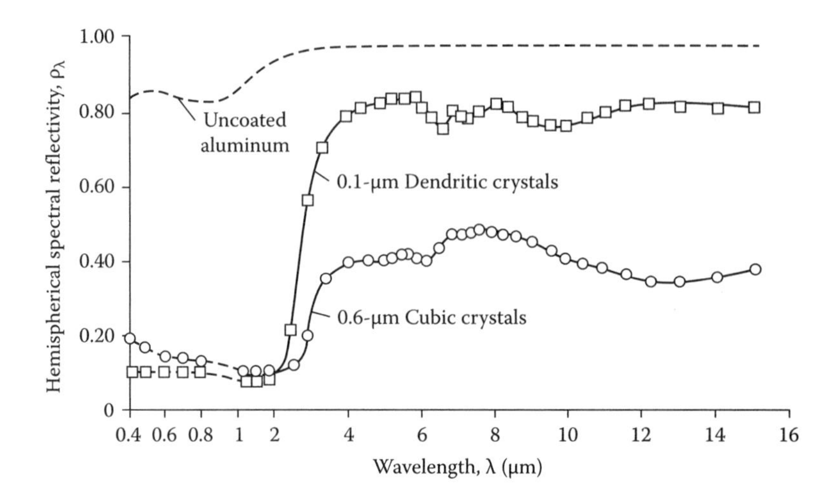

FIGURE 3.44 Hemispherical-spectral reflectivity for normal incident intensity on aluminum coated with lead sulfide. Coating mass per unit surface area, 0.68 mg/cm^2. (From Williams, D.A. et al., *J. Eng. Power*, 85(3), 213, 1963.)

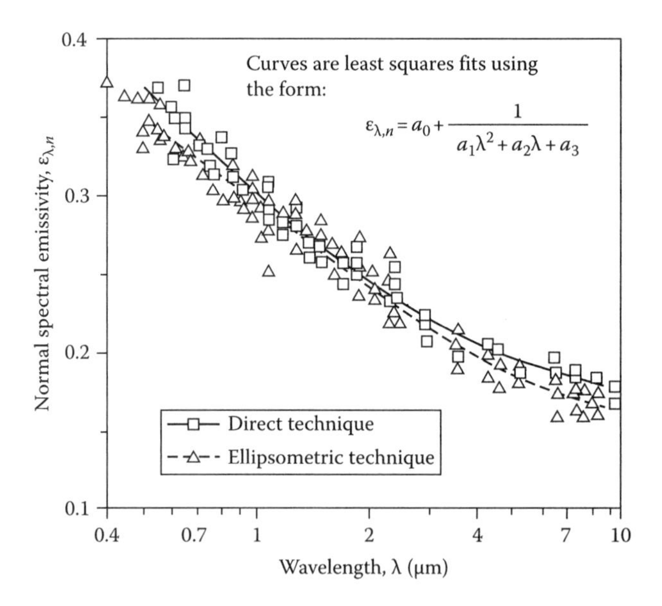

FIGURE 3.45 The normal spectral emissivity of liquid uranium between $T = 1410$ K and $T = 1630$ K. (From Havstad, M.A. et al., *JHT*, 115, 1013, 1993.)

for the liquid state at the melting temperatures of several metals (Bonnnell et al. 1972). The values (Table 3.3) are generally comparable to those for solid metals at high temperatures near their melting points. Measurements of electrical resistivity and Hall coefficients for liquid metals were used by Havstad and Qiu (1996) to predict the spectral normal emissivity of liquid cerium copper alloys. Agreement with published data is good, and extension to copper-, gold-, silver-, and aluminum-rare Earth alloys is also expected to be good on the basis of similarities in property behavior for these alloys (important in laser-cladding technologies).

In Grigoropoulos et al. (1991), the reflectivity of a thin film of silicon was measured as it was melted by incident laser radiation. An approximate increase in reflectivity of 0.1 was noted,

TABLE 3.3
Normal Spectral Emissivities of Liquid Metals at $\lambda = 0.645$ μm

Metal	Melting Point (K)	Spectral Emissivity	Temperature Range (K)
Lanthanum	1193	0.282	1193–1493
Cerium	1071	0.309	1123–1498
Praseodymium	1204	0.294	1203–1498
Cobalt	1767	0.335	1767
Chromium	2133	0.262	2133
Copper	1358	0.147	1358
Iron	1811	0.357	1811
Molybdenum	2895	0.306	2895
Niobium	2744	0.317	2744
Nickel	1728	0.346	1728
Palladium	1827	0.354	1827
Tantalum	3256	0.309	3256
Titanium	1946	0.434	1946
Vanadium	2178	0.343	2178
Zirconium	2128	0.318	2128

Source: Moscowitz, C.M. et al., *High Temp. Sci.*, 4(5), 372, 1972; Bonnell, D.W. et al., The emissivities of liquid metals at their fusion temperatures, *Fifth Symposium on Temperature, June 1971*, in *Temperature, its Measurement and Control in Science and Industry*, vol. 4, pp. 483–487, American Institute of Physics, College Park, MD, 1972.

possibly because melting eliminated the silicon polycrystalline structure. Solid silicon may be quite transparent under some conditions, but liquid silicon has the characteristics of a liquid metal and approaches being opaque in the range $\lambda = 0.4$–1.0 μm at temperatures from 1450°C to 1600°C (Shvarev et al. 1975).

3.6 SELECTIVE AND DIRECTIONAL OPAQUE SURFACES

It is often desirable to tailor the radiative properties of surfaces to increase or decrease their natural ability to absorb, emit, or reflect radiant energy. Such designer materials can be manufactured, particularly with the recent advances in laboratory techniques enhanced with the help of nanotechnology, to provide a desired spectral or directional performance. These new materials are likely to revolutionize the engineering applications in many traditional fields, from furnaces and windows to buildings. For this, we need to understand the characteristics of solar radiation further, which we do next.

3.6.1 Characteristics of Solar Radiation

Solar energy is a major contributor to many radiative heat transfer applications. This section sets out some of the characteristics of solar radiation that we use throughout the text.

3.6.1.1 Solar Constant

The solar constant refers to the solar flux incident on a surface normal to the sun at a distance equal to the Earth's mean radius from the sun. The accepted value by many standards organizations including the American Society for Standards and Measurement (ASTM) is 1366 W/m^2, although the National Oceanographic and Atmospheric Administration (NOAA) uses a value of 1376 W/m^2. The value fluctuates somewhat with time due to changes in the solar energy output, but these fluctuations are small. Larger changes are due to the eccentricity of the Earth's orbit about the sun, causing variations from 1322 W/m^2 at aphelion to 1412 W/m^2 at perihelion. We, however, in this chapter use a value of 1366 W/m^2 unless specified otherwise.

3.6.1.2 Solar Radiating Temperature

The radiation reaching the Earth's surface comes from the outer layers of the sun's atmosphere, which has a much lower temperature than the sun's interior where thermonuclear reaction is occurring. The effective radiating temperature of the sun and its spectral distribution appear to be near to that of a blackbody at a temperature of 5780 K.

3.6.2 Modification of Surface Spectral Characteristics

For surfaces used for collection of radiant energy, such as in solar distillation units, solar furnaces, or solar collectors for energy conversion, it is desirable to maximize the energy absorbed while minimizing the loss by emission. In solar thermionic or thermoelectric devices, it is desirable to maintain the highest possible equilibrium temperature of the surface exposed to the sun. For photovoltaic solar cells, absorption should be maximized within the bandgap of the PV material to maximize electrical output, but absorption should be minimized in the IR spectrum to minimize heating of the cells and minimize cooling requirements (Hajimirza et al. 2011, 2012, Hajimirza and Howell, 2012, 2013a,b, 2014a,b.). For situations where a surface is to be kept cool while exposed to the sun, it is desired to have maximum reflection of solar energy with maximum radiative emission from the surface.

For solar energy collection, a black surface maximizes the absorption of incident solar energy; unfortunately, it also maximizes the emissive losses. However, if a surface could be manufactured that had an absorptivity large in the spectral region of short wavelengths about the peak solar energy, yet small in the spectral region of longer wavelengths where the peak surface emission would occur, it might be possible to absorb nearly as well as a blackbody while emitting very little energy. Such surfaces are called "spectrally selective." One method of manufacture is to coat a thin, nonmetallic layer onto a polished metallic substrate. For radiation with large wavelengths, the thin coating is essentially transparent, and the surface behaves as a metal yielding low values for spectral emissivity and absorptivity. At short wavelengths, however, the radiation characteristics approach those of the nonmetallic coating, so the spectral emissivity and absorptivity are relatively large. Some examples of this behavior are in Figure 3.46 (Shaffer 1958, Hibbard 1961, Long 1965).

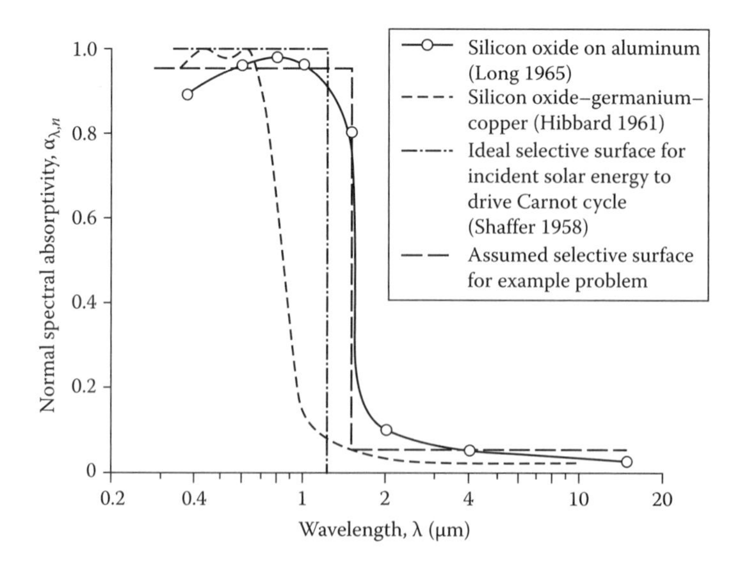

FIGURE 3.46 Characteristics of some spectrally selective surfaces.

An ideal solar-selective surface would absorb a maximum of solar energy while emitting a minimum amount of energy. The surface would thus have an absorptivity of 1 over the range of short wavelengths where the incident solar energy has a large intensity. At longer wavelengths, the absorptivity should drop sharply to zero. The wavelength λ, at which this sharp drop occurs, as in Figure 3.46, is termed the *cutoff wavelength.* The following example demonstrates the use of selective spectral properties in radiative transfer analysis.

Example 3.5

An ideal selective surface is exposed to a normal radiation flux equal to the average solar constant q_{sol} = 1366 W/m^2. The only means of heat transfer to or from the exposed surface is by radiation. Determine the maximum equilibrium temperature T_{eq} corresponding to a cutoff wavelength of λ_c = 1 μm. The solar energy can be assumed to have a spectral distribution proportional to that of a blackbody at 5780 K.

Because the only means of heat transfer is by radiation, the radiant energy absorbed must equal that emitted from the exposed side. For an ideal selective absorber, the hemispherical emissivity and absorptivity are

$$\varepsilon_\lambda = \alpha_\lambda = 1; \quad 0 \le \lambda \le \lambda_c$$
$$\varepsilon_\lambda = \alpha_\lambda = 0; \quad \lambda_c \le \lambda \le \infty$$

The energy absorbed by the surface per unit time is

$$Q_a = \alpha_\lambda F_{0\to\lambda_c}(T_R) q_{sol} A = (1) F_{0\to\lambda_c}(T_R) q_{sol} A$$

where $F_{0\to\lambda_c}(T_R)$ is the fraction of blackbody energy in the range of the wavelengths between zero and the cutoff value, for a radiating temperature T_R. In the case, T_R is the effective solar radiating temperature 5780 K. Similarly, the energy emitted by the selective surface is

$$Q_e = \varepsilon_\lambda F_{0\to\lambda_c}(T_{eq}) \sigma T_{eq}^4 A = (1) F_{0\to\lambda_c}(T_{eq}) \sigma T_{eq}^4 A$$

Equating Q_e and Q_a gives

$$T_{eq}^4 F_{0\to\lambda_c}(T_{eq}) = \frac{q_{sol} F_{0\to\lambda_c}(T_R)}{\sigma}$$

For the chosen value for λ_c all terms on the right are known, and we can solve for T_{eq} by trial and error. The equilibrium temperature for $\lambda_c = 1$ μm is 1337 K. Here are the values of T_{eq} for various λ_c:

Cutoff Wavelength λ_c (μm)	**Equilibrium Temperature T_{eq}, K (°R)**
0.4	2287 (4118)
0.6	1817 (3271)
0.8	1530 (2755)
1.0	1337 (2406)
1.2	1197 (2155)
1.5	1048 (1886)
$\to \infty$	394 (709)

For a blackbody surface ($\lambda_c \to \infty$), the equilibrium temperature is 394 K (709°R); this is the equilibrium temperature of the surface of a black object in space at the Earth's orbit when exposed to normally incident solar radiation and with all other surfaces of the object perfectly insulated. The same equilibrium temperature is reached by a gray body, since a gray emissivity cancels out of the energy-balance equation. As λ_c is decreased, the T_{eq} continues to increase even though less energy is absorbed; this is because it also becomes relatively more difficult to emit energy.

A performance parameter for a solar-selective surface is the ratio of its directional total absorptivity $\alpha(\theta_i, \phi_i, T)$ for incident solar energy to its hemispherical total emissivity $\varepsilon(T)$. The ratio $\alpha(\theta_i, \phi_i, T)/\varepsilon(T)$ for the condition of incident solar energy is a measure of the theoretical maximum temperature that an otherwise insulated surface can attain when exposed to solar radiation. In general, the energy absorbed per unit time by a surface element dA exposed to a directional energy flux q_{sol} [incident from direction (θ_i, ϕ_i)] is

$$dQ_a(\theta_i,\phi_i,T) = \alpha(\theta_i,\phi_i,T)q_{sol}(\theta_i,\phi_i)dA\cos\theta_i \tag{3.46}$$

The total energy emitted per unit time by the surface element is

$$dQ_e = \varepsilon(T)\sigma T^4 dA \tag{3.47}$$

If there are no other means for energy transfer, the emitted and absorbed energies are equated to give

$$\frac{\alpha(\theta_i,\phi_i,T)}{\varepsilon(T_{eq})} = \frac{\sigma T_{eq}^4}{q_{sol}(\theta_i,\phi_i)\cos\phi_i} \tag{3.48}$$

where T_{eq} is the equilibrium temperature that is achieved. Thus, the ratio $\alpha(\theta_i,\phi_i,T)/\varepsilon(T)$ for $T \approx T_{eq}$ is a measure of the equilibrium temperature of the element. Note that the temperature at which the properties α and ε are selected should be the equilibrium temperature that the body attains.

For the collection and utilization of solar energy on Earth or in outer space applications, it is common to have normal solar incidence so that $\alpha = \alpha_n$ and $\cos\theta_i = 1$. Naturally, a high α_n/ε is desired. For the relatively low temperatures of solar collection in ground-based systems without solar concentrators, selective paints on an aluminum substrate have $\alpha_n = 0.92$ and $\varepsilon = 0.10$ (Moore 1985). Low-emittance metallic flakes also mixed in a binder with high-absorptance metallic oxides to yield coatings with $\alpha_n = 0.88$ and $\varepsilon = 0.40$. Note that in a ground-based solar collector, convection must be included in the energy balances. To attain high equilibrium temperatures for space power systems, polished metals attain α_n/ε of 5–7, and specially manufactured surfaces have α_n/ε approaching 20. Coatings with $\alpha_n/\varepsilon \approx 13$ and stability at temperatures up to about 900 K in air are reported by Craighead et al. (1979). Space power systems usually have a concentrator such as a parabolic mirror. This increases the collection area relative to the area for emission and thus effectively increases the absorption-to-emission ratio even further. Economical and durable paints are desired for application to large solar collection areas. Many of these ideas are discussed in reviews by Granqvist (2003) and Wijewardane and Goswami (2012).

Example 3.6

The properties of a real SiO–Al selective surface are approximated by the long dashed curve in Figure 3.46 (it is assumed that this curve can be extrapolated toward $\lambda = 0$ and $\lambda = \infty$). What is the equilibrium temperature of the surface for normally incident solar radiation at Earth orbit when energy transfer is only by radiation? What is α_n/ε for the surface? Describe the spectra of the absorbed and emitted energy at the surface. Assume normal and hemispherical emissivities are equal.

As in the derivation of Equation 3.48, equate the absorbed and emitted energies. The emissivity has nonzero constant values on both sides of the cutoff wavelength, so that

$$Q_a = A\int_{\lambda=0}^{\infty} \alpha_{\lambda,n} q_{\lambda,sol} d\lambda = \left[\varepsilon_{0\to\lambda_c} F_{0\to\lambda_c}(T_R) + \varepsilon_{\lambda_c\to\infty} F_{\lambda_c\to\infty}(T_R)\right] q_{sol} A = \alpha_n q_{sol} A$$

$$Q_e = A\int_{\lambda=0}^{\infty} \varepsilon_\lambda E_{\lambda b}(T_{eq}) d\lambda = \left[\varepsilon_{0\to\lambda_c} F_{0\to\lambda_c}(T_R) + \varepsilon_{\lambda_c\to\infty} F_{\lambda_c\to\infty}(T_{eq})\right] \sigma T_{eq}^4 A = \varepsilon\sigma T_{eq}^4 A$$

where T_R is the temperature of the radiating source. Equating Q_e and Q_a gives

$$\left\{0.95F_{0\to\lambda_c}(T_R) + 0.05\left[1 - F_{0\to\lambda_c}(T_R)\right]\right\} q_{sol}$$
$$= \left\{0.95F_{0\to\lambda_c}(T_{eq}) + 0.05\left[1 - F_{0\to\lambda_c}(T_{eq})\right]\right\} \sigma T_{eq}^4$$

Solving by trial and error, for λ_c = 15 μm, we obtain T_{eq} = 790 K. The small difference in the properties of an ideal selective surface produces a significant change in T_{eq}, which in the previous example was 1048 K for an ideal selective surface with the same λ_c. The spectral curve of incident solar energy is given by $E_{\lambda,i}(T_R) \propto E_{b,\lambda}(T_R)$. It has the shape of the blackbody curve at the solar temperature, T_R = 5780 K, but it is reduced in magnitude so that the integral of E_λ over all λ is equal to q_{sol}, the total incident solar energy per unit area at Earth orbit. Multiplying this curve by the spectral absorptivity of the selective surface gives the spectrum of the absorbed energy. The spectrum of emitted energy is that of a blackbody at 790 K multiplied by the spectral emissivity of the selective surface. The integrated energies under the spectral curves of absorbed and emitted energy are equal.

The energy equation solved in Example 3.6 is a two-spectral-band approximation to the following more general energy-balance equation for a *diffuse* surface:

$$\int_{\lambda=0}^{\infty} \alpha_\lambda(T_{eq}) q_{\lambda,sol}(\lambda) d\lambda = \int_{\lambda=0}^{\infty} \varepsilon_\lambda(T_{eq}) E_{\lambda b}(T_{eq}) d\lambda$$

The $q_{\lambda,sol} d\lambda$ can have any spectral distribution, and by Kirchhoff's law $\alpha_\lambda(T_{eq}) = \varepsilon_\lambda(T_{eq})$ for a diffuse surface. A more general situation is in Figure 3.47, where the absorption of incident energy from the θ_i direction depends on the directional-spectral absorptivity $\alpha_\lambda(\theta_i,\phi_i,T_{eq})$. The emission from the surface depends on its hemispherical-spectral emissivity $\varepsilon_\lambda(T_{eq})$. The q_e is the heat flux supplied to the surface by any other means, such as convection, electrical heating, or radiation to its lower side. The heat balance then becomes

$$\cos\theta_i \int_{\lambda=0}^{\infty} \alpha_\lambda(\theta_i,\phi_i,T_{eq}) q_{\lambda,sol} d\lambda + q_e = \int_{\lambda=0}^{\infty} \varepsilon_\lambda(T_{eq}) E_{\lambda b}(T_{eq}) d\lambda \quad (3.49)$$

Equation 3.49 is readily solved by using integration subroutines and root solvers. This analysis can be used for temperature control of space vehicles, as shown by Furukawa (1992), who determined the solar and Earth radiation fluxes incident on the different surfaces of orbiting vehicle.

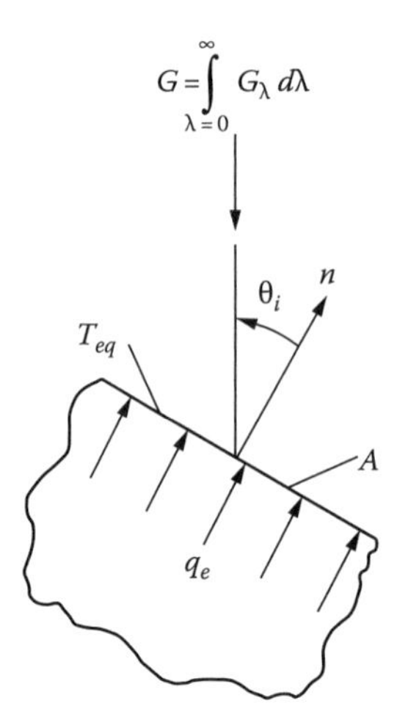

FIGURE 3.47 Radiative energy incident on a selective surface.

Example 3.7

A spectrally selective surface is to be used as a solar energy absorber. The surface that has the same properties as given in Example 3.6 is to be maintained at $T = 394$ K by extracting energy for a power-generating cycle. If the absorber is in orbit around the sun at the same radius as the Earth and is normal to the solar direction, how much energy will a square meter of the surface provide? How does this energy compare with that provided by a black surface at the same temperature? Emitted energy and reflected solar energy by Earth are neglected.

The energy extracted from the surface is the difference between the absorbed and emitted radiation. The absorbed energy flux is calculated as in Example 3.6, where $T_R = 5780$ K,

$$q_a = \int_{\lambda=0}^{\infty} \varepsilon_\lambda (T) q_{\lambda,\text{sol}}(T_R) d\lambda = \left\{0.95F_{0\to\lambda_c}(T_R) + 0.05\left[1 - F_{0\to\lambda_c}(T_R)\right]\right\} q_{\text{sol}}$$

$$= \left[0.95(0.880) + 0.05(1 - 0.880)\right]1366 = 1142.1 \ \text{W/m}^2$$

The emitted flux is

$$q_e = \int_{\lambda=0}^{\infty} \varepsilon_\lambda E_{\lambda b}(T) d\lambda = \left\{0.95F_{0\to\lambda_c}(T) + 0.05\left[1 - F_{0\to\lambda_c}(T)\right]\right\} \sigma T^4$$

$$= \left[0.95 \times (\sim 0) + 0.05(\sim 1)\right]5.6704 \times 10^{-8} \times 394^4 = 70.7 \ \text{W/m}^2$$

Therefore, the energy that can be used for power generation is 1142.1 − 70.7 = 1071.4 W/m^2. For a blackbody or gray body, the equilibrium temperature is 394 K as obtained from Example 3.6 so, for a black or gray absorber, no useful energy can be removed for the stated conditions.

Spectrally selective surfaces can also be useful where it is desirable to cool an object exposed to incident radiation from a high-temperature source. Common situations are objects exposed to the sun, such as a hydrocarbon storage tank, a cryogenic fuel tank in space, or the roof of a building. Equation 3.48 shows that the smaller the value of α/ε that can be reached, the smaller will be the equilibrium temperature. For a cryogenic storage tank exposed to solar flux in the vacuum of outer space, α/ε should be as small as possible in order to reduce losses by heating the stored cryogen. In practice, values of α/ε in the range 0.20–0.25 can be obtained for normal incidence ($\cos\theta_i = 1$). A highly reflecting coating such as a polished metal can also be used for some applications.

This would reflect much of the incident energy, but would be poor for radiating away energy that was absorbed or generated within an enclosure, such as by electronic equipment. This behavior is important for the energy balance in the vacuum of outer space and may not provide a low α/ε, but may not be important when there is appreciable convective cooling that dominates over radiant emission. Some metals may not work well because they have a tendency toward lower reflectivity at the shorter-wavelength characteristic of the incident solar energy; this is shown for uncoated aluminum in Figure 3.44. For some applications, spectrally selective materials are used. White paint is another example of spectrally selective surface (Dunkle 1963). As shown in Figure 3.48, the paint not only reflects the incident solar radiation predominant at short wavelengths but also radiates well at the longer-wavelength characteristic of the relatively low body temperature.

For thermal control in outer space, different spectrally selective surfaces have been defined. Among them, the optical solar reflector (OSR) is a mirror composed of a glass layer silvered on the back side. The glass, being transparent in the short-wavelength region, $\lambda < \sim 2.5$ μm, which includes the visible range, lets the silver reflect incident radiation in this spectral region. The small fraction of short-wavelength energy that is absorbed by the silver and the energy absorbed by the glass at longer wavelengths are radiated away by the glass in the longer-wavelength IR region where glass emits well. Commonly used thin plastic sheets for solar reflection are Kapton, Mylar, and Teflon with silver or aluminum coated on the back side. The long-term radiative performance of these materials after 10-year exposure in geosynchronous orbit is evaluated by Hall and Fote (1992). Fused-silica second-surface mirrors and polished metals are essentially stable in orbit. Metalized Teflon, aluminized Kapton, and some light-colored paints, on the other hand, darken over a long period of time, degrading their performance (Hall and Fote 1992).

Radiative dissipation is vital in outer space applications as there is no other means to eliminate waste energy except to dissipate small quantities by using expendable coolants. For a device on the ground or in the Earth's atmosphere, convection and conduction to the surrounding environment are available. The significance of each of the three heat transfer modes depends on the particular conditions in the energy balance. An interesting example of such an energy balance is discussed by Berdahl et al. (1983) to show that radiative cooling might be useful for cooling buildings to help with air conditioning. Objects exposed to the night sky can cool by radiation to achieve temperatures

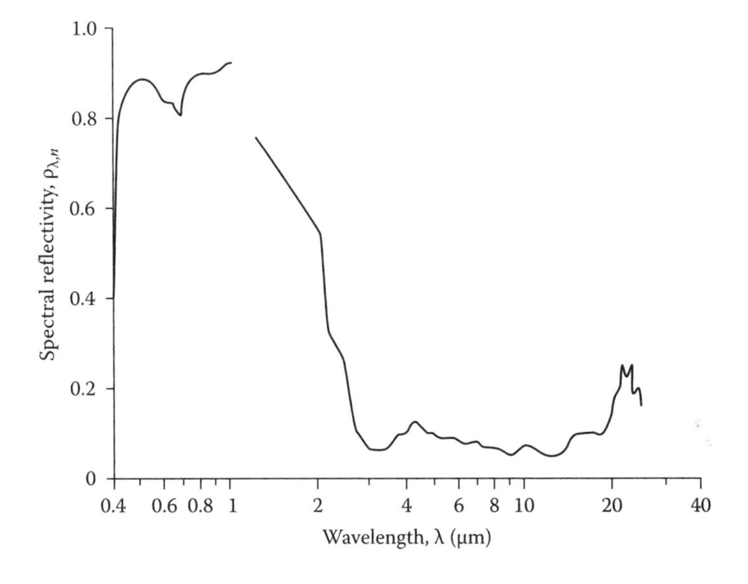

FIGURE 3.48 Reflectivity of white paint coating on aluminum. (Replotted from Dunkle, R.V., Thermal radiation characteristics of surfaces, in J.A. Clark (ed.), *Theory and Fundamental Research in Heat Transfer*, Pergamon Press, New York, pp. 1–31, 1963.)

below the ambient air temperature. This cooling effect can also be utilized during the day if the solar reflectivity of a surface is high (greater than about 0.95) and its emissivity is large in the IR. For this purpose, titanium dioxide white paint is somewhat superior, as an external solar-selective coating, to polyvinyl fluoride film with an aluminized coating 12 μm thick on the underside. These are the same types of materials, as well as many others that are used for spacecraft thermal control. The properties of these types of material are listed by Henninger (1984).

There is also interest in developing selective emitters that emit most of their energy in a narrow spectral region rather than over a broad spectrum as for a blackbody emission. In a thermophotovoltaic converter, the absorbed radiant energy is converted into electrical energy. The conversion is possible only for energy within a small wavelength range. Hence, it is desired to irradiate the converter from a high-temperature surface that radiates primarily in the spectral region that provides effective electrical conversion. This provides good efficiency in the use of the radiated energy and hence efficient utilization of the energy used to heat the radiating surface such as a nuclear power source for applications in outer space. The fabrication of selective emitters by using rare Earth oxides is discussed by Rose et al. (1996). Another approach is discussed by Sentenac and Greffet (1994), where theory is provided for designing a regular microroughness (a "grating") that will provide selective emission. The roughness dimension is of the same size range as the radiation wavelengths. An examination of performance, using EM scattering theory, indicates that wavelength-selective behavior can be obtained, which has been experimentally verified (Greffet and Henkel 2007).

An extended discussion of spectrally selective surfaces for radiative cooling, along with recent references, is in the online Appendix H at www.thermalradiation.net. Optimization of nanogeometries for desired spectral-directional properties is discussed in Chapter 8.

3.6.3 Modification of Surface Directional Characteristics

As discussed previously, surface roughness can have profound effects on radiative properties and can become a controlling factor when roughness is large compared with the radiation wavelength. This leads to the concept of controlling roughness to tailor the directional characteristics of a surface. A surface used as an emitter might be designed to emit strongly in preferred directions, while reducing emission into unwanted directions. Commercial radiant area-heating equipment would operate more efficiently by using such surfaces to direct energy where it is most needed. If the directional surface is primarily an absorber, then, using a solar absorber as an example, it should be strongly absorbing in the direction of incident solar radiation but poorly absorbing in other directions. The surface would, because of Kirchhoff's law for directional properties, emit strongly toward the sun but weakly in other directions. Such a surface then would absorb the same energy as a nondirectional absorber but would emit less than a surface that emits well into all directions.

The characteristics of one such surface are shown in Figure 3.49. Long parallel grooves are of angle 18.2° (Perlmutter and Howell 1963, Brandenberg and Clausen 1965). Each groove is coated with a highly reflecting specular material on the side walls, and a black surface is at the base. The solid line gives the behavior predicted by analysis of such an ideal surface, while the data points show experimental results at $\lambda = 8$ μm for an actual surface. The directional emissivity is very high for θ less than about 30° and decreases rapidly as the angle becomes large. Other such surface configurations exhibit similar characteristics; results for various vee, rectangular, and other grooves are also reported by Demont et al. (1982). EM wave theory was used by Ford et al. (1995) and Cohn et al. (1997) to predict the bidirectional-spectral reflectivity from surfaces with regular microcontours. The microcontour scales for sinusoidal, rectangular, and V-grooves considered are on the order of the wavelength of the incident radiation so that geometric optics is not expected to apply. The theory and measurements found to be in quite good agreement. The reflectivity of a V-groove is shown in Figure 3.50.

The emissivity of a surface with random submicron roughness elements is discussed by Carminati et al. (1998). Through scaling arguments, it is shown that interference effects between radiations emitted by closely spaced elements can be neglected, which is not the case for reflected energy. Based

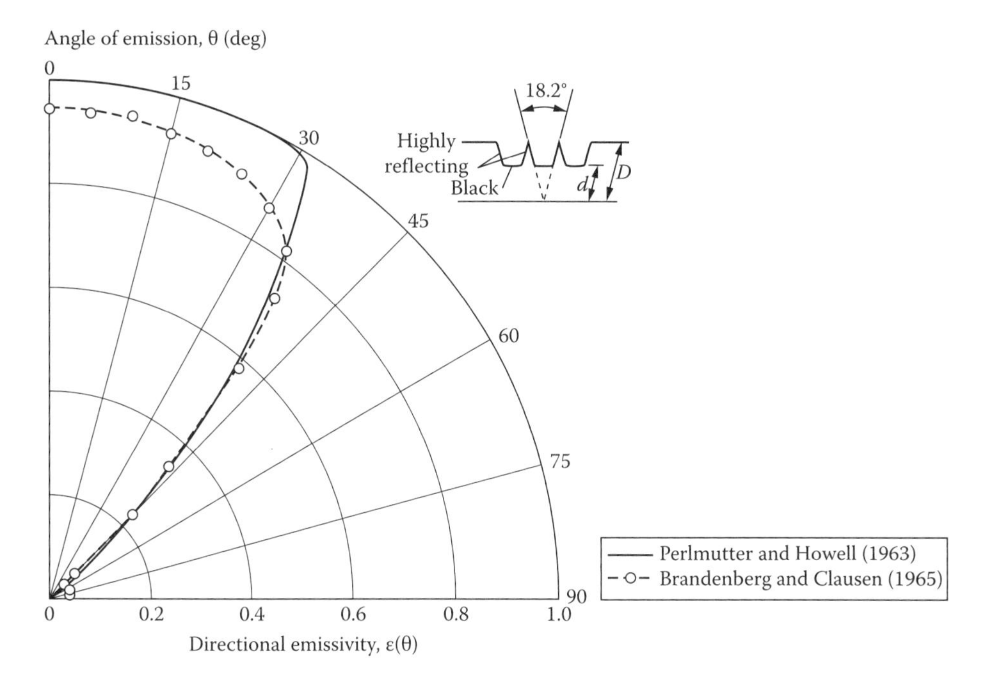

FIGURE 3.49 Directional emissivity of grooved surface with highly reflecting specular side walls and highly absorbing base; $d/D = 0.649$. Results in plane perpendicular to groove direction; data at $\lambda = 8$ μm.

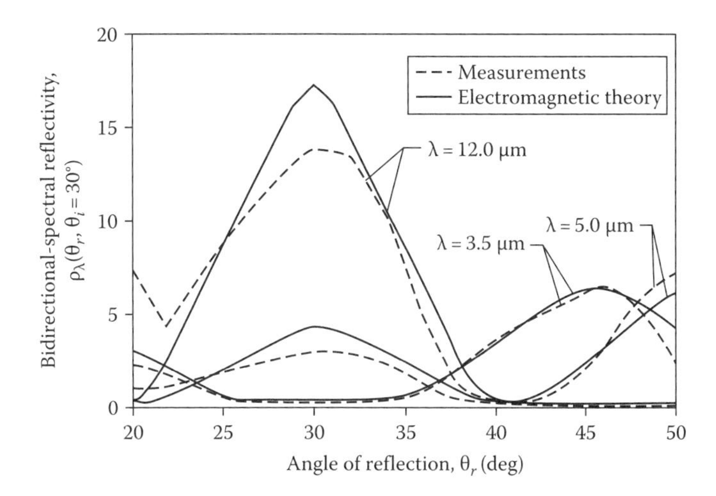

FIGURE 3.50 Bidirectional-spectral reflectivity in plane of incidence for a nickel V-groove cavity; width of cavity opening = 16.7 μm, cavity depth = 1.38 μm, angle of incidence $\theta_i = 30°$. (From Cohn, D.W. et al., *IJHMT*, 40(13), 3223, 1997.)

on this observation, predictions of directional-spectral emissivity from surfaces with random submicron roughness elements can be made through numerical simulation. Because the model initially neglects shadowing and blocking effects, the predictions are restricted to rms height σ_0 to wavelength ratios of $\sigma_0/\lambda < 0.3$, as obtained by comparison with experimental emissivity data for the two components of polarization. Random microscale roughness effects on scattering are analyzed by Fu and Hsu (2006, 2007, 2008). Surfaces with tailored roughness at the submicron scale can produce directional emissivity with extreme directionality (see, e.g., Figures 16.8 and 16.9) (Greffet et al. 2002, Greffet and Henkel 2007), and the theory behind these near-field effects is provided in Chapter 16.

3.7 CONCLUDING REMARKS

The radiative property examples in this chapter have illustrated features that may be encountered when dealing with real surfaces. Based on these data, we can make some useful generalizations, such as that the total emissivities of dielectrics at moderate temperatures are larger than those for metals and the spectral emissivity of metals increases with temperature over a broad range of wavelengths. However, these generalizations can be misleading because of the large property variations that may occur as a result of surface roughness, contamination, oxide coating, grain structure, and so forth. It is usually not possible to predict radiative property values except for surfaces that approach ideal conditions of composition and finish. By coupling analytical trends with observations of experimental trends, it is possible to gain insight into what classes of surfaces would be expected to be suitable for specific applications and how surfaces may be fabricated to obtain certain types of radiative behavior. The latter includes spectrally selective surfaces that are of great value in practical applications such as collection of solar energy and spacecraft temperature control. For spacecraft surfaces, exposure to ultraviolet radiation; cosmic rays; neutron, gamma, and proton bombardment; atomic oxygen; and the solar wind can cause significant changes in radiative properties. These effects are of major concern in the design of devices operating in space (Stevens 1990). Finally, because of the increased interest in solar energy as the major renewable energy source and the development of novel nanopatterning techniques, new designer materials for various applications are likely to surface. Similarly, traditional building materials, such as roof tiles, paints, facades, and windows, can be spectrally optimized to decrease the energy load on the buildings. The spectral emissivity could be designed to provide surfaces that strongly emit during the nighttime to optimize cooling of structures in hot climates (Rephaeli et al. 2013). The developments in spectral radiative transfer analyses may help to develop a wider database for spectral and directional radiative properties of different materials.

HOMEWORK

3.1 An electrical insulator has a refractive index of $n = 1.332$ and has a smooth surface radiating into air. What is the directional emissivity for the direction normal to the surface? What is it for the direction 60° away from the normal?
Answer: 0.9797; 0.9405.

3.2 A smooth hot ceramic dielectric sphere with an index of refraction $n = 1.43$ is photographed with an IR camera. Calculate how bright the image is at locations B and C relative to that at A. (Camera is distant from sphere.)

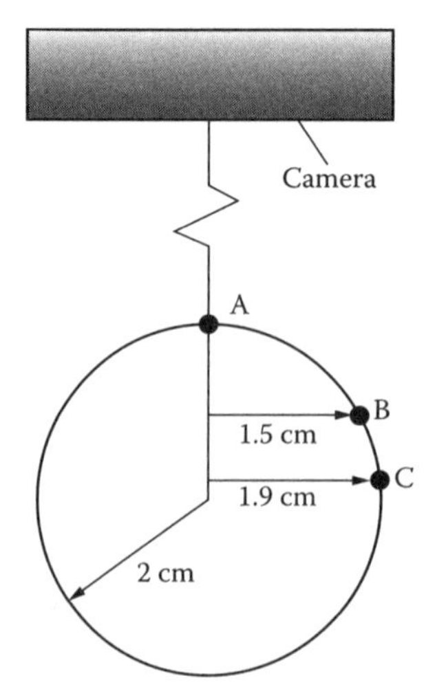

Answer: 0.986; 0.845.

3.3 A particular dielectric material has a refractive index of $n = 1.350$. For a smooth radiating surface, estimate the following:
(a) The hemispherical emissivity of the material for emission into air.
(b) The directional emissivity at $\theta = 60°$ into air.
(c) The directional-hemispherical reflectivity in air for both components of polarized reflectivity. Plot both components for $n = 1.350$ on a graph similar to Figure 3.5. Let θ be the angle of incidence.

Answer: (a) 0.931, (b) 0.937.

3.4 A smooth dielectric material has a normal spectral emissivity of $\varepsilon_{\lambda,n} = 0.725$ at a wavelength in air of 6 μm. Find or estimate values for
(a) The hemispherical-spectral emissivity ε_λ at the same wavelength.
(b) The perpendicular component of the directional-hemispherical-spectral reflectivity $\rho_{\lambda,\perp}(\theta)$ at the same wavelength and for incidence at $\theta = 40°$.

Answer: (a) 0.703; (b) 0.369.

3.5 An inventor wants to use a light source and some Polaroid glasses to determine when the wax finish is worn from her favorite bowling alley. She reasons that the wax will reflect as a dielectric with $n = 1.40$ and that the parallel component of light from the source will be preferentially absorbed and the perpendicular component strongly reflected by the wax. When the wax is worn away, the wood will reflect diffusely. At what height should the light source be placed to maximize the ratio of perpendicular to parallel polarization from the wax as seen by the viewer?

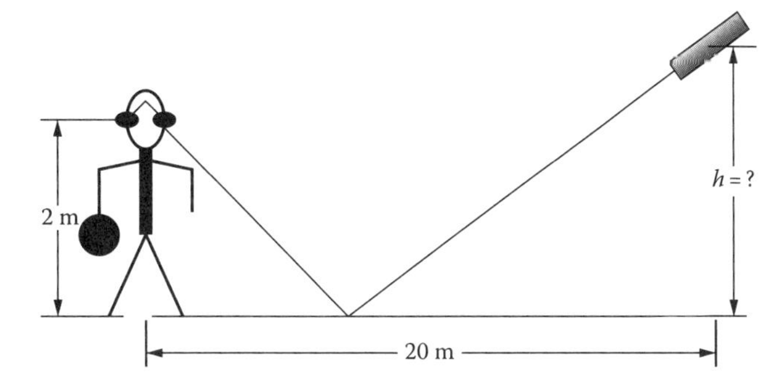

Answer: 12.3 m.

3.6 A smooth ceramic dielectric has an index of refraction $n = 1.58$, which is independent of wavelength. If a flat ceramic disk is at 1100 K, how much emitted energy per unit time is received by the detector when it is placed at $\theta = 0°$ or at $\theta = 60°$? Use relations from the EM theory.

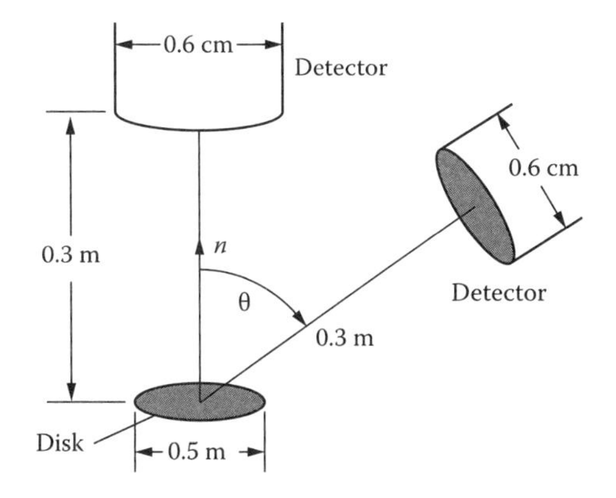

Answer: 15.48×10^{-5} W at $\theta = 0°$; 7.318×10^{-5} W at $\theta = 60°$.

3.7 At a temperature of 300 K, these metals have the following resistivities (Table 3.2):

Copper	1.72×10^{-6} Ohm-cm
Gold	2.27×10^{-6} Ohm-cm
Aluminum	2.73×10^{-6} Ohm-cm

What are the theoretical normal total emissivities and hemispherical total emissivities of these metals, and how do they compare with tabulated values for clean, unoxidized, polished surfaces?
Answer: ε_n 0.0131, 0.0157, 0.0168; ε 0.0170, 0.0202, 0.0217.

3.8 A highly polished metal disk is found to have a measured normal spectral emissivity of 0.095 at a wavelength of 12 μm. What is
(a) The electrical resistivity of the metal (Ohm-cm)
(b) The normal spectral emissivity of the metal at $\lambda = 10$ μm
(c) The refractive index n of the metal at $\lambda = 10$ μm
(Note any assumptions that you make in obtaining your answers.)
Answer: (a) 8.716×10^{-5} Ohm-cm; (b) 0.104; (c) 18.3.

3.9 Show using Equation 3.7 that the parallel component of reflectivity becomes zero when $\theta = \tan^{-1}(n_2/n_1)$.

3.10 A clean metal surface has a normal spectral emissivity of $\varepsilon_{\lambda,n} = 0.06$ at a wavelength of 12 μm. Find the value of the electrical resistivity of the metal.
Answer: 2.271×10^{-4} Ohm-cm.

3.11 Evaluate the normal spectral reflectivity of clean aluminum at 420 K when $\lambda_0 = 6$, 12, and 24 μm. For aluminum, the temperature coefficient of resistivity is 0.0039.
Answer: 0.970; 0.979; 0.985.

3.12 Polished platinum at 300 K is irradiated normally by a gray-body source at 1250 K. Evaluate its normal total absorptivity α_n. (Use the method of Example 3.4.)
Answer: 0.0700.

3.13 A smooth polished gold surface must radiate 350 W/m^2. What is its surface temperature as calculated from Hagen–Rubens relations? At this temperature, what is the normal total emissivity of the gold? What is the ratio of its hemispherical total emissivity to its normal total emissivity?
Answer: 626 K; 0.0313; 1.29.

3.14 The hemispherical total emissive power emitted by a polished metallic surface is 2600 W/m^2 at temperature T_A. What would you expect the emissive power to be if the temperature were doubled? What assumptions are involved in your answer?
Answer: 83,200 W/m^2.

3.15 The following figure gives some experimental data for the hemispherical-spectral reflectivity of polished aluminum at room temperature. Extrapolate the data to $\lambda = 10$ μm. Use whatever method you want, but list your assumptions. Discuss the probable accuracy of your extrapolation. (*Hint:* The electrical resistivity of pure aluminum is about $r_e = 2.73 \times 10^{-6}$ Ohm-cm at 293 K. At 10 μm, take $\bar{n} = 33.6 - 76.4i$. You may use any, all, or none of these data as you wish.)

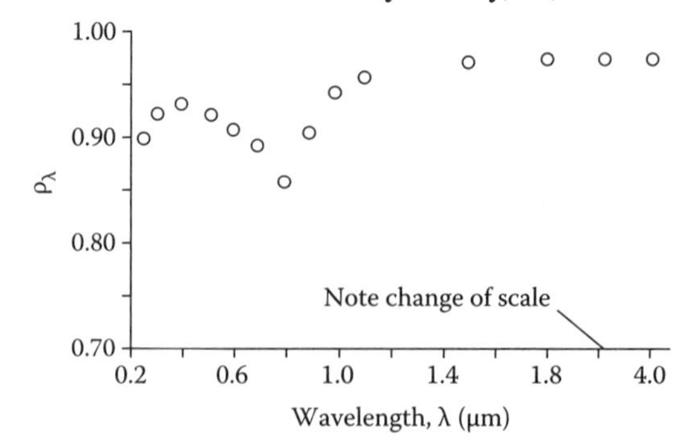

Answer: 0.975.

3.16 An unoxidized titanium sphere is heated until it is glowing red. From a distance, it appears as a red disk. From the EM theory how would you expect the brightness to vary across the disk? What would you expect after looking at Figure 3.29?

3.17 Using Hagen–Rubens emissivity relation, plot the normal spectral emissivity as a function of wavelength for a polished aluminum surface used in a cryogenic application at 55 K. What is the normal total emissivity? (*Note*: Do not use any relations valid only near room temperature.)

3.18 A sample of highly polished platinum has a value of normal spectral emissivity of 0.042 at a wavelength of 8.2 μm at 293 K. What value of normal spectral absorptivity do you expect the sample to have at

(a) A wavelength of 10 μm and at 293 K
(b) A wavelength of 10 μm and at 600 K

Answer: (a) 0.0391; (b) 0.0541.

3.19 Metals cooled to very low temperatures approaching absolute zero become superconducting; that is, the value of $r_e(T \approx 0) \approx 0$. Based on EM theory predictions, what is your estimate of the values of the simple refractive index n, the absorption index k, and the normal spectral and normal total emissivities at such conditions? What assumptions are implicit in your estimates? [The results predicted by Hagen–Rubens relation, and other results from classical EM theory, become inaccurate at $T < 100$ K. Predictions of radiative properties at low absolute temperatures using more exact theoretical approaches are reviewed by Toscano and Cravalho (1976).]

3.20 A smooth polished stainless steel surface must emit a total intensity in the normal direction of 65.0 W/m²·sr. What is its surface temperature as calculated from equations derived from Hagen–Rubens relations? The r_e of the steel is 11.9×10^{-6} Ohm-cm at 293 K.

Answer: 502 K.

3.21 Using Equation 3.31 with data for n and k from Table 3.2, find the hemispherical emissivity of aluminum and titanium at 300 K at 0.589 and 10 μm.

Answer: $\varepsilon_{\lambda,\text{Al}}$ (λ = 0.589 μm) = 0.094; $\varepsilon_{\lambda,\text{Al}}$ (λ = 10 μm) = 0.015; $\varepsilon_{\lambda,\text{Ti}}$ (λ = 0.589 μm) = 0.64; $\varepsilon_{\lambda,\text{Ti}}$ (λ = 10 μm) = 0.051.

3.22 For a selective surface with $\alpha_\lambda = 0.93$ in the range $0 < \lambda < \lambda_c$ and $\alpha_\lambda = 0.12$ at $\lambda > \lambda_c$, plot the curve of equilibrium temperature versus λ_c for the range $2 \le \lambda_c \le 10$ μm.

3.23 For a free convective heat transfer coefficient of 6.0 W/m²·°C, an air temperature of 0°C, and an incident solar flux on the surface of 850 W/m², calculate the equilibrium temperature of the Earth's surface when it is

(a) Covered with fine fresh snow
(b) Plowed soil

Discuss how this might impact global warming in snow-covered regions as the average global air temperature increases.

Answers: (a) 17.4°C; (b) 82.2°C.

3.24 The normal spectral absorptivity of a SiO–Al selective surface can be approximated as shown by the long dashed line in Figure 3.46. The surface receives a flux q from the normal direction. The equilibrium temperature of the surface is 1150 K. Assume the hemispherical-spectral $\varepsilon_\lambda = \alpha_\lambda(\theta = 0)$. What is the value of q if it comes from a gray-body source at 4900 K?

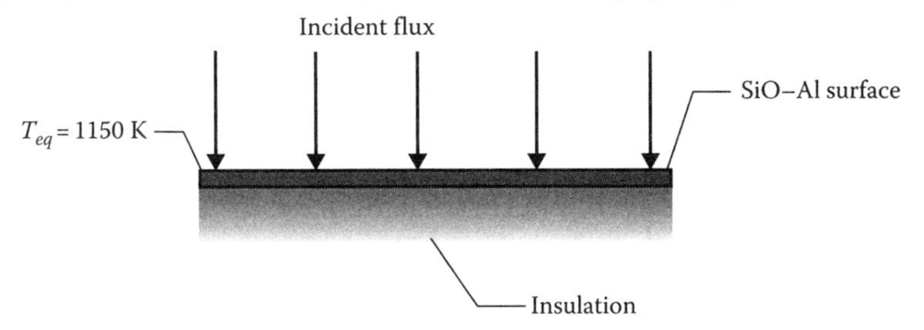

Answer: 9732 W/m².

3.25 A directionally selective gray surface has properties as shown below. The $\alpha(\theta)$ is isotropic with respect to the azimuthal angle ϕ.

(a) What is the ratio $\alpha_n(\theta = 0)/\varepsilon$ (the normal directional absorptivity over the hemispherical emissivity) for this surface?

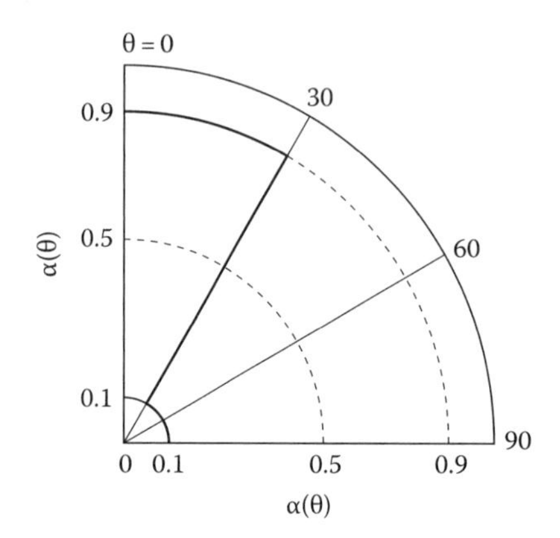

(b) If a thin plate with the aforementioned properties is in Earth orbit around the sun with incident solar flux of 1350 W/m^2, what equilibrium temperature will it reach? Assume that the plate is oriented normal to the sun's rays and is perfectly insulated on the side away from the sun.

(c) What is the equilibrium temperature if the plate is oriented at 40° to the sun's rays?

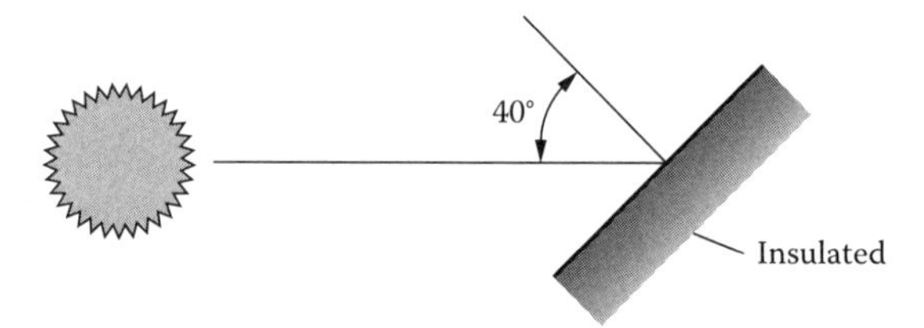

(d) What is the equilibrium temperature if the plate is normal to the sun's rays but is not insulated? Assume the plate is very thin and has the same directional radiation properties on both sides. Neglect radiation emitted by or reflected from the Earth.

Answer: (a) 3.0; (b) 517 K; (c) 297 K; (d) 435 K.

3.26 A flat plate in Earth orbit is insulated on one side, and the other side is facing normal to the solar intensity. The incident solar flux is 1350 W/m^2. A coating on the plate surface facing the sun has a total hemispherical emissivity of 0.235 over a broad range of plate temperatures. Surroundings above the plate are at a very low temperature. Telemetry signals to Earth indicate that the plate temperature is 730 K.

(a) What is the normal solar absorptivity α_{solar} of the plate surface facing the sun?

(b) If α_{solar} is independent of angle, what is the plate temperature if the plate is tilted so that its normal is 40° away from the solar direction?

Answer: (a) 2.803; (b) 683 K.

3.27 A diffuse-spectral coating has the spectral emissivity approximated as follows. The coating is placed on one face of a thin sheet of metal. The sheet is placed in an orbit around the sun where the solar flux is 1350 W/m^2. The other face of the sheet is coated with a diffuse gray coating of hemispherical total emissivity $\varepsilon = 0.510$. What is the temperature (K) of the sheet if the following applies:

(a) The side with the spectral coating is facing normal to the sun.
(b) The gray side is facing normal to the sun.
(c) What is the normal-hemispherical total reflectivity of the diffuse-spectral coating when exposed to solar radiation? Take the effective solar radiating temperature to be 5780 K. Note any necessary additional assumptions.

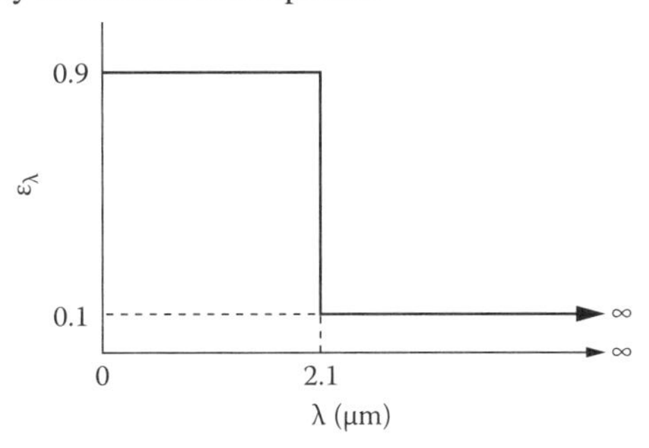

Answer: (a) 428.0 K; (b) 376 K; (c) 0.143.

3.28 The spectral absorptivity of a SiO–Al selective surface can be approximated as shown below. The surface is in Earth orbit around the sun and has the solar flux 1353 W/m^2 incident on it in the normal direction. What is the equilibrium temperature of the surface if the surroundings are very cold?

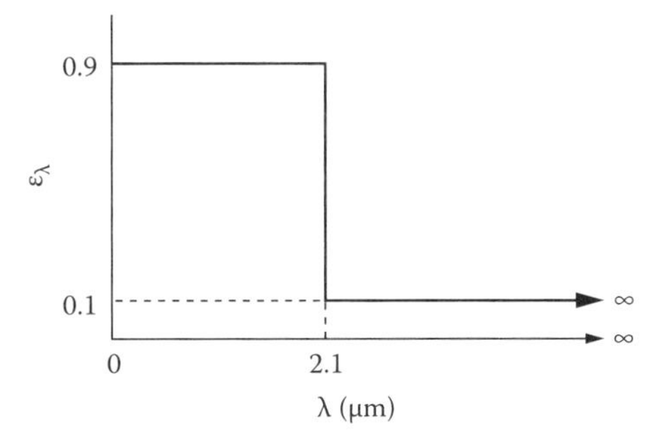

Answer: 662 K.

3.29 A thin plate has a directional-gray surface on one side with the directional emissivity shown below on the left. On the other side of the plate is a coating with diffuse-spectral emissivity shown below on the right. The surroundings are at very low temperature. Find the equilibrium temperature of the plate if it is exposed in vacuum to a normal solar flux of 1353 W/m^2 with a solar spectrum equivalent to that of a blackbody at 5780 K when

(a) The directional-gray side is facing normal to the sun
(b) The diffuse-spectral side is facing normal to the sun

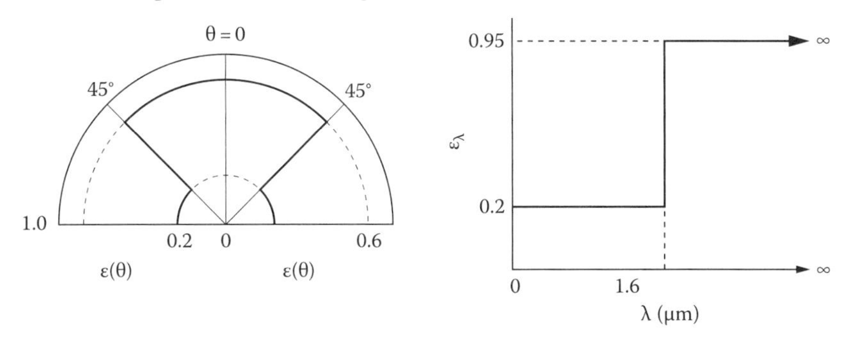

Answer: (a) 376 K; (b) 265 K.

3.30 A gray surface has a directional total absorptivity given by $\alpha(\theta) = 0.82 \cos^4\theta$. This flat surface is exposed to normally incident sunlight of flux 1050 W/m^2. A fluid flows past the back of the thin radiation absorber plate at $T_{fluid} = 325$ K and with a velocity that gives a heat transfer coefficient of $h = 64$ W/m$^2\cdot$K. What is the equilibrium temperature of this flat-plate radiation collector?

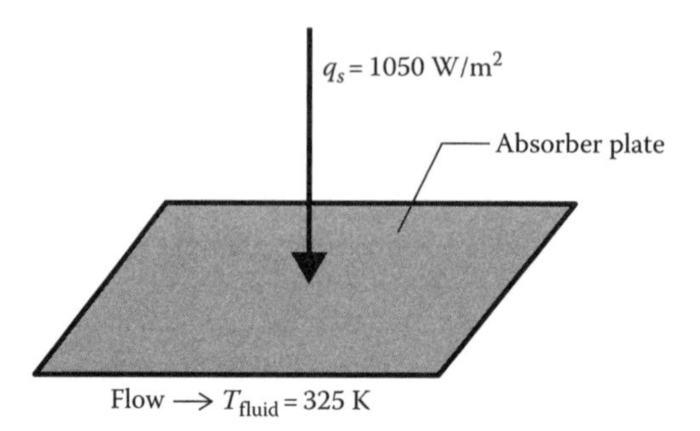

Answer: 335 K.

3.31 A plate is coated with a material combining directional and spectral selectivity so that the plate has a normal total solar absorptivity of 0.92 and an IR hemispherical emissivity of 0.040 at long wavelengths. When placed in sunlight normal to the sun's rays, what temperature will the plate reach (neglecting conduction and convection and with no heat losses from the unexposed side of the plate)? What assumptions did you make in reaching your answer? The incident solar flux ("insolation") is 1050 W/m^2.

Answer: 652 K.

3.32 A solar water heater consists of a sheet of glass 1 cm thick over a black surface that is assumed in perfect contact with the water below it. Estimate the water temperature for normally incident solar radiation. (Assume that Figure 9.22 can be used for the glass properties and that the glass is perfectly transparent for wavelengths shorter than those shown. Take into account approximately the reflections at the glass surfaces; this is treated in more detail in Chapter 17.)

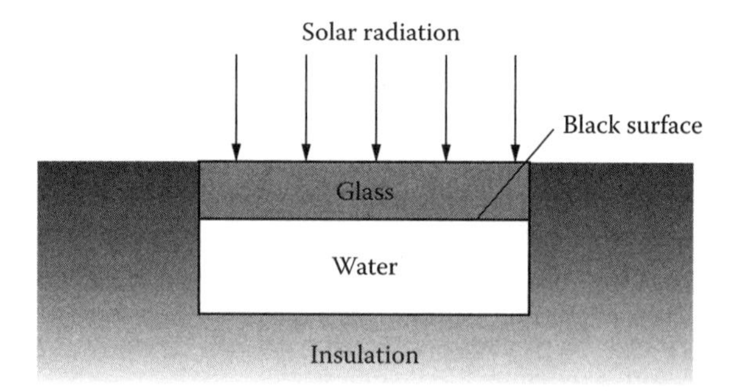

Answer: 389 K.

3.33 A gasoline storage tank is receiving sunlight on a somewhat cloudy day so that the incident radiation normal to the top of the tank is 950 W/m^2. The sides are not receiving solar radiation. The tank top and sides are painted with white paint having the spectral reflectivity in Figure 3.48.

(a) Estimate the average equilibrium temperature that the tank will achieve. (Neglect emitted and reflected radiation from the ground. Do not account for free or forced convection to the air, although this will be appreciable.) The ambient radiating temperature of the surroundings is $T_e = 300$ K.

(b) What is the average tank temperature if the top is painted white as before but the sides have a gray coating with an emissivity of 0.825?

(c) What is the temperature if the entire tank is painted with the gray coating?

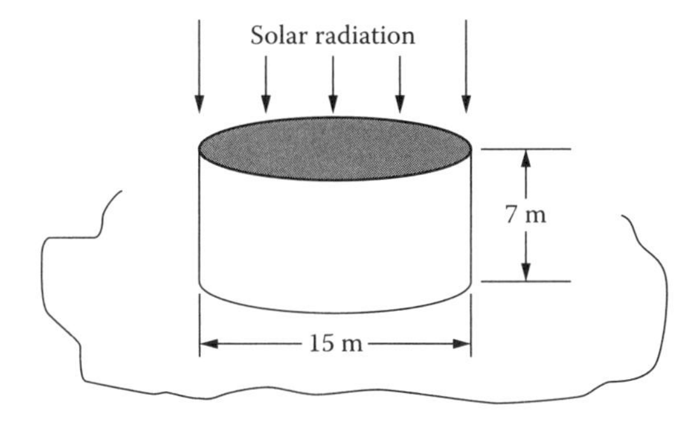

Answer: (a) 308.5 K; (b) 315.3 K; (c) 343.6 K.

3.34 Consider the Earth as a spinning sphere exposed on one hemisphere to solar energy at the solar flux of 1353 W/m^2.

(a) If the average solar absorptivity of the Earth is taken as equal to its emissivity (a gray body), what is the estimated equilibrium temperature of the Earth?

(b) If the solar absorptivity and low-temperature emissivity are taken as the properties of fine snow (Tables B.1, B.2), what will be the Earth's equilibrium temperature?

(c) If the solar absorptivity and low-temperature emissivity are taken as the properties of plowed soil (Tables B.1, B.2), what will be the Earth's equilibrium temperature?

(d) Given the results of parts c and d, what do you see as the impact of the melting of polar ice and glaciers due to global warming? Is there a feedback mechanism that tends to increase or mitigate the effects of warming? (For more in-depth discussion, see Maslowski et al. 2012)

Answer: (a) 4.7°C; (b) −133.5°C; (c) 56.3°C.

3.35 A spinning spherical satellite 0.85 m in diameter is in the Earth's orbit and is receiving solar radiation of 1353 W/m^2. As a consequence of the rotation, the sphere surface temperature is assumed uniform. The sphere exterior is a SiO–Al selective surface as in Example 3.6, with a cutoff wavelength of 2.00 μm. The surface properties do not depend on angle. Neglect energy emitted by the Earth and solar energy reflected from the Earth.

(a) What is the equilibrium surface temperature for heat transfer only by radiation? How does the temperature depend on the sphere diameter?

(b) It is desired to maintain the sphere surface at 815 K. At what rate must energy be supplied to the entire sphere to accomplish this?

Answer: 569 K; 3284 W.

3.36 A flat-plate radiator in space in Earth orbit is oriented normal to the solar radiation. It is receiving direct solar radiation of 1350 W/m^2, radiation emission from the Earth, and solar radiation reflected from the Earth. What must the radiator temperature be to dissipate a total of 3000 W of waste heat from both sides of each 1 m^2 of the radiator?

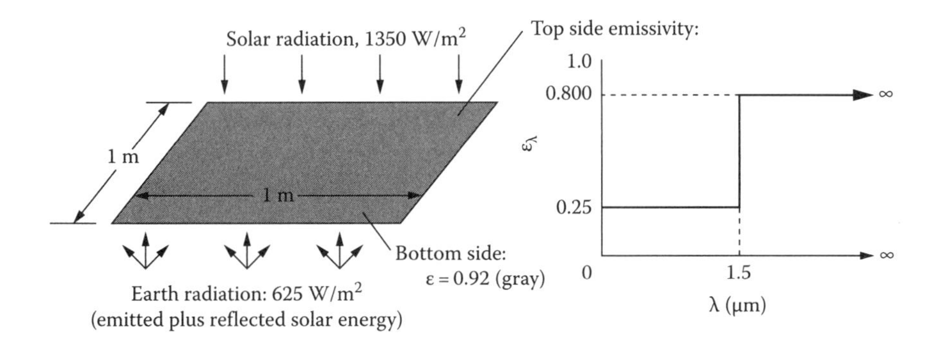

Answer: 435.3 K.

3.37 A cylindrical concentrator (reflector) is long in the direction normal to the cross section shown, so that end effects may be neglected. The concentrator is gray and reflects 97% of the incident solar energy onto the central tube. The central tube receiving the energy is assumed to be at uniform temperature and is coated with a material that has the spectral properties shown. The properties are independent of direction. Emitted energy from the tube that is reflected by the concentrator may be neglected, as may be emission from the concentrator. The surrounding environment is at low temperature.

(a) If the heat exchange is only by radiation, compute the temperature of the central tube.

(b) If the tube is cooled to 390 K by passing a coolant through its interior, how much energy must be removed by the coolant per meter of tube length?

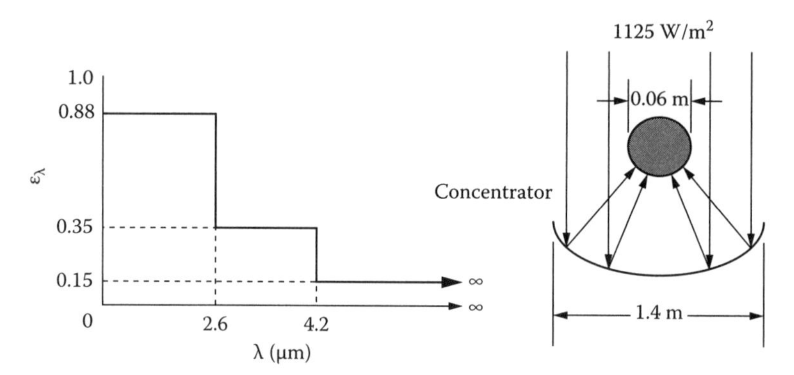

Answer: (a) 822.0 K; (b) 1280.3 W/m of length.

3.38 Specimens of metals are prepared by machining the surfaces so that the surface roughness is composed of parallel elements of rms roughness σ_0 with a correlation distance τ between roughness elements as measured by a profilometer. Determine whether geometric optics or complete EM theory should be applied for determining the reflectivity of the surfaces for the cases below for laser radiation of wavelength λ incident on the surface at angle θ in the plane normal to the roughness elements:

Case	rms Roughness, σ_0, μm	Roughness Correlation Length, τ, μm	Angle of Incidence, θ, Degrees	Laser Wavelength, λ, μm
1	5	5	0	1
2	1	5	0	10.2
3	2	5	60	5
4	10	20	85	10.2
5	10	2	30	1

Answer: Cases 2, 4 and 5, EM theory; Case 1, geometric optics; Case 3 on boundary.

3.39 A nuclear-powered Carnot engine is to produce $W = 110$ kW of work output to drive an electrical generator for a proposed small space station. The temperature of the energy input to the cycle is limited to 1000 K. Find the heat rejection temperature from the cycle (the average temperature of a heat rejection radiator) that will minimize the area of a radiator that rejects the cycle energy.

Answer: 9.20 m^2.

3.40 Suppose that the designer of the nuclear power system in Homework Problem 3.39 insists that the Carnot efficiency of the system be increased to 33%, holding the power output and inlet temperatures at 110 kW and 1000 K. How will this affect the required area of the heat rejection radiator?

Answer: 9.77 m^2.

3.41 For the Carnot system described in Homework Problem 3.37, assume that solar energy is normal to one side of the heat rejection radiator. Use the solar constant of 1353 W/m^2. What radiator area is now required?
Answer: $A = 8.68$ m^2.

3.42 Consider Homework Problem 3.41 for the case when solar energy is normal to the heat rejection radiator. The solar constant on one face of the radiator is 1353 W/m^2. In this case, rather than for a black surface, the radiator has equal spectral normal and spectral hemispherical emissivities on the side of the radiator normal to the sun of $\varepsilon_1 = 0.1$ for $0 \le \lambda_c \le 2$ μm and $\varepsilon_2 = 0.9$ for $2 < \lambda_c \le \infty$ μm and is black on the side facing away from the sun. Compare the result for the radiator area with that calculated for Homework Problems 3.39 and 3.41 (if assigned).
Answer: $A = 8.93$ m.

3.43 A light pipe (refractive index $n_2 = 1.4400$) of diameter 0.1 cm is placed with its flat end 1 cm from a heated semiconductor wafer. What diameter of the wafer surface is viewed by the light pipe?

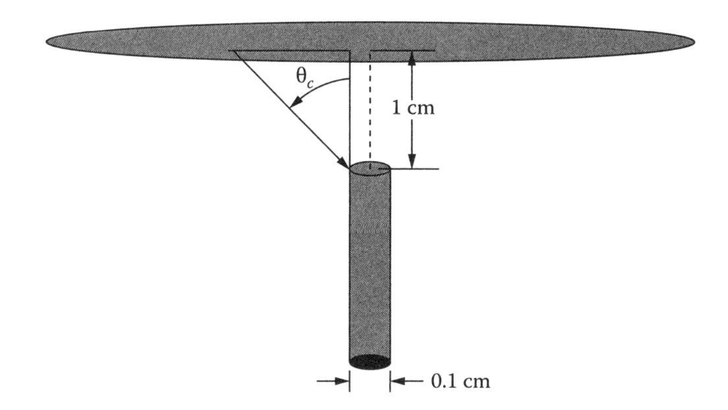

Answer: All values of $0 < \theta < \pi/2$.

3.44 For a small-diameter light pipe with refractive index $n_2 = 1.4400$, plot the reflectivity of the light pipe end versus incident angle θ_1 for radiation incident on the end. Also, plot the angle of refraction χ versus angle of incidence, and determine the range of incident angles θ_1 that will have angles θ_2 above the critical angle when the radiation encounters the cylindrical light pipe wall.

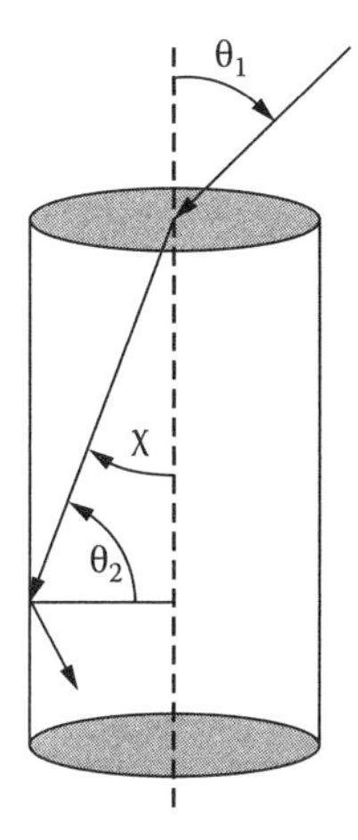

4 Configuration Factors for Diffuse Surfaces with Uniform Radiosity

4.1 RADIATIVE TRANSFER EQUATION FOR SURFACES SEPARATED BY A TRANSPARENT MEDIUM

Radiation interchange among surface areas is required for calculation of heat transfer, illumination engineering, and applied optics. Studies have been conducted for many years, as evidenced by the publication dates of d'Aguillon (1613) and Charle (1888). Since 1960, the study of radiative transfer has been given impetus by technological advances that provided systems in which thermal radiation is very important. Examples abound and include radiant heating, curing, and surface modification systems, satellite temperature control, devices for collection and utilization of solar energy, advanced engines with increased operating temperatures, reduction of energy leakage into cryogenic fuel storage tanks, and space station power systems and thermal control during spacecraft into planetary atmospheres.

The general radiative transfer equation describes the propagation of intensity along a path and was introduced in Sections 1.6 and 1.7. For the important case when the medium separating two surfaces is transparent (i.e., there is no attenuation of intensity along a path by scattering or absorption, and the medium along the path does not emit), then Equation 1.76 shows that the intensity leaving a radiating surface is invariant along a path. Figure 4.1 shows this situation.

As will be shown, under certain assumptions, the geometric relations between surfaces 1 and 2 can be separated from the intensity, allowing the description of the geometric configuration of surface 2 relative to surface 1 to be treated independently of the thermal states of the two surfaces. The factors that contain the geometric relations are called *configuration factors.*

The geometric configuration factors derived in this chapter are an important component for analyzing radiation exchange. Before considering them, some introductory comments are in order about "enclosure" theory that motivates the need for the configuration factors. In Chapters 5 through 8, the theory is developed for radiation exchange in enclosures that are evacuated or contain radiatively nonparticipating media. First, it must be understood what is meant by an *enclosure.* Any surface can be considered as completely surrounded by an envelope of other solid surfaces or open areas. This envelope is the enclosure for the surface; thus, an enclosure accounts for all directions surrounding a surface. By considering the radiation from the surface to all parts of the enclosure and the radiation arriving at the surface from all parts of the enclosure, it is assured that all radiative contributions are included. A convenient enclosure is usually evident from the physical configuration. An opening into a large empty environment can often be considered as an area of zero reflectivity. The opening also acts as a radiation source that can be diffuse or directional, when radiation is entering from the surrounding environment.

In Chapters 1 through 3, we have defined the radiative properties of solid opaque surfaces. For some materials, the properties vary substantially with wavelength, surface temperature, and direction. For radiation computations within enclosures, the geometric aspects of the exchange introduce another challenge in addition to the surface property variations. For simple geometries, it may be

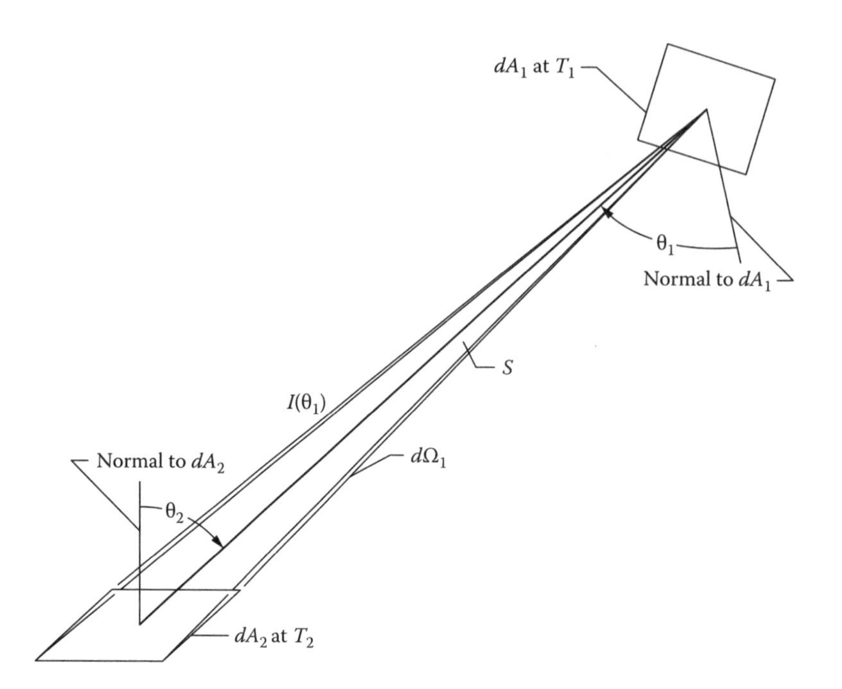

FIGURE 4.1 Intensity leaving surface 1 and transferring energy to surface 2.

possible to account in detail for property variations without the analysis becoming complex. As the geometry becomes more involved, it may be necessary to invoke more idealizations of the surface properties so that solutions can be obtained with reasonable effort, if the resulting decrease in accuracy is acceptable. This development begins with simple situations; successive complexities are then added to build more comprehensive treatments in Chapters 5 through 8.

4.1.1 Enclosures with Diffuse Surfaces

Consider an enclosure where each surface is black and each is isothermal. For black surfaces, there is no reflected radiation, and all emitted energy is diffuse (the intensity leaving a surface is independent of direction). The local energy balance at a surface involves the enclosure geometry, which governs how much radiation leaving a surface will reach another surface. For a black enclosure, the geometric effects are expressed in terms of the diffuse configuration factors developed in this chapter. A configuration factor is the *fraction of uniform diffuse radiation leaving a surface that directly reaches another surface.*

The computation of configuration factors involves analytical or numerical integration over the solid angles by which surfaces can view each other. Some examples are given to demonstrate analytical integrations. Factors are usually evaluated numerically, and numerical routines have been incorporated into many of the computer programs developed for thermal analysis. Because these integrations can be tedious, it is often helpful to use relations that exist between configuration factors. This may make it possible to obtain a desired factor from factors that are already known. These relations, along with various shortcut methods that can be used to obtain configuration factors, are given in this chapter.

The next step in enclosure complexity is to have *gray* rather than black surfaces that emit and reflect *diffusely.* The enclosure surfaces are specified such that it can be assumed that both emitted and reflected energies are *uniform* over each surface. The configuration factors can then be used for radiation leaving a surface by both emission and reflection. For gray surfaces, reflections among surfaces must be accounted for in the analysis; we discuss this in Chapter 5.

4.1.2 Enclosures with Directional (Nondiffuse) and Spectral (Nongray) Surfaces

Sometimes the approximations of black or diffuse–gray surfaces are not sufficiently accurate, and directional and/or spectral effects must be included. The necessity of treating spectral effects was noticed quite early in the field of radiative transfer. Sir William Herschel (1800) published "Investigation of the Powers of the Prismatic Colours to Heat and Illuminate Objects; with Remarks, that prove the Different Refrangibility of Radiant Heat, to which is added, an Inquiry into the Method of Viewing the Sun Advantageously, with Telescopes of Large Apertures and High Magnifying Powers," which includes the following statement:

> In a variety of experiments I have occasionally made, relating to the method of viewing the sun, with large telescopes, to the best advantage, I used various combinations of differently coloured darkening glasses. What appeared remarkable was, that when I used some of them, I felt a sensation of heat, though I had but little light; while others gave me much light, with scarce any sensation of heat. Now, as in these different combinations, the sun's image was also differently coloured, it occurred to me, that the prismatic rays might have the power of heating bodies very unequally distributed among them.

This paper was the first in which the infrared region of the spectrum was defined and the energy radiated as "heat" shown to be of wavelengths different than those for "light." The quotation shows an early awareness that in some instances spectral effects must be included in a radiative analysis. The performance of spectrally selective surfaces, such as those for satellite temperature control and solar collectors, can be understood only by considering wavelength variations of the surface properties. If the spectral surfaces are diffuse, the configuration factors developed here remain applicable for enclosure analysis.

In some instances, the surface properties have significant directional characteristics. In Chapter 3, directionally dependent surface properties were examined. They sometimes differ considerably from the diffuse approximation. Consequently, diffuse configuration factors cannot be used when surfaces emit or reflect in a significantly directional (nondiffuse) manner. For example, a mirror is a special directional surface. Emission from this type of surface is often approximated as being *diffuse*; hence, the emitted energy is treated by using diffuse configuration factors. Reflected energy is followed within the enclosure by using the mirror characteristic that the angles of reflection and incidence are equal in magnitude.

4.2 GEOMETRIC CONFIGURATION FACTORS BETWEEN TWO SURFACES

When calculating radiative transfer among surfaces, geometric relations are needed to determine how the surfaces view each other. In this section, a *geometric configuration factor* is developed to account for geometric effects. Such factors allow computation of radiative transfer in many systems by referring to formulas or tabulated values for the geometric relations between two surfaces. This simplifies a time-consuming portion of the analysis. Radiative exchange in systems with directional properties is analyzed in Chapter 6; first, we discuss the factors for uniformly distributed diffuse energy leaving a surface.

4.2.1 Configuration Factor for Energy Exchange between Diffuse Differential Areas

The radiative transfer from a diffuse differential area element to another area element is used to derive relations for transfer between finite areas. Consider the two differential area elements in

Figure 4.1. The dA_1 and dA_2 are at T_1 and T_2, are arbitrarily oriented, and have their normals at angles θ_1 and θ_2 to the line of length S between them.

If I_1 is the total intensity leaving dA_1, the total energy per unit time leaving dA_1 and incident on dA_2 is

$$d^2Q_{d1-d2} = I_1 dA_1 \cos\theta_1 d\Omega_1 \tag{4.1}$$

where
$d\Omega_1$ is the solid angle subtended by dA_2 when viewed from dA_1
the dash in the subscript of Q means "to"

Equation 4.1 follows directly from the definition of I_1 as the total energy leaving surface 1 per unit time, per unit area projected normal to S, and per unit solid angle. Because I_1 is diffuse, it is independent of the angle at which it leaves dA_1. It may consist of both diffusely emitted and diffusely reflected portions. The d^2Q is a second-order differential as it depends on two differential quantities: dA_1 and $d\Omega_1$.

Equation 4.1 can also be written for radiant energy in a wavelength interval $d\lambda$:

$$d^2Q_{\lambda,d1-d2}d\lambda = I_{\lambda,1}d\lambda dA_1 \cos\theta_1 d\Omega_1 \tag{4.2}$$

The total radiant energy is then found by integrating over all wavelengths:

$$d^2Q_{d1-d2} = dA_1 \cos\theta_1 d\Omega_1 \int_{\lambda=0}^{\infty} I_{\lambda,1} d\lambda \tag{4.3}$$

For a diffuse surface, I_λ does not depend on direction. Because all of the geometric factors are independent of λ, they can be removed from under the integral sign, and the integration over λ is independent of geometry. Thus, the results that follow for diffuse geometric configuration factors involving finite areas apply for *both spectral and total* quantities. For simplicity in notation, the remaining development is carried out for total quantities.

The solid angle $d\Omega_1$ is related to the projected area of dA_2 and the distance between the differential elements by

$$d\Omega_1 = \frac{dA_2 \cos\theta_2}{S^2} \tag{4.4}$$

Substituting this into Equation 4.1 gives the following relation for the total energy per unit time leaving dA_1 that is incident upon dA_2:

$$d^2Q_{d1-d2} = \frac{I_1 dA_1 \cos\theta_1 dA_2 \cos\theta_2}{S^2} \tag{4.5}$$

An analogous derivation for the radiation leaving a diffuse dA_2 that arrives at dA_1 results in

$$d^2Q_{d2-d1} = \frac{I_2 dA_2 \cos\theta_2 dA_1 \cos\theta_1}{S^2} \tag{4.6}$$

Note that both Equations 4.5 and 4.6 contain the same geometric factors.

The *fraction* of energy leaving diffuse surface element dA_1 that arrives at element dA_2 is defined as the *geometric configuration factor* dF_{d1-d2}. The dash in the subscript means "to" (it is not a minus sign). Using Equation 4.5, the definition of dF gives

$$dF_{d1-d2} = \frac{I_1 \cos\theta_1 \cos\theta_2 dA_1 dA_2 / S^2}{\pi I_1 dA_1} = \frac{\cos\theta_1 \cos\theta_2}{\pi S^2} dA_2 \tag{4.7}$$

where $\pi I_1\, dA_1$ is the total diffuse energy leaving dA_1 within the entire hemispherical solid angle over dA_1. Equation 4.7 shows that dF_{d1-d2} *depends only on the size of* dA_2 *and its orientation with respect to* dA_1. By substituting Equation 4.4, Equation 4.7 can also be written as

$$dF_{d1-d2} = \frac{\cos\theta_1 d\Omega_1}{\pi} \tag{4.8}$$

Consequently, elements dA_2 have the same configuration factor if they subtend the same solid angle $d\Omega_1$ when viewed from dA_1 and are positioned along a path at angle θ_1 with respect to the normal of dA_1.

The factor dF_{d1-d2} has various names, being called the *view*, *angle*, *shape*, *interchange*, *exchange*, or *configuration factor*. The notation used here has subscripts designating the areas involved and a derivative consistent with the mathematical description. For the subscript notation, $d1$, $d2$, etc., indicate differential area elements, while 1, 2, etc., indicate finite areas. Thus, dF_{d1-d2} is a factor between two differential elements, and dF_{1-d2} is from finite area A_1 to differential area dA_2. The derivative dF indicates that the factor is for energy to a *differential* element; this keeps the mathematical form of equations such as Equation 4.7 consistent by having a differential quantity on both sides. The F denotes a factor to a *finite* area; thus, F_{d1-2} is from differential element dA_1 to finite area A_2.

4.2.1.1 Reciprocity for Differential Element Configuration Factors

By a derivation similar to that for Equation 4.7, the configuration factor for energy from dA_2 to dA_1 is

$$dF_{d2-d1} = \frac{\cos\theta_1 \cos\theta_2}{\pi S^2} dA_1 \tag{4.9}$$

Multiplying Equation 4.7 by dA_1 and Equation 4.9 by dA_2 gives the *reciprocity relation* between dF_{d1-d2} and dF_{d2-d1}

$$dF_{d1-d2} dA_1 = dF_{d2-d1} dA_2 = \frac{\cos\theta_1 \cos\theta_2}{\pi S^2} dA_1 dA_2 \tag{4.10}$$

4.2.1.2 Sample Configuration Factors between Differential Elements

The derivation of configuration factors in terms of system geometry parameters is now illustrated by a few examples.

Example 4.1

In Figure 4.2, two elemental areas are shown that are located on strips that have parallel generating lines. Derive an expression for the configuration factor between dA_1 and dA_2. The angle β is in the y–z plane.

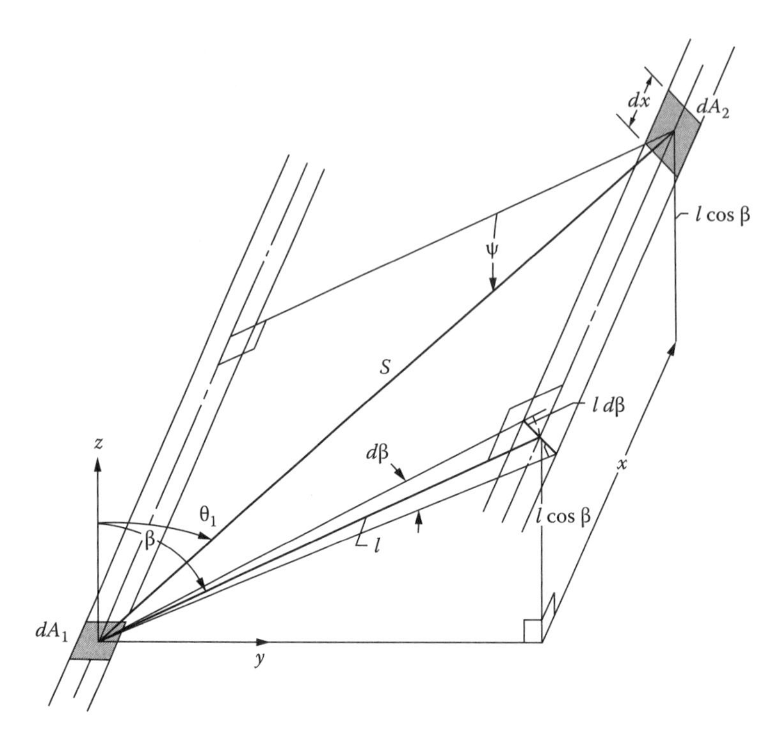

FIGURE 4.2 Geometry for configuration factor between elements on strips formed by parallel generating lines.

The y–z plane is normal to the generating lines. The distance $S = (l^2 + x^2)^{1/2}$, and $\cos\theta_1 = (l\cos\beta)/S = (l\cos\beta)/(l^2 + x^2)^{1/2}$. The angle β is in the y–z plane normal to the two strips. The solid angle subtended by dA_2, when viewed from dA_1, is

$$d\Omega_1 = \frac{\text{Projected area of } dA_2}{S^2} = \frac{(\text{Projected width of } dA_2)(\text{Projected length of } dA_2)}{S^2}$$

$$= \frac{(l d\beta)(dx \cos\psi)}{S^2} = \frac{l d\beta dx}{S^2}\frac{l}{S}$$

Substituting into Equation 4.8 gives

$$dF_{d1-d2} = \frac{\cos\theta_1 d\Omega_1}{\pi} = \frac{l\cos\beta}{(l^2 + x^2)^{1/2}}\frac{1}{\pi}\frac{l^2 d\beta dx}{(l^2 + x^2)^{3/2}} = \frac{l^3 \cos\beta d\beta dx}{\pi(l^2 + x^2)^2}$$

which is the desired configuration factor in terms of parameters that specify the geometry.

Example 4.2

Find the configuration factor between an elemental area and an infinitely long strip of differential width as in Figure 4.3, where the generating lines of dA_1 and $dA_{strip,2}$ are parallel.

Example 4.1 gave the configuration factor between element dA_1 and element dA_2 of length dx. To find the factor when dA_2 becomes an infinite strip, as in Figure 4.3, integrate over all x to obtain

$$dF_{d1-strip,2} = \frac{l^3 \cos\beta d\beta}{\pi}\int_{-\infty}^{\infty}\frac{dx}{(l^2 + x^2)^2} = \frac{l^3 \cos\beta d\beta}{\pi}\frac{\pi}{2l^3} = \frac{1}{2}d(\sin\beta)$$

where β is in the y–z plane normal to the strips.

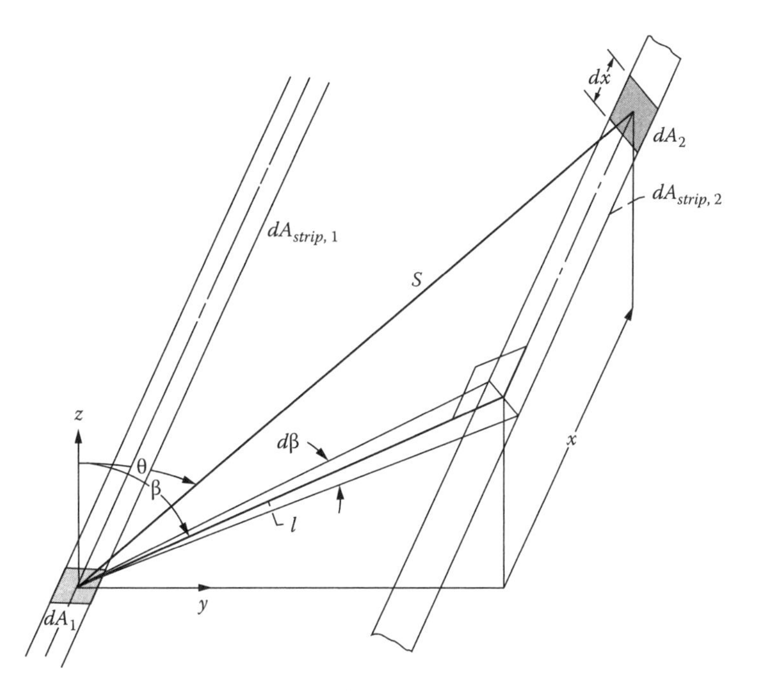

FIGURE 4.3 Geometry for configuration factor between elemental area and infinitely long strip of differential width; area and strip are on parallel generating lines.

Figure 4.3 also shows that since dA_1 lies on an *infinite* strip $dA_{strip,1}$ with elements parallel to $dA_{strip,2}$, the $dF_{d1-strip,2}$ is valid for dA_1 at any location along $dA_{strip,1}$. Then, since any element dA_1 on $dA_{strip,1}$ has the same fraction of its energy reaching $dA_{strip,2}$, it follows that the fraction of energy from the *entire* infinite $dA_{strip,1}$ that reaches $dA_{strip,2}$ is the same as the fraction for each element dA_1. Thus, the configuration factor between two infinitely long strips of differential width and having parallel generating lines must also be the same as for element dA_1 to $dA_{strip,2}$, or $(1/2)d(\sin\beta)$.

Example 4.3

Consider an infinitely long wedge-shaped groove as shown in cross section in Figure 4.4. Determine the configuration factor between the differential strips dx and $d\xi$ in terms of x, ξ, and α.

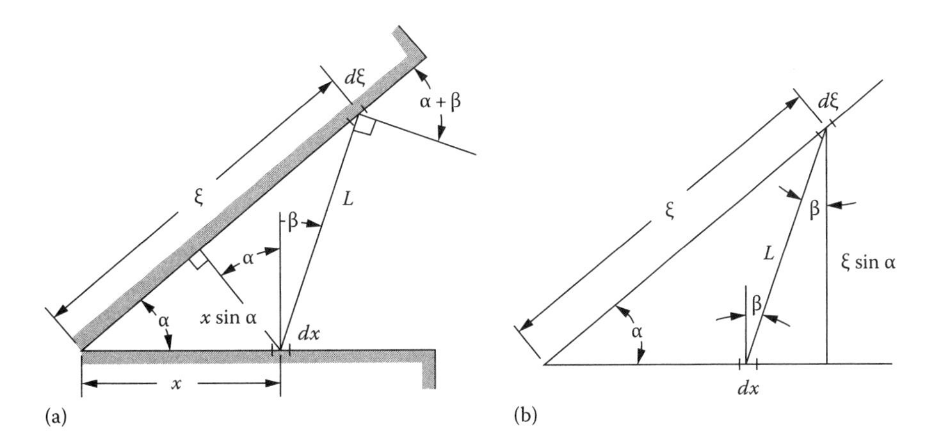

FIGURE 4.4 Configuration factor between two strips on sides of wedge groove: (a) wedge-shaped groove geometry and (b) auxiliary construction.

From Example 4.2, the configuration factor is

$$dF_{dx-d\xi} = \frac{1}{2}d(\sin\beta) = \frac{1}{2}\cos\beta d\beta$$

From the construction in Figure 4.4b, $\cos\beta = (\xi \sin\alpha)/L$. The $d\beta$ is the angle subtended by the projection of $d\xi$ normal to L, that is,

$$d\beta = \frac{d\xi\cos(\alpha+\beta)}{L} = \frac{d\xi}{L}\frac{x\sin\alpha}{L}$$

From the law of cosines, $L^2 = x^2 + \xi^2 - 2x\xi\cos\alpha$. Then

$$dF_{dx-d\xi} = \frac{1}{2}\cos\beta d\beta = \frac{1}{2}\frac{x\xi\sin^2\alpha}{L^3}d\xi = \frac{1}{2}\frac{x\xi\sin^2\alpha}{(x^2+\xi^2-2x\xi\cos\alpha)^{3/2}}d\xi$$

4.2.2 Configuration Factor between a Differential Area Element and a Finite Area

Consider an isothermal diffuse element dA_1 with uniform emissivity at temperature T_1 exchanging energy with a finite area A_2 that is isothermal at temperature T_2 as in Figure 4.5. The relations for exchange between two differential elements must be extended to a finite A_2. Figure 4.5 shows that θ_2 is different for different positions on A_2 and that θ_1 and S will also vary as different differential elements on A_2 are viewed from dA_1.

Two configuration factors need to be considered. The F_{d1-2} is from the differential area dA_1 to the finite area A_2, and dF_{2-d1} is from A_2 to dA_1. Each of these is obtained by evaluating the fraction of energy leaving one diffusely emitting and reflecting area that reaches the second area. Factor F_{d1-2} gives the radiation leaving dA_1 as $\pi I_1\, dA_1$. The energy reaching dA_2 on A_2 is $\pi I_1(\cos\theta_1\cos\theta_2/\pi S^2)dA_1\, dA_2$.

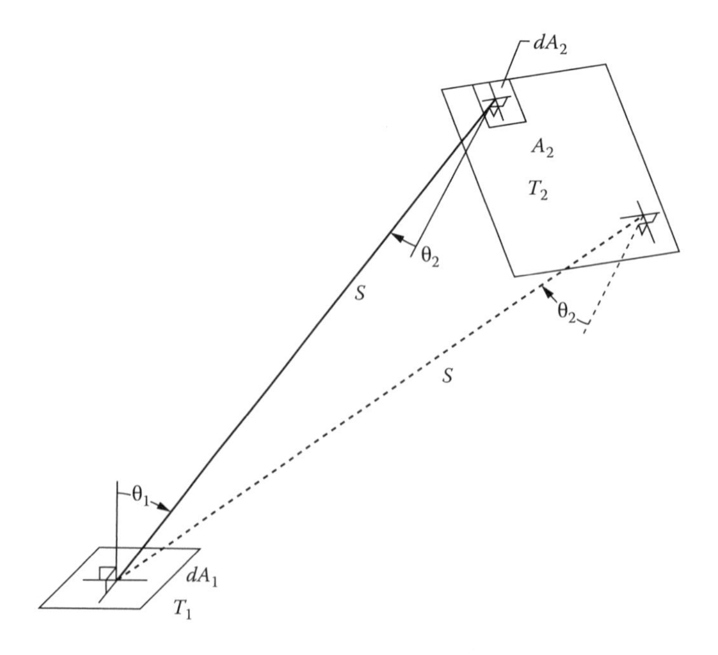

FIGURE 4.5 Radiant interchange between differential element and finite area.

Then, integrating over A_2 to obtain the energy reaching all of A_2 and dividing by the energy leaving dA_1 result in

$$F_{d1-2} = \frac{\int_{A_2} I_1 \cos\theta_1 (\cos\theta_2 dA_1/S^2) dA_2}{\pi I_1 dA_1} = \int_{A_2} \frac{\cos\theta_1 \cos\theta_2}{\pi S^2} dA_2 \tag{4.11}$$

From Equation 4.7, the quantity inside the integral of Equation 4.11 is dF_{d1-d2}, so F_{d1-2} can also be written as

$$F_{d1-2} = \int_{A_2} dF_{d1-d2} \tag{4.12}$$

This expresses the fact that the fraction of the energy reaching A_2 is the sum of the fractions reaching all the elements of A_2.

For obtaining the configuration factor from the finite area A_2 to the element dA_1, the energy leaving A_2 is $\int_{A_2} \pi I_2 dA_2$. The energy reaching dA_1 from A_2 is by integrating Equation 4.6 over $\int_{A_2} I_2(\cos\theta_1 \cos\theta_2/S^2)dA_2$. The configuration factor dF_{2-d1} is then

$$dF_{2-d1} = \frac{dA_1 \int_{A_2} I_2(\cos\theta_1 \cos\theta_2/S^2)dA_2}{\int_{A_2} \pi I_2 dA_2} = \frac{dA_1}{A_2} \int_{A_2} \frac{\cos\theta_1 \cos\theta_2}{\pi S^2} dA_2 \tag{4.13}$$

The last integral on the right was obtained by imposing the condition that A_2 has *uniform emitted plus reflected intensity I_2 over its entire area*. From Equation 4.7, the quantity under the integral sign in Equation 4.13 is dF_{d1-d2}, so the alternative form is obtained:

$$dF_{2-d1} = \frac{dA_1}{A_2} \int_{A_2} dF_{d1-d2} \tag{4.14}$$

4.2.2.1 Reciprocity Relation for Configuration Factors between Differential and Finite Areas

By use of Equation 4.12 to eliminate the integral in Equation 4.14, the *reciprocity relation* is obtained:

$$A_2 dF_{2-d1} = dA_1 dF_{d1-2} \tag{4.15}$$

4.2.2.2 Configuration Factors between Differential and Finite Areas

Certain geometries have configuration factors with closed-form algebraic expressions, while others require numerical integration of Equation 4.11. Two examples are now given to illustrate how factors that have algebraic forms are obtained.

Example 4.4

An element dA_1 is perpendicular to a circular disk A_2 with outer radius r as in Figure 4.6a. Find an equation for the configuration factor F_{d1-2} in terms of h, l, and r.

The first step is to find expressions for the quantities inside the integral of Equation 4.11 in terms of known quantities so that the integration can be carried out. The element $dA_2 = \rho d\rho\, d\phi$. Because

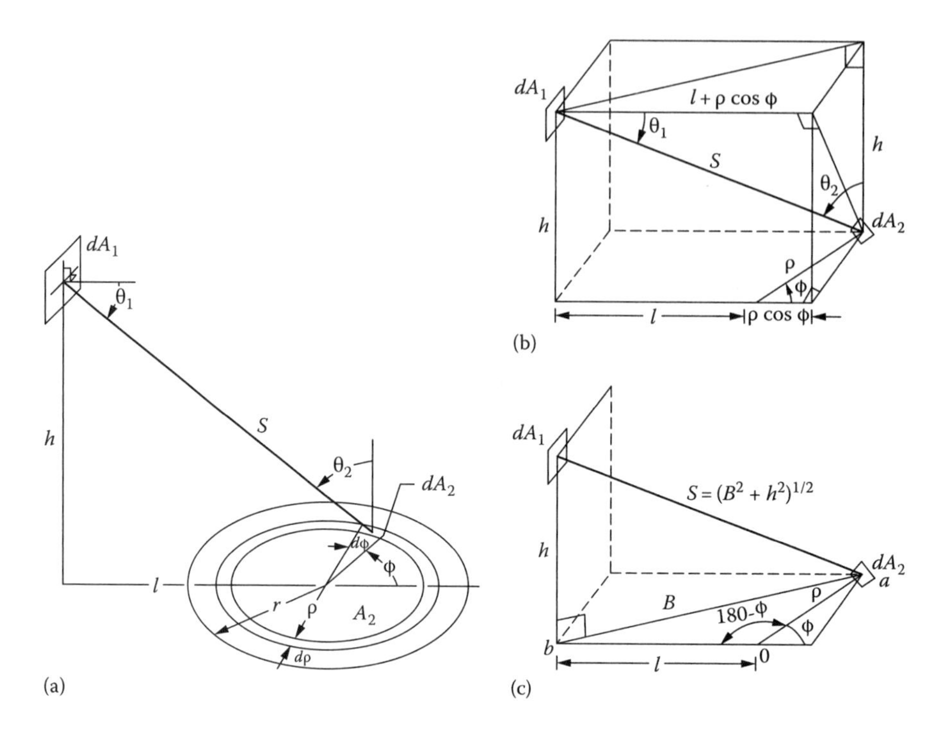

FIGURE 4.6 Geometry for radiative exchange between differential area and circular disk: (a) geometry of problem, (b) auxiliary construction for determining cos θ_1 and cos θ_2, and (c) auxiliary construction for determining S.

the integral in Equation 4.11 must be carried out over all ρ and ϕ, the other quantities in the integral must be put in terms of these variables.

Figure 4.6b is drawn to evaluate cos θ_1 and cos θ_2, which are $\cos\theta_1 = (l + \rho\cos\phi)/S$ and $\cos\theta_2 = h/S$. Figure 4.6c shows that $S^2 = h^2 + B^2$, where B^2 can be evaluated by using the law of cosines on triangle $a0b$. This gives $B^2 = l^2 + \rho^2 - 2l\rho\cos(\pi - \phi) = l^2 + \rho^2 + 2l\rho\cos\phi$. Substituting into Equation 4.11 results in

$$F_{d1-2} = \int_{A_2} \frac{\cos\theta_1\cos\theta_2}{\pi S^2} dA_2 = \int_{A_2} \frac{h(l+\rho\cos\phi)}{\pi S^4}\rho d\rho d\phi$$

$$= \frac{h}{\pi}\int_{\rho=0}^{r}\int_{\phi=0}^{2\pi} \frac{\rho(l+\rho\cos\phi)}{(h^2+l^2+\rho^2+2\rho l\cos\phi)^2} d\phi d\rho$$

This integration is carried out using the symmetry of the configuration and is nondimensionalized to give, after considerable manipulation,

$$F_{d1-2} = \frac{2h}{\pi}\int_{\rho=0}^{r}\int_{\phi=0}^{\pi} \frac{\rho(l+\rho\cos\phi)}{(h^2+\rho^2+l^2+2\rho l\cos\phi)^2} d\phi d\rho$$

$$= \frac{H}{2}\left\{\frac{H^2+R^2+1}{[(H^2+R^2+1)^2-4R^2]^{1/2}} - 1\right\}$$

where

$H = h/l$

$R = r/l$

To avoid the complicated double integration, the analysis can be carried out by contour integration as in Section 4.3.3.

Example 4.5

An infinitely long 2D wedge has an opening angle α. Derive an expression for the configuration factor from one wall of the wedge to a strip element of width dx on the other wall at x as in Figure 4.7a. Such configurations approximate the geometries of long fins and ribs used in radiators for devices in outer space.

The configuration factor between two infinitely long strip elements having parallel generating lines is used from Example 4.2. Note that β is measured clockwise from the normal of dx; Equation 4.5 then gives

$$F_{dx-1} = \int_{\xi=0}^{l} dF_{dx-d\xi} = \int_{\beta=-\pi/2}^{0} d(\sin\beta) + \int_{0}^{\beta'} \frac{1}{2} d(\sin\beta) = \frac{1}{2}(1+\sin\beta')$$

The function $\sin\beta'$ is found by the auxiliary construction in Figure 4.7b to be $\sin\beta' = B/C = (l\cos\alpha - x)/(x^2 + l^2 - 2xl\cos\alpha)^{1/2}$. Then

$$F_{dx-1} = \frac{1}{2} + \frac{l\cos\alpha - x}{2(x^2 + l^2 - 2xl\cos\alpha)^{1/2}}$$

The problem requires dF_{l-dx}. Using the reciprocal relation Equation 4.15 gives

$$dF_{l-dx} = \frac{dx}{l} F_{dx-l} = dX\left[\frac{1}{2} + \frac{\cos\alpha - X}{2(X^2 + 1 - 2X\cos\alpha)^{1/2}}\right]$$

where $X = x/l$. The only parameters are the wedge opening angle and the dimensionless position of dx from the vertex.

4.2.3 Configuration Factor and Reciprocity for Two Finite Areas

Consider the configuration factor for radiation with uniform intensity leaving the diffuse surface A_1 and reaching A_2 as shown in Figure 4.8. The energy leaving A_1 is $\pi I_1 A_1$. The radiation leaving an element dA_1 that reaches dA_2 was given previously as $\pi I_l(\cos\theta_1\cos\theta_2/\pi S^2)dA_1 dA_2$. This is integrated

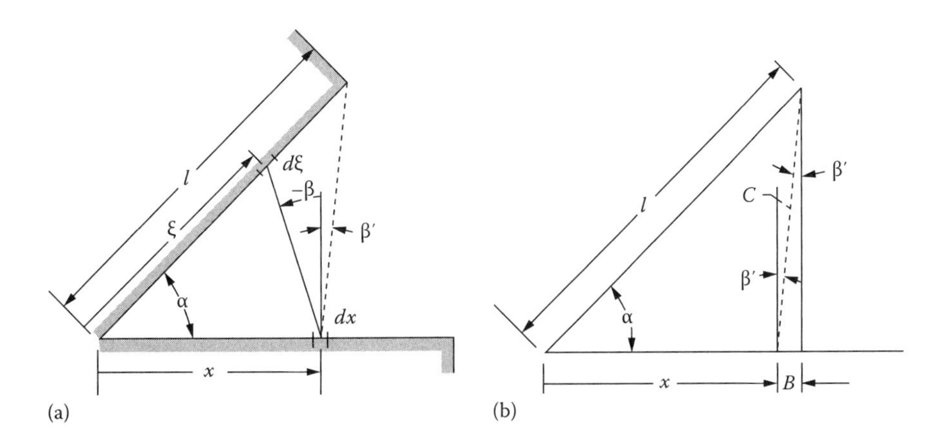

FIGURE 4.7 Configuration factor between one wall and strip on other wall of infinitely long wedge cavity: (a) wedge-cavity geometry and (b) auxiliary construction to determine $\sin\beta'$.

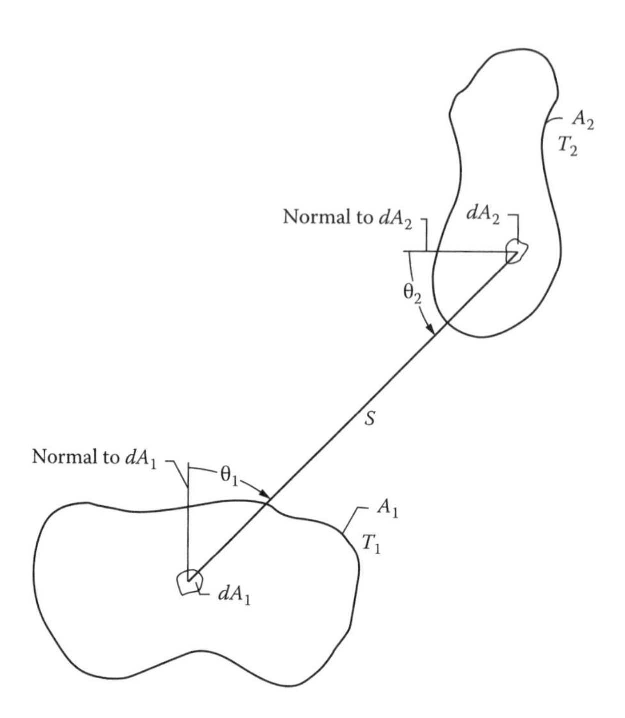

FIGURE 4.8 Geometry for energy exchange between finite areas.

over both A_1 and A_2 to give the energy leaving A_1 that reaches A_2. The configuration factor is then (since I_1 is constant)

$$F_{1-2} = \frac{\int_{A_1}\int_{A_2} \pi I_1(\cos\theta_1 \cos\theta_2 / \pi S^2) dA_2 dA_1}{\pi I_1 A_1} = \frac{1}{A_1}\int_{A_1}\int_{A_2} \frac{\cos\theta_1 \cos\theta_2}{\pi S^2} dA_2 dA_1 \quad (4.16)$$

This expression can be written in terms of configuration factors with differential areas:

$$F_{1-2} = \frac{1}{A_1}\int_{A_1}\int_{A_2} dF_{d1-d2}\, dA_1 = \frac{1}{A_1}\int_{A_1} F_{d1-2} dA_1 \quad (4.17)$$

A derivation similar to Equation 4.16 gives the configuration factor from A_2 to A_1 as

$$F_{2-1} = \frac{1}{A_2}\int_{A_1}\int_{A_2} \frac{\cos\theta_1 \cos\theta_2}{\pi S^2} dA_2 dA_1 \quad (4.18)$$

The *reciprocity relation* for configuration factors between finite areas is found from the identical double integrals in the aforementioned equations:

$$A_1 F_{1-2} = A_2 F_{2-1} \quad (4.19)$$

Further relations among configuration factors are found by using Equation 4.17 in conjunction with the reciprocity relations Equations 4.15 and 4.19:

$$F_{2-1} = \frac{A_1}{A_2} F_{1-2} = \frac{A_1}{A_2}\frac{1}{A_1}\int_{A_1} F_{d1-2} dA_1 = \frac{1}{A_2}\int_{A_1} dF_{2-d1} A_2 = \int_{A_1} dF_{2-d1} \quad (4.20)$$

TABLE 4.1
Summary of Configuration Factor and Reciprocity Relations

Geometry	Configuration Factor	Reciprocity
Elemental area to elemental area	$dF_{d1-d2} = \frac{\cos\theta_1 \cos\theta_2}{\pi S^2} dA_2$	$dA_1 dF_{d1-d2} = dA_2 dF_{d2-d1}$
Elemental area to finite area	$F_{d1-2} = \int_{A_2} \frac{\cos\theta_1 \cos\theta_2}{\pi S^2} dA_2$	$dA_1 F_{d1-2} = A_2 dF_{2-d1}$
Finite area to finite area	$F_{1-2} = \frac{1}{A_1} \int_{A_1}\int_{A_2} \frac{\cos\theta_1 \cos\theta_2}{\pi S^2} dA_2 dA_1$	$A_1 F_{1-2} = A_2 F_{2-1}$

Example 4.6

Two plates of the same finite width and of infinite length are joined along one edge at angle α as in Figure 4.7. Using the same nondimensional parameters as in Example 4.5, derive the configuration factor between the plates.

Example 4.5 gives the configuration factor dF_{l-dx} between one plate and an infinite strip on the other plate. Substituting into Equation 4.20 gives

$$F_{l-l^*} = \int_{x=0}^{l^*} dF_{l-dx} = \int_0^l \left[\frac{1}{2} + \frac{\cos\alpha - X}{2(X^2 + 1 - 2X\cos\alpha)^{1/2}}\right] dX$$

where the width of the side in Figure 4.7 having element dx is designated as l^*, $X = x/l$, and $l^* = l$. Upon integration yields

$$F_{l-l^*} = 1 - \left(\frac{1-\cos\alpha}{2}\right)^{1/2} = 1 - \sin\frac{\alpha}{2}$$

For the present case with equal plate widths, the only parameter is α; hence, the configuration factor is the same for plates of any equal width.

Table 4.1 summarizes the integral definitions of the configuration factors and the configuration factor reciprocity relations.

4.3 METHODS FOR DETERMINING CONFIGURATION FACTORS

4.3.1 Configuration Factor Algebra

Configuration factor algebra is the manipulation of various relations among configuration factors to derive new factors from those already known. The algebra is based on concepts such as reciprocal relations, the F factor being a fraction of energy that is intercepted, and energy conservation for a complete enclosure.

Consider an arbitrary isothermal area A_1 in Figure 4.9 exchanging energy with a second area A_2. The F_{1-2} is the fraction of diffuse energy leaving A_1 that is incident on A_2. If A_2 is divided into A_3 and A_4, the fractions of the energy leaving A_1 that are incident on A_3 and A_4 must add to F_{1-2}:

$$F_{1-2} = F_{1-(3+4)} = F_{1-3} + F_{1-4} \tag{4.21}$$

$$F_{1-3} = F_{1-2} - F_{1-4} \tag{4.22}$$

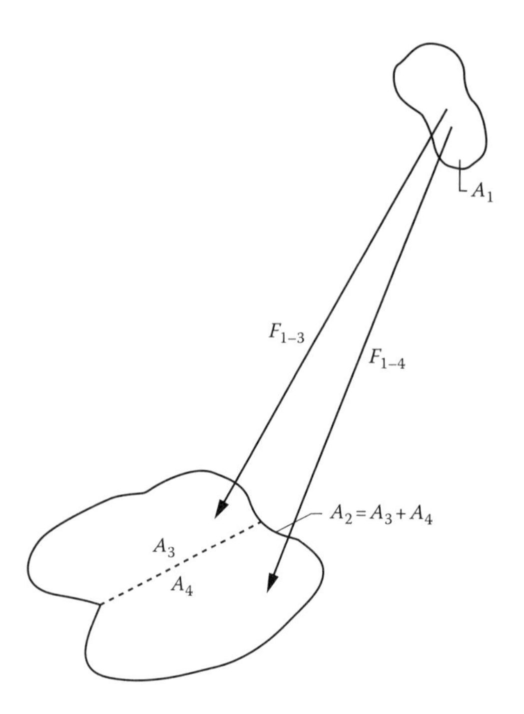

FIGURE 4.9 Energy exchange between two finite areas with one area subdivided: $F_{1-3} + F_{1-4} = F_{1-2}$.

The reciprocity relation, Equation 4.19, gives

$$F_{3-1} = \frac{A_1}{A_3} F_{1-3} = \frac{A_1}{A_3}(F_{1-2} - F_{1-4}) \tag{4.23}$$

In Equations 4.21 through 4.23, the dashes in the subscript mean "to" and are not to be confused with negative signs. Thus, F_{1-2} means from area A_1 to area A_2. Combined areas are grouped with parentheses so that (3 + 4) means the combination of A_3 and A_4. Thus, the notation $F_{1-(3+4)}$ means from A_1 to the combined areas of A_3 and A_4.

The reciprocity relation of Equation 4.23 is a powerful computational tool for obtaining new configuration factors from those available. The following examples demonstrate how this is done.

Example 4.7

An elemental area dA_1 is perpendicular to a ring of outer radius r_o and inner radius r_i as in Figure 4.10. Derive an expression for the configuration factor $F_{d1-ring}$.

From Example 4.4, the configuration factor between dA_1 and the entire disk of area A_2 with outer radius r_o is

$$F_{d1-2} = \frac{H}{2}\left\{\frac{H^2 + R_o^2 + 1}{\left[\left(H^2 + R_o^2 + 1\right)^2 - 4R_o^2\right]^{1/2}} - 1\right\}$$

where

$H = h/l$

$R_o = r_o/l$

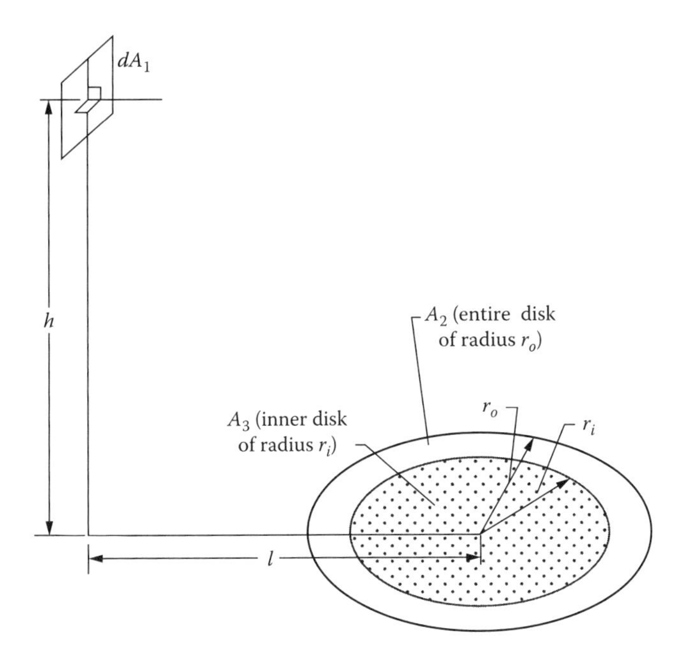

FIGURE 4.10 Energy exchange between elemental area and finite ring.

The configuration factor to the inner disk of area A_3 with radius r_i has the same form, with $R_i = r_i/l$ substituted for R_o. Using configuration factor algebra, the desired configuration factor from dA_1 to the ring $A_2 - A_3$ is $F_{d1-2} - F_{d1-3}$, so that

$$F_{d1-ring} = \frac{H}{2}\left\{\frac{H^2 + R_o^2 + 1}{\left[\left(H^2 + R_o^2 + 1\right)^2 - 4R_o^2\right]^{1/2}} - \frac{H^2 + R_i^2 + 1}{\left[\left(H^2 + R_i^2 + 1\right)^2 - 4R_i^2\right]^{1/2}}\right\}$$

Example 4.8

Suppose the configuration factor is known between two parallel disks of arbitrary size whose centers lie on the same axis. From this, derive the configuration factor between the two rings A_2 and A_3 of Figure 4.11. Give the result in terms of known disk-to-disk factors from disk areas on the lower surface to disk areas on the upper surface.

The factor desired is F_{2-3}. From configuration factor algebra, F_{2-3} is equal to $F_{2-3} = F_{2-(3+4)} - F_{2-4}$. The factor $F_{2-(3+4)}$ can be found from the reciprocal relation $A_2\,F_{2-(3+4)} = (A_3 + A_4)F_{(3+4)-2}$. Applying configuration factor algebra to the right-hand side results in

$$A_2F_{2-(3+4)} = (A_3 + A_4)[F_{(3+4)-(1+2)} - F_{(3+4)-1}]$$
$$= (A_3 + A_4)F_{(3+4)(1+2)} - (A_3 + A_4)F_{(3+4)-1}$$

Applying reciprocity to the right side gives

$$A_2F_{2-(3+4)} = (A_1 + A_2)F_{(1+2)-(3+4)} - A_1F_{1-(3+4)}$$

where the F factors on the right are both disk-to-disk factors from the lower surface to the upper.

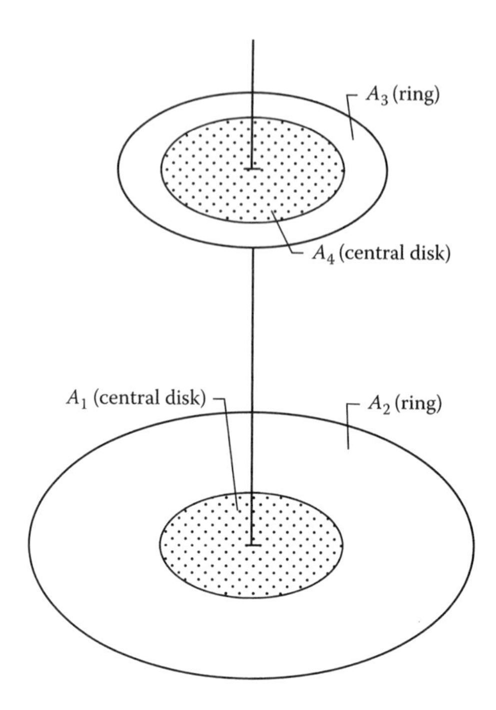

FIGURE 4.11 Energy exchange between parallel ring areas having a common axis.

Next, the factor F_{2-4} is to be determined. Applying reciprocity relations and configuration factor algebra,

$$F_{2-4} = \frac{A_4}{A_2}F_{4-2} = \frac{A_4}{A_2}[F_{4-(1+2)} - F_{4-1}] = \frac{1}{A_2}[(A_1 + A_2)F_{(1+2)-4} - A_1F_{1-4}]$$

Substituting the relations for F_{2-4} and $F_{2-(3+4)}$ into the first equation gives

$$F_{2-3} = \frac{A_1 + A_2}{A_2}[F_{(1+2)-(3+4)} - F_{(1+2)-4}] - \frac{A_1}{A_2}[F_{1-(3+4)} - F_{1-4}]$$

and all configuration factors on the right-hand side are for exchange between two disks from the lower surface to the upper surface.

Because small differences in large numbers can occur in obtaining an F factor by using configuration factor algebra, as might occur on the right side of the last equation of the preceding example, care must be taken that enough significant figures are retained. Feingold (1966) gives one example in which an error of 0.05% in a known factor causes an error of 57% in another factor computed from it by using configuration factor algebra.

Example 4.9

The internal surface of a hollow circular cylinder of radius R is radiating to a disk A_1 of radius r as in Figure 4.12. Express the configuration factor from the internal cylindrical side A_3, to the disk in terms of disk-to-disk factors for the case of $r \leq R$.

From any position on A_1, the solid angle subtended when viewing A_3 is the difference between the solid angle when viewing A_2, or $d\Omega_2$, and that viewing A_4, or $d\Omega_4$. This gives the F_{d1-3} factor from an area element dA_1 on A_1 to area A_3 as $F_{d1-3} = F_{d1-2} - F_{d1-4}$. By integrating over A_1 and

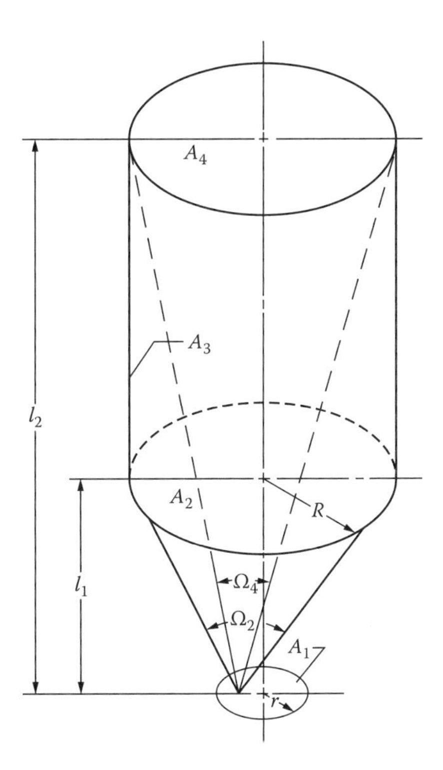

FIGURE 4.12 Geometry for a cylindrical cavity radiating to circular disk A_1 for $r \leq R$.

using Equation 4.17, this can be written for the entire A_1 as $F_{d1-3} = F_{d1-2} - F_{d1-4}$. The factors on the right are between parallel disks. The final result for F from the internal cylindrical side A_3 to the disk A_1 is

$$F_{3-1} = \frac{A_1}{A_3}(F_{1-2} - F_{1-4}) \quad (r \leq R)$$

From symmetry, the configuration factor from A_1 to *any sector* A_s of A_1 is $(A_s/A_1)F_{3-1}$.

In formulating relations between configuration factors, it is sometimes useful to use *energy quantities* rather than fractions of energy leaving a surface that reach another surface. For example, in Figure 4.9, the energy leaving A_2 that arrives at A_1 is proportional to A_2F_{2-1} and is equivalent to the sums of the energies from A_3 and A_4 that arrive at A_1. Thus,

$$(A_3 + A_4)F_{(3+4)-1} = A_3F_{3-1} + A_4F_{4-1} \tag{4.24}$$

This can also be proved by using reciprocity relations:

$$(A_3 + A_4)F_{(3+4)-1} = A_1F_{1-(3+4)} = A_1F_{1-3} + A_1F_{1-4} = A_3F_{3-1} + A_4F_{4-1}$$

4.3.1.1 Configuration Factors Determined by the Use of Symmetry

A reciprocity relation can be derived from the symmetry of a geometry. Consider the opposing areas in Figure 4.13a. From symmetry, $A_2 = A_4$ and $F_{2-3} = F_{4-1}$, so $A_2F_{2-3} = A_4F_{4-1}$. From reciprocity, $A_4F_{4-1} = A_1F_{1-4}$. Hence, the relation

$$A_2F_{2-3} = A_1F_{1-4} \tag{4.25}$$

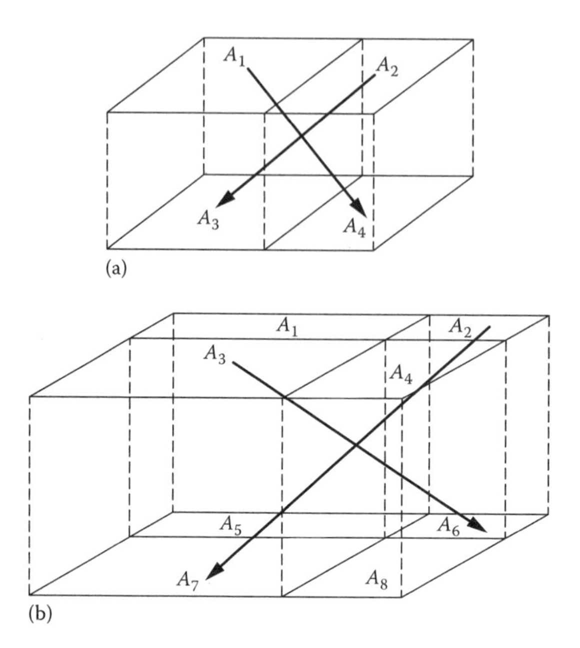

FIGURE 4.13 Geometry to determine reciprocity relations between opposing rectangles: (a) two pairs of opposing rectangles, $A_1F_{1-4} = A_2F_{2-3}$, and (b) four pairs of opposing rectangles, $A_2F_{2-7} = A_3F_{3-6}$.

relates the diagonal directions shown by the arrows. Similarly, the symmetry of Figure 4.13b yields

$$A_2F_{2-7} = A_3F_{3-6} \tag{4.26}$$

Figure 4.14a shows four areas on two perpendicular rectangles having a common edge. Since these areas are of unequal size, there is no apparent symmetry relation. However, a valid relation is

$$A_1F_{1-2} = A_3F_{3-4} \tag{4.27}$$

To prove this, begin with the basic definition, Equation 4.16,

$$A_1F_{1-2} = \frac{1}{\pi}\int_{A_1}\int_{A_2} \frac{\cos\theta_1 \cos\theta_2}{S^2}\, dA_2 dA_1$$

From Figure 4.14b, $S^2 = (x_2 - x_1)^2 + y_1^2 + z_2^2$, $\cos\theta_1 = z_2/S$, and $\cos\theta_2 = y_1/S$ so that

$$A_1F_{1-2} = \frac{1}{\pi}\int_{x_1=0}^{c}\int_{y_1=0}^{a}\int_{x_2=c}^{c+d}\int_{z_2=0}^{b} \frac{y_1 z_2}{\left[(x_2 - x_1)^2 + y_1^2 + z_2^2\right]^2}\, dz_2 dx_2 dy_1 dx_1 \tag{4.28}$$

Similarly, Figure 4.14c reveals that

$$A_3F_{3-4} = \frac{1}{\pi}\int_{A_3}\int_{A_4} \frac{\cos\theta_3 \cos\theta_4}{S^2}\, dA_4 dA_3$$

$$= \frac{1}{\pi}\int_{x_3=c}^{c+d}\int_{y_3=0}^{a}\int_{x_4=0}^{c}\int_{z_4=0}^{b} \frac{y_3 z_4}{\left[(x_3 - x_4)^2 + y_3^2 + z_4^2\right]^2}\, dz_4 dx_4 dy_3 dx_3 \tag{4.29}$$

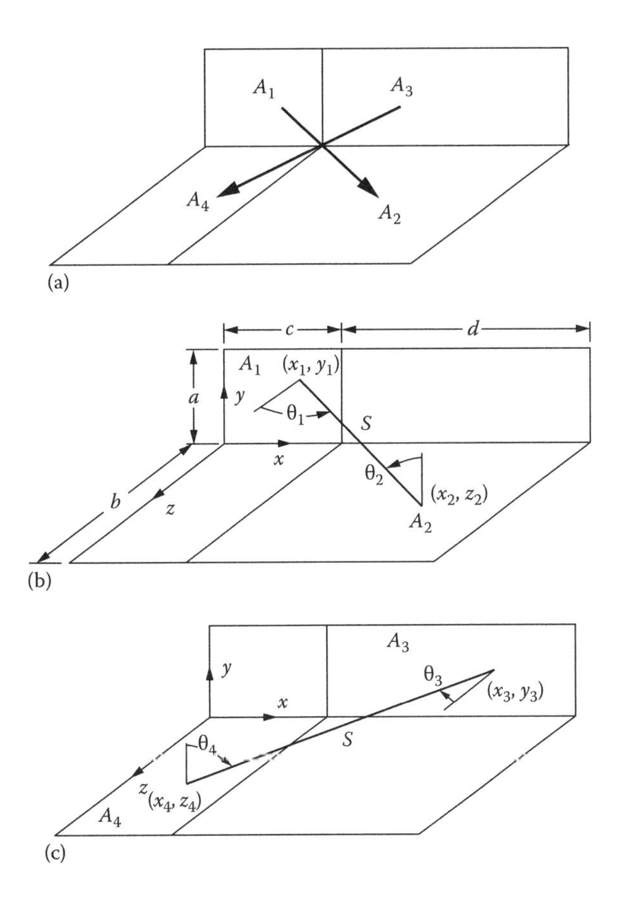

FIGURE 4.14 Geometry for application of reciprocity relations among diagonally opposite pairs of rectangles on two perpendicular planes having a common edge: (a) representation of reciprocity, $A_1F_{1\ 2} = A_3F_{3-4}$, (b) construction for F_{1-2}, and (c) construction for F_{3-4}.

By interchanging the dummy integration variables x_1, y_1, x_2, and z_2 for x_4, y_3, x_3, and z_4, the integrals in Equations 4.28 and 4.29 are identical, thus proving Equation 4.27. Lebedev (1988) extends these diagonal relations to nonplanar surfaces: two infinitely long parallel cylindrical areas of finite width and two coaxial surfaces of rotation.

Example 4.10

If the configuration factor is known for two perpendicular rectangles with a common edge as in Figure 4.15a, derive F_{1-6} for Figure 4.15b.

If F_{1-2} and F_{1-4} are known and F_{3-1} is desired, then first consider the geometry in Figure 4.15c and derive F_{7-6} as follows:

$$F_{(5+6)-(7+8)} = F_{(5+6)-7} + F_{(5+6)-8} = \frac{A_7}{A_5 + A_6}F_{7-(5+6)} + \frac{A_8}{A_5 + A_6}F_{8-(5+6)}$$

$$= \frac{A_7}{A_5 + A_6}(F_{7-5} + F_{7-6}) + \frac{A_8}{A_5 + A_6}(F_{8-5} + F_{8-6})$$

Substitute A_7F_{7-6} for A_8F_{8-5} and solve the resulting relation for F_{7-6} to obtain

$$F_{7-6} = \frac{1}{2A_7}[(A_5 + A_6)F_{(5+6)-(7+8)} - A_7F_{7-5} - A_8F_{8-6}]$$

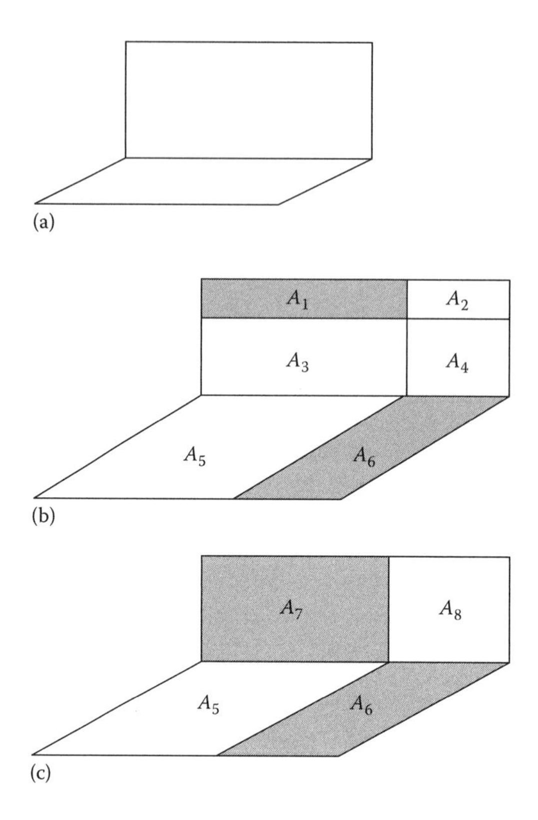

FIGURE 4.15 Orientation of areas for Example 4.10: (a) perpendicular rectangles with one common edge, (b) geometry for F_{1-6}, and (c) auxiliary geometry.

Now, in Figure 4.15b

$$F_{1-6} = \frac{A_6}{A_1}F_{6-1} = \frac{A_6}{A_1}F_{6-(1+3)} - \frac{A_6}{A_1}F_{6-3}$$

The factors $F_{6-(1+3)}$ and F_{6-3} are of the same type as F_{7-6} so F_{1-6} can finally be written as

$$F_{1-6} = \frac{A_6}{A_1}\left\{\frac{1}{2A_6}[(A_1 + A_2 + A_3 + A_4)F_{(1+2+3+4)-(5+6)} - A_6F_{6-(2+4)} - A_5F_{5-(1+3)}] - \frac{1}{2A_6}[(A_3 + A_4)F_{(3+4)-(5+6)} - A_6F_{6-4} - A_5F_{5-3}]\right\}$$

All the F factors on the right side are for two rectangles having one common edge as in Figure 4.15a.

4.3.2 Configuration Factor Relations in Enclosures

So far, only configuration factors between two surfaces have been considered, although subdivision of one or both surfaces into smaller areas has been examined. Consider the very important situation where the configuration factors are between surfaces that form a complete enclosure.

For an enclosure of N surfaces, such as in Figure 4.16 where $N = 8$ as an example, the entire energy leaving any surface k inside the enclosure must be incident on all surfaces of the enclosure.

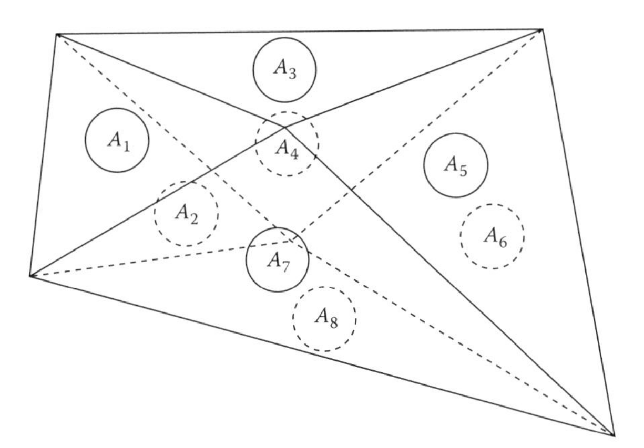

FIGURE 4.16 Geometry for an isothermal enclosure composed of black surfaces.

Thus, all the fractions of energy leaving one surface and reaching the surfaces of the enclosure must total to one:

$$F_{k-1} + F_{k-2} + F_{k-3} + \cdots + F_{k-k} + \cdots + F_{k-N} = \sum_{j=1}^{N} F_{k-j} = 1 \tag{4.30}$$

The factor F_{k-k} is included because when A_k is *concave*, it will intercept a portion of its own outgoing energy.

Example 4.11

Two diffuse isothermal concentric spheres are exchanging energy. Find all the configuration factors for this geometry if the surface area of the inner sphere is A_1 and the surface area of the outer sphere is A_2.

All energy leaving A_1 is incident upon A_2, so $F_{1-2} = 1$. Using the reciprocal relation gives $F_{2-1} = A_1F_{1-2}/A_2 = A_1/A_2$. From Equation 4.30, $F_{2-1} + F_{2-2} = 1$, giving $F_{2-2} = 1 - F_{2-1} = (A_2 - A_1)/A_2$.

Example 4.12

An isothermal cavity of internal area A_1 has a plane opening of area A_2. Derive an expression for the configuration factor of the internal surface of the cavity to itself.

Assume that a black plane surface A_2 replaces the cavity opening; this has no effect on the F_{1-1} factor, which depends only on geometry for diffuse surfaces. Then $F_{2-1} = 1$ and $F_{1-2} = A_2F_{2-1}/A_1 = A_2/A_1$, which is the configuration factor from the entire internal area to the opening. Since A_1 and A_2 form an enclosure, $F_{1-1} = 1 - F_{1-2} = (A_1 - A_2)/A_1$, which is the desired F factor.

Example 4.13

An enclosure of triangular cross section consists of three plane areas, each of finite width and infinite length, thus forming a hollow infinitely long triangular prism. Derive an expression for the configuration factor between any two of the areas in terms of their widths L_1, L_2, and L_3.

For area 1, $F_{1-2} + F_{1-3} = 1$ since $F_{1-1} = 0$. Using similar relations for each area and multiplying through by the respective areas result in

$$A_1F_{1-2} + A_1F_{1-3} = A_1; \quad A_2F_{2-1} + A_2F_{2-3} = A_2; \quad A_3F_{3-1} + A_{3-2} = A_3$$

By applying reciprocal relations to some terms, these equations become

$$A_1F_{1-2} + A_1F_{1-3} = A_1; \quad A_1F_{1-2} + A_2F_{2-3} = A_2; \quad A_1F_{1-3} + A_2F_{2-3} = A_3$$

thus giving three equations for the three unknown F factors. Solving gives

$$F_{1-2} = \frac{A_1 + A_2 - A_3}{2A_1} = \frac{L_1 + L_2 - L_3}{2L_1}$$

When $L_1 = L_2$, this reduces to the factor between infinitely long adjoint plates of equal width separated by an angle α as in Example 4.6:

$$F_{1-2} = \frac{2L_1 - L_3}{2L_1} = 1 - \frac{L_3/2}{L_1} = 1 - \sin\frac{\alpha}{2}$$

The set of three simultaneous equations yielding the final result in Example 4.13 is now examined more closely. The first equation involves two unknowns, F_{1-2} and F_{1-3}; the second has one additional unknown F_{2-3}; and the third has no additional unknowns. Generalizing the procedure from a three-surface enclosure to any N-sided enclosure of plane or convex surfaces shows that, of N simultaneous equations, the first involves N–1 unknowns, the second N–2 unknowns, and so forth. The total number of unknowns U is then

$$U = (N-1) + (N-2) + \cdots 1 = \frac{N(N-1)}{2} \tag{4.31}$$

Thus, $[N(N-1)/2] - N = N(N-3)/2$ factors must be provided. For a four-sided enclosure made up of planar or convex areas, four equations relating $4(4-1)/2 = 6$ unknown configuration factors can be written. Specifying any two factors allows calculation of the rest by solving the four simultaneous equations.

If any surface k can view itself, the factor F_{k-k} must be included in each of the equations. Analyzing this situation shows that an N-sided enclosure provides N equations in $N(N+1)/2$ unknowns. Thus, $[N(N+1)/2] - N = N(N-1)/2$ factors must be specified. For a four-sided enclosure, 4 equations involving 10 unknown F factors could be written. The specification of six factors is required; then the simultaneous relations could be solved for the remaining four factors. If only M surfaces can view themselves, then $[N(N-3)/2] + M$ factors must be specified. For some geometries, the use of symmetry can reduce the number of factors that are required. Use of these relations is discussed further in Section 4.4.

4.3.3 Techniques for Evaluating Configuration Factors

Evaluating configuration factors F_{d1-2} and F_{1-2} requires integration over the finite areas involved. A number of mathematical methods are useful for evaluating configuration factors when analytical integration is too cumbersome or is not possible. A few methods that are especially valuable are discussed next.

4.3.3.1 Hottel's Crossed-String Method

Consider configurations such as long grooves in which all surfaces are assumed to extend infinitely along one coordinate. Such surfaces can be generated by moving a line such that it is always parallel to its original position. A typical configuration is shown in cross section in Figure 4.17. Suppose that the configuration factor is needed between A_1 and A_2 when some blockage of radiant transfer occurs because of other surfaces A_3 and A_4. To obtain F_{1-2}, first consider that A_1 may be concave. In this case, draw the dashed line *agf* across A_1. Then draw in the dashed lines *cf* and *abc* to complete

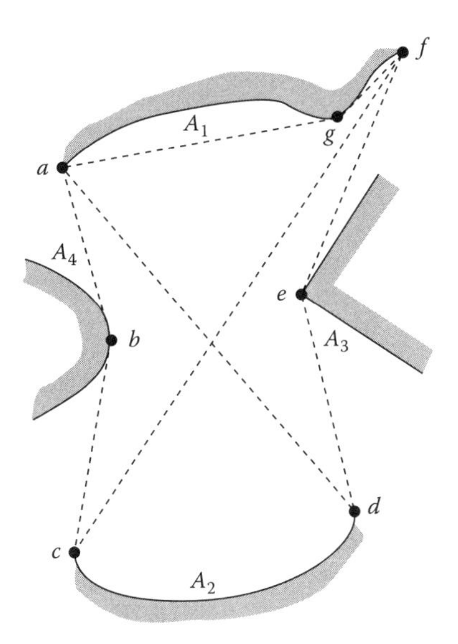

FIGURE 4.17 Geometry for determining configuration factors using Hottel's crossed-string method.

the enclosure *abcfga*, which has three sides that are either convex or planar. The relation found in Example 4.13 for enclosures of this type can be written as

$$A_{agf}F_{agf-abc} = \frac{A_{agf} + A_{abc} - A_{cf}}{2} \tag{4.32}$$

For the three-sided enclosure *adefga*, similar reasoning gives

$$A_{agf}F_{agf-def} = \frac{A_{agf} + A_{def} - A_{ad}}{2} \tag{4.33}$$

Further, note that

$$F_{agf-abc} + F_{agf-2} + F_{agf-def} = 1 \tag{4.34}$$

Substituting Equations 4.32 and 4.33 into Equation 4.34 results in

$$A_{agf}F_{agf-2} = A_{agf}(1 - F_{agf-abc} - F_{agf-def}) = \frac{A_{cf} + A_{ad} - A_{abc} - A_{def}}{2} \tag{4.35}$$

Now $F_{2-agf} = F_{2-1}$ since A_{agf} and A_1 subtend the same solid angle when viewed from A_2. Then, with the additional use of reciprocity, the left side of Equation 4.35 can be written as

$$A_{agf}F_{agf-2} = A_2F_{2-agf} = A_2F_{2-1} = A_1F_{1-2} \tag{4.36}$$

Substituting Equation 4.36 into Equation 4.35 results in

$$A_1F_{1-2} = \frac{A_{cf} + A_{ad} - A_{abc} - A_{def}}{2} \tag{4.37}$$

If the dashed lines in Figure 4.17 are imagined as being lengths of string stretched tightly between the outer edges of the surfaces, then the term on the right of Equation 4.37 is interpreted as *one-half the total quantity formed by the sum of the lengths of the crossed strings connecting the outer edges of A_1 and A_2 minus the sum of the lengths of the uncrossed strings.* This is a very useful way for determining configuration factors, but is limited to a two-dimensional (2D) geometry. It was first described by Hottel (1954).

Example 4.14

Two infinitely long semicylindrical surfaces of radius R are separated by a minimum distance D as in Figure 4.18a. Derive the configuration factor F_{1-2}.

The length of crossed string *abcde* is denoted as L_1 and that of uncrossed string *ef* as L_2. From the symmetry of the geometry, Equation 4.37 gives

$$F_{1-2} = \frac{2L_1 - 2L_2}{2\pi R} = \frac{L_1 - L_2}{\pi R}$$

Then, $L_2 = D + 2R$. The L_1 is twice the length *cde*. The segment of L_1, from *c* to *d*, is found from right triangle 0cd to be

$$L_{1,c-d} = \left[\left(\frac{D}{2}+R\right)^2 - R^2\right]^{1/2} = \left[D\left(\frac{D}{4}+R\right)\right]^{1/2}$$

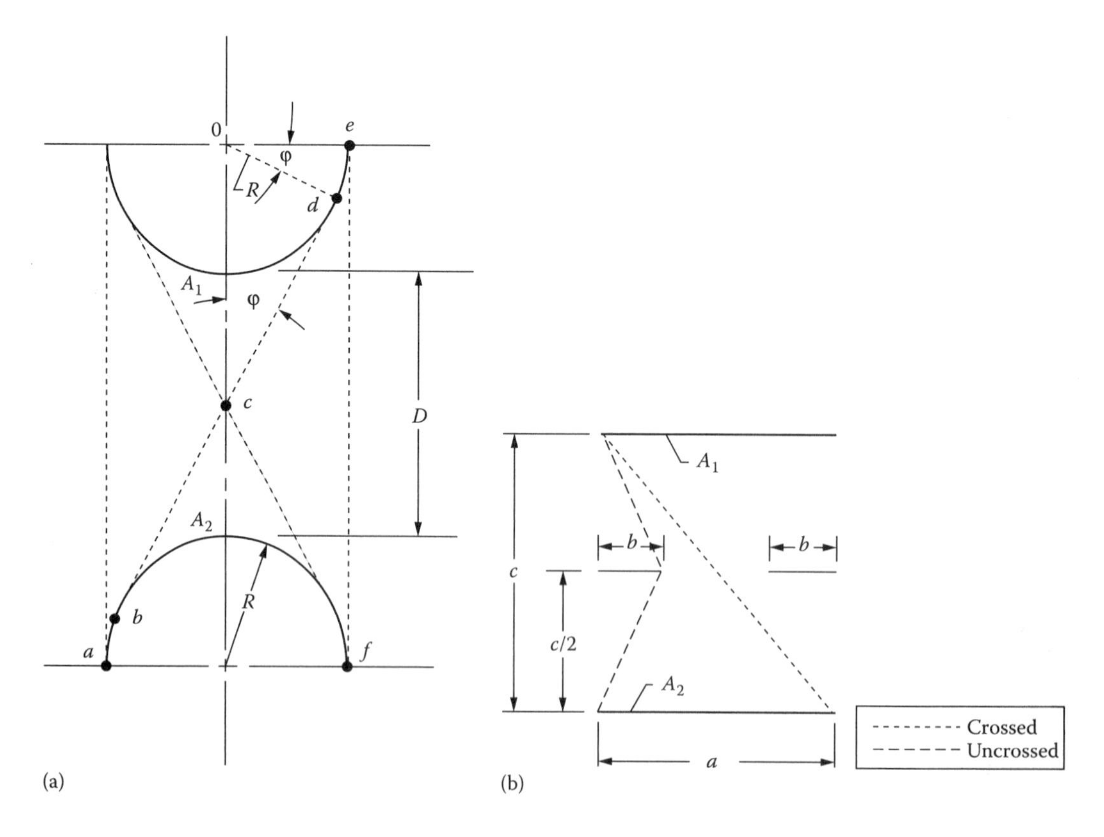

FIGURE 4.18 Examples of applying the crossed-string method: (a) configuration factor between infinitely long semicylindrical surfaces and (b) partially blocked view between parallel strips.

The segment of L_1 from d to e is $L_{1,\,d\text{–}e} = R\varphi$ where, from triangle 0cd, $\varphi = \sin^{-1}[R/(D/2 + R)]$. Then

$$F_{1-2} = \frac{L_1 - L_2}{\pi R} = \frac{2(L_{1,c-d} + L_{1,d-e}) - L_2}{\pi R}$$

$$= \frac{[4D(D/4 + R)]^{1/2} + 2R\sin^{-1}[R/(D/2 + R)] - D - 2R}{\pi R}$$

Letting $X = 1 + D/2R$ gives

$$F_{1-2} = \frac{2}{\pi}\left[(X^2 - 1)^{1/2} + \sin^{-1}\frac{1}{X} - X\right] \quad (4.14.1)$$

Example 4.15

The view between two infinitely long parallel strips of width a is partially blocked by strips of width b as in Figure 4.18b. Obtain the factor F_{1-2}.

The length of each crossed string is $(a^2 + c^2)^{1/2}$, and the length of each uncrossed string is $2[b^2 + (c/2)^2]^{1/2}$. From the crossed-string method the configuration factor is then

$$F_{1-2} = \frac{\sqrt{a^2 + c^2} - 2\sqrt{b^2 + (c/2)^2}}{a} = \sqrt{1 + \left(\frac{c}{a}\right)^2} - \sqrt{\left(\frac{2b}{a}\right)^2 + \left(\frac{c}{a}\right)^2}$$

As expected, $F_{1-2} \to 0$ as b is extended inward so that $b \to a/2$.

Example 4.16

The view between two infinitely long parallel strips of width c and spaced c apart is partially blocked by a thin plate of width $c/2$ as in Figure 4.19. Obtain the configuration factor F_{1-2}.

Following the idea of a partially blocked view as in Figure 4.19, area A_1 can view A_2 either to the right or to the left of the center plate. Figure 4.19a shows the strings drawn to the right side by considering the region to the left of the center plate to be obstructed. Figure 4.19b shows

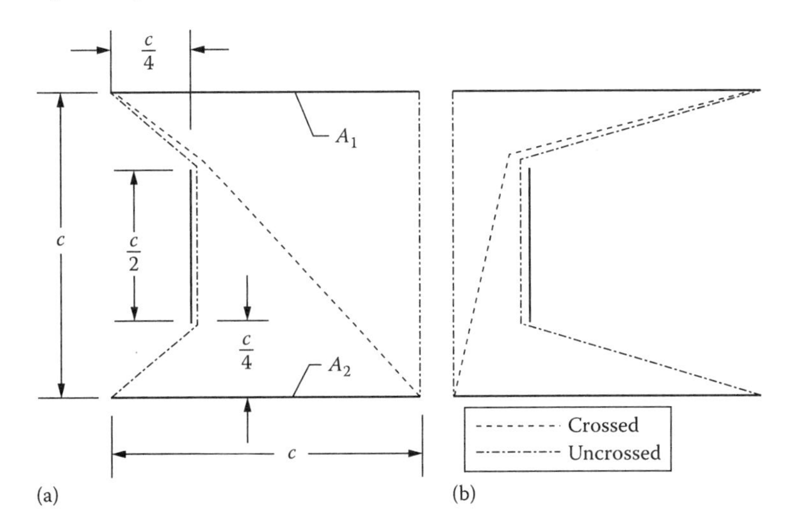

FIGURE 4.19 Partially blocked view between two parallel strips: (a) view through the right side of vertical plate and (b) view through the left side.

the remainder of the view by drawing the strings as if the region to the right of the center plate were obstructed. The final configuration factor is the sum of the results from the two parts. For each part, the crossed strings are of equal length, so for clarity, only one is shown. For part (a), the lengths of the crossed and uncrossed strings are, respectively, $2\sqrt{2}c$ and $c+c/2+2\sqrt{2}(c/4)$. Similarly, for part (b), the lengths are $4c\ [(1/4)^2 + (3/4)^2]^{1/2} = \sqrt{10}c$ and $c+c/2+\sqrt{10}c/2$. Then, from Equation 4.37,

$$cF_{1-2} = \frac{1}{2}\left[2\sqrt{2}c - \left(\frac{3c}{2} + \sqrt{2}\frac{c}{2}\right)\right] + \frac{1}{2}\left[\sqrt{10}c - \left(\frac{3c}{2} + \sqrt{10}\frac{c}{2}\right)\right]$$

This simplifies to

$$F_{1-2} = \frac{1}{4}(3\sqrt{2} + \sqrt{10} - 6) = 0.35123$$

When the center plate is absent the crossed-string method yields $F_{1-2} = \sqrt{2} - 1 = 0.41421$, so the center plate has provided about 15% blockage.

4.3.3.2 Contour Integration

Another useful tool for evaluating configuration factors is to apply Stokes' theorem to reduce the multiple integration over a surface area to a single integration around the boundary of the area. This method is developed by Moon (1961), de Bastos (1961), Sparrow (1963a), and Sparrow and Cess (1978).

Consider a surface area A as in Figure 4.20 with its boundary designated as C (where C is piecewise continuous). An arbitrary point on the area is at position x, y, z. At this point the normal to A is constructed, and the angles between this normal and the x, y, and z axes are α, γ, and δ. Let P, Q, and R be any twice-differentiable functions of x, y, and z. Stokes' theorem in 3D provides the following relation between an integral of P, Q, and R around the boundary C of the area and an integral over the surface A of the area:

$$\oint_C (Pdx + Qdy + Rdz) = \int_A \left[\left(\frac{\partial R}{\partial y} - \frac{\partial Q}{\partial z}\right)\cos\alpha + \left(\frac{\partial P}{\partial z} - \frac{\partial R}{\partial x}\right)\cos\gamma + \left(\frac{\partial Q}{\partial x} - \frac{\partial P}{\partial y}\right)\cos\delta\right] dA \quad (4.38)$$

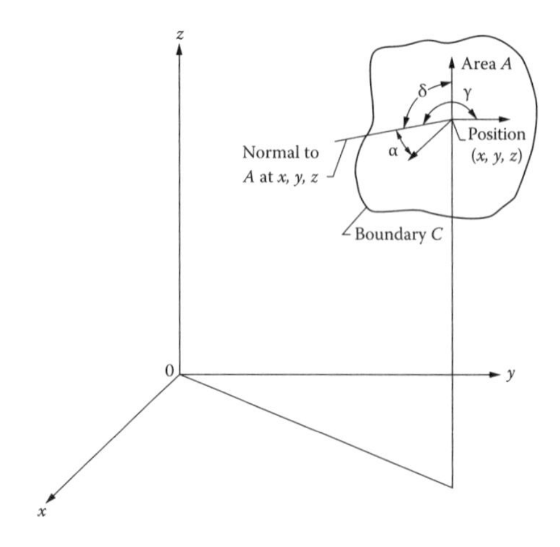

FIGURE 4.20 Geometry for quantities used in Stokes' theorem.

This relation is now applied to express area integrals in configuration factor computations in terms of integrals around the boundaries of the areas.

4.3.3.2.1 Configuration Factor between a Differential and a Finite Area

The integrand in the configuration factor F_{d1-2} is $(\cos\theta_1 \cos\theta_2/\pi S^2)dA_2$. The two cosines can be written (Figure 4.21):

$$\cos\theta_1 = \frac{x_2 - x_1}{S}\cos\alpha_1 + \frac{y_2 - y_1}{S}\cos\gamma_1 + \frac{z_2 - z_1}{S}\cos\delta_1 \tag{4.39}$$

$$\cos\theta_2 = \frac{x_1 - x_2}{S}\cos\alpha_2 + \frac{y_1 - y_2}{S}\cos\gamma_2 + \frac{z_1 - z_2}{S}\cos\delta_2 \tag{4.40}$$

This follows from the relation that for two vectors $\mathbf{V}_1$ and $\mathbf{V}_2$ having direction cosines (l_1, m_1, n_1) and (l_2, m_2, n_2), the cosine of the angle between the vectors is $l_1l_2 + m_1m_2 + n_1n_2$.

Substituting Equations 4.39 and 4.40 into the integral relation for F_{d1-2} gives

$$F_{d1-2} = \int_{A_2} \frac{\cos\theta_1 \cos\theta_2}{\pi S^2} dA_2$$

$$= \frac{1}{\pi}\int_{A_2} \frac{(x_2 - x_1)\cos\alpha_1 + (y_2 - y_1)\cos\gamma_1 + (z_2 - z_1)\cos\delta_1}{S^4}$$

$$\times\left[(x_1 - x_2)\cos\alpha_2 + (y_1 - y_2)\cos\gamma_2 + (z_1 - z_2)\cos\delta_2\right]dA_2 \tag{4.41}$$

Now, let

$$l = \cos\alpha; \quad m = \cos\gamma; \quad n = \cos\delta \tag{4.42}$$

and

$$f = \frac{(x_2 - x_1)l_1 + (y_2 - y_1)m_1 + (z_2 - z_1)n_1}{\pi S^4} \tag{4.43}$$

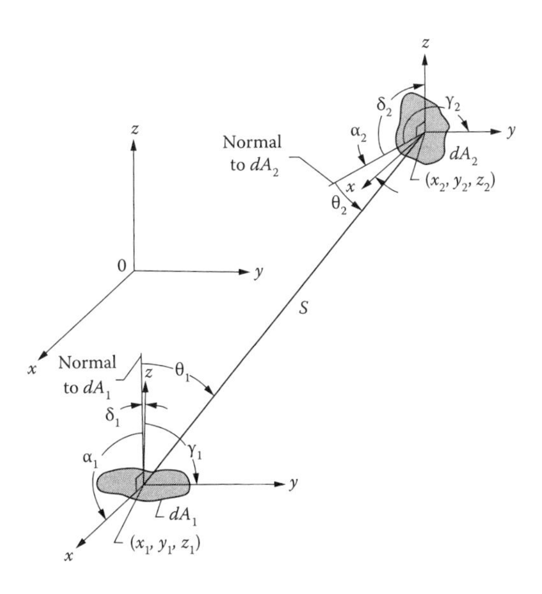

FIGURE 4.21 Geometry for contour integration.

Equation 4.41 becomes

$$F_{d1-2} = \int_{A_2} \left[(x_1 - x_2) f l_2 + (y_1 - y_2) f m_2 + (z_1 - z_2) f n_2\right] dA_2 \tag{4.44}$$

Comparison of Equation 4.44 with the right side of Equation 4.38 shows that Stokes' theorem can be applied if

$$\frac{\partial R}{\partial y_2} - \frac{\partial Q}{\partial z_2} = (x_1 - x_2) f \tag{4.45}$$

$$\frac{\partial P}{\partial z_2} - \frac{\partial R}{\partial x_2} = (y_1 - y_2) f \tag{4.46}$$

and

$$\frac{\partial Q}{\partial x_2} - \frac{\partial P}{\partial y_2} = (z_1 - z_2) f \tag{4.47}$$

Useful solutions to these three equations are (Sparrow 1963a) of the form

$$P = \frac{-m_1(z_2 - z_1) + n_1(y_2 - y_1)}{2\pi S^2} \tag{4.48}$$

$$Q = \frac{l_1(z_2 - z_1) - n_1(x_2 - x_1)}{2\pi S^2} \tag{4.49}$$

$$R = \frac{-l_1(y_2 - y_1) + m_1(x_2 - x_1)}{2\pi S^2} \tag{4.50}$$

Equation 4.38 is used to express F_{d1-2} in Equation 4.44 as a contour integral; that is,

$$F_{d1-2} = \oint_{C_2} (P dx_2 + Q dy_1 + R dz_2) \tag{4.51}$$

Then P, Q, and R are substituted from Equations 4.48, 4.49 and 4.50, and the result is rearranged to obtain

$$F_{d1-2} = \frac{l_1}{2\pi} \oint_{C_2} \frac{(z_2 - z_1)dy_2 - (y_2 - y_1)dz_2}{S^2} + \frac{m_1}{2\pi} \oint_{C_2} \frac{(x_2 - x_1)dz_2 - (z_2 - z_1)dx_2}{S^2} + \frac{n_1}{2\pi} \oint_{C_2} \frac{(y_2 - y_1)dx_2 - (x_2 - x_1)dy_2}{S^2} \tag{4.52}$$

The double integration over area A_2 for F_{d1-2} has been replaced by a set of three line integrals. Sparrow (1963a) discusses the superposition properties of Equation 4.44 that allow additions of the configuration factors of elements aligned parallel to the x, y, and z axes to obtain the factors for arbitrary orientation. A further derivation of the superposition procedure is in Hollands (1995) based on the concepts of the radiation flux vector rather than using contour integration. The numerical integration of the configuration factor for parallel directly opposed squares is discussed by Shapiro (1985). The contour integral method was found to be advantageous in terms of both accuracy and computing time.

Example 4.17

Determine the configuration factor F_{d1-2} from an element dA_1 to a right triangle as in Figure 4.22.

The normal to dA_1 is perpendicular to both the x and y axes and is thus parallel to z. The direction cosines for dA_1 are then $\cos\alpha_1 = l_1 = 0$, $\cos\gamma_1 = m_1 = 0$, and $\cos\delta_1 = n_1 = 1$, and Equation 4.52 becomes

$$F_{d1-2} = \frac{1}{2\pi}\oint_{C_2} \frac{(y_2 - y_1)dx_2 - (x_2 - x_1)dy_2}{S^2}$$

Since dA_1 is situated at the origin of the coordinate system, $x_1 = y_1 = 0$ and F_{d1-2} further reduces to

$$F_{d1-2} = \frac{1}{2\pi}\oint_{C_2} \frac{y_2 dx_2 - x_2 dy_2}{S^2}$$

The distance S between dA_1 and any point (x_2, y_2, z_2) on A_2 is

$$S^2 = x_2^2 + y_2^2 + z_2^2 = x_2^2 + y_2^2 + d^2$$

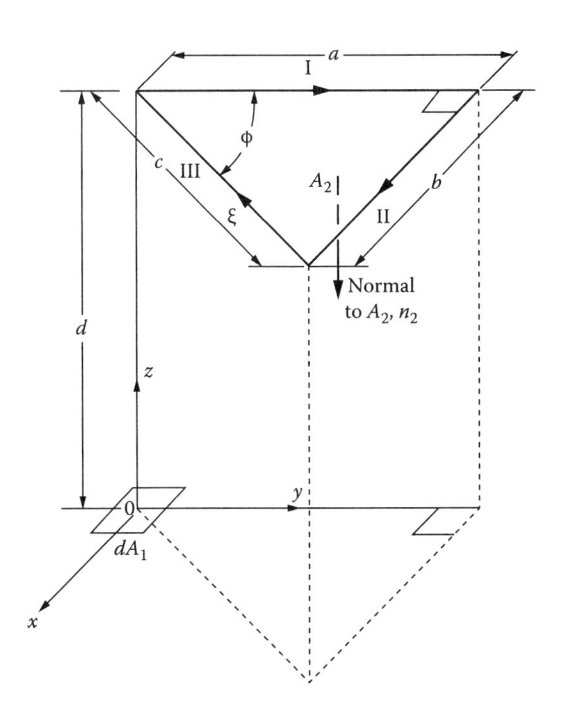

FIGURE 4.22 Geometry to determine the configuration factor between a plane-area element and a right triangle in parallel plane.

The contour integration for F_{d1-2} must now be carried out around the three sides of the right triangle. To keep the sign of F_{d1-2} positive, the integration is performed by traveling around the boundary lines I, II, and III in a particular direction. The correct direction is that of a person walking around the boundary with their head in the direction of the normal n_2 and always keeping A_2 to their left. Along boundary line I, $x_2 = 0$, $dx_2 = 0$, and $0 \le y_2 \le a$. On boundary II, $y_2 = a$, $dy_2 = 0$, and $0 \le x_2 \le b$. On boundary III, the integration is from $\xi = 0$ to c, where ξ is a coordinate along the hypotenuse of the triangle, so that $x_2 = (c - \xi)\sin\varphi$, $y_2 = (c - \xi)\cos\varphi$, and $dx_2 = -\sin\varphi\, d\xi$, $dy_2 = -\cos\varphi\, d\xi$. Substituting into the integral for F_{d1-2} gives

$$2\pi F_{d1-2} = \oint_{\mathrm{I,II,III}} \frac{y_2 dx_2 - x_2 dy_2}{x_2^2 + y_2^2 + d^2}$$

$$= 0 + \int_{x_2=0}^{b} \frac{a dx_2}{x_2^2 + a^2 + d^2} + \int_{\xi=0}^{c} \frac{-(c-\xi)\cos\varphi\sin\varphi d\xi + (c-\xi)\sin\varphi\cos\varphi d\xi}{(c-\xi)^2\sin^2\varphi + (c-\xi)^2\cos^2\varphi + d^2}$$

$$= \int_0^b \frac{a dx_2}{x_2^2 + a^2 + d^2} = \frac{X}{2\pi(1+X^2)^{1/2}} \tan^{-1}\frac{X\tan\varphi}{(1+X^2)^{1/2}}$$

where
$X = a/d$
$\tan\varphi = b/a$

4.3.3.2.2 *Configuration Factor between Finite Areas*

For configuration factors between two finite areas, substituting Equation 4.52 into Equation 4.17 gives

$$A_1F_{1-2} = A_2F_{2-1} = \int_{A_1} F_{d1-2} dA_1$$

$$= \frac{1}{2\pi}\oint_{C_2}\left[\int_{A_1} \frac{(y_2-y_1)n_1 - (z_2-z_1)m_1}{S^2} dA_1\right] dx_2$$

$$+ \frac{1}{2\pi}\oint_{C_2}\left[\int_{A_1} \frac{(z_2-z_1)l_1 - (x_2-x_1)n_1}{S^2} dA_1\right] dy_2$$

$$+ \frac{1}{2\pi}\oint_{C_2}\left[\int_{A_1} \frac{(x_2-x_1)m_1 - (y_2-y_1)l_1}{S^2} dA_1\right] dz_2 \qquad (4.53)$$

where the integrals have been rearranged and dx_2, dy_2, and dz_2 have been factored out since they are independent of the area integration over A_1.

Stokes' theorem is applied in turn to each of the three area integrals. Compare the first of these integrals

$$\int_{A_1} \frac{(y_2-y_1)n_1 - (z_2-z_1)m_1}{S^2} dA_1$$

with the area integral in Stokes' theorem, Equation 4.38. This gives the identities

$$\frac{\partial R}{\partial y_1}-\frac{\partial Q}{\partial z_1}=0;\quad \frac{\partial P}{\partial z_1}-\frac{\partial R}{\partial x_1}=\frac{-(z_2-z_1)}{S^2};\quad \frac{\partial Q}{\partial x_1}-\frac{\partial P}{\partial y_1}=\frac{y_2-y_1}{S^2}$$

A solution to this set of partial differential equations (Sparrow and Cess 1978) is $P = \ln S$, $Q = 0$, and $R = 0$. By use of Equation 4.38 to convert it into a surface integral, the area integral becomes

$$\int_{A_1}\frac{(y_2-y_1)n_1-(z_2-z_1)m_1}{S^2}dA_1=\oint_{C_1}\ln S\,dx_1$$

By applying Stokes' theorem in a similar fashion to the other two integrals in Equation 4.53, that equation becomes

$$A_1F_{1-2}=\frac{1}{2\pi}\oint_{C_2}\left(\oint_{C_1}\ln S\,dx_1\right)dx_2+\frac{1}{2\pi}\oint_{C_2}\left(\oint_{C_1}\ln S\,dy_1\right)dy_2+\frac{1}{2\pi}\oint_{C_2}\left(\oint_{C_1}\ln S\,dz_1\right)dz_2$$

or more compactly as

$$F_{1-2}=\frac{1}{2\pi A_1}\oint_{C_1}\oint_{C_2}(\ln S\,dx_2dx_1+\ln S\,dy_2dy_1+\ln S\,dz_2dz_1)\qquad(4.54)$$

Thus, the integrations over two areas, which would involve integrating over four variables, are replaced by integrations over the two surface boundaries. This allows considerable computational savings when numerical evaluations must be carried out. In addition, it can sometimes facilitate analytical integration.

Example 4.18

Using the contour integration method, formulate the configuration factor for the parallel rectangles in Figure 4.23.

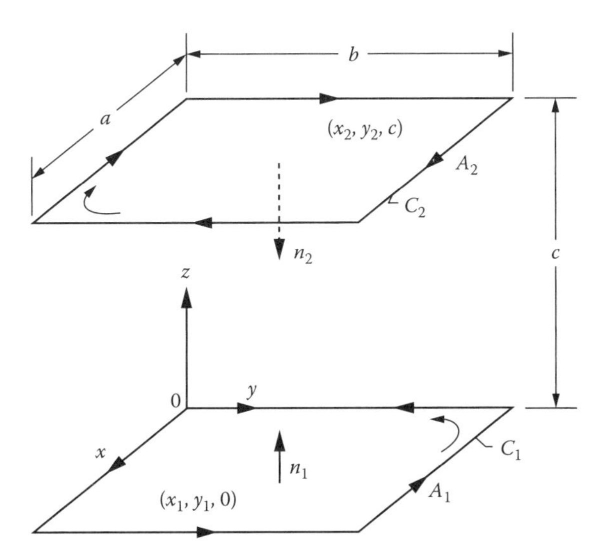

FIGURE 4.23 Contour integration to determine configuration factor between two parallel rectangles.

Note that on both surfaces dz is zero. First, integrate Equation 4.54 around the boundary C_2. The value of S to be used in Equation 4.54 is measured from an arbitrary point $(x_1, y_1, 0)$ on A_1 to a point on the portion of the boundary C_2 being considered. This gives

$$F_{1-2} = \frac{1}{2\pi ab} \oint_{C_1} \left\{ \int_{y_2=0}^{b} \ln\left[x_1^2 + (y_2 - y_1)^2 + c^2\right]^{1/2} dy_2 \right.$$

$$\left. + \int_{y_2=0}^{0} \ln\left[(a - x_1)^2 + (y_2 - y_1)^2 + c^2\right]^{1/2} dy_2 \right\} dy_1$$

$$+ \frac{1}{2\pi ab} \oint_{C_1} \left\{ \int_{x_2=0}^{a} \ln\left[(x_2 - x_1)^2 + (b - y_1)^2 + c^2\right]^{1/2} dx_2 \right.$$

$$\left. + \int_{x_2=0}^{0} \ln\left[(x_2 - x_1)^2 + y_1^2 + c^2\right]^{1/2} dx_2 \right\} dx_1$$

Carrying the integration out over C_1 gives, in this case, eight integrals. The first four, corresponding to the first two integrals of the previous equation, are

$$2\pi ab F_{1-2} = \int_{y_1=0}^{b} \int_{y_2=0}^{b} \ln\left[a^2 + (y_2 - y_1)^2 + c^2\right]^{1/2} dy_2 dy_1$$

$$+ \int_{y_1=b}^{0} \int_{y_2=0}^{b} \ln\left[(y_2 - y_1)^2 + c^2\right]^{1/2} dy_2 dy_1$$

$$+ \int_{y_1=0}^{b} \int_{y_2=0}^{0} \ln\left[(y_2 - y_1)^2 + c^2\right]^{1/2} dy_2 dy_1$$

$$+ \int_{y_1=0}^{0} \int_{y_2=0}^{0} \ln\left[a^2 + (y_2 - y_1)^2 + c^2\right]^{1/2} dy_2 dy_1$$

$$+ (4 \text{ integral terms in } x)$$

$$= \int_{y_1=0}^{b} \int_{y_2=0}^{b} \ln\left[\frac{a^2 + (y_2 - y_1)^2 + c^2}{(y_2 - y_1) + c^2}\right] dy_2 dy_1$$

$$+ \int_{x_1=0}^{a} \int_{x_2=0}^{a} \ln\left[\frac{(x_2 - x_1)^2 + b^2 + c^2}{(x_2 - x_1)^2 + c^2}\right] dx_2 dx_1$$

and the configuration factor is now given by the sum of two double integrals. These can be integrated analytically, and the result is factor 4 in Appendix C.

When the contour integration method is applied to geometries in which two surfaces share a common edge, the contour integrals become indeterminate. This difficulty is overcome for two quadrilaterals by a partial analytical integration, as developed in Mitalas and Stephenson (1966) and Shapiro (1983).

4.3.3.3 Differentiation of Known Factors

An extension of configuration factor algebra is to obtain configuration factors to differential areas by differentiating known factors to finite areas. This technique is useful in certain cases, as is demonstrated by the following example.

Example 4.19

As part of the determination of radiative exchange in a square channel whose temperature varies longitudinally, the configuration factor dF_{d1-d2} is needed between an element dA_1 at one corner of the channel end and a differential length of wall section dA_2 as in Figure 4.24a.

Configuration factor algebra plus differentiation can be used to find the required factor. Referring to Figure 4.24b, the fraction of energy leaving dA_1 that reaches dA_2 is the difference between the fractions reaching the squares A_3 and A_4, so the factor dF_{d1-d2} is the difference between F_{d1-3} and F_{d1-4}. Then

$$dF_{d1-d2} = F_{d1-3} - F_{d1-4} = -\frac{\Delta F_{d1-\square}}{\Delta x}\Delta x\bigg|_{\Delta x\to 0} = -\frac{\partial F_{d1-\square}}{\partial x}dx$$

Thus, if the configuration factor $F_{d1-\square}$ between a corner element and a square in a parallel plane were known, the derivative of this factor with respect to the separation distance could be used to obtain the required factor.

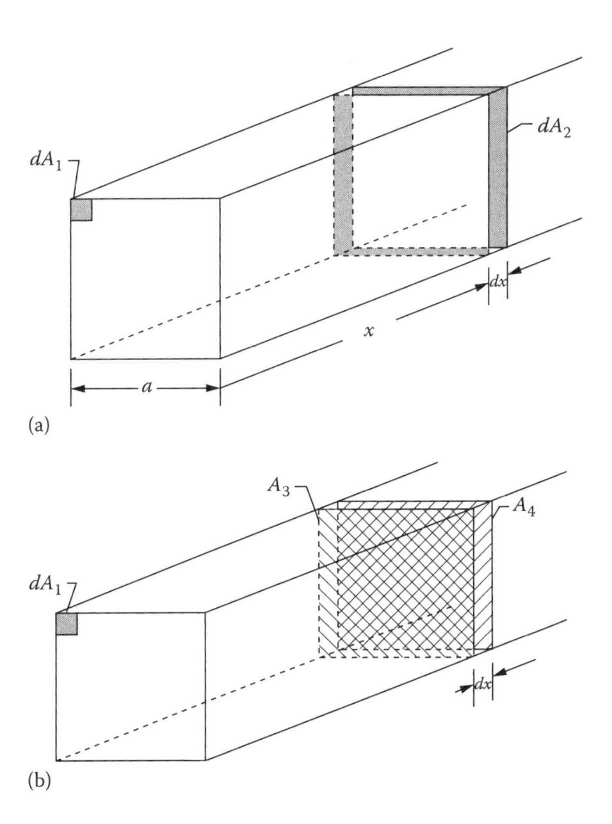

FIGURE 4.24 Geometry for the derivation of configuration factor between differential length of square channel and element at corner of channel end: (a) configuration factor between dA_1 and differential length of channel wall dA_2 and (b) configuration factor between dA_1 and squares A_3 and A_4.

From Example 4.17, the configuration factor between a corner element and a parallel isosceles right triangle is given by setting tan $\varphi = 1$ in the expression for a general right triangle. This yields for the present case, where the distance $d = x$

$$F_{d1-\Delta} = \frac{a}{2\pi(a^2+x^2)^{1/2}}\tan^{-1}\frac{a}{(a^2+x^2)^{1/2}}$$

Inspection shows that, by symmetry, the factor between a corner element and a square is twice the factor $F_{d1-\Delta}$. The dF_{d1-d2} is then

$$dF_{d1-d2} = -\frac{\partial F_{d1-\square}}{\partial x}dx = -\frac{adx}{\pi}\frac{\partial}{\partial x}\left[\frac{1}{(a^2+x^2)^{1/2}}\tan^{-1}\frac{a}{(a^2+x^2)^{1/2}}\right]$$

$$= \frac{axdx}{\pi(a^2+x^2)^{3/2}}\left[\tan^{-1}\frac{a}{(a^2+x^2)^{1/2}} + \frac{a(a^2+x^2)^{1/2}}{x^2+2a^2}\right]$$

$$= \frac{XdX}{\pi(1+X^2)^{3/2}}\left[\tan^{-1}\frac{1}{(1+X^2)^{1/2}} + \frac{(1+X^2)^{1/2}}{2+X^2}\right]; \quad X = \frac{x}{a}$$

More generally, start with F_{1-2} for two parallel areas A_1 and A_2 that are cross sections of a cylindrical channel of arbitrary cross section (Figure 4.25a). This factor depends on the spacing $|x_2-x_1|$ between the two areas and includes blockage due to the channel wall; that is, it is the factor by

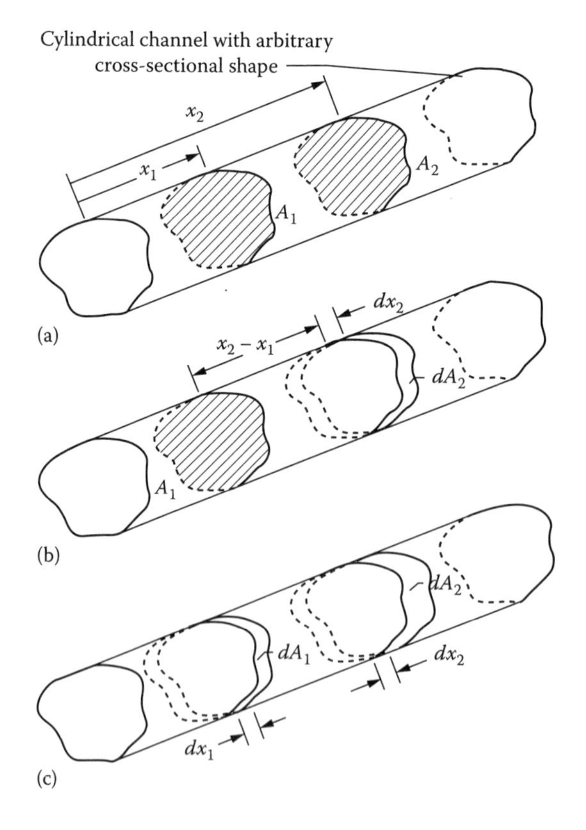

FIGURE 4.25 Geometry for derivation of the configuration factors for differential areas starting from the factors for finite areas: (a) two finite areas, F_{1-2}; (b) finite to differential area, $dF_{1-d2} = -(\partial F_{1-2}/\partial x_2)dx_2$; and (c) two differential areas, $dF_{d1-d2} = -(A_1/dA_1)(\partial^2 F_{1-2}/\partial x_1 \partial x_2)dx_2 dx_1$.

which A_2 is viewed from A_1 with the channel wall present. Note that for simple geometries such as a circular tube or rectangular channel, the wall blockage is zero. The factor between A_1 and dA_2 in Figure 4.25b is then given by

$$dF_{1-d2} = -\frac{\partial F_{1-2}}{\partial x_2} dx_2 \tag{4.19.1}$$

as in Example 4.19. Equation 4.19.1 is now used to obtain dF_{d1-d2}, the configuration factor between the two differential area elements in Figure 4.25c.

By reciprocity, $F_{d2-1} = (-A_1/dA_2)(\partial F_{1-2}/\partial x_2)dx_2$. Then in a fashion similar to the derivation of Equation 4.19.1, $dF_{d2-d1} = (\partial F_{d2-1}/\partial x_1)dx_1$. Substituting F_{d2-1} results in

$$dF_{d2-d1} = -\frac{A_1}{dA_2}\frac{\partial^2 F_{1-2}}{\partial x_1 \partial x_2} dx_2 dx_1 \tag{4.19.2}$$

or after using reciprocity,

$$dF_{d1-d2} = -\frac{A_1}{dA_1}\frac{\partial^2 F_{1-2}}{\partial x_1 \partial x_2} dx_2 dx_1 \tag{4.19.3}$$

Hence, by two differentiations, the factor dF_{d1-d2} can be found from F_{1-2} for the cylindrical configuration.

4.3.4 Unit-Sphere and Hemicube Methods

Experimental determination of configuration factors is possible by the *unit-sphere method* introduced by Nusselt (1928). If a hemisphere of unit radius $r = 1$ is constructed over the area element dA_1 in Figure 4.26, the configuration factor from dA_1 to an area A_2 is, by Equation 4.11,

$$F_{d1-2} = \frac{1}{\pi}\int_{A_2} \cos\theta_1 \frac{\cos\theta_2 dA_2}{S^2} = \frac{1}{\pi}\int_{A_2} \cos\theta_1 \, d\Omega_1$$

The projection of dA_2 onto the surface of the hemisphere is $d\Omega_1$, because

$$d\Omega_1 = \frac{dA_s}{r^2} = dA_s = \frac{\cos\theta_2 dA_2}{S^2}$$

The F_{d1-2} is then

$$F_{d1-2} = \frac{1}{\pi}\int_{A_s} \cos\theta_1 dA_s$$

However, $dA_s \cos\theta_1$ is the projection of dA_s onto the base of the hemisphere. It follows that integrating $\cos\theta_1 \, dA_s$ gives the projection A_b of A_s onto the base of the hemisphere or

$$F_{d1-2} = \frac{1}{\pi}\int_{A_s} \cos\theta_1 dA_s = \frac{A_b}{\pi} \tag{4.55}$$

The relation in Equation 4.55 forms the basis of several graphical and experimental methods for determining configuration factors. In one method, a spherical sector mirrored on the outside is placed over the area element dA_1. A photograph taken by a camera placed above the sector and

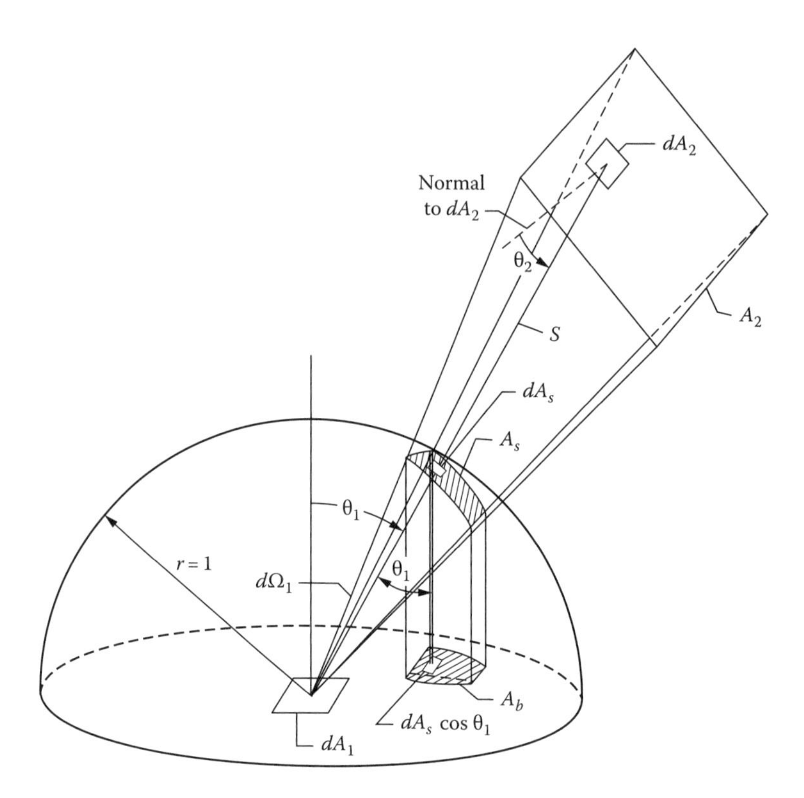

FIGURE 4.26 Geometry of unit-sphere method for obtaining configuration factors.

normal to dA_1 then shows the projection of A_2, which is A_b. The measurement of A_b on the photograph then provides F_{d1-2} from $F_{d1-2} = A_b/\pi r_e^2$, where r_e is the radius of the experimental mirrored sector. A means of optical projection is in Farrell (1976), and numerical techniques are in Lipps (1983) and Rushmeier et al. (1991). A semianalytical computational algorithm based on this method is developed by Mavroulakis and Trombe (1998) that may be useful in complex enclosures where two surfaces can share part of a contour.

The hemicube approach to finding configuration factors is a modified unit-sphere method to meet speed requirements for applications in computer graphics. The method uses the fact that the configuration factor is the same from any object that subtends the same solid angle to the viewer, regardless of the actual shape and orientation of the object. Projection onto the face of one-half of a cube centered over the receiving element dA_1 rather than onto a hemisphere, as shown in Figure 4.27, is more convenient for some purposes. In practice, the five surfaces of the hemicube are divided into small square elements (pixels), and the configuration factors from the element dA_1 to each pixel are precomputed using an element-pixel configuration factor. This can be approximated by Equation 4.7 or, more exactly, by using factors for square finite small areas. The "patch" B projected by A onto the surface of the hemicube is found by available efficient techniques (Cohen and Greenberg 1985, 1988, Watt and Watt 1992), and the configuration factor is found by summing the factors for the pixels contained within the patch. The projection on the unit hemisphere as used in the unit-sphere method is patch C. The precomputation of factors makes this summation very fast, and the method easily handles blocking and shading by intermediate elements (Kramer et al. 2014).

4.3.5 Direct Numerical Integration

It is common to determine configuration factors between a differential area and a finite area, or between two finite areas, by direct numerical integration of Equation 4.11 or 4.16. An example for a differential-to-finite area geometry is carried out in the on-line Appendix G, Example G.1, at

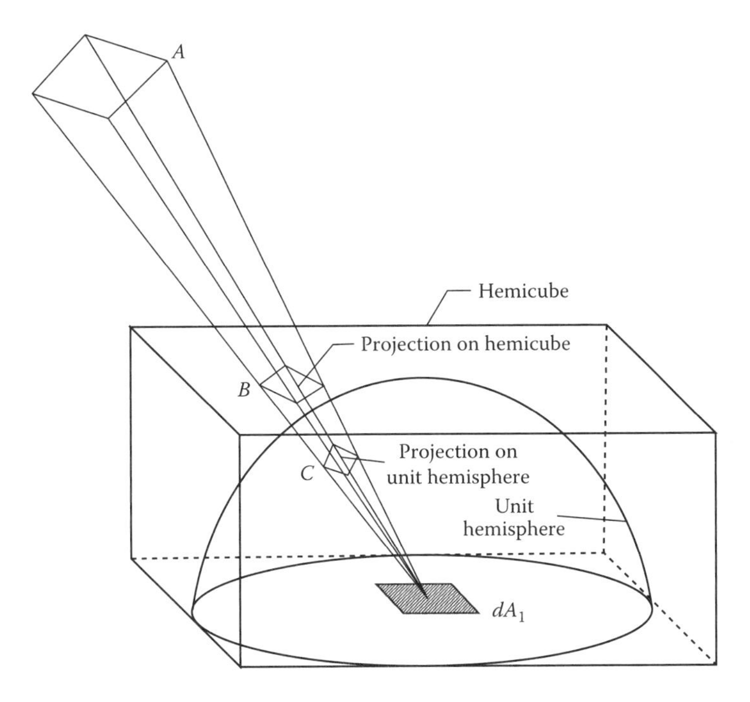

FIGURE 4.27 Geometry for hemicube analysis and relation to unit sphere.

www.thermalradiation.net. Such integrations require care to incorporate the effects of shading or blocking by intervening surfaces, and they can be quite computationally intensive to determine all the factors in a complex system with many surfaces. Methods are available for increasing computational speed and/or accuracy, including adaptive gridding to reduce the number of grid nodes while retaining accuracy (Campbell and Fussell 1990, Saltiel and Kolibal 1993) and partitioning of space by various search algorithms to find and minimize the number of equal-intensity areas that must be summed in the numerical integration (Thibault and Naylor 1987, Cohen et al. 1988, Buckalew and Fussell 1989). A numerical integration method is developed in Krishnaprakas (1998b) by dividing the surfaces into triangular elements and using Gaussian integration. A triangular grid can fit some surface shapes better than a rectangular grid. Contour integration carried out with Gaussian integration is used in Rao and Sastri (1996), and a procedure is derived similar to a finite-element line integral method. Using numerical integration, configuration factors can be calculated between surfaces having common or curved edges.

The discrete ordinates method, described in Chapter 13, was developed to determine radiative transfer in enclosures filled with a medium that absorbs, emits, and scatters radiation. The method involves following radiation along paths between surfaces of an enclosure with the radiative effects of the intervening medium included. By omitting the radiative interaction with the intervening medium, the discrete ordinates method can also be used to evaluate configuration factors as considered in Sanchez and Smith (1992) and Byun and Smith (1997). The radiation exchange for a 3D rectangular box compared very well with those computed from the known formulas for exchange between rectangles. In addition, the effect of an obstruction in the view between the enclosure surfaces in the form of a rectangular solid was introduced and investigated.

Another approach uses a technique common in computer graphics and visualization (Daun and Hollands 2001) based on use of nonuniform rational B-splines to transform a parametrically defined 2D surface into the kernel of the integral equation describing radiative exchange. Some representative 2D radiative exchange solutions are obtained, and the method is shown to be fast and accurate.

4.3.6 Computer Programs for the Evaluation of Configuration Factors

Many computer programs are available that use one or more of the methods outlined in this chapter for numerical calculation of configuration factors. Examples are FACET (Shapiro 1983), which uses area integration and contour integration; VIEW (Emery 1986), which can be used with the NASTRAN thermal analysis code; and a program that relies on the computer-graphical analog to the unit-sphere method (Alciatore et al. 1989). The latter program provides factors between a differential element and an arbitrary 3D object. The program Thermal Simulation System (Chin et al. 1992), developed under NASA sponsorship, incorporates an advanced graphical user interface for displaying configurations. The CHAPARRAL program (Glass 1995) incorporates FACET for 2D factors and uses the hemicube method for computing 3D factors in very large surface element arrays. Many commercially available thermal analysis programs such as COMSOL, FLUENT, FIDAP, and NEVADA also incorporate configuration factor computation using various methods, sometimes with choices among methods.

These and other computer codes provide a means to generate configuration factors for complex geometries and are invaluable for radiative analyses. Their accuracy can be assessed by comparison of computed results with the analytical expressions developed here for simpler geometries that can be used for test cases. Several different numerical methods for calculating configuration factors in complex configurations are compared by Emery et al. (1991) for computing speed, accuracy, and convenience. The geometries range from surfaces almost unobstructed in their view, to highly obstructed intersecting surfaces. The methods compared include double integration, Monte Carlo, contour integration, and projection techniques. If the view is not too complex, methods based on contour integration are found to be successful. The advent of massively parallel computers is making Monte Carlo methods (Chapter 7) particularly attractive for computing configuration factors. Walker et al. (2010, 2012) and Walker (2013) have examined the use of parallel Monte Carlo using either standard central processing units (CPUs) or graphical processing units and find good speed and accuracy in comparison with finite-element-based numerical integration for computing configuration factors for complex geometries. They employ superimposed primitives for fast rendering of many common objects.

4.4 CONSTRAINTS ON CONFIGURATION FACTOR ACCURACY

To obtain accurate results when configuration factors are used for calculating radiative heat transfer in an enclosure, the configuration factors must satisfy their physical constraints. One constraint is that factors for any pair of surfaces satisfy reciprocity. A second constraint provides that the sum of all the factors from a surface to all of the surrounding surfaces in the enclosure, including the view to itself, must equal 1 for energy to be conserved; this is called the "closure" constraint. These constraints may not be accurately satisfied by factors individually evaluated in complex enclosures with many surfaces. Factors may become difficult to evaluate to high precision when enclosures have complex shapes with obstructions and partial views of surfaces. Individual factors are often computed by numerical integrations that introduce approximations or by statistical methods such as Monte Carlo that may be converged to only a certain degree of accuracy. Vujičić et al. (2006a,b) calculated the sensitivity to grid size and sample number for Monte Carlo configuration factor calculation for finite areas for the cases of directly opposed rectangles, hinged rectangles, and concentric parallel disks of equal diameter. Koptelov et al. (2012, 2014) present criteria for choosing the best integration method and number of nodes for numerically computing configuration factors. The latter paper includes criteria for *a priori* choice of number of nodes to achieve a specified level of accuracy.

It was shown in Section 4.3.2 that if a certain number of configuration factors are calculated in an enclosure, the physical constraints can be used to obtain the remaining factors by solving a set of simultaneous equations. Sowell and O'Brien (1972) point out that in an enclosure with many surfaces, the configuration factors most easily specified or computed may not lend themselves to subsequent accurate calculation of the remaining factors. Reciprocity or the energy conservation ("closure") relation, Equation 4.30, may be difficult to apply. They present a computer scheme using matrix algebra that allows calculations of all remaining factors in an N-surface planar or convex-surface enclosure once the required minimum number of configuration factors is specified.

To improve accuracy, some methods are discussed in Larsen and Howell (1986) for smoothing sets of "direct exchange areas" for enclosures, using the constraints imposed by reciprocity and energy conservation. These methods are directly applicable to configuration factors and can be used when some inaccurate or ill-defined configuration factors are present in a set. Another presentation of the compatibility conditions for enclosure configuration factors is given by van Leersum (1989). This guarantees overall energy conservation for the enclosure. An uncertainty analysis in Taylor et al. (1995) shows that strict enforcement of the reciprocity and closure constraints yields an order-of-magnitude reduction in the uncertainty in heat flux results that arises from uncertainties in the areas and configuration factors. Methods for adjusting the factors to provide enforcement are provided in Taylor and Luck (1995) and a least-squares smoothing method is recommended, similar to that in Larsen and Howell and in Vercammen and Froment (1980) for enclosures filled with a gas that is not transparent.

If some of the factors are known to have better accuracy, weighting factors w_{ij} can be used to give them more importance in the least-squares smoothing method; otherwise, the weighting factors are all equal to one. The least-squares optimization for an enclosure with N surfaces is done by minimizing the weighted error, given by E_w, as

$$E_w = \sum_{i=1}^{N}\sum_{j=1}^{N} w_{ij}(F_{c,i-j} - F_{i-j})^2 \tag{4.56}$$

where

the F_{i-j} are the configuration factors that have been calculated with approximations

the $F_{c,i-j}$ are the corrected configuration factors that result from the least-squares smoothing procedure

During the minimization procedure for E_w, constraints are imposed of reciprocity and closure:

$$A_i F_{c,i-j} = A_j F_{c,j-i} \quad i = 1,\ldots,N-1, \quad j = i+1,\ldots,N \tag{4.57}$$

$$\sum_{j=1}^{N} F_{c,i-j} = 1 \quad i = 1,\ldots,N \tag{4.58}$$

$$F_{c,i-j} \geq 0 \quad i = 1,\ldots,N, \quad j = 1,\ldots,N \tag{4.59}$$

The corrected $F_{c,i-j}$ are then used in the enclosure calculations. This provides significantly improved results compared with using configuration factors that have not been optimized to guarantee that the physical constraints are satisfied.

Other approaches are presented by Clarksean and Solbrig (1994), Lawson (1995), and Loehrke et al. (1995). A method that enforces reciprocity and conservation while also enforcing a nonnegativity constraint on all factors using least-squares minimization is given by Daun et al. (2005).

4.5 COMPILATION OF KNOWN CONFIGURATION FACTORS AND THEIR REFERENCES: APPENDIX C AND WEB CATALOG

Many configuration factors for specific geometries have been derived in analytical form or tabulated, and they are spread throughout the literature. Some factors that have convenient analytical forms are given in Appendix C for use in examples and homework problems. An open website giving extensive information on configuration factors is http://www.thermalradiation.net. The website provides references, algebraic relations, tables, and graphical results for over 300 configurations, and a review of the site organization and contents is given by Howell and Mengüç (2011). The most recent version also provides calculation capability to find numerical values for most factors.

HOMEWORK

4.1 Derive the configuration factor F_{d1-2} between a differential area centered above a disk and a finite disk of unit radius parallel to the dA_1.

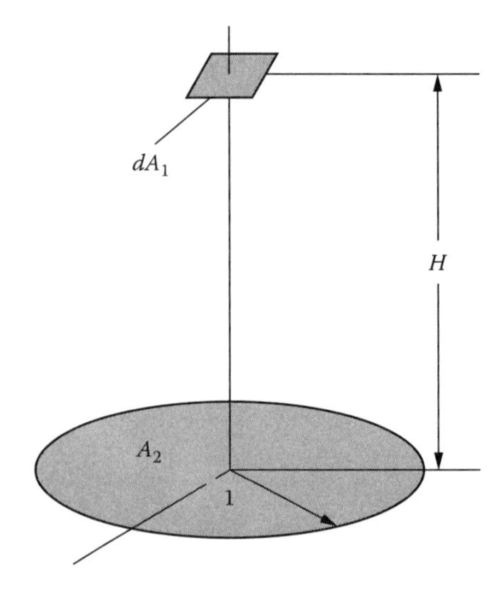

Answer: $1/(H^2 + 1)$.

4.2 Find the configuration factor F_{d1-2} from a planar element to a coaxial parallel rectangle as shown below; the dA_1 is above the center of the rectangle. Use any method except contour integration.

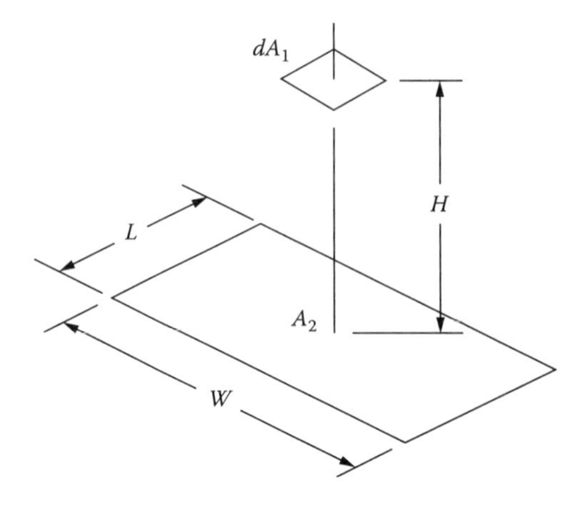

Answer: $F_{d1-2} = \frac{2}{\pi}\left\{\left[\frac{X}{(1+X^2)^{1/2}}\right]\tan^{-1}\left[\frac{Y}{(1+X^2)^{1/2}}\right] + \left[\frac{Y}{(1+Y^2)^{1/2}}\right]\tan^{-1}\left[\frac{X}{(1+Y^2)^{1/2}}\right]\right\},$

where

$X = L/2H$

$Y = W/2H$

4.3 Derive by any three methods, including use of the factors in Appendix C if you choose, the configuration factor F_{1-2} for the infinitely long geometry shown below in cross section.

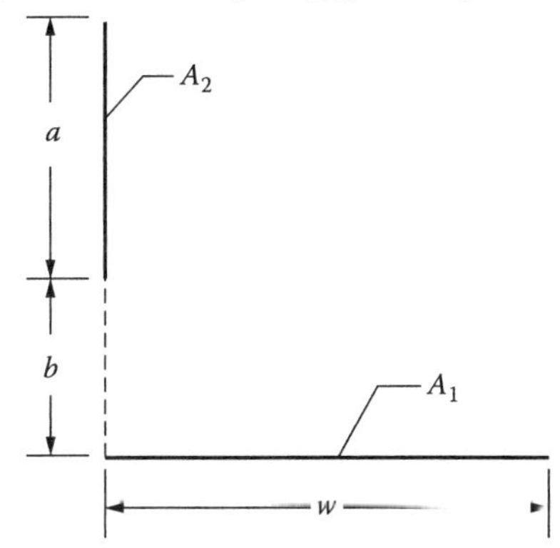

Answer: $F_{1-2} = \{A + (1 + B^2)^{1/2} - [(A + B)^2 + 1]^{1/2}\}/2$,

where

$A = a/w$

$B = b/w$

4.4 Find the configuration factor F_{1-2} for the geometries shown below. Give a numerical value using any method:

(a) Two plates infinitely long normal to the cross section shown

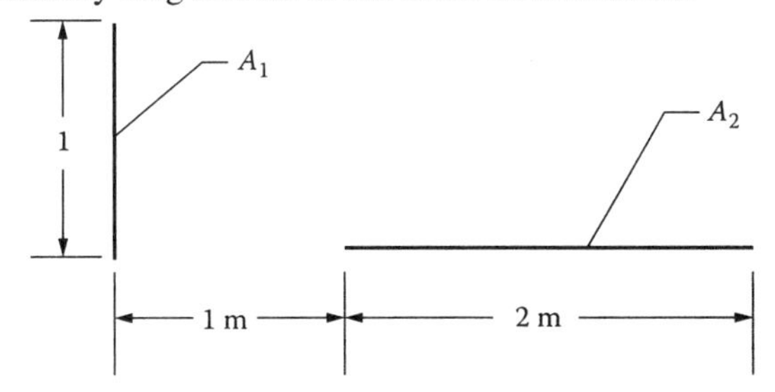

(b) Two plates of finite length as shown

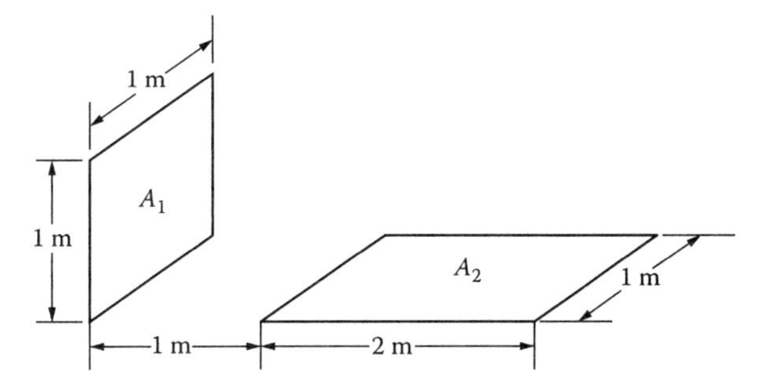

Answer: (a) 0.126; (b) 0.0417.

4.5 Find the configuration factor F_{1-2} for the geometries shown below. Give a numerical value using any method:

(a) Disk to concentric ring

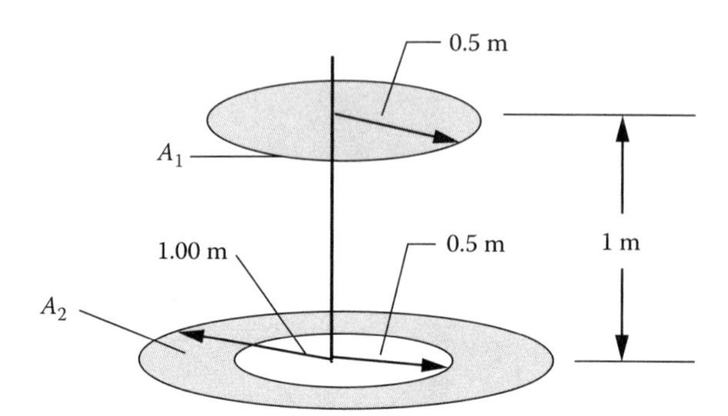

(b) Configuration factor F_{1-2} between the interior conical surface A_1 of a frustum of a right circular cone and a hole in its top

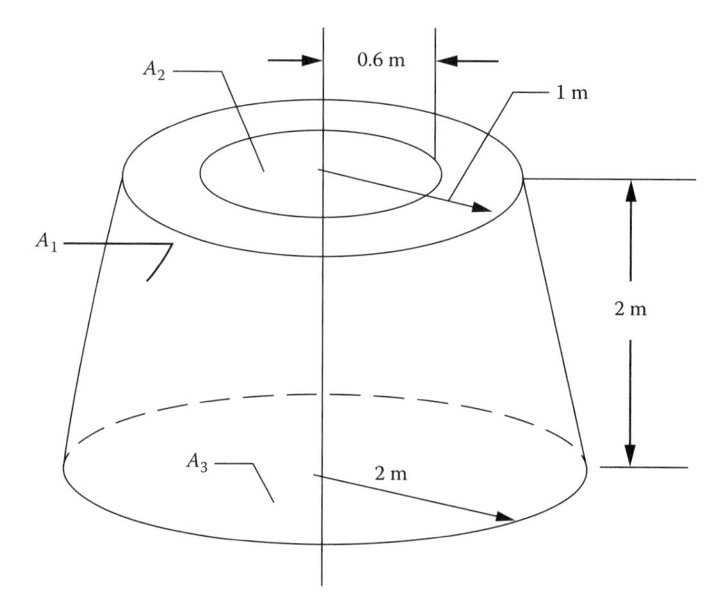

(c) Configuration factor from infinitely long rectangular shape 1 shown in cross section to infinitely long plane 2

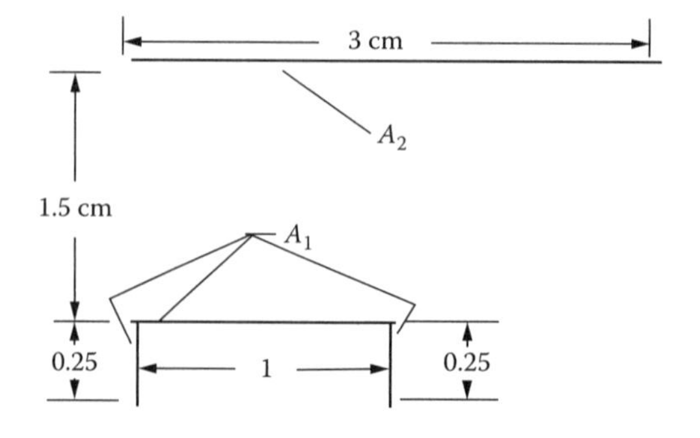

Answer: (a) 0.2973; (b) 0.02743; (c) 0.4164.

4.6 The configuration factor between two infinitely long directly opposed parallel plates of finite width L is F_{1-2}. The plates are separated by a distance D.

(a) Derive an expression for F_{1-2} by integration of the configuration factor between parallel differential strip elements.

(b) Derive an expression for F_{1-2} by the crossed-string method.

Answer: $\left[1+\left(\dfrac{D}{L}\right)^2\right]^{1/2}-\left(\dfrac{D}{L}\right)$.

4.7 The configuration factor between two infinitely long parallel plates of finite width L is F_{1-2} in the configuration shown below in cross section.

(a) Derive an expression for F_{1-2} by the crossed-string method.

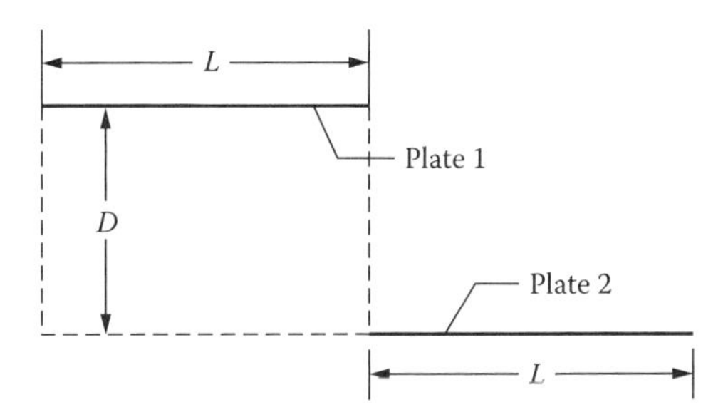

(b) Derive an expression for F_{1-2} by using the results of Homework Problem 4.6 and configuration factor algebra.

(c) Find the configuration factor F_{1-2} for the geometry of infinitely long plates shown below in cross section.

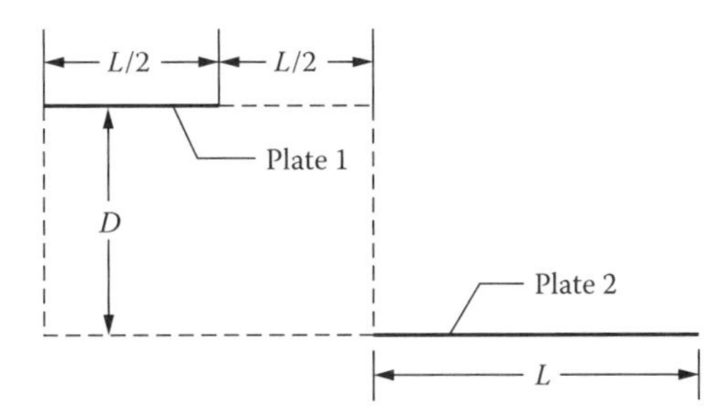

Answer: (a), (b) $\left[1+\left(\dfrac{D}{2L}\right)^2\right]^{1/2}+\dfrac{D}{2L}-\left[1+\left(\dfrac{D}{L}\right)^2\right]^{1/2}$;

(c) $F_{1-2}=2\left[1+\left(\dfrac{D}{2L}\right)^2\right]^{1/2}-\left[1+\left(\dfrac{D}{L}\right)^2\right]^{1/2}-\left[\left(\dfrac{D}{L}\right)^2+\left(\dfrac{3}{2}\right)^2\right]^{1/2}+\left[\left(\dfrac{D}{L}\right)^2+\left(\dfrac{1}{2}\right)^2\right]^{1/2}$.

4.8 (a) For the 2D geometry shown in cross section, derive a formula for $F_{2\text{-}2}$ in terms of r_1 and r_2.

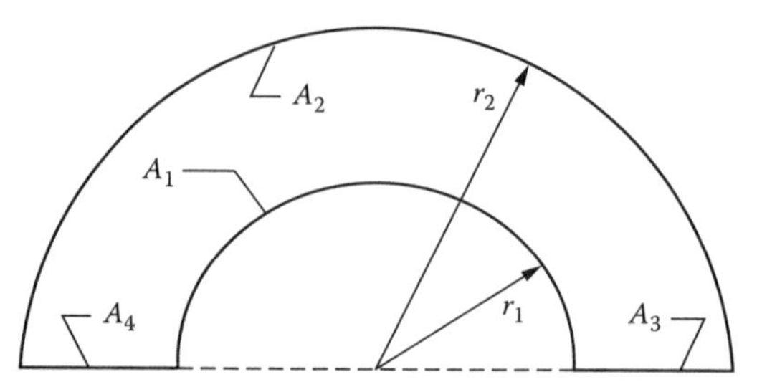

(b) Find $F_{2\text{-}2}$ for the 2D geometry. A_1 is a square (A_1 refers to the total area of the four sides) and A_2 is part of a circle.

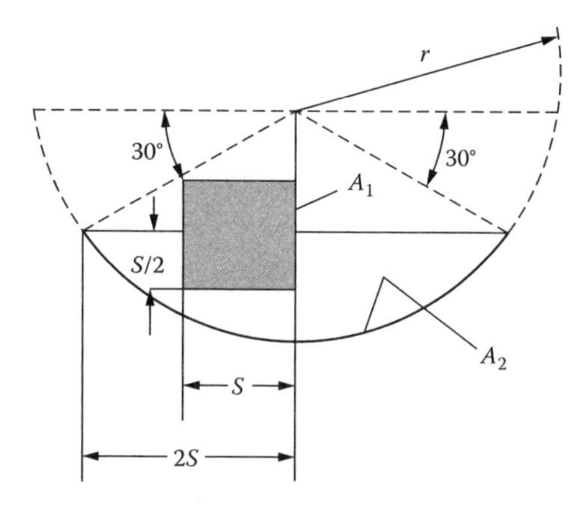

Answer:

(a) $F_{2\text{-}2} = 1 - \dfrac{2}{\pi}\left[1 - \left(\dfrac{r_1}{r_2}\right)^2\right]^{1/2} - \dfrac{2}{\pi}\dfrac{r_1}{r_2}\sin^{-1}\left(\dfrac{r_1}{r_2}\right).$

(b) $F_{2\text{-}2} = 0.1359.$

4.9 Derive a formula for $F_{2\text{-}2}$ in terms of r and α. The A_1 is inside the cone, and A_2 is part of a sphere.

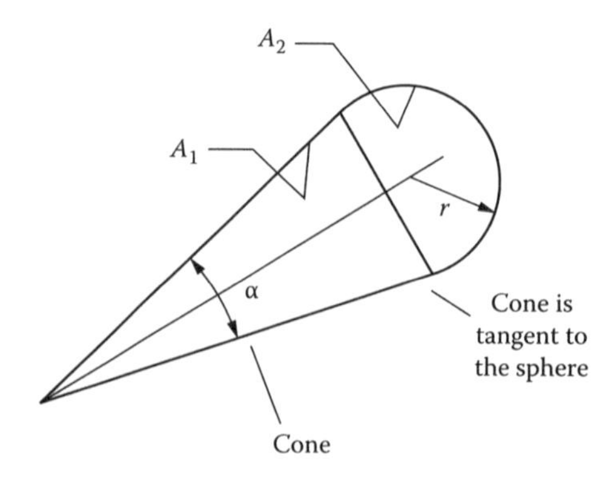

Answer: $F_{2\text{-}2} = \dfrac{1}{2}\left(1 + \sin\dfrac{\alpha}{2}\right).$

4.10 Compute the configuration factor F_{1-2} between faces A_1 and A_2 of the infinitely long parallel plates shown below in cross section when the angle β is equal to (a) 30° and (b) 75°.

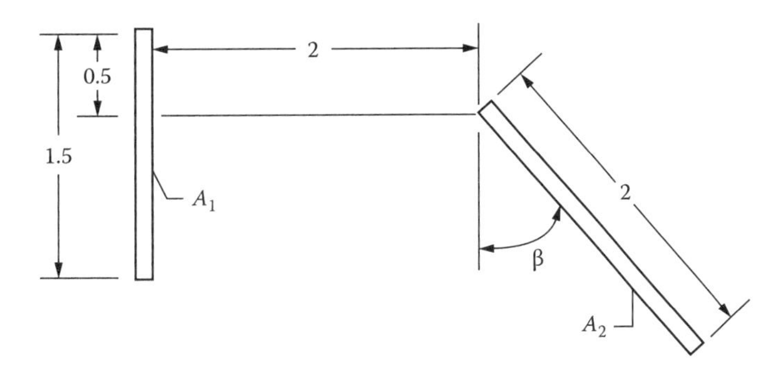

Answer: (a) 0.2752; (b) 0.0915.

4.11 Find the configuration factor between the two infinitely long parallel plates shown below in cross section. Use

(a) The crossed-string method
(b) Configuration factor algebra with factors from Appendix C
(c) Integration of differential strip-to-differential strip factors

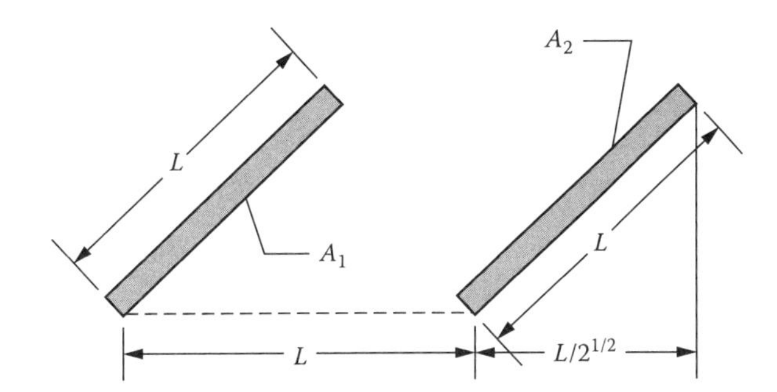

Answer: 0.3066.

4.12 A sphere of radius r is divided into two quarter spheres and one hemisphere. Obtain the configuration factors between all areas inside the sphere, F_{1-2}, F_{2-2}, F_{3-1}, F_{1-1}, etc. From these factors find F_{4-5}, where A_4 and A_5 are equal semicircles that are at right angles to each other.

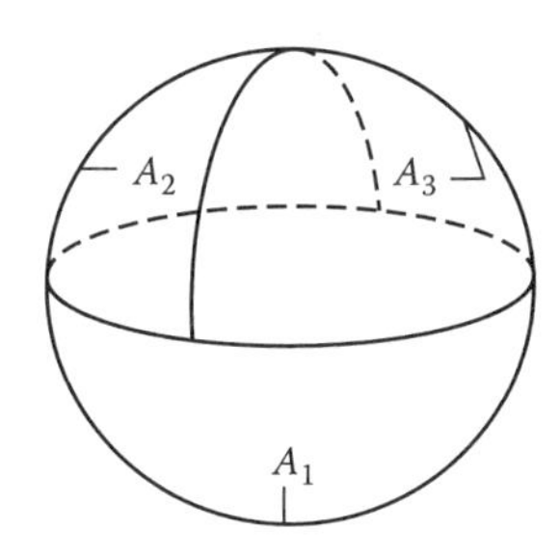

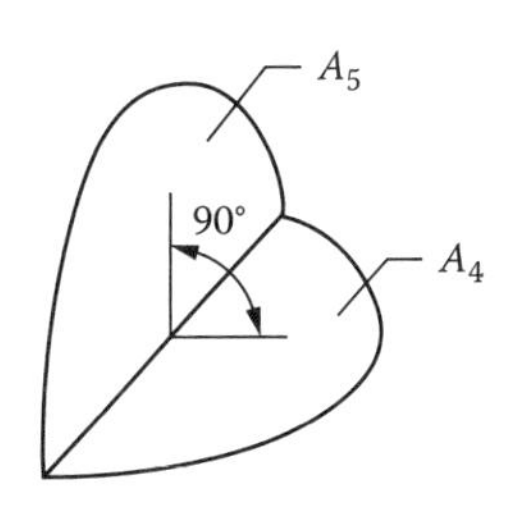

Answer: $F_{4-5} = 0.25$.

4.13 For the 2D geometry shown, the view between A_1 and A_2 is partially blocked by an intervening structure. Determine the configuration factor F_{1-2}.

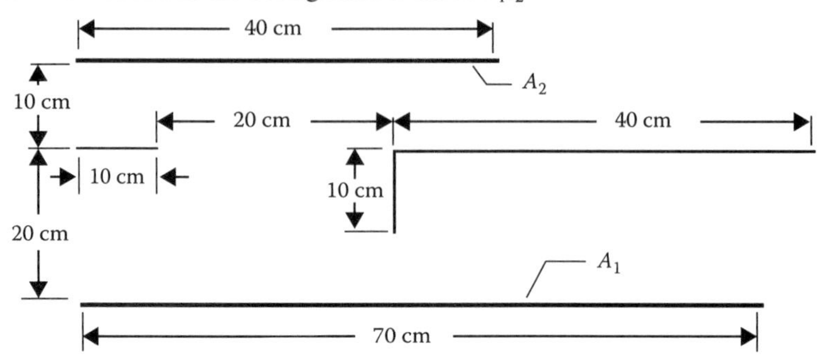

Answer: 0.1815.

4.14 The cylindrical geometry shown in cross section is very long in the direction normal to the plane of the drawing. The cross section consists of two concentric three-quarter circles and two straight lines. Obtain the value of the configuration factor F_{2-2}.

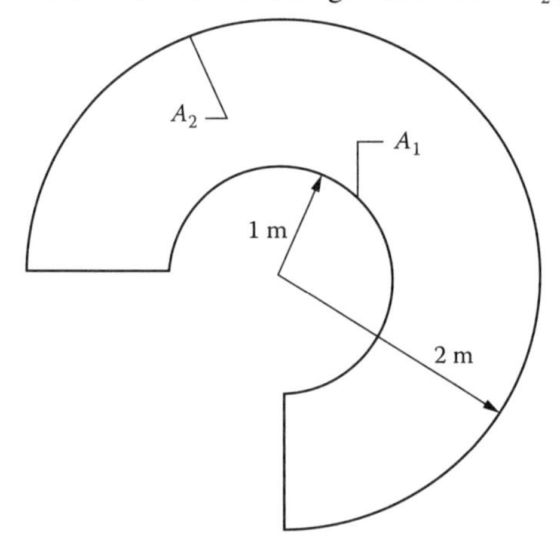

Answer: 0.3547.

4.15 Using the relation $F_{3-(1+2)} = F_{3-1} + F_{3-2}$, show whether the relation $F_{(1+2)-3} = F_{1-3} + F_{2-3}$ is also valid.

4.16 For the cylindrical enclosure of diameter d and length $l = 3d$ shown below, find all configuration factors among the surfaces.

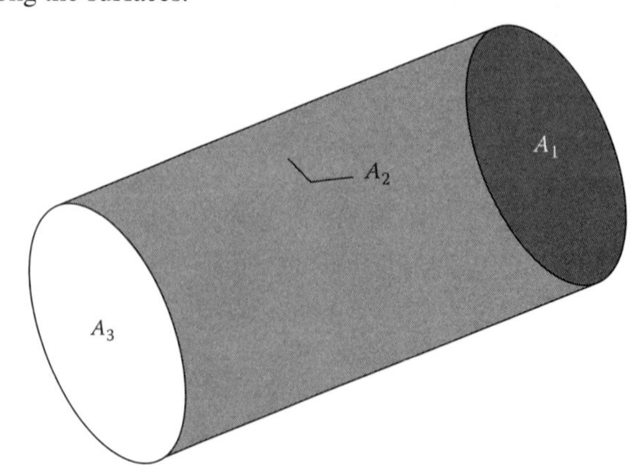

Answer: $F_{2-2} = 0.838$.

4.17 Using the factor for an "infinitely long enclosure formed of three plane areas" derived in Example 4.13, find the factor from one side of an infinitely long enclosure to its base. The enclosure cross section is an isosceles triangle with apex angle α.

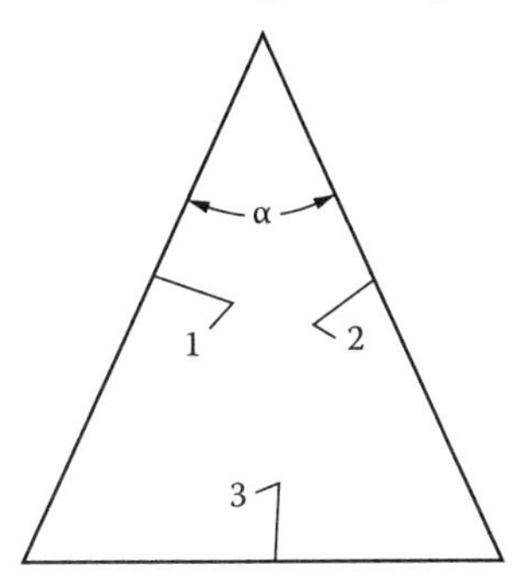

Answer: $F_{1-3} = \sin\dfrac{\alpha}{2}$.

4.18 A four-sided enclosure is formed of three mutually perpendicular isosceles triangles of short side S and hypotenuse H and an equilateral triangle of side H. Find the configuration factor F_{1-2} between two perpendicular isosceles triangles sharing a common short side.

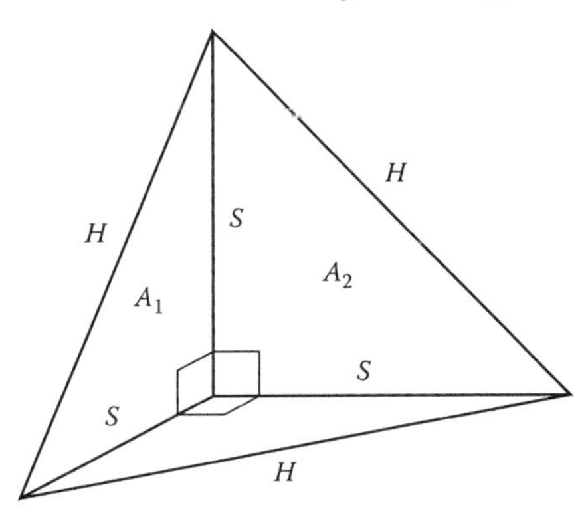

Answer: 0.2113.

4.19 A plate A_1 in the configuration below is to be moved along positions from S = 0 cm to S = 100 cm. Plot the configuration factor F_{1-2} versus S for this configuration.

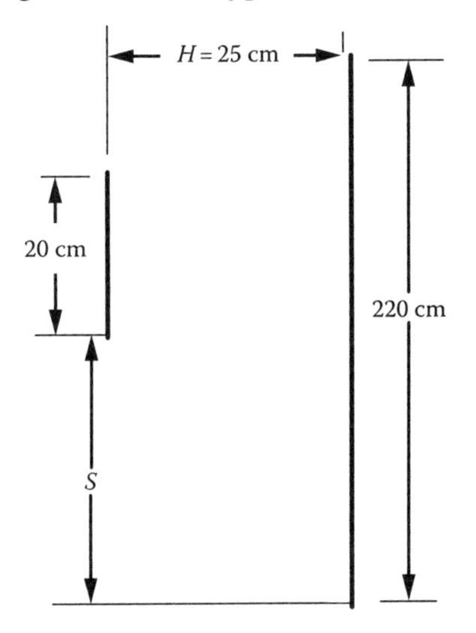

4.20 Derive the configuration factor F_{1-2} from a finite rectangle A_1 to an infinite plane A_2 where the rectangle is tilted at an angle η relative to the plane and one edge of the rectangle is in the infinite plane.

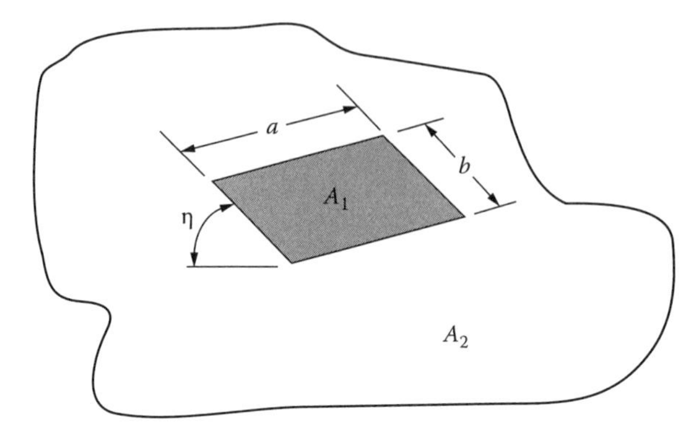

Answer: $F_{1-2} = (1/2)(1 - \cos\eta)$.

4.21 The four flat plates shown in cross section are very long in the direction normal to the plane of the cross section shown. Obtain the value of the configuration factor F_{1-2}. What are the values of F_{2-1} and F_{1-3}?

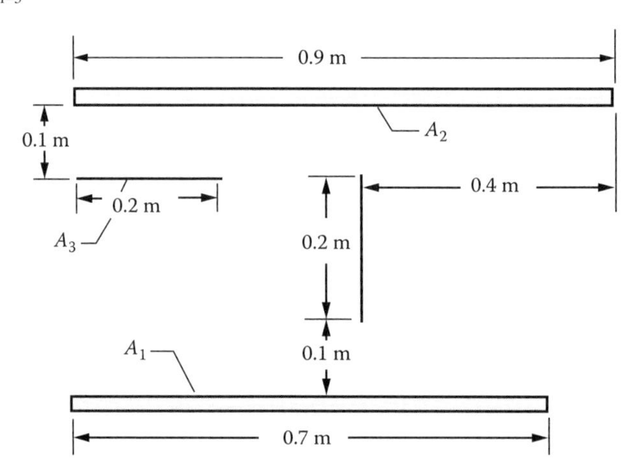

Answer: $F_{1-2} = 0.4260$; $F_{2-1} = 0.3314$; $F_{1-3} = 0.1704$.

4.22 Using the crossed-string method, derive the configuration factor F_{1-2} between the infinitely long plate and the parallel cylinder shown below in cross section. Compare your result with that given for configuration 11 in Appendix C.

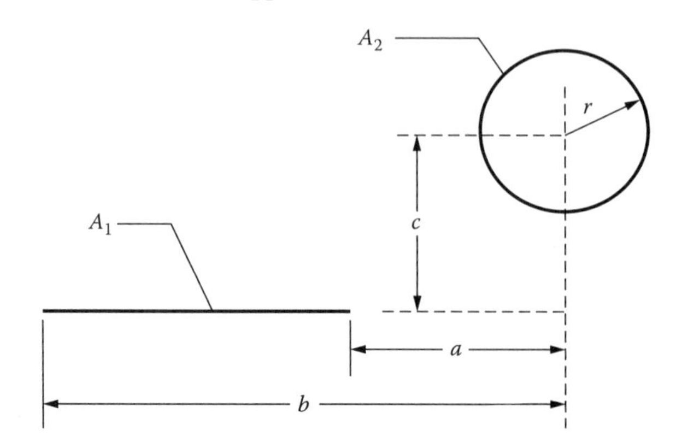

Answer: $F_{1-2} = \left[\dfrac{r}{(b-a)}\right]\left[\tan^{-1}\left(\dfrac{b}{c}\right) - \tan^{-1}\left(\dfrac{a}{c}\right)\right]$.

4.23 A long tube in a tube bundle is surrounded by six other identical equally spaced tubes as shown in cross section below. What is the configuration factor from the central tube to each of the surrounding tubes?

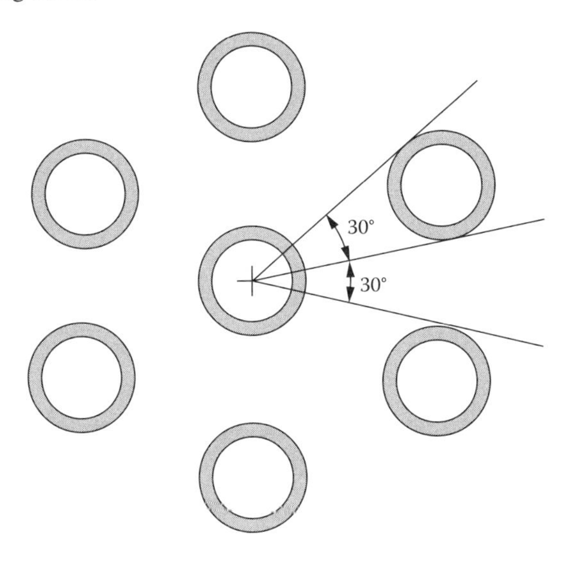

Answer: 0.0844 (>1/12).

4.24 For the 2D geometry shown, the view between A_1 and A_2 is partially blocked by two identical cylinders. Determine the configuration factor F_{1-2}.

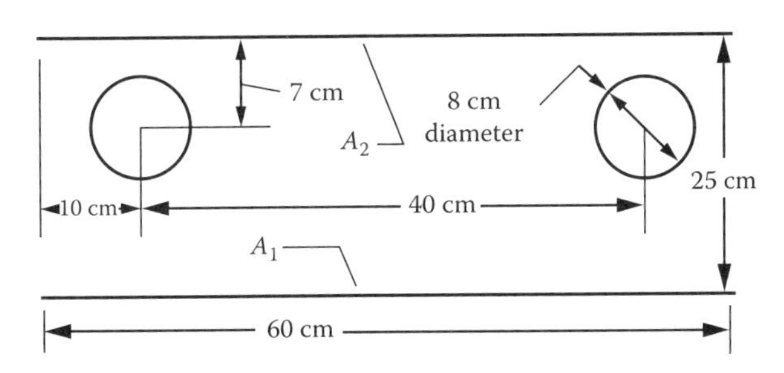

Answer: 0.473.

4.25 Given F_{n-k}, determine the configuration factor between two perpendicular rectangles having a common edge. In terms of this factor, use configuration factor relations to derive the F_{1-8} factor between the two areas A_1 and A_8.

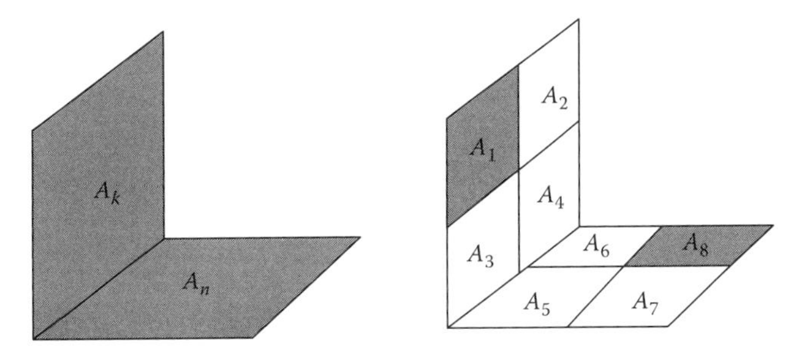

4.26 Find F_{1-2} by any two methods for the 2D geometry shown in cross section.

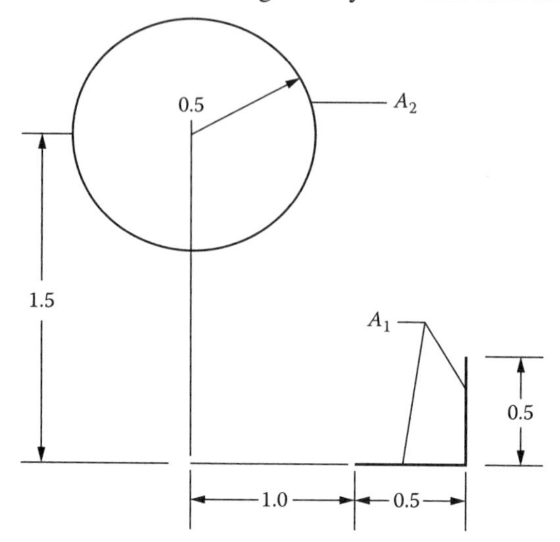

Answer: 0.1974.

4.27 An infinitely long enclosure is shown below in cross section. The outer surfaces form a square in cross section, and the outer surfaces are parallel to the inner circular coaxial cylinder, so the geometry is 2D. Find the configuration factors F_{2-1}, F_{2-3}, and F_{2-4}.

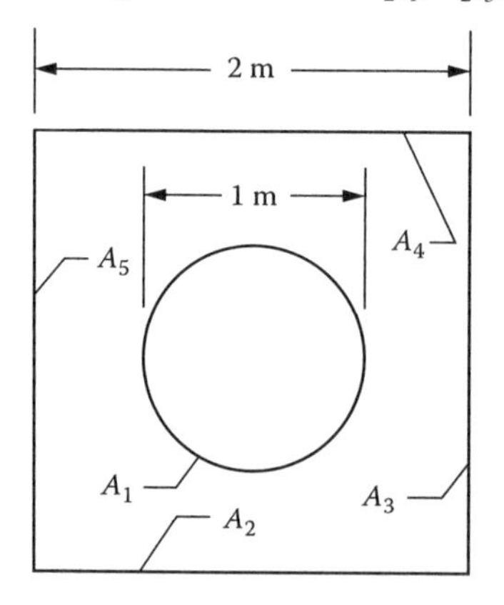

Answer: $F_{2-1} = \pi/8$; $F_{2-3} = 0.2482$; $F_{2-4} = 0.1109$.

4.28 Find the configuration factor F_{1-2} from the quarter disk to the parallel planar ring.

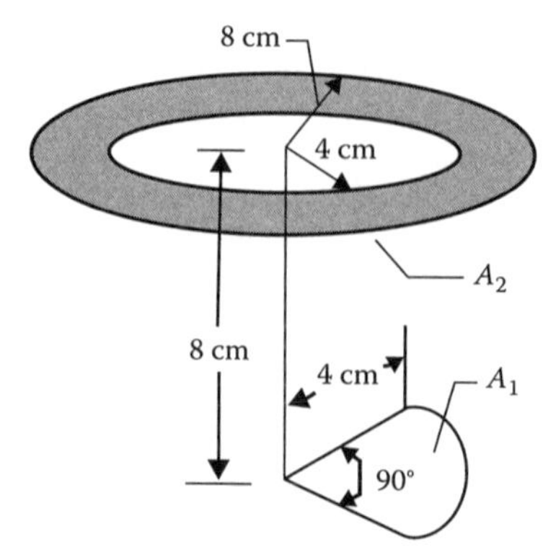

Answer: 0.2973.

4.29 (a) Derive the configuration factor between a sphere of radius R and a coaxial disk of radius r. (Hint: Think of the disk as being a cut through a spherical envelope concentric around the sphere.)

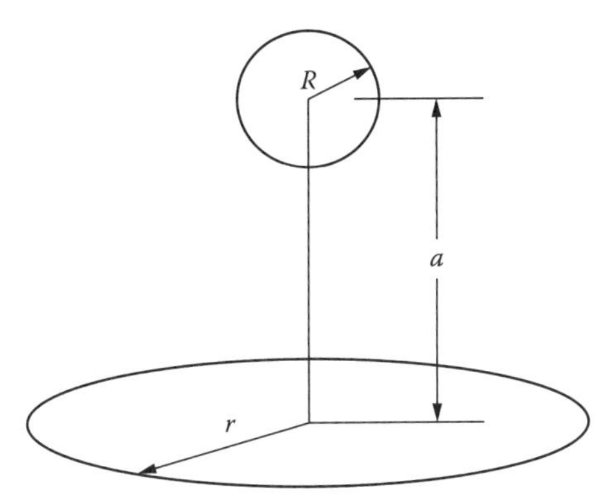

(b) What is the F from a sphere to a sector of a disk?

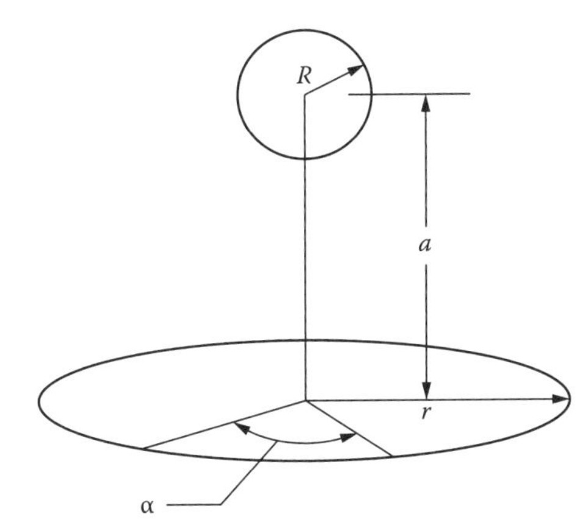

(c) What is the dF from a sphere to a portion of a ring of differential width?

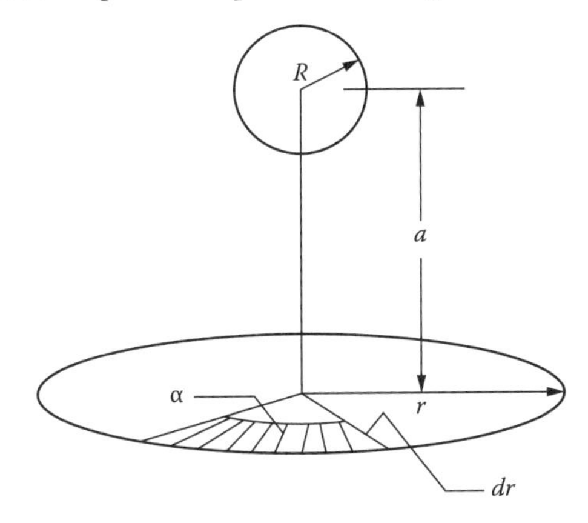

Answer:

(a) $\frac{1}{2}\left[1-\frac{a}{(a^2+r^2)^{1/2}}\right]$; (b) $\frac{\alpha}{4\pi}\left[1-\frac{a}{(a^2+r^2)^{1/2}}\right]$; (c) $\frac{\alpha}{4\pi}\frac{ar}{(a^2+r^2)^{3/2}}dr$.

4.30 Use the crossed-string method to derive the configuration factor between an infinitely long strip of differential width and a parallel infinitely long cylindrical surface.
Answer: $F_{d1-2} = (1/2)(\sin\beta_2 + \sin\beta_1)$.

4.31 Use the disk-to-disk configuration factor 10 of Appendix C to obtain the factor F_{d1-d2} between the interior surfaces of two differential rings on the interior of a right circular cone.

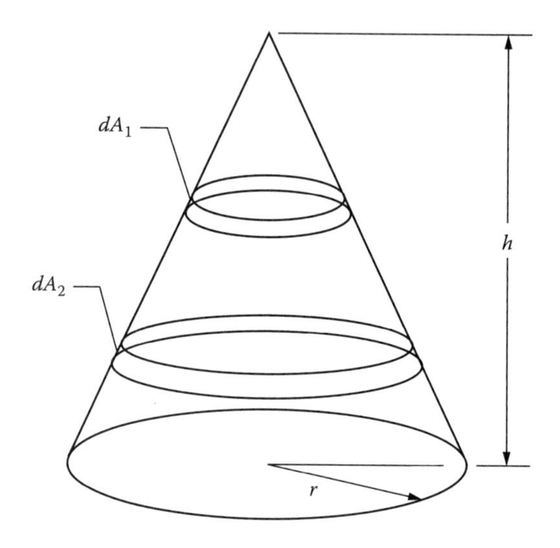

Answer:

$$R \equiv r/h;\ \alpha \equiv \left(R^2+1\right);\ \beta \equiv \left(2R^4 - R^2 - 5\right)$$

$$dF_{d1-d2} = -\frac{\cos\alpha}{2Rx^2}\left\{x - 2y\alpha + \frac{\alpha^{1/2}\left[\alpha x^4 + x^3y\beta + 3x^2y^2(\beta+8) + xy^3\alpha(6R^2-7) + 2y^4Z^2\right]}{\left[\alpha(x+y)^2 - 4xy\right]^{3/2}}\right\}$$

4.32 In terms of disk-to-disk configuration factors, derive the factor between the finite ring A_1 and the finite area A_2 on the inside of a cone.

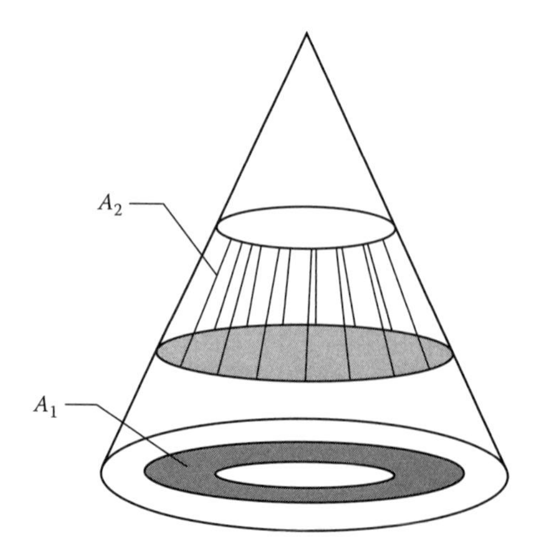

4.33 A closed right circular cylindrical shell with base diameter 1 m and height 1 m is located at the center of a spherical shell 1 m in radius.

(a) Determine the configuration factor between the inside of the sphere and itself.
(b) If the top of the cylindrical shell is removed, determine the configuration factor between the inside of the sphere and the inside of the bottom of the cylindrical shell.

Answer: (a) 0.6250; (b) 0.01072.

4.34 Show by an algebraic derivation whether the configuration factor from the interior curved surface of the frustum of a cone A_1 to its base A_2 as given in configuration factor C-111 of the configuration factor catalog at http://www.thermalradiation.net catalog is equivalent to that given in factor C-112.

4.35 Consider the interior of a black cubical enclosure. Determine the configuration factors between (a) two adjacent walls and (b) two opposite walls. A sphere of diameter equal to one-half the length of a side of the cube is placed at the center of the cube. Determine the configuration factors between (c) the sphere and one wall of the enclosure, (d) one wall of the enclosure and the sphere, and (e) the enclosure and itself.

Answer: (a) 0.20004; (b) 0.19982; (c) 1/6; (d) 0.1309; (e) 0.8691.

4.36 Obtain the configuration factor dF_{d1-2} in Homework Problem 4.2 by using contour integration.

$$\text{Answer: } F_{d1-2}=\left(\frac{2}{\pi}\right)\left\{\left[\frac{X}{(1+X^2)^{1/2}}\right]\tan^{-1}\left[\frac{Y}{(1+X^2)^{1/2}}\right]+\left[\frac{Y}{(1+Y^2)^{1/2}}\right]\tan^{-1}\left[\frac{X}{(1+Y^2)^{1/2}}\right]\right\}$$

where

$X = L/2H$

$Y = W/2H$

4.37 By use of the contour integration method of Section 4.3.3, obtain the final result in Example 4.4 for the configuration factor from an elemental area to a perpendicular circular disk.

$$\text{Answer: } F_{d1-2}=\frac{H}{2}\left[\frac{H^2+R^2+1}{\sqrt{(H^2+R^2+1)^2-4R^2}}-1\right].$$

4.38 Obtain the value of the configuration factor dF_{d1-2} for the geometry shown. The areas dA_1 and A_2 are parallel.

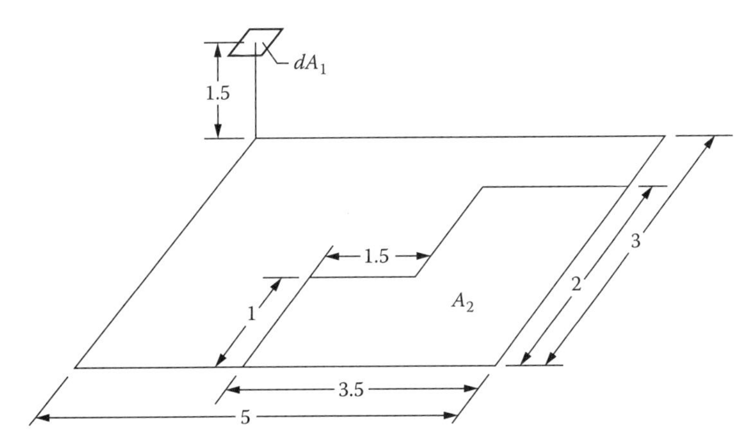

Answer: 0.01261.

4.39 For the geometry below and the special case of $l_1 = l_3$, derive algebraic expressions for the factors F_{1-2}, F_{1-3}, and F_{3-3}. Use relations from factors C-77 and C-97 from the online catalog at http://www.thermalradiation.net.

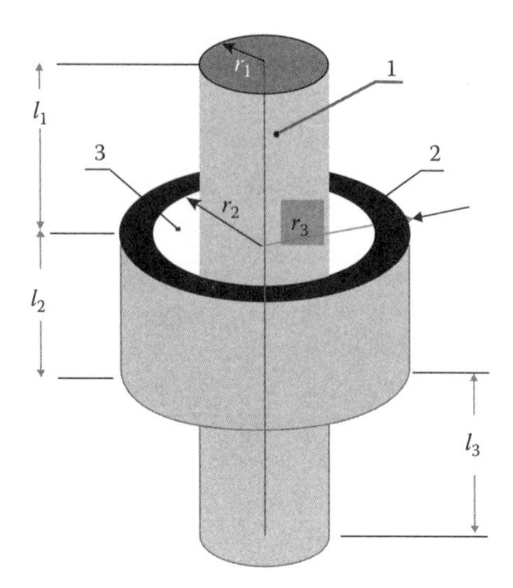

4.40 For the infinite parallel cylinders 1 and 2 with plate 3 forming a "deck" between them, find the factors F_{1-2}, F_{1-3}, and F_{2-2} in algebraic form.

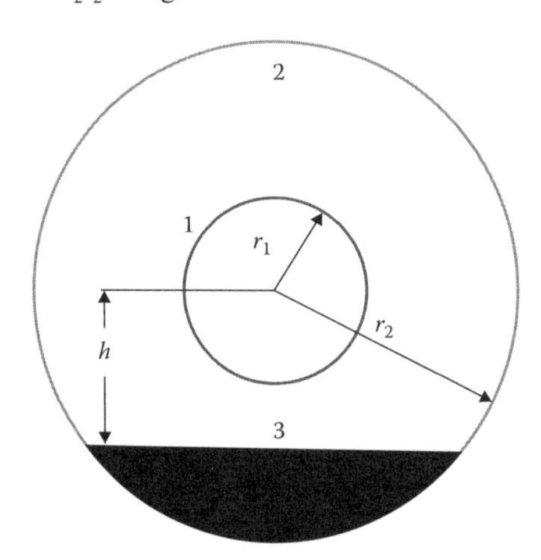

Answer: $F_{1-2} = 1 - F_{1-3} = 1 - \dfrac{\sqrt{r_2^2 - h^2}}{\pi^2 r_1} \tan^{-1}\left(\dfrac{\sqrt{(r_2^2 - h^2)}}{h}\right)$

$$F_{1-3} = \frac{A_3}{A_1} F_{3-1} = \frac{\sqrt{r_2^2 - h^2}}{\pi^2 r_1} \tan^{-1}\left(\frac{\sqrt{(r_2^2 - h^2)}}{h}\right)$$

$$F_{2-2} = 1 - \frac{\pi r_1}{(\pi - cos^{-1}(h/r_2))}\left\{1 - \frac{\sqrt{r_2^2 - h^2}}{\pi^2 r_1} \tan^{-1} \frac{\sqrt{r_2^2 - h^2}}{h}\right\}$$

$$- \frac{\sqrt{r_2^2 - h^2}}{(\pi - cos^{-1}(h/r_2))}\left\{1 - \frac{1}{\pi} \tan^{-1} \frac{\sqrt{r_2^2 - h^2}}{h}\right\}$$

4.41 Interpreting the figure in Homework Problem 4.40 as concentric spheres 1 and 2 with a disk 3 inserted between them, find the factors F_{1-2}, F_{1-3}, and F_{2-2} in algebraic form.

Answer: $R = \frac{b}{h} = \frac{\sqrt{r_2^2 - h^2}}{h}$; $F_{1-2} = 1 - F_{1-3} = 1 - \frac{1}{2}\left[1 - \frac{1}{(1+R^2)^{1/2}}\right]$

$$F_{1-3} = \frac{1}{2}\left[1 - \frac{1}{(1+R^2)^{1/2}}\right];$$

$$F_{2-2} = 1 - \frac{\pi r_1^2}{r_2^2\left[\pi - \cos^{-1}(h/r_2)\right]}\left(1 - \frac{1}{2}\left[1 - \frac{1}{(1+R^2)^{1/2}}\right]\right)$$

$$- \frac{\pi\left(r_2^2 - h^2\right)}{4r_2^2\left[\pi - \cos^{-1}(h/r_2)\right]}\left(1 - \frac{2r_1^2}{\left(r_2^2 - h^2\right)}\left(1 - \frac{1}{(1+R^2)^{1/2}}\right)\right)$$

5 Radiation Exchange in Enclosures Composed of Black and/or Diffuse–Gray Surfaces

5.1 INTRODUCTION

We begin this chapter with a discussion of the analysis of radiation exchange in an enclosure where all surfaces are black, so that no reflections need be considered. Ideal surfaces emit diffusely, so the intensity leaving each surface is *independent of direction*. For emission from an isothermal surface, the configuration factors discussed in Chapter 4 can then be used to calculate how much radiation will reach another surface. The relation for exchange between two surfaces is applied to multiple surfaces, each at a different uniform temperature, in an enclosure of black surfaces.

In the next step toward accounting for real surface properties, the surfaces of the enclosure are assumed to be *diffuse* and *gray* (some surfaces can be black), so the directional spectral emissivity and absorptivity of each surface do not depend on direction or wavelength but can depend on surface temperature. At any surface temperature T, the hemispherical total absorptivity and emissivity are equal and depend only on T, $\alpha(T) = \varepsilon(T)$. Even though this behavior is approached by only a limited number of real materials, the diffuse–gray approximation is often made to greatly simplify enclosure theory.

A specific comment is warranted as to what is meant by the individual "surfaces" or "areas" of an enclosure. Usually, the geometry tends to divide the enclosure into surface areas, such as the individual sides of a rectangular chamber. It also may be necessary to specify surface areas on the basis of heating or cooling conditions; for example, if one side of an enclosure has portions at two different temperatures, that side would be divided into two areas so this difference in boundary condition could be included. An area may also be subdivided on the basis of surface characteristics, such as separate smooth and rough portions with different emissivities. Surface areas in the radiation analysis are then defined as each portion of the enclosure boundary for which an energy balance is formed; these portions are selected on the basis of geometry, imposed heating or temperature conditions, or surface characteristics. A further consideration is solution accuracy. If too few areas are designated, the accuracy may be poor because significant nonuniformity in reflected flux over an area is not accounted for in the analysis. Dividing a surface into too many smaller areas may require excessive computation time. Thus, engineering judgment is required in selecting the size and shape of the enclosure areas and their number.

Surface areas of the enclosure can have various imposed thermal boundary conditions. A surface can be at a specified temperature, have a specified energy input, or be perfectly insulated from external energy transfer. The present analysis requires that each surface area for the enclosure analysis must be at a uniform temperature. If the heating conditions are such that the temperature would vary markedly over an area, the area should be subdivided into smaller, more nearly isothermal portions; if necessary, these portions can be of differential size. From this isothermal area requirement, the emitted energy is uniform over each surface.

A gray surface reflects only a portion of the incident energy. Two assumptions are made for reflected energy: (1) it is diffuse, so the reflected intensity at each position is uniform over all directions and (2) it is uniform over each surface area. If the reflected energy is expected to vary over an area, the area should be subdivided so that the variation is not significant over each surface area considered for the analysis. With these restrictions reasonably met, the reflected energy for each area is assumed to be diffuse and uniform like the emitted energy. Hence, the reflected and emitted energies can be combined into a single diffuse energy flux leaving the area. The geometric configuration factors can then be used for the enclosure analysis. The derivation of the F factors was based on a *diffuse-uniform intensity* leaving the surface; this diffuse-uniform condition must be valid for the sum of *both emitted and reflected* energies.

Radiative energy balances are not limited to steady-state conditions, but can be directly applied when there are transient temperature changes. The net energy flux q in the enclosure theory is the instantaneous net radiative loss from the location being considered on the boundary. For example, if a solid body is cooling only by radiation, q provides the boundary condition for the transient heat conduction solution for the temperature distribution within the solid.

In some instances, an analysis assuming diffuse–gray surface areas may not yield good results. For example, if the temperatures of the individual areas of the enclosure differ considerably from each other, then an area will be emitting predominantly in the range of wavelengths characteristic of its temperature while receiving energy predominantly in different wavelength regions. If the spectral emissivity varies with wavelength, the fact that the incident radiation has a different spectral distribution from the emitted energy makes the gray assumption invalid; that is, $\varepsilon(T) \neq \alpha(T)$. When polished (specular) surfaces are present, the diffuse reflection assumption is invalid, and the directional paths of the reflected energy should be considered. These complications are treated later.

In summary, the enclosure boundary is subdivided into areas so that over each such area the following restrictions are met:

1. All surfaces are opaque.
2. The temperature is uniform.
3. The surface properties are uniform.
4. The ε_λ, α_λ, and ρ_λ are independent of wavelength and direction so that $\varepsilon(T) = \alpha(T) = 1 - \rho(T)$, where ρ is the reflectivity.
5. All energy is emitted and reflected diffusely.
6. The incident and hence reflected energy flux is uniform over each individual area.

5.2 RADIATIVE TRANSFER FOR BLACK SURFACES

Equations 4.5 and 4.6 can be written for black surfaces to give the total energy per unit time leaving dA_1 that is incident upon dA_2 as $I_{b,1}\, dA_1 \cos\theta_1\, dA_2 \cos\theta_2/S^2$ and the total radiation leaving dA_2 that arrives at dA_1 is $I_{b,2}\, dA_2 \cos\theta_2\, dA_1 \cos\theta_1/S^2$. For a black receiving element, all incident energy is absorbed. From Equation 1.32, the blackbody total intensity is related to the blackbody hemispherical total emissive power by $I_b = E_b/\pi = \sigma T^4/\pi$, so the *net* energy per unit time $d^2Q_{d1\leftrightarrow d2}$ transferred from black element dA_1 to black element dA_2 along path S is (Figure 5.1)

$$d^2Q_{d1\leftrightarrow d2} = \sigma\left(T_1^4 - T_2^4\right)\frac{\cos\theta_1\cos\theta_2}{\pi S^2}dA_1 dA_2 \tag{5.1}$$

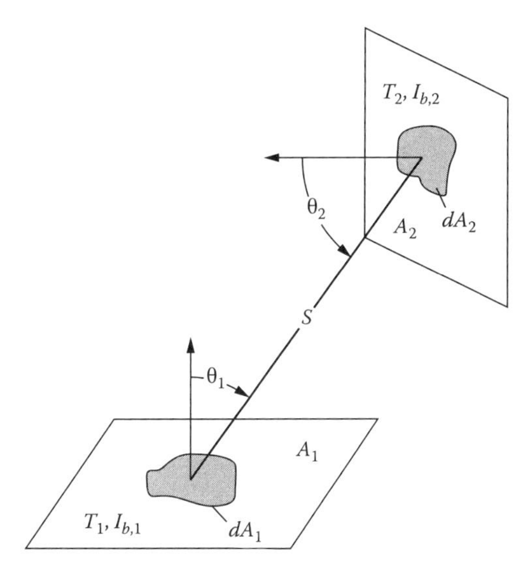

FIGURE 5.1 Radiative exchange between two black isothermal elements.

Example 5.1

The sun emits energy at a rate that can be approximated by a blackbody at T_s = 5780 K. A blackbody area element in orbit around the sun at the mean radius of the Earth's orbit, 1.49×10^{11} m, is oriented normal to the line connecting the centers of the area element and the sun. If the sun's radius is $R_S = 6.95 \times 10^8$ m, what solar flux is incident upon the element?

To the element in orbit, the sun appears as a diffuse isothermal disk element of area $dA_1 = \pi R_S^2 = \pi(6.95\times10^8)^2 = 1.52\times10^{18}\ \text{m}^2$. From the derivation of Equation 5.1, since $\theta_1 = \theta_2 = 0$, the incident energy flux on element dA_2 in orbit is $(\sigma T_s^4/\pi)dA_1(\cos\theta_1\cos\theta_2/S^2) = (\sigma T_s^4/\pi)(dA_1/S^2) = (5.6704\times10^{-8}\times5780^4/\pi)\left[1.52\times10^{18}/(1.49\times10^{11})^2\right] = 1379\ \text{W/m}^2$. This value is consistent with measured values of the solar constant, 1353 to 1394 W/m^2 and the accepted standard value of 1366 W/m^2 (Section 3.6.1).

An alternative procedure is to utilize the fact that radiant energy leaves the sun in a spherically symmetric fashion. The energy radiated from the solar sphere is $\sigma T_s^4 4\pi R_S^2$, and the area of a sphere surrounding the sun and having a radius equal to the Earth's orbit is $4\pi S^2$. Hence, the flux received at the Earth's orbit is $\sigma T_s^4 4\pi R_S^2/4\pi S^2 = \sigma T_s^4 (R_S/S)^2$, as obtained before.

Example 5.2

As shown in Figure 5.2, a black square with side 0.25 cm is at T_1 = 1100 K and is near a tube 0.30 cm in diameter. The tube opening acts as a black surface at 700 K. What is the net radiative transfer from A_1 to A_2 along the connecting path S?

For this geometry, the A_1 and A_2 have small dimensions compared with the length S between them. If Equation 5.1 is used, the cos θ_1, cos θ_2, and S do not vary significantly for positions along A_1 and A_2. Hence, accurate transfer results can be obtained without integrating to obtain an F factor. Then, for Equation 5.1, cos θ_1 is found from the sides of the right triangle A_2-0-A_1 as $\cos\theta_1 = 5/(8^2 + 5^2)^{1/2} = 5/89^{1/2}$. The other factors in the energy exchange equation are given, so the net radiative exchange is

$$\sigma(T_1^4 - T_2^4)\frac{\cos\theta_1\cos\theta_2}{\pi S^2}A_1A_2 = 5.6704\times10^{-8}(1100^4 - 700^4)\frac{5}{89^{1/2}}\frac{\cos 20°}{\pi(89/10^4)}\frac{0.25^2}{10^4}\frac{\pi 0.30^2}{4\times10^4}$$

$$= 5.462\times10^{-5}\ \text{W}$$

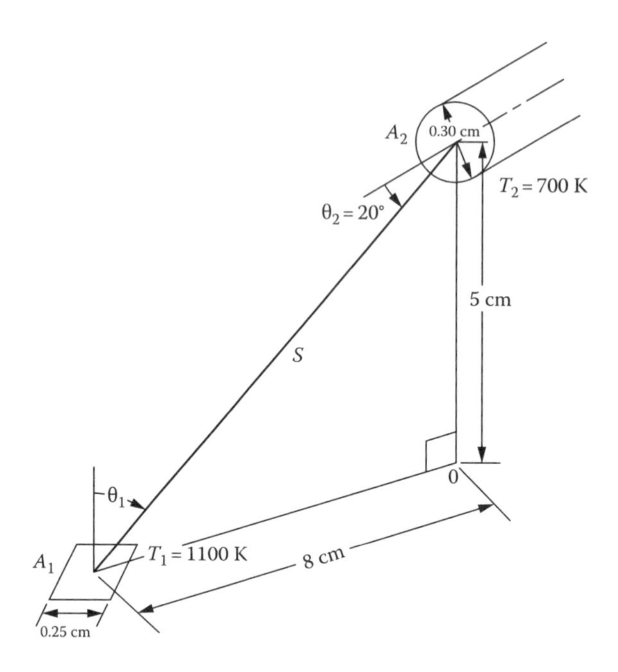

FIGURE 5.2 Radiative transfer between a small square area and a small circular-tube opening.

5.2.1 Transfer between Black Surfaces by the Use of Configuration Factors

Using Equation 4.7, the geometric factors in Equation 5.1 for black area elements can be written in terms of configuration factors to provide a shorter form as

$$d^2Q_{d1\leftrightarrow d2} = \sigma\left(T_1^4 - T_2^4\right)dF_{d1-d2}dA_1 = \sigma\left(T_1^4 - T_2^4\right)dF_{d2-d1}dA_2 \tag{5.2}$$

Similarly, for radiative transfer between a black differential element and a black finite area,

$$dQ_{d1\leftrightarrow 2} = dQ_{d1\to 2} - dQ_{2\to d1} = \sigma T_1^4 dA_1\, F_{d1-2} - \sigma T_2^4 A_2\, dF_{2-d1} \tag{5.3}$$

or, using Equation 4.15, the net transfer is

$$dQ_{d1\leftrightarrow 2} = \sigma\left(T_1^4 - T_2^4\right)dA_1\, F_{d1-2} = \sigma\left(T_1^4 - T_2^4\right)A_2\, dF_{2-d1} \tag{5.4}$$

For two black surfaces with finite areas

$$Q_{1\leftrightarrow 2} = Q_{1\to 2} - Q_{2\to 1} = \sigma T_1^4 A_1 F_{1-2} - \sigma T_2^4 A_2 F_{2-1} \tag{5.5}$$

or, using the reciprocity relation, Equation 4.20,

$$Q_{1\leftrightarrow 2} = \sigma\left(T_1^4 - T_2^4\right)A_1\, F_{1-2} = \sigma\left(T_1^4 - T_2^4\right)A_2\, F_{2-1} \tag{5.6}$$

5.2.2 Radiation Exchange in a Black Enclosure

The derivations so far are for enclosures with black surfaces, which is seldom the case in practical applications. In addition, a surface may not be isothermal, in which case the surfaces must be

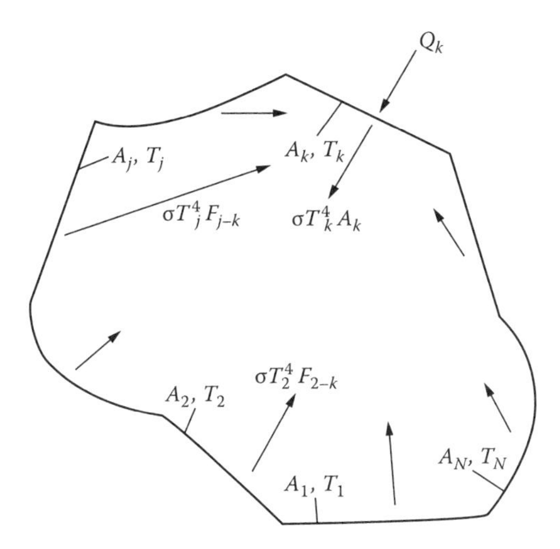

FIGURE 5.3 Enclosure composed of N black isothermal surface areas (shown in cross section for simplicity).

subdivided into smaller ones that meet the isothermal approximation. The analysis here is for a large number of black surfaces in an enclosure.

An energy balance is formed on a typical enclosure surface A_k as in Figure 5.3. The Q_k is the combined energy supplied to A_k by all sources other than by radiation inside the enclosure; this maintains A_k at T_k while A_k exchanges radiation with the enclosure surfaces. The Q_k could be composed of convection heat transfer to the inside of the wall, and/or conduction heat transfer through the wall from an energy source outside, and/or an energy source in the wall itself such as an electrical heater. For wall cooling, such as by cooling channels within the wall, the contribution to Q_k is negative. The emission from A_k to the enclosure is $\sigma T_k^4 A_k$. The radiant energy received by A_k from another surface A_j is $\sigma T_j^4 A_j F_{j-k}$. The energy balance is then

$$Q_k = \sigma T_k^4 A_k - \sum_{j=1}^{N} \sigma T_j^4 A_j F_{j-k} \tag{5.7}$$

where the summation includes energy arriving from all surfaces inside the enclosure including A_k if A_k is concave. Equation 5.7 can be written in alternative forms. Applying reciprocity to the terms in the summation results in

$$Q_k = A_k \left(\sigma T_k^4 - \sum_{j=1}^{N} \sigma T_j^4 F_{k-j} \right) \tag{5.8}$$

For a complete enclosure, from Equation 4.29, $\sum_{j=1}^{N} F_{k-j} = 1$, so that

$$Q_k = A_k \left(\sigma T_k^4 \sum_{j=1}^{N} F_{k-j} - \sigma \sum_{j=1}^{N} T_j^4 F_{k-j} \right) = \sigma A_k \sum_{j=1}^{N} \left(T_k^4 - T_j^4 \right) F_{k-j} \tag{5.9}$$

This is in the form of a sum of the net radiative energy transferred from A_k to each surface within the enclosure.

Example 5.3

The three-sided black enclosure of Example 4.13 has its black surfaces maintained at T_1, T_2, and T_3. To maintain each temperature, determine the amount of energy that must be supplied to each surface by means other than radiation inside the enclosure. This energy equals the net radiative loss from each surface by radiative exchange within the enclosure.

Equation 5.9 is written for each surface as

$$Q_1 = A_1\left[F_{1-2}\sigma\left(T_1^4 - T_2^4\right) + F_{1-3}\sigma\left(T_1^4 - T_3^4\right)\right]$$

$$Q_2 = A_2\left[F_{2-1}\sigma\left(T_2^4 - T_1^4\right) + F_{2-3}\sigma\left(T_2^4 - T_3^4\right)\right]$$

$$Q_3 = A_3\left[F_{3-1}\sigma\left(T_3^4 - T_1^4\right) + F_{3-2}\sigma\left(T_3^4 - T_2^4\right)\right]$$

The configuration factors are in Example 4.13. Thus, all factors on the right sides are known, and the Q_k values can be computed. A check on the numerical results is that, from overall energy conservation, the net Q_k added to the entire enclosure, $\sum_{k=1}^{N} Q_k$, must be zero to maintain steady temperatures. This is also shown by adding all the terms on the right sides of the three equations and using reciprocity such as $A_2F_{2-1} = A_1F_{1-2}$.

Example 5.4

The enclosure of Example 4.13 has two sides maintained at T_1 and T_2. The third side is insulated on the outside, and there is only radiative transfer on the inside so that $Q_3 = 0$. Determine Q_1, Q_2, and T_3.

Equation 5.9 is written for each surface as

$$Q_1 = A_1\left[F_{1-2}\sigma\left(T_1^4 - T_2^4\right) + F_{1-3}\sigma\left(T_1^4 - T_3^4\right)\right]$$

$$Q_2 = A_2\left[F_{2-1}\sigma\left(T_2^4 - T_1^4\right) + F_{2-3}\sigma\left(T_2^4 - T_3^4\right)\right]$$

$$0 = A_3\left[F_{3-1}\sigma\left(T_3^4 - T_1^4\right) + F_{3-2}\sigma\left(T_3^4 - T_2^4\right)\right]$$

The final equation is solved for T_3^4. This is inserted into the first two equations to obtain Q_1 and Q_2.

Example 5.5

A very long black heated tube A_1 of length L is enclosed by a concentric black split cylinder as in Figure 5.4. The diameter of the split cylinder is twice that of the heated tube, and one-half as much energy *flux* is to be removed from the upper area A_3 as from the lower area A_2. If T_1 = 1700 K and a heat flux $Q_1/A_1 = 3 \times 10^5$ W/m² is supplied to the heated tube, what are the values of T_2, T_3, Q_2, and Q_3? Neglect the effect of the tube ends.

Writing Equation 5.9 for each surface gives

$$Q_1 = A_1\left[F_{1-2}\sigma\left(T_1^4 - T_2^4\right) + F_{1-3}\sigma\left(T_1^4 - T_3^4\right)\right]$$

$$Q_2 = A_2\left[F_{2-1}\sigma\left(T_2^4 - T_1^4\right) + F_{2-3}\sigma\left(T_2^4 - T_3^4\right)\right]$$

$$Q_3 = A_3\left[F_{3-1}\sigma\left(T_3^4 - T_1^4\right) + F_{3-2}\sigma\left(T_3^4 - T_2^4\right)\right]$$

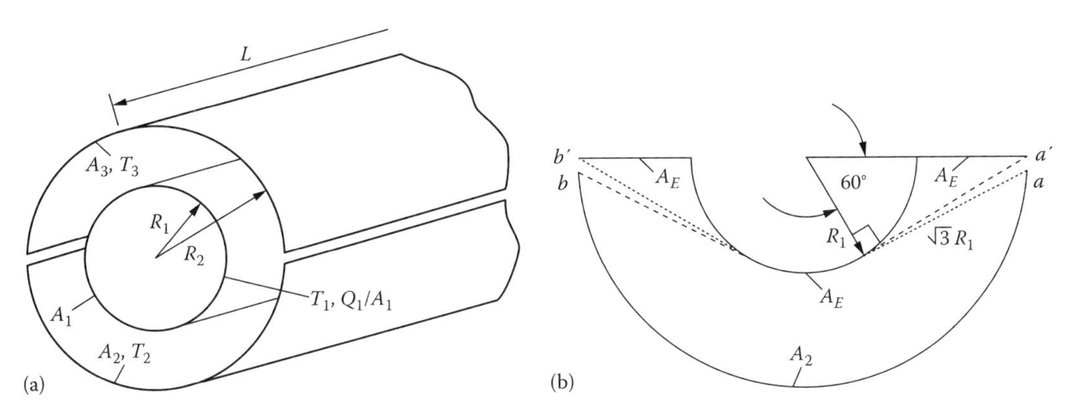

FIGURE 5.4 Radiant energy exchange in split-circular cylinder configuration, $L \gg R_2$. (a) Geometry of enclosure and (b) auxiliary construction to determine F_{2-2}.

From the geometry, $A_1/A_3 = A_1/A_2 = \pi D_1 L/\frac{1}{2}\pi D_2 L = 1$, since $D_2 = 2D_1$. From an energy balance, $Q_1 + Q_2 + Q_3 = 0$ and, since $A_1 = A_2 = A_3$, $(Q_1/A_1) + (Q_2/A_2) + (Q_3/A_3) = 0$. From the statement of the problem, $Q_3/A_3 = Q_2/2A_2$, and this yields

$$\frac{Q_2}{A_2} = -\frac{2}{3}\frac{Q_1}{A_1} = -2.0\times10^5 \text{ W/m}^2, \quad \frac{Q_3}{A_3} = -\frac{1}{3}\frac{Q_1}{A_1} = -1.0\times10^5 \text{ W/m}^2$$

From the symmetry and configuration-factor algebra, $F_{1-2} = F_{1-3} = 1/2$, $F_{2-1} = F_{3-1} = A_1 F_{1-3}/A_3 = 1/2$, and $F_{2-3} = F_{3-2}$. To determine F_{2-3}, it is known that $F_{2-1} + F_{2-2} + F_{2-3} = 1$. Using $F_{2-1} = 1/2$ gives $F_{2-3} = (1/2) - F_{2-2}$. In the auxiliary construction of Figure 5.4b, $F_{2-2} = 1 - F_{2-E}$. The effective area A_E has been drawn in to leave unchanged the view of surface 2 to itself and to simplify the geometry so that the crossed-string method can be used to determine F_{2-E}. The uncrossed strings from a to a' and b to b' have zero length. The crossed strings extend from a to b' and a' to b, and each has length $2\sqrt{3}R_1 + \pi R_1/3$. Then, from Section 4.3.3 and the fact that $A_2 = A_1 = 2\pi R_1$, $F_{2-E} = (2\sqrt{3}R_1 + \pi R_1/3)/2\pi R_1 = (\sqrt{3}/\pi) + (1/6)$. It then follows that $F_{2-3} = (1/2) - F_{2-2} = (1/2) - (1 - F_{2-E}) = (\sqrt{3}/\pi) - (1/3) = 0.2180$. Using this information, the energy exchange equations become

$$3\times10^5 = \frac{\sigma}{2}\left(1700^4 - T_2^4\right) + \frac{\sigma}{2}\left(1700^4 - T_3^4\right)$$

$$-2.0\times10^5 = \frac{\sigma}{2}\left(T_2^4 - 1700^4\right) + 0.2180\sigma\left(T_2^4 - T_3^4\right)$$

$$-1.0\times10^5 = \frac{\sigma}{2}\left(T_3^4 - 1700^4\right) + 0.2180\sigma\left(T_3^4 - T_2^4\right)$$

Adding the second and third equations results in the first, so only two equations are independent. Solving the first and second equations gives $T_2 = 1207$ K and $T_3 = 1415$ K.

5.3 RADIATION BETWEEN FINITE DIFFUSE–GRAY AREAS

5.3.1 Net-Radiation Method for Enclosures

Consider an enclosure of N discrete internal surface areas as in Figure 5.5. To analyze the radiation exchange between the surface areas within the enclosure, two common problems are to find: (1) the required energy supplied to a surface is to be determined when the surface temperature is specified and (2) the temperature that a surface will achieve is to be found when a known heat input is imposed. Some more general boundary conditions are considered later.

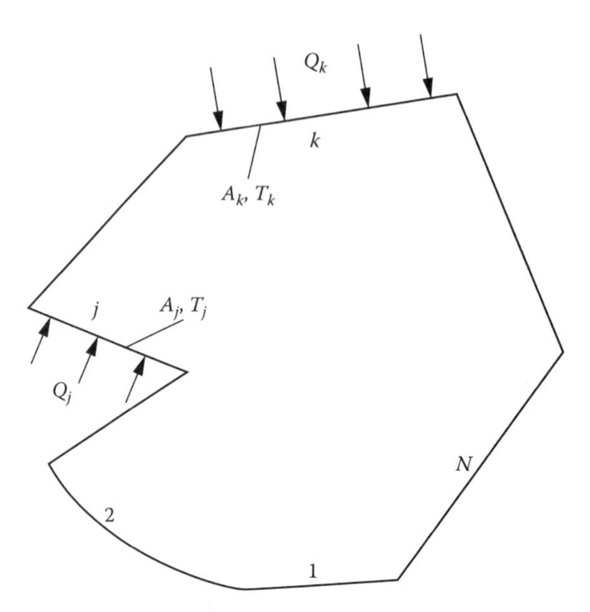

FIGURE 5.5 Enclosure composed of N discrete surface areas with typical surfaces j and k (shown in cross section for simplicity).

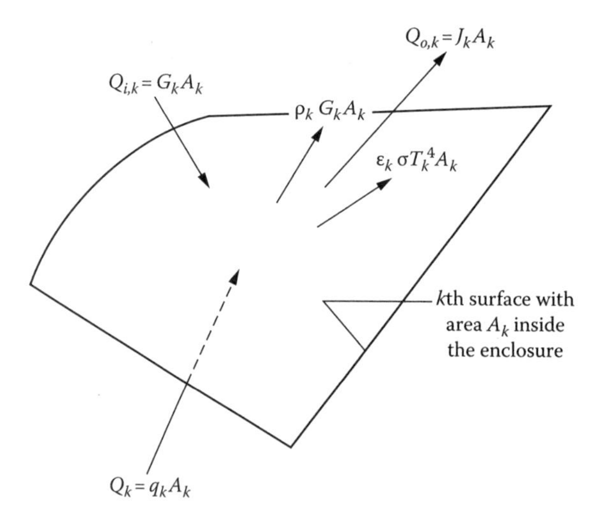

FIGURE 5.6 Energy quantities incident upon and leaving typical surface inside enclosure.

A complex radiative exchange occurs inside the enclosure as radiation leaves a surface, travels to other surfaces, is partially reflected, and is then rereflected many times within the enclosure with partial absorption at each contact with a surface. It is complicated to follow the radiation as it undergoes this process; fortunately, this is not always necessary. A convenient analysis can be formulated by using the *net-radiation method*. This method was first devised by Hottel and later developed in a different manner by Poljak (1935). An alternative approach was given by Gebhart (1961, 1971). All the methods are basically equivalent, as demonstrated by Sparrow (1963b) and shown here in an example. Further, Poljak approach, which is usually convenient, is now given; the other formulations are then presented more briefly.

Consider the kth *inside* surface area A_k of the enclosure in Figures 5.5 and 5.6. The G_k and J_k are the rates of incoming and outgoing *radiant* energy per unit area of A_k. The quantity q_k

is the energy flux supplied to A_k by some means other than the radiation inside the enclosure, to make up for the net radiative gain or loss and thereby maintain the specified inside surface temperature. For example, if A_k is the inside surface of a wall of finite thickness, Q_k could be the heat conducted through the wall from the outside to A_k. An energy balance for A_k provides the relation

$$Q_k = q_k A_k = \left(J_k - G_k\right) A_k \tag{5.10}$$

A second equation results from the energy flux leaving the surface being composed of emitted plus reflected energy. This gives, noting $\alpha_k = \varepsilon_k$ for a gray surface,

$$J_k = \varepsilon_k \sigma T_k^4 + \rho_k G_k = \varepsilon_k \sigma T_k^4 + \left(1 - \alpha_k\right) G = \varepsilon_k \sigma T_k^4 + \left(1 - \varepsilon_k\right) G_k \tag{5.11}$$

where $\rho_k = 1 - \alpha_k = 1 - \varepsilon_k$ has been used for opaque gray surfaces. The term *radiosity* is often used for J_k. The incident flux or *irradiation*, G_k, is derived from the portions of radiant energy leaving the surfaces inside the enclosure that arrive at the kth surface. If the kth surface can view itself (is concave), a portion of its outgoing flux will contribute directly to its incident flux. The incident energy is then equal to

$$A_k G_k = A_1 J_1 F_{1-k} + A_2 J_2 F_{2-k} + \cdots + A_j J_j F_{j-k} + \cdots A_k J_k F_{k-k} + \cdots + A_N J_N F_{N-k} \tag{5.12}$$

From configuration-factor reciprocity, Equation 4.19,

$$A_1 F_{1-k} = A_k F_{k-1}; \quad A_2 F_{2-k} = A_k F_{k-2}; \quad \cdots \quad A_N F_{N-k} = A_k F_{k-N} \tag{5.13}$$

Then Equation 5.12 can be written so the only area appearing is A_k:

$$A_k G_k = A_k J_1 F_{k-1} + A_k J_2 F_{k-2} + \cdots + A_k J_j F_{k-j} + \cdots + A_k J_k F_{k-k} + \cdots + A_k J_N F_{k-N} \tag{5.14}$$

so that the incident flux is

$$G_k = \sum_{j=1}^{N} J_j F_{k-j} \tag{5.15}$$

Equations 5.10, 5.11, and 5.15 are simultaneous relations between q_k, T_k, G_k, and J_k for each surface, and, through Equation 5.15, to the other surfaces. One solution procedure is to note that Equations 5.11 and 5.15 provide two different expressions for G_k. These are each substituted into Equation 5.10 to eliminate G_k and provide two energy balance equations for q_k in terms of T_k and J_k:

$$\frac{Q_k}{A_k} = q_k = \frac{\varepsilon_k}{1 - \varepsilon_k}\left(\sigma T_k^4 - J_k\right) \tag{5.16}$$

$$\frac{Q_k}{A_k} = q_k = J_k - \sum_{j=1}^{N} J_j F_{k-j} = \sum_{j=1}^{N} \left(J_k - J_j\right) F_{k-j} \tag{5.17}$$

The q_k can be regarded as either the energy flux supplied to surface k by means other than internal radiation (such as by convection or conduction to A_k) or as the net radiative loss from A_k by radiation exchange inside the enclosure. Equation 5.16 or 5.17 is the balance between net radiative energy loss and energy supplied by means other than the radiation inside the enclosure.

Equations 5.16 and 5.17 are written for each of the N surfaces in the enclosure. This provides $2N$ equations in $2N$ unknowns. The J_k are N of the unknowns, and the remaining unknowns consist of q and T, depending on what boundary quantities are specified. Later, the J_k are eliminated to give N equations relating the N unknown q and T. Other forms of Equation 5.16 are

$$J_k = \sigma T_k^4 - \frac{1-\varepsilon_k}{\varepsilon_k} q_k \quad \text{or} \quad \sigma T_k^4 = \frac{1-\varepsilon_k}{\varepsilon_k} q_k + J_k \tag{5.18}$$

Equations 5.16 and 5.18 were obtained by eliminating G_k from Equations 5.10 and 5.11. If J_k is eliminated instead, then the results are

$$q_k = \varepsilon_k \sigma T_k^4 - \varepsilon_k G_k \quad \text{or} \quad G_k = \sigma T_k^4 - \frac{q_k}{\varepsilon_k} \tag{5.19}$$

Equation 5.19 shows that the net energy leaving the surface by radiation is the emitted energy minus the absorbed incident energy $\varepsilon_k G_k$. Hence, if q_k and T_k^4 are available from a solution, the incident energy that is absorbed by a gray surface can be found from

$$\alpha_k G_k = \varepsilon_k G_k = \varepsilon_k \sigma T_k^4 - q_k \tag{5.20}$$

The absorbed energy added to the energy supplied by other means is equal to the emitted energy, $\alpha_k G_k + q_k = \varepsilon_k \sigma T_k^4$.

Before continuing with the development, examples are given to illustrate using Equations 5.16 and 5.17 as simultaneous equations for each enclosure surface.

Example 5.6

Derive the expression for the net radiative heat exchange between two infinite parallel flat plates in terms of their temperatures T_1 and T_2 (Figure 5.7).

Since for infinite plates all radiation leaving one plate will arrive at the other plate, the $F_{1-2} = F_{2-1} = 1$. Equations 5.16 and 5.17 for each plate are then

$$q_1 = \frac{\varepsilon_1}{1-\varepsilon_1}\left(\sigma T_1^4 - J_1\right), \quad q_1 = J_1 - J_2 \tag{5.6.1}$$

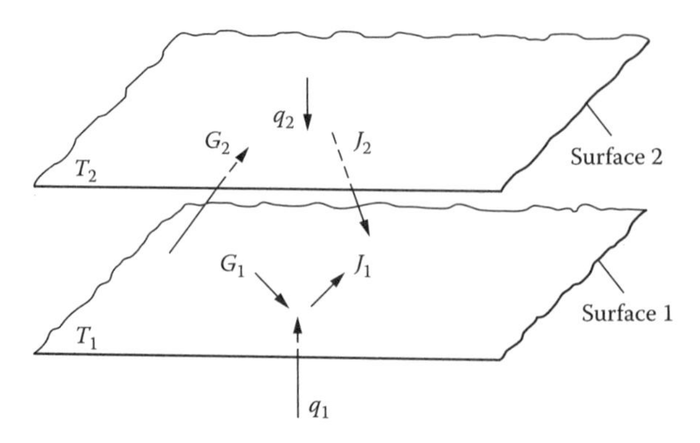

FIGURE 5.7 Heat fluxes for radiant interchange between infinite parallel flat plates.

$$q_2 = \frac{\varepsilon_2}{1-\varepsilon_2}\left(\sigma T_2^4 - J_2\right), \quad q_2 = J_2 - J_1 \tag{5.6.2}$$

Comparing Equations 5.6.1 and 5.6.2, $q_1 = -q_2$, so the energy added to surface 1 is removed from surface 2. The q_1 is thus the net energy transferred from 1 to 2 as requested in the problem statement. Equations 5.6.1 and 5.6.2 yield

$$J_1 = \sigma T_1^4 - \frac{1-\varepsilon_1}{\varepsilon_1} q_1, \quad J_2 = \sigma T_2^4 - \frac{1-\varepsilon_2}{\varepsilon_2} q_2 = \sigma T_2^4 + \frac{1-\varepsilon_2}{\varepsilon_2} q_1$$

These are substituted into Equation 5.6.1, and the result solved for q_1:

$$q_1 = -q_2 = \frac{\sigma\left(T_1^4 - T_2^4\right)}{1/\varepsilon_1\left(T_1\right) + 1/\varepsilon_2\left(T_2\right) - 1} \tag{5.6.3}$$

The functional notation $\varepsilon(T)$ is used to emphasize that ε_1 and ε_2 can be functions of temperature. Since T_1 and T_2 are specified, ε_1 and ε_2 are substituted at their proper temperatures and q_1 is directly calculated.

Example 5.7

For the parallel-plate geometry of the previous example, what temperature will surface 1 reach for a given energy flux input q_1 while T_2 is held at a specified value?

Equation 5.6.3 applies and, when solved for T_1, gives

$$T_1 = \left\{\frac{q_1}{\sigma}\left[\frac{1}{\varepsilon_1\left(T_1\right)} + \frac{1}{\varepsilon_2\left(T_2\right)} - 1\right] + T_2^4\right\}^{1/4} \tag{5.7.1}$$

Since $\varepsilon_1(T_1)$ is a function of T_1, which is unknown, an iterative solution is necessary. A trial T_1 is selected, and ε_1 is chosen at this value. Equation 5.6.3 is solved for T_1, and this value is used to select ε_1 for the next approximation. The process is continued until $\varepsilon_1(T_1)$ and T_1 do not change with further iterations.

Example 5.8

Derive an expression for the net-radiation exchange between two uniform temperature concentric diffuse–gray spheres as in Figure 5.8.

This situation is more complicated than for infinite parallel plates, as the two surfaces have unequal areas and surface 2 can partially view itself. The configuration factors were derived in Example 4.11 as $F_{1-2} = 1$, $F_{2-1} = A_1/A_2$, and $F_{2-2} = 1 - (A_1/A_2)$. Equations 5.16 and 5.17 are written for each of the two sphere surfaces as

$$Q_1 = A_1\frac{\varepsilon_1}{1-\varepsilon_1}\left(\sigma T_1^4 - J_1\right), \quad Q_1 = A_1\left(J_1 - J_2\right) \tag{5.8.1}$$

$$Q_2 = A_2\frac{\varepsilon_2}{1-\varepsilon_2}\left(\sigma T_2^4 - J_2\right) \tag{5.8.2}$$

$$Q_2 = A_2\left[J_2 - \frac{A_1}{A_2}J_1 - \left(1 - \frac{A_1}{A_2}\right)J_2\right] = A_1\left(J_2 - J_1\right) \tag{5.8.3}$$

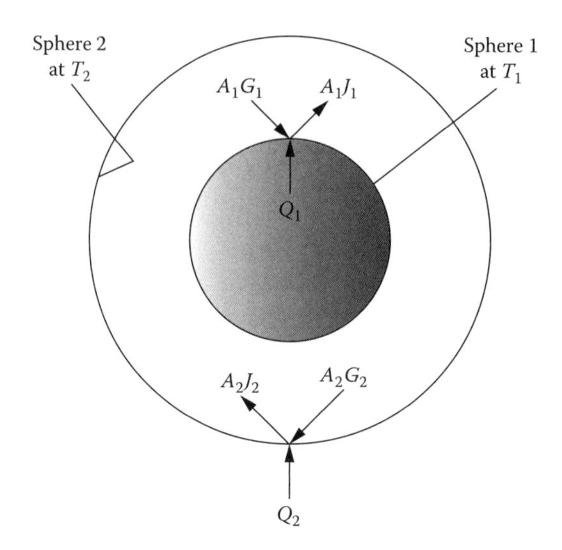

FIGURE 5.8 Energy quantities for radiant interchange between two concentric spheres.

Comparing Equations 5.8.1 and 5.8.3 reveals that $Q_1 = -Q_2$, as expected from an overall energy balance. The four equations (Equations 5.8.1, 5.8.2, and 5.8.3) can be solved for the four unknowns J_1, J_2, Q_1, and Q_2. This yields the net energy transfer (supplied to surface 1 and removed at surface 2):

$$Q_1 = \frac{A_1\sigma\left(T_1^4 - T_2^4\right)}{1/\varepsilon_1(T_1) + (A_1/A_2)\left[1/\varepsilon_2(T_2) - 1\right]} \tag{5.8.4}$$

When the spheres in Example 5.8 are not concentric, all the radiation leaving surface 1 is still incident on surface 2. The configuration factor F_{1-2} is again 1 and, with the use of the same assumptions, the analysis follows as before, leading to Equation 5.8.4. However, when sphere 1 is relatively small (say, one-half the diameter of sphere 2) and the eccentricity is large, the geometry is so different from the concentric case that using Equation 5.8.4 seems intuitively incorrect. The error in using Equation 5.8.4 is that it was derived on the basis that q, G, and J are uniform over each of A_1 and A_2. These conditions are exactly met only for the concentric case. For the eccentric case, the A_1 and A_2 need to be subdivided to improve accuracy.

Example 5.9

A gray isothermal body with area A_1 and temperature T_1 and without any concave indentations is completely enclosed by a much larger gray isothermal enclosure having area A_2. How much energy is being transferred by radiation from A_1 to A_2? The area A_1 cannot see any part of itself, and A_1 is not near A_2.

Since A_1 is completely enclosed and $F_{1-1} = 0$, the configuration factors and analysis are the same as in Example 5.8, which results in Q_1 given by Equation 5.8.4. This is valid, as A_1 is specified as rather centrally located within A_2 and hence the heat fluxes tend to be uniform over A_1. For the present situation, $A_1 \ll A_2$, and Equation 5.8.4 reduces to (unless $\boldsymbol{\varepsilon}_2$ is very small)

$$Q_1 = A_1\varepsilon_1(T_1)\sigma\left(T_1^4 - T_2^4\right) \tag{5.9.1}$$

Note that this result is independent of the emissivity ε_2 of the enclosure (the enclosure acts like a black cavity unless ε_2 is very small so that A_2 is highly reflective).

Example 5.10

Consider a long enclosure made up of three surfaces as in Figure 5.9. The enclosure has a uniform cross section and is long enough that its ends can be neglected in the radiative energy balances. How much energy has to be supplied to each surface (equal to the net radiative energy loss from each surface resulting from exchange within the enclosure) to maintain the surfaces at temperatures T_1, T_2, and T_3?

Write Equations 5.16 and 5.17 for each surface:

$$\frac{Q_1}{A_1} = \frac{\varepsilon_1}{1-\varepsilon_1}\left(\sigma T_1^4 - J_1\right) \tag{5.10.1}$$

$$\frac{Q_1}{A_1} = J_1 - F_{1-1}J_1 - F_{1-2}J_2 - F_{1-3}J_3 \tag{5.10.2}$$

$$\frac{Q_2}{A_2} = \frac{\varepsilon_2}{1-\varepsilon_2}\left(\sigma T_2^4 - J_2\right) \tag{5.10.3}$$

$$\frac{Q_2}{A_2} = J_2 - F_{2-1}J_1 - F_{2-2}J_2 - F_{2-3}J_3 \tag{5.10.4}$$

$$\frac{Q_3}{A_3} = \frac{\varepsilon_3}{1-\varepsilon_3}\left(\sigma T_3^4 - J_3\right) \tag{5.10.5}$$

$$\frac{Q_3}{A_3} = J_3 - F_{3-1}J_1 - F_{3-2}J_2 - F_{3-3}J_3 \tag{5.10.6}$$

The first of each of these three pairs is solved for J, and the J is substituted into the second equation of each pair to obtain

$$\begin{aligned}
&\frac{Q_1}{A_1}\left(\frac{1}{\varepsilon_1} - F_{1-1}\frac{1-\varepsilon_1}{\varepsilon_1}\right) - \frac{Q_2}{A_2}F_{1-2}\frac{1-\varepsilon_2}{\varepsilon_2} - \frac{Q_3}{A_3}F_{1-3}\frac{1-\varepsilon_3}{\varepsilon_3} \\
&\quad = \left(1-F_{1-1}\right)\sigma T_1^4 - F_{1-2}\sigma T_2^4 - F_{1-3}\sigma T_3^4 \\
&\quad = F_{1-2}\sigma\left(T_1^4 - T_2^4\right) + F_{1-3}\sigma\left(T_1^4 - T_3^4\right)
\end{aligned} \tag{5.10.7}$$

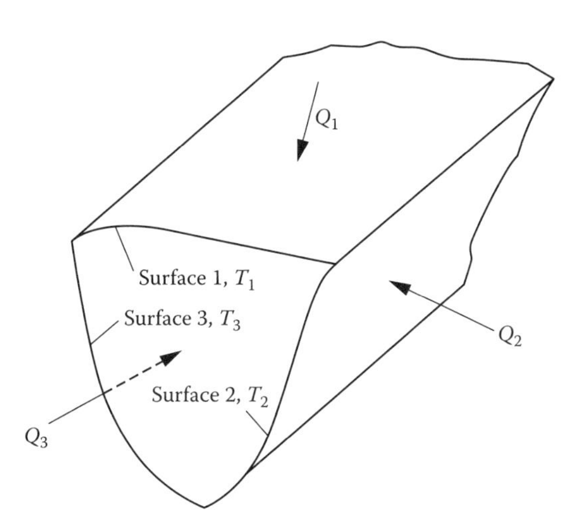

FIGURE 5.9 Long enclosure composed of three surfaces (ends neglected); cross section is uniform.

$$-\frac{Q_1}{A_1}F_{2-1}\frac{1-\varepsilon_1}{\varepsilon_1}+\frac{Q_2}{A_2}\left(\frac{1}{\varepsilon_2}-F_{2-2}\frac{1-\varepsilon_2}{\varepsilon_2}\right)-\frac{Q_3}{A_3}F_{2-3}\frac{1-\varepsilon_3}{\varepsilon_3}$$
$$=-F_{2-1}\sigma T_1^4+(1-F_{2-2})\sigma T_2^4-F_{2-3}\sigma T_3^4$$
$$=F_{2-1}\sigma\left(T_2^4-T_1^4\right)+F_{2-3}\sigma\left(T_2^4-T_3^4\right) \tag{5.10.8}$$

$$-\frac{Q_1}{A_1}F_{3-1}\frac{1-\varepsilon_1}{\varepsilon_1}-\frac{Q_2}{A_2}F_{3-2}\frac{1-\varepsilon_2}{\varepsilon_2}+\frac{Q_3}{A_3}\left(\frac{1}{\varepsilon_3}-F_{3-3}\frac{1-\varepsilon_3}{\varepsilon_3}\right)$$
$$=-F_{3-1}\sigma T_1^4-F_{3-2}\sigma T_2^4+(1-F_{3-3})\sigma T_3^4$$
$$=F_{3-1}\sigma\left(T_3^4-T_1^4\right)+F_{3-2}\sigma\left(T_3^4-T_2^4\right) \tag{5.10.9}$$

Since the T values are known, the ε can be specified from surface-property data at their appropriate T values and the three simultaneous equations solved for the Q supplied to each surface. Note that the results are approximations, because the radiosity leaving each surface is not uniform as assumed by using Equations 5.16 and 5.17. This is because the reflected flux is not uniform as a result of the enclosure geometry. Greater accuracy is obtained by dividing each of the three sides into additional surface areas to yield a larger set of simultaneous equations.

5.3.1.1 System of Equations Relating Surface Heating Q and Surface Temperature T

The form of Equations 5.10.7 through 5.10.9 shows that the Q and T for an enclosure of N surfaces can be related by a system of N equations. It is evident from Equations 5.10.7 through 5.10.9 that the general form for the kth surface is

$$\sum_{j=1}^{N}\left(\frac{\delta_{kj}}{\varepsilon_j}-F_{k-j}\frac{1-\varepsilon_j}{\varepsilon_j}\right)\frac{Q_j}{A_j}=\sum_{j=1}^{N}\left(\delta_{kj}-F_{k-j}\right)\sigma T_j^4=\sum_{j=1}^{N}F_{k-j}\sigma\left(T_k^4-T_j^4\right) \tag{5.21}$$

Corresponding to each surface, k takes on one of the values 1, 2, … N, and δ_{kj} is Kronecker delta, defined as

$$\delta_{kj}=\begin{cases}1 & \text{when } k=j\\ 0 & \text{when } k\neq j\end{cases}$$

When the surface temperatures are specified, the right side of Equation 5.21 is known and there are N simultaneous equations for the unknown Q. In general, the energy inputs to some of the surfaces may be specified and the temperatures of these surfaces are to be determined. There are still a total of N unknown Q and T, and Equation 5.21 provides the necessary number of relations. If the values of ε depend on temperature, it is necessary initially to guess the unknown T. Then, the ε values can be chosen and the system of equations solved. The T values are used to select new ε, and the process repeated until the T and ε values no longer change. Again, note that the results may be approximate because the uniform radiosity assumption is not perfectly fulfilled over each finite area. Division into smaller areas is required to improve accuracy. As a practical example, the solution of an enclosure with mixed boundary conditions to simulate heat-generating electronic modules can be found in Arimilli and Ketkar (1988).

At least one enclosure surface must have a specified temperature as a boundary condition. If all surfaces have a specified heat flux (even if they meet the energy conservation requirement that $\sum_k q_k A_k = 0$), then the left-hand side of Equation 5.21 is a known constant, and the result is that the surface temperatures become indeterminate.

Example 5.11

Consider an enclosure of three sides as in Figure 5.9. Side 1 is maintained at T_1, side 2 is uniformly heated with flux q_2, and the third side is perfectly insulated on the outside. What are the equations to determine Q_1, T_2, and T_3?

The conditions give $Q_2/A_2 = q_2$ and $Q_3 = 0$. Then Equation 5.21 yields three equations, where the unknowns have been gathered on the left sides:

$$\frac{Q_1}{A_1}\left(\frac{1}{\varepsilon_1}-F_{1-1}\frac{1-\varepsilon_1}{\varepsilon_1}\right)+F_{1-2}\sigma T_2^4+F_{1-3}\sigma T_3^4=(1-F_{1-1})\sigma T_1^4+q_2F_{1-2}\frac{1-\varepsilon_2}{\varepsilon_2} \tag{5.11.1}$$

$$-\frac{Q_1}{A_1}F_{2-1}\frac{1-\varepsilon_1}{\varepsilon_1}-(1-F_{2-2})\sigma T_2^4+F_{2-3}\sigma T_3^4=-F_{2-1}\sigma T_1^4-q_2\left(\frac{1}{\varepsilon_2}-F_{2-2}\frac{1-\varepsilon_2}{\varepsilon_2}\right) \tag{5.11.2}$$

$$-\frac{Q_1}{A_1}F_{3-1}\frac{1-\varepsilon_1}{\varepsilon_1}+F_{3-2}\sigma T_2^4-(1-F_{3-3})\sigma T_3^4=-F_{3-1}\sigma T_1^4+q_2F_{3-2}\frac{1-\varepsilon_2}{\varepsilon_2} \tag{5.11.3}$$

Note that for this simple situation, the solution could be shortened by using overall energy conservation to give $Q_1 = -Q_2$. However, it is generally a good idea to solve directly for all the unknowns and use the overall energy balance $\sum_{k=1}^{N} Q_k = 0$ as a check.

Example 5.12

A hollow cylinder is heated on the outside in such a way that the cylinder is maintained at a uniform temperature. The outside of the cylinder is otherwise insulated so the energy must be transferred by radiation from the inside of the cylinder through the cylinder ends. The system is in a vacuum, so radiation is the only mode of energy transfer. As shown in Figure 5.10, there is a disk centered on the cylinder axis and facing normal to the ends of the cylinder. The disk is exposed on both sides, so the side facing away from the cylinder radiates to the surroundings at T_e. The other side receives radiation from the inside of the cylinder. Provide equations to compute the disk temperature, T_d, if the inside of the cylinder is at a uniform temperature T_c.

The heat loss from the surface of the disk facing away from the cylinder can be considered as radiation into a large environment at T_e. Hence, from Equation 5.9.1, the net energy flux leaving this surface is $q_{d,e} = \varepsilon_{d,e}\sigma(T_d^4 - T_e^4)$. The exchange with the cylinder is analyzed as a three-surface enclosure consisting of the inside of the cylinder, the side of the disk facing the cylinder, and the open area between the disk and the cylinder (a frustum of a cone) combined with the open area

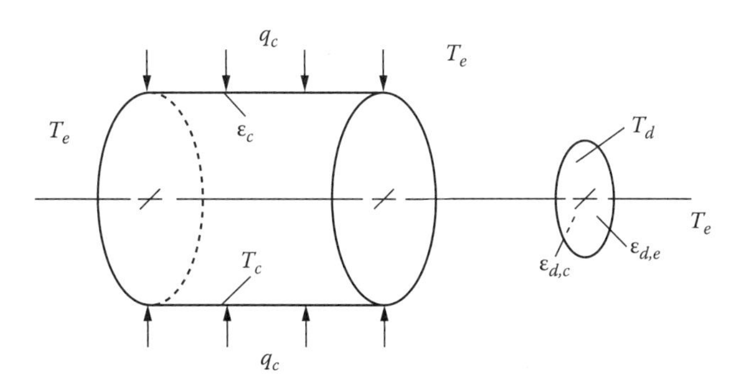

FIGURE 5.10 Hollow cylinder and disk configuration for Example 5.12.

at the left side of the cylinder (this is a first approximation without subdividing the cylinder and disk areas to achieve greater accuracy). In the configuration-factor designations, the subscript *e* designates the combination of the two open areas at T_e. Since the energy lost to the environment is not asked for, only two equations are written, one for the inside of the cylinder and one for the surface of the disk facing the cylinder. The $q_{d,c} = -\varepsilon_{d,e}\sigma\left(T_d^4 - T_e^4\right)$ is the radiative heat flux supplied to the left surface of the disk. From Equation 5.21, the two equations are

$$q_c\left(\frac{1}{\varepsilon_c} - F_{c-c}\frac{1-\varepsilon_c}{\varepsilon_c}\right) - q_{d,c}F_{c-d}\frac{1-\varepsilon_{d,c}}{\varepsilon_{d,c}} = F_{c-d}\sigma\left(T_c^4 - T_d^4\right) + F_{c-e}\sigma\left(T_c^4 - T_e^4\right) \tag{5.12.1}$$

$$-q_cF_{d-c}\frac{1-\varepsilon_c}{\varepsilon_c} + q_d\frac{1}{\varepsilon_{d,c}} = F_{d-c}\sigma\left(T_d^4 - T_c^4\right) + F_{d-e}\sigma\left(T_d^4 - T_e^4\right) \tag{5.12.2}$$

The q_c is eliminated from Equations 5.12.1 and 5.12.2; this yields an analytical relation for T_d.

Example 5.13

As an example with a more complex geometry, consider the long rectangular enclosure shown in cross section in Figure 5.11 (the end walls are neglected). There is a thin baffle along a diagonal dividing the interior into two triangular enclosures. Heat fluxes are specified along two sides, and the other two sides are each cooled to a specified uniform temperature. It is required to find the temperatures T_1 and T_2 of the heated walls and the heat fluxes q_5 and q_6 that must be removed through the cooled walls to maintain their temperatures. For simplicity, the surfaces are not subdivided into smaller areas. Subdivision improves the results (see Example 5.14).

Since T_3, T_4, q_3, and q_4 are also unknown, there are eight unknowns, so eight simultaneous equations are required. Two are the constraints at the baffle wall that yield $T_3 = T_4$ and $q_3 = -q_4$. The

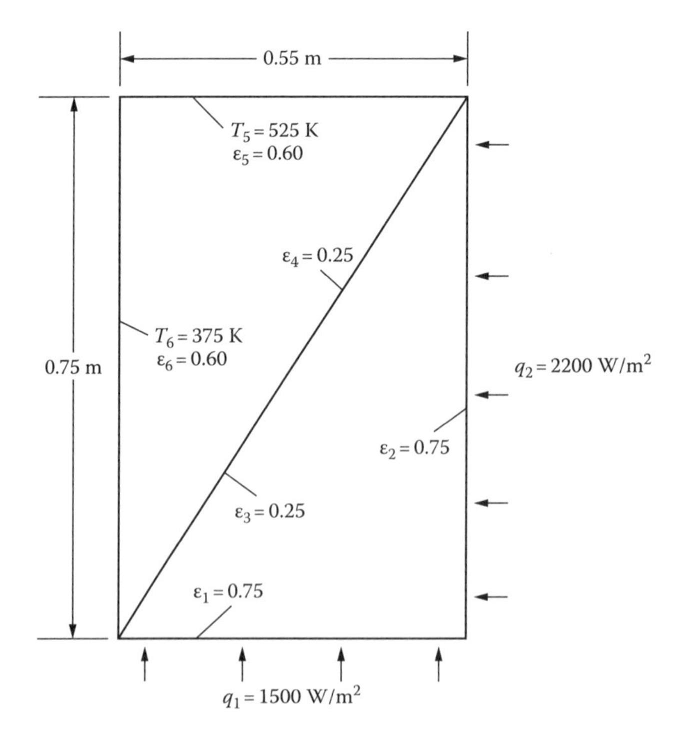

FIGURE 5.11 Rectangular enclosure divided into two triangular enclosures for Example 5.13.

other six equations are found from Equation 5.21 and are written here in abbreviated form using the notation $F_{mn} \equiv F_{m-n}$, $\vartheta \equiv T^4$, $E_n \equiv (1-\varepsilon_n)/\varepsilon_n$:

$$(q_1/\varepsilon_1) - q_2E_2F_{12} - q_3E_3F_{13} = \sigma(\vartheta_1 - F_{12}\vartheta_2 - F_{13}\vartheta_3)$$
$$-q_1E_1F_{21} + (q_2/\varepsilon_2) - q_3E_2F_{23} = \sigma(-F_{21}\vartheta_1 + \vartheta_2 - F_{23}\vartheta_3)$$
$$-q_1E_1F_{31} - q_2E_2F_{32} + (q_3/\varepsilon_3) = \sigma(-F_{31}\vartheta_1 + F_{32}\vartheta_2 + \vartheta_3)$$
$$(q_5/\varepsilon_5) - q_6E_6F_{56} - q_4E_4F_{54} = \sigma(\vartheta_5 - F_{56}\vartheta_6 - F_{54}\vartheta_4)$$
$$-q_5E_5F_{65} + (q_6/\varepsilon_6) - q_4E_4F_{64} = \sigma(-F_{65}\vartheta_5 + \vartheta_6 - F_{64}\vartheta_4)$$
$$-q_5E_5F_{45} - q_6E_6F_{46} + (q_4/\varepsilon_4) = \sigma(-F_{45}\vartheta_5 - F_{46}\vartheta_6 + \vartheta_4)$$

For each three-sided enclosure, the configuration factors are found from Example 4.13. Typical values are $F_{1-2} = F_{5-6} = 0.33632$ and $F_{1-3} = F_{5-4} = 0.66369$. A matrix solver can be used to substitute the values and solve the equations to obtain $T_1 = 817.0$ K, $T_2 = 821.7$ K, $T_3 = 709.1$ K, $q_5 = -559.0$ W/m², and $q_6 = -2890.1$ W/m². Most of the energy is leaving through the more strongly cooled side A_6. The effect of various parameters can easily be examined. If $\varepsilon_1 = \varepsilon_2$ are changed, the only results affected are T_1 and T_2. This is because the same total energy must pass through surface 4, so T_4 remains the same in order to transfer the energy to sides 5 and 6. For $\varepsilon_1 = \varepsilon_2 = 1.0$, the T_1 and T_2 are slightly decreased to $T_1 = 812.9$ K and $T_2 = 815.8$ K. When $\varepsilon_1 = \varepsilon_2 = 0.50$ these temperatures increase to $T_1 = 824.9$ K and $T_2 = 833.1$ K. The results should be checked by determining whether $\sum_{k=0}^{N} q_k A_k = 0$. In this case (0.55 × 1500.0) + (0.75 × 2200.0) – (0.55 × 559.0)–(0.75 × 2890.1) = –0.025 W/m of length, which is within the round-off accuracy.

Example 5.14

A triangular enclosure has long sides normal to the cross section shown in Figure 5.12. The triangular end walls can be neglected in the radiative exchange. Specified quantities are the temperatures of two sides and the heat flux added to the third. The q_1, q_2, and T_3 are required. The solution is obtained first with side 3 being a single area. Then, this side is divided into two equal parts to refine the calculation.

Equation 5.21 is used for analysis. From Example 4.13, $F_{1-2} = (A_1 + A_2 - A_3)/2A_1 = 0.2929 = F_{2-1}$. Then, $F_{1-3} = 1 - F_{1-2} = 0.7071 = F_{2-3}$. From symmetry, $F_{3-1} = F_{3-2} = 0.5$. The self-view factors are zero. The first three equations from Example 5.13 are used. If $\varepsilon_1 = \varepsilon_2 = 0.80$ and $\varepsilon_3 = 0.90$, then the results are $q_1 = -6346$ W/m², $q_2 = 1820$ W/m², and $T_3 = 649.1$ K. Thus, energy must be added

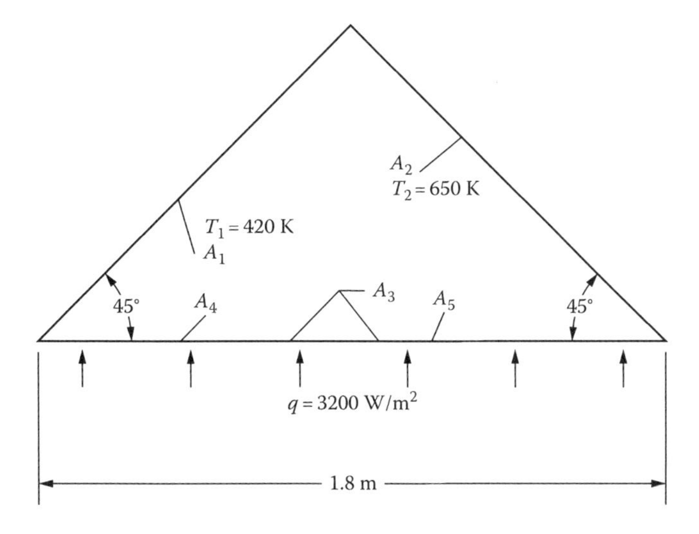

FIGURE 5.12 Triangular enclosure for Example 5.14. Areas A_4 and A_5 are each one-half of A_3.

to maintain the high temperature of A_2. The energy added to A_2 combined with that from A_3 flows out of A_1 at a lower temperature. Now consider what happens if the emissivities are reduced by half, $\varepsilon_1 = \varepsilon_2 = 0.40$ and $\varepsilon_3 = 0.45$. The energy supplied to A_3 has greater difficulty in being transferred to the other walls, and the temperature rises to $T_3 = 733.9$ K. Since T_3 is now larger than both T_1 and T_2, energy is transferred out from both A_1 and A_2: $q_1 = -4101$ W/m² and $q_2 = -424.6$ W/m².

To refine the calculation, A_3 is divided into two equal parts A_4 and A_5. From the geometry, $F_{1-4} = 0.5 = F_{1-5} + F_{1-2} = F_{2-5}$. With F_{1-2} known, this gives $F_{1-5} = 0.2071 = F_{2-4}$. As in the previous examples, four equations are written from Equation 5.21. With T_1, T_2, q_4, and q_5 known, these are solved for q_1, q_2, T_4, and T_5. For $\varepsilon_1 = \varepsilon_2 = 0.80$ and $\varepsilon_3 = 0.90$, this gives $q_1 = -6049$ W/m², $q_2 = 1524$ W/m², $T_4 = 626.3$ K, and $T_5 = 669.7$ K. This reveals the temperature variation along A_3 as compared with the uniform value obtained in the first part of this example. The q_2 is somewhat smaller because A_2 is adjacent to the higher-temperature portion of A_3. By further subdividing A_3, its temperature distribution would be obtained. For $\varepsilon_1 = \varepsilon_2 = 0.40$ and $\varepsilon_3 = 0.45$, $T_4 = 726.8$ K and $T_5 = 740.8$ K, compared with $T_3 = 733.9$ K in the previous calculation. The q are changed somewhat to $q_1 = -4038$ W/m² and $q_2 = -487.2$ W/m². The q values in this example have all been verified to satisfy overall energy conservation (the results given have been rounded off).

5.3.1.2 Solution Method in Terms of Radiosity *J*

An alternative approach for computing the radiative exchange is to solve for J for each surface and then compute q and T. When the surface is viewed with a radiation detector, it is J that is intercepted, that is, the *sum* of *emitted* and *reflected* radiation. Hence, it is sometimes desirable to determine the J as primary quantities. In the previous formulation, the J can be found from the q and T by using Equation 5.18.

When the surface temperatures are all specified, the simultaneous equations for J are obtained by eliminating Q_k from Equations 5.17 and 5.18. This yields either of the following equations for the kth surface:

$$J_k - (1-\varepsilon_k)\sum_{j=1}^{N} J_j F_{k-j} = \sum_{j=1}^{N}\left[\delta_{kj} - (1-\varepsilon_k)F_{k-j}\right]J_j = \varepsilon_k \sigma T_k^4 \tag{5.22}$$

$$J_k + \frac{(1-\varepsilon_k)}{\varepsilon_k}\sum_{j=1}^{N}\left(J_k - J_j\right)F_{k-j} = \sigma T_k^4 \tag{5.23}$$

To illustrate, for a system of three surfaces, Equation 5.22 becomes

$$\left[1-(1-\varepsilon_1)F_{1-1}\right]J_1 - (1-\varepsilon_1)F_{1-2}J_2 - (1-\varepsilon_1)F_{1-3}J_3 = \varepsilon_1\sigma T_1^4 \tag{5.24}$$

$$-(1-\varepsilon_2)F_{2-1}J_1 + \left[1-(1-\varepsilon_2)F_{2-2}\right]J_2 - (1-\varepsilon_2)F_{2-3}J_3 = \varepsilon_2\sigma T_2^4 \tag{5.25}$$

$$-(1-\varepsilon_3)F_{3-1}J_1 - (1-\varepsilon_3)F_{3-2}J_2 + \left[1-(1-\varepsilon_3)F_{3-3}\right]J_3 = \varepsilon_3\sigma T_3^4 \tag{5.26}$$

With the T given, the J can be found. Then, if desired, Equation 5.16 can be used to compute q for each surface.

When q is specified for some surfaces and T for others, Equation 5.22 is used for the surfaces with known T in conjunction with Equation 5.17 for the surfaces with known q to obtain simultaneous equations for the unknown J. Once J is obtained for a surface, it can be combined with the given q (or T) and Equation 5.16 used to determine the unknown T (or q). In general, if an enclosure has surfaces 1, 2, …, m with specified T, and the remaining surfaces $m+1, m+2, \ldots, N$ with specified q, then the system of equations for the J is, from Equations 5.17 and 5.22,

$$\sum_{j=1}^{N}\left[\delta_{kj}-(1-\varepsilon_k)F_{k-j}\right]J_j=\varepsilon_k\sigma T_k^4 \quad 1\le k\le m \tag{5.27}$$

$$\sum_{j=1}^{N}\left[\delta_{kj}-F_{k-j}\right]J_j=\frac{Q_k}{A_k} \quad m+1\le k\le N \tag{5.28}$$

For a black surface with T_k specified, Equation 5.27 gives $J_k=\sigma T_k^4$, so the J_k is known and the number of simultaneous equations is reduced by one. The following example demonstrates obtaining a solution first by using Equation 5.21 and then by using Equations 5.27 and 5.28.

Example 5.15

A frustum of a cone has its base uniformly heated as in Figure 5.13. The top is maintained at 550 K while the side is perfectly insulated on the outside. Surfaces 1 and 2 are diffuse–gray, while surface 3 is black. For energy transfer only by radiation, what is the temperature of side 1? How important is the value of ε_2?

By using the configuration factor for two parallel disks (factor 10 in Appendix C), $F_{3-1}=0.3249$. Then, $F_{3-2}=1-F_{3-1}=0.6751$. From reciprocity, $A_1F_{1-3}=A_3F_{3-1}$ and $A_2F_{2-3}=A_3F_{3-2}$, so that $F_{1-3}=0.1444$ and $F_{2-3}=0.1310$. Then, $F_{1-2}=1-F_{1-3}=0.8556$. From $A_1F_{1-2}=A_2F_{2-1}$, $F_{2-1}=0.3735$. Finally, $F_{2-2}=1-F_{2-1}-F_{2-3}=0.4955$. From Equation 5.21, and by noting that $Q_2=0$ and $1-\varepsilon_3=0$, the three equations are

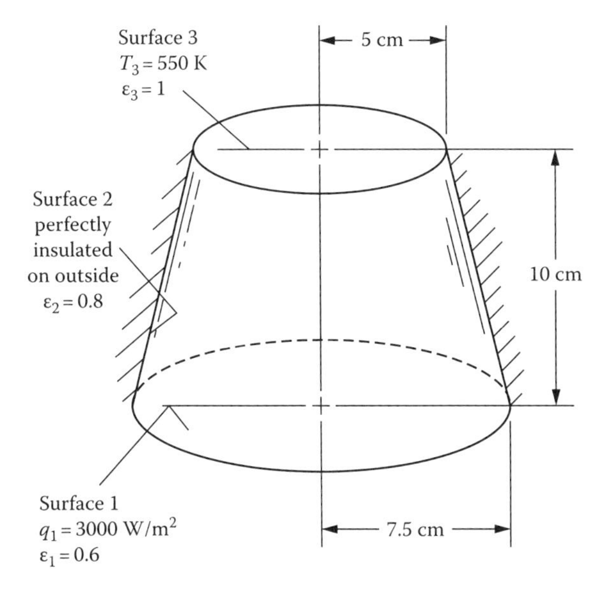

FIGURE 5.13 Enclosure used in Example 5.15.

$$\frac{3000}{0.6} = \sigma\left[T_1^4 - 0.8556T_2^4 - 0.1444(550)^4\right]$$

$$-3000(0.3735)\frac{1-0.6}{0.6} = \sigma\left[-0.3735T_1^4 + (1-0.4955)T_2^4 - 0.1310(550)^4\right]$$

$$-3000(0.3249)\frac{1-0.6}{0.6} + \frac{Q_3}{A_3} = \sigma\left[-0.3249T_1^4 - 0.6751T_2^4 + (550)^4\right]$$

These can be solved for T_1, T_2, and Q_3. (For this particular example, Q_3 can also be obtained from overall energy conservation, $Q_3 = -Q_1$.) The result is $T_1 = 721.6$ K ($T_2 = 667.4$ K). Since $Q_2 = 0$, all the terms involving ε_2 are zero, so ε_2 does not appear in the simultaneous equations, and the emissivity of the insulated surface is not important. Physically, this results from the fact that, for no convection or conduction, all absorbed energy must be reemitted and hence $J_2 = G_2$ independent of ε_2.

This example is now solved using Equations 5.27 and 5.28. Because T_3 is specified and A_3 is black, Equation 5.27 gives $J_3 = \sigma T_3^4$. Equation 5.28 is used at A_1 and A_2 since q_1 and q_2 are specified:

$$J_1 - F_{1-2}J_2 = F_{1-3}\sigma T_3^4 + q_1$$

$$-F_{2-1}J_1 + (1 - F_{2-2})J_2 = F_{2-3}\sigma T_3^4$$

This yields $J_1 = 13{,}370$ W/m^2 and $J_2 = 11{,}250$ W/m^2. From Equation 5.18, $\sigma T_1^4 = J_1 + [(1-\varepsilon_1)/\varepsilon_1]q_1$, and similarly for σT_2^4. This gives the same temperatures as in the first part of this example.

Example 5.16

The second half of Example 5.14 is now calculated by using Equations 5.27 and 5.28. A_1 and A_2 have specified T, while A_4 and A_5 have specified q. Then, Equations 5.27 and 5.28 yields

$$J_1 - (1-\varepsilon_1)\left[F_{1-2}J_2 + F_{1-4}J_4 + F_{1-5}J_5\right] = \varepsilon_1\sigma T_1^4$$
$$J_2 - (1-\varepsilon_2)\left[F_{2-1}J_1 + F_{2-4}J_4 + F_{2-5}J_5\right] = \varepsilon_2\sigma T_2^4$$
$$-F_{4-1}J_1 - F_{4-2}J_2 + J_4 - F_{4-5}J_5 = q_4$$
$$-F_{5-1}J_1 - F_{5-2}J_2 - F_{5-4}J_4 + J_5 = q_5$$

The solution gives J in W/m^2: $J_1 = 3277$, $J_2 = 9741$, $J_4 = 8370$, and $J_5 = 11{,}048$. Then, from Equation 5.16, $q_1 = \left[\varepsilon_1/(1-\varepsilon_1)\right]\left(\sigma T_1^4 - J_1\right)$, and similarly for q_2. From Equation 5.18, $\sigma T_4^4 = \left[\varepsilon_4/(1-\varepsilon_4)\right]q_4 + J_4$, and similarly for T_5. This provides the same results as for Example 5.14.

5.3.2 Enclosure Analysis in Terms of Energy Absorbed at Surface

An alternative procedure to the net-radiation method is to examine the energy *absorbed* at a surface. In the net-radiation method, the energy absorbed at each surface is found from Equation 5.20 as $\alpha G = \varepsilon G$ for a gray surface. A somewhat different viewpoint is briefly presented here; however, additional discussion is in Gebhart (1961, 1976). This formulation yields coefficients that provide the fraction of energy emitted by a surface that is *absorbed* at another surface after reaching that surface by all possible paths. These coefficients are useful in formulating some thermal analyses.

For a typical surface A_k, the net energy loss is the emission from the surface minus the energy that is absorbed by the surface from all incident radiation sources. Let Γ_{jk} be the fraction of the emission from A_j that reaches A_k and is *absorbed*. The Γ_{jk} includes all the paths for reaching A_k, that is, the direct path and paths by means of one or multiple reflections. Thus, $A_j\varepsilon_j\sigma T_j^4\Gamma_{jk}$ is the amount of energy emitted by A_j that is absorbed by A_k. An energy balance on A_k then gives

$$Q_k = A_k\varepsilon_k\sigma T_k^4 - \sum_{j=1}^{N} A_j\varepsilon_j\sigma T_j^4\Gamma_{jk} \tag{5.29}$$

The Γ_{kk} would generally not be zero, since even for a plane or convex surface, some of the emission from the surface is returned to itself by reflection from other surfaces. Equation 5.29 is written for each surface; this relates each Q to the surface temperatures in the enclosure. The Γ factors must now be found.

The Γ_{jk} is the fraction of energy emitted by A_j that reaches A_k and is absorbed. The total emitted energy from A_j is $A_j\varepsilon_j\sigma T_j^4$. The portion traveling directly to A_k and then absorbed is $A_j\varepsilon_j\sigma T_j^4 F_{j-k}\varepsilon_k$, where for a gray surface $\alpha_k = \varepsilon_k$. All other radiation from A_j arriving at A_k will first undergo one reflection. The emission from A_j that arrives at a typical surface A_n and is then reflected is $A_j\varepsilon_j\sigma T_j^4 F_{j-n}\rho_n$. The fraction Γ_{nk} reaches A_k and is absorbed. Then all the energy absorbed at A_k that originated by emission from A_j is

$$A_j\varepsilon_j\sigma T_j^4 F_{j-k}\varepsilon_k + \left(A_j\varepsilon_j\sigma T_j^4 F_{j-1}\rho_1\Gamma_{1k} + A_j\varepsilon_j\sigma T_j^4 F_{j-2}\rho_2\Gamma_{2k} + \cdots \right.$$
$$\left. + A_j\varepsilon_j\sigma T_j^4 F_{j-k}\rho_k\Gamma_{kk} + \cdots + A_j\varepsilon_j\sigma T_j^4 F_{j-N}\rho_N\Gamma_{NK}\right)$$

Dividing this energy by emission from A_j gives the fraction

$$\Gamma_{jk} = F_{j-k}\varepsilon_k + F_{j-1}\rho_1\Gamma_{1k} + F_{j-2}\rho_2\Gamma_{2k} + \cdots$$
$$+ F_{j-k}\rho_k\Gamma_{kk} + \cdots + F_{j-N}\rho_N\Gamma_{Nk}$$

This is rearranged to the form

$$-F_{j-1}\rho_1\Gamma_{1k} - F_{j-2}\rho_2\Gamma_{2k} - \cdots + \Gamma_{jk} - \cdots - F_{j-N}\rho_N\Gamma_{Nk} = F_{j-k}\varepsilon_k$$

By letting j take on all values from 1 to N, the set of equations is

$$\begin{array}{l}
\left(1-F_{1-1}\rho_1\right)\Gamma_{1k} - F_{1-2}\rho_2\Gamma_{2k} - F_{1-3}\rho_3\Gamma_{3k} - \cdots - F_{1-N}\rho_N\Gamma_{Nk} = F_{1-k}\varepsilon_k \\
-F_{2-1}\rho_1\Gamma_{1k} + \left(1-F_{2-2}\rho_2\right)\Gamma_{2k} - F_{2-3}\rho_3\Gamma_{3k} - \cdots - F_{2-N}\rho_N\Gamma_{Nk} = F_{2-k}\varepsilon_k \\
-F_{3-1}\rho_1\Gamma_{1k} - F_{3-2}\rho_2\Gamma_{2k} + \left(1-F_{3-3}\rho_3\right)\Gamma_{3k} - \cdots - F_{3-N}\rho_N\Gamma_{Nk} = F_{3-k}\varepsilon_k \\
\quad\cdot \qquad\qquad \cdot \qquad\qquad\qquad \cdot \qquad \cdot \\
\quad\cdot \qquad\qquad \cdot \qquad\qquad\qquad \cdot \qquad \cdot \\
-F_{N-1}\rho_1\Gamma_{1k} - F_{N-2}\rho_2\Gamma_{2k} - F_{N-3}\rho_3\Gamma_{3k} - \cdots + \left(1-F_{N-N}\rho_N\right)\Gamma_{Nk} = F_{N-k}\varepsilon_k
\end{array} \tag{5.30}$$

Equations 5.30 are solved simultaneously for Γ_{1k}, Γ_{2k}, ..., Γ_{Nk}. This is done for each k where $1 \le k \le N$, and the solution is simplified by the fact that the coefficients of the Γ do not depend on k.

The amount of calculation can be reduced by using some relations between the Γ values. Since all emission by a surface is absorbed by all of the surfaces, the sum of the fractions absorbed must be unity:

$$\sum_{j=1}^{N}\Gamma_{kj} = 1 \tag{5.31}$$

Gebhart also showed that there is the reciprocity relation

$$\varepsilon_k A_k \Gamma_{kj} = \varepsilon_j A_j \Gamma_{jk} \tag{5.32}$$

Equations 5.31 and 5.32 can be substituted into Equation 5.29 to yield

$$Q_k = A_k \varepsilon_k \sigma T_k^4 \sum_{j=1}^{N} \Gamma_{kj} - \sum_{j=1}^{N} A_k \varepsilon_k \Gamma_{kj} \sigma T_j^4 = A_k \varepsilon_k \sum_{j=1}^{N} \Gamma_{kj} \sigma \left(T_k^4 - T_j^4\right) \tag{5.33}$$

Example 5.17 demonstrates using these relations.

5.3.3 Enclosure Analysis by the Use of Transfer Factors

The transfer factors $\mathfrak{F}$ are defined by stating that Q has the form

$$Q_k = A_k \sum_{j=1}^{N} \mathfrak{F}_{kj} \sigma \left(T_k^4 - T_j^4\right) \quad \left(1 \le k \le N\right) \tag{5.34}$$

It is evident by comparing with Equation 5.33 that

$$\mathfrak{F}_{kj} = \varepsilon_k \Gamma_{kj} \tag{5.35}$$

Hence, from Equations 5.31 and 5.32,

$$\sum_{i=1}^{N} \mathfrak{F}_{kj} = \varepsilon_k \tag{5.36}$$

$$A_k \mathfrak{F}_{kj} = A_j \mathfrak{F}_{jk} \tag{5.37}$$

The $\mathfrak{F}$ can obviously be calculated by first obtaining the Γ. An equivalent approach is to let $T_n^4 = 1$ for the nth surface and $T = 0$ for all other surfaces. Then, Equations 5.34 and 5.36 yield

$$\begin{aligned} \mathfrak{F}_{kn} &= -q_k \quad \left(k \ne n\right) \\ \mathfrak{F}_{nn} &= -q_n + \varepsilon_n \end{aligned} \tag{5.38}$$

For the same conditions, Equation 5.21 yields

$$\sum_{j=1}^{N} \left(\frac{\delta_{kj}}{\varepsilon_j} - F_{k-j} \frac{1-\varepsilon_j}{\varepsilon_j} \right) q_j = -F_{k-n} \quad \left(k \ne n\right) \tag{5.39}$$

$$\sum_{j=1}^{N} \left(\frac{\delta_{nj}}{\varepsilon_j} - F_{n-j} \frac{1-\varepsilon_j}{\varepsilon_j} \right) q_j = 1 - F_{n-n} \tag{5.40}$$

If Equations 5.39 and 5.40 are solved for the q, these are obtained as a function of the F and ε. The $\mathfrak{F}$ are found from the q by using Equation 5.38; thus, the $\mathfrak{F}$ are functions of F and ε and do not depend on the T [provided that $\varepsilon \neq \varepsilon(T)$]. Example 5.17 will show how the $\mathfrak{F}$ values are obtained by using matrix inversion to solve Equations 5.39 and 5.40.

5.3.4 Matrix Inversion for Enclosure Equations

When many surfaces are present in an enclosure, a large set of simultaneous equations results such as Equations 5.21, 5.30, or 5.39 and 5.40. The equations can be solved using computation algorithms that accommodate hundreds of simultaneous equations. A set of equations such as Equation 5.21 can be written in a shorter form. Let the right side be C_k and the quantities in parentheses on the left be a_{kj}. Then, the k equations can be written as

$$\sum_{j=1}^{N} a_{kj} q_j = C_k \tag{5.41}$$

where

$$a_{kj} = \frac{\delta_{kj}}{\varepsilon_j} - F_{k-j}\frac{1-\varepsilon_j}{\varepsilon_j}; \quad C_k = \sum_{j=1}^{N} F_{k-j}\sigma\left(T_k^4 - T_j^4\right) \tag{5.42}$$

For an enclosure of N surfaces, the set of equations then has the form

$$\begin{array}{ccccccc}
a_{11}q_1 + a_{12}q_2 + \cdots + a_{1j}q_j + \cdots + a_{1N}q_N = C_1 \\
a_{21}q_1 + a_{22}q_2 + \cdots + a_{2j}q_j + \cdots + a_{2N}q_N = C_2 \\
\vdots \quad \vdots \quad \vdots \quad \vdots \quad \vdots \\
a_{k1}q_1 + a_{k2}q_2 + \cdots + a_{kj}q_j + \cdots + a_{kN}q_N = C_k \\
\vdots \quad \vdots \quad \vdots \quad \vdots \quad \vdots \\
a_{N1}q_1 + a_{N2}q_2 + \cdots + a_{Nj}q_j + \cdots + a_{NN}q_N = C_N
\end{array} \tag{5.43}$$

The array of a_{kj} coefficients is the *matrix of coefficients* $\mathbf{a}$:

$$\mathbf{a} \equiv \left[a_{kj}\right] \equiv \begin{bmatrix} a_{11} & a_{12} & \cdots & a_{1j} & \cdots & a_{1N} \\ a_{21} & a_{22} & \cdots & a_{2j} & \cdots & a_{2N} \\ \vdots & \vdots & \vdots & \vdots & \vdots & \vdots \\ a_{k1} & a_{k2} & \cdots & a_{kj} & \cdots & a_{kN} \\ \vdots & \vdots & \vdots & \vdots & \vdots & \vdots \\ a_{N1} & a_{N2} & \cdots & a_{Nj} & \cdots & a_{NN} \end{bmatrix} \tag{5.44}$$

A method for solving a set of equations such as Equation 5.43 is to obtain a second matrix $\mathbf{a}^{-1}$, called the *inverse* of matrix **a**

$$\mathbf{a}^{-1} \equiv \left[A_{kj}\right] \equiv \begin{bmatrix} A_{11} & A_{12} & \cdots & A_{1j} & \cdots & A_{1N} \\ A_{21} & A_{22} & \cdots & A_{2j} & \cdots & A_{2N} \\ \vdots & \vdots & \vdots & \vdots & \vdots & \vdots \\ A_{k1} & A_{k2} & \cdots & A_{kj} & \cdots & A_{kN} \\ \vdots & \vdots & \vdots & \vdots & \vdots & \vdots \\ A_{N1} & A_{N2} & \cdots & A_{Nj} & \cdots & A_{NN} \end{bmatrix} \quad (5.45)$$

The inverse matrix is comprised of A_{kj} elements corresponding to each a_{kj} in the original matrix. The A_{kj} elements are found by operating on the **a** as follows: The kth row and jth column that contain element a_{kj} in a square matrix **a** are deleted, and the determinant of the remaining square array is called the *minor* of element a_{kj} and is denoted by M_{kj}. The cofactor of a_{kj} is defined as $(-1)^{k+j}M_{kj}$. To obtain the inverse of a square matrix $[a_{kj}]$, each element a_{kj} is first replaced by its cofactor. The rows and columns of the resulting matrix are then interchanged. The elements of the matrix thus obtained are then each divided by the determinant $|a_{kj}|$ of the original matrix $[a_{kj}]$. The elements obtained in this fashion are the A_{kj}. For more detailed information on matrix inversion, the reader should refer to a mathematics text. Most computer mathematics packages such as MATLAB®, Mathcad, COMSOL, and others can also be used to obtain the inverse coefficients A_{kj} from the matrix of a_{kj} values.

After the inverse matrix is obtained, the q_k values in Equation 5.21 are found as the sum of products of A and C:

$$q_k = \sum_{j=1}^{N} A_{kj} C_j \quad (5.46)$$

Thus, the solution for each q_k is in the form of a weighted sum of $\left(T_k^4 - T_j^4\right)$; this is the same form as Equation 5.34.

For a given enclosure, the configuration factors F_{k-j} remain fixed. If in addition the ε_k are constant, then the elements a_{kj}, and hence the inverse elements A_{kj}, remain fixed for the enclosure. The fact that the A_{kj} remain fixed has utility when it is desired to compute radiation quantities within an enclosure for many different T values. The matrix is inverted only once, and then Equation 5.46 is applied for different values of C.

The previous two sections have presented alternative ways of formulating the radiation exchange equations. These are equivalent approaches for obtaining the set of Equations 5.21. For any appreciable number of simultaneous equations, the solution would be found with a computer subroutine or solver. The following example gives the solution to the same problem using three different methods. This illustrates the matrix operations and shows how the various transfer factors are related.

Example 5.17

Figure 5.14 is a two-dimensional (2D) rectangular enclosure that is long in the direction normal to the cross section shown. All the surfaces are diffuse–gray and their temperatures and emissivities are given. For simplicity, the four sides are not subdivided. The configuration factors are obtained by the crossed-string method as $F_{1-3} = F_{3-1} = 0.2770$, $F_{2-4} = F_{4-2} = 0.5662$, $F_{1-2} = F_{1-4} = F_{3-2} = F_{3-4} = 0.3615$,

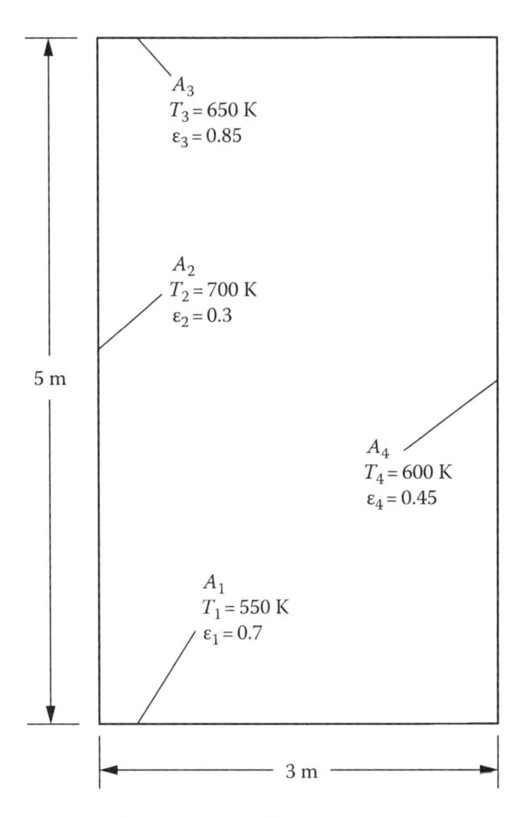

FIGURE 5.14 Rectangular geometry for Example 5.17.

and $F_{4-1} = F_{4-3} = F_{2-1} = 0.2169$. In what follows, an abbreviated notation is used: $E = (1 - \varepsilon)/\varepsilon$, $\vartheta = T^4$, and $F_{kj} = F_{k-j}$. Then, the four equations from Equation 5.21 are

$$\frac{q_1}{\varepsilon_1} - q_2F_{12}E_2 - q_3F_{13}E_3 - q_4F_{14}E_4 = \sigma\left(\vartheta_1 - F_{12}\vartheta_2 - F_{13}\vartheta_3 - F_{14}\vartheta_4\right)$$

$$-q_1F_{21}E_1 + \frac{q_2}{\varepsilon_2} - q_3F_{23}E_3 - q_4F_{24}E_4 = \sigma\left(-F_{21}\vartheta_1 + \vartheta_2 - F_{23}\vartheta_3 - F_{24}\vartheta_4\right)$$

$$-q_1F_{31}E_1 - q_2F_{32}E_2 + \frac{q_3}{\varepsilon_3} - q_4F_{34}E_4 = \sigma\left(-F_{31}\vartheta_1 - F_{32}\vartheta_2 + \vartheta_3 - F_{34}\vartheta_4\right)$$

$$-q_1F_{41}E_1 - q_2F_{42}E_2 - q_3F_{43}E_3 + \frac{q_4}{\varepsilon_4} = \sigma\left(-F_{41}\vartheta_1 - F_{42}\vartheta_2 - F_{44}\vartheta_3 + \vartheta_4\right)$$

All quantities except the q are known. Substitution and solution by a computer software package gives $q_1 = -2876.52$, $q_2 = 1612.53$, $q_3 = 1508.92$, $q_4 = -791.97$ W/m^2.

Now, the method in Section 5.3.2 is used. In Equation 5.30, it is noted that the Γ_{jk} have the same matrix of coefficients for all k. There are four sets of four equations, one set for each k, and within each set j has values 1, 2, 3, and 4. The matrix of the Γ coefficients is (note that the $F_{kk} = 0$ for this example)

$$\mathbf{m} = \begin{bmatrix} 1 & -\rho_2F_{12} & -\rho_3F_{13} & -\rho_4F_{14} \\ -\rho_1F_{21} & 1 & -\rho_3F_{23} & -\rho_4F_{24} \\ -\rho_1F_{31} & -\rho_2F_{32} & 1 & -\rho_4F_{34} \\ -\rho_1F_{41} & -\rho_2F_{42} & -\rho_3F_{43} & 1 \end{bmatrix}$$

When the values are substituted into this matrix, it becomes

$$\mathbf{m}=\begin{bmatrix} 1 & -0.25306 & -0.04155 & -0.19883 \\ -0.06507 & 1 & -0.03254 & -0.31140 \\ -0.08310 & -0.25306 & 1 & -0.19883 \\ -0.06507 & -0.39633 & -0.03254 & 1 \end{bmatrix}$$

The matrix of the four columns of values on the right-hand sides of the four sets of equations is, in symbolic and numerical form,

$$\mathbf{f}=\begin{bmatrix} F_{11}\varepsilon_1 & F_{12}\varepsilon_2 & F_{13}\varepsilon_3 & F_{14}\varepsilon_4 \\ F_{21}\varepsilon_1 & F_{22}\varepsilon_2 & F_{23}\varepsilon_3 & F_{24}\varepsilon_4 \\ F_{31}\varepsilon_1 & F_{32}\varepsilon_2 & F_{33}\varepsilon_3 & F_{34}\varepsilon_4 \\ F_{41}\varepsilon_1 & F_{42}\varepsilon_2 & F_{43}\varepsilon_3 & F_{44}\varepsilon_4 \end{bmatrix}$$

$$\mathbf{f}=\begin{bmatrix} 0 & 0.10845 & 0.23544 & 0.16268 \\ 0.15183 & 0 & 0.18437 & 0.25479 \\ 0.19389 & 0.10845 & 0 & 0.16268 \\ 0.15183 & 0.16986 & 0.18437 & 0 \end{bmatrix}$$

The matrix of Γ_{kj} factors is obtained by matrix inversion as $[\Gamma_{kj}] = \mathbf{m}^{-1}\mathbf{f}$:

$$\Gamma=\begin{bmatrix} 0.13233 & 0.18271 & 0.39315 & 0.29181 \\ 0.25579 & 0.08738 & 0.32299 & 0.33384 \\ 0.32377 & 0.19000 & 0.18278 & 0.30345 \\ 0.27236 & 0.222.56 & 0.34391 & 0.16117 \end{bmatrix}$$

By summing values in each row, it is evident that $\sum_{j=1}^{4}\Gamma_{kj}=1$. The q values are then found from Equation 5.33, and they agree with those given previously.

The solution is now obtained by the method in Section 5.3.3. The coefficient matrix for the set of Equations 5.39 and 5.40 is

$$\mathbf{D}=\begin{bmatrix} 1/\varepsilon_1 & -\rho_2(F_{12}/\varepsilon_2) & -\rho_3(F_{13}/\varepsilon_3) & -\rho_4(F_{14}/\varepsilon_4) \\ -\rho_1(F_{21}/\varepsilon_1) & 1/\varepsilon_2 & -\rho_3(F_{23}/\varepsilon_3) & -\rho_4(F_{24}/\varepsilon_4) \\ -\rho_1(F_{31}/\varepsilon_1) & -\rho_2(F_{32}/\varepsilon_2) & 1/\varepsilon_3 & -\rho_4(F_{34}/\varepsilon_4) \\ -\rho_1(F_{41}/\varepsilon_1) & -\rho_2(F_{42}/\varepsilon_2) & -\rho_3(F_{43}/\varepsilon_3) & 1/\varepsilon_4 \end{bmatrix}$$

and this gives in numerical form

$$\mathbf{D}=\begin{bmatrix} 1.42857 & -0.84352 & -0.04888 & -0.44184 \\ -0.09296 & 3.33333 & -0.03828 & -0.69201 \\ -0.11871 & -0.84352 & 1.17647 & -0.44184 \\ -0.09296 & -1.32111 & -0.03828 & 2.22222 \end{bmatrix}$$

The matrix for the right side coefficients of Equations 5.39 and 5.40 to determine $-q$ is (note that $F_{nm} = 0$ for this example)

$$\mathbf{M} = \begin{bmatrix} -1 & F_{12} & F_{13} & F_{14} \\ F_{21} & -1 & F_{23} & F_{24} \\ F_{31} & F_{32} & -1 & F_{34} \\ F_{41} & F_{42} & F_{43} & -1 \end{bmatrix}$$

To relate the $-q$ and $\mathfrak{F}$ from Equation 5.38, the emissivity matrix is

$$\varepsilon = \begin{bmatrix} \varepsilon_1 & 0 & 0 & 0 \\ 0 & \varepsilon_2 & 0 & 0 \\ 0 & 0 & \varepsilon_3 & 0 \\ 0 & 0 & 0 & \varepsilon_4 \end{bmatrix}$$

Then, according to Equation 5.38, the $\mathfrak{F}$ factors are obtained from the matrix operations

$$\mathfrak{F} = \mathbf{D}^{-1}\mathbf{M} + \varepsilon$$

This yields the matrix of $\mathfrak{F}$ values:

$$\mathfrak{F} = \begin{bmatrix} 0.09263 & 0.12790 & 0.27520 & 0.20427 \\ 0.07674 & 0.02621 & 0.09690 & 0.10015 \\ 0.27520 & 0.16150 & 0.15537 & 0.25793 \\ 0.12256 & 0.10015 & 0.15476 & 0.07253 \end{bmatrix}$$

From these values, it is evident that there is the relation $\mathfrak{F}_{kj} = \varepsilon_k \Gamma_{kj}$ as in Equation 5.35. The q values are computed from Equation 5.34 and are the same as before.

Another procedure proposed for enclosure analysis, which was pointed out in Section 4.3.5, is to use the *discrete ordinates method* where the geometric aspects are included by following rays of radiation. The rays are in a specified number of angular directions from each specified finite area element. This method is described in Chapter 12 for enclosures that contain a gas such as carbon dioxide that is not transparent and hence interacts with radiation. By omitting the radiative interaction with the gas, discrete ordinates can be used for an enclosure that is evacuated or contains a transparent medium. This has been examined by Sanchez and Smith (1992) and may be utilized as further developments are made on the discrete ordinates method. Possible applications would be for complex enclosures that have internal obstructions and shadowing of surfaces. Results are obtained by Sanchez and Smith for a 2D square enclosure with either an internal obstruction or two rectangular protrusions on one wall. The boundary temperatures were all specified and radiative heat fluxes were computed. As demonstrated by comparisons with the net-radiation method described in this chapter, good results can be obtained if a sufficient number of area increments and angular divisions are used for the discrete ordinate integrations.

5.4 RADIATION ANALYSIS USING INFINITESIMAL AREAS

5.4.1 Generalized Net-Radiation Method Using Infinitesimal Areas

In the previous section, the enclosure was divided into finite areas. The accuracy of the results is limited by the assumptions that the temperature and the energy incident on and leaving each surface are uniform over that surface. If these quantities are nonuniform over part of the enclosure

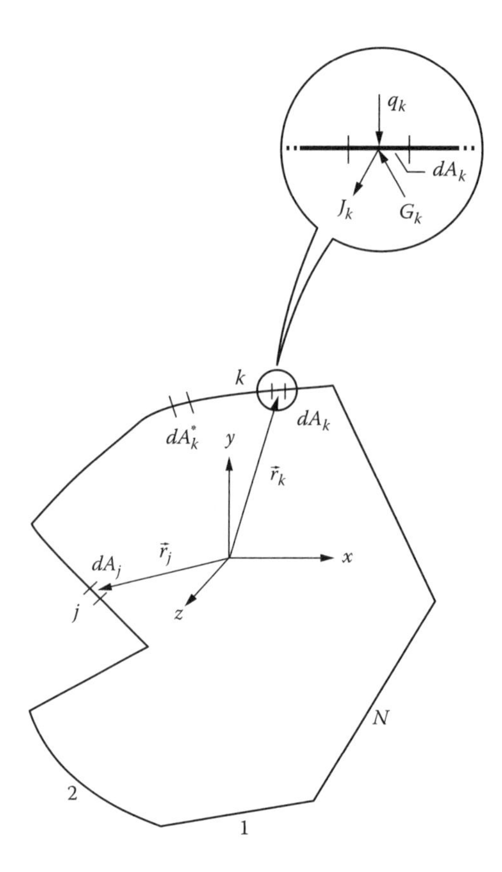

FIGURE 5.15 Enclosure composed of N discrete surface areas with areas subdivided into infinitesimal elements.

boundary, that part must be subdivided until the variation over each area in the analysis is not too large. Several calculations can be made in which successively smaller areas (and hence more simultaneous equations) are used until the results no longer change significantly when the area sizes are further diminished. In the limit, the enclosure boundary, or a portion of it, is divided into infinitesimal parts; this allows large variations in T, q, G, and J to be accounted for in calculations.

The formulation in terms of infinitesimal areas leads to energy balances in the form of integral equations. By using exact or approximate mathematical techniques for integral equations, it may be possible to obtain an analytical solution. Usually, this is not possible, and the integral equations are solved numerically.

Consider an enclosure of N finite areas. The areas would usually be the major geometric divisions of the enclosure or the areas on which a specified boundary condition is held constant. Some or all of these areas are further subdivided into differential area elements as in Figure 5.15. The surfaces are diffuse–gray, and for simplicity, the additional restriction is made that the radiative properties are independent of temperature.

An energy balance on element dA_k at $\mathbf{r}_k$ gives

$$q_k(\mathbf{r}_k) = J_k(\mathbf{r}_k) - G_k(\mathbf{r}_k) \tag{5.47}$$

The outgoing flux is composed of emitted and reflected energy:

$$J_k(\mathbf{r}_k) = \varepsilon_k \sigma T_k^4(\mathbf{r}_k) + (1-\varepsilon_k) G_k(\mathbf{r}_k) \tag{5.48}$$

The incoming flux in Equation 5.48 is composed of portions of the outgoing fluxes from the other area elements of the enclosure. This is a generalization of Equation 5.12 in that an integration is performed over each finite surface to determine the contribution that the local flux leaving that surface makes to G_k:

$$dA_k J_k(\mathbf{r}_k) = \int_{A_1} J_1(\mathbf{r}_1) dF_{d1-dk}(\mathbf{r}_1,\mathbf{r}_k) dA_1 + \cdots$$

$$+ \int_{A_k} J_k(\mathbf{r}_k^*) dF_{dk^*-dk}(\mathbf{r}_k^*,\mathbf{r}_k) dA_k^* + \cdots$$

$$+ \int_{A_N} J_N(\mathbf{r}_N) dF_{dN-dk}(\mathbf{r}_N,\mathbf{r}_k) dA_N \tag{5.49}$$

The second integral is the contribution that other differential elements dA_k^* on surface A_k make to the incident energy at dA_k. By using reciprocity, a typical integral can be transformed to obtain

$$\int_{A_j} J_j(\mathbf{r}_j) dF_{dj-dk}(\mathbf{r}_j,\mathbf{r}_k) dA_j = dA_k \int_{A_j} J_j(\mathbf{r}_j) dF_{dk-dj}(\mathbf{r}_j,\mathbf{r}_k)$$

By operating on all the integrals in Equation 5.49 in this manner, the result becomes

$$G_k(\mathbf{r}_k) = \sum_{j=1}^{N} \int_{A_j} J_j(\mathbf{r}_j) dF_{dk-dj}(\mathbf{r}_j,\mathbf{r}_k) \tag{5.50}$$

Equations 5.48 and 5.50 provide two expressions for $G_k(\mathbf{r}_k)$. These are each substituted into Equation 5.47 to provide two expressions for $q_k(\mathbf{r}_k)$ comparable to Equations 5.16 and 5.17:

$$q_k(\mathbf{r}_k) = \frac{\varepsilon_k}{1-\varepsilon_k}\left[\sigma T_k^4(\mathbf{r}_k) - J_k(\mathbf{r}_k)\right] \tag{5.51}$$

$$q_k(\mathbf{r}_k) = J_k(\mathbf{r}_k) - \sum_{j=1}^{N} \int_{A_j} J_j(\mathbf{r}_j) dF_{dk-dj}(\mathbf{r}_j,\mathbf{r}_k) \tag{5.52}$$

As shown by Equation 4.9, the differential configuration factor dF_{dk-dj} contains the differential area dA_j. To place Equation 5.52 in a form where the variable of integration is explicitly shown, it is convenient to define a quantity $K(\mathbf{r}_j, \mathbf{r}_k)$ by

$$K(\mathbf{r}_j,\mathbf{r}_k) \equiv \frac{dF_{dk-dj}(\mathbf{r}_j,\mathbf{r}_k)}{dA_j} \tag{5.53}$$

Then, Equation 5.52 becomes the integral equation

$$q_k(\mathbf{r}_k) = J_k(\mathbf{r}_k) - \sum_{j=1}^{N} \int_{A_j} J_j(\mathbf{r}_j) K(\mathbf{r}_j,\mathbf{r}_k) dA_j \tag{5.54}$$

The $K(\mathbf{r}_j, \mathbf{r}_k)$ is the *kernel* of the integral equation.

As in the previous discussion for finite areas, there are two paths that can be followed: (1) When the temperatures and imposed heat fluxes are important. Equations 5.51 and 5.52 can be combined to eliminate the J. This gives a set of simultaneous relations relating the surface temperatures T and the heat fluxes q, and the problem is determinate if half of the total T's and q's are specified. Along each surface area, T or q can be specified, and the unknown T and q are found by solving the simultaneous relations. (2) Alternatively, when J is an important quantity, the unknown q can be eliminated by combining Equations 5.51 and 5.52 for each surface that does not have its q specified. For a surface where q is known, Equation 5.52 can be used to directly relate the J's to each other. This yields a set of simultaneous relations for the J in terms of the specified q and T. After solving for the J, Equations 5.51 can be used, if desired, to relate the q and T, where either the q or the T is known at each surface from the boundary conditions. Each of these procedures is now examined.

5.4.1.1 Relations between Surface Temperature T and Surface Heat Flux q

To eliminate the J in the first solution method, Equation 5.51 is solved for $J_k(\mathbf{r}_k)$, giving

$$J_k(\mathbf{r}_k) = \sigma T_k^4(\mathbf{r}_k) - \frac{1-\varepsilon_k}{\varepsilon_k} q_k(\mathbf{r}_k) \tag{5.55}$$

Equation 5.55 in the form shown and with k changed to j is then substituted into Equation 5.52 to eliminate J_k and J_j, which yields

$$\begin{aligned} \frac{q_k(\mathbf{r}_k)}{\varepsilon_k} - \sum_{j=1}^{N} \frac{1-\varepsilon_j}{\varepsilon_j} \int_{A_j} q_j(\mathbf{r}_j)\, dF_{dk-dj}(\mathbf{r}_j, \mathbf{r}_k) &= \sigma T_k^4(\mathbf{r}_k) - \sum_{j=1}^{N} \int_{A_j} \sigma T_j^4(\mathbf{r}_j)\, dF_{dk-dj}(\mathbf{r}_j, \mathbf{r}_k) \\ &= \sum_{j=1}^{N} \int_{A_j} \sigma\left[T_k^4(\mathbf{r}_k) - T_j^4(\mathbf{r}_j)\right] dF_{dk-dj}(\mathbf{r}_j, \mathbf{r}_k) \end{aligned} \tag{5.56}$$

Equation 5.56 relates the surface temperatures to the heat fluxes supplied to the surfaces. It corresponds to Equation 5.21 in the formulation for finite uniform surfaces.

Example 5.18

An enclosure of the general type in Figure 5.9 consists of three plane surfaces. Surface 1 is heated uniformly, and surface 2 has a uniform temperature. Surface 3 is black and at $T_3 \approx 0$. What are the equations needed to determine the temperature distribution along surface 1?

With $T_3 = 0$, $\varepsilon_3 = 1$, and the self-view factors $dF_{dj-dj} = 0$, Equation 5.56 is written for the two plane surfaces 1 and 2 having uniform q_1 and T_2:

$$\frac{q_1}{\varepsilon_1} - \frac{1-\varepsilon_2}{\varepsilon_2} \int_{A_2} q_2(r_2)\, dF_{d1-d2}(r_2, r_1) = \sigma T_1^4(r_1) - \sigma T_2^4 \int_{A_2} dF_{d1-d2}(r_2, r_1) \tag{5.18.1}$$

$$\frac{q_2(r_2)}{\varepsilon_2} - q_1 \frac{1-\varepsilon_1}{\varepsilon_1} \int_{A_1} dF_{d2-d1}(r_1, r_2) = \sigma T_2^4 - \int_{A_1} \sigma T_1^4(r_1)\, dF_{d2-d1}(r_1, r_2) \tag{5.18.2}$$

An equation for surface 3 is not needed since Equations 5.18.1 and 5.18.2 do not involve the unknown $q_3(r_3)$ as a consequence of $\varepsilon_3 = 1$ and $T_3 = 0$. From the definitions of F factors, $\int_{A_2} dF_{d1-d2} = F_{d1-2}$ and $\int_{A_1} dF_{d2-d1} = F_{d2-1}$. Equations 5.18.1 and 5.18.2 simplify to the following relations where the unknowns are on the left:

$$\sigma T_1^4(r_1) + \frac{1-\varepsilon_2}{\varepsilon_2}\int_{A_2} q_2(r_2)\,dF_{d1-d2}(r_2,r_1) = \sigma T_2^4 F_{d1-2}(r_1) + \frac{q_1}{\varepsilon_1} \tag{5.18.3}$$

$$\int_{A_1} \sigma T_1^4(r_1)\,dF_{d2-d1}(r_1,r_2) + \frac{q_2(r_2)}{\varepsilon_2} = \sigma T_2^4 + q_1\frac{1-\varepsilon_1}{\varepsilon_1}F_{d2-1}(r_2) \tag{5.18.4}$$

Equations 5.18.3 and 5.18.4 can be solved simultaneously for the distributions $T_1(r_1)$ and $q_2(r_2)$. Some solution methods are in Section 5.4.2 for these types of simultaneous integral equations.

5.4.1.2 Solution Method in Terms of Outgoing Radiative Flux *J*

Another method results from eliminating the $q_k(\mathbf{r}_k)$ from Equations 5.51 and 5.52 for the surfaces where $q_k(\mathbf{r}_k)$ is unknown. This provides a relation between J and the T specified along a surface:

$$J_k(\mathbf{r}_k) = \varepsilon_k\sigma T_k^4(\mathbf{r}_k) + (1-\varepsilon_k)\sum_{j=1}^{N}\int_{A_j} J_j(\mathbf{r}_j)\,dF_{dk-dj}(\mathbf{r}_j,\mathbf{r}_k) \tag{5.57}$$

When $q_k(\mathbf{r}_k)$, the heat supplied to surface k, is known, Equation 5.52 can be used directly to relate q_k and J. The combination of Equations 5.52 and 5.57 thus provides a complete set of relations for the unknown J in terms of known T and q.

This set of equations for the J is now formulated more explicitly. In general, an enclosure can have surfaces 1, 2, …, m with specified temperature distributions; for these surfaces, Equation 5.57 is used. The remaining N–m surfaces $m + 1$, $m + 2$, …, N have an imposed heat flux distribution specified; for these surfaces, Equation 5.52 is applied. This results in N equations for the unknown J distributions:

$$J_k(\mathbf{r}_k) - (1-\varepsilon_k)\sum_{j=1}^{N}\int_{A_j} J_j(\mathbf{r}_j)\,dF_{dk-dj}(\mathbf{r}_j,\mathbf{r}_k) = \varepsilon_k\sigma T_k^4(\mathbf{r}_k) \quad 1 \le k \le m \tag{5.58}$$

$$J_k(\mathbf{r}_k) - \sum_{j=1}^{N}\int_{A_j} J_j(\mathbf{r}_j)\,dF_{dk-dj}(\mathbf{r}_j,\mathbf{r}_k) = q_k(\mathbf{r}_k) \quad m+1 \le k \le N \tag{5.59}$$

After the J is found from these simultaneous integral equations, Equation 5.51 is applied to determine the unknown q or T distributions:

$$q_k(\mathbf{r}_k) = \frac{\varepsilon_k}{1-\varepsilon_k}\left[\sigma T_k^4(\mathbf{r}_k) - J_k(\mathbf{r}_k)\right] \quad 1 \le k \le m \tag{5.60}$$

$$\sigma T_k^4(\mathbf{r}_k) = \frac{1-\varepsilon_k}{\varepsilon_k}q_k(\mathbf{r}_k) + J_k(\mathbf{r}_k) \quad m+1 \le k \le N \tag{5.61}$$

5.4.1.3 Special Case When Imposed Heating q Is Specified for All Surfaces

An interesting special case is when the imposed energy flux q is specified for all surfaces of the enclosure except one and it is desired to determine the surface temperature distributions. Note that all but one q can be specified independently since $\Sigma_k q_k A_k = 0$ for steady state; however, at least one enclosure surface temperature must be given, or the temperatures become indeterminate. In many cases, the environmental temperature provides this anchor. For this case, the use of the method of the previous section where the J is first determined has an advantage over the method given by Equation 5.56 where the T are directly determined from the specified q. This advantage arises from Equation 5.59 being independent of the radiative surface properties. For a given set of q, the J need be determined only once from Equation 5.59. Then the temperature distributions are found from Equation 5.61, which introduces the emissivity dependence. This has an advantage when it is desired to examine temperature variations for various emissivity values when there is a fixed set of q. A useful relation is obtained by first considering the case in which the surfaces are all black, $\varepsilon_k = 1$. Equation 5.61 shows that $J_k(\mathbf{r}_k) = \sigma T_k^4(\mathbf{r}_k)_{black}$. Since the J_k are independent of the emissivities, the solution in Equation 5.61 can then be written for $\varepsilon_k \neq 1$ as

$$\sigma T_k^4(\mathbf{r}_k) = \frac{1-\varepsilon_k}{\varepsilon_k} q_k(\mathbf{r}_k) + \sigma T_k^4(\mathbf{r}_k)_{black} \quad (5.62)$$

This relates the temperature distributions in an enclosure for $\varepsilon_k \neq 1$ to the temperature distributions in a black enclosure having the same imposed heat fluxes, $q_k(\mathbf{r}_k)$. Thus, once the temperature distributions have been found for the black enclosure, the $\sigma T_k^4(\mathbf{r}_k)$ for gray surfaces are found by simply adding $[(1 - \varepsilon_k)/\varepsilon_k] q_k(\mathbf{r}_k)$.

Example 5.19

A heated enclosure is the circular tube in Figure 5.16, open at both ends and insulated on the outside surface (Usiskin and Siegel 1960). For a uniform heat addition (such as by electrical heating in the tube wall) to the inside surface of the tube wall and a surrounding environment at 0 K, what is the temperature distribution along the tube? If the surroundings are at T_e, how does this influence the temperature distribution?

Since the open ends of the tube are nonreflecting, they can be assumed to act as black disks at the surrounding temperature 0 K. Then with $\varepsilon_1 = \varepsilon_3 = 1$, Equation 5.61 gives $J_3 = \sigma T_1^4 = \sigma T_3^4 = 0$. Consequently, the summation in Equation 5.59 provides only radiation from surface 2 to itself. Since the tube is axisymmetric, the two differential areas dA_k and dA_k^* can be rings located at x and y. Then, Equation 5.59 yields

$$J_2(\xi) - \int_{\eta=0}^{\eta=1} J_2(\eta)\, dF_{d\xi-d\eta}(|\eta-\xi|) = q_2 \quad (5.19.1)$$

where $\xi = x/D$, $\eta = y/D$, $l = L/D$, and $dF_{d\xi-d\eta}(|\eta - \xi|)$ is the configuration factor for two rings a distance $|\eta - \xi|$ apart and is given by factor 15 in Appendix C as

$$dF_{d\xi-d\eta}(|\eta-\xi|) = \left\{1 - \frac{|\eta-\xi|^3 + \frac{3}{2}|\eta-\xi|}{\left[(\eta-\xi)^2+1\right]^{3/2}}\right\} d\eta \quad (5.19.2)$$

Absolute-value signs are used because the configuration factor depends only on the magnitude of the distance between the rings. When $|\eta - \xi| = 0$, $dF = d\eta$, and this is the configuration factor from

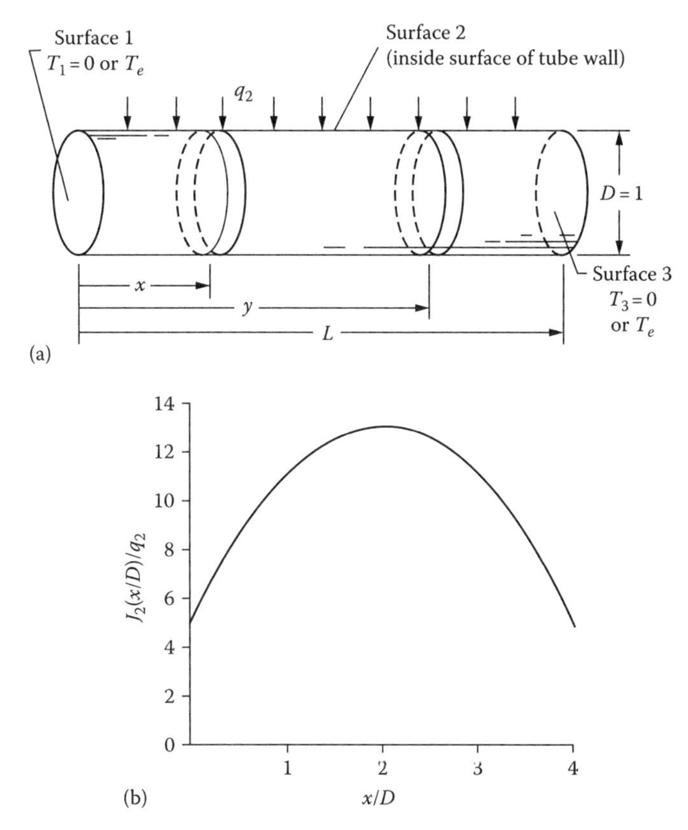

FIGURE 5.16 Uniformly heated tube insulated on outside and open to environment at both ends, (a) geometry and coordinate system and (b) distribution of J on inside of tube for $L/D = 4$.

a differential ring to itself. Equation 5.19.1 can be divided by the constant q_2 and the dimensionless quantity $J_2(\xi)/q_2$ found by numerical or approximate solution methods for linear integral equations as discussed in Section 5.4.2. The $J_2(\xi)/q_2$ distribution is in Figure 5.16b for a tube four diameters in length. From Equation 5.61, the $T_2^4(\xi)$ along the tube is given by

$$\sigma T_2^4(\xi) = \frac{1-\varepsilon_2}{\varepsilon_2} q_2 + J_2(\xi)$$

Since q_2 is constant, the $T_2^4(\xi)$ has the same shape as $J_2(\xi)$. The wall temperature is high in the central region of the tube and low near the ends where energy is radiated more readily to the low-temperature environment.

Now, consider the environment at $T_e \neq 0$. The open ends of the tube can be regarded as perfectly absorbing disks at T_e, and the integral Equation 5.59 yields

$$J_2(\xi) - \int_{\eta=0}^{l} J_2(\eta)\, dF_{d\xi-d\eta}(|\eta-\xi|) - \sigma T_e^4 F_{d\xi-1}(\xi) - \sigma T_e^4 F_{d\xi-3}(l-\xi) = q_2$$

where $F_{d\xi-1}(\xi)$ is the configuration factor from a ring element at ξ to disk 1 at $\xi = 0$, $F_{d\xi-1}(\xi) = \left\{\left(\xi^2 + \frac{1}{2}\right)/\left(\xi^2+1\right)^{1/2}\right\} - \xi$. Since the integral equation is linear in $J_2(\xi)$, let a trial solution be the sum of two parts where, for each part, either $T_e = 0$ or $q_2 = 0$:

$$J_2(\xi) = J_2(\xi)\big|_{T_e=0} + J_2(\xi)\big|_{q_2=0}$$

Substitute the trial solution into the integral equation to get

$$J_2(\xi)\big|_{T_e=0} + J_2(\xi)\big|_{q2=0} - \int_{\eta=0}^{l} J_2(\eta)\big|_{T_e=0}\, dF_{d\xi-d\eta}(|\eta-\xi|)$$

$$-\int_{\eta=0}^{l} J_2(\eta)\big|_{q2=0}\, dF_{d\xi-d\eta}(|\eta-\xi|) - \sigma T_e^4 F_{d\xi-1}(\xi)$$

$$-\sigma T_e^4 F_{d\xi-3}(l-\xi) = q_2$$

For $T_e = 0$, Equation 5.19.1 applies; subtract this equation to obtain

$$J_2(\xi)\big|_{q2=0} - \int_{\eta=0}^{l} J_2(\eta)\big|_{q2=0}\, dF_{d\xi-d\eta}(|\eta-\xi|) - \sigma T_e^4 F_{d\xi-1}(\xi) - \sigma T_e^4 F_{d\xi-3}(l-\xi) = 0$$

The solution is $J_2\big|_{q2=0} = \sigma T_e^4$, and this would be expected physically for an unheated surface in a uniform temperature environment. The tube temperature distribution is found from Equation 5.61 as

$$\sigma T_2^4(\xi) = \frac{1-\varepsilon_2}{\varepsilon_2} q_2 + J_2(\xi)\big|_{T_e=0} + J_2(\xi)\big|_{q2=0}$$
$$= \frac{1-\varepsilon_2}{\varepsilon_2} q_2 + J_2(\xi)\big|_{T_e=0} + \sigma T_e^4$$

where $J_2(\xi)\big|_{T_e=0}$ was found in the first part of this example. The environment has thus added σT_e^4 to the solution for $\sigma T_2^4(\xi)$ found previously for $T_e = 0$.

To arrive at a general conclusion for the effect of T_e in this type of problem, the final result of Example 5.19 is written in the form

$$\sigma\left[T_2^4(\xi) - T_e^4\right] = \frac{1-\varepsilon_2}{\varepsilon_2} q_2 + J_2(\xi)\big|_{T_e=0}$$

Thus, for nonzero T_e, the quantity $T_2^4(\xi) - T_e^4$ equals $T_2^4(\xi)$ for the case when $T_e = 0$. This illustrates a general way of accounting for a finite environment temperature. For energy transfer only by radiation, the governing equations are linear in T^4. As a result, in a cavity, the wall temperature $T_w|_{T_e=0}$ can be calculated for a zero environment temperature. Then, by superposition, the wall temperature for any finite T_e is $T_w^4|_{T_e\neq 0} = T_w^4|_{T_e=0} + T_e^4$. Hence, the thermal characteristics of a cavity having a wall temperature variation T_w and an external environment at T_e are the same as a cavity with wall temperature variation $\left(T_w^4 - T_e^4\right)^{1/4}$ and a zero environment temperature.

Example 5.20

Consider emission from a long cylindrical hole drilled into a material at uniform temperature T_w (Figure 5.17a). The hole is long, so the portion of its internal surface at its bottom end can be neglected in the radiative energy balances. The outside environment is at T_e.

As in the previous discussion, the solution is obtained by using the reduced temperature $T_r = \left(T_w^4 - T_e^4\right)^{1/4}$. If a position is viewed at x on the cylindrical side wall, the energy leaving is $J(x)$. An apparent emissivity is defined as $\varepsilon_a(x) = J(x)/\sigma T_r^4$. The analysis determines how $\varepsilon_a(x)$ is related to the surface emissivity ε, where ε is constant over the cylindrical side of the hole. The integral

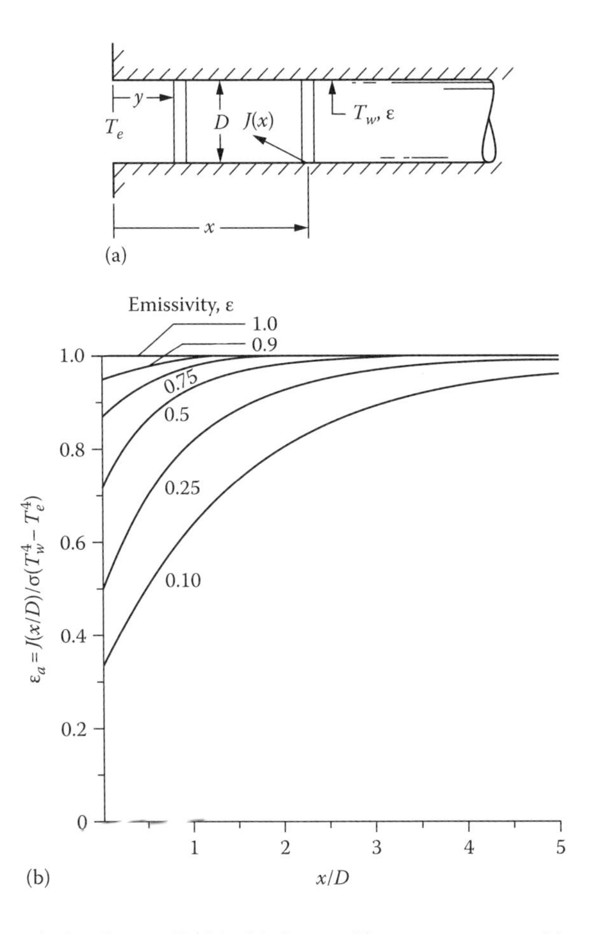

FIGURE 5.17 Radiant emission from cylindrical hole at uniform temperature, (a) geometry and coordinate system and (b) apparent emissivity of cylinder wall.

equation governing the radiation exchange within a hole was first derived by Buckley (1927, 1928) and later by Eckert (1935); both obtained approximate analytical solutions. Results were evaluated numerically by Sparrow and Albers (1960).

Using the reduced temperature, the opening of the hole is approximated by a perfectly absorbing disk at zero reduced temperature. Then, from Equation 5.61 (because $\varepsilon = 1$ and $T_r = 0$ for the opening area), $J = 0$ from the opening into the cavity. Hence, the governing equation for the enclosure is Equation 5.58, written for the cylindrical side wall and including only radiation from the cylindrical wall to itself. As in Example 5.19, the configuration factor is for a ring of differential length on the cylindrical enclosure exchanging radiation with a ring at a different axial location. Equation 5.58 then yields along the surface of the hole

$$J(\xi) - (1-\varepsilon)\int_{\eta=0}^{\infty} J(\eta)\,dF_{d\xi-d\eta}\left(|\eta-\xi|\right) = \varepsilon\sigma T^4 \tag{5.20.1}$$

where $\xi = x/D$, $\eta = y/D$, and $dF_{d\xi-d\eta}(|\eta-\xi|)$ is in Equation 5.19.2. After division by the constant σT_r^4, the apparent emissivity $\varepsilon_a(\xi)$ is governed by the integral equation

$$\varepsilon_a(\xi) - (1-\varepsilon)\int_{\eta=0}^{\infty} \varepsilon_a(\eta)\,dF_{d\xi-d\eta}\left(|\eta-\xi|\right) = \varepsilon \tag{5.20.2}$$

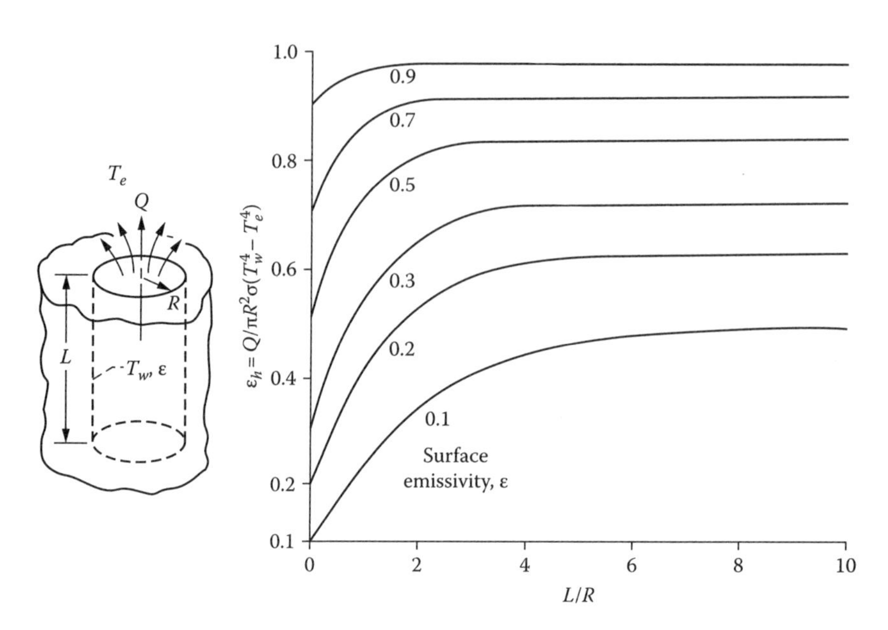

FIGURE 5.18 Apparent emissivity of cavity opening for a cylindrical cavity of finite length with diffuse reflecting walls at constant temperature. (From Lin, S.H. and Sparrow, E.M., *J. Heat Trans.*, 87(2), 299, 1965.)

The solution of Equation 5.20.2 was carried out for various surface emissivities ε, and the results are in Figure 5.17b. The radiation leaving the surface approaches that of a blackbody as the wall position is at greater depths into the hole. At the mouth of the hole $\varepsilon_a \approx \sqrt{\varepsilon}$, as shown in Buckley (1927, 1928).

Radiation from a hole of finite depth was analyzed by Lin and Sparrow (1965) and results are in Figure 5.18. Approximate solutions are in Kholopov (1973). The effective hemispherical emissivity ε_h in Figure 5.18 is a different quantity from that in Figure 5.17b; it gives the total energy leaving the mouth of the cavity ratioed to that emitted from a black-walled cavity. The latter is the same as the energy emitted from a black area across the mouth of the cavity. For each surface emissivity ε, the ε_h increases to a limiting value as the cavity depth increases; the limiting values are less than one. Unless ε is small, a cavity more than only a few diameters deep radiates the same amount as an infinitely deep cavity.

To construct a cavity that provides very close to blackbody emission for use in calibrating measuring equipment, the cavity opening can be partially closed as in Figure 5.19 (Alfano 1972, Alfano and Sarno 1975). The apparent emissivity ε_a at the center of the bottom of the cavity is in Table 5.1 as a function of the diaphragm opening-to-outer-radius ratio and the cavity depth-to-radius ratio. The ε_a increases toward one as the cavity is made deeper and the diaphragm opening decreases. Similar trends are in Heinisch et al. (1973) for partially baffled conical cavities. In Masuda (1973), cavities between radial fins are analyzed. Very detailed results for radiating cavities are in Bedford (1988).

Some information on directional emission from a cylindrical cavity is in Sparrow and Heinisch (1970) where emission normal to the cavity opening is calculated as would be received by a small detector along the cavity centerline and facing the opening. The normal emissivity ε_n of the cavity is defined as the energy received by the detector divided by the energy that would be received if the cavity walls were black. The ε_n depends on the detector distance from the cavity opening; the values in Table 5.2 are for large distances where ε_n no longer changes with distance. The table compares ε_n with ε_h from Figure 5.18 (ε is the emissivity of the cavity walls). The ε_n is larger than ε_h except for small values of cavity L/R.

Example 5.21

What are the integral equations for radiation exchange between two parallel opposed plates finite in 1D and infinite in the other, as in Figure 5.20? Each plate has a specified temperature variation that depends only on the *x*- or *y*-coordinate, and the environment is at T_e.

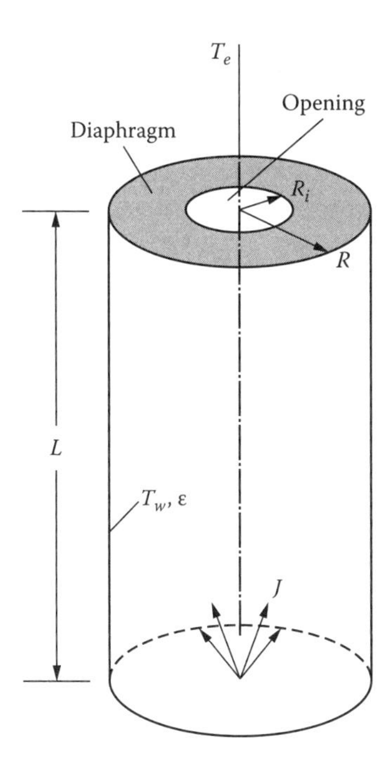

FIGURE 5.19 Cylindrical cavity with annular diaphragm partially covering opening. (From Alfano, G., *Int. J. Heat Mass Trans.*, 15(12), 2671, 1972; Alfano, G. and Sarno, A., *J. Heat Trans.*, 97(3), 387, 1975.)

TABLE 5.1
Values of Apparent Emissivity at the Center of Cavity Bottom, $\varepsilon_a = J/\sigma\left(T_w^4 - T_e^4\right)$

Wall Emissivity		ε_a		
(ε)	R_i/R	$L/R = 2$	$L/R = 4$	$L/R = 8$
0.25	0.4	0.916	0.968	0.990
	0.6	0.829	0.931	0.981
	0.8	0.732	0.888	0.969
	1.0	0.640	0.844	0.965
0.50	0.4	0.968	0.990	0.998
	0.6	0.932	0.979	0.995
	0.8	0.887	0.964	0.992
	1.0	0.839	0.946	0.989
0.75	0.4	0.988	0.997	0.999
	0.6	0.975	0.993	0.998
	0.8	0.958	0.988	0.997
	1.0	0.939	0.982	0.996

From the discussion in Example 4.2, the configuration factors between the infinitely long parallel strips dA_1 and dA_2 are

$$dF_{d1-d2} = \frac{1}{2} d\left(\sin\beta\right) = \frac{1}{2} \frac{a^2}{\left[\left(y-x\right)^2 + a^2\right]^{3/2}} dy$$

$$dF_{d2-d1} = \frac{1}{2} \frac{a^2}{\left[\left(y-x\right)^2 + a^2\right]^{3/2}} dx$$

TABLE 5.2
Comparison of Normal and Hemispherical Emissivity of Cavity

	ε = 0.9		ε = 0.7		ε = 0.5		ε = 0.3	
L/R	ε_h	ε_n	ε_h	ε_n	ε_h	ε_n	ε_h	ε_n
0.5	0.943	0.937	0.815	0.800	0.657	0.638	0.455	0.437
1	0.962	0.960	0.869	0.865	0.742	0.738	0.556	0.554
2	0.972	0.982	0.904	0.934	0.809	0.858	0.657	0.720
4	0.974	0.994	0.914	0.978	0.833	0.950	0.710	0.884
8	0.975	0.999	0.915	0.995	0.836	0.989	0.719	0.973

Source: Sparrow, E.M. and Heinisch, R.P., *Appl. Opt.*, 9, 2569, 1970.

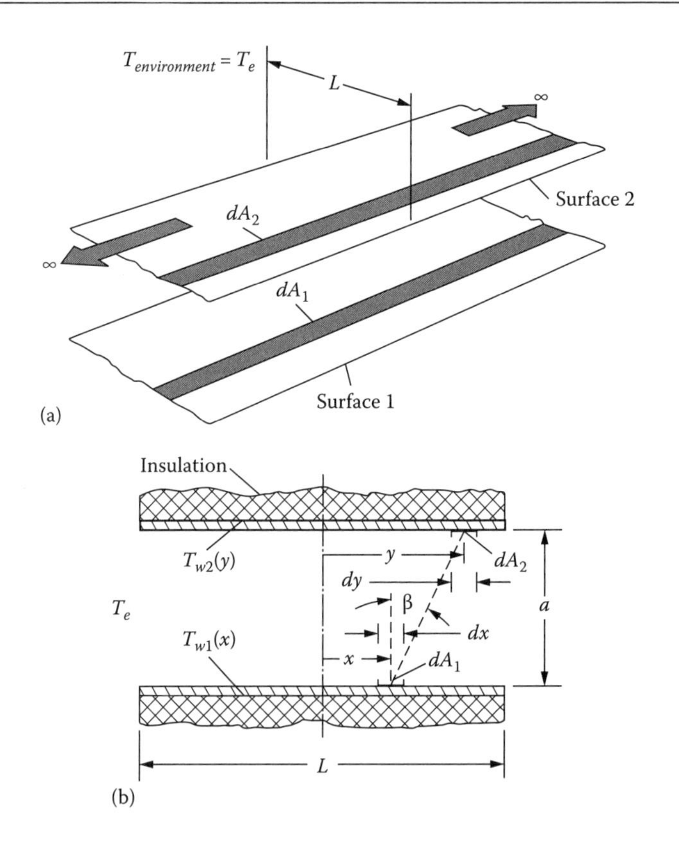

FIGURE 5.20 Geometry for radiation between two parallel plates infinitely long in one direction and of finite width, (a) parallel plates of width L and infinite length and (b) coordinates in cross section of gap between parallel plates.

The distribution of heat flux added to each plate is found by applying Equation 5.56 to each plate. As discussed before Example 5.20, reduced temperatures are used to account for T_e. With $T_1 = \left(T_{w1}^4 - T_e^4\right)^{1/4}$ and $T_2 = \left(T_{w2}^4 - T_e^4\right)^{1/4}$, the governing equations are

$$\frac{q_1(x)}{\varepsilon_1} - \frac{1-\varepsilon_2}{\varepsilon_2}\int_{-L/2}^{L/2} q_2(y)\frac{1}{2}\frac{a^2}{\left[(y-x)^2+a^2\right]^{3/2}}dy = \sigma T_1^4(x) - \int_{-L/2}^{L/2} \sigma T_2^4(y)\frac{1}{2}\frac{a^2}{\left[(y-x)^2+a^2\right]^{3/2}}dy \tag{5.21.1}$$

$$\frac{q_2(y)}{\varepsilon_2} - \frac{1-\varepsilon_1}{\varepsilon_1}\int_{-L/2}^{L/2} q_1(x)\frac{1}{2}\frac{a^2}{\left[(y-x)^2+a^2\right]^{3/2}}dx = \sigma T_2^4(y) - \int_{-L/2}^{L/2}\sigma T_1^4(x)\frac{1}{2}\frac{a^2}{\left[(y-x)^2+a^2\right]^{3/2}}dx \tag{5.21.2}$$

An alternative formulation using Equation 5.57 yields two equations for $J_1(x)$ and $J_2(y)$:

$$J_1(x) - (1-\varepsilon_1)\int_{-L/2}^{L/2} J_2(y)\frac{1}{2}\frac{a^2}{\left[(y-x)^2+a^2\right]^{3/2}}dy = \varepsilon_1\sigma T_1^4(x) \tag{5.21.3}$$

$$J_2(y) - (1-\varepsilon_2)\int_{-L/2}^{L/2} J_1(x)\frac{1}{2}\frac{a^2}{\left[(y-x)^2+a^2\right]^{3/2}}dx = J_2\sigma T_2^4(y) \tag{5.21.4}$$

After the $J_1(x)$ and $J_2(y)$ are found, the desired $q_1(x)$ and $q_1(y)$ are obtained from Equation 5.60 as

$$q_1(x) = \frac{\varepsilon_1}{1-\varepsilon_1}\left[\sigma T_1^4(x) - J_1(x)\right] \tag{5.21.5}$$

$$q_2(y) = \frac{\varepsilon_2}{1-\varepsilon_2}\left[\sigma T_2^4(y) - J_2(y)\right] \tag{5.21.6}$$

5.4.2 Methods for Solving Integral Equations

The unknown wall heat fluxes or temperatures along the surfaces of an enclosure are found from solutions of single or simultaneous integral equations. The integral equations in the formulations up to now are linear; that is, the unknown q, J, or T^4 variables always appear to the first power (note that T^4 is the linear variable rather than T). For linear integral equations, there are various numerical and analytical solution methods; these are discussed in mathematics texts. The use of some of these methods for radiation analysis is now discussed.

5.4.2.1 Numerical Integration

In most instances, the functions inside the integrals of the integral equations are complicated algebraic quantities. This is because they involve a configuration factor. There is usually little chance that an analytical solution can be found, so a numerical solution is used. Consider the simultaneous integral equations in Equations 5.21.1 and 5.21.2. With $T_1(x)$ and $T_2(y)$ specified, the right sides are known functions of x and y. Starting with Equation 5.21.1, a distribution for $q_2(y)$ is assumed as a first trial. Then the integration is carried out numerically for various x values to yield $q_1(x)$ at these x locations. This $q_1(x)$ distribution is inserted into Equation 5.21.2 and a $q_2(y)$ distribution is determined. This $q_2(y)$ is used to compute a new $q_1(x)$ from Equation 5.21.2 and the process is continued until $q_1(x)$ and $q_2(y)$ are no longer changing as the iterations proceed.

To perform the integrations in a computer solution, an accurate integration subroutine is required. Many subroutines are available, and they may require functions such as $q_1(x)$ and $q_2(y)$ at many evenly or unevenly spaced values of the x- and y-coordinates. The values can be obtained by curve fitting the $q_1(x)$ and $q_2(y)$ after each iteration with standard subroutines such as cubic splines. The $q_1(x)$ and $q_2(y)$ are then interpolated at the x and y values called for by the integration subroutine. A precaution should be noted. A quantity such as $J_j\, dF_{dk-dj}$ may go through rapid changes in magnitude because of the geometry involved in the configuration factor; for example, dF_{dk-dj} may decrease rapidly as the distance increases between dA_k and dA_j (Figure 5.21, e.g., for Equation 5.19.2). For small separation distances, there can be a strong peak in the integration kernel. Care should be

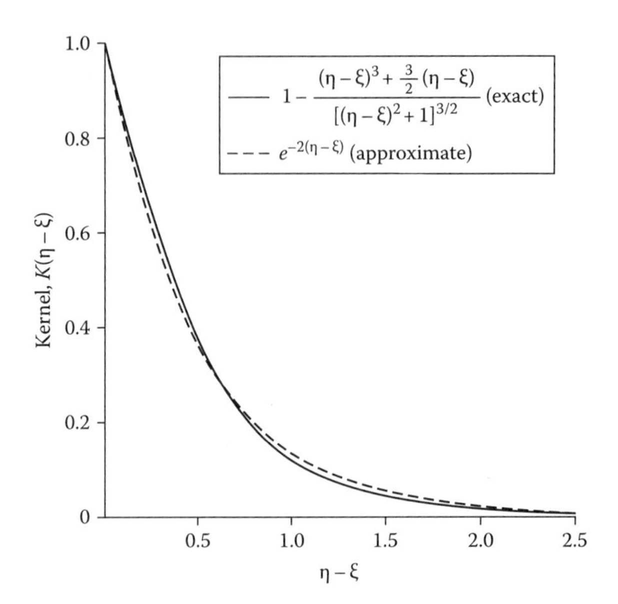

FIGURE 5.21 Exponential approximation to configuration-factor kernel for cylindrical enclosure.

taken that the integration method is accurate for the functions involved. The integration should be done on each side of a sharp peak and not passed through it.

Direct solvers for a set of simultaneous equations can also be used for integral equations. The integrals are expressed in finite-difference form to provide a set of simultaneous equations for the unknowns at each increment position as in Example 5.22.

Example 5.22

For integral Equation 5.19.1, derive a set of simultaneous algebraic equations to determine $J_2(\xi)$ for $l = 4$. For simplicity, divide the length into four equal increments ($\Delta\eta = 1$) and use the trapezoidal rule for integration.

When Equation 5.19.1 is applied at the end of the tube where $\xi = 0$, the relation is obtained:

$$J_2(0) - \left[\frac{1}{2}J_2(0)K(|0-0|) + J_2(1)K(|1-0|) + J_2(2)K(|2-0|) + J_2(3)K(|3-0|) + \frac{1}{2}J_2(4)K(|4-0|)\right](1) = q_2 \quad (5.22.1)$$

The quantity in brackets is the trapezoidal-rule approximation for the integral. The $K(|\eta-\xi|) = dF(|\eta-\xi|)/d\eta$ is the algebraic expression within the braces of Equation 5.19.2. The $J_2(0)$ terms in Equation 5.22.1 are grouped together to provide the first of Equation 5.22.2. The other four equations are the finite-difference equations at the other incremental positions along the enclosure:

$$\begin{aligned}
&J_2(0)\left[1-\frac{1}{2}K(0)\right] - J_2(1)K(1) - J_2(2)K(2) - J_2(3)K(3) - \frac{1}{2}J_2(4)K(4) = q_2\\
&-\frac{1}{2}J_2(0)K(1) + J_2(1)\left[1-K(0)\right] - J_2(2)K(1) - J_2(3)K(2) - \frac{1}{2}J_2(4)K(3) = q_2\\
&-\frac{1}{2}J_2(0)K(2) - J_2(1)K(1) + J_2(2)\left[1-K(0)\right] - J_2(3)K(1) - \frac{1}{2}J_2(4)K(2) = q_2\\
&-\frac{1}{2}J_2(0)K(3) - J_2(1)K(2) - J_2(2)K(1) + J_2(3)\left[1-K(0)\right] - \frac{1}{2}J_2(4)K(1) = q_2\\
&-\frac{1}{2}q_{o,2}(0)K(4) - q_{o,2}(1)K(3) - q_{o,2}(2)K(2) - q_{o,2}(3)K(1) + q_{o,2}(4)\left[1-\frac{1}{2}K(0)\right] = q_2
\end{aligned} \quad (5.22.2)$$

These equations are solved for J_2 at the five surface locations. From symmetry, and with q_2 uniform along the enclosure, it is possible to simplify the solution for this example by using $J_2(0) = J_2(4)$ and $J_2(1) = J_2(3)$.

Equations such as Equation 5.22.2 are first solved for a moderate number of increments along the surfaces. Then the increment size is reduced, and the solution is repeated. This is continued until sufficiently accurate, $J(\xi)$ values are obtained. Equations 5.22.2 use the trapezoidal rule as a simple numerical approximation to the integrals. More accurate numerical integration schemes can be used, which may reduce the number of increments required for sufficient accuracy.

Example 5.22 has only one integral equation. For the situation with two integral equations in Equations 5.21.1 and 5.21.2, each surface can be divided into increments and equations written at each incremental location. This yields N simultaneous equations for the total of N positions on both plates that are solved simultaneously for the $q_1(x)$ and $q_2(y)$. This procedure is an alternative to the iterative solution described previously. The solver for the system of simultaneous equations may actually work by iteration.

5.4.2.2 Analytical Solutions

For some simple geometries and special conditions, the integral equations describing radiative transfer among surfaces may be solved analytically. Such solutions are usually limited to single-surface or two-surface enclosures, so are not described here in detail.

If the kernel of the integral equation is *separable*, that is, $K(\mathbf{r}_j,\mathbf{r}_k) = F_j(\mathbf{r}_j)F_k(\mathbf{r}_k)$, then $F_k(\mathbf{r}_k)$ may be removed from the integral over $\mathbf{r}_j$, possibly simplifying analytical or numerical integration. However, the kernel in radiation problems usually is *not* separable. The general theory of solution of integral equations using separable kernels is in Hildebrand (1992) and an application using a separable exponential approximation to the kernel (Usiskin and Siegel 1960), allowing reduction of the integral equation to a differential equation, is in Buckley (1927, 1928).

The *variational method* (Hildebrand 1992) may be applied if the kernel is *symmetric*, that is, $K(\xi,\eta) = K(\eta,\xi)$. This approach has been used for radiation in a cylindrical tube (Usiskin and Siegel 1960) and for radiative exchange between infinitely long parallel plates of finite width (Sparrow 1960).

An approximate solution may be obtained through a *Taylor series expansion* (Krishnan and Sundaram 1960, Perlmutter and Siegel 1963), which works well if the kernel $K(|\xi-\eta|)$ decays rapidly as $\xi - \eta$ increases as for the cylindrical geometry illustrated in Figure 5.20. The integrand of the integral equations then becomes a series that may be truncated after a few terms and then integrated term by term, reducing the integral equation to a differential equation. Applications are in Choi and Churchill (1985) and Qiao et al. (2000).

The *method of Ambartsumian* can be applied if the temperature or heat flux boundary conditions can be approximately described by an exponential variation or a sum of exponentials, allowing transformation of the integral equation into an initial value problem (Ambartsumian 1942, Kourganoff 1963, Crosbie and Sawheny 1974, 1975).

The problem of finding the intensity leaving a circular opening in a spherical cavity exposed to external uniform heat flux incident on element dA_2, $q_e(dA_2)$, and with a prescribed internal temperature distribution $T(dA_1)$ on the cavity surface has been solved analytically (Jakob 1957, Sparrow and Jonsson 1962). If the internal surface of the cavity has emissivity ε, then the intensity leaving an element dA_1^* through the cavity opening in a particular direction is found to be

$$I\left(dA_1^*\right) = \frac{J_1(dA_1^*)}{\pi} = \frac{\varepsilon\sigma T_1^4\left(dA_1^*\right)}{\pi} + \frac{\dfrac{1}{\pi}\left(\dfrac{1-\varepsilon}{4\pi R^2}\right)\left[\displaystyle\int_{A_1} \varepsilon\sigma T_1^4(dA_1)dA_1 + \int_{A_2} q_e(dA_2)dA_2\right]}{1-\left(1-\varepsilon\right)A_1/4\pi R^2} \tag{5.63}$$

All of these analytical methods become intractable when multiple surfaces are present, and numerical solution techniques are almost always required for more realistic cases.

5.4.2.3 Exact Solution of Integral Equation for Radiation from a Spherical Cavity

Radiation from a spherical cavity as in Figure 5.22a was analyzed by Jakob (1957) and Sparrow and Jonsson (1962). The spherical shape leads to a relatively simple integral-equation solution because there is an especially simple configuration factor between elements on the inside of a sphere. For the two differential elements dA_j and dA_k in Figure 5.22b,

$$dF_{dj-dk} = \frac{\cos\theta_j \cos\theta_k}{\pi S^2} dA_k \tag{5.64}$$

Since the sphere radius is normal to both dA_j and dA_k, the distance between these elements is $S = 2R\cos\theta_j = 2R\cos\theta_k$. Then, Equation 5.64 becomes

$$dF_{dj-dk} = \frac{dA_k}{4\pi R^2} = \frac{dA_k}{A_s} \tag{5.65}$$

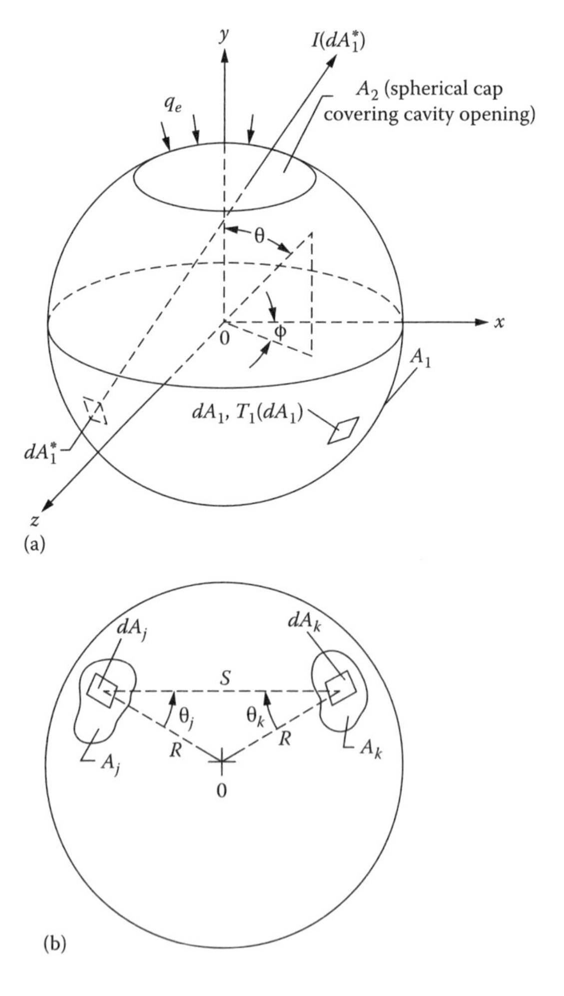

FIGURE 5.22 Geometry for radiation within spherical cavity, (a) spherical cavity with diffuse entering radiation q_e and with surface at variable temperature T_1 and (b) area elements on spherical surface.

where A_s is the surface area of the entire sphere. If dA_j exchanges with the finite area, then Equation 5.65 becomes

$$F_{dj-k} = \frac{1}{4\pi R^2}\int_{A_k} dA_k = \frac{A_k}{4\pi R^2} = \frac{A_k}{A_s} \tag{5.66}$$

Equation 5.66 is *independent of* dA_j, so dA_j can be replaced by any finite area A_j to give

$$F_{j-k} = \frac{A_k}{4\pi R^2} = \frac{A_k}{A_s} \tag{5.67}$$

The configuration factor from any area to any area is simply the fraction of the total sphere area that the *receiving* area occupies.

Consider the spherical cavity in Figure 5.22a with a temperature distribution $T_1(dA_1)$ and a total surface area A_1. The spherical cap that would cover the cavity opening has area A_2. Assume there is diffuse radiative flux q_e (per unit area of A_2) entering through the cavity opening; the q_e can vary over A_2. It is desired to compute the radiation intensity $I(dA_l^*)$ leaving the cavity at a specified location and in a specified direction, as shown by the arrow in Figure 5.22a. The desired intensity results from the diffuse flux leaving dA_l^* and equals $J_1(dA_l^*)/\pi$. The $J_1(dA_l^*)$ is found by applying Equation 5.58:

$$J_1\left(dA_1^*\right) - \left(1-\varepsilon_1\right)\int_{A_1} J_1\left(dA_1\right)dF_{d1^*-d1} - \left(1-\varepsilon_1\right)\int_{A_2} q_e\left(dA_2\right)dF_{d1^*-d2} = \varepsilon_1\sigma T_1^4\left(dA_1^*\right) \tag{5.68}$$

and an exact solution will be found. The F factors from Equation 5.65 are substituted to give

$$J_1\left(dA_1^*\right) - \frac{1-\varepsilon_1}{4\pi R^2}\int_{A_1} J_1\left(dA_1\right)dA_1 = \frac{1-\varepsilon_1}{4\pi R^2}\int_{A_2} q_e\left(dA_2\right)dA_2 + \varepsilon_1\sigma T_1^4\left(dA_1^*\right) \tag{5.69}$$

To solve Equation 5.69, a trial solution $J_1\left(dA_l^*\right) = f\left(dA_l^*\right) + C$ is assumed, where $f(dA_l^*)$ is an unknown function of the location of dA_1^*, and C is a constant. Substituting into Equation 5.69 gives

$$f\left(dA_1^*\right) + C - \frac{1-\varepsilon_1}{4\pi R^2}\int_{A_1} f\left(dA_1\right)dA_1 - \frac{1-\varepsilon_1}{4\pi R^2}CA_1 = \frac{1-\varepsilon_1}{4\pi R^2}\int_{A_2} q_e\left(dA_2\right)dA_2 + \varepsilon_1\sigma T_1^4\left(dA_1^*\right)$$

From the two terms that are functions of local position, $f\left(dA_1^*\right) = \varepsilon_1\sigma T_1^4\left(dA_1^*\right)$. The remaining terms are equated to determine C. This gives the desired result as an exact solution:

$$I\left(dA_1^*\right) = \frac{J_1\left(dA_1^*\right)}{\pi} = \frac{\varepsilon_1\sigma T_1^4\left(dA_1^*\right)}{\pi} + \frac{\frac{1}{\pi}\left[\frac{1-\varepsilon_1}{4\pi R^2}\right]\left[\int_{A_1} \varepsilon_1\sigma T_1^4\left(dA_1\right)dA_1 + \int_{A_2} q_e\left(dA_2\right)dA_2\right]}{1-\left(1-\varepsilon_1\right)A_1/4\pi R^2} \tag{5.70}$$

5.4.3 General Boundary Conditions That Provide Inverse Problems

Throughout this chapter, it has been required that a boundary condition be prescribed on every surface making up the enclosure boundary and that the radiative properties of each surface be known. If this is done, then there are automatically an equal number of equations and unknowns to allow solution of the radiative exchange problem. However, there are classes of problems where two boundary conditions (perhaps heat flux and temperature) are prescribed on one or more surfaces, and the conditions on one or more other surfaces that allow this prescription are to be determined. These problems are *inverse boundary condition problems*. Such problems and others where properties or other conditions are to be found may require special treatment and are covered in Chapter 8.

5.5 COMPUTER PROGRAMS FOR ENCLOSURE ANALYSIS

So far, we have addressed radiative energy exchange within enclosures having black and/or diffuse–gray surfaces. The surface areas can be of finite or infinitesimal size. Each enclosure surface can have a specified net energy flux added to it by some external means such as conduction or convection, or it can have a specified surface temperature. Various methods were presented for solving the array of simultaneous linear equations or the linear integral equations that result from the formulation of these interchange problems. Inverse problems were also discussed that considered other specifications of conditions that are more general and are more difficult to deal with. It was pointed out that most practical problems become so complex that numerical techniques are required for solution. To solve these problems, several computer programs have been developed. For example, the program thermal simulation system (TSS) (Chin et al. 1992), developed under NASA sponsorship, incorporates solvers for enclosures with a very large number of surfaces as used for satellite and space vehicle design. Some other commercially available computer programs for thermal analysis also incorporate solvers for radiative enclosures in addition to including computation of configuration factors (see Section 4.3.6). Computer subroutines such as matrix solvers and integration subroutines can be used and are available in mathematics software packages.

HOMEWORK

5.1 What is the net radiative energy transfer from black surface dA_1 to black surface A_2?

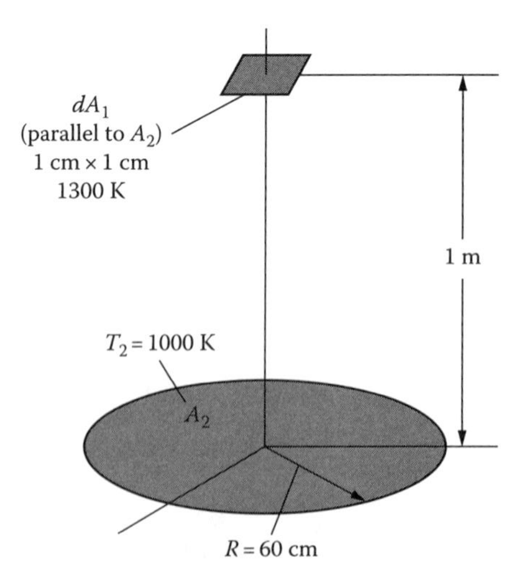

Answer: $dQ_{1-2} = 2.786$ W.

5.2 A person holds her open hand (approximated by a circular disk 10 cm in diameter) 6 cm directly above and parallel to a black heater element in the form of a circular disk 20 cm in diameter (such as an electric range element). The heater element is at 750 K. How much radiant energy from the heater element is incident on the hand?
Answer: 98.1 W.

5.3 A rectangular carbon steel billet 1 × 0.5 × 0.5 m is initially at 1100 K and is supported in such a manner that it transfers heat by radiation from all of its surfaces to surroundings at T_e = 27°C (assume the surroundings are black). Neglect convective heat transfer and assume the billet radiates like a blackbody. Also, assume for simplicity that the thermal conductivity of the steel is infinite. How long will it take for the billet to cool to 410 K? (For carbon steel, let ρ_{cs} = 7800 kg/m^3 and c_{cs} = 470 J/kg · K.)

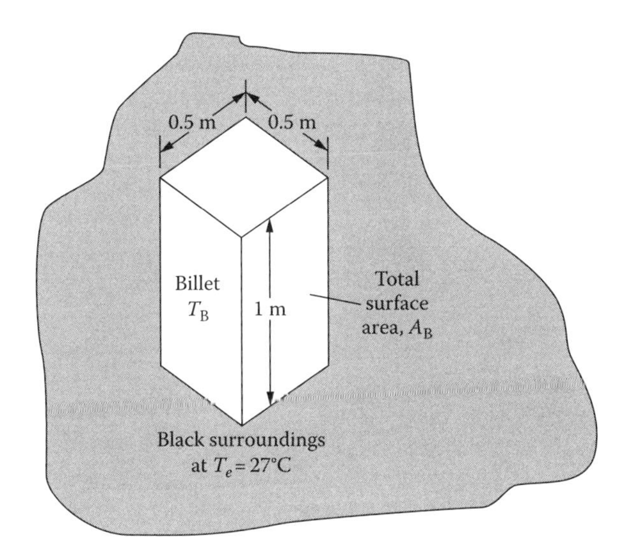

Answer: 9.55 h.

5.4 A black circular disk 0.15 m in diameter and well insulated on one side is electrically heated to a uniform temperature. The electrical energy input is 1200 W. The surroundings are black and are at T_e = 500 K. What fraction of the emitted energy is in the wave number region from 0.2 to 1 μm^{-1}?
Answer: 0.668.

5.5 A black electrically heated rod is in a black vacuum jacket. The rod must dissipate 120 W without exceeding 950 K. Calculate the maximum allowable jacket temperature (neglect end effects).

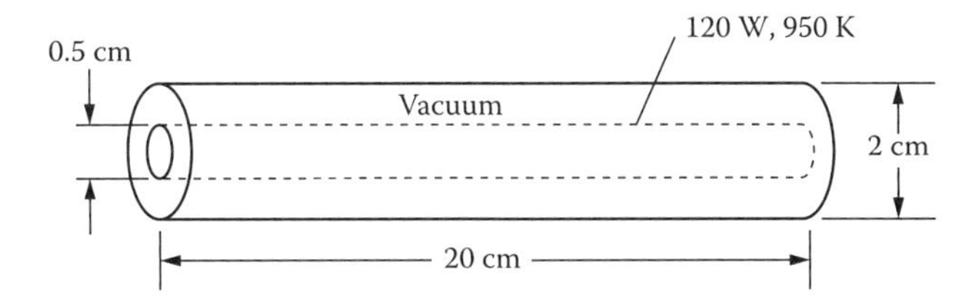

Answer: 613 K.

5.6 A hollow cylindrical heating element is insulated on its outside surface. The element has a 25 cm inside diameter and is 25 cm long. The black internal surface is to be held at 1000 K.

The surroundings are in vacuum and are at 500 K. Both ends of the cylinder are open to the surroundings. Estimate the energy that must be supplied to the element (W).

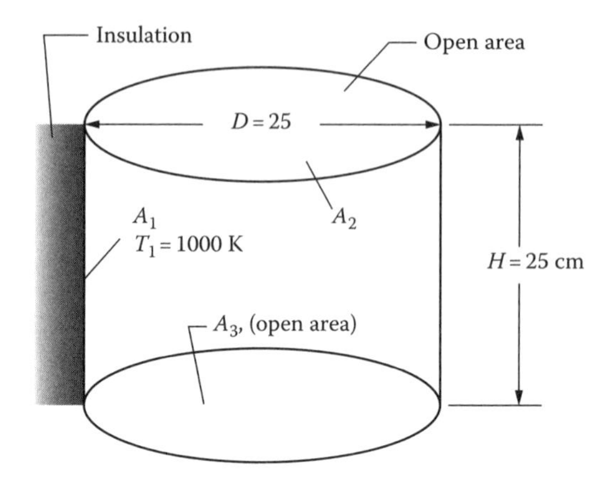

Answer: 4320 W.

5.7 A regular tetrahedron of side length 0.5 m has black internal surfaces with the following characteristics:

Surface	Q(W)	T(K)
1	$Q_1 = 0$	T_1
2	Q_2	600
3	Q_3	1000
4	400	T_4

What are the values of T_1, Q_2, Q_3, and T_4?
Answers: $T_1 = 876$ K; $Q_2 = -3762$ W; $Q_3 = 3362$ W; $T_4 = 894$ K.

5.8 A rectangular solar collector resting on the ground has a black upper surface and dimensions 1.5 × 2.5 m. The lower surface is very well insulated. The collector is tilted at an angle of 40° from the horizontal. Fluid at 95°C is passed through the collector at night. The night sky acts as a blackbody with effective temperature of 210 K and the ground around the collector is at 5°C. Neglecting conduction and convection to the surroundings, at night,

(a) At what rate is energy lost by the collector (W)?
(b) If flow through the collector is stopped, what temperature (K) will the collector attain?

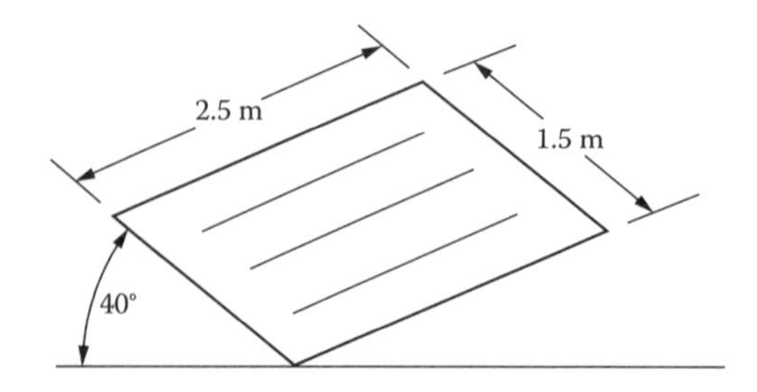

Answers: (a) 3392 W; (b) 222 K.

5.9 The black-surfaced three-sided enclosure shown in the following has infinitely long parallel sides, with the specified temperatures and energy rate additions. Find Q_1, Q_2, and T_3. (Side 3 is a circular arc.)

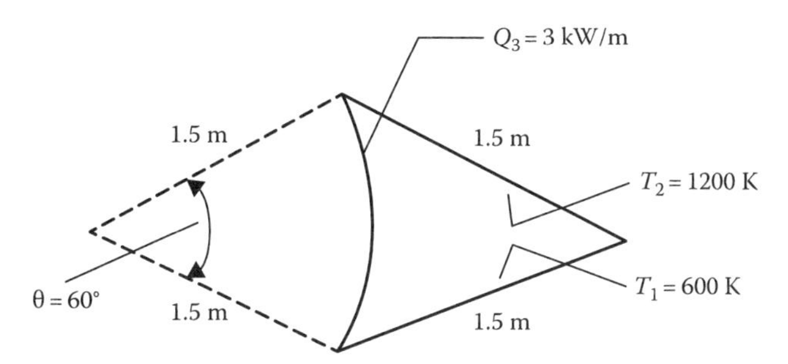

Answers: $Q_1 = -123{,}560$ W/m; $Q_2 = 121{,}560$ W/m; $T_3 = 1032$ K.

5.10 Two enclosures are identical in shape and size and have black surfaces. For one enclosure, the temperatures of the surfaces are $T_1, T_2, T_3, \ldots, T_N$. For the second, the surface temperatures are $(T_1^4 + k)^{1/4}, (T_2^4 + k)^{1/4}, (T_3^4 + k)^{1/4}, \ldots (T_N^4 + k)^{1/4}$, where k is a constant. Show how the heat transfer rates Q_j at any surface A_j are related for the two enclosures.
Answer: They are the same.

5.11 An enclosure with black interior surfaces has one side open to an environment at temperature T_e. The sides of the enclosure are maintained at temperatures of $T_1, T_2, T_3, \ldots, T_N$. How are the rates of energy input to the sides $Q_1, Q_2, Q_3, \ldots, Q_N$ influenced by the value of T_e? How can the results for $T_e = 0$ be used to obtain solutions for other T_e?

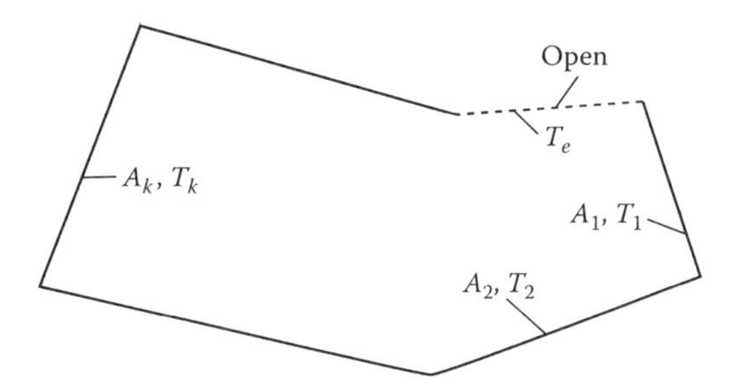

5.12 (a) A circular cylindrical enclosure has black interior surfaces, each maintained at uniform interior temperature as shown. The outside of the entire cylinder is insulated so that the outside does not radiate to the surroundings. How much Q (W) is supplied to each area as a result of the interior radiative exchange? (Perform the calculation without subdividing any surface areas.)

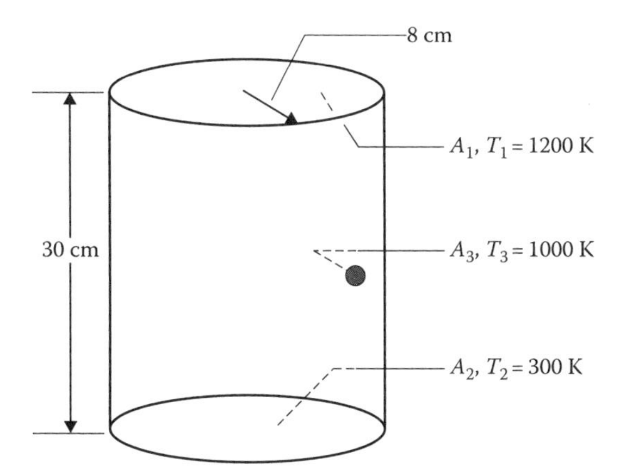

(b) For the same enclosure and the same surface temperatures, divide A_3 into two equal areas A_4 and A_5. What is the Q to each of these two areas, and how do they and their sum compare with Q_3 from part (a)?

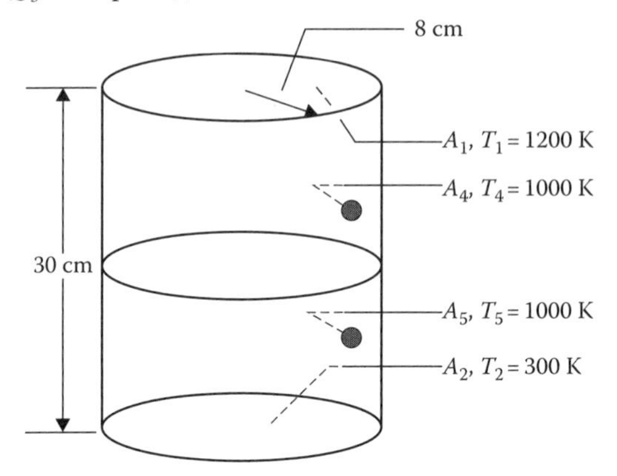

(c) What are Q_6/A_6 and Q_7/A_7 for the same enclosure and surface temperatures? How do they compare with Q_3/A_3 from part (a) and Q_4/A_4 and Q_5/A_5 from part (b)?

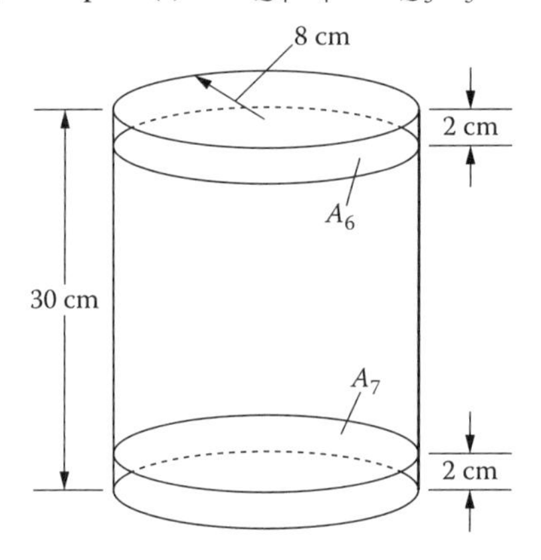

Answer:
(a) $Q_1 = 1295$ W; $Q_2 = -1207$ W; $Q_3 = -87.3$ W.
(b) $Q_1 = 1295$ W; $Q_2 = -1207$ W; $Q_4 = -853$ W; $Q_5 = 765$ W.
(c) $Q_1 = 1295$ W; $Q_2 = -1207$ W; $Q_6 = -261$ W; $Q_7 = 240$ W.

5.13 A black 6 cm diameter sphere at a temperature of 1100 K is suspended in the center of a thin 10 cm diameter partial sphere having a black interior surface and an exterior surface with a hemispherical total emissivity of 0.1. The surroundings are at 300 K. A 7.5 cm diameter hole is cut in the outer sphere. What is the temperature of the outer sphere? What is the Q being supplied to the inner sphere? (For simplicity, do not subdivide the surface areas into smaller zones.)

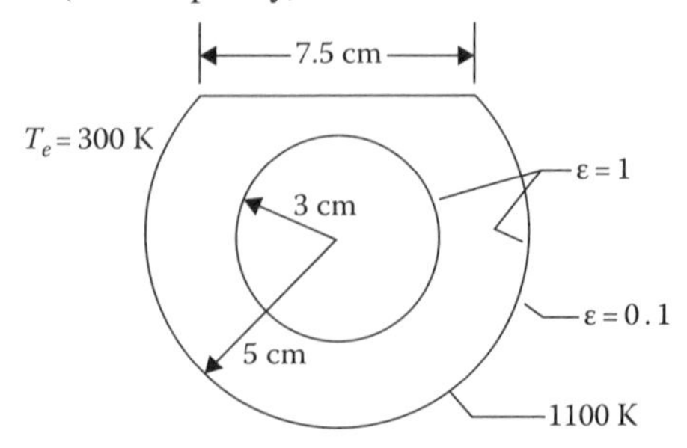

Answer: 988 K; 431 W.

5.14 A 3 cm diameter, very long heater rod at the center of a long box furnace of 10 cm × 10 cm cross section is heated in vacuum to 1000°C, as shown in the following. The top and bottom walls of the box furnace are thermally insulated on the back. The two side walls of the box furnace are covered by metallic panels with the temperature maintained at 700°C. All surfaces are diffuse and gray. The emissivity is 0.8 for the heater rod, 0.5 for the top and bottom walls of the box furnace, and 0.2 for the two metallic panels. Assuming uniform irradiation and temperature on each surface, find the rate of electrical heating applied to the heater rod and the temperature of the top and bottom walls.

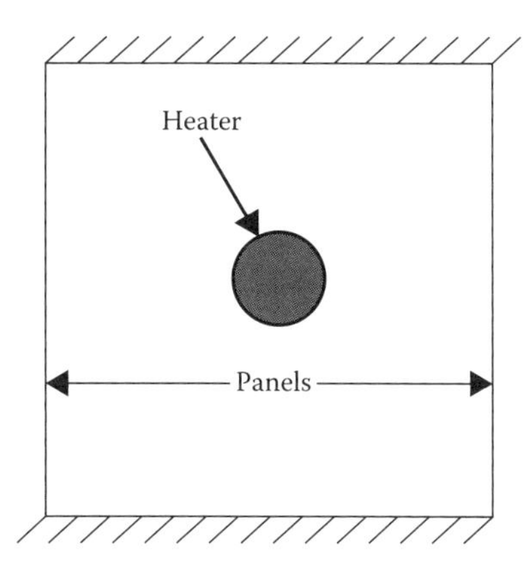

Answer: $q_{heater} = 29.7$ kW/m^2; $q_{wall} = -14.0$ kW/m^2; $T_{top} = 1200$ K (925°C).

5.15 An infinitely long enclosure (normal to the direction shown) is shaped as shown in the following. Assuming that the uniform flux restrictions are met along each surface, find Q_1 and T_2.

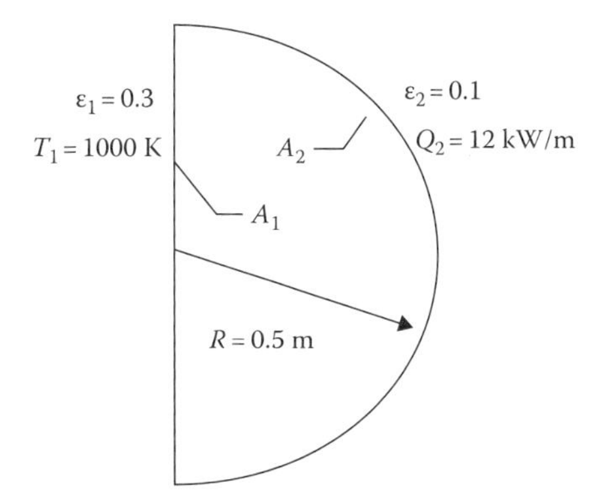

Answers: $Q_1 = -12$ kW/m; $T_2 = 1307$ K.

5.16 Two infinitely long diffuse–gray concentric circular cylinders are separated by two concentric thin diffuse–gray radiation shields. The shields have identical emissivities on both sides.

(a) Derive an expression for the energy transferred between the inner and outer cylinders in terms of their temperatures and the necessary radiative and geometric quantities. (Number the surfaces from the inside out; i.e., the inner surface is number 1, the outer surface is number 4.)

(b) Check this result by showing that in the proper limit it reduces to the correct result for four parallel plates with identical emissivities.

(c) Find the percent reduction in heat transfer when the shields are added if the radii for the surfaces are in the ratio 1:2:4:8, and if $\varepsilon_1 = \varepsilon_4 = 0.6$ and $\varepsilon_2 = \varepsilon_3 = 0.1$.

Answer:

(a) $Q = \dfrac{A_1\sigma\left(T_1^4 - T_4^4\right)}{G_{12} + \dfrac{A_1G_{23}}{A_2} + \dfrac{A_1G_{34}}{A_3}}$ where $G_{ab} \equiv \dfrac{1}{\varepsilon_a} + \dfrac{A_a}{A_b}\left(\dfrac{1}{\varepsilon_b} - 1\right)$.

(b) $Q = \dfrac{A_1\sigma\left(T_1^4 - T_4^4\right)}{3\left(\dfrac{2}{\varepsilon} - 1\right)}$.

(c) $Q_{with}/Q_{without} = 0.109$.

5.17 Consider a diffuse–gray right circular cylindrical enclosure. The diameter is the same as the height of the cylinder, 2.75 m. The top is removed. If the remaining internal surfaces are maintained at 900 K and have an emissivity of 0.7, determine the radiative energy escaping through the open end. Assume uniform irradiation on each surface so that subdivision of the base and curved wall is not included. The outside environment is at $T_e = 0$ K. How do the results compare with Figure 5.18? Explain any difference. If $T_e = 500$ K, what percent reduction occurs in radiative energy loss?
Answer: -2.04×10^5 W; 90.5% reduction in energy rate.

5.18 Consider the gray cylindrical enclosure described in Homework Problem 5.17 with the top in place. A hole 30 cm in diameter is cut in the top. Determine the configuration factors between (a) the base and the hole and (b) the curved wall and the hole. Estimate the radiant energy escaping through the hole. The outside environment is at $T_e \approx 0$ K.
Answer: (a) 0.00238; (b) 0.00238; (c) 2630 W.

5.19 A 2D diffuse–gray enclosure (infinitely long normal to the cross section shown) has each surface at a uniform temperature. Compute the energy added per meter of enclosure length at each surface to account for the radiative exchange within the enclosure, Q_1, Q_2, Q_3. (Assume for simplicity that it is not necessary to subdivide the three areas.) The conditions are $T_1 = 1000$ K, $\varepsilon_1 = 0.6$, $T_2 = 300$ K, $\varepsilon_2 = 0.5$, $T_3 = 800$ K, $\varepsilon_3 = 0.7$.

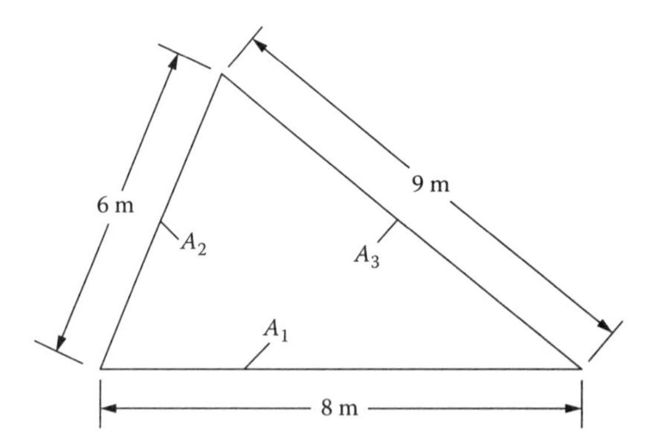

Answer: 159,400 W/m; −95,860 W/m; −63,540 W/m.

5.20 A thin gray disk with emissivity 0.9 on both sides is in Earth's orbit. It is exposed to normally incident solar radiation (neglect radiation emitted or reflected from the Earth.) What is the equilibrium temperature of the disk? A single thin radiation shield having emissivity 0.05 on both sides is placed as shown. What is the disk temperature? What is the effect on both of these results of reducing the disk emissivity to 0.5? (Assume the surroundings are at zero absolute temperature and that for simplicity, it is not necessary to subdivide the areas.)

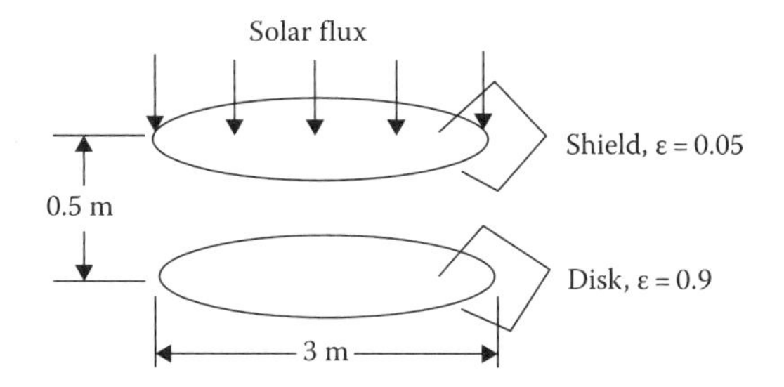

Answer: For unshaded disk, 330.5 K (either value of ε); with shield, 113.8 K (ε = 0.05); 136.1 K (for ε = 0.5).

5.21 For the enclosure with four infinitely long parallel walls shown in the following in cross section, calculate the average heat flux on surface 2 (W/m²). All surfaces are diffuse–gray and are assumed for simplicity to have uniform outgoing flux distributions.

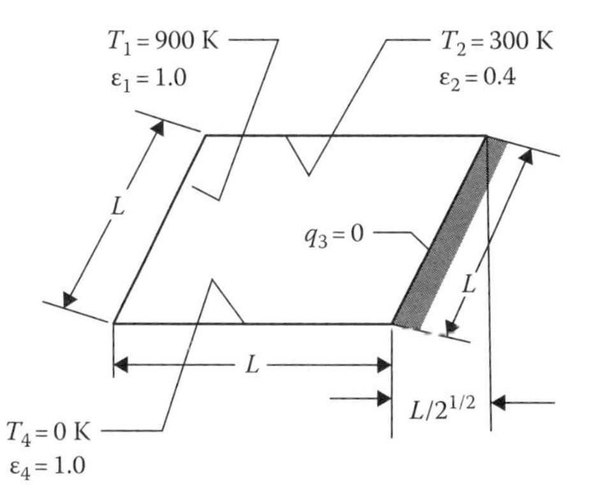

Answer: −4972 W/m².

5.22 Two infinite parallel black plates are at temperatures T_1 and T_2. A perforated thin sheet of gray material of emissivity ε_S is inserted between the plates. Let *SF*, the shading factor, be defined as the ratio of the solid area of the perforated sheet to the total area of the sheet, A_g/A. Determine the temperature of the perforated plate and find the ratio of the energy transfer Q_1 between the black plates when $SF = 1$ (i.e., with a single solid gray radiation shield) to the energy transfer with a perforated shield, Q_{SF}.

Answer: $T_g^4 = \dfrac{T_1^4 + T_2^4}{2}$; $Q_{1,shield}/Q_{1,perforated} = \dfrac{\varepsilon_s}{2} \dfrac{1}{\left[1 + (\varepsilon_s SF/2) - SF\right]}$.

5.23 A very long cylinder at temperature T_1 is coaxial with a long square enclosure shown in the following cross section. The conditions on surfaces 1–5 are shown in the table. Find Q_1 and T_2.

(For simplicity, do not subdivide the surfaces. Also, note that the configuration factors for this geometry were derived in Homework Problem 4.22.)

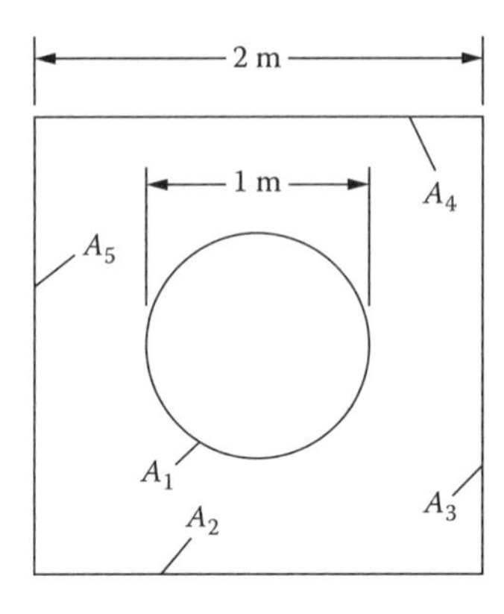

Surface	T(K)	Q (W/m)	ε
1	1100		0.4
2		0	0.5
3	0		1
4	0		1
5	0		1

Answer: $T_2 = 703$ K; $Q_1 = 100$ kW/m.

5.24 A frustum of a cone has its base heated as shown. The top is held at 800 K while the side is perfectly insulated. All surfaces are diffuse–gray. What is the temperature attained by surface 1 as a result of radiative exchange within the enclosure? (For simplicity, do not subdivide the areas.)

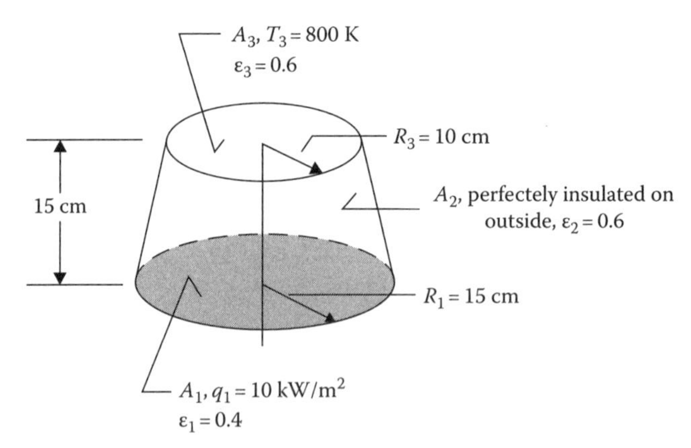

Answer: 1087 K.

5.25 In a metal-processing operation, a metal sphere at uniform temperature is heated in a vacuum to high temperature by radiative exchange with a circular heating element. The surroundings are cool enough that they do not affect the radiative exchange and may be neglected. The surfaces are diffuse–gray. Derive an expression for the net rate of energy absorption by the sphere. The expression should be given in terms of the quantities shown. For simplicity, do not subdivide the surface areas. Discuss whether this is a reasonable approximation for this geometry with regard to the distribution of reflected energy from the sphere.

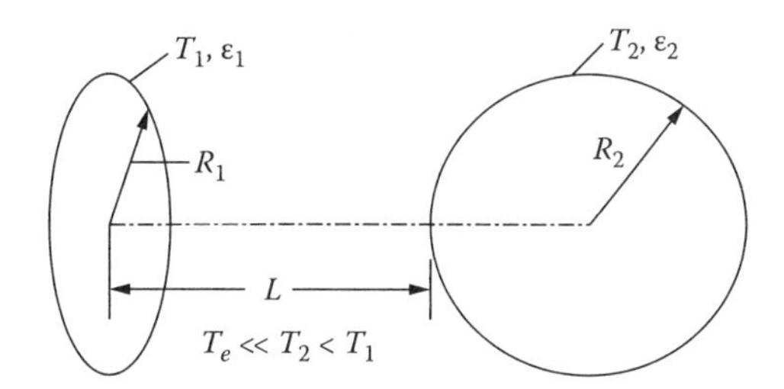

5.26 An enclosure has four sides that are all equilateral triangles of the same size (i.e., it is an equilateral tetrahedron). The sides are of length $L = 4$ m and have conditions imposed as follows:
Side 1 is black and is at uniform temperature $T_1 = 1100$ K.
Side 2 is diffuse–gray and is perfectly insulated on the outside.
Side 3 is black and has a uniform heat flux of 8 kW/m^2 supplied to it.
Side 4 is black and is at $T_4 = 0$ K.
Find q_1, T_2, and T_3. For simplicity, do not subdivide the surface areas.
Answer: 51.4 kW/m^2; 941 K; 972 K.

5.27 A 10 cm diameter hole extends through the wall of a furnace having an interior temperature of 1350 K. The wall is a 20 cm thick refractory brick. Divide the wall thickness into two zones of equal length and compute the net radiation out of the hole into a room at 300 K. (Neglect heat conduction in the wall.)

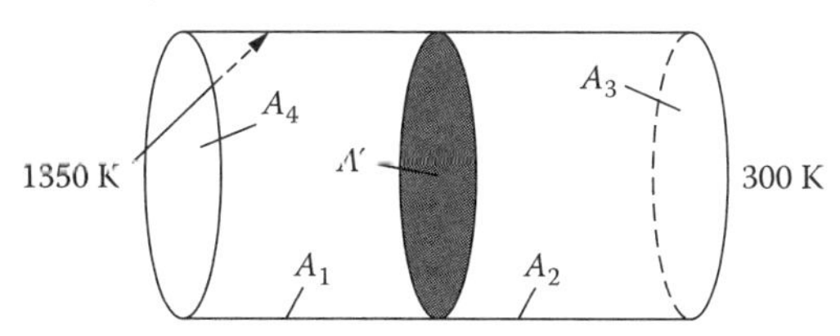

Answer: $T_1 = 1216$ K; $T_2 = 1033$ K; $Q_3 = -560$ W.

5.28 A cubical enclosure with edge length of 6 m has a very small sphere placed at its center ($A_{sphere} \ll A_{side}$). The sphere has emissivity $\varepsilon = 0.4$ and is maintained electrically at $T_s = 1300$ K. The interior walls of the cube have the following properties:

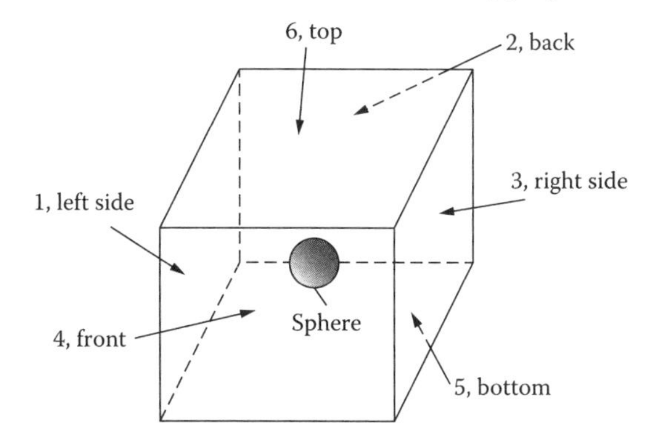

Side	Temperature (K)	Emissivity
1	1100	1.0
2	800	0.5
3	700	0.5
4	400	1.0
5	200	0.2
6	0	1.0

Determine the net q added or removed from each side of the cube and the q added to the sphere. Find the results in kW/m^2. Tabulate all the required configuration factors. (Assume the incident radiation on each surface is uniform.)

Answer: $q_1 = 69.9$ kW/m$_2$; $q_2 = -0.86$ kW/m^2; $q_3 = -6.10$ kW/m^2; $q_4 = -28.0$ kW/m^2; $q_5 = -5.11$ kW/m^2; $q_6 = -29.8$ kW/m^2; $q_s = 54.9$ kW/m^2.

5.29 A rod 2 cm in diameter and 20 cm long is at temperature $T_1 = 1200$ K and has a hemispherical total emissivity of $\varepsilon_1 = 0.30$. It is within a thin-walled concentric cylinder of the same length having a diameter of 8 cm. The emissivity on the inside of the cylinder is $\varepsilon_2 = 0.21$, and on the outside is $\varepsilon_o = 0.15$. All surfaces are diffuse–gray. The entire assembly is suspended in a large vacuum chamber at $T_e = 300$ K. What is the temperature T_2 of the cylindrical shell? For simplicity, do not subdivide the surface areas. (Hint: $F_{2-1} = 0.225$, $F_{2-2} = 0.617$.)

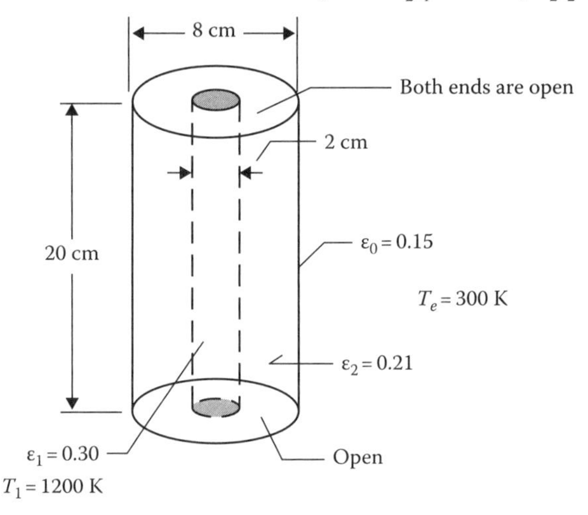

Answer: 722 K.

5.30 A thin diffuse–gray circular disk with emissivity ε_1 on one side and ε_2 on the other side is being heated in a vacuum by a cylindrical electrical heater with a diameter D. The heater has a diffuse–gray interior surface and is open at both ends.

(a) Derive a formula (which can be in terms of configuration factors) for the net radiative energy rate being gained by the disk while it is being heated. The formula should be in terms of the instantaneous disk temperature and the quantities shown.
For the specific case, $T_h = 1200$ K; $T_e = 300$ K; $D = 0.50$ m; $L = 0.80$ m; $d = 0.30$ m; $l = 0.10$ m; $\varepsilon_1 = 0.70$; $\varepsilon_2 = 0.85$; $\varepsilon_h = 0.80$.

(b) What is the net gain (W) when $T_d = 600$ K?

(c) What is the equilibrium disk temperature long after the heater is turned on?

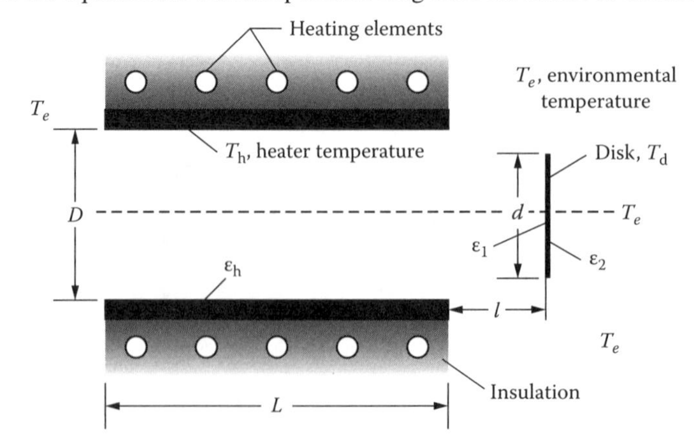

Answer: (a) Net gain =

$$= \varepsilon_2 \sigma \frac{\pi d^2}{4}\left(T_e^4 - T_d^4\right) - \frac{\varepsilon_1 \pi d^2}{4} \frac{\left\{\begin{matrix} \sigma T_d^4\left[1-(1-\varepsilon_3)(F_{3-3}+F_{3-1}F_{1-3})\right]-\varepsilon_3 F_{1-3}\sigma T_3^4 \\ +\sigma T_4^4\left[-F_{1-4}+(1-\varepsilon_3)(F_{3-3}F_{1-4}-F_{3-4}F_{1-3})\right] \end{matrix}\right\}}{1-(1-\varepsilon_3)\left[F_{3-3}+F_{3-1}F_{1-3}(1-\varepsilon_1)\right]}$$

(b) 4010 W; (c) 902 K.

5.31 A hollow satellite in Earth orbit consists of a circular disk and a hemisphere. The disk is facing normal to the direction to the sun. The surroundings are at $T_e = 100$ K. The satellite walls are thin. All surfaces are diffuse. The properties are $\alpha_{1,solar} = 0.95$; $\varepsilon_{1,infrared} = 0.13$; $\varepsilon_2(\text{gray}) = 0.80$; $\varepsilon_3(\text{gray}) = 0.50$; $\varepsilon_4(\text{gray}) = 0.60$. What are the values of T_1 and T_4? (Do not subdivide surfaces. Neglect any emitted or reflected radiation from the Earth.)

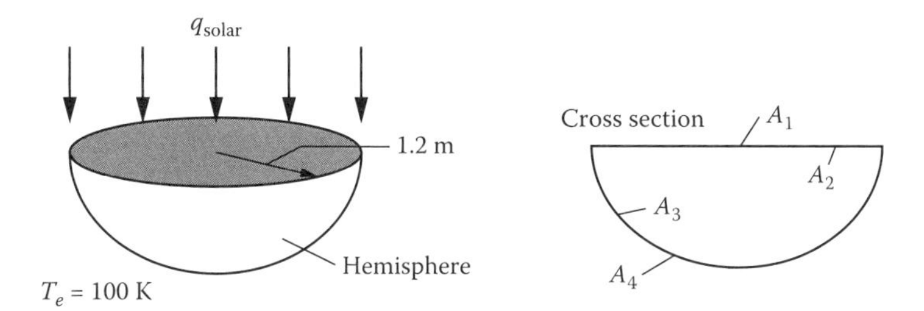

Answer: $T_2 = 458$ K; $T_3 = 345$ K.

5.32 Three parallel plates of finite width are shown in cross section. The plates are maintained at $T_1 = 700$ K and $T_2 = 400$ K. The surroundings are at $T_e = 300$ K. The plates are very long in the direction normal to the cross section shown. What are the values of q_1 and q_2? All plate surfaces are diffuse–gray. (For simplicity, do not subdivide the plate areas.)

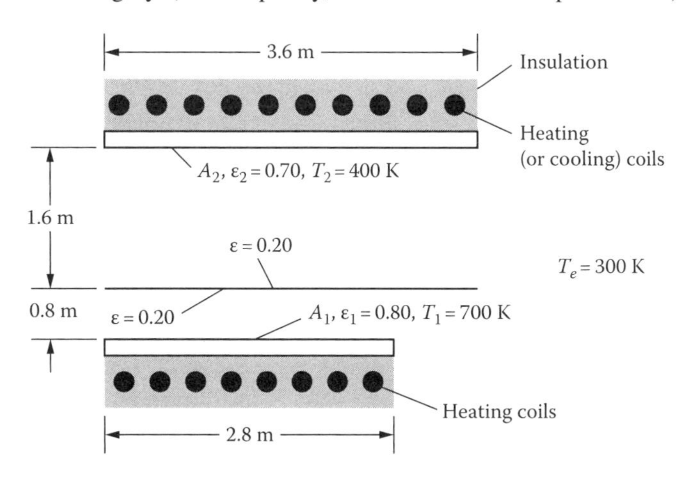

Answer: $q_1 = 5941.2$ W/m²; $q_2 = 97.5$ W/m².

5.33 A space vehicle in orbit around the sun is at the same distance as the Earth. It is a hollow cube with thin walls and is oriented with one side always facing directly toward the sun and the other five sides in the shade. The interior is painted with a coating with $\varepsilon = 0.60$. On the outside, the top is coated with a material with $\alpha_{solar} = 0.93$ and $\varepsilon = 0.80$, the front and back sides are faced with aluminum foil with $\varepsilon = 0.04$, and the two sides and the bottom have white paint with $\varepsilon = 0.80$. The surroundings are at 0 K. Using as simple a radiation model as is reasonable, obtain the temperatures of the six faces of the cube.

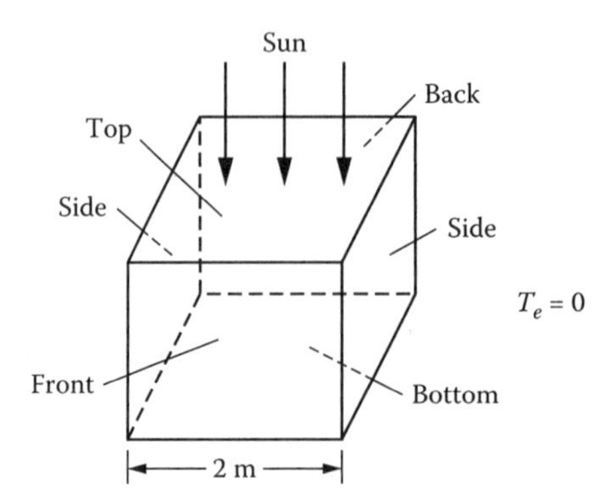

Answer: T_1 (top) = 367 K; T_2 (2 sides, bottom) = 234 K; T_3 (front and back) = 281 K.

5.34 Two infinitely long, directly opposed, parallel diffuse–gray plates of width W have the same uniform heat flux q supplied to them. The environment is at temperature T_e. Set up the governing equation for determining the temperature distribution $T(x)$ on the bottom plate. There is no energy loss from the top side of the upper plate or from the bottom side of the lower plate.

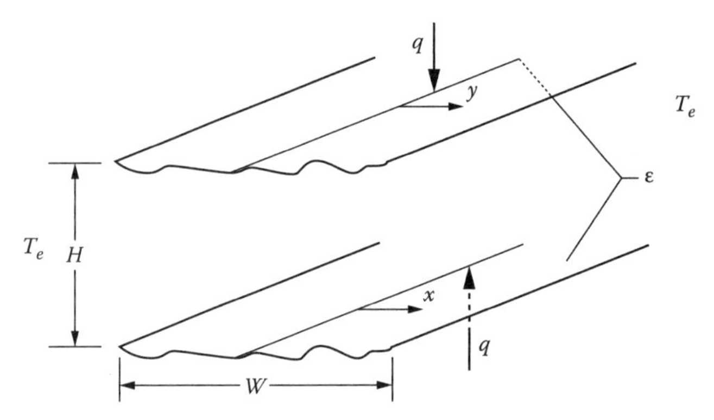

5.35 For the geometry specified in the preceding problem, solve for the temperature distribution $T(x) = T(y)$ on either plate if $q = 3000$ W/m^2, $\varepsilon = 0.25$, $H = 0.25$ m, and $W = 1$ m. A numerical solution is required. Present your results graphically. The environment temperature is $T_e = 300$ K.

Answer: Peak $T(x/W) = 1/2$ of 796 K and a temperature at the plate edge of 712 K.

5.36 Consider two parallel plates of finite extent in one direction. Both plates are perfectly insulated on the outside. Plate 1 is uniformly heated electrically with heat flux q_e. Plate 2 has no external heat input. The environment is at zero absolute temperature.

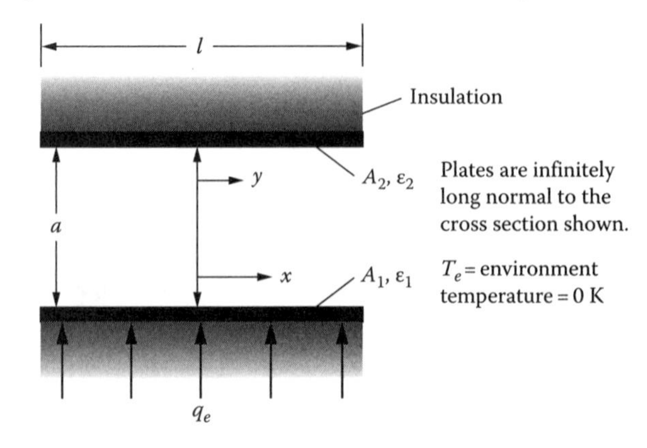

(a) For both plates *black*, show that the integral equations for the surface temperatures are

$$\theta_{b,1}(X) = 1 + \frac{1}{2}\int_{-L/2}^{L/2} \theta_{b,2}(Y)\frac{dY}{\left[(Y-X)^2+1\right]^{3/2}}$$

$$\theta_{b,2}(Y) = \frac{1}{2}\int_{-L/2}^{L/2} \theta_{b,1}(X)\frac{dX}{\left[(X-Y)^2+1\right]^{3/2}}$$

where $X = x/a$, $Y = y/a$, $\theta = \sigma T^4/q_e$, and $L = l/a$.

(b) If both plates are *gray*, show that $\theta_1(X) = \theta_{b,1}(X) + \dfrac{1-\varepsilon_1}{\varepsilon_1}$; $\theta_2(Y) = \theta_{b,2}(Y)$.

5.37 Two diffusely emitting and reflecting parallel plates are of infinite length into and out of the cross section shown in the following. The lower plate has uniform temperature T_1 = 1000 K, and the upper plate has uniform temperature T_2 = 500 K. The surroundings have a temperature of T_e = 300 K. Surface 1 has gray emissivity ε_1 = 0.8 and surface 2 has gray emissivity ε_2 = 0.2.

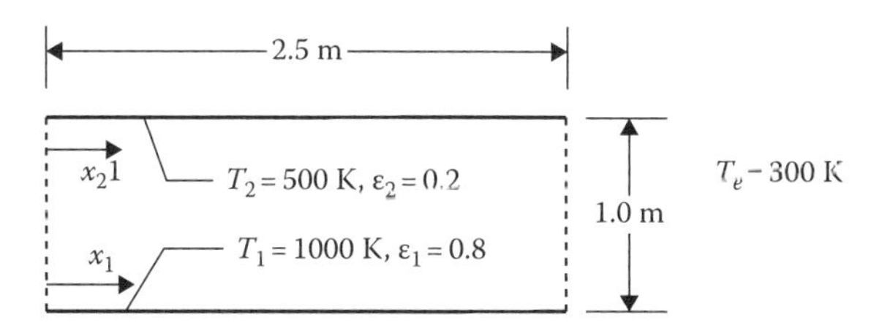

(a) Find the average net radiative heat fluxes q_1 and q_2 on surfaces 1 and 2 assuming uniform irradiation over each surface.
(b) Partition each of the two parallel plates into four equal segments (upper: 1, 2, 3, 4; lower: 5, 6, 7, 8). Find the net radiative heat flux for each segment, assuming uniform irradiation on each segment individually. Compare the results with those of part (a).
(c) Do not assume uniform irradiation on either of the parallel plates and find the distribution of net radiative heat flux $q_1(x_1)$ and $q_2(x_2)$ on each surface. Plot these results and the results of parts (a) and (b) on the same graph.

Answer: (a) 30,390 W/m, –5,970 W/m; (b) $q_1 = q_3$ = 32,190 W/m, q_2 = 28,120 W/m, $q_5 = q_7$ = –5,200 W/m, q_6 = –6,770 W/m.

5.38 A long groove is cut into a metal surface as shown in the following cross section. The groove surface is diffuse–gray and has emissivity ε. The temperature profile along the groove sides, as measured from the apex, is found to be $T(x)$. The environment is at temperature T_e.

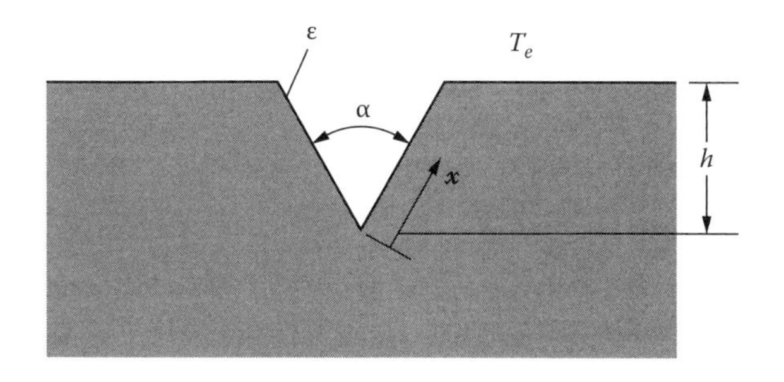

(a) Derive the equations for the heat flux distribution $q(x)$ along the groove surface.
(b) Examine the kernel of the integral equation found in part (a) and show whether it is symmetrical and/or separable.

5.39 A hemispherical cavity is in a block of lightly oxidized copper maintained at 900 K in vacuum (see Figure 3.41a). The surroundings are at 300 K. Use the integral equation method to compute the outgoing heat flux from the cavity surface.
Answer: 27,140 W/m^2.

5.40 A cavity having a gray interior surface S is uniformly heated electrically and achieves a surface temperature distribution $T_{w,0}(S)$ while being exposed to a zero absolute temperature environment, $T_e = 0$. If the environment is raised to T_e and the heating kept the same, what is the surface temperature distribution?

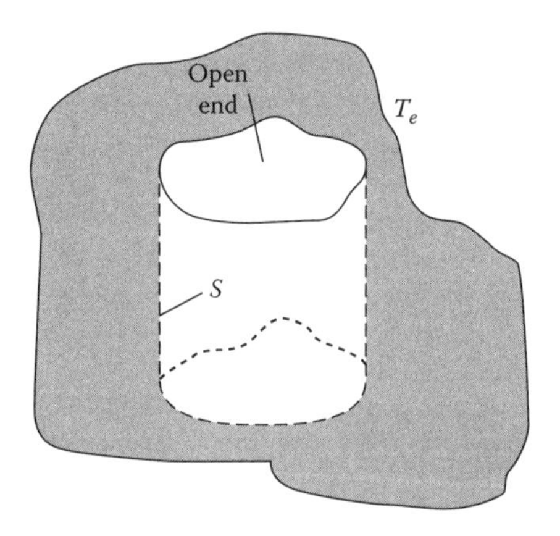

Answer: $T_{w,0}\left(r\right)=\left[T_w^4\left(r\right)-T_e^4\right]^{1/4}$, where $T_{w,0}(S) = T_w(S, T_e = 0)$.

5.41 A gray circular tube insulated on the outside is exposed to an environment at $T_e = 0$ at both ends. The $q(x, T_e = 0)$ has been calculated to maintain the wall temperature at any constant value. Now, let $T_e \approx 0$, and let the wall temperature be uniform at T_w. Show that the $q(x, T_e \neq 0)$ can be obtained as the $q(x, T_e = 0)$ corresponding to the wall temperature $(T_w^4 - T_e^4)^{1/4}$.

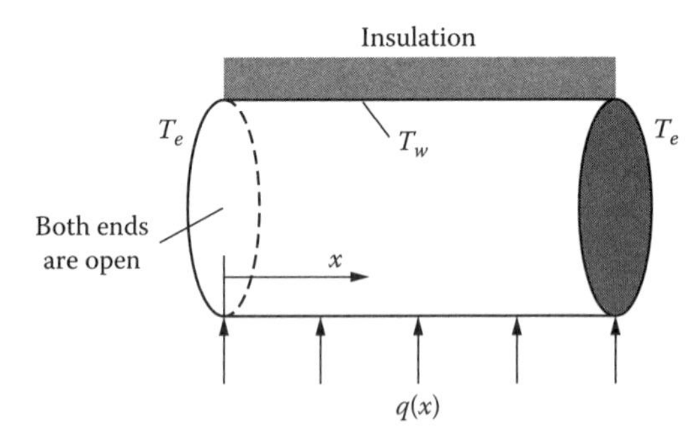

5.42 Two infinitely long plates are joined at right angles along one infinite edge. The plates are each of width W, and both have diffuse–gray surfaces with the same emissivity ε. The heat flux supplied to both surfaces is maintained at a uniform value q. Derive an equation for the temperature distribution on each surface and describe how you would solve for $T(x)$ and $T(y)$. Neglect conduction in the plates and convection to the atmosphere. The surroundings are at temperature T_e.

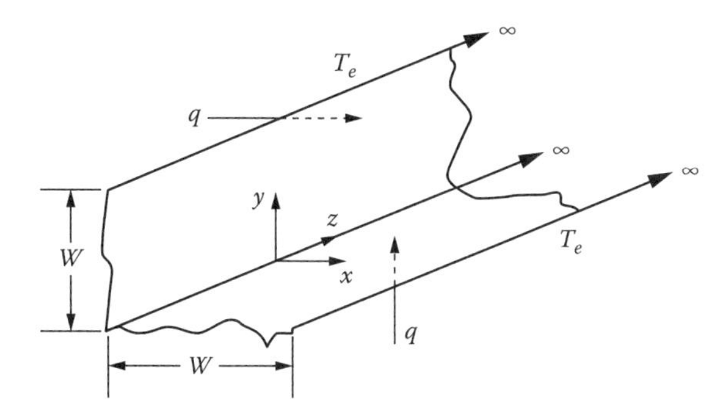

5.43 Two plates are joined at 90° and are very long normal to the cross section shown. The vertical plate (plate 1) is heated uniformly with a heat flux of 300 W/m². The horizontal plate has a uniform temperature of 400 K. Both plates have an emissivity of 0.6. The environment is at T_e = 300 K. The dimensions are shown on the figure.

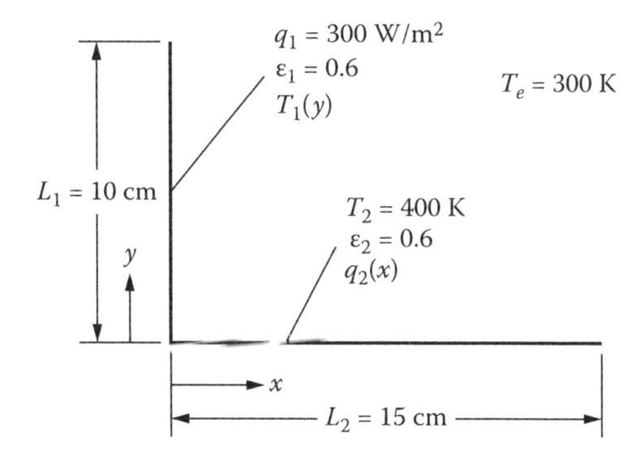

Derive the equations necessary for finding the temperature distribution on surface 1 and the net radiative heat flux on surface 2.

5.44 For the geometry and conditions shown in Homework Problem 5.43, solve for the unknown distributions of temperature $T_1(y)$ and net radiative heat flux $q_2(x)$ and plot the results.

5.45 For the geometry and conditions shown in the following, set up the required integral equations for finding $q_1(y)$, $T_2(z)$, and $T_3(x)$. Put the equations in dimensionless form and discuss how you would go about solving the equations. Which method of Chapter 5 appears most useful?

$T_1(y) = T_1$ = constant
$q_2(z) = 0$ (insulated on the outside); $q_3(x) = q_3$ = constant
$\varepsilon_1(y) = 1.0$; $\varepsilon_2(z) = \varepsilon_3(x) = 0.5$

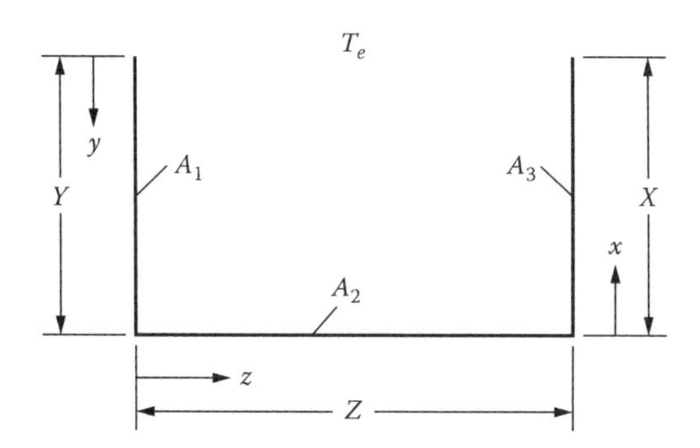

5.46 A three-surface enclosure has the properties shown in the following. Assuming each surface can be treated as having uniform radiosity, find the unknown temperatures and heat fluxes.

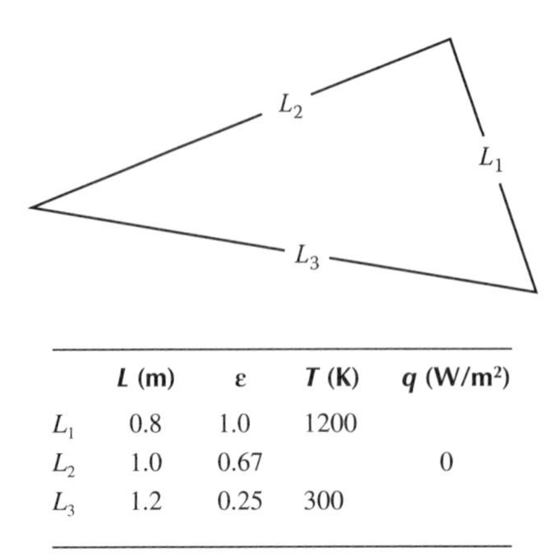

	L (m)	ε	T (K)	q (W/m²)
L_1	0.8	1.0	1200	
L_2	1.0	0.67		0
L_3	1.2	0.25	300	

Answer: $T_2 = 1120$ K; $q_1 = 37.5$ kW/m²; $q_3 = -25.0$ kW/m².

5.47 A four-surface enclosure has the properties and temperatures shown in the following. Find the heat flux for each surface and demonstrate that the radiative heat balance is satisfied. Assume that radiosities can be taken as uniform across each surface.

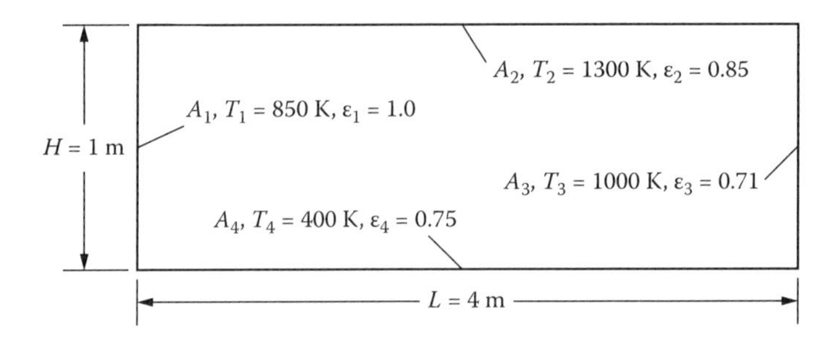

Answer: $q_1 = -54{,}690$ W/m; $q_2 = 108{,}100$ W/m; $q_3 = -16.620$ W/m; $q_1 = -90{,}230$ W/m.

5.48 A very long A-frame grain dryer is built with the cross section of an isosceles triangle with the dimensions shown in the following. For solar heating, the right opaque side is exposed to the sun and is black, the left side is insulated. At a particular time, the conditions for the inside surfaces of the dryer are as shown in the table.

(a) What will be the floor temperature T_3 (K)?

(b) What will be the temperature of the insulated wall T_2?

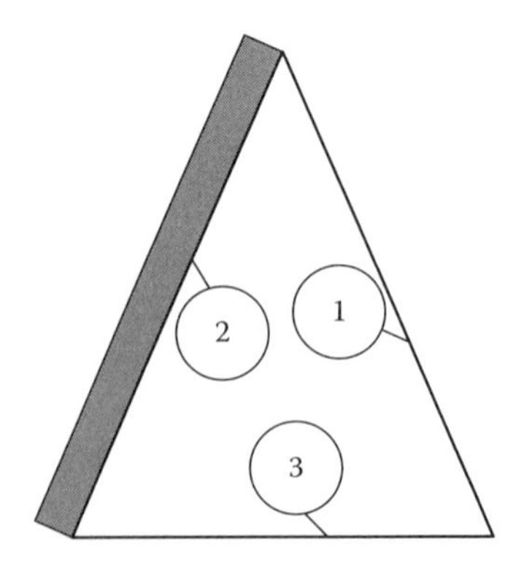

Surface	L (m)	T (K)	q (W/m²)	ε
1	5	450		1.0
2	5		0	0.537
3	3.5		−600	0.35

Answer: (a) 304 K; (b) 437 K.

5.49 A bakery oven will burn biomass for heat to produce organic nonfat sugarless gluten-free whole-grain cookies. The geometry and conditions are shown in the following. The oven is quite long in the dimension into the paper.

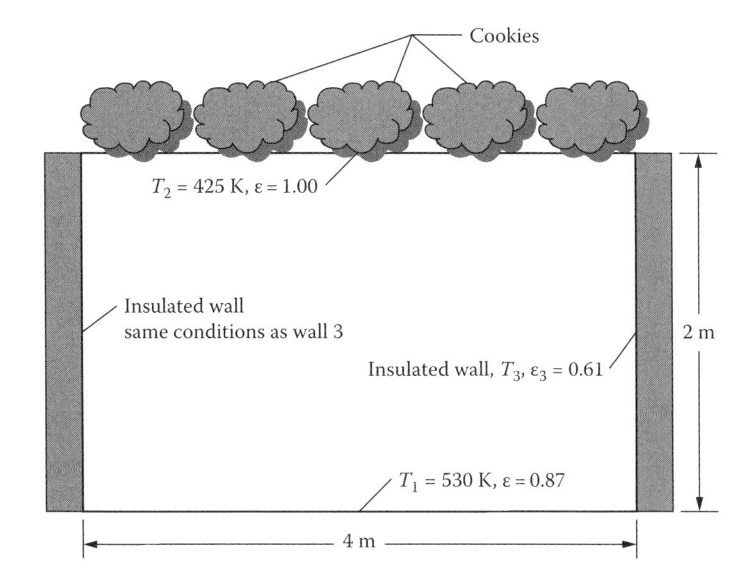

(a) What radiative heat flux q_1 must the biomass fuel produce on the heated wall at T_1 = 530 K if the cookie tray is to be heated to 425 K?
(b) What will the temperature of the sidewalls (T_3) be under these conditions?

Answer: (a) 1895 W/m; (b) 480 K.

5.50 A radiator is planned to provide heat rejection from a nuclear power plant that is to provide electrical power for a lunar outpost. The radiator will itself will be horizontal on the lunar surface, and condensing working fluid in the power cycle will maintain the uniform surface temperature of the radiator at 800 K. The radiator is shielded from the nearby lunar outpost by a 1.5 m high vertical insulated plate (see diagram on the next page). The width of the radiator is limited to 2 m. The radiator has a diffuse–gray emissivity of 0.92, while the insulated shield has an emissivity of 0.37. It is expected that the radiator will be quite long. Conduction within the radiator and heat shield can be neglected. For the situation when the radiator is on the night side of the moon,
(a) Set up the equations for finding the heat flux distribution on the radiator, $q_1(x)$, and the temperature distribution on the shield, $T_2(y)$. Note any assumptions.
(b) Solve the equations for $q_1(x)$ and $T_2(y)$. Show that the increment size chosen for solution is small enough that the solutions are grid independent.
(c) If the total heat rejection from the radiator is required to be 1.0 MW, what must be the length of the radiator (m)?
(d) Discuss how the results would change if the influence of incident solar energy on the radiator during daytime is considered.

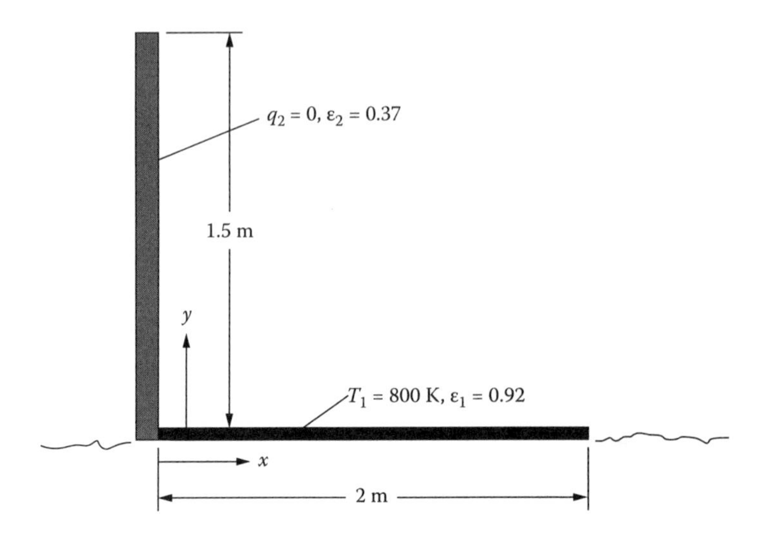

Answers: $T_2(y = 1.5\text{ m}) = 525$ K; $T_2\ (y = 0) = 662$ K; $q_1(x = 0) = 16.5$ kW/m²; $q_1(x = 2\text{ m}) =$ 20.8 kW/m²; length of radiator = 47.2 m.

5.51 For the geometry shown in the following with nonconducting walls,

(a) Provide the final governing equations necessary for finding $T_1(x_1)$ and $q_2(x_2)$ in two forms (1) explicit in the required variables and (2) in nondimensional form using appropriate nondimensional variables.

(b) Find the temperature distribution $T_1(x_1)$ and the heat flux distribution $q_2(x_2)$ and show them on appropriate graphs.

Boundary conditions are $q_1(x_1) = \left[100x_1 - 50x_1^2\right]$(kW/m) where x_1 is in meters, $T_2 = 450$ K, and $T_3 = T_4 = 300$ K. Properties for gray (or black) diffuse surfaces are shown in the figure.

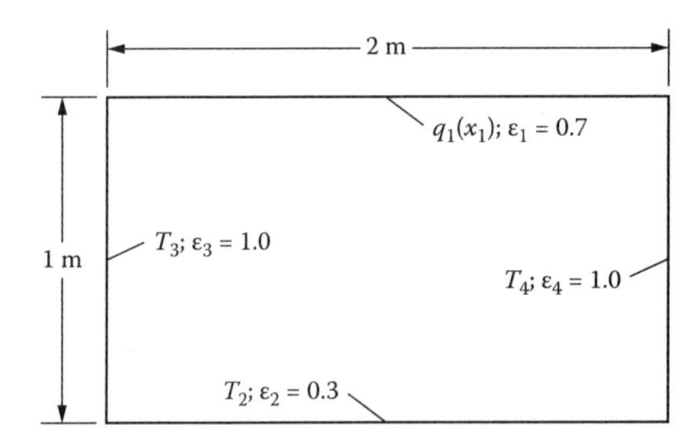

Show that your solution is grid independent and meets overall energy conservation.

Answer: $T_{1,\max} = 1117$ K; $q_{2,\max} = 10.5$ kW/m.

6 Exchange of Thermal Radiation among Nondiffuse Nongray Surfaces

6.1 INTRODUCTION

In Chapter 5, we presented a detailed analysis and solution procedure for radiation exchange analysis within enclosures made up of black or gray diffuse boundary surfaces. One additional restriction was that the radiative properties were independent of temperature. However, as discussed in Chapter 3, most materials deviate from the idealizations of being black, gray, diffuse, or having temperature-independent radiative properties. The assumption of idealized surfaces is made to simplify the computations. This is often reasonable because the radiative properties may not be known to high accuracy, especially their detailed dependence on wavelength and direction; hence, refined computations are not meaningful if only very approximate property data are available. Also, the many reflections and re-reflections in an enclosure tend to average out radiative nonuniformities, for example, emitted plus reflected radiation leaving a directionally emitting surface may be fairly diffuse if it is chiefly composed of reflected energy from radiation incident from many directions.

In some applications more precision is required and the gray and/or diffuse assumption cannot be trusted to be sufficiently accurate. It is desired to carry out exchange computations using as exact a procedure as reasonably possible. The results can be compared with those from simplified methods to provide insight into where simplifying assumptions can be applied and yield reasonable results. To provide techniques for more accurate computations, some methods are examined here for radiative exchange between realistic surfaces. Such analyses are inherently more difficult than for ideal surfaces, and a treatment of real surfaces including all types of variations, while possible in principle, is not usually attempted or justified. For a detailed solution the directional spectral properties must be available. Property variations with wavelength for the normal direction are available for a variety of materials, as given by the references in Chapter 3, but data are usually sparse at short ($\lambda < \sim 0.3$ μm) and long ($\lambda > \sim 15$ μm) wavelengths and for other than the normal direction. For many materials there is little detailed information available. Directional variations for some materials with optically smooth surfaces can be computed using electromagnetic theory, but real materials can deviate from ideal behavior. Certain solutions obviously must include spectral property variations, such as for spectrally selective coatings used for temperature control of space vehicles exposed to solar radiation, and for solar energy collection.

6.2 ENCLOSURE THEORY FOR DIFFUSE NONGRAY SURFACES

By considering diffusely emitting and reflecting surfaces, directional surface effects are eliminated. By assuming diffuse surfaces, spectral effects can be separated from directional ones and understand how spectral property variations need to be accounted for. Under the diffuse assumption, emissivity, absorptivity, and reflectivity are independent of direction, but properties must be available as functions of λ and T to evaluate the radiative exchange.

For diffuse spectral surfaces, configuration factors are valid because they involve only geometry and are for diffuse radiation leaving a surface. The energy-balance equations and methods in

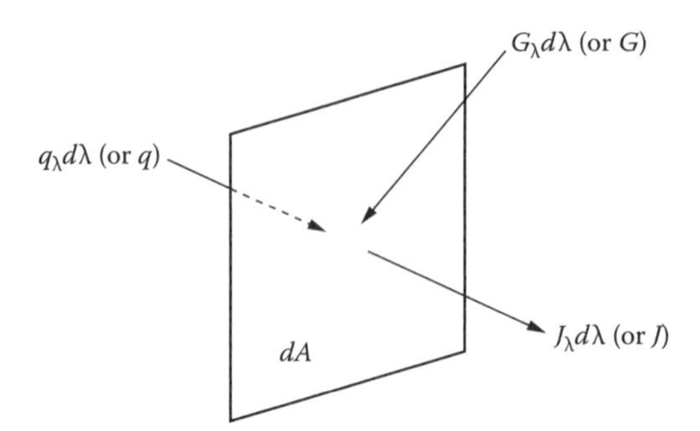

FIGURE 6.1 Spectral (or total) energy fluxes at a surface element.

Chapter 5 are valid if they are written for energy in each wavelength interval $d\lambda$. Usually, however, the boundary conditions involve *total* (including all wavelengths) energy. Total energy boundary conditions cannot be applied to the spectral energies. To illustrate this, consider the surface of Figure 6.1 with locally incident radiation heat flux G, and total radiosity J leaving by combined emission and reflection. If the surface is perfectly insulated, the G and J must be equal,

$$J - G = q = 0 \tag{6.1}$$

However, for an insulated surface with $q = 0$, in each $d\lambda$ the incident and outgoing spectral fluxes are not generally equal so that

$$J_\lambda d\lambda - G_\lambda d\lambda = q_\lambda d\lambda \neq 0 \tag{6.2}$$

An adiabatic surface only has a total radiation gain of zero, or, with Equation 6.1 restated in terms of spectral quantities,

$$q = \int_{\lambda=0}^{\infty} q_\lambda d\lambda = \int_{\lambda=0}^{\infty} \left(J_\lambda - G_\lambda\right) d\lambda = 0 \tag{6.3}$$

The $q_\lambda d\lambda$ is the net energy flux supplied in $d\lambda$ at λ as a result of the radiative exchange process. For an adiabatic surface the $q_\lambda d\lambda$ can vary substantially with λ, depending on the spectral property variations and the spectral distribution of incident energy.

Now consider a nonadiabatic surface where a total energy flux q is supplied by some means such as heat conduction or convection. Then, from Figure 6.1,

$$q = \int_{\lambda=0}^{\infty} q_\lambda d\lambda = \int_{\lambda=0}^{\infty} \left(J_\lambda - G_\lambda\right) d\lambda \tag{6.4}$$

The q may be specified or may be a quantity to be determined so that the surface is locally maintained at a specified temperature. In any $d\lambda$ the net spectral energy $(J_\lambda - G_\lambda)d\lambda$ is unknown and may be positive or negative. The boundary conditions state only that the integral of all such spectral energy values must locally equal the total q.

6.2.1 Parallel-Plate Geometry

To build familiarity while avoiding geometric complexity, these concepts are applied for a geometry of two infinite parallel plates. Then relations for a general geometry are given.

Example 6.1

Two infinite parallel plates of tungsten at specified temperatures T_1 and T_2 ($T_1 > T_2$) are exchanging radiation. Branstetter (1961) determined the temperature-dependent hemispherical spectral emissivity of tungsten by using electromagnetic theory relations to extrapolate limited experimental data, and some of his results are in Figure 6.2. Using these data, compare the net energy exchange between the tungsten plates with that for gray parallel plates using total emissivities.

The solution for gray plates is in Example 5.6, and the present analysis follows in the same way by writing the equations spectrally and solving for the desired quantities. The spectral result is Equation 5.6.3 written for a wavelength interval $d\lambda$,

$$q_{\lambda,1}d\lambda = -q_{\lambda,2}d\lambda = \frac{E_{\lambda b,1}(T_1) - E_{\lambda b,2}(T_2)}{\dfrac{1}{\varepsilon_{\lambda,1}(T_1)} + \dfrac{1}{\varepsilon_{\lambda,2}(T_2)} - 1} d\lambda \tag{6.1.1}$$

Note that for each $d\lambda$ the overall spectral energy balance is zero, that is, $(q_{\lambda,1} + q_{\lambda,2})d\lambda = (q_{\lambda,1} - q_{\lambda,1}) d\lambda = 0$.

The total heat flux exchanged (supplied to surface 1 and removed from surface 2) is found by substituting the property data of Figure 6.2 into Equation 6.1.1 and integrating over all wavelengths:

$$q_1 = -q_2 = \int_{\lambda=0}^{\infty} q_{\lambda,1}d\lambda = \int_{\lambda=0}^{\infty} \frac{E_{\lambda b,1}(T_1) - E_{\lambda b,2}(T_2)}{\dfrac{1}{\varepsilon_{\lambda,1}(T_1)} + \dfrac{1}{\varepsilon_{\lambda,2}(T_2)} - 1} d\lambda \tag{6.1.2}$$

The integration is performed numerically for each set of specified plate temperatures T_1 and T_2.

Some results from Branstetter are in Figure 6.3, where the ratio of gray to nongray exchange is given. The gray results were obtained using Equation 5.6.3 with total emissivities computed from the spectral emissivities of Figure 6.2. In the gray computation, the emissivity of the colder surface 2 was inserted at the mean temperature $\sqrt{T_1T_2}$ rather than at T_2, which is a modification based on electromagnetic theory that is sometimes recommended for metals (Eckert and Drake 1959). Over the range of surface temperatures shown, the gray results deviate up to 25% below the nongray exchange.

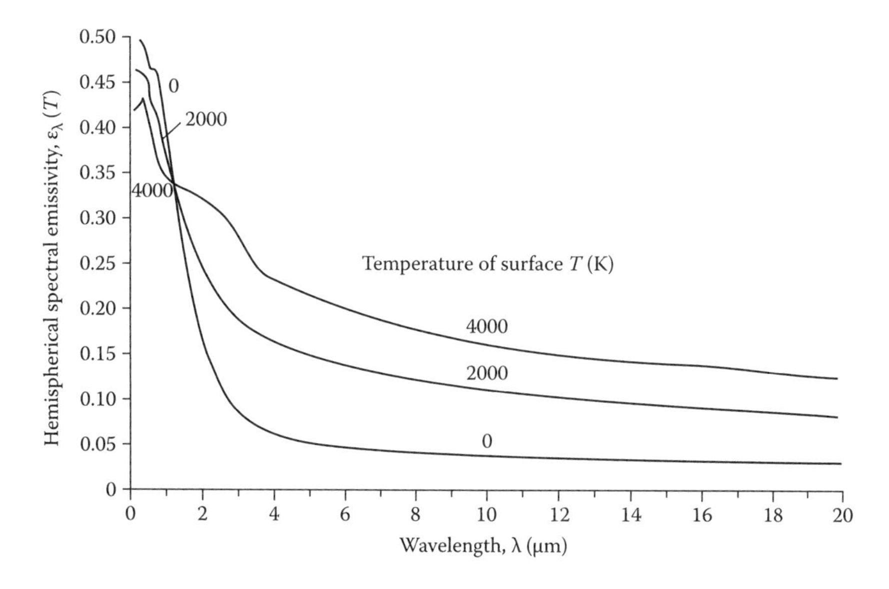

FIGURE 6.2 Hemispherical spectral emissivity of tungsten.

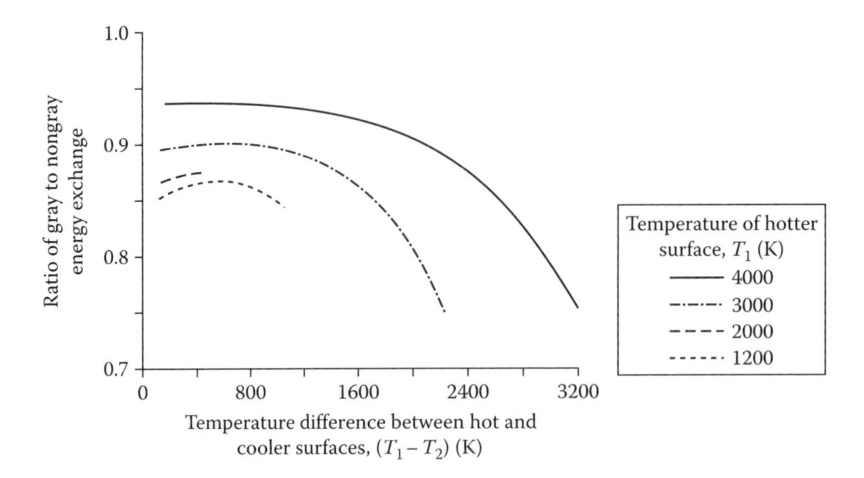

FIGURE 6.3 Comparison of effect of gray and nongray surfaces on computed energy exchange between infinite tungsten plates. (From Branstetter, J.R., Radiant heat transfer between nongray parallel plates of tungsten, NASA TN D-1088, Washington, DC, 1961.)

The following example illustrates how Equation 6.1.2 can be expressed more simply if the spectral properties can be approximated by constant values in a stepwise fashion over a few spectral regions (spectral bands). The evaluation is then done in spectral bands that each extend over a wavelength range; this may be useful for making quick estimates of spectral effects. More generally, the evaluation with variable properties is done by computer integration using numerical integration software.

Example 6.2

Two infinite parallel plates and their approximate spectral emissivities at their respective temperatures are in Figure 6.4. What is the heat flux q transferred across the space between the plates?

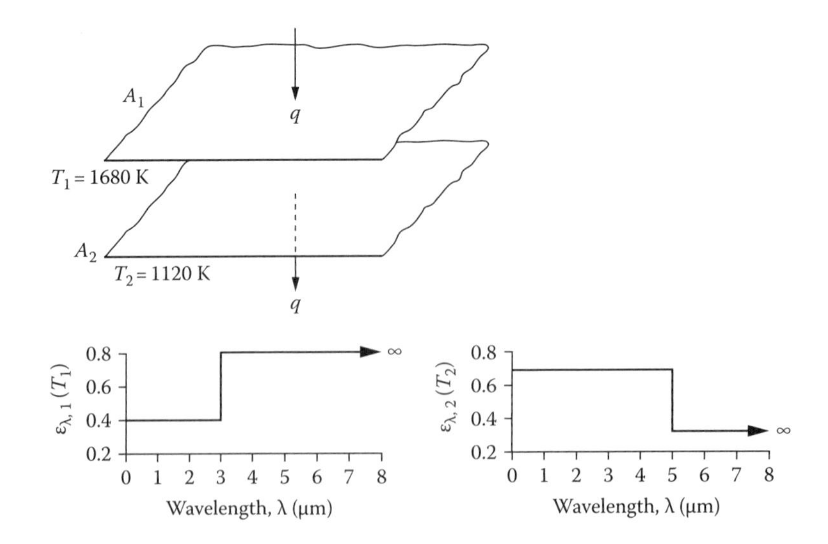

FIGURE 6.4 Example of radiative transfer between infinite parallel plates having spectrally dependent emissivities approximated in spectral bands.

From Equation 6.1.2,

$$q_1 = \int_{\lambda=0}^{3} \frac{E_{\lambda b,1}(T_1) - E_{\lambda b,2}(T_2)}{\dfrac{1}{0.4} + \dfrac{1}{0.7} - 1} d\lambda + \int_{\lambda=3}^{5} \frac{E_{\lambda b,1}(T_1) - E_{\lambda b,2}(T_2)}{\dfrac{1}{0.8} + \dfrac{1}{0.7} - 1} d\lambda + \int_{\lambda=5}^{\infty} \frac{E_{\lambda b,1}(T_1) - E_{\lambda b,2}(T_2)}{\dfrac{1}{0.8} + \dfrac{1}{0.3} - 1} d\lambda$$

This can be directly evaluated by numerical integration, or it can be written as

$$q_1 = \sigma T_1^4 \left[\frac{0.341}{\sigma T_1^4} \int_{\lambda=0}^{3} E_{\lambda b,1}(T_1) d\lambda + \frac{0.596}{\sigma T_1^4} \int_{\lambda=3}^{5} E_{\lambda b,1}(T_1) d\lambda + \frac{0.279}{\sigma T_1^4} \int_{\lambda=5}^{\infty} E_{\lambda b,1}(T_1) d\lambda \right]$$

$$-\sigma T_2^4 \left[\frac{0.341}{\sigma T_2^4} \int_{\lambda=0}^{3} E_{\lambda b,2}(T_2) d\lambda + \frac{0.596}{\sigma T_2^4} \int_{\lambda=3}^{5} E_{\lambda b,2}(T_2) d\lambda + \frac{0.279}{\sigma T_2^4} \int_{\lambda=5}^{\infty} E_{\lambda b,2}(T_2) d\lambda \right]$$

An integral such as $\left(1/\sigma T_1^4\right)\int_{\lambda=3}^{5} E_{\lambda b,1}(T_1) d\lambda$ is the fraction of blackbody radiation at T_1 between λ = 3 and 5 μm, which is $F_{3T_1 \to 5T_1} = F_{5040 \to 8400}$, which can be readily evaluated from Equation 1.36 or 1.37. This yields

$$q = \sigma T_1^4 \left(0.341 F_{0 \to 3T_1} + 0.596 F_{3T_1 \to 5T_1} + 0.279 F_{5T_1 \to \infty}\right) - \sigma T_2^4 \left(0.341 F_{0 \to 3T_2} + 0.596 F_{3T_2 \to 5T_2} + 0.279 F_{5T_2 \to \infty}\right) = 140{,}500 \text{ W/m}^2$$

Compared with gray calculations, the examples illustrate the additional complication of including spectrally dependent surface properties. The enclosure calculations must be carried out in wavelength intervals, and the total energy quantities obtained by integration of the spectral energy values. The integration is done rather easily for parallel plates since the formulation can be algebraically reduced to only one exchange equation, Equation 6.1.2. However, when many surfaces are present in an enclosure, the computations increase substantially since there are multiple exchange equations that need to be evaluated for each wavelength interval. It may be possible to assume that some of the surfaces are gray to provide some simplification.

When specifying wavelength intervals, the most important spectral regions for radiative transfer are where both $E_{\lambda b}$ and ε_λ are large; thus the wavelength range should be divided such that most of the wavelength intervals lie within these regions. If the number of wavelength intervals is increased, the exact results for energy transfer are approached. Dunkle and Bevans (1960) give results that show errors, compared with an exact numerical result, of less than 2% when using a spectral band approximation as compared to about 30% error for the gray-surface approximation. Examples are given for enclosures with specified temperatures or specified net energy fluxes. Additional references providing analyses of energy exchange between spectrally dependent surfaces are Goodman (1957), and Rolling and Tien (1967).

6.2.2 Spectral and Finite Spectral Band Relations for an Enclosure

For an enclosure of N diffuse surfaces as in Figure 5.5, the relation in Equation 5.21 is written spectrally for $d\lambda$ for each of the k surfaces:

$$\sum_{j=1}^{N} \left[\frac{\delta_{kj}}{\varepsilon_{\lambda,j}(T_j)} - F_{k-j} \frac{1 - \varepsilon_{\lambda,j}(T_j)}{\varepsilon_{\lambda,j}(T_j)} \right] q_{\lambda,j} d\lambda = \sum_{j=1}^{N} \left(\delta_{kj} - F_{k-j}\right) E_{\lambda b,j}(T_j) d\lambda \quad (k = 1,2,\ldots,N) \tag{6.5}$$

If the N surface temperatures are all specified, the right sides of these simultaneous equations are known, as are the quantities in the square brackets on the left. The solution yields the $q_{\lambda,k}$ at λ for each of the surfaces. As a check on the calculations, the overall *spectral* energy balance within each $d\lambda$ yields the summation $\sum_{k=1}^{N} A_k q_{\lambda,k} = 0$. For parallel plates this is evident from Equation 6.1.1. The total energy supplied to each surface by other means is then $Q_k = A_k \int_{\lambda=0}^{\infty} q_{\lambda,k} d\lambda$. If some of the q_k are specified, and hence some of the surface temperatures are unknown, an iterative procedure may be required, as discussed following Equation 5.21. This can be especially difficult for a spectrally dependent case because the specified q_k is a *total* energy quantity and represents an integral of the values found from Equation 6.5. The boundary condition does not provide the spectral values of $q_{\lambda,k}$.

Equation 6.5 is usually applied in finite spectral bands. The bands are selected so that the properties for all surfaces are approximated as constant within each band. Equation 6.5 is then integrated over a band $\Delta\lambda$ to yield $\left(\text{after noting } q_{\Delta\lambda,j} \equiv \int_{\Delta\lambda} q_{\lambda,j} d\lambda\right)$

$$\sum_{j=1}^{N}\left[\frac{\delta_{kj}}{\varepsilon_{\Delta\lambda,j}(T_j)} - F_{k-j}\frac{1-\varepsilon_{\Delta\lambda,j}(T_j)}{\varepsilon_{\Delta\lambda,j}(T_j)}\right] q_{\Delta\lambda,j} d\lambda$$
$$= \sum_{j=1}^{N}(\delta_{kj} - F_{k-j}) F_{\lambda T_j \to (\lambda+\Delta\lambda)T_j} E_{\lambda b,j}(T_j)\Delta\lambda \quad (k = 1,2,\ldots,N) \tag{6.6}$$

If the values of T are specified, Equation 6.6 provides N equations for the N unknown $q_{\Delta\lambda,k}$ for each $\Delta\lambda$ band. The solution is carried out for *each* $\Delta\lambda$. Then for each of the k surfaces, the total q_k is found by summing over all wavelength bands

$$q_k = \sum_{\Delta\lambda} q_{\Delta\lambda,k} \tag{6.7}$$

Example 6.3

An enclosure consists of three diffuse surfaces of finite width and infinite length normal to the cross section shown in Figure 6.5. One surface is concave. The radiative properties of each surface depend on wavelength and temperature, and the specified surface temperatures are T_1, T_2, and T_3. Provide a set of band equations for the radiative energy exchange among the surfaces.

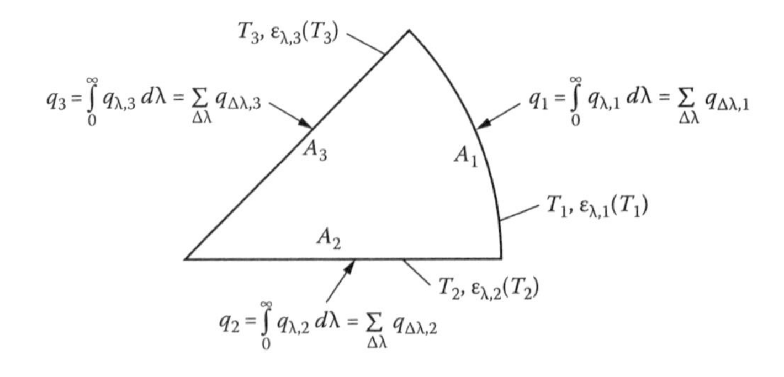

FIGURE 6.5 Radiant interchange in enclosure with surfaces having spectrally varying radiation properties.

From Equation 6.6 for k = 1, 2, and 3, and for wavelength band $\Delta\lambda$, the following equations are written. As a more compact notation let $\bar{E}_{\Delta\lambda,j}(T_j) \equiv [1-\varepsilon_{\Delta\lambda,j}(T_j)]/\varepsilon_{\Delta\lambda,j}(T_j)$ and $F_{\Delta\lambda T_j} \equiv F_{\lambda T_j \to (\lambda+\Delta\lambda)T_j}$ to yield

$$\left[\frac{1}{\varepsilon_{\Delta\lambda,1}(T_1)} - F_{1-1}\bar{E}_{\Delta\lambda,1}(T_1)\right]q_{\Delta\lambda,1} - F_{1-2}\bar{E}_{\Delta\lambda,2}(T_2)q_{\Delta\lambda,2} - F_{1-3}\bar{E}_{\Delta\lambda,3}(T_3)q_{\Delta\lambda,3}$$
$$= (1-F_{1-1})F_{\Delta\lambda T_1}\sigma T_1^4 - F_{1-2}F_{\Delta\lambda T_2}\sigma T_2^4 - F_{1-3}F_{\Delta\lambda T_3}\sigma T_3^4 \quad (6.3.1)$$

$$-F_{2-1}\bar{E}_{\Delta\lambda,1}(T_1)q_{\Delta\lambda,1} + \frac{1}{\varepsilon_{\Delta\lambda,2}(T_2)}q_{\Delta\lambda,2} - F_{2-3}\bar{E}_{\Delta\lambda,3}(T_3)q_{\Delta\lambda,3}$$
$$= -F_{2-1}F_{\Delta\lambda T_1}\sigma T_1^4 + F_{\Delta\lambda T_2}\sigma T_2^4 - F_{2-3}F_{\Delta\lambda T_3}\sigma T_3^4 \quad (6.3.2)$$

$$-F_{3-1}\bar{E}_{\Delta\lambda,1}(T_1)q_{\Delta\lambda,1} - F_{3-2}\bar{E}_{\Delta\lambda,2}(T_2)q_{\Delta\lambda,2} + \frac{1}{\varepsilon_{\Delta\lambda,3}(T_3)}q_{\Delta\lambda,3}$$
$$= -F_{3-1}F_{\Delta\lambda T_1}\sigma T_1^4 - F_{3-2}F_{\Delta\lambda T_2}\sigma T_2^4 + F_{\Delta\lambda T_3}\sigma T_3^4 \quad (6.3.3)$$

These are three simultaneous equations for $q_{\Delta\lambda,1}$, $q_{\Delta\lambda,2}$, and $q_{\Delta\lambda,3}$. The solution is carried out for the $q_{\Delta\lambda,k}$ values in each band $\Delta\lambda$. The $q_{\Delta\lambda,k}$ is the energy supplied to surface k in wavelength interval $\Delta\lambda$ as a result of external heat addition to the surface (e.g., conduction and/or convection) and energy transferred in from other wavelength bands by radiation exchange within the enclosure. Such boundary conditions are discussed in more detail in Chapter 7 for cases including conduction, convection, or imposed heating. Finally, q_k at each surface is found by summing $q_{\Delta\lambda,k}$ for that surface over all wavelength bands as in Equation 6.7. This is the heat flux that must be supplied to surface k by some means other than internal radiation, to maintain its specified surface temperature.

Example 6.4

Consider the geometry of Figure 6.5. Total energy flux is supplied to the three surfaces at rates q_1, q_2, and q_3. Determine the surface temperatures.

The equations are the same as in Example 6.3. Now, however, the prescribed boundary conditions have made the solution much more difficult. Because the surface temperatures are unknown, the $F_{\Delta\lambda T}$ and $\varepsilon_{\Delta\lambda}$ are unknown because they depend on temperature. The solution is carried out as follows: A temperature is assumed for each surface, and $q_{\Delta\lambda,k}(T)$ for each surface is computed. This is done for each of the wavelength bands. The $q_{\Delta\lambda,k}(T)$ values are then summed over all the $\Delta\lambda$ to find q_1, q_2, and q_3, which are compared to the specified boundary values. New temperatures are chosen and the process repeated until the computed q_k values agree with the specified values. The new temperatures for successive iterations are guessed on the basis of the $F_{\Delta\lambda T}$ and $\varepsilon_{\Delta\lambda}$ variations and the trends of how changes in T_k produce changes in q_k throughout the system. As noted in Section 5.4.1.3, temperature must be specified on at least one surface of the enclosure. Thus, there may be many solutions to this problem.

6.2.3 Semigray Approximations

In some practical situations, the radiant energy in an enclosure has two well-defined spectral regions. An important example is an enclosure with an opening that is admitting solar energy. Solar energy is mostly in the short-wavelength spectral region, while emitted energy from the lower-temperature surfaces within the enclosure is at longer wavelengths. One way used to treat this situation is to define, for each surface in the enclosure, a hemispherical total absorptivity for incident solar radiation, and a second hemispherical total absorptivity for incident energy from emission that originates within the enclosure. More generally, N different absorptivities can be defined for each surface A_k, one for incident energy from each of the N enclosure surfaces.

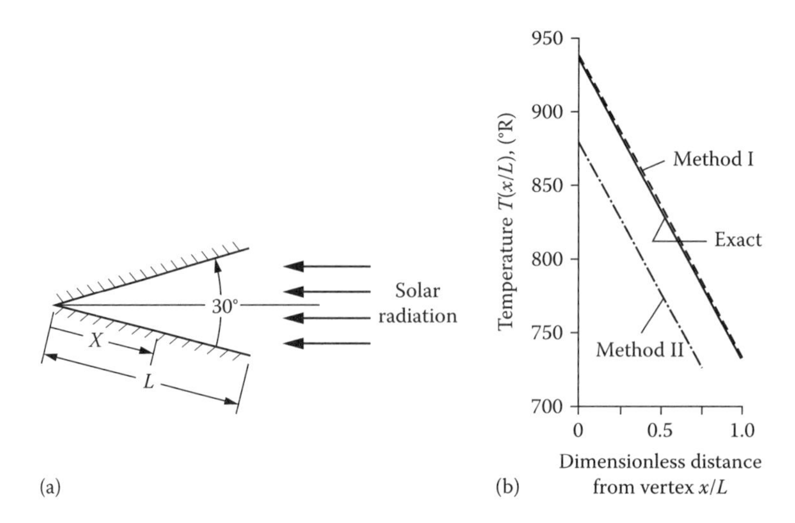

FIGURE 6.6 Effect of semigray approximations on computed temperature distribution in a wedge cavity: (a) geometry of wedge cavity; (b) temperature distribution along wedge, $\alpha_{solar} = 0.220$; $\alpha_{infrared} = 0.099$.

An assumption in these approximate analyses is that each absorptivity $\alpha_k(T_k, T_j)$ is based on an incident *blackbody* spectrum at the temperature of the originating surface T_j. The incident spectrum may actually be quite different from the blackbody form, and this is a weakness of the method. Often the dependence of α_k on T_k is small, and its principal dependence is on T_j through the spectral distribution of the incident energy. Because the absorptivity $\alpha_k(T_k, T_j)$ and emissivity $\varepsilon_k(T_k)$ of surface k are not in general equal, this approach is often called the *semigray enclosure theory*. Bobco (1964) provides the formulation for a general enclosure.

The semigray and exact solutions for the temperature profiles along the surface of a nongray wedge cavity exposed to incident solar radiation are in Figure 6.6 (Plamondon and Landram 1966). The cavity is in vacuum with an environment at ~0 K except for the solar radiation source. The cavity surface properties are independent of temperature and the surfaces are diffuse.

Three solution techniques are given, labeled in Figure 6.6b as Exact, Method I, and Method II. The first is an *exact* solution of the complete integral equations. *Method I* is the semigray analysis that assigns an absorptivity α_{solar} for direct and reflected radiation originating from the solar energy, and a second absorptivity $\alpha_{infrared}$ (equal to the surface emissivity) for radiation originating by emission from the wedge surfaces. *Method II* is a poorer approximation that retains these same two absorptivities but applies α_{solar} only for the incident solar energy and uses $\alpha_{infrared}$ for *all* energy after reflection, regardless of its source. The results in Figure 6.6b are for a polished aluminum surface in a 30° wedge. Method I is in excellent agreement with the exact solution, while method II underestimates the temperatures by about 10%.

6.3 DIRECTIONAL-GRAY SURFACES

Radiant exchange between gray surfaces with directional properties is now considered. Many radiation analyses assume diffuse emitting and reflecting surfaces, and some treatments include the effect of specular reflections with diffuse emission (Section 6.6). Diffuse or specular surface conditions are the most convenient to treat analytically, and in many instances the detailed consideration of directional emission and reflection effects is unwarranted. However, certain materials and special situations require the examination of directional effects.

The difficulty in treating the general case of directionally dependent properties is illustrated by performing an energy balance in a simple geometry: the radiative exchange between two infinitely

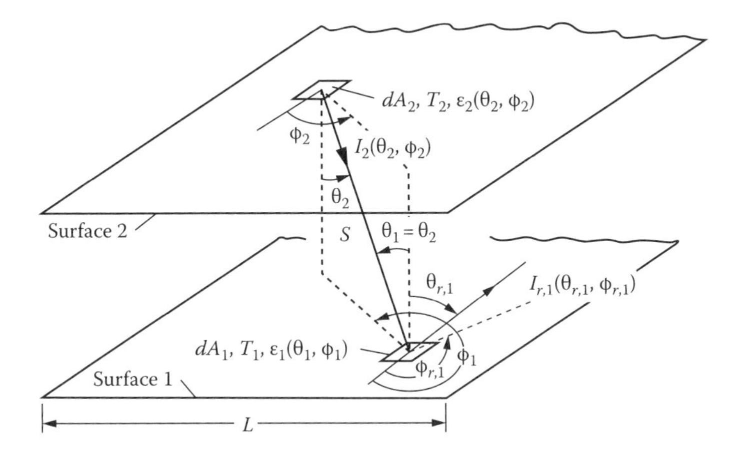

FIGURE 6.7 Radiant interchange between parallel directional surfaces of finite width L that are infinitely long in the direction normal to the plane of the drawing.

long parallel nondiffuse gray surfaces of finite width L (Figure 6.7). The radiation intensity leaving element dA_1 in direction $(\theta_{r,1},\phi_{r,1})$ is composed of emitted and reflected intensities:

$$I_1(\theta_{r,1},\phi_{r,1}) = I_{e,1}(\theta_{r,1},\phi_{r,1}) + I_{r,1}(\theta_{r,1},\phi_{r,1}) \tag{6.8}$$

These two components are given by modifications of Equations 2.4 and 2.40 to yield Equation 6.8 in the form

$$I_1(\theta_{r,1},\phi_{r,1}) = \varepsilon_1(\theta_{r,1},\phi_{r,1}) I_{b,1}(T_1) + \int_{A_2} \rho_1(\theta_{r,1},\phi_{r,1},\theta_1,\phi_1) I_2(\theta_2,\phi_2) \frac{\cos^2\theta_2}{S^2} dA_2 \tag{6.9}$$

In the second term on the right of Equation 6.9, the energy incident on dA_1 from each element dA_2 is multiplied by the bidirectional total reflectivity to give the contribution to the intensity reflected from dA_1 into direction $(\theta_{r,1},\phi_{r,1})$. This is then integrated over A_2 to include all energy incident on dA_1 from A_2.

A similar equation is written for an arbitrary element dA_2 on surface 2. The result is a pair of coupled integral equations to be solved for $I(\theta, \phi)$ at each position and for each direction on the two surfaces. The integral equations are analogous to Equations 5.21.3 and 5.21.4 for diffuse-gray surfaces. Detailed property data for $\varepsilon(\theta, \phi)$ and $\rho(\theta_r, \phi_r, \theta_i, \phi_i)$ are not usually available. For the case when T_1 and T_2 are not known and the temperature dependence of the properties is considerable, the solution for the entire energy-exchange distribution becomes very tedious. Approximations can be made, such as analytically simulating the real properties with simple functions, omitting certain portions of energy deemed negligible, or ignoring directional effects except those expected to provide significant changes from diffuse or specular analyses. Some of these methods are outlined in Bevans and Edwards (1965), Hering (1966a,b), Viskanta et al. (1967), Toor (1967), and Naraghi and Chung (1984, 1986). A method using discrete ordinates was developed by Brockmann (1997) (as in Section 13.4, but without an absorbing gas in the enclosure) to examine directional effects for radiation between the ends of a cylindrical enclosure. The radiating surfaces had the predicted directional behavior for dielectrics or metals as in Figures 3.2 and 3.5, and heat fluxes were compared with those for diffuse surfaces. In most instances, the directional effects were found to be small as a consequence of multiple reflections from the cylindrical sidewall and the radiating ends that tend to diffuse the radiation in the enclosure.

An example is now given illustrating the effect of a directional-gray surface on radiative exchange.

Example 6.5

Two parallel isothermal plates of infinite length and finite width L are arranged as in Figure 6.8a. The upper plate is black, while the lower is a highly reflective gray material with parallel deep grooves of open angle 1° in its surface extending along the infinite direction. Such a surface might be constructed by stacking polished razor blades. The surroundings are at zero temperature. Compute the net energy gain by the directional surface if $T_2 > T_1$, and compare the result to the net energy gain if the directional-gray surface is replaced by a diffuse-gray surface with an emissivity equivalent to the hemispherical emissivity of the directional surface.

In Howell and Perlmutter (1963), the directional emissivity is calculated at the opening of an infinitely long groove with specularly reflecting walls of surface emissivity 0.01. This is given by the dot-dashed line in Figure 6.8b. The angle β_1 is measured from the normal of the opening plane of the grooved surface and is in the cross-sectional plane perpendicular to the length of the groove as in Figure 6.8a. The $\varepsilon_1(\beta_1)$ in Howell and Perlmutter has already been averaged over all circumferential angles for a fixed β_1. Thus, it is an effective emissivity for radiation from a strip on the grooved surface to a parallel infinitely long strip element on an imaginary semicylinder over the groove and with its axis parallel to the grooves. The angle β_1 is different from the usual cone angle θ_1. The actual emissivity $\varepsilon_1(\beta_1)$ of Figure 6.8b is approximated for convenience by the analytical

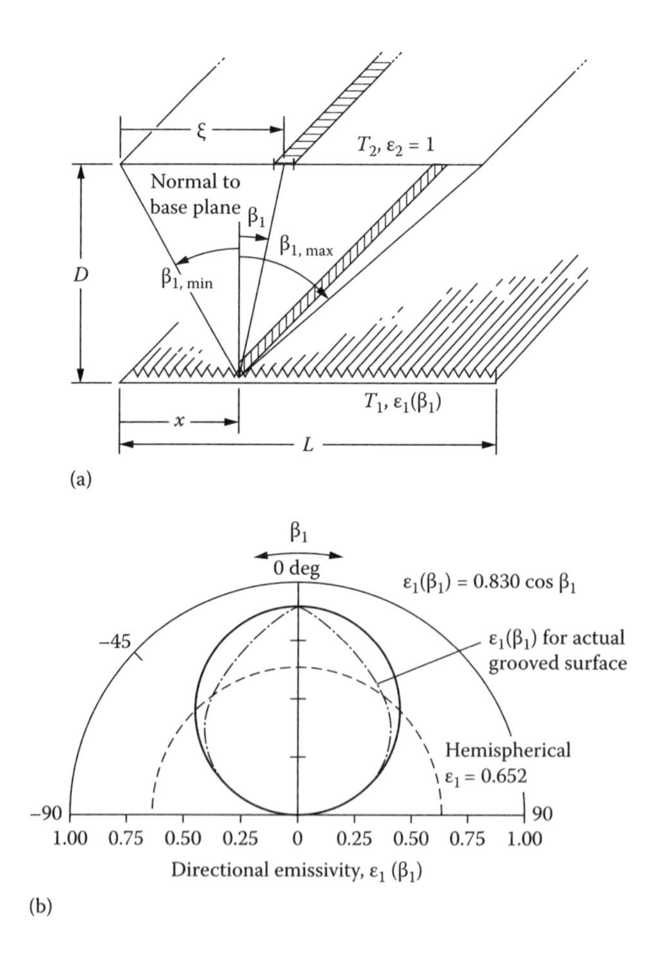

FIGURE 6.8 Interchange between grooved directional-gray surface and black surface: (a) geometry of problem (environment at zero temperature); (b) emissivity of directional surface.

expression $\varepsilon_1(\beta_1) \approx 0.830 \cos \beta_1$. Using cylindrical coordinates to integrate over all β_1, the hemispherical emissivity of this surface is

$$\varepsilon_1 = \frac{\int_{-\pi/2}^{\pi/2} \varepsilon_1(\beta_1)\cos\beta_1 d\beta_1}{\int_{-\pi/2}^{\pi/2} \cos\beta_1 d\beta_1} = 0.830\int_0^{\pi/2} \cos^2\beta_1 d\beta_1 = 0.652$$

and this is the dashed line in Figure 6.8b.

The energy gained by surface 1 will first be determined when surface 2 is black and surface 1 is diffuse with $\varepsilon_1 = 0.652$. The energy emitted by the diffuse surface 1 per unit of the infinite length and per unit time is $Q_{e,1} = 0.652\,\sigma T_1^4 L$. Because surface 2 is black, none of this energy is reflected back to surface 1. The energy per unit length and time emitted by black surface 2 that is absorbed by surface 1 is

$$Q_{a,1} = \alpha_1\sigma T_2^4 \int_{A_2}\int_{A_1} dF_{d2-d1}dA_2 = \varepsilon_1\sigma T_2^4 \int_{A_1}\int_{A_2} dF_{d1-d2}dA_1$$

The configuration factor between infinite parallel strips, from Example 4.2, is $dF_{d1-d2} = d(\sin\beta_1)/2$ so that

$$\int_{A_1}\left(\int_{A_2} dF_{d1-d2}\right)dA_1 = \frac{1}{2}\int_{x=0}^{L}\left(\sin\beta_{1,\max} - \sin\beta_{1,\min}\right)dx$$

From Figure 6.8a, $\sin\beta_1 = (\xi - x)\Big/\left[(\xi - x)^2 + D^2\right]^{1/2}$, which gives

$$\begin{aligned} Q_{a,1} &= \varepsilon_1\sigma T_2^4\,\frac{1}{2}\int_{x=0}^{L}\left[\frac{L-x}{\left(x^2 - 2xL + L^2 + D^2\right)^{1/2}} + \frac{x}{\left(x^2 + D^2\right)^{1/2}}\right]dx \\ &= 0.652\sigma T_2^4\left[\left(L^2 + D^2\right)^{1/2} - D\right] \end{aligned}$$

The net energy gained by surface 1, $Q_{a,1} - Q_{e,1}$, divided by the energy emitted by surface 2, is a measure of the efficiency of the surface as a directional absorber. For surface 1, being diffuse, this ratio is ($l = L/D$)

$$Eff_{diffuse} = \frac{Q_{a,1} - Q_{e,1}}{\sigma T_2^4 L} = \frac{0.652}{l}\left[\left(1 + l^2\right)^{1/2} - 1 - \frac{T_1^4}{T_2^4}l\right]$$

When surface 1 is a directional (grooved) surface, the amount of energy emitted from surface 1 is the same as for a diffuse surface (although it has a different directional distribution) since both have the same hemispherical emissivity. The energy absorbed by the grooved surface is, by using $\alpha_1(\beta_1) = \varepsilon_1(\beta_1)$ for a gray surface,

$$\begin{aligned} Q_{a,1} &= \sigma T_2^4 \int_{A_2}\int_{A_1} \alpha_1(\beta_1)dF_{d2-d1}dA_2 = \frac{0.830\sigma T_2^4}{2}\int_{x=0}^{L}\int_{\beta_{1,\min}}^{\beta_{1,\max}} \cos^2\beta_1 d\beta_1 dx \\ &= \frac{0.830\sigma T_2^4}{4}\int_{x=0}^{L}\left[\frac{D(L-x)}{x^2 - 2xL + L^2 + D^2} + \tan^{-1}\left(\frac{L-x}{D}\right) + \frac{xD}{x^2 + D^2} + \tan^{-1}\frac{x}{D}\right]dx \\ &= \frac{0.830\sigma T_2^4}{2}\tan^{-1}\frac{L}{D} \end{aligned}$$

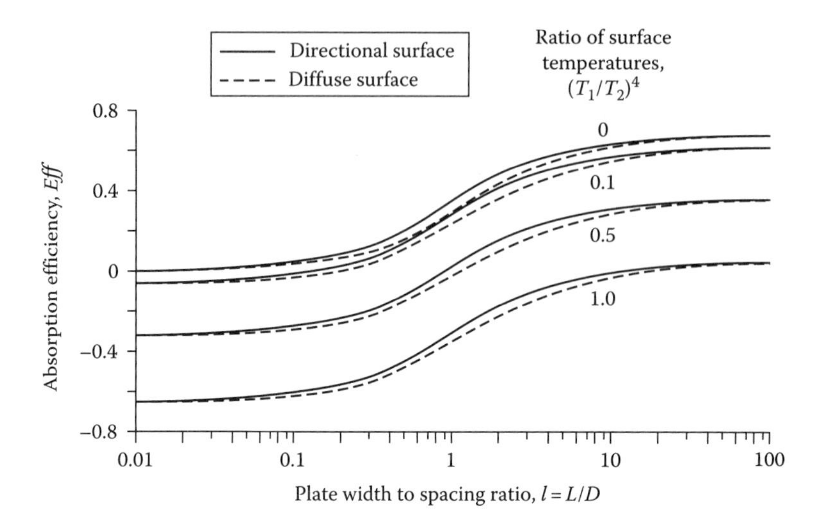

FIGURE 6.9 Effect of directional emissivity on absorption efficiency of surface.

The absorption efficiency of the directional surface is then

$$Eff_{directional} = \frac{Q_{a,1} - Q_{e,2}}{\sigma T_2^4 L} = \frac{0.830}{2}\tan^{-1} l - 0.652\left(\frac{T_1}{T_2}\right)^4$$

The absorption efficiencies of the directional-gray and diffuse-gray surfaces are in Figure 6.9 as a function of l with $(T_1/T_2)^4$ as a parameter. The *Eff* for the directional surface is higher than that for the diffuse surface for all values of l. As l approaches zero, the configuration approaches that of infinite elemental strips, and emission from surface 1 becomes much larger than absorption from surface 2. Thus, $Eff_{directional}$ and $Eff_{diffuse}$ are nearly equal since the surfaces always emit the same amount. As l approaches infinity, the directional effects are lost. At intermediate l a 10% difference in absorption efficiency is attainable.

The effects of directional properties on the local energy loss can be considerable for some geometries. In Figure 6.10 some directional distributions of reflectivity are examined for their influence on local energy loss from the walls of an infinitely long groove. The results are from Viskanta et al. (1967), where for comparison the curves were gathered from original work and various sources (Eckert and Sparrow 1961, Sparrow et al. 1961, Hering 1966a, Toor 1967). The walls of the groove are at 90°, and the surface emissivity distributions are all normalized to give a hemispherical emissivity of 0.1. Curves are presented for diffuse reflectivity ρ, specular reflectivity assumed independent of incident angle ρ_s, specular reflectivity dependent on incident angle $\rho_s(\theta_i)$ based on electromagnetic theory, and three distributions of bidirectional reflectivity $\rho(\theta_r, \theta_i)$. The bidirectional distributions are based on Beckmann and Spizzichino (1963) for rough surfaces having various combinations of the ratio of rms optical-surface roughness amplitude to radiation wavelength, σ_0/λ, and the ratio of roughness autocorrelation distance to radiation wavelength, a_0/λ. Note that the results in Figure 6.10 for the specular and diffuse models do *not* provide upper and lower limits to all the solutions, as is sometimes claimed. Additional information on surface roughness as it affects the directional properties of surfaces is in Section 3.5.3. Energy transfer was studied in a groove with two rough sides, each at uniform temperature. The roughness has a greater influence on the radiation exchange between the two sides than on the net radiation from the groove.

Howell and Durkee (1971) compared analysis and experiment for collimated radiation entering a low-temperature very long three-sided cavity. Because of the low temperature, surface emission

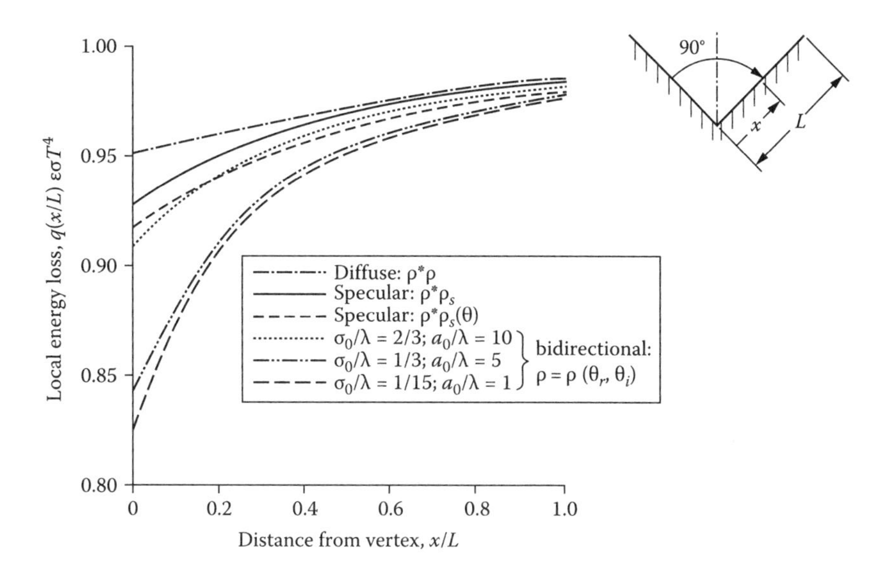

FIGURE 6.10 Local radiative energy loss from surface of isothermal groove cavity. Hemispherical emissivity of surface $\varepsilon = 0.1$.

was not important. The cavity had two surfaces with reflectivities that were diffuse with a specular component, while the third surface was honeycomb material with a strong bidirectional reflectivity. It was found necessary to include all surface characteristics in the analysis to obtain agreement with experiment. This is characteristic of geometries interacting with collimated incident radiation.

Black (1973) analyzed the directional emission characteristics of two types of grooves with some black and some specular sides, as in Figure 6.11a and b. The V-groove tends to emit in the normal direction, while the rectangular groove emits more in the grazing direction $\beta \to 90°$. Some emission results for the rectangular shape are in Figure 6.11c, and they exhibit a strong directional characteristic for small h/w. Grooves and enclosures having specular surfaces are analyzed in Section 6.5.3.

6.4 SURFACES WITH DIRECTIONALLY AND SPECTRALLY DEPENDENT PROPERTIES

The general case of radiative transfer in enclosures where surface properties depend on both direction and wavelength, and where properties can be temperature dependent, is complex to treat fully. When those dependencies must be included, numerical techniques are necessary. Toor (1967) used the Monte Carlo method to study radiation interchange for various simply arranged surfaces with directional properties. Zhang et al. (1997) derived the directional-spectral relation for radiative transfer between parallel plates that is a generalization (with properties independent of angle ϕ) of Equation 6.1.2 (also see Example 6.9):

$$q = 2\int_{\theta=0}^{\pi/2}\int_{\lambda=0}^{\infty} \frac{E_{\lambda b,1}(T_1) - E_{\lambda b,2}(T_1)}{\dfrac{1}{\varepsilon_{\lambda,1}(\theta_1, T_1)} + \dfrac{1}{\varepsilon_{\lambda,2}(\theta_2, T_2)} - 1} d\lambda \sin\theta \cos\theta d\theta \tag{6.10}$$

A difficulty for such an evaluation is in finding the detailed radiative properties to sufficient accuracy. The technology of interest was for evaluating the insulating performance of a double glass window with a vacuum between the two panes. The glass surfaces are opaque in the infrared region so that, for the temperatures involved, Equation 6.10 could be used for transfer across the vacuum space. Comparisons were made with experiment.

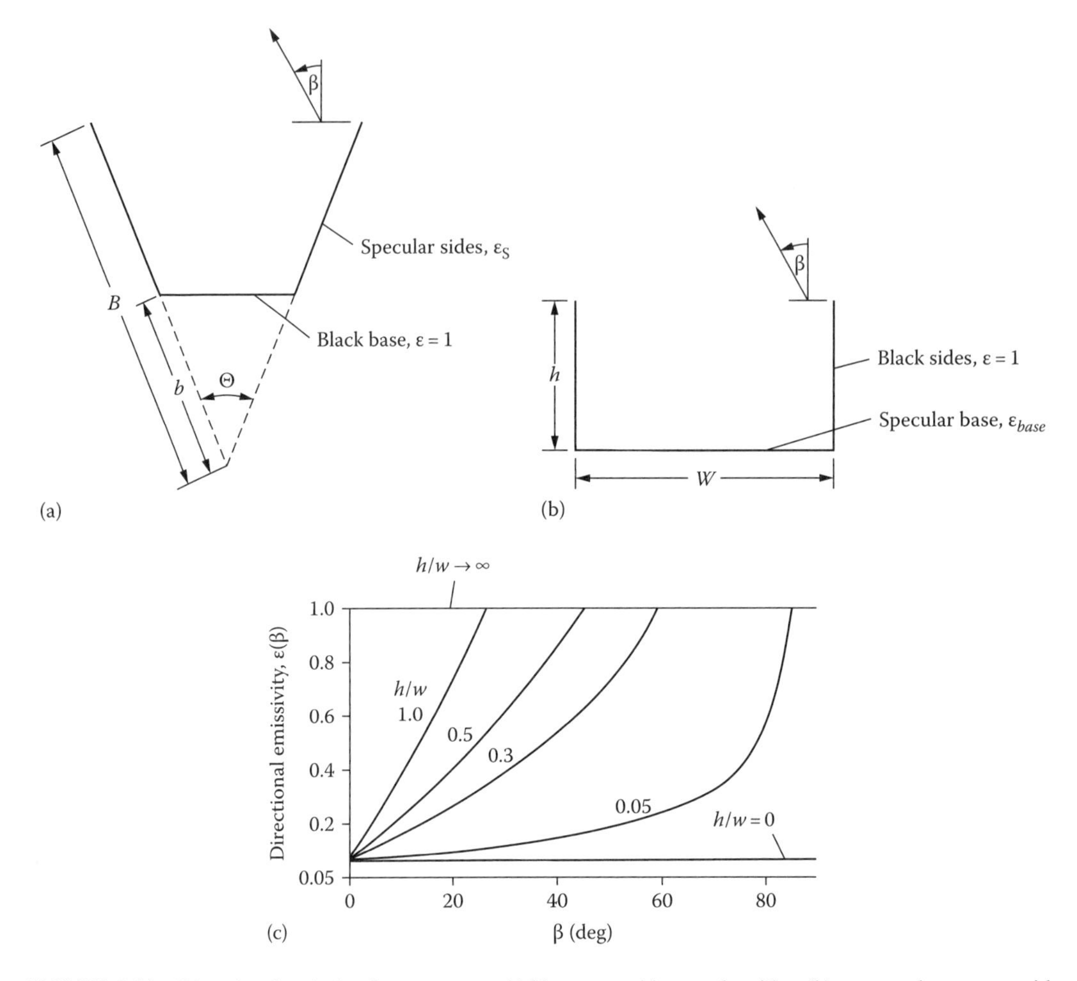

FIGURE 6.11 Directional emission from grooves: (a) V-groove with specular sides; (b) rectangular groove with specular base; and (c) directional emissivity for rectangular groove with specular base ($\varepsilon_{base} = 0.05$) and black sides.

In this section, the general integral equations are formulated for radiation in such systems, and a considerably simplified example problem is solved. The procedure is a combination of the previous diffuse-spectral and directional-gray analyses. The equations are formulated at one wavelength as in Section 6.2 and in terms of intensities for each direction as in Section 6.3; this accounts for both spectral and directional effects. The interaction between two plane surfaces is developed first; this can be generalized to a multisurface enclosure as for gray surfaces in Chapter 5.

The energy balance is now developed for an area element dA at location r in an x, y, z coordinate system as in Figure 6.12. The $I_{\lambda o}(\theta_r, \phi_r, \mathbf{r})$ is the outgoing spectral intensity from dA in the direction θ_r, ϕ_r as the result of both emission and reflection. The spectral intensity emitted by dA in this direction is

$$I_{\lambda e}(\theta_r, \phi_r, \mathbf{r}) = \varepsilon_\lambda(\theta_r, \phi_r, \mathbf{r}) I_{\lambda b}(\mathbf{r}) \tag{6.11}$$

These quantities also depend on T_{dA}, but this functional designation is omitted to simplify the notation. The intensity reflected from dA into the θ_r, ϕ_r direction results from the incident intensity from all directions of a hemisphere above dA. If the spectral intensity incident on dA within $d\Omega_i$ is $I_{\lambda i}(\theta_i, \phi_i, \mathbf{r})$, the intensity reflected from dA into direction θ_r, ϕ_r is

$$I_{\lambda,r}(\theta_r, \phi_r, \mathbf{r}) = \int_{\Omega_i=0}^{2\pi} \rho_\lambda(\theta_r, \phi_r, \theta_i, \phi_i, \mathbf{r}) I_{\lambda,i}\left(\theta_i, \phi_i, \mathbf{r}\right) \cos\theta_i d\Omega_i \tag{6.12}$$

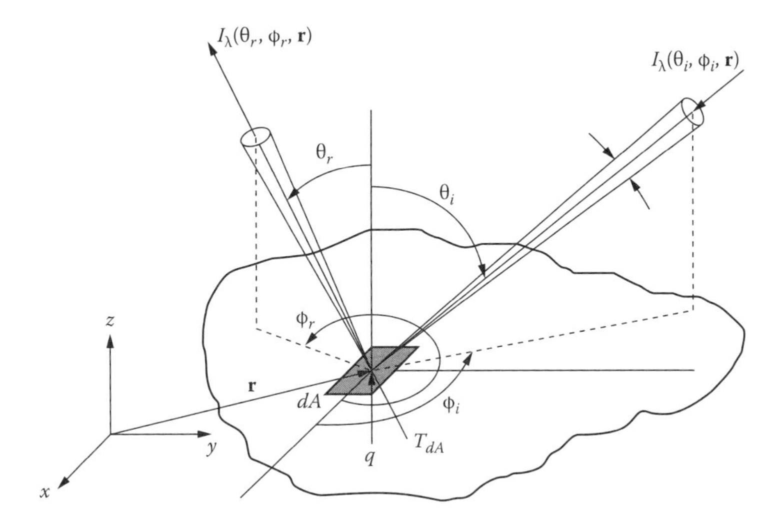

FIGURE 6.12 Geometry for incoming and outgoing intensities at a differential surface area.

The net energy flux supplied to dA for steady state is the difference between the outgoing and incoming radiative fluxes:

$$q(\mathbf{r}) = \int_{\lambda=0}^{\infty} J_\lambda(\mathbf{r})d\lambda - \int_{\lambda=0}^{\infty} G_\lambda(\mathbf{r})d\lambda \tag{6.13}$$

The $J_\lambda(\mathbf{r})d\lambda$ is the angular integration of the emitted and reflected spectral fluxes over all outgoing directions:

$$\begin{aligned} J_\lambda(\mathbf{r})d\lambda = I_{\lambda b}(\mathbf{r})d\lambda \int_{\phi_r=0}^{2\pi}\int_{\theta_r=0}^{\pi/2} \varepsilon_\lambda(\theta_r, \phi_r, \mathbf{r}) \sin\theta_r \cos\theta_r d\theta_r d\phi_r \\ + \int_{\phi_r=0}^{2\pi}\int_{\theta_r=0}^{\pi/2} I_{\lambda,r}(\theta_r, \phi_r, \mathbf{r}) d\lambda \sin\theta_r \cos\theta_r d\theta_r d\phi_r \end{aligned} \tag{6.14}$$

The $G_\lambda(\mathbf{r})d\lambda$ is the result of spectral fluxes incident from all $d\Omega_i$ directions:

$$G_\lambda(\mathbf{r})d\lambda = \int_{\phi_i=0}^{2\pi}\int_{\theta_i=0}^{\pi/2} I_{\lambda,i}(\theta_i, \phi_i, \mathbf{r}) d\lambda \sin\theta_i \cos\theta_i d\theta_i d\phi_i \tag{6.15}$$

Equations 6.11 through 6.15 provide an exact formulation to obtain the heat flux $q(\mathbf{r})$ that must be supplied by other means to area dA to maintain its temperature at T_{dA} in the presence of incident and emitted radiation.

To develop an enclosure theory, various degrees of approximation can be made. If the enclosure is very simple, such as having only two plane surfaces, it may be feasible to include variations of properties and surface temperature across each surface. To develop the required integral equations, consider the two surfaces in Figure 6.13 and let the surrounding environment be at low temperature so that it does not contribute incident radiation. The spectral energy leaving dA_2 at $\mathbf{r}_2$ that reaches dA_1 at $\mathbf{r}_1$ is $I_{\lambda o,2}(\theta_2,\phi_2,\mathbf{r}_2)d\lambda dA_2\cos\theta_2(dA_1\cos\theta_1/S^2)$. In terms of the incident intensity, the

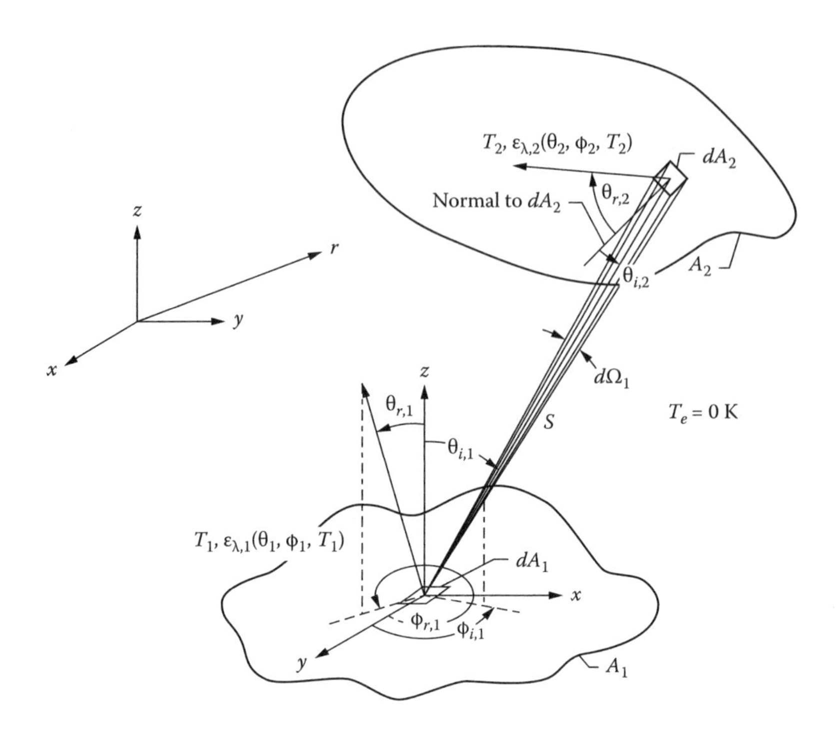

FIGURE 6.13 Interchange between surfaces having directional spectral properties (environment at ~0 K).

incident spectral intensity in $d\Omega_1$ is $I_{\lambda i,1}(\theta_1,\phi_1,\mathbf{r}_1)d\lambda dA_1\cos\theta_1 d\Omega_1$, where $d\Omega_1 = (dA_2\cos\theta_2)/S^2$. Thus, $I_{\lambda i,1}(\theta_1,\phi_1,\mathbf{r}_1) = I_{\lambda o,2}(\theta_2,\phi_2,\mathbf{r}_2)$ and, by using Equations 6.11 and 6.12,

$$I_{\lambda o,1}\left(\theta_{r,1},\phi_{r,1},\mathbf{r}_1\right) = \varepsilon_{\lambda,1}\left(\theta_{r,1},\phi_{r,1},\mathbf{r}_1\right)I_{\lambda b,1}\left(\mathbf{r}_1\right) + \int_{A_2}\rho_{\lambda,1}\left(\theta_{r,1},\phi_{r,1},\theta_{i,1},\phi_{i,1},\mathbf{r}_1\right)I_{\lambda o,2}\left(\theta_2,\phi_2,\mathbf{r}_2\right)\frac{\cos\theta_1\cos\theta_2}{\left|\mathbf{r}_2-\mathbf{r}_1\right|^2}dA_2 \tag{6.16}$$

Similarly, for surface 2, the outgoing intensity is

$$I_{\lambda o,2}\left(\theta_{r,2},\phi_{r,2},\mathbf{r}_2\right) = \varepsilon_{\lambda,2}\left(\theta_{r,2},\phi_{r,2},\mathbf{r}_2\right)I_{\lambda b,2}\left(\mathbf{r}_2\right) + \int_{A_1}\rho_{\lambda,1}\left(\theta_{r,2},\phi_{r,2},\theta_{i,2},\phi_{i,2},\mathbf{r}_2\right)I_{\lambda o,1}\left(\theta_1,\phi_1,\mathbf{r}_1\right)\frac{\cos\theta_2\cos\theta_1}{\left|\mathbf{r}_1-\mathbf{r}_2\right|^2}dA_1 \tag{6.17}$$

Equations 6.16 and 6.17 are both in terms of outgoing intensities, and they provide a set of simultaneous integral equations for $I_{\lambda o,1}$ and $I_{\lambda o,2}$. An iterative numerical solution is required that can be quite complex as both unknowns are functions of position and angle. After $I_{\lambda o,1}(\theta_1, \phi_1, \mathbf{r}_1)$ and $I_{\lambda o,2}$ $(\theta_2, \phi_2, \mathbf{r}_2)$ are obtained, the total energy can be determined that must be supplied to each surface element to maintain the specified local surface temperature. The total energy supplied is the difference between the total emitted and absorbed energies:

$$\frac{dQ_1}{dA_1} = \int_{\lambda=0}^{\infty}\int_{\cap}\varepsilon_{\lambda,1}\left(\theta_{r,1},\phi_{r,1},\mathbf{r}_1\right)I_{\lambda b,1}\left(\mathbf{r}_1\right)\cos\theta_1 d\Omega_1 d\lambda - \int_{\lambda=0}^{\infty}\int_{A_2}\alpha_{\lambda,1}\left(\theta_{i,1},\phi_{i,1},\mathbf{r}_1\right)I_{\lambda o,2}\left(\theta_2,\phi_2,\mathbf{r}_2\right)\frac{\cos\theta_1\cos\theta_2}{\left|\mathbf{r}_2-\mathbf{r}_1\right|^2}dA_2 d\lambda \tag{6.18}$$

Usually, to develop an enclosure theory with more than a few surfaces, the $I_{\lambda o}$, temperature, and surface properties are assumed uniform over each enclosure surface. In addition, a finite number of angular intervals must be specified. If we let A_k and A_j be the kth and jth surfaces of an enclosure with N surfaces, then, by integrating Equation 6.17 over A_k and summing the contributions from all of the A_j surfaces,

$$I_{\lambda o,k}\left(\theta_{r,k},\phi_{r,k}\right)=\varepsilon_{\lambda,k}\left(\theta_{r,k},\phi_{r,k}\right)I_{\lambda b,k}+\frac{1}{A_k}\sum_{j=1}^{N}\int_{A_k}\int_{A_j}\rho_{\lambda,k}\left(\theta_{r,k},\phi_{r,k},\theta_{i,k},\phi_{i,k}\right)I_{\lambda o,j}\left(\theta_j,\phi_j\right)\frac{\cos\theta_k\cos\theta_j}{\left|\mathbf{r}_j-\mathbf{r}_k\right|^2}dA_j dA_k \quad (6.19)$$

When written out for each surface k and for a sufficient number of directions ($\theta_{r,k}$, $\phi_{r,k}$) to obtain acceptable accuracy in the angular integrations that follow, this yields a set of simultaneous equations for $I_{\lambda o,k}(\theta_{r,k},\phi_{r,k})$ for $k = 1,\ldots,N$.

With this degree of approximation, which is characteristic for enclosure analyses, consider the interaction of the two plane surfaces in Figure 6.13. The surroundings are at low temperature relative to the surface temperature, so radiation from the surroundings is neglected; this provides a two-surface enclosure. Then, writing Equation 6.19 for $k = 1$ and 2,

$$I_{\lambda o,1}\left(\theta_{r,1},\phi_{r,1}\right)=\varepsilon_{\lambda,1}\left(\theta_{r,1},\phi_{r,1}\right)I_{\lambda b,1}+\frac{1}{A_1}\int_{A_1}\int_{A_2}\rho_{\lambda,1}\left(\theta_{r,1},\phi_{r,1},\theta_{i,1},\phi_{i,1}\right)I_{\lambda o,2}\left(\theta_2,\phi_2\right)\frac{\cos\theta_1\cos\theta_2}{S^2}dA_j dA_k \quad (6.20)$$

$$I_{\lambda o,2}\left(\theta_{r,2},\phi_{r,2}\right)=\varepsilon_{\lambda,2}\left(\theta_{r,2},\phi_{r,2}\right)I_{\lambda b,2}+\frac{1}{A_2}\int_{A_2}\int_{A_1}\rho_{\lambda,2}\left(\theta_{r,2},\phi_{r,2},\theta_{i,2},\phi_{i,2}\right)I_{\lambda o,1}\left(\theta_1,\phi_1\right)\frac{\cos\theta_2\cos\theta_1}{S^2}dA_1 dA_2 \quad (6.21)$$

Equations 6.20 and 6.21 are in terms of outgoing intensities in each direction $\theta_{r,k}$, $\phi_{r,k}$; they are simultaneous integral equations for $I_{\lambda o,1}$ and $I_{\lambda o,2}$. A numerical solution can be obtained by writing these equations for a number of discrete angular intervals to develop a set of simultaneous equations.

After $I_{\lambda o,1}(\theta_{r,1},\phi_{r,1})$ and $I_{\lambda o,2}(\theta_{r,2},\phi_{r,2})$ are obtained for a sufficient number of angular intervals to yield good accuracy, the total energy can be determined that must be supplied to each surface to maintain its specified temperature. This is the difference between energies carried away from and to the surface; for A_1 this gives

$$\frac{Q_1}{A_1}=\int_{\lambda=0}^{\infty}\int_{\cap}I_{\lambda o,1}\left(\theta_{r,1},\phi_{r,1}\right)\cos\theta_{r,1}d\Omega_{r,1}d\lambda-\frac{1}{A_1}\int_{\lambda=0}^{\infty}\int_{A_1}\int_{A_2}I_{\lambda o,2}\left(\theta_2,\phi_2\right)\frac{\cos\theta_1\cos\theta_2}{S^2}dA_2 dA_1 d\lambda \quad (6.22)$$

and similarly for A_2. For diffuse-gray surfaces, so that $I_{\lambda o,1}$ and $I_{\lambda o,2}$ are independent of λ, θ, and ϕ, this simplifies to $Q_1/A_1 = \pi I_{o,1} - \pi I_{o,2}F_{1-2} = J_1 - J_2F_{1-2}$ as given by Equation 5.17.

If Q_1 rather than T_1 is specified, T_1 must be determined and the solution becomes more difficult. A temperature is assumed for A_1, and the enclosure equations of the form Equations 6.20 and 6.21 are solved to find the $I_{\lambda o}$. The outgoing intensities are substituted into Equation 6.22, and the computed Q_1 is compared to the given value. The T_1 is then adjusted and the procedure repeated until agreement between given and computed Q_1 is attained. If Q is specified for more than one surface, the solution is even more difficult.

Example 6.6

For an example that can be carried out in analytical form, a small area element dA_1 is considered on the axis of, and parallel to, a black circular disk as in Figure 6.14. The element is at T_1, the disk at T_2, and the environment at $T_e \approx 0$ K. The dA_1 has a directional spectral emissivity independent of ϕ and approximated by

$$\varepsilon_{\lambda,1}(\theta_1, T_1) = 0.8\cos\theta_1(1 - e^{-C_2/\lambda T_1})$$

where C_2 is one of the constants in Planck's spectral distribution. As will be evident, this dependence on λ and T_1 was devised to simplify this illustrative example and obtain an analytical result. More generally, numerical integration can be used. Find the energy dQ_1 added to dA_1 to maintain T_1.

The energy balance in Equation 6.18 is emitted energy minus absorbed incident energy. The energy emitted by dA_1 is

$$dQ_{e,1} = dA_1 \int_{\lambda=0}^{\infty} \int_{\cap} \varepsilon_{\lambda,1}(\theta_1) I_{\lambda b,1} \cos\theta_1 d\Omega_1 d\lambda$$

Insert the expressions for $\varepsilon_{\lambda,1} I_{\lambda b,1}$ (Equation 1.15), and $d\Omega_1 = \sin\theta_1 d\phi_1$ to obtain

$$dQ_{e,1} = 0.8 dA_1 \int_{\lambda=0}^{\infty} \int_{\phi_1=0}^{2\pi} \int_{\theta_1=0}^{\pi/2} \cos^2\theta_1 (1 - e^{-C_2/\lambda T_1}) \frac{2C_1}{\lambda^5 (e^{C_2/\lambda T_1} - 1)} \sin\theta_1 d\theta_1 d\phi_1 d\lambda$$

Integrating over ϕ_1 and θ_1 gives

$$dQ_{e,1} = 0.8 dA_1 \frac{2\pi}{3} \int_0^{\infty} \frac{2C_1}{\lambda^5 e^{C_2/\lambda T_1}} d\lambda$$

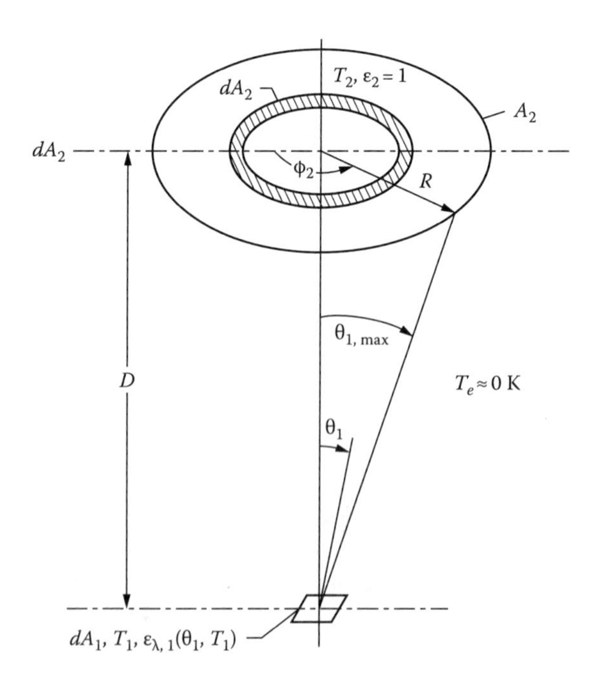

FIGURE 6.14 Energy exchange involving directional spectral surface element (environment at ~0 K).

Using the transformation $\zeta = C_2/\lambda T_1$, the relation $\int_0^\infty \zeta^3 e^{-\zeta} d\zeta = 3!$ from a table of definite integrals, and Stefan–Boltzmann constant σ from Equation 1.31 give

$$dQ_{e,1} = \frac{48}{\pi^4}\sigma T_1^4 dA_1 = 0.493\sigma T_1^4 dA_1$$

Thus, the total hemispherical emission is about half that of a blackbody.

The energy absorbed by dA_1 is

$$dQ_{a,1} = dA_1 \int_{\lambda=0}^{\infty}\int_{A_2} \alpha_{\lambda,1}(\theta_1,\phi_1) I_{\lambda o,2}(\theta_2,\phi_2)\frac{\cos\theta_1\cos\theta_2}{S^2} dA_2 d\lambda$$

From Kirchhoff's law, the directional spectral absorptivity and emissivity are equal. For dA_2 taken as a ring element, the solid angle $\cos\theta_2 dA_2/S^2$ is written as $2\pi\sin\theta_1 d\theta_1$. This is used to write the absorbed energy as (where $I_{\lambda o,2} = I_{\lambda b,2}$ since A_2 is a black surface)

$$dQ_{a,1} = 2\pi(0.8)dA_1 \int_{\lambda=0}^{\infty}\int_{\theta_1=0}^{\theta_{1,max}} (\cos^2\theta_1 \sin\theta_1 d\theta_1) I_{\lambda b,2}(1-e^{-C_2/\lambda T_1}) d\lambda$$

$$= -1.6\pi dA_1 \frac{\cos^3\theta_1}{3}\Bigg|_{\theta_1=0}^{\theta_{1,max}} \int_0^\infty \frac{2C_1(1-e^{-C_2/\lambda T_1})}{\lambda^5(e^{C_2/\lambda T_2}-1)} d\lambda$$

$$= \frac{3.2\pi C_1 dA_1}{3}\left[1-\frac{D^3}{(D^2+R^2)^{3/2}}\right]\int_0^\infty \frac{1-e^{-C_2/\lambda T_1}}{\lambda^5(e^{C_2/\lambda T_2}-1)} d\lambda$$

Using the transformation $\zeta = C_2/\lambda T_1$, this is placed in the form

$$dQ_{a,1} = \frac{48}{\pi^4}\left[1-\frac{1}{(1+r^2)^{3/2}}\right]\sigma T_2^4 dA_1 G\left(\frac{T_2}{T_1}\right)$$

where $r = R/D$ and $G(T_2/T_1) = (1/6)\int_0^\infty \zeta^3 e^{-\zeta}(1-e^{\zeta T_2/T_1})/(1-e^{-\zeta})d\zeta$. This integral was evaluated numerically, giving $G(1.0) = 1.000$, $G(1.5) = 1.045$, and $G(2.0) = 1.063$; hence, the effect of temperature ratio is small. Finally, the energy added to dA_1 to maintain it at T_1 is given by

$$dQ_1 = dQ_{e,1} - dQ_{a,1} = \frac{48\sigma}{\pi^4}\left\{T_1^4 - T_2^4\left[1-\frac{1}{(1+r^2)^{3/2}}\right]G\left(\frac{T_2}{T_1}\right)\right\}dA_1$$

As shown by Example 6.6, it is difficult to devise an analytical function for $\varepsilon_\lambda(\theta, T)$ that can be integrated in closed form over both angle and wavelength. Numerical methods are required to solve problems of this type for $\varepsilon_\lambda(\theta, T)$ functions that represent experimental data.

The development in this section has shown that, although the formulation of radiation-exchange problems involving directional and/or spectral properties is not conceptually difficult, it is usually tedious to solve the resulting integral equations. To simplify the equations, it is usually necessary to make assumptions and approximations that can vary from case to case. Numerical techniques such as iteration are used for directional spectral formulations, since closed-form analytical solutions can rarely be obtained. An alternative numerical technique is the Monte Carlo method presented in Chapter 7. For complicated directional and spectral effects, this is often a better approach than using an integral equation formulation.

Some of the original research is now summarized. A surface with part specular and part diffuse reflectivity and a semigray analysis were used by Shimoji (1977) to find local temperatures in conical and V-groove cavities exposed to incident solar radiation parallel to the cone axis or V-groove bisector plane. Toor and Viskanta (1972) compared with experiment various analytical models using diffuse, specular, semigray, nongray, and combinations of these characteristics. They found, for the particular geometries and materials studied, that spectral effects were less important than directional effects and that the presence of one or more diffuse surfaces in an enclosure made the presence of specularly reflecting surfaces unimportant. Hering and Smith (1970, 1971) and Edwards and Bertak (1971) applied various models of surface roughness in calculating radiant exchange between surfaces. If grooves on a surface have a size that is comparable to or smaller than the wavelength of the incident or emitted radiation, there can be complex interactions of the electromagnetic waves within the grooves. This can produce unusual spectral and directional effects. The radiation behavior of materials with a grooved microstructure was studied by Hesheth et al. (1988), Glass et al. (1982), Wirgin and Maradudin (1985), Sentenac and Greffet (1994), and Hajimirza et al. (2011, 2012). The specularity of spacecraft thermal control materials is reported by Drolen (1992). Innovative treatments of directional properties are required in modeling the illumination of scenes in computer graphic representations. Representative approaches are in Buckalew and Fussell (1989), Immel et al. (1986), Kajiya (1985), Wolff and Kurlander (1990), He et al. (1991), and Sillion et al. (1991). Billings et al. (1991a) give a method using two-parameter Markov chains for treating bidirectional surface properties.

6.5 RADIATION EXCHANGE IN ENCLOSURES WITH SPECULARLY REFLECTING SURFACES

In Chapter 5, the enclosure surfaces are all diffuse emitters and reflectors, and in Sections 6.1 through 6.4 of this chapter, surfaces with general directional and reflection characteristics are analyzed. This section considers a particular class of reflections that was not specifically treated. Except for one special example, all of the surfaces are still assumed to emit diffusely. Some of the surfaces in an enclosure are assumed to reflect diffusely; the remaining surfaces are assumed to be specular, that is, to reflect in a mirror-like manner. From the discussion of roughness in Chapter 3, an important surface parameter for radiative properties is the optical roughness (the ratio of the root-mean-square roughness height to the radiation wavelength.) As the radiation wavelength increases, a smooth surface tends toward being optically smooth, and reflections at larger wavelengths tend to become more specular. Thus, although a surface may not appear mirror-like to the eye for the short wavelengths of the visible spectrum, it may be specular for the longer wavelengths in the infrared.

When reflection is diffuse, the directional history of the incident radiation is lost upon reflection, and the reflected energy has the same directional distribution as if it had been absorbed and diffusely re-emitted. For specular reflection, the reflection angle relative to the surface normal is equal in magnitude to the angle of incidence. Hence, the directional history of the incident radiation is *not* lost upon reflection. When dealing with specular surfaces, it is necessary to consider the specific directional paths that the reflected radiation follows between surfaces.

The specular reflectivities used in this chapter are assumed independent of incidence angle; the same fraction of the incident energy is reflected for any angle of incidence. In addition, all the surfaces are assumed gray (properties do not depend on wavelength). A spectral or nongray analysis can be done as in Section 6.2; heat flow calculations are carried out in each wavelength band that is significant in the radiative transfer, and the results are summed over all bands to obtain total energy quantities.

6.5.1 Some Situations with Simple Geometries

Radiation exchange between infinite parallel plates, concentric cylinders, and concentric spheres, as shown in Figure 6.15, is of practical importance for predicting the heat transfer performance

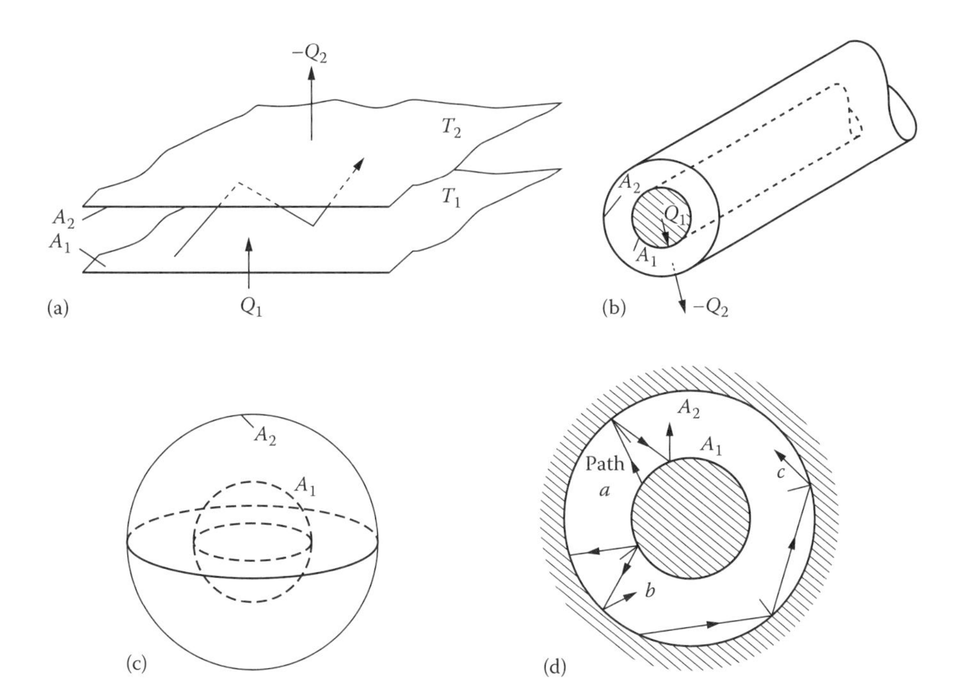

FIGURE 6.15 Radiation exchange for specular surfaces having simple geometries: (a) infinite parallel plates; (b) gap between infinitely long concentric cylinders; (c) gap between concentric spheres; and (d) paths for specular radiation in gap between concentric cylinders or spheres.

of radiation shields, Dewar vessels, and cryogenic insulation. Specular exchange for these geometries is well understood, having been discussed and analyzed by Christiansen (1883) and Saunders (1929).

Consider radiation between two infinite, gray, parallel specularly reflecting surfaces, as in Figure 6.15a. Although the reflections are specular, the emission is diffuse. All emitted and reflected radiation leaving surface 1 directly reaches surface 2; similarly, all emitted and reflected radiation leaving surface 2 directly reaches surface 1. This is true whether the surfaces are specular or diffuse reflectors. For diffuse emission the surface absorptivity is independent of direction, so there are no directional considerations. Hence, for specular reflections and diffuse emission Equation 5.6.3 applies and the net radiative heat transfer from surface 1 to surface 2 is

$$Q_1 = -Q_2 = \frac{A_1\sigma\left(T_1^4 - T_2^4\right)}{\left[1/\varepsilon_1\left(T_1\right)\right] + \left[1/\varepsilon_2\left(T_2\right)\right] - 1} \tag{6.23}$$

Now consider radiation between the concentric cylinders or spheres in Figure 6.15b and c, assuming that the emission is diffuse. Typical radiation paths for specular reflections are in Figure 6.15d. As shown by path a, all radiation emitted by surface 1 will directly reach surface 2. A portion will be reflected from surface 2 back to surface 1, and a portion of this will be re-reflected from 1. This sequence of reflections continues until insignificant energy remains because radiation is partially absorbed on each contact with a surface. From the symmetry of the concentric geometry and the equal incidence and reflection angles for specular reflections, none of the radiation following path a can ever be reflected directly from surface 2 to another location on surface 2. Thus, the exchange process for radiation emitted from surface 1 is the same as though the two concentric surfaces were infinite parallel plates. However, the radiation emitted from the outer surface 2 can travel along either of two types of paths, b or c, as shown in Figure 6.15d. Since emission is assumed diffuse, the fraction F_{2-2} will follow paths of type c (F is a diffuse configuration factor). From the geometry of specular reflections these rays will always be reflected along surface 2 with none reaching surface 1. The fraction $F_{2-1} = A_1/A_2$ is reflected back and forth between the surfaces along paths b in the same fashion as radiation emitted from surface 1. The amount of radiation following this type of path is $A_2\varepsilon_2 F_{2-1}\sigma T_2^4 = A_2\varepsilon_2(A_1/A_2)\sigma T_2^4 = A_1\varepsilon_2\sigma T_2^4$. The fraction of the radiation leaving surface 2 that impinges on surface 1 thus depends on A_1 and not on A_2. Hence, for specular surfaces the exchange behaves as if both surfaces were equal portions of infinite parallel plates with size equal to the area of the inner body. The net radiative heat transfer from surface 1 to surface 2 is thus given by Equation 6.23.

Example 6.7

A spherical vacuum bottle consists of two silvered, concentric glass spheres, the inner being 15 cm in diameter and the evacuated gap between the spheres being 0.65 cm. The emissivity of the silver coating is 0.02. If hot coffee at 368 K is in the bottle and the outside temperature is 294 K, what is the initial radiative heat leakage rate from the bottle?

Equation 6.23 applies for concentric specular spheres. For the small rate of heat leakage expected, it is assumed that the surfaces will be close to 368 and 294 K. This gives

$$Q_1 = \frac{\pi(0.15)^2\, 5.6704\times 10^{-8}\left(368^4 - 294^4\right)}{(1/0.02)+(1/0.02)-1} = 0.440 \text{ W}$$

If, instead of using the specular formulation, both surfaces are assumed diffuse reflectors with ε still 0.02, then Equation 5.8.4 applies. The denominator of the Q_1 equation becomes

$$\frac{1}{\varepsilon_1} + \frac{A_1}{A_2}\left(\frac{1}{\varepsilon_1} - 1\right) = \frac{1}{0.02} + \left(\frac{15}{16.3}\right)^2\left(\frac{1}{0.02} - 1\right) = 91.50$$

instead of 99 as in the specular case. For diffuse surfaces, the heat loss is increased to 0.476 W.

Example 6.8

For the previous example, to illustrate the insulating ability of a vacuum bottle, how long will it take for the coffee to cool from 368 to 322 K if the only heat loss is by radiation?

The heat capacity of the coffee is $\rho_M V_c T_1$. Assuming the coffee is always mixed well enough that it is at uniform temperature, the cooling rate is equal to the *instantaneous* radiation loss. The energy loss by radiation at any time t, given by Equation 6.23, is related to the change of internal energy of the coffee by

$$-\rho_M V c \frac{dT_1}{dt} = \frac{A_1\sigma\left[T_1^4(\tau) - T_2^4\right]}{1/\varepsilon_1 + 1/\varepsilon_2 - 1}$$

where it is assumed that surface 1 is at the coffee temperature and surface 2 is at the outside environment temperature. Then

$$-\int_{T_1=T_I}^{T_1=T_F} \frac{dT_1}{T_1^4 - T_2^4} = \frac{A_1\sigma}{\rho_M V c\left(1/\varepsilon_1 + 1/\varepsilon_2 - 1\right)} \int_0^{\tau} dt$$

where

T_I and T_F are the initial and final temperatures, respectively, of the coffee
ε_1 and ε_2 are assumed independent of temperature

Carrying out the integration and solving for t gives the cooling time from T_I to T_F as

$$t = \frac{\rho_M V c(1/\varepsilon_1 + 1/\varepsilon_2 - 1)}{A_1\sigma}\left[\frac{1}{4T_2^3}\ln\left|\frac{(T_F+T_2)/(T_F+T_2)}{(T_I+T_2)/(T_I+T_2)}\right| + \frac{1}{2T_2^3}\left(\tan^{-1}\frac{T_F}{T_2} - \tan^{-1}\frac{T_I}{T_2}\right)\right]$$

Substituting $\rho_M = 975$ kg/m^3, $V = \frac{1}{6}\pi(0.15)^3$ m^3, $c = 4195$ J/(kg·K), $\varepsilon_1 = \varepsilon_2 = 0.02$, $A_1 = \pi\cdot(0.15)^2$ m^2, $\sigma = 5.6704 \times 10^{-8}$ W/(m^2·K^4), $T_2 = 294$ K, $T_I = 368$ K, and $T_F = 322$ K gives $t = 374.7$ h. The coffee would require about 16 days to cool if heat losses were only by radiation. Conduction losses through the glass wall of the bottle neck usually cause the cooling to be considerably faster.

The following is a summary for exchange between two surfaces that both have diffuse emission. When both surfaces are specular reflectors, Equation 6.23 applies for infinite parallel plates, infinitely long concentric cylinders, and concentric spheres. For infinite parallel plates, Equation 6.23 also applies when both surfaces are diffuse reflectors, or when one surface is diffuse and the other is specular. For cylinders and spheres, Equation 6.23 will apply when the surface of the inner body (surface 1) is a diffuse reflector if the outer body (surface 2) is specular. This is because all radiation leaving surface 1 goes directly to surface 2 for surface 1 either specular or diffuse. When surface 2 is diffuse, Equation 5.8.4 applies and may be used when surface 1 is either specular or diffuse. The relations are summarized in Table 6.1.

The previous development was for surfaces with diffuse emission, which is a common assumption, and it will be used in the more general enclosure theory that will be developed. Before continuing with more general geometries, a special case is considered for two parallel surfaces where it is possible to conveniently examine the effect of directional emission for surfaces that are specularly reflecting.

Example 6.9

Two infinite parallel plates as shown in Figure 6.15a are specular reflectors, but have emissivities (and hence absorptivities) that depend on angle θ (but not on ϕ) such as for polished metals in Section 3.2.2.1. What is the equation for radiative exchange between the surfaces?

The radiation per unit area emitted in solid angle $d\Omega$ from surface 1 in direction θ is $\varepsilon_1(\theta)(\sigma T_1^4/\pi)\cos\theta d\Omega = 2\varepsilon_1(\theta)\sigma T_1^4\cos\theta\sin\theta d\theta$ for ϕ_1 independent of ϕ, and similarly for surface 2. For specular reflections, the directionality of the emitted energy is retained throughout the exchange process for the parallel-plate geometry. Hence, for each direction the energy is exchanged as in Equation 6.23. Then, by integrating over the exchanges within all solid angles, the energy transfer for nondiffuse emission is (all reflections are *specular*)

$$q_{1,nd} = \sigma(T_1^4 - T_2^4)2\int_0^{\pi/2}\frac{\sin\theta\cos\theta}{1/\varepsilon_1(\theta) + 1/\varepsilon_2(\theta) - 1}d\theta$$

For diffuse emission the ε_1 and ε_2 are each independent of θ, so the ratio of transfer for nondiffuse emitting surfaces to diffuse emitting surfaces is

$$\frac{q_{1,nd}}{q_{1,d}} = 2\int_0^{\pi/2}\frac{\sin\theta\cos\theta}{[1/\varepsilon_1(\theta)] + [1/\varepsilon_2(\theta)] - 1}d\theta \bigg/ \frac{1}{(1/\varepsilon_1) + (1/\varepsilon_2) - 1}$$

As a simple illustration, let $\varepsilon_1(\theta) = \varepsilon_2(\theta) = C\cos\theta$ as in Example 2.3. The $C \le 1$ is constant, and from Example 2.3 the hemispherical $\varepsilon_1 = \varepsilon_2 = 2C/3$. The integration yields

$$\frac{q_{1,nd}}{q_{1,d}} = -\left[\frac{8}{C^2}\ln\left(1 - \frac{C}{2}\right) + \frac{4}{C} + 1\right] \bigg/ \left(\frac{C}{3 - C}\right)$$

TABLE 6.1
Radiant Interchange between Some Simply Arranged Surfaces (Emission Is Diffuse)

Geometry	Configuration	Type of Surface Reflection	Energy Transfer Rate, Q_1
Infinite parallel plates	A_2, A_1	A_1 or A_2, either specular or diffuse	$\frac{A_1\sigma(T_1^4 - T_2^4)}{1/\varepsilon_1 + 1/\varepsilon_2 - 1}$
Infinitely long concentric cylinders	A_1, A_2	A_1, specular or diffuse; A_2, diffuse	$\frac{A_1\sigma(T_1^4 - T_2^4)}{1/\varepsilon_1 + (A_1/A_2)(1/\varepsilon_2 - 1)}$
		A_1 specular or diffuse; A_2, specular	$\frac{A_1\sigma(T_1^4 - T_2^4)}{1/\varepsilon_1 + 1/\varepsilon_2 - 1}$
Concentric spheres	A_1, A_2	A_1, specular or diffuse; A_2, diffuse	$\frac{A_1\sigma(T_1^4 - T_2^4)}{1/\varepsilon_1 + (A_1/A_2)(1/\varepsilon_2 - 1)}$
		A_1, specular or diffuse; A_2, specular	$\frac{A_1\sigma(T_1^4 - T_2^4)}{1/\varepsilon_1 + 1/\varepsilon_2 - 1}$

For $C = 0.5$, 0.8, 1.0 this gives, respectively, $q_{1,nd}/q_{1,d} = 1.029$, 1.060, 1.090. The specified $\varepsilon(\theta)$ can increase the heat exchange up to 9%; the effect is smaller when C, the maximum $\varepsilon(\theta)$, is small.

6.5.2 Ray Tracing and the Construction of Images

When there is one or more specular surfaces in an enclosure so that mirror-like reflections occur, the procedures of geometric optics can be applied to the radiative exchange process. The basic ideas are outlined here; more advanced ideas are in Born et al. (1999) and Stone (1963).

An incident ray striking a specular surface is reflected symmetrically about the surface normal, with the angle of reflection equal in magnitude to the angle of incidence. This is used to formulate the concept of *images*. An image is an apparent point of origin for an observed ray. In Figure 6.16a, an observer views an object in a mirror. The object appears to be an image behind the mirror in the position shown by the dotted object. This concept is readily extended to cases in which a series of reflections occur, as in Figure 6.16b. An interesting example of this is the "barber-chair" geometry, where there are planar mirrors on opposite walls of a barber shop. If the mirrors are parallel, a large number of reflections occur and a person receiving a haircut can view many images of himself or herself.

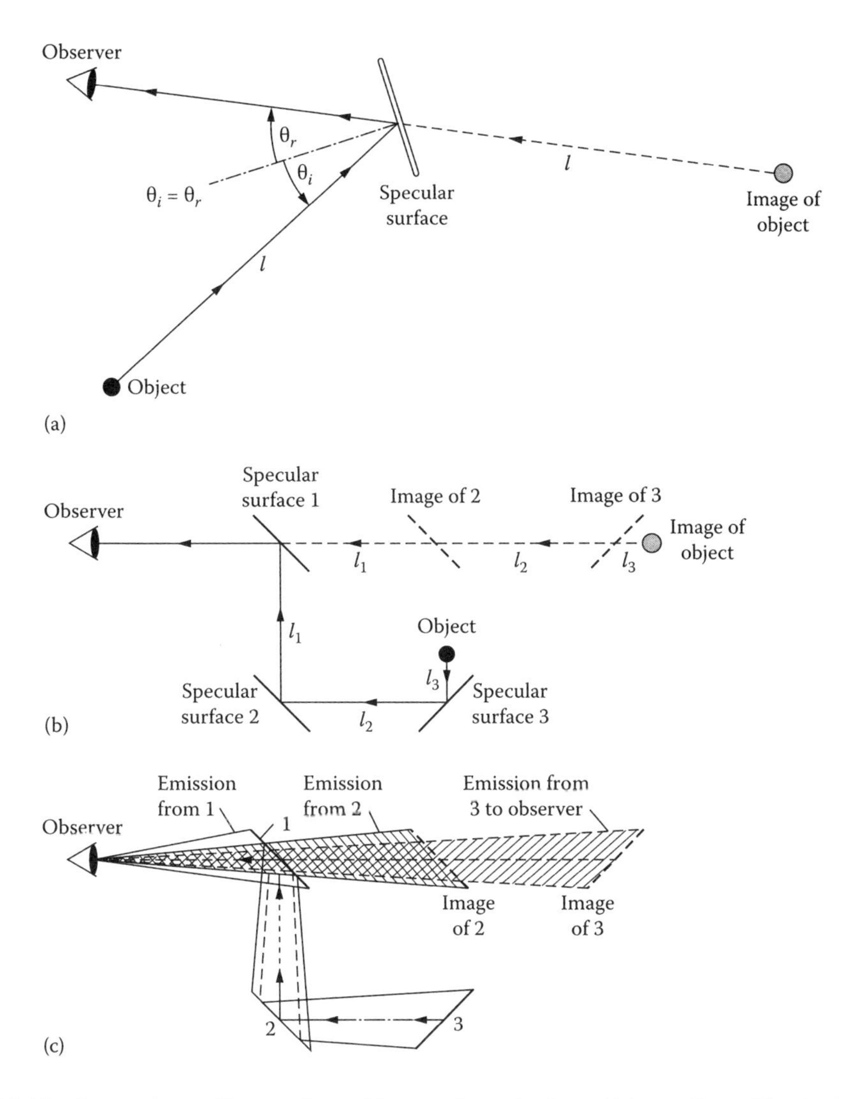

FIGURE 6.16 Ray tracing and images formed by specular reflections: (a) image formed by single reflection; (b) image formed by multiple reflections; and (c) contributions due to emission from specular surfaces.

To this point, mirrors have been discussed only with regard to their ability to change the direction of the rays. For thermal-radiation analysis, the specular surfaces have a finite absorptivity and will thus attenuate the energy of the rays for each reflection. A mirror surface will also emit energy. Emission can be conveniently analyzed using an image system because in the image system all radiation will act along straight lines without the complexity of directional changes at each reflecting surface. The attenuation at each surface is included by multiplying the ray intensity by the specular reflectivity at each reflection. The emission from three surfaces is illustrated in Figure 6.16c. For example, emitted energy reaching the viewer from surface 3 may be considered to be coming directly from the image of 3, with attenuation due to reflections at surfaces 2 and 1.

6.5.3 Radiative Transfer by Means of Simple Specular Surfaces for Diffuse Energy Leaving a Surface

As an introduction to radiative analysis in an enclosure with some specularly reflecting surfaces, a few examples are considered for plane surfaces. The emission from all surfaces is assumed diffuse.

This is a reasonable simplifying assumption in many cases, as indicated by the electromagnetic-theory predictions of the emissivity of specular surfaces in Figures 3.2 and 3.5. The reflected energy is assumed either diffuse (for the nonmirror-like surfaces) or specular. In the enclosure theory that is developed, the diffuse reflected energy is combined with the diffuse emitted energy, and it is then necessary to know how the transfer of this diffuse energy is influenced by the presence of specularly reflecting surfaces. The exchange factors F^s that will be obtained by considering the presence of the specular surfaces will already account for the specularly reflected energy, which has a specific directional behavior in contrast to the diffuse energy. Hence, it is necessary to consider here only the transfer of diffuse energy leaving a surface in the presence of specular surfaces.

Figure 6.17a shows a diffusely emitting and reflecting plane surface A_1 facing a specularly reflecting plane surface A_2. Surface 1 cannot directly view itself; the ordinary configuration factor from any part of A_1 to any other part of A_1 is thus zero. However, with A_2 specular, A_1 can view its image, and a path exists by means of specular reflection for diffuse radiation to travel from the differential area dA_1 to dA_1^*. By ray tracing, the radiation arriving at dA_1^* from dA_1 appears to come from the image $dA_{1(2)}$. Thus, the configuration factor between dA_1 and dA_1^* resulting from one reflection can be obtained as $dF_{d1(2)-d1*}$ for diffuse radiation leaving dA_1. The subscript notation refers to dF from the image of dA_1 (as seen in A_2) to dA_1^*.

There are points of similarity that should be noted when comparing specular and diffuse reflecting cases. When A_1 and A_2 in Figure 6.17a are both diffuse reflectors, the diffuse radiation leaving dA_1 is received at dA_1^* by means of diffuse reflection from A_2 and is governed by dF_{d1-d2} and then dF_{d2-d1} for each element dA_2 on A_2. The portion of the diffuse energy from dA_1 that reaches dA_1^* after one reflection from A_2 is $dA_1 J_1 \rho_2 \int_{A2} dF_{d1-d2} F_{d2-d1*}$ for a diffuse A_2, and is $dA_1 J_1 \rho_{s,2} dF_{d1(2)-d1*}$ for a specular A_2.

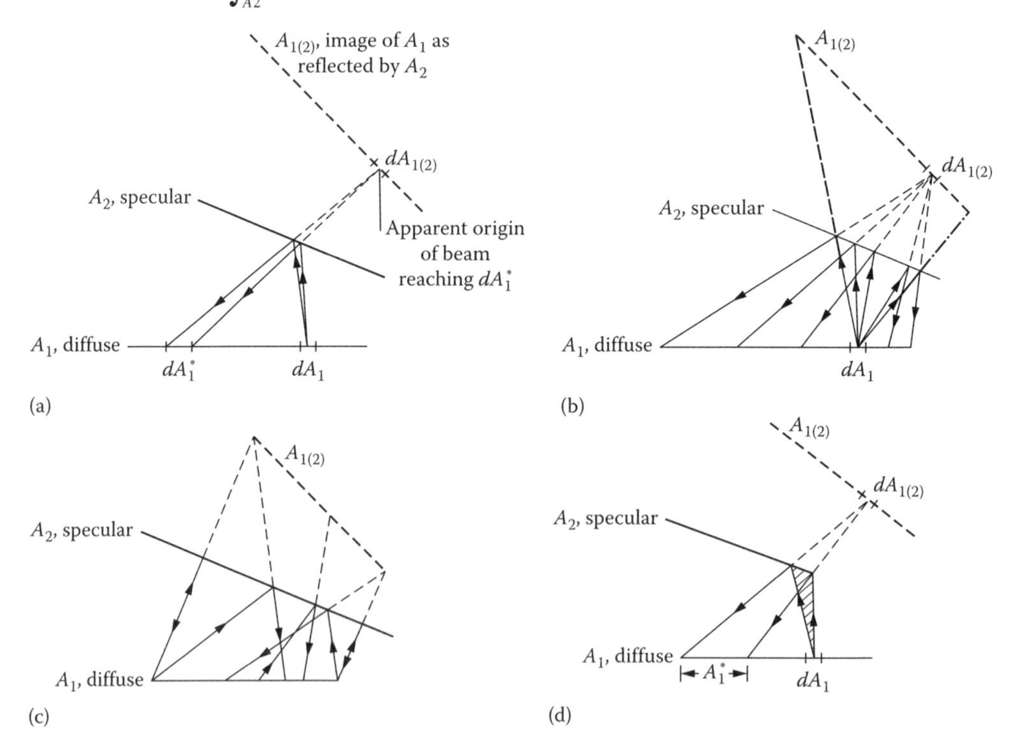

FIGURE 6.17 Radiation between a diffuse surface and itself by means of a specular surface: (a) radiation between two differential areas with one intermediate specular reflection; (b) radiation from differential area to finite area by means of one intermediate specular reflection; (c) radiation from finite area that is reflected back to that area by means of one specular reflection; and (d) radiation from dA_1 can reach only a portion of A_1 by means of specular reflection from A_2.

This reveals that, for $\rho_2 = \rho_{s,2}$, the difference in the two exchanges is incorporated in the configuration factors resulting from the nature of the reflection; this is a purely geometric effect.

Figure 6.17b describes the diffuse radiation from dA_1 that reaches the entire area A_1 by means of one specular reflection. The reflected radiation appears to originate from the image $dA_{1(2)}$. Thus, the geometric configuration factor involved from dA_1 to A_1 is $F_{d1(2)-1}$. From the symmetry revealed by the dot-dash lines in Figure 6.17b, $F_{d1(2)-1} = F_{d1-1(2)}$. Thus, the radiative transfer from dA_1 to A_1 can also be expressed using a configuration factor from the first surface to the image of the second surface.

Figure 6.17c shows typical rays leaving A_1 that are reflected back to A_1; the rays appear to originate from the image $A_{1(2)}$. The configuration factor from A_1 back to itself by means of one specular reflection is then $F_{1(2)-1}$. For the geometry of Figure 6.17c, all of the image $A_{1(2)}$ is visible in A_2 from any position on A_1. In some instances, however, this is not true. For the displaced A_2 in Figure 6.17d the radiation from dA_1 must be within the limited solid angle shown shaded for the radiation to be reflected back to A_1. The configuration factor between dA_1 and A_1 is still $F_{d1(2)-1}$, but this factor is evaluated only over the portion of A_1 that receives reflected rays. The $F_{d1(2)-1}$ is the factor by which $dA_{1(2)}$ views A_1 and the view may be a partial one.

The factor from dA_1 to A_1 has a different value as the location of dA_1 changes along A_1. This means that the energy from A_1 that is reflected back to A_1 will have a nonuniform distribution along A_1. The reflected part of this energy from A_1 will provide a nonuniform J from A_1; this violates the assumption in the enclosure theory of uniform J from each surface. When partial images are present, the area is subdivided into sufficiently small portions that the solution accuracy is adequate.

Now consider the geometry when there are additional specular surfaces involved in the radiation exchange; this provides multiple specular reflections. At each reflection, the radiation is modified by the ρ_s of the reflecting surface. Two specular surfaces are shown in Figure 6.18. Energy is emitted diffusely from A_2 and travels to A_1. The fraction arriving at dA_1 is given by the diffuse factor dF_{2-d1}. This direct path is illustrated in Figure 6.18a. A portion of the energy intercepted by A_1 is reflected back to A_2 and then reflected back to A_1. Hence, A_2 views dA_1 not only directly but also by means of an image formed by two reflections as constructed in Figure 6.18b. First the image $A_{1(2)}$ of A_1 reflected in A_2 is drawn. Then, A_2 is reflected into this image to form $A_{2(1-2)}$. The notation $A_{2(1-2)}$ is read as the image of area 2 formed by reflections in area 1 and area 2 (in that order). The radiation paths and the shaded area in Figure 6.18b reveal that the solid angle within which radiation leaving A_2 reaches dA_1 by means of two reflections is the same as the solid angle by which dA_1 views the image $A_{2(1-2)}$. Thus, the configuration factor involved for two reflections is $dF_{2(1-2)-d1}$. This is interpreted as the factor from the image of surface 2 formed by reflections in surfaces 1 and 2 (in that order) to area element d_1.

Consider the possibility of additional images. The geometric factor involved is always found by viewing dA_1 from the appropriate reflected image of A_2 as seen through A_2 and all intermediate images. In the case of Figure 6.18c, the images of A_2 after four reflections $A_{2(1-2-1-2)}$ cannot view dA_1 by looking through A_2. Hence, there is no radiation leaving A_2 that reaches dA_1 by means of four reflections, and no additional images need be considered.

Example 6.10

The infinitely long (normal to the cross section) groove in Figure 6.19 has specularly reflecting sides that emit diffusely. What fraction of the energy emitted from A_2 reaches the black receiver surface element dA_3 (an infinitely long differential strip)? Express the result in terms of diffuse configuration factors.

Consider first the energy reaching dA_3 directly from A_2 and by means of an even number of reflections. The fraction of emitted radiation that reaches dA_3 directly from A_2 is dF_{2-d3}, as illustrated in Figure 6.19a. A second portion is that emitted from A_2 to A_1, reflected back to A_2, and then reflected to dA_3. From the image diagram in Figure 6.19b, only part of the reflected image $A_{2(1-2)}$ can be viewed by dA_3 through A_2. The fraction of emitted energy reaching dA_3 by this path is the configuration factor evaluated only over the part of $A_{2(1-2)}$ visible to dA_3 multiplied by the two specular reflectivities, $\rho_{s,1}\rho_{s,2}dF_{2(1-2)-d3}$. This is not an ordinary configuration factor but takes into

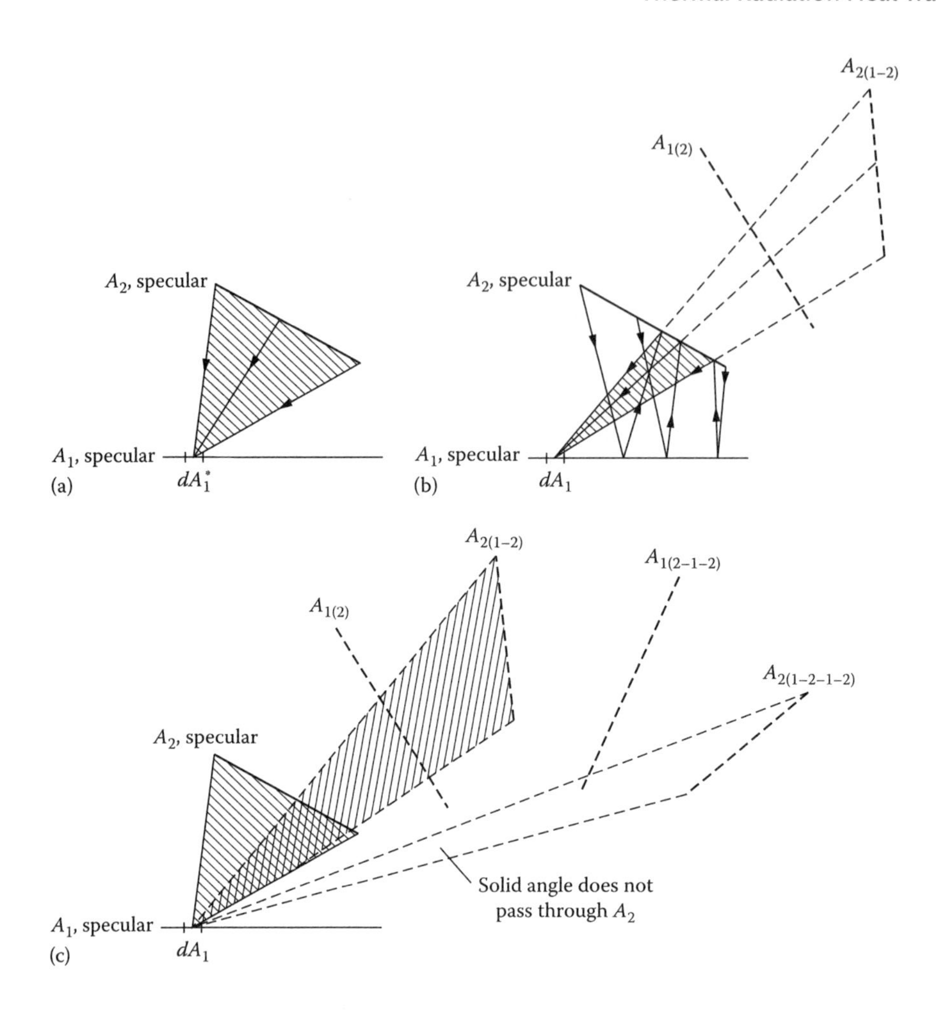

FIGURE 6.18 Radiant interchange between two specularly reflecting surfaces. (a) Energy emitted from A_2 that directly reaches dA_1; (b) energy emitted from A_2 that reaches dA_1 after two reflections; and (c) none of the energy emitted by A_2 reaches dA_1 by means of four reflections.

account the view through the image system. Similarly, there is a contribution after two reflections from each of A_1 and A_2. This is illustrated by the shaded solid angle in Figure 6.19c. The third image of A_2, $A_{2(1-2-1-2-1-2)}$, cannot be viewed by dA_3 through A_2 and hence does not make a contribution. Also, the third image of A_2 cannot view A_1 through A_2 so there are no additional images of A_2. For the energy emitted by A_2, the sum of the fractions that reach dA_3 both directly and by means of the images of A_2 resulting from an even number of reflections is then

$$dF_{2-d3} + \rho_{s,1}\rho_{s,2}dF_{2(1-2)-d3} + \rho_{s,1}^2\rho_{s,2}^2 dF_{2(1-2-1-2)-d3}.$$

Now consider the energy that reaches dA_3 from A_2 by means of an odd number of reflections. Using Figure 6.19d and arguments similar to those for an even number of reflections results in

$$\rho_{s,1}dF_{2(1)-d3} + \rho_{s,1}^2\rho_{s,2}^2 dF_{2(1-2-1)-d3} + \rho_{s,1}^3\rho_{s,2}^2 dF_{2(1-2-1-2-1)-d3}.$$

The first two F factors are evaluated over only the portions of the images of A_2 that can be viewed by dA_3.

The diffuse energy emitted by surface A_2 that reaches dA_3 directly and by all interreflections from both A_1 and A_2 is then governed by the summation

$$dF_{2-d3} + \rho_{s,1}dF_{2(1)-d3} + \rho_{s,1}\rho_{s,2}dF_{2(1-2)-d3} + \rho_{s,1}^2\rho_{s,2}dF_{2(1-2-1)-d3}$$
$$+ \rho_{s,1}^2\rho_{s,2}^2 dF_{2(1-2-1-2)-d3} + \rho_{s,1}^3\rho_{s,2}^2 dF_{2(1-2-1-2-1)-d3}.$$

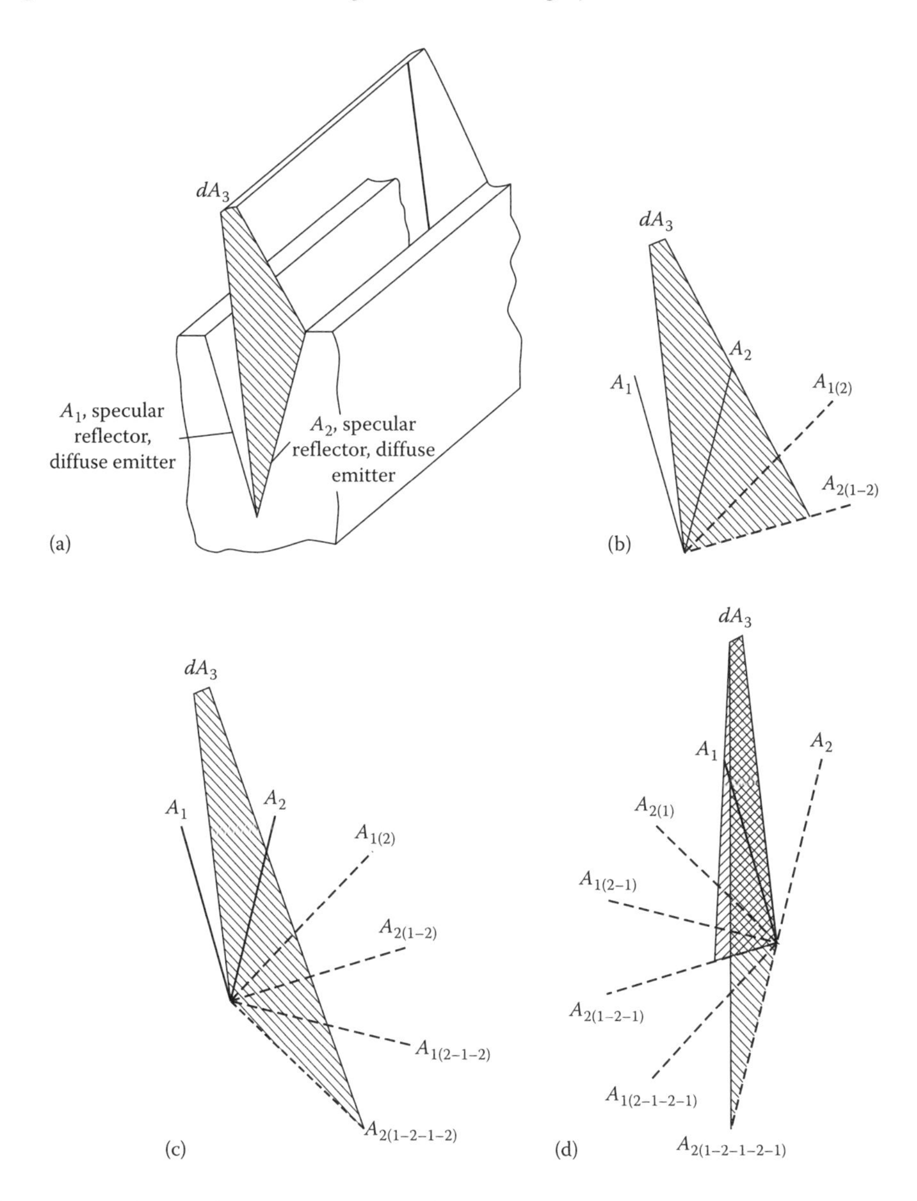

FIGURE 6.19 Diffuse emission from one side of a specularly reflecting groove that reaches a differential-strip receiving area outside the groove opening: (a) geometry of direct exchange from A_2 to dA_3; (b) geometry of exchange for radiation from A_2 that reaches dA_3 by means of one intermediate reflection from each of A_1 and A_2; (c) geometry of exchange for radiation from A_2 that reaches dA_3 by means of two intermediate reflections from each of A_1 and A_2; and (d) geometry of exchange for radiation from A_2 to A_3 by means of an odd number of reflections.

The notation adopted for the specular configuration factors allows a check on the form of the equations for radiant interchange among specular surfaces. The numbers within the subscripted parentheses designate the sequence of reflections from the specular surfaces. The configuration factor is multiplied by a reflectivity for each of these specular surfaces to account for energy absorption at the surfaces. For example, the factor $F_{A(B-C-D)-E}$ is multiplied by $\rho_{s,B}\rho_{s,C}\rho_{s,D}$.

Additional information on the absorption and emission of radiation by specular grooves is in Howell and Perlmutter (1963). Grooves may be useful in the collection and concentration of solar energy. For this purpose an absorbing surface is placed at the bottom of each groove, which can

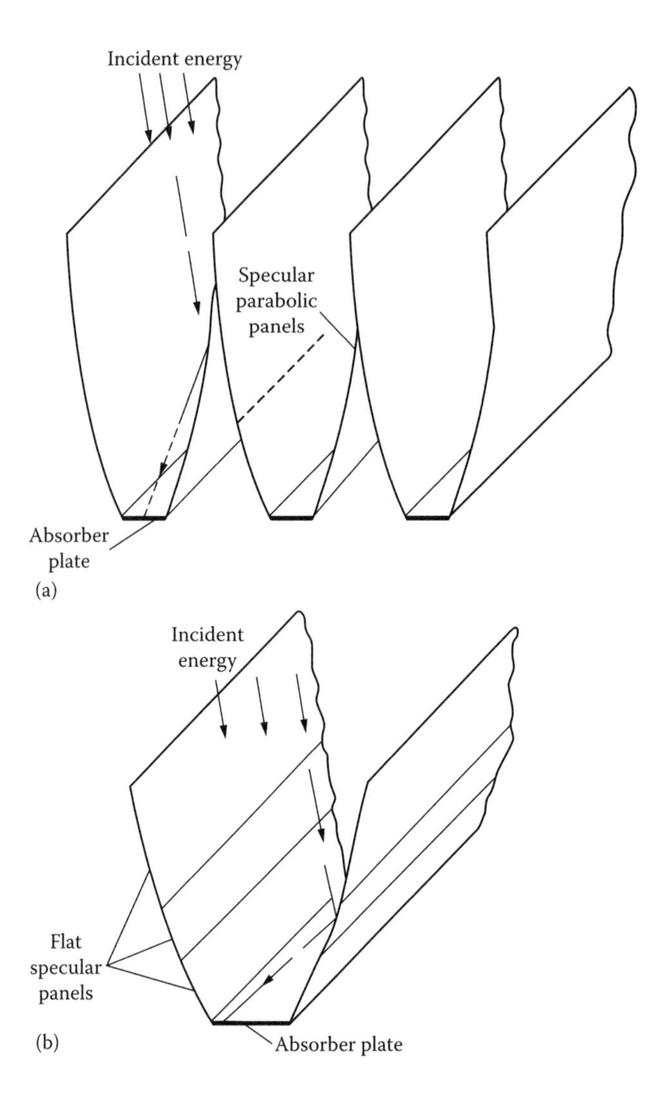

FIGURE 6.20 Concentrating solar collectors: (a) parabolic trough concentrator; (b) concentrator made with flat segments.

have sides in the form of parabolic segments (Winston 1974) as in Figure 6.20a or a series of straight sections (Mannan and Cheema 1977) as in Figure 6.20b. These configurations provide concentration of the incident energy onto the absorber plane, which provides elevated temperatures needed for efficient energy conversion. The groove shapes give good concentration even when incident radiation is not well aligned normal to the absorber plate. This type of solar collector will perform well as the angle of the sun changes throughout the day, even though the collector remains in a fixed position.

6.5.4 Configuration-Factor Reciprocity for Specular Surfaces; Specular Exchange Factors

Reciprocity relations analogous to those for configuration factors between diffuse surfaces apply for the factors involving specular surfaces under certain conditions. Consider a three-sided isothermal enclosure at temperature T, made up of two black surfaces 1 and 2 and a specular surface 3 of

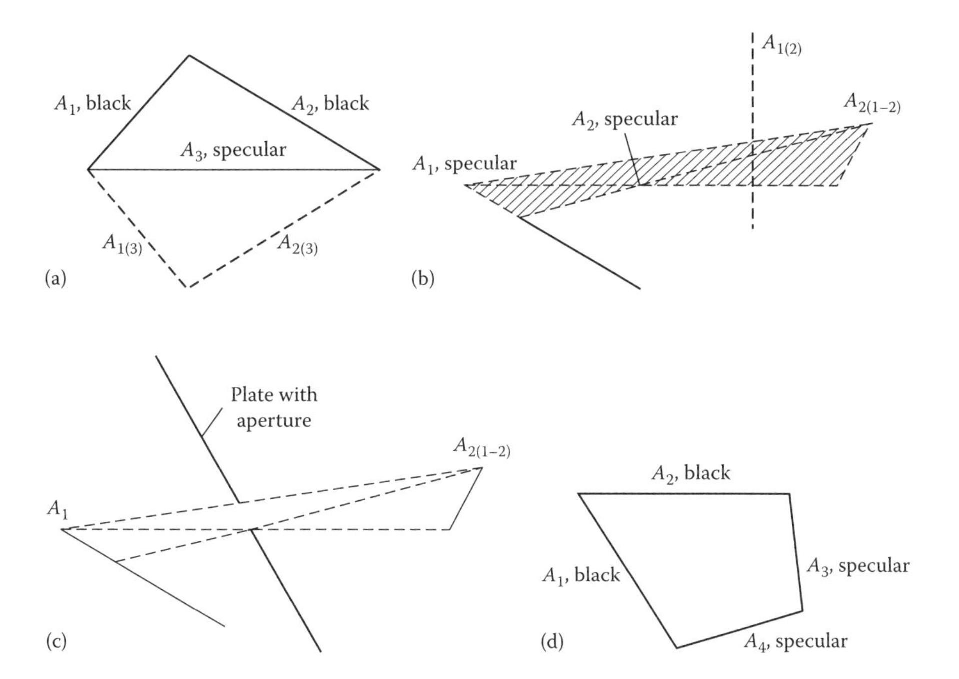

FIGURE 6.21 Reciprocity of configuration factors involving specular surfaces: (a) three-sided enclosure with one specular reflecting surface; (b) system of images of surfaces 1 and 2; (c) energy-exchange analog of image system in (b); and (d) enclosure with two specular and two black surfaces.

reflectivity $\rho_{s,3}$ (Figure 6.21a). The energy emitted by black surface 1 that reaches black surface 2 directly and by reflection from specular surface 3 is

$$Q_{1-2} = \sigma T^4 \left(A_1 F_{1-2} + A_1 \rho_{s,3} F_{1(3)-2} \right) \tag{6.24}$$

The energy leaving surface 2 that reaches surface 1 directly and by specular reflection from surface 3 is

$$Q_{2-1} = \sigma T^4 \left(A_2 F_{2-1} + A_2 \rho_{s,3} F_{1(3)-1} \right) \tag{6.25}$$

Since $A_1 F_{1-2} = A_2 F_{2-1}$ and, for the isothermal enclosure, $Q_{2-1} = Q_{1-2}$, the reciprocity relation is obtained for one specular surface in the enclosure:

$$A_1 F_{1(3)-2} = A_2 F_{2(3)-1} \tag{6.26}$$

A second type of reciprocity relation exists for configuration factors involving specular surfaces. Consider the energy exchange between two surfaces A_1 and A_2 of an isothermal enclosure. If both surfaces are specular, the images in Figure 6.21b can be constructed for radiation from surface 2 to 1 by means of a reflection at surface 1 and at surface 2. An analogous system can be constructed in which a plate with an aperture is substituted for the restraints on the ray paths that are present, as in Figure 6.21c. The aperture allows passage of only those rays that pass through the image system by which $A_{2(1-2)}$ can view at least a portion of A_1 through A_2 and $A_{1(2)}$.

The emitted energy leaving specular surface A_2 in the analog system and absorbed by A_1 is

$$Q_{2(1-2)-1} = Q_{e,2} \rho_{s,1} \rho_{s,2} F_{2(1-2)-1} \alpha_1 = A_{2(1-2)} \varepsilon_2 \sigma T^4 \rho_{s,1} \rho_{s,2} F_{2(1-2)-1} \varepsilon_1 \tag{6.27}$$

for the assumed gray surfaces where $\alpha_1 = \varepsilon_1$, and there are two intermediate specular reflections. The $F_{2(1-2)-1}$ is the diffuse-surface configuration factor for the paths through the aperture (see Example 6.11). These paths are exactly those through the image system, so this is also the specular configuration factor. Similarly, the energy absorbed for the reverse path is

$$Q_{1-2(1-2)} = A_1\varepsilon_1\sigma T^4\rho_{s,2}\rho_{s,1}F_{1-2(1-2)}\varepsilon_2 \tag{6.28}$$

Equating the energy exchanges in either direction between A_1 and $A_{2(1-2)}$ for the isothermal enclosure gives the reciprocity relation

$$A_1F_{1-2(1-2)} = A_{2(1-2)}F_{2(1-2)-1} = A_2F_{2(1-2)-1} \tag{6.29}$$

For nongray surfaces this still applies, as shown by considering the energy in each spectral region. For many intermediate reflections from surfaces A, B, C, D, …, Equation 6.29 can be generalized to

$$A_1F_{1-2(A-B-C-D\ldots)} = A_2F_{2(A-B-C-D\ldots)-1} \tag{6.30}$$

For two-dimensional (2D) areas, the crossed-string method (Section 4.3.3) can be used to obtain the configuration factors. In Figure 6.21b the A_2 and $A_{1(2)}$ are regarded as apertures in the view between A_1 and $A_{2(1-2)}$, the $F_{1-2(1-2)}$ is found by having the crossed and uncrossed strings pass through these apertures.

Example 6.11

A black surface A_1 faces a smaller parallel mirror A_2 as in Figure 6.22. Compute the configuration factor $F_{1-1(2)}$ between A_1 and the image of A_1 formed by one specular reflection in A_2. The surfaces are infinitely long in the direction normal to the plane of the drawing.

The factor is computed from the integral $F_{1-1(2)} = (1/A_1)\int_{A1} F_{d1-1(2)}dA_1$. Consider the element dA_1 at location x on A_1. The factor for radiation from dA_1 to the portion of $A_{1(2)}$ in view through A_2 is (see Example 4.2)

$$F_{d1-1(2)} = \frac{1}{2}(\sin\beta' - \sin\beta'') = \frac{1}{2}\left[\frac{x+a}{\sqrt{(x+a)^2+b^2}} - \frac{x-a}{\sqrt{(x-a)^2+b^2}}\right]$$

This is valid until position $x = l - 2a$ is reached (Figure 6.22b). For larger x, the geometry is as shown in Figure 6.22c. Then

$$F_{d1-1(2)} = \frac{1}{2}(\sin\beta' - \sin\beta'') = \frac{1}{2}\left[\frac{x+l}{\sqrt{(x+l)^2+4b^2}} - \frac{x-a}{\sqrt{(x-a)^2+b^2}}\right]$$

The desired configuration factor is obtained by integrating to give

$$F_{1-1(2)} = \frac{1}{2l}2\int_0^l F_{d1-1(2)}dx = \frac{1}{l}\left\{\frac{1}{2}\int_0^{l-2a}\left[\frac{x+a}{\sqrt{(x+a)^2+b^2}} - \frac{x-a}{\sqrt{(x-a)^2+b^2}}\right]dx\right.$$

$$\left.+\frac{1}{2}\int_{l-2a}^{l}\left[\frac{x+l}{\sqrt{(x+l)^2+4b^2}} - \frac{x-a}{\sqrt{(x-a)^2+b^2}}\right]dx\right\}$$

$$= \sqrt{1+\left(\frac{b}{l}\right)^2} - \sqrt{\left(1-\frac{a}{l}\right)^2+\left(\frac{b}{l}\right)^2}$$

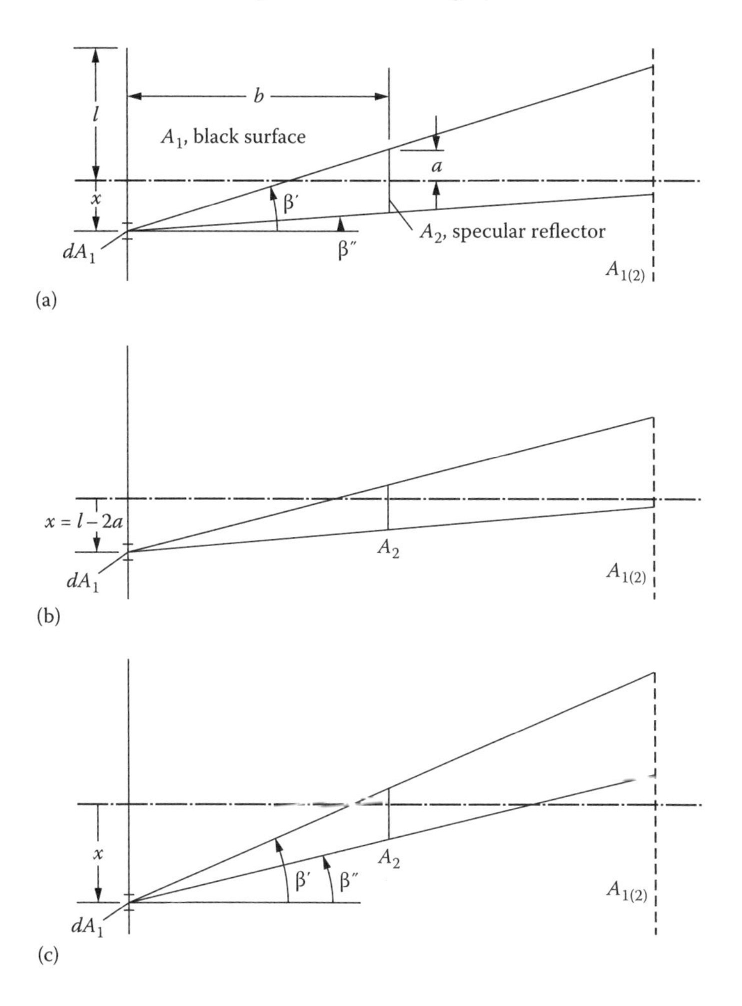

FIGURE 6.22 Configuration-factor computation involving partial views of surface and image (Example 6.11). (a) Portion of $A_{1(2)}$ in view from dA_1 through entire A_2; (b) limiting x for portion of $A_{1(2)}$ to be in view through entire A_2; and (c) portion of $A_{1(2)}$ in view through part of A_2.

This result can be found more easily by the crossed-string method. The surface views its image through the aperture of A_2. Using the crossed-string method (Section 4.3.3.1), the crossed and uncrossed strings are passed through the aperture (Figure 6.23). Then

$$2lF_{1-1(2)} = \frac{1}{2}\left(\sum \text{crossed} - \sum \text{uncrossed}\right)$$

$$F_{1-1(2)} = \frac{1}{4l}\left[4\sqrt{l^2 + b^2} - 4\sqrt{(l-a)^2 + b^2}\right]$$

Reciprocity is now considered when there is more than one specular surface in an isothermal enclosure at temperature T. For simplicity, an enclosure such as Figure 6.21d is used, where there are two specular and two black surfaces. The heat exchange between the two black surfaces by direct exchange and by all specular reflection paths is

$$\frac{Q_{1-2}}{\sigma T^4} = A_1\left(F_{1-2} + \rho_{s,3}F_{1(3)-2} + \rho_{s,4}F_{1(4)-2} + \rho_{s,3}\rho_{s,4}F_{1(3-4)-2} + \cdots + \rho_{s,3}^m\rho_{s,4}^m F_{1(3^m-4^n)-2} + \cdots\right) \quad (6.31)$$

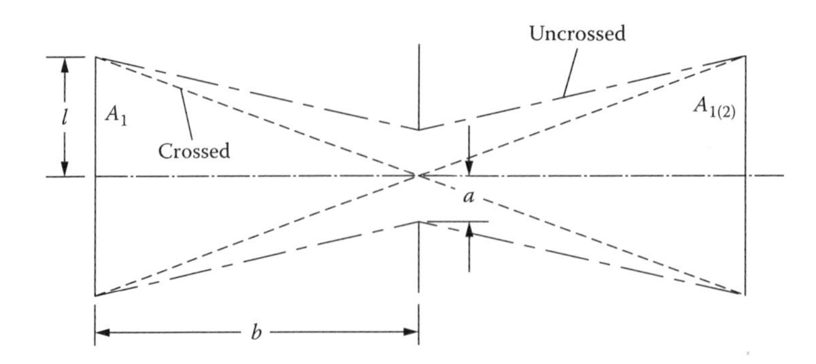

FIGURE 6.23 Geometry for the crossed-string method.

$$\frac{Q_{2-1}}{\sigma T^4} = A_2\left(F_{2-1} + \rho_{s,3}F_{2(3)-1} + \rho_{s,4}F_{2(4)-1} + \rho_{s,3}\rho_{s,4}F_{2(4-3)-1} + \cdots + \rho_{s,3}^m\rho_{s,4}^m F_{2(4^n-3^m)-1} + \cdots\right) \tag{6.32}$$

The shorthand notation ($3^m - 4^n$) means m reflections in 3 and n in 4; hence $F_{1(3m-4n)-2}$ is the configuration factor to area 2 from the image of 1 formed by these m and n reflections. For an isothermal enclosure $Q_{1-2} = Q_{2-1}$, so Equations 6.31 and 6.32 can be written as

$$\frac{Q_{1-2}}{\sigma T^4} = \frac{Q_{2-1}}{\sigma T^4} = A_1 F_{1-2}^s = A_2 F_{2-1}^s \tag{6.33}$$

where the F^s are *specular exchange factors* equal to the quantities in parentheses in Equations 6.31 and 6.32.

$$F_{1-2}^s = F_{1-2} + \rho_{s,3}F_{1(3)-2} + \rho_{s,4}F_{1(4)-2} + \rho_{s,3}\rho_{s,4}F_{1(3-4)-2} + \cdots + \rho_{s,3}^m\rho_{s,4}^n F_{1(3^m-4^n)-2} + \cdots \tag{6.34}$$

The exchange factor expresses how diffuse energy leaving a surface arrives at a second surface directly and by all possible intermediate specular reflections. In contrast to a configuration factor that cannot exceed 1, a specular exchange factor can be larger than 1. Equation 6.34 gives F_{1-2}^s in terms of energy going from images of 1 to surface 2. An alternative form is in terms of radiation from surface 1 directly to surface 2 and by means of reflections to the images of 2; from Equation 6.33,

$$F_{1-2}^s = \frac{A_2}{A_1}F_{2-1}^s = \frac{A_2}{A_1}\left(F_{2-1} + \rho_{s,3}F_{2(3)-1} + \rho_{s,4}F_{2(4)-1} + \rho_{s,3}\rho_{s,4}F_{2(4-3)-1} + \cdots\right)$$

Equation 6.30 is applied to each F factor in the series. The area ratio cancels and the desired result is

$$F_{1-2}^s = F_{1-2} + \rho_{s,3}F_{1-2(3)} + \rho_{s,4}F_{1-2(4)} + \rho_{s,3}\rho_{s,4}F_{1-2(4-3)} + \cdots \tag{6.35}$$

Now, looking at Equations 6.31 and 6.32 in more detail, since $A_1F_{1-2} = A_2F_{2-1}$, and from Equation 6.26 for one reflection $A_1F_{1(3)-2} = A_2F_{2(3)-1}$ and $A_1F_{1(4)-2} = A_2F_{2(4)-1}$, the equality in Equation 6.33 reduces to (after dividing by $\rho_{s,3}\rho_{s,4}$)

$$A_1\left(F_{1(3-4)-2} + \cdots + \rho_{s,3}^{m-1}\rho_{s,4}^{n-1}F_{1(3^m-4^n)-2} + \cdots\right) = A_2\left(F_{2(4-3)-1} + \cdots + \rho_{s,3}^{m-1}\rho_{s,4}^{n-1}F_{2(4^n-3^m)-1} + \cdots\right) \tag{6.36}$$

This equality must hold in the limit as $\rho_{s,3}$ and $\rho_{s,4}$ approach zero so that

$$A_1F_{1(3-4)-2} = A_2F_{2(4-3)-1} \tag{6.37}$$

which is a geometric property of the system. A continuation of this reasoning leads to the general reciprocity relation:

$$A_1F_{1(A-B-C-D\cdots)-2} = A_2F_{2(\cdots D-C-B-A)-1} \tag{6.38}$$

An additional relation is found by combining Equations 6.30 and 6.38 to give

$$A_1F_{1(A-B-C-D\cdots)-2} = A_2F_{2(\cdots D-C-B-A)-1} = A_1F_{1-2(\cdots D-C-B-A)} \tag{6.39}$$

which shows that

$$F_{1(A-B-C-D\cdots)-2} = F_{1-2(\cdots D-C-B-A)} \tag{6.40}$$

This can also be deduced directly from the fact that an image system can be constructed either starting with the real surface 1 and working toward image 2(... $D - C - B - A$), or starting with image 1($A - B - C - D$ ...) and working toward real surface 2; the geometry of the construction is the same in both systems, so the configuration factors between the initial and final surfaces are the same.

A specular exchange factor F^s can be larger than 1 as the infinite number of reflections between two parallel mirrors of infinite extent takes place. This can be shown by considering two mirrors, designated as 1 and 2. For infinite parallel areas, the configuration factor between the mirrors and any of its images is 1, so that the exchange factor is given by

$$F_{1-2}^s = 1 + \rho_{s,2}\rho_{s,1} + (\rho_{s,2}\rho_{s,1})^2 + (\rho_{s,2}\rho_{s,1})^3 + (\rho_{s,2}\rho_{s,1})^4 + \cdots = \frac{1}{1-\rho_{s,1}\rho_{s,2}}$$

which is larger than 1. This does not violate energy conservation.

6.6 NET-RADIATION METHOD IN ENCLOSURES HAVING BOTH SPECULAR AND DIFFUSE REFLECTING SURFACES

6.6.1 Enclosures with Planar Surfaces

In this section, radiation exchange theory is developed for an N-surfaced enclosure, where d surfaces are diffuse reflectors and $N - d$ are specular. All the surfaces emit diffusely and are gray. Let the diffusely reflecting surfaces be 1 through d and the specularly reflecting surfaces be $d + 1$ through N. If there are no external sources of unidirectional energy entering through a window and striking a specular surface, all the energy within the enclosure originates by surface emission that is diffuse. At each diffuse reflecting surface, the reflected diffuse energy is combined with the emitted energy to form J, which is all diffuse. At each specular surface the only diffuse energy leaving is $\varepsilon\sigma T^4$, since reflected energy is directional and cannot be combined with the diffuse emission to give J. The transport between surfaces of the diffuse energies J and $\varepsilon\sigma T^4$ is determined by the specular exchange factors F^s. The $F_{A\text{-}B}^s$ expresses how the diffuse energy leaving surface A arrives at surface B by the direct path and by all possible paths involving specular reflections. The reflected energy from the specular surfaces is already included in the F^s and does not have to be considered after the F^s have been obtained. Hence, by accounting for all paths for J from the diffusely reflecting surfaces and for $\varepsilon\sigma T^4$ from the specularly reflecting surfaces to reach a particular surface, all of the incident energy has been accounted for, including the effects of both diffuse and specular reflections. Then, at any surface, the incident energy is

$$G_k A_k = \sum_{j=1}^{d} J_j A_j F_{j-k}^s + \sigma \sum_{j=d+1}^{N} \varepsilon_j T_j^4 A_j F_{j-k}^s \quad 1 \le k \le N$$

After applying reciprocity (Equation 6.33), the areas cancel and G_k becomes

$$G_k = \sum_{j=1}^{d} J_j F_{k-j}^s + \sigma \sum_{j=d+1}^{N} \varepsilon_j T_j^4 F_{k-j}^s \quad 1 \le k \le N \tag{6.41}$$

A set of enclosure relations is now derived in terms of G and J. For the diffuse reflecting surfaces it is convenient to start with the energy-balance Equations 5.10 and 5.11:

$$Q_k = q_k A_k = \left(J_k - G_k\right) A_k \quad 1 \le k \le d \tag{6.42}$$

$$J_k = \varepsilon_k \sigma T_k^4 + \left(1 - \varepsilon_k\right) G_k \quad 1 \le k \le d \tag{6.43}$$

For the specular reflecting surfaces, although these same energy balances apply, the J_k is eliminated as it consists of diffuse emission $\varepsilon_k \sigma T_k^4$, and specular reflection $(1 - \varepsilon_k)G_k$ and hence has an inconvenient directional character. Eliminating J_k from Equations 6.42 and 6.43 gives

$$Q_k = A_k \varepsilon_k \left(\sigma T_k^4 - G_k\right) \quad d + 1 \le k \le N \tag{6.44}$$

Equations 6.41 through 6.44 can be combined in various ways to obtain equations for the desired unknowns, depending on which quantities are specified. Consider when the temperatures are specified for all the surfaces and it is desired to obtain the net external energy Q_k added to each surface. Equation 6.41 is substituted into Equation 6.43 to eliminate G_k and obtain the following equation for each diffuse surface:

$$J_k - \left(1 - \varepsilon_k\right) \sum_{j=1}^{d} J_j F_{k-j}^s = \varepsilon_k \sigma T_k^4 + \left(1 - \varepsilon_k\right) \sigma \sum_{j=d+1}^{N} \varepsilon_j T_j^4 F_{k-j}^s \quad 1 \le k \le d \tag{6.45}$$

This set of equations is solved for the J for the diffuse reflecting surfaces. This is somewhat simpler than for an enclosure having all diffuse surfaces, as there are only d simultaneous equations rather than N. For each specular surface, the J for the diffuse surfaces are used to obtain G_k from Equation 6.41:

$$G_k = \sum_{j=1}^{d} J_j F_{k-j}^s + \sigma \sum_{j=d+1}^{N} \varepsilon_j T_j^4 F_{k-j}^s \quad d + 1 \le k \le N \tag{6.46}$$

The net external energy added to each diffuse reflecting surface is obtained by eliminating G_k from Equations 6.42 and 6.43,

$$Q_k = A_k \frac{\varepsilon_k}{1 - \varepsilon_k} \left(\sigma T_k^4 - J_k\right) \quad 1 \le k \le d \tag{6.47}$$

and the Q_k to each specular surface is found from Equation 6.44. Equations 6.44 through 6.47 are general energy-interchange relations for enclosures of diffuse and specular reflecting surfaces.

If the kth diffuse surface is black, then $J_k = \sigma T_k^4$ and $1 - \varepsilon_k = 0$, so Equation 6.47 is indeterminate. In this case, from Equation 6.44,

$$Q_k = A_k\left(\sigma T_k^4 - G_k\right) \tag{6.48}$$

where G_k is found from Equation 6.46 with $1 \le k \le d$.

If the energy input Q_k rather than T_k is specified for a diffuse reflecting surface $1 \le k \le d$, then T_k is unknown in Equation 6.45. Equation 6.47 can be used to eliminate this unknown in terms of J_k and the known Q_k. If the heat input Q_k is specified for a specular surface, $d + 1 \le k \le N$, then one of the T_j^4 in the last term of Equation 6.45 is unknown. Equation 6.44 is combined with Equation 6.46 to eliminate G_k, to give

$$\sigma_k T_k^4 - \frac{Q_k}{A_k\varepsilon_k} = \sum_{j=1}^{d} J_j F_{k-j}^s + \sigma \sum_{j=d+1}^{N} \varepsilon_j T_j^4 F_{k-j}^s \quad d+1 \le k \le N \tag{6.49}$$

Since Q_k is known, Equation 6.49 can be combined with Equation 6.45 to yield a simultaneous set of equations to determine the J of the diffusely reflecting surfaces and the T for the specularly reflecting surfaces having specified Q.

If all the surfaces are specular, a simultaneous solution is not required. The G_k are given by Equation 6.46 as

$$G_k = \sigma \sum_{j=1}^{N} \varepsilon_j T_j^4 F_{k-j}^s \quad 1 \le k \le N \tag{6.50}$$

and the Q_k are then found from Equation 6.44. By substituting Equation 6.50 into Equation 6.44, the Q_k are found directly from the specified surface temperatures:

$$Q_k = \sigma A_k \varepsilon_k \left(T_k^4 - \sum_{j=1}^{N} \varepsilon_j T_j^4 F_{k-j}^s \right) \tag{6.51}$$

A useful form for the enclosure equations is found by using Equations 6.47 and 6.44 to eliminate G and J from Equations 6.47 and 6.48. This gives a set of equations, all of the same form, that directly relate the Q and T:

$$\frac{1}{\varepsilon_k}\frac{Q_k}{A_k} - \sum_{j=1}^{d} \frac{Q_j}{A_j}\frac{1-\varepsilon_j}{\varepsilon_j} F_{k-j}^s = \sigma T_k^4 - \sigma \sum_{j=1}^{d} T_j^4 F_{k-j}^s - \sigma \sum_{j=d+1}^{N} \varepsilon_j T_j^4 F_{k-j}^s \quad 1 \le k \le N \tag{6.52}$$

Equation 6.51 is the special case of Equation 6.52 for $d = 0$.

Equation 6.52 can be used to obtain relations between the F^s exchange factors analogous to the relation for diffuse surfaces $\sum_{j=1}^{N} F_{k-j} = 1$. Let the entire enclosure be at uniform temperature. Then there is no net energy exchange and all the Q are zero, so Equation 6.52 reduces to

$$\sum_{j=1}^{d} F_{k-j}^s + \sum_{j=d+1}^{N} \varepsilon_j F_{k-j}^s = 1 \tag{6.53}$$

If all the surfaces in the enclosure are specular ($d = 0$), this further reduces to

$$\sum_{j=1}^{N} \varepsilon_j F_{k-j}^s = \sum_{j=1}^{N} (1-\rho_{s,j}) F_{k-j}^s = 1 \tag{6.54}$$

The set of enclosure Equations 6.52 is not difficult to solve after the exchange factors have been found; determining these factors, however, may not be easy, depending on the complexity of the enclosure geometry. To illustrate the calculations, some specific enclosures are considered.

Figure 6.24a shows an enclosure of three plane surfaces at different uniform specified temperatures; two sides are diffuse reflectors, and the third is specular. In Figure 6.24b, the energy arriving at

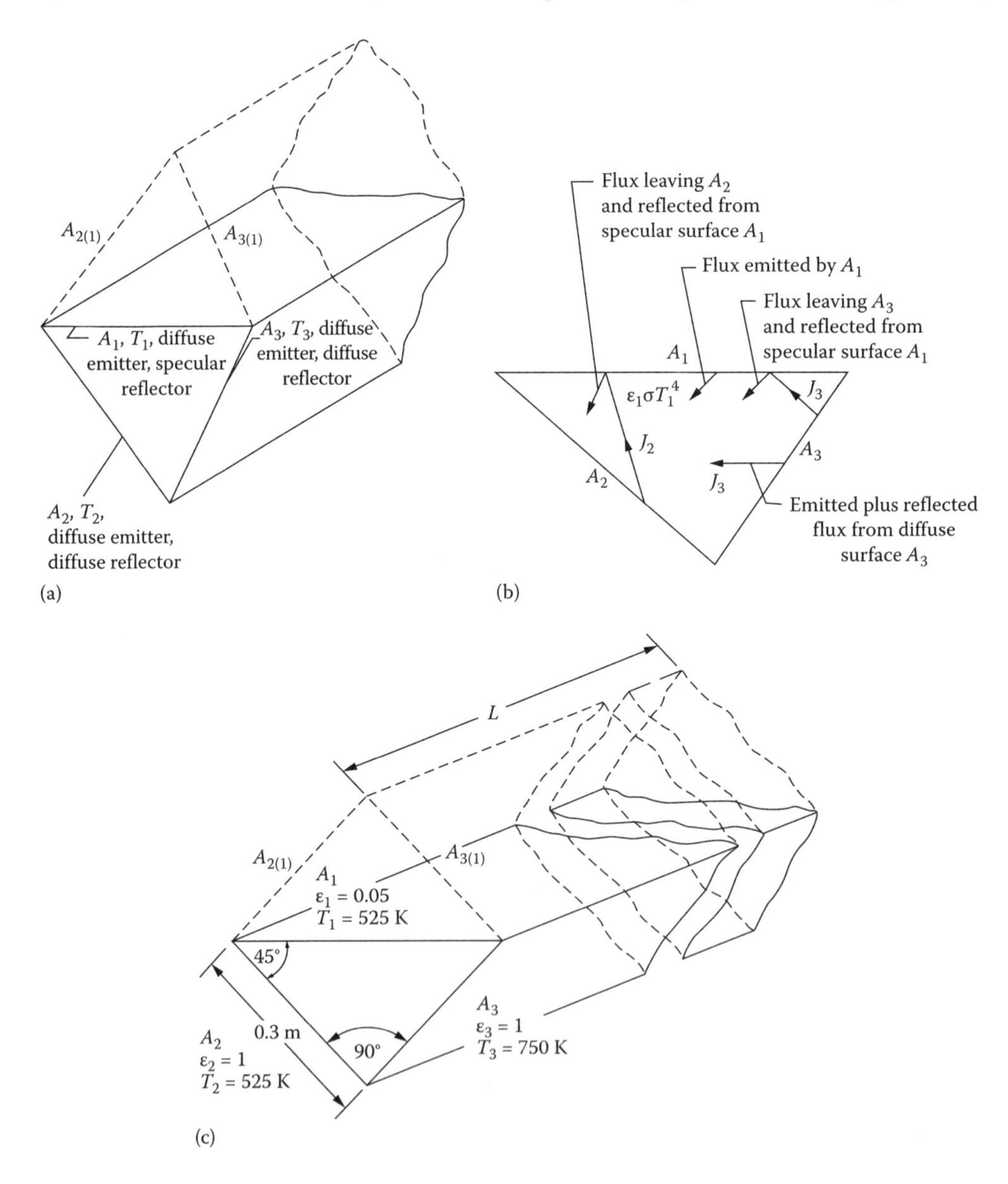

FIGURE 6.24 Enclosure having one specular reflecting surface and two surfaces that are diffuse reflectors: (a) general geometry; (b) energy fluxes to A_1, and energy fluxes that contribute to flux incident upon A_2; and (c) enclosure for Example 6.12.

A_1 comes directly from the diffuse A_2 and A_3, without any intermediate specular reflections. Hence, $F^s_{2-1} = F_{2-1}$ and $F^s_{3-1} = F_{3-1}$. By applying reciprocity, $F^s_{1-2} = F_{1-2}$ and $F^s_{1-3} = F_{1-3}$. For A_2 the incoming radiation is composed of four parts that originate as shown in Figure 6.24b. The first is the diffuse energy originating from A_3 and going directly to A_2, which is $J_3A_3F_{3-2}$. The remaining three parts arrive by means of A_1 and consist of an emitted portion $\varepsilon_1\sigma T_1^4 A_1F_{1-2}$ and two specularly reflected portions. The latter arise from the energy leaving A_2 and A_3 that is specularly reflected to A_2 and appears to come from the images $A_{2(1)}$ and $A_{3(1)}$ in Figure 6.24a. The specularly reflected portions are $J_2\rho_{s,1}A_2F_{2(1)-2} + J_3\rho_{s,1}A_3F_{3(1)-2}$. Multiple specular reflections cannot occur when only one planar specular surface is present. The specular exchange factors are then $F^s_{1-2} = F_{1-2}$, $F^s_{2-2} = \rho_{s,1}F_{2(1)-2}$, and $F^s_{2-3} = F_{2-3} + \rho_{s,1}F_{3(1)-2}$. By using reciprocity, $F^s_{2-1} = F_{2-1}, F^s_{2-2} = \rho_{s,1}F_{2-2(1)}$, and $F^s_{2-3} = F_{2-3} + \rho_{s,1}F_{2-3(1)}$. Similarly, for surface 3, $F^s_{3-1} = F_{3-1}, F^s_{3-2} = F_{3-2} + \rho_{s,1}F_{3-2(1)}$ and $F^s_{3-1} = \rho_{s,1}F_{3-3(1)}$. A numerical example using these factors is now given.

Example 6.12

An enclosure with three sides is in Figure 6.24b. The length L is long enough that the triangular ends can be neglected in the radiative energy balances. Two surfaces are black, and the third is a gray diffuse emitter with $\varepsilon_1 = 0.05$. Determine the energy added per meter of enclosure length to each surface for the two cases: (1) A_1 is a diffuse reflector and (2) A_1 is a specular reflector.

From symmetry, $F_{1-2} = F_{1-3}$. Also $F_{1-2} + F_{1-3} = 1$, so $F_{1-2} = F_{1-3} = 1/2$. From reciprocity, $F_{2-1} = A_1F_{1-2}/A_2 = \sqrt{2}/2 = F_{3-1}$. In addition, $F_{2-1} + F_{2-3} = 1$, so that $F_{2-3} = 1 - \sqrt{2}/2 = F_{3-2} = F_{2-2(1)} = F_{3-3(1)}$. Finally, $F_{3-2(1)} = F_{2-3(1)} = 1 - F_{3-2} = \sqrt{2} - 1$.

For case 1, apply Equations 5.11.1 through 5.11.3 to obtain

$$\frac{Q_1}{0.3\sqrt{2}}\frac{1}{0.05} = \sigma\left(525^4 - \frac{1}{2}525^4 - \frac{1}{2}750^4\right)$$

$$-\frac{Q_1}{0.3\sqrt{2}}\frac{\sqrt{2}}{2}\frac{1-0.05}{0.05} + \frac{Q_2}{0.3} = \sigma\left[-\frac{\sqrt{2}}{2}525^4 + 525^4 - \left(1 - \frac{\sqrt{2}}{2}\right)750^4\right]$$

$$-\frac{Q_1}{0.3\sqrt{2}}\frac{\sqrt{2}}{2}\frac{1-0.05}{0.05} + \frac{Q_3}{0.3} = \sigma\left[-\frac{\sqrt{2}}{2}525^4 - \left(1 - \frac{\sqrt{2}}{2}\right)525^4 + 750^4\right]$$

The solution of these three equations yields (Q's are per meter of enclosure length): $Q_1 = -144.6$ W/m, $Q_2 = -2571.7$ W/m, and $Q_3 = 2716.4$ W/m. The energy supplied to A_3 is removed from A_1 and A_2. The amount removed from A_1 is small because A_1 is a poor absorber.

For case 2, apply Equation 6.45 to compute J_2 and J_3. Since $\varepsilon_2 = \varepsilon_3 = 1$, this yields $J_2 = \sigma T_2^4$ and $J_3 = \sigma T_3^4$ for the black surfaces. Then Equation 6.41 yields the G for each surface as

$$G_1 = \sigma\left[\frac{1}{2}525^4 + \frac{1}{2}750^4\right] = 11{,}125 \text{ W/m}^2$$

$$G_2 = \sigma\left\{0.05(525)^4\frac{\sqrt{2}}{2} + 525^4(1-0.05)\left(1 - \frac{\sqrt{2}}{2}\right) + 750^4\left[1 - \frac{\sqrt{2}}{2} + (1-0.05)(\sqrt{2}-1)\right]\right\}$$

$$= 13{,}666 \text{ W/m}^2$$

$$G_3 = \sigma\left\{0.05(525)^4\frac{\sqrt{2}}{2} + 525^4\left[1 - \frac{\sqrt{2}}{2} + (1-0.05)(\sqrt{2}-1)\right] + 750^4(1-0.05)\left(1 - \frac{\sqrt{2}}{2}\right)\right\}$$

$$= 8{,}101 \text{ W/m}^2$$

Equations 6.44 and 6.48 are applied to find the Q's (per meter of enclosure length) as: $Q_1 = -144.6$ W/m, $Q_2 = -2807.4$ W/m, and $Q_3 = 2952.1$ W/m. By making A_1 specular, the heat transferred from A_3 to A_2 is increased by 9% from 2572 to 2807 W/m.

An alternative for case 2 is to use Equation 6.52 to give

$$Q_1 = \varepsilon_1 A_1 \sigma (T_1^4 - T_2^4 F_{1-2} - T_3^4 F_{1-3})$$
$$Q_2 = A_2 \sigma [\varepsilon_1 T_1^4 F_{2-1} + T_2^4 (1 - \rho_{s,1} F_{2(1)-2}) - T_3^4 (F_{2-3} + \rho_{s,1} F_{2-3(1)})]$$
$$Q_3 = A_3 \sigma [-\varepsilon_1 T_1^4 F_{3-1} - T_2^4 (F_{3-2} + \rho_{s,1} F_{3-2(1)}) + T_3^4 (1 - \rho_{s,1} F_{3-3(1)})]$$

Substituting values yields the same results as before.

To further demonstrate enclosure analysis with some specularly reflecting surfaces, the rectangular geometry in Figure 6.25 is used. All surfaces are diffuse emitters; two surfaces are diffuse reflectors, and two are specular. The reflected images are shown dashed. Reflections continue until all of the outer perimeter enclosing the composite of the original enclosure plus the reflected images consists of either diffuse reflecting (or nonreflecting, such as an opening) surfaces, or images of diffuse surfaces.

For the enclosure in Figure 6.25, the first step is to use Equation 6.45 to obtain J_1 and J_2 for the two diffuse areas. For the exchange factors, consider first F_{1-1}^s. Part of the energy leaving A_1 returns to A_1 by each of three paths: direct reflection from A_3, reflection from A_3 to A_4 and then to A_1, and reflection from A_4 to A_3 and then to A_1. Thus, the energy leaving A_1 that returns to A_1 is expressed by the specular exchange factor $F_{1-1}^s = \rho_{s,3} F_{1(3)-1} + \rho_{s,3}\rho_{s,4} F_{1(3-4)-1} + \rho_{s,4}\rho_{s,3} F_{1(4-3)-1}$. The $F_{1(3-4)-1}$ is the

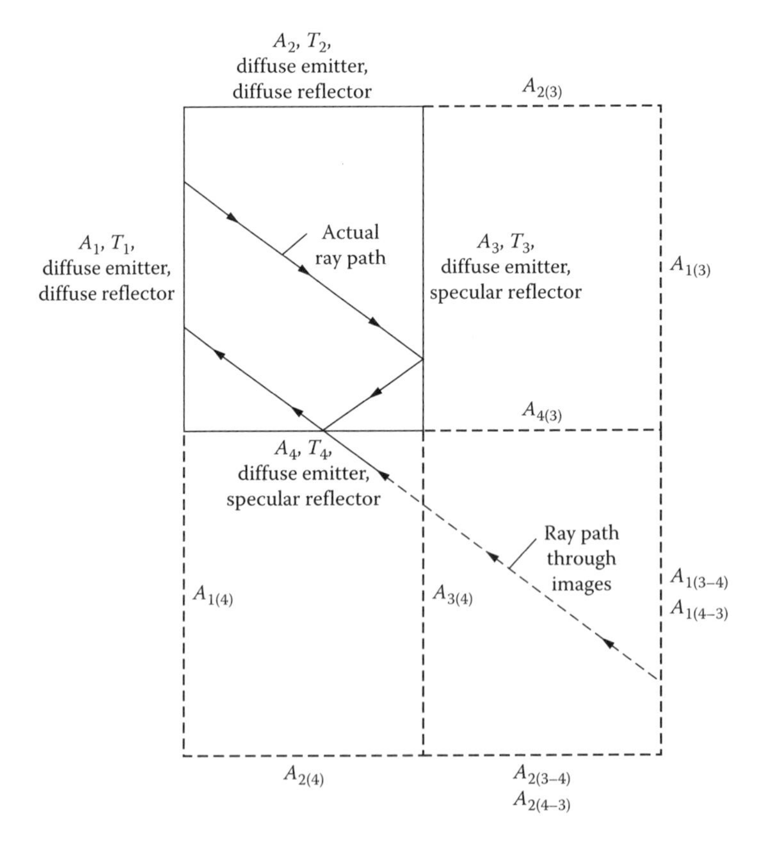

FIGURE 6.25 Rectangular enclosure and reflected images when two adjacent surfaces are specular reflectors and the other two are diffuse reflectors.

configuration factor by which $A_{1(3-4)}$ is viewed from A_1 through A_4 and then $A_{3(4)}$, which are the reflection areas by means of which the $A_{1(3-4)}$ image was formed. Similarly, $F_{1(4-3)-1}$ is the factor by which the same area $A_{1(3-4)}$ is viewed from A_1 through A_3 and then $A_{4(3)}$.

Radiation leaving A_2 reaches A_1 along four paths: direct exchange, reflection from A_3, reflection from A_4, and reflection from A_3 to A_4. No energy from A_2 reaches A_1 by means of reflections from A_4 and then A_3. This is because A_1 cannot view the image $A_{2(4-3)}$ through area A_3. This yields $F^s_{2-1} = F_{2-1} + \rho_{s,3}F_{2(3)-1} + \rho_{s,4}F_{2(4)-1} + \rho_{s,3}\rho_{s,4}F_{2(3-4)-1}$.

The diffuse energy leaving the specular surface A_3 (and similarly for A_4) consists only of emitted energy $\varepsilon_3 A_3 \sigma T_3^4$. There are two paths by which some of this reaches A_1: direct exchange, and by means of specular reflection from A_4. This yields $F^s_{3-1} = F_{3-1} + \rho_{s,4}F_{3(4)-1}$ and $F^s_{4-1} = F_{4-1} + \rho_{s,3}F_{4(3)-1}$.

After applying F factor reciprocity, the factors are substituted into Equation 6.45 to yield:

$$\begin{aligned}
&J_1 - (1-\varepsilon_1)\{J_1[\rho_{s,3}F_{1-1(3)} + \rho_{s,3}\rho_{s,4}(F_{1-1(3-4)} + F_{1-1(4-3)})] \\
&\quad + J_2(F_{1-2} + \rho_{s,3}F_{1-2(3)} + \rho_{s,4}F_{1-2(4)} + \rho_{s,3}\rho_{s,4}F_{1-2(3-4)})\} \\
&= \varepsilon_1\sigma T_1^4 + (1-\varepsilon_1)\sigma[\varepsilon_3 T_3^4(F_{1-3} + \rho_{s,4}F_{1-3(4)}) \\
&\quad + \varepsilon_4 T_4^4(F_{1-4} + \rho_{s,3}F_{1-4(3)})]
\end{aligned} \tag{6.55}$$

Similarly, considering J_2 for surface 2 yields:

$$\begin{aligned}
&J_2 - (1-\varepsilon_2)\{J_1(F_{2-1} + \rho_{s,3}F_{2-1(3)} + \rho_{s,4}F_{2-1(4)} + \rho_{s,3}\rho_{s,4}F_{2-1(4-3)}) \\
&\quad + J_2[\rho_{s,4}F_{2-2(4)} + \rho_{s,3}\rho_{s,4}(F_{2-2(4-3)} + F_{2-2(3-4)})]\} \\
&= \varepsilon_2\sigma T_2^4 + (1-\varepsilon_2)\sigma[\varepsilon_3 T_3^4(F_{2-3} + \rho_{s,4}F_{2-3(4)}) \\
&\quad + \varepsilon_4 T_4^4(F_{2-4} + \rho_{s,3}F_{2-4(3)})]
\end{aligned} \tag{6.56}$$

Equations 6.55 and 6.56 are solved simultaneously for J_1 and J_2.

For the two specular surfaces, the G_3 and G_4 can be found as soon as the J for the diffuse surfaces are known. From Equation 6.46,

$$G_3 = J_1(F_{3-1} + \rho_{s,4}F_{3-1(4)}) + J_2(F_{3-2} + \rho_{s,4}F_{3-2(4)}) + \varepsilon_4\sigma T_4^4 F_{3-4} \tag{6.57}$$

$$G_4 = J_1\left(F_{4-1} + \rho_{s,3}F_{4-1(3)}\right) + J_2\left(F_{4-2} + \rho_{s,3}F_{4-2(3)}\right) + \varepsilon_3\sigma T_3^4 F_{4-3} \tag{6.58}$$

With the J for the diffuse surfaces and the G for the specular surfaces known, the Q to maintain the specified surface temperatures are found from Equations 6.47 and 6.44. The solution can also be found from Equation 6.52, directly relating the Q and T.

Determining the specular exchange factors can become tedious in enclosures with many surfaces. A more direct approach is the Monte Carlo method, wherein the direction of reflected energy can be specified for each incident small "bundle" of energy since it approaches a specular surface from a definite direction (Chapter 7). A similar statistical method involving Markov chain theory has been developed by Naraghi and Chung (1984) and extended by Billings et al. (1991b).

6.6.2 Curved Specular Reflecting Surfaces

When specular reflecting surfaces are curved, the reflected image geometry can become quite complex. To demonstrate some ideas, consider the exchange of radiation within a specular tube (Perlmutter and Siegel 1963), as in Figure 6.26.

It is assumed that the imposed temperature or heating conditions depend only on axial position and are independent of location around the tube circumference. To compute the radiative exchange

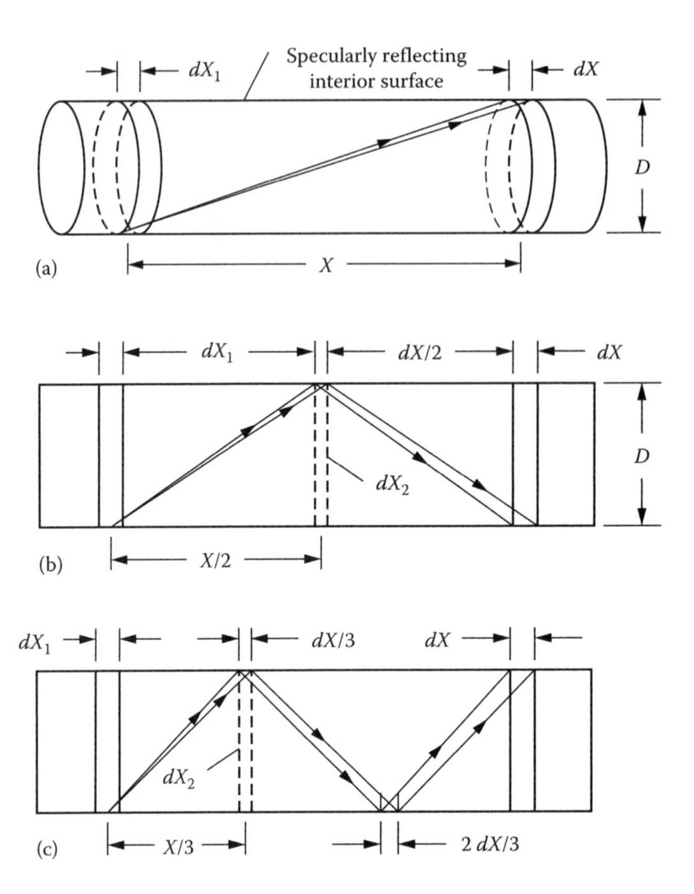

FIGURE 6.26 Radiation exchange within a specularly reflecting cylindrical tube: (a) direct exchange between two ring elements; (b) exchange by one reflection; and (c) exchange by two reflections.

within the tube for axisymmetric heating, the configuration factor between two ring elements on the tube wall is needed. The direct exchange (Figure 6.26a) is governed by the factor (see Appendix C, factor 15):

$$dF_{dX_1-dX} = \left\{1 - \frac{(X/D)^3 + 3X/2D}{[(X/D)^2 + 1]^{3/2}}\right\} dX$$

Figure 6.26b illustrates the configuration factor for one specular reflection. All the radiation from dX_1 that reaches dX by one reflection has been reflected from a ring element halfway between dX_1 and dX. The ring at $X/2$ is only $dX/2$ wide, so the solid angle subtending it will spread to a width dX at the location X. The configuration factor for one reflection is then the factor between dX_1 and the dashed element $dX/2$:

$$dF_{dX_1-dX/2} = \left\{1 - \frac{(X/2D)^3 + 3X/4D}{[(X/2D)^2 + 1]^{3/2}}\right\} \frac{dX}{2}$$

Similarly, the factor for exchange by two reflections (Figure 6.26c) is

$$dF_{dX_1-dX/3} = \left\{1 - \frac{(X/3D)^3 + 3X/6D}{[(X/3D)^2 + 1]^{3/2}}\right\} \frac{dX}{3}$$

and for n reflections

$$dF_{dX_1-dX/(n+1)} = \left(1 - \frac{[X/(n+1)D]^3 + 3X/2(n+1)D}{\{[(X/(n+1)D]^2 + 1\}^{3/2}}\right)\frac{dX}{n+1}$$

In general, the geometric factor for *any* number of reflections is found by considering the exchange between the originating element (dX_1 in this case) and the element (call it dX_2) from which *the first* reflection is made (the dashed element in Figure 6.26b and c). This is because the fraction of energy leaving dX_1 in the solid angle subtended by dX remains the same through the succeeding reflections along the path to dX.

At each reflection the energy is multiplied by the specular reflectivity ρ_s. If all the contributions are summed, the energy leaving dX_1 that reaches dX by direct exchange and all reflection paths is expressed by the specular exchange factor for the inside of a tube that is open at both ends (no reflections from end walls):

$$dF^s_{dX_1-dX} = \sum_{n=0}^{\infty} \rho_s^n \left(1 - \frac{[X/(n+1)D]^3 + 3X/2(n+1)D}{\{[(X/(n+1)D]^2 + 1\}^{3/2}}\right)\frac{dX}{n+1} \qquad (6.59)$$

For a complete heat transfer formulation, the energy entering through the tube ends must be included, which requires the exchange factor from the end openings. This factor is obtained by a derivation similar to Equation 6.59. The energy is considered that leaves an element on the tube wall and travels to the circular disk opening at the end of the tube by a direct path or by one or more reflections along the tube wall between the element and the end opening. The exchange factor was obtained by Perlmutter and Siegel (1963), where it was shown that the energy exchange in the tube can be divided into two separate parts with the complete solution given by superposition. One part is for the tube being heated along its length, such as by electric heaters in the tube wall, but with the end environments at zero temperature (Figure 6.27). The outer surface of the tube is insulated, so the only energy loss is by radiation leaving through the tube ends. The energy balance on an element at x states that the heat addition $q(x)$ equals the emitted energy minus that absorbed as a result of incident energy arriving by specular reflections. Since there is no diffuse reflection in this analysis, all of the arriving specular energy is obtained by using the specular exchange factor, Equation 6.59. The energy equation governing the wall temperature at x is then given by (a gray tube wall is assumed so that $\alpha = \varepsilon$)

$$q(x) = \varepsilon\sigma T_w^4(x) - \varepsilon^2 \left[\int_0^x \sigma T_w^4(y)dF^s(x-y) + \int_x^L \sigma T_w^4(y)dF^s(y-x)\right] \qquad (6.60)$$

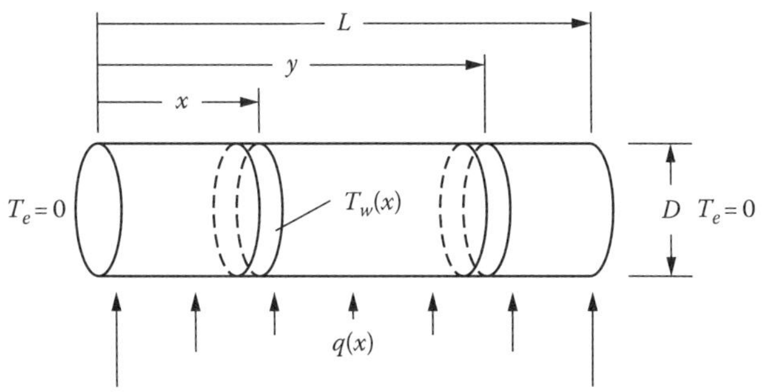

FIGURE 6.27 Heated tube with internal specular reflections; the outside surface of the tube is insulated.

where $dF^s(x-y)$ is given by Equation 6.59 with $X = (x - y)$. Equation 6.60 can be solved numerically as an integral equation for $T_w(x)$ if $q(x)$ is specified, or it can be directly integrated to give $q(x)$ if $T_w(x)$ is specified. Effects of the differing reflectivities of the polarized components on transmission through channels are investigated by Edwards and Tobin (1967) and Miranda (1996). The radiative energy transmission through a polished stainless steel tube is investigated by Miranda, and comparisons are made of analysis with experiment. The tube was cooled to 77 K so radiation from the tube wall could be neglected. One end of the tube was cooled to 77 K, while a blackbody source at 300 K was placed at the other end. For a tube diameter of 0.067 m, measurements were made for tube lengths up to 1.2 m, and agreement with analysis was always within 8%. Additional comparison with experiment is in Qu et al. (2007a).

When the configuration is more involved than the cylindrical geometry, the reflection patterns can become quite complex. Some examples for radiation within a specular conical cavity and a specular cylindrical cavity with a specular end plane are in Lin and Sparrow (1965); a more generalized treatment of nonplanar reflections is in Plamondon and Horton (1967). An approximate analysis of the transfer through specular passages is in Rabl (1977). It is based on estimating the average number of reflections that occur during transmission.

The analyses by Tsai and Strieder (1985, 1986) for radiation within a cylindrical and a spherical enclosure consider a more general reflection behavior obtained by dividing the reflectivity into specular and diffuse components. These components can be treated using the methods in the enclosure theory given in this chapter. This reflection model was for application to radiation within high-temperature porous materials. The results of Tsai and Strieder (1985) show that, for a spherical cavity, the many specular reflections tend to eliminate directional effects and cause the radiation to behave in a diffuse manner.

Example 6.13

A cylindrical cavity open at one end has a specularly reflecting cylindrical wall and base (Figure 6.28a). Determine the fraction of radiation from ring element dX_1 that reaches dX by means of one reflection from the base with reflectivity $\rho_{s,1}$ and one reflection from the cylindrical wall with reflectivity $\rho_{s,2}$.

As shown in Figure 6.28b, for this geometry the reflected radiation from the base can be regarded as originating from an image of dX_1. The second reflection occurs from an element of width $dX/2$ midway between the image dX_1 and dX. The desired radiation fraction is given by the configuration factor from the image dX_1 to the dashed ring area $dX/2$ multiplied by the two reflectivities:

$$\rho_{s,1}\rho_{s,2}dF_{dX_1-dX} = \rho_{s,1}\rho_{s,2}\left(1 - \frac{[(X+X_1)/2D]^3 + 3(X+X_1)/4D}{\{[(X+X_1)/2D]^2+1\}^{3/2}}\right)\frac{dX}{2}$$

Another type of curved specular surface of practical importance is a paraboloidal mirror such as in a solar furnace. The mirror axis is aligned in the direction of the sun and a receiver is placed at the mirror focal plane. It is desired to estimate the receiver temperature. Information on concentrators for solar furnaces is in Cobble (1961) and Kamada (1965). An inverse analysis for designing a mirror to achieve a specified intensity distribution at the receiver is in Zakhidov (1989).

In Maruyama (1991, 1993), the angular distribution of intensity emitted from systems of cylindrical black emitters with various reflector shapes (circular arc, parabolic, and involute) are found by ray tracing. The reflector surfaces are assumed to be either perfect or metallic reflecting surfaces with reflectivity predicted by electromagnetic theory. In Maruyama (1993), there is experimental verification of predictions for circular arc and involute reflectors. The involute is effective for providing a uniformly distributed radiation source.

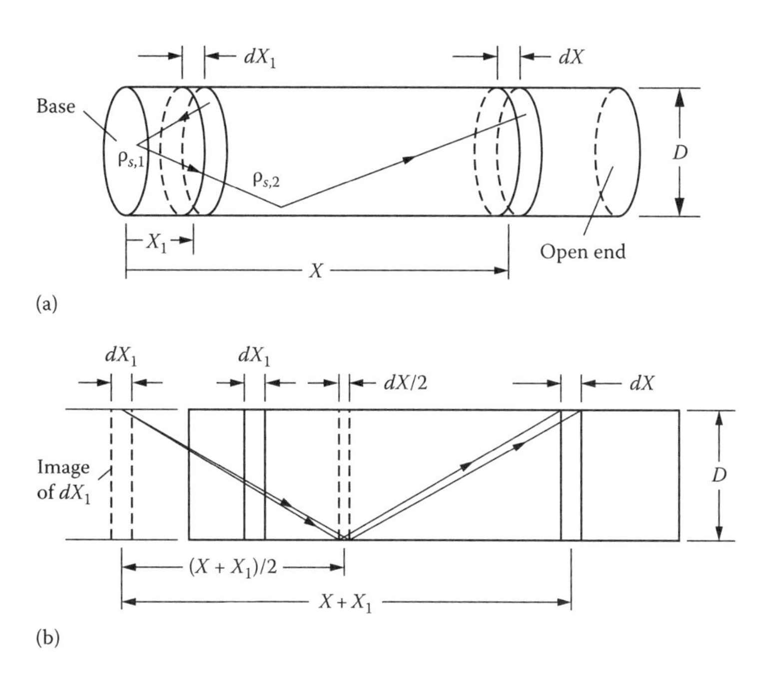

FIGURE 6.28 Reflection in cylindrical cavity with specular curved wall and base: (a) cavity geometry; (b) image of dX_1 formed by reflection in cavity base.

6.7 MULTIPLE RADIATION SHIELDS

The results in Table 6.1 can be extended to obtain the performance of multiple radiation shields, as shown in Figures 6.29 and 6.30. The shields are thin, parallel, highly reflecting sheets, placed between radiating surfaces to reduce energy transfer between them. A highly effective insulation can be obtained by using many sheets separated by vacuum to provide a series of alternate radiation and conduction barriers. One construction approach is depositing highly reflecting metallic films

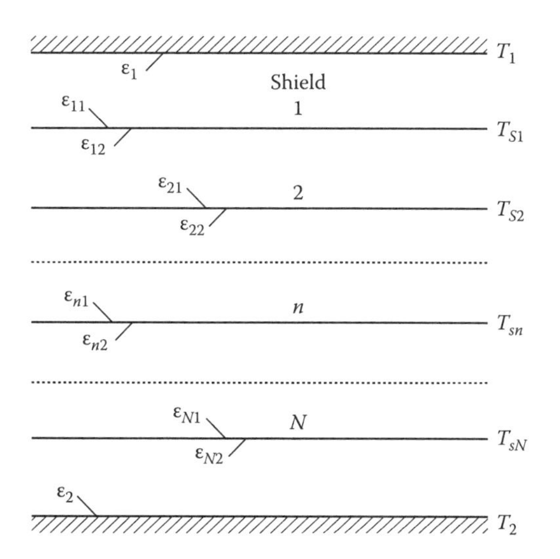

FIGURE 6.29 Parallel walls separated by N radiation shields.

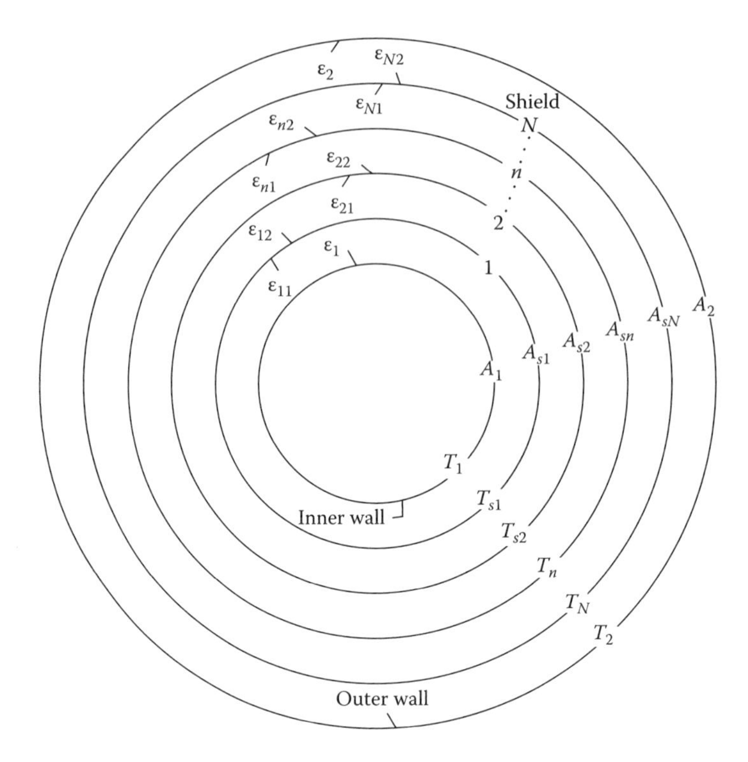

FIGURE 6.30 Radiation shields between concentric cylinders or spheres.

such as aluminum or silver on both sides of thin sheets of plastic spaced apart by placing between them a cloth net having a large open area between its fibers. A stacking of 20 radiation shields per centimeter of thickness can be obtained. An important use of this multilayer insulation is in providing emergency protective blankets for forest-fire fighters and in low-temperature applications such as insulation of cryogenic storage tanks. It is used in satellites and other vehicles operating in outer space (Lin et al. 1996).

For the results here, the spaces between the shields are evacuated so that heat transfer is only by radiation, and all emission is assumed diffuse. To analyze shield performance, consider N radiation shields between two surfaces at temperatures T_1 and T_2 with emissivities ε_1 and ε_2. As a general case, let a typical shield n have emissivity ε_{n1} on one side and ε_{n2} on the other, as in Figure 6.29. As a result of the heat flow, the nth shield will be at temperature T_{sn}. Because for steady state the same q passes through the entire series of shields, Equation 6.23 can be written for each pair of adjacent surfaces as

$$q\left(\frac{1}{\varepsilon_1}+\frac{1}{\varepsilon_{11}}-1\right)=\sigma\left(T_1^4-T_{s1}^4\right)$$

$$q\left(\frac{1}{\varepsilon_{12}}+\frac{1}{\varepsilon_{21}}-1\right)=\sigma\left(T_{s1}^4-T_{s2}^4\right)$$

$$\vdots$$

$$q\left(\frac{1}{\varepsilon_{(N-1)2}}+\frac{1}{\varepsilon_{N1}}-1\right)=\sigma\left(T_{s(N-1)}^4-T_{sN}^4\right)$$

$$q\left(\frac{1}{\varepsilon_{N2}}+\frac{1}{\varepsilon_2}-1\right)=\sigma\left(T_{sN}^4-T_2^4\right)$$

Adding these equations and dividing by the resulting factor multiplying q on the left-hand side gives, after some rearrangement,

$$q = \frac{\sigma\left(T_1^4 - T_2^4\right)}{1/\varepsilon_1 + 1/\varepsilon_2 - 1 + \sum_{n=1}^{N}\left(1/\varepsilon_{n1} + 1/\varepsilon_{n2} - 1\right)} \tag{6.61}$$

In most instances ε is the same on both sides of each shield, and all the shields have the same ε. Let all the shield emissivities be ε_s; then, q becomes

$$q = \frac{\sigma\left(T_1^4 - T_2^4\right)}{1/\varepsilon_1 + 1/\varepsilon_2 - 1 + N\left(2/\varepsilon_s - 1\right)} \tag{6.62}$$

If the wall emissivities are the same as the shield emissivities, $\varepsilon_1 = \varepsilon_2 = \varepsilon_s$, Equation 6.62 further reduces to

$$q = \frac{\sigma\left(T_1^4 - T_2^4\right)}{\left(N+1\right)\left(2/\varepsilon_s - 1\right)} \tag{6.63}$$

In this instance q decreases as $1/(N + 1)$ as the number of shields N increases. When there are no shields, $N = 0$, and Equation 6.62 reduces to $q = \sigma\left(T_1^4 - T_2^4\right)/\left[\left(1/\varepsilon_1\right) + \left(1/\varepsilon_2\right) - 1\right]$ as in Table 6.1. To illustrate shield performance, if in Equation 6.62 $\varepsilon_1 = \varepsilon_2 = 0.8$ and $\varepsilon_s = 0.05$, then $q = \sigma\left(T_1^4 - T_2^4\right)/\left(1.5 + 39N\right)$ and the ratio $q(N \text{ shields})/q(\text{no shields}) = 1.5/(1.5 + 39\,N)$. For various numbers of shields, this yields

	N = 0	N = 1	N = 10	N = 100
$q(N)/q(N = 0)$	1	0.0370	0.00383	0.00038
$1/(N + 1)$	1	0.5	0.0909	0.0099

In the table, the factor $1/(N + 1)$ is the q ratio (independent of ε) when the walls have the same ε value as for the shields, as in Equation 6.63. This illustrates that the fractional reduction in q is larger when the wall ε are large compared to the ε for the shields, since for small wall ε the unshielded heat flow is already low.

As for flat plates, the expressions in Table 6.1 can be used to derive the heat flow through a series of concentric cylindrical or spherical radiation shields, as in Figure 6.30. If the walls A_1 and A_2 and all the shields A_{sn} are diffuse reflectors, the heat flow is (emission is diffuse)

$$Q = \frac{A_1\sigma\left(T_1^4 - T_2^4\right)}{1/\varepsilon_1 + \left(A_1/A_2\right)\left(1/\varepsilon_2 - 1\right) + \sum_{n=1}^{N}\left(A_1/A_{sn}\right)\left(1/\varepsilon_{n1} + 1/\varepsilon_{n2} - 1\right)} \tag{6.64}$$

If the walls are diffuse reflectors and all the shields are specular, then

$$Q = \frac{A_1\sigma\left(T_1^4 - T_2^4\right)}{\left\{1/\varepsilon_1 + 1/\varepsilon_{11} - 1 + \sum_{n=1}^{N-1}\left(A_1/A_{sn}\right)\left(1/\varepsilon_{n2} + 1/\varepsilon_{(n+1)1} - 1\right) + \left(A_1/A_{sN}\right)\left[1/\varepsilon_{N2} + \left(A_{sN}/A_2\right)\left(1/\varepsilon_2 - 1\right)\right]\right\}} \tag{6.65}$$

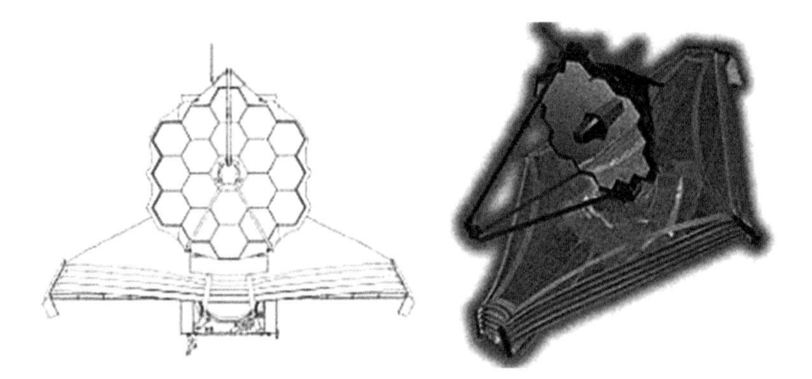

FIGURE 6.31 Conceptual drawings of five-layer metalized polymer radiation shield for the Webb infrared telescope. Shield has roughly the area of a regulation tennis court. (Courtesy of NASA.)

If all the walls and shields are specular, Equation 6.65 applies if A_{sN}/A_2 is replaced by 1 in the last term in the denominator. In this instance, if all shield emissivities are the same and equal to ε_s, the result is

$$Q = \frac{A_1\sigma\left(T_1^4 - T_2^4\right)}{1/\varepsilon_1 + 1/\varepsilon_s - 1 + \sum_{n=1}^{N-1}\left[\left(A_1/A_{sn}\right)\left(2/\varepsilon_s - 1\right)\right] + \left(A_1/A_{sN}\right)\left(1/\varepsilon_s + 1/\varepsilon_{n2} - 1\right)} \tag{6.66}$$

The sum in Equations 6.65 and 6.66 is zero if $N = 1$. When using radiation shields to insulate a cryogenic tank, one or more of the shields can be cooled by using vapor boil-off from the tank. The resulting decrease of temperature within the insulation helps reduce the heat loss from the tank. The optimum utilization of vapor cooling can be examined thermodynamically by minimizing the entropy production resulting from the heat losses and the controlled cryogen boil-off used for vapor cooling of one or more shields. This optimization has been analyzed by Chato and Khodadadi (1984) using the thermodynamic ideas of Schultz and Bejan (1983), and the idea is used in design of the sun shield for the Webb infrared telescope (Figure 6.31).

6.8 CONCLUDING REMARKS

This chapter presents treatments of radiative interchange between directional and specularly reflecting surfaces, and exchange in enclosures containing both specularly and diffusely reflecting surfaces.

In many instances, as in Example 6.10, the interchange of energy in enclosures is modified only a small amount by considering specular in place of diffuse reflecting surfaces. In other applications, however, such as solar concentrators and solar furnaces, and radiative transmission through highly reflecting tubes to guide radiative energy to a detector, specular reflection is a dominant requirement. The design for these types of devices is often done by ray tracing; this has not been treated here, where the emphasis was on analysis of enclosures using configuration factors. A ray-tracing technique by the Monte Carlo method is presented in Chapter 7. For enclosures, Bobco (1964), Sparrow and Lin (1965), Sarofim and Hottel (1966), Mahan et al. (1979), Tsai and Strieder (1986a,b), and Qu et al. (2007a,b) have examined radiative exchange involving surfaces with reflectivities that have both diffuse *and* specular components. Schornhorst and Viskanta (1968) compared experimental and analytical results for radiant exchange among various types of surfaces and found that, *regardless* of the presence of specular surfaces, the diffuse-surface analysis agreed best with experimental results. Bobco and Drolen (1989) provide a model to represent the reflectivity of a surface by combined diffuse and specular components. Jamaluddin and Fiveland (1990) analyzed the effect in a furnace of having walls with diffuse and specular reflectivity components. The enclosures were 2- and 3D in nature and were filled with radiating combustion products. It was found that for some applications the heat transfer performance in furnaces could be improved by using highly reflecting specular walls.

The multiple specular reflections during radiative transmission within a highly reflecting circular tube or similar device can lead to significant polarization effects. As in Figure 6.32, radiation in a given direction can undergo multiple reflections at the same incidence angle. As discussed in connection with the Fresnel reflection equations, the reflectivity differs for the two components of polarization, as in Equations 3.19 and 3.20. This provides different absorption for each of the components, and hence after reflection an initially unpolarized intensity becomes polarized. After multiple reflections at the same incidence angle during transit through a specular tube, for example, the radiation can become strongly polarized (Edwards and Tobin 1967, Miranda 1996, Qu et al. 2007a). In contrast, for diffuse reflections, a multiple reflection process involves many different incidence angles, and polarization effects are not a concern. Polarization is considered in Chapter 17 for the specular reflections at the surfaces of glass windows as in Example 17.1.

It is sometimes implied that the energy transfer between two real surfaces can be bracketed by calculating two limiting magnitudes: (1) interchange between diffuse surfaces of the same total hemispherical emissivities as the real surfaces, and (2) interchange between specularly reflecting surfaces of the same total hemispherical emissivities as the real surfaces. This implication is not generally correct. Consider a surface that has a reflectivity as given by Figure 6.32a where there is strong reflection back into the incident direction. Now consider the radiant interaction between the real surface 2 and black surface 1 as in Figure 6.32b. If surface 2 is specular, it will not return any reflected energy to the black surface (Figure 6.32c). If 2 is diffuse, it will return a portion of the incident energy by reflection (Figure 6.32d). For its real directional properties, however, it will reflect more energy to the black surface than for *either* of the ideal surfaces (Figure 6.32e). The ideal directional surfaces therefore do not constitute limiting cases for energy transfer in general.

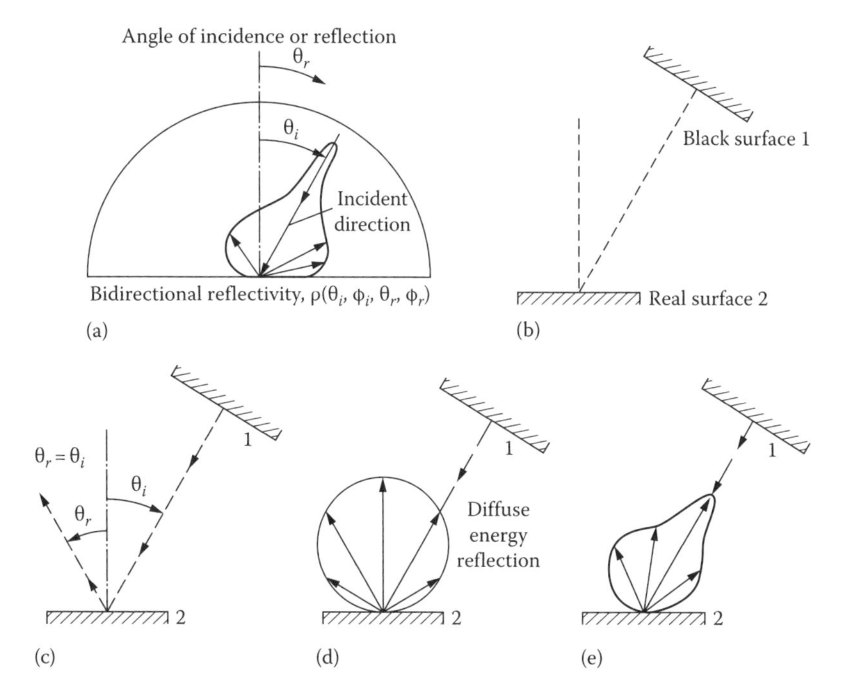

FIGURE 6.32 Radiant energy reflection with various idealizations of surface properties: (a) bidirectional reflectivity of real surface; (b) geometry of surfaces; (c) surface 2 reflects specularly; (d) surface 2 reflects energy diffusely; and (e) surface 2 reflects energy with real properties.

Figure 6.10 demonstrates another case in which diffuse and specular properties do not provide limiting solutions. At best, enclosure calculations based on idealized specular and diffuse assumptions for the surface characteristics give some indication of the possible importance of directional effects. Within enclosures, these directional effects may be small because the multiple reflections between the surfaces tend to make the overall behavior diffuse.

HOMEWORK

6.1 An anti-satellite missile has a total radiation sensor built into its nose. The sensor detects total radiation emitted by a hot satellite, and normally tracks to the satellite, where it impacts. A target satellite that is to be protected from the missile is gray with $\varepsilon = 0.10$, has a surface temperature of $T_s = 450$ K, and is spherical with diameter $D_s = 3$ m.

You are to design a dummy countermeasure spherical satellite that is to be ejected from the real satellite as a decoy. The decoy material has the hemispherical spectral emissivity shown. The diameter of the decoy will be $D_d = 0.45$ m.

(a) What surface temperature T_d should be used for the decoy?

(b) How large a heat source (kW) is necessary in the decoy?

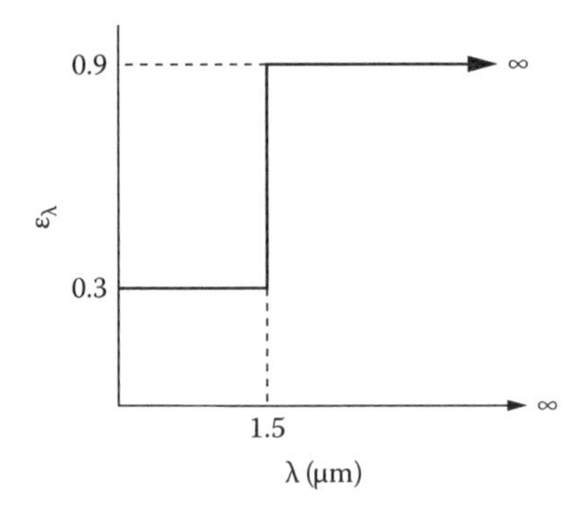

Answer: (a) 671.0 K; (b) 6.58 kW.

6.2 Radiative energy is being transferred across the space between two large parallel plates with hemispherical spectral emissivities approximated as shown. What is the net heat flux q_1 transferred from 1 to 2?

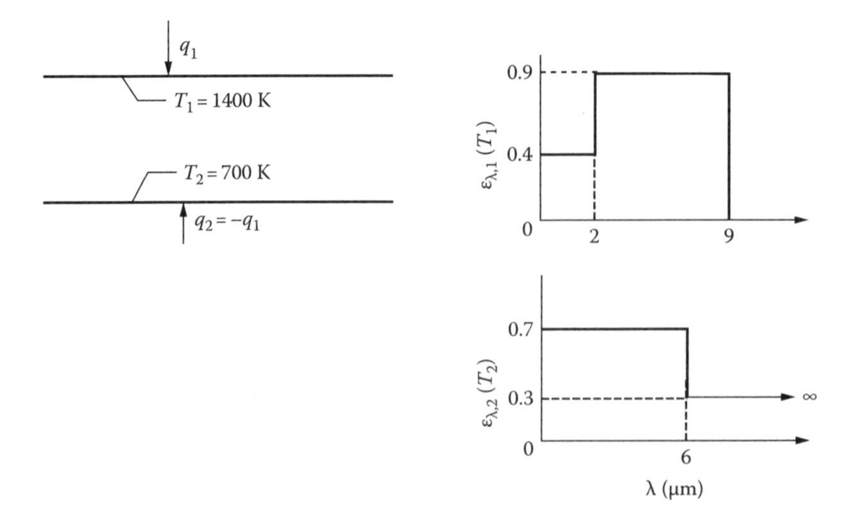

Answer: 107,530 W/m^2.

6.3 The two plates in Homework Problem 6.2 have a flat plate radiation shield placed between them so the geometry is now three large parallel plates. The shield is gray, and has an emissivity of 0.10 on both sides. What is the shield temperature, and what is the heat flux being transferred from plate 1 to plate 2? The spectral emissivities of 1 and 2 are in Homework Problem 6.2.

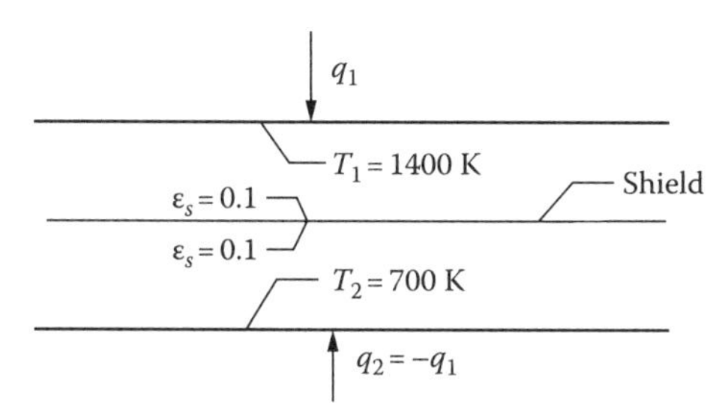

Answer: $T_s = 1194$ K; $q_1 = 9{,}525$ W/m^2.

6.4 Two large diffuse parallel plates are maintained at temperatures $T_1 = 1400$ K and $T_2 = 700$ K. The plates are made from the same metal, and their spectral emissivities as a function of wavelength, ε_λ, are approximated as shown by two constant values joined by a linear decrease with wavelength. Compute the net radiant energy flux being transferred from plate 1 to plate 2. What is the energy flux if both plates are assumed gray with an approximate average emissivity of 0.5 applied over the entire spectral range?

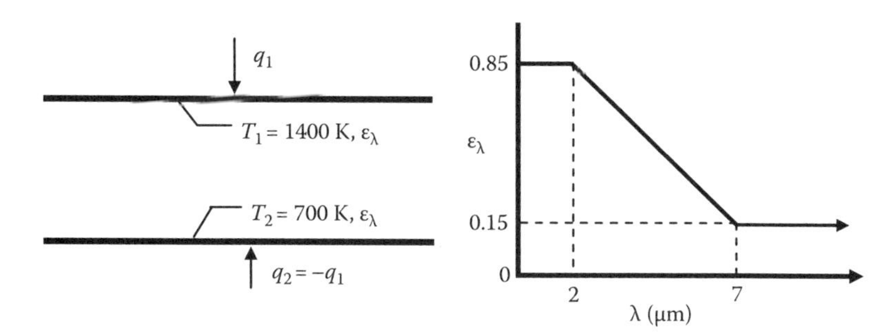

Answer: $q_1 = 106{,}850$ W/m^2; $q_{1,gray} = 68{,}070$ W/m^2.

6.5 Radiation is being transferred across the space between two large parallel plates. Both plates are made of the same metal, and the spectral emissivity of the metal does not depend on temperature. The wavelength variation of the emissivity for both plates is approximated in three steps as shown. The lower plate is maintained at $T_2 = 700$ K, and the upper plate has a uniform heat addition of $q_1 = 60{,}000$ W/m^2. What is the temperature, T_1, of the upper plate?

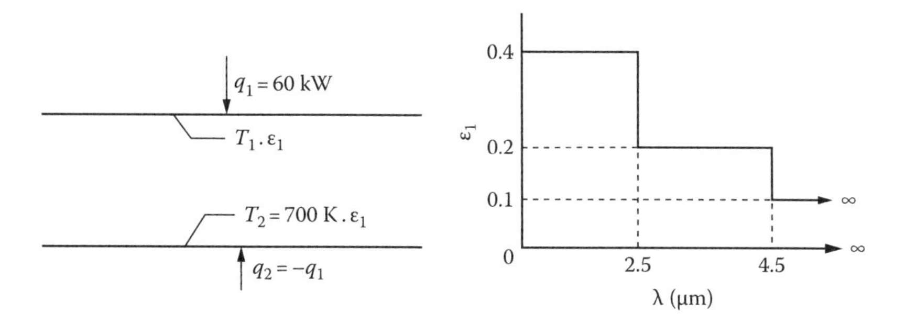

Answer: 1594 K.

6.6 A polished aluminum tank in vacuum in orbit is surrounded by one thin aluminum radiation shield. The shield is polished on the side facing the tank, and is painted with white paint on the outside facing the sun (see Figure 3.48). Treating the geometry as infinite parallel plates, what are the temperature of the shield and the heat flux into the tank for normally incident solar radiation? Assume the tank is maintained at 100 K.

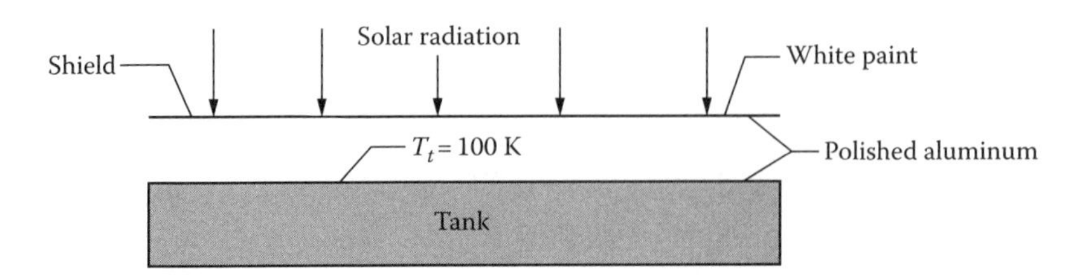

Answer: 266.7 K; 14.0 W/m^2.

6.7 Consider the enclosure of Homework Problem 5.19 with three walls at different specified temperatures. Compute the radiative heat transfer for the following values of hemispherical spectral emissivity. (For simplicity, do not subdivide the three areas into smaller regions.)

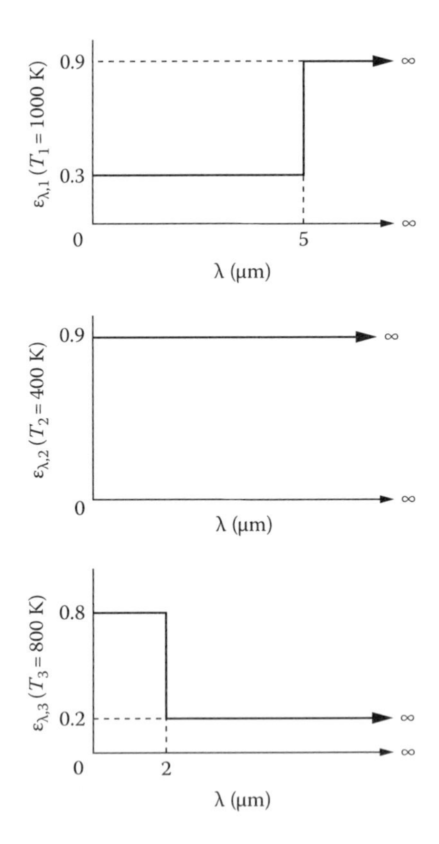

Answer: q_1 = 18,479 W/m^2; q_2 = –24,188 W/m^2; q_3 = –301 W/m^2.

6.8 Estimate the heat flux leaking into a liquid hydrogen container from an adjacent liquid nitrogen container. The glass walls are coated with polished aluminum. Use electromagnetic theory to estimate the radiative properties.

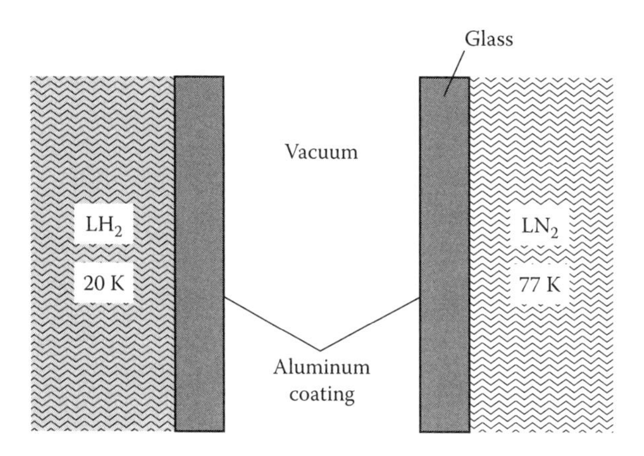

Answer: $q = 0.0030\ \text{W/m}^2$.

6.9 Two parallel plates are of finite width and of infinite length in the direction normal to that shown. The plates are diffuse, but their spectral emissivity varies with wavelength and is approximated by a step function as shown. Both plates are made of the same material, and hence have the same emissivity, which is independent of temperature. The plate edge openings are exposed to the environment at $T_e = 300$ K, and there is energy exchange only from the internal surfaces. Energy is being added to the lower plate, and the upper plate is maintained at a fixed temperature. Obtain values for the temperature of the lower plate, T_1, and for the heat flux q_2 that must be added or removed from the upper plate to maintain its specified temperature. For simplicity, do not subdivide the surface areas.

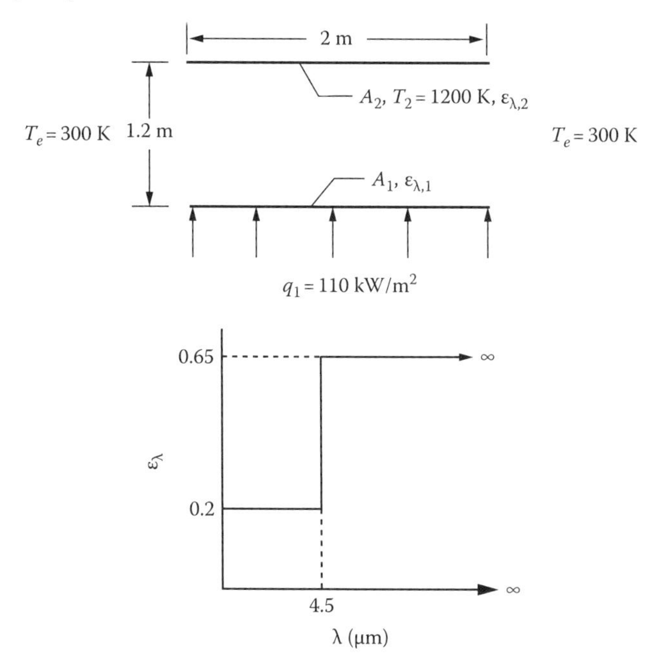

Answer: 1690 K; −8,893 W/m^2.

6.10 Two circular disks are parallel and directly facing each other. The disks are diffuse, but their emissivities vary with wavelength. The properties are approximated with step functions as shown. The disks are maintained at temperatures $T_1 = 1200$ K and $T_2 = 800$ K. The surroundings are at $T_e = 400$ K. Compute the rate of energy that must be supplied to or removed from the disks to maintain their specified temperatures. The outer surfaces of the disks are insulated so there is radiation interchange only from the inner surfaces that are facing each other.

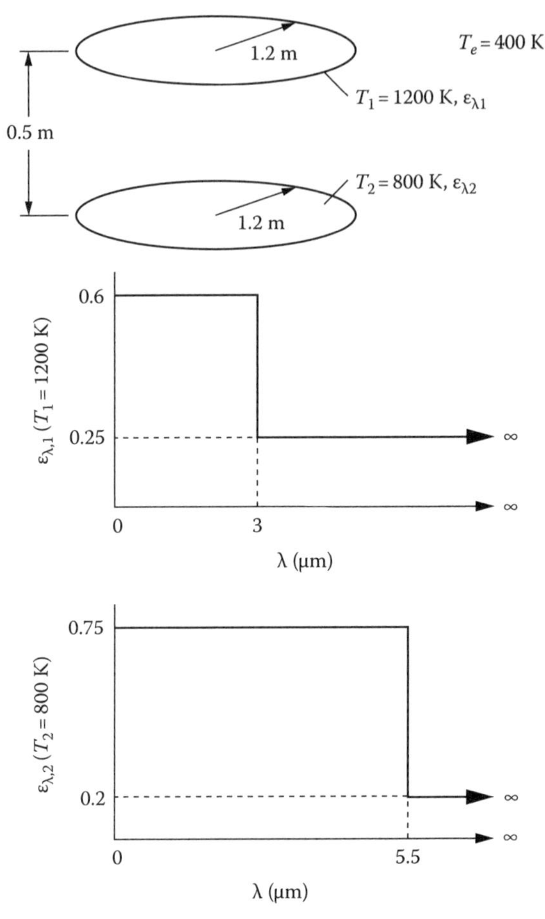

Answer: $Q_1 = 180{,}752$ W; $Q_2 = -58{,}570$ W.

6.11 An area element dA at temperature T is radiating out through a circular opening of radius r in a plate above and parallel to dA. The directional total emissivity of dA is $\varepsilon(\theta) = 0.85 \cos\theta$ as in Figure 2.4. Obtain a relation for the energy emission $Q_{nondiffuse}$ from dA through the opening. From Example 2.3, the hemispherical total emissivity of dA is 0.57. Using this ε and the diffuse configuration factor, compute $Q_{diffuse}$ through the opening. Plot $Q_{nondiffuse}/Q_{diffuse}$ as a function of h/r.

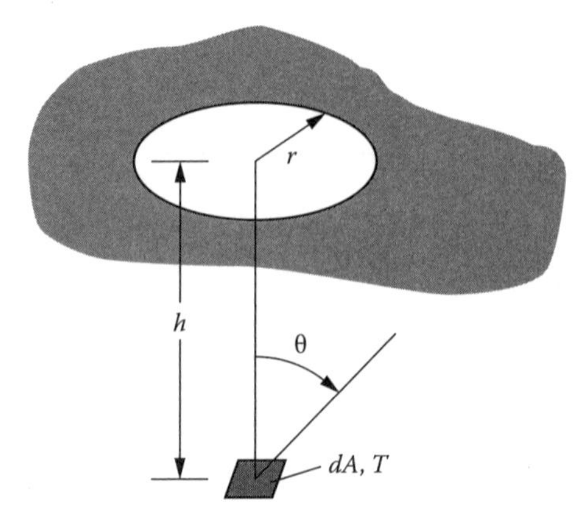

Answer: $\dfrac{Q_{nondiffuse}}{Q_{diffuse}} = \dfrac{\left(H^2+1\right)^{3/2} - H^3}{\left(H^2+1\right)^{1/2}}; \quad H = \dfrac{h}{r}$

6.12 A sphere and area element are positioned as shown. The sphere is gray but has a nondiffuse emissivity $\varepsilon_s = \mu \cos\theta$, where θ is the angle measured from the normal to the sphere surface and μ is a constant. Set up the integral for the direct radiation from the sphere to the area element in terms of the quantities given.

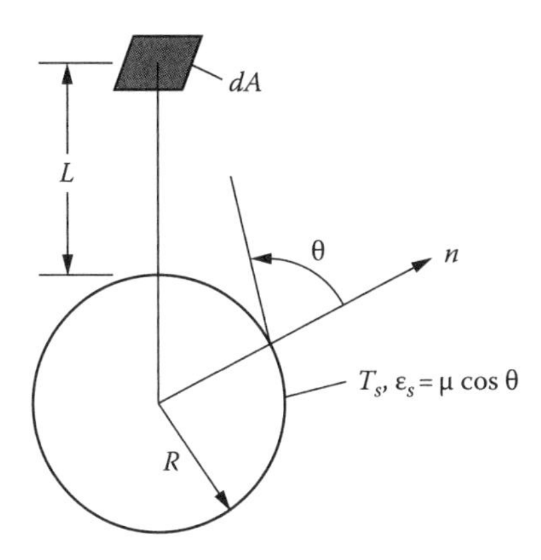

6.13 Two diffusely emitting and reflecting parallel plates are of infinite length normal to the cross-section shown in the figure. The upper plate has uniform temperature $T_1 = 1000$ K, and the lower plate has uniform temperature $T_2 = 500$ K. The surroundings have a temperature of $T_e = 450$ K. The surfaces are nongray with properties shown in the figure. Do NOT assume uniform irradiation on either surface, and find the distribution of net radiative heat flux on each surface.

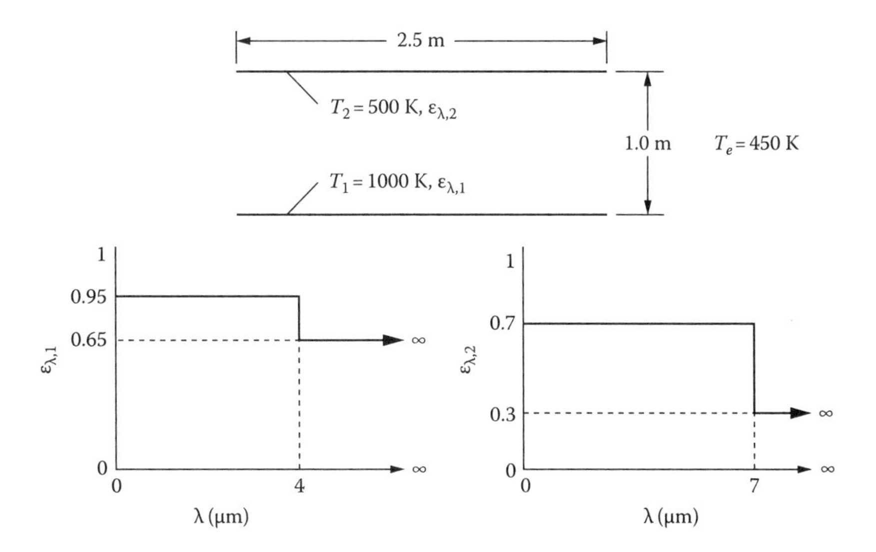

6.14 Obtain the result for $F_{1-1(2)}$ in Example 6.5 by using the crossed-string method. What is $F_{1-1(2)}$ if A_1 in Figure 6.23 is rotated 45° about its center point? (The geometry remains in 2D.)

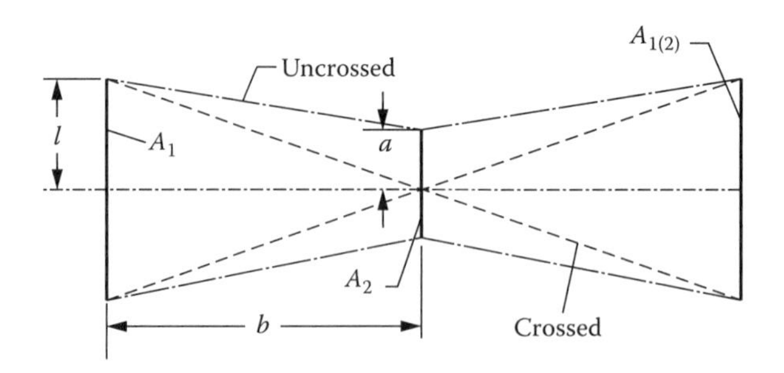

Answer:

First part: $F_{1-1(2)} = \left(\sqrt{1+b^{*2}} - \sqrt{(1-a^*)^2 + b^{*2}}\right)$

Second part: $F_{1-1(2)} = \frac{1}{2}\Big[2\left(b^{*2}+1/2\right)^{1/2}$

$$-\left(\sqrt{a^{*2}+b^{*2}+1-(a^*+b^*)\sqrt{2}}+\sqrt{a^{*2}+b^{*2}+1-(a^*-b^*)\sqrt{2}}\right)\Big]$$

where $a^* = a/\lambda$, $b^* = b/\lambda$.

6.15 In Figure 6.25, the image $A_{2(3-4)}$ is shown as a dotted horizontal line on the lower right hand side of the image diagram. By a suitable construction of rays (actual and through image surfaces) similar to the one shown in the text, show whether $A_{2(3-4)}$ is where Figure 6.25 indicates it to be.

6.16 Obtain an analytical expression for the configuration factor $F_{1-1(2)}$ between the cylinder A_1 and its image $A_{1(2)}$ for each of the two situations shown (the geometries are 2D in nature and are shown in cross section).

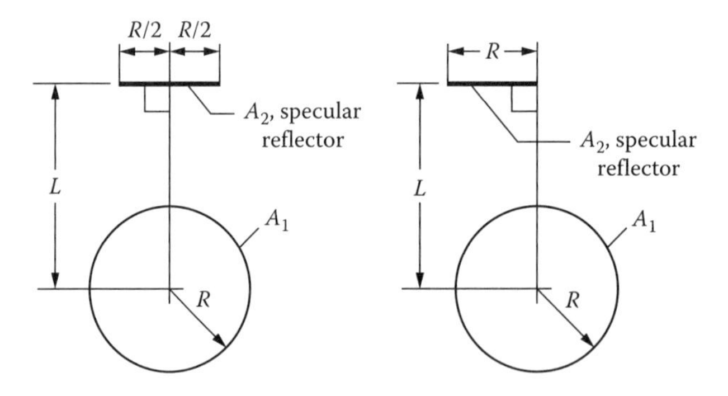

Answer:

$$\text{(a)}\quad F_{1-1(2)} = \frac{1}{\pi}\left\{\sqrt{\left(\frac{L}{R}\right)^2 - 1} + \cos^{-1}\left[\frac{1}{\sqrt{\left(\frac{L}{R}\right)^2 + \frac{1}{4}}}\right] + \tan^{-1}\left(\frac{R}{2L}\right) - \cos^{-1}\left(\frac{R}{L}\right) - \sqrt{\left(\frac{L}{R}\right)^2 - \frac{3}{4}}\right\}$$

$$\text{(b)}\quad F_{1-1(2)} = \frac{1}{2\pi}\left[\sqrt{\left(\frac{L}{R}\right)^2 - 1} + \frac{\pi}{2} - \cos^{-1}\left(\frac{R}{L}\right) - \frac{L}{R}\right].$$

6.17 A long black circular cylinder is partially surrounded by two parallel long plane specular surfaces as shown in cross section (geometry is 2D in nature). What is the rate of heat loss from the cylinder in terms of the quantities shown?

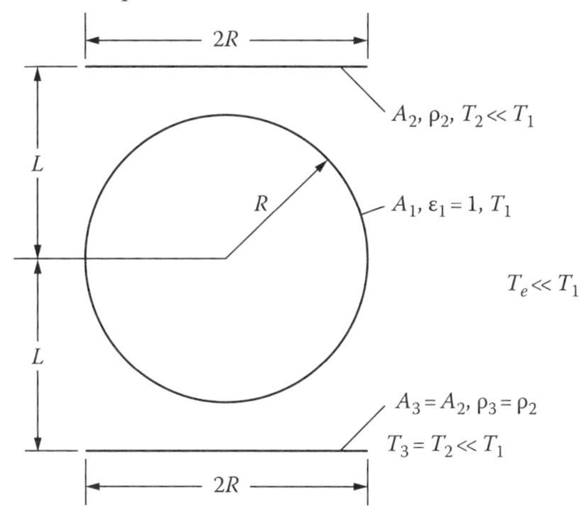

Answer: $\dfrac{Q}{\text{length}} = 2\pi R\sigma T_1^4\left\{1 - \dfrac{2}{\pi}\left[\left(X^2 - 1\right)^{1/2} + \sin^{-1}\left(\dfrac{1}{X}\right) - X\right]\right\}$; $X = L/R$.

6.18 A long enclosure is made up of two specular and two diffuse surfaces as shown in cross section. Draw a diagram of the images that are needed to determine the energy exchange process. Then write the equations for F_{1-2}^s and F_{1-3}^s in terms of the required specular configuration factors and reflectivities (i.e., $F_{1-2}^s = F_{1-2} + \rho_{s,3} F_{1-2(3)} + \cdots$). Now write the set of energy exchange equations for finding Q_1, Q_2, Q_3, Q_4.

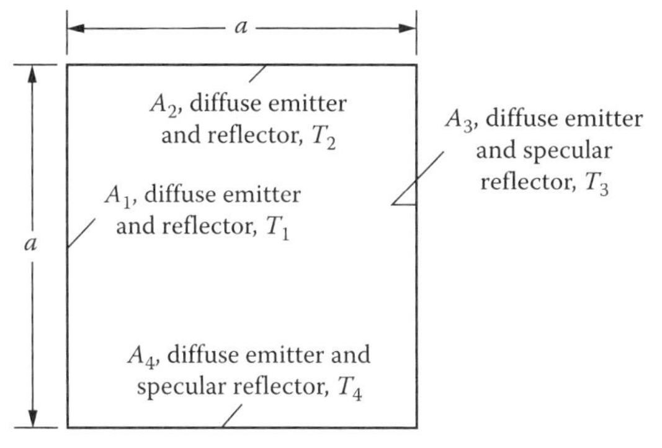

6.19 An equilateral triangular enclosure of infinite length normal to the cross section shown has black surfaces A_1 and A_2 and a specularly reflecting surface A_3 with reflectivity $\rho_{s,3} = 0.6$. Find the values of F_{1-1}^s, F_{1-2}^s, and F_{1-3}^s.

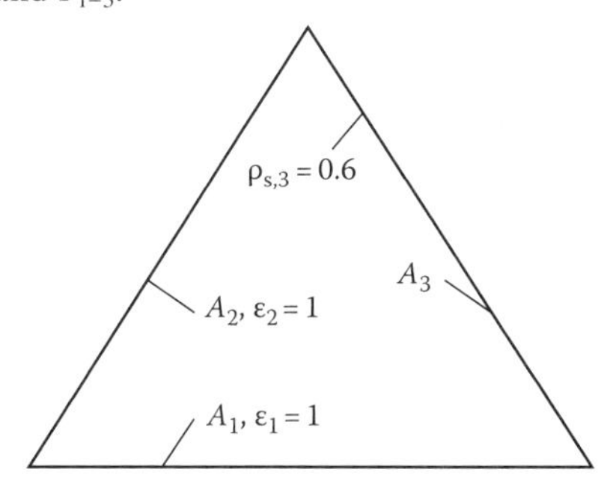

Answer: 0.0804; 0.7196; 0.5000.

6.20 (a) What is the value of the sum of the specular configuration factors $\sum_{j=1}^{N} F_{1-j}^{s}$ for Homework Problem 6.19?

(b) What is the value of the summation $\sum_{j=1}^{N} (1-\rho_{s,j}) F_{1-j}^{s}$ for Homework Problem 6.19?

(c) Explain the results of parts (a) and (b) in terms of the definition of F_{1-j}^{s}. Is the result of part (b) a general relation for all specular enclosures?

Answer: (a) 1.30; (b) 1.00.

6.21 An equilateral triangular enclosure has sides that extend in the normal direction infinitely far into and out of the plane of the cross section shown in the figure. Side 3 is specular and perfectly insulated on the outside. Find:

(a) All necessary exchange factors F^s needed to solve for the unknown temperatures.

(b) The temperatures of surfaces A_2 and A_3.

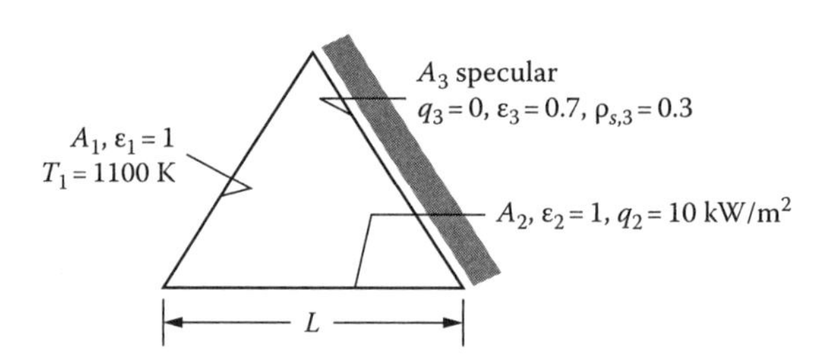

Answer: $T_2 = 1140$ K; $T_3 = 1121$ K.

6.22 An enclosure of equilateral triangular cross section and of infinite length normal to the cross-section shown has two diffuse reflecting interior surfaces and one specularly reflecting interior surface that is perfectly insulated on the outside (i.e., $q_3 = 0$). All surfaces are gray and are diffuse emitters. Compute T_3 and the Q added to each of A_1 and A_2 as a result of radiative exchange within the enclosure for the conditions shown. (For simplicity, do not subdivide the surface areas.)

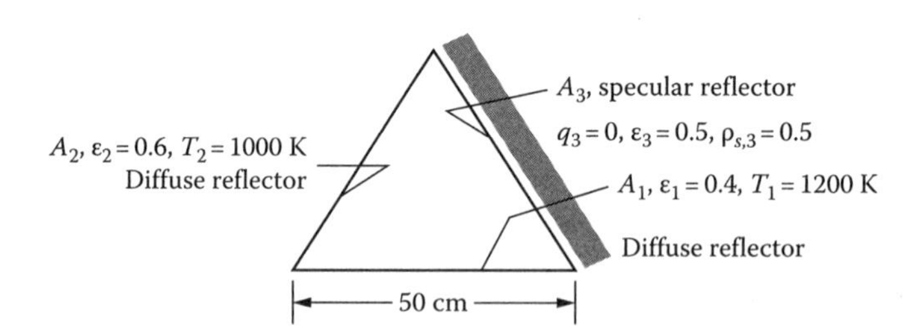

Answer: $T_3 = 1090$ K; $-q_2 = q_1 = 16.9$ kW/m^2.

6.23 An enclosure is made up of three sides as shown. The length L is sufficiently long that the triangular ends can be neglected in the energy balances. Two of the surfaces are black, and the other is a diffuse-gray emitter with emissivity $\varepsilon_1 = 0.10$. What is the energy added, per meter along the length L, to each surface because of radiative exchange within the enclosure for each of the two cases:

(a) Area 1 is a diffuse reflector.

(b) Area 1 is a specular reflector.

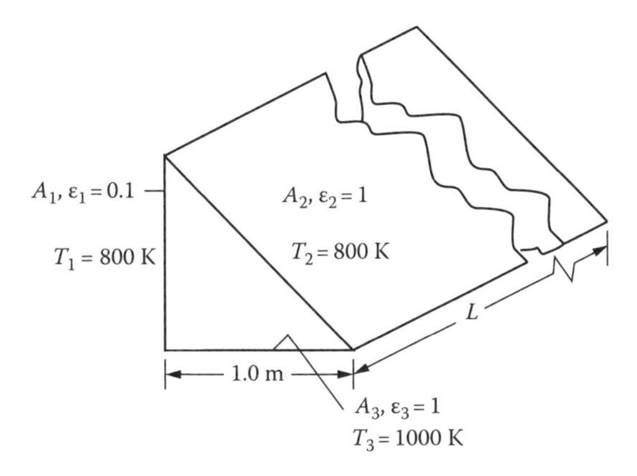

6.24 A triangular enclosure has the conditions shown in the figure. Side 3 is a specular reflector.
(a) Derive relations for T_1^4, q_2, and q_3.
(b) Prove that $q_2A_2 = -q_3A_3$.

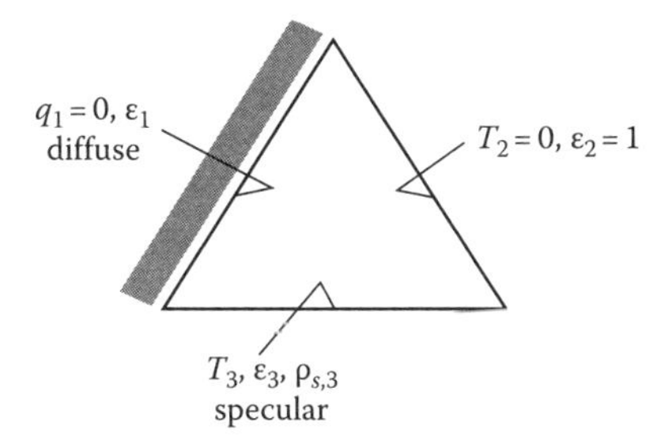

Answer: $T_1^4 = \dfrac{\varepsilon_3 T_3^4 F_{1-3}}{1-\rho_{s,3}F_{1(3)-1}}$; $\quad q_2 = -\varepsilon_3\sigma T_3^4\left(\dfrac{F_{1-3}}{1-\rho_{s,3}F_{1(3)-1}}F_{2-1}^s + F_{2-3}\right)$

$$q_3 = \varepsilon_3\sigma T_3^4\left(1-\frac{\varepsilon_3 F_{1-3}}{1-\rho_{s,3}F_{1(3)-1}}F_{3-1}\right).$$

6.25 Compute the specular exchange factor F_{2-1}^s for the 2D rectangular enclosure shown. All surfaces are gray and are diffuse emitters. Surfaces A_1 and A_2 are diffuse reflectors, while A_3 and A_4 are specular reflectors with $\rho_{s,3} = 0.3$ and $\rho_{s,4} = 0.8$.

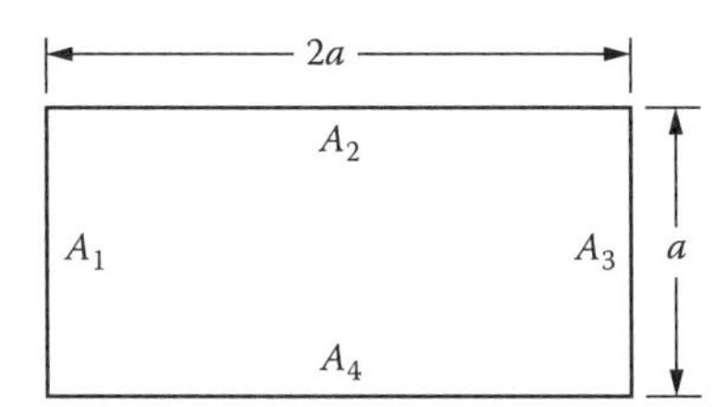

Answer: 0.296.

6.26 An infinitely long square bar of size 2 m on a side is enclosed by an infinitely long concentric circular cylinder of 2 m radius. The temperatures and emissivities of the bar and cylinder are, respectively, T_b, ε_b, and T_c, ε_c. Find the rate at which radiant energy is exchanged between A_1 and A_2 per unit of enclosure length if

(a) Both are diffuse.
(b) Both are specular.
(c) The bar is diffuse, and the cylinder is specular.

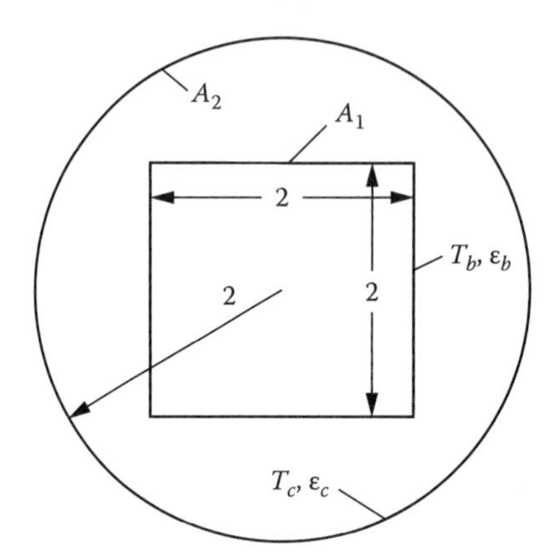

Answer: (a) $Q = \dfrac{8\sigma\left(T_b^4 - T_c^4\right)}{\dfrac{1}{\varepsilon_b} + \dfrac{2}{\pi}\left(\dfrac{1}{\varepsilon_c} - 1\right)}$; (b), (c) $Q = \dfrac{8\sigma\left(T_b^4 - T_c^4\right)}{\dfrac{1}{\varepsilon_b} + \dfrac{1}{\varepsilon_c} - 1}$.

6.27 A 2D rectangular enclosure has gray interior surfaces that are all diffuse emitters. Two opposing surfaces are specular reflectors, while the other two reflect diffusely. Write the set of energy equations to determine Q_1, Q_2, Q_3, and Q_4, including the expressions for the F^s factors in terms of the configuration factors F. (*Note*: Each F^s will consist of an infinite sum.)

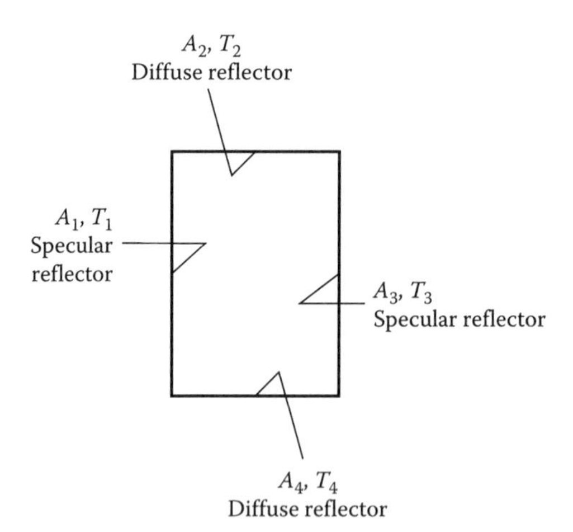

6.28 Two parallel plates of unequal finite width are facing each other. The lower plate is diffuse-gray, and the upper plate is a specular reflector and is gray. The geometry is long in the direction normal to the cross section shown, so the configuration is in 2D. The lower plate is uniformly heated from below so the energy supplied is radiated away from only its upper surface. The upper plate has its upper surface cooled so its lower surface (the specular reflector) is maintained at 600 K. The surroundings are at 400 K. For simplicity, do not subdivide the plate areas.

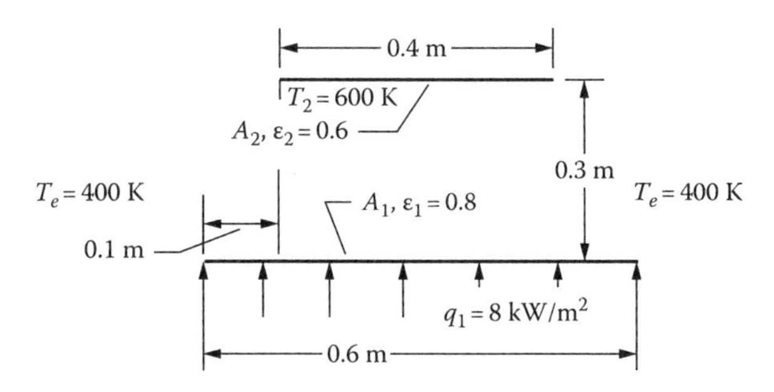

(a) Obtain the temperature of the lower plate and determine what heat flux must be extracted from the upper plate to maintain its specified temperature.

(b) The surface A_2 on the upper plate is now roughened to make it a diffuse reflector, without changing its emissivity. How much will this change the temperature of the lower plate?

Answer:

(a) $T_1 = 712.8$ K, $q_2 = -939.0$ W/m^2;

(b) $T_1 = 708.8$ K (4.0 K decrease), $q_2 = -810.2$ W/m^2.

6.29 A diffuse plate A_1 that is not heated or cooled by external means is facing a larger plate as shown. The spacing between the plates is small. The plates are long in the direction normal to the cross section shown, so the geometry is 2D in nature. The upper plate is gray and is a specular reflector. It is cooled to 300 K. The environment is at a higher temperature, 500 K, so that radiant energy is reflected into the space between the plates. For simplicity, do not subdivide the plate areas.

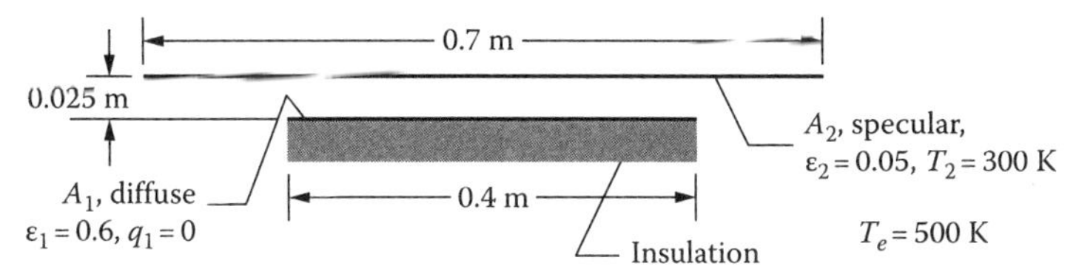

(a) What is the temperature of the lower plate A_1?

(b) If A_2 on the upper plate is made diffuse without changing its emissivity, how much is the temperature of the lower plate changed?

Answer: (a) $T_1 = 462.4$ K; (b) $T_1 = 487.8$ K (a 25 K increase).

6.30 Two parallel plates shown in Homework Problem 5.36 are of finite width (infinite length normal to the cross section). Both plates are perfectly insulated on the outside. Plate 1 is uniformly internally heated electrically with heat flux q_e. Plate 2 has no externally supplied heat input ($q_2 = 0$). The environment is at zero temperature. Plate 1 is black, while plate 2 is a diffuse-gray emitter and specular reflector with emissivity ε_2. Derive an integral equation formulation for the surface temperature distributions. Compare with the formulation in Homework Problem 5.36.

6.31 Derive Equations 6.64 and 6.65 for concentric diffuse cylindrical or spherical surface containing N radiation shields when the shields are either (a) diffuse or (b) specular. Derive an equation for the temperature of the nth shield in each case.

Answer: Q is given by Equations 6.64 and 6.65. For all diffuse shields, the temperature of shield n is

$$T_n^4 = T_1^4 - \frac{\left(T_1^4 - T_2^4\right)\left\{\dfrac{1}{\varepsilon_1} + \dfrac{A_1}{A_{s1}}\left(\dfrac{1}{\varepsilon_{s11}} - 1\right) + \sum_{n=1}^{n-1} \dfrac{A_1}{A_{sn}}\left[\dfrac{1}{\varepsilon_{n2}} + \dfrac{A_{sn}}{A_{s(n+1)}}\left(\dfrac{1}{\varepsilon_{(n+1)1}} - 1\right)\right]\right\}}{\dfrac{1}{\varepsilon_1} + \dfrac{A_1}{A_2}\left(\dfrac{1}{\varepsilon_2} - 1\right) + \sum_{n=1}^{N} \dfrac{A_1}{A_{sn}}\left[\dfrac{1}{\varepsilon_{n1}} + \dfrac{1}{\varepsilon_{n2}} - 1\right]}$$

and for all specular shields with diffuse boundary surfaces, the temperature of any shield n is

$$T_n^4 = T_1^4 - \frac{\left(T_1^4 - T_2^4\right)\left\{\left(\frac{1}{\varepsilon_1} + \frac{1}{\varepsilon_{s11}} - 1\right) + \sum_{n=1}^{n-1} \frac{A_1}{A_{sn}}\left(\frac{1}{\varepsilon_{sn2}} + \frac{1}{\varepsilon_{s(n+1)1}} - 1\right)\right\}}{\left(\frac{1}{\varepsilon_1} + \frac{1}{\varepsilon_{s11}} - 1\right) + \sum_{n=1}^{N-1} \frac{A_1}{A_{sn}}\left(\frac{1}{\varepsilon_{sn2}} + \frac{1}{\varepsilon_{s(n+1)1}} - 1\right) + \frac{A_1}{A_{sN}}\left[\frac{1}{\varepsilon_{sN2}} + \frac{A_{sN}}{A_2}\left(\frac{1}{\varepsilon_2} - 1\right)\right]}.$$

6.32 Two infinite gray parallel plates are separated by a thin gray radiation shield. What is T_s, the temperature of the shield? What radiative energy flux is transferred from plate 2 to plate 1 with the shield in place? What is the ratio of the heat transferred from 2 to 1 with the shield to that transferred without the shield?

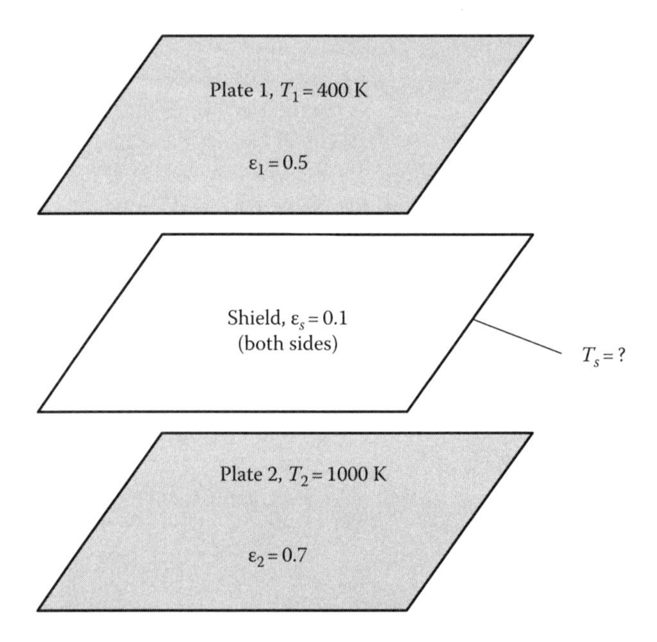

Answer: 852 K; 2,578 W/m²; 0.113.

6.33 A radiative heat flux q_0 is transferred across the gap between two gray parallel plates having the same emissivity ε that are at temperatures T_1 and T_2 (ε is independent of temperature). A single thin radiation shield also having emissivity ε on both sides is placed between the plates. Show that the resulting heat flux is $q_0/2$. Show that adding a second identical shield reduces the heat flux to $q_0/3$. Show that for n shields the heat flux is $q_0/(n + 1)$ when all the surface emissivities are the same.

6.34 (a) What is the effect of a single thin radiation shield on the transfer of energy between two concentric cylinders? Assume the cylinder and shield surfaces are diffuse-gray with emissivities independent of temperature. Both sides of the shield have emissivity ε_s, and the inner and outer cylinders have respective emissivities ε_1 and ε_2.

(b) What is the effect of a single thin radiation shield on the transfer of energy between two concentric spheres? Assume the sphere and shield surfaces are diffuse-gray with emissivities independent of temperature. Both sides of the shield have emissivity ε_s, and the inner and outer spheres have respective emissivities ε_1 and ε_2.

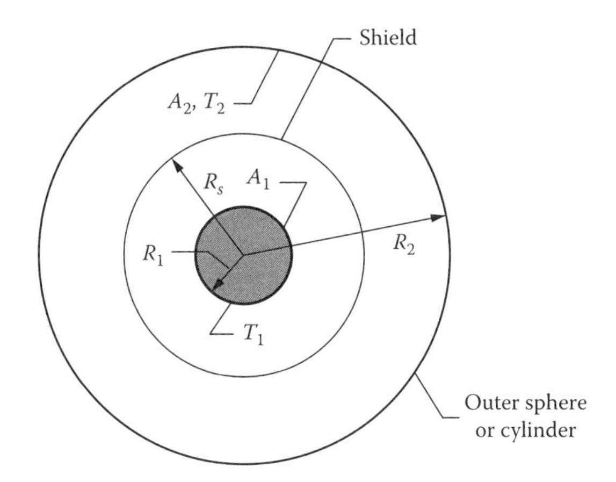

Answer:

(a) $$\frac{Q_{1,shield}}{Q_{1,no\ shield}} = \frac{\Gamma_{12}}{\Gamma_{1s} + \dfrac{A_1\Gamma_{s2}}{A_s}};$$ (b) $$\frac{Q_{1,shield}}{Q_{1,no\ shield}} = \frac{\Gamma_{12}}{\Gamma_{1s} + \dfrac{A_1\Gamma_{s2}}{A_s}}$$

where

$$\Gamma_{1s} \equiv \frac{1}{\varepsilon_1} + \frac{A_1}{A_s}\left(\frac{1}{\varepsilon_s} - 1\right); \quad \Gamma_{s2} \equiv \frac{1}{\varepsilon_s} + \frac{A_s}{A_2}\left(\frac{1}{\varepsilon_2} - 1\right)$$

6.35 Derive a relation for the heat transfer q through an N-layer set of concentric circular cylindrical shields placed between an inner surface at T_1 and an outer layer at T_2. The outer surface has diameter D_1, and the inner surface has diameter D_2. All surfaces are diffuse and have an emissivity of ε.

Answer: $$q' = \frac{\sigma\left(T_1^4 - T_2^4\right)}{\dfrac{1}{\pi D_1}\left[\dfrac{1}{\varepsilon} + \dfrac{D_1}{D_{S1}}\left(\dfrac{1}{\varepsilon} - 1\right)\right] + \sum_{n=1}^{N-1} R_{n-(n+1)} + \dfrac{1}{\pi D_{SN}}\left[\dfrac{1}{\varepsilon} + \dfrac{D_{SN}}{D_2}\left(\dfrac{1}{\varepsilon} - 1\right)\right]}$$

where $$R_{n-(n+1)} = \frac{1}{\pi D_n}\left\{\frac{1}{\varepsilon}\left[1 + \left(\frac{D_n}{D_{n+1}}\right)\right] - \left(\frac{D_n}{D_{n+1}}\right)\right\}$$

6.36 A shield system of N large closely spaced flat-plate thin shields with all shield surfaces of emissivity ε_s is placed between an outer surface at T_1 with emissivity ε_1 and another surface at T_2 with emissivity ε_2. Derive an expression for the heat flux q between surfaces 1 and 2.

Answer: $$q = \frac{\sigma\left(T_1^4 - T_2^4\right)}{\left(\dfrac{1}{\varepsilon_1} + \dfrac{1}{\varepsilon_2} - 1\right) + N\left(\dfrac{2}{\varepsilon_s} - 1\right)}.$$

6.37 Two large gray parallel plates have 20 gray radiation shields between them. The plate temperatures and emissivities are shown in the figure, and all of the shield surfaces have $\varepsilon_s = 0.035$. Determine the temperatures of the 20 shields.

A_1, $\varepsilon_1 = 0.85$, $T_1 = 1050$ K

20 shields

$\varepsilon_s = 0.035$

A_2, $\varepsilon_2 = 0.65$, $T_2 = 525$ K

Answer: $T_n = \left\{T_1^4 - \dfrac{q}{\sigma}\left[\dfrac{1}{\varepsilon_1} + \dfrac{1}{\varepsilon_s} - 1 + (n-1)\left(\dfrac{2}{\varepsilon_s} - 1\right)\right]\right\}^{1/4}$ resulting in $T_1 = 1043$ K and $T_{20} =$ 569.9 K.

6.38 Two large plane parallel surfaces A_1 and A_2 are separated by 80 radiation shields having an $\varepsilon_s = 0.042$ coating on both sides. However, by accident, 15 of the shields were coated on only the upper side. The $\varepsilon_s = 0.85$ for the uncoated sides. What is the rate of heat transfer for the defective system, and how does it compare with the heat flow rate when none of the shields are defective (all are coated on both sides)? All surfaces are gray. Give heat flow rates in W/m².

A_1, $\varepsilon_1 = 0.85$, $T_1 = 800$ K

80 shields

$\varepsilon_s = 0.042$

A_2, $\varepsilon_2 = 0.50$, $T_2 = 400$ K

Answer: 5.84 W/m²; 6.42 W/m²; 10% increase.

6.39 A long rectangular enclosure is shown in cross section (2D geometry). The interior surfaces 1 and 4 are diffuse-gray with specified wall temperatures. A group of N thin diffuse-gray radiation shields is placed on the diagonal. Derive an algebraic expression for the heat flow Q from the surfaces at T_1 to those at T_4 in terms of the quantities given. Evaluate the result for $\varepsilon_1 = 0.75$, $\varepsilon_4 = 0.20$, $\varepsilon_s = 0.10$, $H = 1.20$ m, $W = 0.90$ m, $N = 6$, $T_1 = 600$ K, $T_4 = 400$ K, and for 1 m of length in the third dimension.

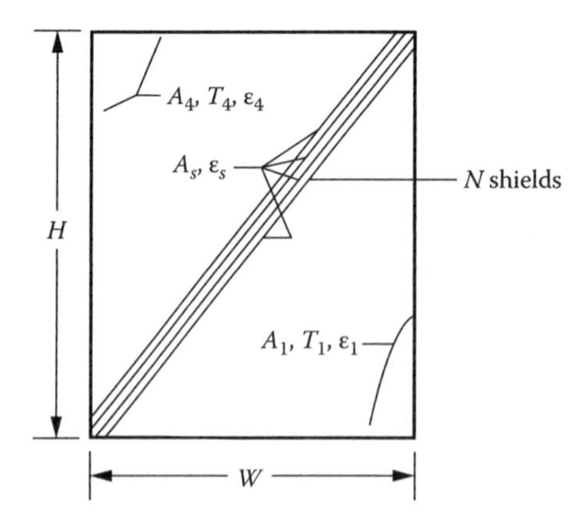

Answer: $Q = \dfrac{A_2\sigma\left(T_1^4 - T_4^4\right)}{\dfrac{A_2}{A_1}\left(\dfrac{1}{\varepsilon_1} + \dfrac{1}{\varepsilon_4} - 2\right) + 1 + N\left(\dfrac{2}{\varepsilon_s} - 1\right)}$; 74.9 *W*.

6.40 Solar radiation with a flux of $q_{solar} = 1353$ W/m^2 is normally incident on a front surface of a shield configuration. The front surface has solar absorptivity $\alpha_{solar} = 0.10$, and with IR emissivity of 0.86. The back of the surface has emissivity 0.015 and faces a 4-layer shield system; each surface of the shields has an emissivity of $\varepsilon_s = 0.015$. The outer face of the last layer of the shield faces space that can be taken as having a temperature of 4 K. What is the temperature of each layer (i.e., the front surface and each of the shields)?
Answer: $T_1 = 228.2$ K; $T_2 = 214.4$ K; $T_3 = 197.4$ K; $T_4 = 174.2$ K; $T_5 = 134$ K; $q = 2.74$ W/m^2.

6.41 It is proposed to add a cryogenic cooling loop to the face of the last layer of the shield system described in Homework Problem 6.40. The cooler is needed to reduce the last shield face temperature to below 30 K. What cooling capacity per unit shield area q_c (mW/m^2) is required to achieve this temperature?
Answer: $q_{cry} = 3090$ mW/m^2.

7 Radiation Combined with Conduction and Convection at Boundaries

7.1 INTRODUCTION

In the preceding chapters, enclosure theory was formulated for radiative exchange between surfaces. The local net radiation loss at a surface was balanced by energy supplied by "some other means" that were not explicitly described. This chapter is concerned with this energy at the surface either by conduction from within the volume interior to the surface (such as from within a wall of an enclosure) or by convection or conduction at the surface from a surrounding medium. At each location along the surface, the radiation, convection, and conduction combine to form a thermal boundary condition. The solution to the energy equations subject to this condition provides the surface temperature and heat flux distributions. The analysis has the same restrictions as in the previous theory: the surfaces are *opaque*, and the *medium* between the radiating surfaces is *perfectly transparent*. The medium between the radiating surfaces may be conducting or convecting energy, but it does not interact with radiation passing through it.

The following situations illustrate combined-mode energy transfer; and additional applications are discussed throughout this chapter. For a vapor-cycle power plant operating in outer space, waste heat must be rejected by radiation. In the space radiator in Figure 7.1a, the vapor of the working fluid in a thermodynamic cycle is condensed, thereby releasing its latent heat. This energy is conducted through the condenser wall and into fins that radiate the energy into outer space. The temperature distribution in the fins and their radiating efficiency depend on combined radiation and conduction. A fin-tube geometry is also commonly used for the absorber in a flat-plate solar collector. Solar energy is incident on the absorber plate through one or more transparent cover glasses. A fluid is heated as it flows through tubes attached to the absorber plate. The collector design requires analysis of combined radiation, conduction, and convection.

In one type of steel-strip cooler in a steel mill (Figure 7.1b), a sheet of hot metal moves past a bank of cold tubes and loses energy to them by radiation. At the same time cooling gas is blown across the sheet. A combined radiation and convection analysis is required to determine the temperature distribution along the steel strip. Controlled radiative and convective cooling is also used in the tempering of sheets of high-strength glass for automobile windows.

In a possible design for a nuclear rocket engine, illustrated by Figure 7.1c, hydrogen gas is heated by flowing through a high-temperature nuclear reactor. The hot gas then passes out through the rocket nozzle. The interior surface of the rocket nozzle receives energy by radiation from the exit face of the reactor core and by convection from the flowing gas. Cooling the nozzle depends on conducting this energy through the nozzle wall and removing it by flowing coolant.

These examples involve energy transfer by two or more modes. The modes can be in series, such as conduction through a wall followed by radiation from its surface. Energy transfer can also be by parallel modes, such as simultaneous conduction and radiation through a transparent material, or simultaneous radiation and convection from a hot surface. Often, both series and parallel modes are present. The interaction of the modes can be simple in some cases. For example, if the amounts of energy transferred from a surface by radiation and convection are independent, they can be

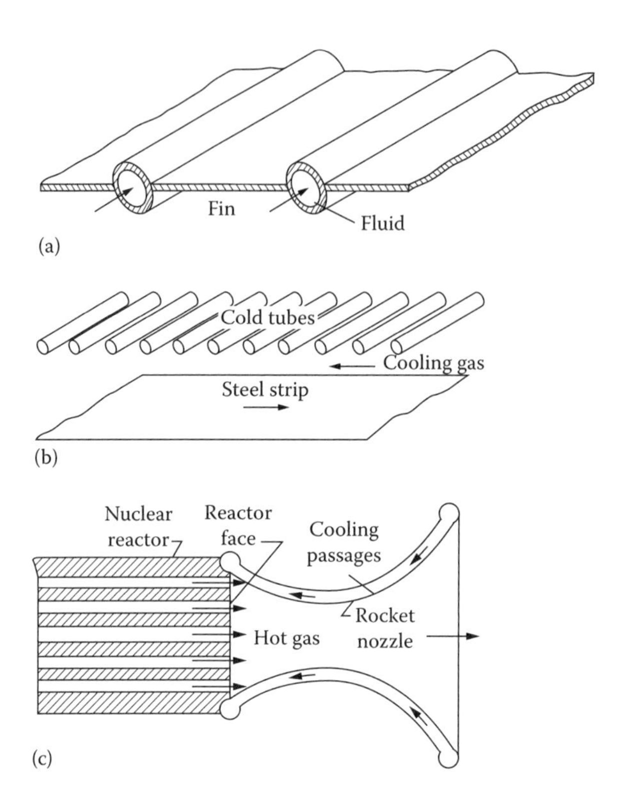

FIGURE 7.1 Heat transfer devices involving combined radiation, conduction, and convection: (a) space radiator or absorber plate of flat-plate solar collector, (b) steel-strip cooler, and (c) nuclear rocket.

computed separately and added. In other instances, the interaction can be complex, such as coupled radiation and free convection.

The various heat transfer modes depend on temperature and/or temperature differences to different powers. When radiation exchange between black surfaces is considered, the energy fluxes depend on surface temperatures to the fourth power. For nonblack surfaces, the temperature dependence may differ somewhat because of emissivity variations with temperature. Heat conduction depends on the local temperature gradient. Convection depends approximately on the first power of the temperature difference. The exact power depends on the type of flow; for example, free convection depends on temperature difference to a power from 1.25 to 1.4. Physical properties that vary with temperature introduce additional temperature dependencies. The various powers and dependencies of temperature provide nonlinear energy transfer relations, and it is usually necessary to use numerical solution techniques. This chapter provides methods for setting up the energy-balance relations, and physical behavior is illustrated and some common solution methods are presented.

7.2 ENERGY RELATIONS AND BOUNDARY CONDITIONS

7.2.1 General Relations

In the analyses developed for enclosures, the net radiative energy flux at any position on the boundary was balanced by the energy flux q supplied by "some other means." The means considered here are conduction, convection, or wall internal energy sources such as electric heaters or nuclear reactions. Since the enclosure walls are assumed opaque, the absorption of radiation is at the surface, and the energy balance provides a boundary condition. Although there can be conduction or

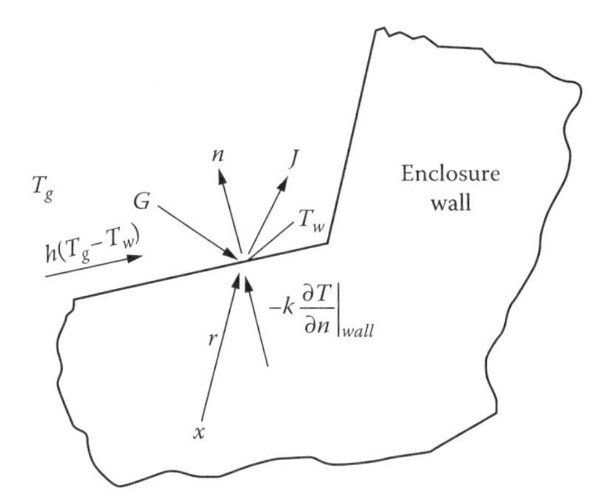

FIGURE 7.2 Boundary condition at location r on the surface of an opaque-walled enclosure.

convection in a medium between radiating surfaces, the medium is assumed here to be perfectly transparent, so radiation passes through with undiminished intensity. The radiation exchange relations developed previously for an enclosure are unchanged. If, for example, convection is expressed in terms of a heat transfer coefficient, Equation 7.1 for q can be written as

$$q = h(T_g - T_w) - k\frac{\partial T}{\partial n}\bigg|_{wall} = J - G \tag{7.1}$$

where all quantities are at r on the surface of the enclosure wall in Figure 7.2. The previous enclosure relations are valid as they are written in terms of q. For example, Equation 5.56 relates T and q along the enclosure boundaries. If the T are given, Equation 5.56 is solved for the q; then Equation 7.1 yields $\partial T/\partial n|_{wall}$. The T and $\partial T/\partial n$ at the wall surface are the boundary conditions for the heat conduction equation within the wall interior:

$$\rho c\frac{\partial T}{\partial t} = \nabla \cdot (k\nabla T) + \dot{q} \tag{7.2}$$

The form of the energy equation inside the enclosure in the space between the radiating surfaces depends on the type of convection, such as forced convection in a channel, a boundary layer flow, or free convection. If the convection depends significantly on the boundary temperatures or heat flux distributions, the solution may require simultaneous solution of the radiation exchange, heat conduction in the wall, and convection relations.

In some problems, the net energy added to the surface by external means is specified more directly than by the normal derivative in Equation 7.1. For example (Figure 7.3), if electric heating generates an energy flux q_e in the wall with insulation on one boundary and negligible heat conduction along the wall, all of the q_e appears at the radiating boundary, and Equation 7.1 becomes, at each location along the boundary,

$$q = h(T_g - T_w) + q_e = J - G \tag{7.3}$$

The q_e may be uniform over the surface area, or it can have a specified variation with location. The heating might be by passage of electric current within the wall, such as for an electrically heated wire.

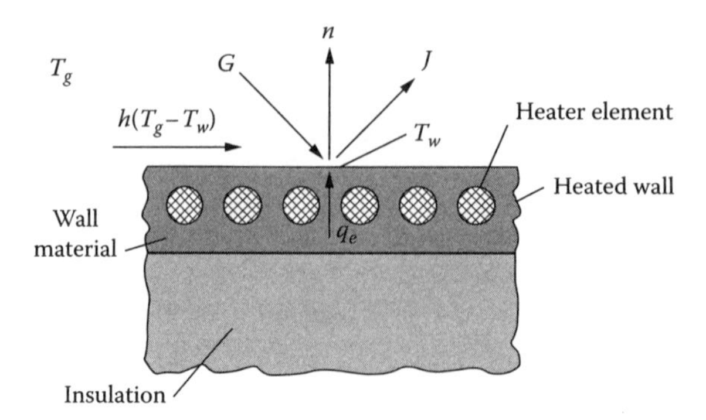

FIGURE 7.3 Boundary condition at surface of opaque wall with specified heat flux.

7.2.2 Uncoupled and Coupled Energy Transfer Modes

In the simplest situations, the radiation, conduction, and convection contributions to an unknown quantity, such as heat flux, are independent. The contributions are computed separately and the results combined. The energy transfer modes are *uncoupled* with regard to the desired quantity.

Example 7.1

Consider the region between two large gray parallel walls with a transparent gas between them (Figure 7.4). The internal surface temperatures T_1 and T_2 are specified. There is free convection in the gas, and the free-convection heat transfer coefficient h_{fc} depends on T_1 and T_2. What is the steady-state energy transfer from wall 1 to wall 2?

The energy transfer is the net radiative exchange and the transfer by free convection. It is equal to the flux q_1 that must be added to wall 1 to maintain it at its specified temperature. Since T_1 and T_2 are given, the h_{fc} can be computed from free-convection correlations and the net energy transfer is, by use of Equation 5.6.3,

$$q_1 = \frac{\sigma\left(T_1^4 - T_2^4\right)}{1/\varepsilon_1\left(T_1\right) + 1/\varepsilon_2\left(T_2\right) - 1} + h_{fc}\left(T_1, T_2\right)\left(T_1 - T_2\right)$$

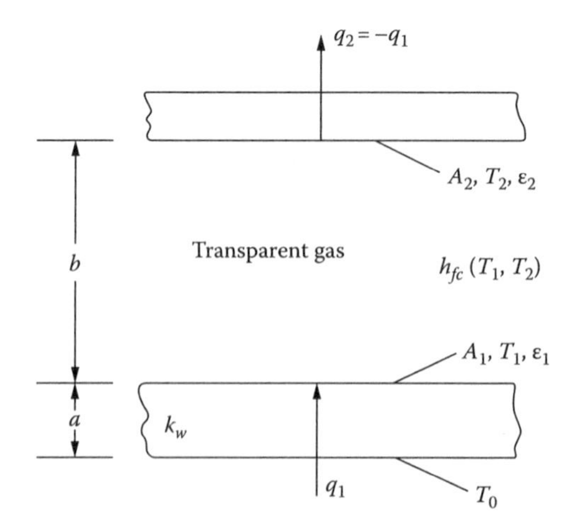

FIGURE 7.4 Parallel wall geometry for Examples 7.1 and 7.2 ($T_1 > T_2$).

The radiative and convective components are *uncoupled*. The q for each mode is computed independently, and the contributions added. The methods of radiative computation developed earlier can be applied without modification.

Coupled problems are more common than uncoupled problems. In coupled problems, the desired unknown quantity cannot be found by adding separate solutions; the energy relations must be solved with the transfer modes simultaneously included.

In some situations, it may be possible to assume that the modes are uncoupled because only weak coupling occurs.

Example 7.2

Let T_0 and T_2 in Figure 7.4 be specified. Since energy must be conserved in crossing surface 1 of the lower wall, the conduction through the lower wall must equal the transfer from surface 1 to surface 2 by combined radiation and free convection. Then, for constant thermal conductivity k_w,

$$q_1 = \frac{k_w}{a}\left(T_o - T_1\right) = \frac{\sigma\left(T_1^4 - T_2^4\right)}{1/\varepsilon_1\left(T_1\right) + 1/\varepsilon_2\left(T_2\right) - 1} + h_{fc}\left(T_1, T_2\right)\left(T_1 - T_2\right)$$

The problem is *coupled* since the unknown T_1 must be found from an equation that simultaneously incorporates all heat transfer processes. The equation for T_1 is highly nonlinear and T_1 can be obtained by iteration or by a computer math package root solver.

These examples demonstrate that the type of boundary condition governs the possibility of uncoupling the calculations. When all temperatures are specified, the energy fluxes can often be uncoupled. If energy fluxes are specified, the entire problem must be treated simultaneously because of nonlinear coupling of the unknown temperatures.

7.2.3 Control Volume Approach for 1D or 2D Conduction along Thin Walls

In some situations the radiating wall is thin and temperature variations are principally along the length and width of the wall rather than across its thickness. An important example is energy dissipation by radiating fins in devices that operate in outer space. Energy is conducted along the fin and radiated from the fin surface. The determination of the fin temperature distribution and performance requires a coupled conduction–radiation solution. The analysis is usually simplified by assuming a uniform temperature across the thin fin thickness at each location. A control volume across the thickness can be used to derive the heat balance equation.

A volume element of area $dx\, dy$ and thickness a is shown in Figure 7.5. The thickness is small, so $T(x, y)$ is considered uniform within the element. Transparent fluids at $T_{m,1}$ and $T_{m,2}$ are flowing across the upper and lower surfaces providing convective heat transfer coefficients h_1 and h_2. The temperature can change with time, and there can be internal heat generation within the element such as by electric heating. An energy balance expresses that the change with time of internal energy of the element equals the energy gains by radiation exchange, conduction, convection, and internal energy sources:

$$\rho c a \frac{\partial T}{\partial t} = G_1 - J_1 + G_2 - J_2 + \frac{\partial}{\partial x}\left(ka\frac{\partial T}{\partial x}\right) + \frac{\partial}{\partial y}\left(ka\frac{\partial T}{\partial y}\right)$$

$$+ h_1\left(T_{m,1} - T\right) + h_2\left(T_{m,2} - T\right) + \dot{q}a \tag{7.4}$$

This is used in the analyses of thin fins that follow.

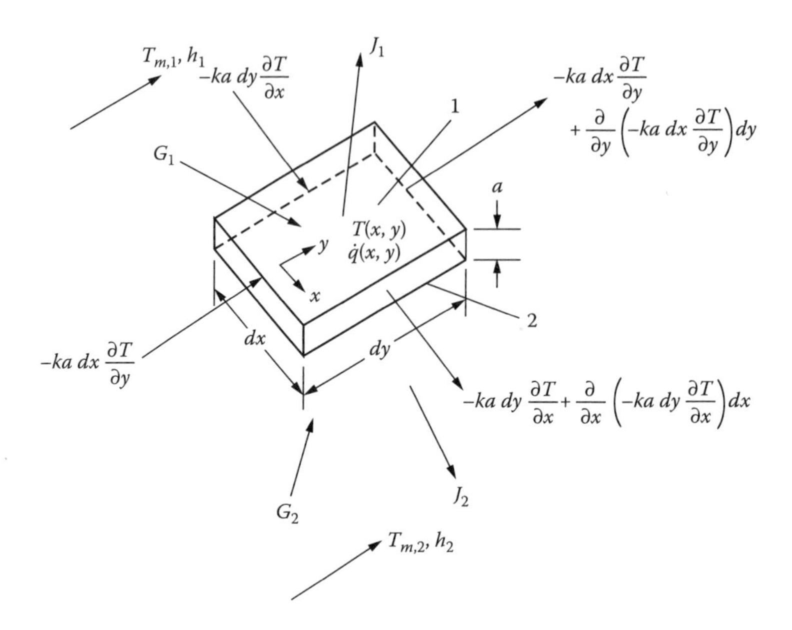

FIGURE 7.5 Element of thin plate for control volume derivation.

7.3 RADIATION TRANSFER WITH CONDUCTION BOUNDARY CONDITIONS

Combined conduction and radiation is fairly common, such as energy losses from radiating fins, energy transfer through the walls of a vacuum Dewar, energy transfer through insulation made of many separated layers of highly reflective material, energy losses from radiators in outer space, and temperature distributions in satellite and spacecraft structures. The sophistication of the radiative portion of the analysis can vary considerably, depending on the accuracy required and the importance of radiation relative to heat conduction. If conduction dominates, approximations can be made in the radiative portion of the analysis, and vice versa.

7.3.1 Thin Fins with 1D or 2D Conduction

7.3.1.1 1D Heat Flow

The heat transfer performance of a thin circular fin is now considered. From circular symmetry, the heat flow is one-dimensional (1D) in the radial direction.

Example 7.3

A thin annular fin in vacuum is embedded in insulation, so it is insulated on one face and around its outside edge (Figure 7.6a,b). The disk has thickness a, inner radius r_i, outer radius r_o, and thermal conductivity k. Energy is supplied to the inner edge from a solid rod of radius r_i that fits the central hole and maintains the inner edge at T_i. The exposed annular surface, which is diffuse–gray with emissivity ε, radiates to the environment at $T_e \sim 0$ K to investigate performance in the cold environment of outer space. Find the temperature distribution as a function of radial position along the disk. The results also apply for the more general annular fin in Figure 7.6c if the heat loss through the end edge of the fin is neglected. There is no heat flow across the symmetry plane of the fin, and hence this plane acts as the insulated boundary in Figure 7.6a.

Assume the disk is thin enough that the local temperature can be considered constant across the thickness a, which is the usual thin fin assumption. For surroundings at zero temperature, there is

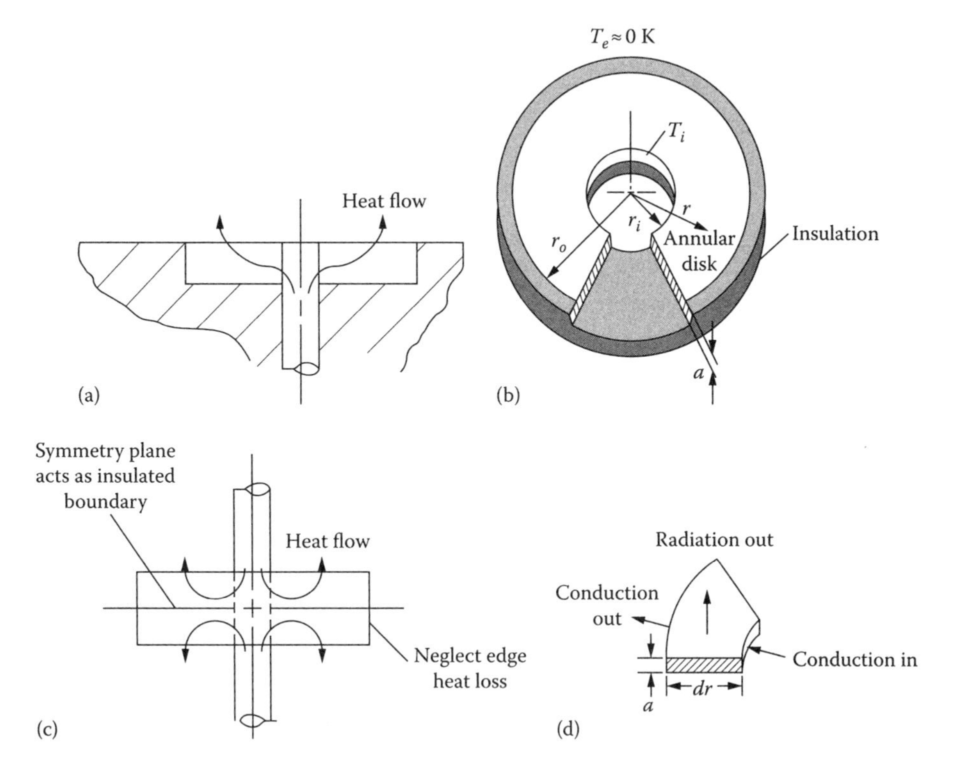

FIGURE 7.6 Geometry for finding temperature distribution in thin radiating annular plate insulated on one side and around outside edge: (a) heat flow path through fin, (b) disk geometry, (c) application to annular fin, and (d) portion of ring element on annular disk.

no incoming radiation. If a and k are constant, the control volume Equation 7.4 for a ring element of width dr (Figure 7.6d) gives

$$ka\frac{1}{r}\frac{d}{dr}\left(r\frac{dT}{dr}\right) - \varepsilon\sigma T^4 = 0 \tag{7.3.1}$$

This is to be solved for $T(r)$ subject to two boundary conditions: at the inner edge $T = T_i$ at $r = r_i$ and at the insulated outer edge where there is no heat flow $dT/dr = 0$ at $r = r_o$. Using dimensionless variables $\vartheta = T/T_i$ and $R = (r - r_i)/(r_o - r_i)$ and two parameters $\delta = r_o/r_i$ and $\gamma = (r_o - r_i)^2 \varepsilon\sigma T_i^3/ka$ results in

$$\frac{d^2\vartheta}{dR^2} + \frac{1}{R + 1/(\delta - 1)}\frac{d\vartheta}{dR} - \gamma\vartheta^4 = 0 \tag{7.3.2}$$

with the boundary conditions $\vartheta = 1$ at $R = 0$ and $d\vartheta/dR = 0$ at $R = 1$. Equation 7.3.2 is a second-order nonlinear differential equation where $\vartheta(R)$ depends on the two parameters δ and γ. Solutions can be obtained by numerical methods, and solvers are available in computer mathematics software packages.

Of interest in the design of cooling fins is the *fin efficiency*, η. This is the energy radiated away by the fin divided by the energy that would be radiated if the entire fin were at the maximum temperature T_i, which would occur for a fin with infinite thermal conductivity. The fin efficiency for the circular radiating fin is then

$$\eta = \frac{2\pi\varepsilon\sigma\int_{r_i}^{r_o} rT^4(r)dr}{\pi\left(r_o^2 - r_i^2\right)\varepsilon\sigma T_i^4} = \frac{2\int_0^1\left[R(\delta - 1) + 1\right]\vartheta^4(R)dR}{\delta + 1}$$

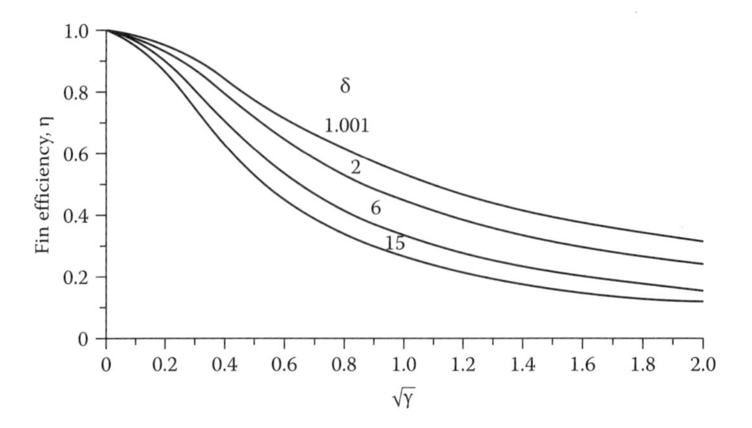

FIGURE 7.7 Radiation fin efficiency for fin of Example 7.3. (From Chambers, R.L. and Somers, E.V., *JHT*, 81(4), 327, 1959.)

and is evaluated after $\vartheta(R)$ has been determined from the differential Equation 7.3.2. The η has been obtained by Chambers and Somers (1959) and is in Figure 7.7. Keller and Holdredge (1970) extended the results to fins of radially varying thickness. The annular fin is a model for a circular foil heat flux sensor, and solutions for that application that also include convection are in Kuo and Kulkarni (1991).

In a more general situation, if the environment is at T_e and the fin is nongray with a total absorptivity α for the incoming radiation spectrum (such as for incident solar radiation in a space application), the energy balance in Equation 7.3.1 becomes (ε is the total emissivity for the spectrum emitted by the fin)

$$ka\frac{1}{r}\frac{d}{dr}\left(r\frac{dT}{dr}\right)-\sigma\left(\varepsilon T^4-\alpha T_e^4\right)=ka\frac{1}{r}\frac{d}{dr}\left(r\frac{dT}{dr}\right)-\varepsilon\sigma\left(T^4-\frac{\alpha}{\varepsilon}T_e^4\right)=0 \quad (7.5)$$

where $(\alpha/\varepsilon)T_e^4$ is an additional parameter. For a gray fin, $\alpha = \varepsilon$; hence, a nongray fin acts like a gray fin in an effective radiating environment of $(\alpha/\varepsilon)T_e^4$. By using this effective environment, results for gray fins can be utilized for nongray fins. Design results for rectangular fins, including incident radiation from the environment, are in Mackay (1963).

For transients where the fin temperature distribution changes with time, the energy storage term in Equation 7.4 must be included. The partial differential equation for $T(r, \tau)$ is then

$$\rho ca\frac{\partial T}{\partial \tau}=ka\frac{1}{r}\frac{\partial}{\partial r}\left(r\frac{\partial T}{\partial r}\right)-\varepsilon\sigma\left(T^4-\frac{\alpha}{\varepsilon}T_e^4\right)=0 \quad (7.6)$$

Results for transient radiating fin behavior are in Eslinger and Chung (1979). Fins of various shapes are treated in Kraus et al. (2001).

Example 7.4

A thin plate of thickness a and length $2L$ is between two tubes in a radiator used to dissipate energy in orbit as in Figure 7.8. The dimension is long in the direction normal to the cross section shown. Both sides of the plate have the same emissivity and are losing energy by radiation.

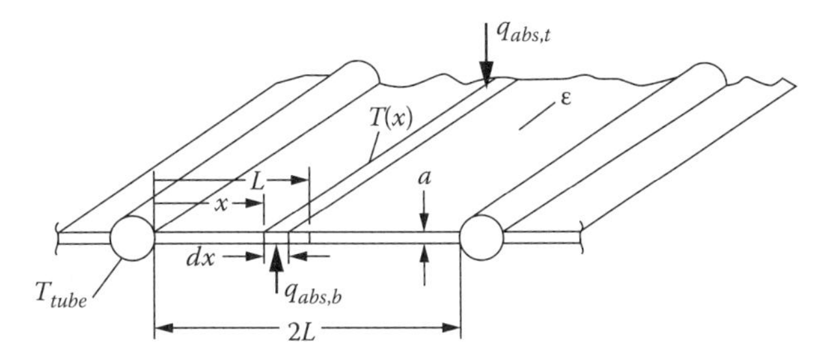

FIGURE 7.8 Flat-plate fin geometry for Example 7.4.

Radiation from the surroundings, such as from the sun, Earth, or a planet, is incident on the plate surfaces, and the fluxes absorbed on the top and bottom sides are $q_{abs,t}$ and $q_{abs,b}$. The plate is diffuse–gray with emissivity ε on both sides and has constant thermal conductivity. Find an expression for the plate temperature distribution in the x direction. Neglect radiative interaction with the tube surfaces.

From the control volume relation Equation 7.4, the energy equation for a plate element of width dx is

$$-ka\frac{d^2T}{dx^2}+2\varepsilon\sigma T^4=q_{abs,t}+q_{abs,b} \tag{7.4.1}$$

The boundary conditions for the thin plate are $T = T_{tube}$ specified at $x = 0$ and, from symmetry, $dT/dx = 0$ at $x = L$. To find $T(x)$, multiply Equation 7.4.1 by dT/dx and integrate to yield

$$-\frac{ka}{2}\left(\frac{dT}{dx}\right)^2+\frac{2}{5}\varepsilon\sigma\left[T^5-T^5(L)\right]=\left(q_{abs,t}+q_{abs,b}\right)\left[T-T(L)\right]$$

where $T(L)$ (which is unknown) is inserted to satisfy the boundary condition at $x = L$. Solve for dT/dx to yield

$$\frac{dT}{dx}=-\left(\frac{4\varepsilon\sigma}{5ka}\right)^{1/2}\left\{T^5-T^5(L)-\frac{5}{2\varepsilon\sigma}\left(q_{abs,t}+q_{abs,b}\right)\left[T-T(L)\right]\right\}^{1/2}$$

The minus sign was chosen for the square root because $T(x)$ must be decreasing with x. Separate variables and integrate again to obtain

$$x=\left(\frac{5ka}{4\varepsilon\sigma}\right)^{1/2}\int_{T}^{T_{tube}}\frac{dT}{\left\{T^5-T^5(L)-(5/2\varepsilon\sigma)\left(q_{abs,t}+q_{abs,b}\right)\left[T-T(L)\right]\right\}^{1/2}} \tag{7.4.2}$$

which satisfies $T = T_{tube}$ at $x = 0$. To obtain $T(L)$, the relation is used that the known length L must equal

$$L=\left(\frac{5ka}{4\varepsilon\sigma}\right)^{1/2}\int_{T(L)}^{T_{tube}}\frac{dT}{\left\{T^5-T^5(L)-(5/2\varepsilon\sigma)\left(q_{abs,t}+q_{abs,b}\right)\left[T-T(L)\right]\right\}^{1/2}} \tag{7.4.3}$$

A numerical root finder in mathematics software packages can be used to obtain $T(L)$ from Equation 7.4.3. The temperature distribution is then found by evaluating the integral in Equation 7.4.2 numerically to find x for various T values (in the lower limit of the integral) between T_{tube} and $T(L)$.

Examples 7.3 and 7.4 considered a single radiating fin. When there are multiple fins that have radiative exchange among them, integral terms are introduced into the energy equations as shown in the next example.

Example 7.5

An infinite array of identical thin fins of thickness a, width W in the x direction, and infinite length in the z direction is attached to a black base maintained at a constant temperature T_b as in Figure 7.9. The fin surfaces are diffuse–gray and are in vacuum. Set up the equation describing the local fin temperature, assuming the environment is at $T_e \approx 0$ K.

Because the fins are thin, their local temperature is assumed constant across the thickness a, and the control volume Equation 7.4 is used for the circled differential element in Figure 7.9. Since there is an infinite array of fins, the surroundings are identical for each fin and are the same on both sides of each fin. From symmetry, only half the fin thickness need be considered. All the fins are the same, so the energy balance need be considered for only one fin, and the fin temperature distributions are all the same.

The net conduction into the element dx per unit time and *per unit length of fin* in the z direction is, for constant thermal conductivity, $(ka/2)(d^2T_f/dx^2)dx$. The radiation relations are formulated from the net-radiation method as given by the enclosure Equation 5.56. Writing this for an element dx along the fin gives

$$\frac{q(x)}{\varepsilon} - \frac{1-\varepsilon}{\varepsilon}\int_{\xi=0}^{W} q(\xi)\,dF_{dx-d\xi} = \sigma T_f^4(x) - \sigma T_b^4\int_{0}^{b} dF_{dx-db} - \int_{\xi=0}^{W} \sigma T_f^4(\xi)\,dF_{dx-d\xi} \tag{7.5.1}$$

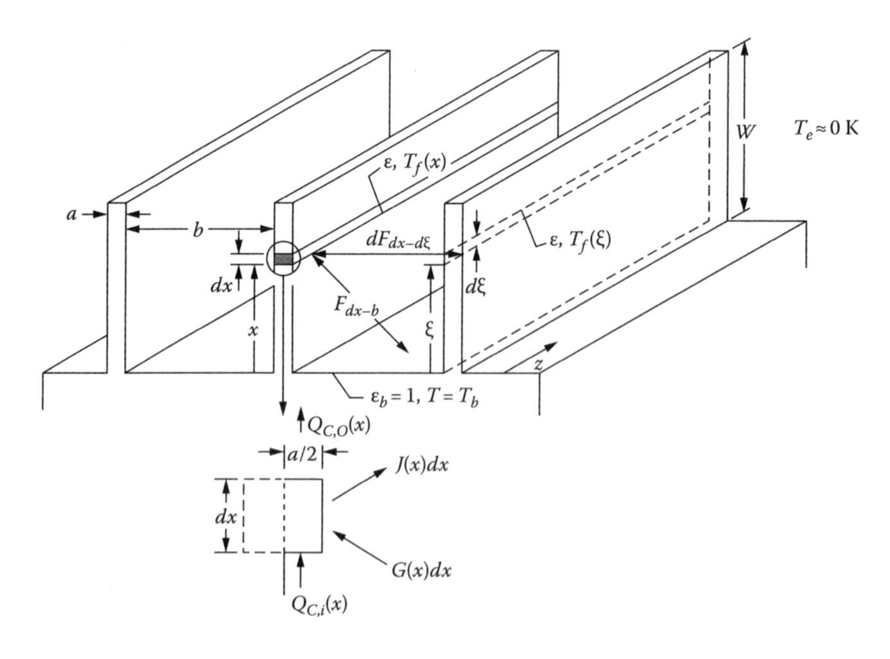

FIGURE 7.9 Geometry for determination of local temperatures on parallel fins.

Using the relations

$$\int_0^b dF_{dx-db} = F_{dx-b}, \quad q(x) = \frac{ka}{2}\frac{d^2T_f(x)}{dx^2}, \quad q(\xi) = \frac{ka}{2}\frac{d^2T_f(\xi)}{d\xi^2}$$

gives after rearrangement

$$-\frac{ka}{2\varepsilon}\frac{d^2T_f(x)}{dx^2} + \sigma T_f^4(x) = \sigma T_b^4 F_{dx-b} + \int_{\xi=0}^{W}\left[-\frac{1-\varepsilon}{\varepsilon}\frac{ka}{2}\frac{d^2T_f(\xi)}{d\xi^2} + \sigma T_f^4(\xi)\right] dF_{dx-d\xi} \tag{7.5.2}$$

In dimensionless form this becomes

$$-\mu\frac{d^2\vartheta(X)}{dX^2} + \vartheta^4(X) = F_{dX-B} + \int_{Z=0}^{1}\left[-\mu(1-\varepsilon)\frac{d^2\vartheta(Z)}{dZ^2} + \vartheta^4(Z)\right] dF_{dX-dZ} \tag{7.5.3}$$

where

$$\vartheta(X) = \frac{T_f(X)}{T_b}, \quad B = \frac{b}{W}, \quad \mu = \frac{ka}{2\varepsilon\sigma T_b^3 W^2}, \quad X = \frac{x}{W}, \quad Z = \frac{\xi}{W}$$

Equation 7.5.3 is a nonlinear integrodifferential equation and can be solved numerically. Two boundary conditions are needed. At the base of the fin, $T_f(x = 0) = T_h$, so

$$\vartheta = 1 \quad \text{at } X = 0 \tag{7.5.4}$$

At the tip of the fin, $x = W$, the conduction to the tip boundary must equal the energy radiated: $-k\partial T_f/\partial x|_{x=W} = \varepsilon\sigma T_f^4(W)$. In terms of ϑ,

$$-\frac{d\vartheta}{dX} = \frac{\varepsilon\sigma T_b^3 W}{k}\vartheta^4 = \frac{1}{2\mu}\frac{a}{W}\vartheta^4 \quad \text{at } X = 1 \tag{7.5.5}$$

and the fin thickness-to-width ratio a/W enters as another parameter. If $(a/W)/2\mu$ is very small, $d\vartheta/dX$ can be approximated as zero (since the maximum ϑ is 1). The configuration factors in Equation 7.5.3 are found by the methods of Examples 4.2 and 4.5.

A black base surface without fins provides the maximum radiative emission. Adding an infinite array of fins to a plane nonblack surface produces a series of radiating cavities that can approach the performance of a black surface. The fins provide additional weight and complexity, and hence it is better to simply use a plane surface with a high emissivity than to use a plane surface with a large array of fins. This conclusion is also reached in the analysis and discussion in Krishnaprakas (1996, 1997). However, a finned surface can provide directional emission or absorption characteristics that make it attractive for some applications, as discussed in Chapter 3. Because of the interest in radiator design for application in power systems operating in the vacuum of Earth orbit or outer space, many conducting–radiating systems have been analyzed. Typical are the previous references in this chapter and Stockman and Kramer (1963), Sparrow et al. (1962), Heaslet and

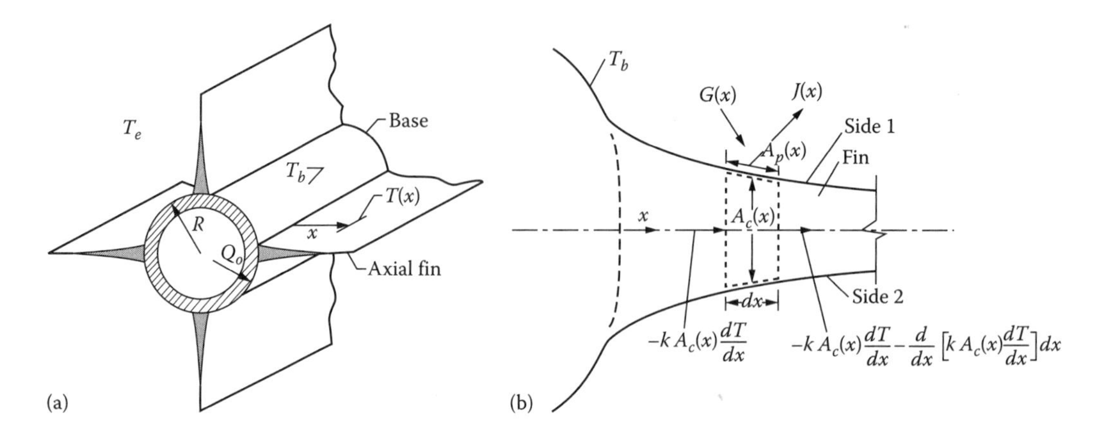

FIGURE 7.10 Axial fin array on a cylindrical pipe: (a) array with four tapered fins and (b) energy balance for tapered fin.

Lomax (1961), Schnurr et al. (1976), Masuda (1973), Frankel and Silvestri (1991), and Chung and Zhang (1991). Many other references are in the literature.

In contrast to a plane surface, if fins are placed on a cylindrical pipe, the radiating area is increased compared with the pipe surface area alone, and radiative dissipation can be substantially improved. The fins can be circular and normal to the pipe axis or axial as in Figure 7.10a. Fins of uniform thickness were analyzed by Frankel and Silvestri (1991), but to save weight, the fins can have a nonuniform cross section that decreases with x (Figure 7.10a) as the remaining energy to be radiated decreases that is flowing outward by conduction within the fin. The optimization of the fin cross-sectional shape has been analyzed by Chung and Zhang (1991) and Krishnaprakas (1997). These include radiative interactions between the fins and between the fins and the base surface.

7.3.1.2 2D Heat Flow

The configuration and boundary conditions for many fin applications is such that the fins can be analyzed as having 1D heat flow. Some applications use fins with 2D heat flow, such as for removing excess heat from electronic equipment in satellites and other space devices. The 2D fin is typically a thin metal plate, such as aluminum, with heat generating equipment in good thermal contact with a portion of the plate area. The heat transferred to the plate is conducted away in 2D and is dissipated by radiation to cooler surroundings. In Badari Narayana and Kumari (1988), the cooling modes are radiation and conduction; in Bobco and Starkovs (1985), convection is also included.

The fin analyzed in Badari Narayana and Kumari is shown in Figure 7.11. It has a uniform thickness and constant thermal properties, and its radiative properties are diffuse–gray. Energy to be dissipated to cool equipment is transferred to the shaded area on one side and is radiated away from both sides. The radiating plate is exposed to the sun on the outside and to surroundings at T_i on the inside. The energy equation within the zone (shown shaded) over which energy from the equipment is being transferred to the plate is given by Equation 7.4 as

$$ka\left(\frac{\partial^2 T}{\partial x^2}+\frac{\partial^2 T}{\partial y^2}\right)+\alpha_s q_{solar}\cos\theta+q_e-\varepsilon_o\sigma T^4(x,y)=0 \tag{7.7}$$

In this region, an element of the plate receives energy by 2D conduction within the plate, by absorption of solar radiation, and by heat addition from electronic equipment and/or other heat sources.

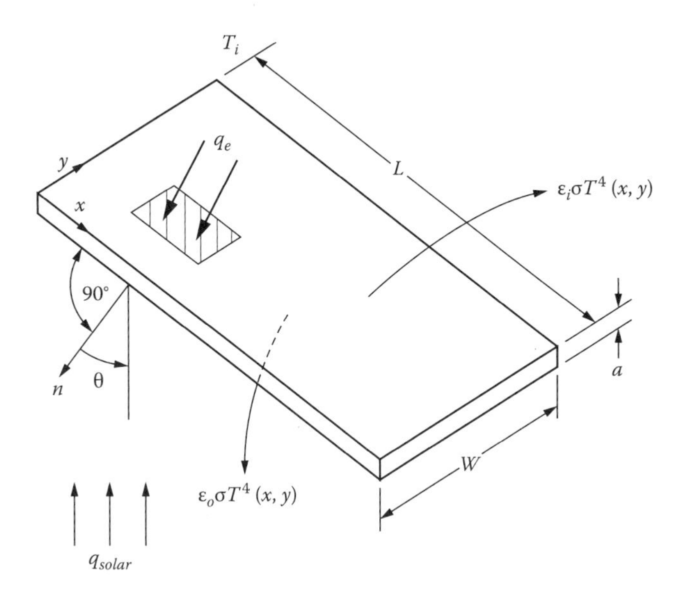

FIGURE 7.11 Two-dimensional radiating fin with heat flux addition q_e to area on one side.

The plate loses energy by radiation from its outside surface. For the other portions of the plate, where $q_e = 0$, the energy equation becomes

$$ka\left(\frac{\partial^2 T}{\partial x^2}+\frac{\partial^2 T}{\partial y^2}\right)+\alpha_s q_{solar}\cos\theta+\sigma\varepsilon_i\left[T_i^4-T^4(x,y)\right]-\varepsilon_o\sigma T^4(x,y)=0 \quad (7.8)$$

where there is a term for the net radiation loss from the surface that is inside the enclosure. Although this was the term used in Badari, Narayana, and Kumari, it is written more generally in Equation 7.4 as $(G - J)$ to account for a more complex heat exchange with a surrounding enclosure. Heat losses from the end edges of the fin were assumed small, so insulated edge boundary conditions are used:

$$\frac{\partial T}{\partial x}=0 \quad \text{at } x=0,L; \quad \frac{\partial T}{\partial y}=0 \quad \text{at } y=(0,W)$$

Along the boundary between the area of heat addition (shaded area) and the remaining area of the plate, there is continuity of fin temperature and heat flow in the x and y directions.

Solutions to other fin problems involving mutual interactions are in Stockman and Kramer (1963), Sparrow and Eckert (1962), Heaslet and Lomax (1961), Nichols (1961), Frankel and Silvestri (1991), Sparrow et al. (1962), and Frankel and Wang (1988). The optimization of a fin array with respect to minimum weight is in Wilkins (1962) and Chung and Zhang (1991). The radiant interchange is analyzed by Masuda (1972) between diffuse–gray external circular fins on cylinders, extending normal to the cylinder axis, and the local heat flux distribution is obtained on the fins and cylinders as well as the fin effectiveness.

7.3.2 Multidimensional and Transient Heat Conduction with Radiation

For a thin radiating fin, the local fin temperature is assumed uniform across the fin thickness, and temperature variations are only in directions along the radiating surfaces. If the conducting solid is thick, however, the temperature will also vary normal to the radiating surface. The conduction can be steady or transient. The surfaces are assumed opaque, so without external convection, the net radiation at the surface is the boundary condition for conduction within the solid. If n is the outward normal from the surface, the conduction heat flux at the surface, flowing outward from within the solid, is $-k(\partial T/\partial n)$. This is the local flux supplied to the surface to balance the net radiative loss. Hence, in the absence of convection, there is the boundary condition, Equation 7.1, $-k(\partial T/\partial n)|_{wall} = (J - G) = q_r$. The governing partial differential equation in the solid is Equation 7.2. The distribution of $(J - G)$ along the conducting surfaces is found from the radiative enclosure methods described previously. For a nongray surface, the radiative q_r is the integral of the spectral fluxes as developed in Chapter 6. Subject to these boundary conditions, the multidimensional heat conduction equations can be solved by finite-difference or finite-element methods (FEM) as in Section 7.7.

When the temperature distributions within the solid are transient as well as spatially dependent, relatively few analytical solutions can be obtained in view of the complexity of the radiative boundary conditions. Some analytical investigations were made by Abarbanel (1960). Transient solutions are more feasible if the geometry is 1D. An example is electric heating of a thin wire where the transient temperature distribution varies only along the wire length (Carslaw and Jaeger 1959). Transient radiative cooling of a wire was used in Masuda and Higano (1988) to obtain the hemispherical total emittance of a metal by measuring the cooling rate. From the control volume approach, the 1D energy equation for a wire of radius r is

$$r^2 \rho c \frac{\partial T}{\partial t} = kr^2 \frac{\partial^2 T}{\partial x^2} - 2r\left[J(x,t) - G(x,t)\right] \tag{7.9}$$

The properties have been assumed constant. The J and G depend on the radiative exchange with the surroundings and are functions of axial position x and time t. The J varies considerably as the wire temperature changes with time and position. The solution for $T(x, t)$ requires an initial temperature distribution and two boundary conditions in x. For example, the boundary conditions could be fixed electrode temperatures at the ends of the wire.

A 2D transient solution was carried out numerically in Sunden (1989). A hollow cylinder, insulated on its internal surface, is heated on its exterior by a time-varying radiation flux from one direction. Absorbed energy is then conducted within the cylinder in radial and circumferential directions. During the transient heating, the outer surface loses energy by radiation and convection. When the cylinder has low thermal conductivity and low thermal diffusivity, the temperature distributions are quite nonuniform and the surface temperatures are high, so radiative cooling is important. For high-conductivity materials, the temperature levels are lower and the temperature distributions are more uniform.

7.4 RADIATION WITH CONVECTION AND CONDUCTION

Interactions of radiation, convection, and conduction are found in a wide variety of situations, such as convective and radiative cooling in air of high-temperature components, furnace and combustion chamber design, cooling of hypersonic and reentry vehicles, interactions of incident solar radiation with the Earth's surface to produce complex free-convection patterns, convection cells and their effect on radiation within and from stars, and marine environment studies for predicting free-convection patterns in oceans with absorption of solar energy. The energy equations and boundary conditions contain temperature differences from convection and temperature derivatives from conduction. Results must usually be obtained by using numerical solution methods. The basic ideas are developed here by using some illustrative examples of practical interest. Additional information and results are in Kuo and Kulkarni (1991), Perlmutter and Siegel (1962), Siegel and Perlmutter (1962),

Cess (1961), Keshock and Siegel (1964), Aziz and Benzies (1976), Siegel and Keshock (1964), Okamoto (1964, 1966a,b), Sohal and Howell (1973, 1974), Campo (1976), Shouman (1965, 1968), and Razzaque et al. (1982). In Hadley et al. (1999), the transient response of a thermocouple with radiation and convection is analyzed using a lumped-parameter model.

7.4.1 Thin Radiating Fins with Convection

Thin fins with multiple heat transfer modes are used extensively for providing effective energy dissipation.

Example 7.6

Examine the performance of the fin in Figure 7.12; it depends on combined conduction, convection, and radiation. A gas at T_e is flowing over the fin and removing energy by convection. The environment to which the fin radiates is also assumed to be at T_e. The fin cross section has area A and perimeter P. The fin is nongray with total absorptivity α for radiation incident from the environment.

Using a control volume, an energy balance on a fin element of length dx yields

$$kA\frac{\partial^2 T}{\partial x^2}dx = \sigma\left[\varepsilon T^4(x) - \alpha T_e^4\right]Pdx + hPdx\left[T(x) - T_e\right] \tag{7.6.1}$$

The term on the left is the net conduction into the element, and on the right are the radiative and convective losses. The radiative exchange between the fin and its base is neglected here but has been included in more detailed analyses. This equation is to be solved for $T(x)$, which can then be used to obtain the energy dissipation. Multiply by $[1/(kA\,dx)]\,dT/dx$, and integrate once to yield

$$\frac{1}{2}\left(\frac{dT}{dx}\right)^2 = \frac{\varepsilon\sigma P}{kA}\left(\frac{T^5}{5} - \frac{\alpha}{\varepsilon}TT_e^4\right) + \frac{hP}{kA}\left(\frac{T^2}{2} - TT_e\right) + C \tag{7.6.2}$$

where C is a constant of integration.

For convection and radiation from the end surface of the fin, or for the end of the fin assumed insulated, the dT/dx in Equation 7.6.2 is expressed in terms of the appropriate end conditions. For a case that leads to an analytical solution, let $T_e \approx 0$ and let the fin be long. For large x, $T(x) \to 0$ and $dT/dx \to 0$, and from Equation 7.6.2, $C = 0$. Solving for dT/dx gives

$$\frac{dT}{dx} = -\left(\frac{2}{5}\frac{P\varepsilon\sigma}{kA}T^5 + \frac{hP}{kA}T^2\right)^{1/2} \tag{7.6.3}$$

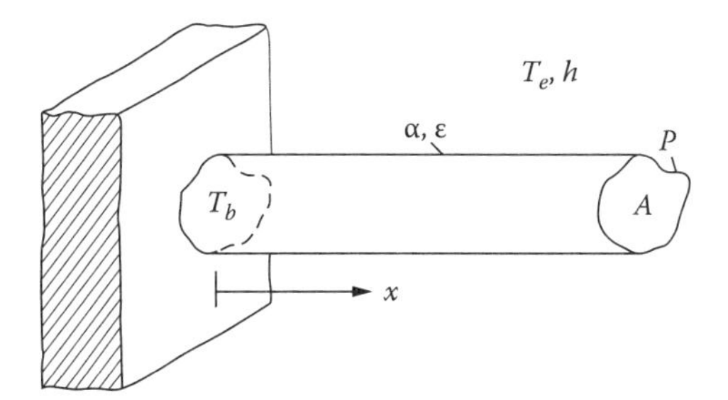

FIGURE 7.12 Fin of constant cross-sectional area transferring energy by radiation and convection. (Flowing gas and environment are both at T_e.)

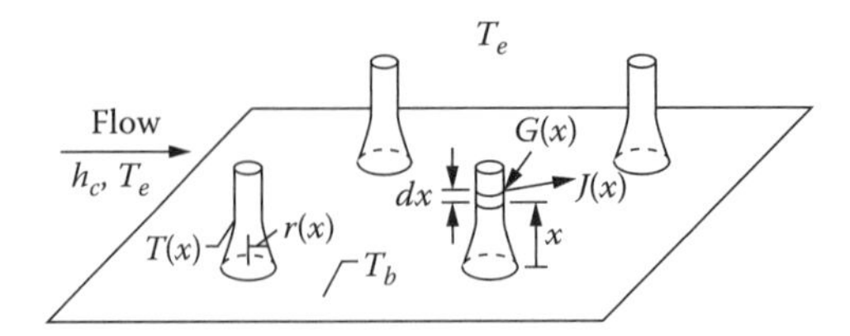

FIGURE 7.13 An array of tapered pin fins with external radiation and convection.

The minus sign is used for the square root since T decreases as x increases. The variables in Equation 7.6.3 are separated and the equation integrated with the condition that $T(0) = T_h$,

$$\int_0^x dx = -\int_{T_b}^{T} \frac{dT}{T\left[\frac{2}{5}(P\varepsilon\sigma/kA)T^3 + hP/kA\right]^{1/2}}$$

$$x = \frac{1}{3}M^{-1/2}\left[\ln\frac{\left(GT_b^3+M\right)^{1/2}-M^{1/2}}{\left(GT_b^3+M\right)^{1/2}+M^{1/2}} - \ln\frac{\left(GT^3+M\right)^{1/2}-M^{1/2}}{\left(GT^3+M\right)^{1/2}+M^{1/2}}\right] \tag{7.6.4}$$

where

$$G = \frac{2}{3}P\varepsilon\sigma/kA$$

$$M = hP/kA$$

For this simplified limiting case, an analytical relation for $T(x)$ is obtained. The solution can be carried somewhat further, as considered in Homework Problem 7.3. A detailed treatment of this type of fin is in Shouman (1965, 1968).

An array of pin fins extending from a surface is useful for heat transfer augmentation, and the optimization of a triangular array is analyzed by Gerencser and Razani (1995). The fins are circular in cross section, and the cross section is variable with length as in Figure 7.13. The convective heat transfer coefficient h_c is assumed constant throughout the array, and for a constant fin thermal conductivity, the energy balance on an element at x of a typical fin is

$$k\frac{d}{dx}\left[\pi r(x)^2\frac{dT(x)}{dx}\right] = 2\pi r(x)\left\{h_c\left[T(x)-T_e(x)\right]+J(x)-G(x)\right\} \tag{7.10}$$

where the differential surface area has been approximated as $\pi r^2\,dx$. The incident radiative flux, $G(x)$, is obtained by the interaction of a fin with the base surface, which is assumed at uniform temperature, and with the surrounding fins. It is assumed in Gerencser and Razani that all of the fins surrounding any one fin may be approximated as a surrounding circular cylinder of constant radius that has a temperature distribution in the x direction that is the same as along the fins. The optimization for minimum fin volume to provide the required energy dissipation yields fins with a curved profile along their length. It is found that good performance can also be achieved by using tapered pin fins with a triangular profile, which are more economical to manufacture.

7.4.2 Channel Flows

Flow of a transparent gas through a heated radiating tube is now considered. Solutions of this type are in Perlmutter and Siegel (1962), Siegel and Perlmutter (1962), Cess (1961), Keshock and Siegel (1964), Aziz and Benzies (1976), Siegel and Keshock (1964), and Razzaque et al. (1982).

The presentation in this chapter is for convective media that are completely transparent to radiation. The enclosure energy-balance equations, such as Equations 5.16 and 5.17, can be used as before, and the q_k at the wall surface will contain convective heat addition to the wall.

Example 7.7

A transparent gas flows through a black circular tube (Figure 7.14). The tube wall is thin, and its outer surface is perfectly insulated. The wall is heated electrically to provide uniform energy input q_e per unit area and time. The wall temperature along the tube length is to be determined. The convective heat transfer coefficient h between the gas and the inside of the tube is assumed constant. The gas has a mean velocity u_m, heat capacity c_p, and density ρ_f. Axial conduction in the thin tube wall is neglected.

If radiation were not considered, the local heat addition to the gas would equal the local electric heating (since the outside of the tube is insulated) and hence would be invariant with x along the tube. The gas temperature and wall temperature would both rise linearly with x. If convection were not considered, the only means for heat removal would be by radiation out of the tube ends as in Example 5.19. In this instance, for equal environment temperatures at the tube ends, the wall temperature is a maximum at the center of the tube and decreases toward each end. The solution for combined radiation and convection is expected to exhibit trends of both limiting solutions.

Consider a ring element dA_x of length dx on the interior of the tube wall at x, as in Figure 7.14. The energy supplied per unit time is composed of electric heating, energy radiated to dA_x by other wall elements of the tube interior (see Example 5.19), and energy radiated to dA_x through the tube inlet and exit:

$$q_e \pi D dx + \int_{z=0}^{l} \sigma T_w^4(z) dF_{dz-dx}\left(|z-x|\right) \pi D dz$$

$$+\sigma T_{r,1}^4 \frac{\pi D^2}{4} dF_{1-dx}(x) + \sigma T_{r,2}^4 \frac{\pi D^2}{4} dF_{2-dx}(l-x)$$

The tube ends are assumed to act as black disks at the inlet and outlet reservoir temperatures, which are assumed equal to the inlet and outlet gas temperatures. The energy leaving the ring element at x by convection and radiation is $\left\{h\left[T_w(x) - T_g(x)\right] + \sigma T_w^4(x)\right\} \pi D dx$. Neglecting axial heat

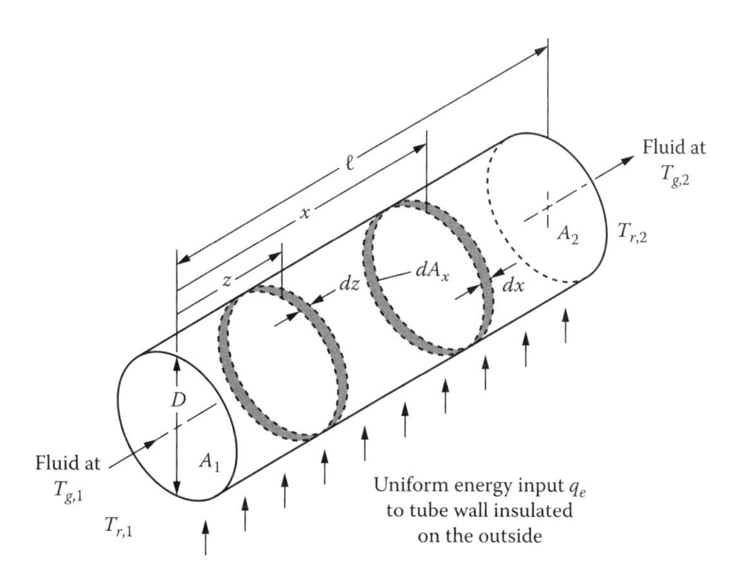

FIGURE 7.14 Flow through tube with uniform internal energy input to wall and outer surface insulated.

conduction in the wall, the energy quantities are equated to yield (reciprocity was used on the F factors so that dx could be divided out)

$$h\left[T_w(x) - T_g(x)\right] + \sigma T_w^4(x) = q_e + \int_{z=0}^{l} \sigma T_w^4(z) dF_{dx-dz}(|x-z|)$$
$$+\sigma T_{r,1}^4 F_{dx-1}(x) + \sigma T_{r,2}^4 F_{dx-2}(l-x) \quad (7.7.1)$$

This has the form of Equation 5.8 with $Q_k/A_k = q_e + h[T_g(x) - T_w(x)]$. Equation 7.7.1 has two unknowns, $T_w(x)$ and $T_g(x)$; a second equation is needed before a solution can be found. This is obtained from an energy balance on a volume element of length dx in the tube. The energy carried into this volume by the gas is $u_m\rho_f c_p T_g(x)(\pi D^2/4)$ and that added by convection from the wall is $\pi D h[T_w(x) - T_g(x)]dx$. The energy carried out by the gas is $u_m\rho_f c_p(\pi D^2/4)\left\{T_g(x) + \left[dT_g(x)/dx\right]dx\right\}$. An energy balance gives

$$u_m\rho_f c_p \frac{D}{4}\frac{dT_g(x)}{dx} = h\left[T_w(x) - T_g(x)\right] \quad (7.7.2)$$

By defining the dimensionless quantities,

$$\mathrm{St} = \frac{4h}{u_m\rho_f c_p} = \frac{4\mathrm{Nu}}{\mathrm{Re\,Pr}}; \quad H = \frac{h}{q_e}\left(\frac{q_e}{\sigma}\right)^{1/4}; \quad \vartheta = T\left(\frac{\sigma}{q_e}\right)^{1/4}$$

and $X = x/D$, $Z = z/D$, and $L = l/D$; the energy balances on the wall and fluid are

$$\vartheta_w^4(x) + H\left[\vartheta_w(x) - \vartheta_g(x)\right] = 1 + \int_0^X \vartheta_w^4(Z) dF_{dX-dZ}(X-Z) + \int_X^L \vartheta_w^4(Z) dF_{dX-dZ}(Z-X)$$
$$+\vartheta_{r,1}^4 F_{dx-1}(X) + \vartheta_{r,2}^4 F_{dx-2}(L-X) \quad (7.7.3)$$

$$\frac{d\vartheta_g(X)}{dX} = \mathrm{St}\left[\vartheta_w(X) - \vartheta_g(X)\right] \quad (7.7.4)$$

The two equations have the unknowns $\vartheta_w(X)$ and $\vartheta_g(X)$ and five parameters: St, H, L, $\vartheta_{r,1}$, and $\vartheta_{r,2}$.

Equation 7.7.4 can be solved by using an integrating factor. The boundary condition is that $\vartheta_g(X)$ has a specified value $\vartheta_{g,1}$ at $X = 0$. The solution is

$$\vartheta_g(X) = \mathrm{St}e^{-\mathrm{St}X}\int_0^X e^{\mathrm{St}Z}\vartheta_w(Z)dZ + \vartheta_{g,1}e^{-\mathrm{St}X} \quad (7.7.5)$$

This is substituted into Equation 7.7.3 to yield an integral equation for $\vartheta_w(X)$

$$\vartheta_w^4(X) + H\vartheta_w(X) - H\mathrm{St}e^{-\mathrm{St}X}\int_{Z=0}^{X} e^{\mathrm{St}Z}\vartheta_w(Z)dZ - H\vartheta_{g,1}e^{-\mathrm{St}X}$$
$$= 1 + \int_{Z=0}^{X} \vartheta_w^4(Z) dF_{dX-dZ}(Z-X) + \int_{Z=X}^{L} \vartheta_w^4(Z) dF_{dX-dZ}(Z-X)$$
$$+\vartheta_{r,1}^4 F_{dX-1}(X) + \vartheta_{r,2}^4 F_{dX-2}(L-X) \quad (7.7.6)$$

Solutions to Equation 7.7.6 were obtained by Perlmutter and Siegel (1962) and representative results calculated by numerical integration are in Figure 7.15. Note that the predicted temperatures

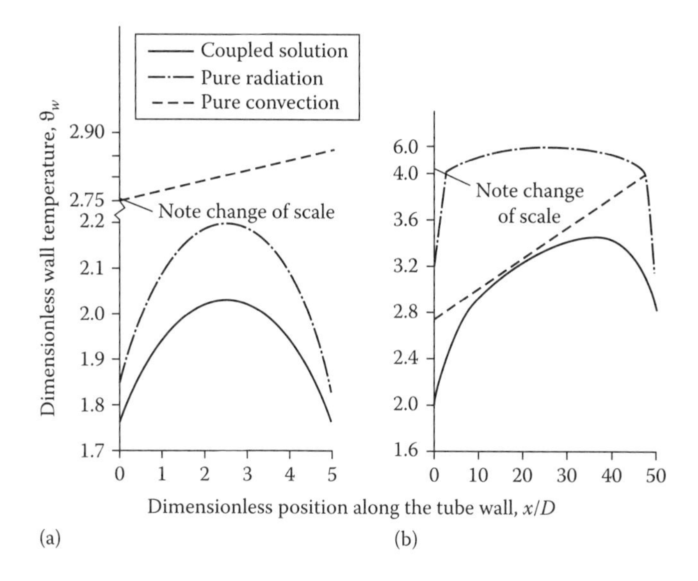

FIGURE 7.15 Tube wall temperatures resulting from combined radiation and convection for transparent gas flowing in uniformly heated black tube for St = 0.02, $H = 0.8$, $\vartheta_{r,1} = \vartheta_{g,1} = 1.5$, and $\vartheta_{r,2} = \vartheta_{g,2}$. (a) Tube length, $l/D = 5$; (b) tube length, $l/D = 50$.

for combined radiation and convection fall below the temperatures predicted for either convection or radiation acting independently. For a short tube, radiation effects are significant over the entire tube length, and for the parameters shown, the combined-mode temperature distribution is similar to that for radiation alone. For a long tube, the combined-mode distribution is close to that for convection alone over the central portion of the tube. The heat transfer resulting from combined convection–radiation is more efficient than by either mode alone. Hence, the wall temperature distribution in the combined problem is below distributions predicted by using either mode alone.

Example 7.8

What are the governing energy equations if the tube interior in Example 7.7 is diffuse–gray with emissivity ε rather than being black?

A convenient derivation is by use of the enclosure Equation 5.56. The energy flux added to the interior surface of the wall by means other than internal radiative exchange is $q_w(x) = h[T_g(x) - T_w(x)] + q_e$. The enclosure equation yields

$$\frac{q_w(x)}{\varepsilon} - \frac{1-\varepsilon}{\varepsilon}\int_{z=0}^{l} q_w(z)\,dF_{dx-dz}(z,x) = \sigma T_w^4(x) - \int_{z=0}^{l} \sigma T_w^4(z)\,dF_{dx-dz}(z,x) - \sigma T_{r,1}^4 F_{dx-1} - \sigma T_{r,2}^4 F_{dx-2} \tag{7.8.1}$$

where $q_w(x)$ can be substituted to yield an equation with $T_w(x)$ and $T_g(x)$. Equation 7.7.5 is unchanged by having the wall gray. Thus, Equations 7.8.1 and 7.7.5 are two equations for the unknowns $T_w(x)$ and $T_g(x)$. Numerical solutions are in Siegel and Perlmutter (1962).

Example 7.9

Consider again the tube in Example 7.7 that is uniformly heated and perfectly insulated on the outside and has a black interior surface. Gas flows through the tube, and the convective heat transfer coefficient is assumed constant. Axial heat conduction in the tube wall is now included.

The tube wall has thermal conductivity k_w, thickness b, and inside and outside diameters D_i and $D_o = D_i + 2b$. The desired result is $T_w(x)$ along the tube length. The wall is assumed sufficiently thin that the local $T_w(x)$ is constant across the wall thickness.

The energy balance in Equation 7.7.1 is modified to include the net gain of energy by an element of the tube wall from axial wall heat conduction. This yields the energy balance

$$h\left[T_w(x) - T_g(x)\right] + \sigma T_w^4(x) = q_e + k_w \frac{D_o^2 - D_i^2}{4D_i} \frac{d^2 T_w(x)}{dx^2} + \int_{z=0}^{l} \sigma T_w^4(z) dF_{dx-dz}(|x - z|)$$

$$+ \sigma T_{r,1}^4 F_{dx-1}(x) + \sigma T_{r,2}^4 F_{dx-2}(l - x) \tag{7.9.1}$$

As in connection with Equation 7.7.1, all lengths are nondimensionalized by dividing by the internal tube diameter, and dimensionless parameters are introduced. The conduction term yields a new parameter

$$N_{CR} = \frac{k_w}{4q_e D_i}\left[\left(\frac{D_o}{D_i}\right)^2 - 1\right]\left(\frac{q_e}{\sigma}\right)^{1/4}$$

For thin walls where $(D_o - D_i)/2 = b \ll D_i$, this reduces to $N_{CR} = \left(k_w b / q_e D_i^2\right)\left(q_e/\sigma\right)^{1/4}$, used in some references.

The dimensionless form of the energy equation is

$$\vartheta_w^4(X) + H\left[\vartheta_w(X) - \vartheta_g(X)\right] = 1 + N_{CR}\frac{d^2\vartheta_w(X)}{dX^2} + \int_{Z=0}^{X} \vartheta_w^4(Z) dF_{dX-dZ}(X - Z)$$

$$+ \int_{Z=X}^{L} \vartheta_w^4(Z) dF_{dX-dZ}(Z - X) + \vartheta_{r,1}^4 F_{dX-1}(X) + \vartheta_{r,2}^4 F_{dX-2}(L - X) \tag{7.9.2}$$

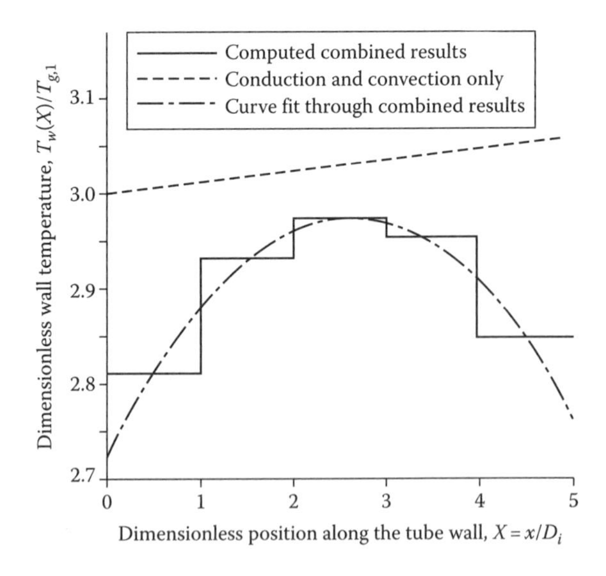

FIGURE 7.16 Wall temperature distribution for flow of transparent fluid through black tube with combined radiation, convection, and conduction for $L = 5$, St = 0.005, $N_{CR} = 0.316$, $H = 1.58$, $\vartheta_{r,1} = \vartheta_{g,1} = 0.316$, and $\vartheta_{r,2} = \vartheta_{g,2}$.

The energy equation for the fluid is still Equation 7.7.5; these equations can be combined as in Equation 7.7.6. Hottel discussed this problem in terms of slightly different parameters. He obtained an early numerical solution for five ring-area intervals on the tube wall, before the common use of computers. Results are in Figure 7.16 in terms of the parameters derived here.

If the formulation includes axial conduction, there are two additional conduction boundary conditions. The solution of Equation 7.9.2 requires two boundary conditions because of the constants introduced by integrating the $d^2\vartheta_w/dX^2$ term. The boundary conditions depend on the physical construction at each end of the tube that determines the amount of conduction. In Siegel and Keshock (1964), some detailed results were obtained where it was assumed for simplicity that the tube end edges were insulated, $(d\vartheta_w/dx)\big|_{x=0} = (d\vartheta_w/dx)\big|_{x=l} = 0$. The extension was also made in Siegel and Keshock to have the convective heat transfer coefficient inside the tube vary with position along the tube as in a thermal entrance region. The FEM (Section 7.7.2) was used by Razzaque et al. (1982) to extend the results to large tube lengths and to include a sinusoidal heat flux addition along the tube length. Ganesan et al. (2015) give experimental results for airflow in a horizontal duct with axial temperature variation and show that surface–surface radiation has a significant effect on total heat transfer.

7.4.3 Natural Convection with Radiation

At moderate temperatures radiative fluxes are small, but in conjunction with natural convection in air that generally produces small convective heat transfer coefficients, the radiative transfer may be comparable to convection. If a single vertical plate in air is internally heated, there can be an interaction of radiation with free convection, depending on the heating condition of the plate. For a specified amount of heating along the plate, such as by electric heating, the local temperatures along the plate for steady state must be such that the local heating is dissipated by radiation, convection, and conduction within the plate to adjacent elements. The behavior of a very thin vertical electrically heated stainless steel foil with negligible lengthwise heat conduction was investigated analytically and experimentally by Webb (1990). With radiative dissipation included with natural convection, the foil tends to have a more uniform temperature distribution than for natural convection alone.

For a fin on a surface, the net heat conduction along the fin is balanced by local radiative dissipation and natural convection. An analysis for a single vertical fin on an isothermal base is in Balaji and Venkateshan (1996) for combined radiation and natural convection. From symmetry about the vertical centerplane of the fin, half of the geometry can be considered for analysis with the centerplane of the fin perfectly insulated to provide the symmetry boundary condition. With the base surface included, this results in a study of combined natural convection and radiation inside an L-shaped corner, with the temperature distribution in the vertical wall depending on heat conduction in the fin, free convection, and radiation exchange with the base surface of finite width and with the surrounding environment. Radiative exchange with adjacent fins was not considered. The results were verified by comparison with experiments from Rodigheiro and de Socio (1983). A proposed method is discussed in Guglielmini et al. (1987) for cooling electronic components attached to a base plate. A series of staggered fins can be attached to a base plate to provide cooling by radiation and free convection. Enclosure theory was used to evaluate the radiative interaction between the fins and the base surface. Predicted results agreed well with an experimental study.

Natural convection inside a closed 2D horizontal rectangular enclosure is analyzed in Dehghan and Behnia (1996) (rectangular cross section in the vertical x–y plane and a large length in the horizontal z direction). The two horizontal boundaries have no external energy supplied to them and are insulated on the outside (adiabatic top and bottom boundaries). The two vertical boundaries are each maintained at a different uniform temperature. The radiative transfer within the enclosure filled with transparent gas was formulated using the net radiation method in Chapter 5. The enclosure was divided into 20 zones on each boundary, and the configuration factors were evaluated with the crossed-string method (Section 4.3.3.1). The resulting simultaneous equations were solved directly by the Gaussian elimination method. For the gas, the steady 2D laminar free-convection flow equations were used with the Boussinesq approximation. The flow and energy equations in the gas were solved by using a finite-volume numerical method given in Gosman et al. (1969).

In a closed rectangular space, as in Balaji and Venkateshan (1994a,b), there is heat flow across the space if the two vertical walls are at different temperatures. Interferometric measurements are given in Ramesh and Venkateshan (1999). In some instances an enclosed air space is used to provide insulation; for this application the heat transfer across the space must be reduced. This can be done by placing vertical partitions in the space, thus providing radiation barriers and suppressing natural convection as has been shown by Nishimura et al. (1988). Radiation was not included, and to examine radiative effects, an analysis was made by Sri Jayaram et al. (1997) of a closed rectangular space with a single vertical partition dividing the interior into two equal rectangular enclosures. The finite-volume numerical method and enclosure theory were used, as in Balaji and Venkateshan (1994a). Including radiation between the surfaces was found to suppress free convection and to augment the total heat transfer compared with the results without radiation. The vertical partition had a strong influence on the radiative transfer, and results in the partitioned enclosure could not be predicted from results in an enclosure without a partition. This arose from the nonlinear behavior of the radiative transfer.

Natural convection induced by heated walls forming a vertical 2D channel was studied in Carpenter et al. (1976) and Moutsoglou et al. (1992). Two parallel walls of length L are spaced b apart as in Figure 7.17, and the bottom and top of the channel are open. The internal surfaces A_1 and A_2 can have specified temperatures or specified heat fluxes. Because the walls are unequally heated, an asymmetric free-convection velocity distribution develops. The radiative exchange tends to equalize the convective heat transfer from the walls, which leads to an improvement in the overall convective cooling. In addition, there is radiative energy dissipation through the end openings of the channel. The radiative exchange can considerably alter the free-convection behavior. Heat transfer away from the heated walls can be augmented by placing an unheated vertical plate between the walls. Energy is radiated from A_1 and A_2 to this additional surface and is then transferred away by natural convection and radiation.

A two-wall geometry as in Figure 7.17 was analyzed by Moutsoglou et al. (1992) to determine the effect of placing one or more vents in one wall with that wall unheated and insulated on the outside. The other wall was uniformly heated on the outside. Radiation is exchanged between the walls, thereby heating the vented insulated wall. The increased temperatures of both walls induce

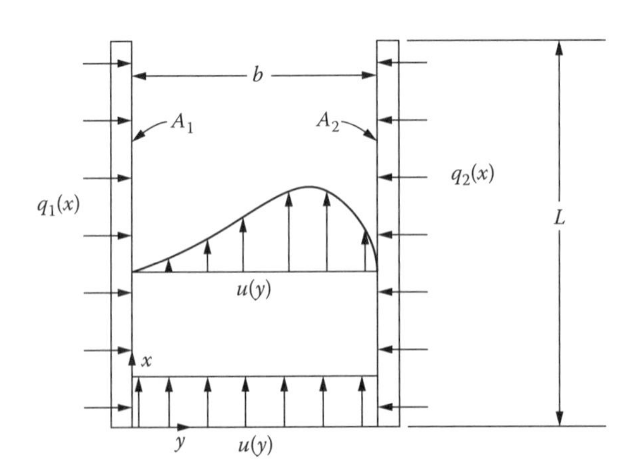

FIGURE 7.17 Natural convection between parallel walls exchanging radiation.

free convection, and air enters through the bottom space between the walls and through the vents in the insulated wall. The radiative boundary conditions are formulated by standard enclosure theory as in Chapter 5. The natural convection flow equations using Boussinesq approximation were solved numerically. It was found that the presence of vents degrades the natural convection cooling process; an unvented channel provides the best performance.

In electronic equipment, there can be isolated heat sources that are cooled by combined radiation and free convection. In Dehghan and Behnia (1996), the heat transfer was analyzed and compared with experiments for a heat source on the boundary of a cavity formed by two parallel walls as in Figure 7.17, but with the opening at the bottom closed by an adiabatic boundary. The two side walls have a finite thickness and are each adiabatic on the outside. There is heat conduction within these walls along their vertical length. A heat source is located at the midheight of one vertical wall. The radiation exchange was formulated by subdividing the vertical walls and using enclosure theory. The natural convection flow is 2D and laminar, and Boussinesq approximation was used. A finite-difference solution was obtained using a pseudo time-dependent iteration with the alternating direction implicit method. Radiation was found to have a significant effect on the flow as it caused a recirculation zone to form, and including radiation in the analysis was necessary to obtain good comparisons with experiment. Lage et al. (1992) consider a cavity formed by two parallel vertical walls with a horizontal adiabatic boundary at the bottom. The vertical walls are at differing uniform temperatures. Based on some previous work that was found to provide sufficiently accurate results, a simplifying approximation is made by obtaining the solution in two steps. First, an analysis for natural convection in the cavity is carried out without radiation. A finite-difference solution is obtained that provides the free-convection heat transfer coefficients at the walls. Then net radiation enclosure analysis is used to add the radiation exchange.

A more complex geometry is considered by Zhao et al. (1992) to determine the cooling behavior of three electrically heated power cables inside a horizontal rectangular conduit. The conduit is a closed rectangular channel that is long, so the geometry can be considered 2D, and the three cables are internally heated horizontal cylinders along the bottom of the rectangular enclosure. The stream function, vorticity, and energy equations were used for the gas in the conduit, and a finite-difference solution was obtained using a pseudo transient convergence method (DeVahl Davis 1986). The radiative portion was formulated by using net radiation enclosure theory (Chapter 5). The effect of radiative transfer was found to be important in the results, and it must be included in the theoretical modeling. The heat transfer behavior for each of the three cylindrical power cables in the conduit was significantly different from that for a single cable in a horizontal conduit.

Natural convection instabilities can be produced or modified by radiation exchange. Instabilities were analyzed by Lienhard (1990) for a plane layer of transparent fluid between two horizontal walls with the lower wall heated; the walls radiate and conduct energy. The radiation exchange between the walls tends to partially equalize temperature nonuniformities and thereby stabilize the fluid against the development of free-convection circulation cells. The stability of confined plane horizontal fluid layers has application to the design of flat-plate solar collectors.

The radiation exchange between surfaces can have significant effects on crystals being grown from the vapor phase within an enclosure. An analysis by Kassemi and Duval (1990) showed that the radiative exchange can induce natural convection for a normally stable heating configuration, thus altering the vapor transport to the crystal growth interface. Another study (de Groh and Kassemi 1993) considered a short circular cylindrical enclosure closed at both the top and bottom with flat plates. The top plate is heated, and both analytical and experimental information is obtained for combined radiation and free convection in the enclosure. If an enclosure is heated on the top and cooled at the bottom, the buoyancy associated with the temperature distribution in the gas tends to make the gas stable. Radiation exchange in the enclosure changes this; energy is radiated to the side walls and free-convection patterns are thereby initiated. The radiative exchange in this analysis was formulated for diffuse–gray surfaces using the net radiation enclosure theory (Chapter 5). The analysis in the gas was carried out with the finite-element computer code FIDAP. It was found by comparison with experiments that radiation effects are important even at temperature levels as low as 300°C. If radiation is not included, the numerical predictions can be far from reality. Radiative effects resulted in a double annular circulation pattern within the enclosure.

Because of the bifurcation/chaos characteristics of natural convection analysis, care should be taken in assuming that 2D solutions are valid, as 3D cell structures can form in apparently 2D geometries (e.g., long channels with asymmetrically heated walls.) These structures can be augmented or retarded by the presence of radiation.

7.5 NUMERICAL SOLUTION METHODS

Numerical solution methods are now presented and illustrated for radiation combined with conduction and/or convection. The methods also apply to pure radiation problems, which are usually easier to solve; for example, for gray surfaces, the equations are linear in temperature to the fourth power when there is only radiative transfer. In the previous chapters, energy equations were derived from an energy balance on each element of the system used to model the real configuration. For some pure radiation solutions, detailed temperature distributions may not be needed and an enclosure may be divided into relatively few elements. With conduction and convection included, there is usually a need for finer detail because these transfer modes depend on local temperature derivatives; accurate and detailed temperature distributions must be obtained so that derivatives can be evaluated accurately. Sometimes the difference between large incoming and outgoing radiation is needed to determine a relatively small amount of conduction and/or convection, and this leads to difficulties with convergence and/or accuracy.

The local energy balance on each element of the system involves the net radiation that is absorbed at the surface, and this depends on the summation of all the contributions from the surroundings. In a detailed formulation, the radiative portion may be set up as integrals involving the temperature distributions and configuration factors from the surrounding surfaces. Some of the solution methods using numerical integration are presented here and are illustrated with a few examples. Integration subroutines are available in computer mathematics software packages.

The benefits of nondimensionalization are discussed for multimode analyses. The relative sizes of the dimensionless parameters may provide insight into the best numerical approach. Some illustrative examples are set up using the finite-difference method and FEM. The result of the numerical formulation is a set of nonlinear algebraic equations, and solution methods are discussed such as by successive substitutions or iteration. Because of the nonlinearity of the combined-mode equations, it is usually necessary to use damping or underrelaxation factors to obtain convergence. Information is presented on the amount of underrelaxation to be used. Problems involving only radiation will sometimes permit overrelaxation to speed convergence. Further, Newton–Raphson method is presented as another useful technique for solving a set of nonlinear algebraic equations.

The Monte Carlo method is also presented to obtain radiative or multimode heat transfer solutions. This is a statistical method in which many small quantities of radiant energy are individually followed along their paths during radiative transfer. This method is relatively easy to set up for complex problems that involve spectral effects and/or directional surfaces. The solutions may require relatively long computer running times, but the Monte Carlo method may be the only reasonable way to attack some complex problems.

7.6 NUMERICAL INTEGRATION METHODS FOR USE WITH ENCLOSURE EQUATIONS

Integration is needed for numerical solution of pure radiation or combined-mode problems. For radiative exchange, the integrals are often functions of two position variables, and integration is over one or both of them. For example, the configuration factor dF_{di-dj} from position $\mathbf{r}_i$ on surface

i to position $\mathbf{r}_j$ on surface j appears in the integral over surface j to obtain F_{di-j} in the form (see Equations 4.11 and 5.53)

$$F_{di-j}(\mathbf{r}_i) = \int_{Aj} dF_{di-dj}(\mathbf{r}_i, \mathbf{r}_j) = \int_{Aj} K(\mathbf{r}_i, \mathbf{r}_j) dA_j \tag{7.11}$$

Many ways can be used to numerically approximate an integral. Because the integrands in radiative enclosure formulations are usually well behaved at the end points, *closed* numerical integration forms are often used that include the end points. *Open* methods do not include the end points and can be used when end-point values are indeterminate, such as for improper integrals that yield finite values when integrated. In analyses including convection and/or conduction, the numerical integration will usually use the grid spacing imposed by the differential terms. In some situations it is sufficient to use numerical integration methods that have regular grid spacing. However, uneven spacings are often advantageous to place more points in regions where functions have large variations or to adequately follow irregular boundaries. *Gaussian quadrature* can be used for variable grid spacing. Simpler schemes such as the *trapezoidal rule* or *Simpson's rule* may be adequate for some problems. These often employ uniform grid spacing and are closed, whereas Gaussian quadrature is open. The trapezoidal rule can readily be used with a nonuniform grid size.

The standard methods are discussed in detail in Appendix G of the online appendices at http://www.thermalradiation.net. Most are included in the standard mathematical packages such as MATLAB® and Mathcad.

7.7 NUMERICAL FORMULATIONS FOR COMBINED-MODE ENERGY TRANSFER

Figures 7.2 and 7.3 illustrated the boundary condition for an opaque surface where the net radiation $(G - J)$ provides the local addition of radiative heat flux. The $(G - J)$ is found by analyzing the radiative enclosure surrounding the surface area. This boundary condition can be applied to obtain the 3D conduction solution within the wall. As a simplified case, Figure 7.5 gave a control volume derivation for a thin wall in which the temperature distribution is 2D, as it is assumed not to vary significantly across the wall thickness. The control volume approach is further developed here to illustrate combined-mode solutions by using thin-walled enclosures as an example.

Consider an enclosure as in Figure 5.15, and let one or more of the walls be thin and heat conducting. There is uniform heat generation within each wall and convection at the inside surfaces, as in Figure 7.18. The outside of the enclosure is assumed well insulated for simplicity, but external heat flows can be added in a similar way to those included here.

A local rectangular coordinate system is positioned along a typical wall A_k (Figure 7.18). For a wall element at $\mathbf{r}_k$, an energy balance is written where the net radiative loss is balanced by the decrease in internal energy and by the energy added by conduction, convection, and internal heat generation:

$$J_k(\mathbf{r}_k) - G_k(\mathbf{r}_k) = q_k(\mathbf{r}_k) = \left[-\rho c a \frac{\partial T_k}{\partial t} + ka\left(\frac{\partial^2 T_k}{\partial x^2} + \frac{\partial^2 T_k}{\partial y^2} \right) + h\left(T_m - T_k\right) + \dot{q}a \right]_{r_k} \tag{7.12}$$

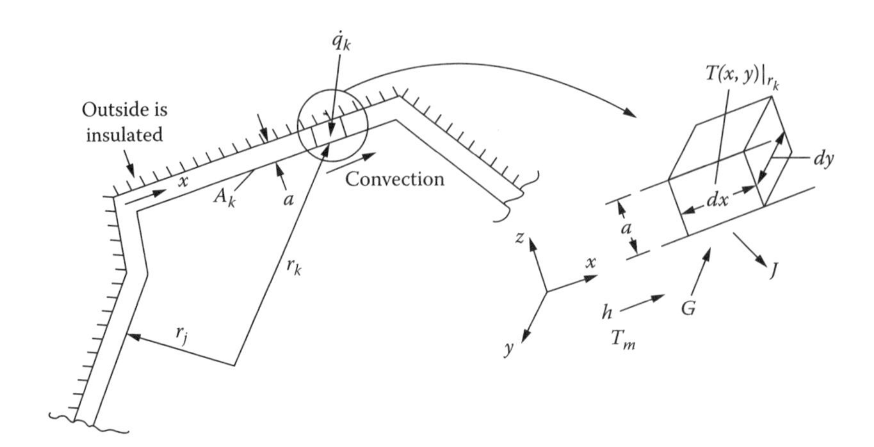

FIGURE 7.18 Radiative enclosure with thin walls in which there is 2D heat conduction.

The local radiative heat loss $q_k(\mathbf{r}_k)$ is found from the enclosure Equation 5.56:

$$\frac{q_k(\mathbf{r}_k)}{\varepsilon_k} - \sum_{j=1}^{N} \frac{1-\varepsilon_j}{\varepsilon_j} \int_{Aj} q_j(\mathbf{r}_j)\, dF_{dk-dj}(\mathbf{r}_j, \mathbf{r}_k) = \sigma T_k^4(\mathbf{r}_k) - \sum_{j=1}^{N} \int_{Aj} \sigma T_j^4(\mathbf{r}_j)\, dF_{dk-dj}(\mathbf{r}_j, \mathbf{r}_k) \tag{7.13}$$

The local temperature of the convecting medium, $T_m(\mathbf{r}_k)$, is obtained from additional convective heat transfer relations, as we illustrate. The equations are placed in dimensionless form to yield

$$\tilde{q}_k(\mathbf{R}_k) = \left[-\frac{\partial \vartheta_k}{\partial \tilde{t}} + N_{\mathrm{CR}} \left(\frac{\partial^2 \vartheta_k}{\partial X^2} + \frac{\partial^2 \vartheta_k}{\partial Y^2} \right) + H(\vartheta_m - \vartheta_k) + \dot{S} \right]_{\mathbf{R}_k} \tag{7.14}$$

$$\frac{\tilde{q}_k(\mathbf{R}_k)}{\varepsilon_k} - \sum_{j=1}^{N} \frac{1-\varepsilon_j}{\varepsilon_j} \int_{A_j} \tilde{q}_j(\mathbf{R}_j)\, dF_{dk-dj}(\mathbf{R}_j, \mathbf{R}_k) = \vartheta_k^4(\mathbf{R}_k) - \sum_{j=1}^{N} \int_{A_j} \vartheta_j^4(\mathbf{R}_j)\, dF_{dk-dj}(\mathbf{R}_j, \mathbf{R}_k) \tag{7.15}$$

where

$$\tilde{q} = q/\sigma T_{ref}^4, \quad N_{\mathrm{CR}} = k/a\sigma T_{ref}^3, \quad H = h/\sigma T_{ref}^3, \quad \dot{S} = \dot{q}a/\sigma T_{ref}^4$$

$$\tilde{t} = \sigma T_{ref}^3/\rho c a, \quad R = r/a, \quad X = x/a, \quad Y = y/a, \quad \vartheta = T/T_{ref}$$

In Equations 7.14 and 7.15, the dimensionless temperature ϑ and the dimensionless net radiative heat flux $\tilde{q}$ are the dependent variables, and X, Y, and $\tilde{t}$ are independent variables. The dimensionless parameters N_{CR}, $\dot{S}$, and H (or slight modifications of them) appear in combined-mode problems involving radiative transfer. They provide a measure of the importance, relative to radiation, of conduction, internal energy generation, and convection, and their relative magnitudes can help indicate the best solution method. If H is large, the problem could be solved as convective heat transfer with a small effect of radiation. The solution will probably converge best if ϑ is chosen as the dependent variable. For small H, the problem might best be solved as a radiative transfer problem using ϑ^4 as the dependent variable. In a transient problem, the variation in temperature level may shift the relative importance of the modes.

Equation 7.15 is the enclosure equation for gray surfaces. If the enclosure has surface properties that vary with wavelength, Equations 6.6 and 6.7 can be used. This is solved for $q_{\Delta\lambda,k}$ in each wavelength band for each surface. The $\tilde{q}_k(\mathbf{R}_k)$ in Equation 7.14 is then found as the summation $\sum_{\Delta\lambda} \tilde{q}_{\Delta\lambda,k}$. Equation 7.14 is otherwise unchanged. Once the best arrangement of the equations is determined, they must be placed in a form for numerical solution. The application of the finite-difference method and FEM are described in some examples.

7.7.1 Finite-Difference Formulation

The solution technique by finite differences is described by two examples.

Example 7.10

The fin temperature distribution in the array of fins in Figure 7.9 is considered in Example 7.5 and is governed by the dimensionless energy Equation 7.5.3

$$-\mu\frac{d^2\vartheta(X)}{dX^2}+\vartheta^4(X)=F_{dX-B}(X)+\int_{Z=0}^{1}\left[-\mu(1-\varepsilon)\frac{d^2\vartheta(Z)}{dZ^2}+\vartheta^4(Z)\right]\cdot dF_{dX\ dZ}(X,Z) \quad (7.10.1)$$

where

$$F_{dX-B}(X)=\frac{1}{2}\left[1-\frac{X}{(B^2+X^2)^{1/2}}\right]$$

$$dF_{dX-dZ}(X,Z)=\frac{1}{2}\frac{B^2}{\left[B^2+(Z-X)^2\right]^{3/2}}dZ$$

$$\mu=ka/2\varepsilon\sigma T_b^3W^2,\quad \text{and}\quad B=b/W.$$

The boundary conditions are $\vartheta = 1$ at the fin base $X = 0$, and $d\vartheta/dX = 0$ at $X = 1$ as it is assumed for simplicity that the end edge of each fin has negligible energy loss. To evaluate the integral on the right-hand side of Equation 7.10.1, the entire distributions of temperature and its second derivative must be known. Hence, if this equation is written at each of a set of X values, each equation will involve the unknown $\vartheta(X)$ at all of the X values, and all of the equations must be solved simultaneously.

To proceed with the solution of Equation 7.10.1, the fin is divided into N small elements, so that $X_i = i\Delta X$ and $Z_j = j\Delta Z$, where $0 \le i \le N$ and $0 \le j \le N$. The second derivative is approximated by

$$\frac{d^2\vartheta}{dX^2}=\frac{\vartheta_{i+1}-2\vartheta_i+\vartheta_{i-1}}{(\Delta X)^2}$$

The integral on the right-hand side of Equation 7.10.1 can be approximated using the trapezoidal rule (see Appendix G of the online appendices at http://www.thermalradiation.net):

$$\int_0^1 f(X,Z)dZ \approx \Delta Z\left[\frac{1}{2}f_0(X) + \sum_{j=1}^{N-1} f_j(X) + \frac{1}{2}f_N(X)\right]$$

where

$$f(X,Z) = \frac{B^2}{2}\left[-\mu(1-\varepsilon)\frac{d^2\vartheta}{dZ^2} + \vartheta^4(Z)\right]\frac{1}{\left[B^2 + (Z-X)^2\right]^{3/2}}$$

Using $\Delta Z = \Delta X$ and substituting the finite-difference form of the second derivative results in

$$f_j(X_i) = \frac{B^2}{2}\left[-\mu(1-\varepsilon)\frac{\vartheta_{j+1} - 2\vartheta_j + \vartheta_{j-1}}{(\Delta X)^2} + \vartheta_j^4\right]\frac{1}{\left\{B^2 + \left[(j-i)\Delta X\right]^2\right\}^{3/2}}$$

The two limits where $j = 0$ and $j = N$ are evaluated by applying the boundary conditions. For $j = i = 0$, $\vartheta_0 = 1$ and $d^2\vartheta/dZ^2 = 0$. The latter condition is needed because the ϑ_{i-1} term would otherwise be undefined; the condition arises because the energy entering the fin at the base is all by conduction, so the temperature gradient is linear at that location. Then, for $j = 0$,

$$f_0(X_i) = \frac{B^2}{2\left[B^2 + (i\Delta X)^2\right]^{3/2}}$$

For $j = N$, $\vartheta_{N+1} = \vartheta_{N-1}$ from $(d\vartheta/dx)_{X=1} = (\vartheta_{N-1} - \vartheta_{N-1})/(2\Delta X) = 0$, where $N + 1$ is a symmetric image point of $N - 1$. Then

$$f_N(X_i) = \frac{B^2}{2}\left[-\mu(1-\varepsilon)\frac{2(\vartheta_{N-1} - \vartheta_{N})}{(\Delta X)^2} + \vartheta_N^4\right]\frac{1}{\left\{B^2 + \left[(N-i)\Delta X\right]^2\right\}^{3/2}}$$

The energy Equation 7.10.1 for element i at $X_i = i\Delta X$ can now be written in finite-difference form as

$$-\mu\frac{\vartheta_{i+1} - 2\vartheta_{i-1}}{(\Delta X)^2} + \vartheta_i^4 = \frac{1}{2}\left\{\frac{1}{\left\{[B/(i\Delta X)]^2 + 1\right\}^{1/2}}\right\} + \Delta X\left[\frac{1}{2}f_0(X_i) + \sum_{j=1}^{N-1} f_j(X_i) + \frac{1}{2}f_N(X_i)\right] \quad 7.10.2$$

This equation is written for each element i for the range $1 \le i \le N$, giving N equations for the N unknown temperatures ϑ_i. Each equation contains every unknown ϑ_i, which appears in the f_j terms. If the resulting set of equations is written in matrix form, the coefficient matrix is full. This is in contrast to 1D pure conduction problems where the coefficient matrix is usually tridiagonal.

This illustrates that in radiative transfer problems, the temperature of every element can be influenced by the temperature of all of the surrounding elements.

Equation 7.10.2 written for element $i = 1$ gives

$$-\mu\frac{\vartheta_2-2\vartheta_1+1}{(\Delta X)^2}+\vartheta_1^4=\frac{1}{2}\left\{1-\frac{1}{[(B/\Delta X)^2+1]^{1/2}}\right\}+\frac{\Delta XB^2}{2}\left(\frac{1}{2}\frac{1}{\left[B^2+(\Delta X)^2\right]^{3/2}}\right.$$

$$-\sum_{j=1}^{N-1}\left[\mu(1-\varepsilon)\frac{\vartheta_{j+1}-2\vartheta_j+\vartheta_{j-1}}{(\Delta X)^2}-\vartheta_j^4\right]\frac{1}{\left\{B^2+\left[(j-1)\Delta X\right]^2\right\}^{3/2}}$$

$$\left.+\frac{1}{2}\left[-\mu(1-\varepsilon)\frac{2\left(\vartheta_{N-1}-\vartheta_N\right)}{(\Delta X)^2}+\vartheta_N^4\right]\frac{1}{\left\{B^2+\left[(N-1)\Delta X\right]^2\right\}^{3/2}}\right) \tag{7.10.3}$$

For the infinite array of fins, $\vartheta_i = f_j$, so that, gathering terms that have ϑ_j and ϑ_j^4 and using $\vartheta_0 = 1$, Equation 7.10.3 is written as (a repeated subscript denotes a summation over the values of that subscript, e.g., $A_{1j}\vartheta_j = \sum_{j=1}^{N} A_{1j}\vartheta_j$).

$$A_{1j}\vartheta_j + B_{1j}\vartheta_j^4 = C_1 \tag{7.10.4}$$

where $\left[\text{let } \tilde{B} \equiv B^2\mu(1-\varepsilon)/2\Delta X)\right]$

$$A_{11}=\tilde{B}\left\{\frac{4}{(1-\varepsilon)\Delta XB^2}-\frac{2}{B^3}+\frac{1}{\left[B^2+(\Delta X)^2\right]^{3/2}}\right\}$$

$$A_{12}=\tilde{B}\left\{\frac{-2}{(1-\varepsilon)\Delta XB^2}+\frac{1}{B^3}-\frac{2}{[B^2+(\Delta X)^2]^{3/2}}+\frac{1}{[B^2+(2\Delta X)^2]^{3/2}}\right\}$$

$$\vdots$$

$$A_{1j}=\tilde{B}\left(\frac{1}{\{B^2+[(j-2)\Delta X]^2\}^{3/2}}-\frac{2}{\{B^2+[(j-1)\Delta X]^2\}^{3/2}}+\frac{1}{[B^2+(j\Delta X)^2]^{3/2}}\right)\qquad 2<j<N-1$$

$$\vdots$$

$$A_{1(N-1)}=\tilde{B}\left(\frac{1}{\{B^2+[(N-3)\Delta X]^2\}^{3/2}}-\frac{2}{\{B^2+[(N-2)\Delta X]^2\}^{3/2}}+\frac{1}{\{B^2+[(N-1)\Delta X]^2\}^{3/2}}\right)$$

$$A_{1N}=\tilde{B}\left(\frac{1}{\left\{B^2+[(N-2)\Delta X]^2\right\}^{3/2}}-\frac{1}{\left\{B^2+[(N-1)\Delta X]^2\right\}^{3/2}}\right)$$

$$B_{1j}=\delta_{1j}-\frac{B^2\Delta X}{\beta_j\{B^2+[(j-1)\Delta X]^2\}^{3/2}};\quad \begin{matrix}\beta_j=2,\ 1\le j\le N\\ \beta_j=4,\ j=N\end{matrix}$$

$$C_1=\frac{1}{2}\left\{1-\frac{\Delta X}{[B^2+(\Delta X)^2]^{1/2}}\right\}+\frac{\mu}{(\Delta X)^2}+\frac{B^2(\Delta X)}{4[B^2+(\Delta X)^2]^{3/2}}-\frac{\mu(1-\varepsilon)}{2B(\Delta X)}$$

This is done for each element i, $1 \leq i \leq N$, and a matrix equation is generated of the form

$$[A_{ij}][\vartheta_j]+[B_{ij}][\vartheta_j^4]=[C_i] \tag{7.10.5}$$

This is a set of nonlinear algebraic equations for the unknown temperatures $\vartheta_i = \vartheta_j$. Solution methods are in Section 7.8.

Example 7.11

A transparent gas with mean velocity u_m and constant physical properties flows through a circular tube of inner diameter D_i and length l (Figure 7.19). A specified heat flux $q_e(x) = q_{max} \sin(\pi x/l)$ is applied along the tube length. The heat transfer coefficient h between the gas and the tube interior surface is assumed independent of x. The tube wall is thin and has thermal conductivity k_n. The gas enters the tube from a large plenum at $T_{g,1}$. The gas leaves the tube at $T_{g,2}$ and enters a mixing plenum that is also at $T_{g,2}$. The tube interior surface is diffuse–gray with emissivity ε. Set up the energy equations and boundary conditions to determine the wall temperature $T(x)$. Put the equations into a finite-difference form for numerical solution.

Following Examples 7.8 and 7.9, the governing energy equations are

$$\begin{aligned}\varepsilon\Bigg\{\vartheta^4(X)&+\int_{Z=0}^{X}\left[\frac{1-\varepsilon}{\varepsilon}\phi(Z)-\vartheta^4(Z)\right]dF_{dX-dZ}(X-Z)\\&+\int_{Z=X}^{L}\left[\frac{1-\varepsilon}{\varepsilon}\phi(Z)-\vartheta^4(Z)\right]dF_{dX-dZ}(Z-X)\\&-\vartheta_{r,1}^4F_{dX-1}(X)-\vartheta_{r,2}^4F_{dX-2}(L-X)\Bigg\}=\phi(X)\end{aligned} \tag{7.11.1}$$

$$\phi(X)=\sin\left(\frac{\pi X}{L}\right)+H[\vartheta_g(X)-\vartheta(X)]+N_{CR}\frac{d^2\vartheta}{dX^2} \tag{7.11.2}$$

$$\vartheta_g(X)=\mathrm{St}\,e^{-\mathrm{St}X}\int_0^X e^{\mathrm{St}Z}\,\vartheta(Z)dZ+\vartheta_{g,1}e^{-\mathrm{St}X} \tag{7.11.3}$$

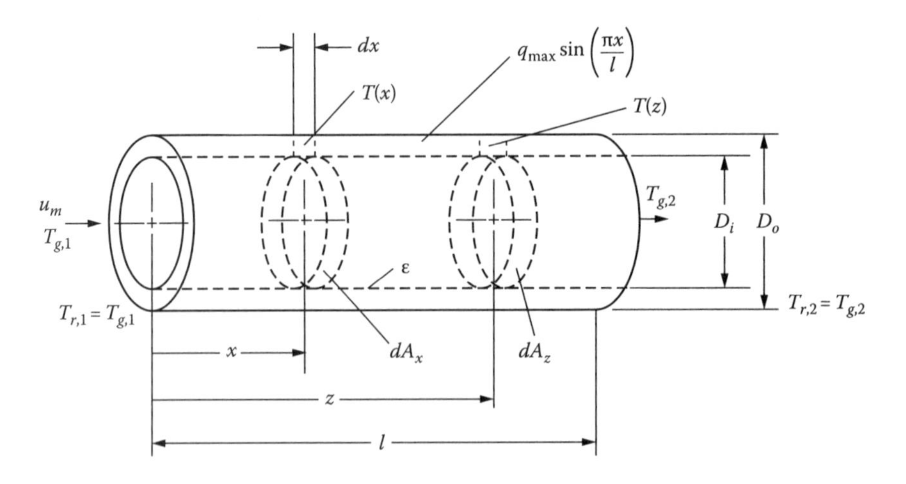

FIGURE 7.19 Cylindrical tube geometry with gas flow and internal radiation exchange.

In this form, $\vartheta = T(\sigma/q_{max})^{1/4}$ and the other parameters are as defined in Examples 7.7 and 7.9 (the reference value of q_e is q_{max} and the reference length is D_i). Substituting the last two equations to eliminate $\Phi(X)$ and $\vartheta_g(X)$ in Equation 7.10.4 results in

$$\varepsilon\vartheta^4(X) + \int_0^L \left[(1-\varepsilon)\left\{\sin\left(\frac{\pi Z}{L}\right) + H\left[\mathrm{Ste}^{-\mathrm{St}Z}\int_0^Z e^{\mathrm{St}\xi}\vartheta(\xi)d\xi + \vartheta_{g,1}e^{-\mathrm{St}Z} - \vartheta(Z)\right]\right.\right.$$
$$\left.\left. + N_{CR}\frac{d^2\vartheta}{dZ^2}\right\} - \varepsilon\vartheta^4(Z)\right]dF_{dX-dZ}(|X-Z|)$$
$$= N_{CR}\frac{d^2\vartheta}{dX^2} + \sin\left(\frac{\pi X}{L}\right) + H\left[\mathrm{Ste}^{\mathrm{St}X}\int_0^X e^{\mathrm{St}Z}\vartheta(Z)dZ + \vartheta_{g,1}e^{-\mathrm{St}X} - \vartheta(X)\right]$$
$$+ \vartheta_{r,1}^4F_{dX-1}(X) + \vartheta_{r,2}^4F_{dX-2}(L-X) \tag{7.11.4}$$

For the boundary conditions required by the $d^2\vartheta/dX^2$ term, both end edges of the tube are assumed to have negligible heat losses, so $(d\vartheta/dX)_{x=0} = (d\vartheta/dX)_{x=L} = 0$. The condition $\vartheta_g(X = 0) = \vartheta_{g,1}$ was used in deriving Equation 7.11.3. To proceed with the numerical solution, define

$$f(X,Z) = \left(\frac{1-\varepsilon}{\varepsilon}\left\{\sin\left(\frac{\pi Z}{l}\right) + H\left[\mathrm{Ste}^{-\mathrm{St}Z}\int_0^Z g(\xi)d\xi + \vartheta_{g,1}e^{-\mathrm{St}Z} - \vartheta(7)\right]\right.\right.$$
$$\left.\left. + N_{CR}\frac{d^2\vartheta}{dZ^2}\right\} - \vartheta^4(Z)\right)\frac{dF_{dX-dZ}(|X-Z|)}{dZ} \tag{7.11.5}$$

where $g(\xi) = e^{\mathrm{St}\xi}\vartheta(\xi)$. Equation 7.11.4 becomes

$$\varepsilon\left[\vartheta^4(X) + \int_0^L f(X,Z)dZ\right] = N_{CR}\frac{d^2\vartheta}{dX^2} + \sin\left(\frac{\pi X}{L}\right) + H\left[\mathrm{Ste}^{-\mathrm{St}X}\int_0^X g(Z)dZ\right.$$
$$\left. + \vartheta_{g,1}e^{-\mathrm{St}X} - \vartheta(X)\right] + \vartheta_{r,1}^4F_{dX-1}(X) + \vartheta_{r,2}^4F_{dX-2}(L-X) \tag{7.11.6}$$

Numerical integration is applied to each integral, and a set of nonlinear algebraic equations is obtained as in Example 7.5.1. At $X_i = i\Delta X$, Equation 7.11.6 becomes, by use of the trapezoidal rule and having $\Delta X = \Delta Z$ where $\Delta X = L/I$ (I is the number of ΔX increments),

$$\varepsilon\left\{\vartheta_i^4 + \Delta X\left[\frac{1}{2}f_0(X_i) + \sum_{j=1}^{I-1}f_i(X_i) + \frac{1}{2}f_I(X_i)\right]\right\} = N_{CR}\frac{\vartheta_{i+1} - 2\vartheta_i + \vartheta_{i-1}}{(\Delta X)^2} + \sin\left(\frac{X_i}{L}\right)$$
$$+ H\left\{\mathrm{Ste}^{-\mathrm{St}X_i}\Delta X\left[\frac{1}{2}g_0 + \sum_{j=1}^{i-1}g_j + \frac{1}{2}g_i\right] + \vartheta_{g,1}e^{-\mathrm{St}X_i} - \vartheta_i\right\}$$
$$+ \vartheta_{r,1}^4F_{dX_i-1}(X_i) + \vartheta_{r,2}^4F_{dX_i-2}(L-X_i) \tag{7.11.7}$$

Note that for the reservoir temperatures for the specified conditions of this example, $\vartheta_{r,1} = \vartheta_{g,1}$ and $\vartheta_{r,2} = \vartheta_{g,2}$. By gathering terms after expansion as in Example 7.5.1, the full set of equations can be written as

$$\begin{array}{llllll}
(A_{00}\vartheta_0 + B_{00}\vartheta_0^4) & +(A_{01}\vartheta_1 + B_{01}\vartheta_1^4) & +\cdots & +\cdots & +(A_{0I}\vartheta_I + B_{0I}\vartheta_I^4) & = C_0 \\
(A_{10}\vartheta_0 + B_{10}\vartheta_0^4) & +(A_{11}\vartheta_1 + B_{11}\vartheta_1^4) & +\cdots & +\cdots & +(A_{1I}\vartheta_I + B_{1I}\vartheta_I^4) & = C_1 \\
\cdots & +\cdots & +\cdots & +\cdots & +\cdots & = \cdots \\
(A_{i0}\vartheta_0 + B_i\vartheta_0^4) & +\cdots & +(A_{ij}\vartheta_j + B_{ij}\vartheta_j^4) & +\cdots & +(A_{iI}\vartheta_I + B_{iI}\vartheta_I^4) & = C_i \\
(A_{I0}\vartheta_0 + B_{I0}\vartheta_0^4) & +\cdots & +(A_{Ij}\vartheta_j + B_{Ij}\vartheta_j^4) & +\cdots & +(A_{II}\vartheta_I + B_{II}\vartheta_I^4) & = C_I
\end{array} \tag{7.11.8}$$

or in matrix form

$$\left[A_{ij}\right]\left[\vartheta_j\right] + \left[B_{ij}\right]\left[\vartheta_j^4\right] = \left[C_i\right] \tag{7.11.9}$$

Once ϑ_j is determined from the solution of Equations 7.11.8 and 7.11.9, the heat flux ϕ_j can be found from Equation 7.11.7 if needed.

7.7.2 FEM Formulation

The FEM has the advantage that the approximation for the temperature in a volume or surface element can vary across the element. Finite-difference formulations assign a single uniform temperature to each element. The temperature variation in the FEM can be specified to increased degrees of approximation (constant, linear, parabolic, etc.) at the cost of increasing the computation time. Temperatures at the boundaries of adjacent elements can be matched, temperature gradients can be forced to match, and, with increased complexity and computer time, the second or higher derivatives can be forced to match.

Various approaches to the FEM formulation are used, but the most prevalent is the Galerkin method. It is only feasible to provide a brief description here. After this brief description oriented toward solving the energy equation, the use of the FEM is illustrated by formulating Example 7.12. Further developments using FEM for other types of radiative energy transfer problems are in Section 13.6.

Consider the energy equation for transient heat conduction and for constant properties, with the local radiative energy source $-\nabla \cdot q_r$ combined with the local source by internal heat generation $\dot{q}$ into a term q_s (see Section 7.2):

$$\rho c_p \frac{\partial T}{\partial t} - k\nabla^2 T - q_s(r,T,t) = 0 \tag{7.16}$$

The volume in which the energy equation is to be solved is divided into finite subregions; these are the *finite elements*. For 1D geometries, the elements are plane, cylindrical, or spherical layers that can have different thicknesses. In two dimensions, triangles are usually used, as in Figure 7.20, or irregular quadrilaterals. In three dimensions, tetrahedrons or rectangular prisms are often used. These are examples of the types of finite elements used to divide geometries of irregular shape; the treatment of irregular volumes is one advantage of using the FEM.

Nodes are assigned to locations in the elements at which the unknown function, such as temperature, is to be determined. Nodes are often placed only at the corners of the elements, but additional nodes can be placed internally or along the element boundaries.

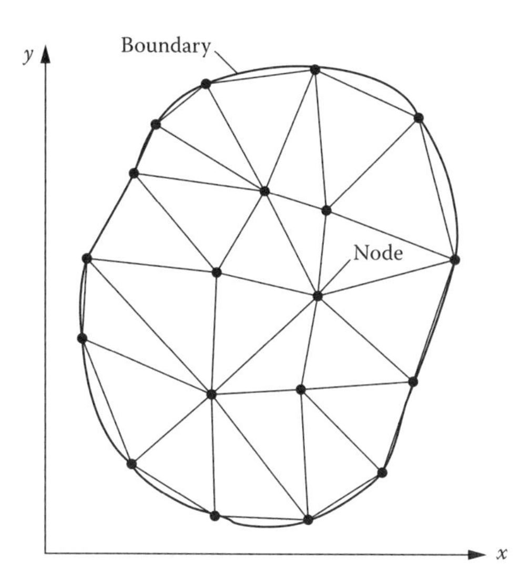

FIGURE 7.20 Two-dimensional region represented by triangular finite elements having nodes at vertexes.

7.7.2.1 Shape Function

The next step in the FEM is to choose *shape* or *interpolation* functions to provide an approximate variation of the dependent variable within each element between the values at the nodes. The simplest shape function is linear, but quadratic and higher-order variations can be used. Each shape function is a *local interpolation function* that is finite only within elements containing a particular node. To describe the shape function in more detail, consider a planar 1D problem with coordinate X. The shape function can be derived by using a series expansion of the unknown function, $T(X) = a_0 + a_1X + a_2X^2 + a_3X^3 + \cdots$. The number of terms used in the series is determined by two considerations: retaining more terms allows a more accurate representation of the temperature distribution within each element, but using fewer terms reduces computation time. The choice of the form of the shape function is a trade-off between accuracy within an element, and thus the number of elements required, and the average computation time per element. Most solutions have used linear or quadratic forms.

For a *linear* shape function in one dimension, there is a node at each end of the element; these nodes are designated here by subscripts 1 and 2 on T and X (note: the a_0 and a_1 are coefficients, and their subscripts do not refer to the nodes). The values of the dependent variable at the nodes are $T_1 = T(X_1) = a_0 + a_1X_1$ and $T_2 = T(X_2) = a_0 + a_1X_2$. Solving for a_0 and a_1 gives $a_0 = (T_1X_2 - T_2X_1)/(X_2 - X_1)$ and $a_1 = (T_2 - T_1)/(X_2 - X_1)$. Then, in the element, $T(X) = \Phi_1(X)T_1(X_1) + \Phi_2(X)T_2(X_2)$, where $\Phi_1(X) = (X_2 - X)/(X_2 - X_1)$ and $\Phi_2(X) = (X - X_1)/(X_2 - X_1)$. The Φ are thus the desired shape or interpolation functions. These functions are each equal to 1 at the node designated by their subscript, and they decrease linearly to zero at the neighboring node. Each shape function is zero outside of the element that contains its particular node (Figure 7.21).

For a quadratic shape function in a 1D geometry, three nodes are used per element. For this and higher-order shape functions, to evaluate the a_m coefficients, the first- and higher-order derivatives of $T(X)$ are matched at the nodes, in addition to the T values. For 2D problems, the form of the linear shape function for triangular elements is found by using the expansion $T(X,Y) = a_0 + a_1X + a_2Y$. For square elements, the biquadratic form can be used: $T(X,Y) = a_0 + a_{1,1}X + a_{1,2}X^2 + a_{2,1}Y + a_{2,2}Y^2$. The shape functions are then derived by substituting the T values at the nodes and solving simultaneously for the unknown a_m or $a_{m,n}$. For the five coefficients in the biquadratic form, there can be a node at each corner of the square and one in the center.

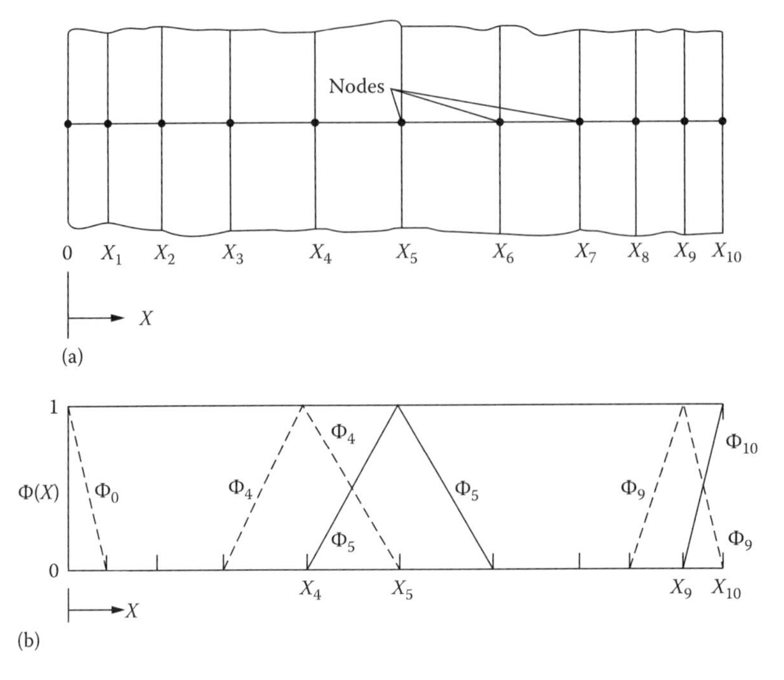

FIGURE 7.21 One-dimensional finite elements and linear shape functions: (a) plane layer with elements of unequal size and (b) linear shape functions for typical elements.

7.7.2.2 Galerkin Form for the Energy Equation

By using the shape functions, an approximate solution $\hat{T}(\mathbf{r},t)$ for $T(\mathbf{r},t)$ is assumed in the form

$$T(\mathbf{r},t) \approx \hat{T}(\mathbf{r},t) = \sum_{j=1}^{N} T_j(t)\Phi_j(\mathbf{r}) \tag{7.17}$$

where

the T_j are the values at the nodes desired from the solution
$\mathbf{\Phi}_j(\mathbf{r})$ are the shape functions that are equal to 1 at each node

When this approximation is substituted into the energy equation, Equation 7.16, there is a residual that depends on $\mathbf{r}$ and t:

$$\rho c_p \frac{\partial \hat{T}}{\partial t} k\nabla^2 \hat{T} - q_s(\mathbf{r},\hat{T},t) = Res(\mathbf{r},t) \tag{7.18}$$

It is desired to obtain a solution that, in an average sense over the entire volume, is as close as possible to the exact solution at each time. Variational principles are applied to minimize the residual. A set $W_j(\mathbf{r})$ of independent weighting functions is applied, and the residual is made orthogonal with respect to *each* of the weighting functions. This provides the following integral, which is evaluated for each of the set of independent weighting functions,

$$\int_V Res(\mathbf{r},t)W_i(\mathbf{r})dV = 0 \tag{7.19}$$

The integration is over the whole volume in which the solution is being obtained. Carrying out the integration for each of the set of weighting functions yields a set of simultaneous equations that can be solved for the $\hat{T}_i$ values at the nodes. In Galerkin method, the weighting functions are chosen to be the same function set as the shape functions. Since each shape function Φ_j is zero except within an element containing $\hat{T}_j$, the resulting matrix for solving the simultaneous equations for $\hat{T}_j$ is banded and sparse. By using the residual from Equation 7.18, Equation 7.17 for $\hat{T}$, and the $\Phi_j(\mathbf{r})$ as the set of weighting functions, Equation 7.19 provides Galerkin form of the energy equation for each of the weighting functions:

$$\int_V \left[\rho c_p \sum_{j=1}^{N} \Phi_j(\mathbf{r}) \frac{\partial T_j(t)}{\partial t} - k \sum_{j=1}^{N} T_j(t) \nabla^2 \Phi_j(\mathbf{r}) - q_s(\mathbf{r}, \hat{T}, t) \right] \Phi_i(\mathbf{r}) dV = 0 \tag{7.20}$$

Evaluating Equation 7.20 for each i provides N simultaneous equations for the $\hat{T}_j$. This method is now illustrated through an example.

Example 7.12

Set up Example 7.11 for numerical solution using the FEM. For simplicity, let the tube interior surface be black.

Following the analysis in Razzaque et al. (1982), it is advantageous to use the independent variable $\omega(X) \equiv \vartheta^4(X)$, so Equation 7.16 becomes, with $\varepsilon = 1$,

$$\omega(X) - \int_0^L \omega(Z) dF_{dX-dZ}(|X-Z|) = \frac{N_{CR}}{4} \frac{d}{dX}\left(\frac{1}{\omega^{3/4}} \frac{d\omega}{dX} \right) + \sin\left(\frac{\pi X}{L} \right)$$
$$+ H \left[\mathrm{Ste}^{-\mathrm{St}X} \int_0^X e^{\mathrm{St}Z} \omega^{1/4}(Z) dZ + \vartheta_{g,1} e^{-\mathrm{St}X} - \omega^{1/4}(X) \right]$$
$$+ \vartheta_{r,1}^4 F_{dX-1}(X) + \vartheta_{r,2}^4 F_{dX-2}(L-X) \tag{7.12.1}$$

with boundary conditions $d\omega/dX = 0$ at $X = 0, L$. If we define $A(\omega) \equiv N_{CR}/4\omega^{3/4}$ and

$$\psi(X,\omega) \equiv \sin\left(\frac{\pi X}{L} \right) + H \left[\mathrm{Ste}^{-\mathrm{St}X} \int_0^X e^{-\mathrm{St}Z} \omega^{1/4}(Z) dZ + \vartheta_{g,1} e^{-\mathrm{St}X} - \omega^{1/4}(X) \right]$$
$$+ \vartheta_{r,1}^4 F_{dX-1}(X) + \vartheta_{r,2}^4 F_{dX-2}(L-X) + \int_0^L \omega(Z) dF_{dX-dZ}(|X-Z|) \tag{7.12.2}$$

then Equation 7.12.1 has the form

$$-\frac{d}{dX}\left[A(\omega) \frac{d\omega}{dX} \right] + \omega = \psi(X,\omega) \tag{7.12.3}$$

Note that $\psi(X, \omega)$ contains nonlinear terms in the variable ω.

To solve Equation 7.12.3 by Galerkin FEM, the $\omega(X)$ is required to satisfy a variational form of Equation 7.12.3 and its boundary conditions, which has the form (using Equation 7.19 with the residual of Equation 7.12.3)

$$\int_0^L \left\{ -\frac{d}{dX}\left[a(\omega)\frac{d\omega}{dX} \right] + \omega \right\} W(X)dX - \int_0^L \psi(X,\omega)W(X)dX = 0 \tag{7.12.4}$$

The $W(X)$ is a weighting function defined by (see Equation 7.17)

$$W(X) = \sum_{i=1}^{N} W_i \Phi_i(X) \tag{7.12.5}$$

The $\Phi_i(X)$ is the *shape function*, and the W_i are coefficients at the nodes. The first term in the first integral in Equation 7.12.4 is integrated by parts, and the boundary conditions of the insulated end edges of the tube wall are used: $d\vartheta/dX = d\omega/dX = 0$ at $X = 0$ and L. This gives

$$\int_0^L \left[A(\omega)\frac{d\omega}{dX}\frac{dW}{dX} + \omega W \right] dX - \int_0^L \psi(X,\omega)W(X)dX = 0 \tag{7.12.6}$$

Now, as in Equation 7.17, an approximate solution is sought of the form

$$\omega(X) \approx \Omega(X) = \sum_{j=1}^{N} \Omega_j \Phi_j(X) \tag{7.12.7}$$

In Galerkin method, the shape functions are the same as in the weighting function, Equation 7.12.5, and Ω_j are the values of Ω at the nodes. Substituting Equations 7.12.5 and 7.12.7 into Equation 7.12.1 results in

$$\sum_{i=1}^{N} W_i \left(\sum_{j=1}^{N} \left\{ \int_0^L \left[A(\Omega)\frac{d\Phi_i}{dX}\frac{d\Phi_j}{dX} + \Phi_i\Phi_j \right] dX \right\} \Omega_j - \int_0^L \psi(X,\Omega)\Phi_i dX \right) = 0 \tag{7.12.8}$$

If we now define

$$K_{ij} = \int_0^L \left[A(\Omega)\frac{d\Phi_i}{dX}\frac{d\Phi_j}{dX} + \Phi_i\Phi_j \right] dX \quad \text{and} \quad \psi_i \int_0^L \psi(X,\Omega)\Phi_i dX$$

Equation 7.12.8 becomes

$$\sum_{i=1}^{N} W_i \left[\sum_{j=1}^{N} K_{ij}\Omega_j - \psi_i \right] = 0 \tag{7.12.9}$$

To satisfy this equation, the quantity inside the square brackets must equal zero, since the W_i are constant coefficients. Then the required Ω_j values can be obtained by solving the equivalent set of equations

$$\begin{aligned} K_{11}\Omega_1 + K_{12}\Omega_2 + \cdots + \cdots + K_{1N}\Omega_N &= \psi_1 \\ K_{21}\Omega_1 + K_{22}\Omega_2 + \cdots + \cdots + K_{2N}\Omega_N &= \psi_2 \\ &\vdots \\ K_{i1}\Omega_1 + \cdots + K_{ij}\Omega_j + \cdots + K_{iN}\Omega_N &= \psi_i \\ &\vdots \\ K_{N1}\Omega_1 + \cdots + K_{Nj}\Omega_j + \cdots + K_{NN}\Omega_N &= \psi_N \end{aligned} \tag{7.12.10}$$

In matrix form, this is

$$[K_{ij}][\Omega_j] = [\psi_i] \tag{7.12.11}$$

Because each shape function (and hence each weighting function in Galerkin method) is usually zero except within elements containing a particular node, the $[K_{ij}]$ is a sparse matrix that is banded along the diagonal. As noted in the introduction to this section, the shape function Φ can have many forms. For example, a linear form could be used for $\Phi_i(X)$ within each element $X_i \le X \le X_{i+1}$ (where $X_{i+1} = X_i + \Delta X$) to yield $\Phi_i(X) = (X_{i+1} - X)/(X_{i+1} - X_i)$. The FEM solution can now be carried out by solving Equations 7.12.10 and 7.12.11 using a numerical technique, as in Section 7.8. The equation for numerical solution in the FEM, Equations 7.12.10 and 7.12.11, has the same form as Equations 7.11.8 and 7.11.9 for finite differences. Numerical methods that work well with finite-difference formulations will usually apply for solving FEM formulations. After the Ω_j are obtained, Equation 7.12.7 is used to find the required temperature values from $\Omega(X) \approx \omega(X) = \vartheta^4(X)$. Note, however, that both K_{ij} and ψ_i are functions of Ω and thus of Ω_j, so the solution is iterative.

Razzaque et al. (1982) applied the FEM to the situation in Examples 7.12 and 7.13. They used a quadratic shape function and were able to extend previous results to a dimensionless tube length of $L = 20$ (limited in other methods to less than 10 and in some cases to less than 5, by numerical instabilities). Some results are in Figure 7.22. The FEM was applied by Altes et al. (1986) to determine the 3D temperature distribution inside a conducting solid exposed to combined conduction and radiation boundary conditions. Complex geometries representing circuit board chips and a magnetoplasmadynamic propulsion system were modeled using this approach. Radiation interaction among surface elements was included in the solutions, which were obtained using a modified version of a commercially available computer code.

7.8 NUMERICAL SOLUTION TECHNIQUES

Examples 7.11 through 7.13 result in a matrix of nonlinear algebraic equations of the form of Equation 7.11.9. It is important to examine the relative values of the elements A_{ij} and B_{ij}. If the A_{ij} are comparatively large, the problem can be treated as linear in ϑ_j; conversely, for large B_{ij}, the problem can be treated as linear in ϑ_j^4. When the coefficients A and B are approximately equal, other treatments are in order.

If we define $A_{ij}^* = A_{ij} + B_{ij}\vartheta_j^3$, Equation 7.11.9 becomes

$$[A_{ij}][\vartheta_j] + [B_{ij}][\vartheta_j^4] = [A_{ij} + B_{ij}\vartheta_j^3][\vartheta_j] = [A_{ij}^*][\vartheta_j] = [C_i] \tag{7.21}$$

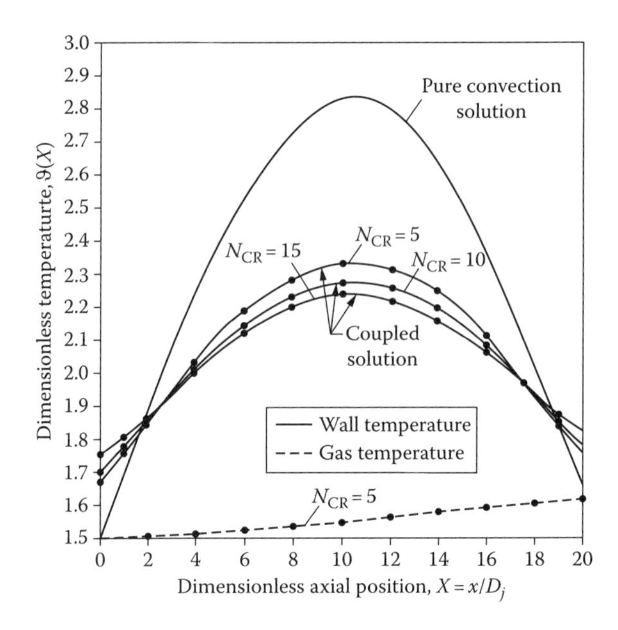

FIGURE 7.22 Comparison of solutions with sinusoidal wall heat flux to illustrate effects of radiation and axial wall conduction: $L = 20$, $St = 0.01$, $\vartheta_{r,1} = \vartheta_{g,1} = 1.5$, $\vartheta_{r,2} = \vartheta_{g,2}$, $H = 0.8$, and N_{CR} is defined in Example 7.9.

This is a set of linear algebraic equations with coefficients A^*_{ij} that are variable and nonlinear. The equations cannot be solved by elimination or direct matrix inversion, because the A^*_{ij} are temperature dependent and thus are not known. Some numerical solution methods are now discussed.

7.8.1 Successive Substitution Methods

7.8.1.1 Simple Successive Substitution

A simple solution method is to assume an initial set of temperatures $\vartheta_j^{(0)}$ and use them to compute $\left[A^*_{ij}\left(\vartheta_j^{(0)}\right)\right]$. This provides values for the elements in the matrix of coefficients, leaving the temperature vector $[\vartheta_j]$ as the unknown. Equation 7.21 or 7.11.9 is then solved for a new set of temperatures $\left[\vartheta_j^{(n+1)}\right]$ from

$$\left[A^*_{ij}\left(\vartheta_j^{(n)}\right)\right]\left[\vartheta_j^{(n+1)}\right] = \left[C_i\right] \tag{7.22}$$

This process is continued until the difference between successive temperature sets is less than an acceptable error, indicating convergence. A difficulty is that this method depends on an accurate initial guess for $[\vartheta_j]$. An inaccurate guess can lead to unstable iterations that may diverge rapidly.

7.8.1.2 Successive Underrelaxation

The simple successive substitution (SSS) method can be modified to obtain convergence in many cases if Equation 7.11.9 or 7.21 is written as

$$\left[A^*_{ij}\left(\vartheta_j^{*(n)}\right)\right]\left[\vartheta_j^{(n-1)}\right] = \left[C_i\right] \tag{7.23}$$

where the $A_{ij}^*\left(\vartheta_j^{*(n)}\right)$ are computed at each iteration by using a modified temperature

$$\vartheta_j^{*(n)} = \alpha\vartheta_j^{(n)} + (1-\alpha)\vartheta_j^{(n-1)} \tag{7.24}$$

The α is a weighting coefficient, or *relaxation parameter*, in the range $0 \le \alpha \le 1$. When $\alpha = 1$, the successive underrelaxation (SUR) method reduces to SSS; when $\alpha < 1$, the new guess is weighted toward the previous guess (i.e., underrelaxed), and oscillations between iterations are damped. If possible, the α should be chosen or found that provides optimized convergence. Values in the range $0.1 < \alpha < 0.4$ give convergence in many cases. Decreasing α usually provides slower convergence, but greater assurance that convergence will occur. Sometimes decreasing α somewhat will increase convergence by reducing oscillatory behavior. Values of $\alpha \approx 0.3$ are reported by Cort et al. (1982) to provide rapid convergence in many cases.

7.8.1.3 Regulated Successive Underrelaxation

Cort et al. (1982) proposed a method of regulated successive underrelaxation (RSUR) that allows the underrelaxation factor α to be chosen and modified for successive iterations. They recommend the following: (1) Initialize $\alpha = 1$; (2) solve Equation 7.24 for $\vartheta_j^{*(n)}$ (for the first iteration, an initial guess $\vartheta_j^{(0)}$ must be provided); (3) solve Equation 7.23 for ϑ_j^{n+1}; (4) calculate

$$\nu^{(n+1)} = \left[\sum_{j=1}^{N}\left(\vartheta_j^{(n+1)} - \vartheta_j^{(n)}\right)^2\right]^{1/2} \tag{7.25}$$

$$R^{(n+1)} = \left[\sum_{j=1}^{N}\left(\vartheta_j^{(n+1)}\right)^2\right]^{1/2} \tag{7.26}$$

and if $\nu^{(n+1)} > \nu^{(n)}$ or if $\nu^{(n+1)} > (1/3)R^{(n+1)}$, reduce α by 0.1; and (5) repeat steps (2) through (4) until convergence.

Equation 7.25 checks for divergence of the solution between iterations, and Equation 7.26 is used to see whether the residual error after each iteration is smaller than a measure of the root-mean-square temperature over the region of the solution. The latter check eliminates slowly oscillating but converging solutions that pass the test of Equation 7.25 but converge very slowly.

Another approach is to rewrite Equation 7.11.9 in the form

$$A_{ii}\vartheta_i^{(n+1)} + B_{ii}\left(\vartheta_i^{(n+1)}\right)^4 = C_i - \sum_{j=1}^{N}(1-\delta_{ij})\left[A_{ij}\vartheta_j^{(n+1)} + B_{ij}\left(\vartheta_j^{(n+1)}\right)^4\right] \equiv D_i \tag{7.27}$$

where δ_{ij} is the Kronecker delta. An initial set of temperatures $\vartheta_j^{(0)}$ is guessed, and D_i is evaluated on the basis of this set. Then the $\vartheta_j^{(1)}$ are found by iterative solution of Equation 7.27 and are used to evaluate the next set of D_i. This process is repeated to solve for $\vartheta_i^{(n)}$ until convergence. Tan (1989) points out that, for a given value of i, Equation 7.27 is a quartic equation with a single real positive root $\vartheta_i^{(n+1)}$ given by

$$\vartheta_i^{(n+1)} = \frac{y^{1/2}}{2}\frac{p-2}{(p-1)^{1/2}+1} \tag{7.28}$$

where

$$p = 2\left(1 + \frac{4D_i}{B_{ii}y^2}\right)^{1/2}, \quad y = \frac{2r}{(s+r)^{2/3} + (s-r)^{1/3}[(s+r)^{1/3} + (s-r)^{1/3}]}$$

and

$$r = \frac{1}{2}\left(\frac{A_{ii}}{B_{ii}}\right)^2, \quad s = \left[r^2 + \left(\frac{4D_i}{3B_{ii}}\right)^3\right]^{1/2}$$

Thus, for each set of D_i, the $\vartheta_i^{(n+1)}$ can be found directly from the nonlinear Equation 7.27 rather than by an inner iteration and then can be used to evaluate new D_i and continue to the next main iteration. This method is quite fast and can be combined with the SUR technique to determine succeeding approximations to provide a method that is *both* stable and fast.

7.8.2 Newton–Raphson-Based Methods for Nonlinear Problems

7.8.2.1 Modified Newton–Raphson

A modified Newton–Raphson (MNR) method is in Ness (1959) for the class of nonlinear problems encountered here. Starting from Equation 7.11.9,

$$\left[A_{ij}\right]\left[\vartheta_j\right] + \left[B_{ij}\right]\left[\vartheta_j^4\right] - \left[C_i\right] = 0 \tag{7.29}$$

an initial approximate temperature $\vartheta_j^{(0)}$ is guessed at each node. A correction factor δ_j is then computed so that $\vartheta_j = \vartheta_j^{(0)} + \delta_j$. This ϑ_j is used to compute a new δ_j, and this process is continued until δ_j becomes smaller than a specified value. The δ_j are found by solving the set of linear equations:

$$[f_{ij}][\delta_j] + [f_i] = 0 \tag{7.30}$$

where

$$f_i = \sum_{j=1}^{N}\left[A_{ij}\vartheta_j^{(0)} + B_{ij}\left(\vartheta_j^{(0)}\right)^4\right] - C_i \tag{7.31}$$

and

$$f_{ij} = A_{ij} + 4B_{ij}\left(\vartheta_j^{(0)}\right)^3 \tag{7.32}$$

The MNR method may not converge if a poor initial temperature set is chosen.

7.8.2.2 Accelerated Newton–Raphson

Cort et al. (1982) proposed a method in which the amount of change in ϑ_i at each iteration is adjusted to accelerate convergence. They recommended that the f_{ij} in the MNR method be replaced by

$$f_{ij} = A_{ij} + \frac{4B_{ij}}{[1-(\beta/3)]}\left(\vartheta_j^{(0)}\right)^{(3-\beta)} \quad \beta \ge 0 \tag{7.33}$$

This effectively modifies the slope of the changes in ϑ_j with respect to iteration number compared with that used in the MNR method. For $\beta = 0$, the accelerated Newton–Raphson (ANR) method reduces to MNR. If β is too large, oscillations and divergence between iterations may occur. For $\beta = 0.175$, the number of iterations to provide a given accuracy for a particular problem was reduced from 28 using MNR to 12 using ANR, and reductions in computer time of up to 80% were obtained. A starting value of $\beta = 0.15$ is recommended by Cort et al.

7.8.3 Applications of the Numerical Methods

Results using the aforementioned methods were compared in Cort et al. (1982) for some typical radiation–conduction problems with temperature-dependent properties and internal energy generation. Consideration was limited to surfaces with radiative exchanges to black surroundings at a single temperature, and the solutions were by finite elements. Because the example problems in this section showed that even complicated radiation–conduction–convection problems with multiple surfaces reduce to the same general form of Equations 7.11.9 and 7.12.10, the conclusions probably apply to a broader class of problems than was studied. In Costello and Shrenk (1966), a linearized solution of Equation 7.12.11 is proposed that speeds convergence over the MNR method. For problems that are either conduction or radiation dominated or where both modes are important, the method performed well, providing a factor-of-ten improvement in solution speed. It was found that the SUR method gave convergence with the fewest iterations and the least computer time; RSUR was useful to find the optimum value of the relaxation parameter α for use in the SUR method. For the Newton–Raphson method, ANR was always faster than MNR, but neither method was as fast as SUR.

In Howell (1992), the convergence ranges and behavior of equations of the form of Equations 7.11.8 or 7.12.11 are discussed, and the various solution methods of this chapter are examined. Nonlinear equations of this type can have behavior characterized by bifurcations and chaos so that steady solutions carried out by SSS, SUR, etc., may not converge. This is true whether the equations are cast as radiation dominated or first-order temperature dominated or the equations used are in mixed form such as Equation 7.21. Decreasing the relaxation factor extends the range of convergence, but often will not yield a solution for some ranges of parameters without unacceptable computer time. For conduction–radiation problems, a bifurcation–chaos behavior results from the numerical method chosen and the equation form and does not imply that multiple physical solutions can exist. However, when there is coupling between radiation and the flow field, as in combined radiation and free convection, multiple physical steady-state solutions may exist. The particular flow configuration reached in steady state may depend on the initial conditions chosen and the set of velocity and temperature fields that are traversed in reaching steady state. In some cases, no steady solution is reached; it may be possible to solve for the steady-state solution by using a fully transient solution that proceeds to the final steady state from physically specified initial conditions.

Numerical solution techniques for steady-state and transient combined-mode problems with surface–surface radiative exchange are examined and discussed by Hogan and Gartling (2008). Three techniques that sequentially solve radiative transfer followed by solution of the energy equation with a radiative source term are compared with a fully coupled solution. For the two example problems studied, the fully coupled method always produced the most accurate solution, although execution time made it unattractive for very large problems. A semi-implicit technique with a Newton type of update appeared to be the best choice for very large problems.

The finite-difference and finite-element numerical procedures that have been described used radiative enclosure theory with finite or infinitesimal areas to obtain a set of simultaneous equations with configuration factors for radiative exchange between surface areas. Convection was specified in terms of a heat transfer coefficient for each area; for example, for radiation exchange inside a tube with a flowing transparent gas, the heat transfer coefficient inside the tube is obtained from available results from tube flow analyses or experimental correlations. For some situations, however, convection is quite dependent on the surface temperatures, such as for free convection, or the geometry is complex so that convective heat transfer correlations are not available with desired accuracy. In these cases, analyses have been made where convection is solved simultaneously with radiation as the flow and surface temperatures are strongly coupled; conduction may also be included, such as for free convection and radiation from a cooling fin as discussed in Section 7.4.3. To solve for the convection heat transfer, the methods of computational fluid mechanics are used.

Another consideration is that the analysis may not use configuration factors. The radiative exchange in an enclosure can be computed directly by a ray-tracing technique such as the Monte Carlo method (Section 7.9). This may be necessary if the surfaces are not diffuse so that configuration factors do not apply. Another solution method where radiation is followed along directions is by use of discrete ordinates. The discrete ordinates method, discussed in Chapter 12, was developed for enclosures filled with a medium that is not transparent, but rather absorbs, emits, and scatters radiation. If radiative participation by the medium is omitted, the method can be applied to enclosures containing a transparent medium such as a convecting nonradiating gas. In this method, the angular directions from each surface element are divided into a finite number, and radiation is followed along these discrete directions to evaluate the radiative exchange. In Tan et al. (1998), discrete ordinates are used in combination with the SIMPLE computer algorithms developed for computational fluid mechanics (Patankar 1980) to simultaneously solve the mass, momentum, and energy equations along with radiation transfer between surfaces.

For natural convection combined with radiation, a variety of computational methods have been used for simultaneously solving the fluid flow and energy equations with radiative exchange. In Zhao et al. (1992), free convection and radiation were analyzed for heated cylinders in a rectangular enclosure. In Dehghan and Behnia (1996), net radiation enclosure analysis was used for the radiative transfer, and the flow and energy equations were placed in finite-difference form and solved with a pseudo transient method to analyze free convection in a cavity with a local heated area on one vertical wall. A vented cavity with a discrete heat source was analyzed by Yu and Joshi (1999) using the numerical methods from Patankar (1980); this study included combined radiation exchange, conduction, and natural convection, with the gas in the cavity being transparent. Free convection of transparent air in a heated vertical channel with one or more vents in one wall was analyzed by Moutsoglou et al. (1992). The flow and energy equations were solved by using finite-difference computational methods as developed by Patankar and Spalding (1972) and van Doormall and Raithby (1983).

7.9 MONTE CARLO METHOD

In Chapter 6, it was found that the enclosure theory analysis becomes very complex when both directional and spectral surface property variations are included. The Monte Carlo method is an alternative approach that can deal with these complexities. Since Monte Carlo is a statistical numerical method, it is first necessary to discuss some concepts of statistical theory. Then the basic procedure is outlined with regard to radiative exchange and two example problems are formulated to demonstrate the method. The straightforward Monte Carlo approach is presented. Some refinements that can shorten computation time by increasing accuracy are discussed briefly.

7.9.1 Definition of Monte Carlo Method

In 1956, Herman Kahn gave a definition of the Monte Carlo method that incorporates the salient ideas: "The expected score of a player in any reasonable game of chance, however complicated, can in principle be estimated by averaging the results of a large number of plays of the game. Such estimation can be rendered more efficient by various devices which replace the game with another known to have the same expected score. The new game may lead to a more efficient estimate by being less erratic, that is, having a score of lower variance, or by being cheaper to play with the equipment on hand. There are obviously many problems about probability that can be viewed as problems of calculating the expected score of a game. Still more, there are problems that do not concern probability but are nonetheless equivalent for some purposes to the calculation of an expected score. The Monte Carlo method refers simply to the exploitation of these remarks" (Kahn 1956).

This definition provides a good outline for the method. What must be done for a specific problem is to set up a game or model that has the same behavior and hence is expected to produce the same outcome as the physical problem that the model simulates, make the game as simple and fast to play as possible, use any available methods to reduce the variance of the average outcome of the game, and then play the game many times and find the average outcome.

Referring to *the* Monte Carlo method is probably meaningless. Any specific problem more likely entails *a* Monte Carlo method, as the label has been placed on a large class of loosely related techniques. General books and monographs are available that detail methods and/or review the literature. A valuable early outline is by Metropolis and Ulam (1949), which is the first work to use the term Monte Carlo for the approach considered here. For clarity and usefulness, both Kahn (1956) and Hammersley and Handscomb (1964) are valuable, as are the general texts by Cashwell and Everett (1959) and Halton (1970) and the survey article by Howell (1998).

No definitive method of predicting computer running time exists for most Monte Carlo problems. The time depends on the computer used and on the ability of the programmer to pick appropriate methods and shortcuts. An example is the use of special subroutines for computing such functions as sine and cosine. These routines sacrifice some accuracy to gain speed. If problem answers accurate to only a small percentage are desired, eight-place accuracy from a slower subroutine is not needed.

7.9.2 Fundamentals of the Method

7.9.2.1 Random Walk

A *Markov chain* is a sequence of events with the condition that the probability of each succeeding event is uninfluenced by prior events. An example is an inebriated gentleman who begins to walk through a city. At each street corner, he becomes confused and chooses completely at random one of the streets leading from the intersection. He may walk up and down the same block several times before he chances to move off down a new street. His walk is a Markov chain, as his decision at any corner is not influenced by where he has been. Because of the randomness of choice at each intersection, it is possible to simulate a sample walk by having a roulette wheel with only four positions, each corresponding to a possible direction. The probability of the gentleman starting from his favorite bar and reaching any point in the city limits could be found by simulating a large number of histories, using the wheel to determine the direction of the walk at each decision point.

The probability of the man reaching intersection (l, m) on a square grid representing the city street map is simply

$$P(l,m) = \frac{1}{4}[P(l+1,m) + P(l-1,m) + P(l,m+1) + P(l,m-1)] \tag{7.34}$$

where the factors in brackets are the probabilities of his being at each of the adjacent four intersections. This is because the probability of reaching $P(l, m)$ from a given adjacent intersection is one-fourth. This type of random walk is a model for processes described by Laplace's equation; Equation 7.34 is the finite-difference analog of Laplace equation.

The probability of a certain occurrence for other processes is usually not as obvious as Equation 7.34. More often, the probability of an event must be determined from physical constraints; then the decision as to what event will occur is made on the basis of this probability.

7.9.2.2 Choosing from Probability Distributions

Consider small packets of radiative energy leaving a differential area and arriving at a disk with an outer radius 10 units in length. For many packets, the number $F(\xi)$ that have arrived at the disk within each small radial increment $\Delta\xi$ about some radius ξ can be represented by a histogram of the frequency function $f(\xi) = F(\xi)/\Delta\xi$, where $0 \le \xi \le 10$. A smooth curve can then be passed through the histogram to give a frequency distribution, as in Figure 7.23. What is needed is a method for simulating additional packets. This method should assign an expected radius ξ to each succeeding packet. The distribution of ξ values should correspond to the distribution that the packets follow according to a known physical process. For a Markov process, the values must be assigned in a random manner so that each decision at each step in the transfer is independent.

To show how this is done, the frequency curve in Figure 7.23 is approximated by

$$f(\xi) = \xi^2 \tag{7.35}$$

in the interval $0 \le \xi \le 10$, and $f(\xi) = 0$ elsewhere, because the only packets being considered for the moment are those directed toward the disk. Equation 7.35 is normalized by dividing by the area under the frequency curve (the total number of packets) to obtain

$$P(\xi) = \frac{f(\xi)}{\int_0^{10} f(\xi)d\xi} = \frac{3\xi^2}{1000} \tag{7.36}$$

If the distribution with which packets have struck the target radii is taken as the basis for estimating the locations the next packets will strike, then the *probability density function* defined by Equation 7.36 is the average distribution that must be satisfied by the ξ values determined by the simulation scheme. The $P(\xi)$ is plotted in Figure 7.24 and is the proportion of values that lie in the region $\Delta\xi$ around ξ.

To determine ξ values, one simulation scheme proceeds as follows: Choose two *random numbers* R_A and R_B from a large set of numbers evenly distributed in the range 0 to 1. The two random numbers are then used to select a point $(P(\xi), \xi)$ in Figure 7.24 by setting $P(\xi) = R_A$ and $\xi = (\xi_{max} - \xi_{min}) R_B = 10\,R_B$. This value of $P(\xi)$ is then compared to the value of $P(\xi)$ computed at ξ from Equation 7.31. If the randomly selected value lies above the computed value of $P(\xi)$, then the randomly selected value of ξ is rejected and two new random numbers are selected. Otherwise, the value of ξ that has been found is listed as the location that the packet will strike. Referring again to Figure 7.24, it is

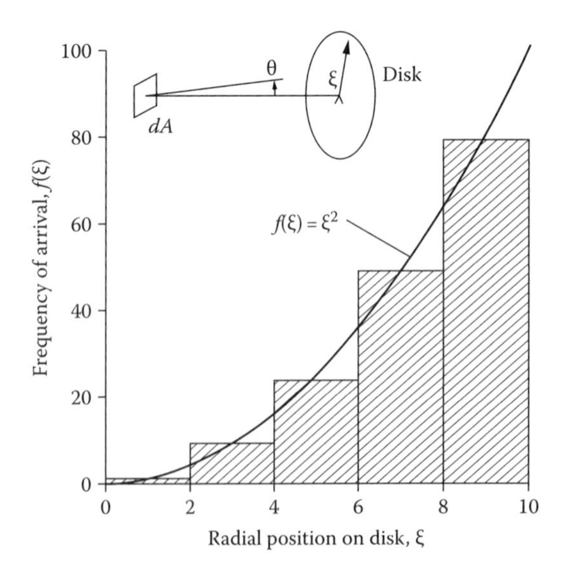

FIGURE 7.23 Distribution of radiation packets arriving at various disk radii.

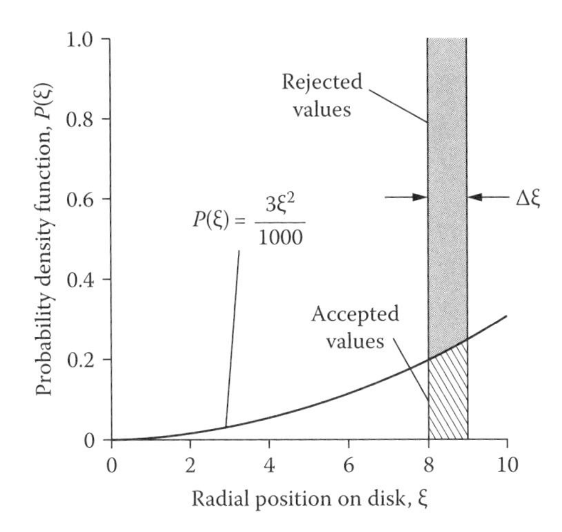

FIGURE 7.24 Probability density function of energy packets hitting disk.

seen that such a procedure ensures that the correct fraction of ξ values selected for use will lie in each increment $\Delta\xi$ after a large number of completely random selections of $P(\xi)$, ξ is made.

The disadvantage of such an event-choosing procedure is that a large portion of the ξ values will be rejected because they lie above the $P(\xi)$ curve. A more efficient method for choosing ξ is therefore desirable. One such method is to integrate the probability density function $P(\xi)$ using the general relation

$$R(\xi) = \int_{-\infty}^{\xi} P(\xi^*)d\xi^* \tag{7.37}$$

where $R(\xi)$ can only have values in the range 0–1 because the integral under the entire $P(\xi)$ curve equals 1, according to Equation 7.36. Equation 7.37 is the general definition of the *cumulative distribution function.* A plot of R against ξ from Equation 7.37 shows the probability of an event occurring in the range $-\infty$ to ξ. For the method given here, the function R is taken to be a random number; each value of ξ is then obtained by choosing an R value at random and using the functional relation $R(\xi)$ to determine the corresponding value of ξ. To show that the probability density of ξ formed in this way corresponds to the required $P(\xi)$, the probability density function of Figure 7.24 is used as an illustrative example. Inserting the example $P(\xi)$ of Equation 7.36 into Equation 7.37 and noting that $P(\xi) = 0$ for $-\infty < \xi < 0$ gives

$$R = \int_{-\infty}^{\xi} P(\xi^*)d\xi^* = \frac{\xi^3}{1000} \quad 0 \le R \le 1 \tag{7.38}$$

Equation 7.38 is plotted in Figure 7.25. Here, * denotes a dummy variable.

Choosing R at random and determining a corresponding value of ξ from Equation 7.38 is equivalent to taking the derivative of the cumulative distribution function. By examination of Equations 7.38 and 7.36, this derivative is $P(\xi)$. Divide the range of ξ into a number of equal increments $\Delta\xi$. Suppose that M values of R are chosen in the range 0–1 at equal intervals along R. There will be M values of ξ corresponding to these M values of R. The fraction of the M values of ξ per given increment $\Delta\xi$ is then $M_{\Delta\xi}/M = \Delta R$, which gives

$$\frac{M\Delta\xi/M}{\Delta\xi} = \frac{\Delta R}{\Delta\xi} \tag{7.39}$$

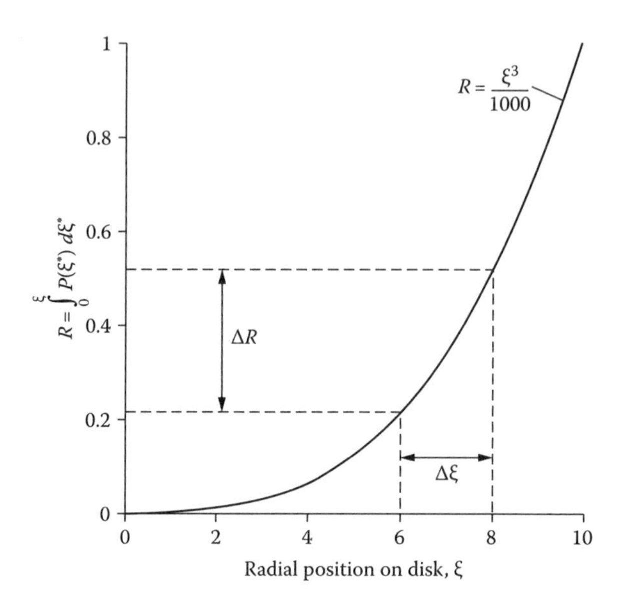

FIGURE 7.25 Cumulative distribution of energy packets on disk.

The quantity $\Delta R/\Delta\xi$ approaches $dR/d\xi$ if a large enough M is used and small increments $\Delta\xi$ are examined. But $dR/d\xi$ can be seen from Equations 7.38 and 7.36 to be simply $P(\xi)$; therefore, by obtaining values of ξ as described preceding Equation 7.38, the required probability distribution is indeed generated.

In physical problems, the frequency distribution often depends on more than one variable. If the interdependence of the variables is such that the frequency distribution can be factored into a product form, the following can be written:

$$f(\xi,\phi) = g(\xi)h(\phi). \tag{7.40}$$

Values of $P(\xi)$ and $P(\phi)$ can be found by integrating out each variable to obtain

$$P(\xi) = \frac{\int_{\phi_{min}}^{\phi_{max}} f(\xi,\phi)d\phi}{\int_{\xi_{min}}^{\xi_{max}}\int_{\phi_{min}}^{\phi_{max}} f(\xi,\phi)d\phi d\xi} = \frac{g(\xi)\int_{\phi_{min}}^{\phi_{max}} h(\phi)d\phi}{\int_{\xi_{min}}^{\xi_{max}} g(\xi)d\xi \int_{\phi_{min}}^{\phi_{max}} h(\phi)d\phi} = \frac{g(\xi)}{\int_{\xi_{min}}^{\xi_{max}} g(\xi)d\xi} \tag{7.41}$$

and similarly,

$$P(\phi) = \frac{\int_{\xi_{min}}^{\xi_{max}} f(\xi,\phi)d\xi}{\int_{\phi_{min}}^{\phi_{max}}\int_{\xi_{min}}^{\xi_{max}} f(\xi,\phi)d\xi d\phi} = \frac{h(\phi)}{\int_{\phi_{min}}^{\phi_{max}} h(\phi)\phi d\phi} \tag{7.42}$$

The methods given previously in this section are used to evaluate ξ and ϕ independently after two random numbers are chosen.

If $f(\xi, \phi)$ cannot be placed in the form of Equation 7.40, then ξ and ϕ values can be determined by choosing two random numbers R_ξ and R_ϕ. Note that

$$P(\xi,\phi) = \frac{f(\xi,\phi)}{\int_{\xi_{min}}^{\xi_{max}} \int_{\phi_{min}}^{\phi_{max}} f(\xi,\phi)d\xi d\xi}$$

Then, ξ and φ are found from the equations

$$R_\xi = \int_{-\infty}^{\xi} \int_{\phi_{min}}^{\phi_{max}} P(\xi^*,\phi)d\phi d\xi^* \tag{7.43}$$

and

$$R_\phi = \int_{-\infty}^{\xi} P(\phi^*,\xi = \text{fixed})d\phi^* \tag{7.44}$$

where ξ in Equation 7.44 is the value obtained from Equation 7.43. This procedure may be extended to any number of variables. Equations 7.43 and 7.44 define the *marginal* and *conditional* distributions of $P(\xi, \phi)$, respectively.

7.9.2.3 Random Numbers

A *random number* is a number chosen without sequence from a large set of numbers spaced at equivalued intervals. For our purposes, the numbers are in the range 0–1. If the numbers 0, 0.01, 0.02, 0.03, …, 0.99, 1.00 are placed on slips of paper and mixed, there would be fair assurance that if a few slips are picked, they would provide random numbers. If many choices are to be made, then smaller intervals (more slips) should be used; after it is drawn, each slip should be replaced and randomly mixed with the others.

For computer solutions, random numbers might be needed for 10^9 or more decisions. It is desirable to have a rapid way of obtaining them and to have the numbers be truly random. A useful method for obtaining random numbers for computer use is a pseudo random number generator. This is a subroutine using the apparent randomness of groups of digits in large numbers. A simple example is to take an 8-digit number, square it, and then choose the middle 8 digits of the resulting 16-digit number as a random number. When a new random number is needed, square the previous random number and take a new random number as the middle 8 digits of the result. This process degenerates after a few thousand cycles by propagating to an all-zero number (Schreider 1964). In a more satisfactory routine, Taussky and Todd (1956), a random number is generated by taking the low-order 36 bits of the product $R_{n-1}K$, where $K = 5^{15}$ and R_{n-1} is the previously computed random number. By always starting a solution with the same R_0, it is possible to check calculations by step-by-step tracing of a few histories. Many subroutines for generating random numbers are based on this approach. The fact that such subroutines generate *pseudorandom* numbers raises the question of whether they are sufficiently random for the problem being treated. Does the sequence repeat, and if so, after how many numbers? Certain standard tests that give partial answers are discussed in Hammersley and Handscomb (1964), Taussky and Todd (1956), and Kendall and Smith (1938). Walker (2013) reviews many contemporary routines and compares them on the basis of period of repetition, serial correlation, and memory requirements when used in computing configuration factors in highly parallel routines. Best choices are recommended for use with either central processing unit (CPU) or graphics processors. No finite set of tests is sufficient to establish randomness, although passing the tests is necessary. Perhaps the safest course is to obtain a standard subroutine whose properties have been established by such tests and use it within its proven limits.

The website random.org provides resources for accessing sets of true random numbers generated from sampling atmospheric noise.

7.9.2.4 Evaluation of Uncertainty

Because solutions obtained by Monte Carlo are averages over results of individual samples, they contain fluctuations about a mean value. The mean can be determined more accurately by increasing the number of values used for determining the mean, as long as there is sufficient computer time. Some *ad hoc* rules of computer use and an estimate of desired accuracy in a given problem can be applied and solutions obtained by trading off within these limits.

To establish the accuracy of solutions, one of several tests can be applied. For example, suppose we want to know the probability of the drunken gentleman reaching a location at the city limits. To determine his success exactly, an infinite number of hypothetical paths need to be followed, and the probability $P(l, m)$ of reaching the boundary point (l, m) determined as

$$P(l,m) = \left[\frac{S(l,m)}{N}\right]_{N\to\infty} \tag{7.45}$$

where

$S(l, m)$ is the number of samples reaching the boundary point
N is the total number of samples

In practice, a probability would be computed based on a finite number, perhaps $N = 10^2$–10^6. Then an estimate is needed of the error μ involved in using this sample size.

For a sample size greater than about $N = 20$, application of the central limit theorem and relations governing normal probability distributions show that the following relation holds whenever the samples S in question can be considered to leave a source and either reach a scoring position with probability P or not reach it with probability $1 - P$. The probability that the average $S(l, m)/N$ for finite N differs by less than some value μ from $[S(l, m)/N]_{N\to\infty}$ is given by

$$P\left[\left|\frac{S}{N} - \left(\frac{S}{N}\right)_{N\to\infty}\right| \le \mu\right] = \frac{2}{\sqrt{\pi}} \int_0^{\eta/\sqrt{2}} e^{-\eta^{*2}}\, d\eta^* = \operatorname{erf}\frac{\eta}{\sqrt{2}} \tag{7.46}$$

where

$$\eta \approx \mu\left[\frac{N}{(S/N)(1 - S/N)}\right]^{1/2} \tag{7.47}$$

Tabulations of the error function (erf), and series expressions for it, are in tables of mathematical functions and are available in mathematics computational software packages.

In many problems, such an error estimation cannot be applied because the samples do not originate from a single source. The radiative energy flux at a location on a surface in an enclosure usually depends on energy arriving from many sources. For such situations, the error may be estimated by subdividing the calculation of the desired statistical mean result into a group of I submeans. The central limit theorem then applies. This states that the statistical fluctuations in the submeans are distributed in a normal or Gaussian distribution about the overall mean. For such a distribution, the variance of the fluctuations in the means can be calculated. For example, if 20,000 samples are examined, a mean result $\bar{P}$ is calculated on the basis of the samples, and 20 submeans $P_1, P_2, \ldots, P_I$ of 1000 samples each are calculated. Then, the variance γ^2 of the mean solution $\bar{P}$ is

$$\gamma^2 = \frac{1}{I-1}\left[\sum_{i=1}^{I}(P_i - \bar{P})^2\right] = \frac{1}{I-1}\left[\sum_{i=1}^{I}P_i^2 - \frac{\sum_{i=1}^{I}P_i}{I}\right] \tag{7.48}$$

This variance is an estimate of the mean-square deviation of the sample mean $\bar{P}$ from the true mean, where the true mean would be obtained by using an infinite number of samples. From the properties of the normal frequency distribution, which the fluctuations in the results computed by Monte Carlo will generally follow, the probability of the sample mean $\bar{P}$ lying within $\pm\gamma$ of the true mean is about 68%, that of its lying within $\pm 2\gamma$ is about 95%, and that of its lying within $\pm 3\gamma$ is 99.7%.

Another measure of the statistical fluctuations in the mean is γ, the standard deviation. Because γ is given by the square root of Equation 7.48, to reduce γ by half, the number of samples used in computing the results must be quadrupled (thereby quadrupling I for constant submean size). This probably means quadrupling the computer time unless the term in brackets can somehow be reduced by decreasing the variance (scatter) of the individual submeans by methods such as stratified sampling, splitting, importance sampling, and energy partitioning. These and other variance-reducing techniques are discussed in the standard references and in Shamsundar et al. (1973) and Haji-Sheikh and Howell (2006), and the saving in computer time is substantial. Baranovski et al. (2001a,b) provide an *a priori* estimation method for determining the number of bundles required for accurate determination of configuration factors.

7.9.3 Application of Monte Carlo Techniques to Thermal Radiative Transfer

The formulation of radiation exchange in enclosures leads to integral equations for the unknown surface temperature or heat flux distributions. By using a probabilistic model of radiative exchange and applying Monte Carlo techniques, it is possible to avoid many of the difficulties inherent in the integral equation formulations. Actions of small parts of the total energy can be examined on an individual basis, in place of solving simultaneously for the entire behavior of the energy involved. A microscopic model for the radiative exchange process is examined; then the solutions of two examples are outlined.

7.9.3.1 Model of the Radiative Exchange Process

In radiation calculations, the usual quantities of interest are local temperatures and energy fluxes. We can model the radiative exchange by following the progress of discrete amounts ("bundles" or "packets") of radiative energy, since local energy flux is then easily computed as the number of these bundles arriving per unit area and time at a location. An obvious bundle is the photon, but this has a disadvantage because the energy of a photon depends on its wavelength, which introduces a needless complication; further, the modern interpretation of the photon and its properties predicts a much more complex picture than implied by this simple model. A more convenient quantity is a bundle that is carrying a given amount of energy w. For spectral problems, the wavelength of the bundle is specified but the energy of the bundle remains equal to w. By assigning equal energies to all bundles, local energy flux is computed by counting the number of bundles arriving at a position of interest per unit time and per unit area and multiplying by w. The bundle paths and histories are computed by Monte Carlo, as now described.

To aid in showing the development of a Monte Carlo radiative model, it is helpful to consider a sample calculation involving directional-spectral surface properties. It is desired to obtain the amount of energy radiated from element dA_1 at temperature T_1 that is absorbed by an infinite plane A_2 at temperature $T_2 = 0$, as shown in Figure 7.26. Let dA_1 have directional-spectral emissivity $\varepsilon_{\lambda,1}(\theta_1, T_1)$ and area 2 have directional-spectral emissivity $\varepsilon_{\lambda,2}(\theta_2, T_2)$, and assume that the directional emissivities of both surfaces are independent of circumferential angle ϕ. For element dA_1, the total

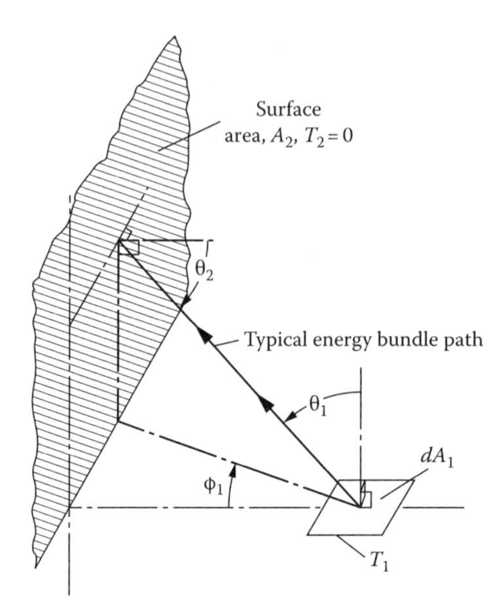

FIGURE 7.26 Radiant interchange between two surfaces.

emitted energy per unit time is $\varepsilon_1(T_1)\sigma T_1^4 dA_1$, where $\varepsilon_1(T_1)$ is the hemispherical total emissivity (Equation 2.8):

$$\varepsilon_1(T_1) = \frac{\left\{\int_0^{\infty}\left[I_{\lambda b,1}(T_1)2\pi\int_{\theta_1=0}^{\pi/2}\varepsilon_{\lambda,1}(\theta_1,T_1)\cos\theta_1\sin\theta_1 d\theta_1\right]d\lambda\right\}}{\sigma T_1^4}$$

If the total energy emitted per unit time by dA_1 is composed of N energy bundles emitted per unit time, the energy assigned to each bundle is $w = \varepsilon_1(T_1)\sigma T_1^4 dA_1/N$. To determine the energy radiated from dA_1 that is absorbed by A_2, follow N bundles of energy after their emission from dA_1 and determine the number S_2 absorbed at A_2. If the energy reflected from A_2 back to dA_1 and then rereflected to A_2 is neglected, the energy transferred per unit time from dA_1 to A_2 is $wS_2 = \left[\varepsilon_1(T_1)\sigma T_1^4 dA_1/N\right]S_2$.

The next question is how to determine the path direction and wavelength assigned to each bundle. This must be done in such a way that the directions and wavelengths of the N bundles conform to the constraints given by the emissivity of the surface and the laws governing radiative processes. For example, if wavelengths are assigned to N bundles, the spectral distribution of emitted energy generated by the Monte Carlo method, consisting of the energy $wN_\lambda\,\Delta\lambda$ for discrete intervals $\Delta\lambda$, must closely approximate the spectrum of the actual emitted energy $\varepsilon_{\lambda,1}(\theta_1)\pi I_{\lambda b,1}d\lambda$. The probability $P_\lambda(\theta_1)d\theta_1 d\lambda$ of emission in a wavelength interval about λ and in an angular interval around θ_1 is the energy in $d\theta_1 d\lambda$ divided by the total emitted energy:

$$P_\lambda(\theta_1)d\theta_1 d\lambda = \frac{2\pi\varepsilon_{\lambda,1}(\theta_1)I_{\lambda b,1}\cos\theta_1\sin\theta_1 d\theta_1 d\lambda}{\varepsilon_1\sigma T_1^4} \tag{7.49}$$

The T_1 in the functional notation is being omitted for simplicity.

It is assumed here that the directional-spectral emissivity is a product function of the variables wavelength and angle:

$$\varepsilon_{\lambda,1}(\theta_1) = \Phi_1(\lambda)\Phi_2(\theta_1) \tag{7.50}$$

This assumption is probably not valid for many real surfaces because, in general, the angular distribution of emissivity depends on wavelength as shown, for example, by Figure 3.17. For the assumed form in Equation 7.50, it follows that the emissivity dependence on either variable may be found by integrating out the other variable (see Equation 7.41). Then, the normalized probability of emission occurring in the interval $d\lambda$ is

$$P_\lambda d\lambda = d\lambda \int_0^{\pi/2} P_\lambda(\theta_1)\,d\theta_1 = \frac{2\pi d\lambda I_{\lambda b,1}\int_{\theta_1=0}^{\pi/2} \varepsilon_{\lambda,1}(\theta_1)\sin\theta_1\cos\theta_1 d\theta_1}{\varepsilon_1\sigma T_1^4} \tag{7.51}$$

Substituting into Equation 7.32 and noting that $P_\lambda\, d\lambda$ is zero in the range $-\infty < \lambda < 0$ gives

$$R_\lambda = \frac{2\pi\int_{\lambda^*=0}^{\lambda}\int_{\theta_1=0}^{\pi/2} \varepsilon_{\lambda^*,1}(\theta_1) I_{\lambda^* b,1}\sin\theta_1\cos\theta_1 d\theta_1 d\lambda^*}{\varepsilon_1\sigma T_1^4} \tag{7.52}$$

where the asterisk denotes a dummy variable of integration. If the number of bundles is very large and this equation is solved for λ each time a random R_λ value is chosen, the computing burden becomes very large. To circumvent this, equations like Equation 7.52 can be numerically integrated once over the range of λ values and a curve fitted to the result. A polynomial approximation

$$\lambda = A + BR_\lambda + CR_\lambda^2 + \cdots \tag{7.53}$$

is often adequate. Equation 7.53 rather than Equation 7.52 is used in the computer program. Alternatively, a table of λ versus R_λ can be put into computer memory and interpolated to obtain λ for chosen R_λ values.

Following a similar procedure for the variable cone angle of emission θ_1 gives

$$R_{\theta_1} = \int_{\theta_1^*=0}^{\theta_1}\int_{\lambda=0}^{\infty} P_\lambda(\theta_1^*)\,d\lambda d\theta_1^* = \frac{2\pi\int_{\theta_1^*=0}^{\theta_1}\int_{\lambda=0}^{\infty} \varepsilon_{\lambda,1}(\theta_1^*) I_{\lambda b,1}\sin\theta_1^*\cos\theta_1^* d\lambda d\theta_1^*}{\varepsilon_1\sigma T_1^4} \tag{7.54}$$

which is curve fit to give

$$\theta_1 = D + ER_{\theta_1} + FR_{\theta_1}^2 + \cdots \tag{7.55}$$

If dA_1 is a *diffuse–gray* surface, then Equation 7.54 reduces to

$$R_{\lambda,diffuse\text{-}gray} = \frac{\pi\int_{\lambda^*=0}^{\lambda} I_{\lambda^* b,1} d\lambda^*}{\sigma T_1^4} = \frac{\int_{(\lambda T)^*=0}^{\lambda T} E_{\lambda^* b,1} d(\lambda T)^*}{\sigma T_1^4} = F_{0\to\lambda T} \tag{7.56}$$

where $F_{0\to\lambda T}$ is the fraction of blackbody emission in the wavelength interval 0–λT. Haji-Sheikh (1988) presents the inverse function $\lambda T = F(R_\lambda)$ for a diffuse–gray surface at temperature T. The relations are in Table 7.1.

For the diffuse–gray case, Equation 7.54 reduces to

$$R_{\theta_1,diffuse\text{-}gray} = 2\int_{\theta_1^*=0}^{\theta_1} \sin\theta_1^*\cos\theta_1^* d\theta_1^* = \sin^2\theta_1 \tag{7.57}$$

TABLE 7.1
Inverse Probability Function for Choosing Wavelength of Emission from a Gray or Black Surface (λT in μm·K)

$$\lambda T = D_1 + D_2 R_\lambda^{1/8} + D_3 R_\lambda^{1/4} + D_4 R_\lambda^{3/8} + D_5 R_\lambda^{1/2} \qquad 0.0 < R_\lambda < 0.1$$

$$\lambda T = D_1 + D_2 R_\lambda + D_3 R_\lambda^2 + D_4 R_\lambda^3 + D_5 R_\lambda^4 \qquad 0.1 < R_\lambda < 0.9$$

$$\lambda T = \left[\frac{0.152886\times 10^{12}}{D_1(1-R_\lambda) + D_2(1-R_\lambda)^2 + D_3(1-R_\lambda)^3 + D_4(1-R_\lambda)^4}\right]^{1/3} \qquad 0.9 < R_\lambda < 1$$

	Coefficients				
Range of R_λ	D_1	D_2	D_3	D_4	D_5
0.0–0.1	503.247	230.243	5,863.85	−10,759.6	8,723.14
0.1–0.4	1,560.84	7,603.61	−15,540.1	31,257.7	−20,844.8
0.4–0.7	2,846.63	−1,430.38	27,936.0	−41,041.9	25,960.9
0.7–0.9	345,197	−1,828,567	3,674,856	−3,284,391	1,108,939
0.9–0.99	1.200	9.476	−44.84	156.9	—
0.99–1.0	1.10064	16.8148	−183.445	890.699	—

Source: Haji-Sheikh, A., Monte Carlo methods, in Minkowycz, W.J., Sparrow, E.M., Schneider, G.E., and Pletcher, R.H. (eds.), Chapter 16, *Handbook of Numerical Heat Transfer*, 1st edn., Wiley Interscience, New York, 1988, pp. 672–723. (Slightly modified for $R_\lambda > 0.9$ as a result of personal communication with A. Haji-Sheikh.)

Note: An alternative formulation accurate within 1% for the range $750 \le \lambda T \le 65 \times 10^3$ ($5.96 \times 10^{-6} \le R_\lambda \le 0.99957$) is given by Haji-Sheikh and Howell (2006):

$$\lambda T = 1 - \exp\left[-1.2\sqrt[3]{R_\lambda/(1-R_\lambda)}\right]$$

$$-\frac{0.12 + 7.0\times 10^{-5}\left[R_\lambda/(1-R_\lambda)\right] - 0.005\sqrt{R_\lambda/(1-R_\lambda)}}{\left\{1+0.30\left[R_\lambda/(1-R_\lambda)\right]^{-3/4}\right\}\left\{1+7.0\times 10^{-6}\left[R_\lambda/(1-R_\lambda)\right]^{3/2}\right\}} + \frac{0.12 + 6.0\times 10^{-4}(1-R_\lambda)^{-2}}{\left\{1+5.0\left[R_\lambda/(1-R_\lambda)\right]^{2/3}\right\}^4}$$

or

$$\sin\theta_1 = \sqrt{R_{\theta_1,\textit{diffuse-gray}}} \qquad (7.58)$$

Computational difficulty is not greatly different in obtaining λ from either Equation 7.53 or 7.56, nor is it much different for obtaining θ_1 from either Equation 7.55 or 7.58. The difference between directional-spectral and diffuse–gray cases is mainly in the auxiliary numerical integrations of Equations 7.52 and 7.54. These integrations are performed once to obtain the curve fits; then, in the main problem-solving computer program, the more difficult case might just as well be handled. Thus, increasing problem complexity leads to only gradual increases in complexity of the Monte Carlo computer program. For a diffuse surface, a possible improvement in selecting the θ and ϕ directions is discussed by Kowsary (1999). For Monte Carlo programs using massively parallel computers, the use of fast and accurate subroutines for trigonometric and other often used functions can greatly speed calculations. Walker (2013) presents examples and shows considerable speedup.

For emission of an individual energy bundle from surface dA_1, a wavelength λ is obtained from Equation 7.53 and a cone angle of emission θ_1 from 7.90b by choosing two random numbers R_λ and R_{θ_1}. To define the bundle path, there remains specification of the circumferential angle ϕ_1. Because of the earlier assumption that emission does not depend on ϕ_1, it is shown by the formalism outlined, and is also fairly obvious, that ϕ_1 is determined by

$$\phi_1 = 2\pi R_{\phi_1} \qquad (7.59)$$

where R_{ϕ_1} is a random number in the range 0–1.

It is not difficult to determine whether a given energy bundle will strike A_2 after leaving dA_1 in direction (θ_1, ϕ_1). As shown in Figure 7.26, it will hit A_2 whenever $\cos \phi_1 \geq 0$. If it misses A_2, another bundle must be emitted from dA_1. If the bundle strikes A_2, it must be determined whether it is absorbed or reflected. To do this, the geometry is used to find the angle of incidence θ_2 of the bundle onto A_2:

$$\cos\theta_2 = \sin\theta_1 \cos\phi_1 \tag{7.60}$$

Knowing the absorptivity of A_2 from Kirchhoff's law, $\alpha_{\lambda,2}(\theta_2) = \varepsilon_{\lambda,2}(\theta_2)$, and having determined the wavelength λ of the incident bundle from Equation 7.53 and the incident angle θ_2 from Equation 7.60, the probability of absorption of the bundle at A_2 can be determined. This is $\alpha_{\lambda,2}(\theta_2)$, since this is the fraction of energy incident on A_2 in a given wavelength interval and within a given solid angle that is absorbed by the surface. The absorptivity is equivalent to the probability density function for absorption of incident energy. To determine whether a given incident bundle is absorbed, the surface absorptivity $\alpha_{\lambda,2}(\theta_2)$ is compared with a random number $R_{\alpha 2}$. If

$$R_{\alpha_2} \leq \alpha_{\lambda,2}(\theta_2) \tag{7.61}$$

the bundle is absorbed and a counter S_2 in the computer memory is increased by 1 to tally the absorbed bundles. Otherwise, the bundle is reflected and not further accounted for. If the bundle path were followed further, rereflections from dA_1 would be considered. Angles of reflection are chosen from known directional reflectivities, and the bundle is followed along its path until it is absorbed by A_2 or lost from the system. The derivation of the necessary relations is similar to that presented. A new bundle is now chosen at dA_1, and its history is followed. This procedure is continued until N bundles are emitted from dA_1. The energy absorbed at A_2 is then $\left[\varepsilon_1(T_1)\sigma T_1^4 dA_1/N\right]S_2$, where S_2 is the total number of bundles absorbed at A_2.

The derivation of the equations needed for the solution is now complete. When making a flowchart (Figure 7.27) to aid in writing a computer program, some methods for shortening computing time can be applied. For example, the angle ϕ_1 is computed first. If the bundle is not going to strike A_2 on the basis of the calculated ϕ_1, there is no point in computing λ and θ_1 for that bundle. Alternatively, because ϕ_1 values are isotropically distributed, it is noted for this geometry that half the bundles must strike A_2. Therefore, the ϕ_1 values can be constrained to the range $-\pi/2 < \phi_1 < \pi/2$.

For this problem, the desired result could be obtained without much difficulty by standard integral methods. However, extension to only slightly more complex problems could cause difficulties for standard treatments, but would not be very difficult for Monte Carlo. For example, consider introducing a third surface with directional properties and account for all interactions among surfaces.

7.9.3.2 Useful Functions

A number of useful relations for choosing angles of emission and assigning a wavelength to bundles were developed in the previous section. These and other functions are summarized in Table 7.2.

7.9.4 Forward Monte Carlo

The most straightforward application of Monte Carlo models the radiative transfer process from the emission of radiation throughout the history of the radiation until it is absorbed by a surface or lost from an enclosure. Such a model is termed forward Monte Carlo or direct simulation Monte Carlo (DSMC). Example Problem 7.14 is an example of the use of DSMC to radiative transfer in an enclosure.

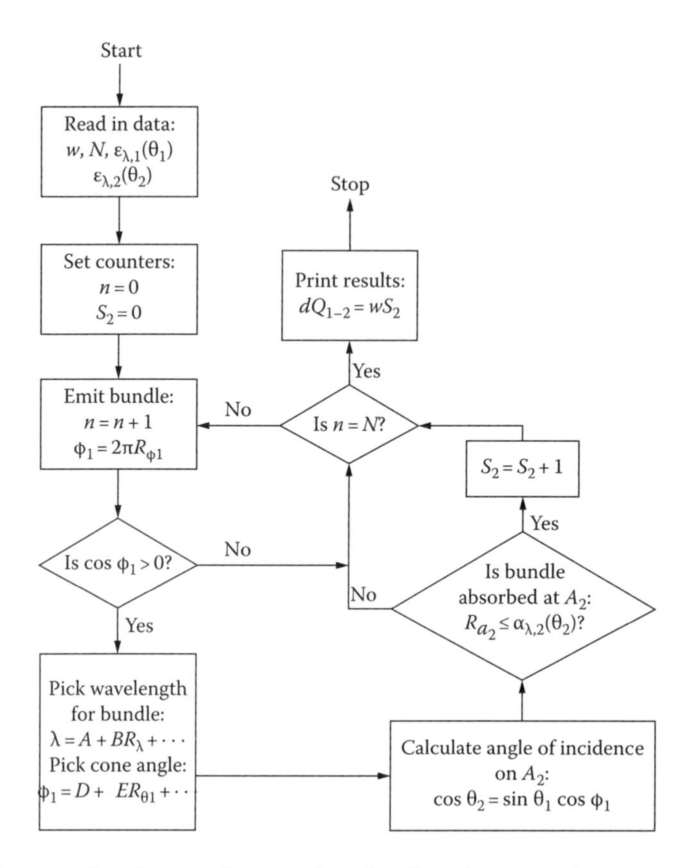

FIGURE 7.27 Computer flow diagram, for example radiant-interchange problem.

Example 7.13

A wedge consists of two very long parallel sides of equal width joined at 90°, Figure 7.28. The surface temperatures are T_1 = 1000 K and T_2 = 2000 K. The effects of the ends may be neglected. Surface 1 is diffuse–gray with ε_1 = 0.5, and surface 2 is directional–gray with directional total emissivity and absorptivity:

$$\varepsilon_2(\theta_2) = \alpha_2(\theta_2) = 0.5\cos\theta_2 \tag{7.13.1}$$

Assume for simplicity that surface 2 reflects diffusely. Set up a Monte Carlo flowchart for determining the energy to be added to each surface to maintain its temperature. Assume that the environment is at T_e = 0 K.

The energy flux emitted by surface 1 is $\varepsilon_1\sigma T_1^4$. If N_1 emitted energy bundles are followed per unit time and area from surface 1, the energy per bundle is $w = \varepsilon_1\sigma T_1^4/N_1$. The energy flux emitted from surface 2 is

$$2\sigma T_2^4\int_{\theta=0}^{\pi/2}\varepsilon_2(\theta)\cos\theta\sin\theta d\theta = \sigma T_2^4\int_{\theta=0}^{\pi/2}\cos^2\theta\sin\theta d\theta = \frac{\sigma T_2^4}{3}$$

If the same amount of energy w is assigned to each bundle emitted by wall 2 as for wall 1, then $wN_2 = \sigma T_2^4/3$. Substituting for w, ε_1 T_1, and T_2 gives

$$N_2 = \frac{\sigma T_2^4}{3}\frac{N_1}{\varepsilon_1\sigma T_1^4} = \frac{32}{3}N_1 \tag{7.13.2}$$

TABLE 7.2
Convenient Functions Relating Random Numbers to Variables for Emission (Assume No Dependence on Circumferential Angle ϕ)

Variable	Type of Emission	Relation
Cone angle, θ	Diffuse	$\sin\theta = R_\theta^{1/2}$
	Directional–gray	$R_\theta = \dfrac{2\int_{\theta^*=0}^{\theta} \varepsilon\left(\theta^*\right)\sin\theta^*\cos\theta^* d\theta^*}{\varepsilon}$
	Directional–nongray	$R_\theta = \dfrac{2\pi\int_{\theta^*=0}^{\theta}\int_{\lambda=0}^{\infty} \varepsilon_\lambda\left(\theta^*\right) I_{\lambda b}\sin\theta^*\cos\theta^* d\lambda d\theta^*}{\varepsilon\sigma T^4}$
Circumferential angle, ϕ	Diffuse	$\phi = 2\pi R_\phi$
Wavelength, λ	Black or gray	$F_{0-\lambda T} = R_{\lambda T}$
	Diffuse-nongray	$R_\lambda = \dfrac{\int_{\lambda^*=0}^{\lambda} \varepsilon_{\lambda^*} E_{\lambda^* b} d\lambda^*}{\varepsilon\sigma T^4}$
	Directional-nongray	$R_\lambda = \dfrac{2\pi\int_{\lambda^*=0}^{\lambda}\int_{\theta=0}^{\pi/2} \varepsilon_{\lambda^*}\left(\theta\right) I_{\lambda^* b}\sin\theta\cos\theta d\theta d\lambda^*}{\varepsilon\sigma T^4}$

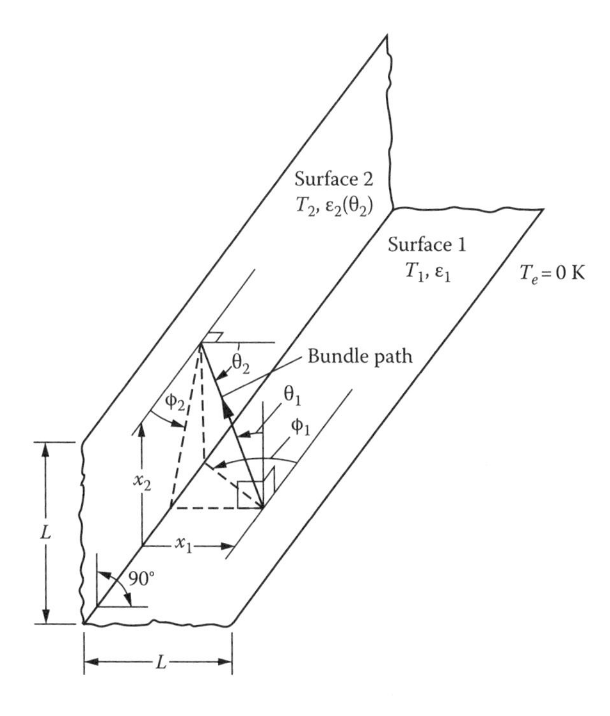

FIGURE 7.28 Geometry of Example 7.14.

Because all bundles have equal energy and 32/3 as many bundles are emitted from surface 2 as from surface 1, it is evident that surface 2 will make the major contribution to the energy transfer.

Now the distributions of directions for emitted bundles from the two surfaces will be derived. Surface 1 emits diffusely, so Equation 7.58 applies. For the directional–gray surface 2, the second line in Table 7.2 is used with $\varepsilon_2(\theta_2)$ from Equation 7.13.1:

$$R_{\theta_2} = \frac{1}{\varepsilon_2} \int_{\theta_2^*=0}^{\theta_2} \cos^2\theta_2^* \sin\theta_2^* d\theta_2^*$$

The $\varepsilon_2(\theta_2)$ is substituted from Equation 2.9 to give

$$R_{\theta_2} = \frac{\int_0^{\theta_2} \cos^2\theta_2^* \sin\theta_2^* d\theta_2^*}{\int_0^{\pi/2} \cos^2\theta_2 \sin\theta_2 d\theta_2} = 1 - \cos^3\theta_2$$

Because R and $1 - R$ are both uniform random distributions in the range $0 \le R \le 1$, this result can be conveniently written as $\cos\theta_2 = R_{\theta_2}^{1/3}$. By similar reasoning, Equation 7.58 can be written as $\cos\theta_1 = R_{\theta_1}^{1/2}$. Since there is no dependence on angle ϕ for either surface, Equation 7.59 applies for both surfaces.

Next, the position must be determined on each surface from which each bundle will be emitted. Because the wedge sides are isothermal, emission is uniform from each side. In such a case, random positions x (Figure 7.28) could be picked on each side as points of emission. This requires generation of a random number. The computer time required to generate a random number can be eliminated by noting that bundle emission is the initial process in each Monte Carlo history; hence, there is no prior history to be removed by using a random number. In this case initial x positions along L can be sequentially chosen as $x = (n/N)L$, where n is the sample-history index for the history being begun, $1 \le n \le N$.

The remaining calculations are to determine whether each emitted bundle will strike the adjacent wall or will leave the cavity. From Figure 7.28, for either surface, when $\pi \le \phi \le 2\pi$ the bundles will leave the cavity for any θ, and when $0 < \phi < \pi$ they will leave if $\sin\theta < (x/\sin\phi)/[(x/\sin\phi)^2 + L^2]^{1/2} = 1/[1 + (L\sin\phi/x)^2]^{1/2}$. The angle of incidence θ_i on a surface is given in terms of the angles θ_δ and ϕ_δ at which the bundle leaves the other surface, by $\cos\theta_i = \sin\theta_\delta \sin\phi_\delta$.

All the necessary relations are now available. A flow diagram is constructed to combine these relations in the correct sequence. Diffuse reflection is assumed from both surfaces. The resulting diagram is in Figure 7.29, showing one way of constructing the flow of events. The indices δ, δ', and δ'' are used to reduce the size of the chart. The index δ always refers to the wall from which the original emission of the bundle occurred, and δ' refers to the wall from which emission or reflection is presently occurring. The δ'' is used to make the emitted distribution of θ angles correspond to either $R_{\theta_1}^{1/2}$ or $R_{\theta_2}^{1/3}$ and have all the reflected bundles correspond to a diffuse distribution.

7.9.5 Reverse Monte Carlo

In certain cases, forward Monte Carlo can be very inefficient in computing radiative transfer. This is easily seen for diffuse surfaces by examining the reciprocity relation, Equation 4.15, $F_{dk-j} = \frac{A_j}{dA_k} dF_{j-dk}$. In surface–surface exchange between diffuse surfaces, the forward Monte Carlo technique in its simplest form is used to determine the factor dF_{j-dk} between a large (finite) surface and small differential area. If the geometry in question makes dF_{j-dk} a small value, then a forward Monte Carlo calculation may take a large number of samples to bring the computed value of dF_{j-dk} into acceptable accuracy. The reciprocity relation may then also produce considerable error in the value of F_{dk-j}. The reverse Monte Carlo technique, in contrast, effectively directly computes F_{dk-j}, and if this is a large factor, accuracy will require fewer samples. Thus, if the incident flux on a small

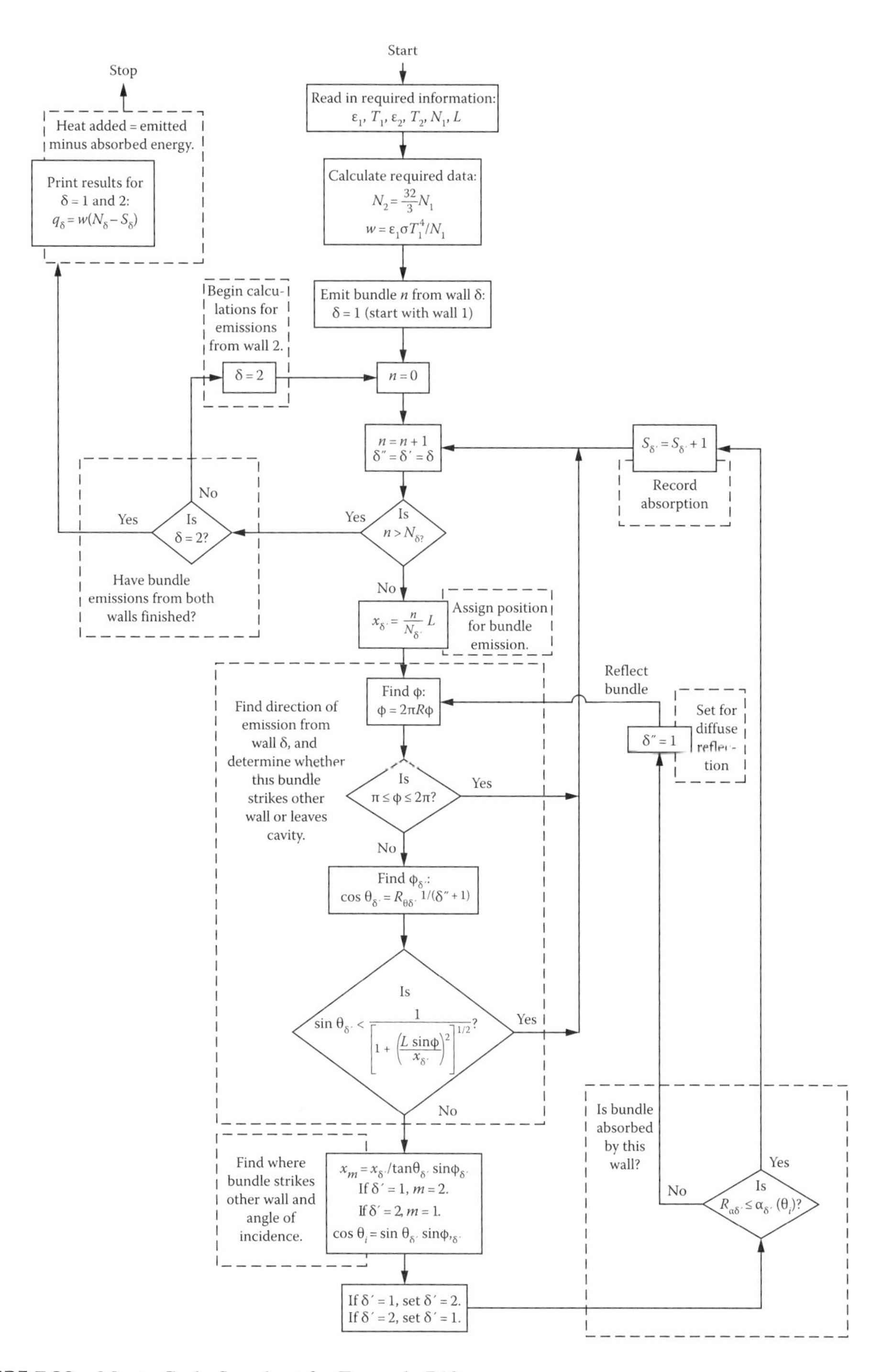

FIGURE 7.29 Monte Carlo flowchart for Example 7.13.

surface in an enclosure is to be determined, many samples may be required from each interacting surface using a forward formulation before a sufficient number of samples strike the small surface to provide acceptable statistical accuracy. In such cases, *reverse* (or *backward*) Monte Carlo methods can be used.

Walters and Buckius (1992, 1994) provide a rigorous derivation showing that the paths of Monte Carlo bundles are reversible, basing the analysis on the radiation reciprocity principle set forth by

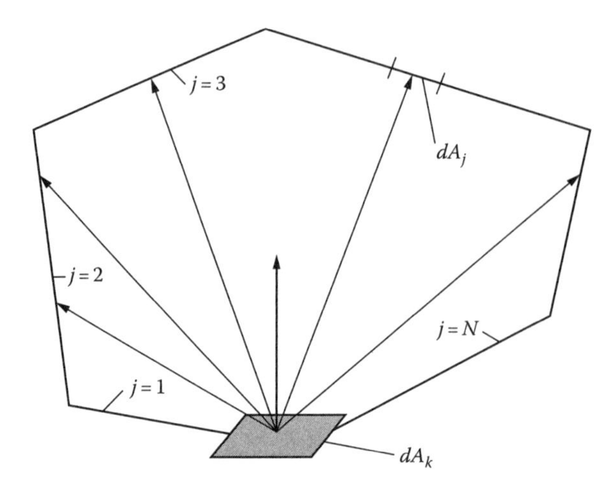

FIGURE 7.30 Enclosure of black surfaces showing reverse Monte Carlo bundle paths.

Case (1957). To find the irradiation on a surface element dA_k, bundle histories can be initiated from that element and followed to their element of origin. Because the angular distribution of irradiation onto dA_k is *a priori* unknown, the reverse paths are initiated using a distribution of angles into the reverse paths prescribed by Equations 7.58 and 7.59. The energy ascribed to each sample is then calculated based on the radiosity at the point of origin reached by the bundle.

Black surface enclosures: Consider first the case of an enclosure composed of black boundaries with known temperatures containing a nonparticipating medium (Figure 7.30).

To find the total radiative flux on the element dA_k, Equation 5.47 is used:

$$q_k(\mathbf{r}_k) = J(\mathbf{r}_k) - G(\mathbf{r}_k) = \sigma T_k^4(\mathbf{r}_k) - G(\mathbf{r}_k) \tag{7.62}$$

Now, N sample bundle reverse paths are originated from dA_k, and their point of intersection with the enclosure surface at location $\mathbf{r}_j$ is found. Each individual bundle n is then assigned energy $w_n = \frac{\left[\sigma T_j^4(\mathbf{r}_j)\right]_n}{N}$. The value of irradiation on the element dA_k is then

$$G(\mathbf{r}_k) = \sum_{n=1}^{N} w_n = \frac{\sigma}{N}\sum_{n=1}^{N}\left[T_j^4(\mathbf{r}_j)\right]_n \tag{7.63}$$

and the local flux $q_k(\mathbf{r}_k)$ is easily found from Equation 7.62.

Diffuse surface enclosures: Now, consider an enclosure with nongray but diffuse surfaces. The spectral radiative flux at any wavelength is found from

$$\begin{aligned} q_{k,\lambda}(\mathbf{r}_k) &= J_{k,\lambda}(\mathbf{r}_k) - G_{k,\lambda}(\mathbf{r}_k) \\ &= \left[\varepsilon_{k,\lambda}E_{\lambda b}(\mathbf{r}_k) + (1-\varepsilon_{k,\lambda})G_{k,\lambda}(\mathbf{r}_k)\right] - G_{k,\lambda}(\mathbf{r}_k) = \varepsilon_{k,\lambda}\left[E_{\lambda b}(\mathbf{r}_k) - G_{k,\lambda}(\mathbf{r}_k)\right] \end{aligned} \tag{7.64}$$

and the total flux is

$$q_k(\mathbf{r}_k) = \int_{\lambda=0}^{\infty} \varepsilon_{k,\lambda}\left[E_{\lambda b}(\mathbf{r}_k) - G_{k,\lambda}(\mathbf{r}_k)\right]d\lambda = \varepsilon_k \sigma T_k^4(\mathbf{r}_k) - \int_{\lambda=0}^{\infty} \varepsilon_{k,\lambda}G_{k,\lambda}(\mathbf{r}_k)d\lambda \tag{7.65}$$

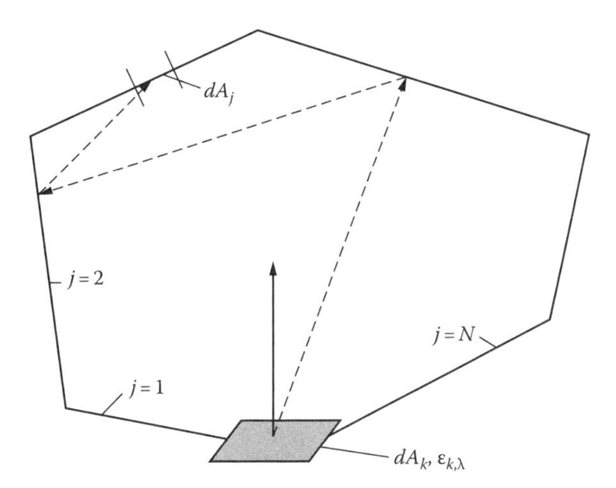

FIGURE 7.31 Enclosure of nongray–diffuse surfaces showing a single reverse Monte Carlo bundle path.

Finding the total radiative flux on nongray–diffuse surface dA_k then reduces to finding the value of the integral in Equation 7.65. In reverse Monte Carlo, this is done by again determining the weighted energy of the incident bundles, assigned by following the reverse bundle paths. Now, however, the energy carried by the bundle is complicated by the interreflections among the diffuse surfaces along the bundle reverse history (Figure 7.31); that is, the radiosity of each surface must be taken into account.

The reverse path is followed by initiating the reverse path as for the black case, except that a wavelength must also be assigned to the bundle through Equation 7.52, or for the nongray–diffuse surface k:

$$R_\lambda = \frac{\int_{\lambda^*=0}^{\lambda} \varepsilon_\lambda E_{\lambda b}(\mathbf{r}_k)\, d\lambda^*}{\varepsilon_k \sigma T_k^4(\mathbf{r}_k)} \tag{7.66}$$

Equation 7.66 can be curve fit as for Equation 7.53. Upon intersection of the bundle with an enclosure surface j, a decision is made as to whether the bundle originated at that surface by emission or was reflected from that surface. This is done by next determining the spectral absorptivity of the intersected surface at the wavelength determined from Equation 7.66, $\alpha_{j,\lambda} = \varepsilon_{j,\lambda}$. A new random number R is chosen, and if $R \le \alpha_{j,\lambda}$, the bundle is assumed to have been emitted by the intersected surface and is assigned the energy $w_n = \left[\varepsilon_{j,\lambda} E_{\lambda b}(\mathrm{r}_j)\right]_n$, and its reverse history is terminated. If, however, $R > \alpha_{j,\lambda}$, the bundle is assumed to have been reflected from the intersected surface, and its history is continued by choosing a further inverse reflected path by choosing the diffuse angles (θ_i, ϕ_i) using the diffuse relations of Equation 7.58 and 7.59. This process is continued through multiple reflections until the location of origin surface j is found. Many reverse bundle paths are then followed from surface k. The integral term in Equation 7.65 is then given by

$$\int_{\lambda=0}^{\infty} \alpha_{k,\lambda} G_{k,\lambda}(\mathrm{r}_k)\, d\lambda = \sum_{n=1}^{N} w_n = \frac{1}{N} \sum_{n=1}^{N} \alpha_{k,\lambda} [\varepsilon_{j,\lambda} E_{\lambda b,j}(\mathrm{r}_j)]_n \tag{7.67}$$

which can be tallied "on the fly" without intermediate storage of w_n for each bundle.

If the enclosure surfaces are not diffuse, then the reciprocity relations shown for reflectivity can be invoked to allow following reverse paths that account for directional surface properties.

If all surfaces are both gray and diffuse, then substituting Equation 7.67 into Equation 7.65 gives

$$q_k(\mathbf{r}_k) = \varepsilon_k \sigma T_k^4(\mathbf{r}_k) - \frac{1}{N}\sum_{n=1}^{N} \alpha_k \left[\varepsilon_j \sigma T_j^4(\mathbf{r}_j)\right]_n \overset{\alpha_k=\varepsilon_k}{\rightarrow} \frac{\varepsilon_k \sigma}{N}\left[\sum_{n=1}^{N}\left(T_k^4(\mathbf{r}_k) - \varepsilon_j T_j^4(\mathbf{r}_j)\right)_n\right] \quad (7.68)$$

Observe that the forward Monte Carlo approach uses equal-energy bundles and follows their paths from origin to point of absorption and gives the correct angular distribution of irradiation onto the absorbing surface. In contrast, the reverse Monte Carlo procedure assumes a uniform distribution of *number* of bundles in the irradiation but determines the correct angular distribution of irradiation through correctly weighting the energy per incident bundle.

7.9.6 Results for Radiative Transfer

7.9.6.1 Literature on Radiation Exchange between Surfaces

A value of Monte Carlo methods is that computer program complexity increases approximately in proportion to problem complexity, while the difficulty of carrying out conventional solutions increases approximately as the square of the complexity because of the matrix form for conventional formulations. Because a Monte Carlo method is somewhat more difficult to apply to the simplest problems, it is most effective when complex geometries and variable properties must be considered. In complex geometries, Monte Carlo has the advantage that simple relations will specify the path of a given energy bundle, whereas most other methods require integrations over surface areas. The integrations become difficult when various curved or skewed surfaces are present.

The fundamental statistical measures needed to determine the accuracy of Monte Carlo solutions, along with a relation for determining the variance in the results as the number of energy bundles is increased, are considered by Haji-Sheikh and Sparrow (1969) so that determination of solution accuracy can be monitored during a solution. Partitioning the energy carried by bundles after each event in the path, and then following the partitioned energy portions individually, is shown by Shamsundar et al. (1973) to reduce the variance in some cases. This is particularly the case for enclosure configurations with open areas. Various techniques for biasing direct Monte Carlo results to reduce variance, along with methods for computing expected variance, are discussed by Haji-Sheikh and Howell (2006). A review of the method is in Howell (1998) and applications specific to radiative exchange between surfaces are in Maltby and Burns (1991), Burns and Pryor (1998), and Zeeb and Burns (1999).

Monte Carlo is used for direct calculation of radiative exchange in complex geometries in Polgar and Howell (1965, 1966), Corlett (1966), Toor and Viskanta (1968a), Howell and Durkee (1971), Blechschmidt (1974), McHugh et al. (1992), Palmer et al. (1996), Antoniak et al. (1996), and Zaworski et al. (1996b). Configuration factors are calculated by Yarbrough and Lee (1985), Bushinskii (1976), Modest (1978), Walker (2013), and Walker et al. (2010, 2012). Some inverse problems are in Wu and Mulholland (1989) and Oguma and Howell (1995). The method has been used to obtain radiative transfer through fixed arrays of objects and in packed beds of spheres or cylinders in Yang et al. (1983), Abbasi and Evans (1982), Kudo et al. (1991a), Kudo et al. (1995), Singh and Kaviany (1991, 1992, 1994), Kaviany and Singh (1993), Argento and Brouvard (1996), and Li et al. (1996). Engineering systems were analyzed by Howell and Bannerot (1976), Omori et al. (1991), Villaneuve et al. (1994), and Mochida et al. (1995).

A probabilistic approach closely related to Monte Carlo relies on *Markov chains*. These usually find a transition matrix that defines the probability of transfer of radiation from one element on an enclosure surface to all other points. The matrix can include the effects of directional surfaces. Repeated matrix multiplications lead to a converged solution for radiative transfer. Surface exchange for various types of surface properties is in Esposito and House (1978), Naraghi and Chung (1984, 1986), and Billings et al. (1991a,b). A hybrid of Monte Carlo and Markov chain approaches is in the discrete probability function method (Sivathanu and Gore 1993, 1994, 1996).

Both Monte Carlo and Markov methods for radiative transfer rely on, or are used to calculate, configuration or exchange factors. The resulting factors may not exactly meet reciprocity and energy conservation constraints because of the statistical fluctuations in the methods. Various methods for improving the factors are discussed in Section 4.4.

Monte Carlo lends itself to parallel systems, because each history may proceed independently. Walker (2013) examined the use of both standard CPUs and graphical processing units (GPUs) in massively parallel computation of configuration factors. Various programming options were investigated. In general, GPU-based systems showed significant speed advantages.

Vujičić et al. (2006a,b) used forward Monte Carlo to compute area–area configuration factors for the cases of directly opposed rectangles, hinged rectangles, and concentric parallel disks of equal diameter. They examined the sensitivity of the results obtained by using the center of triangular surface elements as the points of sample origin to using random points of origin within the elements and generally found small differences in the area–area factors; however, significant differences were found in the element–element factors. In addition, when intervening surfaces caused blocking or shadowing, use of the element center as the origin could cause major errors by making some element–element factors appear to be zero when use of random origin points produced nonzero factors. Methods for accelerating the calculation of ray intersections with boundaries in 3D geometries with and without blocking by intermediate surfaces are compared by Mazumder (2006). Ertürk et al. (2008a) used reverse Monte Carlo to analyze the signal propagation through a light-pipe radiation thermometer, accounting for surfaces with specular-diffuse characteristics that contributed to radiation entering the light pipe.

Many applications of reverse Monte Carlo are for cases involving a participating medium. This extension is discussed in Section 12.6.5. In addition, a detailed discussion of Monte Carlo techniques for photon, electron, and phonon bundles in participating media was recently given by Wong and Mengüç (2002, 2004, 2008, 2010), Vaillon et al. (2004), and Wong et al. (2004, 2011, 2014).

7.9.6.2 Radiative Transmission through the Inside of a Channel

Some measurements of radiation depend on the axial transmission of radiation through a viewing tube or channel from the source to the sensor. The accuracy of the measurements depends on correcting for the transmission characteristics. This may be difficult because details of the surface reflection for a real surface are in terms of a bidirectional reflectivity that can have a complicated form, and uncertainty is amplified by multiple reflections. In Zaworski et al. (1996b), Sivathanu and Gore (1997), and Edwards and Tobin (1967), the transmission is analyzed along the length of a tube and through a channel formed by the gap between parallel plates. The cylindrical tube analyzed by Sivathanu and Gore models a long stainless steel tube used in two-wavelength pyrometry to collimate and guide radiation from a source onto a detector at the other end. The tube surface is modeled as a partially diffuse and a partially specular reflector. The method of analysis used was developed by Sivathanu and Gore (1993) and is similar to Monte Carlo. As a result of wavelength-dependent reflections along the tube wall, the spectral distribution at the detector differs from that at the source. As a consequence, it was found that the temperature measurement deduced from the radiation received can be considerably in error if the interior wall of the collimating tube is highly reflective. Because the reflectivity of the two components of polarization differs greatly for polished surfaces (Section 3.2), one component is propagated with little loss through a passage, while the other component has high losses to absorption after a few reflections. Use of an average specular reflectivity can therefore underpredict transmission through a specular channel or tube. Differences by factors of up to 14 between transmission calculations using an averaged reflectivity, or using the polarized directional components, are predicted Edwards and Tobin for square passages and infinitely wide slots with lengths of 30 times the wall spacing. In Zaworski et al. (1996b), radiative transmission through a channel formed by the gap between two parallel plates was calculated and compared with experiment. The plate surfaces were smooth aluminum coated with a flat white paint having a bidirectional reflectivity that had been previously measured. The bidirectional reflectivity had a specular component that increased considerably as the angle of incidence increased. Two million trajectories were followed for each Monte Carlo simulation. The results of the simulations compared well with experimental intensity distributions along the channel except when the incident radiation was at a small angle to the channel surfaces. The bidirectional reflectivity was not known with good accuracy in this near-tangential angular range. It was found that the assumption of diffuse–gray surfaces can lead to poor results, but the inclusion of the detailed bidirectional reflectivity can be a difficult undertaking. Representing the reflectivity as the sum of a pure diffuse and a pure specular component did not model the system accurately. A broader specular peak was necessary to obtain satisfactory agreement with experimental results.

7.9.6.3 Extension to Directional and Spectral Surfaces

Few references exist that treat problems involving both directionally and spectrally dependent properties. The reasons are twofold. First, accurate and complete directional-spectral properties are not often available. Second, when solutions are obtained, they are often so specialized that little interest exists to warrant their dissemination in the open literature. As pointed out by Dunn et al. (1966), when the radiative properties become available, the methods for handling such surface radiative energy–exchange problems now exist, and Monte Carlo appears to be one of the better-suited techniques. Toor and Viskanta (1968a,b) applied Monte Carlo to some interchange problems involving surfaces with directional and spectral properties. Some of these results were discussed in Chap. 6. Howell and Durkee (1971) obtained good results in a comparison of experimental data with a Monte Carlo analysis of radiative exchange in an enclosure with directional surfaces. Naraghi and Chung (1986) analyzed enclosures with directional properties. A general-purpose Monte Carlo computer code for calculating radiative exchange factors in 3D enclosures is described in Maltby and Burns (1991). This includes mixed specular and diffuse reflection models for surfaces, wavelength-banded spectral surface properties, transmission of radiation through surfaces, and incident beam radiation. For surfaces with bidirectional reflection properties, an improved method of tracking multiple paths for reflected radiation is considered in Shaughnessy and Newborough (1998). The method is applied to a square enclosure with a concentric square insert in its center that provides an obstruction. Good agreement for computed exchange factors was obtained with a standard Monte Carlo solution, and the new method was between 2.6 and 10.4 times greater in speed.

Extensions of Monte Carlo methods are in Howell (1973) and Zigrang (1975) to include property uncertainties in problems with radiation and conduction. In these treatments, mean values and standard deviations of both material properties and dimensional tolerances were used to choose sample systems. Large numbers of such systems were analyzed to predict the average expected performance of the thermal system and the variance in expected performance.

7.9.6.4 Application of Monte Carlo Methods to Combined-Mode Problems

Monte Carlo methods can be used in multimode problems by referring to Equation 7.4, which gives the general energy-balance equation including radiation for a thin (2D) element. The net radiative flux on either surface is given by the term $q_{R,net}(x,y) = G(x,y) - J(x,y)$. Monte Carlo methods can be used as described in Section 7.9.3 to determine the value of $q_{R,net}(x,y)$ on the basis of an assumed initial temperature distribution $T_j^{(0)}$. The resulting $Q_{R,net}^{(0)}(x,y)$ is then substituted into Equation 7.4, which may be solved for a new $T_j^{(n)}$ using the methods in Sections 7.8.1 and 7.8.2. This new $T_j^{(n)}$ is then used in the Monte Carlo routine to find a new $q_{R,net}^{(n)}(x,y)$, and the process is repeated until convergence. This is an iterative approach, and an underrelaxation factor may be used between iterations to aid stability and convergence.

The advantages of using Monte Carlo methods are the ability to treat complex geometries and to include spectral and directional surfaces. A disadvantage is that the statistical fluctuations in the $q_{R,net}^{(n)}(x,y)$ values from Monte Carlo may cause the $T_j^{(n)}$ values computed from Equation 7.4 to fluctuate between iterations, and convergence to an acceptable accuracy may be difficult. Problems in which radiation dominates will converge fairly quickly to the accuracy imposed by the Monte Carlo solution, while those in which convection dominates will be insensitive to the fluctuating values of $q_{R,net}^{(n)}(x,y)$. For problems where both radiation and at least one other heat transfer mode are present and have approximately equal influence, it may be necessary to use care in the Monte Carlo solution so that smooth and accurate $q_{R,net}^{(n)}(x,y)$ values are obtained for the 2D situation discussed here (Equation 7.4). The same considerations apply to more general transient or steady-state 3D problems.

7.10 CONCLUDING REMARKS

This chapter has provided extensive information on modeling and numerical solution methods for radiative transfer either alone or in combined modes with other heat transfer mechanisms.

7.10.1 Verification, Validation, and Uncertainty Quantification

Computer solution of radiative transfer problems, particularly multimode problems, can be quite challenging. In all computational problems, it is necessary to establish *a priori* conditions for the convergence and insensitivity to grid resolution of the quantity of interest (QOI). However, these are not enough. Additionally, the code should be *verified* and *validated*, and the *uncertainty in results should be quantified.*

Simple limiting cases can be checked. It is also possible to compare with available solutions for more complete cases such as in the results presented in Tong and Skocypec (1992) or the benchmark solutions presented in the web appendix for this book at http://www.thermalradiation.net.

To determine the quality of the code, we must first define what it is you really want to know (the QOI). Is it the heat flux at a boundary? The temperature at a location on the boundary? Secondary factors might be allowed to have imprecise prediction as long as the prediction of the QOI is accurate. Usually, for the QOI, the programmer establishes *a priori* measures for acceptable limits on convergence and grid resolution errors. But these are not enough! It should be recognized that any computer model or code ideally should be subjected to three categories of tests: *verification*, *validation*, and *uncertainty quantification* (UQ). This applies to any code, not just those for radiative transfer.

7.10.1.1 Verification

A software engineer needs to ask the following questions to verify his or her code: Is the code bug-free, and is it providing results that reflect the governing equations within the assumptions of the analysis of the physical problem being addressed? To ensure that a code is bug-free and giving correct results within the physical assumptions, one possibility is comparison of code predictions with known benchmark solutions. Benchmarks can originate by comparison with well-accepted solutions from the literature or can be generated. For example, the code can be tested against known simple limiting cases:

1. If all surfaces are isothermal at the same T, is the predicted heat transfer among all elements = 0?
2. If $\varepsilon = 0$ on a surface, is the predicted radiative heat flux = 0 at that surface?
3. For a combined-mode problem, does the code give correct results for each mode independently when the other modes are set to zero?

7.10.1.2 Validation

Even if verified, the numerical code may not reflect reality, since various assumptions are usually implicit in the code (gray and/or diffuse surfaces, temperature-independent properties, etc.) The code should therefore be compared with "reality," that is, usually, with available experimental data. For comparison of predictions with experiment, we must know both the error limits on the experimental data and the limits on the accuracy in the code predictions to determine whether the predictions lie within the range of the experimental measurements. Because perfect agreement between measurement and prediction is seldom obtained, *a priori* limits on what constitutes acceptable validation in the prediction of the QOI should be available. Must the prediction be within 1%, 5%, or 20% before the code predictions are acceptable? The answer will depend on the particular problem and the QOI being predicted.

7.10.1.3 Uncertainty Quantification

What is the uncertainty in the results of the code (the QOI)? A code prediction usually has two types of error source. First, input data to the code such as physical properties, dimensions, and energy source values will have a degree of uncertainty. Second, the model itself has uncertainties due to the assumptions within the model (e.g., angular and spectral discretization errors, gray and diffuse assumption, assumptions as to whether a 1D, 2D, or 3D analysis is acceptable). Having defined possible sources of uncertainty in both input data and the model itself, how do we quantify how these uncertainties propagate through the code so that the uncertainty in QOI prediction can be quantified? This is a quite difficult question to answer and is an ongoing research area. It is, however, a very important question, because unless some bounds are available on the uncertainty in prediction of the QOI, the code is really not useful. For example, if the code is predicting global warming effects but the uncertainty provides predictions that range from an ice age to polar melting, then the code is not of practical benefit.

If we are able to provide a measure of the uncertainty in the QOI and it is so large that the code results are not useful, how do we proceed to reduce prediction uncertainty? It is possible to find the sensitivity of the code to the input data and to the modeling uncertainties and to then improve the data or code to reduce the uncertainty inherent in the most sensitive factors. This is a daunting task when there are many parameters and much data to be considered. The field of verification, validation, and UQ for complex codes is an active research field (Helton 2009, Oden et al. 2010a,b).

HOMEWORK

7.1 An uncovered Styrofoam pan filled with water is placed outdoors on a cloudy night. The air temperature is 5°C. There is almost no wind, so the heat transfer coefficient between the air and water is h = 10 W/(m^2·K). The cloudy night sky acts as blackbody surroundings at T_e = 210 K. Water is opaque for long-wavelength radiation. The index of refraction of water is 1.33.
(a) Will the water start to freeze? Show calculations to prove your answer.
(b) What is the minimum air temperature required to prevent freezing?
Answer: (a) Freezing does begin; (b) 20.5°C.

7.2 A thin 2D fin in vacuum is radiating to outer space, which is assumed at $T_e \approx 0$ K. The base of the fin is at T_b, and the heat loss from the end edge of the fin is negligible. The fin surface is gray with emissivity ε. Write the differential equation and boundary conditions in dimensionless form for determining the temperature distribution $T(x)$ of the fin. (Neglect any radiant interaction with the fin base.) Can you separate variables and indicate the integration necessary to obtain the temperature distribution? (*Hint*: $\int (d^2\theta/dx^2)\,(d\theta/dx) = (1/2)\,(d\theta/dx)^2$ + constant.)

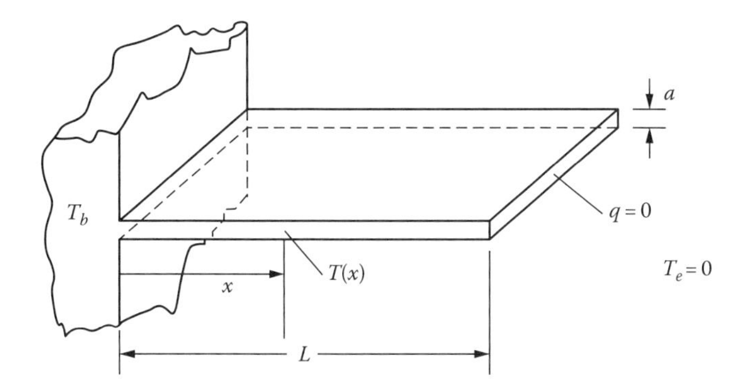

7.3 Consider the fin in Figure 7.12 as analyzed in Example 7.6. The heat transfer coefficient at the tip of the fin is h_L, and the emissivity of the end area is ε as for the rest of the fin surface. Formulate the boundary condition for the end face of the fin, and apply this condition to the general solution of the fin energy equation. Formulate all of the analytical relations, and describe how you would obtain the fin efficiency.

7.4 A very small-diameter pipe is at T_{pipe} = 600 K. The pipe is thin-walled polished copper, has diameter D = 0.2 cm, and is in a large room at T_e = 300 K. The radiative emissivity of the copper is ε_c = 0.04. A cylindrical opaque insulation layer with thickness t and thermal conductivity k = 0.07 W/m·K is added to the surface of the pipe. The emissivity of the outer insulation surface is ε_i = 0.85. The free convective heat transfer coefficient on the surface of the insulation is h = 15 W/m^2·K. For simplicity, h is assumed to be independent of insulation diameter and surface temperature, but a more precise analysis should include these effects. It is found that adding the insulation increases the rate of energy loss from the pipe. Find the thickness of insulation $t = t_{max}$ that *maximizes* the heat loss from the pipe.
Answer: t_{max} = 0.63 mm.

7.5 Consider the fin in Example 7.3 and in the further development in Equation 7.5 that follows the example. The environment temperature is nonzero at T_e. The diffuse fin now has more general spectral properties such that the spectral emissivity is ε_λ. However, it is assumed that ε_λ can be approximated as independent of temperature.
(a) Write the energy balance that now applies of the type in Equation 7.5.
(b) Put the equation in dimensionless form similar to Equation 7.3.2, using the same parameters where possible.
(c) Discuss how to best obtain the solution for $T(x)$ and how to present the results. Discuss whether the results with $T_e \neq 0$ can be related to those for $T_e = 0$.

7.6 Assume the fin in Example 7.6 extends from a black plane surface at T_b that is very large compared with the fin length. How would the fin formulation be modified to account for the interaction between the fin and the base surface?

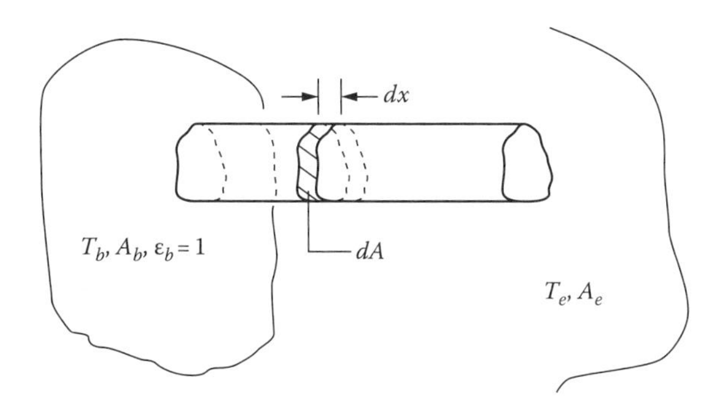

7.7 Consider Homework Problem 6.32. The radiation shield now consists of a layer of opaque plastic 0.10 cm thick coated on each side with a thin layer of metal having the same emissivity, $\varepsilon_s = 0.1$, as in Homework Problem 6.32. The thermal conductivity of the plastic is 0.200 W/(m·K). What is the heat transfer from plate 2 to plate 1, and how does it compare with that for the very thin shield analyzed in Homework Problem 6.32?

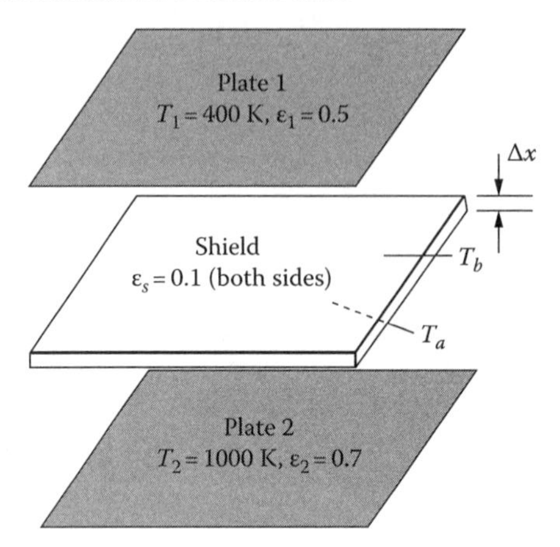

Answer: 2497 W/m^2.

7.8 A space radiator is composed of a series of plane fins of thickness $2t$ between tubes with radius R. The tubes are at uniform temperature T_b. The tubes are black and the fins are gray with emissivity ε. The radiator operates in a vacuum with an environment temperature of $T_e \approx 0$ K. Formulate the differential equation (including the analytical expressions for the configuration factors) and boundary conditions to obtain the temperature distribution $T(x)$ along the fin. Include the interaction between the fin and the tubes.

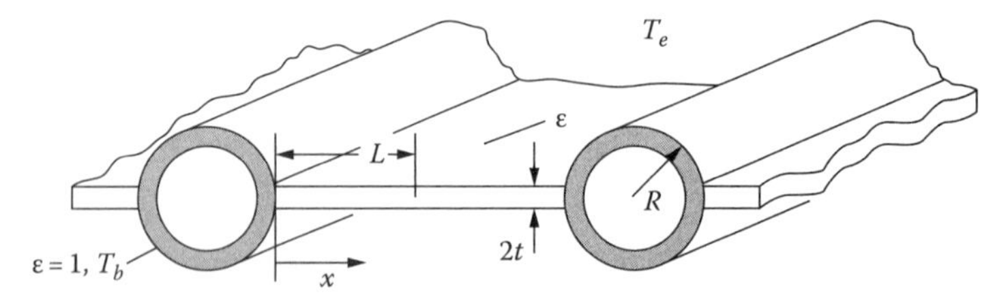

7.9 Two thin vertical posts stand immediately adjacent to a pool of molten material at temperature T_m and are diametrically across the pool from each other. The posts have a square cross section of area A_x, are of length L, and have thermal conductivity k. The entire surface of the posts has emissivity ε. The pool is of radius r and is assumed to be black. A breeze blows across the posts, and the air has temperature T_a. The air motion produces a heat transfer coefficient h between the post surface and the air.

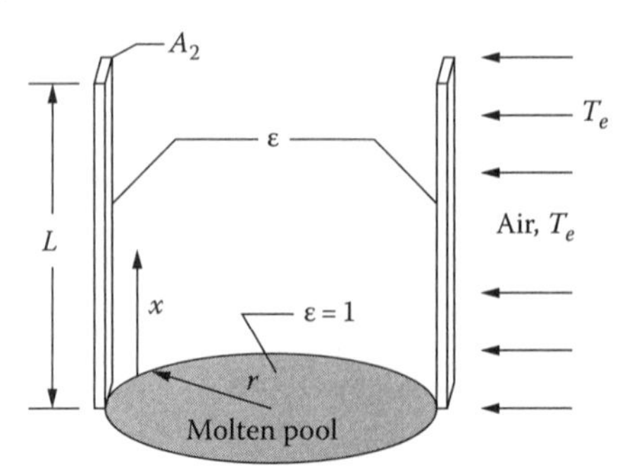

Derive an equation for the temperature distribution $T(x)$ along the posts, including the effect of mutual radiative exchange by the posts. Assume that the temperature at the bottom of the posts is equal to the temperature of the molten pool. Also, assume that the effect of the temperature of the surroundings T_e on radiative transfer can be neglected. Show the necessary boundary conditions for the problem and relations for all of the required configuration factors. (You do not need to substitute the F's into the equation.)

7.10 Steam is condensing inside a thin-walled tube of radius r_i. The tube has a coating of emissivity $\varepsilon_t = 1$ on the outer surface. The saturation temperature of the steam is T_b. Identical annular fins of outer radius r_o and emissivity ε_f are evenly spaced a distance L (between fin faces) along the tube. The fins are of thickness δ and thermal conductivity k. The environment surrounding the fin-tube assembly is at $T_e \approx 0$ K. Convection can be ignored. The configuration factor from a ring element on the tube to a ring element on fin 1 is $dF_{dt-d1}(x, \rho_1)$ and from a ring element on fin 2 to a ring element on fin 1 is $dF_{d2-d1}(L, \rho_1, \rho_2)$.
Set up the governing equation for the temperature distribution of the fin, $T(\rho)$.

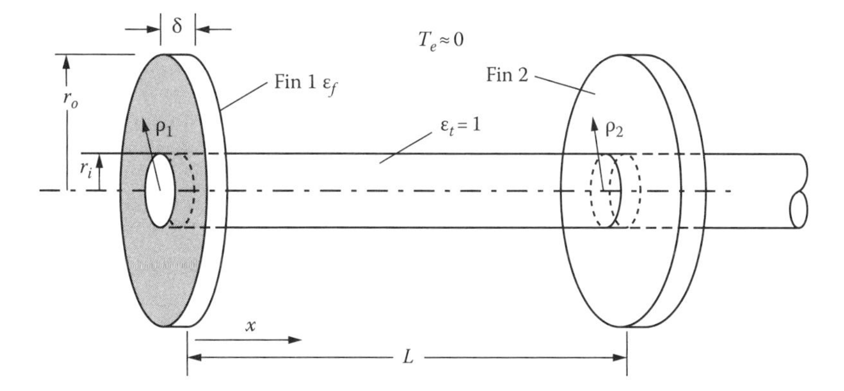

7.11 Two directly opposed parallel diffuse–gray plates of finite width W have a uniform heat flux q_e supplied to each of them. The plates are infinitely long in the direction normal to the cross section shown. They are separated by a distance H. The plates are each of thickness t ($t \ll W$) and thermal conductivity k, and both have emissivity ε. The plates are in vacuum, and the surroundings are at T_e. Set up the governing integral equation for finding $T(x)$, the temperature distribution across both plates. The outer surface of each plate is insulated, so all radiative heat exchange is from the inner surfaces.

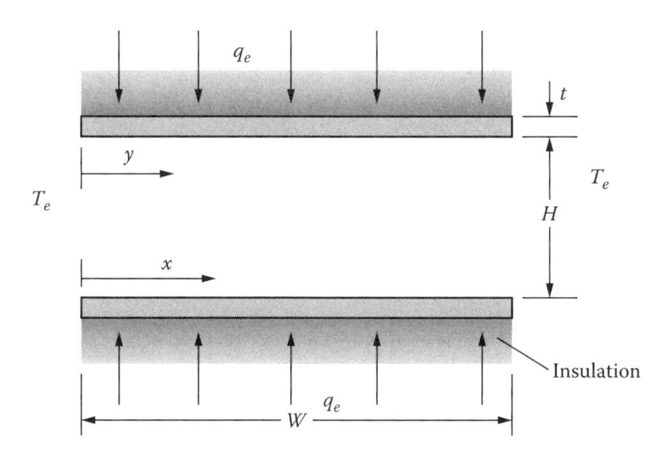

7.12 An infinitely long enclosure is shown in cross section below. It is separated into two compartments by a conducting plate with thermal conductivity $k = 4$ W/m·K. The properties and conditions on the enclosure surfaces are shown in the table. The vertical ends are at specified temperatures, and the horizontal sides are insulated on the outside.

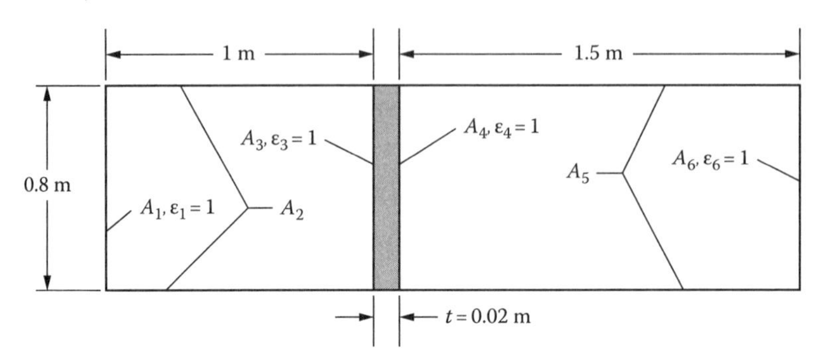

Surface	Emissivity, ε	Net Heat Flux, q (W/m^2)	Temperature (K)
1	1.0		1800
2	0.1	0	
3	1.0		
4	1.0		
5	0.3	0	
6	1.0		200

Determine the values of the missing table entries. Assume for simplicity that the surfaces need not be subdivided.

Answer: $q_1 = q_4 = -q_3 = -q_6 = 87667$ W/m^2, $T_2 = 1749$ K, $T_3 = 1693$ K, $T_4 = 1254$ K, and $T_5 = 1055$ K.

7.13 A copper–constantan thermocouple ($\varepsilon = 0.15$) is in a transparent gas stream at 300 K adjacent to a large blackbody surface at 900 K. The heat transfer coefficient from the gas to the thermocouple is 32 W/(m^2·K). Estimate the thermocouple temperature if it is (a) bare or (b) surrounded by a single polished aluminum radiation shield with $\varepsilon_{shield} = 0.075$ in the form of a cylinder open at both ends. The heat transfer coefficient from the gas to both sides of the shield is 15 W/(m^2·K).

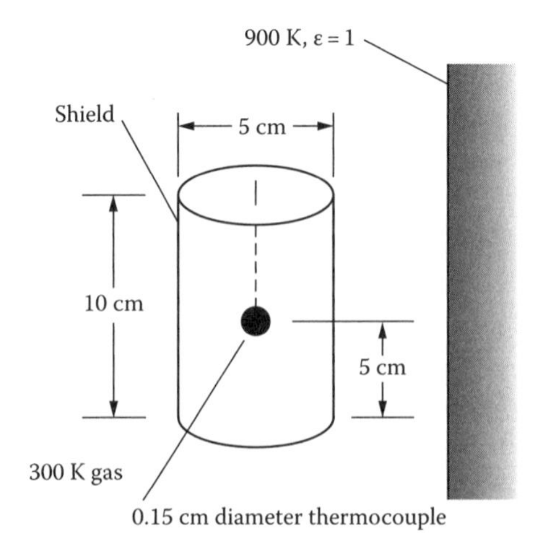

Answer: (a) 383 K; (b) 365 K.

7.14 Thin wire is extruded at fixed velocity through a die at temperature T_0. The wire then passes through air at T_a until its temperature is reduced to T_L. The heat transfer coefficient to the air is h, and the wire emissivity is ε. It is desired to obtain the relation between T_L and T_0 as a function of wire velocity V and distance L. Derive a differential equation for wire temperature as a function of distance from the die, and state the boundary conditions. (*Hint:* Compute the energy balance for flow in and out of a control volume fixed in space.)

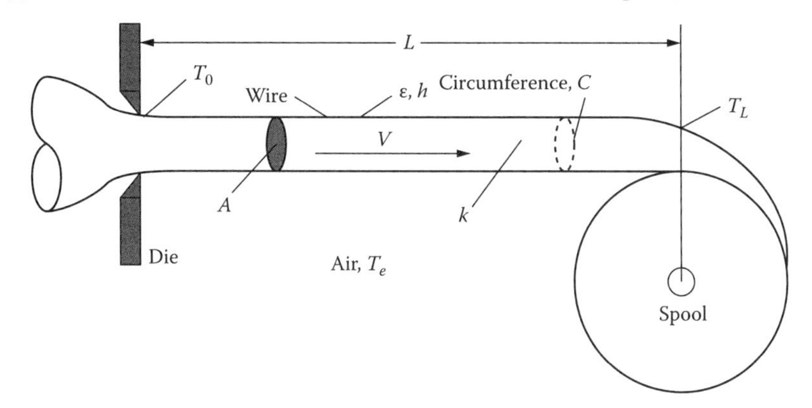

7.15 A single circular fin is to dissipate energy from both sides in a vacuum to surroundings at low temperature. The fin is on a tube with 2 cm outer diameter. The tube wall is maintained at 750 K by vapor condensing on the inside of the tube. The fin has 20 cm outer diameter and is 0.30 cm thick. Estimate the rate of energy loss by radiation from the fin if the fin is made from (a) copper with a polished surface (Figure 3.30), (b) copper with a lightly oxidized surface, and (c) stainless steel (k = 35 W/(m·K) with a clean surface (ε = 0.2). What is the effect on energy dissipation of increasing the fin thickness to 0.60 cm?

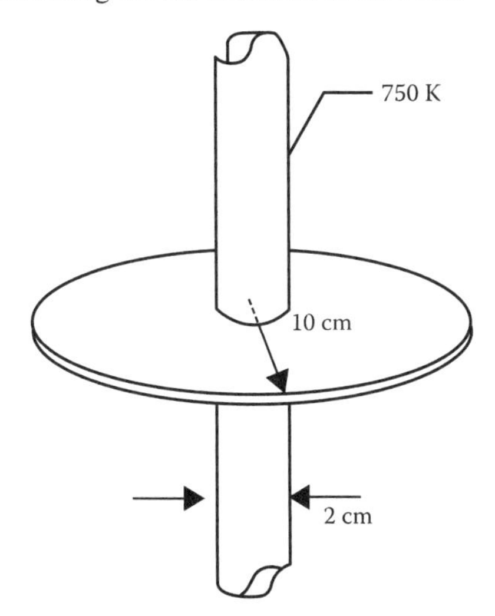

Answer: (c) 69 W; 98 W.

7.16 How would the analysis in Homework Problem 7.10 be modified to include the effect of a nonzero environment temperature?

7.17 A 1 cm diameter thin cryogenic electronic device is glued by epoxy to the bottom surface of a 0.5 cm deep, 1 cm diameter cavity of a high-thermal conductivity ceramic package. The ceramic package is kept at liquid nitrogen temperature of 77 K inside a relatively large vacuum enclosure at the ambient temperature of 300 K. The epoxy provides a thermal interface conductance of 1×10^4 W/m²·K between the electronic device and the ceramic package.

All surfaces are gray with diffuse emission. The emissivity is 0.5 on the device surface and 0.3 on the ceramic surface. Calculate the temperature of the electronic device when the device is turned off.

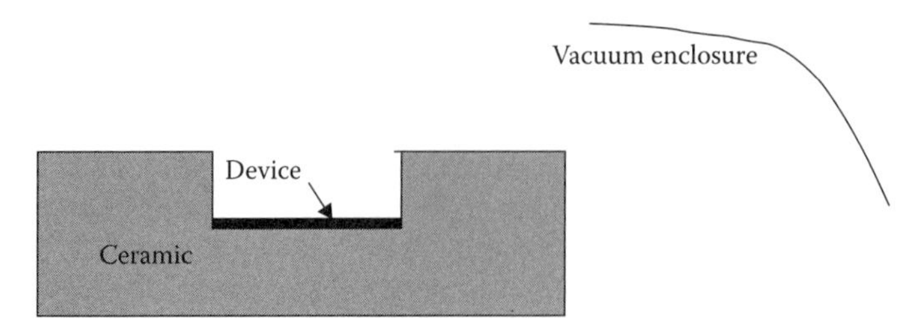

Answer: $T_d = 77.04$ K.

7.18 For the cryogenic device in Homework Problem 7.17, a thin metallic specularly reflecting foil with reflectivity of 0.1 is inserted to cover the cavity as shown in the figure. Calculate the temperatures of the electronic device and the thin metallic foil when the device is turned off. Compare with the results of Homework Problem 7.17.

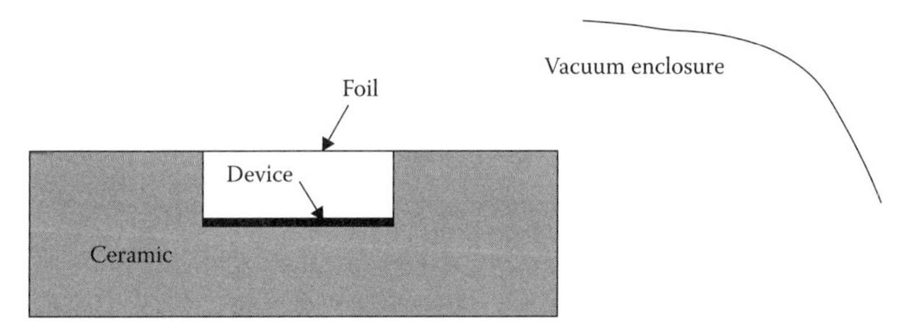

Answer: $T_d = 77.03$ K and $T_f = 295.7$ K.

7.19 The billet in Homework Problem 5.3 has air at 27°C blowing across it that provides an average convective heat transfer coefficient $\bar{h}$ of 24 W/(m$_2$·K). Estimate the cooling time with both radiation and convection included.

Answer: 4.47 h.

7.20 Opaque liquid at temperature $T(0)$ and mean velocity $\bar{u}$ enters a long tube that is surrounded by a vacuum jacket and a concentric electric heater that is kept at uniform axial temperature T_e. The heater is black, and the tube exterior is diffuse–gray with emissivity ε. The convective heat transfer coefficient between the liquid and the tube wall is h, and the tube wall thermal conductivity is k_w. Derive the relations to determine the mean fluid temperature and the tube wall outer surface temperature as a function of distance x along the tube (assume that the liquid properties are constant).

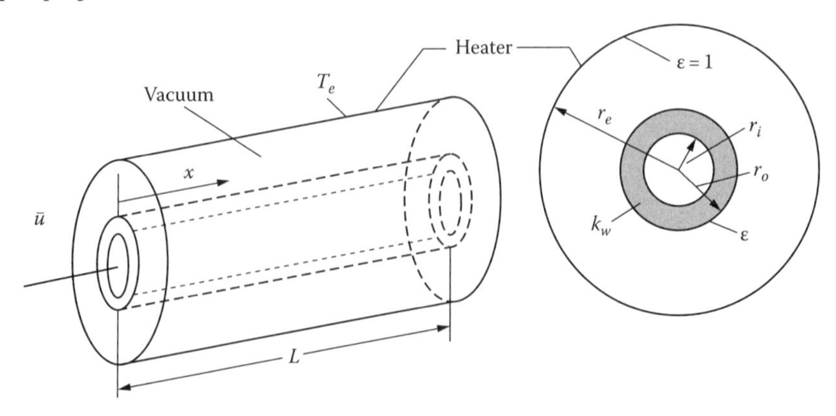

7.21 A solar collector is designed to fit onto the horizontal section of a roof as diagrammed below. Flow is from the right to the left in the tubes of the collector. A tilted white diffuse roof section at the left side helps to reflect additional solar flux onto the collector. Set up the equations for determining the local temperature of the tubes for two cases: (a) no flow in the tubes and (b) flow of water in each tube of 2.00 kg/min. Indicate a possible solution method. Assume that the roof and collector are very long normal to the cross section shown.

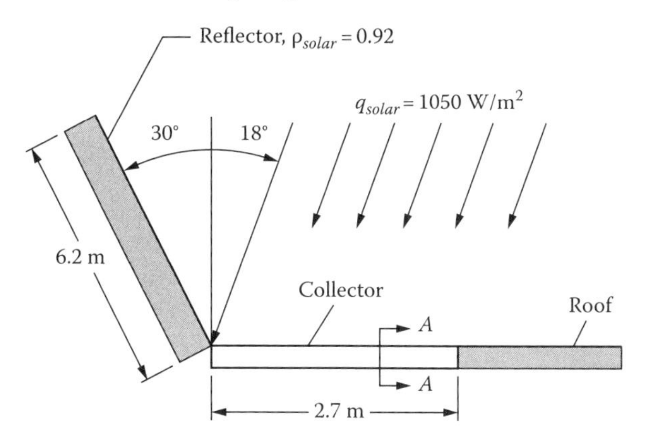

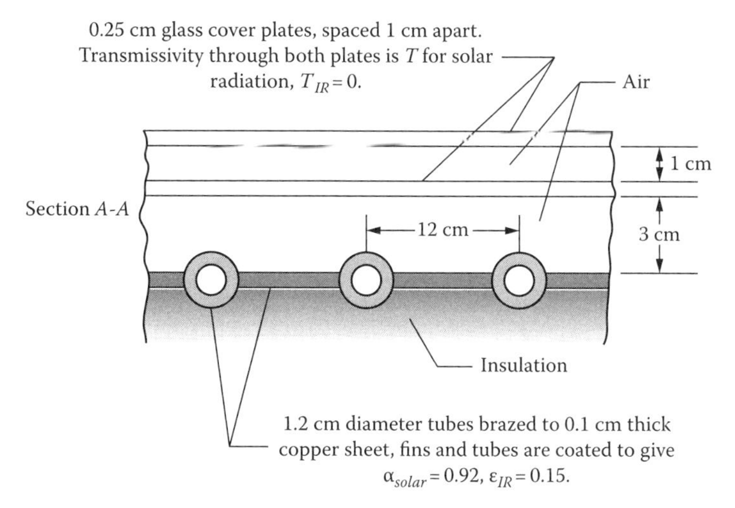

7.22 A rod of circular cross section extends out from a slender space vehicle in Earth orbit into surroundings at T_e. The rod axis is normal to the direction from the sun. The rod is coated so that its infrared emissivity is ε_{IR} and its solar absorptivity is α_s. The base of the rod is at $T_b > T_e$. Derive a differential equation to predict the rod temperature distribution, $T(x)$. State the boundary conditions, including radiation at the circular end face. Neglect temperature variations within the rod cross section at each x. Neglect radiation to the rod from the slender vehicle surface, and neglect any emitted or reflected radiation from the Earth.

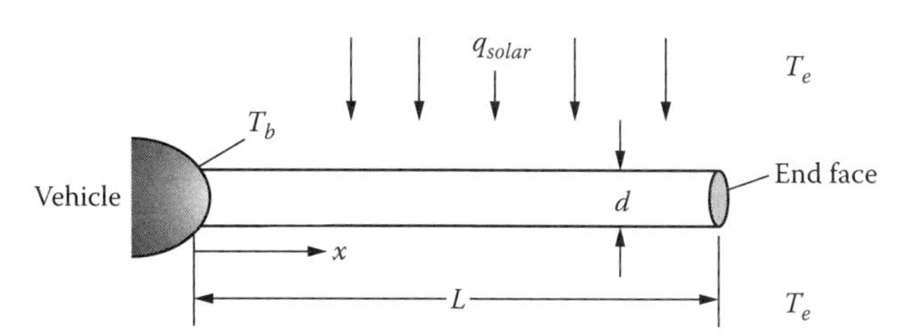

7.23 A radio antenna extends normal to a spacecraft surface as shown in the diagram below. The antenna has a circular cross section with diameter $D = 1$ cm. The spacecraft itself is very large, and its surface can be considered to be black with uniform temperature $T_b = 500$ K. The environment is at $T \approx 0$ K. The antenna material has emissivity = ε (gray, diffuse) and thermal conductivity $k = 40$ W/(m·K).

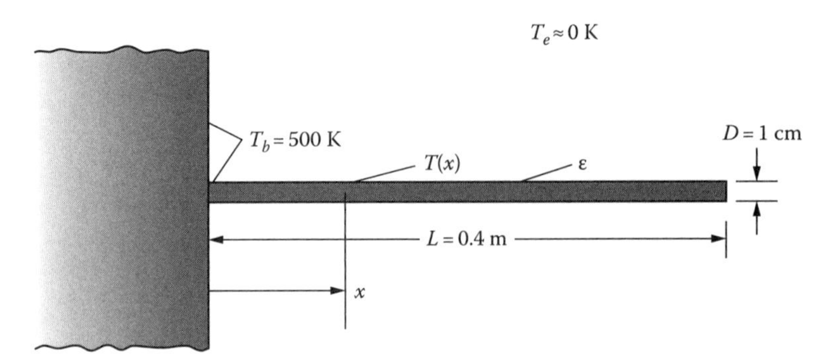

(a) Derive the differential equation for the temperature distribution in the antenna, $T(x)$. Note any assumptions. Include the effect of radiation exchange between the antenna and the spacecraft.
(b) Place the equation in a convenient nondimensional form, using the nondimensional temperature $\vartheta(X) = T(X)/T_b$ and $X = x/L$.
(c) Provide a plot of $\vartheta(X)$ versus X for $\varepsilon = 0.3$, 0.5, and 1.0.

7.24 A_1 and A_2 are diffuse concentric spheres. The inner surface of the inner sphere is heated with nonradiating combustion products at $T_{comb} = 1100$ K with a convective heat transfer coefficient to the surface of 50 W/(m_2·K). Calculate the temperature T_1 of the inner sphere for the surface spectral emissivities shown, where $\varepsilon_{\lambda,1}$ is assumed independent of temperature.

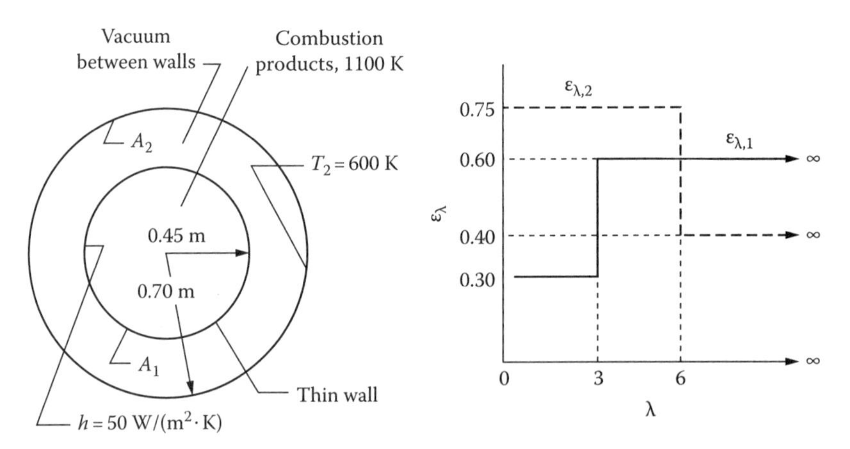

Answer: 869 K.

7.25 A wire between two electrodes is heated electrically with a total of Q_e Watts. The wire resistivity r_e Ohm-cm is constant, and the current is I. One end of the wire is at T_1 and the other is at T_2. The immediate surroundings are a vacuum, and the surroundings have a radiating temperature of T_0. The wire is gray with emissivity ε, diameter D, thermal conductivity k, and length L.

(a) Set up the differential equation for the steady-state temperature distribution along the wire, neglecting radial temperature variations within the wire. Integrate the equation, and find an expression (in the form of an integral) for the wire temperature as a function of x. The final results should contain only the quantities given and should not contain dT/dx. Explain how you would evaluate the expression to find $T(x)$.

(b) Let both the emissivity and the electrical resistivity be proportional to T (see Equations 3.43 and 3.44). Derive the solution for this case, and describe how it can be evaluated for a fixed value of Q_e.

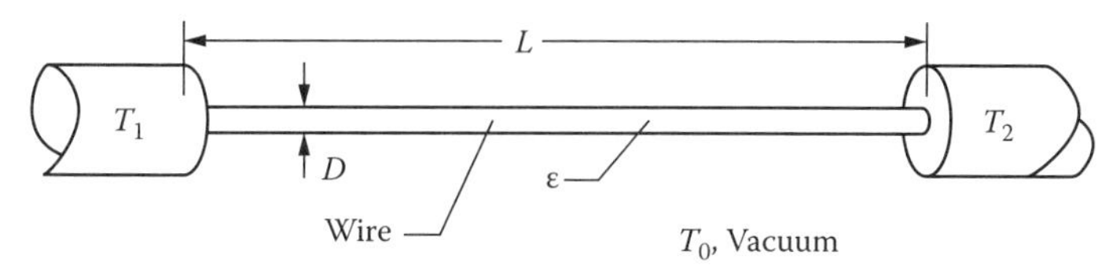

7.26 A spherical temperature sensor, 0.15 cm in diameter, is on the axis of a short pipe, halfway between the open ends. Air at 400 K is flowing through the pipe, and the convective heat transfer coefficient on the sensor is 20 W/(m^2·K). Calculate the sensor temperature. All surfaces are gray. Neglect blockage (shadowing) by the sensor when computing configuration factors between boundaries of the pipe. Do not subdivide surfaces.

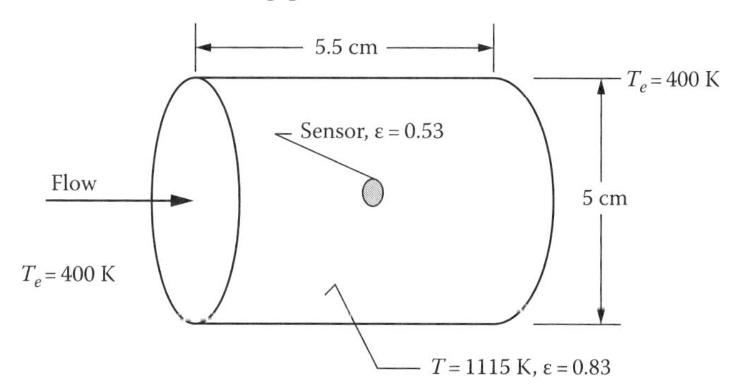

Answer: 921.4 K.

7.27 Two plates are joined at 90° and are very long normal to the cross section shown. The vertical plate (plate 1) is heated uniformly with a heat flux of 300 W/m^2. The horizontal plate has a uniform temperature of 400 K. Both plates have an emissivity of 0.6. The environment is at T_e = 300 K. Both plates are made of tungsten, with thermal conductivity k = 174 W/m·K. The plates are 0.056 cm in thickness. At the edge where the plates join, a thin layer of ideal insulation provides an infinite contact resistance. The dimensions are shown in the figure.

Derive the equations necessary for finding the temperature distribution on surface 1 and the net radiative heat flux on surface 2. Determine reasonable boundary conditions for the plate ends; justify your choice. (This is a continuation of Homework Problem 5.43.)

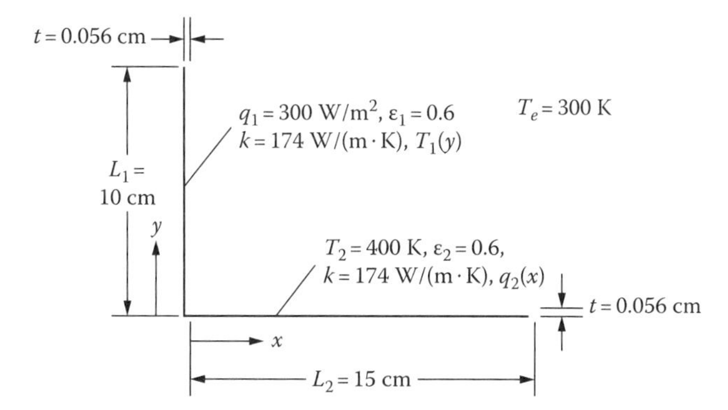

7.28 Evaluate numerically the configuration factor between two infinitely long parallel plates of equal finite width joined along one edge and making an angle of 45°. Use a numerical

integration of the analytical integral form in Example 4.6 using the trapezoidal rule to obtain the result, and compare it with the exact solution.
Answer: 0.61732.

7.29 Evaluate numerically the configuration factor between two infinitely long parallel plates of equal finite width joined along one edge and making an angle of 45°. Use a numerical integration of the analytical integral form in Example 4.6 using Simpson's rule to obtain the result, and compare it with the exact solution.
Answer: 0.61732.

7.30 An electrically heated nickel wire is suspended in vacuum between two water cooled electrodes maintained at 810 K. The wire diameter is 0.15 cm and the wire length is 4.15 cm. The surroundings are at a uniform temperature of $T_e = 900$ K and act as a black environment. Air at a temperature of T_e flows over the wire, causing a heat transfer coefficient between the wire and the air of $h = 35$ W/m^2·K. Set up the combined radiation and conduction relations to determine the wire temperature $T(x)$ assuming that the radial temperature distribution within the wire is uniform at each x. Determine how many Watts must be generated within the wire for its center temperature to be 1050 K. (The wire thermal conductivity is constant with a value of 85 W/(m·K) and the wire emissivity is constant with value $\varepsilon_w = 0.13$.) What is the required wattage if the nickel becomes oxidized so that $\varepsilon_w = 0.47$?

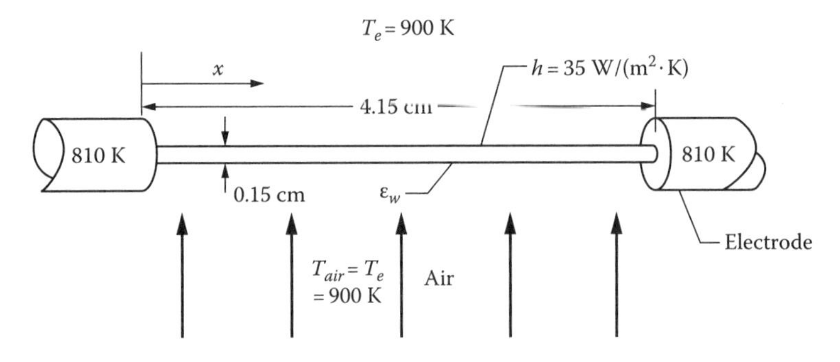

Answer:

ε_w	Q (W)
0.13	14.5
0.47	16.0

7.31 A long solid rectangular region in vacuum has the cross section shown and thermal conductivity k_w. One-half of one of the long sides is heated by contact with an opaque source of uniform flux q_e. The surroundings, which act as a black environment, are at a uniform temperature T_e. The exposed surfaces of the region are gray and have emissivity ε_w. Using the grid shown (for simplicity), set up the finite-difference relations to be solved for the steady temperature distribution in the rectangular solid.

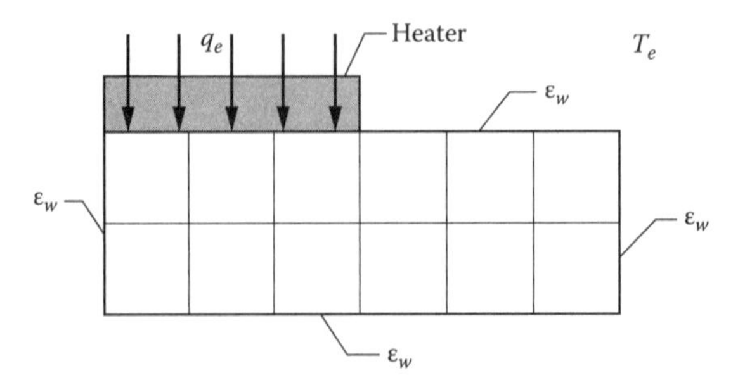

7.32 A long stainless steel tube with a thick wall is filled with a highly insulating material and is near an infinite black hot wall. Divide the circumference of the tube symmetrically about $\theta = 0$ into eight increments and obtain an expression for the radiant energy from the hot wall to each tube increment. Then, using a finite-difference approximation, obtain an approximate temperature distribution around the tube wall. Neglect radial temperature distributions in the tube wall. The tube wall material has thermal conductivity k_w = 34 W/(m·K) and its outer surface is gray with emissivity $\varepsilon_w = 0.187$. The surroundings are at low temperature that can be neglected in the analysis, $T_e \approx 0$ K.

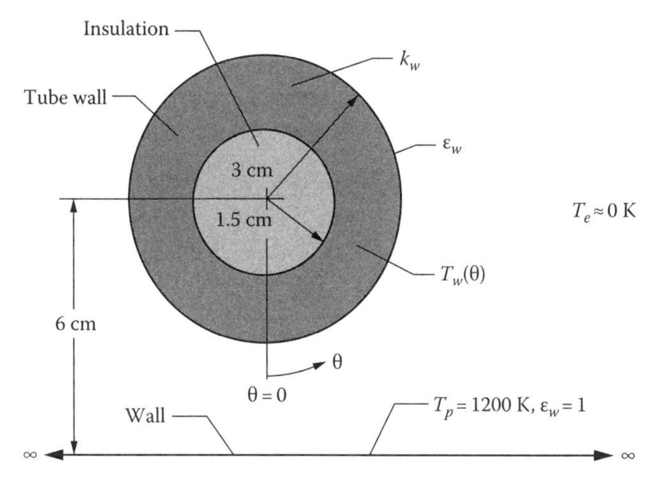

Answer:

T_1	T_2	T_3	T_4
1021.9 K	1014.3 K	1003.5 K	995.9 K

7.33 A long gray empty circular tube is in the vacuum of outer space so that the only external heat exchange is by radiation. The metal tube is coated with a material that has a solar absorptivity α_s and an emissivity in the infrared region of ε_{IR}. The solar flux q_s is incident from a direction normal to the tube axis, and the surrounding environment is at a very low temperature T_e that can be neglected in the radiative energy balances. The geometry is as shown in cross section. The tube is empty so there is internal radiative exchange. Energy is conducted circumferentially within the tube wall. The wall thermal conductivity is k_w.

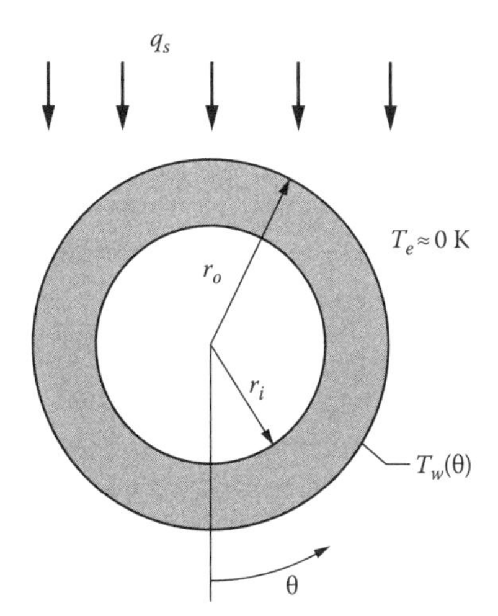

(a) Set up the energy relations required to obtain the temperature distribution around the tube circumference assuming that radial temperature variations within the tube wall can be neglected.
(b) Place the energy relations in finite-difference form and describe how a numerical solution can be obtained.

7.34 The tube in Homework Problem 7.33 is shielded from solar radiation by being in the shadow of a space vehicle, so that it cools to a very low temperature. It is then suddenly exposed to the solar flux. Set up the transient energy relations required to calculate the tube circumferential temperature distribution as a function of time using the same conditions and assumptions as in Homework Problem 7.33. Place the equations in finite-difference form, and describe how a numerical solution can be obtained.

7.35 A tube has length $L = 2.00$ m and has an inside diameter of $D_i = 0.5$ m. The tube has a wall thickness of $b = 1$ cm, and the tube wall material has a thermal conductivity of $k = 300$ W/m·K. An electric heating tape is wrapped around the outside of the tube and is uniformly and carefully insulated on its outer surface. The tape heater imposes a uniform heat flux of $q_e = 6000$ W/m^2 at the inner tube surface. A transparent fluid at $T_{f,\,in} = 300$ K enters the tube from a large plenum at the same temperature. The fluid flows through the tube at a mass flow rate of 0.2 kg/s. The fluid has specific heat $c_p = 4100$ J/(kg·K) and density $\rho = 1000$ kg/m$_3$. At the given flow rate, the heat transfer coefficient between the fluid and the tube surface is given by $h_x = 1000/(x + 0.01)^{1/2}$ W/(m^2·K), where x is the distance from the tube entrance in meters. Four gray–diffuse coatings are available to cover the inside of the tube. These have emissivities of $\varepsilon = 0.05, 0.35, 0.65$, and 0.95.
(a) Derive the energy equations that govern the heat transfer behavior to determine the tube wall and fluid temperatures along the tube length, and place them in dimensionless form.
(b) Find the maximum tube surface temperature, the position of that temperature, $x(T_{\max})$, and the mean fluid temperature at the tube exit, $T_f(x = L)$ for each of the three possible emissivities. Show a plot of the tube inner surface and fluid temperature distributions versus x for each emissivity. These may be in dimensionless form.
(c) Discuss the temperature results, giving some physical discussion of the relative effects of the various heat transfer mechanisms on the shapes of both the tube wall and fluid temperature profiles.

Discuss the numerical accuracy of the results. Are they converged, and are they within acceptable accuracy? Estimate the possible error in your solutions.

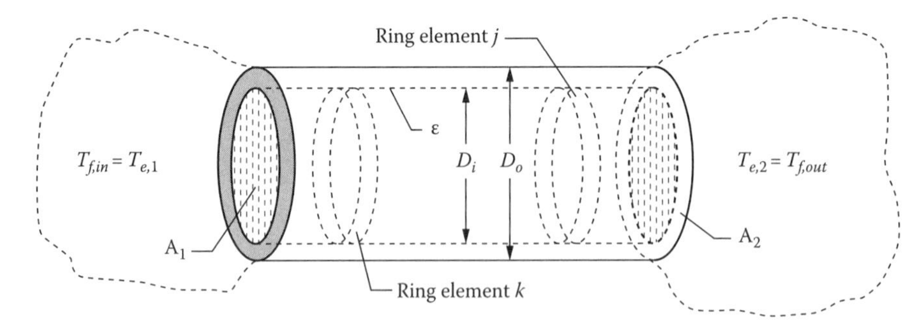

7.36 A thin sheet of copper moves through a radiative–convective oven at a velocity of 0.1 m/s. The sheet and oven are very wide. Air flows at a mass flow rate of 0.2 kg/s/m of oven width over the sheet in counterblow, and the heat transfer coefficient between the air and sheet surface is constant along the sheet at a value of $h = 100$ W/(m^2·K). The back of the sheet is insulated. A black radiant heater at $T_{Heater} = 1200$ K covers the top of the oven as shown. The radiant heater does not interact convectively with the air stream. Louvered curtains at each end of the oven are opaque to radiation but allow airflow. The emissivities of all surfaces are shown.

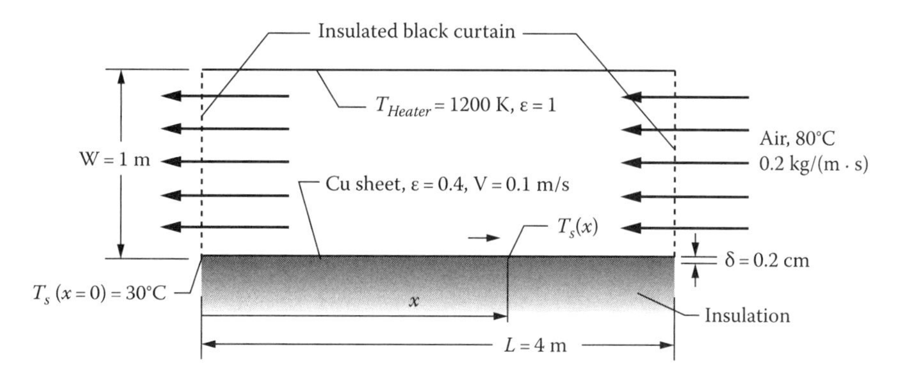

Find the temperature distribution $T_s(x)$ along the copper sheet as a function of position x within the oven, and present the result graphically. Discuss all assumptions made in the solution, and justify them by numerical argument where possible.

Data for copper sheet: k_s = 400 W/(m·K); $c_{p,s}$ = 385 J/(kg·K); ρ_s = 9000 kg/m^3.

7.37 A high-temperature nuclear reactor is cooled by a transparent gas. The gas flows through tubular cylindrical fuel elements. The fuel elements are of length L m and inside diameter D m, and the gas has specific heat c_p kJ/(kg·K), density ρ kg/m^3 (both temperature independent), and a mass flow rate of $\dot{m}$ kg/s through each fuel element. The tube has wall thickness b. Energy is generated in a sinusoidal distribution along the length x of the tube at a rate $q(x) = q_{\max} \sin\left(\frac{\pi x}{L}\right)$ W/m^2 based on the inside tube area. The heat transfer coefficient h between the gas and the tube surface is assumed constant with x and has units of W/(m^2·K). The gas enters the tube at temperature $T_{g,i}$ from a large chamber at temperature $T_{r,i}$. The gas leaves the tube at $T_{g,e}$ and enters a mixing plenum that is at temperature $T_{r,e}$. The tube interior surface is diffuse–gray with emissivity ε. The tube exterior surface is perfectly insulated. The tube wall material has thermal conductivity k W/(m·K).

(a) Set up the governing equations for determining the wall and gas temperature distributions along the tube length (see Examples 7.7 and 7.8 for some help).

(b) Using the dimensionless groups given in Example 7.7 or modifications appropriate for this problem, put the equations in dimensionless form.

(c) For values of the dimensionless parameters St = 0.02, N = 5, H based on $q_{\max}$ = 0.08, $\vartheta_{r,i} = \vartheta_{g,i}$ = 1.5, and $\vartheta_{r,e} = \vartheta_{g,e}$ = 5, solve for the wall temperature distribution along the tube for the cases ε = 1 and ε = 0.5. You may wish to obtain the pure radiation and pure convection results as limiting cases.

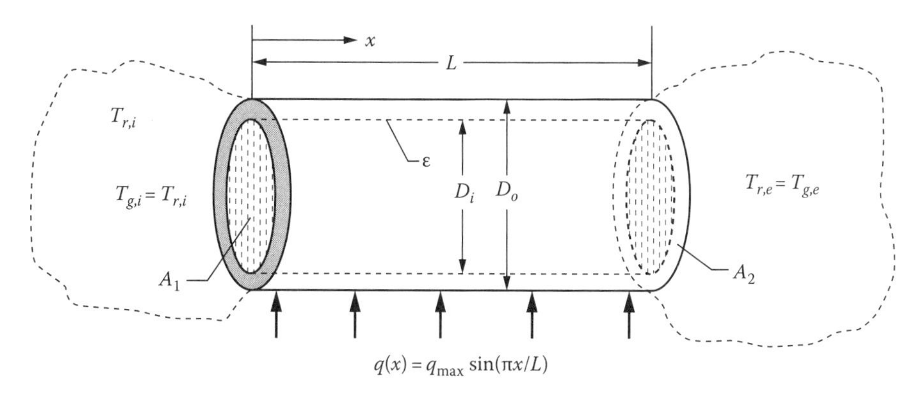

7.38 A square enclosure of length 1 m on a side has the conditions shown below and is filled with a transparent stagnant gas. Water is flowing upward along the right-hand vertical surface and enters the channel at the bottom of that wall at $T_w = 90°C$. The flow rate of the water is large, so that its temperature changes by a negligible amount in passing along surface A_1. There is a heat transfer coefficient between the water and the thin enclosure wall of 50 W/m²·K. Surface 4 is specular and perfectly reflecting, surface 3 is black, and the remaining surfaces are diffuse–gray. Surface 2 has a specified temperature and surface 3 has a specified heat flux.

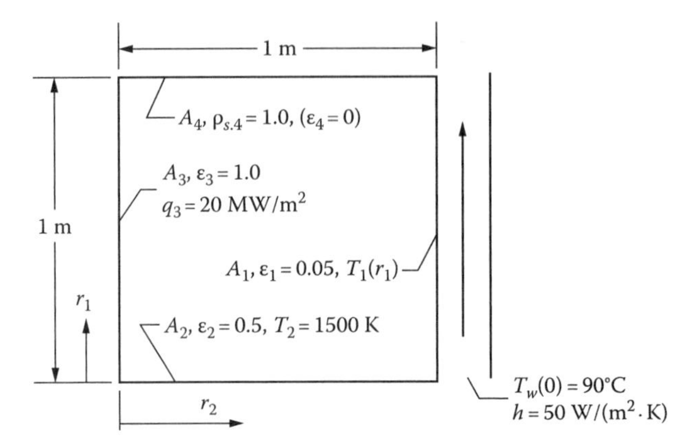

Find the temperature distribution $T_1(r_1)$ along surface 1.

7.39 A 2D problem is to be solved by the FEM in a rectangle that has a width significantly larger than its height. A rectangular finite element is to be used with unequal width and height so the elements will conveniently scale into the problem geometry. Linear shape functions are used, so the temperature distribution is represented as $T(x,y) = a + bx + cy + dxy$. Obtain the shape functions at each of the four nodes for the element and geometry shown.

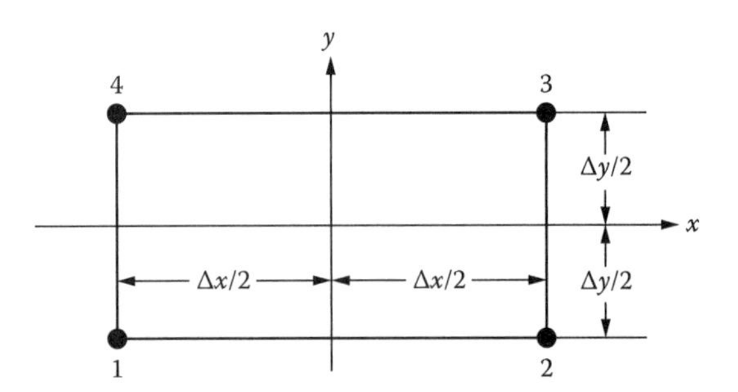

7.40 To provide a more accurate representation of the functional variation, the previous problem for a rectangular element can be extended to use quadratic shape functions. An intermediate node is inserted at the center of each side as shown in the figure. The interpolating polynomial is chosen to have the quadratic form, $T(x, y) = a + bx + cy + dxy + ex^2 + fy^2 + gx^2y + hxy^2$. Obtain the shape functions at the eight nodes. Make a 3D plot of the shape function for a corner node.

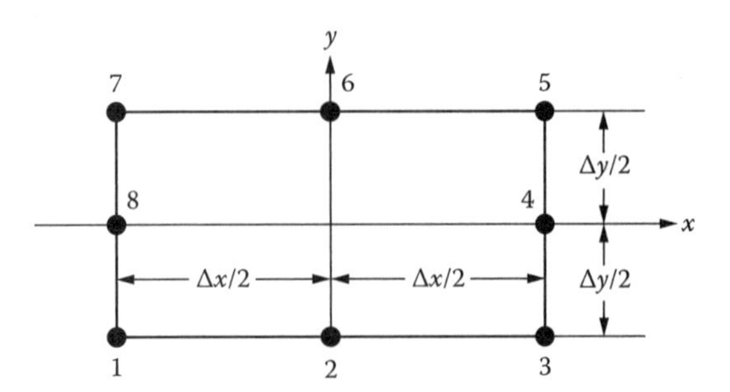

7.41 A Monte Carlo solution has been proposed for a 2D geometry. Rather than using the equations to choose (θ, ϕ) (Equations 7.58 and 7.59) for a diffuse surface, it is proposed to randomly choose the angle β in the plane normal to the two surfaces. This makes the geometric relations for such a problem much simpler.

Derive the relation between β and a random number R for this case.

Answer: $\beta = \sin^{-1}(2R - 1)$.

7.42 The hemispherical-spectral emissivity of a surface is approximated by $\varepsilon_\lambda = 0.157\lambda^3$ for $\lambda \le 5$ μm. Derive a closed-form relation between λ and a random number R for a surface at temperature T for energy emitted from the surface in the range $0 \le \lambda \le 5$ μm.

$$\left[\textit{Hint}: \int \frac{dx}{e^x - 1} = -x + \ln(e^x - 1) = \ln(1 - e^{-x})\right].$$

$$\textit{Answer}: \lambda = \frac{-C_2}{T} \frac{1}{\ln\left\{1 - \left[1 - \exp\left(\frac{-C_2}{5T}\right)\right]^R\right\}}$$

7.43 An area element dA_1 has directional emissivity given by $\varepsilon_1(\theta) = 0.659 \cos\theta$. For the geometry pictured below, find the fraction of energy leaving dA_1 that is absorbed by the black disk A_2. Use the Monte Carlo method. Compare your result with the analytical solution.

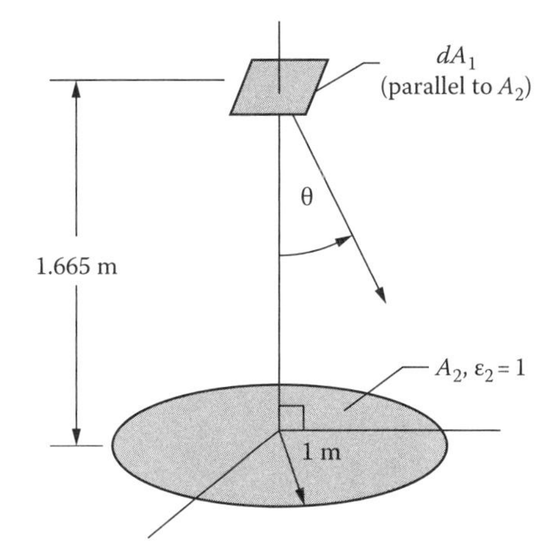

Answer: 0.3700.

7.44 Construct a computer flow chart for the Monte Carlo computation of the configuration factor F_{d1-2} from an area element to a perpendicular disk as shown in Example 4.4.

7.45 Construct a complete computer flow diagram for the Monte Carlo solution of the problem outlined in Homework Problem 5.35 for gray plates.

7.46 Construct a Monte Carlo computer flow diagram to obtain the specular exchange factor F_{1-2}^S in Homework Problem 6.24. (Assume it's an equilateral triangle.)

7.47 Program and solve Homework Problem 7.45 for $L = 1$, $\varepsilon_1 = \varepsilon_2 = 1$, and for $L = 1$, $\varepsilon_1 = \varepsilon_2 = 0.5$. By comparison of the two results, verify the result of Homework Problem 5.36b.

7.48 Obtain a Monte Carlo solution for the nongray heat transfer between infinite parallel plates computed in Example 7.2.

7.49 Obtain a Monte Carlo solution for the second part of Example 6.12 (i.e., for surface A_1 being a specular reflector).

7.50 For the configuration and conditions in Homework Problem 7.27, find $T_1(y)$ and $q(x)$ using a numerical solution, and provide plots of these quantities. Compare with the results for no conduction in the plates found in Homework Problems 5.42 and 5.43. Show grid independence for your results.

7.51 Repeat Homework Problem 7.50 using the Monte Carlo method, and compare the results with those of Homework Problem 7.50. Repeat for enough samples to show independence of the results to the number of samples used in the solution.

7.52 For the geometry shown, find and plot $T(x)$ and $T(\xi)$. Provide a numerical value for $T(x = 0.5\text{ m})$ and $T(\xi = 0.5\text{ m})$ and for $T(x = 0)$ and $T(\xi = 0)$. Use a fine enough grid to show an accurate profile (and reasonable grid independence). Also, show that energy conservation is obeyed. Surfaces 1 and 2 are gray and diffuse.

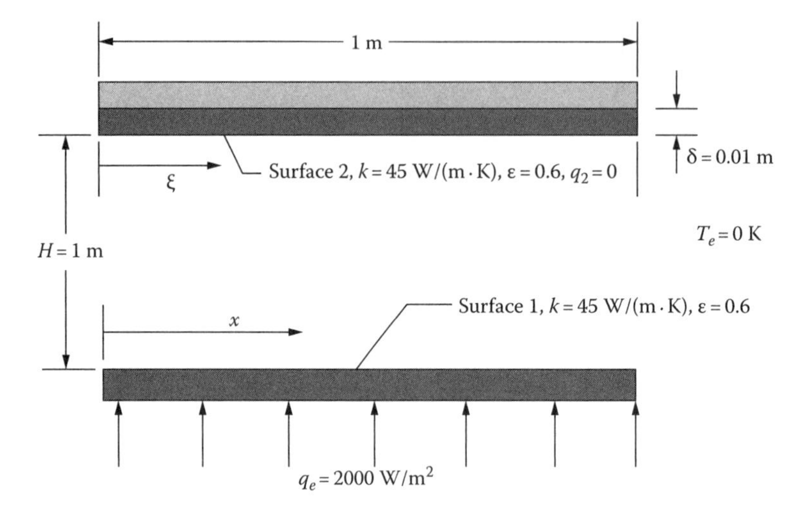

Answer: $T(x = 0, 1) = 500$ K; $T(x = 0.5) = 505.2$ K; $T(\xi = 0, 1) = 355.7$ K; $T(\xi = 0.5) = 360.0$ K.

7.53 For the geometry shown, use the Monte Carlo method to find and plot $T(x)$ and $T(\xi)$. Provide a numerical value for $T(x = 0.5\text{ m})$ and $T(\xi = 0.5\text{ m})$ and for $T(x = 0)$ and $T(\xi = 0)$. Use a fine enough grid and a sufficient number of samples to show an accurate profile (and reasonable grid independence). Also, show that energy conservation is obeyed. Surfaces 1 and 2 are gray and diffuse.

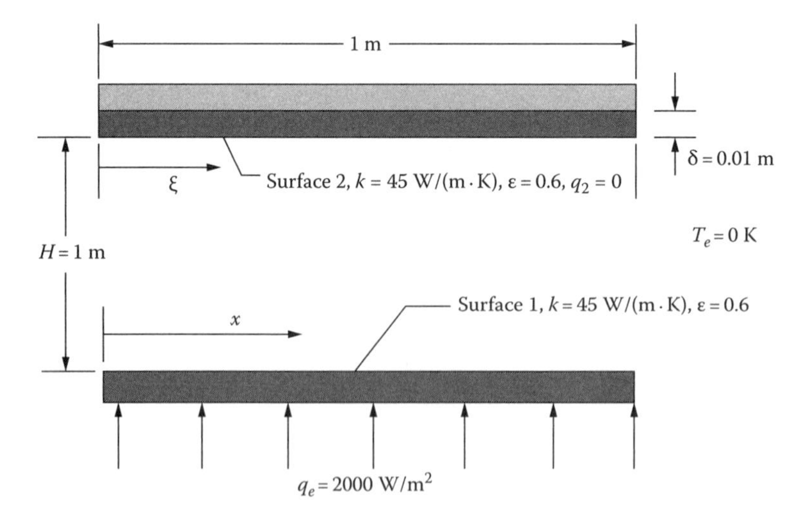

Answer: $T(x = 0, 1) = 500$ K; $T(x = 0.5) = 505.2$ K; $T(\xi = 0, 1) = 355.7$ K; $T(\xi = 0.5) = 360.0$ K.

7.54 A radiator is planned to provide heat rejection from a nuclear power plant that is to provide electrical power for a lunar outpost. The radiator will itself will be horizontal on the lunar surface, and condensing working fluid in the power cycle will maintain the surface temperature of the radiator at a uniform temperature of 800 K. The radiator is shielded from the nearby lunar outpost by a 1.5 m high vertical plate (see diagram). The width of the radiator is limited to 2 m. The radiator has a diffuse–gray emissivity of 0.92, while the shield has a diffuse–gray emissivity of 0.37. The shield is made of a material with thermal conductivity of k = 300 W/m·K and has a thickness of 2 cm. It is backed with a layer of very good insulation. It is expected that the radiator will be quite long. Solar flux of 1360 W/m^2 is incident on the radiator/shield system at an angle of 30° to the normal of the radiator.

(a) Set up the equations for finding the heat flux distribution on the radiator, $q_1(x)$, and the temperature distribution on the shield, $T_2(y)$. Note any assumptions.

(b) Solve the equations for $q_1(x)$ and $T_2(y)$. Compare your solution to the results for the nonconducting case with and without solar energy input.

(c) If the total heat rejection from the radiator is required to be 1 MW, what must be the length of the radiator (m)?

(d) If the radiator itself has a thickness of 1 cm and a thermal conductivity of 287 W/m·K, how would your solution change?

Show that your solution is grid independent and meets overall energy conservation.

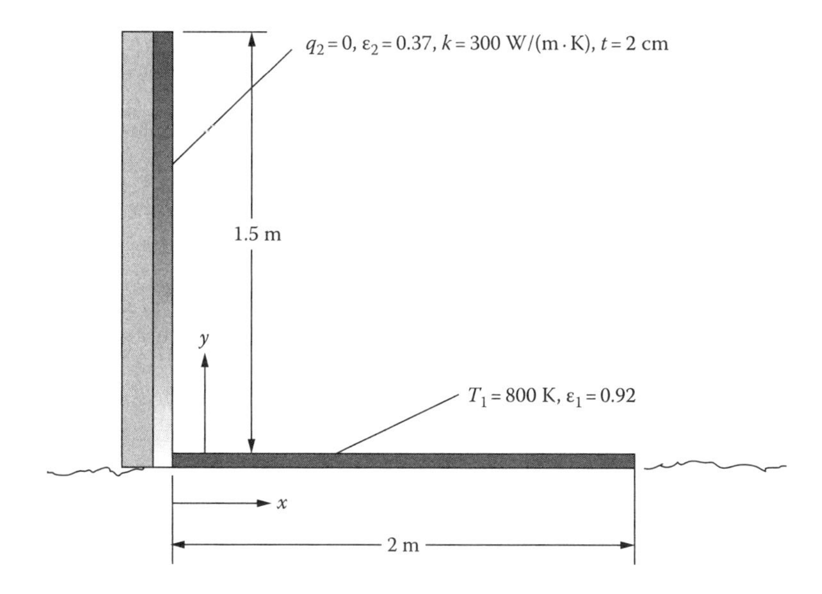

Answer: $q_1(x = 0) = 15.5$ kW/m^2, $q_1(x = 2\text{ m}) = 19.6$ kW/m^2, $T_2(y = 0) = 630$ K; $T_2(y = 1.5\text{ m}) = 588$ K; required length = 49.6 m.

7.55 For the geometry shown below with one conducting wall:

(a) Provide the final governing equations necessary for finding $T_1(x_1)$ and $q_2(x_2)$ in nondimensional form using appropriate nondimensional variables.

(b) Find the temperature distribution $T_1(x_1)$ and the heat flux distribution $q_2(x_2)$, and show them on appropriate graphs.

Boundary conditions are $q_1(x_1) = \left[100x_1 - 50x_1^2\right]$ kW/m^2 where x_1 is in meters, $T_2 = 500$ K, and $T_3 = T_4 = 300$ K. Properties for the diffuse surfaces are shown in the figure.

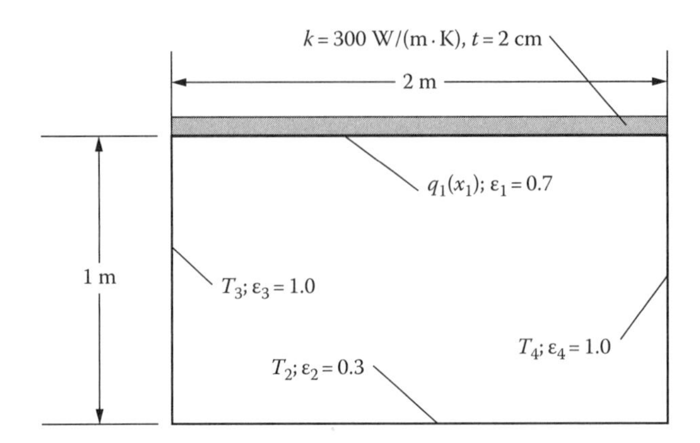

Show that your solution is grid independent and meets overall energy conservation. Compare your solution to the results for the nonconducting case.

Answer: $T_{1,\,max}$ = 1,117.7 K with no conduction and $T_{1,max}$ = 1,102.1 K with conduction and $q_{2,max}$ = ~ −10,100 W/m^2 with no conduction and $q_{2,max}$ = ~ −9,700 W/m^2 with conduction.

7.56 A microelectronic system is to be cooled convectively. The system is shown below.

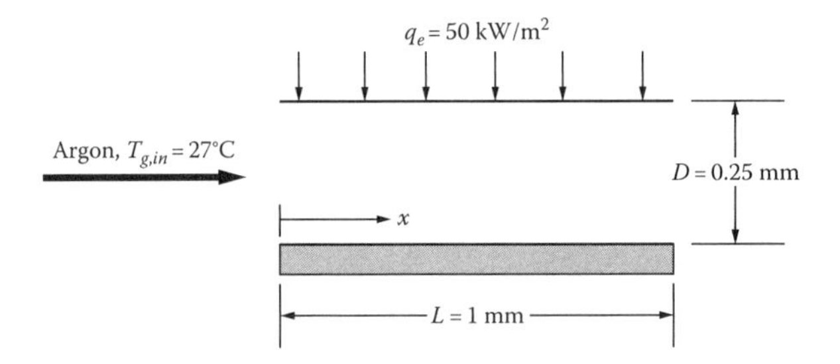

The flow channel is composed of very wide parallel plates separated by a distance D = 0.25 mm, and the channel length is L = 1 mm. Argon enters the channel from a large plenum at $T_{g,in}$ = 27°C. The mass flow rate of the argon is very high. The heat transfer coefficient between the argon and either channel wall is given by $h(x) = 20/[0.1+ (1000x)^{0.2}]$ (W/m^2·K) where x is in meters. The top thin channel wall is heated uniformly by electric heaters at a rate of 50 kW/m^2. The lower channel wall is carefully insulated. The hemispherical–spectral emissivity of the both channel walls is shown in the graph below.

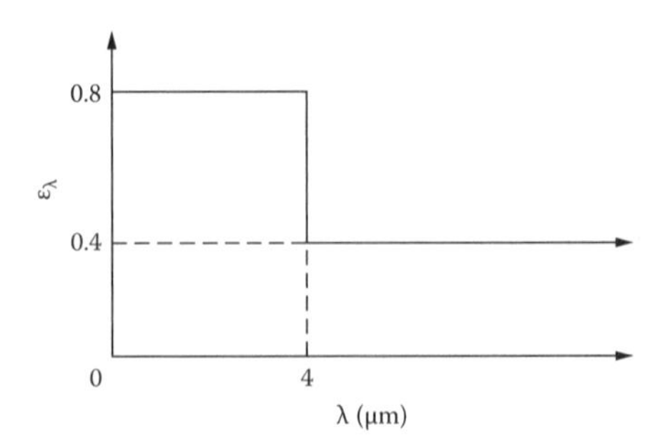

Provide plots of the temperature of each surface versus x, and show the grid independence of your results.

Answer: $T_{top}(x = 0)$ = 550 K, $T_{top,\,max}$ = 995 K, T_{top} $(x = 1)$ = 991 K, $T_{bottom}(x = 0)$ = 340 K, $T_{bottom,\,max}$ = 575 K, T_{bottom} $(x = 1)$ = 530 K.

8 Inverse Problems in Radiative Heat Transfer

8.1 INTRODUCTION TO INVERSE PROBLEMS

Experimental observations of temperature or heat flux profiles may not be available at the physical location where they are needed. Radiative property distributions in a participating medium must often be obtained from remote measurements. These situations belong to the mathematical class known as inverse problems. The solution of these problems is difficult, because the governing equations tend to be mathematically ill posed, and predicting conditions on the remote boundary can result in multiple solutions, physically unrealistic solutions, or solutions that oscillate in space and time. Various methods may be applied for overcoming the ill-posed nature of the governing equations. For problems dominated by conduction, there are texts and monographs available that demonstrate many of these methods (Tikhonov 1963, Alifanov 1994, Alifanov et al. 1995, Beck et al. 1995, Özişik and Orlande 2000). Phillips (1962) and Tikhonov are often credited with developing the first systematic treatment for these types of inverse conduction problems.

For determining radiative properties from remote measurements, various inverse techniques have been employed. This type of problem is required in analyzing flame radiation and is also closely related to problems in X-ray tomography. The radiative properties of porous media have been found using inverse algorithms by many researchers, and a comprehensive literature review is in the paper by Randrianalisoa et al. (2006) and Sacadura (2011). Inverse approaches are also used to characterize the properties of particles and agglomerates from light absorption and scattering techniques. These approaches were carried out by one us over the years as well (see Mengüç, 2011 for the summary of relevant research and the citations). In this chapter, however, we use inverse radiation methods for the design and control of systems.

Aside from determining conditions at an inaccessible boundary or radiative properties, another important class of inverse problems arises in the *design and control* of systems with radiant heat sources. In these problems, the designer specifies the desired output of the thermal system being designed; in most cases, this is a desired temperature and heat flux distribution over a product located on a *design surface*. The designer must then predict the necessary energy inputs to the thermal system that will produce the desired distributions over the design surface. The unknown inputs may be the required steady or transient temperature and energy input distributions to heaters or burners, heater or burner locations, or oven or furnace geometry. In these cases, two boundary conditions are being prescribed on the design surface, and the boundary conditions on the heater surface are to be determined.

For thermal systems dominated by radiative transfer, the problem is complicated because the thermal input at any location on the design surface may be affected by some or all radiant sources in the system, depending on the presence of blocking or shading. The mathematical form of the inverse solution is the same set of integral equations that are found for the forward problem when one boundary condition is set for each surface (Section 5.4). However, when the inverse problem is formulated, some of the equations in the set take the form of Fredholm integral equations of the first kind, which are notoriously ill conditioned (Wing 1991, Hansen 1998). In addition, the number of unknowns in the governing radiative exchange equations may be less than, equal to, or greater than the number of equations that describe the system. These factors imply that design and control of distributed radiative sources may be difficult, especially in problems where both a transient temperature and heat flux distribution are prescribed over the design surface and where other heat transfer modes play a role. The traditional design strategy has been to use trial-and-error solutions with guidance by experience to attain a viable solution. However, this can lead to suboptimal solutions and can be quite time consuming. Some problems are simply not amenable to trial-and-error approaches.

Reviews of inverse methods for design of radiative transfer systems (França et al. 2002, Daun et al. 2003a,b, Daun and Howell 2004) cover the work of many contributors. Various methods can be applied. The truncated singular value method is used in França et al. (2001), França and Howell (2006), and Mossi et al. (2007); the Tikhonov approach in Leduc et al. (2004a,b, 2006); conjugate gradient regularization (CGR) in Ertürk et al. (2001, 2002a–c, 2004), Hosseini-Sarvari et al. (2003a,c), and Pourshghaghy et al. (2006); optimization techniques in Federov et al. (1999), Daun et al. (2003c–e, 2005), Hosseini-Sarvari et al. (2003b), and Daun and Howell (2005); simulated annealing (SA) and Tabu search (TS) in Larsen and Porter (2005) and Porter et al. (2005, 2006); genetic algorithms in Li and Yang (1997), Suram et al. (2004), Kim and Baek (2004), Das et al. (2008), Safavinejad et al. (2005, 2009), Amiri et al. (2011), and Kamkari and Darvishvand (2014); artificial neural networks in Ertürk et al. (2002a) and Deiveegan et al. (2006); and swarm optimization in Yoon et al. (2013) and Yüksel (2013). Experimental validation through matching of inverse solution predictions with experimental results is in Gamba et al. (2002, 2003), Ertürk et al. (2007), and, at the nanoscale, Hajimirza and Howell (2014a).

Two- and three-dimensional (2 and 3D) radiant enclosures enclosing various object shapes are optimized in Safavinejad et al. (2005, 2009), Chopade (2012), and Kamkari and Darvishvand (2014) among others. Enclosures with diffuse-specular surfaces are considered in Safavinejad et al. (2009), Bayat (2010), Kamkari and Darvishvand (2014), and Rukolaine (2015).

8.1.1 Inverse Design and Data Analysis

Analytical techniques used for inverse design and control of radiative systems are similar to those used in inverse data analysis, property determination, and remote measurement, but there are significant differences in how the techniques are implemented. The latter types of problem should have a single solution, that is, we don't expect a material to have multiple properties that give the same measurements, nor multiple temperatures or heat fluxes at a remote boundary that give the same measurements at an accessible location. However, design problems may allow significantly wider tolerances in specification of acceptable results. In design, solving the inverse problem may produce multiple solutions that fall within the allowable tolerances but are very different in form.

Figure 8.1 shows a 2D enclosure. Consider an annealing furnace, where a billet of metal is to be heated to a specified temperature. Once the temperature history is prescribed, a first law energy balance on the billet imposes a second boundary condition, because the required heat flux is specified by the heat capacity, volume, density, and transient temperature trajectory of the billet. The designer may therefore prescribe this required radiative heat flux $q_1(x_1)$ and temperature distribution $T_1(x_1)$ for design surface 1 and seek to find the energy input distributions required on the heater (upper) surface 2 that will provide the desired result. As the allowable tolerances on $q_1(x_1)$ and $T_1(x_1)$ are relaxed, multiple allowable solutions for the heater temperature distribution $T_2(x_2)$ may be found. Multiple solutions are not acceptable in many data analyses, property determination, or remote sensing problems where the particular thermal input or property value that provides a measured signal is sought. For the designer, however, multiple acceptable solutions are desirable, as they allow the designer to choose among the solutions based on considerations such as smoothness and ease of implementation.

It is possible in design problems to specify conditions on the design surface for which no acceptable physical solution for $T_2(x_2)$ exists. The designer may specify design surface characteristics for $q_1(x_1)$ and $T_1(x_1)$ that cannot be obtained (at least within acceptable error limits around the desired distributions) by *any* distribution of heater settings. Just because the designer wants a particular outcome, there is no *a priori* guarantee that it can be obtained. This is in contrast to data analysis problems, where in most cases a feasible solution to the inverse problem is known to exist because some set of physical variables must have produced the observed experimental data.

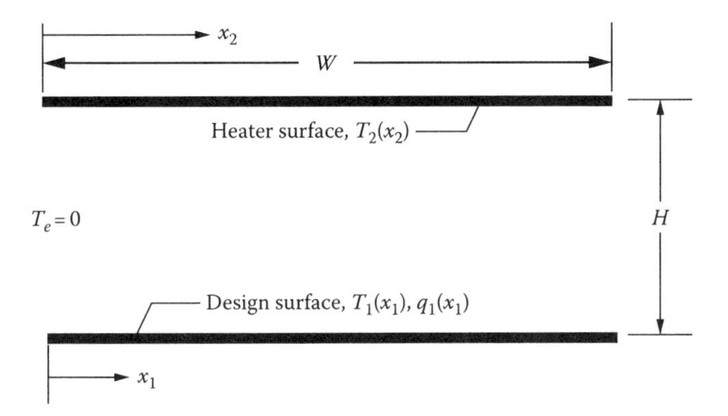

FIGURE 8.1 Simple radiant furnace with design surface 1 [$T_1(x_1)$ and $q_1(x_1)$ prescribed]; designer to find required $T_2(x_2)$.

Designing radiant systems involves inverse solution of integral equations rather than the differential equations that describe applications involving the other heat transfer modes. Since there are few radiation heat transfer problems where conduction and convection can be completely neglected, the governing energy relations are often highly nonlinear integral-differential equations as seen in Chapter 7.

8.1.1.1 Direct Inverse Solutions

A direct or explicit solution to inverse problems requires the use of an inverse formulation. Inverse problems are inherently *ill posed*, meaning that they may be extremely sensitive to the values of boundary conditions and may not possess a mathematical or physically reasonable solution. The corresponding discretized set of equations may be *ill conditioned*, in that the matrix of coefficients defining the solution may be singular or near singular. Ordinary techniques for solving the discretized set of integral equations (e.g., Gauss–Seidel, Gauss elimination or lower upper (LU) decomposition) are likely to either identify nonphysical solutions with high-amplitude fluctuations and/or imaginary absolute temperatures or may completely fail to find a solution.

Before considering direct inverse solution techniques, the ill-conditioned behavior of a system of equations of the type arising in radiative systems can be diagnosed by carrying out a singular value decomposition (SVD) (Hansen 1999) of the matrix of coefficients that arises. The SVD of an arbitrary MxN matrix $\mathbf{A}$ is $\mathbf{A} = \mathbf{USV}^T$, where $\mathbf{U}$ and $\mathbf{V}$ are orthogonal matrices and $\mathbf{S}$ is the diagonal matrix of singular values, and the elements $S_{i,i}$ of the $\mathbf{S}$ matrix are arranged so that $S_{1,1} > S_{2,2} > \ldots > S_{N,N} \geq 0$. The inverse of $\mathbf{A}$ is then given by

$$\mathbf{A}^{-1} = \mathbf{VS'U}^T \tag{8.1}$$

where $S'_{i,i} = 1/S_{i,i}$. If the *condition number* of this matrix ($S_{1,1}/S_{N,N}$) is large, small singular values dominate the inverse matrix and the solution becomes unstable. If $\mathbf{A}$ is rank deficient (which can arise when the number of unknowns and equations is not equal, as often happens in inverse problems), some of the singular values equal zero and conventional matrix inversion methods fail.

Example 8.1

For the geometry in Figure 8.1, set $h = H/W = 0.5$, and let $X = x/W$. Discretize black surfaces 1 and 2 into 30 discrete elements each, and let design surface 1 have uniform dimensionless emissive power $\vartheta_{b1}^4(X_1) = T_1^4/T_{ref}^4 = 1$ with net dimensionless radiative heat flux distribution of $\tilde{q}_1(X_1) = q_1(X_1)/\sigma T_{ref}^4 = 16X_1^2 - 16X_1 - 6$. Using direct matrix inversion, find the required net radiative heat flux distribution on the upper surface, $q_2(x_2)$. Also, plot the singular values for the matrix used in the solution, and determine the condition number.

The governing relation for the dimensionless emissive power distribution on surface 2, $\vartheta_{b2}^4(X_2)$, is then (Equation 5.56)

$$\int_{X_2=0}^{1} \vartheta_{b2}^4(X_2)dF_{dX_1-dX_2} = \sum_{i=1}^{30} \vartheta_{b2,j}^4 dF_{i-j} = \left[1-\tilde{q}_1(X_1)\right] = \left[1-\tilde{q}_{1,i}\right] \quad (8.1.1)$$

The $dF_{i-j} = dF_{dX1-dX2}$ is the configuration factor between two parallel differential strip elements on the surfaces (Example 4.2). Equation 8.1.1 can be written in discretized form as

$$\sum_{j=1}^{m} A_{i,j}\vartheta_{b2,j}^4 = b_i \quad (8.1.2)$$

where

$$A_{i,j} = dF_{dX2-dX1} = \frac{h^2}{2\left\{h^2 + \left[X_{2,i} - X_{1,j}\right]^2\right\}^{3/2}}$$

$$b_j = \left(1-\tilde{q}_{1,j}\right) = \left(7 - 16X_{1,j}^2 + 16X_{1,j}\right),$$

and

$$X_{2,i} = \left[i - \left(\frac{\Delta X_2}{2}\right)\right]\Delta X_2; \quad X_{1,j} = \left[j - \left(\frac{\Delta X_1}{2}\right)\right]\Delta X_1.$$

The resulting linear matrix equation $A\vartheta_{b2}^4 = b$ now needs to be solved. Note that in Equation 8.1.1, the right-hand side is a known function, so this is a Fredholm integral equation of the first kind, well known to have unfortunate characteristics (Wing 1991, Hansen 1998). Using a direct matrix inversion routine (in this case, using Mathcad, which employs a simplex solver) to obtain $\vartheta_{b2}^4 = A^{-1}b$ gives the result shown in Figure 8.2.

If the values shown in Figure 8.2 are introduced as knowns in Equation 8.1.2 using $\vartheta_{b1}^4 = 1$ as the boundary condition in the forward problem to calculate values of $\tilde{q}_1(X_1)$, exact agreement is found with the prescribed $\tilde{q}_1(X_1)$. Thus, the solution shown in Figure 8.2 is in a sense correct: it solves the given problem exactly. However, negative values of dimensionless emissive power are predicted, implying imaginary absolute temperatures on the heater; additionally, the solution is expected on physical grounds to be symmetrical around $j = 15$. Therefore, although the solution in Figure 8.2 mathematically satisfies the problem formulation, it is not a useful physical solution. The reason for this can be seen by viewing the 30 singular values of the **A** matrix for this problem. MATHCAD returns the values shown in Figure 8.3. The singular values range from $S_{1,1} = 0.63$ at $i = 1$ to $S_{30,30} = 1.055 \times 10^{-17}$ at $i = 30$, giving a matrix condition number of $S_{1,1}/S_{30,30} = 6.0 \times 10^{17}$. Such a large condition number implies difficulty in matrix inversion and/or large errors in the solution. Because elements in the inverse of **A** contain the reciprocals of the singular values (Equation 8.1), it is easy to see why direct inversion fails. Alternative solution techniques are necessary. To achieve an accurate and physically useful

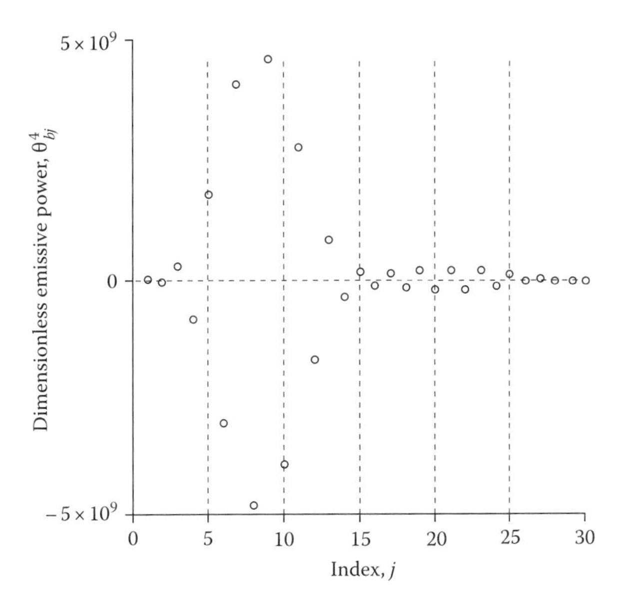

FIGURE 8.2 Predicted dimensionless emissive power of heater surface 2 using direct matrix inversion.

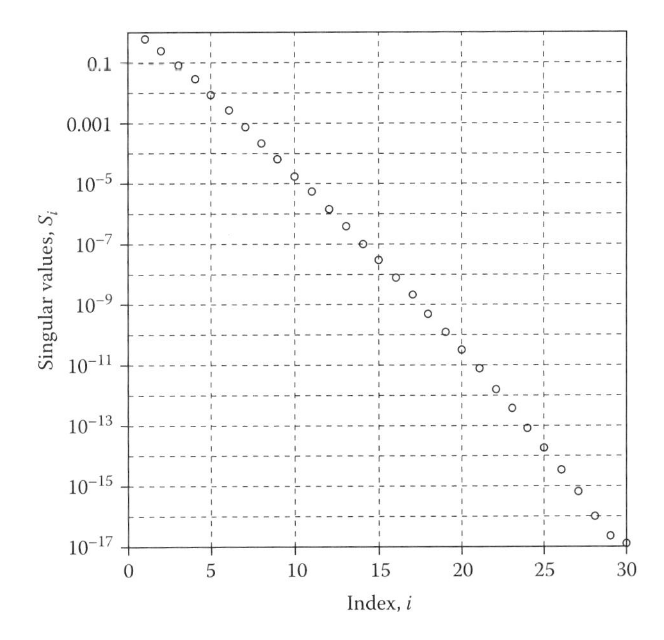

FIGURE 8.3 Singular values of the **A** matrix in Example 8.1.

solution, the explicit system must instead be *regularized* by modifying the ill-conditioned system of equations, or some other approach in formulation must be used. The modified solution will be subject to some error because some information has been deleted in the regularization process, and the level of error must be selected so that the accuracy of the solution satisfies the designer's needs.

In contrast to most finite difference solutions, if the number of increments N on the surfaces is increased, the condition number of the **A** matrix increases and the dimensionless emissive power becomes even more erratic than shown in Figure 8.1.

8.2 GENERAL INVERSE SOLUTION METHODS

For a general transient system, the discretized energy equation for an element on the design surface i (where the desired distributions of T and q are specified) can be written in terms of the emissive power $E = \varepsilon\sigma T^4$ of surface elements elsewhere in the enclosure. Assuming the elements are numbered so that the first N_{DS} elements are on the design surface, at any time t (see Equation 7.2),

$$\rho_i c_{pi}\delta_i A_i \frac{dT_i(t)}{dt} = \left[q_{cond,i}(t) + q_{conv,i}(t)\right]A_i - (J_i - G_i)A_i$$

$$= \left[q_{cond,i}(t) + q_{conv,i}(t)\right]A_i + \sum_{j=1}^{N_{DS}} J_j(t)A_jF_{j\to i} + \sum_{j=N_{DS}+1}^{N} J_j(t)A_jF_{j\to i} - J_i(t)A_i \tag{8.2}$$

Applying reciprocity and rearranging, this becomes

$$\sum_{j=N_{DS}+1}^{N} J_j(t)F_{i\to j} = \rho_i c_{pi}\delta_i \frac{dT_i(t)}{dt} + J_i(t) - \left[q_{cond,i}(t) + q_{conv,i}(t)\right] - \sum_{j=1}^{N_{DS}} J_j(t)F_{i\to j}. \tag{8.3}$$

(The effects of a participating medium between the surfaces are neglected.)

All the specified terms on the right-hand side of Equation 8.3 are for the design surface elements and are therefore known (note that from Equation 5.55, the J_j terms on the design surface are known when T_j and q_j are prescribed); the terms on the left-hand side are for the heaters or other energy sources and are to be determined. Equation 8.3 is thus a discretized version of an ill-posed Fredholm integral equation of the first kind and a generalization of the relation found in Example 8.1. The various techniques for solving design problems of this type can be grouped under the three broad headings of *regularization*, *optimization*, and *metaheuristics*. Vogel (2002) gives numerical techniques for implementing many of these approaches.

The regularization techniques consist of modifying the explicit governing relations to reduce their "ill posedness," accepting some loss of accuracy to gain a useful solution.

Optimization techniques approach the design problem by casting the governing relations in the conventional form with one boundary condition on the design surface fixed and an assumed condition (most often a heat flux distribution) on the heater surface. The assumed condition is varied in a systematic way until the second boundary condition on the design surface is satisfied within acceptable limits.

Metaheuristics is a class of techniques that search for an optimal solution from among a very large set of discrete solutions.

8.2.1 Regularization

A *direct* or *explicit* solution to the inverse design problem requires use of an inverse formulation. As found in Example 8.1, inverse design problems are inherently ill posed, and the corresponding discretized set of equations is ill conditioned. As illustrated in the simple example of Figures 8.1 through 8.3, ordinary techniques (e.g., Gauss–Seidel, Gauss elimination or LU decomposition) are likely to either identify nonphysical solutions with large amplitude fluctuations and/or complex absolute temperatures or completely fail to find a solution.

To achieve an accurate and reasonable solution, the explicit system may be *regularized* by modifying the ill-conditioned system of equations to a nearly equivalent set of well-conditioned equations. The solution is then subject to some error and the level of regularization must be selected so that the accuracy of the solution satisfies the designer's needs. Widely used regularization techniques include *truncated singular value decomposition* (TSVD), *conjugate gradient regularization* (CGR), and *Tikhonov regularization* (TR).

8.2.1.1 Truncated Singular Value Decomposition

This method is based on modifying the SVD of **A**. The solution uses the pseudoinverse matrix that is formed by filtering or truncating small singular values, thus reducing the condition number of the matrix **A**. The solution to $\mathbf{A}x = \mathbf{b}$ using the p largest singular values becomes

$$x_n = \sum_{k=1}^{p} V_{n,k} \frac{b_m U_{m,k}}{S_{k,k}}, \quad n = 1,2,\ldots,N, \tag{8.4}$$

where p has a value less than or equal to the rank of **A** (Hansen 1998). Retaining different numbers of singular values yields alternative solutions. Those with acceptable accuracy provide allowable alternatives. For the case analyzed in Example Problem 8.1 with singular values shown in Figure 8.3, the solution can be regularized by retaining some small subset of the 30 singular values. The predicted heater emissive power distributions found by retaining 1, 3, or 7 singular values are shown in Figure 8.4, illustrating that multiple physically reasonable solutions may be generated. The emissive power distribution for p = 7 requires some negative $\vartheta_{b2,j}^4$ values (which cannot be realized in a physical system), and the results for larger p values become even more unrealistic. The predicted accuracy of the three solutions in predicting the required $\tilde{q}_1(X_1)$ is compared in Figure 8.5.

The solution shown in Figure 8.2, which results from retaining all 30 singular values of the matrix **A**, will produce a result that exactly matches the desired $\tilde{q}_{1,j}$ curve, but as mentioned in the example, requires heater inputs that cannot be physically realized. The RMS error in the solutions is $\left\|\tilde{q}_{1,i,given} - \tilde{q}_{1,i,calculated}\right\|_2$ and has useable values for retained singular values p of 1, 3, and 7 of 0.163, 0.0884, and 0.0397. Thus, the error becomes smaller as more singular values are retained, even though the results tend to become more nonphysical.

8.2.1.2 Conjugate Gradient Regularization

Generally, CGR is based on the incomplete conjugate gradient (CG) minimization of the degenerate functional $F(x) = [\mathbf{A}(x_{exact} - x)] \cdot (x_{exact} - x)]$, which is minimized when $\|\mathbf{A}x - \mathbf{b}\| = 0$. For inverse problems, each minimization step corresponds to a unique solution having distinct accuracy and smoothness characteristics. Small memory requirements, computation economy, robust

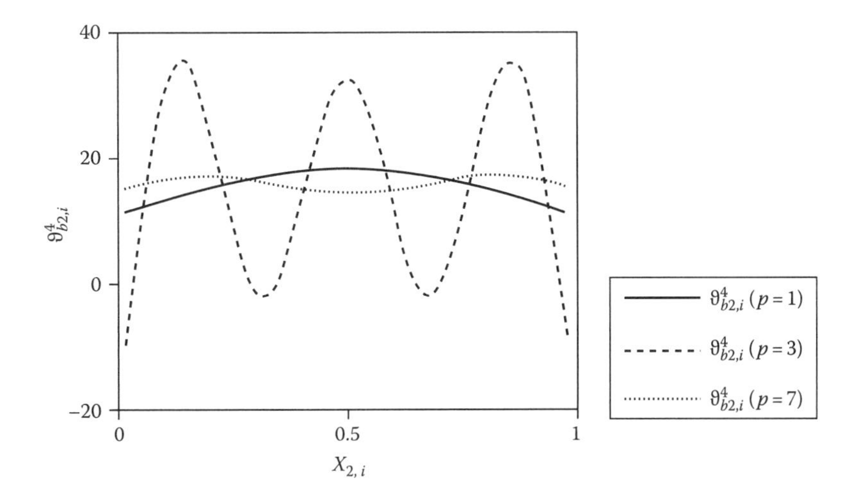

FIGURE 8.4 Regularized TSVD solutions to Example 8.1.

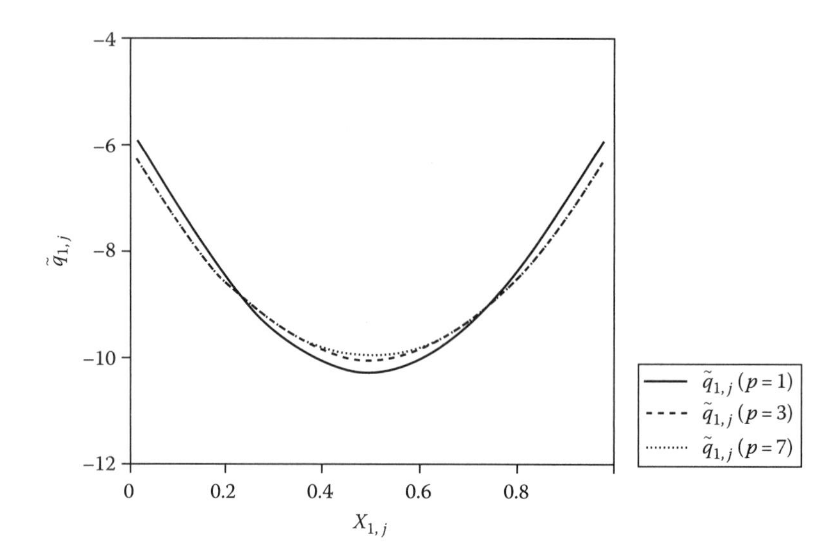

FIGURE 8.5 Predicted heat flux on surface 1 using various TSVD solutions.

convergence characteristics, and the ability to store and use the original matrix often make CGR the method of choice for regularization of large systems of equations (Beckman 1960). Shokouhi et al. (2015) used a CG solution coupled with the finite volume method to determine the heat flux distribution on a heater to achieve defined temperature and total heat flux distribution on the opposite side of a parallel plate channel in the presence of turbulent flow.

8.2.1.3 Tikhonov Regularization

Like CGR, TR is based on minimizing a functional. In this method, the functional is equal to the L_2 norm of the residual vector plus an added constraint on the shape of the solution. The constraint is generally a derivative operator.

$$F(x) = \|\mathbf{A}x - \mathbf{b}\|_2 + \sum_{i=1}^{p} \alpha_i^2 \left[\left\|\mathbf{L}_i (x - x_0)\right\|\right]^2 \tag{8.5}$$

For a pth order scheme, $\mathbf{L}_i$ approximates the discretized ith derivative operator and α_i is the ith order regularization parameter. Using a small regularization parameter results in an accurate solution by emphasizing minimization of the residual norm, but as we have seen, minimization of the residual norm is usually accompanied by widely fluctuating solutions. Using a large regularization parameter, on the other hand, results in a solution with improved smoothness. For the standard or zeroth order TR scheme, $p = 0$ and $\mathbf{L}_i$ becomes the identity matrix $\mathbf{I}$, leading to the set of linear equations $(\mathbf{A}^T\mathbf{A} + \alpha_0^2\mathbf{I})x = \mathbf{A}^T\mathbf{b} + \alpha_0^2\mathbf{I}x_0$. Using the correct regularization parameter results in an optimal solution that is both smooth and sufficiently accurate.

8.2.2 Optimization

Optimization techniques approach the design problem by casting the governing relations in the conventional form with one boundary condition on the design surface fixed and an assumed condition (most often a heat flux distribution) on the heater surface.

Like conventional trial-and-error design, optimization techniques use an iterative process to arrive at the final design configuration. However, in optimization, the assumed solution sequence is adjusted in a systematic way, and the performance of a particular configuration (e.g., IR [infrared] heater power distribution) is evaluated at each iteration. If it does not satisfy the design requirements, the configuration is modified and checked again. This process is repeated until a satisfactory design configuration is identified.

The efficiency of this process and the quality of the final design depend on how much the design performance improves at each iteration. While the trial-and-error technique relies solely on the designer's intuition and experience to improve the design, optimization techniques modify the design configuration systematically, based on sensitivity information and numerical algorithms that maximize the improvement between successive iterations. Consequently, optimization techniques require far fewer iterations than the trial-and-error approach, and the final solution is usually near optimal.

In the design optimization methodology, the design problem is converted into a multivariate minimization problem by first defining an *objective function*, $F(\Phi)$, which quantifies the "goodness" of a particular design in such a way that the minimum of $F(\Phi)$ corresponds to the desired design outcome. For example, the objective function for the examples of this chapter might be defined as the RMS difference between the predicted and required heat flux on the design surface, $F(\Phi) = \left\| \tilde{q}_{1,i,required} - \tilde{q}_{1,i,predicted} \right\|_2$. The objective function is dependent on a set of *design parameters* contained in the vector Φ, which specifies the design configuration. Once these quantities are defined, the design can be optimized using a multivariate minimization algorithm to find the vector Φ^* such that $F(\Phi^*) = Min[F(\Phi)]$. Optimization methods differ in how $F(\Phi)$ is minimized and can be broadly classed as either *deterministic* or *metaheuristic*; deterministic methods work by minimizing $F(\Phi)$ systematically by changing Φ at each step based on the local topography of the objective function, while in metaheuristic algorithms, Φ^* is selected from a large set of candidate solutions generated by a random process. One of the most common optimization methods is the quasi-Newton method, belonging to the deterministic class of methods.

8.2.2.1 Deterministic (Quasi-Newton) Approach

The quasi-Newton method is a gradient-based approach, where the set of design parameters is adjusted systematically at each iteration based on the local objective function curvature; at the kth iteration, the new set of design parameters is found by

$$\Phi^{k+1} = \Phi^k + \alpha^k \boldsymbol{p}^k, \tag{8.6}$$

where

α^k is the step size
$\boldsymbol{p}^k$ is the search direction

The performance of gradient-based approaches is based largely on the choice of search direction (this is, in fact, how they are named). In Newton's method, the search direction at the kth iteration is set equal to Newton's direction, which is found by solving

$$\nabla^2 F\left(\Phi^k\right)\boldsymbol{p}^k = -\nabla F\left(\Phi^k\right), \tag{8.7}$$

where

$\nabla F(\Phi^k)$ is the gradient vector
$\nabla^2 F(\Phi^k)$ is called the Hessian matrix

These contain the first- and second-order objective function sensitivities, respectively. Newton's method usually requires the fewest iterations to minimize the objective function and reaches Φ^* in exactly one step if $F(\Phi)$ is quadratic. Nevertheless, Newton's method is not always the most efficient due to the computational effort required to calculate the Hessian matrix; this is particularly true for large problems where the second-order objective function sensitivities are expensive to calculate. In such cases, it is usually more computationally efficient to use the *quasi-Newton method*. In this approach, the search direction is found by solving

$$\mathbf{B}^k p^k = -\nabla F\left(\Phi^k\right), \tag{8.8}$$

where $\mathbf{B}^k$ approximates the Hessian matrix. Most often, $\mathbf{B}^0$ is set equal to the identity matrix and is updated at each subsequent iteration based on values of $F(\Phi^k)$ and $\nabla F(\Phi^k)$ from previous iterations to improve the approximation.

Various update schemes are described in Kiefer and Wolfowitz (1952), Kushner and Clark (1978), and Bertsekas (1999). Rukolaine (2007) suggests application of level set methods as an efficient technique for representing the surface to be optimized (Burger and Osher 2005, Dorn and Lesselier 2006).

8.2.3 Metaheuristic Approaches

Metaheuristic optimization is an alternative to optimization or regularization in radiation-dominated heat transfer design problems. When all design variables are continuous, regularization and optimization methods are preferred due to the reduced number of iterations required. However, when designs require optimization of a set of discrete variables, the complexity of the problem precludes gradient information, or regularization results in infeasible solutions, metaheuristics are a reliable alternative. Metaheuristic approaches include *simulated annealing* (SA) (Kirkpatrick et al. 1983, Corana et al. 1987, Van Laarhove et al. 1992, Goffe et al. 1994, Chiang and Russell 1996, Porter et al. 2005, 2006), *genetic algorithms* (Li and Yang 1997, Kim and Baek 2004, Suram et al. 2004), *neural networks* (Erturk et al. 2002a, Deivegan et al. 2006), and a more recent method, *tabu search* (TS) (Glover 1989, Barnes and Chambers 1995, Porter et al. 2006).

8.2.3.1 Simulated Annealing

As with deterministic, gradient-based techniques, the SA approach finds Φ^* iteratively, starting from an initial guess Φ^0. The algorithm used for probabilistic acceptance of solutions in SA was developed by Metropolis et al. (1953) as a model for annealing processes in metals and later adapted by Kirkpatrick et al. (1983) and Glover (1986) to be used in metaheuristic optimization.

The algorithm begins by evaluating the objective function for an initial solution that has been randomly selected or chosen by intuition. Subsequent solutions are found by randomly perturbing variables and repeating the objective function calculation. If the new solution is better than the previous one, the new solution automatically replaces the previous solution. If the new solution is worse than the previous solution, a random number is chosen between 0 and 1 and compared with the probability of accepting a worse solution. If the random number is less than this probability, the new (worse) solution is accepted; if not, it is discarded. Occasionally accepting a worse solution allows the progression the chance to escape from local minima in the objective function. After a defined number of new solutions, Φ, have been evaluated, the probability of accepting a worse design is lowered as the algorithm proceeds.

An advantage of SA is that it does not require the calculation of objective function derivatives. Gradient-based methods may converge to local minima faster than metaheuristic methods, since the former class of methods updates approximations of Φ^* based on the objective function topography, while the latter class of methods instead relies on random perturbations. A drawback of all metaheuristic methods is that the heuristic search parameters (initial annealing temperature, rate of annealing temperature decrease, and termination criteria) are not generally apparent at the onset but instead must be discovered by experience accrued from repeated experiment, hence the name *metaheuristic*.

8.2.3.2 Tabu Search

Tabu search (TS) generates a candidate list of solutions at each iteration of the TS optimization. The next solution is forced to be within this list, so there is no mechanism for solution rejection as in SA. Worse designs can be accepted only if worse solutions are present in the candidate list. Populating the candidate list with good solutions becomes a key factor to the success of the algorithm. An intelligent search is needed that deterministically searches the topology of the solution space. This intelligent search is based on the formulation of an intelligent dynamic neighborhood selection strategy. Using *dynamic neighborhood selection* allows for several different neighborhoods to be utilized during the search.

In applying the TS algorithm for radiative transfer problems, a search neighborhood is automatically embedded in the problem formulation in the form of configuration factors. Given an elemental surface area of interest, configuration factors can give an ordinal rank to every other elemental surface area in the problem. It is immediately known which surfaces most strongly influence the surface of interest. This feature of radiation heat transfer can be used to form an intelligent search neighborhood, greatly speeding the search algorithm.

8.2.3.3 Genetic Algorithm

This algorithm, one of the class of evolutionary algorithms, has been successfully applied to radiative transfer problems (Kim and Baek 2004, Safavinejad et al. 2005, 2009, Deiveegan et al. 2006, Darvishvand et al. 2015). A set of initial randomly distributed solutions is generated, and a subset of these is selected using various processes (selection, crossover, and mutation) to breed a new solution set. The new solution sets are compared to see which best approach a minimum objective function, which is based on the design criteria. Those solution sets that best fit are again bred, and this process is continued until a solution is reached that satisfactorily meets the design criteria.

8.3 COMPARISON OF METHODS FOR A PARTICULAR PROBLEM

A design problem that incorporates many attributes of a practical design process for a radiant furnace was proposed and solved by a diverse team to compare the various solution methods (Daun et al. 2006), and some of those results are used here to illustrate the methods. The problem was to determine the radiant heater settings that provide a prescribed transient but spatially uniform heating of the design surface in a 3D enclosure. The design surface is inset slightly from the edges of the bottom enclosure surface.

The design surface (6a) located on the bottom of the enclosure has dimensions of 0.75 × 1.75 m (i.e., it is set in 0.125 m from each boundary). The remainder of the bottom surface around the design surface is a diffuse adiabatic surface with the same thermal properties as the sides. There is no thermal exchange between the design surface and the rest of the bottom surface (i.e., infinite contact resistance). Figure 8.6a shows the enclosure geometry and Figure 8.6b the computational domain, which was formed by inserting imaginary surfaces that are perfect specular reflectors along the lines of symmetry in the original enclosure. Two enclosure geometries were considered with height-to-width ratios of $H/W = 0.5$ m (as shown) and $H/W = 0.25$ m. Table 8.1 gives the enclosure properties that were used in the analyses. Surface 6a is representative of a steel alloy and the remaining surfaces of a refractory brick with a glazed surface that is partially specularly reflecting. All boundaries are assumed to be lumped capacitances, that is, no temperature profiles normal to the boundaries are included.

The computational domain was discretized into a total of 112 equal-area surface elements. (No surface elements are located on the specularly reflecting symmetry surfaces.) However, 16 elements are located on surface 3, while surfaces 4, 5, and 6 each have 32 elements. The design surface has a spatially uniform linearly increasing temperature that increases from 300 o 500 K over a period of 5 h.

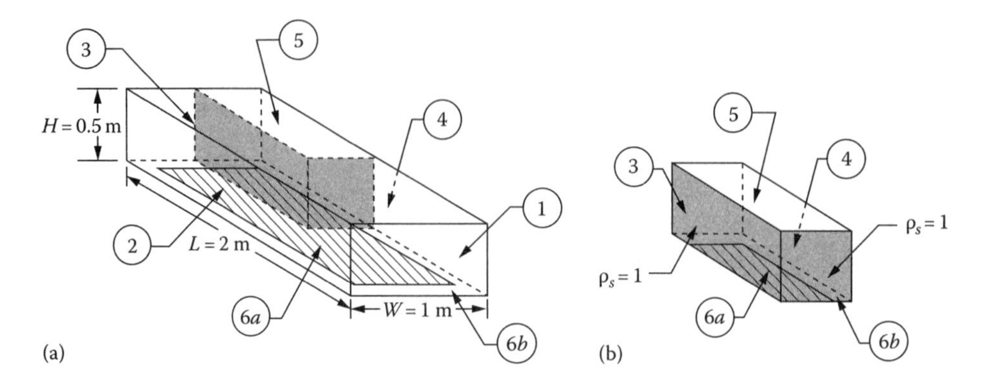

FIGURE 8.6 Basic dimensions of 3D enclosure investigated. (a) enclosure geometry, (b) computational domain. (From Daun, K.J. et al., *JHT*, 128(3), 269, March 2006.)

TABLE 8.1
Enclosure Properties and Specifications

Surface	1 Brick	2 Brick	3 Brick	4 Brick	5 Brick	6a Steel	6b Brick
Emissivity, ε	0.3	0.3	0.3	0.3	0.9	0.6	0.3
Diffuse reflectivity, ρ_d	0.2	0.2	0.2	0.2	0.1	0.3	0.2
Specular reflectivity, ρ_s	0.5	0.5	0.5	0.5	0.0	0.1	0.5
k [W/(m·K)]	1.0	1.0	1.0	1.0	1.0	63.9	1.0
ρ [kg/m^3]	2645	2645	2645	2645	2645	7832	2645
c [J/(kg·K)]	960	960	960	960	960	487	960
δ [m]	0.1	0.1	0.1	0.1	0.1	0.02	0.1

At any time t throughout the process, the energy applied to a given surface element i is given by writing Equation 8.3 in terms of exchange factors (Section 5.3.2) to allow treatment of the specular-diffuse surfaces:

$$(\rho c\delta)_i A_i \frac{dT_i(t)}{dt} = \sum_{j=1}^{N} E_j(t) A_j \Gamma_{j\to i} - E_i(t) A_i + q_{i,cond}\delta(t) \tag{8.9}$$

where $E = \varepsilon\sigma T^4$ is the emissive power of a given element and $\Gamma_{i\to j}$ is the exchange factor, defined as the fraction of energy emitted by surface i that is absorbed by surface j, considering all possible paths of reflection among intervening surfaces. The exchange factor set was found using the Monte Carlo method to accommodate partially specular reflective surfaces. The accuracy of the exchange factor set was enhanced using the smoothing algorithm presented by Daun et al. (2005) outlined in Section 4.4, and the same set of factors was used in all solutions. In Equation 8.9, $q_{i,\,cond}$ accounts for the conductive heat exchanged between surface i and the surrounding elements; convection heat transfer was neglected.

Equation 8.9 is placed in dimensionless form by using the reciprocity relation $\varepsilon_i A_i \Gamma_{i\to j} = \varepsilon_j A_j \Gamma_{j\to i}$:

$$C_i \frac{\partial \vartheta_i(\tilde{t})}{d\tilde{t}} = \sum_{j=1}^{N} \vartheta_j^4(\tilde{t}) \Gamma_{i\to j} - \vartheta_i^4(\tilde{t}) + \frac{\Psi_{i,cond}}{\varepsilon_i} \tag{8.10}$$

where

$$\tilde{t} = \frac{\varepsilon_{DS}\sigma T_{ref}^3}{(\rho c \delta)_{DS}} t; \quad C_i = \left(\frac{\varepsilon_{DS}}{\varepsilon_i}\right)\frac{(\rho c\,\delta)_i}{(\rho c\,\delta)_{DS}}; \quad \vartheta(\tilde{t}) = \frac{T(\tilde{t})}{T_{ref}} \quad \text{and} \quad \psi_i = \frac{q_i \delta}{\sigma T_{ref}^4 A_i}.$$

8.3.1 Solution by Direct Inversion

8.3.1.1 TSVD Solution Method

At $\tilde{t} = 0$ (time step $k = 0$), the dimensionless temperatures on the design surface and wall elements are all at $\vartheta(k = 0)$. A discretized form of Equation 8.10 is applied to find $\vartheta_{HS,j}^4(k = 0)$ for the heater elements. Writing Equation 8.10 for each design surface element i leads to a system having 21 equations and 32 unknowns (the emissive powers in the 32 heating elements). The TSVD method was applied, and the coefficient matrix $\mathbf{A}^{m\times n}$ ($m = 21$, $n = 32$) was decomposed to find the singular values. Since the number of columns of matrix $\mathbf{A}$ is 32, there are 32 singular values. The TSVD is applied so that only the terms related to the largest p ($p \leq 21$) singular values are kept in the solution. The solution provides the emissive powers of the heating elements at time step k, $\vartheta_{HS,j}^{4(k)}$: the radiative flux on each element is then found. These are used to compute the change in temperature of each element using the discretized form of Equation 8.10. The process is repeated at each time step.

8.3.1.2 Tikhonov Solution Method

The Tikhonov method followed the procedure of the TSVD method in deriving the ill-posed equations for elements on the design surface and found the unknown temperatures of the heaters through Equation 8.3. These were then used to find the energy input to the heaters. Both solutions (TSVD and Tikhonov) found the identical set of singular values. For the case of $H/W = 0.25$, the singular values were large enough, and the condition number of the coefficient matrix was small enough, that direct inversion was possible without truncation in TSVD and without use of a smoothing function (i.e., $\lambda_0 = 0$) in the Tikhonov solution.

For the case of $H/W = 0.5$, truncation was necessary in TSVD, and only 3 of 21 singular values were retained to obtain a solution with all positive energy inputs to the heaters. For the Tikhonov solution, a first-order Tikhonov smoothing function with various values of the parameter λ were computed. The appropriate regularization parameter λ needs to be chosen, and after some experimentation, setting $\lambda = 0.2$ provided the best results.

8.3.1.3 CGR Solution

The CGR method followed the procedure used by the TSVD and Tikhonov methods in deriving the equations for elements on the design surface. The system of equations was then solved by the CG algorithm. For the case where $H/W = 0.25$, for which a solution can be found without regularization, the CGR solution converged to the same solution obtained using TSVD and TR. For the case $H/W = 0.50$, the best solution provided by the CGR method was found using two iterative steps. It was possible to obtain physically acceptable solutions (i.e., positive emissive powers and net powers in all heating elements) for as many as 10 CGR steps, leading to solutions that were much more accurate. However, for such cases, it was found that in some heaters the net radiative heat flux was negative. This is because the net power is being computed from the sum of the net radiative heat flux and the increase in the temperature of the adjacent wall, so the power can be positive even if the net radiative heat flux is negative.

8.3.2 Optimization Techniques

For this problem, the quasi-Newton technique was used, and the objective function was set equal to the variance of the design surface temperature from the desired temperature at a given process time, integrated throughout the entire process. If the process time is discretized into N_t discrete time steps, this is equivalent to

$$F(\Phi) = \frac{1}{N_t N_{DS}} \sum_{k=1}^{N_t} \sum_{i=1}^{N_{DS}} \left[\vartheta_i(\Phi, \tilde{t}_k) - \vartheta_i^{\text{target}}(\tilde{t}_k) \right]^2. \tag{8.11}$$

In the problem described here, the optimized heater settings must remain strictly nonnegative for the solution to be practical. The best results were found by simply setting the heater outputs equal to zero if Equation 8.10 predicted a negative value at any time step.

8.3.3 Metaheuristic Results

8.3.3.1 Simulated Annealing

In contrast to the quasi-Newton implementation, the optimal heater settings using the SA approach were found at each time step by minimizing

$$F(\Phi, \tilde{t}) = \frac{1}{N_{DS}} \sum_{j=1}^{N_{DS}} \left[\psi_j(\Phi, \tilde{t}) - \psi_j^{target}(\tilde{t}) \right]^2, \tag{8.12}$$

where $\psi_j^{target}(t)$ is the net radiant heat flux (equal in this case to the conductive flux) needed by each element on the design surface to satisfy the increase in sensible energy during that time step, as defined in Equation 8.10. If each of the 32 heater elements were controlled by a single design parameter, the resulting optimization problem (or sequence of optimization problems in this case) would be too large to be computationally tractible, so the heater surface is divided into six regions and a uniform heat flux over each region (at each time step) is specified by a corresponding Φ_i; thus, Φ has a dimension of 6 instead of 32.

The SA algorithm used was a modification of the one presented by Kirkpatrick et al. (1983) and Glover (1986). Unlike for gradient-based methods, it is comparatively easy to impose bounds on the heater settings in SA by restricting the domain of possible values that Φ_i can assume; in this problem, Φ_i was constrained to be nonnegative. At each time step, optimization was started from the optimal heater settings of the previous time step, with the exception of the second time step. This is because a large amount of energy must be applied by the heaters during the first time step relative to the second time step to overcome the thermal inertia of the enclosure surfaces, since the heating process starts from thermal equilibrium, and the optimized heater settings from the first time step do not provide a good starting point for optimizing the heater settings in the second time step. SA was then applied at each successive time step, using the state of the heaters from the previous step as a good approximation for the current step.

Figure 8.7 shows the dimensionless heater fluxes for the six regions for $H/W = 0.5$ after the initial time increment. Although the method delivered uniform design surface conditions, the lack of smoothness of heater operation (Figure 8.7) would be problematic if the solution were to be implemented in an actual furnace. For $H/W = 0.25$, the system behaved well throughout the transient and all of the heater zones exhibited smooth, monotonically increasing power fluxes.

8.3.4 Comparison of Selected Results

For comparison of the predictions of the various methods for the transient case, results for the design surface at hourly intervals are presented in Table 8.2. The mean value and standard deviation of the dimensionless temperatures on the design surface elements $\vartheta_{DS}(t)$ are given for $H/W = 0.5$, the least accurate of the two cases.

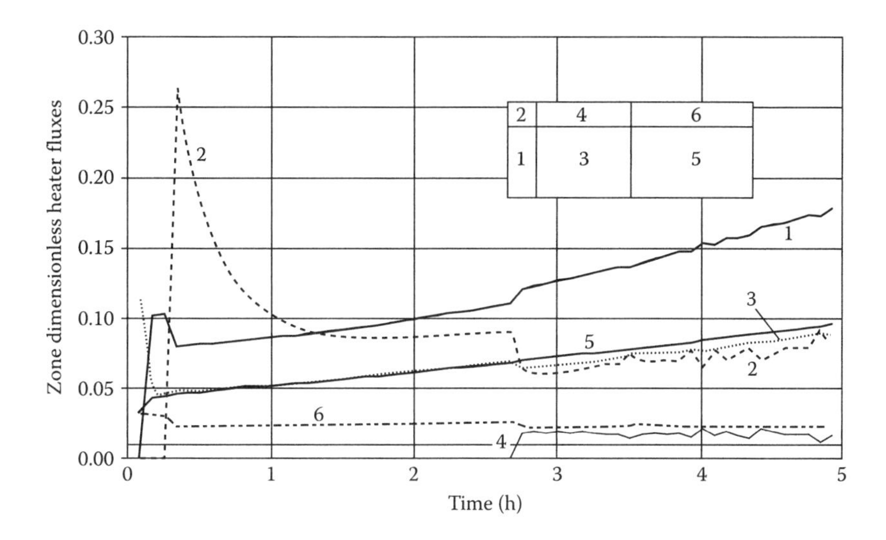

FIGURE 8.7 Predicted power to enclosure roof heater zones 1–6 using simulated annealing; $H/W = 0.5$; $T_{ref} = 1000$ K. (From Daun, K.J. et al., *JHT*, 128(3), 269, March 2006.)

TABLE 8.2
Mean Temperature $[T(t)/T_{ref}]$ and Standard Deviation on Design Surface at Various Times for $H/W = 0.50$, $T_{ref} = 1000$ K

	$t = 1$ h	$t = 2$ h	$t = 3$ h	$t = 4$ h	$t = 5$ h
Target	0.340	0.380	0.420	0.460	0.500
TSVD $p = 3$	0.3400 ± 0.0003	0.3799 ± 0.0004	0.4199 ± 0.0004	0.4598 ± 0.0005	0.4998 ± 0.0006
Tikhonov $L = L_1$, $\lambda = 0.2$	0.3400 ± 0.0002	0.3800 ± 0.0004	0.4199 ± 0.0005	0.4599 ± 0.0006	0.4999 ± 0.0007
CGR ($I = 2$)	0.3399 ± 0.0004	0.3797 ± 0.0005	0.4199 ± 0.0004	0.4599 ± 0.0007	0.4999 ± 0.0008
Constrained optimization	0.3408 ± 0.0033	0.3799 ± 0.0052	0.4199 ± 0.0008	0.4600 ± 0.0018	0.5003 ± 0.0019
Simulated annealing	0.3400 ± 0.0000	0.3801 ± 0.0000	0.4201 ± 0.0001	0.4602 ± 0.0001	0.5003 ± 0.0001

Source: Daun, K.J. et al., *JHT*, 128(3), 269, March 2006.

Each of the methods gives quite good agreement with the target uniform design surface temperatures. For the case of $H/W = 0.25$, the TSVD, Tikhonov, and CGR methods give identical results and are in exact agreement with the target temperatures. The methods based on optimization (quasi-Newton minimization and SA) have minor differences from the target temperatures for this case but are certainly within any reasonable design standard.

For $H/W = 0.5$ (Table 8.2), the results of all of the methods have excellent agreement with the target values at all times; unlike the $H/W = 0.25$, however, the three direct inversion methods (TSVD, Tikhonov, and CGR) now show some variation in the surface temperatures around the target value.

Although each of the methods finds solutions that closely satisfy the desired conditions over the design surface throughout the process, there are significant differences in the predicted heater energy distributions that are necessary to achieve this result. The optimization methods do not encourage spatially smooth solutions, but do guarantee nonnegative heater settings throughout the process. Thus, the solutions for heater power are much less uniform but allow elimination of heaters that require no power input throughout the process.

All of the five techniques predicted solutions that satisfy the design surface requirements within acceptable accuracy. Each solution was verified by taking the predicted heater inputs from the inverse solution, using this distribution as a boundary condition in the forward solution, and verifying that the predicted temperatures on the uniform flux design surface matched the required temperature. The characteristics of the methods used for predicting the required heater power inputs in some cases provide widely different distributions that achieve the same final result, which is characteristic of solutions to inverse problems. The optimization methods lend themselves to predictions that allow a reduction in the number of required heaters, however, in that some heaters were not active at any time throughout the process; on the other hand, the solutions obtained using the regularization methods generally had all heaters active at lower power levels.

8.4 APPLICATION OF METAHEURISTIC METHODS

It is understood that SA is one of a number of metaheuristic approaches that are well suited to handling combinatorial problems. Recently, the method of TS has been applied to radiative furnace design and compared with results by SA (Porter et al. 2006). A problem was posed for a radiant heater that has 63 possible equal-power heater locations that can be used to provide a desired radiative flux distribution on a target. There are 2^{63} potential solutions, so a straightforward search among all possible solutions is not feasible. The inverse problem is to find which of the heaters should be turned on to provide a specified radiative flux on the target design surface. Both SA and TS were used to find the solution. For this problem, TS provided a superior solution, with fewer heaters required, smaller computation time, and smaller value of the final objective function than for SA. The solution of the heater placement problem demonstrated the strengths of metaheuristic optimization when applied to discrete solution spaces with multiple local minima. Cassol et al. (2011) applied optimization to the problem of optimizing the distribution of illumination while minimizing power input to the lamp array.

8.5 UNRESOLVED PROBLEMS

Although a considerable amount of progress has been made in developing methods to solve inverse problems involving radiating systems, there remains significant work to be done. In specifying the conditions on the design surface, for example, the design engineer may choose conditions for which there is no acceptable physical solution, that is, the solution might not be achievable without unacceptable heater conditions (coolers in place of heaters, excessive heater power or temperature requirements, or even imaginary absolute temperatures on the heaters). Such solutions may satisfy the conditions on the design surface mathematically, but they are not useful engineering solutions. An *a priori* determination of the existence of acceptable physical solutions would save a lot of fruitless calculation, but a means to do this is not yet at hand.

Predictions from an inverse solution will not be exact, since assumptions are invariably built into the forward solution, which is being inverted. Such assumptions may include approximated surface properties (diffuse, gray, specular, etc.) or approximated thermophysical properties (conductivity, specific heat, etc.). Thus, to achieve specified conditions on the design surface, at least three additional factors must be addressed: (1) how feedback from the design surface of measured temperature and/or radiative flux can be used to adjust the inverse predictions, (2) how many feedback signals and their locations on the design surface are necessary to adequately provide feedback information, and (3) how errors in properties, geometry, and models affect the results of an inverse analysis. The first of these has been addressed in a preliminary way (Ertürk et al. 2002b) through the use of neural nets. The third question has been addressed in theory by Korolev and Yagola (2012) and Korolev et al. (2013).

8.6 INVERSE PROBLEMS AT THE NANOSCALE

The burgeoning interest in nanoscale heat transfer has opened a new application for inverse methods. Research on near-field interactions of radiation with surface structures has resulted in inverse optimization methods to maximize the spectral absorptivity of solar cells in the bandgap region while minimizing IR absorption to minimize heating effects (Hajimirza et al. 2011, 2012,

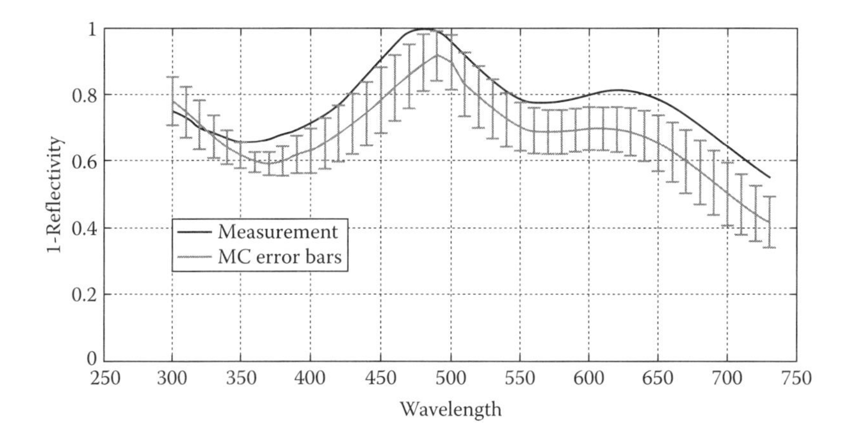

FIGURE 8.8 Spectral absorptivity of optimized multilayer samples: measured values and values predicted using measured rather than handbook values of refractive index. Measured values with error bars showing potential effect of ±manufacturing tolerances. (From Hajimirza, S. and Howell, J.R., *J. Quant. Spectrosc. Radiat. Transf.*, 143, 56, 2014.)

Hajimirza and Howell 2012, 2013a,b, 2014a–c). They showed that the spectral absorptivity of predicted optimized multilayer thicknesses agreed well with experimental measurements (Figure 8.8).

Similarly, optimized surface patterning can allow better spectral control and energy conservation in windows and building materials (Mann et al. 2013). Additional inverse solutions of micro-nanoscale radiation/surface interactions can be found in Drevillon and Ben Abdallah (2007).

8.7 INVERSE PROBLEMS INVOLVING PARTICIPATING MEDIA

When a participating medium is present, other classes of inverse problems may arise, such as determining medium property or temperature distributions from measurements at the boundary. The methods used for such problems are the same as for surface–surface radiative exchange problems.

Hendricks and Howell (1994, 1996) measured the spectral transmittance and reflectance of samples of porous ceramics and used a Levenberg–Marquardt optimization to determine the best set of radiative properties that reproduced the measured radiation at the boundaries. França et al. (2001) examined the laminar flow of a gray gas between parallel walls, with a portion of the lower wall maintained at uniform temperature. The required radiant heater distribution on the upper wall was found by TSVD. This multimode problem involved conduction in the fluid and convective transfer from the heated medium to the design surface. França also examined inverse problems involving a medium with spectrally dependent properties and the determination of the best radiant burner location to obtain a prescribed surface heat flux in a 2D furnace (França et al. 1998, 1999). Hosseini-Sarvari (2003b,c) has examined application of various methods to problems with participating media, including reconstruction of the k-distribution of a medium in a plane-parallel geometry from boundary intensity measurements (2014). Safanivejad et al. (2005) used a genetic algorithm to optimize boundary conditions in a 2D enclosure with an absorbing–emitting medium. Deiveegan et al. (2006) used a genetic algorithm, a Bayesian algorithm, an artificial neural network, and the Levenberg–Marquardt optimization procedure to find surface emissivities and gas properties (scattering and absorption coefficients and simple scattering phase functions) from surface temperature and radiative flux measurements for a gray medium between infinite parallel planes. Pourshagaghy et al. (2006) carried out an inverse boundary design problem for an enclosure containing a participating medium including scattering. Inverse problems to determine the properties of particulate matter in the laboratory and from remote sensing applications is an active research area (Sacadura 2011). Inverse approaches are also used to characterize the properties of particles and agglomerates from light absorption and scattering techniques (Mengüç, 2011). This type of research is presented regularly at the Electromagnetic and Light Scattering Conferences (ELS) and published in the Journal of Quantitative Spectroscopy and Radiative Transfer. In Chapter 15, an introduction is given to these types of studies.

8.8 CONCLUDING REMARKS

Both regularization and optimization techniques have advantages in certain settings. Regularization, particularly TSVD, can provide insight into the reasons for the ill-posed nature of a particular problem and can also indicate how much regularization is required in order to extract an acceptable or practical solution from an ill-posed problem. However, TSVD can be a costly process in terms of computational time and effort, and may not be practical for large systems involving a great number of equations. The CG technique provides similar regularization to TSVD at less cost in computational time and effort, but in transient problems, it can introduce unacceptable fluctuations along time unless the same degree of regularization (i.e., number of steps) is maintained at each time step throughout the process.

Optimization provides an alternative approach that allows easy incorporation of constraints into the formulation and can provide economical solution of complex problems. For problems in which the kernel of the integrals is a variable (problems in which the radiating enclosure geometry is to be optimized), optimization is the only viable solution method.

Although metaheuristic methods are most valuable in combinatorial optimization, they have also found application in continuous problems (cases of this are found in Li and Yang 1997, Kim and Baek 2004). Discretizing the solution space of a continuous optimization problem results in a combinatorial optimization problem. This also reduces the solution space. By using this approach, fully continuous transient radiation problems can be optimized using the powerful memory structure and dynamic neighborhood of TS.

An initial inverse design can be obtained using the methods described here. Even if the design is based on a crude model, it can greatly reduce the complexity of the design and/or increase the reliability of the necessary control system, as the design should produce characteristics close to those for the final system.

Although the obvious application of these methods in macrolevel radiative design is to size heaters when designing a furnace for a particular application, the results may also be incorporated into feedback algorithms that control the heaters during furnace operation. At the microscale, inverse design can greatly reduce the large cost of trial-and-error designs and fabrication.

HOMEWORK

8.1 Four experiments are performed to find the radiative flux on the three walls of a triangular enclosure at measured absolute temperatures T_1, T_2, and T_3. The heat fluxes on the three surfaces are q_1, q_2, and q_3. From the four experiments, the matrix of equations that relate the measurements to the heat fluxes are found by the net radiation method to be

$$\mathbf{A}\cdot\mathbf{q} = \begin{pmatrix} 93.48 & 10.20 & -28.83 \\ 1.96 & 32.82 & 62.41 \\ 26.82 & 36.82 & 57.23 \\ 23.21 & -86.40 & 44.69 \end{pmatrix}\begin{pmatrix} q_1 \\ q_2 \\ q_3 \end{pmatrix} = \begin{pmatrix} 34.72 \\ 70.92 \\ 82.93 \\ -26.22 \end{pmatrix} = \mathbf{b}$$

(a) Believing that only three equations are needed for three unknowns, a grad student deletes the last row of the **A** matrix and the **b** vector and solves for the unknowns. What does he predict? (Make sure the determinant of the reduced 3×3 **A** matrix is nonzero.)

(b) Knowing that measurements of T at high temperatures are generally accurate to three significant figures at best, the student rounds the values in the **b** matrix to three significant figures and again solves for the unknowns using the three-row matrices. Compare with the answer to part (a).

(c) Find the singular values and condition number of the 3×3 **A** matrix, and see whether they might indicate problems in finding the answer to parts (a) and (b).

(d) Now believing that you should use all the data you can get, determine the singular values and condition number of the complete **A** matrix.

(e) Determine the values of **q** using the complete **A** matrix and the four-row **b** vector for both the complete and rounded **b** values.

8.2 In the inverse problem posed in the following, the two black surfaces exchange radiant energy. The surroundings are at $T_\infty = 0$.

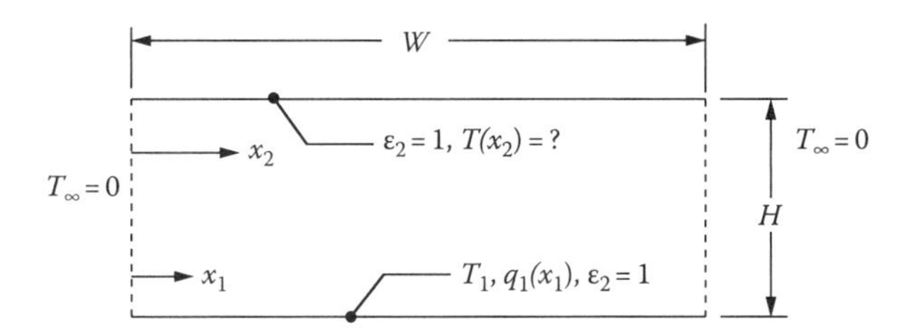

The general problem is as follows. Given $T_1(x_1)$ = constant = T_1 and $q_1(x_1)$ is specified, find the required distribution of temperature $T_2(x_2)$ and net heat flux $q_2(x_2)$ on the upper surface.

Show that at x_1 on the lower surface, the governing equation for this geometry and boundary conditions is $\int_{x_2=0}^{W} E_{b2}(x_2)K(x_1,x_2)dx_2 = E_{b1} - q_1(x_1)$ where $E_{b2}(x_2) = \sigma T_2^4(x_2)$ and that the energy equation can be nondimensionalized using $Q'' = q/\sigma T_2^4$, $X = x/L$, and $\vartheta = E_b/\sigma T_1^4$ to the form $\int_{X_2=0}^{1} \vartheta_2(X_2)K(X_1,X_2)dX_2 = 1 - Q_1''(X_1)$. The kernel of the integral is given by

$$K(X_1,X_2) = \frac{1}{2}\frac{h^2}{\left[h^2 + (X_2 - X_1)^2\right]^{3/2}}$$

where $h = W/L$.

8.3 For the geometry in Homework Problem 8.2, set the heat flux distribution on the lower surface to $Q_1''(X_1) = AX_1^2 - BX_1 - C$, discretize the governing integral equation, and provide the resulting set of algebraic equations in matrix form for determining the unknown values of $\vartheta_2(X_2)$. Note that there are parameters h, (A, B, C), and the number of discrete elements m and n of size ΔX_1 and ΔX_2 on each surface.

8.4 For the geometry in Homework Problem 8.2 and the particular case $Q_1''(X_1) = AX_1^2 - BX_1 - C$, $h = 0.6$, $A = B = 16$ and $C = 6$, and $m = n = 30$, solve for $\vartheta_{2,i}$ for the elements on the top surface using a standard matrix inversion routine (Mathcad, Matlab, Maple, etc.)

8.5 For the problem stated in Homework Problem 8.4, use SVD on the **A** matrix of the problem and show a plot of the singular values.

Answer:

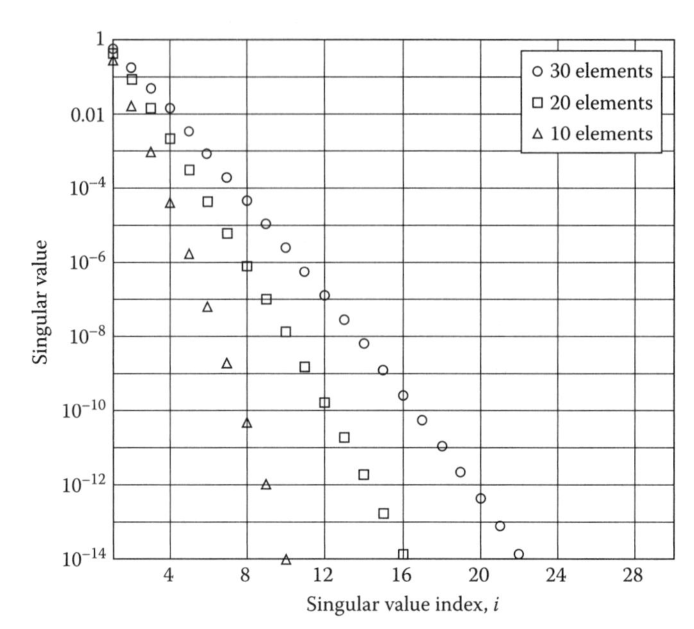

8.6 Choosing an appropriate number of singular values, use TSVD to find a reasonable distribution for $\vartheta(x_2)$ for the parameters of Homework Problem 8.4.

8.7 Show how the solution to Homework Problem 8.2 using the parameters of Homework Problem 8.4 and $p = 3$ is affected if the number of elements $m = n$ on each surface is varied, using $m = 10$, 20, and 30.

Answer:

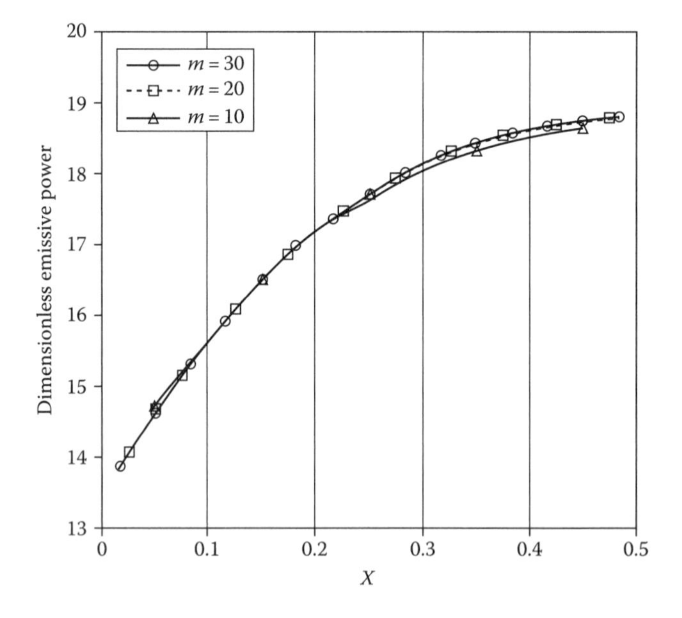

9 Properties of Absorbing and Emitting Media

9.1 INTRODUCTION

Some of the important examples of radiative transfer in participating media are energy transfer through hot gases in engine combustion chambers at high pressures and temperatures, rocket propulsion, glass manufacturing, fibrous insulating layers, nuclear explosions, hypersonic shock layers, plasma generators for nuclear fusion, ablating thermal protection systems, translucent ceramics at high temperature, irradiation of biological systems, and heat transfer in porous materials.

Many possible energy states are present in gases, particularly those with multi-atomic species in their molecules such as water vapor, carbon dioxide, carbon monoxide, methane, and others. Predicting the radiative properties of radiatively participating gases constitutes a challenge in calculating radiative energy transfer. Although some of the applications are relatively recent, the study of radiation in gases has been of interest for over 100 years. An early consideration was absorption and scattering of radiation in the Earth's atmosphere, as this interfered with observations of light from the sun and more distant stars. The solar spectrum received on Earth was recorded by Samuel Langley (1883), and more recent results are in Figure 9.1. The dashed curve is a scaled blackbody emission spectrum at the effective solar temperature of 5780 K. The intermediate curve is the incident solar spectrum outside the Earth's atmosphere. The lowest solid curve with several sharp dips shows the spectrum received at ground level after the solar radiation has passed through the atmosphere along a path normal to the Earth. The dips in the curve show where radiation has been absorbed by various atmospheric constituents, mainly water vapor and carbon dioxide. Absorption occurs in specific wavelength regions, illustrating that gas radiation properties vary considerably with wavelength. Extensive discussions of absorption in the atmosphere are in Kondratyev (1969), Lacis and Hansen (1973), Goody and Yung (1989), Liou (2002), Bohren and Clotiaux (2006) and in Coackley and Yang (2014). The topic is extremely important from the perspective of climate change. Radiation by different gases is also of interest to astrophysicists studying stellar structure. The observed spectrum caused by the emission or absorption of radiation by a gas is characteristic of the specific gas, and it can be used as a diagnostic tool to determine the gas temperature and concentration. Models of stellar atmospheres and the energy transfer processes within them have been constructed and compared with observed stellar behavior.

The importance of gas radiation in industry was recognized in the 1920s for heat transfer in furnaces. Combustion products, chiefly carbon dioxide and water vapor, were found to be significant emitters and absorbers of radiant energy. The energy emitted from flames arises not only from the gaseous emission but also from hot carbon (soot) particles within the flame and from suspended particulate material, as in pulverized-coal combustion. Radiation can also be appreciable in engine combustion chambers, where temperatures can reach a few thousand Kelvins.

The spectral properties of solids and liquids are also of interest. Radiative transfer in a glass-melting furnace was originally studied by Gardon (1958, 1961). He observed that the temperature distribution measured in a deep tank of molten glass was more uniform than expected from heat conduction alone. It was thought that convection might account for the discrepancy, but experimental investigations did not show this. In the late 1940s, it became evident that radiative transfer by absorption and re-emission within the glass is a significant means of energy transfer.

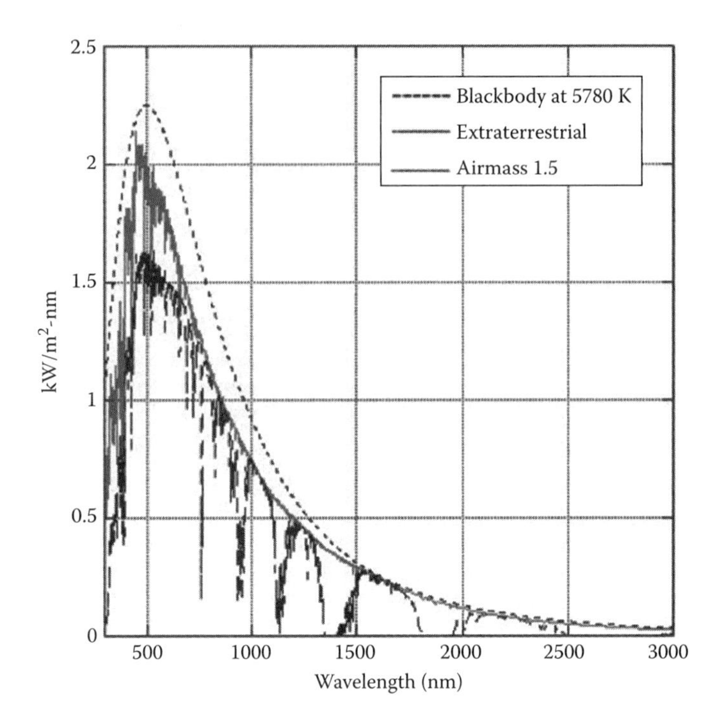

FIGURE 9.1 Attenuation by Earth's atmosphere of incident solar spectral energy flux and comparison with 5780 K scaled blackbody spectrum. (Data extracted from NREL/ASTM G173-03 Tables of Reference Solar Spectral Irradiance.)

Two difficulties make the study of radiation in absorbing, emitting, and scattering media quite challenging. The first is the spatial variation in radiative properties throughout the medium; absorption, emission, and scattering can occur at all locations within the medium at different strengths depending on the concentration of species and local temperature variations. A complete solution for energy exchange requires knowing the radiation intensity, temperature, and physical properties throughout the medium. The mathematics describing the radiative field is inherently complex. A second difficulty is that spectral effects are often much more pronounced in gases, translucent solids, and translucent liquids than for solid surfaces and a detailed spectrally dependent analysis may be required. Most of the simplifications introduced for solving radiation problems in gases and other translucent materials are aimed at decreasing one or both of these complexities.

In this chapter, the definitions of the absorption and emission properties of participating media are given, and information is also presented for finding values for these properties for use in radiative transfer calculations. First, some additional definitions of useful properties are given, and then the physical characteristics governing the radiative emission and absorption properties are described as an aid to understanding the behavior of the radiative properties. Models and correlations of the properties that are useful in engineering radiative transfer are then presented. Scattering properties are covered in Chapter 15.

Because the properties of absorbing-emitting media, particularly gases, are extremely wavelength dependent, they present a major difficulty in carrying out radiative transfer calculations. For this reason, past and present effort seeks to provide useful correlations of the properties that can give accurate results without the extreme computational requirements imposed by detailed spectral calculations. This is an ongoing research field and new methods and improvements of existing methods continue to appear in the literature.

Discussion is limited here to the case of local thermodynamic equilibrium (LTE), where the population of energy states depends on the single parameter of temperature. This is the case for the great majority of engineering applications. There are important exceptions. Such cases are outside the scope of this text, but some introductory material on treating non-LTE effects is in Sections 9.2.2 and 10.6.

Radiant energy intensity (see discussion in Section 3.2.2.1) is attenuated in conducting media according to the relation $I_\lambda(S)/I_\lambda(0) = \exp(-4\pi kS/\lambda)$, where k is the *extinction index* from electromagnetic theory and should not be confused with the κ_λ in Equation 1.44 that is the *absorption coefficient*. The electromagnetic k is related to the magnetic permeability, electrical resistivity, and electrical permittivity of the medium (Equations 3.6 and 3.32). The absorption coefficient κ_λ is related to the electromagnetic extinction index $k(\lambda)$ by

$$\kappa_\lambda = \frac{4\pi k(\lambda)}{\lambda} \tag{9.1}$$

where κ is a function of λ, which is related to the wavelength in vacuum through $\lambda_o = n_\lambda \lambda$. This provides a theoretical basis for Bouguer's law (Equation 1.41), which was originally based on experimental observations. Equation 9.1 provides a means for obtaining the spectral absorption coefficient from optical data available for the electromagnetic extinction index $k(\lambda_0)$, which is a component of the complex refractive index.

The absorption coefficient $\kappa_\lambda(T, P)$ often varies strongly with wavelength and substantially with temperature and pressure since, for a gas, it depends on density; for that reason it can explicitly be denoted as a function of space $\kappa_\lambda(S)$ or a function of T and P as $\kappa_\lambda(T, P)$. For simplicity, we will write it as κ_λ. Considerable analytical and experimental effort has been expended to determine κ_λ for various gases, liquids, and solids. Analytical determinations of κ_λ require detailed quantum-mechanical calculations. Except for the simplest gases, such as atomic hydrogen, the calculations are very tedious and may require simplifying assumptions. Detailed discussions of calculations for κ_λ are in Penner (1959), Bates (1962), Bond et al. (1965), and Herzberg (1992). Recent accounts for different gases are available in the *Journal of Quantitative Spectroscopy and Radiative Transfer* and in *Molecular Spectroscopy*.

Figure 9.2 shows the calculated emission spectrum of hydrogen gas at 40 atm and 11,300 K for a path length through the gas of 50 cm (Garbuny 1967). The presence of "spikes" or strong emission

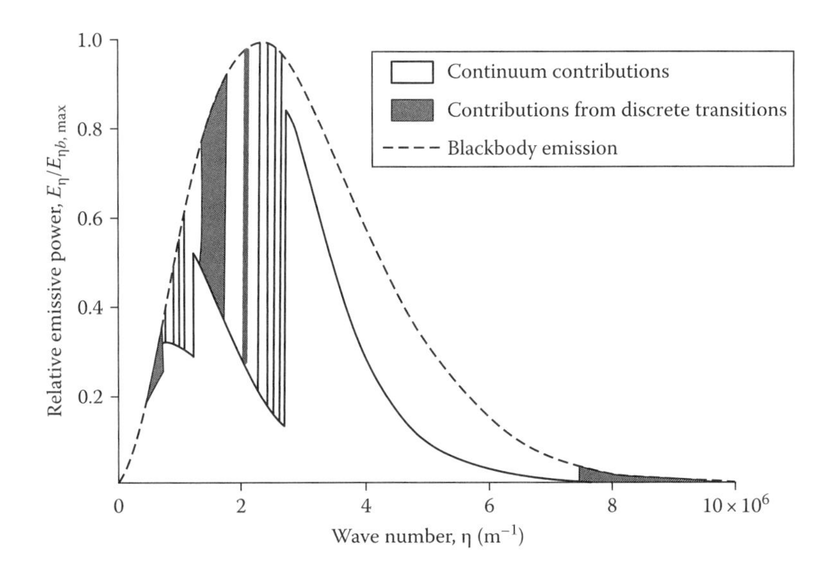

FIGURE 9.2 Normalized emission spectrum of hydrogen at 11300 K and 40 atm, for a path length of 0.5 m. (From Garbuny, M., *Optical Physics*, Academic Press, New York, 1965.)

lines is the result of transitions between bound energy states. The continuous part of the emission spectrum is due to various photodissociations, photoionization, and free electron-atom-photon interactions of other types. The lines and the continuous regions are common features of both emission and absorption spectra at high temperatures, as also shown in Figure 9.3 for high temperature air. When the lines are closely spaced and overlap, a low resolution spectrum makes the lines appear to merge into bands of absorption, as in Figure 9.4.

Note that Figures 9.2 and 9.3 have different abscissas of wavelength or wave number. Often, frequency is used as well. For surfaces considered opaque (Chapter 3), wavelength is often used.

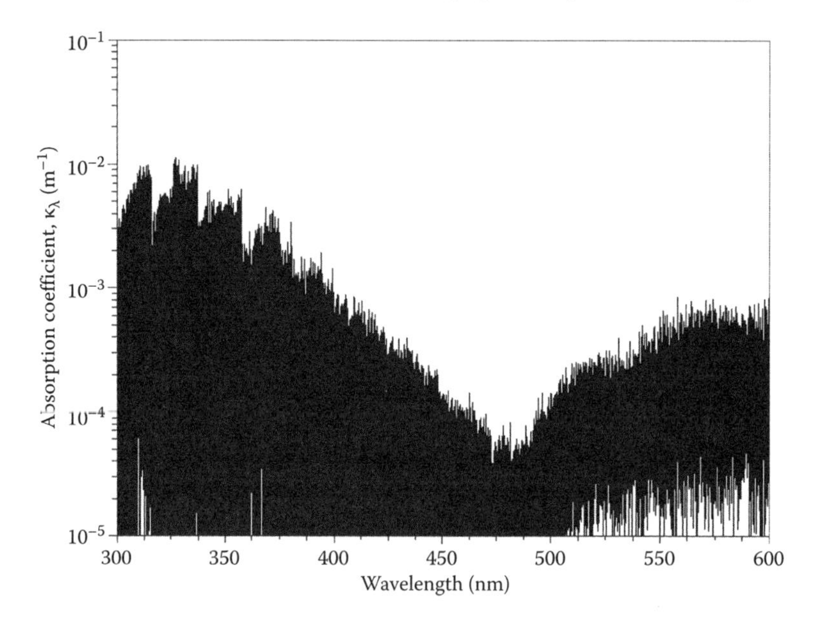

FIGURE 9.3 Emission spectrum of air at 1 atm, 10000 K as calculated from the SPECAIR program. (Program from Laux, C.O., Radiation and nonequilibrium collisional-radiative models, von Karman Institute Lecture Series 2002-07, D. Fletcher, J.-M. Charbonnier, G.S.R. Sarma, and T. Magin (eds.), *Physico-Chemical Modeling of High Enthalpy and Plasma Flows*, Von Karman Institute for Fluid Dynamics, Rhode-Saint-Genèse, Belgium, 2002, available at http://www.specair-radiation.net/.)

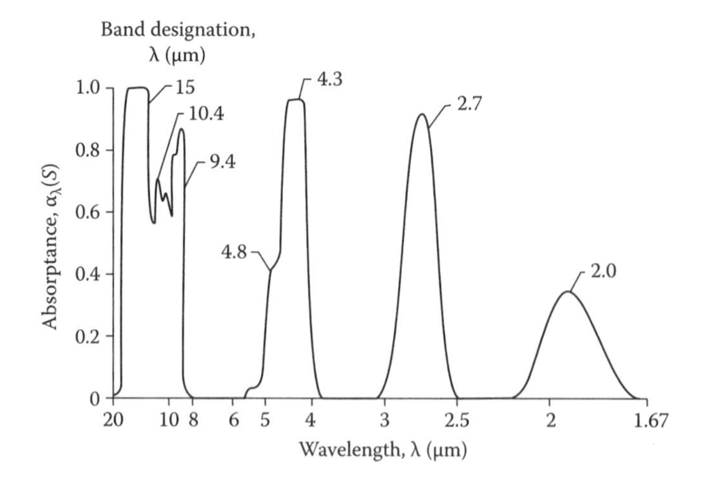

FIGURE 9.4 Low-resolution spectrum of absorption bands for carbon dioxide gas at 830 K, 10 atm, and for path length through gas of 0.388 m. (From Edwards, D.K., Molecular gas band radiation, in T.F. Irvine, Jr. and J.P. Hartnett (eds.), *Advances in Heat Transfer*, vol. 12, Academic Press, New York, 1976, pp. 115–193.)

For radiation within translucent media, frequency is commonly used. Frequency has the advantage that it does not change when radiation passes from one medium into another with a different refractive index. Wavelength does change because of the change in propagation velocity.

9.2 SPECTRAL LINES AND BANDS FOR GAS ABSORPTION AND EMISSION

9.2.1 Physical Mechanisms

Radiation transfer with absorption, emission, and scattering occurs in gases, liquids, and solids. Gases are a very important class of these media. If the radiation properties of gases are compared with solid surfaces that are considered opaque, a difference in spectral behavior is usually quite evident. As shown in Chapter 3, the property variations with wavelength for optically dense solids, which can be treated as opaque, range from fairly smooth to somewhat irregular. Gas properties, however, exhibit very irregular wavelength dependencies. Absorption or emission by gases is significant only in certain wavelength regions, especially when the gas temperature is below a few thousand Kelvins. The absorbing ability of a gas layer as a function of wavelength typically looks like that for carbon dioxide (CO_2), water vapor (H_2O), and methane (CH_4) in Figure 9.5. Note that here "absorptance" in the figure refers to what is not transmitted through a cloud of a given gas species. Its definition is introduced in Equation 9.11.

A radiating gas can be composed of molecules, atoms, ions, and free electrons that can be at various energy levels. In a molecule, the atoms form a dynamic system with vibrational and rotational modes that have specific quantized energy levels. A schematic diagram of the energy levels for an atom or ion is in Figure 9.6. Zero energy is assigned to the ground state (lowest-energy bound state) e_1, with the higher bound states being at positive energy levels. The energy e_I is the ionization potential, which is the minimum energy required to produce ionization from the ground state. For energies above e_I, ionization has taken place and free electrons have been produced.

It is convenient to discuss the radiation process by utilizing a photon or quantum point of view. The photon is the basic unit of radiative energy. Radiative emission releases photons, and absorption

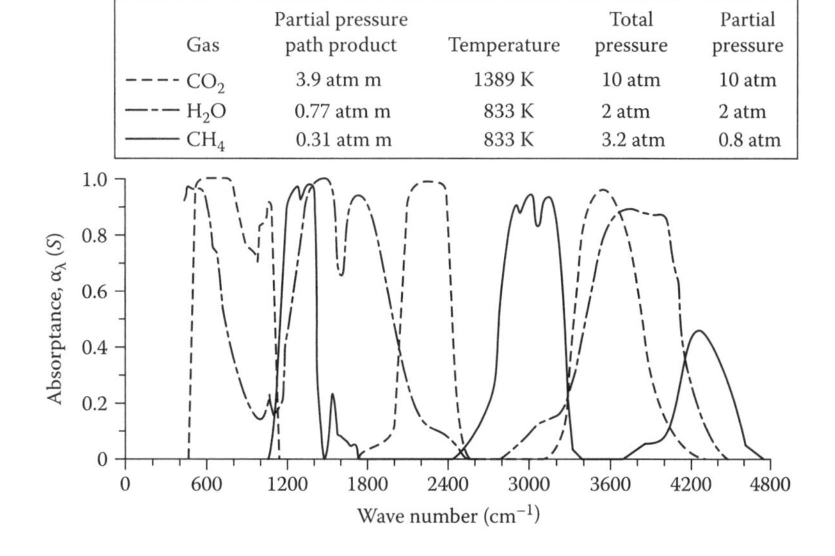

FIGURE 9.5 Low-resolution spectrum of absorption bands at 830 K, 10 atm, and for path length through gas of 0.388 m for carbon dioxide, water vapor, and methane. (From Edwards, D.K., Molecular gas band radiation, in T.F. Irvine, Jr. and J.P. Hartnett (eds.), *Advances in Heat Transfer*, vol. 12, Academic Press, New York, 1976, pp. 115–193.)

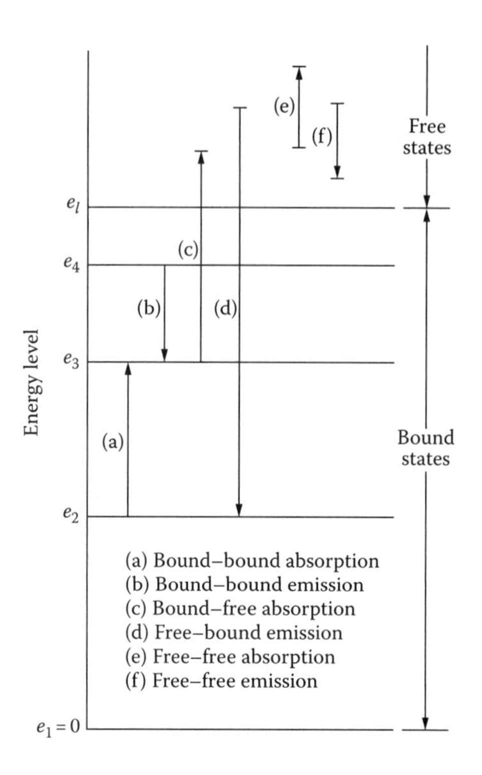

FIGURE 9.6 Schematic diagram of energy states and transitions for atom or ion. e_1 is the ground state and e_l the ionization energy for the gas.

is the capture of photons. When a photon is emitted or absorbed, the energy of the emitting or absorbing particle is correspondingly decreased or increased. Figure 9.6 diagrams three types of transitions that can occur: bound–bound, bound–free, and free–free. A photon can also transfer part of its energy in certain inelastic scattering processes that are of minor engineering importance.

The magnitude of a radiative energy transition is related to the frequency of the emitted or absorbed radiation. The energy of a photon is $h\nu$, where h is Planck's constant and ν is the frequency of the photon energy. For an energy transition from bound state e_3 down to bound state e_2 in Figure 9.6, a photon is emitted with energy $e_3 - e_2 = h\nu$. The frequency of the emitted energy is then $\nu = (e_3 - e_2)/h$, so a *fixed frequency* is associated with the transition from a specific energy level to another. Thus, in the absence of other effects, the spectrum of the emitted radiation is a spectral line at that frequency. Conversely, in a transition between two bound states when a particle absorbs energy, the quantum nature of the process dictates that the absorption is such that the particle can only go to one of the discrete higher energy levels. Consequently, for a photon to be absorbed, the frequency of the photon energy must have one of the certain discrete values. For example, a particle in the ground state in Figure 9.6 can absorb photons with frequencies $(e_2 - e_1)/h$, $(e_3 - e_1)/h$, or $(e_4 - e_1)/h$ and undergo a transition to a higher bound energy level. Photons with other frequencies in the range $0 < \nu < e_l/h$ cannot be absorbed.

When a photon is absorbed or emitted by an atom or molecule and there is no ionization or recombination of ions and electrons, the process is a *bound–bound* absorption or emission (processes (a) and (b) in Figure 9.6). The atom or molecule moves from one quantized bound energy state to another. These states can be rotational, vibrational, or electronic in molecules, and electronic in atoms. Since bound–bound energy changes are associated with specific energy levels, the absorption and emission coefficients are sharply peaked functions of frequency in the form of a series of spectral lines. The lines have a finite spectral width resulting from various line broadening effects discussed in Section 9.2.3.

Vibrational energy modes are always coupled with rotational modes. The rotational spectral lines superimposed about a vibrational line give a band of closely spaced spectral lines. If these lines overlap into a continuous region, a *vibration-rotation band* is formed (see Section 9.2.5). Rotational transitions within a given vibrational state are associated with low energies (long wavelengths, ~8–1000 μm). Vibration-rotation transitions are at infrared energies of about 1.5–20 μm. Electronic transitions are at short wavelengths in the visible region 0.4–0.7 μm and at portions of the ultraviolet and infrared near the visible region. At industrial temperatures, radiation is principally from vibrational and rotational transitions; electronic transitions become important at temperatures above several thousand Kelvins.

Process (c) in Figure 9.6 is a *bound–free* absorption (photoionization). An atom absorbs a photon with energy sufficient to cause ionization. The resulting ion and electron are free to take on any kinetic energy; hence, the bound–free absorption coefficient is a continuous function of photon energy frequency ν as long as the photon energy $h\nu$ is large enough to cause ionization. The reverse (process (d) in Figure 9.6) is free–bound emission (photorecombination). Here an ion and free electron combine, a photon is released, and the energy of the resulting atom drops to that of a discrete bound state. Free–bound emission produces a continuous spectrum, as the combining particles can have any initial energy.

In an ionized gas, a free electron can pass near an ion and interact with its electric field. This can produce a *free–free* transition that is often called *bremsstrahlung*, meaning brake radiation. The electron can absorb a photon (process (e) in Figure 9.6), thereby going to a higher kinetic energy, or it can emit a photon (process (f)) and drop to a lower free energy. Since the initial and final free energies can have any values, a continuous absorption or emission spectrum is produced.

9.2.2 Condition of Local Thermodynamic Equilibrium

It has been assumed in earlier chapters that opaque solids emit energy based solely on their temperature and physical properties. The spectrum of emitted energy was assumed unaffected by the characteristics of any incident radiation. This is usually true because all the absorbed incident energy is quickly redistributed into an equilibrium distribution of internal energy states at the temperature of the solid. However, there are important exceptions. In a gas, the redistribution of absorbed energy occurs by various types of collisions between the atoms, molecules, electrons, and ions that constitute the gas. Under most engineering conditions, this redistribution occurs rapidly, and the energy states of the gas are populated in equilibrium distributions described by the local temperature. When this is true, the Planck spectral distribution along with the spectral absorption coefficient, Equation 1.61, describes the emission from a gas volume element. The assumption that a gas emits according to Equation 1.61 regardless of the spectral distribution of intensity passing through and being absorbed by dV is called "local thermodynamic equilibrium" (LTE).

As discussed in Section 9.1, non-LTE occurs in very rarefied gases, where the rate and/or effectiveness of interparticle collisions in redistributing absorbed radiant energy is low, when rapid transients exist so that the populations of energy states of the particles cannot fully adjust to new conditions during the transient [see Qiu and Tien (1992) for an analysis of the heating response of a metal to a very short laser energy pulse and Tan and Hsu (2002) for the domain of effects in an absorbing–emitting medium subjected to a pulse in boundary conditions], where very sharp gradients occur so that local conditions depend on particles that arrive from adjacent localities at widely different conditions and that may emit before reaching local equilibrium, and where extremely large radiative fluxes exist, so that absorption of energy and therefore population of higher energy states occur so strongly that collisional processes cannot repopulate the lower states to an equilibrium density. For these conditions, the spectral distribution of emitted radiation is not given by Equation 1.61. The populations must be determined by detailed examination of the relation between the collision and radiation processes and their effect on the distribution of energy among the various possible states; this is a formidable undertaking. This is necessary in the examination of shock phenomena and picosecond pulsed laser interactions with matter (sharp gradients and fast transients), stellar atmospheres (extreme energy flux and low density), nuclear explosions (transients, sharp gradients, and extreme fluxes), and high-altitude and interplanetary gas dynamics (very low densities). A gas with small optical thickness can have transmitted within it radiation from regions at widely different conditions and is more likely to depart from LTE than an optically thick gas of the same density.

A prominent non-LTE effect is in the laser, in which a material with a metastable energy state is excited by some external means. Because the excited state is metastable and is chosen so that no competing process is trying to depopulate it, its population can reach a value well above the LTE value; this is a *population inversion*. The material is then exposed to radiation with photons having the same frequency as the transition frequency from the excited metastable state to a lower state. This stimulates the transition to the lower state, and many photons with the transition frequency are emitted, thus amplifying the intensity of the incident radiation. This leads to the acronym *laser*—light amplification by stimulated emission of radiation.

The term *luminescence* covers additional non-LTE mechanisms of radiant emission by transitions from a metastable excited state to a lower energy state, where the original excitation is by means other than thermal agitation. Common excitations are by visible light, ultraviolet radiation, electron bombardment, and biological processes such as decay or those present in some marine creatures and "lightning bugs." Because the transitions are between discrete energy states, the span of wavelengths over which the emission occurs is quite small. Luminescence, therefore, does not add significant energy to the emission spectrum in engineering situations and is neglected in heat transfer calculations. Because luminescence is common to materials at room temperature, it is not predicted by the laws for thermal radiation that would provide negligible visible radiation at such temperatures. This is the origin of the term *cold light* for fluorescent-lamp emission. The quantum-mechanical properties of such luminescent materials must be examined to explain their behavior.

Non-LTE problems are not within the general scope of this work. It is assumed here that LTE exists, the properties defined in this chapter can be computed based on LTE states, and that the spontaneous emission from dV is governed by Equation 1.61. Some cases of non-LTE calculations of radiative transfer are discussed in Section 10.6.

9.2.3 Spectral Line Broadening

As discussed in Section 9.2.1, if a gas is not dissociated or ionized, its internal energy is in discrete vibrational, rotational, and electronic energy states of its atoms or molecules. If the energies of the upper and lower discrete states are e_j and e_i, only photons of energy e_p can cause a transition, where

$$e_p = e_j - e_i = h\nu_{ij} = hc\eta_{ij} \tag{9.2}$$

The discrete transitions result in the absorption of photons of definite frequencies or wave numbers, causing the appearance of dark lines in the transmission spectrum; this is *line absorption*. Equation 9.2 would predict that very little energy could be absorbed from the entire incident spectrum by an absorption line, because only those photons having a single wave number given by Equation 9.2 could be absorbed. Other effects, however, cause the line to be broadened and consequently to have a finite wave number span around the transition wave number η_{ij}. The wave number span of the broadened line, and the variation within it of its absorption ability, depends on the physical mechanism of the broadening. Some of the important mechanisms are natural, Doppler, collision, and Stark broadening. Collision broadening is the most important for most engineering conditions involving infrared radiation.

The variation of the absorption coefficient with wave number within a broadened line is the line *shape*. The shape is important as it is related to the trends of gas absorption with temperature,

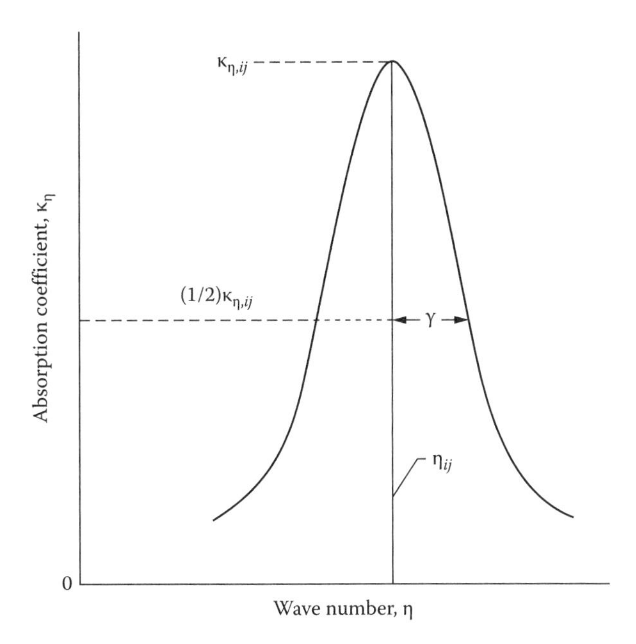

FIGURE 9.7 Absorption coefficient for symmetric broadened spectral line for transition between energy levels i and j.

pressure, and path length through the gas. The shape of a typical line is illustrated by Figure 9.7. The *line intensity* S_{ij} is the integral under the $\kappa_{\eta,ij}$ versus wave number curve,

$$S_{ij} = \int_0^\infty \kappa_{\eta,ij} d\eta = \int_{-\infty}^\infty \kappa_{\eta,ij} d(\eta - \eta_{ij}) \tag{9.3}$$

The $\kappa_{\eta,ij}$ is small except for η close to η_{ij}. The regions away from η_{ij}, where $\kappa_{\eta,ij}$ is small, are the "wings" of the line. The magnitudes of $\kappa_{\eta,ij}$ and S_{ij} depend on the number of molecules in energy level i and hence depend on gas density. Taking the ratio $\kappa_{\eta,ij}/S_{ij}$ tends to cancel the effect of density on magnitude and shows the effect of density in changing the line shape.

One characteristic of the line shape is the *line half-width*, γ. This is one-half of the line width (in units of wave number for the present discussion) at half the maximum line height (Figure 9.7). It provides a definite width to help describe the line. Since $\kappa_{\eta,ij}$ goes to zero asymptotically as $|\eta-\eta_{ij}|$ increases, it is not possible to define a line width in terms of a wave number where $\kappa_{\eta,ij}$ becomes zero. Four mechanisms for line broadening are now discussed, along with their resulting line shapes.

9.2.3.1 Natural Broadening

A stationary emitter unperturbed by any external effects emits energy over a finite spectral interval about each transition wave number. This *natural line broadening* arises from the uncertainty in the exact levels e_i and e_j of the transition energy states, which is related to the Heisenberg uncertainty principle. Natural line broadening produces the line shape

$$\frac{\kappa_{\eta,ij}}{S_{ij}} = \frac{\gamma_n/\pi}{\gamma_n^2 + (\eta - \eta_{ij})^2} \tag{9.4}$$

where γ_η is the line half-width at half-maximum for natural broadening. This shape is a *resonance* or *Lorentz* profile. In units of wave number, it provides a symmetric profile about η_{ij}. In engineering applications, the half-width for natural broadening is usually quite small compared with that for other line-broadening mechanisms; hence, natural broadening is usually neglected.

9.2.3.2 Doppler Broadening

The atoms or molecules of an absorbing or emitting gas are not stationary, but have a distribution of velocities associated with their thermal energy. If an atom or molecule is emitting at wave number η_{ij} and at the same time is moving at velocity v toward an observer, the waves arrive at the observer at an increased η given by $\eta = \eta_{ij}[1 + (v/c)]$. If the emitter is moving away from the observer, v is negative and the observed wave number is less than η_{ij}. In thermal equilibrium, the gas molecules have a Maxwell–Boltzmann distribution of velocities. This velocity distribution results in a spectral line shape with a Gaussian distribution:

$$\frac{\kappa_{\eta,ij}}{S_{ij}} = \frac{1}{\gamma_D}\sqrt{\frac{\ln 2}{\pi}}\exp\left[-\left(\eta-\eta_{ij}\right)^2\frac{\ln 2}{\gamma_D^2}\right] \tag{9.5}$$

The γ_D is the line half-width for Doppler broadening:

$$\gamma_D = \frac{\eta_{ij}}{c}\left(\frac{2kT}{M}\ln 2\right)^{1/2} \tag{9.6}$$

The dependence of γ_D on $T^{1/2}$ shows that Doppler broadening becomes important at high temperatures.

9.2.3.3 Collision Broadening and Narrowing

As the pressure of a gas increases, the collision rate experienced by an atom or molecule with surrounding particles is increased. Collisions can perturb the energy states of the atoms or molecules, resulting in collision broadening of the spectral lines. For noncharged particles, the line has a Lorentz profile (Kondratyev 1969), which is the same shape as for natural broadening:

$$\frac{\kappa_{\eta,ij}}{S_{ij}} = \frac{\gamma_c/\pi}{\gamma_c^2+(\eta-\eta_{ij})^2} \tag{9.7}$$

The collision half-width γ_c is determined by the collision rate. An approximate value from kinetic theory is

$$\gamma_c = \frac{1}{2c}\frac{4D^2P}{(\pi MkT)^{1/2}} \tag{9.8}$$

where

D is the diameter of the atoms or molecules
P is the gas pressure for the single-component gas
M is the mass of an individual molecule

Equation 9.8 shows that collision broadening becomes important at high pressures and low temperatures or, from the perfect gas law, high pressures and densities.

Collision broadening is often the chief contributor to line broadening for engineering infrared conditions, and the other line-broadening mechanisms can usually be neglected. The shapes of Doppler and Lorentz broadened lines are compared in Figure 9.8 for the same half-width and area

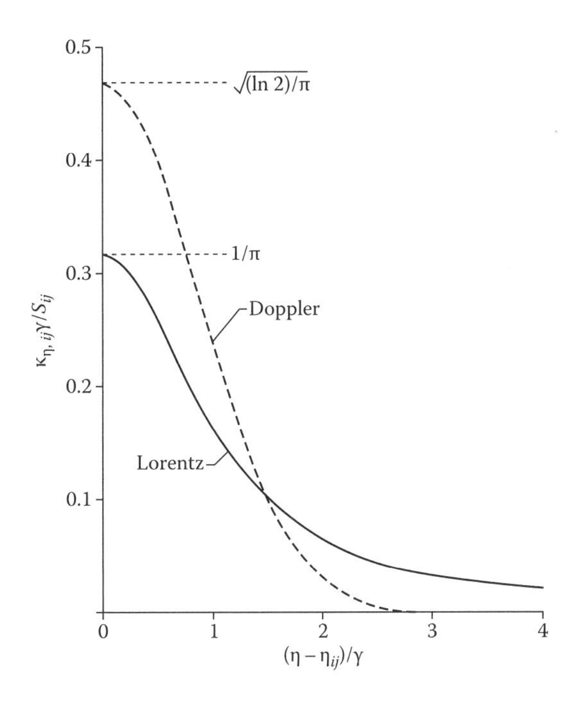

FIGURE 9.8 Line-shape parameter for equal intensity Doppler and Lorentz broadened spectral lines (areas under two curves are equal).

under the curves. The Lorentz profile is lower at the line center but remains appreciable farther out in the wings of the line. Even when Doppler broadening is dominant near the line center, collision broadening is often the important mechanism far from the center. At low pressures and high temperatures, Doppler broadening becomes important.

When both collision and Doppler broadening are important, a convolution of the collision and Doppler profiles results, called the Voigt profile (Varghese and Hanson 1984). This depends on two parameters, the normalized frequency separation from the line center, $[(\nu - \nu_{ij})(\ln 2)^{1/2}]/(\pi^{1/2}\gamma_D)$, and the broadening parameter, $\gamma_c(\ln 2)^{1/2}/\gamma_D$. Here, the half-widths are in terms of frequency rather than wave number.

Collision narrowing refers to the reduction of Doppler broadening because of collisions that limit the scale of molecular motion. Such narrowing is normally important when both Doppler and collision broadening are present and a more complex line shape is produced. The shape is given by Galatry profile (Galatry 1961, Ouyang and Varghese 1989), which is a function of the two parameters of the Voigt profile and an additional narrowing parameter $kT(\ln 2)^{1/2}/(2M\pi c\gamma_D D_d)$. The definitions are as for the Voigt profile, and D_d is the optical diffusion coefficient, usually approximated as equal to the mass diffusion coefficient.

9.2.3.4 Stark Broadening

When strong electric fields are present, the energy levels of the radiating gas particles can be greatly perturbed. This is *Stark effect*, which can produce very large line broadening. It is often observed in partially ionized gases where interactions of radiating particles with electrons and protons give large Stark effects. Calculation of the line shapes must be by quantum mechanics, and the shapes are quite unsymmetrical and complicated.

Stark and collision broadening are often lumped under the general heading of pressure broadening. Both depend on the pressure of the broadening component of the gas. When two or more broadening effects contribute simultaneously, calculation of the resulting line shape is more difficult. Additional information is in Penner (1959), Bond et al. (1965), Breene (1981), Griem (2005), and Oks (2006).

Broadening has been discussed here under the assumption that only one atomic or molecular species is present. If the gas has more than one component, collision broadening is caused by collisions with like molecules (self-broadening) and with other species. Both collision processes must be included in calculating line shapes.

9.2.4 Absorption or Emission by a Single Spectral Line

9.2.4.1 Property Definitions for a Path in a Uniform Absorbing and Emitting Medium

The variation of intensity along a path is described by the radiative transfer equation (RTE), Equation 1.73. It is useful for what follows to consider a limited form of the RTE and some definitions that result from it. For the present, scattering effects are not included. Then $\beta_\lambda = \kappa_\lambda$ and, from Equation 1.73, the spectral intensity along a path is attenuated by absorption and augmented by emission to yield the radiative transfer equation without scattering:

$$\frac{\partial I_\lambda}{\partial S} = -\kappa_\lambda(S)I_\lambda(S) + \kappa_\lambda(S)I_{\lambda b}(S) \tag{9.9}$$

If the refractive index $n_\lambda \neq 1$, the $I_{\lambda b}$ contains an n^2 factor as in Equation 1.20.

A gas with uniform temperature and uniform composition is now considered, such as in a well-mixed furnace or combustion chamber. The κ_λ and $I_{\lambda b}$ are then constant throughout the volume, and Equation 9.9 is integrated from $S = 0$ to S starting from $I_\lambda(0)$ at $S - 0$ to give

$$I_\lambda(S) = I_\lambda(0)e^{-\kappa_\lambda S} + I_{\lambda b}\left[1 - e^{-\kappa_\lambda S}\right] \tag{9.10}$$

The $e^{-\kappa_\lambda S}$ is the spectral *transmittance* (fraction transmitted), $t_\lambda(S)$, of the initial intensity. Then $1\text{–}1 - e^{-\kappa_\lambda S}$ is the fraction of $I_\lambda(0)$ that was absorbed; this is the spectral *absorptance* $\alpha_\lambda(S)$ along the path. By virtue of Kirchhoff's law, this quantity appears in the spectral emission along the path as given by the last term of Equation 9.10, which becomes

$$I_\lambda(S) = I_\lambda(0)t_\lambda(S) + I_{\lambda b}\alpha_\lambda(S) \tag{9.11}$$

and

$$\alpha_\lambda(S) = 1 - t_\lambda(S) = 1 - e^{-\kappa_\lambda S}$$

By integrating over all λ, the *total absorptance* along a path in a uniform gas (uniform composition, temperature, and pressure) is

$$\alpha(S) = \frac{\int_0^\infty I_\lambda(0)\alpha_\lambda(S)d\lambda}{\int_0^\infty I_\lambda(0)d\lambda} = \frac{\int_0^\infty I_\lambda(0)[1 - e^{-\kappa_\lambda S}]d\lambda}{\int_0^\infty I_\lambda(0)d\lambda} \tag{9.12}$$

Similarly, the total emittance along a path in a uniform gas is

$$\varepsilon(S) = \frac{\int_0^\infty I_{\lambda b}\alpha_\lambda(S)d\lambda}{\int_0^\infty I_{\lambda b}d\lambda} = \frac{\pi\int_0^\infty I_{\lambda b}[1 - e^{-\kappa_\lambda S}]d\lambda}{\sigma T^4} \tag{9.13}$$

where T is the uniform gas temperature. The *total transmittance* along the gas path is

$$t(S)=\frac{\int_0^\infty I_\lambda(0)t_\lambda(S)d\lambda}{\int_0^\infty I_\lambda(0)d\lambda}=\frac{\int_0^\infty I_\lambda(0)e^{-\kappa_\lambda S}d\lambda}{\int_0^\infty I_\lambda(0)d\lambda} \tag{9.14}$$

It follows that

$$1-\alpha(S)=t(S) \tag{9.15}$$

As a step toward the evaluation of gas absorptance and emittance, the integrals in Equations 9.12 and 9.13 are now considered over the spectral region of a single broadened line. For a line centered about η_{ij}, the absorption coefficient $\kappa_{\eta,ij}$ is essentially zero except in a narrow wave number range surrounding η_{ij}. Unless the path length S is large, the integrands in the numerators of Equations 9.12 and 9.13 are appreciable only within this range. The $I_\eta(0)$ or $I_{\eta b}$ remain essentially constant within this range, and since the largest absorption is at η_{ij}, the $I_\eta(0)$ and $I_{\eta b}$ are ordinarily evaluated at η_{ij}. Then Equations 9.12 and 9.13 become, for a spectral line (in terms of wave number),

$$\alpha_{ij}(S)=\frac{I_{\eta_{ij}}(0)\int_{-\infty}^{\infty}\{1-\exp[-\kappa_{\eta,ij}S]\}d(\eta-\eta_{ij})}{\int_0^\infty I_\eta(0)d\eta} \tag{9.16}$$

$$\varepsilon_{ij}=\frac{\pi I_{\eta_{ij}b}\int_{-\infty}^{\infty}\{1-\exp[-\kappa_{\eta,ij}S]\}d(\eta-\eta_{ij})}{\sigma T^4} \tag{9.17}$$

The absorbed and emitted energies for the spectral line both contain the same integral. This is the *equivalent line width* $\bar{A}_{ij}(S)$,

$$\bar{A}_{ij}(S)\equiv\int_{-\infty}^{\infty}\{1-\exp[-\kappa_{\eta,ij}S]\}d(\eta-\eta_{ij}) \tag{9.18}$$

which has units of the spectral variable (η in this instance). By considering a spectral line within which the gas is perfectly absorbing ($\kappa_{\eta,ij}\to\infty$), and having no absorption outside this line, Equation 9.18 shows that $\bar{A}_{ij}$ can be interpreted as *the spectral width of a black line centered about* η_{ij} *that produces the same absorption or emission as the actual line.* The evaluation of $\bar{A}_{ij}$ is now considered for some important limiting cases.

9.2.4.2 Weak Lines

First let the optical path length be small, $\kappa_{\eta,ij}S \ll 1$ so the exponential term in Equation 9.18 can be approximated by $1-\exp[-\kappa_{\eta,ij}S]\approx\kappa_{\eta,ij}S$. Using the line intensity S_{ij} from Equation 9.3,

$$\bar{A}_{ij}(S)=S\int_{-\infty}^{\infty}\kappa_{\eta,ij}d(\eta-\eta_{ij})=SS_{ij} \tag{9.19}$$

The line intensity S_{ij} should not be confused with the path length S. For a small optical path, the equivalent line width $\bar{A}_{ij}(S)$ is *linear* with path length S regardless of the line shape. A line with linear behavior of $\bar{A}_{ij}(S)$ with S is called a *weak* line.

9.2.4.3 Lorentz Lines

Some useful relations for the Lorentz line shape for collision broadening (Equation 9.7) are examined, as this is the most important broadening for engineering applications in the infrared. Substituting Equation 9.7 into Equation 9.18 gives

$$\bar{A}_{ij}(S) = \int_{-\infty}^{\infty} \left\{ 1 - \exp\left[-\frac{S_{ij}}{\pi} \frac{\gamma_c S}{\gamma_c^2 + (\eta - \eta_{ij})^2} \right] \right\} d(\eta - \eta_{ij}) \tag{9.20}$$

This is integrated to yield

$$\bar{A}_{ij}(S) = 2\pi\gamma_c \xi e^{-\xi}[I_0(\xi) + I_1(\xi)] \equiv 2\pi\gamma_c £(\xi) \tag{9.21}$$

where

$\xi = SS_{ij}/2\pi\gamma_c$
I_n is the imaginary Bessel function of order n
$£(\xi)$ is Ladenberg–Reiche function tabulated in Goody and Yung (1989)

Some relations that approximate Equation 9.21 are in Tien (1968). One useful approximation for $£(\xi)$ is

$$£(\xi) \approx \xi \left[1 + \left(\frac{\pi\xi}{2} \right)^{5/4} \right]^{-2/5} \tag{9.22}$$

which is within 1% of the exact function.

For a weak line, $\xi \ll 1$, Equation 9.21 reduces to Equation 9.19. For a strong Lorentz line, $\xi \gg 3$, the asymptotic form of Equation 9.21 is

$$\bar{A}_{ij}(S) = 2\sqrt{S_{ij}\gamma_c S} \tag{9.23}$$

This shows that, for a *strong Lorentz line*, the equivalent line width and thus absorptance varies as the square root of the path length S. This is in contrast to the behavior for any weak line, Equation 9.19, where the absorptance varies linearly with path length. Experimental results bear out these dependencies.

Numerical values of $\bar{A}_{ij}(S)$ can be calculated if γ_c and S_{ij} are known. An expression for γ_c was given in Equation 9.8, and it depends on gas pressure and temperature. The S_{ij} is also a function of these variables, as it depends on the number of gas molecules occupying an energy level and on the probability of a transition occurring. This gives the line intensity S_{ij} in the form

$$S_{ij} \propto \rho e^{-K_{ij}/T} \propto \frac{P}{T} e^{-K_{ij}/T} \tag{9.24}$$

where K_{ij} is a coefficient depending on the particular quantum state involved in the transition. As a result, in Equation 9.24, for a fixed gas temperature the $\bar{A}_{ij}$ depends on ρS or PS (from $S_{ij}S$) and on ρ or P alone (from γ_c). There are thus two effects of increasing gas density or pressure. One is the increase in absorption because there are more molecules along the radiation path, and the second is the increased width of the spectral line because of greater collisional line broadening. These trends help guide the correlation of absorption and emission behavior when many lines are superposed into an absorption band.

9.2.5 Band Absorption

9.2.5.1 Band Structure

The common absorbing/emitting gases in engineering calculations are diatomic or polyatomic and possess vibrational and rotational energy states that are absent in monatomic gases. The transitions between vibrational and rotational states usually provide the main contribution to the absorption coefficient in the significant thermal radiation spectral regions at moderate temperatures. As the temperature is raised, dissociation, electron transitions, and ionization become more probable, and their contributions to the absorption coefficient must be included. When the absorption coefficient of a gas is determined experimentally, contributions of all the line and continuum processes are superimposed. In computing such coefficients, each absorption process must be analyzed, and the complete coefficient obtained by combining contributions.

Since vibration-rotation bands usually occur in the most important spectral regions in engineering radiation calculations, their structure is now examined and band models are discussed for analysis of band-absorption features. The correlation of experimental band absorptance data is illustrated for application in engineering problems.

A vibration-rotation band consists of groups of very closely spaced spectral lines resulting from transitions between vibrational and rotational energy states. Consider the vibrational-rotation transitions governing the absorption coefficient of most polyatomic gases up to $T \approx 3000$ K. The transitions depend strongly on wave number, so the absorption coefficient is strongly spectrally dependent. An example is Figure 9.9 for a portion of the carbon dioxide spectrum. The absorption

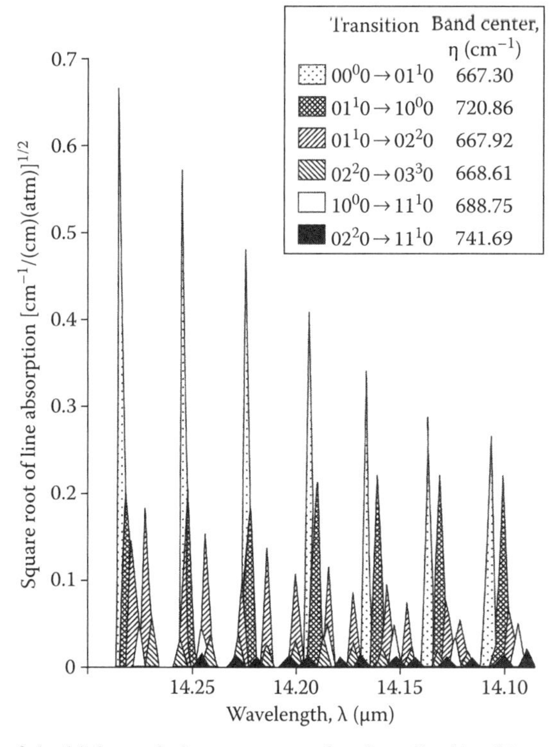

FIGURE 9.9 A portion of the high-resolution spectrum of carbon dioxide. (From Goody, R.M., and Yung, Y.L., *Atmospheric Radiation*, 2nd edn., Oxford University Press, New York, 1989.) The notation (01^10), etc., is a designation used to show the quantum state of a harmonic oscillator. In the general case $(v_1v_2v_3)$, the v_i are the vibrational quantum numbers, and l is the quantum number for angular momentum. Transitions between two energy states, such as those denoted by $(00^00) \rightarrow (01^10)$, give rise to absorption lines. Certain selection rules govern the allowable transitions.

lines are so closely spaced in certain spectral regions that individual lines are not fully resolved by experimental measurements. As a consequence of broadening, the lines may overlap to form absorption bands. Figure 9.4 shows an example of carbon dioxide absorption bands observed with low spectral resolution.

The large number of possible energy transitions that can produce spectral lines is illustrated by the many energy levels and transition arrows in Figure 9.10. This shows the potential energy for a diatomic molecule as a function of the separation distance between its two atoms. The two curves are each for a different electronic energy state where the electron may be shared by the two atoms. The distance R_e is the mean interatomic distance corresponding to each of the electronic states. The long-dashed horizontal lines denote vibrational energy levels, while the short-dashed lines are rotational states superimposed on the vibrational states. Transitions between rotational levels of the *same* vibrational state involve small $e_j - e_i$. Hence, from Equation 9.2, these transitions give lines at low wave numbers in the far infrared. Transitions between rotational levels in *different* vibrational states of the same electronic state give vibration-rotation bands at wave numbers in the near infrared. If transitions occur from a rotational level of an electronic and vibrational state to a rotational level in a different electronic and vibrational state, then large $e_j - e_i$ are involved and a band system can be formed in the high wave number visible and ultraviolet regions.

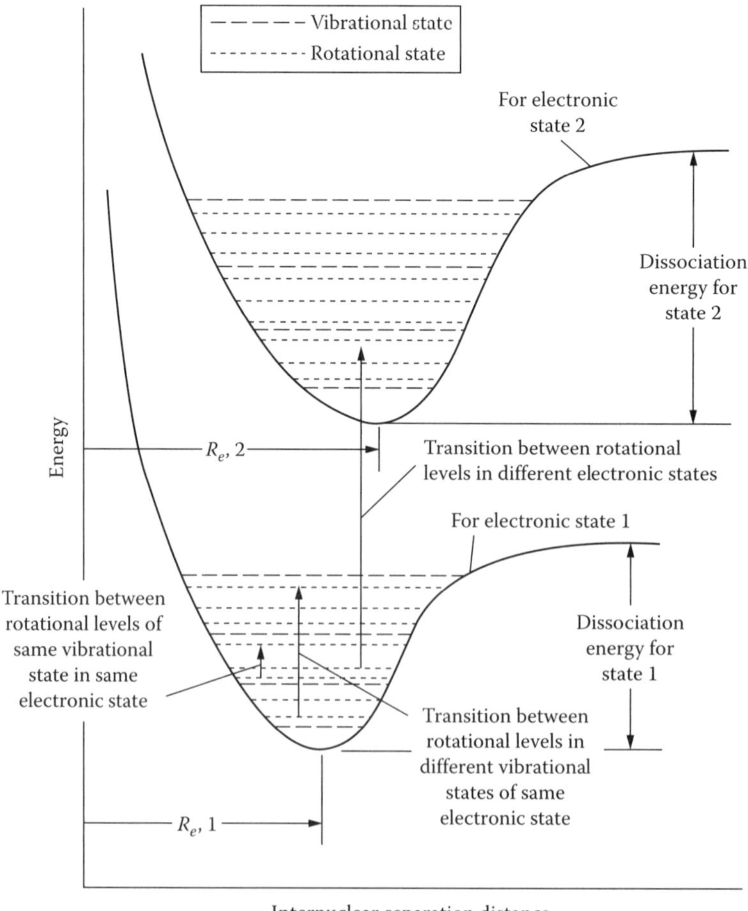

FIGURE 9.10 Potential energy diagram and transitions for a diatomic molecule.

9.2.5.2 Types of Band Models

Two categories of models are employed for the absorption and emission of bands of closely spaced lines. *Narrow-band* models use characterizations of individual line shapes, widths, and spacings to derive band characteristics within a specified wave number interval. *Wide-band* models provide correlations of band characteristics over the entire wave number region of the band, and account for the increasing importance of weakly absorbing lines in the wings of the band as the radiation path length becomes large. General discussions of earlier wide-band models are in Edwards (1976), Tiwari (1976), and Morizumi (1979). More accurate contemporary wide-band models use the k-distribution or spectrally line-weighted models (Section 9.3.2.1). In a detailed radiation exchange calculation, the absorbed and emitted energy are needed in each band region, such as in the four main CO_2 bands of Figure 9.4. These bands are separated by spectral regions that are nearly transparent. For the total absorbed or emitted energy in a *uniform* gas, Equations 9.12 and 9.13 are used; both involve the same type of integral. For an absorption band that occupies a narrow spectral region, an average value of $I_\eta(0)$ or $I_{\eta b}$ can be taken out of the integral for each band. For example, the total emittance in Equation 9.13 is evaluated as

$$\varepsilon(S) = \frac{\pi \sum_l I_{\eta b,l} \int_l \left[1 - e^{-\kappa_\eta S}\right] d\eta}{\sigma T^4} \tag{9.25}$$

where the subscript l denotes a band, the integral is over each band, and the summation is over all the bands.

Similar to the equivalent line width in Equation 9.18, the integral in Equation 9.25 is defined as the *effective bandwidth* $\bar{A}_l(S)$:

$$\bar{A}_l(S) \equiv \int_{\substack{absorption\\bandwidth}} [1 - e^{-\kappa_\eta S}] d\eta \tag{9.26}$$

and Equation 9.25 becomes

$$\varepsilon(T,P,S) = \frac{\pi}{\sigma T^4} \sum_l I_{\eta b,l} \bar{A}_l(S) \tag{9.27}$$

The $\bar{A}_l(S)$ has units of the spectral variable, which is η for Equation 9.26. The span of the absorption band that provides the upper and lower limits of the integral in Equation 9.26 does not have a specific value that applies to all conditions. It can be defined as the spectral interval beyond which there is only a specified small fractional contribution to $\bar{A}_l(S)$. The width of this interval increases slowly with path length as absorption becomes more important in the wings of the band.

The total emittance in Equation 9.27 can be used for engineering calculations of radiation from an isothermal uniform gas to an enclosure boundary. The $\bar{A}_l$ can also be used to obtain each band absorptance α_l (T, P, S) for detailed spectral-exchange calculations in enclosures. To calculate quantities such as ε_l and α_l, (and, from Equation 9.15, t_l) for various conditions, correlations for $\bar{A}_l$ are needed as a function of path length, pressure, temperature, etc., for the important bands of each important radiating gas. Correlations for $\bar{A}_l$ are now discussed.

By comparing Equations 9.26 and 9.18, the effective bandwidth $\bar{A}_l$ for a band is the sum of the $\bar{A}_{ij}$ for all the spectral lines in the band if each $\bar{A}_{ij}$ acts independently. Generally, spectral lines overlap, and each line does not absorb as much energy as it would if it acted independently. An absorption band typically includes many broadened lines. Hence, κ_η in Equation 9.26 is a complicated irregular function of wave number, and the integration for $\bar{A}_l$ is difficult as it requires that the detailed shape of each broadened line be known. This motivates the need for detailed spectral line databases for gases that are important in engineering radiative heat transfer.

9.2.5.3 Databases for the Line Absorption Properties of Molecular Gases

Compilations of the line characteristics of common molecular gases are available for download. These line-by-line data sources generally provide the wave number of the line center, Lorentz air-broadened line half-width, and line intensity for all important lines of common molecular gases. This is the basic data required for performing detailed line-by-line spectral analyses. Their high resolution makes them impractical for engineering calculations, but they can be used to provide benchmark calculations for determining the accuracy of less detailed engineering models.

The HITRAN2012 compilation* provides 7,400,447 spectral lines for 47 different molecules at 297 K, and details of how the data are compiled is in Rothman (1996) and Rothman et al. (2005, 2010). At higher temperatures, certain lines ("hot lines") that are unimportant and are neglected near room temperature become much more prominent, and the low-temperature database is not adequate for determining spectral properties. Approximate databases that include the hot lines are available (Scutaru et al. 1993, Rivière et al. 1995, 2012). They include appropriate hot lines for use to 2500 K. The HITEMP database extends the HITRAN database to include hot lines (Varanasi 2001, Rothman et al. 2005, 2010, 2012, Gordon et al. 2007). The HITEMP 2010 version extends the HITRAN data for CO_2, H_2O, CO, NO, and OH. The HITEMP 2010 data for water vapor has been validated up to 4000 K. HITRAN 2010 absorption cross section data for a CO_2 band are shown in Figure 9.11. The HITRAN database is published (Rothman et al. 2013) within a special issue of the *Journal of Quantitative Spectroscopy and Radiative Transfer* (see Ioniu et al. [2013] for the preface and the details).

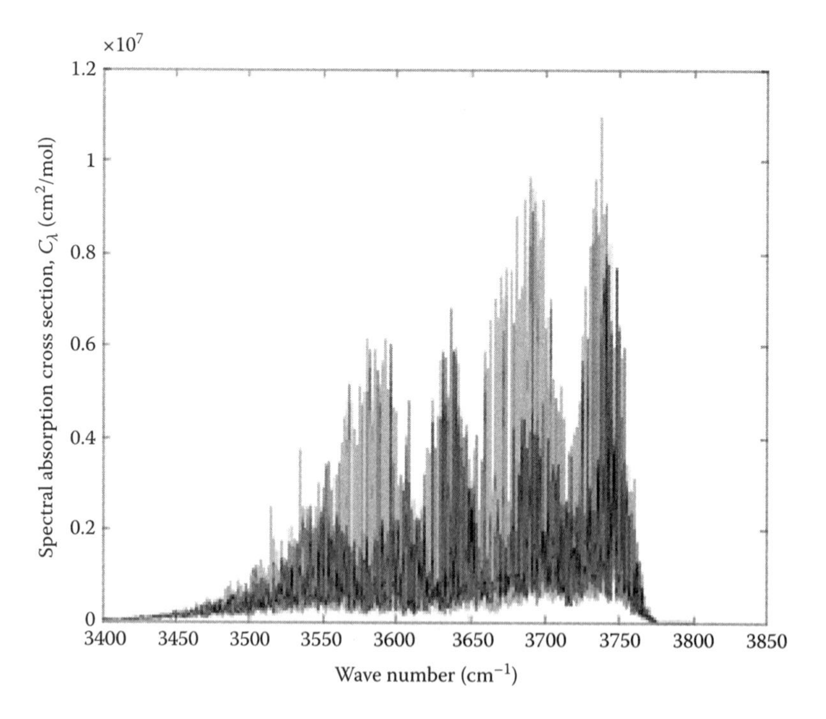

FIGURE 9.11 Example spectral absorption cross section (κ_λ/N) for lines in the 2.7 μm band of homogeneous CO_2 with uniform temperature of $T = 1000$ K, 1 atm. (Generated from program HITRAN 2010.)

* These data can be accessed at the website http://www.cfa.hitran.com/. The advances in the field are periodically reported in *JQSRT*. A similar database that includes data for 50 Earth and other planetary atmospheric gases at 296 K is at http://ara.lmd.polytechnique.fr/htdocs-public/products/GEISA/HTML-GEISA, home of the GEISA line-by-line database (Gestion et Etude des Informations Spectroscopiques Atmospheriques). (Jacquinet-Husson, N. et al., The 2009 edition of the GEISA spectroscopic database, *JQSRT*, 112, 2395–2445 2011). The 2011 version of GEISA is accessible at http://www.pole-ether.fr.

More specialized line-by-line databases are available, but access may require security clearance and/or payment. For example, databases for the very high temperatures and species experienced during spacecraft entry into planetary atmospheres can be accessed through the programs NEQAIR (Whiting et al. 1996); SPECAIR (Laux 2002, 2006); HARA (Johnston et al. 2008a,b), and, for atomic spectra, the Atomic Spectra Database (ASD) available from the National Institutes of Standards and Testing (Kramida et al. 2012). A new code, HyperRad, is being developed at NASA Ames Research Center. It is meant for incorporation into hypersonic flow codes and includes line spectra based on ab initio calculation from basic quantum physics to the extent possible, and includes non-LTE effects. Comparisons of predictions of line spectra from NEQAIR, HARA, and HyperRad with experimental shock-tube data are in Brandis et al. (2011, 2013) and Cruden et al. (2012).

9.3 BAND MODELS AND CORRELATIONS FOR GAS ABSORPTION AND EMISSION

Because of the difficulty of computing radiative transfer by integrating over the detailed line-by-line data, various models have been constructed to approximate the line structure over a spectral interval. When this spectral interval is a portion of a complete vibration-rotation band, the model is a *narrow-band model.* When the model covers an entire vibration-rotation band, it is a *wide-band model.*

9.3.1 Narrow-Band Models

The radiative absorption within a narrow spectral bandwidth $\Delta\eta$ can be determined by integrating over all lines with contributions within the band. This line-by-line calculation can use a database such as HITRAN for individual line intensities, wave numbers, and shapes that occur within the narrow band at the specified temperature, total pressure, and species partial pressure. Although the evaluation can be done, it is tedious.

To simplify the calculation of band behavior, models of the line structure typical of a narrow band have been proposed. These models use the observation that the large variations of line intensity provide a much larger effect than the variation in line widths. The common models then assume that all lines have the same half-width. The models vary in their assumed distributions of line spacing, shape, and intensity. For engineering applications at or above atmospheric pressure, the lines can be assumed to be pressure-broadened and thus all to have a Lorentz profile. At lower total pressures, the Voigt or Doppler profile may be appropriate (Goody and Yung 1989). Three commonly used narrow-band models are now discussed.

9.3.1.1 Elsasser Model

The lines within the narrow band are assumed to be overlapping, but have the same intensity $S_{ij} = S_c$, have equal spacing δ within the band (Figure 9.12a), and all have a Lorentz shape with the same half-width γ_c. The absorption coefficient for the band is found by summing Equation 9.7 over the identical lines:

$$\kappa_\eta = \frac{S_c}{\pi}\sum_{n=-\infty}^{\infty}\frac{\gamma_c}{\gamma_c^2+(\eta-n\delta)^2} = \frac{S_c}{\delta}\frac{\sinh(2\beta)}{\cosh(2\beta)-\cos[2\pi(\eta-\eta_{ij})/\delta]} \tag{9.28}$$

where $(\eta - \eta_{ij})$ varies between $-\delta/2$ and $\delta/2$ over each periodic line interval and $\beta = \pi\gamma_c/\delta$. Defining $\mu = SS_c/\delta$ and $z = 4(\eta - \eta_{ij})/\delta$, the effective bandwidth for a band of m lines is then

$$\frac{\bar{A}_l}{m\delta} = 1 - \frac{1}{2}\int_0^2 \exp\left\{-\frac{\mu\sinh(2\beta)}{\cosh(2\beta)-\cos(\pi z/2)}\right\}dz \tag{9.29}$$

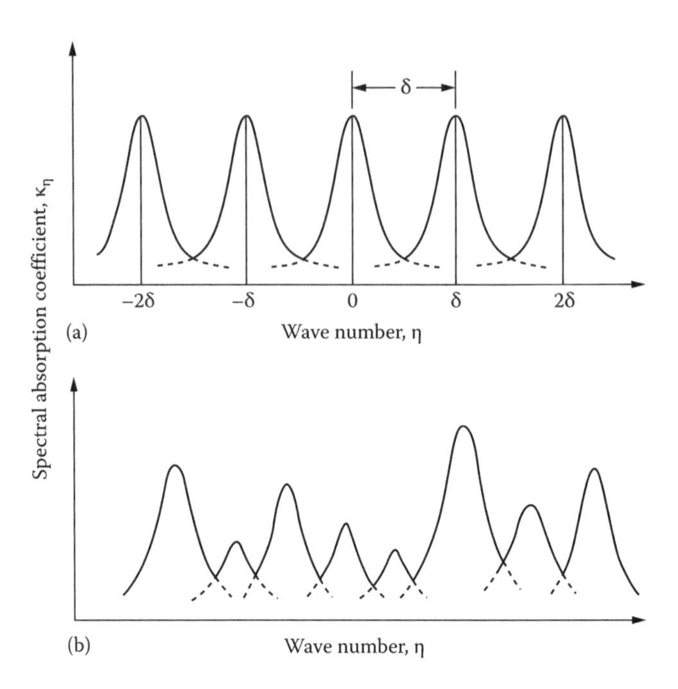

FIGURE 9.12 Models of absorption lines forming an absorption band, (a) Elsasser model that uses equally spaced Lorentz lines of equal line intensity; (b) statistical model.

Although the integral in Equation 9.29 cannot be evaluated in closed form, some informative limiting results are found. For *weak nonoverlapping lines*, $\mu \ll 1$, each line acts independently, and $\bar{A}_l/m\delta$ for the band of m lines is the same as for an individual line. Using Equation 9.19 with $S_{ij} = S_c$ and $\omega = m\delta$ in this band model gives

$$\frac{\bar{A}_l}{m\delta} = \frac{\bar{A}_l}{\omega} = \frac{\bar{A}_{ij}}{\delta} = \mu \quad \mu \ll 1; \quad \mu = \frac{SS_c}{\delta} \tag{9.30}$$

For a *long radiation path*, $\mu \gg 1$, and, letting the overall bandwidth of m lines be $\omega = m\delta$, Equation 9.29 yields

$$\frac{\bar{A}_l}{\omega} = 1 \quad \mu \gg 1 \tag{9.31}$$

This represents total absorption within the entire width of the band. Three other limits and the conditions giving them are summarized in Tiwari (1976). For strong non-overlapping lines, Equation 9.29 yields

$$\frac{\bar{A}_l}{\omega} = \operatorname{erf}\left(\sqrt{\beta\mu}\right) \quad \beta \ll 1, \quad \frac{\mu}{\beta} \gg 1 \tag{9.32}$$

If in addition $\beta\mu$ is small so the lines are thin and spaced well apart,

$$\frac{\bar{A}_l}{\omega} = 2\sqrt{\frac{\beta\mu}{\pi}} \quad \beta \ll 1, \quad \frac{\mu}{\beta} \gg 1, \quad \beta\mu \ll 1 \tag{9.33}$$

This is called the *square-root limit*. If β is very large, $\gamma_c \gg \delta$, the lines are very broad compared with the spacing between them, and the lines are spread out over the band. For this case,

$$\frac{\bar{A}_l}{\omega} = 1 - e^{-\mu} \tag{9.34}$$

where, with the line structure spread out, the S_c/δ in $\mu = SS_c/\delta$ is the average absorption coefficient.

9.3.1.2 Goody Model

The lines are assumed to have a random distribution of wave number positions within the band as in Figure 9.12b, but have an exponential distribution of intensities. Again, all lines are assumed to have a Lorentz shape and to have the same half-width.

9.3.1.3 Malkmus Model

In this model, the lines are assumed to have a random distribution of wave numbers within the band, and to have a line intensity, S_{ij}, distribution proportional to the inverse of the intensity for low-intensity lines, and a distribution that decreases exponentially for high-intensity lines (the "exponential tailed S_{ij}^{-1} distribution"). In this model, the intensity distribution is cut off above a maximum intensity and below a minimum line intensity. The inverse distribution for low-intensity lines has been shown to be in better agreement with actual line intensity distributions than is the exponential distribution (Malkmus 1967). The Malkmus model has been shown in a number of studies to provide the most accurate prediction of band absorption behavior in comparison with detailed line-by-line calculations.

Many other narrow band models have been proposed, some of them more useful in certain cases than the Elsasser, Goody, or Malkmus models (Edwards and Menard 1964, Ludwig et al. 1973, Goody and Yung 1989). Modifications to the Elsasser model have been made by using evenly spaced lines with a Doppler profile (Golden 1967, 1968, Kyle 1967), and by using a Voigt profile (Golden 1969).

9.3.1.4 Wide Band Models

Models have been proposed that provide correlations of the absorption properties of an entire vibration-rotation band. The most complete such model is the exponential wide band model developed by Edwards and coworkers, who provide correlations for the bands found for many gases of practical interest. The correlations are based on a combination of the analytical predictions of line and band behavior coupled with experimental measurements. The details are in Appendix A in the online website at http://www.thermalradiation.net. The models are not discussed here, because they have been largely superseded by the *k*-distribution method, which can provide much more accurate treatment of spectral effects. However, the wide-band models can still provide useful correlations for some gases of industrial interest such as NO_x and CH_4.

9.3.2 Contemporary Band Correlations

Recent approaches to band correlations have led to helpful simplifications. Most of these methods involve transforming the spectral distribution of the absorption coefficient into a probability density versus absorption coefficient space for a given narrow or wide band, and in some cases integrating the probability density function (PDF) to give a cumulative distribution function (CDF) of the absorption coefficient. The smoothly varying PDF or CDF is then used to provide spectral property dependence in the radiative transfer relations. The PDFs and CDFs must be developed from fundamental data on the spectral line absorption behavior in each band from the HITRAN database for low temperatures (Rothman et al. 2005, 2010, 2012), and in HITEMP for higher temperature data (Rothman et al. 2010). They can also be developed from summing narrow-band models (such as the Elsasser, Goody, or Malkmus models) over the entire wide band, or from the parameters developed for other wide-band models. The *k*-distribution base models are discussed in the next section.

9.3.2.1 *k*-Distribution Method

To find solutions for total radiative energy transfer, the transfer equations described in Chapter 1 can be solved in many spectral intervals and integration is then carried out to determine the spectral energy. Since κ_η varies in a detailed way with η, this integration requires solving the transfer equations at many very closely spaced spectral intervals, followed by integration across the spectrum to obtain the total flux or flux divergence. For complex line spectra, such a line-by-line solution and integration is time consuming and, in most cases, impractical for engineering calculations.

The *k-distribution method* uses a transformation of variables to reduce the extent of the spectral calculations. It was introduced for astrophysical applications and was originally limited to homogeneous absorbing media. Many approaches to improving the method for engineering applications have been proposed. Modest (2013) provides a review of developments of the method.

The method was first applied to a narrow band of the complete spectrum. In this case, the assumption is made that the local blackbody spectral intensity does not vary appreciably over the narrow absorption band and can be treated as constant over the band width. The spectral intensity within the band as obtained from the radiative transfer equation is then only a function of the spectral absorption coefficient κ_η (the name for the method arises from the use in atmospheric radiation of the symbol *k* for the absorption coefficient). The local radiative flux within the band *l* is expressed, using $\mu = \cos\theta$ and integrating over all incidence angles over a sphere surrounding the direction of propagation (derivation is shown in Equation 10.28),

$$q_l = \int_{\Delta\eta_l} q_{\eta,l}(\kappa_\eta)d\eta = 2\pi \int_{\Delta\eta_l} \left[\int_{\mu=-1}^{1} I_\eta(\kappa_\eta,\mu)\mu d\mu \right] d\eta \tag{9.35}$$

Most of the line-by-line databases use wave number η as the spectral variable.

Figure 9.11 is typical of the line spectra of molecular gases. To eliminate the line-by-line integration of such properties over the spectrum, we observe that the integration of an arbitrary function $H(\kappa_\eta)$ over wave number can be replaced by integration over absorption coefficient κ_η by using the transformation

$$\frac{1}{\Delta\eta_l} \int_{\Delta\eta_l} H(\kappa_\eta)d\eta = \int_{\kappa_\eta=0}^{\infty} H(\kappa_\eta) f(\kappa_\eta) d\kappa_\eta \tag{9.36}$$

where $f(\kappa_\eta)d\kappa_\eta$ is the PDF of the spectral absorption coefficient within the band (or *k-distribution*), which must satisfy

$$\int_{\kappa_\eta=0}^{\infty} f(\kappa_\eta)d\kappa_\eta = 1 \tag{9.37}$$

Because of the rapid variation of line-by-line absorption coefficients with wave number, it is more accurate and takes less computational effort to replace the integral on the left of Equation 9.36 with the integration over the smoothly varying PDF of the band absorption coefficient. By examining Figure 9.12b, which is typical of absorption band line structure, the wave number integration across the band can be replaced by a summation of individual integrations between the peaks and valleys of the absorption coefficient. Each integration is then over a monotonically increasing or decreasing κ_η in a portion of the band $\Delta\eta_j$, and Equation 9.35 becomes

$$q_l = 2\pi \sum_{j=1}^{N_j} \int_{\Delta\eta_j} \left[\int_{\mu=-1}^{1} I_{\eta,j}(\kappa_\eta, \mu)\mu d\mu \right] d\eta \tag{9.38}$$

Equation 9.38 is now transformed into a sum of integrations over absorption coefficient rather than over wave number by using Equation 9.36. The first step is to write

$$q_l = 2\pi \sum_{j=1}^{N_j} \int_{\Delta\eta_j} \left[\int_{\mu=-1}^{1} I_{\eta,j}(\kappa_\eta, \mu)\mu d\mu \right] d\eta$$

$$= 2\pi\Delta\kappa_l \sum_{j=1}^{N_j} \int_{\kappa_{j,\min}}^{\kappa_{j,\max}} \left[\int_{\mu=-1}^{1} I_{\eta,j}(\kappa_\eta, \mu)\mu d\mu \right] \frac{1}{\Delta\kappa_l} \left| \frac{d\eta}{d\kappa_\eta} \right|_j d\kappa_\eta \tag{9.39}$$

The PDF of the absorption coefficient within the band l is

$$f(\kappa_\eta) d\kappa_\eta = \sum_{j=1}^{N_j} \frac{1}{\Delta\kappa_l} \left| \frac{d\eta}{d\kappa_\eta} \right|_j \left[h(\kappa_\eta - \kappa_{\eta,\min}) - h(\kappa_\eta - \kappa_{\eta,\max}) \right] d\kappa_\eta \tag{9.40}$$

where h is the step function and is used to define the variation in κ_η between peaks and valleys or vice versa. Equation 9.39 has yielded the form of the right side of Equation 9.36 so that

$$q_l = 2\pi\Delta\eta_l \int_{\kappa_\eta=0}^{\infty} \left[\int_{\mu=-1}^{1} I_\eta(\kappa_\eta, \mu)\mu d\mu \right] f(\kappa_\eta) d\kappa_\eta = 2\pi\Delta\eta_l \int_{\mu=-1}^{1} I_l(\mu)\mu d\mu \int_{\kappa_\eta=0}^{\infty} f(\kappa_\eta) d\kappa_\eta \tag{9.41}$$

where it is assumed that I_l does not vary over the band width of band l.

The integration over wave number has now been transformed into integration over the PDF of κ_η within the band. The PDF must be generated for the given narrow band. This may be done by using a standard narrow-band model or a line-by-line database. Domoto (1974) derived an expression for the k-distribution using the Malkmus narrow band model. Tang and Brewster (1994) used the Elsasser narrow-band model with coefficients from Edwards' wide-band model (see Appendix A in the on-line Appendices at www.thermalradiation.net) to derive the k-distribution for six CO_2 bands. Most contemporary k-distributions are generated from direct line-by-line integration using online databases (Section 9.2.5.3).

Difficulties with the standard k-distribution method for engineering applications include accounting for (1) the overlap of lines and bands in mixtures of participating gases; and (2) the effects of inhomogeneities in gas properties due to variations in temperature, concentration, and possibly pressure. These are concerns with all attempts to construct an easily used technique for accurately incorporating spectral characteristics of gases.

9.3.2.2 Correlated-*k* Assumption

If properties vary with position, the *correlated-k* or *c-k assumption* can be implemented, resulting in the *c-k* method (Goody and Yung 1989, Goody et al. 1989, Liu et al. 2000a). The assumption is based on the observation that over a narrow spectral interval the radiative transfer is insensitive to the exact placement of spectral lines within the interval. Thus, reordering the lines within the interval should not affect the radiative transfer. The *c-k* method can be applied to either a narrow- or wide-band model if Planck distribution does not vary significantly across the band, and with scattering, if the scattering coefficient and phase function are invariant across the spectral width of the band.

The absorption coefficient PDF, $f(\kappa_\eta)d\kappa_\eta$, can then be used to compute a CDF, $g(\kappa_\eta)$, through the relation

$$g(\kappa_\eta) = \int_{\kappa_\eta^*=0}^{\kappa_\eta} f(\kappa_\eta^*)d\kappa_\eta^* \tag{9.42}$$

The advantage of having the CDF is that the average transmission within an absorption band l over a path length S is

$$\bar{T}_l(S) = \int_{g=0}^{1} \exp\left[-\kappa_\eta(g_l)S\right]dg_l \tag{9.43}$$

where $\kappa_n(g_l)$ is the spectral absorption coefficient evaluated at the value g_l of the CDF. If the absorption properties vary with position, this relation is modified to

$$\bar{T}_l(S) = \int_{g=0}^{1} \exp\left[-\int_0^S \kappa_\eta(g_l, S^*)dS^*\right]dg_l \tag{9.44}$$

Closed-form relations for $f(\kappa_\eta)$ are in Marin and Buckius (1996) for narrow bands based on the Malkmus and Goody narrow-band models using Edwards' exponential wide-band model (see the details of Edwards' model in the online Appendix A website http://www.thermalradiation.net) to provide the line-spacing and pressure-broadening parameters. The narrow-band models are then used to generate wide-band absorptances that are shown to be within ±12% of experimental values. The Malkmus closed-form PDF relations are then integrated to provide closed-form CDF relations. This work shows that the *c-k* method can be applied to provide accurate wide-band property relations.

Marin and Buckius (1998a,b) further developed the wide-band $g(\kappa_\eta)$ correlations $[\equiv g_{wb}(\kappa_\eta)]$ by using line-by-line data for six CO_2 bands for temperatures up to 1000 K, and four H_2O rotational-vibrational bands for T to 2900 K, and the rotational H_2O band for T to 1900 K. Simplifying assumptions provided closed-form expressions for $g_{wb}(\kappa_\eta)$ as

$$g_{wb}(\kappa^*) = \begin{cases} 0 \quad \text{for } \kappa^* \le \exp\left[2\exp\left(-\dfrac{B}{4\pi}\right) - W(B,R)\ln R\right] \\[2ex] W(B,R) + \dfrac{1}{\ln R}\left[\ln\kappa^* - 2\exp\left(-\dfrac{B}{4\pi}\right)\right] \quad \text{for } \exp\left[2\exp\left(-\dfrac{B}{4\pi}\right) - W(B,R)\ln R\right] < \kappa^* < R \\[2ex] 1 + W(B,R) - \dfrac{2\exp(-B/4\pi)}{\ln R}\sqrt{\dfrac{R}{\kappa^*}} \quad \text{for } R < \kappa^* \le \kappa_{\max}^* \\[2ex] 1.0 \quad \text{for } \kappa^* > \kappa_{\max}^* \end{cases} \tag{9.45}$$

where

$$\kappa^* = \kappa_\eta \omega R / \{\rho\alpha[1-\exp(-B/\pi)]\}; \quad R = \exp\left(\frac{\Delta\eta}{\omega}\right); \quad \Delta\eta = \eta_u - \eta_l;$$

$$W(B,R) = \frac{2\exp(-B/4\pi)}{\ln R}\sqrt{\frac{[1-\exp(-B/\pi)][\cosh(2B)-1]}{\sinh(2B)}}$$

$$\kappa^*_{\max} = \{R/[1-\exp(-B/\pi)]\}\sinh(2B)/[\cosh(2B)-1]$$

Values for B, P_e, $\alpha(T)$, $\beta(T)$, and $\omega(T)$ are obtained from relations in Edward's wide-band models (see http://www.thermalradiation.net) as in the wide-band correlation Example A.1 in the on-line appendix, but the wide band correlation parameters to use in the simplified model are in Tables 9.1 and 9.2 for CO_2 and H_2O. As in the exponential wide band correlation, P is the total pressure of the gas and ρ is the density of the radiating constituent. The relations presented here differ somewhat from the original references so that the definition of B is kept consistent.

The form of Equation 9.45 was chosen so that it can be inverted to provide expressions for the normalized absorption coefficient $\kappa^*(g_{wb})$. The result is

$$\left.\begin{aligned}
&\kappa^*(g_{wb}) = \exp\left\{[g_{wb}(\kappa^*) - W(B,R)]\ln R + 2e^{-B/4\pi}\right\} \\
&\text{for } 0 \le g_{wb}(\kappa^*) \le 1 + W(B,R) - \frac{2e^{-B/4\pi}}{\ln R} \\
&\kappa^*(g_{wb}) = R\left\{[1 + W(B,R) - g_{wb}(\kappa^*)]\frac{\ln R}{2(e^{-B/4\pi})}\right\}^{-2} \\
&\text{for } \left[1 + W(B,R) - \frac{2e^{-B/4\pi}}{\ln R}\right] < g_{wb}(\kappa^*) \le 1
\end{aligned}\right\} \quad (9.46)$$

TABLE 9.1
Band Parameters for Carbon Dioxide[a]

Band (μm)	η_l (cm^{-1})	η_u (cm^{-1})	Transition δ	α_0 (m^2/cm·g)	β_0/π	ω_0 (cm^{-1})
2.0	4700	5250	2, 0, 1	0.084	0.017	61.0
2.7	3300	3800	1, 0, 1	4.3	0.040	22.1
			$\Delta\eta_1^b$	4.3[b]	0.040[b]	22.1[b]
			$\Delta\eta_2^b$	0.2[b]	1×10^{4} [b]	40.0[b]
4.3	1950	2400	0, 0, 1	117.0	0.102	11.7
			$\Delta\eta_2^b$	105.3[b]	0.102[b]	10.5[b]
			$\Delta\eta_2^b$	0.045[b]	0.0030[b]	10.8[b]
9.4	1000	1125	0, −2, 1[c]	3.4×10^{-9}	0.025	20.0
10.4	850	1000	−1, 0, 1	2.58×10^{-9}	0.0210	30.0
15.0	450	850	0, 1, 0	12.3	0.042	15.2

Source: Marin, O. and Buckius, R., *IJHMT*, 41, 3881, 1998b.

[a] The $b = 1.3$ for all bands; $n = 0.65$ for the 2.0 and 2.7 μm bands, $n = 0.7$ for the 15 μm band, and $n = 0.8$ for the other bands; $T_0 = 100$ K, $P_0 = 1$ atm.

[b] For use when bands overlap: see Marin and Buckius (1998b).

[c] Use values for the 10.4 μm band instead of those for the 9.4 μm band.

TABLE 9.2
Band Parameters for Water Vapor[a]

Band (μm)	η_l (cm^{-1})	η_u (cm^{-1})	Transition δ	α_0 (m^2/cm·g)	β_0/π	ω_0 (cm^{-1})
1.38	6000	8000	1, 0, 1	2.17	0.1180	78.0
1.87	4400	6000	0, 1, 1	2.64	0.0850	70.6
2.7	2600	4400	0, 2, 0	0.19[b]	0.0360[c]	83.0
			1, 0, 0	2.30[b]		
			0, 0, 1	23.3[b]		
6.3	1000	2600	0, 1, 0	39.6	0.0566	72.8
Rotational	150	1000	0, 0, 0	$420.0\exp(-3\sqrt{100/T})$	0.1140	37.0

Source: Marin, O. and Buckius, R., *IJHMT*, 41, 2877, 1998a.

[a] For all bands, $b = 8.6\sqrt{(T/T_0)} + 0.5, n = 1, T_0 = 100\,K, P_0 = 1\,\text{atm}$.

[b] $\alpha_{2.7} = \Sigma_{j=1}^{3}\alpha_j$.

[c] $\gamma_{2.7} = (1/\alpha_{2.7})(\Sigma_{j=1}^{3}\sqrt{\alpha_j\gamma_j})^2$.

Example 9.1

Find and plot a graph of the CDF versus κ_η relation for the 9.4 μm band of pure CO_2 at 500 K, and find the effective bandwidth for a path length of $S = 0.364$ m.

Relations from the exponential wide-band model for α, β, and ω, and the transitions (the −1, 0, 1 are used). These are given in Example A.1 for the same parameters as in this example in web Appendix A at http://www.thermalradiation.net. However, the coefficients α_0, β_0, and ω_0 for these relations differ somewhat, and the values from Table 9.1 must be used for best agreement of Marin–Buckius relations with experimental data. From the web example, the resulting values are $\alpha = 0.0185$ m^2/g·cm, $\beta = 0.1537$, $B = 0.1896$, and $\omega = 44.72$ cm^{-1}. The necessary quantities given below Equation 9.45 are now calculated for using the correlations. This gives $R = 16.36$, $W = 0.0738$, and $\kappa^*_{max} = 1492$. Substituting these values into the correlation Equation 9.45 results in the CDF having the values

$$g_{wb}(\kappa^*) = 0 \quad \text{if } \kappa^* \le 5.834$$

$$g_{wb}(\kappa^*) = W(B,R) + \frac{1}{\ln R}\left[\ln\kappa^* - 2e^{-B/4\pi}\right] \quad \text{if } 5.834 < \kappa^* < 16.36$$

$$g_{wb}(\kappa^*) = 1 + W(B,R) - \frac{2e^{-B/4\pi}}{\ln R}\sqrt{\frac{R}{\kappa^*}} \quad \text{if } 16.36 < \kappa^* \le 1492$$

$$g_{wb}(\kappa^*) = 1 \quad \text{if } \kappa^* > 1492$$

Similarly, the relation for $\kappa^*(g_{wb})$ are, from Equation 9.46,

$$\kappa_1^*(g_{wb}) = \exp\{[g_{wb}(\kappa^*) - W(B,R)]\ln R + 2e^{-B/4\pi}]\} \quad \text{for } 0 \le g_{wb}(\kappa^*) \le 0.369$$

$$\kappa_2^*(g_{wb}) = R\left\{[1 + W(B,R) - g_{wb}(\kappa^*)]\frac{\ln R}{2e^{-B/4\pi}}\right\}^{-2} \quad \text{for } 0.369 < g_{wb}(\kappa^*) \le 1$$

From the definition of κ^* below Equation 9.46, the actual absorption coefficient κ_η is found from $\kappa_\eta = \kappa^*\,\rho\alpha[1-\exp(-B/\pi)]/\omega R$. The plots of $g_{wb}(\kappa_\eta)$ versus κ_η and the inverted function $\kappa_\eta(g_{wb})$ versus $g_{wb}(\kappa_\eta)$ are in Figure 9.13.

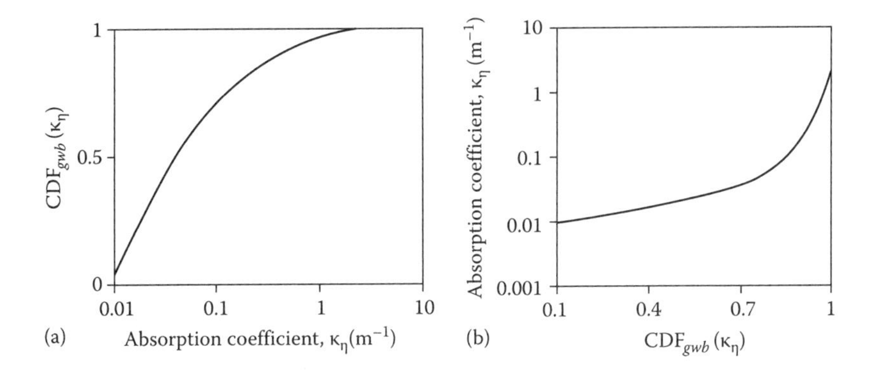

FIGURE 9.13 Results for *c-k* distribution for 9.4 μm band of pure CO_2 at 1 atm, 500 K.

Given the functions for $\kappa^*(g_{wb})$, Equation 9.45 can be used to find the effective bandwidth. Using a path length of $S = 0.364$ m, and $\rho = pM/RT = 1$ atm × 44 (kg/kg·mol)/[0.08206(atm·m³/kg·mol·K) × 500 K] = 1.072 kg/m³ = 1072 g/m³. The transmittance is

$$T_{9,4\mu m}(S) = \int_{g=0}^{0.369} \exp\left[-\frac{\rho\alpha}{\omega R}(1-e^{-B/\pi})\kappa_1^*(g_{wb})S\right]dg_{wb}$$

$$+\int_{g=0.369}^{1} \exp\left[-\frac{\rho\alpha}{\omega R}(1-e^{-B/\pi})\kappa_2^*(g_{wb})\right]dg_{wb} = 0.949$$

The effective bandwidth is then, using η_u and η_l from Table 9.1,

$$\frac{\bar{A}_{9.4\ \mu m}}{\eta_u - \eta_l} = 1 - T_{9.4\ \mu m}$$

giving

$$\bar{A}_{9.4\ \mu m} = (1-0.949)125 = 6.40\ \ cm^{-1}.$$

Marin and Buckius correlations use a constant bandwidth of $\eta_u - \eta_l$ in their calculations rather than the calculated bandwidth ω used in the exponential wide-band models. The answer is in reasonable agreement with the experimental result of 5.9 cm⁻¹.

The statistical narrow-band correlated-*k* method was tested in two-dimensional (2D) applications (Liu et al. 2000a, Goutiere et al. 2001) and was extended to formulate a wide-band model by lumping 5–20 narrow bands. A number of discrete-ordinates-based solutions were presented for various assumed gas temperature distributions. Comparisons of wall heat flux and radiative source distributions using lumped bands were within 2%–3% of the statistical narrow-band solutions, but showed computational speedup factors of 6 or more. Park and Kim (2005) carried out similar calculations in 3D, and showed similar gains in efficiency.

9.3.2.3 Full Spectrum *k*-Distribution Methods

Full-spectrum methods abandon the band-by-band approach of the previous section, and provide *k*-distributions valid over the entire spectrum. The *k*-distributions are weighted by the blackbody spectrum evaluated at a reference temperature. For a homogeneous gas, the CDF for the full-spectrum k-distribution (FSK) is defined by

$$g(\kappa) = \frac{\pi}{\sigma T^4} \int_{\lambda \text{ s.t. } \kappa_\lambda \le \kappa} I_{\lambda,b}\, d\lambda \tag{9.47}$$

This is also known as the absorption line blackbody distribution function (ALBDF).

Figure 9.14 illustrates the regions over which the CDF is generated for a particular value of κ for a region of the entire spectrum. As noted earlier, for an actual molecular spectrum, solution of the RTE at each spectral interval followed by integration of intensity over the spectrum requires extremely small increment size for accurate integration (e.g., see Figure 9.11). Generation of the CDF at a particular temperature is also time consuming and requires extremely small increments in wave number, but is done prior to solution of the RTE; the RTE then need be solved over only a relatively few increments of the CDF (Chapter 10). The presence of the HITRAN, HITEMP, and GEISA databases for molecular lines (Section 9.2.5.3) makes possible an accurate generation of the FSK CDF across the entire spectrum.

Figure 9.15 shows the FSK CDF generated from HITRAN2010 data for CO_2 at T = 1000 K. Rather than absorption coefficient κ, the CDF is shown in terms of the absorption cross section C, where $C = \kappa/N$ and N is the number density (molecules/volume) of the CO_2 found for an ideal gas from $N = p/RT$. Here, p is the partial pressure of the CO_2. The range of wavenumber from 0 to 14000 encompasses over 99.9% of the energy in the blackbody spectrum at 1000 K (Table A.5).

When the CDF $g(C)$ is generated, the RTE for a nonscattering medium (Equation 1.73) can be rewritten for each value of $g(C)$ as

$$\frac{dI_g}{ds} = \kappa_g(I_b - I_g) \tag{9.48}$$

After solving for I_g, the total intensity is found by integrating over g as

$$I = \int_{g=0}^{1} I_g\, dg \tag{9.49}$$

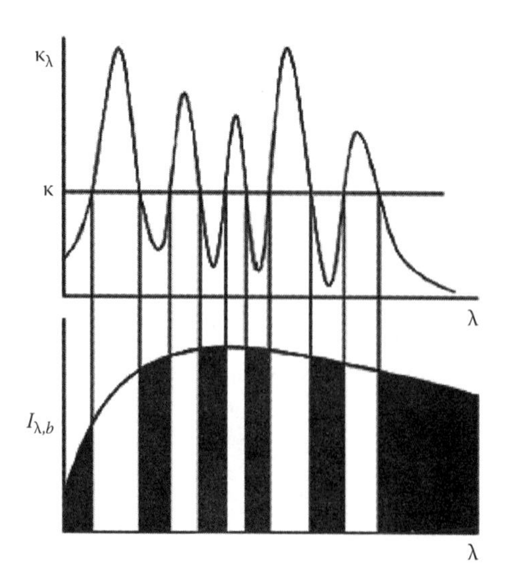

FIGURE 9.14 Illustration of generation of the CDF for the full spectrum k-distribution. The black sections of the blackbody distribution add to the CDF as the integration is carried out with κ varying from $\kappa_\lambda = 0$ to $\kappa_\lambda = \kappa_{\lambda\max}$.

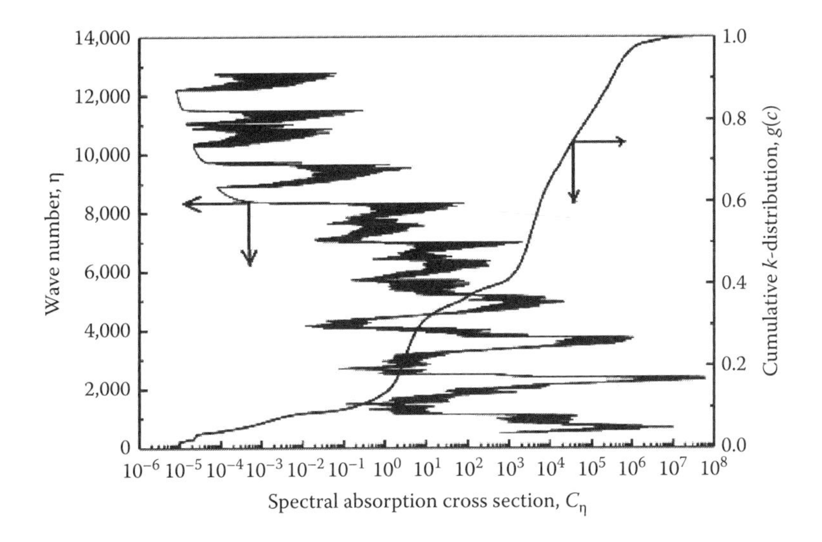

FIGURE 9.15 CDF of k-distribution for 10% CO_2 in air at 1000 K, 1 atm. (Data calculated from program HITRAN 2010.)

Because g is a monotonically and smoothly increasing function, it generally suffices to solve for I_g for a relatively few g intervals and then to use a simple quadrature to carry out the integration. This is much simpler as compared with a line-by-line integration over κ_η; that is the great advantage of the FSK approach.

André and Vaillon (2012) show that moments of the FSK (Equation 9.47) can be related to the Planck and Rosseland mean absorption coefficients (Equations 9.71 and 12.29), and also can be used to show the variance of absorption coefficients relative to the mean absorption coefficient for an arbitrary band or full spectrum distribution. Their generalized k-moment method is used to generate high-temperature FSK distribution for H_2O (André et al. 2014b) and is proposed as a new basis for k-distribution analysis in André et al. 2014a.

9.3.2.3.1 *Effect of Temperature and Concentration Gradients in the Medium*

As described in Equation 9.47, the $g(C)$ was derived based on a single reference temperature. However, for a nonisothermal or nonhomogeneous medium, the $g(C)$ will vary across the medium, introducing error into the solution.

Tencer and Howell (2014a) present a method for estimating the error introduced by applying the FSK method based on a reference temperature to multi-D problems with temperature-dependent properties. (The error introduced by using a single $g(C)$ distribution is generally much larger than the uncertainty in line intensity, line width, and line spectral location that are given in the line-by-line databases.)

Some of the ongoing attempts to treat the FSK approximations for CO_2 and water vapor mixtures are described by Wang and Modest (2005a,b) and Modest (2013), where various techniques are examined. The multigroup full-spectrum k-distribution (MGFSK) method arranges the spectral positions into some number M of separate groups, with the groups determined from the temperature and partial pressure dependencies of the spectral positions. This works well for individual gas species, but cannot handle mixtures of gases or inhomogeneous spatial property variations. Zhang and Modest (2003a) provide a MGFSK database for water vapor (available from Modest on request) and full-spectrum k-distribution correlations based on the 2000 HI TEMP line-by-line data are in Modest and Singh (2005).

The multiscale full-spectrum k-distribution (MScFSK) method groups the spectral lines into separate scales that depend on their temperature dependence. The MScFSK method uses the RTE for an absorbing gas (Equation 9.9). If M participating gases are present in a nonscattering mixture, then at a particular wavenumber η, the local absorption coefficient for the mixture is

$$\kappa_\eta = \sum_{m=0}^{M} \kappa_{\eta,m} \tag{9.50}$$

and the local intensity is

$$I_\eta = \sum_{m=0}^{M} I_{\eta,m} \tag{9.51}$$

The RTE (Equation 9.9) written in wavenumber form for each nonscattering "scale" (in this case, each participating molecular species) is then

$$\frac{\partial I_{\eta,m}}{\partial S} = \kappa_{\eta,m} I_{b,\eta} - \kappa_\eta I_{\eta,m} \tag{9.52}$$

and Equation 9.52 is written for each component gas m. Note that Equation 9.52 indicates that the change in intensity of medium m at wavenumber η is due to emission from individual medium m less absorption at that wavenumber by *all* components of the mixture.

The spectrum is now divided into N narrow bands, and the correlated-k method (Equations 9.42 through 9.44) is used for solution within each band. An approximate overlap correction must be determined for the overlap in a given band among the M species. This approach requires the solution of $M \times N$ equations with the appropriate boundary conditions. Narrow-band k-distributions for water vapor and CO_2 for use in constructing MScFSK solutions are discussed by Wang and Modest (2005a,b), along with discussion of the magnitude of the overlap correction for mixtures of these gases. The MScFSK has been extended to include the effect of gray-wall boundaries in Wang and Modest (2007), and to handle inhomogeneous mixtures through a hybrid of the multiscale and multigroup approaches in Pal et al. (2008). The use in Monte Carlo solvers is in Wang et al. (2007).

Caliot et al. (2008) examine the application of the c-k distribution method for mixtures at high temperature in remote sensing applications, applying a reordering of spectral lines to determine the properties of a mixture of fictitious gases with properties that mimic those of the real gases. Tencer and Howell (2013a) present an exact method for using FSK in 1D cases with prescribed temperature variations in the medium [the multisource full spectrum k-distribution (MSoFSK) method]. They show precise agreement with line-by-line solutions for test cases. Solovjov et al. (2014) present a similar approach. Hosseini Sarvari (2014) uses an inverse method to recover the k-distribution variation throughout a planar medium based on measurement of the spectral intensity at the medium boundaries. Cai et al. (2014) present k-distributions and mixing models for combustion gases for use in various PDE solution methods, including P_N and SP_N (Chapter 12).

9.3.3 Weighted Sum of Gray Gases

The weighted-sum-of-gray-gases (WSGG) approach seeks to replace the integration of spectral properties with a summation over a small set of J gray gases to simulate the properties of the nongray gas, and is thus closely related to the FSK method. Hottel (1954) introduced the method and proposed that the emittance of an isothermal gas could be approximated by

$$\varepsilon(S) = \sum_{j=1}^{J} w_j (1 - e^{-\kappa_j S}) \tag{9.53}$$

where

κ_j is the absorption coefficient
w_j is the corresponding weight for the jth gray gas

Modest (1991) showed that the WSGG method could be used with any solution technique for the radiative transfer equation. Denison and Webb (1993a,b) further developed the method. In their approach, line-by-line data weighted by the blackbody spectral distribution were integrated across narrow bands one time. The number, weights, and effective absorption coefficients of a set of gray gases were then obtained. This approach is the "spectral-line-weighted sum-of-gray-gases" (SLWSGG) method, and it allows for windows in the spectrum, as one of the gray gases can be nonabsorbing. An extension to include gray particles in a nongray gas is in Yu et al. (2000). Solovjov and Webb (2000, 2002, 2005, 2007) further extended the method to handle multicomponent mixtures, nonuniform media, the presence of soot, and multilayer media.

The RTE for use with the SLWSGG approach in a nonscattering medium is that proposed by Modest (1991):

$$\frac{dI_j(S)}{dS} = \kappa_j\left(w_j\frac{\sigma T^4(S)}{\pi} - I_j(S)\right) \tag{9.54}$$

This equation is solved for I_j for each gray gas, and the results summed over the J gray gases to obtain the total intensity. Denison and Webb (1993a) showed that Equation 9.54, which is independent of the geometry of the participating medium, can be applied to systems with weights and gas spectral absorption coefficients that vary across the participating medium, and to systems with gray boundaries.

The weights and absorption coefficients are obtained using the SLWSGG method as follows (Denison and Webb 1993a,b, Pearson et al. 2014a): An absorption-line blackbody distribution function (ALBDF) (equivalent to the k-distribution) is defined as *the fraction of the blackbody energy in the portions of the total spectrum where the high-resolution spectral molar absorption cross section of the gas* $\bar{C}_{abs,n}$ *is less than a prescribed value* $\bar{C}_{abs}$. This distribution function is expressed for an arbitrary absorbing species (e.g., H_2O, CO_2) as

$$F(\bar{C}_{abs},T_b,T_g,P,Y_s) = \frac{\pi}{\sigma T_b^4}\int\limits_{(\eta,\bar{C}_{abs},T_g,P,Y_s)} I_{b\eta}(T_b,\eta)d\eta \tag{9.55}$$

The equivalence to the FSK method is obvious by comparing Equation 9.55 with Equation 9.47. In fact, Chu et al. (2014) showed an exact mathematical equivalence between the FSK and the SLWSGG methods, although there are differences in the numerical methods used to generate the absorption line blackbody distribution functions used in the methods.

The fractional function F in Equation 9.55 has a monotonic increase between 0 and 1 with increasing absorption cross section. The function is dependent on the molar absorption cross section $\bar{C}_{abs}$, blackbody source temperature T_b, gas temperature T_g, total pressure P, and mole fraction of broadening species Y_s. The dependence of the function on the spectrum is through the spectral interval of integration, η. The function F depends on both T_g and T_b. For total emissivity calculations of isothermal gaseous media, these temperatures are equal. However, for gas total absorptivity calculations, the gas and the source temperatures may differ. The correlations are for radiation incident from a blackbody at T_b.

To simplify using the SLWSGG method, Denison and Webb (1993a,b) proposed a correlation of the absorption-line blackbody distribution function for common absorbing/emitting gases at atmospheric pressure. For CO_2 and water vapor, the correlation relation is

$$F\left(\bar{C},T_g,T_b,Y\to 0\right) = \frac{1}{2}\tanh[P_F(\bar{C},T_g,T_b,Y)] + \frac{1}{2} \tag{9.56}$$

where the function P_F is

$$P_F(\bar{C},T_g,T_b,Y \to 0) = \sum_{l=0}^{3}\sum_{m=0}^{3}\sum_{n=0}^{3} b_{lmn}\left(\frac{T_g}{2500}\right)^n\left(\frac{T_b}{2500}\right)^m (\xi-\xi_{sb})^l \tag{9.57}$$

$$\xi = \ln(\bar{C}_{abs}) \tag{9.58}$$

and

$$\xi_{sb} = \sum_{l=0}^{3}\sum_{m-0}^{3}\sum_{n=0}^{2} c_{lmn}\left(\frac{T_b}{2500}\right)^n \xi^m (Y)^{l+1} \tag{9.59}$$

The T_g and T_b are in Kelvins, and $\bar{C}_{abs}$ has the units m²/mol. The ξ is given by Equation 9.58, and Y is the mole fraction of the absorbing/emitting species. Equation 9.59 is used only for H_2O; for CO_2 self-broadening is independent of Y, and $\xi_{sb} = 0$.

Kangwanpongpan et al. (2012) and Dorigon et al. (2013) provide WSGG correlations based on the HITRAN 2010 database suitable for combustion calculations. Pearson et al. (2014a,b) have also updated calculation of the ALBDF based on the HITRAN 2010 database, along with the coefficients needed for computing the correlations of Equations 9.56 through 9.59. They included CO_2, H_2O, and CO for temperatures from 300 to 3000 K, pressures from 0.1 to 50 atm, and mole fractions between 0 and 1.

For H_2O the coefficients b_{lmn} of the correlation are in Table 9.3. In Equations 9.56 through 9.59 where there is only air broadening (negligible self-broadening), the $Y_{H_2O} \to 0$. The coefficients c_{lmn}

TABLE 9.3
Values for the Coefficients b_{lmn} for H_2O

	$n = 0$	$n = 1$	$n = 2$	$n = 3$
$l = 0$				
$m = 0$	0.978924	−0.207426	−2.80673	1.38942
$m = 1$	2.43339	−5.54212	16.9379	−7.31920
$m = 2$	−2.27043	9.80203	−24.4373	10.2661
$m = 3$	0.968416	−4.41664	10.2661	−4.20383
$l = 1$				
$m = 0$	0.231177	−0.0946050	0.908440	−0.538832
$m = 1$	0.253813	2.08204	−4.47671	2.44379
$m = 2$	−0.337419	−2.30864	4.63425	−2.47350
$m = 3$	0.179163	0.787972	−1.52075	0.829104
$l = 2$				
$m = 0$	0.166909	−0.795281	0.939269	−0.304703
$m = 1$	−0.669061	3.93941	−4.43049	1.19632
$m = 2$	0.837513	−4.93114	5.02758	−0.976627
$m = 3$	−0.321007	1.90733	−1.76314	0.215567
$l = 3$				
$m = 0$	0.0189884	−0.0755879	0.0692804	2.47619E−06
$m = 1$	−0.0794413	0.346149	−0.231489	−0.0657353
$m = 2$	0.0969300	−0.404604	0.151771	0.177646
$m = 3$	−0.0367609	0.147107	−0.0148638	−0.0984692

Source: Pearson, J.T. et al., *JQSRT*, 138, 82, 2014a.

TABLE 9.4
Self-Broadening Shift Parameters c_{lmn} for H_2O

	$n = 0$	$n = 1$	$n = 2$
$l = 0$			
$m = 0$	1.68529	−2.91926	1.41928
$m = 1$	0.160822	−5.18188	4.53813
$m = 2$	0.593801	−2.66943	2.02036
$m = 3$	0.0616066	−0.227721	0.166014
$l = 1$			
$m = 0$	−1.19664	3.13626	−1.97238
$m = 1$	−1.47806	17.7436	−15.5388
$m = 2$	−2.40775	9.54136	−7.05450
$m = 3$	−0.247551	0.835499	−0.589727
$l = 2$			
$m = 0$	−0.156309	−1.63344	1.49051
$m = 1$	2.15385	−23.6019	20.8035
$m = 2$	3.34190	−12.9818	9.57359
$m = 3$	0.344856	−1.14900	0.807114
$l = 3$			
$m = 0$	0.408081	0.218904	−0.421655
$m = 1$	−0.988248	10.5424	−9.32335
$m = 2$	−1.51699	5.85632	−4.31687
$m = 3$	−0.156857	0.520449	−0.365078

Source: Pearson, J.T. et al., *JQSRT*, 138, 82, 2014a.

in Table 9.4 were obtained from a least-squares fit, with the shift from ξ to $\xi - \xi_{sb}$ by self-broadening evaluated for H_2O mole fractions between 0 and 1.0.

For CO_2, the coefficients b_{lmn} are in Table 9.5. These were determined from a least-squares fit of the distribution function for $\bar{C}_{abs}$ for equal source and gas temperatures between 400 and 3000 K. Pressures from 0.1 to 50 atm can be treated by referring to the references. Unlike H_2O, which has strong self-broadening, the absorption-line distribution function for CO_2 is essentially independent of Y_{CO2}, and the self-broadening ξ_{sb} in Equation 9.56 is neglected.

Pressure effects on the ALBDF for pressures 0.1–50 atm are given in Pearson et al. (2014b). The ALBDFs for H_2O, CO_2, and CO at various pressures were generated from the HITEMP 2010 database for temperatures up to 3000 K and pressures from 0.1 to 50 bar. Voigt line shapes were used because Doppler line broadening becomes significant at the lower pressures. Pearson et al. (2014b) integrated SLWSGG results to calculate total emissivity and compared well with the correlations of Leckner (1972) (Section 9.4).

An alternative method for computing the absorption line blackbody distribution function is to use the CDF $g(\kappa_\eta)$ from the cumulative k-distribution in the relation

$$F = \frac{\pi}{\sigma T_b^4} \int_0^\infty g_\eta(\kappa_\eta) I_{b,\eta}(T_b) d\eta \tag{9.60}$$

This form allows including spectrally dependent scattering effects and nongray boundary conditions (Denison and Webb 1996b).

After the function F is found for a particular gas at known conditions, the weights w_j and absorption coefficients κ_j for use in the WSGG Equation 9.54 are found by segmenting the spectrum into J

TABLE 9.5
Values for the Coefficients b_{lmn} for CO_2

	$n = 0$	$n = 1$	$n = 2$	$n = 3$
$l = 0$				
$m = 0$	1.83492	−1.82218	1.27833	−0.209800
$m = 1$	−1.54969	1.11312	−2.42581	0.919086
$m = 2$	3.51788	−1.14955	3.54657	−1.51642
$m = 3$	−1.50580	0.522032	−1.65400	0.729684
$l = 1$				
$m = 0$	0.202157	0.885000	−0.982475	0.375315
$m = 1$	−0.531916	−2.21448	2.29322	−0.867072
$m = 2$	1.01979	1.91290	−1.68833	0.600486
$m = 3$	−0.506065	−0.526424	0.340073	−0.101266
$l = 2$				
$m = 0$	0.054472216	−0.158181	0.290036	−0.137608
$m = 1$	−0.172337	0.538061	−0.853519	0.383656
$m = 2$	0.233086	−0.702110	1.04692	−0.458833
$m = 3$	−0.101433	0.291151	−0.424021	0.184048
$l = 3$				
$m = 0$	6.30156E-03	−0.0300971	0.0445779	−0.0194831
$m = 1$	−0.0182273	0.0950925	−0.122626	0.0501599
$m = 2$	0.0209971	−0.109841	0.134897	−0.0540501
$m = 3$	−8.18055E-03	0.0416371	−0.0502137	0.0200237

Source: From Pearson, J.T. et al., *JQSRT*, 138, 82, 2014a.

regions. The weights are given for any region j by choosing values for the absorption cross sections at the limits of each region, $\bar{C}_j$. Then

$$w_j = F(\bar{C}_{j+1}, T_b, T_g, Y) - F(\bar{C}_j, T_b, T_g, Y) \tag{9.61}$$

and the corresponding absorption coefficients $\kappa_j = N\bar{C}_j^*$ are then found from appropriate mean values of $\bar{C}_j^*$ within each interval (these will generally not be equal to $\bar{C}_j$) and from N, the molar concentration of the absorbing–emitting gas in g·mol/m³. For $\bar{C}_j^*$, a mean given by

$$\bar{C}_j^* = \exp\left[\frac{\ln(\bar{C}_j) + \ln(\bar{C}_{j+1})}{2}\right] \tag{9.62}$$

is suggested. The weights and absorption coefficients are now used in Equation 9.54, and the resulting values for intensity are then summed over the J spectral regions to obtain total intensity values. As few as $J = 3$ regions have been found to provide good accuracy for some problems, but $J = 10$–20 is more commonly used.

Example 9.2

Develop and plot the absorption line blackbody distribution function for pure CO_2 at equal gas and blackbody source temperatures of 1000 K. Discuss the choice of weights and absorption coefficients for the SLWSGG method for this case.

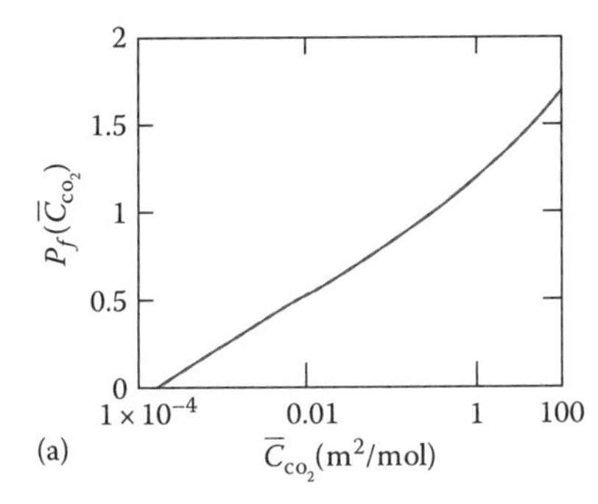

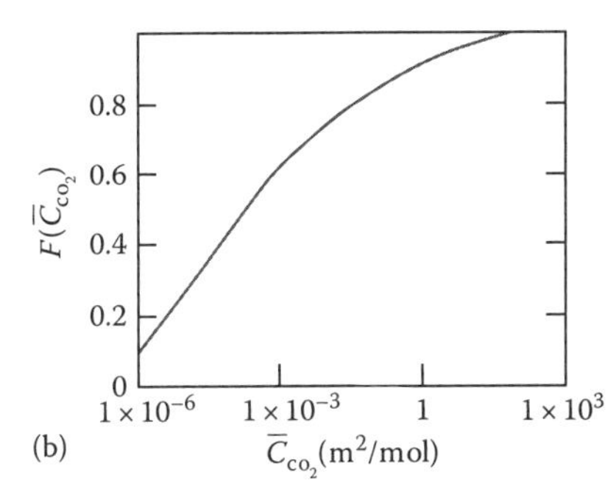

FIGURE 9.16 Results for pure CO_2 at 1 atm pressure and $T_g = T_h = 1000$ K. (a) The function $P_F\left(\bar{C}_{CO_2}\right)$ versus molar absorption cross section $\bar{C}_{CO_2}$. (b) The absorption line blackbody function $F\left(\bar{C}_{CO_2}\right)$ versus molar absorption cross section $\bar{C}_{CO_2}$.

For CO_2, set the self-broadening $\xi_{sb} = 0$. Using the updated coefficients in Table 9.3 and $T_g = T_b = 1000$ K, the $P_F(\bar{C}_{CO_2})$ is calculated from Equation 9.57. A plot of the result is in Figure 9.16a. The absorption line blackbody function is then calculated from Equation 9.56, and the result is in Figure 9.16b. The absorption coefficient is now found from the molar absorption cross section $\bar{C}_{CO_2}$ as $\kappa = \bar{C}_{CO_2}N$ where the molar volumetric concentration

$$
\begin{aligned}
N &= n/V = P/RT \\
&= 101.2(\text{kPa})/[8.314(\text{kJ/kmol}\cdot\text{K})\times 1000(\text{K})] \\
&= 0.0122\ \text{kmol/m}^3 \\
&= 12.2\ \text{mol/m}^3.
\end{aligned}
$$

The absorption coefficients are generally quite small over much of the range of significant values of $F\left(\bar{C}_{CO_2}\right)$. For example, for all κ less than $\bar{C}_{CO_2}N = 1\times 12.2\ \text{m}^{-1}$, the $F\left(\bar{C}_{CO_2}\right)$ has already reached about 0.9, that is, 90% of the blackbody energy passes through CO_2 with absorption coefficient less than 12.2 m^{-1}.

Note that the path length has not yet entered the calculation. Determining the weights and absorption coefficients requires considering path lengths that are important in a particular problem.

Line-by-line and *k*-distribution solutions in 1D have been compared by Tang and Brewster (1999) with excellent agreement if sufficient resolution was used in generating the *k*-distribution. However, errors greater than 10% were observed in the local total heat flux and flux divergence. Galarça et al. (2011) indicate that the SLWSGG formulation for radiative transfer is not conservative, and propose a modification.

The statistical narrow band correlated-*k*, cumulative-*k* distribution, SLWSGG, exponential wide band, and some hybrid methods were compared in 2D enclosure problems using uniform and non-homogeneous-nonisothermal pure and mixed gases by Liu et al. (2000a). Discrete ordinates and ray-tracing solution methods (Chapter 12) were used. It was reported that the *c-k* method and one of the hybrids based on lumping of narrow bands provided the most accurate solutions. Errors in computed radiative flux divergence by both methods compared with statistical narrow-band model results were within 4%, but wall flux values had errors of 11%–13% for the *c-k* method. The SLWSGG method had somewhat lower accuracy, but had much lower computational costs, and was reported to have the best combination of error and computing time among the methods. The exponential wide-band results had greater errors. Maurente et al. (2008) implemented the SLWSGG method in a Monte Carlo analysis of a cylindrical combustion chamber. Cassoll et al.

(2014) propose an efficient method for calculating the ALBDF for various H_2O-CO_2 gas mixtures including soot in nonhomogeneous media. They use ALBDFs generated from the HITEMP 2010 database and combine them using a mixture rule. Comparison with line-by-line calculations for various 1D test cases with nonhomogeneous temperatures and concentrations show good agreement (generally, average errors in predicted surface heat flux were less than 6% and maximum errors within 10%).

9.4 GAS TOTAL EMITTANCE CORRELATIONS

The total emittance was defined in Equation 9.13 and when it is obtained from κ_λ it involves an integration of the significant radiation contributions over the spectrum. Graphical presentations of the total emittance for the important radiating gases have been developed from total radiation measurements, by integration using spectral measurements of absorption lines and bands, and from the theories of line and band absorption. At industrial furnace and combustion chamber temperatures, it is only heteropolar gases that absorb and emit significantly, such as CO_2, H_2O, CO, SO_2, NO, and CH_4. Gases with symmetric diatomic molecules, such as N_2, O_2, and H_2, are transparent to infrared radiation and do not emit significantly; however they become important absorbing/emitting contributors at very high temperatures. Charts for the total ε_g were originally developed by Hottel (1954) from experimental measurements. The thickness of the gas enters through the parameter L_e, which is an average path length through a uniform gas volume; values for various volume shapes are in Chapter 10. The gas pressure enters as a parameter because κ_λ depends on gas density.

If the gas is in a mixture, both the mixture pressure and the partial pressures of the radiating constituent are parameters. Charts for water vapor and CO_2 are in Figures 9.17 and 9.18. Additional charts for sulfur dioxide, ammonia, carbon monoxide, methane, and a few other gases are in Hottel and Sarofim (1967). The discussion here is for CO_2 and water vapor.

Before proceeding to describe the use of total emittance, a simple example is given to show how the total properties are developed by an integration using the spectral properties. The example shows how the transparent spectral regions between the absorbing parts of the spectrum are important in limiting the size of total emittance values. The example illustrates the limit for a thick layer of absorbing gas.

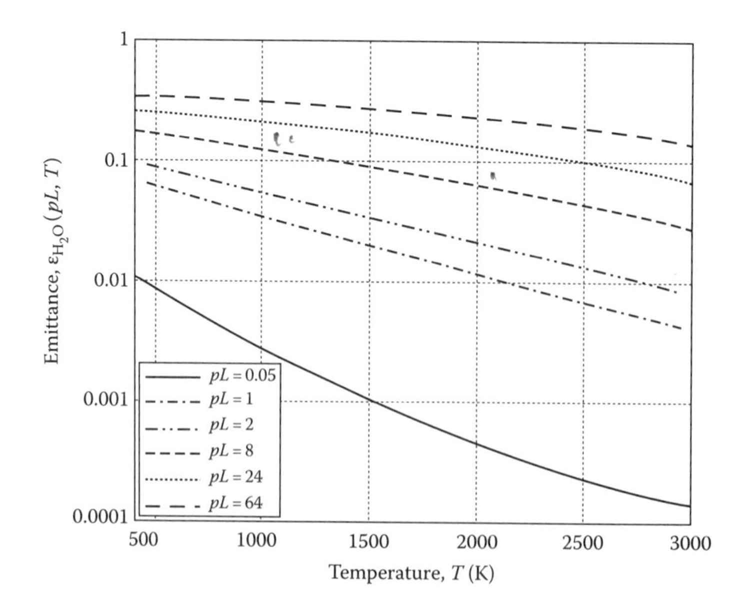

FIGURE 9.17 Computed emittance of water vapor from Equation 9.64 using the coefficients of Table 9.9. Units of pL are bar-cm.

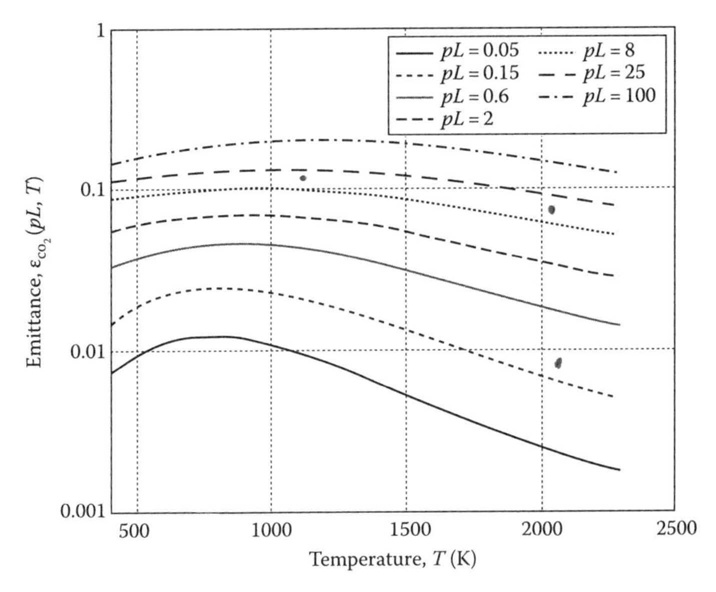

FIGURE 9.18 Computed emittance of CO_2 from Equation 9.64 using the coefficients of Table 9.9. Units of pL are bar-cm.

Example 9.3

As a rough approximation, idealize the absorptance of CO_2 at $T_g = 830$ K and 10 atm as in Figure 9.4 so that it consists of four bands having vertical boundaries at $\lambda = 1.8$ and 2.2, 2.6 and 2.8, 4.0 and 4.6, and 9 and 19 μm. What is the total emittance of a thick layer of gas at this temperature?

For a thick layer (large L_e), Equation 9.13 shows that $\varepsilon_\lambda = 1 - e^{-\kappa_\lambda L_e} = 1$ in the absorbing regions. Hence, the thick gas emits like a blackbody in the four absorption bands. In the nonabsorbing regions between the bands, ε_λ is small and is neglected in this simplified model. The total emittance becomes

$$\varepsilon(T_g, P, S) = \frac{\int_0^\infty \varepsilon_\lambda(T_g, P, S) E_{\lambda b,g} d\lambda}{\sigma T_g^4} = \frac{\int_{\substack{absorbing\\bands}} E_{\lambda,b,g} d\lambda}{\sigma T_g^4}$$

The emittance is thus the fractional emission of a blackbody in the wavelength intervals of the absorbing bands. The required values are:

λ (μm)	λT_g (μm · K)	$F_{0\to\lambda T_g}$	λ (μm)	λT_g (μm · K)	$F_{0\to\lambda T_g}$
1.8	1494	0.01250	4.0	3320	0.34446
2.2	1826	0.04248	4.6	3818	0.44684
2.6	2158	0.09323	9	7470	0.83292
2.8	2324	0.12479	19	15,770	0.97275

Then, the emittance is

$$\varepsilon(T_g, P, S) = \sum_{\substack{absorbing\\bands}} (F_{0\to(\lambda T_g)_{upper}} - F_{0\to(\lambda T_g)_{lower}})_{band}$$

$$\varepsilon = (0.04248 - 0.01250) + (0.12479 - 0.09323) + (0.44684 - 0.34446) + (0.97275 - 0.83292) = 0.304$$

Even for an optically thick CO_2 volume, the emittance is much less than for a blackbody.

Example 9.4

What fraction of incident solar radiation is absorbed by a thick layer of CO_2 at 10 atm and 830 K? Use the approximate absorption bands of Example 9.3.

The effective radiating temperature of the sun is T_s = 5780 K. The desired result is the fraction of the solar spectrum that lies within the four CO_2 bands, as this is the only portion of the incident radiation that will be absorbed. Using the $F_{0\to\lambda T_s}$, factors obtained using T_s gives

λ (μm)	λT_g (μm · K)	$F_{0\to\lambda T_s}$	λ (μm)	λT_s (μm · K)	$F_{0\to\lambda T_s}$
1.8	10,400	0.92195	4.0	23,120	0.99028
2.2	12,720	0.95251	4.6	26,590	0.99340
2.6	15,030	0.96909	9	52,020	0.99902
2.8	16,180	0.97455	19	109,800	0.99989

The fraction absorbed is then

$$\alpha = \sum_{\substack{absorbing\\bands}} (F_{0\to(\lambda T_s)upper} - F_{0\to(\lambda T_s)lower})_{band}$$

$$= (0.95251 - 0.92195) + (0.97455 - 0.96909) + (0.99340 - 0.99028)$$

$$+(0.99989 - 0.99902) = 0.040$$

Even though the gas layer is thick, only 4.0% of the incident energy is absorbed since the gas transmits well in the λ regions between the absorption bands and much of the solar energy is at shorter wavelengths than the absorbing bands.

Hottel's curves of the total emittance $\varepsilon(pL_e,T)$ for CO_2 and H_2O have been widely used for total gas emittance. The curves are based on experimental data with extrapolations to high temperatures and large L_e-partial pressure regions based on theory. These curves have been updated based on more recent data. Rather than using Hottel's original curves, it is useful to have analytical expressions for the emittance for use in computation.

For water vapor in air, Cess and Lian (1976) propose the correlation

$$\varepsilon_{H_2O}(X,T) = a_0\left[1 - \exp(-a_1\sqrt{X})\right] \tag{9.63}$$

where $X = p_{H_2O}L_e\left(p_{air} + bp_{H_2O}\right)\left(300/T\right)$ and $b = 5.0(300/T)^{1/2} + 0.5$. In these relations, T is in K, p in atm, and L_e in m. The constants for use in Equation 9.63 are in Table 9.6.

Leckner (1972) gives empirical correlations for the total emittance derived from calculations summing narrow band behavior over the spectrum for both water vapor and CO_2. The most accurate expressions from Leckner agree within 5% to values calculated from spectral data for $T > 400$ K,

TABLE 9.6
Constants for Use in Equation 9.57

T (K)	a_0	a_1 ($m^{-1/2}$ atm^{-1})
300	0.683	1.17
600	0.674	1.32
900	0.700	1.27
1200	0.673	1.21
1500	0.624	1.15

TABLE 9.7
Coefficients c_{ij} for Equation 9.58 to Calculate Water Vapor and CO_2 Emittance

j	c_{0j}	c_{1j}	c_{2j}	c_{3j}	c_{4j}
Water vapor, $T > 400$ K, $M = 2$, $N = 2$					
0	−2.2118	−1.1987	0.035596		
1	0.85667	0.93048	−0.14391		
2	−0.10838	−0.17156	0.045915		
Carbon dioxide, $T > 400$ K, $M = 3$, $N = 4$					
0	−3.9781	2.7353	−1.9822	0.31054	0.015719
1	1.9326	−3.5932	3.7247	−1.4535	0.20132
2	−0.35366	0.61766	−0.84207	0.39859	−0.063356
3	−0.080181	0.31466	−0.19973	0.046532	−0.0033086

and are in close agreement with Hottel charts for ranges where Hottel based the charts on experimental data. Docherty (1982) compares Leckner's predictions with those of Hottel as well as with more recent experimental data, and concludes that Leckner's predictions are more accurate than Hottel charts in the regions where Hottel extrapolated outside the range available to him.

In Leckner's correlations and equations, p is in bar, and L_e is in cm. The correlation equation is

$$\varepsilon(T, pL_e) = \exp\left\{ a_0 + \sum_{j=1}^{M} a_j \left[\log(pL_e)\right]^j \right\} \tag{9.64}$$

where

$$a_j = c_{0j} + \sum_{i=1}^{N} c_{ij} \left(T/1000\right)^i$$

and the values of c_{ij} are in Table 9.7 for water vapor and for CO_2.

A plot of the emittance of water vapor predicted by Equation 9.64 is in Figure 9.17. Observe that the emittance increases with the pressure-path length product as expected. The trend with temperature is that emittance generally decreases with increasing temperature for water vapor. In contrast, CO_2 tends to go through a peak in emittance at about 1200 K (Figure 9.18).

The correlations are for properties of the absorbing gases at essentially zero pressure mixed with air at a total pressure of one bar. If the total pressure differs considerably from one bar, then a pressure correction must be applied to the predicted one bar emittance of the individual gases because of increased pressure broadening of the individual lines that make up the bands that are summed to obtain the total emittance. Hottel (1954) presented graphs for this correction, and Leckner (1972) has provided algebraic expressions. For water vapor, the pressure correction from Leckner is

$$C_{H_2O} = 1 + (\Lambda_{H_2O} - 1)\Xi_{H_2O} \tag{9.65}$$

where

$$\Lambda_{H_2O} = \frac{\left[1.888 - 2.053\log_{10}(T/1000)\right] P_{E,H_2O} + 1.10(T/1000)^{-1.4}}{P_{E,H_2O} + \left[1.888 - 2.053\log_{10}(T/1000)\right] + 1.10(T/1000)^{-1.4} - 1}$$

$$\Xi_{H_2O} = \exp\left(-\frac{\left\{\log_{10}[13.2(T/1000)^2] - \log_{10}(p_{H_2O}L_e)\right\}}{2} \right)$$

The effective pressure is given by $P_{E,H_2O} = P_t\left[1 + 4.9(p_{H_2O}/P_t)(273/T)^{1/2}\right]$ and P_t is the total pressure of the air–H_2O mixture. In the expression for Λ, the T in the expression in square brackets is replaced by 750 if $T < 750$ K.

The pressure correction for CO_2 from Leckner is given by

$$C_{CO_2} = 1 + (\Lambda_{CO_2} - 1)\Xi_{CO_2} \tag{9.66}$$

where

$$\Lambda_{CO_2} = \frac{\left[1.00 + 0.10(T/1000)^{-1.45}\right]P_{E,CO_2} + 0.23}{P_{E,CO_2} + \left[1.00 + 0.10(T/1000)^{-1.45}\right] - 0.77}$$

$$\Xi_{CO_2} = \exp\left\{-1.47\left[\mu - \log_{10}(p_{CO_2}L_e)\right]\right\}$$

The effective pressure is given by $P_{E,CO_2} = P_t\left[1 + 0.28(p_{CO_2}/P_t)\right]$, where P_t is the total pressure of the air–CO_2 mixture. In the expression for Ξ, $\mu = \log_{10}[0.225(T/1000)^2]$ if $T > 700$ K, and $\mu = \log_{10}[0.054(T/1000)^{-2}]$ if $T < 700$ K.

The individual emittances for H_2O and CO_2 in air must be modified when both gases are present in a mixture, which is commonly the case. This is because the individual spectral lines and absorption bands for the two gases overlap in some spectral regions, and simple addition of the individual emittances will overpredict the emittance of the mixture. In some cases, a simple addition predicts a gas absorptance and emittance that are greater than unity at certain wavelengths. This can be seen as follows: If two absorbing gases have spectral absorption coefficients $\kappa_{\lambda,1}$ and $\kappa_{\lambda,2}$ then from Equation 9.13,

$$\varepsilon = \frac{1}{\sigma T_g^4}\int_{\lambda=0}^{\infty}\left[1 - e^{-(\kappa_{\lambda,1}+\kappa_{\lambda,2})}\right]E_{\lambda b}(T_g)d\lambda$$

$$= \frac{1}{\sigma T_g^4}\int_{\lambda=0}^{\infty}\left[1 - e^{-\kappa_{\lambda,1}} + 1 - e^{-\kappa_{\lambda,2}} - \left(1 - e^{-\kappa_{\lambda,1}}\right)\left(1 - e^{-\kappa_{\lambda,2}}\right)\right]E_{\lambda b}(T_g)d\lambda \tag{9.67}$$

The first four terms integrate to give the total emittances of the two individual gases, resulting in

$$\varepsilon = \varepsilon_1 + \varepsilon_2 - \frac{1}{\sigma T_g^4}\int_{\lambda=0}^{\infty}\left(1 - e^{-\kappa_{\lambda,1}}\right)\left(1 - e^{-\kappa_{\lambda,2}}\right)E_{\lambda b}(T_g)d\lambda = \varepsilon_1 + \varepsilon_2 - \Delta\varepsilon \tag{9.68}$$

The final term on the right is the band overlap correction. Hottel (1954) presents a graph of the approximate band overlap correction. An empirical expression for the band overlap correction that is in good agreement with Hottel chart (Leckner 1972) valid for $1000 < T < 2200$ K and all pressures is

$$\Delta\varepsilon = \left(\frac{\zeta}{10.7 + 101\zeta} - 0.0089\zeta^{10.4}\right)\left[\log_{10}(pL_e)\right]^{2.76} \tag{9.69}$$

where $\zeta = p_{H_2O}/\left(p_{H_2O} + p_{CO_2}\right)$, $p = \left(p_{H_2O} + p_{CO_2}\right)$ is in bars, and L_e is in cm.

The final emittance equation including the pressure corrections and overlap correction $\Delta\varepsilon$ is

$$\varepsilon(pL_e) = C_{H_2O}\varepsilon_{H_2O}(p_{H_2O}L_e) + C_{CO_2}\varepsilon_{CO_2}(p_{CO_2}L_e) - \Delta\varepsilon \tag{9.70}$$

Example 9.5

A container with effective radiation thickness of $L_e = 2.4$ m contains a mixture of 15 volume percent of CO_2, 20% H_2O vapor, and the remainder air. The total pressure of the gas mixture is 1 atm, and the gas temperature is 1200 K. What is the emittance of the gas?

The partial pressures of the gases are equal to the mole fraction of each times the total pressure. The mole fraction in an ideal gas mixture is equal to the volume fraction, so the partial pressures are $p_{CO_2} = 0.15$, $p_{H_2O} = 0.20$, and $p_{air} = 0.65$ atm. For water vapor, the a_j values are $a_j = c_{0j} + \sum_{i=1}^{N} c_{ij}\left(T/1000\right)^i$, giving $a_1 = -3.599$, $a_2 = 1.766$, and $a_3 = -0.248$. Using Equation 9.64 (remembering to convert the pressures to bars), $\varepsilon_{H_2O}(T, pL_e) = \exp\left\{a_0 + \sum_{j=1}^{M} a_j\left[\log(pL_e)\right]^j\right\} = 0.266.$

A similar calculation for CO_2 gives $\varepsilon(CO_2) = 0.146$. No pressure correction is necessary because the total pressure is 1 atm. The overlap correction is calculated using

$$\zeta = \frac{p_{H_2O}}{p_{H_2O} + p_{CO_2}} = \frac{0.20}{0.15 + 0.20} = 0.571$$

and

$$\left(p_{H_2O} + p_{CO_2}\right)L_e = \left(0.15 + 0.20\right)(\text{atm}) \times 1.01325\ (\text{bar/atm}) \times 240\ (\text{cm})$$
$$= 85.1\ (\text{bar} \times \text{cm}).$$

Substituting results in

$$\Delta\varepsilon = \left(\frac{\zeta}{10.7 + 101\zeta} - 0.0089\zeta^{10.4}\right)\left[\log_{10}(pL_e)\right]^{2.76} = 0.051$$

The total emittance of the gas mixture is then

$$\varepsilon(pL_e) = \varepsilon_{H_2O}(p_{H_2O}L_e) + \varepsilon_{CO_2}(p_{CO_2}L_e) - \Delta\varepsilon = 0.266 + 0.146 - 0.051 = 0.360$$

Modest and Singh (2005) compare predictions of the HITEMP line-by-line and full-spectrum cumulative-*k* distribution predictions of H_2O total emittance, and show reasonable agreement with Hottel and Leckner charts. Pearson et al. (2013) used the updated ALBDFs based on the HITEMP 2010 database and also show excellent agreement with Leckner's correlations.

For some calculations, it is convenient to use mean absorption coefficients that are averaged over some or all of the wavelength spectrum. Absorption coefficients can be averaged in many ways, depending on the particular way they are to be used. For example, when computing emission from a volume element, a useful mean is *Planck mean absorption coefficient*, κ_P, defined by

$$\kappa_P(T,P) \equiv \frac{\int_{\lambda=0}^{\infty} \kappa_\lambda(T,P)E_{\lambda b}\left(T\right)d\lambda}{\int_{\lambda=0}^{\infty} E_{\lambda b}\left(T\right)d\lambda} = \frac{\int_{\lambda=0}^{\infty} \kappa_\lambda(T,P)E_{\lambda b}\left(T\right)d\lambda}{\sigma T^4} \tag{9.71}$$

The κ_P is the mean of the spectral coefficient weighted by the blackbody (Planck distribution) emission spectrum. Planck mean is convenient since it depends only on the local properties at dV and can be tabulated. Other weighting factors and particular mean absorption coefficients are discussed in depth in Section 12.11.

9.5 TRUE ABSORPTION COEFFICIENT

Equation 1.44 gives the attenuation of intensity passing through an absorbing, nonemitting, nonscattering medium as would be observed by detectors of incident and emerging radiation. Such information could be used in determining κ_λ. Actually, as radiative energy passes through a translucent medium, not only is it absorbed, but there is an additional phenomenon where the radiation field stimulates some of the atoms or molecules to emit energy. This is not *ordinary* or *spontaneous* emission caused by the temperature of the medium, discussed in Section 1.6.4. The spontaneous emission is the result of an excited energy state of the medium being unstable and decaying spontaneously to a lower energy state. Emission resulting from the presence of the radiation field is termed *stimulated* or *induced* emission and acts like negative absorption.

For induced emission, a photon from the radiation field encounters a particle, such as an atom or molecule, that is in an excited energy state. There is a probability that the incident photon will trigger a return of the particle to a lower energy state. If this occurs, the particle emits a photon at the *same frequency* and *in the same direction* as for the incident photon. Thus, the incident photon is not absorbed but is joined by a second identical photon.

Induced emission constitutes a portion of the intensity that is emerging from a translucent volume. Consequently, the energy actually absorbed by the medium is greater than the difference between the entering and leaving intensities. This is because the observed emerging intensity is the result of the actual absorption modified by induced emission added along the path. The actual absorbed energy depends on the *true absorption coefficient* $\kappa_\lambda^+(T,P)$, which is larger than the absorption coefficient $\kappa_\lambda(T, P)$ calculated from observed attenuation data.

Statistical-mechanical considerations give the relation between $\kappa_\lambda(T, P)$ and $\kappa_\lambda^+(T,P)$ for a gas with refractive index $n = 1$ as

$$\kappa_\lambda(T,P) = \left[1 - \exp\left(-\frac{hc_0}{k\lambda T}\right)\right]\kappa_\lambda^+(T,P) = \left[1 - \exp\left(-\frac{C_2}{\lambda T}\right)\right]\kappa_\lambda^+(T,P) \tag{9.72}$$

Because of the negative exponential term, κ_λ^+ is always larger than κ_λ (hence the use of the superscript +). Because induced emission depends on the incident radiation field, it is usually combined, as is done here, with the true absorption, thereby yielding the absorption coefficient κ_λ, and it is κ_λ that is used in the radiative energy calculations. The emission term in the equation of radiative transfer is then *only* spontaneous emission and consequently depends only on the local conditions of the medium; it is then not necessary to include directional effects in the emission term. The exponential term in Equation 9.72 is small unless λT is large. Thus κ_λ and κ_λ^+ are nearly equal except at large λT (long wavelengths and/or high temperatures). The values are within 1% for λT less than 3120 μm·K, and within 5% for λT less than 4800 μm·K. For calculations of radiative transfer in absorbing-emitting media, the properties in the literature give κ_λ with few exceptions; hence, the true absorption coefficient κ_λ^+ does not need to be considered further in this development.

9.6 RADIATIVE PROPERTIES OF TRANSLUCENT LIQUIDS AND SOLIDS

Figure 9.19 shows κ_λ for diamond (Garbuny 1965). Strong absorption peaks exist due to crystal-lattice vibrational energy states at certain wavelengths. However, the spectral variations are more regular than for gases. This is true for most solids and liquids that are translucent in various spectral regions.

Now consider the selective transmission and absorption behavior of glass and water. Figure 9.20 shows the overall spectral transmittance of glass plates for normally incident radiation. As shown in Section 17.2.1, the overall transmittance T_λ includes the effect of absorption (related to the absorption coefficient κ_λ) within the glass of thickness d, and multiple surface reflections (related to the surface

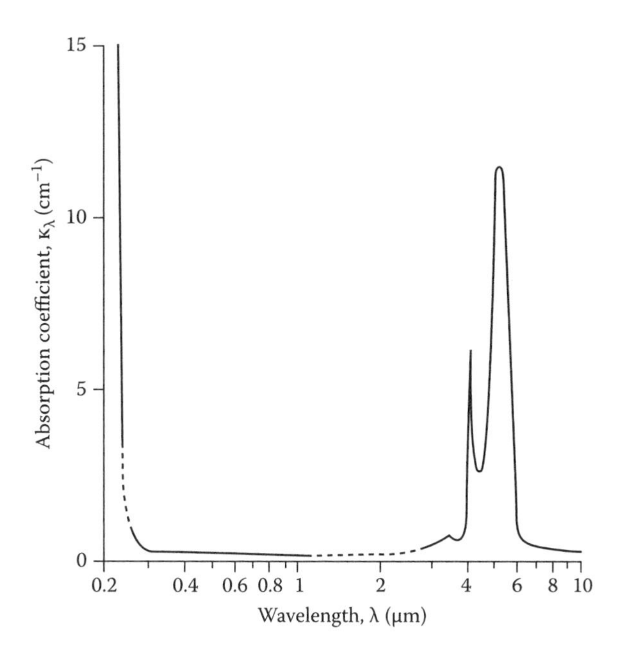

FIGURE 9.19 Spectral absorption coefficient of diamond. (From Garbuny, M., *Optical Physics*, Academic Press, New York, 1965.)

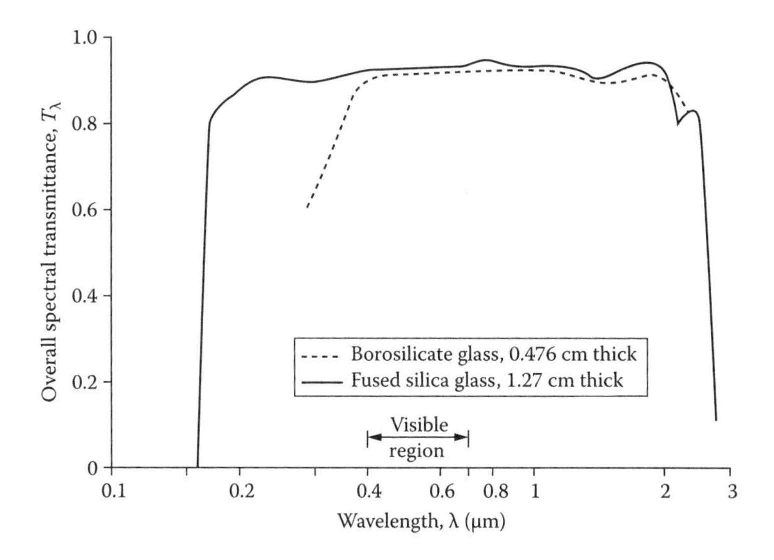

FIGURE 9.20 Normal overall spectral transmittance of a borosilicate or fused silica glass plate (includes surface inter-reflections) at 298 K. (Replotted from Touloukian, Y.S. and Ho, C.Y. (eds.), Thermophysical properties of matter, TRPC data services, *Thermal Radiative Properties: Nonmetallic Solids*, vol. 8, Touloukian, Y.S. and DeWitt, D.P., Plenum Press, New York, 1972a.)

reflectivity, ρ_λ); it is given by Equation 17.2 as $T_\lambda = \tau_\lambda\left(1-\rho_\lambda\right)^2/\left(1-\rho_\lambda^2\tau_\lambda^2\right)$ where $\tau_\lambda = \exp(-\kappa_\lambda d)$ and $\rho_\lambda = [(n-1)/(n+1)]^2$. For small $\kappa_\lambda d$ this reduces to $T_\lambda = (1-\rho_\lambda)/(1+\rho_\lambda)$. Typically for glass $n \approx 1.5$, so $\rho_\lambda = (0.5/2.5)^2 = 0.04$. Then including only reflection losses gives $T_\lambda = (1-0.04)/(1+0.04) = 0.92$. In Figure 9.21 the fused silica has very low absorption in the range $\lambda = 0.2$–2 μm, and $T_\lambda \approx 0.9$ in this region as a result of surface reflections. Ordinary glasses typically have two strong cutoff wavelengths beyond which the glass becomes highly absorbing and T_λ decreases rapidly to near zero except for very

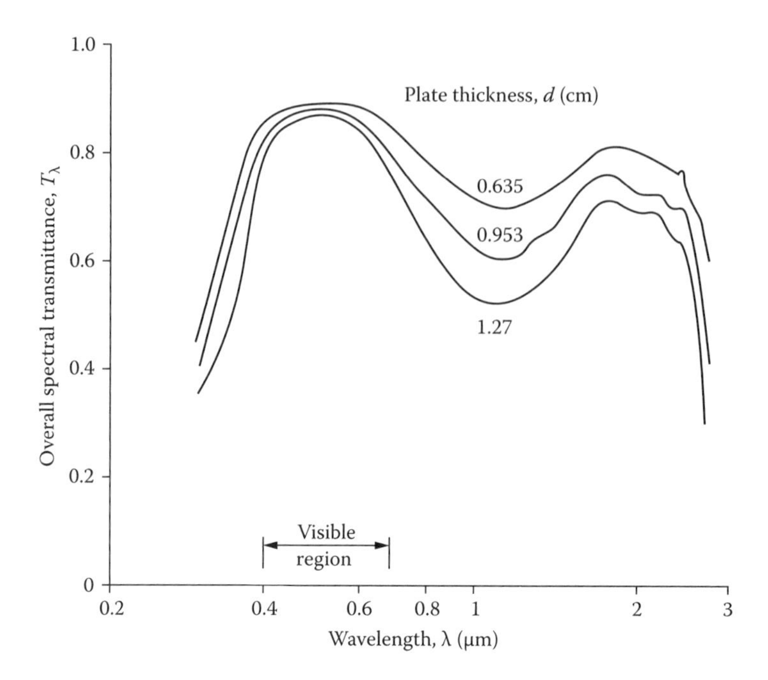

FIGURE 9.21 Effect of plate thickness on normal overall spectral transmittance of soda-lime glass (includes surface inter-reflections) at 298 K. (From Hsieh, C.K. and Su, K.C., *Sol. Energy*, 22(1), 37, 1979.)

thin plates. The measured curve for fused silica in Figure 9.20 shows this. There are strong cutoffs in the far ultraviolet at $\lambda \approx 0.17$ μm and in the near infrared at $\lambda \approx 2.5$ μm. The glass is therefore a strong absorber or emitter for $\lambda < 0.17$ μm and $\lambda > 2.5$ μm. Figure 9.21 shows the overall transmittance for various thicknesses of soda-lime glass, which is more absorbing than fused silica. The effect of absorption is illustrated quite well as the thickness increases. Typical optical constants for glass are in Hsieh and Su (1979). Nicolau and Maluf (2001) provide measured values of tinted commercial glass. A calculator is available on line to determine glass radiative properties: http://www.guardianglass.co.uk/industry.

For windows in high-temperature devices, such as furnaces or solar-cavity receivers, emission from within the windows can be significant. From Kirchhoff's law the overall spectral emittance E_λ that includes the effect of surface reflections is equal to the spectral absorptance. Hence, from Equation 17.3 for an isothermal window, $E_\lambda = (1-\rho_\lambda)(1-\tau_\lambda)/(1-\rho_\lambda\tau_\lambda)$. For a thick window, beyond the cutoff wavelength, $\tau_\lambda \to 0$ and $E_\lambda \approx 1 - \rho_\lambda$. In this instance, reflection from only one surface is significant because all the radiation passing through the first surface is absorbed before it can be transmitted to the second surface of the window. If the n for glass is 1.5, then $\rho_\lambda \sim 0.04$ for incidence from the normal direction; hence, $E_\lambda = 0.96$ in the normal direction for the highly absorbing spectral regions of the glass. In a fashion similar to Example 3.3, the hemispherical value is found as $E_\lambda = 0.90$. This is the upper value in Figure 9.22, which shows the hemispherical emittance of window-glass sheets of various thicknesses (Gardon 1956).

The transmission behavior in Figures 9.21 and 9.22 provides glass windows with the important ability to trap solar energy. The sun radiates a spectral energy distribution very much like a blackbody at 5780 K (10,400°R). Considering the range $0.3 < \lambda < 2.7$ μm as being between the cutoff wavelengths in Figure 9.21, the blackbody characteristics show that 95% of the solar energy is in this range. Hence, glass has a low absorptance for solar radiation, and most of the solar radiation spectrum passes readily through a glass window. The emission from objects at the ambient temperature inside the enclosure behind the glass is at long wavelengths and is trapped because of the high absorptance (poor transmission) of the glass in this long-wavelength spectral region. This trapping behavior is the well-known "greenhouse effect," which also occurs because of the strong infrared (IR) absorption of various gases (particularly CO_2) in the Earth's atmosphere.

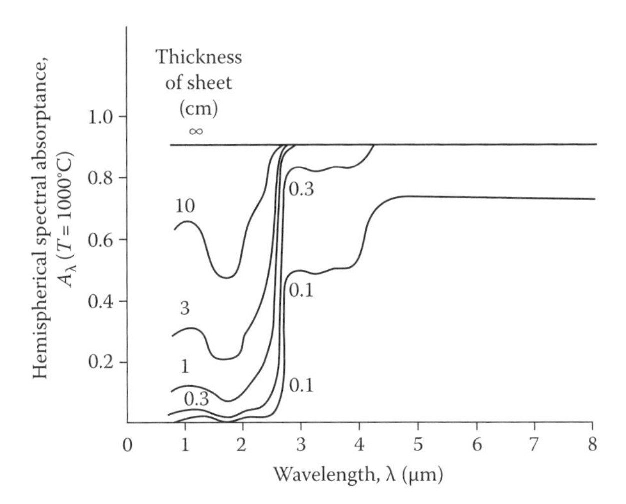

FIGURE 9.22 Absorptance (emittance) of sheets of window glass at 1000°C. (Data from Gardon, R., *J. Am. Ceram. Soc.*, 39(8), 278, 1956.)

An interesting application of a selectively transmitting layer is the transparent heat mirror. As mentioned in Fan and Bachner (1976), the transparent heat mirror has been used to construct transparent metallurgical furnaces for observing the growth of crystals at temperatures up to 1300 K. The thermal insulation for this furnace is provided by a gold film about 0.02 μm thick deposited on the inside of a Pyrex tube that encloses the heated region. These films have a high reflectance in the infrared and are equivalent to several inches of fibrous insulation in preventing radiative heat loss. The films, however, have a transmittance of about 0.2 in the visible region, which is adequate for observation into the high-temperature furnace.

A selectively transparent coating may also be useful for collecting solar energy. It can allow the short-wavelength solar energy to pass into a solar collector and prevent the escape of long-wavelength radiation re-emitted by the energy receiver. Possible coating materials are indium trioxide (In_2O_3), magnesium oxide (MgO), tin dioxide (SnO_2), and zinc oxide. Thin films of these materials are transparent for much of the solar spectrum and have a rapid increase of reflectance in the infrared. The measured transmittance and reflectance of a 0.35-μm-thick layer of Sn-doped In_2O_3 deposited on Corning 7059 glass are in Figure 9.23. The application to a solar-energy receiver is discussed by Jarvinen (1977).

Another option to control the spectral reflectivity is to imprint nanoscale patterns into the coating on a glass surface. Careful control of the pattern geometry can provide highly selective spectral effects (Mann et al. 2013), allowing the reflectivity spectrum to be adjusted for improved absorption or reflection in the IR portion of solar spectrum. Figure 9.24 shows the reflected solar intensity for a particular nanopattern in a silver layer compared with conventional low emissivity glass.

Water is also a selective absorber for radiant transmission. Figure 1.2 and Table 9.8 give the spectral absorption coefficient of water (Hale and Querry 1973) (see Irvine and Pollack (1968), Pinkley et al. (1977), and Rabl and Nielsen (1975) for additional information). The absorption coefficient determines the exponential attenuation of the radiant intensity as given by Equation 1.44. The absorption coefficient is quite small in the visible region (λ = 0.4–0.7 μm). In the vicinity of 1 μm, κ_λ begins to increase, and at longer wavelengths in the near-infrared, absorption is quite large. In the visible region, κ_λ is especially low in the blue-green region (0.5–0.55 μm). This accounts for the green appearance of sunlight penetrating under water to large depths; only the blue-green light has not been absorbed. Salt water has a higher absorption coefficient in the infrared than pure water (Smith and Baker 1981, Pelevin and Rostovtseva 2001).

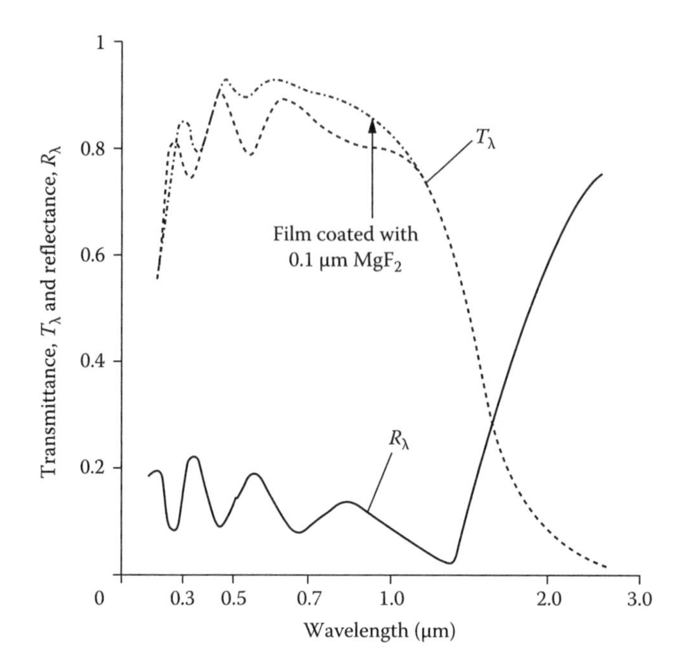

FIGURE 9.23 Transmittance and reflectance of 0.35 μm thick film of Sn-doped In_2O_3 film on Corning 7059 glass. Also shown is the effect on T_λ of an antireflection coating of MgF_2. (From Fan, J.C.C. and Bachner, F.J., *Appl. Opt.*, 15(4), 1012, 1976.)

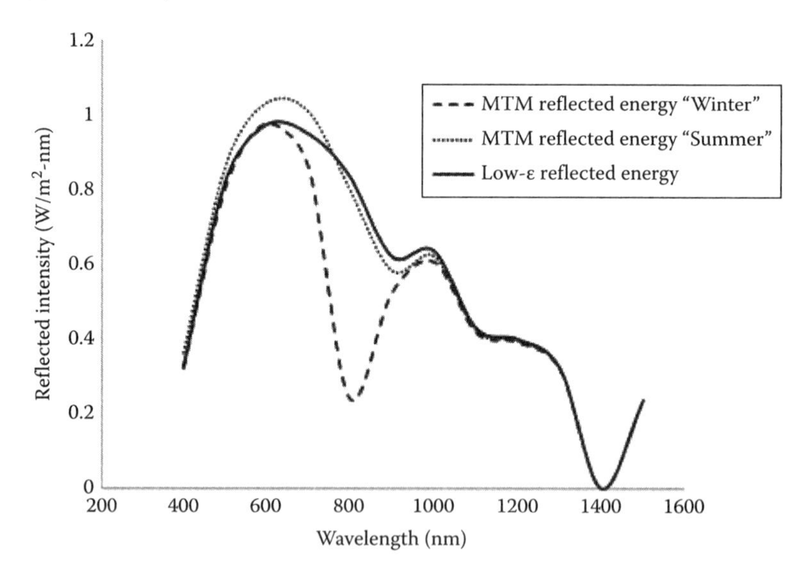

FIGURE 9.24 Reflected radiant intensity as a function of wavelength for low-ε and MTM glass configurations. (From Mann, T. et al., Metamaterial window glass for adaptable energy efficiency, *Proceedings of the ASME 2013 Summer Heat Transfer Conference*, Minneapolis, MN, July 14–19, 2013, paper HT2013-17511.)

Application of the values of the absorption coefficient in Table 9.8 to the solar spectrum results in the energy penetrations in Table 9.9. The second column shows the portions of the solar energy spectrum that are in various wavelength intervals. The successive columns demonstrate the high transparency for visible radiation as compared with very strong energy absorption in most of the near-infrared region. A similar table is in Rabl and Nielsen (1975) for the capture of solar radiation by shallow bodies of water called "solar ponds."

TABLE 9.8
Absorption Coefficient of Water

λ (μm)	κ_λ (cm^{-1})	λ (μm)	κ_λ (cm^{-1})
0.20	0.0691	2.4	50.1
0.25	0.0168	2.6	153
0.30	0.0067	2.8	5160
0.35	0.0023	3.0	11400
0.40	0.00058	3.2	3630
0.45	0.00029	3.4	721
0.50	0.00025	3.6	180
0.55	0.000045	3.8	112
0.60	0.0023	4.0	145
0.65	0.0032	4.2	206
0.70	0.0060	4.4	294
0.75	0.0261	4.6	402
0.80	0.0196	4.8	393
0.85	0.0433	5.0	312
0.90	0.0679	5.5	265
0.95	0.388	6.0	2240
1.0	0.363	6.5	758
1.2	1.04	7.0	574
1.4	12.4	7.5	546
1.6	6.72	8.0	539
1.8	8.03	8.5	543
2.0	69.1	9.0	557
2.2	16.5	9.5	587
		10.0	638

Source: Hale, G.M. and Querry, M.R., *Appl. Opt.*, 12(3), 555, 1973.

TABLE 9.9
Fractions of Solar Radiation Spectrum Transmitted through Various Thicknesses of Water

Spectral Interval	Incident Solar	Transmitted Energy Distribution for Water-Layer Thickness (cm)							
λ (μm)	Energy Distribution	0.001	0.01	0.1	1	10	100	1,000	10,000
0.3–0.6	0.237	0.237	0.237	0.237	0.237	0.236	0.229	0.173	0.014
0.6–0.9	0.360	0.360	0.360	0.359	0.353	0.305	0.129	0.010	
0.9–1.2	0.179	0.179	0.178	0.172	0.123	0.008			
1.2–1.5	0.087	0.086	0.082	0.063	0.017				
1.5–1.8	0.080	0.078	0.064	0.027					
1.8–2.1	0.025	0.023	0.011						
2.1–2.4	0.025	0.025	0.019	0.001					
2.4–2.7	0.007	0.006	0.002						
Totals	1.000	0.994	0.953	0.859	0.730	0.549	0.358	0.183	0.014

Source: From Kondratyev, Ya.K., *Radiation in the Atmosphere*, Academic Press, New York, 1969.

Clear ice also has a low absorption coefficient in the visible range, and the absorption coefficient increases by a factor of the order of 10^3 as the radiation wavelength increases from about 0.55 to 1.2 μm. Radiation in the visible and near-visible range can therefore be passed through ice that is not cloudy due to impurities or air bubbles. If an ice layer has adhered to a surface, visible and near-infrared radiation passes through the ice, thereby heating the substrate surface and providing a means for ice removal (Gilpin et al. 1977, Seki et al. 1979, Song and Viskanta 1990). Information on melting of semitransparent materials is in Diaz and Viskanta (1986) and Shih et al. (1986b).

A semitransparent material important in modern technology is silicon. Pure solid silicon is used in manufacturing semiconductor chips, and the rapid thermal processing of silicon wafers may entail heating rates yielding temperature increases greater than 200 K/s. Intense radiant heaters are used to provide the energy for this process. In the required spectral and temperature range for processing, the wavelength-dependent complex refractive index of silicon is very temperature dependent and hence, as expected from electromagnetic theory, so are the spectral reflectivity and transmissivity of the silicon wafers. At low temperatures, the wafers are fairly transparent in the infrared region, becoming more opaque as the wafer temperature increases. Design of wafer heaters requires careful consideration of these factors. Fairly complete data for absorption coefficients, including curve fits to the data at elevated temperatures for various wavelengths for T = 300–1600 K, are in Sin et al. (1984). More comprehensive data (spectral complex refractive index, coefficients of temperature dependence, normal spectral reflectivity, spectral absorption coefficient, complex dielectric constant) are in Sun et al. (1997) for λ = 0.25–1.1 μm for temperatures up to 1527 K. Jellison and Modine (1994) provide the measured components of the complex refractive index between 300 and 490 K for wavelengths in the range 400–780 nm. Zhou and Zhang (2003) use a Monte Carlo analysis to determine the spectral reflectance, absorptance, and transmittance of silicon wafers with prescribed surface roughness. Molton silicon is opaque at all wavelengths.

HOMEWORK

9.1 Compute the half-width for Doppler broadening, γ_D, of neon at a wavelength of 0.75 μm and for $T = 400$ K.
Answer: $\gamma_D = 0.0214$ cm^{-1}.

9.2 Two absorption lines have the same transition (centerline) wave number, η_{ij} = 550 cm^{-1}. Both have the same half-width, 0.15 cm^{-1}. One line has the Doppler profile; the other has the Lorentz profile. Draw the two line shapes, $\kappa_{\eta,ij}(\eta)/S_{ij}$, as a function of η, on the same plot.

9.3 A gas is composed of pure atomic hydrogen at a temperature of 1000 K. Calculate the half-width of the hydrogen Lyman alpha line (transition centerline frequency = 2.4675×10^{15} Hz) for the case of Doppler broadening. Then plot the line shape $\kappa_{\eta,ij}/S_{ij}$ for this line as a function of wave number. The mass of the hydrogen atom is 1.66×10^{-24} g.
Answer: $\gamma_D = 0.932$ cm^{-1}.

9.4 For the same gas and temperature as Homework Problem 9.3, compute the half-width of the line for collision broadening at a pressure of 1 atm. Assume the diameter of the hydrogen atom is about 1.06×10^{-8} cm. Plot $\kappa_{\eta,ij}(\eta)/S_{ij}$ for collision broadening on the same wave number plot as for Homework Problem 9.3.
Answer: $\gamma_c = 0.0089$ cm^{-1}

9.5 Prove that for weak, nonoverlapping Lorentz lines, the Elsasser band model predicts the weak absorption band relation $\bar{A}_\ell = C_\ell S$ where C_ℓ is a constant for the particular band.

9.6 From the spectral absorptance α_λ in Figure 9.4, estimate the total emittance of CO_2 for the temperature, pressure, and path length given in the figure caption. Do not assume that the spectral bands are black. Compare the result with the value obtained from Leckner's correlation, Equation 9.64.
Answer: $\varepsilon = 0.259$; $\varepsilon_{Lechner} = 0.31$.

9.7 Carbon dioxide is in a mixture with air. The CO_2 has a mole fraction of 0.4, and the gas mixture is at a temperature of 1250 K and a total pressure of 1 atm. Using the correlated-*k* method, determine the emittance of the gas in each band and the total emittance for a path length of 2 m.
Answer: $\varepsilon(X = pS = 0.8 \text{ atm·m}, T = 1250 \text{ K}) = 0.161$.

9.8 Carbon dioxide is in a mixture with air. The CO_2 has a mole fraction of 0.4, and the gas mixture is at a temperature of 1250 K and a total pressure of 1 atm. Using the spectral-line-weighted sum-of-gray-gases (SLWSGG) method, determine the total emittance for a path length of 2 m. Take the source and gas temperatures both as 1250 K for the gas absorptance calculation, so the emittance is then equal to the absorptance.
Answer: $\varepsilon(X = pS = 0.8 \text{ atm·m}, T = 1250 \text{ K}) = 0.190$.

9.9 For water vapor at 1200 K with a mean beam length of 3 m and a partial pressure of 0.2 atm when mixed with air at a total pressure of 1 atm, compare the water vapor total emittance $\varepsilon(T, pL_e)$ using the predictions of Cess and Lian, Equation 9.63 and Leckner, Equation 9.64.
Answer: Cess and Lian—0.301; Leckner—0.290.

9.10 Use Leckner correlation to find the emittance of the Earth's atmosphere for a temperature of 300 K, a CO_2 concentration of 350 ppm by volume, and a mean beam length through the atmosphere of 60 km.
Answer: $\varepsilon = 0.113$.

10 Fundamental Radiative Transfer Relations

10.1 INTRODUCTION

For a translucent medium that is hot and/or is subjected to external radiation, energy is transferred internally by radiation in addition to conduction and convection. The energy equation in a material expresses a local balance of energy arriving by all modes of energy transfer, internal energy stored, energy generated by local sources, and energy leaving by all modes of transfer. For radiative transfer in a translucent material, energy is deposited locally by absorption and leaves by local emission. The net energy deposited by all of the radiative effects can be viewed as a local energy source for convection and conduction transfer in the same manner as an energy source provided by electrical dissipation or by nuclear or chemical reactions. The energy equation including the radiative energy source is provided in this chapter, and it is to be solved to provide the temperature distribution in the translucent material and other heat transfer characteristics such as local heat fluxes. Chapters 11 and 12 present methods for determining the radiative flux divergence (the radiative source term in the energy equation). This is the quantity of interest in most heat transfer problems requiring consideration of radiation transfer. Chapter 13 then examines solution of the energy equation including the thermal radiation effects.

To obtain the radiative contribution to the energy equation, radiative transfer relations must be provided. Because some of the radiative terms depend on temperature, the radiative transfer equation (RTE) must be solved simultaneously with the energy equation to determine the temperature distribution and radiating characteristics. Basic concepts in Chapter 1 showed that radiation traveling along a path is attenuated by absorption and scattering and is enhanced by emission and by scattering coming in from other directions. These concepts are expanded here to develop an equation for the radiation intensity along a typical path through a translucent medium. This equation contains a first derivative of intensity with respect to the path coordinate, so a solution requires one boundary condition. This is usually the intensity at the origin of the radiation path being considered. Because the path usually begins at a boundary of the radiating medium, the radiation at the boundary is thereby incorporated into the radiation distribution within the medium. The radiative boundary conditions for solving the RTE throughout the medium are in addition to the thermal boundary conditions that must be specified to solve the energy equation.

The intensity obtained from the RTE is the local radiative energy traveling in a single direction per unit solid angle and wavelength interval and crossing a unit area element within a medium normal to the path direction. To determine the net radiative energy crossing an area element then requires integrations to include energy contributions by intensities crossing in all directions and for all wavelengths. This results in the local radiative flux and the radiative flux divergence that is used to obtain the radiative internal heat source distribution in the energy equation.

If the medium is scattering, the local radiative intensity is affected by radiation emitted and scattered throughout the medium. The scattering into a direction along a path is often combined with the local emission into that direction to form the *source function* in the RTE, which is different from the radiative source term in the conservation of energy equation. For a solution of the RTE including scattering, the equation for the source function is solved simultaneously with the energy equation to determine both the source function and temperature distributions. The formulations provided here do not consider interface reflection and refraction effects for translucent materials with a refractive index of $n > 1$. Additional analytical relations are provided in Chapter 17 for materials such as glass, water, translucent plastics, or translucent ceramics that have $n > 1$.

10.2 ENERGY EQUATION AND BOUNDARY CONDITIONS FOR A PARTICIPATING MEDIUM

An energy balance on a volume element within a material includes contributions by conduction, convection, internal heat sources such as by electrical dissipation and combustion, compression work, viscous dissipation, energy storage during transients, and the contribution by radiative transfer in a translucent material. Compared with internal energy in the form of heat capacity, the storage of radiant energy by the increase of photons within a volume element is usually negligible except for some special transients such as are considered in Kumar and Mitra (1995) and Longtin and Tien (1997) where this additional transient term is provided; hence, no modification of the usual transient heat capacity term in the energy equation is considered in this chapter as a result of the thermal radiation. In most problems, radiation pressure is negligible relative to fluid pressure and is not included in the compression work. For heat conduction, the net contribution to a volume element can be written as the negative divergence of a conduction flux vector $-\nabla \cdot q_c = \nabla \cdot (k\nabla T)$. Similarly, the net contribution by radiant energy per unit volume within a translucent medium can be written as the negative of the divergence of the radiant flux vector $\mathbf{q}_r$, and expressions for this term will be obtained in the subsequent development of the radiative transfer relations. Thus, the conventional energy equation for a single-component translucent fluid can be modified for the effect of radiative transfer by adding $-\mathbf{q}_r$ to the $k\nabla T$ to yield

$$\rho c_p \frac{DT}{Dt} = \beta T \frac{DP}{Dt} + \nabla \cdot (k\nabla T - \mathbf{q}_r) + \dot{q} + \Phi_d \tag{10.1}$$

where

- D/Dt is the substantial derivative
- The β is the thermal coefficient of volume expansion of the fluid
- $\dot{q}$ is the local energy source (electrical, chemical, nuclear) per unit volume and time
- Φ_d is the energy production by viscous dissipation

Although traditionally referred to as the energy equation, the terms are each energy rates (or power) per unit volume. An alternative form in terms of enthalpy is

$$\rho \frac{Dh}{Dt} = \frac{DP}{Dt} + \nabla \cdot (k\nabla T - \mathbf{q}_r) + \dot{q} + \Phi_d \tag{10.2}$$

For an incompressible fluid with constant properties in rectangular coordinates,

$$\frac{\partial T}{\partial t} + u\frac{\partial T}{\partial x} + v\frac{\partial T}{\partial y} + w\frac{\partial T}{\partial z} = \frac{k}{\rho c}\left(\frac{\partial^2 T}{\partial x^2} + \frac{\partial^2 T}{\partial y^2} + \frac{\partial^2 T}{\partial z^2}\right) - \frac{1}{\rho c}\left(\frac{\partial q_{r,x}}{\partial x} + \frac{\partial q_{r,y}}{\partial y} + \frac{\partial q_{r,z}}{\partial z}\right)$$
$$+ \frac{1}{\rho c}\dot{q}(x,y,z,t) + \frac{1}{\rho c}\Phi_d \tag{10.3}$$

To obtain the temperature distribution in the medium by solving Equation 10.1, an expression for $\nabla \cdot \mathbf{q}_r$ is needed in terms of the temperature distribution. In addition, the $\nabla \cdot \mathbf{q}_r$ is also affected by scattered radiation, which will need to be determined within the medium. One approach for obtaining the required relations is to derive the $\mathbf{q}_r$ and then differentiate to obtain $\nabla \cdot \mathbf{q}_r$; this is demonstrated in Chapter 12 for the specific geometries of a plane layer and a long circular cylinder.

Another approach is to obtain $\nabla \cdot \mathbf{q}_r$ directly by considering the local radiative interaction within a differential volume in the medium. The $\nabla \cdot \mathbf{q}_r$ is found from the difference between the radiation emitted from the same volume element and the absorption in that same element. This depends on the absorption coefficient and the intensity incident from all directions (note that for a medium with refractive index $n > 1$, an n^2 factor is in the emission term, as given in Chapter 17). The radiative energy required for the energy equation is the total radiation (includes all wavelengths, wave numbers, or frequencies). Hence, as will be developed in detail, relations are required so that the quantities can be evaluated to obtain the difference between total radiation incident from all solid angles Ω_i that is locally absorbed and the locally emitted radiation. These two quantities (absorbed and emitted radiation) are expressed by the two integral terms over all wavelengths on the right side of the following relation (see Section 10.4.2):

$$-\nabla \cdot \mathbf{q}_r = \int_{\lambda=0}^{\infty} \kappa_\lambda \left[\int_{\Omega_i=0}^{4\pi} I_\lambda(\Omega_i) d\Omega_i \right] d\lambda - 4\pi \int_{\lambda=0}^{\infty} \kappa_\lambda I_{\lambda b} d\lambda \tag{10.4}$$

The RTE is developed in the next section as needed to obtain the local intensity throughout the material volume in the $\int_{\Omega_i=0}^{4\pi} I_\lambda(\Omega_i) d\Omega_i$ term.

A special case of the energy equation is when radiation dominates over both conduction and convection. Then, for steady state, the equilibrium temperature distribution is achieved only by radiative effects with or without heat sources, $\dot{q}$, present. This condition is called *radiative equilibrium*. This limit provides useful results for some high-temperature applications, and solutions for radiative equilibrium provide limiting results for comparison with solutions including conduction and/or convection combined with radiation. If a medium is considered with internal energy sources $\dot{q}$, such as chemical or electrical energy release, the energy equation for radiative equilibrium is

$$\nabla \cdot \mathbf{q}_r(\mathbf{r}) = \dot{q}(\mathbf{r}) \tag{10.5}$$

where $\mathbf{r}$ is a position vector in the participating medium.

The solution of the energy equation requires boundary conditions. For the general energy relations given by Equations 10.1 through 10.3, the dependent variable is a temperature or a temperature-dependent property such as enthalpy. Because the equation is second order in the space coordinates and first order in time, the equation requires two boundary conditions for each independent space variable such as x, y, and z in Cartesian coordinates and an initial condition. When radiation is present, there are additional boundary conditions for the radiative intensities. The radiative boundary conditions enter the results through the solution of the RTE that is used to evaluate the $\nabla \cdot \mathbf{q}_r$ term in the energy equation. Thus, the radiative boundary conditions are not explicitly stated for the energy equation, but are incorporated in the solution for the radiative flux divergence, $\nabla \cdot \mathbf{q}_r$.

In this chapter and in Chapters 11 and 12, discussion centers on cases with either radiative equilibrium or a prescribed internal source $\dot{q}(\mathbf{r})$. These cases are generally solved as linear equations. Cases with conduction and/or convection become highly nonlinear in temperature and are discussed in Chapter 13.

10.3 RADIATIVE TRANSFER AND SOURCE-FUNCTION EQUATIONS

To obtain the radiative intensity throughout the translucent material, the radiative transfer equation is used. This is the differential equation that describes the radiation intensity along a path in a fixed direction through an absorbing, emitting, and scattering medium. Bouguer's law, Equation 1.41, accounts for attenuation of intensity by absorption and scattering. The RTE also includes augmentation of the radiation intensity by emission and scattering into the path direction. We now discuss the details.

10.3.1 Radiative Transfer Equation

As derived in Chapter 1 (Equation 1.72), the RTE is

$$\begin{aligned} &I_\lambda(r+dr,\ \Omega,\ t+dt)-I_\lambda(r,\Omega,\ t) \\ &= \kappa_\lambda I_{\lambda b}(r,t)ds-\kappa_\lambda I_\lambda(r,\Omega,t)ds-\sigma_{s,\lambda}I_\lambda(r,\Omega,t)ds \\ &\quad +\frac{\sigma_{s,\lambda}}{4\pi}\int_{4\pi} I_\lambda(r,\Omega,t)\Phi_\lambda(\Omega',\Omega)d\Omega' \end{aligned} \tag{10.6}$$

We can combine the attenuation terms and define the *extinction coefficient* $\beta_\lambda = \kappa_\lambda + \sigma_{s,\lambda}$. Equation 10.6 is written for a particular propagation direction S and for steady conditions (e.g., over a time interval in which the radiation intensity varies insignificantly due to photon time-of-flight effects, $\partial I/c\partial t \approx 0$):

$$\frac{dI_\lambda}{dS} = -\beta_\lambda(S)I_\lambda(S)+\kappa_\lambda(S)I_{\lambda b}(S)+\frac{\sigma_{s,\lambda}}{4\pi}\int_{\Omega_i=0}^{4\pi} I_\lambda(S,\Omega_i)\Phi_\lambda(\Omega,\Omega_i)d\Omega_i \tag{10.7}$$

In Equation 10.7 and the development that follows, $\sigma_{s,\lambda}$ is assumed to be independent of direction of incidence, Ω_i; this assumption is generally valid except for a few special cases such as propagation through filament wound structures and other layered scattering materials.

We define the *single scattering albedo*, ω_λ, as the ratio of scattering to extinction coefficients,

$$\omega_\lambda = \frac{\sigma_{s,\lambda}}{\kappa_\lambda+\sigma_{s,\lambda}} = \frac{\sigma_{s,\lambda}}{\beta_\lambda};\quad 1-\omega_\lambda = \frac{\kappa_\lambda}{\kappa_\lambda+\sigma_{s,\lambda}} = \frac{\kappa_\lambda}{\beta_\lambda} \tag{10.8}$$

If the medium is not absorbing, but scattering alone, $\omega_\lambda \to 1$. For a purely absorbing medium, $\omega_\lambda \to 0$. The *optical thickness* or *opacity* was defined in Equation 1.53 as

$$\tau_\lambda(S) = \int_{S^*=0}^{S} \beta_\lambda(S^*)dS^* = \int_{S^*=0}^{S} \left[\kappa_\lambda(S^*)+\sigma_{s,\lambda}(S^*)\right]dS^* \tag{10.9}$$

The RTE (Equation 10.7) in terms of albedo and optical thickness becomes

$$\frac{dI_\lambda}{d\tau_\lambda} = -I_\lambda(\tau_\lambda)+(1-\omega_\lambda)I_{\lambda b}(\tau_\lambda)+\frac{\omega_\lambda}{4\pi}\int_{\Omega_i=0}^{4\pi} I_\lambda(\tau_\lambda,\Omega_i)\Phi_\lambda(\Omega,\Omega_i)d\Omega_i \tag{10.10}$$

where

$d\tau_\lambda = \beta_\lambda(S)dS$ is the *optical differential thickness*

$\tau_\lambda = \tau_\lambda(S)$

$\omega_\lambda = \omega_\lambda(S)$

The final two terms in Equation 10.10 act to increase the intensity along the direction S by absorption and in scattering. They are often combined into the *source function*, $\hat{I}_\lambda(\tau_\lambda,\Omega)$,

$$\hat{I}_\lambda(\tau_\lambda,\Omega) = (1-\omega_\lambda)I_{\lambda b}(\tau_\lambda) + \frac{\omega_\lambda}{4\pi}\int_{\Omega_i=0}^{4\pi} I_\lambda(\tau_\lambda,\Omega_i)\Phi_\lambda(\Omega,\Omega_i)\,d\Omega_i \tag{10.11}$$

For anisotropic scattering, $\hat{I}_\lambda(\tau_\lambda,\Omega)$ is a function of Ω (i.e., of the direction S). The RTE (Equation 10.10) then becomes

$$\frac{dI_\lambda}{d\tau_\lambda} + I_\lambda(\tau_\lambda) = \hat{I}_\lambda(\tau_\lambda,\Omega) \tag{10.12}$$

where $\tau_\lambda = \tau_\lambda(S)$. This appears to be a differential equation; however, because I_λ is within the source function in Equation 10.11, Equation 10.12 is actually an integro-differential equation. It can be formally integrated after multiplying it by the integrating factor e^{τ_λ}:

$$\frac{dI_\lambda}{d\tau_\lambda}e^{\tau_\lambda} + I_\lambda(\tau_\lambda)e^{\tau_\lambda} = \frac{d}{d\tau_\lambda}\left[I_\lambda(\tau_\lambda)e^{\tau_\lambda}\right] = \hat{I}_\lambda(\tau_\lambda,\Omega)e^{\tau_\lambda} \tag{10.13}$$

Integrating over the optical path from $\tau_\lambda = 0$ to $\tau_\lambda = \tau_\lambda(S)$ and rearranging gives

$$I_\lambda(\tau_\lambda,\Omega) = I_\lambda(0,\Omega)e^{-\tau_\lambda} + \int_{\tau_\lambda^*=0}^{\tau_\lambda} \hat{I}_\lambda(\tau_\lambda^*)\,e^{-(\tau_\lambda-\tau_\lambda^*)}\,d\tau_\lambda^*;\quad \tau_\lambda(S) = \int_{S^*=0}^{S}\beta_\lambda(S^*)\,dS^* \tag{10.14}$$

where

τ_λ^* is a dummy optical variable of integration along S (see Equation 10.9)
$I_\lambda(0,\Omega)$ is the intensity in the direction of S at the boundary or location where $S = 0$

Equation 10.14 is the *integrated form of the RTE*. Equation 10.14 is interpreted physically as the intensity at optical depth τ_λ, being composed of two terms. The first is the attenuated initial intensity that arrives at τ_λ. The second is the intensity at τ_λ resulting from emission and incoming scattering in the S direction by all thickness elements along the path from 0 to S, reduced by exponential attenuation between each location of emission and incoming scattering τ_λ^* and the location τ_λ.

Without scattering (i.e., for absorption only) the source function defined in Equation 10.11, $\hat{I}_\lambda(\tau_\lambda) = I_{\lambda b}(\tau_\lambda)$, reduces to the local blackbody intensity that is isotropic, and the RTE (Equation 10.14) becomes

$$I_\lambda(\tau_\lambda,\Omega) = I_\lambda(0,\Omega)e^{-\tau_\lambda} + \int_{\tau_\lambda^*=0}^{\tau_\lambda} I_{\lambda b}(\tau_\lambda^*)\,e^{-(\tau_\lambda-\tau_\lambda^*)}\,d\tau_\lambda^*;\quad \tau_\lambda(S) = \int_{S^*=0}^{S}\kappa_\lambda(S^*)\,dS^* \tag{10.15}$$

For a purely scattering medium, the single scattering albedo $\omega_\lambda = 1$, and the source function Equation 10.11 becomes

$$\hat{I}_\lambda(\tau_\lambda,\Omega) = \frac{1}{4\pi}\int_{\Omega_i=0}^{4\pi} I_\lambda(\tau_\lambda,\Omega_i)\Phi_\lambda(\Omega,\Omega_i)\,d\Omega_i;\quad \tau_\lambda(S) = \int_{S^*=0}^{S}\sigma_{s,\lambda}(S^*)\,dS^* \tag{10.16}$$

The RTE becomes Equation 10.14 with this substitution for the source function. The local intensity is then no longer a function of medium temperature except through the possible temperature dependence of the properties.

For isotropic scattering, $\Phi_\lambda = 1$, so the source function Equation 10.11 including emission becomes

$$\hat{I}_\lambda(\tau_\lambda) = (1-\omega_\lambda) I_{\lambda b}(\tau_\lambda) + \frac{\omega_\lambda}{4\pi} \int_{\Omega_i=0}^{4\pi} I_\lambda(\tau_\lambda, \Omega_i)\, d\Omega_i \tag{10.17}$$

The source function is now independent of direction (isotropic). In the limit of no absorption ($\omega_\lambda = 1$) and isotropic scattering, Equation 10.17 shows that the source function reduces to the local mean incident intensity,

$$\hat{I}_\lambda(\tau_\lambda) = \frac{1}{4\pi} \int_{\Omega_i=0}^{4\pi} I_\lambda(\tau_\lambda, \Omega_i)\, d\Omega_i \equiv \bar{I}_\lambda(\tau_\lambda) \tag{10.18}$$

The boundary condition $I_\lambda(0, \Omega)$ needed for the solution of the radiative transfer equation in Equation 10.14 is obtained as in the usual enclosure theory. At an opaque solid boundary, the intensity consists of emitted and reflected intensities. At a translucent boundary, the $I_\lambda(0, \Omega)$ depends on radiation entering from the exterior surroundings of the boundary. Obtaining and applying the boundary condition is illustrated by examples and solutions in succeeding chapters where the equation of transfer is applied; translucent boundaries are considered in Chapter 17 for materials with $n > 1$.

10.3.2 Source-Function Equation

The source function was defined in Equation 10.11. Eliminate I_λ by using Equation 10.14 in Equation 10.11 to obtain an equation for $\hat{I}_\lambda$; the result is the *source-function equation.* As shown in Figure 10.1, the location τ_λ in Equation 10.11 is along path S, and the source function is being considered along this path. At each location along S, radiation is augmented by scattering coming in from all incident paths such as S_i. The optical length along S is $\tau_{\lambda,i}$ from its origin $\tau_{\lambda,i} = 0$ to the location τ_λ along S. Then, from Equation 10.14, the incident intensity in $d\Omega_i$ at τ_λ in Figure 10.1 provided by radiation along S_i is

$$I_\lambda(\tau_{\lambda,i}, \Omega_i) = I_\lambda(\tau_{\lambda,i} = 0, \Omega_i) e^{-\tau_{\lambda,i}} + \int_{\tau_{\lambda,i}^*=0}^{\tau_{\lambda,i}} \hat{I}_\lambda(\tau_{\lambda,i}^*, \Omega_i)\, e^{-(\tau_{\lambda,i}-\tau_{\lambda,i}^*)}\, d\tau_{\lambda,i}^* \tag{10.19}$$

Substituting this into Equation 10.11 yields the source-function equation as

$$\hat{I}_\lambda(\tau_\lambda, \Omega) = (1-\omega_\lambda) I_{\lambda b}(\tau_\lambda) + \frac{\omega_\lambda}{4\pi} \int_{\Omega_i=0}^{4\pi} \left[I_\lambda(\tau_{\lambda,i} = 0, \Omega_i) e^{-\tau_{\lambda,i}} + \int_{\tau_{\lambda,i}^*=0}^{\tau_{\lambda,i}} \hat{I}_\lambda(\tau_{\lambda,i}^*, \Omega_i)\, e^{-(\tau_{\lambda,i}-\tau_{\lambda,i}^*)}\, d\tau_{\lambda,i}^* \right] \Phi_\lambda(\Omega, \Omega_i)\, d\Omega_i \tag{10.20}$$

This integral equation shows how the radiative source function along a path depends on the source function along all of the other intersecting paths arriving from throughout the medium volume. The solution requires obtaining results for the entire radiation field. Because of the complexity of

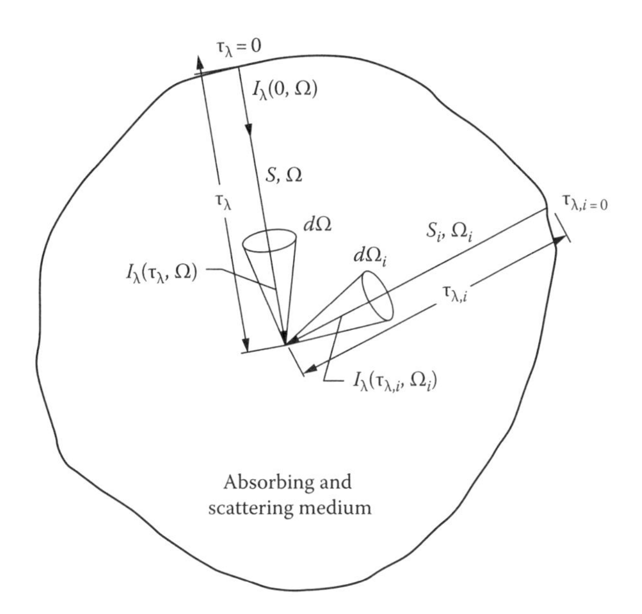

FIGURE 10.1 Path direction S for intensity and typical path S_i.

the source-function equation, simplifying approximations are often made that are reasonable for various physical situations. Some of these are now examined.

A common assumption is that the scattering is isotropic (in addition to $\sigma_{s,\lambda}$ being independent of incidence angle as assumed earlier). This is a limiting approximation that is only valid for random distributions of scattering particles and for conditions without strong directionally incident radiation so that the radiation interacting with the particles tends to be isotropic. Under this assumption, the phase function $\Phi = 1$ and $\hat{I}_\lambda$ is isotropic (see Equation 10.17) so that Equation 10.20 reduces to

$$\hat{I}_\lambda(\tau_\lambda) = (1-\omega_\lambda)I_{\lambda b}(\tau_\lambda)$$

$$+\omega_\lambda \int_{\Omega_i=0}^{4\pi} \left[\frac{1}{4\pi} I_\lambda(\tau_{\lambda,i}=0,\Omega_i)e^{-\tau_{\lambda,i}} + \int_{\tau^*_{\lambda,i}=0}^{\tau_{\lambda,i}} \hat{I}_\lambda\left(\tau^*_{\lambda,i}\right) e^{-(\tau_{\lambda,i}-\tau^*_{\lambda,i})}\, d\tau^*_{\lambda,i} \right] d\Omega_i \tag{10.21}$$

For a gray medium with constant properties and isotropic scattering, the equation for $\hat{I}$ in terms of the physical position S (rather than the optical length), the incident path length S_i, and the extinction coefficient β is

$$\hat{I}(S) = (1-\omega)I_b(S)$$

$$+\omega \left[\frac{1}{4\pi} \int_{\Omega_i=0}^{4\pi} I(S_i=0,\Omega_i)e^{-\beta S}\, d\Omega_i + \beta \int_{S^*=0}^{S_i} \hat{I}\left(S_i^*\right) e^{-\beta(S_i-S_i^*)}\, dS_i^* \right] \tag{10.22}$$

Without scattering, $\omega_\lambda = 0$ or $\omega = 0$, the source function is simply the blackbody intensity, $\hat{I}_\lambda(\tau_\lambda) = I_{\lambda b}(\tau_\lambda)$ or $\hat{I}(S) = I_b(S)$. For a medium with only scattering and no absorption, $\omega_\lambda = 1$ or $\omega = 1$ in Equations 10.21 and 10.22, or the source function is given by Equations 10.16 and 10.18 in terms of local mean incident intensity.

Equations 10.19 and 10.20 often cannot be solved simultaneously for the source function and intensity distributions throughout the translucent medium because they contain the unknown temperature distribution within the medium that affects the blackbody intensity $I_{\lambda b}(T)$. For temperature-dependent properties, the temperature distribution is also needed to determine the absorption and scattering coefficients; the local optical depth $\tau_\lambda(S)$ along a path can then be computed from Equation 10.9 and the physical path length S related to the optical length τ_λ. The temperature distribution depends on energy conservation within the medium, Equation 10.1, which in turn depends on the total radiative source (radiative flux divergence) in each volume element, $-\nabla \cdot \mathbf{q}_r$, that is obtained from the intensities and the source function. As will be shown as the development continues, the energy equation is solved along with the intensity and source-function equations to yield the temperature distribution and the angular and spatial distributions of intensity. The equations are sufficiently complex that numerical solutions are almost always required. In a few instances analytical techniques have been used to obtain closed-form solutions for a limited range of geometries and conditions. The energy, intensity, and source-function equations provide the set of basic equations required for analysis.

Example 10.1

A black surface element dA is 10 cm from an element of gas dV (Figure 10.2). The gas element is a part of a gas volume V that is isothermal and at the same temperature T as dA. If the gas has an absorption coefficient $\kappa_\lambda = 0.1$ cm^{-1} at $\lambda = 1$ μm and there is no scattering, what is the spectral intensity at $\lambda = 1$ μm that arrives at dV along the path S from dA to dV?

Because dA is black and at temperature T, the intensity at $S = 0$ is $I_\lambda(0) = I_{\lambda b}(T)$. Since the gas is isothermal, the emitted blackbody intensity in the gas is $I_{\lambda b}(\tau_\lambda) = I_{\lambda b}(T)$. Substituting into the integrated RTE, Equation 10.15, that is, for $\sigma_{s,\lambda} = 0$, gives

$$I_\lambda(\tau_\lambda) = I_{\lambda b}(T)e^{-\tau_\lambda} + I_{\lambda b}(T)e^{-\tau_\lambda} \int_{\tau_\lambda^* = 0}^{\tau_\lambda} e^{\tau_\lambda^*} d\tau_\lambda^*$$

After integration, this reduces to $I_{\lambda b}(\tau_\lambda) = I_{\lambda b}(T)$. The intensity arriving at dV along an isothermal path through the gas from a black surface element at the same temperature as the gas is thus equal to the blackbody spectral intensity emitted by the wall and does not depend on κ_λ or S. The attenuation by gas absorption of the spectral intensity emitted by the wall is exactly compensated by emission from the gas along the path from dA to dV. This is true at each wavelength so, from Equation 10.4, the radiative source is $-\nabla \cdot \mathbf{q}_r = 0$. If T is specified, the spectral intensity arriving at dV can be calculated using Equation 1.13, since $n_{gas} \approx 1$.

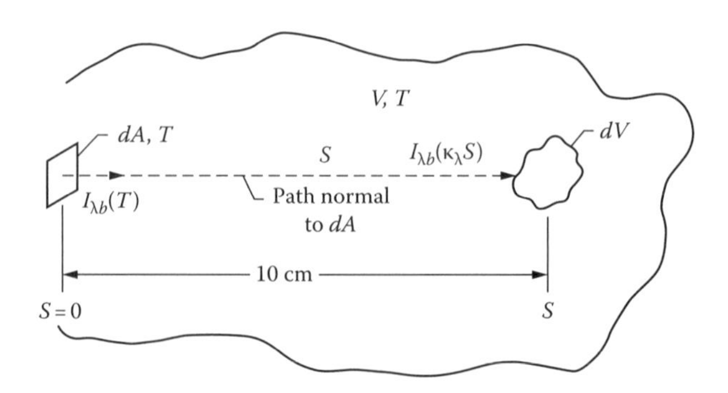

FIGURE 10.2 Geometry for Example 10.1.

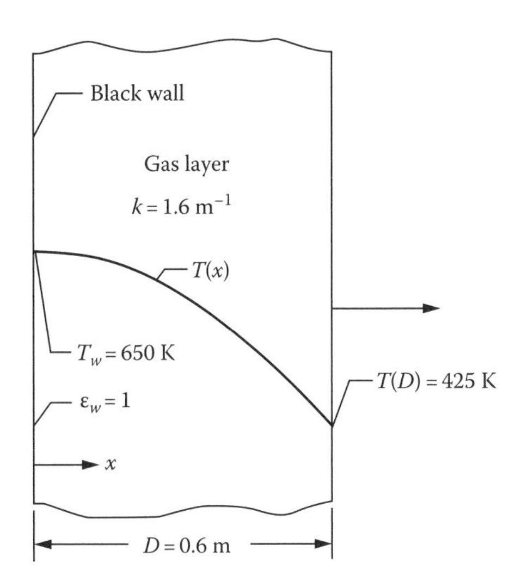

FIGURE 10.3 Conditions for Example 10.2.

Example 10.2

An absorbing–emitting layer of gray gas with $\kappa = 1.6\ m^{-1}$ is adjacent to a black wall. As a result of internal energy generation, the medium has a parabolic temperature distribution decreasing from 650 K at the wall to 425 K at the boundary $x = D$ (Figure 10.3). What is the total intensity $I(D)$ in the direction normal to the wall?

The parabolic temperature distribution is given by $T(x) = T_w - (T_w - 425)(x/D)^2$. From Equation 10.15, the intensity normal to the wall at $x = D$ is given for a gray gas by

$$I(D) = \frac{\sigma T_w^4}{\pi} e^{-\kappa D} + \int_{\kappa x=0}^{\kappa D} \frac{\sigma T^4(x)}{\pi} e^{-\kappa(D-x)} \kappa dx$$

If we insert the $T(x)$ and other numerical values, numerical integration yields $I(D) = 2510\ W/(m^2 \cdot sr)$.

10.4 RADIATIVE FLUX AND ITS DIVERGENCE WITHIN A MEDIUM

The radiative flux vector $\mathbf{q}_r$ and its divergence $-\nabla \cdot \mathbf{q}_r$ are now considered as required for solving the energy Equation 10.1.

10.4.1 Radiative Flux Vector

For an energy balance on a volume element dV, the net radiative energy supplied to dV is needed. For a volume element $dx\, dy\, dz$, the radiative energies in and out of the $dx\, dz$ faces are shown in Figure 10.4. The energies are similarly written for the other faces, the outgoing energies are subtracted from the incoming, and the result divided by $dx\, dy\, dz$. The result is the net rate of radiative energy supplied to dV per unit volume:

$$-\left[\frac{\partial q_{r,x}}{\partial x} + \frac{\partial q_{r,y}}{\partial y} + \frac{\partial q_{r,z}}{\partial z}\right] = -\nabla \cdot \mathbf{q}_r \tag{10.23}$$

This is the negative of the divergence of the radiative flux vector $\mathbf{q}_r$,

$$\mathbf{q}_r = \mathbf{i} q_{r,x} + \mathbf{j} q_{r,y} + \mathbf{k} q_{r,z} \tag{10.24}$$

The vector form $-\nabla \cdot \mathbf{q}_r$ can be used for any coordinate systems.

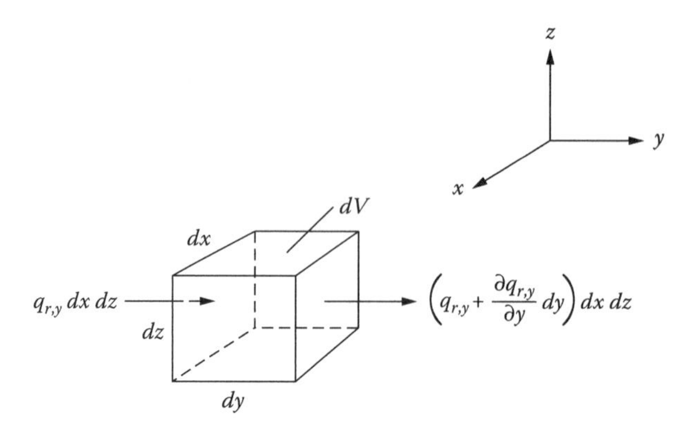

FIGURE 10.4 Radiative fluxes for volume element.

The flux vector is now related to the intensity from the equation of transfer. Consider the area dA in Figure 10.5. This could be one of the faces of the volume element in Figure 10.4, and **n** is then along a coordinate direction. In general, **n** is any direction, and the flux $q_{r,n}$ is through an area normal to the **n** direction. Let **s** be a unit vector in the S direction, which is the direction for the intensity I. The direction cosines for S in the rectangular coordinate system are α, δ, and γ. Then, $\cos\theta = \mathbf{s}\cdot\mathbf{n}$, and in Figure 10.5, the intensity is the energy rate per unit solid angle crossing dA per unit area normal to the direction of $I = I(\alpha, \delta, \gamma)$. Hence, the energy rate crossing dA as a result of I is $I = I(\alpha, \delta, \gamma)\, dA \cos\theta\, d\Omega$. The radiative flux crossing dA as a result of intensities incident from all directions is then

$$q_{r,n} = \int_{\Omega=0}^{4\pi} I(\alpha,\delta,\gamma)\cos\theta d\Omega = \int_{\Omega=0}^{4\pi} I(\alpha,\delta,\gamma)\mathbf{s}\cdot\mathbf{n}d\Omega \tag{10.25}$$

where θ is the angle from the normal of dA to the direction of $I(\alpha, \delta, \gamma)$. The $q_{r,n}$ depends on the direction of **n** relative to **s**. The cosθ becomes negative for $\theta > \pi/2$, so the sign of the portion of the energy flux traveling in the direction opposite to the positive **n** direction is automatically included. The $q_{r,n}$ is the component in the **n** direction of the radiative flux vector given by

$$\mathbf{q}_r = \int_{\Omega=0}^{4\pi} I(\alpha,\delta,\gamma)\mathbf{s}d\Omega \tag{10.26}$$

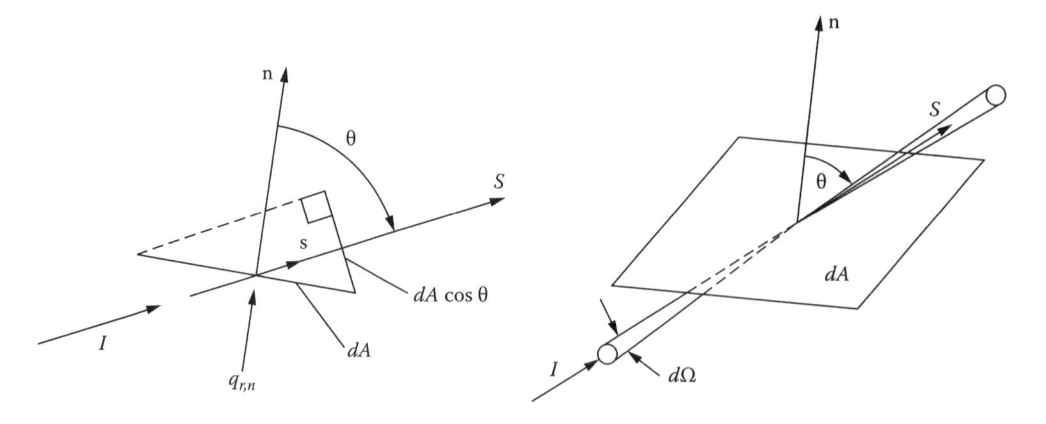

FIGURE 10.5 Quantities in derivation of radiative flux vector.

that is, $q_{r,n} = \mathbf{n}\cdot\mathbf{q}_r$. If the direction cosines of **n** in the rectangular coordinate system are α', δ', and γ' and those for **s** (the direction of I) are α, δ, and γ, then $\mathbf{n} = \mathbf{i}\alpha' + \mathbf{j}\delta' + \mathbf{k}\gamma'$ and $\mathbf{s} = \mathbf{i}\alpha + \mathbf{j}\delta + \mathbf{k}\gamma$ so that $\mathbf{s}\cdot\mathbf{n} = \cos\theta = \alpha\alpha' + \delta\delta' + \gamma\gamma'$ and, from Equation 10.25,

$$q_{r,n} = \int_{\Omega=0}^{4\pi} I(\alpha,\delta,\gamma)(\alpha\alpha' + \delta\delta' + \gamma\gamma')d\Omega.$$

Thus,

$$q_{r,n} = \alpha' q_{r,x} + \beta' q_{r,y} + \gamma' q_{r,z} \quad (10.27)$$

The $q_{r,x}$, $q_{r,y}$, and $q_{r,z}$ are fluxes across areas normal to the x, y, and z directions and are the components of the radiative flux vector. Each component is obtained from the integral in Equation 10.25 with **n** oriented in that coordinate direction. For example,

$$q_{r,x} = \int_{\Omega=0}^{4\pi} I(\alpha,\delta,\gamma)\,\alpha d\Omega,$$

where α is the cosine of the angle between the x axis and the direction of I that is in the direction of **s**.

The radiative flux vector can also be written by using Equation 10.26 with a spherical coordinate system as in Figure 10.6. The unit vector **s** is then

$$\mathbf{s} = \mathbf{i}\cos\phi\sin\theta + \mathbf{j}\sin\phi\sin\theta + \mathbf{k}\cos\theta \quad (10.28)$$

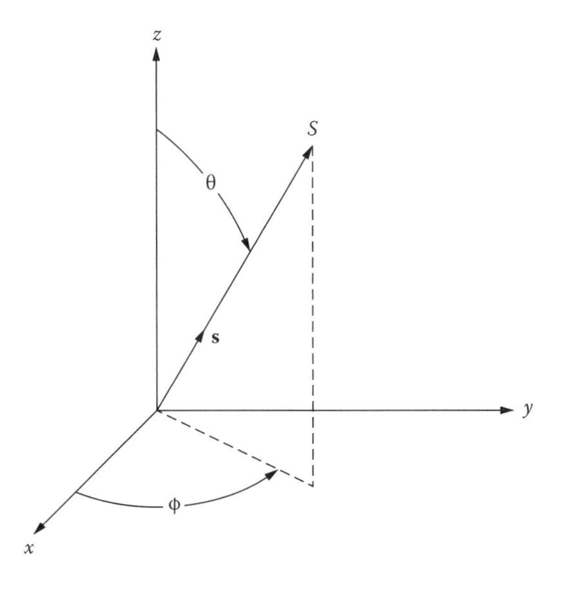

FIGURE 10.6 Spherical coordinate system for radiative flux vector.

Substituting **s** and $d\Omega = \sin\theta d\theta d\phi$ into Equation 10.26 gives the vector $\mathbf{q}_r$ in terms of its three components:

$$\mathbf{q}_r = \mathbf{i}q_{r,x} + \mathbf{j}q_{r,y} + \mathbf{k}q_{r,z} = \mathbf{i}\int_{\phi=0}^{2\pi}\int_{\theta=0}^{\pi} I(\theta,\phi)\cos\phi\sin^2\theta d\theta d\phi$$

$$+\mathbf{j}\int_{\phi=0}^{2\pi}\int_{\theta=0}^{\pi} I(\theta,\phi)\sin\phi\sin^2\theta d\theta d\phi$$

$$+\mathbf{k}\int_{\phi=0}^{2\pi}\int_{\theta=0}^{\pi} I(\theta,\phi)\cos\theta\sin\theta d\theta d\phi \tag{10.29}$$

The negative divergence of the radiant flux vector, $-\nabla\cdot\mathbf{q}_r$, is considered in Sections 10.4.2 and 10.4.3 and is used in the energy equation as discussed in Section 10.2.

In some situations, such as extremely rapid transients, it is necessary to account for the radiative energy contained within a volume element (Kumar and Mitra 1995, Longtin and Tien 1997). This is related to the local intensity averaged over all solid angles, as will now be shown. Consider radiation as a collection of photons, and the conditions at any location in a medium are given by a photon energy distribution function f. Since photon energy is related more directly to frequency than to wavelength, frequency is used here as the spectral variable. In volume dV at position **r**, let $f(\nu, \mathbf{r}, S)\, d\nu\, dV\, d\Omega$ be the number of photons traveling in the direction of S in frequency interval $d\nu$ centered about ν and within solid angle $d\Omega$ about the direction S (see Figure 10.7). Each photon has energy $h\nu$. The energy per unit volume per unit frequency interval is then $h\nu f(\nu, \mathbf{r}, S)\, d\Omega$ integrated over all solid angles. This is the *spectral radiant energy density:*

$$U_\nu(\nu,\mathbf{r}) = h\nu\int_{\Omega=0}^{4\pi} f(\nu,\mathbf{r},S)d\Omega \tag{10.30}$$

To obtain the intensity and show its relation to U_ν, the energy flux in the S direction is needed across area dA normal to the S direction, Figure 10.7. The photons have velocity c, and the number density

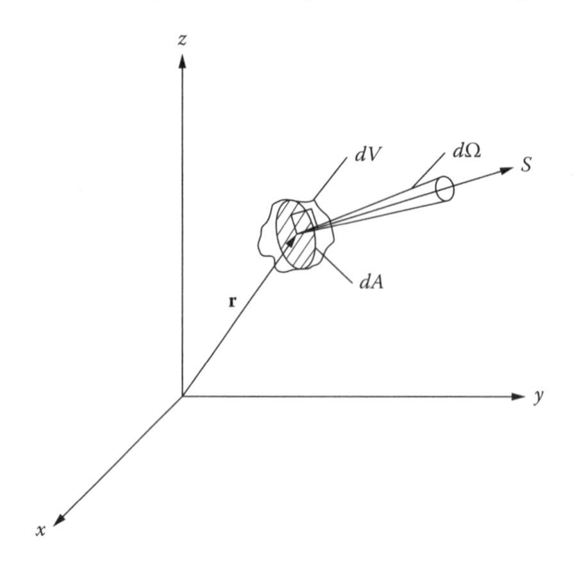

FIGURE 10.7 Schematic for derivation of radiative energy density.

traveling in the normal direction across dA is $f\,d\nu\,d\Omega$. The number of photons crossing dA in the S direction per unit time is then $cf\,d\nu\,d\Omega\,dA$. The spectral radiative energy carried by these photons is $h\nu\,cf(\nu, \mathbf{r}, S)\,d\nu\,d\Omega\,dA$. The spectral intensity is the energy in a single direction per unit time, unit frequency interval, and unit solid angle crossing a unit area normal to that direction. This gives the spectral intensity at location r and in direction S as $I_\nu = h\nu cf(\mathbf{r},S)$. The *relation between energy density and the volume integrated intensity* is then obtained by using I_ν to eliminate f in Equation 10.30,

$$U_\nu(\nu,\mathbf{r}) = \frac{1}{c}\int_{\Omega=0}^{4\pi} I_\nu(\mathbf{r},S)d\Omega = \frac{4\pi}{c}\left[\frac{1}{4\pi}\int_{\Omega=0}^{4\pi} I_\nu(\mathbf{r},S)d\Omega\right] = \frac{4\pi}{c}\bar{I}_\nu(\mathbf{r}) \tag{10.31}$$

where $\bar{I}_\nu(\mathbf{r})$ is the mean incident spectral intensity, Equation 10.18. This is the relation given in Equation 1.64, followed by expressions for radiative pressure in Equations 1.65 and 1.66.

10.4.2 Divergence of Radiative Flux without Scattering (Absorption Alone)

In some important applications, such as for water vapor or CO_2 gas without suspended particles, scattering can be neglected, so $\omega_\lambda \to 0$ and the spectral source function Equation 10.11 reduces to $\hat{I}_\lambda = I_{\lambda b}$. Equation 10.15 for the spectral intensity along a path within a medium (as in Figure 10.8) then yields the intensity incident at τ_λ in solid angle $d\Omega_i$,

$$I_\lambda(\tau_\lambda,\Omega_i) = I_\lambda(0,\Omega_i)e^{-\tau_\lambda} + \int_{\tau^*=0}^{\tau_\lambda} I_{\lambda b}(\tau_\lambda^*)e^{-(\tau_\lambda-\tau_\lambda^*)}d\tau_\lambda^* \tag{10.32}$$

To obtain the net rate of radiative energy supplied to a volume element dV, consider the energy absorbed and emitted by dV. The rate of energy absorbed from the incident spectral intensity $I_\lambda(\tau_\lambda, \Omega_i)$ arriving within solid angle $d\Omega_i$ in Figure 10.8 is, as in the development of Equation 1.61, $\kappa_\lambda(dV)I_\lambda(\tau_\lambda, \Omega_i)\,dV\,d\lambda\,d\Omega_i$. The incident intensity $I_\lambda(\tau_\lambda, \Omega_i)$ is given by Equation 10.32 where $I_\lambda(0, \Omega_i)$ is the spectral intensity directed toward dV from the boundary. The rate of energy absorbed by dV from all incident directions is the integral over all Ω_i, $\kappa_\lambda(dV)d\lambda\int_{\Omega_i=0}^{4\pi} I_\lambda(\tau_\lambda,\Omega_i)d\Omega_i$. The mean incident intensity $\bar{I}_\lambda(\tau_\lambda)$ at dV as in Equation 10.31 is used for convenience, $\bar{I}_\lambda(\tau_\lambda) \equiv (1/4\pi)\int_{\Omega_i=0}^{4\pi} I_\lambda(\tau_\lambda,\Omega_i)d\Omega_i$,

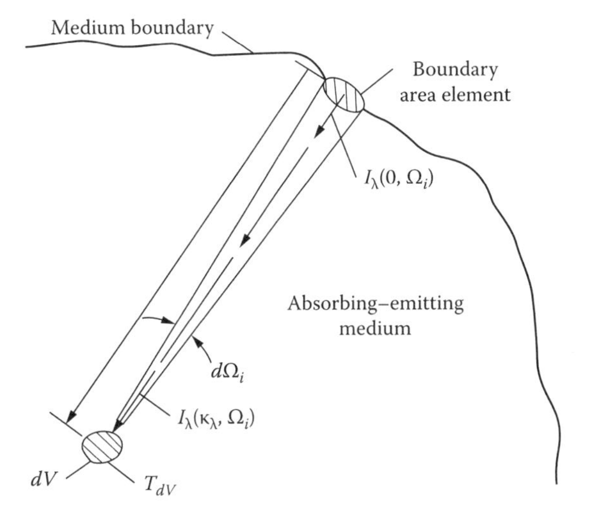

FIGURE 10.8 Geometry and incident intensity for derivation of energy-conservation relation.

and the spectral energy absorbed in dV is then $4\pi\kappa_\lambda(dV)\bar{I}_\lambda(\tau_\lambda)dV\,d\lambda$. By integrating over all wavelengths, the rate of total energy absorbed by dV from the radiation field is $4\pi dV\int_{\lambda=0}^{\infty}\kappa_\lambda(dV)\bar{I}_\lambda(\tau_\lambda)\,d\lambda$.

The rate of total energy emission by dV is obtained from Equation 1.61 by integrating over all wavelengths to give $4\pi dV\int_{\lambda=0}^{\infty}\kappa_\lambda(dV)I_{\lambda b}(T,dV)d\lambda$. Using Planck mean absorption coefficient, Equation 9.71, this is also $4dV\kappa_P\sigma T^4(S)$. Since dV is very small, the energy emitted by dV escapes without reabsorption within dV.

The net outflow of radiant energy per unit volume, which is the desired divergence of the radiant flux vector, is then the emitted energy rate minus the absorbed energy rate so that, at any location S,

$$\nabla\cdot\mathbf{q}_r(S)=4\pi\int_{\lambda=0}^{\infty}\kappa_\lambda(S)\left[I_{\lambda b}(S)-\frac{1}{4\pi}\int_{\Omega_i=0}^{4\pi}I_\lambda(S,\Omega_i)d\Omega_i\right]d\lambda$$

$$=4\pi\int_{\lambda=0}^{\infty}\kappa_\lambda(S)\left[I_{\lambda b}(S)-\bar{I}_\lambda(S)\right]d\lambda=4\left[\kappa_P\sigma T^4(S)-\pi\int_{\lambda=0}^{\infty}\bar{I}_\lambda(S)d\lambda\right] \tag{10.33}$$

This is the relation given earlier by Equation 10.4 when the concept of radiative flux divergence was being introduced.

10.4.3 Divergence of Radiative Flux Including Scattering

The relations in the previous section to obtain $\nabla\cdot\mathbf{q}_r(S)$ are now extended to include scattering for media such as gases with suspended particles. Begin with the equation of transfer, Equation 10.7, which applies for a phase function that is anisotropic. The first term of Equation 10.7 is expanded as

$$\frac{dI_\lambda}{dS}=\frac{\partial I_\lambda}{\partial x}\frac{dx}{dS}+\frac{\partial I_\lambda}{\partial y}\frac{dy}{dS}+\frac{\partial I_\lambda}{\partial z}\frac{dz}{dS}=\alpha\frac{\partial I_\lambda}{\partial x}+\delta\frac{\partial I_\lambda}{\partial y}+y\frac{\partial I_\lambda}{\partial z} \tag{10.34}$$

where α, δ, γ are the direction cosines for I_λ in the S direction in an x, y, z coordinate system. In a small number of situations, such as in Chern et al. (1995a,b,c), a layered embedded structure that is producing the scattering in the translucent material is modeled as an anisotropic scattering medium. For scattering behavior in this type of application, the amount of energy that is scattered from an incident direction can also depend on that incident direction. After the energy is scattered, its directional distribution can be anisotropic. To include these more realistic effects, Equation 10.10, modified to have $\sigma_{s,\lambda}$ depend on incidence direction, is integrated at location S over all Ω,

$$\int_{\Omega=0}^{4\pi}\frac{dI_\lambda}{dS}d\Omega=\int_{\Omega=0}^{4\pi}\left(\alpha\frac{\partial I_\lambda}{\partial x}+\delta\frac{\partial I_\lambda}{\partial y}+\gamma\frac{\partial I_\lambda}{\partial z}\right)d\Omega$$

$$=-\kappa_\lambda(S)\int_{\Omega=0}^{4\pi}I_\lambda(S,\Omega)\,d\Omega-\int_{\Omega=0}^{4\pi}\sigma_{s,\lambda}(S,\Omega)I_\lambda(S,\Omega)\,d\Omega+\kappa_\lambda(S)\int_{\Omega=0}^{4\pi}I_{\lambda b}(S)d\Omega$$

$$+\frac{1}{4\pi}\int_{\Omega=0}^{4\pi}\int_{\Omega_i=0}^{4\pi}\sigma_{s,\lambda}(S,\Omega_i)I_\lambda(S,\Omega_i)\Phi_\lambda(\Omega,\Omega_i)\,d\Omega_i\,d\Omega \tag{10.35}$$

The modeling of the macroscopic anisotropic structure might also lead to directional internal absorption and emission (i.e., $\kappa_\lambda(S) = \kappa_\lambda(S, \Omega)$), but this effect is not included here. From Section 10.4.1, $\int_{\Omega=0}^{4\pi} \alpha I_\lambda(S,\Omega)d\Omega$ is the radiative flux $q_{r\lambda,x}(S)$ and similarly in the y and z directions. The $I_{\lambda b}$ is isotropic, so it is independent of angular direction Ω. Then, Equation 10.35 becomes

$$\int_{\Omega=0}^{4\pi} \frac{dI_\lambda}{dS} d\Omega = \frac{\partial q_{r\lambda,x}}{\partial x} + \frac{\partial q_{r\lambda,y}}{\partial y} + \frac{\partial q_{r\lambda,z}}{\partial z} = \nabla \cdot \mathbf{q}_{r\lambda}$$

$$= -\kappa_\lambda(S) \int_{\Omega=0}^{4\pi} I_\lambda(S,\Omega)d\Omega - \int_{\Omega=0}^{4\pi} \sigma_{s,\lambda}(S,\Omega) I_\lambda(S,\Omega)d\Omega + 4\pi\kappa_\lambda(S) I_{\lambda b}(S)$$

$$+ \int_{\Omega_i=0}^{4\pi} \sigma_{s,\lambda}(S,\Omega_i) I_\lambda(S,\Omega_i) \left(\frac{1}{4\pi} \int_{\Omega=0}^{4\pi} \Phi_\lambda(\Omega,\Omega_i) d\Omega \right) d\Omega_i \tag{10.36}$$

The phase function Φ_λ can be normalized as given by Equation 1.72, $(1/4\pi)\int_{\Omega=0}^{4\pi} \Phi_\lambda(\Omega,\Omega_i)d\Omega = 1$. The two integrals of the local $\sigma_{s,\lambda}(\Omega)I_\lambda(\Omega)$ and $\sigma_{s,\lambda}(\Omega_i)I_\lambda(\Omega_i)$ are both over all solid angles, so they combine to zero. For use in the energy Equation 10.1, the radiative source must include the energy for all wavelengths, so that spectrally integrating the remaining terms of Equation 10.36 yields the following result that is equivalent to the relation in Equation 10.33:

$$\nabla \cdot \mathbf{q}_r(S) = 4\pi \int_{\lambda=0}^{\infty} \kappa_\lambda(S) \left[I_{\lambda b}(S) - \frac{1}{4\pi} \int_{\Omega=0}^{4\pi} I_\lambda(S,\Omega) d\Omega \right] d\lambda$$

$$= 4\pi \int_{\lambda=0}^{\infty} \kappa_\lambda(S) \left[I_{\lambda b}(S) - \bar{I}_\lambda(S) \right] d\lambda = 4\kappa_P \left[\sigma T^4(S) - \pi \int_{\lambda=0}^{\infty} \bar{I}(S)\, d\lambda \right] \tag{10.37}$$

This relation for $\nabla \cdot \mathbf{q}_r$ is valid for both anisotropic and isotropic scattering.

Although the scattering coefficient is not explicitly evident in Equation 10.37, I_λ is a function of the scattering as it is obtained by integrating Equation 10.14 over Ω and λ, where $\hat{I}_\lambda$ is from Equation 10.11 and depends on ω_λ. The $\nabla \cdot \mathbf{q}_r$ in Equation 10.37 is in terms of the mean incident intensity, but it can be written in terms of the source function, which makes the scattering albedo evident. For isotropic scattering (and no dependence of $\sigma_{s,\lambda}$ on incidence angle), Equation 10.11 relates $\hat{I}_\lambda(\tau_\lambda)$ and $\bar{I}_\lambda(\tau_\lambda)$ by

$$\hat{I}_\lambda(\tau_\lambda) = (1-\omega_\lambda) I_{\lambda b}(\tau_\lambda) + \omega_\lambda \bar{I}_\lambda(\tau_\lambda); \quad \omega_\lambda = \frac{\sigma_{s,\lambda}}{\kappa_\lambda + \sigma_{s,\lambda}} = \omega_\lambda(\tau_\lambda) \tag{10.38}$$

The $\bar{I}_\lambda(\tau_\lambda)$ is then eliminated by combining Equations 10.37 and 10.38 to obtain $\nabla \cdot \mathbf{q}_r$ in terms of $\hat{I}_\lambda(\tau_\lambda)$,

$$\nabla \cdot \mathbf{q}_r(S) = 4\pi \int_{\lambda=0}^{\infty} \frac{\kappa_\lambda(\tau_\lambda)}{\omega_\lambda(\tau_\lambda)} \left[I_{\lambda b}(\tau_\lambda) - I_\lambda(\tau_\lambda) \right] d\lambda; \quad \tau_\lambda(S) = \int_{S^*=0}^{S} \beta_\lambda(S^*) dS^* \tag{10.39}$$

where the source function $\hat{I}_\lambda(\tau_\lambda)$ is obtained by solving the integral Equation 10.20.

Consider the special case of radiative propagation in a purely scattering medium that does not absorb or emit. Then, $\kappa_\lambda = 0$, and Equations 10.37 and 10.39 show that for any type of scattering, in the absence of absorption,

$$\nabla \cdot \mathbf{q}_r = 0 \tag{10.40}$$

Without absorption there is no energy being locally absorbed or emitted so there is no radiative heat source. Energy is being redirected only by scattering. In this limit, Equation 10.5 cannot apply; there must be conduction and/or convection to yield a steady temperature distribution. Also, in the limit of no scattering ($\omega_\lambda = 0$), Equation 10.39 is indeterminate, and Equation 10.37 applies.

10.5 SUMMARY OF RELATIONS FOR RADIATIVE TRANSFER IN ABSORBING, EMITTING, AND SCATTERING MEDIA

10.5.1 Energy Equation

The energy equation for a single-component translucent medium with a radiative internal energy source $-\nabla \cdot \mathbf{q}_r$, and other internal heat sources $\dot{q}$, is

$$\rho c_p \frac{DT}{Dt} = \beta T \frac{DP}{Dt} + \nabla \cdot (k\nabla T - \mathbf{q}_r) + \dot{q} + \Phi_d \tag{10.41}$$

10.5.2 Radiative Energy Source

For a medium with or without scattering, the radiative source term for the energy equation is absorbed minus emitted energy,

$$-\nabla \cdot \mathbf{q}_r(S) = 4\pi \int_{\lambda=0}^{\infty} \kappa_\lambda(\tau_\lambda) \left[\frac{1}{4\pi} \int_{\Omega_i=0}^{4\pi} I_\lambda(\tau_\lambda, \Omega_i) d\Omega_i - I_{\lambda b}(\tau_\lambda) \right] d\lambda;$$
$$\tau_\lambda(S) = \int_{S^*=0}^{S} \beta_\lambda(S^*) dS^* \tag{10.42}$$

For pure scattering of any type (no absorption), the radiative energy source is zero,

$$-\nabla \cdot \mathbf{q}_r(S) = 0 \tag{10.43}$$

For isotropic scattering with absorption, the radiative energy source can be obtained from the temperature and source-function distributions by evaluating

$$-\nabla \cdot \mathbf{q}_r(S) = 4\pi \int_{\lambda=0}^{\infty} \frac{\kappa_\lambda(\tau_\lambda)}{\omega_\lambda(\tau_\lambda)} \left[\hat{I}_\lambda(\tau_\lambda) - I_{\lambda b}(\tau_\lambda) \right] d\lambda; \quad \tau_\lambda(S) = \int_{S^*=0}^{S} \beta_\lambda(S^*) dS^* \tag{10.44}$$

For absorption only without scattering, $\omega_\lambda = 0$, and Equation 10.44 becomes singular, so Equation 10.42 should be used.

10.5.3 Source Function

The source function for isotropic scattering with $\sigma_{s,\lambda}$ independent of Ω_i is obtained by solving an integral equation that requires the temperature distribution and the radiative boundary condition $I_\lambda(\tau_{\lambda,i} = 0, \Omega_i)$ for all paths $\tau_{\lambda,i}$ in the radiation field (see Figure 10.1),

$$\hat{I}_\lambda(\tau_\lambda) = (1-\omega_\lambda) I_{\lambda b}(\tau_\lambda) + \omega_\lambda \left[\frac{1}{4\pi} \int_{\Omega_i=0}^{4\pi} I_\lambda(\tau_{\lambda,i}=0,\Omega_i) e^{-\tau_{\lambda,i}} d\Omega_i + \int_{\tau_{\lambda,i}^*=0}^{\tau_{\lambda,i}} \hat{I}_\lambda(\tau_{\lambda,i}^*) e^{-(\tau_{\lambda,i}-\tau_{\lambda,i}^*)} d\tau_{\lambda,i}^* \right] \tag{10.45}$$

where $\omega_\lambda = \omega_\lambda(\tau_\lambda)$ and τ_λ and $\tau_{\lambda,i}$ are optical paths to the same location as shown in Figure 10.1.

The $\tau_{\lambda,i}$ account for all paths to τ_λ through the medium. For absorption only (no scattering), $\omega_\lambda = 0$, and the source function is the local blackbody intensity

$$\hat{I}_\lambda(\tau_\lambda) = I_{\lambda b}(\tau_\lambda) \tag{10.46}$$

For scattering only (no absorption), $\omega_\lambda = 1$, and Equation 10.45 becomes

$$\hat{I}_\lambda(\tau_\lambda) = \frac{1}{4\pi} \int_{\Omega_i=0}^{4\pi} I_\lambda(\tau_{\lambda,i}=0,\Omega_i) e^{-\tau_{\lambda,i}} d\Omega_i + \int_{\tau_{\lambda,i}^*=0}^{\tau_{\lambda,i}} \hat{I}_\lambda(\tau_{\lambda,i}^*) e^{-(\tau_{\lambda,i}-\tau_{\lambda,i}^*)} d\tau_{\lambda,i}^* \tag{10.47}$$

10.5.4 Radiative Transfer Equation

If scattering is independent of incidence direction, the change of intensity along a path is given by the RTE in the form

$$\frac{dI_\lambda}{dS} = -\beta_\lambda(S) I_\lambda(S) + \kappa_\lambda(S) I_{\lambda b}(S) + \frac{\sigma_{s,\lambda}(S)}{4\pi} \int_{\Omega_i=0}^{4\pi} I_\lambda(S,\Omega_i) \Phi_\lambda(S,\Omega,\Omega_i) d\Omega_i \tag{10.48}$$

An integrated form of the RTE is then

$$I_\lambda(\tau_\lambda,\Omega) = I_\lambda(0,\Omega) e^{-\tau_\lambda} + \int_{\tau_\lambda^*=0}^{\tau_\lambda} \hat{I}_\lambda(\tau_\lambda^*,\Omega) e^{-(\tau_\lambda-\tau_\lambda^*)} d\tau_\lambda^* \tag{10.49}$$

10.5.5 Relations for a Gray Medium

In addition to isotropic scattering, if the simplification is made that the medium is gray, the radiative energy source Equation 10.44 and source function Equation 10.45 reduce to

$$\begin{aligned} -\nabla \cdot \mathbf{q}_r(S) &= 4\pi \frac{\kappa(\tau)}{\omega(\tau)} \left[\hat{I}(\tau) - I_b(\tau) \right] \\ &= 4 \frac{\kappa(S)}{\omega(S)} \left[\pi \hat{I}(S) - \sigma T^4(S) \right] \end{aligned} \tag{10.50}$$

where

$$\tau(S) = \int_{S^*=0}^{S} \beta(S^*)\, dS^*$$

$$\hat{I}(\tau) = (1-\omega) I_b(\tau) + \omega \left[\frac{1}{4\pi} \int_{\Omega_i=0}^{4\pi} I(\tau_i = 0, \Omega_i) e^{-\tau_i} d\Omega_i + \int_{\tau^*=0}^{\tau_i} \hat{I}(\tau_i^*) e^{-(\tau_i - \tau_i^*)} d\tau_i^* \right] \quad (10.51)$$

where $\omega = \omega(\tau)$. For $\omega \to 0$ (no scattering), Equation 10.50 is singular, so Equation 10.42 (which is valid for $\omega \geq 0$) can be used:

$$-\nabla \cdot \mathbf{q}_r(S) = 4\pi\kappa(\tau) \left[\frac{1}{4\pi} \int_{\Omega_i=0}^{4\pi} I(\tau, \Omega_i) d\Omega_i - I_b(\tau) \right] = 4\kappa(\tau)\left[\pi \bar{I}(S) - \sigma T^4(S)\right] \quad (10.52)$$

where $\tau(S) = \int_{S^*=0}^{S} \kappa(S^*) dS^*$. For a uniform extinction coefficient β, Equation 10.51 can be written in terms of the physical position S along a path as

$$\hat{I}(S) = \frac{(1-\omega)}{\pi} \sigma T^4(S) + \omega \left[\frac{1}{4\pi} \int_{\Omega_i=0}^{4\pi} I(S_i = 0, \Omega_i) e^{\beta S_i} d\Omega_i + \beta \int_{S^*=0}^{S_i} \hat{I}(S_i^*) e^{-\beta(S_i - S_i^*)} dS_i^* \right] \quad (10.53)$$

where the S_i are all incident paths to location S.

In Chapter 11, the basic equations are further developed for specific geometries such as a plane layer, a two-dimensional (2D) rectangular volume, and a cylinder. Some solutions are obtained. This clarifies the application of the relations given here and their simultaneous solution subject to both radiative (for determining I and $\hat{I}$) and thermal (for determining T) boundary conditions for physical situations.

10.6 TREATMENT OF RADIATION TRANSFER IN NON-LTE MEDIA

Section 9.2.2 has a discussion of the conditions when local thermodynamic equilibrium (LTE) conditions might not occur.

In this book, we generally assume that LTE exists and that the emission from dV is governed by Equation 1.61. However, the remainder of this section briefly examines some engineering approximations that allow basic non-LTE radiation problems to be approached.

Non-LTE cases have been treated by using gray-gas and emission-dominated approximations (Vincenti and Kruger 1986). A compromise between complete non-LTE calculations (which require detailed balancing among all energy states based on transition rates between them) and the assumption that LTE exists is that a different local equilibrium temperature is applied to each energy mode (translational, vibrational, rotational, electronic, and electron plasma). Then, local properties can be approximately calculated for each mode of the various chemical species present in the gas. This is still a formidable computational task when multiple chemical species and their forms (atoms, molecules, ions, and electrons) are present. To further reduce the complexity, a two-temperature model of local states is often used. Usually, $T_e(\vec{\mathbf{r}})$, the electron temperature, is assumed to apply to the vibrational and electronic energy modes of heavy particles (atoms, molecules, and ions) as well as to the translational energy distribution of the local electron plasma, and $T_r(\vec{\mathbf{r}})$, the rotational (and translational) temperature, applies to all rotational line transitions of heavy particles. This assumption allows k-distribution calculation of spectral properties based on only two local temperature states and greatly reduces the computational load. This model reflects a physical observation that energy distributions controlling emission from a lower temperature medium are governed by the translational temperature and the associated closely spaced rotational–vibrational states; these are chiefly influenced by local collisions among molecules. This temperature thus affects emission from the medium. In a non-LTE case, the electronic and bound-free (ionization) states are assumed to be governed by the electron temperature T_e, which is influenced by high-energy radiation from more remote locations, causing a higher effective temperature for these distributions, chiefly influenced by radiation absorption. Such a model is clearly not exact, but does capture first-order non-LTE effects.

For the particular case of reentry radiation, this or a similar approach is used by Lamet et al. (2008, 2010), Bansal et al. (2009a,b, 2010a,b, 2011a,b), Sohn et al. (2010), and Maurente et al. (2012). Wilbers et al. (1991) and Trelles et al. (2007) apply a two-temperature model to analyze radiation in an argon arc plasma torch.

Bansal (2011b) and Maurente present alternative formulations of the non-LTE RTE applicable to spacecraft re-entry conditions. Both assume that a multiscale full spectrum k-distribution (see Section 9.3.2.3) can be used for the medium properties (Howell 2014).

In the Bansal approach, a separate k-distribution-based RTE is written for each species m:

$$\frac{dI_{g_{b,m}^{ne}}}{ds} = NC\left(I_{b,m}^{ne} - I_{g_{b,m}^{ne}}\right) \quad m = 1,2,3,\ldots,M \tag{10.54}$$

where

N is the number density of the absorbing species
C is the absorption cross section (and thus $NC = \kappa$)
$I_{g_{b,m}^{ne}}$ is the intensity for species m at the cumulative distribution function (CDF) value g of the k-distribution
$I_{b,m}^{ne}$ is the non-LTE black intensity for species m, based on the temperature of the individual species

In this approach, the nonequilibrium effects are contained in the source term intensity, $I_{b,m}^{ne}$. If there is no spectral overlap between the species and the absorption characteristics of a given species are unaffected by the presence of the others, then the total intensity is obtained after solution of the RTE for each species by summation of $I_{g_{b,m}^{ne}}$ over all m and g.

Maurente et al. proposed an alternative form of the RTE for non-LTE atmospheric reentry conditions, which after some manipulation becomes

$$\frac{dI_{g_j}}{ds} = NC\left[\sum_{m=1}^{M}\left(\gamma_{g_{b,m}^{ne}} I_{b,m}^{ne}\right) - I_{g_j}\right] \quad m = 1,2,3,\ldots,M \tag{10.55}$$

Here, the $\gamma_{g_{b,m}^{ne}}$ is a factor that relates the CDF of the k-distribution for a given species to a reference CDF, and the j subscript indicates evaluation over the joint cumulative k-distribution for *all* species (rather than for a single species). This equation need be solved only once rather than for each species as for the Bansal approach, and I is then found by integrating over all g. Both Bansal and Maurente compare their method with line-by-line calculations for test cases. Further details can be found in the references, as they are outside the scope of this book.

10.7 NET-RADIATION METHOD FOR ENCLOSURES FILLED WITH AN ISOTHERMAL MEDIUM OF UNIFORM COMPOSITION

The radiation-exchange equations were developed in Section 5.2 for an enclosure that does not contain an absorbing–emitting medium and has diffuse surfaces with spectrally dependent properties. This section expands that treatment to include an isothermal absorbing medium between the surfaces.

Since the absorption properties of gases and other absorbing media are almost always strongly wavelength dependent, the present development is carried out for a differential wavelength interval. Integrations over all wavelengths or over a k-distribution then yield the total radiative behavior. It is assumed that surface directional property effects are sufficiently unimportant in the desired results that surfaces can be treated as diffuse emitters and reflectors.

In gas-filled enclosures, such as in industrial furnaces or engine combustion chambers, there is often sufficient mixing that the entire gas is essentially isothermal and of uniform composition. In this instance, the analysis is simplified, as it is not necessary to compute or specify the gas temperature distribution. Sometimes, the uniform gas temperature can be found from the governing energy balances. Even with this uniform gas simplification, a detailed spectral radiation exchange computation between the gas and bounding surfaces can become quite involved.

Consider an enclosure of N surfaces, each at a uniform temperature, as in Figure 10.9. Typical surfaces are designated by subscripts j and k. Some of the surfaces can be open boundaries such as a window or hole; these are usually modeled as a perfectly absorbing (black) surface at the temperature of the surroundings outside the opening. The enclosure is filled with an absorbing–emitting medium such as a radiating gas at uniform temperature T_g. The quantity Q_g is the energy rate that it is necessary to supply by means other than radiation to the entire absorbing–emitting medium to maintain its uniform temperature. A common source for Q_g is combustion. If in a problem solution the Q_g is found to be negative, the medium is gaining a net amount of radiative energy from the enclosure boundaries, and energy must be removed from the medium to maintain it at its steady temperature T_g. The Q_g is analogous to the Q_k at a surface, which is the energy rate supplied to area A_k by means other than radiation inside the enclosure.

Enclosure theory yields equations relating Q_k and T_k for each surface to Q_g and T_g for the gas or other absorbing uniform isothermal medium filling the enclosure. Considering all the surfaces and the medium, if half of the Q and T are specified, there are sufficient radiative heat balance equations to solve for the remaining unknown Q and T values. Either Q or T must be specified for each boundary surface, or the final matrix of equations may become singular. Further, at least one boundary temperature or the medium temperature must be specified to anchor the solution. Problems where an unspecified boundary condition is to be found are addressed in Chapter 8 and in Sections 12.10 and 13.7. For some applications, the energy input to the gas from external sources Q_g is given, and

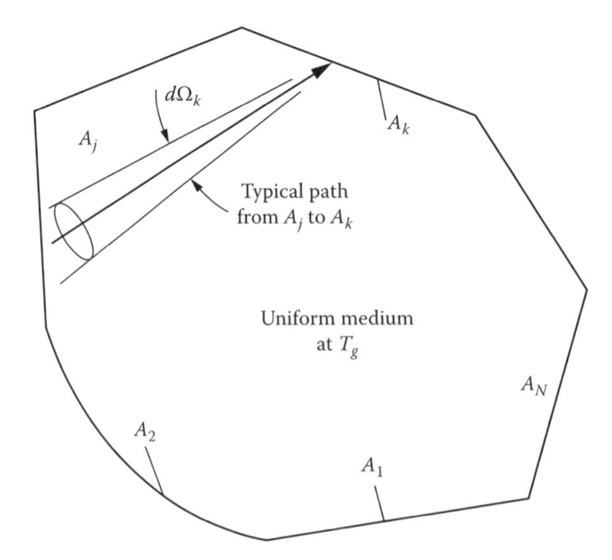

FIGURE 10.9 Enclosure composed of N discrete surface areas and filled with uniform medium at T_g (enclosure shown in cross section for simplicity).

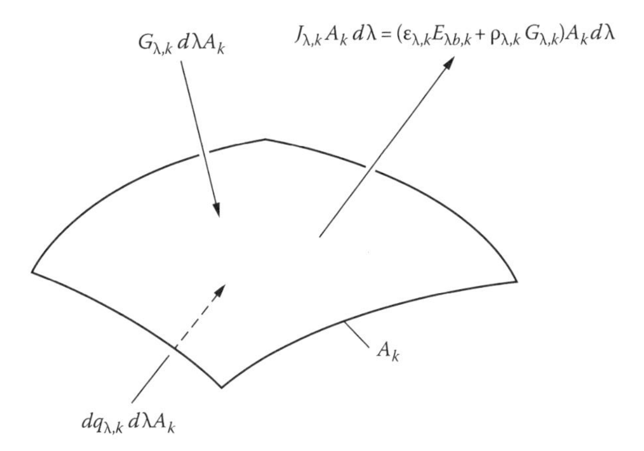

FIGURE 10.10 Spectral energy quantities incident on and leaving typical surface area of enclosure.

the analysis yields the steady gas temperature T_g. Conversely, if a desired T_g is specified, the analysis yields the energy that must be supplied to the gas to maintain its temperature.

The net radiation method in Chapters 5 and 6 is now extended to include radiation exchange with the medium. At the kth surface of an enclosure (Figure 10.10), an energy balance, as in Equation 7.1, gives

$$Q_{\lambda,k} d\lambda = q_{\lambda,k} A_k d\lambda = (J_{\lambda,k} - G_{\lambda,k}) A_k d\lambda \tag{10.56}$$

The $J_{\lambda,k}\, d\lambda$ and $G_{\lambda,k}\, d\lambda$ are, respectively, the outgoing and incoming radiative energy fluxes (radiosity and irradiation) in wavelength interval $d\lambda$. The $Q_{\lambda,k}\, d\lambda$ is the energy rate supplied to the surface A_k in the wavelength region $d\lambda$. The external energy supplied to A_k by some means such as conduction and/or convection is equal to $\int_{\lambda=0}^{\infty} Q_{\lambda,k} d\lambda$.

The radiosity $J_{\lambda,k}$ is composed of emitted and reflected energy, as in Equation 5.11, written spectrally

$$J_{\lambda,k} d\lambda = \varepsilon_{\lambda,k}(T_k) E_{\lambda b,k}(T_k) d\lambda + \rho_{\lambda,k}(T_k) G_{\lambda,k} d\lambda \tag{10.57}$$

The functional notation is often omitted in what follows, to shorten the equations. The $E_{\lambda b,k}(T_k)\, d\lambda$ is the blackbody spectral emission at T_k in wavelength region $d\lambda$ about wavelength λ.

The $G_{\lambda,k}\, d\lambda$ in Equation 10.56 is the incoming spectral flux to A_k. It is the sum of the contributions, from all the surfaces, that reach the kth surface after partial absorption while passing through the intervening medium, plus the contribution by emission from the medium. The equation of transfer allows for both attenuation and emission as radiation passes along a path through a medium. A typical path from A_j to A_k within incident solid angle $d\Omega_k$ is in Figure 10.9. If all such paths and solid angles are accounted for by which radiation can pass from all the surfaces (including A_k itself if it is concave) to A_k, the solid angles $d\Omega_k$ will encompass all of the medium that can radiate to A_k. Thus, by using the equation of transfer to compute the energy transported along all paths between surfaces, the emission by the medium is included.

The radiation passing from one surface to another through a nonscattering medium is now considered, including emission and absorption by the intervening medium. In enclosure theory, J_λ is assumed uniform over each surface. Since the surfaces are assumed diffuse, the spectral intensity leaving dA_j is $I_{\lambda o,j} = J_{\lambda,j}/\pi$. From the RTE given in Equation 10.15, the intensity arriving at dA_k after traversing path S is, for a nonscattering medium with uniform temperature and composition (constant κ_λ throughout the volume),

$$I_{\lambda,i,j-k} = I_{\lambda,o,j} e^{-\kappa_\lambda S} + I_{\lambda b,g}\left(1 - e^{-\kappa_\lambda S}\right) \tag{10.58}$$

where the subscript j–k means j to k (not a minus sign). Using the definitions $t_\lambda(S) \equiv e^{-\kappa_\lambda S}$ (spectral transmittance of the medium along path length S) and $\alpha_\lambda(S) \equiv 1 - e^{-\kappa_\lambda S}$ (spectral absorptance along the path),

$$I_{\lambda,i,j-k} = t_\lambda(S) I_{\lambda,o,j} + \alpha_\lambda(S) I_{\lambda b,g}(T_g) \tag{10.59}$$

This intensity arriving at dA_k in solid angle $d\Omega_k$ provides the energy $I_{\lambda,i,j-k} dA_k \cos\theta_k d\Omega_k d\lambda$. Using $d\Omega_k = dA_j \cos\theta_j / S^2$, the arriving spectral energy is

$$d^2 Q_{\lambda i,j-k} d\lambda = \left[t_\lambda(S) I_{\lambda,o,j} + \alpha_\lambda(S) I_{\lambda b,g}(T_g)\right] \frac{dA_k dA_j \cos\theta_k \cos\theta_j}{S^2} d\lambda \tag{10.60}$$

For a diffuse surface, $J_{\lambda,j} = \pi I_{\lambda,o,j}$ and $E_{\lambda b,g} = \pi I_{\lambda b,g}$, so

$$d^2 Q_{\lambda i,j-k} d\lambda = \left[t_\lambda(S) J_{\lambda,j} + \alpha_\lambda(S) E_{\lambda b,g}(T_g)\right] d\lambda \frac{\cos\theta_k \cos\theta_j}{\pi S^2} dA_k dA_j \tag{10.61}$$

Equation 10.61 is integrated over both A_j and A_k to give the spectral energy transmitted and emitted along all paths from A_j that are incident on A_k:

$$Q_{\lambda i,j-k} d\lambda = \int_{A_k}\int_{A_j} \left[t_\lambda(S) J_{\lambda,j} + \alpha_\lambda(S) E_{\lambda b,g}(T_g)\right] d\lambda \frac{\cos\theta_k \cos\theta_j}{\pi S^2} dA_k dA_j \tag{10.62}$$

10.7.1 Definitions of Spectral Geometric-Mean Transmission and Absorption Factors

The double integral in Equation 10.62 is similar to the double integral in Equation 4.16 for the configuration factor between two surfaces. By analogy, define $\bar{t}_{\lambda,j-k}$ such that

$$\bar{t}_{\lambda,j-k} \equiv \frac{1}{A_j F_{j-k}} \int_{A_k}\int_{A_l} \frac{t_\lambda(S)\cos\theta_k \cos\theta_j}{\pi S^2} dA_j dA_k \tag{10.63}$$

where F_{j-k} is the configuration factor. For no absorbing medium, $t_\lambda(S) = 1$, and the double integral on the right side of Equation 10.63 becomes F_{j-k} so, for no absorption, $\bar{t}_{\lambda,j-k} = 1$. For complete absorption in the medium between A_j and A_k, $\bar{t}_{\lambda,j-k} = 0$. $\bar{t}_{\lambda,j-k}$ is the *geometric-mean transmittance* from A_j to A_k. From the second quantity in brackets in Equation 10.62, a *geometric-mean absorptance* $\bar{\alpha}_{\lambda,j-k}$ is defined,

$$\bar{\alpha}_{\lambda,j-k} \equiv \frac{1}{A_j F_{j-k}} \int_{A_k}\int_{A_j} \frac{\alpha_\lambda(S)\cos\theta_k \cos\theta_j}{\pi S^2} dA_j dA_k \tag{10.64}$$

For a nonabsorbing medium, $\bar{\alpha}_{\lambda,j-k} = 0$, while for complete absorption, $\bar{\alpha}_{\lambda,j-k} = 1$. From the definitions of t_λ and α_λ, $\bar{t}_\lambda$ and $\bar{\alpha}_\lambda$ are related by

$$\bar{\alpha}_{\lambda,j-k} = 1 - \bar{t}_{\lambda,j-k} \tag{10.65}$$

Equation 10.62 becomes

$$Q_{\lambda i,j-k}d\lambda = \left[\bar{t}_\lambda(S)J_{\lambda,j} + \bar{\alpha}_\lambda(S)E_{\lambda b,g}(T_g)\right]A_jF_{j-k}d\lambda \tag{10.66}$$

An alternative terminology is also used in which $A_jF_{j-k}\bar{t}_{\lambda,j-k}$ is the *geometric transmission factor* and $A_jF_{j-k}\bar{\alpha}_{\lambda,j-k}$ the *geometric absorption factor*. To compute the heat exchange in an enclosure, it is necessary to determine each $\bar{t}_\lambda$ and $\bar{\alpha}_\lambda$; only one double integration is needed because of relation Equation 10.65.

10.7.2 Matrix of Enclosure: Theory Equations

For an enclosure with N surfaces bounding a uniform isothermal medium at T_g, the incident spectral energy on any surface A_k is that arriving from all the surrounding surfaces and the enclosed isothermal medium:

$$G_{\lambda,k}A_k = \sum_{j=1}^{N}\left[\bar{t}_\lambda(S)J_{\lambda,j} + \bar{\alpha}_\lambda(S)E_{\lambda b,g}(T_g)\right]A_jF_{j-k} \tag{10.67}$$

From reciprocity, $A_jF_{j-k} = A_kF_{k-j}$; so, A_k is eliminated to give

$$G_{\lambda,k} = \sum_{j=1}^{N}\left[\bar{t}_\lambda(S)J_{\lambda,j} + \bar{\alpha}_\lambda(S)E_{\lambda b,g}(T_g)\right]F_{k-j} \tag{10.68}$$

Equations 10.56, 10.57, and 10.68 relate J_λ, G_λ, and q_λ for each of the surfaces in the enclosure to $E_{\lambda b}$ for that surface and to $E_{\lambda b,g}$. The G_λ is eliminated by combining Equations 10.56 and 10.57 and by substituting Equations 10.56 and 10.68. This yields two equations for $J_{\lambda,k}$ and $q_{\lambda,k}$ for each surface in terms of $E_{\lambda b,k}$ and $E_{\lambda b,g}$,

$$q_{\lambda,k} = \frac{\varepsilon_{\lambda,k}}{1-\varepsilon_{\lambda,k}}(E_{\lambda b,k} - J_{\lambda,k}) \tag{10.69}$$

$$q_{\lambda,k} = J_{\lambda,k} - \sum_{j=1}^{N}\left[\bar{t}_{\lambda,j-k}(S)J_{\lambda,j} + \bar{\alpha}_{\lambda,j-k}(S)E_{\lambda b,g}(T_g)\right]F_{k-j} \tag{10.70}$$

Equation 10.69 is the same as for an enclosure without an absorbing medium; if $\varepsilon_{\lambda,k} = 1$, Equation 10.69 becomes $J_{\lambda,k} = E_{\lambda b,k}$. Equations 10.69 and 10.70 are analogous to Equations 5.16 and 5.17 for a gray enclosure with no absorbing medium. From the symmetry of the integrals in Equations 10.63 and 10.64, and from $A_jF_{j-k} = A_kF_{k-j}$, it is found that

$$\bar{t}_{\lambda,j-k} = \bar{t}_{\lambda,k-j} \tag{10.71}$$

and

$$\bar{\alpha}_{\lambda,j-k} = \bar{\alpha}_{\lambda,k-j} \tag{10.72}$$

Then, Equation 10.70 can also be written as

$$q_{\lambda,k} = J_{\lambda,k} - \sum_{j=1}^{N} \left[\bar{t}_{\lambda,k-j}(S) J_{\lambda,j} + \bar{\alpha}_{\lambda,k-j}(S) E_{\lambda b,g}(T_g) \right] F_{k-j} \tag{10.73}$$

As in Section 5.3.1, Equations 10.69 and 10.70 can be further reduced by solving Equation 10.69 for J_λ and inserting it into Equation 10.70. This gives

$$\sum_{j=1}^{N} \left(\frac{\delta_{kj}}{\varepsilon_{\lambda,k}} - F_{k-j} \frac{1-\varepsilon_{\lambda,k}}{\varepsilon_{\lambda,k}} \bar{t}_{\lambda,k-j} \right) q_{\lambda,j} = \sum_{j=1}^{N} \left[(\delta_{kj} - F_{k-j} \bar{t}_{\lambda,k-j}) E_{\lambda b,j} - F_{k-j} \bar{\alpha}_{\lambda,k-j} E_{\lambda b,g} \right] \tag{10.74}$$

Kronecker delta is $\delta_{kj} = 1$ when $k = j$, and $\delta_{kj} = 0$ when $k \neq j$. This equation is analogous to Equation 5.21. If Equation 10.74 is written for each k from 1 to N, a set of N equations is obtained relating the $2N$ quantities q_λ and $E_{\lambda b}$ for all the surfaces. If the medium temperature (and hence $E_{\lambda b,g}$) is known, then one-half of the q_λ and $E_{\lambda b}$ values need to be specified, and the equations can be solved. To determine total energy quantities, the equations must be solved in a number of wavelength intervals, and integration of each energy quantity must then be performed over all wavelengths. This is the same as for the band equations in Section 6.2.2. If T_g is unknown, an additional equation is needed, as given in the next section.

10.7.3 Energy Balance on a Medium

An energy balance on the medium relates the medium temperature, T_g, and the energy that is supplied to the medium, Q_g, by means other than radiative exchange within the enclosure. From an energy balance on the entire enclosure, the energy that must be supplied to the medium, for example, by combustion, is equal to the net energy escaping from all the N boundary surfaces,

$$Q_g = -\sum_{k=1}^{N} A_k \int_{\lambda=0}^{\infty} q_{\lambda,k} d\lambda \tag{10.75}$$

This can be evaluated after the $q_{\lambda,k}$ are found for each surface from Equation 10.74 in a sufficient number of wavelength intervals. These equations can be solved if T_g and hence $E_{\lambda b,g}$ is known. If T_g is unknown and instead Q_g is specified, a less direct solution is required. The T_g is guessed and Q_g is found. The solution is iterated on T_g and the resulting relation between Q_g and T_g yields the T_g corresponding to the specified Q_g.

Example 10.3

Consider a well-mixed gas at T_g completely surrounded by a single wall at uniform temperature T_1 such as in a simple cooled combustion chamber. The heat transfer from the gas to the wall is to be found in terms of T_1 and T_g.

For an enclosure with a single wall, Equation 10.74 yields

$$\left(\frac{1}{\varepsilon_{\lambda,1}} - F_{1-1} \frac{1-\varepsilon_{\lambda,1}}{\varepsilon_{\lambda,1}} \bar{t}_{\lambda,1-1} \right) q_{\lambda,1} = (1 - F_{1-1} \bar{t}_{\lambda,1-1}) E_{\lambda b,1} - F_{1-1} \bar{\alpha}_{\lambda,1-1} E_{\lambda b,g}$$

Using $F_{1-1} = 1$ and $1-\bar{t}_{\lambda,1-1} = \bar{\alpha}_{\lambda,1-1}$, this equation is simplified and then integrated over all λ to obtain the energy rate added to the gas that is transferred to the wall,

$$Q_g = -Q_1 = -A_1 \int_{\lambda=0}^{\infty} q_{\lambda,1} d\lambda = A_1 \int_0^{\infty} \frac{E_{\lambda b,g} - E_{\lambda b,1}}{1/\varepsilon_{\lambda,1} + 1/\bar{\alpha}_{\lambda,1-1} - 1} d\lambda \tag{10.3.1}$$

Example 10.4

Obtain relations for the energy transfer in an enclosure of two infinite parallel plates at temperatures T_1 and T_2 bounding a well-mixed gas at uniform temperature T_g.

Equation 10.74, applied to a two-surface enclosure, gives for $k = 1$ and 2 (note that $F_{1-1} = F_{2-2} = 0$):

$$\frac{1}{\varepsilon_{\lambda,1}} q_{\lambda,1} - F_{1-2} \frac{1-\varepsilon_{\lambda,2}}{\varepsilon_{\lambda,2}} \bar{t}_{\lambda,1-2} q_{\lambda,2} = E_{\lambda b,1} - F_{1-2}\bar{t}_{\lambda,1-2} E_{\lambda b,2} - F_{1-2}\bar{\alpha}_{\lambda,1-2} E_{\lambda b,g} \tag{10.4.1}$$

$$-F_{2-1} \frac{1-\varepsilon_{\lambda,1}}{\varepsilon_{\lambda,1}} \bar{t}_{\lambda,2-1} q_{\lambda,1} + \frac{1}{\varepsilon_{\lambda,2}} q_{\lambda,2} = -F_{2-1}\bar{t}_{\lambda,2-1} E_{\lambda b,1} - F_{2-1}\bar{\alpha}_{\lambda,2-1} E_{\lambda b,2} + E_{\lambda b,g} \tag{10.4.2}$$

For infinite parallel plates, $F_{1-2} = F_{2-1} = 1$, and from Equations 10.71 and 10.72, $\bar{t}_{\lambda,2-1} = \bar{t}_{\lambda,1-2}$ and $\alpha_{\lambda,2-1} = \bar{\alpha}_{\lambda,1-2}$. For simplicity, the numerical subscripts on $\bar{t}$ and $\bar{\alpha}$ are omitted. Then, Equations 10.4.1 and 10.4.2 become

$$\frac{1}{\varepsilon_{\lambda,1}} q_{\lambda,1} - \frac{1-\varepsilon_{\lambda,2}}{\varepsilon_{\lambda,2}} \bar{t}_{\lambda} q_{\lambda,2} = E_{\lambda b,1} - \bar{t}_{\lambda} E_{\lambda b,2} - \bar{\alpha}_{\lambda} E_{\lambda b,g} \tag{10.4.3}$$

$$-\frac{1-\varepsilon_{\lambda,1}}{\varepsilon_{\lambda,1}} \bar{t}_{\lambda} q_{\lambda,1} + \frac{1}{\varepsilon_{\lambda,2}} q_{\lambda,2} = \bar{t}_{\lambda} E_{\lambda b,1} + E_{\lambda b,2} - \bar{\alpha}_{\lambda} E_{\lambda b,g} \tag{10.4.4}$$

Equations 10.4.3 and 10.4.4 are solved for $q_{\lambda,1}$ and $q_{\lambda,2}$. After using the relation $\bar{\alpha}_{\lambda} = 1 - \bar{t}_{\lambda}$, this results in

$$q_{\lambda,1} = \frac{1}{1-(1-\varepsilon_{\lambda,1})(1-\varepsilon_{\lambda,2})\bar{t}_{\lambda}^2} \left\{ \varepsilon_{\lambda,1}\varepsilon_{\lambda,2}\bar{t}_{\lambda}(E_{\lambda b,1} - E_{\lambda b,2}) + \varepsilon_{\lambda,1}(1-\bar{t}_{\lambda})\left[1+(1-\varepsilon_{\lambda,2})\bar{t}_{\lambda}\right](E_{\lambda b,1} - E_{\lambda b,g}) \right\} \tag{10.4.5}$$

$$q_{\lambda,2} = \frac{1}{1-(1-\varepsilon_{\lambda,1})(1-\varepsilon_{\lambda,2})\bar{t}_{\lambda}^2} \left\{ \varepsilon_{\lambda,1}\varepsilon_{\lambda,2}\bar{t}_{\lambda}(E_{\lambda b,2} - E_{\lambda b,1}) + \varepsilon_{\lambda,2}(1-\bar{t}_{\lambda})\left[1+(1-\varepsilon_{\lambda,1})\bar{t}_{\lambda}\right](E_{\lambda b,2} - E_{\lambda b,g}) \right\} \tag{10.4.6}$$

The total energy fluxes added to surfaces 1 and 2 are

$$q_1 = \int_{\lambda=0}^{\infty} q_{\lambda,1} d\lambda \quad \text{and} \quad q_2 = \int_{\lambda=0}^{\infty} q_{\lambda,2} d\lambda \tag{10.4.7}$$

The total energy added to the gas to maintain its temperature T_g is equal to the net energy leaving the parallel plates. Hence, per unit area of the plates,

$$q_g = -(q_1 + q_2) \tag{10.4.8}$$

When the medium between the plates does not absorb or emit radiation, then $\bar{t}_\lambda = 1$, and Equations 10.4.5 and 10.4.6 reduce to Equation 5.5. With an absorbing radiating gas, the numerical integration of Equations 10.4.5 and 10.4.6 over all λ to obtain q_1 and q_2 is difficult because of the very irregular variations of the gas absorption coefficient with λ; some methods for avoiding the detailed integrations were discussed in Chapter 9.

10.7.4 Spectral Band Equations for an Enclosure

One approach for integrating over λ is to divide the spectrum into bands in which the gas is either absorbing or essentially nonabsorbing. k-Distribution methods or the spectral line-based weighted-sum-of-gray-gases approaches for computing band properties (Chapter 9) can be used. For situations where both the geometry and thermal boundary conditions are simple, the enclosure equations can be solved in closed form as in Examples 10.3 and 10.4. In Equation 10.3.1, for example, if $\bar{\alpha}_\lambda \approx 0$ within a wavelength region, this nonabsorbing band does not contribute to Q_g. Then, if l designates an absorbing band,

$$\frac{Q_g}{A_1} = \sum_l \left(\frac{E_{\lambda b,g} - E_{\lambda b,1}}{1/\varepsilon_{\lambda,1} + 1/\bar{\alpha}_{1-1} - 1} \right)_l \Delta\lambda_l \tag{10.76}$$

The $\bar{\alpha}_{1-1}$ for each band is found from one of the methods in Chapter 9. Usually, the bandwidth is small, so that $E_{\lambda b,g}$, $E_{\lambda b,1}$, and $E_{\lambda,1}$ can be considered constant over each band. If the $E_{\lambda b}$ variation is significant over a band, the integrated blackbody functions can be used, as in Example 6.2.

For more complex enclosure geometries, Equation 10.74 is used. For a band of width $\Delta\lambda$, the integration of Equation 10.74 over the band gives

$$\int_{\Delta\lambda} \sum_{j=1}^{N} \left(\frac{\delta_{kj}}{\varepsilon_{\lambda,j}} - F_{k-j} \frac{1-\varepsilon_{\lambda,j}}{\varepsilon_{\lambda,j}} \bar{t}_{\lambda,k-j} \right) q_{\lambda,j} d\lambda = \int_{\Delta\lambda} \sum_{j=1}^{N} \left[\left(\delta_{kj} - F_{k-j}\bar{t}_{\lambda,k-j} \right) E_{\lambda b,j} - F_{k-j}\bar{\alpha}_{\lambda,k-j} E_{\lambda b,g} \right] d\lambda \tag{10.77}$$

It is assumed that the bands are sufficiently narrow that $q_{\lambda,j}$, $\varepsilon_{\lambda,j}$, $\bar{t}_{\lambda,k-j}$, $\bar{\alpha}_{\lambda,k-j}$, $E_{\lambda b,j}$, and $E_{\lambda b,g}$ can be regarded as constants over the bandwidth, being characteristic of some mean wavelength within the band or, in the case of $\bar{t}$ and $\bar{\alpha}$, being averaged over the band. Then, Equation 10.77 is written for band l as

$$\sum_{j=1}^{N} \left(\frac{\delta_{kj}}{\varepsilon_{l,j}} - F_{k-j} \frac{1-\varepsilon_{l,j}}{\varepsilon_{l,j}} \bar{t}_{l,k-j} \right) q_{l,j} = \sum_{j=1}^{N} \left[\left(\delta_{kj} - F_{k-j}\bar{t}_{l,k-j} \right) E_{lb,j} - F_{k-j}\bar{\alpha}_{l,k-j} E_{lb,g} \right] \tag{10.78}$$

In a spectral region where the gas can be approximated as nonabsorbing, $\bar{t}_l = 1$ and $\bar{\tau}$, and Equation 10.78 reduces to

$$\sum_{j=1}^{N} \left(\frac{\delta_{kj}}{\varepsilon_{l,j}} - F_{k-j} \frac{1-\varepsilon_{l,j}}{\varepsilon_{l,j}} \right) q_{l,j} = \sum_{j=1}^{N} (\delta_{kj} - F_{k-j}) E_{lb,j} \tag{10.79}$$

Equations 10.78 and 10.79 provide simultaneous equations for q_l in each band at each boundary. The $\bar{\alpha}_{l,k-j}$ in Equation 10.78 is found from Equation 10.64 by taking an integrated average over the band:

$$\bar{\alpha}_{l,k-j} = \frac{1}{A_k F_{k-j}} \int_{A_l}\int_{A_k} \frac{\left[(1/\Delta\lambda_l)\int_{\Delta\lambda_l} \alpha_\lambda(S)d\lambda\right]\cos\theta_j \cos\theta_k}{\pi S^2} dA_k dA_j \tag{10.80}$$

For each band, $\bar{t}_{l,k-j} = 1 - \bar{\alpha}_{l,k-j}$, and only a single evaluation is needed:

$$\bar{\alpha}_{l,k-j} = \frac{1}{A_k F_{k-j}} \int_{A_l}\int_{A_k} \frac{\alpha_l(S)\cos\theta_j \cos\theta_k}{\pi S^2} dA_k dA_j \tag{10.81}$$

where $\alpha_l(S)$ is the integrated band absorption,

$$\alpha_l(S) = \frac{1}{\Delta\lambda_l}\int_{\Delta\lambda_l} \alpha_\lambda(S)d\lambda = \frac{1}{\Delta\lambda_l}\int_{\Delta\lambda_l}\left(1 - e^{-\kappa_\lambda S}\right)d\lambda \tag{10.82}$$

If desired, the α_l can be expressed from Equation 9.26 in terms of the effective bandwidth as

$$\alpha_l(S) = \frac{\bar{A}_l(S)}{\Delta\lambda_l} \tag{10.83}$$

A detailed development is given by Nelson (1984) of the band relations for a nongray isothermal gas in an enclosure with diffuse walls. The net radiation enclosure equations are examined in detail in Stasiek (1998) for an enclosure filled with a uniform gas. An extension is made for some of the walls being windows so that the enclosure interior can interact directly with its surrounding environment.

10.7.5 Gray Medium in a Gray Enclosure

If a gas contains many suspended particles or droplets, it may be reasonable to neglect spectral variations of the properties of the suspension. In addition, the walls in a furnace or combustion chamber may be partially soot covered, so an examination of the radiation transfer with gray boundaries may yield accurate results.

From Example 10.3, if a gray gas at T_g is bounded by a chamber consisting of one gray wall with area A_1 at T_1, the energy supplied to the gas by combustion or other nonradiative means and transferred to the wall is

$$Q_g = \frac{A_1\sigma\left(T_g^4 - T_1^4\right)}{1/\varepsilon_1 + 1/\bar{\alpha}_{1-1} - 1} = \frac{\varepsilon_1\bar{\alpha}_{1-1}}{1-\left(1-\varepsilon_1\right)\left(1-\bar{\alpha}_{1-1}\right)} A_1\sigma\left(T_g^4 - T_1^4\right) \tag{10.84}$$

From Example 10.4, for a gray medium between infinite gray parallel plates, the q_1 from Equation 10.4.5 becomes

$$q_1 = \frac{\varepsilon_1\varepsilon_2\bar{t}\sigma\left(T_1^4 - T_2^4\right) + \varepsilon_1\left(1-\bar{t}\right)\left[1 + (1-\varepsilon_2)\bar{t}\right]\sigma\left(T_1^4 - T_g^4\right)}{1-(1-\varepsilon_1)(1-\varepsilon_2)\bar{t}^2} \tag{10.85}$$

and similarly for q_2 from Equation 10.4.6.

Now, consider a chamber made up of two finite gray surfaces that enclose a gray medium. From Equation 10.78,

$$\left(\frac{1}{\varepsilon_1} - F_{1-1}\frac{1-\varepsilon_1}{\varepsilon_1}\bar{t}_{1-1}\right)q_1 - F_{1-2}\frac{1-\varepsilon_2}{\varepsilon_2}\bar{t}_{1-2}q_2$$
$$= \left(1 - F_{1-1}\bar{t}_{1-1}\right)\sigma T_1^4 - F_{1-2}\bar{t}_{1-2}\sigma T_2^4 - \left(F_{1-1}\bar{\alpha}_{1-1} + F_{1-2}\bar{\alpha}_{1-2}\right)\sigma T_g^4 \tag{10.86}$$

$$-F_{2-1}\frac{1-\varepsilon_1}{\varepsilon_1}\bar{t}_{2-1}q_1 + \left(\frac{1}{\varepsilon_2} - F_{2-2}\frac{1-\varepsilon_2}{\varepsilon_2}\bar{t}_{2-2}\right)q_2$$
$$= -F_{2-1}\bar{t}_{2-1}\sigma T_1^4 + \left(1 - F_{2-2}\bar{t}_{2-2}\right)\sigma T_2^4 - \left(F_{2-1}\bar{\alpha}_{2-1} + F_{2-2}\bar{\alpha}_{2-2}\right)\sigma T_g^4 \tag{10.87}$$

If T_1, T_2, and T_g are specified, these equations can be solved for q_1 and q_2. Then, the energy supplied to the medium by some means other than radiation within the enclosure is $Q_g = -(q_1A_1 + q_2A_2)$.

If one of the walls (wall 2) is adiabatic (e.g., an uncooled refractory wall), then $q_2 = 0$. The T_2 can be eliminated from Equations 10.86 and 10.87 and the result solved for q_1 in terms of T_1 and T_g to yield

$$\frac{q_1A_1}{\sigma\left(T_1^4 - T_g^4\right)} = \left(\frac{1-\varepsilon_1}{A_1\varepsilon_1} + \frac{1}{1/f_{1g} + 1/(f_{12} + f_{2g})}\right)^{-1} \tag{10.88}$$

where

$$f_{1g} = \left[A_1\left(F_{1-1}\bar{\alpha}_{1-1} + F_{1-2}\bar{\alpha}_{1-2}\right)\right]^{-1}$$

$$f_{12} = \left(A_1F_{1-2}\bar{t}_{1-2}\right)^{-1}$$

$$f_{2g} = \left[A_2\left(F_{2-1}\bar{\alpha}_{2-1} + F_{2-2}\bar{\alpha}_{2-2}\right)\right]^{-1}$$

Returning to Equations 10.86 and 10.87, if the medium is adiabatic ($Q_g = 0$), then the two walls are exchanging energy through a well-mixed medium that will come to an equilibrium temperature. Then $Q_2 = -Q_1$, so that $q_2 = -q_1A_1/A_2$. q_2 is then eliminated from Equations 10.86 and 10.87 and the two equations are combined to eliminate T_g. The result is

$$\frac{q_1A_1}{\sigma\left(T_1^4 - T_2^4\right)} = \left(\frac{1-\varepsilon_1}{A_1\varepsilon_1} + \frac{1}{1/f_{12} + 1/\left(f_{1g} + f_{2g}\right)} + \frac{1-\varepsilon_2}{A_2\varepsilon_2}\right)^{-1} \tag{10.89}$$

where the f quantities are defined in Equation 10.88. With $Q_1 = q_1A_1$ known and $q_2 = -q_1A_1/A_2$, Equations 10.86 and 10.87 can be solved for the temperature of the medium if desired.

For the general case of a well-mixed gray medium in a gray enclosure where there are more than a few surfaces, a closed-form algebraic equation often cannot be obtained. Rather, the enclosure simultaneous equations would be solved numerically. For the general situation, Equation 10.78 is written for the gray case (one spectral band) as

$$\sum_{j=1}^{N}\left(\frac{\delta_{kj}}{\varepsilon_j} - F_{k-j}\frac{1-\varepsilon_j}{\varepsilon_j}\bar{t}_{k-j}\right)q_j = \sum_{j=1}^{N}(\delta_{kj} - F_{k-j}\bar{t}_{k-j})\sigma T_j^4 - F_{k-j}\bar{\alpha}_{k-j}\sigma T_g^4 \tag{10.90}$$

10.8 EVALUATION OF SPECTRAL GEOMETRIC-MEAN TRANSMITTANCE AND ABSORPTANCE FACTORS

To compute thermal variable values from the enclosure equations, the property and geometric quantities $\bar{\tau}$ and $\bar{\alpha}$ or $AF\bar{\tau}$ and $AF\bar{\alpha}$ must be evaluated. These quantities depend on both geometry and wavelength. The integrations over the surface areas of certain geometries can be carried out analytically using the defining Equations 10.63, 10.64, and the relation of Equation 10.65. The derivation of some of the resulting algebraic relations is provided in Appendix B, *Derivation of Geometric Mean Beam Length Relations*, contained on the book website http://www.thermalradiation.net. Factors are evaluated for hemisphere to differential area at center of its base, top of right circular cylinder to center of its base, side of cylinder to center of its base, entire sphere to any area on its surface or to its entire surface, infinite plate to any area on parallel plate, and rectangle to directly opposed parallel rectangle. The results are summarized in Table 10.1.

Also in the web Appendix B, Section B.7 *Geometric-Mean Beam Length for Spectral Band Enclosure Equations*, the derivation is given for the integrated average mean beam length between pairs of surfaces assuming that the integrated band absorption $\alpha_l(S)$ varies linearly between the surfaces. The geometric-mean beam length can be used as the length in the effective bandwidth correlations; the methodology for this approach is in web Appendix B. Algebraic equations and tables of values for geometric-mean beam lengths from Dunkle (1964) are given for directly opposed parallel equal rectangles and for rectangles at right angles sharing a common edge, allowing computation for radiative transfer in rectangular parallelepipeds containing an absorbing/emitting isothermal medium.

TABLE 10.1
Spectral Geometric-Mean Transmittance for Some Simple Geometries

Geometry	Relation
Hemisphere of radius R to differential area at center of base	$\bar{t}_{\lambda,j-dk} = e^{-\kappa_\lambda R}$
Top of right circular cylinder of radius R and height h to center of its base	$A_j dF_{j-dk}\bar{t}_{\lambda,j-dk} = 2dA_k\left\{E_3(\kappa_\lambda h) - \frac{1}{(R/h)^2+1}E_3\left[\kappa_\lambda h\cdot\sqrt{(R/h)^2+1}\right]\right\}$
Side of right circular cylinder of radius R and height h to center of its base	$A_j dF_{j-dk}\bar{t}_{\lambda,j-dk} = 2dA_k\left\{E_3(\kappa_\lambda R) - \frac{(R/h)^2}{(R/h)^2+1}E_3\left[\kappa_\lambda h\cdot\sqrt{(R/h)^2+1}\right]\right\}$
Entire sphere of radius R or any area on its surface to its entire surface	$\bar{t}_{\lambda,j-k} = \frac{2}{(2\kappa_\lambda R)^2}\left[1-(2\kappa_\lambda R+1)e^{-2\kappa_\lambda R}\right]$
Infinite plate to any area on a parallel plate separated by distance D	$\bar{t}_{\lambda,j-k} = 2E_3(\kappa_\lambda D)$

Note: E_3 is the exponential integral function. Values are in Appendix D.

10.9 MEAN BEAM LENGTH APPROXIMATION FOR SPECTRAL RADIATION FROM AN ENTIRE VOLUME OF A MEDIUM TO ALL OR PART OF ITS BOUNDARY

Some practical situations require evaluating the energy radiated from a volume of isothermal medium with uniform composition to all or part of its boundaries, without considering emission and reflection from the boundaries. Such an approach allows assessing the relative importance of radiation in a complex problem before embarking on a more comprehensive computation.

An example is radiation from hot furnace gases to walls that are cool enough for their emission to be small and that are rough and soot covered so they are essentially nonreflecting. In this section, the energy is considered in $d\lambda$; in the next section, an integration will include all λ to provide the total radiant energy. For the specified conditions, the $J_{\lambda,j}\, d\lambda$ in Equation 10.67, which is the spectral outgoing energy flux from A_j, is zero. The incoming spectral energy at A_k is then

$$A_k G_{\lambda,k} d\lambda = \sum_{j=1}^{N} E_{\lambda b,g} d\lambda A_j F_{j-k} \bar{\alpha}_{\lambda,j-k} \tag{10.91}$$

If the geometry is a hemisphere of medium with radius R radiating to an area element dA_k at the center of its base, Equation 10.91 has an especially simple form. Since the hemispherical boundary is the only surface in view of dA_k, Equation 10.91 reduces to

$$dA_k G_{\lambda,k} = E_{\lambda b,g} A_j dF_{j-dk} \bar{\alpha}_{\lambda,j-dk} = E_{\lambda b,g} dA_k F_{dk-j} \bar{\alpha}_{\lambda,j-dk} \tag{10.92}$$

where

$$\bar{\alpha}_{\lambda,j-dk} = 1 - \bar{t}_{\lambda,j-dk} = 1 - e^{-\kappa_\lambda R}$$

For radiation between dA_k and the surface of a hemisphere, $F_{dk-j} = 1$, so Equation 10.92 reduces to the following simple expression giving the *incident* energy flux from a hemisphere of medium emitting to the center of the hemisphere base:

$$G_{\lambda,k} d\lambda = \left(1 - e^{-\kappa_\lambda R}\right) E_{\lambda b,g} d\lambda \tag{10.93}$$

The term $(1-e^{-\kappa_\lambda R})$ is the *spectral emittance* of the gas $\varepsilon_k(T, P, R)$ for path length R. Then, Equation 10.93 becomes

$$G_{\lambda,k} d\lambda = \varepsilon_\lambda(\kappa_\lambda R) E_{\lambda b,g} d\lambda \qquad \varepsilon_\lambda(\kappa_\lambda R) = 1 - e^{-\kappa_\lambda R} \tag{10.94}$$

The gas emittance and thus the incident spectral energy depend on the optical radius of the hemisphere, $\kappa_\lambda R$.

It would be very convenient if a relation having the simple form of Equation 10.94 could be used to determine the value of $G_{\lambda,k}\, d\lambda$ on A_k for *any* geometry of radiating medium volume and for A_k being all or part of its boundary. Because the geometry of the medium enters Equation 10.93 only through $\varepsilon_\lambda(\kappa_\lambda R)$, we can define a fictitious equivalent value of R, say L_e, that would yield a value of $\varepsilon_\lambda(\kappa_\lambda L_e)$ such that Equation 10.94 would give the correct $G_\lambda\, d\lambda$ for a given geometry. This fictitious length L_e is called the *mean beam length*. Then, for an arbitrary geometry of gas, let

$$G_{\lambda,k} d\lambda = \varepsilon_\lambda(\kappa_\lambda L_e) E_{\lambda b,g} d\lambda = \left(1 - e^{-\kappa_\lambda L_e}\right) E_{\lambda b,g} d\lambda \tag{10.95}$$

The mean beam length is the required radius of an equivalent hemisphere of a medium that radiates a flux to the center of its base equal to the average flux radiated to the area of interest by the actual volume of the medium.

10.9.1 Mean Beam Length for a Medium between Parallel Plates Radiating to Area on Plate

Consider two black infinite parallel plates at $T_1 = T_2 = 0$ separated by a distance D. The plates enclose a uniform medium at T_g with absorption coefficient κ_λ. The rate at which spectral energy is incident upon A_k on one plate is, from Equation 10.91,

$$G_{\lambda,k}A_k d\lambda = E_{\lambda b,g}d\lambda A_j F_{j-k}\bar{\alpha}_{\lambda,j-k} = E_{\lambda b,g}d\lambda A_j F_{j-k}\left[1-2E_3\left(\kappa_\lambda D\right)\right] \quad (10.96)$$

where the expression $\bar{\alpha}_{\lambda,j-k} = [1-2E_3(\kappa_\lambda D)]$ for infinite parallel plates is from Table 10.1 (This is derived in web Appendix B, Equation B.13, available on the book website at http://www.thermalradiation.net). Here, E_3 is the exponential integral function, tabulated in Appendix D of this book. For infinite parallel plates, $F_{j-k} = 1$, and by reciprocity, $F_{j-k} = A_k/A_j$, so Equation 10.96 reduces to

$$G_{\lambda,k}d\lambda = \left[1-2E_3\left(\kappa_\lambda D\right)\right]E_{\lambda b,g}d\lambda \quad (10.97)$$

Comparing Equations 10.94 and 10.97 provides the mean beam length for infinite parallel plates as

$$L_e = -\frac{1}{\kappa_\lambda}\ln[2E_3(\kappa_\lambda D)] \quad (10.98)$$

or, in terms of optical thickness $\tau_\lambda = \kappa_\lambda D$,

$$\frac{L_e}{D} = -\frac{1}{\tau_\lambda}\ln[2E_3(\tau_\lambda)] \quad (10.99)$$

10.9.2 Mean Beam Length for the Sphere of a Medium Radiating to Any Area on Its Boundary

Consider a medium within a nonreflecting sphere of radius R with the sphere boundary A_j at $T_j = 0$. From Equation 10.91, the spectral radiation flux incident on an element dA_k is $G_{\lambda,k} = E_{\lambda b,g}(A_j/dA_k)F_{j-k}\bar{\alpha}_{\lambda,j-k}$. For a sphere, Equation 5.65 gives $dF_{j-dk} = dA_k/A_j$. Substituting into Equation 10.92 for $\bar{\alpha}_{\lambda,j-k}$ from Table 10.1 for this geometry gives

$$G_{\lambda,k} = E_{\lambda b,g}\left\{1-\frac{2}{(2\kappa_\lambda R)^2}\left[1-(2\kappa_\lambda R+1)e^{-2\kappa_\lambda R}\right]\right\} \quad (10.100)$$

Equating Equations 10.95 and 10.100 gives the mean beam length for this geometry as

$$\frac{L_e}{2R} = -\frac{1}{2\kappa_\lambda R}\ln\left\{\frac{2}{(2\kappa_\lambda R)^2}\left[1-(2\kappa_\lambda R+1)e^{-2\kappa_\lambda R}\right]\right\} \quad (10.101)$$

Because the expression used for $\bar{\alpha}_{\lambda,j-k}$ is quite general, Equation 10.101 gives the correct mean beam length for the entire sphere of medium radiating to any portion of its boundary. Additional results for spheres are in Koh (1965).

10.9.3 Radiation from the Entire Medium Volume to Its Entire Boundary for Optically Thin Media

Because of the integrations involved, the mean beam length for an entire medium volume radiating to all or part of its boundary is usually difficult to evaluate except for the simplest shapes. It is fortunate that some practical approximations can be found by looking first at the optically thin limit. For a small optical path length $\kappa_\lambda S$, the transmittance becomes

$$\lim_{\kappa_\lambda S\to 0} t_\lambda = \lim_{\kappa_\lambda S\to 0} e^{-\kappa_\lambda S} = \lim_{\kappa_\lambda S\to 0}\left[1-\kappa_\lambda S+\frac{(\kappa_\lambda S)^2}{2!}-\cdots\right]\to 1$$

Each differential volume of the uniform temperature medium emits spectral energy $4\kappa_\lambda E_{\lambda b,g} d\lambda dV$. Since $t_\lambda = 1$, there is no attenuation of emitted radiation, and it all reaches the enclosure boundary. For the entire radiating volume, the energy reaching the boundary is $4\kappa_\lambda E_{\lambda b,g} d\lambda\, V$, so the average spectral flux received at the entire boundary of area A is, in the optically thin limit,

$$G_\lambda = 4\kappa_\lambda E_{\lambda b,g}\frac{V}{A} \tag{10.102}$$

By use of the mean beam length, the average spectral flux reaching the boundary is given by Equation 10.93. For the optically thin case, let L_e be designated by $L_{e,0}$. Then, from Equation 10.93, for small $\kappa_\lambda L_{e,0}$,

$$G_\lambda = \left\{1-\left[1-\kappa_\lambda L_{e,o}+\frac{(\kappa_\lambda L_{e,o})^2}{2!}-\cdots\right]\right\}E_{\lambda b,g} = \kappa_\lambda L_{e,0}E_{\lambda b,g} \tag{10.103}$$

Equating this to G_λ in Equation 10.102 gives the desired result for the mean beam length of an *optically thin* medium radiating to its entire boundary:

$$L_{e,0} = \frac{4V}{A} \tag{10.104}$$

To give a few examples, for a sphere of diameter D,

$$L_{e,0} = \frac{4V_s}{A_s} = \frac{4\pi D^3/6}{\pi D^2} = \frac{2}{3}D \tag{10.105}$$

For an infinitely long circular cylinder of diameter D,

$$L_{e,0} = \frac{4\pi D^2/4}{\pi D} = D \tag{10.106}$$

For a medium between infinite parallel plates spaced D apart,

$$L_{e,0} = \frac{4V_g}{A_s} = \frac{4DA}{2A} = 2D \tag{10.107}$$

For an infinitely long rectangular parallelepiped with cross-sectional dimensions h and w,

$$L_{e,0} = \frac{4hw}{2(h+w)} = \frac{2hw}{h+w} \tag{10.108}$$

10.9.4 Correction to Mean Beam Length When a Medium Is Not Optically Thin

For a medium that is not optically thin, it would be very convenient if L_e could be obtained by applying a simple correction factor to the $L_{e,0}$ from Equation 10.104. A useful technique is to introduce a correction coefficient C so that L_e is given by

$$L_e = CL_{e,0} \tag{10.109}$$

Then, the incoming spectral flux in Equation 10.95 is

$$G_\lambda d\lambda = \left(1 - e^{-\kappa_\lambda CL_{e,0}}\right) E_{\lambda b,g} d\lambda \tag{10.110}$$

Consider a radiating medium between infinite parallel plates spaced D apart. Using Equation 10.107 in Equation 10.110 gives $G_\lambda d\lambda = [1-\exp(-\kappa_\lambda C2D)]E_{\lambda b,g}d\lambda$. From Equation 10.97, the actual flux received is $G_\lambda d\lambda = [1-2E_3(\kappa_\lambda D)]E_{\lambda b,g}d\lambda$. To compare these fluxes, the ratio $[1-2E_3(\kappa_\lambda D)]/[1-\exp(-2C\kappa_\lambda D)]$ is shown in Figure 10.11 for a range of $\kappa_\lambda D$ using $C = 0.9$. This value of C was found to yield a ratio close to 1 for all $\kappa_\lambda D$ and hence is a valid correction coefficient for this geometry.

Table 10.2 gives the mean beam length $L_{e,0}$ for various geometries, along with L_e values that provide reasonably accurate radiative fluxes for nonzero optical thicknesses. The values of C are found to be in a range near 0.9 (Hottel 1954, Eckert and Drake 1959, Hottel and Sarofim 1967). Hence, it is historically recommended that, for a geometry for which exact L_e values have not been calculated, the approximation

$$L_e \approx 0.9L_{e,0} \approx 0.9\frac{4V}{A} \tag{10.111}$$

be used for an entire uniform isothermal medium volume radiating to its entire boundary.

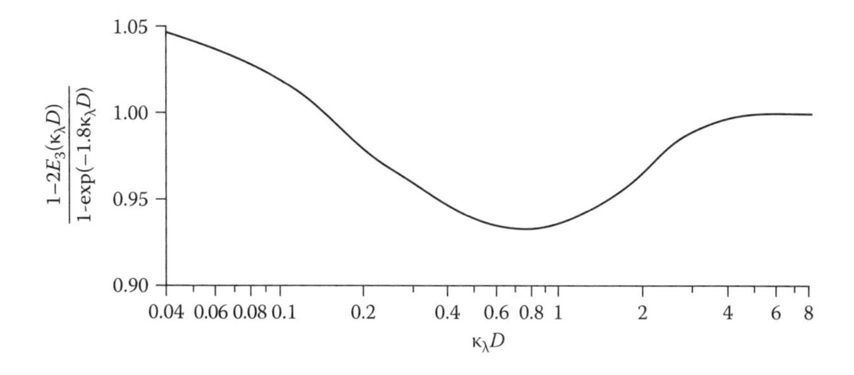

FIGURE 10.11 Ratio of emission by layer of medium to that calculated using a mean beam length, $L_e = 1.8D$. Deviation from 1.00 corresponds to error in the approximation.

TABLE 10.2
Mean Beam Lengths for Radiation from Entire Medium Volume

Geometry of Radiating System	Characterizing Dimension	Mean Beam Length for Optical Thickness, $\kappa_\lambda L_e \to 0$, $L_{e,0}$	Mean Beam Length Corrected for Finite Optical Thickness,[a] L_e	$C = L_e/L_{e,0}$
Hemisphere radiating to element at center of base	Radius R	R	R	1
Sphere radiating to its surface	Diameter D	$\frac{2}{3}D$	$0.65D$	0.97
Circular cylinder of infinite height radiating to concave bounding surface	Diameter D	D	$0.95D$	0.95
Circular cylinder of semi-infinite height radiating to:				
Element at center of base	Diameter D	D	$0.90D$	0.90
Entire base	Diameter D	$0.81D$	$0.65D$	0.80
Circular cylinder of height equal to diameter radiating to:				
Element at center of base	Diameter D	$0.77D$	$0.71D$	0.92
Entire surface	Diameter D	$\frac{2}{3}D$	$0.60D$	0.90
Circular cylinder of height equal to two diameters radiating to:				
Plane end	Diameter D	$0.73D$	$0.60D$	0.82
Concave surface	Diameter D	$0.82D$	$0.76D$	0.93
Entire surface	Diameter D	$0.80D$	$0.73D$	0.91
Circular cylinder of height equal to one-half the diameter radiating to:				
Plane end	Diameter D	$0.48D$	$0.43D$	0.90
Concave surface	Diameter D	$0.52D$	$0.46D$	0.88
Entire surface	Diameter D	$0.50D$	$0.45D$	0.90
Cylinder of infinite height and semicircular cross section radiating to element at center of plane rectangular face	Radius R		$1.26R$	
Infinite slab of medium radiating to:				
Element on one face	Slab thickness D	$2D$	$1.8D$	0.90
Both bounding planes	Slab thickness D	$2D$	$1.8D$	0.90
Cube radiating to a face	Edge X	$\frac{2}{3}X$	$0.6X$	0.90
Rectangular parallelepipeds 1 × 1 × 4 radiating to:				
1 × 4 face	Shortest edge X	$0.90X$	$0.82X$	0.91
1 × 1 face	Shortest edge X	$0.86X$	$0.71X$	0.83
All faces	Shortest edge X	$0.89X$	$0.81X$	0.91
1 × 2 × 6 radiating to:				
2 × 6 face	Shortest edge X	$1.18X$		
1 × 6 face	Shortest edge X	$1.24X$		
1 × 2 face	Shortest edge X	$1.18X$		
All faces	Shortest edge X	$1.20X$		

(Continued)

TABLE 10.2 (*Continued*)
Mean Beam Lengths for Radiation from Entire Medium Volume

Geometry of Radiating System	Characterizing Dimension	Mean Beam Length for Optical Thickness, $\kappa_\lambda L_e \to 0$, $L_{e,0}$	Mean Beam Length Corrected for Finite Optical Thickness,[a] L_e	$C = L_e/L_{e,0}$
Medium between infinitely long parallel concentric cylinders	Radius of outer cylinder R and of inner cylinder r	$2(R-r)$	See Anderson and Handvig (1989)	
Medium volume in the space between the outside of the tubes in an infinite tube bundle and radiating to a single tube:				
Equilateral triangular array:	Tube diameter			
$S = 2D$	D and spacing	$3.4(S-D)$	$3.0(S-D)$	0.88
$S = 3D$	between tube	$4.45(S-D)$	$3.8(S-D)$	0.85
Square array:	centers, S			
$S = 2D$		$4.1(S-D)$	$3.5(S-D)$	0.85

[a] Corrections are those suggested by Hottel (1954), Hottel and Sarofim (1967) or Eckert and Drake (1959). Corrections were chosen to provide maximum L_e where these references disagree.

It is shown in Cartigny (1986) that the concept of mean beam length can also be used for *scattering* media, and Equation 10.111 applies in the limit of pure scattering and small optical thickness.

10.10 EXCHANGE OF TOTAL RADIATION IN AN ENCLOSURE BY USE OF MEAN BEAM LENGTH

The mean beam length was obtained in the previous section at one wavelength. This concept is now applied to obtain the exchange of *total* energy within an enclosure. The use of the mean beam length simplifies the geometric considerations, but it remains to integrate the spectral relations to obtain total energy transfer.

10.10.1 Total Radiation from the Entire Medium Volume to All or Part of Its Boundary

The mean beam length was found to be approximately independent of κ_λ, as evidenced by Equation 10.111. This means that L_e can be used as a characteristic dimension of the gas volume and approximated as constant during integration over wavelength. The total heat flux from the medium that is incident on a surface is found by integrating Equation 10.95 over all λ:

$$G = \int_{\lambda=0}^{\infty} (1 - e^{-\kappa_\lambda L_e}) E_{\lambda b,g} d\lambda \tag{10.112}$$

where L_e is independent of λ. Now, define a *total emittance* ε_g for the medium such that

$$G = \varepsilon_g \sigma T_g^4 \tag{10.113}$$

Equating the last two relations gives

$$\varepsilon_g = \frac{\int_{\lambda=0}^{\infty} \left(1 - e^{-\kappa_\lambda L_e}\right) E_{\lambda b,g} d\lambda}{\sigma T_g^4} \tag{10.114}$$

The ε_g in Equation 10.114 is a convenient quantity that can be provided for each medium in terms of L_e and T_g. Values of ε_g are available for the important radiating gases, and also analytical forms are available that are convenient for computer use; these results for ε_g are in Section 9.4. Then, for a particular geometry and medium temperature and pressure, the ε_g is applied by use of Equation 10.113. An example illustrates how ε_g is obtained and used.

Example 10.5

A cooled right cylindrical tank 4 m in diameter and 4 m long has a black interior surface and is filled with hot gas at a total pressure of 1 atm. The gas is composed of CO_2 mixed with a transparent gas that has a partial pressure of 0.75 atm. The gas is uniformly mixed at $T_g = 1100$ K. Compute how much energy must be removed from the tank surface to keep it cool if the tank walls are all at low temperature so that only radiation from the gas is significant.

The geometry is a finite circular cylinder of gas, and the radiation to its walls will be computed. Using Table 10.2, the corrected mean beam length is $L_e = 0.60D = 2.4$ m. The partial pressure of the CO_2 is 0.25 atm, so that $p_{CO_2}L_e = 0.25 \times 2.4 = 0.6$ atm·m. From Equation 9.64 and Table 9.7, remembering to convert $p_{CO_2}L_e$ into units of bar–cm, $\varepsilon_{CO_2}(p_{CO_2}L_e, T_g) = 0.170$, and C_{CO2} from Equation 9.65 is 1.0, since the mixture total pressure is 1. From Equation 10.113, the energy to be removed is

$$Q_i = GA = \varepsilon_{CO_2}\sigma T_g^4 A = 0.170 \times 5.6704 \times 10^{-8}(1100)^4 24\pi = 1064 \text{ kW}$$

10.10.2 Exchange between the Entire Medium Volume and the Emitting Boundary

In the previous section, the temperature of the black enclosure wall was small enough that emission from the wall could be neglected. If wall emission is significant, the average heat flux removed at the wall is the emission of the medium to the wall, which is all absorbed because the wall is black, minus the average flux emitted from the wall that is absorbed by the medium. For steady state, the total energy removed from the entire boundary equals the energy supplied to the medium by some other means, such as by combustion in a gas. An energy balance gives, for an enclosure with its entire boundary black and at T_w,

$$-\frac{Q_w}{A} = \frac{Q_g}{A} = \sigma\left[\varepsilon_g(T_g)T_g^4 - \alpha_g(T_w)T_w^4\right] \tag{10.115}$$

This also follows from integrating Equation 10.3.1 when the wall is black ($\varepsilon_{\lambda,1} = 1$). The $\alpha_g(T_w)$ is the total absorptance by the medium for radiation emitted from the black wall at temperature T_w. The $\alpha_g(T_w)$ depends on the spectral properties of the medium and on T_w, as this determines the spectral distribution of the radiation received by the medium.

For radiative exchange calculations in furnaces, an approximate procedure for determining α_g is in Hottel and Sarofim (1967). The α_g is obtained from the gas total emittance values by using

$$\alpha_g = \alpha_{CO_2} + \alpha_{H_2O} - \Delta\alpha \tag{10.116}$$

where

$$\alpha_{CO_2} = C_{CO_2}\varepsilon^+_{CO_2}\left(\frac{T_g}{T_w}\right)^{0.5} \quad (10.117)$$

$$\alpha_{H_2O} = C_{H_2O}\varepsilon^+_{H_2O}\left(\frac{T_g}{T_w}\right)^{0.5} \quad (10.118)$$

$$\Delta\alpha = (\Delta\varepsilon)_{\text{at } T_w} \quad (10.119)$$

The $\varepsilon^+_{CO_2}$ and $\varepsilon^+_{H_2O}$ are, respectively, ε_{CO_2} and ε_{H_2O} obtained from Equation 9.64 evaluated at T_w and at the respective parameters $p_{CO_2}L'_e$ and $p_{H_2O}L'_e$ where $L'_e = L_eT_w/T_g$. It is pointed out in Edwards and Matavosian (1984) that the exponent 0.5 is now becoming more accepted to replace the values 0.65 and 0.45 originally used in Equations 10.117 and 10.118. At high temperatures and pressures when there is overlapping of absorption lines in the infrared spectrum, $L'_e = L_e\left(T_w/T_g\right)^{3/2}$, which is discussed by Edwards and Matavosian,

Section 10.10.1 considered a uniform isothermal gas in a black enclosure. If the bounding walls are not black and hence are reflecting, radiation can pass through the gas by means of multiple reflections from the boundary. For an enclosure with a single wall, this can be included by integrating Equation 10.114 over all wavelengths. In Edwards and Matavosian, a procedure is proposed to use total emittance values for situations with multiple reflections. The solution of enclosure heat transfer problems with reflecting walls was treated in Section 10.7.4 by integrating spectral relations over the wavelength absorption bands.

Yuen (2015) extends the mean beam length to three-dimensional enclosures with a nongray $N_2/CO_2/H_2O$ gas, and includes the presence of soot.

An example that considers a black enclosure with walls at differing temperatures is in Example B.2 in Appendix B at the book website http://www.thermalradiation.net.

10.11 OPTICALLY THIN AND COLD MEDIA

The spectral intensity along a path depends on attenuation by absorption and scattering and on augmentation by emission and incoming scattering. From the radiative transfer Equation 10.14, we can write the intensity as

$$I_\lambda(S) = I_\lambda(0)e^{-\int_{S^*=0}^{S}\beta_\lambda(S^*)\,dS^*} + \int_0^S \beta_\lambda(S^*)\hat{I}_\lambda(S^*)e^{-\int_{S^{**}=S^*}^{S}\beta_\lambda(S^{**})\,dS^{**}}dS^* \quad (10.120)$$

where the spectral properties $\beta_\lambda = \kappa_\lambda + \sigma_{s,\lambda}$ may vary along the path S. The source function $\hat{I}_\lambda\left(S\right)$ depends on local blackbody emission and scattering and is found from an integral equation such as Equation 10.10. Since $\hat{I}_\lambda\left(S\right)$ depends on $I_{\lambda b}(S)$, the temperature distribution must be found to evaluate $I_{\lambda b}(S)$, so a simultaneous solution is required with the energy equation. Various limiting cases can be examined, such as an optically thin or thick medium, weak or strong scattering relative to absorption, and weak or strong internal emission relative to radiation incident at the boundaries.

10.11.1 Nearly Transparent Medium

When the optical depth along a path is small, $\int_{S^*=0}^{S}\beta_\lambda(S^*)dS^* \to 0$, Equation 10.120 is simplified as the exponential attenuation terms each approach one, so that

$$I_\lambda(S) = I_\lambda(0) + \int_{S^*=0}^{S}\beta_\lambda(S^*)\hat{I}_\lambda(S^*)dS^* \quad (10.121)$$

TABLE 10.3
Approximations to Radiative Transfer Equation

Approximation	Form of Radiative Transfer Equation	Conditions
Nearly transparent, optically thin medium	$I_\lambda(S) = I_\lambda(0)$	The medium has such a low extinction coefficient that intensity does not change by absorption, emission, or scattering along a path within the medium.
Optically thin medium with cold nonreflecting boundaries or small incident radiation (emission approximation)	$I_\lambda(S) = \int_{S^*=0}^{S} \kappa_\lambda(S^*) I_{\lambda b}(S^*) dS^*$	No energy is incident from the boundaries, and the medium is relatively transparent so that emitted energy from the medium passes within the system without significant attenuation.
Cold medium with weak scattering	$I_\lambda(S) = I_\lambda(0)\exp\left[-\int_{S^*=0}^{S} \beta_\lambda dS^*\right]$	Radiation emitted and scattered by the medium is negligible compared to that originating from boundaries or external sources.
Optically thick medium (diffusion approximation, see Chapter 12)	$q_{r\lambda}(S) = -\frac{4\pi}{3\beta_\lambda}\frac{dI_{\lambda b}(S)}{dS}$	The optical depth of the medium is sufficiently large and the temperature gradients sufficiently small, so the local intensity and radiative flux depend only on the local blackbody intensity.

The intensity that enters at $S = 0$ is not appreciably attenuated, and there is no attenuation along the path of the intensity that is locally emitted or scattered. In the special case when the optical depth is small and the $I_\lambda(0)$ is not small, so that the integral in Equation 10.121 is small relative to $I_\lambda(0)$, Equation 10.121 further reduces to

$$I_\lambda(S) = I_\lambda(0) \tag{10.122}$$

as in Table 10.3.

The incident intensity is dominant and is essentially unchanged along its path through the medium. This approximation is now used to examine a limiting case where conduction is very small compared with radiation and convection is absent.

Example 10.6

Two infinite parallel black walls at T_1 and T_2 are a distance D apart, and the space between them is filled with an absorbing, emitting, and scattering gas with absorption coefficient $\kappa_\lambda(x)$. Assuming the nearly transparent approximation is valid, derive an expression for the gas temperature distribution for the limit where radiation is dominant so that heat conduction can be neglected.

For steady state and no heat conduction, the medium is in *radiative equilibrium* without internal energy sources, so $\mathbf{q}_r$ is constant and Equation 10.37 gives (this includes scattering)

$$\int_{\lambda=0}^{\infty} \kappa_\lambda(x) I_{\lambda b}\left[T(x)\right] d\lambda = \int_{\lambda=0}^{\infty} \kappa_\lambda(x) \bar{I}_{\lambda,i}(x) d\lambda \tag{10.6.1}$$

where the x coordinate is normal to the walls. The $\bar{I}_{\lambda,i}$ is obtained from the intensities reaching a volume element along paths in the positive and negative coordinate directions,

$$4\pi\bar{I}_{\lambda,i}(x) = \int_{\cap} I_\lambda^+(x,\Omega_i)d\Omega_i + \int_{\cap} I_\lambda^-(x,\Omega_i)d\Omega_i \tag{10.6.2}$$

Since the walls are black, the nearly transparent approximation gives, at any location between the walls, $I_\lambda^+(\Omega_i) = I_{\lambda b}(T_1)$ and $I_\lambda^-(\Omega_i) = I_{\lambda b}(T_2)$. Then, since the blackbody intensity is independent of angle, Equation 10.6.2 reduces to

$$4\pi\bar{I}_{\lambda,i} = 2\pi\left[I_{\lambda b}(T_1) + I_{\lambda b}(T_2)\right] \tag{10.6.3}$$

Substituting Equation 10.6.3 into Equation 10.6.1 gives, at any x position between the walls,

$$\int_{\lambda=0}^{\infty} \kappa_\lambda(x) I_{\lambda b}\left[T(x)\right]d\lambda = \frac{1}{2}\int_{\lambda=0}^{\infty} \kappa_\lambda(x)\left[I_{\lambda b}(T_1) + I_{\lambda b}(T_2)\right]d\lambda \tag{10.6.4}$$

If $\kappa_\lambda(x)$ depends on local temperature, Equation 10.6.4 is solved iteratively for $T(x)$. If κ_λ can be assumed independent of gas temperature,

$$\int_{\lambda=0}^{\infty} \kappa_\lambda(x) I_{\lambda b}\left[T(x)\right]d\lambda = \frac{1}{2}\int_{\lambda=0}^{\infty} \kappa_\lambda\left[I_{\lambda b}(T_1) + I_{\lambda b}(T_2)\right]d\lambda \tag{10.6.5}$$

The right side is evaluated for the specified T_1 and T_2. Then, the gas temperature T, which is independent of x for the specified conditions, can be found to satisfy Equation 10.6.5 by using a root solver. For a gray gas with temperature-independent properties, κ is constant and Equation 10.6.5 gives $T^4 = (T_1^4 + T_2^4)/2$. In this optically thin limit without heat conduction, the entire medium approaches a fourth-power temperature, that is, the average of the fourth powers of the black boundary temperatures.

Le Dez and Sadat (2015) present the solution for an optically thin gray medium between concentric cylinders.

10.11.2 Optically Thin Media with Cold Boundaries or Small Incident Radiation: The Emission Approximation

In the nearly transparent approximation, the gas is optically thin, and the local intensity within the medium is dominated by intensities from the boundaries. In the *emission approximation*, the gas is again optically thin, but there is negligible energy from the boundaries. There is only energy emission within the medium and no attenuation by either absorption or scattering. The radiative transfer Equation 10.120, simplified for these conditions and integrated over all wavelengths, becomes

$$I(S) = \int_{S^*=0}^{S}\left[\int_{\lambda=0}^{\infty} \kappa_\lambda(S^*) I_{\lambda b}(S^*)d\lambda\right]dS^* \tag{10.123}$$

The $I(S)$ is the integrated contribution by all the emission along the path as the emitted energy travels through the medium without attenuation.

Example 10.7

Use the emission approximation to find the energy flux emerging from an isothermal gas layer at T_g with an integrated mean absorption coefficient weighted by the blackbody spectrum at T_g (*Planck mean*, Equation 9.71) of $\kappa_P = 0.010$ cm^{-1} and thickness $D = 1.5$ cm, if the layer is bounded by transparent nonradiating walls and cold surroundings (Figure 10.12a).

If $I(\theta)$ is the total intensity emerging from the layer in direction θ, the emerging flux is $q = 2\pi\int_{\theta=0}^{\pi/2} I(\theta)\cos\theta\sin\theta\, d\theta$. The layer is isothermal with constant temperature T_g, so for the inner integral of Equation 10.123, the quantities are independent of S^*, and the Planck mean of $\kappa_\lambda(T_g)$ gives $\pi\int_{\lambda=0}^{\infty}\kappa_\lambda(T_g)I_{\lambda b}(T_g)d\lambda = \kappa_P(T_g)\sigma T_g^4$. Then, Equation 10.123 can be integrated over any path $S = D/\cos\theta$ through the layer to yield $I(\theta) = (1/\pi)\kappa_p(T_g)\sigma T_g^4 D/\cos\theta$. This gives the radiative flux from each boundary as

$$q = 2\int_{\theta=0}^{\pi/2}\kappa_P(T_g)\sigma T_g^4\, D\sin\theta d\theta = 2\kappa_P(T_g)\sigma T_g^4 D \qquad (10.7.1)$$

which gives $q = 0.03\,\sigma T_g^4$ for the specified numerical values. This is not a precise result, even though the layer thickness is optically thin: $\kappa_P D = 0.015 \ll 1$. This is because the radiation reaching the layer boundary along each path has passed along a thickness $D/\cos\theta$. For θ approaching $\pi/2$, the optical path length becomes very large, so the emission approximation cannot hold. A more accurate solution including effects of the proper path lengths gives $q = 1.8\,\kappa_P(T_g)\sigma T_g^4 D$ (see Section 10.9.4), which is a 10% decrease compared with Equation 10.7.1.

Example 10.8

A plane layer of gray medium with thickness D and constant properties is initially at uniform temperature T_0. It has attenuation coefficient β, density ρ, and specific heat c_v. At time $t = 0$, the layer is subjected to a very cold environment. Consider only radiation transfer and obtain the transient temperature of the layer by using the emission approximation.

From the results of Example 10.7, the instantaneous transient heat flux emerging from both boundaries of the layer is $q(t) = 4\beta\sigma T^4(t)D$. The energy equation for the layer then becomes

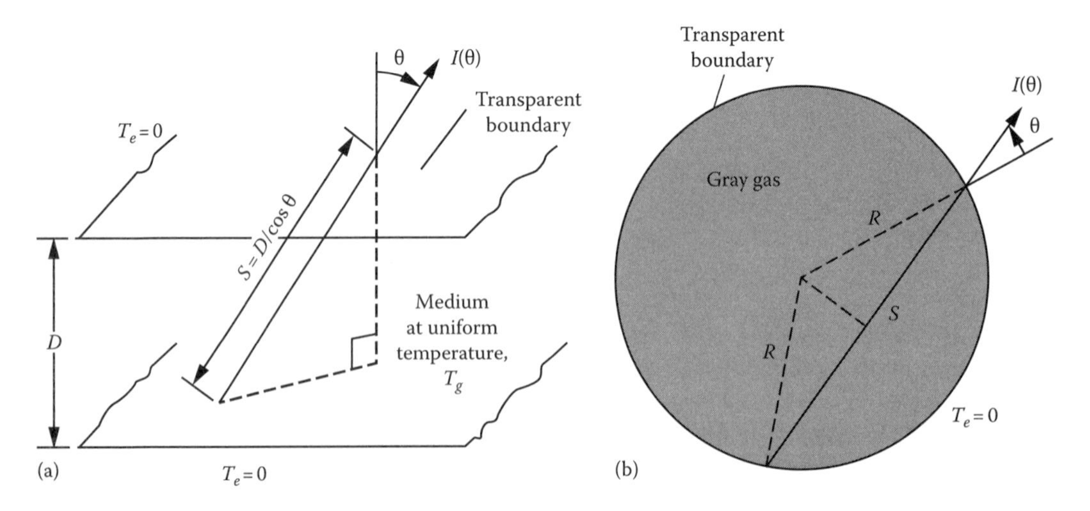

FIGURE 10.12 Examples for emission approximation, (a) layer geometry for Example 10.7, (b) emission from spherical gas-filled balloon with transparent skin.

$\rho c_v dT/dt = -4\beta\sigma T^4(t)$ or, in dimensionless form, $d\vartheta/d\bar{t} = -4\vartheta^4(\bar{t})$ where $\vartheta = T/T_0$ and $\bar{t} = (\beta\sigma T_0^3/\rho c_v)t$. Integrating with the condition that $\vartheta = 1$ at $\bar{t} = 0$ gives the transient uniform temperature throughout the layer for the emission approximation as

$$\vartheta(\bar{t}) = \frac{1}{(1+2\bar{t})^{1/3}}$$

Example 10.9

A spherical balloon of radius R is in orbit around the Earth and enters the Earth's shadow. The balloon has a perfectly transparent wall and is filled with a gray gas with constant absorption coefficient κ, such that $\kappa R = 1$. Neglecting radiant exchange with the Earth, derive a relation for the initial rate of radiant energy loss from the balloon if the initial temperature of the gas is T_0.

Using the emission approximation, Equation 10.123, Figure 10.12b, shows that the intensity at the surface for a typical path S is $I(\theta) = (\kappa\sigma T_0^4/\pi)S = (\kappa\sigma T_0^4/\pi)2R\cos\theta$. The radiative flux leaving the surface is

$$q = 2\pi\int_{\theta=0}^{\pi/2} I(\theta)\cos\theta\sin\theta d\theta = 4\kappa\sigma T_0^4 R\int_{\theta=0}^{\pi/2}\cos^2\theta\sin\theta d\theta = \frac{4}{3}\kappa\sigma T_0^4 R$$

The initial rate of energy loss from the entire sphere is then $Q = (4/3)\kappa\sigma T_0^4 R(4\pi R^2) = 4\kappa\sigma T_0^4 V_s$, where $V_s = (4/3)\pi R^3$ is the sphere volume. This is what is expected, as it was found in Section 1.6.4 that *any* isothermal gas volume with negligible internal absorption radiates according to this relation.

10.11.3 Cold Medium with Weak Scattering

This approximation applies when both the local blackbody emission and scattering within a medium are small. Scattering is small enough that scattering *into* the S direction can be neglected. The radiative transfer Equation 10.120 reduces to

$$I_\lambda(S) = I_\lambda(0)e^{-\int_{S^*=0}^{S}\beta_\lambda(S^*)dS^*} \tag{10.124}$$

so along each path the local intensity consists only of the attenuated incident intensity such as $I_\lambda(0)$ being provided by a heated boundary.

Example 10.10

Radiant energy of 100 W leaves a frosted glass spherical light bulb 10 cm in diameter enclosed in a fixture having a flat glass plate as in Figure 10.13. If the glass is 2 cm thick and has a gray extinction coefficient of 0.05 cm^{-1}, find the intensity directly from a location on the bulb surface leaving the fixture at an angle of $\theta = 60°$ as shown in the figure. Neglect interface-interaction effects resulting from the difference in refractive index between the glass and surrounding air; these effects are included in Chapter 17.

Integrating Equation 10.124 over λ and S results in the total intensity $I(S,\theta) = I(0,\theta)e^{-(\kappa+\sigma_s)S(\theta)}$. To obtain $I(0,\theta)$, consider the bulb to be a diffuse sphere. The emissive power at the sphere surface is 100 W divided by the sphere area. The intensity is this diffuse emissive power divided by π, $I(0,\theta) = 100\,\text{W}/(\pi 10^2\,\text{cm}^2 \times \pi\,\text{sr}) = 0.101\,\text{W}/(\text{cm}^2\cdot\text{sr})$ Then, $I(S,\theta) = 0.101e^{-0.05(2/\cos 60°)} = 0.0827\,\text{W}/(\text{cm}^2\cdot\text{sr})$.

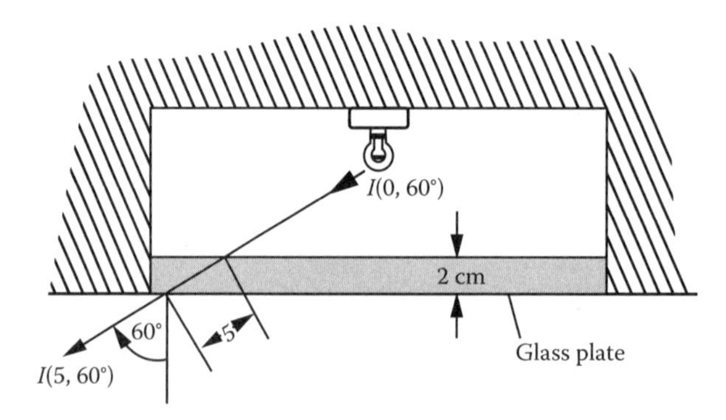

FIGURE 10.13 Directional intensity of radiation from light fixture (Example 10.10).

HOMEWORK

10.1 A layer of isothermal gray gas with T_g = 1000 K and with uniform properties is 1 m thick. The layer boundaries are perfectly transparent. Intensity $I(x = 0)$ is normally incident on the left boundary; there is no incident intensity from the environment on the boundary at x = 1. For each set of conditions shown in the following figures, what is the value of the intensity in the x-direction normal to the layer at x = 1 m?

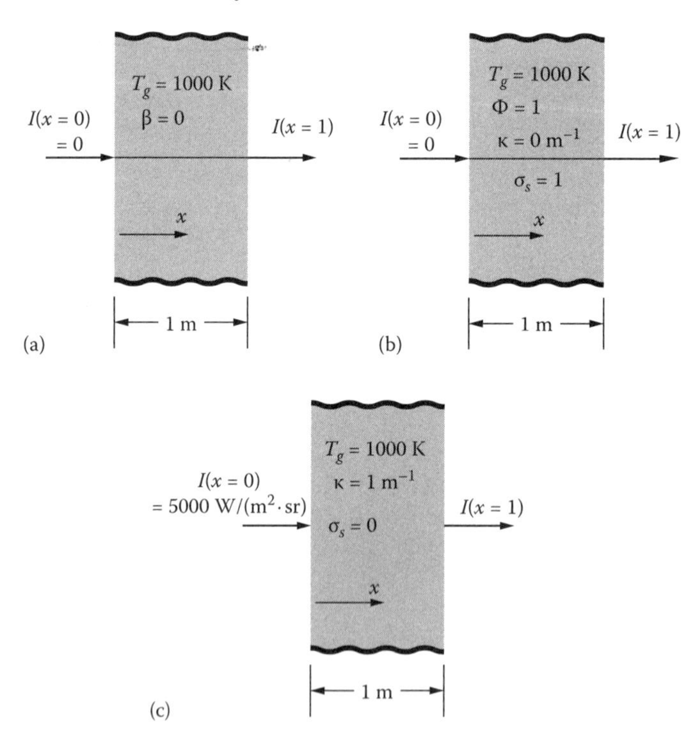

Answer: 0, 0, 13,249 W/(m$^2\cdot$sr)

10.2 A slab of nonscattering solid material has a gray absorption coefficient of κ = 0.2 cm^{-1} and refractive index $n \approx 1$. It is 5 cm thick and has an approximately linear temperature distribution within it as established by thermal conduction.

1. For the condition shown, what is the emitted intensity normal to the slab at $x = D$? What average slab temperature would give the same emitted normal intensity?

2. If the temperature profile is reversed (i.e., $T(x = 0) = 800$ K, $T(x = D) = 300$ K), what is the normal intensity emitted at $x = D$?

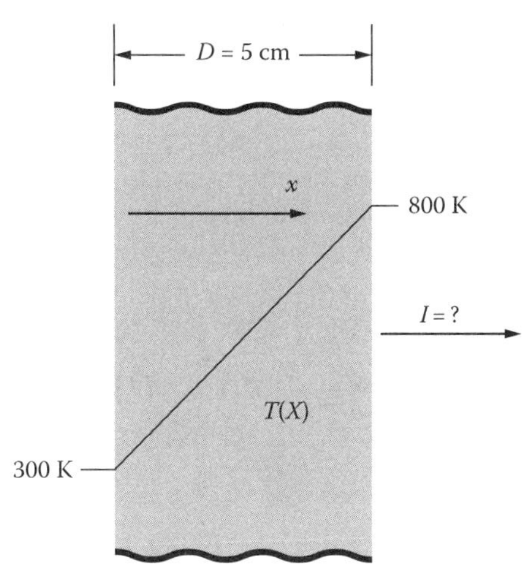

Answer: (a) 1849 W/(m$^2 \cdot$ sr); 634 K (b) 1149 W/(m$^2 \cdot$ sr)

10.3 A plane layer of semitransparent medium without scattering is at a uniform temperature of $T_m = 950$ K. The layer is 0.28 m thick. The medium has three absorption bands with constant absorption coefficients $\kappa_{\lambda,1} = 5.5$ m^{-1}, $\kappa_{\lambda,2} = 4.6$ m^{-1}, and $\kappa_{\lambda,3} = 3.8$ m^{-1} in the wavelength bands 1.3–3.1 μm, 3.65–5.05 μm, and 5.95–8.5 μm. For the remainder of the spectrum, the medium is perfectly transparent. One boundary of the layer is in contact with a black source at $T_w = 1030$ K. Calculate the intensities leaving the layer at the other boundary in the normal direction, 30° from the normal, and 60° from the normal.

Answer: $I(x = 0.28\text{ m}, \theta = 0°) = 10{,}458$ W/(m$^2 \cdot$ sr)

$I(x = 0.28\text{ m}, \theta = 30°) = 10{,}261$ W/(m$^2 \cdot$ sr)

$I(x = 0.28\text{ m}, \theta = 60°) = 9{,}699$ W/(m$^2 \cdot$ sr)

10.4 The radiation property of a gas is measured with the use of a blackbody radiation source at temperature 2000 K and an optical detector, as shown in the figure. The gas at temperature 300 K fills the 2 m spacing between the 1 cm diameter aperture of the radiation source and the 1.5 cm diameter detector surface. Scattering by the gas can be ignored. Derive an expression to relate the spectral radiation ($Q_{\lambda,\,abs}$) absorbed by the opaque detector with the spectral absorption coefficient (κ_λ) of the gas, the directional-hemispherical spectral reflectivity [$\rho_\lambda(\theta_i)$] of the detector surface, the angle θ_1 measured from the normal to the aperture, and the angle θ_2 measured from the normal to the detector surface.

Blackbody
θ_1
$S = 2$ m
θ_2
Detector

Answer: $dQ_\lambda = \left[1 - \rho_\lambda\left(\theta_2\right)\right] I_\lambda(S)\cos(\theta_1)dA_1 \dfrac{dA_2\cos(\theta_2)}{S^2}$

10.5 An isothermal enclosure is filled with a nonscattering absorbing and emitting medium. The medium and enclosure are at the same temperature. Show that $\nabla \cdot \mathbf{q}_r$ must be zero for this condition.

10.6 A large absorbing semitransparent medium without scattering has a single plane boundary. A black plate with constant temperature of $T = 1250$ K is suddenly placed in contact with that boundary. The plate radiates into the medium, which is cool enough that it does not radiate

significantly. Determine the radiative energy source $-\nabla \cdot \mathbf{q}_r$ at a location $x = 0.82$ m into the medium from the plane boundary if the absorption coefficient of the medium is $\kappa = 0.75$ m^{-1}.
Answer: 55,960 W/m^3

10.7 A semi-infinite medium is absorbing, emitting, and isotropically scattering. It is gray, has $n = 1$, and has absorption coefficient κ and scattering coefficient σ_s. The medium is initially at uniform temperature T_i. The transparent plane surface of the medium is suddenly subjected to radiative exchange with a large environment at a lower uniform temperature T_e. It is proposed to carry out a numerical solution to obtain the transient temperature distributions in the medium as it cools. Provide the energy and scattering equations in a convenient dimensionless form that are then to be placed in numerical form for solution. Heat conduction is included and the medium is stationary. The density ρ, specific heat c, and thermal conductivity k of the medium are assumed constant.

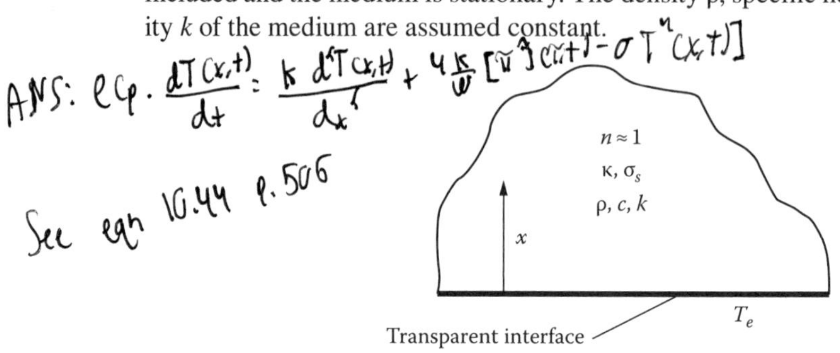

10.8 Pure carbon dioxide at 1 atm and 1955 K is contained between parallel plates 0.3 m apart. What is the radiative flux received at the plates as a result of radiation by the gas? (Use Leckner relations for CO_2 total emittance.)
Answer: 102 kW/m^2

10.9 A furnace at atmospheric pressure with interior in the shape of a cylinder with height equal to 2 times its diameter is filled with a 50:50 mixture by volume of CO_2 and N_2. The furnace volume is 0.689 m^3. The gas temperature is uniform at 1825 K and the walls are cooled. The interior surfaces are black. At what rate is energy being supplied to the gas (and removed from the walls) to maintain these conditions? Use Leckner correlations for gas properties.
Answer: 303 kW

10.10 A rectangular furnace of dimensions 0.6 × 0.5 × 2.2 m has soot-covered interior walls that can be considered black and cold. The furnace is filled with well-mixed combustion products at a temperature of 2220 K composed of 40% by volume CO_2, 30% by volume water vapor, and the remainder N_2. The total pressure is 1 atm. Compute the radiative flux and the total energy rate to the walls using Leckner correlations for the CO_2 and H_2O mixture.
Answer: 975 kW

10.11 Pure carbon dioxide at 1 atm and 2950 K is contained between parallel plates 0.3 m apart. What is the radiative flux received at the plates as a result of radiation by the gas? (Use Leckner correlations for CO_2.)
Answer: 555 kW/m^2

10.12 A pipe 15 cm in diameter is carrying superheated steam at 1.2 atm pressure and at a uniform temperature of 1300 K. What is the radiative flux from the steam received at the pipe wall?
Answer: 34.4 kW/m^2

10.13 A furnace at atmospheric pressure with interior in the shape of a cube having an edge dimension of 1.0 m is filled with a 50:50 mixture by volume of CO_2 and N_2. The gas temperature is uniform at 1825 K and the walls are cooled to 1100 K. The interior surfaces are black. At what rate is energy being supplied to the gas (and removed from the walls) to maintain these conditions? Use the method in Section 10.10.2.
Answer: 333 kW

10.14 Consider the same conditions and furnace volume as in Homework Problem 10.13. The furnace is now a cylinder with height equal to 2 times its diameter. What is the energy rate supplied to the gas?
Answer: 327 kW

10.15 A furnace being designed by a chemical company will be used to burn toxic waste composed of hydrocarbons. Complete elimination of the hydrocarbons with oxygen requires that the temperature of the combustion products in the furnace (60% CO_2, 40% H_2O by volume) be maintained at 1500 K. To prevent leakage of toxic waste or combustion products to the surroundings, the furnace interior is kept at 0.5 atm. The furnace is in the shape of a right circular cylinder of height equal to its diameter of 4 m. What average radiative flux is incident on the interior surface of the furnace? Comment on any assumptions used in obtaining the result.
Answer: 69.4 kW/m^2

10.16 Estimate the maximum radiative flux that is incident on any local area on the interior surface of the furnace described in Homework Problem 10.15.
Answer: 94.1 kW/m^2

10.17 A cubical black-walled enclosure with 2 m edges contains a mixture of gases at $T = 1500$ K and a pressure of 2 atm. The gas has volume fractions of 0.4/0.4/0.2 for $CO_2/H_2O/N_2$. Cooling water is passed over one face of the cube. If the water is initially at 25°C and has a maximum allowable temperature rise of 10°C, what mass flow rate of water (kg/s) is required? (Neglect reradiation from the wall, and assume all other walls are cool.)
Answer: 19.2 kg/s

10.18 A large furnace has within it a large number of parallel tubes arranged in an equilateral triangular array. The tubes are of 2.5 cm outside diameter, and the tube centers are spaced 7.5 cm apart. The furnace gas is composed of 75% CO_2 and 25% H_2O by volume, and the combustion process in the gas maintains the gas temperature at 1300 K and a total pressure of 0.9 atm. What is the radiative flux incident on a tube in the interior of the bundle (i.e., completely surrounded by other tubes) per meter of tube length?
Answer: 2.23 kW/m

10.19 Evaluate the geometric-mean transmittance $\bar{\tau}_{d1-2}$ from the element dA_1 to the area A_2 in Homework Problem 5.1. The region between the areas is filled with a gray medium at uniform temperature having an absorption coefficient $\kappa = 0.035$ cm^{-1}.
Answer: 0.0234

10.20 A thin black plate 1 × 1 cm is at the center of a sphere of CO_2–air mixture at a uniform temperature of 2000 K and 1 atm total pressure. The partial pressure of the CO_2 is 0.8 atm, and the sphere diameter is 1 m. How much energy is absorbed by the plate? What will the plate temperature be? (Assume the boundary of the sphere is black and kept cool so that it does not enter into the radiative exchange.)

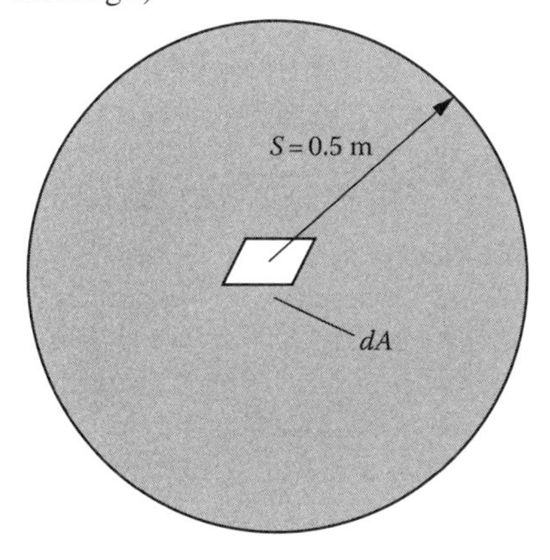

Answer: 19.6 W; 1147 K.

10.21 A spherical cavity is filled with an isothermal gray medium having an absorption coefficient κ. Set up the relations needed to determine the geometric-mean transmittance $\bar{t}_{1-d2}$ from the cavity surface A_1 to the area element dA_2 at the center of the opening A_2. (Note that a similar situation is considered by Koh [1965].)

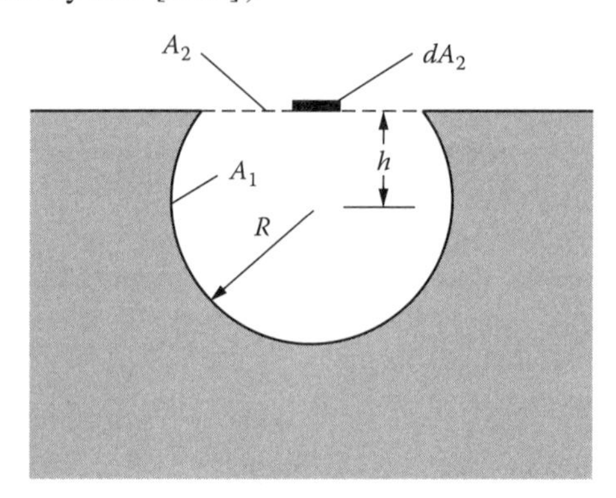

10.22 A sphere of gray gas at uniform temperature is situated above a surface with the region between the sphere and surface being nonabsorbing. Derive a relation for the radiative energy incident on the circular area as shown. (*Hint*: Consider the circular area as being a cut through a concentric sphere surrounding the gas sphere, and make use of the symmetry of the geometry.)

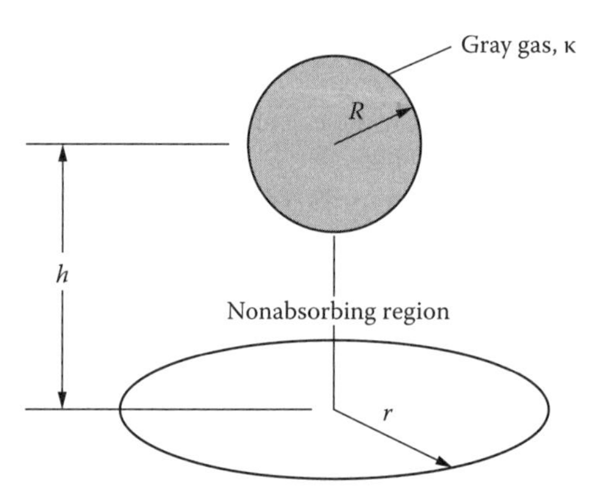

Answer: $Q_i = 2\pi R^2 \sigma T_g^4 \left(1 - \dfrac{h}{\sqrt{h^2+r^2}}\right)\left\{1 - \dfrac{2}{(2\kappa R)^2}\left[1-(2\kappa R+1)e^{-2\kappa R}\right]\right\}$

10.23 For the radiating sphere of gas in Homework Problem 10.22, derive an expression for the local energy flux incident along the plane surface as a function of distance r.

Answer: $q_i(r) = \dfrac{dQ_i}{dr} = 4\pi R^2 \sigma T_g^4 \dfrac{1}{2}\dfrac{hr}{(h^2+r^2)^{3/2}}\left\{1 - \dfrac{2}{(2\kappa R)^2}[1-(2\kappa R+1)e^{-2\kappa R}]\right\}.$

10.24 Two infinite diffuse-gray parallel plates at temperatures T_1 and T_2, with respective emissivities ε_1 and ε_2, are separated by a distance D. The space between them is filled with a gray-absorbing and gray-scattering medium having a constant extinction coefficient β. Obtain expressions for the net radiative heat flux being transferred between the plates and the temperature distribution in the medium, by using the nearly transparent approximation. Heat conduction can be neglected compared with radiative transfer.

Answer: $q = \sigma\left(T_1^4 - T_2^4\right) \Big/ \left(\dfrac{1}{\varepsilon_1} + \dfrac{1}{\varepsilon_2} - 1\right); \; \dfrac{T^4 - T_2^4}{T_1^4 - T_2^4} = \left(\dfrac{1}{\varepsilon_2} - \dfrac{1}{2}\right) \Big/ \left(\dfrac{1}{\varepsilon_1} + \dfrac{1}{\varepsilon_2} - 1\right)$

10.25 A gray isotropically scattering gas is contained between large diffuse-gray parallel plates. The plates both have emissivity $\varepsilon = 0.30$. Plate 1 is maintained at temperature $T_1 = 1150$ K, and plate 2 is at $T_2 = 525$ K. The medium between the plates has a uniform extinction coefficient of $\beta = 0.75\ \text{m}^{-1}$. The plate geometry is shown in the succeeding text.

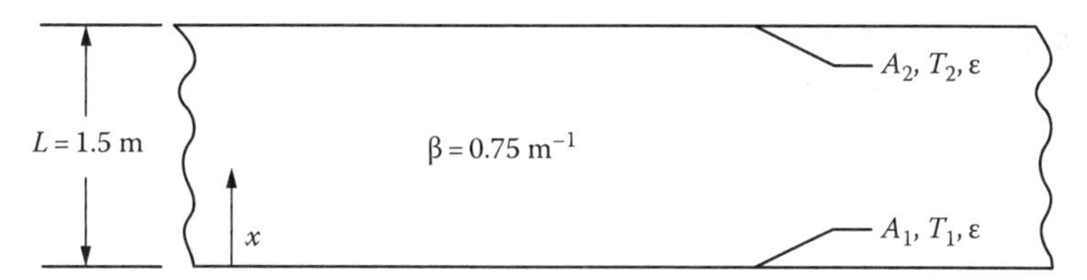

Predict the net radiative heat flux transferred between the surfaces (W/m^2) and plot the temperature distribution $[T^4(k) - T_2^4] - (T_1^4 - T_2^4)$ in the gas, where $\tau = \beta x$. Heat conduction is small. Solve the problem using the nearly transparent optically thin approximation.
Answer: $q = 16{,}740$ W/m^2: $\theta(\kappa) = 0.500$ (constant)

10.26 Two gases have constant extinction coefficients β_1 and β_2 that are both much smaller than 1. The gases are separated by a thin metal barrier with low emissivity ε_s. The gases are in the form of two layers bounded by large parallel diffuse walls as shown. Radiation exchange is large compared with heat conduction that can be neglected. Using the nearly transparent approximation, obtain relations for the temperature distributions within the gases and the net radiative energy transfer from wall 1 to wall 2. Evaluate the results for $\varepsilon_1 = 0.78$, $\varepsilon_s = 0.04$, $\varepsilon_2 = 0.88$, $T_1 = 980$ K, and $T_2 = 765$ K. Determine the temperature jump across the thin metal barrier and determine the size of the jump as a function of ε_s.

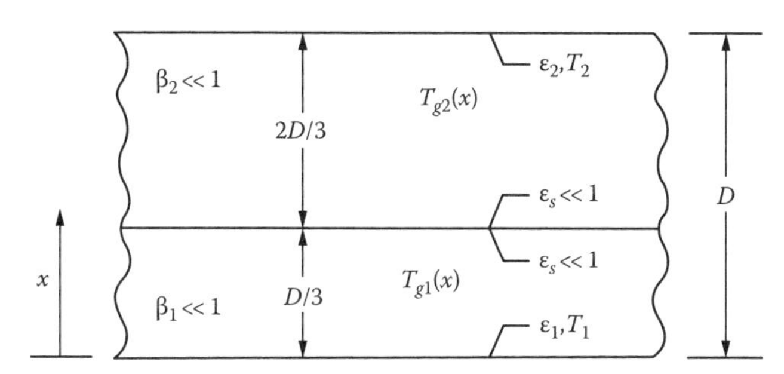

Answer: $q = 652.2$ W/m^2, $T_{g1} = 977.6$ K, $T_{g2} = 769.1$ K, $\Delta T_{jump} = 208.5$ K.

10.27 A sphere of high-temperature optically thin gray gas of fixed volume is being cooled by radiative loss to cool black surroundings (neglect emission from the surroundings). At any instant the entire gas may be considered isothermal and the emission approximation used to compute the radiative loss. Heat conduction is neglected. Write the transient energy equation and solve to obtain the gas temperature as a function of time, starting from an initial temperature T_i.

Answer: $T = \left(\dfrac{12\sigma\kappa}{\rho c_v} t + \dfrac{1}{T_i^3} \right)^{-1/3}$.

10.28 The sphere in Homework Problem 10.27 now consists of a gas that radiates spectrally in three wavelength bands. The sphere is 0.6 m in diameter and has an initial uniform temperature of $T_i = 1150$ K. The three absorption bands have absorption coefficients $\kappa_{\lambda 1} = 1.1\ \text{m}^{-1}$, $\kappa_{\lambda 2} = 1.6\ \text{m}^{-1}$, and $\kappa_{\lambda 3} = 0.9\ \text{m}^{-1}$ in the wavelength bands from $\lambda = 0.9$–2.0 μm, 2.9–4.0 μm, and 6.0–9.0 μm. In the other portions of the spectrum, the gas is transparent. Write the transient energy equation and calculate the transient gas temperature for cooling of the sphere in cool black surroundings. Assume that the emission approximation can be used.

10.29 A spherical cavity 15 cm in diameter is filled with a gray medium having an absorption coefficient of 0.1 cm^{-1}. The cavity surface is black and is at a uniform temperature of 550 K. When the medium is first placed in the cavity, the medium is cold. For this condition, use the cold medium approximation to estimate the heat flux radiated from the small opening as shown.

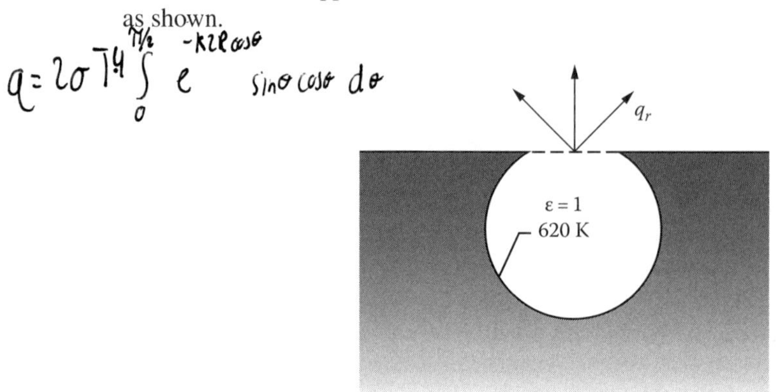

Answer: 2039 W/m^2

10.30 An optically thin gray gas with constant absorption coefficient κ is contained in a long transparent cylinder of diameter D. The surrounding environment is at low temperature that can be considered zero. Initially, the cylinder is at the environment temperature. Then, an electrical discharge is passed through the gas, continuously producing in the gas a uniform energy source $\dot{q}$ per unit volume and time. Derive a relation for the transient gas temperature variation if radiation is assumed to be the only significant mode of heat transfer. What is the maximum temperature T_{max} that the gas will achieve?

Answer: $\dfrac{8\kappa\sigma T_{\max}^3}{\rho c_v} t = \dfrac{1}{2} \ln \dfrac{1+\theta}{1-\theta} + \tan^{-1}\theta;\quad T_{\max} = \left(\dfrac{\dot{q}}{4\kappa\sigma}\right)^{1/4};\quad \theta = \dfrac{T}{T_{\max}}$

11 Radiative Transfer in Plane Layers and Multidimensional Geometries

11.1 INTRODUCTION

The energy equation was introduced in Chapter 10 (Equations 10.1 through 10.3) along with relations for the associated radiative flux divergence (Equation 10.4) and the radiative transfer equation (Equations 10.6 through 10.10). In this chapter, these relations are applied to the cases of radiative transfer in plane layers and multidimensional geometries, with and without internal generation. Solution techniques are covered in Chapter 12, and incorporation of additional energy transfer modes (conduction, convection) is considered in Chapter 13.

11.2 RADIATIVE INTENSITY, FLUX, FLUX DIVERGENCE, AND SOURCE FUNCTION IN A PLANE LAYER

To evaluate the influence of some of the many variables for heat transfer in a radiating gas or other translucent medium, it is helpful to use a simple geometry such as a plane layer of large extent relative to its thickness and with uniform conditions along each boundary. There is considerable literature for this one-dimensional (1D) geometry for engineering applications, atmospheric physics, and astrophysics since the Earth's atmosphere and the outer radiating layers of the sun can be approximated as plane layers (Kourganoff 1963, Chandrasekhar 1960, Goody and Yung 1989, Bohren and Clothiaux 2006, Petty 2006).

11.2.1 RADIATIVE TRANSFER EQUATION AND RADIATIVE INTENSITY FOR A PLANE LAYER

A plane layer between two boundaries is shown in Figure 11.1. It is a 1 D system, so the temperature and properties of the medium vary only in the x direction. The boundaries of the plane layer form a two-surface enclosure, and referring to Figures 10.1 and 10.2, all paths S and S_i for the radiation intensity originate at the lower or upper surface. As shown in Figure 11.1, the path direction is given by the angle θ measured from the positive x direction. The superscript notation, + or −, indicates, respectively, the directions with positive or negative cos θ, so that I_λ^+ corresponds to $0 \le \theta \le \pi/2$ and I_λ^- to $\pi/2 \le \theta \le \pi$.

The optical depth $\tau_\lambda(x)$ within the layer is defined *along the x coordinate across the thickness* as

$$\tau_\lambda(x) = \int_{x^*=0}^{x} \left[\kappa_\lambda(x^*) + \sigma_{s,\lambda}(x^*)\right] dx^* = \int_{x^*=0}^{x} \beta_\lambda(x^*)\, dx^* \tag{11.1}$$

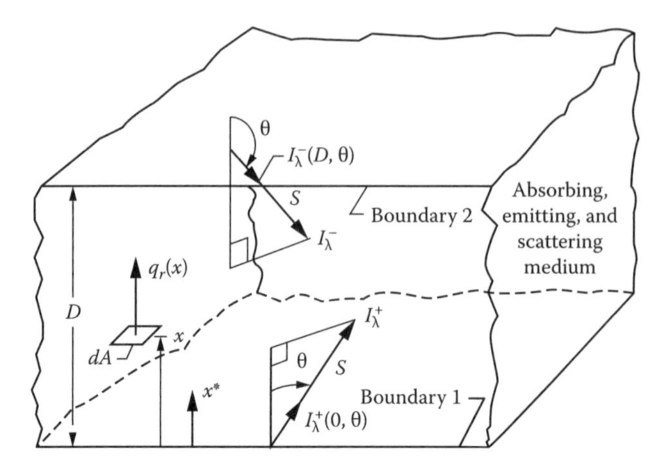

FIGURE 11.1 Plane layer between infinite parallel boundaries.

which accounts for variable properties, $\beta_\lambda(x)$. For constant properties, $\tau_\lambda(x) = \beta_\lambda x$. For the positive x directions, the relation between optical positions along the S and x directions is

$$\tau_\lambda(S) = \int_{S^*=0}^{S} \beta_\lambda(S^*)dS^* = \frac{1}{\cos\theta}\int_{x^*=0}^{x} \beta_\lambda(x^*)dx^* = \frac{\tau_\lambda(x)}{\cos\theta} \tag{11.2}$$

For the minus direction, Equation 11.2 still applies because $dS = -dx/\cos(\pi - \theta) = dx/\cos\theta$. With $d\tau_\lambda(S) = d\tau_\lambda(x)/\cos\theta$, the radiative transfer Equation 10.12 becomes, for I_λ^+ and I_λ^-,

$$\cos\theta\frac{\partial I_\lambda^-}{\partial \tau_\lambda(x)} + I_\lambda^-\left[\tau_\lambda(x),\theta\right] = \hat{I}_\lambda\left[\tau_\lambda(x),\theta\right] \quad \pi/2 \le \theta \le \pi \tag{11.3}$$

$$\cos\theta\frac{\partial I_\lambda^+}{\partial \tau_\lambda(x)} + I_\lambda^+\left[\tau_\lambda(x),\theta\right] = \hat{I}_\lambda\left[\tau_\lambda(x),\theta\right] \quad 0 \le \theta \le \pi/2 \tag{11.4}$$

The source function $\hat{I}_\lambda$, which consists of emission and in-scattering terms, depends on θ because of the angular dependence of the phase function Φ for anisotropic scattering, as in Equation 10.11. The partial derivatives in Equations 11.3 and 11.4 emphasize that I_λ^+ and I_λ^- depend on both $\tau_\lambda(x)$ and θ. It is convenient to let $\mu \equiv \cos\theta$; then Equations 11.3 and 11.4 become

$$\mu\frac{\partial I_\lambda^+}{\partial \tau_\lambda(x)} + I_\lambda^+\left[\tau_\lambda(x),\mu\right] = \hat{I}_\lambda\left[\tau_\lambda(x),\mu\right] \quad 1 \ge \mu \ge 0 \tag{11.5}$$

$$\mu\frac{\partial I_\lambda^-}{\partial \tau_\lambda(x)} + I_\lambda^-\left[\tau_\lambda(x),\mu\right] = \hat{I}_\lambda\left[\tau_\lambda(x),\mu\right] \quad 0 \ge \mu \ge -1 \tag{11.6}$$

Using an integrating factor as in Equation 10.13, Equations 11.5 and 11.6 are integrated subject to the following boundary conditions that need to have their values specified:

$$I_\lambda^+\left[\tau_\lambda,\mu\right] = I_\lambda^+\left[0,\mu\right] \quad \text{at } \tau_\lambda = 0 \tag{11.7}$$

$$I_\lambda^-\left[\tau_\lambda,\mu\right] = I_\lambda^-\left[\tau_{\lambda,D},\mu\right] \quad \text{at } \tau_\lambda = \tau_{\lambda,D} \tag{11.8}$$

where $\tau_{\lambda,D} = \int_{x^*=0}^{x} \beta_\lambda(x)dx = \int_{x=0}^{D} [\kappa_\lambda(x) + \sigma_{s,\lambda}(x)]dx$ is the layer optical thickness. The integration gives the intensities as a function of location and angle in the plane layer as

$$I_\lambda^+(\tau_\lambda,\mu) = I_\lambda^+(0,\mu)e^{-\tau_\lambda/\mu} + \frac{1}{\mu}\int_{\tau_\lambda^*=0}^{\tau_\lambda} \hat{I}_\lambda(\tau_\lambda^*,\mu)e^{-(\tau_\lambda-\tau_\lambda^*)/\mu}d\tau_\lambda^*; \quad 1 \ge \mu \ge 0 \tag{11.9}$$

$$I_\lambda^-(\tau_\lambda,\mu) = I_\lambda^-(\tau_{\lambda,D},\mu)e^{-\tau_\lambda/-\mu} - \frac{1}{\mu}\int_{\tau_\lambda^*=0}^{\tau_{\lambda,D}} \hat{I}_\lambda(\tau_\lambda^*,\mu)e^{-(\tau_\lambda^*-\tau_\lambda)/-\mu}d\tau_\lambda^*; \quad 0 \ge \mu \ge -1 \tag{11.10}$$

In Equation 11.10 the θ is between $\pi/2$ and π so that $\mu = \cos\theta$ is negative.

The use of the source function $\hat{I}_\lambda$ in Equations 11.9 and 11.10 applies for both scattering and nonscattering media; however, without scattering, Equation 10.45 gives $\hat{I}_\lambda(\tau_\lambda,\mu) = I_{\lambda b}(\tau_\lambda)$, which is isotropic, so Equations 11.9 and 11.10 simplify for a medium with absorption only to give

$$I_\lambda^+(\tau_\lambda,\mu) = I_\lambda^+(0,\mu)e^{-\tau_\lambda/\mu} + \frac{1}{\mu}\int_{\tau_\lambda^*=0}^{\tau_\lambda} I_{\lambda b}(\tau_\lambda^*,\mu)e^{-(\tau_\lambda-\tau_\lambda^*)/\mu}d\tau_\lambda^*; \quad 1 \ge \mu \ge 0 \tag{11.11}$$

$$I_\lambda^-(\tau_\lambda,\mu) = I_\lambda^-(\tau_{\lambda,D},\mu)e^{-\tau_\lambda/-\mu} - \frac{1}{\mu}\int_{\tau_\lambda^*=0}^{\tau_{\lambda,D}} I_{\lambda b}(\tau_\lambda^*,\mu)e^{-(\tau_\lambda^*-\tau_\lambda)/-\mu}d\tau_\lambda^*; \quad 0 \ge \mu \ge -1 \tag{11.12}$$

11.2.2 Local Radiative Flux in a Plane Layer

At each x location the total radiative energy flux in the positive x direction is the integral of the spectral flux over all λ. With the notation that $q_{r\lambda}\,d\lambda$ is the spectral flux in $d\lambda$,

$$q_r(x) = \int_{\lambda=0}^{\infty} q_{r\lambda}(x)d\lambda = \int_{\lambda=0}^{\infty} q_{r\lambda}(\tau_\lambda)d\lambda \tag{11.13}$$

where $\tau_\lambda(x)$ is the optical depth at x. For a fixed x, the optical coordinate τ_λ depends on λ since β_λ is a function of λ.

The net spectral flux in the positive x direction crossing dA in the plane at x in Figure 11.1 is obtained in two parts, one from I_λ^+ and one from I_λ^-. Since intensity is energy per unit solid angle crossing an area normal to the direction of I, the projection of dA must be considered normal to either I_λ^+ or I_λ^-. The spectral energy flux in the positive x-direction from I_λ^+ is then (using $d\Omega = 2\pi \sin\theta\, d\theta$ and $\mu = \cos\theta$)

$$q_{r\lambda}^+(\tau_\lambda)d\lambda = 2\pi d\lambda \int_{\theta=0}^{\pi/2} I_\lambda^+(\tau_\lambda,\theta)\cos\theta\sin\theta d\theta = 2\pi d\lambda \int_{\mu=0}^{1} I_\lambda^+(\tau_\lambda,\mu)\mu d\mu \tag{11.14}$$

which agrees with the z component in Equation 10.29. The $q^+_{r\lambda}(\tau_\lambda)$ is equivalent to the local spectral radiosity in the positive x-direction, $J^+_\lambda(\tau_\lambda)$. The total flux from I^+_λ in the positive x-direction is

$$q^+_r(x) = J^+(x) = \int_{\lambda=0}^{\infty} q^+_{r\lambda}(\tau_\lambda)\,d\lambda \tag{11.15}$$

Similarly, in the negative x-direction,

$$q^-_{r\lambda}(\tau_\lambda)\,d\lambda = J^-_\lambda(\tau_\lambda)\,d\lambda = 2\pi d\lambda \int_{\pi-\theta=0}^{\pi/2} I^-_\lambda(\tau_\lambda,\theta)\cos(\pi-\theta)\sin(\pi-\theta)\,d(\pi-\theta)$$

$$= -2\pi d\lambda \int_{\theta=\pi/2}^{\pi} I^-_\lambda(\tau_\lambda,\theta)\cos\theta\sin\theta\,d\theta = 2\pi d\lambda \int_{\mu=0}^{1} I^-_\lambda(\tau_\lambda,-\mu)\,\mu\,d\mu \tag{11.16}$$

The net spectral flux in the positive x-direction is then

$$q_{r\lambda}(\tau_\lambda)\,d\lambda = \left[q^+_{r\lambda}(\tau_\lambda) - q^-_{r\lambda}(\tau_\lambda)\right]d\lambda = \left[J^+_\lambda - J^-_\lambda\right]d\lambda$$

$$= 2\pi d\lambda \int_{\mu=0}^{1} \left[I^+_\lambda(\tau_\lambda,\mu) - I^-_\lambda(\tau_\lambda,\mu)\right]\mu\,d\mu \tag{11.17}$$

The intensities can be substituted from Equations 11.9 and 11.10 or from Equations 11.11 and 11.12 when there is no scattering. Using Equations 11.9 and 11.10 that include scattering gives

$$q_{r\lambda}(\tau_\lambda)\,d\lambda = 2\pi\left[\int_{\mu=0}^{1} I^+_\lambda(0,\mu)e^{-\tau_\lambda/\mu}\mu\,d\mu - \int_{\mu=0}^{1} I^-_\lambda(\tau_{\lambda,D},-\mu)e^{-(\tau_{\lambda,D}-\tau_\lambda)/\mu}\mu\,d\mu\right.$$

$$+\int_{\mu=0}^{1}\int_{\tau^*_\lambda=0}^{\tau_\lambda} \hat{I}_\lambda(\tau^*_\lambda,\mu)e^{-(\tau_\lambda-\tau^*_\lambda)/\mu}\,d\tau^*_\lambda\,d\mu$$

$$\left.+\int_{\mu=0}^{1}\int_{\tau^*_\lambda=\tau_\lambda}^{\tau_{\lambda,D}} \hat{I}_\lambda(\tau^*_\lambda,-\mu)e^{-(\tau^*_\lambda-\tau_\lambda)/\mu}\,d\tau^*_\lambda\,d\mu\right]d\lambda \tag{11.18}$$

The total net radiative flux at x is found by integrating Equation 11.18 from $\lambda = 0$ to $\lambda = \infty$ using the τ_λ at each λ corresponding to the fixed x.

11.2.3 Divergence of the Radiative Flux: Radiative Energy Source

The radiative flux divergence $\nabla\cdot\mathbf{q}_r(x)$ is needed for the energy Equation 10.1. For a plane layer with uniform conditions over each boundary, the $\mathbf{q}_r(x)$ depends only on x so that

$$\nabla\cdot\mathbf{q}_r(x) = \frac{dq_r(x)}{dx} = \frac{d}{dx}\int_{\lambda=0}^{\infty} q_{r\lambda}(x)\,d\lambda = \int_{\lambda=0}^{\infty}\frac{dq_{r\lambda}(x)}{dx}\,d\lambda = \int_{\lambda=0}^{\infty}\beta_\lambda(\tau_\lambda)\frac{dq_{r\lambda}(\tau_\lambda)}{d\tau_\lambda}\,d\lambda \tag{11.19}$$

where $d\tau_\lambda = \beta_\lambda(x)\,dx$ and, throughout the integration over λ, the τ_λ corresponds to the specified x. Differentiating Equation 11.18 with respect to τ_λ yields the quantity in the integral of Equation 11.19 as

$$\frac{dq_{r\lambda}(x)}{dx}d\lambda = \beta_\lambda(\tau_\lambda)\frac{dq_{r\lambda}(\tau_\lambda)}{d\tau_\lambda}$$

$$= -2\pi\beta_\lambda(\tau_\lambda)\left[\int_{\mu=0}^{1} I_\lambda^+(0,\mu)e^{-\tau_\lambda/\mu}d\mu + \int_{\mu=0}^{1} I_\lambda^-(\tau_{\lambda,D},-\mu)e^{-(\tau_{\lambda,D}-\tau_\lambda)/\mu}d\mu\right.$$

$$+\int_{\mu=0}^{1}\frac{1}{\mu}\int_{\tau_\lambda^*=0}^{\tau_\lambda}\hat{I}_\lambda(\tau_\lambda^*,\mu)e^{-(\tau_\lambda-\tau_\lambda^*)/\mu}d\tau_\lambda^*\,d\mu - \int_{\mu=0}^{1}\hat{I}_\lambda(\tau_\lambda,\mu)d\mu$$

$$\left.+\int_{\mu=0}^{1}\frac{1}{\mu}\int_{\tau_\lambda^*=\tau_\lambda}^{\tau_{\lambda,D}}\hat{I}_\lambda(\tau_\lambda^*,-\mu)e^{-(\tau_\lambda^*-\tau_\lambda)/\mu}d\tau_\lambda^*\,d\mu - \int_{\mu=0}^{1}\hat{I}_\lambda(\tau_\lambda,-\mu)d\mu\right] \quad (11.20)$$

Another form is to use Equation 10.42 to obtain

$$-\frac{dq_r(x)}{dx} = 4\pi\int_{\lambda=0}^{\infty}\kappa_\lambda(x)\left[\bar{I}_\lambda(x) - I_{\lambda b}(x)\right]d\lambda \quad (11.21)$$

which applies with or without scattering. The forms in Equations 11.19 and 11.20 or in Equation 11.21 are equivalent because the $\bar{I}_\lambda(\lambda,x)$ in Equation 11.21 must be found from Equations 11.9 and 11.10 by using $\bar{I}_\lambda(x) = \frac{1}{2}\int_0^1\left[I_\lambda^+(x,\mu) + I_\lambda^-(x,-\mu)\right]d\mu$.

In the limit of *absorption only* (no scattering), $\hat{I}_\lambda(\tau_\lambda,\mu) \to I_{\lambda b}(\tau_\lambda)$ and Equations 11.11 and 11.12 apply for I_λ^+ and I_λ^-. In the limit of *scattering only* (no absorption), Equation 10.43 gives $dq_r(x)/dx = 0$. This is valid for any type of scattering. This means that radiation is decoupled from other modes, that is, conduction and convection heat transfer.

11.2.4 Equation for the Source Function in a Plane Layer

The equation for the source function is obtained from Equation 10.20. The substitutions are made that $\tau_\lambda(S) = \beta_\lambda(x)/\cos\theta$ and $d\Omega_i = 2\pi\sin\theta_i d\theta_i$. Then, following the procedure in the derivation of the previous plane layer relations, the source-function equation becomes (where $\omega_\lambda = \omega_\lambda(\tau_\lambda)$)

$$\hat{I}_\lambda(\tau_\lambda,\Omega) = (1-\omega_\lambda)I_{\lambda b}(\tau_\lambda) + \frac{\omega_\lambda}{2}\left[\int_{\mu_i=0}^{1} I_\lambda^+(0,\mu_i)e^{-\tau_\lambda/\mu_i}\Phi_\lambda(\mu,\mu_i)d\mu_i\right.$$

$$+\int_{\mu_i=0}^{1} I_\lambda^+(\tau_{\lambda,D},\mu_i)e^{-(\tau_{\lambda,D}-\tau_\lambda)/\mu_i}\Phi_\lambda(\mu,-\mu_i)d\mu_i$$

$$+\int_{\mu_i=0}^{1}\frac{1}{\mu_i}\int_{\tau_\lambda^*=0}^{\tau_\lambda}\hat{I}_\lambda(\tau_\lambda^*,\mu_i)e^{-(\tau_\lambda-\tau_\lambda^*)/\mu_i}d\tau_\lambda^*\,\Phi(\mu,\mu_i)d\mu_i$$

$$\left.+\int_{\mu_i=0}^{1}\frac{1}{\mu_i}\int_{\tau_\lambda^*=\tau_\lambda}^{\tau_{\lambda,D}}\hat{I}_\lambda(\tau_\lambda^*,-\mu_i)e^{-(\tau_\lambda^*-\tau_\lambda)/\mu_i}d\tau_\lambda^*\Phi(\mu,-\mu_i)d\mu_i\right] \quad (11.22)$$

11.2.5 Relations for Isotropic Scattering

Isotropic scattering is an assumption to provide a substantial reduction in the complexity of the radiative transfer relations. For isotropic scattering the source function is given by Equation 10.21 as

$$\hat{I}_\lambda(\tau_\lambda) = (1-\omega_\lambda) I_{\lambda b}(\tau_\lambda) + \frac{\omega_\lambda}{4\pi} \int_{\Omega_i=0}^{4\pi} I_\lambda(\tau_\lambda, \Omega_i)\, d\Omega_i \tag{11.23}$$

where $\omega_\lambda = \omega_\lambda(\tau_\lambda)$. The source function $\hat{I}_\lambda(\tau_\lambda)$ consists of isotropic scattering and isotropic emission, so it is independent of direction.

For isotropic scattering the spectral radiative flux Equation 11.18 reduces to

$$\begin{aligned} q_{r\lambda}(\tau_\lambda)\, d\lambda = 2\pi \Bigg[& \int_{\mu=0}^{1} I_\lambda^+(0,\mu) e^{-\tau_\lambda/\mu} \mu\, d\mu - \int_{\mu=0}^{1} I_\lambda^-(\tau_{\lambda,D}, -\mu) e^{-(\tau_{\lambda,D}-\tau_\lambda)/\mu} \mu\, d\mu \\ & + \int_{\tau_\lambda^*=0}^{\tau_\lambda} \hat{I}_\lambda(\tau_\lambda^*) \int_{\mu=0}^{1} e^{-(\tau_\lambda - \tau_\lambda^*)/\mu}\, d\mu\, d\tau_\lambda^* \\ & + \int_{\tau_\lambda^*=\tau_\lambda}^{\tau_{\lambda,D}} \hat{I}_\lambda(\tau_\lambda^*) \int_{\mu=0}^{1} e^{-(\tau_\lambda^* - \tau_\lambda)/\mu}\, d\mu\, d\tau_\lambda^* \Bigg] d\lambda \end{aligned} \tag{11.24}$$

This relation and others that follow are written much more conveniently by using the *exponential integral function*

$$E_n(\xi) \equiv \int_0^1 \mu^{n-2} e^{-\xi/\mu}\, d\mu \tag{11.25}$$

for which numerical values are available in handbooks and in computational software packages. The spectral radiative flux for isotropic scattering then becomes

$$\begin{aligned} q_{r\lambda}(\tau_\lambda)\, d\lambda = 2\pi \Bigg[& \int_{\mu=0}^{1} I_\lambda^+(0,\mu) e^{-\tau_\lambda/\mu} \mu\, d\mu - \int_{\mu=0}^{1} I_\lambda^-(\tau_{\lambda,D}, -\mu) e^{-(\tau_{\lambda,D}-\tau_\lambda)/\mu} \mu\, d\mu \\ & + \int_{\tau_\lambda^*=0}^{\tau_\lambda} \hat{I}_\lambda(\tau_\lambda^*) E_2(\tau_\lambda - \tau_\lambda^*)\, d\tau_\lambda^* \int_{\tau_\lambda^*=\tau_\lambda}^{-\tau_{\lambda,D}} \hat{I}_\lambda(\tau_\lambda^*) E_2(\tau_\lambda^* - \tau_\lambda) \Bigg] d\lambda \end{aligned} \tag{11.26}$$

The $E_n(\xi)$ functions are discussed in detail by Kourganoff (1963) and Chandrasekhar (1960), and the important relations for radiative transfer are in Appendix D.

Similarly, by use of $E_n(\xi)$, Equation 11.20 for the flux divergence for isotropic scattering (the negative of the radiative heat source) reduces to

$$\frac{dq_{r\lambda}(\tau_\lambda)}{dx} = -2\pi\beta_\lambda(\tau_\lambda)\left[\int_{\mu=0}^{1} I_\lambda^+(0,\mu)e^{-\tau_\lambda/\mu}\,d\mu + \int_{\mu=0}^{1} I_\lambda^-(\tau_{\lambda,D},-\mu)e^{-(\tau_{\lambda,D}-\tau_\lambda)/\mu}\,d\mu\right.$$
$$\left. + \int_{\tau_\lambda^*=0}^{\tau_{\lambda,D}} \hat{I}_\lambda(\tau_\lambda^*)E_1\left(\left|\tau_\lambda^* - \tau_\lambda\right|\right)d\tau_\lambda^*\right] + 4\pi\beta_\lambda(\tau_\lambda)\hat{I}_\lambda(\tau_\lambda) \tag{11.27}$$

For isotropic scattering the source-function Equation 11.21 reduces to

$$\hat{I}_\lambda(\tau_\lambda) = (1-\omega_\lambda)I_{\lambda b}(\tau_\lambda) + \frac{\omega_\lambda}{2}\left[\int_{\mu=0}^{1} I_\lambda^+(0,\mu)e^{-\tau_\lambda/\mu}d\mu + \int_{\mu=0}^{1} I_\lambda^-(\tau_{\lambda,D},-\mu)e^{-(\tau_{\lambda,D}-\tau_\lambda)/\mu}d\mu\right.$$
$$\left. + \int_{\tau_\lambda^*=0}^{\tau_\lambda} \hat{I}_\lambda(\tau_\lambda^*)E_1\left(\left|\tau_\lambda^* - \tau_\lambda\right|\right)d\tau_\lambda^*\right] \tag{11.28}$$

The expressions in square brackets are the same in Equations 11.27 and 11.28. The two equations are combined to obtain, as in Equation 10.39,

$$-\frac{dq_{r\lambda}}{dx} = 4\pi\frac{\kappa_\lambda(\tau_\lambda)}{\omega_\lambda(\tau_\lambda)}\left[\hat{I}(\tau_\lambda) - I_{\lambda b}(\tau_\lambda)\right] \tag{11.29}$$

Equation 11.21, which is in terms of $\bar{I}_\lambda$, can also be used. This applies with or without scattering. For absorption only, $\omega_\lambda \to 0$ and $\hat{I}_\lambda \to I_{\lambda b}$, so Equation 11.29 becomes singular. Equation 11.21 can then be used so that

$$-\frac{dq_r(x)}{dx} = 4\pi\int_{\lambda=0}^{\infty}\kappa_\lambda(x)\left\{\frac{1}{2}\int_{\mu=0}^{1}\left[I_\lambda^+(x,\mu) + I_\lambda^-(x,-\mu)\right]d\mu - I_{\lambda b}(x)\right\}d\lambda \tag{11.30}$$

11.2.6 Diffuse Boundary Fluxes for a Plane Layer with Isotropic Scattering

An absorbing and emitting gas would usually be confined by an enclosure of solid boundaries. If a plane layer has opaque boundaries that can be assumed diffuse, the intensities leaving the boundaries $I_\lambda^+(0,\mu)$ and $I_\lambda^-(\tau_{\lambda,D},-\mu)$ do not depend on angle (are independent of μ) and can be expressed in terms of diffuse outgoing fluxes

$$I_\lambda^+(0,\mu) = I_\lambda^+(0) = \frac{J_{\lambda,1}}{\pi}; \quad I_\lambda^-(\tau_{\lambda,D},-\mu) = I_\lambda^-(\tau_{\lambda,D}) = \frac{J_{\lambda,2}}{\pi} \tag{11.31}$$

where 1 and 2 correspond to boundaries at $\tau_\lambda = 0$ and $\tau_{\lambda,D}$. Then, in Equations 11.26 through 11.28, $I_\lambda^+(0)$ and $I_\lambda^-(\tau_{\lambda,D})$ are taken out of the integrals over μ. The integrals are expressed in terms of exponential integral functions to give

$$\int_{\mu=0}^{1} I_\lambda^+(0,\mu)e^{-\tau_\lambda/\mu}\mu d\mu = \frac{J_{\lambda,1}}{\pi}E_3(\tau_\lambda) \tag{11.32}$$

$$\int_{\mu=0}^{1}\left[I_\lambda^-(\tau_{\lambda,D},-\mu)e^{-(\tau_{\lambda,D}-\tau_\lambda)/\mu}\right]\mu d\mu = \frac{J_{\lambda,2}}{\pi} \tag{11.33}$$

$$\int_{\mu=0}^{1} I_\lambda^+(0,\mu)e^{-\tau_\lambda/\mu}d\mu = \frac{J_{\lambda,1}}{\pi}E_2(\tau_\lambda) \tag{11.34}$$

$$\int_{\mu=0}^{1} I_\lambda^-(\tau_{\lambda,D},-\mu)e^{-(\tau_{\lambda,D}-\tau_\lambda)/\mu}d\mu = \frac{J_{\lambda,2}}{\pi}E_2(\tau_{\lambda,D}-\tau_\lambda) \tag{11.35}$$

To obtain equations for $J_{\lambda,1}$ and $J_{\lambda,2}$ in terms of boundary emissivities, Equation 11.26 is evaluated at surface 1, $\tau_\lambda = 0$, to give (note that $E_3(0) = 1/2$)

$$q_{r\lambda,1} = J_{\lambda,1} - \left[2J_{\lambda,2}E_3(\tau_{\lambda,D}) + 2\pi\int_{\tau_\lambda^*=0}^{\tau_{\lambda,D}}\hat{I}_\lambda(\tau_\lambda^*)E_2(\tau_\lambda^*)d\tau_\lambda^*\right]$$

By comparison with Equation 6.2, which gives $q_{r\lambda,1} = J_{\lambda,1} - G_{\lambda,1}$, the quantity in the square brackets is $G_{\lambda,1}$. Using the emitted plus reflected incident energy for the outgoing energy (radiosity) gives the equation for $J_{\lambda,1}$ as

$$\begin{aligned} J_{\lambda,1} &= \varepsilon_{\lambda,1}\pi I_{\lambda b,1} + 2(1-\varepsilon_{\lambda,1})\left[J_{\lambda,2}E_3(\tau_{\lambda,D}) + \pi\int_{\tau_\lambda^*=0}^{\tau_{\lambda,D}} I_\lambda(\tau_\lambda^*)E_2(\tau_\lambda^*)d\tau_\lambda^*\right] \\ &= \varepsilon_{\lambda,1}E_{\lambda b,1} + (1-\varepsilon_{\lambda,1})G_{\lambda,1} \end{aligned} \tag{11.36}$$

Similarly, at surface 2 (by use of Equation 11.26 evaluated at $\tau_{\lambda,D}$)

$$\begin{aligned} J_{\lambda,2} &= \varepsilon_{\lambda,2}\pi I_{\lambda b,2} + 2(1-\varepsilon_{\lambda,2})\left[J_{\lambda,1}E_3(\tau_{\lambda,D}) + \pi\int_{\tau_\lambda^*=0}^{\tau_{\lambda,D}} I_\lambda(\tau_\lambda^*)E_2(\tau_{\lambda,D}-\tau_\lambda^*)d\tau_\lambda^*\right] \\ &= \varepsilon_{\lambda,2}E_{\lambda b,2} + (1-\varepsilon_{\lambda,2})G_{\lambda,2} \end{aligned} \tag{11.37}$$

11.3 GRAY PLANE LAYER OF ABSORBING AND EMITTING MEDIUM WITH ISOTROPIC SCATTERING

A gray medium has absorption and scattering coefficients that are independent of wavelength. From the discussion of gas-property spectral variations, Section 9.2.5, it is evident that gases are usually far from being gray. However, in some instances gases may be considered gray over all or a portion of the spectrum. If the temperatures are such that this spectral region contains an appreciable portion of the energy being exchanged, the approximation is reasonable. When particles of soot or other materials are present in the medium, the gas–particle mixture may act nearly gray. Examination of the radiative behavior of a gray medium provides an understanding of some features of a real medium without the complications that real media introduce.

The equations for local flux and the source function are now written for a gray medium with isotropic scattering (properties can vary with x). The equations are expressed in terms of the total quantities $I = \int_{\lambda=0}^{\infty} I_\lambda \, d\lambda$, and ω_λ and τ_λ become ω and τ. Integrating Equation 10.17 over all λ gives

$$\hat{I}(\tau) = (1-\omega)\frac{\sigma T^4(\tau)}{\pi} + \frac{\omega}{4\pi}\int_{\Omega_i=0}^{4\pi} I(\tau,\Omega_i)d\Omega_i = (1-\omega)\frac{\sigma T^4(\tau)}{\pi} + \omega\bar{I}(\tau) \tag{11.38}$$

In the flux Equation 11.26, the boundary values $I_\lambda^+(0,\mu)$ and $I_\lambda^-(\tau_D,-\mu)$ can still be spectrally dependent since the boundaries have not been assumed gray. Integrating over all λ gives

$$q_r(\tau) = 2\pi\left[\int_{\mu=0}^{1} I^+(0,\mu)e^{-\tau/\mu}\mu d\mu - \int_{\mu=0}^{1} I^-(\tau_D,-\mu)e^{-(\tau_D-\tau)/\mu}\mu d\mu \right.$$
$$\left. + \int_{\tau^*=0}^{\tau} \hat{I}(\tau^*)E_2(\tau-\tau^*)d\tau^* - \int_{\tau^*=\tau}^{\tau_D} \hat{I}(\tau^*)E_2(\tau^*-\tau)d\tau^*\right] \tag{11.39}$$

If the boundaries are diffuse and gray, $I^+(0,\mu) = J_1/\pi$ and $I^-(\tau_D,-\mu) = J_2/\pi$, and the first two integrals in Equation 11.39 become exponential integral functions. This gives

$$q_r(\tau) = 2\left[J_1E_3(\tau) - J_2E_3(\tau_D-\tau)\right]$$
$$+2\pi\left[\int_{\tau^*=0}^{\tau} \hat{I}(\tau^*)E_2(\tau-\tau^*)d\tau^* - \int_{\tau^*=\tau}^{\tau_D} \hat{I}(\tau^*)E_2(\tau^*-\tau)d\tau^*\right] \tag{11.40}$$

For a gray medium, Equation 11.27 for the divergence of the radiative flux becomes

$$\frac{dq_r(\tau)}{dx} = -2\pi\beta(\tau)\left[\int_{\mu=0}^{1} I^+(0,\mu)e^{-\tau/\mu}d\mu + \int_{\mu=0}^{1} I^-(\tau_D,-\mu)e^{-(\tau_D-\tau)/\mu}d\mu \right.$$
$$\left. + \int_{\tau^*=0}^{\tau_D} \hat{I}(\tau^*)E_1\left(\left|\tau^*-\tau\right|\right)d\tau^*\right] + 4\pi\beta(\tau)\hat{I}(\tau) \tag{11.41}$$

If the boundaries are diffuse and gray, the flux divergence is

$$\frac{dq_r(\tau)}{dx} = -2\beta(\tau)\left[J_1E_2(\tau)+J_2E_2(\tau_D-\tau)+\pi\int_{\tau^*=0}^{\tau_D}\hat{I}(\tau^*)E_1(|\tau^*-\tau|)d\tau^*\right]+4\pi\beta(\tau)\hat{I}(\tau) \tag{11.42}$$

For a gray medium, the integral Equation 11.28 for the source function is (where $\omega = \omega(\tau)$)

$$\hat{I}(\tau)=(1-\omega)\frac{\sigma T^4(\tau)}{\pi}+\frac{\omega}{2}\left[\int_{\mu=0}^{1} I^+(0,\mu)e^{-\tau/\mu}d\mu+\int_{\mu=0}^{1} I^-(\tau_D,-\mu)e^{-(\tau_D-\tau)/\mu}d\mu\right.$$
$$\left.+\int_{\tau^*=0}^{\tau_D}\hat{I}(\tau^*)E_1(|\tau^*-\tau|)d\tau^*\right] \tag{11.43}$$

If the boundaries are diffuse and gray, the integral equation becomes

$$\hat{I}(\tau)=(1-\omega)\frac{\sigma T^4(\tau)}{\pi}+\frac{\omega}{2\pi}\left[J_1E_2(\tau)+J_2E_2(\tau_D-\tau)+\pi\int_{\tau^*=0}^{\tau_D}\hat{I}(\tau^*)E_1(|\tau^*-\tau|)d\tau^*\right] \tag{11.44}$$

By eliminating the integrals from Equations 11.42 and 11.44, or by using Equation 11.29, the flux divergence has the convenient form (as shown by Equation 10.50)

$$\frac{dq_r}{dx}=4\frac{\kappa(\tau)}{\omega(\tau)}\left[\sigma T^4(\tau)-\pi\hat{I}(\tau)\right]=4\frac{\kappa(x)}{\omega(x)}\left[\sigma T^4(x)-\pi\hat{I}(x)\right] \tag{11.45}$$

where $\tau(x)=\int_0^x \beta(x^*)dx^*$. By use of Equation 11.38, $\hat{I}(\tau)$ can be eliminated to provide the form corresponding to Equation 11.21,

$$\frac{dq_r}{dx}=4\kappa(x)\left[\sigma T^4(x)-\pi\bar{I}(x)\right] \tag{11.46}$$

In the limit of zero absorption, so there is only scattering of any type, $dq_r/dx = 0$, as is evident from Equation 11.46 (see also Equation 10.43).

To illustrate the use of the radiative transfer equation for a plane layer, the dimensionless emission Q is calculated for a plane layer of gray medium at uniform temperature T_g with emission, absorption, and isotropic scattering, in surroundings at T_e, as in Figure 11.2. This has application to the dissipation of waste energy in outer space by radiation from sheets of hot liquid droplets traveling through space (Taussig and Mattick 1986, Siegel 1987b,c). The absorption and scattering coefficients are constant throughout the layer, and $\tilde{Q}=q_r/\left[\sigma(T_g^4-T_e^4)\right]$. This definition is used as it gives $\tilde{Q}$ independent of T_e. The q_r can be found by evaluating Equation 11.40 at $\tau = \tau_D$ and assuming that the surroundings act like a black background at T_e so that $J_1(0)=J_2(\tau_D)=\sigma T_e^4$. The result is arranged into the form

$$\tilde{Q}=2\int_{\tau=0}^{\tau_D}\frac{\pi\hat{I}(\tau)-\sigma T_e^4}{\sigma(T_g^4-T_e^4)}E_2(\tau_D-\tau)d\tau \tag{11.47}$$

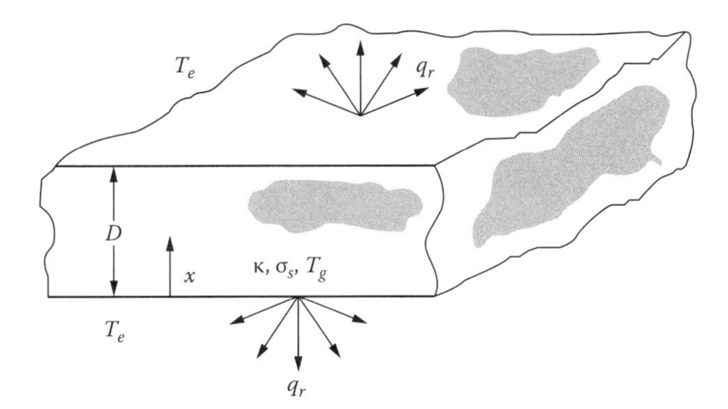

FIGURE 11.2 Plane layer of emitting, absorbing, and scattering medium at uniform temperature.

The $\hat{I}(\tau)$ is obtained from the integral Equation 11.44, which becomes

$$\hat{I}(\tau) = (1-\omega)\frac{\sigma T_g^4}{\pi} + \frac{\omega}{2}\left\{\frac{\sigma T_e^4}{\pi}\left[E_2(\tau) + E_2(\tau_D - \tau)\right] + \int_{\tau^*=0}^{\tau_D} \hat{I}(\tau^*) E_1\left(\left|\tau^* - \tau\right|\right) d\tau^*\right\}$$

This is arranged into

$$\frac{\pi\hat{I}(\tau) - \sigma T_e^4}{\sigma\left(T_g^4 - T_e^4\right)} = 1 - \omega + \frac{\omega}{2}\int_{\tau^*=0}^{\tau_D} \frac{\pi\hat{I}(\tau^*) - \sigma T_e^4}{\sigma\left(T_g^4 - T_e^4\right)} E_1\left(\left|\tau^* - \tau\right|\right) d\tau^* \tag{11.48}$$

This can be solved numerically by iteration for the dimensionless variable $[\pi\hat{I}(\tau) - \sigma T_e^4]/[\sigma(T_g^4 - T_e^4)]$ The results are inserted into Equation 11.47 to obtain $\tilde{Q}$. The $\tilde{Q}$ is a function of τ_D and ω, where $\tau_D = (\kappa + \sigma_s)D$. When $\omega = 0$, Equation 11.48 shows that $\hat{I}(\tau) = \sigma T_g^4/\pi$, so $\tilde{Q} = 2\int_0^{\kappa_D} E_2(\tau_D - \tau)d\tau = 1 - 2E_3(\tau_D)$. Some results from Siegel (1987b) are in Table 11.1. For each ω the $\tilde{Q}$ values increase with κ_D asymptotically to a maximum value that decreases as ω increases.

TABLE 11.1

$\tilde{Q}$ Values for Plane Layer at Uniform Temperature

Optical	Scattering Albedo (ω)					
Thickness (τ_D)	0	0.30	0.60	0.80	0.90	0.95
0.2	0.296	0.225	0.140	0.0748	0.0386	0.0197
0.5	0.557	0.449	0.303	0.172	0.0926	0.0481
1.0	0.781	0.667	0.490	0.304	0.173	0.0926
2	0.940	0.846	0.681	0.475	0.297	0.170
3	0.982	0.900	0.757	0.566	0.382	0.233
4	0.994	0.918	0.786	0.612	0.436	0.281
5	0.998	0.924	0.798	0.637	0.470	0.317
10	1.000	0.933	0.808	0.659	0.518	0.389

Source: Siegel, R., *J. Heat Trans.*, 109(1), 159, 1987b.

Another situation of interest is transient radiative cooling of a layer such as in Figure 11.2. This shows how the layer emissive ability decreases as the outer portions of the layer become cool and then do not radiate as well as the inner regions. If the layer consists of a dispersion of particles or drops, heat conduction in the medium is small, and if convection is also small, radiative transfer dominates. From the energy Equation 10.1,

$$\rho c\frac{\partial T}{\partial t}=-\frac{\partial q_r}{\partial x}=-\beta\frac{\partial q_r}{\partial \tau} \tag{11.49}$$

Using Equation 11.45 for $\partial q_r/\partial\tau$ at each t yields

$$\frac{\rho c}{\beta}\frac{\partial T(\tau,t)}{\partial t}=4\frac{1-\omega}{\omega}\left[\pi\hat{I}(\tau,t)-\sigma T^4(\tau,t)\right] \tag{11.50}$$

The $\hat{I}(\tau)$ is related to $T^4(\tau)$ by the integral Equation 11.44

$$\hat{I}(\tau,t)=(1-\omega)\frac{\sigma T^4(\tau,t)}{\pi}+\frac{\omega}{2}\left\{\frac{\sigma T_e^4}{\pi}\left[E_2(\tau)+E_2(\tau_D-\tau)\right]+\int_{\tau^*=0}^{\tau_D}\hat{I}(\tau^*)E_1\left(\left|\tau^*-\tau\right|\right)d\tau^*\right\} \tag{11.51}$$

Equations 11.50 and 11.51 can be solved starting from a specified initial temperature distribution $T(\tau, 0)$. This is inserted into Equation 11.51, which is solved numerically by iteration for $\hat{I}(\tau, 0)$. By using this along with $T^4(\tau, 0)$, Equation 11.50 is used to extrapolate to $T(\tau, t + \Delta t)$. This is inserted into Equation 11.51 to find $\hat{I}(\tau, t+\Delta t)$ and thereby continue the transient solution. Some numerical techniques are discussed in Section 13.5.

A similarity solution is found when the surroundings are at low temperature, $(T_e/T)^4 \ll 1$. This solution yields constant layer emittance values that are ultimately achieved during transient radiative cooling. A solution is tried of the form

$$\left[\sigma T^4(\tau,t)\right]^{1/4}=\Theta(t)F(\tau) \tag{11.52}$$

$$\left[\pi\hat{I}(\tau,t)\right]^{1/4}=\Theta(t)\Gamma(\tau) \tag{11.53}$$

Then, Equations 11.50 and 11.51 become

$$\frac{\rho c}{\sigma^{1/4}\beta}\frac{1}{\Theta^4(t)}\frac{d\Theta}{\partial t}=4\frac{1-\omega}{\omega}\left[\frac{\Gamma^4(\tau)-F^4(\tau)}{F(\tau)}\right] \tag{11.54}$$

$$\Gamma^4(\tau)=(1-\omega)F^4(\tau)+\frac{\omega}{2}\int_{\tau^*=0}^{\tau_D}\Gamma^4(\tau^*)E_1\left(\left|\tau^*-\tau\right|\right)d\tau^* \tag{11.55}$$

In Equation 11.54 the functions of τ and t have been separated, so the functions on each side of the equation must be a constant. Then, from the right side of Equation 11.54,

$$\frac{\Gamma^4(\tau)-F^4(\tau)}{F(\tau)}=\frac{\Gamma^4(0)-F^4(0)}{F(0)} \tag{11.56}$$

This is solved simultaneously with Equation 11.55 to obtain $F(\tau)$ and $\Gamma(\tau)$; the numerical solution details are in Siegel (1987b). The emittance reached in this "fully developed" transient region is defined as $\varepsilon_{fd} = q_r(\tau = \tau_D, t)/\sigma T_m^4(t)$, where $T_m(t)$ is the integrated mean temperature across the layer. ε_{fd} is independent of time and is a function only of τ_D and ω. ε_{fd} is lower than ε for uniform temperature, which are equal to $\tilde{Q}$ in Table 11.1, because of the relatively larger cooling of the outer portions of the layer during transient cooling. This uneven cooling results in poor radiative dissipation compared with that expected from the mean temperature of the layer.

11.4 GRAY PLANE LAYER IN RADIATIVE EQUILIBRIUM

11.4.1 Energy Equation

In some instances radiation dominates over other means of energy transfer. This condition also provides a limiting case to compare with energy transfer by combined modes, where heat conduction and/or convection is present with radiation. In this section the special case is considered of *steady state* without significant heat conduction, convection, viscous dissipation, or internal heat sources. For simplicity, the development is for a gray medium. When all energy sources and transfer mechanisms are negligible compared with radiation, the total energy emitted from each volume element must equal its total absorbed energy. This is termed *radiative equilibrium* and is steady-state energy conservation in the absence of any transfer but radiation. With only radiation present, Equation 10.1 becomes, for steady state without energy sources ($\dot{q} = 0$),

$$\nabla \cdot q_r = \frac{dq_r(x)}{dx} = 0 \quad \text{or} \quad \frac{dq_r(\tau)}{d\tau} = 0 \tag{11.57}$$

q_r is the total heat flux being transferred, since radiation is the only means of transfer.

11.4.2 Absorbing Gray Medium in Radiative Equilibrium with Isotropic Scattering

With $dq_r/d\tau = 0$, Equation 11.45 shows that

$$\hat{I}(\tau) = \frac{\sigma T^4(\tau)}{\pi} \tag{11.58}$$

so for radiative equilibrium with a nonzero absorption coefficient, the source function equals the local blackbody intensity. With this relation used to eliminate $\hat{I}(\tau)$ and with $dq_r/d\tau = 0$, the same integral equation for $T^4(\tau)$ results from either Equation 11.41 or 11.43:

$$4\sigma T^4(\tau) = 2\left[\pi \int_{\mu=0}^{1} I^+(0,\mu)e^{-\tau/\mu}d\mu + \pi \int_{\mu=0}^{1} I^-(\tau_D,-\mu)e^{-(\tau_D-\tau)/\mu}d\mu \right.$$
$$\left. + \int_{\tau^*=0}^{\tau_D} \sigma T^4(\tau^*)E_1(|\tau^* - \tau|)d\tau^*\right] \tag{11.59}$$

The radiative flux Equation 11.39 becomes, with $\hat{I}(\tau) = \sigma T^4(\tau)/\pi$,

$$q_r = 2\left[\pi \int_{\mu=0}^{1} I^+(0,\mu) e^{-\tau/\mu} \mu d\mu - \pi \int_{\mu=0}^{1} I^-(\tau_D,-\mu) e^{-(\tau_D-\tau)/\mu} \mu d\mu \right.$$
$$\left. + \int_{\tau^*=0}^{\tau} \sigma T^4(\tau^*) E_2(\tau-\tau^*) d\tau^* - \int_{\tau^*=\tau}^{\tau_D} \sigma T^4(\tau^*) E_2(\tau^*-\tau) d\tau^* \right] \tag{11.60}$$

Thus, for a gray medium with $\dot{q} = 0$ in radiative equilibrium with isotropic scattering or without scattering, and in which the absorption coefficient is nonzero, a single integral Equation 11.59 governs the temperature distribution. After the temperature distribution has been obtained, it is inserted into Equation 11.60 to obtain q_r. Since $dq_r/d\tau = 0$, q_r is constant (does not depend on τ). Hence Equation 11.60 can be evaluated at any convenient τ, such as $\tau = 0$. Now the special case is considered in which there is scattering, but absorption is zero.

11.4.3 Isotropically Scattering Medium with Zero Absorption

Consider an isotropically scattering medium with zero absorption coefficient. Although this case is included here because $dq_r/d\tau = 0$, as given by Equation 11.60, it does not require radiative equilibrium. If $\omega \to 1$, then from Equation 11.38, $\hat{I}(\tau) = \bar{I}(\tau)$, where $\bar{I}(\tau)$ is the mean scattered intensity at τ, and $\tau = \int_0^x \sigma_s dx$. Equation 11.43 reduces to an integral equation for $\bar{I}(\tau)$:

$$\bar{I}(\tau) = \frac{1}{2}\left[\int_{\mu=0}^{1} I^+(0,\mu) e^{-\tau/\mu} d\mu + \int_{\mu=0}^{1} I^-(\tau_D,-\mu) e^{-(\tau_D-\tau)/\mu} d\mu + \int_{\tau^*=0}^{\tau_D} \bar{I}(\tau^*) E_1(|\tau^*-\tau|) d\tau^*\right] \tag{11.61}$$

The radiative flux Equation 11.39 becomes

$$q_r = 2\pi\left[\int_{\mu=0}^{1} I^+(0,\mu) e^{-\tau/\mu} \mu d\mu - \int_{\mu=0}^{1} I^-(\tau_D,-\mu) e^{-(\tau_D-\tau)/\mu} \mu d\mu \right.$$
$$\left. + \int_{\tau^*=0}^{\tau} \bar{I}(\tau^*) E_2(\tau-\tau^*) d\tau^* - \int_{\tau^*=\tau}^{\tau_D} \bar{I}(\tau^*) E_2(\tau^*-\tau) d\tau^* \right] \tag{11.62}$$

Equation 10.40 shows that the divergence of the radiative flux is zero, $dq_r/d\tau = 0$, so q_r in Equation 11.62 is constant in space. This result for pure scattering does not require radiative equilibrium; it does not rely on the absence of other heat transfer modes. Within a nonabsorbing medium, the radiative transfer by scattering does not interact with the energy equation.

11.4.4 Gray Medium with $dq_r/dx = 0$ between Opaque Diffuse–Gray Boundaries

For diffuse–gray boundaries, Equation 11.40 applies. For an absorbing medium with or without scattering, $\hat{I}(\tau) \to \sigma T^4(\tau)/\pi$ for $dq_r/dx = 0$ as in Equation 11.58. For a scattering medium without absorption, $\hat{I}(\tau) \to \bar{I}(\tau)$ as for Equation 11.62. The net radiative flux in the x direction is then

$$q_r = 2\left[J_1 E_3(\tau) - J_2 E_3(\tau_D - \tau) + \int_{\tau^*=0}^{\tau} \Gamma(\tau^*) E_2(\tau-\tau^*) d\tau^* - \int_{\tau^*=\tau}^{\tau_D} \Gamma(\tau^*) E_2(\tau^*-\tau) d\tau^*\right] \tag{11.63}$$

where

$$\Gamma(\tau)=\begin{Bmatrix}\sigma T^4(\tau)\\ \pi \bar{I}(\tau)\end{Bmatrix} \quad \tau=\begin{Bmatrix}\int_{x^*=0}^{x} \beta(x^*)dx^* & \text{when } \kappa>0, \sigma_s \ge 0\\ \int_{x^*=0}^{x} \sigma_s(x^*)dx^* & \text{when } \kappa=0, \sigma_s>0\end{Bmatrix} \tag{11.64}$$

q_r is constant across the layer, so it can be evaluated at any convenient location such as $\tau = 0$. $\Gamma(\tau)$ is found from the integral equation

$$\Gamma(\tau)=\frac{1}{2}\left[J_1E_2(\tau)+J_2E_2(\tau_D-\tau)+\int_{\tau^*=0}^{\tau_D}\Gamma(\tau^*)E_1\left(\left|\tau^*-\tau\right|\right)d\tau^*\right] \tag{11.65}$$

J_1 and J_2 are found starting with Equation 11.36 and 11.37

$$J_1=\varepsilon_1\sigma T_1^4+2(1-\varepsilon_1)\left[J_2E_3(\tau_D)+\pi\int_{\tau^*=0}^{\tau_D}\Gamma(\tau^*)E_2(\tau^*)d\tau^*\right] \tag{11.66}$$

$$J_2=\varepsilon_2\sigma T_2^4+2(1-\varepsilon_2)\left[J_1E_3(\tau_D)+\int_{\tau^*=0}^{\tau_D}\Gamma(\tau^*)E_2(\tau_D-\tau^*)d\tau^*\right] \tag{11.67}$$

By using Equation 11.63 evaluated at $\tau = 0$ for Equation 11.66 and at $\tau = \tau_D$ for Equations 11.36 and 11.37, Equations 11.66 and 11.67 become

$$J_1=\varepsilon_1\sigma T_1^4-(1-\varepsilon_1)\left[q_r-J_1\right];\quad J_2=\varepsilon_2\sigma T_2^4+(1-\varepsilon_2)\left[q_r+J_2\right]$$

Solving for J_1 and J_2 gives

$$J_1=\sigma T_1^4-\frac{(1-\varepsilon_1)}{\varepsilon_1}q_r;\quad J_2=\sigma T_2^4+\frac{(1-\varepsilon_2)}{\varepsilon_2}q_r \tag{11.68}$$

11.4.5 Solution for Gray Medium with $dq_r/dx = 0$ between Black or Diffuse–Gray Boundaries at Specified Temperatures

A layer of gray medium with absorption and scattering coefficients $\kappa(T)$ and $\sigma_s(T)$ is between opaque infinite parallel boundaries at specified temperatures T_1 and T_2, as in Figure 11.3. It is desired to obtain the temperature distribution in the medium and the energy transfer q_r between boundaries for energy transfer by radiation only.

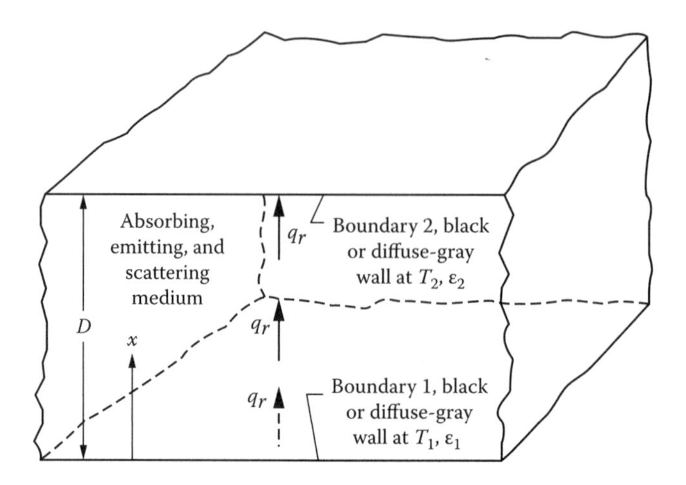

FIGURE 11.3 Plane layer of medium in radiative equilibrium between infinite parallel diffuse surfaces.

11.4.5.1 Gray Medium between Black Boundaries

For black boundaries $\varepsilon_1 = \varepsilon_2 = 1$, $J_1 = \sigma T_1^4$ and $J_2 = \sigma T_2^4$, from Equation 11.68. Evaluating Equation 11.63 at $\tau = 0$ gives the net heat flux from boundary 1 to boundary 2 as

$$q_r = \sigma T_1^4 - 2\sigma T_2^4 E_3(\tau_D) - 2\int_{\tau^*=0}^{\tau_D} \Gamma(\tau^*) E_2(\tau^*)d\tau^* \tag{11.69}$$

where, from Equation 11.65, $\Gamma(\tau)$ is found by solving the integral equation

$$\Gamma(\tau) = \frac{1}{2}\left[\sigma T_1^4 E_2(\tau) + \sigma T_2^4 E_2(\tau_D - \tau) + \int_{\tau^*=0}^{\tau_D} \Gamma(\tau^*) E_1(|\tau^* - \tau|) d\tau^*\right] \tag{11.70}$$

Equations 11.69 and 11.70 are placed in dimensionless form by defining

$$\psi_b \equiv \frac{q_r}{\sigma(T_1^4 - T_2^4)}, \quad \phi_b(\kappa) \equiv \frac{\Gamma(\tau)/\sigma - T_2^4}{T_1^4 - T_2^4} \tag{11.71}$$

where the b subscript emphasizes that this is for black boundaries. This yields

$$\phi_b(\tau) = \frac{1}{2}\left[E_2(\tau) + \int_{\tau^*=0}^{\tau_D} \phi_b(\tau^*) E_1(|\tau - \tau^*|) d\tau^*\right] \tag{11.72}$$

$$\psi_b = 1 - 2\int_{\tau^*=0}^{\tau_D} \phi_b(\tau^*) E_2(\tau^*) d\tau^* \tag{11.73}$$

The $\phi_b(\tau)$ is obtained by solving Equation 11.72 and is then used in Equation 11.73 to obtain ψ_b.

For the limiting case as both absorption and scattering in the medium become very small, $\tau_D \to 0$ and $E_3(\tau_D) \to 1/2$, so Equation 11.69 reduces to

$$q_r\big|_{\tau_D \to 0} = \sigma\left(T_1^4 - T_2^4\right) \tag{11.74}$$

(or $\psi_b = 1$), which is the solution for black infinite parallel plates with a transparent medium between them. For this limit, since $E_2(0) = 1$, Equation 11.70 yields

$$\left.\frac{\Gamma(\tau)}{\sigma}\right|_{\tau_D \to 0} = \frac{T_1^4 + T_2^4}{2} \tag{11.75}$$

(or $\phi_b = 1/2$). For a nearly transparent gray medium with $\kappa > 0$, $\Gamma(\kappa) = \sigma T^4(\kappa)$, and the temperature of the medium to the fourth power approaches the average of the fourth powers of the boundary temperatures. In an isotropically scattering medium without absorption, the $[\Gamma(\tau)/\sigma]_{\tau_D \to 0}$ in Equation 11.75 equals $\pi \bar{I}(\tau)/\sigma$ within the medium, Equation 11.64.

In the following discussion, let $\kappa > 0$ so that the results can be interpreted in terms of the medium temperature. Numerical results from Equations 11.72 and 11.73 for the temperature distribution and heat flux in a gray gas with constant properties between infinite black parallel boundaries have been obtained by many investigators. Solution methods are discussed in succeeding chapters. Heaslet and Warming (1965) presented results accurate to four significant figures. These are in Figure 11.4 and values for ψ_b are also in Table 11.2.

Figure 11.4a shows there is a discontinuity between each specified boundary temperature and the temperature of the medium at the boundary. This is called a temperature "jump" or "slip." If the jump were not present, the curves would all go to 1 at $\tau/\tau_D = 0$ and to 0 at $\tau/\tau_D = 1$. There is no jump when heat conduction is present; the jump is a limit as transfer only by radiation is approached. To determine the magnitude of the jump, the temperature of the medium is evaluated at $\tau = 0$. Using Equation 11.70,

$$\frac{T_1^4 - T^4(\tau=0)}{T_1^4 - T_2^4} = \frac{1}{2}\left[\frac{T_1^4}{T_1^4 - T_2^4} - \frac{T_2^4}{T_1^4 - T_2^4} E_2(\tau_D) - \int_0^{\tau_D} \frac{T^4(\tau^*)}{T_1^4 - T_2^4} E_1\left(\tau^*\right) d\tau^*\right] \tag{11.76}$$

As $\tau_D \to 0$ the integral vanishes and $E_2(\tau_D) \to 1$, so this limit gives

$$\left.\frac{T_1^4 - T^4(\tau=0)}{T_1^4 - T_2^4}\right|_{\tau_D \to 0} = \frac{1}{2} \tag{11.77}$$

The magnitude of the jump for a gray medium with constant absorption coefficient is in Figure 11.5 as a function of the layer optical thickness. From symmetry, $T_1^4 - T^4(\tau=0) = T^4(\tau=\tau_D) - T_2^4$.

11.4.5.2 Gray Medium between Diffuse–Gray Boundaries

An extension can now be made for gray rather than black boundaries. From Equations 11.63 and 11.65, the equations have the same form as Equations 11.69 and 11.70, except that the outgoing fluxes J_1 and J_2 replace σT_1^4 and σT_2^4. Hence, as in Equation 11.71, for gray boundaries let

$$\psi = \frac{q_r}{J_1 - J_2}, \quad \phi(\tau) = \frac{\Gamma(\tau) - J_2}{J_1 - J_2}$$

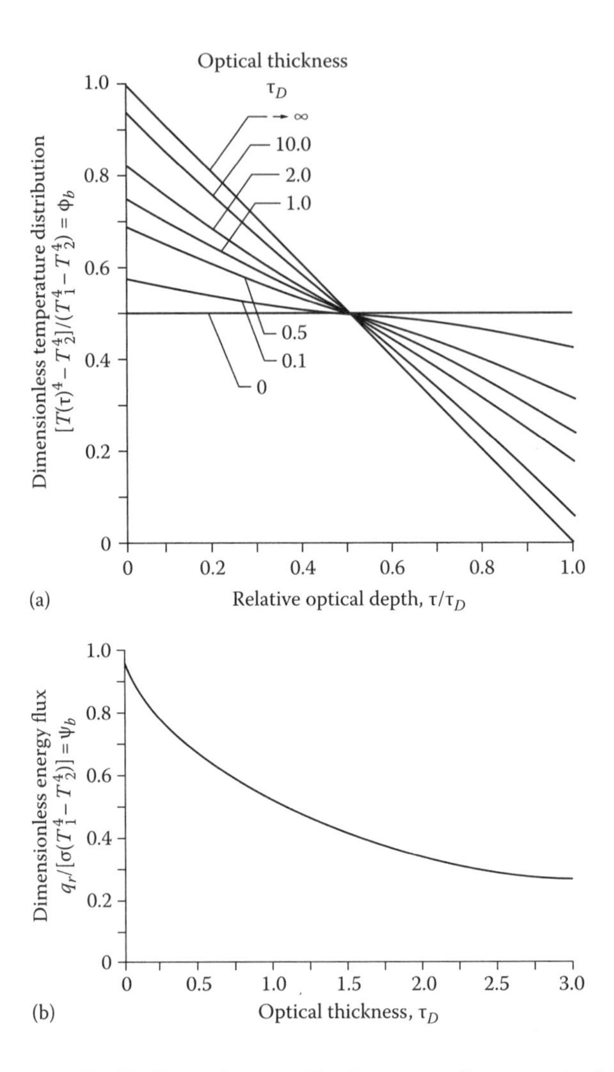

FIGURE 11.4 Temperature distribution and energy flux in gray medium contained between infinite black parallel boundaries. (a) Temperature distribution; (b) energy flux. (From Heaslet, M.A. and Warming, R.F., *Int. J. Heat Mass Trans.*, 8(7), 979, 1965.)

TABLE 11.2
Dimensionless Energy Flux $q_r/\sigma(T_1^4 - T_2^4) = \psi_b$[a]

Optical Thickness (τ_D)	Ψ_b	Optical Thickness (τ_D)	Ψ_b
0	1	0.8	0.6046
0.1	0.9157	1.0	0.5532
0.2	0.8491	1.5	0.4572
0.3	0.7934	2.0	0.3900
0.4	0.7458	2.5	0.3401
0.5	0.7040	3.0	0.3016
0.6	0.6672	4.0	0.2460

Source: Heaslet, M.A. and Warming, R.F., *Int. J. Heat Mass Trans.*, 8(7), 979, 1965.
[a] For $\tau_D = 1$, $\Psi_b = (4/3)/(1.42089 + \tau_D)$.

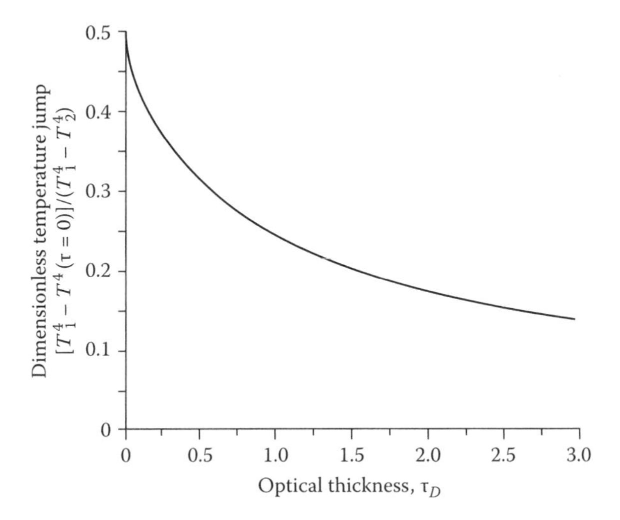

FIGURE 11.5 Discontinuity at boundary between gray gas and black boundary temperatures for radiative equilibrium. (From Heaslet, M.A. and Warming, R.F., *Int. J. Heat Mass Trans.*, 8(7), 979, 1965.)

where

$$\Gamma(\tau) = \begin{cases} \sigma T^4(\tau) & \text{when } \kappa > 0, \sigma_s \ge 0 \\ \pi \bar{I}(\tau) & \text{when } \kappa = 0, \sigma_s > 0 \end{cases}$$

The equations for $\phi(\tau)$ and ψ are then the same as in Equations 11.72 and 11.73, so $\phi = \phi_b$, and $\psi = \psi_b$, and for gray boundaries the $\Gamma(\tau)$ and q_r are given by

$$\Gamma(\tau) = \phi_b(\tau)(J_1 - J_2) + J_2 \tag{11.78}$$

$$q_r = \psi_b(J_1 - J_2) \tag{11.79}$$

Hence, assuming that ϕ_b and ψ_b have been obtained for black boundaries, only J_1 and J_2 are needed to solve the gray boundary case. From Equations 11.68,

$$J_1 = \sigma T_1^4 - \frac{1-\varepsilon_1}{\varepsilon_1} q_r \tag{11.80}$$

$$J_2 = \sigma T_2^4 + \frac{1-\varepsilon_2}{\varepsilon_2} q_r \tag{11.81}$$

Substitute these relations into Equation 11.79 and solve for q_r to obtain

$$\frac{q_r}{\sigma(T_1^4 - T_2^4)} = \frac{\psi_b}{1 + \psi_b(1/\varepsilon_1 + 1/\varepsilon_2 - 2)} \tag{11.82}$$

Then substitute J_1 and J_2 from Equations 11.80 and 11.81 into Equation 11.78 and eliminate q_r by using Equation 11.82 to obtain

$$\frac{\Gamma(\tau)/\sigma - T_2^4}{T_1^4 - T_2^4} = \frac{\phi_b(\tau) + \left[(1-\varepsilon_2)/\varepsilon_2\right]\psi_b}{1 + \psi_b\left(1/\varepsilon_1 + 1/\varepsilon_2 - 2\right)} \tag{11.83}$$

Thus, the solutions for plane layers with any combination of gray boundaries can be found conveniently from the black boundary solutions.

11.4.5.3 Extended Solution for Optically Thin Medium between Gray Boundaries

Section 10.11 provides some solutions for optically thin media contained within black or transparent boundaries. Here, we provide more comprehensive relations for a nearly transparent medium between infinite parallel boundaries.

Consider the heat flux Equation 11.26 used in conjunction with Equations 11.36 and 11.37 for diffuse boundaries that are not black. From the series expansions in Appendix D, the exponential integrals are approximated for small arguments by $E_2(x) = 1 + O(x)$ and $E_3(x) = \frac{1}{2} - x + O(x^2)$. Then Equation 11.26 becomes, after substituting Equations 11.32 and 11.33 for diffuse boundaries in the first two integrals on the right,

$$q_{r\lambda}\left(\tau_\lambda\right) = \left(1 - 2\tau_\lambda\right)J_{\lambda,1} - \left(1 - 2\tau_{\lambda,D} + 2\tau_\lambda\right)J_{\lambda,2} + 2\pi\int_{\tau^*_\lambda=0}^{\tau_\lambda}\hat{I}_\lambda\left(\tau_\lambda^*\right)\left[1 + O\left(\tau_\lambda - \tau_\lambda^*\right)\right]d\tau_\lambda^*$$

$$-2\pi\int_{\tau^*_\lambda=\tau_\lambda}^{\tau_{\lambda,D}}\hat{I}_\lambda\left(\tau_\lambda^*\right)\left[1 + O\left(\tau_\lambda^* - \tau_\lambda\right)\right]d\tau_\lambda^* \tag{11.84}$$

If $\tau_{\lambda,D} \ll 1$, the terms of order τ_λ are neglected and this reduces to $q_{r\lambda}(\tau_\lambda) = J_{\lambda,1} - J_{\lambda,2}$. The local spectral flux in the medium is the difference between the fluxes leaving the boundaries; the fluxes are not attenuated by the medium in the nearly transparent approximation.

In a similar fashion, the source-function Equation 11.28 reduces to

$$\hat{I}_\lambda\left(\tau_\lambda\right) = (1 - \omega_\lambda)I_{\lambda b}\left(\tau_\lambda\right) + \frac{\omega_\lambda}{2\pi}\left(J_{\lambda,1} + J_{\lambda,2}\right) \tag{11.85}$$

The flux derivative Equation 11.27 simplifies to

$$\frac{dq_{r\lambda}}{d\tau_\lambda} = -2\left(J_{\lambda,1} + J_{\lambda,2}\right) + 4\pi\hat{I}_\lambda\left(\tau_\lambda\right)$$

$$\frac{dq_{r\lambda}(x)}{dx} = 2\kappa_\lambda(x)\left\{4\pi I_{\lambda b}(x) - \left(J_{\lambda,1} + J_{\lambda,2}\right)\right\} \tag{11.86}$$

The diffuse spectral fluxes at the boundaries, Equations 11.36 and 11.37, reduce to

$$J_{\lambda,1} = \varepsilon_{\lambda,1}E_{\lambda b,1} + \left(1 - \varepsilon_{\lambda,1}\right)J_{\lambda,2}$$
$$J_{\lambda,2} = \varepsilon_{\lambda,2}E_{\lambda b,2} + \left(1 - \varepsilon_{\lambda,2}\right)J_{\lambda,1}$$

Solving simultaneously for the spectral fluxes leaving the boundaries yields, for the nearly transparent approximation,

$$J_{\lambda,1} = \frac{\varepsilon_{\lambda,1}E_{\lambda b,1} + \varepsilon_{\lambda,2}E_{\lambda b,2}\left(1-\varepsilon_{\lambda,1}\right)}{1-\left(1-\varepsilon_{\lambda,1}\right)\left(1-\varepsilon_{\lambda,2}\right)} \tag{11.87}$$

$$J_{\lambda,2} = \frac{\varepsilon_{\lambda,2}E_{\lambda b,2} + \varepsilon_{\lambda,1}E_{\lambda b,1}\left(1-\varepsilon_{\lambda,2}\right)}{1-\left(1-\varepsilon_{\lambda,1}\right)\left(1-\varepsilon_{\lambda,2}\right)} \tag{11.88}$$

These relations are now applied in an example.

Example 11.1

A nearly transparent medium with extinction coefficient β_λ is between two diffuse parallel boundaries separated by a distance D and temperatures T_1 and T_2. What total heat flux is being transferred between the boundaries in the absence of heat conduction and convection?

For the nearly transparent approximation, $q_{r\lambda} = J_{\lambda,1} - J_{\lambda,2}$. Substituting $J_{\lambda,1}$ and $J_{\lambda,2}$ from Equation 11.36 yields

$$q_{r\lambda}d\lambda = \frac{\varepsilon_{\lambda,1}E_{\lambda b,1} + \varepsilon_{\lambda,2}E_{\lambda b,2}\left(1-\varepsilon_{\lambda,1}\right) - \varepsilon_{\lambda,2}E_{\lambda b,2} + \varepsilon_{\lambda,1}E_{\lambda b,1}\left(1-\varepsilon_{\lambda,2}\right)}{1-\left(1-\varepsilon_{\lambda,1}\right)\left(1-\varepsilon_{\lambda,2}\right)}d\lambda$$

Simplifying and integrating with respect to λ gives the required result,

$$q_r = \int_{\lambda=0}^{\infty} \frac{E_{\lambda b,1} - E_{\lambda b,2}}{1/\varepsilon_{\lambda,1} + 1/\varepsilon_{\lambda,2} - 1} d\lambda \tag{11.1.1}$$

This is the same as Equation 6.1.2; the radiant energy flux transferred is uninfluenced by the nearly transparent medium between the boundaries. The medium temperature is obtained in the next example.

Example 11.2

What is the temperature distribution in the medium for the conditions in Example 11.1?

For radiative equilibrium without internal heat sources (no conduction or convection), $\nabla \cdot q_r = \int_0^\infty (dq_{r\lambda}/dx)d\lambda = 0$. Then, from Equation 11.86,

$$\int_{\lambda=0}^{\infty} \kappa_\lambda(x)\left[2\pi I_{\lambda b}(x) - \left(J_{\lambda,1} + J_{\lambda,2}\right)\right]d\lambda = 0$$

Substituting Equations 11.87 and 11.88 yields

$$2\pi\int_{\lambda=0}^{\infty} \kappa_\lambda(x) I_{\lambda b}\left[T(x)\right]d\lambda = \int_{\lambda=0}^{\infty} \kappa_\lambda(x) \frac{2\left(\varepsilon_{\lambda,1}E_{\lambda b,1} + \varepsilon_{\lambda,2}E_{\lambda b,2}\right) - \varepsilon_{\lambda,1}\varepsilon_{\lambda,2}\left(E_{\lambda b,1} + E_{\lambda b,2}\right)}{\varepsilon_{\lambda,1} + \varepsilon_{\lambda,2} - \varepsilon_{\lambda,1}\varepsilon_{\lambda,2}} d\lambda \tag{11.2.1}$$

This can be solved for $T(x)$ by iteration, noting that $\kappa_\lambda(x)$ can be a function of T and x; Equation 11.2.1 reduces to Equation 10.6.4 when $\varepsilon_{\lambda,1} = \varepsilon_{\lambda,2} = 1$. If all properties are independent of both wavelength and temperature, Equation 11.2.1 reduces to the uniform temperature

$$T^4 = \frac{1}{2}\frac{2\left(\varepsilon_1 T_1^4 + \varepsilon_2 T_2^4\right) - \varepsilon_1\varepsilon_2\left(T_1^4 + T_2^4\right)}{\varepsilon_1 + \varepsilon_2 - \varepsilon_1\varepsilon_2} \tag{11.2.2}$$

11.5 MULTIDIMENSIONAL RADIATION IN A PARTICIPATING GRAY MEDIUM WITH ISOTROPIC SCATTERING

Sections 11.2 through 11.4 developed radiative relations for plane layers. Other geometries are considered next, including rectangular regions. Since the relations become more complicated, the development is for a gray medium, with constant properties, and isotropic scattering. With these simplifications, Equations 10.17, 10.26, 10.38, 10.45, and 10.52 become

$$\mathbf{q}_r = \int_{\Omega=0}^{4\pi} I\mathbf{s}\,d\Omega \tag{11.89}$$

$$\nabla\cdot\mathbf{q}_r = 4\pi\kappa\left(\frac{\sigma T^4}{\pi} - \bar{I}\right) = 4\pi\beta\left(I - \bar{I}\right) = 4\pi\beta\frac{\kappa}{\sigma_s}\left(\frac{\sigma T^4}{\pi} - \hat{I}\right) \tag{11.90}$$

$$\hat{I} = \frac{1}{\beta}\left(\kappa\frac{\sigma T^4}{\pi} + \sigma_s \bar{I}\right) = (1-\omega)\frac{\sigma T^4}{\pi} + \omega\bar{I} \tag{11.91}$$

$$\bar{I} = \frac{1}{4\pi}\int_{\Omega=0}^{4\pi} I\,d\Omega = \frac{1}{4\pi}\int_{\Omega=0}^{4\pi}\left[I(0)e^{-\beta S} + \beta\int_{S^*=0}^{S}\hat{I}(S^*)e^{-\beta(S-S^*)}dS^*\right]d\Omega \tag{11.92}$$

where

κ and σ_s are constants

$\beta = \kappa + \sigma_s$

The temperature is a function of position within the radiating medium and is found from the energy equation that requires $\nabla\cdot\mathbf{q}_r$.

11.5.1 Radiation Transfer Relations in Three Dimensions

A 3D geometry is shown in Figure 11.6. Location vectors $\mathbf{r}_0$, $\mathbf{r}$, and $\mathbf{r}^*$ in Figure 11.6b give the positions of dA, dV, and dV^*. Then the path length in Figure 11.6a is $S = |\mathbf{r} - \mathbf{r}_0|$, and the unit vector along S is $\mathbf{s} = (\mathbf{r} - \mathbf{r}_0)/|\mathbf{r} - \mathbf{r}_0|$ or $(\mathbf{r} - \mathbf{r}^*)/|\mathbf{r} - \mathbf{r}^*|$. The cos θ is the dot product of the unit vector $\mathbf{n}$ and the unit vector $\mathbf{s}$ along S, so $\cos\theta = \mathbf{n}\cdot(\mathbf{r} - \mathbf{r}_0)/|\mathbf{r} - \mathbf{r}_0|$. The solid angle that dA subtends when viewed from dV is $d\Omega = dA\cos\theta/S^2 = dA\,\mathbf{n}\cdot(\mathbf{r} - \mathbf{r}_0)/|\mathbf{r} - \mathbf{r}_0|^3$. In the integral of $\hat{I}(S^*)$ in Equation 11.92, the integration over all $d\Omega$ includes all volume elements of the medium $dV^* = dS^* d\Omega(S - S^*)^2$. By using these vector relations, and $dS^*\,d\Omega = dV^*/|\mathbf{r} - \mathbf{r}^*|^2$, Equation 11.92 becomes

$$\bar{I}(\mathbf{r}) = \frac{1}{4\pi}\int_A I(\mathbf{r}_0)\frac{\mathbf{n}\cdot(\mathbf{r}-\mathbf{r}_0)}{|\mathbf{r}-\mathbf{r}_0|^3}e^{-\beta|\mathbf{r}-\mathbf{r}_0|}dA + \frac{\beta}{4\pi}\int_V \hat{I}(\mathbf{r}^*)\frac{e^{-\beta|\mathbf{r}-\mathbf{r}^*|}}{|\mathbf{r}-\mathbf{r}^*|^2}dV^* \tag{11.93}$$

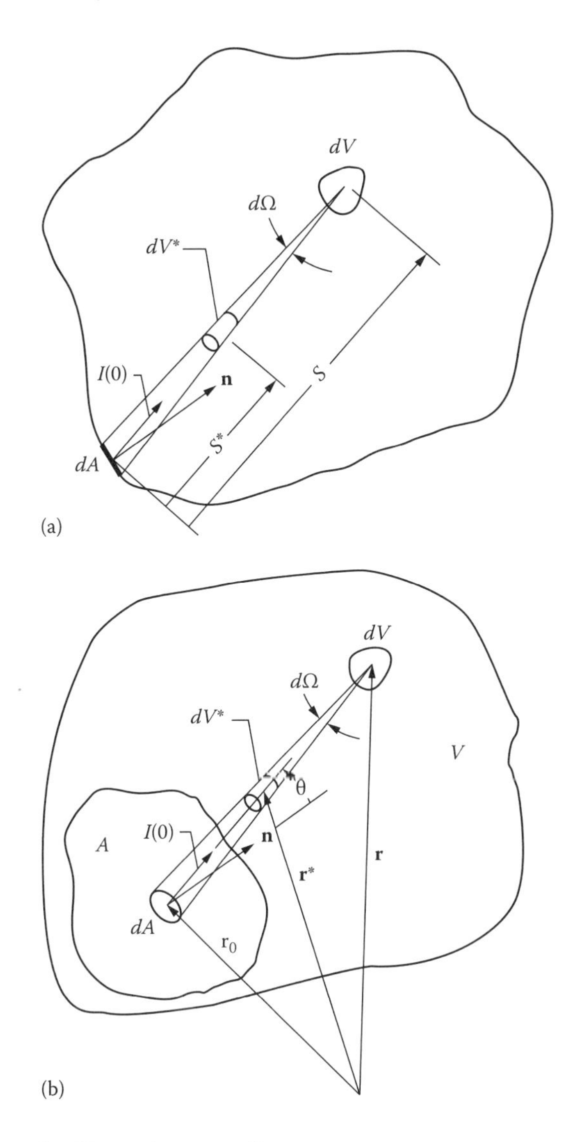

FIGURE 11.6 Geometry for 3D translucent medium.

The expression for $\mathbf{q}_r$ is obtained in the same manner. Equation 11.89 contains the unit vector **s** and is otherwise similar to the definition of $\bar{I}$. Using Equation 11.93 and inserting the additional **s**, $\mathbf{q}_r$ is

$$\mathbf{q}_r(\mathbf{r}) = \int_A I(\mathbf{r}_0)\left[\mathbf{n}\cdot(\mathbf{r}-\mathbf{r}_0)\right]\frac{\mathbf{r}-\mathbf{r}_0}{|\mathbf{r}-\mathbf{r}_0|^4}e^{-\beta|\mathbf{r}-\mathbf{r}_0|}dA + \beta\int_V \hat{I}(\mathbf{r}^*)\frac{\mathbf{r}-\mathbf{r}^*}{|\mathbf{r}-\mathbf{r}^*|^3}e^{-\beta|\mathbf{r}-\mathbf{r}^*|}dV^* \tag{11.94}$$

Relations of this type are given in Lin (1988).

11.5.2 Two-Dimensional Transfer in an Infinitely Long Right Rectangular Prism

To illustrate the geometrical manipulations for applying Equations 11.93 and 11.94, radiative transfer is considered in a rectangular region with uniform conditions along the axial, z, direction (Figure 11.7). The length in the z direction is large compared with dimensions b and d, and

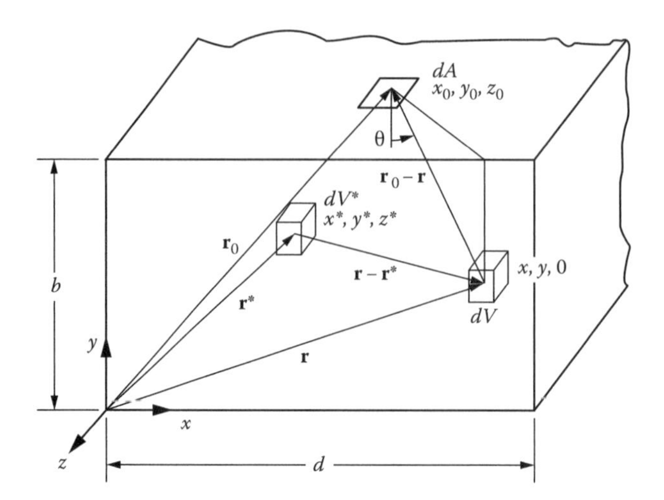

FIGURE 11.7 Geometry for radiation in 2D translucent rectangular medium.

the axial location of dV is arbitrary; hence it is specified to be at $z = 0$. Then, the volume integral in Equation 11.94 becomes

$$\int_V \hat{I}(r^*) \frac{e^{-\beta|r-r^*|}}{|r-r^*|^2} dV^* = 2 \int_{x^*=0}^{d} \int_{y^*=0}^{b} \hat{I}(x^*, y^*) \int_{z^*=0}^{\infty} \frac{e^{-\beta[(x-x^*)^2+(y-y^*)^2+z^{*2}]^{1/2}}}{(x-x^*)^2 + (y-y^*)^2 + z^{*2}} dz^* \, dx^* \, dy^* \quad (11.95)$$

Let $\xi^* = z^*/[(x-x^*)^2 + (y-y^*)^2]^{1/2} = z^*/\rho^*$ to obtain the z^* integral as

$$\int_{\xi=0}^{\infty} \frac{e^{-\beta\rho^*(1+\xi^{*2})^{1/2}}}{\rho^{*2}(1+\xi^{*2})} \rho^* \, d\xi^* = \frac{1}{\rho^*} \int_{t=1}^{\infty} \frac{e^{-\beta\rho^* t}}{t(t^2-1)^{1/2}} dt = \frac{1}{\rho^*} \frac{\pi}{2} S_1(\beta\rho^*) \quad (11.96)$$

The transformation $t = (1+\xi^{*2})^{1/2}$ has been used, and the function S_1 is one of the class of functions

$$S_n(x) \equiv \frac{2}{\pi} \int_1^{\infty} \frac{e^{-xt}}{t^n (t^2-1)^{1/2}} dt = \frac{2}{\pi} \int_0^{\pi/2} e^{-x/\cos\theta} \cos^{n-1}\theta \, d\theta \quad n = 0, 1, 2, \ldots \quad (11.97)$$

S_n are similar to the exponential integral functions E_n that were used for plane layers. Their mathematical properties are examined by Yuen and Wong (1983) and Altaç (1996), and some of their characteristics are in Appendix D along with a table of values.

The first (area) integral on the right side of Equation 11.93 consists of four parts corresponding to the four sides of the rectangle. Consider the top side. From the triangle in Figure 11.7, $\cos\theta = \mathbf{n} \cdot (\mathbf{r} - \mathbf{r}_0) / |\mathbf{r} - \mathbf{r}_0| = (b-y) / (\rho_0^2 + z_0^2)^{1/2}$, where $\rho_0(x_0, y_0) = [(x-x_0)^2 + (y-y_0)^2]^{1/2}$. Then for the upper surface, the area integral becomes, after integration over the z variable,

$$2 \int_{x_0=0}^{d} I(x,b) \int_{z_0=0}^{\infty} e^{-\beta(\rho_0^2+z_0^2)^{1/2}} \frac{b-y}{(\rho_0^2+z_0^2)^{3/2}} dz_0 \, dx_0 = \pi(b-y) \int_{x_0=0}^{d} \frac{I(x_0,b)}{\rho_0^2} S_2(\beta\rho_0) \, dx_0 \quad (11.98)$$

The $\bar{I}(\mathbf{r})$ in Equation 11.93 is then given by

$$4\bar{I}(x,y)=(b-y)\int_{x_0=0}^{d} I(x_0,b)\frac{S_2\left[\beta\rho_0(x_0,b)\right]}{\rho_0^2(x_0,b)}dx_0+(d-x)\int_{y_0=0}^{b} I\left(d,y_0\right)\frac{S_2\left[\beta\rho_0(d,y_0)\right]}{\rho_0^2(d,y_0)}dy_0$$

$$+y\int_{x_0=0}^{d} I\left(x_0,0\right)\frac{S_2\left[\beta\rho_0(x_0,0)\right]}{\rho_0^2(x_0,0)}dx_0+x\int_{y_0=0}^{b} I\left(0,y_0\right)\frac{S_2\left[\beta\rho_0\left(0,y_0\right)\right]}{\rho_0^2\left(0,y_0\right)}dy_0$$

$$+\beta\int_{x^*=0}^{d}\int_{y^*=0}^{b}\hat{I}\left(x^*,y^*\right)\frac{S_1\left(\beta\rho^*\right)}{\rho^*}dx^*\,dy^* \tag{11.99}$$

When calculating results from these equations, it is sometimes advantageous to transform them into cylindrical coordinates. Then a $dx\,dy$ increment becomes $\rho\,d\rho\,d\theta$, and this cancels the ρ that is in the denominator. This eliminates any computational difficulty with the denominator becoming zero or close to zero; this is illustrated by Yuen and Ho (1985) for a rectangular gray radiating medium with internal heat generation.

Some examples are now considered for radiative equilibrium. Radiation combined with other modes is considered in Chapter 13.

Example 11.3

Consider radiative cooling from within a translucent material in the limit when heat conduction is negligible compared with radiation. A hot rectangular bar of emitting, absorbing, and scattering material with nonreflecting boundaries such as in Figure 11.7 is initially at uniform temperature T_i. It is suddenly placed in a very low temperature environment. The medium is gray with constant radiative properties, scatters isotropically, and has density ρ and specific heat c. Derive the energy relation to obtain the transient temperature distribution.

Using Equation 11.90 for $\nabla\cdot q_r$, the energy equation for $T(x, y, t)$ is

$$\rho c\frac{\partial T(x,y,t)}{\partial t}=-4\beta\frac{\kappa}{\sigma_s}\left[\sigma T^4(x,y,t)-\pi\hat{I}\,(x,y,t)\right]\quad T(x,y,0)=T_i \tag{11.3.1}$$

This is solved simultaneously with the integral equation found by substituting $\bar{I}$ from Equation 11.99 into Equations 11.91 and 11.92. The incoming intensities at the boundaries are zero as the surroundings are at low temperature and the boundaries are nonreflecting. Then

$$\hat{I}(x,y,t)=\frac{\kappa}{\beta}\frac{\sigma T^4(x,y,t)}{\pi}+\frac{\sigma_s}{4}\int_{x^*=0}^{d}\int_{y^*=0}^{b}\hat{I}\left(x^*,y^*,t\right)\frac{S_1\left(\beta\rho^*\right)}{\rho^*}dx^*\,dy^* \tag{11.3.2}$$

where

$$\rho^*=\left[\left(x-x^*\right)^2+\left(y-y^*\right)^2\right]^{1/2}.$$

Equations 11.3.1 and 11.3.2 are placed in dimensionless form and solved numerically. Starting at $t = 0$, the $T(x, y, 0)$ in Equation 11.3.2 is set equal to T_i. Then the integral equation is solved numerically for $\hat{I}\left(x,\ y,\ 0\right)$. $T(x, y, 0)$ and $\hat{I}\left(x,\ y,\ 0\right)$ are substituted into the right side of Equation 11.3.1, and the resulting temperature derivative is used to extrapolate forward in time to obtain $T(x, y, \Delta t)$. With this temperature distribution, $\hat{I}(x, y, \Delta t)$ distribution is found by iteration of Equation 11.3.2, and the process is continued to move forward in time. The numerical method is discussed in Section 13.3.

Example 11.4

For the transient cooling problem in Example 11.3, derive an expression for the instantaneous emittance of the rectangular region based on its instantaneous heat loss and mean temperature.

The emittance is obtained from the local heat fluxes leaving the region boundaries. For this example the externally incident intensities are zero, so only the volume integral in Equation 11.94 is needed. Along the upper boundary,

$$q_r(x,b) = 2\beta \int_{x^*=0}^{d} \int_{y^*=0}^{b} \int_{z^*=0}^{\infty} \hat{I}(x^*,y^*) \frac{(b-y^*)e^{-\beta[(x-x^*)^2+(b-y^*)^2+z^{*2}]^{1/2}}}{[(x-x^*)^2+(b-y^*)^2+z^{*2}]^{3/2}} dz^*\,dx^*\,dy^* \quad (11.4.1)$$

Using the same transformation as in Equation 11.96, the integration over z^* is expressed as an S_2 function,

$$q_r(x,b) = \pi\beta \int_{x^*=0}^{d} \int_{y^*=0}^{b} \hat{I}(x^*,y^*) \frac{b-y^*}{\rho_1^{*2}} S_2\left(\beta\rho_1^*\right) dx^*\,dy^* \quad (11.4.2)$$

where $\rho_1^* = \left[(x-x^*)^2 + (b-y^*)^2\right]^{1/2}$. Similarly, along boundary $x = d$,

$$q_r(d,y) = \pi\beta \int_{x^*=0}^{d} \int_{y^*=0}^{b} \hat{I}(x^*,y^*) \frac{d-x^*}{\rho_2^{*2}} S_2(\beta\rho_2^*)\,dx^*\,dy^* \quad (11.4.3)$$

where $\rho_2^* = \left[(d-x^*)^2 + (y-y^*)^2\right]^{1/2}$. The overall emittance of the rectangle will be based on its instantaneous mean temperature

$$T_m(t) = \left[\frac{1}{bd} \int_{x=0}^{d} \int_{y=0}^{b} T^4(x,y,t)\,dx\,dy\right]^{1/4} \quad (11.4.4)$$

The total heat loss $Q(t)$ is obtained by integrating the local q_r over their respective boundaries around the rectangle and using symmetry of the upper and lower boundaries and the two vertical sides. Then, the overall emittance is

$$\varepsilon(t) = \frac{Q(t)}{\sigma T_m^4(t)} = \frac{2}{\sigma T_m^4(t)} \left[\int_0^d q_r(x,b)\,dx + \int_0^d q_r(d,y)\,dy\right] \quad (11.4.5)$$

A useful case for testing the accuracy of numerical solutions is to evaluate the radiation to cold black boundaries by a 2D translucent rectangular region at a uniform temperature T_m. The region contains an absorbing–emitting medium without scattering, and with uniform properties. The fluxes at the boundary can be found from Equations 11.4.2 and 11.4.3 by using the source function

$\hat{I} = \sigma T_m^4/\pi$, which is constant within the volume for this case, and, without scattering, $\beta = \kappa$. Then, Equations 11.4.2 and 11.4.3 become

$$\frac{q_r(x,b)}{\sigma T_m^4} = \kappa \int_{x^*=0}^{d} \int_{y^*=0}^{b} \frac{b-y^*}{\rho_1^{*2}} S_2\left(\kappa\rho_1^*\right) dx^*\, dy^* \tag{11.100}$$

$$\frac{q_r(d,y)}{\sigma T_m^4} = \kappa \int_{x^*=0}^{d} \int_{y^*=0}^{b} \frac{d-x^*}{\rho_2^{*2}} S_2\left(\kappa\rho_2^*\right) dx^*\, dy^* \tag{11.101}$$

In Siegel (1991a) a series of transformations was used, and the integrations in Equations 11.100 and 11.101 were carried out to yield the following readily evaluated forms:

$$\frac{q_r(x,b)}{\sigma T_m^4} = 1 - S_1\left(BX\right) + S_3\left(BX\right) - S_1\left[B(R-X)\right] + S_3\left[B(R-X)\right]$$

$$-B \int_{X^*=0}^{R} \left\{ S_0\left(B\left[\left(X-X^*\right)^2 + 1\right]^{1/2} \right) - S_2\left(B\left[\left(X-X^*\right)^2 + 1\right]^{1/2} \right) \right\} dX^* \tag{11.102}$$

$$\frac{q_r(d,y)}{\sigma T_m^4} = 1 - S_1(BY) + S_3(BY) - S_1\left[B(1-Y)\right] + S_3\left[B(1-Y)\right]$$

$$-B \int_{Y^*=0}^{1} \left\{ S_0\left(B\left[R^2 + \left(Y-Y^*\right)^2\right]^{1/2} \right) - S_2\left(B\left[R^2 + (Y-Y^*)^2\right]^{1/2} \right) \right\} dY^* \tag{11.103}$$

where

$B = ab$ (optical dimension)
$R = d/b$ (aspect ratio)
$X = x/b$, $Y = y/b$

Tables of local heat flux along the boundary are in Siegel (1991a) for a wide range of aspect ratios and optical thicknesses. Figure 11.8 shows results for a square region. For large optical thicknesses, the dimensionless heat flux is equal to 1 along the boundary away from the corners and equals 1/2 at the corners. An extension for a nongray medium is in Siegel (1992a). References that provide solutions for radiative transfer in rectangular geometries are Thynell and Özişik (1987), Fiveland (1984), Yuen and Wong (1984), Razzaque et al. (1984), Siegel (1989b, 1990, 1991a, 1992a, 1993), Yuen and Takara (1990b), Güven and Bayazitoglu (2003), and Tencer (2014).

11.5.3 One-Dimensional Transfer in a Cylindrical Region

A nonplanar geometry that has been extensively studied is the cylindrical shape, which is of interest for radiation in tubular furnaces, cylindrical combustion chambers, and radiating hot gas flows in pipes. Studies in the literature include infinitely long cylinders (Men 1973, Thynell 1990b), concentric cylindrical regions (Greif and Clapper 1966, Pandey 1989, Wu and Wu 1993), and 2D analyses in cylinders of finite length (Reguigui and Dougherty 1992, Hsu and Ku 1994, Pessoa-Filho and Thynell 1996, Zhang and Sutton 1996). The analysis in Fernandes and Francis (1982)

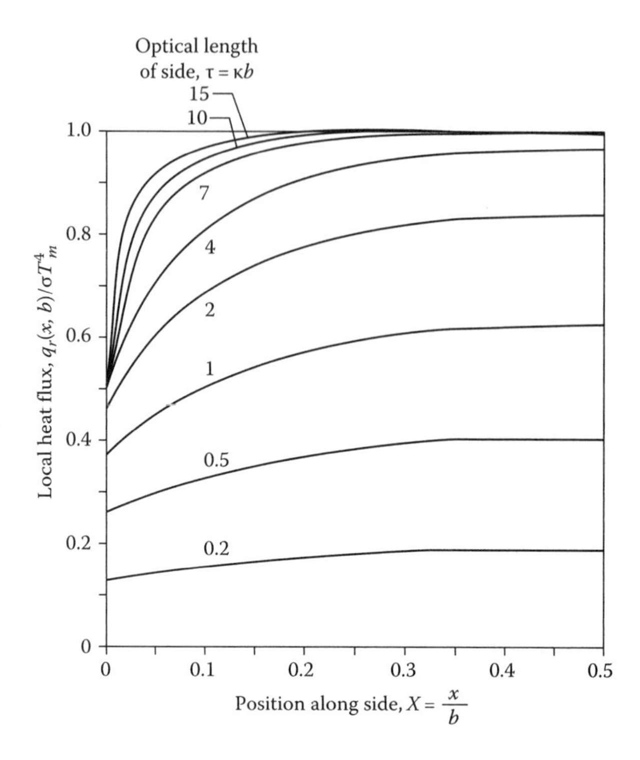

FIGURE 11.8 Local radiative flux along boundary of square absorbing–emitting region as a function of optical length of side. (From Siegel, R., *J. Heat Trans.*, 113(1), 258, 1991a.)

is for an infinite cylinder with isotropic scattering (some transient cases are also included), and Thynell (1992) and Azad and Modest (1981a,b) include models for anisotropic scattering. An inverse problem based on the forward solution for a cylindrical system is given by Mengüç and Manickavasagam (1993). For simplicity, the development here is for 1D behavior in an infinitely long axisymmetric cylinder of a gray medium, in which isotropic scattering is included. The energy equation requires a value for $\nabla \cdot \mathbf{q}_r$. The radiative heat flow at or across a boundary is obtained from $\mathbf{q}_r$. These quantities can be found by integrating, at any location, the contributions supplied by the radiation intensity as a result of radiation leaving boundary and volume elements. For a cylinder, the integrations become geometrically complex. This derivation is limited to axisymmetric conditions with no variations along the cylinder axis and is for constant radiative properties. The results are consequently 1D, and $\mathbf{q}_r$ depends only on radius. However, the required area and volume integrations are 3D.

The geometric aspects of the cylindrical shape are given by Heaslet and Warming (1966) and Kesten (1968) and the results have been used in various forms to obtain solutions in Thynell (1992), Fernandes and Francis (1982), Azad and Modest (1981a,b), and Siegel (1988, 1989a,c). The approach in Kesten is used to illustrate the derivation of the radial radiative heat flux. A typical path *AOB* is considered, Figure 11.9, and the intensities along this path are integrated at location *O* in the required manner to obtain the radiative flux, $q_r(r)$. $d\tau$ along the path is related to the distance dx projected on a cross section normal to the cylinder axis by $d\tau = dx/\cos\alpha = dx/\mu$, where $\mu = \cos\alpha$ and $d\tau$ and dx, and also r and R, are *optical* coordinates (physical coordinates multiplied by the attenuation coefficient β). From this point on in the derivation, the β in the equations is the angle shown in Figure 11.9, and *not* the attenuation coefficient. Then, from Equation 10.12, the radiative transfer equation is

$$\mu \frac{dI}{dx} + I(x) = \hat{I}(x) \tag{11.104}$$

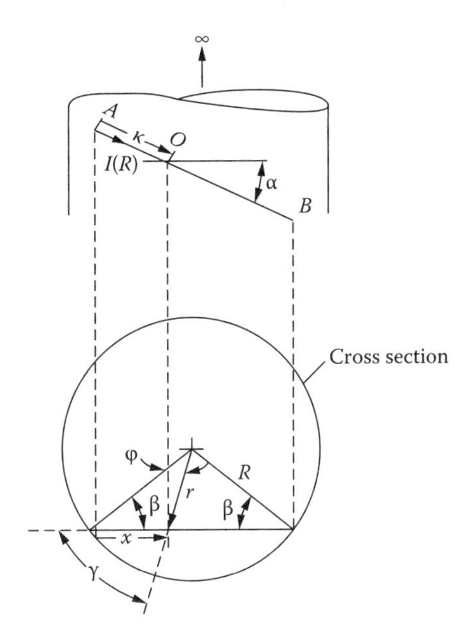

FIGURE 11.9 Geometry for radiative energy transfer in an infinitely long cylinder.

The law of cosines is used to relate x to the radial coordinate r,

$$r^2 = x^2 + R^2 - 2xR\cos\beta \tag{11.105}$$

Angle β remains constant along the path x, so, by differentiating Equation 11.105, dx and dr are related along the path by

$$rdr = (x - R\cos\beta)dx \tag{11.106}$$

Equations 11.105 and 11.106 are combined to eliminate x, with the result that

$$dx = \pm\frac{rdr}{\left(r^2 - R^2\sin^2\beta\right)^{1/2}}\begin{Bmatrix} -\text{for} & 0 \le x < R\cos\beta \\ +\text{for} & R\cos\beta < x \le 2R\cos\beta \end{Bmatrix} \tag{11.107}$$

Letting $F(r,\beta) = \left(r^2 - R^2\sin^2\beta\right)^{1/2}/r$, Equation 11.104 becomes, in terms of r,

$$\frac{dI}{dr} \pm \frac{1}{\mu F(r,\beta)} I = \pm\frac{\hat{I}}{\mu F(r,\beta)} \tag{11.108}$$

with the signs used as previously defined. Equation 11.108 can be integrated as for Equation 10.14. It is assumed that a diffuse intensity $I(R)$, in the direction of τ, is leaving the inside of the boundary at location A. To shorten the relations, define $\Gamma(a,b) \equiv \int_a^b d\xi/F(\xi,\beta)$. Then, from Equation 10.14, I along the path x is given by

$$I(r) = I(R)e^{(1/\mu)\Gamma(R,r)} - \frac{1}{\mu}e^{(1/\mu)\Gamma(R,r)}\int_R^r \frac{e^{-(1/\mu)\Gamma(R,r^*)}\hat{I}\left(r^*\right)}{F(r^*,\beta)}dr^* \quad (0 \le x < R\cos\beta) \tag{11.109}$$

$$I(r) = I(R\sin\beta)e^{-(1/\mu)\Gamma(R\sin\beta,r)} + \frac{1}{\mu}e^{-(1/\mu)\Gamma(R\sin\beta,r)}\int_{R\sin\beta}^{r}\frac{e^{(1/\mu)\Gamma(R\sin\beta,r^*)}\hat{I}(r^*)}{F(r^*,\beta)}dr^* \tag{11.110}$$

$$(R\cos\beta < x \le 2R\cos\beta)$$

where $I(R\sin\beta)$ is Equation 11.109 evaluated at $r = R\sin\beta$.

The heat flux in the radially outward direction is obtained by integrating the intensities passing through a cylindrical surface with constant r, with each intensity weighted by the projected area of the surface normal to it, $q_r(r) = \int_{\Omega=0}^{4\pi} I(r)\cos\theta d\Omega$. The angle θ is between $I(r)$ and the outward normal to the cylindrical surface. Using $\cos\theta = -\cos\alpha\cos\gamma$ and the solid angle as obtained by projecting an element of surface area onto a unit sphere, $d\Omega = \sin(\pi/2-\alpha)d\gamma d(\pi/2-\alpha) = -\cos\alpha d\gamma d\alpha$, $q_r(r)$ becomes

$$q_r(r) = -4\int_{\alpha=0}^{\pi/2}\int_{\gamma=0}^{\pi} I(r)\cos^2\alpha\cos\gamma d\gamma d\alpha \tag{11.111}$$

$q_r(r)$ can be expressed in terms of β rather than γ by using the relation from Figure 11.9,

$$r\sin\gamma = R\sin\beta\begin{Bmatrix} 0 \le \gamma < \dfrac{\pi}{2}, & 0 \le x < R\cos\beta \\ \dfrac{\pi}{2} < \gamma \le \pi, & R\cos\beta < x \le 2R\cos\beta \end{Bmatrix} \tag{11.112}$$

Let I^- and I^+ be the intensities corresponding to Equations 11.109 and 11.110. Then

$$\begin{aligned} q_r(r) &= -4\frac{R}{r}\int_{\alpha=0}^{\pi/2}\left(\int_0^{\sin^{-1}(r/R)} I^-\cos\beta d\beta + \int_{\sin^{-1}(r/R)}^{0} I^+\cos\beta d\beta\right)\cos^2\alpha d\alpha \\ &= 4\frac{R}{r}\int_{\alpha=0}^{\pi/2}\left[\int_0^{\sin^{-1}(r/R)}(I^+ - I^-)\cos\beta d\beta\right]\cos^2\alpha d\alpha \end{aligned} \tag{11.113}$$

Equations 11.109 and 11.110 are now substituted for I^+ and I^-. The integration over α can be carried out as previously done for the rectangular cross section. This accounts for all axial locations along the cylinder and results in S_n functions as defined in Equations 11.97. The expression for $q_r(r)$ becomes

$$\begin{aligned} q_r(r) = 2\pi\frac{R}{r}\int_{\beta=0}^{\sin^{-1}(r/R)} &\Big\{I(R)\Big(S_3\big[2\Gamma(R\sin\beta,r)+\Gamma(r,R)\big] - S_3\big[\Gamma(r,R)\big]\Big) \\ &+ \int_{r^*=R\sin\beta}^{r}\frac{\hat{I}(r^*)}{F(r^*,\beta)}S_2\big[\Gamma(r^*,r)\big]dr^* - \int_{r^*=r}^{R}\frac{\hat{I}(r^*)}{F(r^*,\beta)}S_2\big(\big[\Gamma(r,r^*)\big]\big)dr^* \\ &+ \int_{r^*=R\sin\beta}^{R}\frac{\hat{I}(r^*)}{F(r^*,\beta)}S_2\big[\Gamma(R\sin\beta,r)+\Gamma(R\sin\beta,r^*)\big]dr^*\Big\}\cos\beta d\beta \end{aligned} \tag{11.114}$$

The physical interpretation of the four terms in Equation 11.114 can be visualized by considering the effect on the heat flux at r of radiation traveling along a diameter of the cylinder. The first term is the heat flux from the cylinder boundary attenuated as it passes through the medium. The positive portion is the radiation passing from the boundary through the center region of the cylinder to r, while the negative portion is from the boundary on the opposite side. The second term is the positive contribution by the emitting and scattering medium between the center of the cylinder and r, while the third term is the negative heat flux contribution by the medium between the boundary and r. The fourth term is the radiation emitted and scattered by the medium between the boundary and the center of the cylinder and then attenuated as it passes from the source location through the center of the cylinder to r.

$\nabla \cdot \mathbf{q}_r$ for substitution in the energy equation can then be obtained in cylindrical coordinates for the 1D axisymmetric case by carrying out the differentiation $(1/r)d(rq_r)/dr$. $\nabla \cdot \mathbf{q}_r$ can also be derived by carrying out the integral form for $\bar{I}$ in Equation 11.93 and then using Equation 11.90. This approach was used by Heaslet and Warming (1966). Various forms for $\mathbf{q}_r$ and $\nabla \cdot \mathbf{q}_r$ are also given by Fernandes and Francis (1982), Azad and Modest (1981a,b), and Siegel (1988).

11.5.4 Additional Information on Nonplanar and Multidimensional Geometries

The quantities for the 1D axisymmetric cylindrical case in the previous section depend only on radius. An extension to another radially dependent case is the axisymmetric cylindrical annulus (Habib 1973, Viskanta and Anderson 1975, Wu and Wu 1993). In Wu and Wu, there are two concentric regions around an opaque cylinder. Anisotropic scattering in an annular region is treated in Harris (1989) by using the P_1 and P_3 approximations. A cylindrical region with variable radius is analyzed by Song et al. (1998). The more complex 2D analysis of a cylinder of finite length is considered by Yucel and Williams (1987), Reguigui and Dougherty (1992), Hsu and Ku (1994), Zhang and Sutton (1996), and Pessoa-Filho and Thynell (1996). Mengüç and Manickavasagam (1993) presented a formulation for an axisymmetric cylindrical system and then used it for an inverse solution.

Another radially dependent geometry is a sphere (Thynell and Özişik 1985, Dombrovsky 2000, Liu et al. 2002). The region between two uniform concentric spheres is analyzed by Viskanta and Merriam (1968), Tong and Swathi (1987), and El-Wakil and Abulwafa (2000). In Viskanta and Merriam, the bounding spheres are diffuse–gray, and the medium can emit, absorb, and scatter isotropically. The same geometry is considered in Tong and Swathi and in El-Wakil and Abulwafa anisotropic scattering is included; additional references are also given for spherical geometries.

Radiative transfer within a 2D corner region was analyzed by Wu and Fu (1990). The geometry is a layer of fixed thickness that extends around the corner along the exterior surface of a rectangle. An application is the radiative performance of a coating placed on the outside of a corner.

For 3D situations, a cylinder can be considered with both circumferential and axial variations. In Mengüç et al. (1985) an axisymmetric cylinder of finite length was analyzed, including anisotropic scattering. The long rectangular region discussed previously is further generalized to a 3D rectangular region in Mengüç and Viskanta (1985) and Fiveland (1988), and some exact radiative transfer relations for 3D rectangular geometries are in Crosbie and Schrenker (1982) and Lin (1987). Modeling of more general multidimensional geometries is developed by Carvalho et al. (1993) and Martynenko et al. (1998).

An extensive review of transient solutions in translucent materials including radiation is in Siegel (1998) and further discussion is in Chapter 17. A solution for the penetration of radiation is in Manohar et al. (1995). Transient heat transfer in a layer with Arrhenius heat generation is studied by Crosbie and Pattabongse (1987). Numerical solutions for spherical geometries are found in Viskanta and Lall (1965, 1966) and Tsai and Özişik (1987), and results for the region between coaxial cylinders are in Chang and Smith (1970). Wendlandt (1973) derives expressions for the transient temperature distribution in a semi-infinite absorbing medium exposed to a laser pulse. An analysis for a very short laser pulse is in Guo et al. (2000) including the effect of the speed of radiant propagation. The transient heating of a number of semitransparent solid geometries is reviewed by

Viskanta and Anderson (1975). Guo and Kumar (2002) use the discrete ordinates method to treat transient radiation effects in 3D systems and apply it to a rectangular enclosure. Tan et al. (2006) apply a meshless least-squares collocation technique to transient radiative transfer.

The body of the literature is growing rapidly in both journal articles and conference proceedings. More recent papers are summarized at the web site of the book, www.thermalradiation.net at regular intervals.

HOMEWORK

11.1 A semi-infinite, isotropically scattering, absorbing–emitting medium maintained at uniform temperature T_g is in contact with a black boundary at T_w. The medium is gray with constant scattering and absorption coefficients σ_s and κ. The medium is not moving, and heat conduction is neglected. Show that the heat flux transferred to the boundary can be expressed in the form $q = H\sigma(T_g^4 - T_w^4)$. Provide the relations needed to obtain values for H.

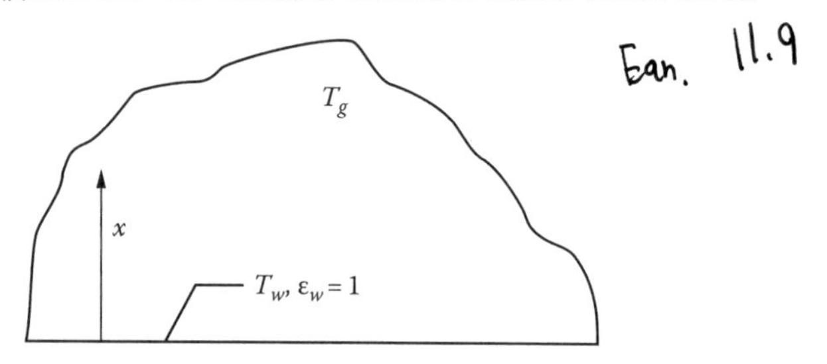

11.2 In Homework Problem 11.1, the plane boundary at $x = 0$ is modified to be diffuse–gray instead of black. Derive the integral equation relations that can be numerically evaluated to determine the heat flux transferred to the boundary. Place the equations in a convenient dimensionless form.

11.3 A nonscattering stagnant gray medium with absorption coefficient $\kappa = 0.1$ cm^{-1} is contained between large black parallel plates 20 cm apart as shown. Assume the medium has constant density, negligible heat conduction, and $n \approx 1$. What is the net energy flux being transferred by radiation from the lower to the upper plate? If the black plates are replaced with gray surfaces with $\varepsilon_1 = 0.6$ and $\varepsilon_2 = 0.1$, what is the energy flux being transferred? Plot the temperature distribution $T(x)$ for black boundaries and for gray boundaries.

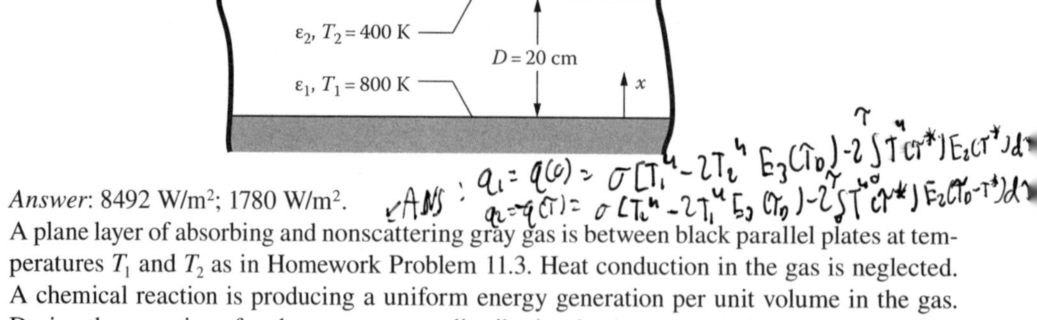

Answer: 8492 W/m^2; 1780 W/m^2.

11.4 A plane layer of absorbing and nonscattering gray gas is between black parallel plates at temperatures T_1 and T_2 as in Homework Problem 11.3. Heat conduction in the gas is neglected. A chemical reaction is producing a uniform energy generation per unit volume in the gas. Derive the equations for the temperature distribution in the gas and the local energy flux in the x direction. What is the equation to obtain the net energy flux q_1 and q_2 supplied to each plate? What does the temperature distribution become for the limiting case when $\tau_D = \kappa_D \to 0$ (optically thin layer)?

11.5 A scattering, nonabsorbing, nonconducting medium is contained between large parallel plates 8 cm apart. The scattering is assumed to be isotropic and independent of wavelength, and the scattering coefficient is $\sigma_s = 0.2$ cm^{-1}. The plate temperatures are $T_1 = 950$ K and $T_2 = 630$ K. Compute the net radiative heat flux transferred from plate 1 to plate 2 if the plates are black or if the plates are gray with $\varepsilon_1 = 0.85$ and $\varepsilon_2 = 0.32$. How do these transfers compare with the values for the plates in a vacuum?

Answer: 37250 W/m^2 (black, vacuum); 11280 W/m^2 (gray, vacuum); 16530 W/m^2 (black, medium); 8179 W/m^2 (gray, medium).

11.6 A gray absorbing and scattering medium is contained between gray parallel plates 8 cm apart with $T_1 = 950$ K, $\varepsilon_1 = 0.85$, $T_2 = 630$ K, $\varepsilon_2 = 0.32$. The scattering is isotropic and independent of wavelength, and heat conduction is neglected. Compute the net energy transfer from plate 1 to plate 2 if the scattering and absorption coefficients are, respectively, $\sigma_s = 0.2$ cm^{-1} and $\kappa = 0.1$ cm^{-1}. Compare the result with that of Homework Problem 11.5 to show the effect of adding absorption in the medium.

Answer: 7210 W/m^2.

11.7 An enclosed rectangular region is filled with an absorbing, emitting, and isotropically scattering medium. The medium is gray and has absorption coefficient κ and scattering coefficient σ_s. The geometry is long in the z direction, and the side dimensions are b and d. The four boundaries are black. The vertical boundaries are both at uniform temperature T_v, and the horizontal boundaries are at T_h. The medium is heated uniformly throughout its volume with a volumetric energy rate $\dot{q}$. The effect of heat conduction is neglected to obtain the limit where radiation is dominating, and the medium is not moving. Provide the energy equation and radiative transfer relations including scattering that are required to solve the steady-state temperature distribution in the rectangular cross section. Place the equations in convenient dimensionless forms for use in a numerical solution.

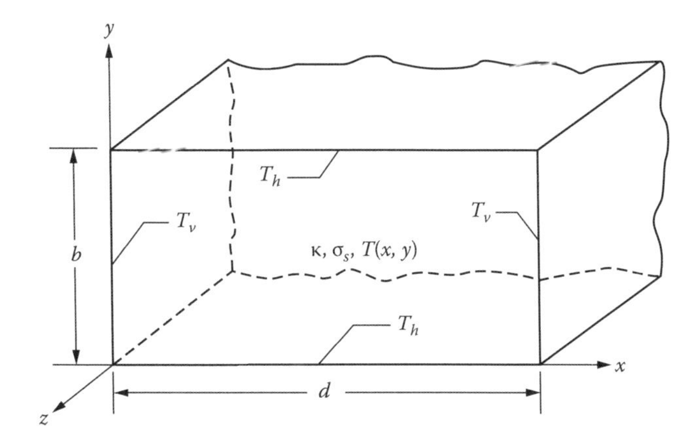

11.8 A rectangular enclosure that is very long normal to the cross section shown has diffuse–gray boundaries at conditions shown in the following and encloses a uniform gray gas at $T_g = 1500$ K. The gas has an absorption coefficient of 0.25 m^{-1}. Find the average net radiative flux at each surface and the energy necessary to maintain the gas at 1500 K. (Some necessary relations are given in Appendix B at the book website http://www.thermalradiation.net.)

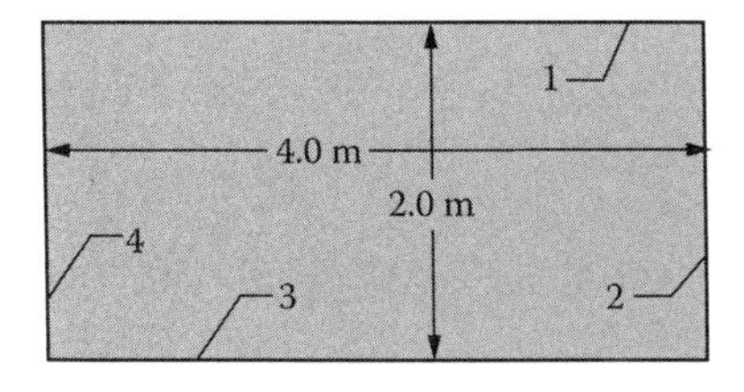

Surface	ε	T_w (K)
1	1.0	2000
2	0.5	1500
3	0.1	1000
4	0	500

Answer: $q_1 = 607$ kW/m^2; $q_2 = -75.1$ kW/m^2; $q_3 = -39.3$ kW/m^2; energy added = 2121 kW/m.

11.9 Two infinite diffuse–gray parallel plates at temperatures T_1 and T_2, with respective emissivities ε_1 and ε_2, are separated by a distance D. The space between them is filled with a gray absorbing and scattering medium having a constant extinction coefficient β. Obtain expressions for the net radiative heat flux being transferred between the plates and the temperature distribution in the medium, by using the nearly transparent approximation. Heat conduction can be neglected compared with radiative transfer.

Example 11.1 p. 569

Answer: $q = \sigma\left(T_1^4 - T_2^4\right)\Big/\left(\dfrac{1}{\varepsilon_1} + \dfrac{1}{\varepsilon_2} - 1\right)$; $\dfrac{T^4 - T_2^4}{T_1^4 - T_2^4} = \dfrac{\dfrac{1}{\varepsilon_2} - \dfrac{1}{2}}{\dfrac{1}{\varepsilon_1} + \dfrac{1}{\varepsilon_2} - 1}$.

11.10 Two gases have constant extinction coefficients β_1 and β_2 that are both much smaller than 1. The gases are separated by a thin metal barrier with low emissivity ε_s. The gases are in the form of two layers bounded by large parallel diffuse boundaries as shown. Radiation exchange is large compared with heat conduction that can be neglected. Using the nearly transparent approximation, obtain relations for the temperature distributions within the gases and the net radiative energy transfer from boundary 1 to boundary 2. Evaluate the results for $\varepsilon_1 = 0.78$, $\varepsilon_s = 0.04$, $\varepsilon_2 = 0.88$, $T_1 = 980$ K, and $T_2 = 550$ K. Determine the temperature jump across the thin metal barrier and determine the size of the jump as a function of ε_s.

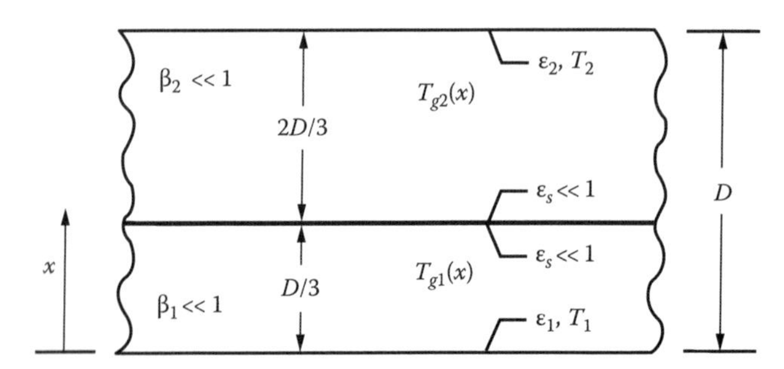

Answer: $q = 934.4$ W/m^2, $T_{g1} = 977$ K, $T_{g2} = 567$ K, $\Delta T_{jump} = 410$ K.

11.11 A gray medium is contained between parallel black boundaries. The optical thickness based on plate spacing is $\tau_D = 1.0$. Find the dimensionless radiative heat flux between the plates, Ψ_b, and compare the result with the value computed by Heaslet and Warming (1965) in Table 11.2. Use the exact relations in Section 11.4.5.2.
Answer: 0.5532.

12 Solution Methods for Radiative Transfer in Participating Media

12.1 INTRODUCTION

In the previous chapters, governing relations were presented for analyzing radiative transfer and energy conservation. These relations were applied to one- and two-dimensional (1- and 2D) systems, along with some approximate methods for optically thin media. In this chapter, methods are given for treating more general cases of radiative transfer along with numerical methods for carrying out solutions.

The presence of computers based on massively parallel architectures now makes solution of very complex problems possible. Some methods outlined in earlier editions of this text have been superseded by more accurate techniques. In particular, the Curtis–Godson method, which seeks to handle spectral properties by substituting a fictitious medium that obeys the radiative transfer equation (RTE) at optically thin and thick limits, and the exponential kernel approximation, which replaces the exponential integral functions E_n with simpler functions to allow analytical solution in some cases, have both been removed from the book. However, they are available in Appendices C and D at http://www.thermalradiation.net.

As noted in Chapter 1, the radiation intensity is dependent on seven independent variables: time, a spectral variable, three spatial variables, and two angular variables. In this text, we assume that the dependence of intensity on time can be neglected, as the time of flight of photons at the speed of light is usually negligibly small in comparison with other time-varying changes encountered through coupling with the energy equation. (For applications where the time-dependent term in the RTE cannot be neglected, see Olson et al. [2000], Frank [2007], for discussions of solutions where temporal effects are important.) Solution of the RTE for the local intensity thus requires treatment of six variables: $I(\lambda, x, y, z, \theta, \phi)$. The spectral variable and methods for finding accurate spectral properties are discussed in Chapter 9. In this chapter, we discuss general methods for solving the RTE to determine the local spectral intensity, the local radiative flux, and the flux gradient. Chapter 13 extends the discussion to combined-mode problems that include radiation.

Each of the methods in this chapter is applied to solution of the general RTE (Equation 10.12) repeated here:

$$\frac{\partial I_\lambda(S,\Omega)}{\beta_\lambda \partial S} = \frac{\partial I_\lambda(S,\Omega)}{\partial \tau_\lambda(S)} = \hat{I}_\lambda(S) - I_\lambda(S,\Omega) \tag{12.1}$$

12.2 SERIES EXPANSION AND MOMENT METHODS

Many methods for the solution of the differential form of the RTE are based on moment methods, and these are often combined with using a series expansion of the intensity along a particular direction vector. The series is then truncated to some small number of terms, resulting in final formulations that are mathematically and numerically tractable. Five such methods are discussed in this section: the diffusion solution, the Milne–Eddington approximation, the P_N (spherical harmonics expansion) method, the SP_N (simplified P_N) method, and the M_N method. Further discussion is in Frank (2007) and Tencer (2013).

First, define the *radiative moment equations*. These equations are developed by multiplying the local intensity at S by powers of the direction cosines $l_i(i = 1, 2, 3)$, either individually or in combination, and then integrating over all solid angles. This results in the zeroth, first, second, and general moments as

$$I^{(0)}(S) = \int_{\Omega=0}^{4\pi} I(S,\Omega)d\Omega = 4\pi\bar{I}(S) = cU_\lambda(S) \quad (12.2)$$

$$I^{(i)}(S) = \int_{\Omega=0}^{4\pi} l_i I(S,\Omega)d\Omega = q_i(S) \quad (i = 1,2,3) \quad (12.3)$$

$$I^{(ij)}(S) = \int_{\Omega=0}^{4\pi} l_i l_j I(S,\Omega)d\Omega = cP_{ij}(S) \quad (i, j = 1,2,3) \quad (12.4)$$

$$I^{(ijkl\ldots)}(S) = \int_{\Omega=0}^{4\pi} l_i l_j l_k l_l \ldots I(S,\Omega)d\Omega \quad (i, j, k, l, \ldots, = 1,2,3) \quad (12.5)$$

The zeroth moment is 4π times the average local intensity or, if divided by c, is the radiative energy density (Equation 10.31). The first moment is the radiative energy flux in the i coordinate direction (Equation 10.26), and the second moment divided by c is the local radiation stress and pressure tensor. Higher-order moments have no physical meaning.

Taking moments of the RTE by integrating with respect to the direction cosines gives the RTE moment equations:

Zeroth moment RTE is

$$\int_{\Omega=0}^{4\pi} l_i \frac{\partial I_\lambda(S,\Omega)}{\partial\tau_\lambda} d\Omega = \int_{\Omega=0}^{4\pi} \hat{I}_\lambda(S)d\Omega - \int_{\Omega=0}^{4\pi} I_\lambda(S,\Omega)d\Omega \quad (12.6)$$

or, using the moment definitions,

$$\frac{\partial I_\lambda^{(1)}(S)}{\partial\tau_\lambda} = \int_{\Omega=0}^{4\pi} \hat{I}_\lambda(S)d\Omega - I_\lambda^{(0)}(S) \quad (12.7)$$

or

$$\frac{\partial q_{i,\lambda}(S)}{\partial\tau_\lambda} = \int_{\Omega=0}^{4\pi} \hat{I}_\lambda(S,\Omega)d\Omega - 4\pi\bar{I}_\lambda(S) \quad (12.8)$$

For the special case of isotropic scattering, the integral of the source function over all directions is zero, and Equation 12.8 reduces to

$$\frac{1}{4\pi}\frac{\partial q_{i,\lambda}(S)}{\partial\tau_\lambda} = -\bar{I}_\lambda(S) \quad (12.9)$$

The first moment RTE is

$$\int_{\Omega=0}^{4\pi} \frac{l_i l_j}{\beta_\lambda} \frac{\partial I_\lambda(S,\Omega)}{\partial S} d\Omega = \int_{\Omega=0}^{4\pi} l_j \hat{I}_\lambda(S) d\Omega - \int_{\Omega=0}^{4\pi} l_j I_\lambda(S,\Omega) d\Omega \tag{12.10}$$

or

$$I_\lambda^{(2)}(S) = \int_{\Omega=0}^{4\pi} l_j \hat{I}_\lambda(S) d\Omega - I_\lambda^{(1)}(S) \tag{12.11}$$

$$cP_{\lambda,ij}(S) = \int_{\Omega=0}^{4\pi} l_j \hat{I}_\lambda(S) d\Omega - q_{\lambda,i}(S) \tag{12.12}$$

Further moment equations are generated in a similar way, but the zero and first moment equations are sufficient for most solution methods. Equations 12.1, 12.8, and 12.11 form a set of three partial differential equations involving the local intensity $I_\lambda(S)$ and the first three moments $I_l^{(0)}$, $I_l^{(1)}$, and $I_l^{(2)}$ (or alternatively, $I_\lambda(S)$, $q_{\lambda,i}(S)$, $\bar{I}_\lambda(S)$, and $P_{\lambda,ij}(S)$). Thus, some approximation must be made to provide a fourth equation to close the set, called a closure condition. Usually, this is in the form of some approximation for the radiative pressure tensor, $P_{\lambda,ij}(S)$.

12.2.1 Optically Thick Media, Radiative Diffusion

In an optically thick medium, radiation travels only a short distance before being scattered or absorbed. The local intensity results from radiation originating from the nearby surroundings where emission and scattering are similar to that of the location under consideration. Radiation from locations where conditions are appreciably different is greatly attenuated in an optically thick medium before reaching the location being considered. For this situation, it is possible to transform the expressions for radiative energy into a diffusion relation similar to that for heat conduction. Energy transfer depends only on the conditions in the immediate vicinity of the position being considered and is expressed in terms of the gradient of the conditions at that position. The diffusion approximation provides a substantial simplification. Standard techniques such as finite-difference schemes can be used for solving the radiative diffusion differential equation.

Real gases are usually transparent in some wavelength regions. The diffusion approach can be applied only in wavelength regions where the optical thickness of the medium is greater than about 10; this value depends on the geometry and conditions of the problem and may need to be larger in some instances. The fact that some *mean* optical thickness meets this criterion is not sufficient. Sometimes good results for some quantities are predicted for much smaller optical thicknesses, as we show in Figure 12.4. Wavelength-band applications of the diffusion method can be made to use the method in the optically thick spectral regions (Siegel and Spuckler 1994a). In a high-temperature gas core nuclear reactor (Slutz et al. (1994), the gaseous uranium fuel has a high opacity, so the diffusion approximation is adequate for the desired thermal radiation heat transfer calculations.

The diffusion approximation requires the intensity to be nearly isotropic within the medium. This can occur in the interior of an optically thick medium with small temperature gradients, but is not valid near boundaries for certain types of conditions. For example, at a boundary of a hot medium adjacent to much lower or much higher temperature surroundings, radiation leaving and entering the medium will be much different. As a result of this large anisotropy, the diffusion approximation loses accuracy near the boundary, and a temperature jump boundary condition may be used to improve accuracy. If heat conduction is present and dominant near a boundary, the inaccuracy of the radiative diffusion can be less important.

12.2.1.1 Simplified Derivation of the Radiative Diffusion Approximation

A simplified derivation for a 1D layer is used to show the basic ideas in the diffusion approximation for an absorbing–emitting medium with isotropic scattering. A more general derivation is then discussed. Let H be a path length along a path direction S over which the radiant energy density changes appreciably, and let the extinction mean free path (Equation 1.50) be $l_{m,\lambda} = 1/(\kappa_\lambda + \sigma_{s\lambda}) = 1/\beta_\lambda$. For the diffusion approximation to apply, $l_{m,\lambda}/H = 1/\beta_\lambda H \ll 1$. In Figure 12.1a, using $dS = dx/\cos\theta$ in Equation 12.1, the RTE for the change of I_λ with x for a fixed θ direction is

$$-\frac{\cos\theta}{\beta_\lambda}\frac{\partial I_\lambda(x,\theta)}{\partial x} = I_\lambda(x,\theta) - \hat{I}_\lambda(x) \tag{12.13}$$

or, defining $\mu = \cos\theta$, using H to nondimensionalize so $X = x/H$, and defining the optical thickness as $\tau_\lambda = \beta_\lambda H$

$$-\frac{\mu}{\tau_\lambda}\frac{\partial I_\lambda(X,\mu)}{\partial X} = I_\lambda(X,\mu) - \hat{I}_\lambda(X) \tag{12.14}$$

where the source function $\hat{I}_\lambda$ does not depend on angle for isotropic scattering.

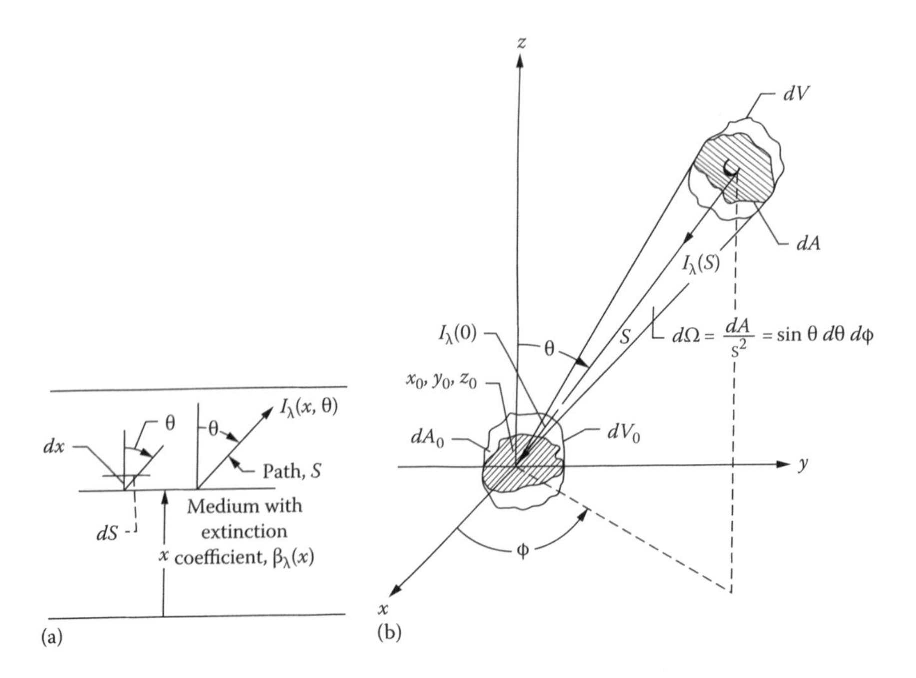

FIGURE 12.1 Geometry for derivation of diffusion equations: (a) 1D plane layer of medium and (b) general 3D region.

Now, solve Equation 12.14 by expanding the intensity in a series of unknown functions $I_\lambda^{(n)}$, $n = 0,1,2,\ldots$ multiplied by powers of $\frac{1}{\tau_\lambda} \ll 1$:

$$I_\lambda = I_\lambda^{(0)} + \frac{1}{\tau_\lambda} I_\lambda^{(1)} + \left(\frac{1}{\tau_\lambda}\right)^2 I_\lambda^{(2)} + \cdots \tag{12.15}$$

where the (n) superscript denotes the terms of the expanded intensity function, not to be confused with superscripts for moments.

Insert Equation 12.15 into 12.14 to obtain ($\hat{I}_\lambda$ is given by Equation 10.17)

$$-\mu \frac{1}{\tau_\lambda}\left[\frac{\partial I_\lambda^{(0)}}{\partial X} + \frac{1}{\tau_\lambda}\frac{\partial I_\lambda^{(1)}}{\partial X} + \cdots\right] = I_\lambda^{(0)} + \frac{1}{\tau_\lambda} I_\lambda^{(1)} + \cdots - (1-\omega_\lambda) I_{\lambda b}$$
$$- \omega_\lambda \left[\frac{1}{4\pi}\int_{\Omega_i=4\pi}\left(I_\lambda^{(0)} + \frac{1}{\tau_\lambda} I_\lambda^{(1)} + \cdots\right) d\Omega_i\right] \tag{12.16}$$

where $\omega_\lambda = \sigma_{s\lambda}/\beta_\lambda$. In addition to the expansion parameter $1/\tau_\lambda$, there is the quantity ω_λ, the scattering albedo, which characterizes extinction by scattering relative to total extinction. For small ω_λ, there is diffusion by absorption alone, as in a dense gas containing few scattering particles. For $\omega_\lambda \to 1$, there is scattering alone, as in a poorly absorbing gas with many suspended scattering particles.

In Equation 12.16, collect the terms of zero order in $1/\tau_\lambda$ to obtain

$$I_\lambda^{(0)} = (1-\omega_\lambda) I_{\lambda b} + \frac{\omega_\lambda}{4\pi}\int_{\Omega_i=0}^{4\pi} I_\lambda^{(0)} d\Omega_i \tag{12.17}$$

The terms $I_{\lambda b}$ and the first moment of $I_\lambda^{(0)}$, $\int_{\Omega_i=0}^{4\pi} I_\lambda^{(0)} d\Omega_i$, on the right do not depend on incidence angle $d\Omega_i$. Hence, $I_\lambda^{(0)}$ on the left cannot depend on Ω_i. Using this fact, the first moment (integral) term in Equation 12.17 reduces to $\omega_\lambda I_\lambda^{(0)}$ so that Equation 12.17 becomes

$$I_\lambda^{(0)} = I_{\lambda b} \tag{12.18}$$

Now collect the terms in Equation 12.16 of first order in $1/\tau_\lambda$ and substitute Equation 12.18 for $I_\lambda^{(0)}$ to obtain

$$-\mu \frac{dI_{\lambda b}}{dX} = I_\lambda^{(1)} - \frac{\omega_\lambda}{4\pi}\int_{\Omega_i=0}^{4\pi} I_\lambda^{(1)} d\Omega_i \tag{12.19}$$

To find $I_\lambda^{(1)}$, multiply by $d\Omega_i = 2\pi \sin\theta d\theta = -2\pi\mu d\mu$ and integrate over all $d\Omega_i$:

$$\frac{dI_{\lambda b}}{dX}\int_{\mu=-1}^{1} 2\pi\mu d\mu = \int_{\Omega_i=0}^{4\pi} I_\lambda^{(1)} d\Omega_i - \frac{\omega_\lambda}{4\pi}\left(\int_{\Omega_i=0}^{4\pi} I_\lambda^{(1)} d\Omega_i\right)\int_{\Omega_i=0}^{4\pi} d\Omega_i$$

The integral on the left is zero, so

$$0 = \int_{\Omega_i=0}^{4\pi} I_\lambda^{(1)} d\Omega_i - \omega_\lambda \int_{\Omega_i=0}^{4\pi} I_\lambda^{(1)} d\Omega_i$$

For any $\omega_\lambda \neq 1$, the $\int_{\Omega_i=0}^{4\pi} I_\lambda^{(1)} d\Omega_i = 0$, and Equation 12.19 reduces to

$$I_\lambda^{(1)} = -\mu \frac{dI_{\lambda b}}{dX} \tag{12.20}$$

Substitute Equations 12.18 and 12.20 into Equation 12.15 to obtain

$$I_\lambda = I_{\lambda b} - \frac{\mu}{\tau_\lambda} \frac{dI_{\lambda b}}{dX} \tag{12.21}$$

This reveals the important feature that *in the diffusion limit, the local intensity depends only on the magnitude and the gradient of the local blackbody intensity.* Since temperature gradients are small and τ_λ is large, the last term on the right is small and I_λ is nearly isotropic, like $I_{\lambda b}$.

The local spectral energy flux at X flowing in the X direction is found by multiplying I_λ by $\mu\, d\lambda$ and integrating over all solid angles:

$$q_\lambda(X) d\lambda = 2\pi d\lambda \int_{\mu=-1}^{1} I_\lambda(\mu) \mu d\mu \tag{12.22}$$

Because the blackbody intensity $I_{\lambda b}$ does not depend on μ, using Equation 12.21 in Equation 12.22 gives

$$q_\lambda(X) d\lambda = 2\pi I_{\lambda b}(X) d\lambda \int_{\mu=-1}^{1} \mu d\mu - \frac{2\pi d\lambda}{\tau_\lambda(X)} \frac{dI_{\lambda b}}{dX} \int_{\mu=-1}^{1} \mu^2 d\mu = -\frac{4\pi}{3\tau_\lambda(X)} \frac{dI_{\lambda b}}{dX} d\lambda \tag{12.23}$$

Equation 12.23 is the *Rosseland diffusion equation*. The local radiative energy flux depends only on local conditions. In a gray medium, since $\pi I_{\lambda b}\, d\lambda$ integrated over all λ is σT^4, Equation 12.23 gives the total radiative flux as

$$q(X) = -\frac{4\sigma}{3\tau(X)} \frac{dT^4}{dX} = -\frac{16\sigma T^3}{3\tau(X)} \frac{dT}{dX} \tag{12.24}$$

For $\kappa = 0$ ($\omega = 1$, pure scattering), energy must be supplied by an external source since the medium does not emit radiation; then σT^4 in Equation 12.24 is replaced by $\pi \bar{I}$ *as* in Section 11.4.3.

12.2.1.2 General Radiation–Diffusion Relations in a Medium

In the previous section, to obtain the diffusion equation, only first-order terms were retained in the series Equation 12.15, and boundaries were not considered. More general equations, including second-order terms, were considered in Deissler (1964) and boundary conditions were introduced so that the diffusion equations could be applied to finite regions. The intermediate equations in the derivation become somewhat complex because of their general form in 3D Taylor series expansions. The final results are provided here, and Deissler can be referred to for details.

12.2.1.2.1 Rosseland Diffusion Equation for Local Radiative Flux

A control volume is used in a 3D region, as in Figure 12.1b. The energy from dV to dV_0 is derived from the equation of transfer. In the general diffusion approximation, the radiation at dV_0 originates only from locations close to dV_0. The intensities are expanded about dV_0 in a 3D Taylor series and then truncated after second-order terms; this is sufficient in an optically dense medium where the diffusion approximation is assumed to apply. The intensities are used to integrate over all directions for energy approaching dV_0 from the positive and negative coordinate directions.

The second-order terms cancel when the diffusion approximation is imposed. Hence, the Rosseland diffusion equation obtained in the general second-order derivation is found to be the same as in Equation 12.24. In a 3D geometry (Figure 12.1b) the general relation for the local radiative energy flux (at any position **r**) in each coordinate direction including isotropic scattering (the z direction is used here) is given by

$$q_{\lambda,z}\Big|_{\mathbf{r}} d\lambda = -\frac{4\pi}{3\beta_\lambda(\mathbf{r})}\left(\frac{\partial I_{\lambda b}}{\partial z}\right)_{\mathbf{r}} d\lambda = -\frac{4}{3\beta_\lambda(\mathbf{r})}\left(\frac{\partial E_{\lambda b}}{\partial z}\right)_{\mathbf{r}} d\lambda \tag{12.25}$$

This is the general relation for local spectral radiative energy flux in terms of the local emissive power gradient. Equation 12.25 has the same form as the Fourier heat conduction law, permitting solution of some radiation problems by solution methods similar to those used for heat conduction.

A mean of the attenuation coefficient β_λ may be defined for the medium. To obtain the energy flux in a wavelength range, integrate Equation 12.25 over $\Delta\lambda$ (the parentheses and **r** subscript are omitted for simplicity):

$$\begin{aligned} q_{\Delta\lambda,z} &= -\int_{\Delta\lambda} \frac{4}{3\beta_\lambda}\left(\frac{\partial E_{\lambda b}}{\partial z}\right) d\lambda \equiv -\frac{4}{3\beta_{R,\Delta\lambda}} \int_{\Delta\lambda} \frac{\partial E_{\lambda b}}{\partial z} d\lambda \\ &= -\frac{4}{3\beta_{R,\Delta\lambda}} \frac{\partial}{\partial z} \int_{\Delta\lambda} E_{\lambda b} d\lambda = -\frac{4}{3\beta_{R,\Delta\lambda}} \frac{\partial E_{\Delta\lambda b}}{\partial z} \end{aligned} \tag{12.26}$$

This defines the mean attenuation coefficient $\beta_{R,\Delta\lambda}$ as

$$\frac{1}{\beta_{R,\Delta\lambda}} \equiv \frac{\int_{\Delta\lambda} (1/\beta_\lambda)(\partial E_{\lambda b} / \partial z) d\lambda}{\int_{\Delta\lambda} (\partial E_{\lambda b}/\partial z) d\lambda} = \frac{\int_{\Delta\lambda} (1/\beta_\lambda)(\partial E_{\lambda b}/\partial E_b) d\lambda}{\int_{\Delta\lambda} (\partial E_{\lambda b}/\partial E_b) d\lambda} \tag{12.27}$$

For the entire λ range, there is the important result that the local total radiative flux is

$$q_z = -\frac{4}{3\beta_R}\frac{\partial E_b}{\partial z} = -\frac{4}{3\beta_R}\frac{\partial(\sigma T^4)}{\partial z} = -\frac{16}{3\beta_R}\sigma T^3 \frac{\partial T}{\partial z} \tag{12.28}$$

where

$$\frac{1}{\beta_R} \equiv \int_{\lambda=0}^{\infty} \left(\frac{1}{\beta_\lambda}\right)\left(\frac{\partial E_{\lambda b}}{\partial E_b}\right) d\lambda \tag{12.29}$$

The β_R is the local *Rosseland mean attenuation coefficient*, named after S. Rosseland, who was the first to use diffusion theory in studying radiation effects in astrophysics (Rosseland 1936). The $\partial E_{\lambda b}/\partial E_b$ is found by differentiating Planck's law (Equation 1.15) after letting $T = (E_b/\sigma)^{1/4}$:

$$\frac{\partial E_{\lambda b}}{\partial E_b} = \frac{\partial}{\partial E_b}\left(\frac{2\pi C_1}{\lambda^5\left\{\exp\left[(C_2/\lambda)(\sigma/E_b)^{1/4}\right]-1\right\}}\right) = \frac{\pi}{2}\frac{C_1 C_2}{\lambda^6}\frac{\sigma^{1/4}}{E_b^{5/4}}\frac{\exp\left[(C_2/\lambda)(\sigma/E_b)^{1/4}\right]}{\left\{\exp\left[(C_2/\lambda)(\sigma/E_b)^{1/4}\right]-1\right\}^2}$$

$$= \frac{\pi}{2}\frac{C_1 C_2}{\lambda^6}\frac{1}{\sigma T^5}\frac{\exp(C_2/\lambda T)}{\left[\exp(C_2/\lambda T)-1\right]^2}$$

Then,

$$\frac{1}{\beta_R} \equiv \int_{\lambda=0}^{\infty}\left(\frac{1}{\beta_\lambda}\right)\left(\frac{\partial E_{\lambda b}}{\partial E_b}\right)d\lambda = \frac{\pi}{2}\frac{C_1 C_2}{\sigma T^5}\int_{\lambda=0}^{\infty}\left(\frac{1}{\beta_\lambda}\right)\left(\frac{\exp(C_2/\lambda T)}{\lambda^6\left[\exp(C_2/\lambda T)-1\right]^2}\right)d\lambda \tag{12.30}$$

If the spectral absorption coefficient β_λ has spectral regions of transparency, this definition breaks down and special steps must be taken.

12.2.1.2.2 Emissive Power Jump Boundary Condition

To this point, the location in the medium was considered to be far enough (in optical thickness) from any boundary that the boundary conditions did not enter the diffusion relations. Now the interaction with a diffuse wall is considered for the limit when radiation is the only means of energy transfer (no conduction or convection). Unlike the cases where conduction or convection is present, the boundary temperature and the medium temperature adjacent to the boundary are not taken as equal. Let the wall bounding the medium from earlier, Figure 12.2, have a hemispherical spectral emissivity $\varepsilon_{\lambda w2}$. All quantities evaluated on the wall have the subscript w to distinguish them from quantities in the medium at the wall; these do not have a w subscript. Consider area dA_2 in the medium parallel and immediately adjacent to the wall. The spectral energy quantities passing through dA_2 are shown in Figure 12.2, so the net spectral flux across dA_2 in the positive z direction is

$$(q_{\lambda,z})_2\,d\lambda = \varepsilon_{\lambda,w2}\left[G_{\lambda,2} - E_{\lambda,bw2}\right]d\lambda \tag{12.31}$$

Rearranging gives

$$-E_{\lambda,bw2} = \frac{(q_{\lambda,z})_2}{\varepsilon_{\lambda,w2}} - G_{\lambda,2} \tag{12.32}$$

The incident radiative flux (irradiation) $G_{2,\lambda}$ is expanded in a Taylor series and terms through second order are retained (Deissler 1964). The diffusion equation (Equation 12.24) is substituted for the first-order terms, and the result is the following relation for the discontinuity ("jump") in emissive power at dA_2 in the limit of only radiative transfer:

$$E_{\lambda,b2} - E_{\lambda,bw2} = \left(\frac{1}{\varepsilon_{\lambda,w2}} - \frac{1}{2}\right)(q_{\lambda,z})_2 - \frac{1}{2\beta_\lambda^2}\left(\frac{\partial^2 E_{\lambda,b}}{\partial z^2} + \frac{1}{2}\frac{\partial^2 E_{\lambda,b}}{\partial y^2} + \frac{1}{2}\frac{\partial^2 E_{\lambda,b}}{\partial x^2}\right)_2 \tag{12.33}$$

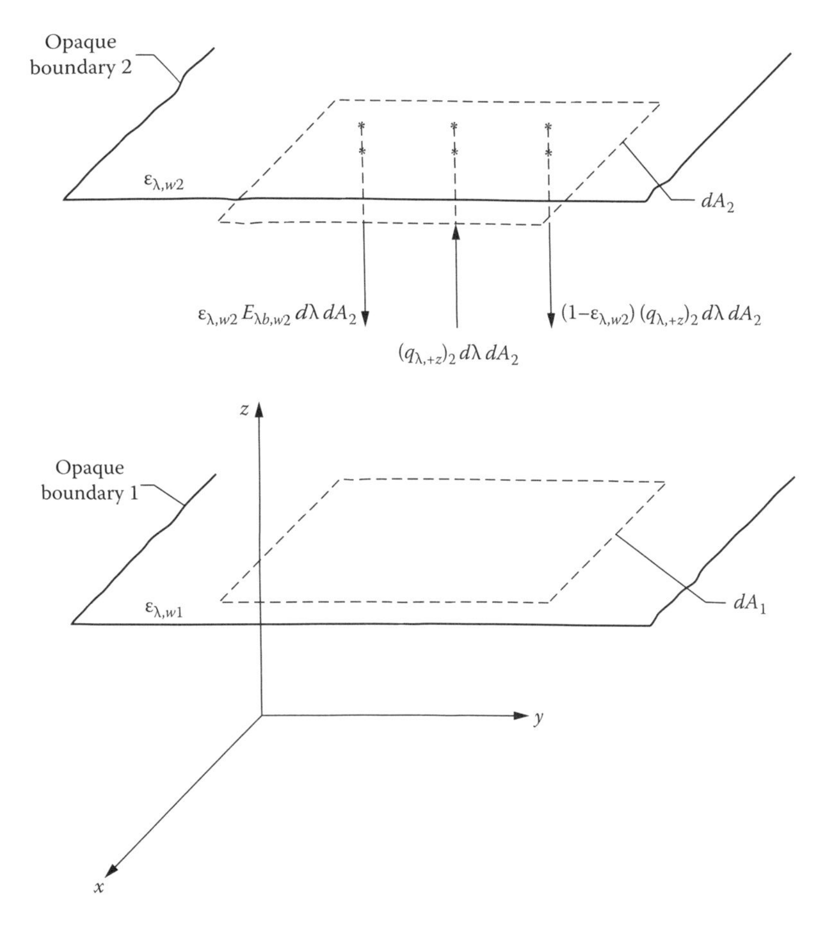

FIGURE 12.2 Geometry for derivation of energy-jump condition at opaque boundary.

All quantities without a w subscript are evaluated at dA_2 in the medium adjacent to the wall. The quantities with a w subscript are on wall 2, and $q_{\lambda,z}d\lambda$ is the net flux in $d\lambda$ in the positive z direction. Similarly, the discontinuity in emissive power at dA_1 in Figure 12.2 is

$$E_{\lambda,bw1}-E_{\lambda,b1}=\left(\frac{1}{\varepsilon_{\lambda,w1}}-\frac{1}{2}\right)\left(q_{\lambda,z}\right)_1+\frac{1}{2\beta_\lambda^2}\left(\frac{\partial^2 E_{\lambda,b}}{\partial z^2}+\frac{1}{2}\frac{\partial^2 E_{\lambda,b}}{\partial y^2}+\frac{1}{2}\frac{\partial^2 E_{\lambda,b}}{\partial x^2}\right)_1 \tag{12.34}$$

where quantities with a w subscript are on wall 1 and those without w are *in the medium* adjacent to wall 1.

Equations 12.33 and 12.34 are boundary conditions relating the emissive power $E_{\lambda,b}$ in the medium adjacent to the wall, to the wall emissive power $E_{\lambda,bw}$. In the limit when radiation is the only means of energy transport in the medium, there is a discontinuity in emissive power in passing from the medium to each wall. The use of the diffusion approximation in the derivation of these boundary relations implies that the proportionality between local radiative flux and emissive power gradient in the medium is valid in the medium very near a bounding surface. Although this is not strictly true, the use of jump boundary conditions corrects, to a good approximation, for wall effects if radiation dominates over conduction and/or convection.

The general radiation–diffusion equation has been provided at λ as Equation 12.25 and, for a wavelength band, as Equations 12.26 and 12.27. The boundary conditions at solid boundaries with normals into the medium in the negative and positive coordinate directions are given at λ by

Equations 12.33 and 12.34. When the diffusion equation is used, it is assumed to apply throughout the entire medium including the region adjacent to a boundary. The boundary effect is accounted for by using a jump boundary condition.

12.2.1.2.3 Gray Stagnant Medium between Parallel Gray Walls

Many molecular gases have strong property variations with wavelength, and it is necessary to solve the diffusion equation in a number of wavelength regions. For some situations, such as soot-filled flames and high-temperature uranium gas, a gray-medium approximation can be made. The equations reduce considerably in this case. For illustration, consider a gray medium between infinite parallel gray walls at unequal temperatures (Figure 12.3) for the limit where radiation is strongly dominant over conduction and/or convection.

For a gray medium, the absorption and scattering coefficients are independent of wavelength and Equation 12.27 gives $q_z = -4/[3\beta(z)](dE_b/dz)$. This can be integrated directly because, with no conduction, convection, or heat sources in the medium, the radiative flux q_z is constant in a 1D planar system. With the additional assumption that β does not depend on temperature and is therefore independent of z, integrating from 0 to z gives

$$E_b(z) - E_{b1} = -\frac{3\beta}{4} q_z z \tag{12.35}$$

Evaluating Equation 12.35 at $z = D$ yields

$$\frac{E_{b2} - E_{b1}}{q_z} = -\frac{3\beta D}{4} \tag{12.36}$$

The E_{b1} and E_{b2} are in the medium adjacent to the walls. To connect the unknown E_{b1} and E_{b2} with the specified wall conditions, the jump boundary conditions are applied. Differentiating Equation 12.35 twice with respect to z shows that the second derivative terms are zero in the boundary condition Equations 12.33 and 12.34, so they become, for a gray medium with gray walls,

$$\frac{E_{b2} - E_{bw2}}{q_z} = \frac{1}{\varepsilon_{w2}} - \frac{1}{2} \tag{12.37}$$

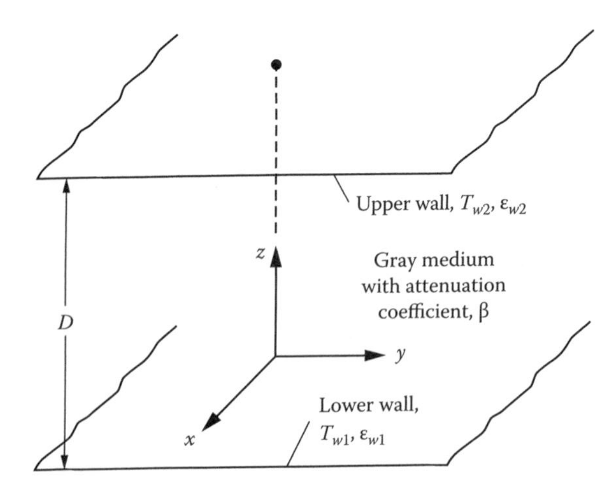

FIGURE 12.3 Schematic for radiative interchange between infinite gray boundaries enclosing a gray medium with absorption and isotropic scattering.

$$\frac{E_{bw1}-E_{b1}}{q_z}=\frac{1}{\varepsilon_{w1}}-\frac{1}{2} \quad (12.38)$$

To eliminate the unknown medium emissive powers at the bounding surfaces, E_{b1} and E_{b2}, add Equations 12.37 and 12.38, and then substitute $E_{b2}-E_{b1}$ from Equation 12.36 to yield

$$\frac{q_z}{E_{bw1}-E_{bw2}}=\frac{1}{3\beta D/4+1/\varepsilon_{w1}+1/\varepsilon_{w2}-1} \quad (12.39)$$

Equation 12.39 gives the radiative energy transfer (for radiative equilibrium without internal heat sources) through a layer of gray medium as a function of the optical thickness βD and the wall emissivities. It is relative to the difference in the black emissive powers of the walls, which is the maximum possible energy transfer. A comparison of this diffusion solution with the solution of the exact integral equations (Heaslet and Warming 1965) is in Figure 12.4 for equal wall emissivities. Agreement is excellent for all optical thicknesses. The distribution of emissive power $E_b(z)$ across the layer is found from Equation 12.35 by eliminating the unknown E_{b1} by use of Equation 12.38 or in another form by eliminating E_{b1} and E_{b2} from Equations 12.35 through 12.38. This result, which is for radiative equilibrium without internal heat sources, is shown in Table 12.1.

12.2.1.2.4 Other Radiative Diffusion Solutions for Gray Media

Table 12.1 provides diffusion solutions for the temperature distribution and energy transfer in simple geometries for gray media with constant properties between gray walls in the limit of radiative equilibrium (see Example 12.1 for further analytical details). Media such as real gases are usually not gray and are not optically thick in all wavelength regions, so caution is advised. For cylindrical or spherical geometries, agreement with exact solutions is often not as good as for infinite parallel plate boundaries. Good agreement has been found in cylindrical and spherical geometries if the optical thickness is greater than about 10, with improved agreement as wall emissivities become lower and diameter ratios D_{inner}/D_{outer} approach unity. A comparison for the cylindrical geometry is discussed later in connection with Figure 12.9.

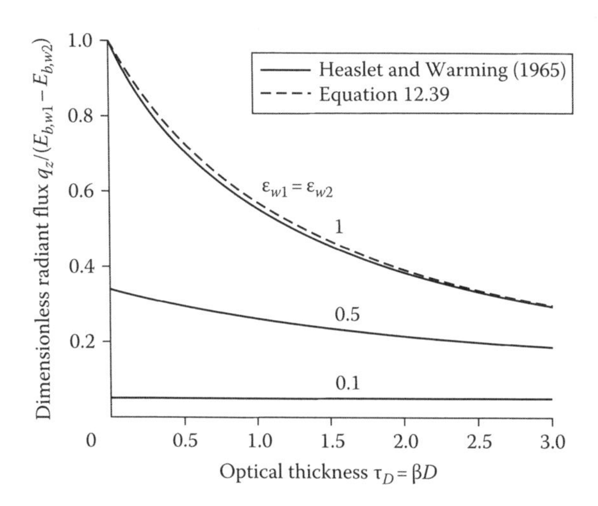

FIGURE 12.4 Validity of diffusion solution for energy transfer through gray medium with absorption and isotropic scattering between parallel gray walls for the limit of radiative equilibrium without internal heat sources.

TABLE 12.1
Diffusion Theory Predictions of Energy Transfer and Temperature Distribution for an Absorbing–Emitting Gray Gas with Isotropic Scattering in Radiative Equilibrium between Gray Walls and without Internal Heat Sources

Geometry	Relations[a]
Infinite parallel plates	$\psi = \dfrac{1}{(3\beta D/4) + \bar{E}_1 + \bar{E}_2 + 1}$
	$\phi(z) = \psi\left[\dfrac{3\beta}{4}(D-z) + \bar{E}_2 + \dfrac{1}{2}\right]$
Infinitely long concentric cylinders	$\psi = \dfrac{1}{\dfrac{3}{8}\left[\beta D_1 \ln\left(\dfrac{D_2}{D_1}\right) + \dfrac{1-(D_1/D_2)^2}{\kappa D_1}\right] + \left(\bar{E}_1 + \dfrac{1}{2}\right) + \dfrac{D_1}{D_2}\left(\bar{E}_2 + \dfrac{1}{2}\right)}$
	$\phi(r) = \psi\left\{-\dfrac{3}{8}\left[\beta D_1 \ln\left(\dfrac{D}{D_2}\right) + \dfrac{D_1}{\kappa D_2^2}\right] + \left(\bar{E}_2 + \dfrac{1}{2}\right)\dfrac{D_1}{D_2}\right\}$
Concentric spheres	$\psi = \dfrac{1}{\dfrac{3}{8}\left[\beta D_1\left(1 - \dfrac{D_1}{D_2}\right) + 2\dfrac{1-(D_1/D_2)^3}{\beta D_1}\right] + \left(\bar{E}_1 + \dfrac{1}{2}\right) + \dfrac{D_1^2}{D_2^2}\left(\bar{E}_2 + \dfrac{1}{2}\right)}$
	$\phi(r) = \psi\left\{-\dfrac{3}{8}\left[\beta D_1\left(\dfrac{D_1}{D_2} - \dfrac{D_1}{D}\right) + \dfrac{2D_1^2}{\beta D_2^3}\right] + \left(\bar{E}_2 + \dfrac{1}{2}\right)\dfrac{D_1^2}{D_2^2}\right\}$

[a] Definitions $\bar{E}_N = (1-\varepsilon_{wN})/\varepsilon_{wN}, \psi = Q_1/\sigma\left(T_{w1}^4 - T_{w2}^4\right), \phi(\xi) = \left[T^4(\xi) - T_{w2}^4\right]/\left(T_{w1}^4 - T_{w2}^4\right), D = 2r, \beta = \kappa + \sigma_s$.
Note: ϕ is defined only if $\kappa > 0$.

Example 12.1

The space between two diffuse–gray spheres (Figure 12.5) is filled with an optically dense stagnant medium having constant attenuation coefficient β. For the limiting condition of radiative equilibrium, compute the radiative energy flow Q_1 across the gap from sphere 1 to sphere 2 and the temperature distribution $T(r)$ in the medium, using the diffusion method with jump boundary conditions.

For a gray medium with constant β, Equation 12.28 gives the net heat flux in the positive r direction as $q_r = -(4/3\beta)\, dE_b/dr$. From energy conservation, with no internal heat sources, q_r varies with r as $q_r = Q_1/4\pi r^2$. Combine these relations and integrate from R_1 to R_2 to obtain

$$\frac{Q_1}{4\pi}\int_{r=R_1}^{R_2}\frac{dr}{r^2} = -\frac{4}{3\beta}\int_{E_b=E_{b1}}^{E_{b2}} dE_b \tag{12.1.1}$$

$$\frac{Q_1}{4\pi}\left(\frac{1}{R_2} - \frac{1}{R_1}\right) = \frac{4}{3\beta}\left(E_{b2} - E_{b1}\right) \tag{12.1.2}$$

or

$$\frac{Q_1}{\left(E_{b1} - E_{b2}\right)} = \frac{16\pi}{3\beta}\frac{1}{\left(\dfrac{1}{R_1} - \dfrac{1}{R_2}\right)} \tag{12.1.3}$$

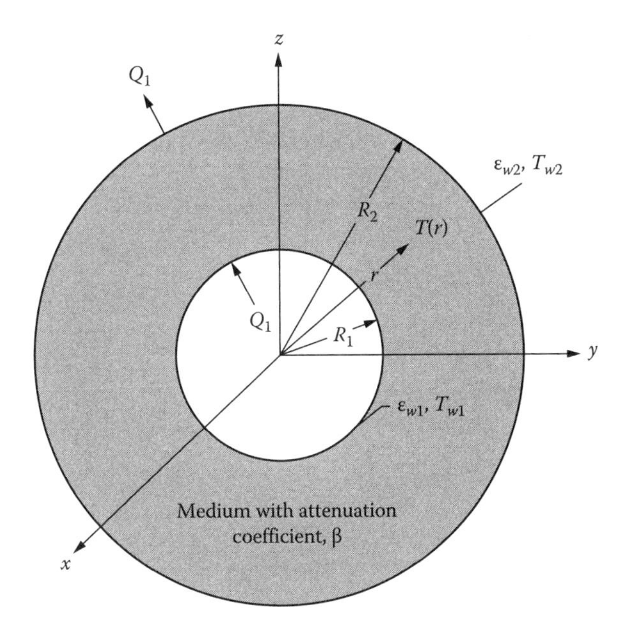

FIGURE 12.5 Radiation across gap between concentric spheres with intervening medium of constant attenuation coefficient.

The E_{b1} and E_{b2} are *in the gas* adjacent to the boundaries, and jump boundary conditions express these quantities in terms of wall values. The jump boundary conditions are in Equations 12.33 and 12.34 and involve second derivatives that are now found. By integrating Equation 12.1.1 from R_1 to r,

$$E_b(r) - E_{b1} = \frac{3\beta Q_1}{16\pi}\left(\frac{1}{r} - \frac{1}{R_1}\right) \tag{12.1.4}$$

Substitute $r = (x^2 + y^2 + z^2)^{1/2}$ and differentiate twice with respect to x to obtain

$$\frac{\partial^2 E_b(r)}{\partial x^2} = -\frac{3\beta Q_1}{16\pi}\frac{\left(x^2+y^2+z^2\right)^{3/2} - 3x^2\left(x^2+y^2+z^2\right)^{1/2}}{\left(x^2+y^2+z^2\right)^3} \tag{12.1.5}$$

and similarly for the y and z directions.

In the boundary condition Equation 12.33, point 2 can be conveniently taken in Figure 12.5 at $x = y = 0$ and $z = R_2$. This gives

$$\left[\frac{\partial^2 E_b(r)}{\partial x^2}\right]_2 = \left[\frac{\partial^2 E_b(r)}{\partial y^2}\right]_2 = -\frac{3\beta Q_1}{16\pi}\frac{1}{R_2^3}; \quad \left[\frac{\partial^2 E_b(r)}{\partial z^2}\right]_2 = \frac{3\beta Q_1}{8\pi}\frac{1}{R_2^3}$$

Also, $q_{z,2} = Q_1/4\pi R_2^2$. Substituting into Equation 12.33 gives

$$E_{b2} - E_{bw2} = \left(\frac{1}{\varepsilon_{w2}} - \frac{1}{2}\right)\frac{Q_1}{4\pi R_2^2} - \frac{3Q_1}{32\beta\pi}\frac{1}{R_2^3} = \frac{Q_1}{4\pi R_2^2}\left[\left(\frac{1}{\varepsilon_{w2}} - \frac{1}{2}\right) - \frac{3}{8\beta R_2}\right] \tag{12.1.6}$$

Similarly, at the inner sphere boundary, from Equation 12.34,

$$E_{bw1} - E_{b1} = \frac{Q_1}{4\pi R_1^2}\left[\left(\frac{1}{\varepsilon_{w1}} - \frac{1}{2}\right) + \frac{3}{8\beta R_1}\right] \quad (12.1.7)$$

Adding Equations 12.1.6 and 12.1.7 gives

$$E_{b2} - E_{b1} = E_{bw2} - E_{bw1} + \frac{Q_1}{4\pi}\left[\left(\frac{1}{\varepsilon_{w2}} - \frac{1}{2}\right)\frac{1}{R_2^2} + \left(\frac{1}{\varepsilon_{w1}} - \frac{1}{2}\right)\frac{1}{R_1^2} + \frac{3}{8\beta}\left(\frac{1}{R_1^3} - \frac{1}{R_2^3}\right)\right]$$

After substituting this into the right side of Equation 12.1.2, the result is solved for Q_1 to give the ψ in the last entry in Table 12.1.

To obtain the temperature distribution, integrate Equation 12.1.1 from R_2 to r to obtain $E_b(r)-E_{b2} = (3\beta Q_1/16\pi)(1/r-1/R_2)$. Add Equation 12.1.6 to eliminate E_{b2}:

$$E_b(r) - E_{bw2} = \frac{3\beta Q_1}{16\pi}\left(\frac{1}{r} - \frac{1}{R_2}\right) + \left(\frac{1}{\varepsilon_{w2}} - \frac{1}{2}\right)\frac{Q_1}{4\pi R_2^2} - \frac{3Q_1}{32\beta\pi}\frac{1}{R_2^3} \quad (12.1.8)$$

This gives the last expression for ϕ in Table 12.1. The ϕ is valid only if $\kappa > 0$, since temperatures are indeterminate in the limit of pure scattering when there is only radiative transfer.

Example 12.2

A plane layer of gray medium with constant properties is originally at a uniform temperature T_0. The attenuation coefficient is β, and the layer thickness is D. The heat capacity of the medium is c_V and its density is ρ. At time $t = 0$, the layer is placed in surroundings at zero temperature. Neglecting conduction and convection, discuss the solution for the transient temperature profiles for cooling only by radiation when β is very large. Contrast this with the solution for β being very small as in Section 10.11.

At the layer center, $x = D/2$ (boundaries are at $x = 0$ and D), the symmetry condition provides that at any time, $(\partial T/\partial x)_t = 0$ for $x = D/2$. At $t = 0$ for any x, $T = T_0$. For only radiation being included, there is a temperature jump at the boundaries $x = 0$, D, so the boundary temperatures are finite rather than being equal to the zero outside temperature.

For large β the diffusion approximation can be employed, and from Equation 12.24 the heat flux in the x direction is

$$q(x,t) = -\frac{4}{3\beta}\frac{\partial E_b(x,t)}{\partial x} = -\frac{4\sigma}{3\beta}\frac{\partial T^4(x,t)}{\partial x}$$

Energy conservation gives $-\partial q(x,t)/\partial x = \rho c_V \partial T/\partial t$. Combining these two equations to eliminate q gives the transient energy diffusion equation for the temperature distribution in the layer with constant absorption coefficient:

$$\rho c_V \frac{\partial T}{\partial t} = \frac{4\sigma}{3\beta}\frac{\partial^2 T^4(x,t)}{\partial x^2}$$

Defining dimensionless variables as $\tilde{t} = \beta\sigma T_0^3 t/\rho c_V$, $\tau = \beta x$, and $\vartheta = T/T_0$ gives

$$\frac{\partial \vartheta}{\partial \tilde{t}} = \frac{4}{3}\frac{\partial^2 \vartheta^4(\tau,\tilde{t})}{\partial \tau^2}$$

The initial condition and the symmetry condition at $x = D/2$ are $\vartheta(\tau,0) = 1$ and $\left.\frac{\partial \vartheta}{\partial \tilde{t}}\right|_{\tau=\kappa D,\tilde{t}} = 0$. At the boundary $\tau = \beta D$, a slip condition must be used. For surroundings that are empty space at zero temperature, $E_{bw} = 0$ and $\varepsilon_w = 1$, so that at the exposed boundary of the medium for any time Equation 12.33 gives

$$\sigma T^4\Big|_{x=D} = \frac{1}{2}\left(-\frac{4\sigma}{3\beta}\frac{\partial T^4}{\partial x}\right)_{x=D} - \frac{\sigma}{2\beta^2}\frac{\partial^2 T^4}{\partial x^2}\Big|_{x=D}$$

or

$$0 = \left(2\vartheta^4 + \frac{4}{3}\frac{\partial \vartheta^4}{\partial \tau} + \frac{4}{3}\frac{\partial^2 \vartheta^4}{\partial \tau^2}\right)_{\tau=\beta D}$$

Similar relations apply at $x = 0$. For these conditions, a numerical solution is necessary. For comparison, for the limit of a small β, an analytical solution is obtained in Example 10.6.

Viskanta and Bathla (1967) obtained numerical solutions to the transient energy equation, along with limiting solutions. Results for an optical thickness of $\beta D = 2$ are in Figure 12.6 for transient profiles in one-half of the symmetric layer.

The results in this section have been for the important limit where radiation is the dominant mode of energy transfer; conduction and convection have been neglected. Combining heat conduction and convection with radiation using the diffusion solution is considered in Chapter 13.

12.2.2 Moment-Based Methods

Moment-based methods are developed by multiplying the differential equation of radiative transfer by various powers of the direction cosines of the intensity to form a simultaneous set of *moment equations* (Equations 12.1, 12.8, and 12.11) that are then solved. The general procedure provides one less equation than the number of unknowns generated. To overcome this difficulty, the local intensity is approximated by a series expansion.

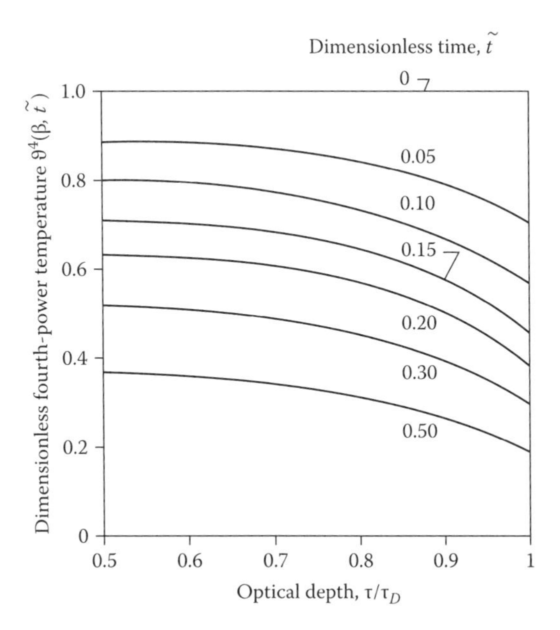

FIGURE 12.6 Dimensionless temperature profiles as a function of time for radiative cooling of a gray slab; optical thickness $\tau(x = D) = 2$. (From Viskanta, R. and Bathla, P.S., *Z. Angew. Math. Phys.*, 18(3), 353, 1967.)

If the series is in terms of *spherical harmonics*, denoted by P, and the series is truncated after a selected number of terms, *N*, this procedure provides a closure relation so that a solution can proceed. When the series is truncated after one or three terms, the method is called P_1 or P_3; in general, it is the P_N method. It is also referred to as the *moment method* and the *differential approximation*. This will serve as an introduction to moment methods.

First, a plane layer is considered. The first two RTE moment equations (zeroth and first) are solved simultaneously. Then, a general development is provided.

12.2.2.1 Milne–Eddington (Differential) Approximation

This approximate method is developed by starting with the RTE Equation 11.3 in a 1D plane layer with scattering that is assumed isotropic. The optical coordinate τ is derived in Equation 11.1. To simplify the notation, the symbols for wavelength are omitted by carrying out the derivation for a gray medium. The Milne–Eddington equations that are obtained also apply to spectral calculations if the same forms are written using spectral quantities. The RTE Equation 11.3, with the source function substituted from Equation 11.23, gives

$$\mu\frac{dI}{d\tau} = -I(\tau) + (1-\omega)I_b(\tau) + \omega\bar{I}(\tau) \tag{12.40}$$

where $\bar{I}(\tau) = (1/4\pi)\int_{\Omega=0}^{4\pi} I(\tau,\Omega_i)d\Omega_i$ and $\tau(x) = \int_{x^*=0}^{x} \beta(x^*)dx^*$. To obtain the zeroth moment equation (equivalent to Equation 12.6), Equation 12.40 is multiplied by $d\Omega$ and integrated over all solid angles to give

$$\int_{\Omega=0}^{4\pi} \mu\frac{dI}{d\tau}d\Omega = \frac{d}{d\tau}\int_{\Omega=0}^{4\pi} \mu I d\Omega = -\int_{\Omega=0}^{4\pi} I d\Omega + (1-\omega)I_b \int_{\Omega=0}^{4\pi} d\Omega + \omega\bar{I}\int_{\Omega=0}^{4\pi} d\Omega \tag{12.41}$$

Because $\int_{\Omega=0}^{4\pi} \mu I d\Omega$ is the radiative flux, and I_b and $\bar{I}$ do not depend on direction, this integrates to become the Milne–Eddington equation:

$$\frac{dq_r(\tau)}{d\tau} = -4\pi\bar{I} + 4\pi(1-\omega)I_b + 4\pi\omega\bar{I} = 4\pi(1-\omega)\left[I_b(\tau) - \bar{I}(\tau)\right] \tag{12.42}$$

To obtain the first moment equation (equivalent to Equation 12.10), the RTE equation (Equation 12.40) is multiplied by $\mu d\Omega$ and integrated over all solid angles to give

$$\int_{\Omega=0}^{4\pi} \mu^2\frac{dI}{d\tau}d\Omega = -\int_{\Omega=0}^{4\pi} I\mu d\Omega + (1-\omega)I_b\int_{\Omega=0}^{4\pi} \mu d\Omega + \omega\bar{I}\int_{\Omega=0}^{4\pi} \mu d\Omega \tag{12.43}$$

Because $d\Omega = -2\pi d\mu = 2\pi sin\theta d\theta$, the integrals involve products of the orthogonal functions $\sin\theta$ and $\cos\theta$, which simplifies their evaluation. Orthogonal functions will also be used in the more general formulation that follows. To evaluate Equation 12.43, the approximation made independently by Eddington (1926) and Milne (1930) is that, for radiation crossing a unit area oriented normal to the *x* direction, all intensities with positive directional components in *x* have a value independent of angle, and all intensities with a negative *x* directional component have a different constant value. Hence, the local radiation in each coordinate direction is assumed to be isotropic, as in Figure 12.7.

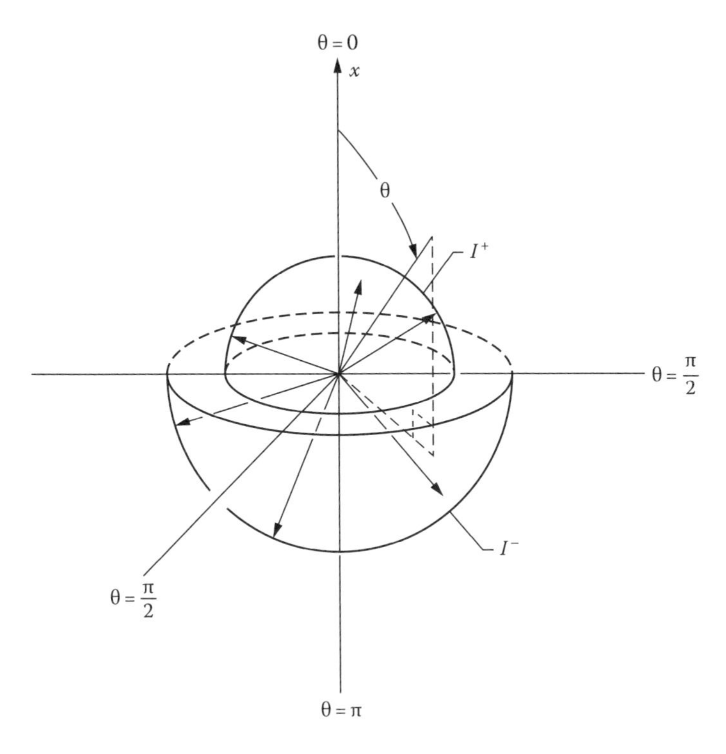

FIGURE 12.7 Approximation of intensities being isotropic in positive and in negative directions as used in the Milne–Eddington and two-flux methods.

After making this assumption, Equation 12.43 is integrated to yield the second Milne–Eddington equation:

$$q_r(\tau) = -\frac{2\pi}{3}\left(I^+ + I^-\right) = -\frac{4\pi}{3}\frac{d\bar{I}(\tau)}{d\tau} \tag{12.44}$$

The I^+ and I^- are isotropic in each of the $+x$ and $-x$ directions, so there are the following relations for the radiative flux q_r and the mean intensity $\bar{I}$ at each τ location:

$$q_r = \pi\left(I^+ - I^-\right) = q_r^+ - q_r^- \tag{12.45}$$

$$\bar{I} = \frac{I^+ + I^-}{2} = \frac{q_r^+ + q_r^-}{2} \tag{12.46}$$

Solving Equations 12.45 and 12.46 for q_r^+ and q_r^- yields the auxiliary relations:

$$q_r^+ = \pi\bar{I} + \frac{q_r}{2} \tag{12.47}$$

$$q_r^- = \pi\bar{I} - \frac{q_r}{2} \tag{12.48}$$

The two Milne–Eddington flux equations (Equations 12.42 and 12.44) can be combined to eliminate either $\bar{I}(\tau)$ or $q_r(\tau)$. This yields two differential equations for $q_r(\tau)$ and $\bar{I}(\tau)$ as

$$\frac{d^2q_r}{d\tau^2} - 3(1-\omega)q_r(\tau) = 4(1-\omega)\frac{dE_b(\tau)}{d\tau} \tag{12.49}$$

$$\frac{d^2\bar{I}}{d\tau^2} - 3(1-\omega)\bar{I}(\tau) = 4(1-\omega)E_b(\tau) \tag{12.50}$$

These relations also apply in spectral form. For example, for radiative properties that depend on λ but not on x, the optical coordinate τ becomes $\tau_\lambda = (\kappa_\lambda + \sigma_{s\lambda})x = \beta_\lambda x$, and Equation 12.49 can be written as (because the relation between τ_λ and x varies with λ, it is sometimes easier to use the physical coordinate)

$$\frac{d^2q_{r\lambda}}{dx^2} - 3\kappa_\lambda(\kappa_\lambda + \sigma_{s\lambda})q_{r\lambda}(x) = 4\kappa_\lambda\frac{dE_{\lambda b}(x)}{dx} \tag{12.51}$$

To obtain a solution, the two equations, Equations 12.42 and 12.44, or Equations 12.49 and 12.51, are used with the energy equation to provide three equations with three unknowns, q_r, $\bar{I}$, and T, that is in the blackbody function $E_b = \pi I_b$. For the special case of steady-state energy transfer only by radiation, the energy equation (Equation 10.5) without convection or conduction (radiative equilibrium) gives, for a 1D plane layer and a gray medium,

$$\frac{dq_r(x)}{dx} = \dot{q}(x) \quad \text{or} \quad \frac{dq_r(\tau)}{d\tau} = \frac{1}{\kappa+\sigma_s}\dot{q}(\tau) = \frac{1}{\beta}\dot{q}(\tau) \tag{12.52}$$

If, in addition, there are no internal heat sources, so $\dot{q} = 0$, the radiative flux q_r is a constant, and Equation 12.49 simplifies to

$$q_r = -\frac{4}{3}\frac{dE_b(\tau)}{d\tau} \quad \text{or} \quad q_r = -\frac{4}{3\beta(x)}\frac{dE_b(x)}{dx} \tag{12.53}$$

where for the total flux, $E_b = \sigma T^4$. For this special case, Equation 12.53 for the radiative heat flux is the same as for the diffusion approximation (Equation 12.28). In spectral form, $q_{r,\lambda} = -\left[4/3\beta_\lambda(x)\right]\left[dE_{\lambda b}(x)/dx\right]$ for radiative equilibrium without internal heat sources.

Two sets of boundary conditions are needed: one set for solving the RTE and the second for solving the energy equation. At an opaque solid boundary for the translucent medium, the outgoing radiation (radiosity) consists of emission by the boundary, and reflected incoming radiation. Hence, for an opaque gray surface at each boundary of a gray translucent layer, $0 \le x \le D$ as in Figure 12.3,

$$J(0) = \varepsilon_{w1}E_{bw1} + (1-\varepsilon_{w1})G(0) \tag{12.54}$$

$$J(D) = \varepsilon_{w2}E_{bw2} + (1-\varepsilon_{w2})G(D) \tag{12.55}$$

For use with the Milne–Eddington equations, these boundary conditions are placed in terms of $\bar{I}$ and q_r by use of Equations 12.47 and 12.48 noting that $J(0) = q_r^+(0)$ and $J(D) = q_r^-(D)$ with the result that, at $x = 0$ and $x = D$,

$$\bar{I}(0) + \frac{1}{\pi}\left(\frac{1}{\varepsilon_{w1}} - \frac{1}{2}\right) q_r(0) = \frac{1}{\pi} E_{bw1} \tag{12.56}$$

$$\bar{I}(D) - \frac{1}{\pi}\left(\frac{1}{\varepsilon_{w2}} - \frac{1}{2}\right) q_r(D) = \frac{1}{\pi} E_{bw2} \tag{12.57}$$

Example 12.3

A plane layer of gray gas with constant absorption and scattering coefficients and isotropic scattering is between two opaque parallel walls spaced D apart, which are at specified temperatures T_1 and T_2 and have gray emissivities ε_1 and ε_2. Uniform energy generation is occurring throughout the gas at a rate $\dot{q}$ per unit volume. The gas is not moving and heat conduction in the gas is neglected to obtain the limiting result for only radiative transfer. Use Milne–Eddington relations to find the distributions of radiative transfer and temperature within the layer.

Integrating the energy Equation 12.52, the distribution of radiative flux in the layer is $q_r(\tau) = \dot{q}\tau/\beta + C_1$, where C_1 is an integration constant. With the functional form of $q_r(\kappa)$ known, the second Milne–Eddington equation (Equation 12.44) can be integrated to obtain $\bar{I}(\tau) = -3\dot{q}\tau^2/8\pi\beta - 3C_1\tau/4\pi + C_2$. The two integration constants are obtained by substituting the $q_r(\kappa)$ and $\bar{I}(\tau)$ evaluated at the boundaries into the two boundary conditions in Equation 12.54 and solving the results simultaneously for C_1 and C_2. The C_1 needed in the distribution $q_r(\tau) = \dot{q}\tau/\beta + C_1$ is found as

$$C_1 = \frac{-(\dot{q}\tau_D/2\beta)\left[(3\tau_D/4) + (2/\varepsilon_2) - 1\right] + E_{b1} - E_{b2}}{(3\tau_D/4) + (1/\varepsilon_1) + (1/\varepsilon_2) - 1} \tag{12.3.1}$$

where $\tau_D = \beta D$. Then from Equation 12.56, $\bar{I}(0) = C_2$ and $q_r(0) = C_1$, $C_2 = -(1/\pi)(1/\varepsilon_1 - 1/2)C_1 + E_{b1}/\pi$. With $\bar{I}(\tau)$ and $q_r(\tau)$ known, the temperature distribution is then obtained from the first Milne–Eddington equation (Equation 12.42) as

$$\sigma T^4(\tau) = \pi I_b(\tau) = E_{b1} + \frac{\dot{q}}{2}\left(\frac{1}{2\kappa} - \frac{3\tau^2}{4\beta}\right) - \left(\frac{3\tau}{4} + \frac{1}{\varepsilon_1} - \frac{1}{2}\right) C_1 \tag{12.3.2}$$

12.2.2.2 General Spherical Harmonics (P_N) Method

In the general P_N method, the integral equations of radiative transfer are reduced to a set of differential equations by approximating the transfer relations by a finite set of moment equations. As discussed in Section 12.2, the moments are generated by multiplying the equation of transfer by powers of the cosine of the angle between the coordinate direction and the direction of the intensity and then integrating over all solid angles. This is a generalization of Milne–Eddington method in which Equations 12.42 and 12.44 are from the equation of transfer multiplied by $(\cos\theta)^0$ and $(\cos\theta)^1$ and then integrated over all $d\Omega$. These are equivalent to the more general moment Equations 12.8 and 12.11. The P_N development is in 3D so that general geometries can be treated. The treatments in Higenyi (1979), Bayazitoglu and Higenyi (1979), Ratzel and Howell (1982), Liu et al. (1992a,b), and Mengüç and Viskanta (1985, 1986) are followed; other pertinent references are Mark (1945), Marshak (1947), Stone and Gaustad (1961), Adrianov and Polyak (1963), Traugott and Wang (1964), Rhyming (1966), Chou and Tien (1968), Dennar and Sibulkin (1969), Cheng (1972), Selçuk and Siddall (1976), Shvartsburg (1976), Modest (1990), and Tencer (2013). The formulation of the P_N approximations is addressed in Modest et al. (2014) and Ge et al. (2015).

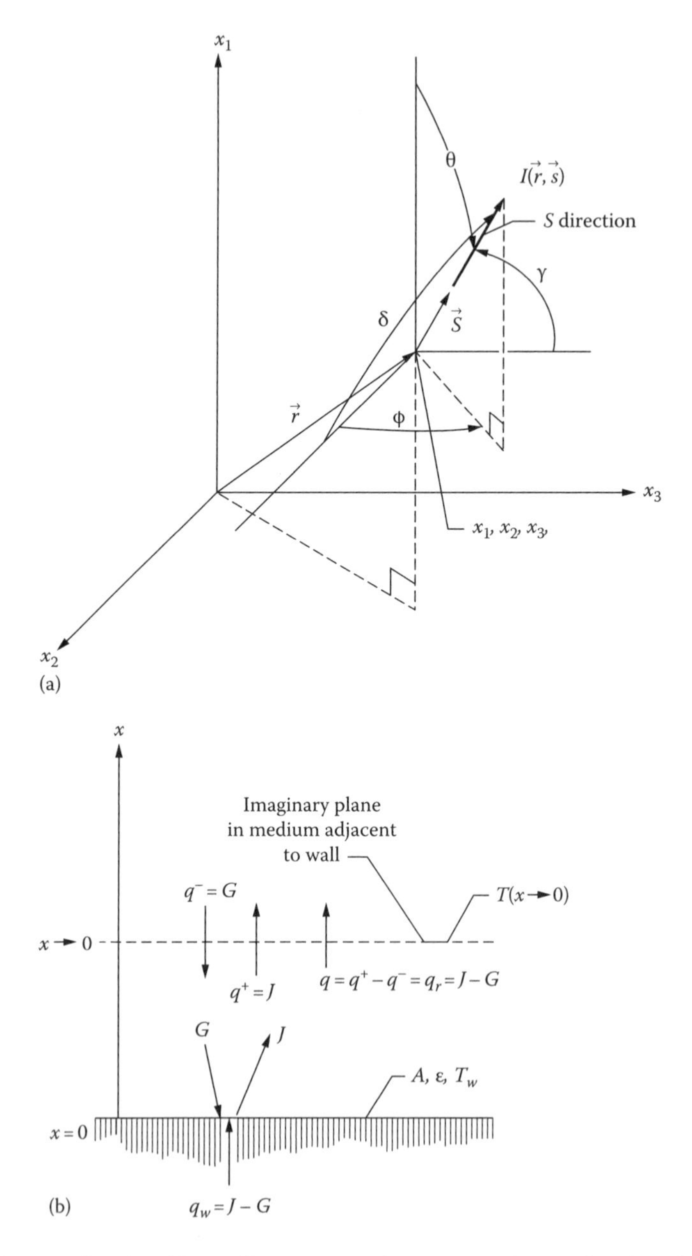

FIGURE 12.8 P_N approximation: (a) coordinate system showing intensity as a function of position and angle for P_N approximation and (b) heat fluxes in boundary condition.

A rectangular coordinate system x_1, x_2, x_3 is shown in Figure 12.8a. The variation of intensity at position **r** along the S direction in the direction of the unit vector **s** is given by the RTE equation (Equation 10.10). If the medium is assumed gray with uniform scattering and absorption coefficients, and assumed to scatter isotropically, Equation 10.10 is integrated over all λ to give

$$\frac{dI}{dS} = \kappa I_b - (\kappa + \sigma_s) I + \frac{\sigma_s}{4\pi} \int_{\Omega_i = 0}^{4\pi} I(S, \Omega_i) \, d\Omega_i \tag{12.58}$$

where $I = I(S, \Omega)$ and $I_b = I_b(S)$. At each location in a medium, radiation is traveling in all directions. It is useful to express the direction S of the intensity I in terms of the angles θ and ϕ in a spherical coordinate system or in terms of direction cosines l_i (i = 1, 2, 3) of the coordinate system (Figure 12.8a), so that

$$\frac{dI}{dS} = \cos\theta\frac{\partial I}{\partial x_1} + \sin\theta\cos\phi\frac{\partial I}{\partial x_2} + \sin\theta\sin\phi\frac{\partial I}{\partial x_3} = l_1\frac{\partial I}{\partial x_1} + l_2\frac{\partial I}{\partial x_2} + l_3\frac{\partial I}{\partial x_3} \tag{12.59}$$

where $l_1 = \cos\theta$, $l_2 = \cos\delta$, and $l_3 = \cos\gamma$. Using the optical coordinate $d\tau_i = (\kappa+\sigma_s)dx_i$, and albedo ω, the RTE, Equation 12.59, becomes

$$\sum_{i=1}^{3} l_i\frac{\partial I}{\partial \tau_i} + I = (1-\omega)I_b + \frac{\omega}{4\pi}\int_{\Omega_i=0}^{4\pi} I(S,\Omega_i)\,d\Omega_i \tag{12.60}$$

where I is a function of location S and direction Ω.

To develop the P_N method, the intensity at each S is expressed as an expansion in a series of orthogonal harmonic functions:

$$I(S,\Omega) = \sum_{l=0}^{\infty}\sum_{m=-l}^{l} A_l^m(S)Y_l^m(\Omega) \tag{12.61}$$

In the limit of an infinite number of terms in the series, $l \to \infty$, the spherical harmonics approximation is exact. The $A_l^m(S)$ are position-dependent coefficients to be determined by the solution, and $Y_l^m(\Omega)$ are the angularly dependent normalized spherical harmonics:

$$Y_l^m(\Omega) = \left[\frac{2l+1}{4\pi}\frac{(l-|m|)!}{(l+|m|)!}\right]^{1/2} e^{jm\phi}P_l^{|m|}(\cos\theta) \tag{12.62}$$

where $j = \sqrt{-1}$ so that $e^{jm\phi}$ provides the harmonics $\cos m\phi$ and $\sin m\phi$.* The $P_l^{|m|}(\cos\theta)$ are associated Legendre polynomials of the first kind, of degree l and order m,

$$P_l^{|m|}(\mu) = \frac{(1-\mu^2)^{|m|/2}}{2^l l!}\frac{d^{l-|m|}}{d\mu^{l-|m|}}(\mu^2-1)^l \tag{12.63}$$

where $\mu = \cos\theta$*. The $P_l^{|m|}(\mu) \equiv 0$ for $|m| > l$ and $P_l^0(\mu) \equiv P_l(\mu)$. Values $P_l^m(\cos\theta)$ for $0 \le l \le 3$ are in Table 12.2.

TABLE 12.2
Associated Legendre Polynomials, $P_l^m(\cos\theta)$

L	*m* = 0	*m* = 1	*m* = 2	*m* = 3
0	1.0	0	0	0
1	$\cos\theta$	$\sin\theta$	0	0
2	$(3\cos^2\theta-1)/2$	$3\cos\theta\sin\theta$	$3\sin^2\theta$	0
3	$(1/2)(5\cos^2\theta-3)\cos\theta$	$(3/2)(5\cos^2\theta-1)\sin\theta$	$15\cos\theta\sin^2\theta$	$15\sin^3\theta$

* Two definitions of normalized spherical harmonics are in common use. One definition (not used here) includes a factor of $(-1)^m$. The factor introduces an alternating sign in spherical harmonics with positive m. If this definition is used, Equation 12.63 must also correctly incorporate this factor.

To apply the P_N method, Equation 12.61 is truncated after a finite number of N terms. Generally, in engineering radiative transfer problems, terms are retained for $l = 0$ and 1 (P_1 approximation) or for $l = 0, 1, 2,$ and 3 (P_3 approximation). It is possible to retain higher-order terms, but for thermal radiation problems, the P_3 approximation usually has been found adequate. A higher-order approximation adds considerable complexity and may not be practical. Even-order approximations (P_2, P_4, etc.) give little increase in accuracy over the next lower-order odd-numbered expansion, and they are difficult to apply for boundary conditions of specified temperature or energy flux. Thus, odd-order expansions are most often used for radiative transfer.

To determine $I(S, \Omega)$, the coefficients $A_l^m(S)$ in the series expansion Equation 12.61 must be evaluated. To do this, the local intensity in Equation 12.61 is substituted into the integrals in the moment equations (Equations 12.2 through 12.4), the series is truncated at the desired approximation, and the integrations are carried out. For the P_3 approximation, this gives 20 coupled algebraic equations in 20 moments of intensity. Ratzel and Howell (1982) solved this set and present forms for the $A_l^m(S)$ coefficients in terms of the moments. Substituting these expressions for $A_l^m(S)$ into Equation 12.61 gives the relation for the P_3 approximation for the local intensity $I(S, \theta, \phi)$ in terms of its moments as

$$\begin{aligned}
4\pi I(S,\theta,\phi) &= I^{(0)} + 3I^{(1)}\cos\theta + 3I^{(2)}\sin\theta\cos\phi + 3I^{(3)}\sin\theta\sin\phi \\
&+ \frac{5}{4}\left(3I^{(11)} - I^{(0)}\right)\left(3\cos^2\theta - 1\right) + 15\left(I^{(12)}\cos\phi + I^{(13)}\sin\phi\right)\cos\theta\sin\theta \\
&+ \frac{15}{4}\left[\left(I^{(22)} - I^{(33)}\right)\cos 2\phi + 2I^{(23)}\sin 2\phi\right]\sin^2\theta \\
&+ \frac{7}{4}\left(5I^{(111)} - 3I^{(1)}\right)\left(5\cos^3\theta - 3\cos\theta\right) \\
&+ \frac{21}{8}\left[\left(5I^{(211)} - I^{(2)}\right)\cos\phi + \left(5I^{(311)} - I^{(3)}\right)\sin\phi\right]\left(5\cos^2\theta - 1\right)\sin\theta \\
&+ \frac{105}{4}\left[\left(I^{(122)} - I^{(133)}\right)\cos 2\phi + 2I^{(123)}\sin 2\phi\right]\cos\theta\sin^2\theta \\
&+ \frac{35}{8}\left[\left(I^{(222)} - 3I^{(233)}\right)\cos 3\phi - \left(I^{(333)} - 3I^{(322)}\right)\sin 3\phi\right]\sin^3\theta
\end{aligned} \tag{12.64}$$

along with the identities

$$I^{(0)} = I^{(11)} + I^{(22)} + I^{(33)}; \quad I^{(1)} = I^{(111)} + I^{(222)} + I^{(333)}; \quad I^{(2)} = I^{(211)} + I^{(222)} + I^{(233)}$$

$$I^{(3)} = I^{(311)} + I^{(322)} + I^{(333)}$$

The P_1 approximation is much simpler; for the local intensity, only the first four terms in Equation 12.64 appear:

$$I(S,\theta,\phi) = \frac{1}{4\pi}\left(I^{(0)} + 3I^{(1)}\cos\theta + 3I^{(2)}\sin\theta\cos\phi + 3I^{(3)}\sin\theta\sin\phi\right) \tag{12.65}$$

This illustrates why the P_1 approximation has been used much more than the P_3.

To continue the solution, expressions for the moments of intensity must be developed so that explicit relations for intensity for the P_3 and P_1 approximations can be obtained from Equation 12.64 or 12.65. This is done by generating moment differential equations from the differential RTE, Equation 12.60. The integral in Equation 12.60 is the zeroth moment $I^{(0)}$, so the RTE, Equation 12.60, can be written as

$$\sum_{i=1}^{3} l_i \frac{\partial I}{\partial \tau_i} + I = (1-\omega) I_b + \frac{\omega}{4\pi} I^{(0)} \tag{12.66}$$

Equation 12.66 is multiplied by powers of the direction cosines individually and in combination, and the results are integrated over all solid angles, resulting in the first- and higher-order RTE moment equations (such as Equation 12.10 for the first-order moment equation). Because the derivative terms on the left of Equation 12.66 are with respect to coordinate position while the integrals are over solid angle, the derivative and integral of each term can be interchanged. The results of carrying out these operations for the P_3 approximation are the differential equations:

$$\sum_{i=1}^{3} \frac{\partial I^{(i)}}{\partial \tau_i} = (1-\omega)\left(4\pi I_b - I^{(0)}\right) \tag{12.67}$$

$$\sum_{i=1}^{3} \frac{\partial I^{(ij)}}{\partial \tau_i} = -I^{(j)} \quad \text{(3 equations: } j = 1,2,3\text{)} \tag{12.68}$$

$$\sum_{i=1}^{3} \frac{\partial I^{(ijk)}}{\partial \tau_i} = -I^{(jk)} + \frac{4\pi}{3}\delta_{jk}\left[(1-\omega) I_b + \frac{\omega}{4\pi} I^{(0)}\right] \quad \text{(9 equations: } j,k = 1,2,3\text{)} \tag{12.69}$$

$$\sum_{i=1}^{3} \frac{\partial I^{(ijkl)}}{\partial \tau_i} = -I^{(jkl)} \quad \text{(27 equations: } j,k,l = 1,2,3\text{)} \tag{12.70}$$

The δ_{jk} is the Kronecker delta, $\delta_{jk} = 1$ for $j = k$, $\delta = 0$ for $j \neq k$. For the P_1 approximation, only Equations 12.67 and 12.68 are used.

Note that Equation 12.70 for the P_3 approximation contains a new set of unknowns, the fourth-order moments $I^{(ijkl)}$. For the P_1 approximation, the second-order moments $I^{(ij)}$ are present. To close the set of equations, values for these moments must be found. This is done by substituting Equation 12.64 (or Equation 12.65 for the P_1 approximation) into the general moment equation (Equation 12.63) to generate a relation for the fourth moment $I^{(ijkl)}$ (or $I^{(ij)}$ for P_1). This relation is not exact because Equations 12.64 and 12.65 were truncated to $N = 3$ and $N = 1$, respectively. However, an approximate closure condition is generated. For P_3, the result is

$$I^{(ijkl)} = \frac{1}{7}\left(I^{(ij)}\delta_{kl} + I^{(ik)}\delta_{jl} + I^{(jk)}\delta_{il} + I^{(il)}\delta_{jk} + I^{(kl)}\delta_{ij} + I^{(jl)}\delta_{ik}\right) - \frac{1}{35}\left(\delta_{ij}\delta_{kl} + \delta_{il}\delta_{jk} + \delta_{ik}\delta_{jl}\right) \tag{12.71}$$

For P_1,

$$I^{(ij)} = \frac{1}{3}\delta_{ij} I^{(0)} \tag{12.72}$$

so that $I^{(11)} = I^{(22)} = I^{(33)} = I^{(0)}/3$ (which is consistent with the first identity following Equation 12.64) and all the other $I^{(ij)} = 0$.

The formulation is now complete, in that a number of equations equal to the number of unknowns are available for the moments of intensity. Once these are determined, Equation 12.64 or 12.65 provides the local angular values of the intensity and the problem is, in principle, solved. The boundary conditions needed to solve the moment differential equations are now formulated, and applications of the P_N method are then presented to illustrate the use of these relations.

12.2.2.2.1 Boundary Conditions for the P_N Method

Useful boundary conditions for engineering applications are due to Marshak (1947): they work well for odd-order expansions. Other discussions of boundary conditions are in Mark (1945), Shokair and Pomraning (1981), and Liu et al. (1992b).

For an opaque directional-gray surface with emissivity $\varepsilon(\Omega)$ in the direction of solid angle $d\Omega$, the outgoing intensity $I_o(\Omega)$ from a boundary is produced by emission and by reflection of incident intensities from Ω_i as

$$I_o(\Omega) = \varepsilon(\Omega) I_{b,w} + \frac{1}{\pi} \int_{\Omega_i=0}^{2\pi} \rho(\Omega,\Omega_i) I(\Omega_i) l_i d\Omega_i \tag{12.73}$$

where l_i is the direction cosine of $I(\Omega_i)$ relative to the surface normal, the integration for the reflected intensity is over the hemisphere of all incident solid angles Ω_i, and $\rho(\Omega, \Omega_i)$ is the bidirectional reflectivity. For a nondiffuse surface, the general Marshak boundary condition is

$$\int_{\Omega=0}^{2\pi} I_o(\Omega) Y_l^m(\Omega) d\Omega = \int_{\Omega=0}^{2\pi} \left[\varepsilon(\Omega) I_{b,w} + \frac{1}{\pi} \int_{\Omega_i=0}^{2\pi} \rho(\Omega,\Omega_i) I(\Omega_i) l_i d\Omega_i \right] d\Omega \tag{12.74}$$

For 1D cases (no dependence on ϕ in Equation 12.62), Y_l^m may be replaced by $Y_l^m = l_i^{2n-1} (n = 1,2,3,\ldots)$. In this notation $2n - 1 = N$. For multidimensional rectangular geometries, $Y_l^m = l_i (i = 1,2,3)$ and $Y_3^m = l_i l_j l_k (i, j, k = 1,2,3)$. These equations provide more relations than are needed, and it is suggested in Ratzel and Howell (1982) that combinations be chosen with moments of the intensity normal to the boundary. Some suggestions for removing ambiguity in the boundary conditions and improving accuracy are in Liu et al. (1992)

For the 1D case with $n = 1$, corresponding to the P_1 approximation, the left side of Equation 12.74 reduces to

$$\int_{\Omega=0}^{2\pi} I_o(\Omega) Y_l^m(\Omega) d\Omega = \int_{\Omega=0}^{2\pi} I_o(\Omega) l_i^{2n-1} d\Omega = \int_{\Omega=0}^{2\pi} I_o(\Omega) l_i d\Omega = J \tag{12.75}$$

so this condition gives the radiosity J (Figure 12.8b). The other boundary terms do not have a physical interpretation.

The use of the P_N relations is now demonstrated in an example for conditions where radiation dominates so that conduction and/or convection can be neglected. The P_1 relations are used, but the P_3 relations can be applied.

Example 12.4

Using the P_1 approximation, derive relations for the temperature distribution and energy transfer between parallel plane walls at T_{w1} and T_{w2} for the limit of radiative equilibrium without internal heat sources. Each wall has the same diffuse–gray emissivity ε_w, and they are separated by an emitting, absorbing, and isotropically scattering medium with albedo ω and optical thickness τ_D based on the spacing between the walls.

Because the geometry requires the radiative energy flux to be only in the τ_1 direction, it follows that the first moments (equal to the fluxes) $I^{(j)} = 0$ for $j = 2, 3$. Then, Equation 12.65 for the P_1 approximation becomes

$$I(S,\theta,\phi) = \frac{1}{4\pi}\left(I^{(0)} + 3I^{(1)}\cos\theta\right) \tag{12.4.1}$$

and the moment equations (Equations 12.67 and 12.68) used for the P_1 approximation become

$$\frac{dI^{(1)}}{d\tau_1} = (1-\omega)\left(4\pi I_b - I^{(0)}\right) \tag{12.4.2}$$

$$\frac{dI^{(11)}}{d\tau_1} = -I^{(1)}, \quad \frac{dI^{(12)}}{d\tau_1} = 0, \quad \frac{dI^{(13)}}{d\tau_1} = 0 \tag{12.4.3}$$

The closure condition for P_1, Equation 12.72, gives $I^{(11)} = (1/3)I^{(0)}$ and $I^{(12)} = I^{(13)} = 0$. Substituting into the second moment differential equation (Equation 12.4.3) gives

$$\frac{1}{3}\frac{dI^{(0)}}{d\tau_1} = -I^{(1)} \tag{12.4.4}$$

Now, because $I^{(1)} = q_r$ and q_r is constant in this geometry for radiative equilibrium with $\dot{q} = 0$, it follows from Equation 12.4.2 that $I^{(0)} = 4\pi I_b = 4\sigma T^4(\tau_1)$. Substituting this into Equation 12.69 gives

$$q_r = -\frac{4\sigma}{3}\frac{dT^4}{d\tau_1} \tag{12.4.5}$$

This is the same relation as for the diffusion solution Equation 12.24. However, for other geometries, the P_1 solution does not generally provide the diffusion result. Integrating Equation 12.4.5 results in a linear T^4 distribution in the medium:

$$\sigma T^4(\tau_1) = -\frac{3q_r}{4}\tau_1 + C \tag{12.4.6}$$

The boundary conditions are applied to relate the temperature distribution to the known temperatures and obtain the integration constant C. Measuring τ_1 from the wall at T_{w1}, Equation 12.74 at this boundary becomes (using Equation 12.75 and $Y_1^0 = (3/4\pi)^{1/2}\cos\theta$)

$$J(\tau_1 = 0) = 2\pi\int_{\theta=0}^{\pi/2} I_o(\tau_1 = 0)\cos\theta\sin\theta d\theta = 2\varepsilon_w\sigma T_{w,1}^4\int_{\theta=0}^{\pi/2}\cos\theta\sin\theta d\theta$$

$$+2(1-\varepsilon_w)\int_{\theta=0}^{\pi/2}\left[2\pi\int_{\theta_i=0}^{\pi/2} I(\tau_1 = 0,\theta_i)\cos\theta_i\sin\theta_i d\theta_i\right]\cos\theta\sin\theta d\theta \tag{12.4.7}$$

The incident intensity $I(\tau_1 = 0, \theta)$ in Equation 12.4.7 is expressed by Equation 12.4.1, so that, from the moments that have been found,

$$I(\tau_1 = 0,\theta_i) = \frac{1}{4\pi}\left(I^{(0)} + 3I^{(1)}\cos\theta_i\right) = \frac{1}{4\pi}\left(4\sigma T^4(\tau_1 = 0) + 3q_r\cos\theta_i\right)$$

Substituting into the boundary condition Equation 12.4.7 gives for the diffuse boundary

$$J(\tau_1 = 0) = \varepsilon_W \sigma T_{w1}^4 + \frac{(1-\varepsilon_W)}{2} \int_{\theta_i=\pi}^{\pi/2} \left[4\sigma T^4(\tau_1 = 0) + 3q_r \cos\theta_i\right] \cos\theta_i \sin\theta_i d\theta_i \qquad (12.4.8)$$

From the net-radiation equation (Equation 5.18) at boundary 1 (*note:* $q_r = J - G = q_1$),

$$J(\tau_1 = 0) = \sigma T_{w1}^4 - \frac{(1-\varepsilon_W)}{\varepsilon_W} q_r \qquad (12.4.9)$$

Integrating Equation 12.4.8 and using Equation 12.4.9 to eliminate J results in

$$\sigma T^4(\tau_1 = 0) = \sigma T_{w1}^4 - \left(\frac{1}{\varepsilon_W} - \frac{1}{2}\right) q_r \qquad (12.4.10)$$

A similar analysis applied at wall 2 gives

$$\sigma T^4(\tau_1 = \tau_D) = \sigma T_{w2}^4 + \left(\frac{1}{\varepsilon_W} - \frac{1}{2}\right) q_r \qquad (12.4.11)$$

Equations 12.4.10 and 12.4.11 are the same as for the jump boundary conditions for diffusion, Equations 12.37 and 12.38. Equation 12.4.10 is used to evaluate the constant in Equation 12.4.6 where $\sigma T^4(\tau_1 = 0) = C$, resulting in

$$\sigma T^4(\tau_1) = \sigma T_{w1}^4 - \left(\frac{1}{\varepsilon_W} - \frac{1}{2}\right) q_r - \frac{3q_r}{4} \tau_1 \qquad (12.4.12)$$

Evaluating Equation 12.4.12 at $\tau_1 = \tau_D$, substituting Equation 12.4.11 to eliminate $T^4(\tau_1 = \tau_D)$, and solving the result for q_r yields

$$\frac{q_r}{\sigma\left(T_{w1}^4 - T_{w2}^4\right)} = \frac{1}{3\tau_D/4 + 2/\varepsilon_W - 1} \qquad (12.4.13)$$

This expression for q_r can be substituted into Equation 12.4.12 to obtain $T^4(\tau_1)$ in terms of the known boundary conditions (see Table 12.3). This completes the solution.

Although the P_1 approximation provides the same result as the diffusion solution for a translucent medium between infinite parallel diffuse–gray walls, for other geometries, it generally provides a different and more accurate solution than the diffusion result. Table 12.3 has P_1 predictions for a medium between parallel plane walls, concentric cylinders, and concentric spheres. Figures 12.9 and 12.10 compare the exact solution for energy transfer with the diffusion, P_1, and P_3 approximations for concentric cylinders and concentric spheres (Bayazitoglu and Higenyi 1979). Physically, ψ cannot be larger than 1 and should approach 1 for small optical thicknesses when the bounding surfaces are black. This is the limit for the Monte Carlo solution (Perlmutter and Howell 1964) in Figure 12.9 and the exact solution (Rhyming 1966) in Figure 12.10. The diffusion approximation is based on the assumption of an optically thick medium and is not expected to give good results for small optical thicknesses. It gives good results for optical thicknesses greater than 1 for $D_{inner}/D_{outer} = 0.5$. For smaller D_{inner}/D_{outer}, the diffusion results are not as good, especially for spheres; larger optical thicknesses are required for good agreement with the exact solution. The P_3 approximation

TABLE 12.3
P_1 Approximations for Energy Transfer and Temperature Distribution for a Gray Translucent Medium in Radiative Equilibrium (Absorption, Emission, Isotropic Scattering) between Gray Surfaces, $\dot{q} = 0$

Geometry	Relations[a]
Infinite parallel plates	$\psi = \dfrac{1}{(3\beta D/4) + E_1 + E_2 + 1}$
	$\phi(z) = \psi\left[\dfrac{3\beta}{4}(D - z) + E_2 + \dfrac{1}{2}\right]$
Infinitely long concentric cylinders	$\psi = \dfrac{1}{\dfrac{3}{8}\beta D_1 \ln\left(\dfrac{D_2}{D_1}\right) + \left(E_1 + \dfrac{1}{2}\right) + \dfrac{D_1}{D_2}\left(E_2 + \dfrac{1}{2}\right)}$
	$\phi(r) = \psi\left[-\dfrac{3}{8}\beta D_1 \ln\left(\dfrac{D}{D_2}\right) + \left(E_2 + \dfrac{1}{2}\right)\dfrac{D_1}{D_2}\right]$
Concentric spheres	$\psi = \dfrac{1}{\dfrac{3}{8}\beta D_1\left(1 - \dfrac{D_1}{D_2}\right) + \left(E_1 + \dfrac{1}{2}\right) + \dfrac{D_2}{D_1}\left(E_2 + \dfrac{1}{2}\right)}$
	$\phi(r) = \psi\left[-\dfrac{3}{8}\beta D_1\left(\dfrac{D_1}{D_2} - \dfrac{D_1}{D}\right) + \left(E_2 + \dfrac{1}{2}\right)\dfrac{D_1^2}{D_2^2}\right]$

[a] Definition: $E_N = (1 - \varepsilon_{wN})/\varepsilon_{wN}$, $\psi = Q_1/A_1\sigma\left(T_{w1}^4 - T_{w2}^4\right)$, $\phi(\xi) = \left[T^4(\xi) - T_{w2}^4\right]/\left(T_{w1}^4 - T_{w2}^4\right)$, $D = 2r$, $\beta = \kappa + \sigma_s$; and for $\phi(r)$ the $k > 0$.

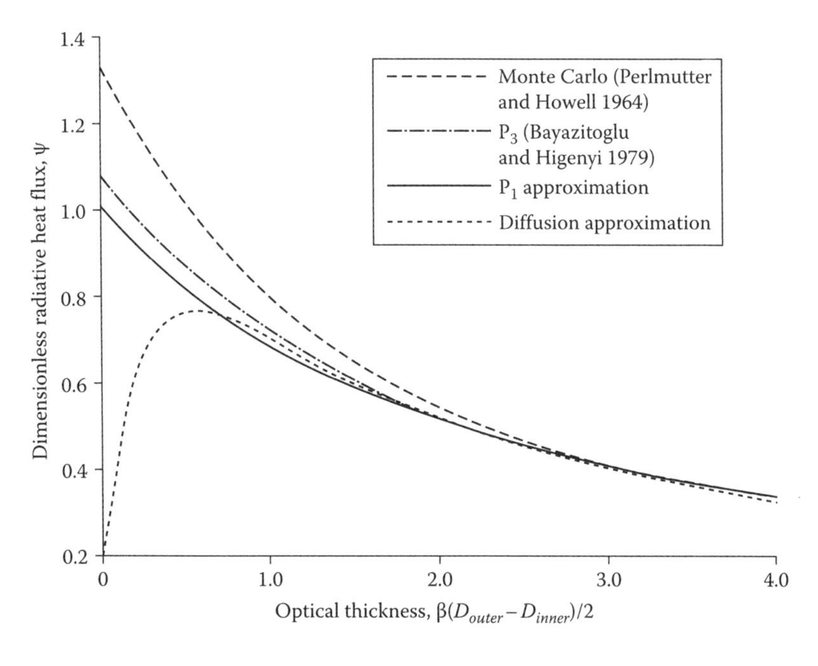

FIGURE 12.9 Comparison of solutions of energy transfer between infinitely long concentric black cylinders enclosing gray medium in radiative equilibrium with $\dot{q} = 0$; $D_{inner}/D_{outer} = 0.5$.

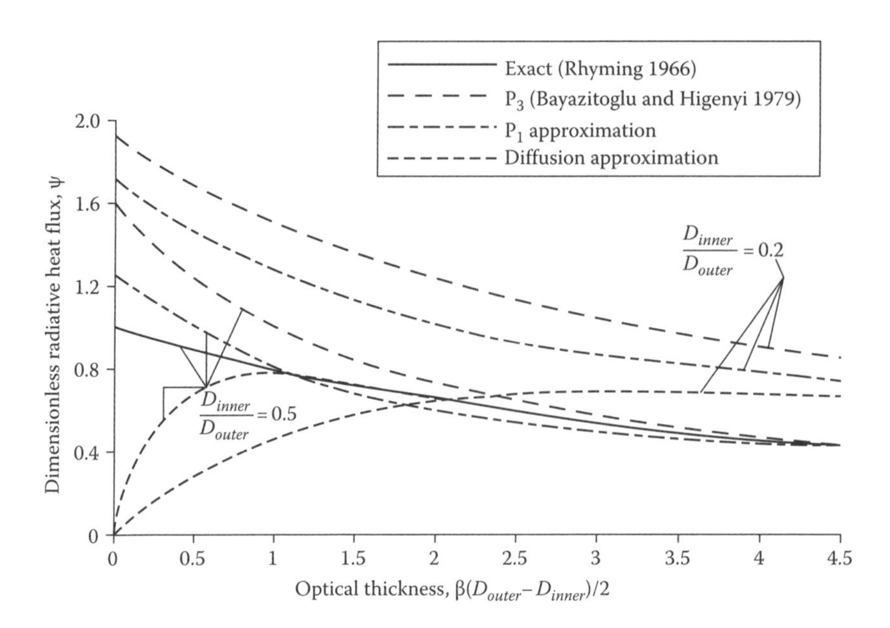

FIGURE 12.10 Comparison of solutions of energy transfer between black concentric spheres enclosing a gray medium.

is better than the P_1 and provides good results for $D_{inner}/D_{outer} = 0.5$. However, the P_N results are poor for smaller diameter ratios, as shown in Figure 12.10. The P_N approximations are based on series expansions about each location and hence would be expected to give less accurate results as the optical thickness decreases, and each location is influenced to an increasing extent by distant surroundings and boundaries. Additional information is in Heaslet and Warming (1966), Kesten (1968), Loyalka (1969), Schmid-Burgk (1974), Dua and Cheng (1975), Modest (1979, 1989), Mengüç and Viskanta (1985, 1986), Liu et al. (1993a), Tencer (2013), Modest et al. (2014), and Ge et al. (2015).

For radiative equilibrium with $\dot{q} = 0$, an engineering solution for ψ can sometimes be obtained by using the optically thin solution, which is exact in the limit of small optical dimension and either the diffusion or the differential approximation at large optical dimension. A curve faired between these solutions may provide acceptable accuracy for ψ over the entire range of optical thicknesses. However, this approach will not provide the temperature distribution.

P_1 and P_3 solutions are in Ratzel and Howell (1982) for radiative transfer in a rectangular enclosure with diffuse–gray or specular walls at differing isothermal temperatures containing a gray isotropically scattering gas. Profiles of wall heat flux distributions and gas temperature distributions are given.

Figure 12.11 shows the error in wall heat flux on the wall opposite the hot wall in the P_1 solution relative to the exact analytical solution (Crosbie and Schrenker 1984) for the case of a square black-bounded enclosure with one hot boundary enclosing a cold medium as a function of the medium optical thickness and albedo. The results are from Tencer (2013) and Tencer and Howell (2013b). Good agreement is present for all values of albedo if the optical thickness is less than 1 and for all values of optical thickness if the albedo exceeds 0.9. Errors will be smaller if gray diffuse rather than black boundaries are present, as this will tend to reduce the gradients in intensity.

12.2.2.3 Simplified P_N (SP_N) Method

A modified P_N method known as the simplified P_N (SP_N) method shows superior accuracy and computational efficiency over the conventional P_N method for multidimensional problems at low orders. It was proposed on a heuristic basis by Gelbard (1961) and was originally applied to neutron diffusion. Larsen et al. (1993) provided a theoretical justification for the simplification and showed that the SP_N approach can be derived from an asymptotic expansion of the intensity in the transport

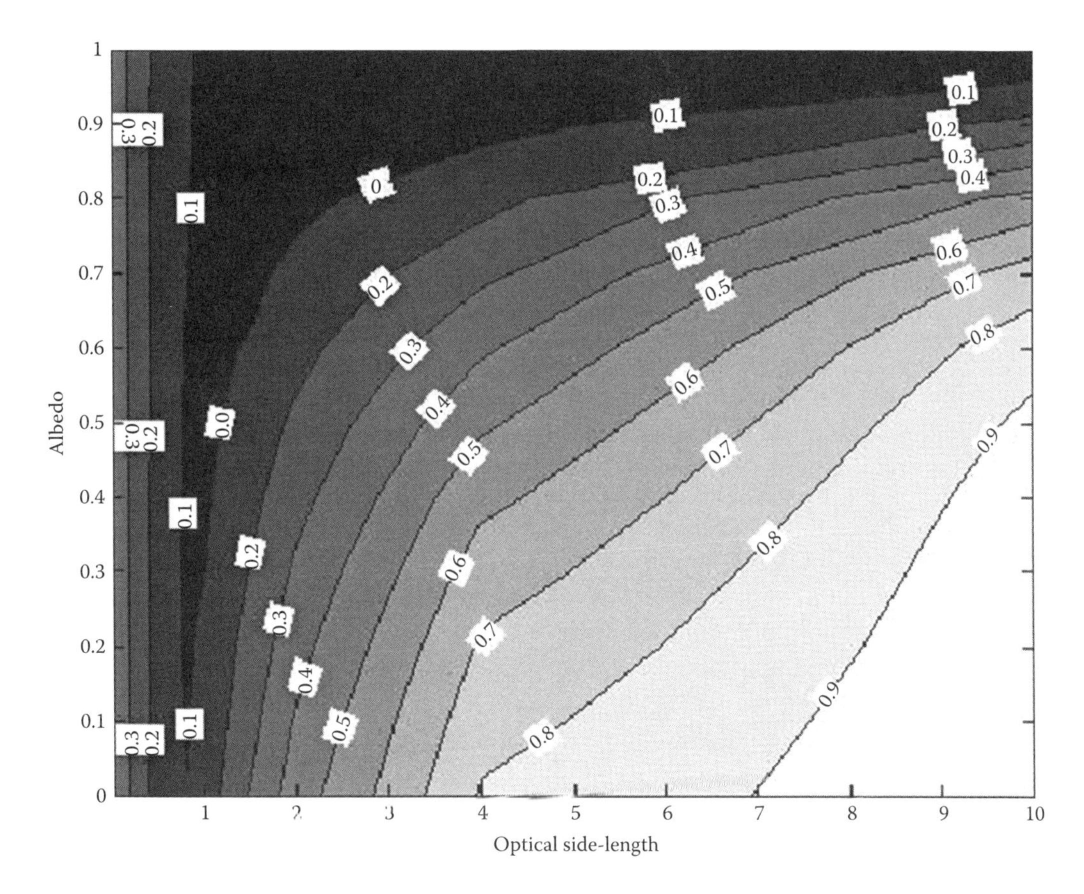

FIGURE 12.11 Error in P_1 solution relative to exact solution for a square enclosure with one hot and three cold black boundaries and a cold medium. (From Tencer, J. and Howell, J.R., A parametric study of the accuracy of several radiative transport solution methods for a set of 2-D benchmark problems, *Proceedings of the ASME 2013 Summer Heat Transfer Conference*, Minneapolis, MN, July 14–19, 2013.)

equation. The expansion results in the diffusion solution as the leading order term, with the SP_N terms providing higher-order approximations.

To obtain the SP_N solution, following the derivation in Larsen et al. (2002), the local intensity is expanded in a Neumann series around the local blackbody intensity for the nonscattering case as

$$I_\lambda = \left[1 - \frac{\delta}{\kappa_\lambda}\mu\cdot\nabla + \left(\frac{\delta}{\kappa_\lambda}\right)^2(\mu\cdot\nabla)^2 - \left(\frac{\delta}{\kappa_\lambda}\right)^3(\mu\cdot\nabla)^3 + \left(\frac{\delta}{\kappa_\lambda}\right)^4(\mu\cdot\nabla)^4 - \cdots\right]I_{\lambda b} \tag{12.76}$$

Here δ is a small expansion parameter defined by $\delta = 1/\kappa_{ref}L_{ref} = 1/\tau_{ref}$, so the derivation assumes an optically thick medium with $\tau_{ref} \gg 1$ at all wavelengths. Using this scaling, the radiative flux divergence Equation 10.4 in dimensionless terms becomes

$$-\nabla\cdot\mathbf{q}_r = \frac{1}{\delta^2}\left[\int_{\lambda=0}^{\infty}\kappa_\lambda\int_{\phi} I_\lambda(\Omega_i)d\Omega_i d\lambda - 4\pi\int_{\lambda=0}^{\infty}\kappa_\lambda I_{\lambda b}d\lambda\right] \tag{12.77}$$

Equation 12.76 is now integrated over the sphere of solid angles, noting that $\int_{\phi}(\mu\cdot\nabla)^n d\Omega = \left[1+(-1)^n\right]\frac{2\pi}{n+1}\nabla^n$. This results in

$$4\pi\bar{I}_\lambda = \int_{\phi} I_\lambda d\Omega = 4\pi\left[1+\frac{\delta^2}{3\kappa_\lambda^2}\nabla^2+\frac{\delta^4}{5\kappa_\lambda^4}\nabla^4+\frac{\delta^6}{7\kappa_\gamma^6}\nabla^6+\cdots\right]I_{\lambda b}+O(\delta^8) \tag{12.78}$$

where $\nabla^2 = \nabla\cdot\nabla$, etc. The expansion for $I_{\lambda b}$ then becomes

$$\begin{aligned} I_{\lambda b} &= \left[1-\frac{\delta}{\kappa_\lambda}\mu\cdot\nabla+\left(\frac{\delta}{\kappa_\lambda}\right)^2(\mu\cdot\nabla)^2-\left(\frac{\delta}{\kappa_\lambda}\right)^3(\mu\cdot\nabla)^3+\left(\frac{\delta}{\kappa_\lambda}\right)^4(\mu\cdot\nabla)^4-\cdots\right]^{-1}\left[\bar{I}_\lambda-O(\delta^8)\right] \\ &= \left\{1-\left[\frac{\delta^2}{3\kappa_\lambda^2}\nabla^2+\frac{\delta^4}{5\kappa_\lambda^4}\nabla^4+\frac{\delta^6}{7\kappa_\lambda^6}\nabla^6+\cdots\right]+\left[\frac{\delta^2}{3\kappa_\lambda^2}\nabla^2+\frac{\delta^4}{5\kappa_\lambda^4}\nabla^4+\frac{\delta^6}{7\kappa_\lambda^6}\nabla^6+\cdots\right]^2\right. \\ &\quad \left.-\left[\frac{\delta^2}{3\kappa_\lambda^2}\nabla^2+\frac{\delta^4}{5\kappa_\lambda^4}\nabla^4+\frac{\delta^6}{7\kappa_\lambda^6}\nabla^6+\cdots\right]^3\cdots\right\}\left[\bar{I}_\lambda-O(\delta^8)\right] \end{aligned} \tag{12.79}$$

Neglecting terms of order greater than $O(\delta^6)$ and gathering the remaining terms gives the SP_N expansion for intensity:

$$I_{\lambda b} = \left(1-\frac{\delta^2}{3\kappa_\lambda^2}\nabla^2-\frac{4\delta^4}{45\kappa_\lambda^4}\nabla^4-\frac{44\delta^6}{945\kappa_\lambda^6}\nabla^6\right)\bar{I}_\lambda \tag{12.80}$$

12.2.2.3.1 SP_1 Solution

Restricting Equation 12.80 to terms of order lower than δ^4 gives

$$\kappa_\lambda\left(I_{\lambda b}-\bar{I}_\lambda\right) = -\delta^2\nabla\cdot\frac{1}{3\kappa_\lambda}\nabla\bar{I}_\lambda+O(\delta^4) \tag{12.81}$$

Integrating Equation 12.81 over wavelength and inserting into Equation 12.78 gives

$$\nabla\cdot\mathbf{q}_r = \frac{4\pi}{\delta^2}\int_{\lambda=0}^{\infty}\kappa_\lambda\left(I_{\lambda b}-\bar{I}_\lambda\right)d\lambda = -\int_{\lambda=0}^{\infty}\nabla\cdot\frac{4\pi}{3\kappa_\lambda}\nabla\bar{I}_\lambda d\lambda \tag{12.82}$$

Equation 12.82 is the SP_1 approximation for the local radiative flux divergence, accurate to order δ^2. To order δ^2, Equation 12.79 indicates that $\bar{I}_\lambda = I_{\lambda b}$, so Equation 12.82 predicts the local radiative flux divergence as

$$\nabla\cdot\mathbf{q}_r = -\int_{\lambda=0}^{\infty}\nabla\cdot\frac{4\pi}{3\kappa_\lambda}\nabla I_{\lambda b}d\lambda = -\nabla\cdot\int_{\lambda=0}^{\infty}\frac{4\pi}{3\kappa_\lambda}\nabla I_{\lambda b}d\lambda = 0 \tag{12.83}$$

indicating that the local radiative flux is

$$q_r = -\int_{\lambda=0}^{\infty}\frac{4}{3\kappa_\lambda}\mathbf{n}\cdot\nabla E_{\lambda b}d\lambda = -\frac{4\sigma}{3\kappa_R}\mathbf{n}\cdot\nabla T^4 \tag{12.84}$$

which is the same as the Rosseland diffusion result of Equation 12.24 as well as the P_1 solution of Equation 12.4.5.

12.2.2.3.2 SP_1 Boundary Conditions

Larsen et al. (2002) note that the boundary conditions for the SP_1 solution in general geometries reduce to the same Marshak conditions found for the P_1 solution (Equations 12.4.9 through 12.4.12), which are in turn the same as those found for the diffusion solution (Equations 12.33 and 12.34 or 12.37 and 12.38). The SP_1 solution and boundary conditions for radiative transfer in general geometries are seen to be equivalent to the P_1 and diffusion solutions. This is not the case for higher-order SP_N solutions.

12.2.2.3.3 Higher-Order Solutions

Retaining one more term in Equation 12.80 than for the SP_1 solution gives the SP_2 relation accurate to order δ^4:

$$I_{\lambda b} = \left(1 - \frac{\delta^2}{3\kappa_\lambda^2}\nabla^2 - \frac{4\delta^4}{45\kappa_\lambda^4}\nabla^4\right)\bar{I}_\lambda$$

$$= \bar{I}_\lambda - \delta^2\nabla\cdot\frac{1}{3\kappa_\lambda^2}\left[\bar{I}_\lambda + \delta^2\nabla\cdot\frac{4}{15\kappa_\lambda^2}\nabla\bar{I}_\lambda\right] \tag{12.85}$$

or

$$\bar{I}_\lambda = I_{\lambda b} + \delta^2\nabla\cdot\frac{1}{3\kappa_\lambda^2}\nabla\bar{I}_\lambda + O(\delta^4) \tag{12.86}$$

Substituting Equation 12.86 into Equation 12.85 results in

$$I_{\lambda b} = \bar{I}_\lambda - \delta^2\nabla\cdot\frac{1}{3\kappa_\lambda^2}\nabla\bar{I}_\lambda - \delta^2\nabla\cdot\frac{4}{15\kappa_\lambda^2}\nabla\left(\bar{I}_\lambda - I_{\lambda b}\right) = \bar{I}_\lambda - \delta^2\nabla\cdot\frac{1}{3\kappa_\lambda^2}\nabla\left[\bar{I}_\lambda + \frac{4}{5}\left(\bar{I}_\lambda - I_{\lambda b}\right)\right] \tag{12.87}$$

Integrating as in Equation 12.82 gives

$$\nabla\cdot\mathbf{q}_r = \frac{4\pi}{\delta^2}\int_{\lambda=0}^{\infty}\kappa_\lambda\left(I_{\lambda b} - \bar{I}_\lambda\right)d\lambda = -\int_{\lambda=0}^{\infty}\nabla\cdot\frac{4\pi}{3\kappa_\lambda}\nabla\left[\bar{I}_\lambda + \frac{4}{5}\left(\bar{I}_\lambda - I_{\lambda b}\right)\right]d\lambda + O(\delta^4) \tag{12.88}$$

The kernel of the integral in the SP_2 result is the same as for the SP_1 solution (Equation 12.82) modified by the addition of the term $4/5\left(\bar{I}_\lambda - I_{\lambda b}\right)$, and the local radiative flux is

$$\mathbf{n}\times\mathbf{q}_r = -\frac{4\pi}{3}\int_{\lambda=0}^{\infty}\frac{1}{\kappa_\lambda}\mathbf{n}\cdot\nabla\left[\bar{I}_\lambda + \frac{4}{5}\left(\bar{I}_\lambda - I_{\lambda b}\right)\right]d\lambda = -\frac{4\pi}{3}\int_{\lambda=0}^{\infty}\frac{1}{\kappa_\lambda}\mathbf{n}\cdot\nabla\left[\frac{9}{5}\bar{I}_\lambda - \frac{4}{5}I_{\lambda b}\right]d\lambda \tag{12.89}$$

Noting that Equation 12.89 gives the local spectral radiative flux as

$$q_{r,\lambda} = -\frac{4\pi}{15\kappa_\lambda}\nabla\left[9\bar{I}_\lambda - 4I_{\lambda b}\right] \tag{12.90}$$

then Equation 12.87 gives the local value for $\bar{I}_\lambda$ as

$$\bar{I}_\lambda = I_{\lambda b} + \delta^2 \nabla \cdot \frac{1}{15\kappa_\lambda^2} \nabla\left[9\bar{I}_\lambda - 4I_{\lambda b}\right] = I_{\lambda b} - \frac{\delta^2}{4\pi\kappa_\lambda} \mathbf{n} \cdot \nabla \mathbf{q}_{r,\lambda} \tag{12.91}$$

Retaining one additional term in Equation 12.80 than for the SP_1 result of Equation 12.81 gives the SP_3 relation accurate to order δ^6:

$$\begin{aligned} I_{\lambda b} &= \left(1 - \frac{\delta^2}{3\kappa_\lambda^2}\nabla^2 - \frac{4\delta^4}{45\kappa_\lambda^4}\nabla^4 - \frac{44\delta^6}{945\kappa_\lambda^6}\nabla^6\right)\bar{I}_\lambda \\ &= \bar{I}_\lambda - \frac{\delta^2}{3\kappa_\lambda^2}\nabla^2\left[\bar{I}_\lambda + \frac{4\delta^2}{15\kappa_\lambda^2}\nabla^2\bar{I}_\lambda + \frac{44\delta^4}{315\kappa_\lambda^4}\nabla^4\bar{I}_\lambda\right] \\ &= \bar{I}_\lambda - \frac{\delta^2}{3\kappa_\lambda^2}\nabla^2\left[\bar{I}_\lambda + \left(1 + \frac{11\delta^2}{21\kappa_\lambda^2}\nabla^2\right)\left(\frac{4\delta^2}{15\kappa_\lambda^2}\nabla^2\bar{I}_\lambda\right)\right] \end{aligned} \tag{12.92}$$

Defining $\bar{I}_\lambda^{(2)} = \left(1 + \frac{11\delta^2}{21\kappa_\lambda^2}\nabla^2\right)\left(\frac{4\delta^2}{15\kappa_\lambda^2}\nabla^2\bar{I}_\lambda\right)$, Equation 12.92 becomes

$$I_{\lambda b} = \bar{I}_\lambda - \delta^2 \nabla \cdot \frac{1}{3\kappa_\lambda^2} \nabla\left[\bar{I}_\lambda + \bar{I}_\lambda^{(2)}\right] \tag{12.93}$$

Integrating Equation 12.93 over wavelength and the sphere of solid angles gives

$$\int_{\phi}\int_{\lambda=0}^{\infty} \kappa_\lambda \left(I_{\lambda b} - \bar{I}_\lambda\right) d\lambda d\Omega = -\delta^2 \int_{\phi}\int_{\lambda=0}^{\infty} \nabla \cdot \frac{1}{3\kappa_\lambda} \nabla\left[\bar{I}_\lambda + \bar{I}_\lambda^{(2)}\right] d\lambda d\Omega + O\left(\delta^6\right) \tag{12.94}$$

Substituting into Equation 12.82 to determine the radiative flux divergence gives

$$\nabla \cdot \mathbf{q}_r = \frac{4\pi}{\delta^2} \int_{\lambda=0}^{\infty} \kappa_\lambda \left(I_{\lambda b} - \bar{I}_\lambda\right) d\lambda = -\int_{\lambda=0}^{\infty} \nabla \cdot \frac{4\pi}{3\kappa_\lambda} \nabla\left[\bar{I}_\lambda + \bar{I}_\lambda^{(2)}\right] d\lambda \tag{12.95}$$

Comparison with Equation 12.83 for the SP_1 (diffusion) prediction of radiative flux divergence shows that the additional term $\bar{I}_\lambda^{(2)}$ has now entered into the radiative flux divergence. Equations 12.93 and 12.95 and associated boundary conditions comprise the SP_3 solution.

12.2.2.3.4 Boundary Conditions for Higher-Order SP_N Solutions

The emissive power jump boundary conditions of Marshak (Equations 12.35 through 12.38) are used with the SP_N equations to generate appropriate boundary conditions for the higher-order solutions. Tomašević and Larsen (1975) and Brantley and Larsen (2000) derive the boundary conditions for black boundaries, and Larsen et al. (2003) extend these to nongray boundaries with directional reflectivity. For a gray diffuse surface at $x = 0$, the SP_2 boundary relations reduce to

$$5\bar{I}_\lambda(0) + \frac{2-\varepsilon}{\varepsilon}\frac{4}{\kappa_\lambda} \mathbf{n} \cdot \nabla \bar{I}_\lambda(0) = \left[6I_{\lambda b,w} - I_{\lambda b}(0)\right] \tag{12.96}$$

Substituting Equation 12.91 for $\bar{I}_\lambda(0)$ and rearranging gives

$$\frac{E_{\lambda b,w}-E_{\lambda b}(0)}{\left(\dfrac{1}{\varepsilon_\lambda}-\dfrac{1}{2}\right)}=\frac{-5}{24\kappa_\lambda\left(\dfrac{1}{\varepsilon_\lambda}-\dfrac{1}{2}\right)}\mathbf{n}\cdot\nabla\mathbf{q}_{r,\lambda}+\frac{4}{3\kappa_\lambda}\mathbf{n}\cdot\nabla E_{\lambda b}(0)-\frac{1}{\kappa_\lambda}\nabla\cdot\left(\frac{1}{3\kappa_\lambda}\nabla\mathbf{q}_{r,\lambda}\right) \tag{12.97}$$

For the case of no sources in the medium, $\nabla\mathbf{q}_{r,\lambda}=0$ and Equation 12.97 reduces to Equation 12.37.

The boundary conditions for the SP_3 relations result in a pair of weakly coupled differential equations that are derived in Larsen et al. (2002). These authors observe that the SP_1 equation and boundary conditions are identical with P_1 solutions for all geometries and that for planar geometries the SP_N and P_N equations and boundary conditions are also identical. However, for multidimensional geometries, the SP_2 and higher-order SP_N solutions differ from the P_N solutions, and the SP_2 and SP_3 relations are found to be more accurate and easier to program and implement than the corresponding P_N equations. Morel et al. (1979) have implemented the SP_N method into a general unstructured grid code. Cai et al. (2012) employ various SP_N orders with *k*-distributions to study spectral effects in hydrogen–air diffusion flames. Zhang et al. (2013b) compare the convergence rate among various formulations of the SP_N method.

12.2.2.4 M_N Method

The M_N method for solving the RTE uses entropy maximization as the closure principle for the first two (zeroth and first) RTE moment equations (Equations 12.6 and 12.10) (Minerbo, 1978). The derived closure relation is exact in the diffusion and transport limits; however, the method itself is not exact, in that it uses the first two moment equations plus closure rather than the RTE itself (Ripoll and Wray 2003, Berthon et al. 2007, González et al. 2008, 2009). The maximum entropy closure condition gives a nonlinear coefficient on the radiative flux equation. The M_N method is quite useful in neutron transport solutions, where the time of flight of neutrons in transient problems requires that the transient terms in the neutron transport equation be retained. Because this term is generally negligible in photon transport, the M_N solutions have been found to be less useful.

In the first moment equation, Equation 12.12, for isotropic or no scattering, or in fact for any source function that is symmetric around the propagation direction, the integral term is zero. With this assumption, the closure condition is given by assuming the radiative pressure term to be proportional to the gradient in energy density with a variable coefficient $\zeta(R)$. Substituting into Equation 12.12 with the assumption noted and treating the medium as gray,

$$q_r(S)=cP_{ij}(S)=-\frac{4}{\beta}\zeta(R)c\nabla U(S)=-\frac{4\pi}{\beta}\zeta(R)\nabla\bar{I}(S) \tag{12.98}$$

For no scattering, this becomes

$$q_r(S)=-\frac{4}{\kappa}\zeta(R)\nabla E_b(S) \tag{12.99}$$

Various forms have been suggested for the variable coefficient $\zeta(R)$, called the *flux limiter* (Pomraning 1982, Levermore 1984, Brunner and Holloway 2001, Frank 2007, Fan 2012). Flux limiters are required to avoid nonphysical solutions from arising in diffusion-based solutions for certain combinations of extinction coefficient and any constant coefficient on the diffusion relation such as on the Rosseland equation, Equation 12.25. The basic constraint can be seen by examining the first and second moment equations, Equations 12.2 and 12.3, which require that $q_r(S) < 4\pi\bar{I}(S)$. Using entropy maximization and this constraint for 1D M_1 solutions, the value of the variable flux limiter is found by solving the transcendental equations (Minerbo 1978, Tencer 2013a):

$$\zeta(R) = \frac{1}{Z^2} - \text{csch}^2(Z)$$
$$R = \frac{\coth(Z) - (1/Z)}{\zeta(R)} \tag{12.100}$$

where R is a function of the *effective albedo* $\tilde{\omega}$ and R is given by Pomraning (1982) as

$$R = \left[\left(\frac{\partial I^{(i)}(S)}{\partial S}\right)\Big/\beta\tilde{\omega}I^{(0)}(S)\right] = \frac{-\nabla\cdot q_r}{4\pi\beta\tilde{\omega}\bar{I}(S)} = \frac{1-\tilde{\omega}}{\tilde{\omega}} \tag{12.101}$$

The parameter R is thus in the range $0 < R < \infty$, giving values of $\zeta(R)$ in the range from $1/3 < \zeta(R) < 0$ (Figure 12.12).

If the value of R is taken as zero, then $\zeta(R)$ becomes 1/3, and Equations 12.98 and 12.99 reduce to the first-order diffusion and P_1 solutions. For larger effective albedo, the flux limiter gets smaller, approaching zero in the transport limit.

The curve in Figure 12.12 is very closely approximated (Tencer 2013) by the relation $\zeta(R) = \frac{2}{3\pi}\cot^{-1}(0.2673|R|)$. Use of this relation eliminates iterative solution of Equation 12.100, but because q_r and $\bar{I}$ are still unknowns in Equation 12.101, solution using the M_1 method requires iteration among q_r, R, and $\bar{I}$.

The diffusion approximation, P_N, SP_N, and M_N methods suffer from the difficulty that the effects of the boundaries enter only through the boundary conditions. In a translucent medium, however, the radiative energy from the boundary can propagate to the interior of the medium, and solutions for the radiative flux by the P_N, SP_N, M_N, and diffusion methods do not account for this. They are

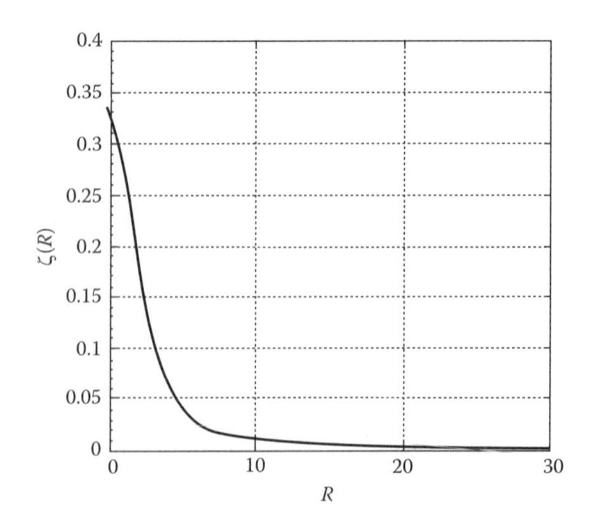

FIGURE 12.12 Flux limiter $\zeta(R)$ versus the parameter R.

thus inaccurate within a few radiative mean free paths from a boundary. The use of slip boundary conditions partially corrects for this. Modest (1989) also partially remedies this for the P_1 method in the "modified P_1" method by dividing the local intensity into a portion originating from the medium, which can be treated through the usual P_1 approximation, plus another portion that originates at the boundary. This approach makes the P_1 results much more accurate, particularly in systems with small to moderate optical thicknesses.

For planar geometries, Mengüç and Iyer (1988) propose the "double P_1" or DP_1 approximation, which is a hybrid of the P_1 and two-flux approximations. The method incorporates linear anisotropic scattering and includes with greater accuracy the variation of intensity with angle for this geometry than the standard P_1 formulation. Comparisons with a P_9 solution using a nine-term anisotropic scattering phase function show excellent agreement for uniform and nonuniform properties in planar geometries. Ge et al. (2015) compare higher-order P_N and SP_N for some simple geometries with varying radiative properties, and compare results with Monte Carlo solutions. Iyer and Mengüç (1989) later provided the formulations for the quadruple (QP_1) and octopule (OP_1) P_1 approximations.

Siewert (1978) discusses the F_N method, which is related to the P_N method. The intensity in a given direction is approximated by a series expansion, but, rather than using spherical harmonics, a Fourier expansion in the direction cosines is used. Kumar and Felske (1986) use the F_N method to obtain a set of coupled equations from the differential equation of transfer and then express the solution as a singular eigenfunction expansion. They solved for the energy flux and intensity distributions at the boundaries of a slab of absorbing–anisotropically scattering medium exposed to collimated incident radiation.

12.3 DISCRETE ORDINATES (S_N) METHOD

The discrete ordinates method is an extension of a more restricted method (the *two-flux method*) proposed independently by Schuster (1905) and Schwarzschild (1906) for studying radiative transfer in stellar atmospheres. Chandrasekhar (1960) extended the two-flux method to include anisotropic scattering and made it applicable to multidimensional geometries. When the solid angle about a location is divided into more than the two hemispheres, each assumed with uniform intensities as used in the two-flux method (Section 12.3.1), the method is known as the *discrete ordinates*, S_N, or *multiflux* method (Chandrasekhar 1960, Lathrop 1966). Fiveland (1984, 1987, 1988, 1994) developed and implemented the discrete ordinates method for analyses of coal-fired furnaces. Viskanta and Mengüç (1983, 1987) presented a review of work using the method. Krook (1955) compared the discrete ordinates method with the P_N method and showed that, in the limit of many terms, they become mathematically equivalent.

12.3.1 Two-Flux Method: The Schuster–Schwarzschild Approximation

As an introduction to the discrete ordinates method, the two-flux or S_2 approximation is discussed first. This is the simplest multiflux approximation. It is assumed that the energy transfer is 1D and that the intensity is isotropic for all radiation with components in the positive coordinate direction and is also isotropic in the negative direction with a different value, as in Figure 12.7. This yields relations similar to Milne–Eddington or P_1 approximation, as pointed out by Krook (1955). In that development, the same isotropy assumption was used to evaluate some of the integral terms *after* taking moments of the transfer equation. Here the isotropic assumption is used at the *start* of the development. To simplify the notation, the derivation is carried out for a gray medium so the wavelength symbols are omitted. The two-flux equations can also be used for spectral calculations if the same equation forms are written using spectral quantities.

Starting with Equation 12.1 that includes isotropic scattering, and making the isotropic assumption for intensity in each + or − direction, we obtain the mean incident intensity as $\bar{I}(\tau) = [I^+(\tau) + I^-(\tau)]/2$. The transfer equation in the + and in the − direction is then multiplied by $d\Omega$ and integrated over each hemisphere of solid angles ($0 \le \theta \le \pi/2$ and $\pi/2 \le \theta \le \pi$) as in Figure 12.7 to obtain for each direction (using $\mu = \cos\theta$):

$$\frac{dI^+}{d\tau}\int_{\mu=0}^{1}\mu d\mu = -I^+\int_{\mu=0}^{1}d\mu + (1-\omega)I_b\int_{\mu=0}^{1}d\mu + \frac{\omega}{2}(I^+ + I^-)\int_{\mu=0}^{1}d\mu \quad (12.102)$$

$$-\frac{dI^-}{d\tau}\int_{\mu=0}^{1}\mu d\mu = -I^-\int_{\mu=0}^{1}d\mu + (1-\omega)I_b\int_{\mu=0}^{1}d\mu + \frac{\omega}{2}(I^+ + I^-)\int_{\mu=0}^{1}d\mu \quad (12.103)$$

The integrations are carried out and the results multiplied by π (since, for isotropic intensities, $q_r^+ = \pi I^+$ and $q_r^- = \pi I^-$) to yield relations in terms of fluxes:

$$\frac{dq_r^+}{d\tau} = -(2-\omega)q_r^+ + \omega q_r^- + 2(1-\omega)E_b \quad (12.104)$$

$$-\frac{dq_r^-}{d\tau} = \omega q_r^+ - (2-\omega)q_r^- + 2(1-\omega)E_b \quad (12.105)$$

These are forms of Schuster–Schwarzschild equations. The same functional form as for Milne–Eddington approximation is obtained by substituting Equations 12.49 and 12.50 into Equations 12.104 and 12.105 to eliminate the q_r^+ and q_r^-. The resulting two equations are added, and subtracted, to yield the two relations:

$$\frac{dq_r(\tau)}{d\tau} = 4(1-\omega)\left[E_b(\tau) - \pi\bar{I}(\tau)\right] \quad (12.106)$$

and

$$q_r(\tau) = -\pi\frac{d\bar{I}(\tau)}{d\tau} \quad (12.107)$$

where $q_r = q_r^+ - q_r^-$. These have the same form as for Milne–Eddington approximation (Equations 12.42 and 12.44), except that the coefficient in Equation 12.107 has changed to 1 from 4/3 in Equation 12.44. These equations also apply when written with spectral quantities.

For 1D steady-state transfer between parallel walls without convection, conduction, or internal heat sources, the q_r is constant (radiative equilibrium with $\dot{q} = 0$) and, from Equation 12.106, $\bar{I} = E_b/\pi$. Then, by substituting into Equation 12.107,

$$q_r(\tau) = -\frac{dE_b(\tau)}{d\tau} = -\frac{1}{\left[\kappa(x)+\sigma_s(x)\right]}\frac{dE_b(x)}{dx} = -\frac{\sigma}{\beta(x)}\frac{dT^4(x)}{dx} \quad (12.108)$$

which has the same form as the diffusion approximation equation, Equation 12.24, but has a different numerical coefficient.

As discussed for Milne–Eddington approximation, Equations 12.106 and 12.107 are combined with the energy equation to obtain three equations for $q_r(\tau)$, $\bar{I}(\tau)$ and $E_b(\tau)$. The same boundary conditions can be used as discussed for Milne–Eddington approximation in Section 12.2.2.1.

By using Schuster–Schwarzschild equations, the results in Example 12.3 have the same form except that the 3/4 is replaced by 1 in two places in Equation 12.3.1 and two places in Equation 12.3.2. If there is no internal heat generation, $\dot{q} = 0$, the radiative energy flux does not depend on position τ and is given by

$$q_r = \frac{\sigma\left(T_1^4 - T_2^4\right)}{\tau_D + 1/\varepsilon_1 + 1/\varepsilon_2 - 1} \tag{12.109}$$

The temperature distribution is

$$\sigma T^4(\tau) = \sigma T_1^4 - \left(\tau + \frac{1}{\varepsilon_1} - \frac{1}{2}\right) q_r \tag{12.110}$$

The two-flux equations given here are for isotropic scattering. Some materials used for high-temperature heat shields can have strong backscattering. In this instance, modified two-flux equations have been used. They are given in Matthews et al. (1985) and Cornelison and Howe (1992) where they are applied to a medium subjected to an intense concentrated solar flux and to a backscattering heat shield for thermal protection.

Example 12.5

A plane layer of gray absorbing medium with isotropic scattering is not confined by solid boundaries. It has an optical thickness τ_D and refractive index $n = 1$ and is exposed on each side to a blackbody environment; the side at $\tau = (\kappa + \sigma_s)x = 0$ is exposed to T_{e1} and the side at $\tau = \tau_D$ to T_{e2}. For the limit where there is no conduction, convection, or internal heat sources, determine the radiative transfer through the gray layer and its temperature distribution.

From the energy equation, for radiative equilibrium with $\dot{q} = 0$, the radiative heat flux within the layer is independent of location so, from Equation 12.106, $dq_r/d\tau = 0$ and $\bar{I}(\tau) = E_b(\tau)/\pi$. By integrating Equation 12.108 with q_r constant, the temperature distribution within the layer has the form $E_b(\tau) = -q_r\tau + E_b(0)$, so that $\bar{I}(\tau) = (1/\pi)[-q_r\tau + E_b(0)]$ and $q_r = [E_b(0) - E_b(\tau_D)]/\tau_D$. The radiative boundary conditions are now considered. Since the medium has a refractive index $n = 1$, there are no reflections at the boundaries for radiation entering the layer from the surroundings. Then, from Equations 12.47 and 12.48,

$$q_r^+(0) = \sigma T_{e1}^4 = \pi\bar{I}(0) + \frac{q_r}{2} \tag{12.5.1}$$

$$q_r^-(\tau_D) = \sigma T_{e2}^4 = \pi\bar{I}(\tau_D) - \frac{q_r}{2} \tag{12.5.2}$$

$\bar{I}(0) = E_b(0)/\pi$, $\bar{I}(\tau_D) = E_b(\tau_D)/\pi$, and $q_r = \left[E_b(0) - E_b(\tau_D)\right]/\tau_D$ are substituted into Equation 12.110 to give

$$\sigma T_{e1}^4 = E_b(0) + \frac{E_b(0) - E_b(\tau_D)}{2\tau_D} \quad \text{and} \quad \sigma T_{e2}^4 = E_b(\tau_D) - \frac{E_b(0) - E_b(\tau_D)}{2\tau_D} \tag{12.5.3}$$

These two equations are solved simultaneously for $E_b(0)$ and $E_b(\tau_D)$. Then, by substituting these values, the radiative heat flux and the temperature distribution in the layer are obtained as

$$q_r = \frac{E_b(0) - E_b(\tau_D)}{\tau_D} = \frac{\sigma\left(T_{e1}^4 - T_{e2}^4\right)}{\tau_D + 1} \tag{12.5.4}$$

$$\sigma T^4(\tau) = E_b(\tau) = -q_r\tau + E_b(0) = \frac{1}{2(\tau_D+1)}\left\{\left[1+2(\tau_D-\tau)\right]\sigma T_{e1}^4 + (1+2\tau)\sigma T_{e2}^4\right\} \tag{12.5.5}$$

For very small τ_D, the layer does not offer any resistance to the energy exchange between the two black surroundings, and the energy flux becomes $q_r = \sigma(T_{e1}^4 - T_{e2}^4)$ with the layer temperature of $T(\tau)^4 = (T_{e1}^4 + T_{e2}^4)/2$; these are the transparent limits. For very large τ_D, $q_r \to 0$ and $T(\tau)^4$ varies linearly from T_{e1}^4 to T_{e2}^4. These results are for the limit of no heat conduction, convection, or internal heat sources in a medium with refractive index $n = 1$.

Example 12.6

Let the layer in Example 12.5 have spectral properties that can be approximated by two spectral bands. Provide a method to obtain the layer temperature distribution.

Let λ_c be the separating wavelength between the two spectral bands, and let the subscripts S and L designate the bands with short ($0 < \lambda < \lambda_c$) and long ($\lambda_c < \lambda < \infty$) wavelengths. Then, for radiative equilibrium without internal heat sources, the derivative of the total radiative heat flow is zero so that $q_r = q_{rS}(x) + q_{rL}(x)$ is constant. This condition will need to be satisfied by the final solution obtained by iteration; the flux is not constant in each wavelength band. For the blackbody surroundings, the blackbody function $F_{0\to\lambda}(T)$ can be used to determine the amount of energy incident at each boundary that is in each spectral band. Then, using Equations 12.5.1 and 12.5.2 for one of the bands,

$$E_{bS}(T_{e1}) = \sigma T_{e1}^4 F_{0\to\lambda_c}(T_{e1}) = \pi \bar{I}_S(0) + \frac{q_{rS}(0)}{2} \tag{12.6.1}$$

$$E_{bS}(T_{e2}) = \sigma T_{e2}^4 F_{0\to\lambda_c}(T_{e2}) = \pi \bar{I}_S(D) - \frac{q_{rS}(D)}{2} \tag{12.6.2}$$

and similarly for the other band where $E_{bL}(T) = \sigma T^4[1 - F_{0\to\lambda_c}(T)]$. For each spectral band, Equations 12.106 and 12.107 can be combined in spectral form and then integrated over the spectral band to yield the differential equation for $\bar{I}_S$:

$$-\frac{\pi}{4}\frac{d^2\bar{I}_S}{d^2x} = \kappa_S\left[E_{bS}(x) - \pi\bar{I}_S(x)\right] \tag{12.6.3}$$

and similarly for $\bar{I}_L$. To start the iterative solution, the two solutions for a gray medium evaluated at the properties for each of the spectral bands are used to estimate values for $T(x)$, $\bar{I}_S(0)$ and $\bar{I}_L(0)$. From Equation 12.6.1, written for each band, the $q_{rS}(0)$ and $q_{rL}(0)$ are then obtained, and from Equation 12.107, this gives an estimated $d\bar{I}/dx$ at $x = 0$ for each spectral band. Starting with these initial values for $\bar{I}_S(0)$ and $\bar{I}_L(0)$ and their first derivatives, the differential Equation 12.6.3 is integrated numerically for each band to reach $x = D$, and the values at $x = 0$ are adjusted iteratively to satisfy the boundary conditions at $x = D$ given by Equation 12.6.2 written for each band. From the $\bar{I}_s(x)$ and $\bar{I}_L(x)$, the $q_r(x)$ can then be found by using Equation 12.107 integrated over each band, and the condition for radiative equilibrium can be checked, $q_{rS}(x) + q_{rL}(x) =$ constant. From the $q_{rS}(x)$ and $q_{rL}(x)$, a new temperature distribution is estimated by using Equation 12.106 integrated over each band and summing the results to find $E_b(x)$. An iterative scheme must be devised to converge toward $q_{rS}(x) + q_{rL}(x) = q_r$ (a constant), and this yields the correct temperature distribution.

This example reveals the complexity of the iterative procedure needed to obtain spectral solutions by the two-flux method. Some examples of spectral solutions that illustrate the numerical techniques that have been used are in Matthews et al. (1985), Cornelison and Howe (1992), Siegel and Spuckler (1994a), and Siegel (1996). Convergence can be difficult for optically thick bands. A combination of the two-flux method with the diffusion method used for the thick bands is developed by Siegel and Spuckler.

Viskanta (1982) reviewed the literature on the two-flux model through 1982. The method works well in 1D systems, and a solution with combined radiation and conduction is in Tremante and Malpica (1994). The two-flux method has been used for analysis of packed beds (Brewster and Tien 1982), radiative transfer through fibers and powders (Tong and Tien 1983, Wang and Tien 1983), and porous layers with penetrating flow and an external radiation source (Matthews et al. 1985, Lee and Howell 1986). Anisotropic scattering can be considered, but is used as an integrated average for forward and backward scattering. The size of the backward and forward scattering fractions to be used in a two-flux model is analyzed by Matthews et al. (1985), Koenigsdorff et al. (1991), and Cornelison and Howe (1992) and compared with other approximations in the literature. Solutions for a composite of plane layers are in Spuckler and Siegel (1996).

DeMarco and Lockwood (1975) modified the two-flux model into a four-flux or six-flux model. This is a hybrid of the discrete ordinates method and the P_1 differential approximation. It has certain computational advantages over the more complete models and generally predicts fluxes in reasonable agreement with other models. An evaluation of multiflux models is in Selçuk (1989) for radiative transfer in cylindrical furnaces. The discrete transfer method in Section 12.4.1 was developed to provide further improvements.

12.3.2 Radiative Transfer Equation with S_N Method

To develop the discrete ordinates method, the general transfer relations are represented by a set of equations for an intensity that is angularly averaged over each of a finite number of ordinate directions. Integrals over a range of solid angles are replaced by sums over the ordinate directions within that range. Depending on the angular quadrature method employed, the method is referred to as the *discrete ordinates* (or S_N method) (Kumar et al. 1990, Fiveland and Jessee 1995, Selçuk and Kayakol 1997), or the *multiflux* method (Sasse et al. 1995). The history of the S_N model in nuclear applications is in Lathrop (1992). Fiveland (1984, 1987, 1988), Kumar et al. (1990), and Tencer (2013a) present outlines of the method as applied to radiative transfer, and Coelho (2014) provides a comprehensive review of advances in the method since the year 2000.

The RTE equation (Equation 10.48) along a path S in the direction of $d\Omega$ is (l_1, l_2, and l_3 are direction cosines relative to the x, y, and z directions)

$$\frac{dI}{dS}=\frac{\partial I}{\partial x}\frac{dx}{dS}+\frac{\partial I}{\partial y}\frac{dy}{dS}+\frac{\partial I}{\partial z}\frac{dz}{dS}=l_1\frac{\partial I}{\partial x}+l_2\frac{\partial I}{\partial y}+l_3\frac{\partial I}{\partial z}$$

$$=\kappa(S)I_b(S)-\left[\kappa(S)+\sigma_s(S)\right]I(S,\Omega)+\frac{\sigma_s(S)}{4\pi}\int_{\Omega_i=0}^{4\pi}I(S,\Omega_i)\Phi(S,\Omega,\Omega_i)d\Omega_i \qquad (12.111)$$

The integral over the incident angular directions is approximated by a weighted sum of the angular quantities. Let m and m' correspond to outgoing and incoming angular directions that each represent a finite solid angle, and the w are weighting factors to provide an accurate representation of the integral when converted into a summation. Then, the equation of transfer in the m-direction is written as

$$l_{1,m}\frac{\partial I_m}{\partial x}+l_{2,m}\frac{\partial I_m}{\partial y}+l_{3,m}\frac{\partial I_m}{\partial z}=\kappa(S)I_b(S)-\left[\kappa(S)+\sigma_s(S)\right]I_m(S)+\frac{\sigma_s(S)}{4\pi}\sum_{m'}w_{m'}I_{m'}(S)\Phi_{m'm}(S) \qquad (12.112)$$

The $l_{1,m}$, $l_{2,m}$, and $l_{3,m}$ are the direction cosines of $I_m(S)$ for $\Omega = \Omega_m$ in the mth direction. For the summation, Gaussian quadrature is often used; various quadrature schemes are discussed in

Fiveland (1988), Koch et al. (1995), and Tencer (2013), and more information is provided in Section 12.3.5 about coordinate pairs, weighting factors, and quadrature schemes.

12.3.3 Boundary Conditions for the S_N Method

When obtaining the intensity along a path in direction m (see Figure 12.13a), the initial value at the boundary at location $\mathbf{r}_w$ must be specified or obtained from the radiation field surrounding the medium being analyzed. Boundary conditions expressing the intensity leaving a solid surface along each ordinate direction m or m' can be expressed as the sum of emitted intensity and the reflected intensity resulting from the reflected incident energy that is written here in terms of incident intensities. For a diffuse–gray surface, this gives, at the origin of a path $S = 0$ that originates at location $\mathbf{r}_w$ on the wall (see Figure 12.13a),

$$I_m\left(\mathbf{r}_w, S=0\right) = \varepsilon_w\left(\mathbf{r}_w\right) I_{b,w}\left(\mathbf{r}_w\right) + \frac{1-\varepsilon_w\left(\mathbf{r}_w\right)}{\pi} \sum_{m'} l_{i,m'} w_{m'} I_{m'}\left(\mathbf{r}_w\right) \tag{12.113}$$

where $l_{i,m'}$ is the direction cosine between the m' direction and the i coordinate direction that is normal to the surface.

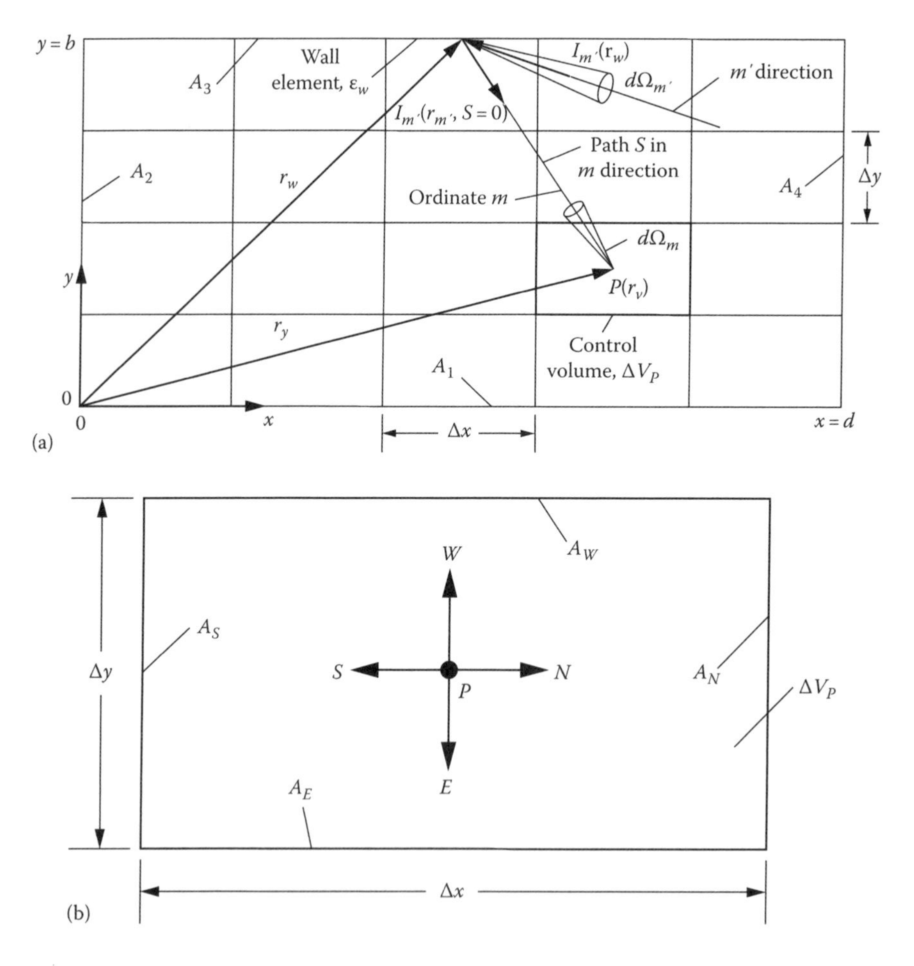

FIGURE 12.13 2D radiating medium and typical control volume: (a) region divided into control volumes and (b) typical control volume.

The use of Equation 12.113 causes a difficulty in obtaining accurate solutions from the discrete ordinates method. The location of $\mathbf{r}_w$, on the boundary, where $S = 0$ for a given ordinate, is generally taken as being at the midpoint of the angular increment that surrounds the m discrete ordinate direction (extending from P in Figure 12.13a). The location of $\mathbf{r}_w$ may be anywhere on a boundary increment. A small change in the location of the volume element (such as P in Figure 12.13a) for which Equation 12.112 is evaluated for a path along a given ordinate can cause $\mathbf{r}_w$ to move from one boundary increment into an adjacent one with possibly very different boundary properties. This may cause apparent discontinuities in temperature and radiative flux to occur in the solutions and is known as the *ray effect* (Lathrop 1968, 1971, Chai et al. 1993, Coelho 2001, 2014). The effect can be reduced by increasing the order of the solution (this is discussed in Section 12.3.4); however, when discontinuous boundary conditions are present (such as adjacent surfaces with different temperatures or emissivities, or even the presence of corners), the effect may persist even to high-order solutions.

12.3.4 Control Volume Method for S_N Numerical Solution

For a numerical solution, the medium is divided into finite volume elements, and the boundary into finite area elements. The equation of transfer is written in each ordinate direction for each volume element, and the boundary condition specified in each direction at each wall element. This provides a set of linear differential equations and boundary conditions sufficient to solve for the unknown intensities I_m along the ordinates at every location in the medium. Once the I_m are known, the local radiative energy flux in the i coordinate direction at location S in the medium is found from

$$q_{r,i}(S) = \int_{\Omega=0}^{4\pi} I(S,\Omega)\cos\theta d\Omega \approx \sum_m l_{i,m} w_m I_m(S) \tag{12.114}$$

where $\cos\theta$ or $l_{i,m}$ is the direction cosine between the m ordinate direction (in direction S) and the i coordinate direction. At a diffuse boundary, the net radiative flux leaving the wall (emitted flux minus absorbed flux) is

$$q_r(\mathbf{r}_w) = \varepsilon_w(\mathbf{r}_w)\pi I_{b,w}(\mathbf{r}_w) - \alpha_w(\mathbf{r}_w)\sum_{m'} l_{i,m'} w_{m'} I_{m'}(\mathbf{r}_w) \tag{12.115}$$

and the summation is for the hemisphere of incident directions m' onto the boundary location.

The numerical methods for multidimensional discrete ordinates solutions were originally developed for neutron transport calculations in nuclear reactor physics (Carlson and Lee 1961, Lee 1962, Lathrop and Carlson 1965, Lathrop 1966, Carlson and Lathrop 1968, Carlson 1970, 1971). In the past 20 years, the methods have been applied to multidimensional radiative transport problems (Fiveland 1984, 1988, Fiveland et al. 1984, Truelove 1987, Yuçel and Williams 1987, Fiveland and Jamaluddin 1991, Coelho 2001, 2014, Tencer 2013). In a numerical solution by the discrete ordinates (S_N) method, the intensity is obtained in discrete directions covering the 4π solid angle about each location in the volume of radiating medium. Because intensities must be found along each positive and negative ordinate direction around a grid point, an even number of simultaneous equations must be solved at each grid point. The solution is denoted S_2, S_4, S_6... S_N, where the subscript N gives the *order* of the solution. The number of discrete ordinates that is used is related to the order and the number of dimensions. The N indicates the number of different direction cosines used for each principal direction. In a 3D problem, each octant of a sphere of solid angles around a grid point contains $(N/2) + [(N/2) - 1] + [(N/2) - 2] + \cdots + 1 = N(N + 2)/8$ ordinates (quadrature points), requiring solution of $N(N + 2)$ simultaneous equations at each grid point to cover the entire range of solid angles. The S_2 discrete ordinates formulation ($N = 2$) thus requires solving 2(2 + 2) = 8 simultaneous equations in a 3D problem, while S_4 requires 4(4 + 2) = 24 equations. For a D-dimensional problem (D = 2, 3), the number of simultaneous equations is $2^D N(N + 2)/8$. When D = 1, there are N simultaneous equations. Hence, an S_4 analysis in a 2D problem requires solving $2^2 4(4 + 2)/8 = 12$ simultaneous equations, whereas for 1D, four equations are required. The angular integrals of intensity are approximated as a weighted finite sum in terms of these directions, as indicated by Equation 12.112. In the numerical procedure, the discrete ordinates equations for the intensities are applied locally within the medium by using a control volume technique, such as is used for convection problems (Patankar 1980). By this technique, the discrete ordinates method can be incorporated into existing computer codes based on using control volumes.

12.3.4.1 Relations for 2D Rectangular Coordinates

To illustrate a numerical method, consider a 2D medium as in Figure 12.13a. The volume is divided into rectangular regions (control volumes), with a typical control volume centered at $\mathbf{r}_v$. A typical wall element is centered at $\mathbf{r}_w$. For simplicity, the walls are assumed diffuse and gray so that the boundary condition Equation 12.113 applies. In Figure 12.13a, the $I_m(\mathbf{r}_w, S = 0)$ is the outgoing intensity from a typical wall element at location $\mathbf{r}_w$, in the direction toward the solid angle $d\Omega_m$ at the typical control volume that has its center at $P(\mathbf{r}_v)$. The S direction is the direction of the discrete ordinate m. The intensity that arrives at the control volume depends, according to the equation of transfer Equation 12.112, on the emission and scattering along the path from $\mathbf{r}_w$ to $\mathbf{r}_v$.

A typical control volume in a radiating 2D medium is shown in Figure 12.13b. The four sides are labeled with subscripts N, E, S, W (north, east, south, west), where north is in the x direction. To set up the numerical solution, a control volume form for the equation of transfer is derived by multiplying the transfer equation (Equation 12.112) written in 2D by $dx\, dy$ and integrating over the control volume to obtain

$$\alpha_m\left(I_{mN}A_N - I_{mS}A_S\right) + \delta_m\left(I_{mW}A_W - I_{mE}A_E\right)$$
$$= \kappa\Delta V_P I_{b,P} - \left(\kappa + \sigma_s\right)\Delta V_P I_{mP} + \Delta V_P \frac{\sigma_s}{4\pi}\sum_{m'} w_{m'} I_{m'P}\Phi_{m'm} \tag{12.116}$$

where uniform radiative properties have been assumed.

The scattering source term for the ordinate directions is computed using the scattering phase function $\Phi_{m'm}$. If the scattering is highly anisotropic, a high-order S_N solution may be required for accuracy. The simultaneous numerical solution of the equation of transfer for each of the m directions is usually done by iteration because the volume emission, scattering source terms, and boundary conditions depend on the intensities. The calculations can be started by assuming the boundaries are black (to eliminate the need for reflected intensities for the first iteration), the medium is at a known uniform temperature (internal emission may be neglected for the first iteration), and the scattering terms are zero. The radiant intensities are then computed along the ordinates as initiated by wall emission. For subsequent iterations, the full boundary conditions and radiative source terms are used. During each iteration, a solution of the control volume equations together with the boundary conditions is found for each x and y by traversing from point to point at the center of each control volume. The solution proceeds by computing intensities at all the x values for a given y and then advancing to a new y. The intensities are obtained for all of the m directions before going on to the next iteration. Values for all the control volumes are recalculated until the intensity values converge, and the converged temperature distribution and wall heat fluxes are then obtained.

Consider the iterative process at a location and for directions for which both direction cosines are positive (to the right and upward in Figure 12.13a). Before using the control volume equation, the number of unknowns is reduced by relating the radiant fluxes at the sides of the control volume to the radiant flux at the center location P of the control volume. For this purpose, a spatially weighted approximation is written as

$$I_{mP} = \eta I_{mN} + (1-\eta) I_{mS} = \eta I_{mW} + (1-\eta) I_{mE} \tag{12.117}$$

where η is the weighting factor. For simplicity, a value of η = 1/2 may be used, corresponding to the "diamond difference" relations by Carlson and Lathrop (1968). For η = 1/2,

$$I_{mP} = \frac{1}{2}(I_{mN} + I_{mS}) = \frac{1}{2}(I_{mW} + I_{mE}) \tag{12.118}$$

If the calculation is going in the direction of positive direction cosines, the intensities I_{mS} and I_{mE} (at the "incoming" faces) are assumed known. Then Equations 12.117 and 12.118 are used to eliminate the I_{mN} and I_{mW} (at the "outgoing" faces) in Equation 12.116. Solving for I_{mP} at the center of the control volume yields

$$I_{mP} = \frac{\left\{ l_{1,m}\left[A_N(1-\eta) + A_S\eta\right] I_{mS} + l_{2,m}\left[A_W(1-\eta) + A_E\eta\right] I_{mE} + \eta\kappa\Delta V_P I_{b,P} + \eta\Delta V_P \frac{\sigma_s}{4\pi}\sum_{m'} w_{m'} I_{m'P}\Phi_{m'm} \right\}}{l_{1,m}A_N + l_{2,m}A_W + \eta(\kappa+\sigma_s)\Delta V_P} \tag{12.119}$$

For η = 1/2, this simplifies to

$$I_{mP} = \frac{\left\{ l_{1,m}\left[A_N + A_S\right] I_{mS} + l_{2,m}\left[A_W + A_E\right] I_{mE} + \kappa\Delta V_P I_{b,P} + \Delta V_P \frac{\sigma_s}{4\pi}\sum_{m'} w_{m'} I_{m'P}\Phi_{m'm} \right\}}{2l_{1,m}A_N + 2l_{2,m}A_W + (\kappa+\sigma_s)\Delta V_P} \tag{12.120}$$

These relations have been written for a direction with positive direction cosines; referring to Figure 12.13b, this is for an intensity moving upward and to the right. The radiation enters through the A_S and A_E faces and exits through the A_N and A_W faces; to generalize the relations, the entering and exiting faces are designated as "in" and "out." Then, for intensities traveling in any direction, such as originating from the vertical sides or top of the enclosure in Figure 12.13a, Equations 12.117 and 12.119 are written to apply for any direction across the enclosure:

$$I_{mP} = \eta I_m(x_{out}) + (1-\eta) I_m(x_{in}) = \eta I_m(y_{out}) + (1-\eta) I_m(y_{in}) \tag{12.121}$$

$$I_{mP} = \frac{\left\{ \begin{array}{c} |l_{1,m}|\left[A(x_{out})(1-\eta) + A(x_{in})\eta\right] I_m(x_{in}) + |l_{2,m}|\left[A(y_{out})(1-\eta) + A(y_{in})\eta\right] I_m(y_{in}) \\ + \eta\kappa\Delta V_P I_{b,P} + \eta\Delta V_P \frac{\sigma_s}{4\pi}\sum_{m'} w_{m'} I_{m'P}\Phi_{m'm} \end{array} \right\}}{|l_{1,m}| A(x_{out}) + |l_{2,m}| A(y_{out}) + \eta(\kappa+\sigma_s)\Delta V_P} \tag{12.122}$$

12.3.4.2 Relations for 3D Rectangular Coordinates

The relations in Equations 12.122 and 12.123 can be extended to 3D rectangular control volumes by adding terms in the z direction with a direction cosine γ_m. This gives

$$I_{mP} = \eta I_m(x_{out}) + (1-\eta) I_m(x_{in}) = \eta I_m(y_{out}) + (1-\eta) I_m(y_{in})$$
$$= \eta I_m(z_{out}) + (1-\eta) I_m(z_{in}) \tag{12.123}$$

$$I_{mP} = \frac{\left\{\begin{aligned}&|l_{1,m}|\left[A(x_{out})(1-\eta) + A(x_{in})\eta\right] I_m(x_{in}) + |l_{2,m}|\left[A(y_{out})(1-\eta) + A(y_{in})\eta\right] I_m(y_{in})\\ &+ |l_{3,m}|\left[A(z_{out})(1-\eta) + A(z_{in})\eta\right] I_m(z_{in}) + \eta\kappa\Delta V_P I_{b,P} + \eta\Delta V_P \frac{\sigma_s}{4\pi}\sum_{m'} w_{m'} I_{m'P}\Phi_{m'm}\end{aligned}\right\}}{|l_{1,m}| A(x_{out}) + |l_{2,m}| A(y_{out}) + |l_{3,m}| A(z_{out}) + \eta(\kappa+\sigma_s)\Delta V_P} \tag{12.124}$$

The application of these relations is demonstrated by an example that shows how the intensities are calculated iteratively in a 2D enclosure.

Example 12.7

A 2D (no variations normal to the x–y plane) square enclosure (Figure 12.14) is filled with a gray medium with $n = 1$ that absorbs and emits radiation but has negligible scattering. Radiative energy transfer dominates so that conduction and convection can be neglected, and there are no internal heat sources. The horizontal side at $y = 0$ is at temperature T_{w1}, and the vertical side at $x = 0$ is at T_{w2}. The other two boundaries are at temperatures low enough that radiation emitted from them can be neglected. All of the boundaries are black. The medium has a constant absorption coefficient, and the optical length of each side is $\kappa D = 2$. Using discrete ordinates with the

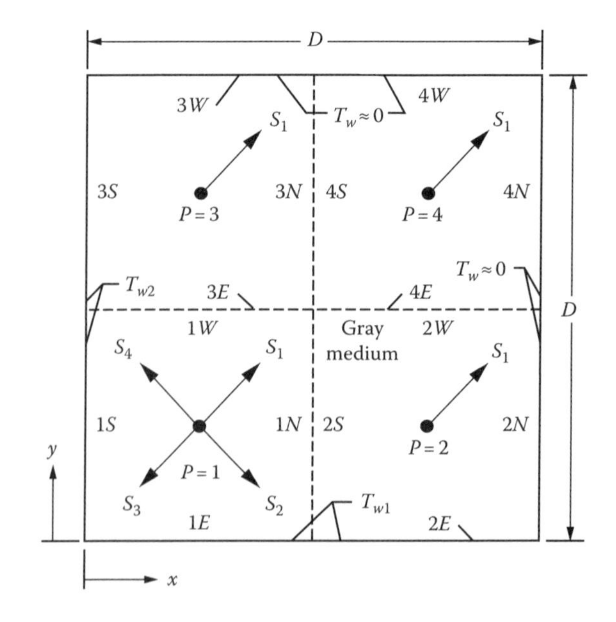

FIGURE 12.14 Square enclosure with black boundaries divided into four control volumes for the discrete ordinates solution in Example 12.7.

TABLE 12.4
Weights (w) and Ordinate Directions ($\mu = \cos\theta$) for S_n Method for 1D Geometry

S_n	w	$\pm\mu$	S_n	w	$\pm\mu$
S_2	1	0.500000	S_{10}	1/5	0.083752
				1/5	0.312729
S_4	1/2	0.211325		1/5	0.500000
	1/2	0.788675		1/5	0.687270
				1/5	0.916248
S_6	1/3	0.146446			
	1/3	0.500000	S_{12}	1/6	0.066877
	1/3	0.853554		1/6	0.366693
				1/6	0.288732
S_8	1/4	0.102672		1/6	0.711267
	1/4	0.406205		1/6	0.633307
	1/4	0.593795		1/6	0.933123
	1/4	0.897327			

Source: Fiveland, W.A., *JHT*, 109(3), 809, 1987.

S_2 approximation, show how the intensities can be calculated that can be used to provide the medium temperature distribution and the wall heat fluxes.

As an illustration of the calculation procedure, the enclosure is divided into four control volumes as in Figure 12.14, and four ordinate directions are used for the path directions $S_1 \ldots S_4$; these S_n directions are projections in the x–y plane, but the effect of radiating directions extending in the third dimension is included in the selection of the direction cosines. As indicated in Table 12.4, for the similar situation of a 1D geometry where there are also no variations normal to the cross section, the direction cosines for the S_2 approximation are recommended to be 1/2 so that, for evaluating the relations that follow, $|l_{1,m}| = |l_{2,m}| = 1/2$. Using a unit depth normal to the cross section in Figure 12.14, the areas of the side faces of the control volumes are all $D/2$, each volume is $D^2/4$, and κ times the volume is $\kappa D \times D/4 = 2 \times D/4 = D/2$. The weighting factors are all π so that quantities in the four ordinate directions will properly sum to the entire 4π solid angle, and the value of the weighting factor $\eta = 1/2$ is used as discussed for Equation 12.124.

The calculation proceeds by using the intensities at two side faces of each control volume to calculate the intensity at the center point; Equation 12.122 is used and, for $\sigma_s = 0$, it becomes

$$I_{mP} = \frac{(1/4)DI_m(x_{in}) + (1/4)DI_m(y_{in}) + (1/4)DI_{b,P}}{(1/4)D + (1/4)D + (1/4)D}$$

$$= \frac{1}{3}\left[I_m(x_{in}) + I_m(y_{in}) + I_{b,P}\right] \tag{12.7.1}$$

With I_{mP} known, the $I_m(x_{out})$ and $I_m(y_{out})$ are found from Equation 12.121:

$$I_m(x_{out}) = 2I_{mp} - I_m(x_{in}); \quad I_m(y_{out}) = 2I_{mp} - I_m(y_{in}) \tag{12.7.2}$$

The $I_m(x_{out})$ and $I_m(y_{out})$ of a control volume provide the $I_m(x_{in})$ and $I_m(y_{in})$ of the adjacent control volumes, so the process can continue.

As a concrete numerical example, consider the S_1 direction for $m = 1$ in Figure 12.14. To begin, let the $I_{bP} = 0$ for $P = 1 \ldots 4$. Then, for $m = 1$ and $P = 1$ $(I_{mP} \rightarrow I_{11})$ Equation 12.7.1 gives

$$I_{11} = \frac{1}{3}\left[I_1(1S) + I_1(1E)\right] = \frac{1}{3}\frac{\sigma}{\pi}\left(T_{w2}^4 + T_{w1}^4\right)$$

Then with I_{11} known, Equation 12.7.2 is used to obtain

$$I_1(1N) = I_1(2S) = 2I_{11} - I_1(1S) = 2I_{11} - \frac{\sigma T_{w2}^4}{\pi}$$

$$I_1(1W) = I_1(3E) = 2I_{11} - I_1(1E) = 2I_{11} - \frac{\sigma T_{w1}^4}{\pi}$$

With $I_1(2S)$ now known, and $I_1(2E) = \sigma T_{w1}^4/\pi$, Equations 12.7.1 and 12.7.2 are used to obtain $I_{12}(m = 1, P = 2)$, and then $I_1(2N)$ and $I_1(2W) = I_1(4E)$. Similarly, with $I_1(3E)$ now known, and $I_1(3S) = \sigma T_{w2}^4/\pi$, there are obtained $I_{13}(m = 1, P = 3)$, $I_1(3W)$, and $I_1(3N) = I_1(4S)$. Since the $I_1(4E)$ and $I_1(4S)$ are now known, the $I_{14}(m = 1, P = 4)$, $I_1(4N)$, and $I_1(4W)$ are calculated. Thus for the $m = 1$ direction, the average intensity at the center of each control volume and the incident intensity at the center of each side face for each of the control volumes have been obtained.

This procedure is now repeated for the $m = 2$, 3, and 4 directions so that all of the intensities are obtained. Then, using the weighting factors, the I_{bP} is calculated for $P = 1 \ldots 4$. For radiative equilibrium without internal heat sources,

$$I_{b,P} = \frac{1}{4}\left(I_{1P} + I_{2P} + I_{3P} + I_{4P}\right) \quad P = 1,2,3,4$$

These $I_{b,P}$ are used in the next iteration.

After the converged intensities are found by iterating this procedure, the temperature at the center of each control volume is obtained as $T_P = (\pi I_{b,P}/\sigma)^{1/4}$. The local energy flux from the wall is found by subtracting the incident energy, all absorbed by the black wall, from the emitted energy. For example, the value at $x = D/4$ and $y = 0$ (the 1E face) is

$$q_w = \sigma T_{w1}^4 - \sum_m \left|l_{1,m}\right| w_m I_m = \sigma T_{w1}^4 - \frac{\pi}{2}\left[I_2(1E) + I_3(1E)\right]$$

Recommendations in the literature help obtain physically valid solutions and minimize errors. Because Equation 12.7.2 is an extrapolation across a control volume, negative intensities can result when the extinction coefficient is large and the control volume is not sufficiently small. Negative intensities are then replaced by zero. Rather than use smaller control volumes, various improved differencing schemes have been proposed that might yield improved accuracy and avoid negative intensities as compared with diamond differencing. In Mohamad (1996) and Chai et al. (1994), the equation of transfer is used to guide the differencing within a control volume. The equation of transfer is solved analytically in a small region to relate nearby intensities. An exponential relation is obtained, and in Chai et al., this is called the *modified-exponential scheme*; in Mohamad, it is the *local analytical method*. Fiveland and Jessee (1994) used a spatial discretization done by using the finite-element method (FEM). Jessee and Fiveland (1997) investigated a *high-resolution* method that uses a high-order interpolation scheme. Coelho (2014) discusses the applications and relative merits of various contemporary angular quadrature methods and spatial discretization methods.

12.3.5 Ordinate and Weighting Pairs

Substantial research effort has been applied to the development and analysis of quadrature rules for integration over the unit sphere. Level symmetric quadrature rules are often recommended for general applications because of their rotational symmetry (Carlson and Lee 1962, Lee 1962), Lathrop and Carlson 1965, Carlson 1970, 1971, Lewis and Miller 1993). Other ordinate and weighting sets have been recommended to improve accuracy, and to promote convergence, particularly in multimode problems (Fiveland 1987, Truelove 1987, Thurgood et al. 1990, Koch et al. 1995, Li et al. 1998, Rukolaine and Yuferev 2001, Koch and Becker 2004, Jarrell 2010). Longoni (2004) discusses appropriate quadrature sets for use in parallel computing.

A recommended set of weights and ordinate directions for a 1D geometry is provided by Fiveland (1987) and in Table 12.4 for S_2, S_4, S_6, S_8, S_{10}, and S_{12} approximations. In Truelove (1987), a set of ordinates is recommended to improve accuracy for the S_2 and S_4 approximations. The ordinates have been arranged to satisfy a half-range flux condition that states that, for a uniform intensity, the weighting factors must obey the sums $\sum_{a_m>0} w_m l_{1,m} = \pi$ and $\sum_{\delta_m>0} w_m l_{2,m} = \pi$. A set of ordinate directions and weights for 3D transfer is shown in Table 12.5. An improved spatial differencing scheme is proposed in Kim and Kim (2001) for discrete ordinates in 2D rectangular enclosures. To aid solutions in media with optical dimensions larger than one, computational acceleration schemes are provided by Fiveland and Jessee (1996). A convergence acceleration multigrid procedure called the "coupled ordinates method" is developed in Mathur and Murthy (1999). This is to

TABLE 12.5
Ordinate Directions and Weights for 3D Solutions: S_n Quadrature for the First Quadrant

		Ordinates			
Approx.	S_n Point	l_1	l_2	l_3	Weights w
S_2	1	0.5773503	0.5773503	0.5773503	1.5707963
S_4	1	0.2958759	0.2958759	0.9082483	0.5235987
	2	0.9082483	0.2958759	0.2958759	0.5235987
	3	0.2958759	0.9082483	0.2958759	0.5235987
S_6	1	0.1838670	0.1838670	0.9656013	0.1609517
	2	0.6950514	0.1838670	0.6950514	0.3626469
	3	0.9656013	0.1838670	0.1838670	0.1609517
	4	0.1838670	0.6950514	0.6950514	0.3626469
	5	0.6950514	0.6950514	0.1838670	0.3626469
	6	0.1838670	0.9656013	0.1838670	0.1609517
S_8	1	0.1422555	0.1422555	0.9795543	0.1712359
	2	0.5773503	0.1422555	0.8040087	0.0992284
	3	0.8040087	0.1422555	0.5773503	0.0992284
	4	0.9795543	0.1422555	0.1422555	0.1712359
	5	0.1422555	0.5773503	0.8040087	0.0992284
	6	0.5773503	0.5773503	0.5773503	0.4617179
	7	0.8040087	0.5773503	0.1422555	0.0992284
	8	0.1422555	0.8040087	0.5773503	0.0992284
	9	0.5773503	0.8040087	0.1422555	0.0992284
	10	0.1422555	0.9795543	0.1422555	0.1712359

Source: Fiveland, W.A., *JTHT*, 2(4), 309, 1988.

improve iterative solution methods used for the discrete ordinates and finite volume (Section 12.4.2) formulations that often converge slowly for optical thicknesses greater than 10.

Tencer (2013) discusses and compares various quadrature schemes for the 2D test problem solved analytically by Crosbie and Schrenker (1984). The case of black walls was chosen to accentuate the ray effects. The quantity of interest is the spatially resolved heat flux on the wall opposite the heated surface that accentuates the ray-effect error. Tencer found that the sets proposed by Koch and Becker (2004) for S_6 and S_8 provided better accuracy than level set solutions with the same number of ordinates and that the sets proposed by Thurgood et al. (1990) were less accurate and converged more slowly than either the Koch et al. (1995) or level sets.

The orientation of the quadrature set relative to a particular problem geometry can greatly affect the accuracy of an S_N solution and the influence of the ray effect. There is presently no method to select the best orientation. Based on this observation, Tencer (2014) proposed averaging S_N solutions using a particular S_N quadrature set that is rotated about an initial orientation. He shows considerable improvement in accuracy over solutions using a single quadrature orientation. Tencer (2014) describes the improvements in S_N solutions that are obtained when the S_N coordinate system is rotated relative to the coordinate system describing the geometry. The results from averaging various rotation sets mitigate ray effects, and result in smooth solutions. Generation of the solution set was found to be easily parallelized.

12.3.6 Results Using Discrete Ordinates

Various solution methods for discrete ordinates have been presented in the literature, and results for basic cases have been compared with solutions by other methods. Fiveland (1984, 1987, 1988) presents an iterative method for solving the discrete ordinates equations, while Kumar et al. (1990) use available linear differential equations solvers. The weights w_m used in the equations depend on the quadrature scheme for approximating the integrals. Fiveland (1984) uses equal weights for all ordinate directions.

Gaussian quadrature that uses adjustable weights for best approximation of integrals over the full range of solid angles, as is needed in the equation of transfer, may not be appropriate for the boundary condition equations that require integration over the half range. Kumar et al. (1990) discuss various weighting options and their relative accuracy for the solution of a 1D highly anisotropic scattering problem. They show that, for the cases studied, Gaussian quadrature provided convergence more quickly than Fiveland's scheme, but Fiveland's results using equal weights for all ordinate directions tended to be more accurate. For multidimensional problems in which highly anisotropic scattering is present, the full-range integrals are more important for solution accuracy than the half-range integrals in the boundary conditions; Gaussian quadrature may then provide better overall accuracy. Koch et al. (1995) discuss formulation of quadrature schemes in depth and propose a weighting scheme based on "double cyclic triangles," which they show to be rotationally invariant with respect to the even moments and show improved accuracy over traditional S_N moments.

Fiveland (1984) compared discrete ordinates results with the P_1 and P_3 results of Ratzel and Howell (1982) for a rectangular enclosure with one wall at T_W and the other three at $T = 0$ K and all walls gray–diffuse emissivity, ε_w. He found that both S_6 results were very good, but even S_4 was more accurate than the P_3.

Kim and Lee (1989) used an S_{14} solution for analyzing a general Mie scattering medium in a 2D enclosure exposed to a collimated incident radiation source on one face. The higher-order discrete ordinates solution was necessary to deal with scattering phase functions that are highly anisotropic. An integral formulation is in Wu (1990) in terms of moments of the intensity. The analysis is for a 3D medium that is inhomogeneous and has anisotropic scattering. An example is given for a 2D rectangular medium exposed to collimated radiation. A nonaxisymmetric cylindrical enclosure is analyzed by Jamaluddin and Smith (1992), and the effect of Fresnel reflecting boundaries in a cylindrical enclosure is analyzed by Wu and Liou (1997).

Mengüç and Iyer (1988) combined the P_N and discrete ordinates methods. They expressed the intensity in each ordinate direction in terms of a spherical harmonics expansion. Using two-flux and eight-flux discrete ordinates with P_1 expansions and Marshak boundary conditions, they solved problems with anisotropically scattering media in 1- and 2D.

Results for the S_4, S_6, and S_8 discrete ordinates method are in Fiveland (1988) for an anisotropically scattering gray participating medium in an idealized 3D furnace. The results are for radiative equilibrium. Fiveland made comparisons among the discrete ordinates, P_3, and zone method results for the same problem except with isotropic scattering. Combined radiation and conduction in a 2D rectangular enclosure was analyzed by Kim and Baek (1991) using the S_4 method. A description of the discrete ordinates method was given, and linear anisotropic scattering was included. Three walls of the rectangular enclosure were at one temperature and the fourth wall was at a different uniform temperature. Comparisons were made with results from other methods given by Razzaque et al. (1984), Yuen and Takara (1988), and Tan (1989a), and good agreement was obtained. Only small amounts of computer time were required. Comparisons of various discrete ordinates formulations are in Fiveland and Jessee (1995) for 2D enclosures with gray walls that contain an absorbing–emitting and isotropically scattering medium.

Application of discrete ordinates for a nongray medium between diffusely reflecting walls in a plane layer is in Menart et al. (1993). A nongray medium introduces the complexity of having spectral regions that are optically thin in the same calculation as spectral regions that are optically thick, making convergence difficult.

The discussion here has been for rectangular, cylindrical, or spherical geometries. Interpolation methods for irregular 2D geometries are in Koo et al. (1997). Results are obtained for an absorbing–emitting medium in a "J"-shaped region, and very good comparisons are obtained with results using the zone method. Pincus and Evans (2009) compare results for a parallelized 3D spherical harmonics S_N code with those from a Monte Carlo code. The application is to an atmospheric science domain with various cloud configurations. Hunter and Guo (2015) examine the trade-off among errors in S_N solutions due to grid resolution, errors due to ray effects, errors caused by false scattering effects, and errors due to anisotropic scattering. They present conclusions as to which errors dominate in particular parameter ranges.

12.4 OTHER METHODS THAT DEPEND ON ANGULAR DISCRETIZATION

A number of methods have been devised that offer simpler integration into general heat transfer analysis codes than offered by the discrete ordinates approach. These are briefly reviewed in this section.

12.4.1 Discrete Transfer Method

The discrete transfer method is an attempt to provide a fast and accurate algorithm for incorporating radiative transfer into codes for combustion and flow in complex geometries (Lockwood and Shah 1981). The method as originally proposed consists of determining the intensity for each of N rays arriving at each surface element P in an enclosure as in Figure 12.15. The rays and their associated solid angles $\Delta\Omega_k$ are equally distributed over the surface of a hemisphere centered over the receiving element, rather than being chosen and weighted through a quadrature technique as for the discrete ordinates method. The point of origin Q of each ray is determined by finding the

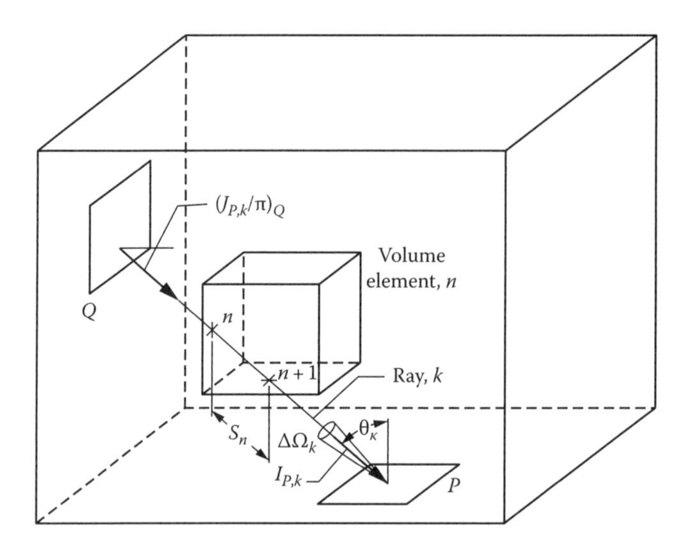

FIGURE 12.15 Ray path for discrete transfer method.

intersection of the ray with the enclosure boundary (Figure 12.15). Note that the origin point Q will usually not be at the center of a boundary area element.

The source function, temperature, and properties are usually assumed to be constant within a given volume element. The RTE equation (Equation 10.15) along each ray, as from Q to P, can then be placed in the form of a recurrence relation:

$$I_{n+1} = I_n e^{-\beta S_n} + \hat{I}_n(1 - e^{-\beta S_n}) \tag{12.125}$$

where I_{n+1} is the ray intensity leaving boundary $n + 1$ of a volume element (and entering the next element) and is equal to the ray intensity I_n crossing boundary n and attenuated along the path S_n within the element, plus the increase from the source function $\hat{I}_n$ within the element.

Because the intensity of the ray k leaving point Q depends on the radiosity J_Q at that point and the radiosity is initially unknown, the intensity $I_{o,Q} = J_Q/\pi$ (that assumes diffuse boundaries) needed to begin the recurrence procedure must be assumed, so the procedure is inherently iterative. After solving for the recurrence relation for each ray, the incident flux for N rays on each surface element P is

$$q_{i,P} = \sum_{k=1}^{N} I_{P,k} \cos\theta_k \Delta\Omega_k \tag{12.126}$$

Once $q_{i,P}$ is known for each surface element, a new value of J_Q for each boundary element is found from the radiosity relation, Equation 5.11. A new iteration can then be carried out. The isotropic scattering portion of the source function Equation 10.20 for each element is found from the summed intensity of the rays passing through the element times the single scattering albedo. For anisotropic scattering, no efficient approach has been proposed for use in the discrete transfer method. If the medium is nonscattering, then the source function with $\omega = 0$ for a gray medium becomes I_b, Equation 10.51, which is found from the energy equation for the particular problem being solved. Because the discrete transfer method is usually a part of a fluid flow or combustion code and the fluid mechanics and combustion solutions are themselves iterative, the iterative nature of the method may not pose a significant penalty on performing the calculations.

Shortcomings of the discrete transfer method in its original formulation are that it suffers from the ray effect for the same reasons as for the discrete ordinates method (Section 12.3.3); it does not obey energy conservation; and it is difficult to include anisotropic scattering in the formulation. Modifications to ensure energy conservation are proposed by Coelho and Carvalho (1997). Accurate quadrature formulas, treatment of linear rather than constant temperatures within individual volume elements, and variations of ray intensity within the associated $\Delta\Omega_k$ are incorporated in Cumber (1995). A modification of the method that initiates ray directions from volume elements in the medium rather than from surface elements provides a more accurate determination of the source terms in the RTE at the expense of additional computational time (Selçuk and Kayakkol 1998). This approach provides improved accuracy near enclosure corners, where the usual method generates the largest errors.

Comparisons of discrete transfer results with benchmark solutions for 3D L-shaped enclosures and other cases (Henson et al. 1996) show accuracy within 1.2% in surface heat flux and divergence of radiative flux. The speed of the method is useful, but its restriction to diffuse boundaries and isotropic scattering is a limitation for analyzing some real systems.

12.4.2 Finite Volume Method

The finite volume method for radiative energy transfer is based on the same ideas as the finite volume analysis for fluid flow and convective energy transfer. In most applications of the method for radiative transfer, the sphere of solid angles surrounding a volume element is divided into equal solid angles or into equal circumferential and equal polar angular increments without the weighting factors as used in the discrete ordinates method. The differential form of the equation of radiative transfer is used to relate the intensity entering the control volume from a given direction to that leaving an adjacent volume element. By summing over all directions, a total energy balance is written for each volume element, thus providing local and overall energy conservation. Anisotropic scattering is incorporated by applying the phase function to each incoming intensity; the accuracy of treating anisotropic scattering is related to the number of discrete directions used.

The finite volume method has been extended to cylindrical geometries (Chui et al. 1992), irregular geometries by use of body-fitted elements (Chai et al. 1995), and geometries with unstructured meshes (Murthy and Mathur 1998). An angular discretization scheme for the finite volume method in 3D is given by Kim et al. (2000). The polar angle θ is divided uniformly into an equal number of increments, while the circumferential (azimuthal) angle ϕ has a sequence of increments that depend on the polar angle. Good agreement was obtained with Monte Carlo benchmark solutions. Chai and Patankar (2000) and Coelho (2014) give reviews of the method. Anisotropic scattering (Trivic et al. 2004), incorporation of the weighted sum of gray gases to treat spectral effects (Trivic 2004, Cai et al. 2007), and blocking effects through the use of curvilinear coordinates (Talukdar et al. 2005) have been studied.

Hassanzadeh et al. (2008) extend the finite volume method to an approach they term the Q_L method and show better accuracy, faster convergence, and lower cost than for the finite volume method for most benchmark 1- and 2D cases. Coelho (2014) presents a comprehensive discussion of this approach. Roger et al. (2014) proposed a multiscale method which is efficient when the radiation solution requires extra computation time.

12.4.3 Boundary Element Method

The boundary element method transforms the integrals in the integrated form of the RTE into integrals over the enclosure boundary. This is done by expressing the integration over the enclosure volume as an integration along a radiative path times an integration over the enclosure boundary. For certain classes of problems, the integration along the path can be carried out analytically or by a simple summation. The RTE is then solved by dividing the boundary into elements and using the method of weighted residuals to solve the resulting set of equations. Because integrations are over the boundary of the enclosure rather than over its volume, the degree of integration is effectively reduced by one, resulting in an efficient numerical scheme. Various radiative transfer problems have been solved using the method, which is shown to be closely related to the zonal method, but the boundary element method does not require explicit evaluation of the zonal exchange areas and thus avoids the various volume integrations necessary in the zonal approach (Bialecki 1992, Sun et al. 1998, Mbiock and Weber 2000). The method shares characteristics with the YIX technique (Tan and Howell 1989a, 1990a).

12.5 ZONAL METHOD

12.5.1 Exchange Area Relations

The zonal method has been used extensively for engineering problems with radiating gases. In this method, a nonisothermal enclosure filled with nonisothermal gas or other translucent medium (with $n \approx 1$ for this development) is subdivided into areas and volumes (called "zones") that are each approximated as isothermal. An energy balance is written for each zone. This provides a set of simultaneous equations for the unknown heat fluxes or temperatures. The method is somewhat an extension of enclosure theory, and Hottel and Sarofim (1967) and Hottel et al. (2007) discuss it in detail. Multidimensional applications have been carried out by Hottel and Cohen (1958), Einstein (1963a,b), Hottel and Sarofim (1965), and França et al. (2001). Scattering can be included, as shown by Noble (1975) and Yuen and Takara (1990).

The basic concepts of the zonal method are developed next for a nonisothermal gas with uniform composition and a uniform absorption coefficient. Consider a finite volume V_γ in Figure 12.16, and

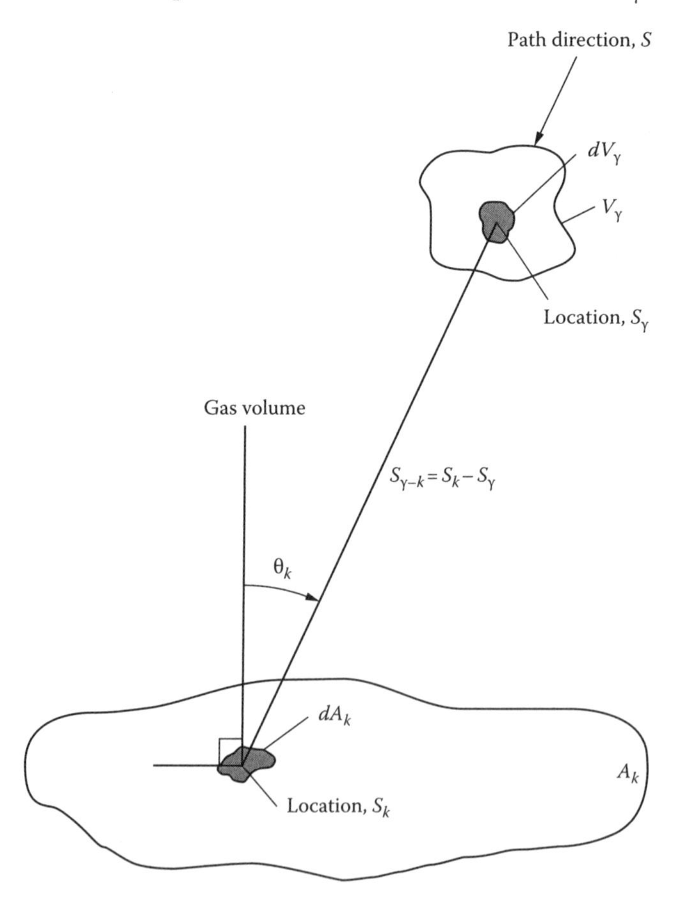

FIGURE 12.16 Radiation from gas volume V_γ to area A_k.

surface A_k. From Equation 1.61, the spectral emissive power from a volume element of gas dV_γ is $4\pi\kappa_\lambda I_{\lambda b} dV_\gamma\, d\lambda$, or, per unit solid angle around dV_γ, it is $\kappa_\lambda I_{\lambda b}\, dV_\gamma d\lambda$. The surface element dA_k subtends the solid angle $dA_k \cos\theta_k / S^2_{\gamma-k}$ when viewed from dV_γ. The fraction of radiation transmitted through the path length $S_{\gamma-k}$ is $\exp\left[-\int_{S_\gamma}^{S_k} \kappa_\lambda(S^*)dS^*\right]$. Multiplying these factors and integrating over V_γ and A_k gives the spectral energy arriving at A_k from gas finite volume V_γ as

$$G_{\lambda,\gamma-k} d\lambda A_k = d\lambda \int_{V_\gamma}\int_{A_k} \frac{\kappa_\lambda(\gamma) I_{\lambda b}(\gamma)\cos\theta_k}{S^2_{\gamma-k}} \exp\left[-\int_{S_\gamma}^{S_k} \kappa_\lambda(S^*)dS^*\right] dA_k dV_\gamma \tag{12.127}$$

Since $\kappa_\lambda(\gamma) = \kappa_\lambda$ is constant, the exponential factor is $\exp\left[-\kappa_\lambda(S_k - S_\gamma)\right] = t_\lambda(S_{\gamma-k})$, where t_λ is the transmissivity. In the zone method, the entire gas volume is divided into finite subvolumes V_γ with uniform conditions assumed over each V_γ. Then $I_{\lambda b}(\gamma)$ need not be included in the integral over dV_γ and Equation 12.127 becomes

$$G_{\lambda,\gamma-k} d\lambda A_k = d\lambda \kappa_\lambda I_{\lambda b}(\gamma) \int_{V_\gamma}\int_{A_k} \frac{\cos\theta_k}{S^2_{\gamma-k}} t_\lambda(S_{\gamma-k}) dA_k dV_\gamma \tag{12.128}$$

For a gray gas, Equation 12.128 integrated over all wavelengths gives the total energy incident on A_k as

$$G_{\gamma-k} A_k = \kappa \frac{\sigma T_\gamma^4}{\pi} \int_{V_\gamma}\int_{A_k} \frac{\cos\theta_k}{S^2_{\gamma-k}} \tau(S_{\gamma-k}) dA_k dV_\gamma \tag{12.129}$$

Now, define the *gas–surface direct-exchange area* $\overline{g_\gamma s_k}$ as

$$\overline{g_\gamma s_k} \equiv \frac{\kappa}{\pi} \int_{V_\gamma}\int_{A_k} \frac{\cos\theta_k}{S^2_{\gamma-k}} t(S_{\gamma-k}) dA_k dV_\gamma \tag{12.130}$$

Equation 12.129 can then be written as

$$G_{\gamma-k} A_k = \overline{g_\gamma s_k} \sigma T_\gamma^4 \tag{12.131}$$

Thus the energy $G_{\gamma-k} A_k$ arriving at A_k is regarded as the blackbody emissive power σT_γ^4 of the gas in V_γ that is radiated from an effective area $\overline{g_\gamma s_k}$. The entire gas volume is divided into Γ finite regions, so the energy flux incident upon A_k from all of the gas is

$$(G_k)_{from\ gas} = \frac{1}{A_k} \sum_{\gamma=1}^{\Gamma} \overline{g_\gamma s_k} \sigma T_\gamma^4 \tag{12.132}$$

The energy leaving surface area A_j of the enclosure that reaches A_k is, for a uniform gray gas in the enclosure,

$$G_{j-k}A_k = \frac{J_j}{\pi}\int_{A_k}\int_{A_j} t(S_{j-k})\frac{\cos\theta_j\cos\theta_k dA_j dA_k}{S_{j-k}^2} \tag{12.133}$$

where J_j is assumed uniform over A_j as in the usual enclosure theory. We can define a *surface–surface direct-exchange area* as

$$\overline{s_j s_k} = \int_{A_k}\int_{A_j} t(S_{j-k})\frac{\cos\theta_j\cos\theta_k dA_j dA_k}{\pi S_{j-k}^2} \tag{12.134}$$

Equation 12.133 is then written as

$$G_{j-k}A_k = \overline{s_j s_k} J_{Sj} \tag{12.135}$$

Thus the energy $G_{j-k}A_k$ from A_j arriving at A_k is considered as the flux J_j leaving A_j times an *effective* area $\overline{s_j s_k}$. The flux incident upon A_k by the fluxes leaving all N enclosure surfaces is

$$(G_k)_{from\ surfaces} = \frac{1}{A_k}\sum_{j=1}^{N}\overline{s_j s_k}J_j \tag{12.136}$$

The total radiative flux incident upon A_k from all of the surfaces and the gas is then

$$G_k = \frac{1}{A_k}\left(\sum_{j=1}^{N}\overline{s_j s_k}J_j + \sum_{\gamma=1}^{\Gamma}\overline{g_\gamma s_k}\sigma T_\gamma^4\right) \tag{12.137}$$

The usual net-radiation Equations 5.10 and 5.11 also apply at surface A_k:

$$q_k = J_k - G_k \tag{12.138}$$

$$J_k = \varepsilon_k\sigma T_k^4 + (1-\varepsilon_k)G_k \tag{12.139}$$

12.5.2 Zonal Formulation for Radiative Equilibrium

In some furnaces, radiation is the dominant means of energy transfer. As an important limiting case, consider the formulation when the contributions by conduction and convection within the gas can be neglected in influencing the radiative calculations. If the gas temperature distribution can be estimated or is known from measurements, the analysis given here will provide the radiative heat fluxes incident from the gas to the wall, and this can then be added to the convection at the wall. As in enclosure calculations in Chapter 5, if the temperatures T_γ are given for all gas elements V_γ, Equations 12.137 through 12.139 are sufficient to solve for N unknown values of T_k and q_k; the other N values of T_k and q_k must be provided as known boundary conditions. The enclosure calculation methods of Section 5.3 can be applied directly.

When the T_γ of the Γ gas elements are unknown, then Γ additional equations are required. These are obtained by taking an energy balance on each isothermal gas zone V_γ. For each V_γ, the emission must equal the absorption of energy plus the local volumetric heat source in the gas, $\dot{q}_\gamma$. For a gray gas with constant properties, an energy balance on the volume region V_γ gives

$$4\kappa\sigma T_\gamma^4 V_\gamma = \kappa^2 \sum_{\gamma^*=1}^{\Gamma} \sigma T_{\gamma^*}^4 \int_{V_\gamma}\int_{V_{\gamma^*}} \frac{t(S_{\gamma^*-\gamma})dV_{\gamma^*}dV_\gamma}{\pi S_{\gamma^*-\gamma}^2}$$

$$+\kappa \sum_{k=1}^{N} J_k \int_{V_\gamma}\int_{A_k} \frac{\cos\theta_k}{\pi S_{k-\gamma}^2} t(S_{k-\gamma})dA_k dV_\gamma + \dot{q}_\gamma V_\gamma \tag{12.140}$$

Now, define the surface–gas direct-exchange area as

$$\overline{s_k g_\gamma} \equiv \frac{\kappa}{\pi} \int_{V_\gamma}\int_{A_k} \frac{\cos\theta_k}{S_{\gamma-k}^2} t(S_{k-\gamma})dA_k dV_\gamma \tag{12.141}$$

Comparing Equations 12.130 and 12.141 shows that there is reciprocity between the surface–gas and gas–surface direct-exchange areas:

$$\overline{s_k g_\gamma} = \overline{g_\gamma s_k} \tag{12.142}$$

Define the gas–gas direct-exchange area as

$$\overline{g_{\gamma^*} g_\gamma} = \overline{g_\gamma g_{\gamma^*}} \equiv \frac{\kappa^2}{\pi} \int_{V_\gamma}\int_{V_{\gamma^*}} \frac{t(S_{\gamma^*-\gamma})dV_{\gamma^*}dV_\gamma}{S_{\gamma^*-\gamma}^2} \tag{12.143}$$

Substituting Equations 12.141 through 12.143 into Equation 12.140 gives

$$4\kappa\sigma T_\gamma^4 V_\gamma = \sum_{\gamma^*=1}^{\Gamma} \sigma T_{\gamma^*}^4 \overline{g_{\gamma^*} g_\gamma} + \sum_{k=1}^{N} J_k \overline{g_\gamma s_k} + \dot{q}_\gamma V_\gamma \tag{12.144}$$

The $\overline{g_{\gamma^*} g_\gamma}$ are universal, and are tabulated by Hottel and Cohen (1958). Equation 12.144 can be written for each V_γ to provide the additional set of Γ equations required to obtain the T_γ distribution in the gas.

It is possible to make approximate allowances for spectral variations in gas properties. Property variations with position in the enclosure are included by defining a suitable mean absorption coefficient between each set of zones. Einstein (1963a,b) modified the $\overline{gs}$ and $\overline{gg}$ factors to give better accuracy when strong temperature gradients are present. These approximations become difficult to carry through if the absorption coefficient is a strong function of temperature. Noble (1975) shows how matrix theory can be used as a computational aid.

Values of $\overline{s_j s_k}$ and $\overline{g_\gamma s_k}$ are tabulated by Hottel and Cohen (1958) for cubical isothermal volumes and square isothermal boundary elements. Exchange areas are given for elements in rectangular (Nelson 1974, Scholand and Schenkel 1986), cylindrical (Einstein 1963b), and conical (Bannerot and Wierum 1974) enclosures. Hottel and Sarofim (1967) provide a tabulation for the cylindrical geometry and a table of references for several other geometries. In Mihail and Maria (1983), numerical integration was used to evaluate direct-exchange areas for examples involving squares, cubes, circular cylinders, and elliptic cylinders. Some relations for a cylindrical enclosure are in Sika (1991).

Useful results are given by Tucker (1986) for application in rectangular furnaces. Since many furnaces can be conveniently divided into cubic gas zones and square surface zones, the $\overline{ss}$, $\overline{gs}$, and $\overline{gg}$ factors were evaluated for these geometries and the results extended to a value of 18 for the optical dimension of the side of the square or cube. The factors are given on charts and are curve fitted with exponential functions that can be used for computer calculations. Results are given for squares and cubes that are close to each other. For larger separation distances, it is a good approximation to assume that the view and path length for absorption are the same for all points within each zone and thus to treat the zones as differential elements. Then differential forms can be used as follows, with θ and S based on the center-to-center orientation and separation distance:

$$\overline{s_j s_k} = \overline{s_k s_j} = \frac{t(S_{j-k})\cos\theta_j \cos\theta_k dA_j dA_k}{\pi S_{j-k}^2} \tag{12.145}$$

$$\overline{s_k g_\gamma} = \overline{g_\gamma s_k} = \frac{\kappa t(S_{k-\gamma})\cos\theta_k dA_k dV_\gamma}{\pi S_{k-\gamma}^2} \tag{12.146}$$

$$\overline{g_{\gamma^*} g_\gamma} = \overline{g_\gamma g_{\gamma^*}} = \frac{\kappa^2 t(S_{\gamma^*-\gamma}) dV_{\gamma^*} dV_\gamma}{\pi S_{\gamma^*-\gamma}^2} \tag{12.147}$$

In Edwards and Balakrishnan (1972b), exchange areas are derived for cubes in nonhomogeneous media. Values are presented for narrow spectral bands with overlapped lines and with nonoverlapped lines for the cases of adjacent and remote pairs of cubes, assuming a linear temperature distribution between the cubes.

The zone method has been modified by Larsen and Howell (1985), using the same assumptions inherent in the original form of the zone method of Hottel (1954) but deriving the transfer equations in terms of *exchange factors*. These factors are defined in terms of the fraction of radiative energy leaving one element that is absorbed by a receiving element, including all possible paths of intermediate scattering as well as intermediate absorption and reemission in a medium in radiative equilibrium. The exchange areas of the Hottel method can be calculated from the exchange factors and vice versa, so the methods are interchangeable. Exchange factors have the advantage that they are measurable, and Liu and Howell (1987) report measured values for a rectangular enclosure. Measuring exchange factors in scale models of furnaces eliminates the restrictions that limit zone analyses to cases where the exchange areas are known or can be calculated. Exchange factors defined somewhat differently were used to analyze a rectangular enclosure in Naraghi and Kassemi (1989).

12.5.3 Developments for the Zone Method

12.5.3.1 Smoothing of Exchange Area Sets

To achieve accurate numerical results with the zone method, the calculated exchange areas must satisfy conditions of reciprocity and energy conservation. Generally, the exchange areas must satisfy the reciprocity constraints

$$\overline{s_i s_j} = \overline{s_j s_i}, \quad \overline{s_i g_\gamma} = \overline{g_\gamma s_i}, \quad \overline{g_\gamma g_\mu} = \overline{g_\mu g_\gamma} \tag{12.148}$$

as well as the conservation relations

$$(4\kappa V)_\gamma = \sum_{i=1}^{N} \overline{g_\gamma s_i} + \sum_{\mu=1}^{\Gamma} \overline{g_\gamma g_\mu}, \quad A_i = \sum_{j=1}^{N} \overline{s_i s_j} + \sum_{\gamma=1}^{\Gamma} \overline{s_i g_\gamma} \tag{12.149}$$

where N is the number of surface elements in the enclosure and Γ is the number of volume elements. However, if the various exchange areas are computed independently, there is no guarantee that Equations 12.148 and 12.149 will be exactly satisfied.

In general, exchange areas between every pair of surface and volume elements must be known in order to carry through a complete analysis by the zone method. If reciprocity is used to compute as many factors as possible, there may remain as many as $M(M + 1)/2$ independent exchange areas to be evaluated, where $M = N + \Gamma$. Symmetry may reduce this number for some situations. Sowell and O'Brien (1972) use Equation 12.149 to evaluate M additional exchange areas, leaving $M(M - 1)/2$ independent areas to be evaluated. However, as pointed out by Sowell and O'Brien, this approach may lump all the errors of the independently evaluated exchange areas into the M areas found by applying Equation 12.149. Vercammen and Froment (1980) obtained exchange areas by a Monte Carlo approach and found the usual statistical scatter in the results, so the constraints of Equation 12.148 and 12.149 were not met exactly. They present a regression method for smoothing all unique and nonzero factors.

In Larsen and Howell (1986), a method of least-squares smoothing uses Lagrangian multipliers with Equations 12.148 and 12.149 as constraints. This method ensures that constraints are met with a minimum disturbance to the original factors. The tendency of this method is to adjust each exchange area in proportion to its original magnitude. This method can be incorporated into general zone computer codes to assure the "best" set of exchange areas is provided. The method appears to work best for large M; for small M, the methods of Sowell and O'Brien or Vercammen and Froment may be more appropriate. The use of these smoothing methods, however, was found to have little effect on the zone predictions for two problems carried out in Murty and Murty (1991). Daun et al. (2005) propose a constrained least-squares optimization method that satisfies Equations 12.148 and 12.149 but precludes negative factors

12.5.3.2 Other Formulations of the Zone Method

A set of explicit matrix relations is in Noble (1975) for the calculation of total exchange areas from the direct-exchange areas, reducing the time required for solution. Naraghi and Chung (1986) use a stochastic approach to recast the zone equations into a third basis (counting the Hottel and Sarofim (1958) and Larsen and Howell (1985) approaches as two others). They claim increased coding efficiency for the method that uses exchange factors defined somewhat differently from those of Larsen and Howell.

The imaginary planes method is a technique directed toward decreasing the computation time required for the zone method. Considerable reductions in computer time are reported in Charette et al. (1990) where an outline of the method is given for 3D geometries. Each volume zone is linked only to the adjacent zones by the net radiative heat fluxes passing through its zone boundaries (imaginary planes). The radiative transfer is therefore modeled in terms of only the interactions from the immediately adjacent zones, as opposed to using direct interactions with more distant zones as in the classical zone method. Although each volume zone has a direct view of only its own boundaries, the transfer with all other zones is linked in a chain fashion through the radiative heat fluxes crossing the imaginary planes. Hence, the interaction between all zones is included. This technique in formulation provided appreciable savings in computation time for the demonstration cases in Charette et al.

Yuen (2006) and Ghannam et al. (2010, 2012) use a modified zonal method (the multiple absorption coefficient zonal method). This approach uses modified generic exchange factors that can better account for variations in absorption coefficient across an enclosure with inhomogeneous nonisothermal media.

12.5.3.3 Numerical Results from the Zone Method

Larsen and Howell (1985) calculated temperature distributions in the medium, and surface heat fluxes, in 2- and 3D enclosures with gray and black bounding walls containing an absorbing, emitting, and isotropically scattering medium with and without heat conduction. The exchange-factor method was used with exchange-factor smoothing techniques developed to ensure energy conservation (Larsen and Howell 1986). An 11 × 11 set of volume elements and corresponding surface elements was used for the 2D solution and a 5 × 5 × 5 set was used for the 3D solution.

Comparisons of Larsen and Howell zone results with the 2D exact solution by Crosbie and Schrenker (1984) for pure radiation (Table 12.6) are quite good. Errors are within 1.5%, and most values of emissive power and surface flux are within 0%–0.5%. Some comparisons with the P_3 and finite-element solutions are in Figure 12.17 for a square 2D enclosure with black surfaces. Some results for a cubic enclosure (Figure 12.18) with and without conduction are in Tables 12.7 and 12.8 (Larsen and Howell 1985).

The zone method was originally developed for an absorbing and emitting gas without scattering, and further developments were made by Noble (1975) to incorporate scattering. In Yuen and Takara (1990a), a method is given for including scattering that is anisotropic. Results are obtained for a medium with linear anisotropic scattering in a cubic enclosure, but no verification is made by comparisons with other methods. To obtain some verification, this method is applied by Ma (1995) for some conditions in a 1D plane layer so that comparisons can be made with results obtained with the Monte Carlo method. A hot isothermal medium is analyzed between two cold black walls, and the wall heat flux is obtained for isotropic scattering and for linear anisotropic scattering. As a second example, a gas in radiative equilibrium is analyzed between a hot and a cold black wall with linear isotropic scattering in the gas. The temperature distribution in the gas and the heat fluxes at the walls are calculated. Very good comparisons are obtained

TABLE 12.6
Comparison of Dimensionless Surface Heat Flux and Dimensionless Centerplane Emissive Power in a 2D Square Enclosure as Computed by Zonal Analysis and Exact Formulation[a] Radiation Only (Optical Side Length = 1)

	$\tilde{q}_{side}$		$\tilde{q}_{top}$		$\tilde{q}_{bottom}$		$\bar{E}_{centerplane}$	
X or *Y*	Zonal	Exact	Zonal	Exact	Zonal	Exact	Zonal	Exact
0.1	0.524	0.518 (1.2)	0.832	0.827 (0.6)	0.189	0.190 (0.5)	0.521	0.519 (0.4)
0.2	0.437	0.431 (1.4)	0.798	0.796 (0.3)	0.212	0.213 (0.5)	0.434	0.433 (0.2)
0.3	0.368	0.366 (0.5)	0.778	0.777 (0.1)	0.229	0.230 (0.4)	0.361	0.361 (0.0)
0.4	0.310	0.308 (0.6)	0.768	0.767 (0.1)	0.240	0.240 (0.0)	0.300	0.299 (0.3)
0.5	0.260	0.259 (0.4)	0.764	0.764 (0.0)	0.243	0.244 (0.4)	0.250	0.250 (0.0)
0.6	0.218	0.217 (0.5)	0.768	0.767 (0.1)	0.240	0.240 (0.0)	0.208	0.208 (0.0)
0.7	0.182	0.181 (0.6)	0.778	0.777 (0.1)	0.229	0.230 (0.4)	0.173	0.173 (0.0)
0.8	0.149	0.149 (0.0)	0.798	0.796 (0.3)	0.212	0.213 (0.5)	0.142	0.142(0.0)
0.9	0.119	0.119 (0.0)	0.832	0.827 (0.6)	0.189	0.190 (0.5)	0.114	0.115 (0.9)

Source: Crosbie, A.L. and Schrenker, R.G. *JQSRT*, 31(4), 339, 1984; Larsen, M.E. and Howell, J.R., *JHT*, 107(4), 936, 1985.
[a] Numbers in parentheses are percentage differences; $\tilde{q}$ = dimensionless heat flux, $\bar{E}$ = dimensionless emissive power.

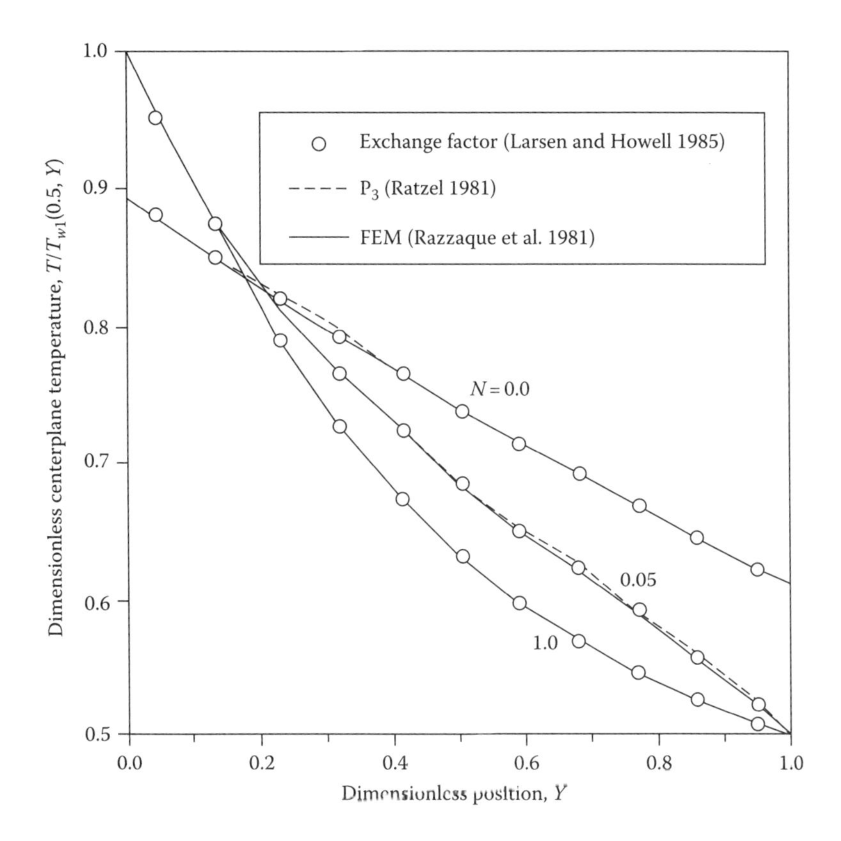

FIGURE 12.17 Centerplane temperature profiles for an infinitely long enclosure of square cross section with black walls; optical side length $\tau_D = 1.0$, $\vartheta_{w1} = 1.0$, $\vartheta_{wi} = 0.5(i = 2, 3, 4)$, at various conduction–radiation parameters, N.

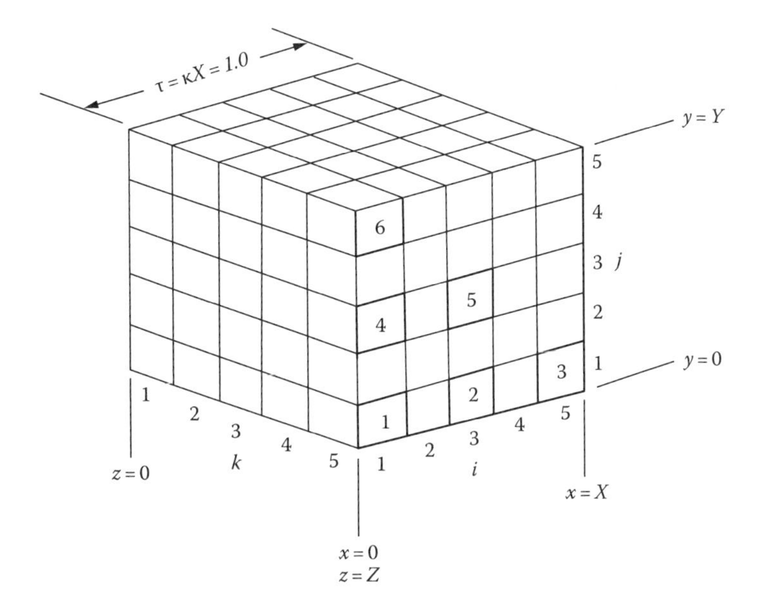

FIGURE 12.18 Black-walled cube with five zones in each direction; optical length of side = 1.

TABLE 12.7
3D Zonal Analysis Results for a Cube, Pure Radiation[a]

		Dimensionless Emissive Power, $[T_{i,j,k}/T_w(\kappa X = 1)]^4$				Dimensionless Heat Flux,
k	j	$i = 1$	$i = 3$	$i = 5$	Zone	$q_w/\sigma T_w^4(\kappa X = 1)$
5	5	0.105	0.177	0.368	1	0.499
5	3	0.177	0.298	0.500	2	0.695
5	1	0.368	0.500	0.632	3	0.821
3	5	0.177	0.298	0.500	4	0.292
3	3	0.298	0.500	0.702	5	0.483
3	1	0.500	0.702	0.823	6	0.177
1	5	0.368	0.500	0.632		
1	3	0.500	0.702	0.823		
1	1	0.632	0.823	0.895		

Source: Larsen, M.E. and Howell, J.R., *JHT*, 107(4), 936, 1985.

[a] Surfaces for which $z = 0$, $y = 0$, or $\kappa X = 1$ have unit emissive power; others are cold. All walls are black, and no internal energy source is present; $N = 0$, optical side length = 1.

TABLE 12.8
Useful Relations for Monte Carlo Solution of Radiation Problems in a Medium

Phenomena	Variables	Relations
Emission from a volume element with absorption coefficient, κ_λ	Cone angle, θ	$\cos\theta = 1 - 2R\theta$
	Circumferential angle, ϕ	$\phi = 2\pi R_\phi$
	Wavelength λ gray medium	$F_{0-l} = R_\lambda$
	Nongray medium	$\frac{\int_0^\lambda \kappa_\lambda I_{\lambda b} d\lambda^*}{\int_0^\infty \kappa_\lambda I_{\lambda b} d\lambda^*} = R_\lambda$
Attenuation by medium with extinction coefficient, β_λ	Path length, l	
	Uniform medium properties	$l = -\frac{1}{\beta_\lambda}\ln R_l$
	Nonuniform medium properties	$-\int_0^l \beta_\lambda(S)dS = \ln R_l$
Isotropic scattering from a volume element	Cone angle, θ	$\cos\theta = 1 - 2R\theta$
	Circumferential angle, ϕ	$\phi = 2\pi R_\phi$
Anisotropic scattering in a gray medium with phase function Φ independent of incidence angle and circumferential scattering angle	Cone angle, θ	$R_\theta = \frac{1}{2}\int_0^\theta \Phi(\theta^*)\sin\theta^* d\theta^*$
	Circumferential angle, ϕ	$\phi = 2\pi R_\phi$

with results from Monte Carlo calculations and from the literature. Tian and Chiu (2003) use a variable transformation to reduce the up to sextuple integrals in direct-exchange area calculations to quadruple integrations, considerably reducing the computational load, and apply the method to systems with nonuniform zones. Hottel et al. (2008) present the most complete embodiment of the zone method, including treatment of specular boundaries and particulate scattering in the formulation.

12.6 MONTE CARLO TECHNIQUE FOR RADIATIVELY PARTICIPATING MEDIA

Monte Carlo utilizes a statistical simulation to determine the behavior of a system. It was applied in Chapter 7 to radiative transfer among surfaces without an intervening participating medium. The information in that chapter is necessary to the development here. Based on the transfer model in Chapter 7, extensions are made here to include absorbing, emitting, and scattering media. The model consists of following a finite number of energy bundles through their transport histories. The average behavior of the bundles provides the radiative performance.

Monte Carlo techniques can be very useful for radiative transfer through absorbing, emitting, and scattering media. The local radiation balance in the medium requires integrating the incoming radiation from the surrounding surfaces and from the volume elements of the surrounding medium. By extending the Monte Carlo model developed for surface-to-surface exchange, it is possible to account for a large variety of effects in radiating media. This can be done without resorting to the simplifying assumptions that are often necessary for numerical solutions based on direct numerical solution of the RTE.

Reviews of the Monte Carlo method applied to participating media are in Farmer and Howell (1998), Howell (1998), and Haji-Sheikh and Howell (2006); these contain references to many applications of the method. Scattering computations by use of Monte Carlo are reviewed by Walters and Buckius (1994), and additional treatments of scattering are in Gupta et al. (1983), Farmer and Howell (1994c), and Ambirajan and Look (1996). Coupled radiation and conduction is analyzed by Al Abed and Sacadura (1983). Inhomogeneous media are treated in Kobiyama (1989) and Hensen et al. (1996), and applications to gases with spectrally dependent properties, using various models, are in Modest (1992), Kudo et al. (1993), Liu and Tiwari (1994), and Cherkaoui et al. (1996). Analysis of thermal treatment of skin cancer using Monte Carlo analysis of laser interaction with embedded nanoparticles is in Bayazitoglu et al. (2013) and Randrianalisoa et al. (2014). For large optical thicknesses, because of the short penetration path lengths and resulting long simulation times, hybrids of Monte Carlo (used for spectral regions providing moderate or long mean free path lengths) with a diffusion solution (for spectral regions with large optical thicknesses) were combined with success by Farmer and Howell (1994a,b). Radiation in multidimensional enclosures and geometries with unusual shapes are in Kaminski (1989), Kudo et al. (1991), and Malalasekera and James (1995). Random number generators, simplified relations for fast processing such as trigonometric functions, and considerations for efficient massively parallel processing using Monte Carlo are discussed by Walker (2013). Eymet et al. (2013) proposed a novel algorithm for Monte Carlo where no volumetric grid is needed; therefore, they eliminate the requirement for numerical inversion. Delatorre et al. (2014) discuss the application of advances on MC techniques to concentrated solar applications. Inverse solutions and reverse Monte Carlo methods are discussed in Section 12.6.5.

12.6.1 Computational Method for Participating Media

An additional factor required in the model in Chapter 7 is the path length traveled in the translucent medium by an individual energy bundle before it is absorbed or scattered, or leaves the system. The required relations between variables and random numbers are in Table 12.8. It is possible to allow for variations in medium properties along the path. In principle, it is even possible to account for variations in refractive index of the medium by having the bundles travel curved paths.

The functions required to determine the angles and wavelengths of emission from a volume element in the medium are in Table 12.8. If a problem is being solved in which there is only steady-state radiative transfer (no conduction or convection) in the medium (radiative equilibrium without internal heat sources), when an energy bundle is absorbed in the medium, a new bundle must be emitted from the same location to ensure no accumulation of energy. The emitted bundle in the medium may be considered as the continuation of the history of the absorbed bundle, and the history continues until the energy reaches a bounding surface.

The total energy emitted by a volume element dV is given by Equation 1.61 integrated over all λ to obtain $4dV\int_0^\infty \kappa_\lambda E_{\lambda b} d\lambda$. For radiative equilibrium without internal heat sources, the energy in the bundles emitted by a volume must equal the energy in the bundles absorbed, wS_{dV}, where w is the energy per bundle and S_{dV} is the number of bundles absorbed per unit time in dV. Then, if we note from Equation 9.71 that $\kappa_P \equiv \int_0^\infty \kappa_\lambda E_{\lambda b}(T_{dV})d\lambda/\sigma T_{dV}^4$, Planck mean absorption coefficient κ_P can be substituted to eliminate the integral of $\kappa_\lambda E_{\lambda b}(T_{dV})$. Equating the two energy terms gives

$$T_{dV} = \left(\frac{wS_{dV}}{4\kappa_p \sigma dV}\right)^{1/4} \tag{12.150}$$

This determines the local temperature in the medium from the Monte Carlo quantities found in the solution and from the medium properties. If κ_P depends on local temperature T_{dV}, iteration is required. A temperature distribution is assumed for a first iteration to obtain the bundle histories. The Monte Carlo quantities are used in Equation 12.150 to obtain a new temperature distribution, which is then used for the second iteration. The process is repeated until the temperatures converge.

There are many variations on the Monte Carlo model for trying to increase efficiency. One most frequently suggested is the fractional absorption of energy when a bundle reaches a surface of known absorptivity. The bundle energy is reduced after each reflection. The bundle history is followed until a sufficient number of reflections have occurred to reduce the bundle energy below some predetermined level where the effect of the bundle in succeeding reflections would be negligible. Such a procedure leads to better accuracy for many solutions because a bundle history extends on the average through many more events, and a given number of bundles thus provide a larger number of events for compiling averages. Other shortcuts for reducing programming difficulties involved with having spectral and directional properties are in Haji-Sheikh and Sparrow (1969) and Farmer and Howell (1998).

Example 12.8

A gray gas with constant absorption coefficient κ is between two infinite parallel black walls spaced D apart. Wall 1 is at T_{w1}, and wall 2 is at $T_{w2} = 0$. Construct a Monte Carlo flowchart for determining the energy transfer and the gas temperature distribution in the limit without conduction or convection (radiative equilibrium, with $\dot{q} = 0$).

The emission per unit time and area from surface 1 is σT_{w1}^4. If N bundles are emitted per unit time, then each carries energy $w = \sigma T_{w1}^4/N$. The bundles are emitted at cone angles θ given by the first line of Table 7.2, $\sin\theta = \sqrt{R_\theta}$, where R_θ is a random number in the range 0–1. A typical bundle will travel path length l after emission. The probability of traveling a given distance S before absorption in a medium of constant absorption coefficient κ is $P(S) = e^{-\kappa S}\Big/\int_0^\infty e^{-\kappa S} dS = \kappa e^{-\kappa S}$. Using Equation 7.48, this is put in the form of a cumulative distribution:

$$R_l = \frac{\int_0^l \kappa e^{-\kappa S} dS}{\int_0^\infty \kappa e^{-\kappa S} dS} = 1 - e^{-\kappa l} \quad \text{or} \quad l = -\frac{1}{\kappa}\ln(1 - R_l)$$

Because R_l is uniformly distributed between 0 and 1, this relation may be written as $l = -(1/\kappa)\ln R_l$ or $L = -(1/\tau_D)\ln R_l$, where $L = l/D$ and $\tau_D = \kappa D$.

The dimensionless distance normal to the wall $X = x/D$ that a bundle will travel when moving through a path L is $X = L\cos\theta = -(\cos\theta/\tau_D)\ln R_l$. Divide the distance D between the walls into k equal increments of dimensionless width $\Delta X = \Delta x/D$, and number the increments $j = 1, 2, 3, \ldots, k$. The increment number at which absorption occurs is $j = \text{TRUNC}(X/\Delta X) + 1$, where TRUNC denotes truncating the value of $X/\Delta X$ to its integer. At each absorption, a tally is kept of j by increasing a counter S_j in the computer memory by one, $S_j = S_j + 1$.

If the bundle is absorbed in an element of the medium, it is immediately emitted from the same element to conserve energy for a medium in radiative equilibrium. This is done by choosing an angle of emission θ from the probability for emission into all cone angles of a unit sphere surrounding dV: $P(\theta) = \sin\theta \Big/ \int_0^\pi \sin\theta d\theta$. Using the cumulative distribution function (CDF), $R_\theta = \int_0^\theta P(\theta^*)d\theta^* = (1-\cos\theta)/2$ gives the emission angle in terms of a random number as $\theta = \cos^{-1}(1 - 2R\theta)$. The distance from the wall to the next absorption point is then $X = X_0 - (\cos\theta/\kappa_D)\ln R_l$, where X_0 is the position of the previous absorption.

The process of absorptions and emissions is continued until the energy bundle reaches a black boundary. This occurs when $X \geq 1$ or $X \leq 0$, and a counter S_{w1} or S_{w2} is then increased by one unit to record the absorption at the surface.

A new bundle is emitted, and the process repeated until all N bundles have been emitted. The dimensionless net energy flux leaving surface 1 is then found from the total bundles emitted minus those reabsorbed at surface 1:

$$\frac{q_1}{\sigma T_{wl}^4} = \frac{\sigma T_{wl}^4 - wS_{w1}}{\sigma T_{w1}^4} = 1 - \frac{S_{w1}}{N}$$

The net energy flux arriving at surface 2, $-q_2$, is given by

$$-\frac{q_2}{\sigma T_{wl}^4} = \frac{wS_{w2}}{wN} = \frac{S_{w2}}{N}$$

The temperature in each increment of the medium is found from Equation 12.150 as

$$\Theta_j = \frac{T_j}{T_{w1}} = \left(\frac{wS_j}{4\tau_D \sigma \Delta X T_{w1}^4}\right)^{1/4} = \left(\frac{S_j}{4\tau_D N \Delta X}\right)^{1/4}$$

The flowchart is shown in Figure 12.19. Note that $S_{w1} + S_{w2} = N$, $q_1/\sigma T_{w1}^4 = 1 - S_{w1}/N = S_{w2}/N = -q_2/\sigma T_{w1}^4$ and, as expected, $q_1 = -q_2$. The only reason for obtaining both radiative flux quantities is to check the results. By noting the linearity of this problem with T^4, it is possible to obtain solutions for any combination of surface temperatures by use of this flowchart (Howell and Perlmutter 1964b). By use of the relations between solutions for black and gray surfaces in Equation 11.82, solutions can be obtained for any combination of gray boundary emissivities.

This example illustrates the power of the Monte Carlo method. Figure 12.19 gives a fairly complete diagram of the logic required for programming the solution for energy transfer through a nonisothermal gray medium between infinite parallel black walls at different temperatures. A comparison of this diagram with other methods for solving the RTE and energy equation in Chapter 12 illustrates the simplifications in both concept and formulation that may be inherent in the Monte Carlo method. This is even more evident in a 2- or 3D geometry.

The computational element size sufficient to yield statistical accuracy in the Monte Carlo radiative source evaluation may be larger than the grid size required for an accurate numerical solution of the energy equation for multimode problems. This grid difference and matching difficulty can be eliminated by taking a sufficiently small Monte Carlo computation element size to match other grid size requirements, with a resulting, possibly large, increase in computer time needed for the increased number of statistical simulations. To conserve computer time, an alternative approach that is usually accurate is to use a grid size compatible with the Monte Carlo statistical requirements and then interpolate intermediate values where the radiative source is needed for grid matching for solving the energy equation. This may not introduce significant error, as the radiative source is often

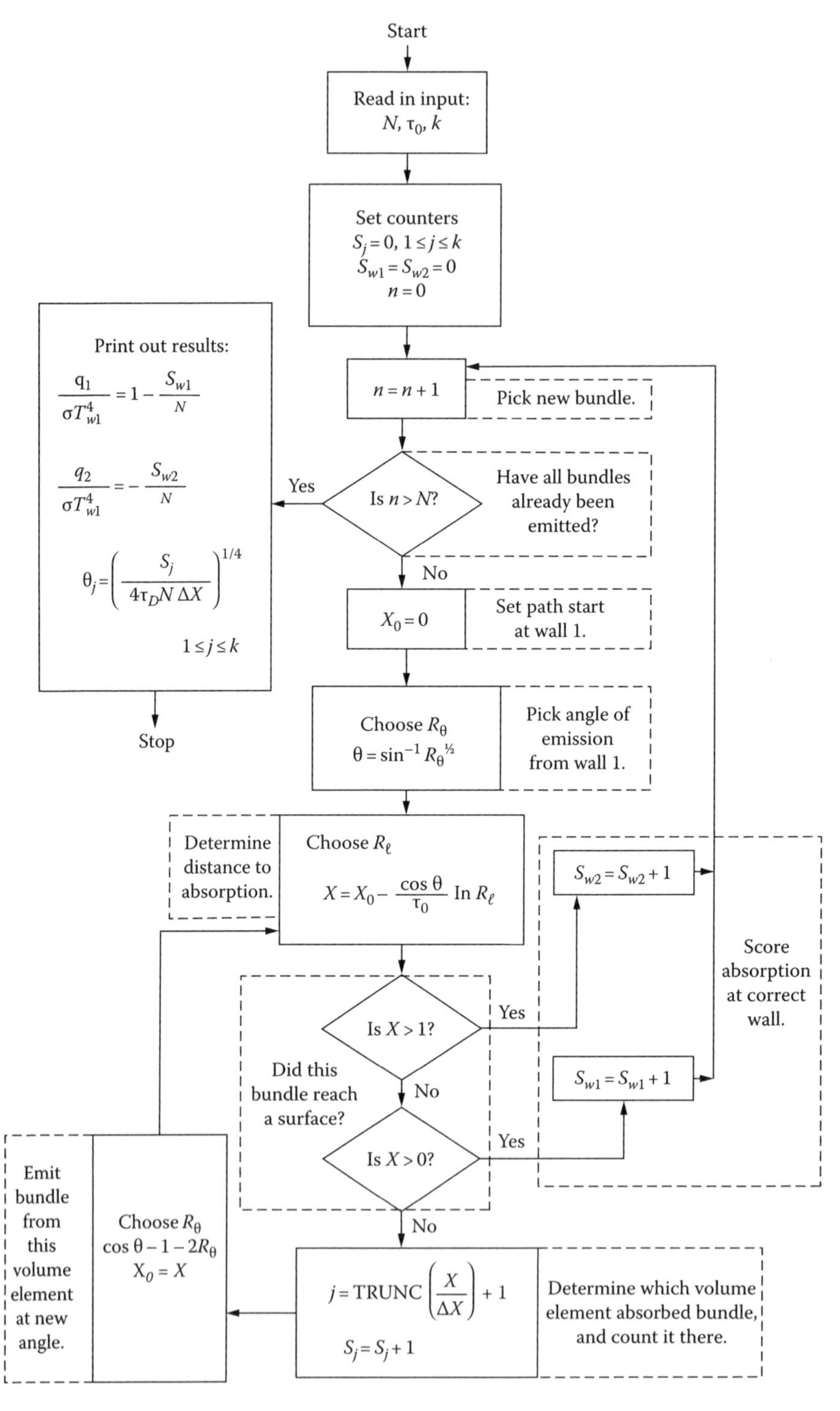

FIGURE 12.19 Flowchart for Monte Carlo solution of radiant transfer between infinite parallel black walls.

a slowly varying function compared with the temperature and its derivatives, and large grid spacing may be adequate for evaluating the local radiative source. Care must be taken, however, because there are situations, such as near boundaries with absorption of large incident external radiation, where the radiative heat source may change rapidly with position, depending on the optical density of the medium.

12.6.2 Monte Carlo Results for Radiation through Gray Gases

12.6.2.1 Infinite Parallel Boundaries

Some results by Monte Carlo methods are now examined. Because accurate solutions are available in the literature for a gray gas between infinite parallel walls, almost every solution method is tried for this situation and compared with the results of analytical approaches typified by Heaslet and Warming (1965). In Howell and Perlmutter (1964b), the local gas emissive power and the net energy transfer between diffuse–gray walls are calculated in a manner similar to Example 12.8. Parameters are the wall emissivity ε_w (specified as equal for both walls) and the gas-layer optical thickness $\tau_D = \kappa D$, where κ is a constant. Two cases are examined without conduction or convection; the first is a gas with no internal energy generation between walls at different temperatures, and the second is for a gas with uniformly distributed internal energy sources between walls at equal temperatures. Figure 12.20 indicates the accuracy that can be obtained by Monte Carlo solutions for such idealized situations; the diffusion solution (Deissler 1964) is also shown. The calculated energy transfer values have a 99.99% probability of lying within ±5% of the midpoints shown.

In Figure 12.21, the emissive power distribution within the gas is shown. Comparison with the exact solutions of Heaslet and Warming is quite good; however, some trends common to all straightforward Monte Carlo solutions in gas radiation problems are evident. First, the calculated individual points in Figure 12.21 reveal increasing error with decreasing optical thickness. This reflects the smaller fraction of energy bundles being absorbed in a given volume element as the optical thickness of the gas decreases. As the number of absorbed bundles decreases, the accuracy of the local emissive power also decreases. Second, the computing time required for problems involving large optical thickness, say larger than 10, becomes large. This is because the free path of an energy bundle, $L = -(1/\tau_D)\ln R_l$, is short for large optical thickness; therefore, many absorptions occur during a typical bundle history.

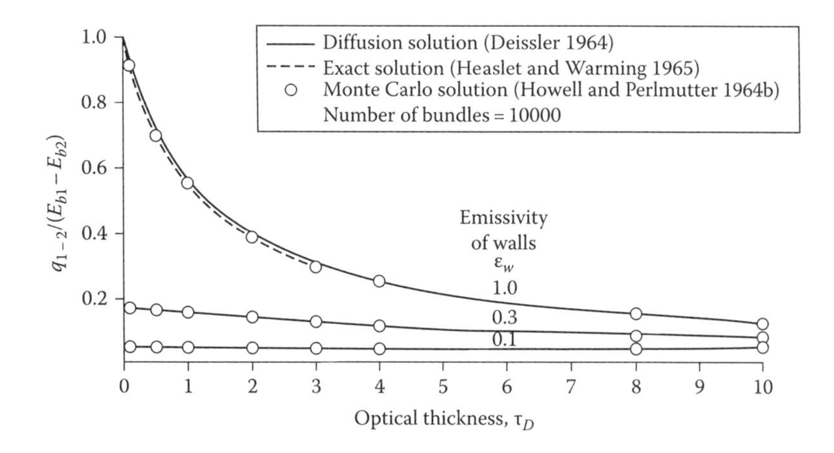

FIGURE 12.20 Net radiative heat transfer between infinite gray parallel walls separated by gray gas without heat conduction.

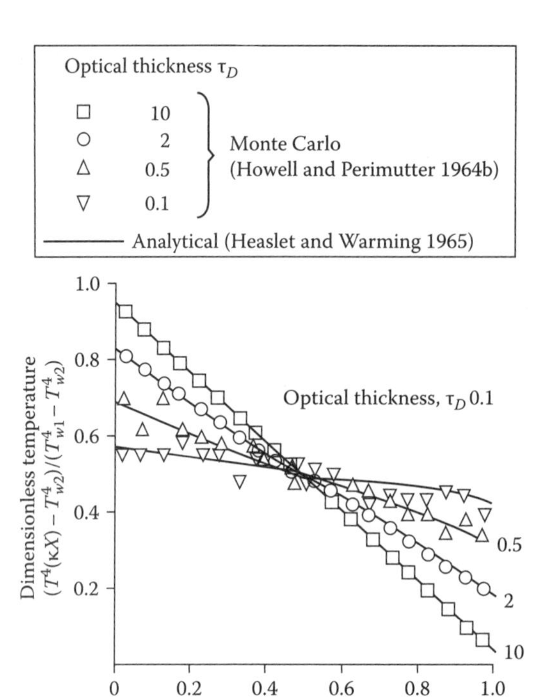

FIGURE 12.21 Emissive power distribution in gray gas without heat conduction between infinite parallel black walls.

Hence, for small optical thickness, accuracy decreases; for large optical thickness, computing time becomes large. These are not serious limitations, as the transparent and diffusion approximations become valid for these conditions. In addition, the range of optical thickness over which a Monte Carlo solution can be effectively utilized can be extended by various techniques such as biasing, splitting, Russian roulette, and a large number of specialized schemes for specific solutions. Many of these involve biasing the path length to increase the number of bundles absorbed in otherwise weakly absorbing regions.

12.6.2.2 Cylindrical Geometry

A more difficult problem to treat analytically is to determine the emissive power distribution and local energy flux in a gray gas between concentric cylinders. The Monte Carlo approach, however, differs only slightly from that for parallel planes. The only additional complication is in determining the bundle position in terms of cylindrical coordinates. Some results from Perlmutter and Howell (1964) for an annular region are in Figure 12.22 and are compared with the modified diffusion solution (Deissler 1964) from Section 12.2.1.2; there is no heat conduction. Trends in accuracy are similar to those for infinite parallel walls. In Avery et al. (1969), finite cylinders of homogeneous nongray gas were analyzed to obtain the local source function and the profiles of the spectral lines for radiation emerging from the cylinder.

12.6.3 Consideration of Radiative Property Variations

For radiative transfer in gases, it is difficult to account accurately for the strong spectral, temperature, and pressure dependence of the radiative absorption coefficient. Many analyses are for gray gases or use various mean absorption coefficients. Monte Carlo is well suited for considering property variations with many variables. It requires relatively little extra effort to assign wavelengths

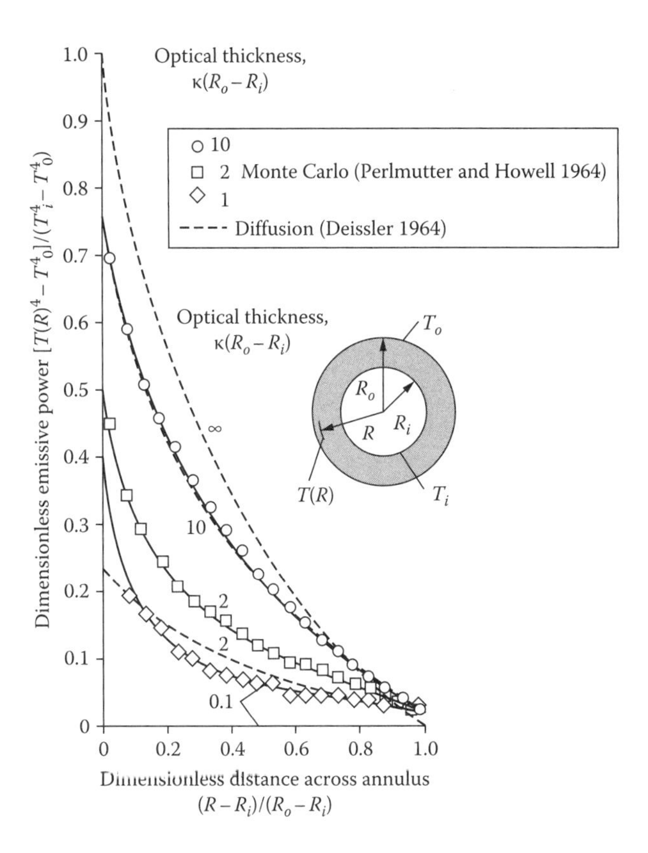

FIGURE 12.22 Dimensionless emissive power distribution in gray gas without heat conduction in annulus between black concentric cylinders of radius ratio $R_i/R_o = 0.1$.

to individual energy bundles and to allow the paths of the bundles to depend on the local spectral absorption coefficient. The required relations are in Table 12.8.

If property variations with temperature are considered, an iterative solution is required because the temperature distribution within the medium is not generally known *a priori*. Determining the path length to absorption becomes more difficult because the absorption coefficient varies with position. By applying the formalism outlined in Section 7.9.2, the path length l is given by

$$\ln R_l = -\int_0^l \kappa_\lambda(S)dS \tag{12.151}$$

To evaluate this integral to determine l along a fixed line after choosing a random number R_l is time consuming but is feasible. Howell and Perlmutter (1964a) used this approach for including temperature- and wavelength-dependent absorption coefficients of hydrogen between infinite parallel walls. They considered energy transfer through a gas between walls at different temperatures, and for a parabolic distribution of internal energy generation in the gas. To evaluate the path length, Equation 12.151 was approximated by dividing the gas into plane increments of thickness Δx. The path length through a given increment was then $\Delta l = \Delta x/\cos\theta$, where θ is the angle between the bundle path and the perpendicular to the walls. Equation 12.151 was replaced by $\ln R_l + \Delta l \sum_{j=1}^{p} \kappa_{\lambda,j} > 0$, and the summation carried out until a value of the integer p was reached that satisfied the inequality. The p is related to the increment number in which absorption occurs. Values of $\kappa_{\lambda,j}$ were assumed for the first iteration and then recalculated in successive iterations on the basis of the newly computed local temperatures.

12.6.4 Parallel Processing and Other Computational Improvements

Monte Carlo is particularly suited to parallel computation. Each energy bundle can be treated individually, so individual bundle histories can be computed on a single processor; with multiple processors, multiple bundles can be followed simultaneously, thus reducing the computation time to obtain the solution.

The usual measures of the efficacy of parallel computing for a particular application are the *speedup* and the *efficiency*, defined by

$$\text{speedup} \equiv \frac{(\text{CPU [central processing unit] time})_{serial}}{(\text{CPU time})_{parallel}} \tag{12.152}$$

$$\text{efficiency} \equiv \frac{\text{speedup}}{\text{number of parallel processors}} \tag{12.153}$$

where the CPU times are those obtained by using a single processor (serial) or multiple parallel processors (parallel) on a computer of the same size. Because there is little or no required communication among the parallel processors during bundle history generation, the speedup in the Monte Carlo calculation is very nearly directly proportional to the number of processors and the efficiency approaches 1. (This is sometimes referred to as being "embarrassingly parallelizable.") This computational advantage of the Monte Carlo method has been recognized and demonstrated by Al-Bahadili and Wood (1993), Burns and Pryor (1989), Farmer and Howell (1998), and Walker (2013). Pincus and Evans (2009) show advantages for a highly parallelized Monte Carlo code for many cases of modeling a cloudy atmosphere, but a parallelized P_N–S_N code was more efficient for some cases. They note that memory limitations may give Monte Carlo an advantage in very large meshes.

Farmer and Howell used an isothermal gray plane layer between black boundaries as a test case. Using the performance measure, $\gamma^2 \times$ (CPU time), where γ is the variance in the results, and then running cases to the same variance level on sequential and parallel processors, the speedup can be defined as

$$\text{speedup} \equiv \frac{\gamma^2 \times (\text{CPU time})_{serial}}{\gamma^2 \times (\text{CPU time})_{parallel}} \equiv \frac{(\text{CPU time})_{serial}}{(\text{CPU time})_{parallel}} \tag{12.154}$$

which gives the same definition as Equation 12.152. The optical thickness was varied, and the speedup was measured for various numbers of parallel processors up to 75. The performance results for an optical thickness of 10 are in Figure 12.23 and are very close to ideal. Ma et al. (2015) used Monte Carlo in an inverse analysis based on particle swarm optimization using graphical programming units (GPUs) for massive parallelization, and showed speedup of over 160× the time using parallel processing based on CPUs.

Alternatives to complete simulation by Monte Carlo have been proposed, combining the advantages of Monte Carlo with other techniques (Farmer and Howell 1994a,b, 1998, Yang et al. 1995, Maruyama and Aihara 1996, Baek et al. 2000, Guo and Maruyama 1999, 2000). Monte Carlo is used to compute the geometric exchange factors, and standard matrix methods are then used to solve for radiative exchange. Once the exchange factors are computed, different sets of boundary temperature or fluxes can be imposed, and results obtained without recomputing the exchange factors. For combined-mode solutions, this has advantages in computer time and convergence. However, if radiative properties are temperature dependent, if nongray gases are considered, or if parametric studies of the effect of properties are required, then the exchange factors must be recomputed for various properties and spectral intervals and these methods may not be practical. Meshless Monte Carlo methods for handling heterogeneous media have been proposed by Gauthier et al. (2013a,b). Other stochastic methods closely related to direct physical simulation as used in Monte Carlo have been proposed, as in Sivathanu and Gore (1993).

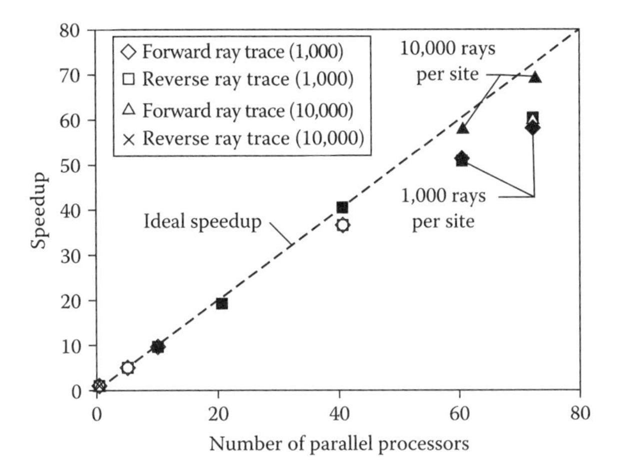

FIGURE 12.23 Speedup versus number of parallel processors for isothermal optically thick medium (optical thickness = 10) between parallel cold black surfaces (Farmer and Howell 1998), using conventional (forward) Monte Carlo or reverse ray tracing by the method of Walters and Buckius (1994).

12.6.5 Reverse Monte Carlo in Participating Media

The reverse Monte Carlo method introduced in Section 7.9.5 can be quite valuable when extended to problems involving a participating medium, especially for media with large optical thickness. Case (1958) and Walters and Buckius (1992, 1994) provide the fundamental framework to show that reversing the Monte Carlo paths from an element of interest to the sources of radiation is a valid concept; this includes a proof of the reciprocity of any path, which may include an anisotropic phase function. Walters and Buckius developed a reverse Monte Carlo method that considers an anisotropically scattering inhomogeneous absorbing–emitting medium, with boundaries that incorporate a bidirectional reflectivity and spectrally dependent properties. Consider a reverse bundle path propagating through an absorbing–emitting–scattering medium (Figure 12.24). The surfaces are assumed to have bidirectional spectral reflectivity $\rho_\lambda(\theta_i,\phi_i,\theta,\phi)$.

The reverse paths from surface dA_k are initiated as for the nonparticipating medium case (see Table 7.2 for angles and wavelength of emission). The method of Walters and Buckius uses attenuation between the point of initiation and a point of scattering to track sample energy by attenuation; a scattering location is then used to choose a change in direction from the scattering coefficient and scattering phase function. The history is followed until the point of origin on a boundary is reached.

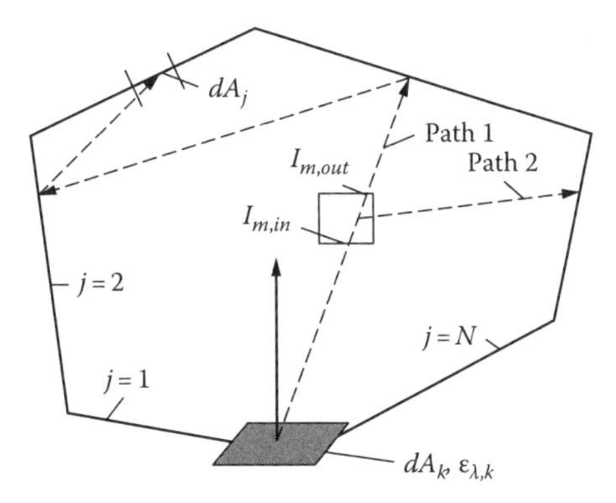

FIGURE 12.24 Enclosure containing inhomogeneous spectrally absorbing–emitting–anisotropically scattering medium.

If the total path length including all direction changes by scattering L includes M segments through cells with homogeneous temperature and properties and it reaches a black boundary at temperature T_w, then the intensity reaching dA_k from sample n is

$$I_{\lambda,i,k} = I_{\lambda b}(T_w)\exp\left[-\int_{l=0}^{L}\kappa_\lambda(l)dl\right] + \sum_{m=1}^{M}\left(I_{\lambda b}(T_m)\left\{\exp\left[-\int_{l=l_{m,out}}^{L}\kappa_\lambda(l)dl\right]-\exp\left[-\int_{l=l_{m,in}}^{L}\kappa_\lambda(l)dl\right]\right\}\right) \quad (12.155)$$

This can be further simplified using summations over the M homogeneous path elements to replace the integrals, resulting in

$$I_{\lambda,i,k} = I_{\lambda b}(T_w)\exp\left[-\sum_{m=1}^{M}\kappa_{\lambda,m}\left(l_{m,out}-l_{m,in}\right)\right] + \sum_{m=1}^{M}\left(I_{\lambda b}(T_m)\left\{\exp\left[-\sum_{p=m,out}^{M}\kappa_{\lambda,p}\left(l_p-l_{p-1}\right)\right]-\exp\left[-\sum_{p=1}^{m,in}\kappa_{\lambda,p}\left(l_p-l_{p-1}\right)\right]\right\}\right) \quad (12.156)$$

The terms in the summation account for the emission from a cell m between the points that the ray enters the cell, m,in, and the point it leaves, m,out. Note that the overall path length L is made up of the various segments of length $l_{m,out} - l_{m,in}$, and could be viewed as a straightened path of length L, where L is determined by the effects of both scattering and absorption. The process is continued for N total samples, and G_k on dA_k is found from

$$G_k = \frac{\pi}{N}\sum_{n=1}^{N} I_{\lambda,i,k,n} \quad (12.157)$$

The heat flux or temperature of dA_k is then found using the net-radiation equations (Equations 5.10 and 5.11).

If the boundary is not black, then boundary reflection is handled as for the nonparticipating medium case in Section 7.9.5; the path is either continued at a randomly chosen angle of reflection when a chosen random number is greater than the spectral absorptance of the boundary, or terminated otherwise.

If the initiating element for the reverse path is a volume element of medium dV_γ rather than a boundary element, the methodology is the same, except that the initial sample angles are chosen as for volume emission.

Propagation of radiation using reverse Monte Carlo or reverse ray tracing is used by Lu and Hsu (2005) and Katika and Pilon (2006).

12.6.6 Expanded Monte Carlo Treatments

Recent applications have extended the Monte Carlo method to cases using a full-spectrum k-distribution (FSK)-related approach (Tessé et al. 2002). Wang et al. (2007) used the FSK CDF in treating spectral properties in gases in Monte Carlo solutions, while Maurente et al. (2007, 2008) invoked the spectral-line-weighted sum-of-gray–gas (SLWSGG) absorption line blackbody distribution function (Chapter 9) for treating spectral properties. Other cases include collimated beams or point sources (Ambirajan and Look 1996, Modest 2003), the existence of very fast transients (Lu and Hsu 2005), the directional emittance of a nonisothermal slab (Li et al. 2005), or the medium having nonhomogeneous optical thickness (Dupoirieux et al. 2006) or anisotropic scattering (Subramanian and Mengüç 1991). Byun et al. (2004) combine Monte Carlo with a finite volume method to avoid the ray effect of the discrete ordinates approach, while Joseph et al. (2009) combine the discrete ordinates (S_N) method and Monte Carlo approaches, using Monte Carlo results to check the accuracy of S_N at controlling points in the geometry. Evans et al. (2003) invoke a residual approach that can greatly speed transient solutions to radiative diffusion problems. De la Torre et al. (2009) modify the RTE to find a simplified way to handle a class of inverse problems for geometrical optimization using Monte Carlo. Zhao et al. (2015) apply Monte Carlo to study radiative transfer in planar media with gradients in refractive index, including the effects of polarization and scattering. Wong and Mengüç (2002) simulated a propagation of collimated beam in a medium via Monte Carlo Approach. They expanded the approach to electron beams (Wong and Mengüç, 2004, 2010), as well as phonons (Wong and Mengüç, 2010, Wong et al. 2011, 2014).

12.7 ADDITIONAL SOLUTION METHODS

12.7.1 Reduction of the Integral Order

To efficiently solve the integral equation of radiative transfer and obtain the radiative source term for the energy equation, it is useful to reduce the order of the integrations required. If the integration over a volume is reduced from triple to double, computation time is greatly decreased. At least two approaches are available: the *point allocation method* (Yuen and Wong 1984) that reduces the order of integrals over volume by one and the *product integration method* (PIM) applied to radiative transfer in Baker (1977) and Tan (1989).

In the PIM, the medium emissive power, the surface radiosity, and the radiative flux divergence are found at prescribed points. Their forms are assumed to be given by a series with unknown interpolation functions as coefficients. The series is substituted into the RTE, which then has the form of a matrix of discretized equations that is solved for the interpolating functions. The matrix elements involve multiple integrals that must be evaluated before solution, but they do not contain the unknowns and therefore need to be evaluated only once. The method is closely related to the FEM but reduces the dimension of the required integrations, thus reducing computation time. For a given degree of approximation of the temperature distribution within each of N finite volume or surface elements, the number of calculations in zonal or FEMs increases as N^2, while in the point allocation or PIM, it increases as N. In Tan (1989b), results using product integration include linear anisotropic scattering in a 2D emitting, absorbing, and scattering medium. A 2D square enclosure divided into 8 × 8 volume elements was analyzed. Results including a uniform internal energy source compared well with the zone method (Figure 12.25).

12.7.2 YIX Method

The YIX method (Tan and Howell 1990a, Tan et al. 2000) is a numerical approach that reduces the order of the multiple integrations and has other important attributes. The YIX name is from the shape of the pattern of the integration points for three, two, and four angular directions in a 2D geometry. The integrals over distance are constructed such that results are stored for use in subsequent integrations, allowing integrals to be computed as simple sums.

Subdivide the local intensity integral from the 1D transfer equation so that

$$I = \int_{x=0}^{L} f(x)E_1(x)dx = \sum_{i=1}^{n} \int_{x=x_{i-1}}^{x_i} f(x)E_1(x)dx + \int_{x=x_n}^{L} f(x)E_1(x)dx \tag{12.158}$$

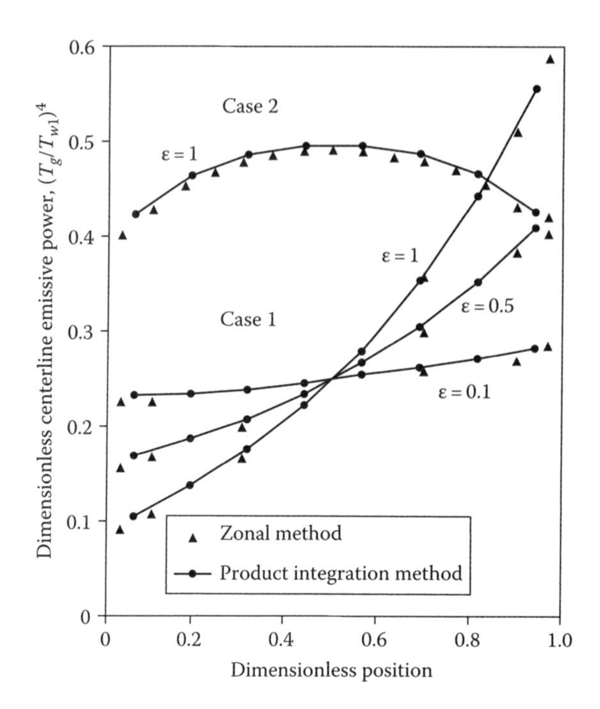

FIGURE 12.25 Centerline dimensionless emissive power in a square enclosure; optical side length $\tau_D = 1$. Case 1, effect of surface emissivity for pure radiation with surfaces at $\vartheta_{w1} = 1$, $\vartheta_{wi} = 0$ for $i = 2, 3, 4$, and case 2, uniform source in medium with $\vartheta_{wi} = 0$ for $i = 1$–4. (From Tan, Z., *JHT*, 111(1), 141, 1989b.)

where $0 = x_0 < x_1 < \ldots < x_n \le L$ and the individual values of x_i are to be found. The sum accounts for the contribution to I from the subregions from $x = 0$ to x_n. The final term is the contribution in the final interval x_n to L. Each of the integrals over a subregion, $x_{i-1} \le x \le x_i$, is expressed as a two-point approximation so that

$$I_i = \int_{x=x_{i-1}}^{x_i} f(x)E_1(x)dx \approx a_i f(x_{i-1}) + b_i f(x_i) \tag{12.159}$$

For $f(x) = 1$ and for $f(x) = x$, Equation 12.159 gives the two equations

$$a_i + b_i = \int_{x=x_{i-1}}^{x_i} E_1(x)dx = E_2(x_{i-1}) - E_2(x_i) \tag{12.160}$$

$$a_i x_{i-1} + b_i x_i = \int_{x=x_{i-1}}^{x_i} xE_1(x)dx = x_{i-1}E_2(x_{i-1}) - x_i E_2(x_i) + E_3(x_{i-1}) - E_3(x_i) \tag{12.161}$$

Equations 12.160 and 12.161 are solved for a_i and b_i, to give

$$a_i = E_2(x_{i-1}) - D(x_i), \quad b_i = D(x_i) - E_2(x_i), \quad \text{and} \quad D(x_i) = \frac{E_3(x_{i-1}) - E_3(x_i)}{x_i - x_{i-1}} \tag{12.162}$$

This approach can be extended to use higher-order approximations of the integrals as provided in Hsu and Tan (1996), but if the intervals $x_i - x_{i-1}$ are small, the two-point approximation is adequate. Substituting Equation 12.162 into 12.159 and the result into Equation 12.158 gives

$$I \approx \left[1-D(x_1)\right]f(0)+\sum_{i=1}^{n-1}\left[D(x_i)-D(x_{i+1})\right]f(x_i)$$
$$+\left[D(x_n)-D(L)\right]f(x_n)+\left[D(L)-E_2(L)\right]f(L) \quad (12.163)$$

The final simplification that makes this method useful is to let each spatial increment provide the same contribution to the summation in Equation 12.163, or $[1 - D(x_1)] = [D(x_i) - D(x_{i+1})] \equiv \beta =$ constant. Substituting into Equation 12.163 gives

$$I \approx \beta\left[f(0)+\sum_{i=1}^{n-1}f(x_i)\right]+\left[D(x_n)-D(L)\right]f(x_n)+\left[D(L)-E_2(L)\right]f(L) \quad (12.164)$$

Note that the term $[D(x_n) - D(L)]$ is not equal to β except in the special case when x_{n+1} falls exactly on the boundary at L. This term must be treated separately.

The number of evaluations of the kernel necessary in the original form Equation 12.158 has been reduced; only a summation over $f(x_i)$ is necessary over most of the increments in the integration. The kernel evaluations are further reduced by approximating the contribution of the final element where $x_n < x < L < x_{n+1}$ by

$$\int_{x=x_n}^{L} E_1(x)f(x)dx \approx \frac{L-x_n}{x_{n+1}-x_n}\int_{x=x_n}^{x_{n+1}} E_1(x)f(x)dx$$
$$\approx \frac{L-x_n}{x_{n+1}-x_n}\left[E_2(x_n)-E_2(x_{n+1})\right]f(x_n), \quad x_n \le L < x_{n+1} \quad (12.165)$$

This provides an estimate of the contribution of the element lying next to the boundary $x_n < x \le L$, in terms of the more easily calculated contribution of the whole fictitious element $x_n < x < x_{n+1}$. Equation 12.164 then becomes

$$I \approx \beta\left[f(0)+\sum_{i=1}^{n-1}f(x_i)\right]+\frac{L-x_n}{x_{n+1}-x_n}\left[E_2(x_n)-E_2(x_{n+1})\right]f(x_n) \quad (12.166)$$

Now, define $Q_i \equiv [E_2(x_{i-1}) - E_2(x_i)]/(x_i - x_{i-1})$ and $P_i \equiv D(x_{i-1}) - E_2(x_{i-1}) - x_{i-1}Q_i - \beta$, and note that $[E_3(x_{n+1}) - E_3(x_n)]/(x_{n+1} - x_n) \approx dE_3(x_n)/dx = -E_2(x_n)$. These relations are substituted into Equation 12.166, which becomes, after some algebra using $\xi \equiv D(x_n) - D(x_{n+1})$,

$$I \approx \xi\left[f(0)+\sum_{i=1}^{n-1}f(x_i)\right]+\left[P_{n+1}+LQ_{n+1}\right]f(x_n) \quad (12.167)$$

Once x_1 is specified, the constants D, P, and Q are computed and stored; evaluation of the integral I is then a straightforward summation. The grid spacing is nonuniform and is chosen so that the contribution to the integral from each increment is roughly the same. The integration grid is thus uncoupled from the choice of increment spacing.

An advantage of this method is that $f(x)$ contains the local properties of the medium; if the properties are nonhomogeneous but temperature independent, they are readily incorporated in the solution. Some cases of this type are treated in Tan and Howell (1990). If the properties are temperature dependent, the solution is iterative. The method is readily extended to multidimensional geometries; the exponential integral function E_n is replaced by S_n (Appendix D) in the 2D formulation, as discussed by Tan and Howell. A 2D enclosure with an internal partition is treated by Tan and Howell (1989b), and an anisotropically scattering square medium exposed to a collimated source is analyzed by Tan and Howell (1990b). The method has been applied to radiation with free convection in Tan and Howell (1991), and solutions of the resulting set of equations were obtained using standard linear equation solvers. The method requires precomputation of a number of coefficients in the solution, but greatly reduces the time for computing the integrals in the radiative source.

12.7.3 Spectral Methods

Spectral methods have been used in fluid mechanics to discretize the spatial domain using Chebyshev collocation points. This gives nonuniform spacing of points, with weighting toward nearby points (Canuto et al. 2006, Shen and Tang 2006). For radiation problems, such a weighting scheme can be beneficial because the exponential decay characteristic means that radiation from nearby points is more important, and thus a method providing more nearby locations should give better accuracy than a method using evenly spaced grids. The method provides high-order (exponential) convergence.

For radiation problems, a discrete ordinates approach is generally used to discretize the angular domain, and then spatial discretization is provided by a spectral method (Sun et al. 2012, Li et al. 2013).

Chebyshev–Gauss–Lobatto collocation points along a given ordinate direction S are given by

$$\frac{S_i}{S} = -\cos\left[\frac{\pi(i-1)}{N-1}\right], \quad i = 1, 2, \ldots, N \tag{12.168}$$

The method has been used by Li et al. (2008) for 1D radiative transfer. Ma et al. (2014) solve a set of 2D radiation problems incorporating spectral radiative properties using FSKs.

12.7.4 FEM for Radiative Equilibrium

In this section, the approach in Razzaque et al. (1983) is followed for problems with radiation as the only transfer mode; the addition of other means of energy transfer is in Section 13.5.2.1. The FEM was applied by Razzaque et al. to obtain results for 2D rectangular geometries. The formulation starts with the energy Equation 10.4, and without conduction, convection, or internal heat sources, $\nabla \cdot \mathbf{q}_r = 0$, so that for a gray medium,

$$4\sigma T^4(x, y) = \int_{\Omega_i=0}^{4\pi} I_i(x, y)\, d\Omega_i \tag{12.169}$$

The incoming intensity $I_i(x, y)$ is evaluated from the equation of radiative transfer, so the integration in Equation 12.169 includes conditions over the entire region. The value of the integral is a function of (x, y, E_b) in a 2D Cartesian system, where $E_b = \sigma T^4(x, y)$. Defining $F(x, y, E_b)$ as the integral on the right-hand side, Equation 12.169 becomes

$$E_b(x, y) = \frac{1}{4} F(x, y, E_b) \tag{12.170}$$

where F, through its dependence on E_b, depends on the radiation field surrounding each x, y. As in Chapter 7, the Galerkin FEM is applied. Equation 12.170 is multiplied by a weighting function $W(x, y)$ as in Equation 7.12.4, and the resulting equation is integrated over the surrounding volume V to give

$$\int_V E_b(x,y)W(x,y)dV = \frac{1}{4}\int_V F(x,y,E_b)W(x,y)dV \tag{12.171}$$

As in Equation 7.12.7, to solve for the temperature distribution, an approximate solution $U(x, y)$ is assumed in the form

$$E_b(x,y) \approx U(x,y) = \sum_{j=1}^{N} U_j\Phi_j(x,y) \tag{12.172}$$

where the U_j are coefficients to be determined. The weighting function $W(x, y)$ is assumed to have the following form, where, for Galerkin's method, the same $\Phi(x, y)$ functions as in $U(x, y)$ are used:

$$W(x,y) = \sum_{i=1}^{N} W_i\Phi_i(x,y) \tag{12.173}$$

As discussed in Section 7.7.2.1, the $\Phi(x, y)$ is the *shape function.* It describes the nondimensional variation of the emissive power assumed over each finite element, and it can be constant, linear, or curvilinear. Its particular form is chosen as a compromise between accuracy and computational effort. Note that for radiative equilibrium without heat sources, the shape functions describe the emissive power $E_b(x, y)$ rather than the temperature.

Substituting Equations 12.172 and 12.173 into 12.171 and rearranging gives

$$\sum_{i=1}^{N} W_i\left[\sum_{j=1}^{N}\left(\int_V \Phi_j(x,y)\Phi_i(x,y)dV\right)U_j - \frac{1}{4}\int_V F(x,y,U)\Phi_i(x,y)dV\right] = 0 \tag{12.174}$$

Defining

$$K_{ij} = \int_V \Phi_j(x,y)\Phi_i(x,y)dV \tag{12.175}$$

and

$$f_i = \frac{1}{4}\int_V F(x,y,U)\Phi_i(x,y)dV \tag{12.176}$$

Equation 12.174 becomes

$$\sum_{i=1}^{N} W_i\left[\sum_{j=1}^{N} K_{ij}U_j - f_i\right] = 0 \tag{12.177}$$

Note that K_{ij} and f_i are integrals over the region surrounding each x, y. The quantities under the integral signs in Equations 12.175 and 12.176 depend on the x, y location and on an integration variable over the region. Because the W_i are arbitrary, Equation 12.177 represents N equations to be satisfied by the N values of U_j. Thus, Equation 12.177 provides the following simultaneous relations:

$$\sum_{j=1}^{N} K_{ij}U_j = f_i \quad i = 1,2,3,\ldots,N \tag{12.178}$$

The K_{ij} in this instance is associated with local emission; the f_i is associated with locally incident intensity. After simultaneous solution of Equation 12.178, the U_j are used in Equation 12.172 to obtain $E_b(x, y)$.

For pure radiation problems, finite-element calculations require solving the integral equation of transfer to determine the $I_i(x, y)$ needed in Equation 12.169 that are in the f_i of Equation 12.177. Numerical accuracy can in principle be improved by increasing the number of finite elements N. Kang and Song (2008) implement the discrete ordinates method on a 3D FEM grid. Zhang et al. (2013) present a meshless FEM for solving the RTE in complex 3D geometries and give application to radiation–conduction problems.

In Razzaque et al. (1983), FEM was used to analyze an absorbing, emitting, and isotropically scattering medium in a rectangular enclosure with gray walls. The shape function $\Phi(x, y)$ was biquadratic in x and y and had nine nodes for each quadrilateral element. The node distribution and some possible shapes of $\Phi(x, y)$ for an element are in Figure 12.26. Curvilinear shape functions are usually necessary to achieve good accuracy in combined-mode solutions.

Results are in Figure 12.27 for the radiative flux at the hot surface of a square enclosure with black walls containing a translucent medium of given optical thickness. The sets of curves are for various optical lengths of the enclosure side. The dimensionless temperature equals one on the hot surface and zero on the other three walls. Comparisons with the zone method (Larsen and Howell 1985) are excellent. The P_3 results of Ratzel (1981) do not compare well, especially for small optical thicknesses where the P_N type of expansion would be expected to decrease in

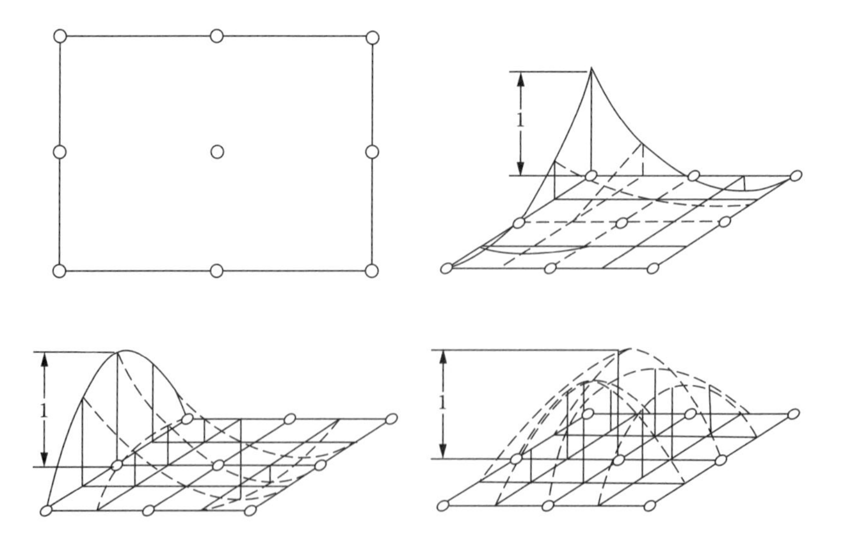

FIGURE 12.26 Node distribution and shape functions for a quadrilateral element. (From Razzaque, M.M. et al., *JHT*, 105(4), 933, 1983.)

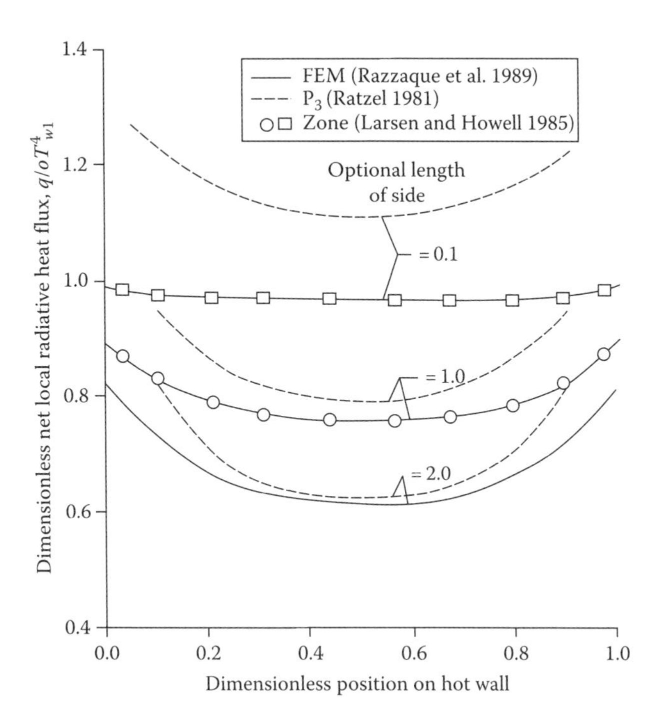

FIGURE 12.27 Dimensionless net local radiative heat flux at the hot wall in square enclosures with various optical thicknesses and black walls. Hot wall at dimensionless temperature $\vartheta = T/T_{w1} = 1.0$ and other walls at $\vartheta = 0$.

accuracy (results obtained are not physically possible). For the FEM results, an array was used of 2 × 2 elements of nine nodes each. The accuracy was checked by using an array of 4 × 4 elements, which increased computational time by a factor of about 4, and the results changed by less than 1%. Thus, for pure radiation solutions, FEM required only a small element array to provide good accuracy. When additional means of energy transfer are present, more elements are required.

12.7.5 Additional Information on Numerical Methods

Considerable effort is still needed to improve numerical methods for solving the radiative transfer relations for radiation only and for radiation combined with other energy transfer modes. In multimode problems, radiation is often the greatest driver in computer time because of its integral nature, in contrast to the differential nature of conduction and convection terms in the energy equation. Difficulties result from the multiple integrations required for the radiative source in multidimensional geometries and from the integration over the spectrum to obtain total energy quantities. The nonlinearity of combined-mode formulations can cause convergence difficulties. Local regions having large temperature changes require small grid sizes to obtain good accuracy in the radiation terms. Reacting flows such as combustion problems present stability difficulties because of the strong dependence of the reaction rates on temperature. In some cases, the radiation terms can lead to either slow or a lack of convergence. Increases in computer size and speed may change the emphasis on the various solution methods. Multimode problem solution techniques are covered in Chapter 13.

The increasing ability of computers to use vector and parallel processing will continue to have an influence (Denning 1985, Shih et al. 1986a, Walker 2013). The best methodology will need to be developed to implement parallel processing for multimode heat transfer calculations. One method is to solve the RTE using an initial assumed temperature field on one processor while simultaneously solving the energy equation on another to compute the temperature field from an initial assumed radiation field. Then the calculated radiation field and temperature information from each parallel path are traded and used as new guesses, and iteration proceeds until convergence. Ghannam (2012) and Ghannam et al. (2012) compared the use of graphical programming units (GPUs) rather than CPUs as the basic computational processors in the zonal method and showed significant speedup using GPUs.

Monte Carlo methods in particular stand to improve significantly in reduced computation time. A vectorized Monte Carlo program in Burns and Pryor (1989) provided a speedup factor of approximately 16 (Section 12.6.4), while Farmer and Howell (1998) showed speedup for parallel processing in proportion to the number of processors. Walker (2013) investigated speedup of Monte Carlo on massively parallel machines and in particular compared the characteristics of various random number generators, different programming methods, optimization of repetitively used elementary functions, and also the use of GPUs rather than CPUs. Although Walker's work was aimed at fast computation of configuration factors, many of the conclusions should apply to general Monte Carlo algorithms. Factors of 10 to over 30 were found through the use of GPUs rather than CPUs.

A finite-volume method (Section 12.4.2) is developed by Raithby and Chui (1990). The method can be applied on the same grid that is used to compute fluid flow and heat transfer. A Taylor series expansion is used to relate the local intensity to that at adjacent grid points. Good agreement was obtained with results for surface fluxes by emission of a square region of radiating medium at uniform temperature in Fiveland (1984) and the exact solution in Siegel (1991a). Good agreement was also obtained with the solution in Crosbie and Schrenker (1984) for a rectangular cavity having three walls at a single temperature and the fourth wall at a higher temperature containing a medium in radiative equilibrium. Mathur and Murthy (2000) discuss the use of unstructured finite volume grids for multimode problems.

12.8 COMPARISON OF RESULTS FOR THE METHODS

The accuracy of the various methods presented in this section depends on the values of the important parameters that describe a particular problem, chiefly the medium albedo and optical thickness.

Gonzalez et al. (2008) compared S_N and M_1 methods for a particular 2D square domain with and without coupling to other heat transfer modes. Further, Planck mean opacity was modeled for an inhomogeneous aluminum plasma. They show that for this particular case, both methods were quite accurate near the source in comparison with an analytical solution for the radiative equilibrium problem using an S_{24} solution. Far from the source, the S_{24} solution suffered from ray effects. In the coupled problems, the M_1 method overpredicted the flux far from a single or distributed boundary heat source, while S_8 results also suffered from ray effects.

Tencer and Howell (2013b) and Tencer et al. (2014) have carried out comparisons of predictions from S_N, P_N, SP_N, and M_N solutions over a wide range of scattering albedo and optical thickness for various 1- and 2D geometries. The general conclusions are that the discrete ordinates methods are highly accurate in problems without ray-effect-inducing discontinuities. For problems with such features, the large number of ordinate directions required to mitigate the ray-effect oscillations may make the discrete ordinates methods computationally intractable. Of the low-order angular approximations considered, the SP_3 method outperformed the M_1, P_1, and P_3 methods in situations when the solution was approximately locally 1D. When the solution was fully 2D, the P_1 method appeared to provide the most effective trade-off between accuracy and computational expense although there were regions where the M_1 and P_3 methods were more accurate than the P_1 method. Maps of solution accuracy for each method as a function of scattering albedo and optical thickness are presented for

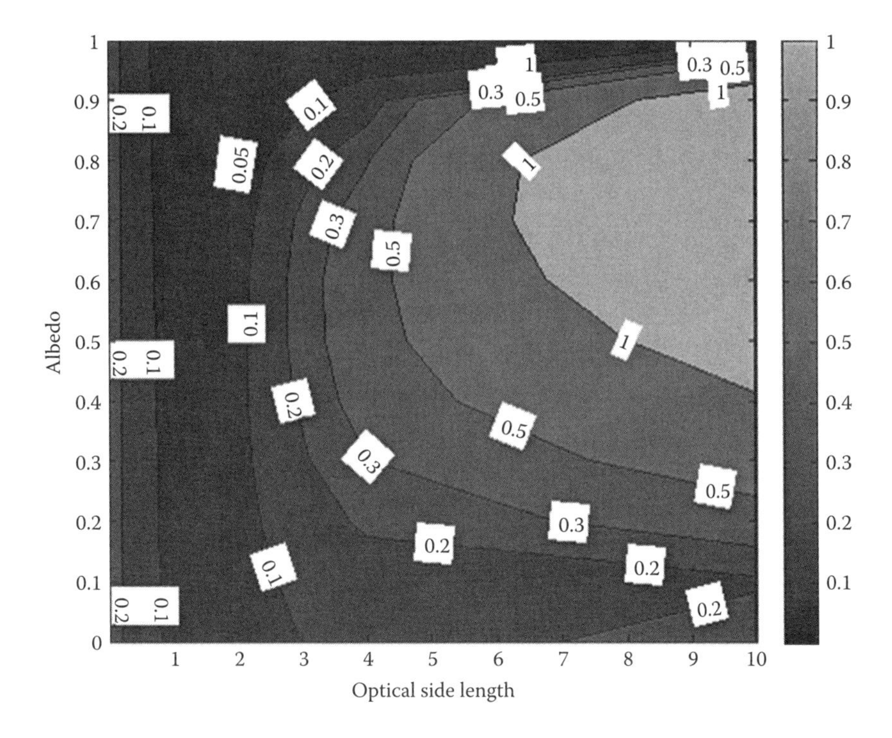

FIGURE 12.28 Relative error (percent) in the SP_3 prediction in a square enclosure with single hot wall and homogeneous cold participating medium. Error is the L^2 norm of the error in the mean heat flux prediction on the hot wall. Reference: analytical solution of Crosbie and Schrenker (1984). (From Tencer, J., Error analysis for radiation transport, PhD Dissertation, Department of Mechanical Engineering, The University of Texas at Austin, Austin, TX, December 2013a.)

various test cases. An analytical solution or a very-high-order S_N solution is used as the benchmark for comparison. Such maps are in Figures 12.11 and 12.28 for the P_1 and P_3 solutions, respectively (Tencer 2013).

12.9 BENCHMARK SOLUTIONS FOR COMPUTATIONAL VERIFICATION

The choice of the best solution method for the radiative transfer and energy relations that model a particular problem is not usually clear. Each method has disadvantages as well as positive attributes. In an effort to evaluate the choice of methods, the American Society of Mechanical Engineers sponsored a series of workshops for which a set of radiative transfer problems was proposed (Tong and Skocypek 1992) for 1-, 2-, and 3D simple geometries. A radiatively participating medium was specified with given spectrally dependent anisotropic scattering properties, and a model was specified for its spectrally dependent band absorptance. A temperature distribution within the medium was given, and the bounding surfaces were specified as cold and black. Researchers were asked to apply their favorite solution method to these benchmark problems and provide numerical values of boundary heat flux distributions and the divergence of the radiative flux at various locations within the medium.

The investigators who participated provided solution results from a generalized zonal method, from three Monte Carlo solutions, from the YIX method, and from two specialized approaches. The boundary heat fluxes from the various methods were found to be in better agreement than the values for the radiative flux divergence, but even the boundary fluxes varied by as much as 40% among investigators for 2D geometries and as much as 100% for 3D geometries.

The poor agreement in these comparisons led to continued computational efforts, and solution methods have been reexamined and modified. In Hsu and Farmer (1997), some benchmark problems were solved using the Monte Carlo and YIX methods, and agreement between these methods is within 5% for all problems. Disagreement was ascribed to differences in spectral integration approaches for the nongray cases. Solutions for gray media were also obtained in Hsu and Farmer by both methods to provide benchmark results for comparison with other methods. For simple geometries and conditions, agreement in wall heat flux and flux divergence at the wall are within 1%. For more complex cases such as variable absorption properties, anisotropic scattering, or an optically dense core region, differences of up to 8% were found and are ascribed to the ray effect in the YIX solutions that uses an S_{16} quadrature on $9 \times 9 \times 9$ and $27 \times 27 \times 27$ uniform grids. Similar cases are solved by an accurate finite-element technique in Burns et al. (1995c) and compared with the results of Hsu and Farmer. The solutions in Hsu and Farmer are further improved by Tan et al. (2000) where modifications to the YIX method are introduced to include linear and quadratic elements in the piecewise integrations. Comparisons are presented for exact, YIX, improved YIX, and FEMs.

In Wu et al. (1996) solutions are generated using quadrature of the RTE for a finite parallelepiped with collimated incident radiation. The volume is homogeneous or nonhomogeneous, is absorbing and emitting, and has anisotropic scattering. The results are compared with YIX solutions (Table 12.9). The quadrature solutions are essentially exact, and results are presented for a wide variety of cases that serve as benchmark solutions for this geometry. Similarly, solutions are in Hsu et al. (1999) for finite-length circular cylinders with anisotropic scattering but without absorption or emission. The YIX solutions are in excellent agreement with the exact solutions, but showed some errors due to ray effects for cases with step changes in

TABLE 12.9

Emissive Power Distribution in Cube of Side Length 2*c* with Three Adjacent Black Surfaces at T_1; Other Surfaces Black at $T_2 = 0$, with Gray Extinction Coefficient 0.5/*c*; and Isotropic Scattering–Emissive Power $E_b(x)$ along Line x, $y = z = 0.25c$

	$E_b(x)/E_{b,w1}$	
$x/(2c)$	QM17[a]	PIM[b]
−0.45833	0.74027	0.74022
−0.37500	0.68285	0.68293
−0.29167	0.63340	0.63355
−0.20833	0.59040	0.59055
−0.12500	0.55223	0.55240
−0.04167	0.51708	0.51725
0.04167	0.48291	0.48309
0.12500	0.44776	0.44793
0.20833	0.40959	0.40977
0.29167	0.36658	0.36676
0.37500	0.31716	0.31736
0.45833	0.25981	0.26004

Source: Wu, S.-H. et al., *ASME HTD*, 332(1), 101, 1996.

[a] Results by interpolation.

[b] Results by numerical quadrature using 17 quadrature points (QM17) and by the PIM (origin of coordinates is at cube center).

the scattering coefficient. This was remedied with an adaptive choice of ray direction and weighting, and further work on reducing ray effects is in Tan et al. (2000). A more complex 3D L-shaped geometry with homogeneous and inhomogeneous mixtures of carbon particles, CO_2, and N_2 is analyzed for heat transfer by combined radiation and conduction in Hsu and Tan (1997). Solutions by the YIX method are compared with discrete transfer solutions for radiation (Malalasekara and James 1995, Henson and Malalasekara 1998) and show excellent agreement to three or four significant figures for gray properties, and agreement within a few percent for nongray analyses. Several solutions for axisymmetric geometries are in Nunes et al. (2000) for radiative exchange in a medium with anisotropic scattering. Comparisons are made with P_1, Monte Carlo, and discrete ordinates methods in a truncated cone, nozzle-shaped enclosure, and a long cylinder.

One benchmark solution that has been used for many years is for radiative equilibrium in a gray gas between infinite parallel black walls at differing uniform temperatures. Table 11.2 provides accurate net radiative heat fluxes as a function of optical spacing between the walls. Examples of benchmark solutions believed to be accurate to four significant figures are in Tables E.1 and E.2 of the online Appendix E on website for this book at http://www.thermalradiation.net; these illustrate the detailed benchmark results available in the references.

12.10 INVERSE PROBLEMS INVOLVING PARTICIPATING MEDIA

Inverse problems with participating media have been examined. Ruperti et al. (1996) used measurements of temperature within a participating medium to predict surface temperatures and fluxes. Linhua et al. (1998) deduce the temperature distribution in a rectangular furnace from measured quantities at the boundary. Orlande et al. (2006) compared three parameter estimation methods to determine the thermophysical properties of a conducting semitransparent medium. De la Torre et al. (2009) use Monte Carlo for optimizing the shape of an enclosure containing an isotropically scattering medium. Le Dez et al. (2009) developed an inverse technique related to singular value decomposition (see Chapter 8) for determining the temperature field within an optically thick cylindrical medium with nonunity refractive index. Inverse multimode problems are discussed in Section 13.7.

12.11 USE OF MEAN ABSORPTION COEFFICIENTS

This chapter has considered various methods for analyzing radiative transfer. Only passing reference has been made to the treatment of the spectral dependence of the radiating medium treated in Chapter 9. As shown there, the properties of many translucent materials can vary considerably with the spectral variable. To try to overcome the need to include detailed property variations in an analysis, there have been attempts to use mean property values over a spectral range. A mean absorption coefficient that is spectrally averaged over all wavelengths provides a simplification that sidesteps the exact but tedious procedure of carrying out a spectral analysis and then integrating the spectral energy over all wavelengths or using a k-distribution or SLWSGG method (Chapter 9) to obtain the total energy. A difficulty is whether it is possible to decide in advance for a particular situation if using a mean absorption coefficient will yield a sufficiently accurate solution.

12.11.1 Definitions of Mean Absorption Coefficients

Planck mean absorption coefficient was briefly discussed in Section 9.4, and the Rosseland mean was introduced in Section 12.1.2.1. Here, we expand the discussion to other mean coefficients and give some applications.

Local total emission by a volume element dV in a material is given by the integral $4dV\int_0^\infty \kappa_\lambda(T,P)\pi I_{\lambda b}(T)d\lambda = 4dV\int_0^\infty \kappa_\lambda(T,P)E_{\lambda b}(T)d\lambda$. For the emission integral, it is convenient to define *Planck mean absorption coefficient* $\kappa_P(T, P)$ (Equation 9.71) as

$$\kappa_p(T,P) \equiv \frac{\int_0^\infty \kappa_\lambda(T,P)E_{\lambda b}(T)d\lambda}{\int_0^\infty E_{\lambda b}(T)d\lambda} = \frac{\int_0^\infty \kappa_\lambda(T,P)E_{\lambda b}(T)d\lambda}{\sigma T^4} \tag{12.179}$$

The κ_P is the mean of the spectral coefficient weighted by the blackbody (Planck distribution) emission spectrum. It is useful in considering *emission* from a volume and for certain special cases of radiative transfer. Further, Planck mean κ_P is convenient since it depends only on the properties at dV. It can be tabulated and is especially useful where the pressure is constant over the volume of a gaseous medium.

The local absorption in a material depends on the integral $\int_0^\infty \kappa_\lambda(T,P)\bar{I}_\lambda d\lambda$. For this absorption integral, an *incident mean* absorption coefficient $\kappa_i(T, P)$ can be defined

$$\kappa_i(T,P) = \frac{\int_0^\infty \kappa_\lambda(T,P)\bar{I}_\lambda d\lambda}{\int_0^\infty \bar{I}_\lambda d\lambda} \tag{12.180}$$

This has little value for general use. A tabulation of κ_i would be needed for many combinations of incident spectral distributions and spectral variations of local absorption coefficients. Except in very limited special cases, this would not be warranted. The incident mean can be used when the incident intensity has a spectral form that remains fixed so that κ_i can be evaluated and tabulated. For example, incident energy having a solar spectrum occurs sufficiently often that κ_i can be tabulated for the solar distribution. The κ_i is useful in the transparent approximation when the incident spectral intensity is known, as this spectrum remains unchanged while radiation travels through the medium. If the mean intensity $\bar{I}_{\lambda,i}$ is proportional to a blackbody spectrum *at the temperature of the position for which* $\kappa_\lambda(T, P)$ *is evaluated*, that is, $\bar{I}_{\lambda,i} \propto I_{\lambda b}(T)$, then the incident mean is equal to κ_P:

$$\kappa_i(T,P) = \frac{\int_0^\infty \kappa_\lambda(T,P)I_{\lambda b}(T)d\lambda}{\int_0^\infty I_{\lambda b}(T)d\lambda} = \kappa_P(T,P) \tag{12.181}$$

The Rosseland mean attenuation coefficient arises from treating radiation transfer in optically thick media, where it acts as a diffusion process. It is considered in Section 12.2.1.2.1. It is defined in Equation 12.27, and for only absorption (no scattering),

$$\kappa_R(T,P) = \left[\int_0^\infty \frac{1}{\kappa_\lambda(T,P)}\frac{\partial E_{\lambda b}(T)}{\partial E_b(T)}d\lambda\right]^{-1} \tag{12.182}$$

At first glance, the Rosseland mean appears to be entirely different from κ_P and κ_i, which are weighted by spectral distributions of energy or intensity. However, for 1D diffusion, radiative spectral flux is found to depend only on the local blackbody emissive power gradient and absorption coefficient:

$$q_{\lambda,z}d\lambda = -\frac{4}{3\kappa_\lambda}\frac{dE_{\lambda b}(T)}{dz}d\lambda = -\frac{4}{3\kappa_\lambda}\frac{\partial E_b}{\partial z}\frac{\partial E_{\lambda b}}{\partial E_b}d\lambda \tag{12.183}$$

Then

$$\int_0^\infty \kappa_\lambda q_{\lambda,z}d\lambda = -\frac{4}{3}\frac{\partial E_b}{\partial z} \quad \text{and} \quad \int_0^\infty q_z d\lambda = -\frac{4}{3}\frac{\partial E_b}{\partial z}\int_0^\infty \frac{1}{\kappa_\lambda}\frac{dE_{\lambda b}}{dE_b}d\lambda$$

Substituting into Equation 12.182 gives for the *diffusion case*

$$\kappa_R(T,P) = \frac{\int_0^\infty \kappa_\lambda q_{\lambda,z}d\lambda}{\int_0^\infty q_{\lambda,z}d\lambda} \tag{12.184}$$

The Rosseland absorption coefficient is thus a mean value of κ_λ weighted by the local spectral energy flux $q_{\lambda,z}d\lambda$ through the assumption that the local flux depends only on the local gradient of emissive power and the local κ_λ.

For a gray gas, the absorption coefficient is independent of wavelength, $\kappa_\lambda(T,P) = \kappa(T,P)$, and the mean values reduce to $\kappa_P(T,P) = \kappa_i(T,P) = \kappa_R(T,P) = \kappa(T,P)$.

Determining any of the mean coefficients from spectral absorption coefficients usually requires detailed line-by-line numerical integrations or use of the *k*-distribution or SLWSGG methods of Chapter 9. Even so, if appropriate spectral mean values can be applied to yield reasonably accurate solutions, the solution effort is often considerably decreased.

12.11.2 Approximate Solutions of the Radiative Transfer Equations Using Mean Absorption Coefficients

Some references incorporate mean absorption coefficients in radiative transfer calculations. Solving the transfer and energy equations is then considerably simplified because integrations over wavelength are not needed. The most common approximation is that gray–gas relations are applied to a real gas by substituting an appropriate mean absorption coefficient in place of the κ in the gray solution. By examining 40 cases, Patch (1967a,b) showed that simple substitution of Planck mean in gray–gas solutions leads to errors in total intensities that varied from −43% to 881% from the solutions obtained by using spectral properties in the transfer equations and integrating the spectral results. Reductions in error were found by dividing the intensity into two or more spectral bands and using an individual Planck mean for each band.

In an effort to provide improvements, a number of other mean absorption coefficients have been introduced. Sampson (1965) synthesized a coefficient that varies from Planck mean to the Rosseland mean as the optical depth increases along a path. Agreement was obtained within a factor of two, with exact solutions for various problems. Abu-Romia and Tien (1967) applied a weighted Rosseland mean over optically thick portions of the spectrum and a Planck mean over optically thin regions and obtained relations for energy transfer between bounding surfaces. Planck and Rosseland mean absorption coefficients for carbon dioxide, carbon monoxide, and water vapor are given to aid such computations.

Patch (1967a,b) defined an *effective mean absorption* coefficient as

$$\kappa_e(S,T,P) = \frac{\int_0^\infty \kappa_\lambda(T,P) I_{\lambda b}(T) \exp[-\kappa_\lambda(T,P)S] d\lambda}{\int_0^\infty I_{\lambda b}(T) \exp[-\kappa_\lambda(T,P)S]\, d\lambda} \tag{12.185}$$

The values of $\kappa_e(S, T, P)$ can be tabulated as a function of temperature and pressure; in addition, κ_e depends on the path length and must be tabulated as a function of S. For small S, κ_e approaches κ_P. For large S, the exponential term in the integrals causes κ_e to approach the minimum value of κ_λ in the spectrum considered. In radiative transfer calculations, the approximation is made that the real gas, with T and P known variables along S, is replaced along any path by an effective uniform gas with absorption coefficient κ_e. Computations are then performed using κ_e in the gray–gas equation of transfer. The κ_e value used is found by equating $\kappa_e S$ at the T and P of the point to which S is measured, to the optical depth of that point in the real gas. For 40 cases, Patch shows agreement of total intensities within −25% to 28% of the integrated spectral solutions. Other methods of using mean coefficients are in Lick (1963), Stewart (1964), Thomas and Rigdon (1964), Grant (1965), and Howe and Scheaffer (1967).

12.12 SOLUTION USING COMMERCIAL CODES

Most of the major commercially available computational fluid dynamics (CFD) codes employ one or more choices of methods for handling radiative transfer within a participating medium. For example, the ANSYS CFD code packages FLUENT and CFX between them provide choice from among surface–surface, diffusion, P_1, discrete transfer, discrete ordinates, and Monte Carlo solvers. Various models are included for treating anisotropic scattering and spectral medium property variations, although these features are not available for all solvers. These and competing codes continue to add features and capabilities, and careful comparison is warranted of the required capabilities for a particular problem or application. More discussion of codes for treating multimode problems is in Chapter 13.

HOMEWORK

12.1 The diffusion solution predicts that the zeroth moment $I^{(0)}(X) = I_{\lambda b}(X)$. Comparing this result with the zeroth moment Equation 12.2, what do you conclude about the local mean intensity in a medium subject to the restrictions of the radiation diffusion assumptions?

12.2 The space between two concentric diffuse–gray cylinders is filled with an optically dense stagnant medium having a constant absorption coefficient κ and isotropic scattering with scattering coefficient σ_s. Compute the radiative heat transfer across the gap from the inner to the outer cylinder and the radial temperature distribution in the medium in the limit of negligible heat conduction by using the diffusion method with jump boundary conditions.

12.3 Two large parallel gray plates are 5.65 cm apart. Their temperatures and emissivities are $T_1 = 800$ K, $\varepsilon_1 = 0.70$; $T_2 = 512$ K, $\varepsilon_2 = 0.31$. Compute the heat transferred by radiation across the space between the plates when the space is a vacuum and when the space is filled with a gray medium that has absorption and isotropic scattering with extinction coefficient $\beta = 0.63$ cm^{-1}. (Use the diffusion method as an approximation and neglect heat conduction.)
Answer: 5.29 kW/m^2; 3.06 kW/m^2.

12.4 Two large gray parallel plates at temperatures T_1 and T_2 and with emissivities ε_1 and ε_2 are separated by an optically thick gray medium with absorption and isotropic scattering. The medium has within it a uniform volumetric energy source of $\dot{q}$ W/m^3. Compute the temperature distribution in the medium by use of the diffusion method with jump boundary conditions for the limit of negligible heat conduction.

12.5 A large plate of translucent glass is laid upon a sheet of polished aluminum. The aluminum is kept at a temperature of 500 K and has an emissivity of 0.03. The glass is 2 cm thick and has a Rosseland mean absorption coefficient of $\kappa_R = 3.6\ \text{cm}^{-1}$. A transparent liquid flows over the exposed face of the glass and maintains that face at a temperature of 270 K.

(a) What is the heat flux through the glass plate?
(b) What is the temperature distribution in the glass plate?

Neglect heat conduction in your calculations, and for simplicity, assume that the refractive indices of the glass and liquid are both one.

Answer: $q = 84.8\ \text{W/m}^2$; $T(\tau) = 100(133.7-11.2\tau)^{1/4}$(K).

12.6 A gray gas is contained between infinite gray parallel plates. The plates both have emissivity $\varepsilon = 0.30$. Plate 1 is maintained at temperature $T_1 = 1150$ K, and plate 2 is at $T_2 = 525$ K. The gray absorbing and isotropically scattering medium between the plates has a uniform extinction coefficient of $\beta = 0.75\ \text{m}^{-1}$, and heat conduction is neglected. The plate geometry is shown as follows.

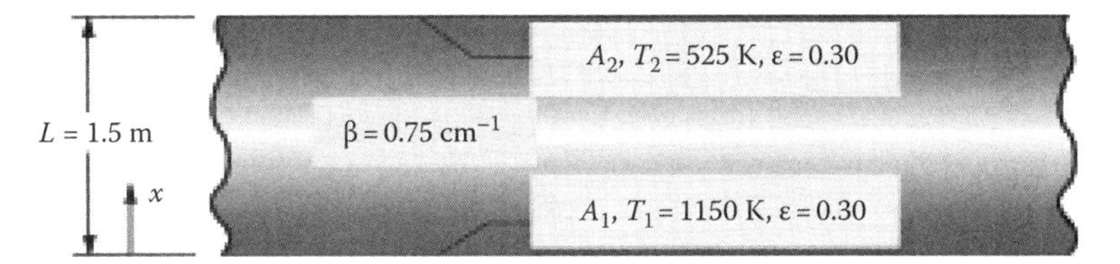

Predict the net radiative heat flux transferred between the surfaces (W/m²), and plot the dimensionless temperature distribution $\phi(\tau) = [T^4(k) - T_2^4]/(T_1^4 - T_2^4)$ in the gas, where $\tau = \beta x$. Solve the problem using

(a) The P_1 approximation (differential approximation)
(b) Milne–Eddington approximation

Compare the results with those of Homework Problem 12.5.

Answer:

	(a)	(b)
q (W/m²)	14,570	14,570
$\phi(\tau)$	0.5648–0.1152 τ	0.5648–0.1152 τ

12.7 A gray isotropically scattering gas is contained between large diffuse–gray parallel plates. The plates both have emissivity $\varepsilon = 0.30$. Plate 1 is maintained at temperature $T_1 = 1150$ K, and plate 2 is at $T_2 = 525$ K. The medium between the plates has a uniform extinction coefficient of $\beta = 0.75\ \text{m}^{-1}$. The plate geometry is shown as follows.

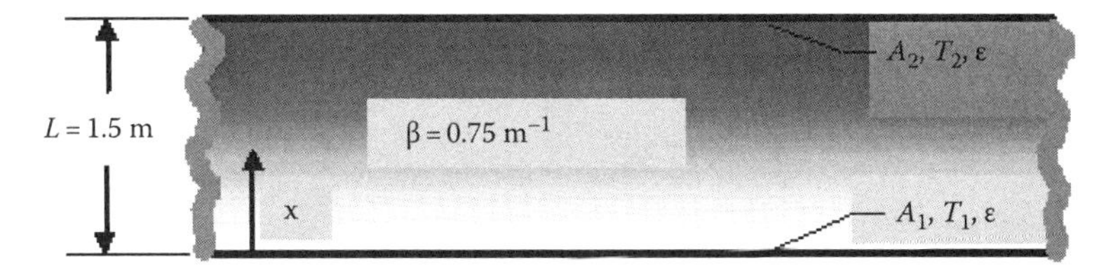

Predict the net radiative heat flux transferred between the surfaces (W/m²) and plot the temperature distribution $[T^4(k) - T_2^4]/(T_1^4 - T_2^4)$ in the gas, where $\tau = \beta x$. Heat conduction is small. Solve the problem using the first-order diffusion method and the nearly transparent optically thin approximation.

Answer: Diffusion: $q = 14{,}570\ \text{W/m}^2$; $\Phi(\kappa) = 0.5648 - 0.1152$ K. Thin: 16,740 W/m²; $\Phi(\kappa) = 0.500$.

12.8 As in Homework Problem 12.7, a gray absorbing and isotropically scattering gas is contained between large gray parallel plates. The plates both have emissivity $\varepsilon = 0.30$. Plate 1 is maintained at temperature $T_1 = 1150$ K, and plate 2 is at $T_2 = 525$ K. The gray medium between the plates has a uniform extinction coefficient of $\beta = 0.75$ m^{-1}, and heat conduction is neglected. The plate geometry is shown as follows and is the same as in Homework Problem 12.6.

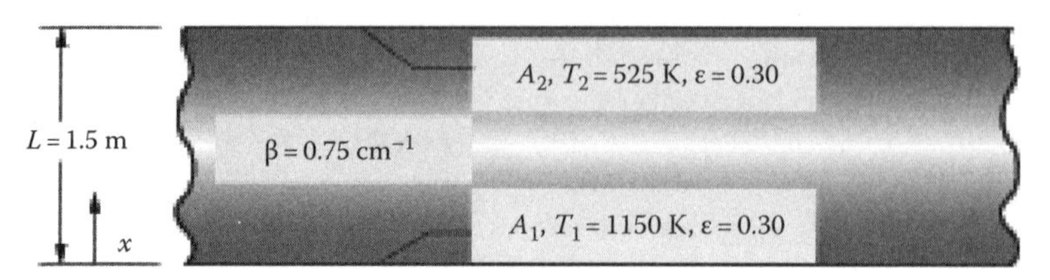

Predict the net radiative heat flux transferred between the surfaces (W/m^2) and plot the dimensionless temperature distribution $\phi(\tau) = [T^4(k) - T_2^4]/(T_1^4 - T_2^4)$ in the gas, where $\tau = \beta x$. Obtain the solution using the two-flux method, and compare the results with those of Homework Problems 12.6 and 12.7.

Answer: 13,970 W/m^2; $\phi(\tau) = 0.5828 - 0.147\tau$.

12.9 A long cylinder 12 cm in diameter is surrounded by another cylinder 24 cm in diameter. The surfaces are gray, the inner cylinder is at $T_1 = 910$ K with $\varepsilon_1 = 0.42$, and the outer cylinder is at $T_2 = 1075$ K with $\varepsilon_2 = 0.83$. What is the heat transfer from the outer cylinder to the inner cylinder per unit length for vacuum between the cylinders? If the space between the cylinders is filled with a gray medium having absorption and isotropic scattering with extinction coefficient $\beta = \kappa + \sigma_s = 0.41$ cm^{-1}, compute the energy transfer using the P_1 method and the diffusion method (heat conduction is neglected).

Answer: 5.59 kW/m; 3.94 kW/m; 3.89 kW/m.

12.10 A fusion reactor core is contained within a spherical shell of radius R_1 blanketed by a concentric layer of gray absorbing and isotropically scattering gas of thickness L so that the outer radius is $R_2 = R_1 + L$. The spherical surfaces that contain the gray–gas blanket are black. The outer spherical surface is at T_{w2}, and the gas has extinction coefficient $\beta = \kappa + \sigma_s$. If the reactor generates power Q, find expressions for the inner surface temperature T_{w1} using the diffusion and the P_1 differential methods (heat conduction is neglected). Plot the ratio $\dfrac{A_1\sigma\left(T_{w1}^4 - T_{w2}^4\right)}{Q} = \dfrac{1}{\Psi}$ as a function of $\tau = \beta L$ and R_1/R_2 for both solutions.

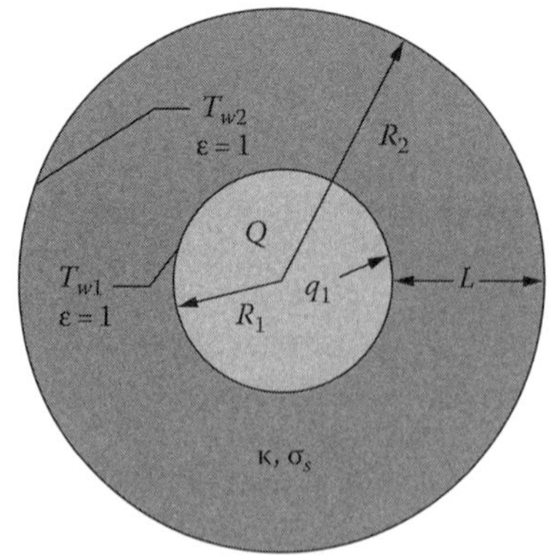

12.11 A fusion reactor core is contained within a spherical shell of radius R_1 blanketed by a concentric layer of gray absorbing and isotropically scattering gas of thickness L so that the outer radius is $R_2 = R_1 + L$. The spherical surfaces that contain the gray–gas blanket are black. The outer spherical surface is at T_{w2}, and the gas has extinction coefficient $\beta = \kappa + \sigma_s$. If the reactor generates power Q, find expressions for the inner surface temperature T_{w1} using the SP_1 differential methods (heat conduction is neglected). Plot the ratio $\frac{A_1\sigma\left(T_{w1}^4 - T_{w2}^4\right)}{Q} = \frac{1}{\Psi}$ as a function of $\tau = \beta L$ and R_1/R_2. Compare with the results of Homework Problem 12.10 if assigned.

Answer: $T_{w1}^4 = T_{w2}^4 + \frac{Q}{A_1\sigma\psi}$.

12.12 An infinitely long rectangular 2D enclosure is shown as follows in cross section. It is filled with an absorbing–emitting but nonscattering gray medium with absorption coefficient $\kappa = 2\ \text{m}^{-1}$. There is no convection or internal heat generation, and conduction is neglected. Using the P_1 method, find and plot the net radiative heat flux distribution along surface 1, $q_r(x_1)$, and also plot representative isotherms, $T_g(x,y)$, in the medium.

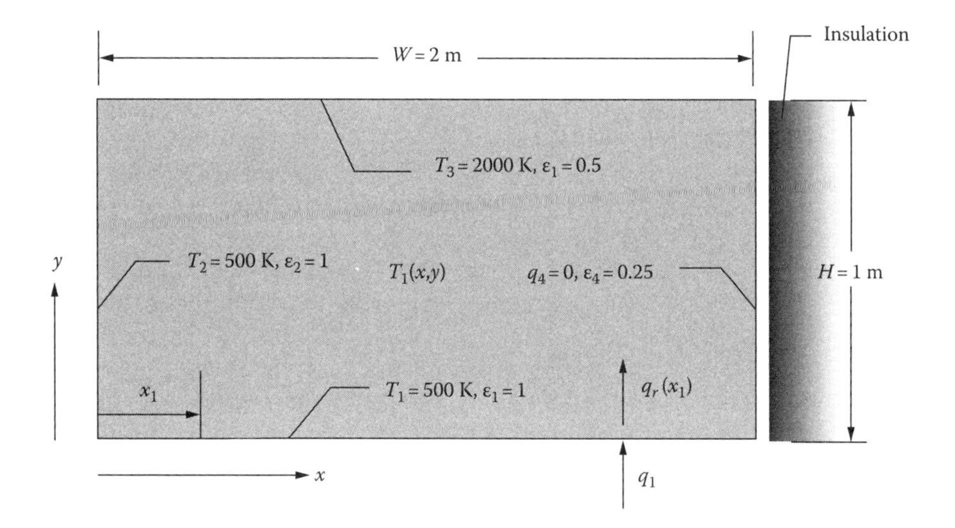

12.13 A 2D square enclosure (long in the direction normal to the cross section) is filled with gray medium that absorbs and emits radiation, but has negligible scattering. The medium is not moving, heat conduction can be neglected relative to radiative transfer, and the refractive index of the medium is $n \approx 1$. As shown, the lower horizontal side is at $T_1 = 950$ K and the vertical side at $x = 0$ is at $T_2 = 825$ K. At the other two boundaries, T_3 and T_4 have small values that can be neglected in the radiative exchange. All of the boundaries are black. The absorption coefficient is constant throughout the medium, $\kappa = 1.4\ \text{m}^{-1}$, and the side length of the enclosure is 0.9 m. Using the development in Example 12.7, set up a discrete ordinates procedure for the S_4 quadrature using four volume elements and carry out a few iterations for the intensities and local temperatures within the volume.

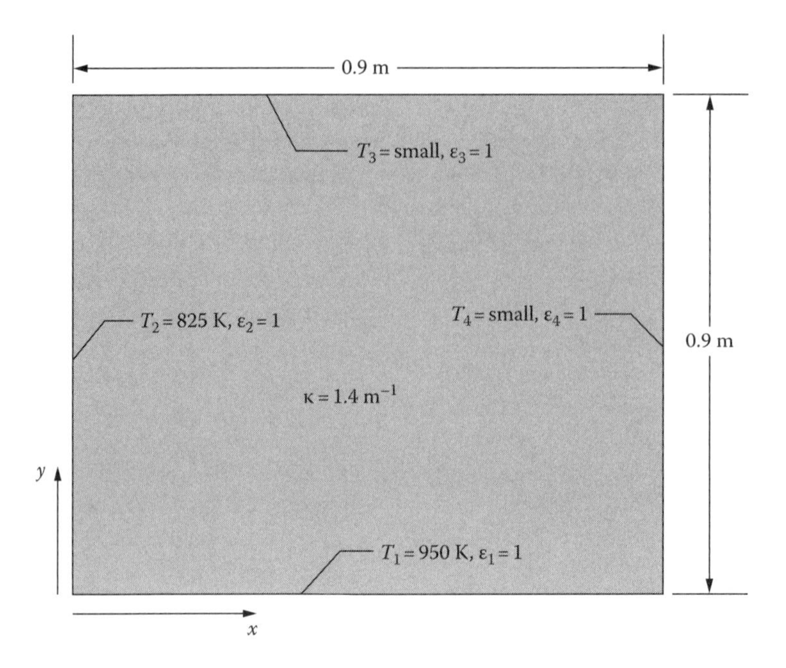

12.14 A rectangular enclosure that is very long normal to the cross section shown has the surface conditions and properties listed in the table. The enclosure is filled with an absorbing, emitting, nonscattering gray gas in radiative equilibrium (no heat conduction, convection, or internal energy sources). Find the heat flux that must be supplied to each surface to maintain the specified temperatures. Use a finite-difference numerical solution of the energy equation including radiative transfer.

Surface	Type	Emissivity	*T*, K	q
1	Specular	0		0
2	Diffuse	0.713	810	
3	Specular	0		0
4	Diffuse	0.305	1300	

Answer: 33.5 kW/m^2.

12.15 A plane layer of thickness D has its boundaries maintained at fixed temperatures T_0 and T_D. It is heated internally with a volumetric heat source distribution $\dot{q}(x)$ that depends on the x location within the layer, and its temperature distribution is needed. As an approximate trial solution by finite elements, it is decided to use only two elements, but to use quadratic shape functions so that the temperature distribution can better adjust to the $\dot{q}(x)$ distribution.

To use quadratic shape functions, an additional node is placed at the center of each element. The two elements are of equal width Δx_e and the nodes are all equally spaced with spacing Δx. Obtain expressions for the shape functions at nodes 1, 2, and 3, and make a plot of these functions.

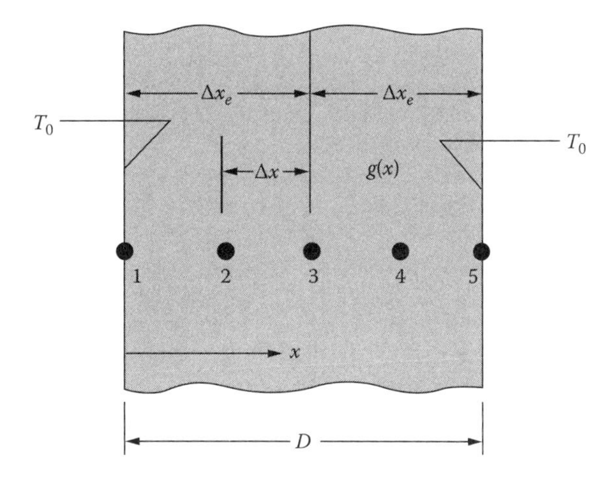

Answer:

$$\Phi_1(x) = 1 - \frac{3}{2}\frac{x - x_1}{\Delta x} + \frac{1}{2}\left(\frac{x - x_1}{\Delta x}\right)^2; \quad \Phi_2(x) = 1 - \left(\frac{x - x_2}{\Delta x}\right)^2; \quad \Phi_3(x) = 1 - \frac{3}{2}\frac{x_3 - x}{\Delta x} + \frac{1}{2}\left(\frac{x_3 - x}{\Delta x}\right)^2.$$

12.16 For use in a finite-element solution in a 2D region, consider a three-node triangular element. The dimensionless temperature distribution $t(X,Y)$ is to be obtained. A linear interpolation function for $t(X,Y)$ is to be used in the form $t(X,Y) = aX + bY + c$. Find relations for the shape functions at each of the nodes.

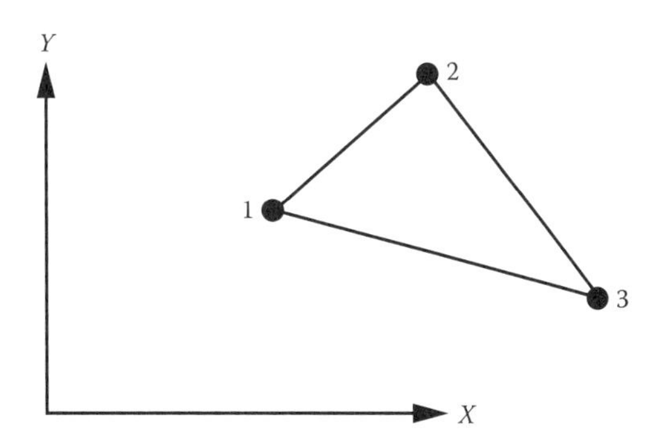

12.17 Obtain the solution to Homework Problem 12.14 by using a Monte Carlo numerical method.

12.18 For a nongray nonscattering gas with nonuniform properties, derive the following relations for the wavelength of emission λ, and the path length to absorption ℓ:

$$R_\lambda(S) = \frac{\int_{\lambda^*=0}^{\lambda} \kappa_\lambda(S) I_{\lambda b}(S)\, d\lambda^*}{\int_{\lambda=0}^{\infty} \kappa_\lambda(S) I_{\lambda b}(S)\, d\lambda}; \quad \ln R_{\ell,\lambda} = -\int_{S=0}^{\ell} \kappa_\lambda(S)\, dS$$

12.19 In Example 12.8, let both plate temperatures T_1 and T_2 be nonzero. Plate 1 is black but plate 2 has a spectrally varying hemispherical emissivity $\varepsilon_{\lambda,2}(T_2)$. Modify the flowchart in Figure 12.19 to account for this in determining the energy transfer between the plates and the gas temperature distribution.

12.20 A gray nonscattering gas with constant absorption coefficient κ is within a rectangular enclosure of finite width L, height D, and infinite length normal to the cross section shown. All of the boundaries are black. The lower and upper boundary temperatures are T_1 and T_2, the gas is at a constant temperature T_g, and the side boundaries are at zero temperature as shown. Construct a Monte Carlo flowchart to obtain the radiative energy transfer to the upper boundary at T_2. Heat conduction is neglected.

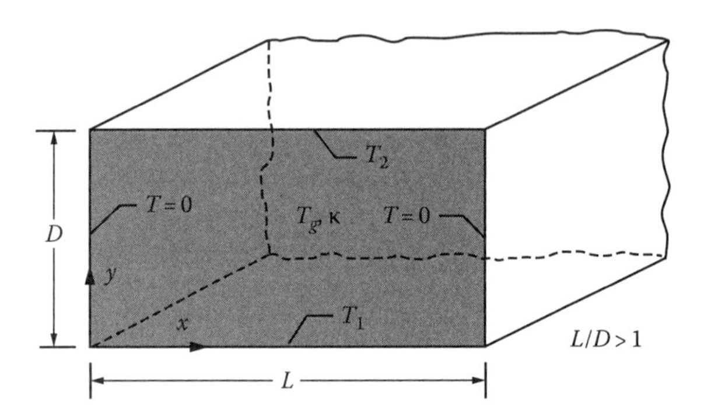

12.21 Derive the CDF for the angle of scatter for isotropic scattering and for Rayleigh scattering. (For Rayleigh scattering, the phase function is $\Phi(\theta) = (3/4)\,(1 + \cos^2\theta)$.)

12.22 Derive the CDF for the angle of scatter for a gas that scatters according to Henyey–Greenstein phase function given by $\Phi(\mu) = (1-g^2)/(1+g^2-2g\mu)^{3/2}$ where g is a constant and $\mu = \cos\theta$.

Answer: $R_\theta = \dfrac{1+g}{2g} + \dfrac{g^2-1}{2g\left(1+g^2-2g\cos\theta\right)^{1/2}}$.

12.23 Two infinite parallel black plates are spaced a distance D apart. The plate temperatures are T_1 and T_2, and the space between them is filled with a gray isotropically scattering medium. The medium does not absorb or emit radiation. Draw a Monte Carlo flowchart to obtain the energy transfer from plate 1 to plate 2 through the scattering medium. Heat conduction is neglected.

12.24 Generalize Homework Problem 12.23 to account for an anisotropic phase function that is independent of both incidence angle and circumferential scattering angle.

12.25 Two diffusely emitting and reflecting parallel plates are of finite width and infinite length normal to the cross section shown. The lower plate has uniform temperature $T_1 = 1000$ K, and the upper plate is at $T_2 = 500$ K. The plate emissivities are $\varepsilon_1 = 0.8$ and $\varepsilon_2 = 0.2$. The surroundings have a temperature of 450 K. An absorbing–emitting medium with absorption coefficient $\kappa = 0.5$ m^{-1} is between the plates. The medium is in radiative equilibrium with its surroundings (heat conduction is negligible).

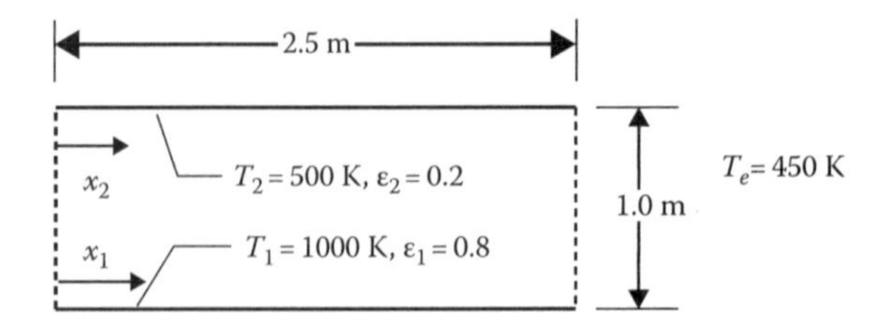

Using the Monte Carlo method, find the distribution of net radiative heat flux on each surface. Show the dependence of the results on the number of samples used in the Monte Carlo solution. Plot the results, and compare them with the results of Homework Problem 5.37.

12.26 The medium in the following enclosure is isotropically scattering and has attenuation coefficient $\beta = 4\ m^{-1}$. The boundary conditions on the enclosure are shown. The medium is in radiative equilibrium.

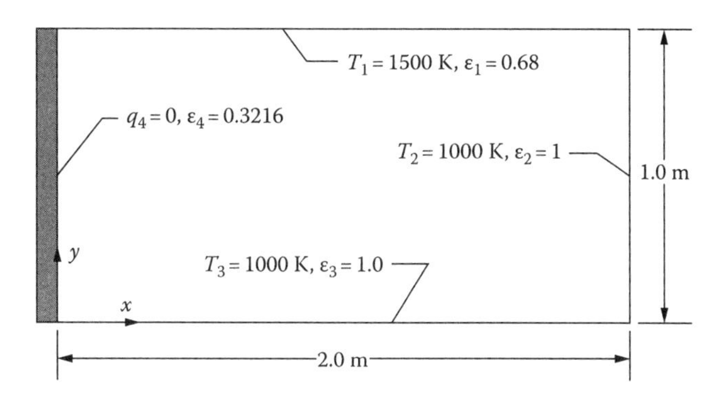

(a) Provide expressions for $E_b(x,y)$ and $q_3(x)$ using the P_1 method.
(b) Provide plots of $E_b(x,y = 0.25$ m$)$, $E_b(x,y = 0.50$ m$)$, and $E_b(x,y = 0.75$ m$)$ on the same graph.
(c) Provide a plot of $q_3(x)$ on surface 3.

Do the problem numerically or analytically.

12.27 Use the Monte Carlo method to determine the temperature distribution in the 2D enclosure shown below. The medium in the enclosure is nonscattering and has absorption coefficient $\kappa = 10\ m^{-1}$.

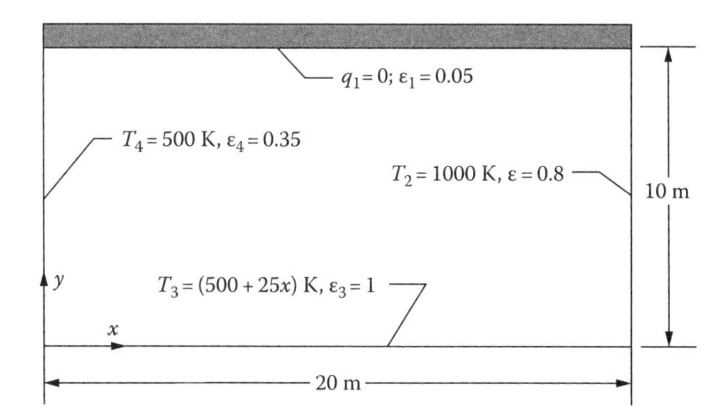

Provide plots of $q_3(x)$ on surface 3, $T_1(x)$ on surface 1, and $T_g(x = 10$ m$,y)$ and $T_g(x, y = 5$ m$)$. Report the number of samples used in the solution.

12.28 A common inverse application is in computational tomography. Typically, one is attempting to determine a distribution of a property or variable using a line-of-sight measurement. A common tomographic reconstruction problem uses a laser to probe the distribution of gases or smoke in a flame. A flame can be thought of as a cylindrical column with a distribution of temperature, gases, and solid particles. Assume that the temperature is constant. By probing the flame using a "laser extinction" measurement system, it is possible to find the distribution of radiatively absorbing particles.

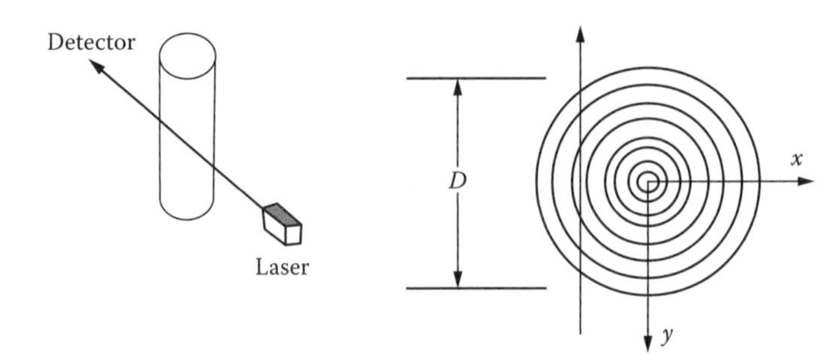

The two figures shown above present a schematic of a laser beam piercing a cylindrical object of diameter $D = 0.25$ cm for interrogation. There are eight subregions within the cylinder. The diameters of the subregions are $nD/8$ where n goes from 1 to 8. Laser extinction obeys the following rule $\frac{I}{I_0} = \exp\left(-\int_0^L \kappa \, dy\right)$ that for this discretized system can be written as $\frac{I}{I_0} = \exp\left(-\sum \kappa_i L_i\right)$. In this equation, κ_i is the absorption coefficient and is proportional to the concentration of absorbing matter in element i. For circles 1 to 8 the $\kappa_i = i^2$ which gives the distribution 1, 4, 9, 16, 25, 36, 49, 64, 88 [1/m].

(a) Calculate I/I_0 at the x values shown as follows.
(b) Set up the linear algebra equations to solve for κ_i.
(c) Invert the system of equations to find κ_i.
(d) Compare results with the distributions used to find I/I_0.
(e) Comment on the methods used to solve this problem.
(f) Comment on the extension of this approach to scattering tomography in Mengüç, and Dutta (1994).

Measurement 1: $x = 0, 0.4D$.
Measurement 2: $x = 0, 0.2D, 0.6D, 0.8D$.
Measurement 3: $x = 0, 0.1D, 0.2D, 0.3D, 0.4D, 0.5D, 0.6D, 0.7D$.
Measurement 4: add to measurement 3 the points $0.05D$, $0.15D$, $0.35D$, $0.45D$, $0.65D$, $0.75D$, $0.8D$, $0.9D$.

13 Conjugate Heat Transfer in Participating Media

13.1 INTRODUCTION

We introduced energy and radiative transfer equations (RTEs) in Chapter 10. Application of the RTE to two-dimensional (2D) and three-dimensional (3D) media are given in Chapter 11, and general RTE solution techniques are provided in Chapter 12.

The energy Equation 10.1 requires the term $-\nabla \cdot \mathbf{q}_r$, which is the energy source per unit volume supplied locally by radiation to each volume element dV. This term can be obtained by using Equation 10.33. The right side of this equation contains the average local intensity received from all directions. Without scattering, the incident intensities for evaluating this integral are obtained by solving the RTE, Equation 10.15, that gives the spectral intensity at each location and in each direction in terms of the intensity at $S = 0$ and the blackbody intensity along each path to dV. The intensity at $S = 0$ accounts for the radiative boundary conditions. Since the temperature is in the local blackbody spectral intensity, a simultaneous solution is required with the energy equation to obtain the temperature distribution. The energy equation has its own thermal boundary conditions that are separate from those used for the RTEs. When scattering is included, the source-function equation such as Equation 10.16 or 10.17 must be solved throughout the medium. The source function can be used in a relation such as Equation 10.39 to obtain the radiative heat source for simultaneous solution with the energy Equation 10.1.

The intensities and the source function throughout the medium depend on the temperature distribution. If the temperature distribution is given, the solution is greatly simplified as the radiative energy fluxes are the only quantities to be determined. The local blackbody spectral energies are known, and the spectral temperature-dependent properties can be specified. With scattering included, the source function must still be obtained from the solution of an integral equation that usually requires solution by iteration. The source function must be obtained at various wavelengths in the spectral regions where radiation is important, and then integration must be performed over all wavelengths to obtain the total energy quantity needed for the energy equation.

If the temperature distribution is unknown, the radiative transfer relations and the energy equation must be solved simultaneously. An iterative procedure is to estimate the temperature distribution and then solve the transfer relations. If scattering is present, this involves solving the integral equation for the source function. The radiative energy source is then evaluated and the energy equation is solved to obtain an improved temperature distribution with which to continue the iteration. Mazumder (2006) examines the possibility of substituting the nonlinear radiative terms directly into the energy equation and solving the resulting nonlinear integrodifferential equation directly rather than using the iterative procedure.

Substantial simplification results if the medium is nonscattering, as this eliminates solving the integral equation for the source function. Another substantial simplification is to assume that spectral and/or temperature-dependent property variations can be neglected. Neglecting spectral variations, however, may lead to inaccurate or unrealistic results.

Generally, the intensity must be obtained in each direction at each location and in each significant spectral interval. These results are obtained by solving the RTE, and extensive computations are usually required because of the dependence of intensity on S, θ, ϕ, and λ. Approaches to solving the RTE are summarized in Chapter 12. Next, we consider cases where the RTE and the energy equation are considered simultaneously.

13.2 RADIATION COMBINED WITH CONDUCTION

Sections 12.2.1.2.2, 12.2.2.2.1, and 12.2.2.3.4 consider boundary conditions for the limiting situation of radiation being dominant so that conduction and convection effects are neglected. In this limit, a temperature discontinuity ("jump") is obtained at a solid boundary. When conduction and/or convection are present, the temperature is continuous at the boundaries, although the gradient in temperature is generally not. For very small conduction and/or convection relative to radiation, the temperature gradients may be steep near a wall and the solution approaches the temperature jump condition. In this section, conduction is included with radiation; however, the addition of convection is in subsequent sections.

There are several cases where energy is transferred within a translucent medium by only radiation and conduction. These usually involve solid or highly viscous media, so convection in the medium is not important. Glass can absorb significant amounts of radiation in certain wavelength regions (see Figures 9.22 through 9.24). At elevated temperatures, there can be appreciable emission within glass. Glass is optically dense in the infrared region, and absorption and emission provide a radiative transport traveling through it layer by layer. In an optically dense material, molecular energy conduction is thus augmented by *radiative conduction*. Radiative effects are quite important for the temperature distribution within molten glass in a furnace, for the process of drawing optical fibers, and for the heat treatment of glass plates and annealing of silicon wafers. Similar effects are found in thermoplastics and thermosets. These effects are analyzed in Gardon (1956), Condon (1968), Eryou and Glicksman (1972), Anderson et al. (1973), Kuriyama et al. (1976), Wendlandt (1973), Lee and Jaluria (1996), Song et al. (1998), Golchert et al. (2002), Nakouzi (2012), Walker (2013), and Siedow et al. (2015). A detailed study of radiation and conduction interactions, including the effects of electrons during micro and nanomachining is provided by Wong and Mengüç (2008).

Another application involves radiative behavior within translucent coatings on surfaces. Glassy materials are sometimes used as sacrificial ablating coatings to protect the interior of a body from high external temperatures. The radiation–conduction process is important in regulating the temperature distribution within the ablating layer (Kadanoff 1961, Boles and Özişik 1972, Nelson 1973, Johnston and Gnoffo 2008, Bauman et al. 2011). The temperature distribution is influential in determining how the ablating material will soften, melt, or vaporize. These processes ultimately govern how efficient the coating is for protecting the surface. Nonablating ceramic coatings that may be translucent are also used for surface protection, such as zirconia coatings on turbine blades (Siegel 1996, 1977a), in diesel engines (Wahiduzzaman and Morel 1992), and in combustion chambers (Siegel 1997b). Another type of translucent coating is formed by cryodeposits of solidified gas on a very cold surface. The surface may be on a space vehicle orbiting at the upper fringe of the atmosphere or may be part of a cryopump used to produce high vacuum by condensing the gas within a chamber. The cryodeposit coating changes the radiative properties of the cold surface and can significantly influence radiative exchange with this surface (McConnell 1966, Merriam and Viskanta 1968, Gilpin et al. 1977).

Radiation can be a significant part of the energy transfer in fibrous insulation materials (Tong and Tien 1983, Mathes et al. 1990, Tong et al. 1983, Gorthala et al. 1994, Cunnington and Lee 1996, Petrov 1997), foam insulations (Glicksman et al. 1987), high-temperature porous insulating materials (Matthews et al. 1985), silica aerogels (Heinemann et al. 1996), gas-fluidized beds (Shafey et al. 1993), and radiation-induced curing of thermoset filament-wound composites (Chern et al. 2002a,b). For temperatures of 300–400 K, conduction in the air and radiation are the significant modes of transfer in lightweight fibrous insulation (Tong and Tien 1983). Radiation can also be significant in polyurethane foam insulation because the cell walls in the foam are partially transparent to infrared radiation.

The foam scatters radiation anisotropically, but scattering is more significant in fiberglass than in foam insulation. Low-density, high-temperature insulations such as hafnia, thoria, and zirconia are used for high-technology applications such as the space shuttle (Cunnington and Lee 1996), reentry vehicles, and solar central receivers. These materials are semitransparent to near-infrared and visible radiation, and radiative transfer in them can equal or exceed conduction in these applications. Solids such as glass have a refractive index significantly larger than 1, such as $n \approx 1.55$ for window glass.

Throughout Section 13.2, the theory is for materials with $n \approx 1$, such as a radiating gas in a chamber or a gas containing suspended particles in a furnace or a hot exhaust plume (Thynell 1992). Index of refraction effects are considered in Chapter 17 for glass windows, translucent ceramic coatings, thin films, and other applications.

13.2.1 Energy Balance

For combined radiation and conduction in an absorbing–emitting and scattering medium, the energy Equation 10.1 is used. This is solved subject to the boundary conditions to obtain the temperature distribution in the medium; heat flows can then be found. The convection, viscous dissipation, and volume expansion terms are omitted from Equation 10.1 to yield

$$\rho c \frac{\partial T}{\partial t} = \nabla \cdot (k\nabla T - \mathbf{q}_r) + \dot{q} \tag{13.1}$$

If the $\nabla \cdot \mathbf{q}_r$ is substituted from Equation 10.4, the local energy balance is

$$\rho c \frac{\partial T}{\partial t} = \nabla \cdot (k\nabla T) + \dot{q} - 4\pi \int_{\lambda=0}^{\infty} \kappa_\lambda(T) I_{\lambda b}(T)\, d\lambda + \int_{\lambda=0}^{\infty} \kappa_\lambda(T) \left[\int_{\Omega_i=0}^{4\pi} I_\lambda(\Omega_i)\, d\Omega_i \right] d\lambda \tag{13.2}$$

The radiation emission and absorption terms in Equation 13.2 depend on both the local temperature and on the surrounding radiation field.

13.2.2 Plane Layer with Conduction and Radiation

13.2.2.1 Absorbing–Emitting Medium without Scattering

A layer of translucent conducting–radiating medium is between parallel black walls at temperatures T_1 and T_2, as in Figure 11.3. The medium is gray and has a constant thermal conductivity k and a constant absorption coefficient κ. The steady energy transfer relations will be developed without scattering.

For 1D heat conduction and constant k, the $\nabla \cdot (k\nabla T)$ reduces to $k(d^2T/dx^2)$, and $\nabla \cdot \mathbf{q}_r$ becomes $dq_{r,x}/dx$. The temperature distribution is steady, $\partial T/\partial t = 0$, and there is no internal heat generation, $\dot{q} = 0$. Then, with $\tau = \kappa x$, Equation 13.1 reduces to

$$k\kappa \frac{d^2T}{d\tau^2} = \frac{dq_r(\tau)}{d\tau} \tag{13.3}$$

For a plane layer with diffuse–gray boundaries, $dq_r/d\tau$ is given by Equation 11.42. For black walls, the boundary fluxes are $J_1 = \sigma T_1^4/\pi$ and $J_2 = \sigma T_2^4/\pi$. For zero scattering in a gray medium, $\hat{I}(\tau) = \sigma T^4(\tau)/\pi$ from Equation 11.38. Then Equation 13.3 becomes, with $\tau_D = \kappa D$,

$$k\kappa \frac{d^2T}{d\tau^2} = -2\sigma T_1^4 E_2(\tau) - 2\sigma T_2^4 E_2(\tau_D - \tau) - 2\int_{\tau^*=0}^{\tau_D} \sigma T^4(\tau^*) E_1(|\tau - \tau^*|)\, d\tau^* + 4\sigma T^4(\tau) \tag{13.4}$$

The boundary conditions for T are $T(\tau) = 0 = T_1$ and $T(\tau = \kappa D = \tau_D) = T_2$. Define the dimensionless quantities $\vartheta = T/T_1$, $\vartheta_2 = T_2/T_1$, and $N_{CR} = k\kappa/4\sigma T_1^3$ to give

$$N_{CR}\frac{d^2\vartheta(\tau)}{d\tau^2} = \vartheta^4(\tau) - \frac{1}{2}\left[E_2(\tau) + \vartheta_2^4 E_2(\tau_D - \tau) + 2\int_{\tau^*=0}^{\tau_D}\vartheta^4(\tau^*)E_1(|\tau - \tau^*|)d\tau^*\right] \tag{13.5}$$

This is the energy equation for $\vartheta(\tau)$; the boundary conditions are $\vartheta(0) = 0$ and $\vartheta(\tau_D) = \vartheta_2$. The solution depends on the parameters N_{CR}, τ_D, and ϑ_2.

The $N_{CR} \equiv k\kappa/4\sigma T_j^3$ is the *conduction–radiation parameter* (or *Stark number*) for a nonscattering medium based on the jth temperature. The N_{CR} does *not* directly give the relative values of conduction to emission because the ratio of these values depends on both temperature difference and absolute temperature level. When scattering is included, the N_{CR} includes the scattering coefficient to become $N_{CR} \equiv k(\kappa + \sigma_s)/4\sigma T_j^3$.

The combined radiation and conduction energy transfer across the translucent layer can be obtained from the temperature distribution. From energy conservation, q is independent of location τ for the conditions considered where $\dot{q} = 0$, so the evaluation for q can be done at any τ location. Equation 11.69 with $\Gamma = \sigma T^4$ gives the net radiative heat flux in terms of the temperature distribution across a gray gas between black walls. This radiative flux relation was obtained for convenience at $\tau = 0$. In addition, at the same location, there is now a conduction flux $-k(dT/dx)|_{x=0} = -k\kappa(dT/d\tau)|_{\tau=0}$, so the heat flux relation becomes, in terms of the unknown temperature distribution,

$$q = -k\kappa\left.\frac{dT}{d\tau}\right|_{\tau=0} + \sigma T_1^4 - 2\sigma T_2^4 E_3(\tau_D) - 2\int_0^{\tau_D}\sigma T^4(\tau^*)E_2(\tau^*)d\tau^* \tag{13.6}$$

On the right, the first term is the conduction away from wall 1 by the medium, the second is the radiation leaving black wall 1, the third is the radiation leaving wall 2 that is then attenuated by the translucent medium and reaches wall 1, and the last term is the radiation from the medium to wall 1. In dimensionless form,

$$\frac{q}{\sigma T_1^4} = -4N_{CR}\left.\frac{d\vartheta}{d\tau}\right|_{\tau=0} + 1 - 2\left[\vartheta_2^4 E_3(\tau_D) + \int_{\tau^*=0}^{\tau_D}\vartheta^4(\tau^*)E_2(\tau^*)d\tau^*\right] \tag{13.7}$$

Viskanta and Grosh (1962b,c) obtained solutions for $\vartheta(\tau)$ from Equation 13.5 by numerical integration and iteration, and temperature distributions are in Figure 13.1. For $N_{CR} \to \infty$, conduction dominates, and the solution reduces to the linear profile for conduction through a plane layer. For $N_{CR} = 0$, there is no conduction, and there is a temperature discontinuity (jump) at each wall as discussed in the previous section for radiation alone. Heat flux results from Viskanta and Grosh (1962c) are in Table 13.1 as obtained from Equation 13.7. Reevaluations of the results by various numerical methods, such as in Nice (1983) and Burns et al. (1995b), confirmed these values within small variations. Results for cylindrical and spherical geometries are in Thynell (1990a,b, 1992), Greif and Clapper (1966), Men (1973), Viskanta and Merriam (1968), and Tuntomo and Tien (1992). In Habib (1973), the effect of radiation is studied on solidification of an annular region bounded by black surfaces. The results are for a constant absorption coefficient and no scattering. Solutions for a sphere with conduction and radiation are in Thynell (1990a) and Tuntomo and Tien (1992). Transient conditions are included in Tuntomo and Tien. Zhao and Liu (2007) implement a spectral element method for combined conduction/radiation and apply it to four test problems in 3D rectangular geometries.

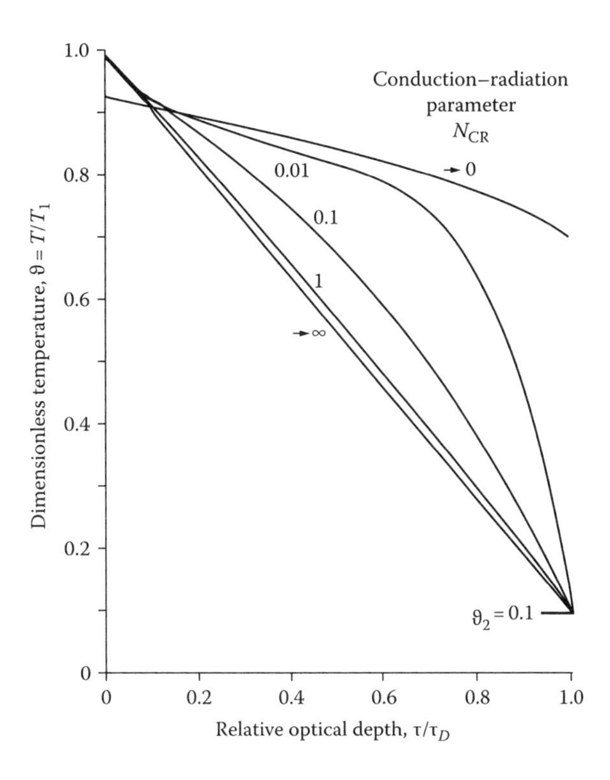

FIGURE 13.1 Dimensionless temperature distribution in gray gas between infinite parallel black plates with conduction and radiation. Plate temperature ratio $\vartheta_2 = 0.1$; optical spacing $\tau_D = 1.0$. (From Viskanta, R. and Grosh, R.J., *J. Heat Trans.*, 84(1), 63, 1962b.)

Example 13.1

Add heat conduction with a constant thermal conductivity to the Milne–Eddington example (Example 12.3) and show how the temperature distribution and heat flows can be determined.

With the addition of heat conduction, the energy Equation 12.52 becomes (see Equation 13.1) $-k(d^2T/dx^2)+dq_r/dx=\dot{q}$. This is integrated with respect to x, and then Equation 12.44 is used to substitute $d\bar{I}/dx$ for q_r. The equation is then integrated again and the result, with two constants of integration, is

$$-kT(x)-\frac{4\pi}{3\beta}\bar{I}(x)=\dot{q}\frac{x^2}{2}+C_3x+C_4 \tag{13.1.1}$$

This is evaluated at $x = 0$ and $x = D$ using the specified boundary temperatures T_1 and T_2, which gives the two constants of integration as

$$C_3=\frac{k}{D}(T_1-T_2)+\frac{4\pi}{3D\beta}\left[\bar{I}(0)-\bar{I}(D)\right]-\dot{q}\frac{D}{2} \tag{13.1.2}$$

$$C_4=-kT_1-\frac{4\pi}{3\beta}\bar{I}(0) \tag{13.1.3}$$

To evaluate the temperature distribution from Equation 13.1.1, the $\bar{I}(\tau)$ must be determined. This requires an iterative numerical procedure, and problems of this type have been solved in Siegel and Spuckler (1994a) and Siegel (1996, 1997a). The $\bar{I}(\tau)$ is found by solving the Milne–Eddington

TABLE 13.1
Heat Flux between Parallel Black Plates by Combined Radiation and Conduction through a Gray Medium

Optical Thickness, τ	Plate-Temperature Ratio, ϑ_2	Conduction–Radiation Parameter, N_{CR}	Dimensionless Energy Flux, $q/\sigma T_1^4$
0.1	0.5	0	0.859
		0.01	1.074
		0.1	2.880
		1	20.88
		10	200.88
1.0	0.5	0	0.518
		0.01	0.596
		0.1	0.798
		1	2.600
		10	20.60
1.0	0.1	0	0.556
		0.01	0.658
		0.1	0.991
		1	4.218
		10	36.60
10	0.5	0	0.102
		0.01	0.114
		0.1	0.131
		1	0.315
		10	2.114

Source: Enoch, I.E. et al., *Num. Heat Trans.*, 5, 353, 1982.

equation (Equation 12.50) numerically. Since this equation has the temperature distribution in $E_b(x)$ on the right side, a $T(x)$ distribution is assumed to start the iterative procedure. A starting estimate can be obtained from the opaque solution (heat conduction only) for a layer with uniform internal heat generation, constant k, and specified boundary temperatures, which is

$$T(x) = T_1 - \frac{x}{D}(T_1 - T_2) + \frac{\dot{q}}{2\beta}x(D - x) \tag{13.1.4}$$

To start the integration of Equation 12.50, an $\bar{I}(0)$ is assumed and $q_r(0)$ is obtained from the boundary condition Equation 12.56; this gives $d\bar{I}/dx$ at $x = 0$ by use of Equation 12.44. The integration can then be done, and after integrating to $x = D$, the second boundary condition in Equation 12.57 is checked by using $d\bar{I}/dx$ at $x = D$ in Equation 12.44 to find $q_r(D)$. The $\bar{I}(0)$ is adjusted to try to satisfy the boundary conditions at both boundaries, which is possible since the $T(x)$ being used is only a guess toward the final solution. The $\bar{I}(x)$ is then used in Equations 13.1.1 and 13.1.2 to determine the next approximation to the temperature distribution. During the iteration, it is usually necessary to use damping factors between successive changes in the functions to keep the iterative process stable. After convergence, the local radiative flux $q_r(x)$ can be found from Equation 12.44 and the heat conduction from $-k\, dT/dx$, as a function of x.

13.2.2.2 Absorbing–Emitting Medium with Scattering

Scattering is now added to the absorbing and emitting plane layer with heat conduction. The medium is gray and scattering is isotropic. Scattering is conveniently included by using the radiative source function. For isotropic scattering the phase function $\Phi(\lambda, \Omega) = 1$, so for gray properties Equation 10.50 applies, and with $\tau = (\kappa + \sigma_s)x$ the energy Equation 13.3 becomes

$$k\beta\frac{d^2T}{d\tau^2} = 4\frac{(1-\omega)}{\omega}\left[\sigma T^4(\tau) - \pi\hat{I}(\tau)\right] \tag{13.8}$$

The $\pi\hat{I}(\tau)$ is found by using the integral Equation 11.43 that gives, for a plane layer with black boundaries at uniform temperatures T_1 and T_2,

$$\pi\hat{I}(\tau) = (1-\omega)\sigma T^4(\tau) + \frac{\omega}{2}\left[\sigma T_1^4 E_2(\tau) + \sigma T_2^4 E_2(\tau_D - \tau) + \pi\int_{\tau^*=0}^{\tau_D} \hat{I}(\tau^*)E_1\left(\left|\tau^* - \tau\right|\right)d\tau^*\right] \tag{13.9}$$

Equations 13.8 and 13.9 are placed in dimensionless form by using the same quantities as in Equation 13.5 and by using $\pi\hat{I}/\sigma T_1^4$ as a dimensionless source function. The $N_{CR} = k(a + \sigma_s)/4\sigma T_1^3$ includes the scattering coefficient, and the scattering albedo ω is an additional parameter. The boundary conditions are the same as given after Equation 13.4. The solution is obtained by numerical integration and iteration, and results are in Viskanta (1965).

Figure 13.2 shows the effect of scattering on some temperature profiles. For $\omega = 0$, the results correspond to those in Figure 13.1. For a fixed τ_D, increased scattering, and hence reduced absorption, tends to make the profiles linear as for heat conduction only. When $\omega \rightarrow 1$, there is no absorption or emission by the medium, and hence there is no mechanism that can convert radiant energy into internal energy of the medium. The temperature then has the linear distribution for conduction alone. Energy is then transferred independently by conduction and by radiant scattering through the medium. The effect of partially reflecting walls (ε_1 and/or $\varepsilon_2 < 1$) is not given here, but is included in Viskanta (1965) and Menart and Lee (1993). Typical heat fluxes are in Table 13.2, from Enoch et al. (1982), for a layer between diffuse–gray walls with isotropic scattering in an absorbing and emitting medium; the results demonstrate the influence of the parameters. In Yuen and Wong (1980), the effect was investigated of having strong forward or backward scattering.

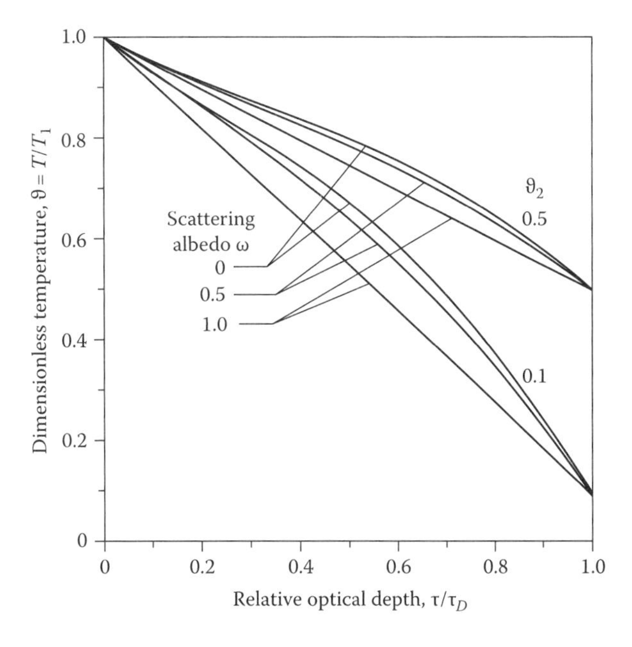

FIGURE 13.2 Effect of scattering albedo on temperature distribution in translucent gray medium between infinite parallel black plates. Conduction–radiation parameter $N_{CR} = 0.1$, optical spacing $\tau_D = 1.0$, and plate temperature ratios $\vartheta_2 = 0.1$ and 0.5. (From Viskanta, R., *J. Heat Trans.*, 87(1), 143, 1965.)

TABLE 13.2
Dimensionless Heat Fluxes across a Translucent Plane Layer by Combined Radiation and Conduction with Isotropic Scattering, $T_1/T_2 = 2$

$\tau_D = \beta D$	ε_1	ε_2	N_{CR}	ω	$(q_r + q_c)/\sigma T_1^4$
1.0	0.1	0.1	0.1	0	0.461
1.0	0.1	0.1	0.1	0.5	0.350
1.0	0.1	0.1	0.1	1.0	0.248
1.0	0.5	0.5	0.1	0	0.571
1.0	0.5	0.5	0.1	0.5	0.513
1.0	0.5	0.5	0.1	1.0	0.449
1.0	1	0	1.0	0.5	2.303
1.0	0	1	1.0	0.5	2.210
1.0	0.1	0	1.0	0.5	2.143
1.0	0	0.1	1.0	0.5	2.132

Source: Enoch, I.E. et al., *Num. Heat Trans.*, 5, 353, 1982.

Some experimental results are given by Schimmel et al. (1970) for layers of CO_2, N_2O, and mixtures CO_2–CH_4 and CO_2–N_2O. The conduction heat fluxes at the walls were compared with analytical results using gray gas, uniform absorption coefficient within each effective spectral bandwidth, and wideband gas property models. The wideband model of Edwards and Menard (see Appendix A at the book website, http://www.thermalradiation.net) gave the best results. An analysis is in Smith et al. (1987) for a nongray medium of CO_2, H_2O vapor, and soot between gray–diffuse walls with different emissivities.

Example 13.2

A purely scattering medium between diffuse–gray walls is nonabsorbing, $\kappa = 0$, $\sigma_s > 0$, and conducts heat with a constant thermal conductivity k. The wall temperatures are T_1 and T_2, and the spacing between the walls is D. Determine the energy transfer from wall 1 to wall 2.

For pure scattering, the temperature distribution within the layer does not enter into the radiative solution as given by Equations 11.82 and 11.83, and $\nabla \cdot \mathbf{q}_r = 0$ as shown by Equation 10.40. Hence, in the absence of absorption, the energy equation that determines the heat conduction is independent of the scattering process. The energy transfer is then found by adding the heat conduction to Equation 11.82 as if scattering were not present:

$$q = \frac{k(T_1 - T_2)}{D} + \frac{\sigma(T_1^4 - T_2^4)\psi_b}{1 + \psi_b(1/\varepsilon_1 + 1/\varepsilon_2 - 2)} \tag{13.2.1}$$

The ψ_b is in Figure 11.4b or Table 11.2, where $\tau_D = \int_0^D \sigma_s(x)dx$.

Example 13.3

For the geometry in Example 13.2, determine the energy transfer from wall 1 to wall 2 by heat conduction combined with absorption and scattering in the optically thin limit.

From Figure 11.4b, as $\tau_D \to 0$, $\psi_b \to 1$, so Equation 11.82 reduces to $q = \sigma(T_1^4 - T_2^4)/(1/\varepsilon_1 + 1/\varepsilon_2 - 1)$, which is the same as the result without a radiating medium between the walls. Equation 11.75 shows that, in the optically thin limit, the radiative flux does not produce a temperature gradient in the medium and hence does not influence heat conduction. Ordinary heat conduction can thus be added to the radiation to give the combined flux in the optically thin limit:

$$q = \frac{k(T_1 - T_2)}{D} + \frac{\sigma(T_1^4 - T_2^4)}{1/\varepsilon_1 + 1/\varepsilon_2 - 1} \tag{13.3.1}$$

13.2.3 Rectangular Region with Conduction and Radiation

The steady-state energy transfer in a rectangular region by radiation with heat conduction included has been analyzed by Razzaque et al. (1984a) and Yuen and Takara (1988) for a gray absorbing–emitting medium without scattering. The solutions are 2D and depend only on the x and y in Figure 11.7. Internal heat generation and gray boundaries are included in Razzaque et al., while in Yuen and Takara, the boundaries are black. From Equation 13.1, the steady-state energy equation for constant properties is

$$k\nabla^2 T - \nabla\cdot\mathbf{q}_r + \dot{q} = 0 \tag{13.10}$$

and for a gray medium, Equation 11.90 gives $\nabla\cdot\mathbf{q}_r = 4\kappa\sigma T^4 - 4\pi\kappa\bar{I}$. For a 2D rectangular region, $\bar{I}(x,y)$ is given by Equation 11.99. If, for simplicity, an enclosure with black walls is considered, the energy equation has the form, since $I_b = \sigma T_b^4/4/\pi$,

$$\begin{aligned} -\frac{k}{\kappa}\left(\frac{\partial^2 T}{\partial x^2} + \frac{\partial^2 T}{\partial y^2}\right) + 4\sigma T^4 = \frac{\dot{q}}{\kappa} &+ (b-y)\int_{x_0=0}^{d} \sigma T_b^4(x_0,b)\frac{S_2[\kappa\rho_0(x_0,b)]}{\rho_0^2(x_0,b)}dx_0 \\ &+ (d-x)\int_{y_0=0}^{b} \sigma T_b^4(d,y_0)\frac{S_2[\kappa\rho_0(d,y_0)]}{\rho_0^2(d,y_0)}dy_0 \\ &+ y\int_{x_0=0}^{d} \sigma T_b^4(x_0,0)\frac{S_2[\kappa\rho_0(x_0,0)]}{\rho_0^2(x_0,0)}dx_0 \\ &+ x\int_{y_0=0}^{b} \sigma T_b^4(0,y_0)\frac{S_2[\kappa\rho_0(0,y_0)]}{\rho_0^2(0,y_0)}dy_0 \\ &+ \kappa\int_{x^*=0}^{d}\int_{y^*=0}^{b} \sigma T^4(x^*,y^*)\frac{S_1(\kappa\rho^*)}{\rho^*}dx^*\,dy^* \end{aligned} \tag{13.11}$$

where $\rho^* = [(x-x^*)^2 + (y-y^*)^2]^{1/2}$, and $\rho_0(x_0,y_0) = [(x-x_0)^2 + (y-y_0)^2]^{1/2}$. The S_n functions are defined in Equation 11.97.

Consider a specific situation from Yuen and Takara (1988). The $\dot{q}=0$ in the radiating medium, and three walls of the enclosure are at the same temperature T_2 with the remaining wall at a different temperature T_1 as in Figure 13.3. With $\dot{q}=0$, Equation 13.11 is written for all of the walls specified at T_2 and is then subtracted from Equation 13.11 to yield, after letting $\vartheta = T/T_1$ and $\vartheta_2 = T_2/T_1$,

$$\begin{aligned} -\frac{k}{\kappa\sigma T_1^3}\left[\frac{\partial^2\vartheta}{\partial x^2} + \frac{\partial^2\vartheta}{\partial y^2}\right] + 4(\vartheta^4 - \vartheta_2^4) = y\int_{x_0=0}^{d}(1-\vartheta_2^4)\frac{S_2[\kappa\rho_0(x_0,0)]}{\rho_0^2(x_0,0)}dx_0 \\ + a\int_{x^*=0}^{d}\int_{y^*=0}^{b}\left(\vartheta^4 - \vartheta_2^4\right)\frac{S_1(\kappa\rho^*)}{\rho^*}dx^*\,dy^* \end{aligned} \tag{13.12}$$

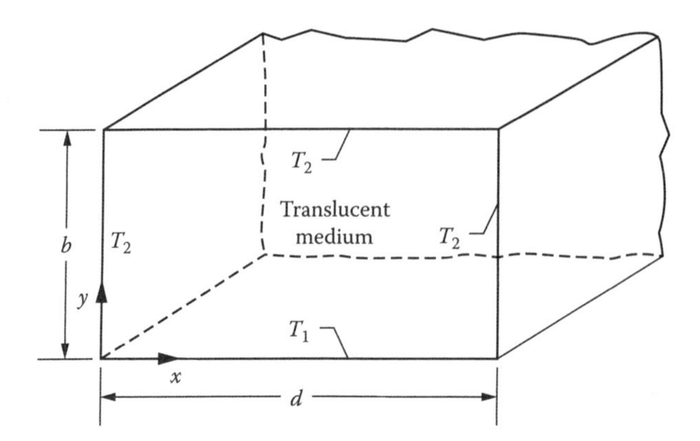

FIGURE 13.3 Two-dimensional (2D) rectangular region of a translucent medium with black interior boundaries.

This was further nondimensionalized in terms of optical dimensions in Yuen and Takara and solved with a numerical technique utilizing a Taylor series expansion of the temperature about each local temperature value. A transient solution is in Mitra et al. (1997) and an inverse problem is considered in Li (1993). Numerical methods are provided in Sections 13.5.

13.2.4 P_N Method for Radiation Combined with Conduction

The P_N method provides an expression for the local radiative source that is *differential* in form and can be incorporated into the energy equation in differential form that includes convection and/or conduction. The P_N method thus can fit into whatever grid size is used for numerically solving the energy equation. The procedure for a combined radiation and conduction solution is in the following example.

Example 13.4

A plane layer of radiating and isotropically scattering medium with constant thermal conductivity k, constant radiative properties, optical thickness τ_D, and albedo ω is between infinite parallel diffuse–gray walls of emissivity ε_w at temperatures T_{w1} and T_{w2}. The medium has a uniform internal heat source $\dot{q}$. Derive the relations needed to obtain the heat flux to each bounding wall and the temperature distribution in the medium using the P_1 approximation.

With conduction, radiation, and internal heat generation, the energy Equation 13.1 is $\nabla \cdot (k\nabla T - \mathbf{q}_r) + \dot{q} = 0$ or, using a summation form for the radiation and conduction fluxes,

$$\sum_{i=1}^{3} \frac{\partial q_{r,i}}{\partial x_i} + \sum_{i=1}^{3} \frac{\partial q_{c,i}}{\partial x_i} = \dot{q} \tag{13.4.1}$$

For the present 1D problem, since $q_r = I^{(1)}$, where $I^{(1)}$ is the first moment of the intensity, the energy equation can be put in the dimensionless form

$$\frac{d\tilde{I}^{(1)}}{d\tau_1} = 4N_{CR}\frac{d^2\vartheta}{d\tau_1^2} + \frac{\dot{S}}{\tau_D} \tag{13.4.2}$$

where

$$N_{CR} = \frac{k(\kappa + \sigma_s)}{4\sigma T_{w1}^3}$$

$$\dot{S} = \frac{\dot{q}D}{\sigma T_{w1}^4}$$

$$\tau_1 = (\kappa + \sigma_s)x_1$$

$$\tau_D = (\kappa + \sigma_s)D$$

$$\tilde{I}^{(1)} = I^{(1)}/\sigma T_{w1}^4$$

$$\vartheta = T/T_{w1}$$

This is the defining equation for the derivative of $\tilde{I}^{(1)}$. In the pure-radiation solution of Example 12.4 without internal heat sources, the derivative of the first moment in Equation 13.4.2 was set equal to zero, because for these conditions $\tilde{I}^{(1)} = q_r/\sigma T_{w1}^4$ is constant and $d\tilde{I}^{(1)}/d\tau_1 = 0$. This is not the case here with conduction and an internal heat source included. The presence of a second derivative of dimensionless temperature requires two boundary conditions for solving the energy equation: $\vartheta(\tau_1 = 0) = 1$ and $\vartheta(\tau_1 = \tau_D) = T_{w2}/T_{w1}$.

To proceed with the solution, two coupled second-order differential equations are derived for $\tilde{I}^{(0)}$ and $\tilde{I}^{(1)}$ (the nondimensional intensity integrated over all solid angles and the radiative flux). The first is obtained by equating Equation 13.4.1 and the first moment differential Equation 12.4.2 in the P_N method. For the present 1D problem, this yields

$$\frac{4N_{CR}}{1-\omega}\frac{d^2\vartheta}{d\tau_1^2} + \frac{\dot{S}}{\tau_D(1-\omega)} - 4\vartheta^4 = -\tilde{I}^{(0)} \tag{13.4.3}$$

where $\tilde{I}^{(0)} = I^{(0)}/\sigma T_{w1}^4$. The second equation is found by substituting the closure equation (Equation 12.72) into the second-moment differential equation (Equation 12.68) to obtain the relation between $\tilde{I}^{(0)}$ and $\tilde{I}^{(1)}$:

$$\frac{d\tilde{I}^{(11)}}{d\tau_1} = \frac{1}{3}\frac{d\tilde{I}^{(0)}}{d\tau_1} = -\tilde{I}^{(1)} \tag{13.4.4}$$

Now, Equation 13.4.4 is differentiated with respect to τ_1, and the result is substituted into Equation 13.4.2 to yield

$$\frac{d^2\tilde{I}^{(0)}}{d\tau_1^2} + 12N_{CR}\frac{d^2\vartheta}{d\tau_1^2} + \frac{3\dot{S}}{\tau_D} = 0 \tag{13.4.5}$$

Equations 13.4.3 and 13.4.5 can be solved simultaneously for $\tilde{I}^{(0)}$ and $\tilde{t}$. They can be combined into a single fourth-order equation in $\tilde{t}$ by differentiating Equation 13.3.3 twice with respect to τ_1 and the result is substituted into Equation 13.4.5 to eliminate the second derivative of $\tilde{I}^{(0)}$. The resulting fourth-order equation, or the two second-order equations, requires two boundary conditions in addition to the known boundary surface temperatures. These are generated from Equations 12.74 and 12.75 using Equation 12.4.9 to eliminate J, which results in

$$\begin{aligned}\frac{1}{4}\tilde{I}^{(0)}(\tau_1 = 0) &= 1 - \left(\frac{1}{\varepsilon_w} - \frac{1}{2}\right)\tilde{I}^{(1)} \quad (\tau_1 = 0)\\ \frac{1}{4}\tilde{I}^{(0)}(\tau_1 = \tau_D) &= \vartheta_{w2}^4 + \left(\frac{1}{\varepsilon_w} - \frac{1}{2}\right)\tilde{I}^{(1)} \quad (\tau_1 = \tau_D)\end{aligned} \tag{13.4.6}$$

or

$$\frac{1}{4}\tilde{I}_i^{(0)} = \vartheta_{wi}^4 \pm \left(\frac{1}{\varepsilon_w} - \frac{1}{2}\right)\tilde{I}_i^{(1)} \tag{13.4.7}$$

where the i subscript denotes walls 1 or 2 and the positive sign applies at wall i = 2. Inserting Equation 13.4.4 to eliminate $\tilde{I}^{(1)}$ results in the final boundary relations for $\tilde{I}^{(0)}$:

$$\pm\left(\frac{d\tilde{I}^{(0)}}{d\tau_1}\right)_i = -\frac{3}{4\left(\dfrac{1}{\varepsilon_w} - \dfrac{1}{2}\right)}\left[\tilde{I}_i^{(0)} - 4\vartheta_{wi}^4\right] \quad (13.4.8)$$

These can be directly applied as the boundary conditions for Equation 13.4.5. The problem is completely specified with two second-order differential Equations 13.4.3 and 13.4.5 and the four boundary conditions, the specified boundary temperatures for Equation 13.4.3, and the conditions (i = 1, 2) in Equation 13.4.8 for Equation 13.4.5. An iterative numerical solution can be used to obtain $\vartheta(\tau_1)$ and $\tilde{I}_i^{(0)}(\tau_1)$. Then $\tilde{I}_i^{(1)}$ at the boundaries (i = 1, 2) is found from Equations 13.4.6, which gives the desired radiative fluxes at the boundaries.

The total energy transfer also requires the amount of heat conduction. Since the temperature distribution has been determined, this can be found by evaluating $-kdT/dx$ at the boundaries.

Example 13.4 was solved by Ratzel and Howell (1982) as a limiting case for a P_3 formulation. The two coupled differential equations were solved by the collocation method (Ascher et al. 1979) for sets of mixed-order boundary value problems in ordinary differential equations, which automatically refines the grid size until a specified tolerance is met in solution accuracy. A modified successive overrelaxation method was used that applies to nonlinear equations. It was found that the particular commercial solver that accommodates equations up to fourth order and should solve the single P_1 equation would not converge for large N_{CR} to a specified accuracy on ϑ of $\approx 10^{-3}$. When the relations were expressed as a set of two second-order equations rather than a single fourth-order equation, the solution was obtained satisfactorily. For the P_3 solution of the same problem, the P_3 equations were used as six coupled nonlinear equations, and convergence was obtained.

Figure 13.4 compares the P_1 and P_3 results (Ratzel and Howell 1982) with a numerical solution (Larsen and Howell 1985) for the centerline temperature profile (with the conduction/radiation number N_{CR} as a parameter) in a square enclosure with one hot surface and three surfaces at one-half the absolute temperature of the hot side. The enclosure contained a gray medium without internal heat generation (i.e., parameter $\dot{S} = 0$). Figure 13.5 shows the P_3 results for the temperature distribution in the entire medium for the same conditions with $N_{CR} = 0.01$.

A significant increase in accuracy is reported for the P_3 compared with the P_1 approximation in Marshak (1947), Mengüç and Viskanta (1983, 1985, 1986), and Mengüç (1985); heat conduction is not included. Although for P_3 there is an increase in the complexity of the solutions, the equations remain algebraic and of closed form for parallel flat plates. For concentric cylinders and spheres, when the P_3 method is used, a numerical solution of the resulting fourth-order linear ordinary differential equations is necessary. Stone and Gaustad (1961) give a formulation for nongray gases for the astrophysical boundary condition of zero incident flux at one boundary. For radiative transfer in a medium within a cylinder of finite length bounded by gray walls at known temperature, a finite-element solution took up to 10 times more computer time than a finite-difference solution for the same accuracy. The results indicate that the P_3 approximation is sufficiently accurate for most applications. Further, 3D rectangular geometries and 2D axisymmetric geometries of finite length with anisotropic scattering were analyzed by Mengüç (1985) and Mengüç and Viskanta (1985, 1986a). Recently, Modest et al. (2014) and Ge et al. (2015) extended the P_N solutions to 2D axisymmetric and 3D cases.

The P_1 approximation was used by Park and Kim (1993) to analyze the effects of combined radiation and conduction. The P_1 approximation was also used by Hartung and Hassan (1993) to analyze radiation effects in a high-temperature region formed in front of a blunt body during reentry of a space vehicle into a planetary atmosphere. Usually, a 1D approximation is made for the layer in

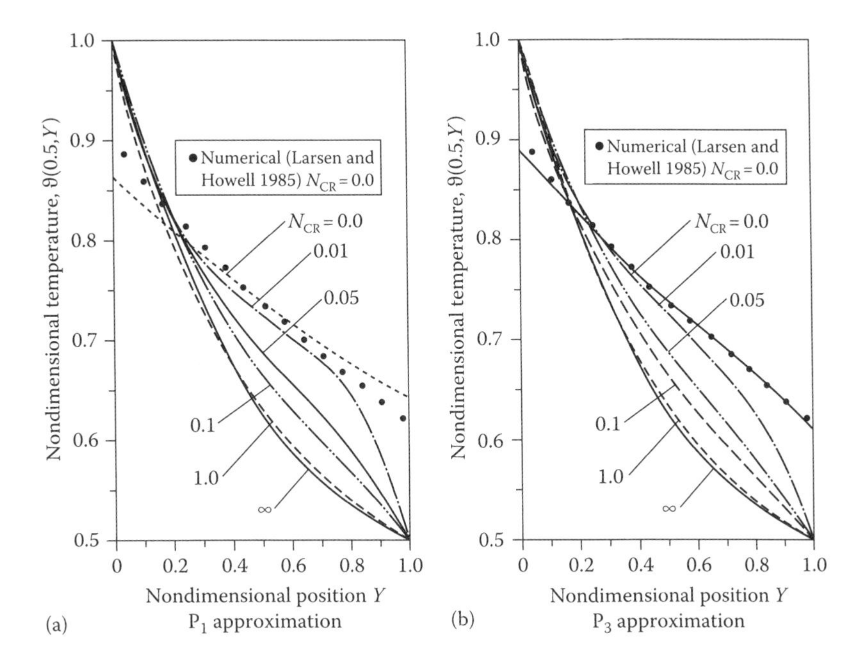

FIGURE 13.4 Comparison of P_1 and P_3 approximation results for the nondimensional centerline temperature distribution in a square enclosure of a gray medium for various values of the conduction–radiation parameter N_{CR}: optical length of side $\tau_D = 1.0$, $\vartheta_{w1} = 1.0$, $\vartheta_{wi} = 0.5$ ($i = 2, 3, 4$), $\varepsilon_{wi} = 1.0$ ($i = 1$–4), $\dot{S} = 0$. (a) P_1 approximation and (b) P_3 approximation. (From Ratzel, A. and Howell, J.R., Two-dimensional energy transfer in radiatively participating media with conduction by the P–N approximation, *Proceedings of 1982 International Heat Transfer Conference*, vol. 2, Munich, Germany, September, 1982, pp. 535–540.)

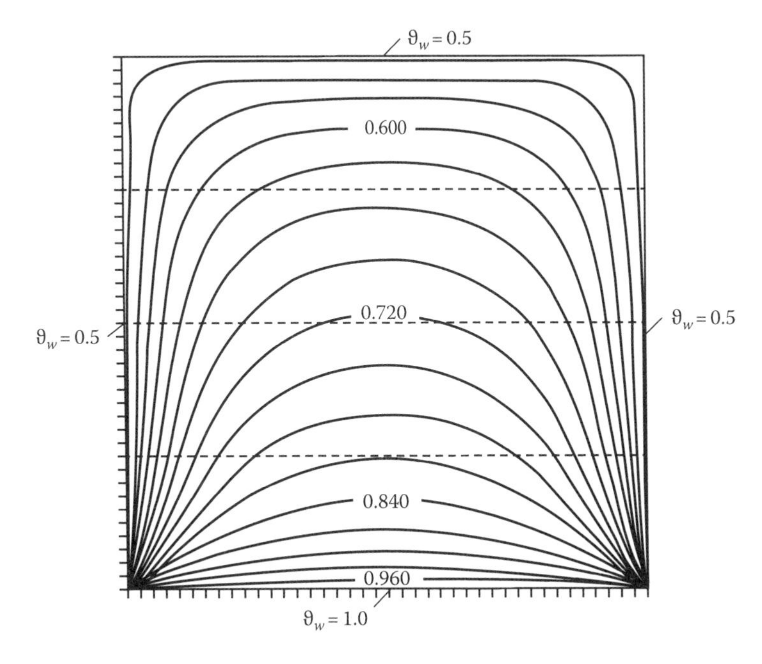

FIGURE 13.5 P_3 approximation results for the nondimensional temperature distribution in a square enclosure of a gray medium for $N_{CR} = 0.01$: optical length of side $\tau_D = 1.0$, $\vartheta_{w1} = 1.0$, $\vartheta_{wi} = 0.5$ ($i = 2, 3, 4$), $\varepsilon_{wi} = 1.0$ ($i = 1$–4), $\dot{S} = 0$. (From Ratzel, A. and Howell, J.R., *Proceedings of 1982 International Heat Transfer Conference*, vol. 2, pp. 535–540, Munich, Germany, September, 1982.)

TABLE 13.3
Three-Dimensional Zonal Analysis Results for a Cube, Combined Radiation, and Conduction[a]

		Dimensionless Emissive Power, $[T_{i,j,k}/T_w(\kappa X = 1)]^4$				Dimensionless Heat
k	j	$I = 1$	$I = 3$	$I = 5$	Zone	Flux, $q_w/\sigma T_w^4(\kappa X = 1)$
5	5	0.069	0.181	0.479	1	0.663
5	3	0.181	0.431	0.679	2	0.959
5	1	0.479	0.679	0.779	3	1.165
3	5	0.181	0.431	0.679	4	0.298
3	3	0.431	0.807	0.917	5	0.589
3	1	0.679	0.917	0.960	6	0.158
1	5	0.479	0.679	0.779		
1	3	0.679	0.917	0.960		
1	1	0.779	0.960	0.980		

Source: Larsen, M.E. and Howell, J.R., *J. Heat Trans.*, 107(4), 936, 1985.
[a] Surface conditions are as in Table 12.7; $N_{CR} = 0.01$, optical side length = 1.

front of a blunt body and behind the shock wave that is formed. This has been found inaccurate for some applications, so a 3D analysis was developed. The P_1 approximation was used in Ezekoye and Zhang (1997) to analyze combined radiation and conduction in a diffusion flame. Effects of soot radiation and gas chemistry are included.

Larsen and Howell (1985) calculated temperature distributions and surface heat fluxes using a modified zonal formulation in 2D and 3D enclosures with gray and black bounding walls containing an absorbing, emitting, and isotropically scattering medium with heat conduction. Results were obtained for various values of the conduction–radiation parameter N_{CR}. The exchange-factor method was used with exchange-factor smoothing techniques developed to ensure energy conservation (Larsen and Howell 1986). The resulting set of nonlinear algebraic equations was solved for temperatures within the medium. The medium temperature distribution was then used to compute radiative fluxes at the walls, and this was added to the conductive flux computed from the temperature gradient at the boundary to obtain the total boundary energy flux (Table 13.3).

13.2.5 Approximations for Combined Radiation and Conduction

13.2.5.1 Addition of Energy Transfer by Radiation and Conduction

Before the diffusion theory for combined radiation and conduction is considered, a relatively simple idea is discussed for obtaining energy transfer by the combination of these two transfer processes. This approximation is to assume that the interaction between the two transfer processes is so weak that, for computing energy transfer, each process can be considered to act independently; the conduction and radiation energy transfers are each formulated as if the other mechanism were absent. Einstein (1963a) and Cess (1964) investigated this approximation for an absorbing–emitting gray medium between infinite parallel walls. When the walls are black, the energy transfer is within 10% of the exact solution. Exact results are approached in the optically thin and thick limits. Larger errors are possible if highly reflecting surfaces are present. Yuen and Wong (1980) showed that, for an absorbing–emitting medium with scattering, the presence of low-emissivity boundaries increases the error of the additive approximation. In Yuen and Takara (1988), the additive solution was examined for a square enclosure with black walls filled with a gray absorbing–emitting medium. Three of the enclosure boundaries are at a single temperature, and the remaining boundary is at a different temperature. The additive solution gave good results for energy transfer. Howell (1965) showed that the additive solution is fairly accurate for a gray medium between black concentric cylinders.

An additive solution cannot be used to predict temperature profiles. It is a simple approximation for estimating energy transfer by combined modes, although the accuracy obtained becomes doubtful for many situations. The use of the additive method is not advised for situations where the accuracy for energy transfer has not been established by some comparisons with more exact solutions.

Example 13.5

By using the additive approximation, obtain a relation for the energy transfer from a gray infinite wall at T_1 with emissivity ε_1 to a parallel infinite gray wall at T_2 with emissivity ε_2. The spacing between the walls is D, and the region between the walls is filled with gray medium having a constant absorption coefficient κ, isotropic scattering coefficient σ_s, and thermal conductivity k. Use the diffusion approximation for the radiative transfer.

The energy flux by only conduction from surface 1 to surface 2 is $q_c = k(T_1 - T_2)/D$. The diffusion solution for only radiation from 1 to 2 is (Table 12.1) $q_r = \sigma(T_1^4 - T_1^4)/(3\beta D/4 + 1/\varepsilon_1 + 1/\varepsilon_2 - 1)$. Since the two energy transfers are assumed independent, the additive solution gives $q = q_c + q_r$. Using the dimensionless quantities $N_{CR} = k\beta/4\sigma T_1^3, \vartheta = T/T_1$, and $\tau_D = (\kappa + \sigma_s)D = \beta D$ gives the combined flux as

$$\frac{q}{\sigma T_1^4} = \frac{4N_{CR}(1-\vartheta_2)}{\tau_D} + \frac{1-\vartheta_2^4}{3\tau_D/4 + 1/\varepsilon_1 + 1/\varepsilon_2 - 1} \tag{13.5.1}$$

Equation 13.5.1 is correct for $N_{CR} = 0$ (radiation only) within the accuracy of the diffusion solution, and for $N_{CR} \to \infty$ (conduction only), because the solution adds these two limiting results. Examples 13.2 and 13.3 show that superposition provides the exact solution for the limits of a medium with only scattering or an optically thin medium.

A comparison of $q/\sigma T_1^4$ from Equation 13.5.1 with exact numerical solutions for $\varepsilon_1 = \varepsilon_2 = 1$, $\vartheta_2 = 0.5$, and $\sigma_s = 0$ from Viskanta and Grosh (1962a,b) is in Figure 13.6. For this geometry and black surfaces, the additive results are very accurate. The accuracy of the additive method is improved here because the diffusion solution gives a pure-radiation heat transfer that is slightly above the exact pure-radiation solution (Figure 13.6), whereas the pure-conduction result is too low. This is because the conduction solution is based on the linear gradient of T, whereas the actual gradients at the boundaries are larger when radiation is present (Figure 13.1). The errors in the two solutions compensate in this geometry. Nelson (1975) shows that superposition provides good results for a nongray gas with a single absorption band.

13.2.5.2 Diffusion Method for Combined Radiation and Conduction

This approximate method is a substantial improvement over the additive approximation since the energy equation is solved with the coupled energy transfers by conduction and radiation; radiative diffusion is included simultaneously with diffusion by heat conduction. This is expected to give improved results for combined energy transfer and also provides the temperature distribution that cannot be calculated with the additive approximation. Spectral variations in properties can also be included, as shown by Siegel and Spuckler (1994a). As shown in the derivation in Chapter 12, the diffusion heat flux relation for radiative transfer has the same form as the Fourier conduction law. By using the Rosseland mean attenuation coefficient defined in Equation 12.27, the radiative

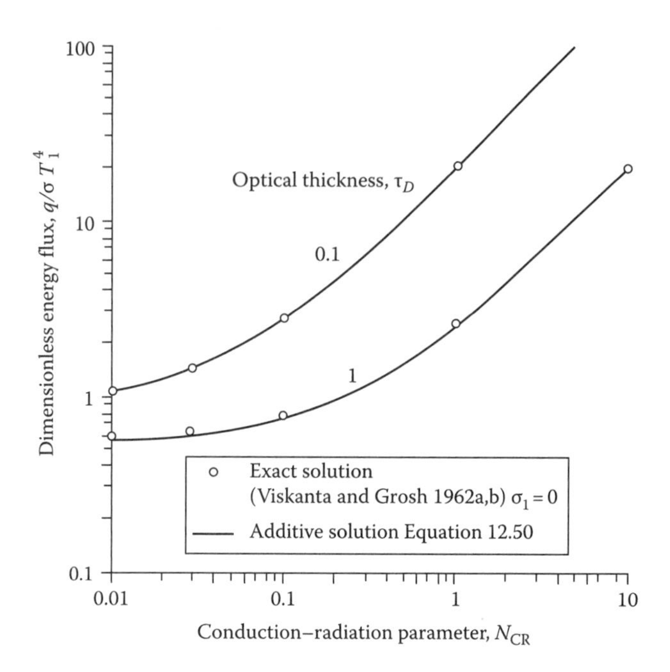

FIGURE 13.6 Comparison of simple additive and exact numerical solutions of combined conduction–radiation energy transfer between black parallel plates. Plate temperature ratio $T_2/T_1 = 0.5$.

flux vector for an absorbing, emitting, and isotropically scattering medium can be written from Equation 12.28 as

$$\mathbf{q}_r = -\frac{4}{3\beta_R}\nabla E_b = -\frac{16\sigma T^3}{3\beta_R}\nabla T \tag{13.13}$$

where β_R can be a function of position. Then, the local energy flux vector by combined radiation and conduction is

$$\mathbf{q} = \mathbf{q}_r + \mathbf{q}_c = -\left(\frac{16\sigma T^3}{3\beta_R} + k\right)\nabla T \tag{13.14}$$

This can be used in energy Equation 13.1. For example, in 2D rectangular coordinates, with internal heat sources, the transient energy equation is

$$\rho c_P \frac{\partial T}{\partial t} = \frac{\partial}{\partial x}\left[\left(\frac{16\sigma T^3}{3\beta_R} + k\right)\frac{\partial T}{\partial x}\right] + \frac{\partial}{\partial y}\left[\left(\frac{16\sigma T^3}{3\beta_R} + k\right)\frac{\partial T}{\partial y}\right] + \dot{q}(x, y) \tag{13.15}$$

The energy transfer is similar to heat conduction, with a thermal conductivity that depends on temperature.

To obtain the temperature distribution in the medium, an equation such as Equation 13.15 is solved subject to the initial and boundary conditions. The boundary conditions would often be specified temperatures of the enclosure surfaces. However, as discussed earlier, near a boundary the diffusion approximation may not be accurate as the radiation is not isotropic. If the solution is in error near the wall it cannot be matched directly to the boundary conditions. To overcome this difficulty, the boundary condition at the edge of the absorbing–emitting medium is modified so that the resulting solution to the diffusion equation using this effective boundary condition will be correct in the region away from the boundaries where the diffusion approximation is valid.

For pure radiation, a temperature jump was introduced to join the diffusion solution in the medium to the wall temperature. For combined conduction–radiation, a similar concept was introduced by Goldstein and Howell (1968) and Howell and Goldstein (1969). By using asymptotic expansions to match linearized solutions for intensity, flux, and temperature near the wall with the diffusion solution for these quantities far from the wall, an effective jump condition was derived. As shown in Figure 13.7, the jump gives the boundary condition $T(x \to 0)$ that the diffusion solution must have if the diffusion solution is to extend to the wall. The jump is given in terms of the jump coefficient ψ, which is a function only of the conduction–radiation parameter $N_{CR} = k\beta/4\sigma T_1^3$. In terms of quantities at wall 1, ψ_1 is given by

$$\psi_1 = \frac{\sigma\left[T_1^4 - T^4(x \to 0)\right]}{q_{r,1}} \tag{13.16}$$

where

$q_{r,1}$ is the radiative flux at the boundary as evaluated by the diffusion approximation
T_1 is the wall temperature
$T(x \to 0)$ is the extrapolated temperature *in the medium* at the wall, which is the effective jump temperature to be used in the diffusion solution

The ψ_1 is computed from the relations of Goldstein and Howell as

$$\psi_1 = \frac{3}{4\pi}\int_0^1 \tan^{-1}\frac{1}{\psi(\gamma)}d\gamma \quad \text{where } \psi(\gamma) = \frac{1}{\pi}\left(\frac{N_{CR}}{2\gamma^3} - \frac{2}{\gamma} - \ln\frac{1-\gamma}{1+\gamma}\right) \tag{13.17}$$

For large N_{CR}, the jump effect can be neglected, as heat conduction dominates over radiation effects near the wall.

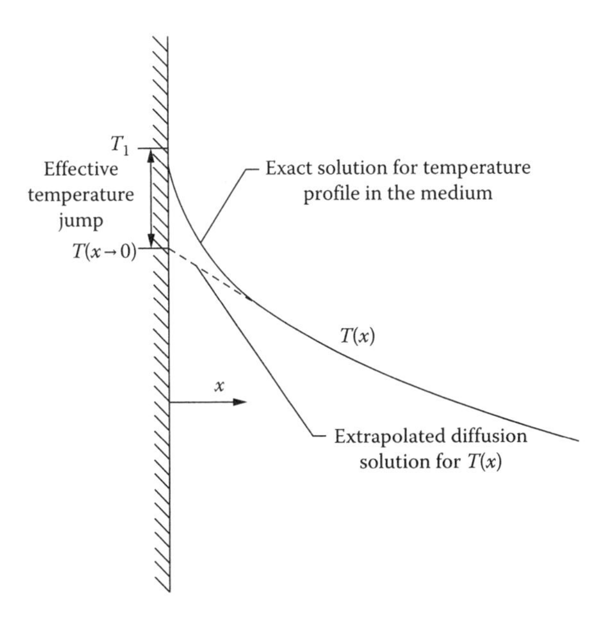

FIGURE 13.7 Use of effective temperature jump as boundary condition for diffusion solution in combined conduction and radiation.

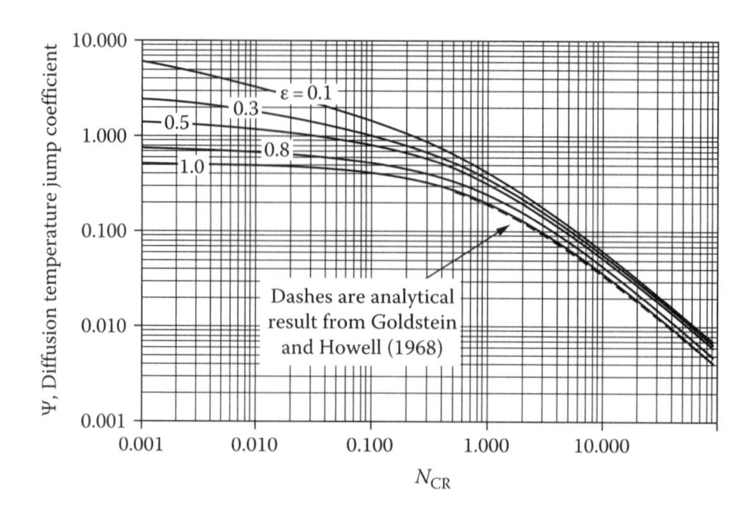

FIGURE 13.8 Temperature jump coefficient for combined conduction–radiation solutions by the diffusion method. (From Goldstein, M.E. and Howell, J.R., Boundary conditions for the diffusion solution of coupled conduction-radiation problems, NASA TN D-4618, 1968 (for $\varepsilon = 1$); Larsen, M.E., Use of contact resistance algorithm to implement jump boundary conditions for the radiation diffusion approximation, *Proceedings of HT2005: 2005 ASME Summer Heat Trans Conference,* Paper HT2005–72561, San Francisco, CA, July, 2005 [for $\varepsilon \neq 1$].)

Larsen (2005) extended the conduction/radiation slip condition to boundary conditions for gray opaque surfaces by numerically solving a range of conduction/radiation cases using the zone method (Section 12.5) and determining the slip condition that results. The numerical predictions for $\varepsilon = 1$ agree well with the analytical solution of Equation 13.17. The resulting slip coefficient Ψ versus N_1 from the analytical solution of Goldstein and Howell for black walls and for gray boundary emissivity from Larsen is shown in Figure 13.8.

With the diffusion approximation, results for combined radiation and conduction can be obtained for both energy transfer and temperature profiles, as in the following example. Other solutions of this general type are in Wang and Tien (1967) and Taitel and Hartnett (1968).

Example 13.6

Using the diffusion method, find the steady temperature profile in a medium with constant attenuation coefficient β and thermal conductivity k, with $\dot{q} = 0$, and contained between infinite parallel black walls at T_1 and T_2 spaced D apart with the lower wall 1 at $x = 0$. What is the heat transfer across the layer?

For this geometry, Equation 13.14 becomes, in dimensionless form,

$$\frac{q}{\sigma T_1^4} = -\left(\frac{4}{3}\frac{d\vartheta^4}{d\tau} + 4N_{CR}\frac{d\vartheta}{d\tau}\right) \tag{13.6.1}$$

From energy conservation, with no internal heat sources, the q is constant across the layer between the walls. Equation 13.6.1 is then integrated from 0 to τ_D to yield

$$\frac{q}{\sigma T_1^4}\tau_D = -\left\{\frac{4}{3}\left[\vartheta^4(\tau_D) - \vartheta^4(0)\right] + 4N_{CR}\left[\vartheta(\tau_D) - \vartheta(0)\right]\right\} \tag{13.6.2}$$

where $\vartheta(0)$ and $\vartheta(\tau_D)$ are *in the medium* at the boundaries. These two temperatures are eliminated by using the jump boundary conditions to relate them to the specified wall temperatures T_1 and T_2.

At wall 1, for the particular N_1 of the problem, the ψ_1 is found from Figure 13.8 and set equal to $\psi_1 = \sigma[T_1^4 - T^4(0)]/q_{r,1}$. From Equations 13.13 and 13.14, the radiative flux $q_{r,1}$ at the wall is

$$q_{r,1} = -\frac{16\sigma T_1^3}{3\beta_R}\frac{dT}{dx}\bigg|_1 = \frac{16\sigma T_1^3}{3\beta_R}\frac{q}{\left(16\sigma T_1^3/3\beta_R\right)+k} = \frac{4q}{3\beta_R\left[\left(4\sigma/3\beta_R\right)+\left(k/4T_1^3\right)\right]}$$

Then, ψ_1 from Equation 13.16 becomes

$$\psi_1 = \frac{\sigma\left[T_1^4 - T^4(0)\right]}{\left(4\sigma/3\beta_R\right)q\Big/\left[\left(4\sigma/3\beta_R\right)+\left(k/4T_1^3\right)\right]}$$

This is rearranged into

$$\frac{4}{3\beta_R}q_1 = \frac{1}{\psi_1}\left\{\frac{4\sigma}{3\beta_R}\left[T_1^4 - T^4(0)\right] + \frac{k}{4T_1^3}\left[T_1^4 - T^4(0)\right]\right\} \tag{13.6.3}$$

As shown in the derivation of ψ_1, the conditions for which the diffusion solution is valid lead to the jump $T_1 - T(0)$ being small. For convenience, a portion of Equation 13.6.3 can then be linearized. With $T_1 - T(0) = \delta$, where δ is small,

$$\frac{T_1^4 - T^4(0)}{4T_1^3} = \frac{T_1^4 - (T_1-\delta)^4}{4T_1^3} \approx \frac{T_1^4 - T_1^4 + 4T_1^3\delta}{4T_1^3} = \delta = T_1 - T(0)$$

Then, Equation 13.6.3 becomes $\frac{4}{3\beta_R}q_1 \approx \frac{1}{\psi_1}\left\{\frac{4\sigma}{3\beta_R}\left[T_1^4 - T^4(0)\right] + k\left[T_1 - T(0)\right]\right\}$, or, in dimensionless form,

$$\frac{4}{3}\psi_1\frac{q_1}{\sigma T_1^4} = \frac{4}{3}\left[1-\vartheta^4(0)\right] + 4N_{CR}\left[1-\vartheta(0)\right] \tag{13.6.4}$$

Similarly, at wall 2 (note that ψ_2 is found by using the value of $N_{CR,2} = k\beta_R/4\sigma T_2^3$ as the abscissa in Figure 13.8),

$$\frac{1}{3}\psi_2\frac{q_1}{\sigma T_1^4} = \frac{1}{3}\left[\vartheta^4(\tau_D) - \vartheta_2^4\right] + N_{CR}\left[\vartheta(\tau_D) - \vartheta_2\right] \tag{13.6.5}$$

Now add Equations 13.6.2, 13.6.4, and 13.6.5 to eliminate the unknown temperatures in the medium $\vartheta(0)$ and $\vartheta(\tau_D)$. This yields the energy flux transferred across the layer as

$$\frac{q_1}{\sigma T_1^4} = \frac{\left[1-\vartheta_2^4\right] + 3N_{CR}\left[1-\vartheta_2\right]}{3\tau_D/4 + \psi_1 + \psi_2} \tag{13.6.6}$$

The results of Equation 13.6.6 are plotted in Figure 13.9 and compared with the exact and additive solutions. For $\tau_D = 1$, the results compare very well with the exact solution, and Equation 13.6.6 will provide good results for larger τ_D. For $\tau_D = 0.1$, a simple additive solution provides better energy transfer values.

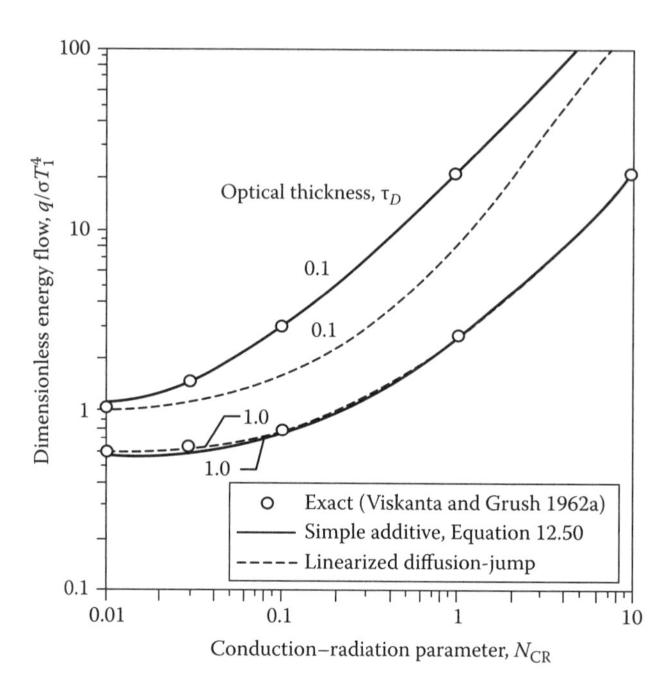

FIGURE 13.9 Comparison of various methods for predicting energy transfer by conduction and radiation across a layer between parallel black walls. Wall temperature ratio $T_2/T_1 = 0.5$.

An advantage of the combined-mode diffusion solution is that it yields the temperature distribution in the medium. Temperature profiles are predicted by integrating Equation 13.6.1 from 0 to τ (note that q is constant) and then using Equations 13.6.5 and 13.6.6 to eliminate $\vartheta(0)$ and q. This yields

$$\frac{1-\vartheta^4(\tau)+3N_{CR}\left[1-\vartheta_2(\tau)\right]}{1-\vartheta_2^4+3N_{CR}(1-\vartheta_2)}=\frac{3\tau/4+\psi_1}{3\tau_D/4+\psi_1+\psi_2} \tag{13.18}$$

Temperature profiles are in Figure 13.10. For $\tau_D = 1$ (Figure 13.10a), the profiles are poor except for the largest N_{CR} shown. Better results are obtained for all N_{CR} for $\tau_D = 10$ as illustrated in Figure 13.10b, because the assumptions in the diffusion solution have greater validity at larger τ_D. For $N_{CR} \to 0$ and $N_{CR} \to \infty$, the diffusion–jump method goes to the correct limiting solutions. The diffusion method provides accurate temperature distributions when the layer is optically thick and there is sufficient heat conduction to minimize temperature jump effects at the boundaries.

Within their limits of applicability, diffusion methods provide a useful interpretation of the conduction–radiation parameter. The ratio of molecular conductivity to radiative conductivity is $k/(16\sigma T^3/3\beta_R) = (3/4)(k\beta_R/4\sigma T^3) = (3/4)N_{CR}$. Therefore, in the diffusion limit, N_{CR} is a direct measure of the conductivity ratio and consequently in this limit is also a direct measure of the ratio of the energy transferred by conduction and radiation.

The diffusion method allows solving of difficult problems by standard analytical and numerical techniques and can be used when the assumptions in its derivation are justified. The most stringent assumption is that of optically thick conditions. Because most gases have band spectra, the optically thick regions arise within band limits. If the absorption mean free path is small, the assumption that only local conditions affect the spectral radiant flux is quite good. At other wavelengths, and in the absence of significant scattering such as by suspended particles, the gas is usually much more transmitting, and diffusion methods are not justified. Care must be taken to apply the diffusion equation only in geometric and spectral regions where the optically thick assumption is valid.

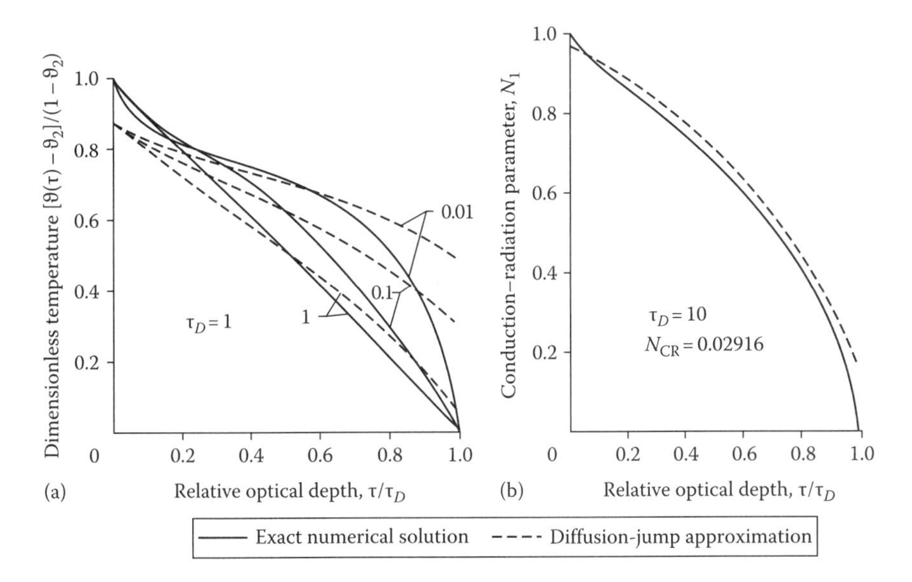

FIGURE 13.10 Comparison of temperature profile by exact solution with diffusion-jump approximation. Wall temperature ratio $T_2/T_1 = 0.5$; wall emissivities $\varepsilon_1 = \varepsilon_2 = 1.0$. (a) Optical thickness $\tau_D = 1$; (b) optical thickness $\tau_D = 10$, conduction-radiation parameter $N_{CR} = 0.02916$. (From Viskanta, R. and Grosh, R.J., *Int. J. Heat Mass Trans*., 5, 729, 1962a.)

The combination of optically thick spectral bands using the diffusion approximation, with bands that are not thick, is in the analysis by Siegel and Spuckler (1994a). The Rosseland mean attenuation coefficient for the entire range of λ should not be used as the criterion for optical thickness. It may have a large value, but the spectral attenuation coefficient may be small in certain spectral regions that allow significant radiant transmission. The use of Rosseland mean coefficient in such cases may lead to large errors. An approach for the optically thick regions is to define the wavelength bands in which the spectral attenuation coefficient is everywhere large and evaluate a Rosseland mean for each of these spectral regions. Howell and Perlmutter (1964a) applied the diffusion method to a real gas and compared the results to an exact solution by the Monte Carlo method. The agreement was not as good as when comparing results for gray gases.

Bobco (1967) used a modified diffusion solution to find the directional emissivity of a semi-infinite layer of isothermal gray scattering–absorbing medium with isotropic scattering. The directional emissivities were found to differ considerably from a diffuse distribution. By comparison with other methods, the diffusion approximation was found in Petrov (1997) to be the preferable approach for studying transient combined radiation and conduction in high-temperature fibrous thermal insulation such as for the space shuttle reentering the Earth's atmosphere.

13.3 TRANSIENT SOLUTIONS INCLUDING CONDUCTION

A few examples (Examples 11.3 and 11.4) have already been given for transient solutions in the limit where radiation was dominant and heat conduction could be neglected. In a transient, there are two types of internal energy storage per unit volume and time: One is the local variation with time of the radiant energy density and the second is because of the ordinary heat capacity of the material as encountered in conventional heat conduction and convection solutions. From Equation 10.30, the first of these is

$$\frac{\partial}{\partial t}\int_0^\infty U_\lambda d\lambda = \frac{\partial}{\partial t}\left(\frac{1}{c}\int_0^\infty\int_0^{4\pi} I_\lambda d\Omega d\lambda\right)$$

The second is $\rho c_v \partial T/\partial t$. Because of the large value of the electromagnetic propagation speed c in a medium, the storage of radiant energy is usually neglected. However, some analyses dealing with the effects of nuclear weapons, high-energy lasers, and some situations in astrophysics require consideration of transient variations in the radiant energy density. Because of the high propagation speeds for electromagnetic energy, the radiation field can adjust very rapidly to temperature changes. Generally, the transient temperature change of a medium would be governed by the heat capacity of the medium, and consequently, transient temperature changes would be much slower than the radiation relaxation time. Hence, when used with the transient energy conservation equation containing the heat capacity term, the unsteady radiation term in the equation of transfer is usually negligible. This is why the steady form of the equation of transfer, as derived in Chapter 10, can be instantaneously applied during almost all transient heat transfer processes involving radiation. In Kumar and Mitra (1995) and Longtin and Tien (1997), analyses are given that include the transient radiative energy density term. A few examples of transient analyses are now considered for combined radiation and conduction. Additional references are in Chapter 17, where translucent materials with $n > 1$ are treated, and in Chapter 16.

Example 13.7

A plane layer with thickness from $x = 0$ to D is initially at $T_i(x)$ and is in vacuum. The layer is gray and is composed of a translucent porous solid material that emits, absorbs, and scatters radiation. Scattering is isotropic, and the thermal conductivity of the layer is a constant, k. The absorption and scattering coefficients, κ and σ_s, are uniform within the layer. The surfaces of the porous layer are assumed nonreflecting. For time $t > 0$, an internal heat generation $\dot{q}(x, t)$ is applied. The surrounding environment for $x < 0$ is at $T_{e,1}$, and for $x > D$, it is at $T_{e,2}$. Provide the energy relations to solve for the transient temperature distribution within the layer, $T(x, t)$.

From Equation 10.1, the energy equation is

$$\rho c \frac{\partial T}{\partial t} = k \frac{\partial^2 T}{\partial x^2} - \frac{\partial q_r}{\partial x} + \dot{q}(x,t)$$

The $\partial q_r/\partial x$ is found from Equation 10.50 for isotropic scattering and, for a gray material,

$$\frac{\partial q_r}{\partial x} = 4\pi \frac{\kappa}{\omega}\left[\frac{\sigma T^4}{\pi}(x,t) - \hat{I}(x,t)\right]$$

so the energy equation becomes

$$\rho c \frac{\partial T}{\partial t} = k \frac{\partial^2 T}{\partial x^2} - 4\frac{\kappa}{\omega}[\sigma T^4(x,t) - \pi \hat{I}(x,t)] + \dot{q}(x,t)$$

Let $T_{i,0}$ be an arbitrary reference temperature such as $T_i(0, 0)$ and $\vartheta = T/T_{i,0}$, $\tau = (\kappa + \sigma_s)x$, and $\tau_D = (\kappa + \sigma_s)D$. The energy equation in dimensionless form is then

$$\frac{\partial \vartheta}{\partial \tilde{t}} = 4N_{CR} \frac{\partial^2 \vartheta}{\partial \tau^2} - \frac{4(1-\omega)}{\omega}[\vartheta^4(\tau,\tilde{t}) - \tilde{I}(\tau,\tilde{t})] + \tilde{q}(\tau,\tilde{t}) \tag{13.7.1}$$

where

$$\tilde{t} = \frac{(\kappa+\sigma_s)\sigma T_{i,0}^3}{\rho c} t, \quad N_{CR} = \frac{k(\kappa+\sigma_s)}{4\sigma T_{i,0}^3}, \quad \tilde{I} = \frac{\pi \hat{I}}{\sigma T_{i,0}^4}, \quad \tilde{q} = \frac{\dot{q}}{(\kappa+\sigma_s)\sigma T_{i,0}^4}$$

The equation for $\tilde{I}$ is obtained from Equation 11.22, where for nonreflecting boundaries $I^+(0, \mu) = \sigma T_{e,1}^4/\pi$ and $I^-(\tau_D, -\mu) = \sigma T_{e,2}^4/\pi$, with the surroundings assumed to act as black radiation sources; this gives

$$\tilde{I}(\tau,\tilde{t}) = (1-\omega)\vartheta^4(\tau,\tilde{t}) + \frac{\omega}{2}\left[\vartheta_{e,1}^4(\tilde{t})E_2(\tau) + \vartheta_{e,2}^4(\tilde{t})E_2(\tau_D - \tau) + \int_0^{\tau_D} \tilde{I}(\tau^*,\tilde{t})E_1(|\tau^* - \tau|)d\tau^*\right] \quad (13.7.2)$$

Equations 13.7.1 and 13.7.2 are solved numerically subject to the boundary conditions $\partial\vartheta(0,\tilde{t})/\partial\tau = 0, \partial\vartheta(\tau_D,\tilde{t})/\partial\tau = 0$ (no heat conduction into surrounding vacuum), and the specified initial condition $\vartheta(\tau,0) = T_i(x)/T_{i,0}$. Numerical solution methods are in Section 13.5.

Example 13.8

Develop an analysis for the onset of transient heating by a unidirectional flux q_i incident at angle θ on a semi-infinite gray solid, as in Figure 13.11a. The absorption coefficient of the solid is κ and there is no scattering. Include partial reflection at the layer surface.

A fraction $\hat{t}_0(\theta)$ of the incident energy is transmitted through the interface, and within the material, the energy is refracted into direction χ (an optically smooth interface has been assumed). The fraction of entering radiation that reaches depth x is $e^{-\kappa x/\cos\chi}$, and the amount $(\kappa/\cos\chi)\,e^{-\kappa x/\cos\chi}$ is absorbed

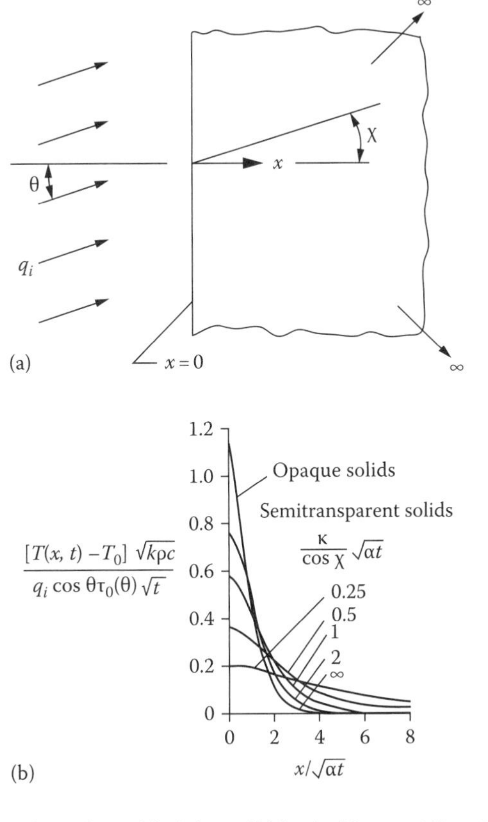

FIGURE 13.11 Transient heating of semi-infinite solid by incident unidirectional radiative flux, (a) flux incident on semi-infinite slab and (b) dimensionless temperature distribution.

at x per unit volume. The solid is assumed to be in its initial state of heating so that temperatures are low, and emission within the solid and heat loss from its surface can be neglected. The energy equation, initial condition, and boundary conditions are then

$$\rho c \frac{\partial T}{\partial t} = k \frac{\partial^2 T}{\partial x^2} + q_i \cos\theta\, \hat{t}_0(\theta) \frac{\kappa}{\cos\chi} e^{-\kappa x/\cos\chi} \tag{13.8.1}$$

$$T(x,0) = T_0 \text{ (initial condition)}$$

$$\frac{\partial T}{\partial x}(0,t) = 0; \quad \lim_{x\to\infty} T(x,t) = T_0$$

The solution is in Carslaw and Jaeger (1959) and can be placed in the form

$$\frac{[T(x,t) - T_0]\sqrt{k\rho c}}{q_i \cos\theta\, \hat{t}_0(\theta)\sqrt{t}} = \frac{2}{\sqrt{\pi}} e^{-X^2} - 2X \operatorname{erfc} X - \frac{1}{A} e^{-2AX} + \frac{1}{2A}\left[e^{A(A-2X)} \operatorname{erfc}(A - X) + e^{A(A+2X)} \operatorname{erfc}(A + X)\right] \tag{13.8.2}$$

where

$$X = x/2\sqrt{\alpha t}$$

$$A = (\kappa/\cos\chi)\sqrt{\alpha t} \quad \text{and} \quad \alpha = k/\rho c \text{ (note that } t = \text{time, } \hat{t}_0 = \text{surface transmissivity)}$$

Some results are in Figure 13.11b. Since the parameter on the curves depends on the absorption coefficient κ, an increase of κ increases the surface temperature reached at any time. A small κ increases the penetration of the temperature distribution into the solid. As $(\kappa/\cos\chi)\sqrt{\sigma t} \to \infty$, the temperature distribution approaches that for an opaque solid.

Spectral methods (Section 12.7.3) have been used for solution of radiation–conduction problems in Tan et al. (2006) and for 3D transient radiation/conduction problems in Sun et al. (2012).

13.4 COMBINED RADIATION, CONDUCTION, AND CONVECTION IN A BOUNDARY LAYER

The laminar boundary layer including radiative transfer has been examined using the optically thin and thick approximations and is a good illustration to present here (Figure 13.12). The flowing medium absorbs and emits radiation, but scattering is not included; the wall at T_1 is black.

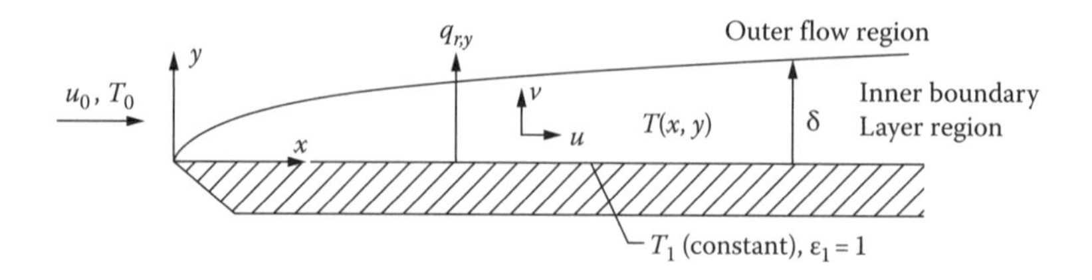

FIGURE 13.12 Boundary layer flow over a flat plate, with free-stream velocity, u_0.

13.4.1 Optically Thin Thermal Layer

To analyze laminar-flow heat transfer on a flat plate, an expression is needed for the radiative source term $-\partial q_{r,y}/\partial y$ in the energy equation. Within the boundary layer, it is assumed that the thermal conditions are changing slowly enough in the x direction, as compared with the y direction, so the conditions contributing to $q_{r,y}$ at a specific x, say x^+, are all at that x^+ and hence are at the temperature distribution $T(x^+, y)$. Then $\partial q_{r,y}/\partial y$ can be evaluated using 1D relations such as Equations 13.2 and 13.3. For only one bounding wall in Equation 13.2, there is only a T_1 term, and the upper limit of the integral is extended to infinity. Also, the $T^4(\tau^*)$ is replaced by $T^4(x, \tau^*)$ to emphasize the approximation for the radiation term that the temperatures surrounding any position x^+ are all assumed at $T(x = x^+, y)$. Then, for flow over a black wall, the laminar boundary layer energy equation for $T(x, y)$ becomes, with the addition of the radiative energy terms,

$$\rho c_P\left(u\frac{\partial T}{\partial x}+v\frac{\partial T}{\partial y}\right)=k\frac{\partial^2 T}{\partial y^2}-4\kappa\sigma T^4+2\kappa\sigma\left[T_1^4E_2(\tau)+\int_{\tau^*=0}^{\infty}T^4(x,\tau^*)E_1(|\tau-\tau^*|)d\tau^*\right] \quad (13.19)$$

where $\tau = \kappa y$ and there is no scattering and c_P is the specific heat of the gas.

The temperature field is considered as composed of two regions. Near the wall in the usual thermal boundary layer of thickness δ that would be present in the absence of radiation, there are large temperature gradients, and heat conduction is important. This layer thickness is usually small; hence, for the formulation in this section, it is assumed optically thin so that radiation passes through it without attenuation. For larger y than in this layer, temperature gradients are small and heat conduction is neglected compared with radiative transfer. The approximate analysis now proceeds along the path used by Cess (1964).

In the outer region, the velocity in the x direction has the free-stream value u_0, and with the neglect of heat conduction in this region, the boundary layer energy equation reduces to

$$\rho c_P u_0\frac{\partial T}{\partial x}=-4\kappa\sigma T^4+2\kappa\sigma\left[T_1^4E_2(\tau)+\int_{\tau^*=0}^{\infty}T^4(x,\tau^*)E_1(|\tau-\tau^*|)d\tau^*\right] \quad (13.20)$$

To obtain an approximate solution by iteration, substitute the incoming free-stream temperature T_0 for the temperature on the right side as a first approximation and then carry out the integral to obtain a second approximation. This yields, for the outer region, to first-order terms,

$$T(x,y)=T_0+\sigma\left(T_1^4-T_0^4\right)E_2(\kappa y)\frac{2\kappa x}{\rho c_P u_0}+\cdots \quad (13.21)$$

where $T = T_0$ at $x = 0$.

At the edge of the thermal layer, $\kappa y = \kappa\delta$, which is small, so that $E_2(\kappa\delta) \approx E_2(0) = 1$. Hence, at $y = \delta$, Equation 13.21 becomes

$$T(x,\delta)=T_0+\sigma\left(T_1^4-T_0^4\right)\frac{2\kappa x}{\rho c_p u_0}+\cdots \quad (13.22)$$

Equation 13.22 is the edge boundary condition that the outer radiation layer imposes on the inner thermal layer. The $T(x, \delta)$ to this approximation is increasing linearly with x. This is the result of the flowing medium absorbing a net radiation from the plate in proportion to the difference $T_1^4 - T_0^4$ and the absorption coefficient κ.

To solve the boundary layer equation in the inner thermal layer region, the last integral in Equation 13.19 is divided into two parts, from $\tau = 0$ to $\kappa\delta$ and from $\tau = \kappa\delta$ to ∞. The first portion is neglected as the thermal layer is optically thin, and the second is evaluated by using the outer solution Equation 13.21. By retaining only first-order terms, the boundary layer energy equation is reduced to

$$u\frac{\partial T}{\partial x}+v\frac{\partial T}{\partial y}=\alpha\frac{\partial^2 T}{\partial y^2}+\frac{2\kappa\sigma}{\rho c_p}\left(T_1^4+T_0^4-2T^4\right) \tag{13.23}$$

The boundary conditions are given by Equation 13.21 at $y = \delta$, and the specified wall temperature $T = T_1$ at $y = 0$. The solution becomes rather complex and is not developed further here. The reader is referred to Cess (1964, 1966a) for additional information.

13.4.2 Optically Thick Thermal Layer

At the opposite limit from the previous section, if the thermal layer has become very thick or the medium is highly attenuating, the boundary layer can be optically thick. The analysis is then simplified, as the diffusion approximation can be employed. From Equation 13.15, radiative diffusion adds a radiative conductivity to the ordinary thermal conductivity. Then, the laminar boundary layer energy equation becomes

$$\rho c_p\left(u\frac{\partial T}{\partial x}+v\frac{\partial T}{\partial y}\right)=\frac{\partial}{\partial y}\left[\left(\frac{16\sigma T^3}{3\beta}+k\right)\frac{\partial T}{\partial y}\right] \tag{13.24}$$

With the assumption of constant fluid properties, the momentum and continuity equations do not depend on temperature; consequently, the flow is unchanged by heat transfer. The velocity distribution is given by Blasius solution (Schlichting 1960) in terms of a similarity variable $\eta = y\sqrt{u_0/\nu x}$. The ν under the square roots is the kinematic viscosity. The stream-function and velocity components are

$$\psi=\sqrt{\nu x u_0 f(\eta)},\quad u=\frac{\partial\psi}{\partial y}=u_0\frac{df}{d\eta},\quad v=-\frac{\partial\psi}{\partial x}=\frac{1}{2}\sqrt{\frac{\nu u_0}{x}}\left(\eta\frac{df}{d\eta}-f\right) \tag{13.25}$$

where the function $f(\eta)$ is in Lee et al. (1990). These quantities are substituted into Equation 13.24, which is then placed in the form

$$-\frac{\mathrm{Pr}}{2}f\frac{d\vartheta}{d\eta}=\frac{d}{d\eta}\left[\left(\frac{4\vartheta^3}{3N_{CR}}+1\right)\frac{d\vartheta}{d\eta}\right] \tag{13.26}$$

where

$$\vartheta = T/T_0$$

$$N_{CR}=k\beta/4\sigma T_0^3$$

and N_{CR} is the conduction–radiation parameter. The boundary conditions that were used in the numerical solution are $\vartheta = \vartheta_1 = T_1/T_0$ at $\eta = 0$, and $\vartheta = 1$ at $\eta = \infty$. To be more precise, a temperature jump condition from the use of radiative diffusion should be used at the wall, but this has not been formulated for combined radiation, convection, and conduction.

Numerical solutions were carried out by Viskanta and Grosh (1962c) and typical temperature profiles are in Figure 13.11. For $N_{CR} = 10$, the profile was found to be within 2% of the opaque limit

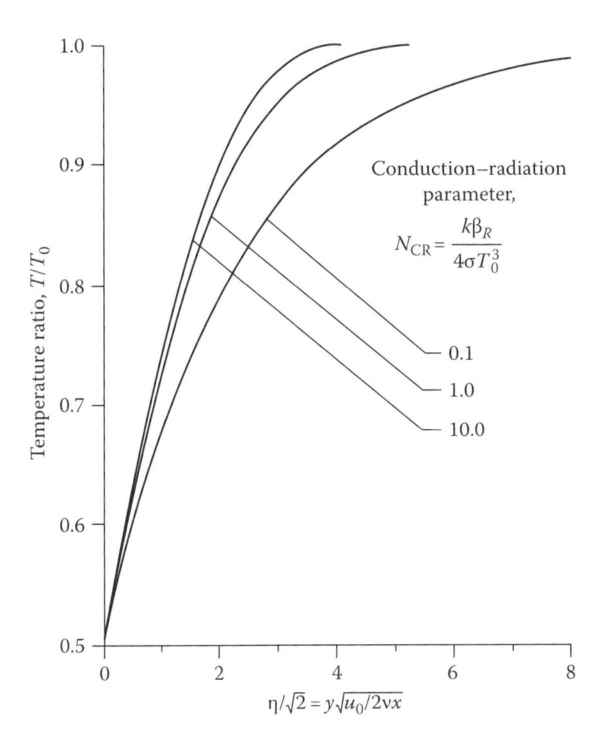

FIGURE 13.13 Boundary layer temperature profiles for laminar flow on flat plate with radiation Prandtl number Pr = 1.0 and temperature ratio $T_1/T_0 = 0.5$. (From Viskanta, R. and Grosh, R.J., *Int. J. Heat Mass Trans.*, 5, 795, 1962.)

$N_{CR} \to \infty$ for conduction and convection alone. The effect of radiation is found to thicken the thermal boundary layer, similar to the effect of decreasing the Prandtl number. This is expected, since the Prandtl number is the ratio of viscous to thermal diffusion ν/α. Radiation supplies an additional means for thermal diffusion, thereby effectively increasing the thermal diffusivity, α (Figure 13.13).

13.5 NUMERICAL SOLUTION METHODS FOR COMBINED RADIATION, CONDUCTION, AND CONVECTION IN PARTICIPATING MEDIA

Solution methods for the RTE were outlined in Chapter 12. In the following sections, numerical methods are described for solving the radiative transfer and energy equations in participating media when energy transfer modes and/or internal heat sources are present in addition to radiation. This includes energy transfer in a translucent medium by radiation, conduction, forced convection, and free convection, and with temperature-dependent volumetric heat sources such as by chemical reactions. Radiative intensities are usually spectrally dependent, and integrating the radiant energy over the spectrum is required to determine the total radiative energy source for the energy equation. For the local radiative source, the contributions by the local incident intensities from the equation of transfer must also be integrated over all directions surrounding each location. Thus, both spectral and directional integrations may be required. With scattering, the source-function distribution must be obtained by solving the source-function equation simultaneously with the energy equation. The two unknowns are the temperature and source-function distributions. The intensities that are obtained for the source function provide local energy fluxes.

Solutions become difficult when both spectrally and spatially dependent variations are included. Some numerical techniques that can be used are considered here. The combination of radiation with conduction, convection, and/or temperature-dependent volumetric energy generation provides a nonlinear energy equation, and some of the standard numerical methods may not work well. For example, the net heat conduction may be the result of a difference between large incoming and outgoing radiation terms; this can result in loss of accuracy for computing the conduction terms. Strongly temperature-dependent volumetric sources, as in the chemical kinetics of combustion processes, are particularly sensitive to small differences in local temperatures, and solutions are difficult to bring to convergence. Extra care is generally required to obtain a method that gives good accuracy, converges well, and can be carried out with a reasonable computational effort.

An application that incorporates all of these difficulties is high-speed atmospheric reentry of space vehicles. Here, the shock layer incorporates non-LTE effects near the shock, hypersonic flow, highly spectral line radiation emission and absorption (and possibly scattering in the boundary layer near the ablating heat shield), and strongly coupled chemical reactions. At velocities expected for Lunar mission returns to Earth, radiation and convection to the spacecraft thermal protection system are roughly equal. For a Mars mission return to the Earth, it is expected that over 90% of the heat flux incident on an ablating heat shield will be from radiation. The flux levels in this case are predicted to be so high that the present generation of heat shield materials is not adequate.

Evaluating the radiative source distribution in the energy equation requires integrating energy contributions from the medium surrounding each location in the translucent material. This can be done by any of the methods of Chapter 12. The integration must be carried out over a sufficient number of locations within the medium to define the radiative source distribution accurately within the volume; these integrations can be very time consuming.

With the radiative source evaluated, the energy equation must be solved including conduction and/or convection. This can be done by finite-difference, finite-element, or other methods. Solutions can involve over- or underrelaxation techniques, implicit methods, or matrix solutions of the set of simultaneous equations developed for the grid locations within the medium. If an iterative solution method is used, it can depend on whether a steady-state or transient solution is required. It may be necessary to use two sets of grid points. A coarse grid can be used for the locations at which the radiative source term is evaluated and intermediate values of the source term within the grid can then be obtained by interpolation. The conduction and convection portions may require a locally finer grid to resolve large temperature variations in regions such as near boundaries. It is usually helpful to use a variable grid size.

Each of the methods for obtaining the local radiative source distribution has advantages and disadvantages when applied to multimode problems. The discretization schemes for the finite-element and finite-difference methods can be made to mesh with discretization schemes for other energy transfer modes. The zone method using a small number of zones and the Monte Carlo method are not as easy to match with other energy transfer modes. However, the Monte Carlo scheme can be easier to formulate for a more complete treatment of the effects of nonuniform and spectral property variations when they are important. The zone technique can be applied to a broad variety of practical problems because its long history of development has provided numerical values for various zone exchange factors.

Some analyses and results from the literature for radiation combined with conduction and/or convection are discussed later in the chapter. These illustrate how the methods presented here such as finite differences, finite elements, and Monte Carlo have been applied. For some solutions, the radiative analyses have been combined with solvers for computational fluid mechanics to obtain the required details of convection processes for irregular geometries and for combined forced and free convection where buoyancy must be included. Mazumder (2005) examines the advantage of directly solving the closely coupled nonlinear energy equation for combined modes as opposed to the loosely coupled approach of solving the energy equation for temperature, using this in the RTE to find the radiative flux divergence, substituting this into the energy equation, and iterating to convergence. Ghannam et al. (2010, 2012) and Ghannam (2012) uses the zonal method for combined radiation and conduction in a complex geometry with an absorbing medium, examining various methods for iterating to convergence. He places special emphasis on the use of graphical processing units (GPUs) as the computing platform for parallel computing rather than the more conventional central processing units (CPUs) and shows considerable advantages.

13.5.1 Finite-Difference Methods

13.5.1.1 Energy Equation for Combined Radiation and Conduction

For simplicity only combined radiation and conduction are considered here; including convection is discussed later with reference to some applications. Solution is now considered of the transient energy equation that contains a radiative heat source. For this development, the medium is gray with constant properties, and there is no internal heat generation. The energy Equation 10.3 is placed in dimensionless form by letting $X = (\kappa + \sigma_s)x = \beta x$, $Y = (\kappa + \sigma_s)y = \beta y$, and $Z = (\kappa + \sigma_s)z = \beta z$. The ∇ in this development is then with respect to optical coordinates X, Y, Z. The $S_r \equiv (1/4\sigma T_{ref}^4)\nabla \cdot \mathbf{q}_r$ where T_{ref} is a convenient reference temperature. The energy equation becomes (without $\dot{q}$ or Φ_d)

$$\frac{\partial \vartheta}{\partial \tilde{t}} = N_{CR}\nabla^2\vartheta - S_r\left(X,Y,Z,\vartheta\right) \tag{13.27}$$

$$N_{CR} = k\beta/4\sigma T_{ref}^3$$

$$\tilde{t} = 4\tau\beta\sigma T_{ref}^3/\rho c$$

$$\vartheta = T/T_{ref}$$

The X, Y, and Z are optical dimensions in the x, y, and z-directions, respectively. An implicit finite-difference method is developed for transient and steady-state solutions. This is an example of a numerical procedure that is stable for forward integration in time. A subscript and superscript notation is used, $\vartheta_{i,j,k}^n$, where n refers to the nth dimensionless time interval and i, j, k are the X, Y, Z grid locations.

In the solution, two operations are necessary: a forward integration in time or iteration for a steady solution and a spatial integration to evaluate the local radiative energy source $-S_r$. To integrate forward in time, the relation $\vartheta = \int (\partial\vartheta/\partial\tilde{t})d\tilde{t}$ can be approximated in various ways. The trapezoidal rule is convenient, and this gives

$$\begin{aligned}\vartheta^{n+1} &= \vartheta^n + \frac{\Delta\tilde{t}}{2}\left(\frac{\partial\vartheta^{n+1}}{\partial\tilde{t}} + \frac{\partial\vartheta^n}{\partial\tilde{t}}\right) \\ &= \vartheta^n + \frac{\Delta\tilde{t}}{2}\left(N_{CR}\nabla^2\vartheta^{n+1} - S_r^{n+1} + N_{CR}\nabla^2\vartheta^n - S_r^n\right)\end{aligned} \tag{13.28}$$

as described by Warming and Beam (1978). The S_r^{n+1} is now expressed by a linearized expansion in time away from S_r^n:

$$S_r^{n+1} = S_r^n + \frac{\partial S_r^n}{\partial\vartheta}\Delta\vartheta, \quad \text{where } \Delta\vartheta = \vartheta^{n+1} - \vartheta^n \tag{13.29}$$

The $\nabla^2\vartheta^{n+1}$ is rewritten in terms of by ϑ^n using the identity

$$\nabla^2\vartheta^{n+1} = \nabla^2\left(\vartheta^{n+1} - \vartheta^n\right) + \nabla^2\vartheta^n = \nabla^2\left(\Delta\vartheta\right) + \nabla^2\vartheta^n \tag{13.30}$$

Equations 13.29 and 13.30 are substituted into Equation 13.28, and the result simplifies to

$$\left(1+\frac{\Delta\tilde{t}}{2}\frac{\partial S_r^n}{\partial\vartheta}-\frac{\Delta\tilde{t}}{2}N_{\mathrm{CR}}\nabla^2\right)\Delta\vartheta=\Delta\tilde{t}\left(N_{\mathrm{CR}}\nabla^2\vartheta^n-S_r^n\right) \tag{13.31}$$

Using the values at time increment n, Equation 13.31 yields a matrix of equations for each $\Delta\vartheta$ at each grid point. Then, the new ϑ at each point is $\vartheta^{n+1} = \vartheta^n + \Delta\vartheta$.

Before giving some illustrative examples, the equation corresponding to Equation 13.31 is given that can be used to obtain a steady-state solution by iteration from an initial guessed temperature distribution. The steady energy equation is

$$N_{\mathrm{CR}}\nabla^2\vartheta=S_r\left(X,Y,Z,\vartheta\right) \tag{13.32}$$

The linearized expansion for the $(n + 1)$th iteration in terms of the nth iteration gives (n now designates the iteration)

$$N_{\mathrm{CR}}\nabla^2\vartheta^{n+1}=S_r\left(\vartheta^n\right)=\frac{\partial S_r^n}{\partial\vartheta}\Delta\vartheta \tag{13.33}$$

Equating the $\nabla^2\vartheta^{n+1}$ given by Equations 13.30 and 13.33 gives the desired equation to solve for $\Delta\vartheta$ as

$$\left(\frac{\partial S_r^n}{\partial\vartheta}-N_{\mathrm{CR}}\nabla^2\right)\Delta\vartheta=N_{\mathrm{CR}}\nabla^2\vartheta^n-S_r^n \tag{13.34}$$

13.5.1.2 Radiation and Conduction in a Plane Layer

To illustrate the application of Equation 13.33, consider a plane layer of gray radiating and conducting medium without scattering between opaque parallel walls, as in Figure 13.14. Initially, the layer temperature is uniform at T_0. Let this be the reference temperature so that $\vartheta = T/T_0$ and, at $t = 0$, $\vartheta_o = 1$. The lower and upper wall temperatures are then suddenly changed to $\vartheta_w(0)$ and $\vartheta_w(X_D)$ where $X_D = \beta D$. The transient temperature distribution is to be obtained.

From Equation 13.5, the radiation source at $X = \beta x$ is related to $\vartheta(X)$ by

$$S_r\left(X,\vartheta\right)=\vartheta^4\left(X\right)-\frac{1}{2}\Bigg[\vartheta_1^4E_2\left(X\right)+\vartheta_2^4E_2\left(X_D-X\right)+\int_{X^*=0}^{X}\vartheta^4\left(X^*\right)E_1\left(X-X^*\right)dX^*+\int_{X^*=X}^{X_D}\vartheta^4\left(X^*\right)E_1\left(X^*-X\right)dX^*\Bigg] \tag{13.35}$$

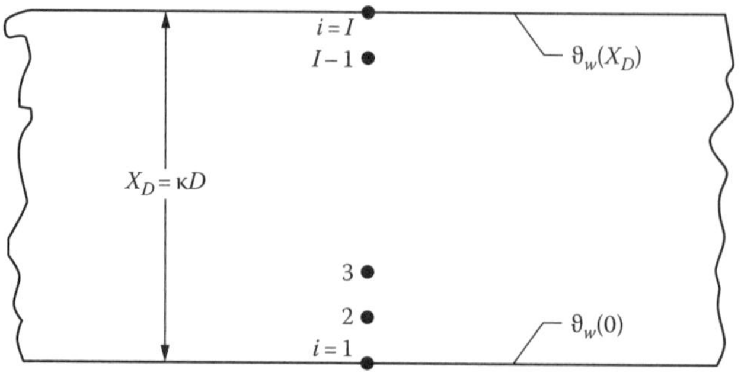

FIGURE 13.14 Plane layer of translucent material for transient thermal analysis using finite differences.

and, by differentiation,

$$\left.\frac{\partial S_r}{\partial \vartheta}\right|_X = 4\left[\vartheta^3(X) - \frac{1}{2}\left\{\int_{X^*=0}^{X} \vartheta^3(X^*)\left(\partial\vartheta/\partial\tilde{t}\right)\Big|_{X^*} E_1(X - X^*)\,dX^* \right.\right.$$
$$\left.\left. + \int_{X^*=X}^{X_D} \vartheta^3(X^*)\left(\partial\vartheta/\partial\tilde{t}\right)\Big|_{X^*} E_1(X^* - X)\,dX^*\right\} \Big/ \left(\partial\vartheta/\partial\tilde{t}\right)\Big|_X\right] \tag{13.36}$$

The finite-difference form for the second derivative in Equation 13.31 is $\partial^2\vartheta_i/\partial X^2 = (\vartheta_{i-1} - 2\vartheta_i + \vartheta_{i+1})/(\Delta X)^2$. The boundary conditions give $\vartheta_1 = \vartheta_w(0)$ and $\vartheta_I = \vartheta_w(X_D)$. Because these values are fixed, $\Delta\vartheta_1 = \Delta\vartheta_I = 0$. Equation 13.31 is then applied at each interior point $i = 2,3, \ldots, I-1$. For example, at $i = 3$,

$$\left(1 + \frac{\Delta\tilde{t}}{2}\frac{\partial S_{r,3}^n}{\partial\vartheta}\right)\Delta\vartheta_3^n - \frac{\Delta\tilde{t}}{2}N_{CR}\frac{\left(\Delta\vartheta_2^n - 2\Delta\vartheta_3^n + \Delta\vartheta_4^n\right)}{(\Delta X)^2} = \Delta\tilde{t}\left(N_{CR}\frac{\left(\vartheta_2^n - 2\vartheta_3^n + \vartheta_4^n\right)}{(\Delta X)^2} - S_{r,3}^n\right) \tag{13.37}$$

where, without scattering, $N_{CR} = k\kappa/4\sigma T_0^3$ and $\tilde{t} = 4t\kappa\sigma T_0^3/\rho c$. By writing similar equations for all interior points, the following tridiagonal matrix is obtained for $\Delta\vartheta_i$:

$$\begin{bmatrix} b_2 & c_2 & & \\ a_3 & \ddots & \ddots & \\ & \ddots & \ddots & c_{I-2} \\ & & a_{I-1} & b_{I-1} \end{bmatrix}\begin{bmatrix} \Delta\vartheta_2^n \\ \Delta\vartheta_3^n \\ \vdots \\ \Delta\vartheta_{I-1}^n \end{bmatrix} = \begin{bmatrix} s_2 \\ s_3 \\ \vdots \\ s_{I-1} \end{bmatrix} \tag{13.38}$$

$$a_i = c_i = -\frac{\Delta\tilde{t}}{2}\frac{N_{CR}}{(\Delta X)^2}, \quad b_i = 1 + \frac{\Delta\tilde{t}}{2}\frac{\partial S_{r,i}^n}{\partial\vartheta} + \Delta\tilde{t}\frac{N_{CR}}{(\Delta X)^2}$$

$$s_i = \Delta\tilde{t}\left(N_{CR}\frac{\vartheta_{i-1}^n - 2\vartheta_i^n + \vartheta_{i+1}^n}{(\Delta X)^2} - S_{r,i}^n\right) \quad 2 \le i \le I-1$$

where for $i = 2$, $\vartheta_{i-1} = \vartheta_1 = \vartheta_w(0)$ and for $i = I-1$, $\vartheta_{i+1} = \vartheta_I = \vartheta_w(X_D)$. The tridiagonal matrix Equation 13.38 is solved by the following well-known algorithm, where f and g are first computed from

$$f_2 = \frac{c_2}{b_2} \quad g_2 = \frac{s_2}{b_2}$$

$$f_i = \frac{c_i}{b_i - a_i f_{i-1}} \quad g_i = \frac{s_i - a_i g_{i-1}}{b_i - a_i f_{i-1}} \quad 3 \le i \le I-1$$

and the $\Delta\vartheta_i$ values are then obtained recursively from the relations

$$\Delta\vartheta_{I-1} = g_{I-1}, \quad \Delta\vartheta_i = g_i - f_i\Delta\vartheta_{i+1}, \quad 2 \le i \le I-2$$

The values of ϑ at the next time step are $\vartheta^{n+1} = \vartheta_i^n + \Delta\vartheta_i$ for $2 \le i \le I-2$.

For a steady situation, the transient solution can be continued to steady state, or Equation 13.34 can be used. For the latter, the iteration is started by making a guess of the steady temperature distribution. The implicit iteration of Equation 13.34 yields a tridiagonal matrix of the same form as Equation 13.38. The coefficients are

$$a_i = c_i = -\frac{N_{CR}}{(\Delta X)^2}, \quad b_i = \frac{\partial S_{r,i}^n}{\partial\vartheta} + \frac{2N_{CR}}{(\Delta X)^2}$$

$$s_i = N_{CR}\frac{\vartheta_{i-1}^n - 2\vartheta_i^n + \vartheta_{i+1}^n}{(\Delta X)^2} - S_{r,i}^n$$

At each step of the forward integration or iteration in these solutions, the S_r and $\partial S_r/\partial\vartheta$ must be obtained at each grid location. These can be evaluated by numerical integration of Equations 13.35 and 13.36, or other techniques such as finite elements or discrete ordinates can be used. Evaluation by numerical integration requires an accurate integration technique. Since $E_1(0) = \infty$, care must be taken as X^* approaches X. Since the integral of E_1 is $-E_2$, and $E_2(0) = 1$, the integration can be carried out analytically over a very small region near the singularity, with ϑ constant over this region. A 1D transient analysis using a Crank–Nicolson procedure and trapezoidal integration is in Yoshida et al. (1990). A transient cooling analysis for a plane layer in Siegel (1992b) used Gaussian integration to evaluate the local radiative source and a finite-difference procedure with variable space and time increments to solve the transient energy equation.

13.5.1.3 Radiation and Conduction in a 2D Rectangular Region

As another example, consider a 2D transient analysis of a gray absorbing–emitting rectangular region, as in Figure 13.15a, that is initially at uniform temperature T_0. The rectangle has height D, width W, optical height κD, and optical width κW. To illustrate the transient, the region is placed into vacuum at a much lower temperature. For these conditions, energy can leave the rectangle only by radiation because there is no surrounding cooling medium. During radiative cooling, heat conduction partially equalizes the transient temperature distribution.

The transient energy equation including conduction and radiation without scattering is obtained from Equations 13.27, 11.89, and 11.98 as

$$\frac{\partial\vartheta}{\partial\tilde{t}} = N_{CR}\left(\frac{\partial^2\vartheta}{\partial X^2} + \frac{\partial^2\vartheta}{\partial Y^2}\right) - \vartheta^4(X,Y,\tilde{t})$$
$$+\frac{1}{4}\int_{X^*=0}^{\kappa W}\int_{Y^*=0}^{\kappa D}\vartheta^4(X^*,Y^*,\tilde{t})\frac{S_1(R)}{R(X,Y,X^*,Y^*)}dX^*dY^* \qquad (13.39)$$

where

$$\vartheta = T/T_0$$

$R = [(X - X^*)^2 + (Y - Y^*)^2]^{1/2}$ and the S_1 function is in Appendix D.

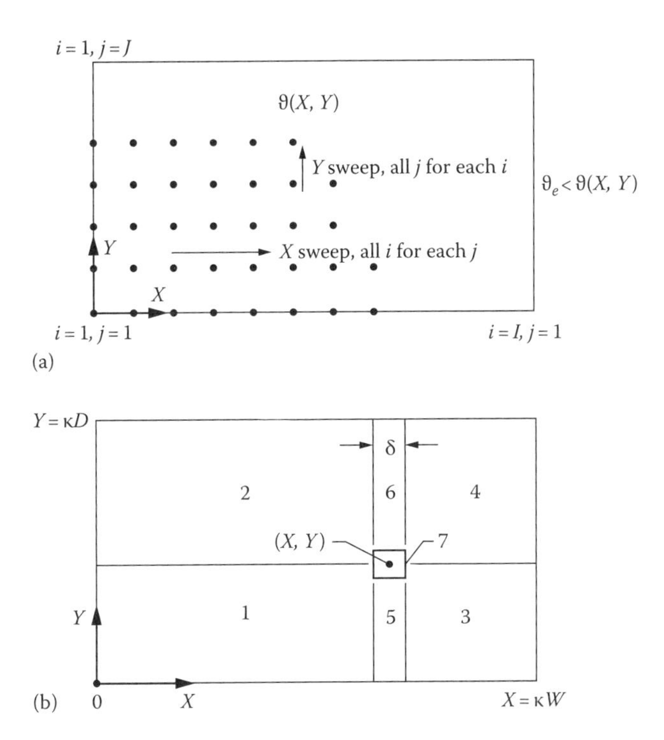

FIGURE 13.15 2D rectangular region for cooling analysis by finite differences of combined radiation and conduction, (a) grid for alternating direction implicit [ADI] method and (b) regions for double integration in energy equation.

From the form of Equation 13.27, the radiative heat source is

$$S_r\left(X,Y,\tilde{t}\right)=\vartheta^4\left(X,Y,\tilde{t}\right)-\frac{1}{4}\int_{X^*=0}^{\kappa W}\int_{Y^*=0}^{\kappa D}\vartheta^4\left(X^*,Y^*,\tilde{t}\right)\frac{S_1(R)}{R}dX^*dY^*$$

Equation 13.31 can now be directly applied to find $\Delta\tilde{t}=\tilde{t}^{n+1}-\tilde{t}^{n}$ at each grid point within the region. The alternating direction implicit (ADI) method is used. This is described by Press et al. (1989); the method here differs somewhat from that given by Warming and Beam (1978). The ∇^2 operator is split into each of the coordinate directions, and Equation 13.31 is approximated by

$$\left(1+\frac{\Delta\tilde{t}}{2}\frac{\partial S_r^n}{\partial\vartheta}-\frac{\Delta\tilde{t}}{2}N_{CR}\frac{\partial^2}{\partial X^2}\right)\Delta\varphi=\Delta\tilde{t}\left(N_{CR}\nabla^2\vartheta^n-S_r^n\right) \tag{13.40}$$

$$\left(1-\frac{\Delta\tilde{t}}{2}N_{CR}\frac{\partial^2}{\partial Y^2}\right)\Delta\vartheta=\Delta\varphi \tag{13.41}$$

To move ahead one time increment, the first equation is solved for $\Delta\varphi$; this is then used on the right side of the second equation to solve for $\Delta\vartheta$. The rectangle is covered with a square grid as in Figure 13.15a. A sweep is made in the X direction for each j to obtain the $\Delta\varphi_{i,j}^n$ for all i. The boundary temperatures are unknown; radiation from the volume passes through the boundaries but cannot interact with conduction exactly at the boundary because the boundary itself has no volume. Hence, conduction, and consequently the normal temperature derivative, is zero at the boundaries; this boundary condition is included in the finite-difference representation. The matrix for each X sweep for $\Delta\varphi^n_{i,j}$, $1 \le i < I$, has the same form as Equation 13.38. A few coefficients are different because Equation 13.38 was written for known boundary temperatures. The values are

$$a_i = -\frac{\Delta\tilde{t}}{2}\frac{N_{CR}}{(\Delta X)^2} \quad 2 \le i \le I-1 \quad a_I = -\Delta\tilde{t}\frac{N_{CR}}{(\Delta X)^2}$$

$$b_i = 1 + \frac{\Delta\tilde{t}}{2}\frac{\partial S^n_{r,i}}{\partial \vartheta} + \Delta\tilde{t}\frac{N_{CR}}{(\Delta X)^2} \quad 1 \le i \le I$$

$$c_1 = -\Delta\tilde{t}\frac{N_{CR}}{(\Delta X)^2} \quad c_i = -\frac{\Delta\tilde{t}}{2}\frac{N_{CR}}{(\Delta X)^2} \quad 2 \le i \le I-1$$

The $S_{i,j}$ on the right side are of the form given by Equation 13.38:

$$s_{1,j} = \Delta\tilde{t}\left(2N_{CR}\frac{\vartheta^n_{2,j} - \vartheta^n_{1,j}}{(\Delta X)^2} + N_{CR}\frac{\vartheta^n_{1,j-1} - 2\vartheta^n_{1,j} + \vartheta^n_{1,j+1}}{(\Delta Y)^2} - S^n_{r,1,j}\right), \quad 2 \le j \le J-1$$

$$s_{i,j} = \Delta\tilde{t}\left(N_{CR}\frac{\vartheta^n_{i-1,j} - 2\vartheta^n_{i,j} + \vartheta^n_{i+1,j}}{(\Delta X)^2} + N_{CR}\frac{\vartheta^n_{i,j-1} - 2\vartheta^n_{i,j} + \vartheta^n_{i,j+1}}{(\Delta Y)^2} - S^n_{r,i,j}\right), \quad 2 \le i \le I-1 \quad 2 \le j \le J-1$$

$$s_{I,j} = \Delta\tilde{t}\left(2N_{CR}\frac{\vartheta^n_{I-1,j} - \vartheta^n_{I,j}}{(\Delta X)^2} + N_{CR}\frac{\vartheta^n_{I,j-1} - 2\vartheta^n_{I,j} + \vartheta^n_{I,j+1}}{(\Delta Y)^2} - S^n_{r,I,j}\right), \quad 2 \le i \le I-1$$

For $j = 1$, replace $\vartheta^n_{i,j-1} - 2\vartheta^n_{i,j} + \vartheta^n_{i,j+1}$ by $2\left(\vartheta^n_{i,2} - \vartheta^n_{i,1}\right)$. For $j = J$, replace $\vartheta^n_{i,j-1} - 2\vartheta^n_{i,j} + \vartheta^n_{i,j+1}$ by $2\left(\vartheta^n_{i,J-1} - \vartheta^n_{i,J}\right)$.

After the $\Delta\varphi^n_{i,j}$ are obtained, they are used to sweep along each set of grid points in the Y direction by use of Equation 13.41. The matrix form for the Y sweep is

$$\begin{bmatrix} 1 + N_{CR}\frac{\Delta\tilde{t}}{(\Delta Y)^2} & -N_{CR}\frac{\Delta\tilde{t}}{(\Delta Y)^2} & \cdots & & \\ -\frac{N_{CR}}{2}\frac{\Delta\tilde{t}}{(\Delta Y)^2} & 1 + N_{CR}\frac{\Delta\tilde{t}}{(\Delta Y)^2} & -\frac{N_{CR}}{2}\frac{\Delta\tilde{t}}{(\Delta Y)^2} & & \\ \vdots & \vdots & \vdots & \vdots & \\ & \cdots & -N_{CR}\frac{\Delta\tilde{t}}{(\Delta Y)^2} & & 1 + N_{CR}\frac{\Delta\tilde{t}}{(\Delta Y)^2} \end{bmatrix} \begin{bmatrix} \Delta\vartheta_{i,1} \\ \Delta\vartheta_{i,2} \\ \vdots \\ \Delta\vartheta_{i,J} \end{bmatrix} = \begin{bmatrix} \Delta\varphi_{i,1} \\ \Delta\varphi i,2 \\ \vdots \\ \Delta\varphi_{i,J} \end{bmatrix}$$

This yields all the $\Delta\vartheta_{i,j}$; the ϑ at $\tilde{t} + \Delta\tilde{t}$ are then $\vartheta^{n+1}_{i,j} = \vartheta^n_{i,j} + \Delta\vartheta^n_{i,j}$.

Since the function $S_1(R)$ in the integrands for the local radiative source S_r is well behaved as $R \to 0$ (see Appendix D.2), the integrands appear to be singular because of the $1/R$ factor when the integration variables X^*, Y^* approach grid point X, Y. The integrands are actually not singular, as is evident by using cylindrical coordinates R, θ about X, Y; the $dX\,dY$ becomes $R\,dR\,d\theta$, and the $1/R$ is thus removed. However, when using rectangular coordinates for numerical integration, the apparent singularity must be dealt with for small R. The integration about each grid point X, Y is divided into seven regions as in Figure 13.15b. Region 7 is a small square of width less than one grid spacing. The integration in this region is carried out in cylindrical coordinates over a circle with area equal to that of region 7. For the other integrations, a grid of points can be used with 2D integration subroutines. The advantage of using direct numerical integration for the radiative source is that high accuracy can be obtained from the energy equation for detailed features such as temperature distributions. This is very important for transients, where the solution will otherwise tend to drift into inaccuracy as time proceeds. Methods for increasing the speed for multiple integration in multidimensional media by using parallel processors are discussed by Genz (1982) and Shih et al. (1986a).

During some transients, there can be larger temperature gradients near boundaries than in the medium interior. This would occur during transient cooling by radiation and conduction in a translucent solid with convective cooling at its boundaries or in a convective boundary layer in a radiating fluid. Additional grid points are necessary in the large-gradient regions for an accurate finite-difference representation of the conduction or convection terms. The radiative source term can be evaluated using a coarser grid with intermediate values and then interpolated at finer grid point locations for solving the energy equation. It is often helpful to use a variable grid size (Siegel and Molls 1992) having relatively more points in regions where the temperature is varying most rapidly.

Yuen and Takara (1988) obtained solutions for combined conduction and radiation in 2D enclosures, using an iterative solution to a finite-difference form of the energy equation. The method is based on an earlier method for radiation analysis without conduction (Yuen and Wong 1984). They use a numerical approximation to the integral form of the equation of transfer for determining the local radiative heat source in the translucent medium. The approximation assumes a linear variation of the emissive power distribution within each difference element and approaches an exact solution for small elements. The only physical approximation is in the radiative boundary condition, which is generated by extrapolation of the linear function normal to the boundary. Good accuracy was found for an 11 × 11 grid in a square enclosure, and comparisons were excellent with the finite-element results of Razzaque et al. (1984). Comparisons with the P_3 results of Ratzel (1981) were good for emissive power in the medium but were poor for surface energy flux in some cases; this is likely due to inaccuracy in the P_3 method. Yuen and Takara (1988) also compared results with the diffusion solution in 2D and found deviations up to 17% in local emissive power of the medium and very large errors in surface energy flux. This was for optical dimensions of the medium from 0.1 to 5 that may be too small for diffusion to be an accurate approximation. Much better predictions of total surface energy flux (within 10% for all cases reported, and within 1% in most cases) were obtained by simply adding the energy flux for a pure-conduction solution to that for a pure-radiation solution; this was for optical dimensions of the rectangle from 0.1 to 5.

A 2D rectangular region was analyzed by Wu and Ou (1994) for transient behavior including isotropic scattering. The translucent medium is gray and the material is enclosed by black boundaries so there are no reflections at the walls. The transient energy equation was solved with a finite-difference method with the radiative heat source obtained with a modified differential approximation. Dimensionless temperature distributions are in Figure 13.16 for a square cross section and for a rectangular region ($0 \leq x \leq W$, $0 \leq y \leq D$) with an aspect ratio of $D/W = 0.2$. The transient starts with the translucent region at uniform temperature T_i and the temperature of the wall at $y = 0$ is then raised to T_1, with the temperature at the other three walls remaining unchanged. The temperatures shown in Figure 13.16 extend across the short dimension $0 \leq y \leq D$ for the rectangular region and are halfway between the short sides. The temperatures are nondimensionalized, so they extend from 1 to 0 over the height of the region. For short times, there is little difference in temperatures between the square and rectangular geometries. At long times, the temperature profile across the short dimension becomes somewhat like that for a translucent layer at steady state between plane walls at unequal temperatures. This limit is approached as the height to width of the rectangle becomes very small but is already being approached for an aspect ratio of 0.2.

A finite-difference solution including the effects of real gas properties is in Kamiuto (1996), where carbon dioxide is between parallel walls at differing constant temperatures. A wideband

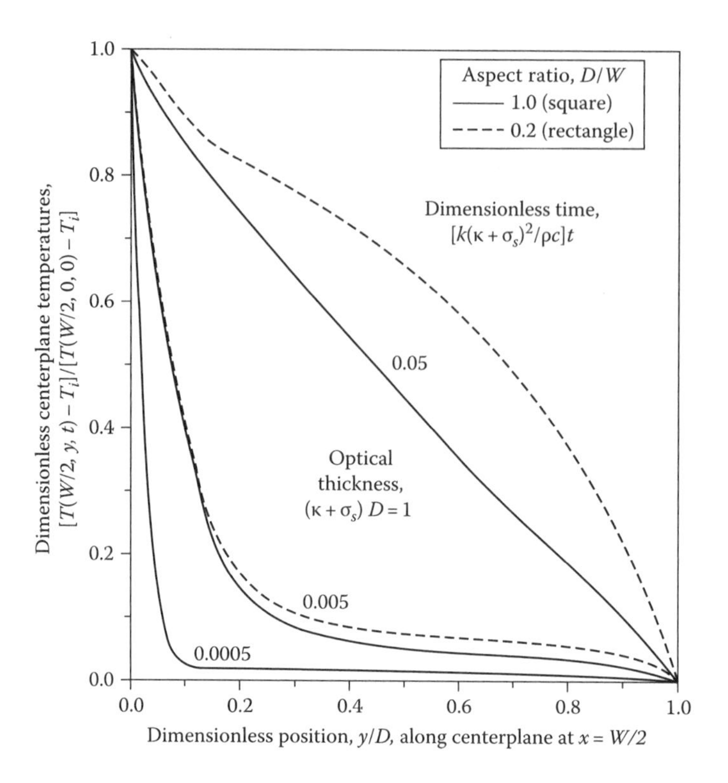

FIGURE 13.16 Transient temperatures along the center plane of a square or rectangular region (across the short dimension), following a temperature increase from T_i to T_1 at $y = 0$, $N_{CR} = k(\kappa + \sigma_s)/4\sigma T_1^3 = 0.01$, $\omega = 0.5$. (Replotted from Wu, C.-Y. and Ou, N.-R., *Int. J. Heat Mass Trans.*, 37(17), 2675, 1994.)

spectral model was used for combined radiation and conduction. The conduction term in the energy equation was represented by a central difference, and 250 uniform increments were used across the thickness of the layer. The radiative heat source was obtained from an equation similar to Equation 13.4 but including spectral property variations, so the energy equation became, with variable thermal conductivity,

$$\frac{d}{dy}\left(k\frac{dT}{dy}\right) = \int_{\lambda=0}^{\infty} 2\pi\kappa_\lambda(y)\Big\{2I_{\lambda b} - I_\lambda^+(0)E_2(\kappa_\lambda y)$$

$$- I_\lambda^-(D)E_2\left[\kappa_\lambda(D-y)\right] - \int_{\xi=0}^{D} \kappa_\lambda(\xi)I_{\lambda b}(\xi)E_1\left(\kappa_\lambda|y-\xi|\right)\Big\}d\lambda \quad (13.42)$$

The radiative integrals were evaluated with the trapezoidal rule. The set of simultaneous equations that resulted from the finite-difference representation was solved by Gaussian elimination.

Some examples using finite-difference procedures for solving the energy equation in geometries other than plane or rectangular regions are in Krishnaprakas (1998) and Chu and Weng (1992) for a cylinder and a sphere. In Krishnaprakas, the derivatives in the energy equation were approximated by second-order finite differences that provided a set of nonlinear algebraic equations for the unknown temperatures at the nodes. The radiative transfer relations were solved with discrete ordinates (Section 12.3). The algebraic equations were solved by Newton–Raphson method (Section 7.8.2). A transient analysis was performed by Chu and Weng (1992) with the radiative contribution to the energy equation obtained from the P_3 approximation. The transient term in the energy equation was represented by a three-term backward difference relation, and the result was a set of ordinary differential equations that were solved with a computer library subroutine.

Another application of finite differences is for solving an inverse problem where surface temperatures and fluxes are to be estimated from temperature measurements made inside a semitransparent plane layer (Ruperti et al. 1996). A backward second-order approximation was used for the time derivative in the energy equation and central differences used for the second-order heat conduction term. The exact form of the radiative transfer relation was used for a gray medium between black boundaries, as in Equation 13.4, and the integrals were evaluated by Gauss–Legendre quadrature.

Various radiative transfer solvers have been combined with lattice Boltzmann solutions of the energy equation. Mishra and Landanasu (2005) use the discrete transfer method for radiation, Mishra et al. (2011) combine the discrete ordinates method, Sakurai et al. (2010) use the radiation element method (REM), and Mishra et al. (2014) invoke the finite volume method for radiation in a planar medium. Das et al. (2008) use a coupled finite volume radiation solver with a lattice Boltzmann conduction solver to solve for multiple parameters via genetic algorithms.

13.5.1.4 Boundary Conditions for Numerical Solutions

The boundary conditions for numerical solution of combined-mode problems are conveniently formulated by writing energy balances on each volume element that is adjacent to a boundary. For example, consider an element $1, j$ within a participating gray medium with refractive index $n \approx 1$ subject to a diffuse external radiative flux $q_r(x=0)$, Figure 13.17. The boundary at $x=0$ is transparent. The medium is 2D and is shown in cross section. There is conduction within the medium, which has uniform thermal conductivity k and absorption coefficient κ. There is convective transfer at the surface to surroundings at T_e with a convective coefficient h, and there is a uniform volumetric energy source of $\dot{q}(x,y)$.

To include the boundary temperature in the boundary condition, a half-width node element is used at the boundary. A steady-state energy balance on element $1, j$ at the boundary requires including conduction crossing each face of the element within the medium, the net radiative flux to the element, the convective transfer at the left-hand face, and the internal energy generation. It is often useful to write each temperature difference from the node temperature $T_{i,j}$ to an adjacent temperature; the sign of the energy transfer is then automatically correct regardless of the temperature gradients. The boundary condition equation for this example is then

$$-k\Delta y\frac{T_{1,j}-T_{2,j}}{\Delta x}-k\frac{\Delta x}{2}\frac{T_{1,j}-T_{1,j+1}}{\Delta y}-k\frac{\Delta x}{2}\frac{T_{1,j}-T_{1,j-1}}{\Delta y}$$

$$-\nabla\cdot\mathbf{q}_{r,i,j}\frac{\Delta x}{2}\Delta y-h\Delta y(T_{1,j}-T_e)+\dot{q}\Delta y\frac{\Delta x}{2}=0 \qquad (13.43)$$

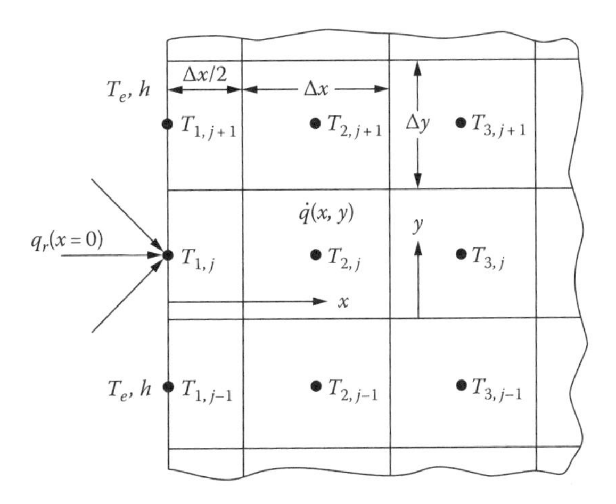

FIGURE 13.17 Treatment of numerical boundary conditions.

The radiative flux divergence $\nabla \cdot \mathbf{q}_r$ is evaluated by using the particular method chosen for treating radiative transfer.

In the limit as $\Delta x \to 0$, the boundary condition reduces to

$$-k \lim_{\Delta x \to 0}\left(\frac{T_{1,j} - T_{2,j}}{\Delta x}\right) - h(T_{1,j} - T_e) = 0$$

which gives, in the limit,

$$k\frac{\partial T}{\partial x}\bigg|_{x=0,y} = h\left(T_{x=0,y} - T_e\right) \tag{13.44}$$

which is the expected analytical boundary condition. The radiative flux divergence, the y direction conduction terms, and the volumetric source do not enter the boundary condition in the analytical form, Equation 13.44, but must be included in the numerical boundary condition. This is because the numerical boundary condition is applied to a finite volume element, whereas the analytical boundary condition is applied at a zero-thickness planar boundary. It is important to realize that simply placing the analytical boundary condition of Equation 13.44 into finite-difference form does not result in the required complete numerical boundary condition for a finite volume element adjacent to a boundary.

Some solutions for combined radiation and conduction were obtained by Ho and Özişik (1987a,b, 1988) and Tsai and Özişik (1987) that include transient conditions, scattering, spherical geometry, and a two-layer planar region. The conduction part of the solution was obtained with finite differences. The radiation source term was obtained by expanding the locally incident radiation in a power series and substituting into the integral equation of transfer. A collocation or Galerkin procedure was used to obtain the unknown coefficients in the power series expansion. The source function is then known in terms of the incident intensities. Li et al. (2009) applied spectral methods to solution of coupled radiation–conduction in the region between concentric spheres.

13.5.2 Finite-Element Method

The FEM holds promise for solving radiative transfer problems with combined energy transfer modes because it can provide a match with the computational grid used when other transfer modes are present (Nice 1983, Razzaque et al. 1983). It can also provide an exact solution in the sense that the exact transfer equations can be solved. The medium is divided into convenient subvolumes (finite elements). As described in Section 12.7.4, nodes are placed along element boundaries, and some can also be located within each element. The values of temperature, its fourth power, or other dependent variable, at the nodes are to be obtained from the solution. The distribution of temperature (or other dependent variable) is described, to the degree of accuracy desired, by using interpolation functions based on the expected form of the solution. If each element is taken to be isothermal, the method is equivalent to a zoning technique (Section 12.5) and, for small elements, the finite-difference method. In 2D problems, the temperature distributions in the elements are usually described by biquadratic functions. This allows a continuous temperature profile to be prescribed in the medium by matching the element boundary node temperatures to the temperatures at the boundary nodes of adjacent elements. The use of higher-order functions for the element temperatures allows additional matching of temperature derivatives at the element boundary nodes.

13.5.3 FEM for Radiation with Conduction and/or Convection

The FEM has been applied to various combined-mode solutions with radiative transfer. Introductory information is in Chung (1988) and Jaluria and Torrance (1986) and a further development is in Altes et al. (1986). Most of the solutions have used Galerkin's method.

The discrete ordinates method (Section 12.3) is often used for computing the radiative flux divergence for use in Equation 10.4. Galerkin FEM is known to be unstable if the element Péclet number (Pe) and Damköhler number (Da) product is larger than one. If this product is less than or equal to unity, the method may or may not be stable. Tencer (2013) points out that this instability condition is trivially satisfied when diffusion is absent from the equation to be solved implying that the Galerkin finite-element solution to the first-order discrete ordinates equation is inherently unstable. He shows possible stabilization techniques based on modifying the local intensity vector or using even- and odd-parity forms for the intensity.

Finite-element numerical codes can be used for solving sets of simultaneous equations from the discretized finite-element formulation. Many codes are limited to linear systems, and an iterative solution may be necessary for nonlinear formulations resulting from combined modes. This can be done by assuming an initial temperature distribution to determine the unknown temperature-dependent coefficients in a linearized formulation and then iterating until convergence. The method for carrying out such a procedure is outlined here.

For an illustrative case with heat conduction and an energy source in the translucent medium, the energy equation for $T(x, y)$ is written from Equations 10.1 and 10.4, for a 2D gray medium with no scattering, convection, or viscous dissipation, as

$$4\kappa\sigma T^4(x,y) - k\nabla^2 T = F(x,y,T) \tag{13.45}$$

The $F(x, y, T)$ includes the local absorption of incident radiation and the specified local internal energy generation in the volume. The energy equation is written in dimensionless form as (N_c is used here for the conduction–radiation parameter to avoid confusion with N, which is the number of elements)

$$(4\vartheta^3)\vartheta - 4N_{\text{CR}}\nabla^2\vartheta = F(\tau_1,\tau_2,\vartheta) \tag{13.46}$$

where $N_{\text{CR}} = k\kappa/4\sigma T_{ref}^3$ and ∇ in Equation 13.46 is with respect to optical coordinates $\tau_1 = \kappa x$ and $\tau_2 = \kappa y$. Following the method of Section 12.7.4, Equation 13.46 is multiplied by a weighting function $W(\tau_1, \tau_2)$ and integrated over the entire volume of the region surrounding each τ_1, τ_2 location. As in the derivation of Equation 12.175, an approximate solution $\vartheta(\tau_1, \tau_2) \approx U(\tau_1, \tau_2)$ and the assumed form for $W(\tau_1, \tau_2)$, as in Equation 12.173, using the same shape functions as in $U(\tau_1, \tau_2)$, are then substituted to obtain as in Equation 12.177:

$$\sum_{i=1}^{N} W_i \left\{ \sum_{j=1}^{N} \left[\int_V (-N_{\text{CR}}\nabla\Phi_j\nabla\Phi_i + \vartheta^3\Phi_j\Phi_i)dV \right] U_j - \frac{1}{4}\int_V F(\tau_1,\tau_2,U)\Phi_i(\tau_1,\tau_2)dV \right\} = 0 \tag{13.47}$$

Note that the Φ_j used here describe the *temperature*, rather than in Equation 12.178 where they describe the *emissive power*. The effect of using the same basis functions Φ_j for both ϑ and ϑ^4 within a solution is discussed by Burns et al. (1995b). The ϑ^3 remains in Equation 13.46 as a result of the linearized iterative method being used for a solution with combined radiation and conduction. As was done for Equations 13.39 and 13.43, define

$$K_{ij} = \int_V \left(-N_{CR}\nabla\Phi_j\nabla\Phi_i + \vartheta^3\Phi_j\Phi_i\right)dV \tag{13.48}$$

and, as in Equation 12.176,

$$f_i = \frac{1}{4}\int_V F(\tau_1,\tau_2,U)\Phi_i\left(\tau_1,\tau_2\right)dV \tag{13.49}$$

Then, Equation 13.47 is written as

$$\sum_{i=1}^{N} W_i\left[\sum_{j=1}^{N} K_{ij}U_j - f_i\right] = 0 \tag{13.50}$$

where the K_{ij} in Equation 13.48 are different functions from those in Equation 12.175. The quantities under the integrals in Equation 12.175 and 12.176 depend on τ_1, τ_2, and on dummy integration coordinates over the 2D region surrounding the location τ_1, τ_2. Comparing with Equation 12.177, Equation 13.50 provides N simultaneous equations of the same form as Equation 12.178 to be satisfied by the values of the coefficients U_j for $j = 1, \ldots, N$:

$$\sum_{j=1}^{N} K_{ij}U_j = f_i \quad i = 1,2,3,\ldots,N \tag{13.51}$$

The solution to the set of equations (represented by Equation 13.51) can be obtained using an available finite-element computer routine, as the $[K_{ij}]$ matrix and the $\mathbf{f}$ vector for the f_i terms are defined. Note that, compared with the pure-radiation case in Equation 12.178, the $[K_{ij}]$ matrix from Equation 13.48 now contains the unknown $\vartheta^3(\tau_1,\tau_2)$ values; thus, the solution is iterative. In the iteration, the $\vartheta(\tau_1,\tau_2)$ from the previous iteration is used in the evaluation of the K_{ij}. The numerical solution methods outlined in Section 7.7.2 can be used.

13.5.3.1 Results from Finite-Element Analyses

For conduction and/or convection combined with radiation, finite-element solutions have been obtained including scattering and for boundary conditions of specified temperature or heat flux. Further, 1D geometries of a plane layer or cylinder were analyzed by Razzaque et al. (1982), Fernandes et al. (1981), Wu et al. (1981), and Fernandes and Francis (1982) for combined conduction or convection and radiation. The latter includes scattering and transient conditions. Moreover, 2D rectangular geometries were analyzed by Razzaque et al. (1983) for radiation only, in a gray, nonscattering medium; conduction was included in Razzaque et al. (1984), and in Sokman and Razzaque (1987), isotropic scattering was incorporated. A boundary layer flow was analyzed in Utreja and Chung (1989) using a 1D radiation model. The medium was gray, and scattering was not included. Another analysis combining radiation with convection is for a 2D flow in a diverging or converging channel with scattering included (Chung and Kim 1984). Kang and Song (2008) use the discrete ordinates method to evaluate the radiative transfer in a finite-element analysis.

The solutions obtained by Razzaque et al. (1984) are for combined radiation and conduction in a medium in a rectangular enclosure with prescribed boundary temperatures and emissivities. For the particular conditions studied, the FEM required modification. The square enclosure had a fixed

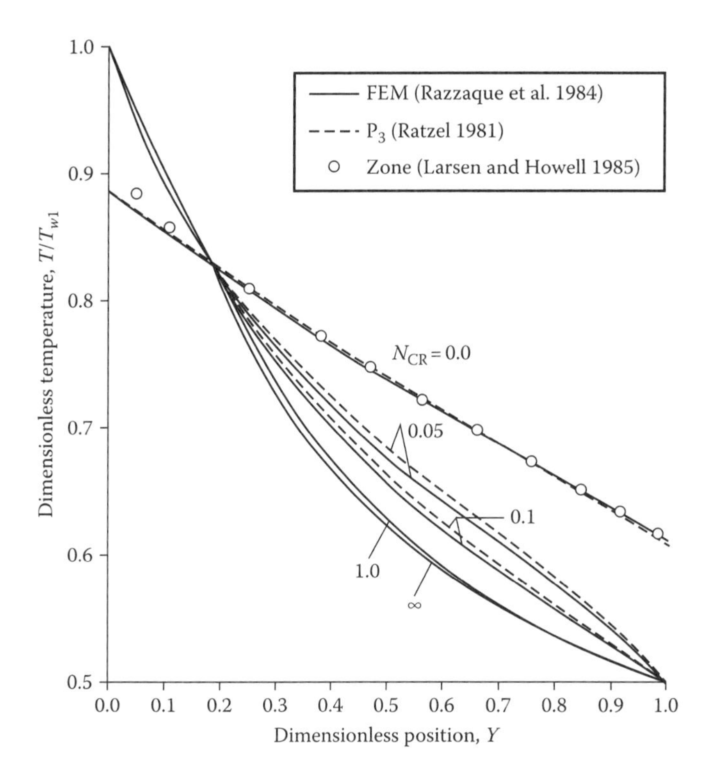

FIGURE 13.18 Nondimensional centerline temperature profiles in a square enclosure for various conduction–radiation parameters N_{CR} and with black walls; optical length of side $\tau_D = 1.0$; hot wall nodes at dimensionless temperature $\vartheta_{w1} = 1.0$, corner nodes at hot–cold intersections at $\vartheta_c = 0.75$, and other wall nodes at $\vartheta_{w2} = 0.5$. (From Razzaque, M.M. et al., *J. Heat Trans.*, 106(3), 613, 1984.)

temperature on one wall and a smaller fixed temperature on the other three walls. These conditions were chosen because solutions for the pure-radiation limit by the P_N and zone methods are available for comparison (Ratzel 1981, Larsen and Howell 1985). However, the FEM requires continuous boundary conditions, and in this instance, the conditions are discontinuous at the enclosure corners. The boundary conditions were modified so that, in the finite elements containing corners, the temperatures on the element boundary were forced to approach the average of the two wall temperatures. This is more realistic than for the original boundary conditions, because with conduction in the medium, it is not possible to have a temperature discontinuity on the boundary. This change in boundary conditions should be noted when comparing these solutions with other work. The solutions are exact within the accuracy of the numerical method, because the energy equation contained the exact integral for the local energy source. Results for centerline temperatures in the medium with N_c as a parameter are in Figure 13.18. Comparisons with the P_3 and zone results indicate that the change in corner conditions had little effect on the centerline profiles for the parameters shown.

Isotropic scattering and convection were included in an analysis of heat transfer in convergent and divergent channels (Chung and Kim 1984). The finite-element scheme parallels that outlined here but includes convection. Shape functions with four nodes were used. A Newton–Raphson method was used to solve the nonlinear matrix of the form in Equation 13.62. Solutions are presented for a set of cases with specified velocity.

A compressible boundary layer with combined conduction, convection, and radiation was studied by Utreja and Chung (1989) using an optimal control penalty method to speed convergence. This method reduces the order of the differential equations by replacing them with a set of lower-order equations and a set of constraint equations, which can be solved in finite-element form. For boundary layer flows, this method produces a well-conditioned positive-definite matrix of the K_{ij} coefficients.

13.5.4 Monte Carlo in Combined-Mode Problems

The Monte Carlo approach in combined-mode problems can be carried out by either combining the Monte Carlo formulation for radiative flux divergence directly into the energy Equation 10.2 or by solving for the radiative flux divergence by Monte Carlo using an assumed temperature distribution in the medium, substituting the local values into the energy equation, solving for a new temperature distribution, and iterating until convergence. The methods discussed in Sections 7.8 and 13.5 apply to Monte Carlo as well as to other techniques for solving for radiative flux divergence, $\nabla \cdot \mathbf{q}_r$. When using Monte Carlo to determine $\nabla \cdot \mathbf{q}_r$, the statistical nature of the results may hinder convergence of the temperature field.

For radiation combined with conduction and/or convection, an iterative solution procedure is to use a Monte Carlo method to evaluate the local radiative energy source on the basis of an initial assumption of the temperature distribution in the medium. This source distribution is then used in the energy equation, which is solved by any convenient numerical method for a new temperature distribution. This temperature distribution is used to repeat the Monte Carlo calculation for the radiative source, and the procedure is continued until convergence. A difficulty with this method is that the statistical fluctuations inherent in Monte Carlo evaluations can produce irregular behavior in the spatial variations of the radiative source. This may produce numerical instabilities in the solution of the energy equation and errors in the resulting conduction and/or convection terms.

13.6 COMBINED RADIATION, CONVECTION, AND CONDUCTION HEAT TRANSFER

Examples of combined convection, conduction, and radiation in absorbing, emitting, and scattering media are found in combustion chambers, atmospheric phenomena, shock-wave problems, rocket nozzles, high-temperature heat exchangers, and industrial furnaces. As a consequence, a large literature is available; review articles and comprehensive books are in Cess (1964), Viskanta (1963, 1966, 1982), Zel'dovich and Raizer (1966), and Edwards (1983). Mengüç et al. solved the coupled conduction, convection and radiation problem between two parallel plates using the F_N method. Kumar et al. (1990) provide a scheme for coupling the solution of the RTE by the discrete ordinates method to the energy equation by use of the local emissive power of the medium. Effectively, the energy differential equation is solved simultaneously with the set of first-order differential equations that result from the discrete ordinates representation of the equation of transfer in the region. Khalil (1982) applied the S_4 discrete ordinates method to predict incident radiative fluxes in a rectangular test furnace 2 × 2 × 6 m in size, with flow in the long wall direction and with a known temperature profile in the flowing gas. Good agreement was obtained with experimental heat fluxes and with calculations using the zone method. Another comparison gave good agreement for an axisymmetric furnace for heat fluxes calculated from known temperature distributions. When the S_4 model was used to predict both the temperature distribution and surface fluxes in a 4 × 4 × 11 m test furnace, comparisons with experiments were less accurate. For this comparison, the gas was assumed to be gray with an absorption coefficient $\kappa = 0.3\ \text{m}^{-1}$. Additional comparisons for S_4 models in rectangular furnaces are in Selçuk and Kayakol (1997).

Modeling of radiation with turbulent combustion in a coal-fired furnace was carried out by Varma and Mengüç (1989). They used the P_1 approximation, and considered different particle types and loadings. Turbulent flow of an absorbing, emitting, and scattering medium in a tube is investigated by Krishnaprakas et al. (1999a). The results showed that heat transfer was increased by having strongly forward scattering. A specular tube wall decreased heat transfer in comparison with a diffuse wall. The interaction of radiation with free convection from a vertical flat plate is in Krishnaprakas et al. (1999b). An inverse problem is considered in Linhua et al. (1998) where the temperature distribution in a rectangular furnace is to be deduced from measurements of other radiative quantities. Bauman et al. (2011) discuss numerical approaches to including radiation with other modes in the context of hypersonic flows during spacecraft reentry.

A boundary layer flow with internal radiative effects was considered in Section 13.4 where it was analyzed by using optically thin and optically thick approximations. Channel flows with radiation are now considered.

13.6.1 Forced Convection Channel Flows

Combined radiative and convective energy transfer for flow of an absorbing–emitting gas in a channel is of engineering interest in some high-temperature heat exchange devices. For laminar flow, the energy Equation 13.15, with $v = 0$ for fully developed flow, applies within the flow with the appropriate radiative energy source. Solutions with varying degrees of approximation are in Einstein (1963b), Viskanta (1963), Chiba and Greif (1973), Greif and Willis (1967), Martin and Hwang (1975) and Mengüç et al. (1983). Approximate numerical solutions for the flow of a gray absorbing–emitting medium in a parallel-plate channel were first discussed by Viskanta (1963). However, here temperature-independent properties are assumed. In addition to τ_D, N, and temperature ratios, a new parameter enters; it is a Nusselt number that can be defined to include a radiative contribution, thus differing from the usual convective parameter. The gas-to-surface and gas-to-gas direct-exchange areas of the zonal method were applied in Einstein (1963a,b) for solving the energy equation in a finite-length parallel-plate channel and circular tube with internal heat generation in the gas. Comparisons were made with Adrianov and Shorin (1961), where the cold-material approximation was used (Section 10.11.3), so that absorption but not emission from the gas was included. The zonal method was also used by Stasiek and Collins (1993) for flow of a radiating gas in a tube. The flow in Kim and Baek (1996b) is laminar and was assumed fully developed throughout the tube, so the flow equation did not need to be solved. The external surface of the tube was cooled by convection that varied around the tube circumference, and the interior tube wall was black. The flowing absorbing gas was gray with isotropic scattering. The radiative portion of the analysis was solved with discrete ordinates, and a finite-difference control volume method was used to discretize the energy equation; the resulting equations were solved by a tridiagonal matrix algorithm. The results illustrate the augmentation of cooling when radiation is significant compared with convection.

The preceding analyses are for gray gases flowing in channels with gray or black walls and for temperature-independent properties. A tube-flow heat transfer analysis is in De Soto and Edwards (1965), which accounts for nongray gases with temperature-dependent radiative properties. An exponential wideband model (book web appendix at http://www.thermalradiation.net) was used to account for the gas spectral effects, and entrance region flows were included. A similar study by Pearce and Emory (1970) employed a box model for the nongray absorption properties where an absorption band is approximated by having a constant absorption coefficient within an effective bandwidth and has zero absorption elsewhere. Jeng et al. (1976) also analyzed the laminar flow of a radiating gas in a tube with uniform wall temperature for a nongray medium. They used an expression for band radiation suggested by Tien and Lowder (1966) and examined the optically thin and thick limits for real gases. In the thick limit, logarithmic behavior suggested by Goody and Belton (1967) was used for a vibration–rotation band. In Soufiani and Taine (1989), a narrowband model was used to analyze emitting and absorbing laminar gas flow between two parallel walls at uniform temperature, and the results are compared with experiment. A water vapor–air mixture was studied for conditions with strong temperature gradients. Variable properties were included by coupling the continuity, momentum, and energy equations. The interaction of convection and radiation for steam has been studied by Kim and Viskanta (1984) and Chiou (1993). For high-pressure steam at 68 atm, it was found by Kim and Viskanta that radiation reduced the convective Nusselt number. In Chiou, for lower pressures of 1.7–5.1 atm, it was found that heat transfer could be computed by adding the radiative transfer to the usual convective transfer, and radiation contributed up to 27% of the total transfer. The calculations included the turbulent conductivity and the effect of spectral property variations of the steam. In Hirano (1988), a nongray analysis was made of a laminar channel flow of high-temperature CO_2 and combustion gases. The effect of installing an additional plate in the channel was studied. For some

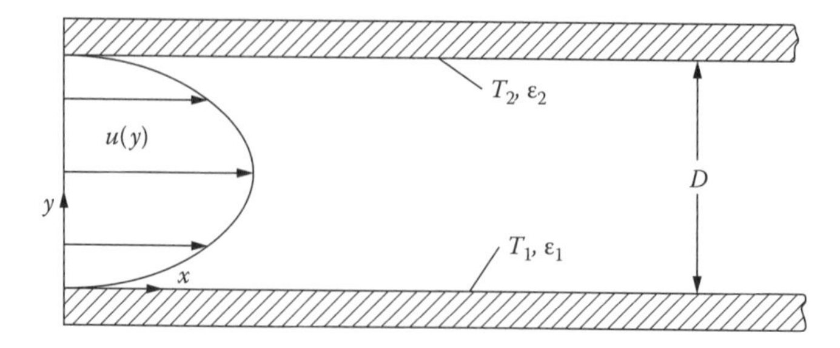

FIGURE 13.19 Flow of radiating and scattering medium between parallel walls at unequal temperatures.

conditions, this serves as a means of heat transfer enhancement by absorption and reradiation. The effect of anisotropic scattering on turbulent flow of a radiating medium in a tube was analyzed by Krishnaprakas and Narayana (1999). França et al. (2001) used an inverse analysis for determining the required heat flux and temperature distribution on one channel wall containing a medium in laminar flow to give a required temperature and heat flux on the opposite boundary.

An example of a channel flow analysis is heat transfer by an absorbing, emitting, and scattering medium flowing between parallel walls at different uniform temperatures (Figure 13.19). The medium and walls are gray, scattering in the medium is isotropic, and the walls are diffuse and opaque. All physical properties are constant. The velocity distribution is specified as fully developed when the flow enters the heat transfer region at $x = 0$; hence, it is only the thermal behavior that develops as a function of x. This is investigated by Chawla and Chan (1980). The energy equation is, from Equation 10.3 with viscous dissipation neglected,

$$\rho c_p u(y)\frac{\partial T}{\partial x} = k\left(\frac{\partial^2 T}{\partial x^2} + \frac{\partial^2 T}{\partial y^2}\right) - \left(\frac{\partial q_r}{\partial x} + \frac{\partial q_r}{\partial y}\right) + \dot{q}(x, y) \tag{13.52}$$

Axial heat conduction is usually small, and in addition, it is assumed that temperatures change slowly enough in the axial direction that $\partial q_r/\partial x \ll \partial q_r/\partial y$. For this example, there are no internal heat sources such as combustion in the flow. The $\partial q_r/\partial y$ is found from Equation 10.50 as $\partial q_r/\partial y = 4\pi(\kappa/\omega)(\sigma T^4/\pi - \hat{I})$, so the energy equation becomes

$$\rho c_p u(y)\frac{\partial T}{\partial x} = k\frac{\partial^2 T}{\partial y^2} - 4\frac{\kappa}{\omega}\left[\sigma T^4(x, y) - \pi\hat{I}(x, y)\right] \tag{13.53}$$

The source function $\hat{I}(x, y)$ is found as a function of y at each x from Equation 11.43 for a gray medium where $\tau = (\kappa + \sigma_s)y$:

$$\pi\hat{I}(\tau) = (1-\omega)\sigma T^4(\tau) + \omega\left[J_1 E_2(\tau) + J_2 E_2(\tau_D - \tau) + \pi\int_0^{\tau_D} I(\tau^*)E_1\left(\left|\tau^* - \tau\right|\right)d\tau^*\right] \tag{13.54}$$

The outgoing fluxes J_1 and J_2 are each equal to the sum of emitted and reflected energies and are given by Equations 11.66 and 11.67 for gray surfaces as

$$J_1 = \varepsilon_1\sigma T_1^4 + 2(1-\varepsilon_1)\left[J_2 E_3(\tau_D) + \pi\int_0^{\tau_D}\hat{I}(\tau^*)E_2(\tau^*)d\tau^*\right] \tag{13.55}$$

$$J_2 = \varepsilon_2 \sigma T_2^4 + 2(1-\varepsilon_2)\left[J_1 E_3(\tau_D) + \pi\int_0^{\tau_D} \hat{I}(\tau^*) E_2(\tau_D - \tau^*)\,d\tau^*\right] \tag{13.56}$$

Equations 13.53 and 13.54 are placed in dimensionless form and then solved numerically with the J_1 and J_2 used from Equations 13.55 and 13.56. For Poiseuille flow, $u(y) = 6\bar{u}[y/D - (y/D)^2]$, and at $x = 0$, $T(0, y) = T_i$, a uniform initial value; other initial temperature distributions could be used. Using $T(0, y) = T_i$, Equation 13.54 is solved by iteration for $\hat{I}(0, y)$ subject to the constraints of Equations 13.55 and 13.56. The $T(0, y)$ and $\hat{I}(0, y)$ are substituted into the right side of Equation 13.53 and the resulting values of $\partial T/\partial x$ as a function of y are used to obtain new T values for a grid of y points a small Δx down the channel. The $T(\Delta x, y)$ are then used in Equations 13.55 and 13.56 to obtain $\hat{I}(\Delta x, y)$, and the process is continued to move forward in the x direction.

Numerical solutions were obtained by Chawla and Chan (1980) for the flow initially at uniform temperature, with both walls at the same uniform temperature T_w and with the same emissivity ε_w. For this case, the local Nusselt number is $\mathrm{Nu}(x) = q_w(x)2D/k[T_w - T_m(x)]$, where $T_m(x)$ is the local mean fluid temperature. The $q_w(x)$ is the local heat flux that must be added at the walls to maintain their uniform temperature. The heat loss from the wall interior surfaces is composed of conduction and radiation components. Using Equation 5.16 for the radiative flux, with J_w being the outgoing radiative flux,

$$q_w = -k\left(\frac{\partial T}{\partial y}\right)_{y=0} + \frac{\varepsilon_w}{1-\varepsilon_w}\left(\sigma T_w^4 - J_w\right) \tag{13.57}$$

Results are in Chawla and Chan (1980) for the variation of bulk mean temperature $T_m(x)$ and $\mathrm{Nu}(x)$ with axial distance as a function of scattering albedo, wall emissivity, and conduction–radiation parameter. For pure convection, $\mathrm{Nu}(x)$ approaches a constant value for large x. When radiation is present, $\mathrm{Nu}(x)$ was found to pass through a minimum and then increase farther downstream. The increase in $\mathrm{Nu}(x)$ became more significant as the optical thickness of the channel increased and as the conduction–radiation parameter decreased. In Kim and Lee (1990) and Huang and Lin (1991a,b), anisotropic scattering was included and a 2D radiative analysis was used to determine heat transfer for Poiseuille flow in a pipe. A significant difference was found from the result using a 1D radiative analysis; including radiative preheating of the entering fluid caused the difference.

A special case of the previous problem was examined by Siegel (1987b) with application to a liquid drop radiator for outer-space applications. This radiator consists of a layer of small hot liquid drops traveling through space and cooling by radiative loss. In this instance, because of the vacuum of space, there is no heat conduction within the layer and the layer has no solid boundaries. The surroundings are assumed to be at very low temperature. The convective velocity originates with a uniform distribution at $x = 0$ and remains that way in the absence of wall friction. Equations 13.53 and 13.54 then have the form

$$\rho c_p \bar{u}\frac{\partial T}{\partial x} = -4\frac{\kappa}{\omega}\left[\sigma T^4(x,y) - \pi\hat{I}(x,y)\right] \tag{13.58}$$

$$\pi\hat{I}(x,\tau) = (1-\omega)\sigma T^4(x,\tau) + \omega\pi\int_0^{\tau_D} I(x,\tau^*) E_1(|\tau^* - \tau|)\,d\tau^* \tag{13.59}$$

The numerical solution of these equations with $T(0, y) = T_i$ (a constant) revealed that, for large x, the local heat flux being radiated away, $q_r(x, 0)$, is proportional to the fourth power of the local mean temperature of the droplet layer, that is, $q_r(x,0)/T_m^4(x)$ is constant. In Siegel (1987a), this limit was examined further and it was found that Equations 13.58 and 13.59 could be solved by a separation of variables solution, which proved analytically that the emittance $q_r/\sigma T_m^4$ of the layer becomes constant as x is increased. An extension of the analysis, in Siegel (1987a), allowed for solidification of the medium within the radiating layer. Since radiation can escape from throughout the medium, there is the interesting feature that solidification occurs in a distribution throughout the volume rather than at a well-defined interface.

In some tube-flow analyses such as deSoto and Edwards (1965), the radiative terms in the energy Equation 13.52 are simplified by assuming that the chief radiative contribution to the gas at an axial location is the result of temperatures only in the immediate surroundings. The axial temperature variation is thus neglected in determining the radiative terms. The radiative fluxes are calculated as those from an infinitely long cylinder of gas having a radial temperature distribution the same as that at the axial location for which the flux is being evaluated.

The neglect of the axial radiation flux component is not valid for a short channel or tube or when conditions are changing rapidly in the flow direction. In this instance, a zone method could be used to account for radiation in both the transverse and axial directions. A specific situation is analyzed by Einstein (1963b) for flow of an absorbing–emitting gas without scattering in a black tube of diameter D, as in Figure 13.20a. Gas enters the tube at temperature T_i and leaves at T_0. The tube wall temperature is constant at T_w. The surrounding environments at the inlet and exit ends of the tube are assumed to be at the inlet and exit gas temperatures, respectively. The governing energy equation at position $\mathbf{r}$ in the tube for laminar flow is

$$\rho c_p u \left.\frac{\partial T}{\partial x}\right|_r = \frac{k}{r}\frac{\partial}{\partial r}\left.\left(r\frac{\partial T}{\partial r}\right)\right|_r - 4\kappa\sigma T^4(\mathbf{r})$$
$$+\kappa\left[\iiint_V \sigma T^4(\mathbf{r}^*) f(\mathbf{r}^*-\mathbf{r})\,dV + \iint_S \sigma T_s^4(\mathbf{r}^*) g\left((\mathbf{r}^*-\mathbf{r})\right) dS\right] \quad (13.60)$$

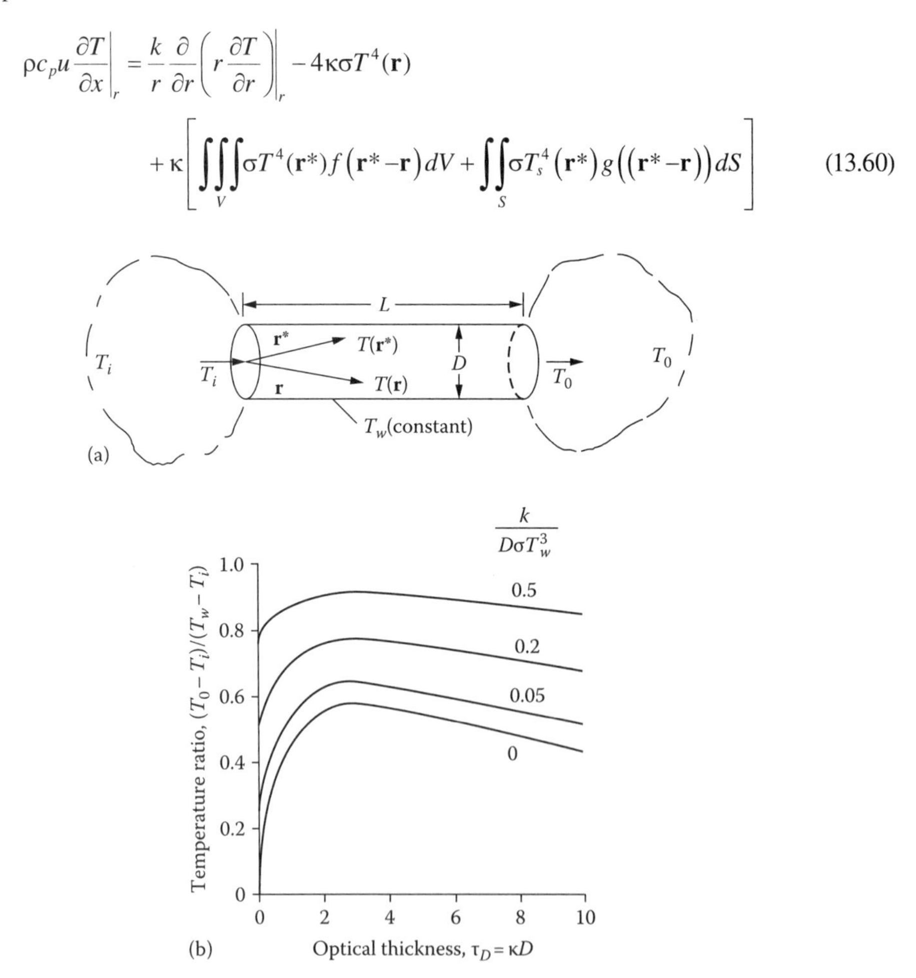

FIGURE 13.20 Combined radiation and convection for absorbing gas flowing in black tube with constant wall temperature. (a) Tube geometry and boundary conditions and (b) exit temperature for $T_i/T_w = 0.4$, $l/D = 5$, and $\rho\bar{u}c_p/\sigma T_w^3 = 33$. (From Einstein, T.H., *Radiant Heat Transfer to Absorbing Gases Enclosed in a Circular Pipe with Conduction, Gas Flow, and Internal Heat Generation*, NASA TR R-156, Washington, DC, 1963b.)

The triple integral is the radiation absorbed at r as a result of emission from all the gas in the tube. The $f(\mathbf{r}^* - \mathbf{r})$ is a gas-to-gas exchange factor from position $\mathbf{r}^*$ to position $\mathbf{r}$. The double integral is the radiation absorbed at $\mathbf{r}$ as a result of emission from the boundaries, which include the tube wall and the end planes of the tube. The $g(\mathbf{r}^* - \mathbf{r})$ is a surface-to-gas exchange factor; the f and g are given in Einstein (1963b).

Typical results of the numerical solution are in Figure 13.20b for a Poiseuille flow velocity distribution. These results show how well the gas obtains energy from the wall, since the ordinate is a measure of how close the exit gas temperature approaches the wall temperature. The results are in terms of gas optical thickness based on tube diameter and a conduction–radiation parameter based on wall temperature. As the optical thickness increases from zero, the amount of radiated energy from the wall that is absorbed by the gas increases to a maximum. Then, for large κ_D, the energy absorbed by the gas decreases. The decrease is caused by the self-shielding of the gas, which means that for high κ_D, most of the direct radiation from the tube wall is absorbed in a thin gas layer near the wall. Since gas emission is isotropic, about one-half the energy reemitted by this thin layer goes back toward the wall. Thus, the gas in the center of the tube is shielded from direct radiation, and the heat transfer efficiency decreases.

A 2D (radial and axial) analysis was made by Huang and Lin (1991a,b) for laminar flow in a circular tube. Uniform heating was imposed along a portion of the tube length, while the remainder of the tube wall was insulated. An absorbing–emitting gray medium was flowing through the tube, which had a black internal surface. Temperature distributions were obtained by solving the energy equation by iteration using Crank–Nicolson finite-difference method. A finite-element node approximation was used to evaluate the radiative source term in the energy equation.

The axial radiative transfer component can be important if a channel is significantly diverging or converging. A 2D analysis is in Chung and Kim (1984) for a radiating medium flowing between diverging or converging flat plates. Scattering is included, the walls are black, and they are at unequal uniform temperatures. The inlet and outlet boundary planes are treated by assuming they act like porous black surfaces. This may not always be a good approximation because of scattering back into the channel by the medium in the end reservoirs. The numerical solution showed that including radiation effects had a more significant effect on temperature profiles in a diverging channel than a converging one.

For fully developed turbulent flow in the tube, the energy equation using a turbulent eddy diffusivity ε_h is

$$\rho c_p u \frac{\partial T}{\partial x} = \frac{1}{r}\frac{\partial}{\partial r}\left[\left(k + \rho c_p \varepsilon_h\right) r \frac{\partial T}{\partial r}\right] - \nabla \cdot \mathbf{q}_r \tag{13.61}$$

The optically thin limit was used by Landram et al. (1969) to approximate $-\nabla \cdot \mathbf{q}_r$. It was evaluated for a volume element as equal to the energy absorbed from the wall by use of an incident mean absorption coefficient minus the energy emitted by use of a Planck mean absorption coefficient (see Section 12.11). If axial diffusion of radiation is neglected, then $\nabla \cdot \mathbf{q}_r = (1/r)\ \partial(rq_{r,r})/\partial r$. Expressions for $q_{r,r}$ (radiative flux in the radial direction) are in Wassel and Edwards (1976b) and Greif (1978), and detailed temperature-distribution and Nusselt-number results are in Wassel and Edwards for heat transfer to turbulent flow in a tube. Detailed calculations using an exponential band absorption model are in Balakrishnan and Edwards (1979) for flow in the thermal entrance region of a flat-plate duct downstream of a step in wall temperature. Laminar and turbulent flow are considered.

There are various applications where solid particles or liquid drops are suspended in gas streams in tubes or channels with conditions providing significant radiation. These include pulverized-coal combustion, steam–water droplet mixtures, and seeded gases in magnetohydrodynamic generators. Radiative analyses for turbulent particulate flow in a tube are in Azad and Modest (1981a) and Tabanfar and Modest (1987). The first of these includes anisotropic scattering by the particles, which are gray emitters and absorbers. The gas, however, is transparent and transfers energy to or from the particles by convection. The gas-phase energy equation has no radiation terms but has a term for convection to the particles. The particulate energy equation contains this convection term as well as the divergence of the radiative flux in cylindrical coordinates. In Tabanfar and Modest, the gas–particulate suspension is nonscattering, but the effect of a nongray gas is included. The gas and particle temperatures are assumed to be locally equal, so only one energy equation is used ($N_P V_P \ll 1$):

$$\left(N_p V_p \rho_p c_{p,p} + \rho_g c_{p,g}\right) u(r) \frac{\partial T}{\partial x} = \frac{1}{r} \frac{\partial}{\partial r}\left[\left(k_g + \rho_g c_{p,g} \varepsilon_{h,g}\right) r \frac{\partial T}{\partial r}\right] - \nabla \cdot \mathbf{q}_r(r) \tag{13.62}$$

with boundary conditions $\partial T/\partial r = 0$ at $r = 0$, $T = T_w$ at $r = D/2$, and $T = T_i$ at $x = 0$. The $\mathbf{q}_r$ is for a cylinder as given in Section 11.5.3. As briefly discussed earlier, the presence of radiation causes the Nusselt number, as a function of axial length, to pass through a minimum and then increase. In contrast, for convection only in a hot tube at uniform wall temperature, the Nu(x) approaches a constant asymptotic value. This behavior is for the wall being at a higher temperature than the gas. As the particulate radiation is increased, such as by having a larger particle density, the minimum Nu moves toward the tube entrance. In Tabanfar and Modest, the situation of a hot gas and cool wall was also studied. In this instance the fluid temperature decreases along the tube length and the importance of radiation is thereby reduced with axial distance. The Nu values are higher than for convection only, but Nu(x) does not pass through a minimum along the axial length.

13.6.2 Free-Convection Flow, Heat Transfer, and Stability

When free convection is significant, buoyancy appears in the momentum equation while the continuity and energy equations are unchanged. A review of literature upto 1986 is in Yang (1986). The effect of free convection on laminar forced upward flow of carbon dioxide in a vertical tube was studied analytically and experimentally by Greif (1978). For fully developed tube flow, the momentum equation is

$$\frac{dP}{dx} + \rho g = \mu \frac{1}{r} \frac{d}{dr}\left(r \frac{du}{dr}\right) \tag{13.63}$$

This must be solved in conjunction with the energy Equation 13.61, as the buoyancy term ρg is temperature dependent; the solution is in Greif. As is usual in free convection, the density was linearized by letting $\rho = \rho_w \cdot [1 + \beta(T_w - T)]$. In the same spirit, the spectral blackbody function was linearized by letting $E_{\lambda b}(T) = E_{\lambda b}(T_w) - (\partial E_{\lambda b}/\partial T)_w(T_w - T)$. Good agreement was obtained between predicted and measured temperature distributions. A more complex channel flow problem was studied analytically and experimentally by Yamada (1988). The geometry is parallel vertical walls with unequal uniform temperatures and with unequal heat addition to them. The heating produces upward free-convection flow between the walls. Spectral variations are included for the wall emissivities and for the radiation properties of the flowing medium.

The complex situation of combined forced and free convection of a radiating gas was analyzed by Yan and Li (2001) for flow in a vertical square duct. The radiative portion was solved by discrete ordinates (Section 12.3), and the 3D flow equations were solved by a vorticity–velocity method developed by Yan in earlier journal articles on convection. Radiation enhances the total heat transfer and speeds up the approach to fully developed flow. Using the same methods, another complex situation is analyzed by Yan et al. (1999) for a gray gas. This is convection in a radially rotating duct with a square cross section. Rotation induces buoyancy effects, and radiation increases both the total heat transfer and the thermal development of the flow. A combination of free convection and radiation heat transfer was studied by Guglielmini et al. (1987) for a vertical surface with a staggered array of fins attached to it. The analytical results were verified by an experimental study.

The effect of radiation on boundary layer development in free convection of an absorbing–emitting gas on a vertical plate is examined by Hasegawa et al. (1973), Cess (1966b), Arpaci and Gözüm (1973), Arpaci and Bayazitoglu (1973), and Pantokratoras (2014). Novotny and Kelleher (1967) consider layer development on a horizontal cylinder. The boundary layer heat transfer is studied by Yih (2000) along a vertical cone with the vertex pointed downward. The diffusion approximation was used for the radiative source. In Hasegawa et al. (1973) and Gille and Goody (1964), the onset of free convection was experimentally determined in a gas exposed to thermal radiation.

Numerical techniques have been applied to analyze combined radiation and free convection in enclosed rectangular regions filled with an absorbing–emitting medium. In Chang et al. (1983) and Kassemi and Naraghi (1993), an enclosure with a square cross section was studied. The 3D is large, so the geometry is 2D. The top and bottom horizontal boundaries of the square are insulated, and the two vertical side walls are at unequal uniform temperatures. Some results were also obtained by Chang et al. (1983) when partial vertical partitions were within the square, partially obstructing the view between the two vertical sides. For an enclosure without internal partitions, radiation increased the temperature of the gas within the enclosure except in the immediate vicinity of the cold wall and in a small region at the lower corner close to the cold wall. The net effect was that natural convection was somewhat reduced by the interaction of radiation with the gas. However, with regard to overall heat transfer, this effect was considerably overcompensated by the increase in heat transfer by radiation. The analysis by Webb and Viskanta (1987) was for a similar 2D rectangular geometry. A semitransparent fluid is in the form of a vertical layer that is bounded by four sides. The two horizontal boundaries are adiabatic. One vertical boundary is opaque and is maintained at uniform temperature in order to cool the fluid. The other vertical boundary is transparent and serves to confine the fluid and to transmit incident external radiation. The fluid is cold, so it absorbs incident radiation but does not radiate. The analysis provided the free-convection flow patterns. An increase in fluid layer opacity promotes the formation of a boundary layer region near the transmitting boundary because of the high volumetric absorption of energy in the region. The effects of internal radiation and free convection on crystal growth were analyzed by Matsushima and Viskanta (1990). A numerical investigation was made by Fusegi and Farouk (1989, 1990) of the radiative interaction with laminar and turbulent free convection in a square enclosure filled with nongray gas. In Fusegi and Farouk (1989), the radiation calculations were made with a P_1 model and a weighted sum of gray gases model. A control-volume-based finite-difference method was used by Fusegi and Farouk (1990) and comparisons were made with experimental data. In Tan and Howell (1991), the product integration method (Section 12.7.1) was used to find temperature and velocity profiles in an absorbing–emitting and isotropically scattering medium in a gray-walled square enclosure with free convection. A 2D radiative analysis was necessary even for the case where both horizontal boundaries are adiabatic; otherwise, the variation in radiative flux on the boundaries and in the medium was not accurately included, and significant changes in streamlines and energy fluxes on the constant temperature boundaries were found. Mahapatra (2014) examined the case of 2D mixed laminar convection with an absorbing, emitting, and scattering gray medium in a rectangular cell driven by a moving upper boundary and with unequal wall temperatures.

A participating medium between concentric horizontal cylinders with free convection is analyzed in Tan and Howell (1989a), Burns et al. (1995a), and Morales and Campo (1992) using the YIX, finite-element, and P_1 methods, respectively. Flow streamlines and heat transfer are presented. All of these analyses assume that a vertical line of symmetry exists that is perpendicular to the horizontal cylinder axis. More recent models (Kuo et al. 1999) are based on analysis of the transient development to steady state and include the effects of flow bifurcation/chaos and do not impose the symmetry condition. These results show that the earlier solutions, while correct within their assumptions, provide only one rather unlikely final steady solution. Other long-term solutions are more likely to occur from plausible initial conditions when symmetry breaking is allowed. Other analyses for free convection in an absorbing–emitting medium between infinite concentric cylinders are in Borjini et al. (1998), Tian and Chiu (2006), Cassell and Williams (2007), and Le Dez and Sadat (2012a, 2012b).

For square and rectangular enclosures, the effect of radiation on the development of 3D free convection in a participating medium has been shown, using the P_1 approximation (Lan et al. 1999) and the complete equation of transfer (Lan et al. 2000). Both solutions used an accurate spectral method for numerical solution. Radiation was shown to suppress the first bifurcation in the flow to higher Rayleigh numbers, while augmenting the second transition to occur at lower Rayleigh numbers.

Another aspect of free convection is to know the stability conditions for which the onset of free convection can occur. The P_1 approximation was used by Arpaci and Gözüm (1973) to include radiation in an analysis of the thermal convective stability of a nongray fluid between horizontal parallel confining boundaries (the Bénard problem with radiation). A combined experimental and analytical study for CO_2 gas is in Hutchinson and Richards (1999). From Arpaci and Gözüm, the radiative flux equation for a nongray fluid is

$$\int_0^\infty \left[\frac{\partial}{\partial x_\kappa} \left(\frac{1}{\kappa_\lambda} \sum_{j=1}^{3} \frac{\partial q_{r\lambda,j}}{\partial x_j} \right) - 4 \frac{\partial E_{\lambda b}}{\partial x_k} - 3\kappa_\lambda q_{r\lambda,k} \right] d\lambda = 0 \tag{13.64}$$

In the limit $\kappa_\lambda x \to 0$, Equation 13.64 reduces to the thin-gas approximation with the term $-3\kappa_\lambda q_{r\lambda,k} \to 0$ so that

$$\int_0^\infty \left[\frac{\partial}{\partial x_k} \left(\frac{1}{\kappa_\lambda} \sum_{j=1}^{3} \frac{\partial q_{r\lambda,j}}{\partial x_j} \right) - 4 \frac{\partial E_{\lambda b}}{\partial x_k} \right] d\lambda = 0$$

By setting the integrand equal to zero, integrating with respect to x_k, and then integrating with respect to λ, this is transformed to

$$\sum_{j=1}^{3} \frac{\partial q_{r,j}}{\partial x_j} = \int_0^\infty 4\kappa_\lambda E_{\lambda b} d\lambda + C = 4\kappa_p \sigma T^4 + C \tag{13.65}$$

The $\kappa_p = \int_0^\infty \kappa_\lambda E_{\lambda b} d\lambda / \sigma T^4$ is from Equation 12.179. For the optically thick limit, Equation 13.64 reduces to (as in Equation 12.28 for no scattering)

$$q_{r,k} = -\frac{4\sigma}{3\kappa_R} \frac{\partial T^4}{\partial x_k} \tag{13.66}$$

where Equation 12.184 has been used for κ_R, the Rosseland mean absorption coefficient.

Because Equation 13.62 applies at both the optically thin and thick limits, Equation 13.65 and 13.64 can be substituted into Equation 13.62 to yield

$$\frac{\partial}{\partial x_k} \left(\sum_{j=1}^{3} \frac{\partial q_{r,j}}{\partial x_j} \right) = -4\kappa_p \sigma \frac{\partial T^4}{\partial x_k} - 3\kappa_p \kappa_R q_{r,k} = 0 \tag{13.67}$$

where, from Equation 13.65, the Planck mean was used to approximate κ_λ in the first term of Equation 13.64. For the 1D case,

$$\frac{d^2 q_{r,x}}{dx^2} - 3\kappa_p \kappa_R q_{r,x} = 4\kappa_p \sigma \frac{dT^4}{dx} \tag{13.68}$$

For the problem of an initial steady temperature profile in the fluid (no free convection, heat generation or dissipation), the general energy Equation 10.1 becomes

$$\frac{d}{dx}\left(k\frac{dT}{dx}\right)-\frac{dq_{r,x}}{dx}=0 \tag{13.69}$$

Equations 13.67 and 13.68 and the boundary conditions on $q_{r,x}$ specified by Equations 12.4.10 and 12.4.11, along with the prescribed wall temperatures, constitute the formulation for the initial fluid-temperature profile. This is used in the stability problem in Arpaci and Gözüm (1973), which requires only the temperature gradient in the initial state to determine whether the fluid will become unstable.

The effect of radiation on fluid stability in enclosed 2D slender vertical cavities (height/width $\approx$ 10) is analyzed by Desrayaud and Lauriat (1988). The fluid is emitting, absorbing, and nonscattering, and the side walls are each isothermal but at unequal temperatures. The stability of the convective region was investigated. The stability of the conduction regime before the onset of convection was first studied by Hassab and Özişik (1979). The instability of flow induced by radiation in a semitransparent medium was analyzed by Yang and Leu (1993) in a slender slot that is inclined to various angles between vertical and horizontal. The Eddington approximation was used for the radiative portion of the analysis where heating is by irradiation on one boundary of the layer. For small angles from the horizontal, the instability starts as stationary longitudinal rolls. The inclination angle for transition to instability in a gas was found to be a minimum when the optical thickness is about 1. For a larger optical thickness, radiation does not penetrate as well and the configuration is more stable.

Lan et al. (2003) solved for the 3D flow distribution in a participating medium in an enclosure heated from below and cooled from above using exact solution of the radiation integrals coupled with the spectral method solution for the flow. They also performed a stability analysis in this situation and found that radiation tended to stabilize the flow (i.e., retard the beginning of free convection to higher Rayleigh numbers).

Another type of stability problem is in solar ponds (Giestas et al. 1996). This is a layer of water in which a gradient in salt concentration is created to stabilize the flow while transmitted solar radiation is being absorbed at the bottom of the pond. This provides an unstable temperature gradient with the largest temperatures at the bottom of the layer. The larger salt concentration at the bottom of the pond stabilizes the water in the presence of the increased temperature tending to decrease water density near the bottom. The results show that decreased transparency in the pond promotes instability by direct absorption of solar energy within the water instead of at the bottom of the pond, and a larger salt gradient is then required for stable operation.

13.6.3 Radiative Transfer in Porous Media and Packed Beds

Radiative transport in porous media and packed beds involves radiative interaction with both conduction and convection in various types of irregular structures. Applications are in combustion, high-temperature heat exchangers, regenerators and recuperators, insulation systems, packed and circulating bed combustors and reactors, and proposed energy storage and energy conversion devices. In many cases, the porous material, which may be translucent such as some ceramics, is absorbing and emitting radiant energy while interacting by convection with a fluid passing through the structure. At high temperatures, the fluid is usually a gas, which can often be assumed transparent to radiation because the dimensions between the elements of the solid structure in the porous medium are usually much less than the radiative mean free path for absorption or scattering in the fluid. When no fluid is present, the energy transfer within the porous structure is by combined radiation and conduction, with radiation being increasingly important for high-temperature applications. Because of the rich literature and applications of radiation in porous media, an extended discussion is included in the book website at http://www.thermalradiation.net.

13.6.4 Radiation Interactions with Turbulence

Radiation is coupled with turbulence effects in absorbing–emitting media. This was first recognized by Townsend (1958). The coupling is through the radiative flux divergence in the energy equation, which affects the local temperature and, importantly for combustion problems, the local species concentrations, which are quite temperature dependent. The turbulent fluctuations in these quantities also affect the local radiative properties. In addition, the fluctuation in local temperatures around the local mean can cause a very distorted distribution of local fourth-power temperatures, which in turn affects local emission of radiation and causes radiative emission to be greater than would be predicted using a local fourth-power temperature based solely on the local mean.

Most research in this area has been devoted to problems in turbulent combustion (Song and Viskanta 1987, Kounalakas et al. 1988, Mazumdar and Modest 1998, Coelho 2012a,b, Cumber and Onokpe 2013; Centeno et al. 2014).

13.6.5 Additional Topics with Combined Radiation, Conduction, and Convection

A large body of literature deals with reentry of bodies into the atmosphere and radiation within and from hypersonic shock waves. A rigorous treatment of these problems is difficult because of the nonequilibrium chemical reactions that are coupled with the radiation effects. Good introductions and discussions of shock problems are in Zel'dovich and Raizer (1966) and Penner and Olfe (1968), radiation effects in rocket exhaust plumes have been examined in deSoto (1965), the influence of radiation in ablating bodies is examined in Nelson (1973) and Boles and Özişik (1972), and recent research includes work by Laux (2002), Lamet et al. (2008), Bourdon and Bultel (2008), Bansal et al. (2010a), Bauman et al. (2011), and Maurente et al. (2012). Radiation interaction within a layer of gas, including transpiration, is treated in Viskanta and Merriam (1967). Investigations are in Kaminski et al. (1995) and Kim and Baek (1996a) of external convection combined with radiation in a semitransparent gas. In Kaminski et al., heat transfer was investigated experimentally and numerically for flow of propane combustion products over a horizontal cylinder. The equation of radiative transfer was solved simultaneously with the continuity, momentum, and energy equations by an FEM. A wideband model was used in each of the important spectral bands, and a P_1 method applied for radiation in each band. For low Reynolds number flow, radiation makes an important contribution to the total heat transfer.

The complex subject of radiative transfer in a chemically reacting flow in a 2D nozzle was analyzed by Liu and Tiwari (1996); this is of interest in combustion for hypersonic propulsion. The 2D Navier–Stokes equations were solved with the energy equation including a chemical reaction model for hydrogen–air combustion and radiative transfer. The radiative transfer was simulated with the Monte Carlo method. The calculations were nongray, and a statistical narrowband model was used for the water vapor resulting from combustion. The calculations showed the significance of radiation on the heat flux along the nozzle wall.

13.7 INVERSE MULTIMODE PROBLEMS

Inverse problems of multimode energy transfer are of interest in the analysis and design of systems with participating media. Ruperti et al. (1996) used measurements of temperature within a participating medium to predict surface temperatures and fluxes. Linhua et al. (1998) deduce the temperature distribution in a rectangular furnace from measured quantities at the boundary. In França et al. (2001), a problem similar to that in Figure 13.19 (fully developed laminar flow of a conducting and radiating medium) is analyzed, except that both $T_1(x_1)$ and $q_1(x_1)$ are prescribed on surface 1, and it is desired to determine the conditions on surface 2, $T_2(x_2)$ and $q_2(x_2)$, that will provide these conditions. The problem is described by Equation 13.52 through 13.56; however, the ill-conditioned nature of the highly nonlinear problem with these prescribed boundary conditions makes the inverse solution difficult. Solutions by two inverse techniques are in França et al. for various values of the governing parameters. A general discussion of solution approaches to nonlinear inverse problems with participating and nonparticipating media is in França et al. (2002). Orlande et al. (2006) compared three parameter estimation methods to determine the thermophysical properties of a conducting semitransparent medium.

A plane layer was considered by Liu et al. (2001) to obtain an inverse solution. The observed data are the characteristics of the radiation leaving the layer and the boundary temperatures. The solution is to obtain the radiative source distribution in the radiating layer and the layer scattering behavior. This has application to remote sensing and for material property measurements. The solution of an inverse problem is used by Li (1999) to determine radiative properties in a medium and the conduction–radiation parameter. Inverse solutions are also in Li (1993) and Mengüç and Manickavasagam (1993) for 2D rectangular and cylindrical geometries. Further details on inverse problems are in Chapter 8.

13.8 VERIFICATION, VALIDATION, AND UNCERTAINTY QUANTIFICATION

The discussion in Section 7.10 discusses the need for computer code verification, validation, and uncertainty quantification for problems of radiative exchange among surfaces. That discussion applies as well for multimode problems with participating media and is worth referring to here.

Given that the code is verified and validated, what is the uncertainty in the results of the code (the quantity of interest, or QOI)? The prediction results may have uncertainty because of uncertainties in input data quantities (absorption and scattering coefficients, phase functions, emissivities, thermal conductivity, etc.). Can distributions of these quantities be known or predicted? Additional prediction uncertainties occur because of uncertainties in the model itself (approximations in geometry, in level of model accuracy [Monte Carlo, P_N, SP_N, S_N, M_N, diffusion, etc.]). Having defined possible sources of uncertainty, how do we quantify the uncertainty in predictions? Brute force methods such as multiple code runs using Monte Carlo choices of properties from their statistical distributions can be done but is computationally expensive. Approaches using Bayesian statistics are under study (Oden et al. 2010a,b). Once the sensitivity of the code to the various uncertain parameters is established, then the accuracy in factors causing the most uncertainty in predictions can be improved (Helton 2009).

Many commercial CFD codes include models of radiation in the multimode solution algorithms. The codes are usually well verified, although occasional bugs are still identified. However, because the codes seek to numerically analyze a very wide variety of physical problems and use a choice of models for radiation, turbulence, properties, etc., it can't be taken for granted that the codes have been validated against appropriate experiments. The treatment of spectral and directional surface properties, absorption and scattering coefficients, and spectral properties is often simplistic. Thus, the uncertainty in the predictions of the codes do not have identifiable uncertainty limits, and the results of these codes need to be used with great caution.

HOMEWORK

13.1 A plane layer of absorbing, emitting, isotropically scattering material with $n \approx 1$ is being uniformly heated throughout its volume with a volumetric heating rate $\dot{q}$. The layer is of thickness D and has absorption coefficient κ, scattering coefficient σ_s, and thermal conductivity k. The layer is confined between opaque walls that are diffuse–gray with emissivities ε_1 and ε_2 and temperatures T_1 and T_2. Derive the energy and RTEs including scattering required to perform a numerical solution for the steady temperature distribution across the layer from $x = 0$ to D with heat conduction included.

13.2 A plane layer of gray isotropically scattering, absorbing, emitting, and heat-conducting medium is being heated uniformly throughout its volume by a chemical reaction that starts at time zero. The volumetric energy generation rate is $\dot{q}$. The medium is in contact with black walls at $x = 0$ and D that are cooled to maintain them at T_i throughout the transient heating. Initially, the entire medium is at T_i. The medium has absorption coefficient κ, scattering coefficient σ_s, thermal conductivity k, density ρ, and specific heat c. Derive the transient energy equation and the integral equation for the source function that could be placed in finite-difference form to solve numerically for the transient temperature distributions as a function of position between the walls. Place the equations in a convenient dimensionless form that has a parameter containing the thermal conductivity.

13.3 Modify the formulation in Homework Problem 13.2 to include the effect of both boundaries being diffuse–gray with the same emissivity $\varepsilon < 1$ rather than being black.

13.4 In Homework Problem 11.5, for the situation of a scattering medium between gray plates, it is desired to double the amount of energy being transferred by having the scattering (nonabsorbing) particles suspended in a thermally conducting but nonabsorbing medium. What thermal conductivity of the medium is required to accomplish this?
Answer: 2.04 W/m·K.

13.5 The region between two large black parallel plates is filled with a stationary gray medium. The plate temperatures are 1000 and 500 K, the spacing between the plates is 10 cm, and the absorption coefficient of the medium is $\kappa = 0.1$ cm^{-1}. If the medium thermal conductivity is 0.045 W/(cm·K), what is the net energy flux being transferred between the plates?
Answer: 29.6 kW/m^2.

13.6 A partially transparent solid absorbing sphere is placed in a black cavity at T_b. The sphere temperature is initially low. The sphere has surface transmissivity τ_0, thermal conductivity k, and absorption coefficient κ, and scattering is negligible. Transparent gas at T_g is circulating in the cavity, providing a heat transfer coefficient h at the sphere surface. Give the equations and boundary conditions to compute the transient temperature distribution in the sphere for the initial period during which emission by the sphere is small. Neglect any refraction effects and any dependence of τ_0 on angle.

13.7 A long chamber with a rectangular cross section has opposite sides at T_1 and T_2, and the other two sides are insulated on the outside. The chamber is filled with a stationary medium that is absorbing and scattering and is optically thin. The medium has thermal conductivity k. The surfaces are diffuse–gray, with emissivities as shown; the insulated walls both have the same emissivity. Develop an approximate expression for the heat transfer from A_1 to A_2. (*Hint*: see Example 13.3.)

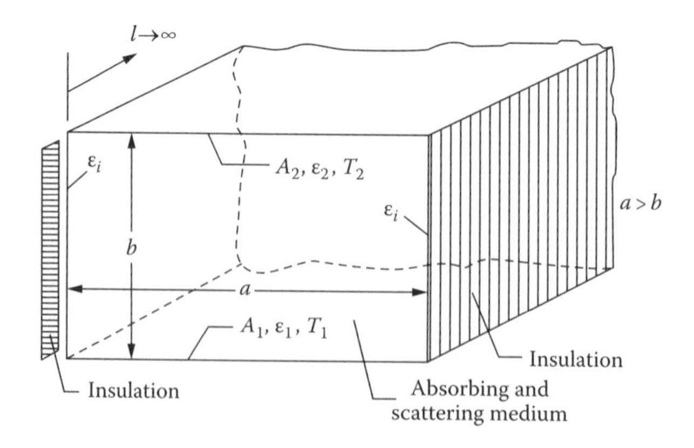

Answer: $\dfrac{Q}{al} = \dfrac{k}{b}(T_1 - T_2) + \dfrac{\sigma\left(T_1^4 - T_2^4\right)}{\dfrac{1}{\varepsilon_1} + \dfrac{1}{\varepsilon_2} - \dfrac{2F_{1-2}}{1+F_{1-2}}}$; where $F_{1-2} = \sqrt{1+B^2} - B$; $B = \dfrac{b}{a}$.

13.8 An enclosed rectangular region is filled with an absorbing, emitting, and heat-conducting medium without scattering. The medium is gray and has absorption coefficient κ and constant thermal conductivity k. The geometry is long in the z direction normal to the cross section shown and the side dimensions in the x and y directions are d and b. The four bounding walls are black. The vertical walls are both at uniform temperature T_v, and the horizontal walls are at T_h. The medium is heated uniformly throughout its volume with a volumetric energy rate $\dot{q}(x, y)$, and the medium is not moving. Provide the energy equation and radiative transfer relations necessary to solve for the steady-state temperature distribution in the rectangular cross section. Place the equations in convenient dimensionless forms for use in a numerical solution.

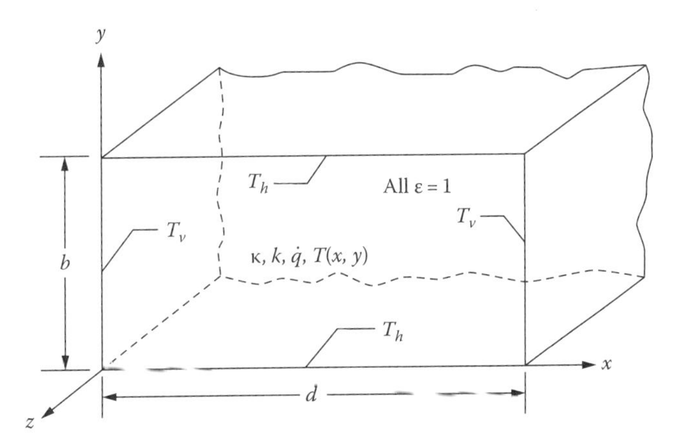

13.9 An absorbing liquid with absorption coefficient κ and thermal conductivity k_ℓ is flowing down a vertical flat plate. The plate is gray and diffusely reflecting with absorptivity α independent of incident angle. There is insulation with thickness B and thermal conductivity k_i on the back of the plate as shown. A solar flux q_s is incident on the plate at angle θ. Formulate the relations necessary to determine the mean temperature of the liquid at the bottom of the plate. Neglect the effects of refraction by the liquid. The result should be in terms of the liquid mass flow rate, plate length, plate absorptivity, and the insulation thickness and conductivity. The temperatures are low enough that emission from the plate and from the liquid may be neglected. The back side of the insulation layer is subjected to a heat transfer coefficient h_a.

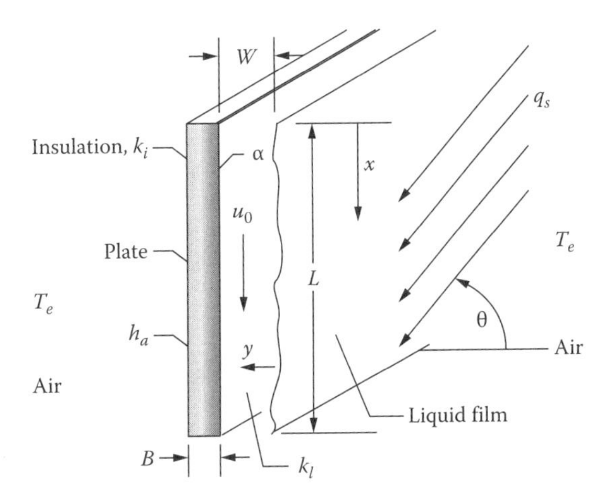

13.10 A plane layer of absorbing, emitting, isotropically scattering, and heat-conducting material is being heated throughout its volume with a parabolic volumetric heating rate $\dot{q}(x) = \dot{q}_{\min} + 4A\left[\frac{x}{D} - \left(\frac{x}{D}\right)^2\right]$ where A is the amplitude $\dot{q}_{\max} - \dot{q}_{\min}$. The layer is of thickness D and has constant absorption and scattering coefficients κ and σ_s and a constant thermal conductivity k. The layer is confined between surfaces that are diffuse–gray. The walls at $x = 0$ and D have emissivities ε_1 and ε_2 and temperatures T_1 and T_2.

1. By using the two-flux method, obtain a solution for the radiative heat transfer and the temperature distribution for the limit of zero heat conduction.
2. Provide the two-flux relations that can be solved for $q_r(x)$ and $T(x)$ with heat conduction included.

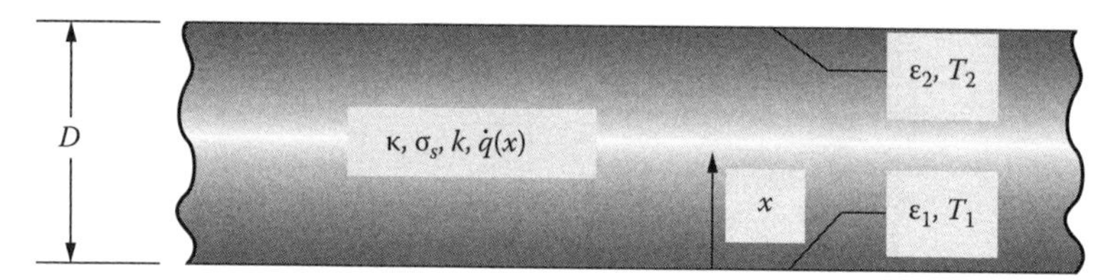

13.11 A gray nonscattering medium is contained between infinite parallel black walls at temperatures $T_1 = 1490$ K, $T_2 = 885$ K. The medium has absorption coefficient $\kappa = 0.4$ m^{-1} and thermal conductivity $k = 1.78$ W/(m·K). The walls are separated by a distance of 1.8 m, and there is uniform volumetric energy generation in the medium of 2.87 kW/m^3. The medium is not moving. Find the fluxes q_1 and q_2 at the walls and plot the temperature distribution in the medium. Solve the problem by a numerical finite-difference solution of the complete integrodifferential equation that applies.

$\varepsilon_2 = 1, T_2 = 885$ K
$\dot{q} = 2.87$ kW/m^3
$D = 1.8$ m
$\kappa = 0.4$ m^{-1}
$k = 1.78$ W/(m·K)
$\varepsilon_2 = 1, T_1 = 1490$ K
x

13.12 The finite-difference formulation for a plane layer in Section 13.5.2 is for a grid of X points with a uniform increment size, ΔX. An improvement is to use a variable grid size so that the distribution of grid points can be modified to place more of the points in regions where the temperature gradients are large. For unequal ΔX increments on either side of an X location, the finite-difference representation for the second derivative of temperature is

$$\left.\frac{\partial^2\theta}{\partial X^2}\right|_i = \frac{2\theta_{i+1}}{\Delta X^+\left(\Delta X^+ + \Delta X^-\right)} - \frac{2\theta_i}{\Delta X^+\Delta X^-} + \frac{2\theta_{i-1}}{\Delta X^-\left(\Delta X^+ + \Delta X^-\right)}$$

where

$\Delta X^+ = X_{i+1} - X_i$

$\Delta X^- = X_i - X_{i-1}$

Using this representation, modify the matrix coefficients in Equation 13.38 to incorporate a variable grid size.

13.13 Using the solution ideas for the 2D transient analysis in Section 13.5.2 and those in the steady 1D analysis at the end of Section 12.7.4, set up the FEM for the ADI solution for the following situation with an emitting, absorbing, and conducting gray medium in a 2D rectangular enclosure that is infinitely long normal to the cross section shown. There is no scattering in the medium. The rectangular region has a uniform heating $\dot{q}$ W/m^3 throughout its volume. The steady 2D temperature distribution is to be determined. For simplicity, use a grid having the same increment size in both the x and y directions. All of the boundary walls are black and are at the same temperature T_w.

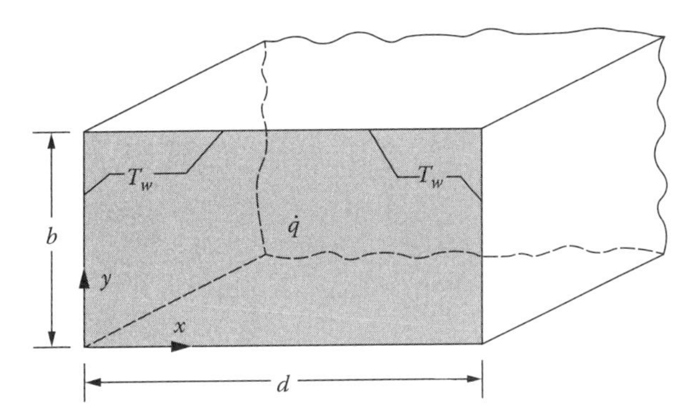

13.14 Modify Homework Problem 13.13 to incorporate a variable grid size into the solution for the steady temperature distribution within a square cross section. Use the ideas in Homework Problem 13.12 extended into 2D.

13.15 A plane layer of thickness D has a constant thermal conductivity k and has uniform internal heat generation $\dot{q}$ W/m^3 throughout its volume. The boundaries are maintained at fixed temperatures T_0 and T_D. It is desired to set up the solution method by finite elements for the limiting situation of pure conduction. Linear shape functions are to be used. The layer is divided into four equal elements as shown. Derive the required finite-element equations to be solved for the temperatures at the interior nodes.

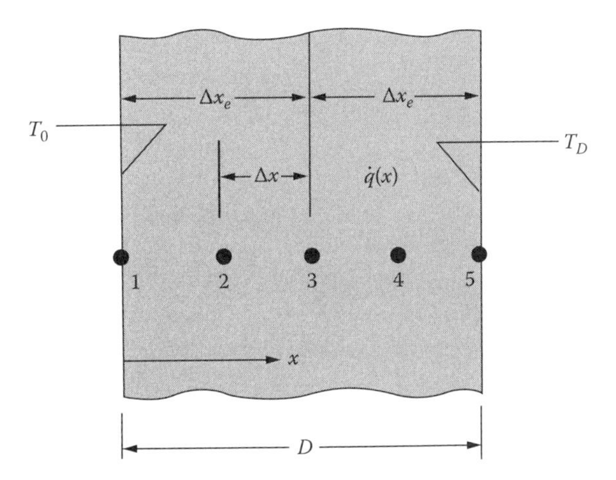

Answer: $T_0 - 2T_2 + T_3 = -\dot{q}(\Delta x)^2/k$

$T_2 - 2T_3 + T_4 = -\dot{q}(\Delta x)^2/k$

$T_3 - 2T_4 + T_D = -\dot{q}(\Delta x)^2/k.$

13.16 Modify Homework Problem 13.15 to have the Δx increments be of arbitrary unequal sizes across the width of the layer.

13.17 The plane layer of optically thin gray gas with heat conduction in Homework Problem 13.15 now undergoes a transient heating process. The layer is initially unheated and is at a low temperature in equilibrium with its surroundings. Then, the uniform internal heat generation $\dot{q}$ is suddenly applied. Develop the finite-element equations required to solve for the transient temperatures at the internal nodes 2, 3, and 4 as shown in the figure for Homework Problem 13.15. The thermal properties are assumed constant.

13.18 Reformulate Homework Problem 13.15 using the FEM with quadratic shape functions. Use the shape functions derived in Homework Problem 12.15, and the 2 elements and 5 nodes in that problem. For this problem, the $\dot{q}(x)$ distribution is specified as

$$\dot{q}(x) = 2 + \cos(\pi x/D).$$

13.19 Two large infinite parallel plates are separated by an optically thick absorbing–emitting, isotropically scattering, and conducting gas in which a chemical reaction is occurring that produces a uniform energy generation rate of $\dot{q}$ per unit volume. The plates are gray and have temperatures and emissivities T_1, ε_1 and T_2, ε_2. Determine the heat fluxes q_1 and q_2 that must be supplied to each of the plates as a result of radiation exchange combined with heat conduction between them. Use the radiative diffusion approximation, and assume that the gas is stationary and that Rosseland mean attenuation coefficient β_R and thermal conductivity k are both constant. Show how to implement the slip coefficient of Figure 13.8.

13.20 A chamber wall is constructed of three opaque parallel plates as shown. The surfaces are all gray, and plate 2 is sufficiently thin so that it can be considered to have uniform temperature across its thickness. What is the net heat flux from plate 1 to 3 if the regions between the plates are both a vacuum? An absorbing–emitting, isotropically scattering, and heat-conducting medium is now introduced between plates 2 and 3. Use the additive approximation with the diffusion method for the radiative transfer to estimate the net energy flux from 1 to 3 for this situation.

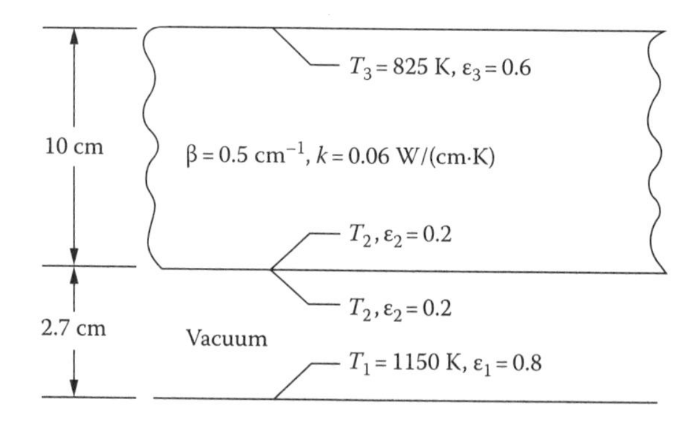

Answer: 6.68 kW/m², 4.97 kW/m².

13.21 Two large parallel plates are separated by a distance D. The plates both have a uniform temperature T_0 and have emissivity ε. A gas with Rosseland mean absorption coefficient κ_R is flowing between the plates at a uniform velocity U parallel to the plates. The gas has density ρ and specific heat c_p. Energy is added to each plate at a rate q_w per unit area. Neglecting heat conduction in the gas, derive an expression for the temperature distribution in the gas using the diffusion method. A fully developed temperature profile may be assumed as an approximation.

13.22 Repeat Homework Problem 13.21, but include heat conduction in the gas with thermal conductivity k. In addition, the absorption coefficient now depends on temperature as $\kappa_R(T) = \kappa_{R,0}(T/T_0)^3$.

13.23 A stagnant gray gas with attenuation coefficient $\kappa + \sigma_s = \beta = 2$ m^{-1} is contained between parallel plates spaced 0.45 m apart. The plates are nongray and have hemispherical spectral emissivities and temperatures as shown. The gas thermal conductivity is $k = 0.42$ W/(m·K).

Compute the energy flux transferred from plate 1 to plate 2. Use the additive approximation and the diffusion solution for the radiative transfer.

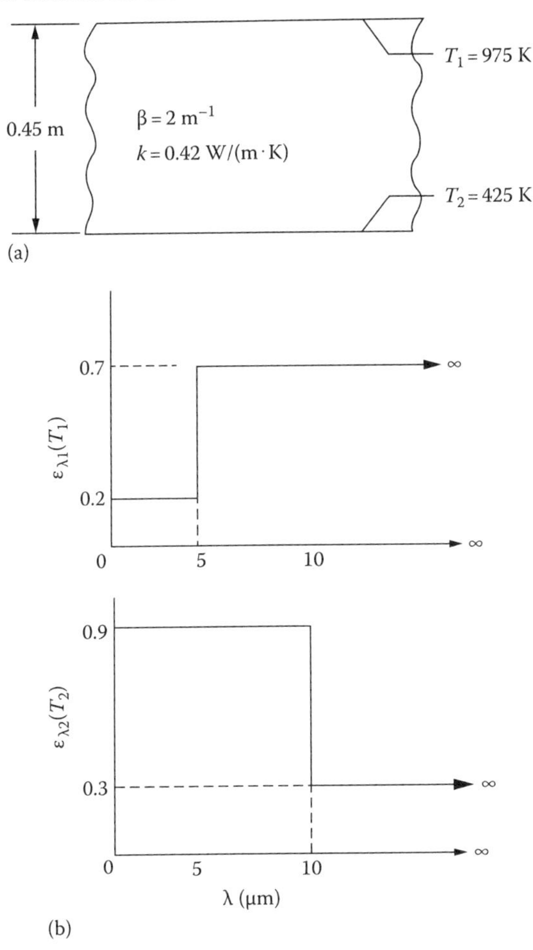

Answer: 13,180 W/m^2.

13.24 Two gray parallel plates with emissivities $\varepsilon_1 = 0.45$ and $\varepsilon_2 = 0.9$ are spaced 0.15 m apart. The plate temperatures are $T_1 = 515$ K and $T_2 = 297$ K. A stagnant nongray gas with spectral absorption coefficient as shown is in the space between the plates. The thermal conductivity of the gas is $k = 0.285$ W/(m·K). Using the additive method with the radiative diffusion solution, find the heat transfer between the plates. How significant is the heat conduction?

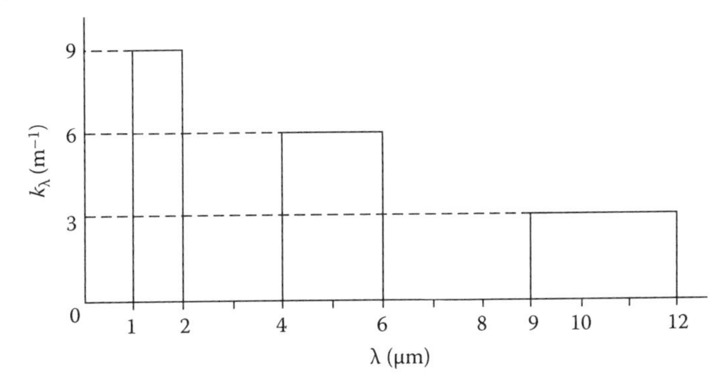

Answer: 1821 W/m^2; 23%.

13.25 The layer with uniform internal energy generation and constant thermal conductivity in Homework Problem 13.15 is now a gray gas without scattering that is optically thin and has absorption coefficient κ. Set up the finite-element equations to be solved for the temperatures at the internal nodes. The boundaries are transparent and do not radiate into the gas. The external surroundings are at a very low temperature.

13.26 A plane layer of semitransparent absorbing, emitting, and heat-conducting medium is between two thin metal (highly conducting) walls. The medium is not moving, and it is gray with absorption coefficient κ and thermal conductivity k. The wall surfaces are diffuse–gray with emissivities ε_1 on both sides of the wall at $x = 0$ and ε_2 on both sides of the wall at $x = D$. The large isothermal surroundings below the lower wall are filled with gas at T_{g1}, and there is a heat transfer coefficient h_1 between the gas and the outside of the lower wall. Outside the upper wall, there is a large isothermal reservoir of gas at T_{g2} that provides a convection coefficient h_2. Set up the necessary finite-difference relations to determine the wall temperatures T_{w1} and T_{w2} and the heat flow q from T_{g1} to T_{g2}.

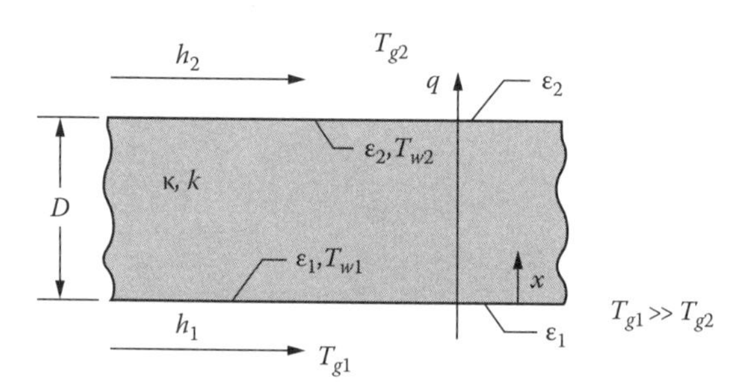

13.27 A parallel-plate channel is heated with a uniform heat flux q along the outside of each of its walls. A semitransparent absorbing, emitting, heat-conducting medium is flowing between the walls with fully developed Poiseuille flow having a parabolic velocity distribution $u(y)$. The medium has absorption coefficient κ and thermal conductivity k. Thermal properties are assumed constant. The channel wall interior surfaces are black. The region being considered is downstream from the channel entrance region. Axial heat conduction and axial radiative transfer are to be neglected in the formulation, although axial radiation can be considerable for some conditions. Set up a numerical procedure to determine the temperature distribution across the channel within the medium. List the assumptions made. Note that the mean temperature of the medium will rise linearly along the channel length as a result of the uniform wall heat addition and the assumptions of negligible axial conduction and axial radiation.

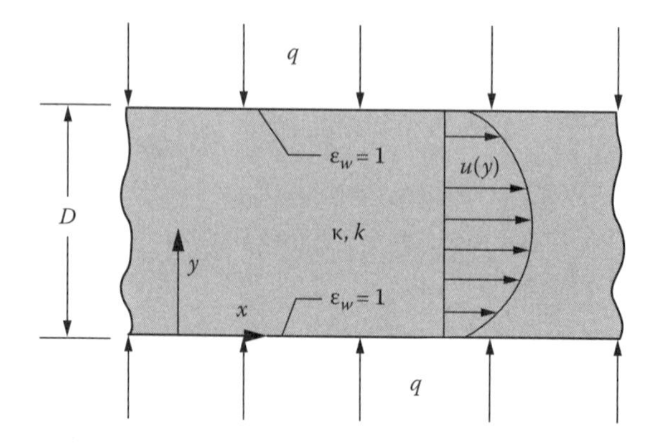

13.28 The gray medium in the following enclosure is nonscattering and has an absorption coefficient of $\kappa = 4\ \text{m}^{-1}$ and a thermal conductivity of $k = 20$ W/m·K.

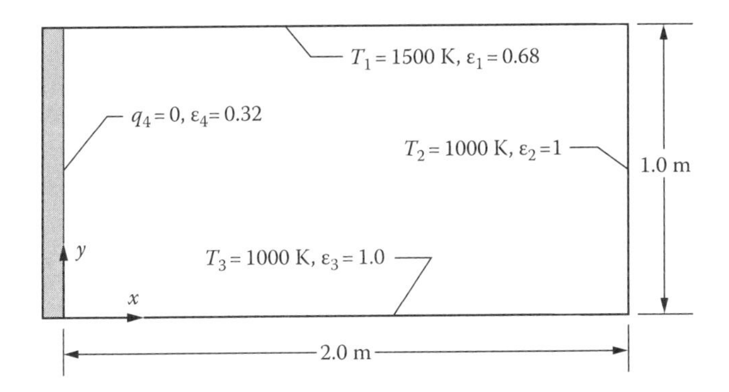

Using the diffusion method for the radiative transfer,

1. Show the governing equation(s) and boundary condition for the problem in specific and dimensionless form. Indicate any simplifying assumptions that are necessary.
2. Provide a plot of $q_3(x)$ on surface 3 and profiles of the medium temperature at $T(x = 0.5, y)$ and $T(x, y = 0.25)$.
3. Show limiting solutions for the cases of $N_{CR} = 0$ and $N_{CR} = \infty$.
4. Show that the solution is grid independent.
5. Show that energy is conserved on the system boundaries.

13.29 The gray medium in the following enclosure is nonscattering and has an absorption coefficient of $\kappa = 4\ \text{m}^{-1}$ and a thermal conductivity of $k = 20$ W/m·K.

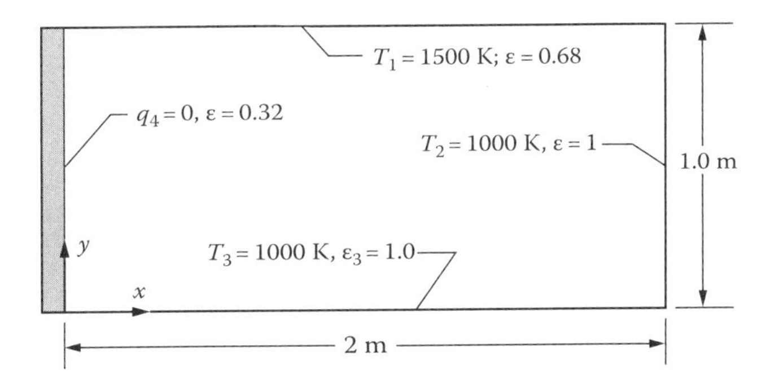

Using the Monte Carlo method for the radiative transfer,

1. Show the governing equation(s) and boundary condition for the problem in specific and dimensionless form. Indicate any simplifying assumptions that are necessary.
2. Provide a plot of $q_3(x)$ on surface 3 and profiles of the medium temperature at $T(x = 0.5, y)$ and $T(x, y = 0.25)$.
3. Show limiting solutions for the cases of $N_{CR} = 0$ and $N_{CR} = \infty$.
4. Show that the solution is grid independent.
5. Show that energy is conserved on the system boundaries.

13.30 Use the Monte Carlo method to determine the temperature distribution in the following 2D enclosure. The medium in the enclosure is nonscattering and has an absorption coefficient $\kappa = 10\ \text{m}^{-1}$ and thermal conductivity of 1.135 W/m·K.

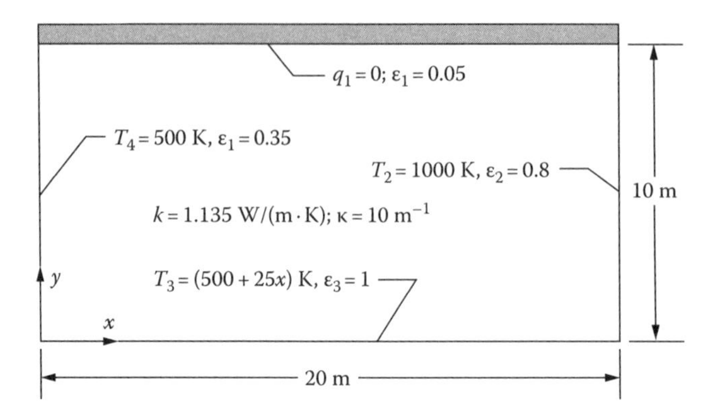

Provide plots of $q_3(x)$ on surface 3, $T_1(x)$ on surface 1, and $T_g(x = 10\text{ m}, y)$ and $T_g(x, y = 5\text{ m})$. Show that your solution is correct in the limit of no radiation and no conduction.

13.31 Use the Monte Carlo method to determine the temperature distribution in the following 2D enclosure. The medium in the enclosure has isotropic scattering coefficient of $\sigma_s = 5\ \text{m}^{-1}$ and absorption coefficient $\kappa = 10\ \text{m}^{-1}$ and thermal conductivity of 1.135 W/(m·K).

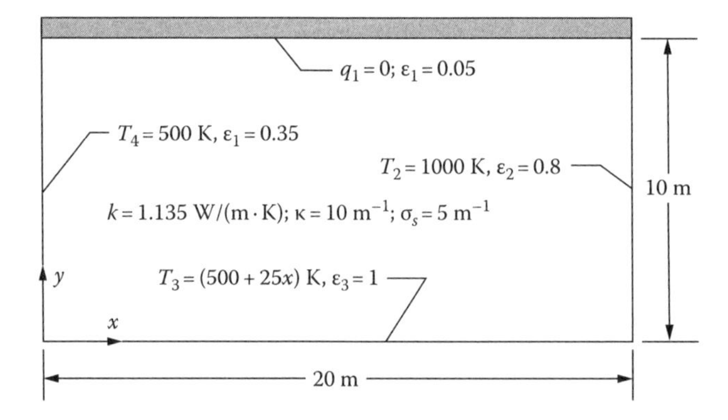

Provide plots of $T_g(x = 10\text{ m}, y)$ and $T_g(x, y = 5\text{ m})$. Show that your solution is correct in the limit of no radiation and no conduction.

14 Electromagnetic Wave Theory

14.1 INTRODUCTION

Light is an electromagnetic (EM) wave, and so is thermal radiation. As we have discussed in Chapter 1, radiation transfer takes place within the EM spectrum between 0.1 and 100 μm, although for most practical engineering applications, we consider the range between 0.3 and 20 μm, including the visible spectrum from 0.4 to 0.7 μm. These wavelengths are quite small length scales compared to many physical objects we study. For example, the thickness of a hair strand is about 100 μm, and most engineered objects are larger than that. Given this, the wave nature of radiation transfer is not needed for the solution of many engineering problems; instead, the radiative transfer equation (RTE) as discussed in the previous chapters can be suitably employed. It should be understood, however, that the wave nature of radiation transfer is crucial for determining the spectral properties of different dielectric and metallic media.

Particles and gases absorb and scatter radiation spectrally, and without detailed understanding of the wave nature of light, it will be impossible to determine their optical properties and their contributions to the radiative energy balance. In addition, as the object sizes get smaller, as is the case of micro- and nanoscale objects, the EM wave propagation needs to be considered in detail to calculate the energy exchange between them. Only with the use of EM theory can we consider polarization, coherence, and radiation tunneling, along with the propagation of radiative energy between such structures. In this chapter, we provide the general expressions for the EM wave theory and consider their use for determining surface properties. Finally, we briefly describe how the EM theory can be used to derive the RTE.

Sir James Clerk Maxwell is given credit for establishing the coherent concept of EM waves with his seminal paper published in 1864. This paper, which is considered to be a crowning achievement of classical physics, shows the relation between electric and magnetic fields and concludes that EM waves propagate with the speed of light, indicating that light itself is in the form of an EM wave (Maxwell 1890). Even though later studies have shown that quantum mechanics is a more general theory for the explanation of energy transfer at all scales, the EM wave theory is sufficient for understanding the light and radiative energy propagation and its interaction with matter at most size ranges. The so-called Maxwell equations can properly explain all the fundamental physics from nanoscales to stellar systems (Jackson 1998, Mishchenko et al. 2006). The concepts of reflection, refraction, transmission, and scattering can be explained with the use of these equations. In addition, values for these parameters, as well as absorptivity of materials can, in certain cases, be calculated from their optical and electrical properties, as shown for the applications in Chapter 3. However, the Maxwell equations do not account for the emission term, which needs to be considered separately and is discussed in Chapter 16.

Here, the relations between radiative, optical, and electrical properties are developed by considering wave propagation in a medium and the interaction between the EM wave and matter. An ideal interaction is considered for optically smooth, clean surfaces that reflect, refract, and transmit the incoming wave in an ideal manner. In its most simple case, these interactions assume specular, that is, mirror-like behavior, as such behavior is mathematically tractable and can be solved using Maxwell equations. Most real surfaces, however, are not mirrorlike and almost all have surface roughness, contamination, impurities, and crystal-structure imperfections, requiring the development of different approximations to account for diffuse or diffuse–specular reflection, refraction, and transmission characteristics. The departures of real materials from the ideal conditions assumed in the theory can produce large variations of measured property values from theoretical predictions.

Although the ideal theory based on the Maxwell equations cannot completely represent the actual physics involving real surfaces, it serves a number of useful purposes. For example, it provides an understanding of why there are basic differences in the radiative properties of insulators and electrical conductors and reveals general trends that help unify the presentation of experimental data. These trends become crucial for engineering calculations when extrapolation of limited experimental data is needed into other spectral and thermal ranges. The classical theory of EM waves also has utility in understanding the angular behavior of the directional reflectivity, absorptivity, and emissivity. Since the theory applies to pure substances with ideally smooth surfaces, it provides a means for computing the limits of attainable properties, such as the maximum reflectivity or minimum emissivity of a metallic surface.

In the following, first, we discuss the EM wave equations, that is, Maxwell equations. After that, we introduce a number of simplifications of these equations.

The derivation of radiative property relations from classical theory is in Sections 14.2 through 14.4. Readers interested only in using the results for property predictions can refer directly to Chapter 3, where we have outlined the property predictions obtained from the EM wave theory. Finally, in Section 14.5, we discuss the relationship between the RTE and the EM wave theory, as outlined by Mishchenko et al. (2006) and Mishchenko (2014).

14.2 EM WAVE EQUATIONS

Maxwell equations can be used to describe the propagation of EM waves within any medium. In most general form, they can be written as

$$\nabla \times \mathbf{E} = -\frac{\partial \mathbf{B}}{\partial t} \quad \text{(Faraday's law)} \tag{14.1}$$

$$\nabla \times \mathbf{H} = \mathbf{J} + \frac{\partial \mathbf{D}}{\partial t} \quad \text{(Ampère s law)} \tag{14.2}$$

$$\nabla \cdot \mathbf{D} = \rho_e \quad \text{(Gauss's law)} \tag{14.3}$$

$$\nabla \cdot \mathbf{B} = 0 \quad \text{(Gauss's law)} \tag{14.4}$$

where

E, **D**, **H**, and **B** are, respectively, the electric field, electric displacement, magnetic field, and magnetic induction

ρ_e is the free charge density

J is the current density

t represents the time

The local conservation of charge is given by the following continuity equation:

$$\frac{\partial \rho_e}{\partial t} + \nabla \cdot \mathbf{J} = 0 \tag{14.5}$$

For linear and isotropic media, the electric displacement and magnetic induction are related to the electric and magnetic fields. Here, it may be desirable to give some introductory explanation for some of these concepts. In a dielectric material, the presence of an electric field **E** causes the atomic nuclei and their electrons, that is, the bound charges in the material, to slightly separate. This causes induction of a local electric dipole moment. The displacement field **D** is then linearly related to electric field and the dipole moment. On the other hand, Faraday's law of induction tells us that the induced electromotive force in any closed circuit is equal to the rate of change of the magnetic flux through the circuit. This law clearly shows that the time-varying electric field is giving us a time-varying magnetic field.

Two constitutive relations can be used: $\mathbf{D} = \gamma\mathbf{E} + \mathbf{P}$ and $\mathbf{B} = \mu\mathbf{H}$, where γ is the electric permittivity and μ the magnetic permeability. The current density **J** is also related to the electric field via the relation $\mathbf{J} = \mathbf{E}/r_e$, where r_e is the electrical resistivity. Here, for our discussion of propagation of EM waves, we drop the so-called polarization density **P**. From these and assuming that there is no accumulation of charge (i.e., $\rho_e = 0$), Maxwell equations are simplified to

$$\nabla \times \mathbf{E} = -\mu \frac{\partial \mathbf{H}}{\partial t} \quad \text{(Faraday's law)} \tag{14.6}$$

$$\nabla \times \mathbf{H} = \gamma \frac{\partial \mathbf{E}}{\partial t} + \frac{\mathbf{E}}{r_e} \quad \text{(Ampère s law)} \tag{14.7}$$

$$\nabla \cdot \mathbf{E} = 0 \quad \text{(Gauss's law)} \tag{14.8}$$

$$\nabla \cdot \mathbf{H} = 0 \quad \text{(Gauss's law)} \tag{14.9}$$

In these equations, SI units are used, where the corresponding quantities are listed in Table 14.1. When referring to vacuum conditions, we use zero subscripts to specify these properties.

The solutions to these equations reveal how EM waves travel within a medium and how the electric and magnetic fields interact with the matter. By knowing how waves move in each of two adjacent media and applying coupling relations at the interface, relations governing expressions for

TABLE 14.1
Quantities for Use in Electromagnetic Equations in SI Units

Symbol	Quantity	Units	Value
c	Speed of electromagnetic wave propagation	m/s	$c_0 = 2.9979246 \times 10^8$
B	Magnetic induction	V·s/m^2	
D	Electric displacement	C/m^2	
E	Electric field	N/C (newtons/coulomb)	
H	Magnetic field	C/(m·s)	
J	Current density	A/m^2	
K	Dielectric constant, γ/γ_0		
r_e	Electrical resistivity	Ohm·m; $\text{N·m}^2\text{·s/C}^2$	
S	Instantaneous rate of energy transport per unit area	$\text{N·m/(s·m}^2)$; W/m^2	
$x,y,z; x',y',z'$	Cartesian coordinate position	m	
γ	Electrical permittivity	$\text{C}^2/(\text{N·m}^2)$	$\gamma_0 = (1/\mu_0 c_0^2) = 8.854188 \times 10^{-12}$
μ	Magnetic permeability	$\text{N·s}^2/\text{C}^2$	
ρ_e	Free charge density	C/m^3	$\mu_0 = 4\pi \times 10^{-7}$

Note: A quantity with a zero subscript refers to evaluation in vacuum. Any value written in bold refers to vector quantity, such as **B**, **D**, **E**, **H**, **J**, and **S**.

reflection and transmission are obtained. Using these expressions, we can obtain the relations for absorption, which can be correlated with the emission via the use of Kirchhoff's law.

14.3 WAVE PROPAGATION IN A MEDIUM

Here, we discuss EM wave propagation within an infinite, homogeneous, isotropic medium. A *perfect dielectric* (defined as nonconductor or perfect insulator) is considered in Section 14.3.1, where the energy of the propagating EM waves is not attenuated. A medium with finite electrical conductivity is analyzed next in Section 14.3.2; examples of such media are the *imperfect dielectrics* (poor conductors), *metals* (good conductors), or *semiconductors* (materials with intermediate conductivity). In these media, the EM waves are attenuated as their energy is absorbed by the medium itself.

14.3.1 EM Wave Propagation in Perfect Dielectric Media

Generally, EM waves propagating in a perfect dielectric medium do not lose any of their energy, as perfect dielectrics are by definition nonabsorbing (nonattenuating). Wave propagation in such media is similar to that in vacuum except the speed of propagation is reduced in proportion to the index of refraction. Here, we first consider a vacuum or a dielectric material (or insulator) with large electrical resistivity such that the $\mathbf{E}/r_e$ term can be neglected in Equations 14.1 through 14.5. For this condition, Equations 14.1 and 14.9 can be written in Cartesian coordinates to provide three equations relating the x, y, and z components of the electric and magnetic fields:

$$\frac{\partial H_z}{\partial y} - \frac{\partial H_y}{\partial z} = \gamma \frac{\partial E_x}{\partial t} \tag{14.10}$$

$$\frac{\partial H_x}{\partial z} - \frac{\partial H_z}{\partial x} = \gamma \frac{\partial E_y}{\partial t} \tag{14.11}$$

$$\frac{\partial H_y}{\partial x} - \frac{\partial H_x}{\partial y} = \gamma \frac{\partial E_z}{\partial t} \tag{14.12}$$

Using Equations 14.8 and 14.9, along with Equations 14.10 through 14.12, we obtain

$$\frac{\partial E_x}{\partial x} + \frac{\partial E_y}{\partial y} + \frac{\partial E_z}{\partial z} = 0 \tag{14.13}$$

$$\frac{\partial H_x}{\partial x} + \frac{\partial H_y}{\partial y} + \frac{\partial H_z}{\partial z} = 0 \tag{14.14}$$

As shown in Figure 14.1, the coordinate system x,y,z is defined following the path of a wave propagating in the x direction. For simplicity, a plane wave is considered where all the quantities concerned with the wave are constant over any yz plane at any time. Hence $\partial/\partial y = \partial/\partial z = 0$, and Equations 14.10 through 14.14 reduce to

$$0 = \gamma \frac{\partial E_x}{\partial t} \tag{14.15}$$

$$-\frac{\partial H_z}{\partial x} = \gamma \frac{\partial E_y}{\partial t} \tag{14.16}$$

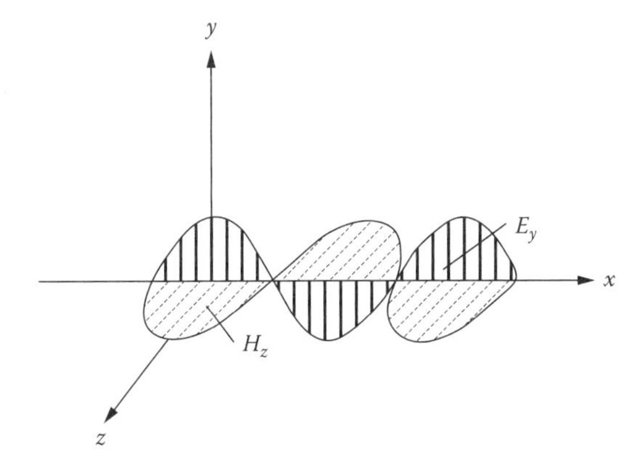

FIGURE 14.1 Wave propagation in homogeneous isotropic material. Shown are an electric field wave polarized in xy plane, traveling in x direction with companion magnetic field wave.

$$\frac{\partial H_y}{\partial x} = \gamma \frac{\partial E_z}{\partial t} \tag{14.17}$$

$$0 = -\mu \frac{\partial H_x}{\partial t} \tag{14.18}$$

$$-\frac{\partial E_z}{\partial x} = -\mu \frac{\partial H_y}{\partial t} \tag{14.19}$$

$$\frac{\partial E_y}{\partial x} = -\mu \frac{\partial H_z}{\partial t} \tag{14.20}$$

$$\frac{\partial E_x}{\partial x} = 0 \tag{14.21}$$

$$\frac{\partial H_x}{\partial x} = 0 \tag{14.22}$$

After differentiating Equation 14.16 with respect to time t and (14.20) with respect to space x, the results are combined to eliminate the H_z component. Similarly, from (14.17) and (14.6d) the H_y is eliminated. This yields two second-order partial differential equations for the electric field components E_y and E_z propagating in the x direction:

$$\mu\gamma \frac{\partial^2 E_y}{\partial t^2} = \frac{\partial^2 E_y}{\partial x^2} \tag{14.23}$$

$$\mu\gamma \frac{\partial^2 E_z}{\partial t^2} = \frac{\partial^2 E_z}{\partial x^2} \tag{14.24}$$

We can further simplify these equations by considering only the E_y component, which is equivalent to assuming that the EM waves are polarized such that the electric field vector E is only in the xy plane (Figure 14.1). Then, E_z and its derivatives are zero and Equation 14.24 does not need to be considered.

Note that the x components of E and H, from Equations 14.15, 14.18, 14.21, and 14.22, are related as $\partial E_x/\partial t = \partial E_x/\partial x = \partial H_x/\partial t = \partial H_x/\partial x = 0$. Hence, the electric and magnetic field components in the direction of propagation are steady and independent of the propagation direction x. The only time-varying component of **E** is E_y as governed by Equation 14.23. Since this component is normal to the x direction of propagation, the wave is called a *transverse* wave. Equation 14.23 is the *wave equation* for propagation of E_y in the x direction. The general solution of this wave equation is expressed as

$$E_y = f\left(x - \frac{t}{\sqrt{\mu\gamma}}\right) + g\left(x + \frac{t}{\sqrt{\mu\gamma}}\right) \tag{14.25}$$

where f and g are *any* differentiable functions. The f provides propagation in the positive x direction, while g accounts for propagation in the negative x direction. The present discussion is for a wave moving in the positive direction, so only the f function is present.

We can obtain the speed of the propagating wave in dielectric medium starting from the wave equation. For this, we first consider an observer moving with the wave such that the observer is always at a fixed E_y location on Figure 14.1. The x location of the observer must then vary with time, which means that the argument of f as given in Equation 14.25 is to be a constant. The speed of the propagating wave is expressed as dx/dt suggesting $dx/dt = 1/\sqrt{\mu\gamma}$. Then, the relation

$$E_y = f\left(x - \frac{t}{\sqrt{\mu\gamma}}\right) \tag{14.26}$$

represents a wave with y component E_y, propagating in the positive x direction with speed $1/\sqrt{\mu\gamma}$. In vacuum, light propagates with the speed of c_0; therefore, $c_0 = 1/\sqrt{\mu_0\gamma_0}$, as given in Table 14.1. Independent measurements of μ_0, γ_0, and c_0 validate this result. Indeed, this basic deduction led Maxwell to conclude that light must be EM wave.

During the propagation of an EM wave, the electric field component E_y is accompanied by a magnetic field component H_z. If Equation 14.16 is differentiated with respect to x and Equation 14.20 with respect to t, the results can be combined to yield

$$\mu\gamma\frac{\partial^2 H_z}{\partial t^2} = \frac{\partial^2 H_z}{\partial x^2} \tag{14.27}$$

which is the same wave equation as Equation 14.23. Hence, the H_z component of the magnetic field propagates along with E_y as in Figure 14.1.

The f function given in Equation 14.26 can have any complex and arbitrary-shaped waveform. Such a function can be represented using a Fourier series as a superposition of harmonic waves, each having a different fixed wavelength. We consider only one such spectral wave, as more complex wave forms can be constructed using a number of others using superposition. Starting here, we use complex algebra in expressing these waves to simplify the mathematical analysis. With this idea in mind, we write the expression for the E_y component of the electric field at the origin $x = 0$ as

$$E_y = E_{yM}\exp(i\omega t) = E_{yM}(\cos\omega t + i\sin\omega t) \tag{14.28}$$

where the M subscript indicates the maximum amplitude of the wave. An observer riding with this wave and leaving the origin at time t_1 will arrive at location x after a time increment of x/c, where

c is the wave speed in the medium. Hence, the time of arrival is $t = t_1 + (x/c)$, so that the time of leaving the origin becomes $t_1 = t - (x/c)$. A wave traveling in the positive x direction is then given by

$$E_y = E_{yM} \exp\left[i\omega\left(t - \frac{x}{c}\right)\right] = E_{yM} \exp[i\omega(t - \sqrt{\mu\gamma}x)] \tag{14.29}$$

This is a solution to the governing wave Equation 14.23, as is shown by comparison with Equation 14.26. Other forms of the solution can be obtained by using relations between the angular frequency ω, the linear frequency ν, or the wavelength, λ: $\omega = 2\pi\nu = 2\pi c/\lambda_m = 2\pi c_0/\lambda_0$, where λ_m and λ_0 are the wavelengths in the medium and in vacuum.

A medium other than vacuum is characterized by its refractive index n, where it is defined as the ratio of the wave speed in vacuum c_0 to the speed in the medium $c = 1/\sqrt{\mu\gamma}$. Hence, $n = c_0/c = c_0\sqrt{\mu\gamma} = \sqrt{\mu\gamma/\mu_0\gamma_0}$, and Equation 14.29 can be written as

$$E_y = E_{yM} \exp\left[i\omega\left(t - \frac{n}{c_0}x\right)\right] \tag{14.30}$$

As shown by Equations 14.29 and 14.30, the amplitude of the wave does not decrease in vacuum or in a *perfect dielectric* medium, which has zero electrical conductivity. In many real materials the electrical conductivity is significant; therefore, the last term on the right of Equation 14.6 cannot be neglected; this causes the propagating wave to get attenuated.

14.3.2 Wave Propagation in Isotropic Media with Finite Electrical Conductivity

In this section we discuss EM wave propagation in conducting media, including imperfect insulators (*imperfect dielectrics*) that have a low electrical conductivity, in semiconductors and in good conductors (*metals*). In these materials the traveling waves are attenuated. Attenuation in the medium is expressed by replacing the index of refraction of the medium with a *complex refractive index* $\bar{n}$:

$$\bar{n} = n - ik \tag{14.31}$$

where the imaginary part k is called as the *extinction index*. It should be understood, however, this is not the same extinction coefficient used in RTE. This extinction index k is related to the absorption coefficient κ of the RTE (see Equation 14.46). Here, both n and k are spectral, although we omit the λ subscript for the sake of clarity. Again, we consider a single plane wave in the following analysis; more complex wave forms can be constructed using superposition techniques and Fourier series expansion. If we substitute Equation 14.31 into 14.30 for n, the electric field takes the form

$$E_y = E_{yM} \exp\left\{i\omega\left[t - (n - ik)\frac{x}{c_0}\right]\right\} \tag{14.32}$$

or after some manipulation,

$$E_y = E_{yM} \exp\left[i\omega\left(t - \frac{n}{c_0}x\right)\right] \exp\left(-\frac{\omega}{c_0}kx\right) \tag{14.33}$$

The last exponential term indicates the attenuation or the energy absorption from the wave as it travels through the medium. Indeed, Equation 14.33 is a solution of the governing equations with Equation 14.6 included. With this term retained, Equation 14.23 takes the form

$$\mu\gamma\frac{\partial^2 E_y}{\partial t^2} = \frac{\partial^2 E_y}{\partial x^2} - \frac{\mu}{r_e}\frac{\partial E_y}{\partial t} \tag{14.34}$$

If the wave amplitude in Equation 14.33 is substituted into Equation 14.34, we obtain

$$c_0^2\mu\gamma = (n - ik)^2 + \frac{i\mu\lambda_0 c_0}{2\pi r_e} \tag{14.35}$$

which is needed to satisfy Maxwell equations. Equating the real and imaginary parts of Equation 14.35 yields

$$n^2 - k^2 = \mu\gamma c_0^2 \tag{14.36}$$

$$nk = \frac{\mu\lambda_0 c_0}{4\pi r_e} \tag{14.37}$$

These equations are solved for the components of the complex refractive index

$$n^2 = \frac{\mu\gamma c_0^2}{2}\left\{1 \pm \left[1 + \left(\frac{\lambda_0}{2\pi c_0 r_e \gamma}\right)^2\right]^{1/2}\right\} \tag{14.38}$$

$$k^2 = \frac{\mu\gamma c_0^2}{2}\left\{-1 \pm \left[1 + \left(\frac{\lambda_0}{2\pi c_0 r_e \gamma}\right)^2\right]^{1/2}\right\} \tag{14.39}$$

The positive sign is chosen for most cases in Equations 14.38 and 14.39; however, certain materials have negative electrical permittivity, requiring choice of the negative sign. Again note that Equation 14.29 for perfect insulating media is identical to Equation 14.33 for conducting media with the exception that the simple refractive index n is replaced by the complex refractive index $n - ik$. This also suggests that most relations for perfect dielectrics can be used for conductors, provided $n - ik$ is substituted for n.

14.3.3 Energy of an EM Wave

The energy carried by an EM wave per unit time and per unit area is expressed by the cross product of the electric and magnetic field vectors, which are orthogonal to each other:

$$\mathbf{S} = \mathbf{E} \times \mathbf{H} \tag{14.40}$$

where **S** is known as the *Poynting* vector, and it propagates perpendicularly to both of the oscillating **E** and **H** fields. The direction of energy propagation can be determined starting from the **E** and **H** vectors and by using the right-hand rule. For the plane wave shown in Figure 14.1, the electric field is in the y direction; therefore, the corresponding magnetic field is in z direction. Consequently, the EM wave propagation is in the positive x direction, and its magnitude S is

$$S = |\mathbf{S}| = E_y H_z \tag{14.41}$$

If E_y is given by Equation 14.33, then Equation 14.20, which applies for conductors as well as perfect insulators, can be used to find H_z as

$$-\mu \frac{\partial H_z}{\partial t} = \frac{\partial E_y}{\partial x} = -\frac{i\omega \bar{n}}{c_0} E_y = \frac{-i\omega}{c_0}(n - ik)E_y \tag{14.42}$$

Then, noting the t dependence of E_y in Equation 14.33 and integrating yield

$$H_z = \frac{\bar{n}}{\mu c_0} E_y \tag{14.43}$$

where the integration constant is set equal to zero. This constant would correspond to the presence of a steady magnetic intensity in addition to that induced by E_y and is zero for the conditions of the present discussion.

When H_z is substituted into Equation 14.41, the magnitude of Poynting vector is written as

$$|\mathbf{S}| = \frac{\bar{n}}{\mu c_0} E_y^2 \tag{14.44}$$

or in more general vector form as

$$|\mathbf{S}| = \frac{\bar{n}}{\mu c_0} |\mathbf{E}|^2 \tag{14.45}$$

The energy transmitted by the wave, per unit time and area, is proportional to the square of the amplitude of the electric field. The $|\mathbf{S}|$ is a spectral quantity that is proportional to spectral radiation intensity, I_λ, which we have introduced in Chapter 1. Radiation intensity passing through an attenuating medium decreases exponentially. This decay can be determined by examining Equations 14.32, 14.33 and 14.44, which shows the corresponding decay factor as $\exp(-2\omega k_\lambda x/c)$ or $\exp(-4\pi k_\lambda x/\lambda)$. As a result of absorption, the spectral intensity I_λ decays as $\exp(-\kappa_\lambda x)$, so the relation between the *absorption coefficient* κ_λ and the *extinction index* k_λ is

$$\kappa_\lambda = \frac{4\pi k_\lambda}{\lambda} \tag{14.46}$$

This equation provides a procedure to determine the spectral absorption coefficient κ_λ from optical data for the extinction index k_λ or the complex index of refraction as a function of wavelength.

14.4 LAWS OF REFLECTION AND REFRACTION

So far we have discussed the wave nature of propagating EM waves. Next, we can focus on the interaction of EM waves with matter. For this, we consider an interface and outline the laws of reflection and refraction in terms of the real and imaginary parts of the complex index of refraction, which are indeed related to the electric and magnetic properties of the two media via Equations 14.38 and 14.39. Here, our focus is on a single ideal planar interface, which is important for many engineering applications. These ideas can be extended to absorption and scattering of EM waves by finite size particles and agglomerates, which are the subject of Chapter 15.

14.4.1 Reflection and Refraction at the Interface between Perfect Dielectrics ($k \to 0$)

Next, let us consider the interaction at an optically smooth interface between two nonattenuating (*perfect dielectric*) materials. For simplicity, a simple cosine wave is used as given by the real part of Equation 14.28. An x', y', z', coordinate system is fixed to the path of the incident wave that is moving in the x' direction. The wave strikes the interface as in Figure 14.2, where the interface is in the yz plane of the x,y,z coordinates attached to the media. *The plane of incidence* is defined by containing both the normal to the interface and the incident direction x'. The coordinate system has been drawn so the y' direction is in the plane of incidence. The interaction of the wave with the interface depends on the wave orientation relative to the plane of incidence. If the amplitude vector of the incident wave is in the plane of incidence (amplitude vector in y', direction), then the amplitude vector is at an angle to the interface. If the amplitude vector is normal to the plane of incidence (amplitude vector in z', direction), then the incident wave vector is parallel to the interface.

Consider an incident wave $E_{\parallel,i}$ polarized so that it has amplitude only in the x',y', plane (Figure 14.3) and hence is parallel to the plane of incidence. From Equation 14.28, retaining only the real part (cosine term), the wave propagating in the x' direction has its electric field written as

$$E_{\parallel,i} = E_{M\parallel,i} \cos\left[\omega\left(t - \frac{n_1 x'}{c_0}\right)\right] \quad (14.47)$$

From Figure 14.3a, the components of the incident wave in the x, y, z coordinate system are (components are taken to be positive in the positive coordinate directions)

$$E_{x,i} = -E_{\parallel,i} \sin\theta_i \quad (14.48)$$

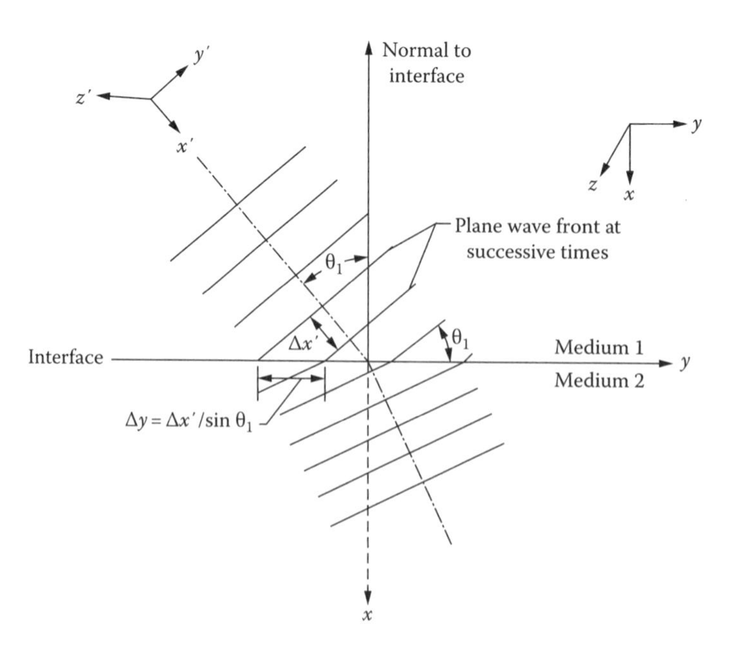

FIGURE 14.2 Plane wave incident upon interface between two media.

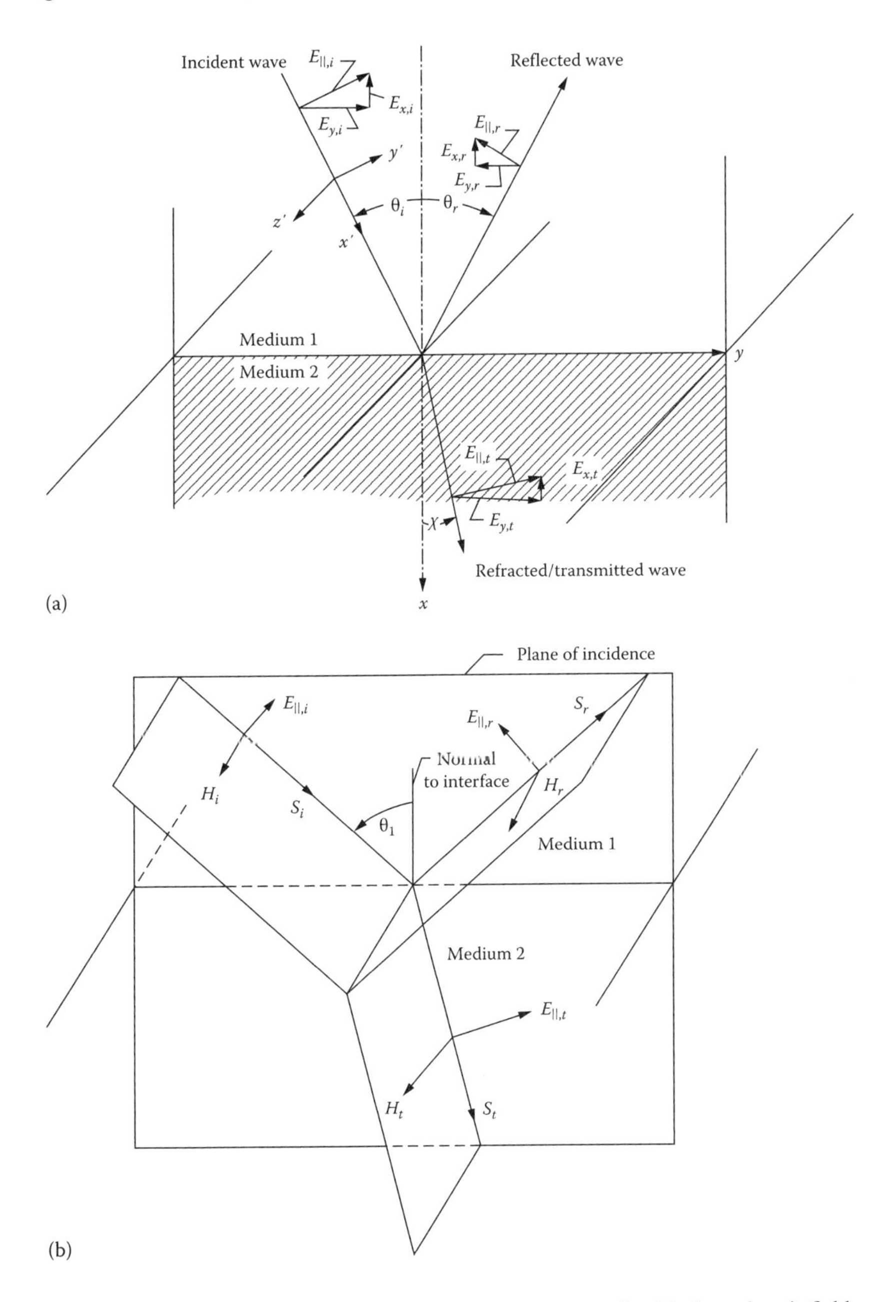

FIGURE 14.3 Interaction of EM wave with boundary between two media: (a) plane electric field wave polarized in xy plane striking intersection of two media and (b) electric intensity, magnetic intensity, and Poynting vectors for incident wave polarized in plane of incidence.

$$E_{y,i} = E_{\|,i} \cos \theta_i \tag{14.49}$$

$$E_z = 0 \tag{14.50}$$

Substituting Equation 14.47 into Equations 14.49 and 14.50 and noting that x', the distance the wave front travels in a given time, is related to the y distance the front travels along the interface by (see Figure 14.2) $x' = y \sin \theta_i$, we obtain for the incident components

$$E_{x,i} = -E_{M\|,i} \sin\theta_i \cos\left[\omega\left(t - \frac{n_1 y \sin\theta_i}{c_0}\right)\right] \tag{14.51}$$

$$E_{y,i} = E_{M\|,i} \cos\theta_i \cos\left[\omega\left(t - \frac{n_1 y \sin\theta_i}{c_0}\right)\right] \tag{14.52}$$

$$E_{z,i} = 0 \tag{14.53}$$

Upon striking the interface, the wave separates into $E_{\|,r}$ reflected at angle θ_r and $E_{\|,t}$ refracted at angle χ and transmitted into medium 2. From Figure 14.3, the components of the reflected ray at the interface in the positive coordinate directions are

$$E_{x,r} = -E_{M\|,r} \sin\theta_r \cos\left[\omega\left(t - \frac{n_1 y \sin\theta_r}{c_0}\right)\right] \tag{14.54}$$

$$E_{y,r} = -E_{M\|,r} \cos\theta_r \cos\left[\omega\left(t - \frac{n_1 y \sin\theta_r}{c_0}\right)\right] \tag{14.55}$$

$$E_{z,r} = 0 \tag{14.56}$$

The direction of $E_{\|,r}$ is such that $E_{\|,r}$, H_r, and S_r are consistent with the right-hand rule connecting Poynting vector with the E and H fields. Similarly, from Figure 14.3, the components of the refracted wave are

$$E_{x,t} = -E_{M\|,t} \sin\chi \cos\left[\omega\left(t - \frac{n_2 y \sin\chi}{c_0}\right)\right] \tag{14.57}$$

$$E_{y,t} = E_{M\|,t} \cos\chi \cos\left[\omega\left(t - \frac{n_2 y \sin\chi}{c_0}\right)\right] \tag{14.58}$$

$$E_{z,t} = 0 \tag{14.59}$$

Boundary conditions must be satisfied at the interface of the two media following the physics of the problem in hand. The sum of the components of the electric fields of the reflected and incident waves parallel to the interface must be equal to the electric field of the refracted wave in the same plane. This is because the electric field in medium 1 is the superposition of the incident and reflected electric fields. For the polarized wave considered here, this condition gives the following equality for the y components that are parallel to the interface:

$$\left\{E_{M\|,i} \cos\theta_i \cos\left[\omega\left(t - \frac{n_1 y \sin\theta_i}{c_0}\right)\right] - E_{M\|,r} \cos\theta_r \cos\left[\omega\left(t - \frac{n_1 y \sin\theta_r}{c_0}\right)\right]\right.$$

$$\left. = E_{M\|,t} \cos\chi \cos\left[\omega\left(t - \frac{n_2 y \sin\chi}{c_0}\right)\right]\right\}_{x=0} \tag{14.60}$$

Since Equation 14.60 must hold for arbitrary t and y and the angles θ_i, θ_r, and χ are independent of t and y, the cosine terms involving time must be equal. As discussed in Chapter 3, this can be true only if

$$n_1 \sin\theta_i = n_1 \sin\theta_r = n_2 \sin\chi \tag{14.61}$$

which results in $\theta_i = \theta_r$ (same as Equations 3.1 and 3.2). Again, as discussed in Chapter 3, the angle of reflection of an EM wave from an ideal interface is equal to its angle of incidence rotated about the normal to the interface through a circumferential angle of $\theta_r = \theta_i + \pi$, which provide the relations that define mirrorlike or specular reflections.

From Equation 14.61, we obtain and rewrite Equation 3.3, which is *Snell's law*;

$$\frac{\sin\chi}{\sin\theta_i} = \frac{n_1}{n_2} \tag{14.62}$$

With the cosine terms involving time equal and with the use of Equation 14.47, there also follows from Equations 14.57 through 14.59

$$(E_{M\|,i}\cos\theta_i - E_{M\|,r}\cos\theta_i = E_{M\|,t}\cos\chi)_{x=0} \tag{14.63}$$

This can be used to relate the reflected electric field to the incident value. The refracted component $E_{M\|,t}$ is eliminated by considering the magnetic field.

The magnetic field parallel to the boundary must be continuous at the boundary plane. The magnetic field vector is perpendicular to the electric field; since the electric field being considered is in the plane of incidence, the magnetic field is parallel to the boundary. Continuity at the boundary provides that

$$(H_i + H_r = H_t)_{x=0} \tag{14.64}$$

The relation between electric and magnetic components is given by Equation 14.42. Although for simplicity this relation was derived for only the specific components H_z and E_y, it is true more generally, so the magnitudes of the E and H vectors are related by

$$|\mathbf{H}| = \frac{\bar{n}}{\mu c_0}|\mathbf{E}| \tag{14.65}$$

For both dielectrics and metals, the magnetic permeability is very close to that of a vacuum, so that $\mu \approx \mu_0$. Then, Equation 14.64 can be written as

$$(\bar{n}_1 E_{M\|,i} + \bar{n}_1 E_{M\|,r} = \bar{n}_2 E_{M\|,t})_{x=0} \tag{14.66}$$

Equations 14.63 and 14.66 are combined to eliminate $E_{M\|,t}$ and give the reflected electric field in terms of the incident intensity for nonattenuating materials ($\bar{n} = n$ since $k = 0$)

$$\frac{E_{M\|,r}}{E_{M\|,i}} = r_\| = \frac{\cos\theta_i/\cos\chi - n_1/n_2}{\cos\theta_i/\cos\chi + n_1/n_2} \tag{14.67}$$

where $r_\|$ is the reflection coefficient for the electric field amplitude. If the derivation is repeated for an incident plane electric wave polarized perpendicular to the incident plane, the relation between reflected and incident components is written after using the corresponding equations similar to Equations 14.63 and 14.66 for perpendicular components (see Homework Problem 14.5):

$$\frac{E_{M\perp,r}}{E_{M\perp,i}} = r_{\perp} = -\frac{\cos\chi/\cos\theta_i - n_1/n_2}{\cos\chi/\cos\theta_i + n_1/n_2} \tag{14.68}$$

Equation 14.62 is used in Equations 14.67 and 14.68 to eliminate n_1/n_2 in terms of $\sin\chi/\sin\theta_i$ to yield

$$\frac{E_{M\|,r}}{E_{M\|,i}} = r_{\|} = \frac{\tan(\theta_i - \chi)}{\tan(\theta_i + \chi)} \tag{14.69}$$

$$\frac{E_{M\perp,r}}{E_{M\perp,i}} = r_{\perp} = -\frac{\sin(\theta_i - \chi)}{\sin(\theta_i + \chi)} \tag{14.70}$$

The energy carried by a wave is proportional to the square of the wave amplitude, as given by Equations 14.44 and 14.45. Then if we take the square of reflection coefficients (or $E_{M,r}/E_{M,i}$), we obtain the ratio of energy reflected from a surface to energy incident from a given direction. For the ideal conditions examined here, the reflection can be considered specular (mirrorlike). The spectral dependence arises from the variation of optical constants with wavelength. The values of $\rho_{\lambda,s}(\theta_i,\phi_i)$ for incident parallel and perpendicular polarized components are then obtained as (note that spectral dependence is omitted from now on for the sake of clarity but implied; the subscript s denotes specular reflectivity, as briefly outlined in Chapter 3)

$$\rho_{\|,s}(\theta_i,\phi_i) = r_{\|}^2 = \left(\frac{E_{M\|,r}}{E_{M\|,i}}\right)^2 \tag{14.71}$$

$$\rho_{\perp,s}(\theta_i,\phi_i) == r_{\perp}^2 = \left(\frac{E_{M\perp,r}}{E_{M\perp,i}}\right)^2 \tag{14.72}$$

Because all reflectivities predicted by the EM theory are specular, the subscript s will not be carried from this point on, to simplify the notation. For the ideal surfaces considered, there is no dependence on the azimuthal angle φ, which is also dropped from the notation for simplicity.

For unpolarized incident radiation, the electric field has no definite orientation relative to the incident plane and has equal parallel and perpendicular components. Then, the reflectivity is the average of $\rho_{\|}(\theta_i)$ and $\rho_{\perp}(\theta_i)$:

$$\rho(\theta_i) = \frac{\rho_{\|}(\theta_i) + \rho_{\perp}(\theta_i)}{2} = \frac{1}{2}\left[\frac{\tan^2(\theta_i - \chi)}{\tan^2(\theta + \chi)} + \frac{\sin^2(\theta_i - \chi)}{\sin^2(\theta + \chi)}\right]$$

$$= \frac{1}{2}\frac{\sin^2(\theta_i - \chi)}{\sin^2(\theta_i + \chi)}\left[1 + \frac{\cos^2(\theta_i + \chi)}{\cos^2(\theta_i - \chi)}\right] \tag{14.73}$$

Equation 14.73 is the well-known *Fresnel equation* and gives the reflectivity for an unpolarized ray incident upon an interface between two perfect (nonattenuating) dielectric media. The relation between χ and θ_i is given by Equation 14.62, which later is used to eliminate χ to obtain and expression in terms of θ_i only (see Equation 14.88).

When the incident radiation is normal to the interface, $\cos\theta_i = \cos\chi = 1$, and Equations 14.67 and 14.68 yield

$$\frac{E_{M\|,r}}{E_{M\|,i}} = r_\| = -\frac{E_{M\perp,r}}{E_{M\perp,i}} = r_\perp = \frac{1 - n_1/n_2}{1 + n_1/n_2} = \frac{n_2 - n_1}{n_2 + n_1} \tag{14.74}$$

The normal reflectivity is then

$$\rho_n = \rho(\theta_i = \theta_r = 0) = \left(\frac{n_2 - n_1}{n_2 + n_1}\right)^2 = \left[\frac{(n_2/n_1) - 1}{(n_2/n_1) + 1}\right]^2 \tag{14.75}$$

The reflectivities are spectral because n_1 and n_2 are always functions of λ, although notation for this dependence is omitted, yet implied.

14.4.2 Reflection and Refraction at the Interface of an Absorbing Medium ($k \neq 0$)

As was shown by Equation 14.33, the propagation of a wave in an infinite medium that attenuates the wave is governed by the same relations as in a nonattenuating medium if the refractive index n for the latter case is replaced by $\bar{n} = n - ik$. When the interaction of a wave with a boundary is considered, the theoretical expressions for reflected wave amplitudes derived for nonattenuating media ($k = 0$) also apply for attenuating media if $\bar{n}$ is used instead of n, but this leads to some complicated interpretations. For example, Snell's law (Equation 14.62) becomes

$$\frac{\sin\chi}{\sin\theta_i} = \frac{\bar{n}_1}{\bar{n}_2} = \frac{n_1 - ik_1}{n_2 - ik_2} \tag{14.76}$$

Because this relation is complex, $\sin\chi$ is complex, and the angle χ can no longer be interpreted physically as a simple angle of refraction for propagation into the material. Except for the special case of normal incidence, n is no longer directly related to the propagation velocity.

Now, consider oblique incidence on an attenuating medium. Figure 14.4 shows a plane wave incident from vacuum on an absorbing material with complex index of refraction $\bar{n} = n - ik$. After refraction, the planes of equal phase are still normal to the direction of propagation, and these planes move with the *phase velocity*, which is equal to c_0/a, where $a = n$ for normal incidence. For other directions, a is related to n and k. For a nonattenuating medium, $a = n$, and the phase velocity is defined as c_0/n. The attenuation of the wave depends on the distance traveled within the medium, and hence the planes of constant amplitude stay parallel to the interface. In a material, the wave front may be *inhomogeneous* as the planes of constant amplitude and constant phase are not necessarily along the same direction. Only for normal incidence, the planes of constant amplitude and constant phase are parallel, so that for this special case, the waves can be considered homogeneous. The *extinction index* depends on the direction, and here it is designated with the symbol b. The *extinction index* b in the x direction is k; also for the normal incidence $b = k$. We will now discuss how a and b are related to n and k for general direction of propagation of incident wave.

By analogy with Equation 14.32, the $E_{y'}$ of the wave in the medium is

$$E_{y'} = E_{y'M} \exp\left[i\omega\left(t - \frac{a}{c_0}x'\right)\right]\exp\left(-\frac{\omega}{c_0}bx\right) \tag{14.77}$$

where propagation is in the x' direction, while attenuation depends on the x direction. The x coordinate can be written as $x = x'$, $\cos\delta = y'$, $\sin\delta$, where $\delta = \sin^{-1}[(1/a)\sin\theta]$ as determined

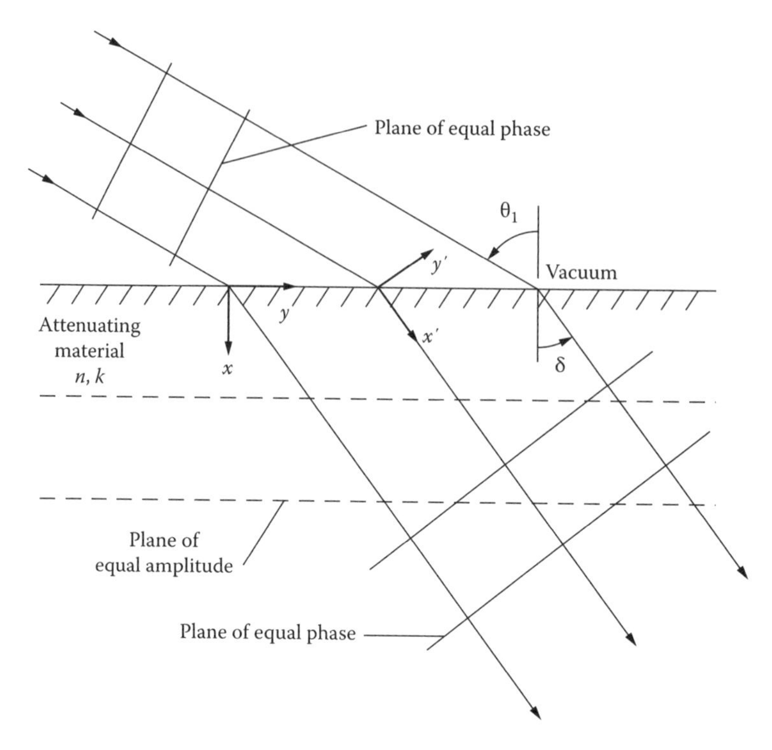

FIGURE 14.4 Planes of equal phase and amplitude for propagation into an attenuating material.

from the phase velocity. The propagation angle δ is not equal to the χ in Equation 14.76 as χ is complex. Then, $E_{y'}$ becomes

$$E_{y'} = E_{y'M} \exp(i\omega t) \exp\left[-x'\left(\frac{i\omega a}{c_0} + \frac{\omega}{c_0} b\cos\delta\right)\right] \exp\left(y'\frac{\omega}{c_0} b\sin\delta\right) \tag{14.78}$$

Since $E_{y'}$ is a function of t, x', and y', Equation 14.34 is written in two space dimensions as

$$\mu\gamma\frac{\partial^2 E_{y'}}{\partial t^2} + \frac{\mu}{r_r}\frac{\partial E_{y'}}{\partial t} = \frac{\partial^2 E_{y'}}{\partial x'^2} + \frac{\partial^2 E_{y'}}{\partial y'^2} \tag{14.79}$$

Substituting Equation 14.78 into 14.79 yields

$$-\mu\gamma\omega^2 + \frac{\mu}{r_r} i\omega = \left(\frac{i\omega a}{c_0} + \frac{\omega}{c_0} b\cos\delta\right)^2 + \left(\frac{\omega}{c_0} b\sin\delta\right)^2 \tag{14.80}$$

Equating real and imaginary parts of Equation 14.80 and using Equations 14.36 and 14.37 yields relations between a and b and n and k: $a^2 - b^2 = n^2 - k^2$ and $ab \cos \delta = nk$, and, as given earlier, $\delta = \sin^{-1} [(1/a) \sin \theta_i]$. This provides three simultaneous equations from which b, a, and δ can be calculated from θ, n, and k. This process yields the attenuation, propagation velocity (phase velocity), and direction of propagation within the material. It is evident that the propagation velocity c_0/a depends on δ; that is, the velocity depends on direction within the material even though the material is isotropic. In the case of normal incidence, $\delta = 0$, and $a = n$ and $b = k$. Hence, it is only for normal incidence that n is directly related to the propagation velocity by $c = c_0/n$ and k is a direct measure of the attenuation with depth in the material.

For the reflection laws, we first consider the case of normal incidence. From Equation 14.66, with $\bar{n}$ replacing n,

$$\frac{E_{M\|,r}}{E_{M\|,i}} = r_{\|} = -\frac{E_{M\perp,r}}{E_{M\perp,i}} = r_{\perp} = \frac{\bar{n}_2 - \bar{n}_1}{\bar{n}_2 + \bar{n}_1} = \frac{n_2 - ik_2 - (n_1 - ik_1)}{n_2 - ik_2 + n_1 - ik_1} \tag{14.81}$$

It is clear from Equations 14.71 and 14.72 that the EM wave energy can be calculated from $|\mathbf{E}|^2$. For a complex quantity z, $|\mathbf{z}|^2 = zz^*$, where z^* is the complex conjugate. Then because the relations for $\|$ and $\perp$ polarizations are the same for normal incidence, the reflectivity is given as (as in Equation 3.20, which is repeated for reference)

$$\rho_n = \left[\frac{n_2 - ik_2 - (n_1 - ik_1)}{n_2 - ik_2 + n_1 - ik_1}\right]\left[\frac{n_2 + ik_2 - (n_1 + ik_1)}{n_2 + ik_2 + n_1 + ik_1}\right]$$

$$= \frac{(n_2 - n_1)^2 + (k_2 - k_1)^2}{(n_2 + n_1)^2 + (k_2 + k_1)^2} \tag{14.82}$$

If the first medium is air, $n_1 \approx 1$ and $k_1 \approx 1$. When the material is also nonattenuating (perfectly transparent), $k_2 \to 0$.

For oblique incidence, the reflectivity can be derived by starting from Equations 14.67 and 14.68 and using the complex index of refraction. For incident rays polarized parallel or perpendicular to the plane of incidence, Equations 14.67 and 14.68 give the complex ratios

$$\frac{E_{M\|,r}}{E_{M\|,i}} = r_{\|} = \frac{\cos\theta_i/\cos\chi - (n_1 - ik_1)/(n_2 - ik_2)}{\cos\theta_i/\cos\chi + (n_1 - ik_1)/(n_2 - ik_2)} \tag{14.83}$$

$$\frac{E_{M\perp,r}}{E_{M\perp,i}} = r_{\perp} = \frac{\cos\chi/\cos\theta_i - (n_1 - ik_1)/(n_2 - ik_2)}{\cos\chi/\cos\theta_i + (n_1 - ik_1)/(n_2 - ik_2)} \tag{14.84}$$

The real and imaginary parts of Equations 14.83 and 14.84 correspond, respectively, to the changes in amplitude and phase. The reflectivity for the parallel or perpendicular component is found by multiplying with its complex conjugate. This requires considerable manipulation, as $\cos\chi$ is a complex number.

The important case considered next is of radiation incident in air or vacuum on a material with a complex index of refraction. Then from Equations 14.83, 14.84, and 14.76 with $\bar{n} = n - ik$,

$$\frac{E_{M\|,r}}{E_{M\|,i}} = \frac{\bar{n}\cos\theta_i - \cos\chi}{\bar{n}\cos\theta_i + \cos\chi} \tag{14.85}$$

$$\frac{E_{M\perp,r}}{E_{M\perp,i}} = \frac{\bar{n}\cos\chi - \cos\theta_i}{\bar{n}\cos\chi + \cos\theta_i} \tag{14.86}$$

$$\frac{\sin\chi}{\sin\theta_i} = \frac{1}{n-ik} = \frac{1}{\bar{n}} \tag{14.87}$$

The $\bar{n}$ cos χ in Equation 14.87 is then

$$\bar{n}\cos\chi = \bar{n}(1-\sin^2\chi)^{1/2} = (\bar{n}^2 - \sin^2\theta_i)^{1/2} \tag{14.88}$$

The results are presented more conveniently by letting $a - ib = (\bar{n}^2 - \sin^2\theta_i)^{1/2}$. By squaring and equating real and imaginary parts, the resulting simultaneous equations are solved for a and b to obtain the set of equations provided in Chapter 3 (see reflectivity relations, Equations 3.14 through 3.23).

14.5 AMPLITUDE AND SCATTERING MATRICES

One of the most important aspects of the EM wave theory is that it allows predictions of the electric field distributions within a medium and its surroundings for different polarization settings. However, electric fields are not measured in engineering applications. The electric field oscillates with the frequency ω of the wave, which is in the order of 10^{12}–10^{14} Hz for the visible to infrared part of the spectrum. What we measure is the intensity of the radiation, which is associated with the electric and magnetic fields of the propagating wave. Therefore, it is important to provide expressions that can also include polarization information. To do this, we start our formulation from Maxwell equation given in Equation 14.1. Then the relation between the incident and the scattered fields in the spherical coordinate system is

$$\begin{pmatrix} E_{\parallel s} \\ E_{\perp s} \end{pmatrix} = \frac{e^{ik(r-z)}}{-ikr}\begin{pmatrix} S_2 & S_3 \\ S_4 & S_1 \end{pmatrix}\begin{pmatrix} E_{\parallel i} \\ E_{\perp i} \end{pmatrix} \tag{14.89}$$

As discussed, $E_\parallel$ and $E_\perp$ are the parallel (transverse magnetic [TM]) and perpendicular (transverse electric [TE]) components of the electric field, respectively. The subscript i denotes the incident component of the field and s for the scattered electric field. The k is the wave vector for the wave and r is the distance between the detector and the reflecting surface or a scatterer. Because of the linearity of Maxwell equations, the scattered and incident fields are related by a simple matrix called the amplitude scattering matrix. The S_i elements are for the scattering amplitude matrix and dependent on both the polar and the azimuthal angles (Bohren and Huffman 1983, Mishchenko 2014). These expressions can also be used for reflections, where a beam is reflected from a surface element. Then, the subscript s is replaced by the subscript r.

The measurable intensity is expressed in terms of Poynting vector, which is the square of the amplitude of the electric field (Equation 14.41). If we are interested only in one polarization component of the wave, we can use a polarizer in front of the beam path. If a horizontal polarizer is used (a polarizer that filters the perpendicular or the TE component of the wave and only allows the parallel [TM] component to go through), the detected intensity is

$$I_\parallel = E_\parallel E_\parallel^* \tag{14.90}$$

where we drop the multiplier $(k/2\omega\mu_0)$ from now on for the sake of simplicity. In these expressions, the superscript * denotes the complex conjugate of a parameter. Similarly, if the beam goes through a vertical polarizer (filtering out the TM component and allowing only the TE component of the wave to pass through), the intensity detected is

$$I_{\perp} = E_{\perp} E_{\perp}^{*} \tag{14.91}$$

Stokes (1852) defined a set of parameters that are related to intensity and polarization of a wave, which serve as an equivalent description of a polarized wave. With this idea, the intensity of unpolarized beam of light scattered by a particle is defined as

$$I_s = \left\langle E_{\parallel s} E_{\parallel s}^{*} + E_{\perp s} E_{\perp s}^{*} \right\rangle \tag{14.92}$$

The difference between these two irradiances as measured by the detector is

$$Q_s = \left\langle E_{\parallel s} E_{\parallel s}^{*} - E_{\perp s} E_{\perp s}^{*} \right\rangle \tag{14.93}$$

If a polarizer is oriented at +45° (a polarizer that filters out the component of all electric field oscillations in the +45° angle of the plane), or at −45°, the difference of intensity measured by a detector from these two measurements is

$$U_s = \left\langle E_{\parallel s} E_{\perp s}^{*} - E_{\perp s} E_{\parallel s}^{*} \right\rangle \tag{14.94}$$

Another case is considered for a beam going through two circular polarizers. If the beam passes through a right-handed polarizer (a polarizer that only transmits electric field oscillations rotating in an anticlockwise direction when viewed toward the source) or a left-handed polarizer (a similar polarizer, but for oscillations in the clockwise direction), the difference of measured intensities is

$$V_s = i\left\langle E_{\parallel s} E_{\perp s}^{*} - E_{\perp s} E_{\parallel s}^{*} \right\rangle \tag{14.95}$$

Equations 14.92 through 14.95 represent the four ellipsometric parameters that describe a plane wave in terms of the four Stokes' parameters I_s, Q_s, U_s, and V_s. These four Stokes' parameters are represented in a column vector called Stokes' vector. The scattered and incident Stokes' vectors are related by a matrix that follows from the amplitude scattering matrix as

$$\begin{pmatrix} I_s \\ Q_s \\ U_s \\ V_s \end{pmatrix} = \frac{1}{k^2 r^2} \begin{bmatrix} S_{11} & S_{12} & S_{13} & S_{14} \\ S_{12} & S_{22} & S_{23} & S_{24} \\ S_{31} & S_{32} & S_{33} & S_{34} \\ S_{41} & S_{42} & S_{43} & S_{44} \end{bmatrix} \begin{pmatrix} I_i \\ Q_i \\ U_i \\ V_i \end{pmatrix} \tag{14.96}$$

The 4 × 4 matrix consisting of 16 S_{ij} (i = 1,2,3,4 and j = 1,2,3,4) elements, which are called the scattering matrix elements, is directly related to the elements of the amplitude scattering matrix in Equation 14.89. This matrix is also called Mueller matrix when the scattering is by a single particle. Reflection from a surface is just a specific form of scattering.

Stokes' vector and the scattering matrix are the most important parameters required for obtaining the complete scattering profile of a scatterer. The relationship between the scattering matrix and the scattering amplitude matrix elements is

$$\begin{aligned}
S_{11} &= \frac{1}{2}\left(|S_1|^2 + |S_2|^2 + |S_3|^2 + |S_4|^2\right) & S_{12} &= \frac{1}{2}\left(|S_2|^2 - |S_1|^2 + |S_4|^2 - |S_3|^2\right) \\
S_{13} &= \mathrm{Re}\left(S_2 S_3^* + S_1 S_4^*\right) & S_{14} &= \mathrm{Im}\left(S_2 S_3^* - S_1 S_4^*\right) \\
S_{21} &= \frac{1}{2}\left(|S_2|^2 - |S_1|^2 - |S_4|^2 + |S_3|^2\right) & S_{22} &= \frac{1}{2}\left(|S_2|^2 + |S_1|^2 - |S_4|^2 - |S_3|^2\right) \\
S_{23} &= \mathrm{Re}\left(S_2 S_3^* - S_1 S_4^*\right) & S_{24} &= \mathrm{Im}\left(S_2 S_3^* + S_1 S_4^*\right) \\
S_{31} &= \mathrm{Re}\left(S_2 S_4^* + S_1 S_3^*\right) & S_{32} &= \mathrm{Re}\left(S_2 S_4^* - S_1 S_3^*\right) \\
S_{33} &= \mathrm{Re}\left(S_1 S_2^* + S_3 S_4^*\right) & S_{34} &= \mathrm{Im}\left(S_2 S_1^* + S_4 S_3^*\right) \\
S_{41} &= \mathrm{Im}\left(S_4 S_2^* + S_1 S_3^*\right) & S_{42} &= \mathrm{Im}\left(S_4 S_2^* - S_1 S_3^*\right) \\
S_{43} &= \mathrm{Im}\left(S_1 S_2^* - S_3 S_4^*\right) & S_{44} &= \mathrm{Re}\left(S_1 S_2^* - S_3 S_4^*\right)
\end{aligned} \tag{14.97}$$

Here, S_{11} specifies the angular distribution of the scattered light given unpolarized incident light. This scattered light is, in general, partially polarized with its degree of polarization given as

$$\sqrt{\frac{\left(S_{21}{}^2 + S_{31}{}^2 + S_{41}{}^2\right)}{S_{11}{}^2}}$$

This relation demonstrates a general aspect of scattering by particles regardless of their nature: scattering is a mechanism for polarizing light. The S_{ij} depend on the scattering direction, and therefore, so does the degree of polarization.

These 16 scattering matrix elements for a single particle are not all independent; only 7 of them can be independent, corresponding to the 4 moduli $|S_j|$ (j = 1,2,3,4) and the 3 differences in phase between the S_j. Thus, there must be nine independent relations among the S_{ij}; these are available in Abhyankar and Fymat (1969). A cloud or suspension of scattering particles may contain many identical particles with different orientations. If the scattering matrix of a particle in a given orientation is known, the scattering matrix of the same particle, or its mirror image, in other orientations is known. Since the scattering matrix of a cloud of particles is the sum of the scattering matrices of the individual scatterers, certain nondiagonal elements in the scattering matrix add up to zero or equal another element, depending on the distribution (Bohren and Huffman 1983, Govindan et al. 1996). For a large cloud of randomly oriented particles, the scattering matrix reduces to

$$\begin{bmatrix} S_{11} & S_{12} & 0 & 0 \\ S_{12} & S_{11} & 0 & 0 \\ 0 & 0 & S_{33} & S_{34} \\ 0 & 0 & -S_{34} & S_{33} \end{bmatrix} \tag{14.98}$$

with only four significant terms, S_{11}, S_{12}, S_{33}, S_{34}, which are sufficient to characterize a particle or a cloud of assembly-averaged agglomerates (Bohren and Huffman 1983, Mengüç and Manickavasagam 1998). Only three of these four matrix elements are independent, such that $(S_{11})^2 = (S_{12})^2 + (S_{33})^2 + (S_{34})^2$. Note that in Equation 14.98 we omit the factor $1/(kr)^2$ for convenience.

If the incident light is 100% parallel polarized (||), Stokes' parameters are defined as (Bohren and Huffman 1983)

$$I_s = (S_{11} + S_{12})I_i \quad Q_s = I_s \quad U_s = V_s = 0 \tag{14.99}$$

On the other hand, if the incident light is perpendicularly polarized to the scattering plane, we obtain

$$I_s = (S_{11} - S_{12})I_i \quad Q_s = -I_s \quad U_s + V_s = 0 \tag{14.100}$$

For unpolarized incident light, we obtain

$$I_s = S_{11}I_i \quad Q_s = S_{12}I_s \quad U_s = V_s = 0 \tag{14.101}$$

Stokes' parameters of the light scattered by a collection of randomly separated particles are the sum of those of the individual particles. Therefore, the scattering matrix for such a collection is merely the sum of the individual particle scattering matrices. With this formulation, one can measure scattered intensity using different polarizers and then determine the properties of the scattering medium (a single particle, a cloud of particles, or a surface) by an inverse analysis (Bohren and Huffman 1983, Mengüç and Manickavasagam 1998, Mishchenko et al. 2000, 2002, 2014).

14.6 EM WAVE THEORY AND THE RADIATIVE TRANSFER EQUATION

So far, in this book, we have considered the RTE obtained via a phenomenological approach. While deriving the RTE, we have not accounted for the wave nature of the energy propagation, and we have neglected both polarization and coherence. Even though classical formulation of the RTE is sufficient for most applications related to engineering, in many modern applications, the wave approach may need to be considered. As discussed by Mishchenko (2006, 2014), the EM wave theory can be used to derive the RTE and to determine the radiation intensity leaving a medium at a given direction. Alternatively, the EM wave theory can be used for determining energy exchange between objects close to each other. That is widely referred to as the *near-field radiative transfer* and is discussed more specifically for nanoscale applications in Chapter 16.

If the EM wave theory is applied to a discrete random media, we obtain the same expressions for the conservation of radiative energy (i.e., RTE), while all the fundamental properties of energy exchange are kept intact. By doing so, the RTE becomes a limiting case of the EM theory. For the derivation, the medium is assumed to be comprised of many discrete absorbers and scatterers. If an N-particle system is considered, then the total electric field at an observation point is the summation of the incident field and spherical waves originating from each of N-particles. These are called Foldy–Lax equations, which can be coupled with additional approximations when the particle locations are completely random, while N goes to infinity. The derivation of the RTE from the EM theory is a lengthy one, as discussed by Mishchenko (2014). The treatment provided is mostly related to scattering media where emission is not considered. Therefore, in the context of the thermal radiative transfer, which is the concern of this book, that derivation is not given. However, we encourage the reader to refer to the publications by Mishchenko et al. (2002, 2006, 2014) for better understanding of the EM theory-based derivation, as it is fundamental in nature and carries a number of important messages: (1) radiative transfer theory is a direct extension of Maxwell equations, (2) the physical meaning and limitations of all parameters and their relationships to fundamental physical quantities become clear with respect to the wavelength of radiation and wave propagation, (3) the range of applicability of the RTE applicable to macroscopic systems is well characterized, and (4) coherent backscattering is related with the radiative transfer theory and opens the way to application of the concept to different diagnostic applications.

HOMEWORK

14.1 Show the derivation of Equations 14.13 and 14.14 with all details. Then obtain the governing equations (Equations 14.15 through 14.22) for a wave propagating along a 45° oblique direction to both x and y directions.

14.2 Show the details of the derivation of Equations 14.38 and 14.39 starting from Equations 14.35 through 14.37.

14.3 Derive Equations 3.14 through 3.17 starting from Equations 14.85 through 14.88, showing all the details. After that, plot them for $n_1 = 1.0$, $n_2 = 1.3 - i0.1$ and comment on trends.

14.4 Assume a gold plate is submerged in olive oil. Find the spectral index of refraction data for gold from the literature. The oil index of refraction is not wavelength dependent and has value $n = 1.4670$. Discuss which equations need to be used under which assumptions. Obtain the angular reflection from this gold plate if you use; (a) a green laser ($\lambda = 0.514$ μm); (b). Plot the results spectrally for afternoon sun (45° oblique incidence, 400–700 nm wavelength range).

14.5 Derive Equation 14.68 for the perpendicular reflection coefficient $r_\perp$ using Equations 14.63 and 14.66.

14.6 Discuss which parameters have the most effect on the energy of an EM wave between two parallel plates. Discuss the importance of local, thermodynamic, spectral, and directional effects.

14.7 Determine the expressions for B and E for a TM EM wave propagating in yz coordinates between two perfect conductor plates at $y = 0$ and $y = b$.

15 Absorption and Scattering by Particles and Agglomerates

15.1 OVERVIEW

In previous chapters, we discussed a number of methods to calculate radiative heat exchange between different surfaces and structures separated by participating or nonparticipating media. If the medium between such structures is not participating, that is, not absorbing, scattering, or emitting, then a relatively simple view factor analysis can be used to determine the net radiation flux on each surface (see Chapters 4 through 7). On the other hand, if the medium is participating, then determination of radiative exchange between surfaces becomes quite complicated. That case requires the solution of the integro-differential radiative transfer equation (RTE) for the corresponding geometry with proper boundary conditions (see Chapters 10 through 13). For this, we need to know the radiative properties of different particles and gases that constitute the medium. These properties may vary locally as they would depend on gas partial pressures, particle volume fraction profiles, shape and size distributions, and material properties. All these properties are wavelength dependent and may be strong functions of temperature; therefore, they need to be predicted together with the solutions of the RTE and the conservation of energy equation.

In most of the analyses presented so far, we assumed the radiative properties were provided *a priori*. In Chapter 9, we outlined different models to determine the absorption and emission properties of different gases as a function of temperature and concentration. In this chapter, we discuss how particle properties can be determined starting from first principles, which requires the solution of Maxwell equations with the corresponding optical properties. We provide a brief, albeit general, framework for the discussion of absorption and scattering phenomena, and after that we outline different solution methodologies. A more comprehensive list of relevant literature and the classical papers in the field can be obtained from the Scattport portal (http://www.scattport.org/index.php/classic-papers). In addition, a recent special issue of the *Journal of Quantitative Spectroscopy and Radiative Transfer* (JQSRT) was devoted to the most recent advances in electromagnetic light scattering by nonspherical particles (Dubovik et al. 2014).

First, we consider two limiting cases: The geometric optics (GO) approximation is for large, transparent spheres and Rayleigh approximation (RA) is for particles much smaller than the wavelength of an incident electromagnetic wave. These two extreme cases correspond to a significant range of particles important for radiative transfer calculations; however, more general solution methodologies are needed to cover the entire size range of particles with different shapes and structures. For this, we discuss Lorenz–Mie theory (LMT) for arbitrary sizes of homogeneous and coated spherical particles. Even though the LMT calculations can be used to obtain the absorption and scattering profiles at GO and RA size ranges, it may yield faster computation and may provide a better physical insight for some engineering applications. We extend discussion to axisymmetric particles where T-matrix approaches (TMA) can be used. Then, we outline the discrete dipole approximations (DDA) where arbitrary structures of particles and agglomerates can be modeled as comprised of a compendium of small dipoles. In addition, a brief overview is provided of the finite element method (FEM) and the finite difference time-domain (FDTD) method, and list relevant literature. We briefly discuss the available computational codes and provide a number of phase function approximations that can be used in radiative transfer calculations.

The discussion provided here for different approximations and models is the starting point for the calculation of more complicated cases. References provided here should assist the reader to find other sources that can be used for various calculations. For a detailed analysis of many of the cases studied here, refer to the recent text by Mishchenko (2014).

In most analyses involving particles, we make the *single scattering* approximation. This means that the scattering by a cloud of particles can be determined as the summation of scattering by the individual particles. To model scattering by a single particle, we assume a planar electromagnetic (EM) wave is incident on a particle; then, the far-field electric field around the particle is determined (see Figure 15.1). For a cloud of particles, as shown in Figure 15.2, the net absorption and scattering can be calculated following a linear summation rule. For radiative transfer calculations, we almost always deal with a cloud of particles within a small control volume along the path of an incident wave. On the other hand, if particles are in close proximity to each other, the net absorption and scattering deviate from the linear approach. One obvious reason is that the field incident on each and every particle within the cloud would not be planar anymore; the incident field on a given particle would be the combination of the incident wave and the scattered EM fields from other close particles. This requires detailed modeling of scattering by all particles at once. In essence, this is also the case in DDA; however, since dipoles are considered in DDA the

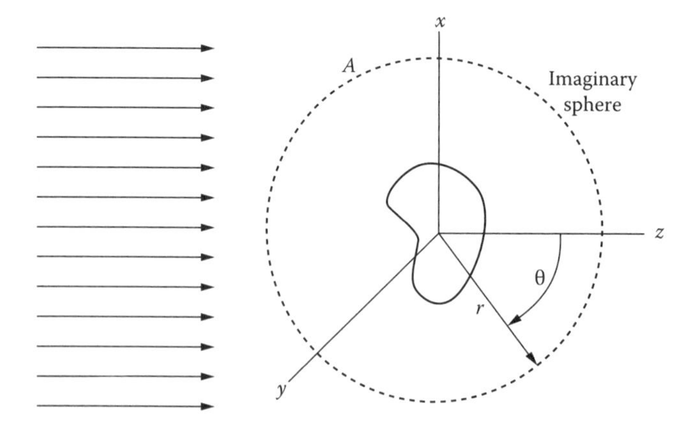

FIGURE 15.1 Plane electromagnetic wave incident on a spherical particle.

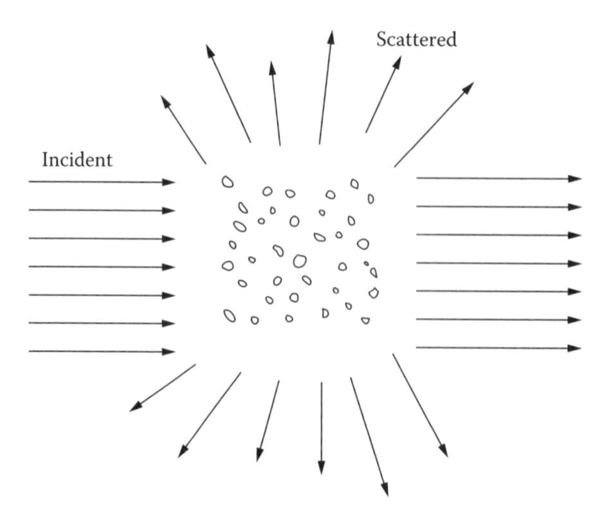

FIGURE 15.2 Plane electromagnetic wave incident on a particle cloud.

formulation is relatively easy. If arbitrary size particles need to be considered, then the rigorous solution of the governing equations is quite challenging beyond a few cases of a finite number of simple shapes. Therefore, these *dependent scattering* cases need to be handled using different approximations, which are the focus of Section 15.6.

15.2 ABSORPTION AND SCATTERING: DEFINITIONS

15.2.1 Background

Absorption and scattering take place when light interacts with a droplet or particle, be it a rain drop, an agglomerate of soot, or a strand of cotton fiber. During this interaction, incident electromagnetic wave energy is either converted into internal thermal energy of the particle via absorption or is scattered to all directions. Once absorbed, the radiative energy may cause the temperature of a particle to increase during the process. This means that after absorption, particle properties may change, and the particle may even go through a phase change. On the other hand, scattering simply refers to the redirection of the incident energy into directions around the particle, and therefore does not affect the internal energy of particles directly. Redirected energy may be scattered back to the particle and eventually be absorbed.

The word "scattering" encompasses optical phenomena such as reflection, refraction, transmission, and diffraction. A radiation scattering event is called *elastic* if the frequency of the incident radiation does not change during the process. Most of the problems we face in engineering radiative transfer fall into this category. On the other hand, if the frequency of the incident and scattered radiation is different (i.e., if incident radiation is reemitted as in the case of fluorescence or Raman scattering), then we call the process *inelastic scattering*. We do not cover inelastic scattering applications.

Depending on the size and shape of the particle and the wavelength of the incident radiation, the degree of absorption and scattering will vary both spectrally and directionally. Of course, if there are more particles along the path of an incident beam, absorbed and scattered radiation increase accordingly. This increase is usually linear in nature, particularly if particles are far from each other, at distances at least several wavelengths long. However, when the particles are in close proximity to each other or to a surface, then absorption and scattering depends on particle density in a nonlinear fashion. Then, the calculation of interaction of incident radiation with particles can be done by considering the effects of other particles and agglomerates simultaneously (see Section 15.6.3 for DDA discussions).

15.2.2 Absorption and Scattering Coefficients, Cross Sections, Efficiencies

The absorption and scattering coefficients, κ_λ and σ_λ, are directly related to the intensity of the beam incident on a control volume along its path. They can be considered as proportionality constants indicating how much of the incident radiative intensity, I_λ, is reduced during the propagation along a small path interval dS (as repeated from Equations 1.42 and 1.45; see Figure 15.2):

$$\frac{dI_\lambda}{dS} = -\kappa_\lambda I_\lambda \quad \text{or} \quad \frac{dI_\lambda}{dS} = -\sigma_\lambda I_\lambda \tag{15.1}$$

If a medium is both absorbing and scattering, then we use the *extinction coefficient* as the proportionality constant of attenuation: $\beta_\lambda = \kappa_\lambda + \sigma_\lambda$. It should be understood that all these proportionality constants are volumetric quantities and reflect on what happens within an elemental control volume along the beam path. They all have the units of inverse length, m^{-1}. The control volume may include different particles, agglomerates, and gases. The required properties are determined by accounting for their individual contributions in a linear fashion:

$$\left.\begin{aligned}
\kappa_\lambda &= \sum_{i,particles} \kappa_{\lambda,i} + \sum_{j,gases} \kappa_{\lambda,j} \\
\sigma_\lambda &= \sum_{i,particles} \sigma_{\lambda,i} \\
\beta_\lambda &= \sigma_\lambda + \kappa_\lambda = \sum_{i,particles} \sigma_{\lambda,i} + \sum_{i,particles} \kappa_{\lambda,i} + \sum_{j,gases} \kappa_{\lambda,j} \\
\omega_\lambda &= \frac{\sigma_\lambda}{\beta_\lambda}
\end{aligned}\right\} \tag{15.2}$$

where

i and j refer to different particle types and gas species, respectively
ω_λ is the spectral single scattering albedo

It is clear from Equation 15.2 that absorption and scattering coefficients required for the solution of the RTE do not necessarily correspond to those of individual particles or gas species, but the combined effect of all of those within a volume element. For most radiative heat transfer calculations, scattering by gas molecules is negligible. Yet, without a detailed understanding of scattering by molecules and small particles, we cannot explain the blue color of the sky or the red color of sunsets/moonsets. In these cases, the path length for radiation is so large that molecular and small particle scattering effects become significant. This is discussed in Section 15.4.

Intuitively one can deduce that how each particle absorbs or scatters radiation must depend on its shape, size, as well as to its optical properties. If these parameters are well defined, then one can determine how much of the incident energy will be absorbed and scattered by an *individual particle* by solving Maxwell equations. The geometry of the problem considered for these solutions is given in Figure 15.1 when a particle is illuminated by a planar EM wave. Corresponding nomenclature for the incident and scattered intensity directions in terms of polar (θ) and azimuthal (ϕ) angles is shown in Figure 15.3. Note that here we use prime (′) to indicate the incident direction.

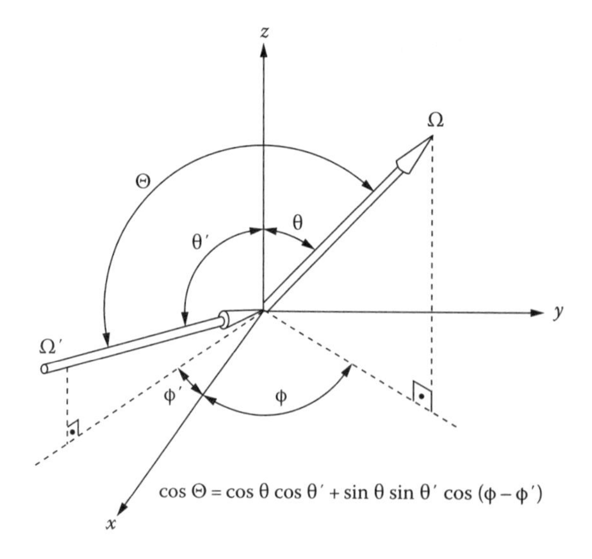

FIGURE 15.3 Nomenclature used for the incident and scattered intensities in Cartesian coordinates. Here, Θ is the scattering angle.

For simple shapes like spheres, the solution of Maxwell equations is relatively easy, and it can be obtained analytically. For complicated structures more detailed numerical algorithms are needed. These solutions give us the so-called absorption and scattering cross sections, $C_{\lambda,a}$ or $C_{\lambda,s}$, respectively, which refer to an "effective" particle area that removes the EM energy from the path of the incident radiation; they have the units of area, m^2. Following this, *efficiency factors* $Q_{\lambda,a}$ or $Q_{\lambda,s}$ are obtained by dividing the absorption and scattering cross sections with the actual geometric cross section of the particle; therefore, efficiency factors have no units.

Particles in many practical applications have a wide range of sizes, shapes, and structures. They are called polydispersed particles. If it is assumed that particles are the same size, shape, and structure, they are called monodispersed. For a cloud of monodispersed spherical particles within a given control volume, the absorption and scattering coefficients are:

$$\kappa_\lambda = NC_{\lambda,a} = NQ_{\lambda,a}\pi\frac{D^2}{4} \qquad \sigma_\lambda = NC_{\lambda,s} = NQ_{\lambda,s}\pi\frac{D^2}{4} \tag{15.3}$$

where

D is particle diameter

N is number of particles per unit volume (with units of m^{-3})

N can be replaced by the volume fraction, f_v, of particles:

$$f_v = NV_{particle} = N\pi\frac{D^3}{6} \tag{15.4}$$

Here f_v has no units (m^3/m^3). These expressions are valid for a homogenous cloud of N *monodispersed* particles per volume with the same diameter D. Similar expressions can be written for a *polydispersed* particle cloud with different size particles, which will require the definition of size distributions. For the sake of simplicity, a monodispersed particle cloud approximation is sufficient for our present discussion.

Extinction cross sections and extinction efficiency factors can also be obtained simply following the addition rule we used in determining the extinction coefficient:

$$C_{\lambda,e} = C_{\lambda,a} + C_{\lambda,s} \qquad Q_{\lambda,e} = Q_{\lambda,a} + Q_{\lambda,s} \tag{15.5}$$

This rule is valid for most applications considered in radiative transfer, where absorption and scattering are *independent* in nature. We discuss dependent absorption and scattering behavior in Section 15.6.

15.2.3 Scattering Phase Function

In addition to the values of the absorption and scattering coefficients, the solution of the radiative transfer equation (RTE) depends on the scattering phase function for different particles or agglomerates. The phase function gives the probability that radiation incident on a particle in the direction of (θ', ϕ') will be scattered into a direction of (θ, ϕ) within solid angle Ω (See Figure 15.3). Then, as shown in Figure 15.3, the scattering angle Θ is defined as the angle between the incident and scattered directions:

$$\cos\Theta = \cos\theta\cos\theta' + \sin\theta\sin\theta'\cos(\phi - \phi') \tag{15.6}$$

If the incident beam is propagating in the z-direction, then the scattering is given in terms of polar angle: $\Theta = \theta$. For the sake of clarity, we assume the incident radiation is in z-direction, and then give the expressions only in terms of the direction of scattered beam, that is, (θ, ϕ).

The intensity of the scattered light in the direction (θ, ϕ) at a far-field distance r is expressed using the differential scattering cross section $dC_s/d\Omega$ as (Bohren and Huffman 1983):

$$\frac{dC_s}{d\Omega} = I_i \frac{S_{11}}{k^2 r^2}; \quad I(\theta,\phi) = I_i \frac{1}{r^2} \tag{15.7}$$

where
I_i is the incident radiation intensity (in z-direction)
k is the wavenumber, which is the inverse of the wavelength of radiation

Here, S_{11} is the first term of the scattering matrix element (see Equation 14.98). The phase function is related to the scattering cross section by

$$\Phi(\theta,\phi) = \frac{1}{C_s} \frac{dC_s}{d\Omega} \tag{15.8}$$

Here, $dC_s/d\Omega$ does not mean a derivative, but is written to show the variation of scattering cross section in the angular domain. The scattering cross section is obtained by integrating the S_{11}:

$$C_s = \frac{1}{k^2} \int_{\phi=0}^{2\pi} \int_{\theta=0}^{\pi} S_{11}(\theta,\phi) \sin\theta \, d\theta \, d\phi \tag{15.9}$$

Because the phase function is a probability function, it is normalized such that:

$$\frac{1}{4\pi} \int_{\phi=0}^{2\pi} \int_{\theta=0}^{\pi} \Phi(\theta,\phi) \sin\theta \, d\theta \, d\phi = 1 \tag{15.10}$$

There are two terms extensively used in the radiative transfer literature to define the angular profile of scattering. *Isotropic scattering* means that scattered intensity is uniformly distributed in all directions. The integral term in the integro-differential RTE is significantly simplified by using the isotropic scattering assumption, which corresponds to Φ(θ, ϕ) = 1. This approximation was extensively used during the early development of radiative transfer theory as it allowed analytical modeling of the radiative transfer problems when powerful computers were not available (see Ozisik 1973, Sparrow and Cess 1978, Viskanta 2005). In nature, however, no object scatters light isotropically; therefore, this simplification should be considered as no more than a mathematical convenience. On the other hand, some packed media, including sand and snow, have a tendency to scatter light more uniformly in all directions; depending on the formulation of a problem, isotropic scattering can be employed for the solution of the RTE in such media. It should be understood, however, that the isotropic scattering in this case corresponds to that of medium rather than the individual sand or snow particles.

The second adjective used to describe scattering events is *anisotropic*, which actually applies to all scattering profiles in nature. As given in Equation 15.7, the angular profile of scattered light by matter varies with respect to both the polar and azimuthal angles. For many applications, azimuthal symmetry can be assumed; therefore, only the polar angle dependence may be needed. Then, we write Φ(θ, ϕ) = Φ(θ).

In general, the scattering phase function, along with values for the absorption and scattering cross sections, is obtained from the solution of Maxwell equations. Several different solution methodologies and algorithms for different shapes and structures are available in the literature (see the books by van de Hulst 1980, Bohren and Huffman 1983, Dombrovsky 1996, Mishchenko 2006, 2014, Doicu et al. 2006, and the website Scattport for the most recent papers [http://www.scattport.org/index.php/]). Once we obtain the cross section of different particles from any of these solution techniques, we can determine the absorption and scattering coefficients using the relations given earlier.

The solution of Maxwell equations can be obtained using three parameters: (1) the wavelength of incident radiation, (2) the physical size/shape/structure of the scatterer, and (3) the complex index of refraction of the particle with respect to the surrounding medium. If particles are spherical in shape, a dimensionless scaling parameter, called the *size parameter*, can be used to combine the first two parameters into a single one. The size parameter is indeed a natural scaling parameter defined as the ratio of the perimeter of a spherical particle with diameter D to the wavelength of incident radiation:

$$\xi = \frac{\pi D}{\lambda_m} \tag{15.11}$$

Here, the wavelength, λ_m, corresponds to that within the medium, such that, $\lambda_m = \lambda_0/n$ with n being the real part of the complex index of refraction of the medium.

For homogeneous spheres, the solution of Maxwell equations can be obtained analytically. In 1891, Lorenz presented a formulation of this problem for the first time, in Danish, which was lost in the annals of scientific literature for years. In 1908, Gustav Mie reported his version of the formulation in German, and his paper became one of the most cited papers of all time (Mie 1908, see also Horvath 2009a,b). During the last 100 years, the theory has been applied to many fields from carbonaceous soot in the atmosphere to nanoparticles for advanced materials synthesis. Lorenz–Mie theory has been expanded to cylindrical particles and ellipsoids (Bohren and Huffman 1983, Mishchenko 2006, 2014) and is commonly used for calculations in all computation platforms (Wriedt, 2008).

Further, Lorenz–Mie theory can be used for all sizes of spherical particles; therefore, it encompasses all values of ξ. Its formulation leads to a series of analytical expressions in terms of complex functions of the size parameter ξ (see Equation 15.11), so it still needs to be solved numerically. With values of ξ beyond 1000 or so, both the solution time and convergence may be problematic. For that reason, at the small and large ξ limits, relatively simple approximations may be used. If ξ is smaller than 0.1, the Rayleigh theory can be adapted. For particles much larger than the wavelength of incident radiation ($\xi \gg 1$), diffraction theory or geometric optics approximations can be used.

In Figure 15.4, two simple phase functions are depicted, where a Rayleigh phase function (see Section 15.4.1) is compared against the isotropic scattering phase function. Of course, a phase function can have much more complicated angular pattern, as shown in four examples of Figure 15.5.

We start our discussion with an intuitive approach for large spherical particles. After that we provide simplifications for Rayleigh spheres. This is followed with a discussion for the general case of arbitrary size spherical particles, where Lorenz–Mie theory is outlined. After that, different models will be presented for scattering by irregularly shaped particles, including a brief discussion of the T-matrix approach and the DDA.

If the scattering behavior is determined from any of these models for a single particle, then scattering for a cloud of particles can be calculated using a simple summation rule. Again, this is possible assuming *independent scattering* applies when the clearance between particles is sufficiently large relative to both the radiation wavelength and the particle diameter D. If particles are close together, such as in a packed bed, foams, granulated sugar, flour, sand, or snow, the scattering can differ from predictions using the addition of independent particle scattering. This *dependent*

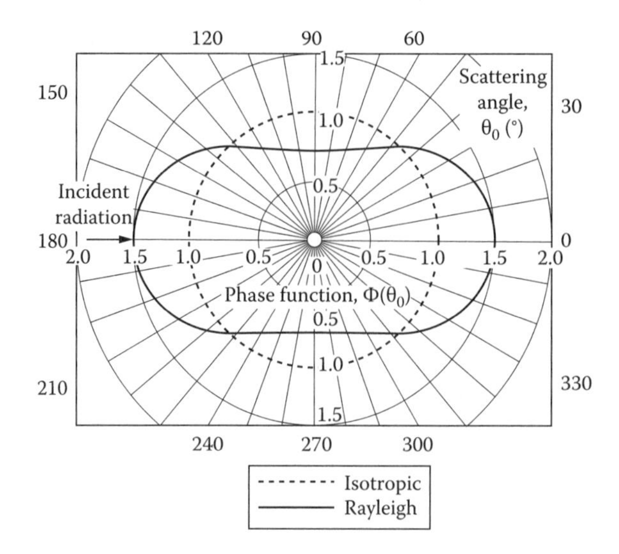

FIGURE 15.4 Comparison of isotropic scattering phase function against a Rayleigh phase function. Note that Rayleigh phase function is for small particles where the size parameter $\xi = \pi D/\lambda$ is much smaller than unity, whereas isotropic phase function is a mathematical idealization.

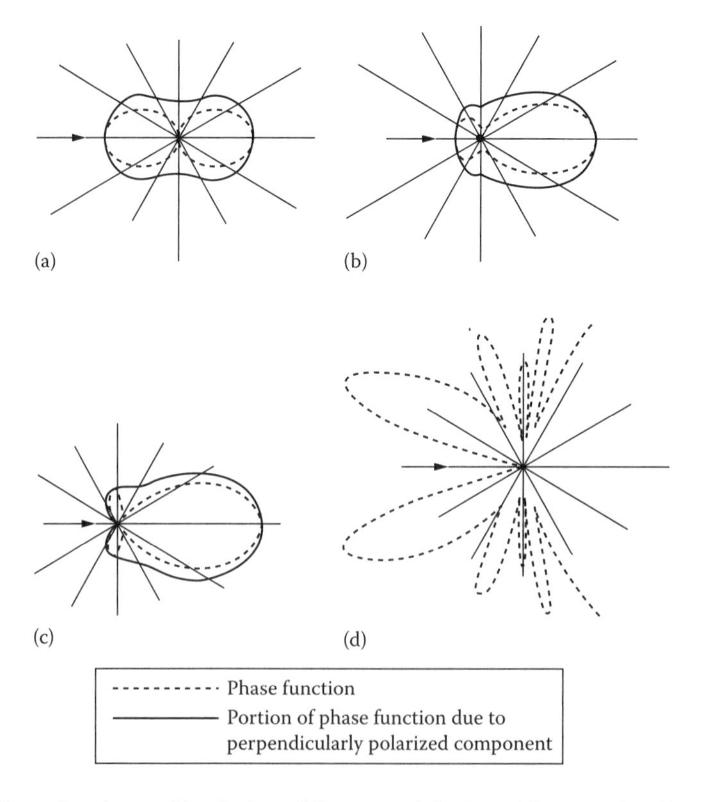

FIGURE 15.5 Phase function profiles for four different particles (on arbitrary scales). Lorenz–Mie profiles, for the corresponding size parameter and complex index of refraction values. (a) $\xi = \pi D/\lambda \to \infty$, metallic sphere, $\bar{n} = 0.57 - 4.29i$; (b) $\xi = 9.15$, metallic sphere, $\bar{n} = 0.57 - 4.29i$; (c) $\xi = 10.3$, metallic sphere, $\bar{n} = 0.57 - 4.29i$; and (d) $\xi = 8$, dielectric sphere, $n = 1.25$.

absorption and scattering regime is discussed in Section 15.7 and in Appendix E on the website for this book at www.thermalradiation.net.

15.3 SCATTERING BY SPHERICAL PARTICLES

To explain scattering and the scattering phase function, consider a simplified case of a large spherical particle illuminated by a collimated light beam, as shown in Figure 15.6. When incident radiation encounters the particle, some of the radiation may be reflected by its surface; the remainder penetrates into the particle medium, where it can be partially absorbed. If the particle does not completely absorb the penetrated energy, some of this radiation leaves the particle from all directions after multiple internal *reflections* and *refractions*. When interacting with the particle boundary, radiation is refracted and its direction is also changed by many internal reflection events. Scattering also occurs by *diffraction* that results from a slight bending of the paths for radiation passing near edges of an obstruction. Scattering is the cumulative effect of all these physical phenomena that alter the path of the original beam. If the particle is not absorbing, there will be no energy exchange between the incident radiation and the particle; then the thermodynamic properties of the medium are not affected. This means that for a completely scattering medium, there is no coupling between the RTE and the energy equation. An extensive review of the literature for large particles, particularly for atmospheric ice, is in Bi et al. (2014).

15.3.1 Scattering by a Specularly Reflecting Sphere

For large strongly internally absorbing (equivalent to strongly reflecting) particles such as metallic spheres with particle size $\xi > 25$, scattering is mostly by reflection. Then, the efficiency factors can be calculated from relatively simple reflection relations. To study these types of particles, let us consider large spherical particles with *specularly* reflecting surfaces. These types of particles are common in engineering and their scattering patterns can be used for diagnostic purposes to monitor different processes. For particles with size parameter ξ larger than about 25, visible light (λ = 400–700 nm) requires a corresponding particle diameter to be larger than about 5 μm. In this case, we can follow a ray tracing approach and determine reflection profiles as a function of the beam direction and surface orientation (see the geometry in Figure 15.7). It is assumed that radiation with a uniform intensity is incident on the particle in the direction (θ_i). The energy intercepted by the projected area of a band of width $R\,d\theta$ and circumference $2\pi R\sin\theta$ on the surface of the sphere

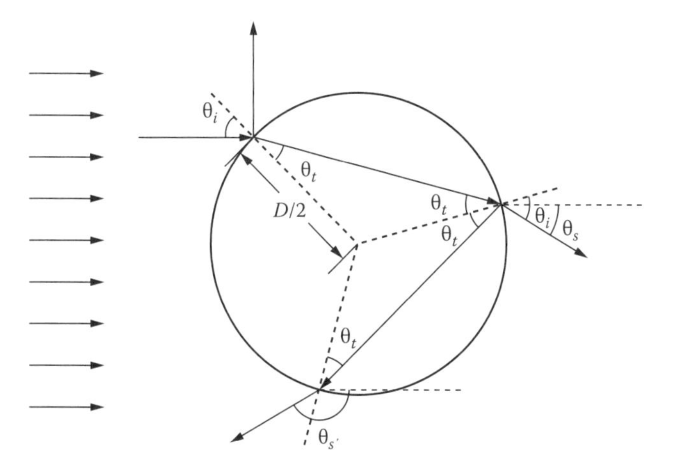

FIGURE 15.6 Geometric optics (Ray tracing) approach through a transparent sphere.

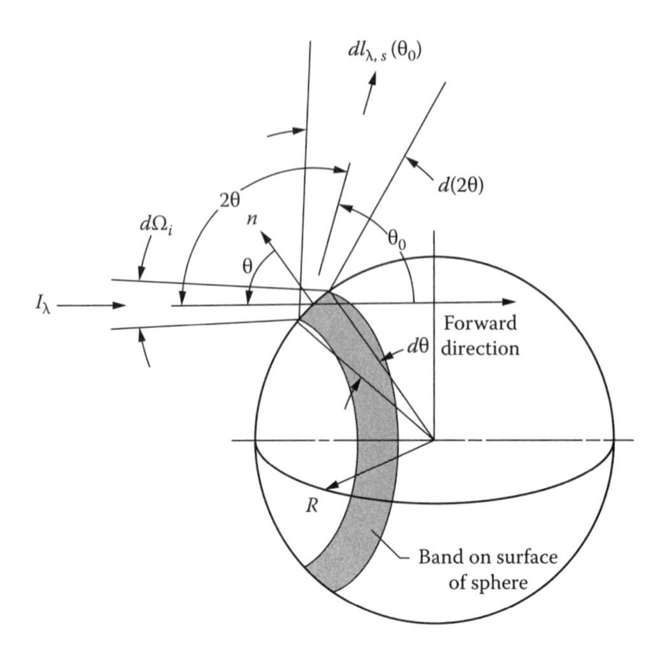

FIGURE 15.7 Reflection of incident radiation by surface of specular sphere.

is $I_\lambda d\Omega_i d\lambda 2\pi R^2 \sin\theta\cos\theta d\theta$. This corresponds to $\rho_{\lambda,s}(\theta)$, the directional specular reflectivity for incidence at θ_i. The energy reflected from the entire sphere is found by integrating over the sphere area:

$$\text{Reflected energy} = I_\lambda d\Omega_i d\lambda \pi R^2 \int_0^{\pi/2} 2\rho_{\lambda,s}(\theta)\sin\theta d(\sin\theta) \tag{15.12}$$

The integral of Equation 15.12 is the hemispherical reflectivity ρ_λ for isotropic radiation incident on the specular surface. Hence, the energy scattered by reflection from the entire sphere is $I_\lambda d\Omega_i d\lambda(\pi R^2)\rho_\lambda$. Using the scattering cross section σ_λ for the particle, the scattered energy is expressed as $\sigma_\lambda I_\lambda d\Omega_i d\lambda$. Hence, the particle scattering cross section is

$$C_{s,\lambda}(D) = \frac{\pi D^2}{4} Q_{s,\lambda} = \pi R^2 \rho_\lambda \tag{15.13}$$

and the scattering efficiency becomes

$$Q_s = \rho_\lambda \tag{15.14}$$

The scattering cross section is therefore equal to the particle projected area times the hemispherical reflectivity. For a cloud of independently scattering specular spheres with the same diameter D, the scattering coefficient σ_λ is

$$\sigma_\lambda = \rho_\lambda \frac{\pi D^2}{4} N \tag{15.15}$$

In a similar fashion, the absorption coefficient for a cloud of specularly reflecting large spheres is

$$\kappa_\lambda = (1-\rho_\lambda)\frac{\pi D^2}{4} N \tag{15.16}$$

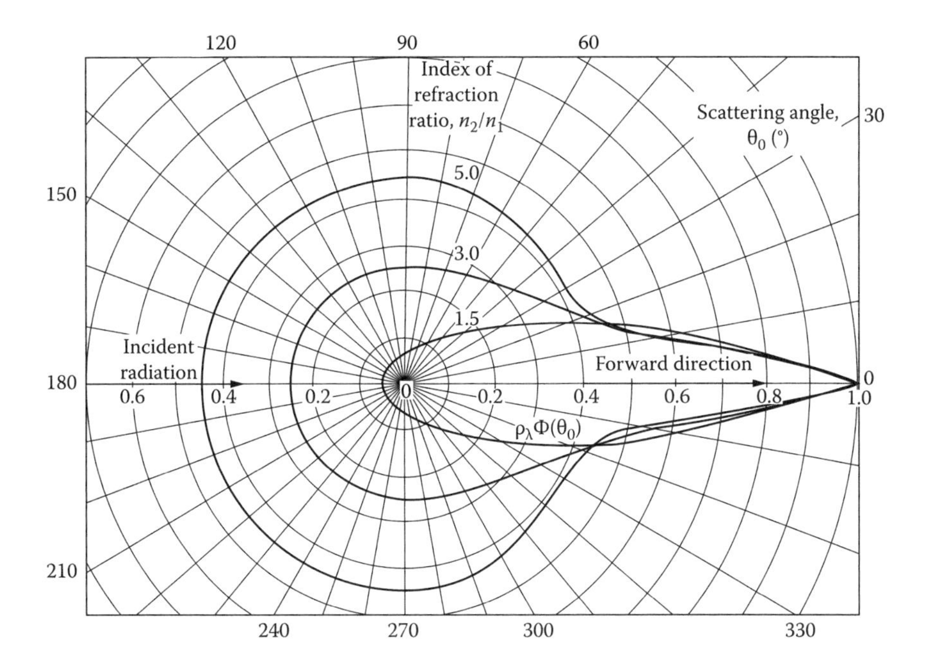

FIGURE 15.8 Scattering diagram for specularly reflecting dielectric sphere that is large compared with radiation wavelength within the sphere, $\xi \gg 1$, (n_2 for particle, n_1 for surrounding medium).

Figure 15.8 shows that the energy specularly reflected from the band of the sphere at angle θ is reflected into direction 2θ and into a solid angle $d\Omega_s = 2\pi\sin 2\theta d(2\theta) = 8\pi\sin\theta\cos\theta d\theta$. The intensity scattered from the incident radiation is the reflected energy per unit incident solid angle, projected area, and wavelength:

$$I_{\lambda,s} = \frac{I_\lambda d\Omega_i d\lambda(\pi D^2/4)\rho_\lambda}{d\Omega_i(\pi D^2/4)d\lambda} = I_\lambda \rho_\lambda \tag{15.17}$$

The energy scattered by a particle into $d\Omega_s$ is $I_\lambda d\Omega_i d\lambda(\pi D^2/4)2\sin\theta\cos\theta d\theta\rho_{\lambda,s}(\theta)$. The intensity scattered into direction 2θ is then

$$I_{\lambda,s}(2\theta) = \frac{I_\lambda d\Omega_i d\lambda(\pi D^2/4)2\sin\theta\cos\theta d\theta\rho_{\lambda,s}(\theta)}{d\Omega_i(\pi D^2/4)d\Omega_s d\lambda} = \frac{I_\lambda\rho_{\lambda,s}(\theta)}{4\pi} = \frac{I_{\lambda,s}}{4\pi}\frac{\rho_{\lambda,s}(\theta)}{\rho_\lambda} \tag{15.18}$$

Inserting this equation into Equation 15.12 yields

$$\Phi(2\theta) = \frac{\rho_{\lambda,s}(\theta)}{\rho_\lambda} \tag{15.19}$$

The angle 2θ is related to the angle θ_0 in Figure 15.7 by $\theta_0 = \pi - 2\theta$ so that, relative to the forward scattering direction

$$\Phi(\theta_0) = \frac{\rho_{\lambda,s}((\pi-\theta_0)/2)}{\rho_\lambda} \tag{15.20}$$

For unpolarized incident radiation, the specular reflectivity $\rho_{\lambda,s}(\theta_i)$ for a dielectric sphere is obtained from Equation 3.7. Also, the directional-hemispherical reflectivity is equal to one minus the emissivity. The quantity $\rho_\lambda\Phi(\theta_0)$ is shown in Figure 15.8 for various refractive index ratios n_2/n_1 of particle to surroundings. For a dielectric the $\rho_{\lambda,s}(\theta_i)$ for normal incidence is usually small compared to that at grazing angles$\rho_\lambda(\theta_i)I_{\lambda,i}d\Omega_i d\lambda dA\cos\theta_i$. Consequently, in Figure 15.8 the forward scatter from the sphere (at $\theta_0 = 0$) is one, and the backward scatter (at $\theta_0 = \pi$) is small.

Similar analysis can be carried out for any particle shape. The accuracy of such approaches was assessed by Bi et al. (2014) for large ice particles. Recent literature can be obtained from this paper.

15.3.2 Reflection from a Large Diffuse Sphere

If a large sphere has a *diffusely reflecting* surface rather than specularly reflecting, then each surface element that receives incident radiation will reflect it into the entire 2π solid angle above that element. Figure 15.9 shows the geometry considered for this case. The shaded portion of the sphere does not contribute any radiative energy in the direction of the observer because it either does not receive radiation or is hidden from the direction of the observer.

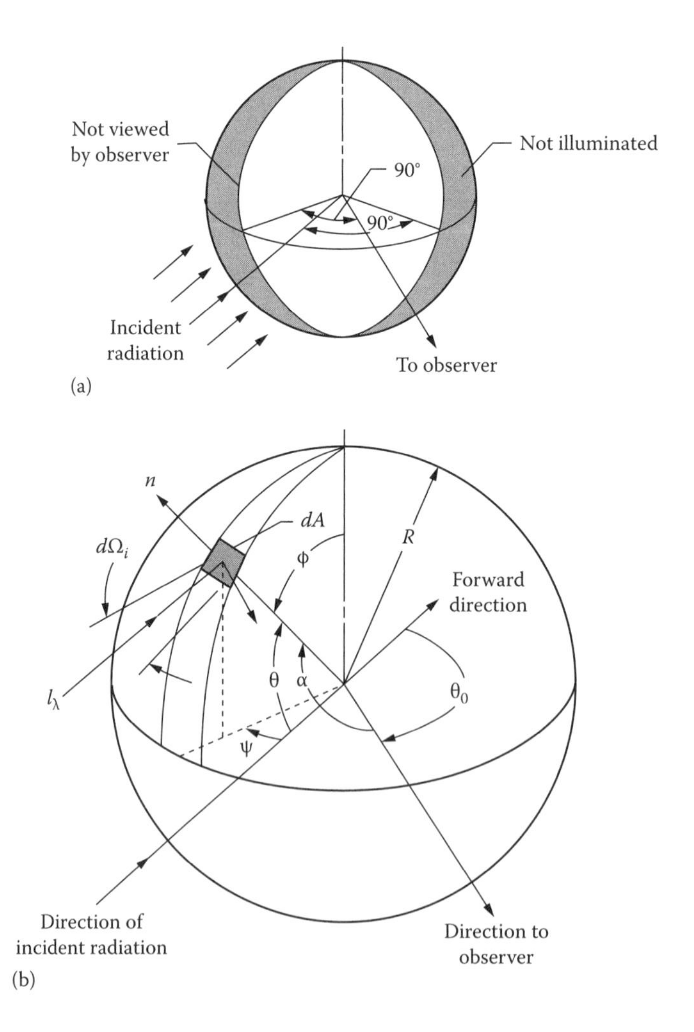

FIGURE 15.9 Scattering by reflection from diffuse sphere. (a) Illuminated region visible to observer; (b) geometry on sphere.

For a simple analysis, let us consider a sphere of radius R as shown in Figure 15.9. A typical surface-area element dA is defined with the angles ψ and ϕ. The observer is at angle θ_0 from the forward direction. The normal to dA is at θ_0 and α relative to the directions of incidence and observation, respectively. The incident spectral energy flux within the incident solid angle $d\Omega_i$ is $I_\lambda d\Omega_i d\lambda$. The projected area of dA normal to the incident direction is $dA \cos\theta$, so the energy received by dA is $I_\lambda d\Omega_i d\lambda dA\cos\theta$. The amount reflected by the sphere would be $\rho_\lambda(\theta) I_\lambda d\Omega_i d\lambda dA \cos\theta$, where $\rho_\lambda(\theta)$ is the diffuse directional-hemispherical spectral reflectivity. In this case, $\rho_\lambda(\theta)$ is assumed independent of incidence angle and hence is equal to the hemispherical reflectivity ρ_λ. Using the cosine-law dependence for diffuse reflection gives the reflected energy per unit solid angle $d\Omega_s$ in the direction of the observer $\rho_\lambda I_\lambda d\Omega_i d_\lambda dA\cos\theta\cos\alpha/\pi$. To integrate the reflected contributions that are received by the observer from all elements on the sphere, the dA, $\cos\theta$, and $\cos\alpha$ are expressed in terms of R, θ, and ϕ, as $dA = R^2 \sin\phi d\phi d\psi$, $\cos\theta = \sin\phi\cos\psi$, and $\cos\theta = \sin\phi \cos(\psi + \pi - \theta_0)$. Then the energy scattered by reflection into the θ_0 direction per unit solid angle $d\Omega_s$ about that direction is found by integrating,

$$\frac{\rho_\lambda I_\lambda d\Omega_i d\lambda R^2}{\pi} \int_{\phi=0}^{\pi} \int_{\psi=-\pi/2}^{\theta_0-(\pi/2)} \sin^3\phi \cos\psi \cos(\psi + \pi - \theta_0)\, d\psi\, d\phi$$

$$= \frac{\rho_\lambda I_\lambda d\Omega_i d\lambda R^2}{\pi} \frac{2}{3} (\sin\theta_0 - \theta_0 \cos\theta_0) \tag{15.21}$$

Dividing the scattered energy by $(\pi D^2/4) d\Omega_i d\lambda$ gives the scattered intensity

$$I_{\lambda,s}(\theta_0) = \frac{\rho_\lambda I_\lambda}{\pi^2} \frac{2}{3} \left(\sin\theta_0 - \theta_0 \cos\theta_0\right)$$

The entire amount of incident intensity that is scattered is $I_{\lambda,s} = \rho_\lambda I_\lambda$. Then, the directional magnitude of the scattered intensity is related to the entire scattered intensity times the phase function by $(\rho_\lambda I_\lambda/\pi^2)\frac{2}{3}(\sin\theta_0 - \theta_0\cos\theta_0) = \rho_\lambda I_\lambda[\Phi(\theta_0)/4\pi]$ so the phase function for a *diffuse* sphere is

$$\Phi(\theta_0) = \frac{8}{3\pi} (\sin\theta_0 - \theta_0 \cos\theta_0) \tag{15.22}$$

The corresponding scattering profiles are shown in Figure 15.10. The largest scattering is for $\theta_0 = 180°$, that is, toward an observer back in the same direction as the source of the incident radiation. For this θ_0 the entire illuminated surface of the sphere is observed.

15.3.3 Large Ideal Dielectric Sphere with $n \approx 1$

For a large nonattenuating dielectric ($k = 0$) sphere with refractive index $n \approx 1$, the reflectivity of the particle surface is almost zero if the index of refraction of the surrounding medium is $n = 1$ and $k = 0$. In this case, the incident radiation passes through the sphere without any change in its electric field amplitude. Therefore, there is no scattering. However, the speed of electromagnetic wave within the sphere, $c = c_0/n$, is slightly smaller than that in the surrounding medium. This means that radiation traveling through different portions of the sphere with different thicknesses would

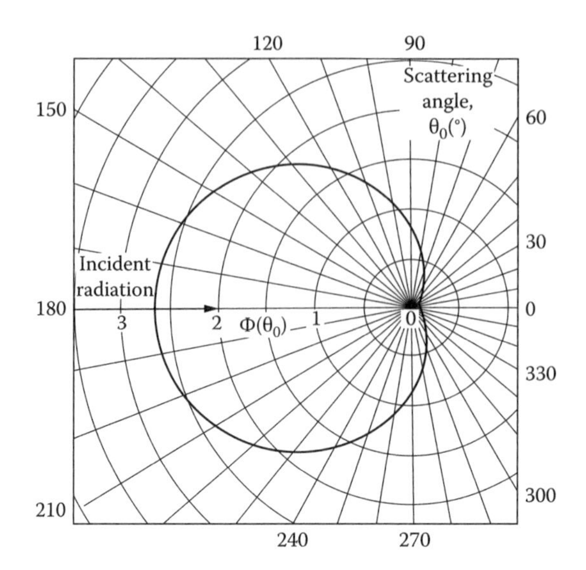

FIGURE 15.10 Scattering phase function for diffusely reflecting sphere, large compared with wavelength of incident radiation and with uniform reflectivity.

have different phase lags. The resulting interference of the waves passing out of the sphere yields a scattering cross section given as

$$C_{s,\lambda} = \frac{\pi D^2}{4}\left[2 - \frac{4}{W}\sin W + \frac{4}{W^2}(1-\cos W)\right] \tag{15.23}$$

where $W = 2(\pi D/\lambda)(n-1)$. The scattering efficiency in this case is

$$Q_{s,\lambda} = \frac{C_{s,\lambda}}{\pi D^2/4} = 2 - \frac{4}{W}\sin W + \frac{4}{W^2}(1-\cos W) \tag{15.24}$$

Additional information for this case is given by Van de Hulst (1957, 1981).

15.3.4 Diffraction from a Large Sphere

In general, the scattering profile for any type and size particle needs to be obtained by solving Maxwell equations, as discussed in Chapter 14. Such a solution is given in terms of the electric field vectors and accounts for interference, reflection, refraction, and diffraction. When the analysis is carried out for large spheres as we have shown earlier, the diffraction is neglected. Diffraction is predominantly in the forward scattering direction. This means that it can be included in the radiative transfer calculations as if it were transmitted past the particle without any interaction. Hence, diffraction can *often be neglected* for energy exchange within a scattering medium. However, if we want to have the total scattering pattern by a large particle, diffraction and reflection must be added.

The most familiar form of diffraction is when light passes through a small hole or slit. As shown in Figure 15.11a, the result is a diffraction pattern of alternate illuminated and dark rings or strips. If a large spherical particle is in the path of incident radiation, Babinet's principle states that the diffracted intensities are the same as for a hole with the same diameter as the sphere. This is because a hole and a particle produce complementary disturbances in the amplitude of an incident electromagnetic wave. The entire projected area of a sphere takes part in the diffraction, and the scattering cross section for diffraction is equal to the projected area $\pi D^2/4$. Because diffraction and reflection occur simultaneously, the total scattering cross section can approach $2\pi D^2/4$ when a sphere is highly reflecting.

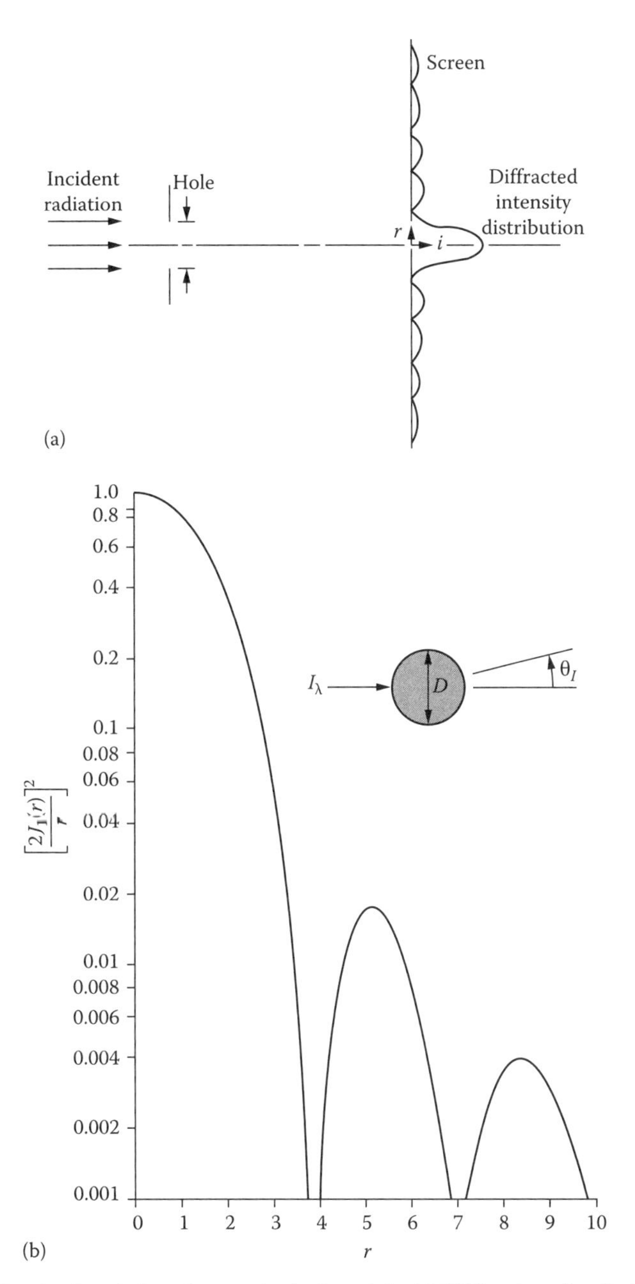

FIGURE 15.11 Diffraction by a hole or large spherical particle. (a) Diffraction of radiation by hole; (b) phase function for diffraction from large sphere.

This is the case for large particles in general, as the scattering efficiency Q_e asymptotically approaches to two as the size parameter ξ goes to infinity. This is known as the *extinction paradox*. This limit suggests that a large particle removes two times the incident energy from the incident beam compared to that cast by its own shadow. The reason for this additional extinction is due to the diffraction of the plane parallel beam around the particle beyond what is actually incident on its physical structure.

The profile of energy diffracted by a large sphere onto a vertical screen is expressed as $[2J_1(r)/r]^2$, where J_1 is a Bessel function of the first kind of order one (Van de Hulst 1957, 1981). This profile is plotted in Figure 15.11b. The function is normalized such that $2\int_{r=0}^{\infty}\{[J_1(r)]^2/r\}dr = 1$ (Born and Wolf 1981).

From the diffraction expression, the distance r is related to the angle θ_0 by $r = \xi \tan \theta_0$. The amount of energy diffracted depends on the particle size. For particles with large ξ the diffracted radiation lies within a narrow angular region in the forward scattering direction. For smaller particles where $\xi \approx 1$, simple relationships would not hold and the general Lorenz–Mie theory must be used to determine the profile. Similarly, diffraction relationships can be extended to other geometries. For example, Yamada and Kurosaki (2000) showed that for a single cylindrical fiber with a large size parameter, scattering can be predicted by diffraction added to reflection from the fiber surface. For large particles of arbitrary sizes, the characterization techniques are usually based on diffraction calculations (Xu 2001).

15.3.5 Geometric Optics Approximation

In many applications, scattering by large spherical particles need to be determined. For example, summer snow cover on Antarctic sea ice is characterized as grains almost in the order of 1 cm, which corresponds to size parameter ξ of 10,000 or so (Zhou et al. 2003). The scattering behavior of spherical and even arbitrarily shaped particles can be determined using the geometric optics approximation (GOA), following a ray tracing approach as depicted in Figure 15.6. For this, an incident plane wave is assumed to be a collection of parallel rays, each with independent path. Each of these rays travel through the media undergoing several reflection/refraction/transmission events before they exit the particle. Exactly where a beam hits on the particle surface and the surface orientation at that point must be determined *a priori*. This requires precise mathematical definition of the surface geometry. In addition, the index of the particle and the surrounding media need to be known. If their path histories are traced from the point of incidence to exit, the angular variation in the scattered light can be computed. If the medium is absorbing, then the change in intensity and in polarization after each reflection/refraction/transmission can be calculated using Fresnel equations (Hecht 2002).

Such a GOA (ray tracing) algorithm can also be constructed for a polarized radiation beam, considering both the parallel and perpendicular component of the electric field. Then, one can use Stoke's formulation provided in Equation 14.96 to determine the reflected and transmitted polarized intensities after the beam hits a surface element. The scattering matrix for the case of reflection and refraction are given following the scattering matrix, Equations 14.96 and 14.98 (and omitting the multiplier $1/(k^2r^2)$):

$$\begin{pmatrix} I_s \\ Q_s \\ U_s \\ V_s \end{pmatrix} = \begin{bmatrix} S_{11} & S_{12} & 0 & 0 \\ S_{12} & S_{11} & 0 & 0 \\ 0 & 0 & S_{33} & S_{34} \\ 0 & 0 & -S_{34} & S_{33} \end{bmatrix} \begin{pmatrix} I_i \\ Q_i \\ U_i \\ V_i \end{pmatrix} \tag{15.25}$$

Specifically for a reflecting surface, it is (Wong and Mengüç 2002, Swamy 2007)

$$S^R = \begin{bmatrix} \frac{1}{2}\left(|r_\parallel|^2 + |r_\perp|^2\right) & \frac{1}{2}\left(|r_\parallel|^2 - |r_\perp|^2\right) & 0 & 0 \\ \frac{1}{2}\left(|r_\parallel|^2 - |r_\perp|^2\right) & \frac{1}{2}\left(|r_\parallel|^2 + |r_\perp|^2\right) & 0 & 0 \\ 0 & 0 & \mathrm{Re}\left(r_\parallel r_\perp^*\right) & \mathrm{Im}\left(r_\parallel r_\perp^*\right) \\ 0 & 0 & -\mathrm{Im}\left(r_\parallel r_\perp^*\right) & \mathrm{Re}\left(r_\parallel r_\perp^*\right) \end{bmatrix} \tag{15.26}$$

$$S^T = \begin{bmatrix} 1-\frac{1}{2}\left(|r_{\parallel}|^2+|r_{\perp}|^2\right) & \frac{1}{2}\left(|r_{\parallel}|^2+|r_{\perp}|^2\right) & 0 & 0 \\ \frac{1}{2}\left(|r_{\parallel}|^2+|r_{\perp}|^2\right) & 1-\frac{1}{2}\left(|r_{\parallel}|^2+|r_{\perp}|^2\right) & 0 & 0 \\ 0 & 0 & \frac{1}{m}\mathrm{Re}\left[(1+r_{\parallel})(1+r_{\perp}^*)\right] & \frac{1}{m}\mathrm{Im}\left[(1+r_{\parallel})(1+r_{\perp}^*)\right] \\ 0 & 0 & -\frac{1}{m}\mathrm{Im}\left[(1+r_{\parallel})(1+r_{\perp}^*)\right] & \frac{1}{m}\mathrm{Re}\left[(1+r_{\parallel})(1+r_{\perp}^*)\right] \end{bmatrix} \tag{15.27}$$

where superscripts R and T refer to either reflection or transmission. In these expressions, $r_{\parallel}$ and $r_{\perp}$ correspond to the electric field reflection coefficients similar to those in Equations 14.69 and 14.70 (and later to Equations 14.81 and 14.83 through 14.86), and the superscript * is as explained there. More specifically they can be expressed as

$$r_{\parallel} = \frac{n^2\cos(\theta_i) - \sqrt{n^2+\cos^2\theta_i - 1}}{n^2\cos(\theta_i) + \sqrt{n^2+\cos^2\theta_i - 1}} \tag{15.28}$$

$$r_{\perp} = \frac{\cos(\theta_i) - \sqrt{n^2+\cos^2\theta_i - 1}}{\cos(\theta_i) + \sqrt{n^2+\cos^2\theta_i - 1}} \tag{15.29}$$

$$n = \frac{n_t}{n_i} \tag{15.30}$$

$$\theta_t = \sin^{-1}\left(\frac{\mathrm{Re}(n_i)}{\mathrm{Re}(n_t)}\sin\theta_i\right) \tag{15.31}$$

Here
- the subscripts i and t are for the incident and transmitted side of the surface
- Re refers to the real value of the argument

The angle θ is measured with respect to the normal of a given surface element. This formulation can be extended to a collection of particles/scatterers by ray tracing for different geometries. The geometric optics approximation provides a useful tool to study multiple scattering media (Wong and Mengüç 2002). Once all Stokes vectors are determined, the scattering phase function is determined using S_{11}, and the scattering cross section is from the integration of the phase function over all angles. A similar approach was applied to liquid foams where the scattering and depolarization of incident radiation on a cylindrical column of foams were explored (see Swamy 2007, Swamy et al. 2007, 2009 and the corresponding references therein). In addition, a GOA algorithm was discussed for large dielectric spheres such as snow grains and compared against the LMT results up to size a parameter of 10,000.

The geometric optics approximation can be extended to absorbing irregularly shaped particles, for example, to ice crystals for applications to remote sensing and climate radiative forcing. These crystals are usually large and absorbing at longer wavelengths, such as $\lambda = 10\ \mu m$, which is

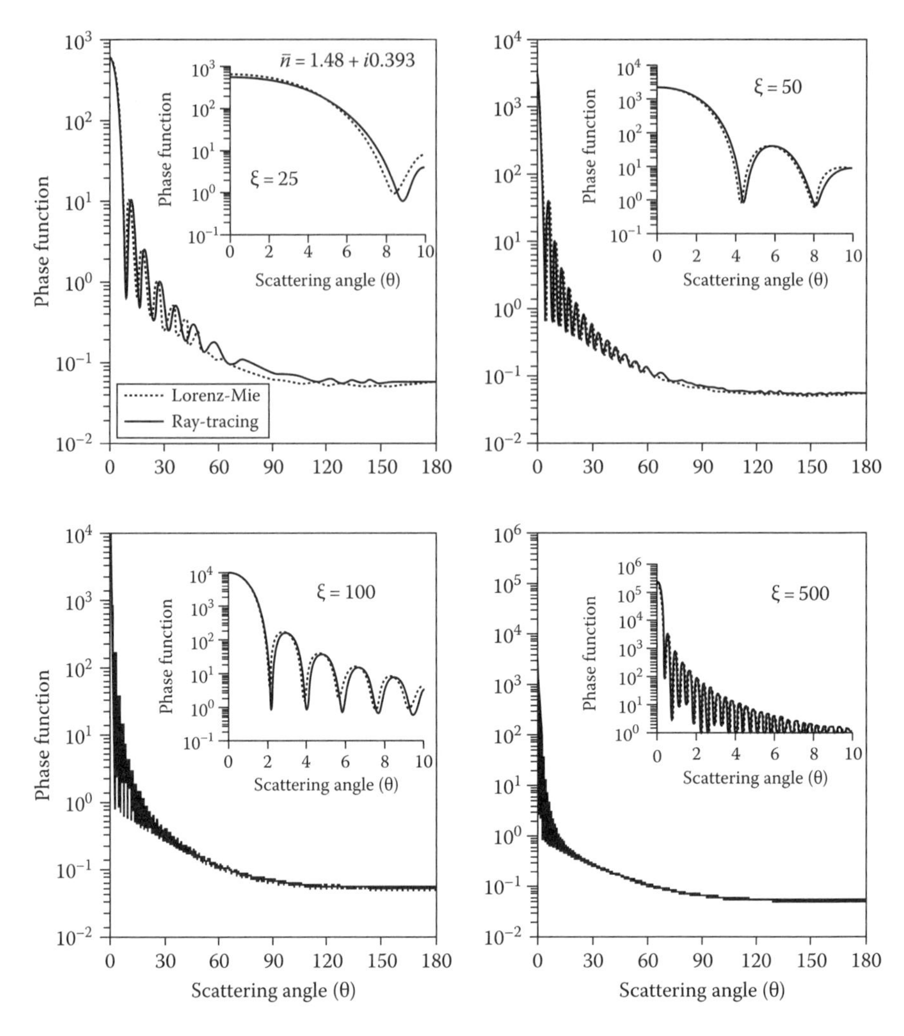

FIGURE 15.12 The angular distribution of scattering phase function for large absorbing spheres; comparisons of GOA against the LMT results. (Adapted from Yang, P. and Liou, K.N., *J. Quant. Spectrosc. Radiative Trans.*, 110, 1162, 2009.)

important for these applications. Yang and Liou (2009) developed a new GOA algorithm to model such particles. They use Maxwell equations to derive the generalized Fresnel reflection and refraction coefficients, and then compute the scattering phase matrix. They compared their predictions against those from the LMT for size parameters of $\xi = 25$, 50, 100, and 500. As shown in Figure 15.12, the agreement between these two approaches is quite good, and the model can be extended to other geometries with confidence.

15.4 SCATTERING BY SMALL PARTICLES

15.4.1 Rayleigh Scattering by Small Spheres

We now introduce simple expressions to determine the absorption and scattering behavior of small particles. *Rayleigh scattering* is attributed to Lord Rayleigh (John William Strutt) who was the first to explore molecular scattering to explain the color of the sky. Although the formulation for small particles (for size parameter ξ much smaller than 1) can be obtained starting from Lorenz–Mie theory, the Rayleigh approximation provides a concise and intuitive methodology to understand scattering phenomena (Rayleigh 1871, 1897). The scattering cross section for Rayleigh scatterers is proportional to the fourth power of size parameter, or proportional to the inverse fourth power of the radiation wavelength. This suggests that when the incident radiation covers a wavelength spectrum, the shorter-wavelength radiation will be Rayleigh scattered with a strong preference. For example, for the visible spectrum of the solar radiation where the wavelength changes from 400 nm (blue) to 700 nm (red), blue light will be scattered strongly by almost the ratio of $(700/400)^4$, that is, nearly 10 times. This means that molecules in the atmosphere preferentially scatter the blue color of the spectrum to all directions, giving the blue background of the sky. Without molecular scattering, the sky would appear black except for the direct view of the sun. As the sun is setting, the path length for direct radiation through the atmosphere becomes much longer than during the middle of the day. In traversing this longer path, proportionately more of the short-wavelength part of the visible spectrum is scattered away from the direct path of the sun's rays. As a result, at sunset the sun takes on a red color as the longer-wavelength red rays are able to penetrate the atmosphere along the path to the observer with less attenuation than for the rest of the visible spectrum. If many dust particles are present, the sunset may be even a deeper red due to the forward scattering of longer wavelength radiation.

If particles with a very limited range of sizes are present in the atmosphere, unusual scattering effects may be observed. The extensive scattering by the ash cloud may affect the radiation balance of the Earth. For example, the 1815 eruption of Mount Tambora on the island of Sumbawa in Indonesia was one of the most powerful in recorded history. The event lowered global temperatures and some experts claim it led to global cooling and harvest failures across the world. Following the eruption of Krakatoa in Indonesia in 1883, the occurrence of blue and green suns and moons was also noted for many years. This was attributed to particles in the atmosphere of such a size range as to scatter only the red portion of the visible spectrum. On September 26, 1950, a blue sun and moon were observed in Europe believed due to finely dispersed smoke particles of uniform size carried from forest fires in Canada. A green moon was observed following the 1982 eruption of El Chichon in Mexico. Indeed, recent research conducted by C. Zerefos found correlation between artists' rendering of sky color and the atmospheric events during 1500s–2000s (Zerefos et al. 2014).

15.4.2 Scattering Cross Section for Rayleigh Scattering

An approximate size limit for Rayleigh scattering was recommended by Kerker (1961) is that the ratio of particle radius to the wavelength λ of radiation within the medium be less than 0.05; hence $(\xi = \pi D/\lambda) < (\sim 0.3)$. However, light scattering studies conducted to show the phase function profiles suggest that this limit is about 0.1 for spherical particles (Ku and Felske 1984, Agarwal and Mengüç 1991). The limits depend on both ξ and the refractive indices of the particle and surrounding medium. Here, let us consider scattering of light from small nonabsorbing spherical particles within a nonabsorbing medium. Assume the particle properties are designated by subscript 2 and the medium by subscript 1 so that $k_2 = 0$ and $k_1 = 0$. Let n be the relative refractive index n_2/n_1. Then the Rayleigh scattering cross section for unpolarized incident radiation is given as

$$C_{s,\lambda} = \frac{24\pi^3 V^2}{\lambda^4}\left(\frac{n^2-1}{n^2+2}\right)^2 = \frac{8}{3}\frac{\pi D^2}{4}\xi^4\left(\frac{n^2-1}{n^2+2}\right)^2 \quad n = \frac{n_2}{n_1} \tag{15.32}$$

where

λ is the wavelength in the medium surrounding the particle
V is the volume of the spherical particle

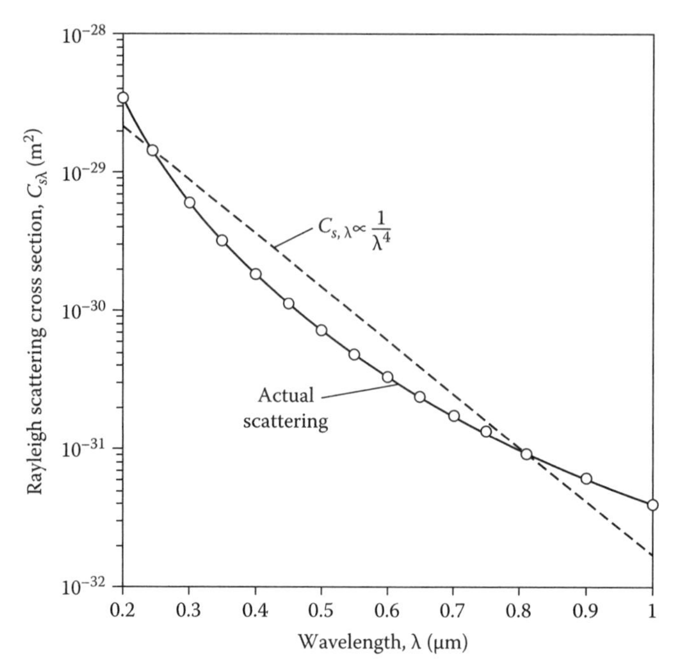

FIGURE 15.13 Comparison of actual Rayleigh scattering cross section for air at standard temperature and pressure with $1/\lambda^4$ variation. (From Goody, R.M. and Yung, Y.L., *Atmospheric Radiation*, 2nd edn., Oxford University Press, New York, 1989.)

The scattering efficiency for Rayleigh scattering is

$$Q_{s\lambda} = \frac{C_{s,\lambda}}{\pi D^2/4} = \frac{8}{3}\xi^4\left(\frac{n^2-1}{n^2+2}\right)^2 \tag{15.33}$$

The actual scattering cross section for particles in a medium may vary with λ, therefore Rayleigh scattering may be somewhat different from the $1/\lambda^4$ dependence. In air at standard temperature and pressure, the restrictions are satisfied and Rayleigh scattering from the gas molecules governs. This is shown in Figure 15.13, where the actual scattering dependence on wavelength is compared with $1/\lambda^4$.

When the particles are of a conducting material with complex refractive index $\bar{n}_2 = n_2 - ik_2$, and the surrounding medium is nonabsorbing so that $\bar{n}_1 = n_1$, the scattering cross section and efficiency factor have the following more general forms:

$$C_{s,\lambda} = \frac{24\pi^3V^2}{\lambda^4}\left|\frac{\bar{n}^2-1}{\bar{n}^2+2}\right| = \frac{8}{3}\frac{\pi D^2}{4}\xi^4\left|\frac{\bar{n}^2-1}{\bar{n}^2+2}\right|^2 \qquad \bar{n} = \frac{n_2 - ik_2}{n_1} \tag{15.34}$$

$$Q_{s\lambda} = \frac{8}{3}\xi^4\left|\frac{\bar{n}^2-1}{\bar{n}^2+2}\right|^2 \tag{15.35}$$

Inserting $\bar{n} = n - ik$ and taking the square of the absolute value as indicated yields

$$\left.\begin{aligned} C_{s\lambda} &= \frac{24\pi^3V^2\left[(n^2-k^2-1)(n^2-k^2+2)+4n^2k^2\right]^2+36n^2k^2}{\lambda^4\left[(n^2-k^2+2)^2+4n^2k^2\right]^2} \\ n &= \frac{n_2}{n_1} \quad k = \frac{k_2}{n_1} \end{aligned}\right\} \tag{15.36}$$

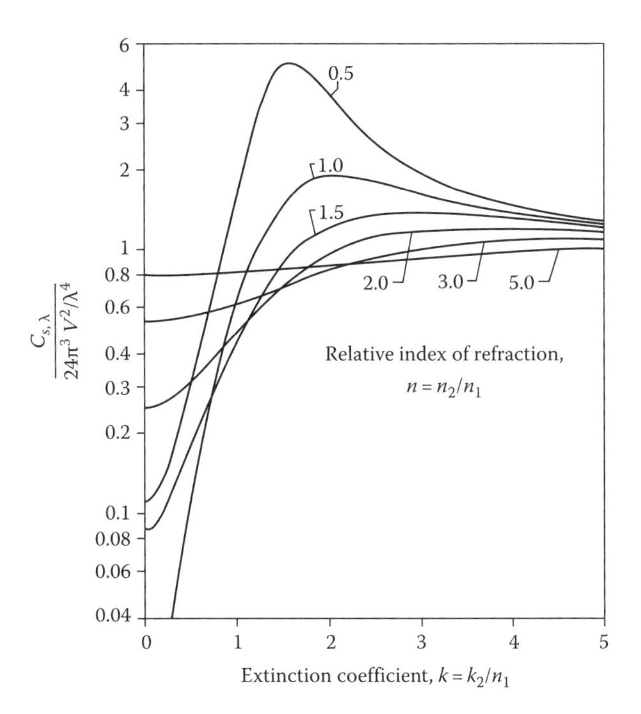

FIGURE 15.14 Rayleigh scattering cross section as a function of relative index of refraction and extinction coefficient.

For $k = 0$ this reduces to Equation 15.34. The quantity $C_{s,\lambda}/(24\pi^3V^2/\lambda^4)$ from Equation 15.36 is given in Figure 15.14 for various n and k values.

The absorption efficiency factor for soot particles can be calculated following a similar approach. Soot particles are often agglomerated; however, before that agglomeration starts, the individual soot monomers are about 15–20 nm size spheres. Then, for small absorbing spheres with $k_2 \neq 0$ in a nonabsorbing medium ($k_1 \neq 0$), we can obtain:

$$\left.\begin{aligned} Q_{a\lambda} &= -4\xi \,\mathrm{Im}\left(\frac{\bar{n}^2 - 1}{\bar{n}^2 + 2}\right) \quad \bar{n} = \frac{n_2}{n_1} - i\frac{k_2}{n_1} = n - ik \\ Q_{a\lambda} &= 24\xi \frac{nk}{(n^2 - k^2 + 2)^2 + 4n^2k^2} \end{aligned}\right\} \tag{15.37}$$

where λ is the wavelength in the medium surrounding the sphere.

15.4.3 Phase Function for Rayleigh Scattering

For incident unpolarized radiation, the phase function can be obtained from the EM theory. This Rayleigh scattering approximation for particles much smaller than the wavelength is

$$\Phi(\theta,\phi) = \frac{3}{4}(1 + \cos^2\theta) \tag{15.38}$$

Note that this expression is independent of the azimuthal angle ϕ. The phase functions for Rayleigh scattering and for *isotropic* scattering (a circle of unit radius) are already shown graphically in Figure 15.4. For Rayleigh scattering, the scattered energy is symmetric with respect to the direction of the incident radiation and has both the forward and backward scattering lobes.

15.5 LORENZ–MIE THEORY FOR SPHERICAL PARTICLES

One of the most celebrated theories in modern day physics is Lorenz–Mie theory, which describes the interaction of a plane-parallel electromagnetic wave with a homogenous spherical particle. This approach was known simply as Mie theory for years, based on the formulation published by Gustav Mie in 1908 (Mie 1908, Lilienfeld 1991, also see Mie 1978 for translation); as of late 2014, Mie's original paper was cited over 4000 times. Recently, the *Journal of Quantitative Spectroscopy and Radiative Transfer* published a compendium of papers discussing the developments since then and the variations of different solution techniques (Horvarth 2009b). As noted earlier, another paper reporting a very similar formulation was published almost two decades earlier, in 1891, albeit in Danish. This paper by the Danish scientist, Ludvig Lorenz, was lost in the archives of scientific literature for years and was discovered much later than Mie's work became famous (Lorenz 1890, 1898). It is fair to call the theory as Lorenz–Mie theory to celebrate both of these scientists. Most of the classical papers related to this development are available at the Scattport site, http://www.scattport.org/index.php/classic-papers.

Lorenz–Mie theory is quite general in the sense that it can be used for any size of spherical particle. Mie originally applied electromagnetic theory to derive the resonance properties of the electromagnetic field when a plane spectral wave is incident on a small spherical surface (i.e., a particle) across which the optical properties n and k change abruptly. For small particles, Rayleigh approach is less computationally intense and can yield quicker predictions, therefore it is sufficient for many applications. If the particles are large and transparent, like rain drops, then the geometric ray tracing approach discussed in Section 15.3 is perhaps more intuitive to use. However, with the increasing availability of computing power, use of the rigorous Lorenz–Mie theory is feasible and more preferable. The theory shows that the scattering profiles obtained may display complicated angular variations depending on the size of the particle, the wavelength of the incident radiation, and the complex index of refraction of the particle at that wavelength. Four different scattering profiles obtained from Lorenz–Mie theory are shown in Figure 15.5. Also, for large absorbing spheres, the LMT results are compared against the GOA profiles in Figure 15.12 (Yang and Liou 2009).

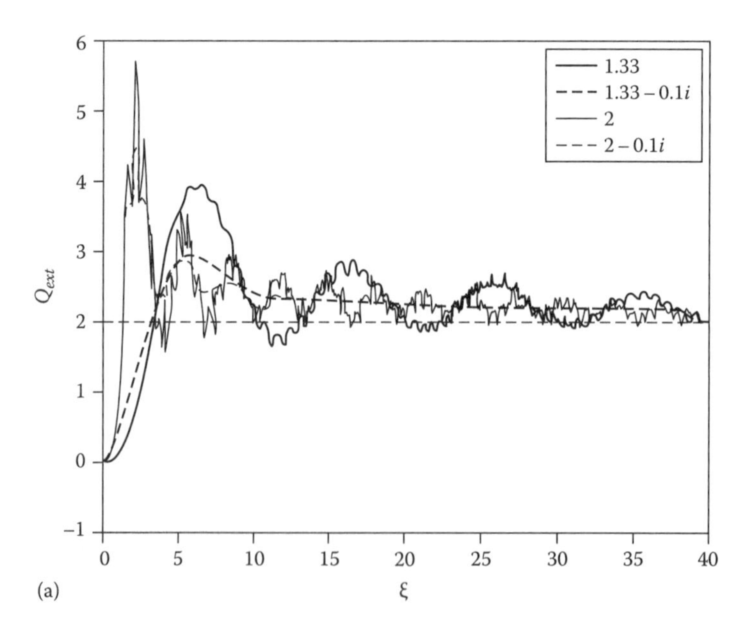

FIGURE 15.15 Extinction, absorption, and scattering efficiency factors for spheres as a function of size parameter and for four different index of refraction values: (a) extinction efficiency. *(Continued)*

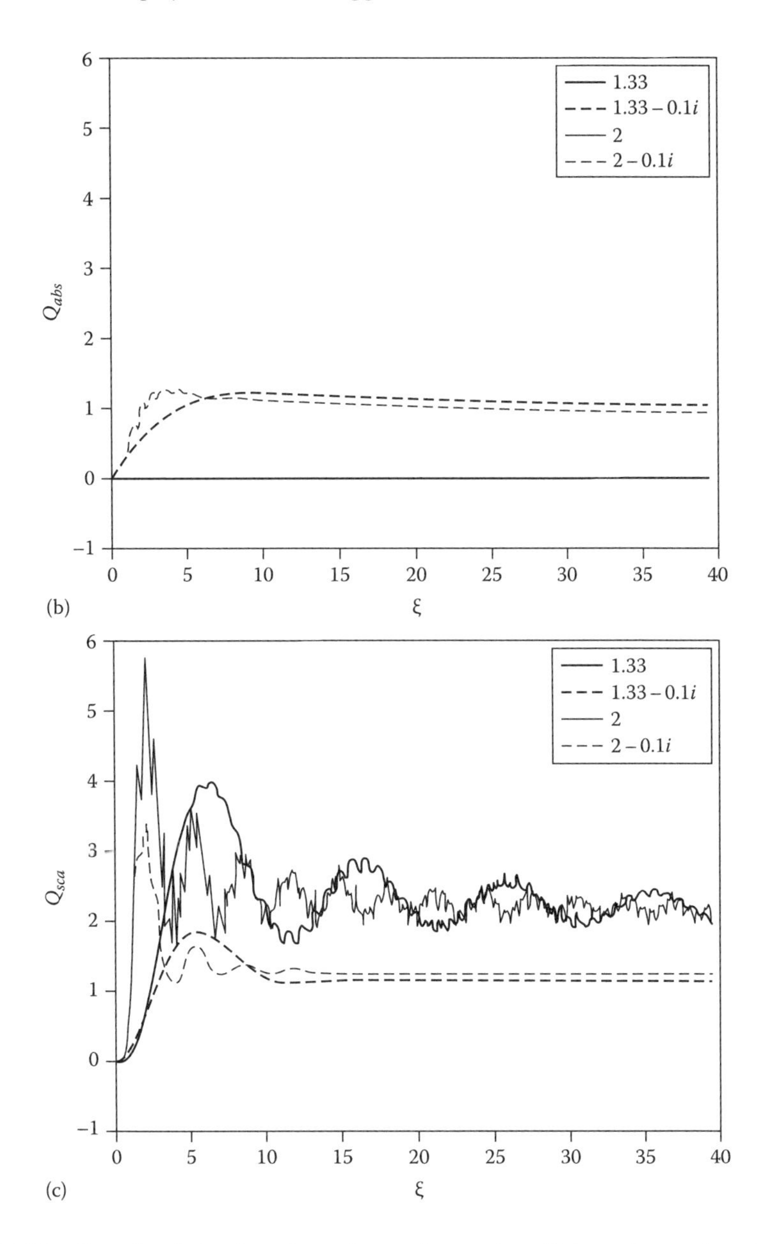

FIGURE 15.15 (*Continued*) Extinction, absorption, and scattering efficiency factors for spheres as a function of size parameter and for four different index of refraction values: (b) absorption efficiency, (c) scattering efficiency.

Discussions of Lorenz–Mie theory was provided by several researchers over the years, including Van de Hulst (1957, 1981), Bohren and Huffman (1983), Barber and Hill (1990), Mishchenko et al. (2006), and Mishchenko (2014). Different computer algorithms were listed by Wriedt and Hellmers (2008) and Wriedt (2010a) (see also the website: http://www.scattport.org/index.php/). General reviews of recent developments in the field are given by Mishchenko and Travis (2008), Horvath (2009a,b), and Mishchenko (2014). It is impossible to outline all the details of Lorenz–Mie theory in a single section; for that reason we invite the reader to examine these books as well as the articles available in the literature. Here, we provide a short discussion of the concept and the formulation.

15.5.1 Formulation for Homogeneous and Stratified Spherical Particles

The original formulation of Lorenz–Mie theory is for homogeneous spherical particles. It can readily be applied to coated spheres, which are important for many engineering systems. If all the layers have the same optical properties, then the solution of the coated sphere model would be identical to that of a homogeneous sphere. The stratified sphere model was first made available by Bohren and Huffman (1983). Mackowski et al. (1990) reported a formulation for the internal cross sections for a stratified sphere, which reduces to the homogeneous sphere model if the properties of each layer are taken as equal. That formulation was used for combustion (Mackowski et al. 1989a,b) and diagnostic calculations (Bhanti et al. 1996, Crofcheck et al. 2002). Other models for coated particles and spheroids are also available (Voshchinnikov 1993, 1996). Here, we provide the analysis for the stratified sphere model.

For the presentation of this general case, let us consider a "layered spherical particle" or a stratified sphere divided into a series of concentric spheres. Each layer is identified with its diameter d_i, corresponding size parameter ($\xi_i = \pi d_i/\lambda$), and a complex refractive index ($\bar{n}_i = n_i - ik_i$), $i = 1,2,\ldots, L$ with L being the total number of concentric spheres (Bohren and Huffman 1983). This general formulation becomes equal to the single homogenous particle case if $L = 1$. The internal and scattered electric fields for the lth sphere are expressed using vector spherical harmonics:

$$E_{lr} = \frac{\cos\phi\sin\theta}{\xi^2}\sum_{n=1}^{\infty} n(n+1)E_n\{\pi_n(\theta)[b_{\text{ln}}\psi_n(\xi) + d_{\text{ln}}\chi_n(\xi)]\} \tag{15.39}$$

$$E_{l\theta} = \frac{\cos\phi}{\xi}\sum_{n=1}^{\infty} E_n\left[\pi_n(\theta)[a_{\text{ln}}\psi_n(\xi) + c_{\text{ln}}\chi_n(\xi)] - i\tau_n(\theta)[b_{\text{ln}}\psi_n'(\xi) + d_{\text{ln}}\chi_n'(\xi)]\right] \tag{15.40}$$

$$E_{l\varphi} = \frac{\sin\phi}{\xi}\sum_{n=1}^{\infty} E_n\left[-\tau_n(\theta)[a_{\text{ln}}\psi_n(\xi) + c_{\text{ln}}\chi_n(\xi)] + i\pi_n(\theta)[b_{\text{ln}}\psi_n'(\xi) + d_{\text{ln}}\chi_n'(\xi)]\right] \tag{15.41}$$

The scattered field from the stratified particle is

$$E_{sr} = \frac{i\cos\phi\sin\theta}{\xi^2}\sum_{j=1}^{\infty} j(j+1)E_j[a_j\pi_j(\theta)\zeta_j(\xi)] \tag{15.42}$$

$$E_{s\theta} = \frac{\cos\phi}{\xi}\sum_{j=1}^{\infty} E_j\left[ia_j\pi_j(\theta)\zeta_j'(\xi) - b_j\pi_j(\theta)\zeta_j(\xi)\right] \tag{15.43}$$

$$E_{s\phi} = -\frac{\sin\phi}{\xi}\sum_{j=1}^{\infty} E_j\left[ia_j\tau_j(\theta)\zeta_j' - b_j\tau_j(\theta)\zeta_j(\xi)\right] \tag{15.44}$$

where

$$\xi = \frac{\pi n_i d}{\lambda} \tag{15.45}$$

$$E_j = \frac{i^j(2j+1)E_o}{j(j+1)} \tag{15.46}$$

$$\pi_j = \frac{dP_j^l(\cos\theta)}{d\theta} \tag{15.47}$$

$$\tau_j = \frac{P_j^l(\cos\theta)}{\sin\theta} \tag{15.48}$$

Here P_j are the Legendre coefficients of the jth order (Ozisik 1973, Born and Wolf 1981):

$$P_j(\cos\theta) = \sum_{i=0}^{[j/2]} (-1)^i \frac{(2j-2i)!}{2^k i!(j-i)!(j-2i)!} (\cos\theta)^{j-2i} \tag{15.49}$$

Lorenz–Mie expansion coefficients are then calculated from

$$a_j = \frac{n\psi_j(\xi)[\psi'_j(n\xi)/\psi_j(n\xi)] - \psi'_j(\xi)}{n\zeta_j(\xi)[\psi'_j(n\xi)/\psi_j(n\xi)] - \zeta'_j(\xi)} \tag{15.50}$$

$$b_j = \frac{\psi_j(\xi)[\psi'_j(n\xi)/\psi_j(n\xi)] - n\psi'_j(\xi)}{\zeta_j(\xi)[\psi'_j(n\xi)/\psi_j(n\xi)] - n\zeta'_j(\xi)} \tag{15.51}$$

These coefficients can be determined by developing a system of boundary conditions and by matching the tangential components of the electric and magnetic fields at each interface. Here, ψ_j and ζ_j are Riccati–Bessel functions that need to be solved iteratively with the field equations until the scattered and the internal energies of the tangential calculations converge. After Mie coefficients are determined, scattering, absorption, and extinction cross sections are obtained:

$$C_s = \frac{2\pi}{k^2} \sum_{j=1}^{\infty} (2j+1)\left(a_j^2 + b_j^2\right) \tag{15.52}$$

$$C_e = C_s + C_a \tag{15.53}$$

$$C_e = \frac{2\pi}{k^2} \sum_{j=1}^{\infty} (2j+1)\,\mathrm{Re}\left\{a_j^2 + b_j^2\right\} \tag{15.54}$$

By dividing these expressions with the geometrical cross section of the particle, we can obtain absorption, scattering, and extinction cross sections (see Equation 15.3).

The elements of the amplitude scattering matrix are of interest as well:

$$S_1 = \sum_{n=1}^{\infty} \frac{2n+1}{n(n+1)}(a_n\pi_n + b_n\tau_n) \tag{15.55}$$

$$S_2 = \sum_{n=1}^{\infty} \frac{2n+1}{n(n+1)}(a_n\tau_n + b_n\pi_n) \tag{15.56}$$

$$S_{11} = \tfrac{1}{2}\left[|S_2|^2 + |S_1|^2\right] \tag{15.57}$$

The phase function in terms of scattering angle is then obtained as

$$\Phi(\theta) = \frac{1}{C_{sca}} \frac{S_{11}(\theta)}{k^2} \tag{15.58}$$

where
- θ is the scattering angle
- S_{11} is the differential scattering cross section of the scatterer and its magnitude changes with the scattering angle

For axisymmetric particles, it is given as

$$\Phi(\theta) = 1 + \sum_{j=1}^{\infty} A_j P_j(\cos\theta) \tag{15.59}$$

Here, again, P_j are Legendre coefficients of the jth order. Note that in many applications, both the azimuthal and polar angle dependence of the scattering phase function is required, especially if irregular shaped particles are considered. However, for many radiative heat transfer calculations this formulation is sufficient.

Extinction, absorption, and scattering cross sections as calculated from Lorenz–Mie theory are shown in Figure 15.15. These results are for four different phase functions that correspond to different size parameters and indices of refraction. Note that with increasing ξ, the extinction efficiency factor approaches 2, which is known as the extinction paradox as noted in the section for large particles. This suggests that a large particle can attenuate two times more energy than its own physical cross section. The reason for this is the bending of EM waves around the particle due to diffraction.

Lorenz–Mie theory discussed so far is applicable for incident plane-parallel electromagnetic waves. For most radiative transfer analysis, this formulation would be sufficient. However, for many laser diagnostics and laser processing application, the laser beam has a Gaussian profile, rather than being plane parallel. For this case, a modified version of Lorenz–Mie theory needs to be used. Interested readers can find the discussion of the literature for these applications in the reviews by Lock and Gouesbet (2009) and Gouesbet and Lock (2013).

Different solution techniques are provided for Lorenz–Mie formulation over the years. Many of them are listed at the Scattport portal http://www.scattport.org/. There are many other applications of Mie Code on the web, including a simple-to-use Mie scattering calculator, http://omlc.ogi.edu/calc/mie_calc.html (Prahl 2009).

15.5.2 Cross Sections for Specific Cases

One of the simpler results from Lorenz–Mie theory is for small spheres somewhat larger than Rayleigh limit. The general Mie solutions are expanded into a power series in size parameter $\xi = \pi D/\lambda$, giving the scattering cross section as

$$C_{s,\lambda} = Q_{s,\lambda}\frac{\pi D^2}{4} = \frac{8}{3}\frac{\pi D^2}{4}\xi^4\left|\frac{\bar{n}^2-1}{\bar{n}^2+2}\left(1+\frac{3}{5}\frac{\bar{n}^2-2}{\bar{n}^2+2}\xi^2+\cdots\right)\right|^2 \tag{15.60}$$

where

$\bar{n} = n - ik$ is the complex index of refraction
D is the particle diameter

The first term in the parenthesis is identical to the formulation provided earlier for Rayleigh theory (see Equation 15.32). The second term in the parentheses is then the first correction to Rayleigh scattering relation.

In a similar fashion the *absorption* cross section in Equation 15.37 becomes, with the second-order term

$$C_{a,\lambda} = Q_{a,\lambda}\frac{\pi D^2}{4}$$
$$= -4\xi\frac{\pi D^2}{4}\,\mathrm{Im}\left\{\left(\frac{\bar{n}^2-1}{n^2+2}\right)\left[1+\frac{\xi^2}{15}\left(\frac{\bar{n}^2-1}{\bar{n}^2+2}\right)\left(\frac{\bar{n}^4+27\bar{n}^2+38}{2\bar{n}^2+3}\right)+\cdots\right]\right\} \tag{15.61}$$

For small spheres, Penndorf (1962) derived from the exact Lorenz–Mie theory the following series expansions for the scattering and extinction (scattering plus absorption) efficiencies are expressed as:

$$Q_{s\lambda} = \frac{8\xi^4}{3z_1^2}\left\{\left[(n^2+k^2)^2+n^2-k^2-2\right]^2+36n^2k^2\right\}\left\{1+\frac{6}{5z_1}\left[(n^2+k^2)^2-4\right]\xi^2-\frac{8nk}{z_1}\xi^3\right\} \tag{15.62}$$

$$Q_{e\lambda} = Q_{s\lambda}+Q_{a\lambda} = \frac{24nk}{z_1}\xi+\left\{\frac{4}{15}+\frac{20}{3z_2}+\frac{4.8}{z_1^2}[7(n^2+k^2)^2+4(n^2-k^2-5)]\right\}nk\xi^3$$
$$+\frac{8}{3z_1^2}\left\{[(n^2+k^2)^2+n^2-k^2-2]^2-36n^2k^2\right\}\xi^4 \tag{15.63}$$

where

$$z_1 = (n^2+k^2)^2+4(n^2-k^2)+4 \tag{15.64}$$

$$z_2 = 4(n^2+k^2)^2+12(n^2-k^2)+9 \tag{15.65}$$

The lowest-order terms in Equations 15.62 and 15.63 are the same as in Equations 15.32 and 15.37. Equations 15.62 and 15.63 can be used up to $\xi = 0.8$ in the range $n = 1.25$–1.75 and $k < 1$. For larger particles the exact Lorenz–Mie solutions are required.

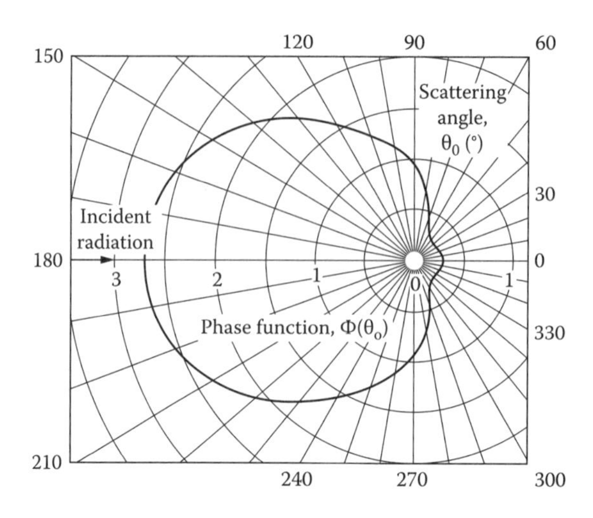

FIGURE 15.16 Phase function for scattering of unpolarized incident radiation from small nonabsorbing sphere with $n \to \infty$.

A cloud of small dielectric spheres with large n becomes highly reflecting. The scattering cross section for this case cannot be obtained directly from Equation 15.60 by considering an asymptotic case where $\bar{n}$ approaches $\infty + i0$. As n becomes large, part of the incident radiation that penetrates the particle is internally reflected. This creates evanescent waves within the particle that produce resonance peaks in the scattering profile. The expansion used to obtain Equation 15.60 does not account for this behavior. In the limit for $n \to \infty$ the scattering cross section for small spheres is expressed as:

$$C_{s,\lambda} = Q_{s,\lambda}\frac{\pi D^2}{4} = \frac{\pi D^2}{4}\left(\frac{10}{3}\xi^4 + \frac{4}{5}\xi^6 + \cdots\right) \tag{15.66}$$

If, in addition to $n \to \infty$, the particles are small enough that only the first term within the parentheses of Equation 15.66 is significant (the use of only the first term is accurate within 2% for $\xi < 0.2$), then the phase function for unpolarized radiation is expressed as

$$\Phi(\theta_0) = \frac{3}{5}\left[\left(1 - \frac{1}{2}\cos\theta_0\right)^2 + \left(\cos\theta_0 - \frac{1}{2}\right)^2\right] \tag{15.67}$$

A polar diagram of this function is shown in Figure 15.16. The highly reflecting particles produce strong scattering back toward the source.

15.6 PREDICTION OF PROPERTIES FOR IRREGULARLY SHAPED PARTICLES

In nature and in most engineering applications, particles are far from being spherical or homogeneous. The shape and structures of particles affect their absorption and scattering significantly; therefore without accounting for the shape, size, and size distribution of particles properly, their effect on radiative transfer cannot be predicted accurately. This necessitates the use of other models to study the interaction of EM waves with particles. In many cases, these models are numerical in nature and require extensive use of computer resources. In the following text, we provide a short review of the literature and then highlight some of the algorithms and computer codes available.

A detailed account of nonspherical particles was first published by Schuerman (1979). A compendium of papers edited by Wriedt (1999) and Mishchenko et al. (2000) covered both the theory and discussions on various computer algorithms. Later, a detailed discussion of different numerical methods was given by Kahnert (2003). In 2006, Mishchenko et al. published a comprehensive monograph. Two recent reviews by Mishchenko et al. (2013) and Wriedt (2009) provided a state-of-the-art review of the subject with more than 100 references each. A compendium of all these studies was reported by Mishchenko (2014). Given this extensive body of information, we limit ourselves only to the general discussion of the subject area.

15.6.1 Integral and Differential Formulations

As we discussed in the previous section, Lorenz–Mie theory provides an analytical formulation for the solution of the EM wave–spherical particle interaction. If the shape of the particle is not spherical, then one can use other semi-analytical techniques, where the electric field is represented as linear combinations of spherical vector wave functions, spheroidal wave functions (Wriedt 2009, Mishchenko et al. 2013, 2014). They are classified as T-matrix approaches, separation of variables methods, the generalized multipole technique, and the null-field method with discrete sources. Among these, the T-matrix approach is the most important and is discussed in the next section.

It is possible to solve Maxwell equations numerically with the proper boundary conditions. These solutions are categorized as the *integral* or *differential equation methods*. Integral equation methods include the method of moments (MoM) and the volume integral equation (VIE) methods, including the DDA. Because of its flexible nature and potential application in many engineering problems, we discuss the DDA formulation in more detail in the following text. Among the differential equation methods are the finite difference time domain (FDTD) method and the FEM. Both of these approaches are available commercially and are highlighted in the following text. Many of the commercial approaches still must be customized to achieve the desired accuracy in the results.

15.6.2 T-Matrix Approach

The T-matrix approach is extensively used to obtain the scattering profiles of rotationally symmetric particles. It is known by different names including the extended boundary condition method (EBCM) and the null-field method (NFM). Its origin goes to Waterman who published the formulation in 1965 (Waterman 1965, 1971). In 1986, Mugnai and Wiscombe provided a compendium of results showing the application of the formulation for axially symmetric particles. A detailed review of the methodology was given by Mishchenko et al. (1996). Again, Mishchenko et al. have regularly published annual lists of T-matrix related papers since 2004 (2004, 2007, 2008, 2010, 2013, 2014). A recent book by Rother (2009) discusses in detail the theory and application of T-matrix method to different shape and size particles. In addition, a special issue of JQSRT was dedicated to Waterman and several papers related to T-matrix were collected. The preface of the special issue discusses the developments and advances (Mishchenko et al. 2013).

In the original T-matrix methods (TMMs), an axisymmetric particle is modeled using Chebyshev polynomials, which allows the formulation for three-dimensional (3D), axisymmetric, structure. These are known as *Chebyshev particles* and their shape function is given in the spherical coordinate system as (Mishchenko et al. 2000):

$$d(\theta,\phi) = d_o[1+\varepsilon_d\, T_a(\cos\theta)] \tag{15.68}$$

where

θ and ϕ are the polar and azimuth angles, respectively

d_o is the diameter of unperturbed sphere

ε_d is the deformation parameter to determine the change of particle from spherical to any other shape

$T_a(\cos\theta)$ is the Chebyshev polynomial of degree a

Usually the aspect ratio of four-to-one can be modeled accurately with the standard approaches before facing instability issues.

In the T-matrix formulation, the incident field is expanded in vector spherical wave functions (VSWF) that are regular at the origin, and the scattered field in VSWFs, which are regular at infinity in the medium outside the particle. After that the incident field expansion coefficients are calculated analytically. A particle response matrix called the T-matrix transforms the incident field expansion coefficients to the scattered field coefficients. The T-matrix depends only on the particle shape, size, refractive index, and the orientation of the particle with respect to a reference frame and is independent of the incident and scattered fields. Because of this, the approach can be used effectively, as once the T-matrix is known for a particle or a configuration of particles, the scattered field can be calculated for any incident field from any direction. The method can be applied to particles of any shape and size although the computations become quite complex for particles that have no symmetry. The computation of the T-matrix for a cluster makes use of the translational addition theorems for VSWFs and uses the T-matrices for all individual spheres. The drawback of the method early in its development was that it was difficult to use it for particles with large aspect ratios or with particle shapes lacking axial symmetry. However, the procedure was extended to a cluster of spheres by Mackowski and Mishchenko (1996) and by Doicu and Wriedt (2001) to nonspherical particles. The analytical orientation averaging method was developed by Mishchenko (1991) for randomly oriented, rotationally symmetric particles. Petrov et al. (2006) considered ensembles of particles with different sizes and random orientations using the T-matrix approach. Recently, Loke et al. (2007, 2009) discussed hybrid formulations of the T-matrix method with FDTD or DDA techniques and Schmidt and Wriedt (2009) applied the T-matrix method to biaxial anisotropic particles.

The T-matrix method was also tested against experimental data by Aslan et al. (2006b) who considered Chebyshev polynomials [$T_6(0.1)$ and $T_4(0.4)$] to model surface irregularities of different metallic MgBaFeO particles. Also, the spherical shape results from the T-matrix method were compared against Lorenz–Mie results. It was shown that the effective sphere model is not sufficient to model the scattering behavior of metallic agglomerates, but the T-matrix approach provides much better agreement with the experimental data. Given this, the measured elliptically polarized scattered light profiles can be used to determine particle shape and structure quite accurately.

Because of its high numerical accuracy, the T-matrix method has gained more interest in the past few years. Mishchenko et al. (2009, 2011, 2013, 2014) have listed several papers that used T-matrix codes and a tutorial for irregularly shaped particles was provided by Mishchenko (2009). One of the widely used formulations is also made available by Mishchenko et al. (2013) and on the web (http://www.giss.nasa.gov/staff/mmishchenko/t_matrix.html). Another T-matrix method (based on Mishchenko's code) was included in a comprehensive toolbox, which is implemented in MATLAB® for modeling mesoscale optical micro-manipulation and other coherent light interaction (Nieminen et al. 2007). It contains functions for generating plane waves, Gaussian beams or beams with arbitrary Laguerre–Gauss modes to be used in conjunction Lorenz–Mie theory or T-matrix approaches. The package can be downloaded from http://www.physics.uq.edu.au/people/nieminen/software.html.

There is an exponential growth in the use of the method; therefore it is not possible to list all relevant papers here. T-matrix papers are regularly reviewed by Mishchenko et al. (2004, 2007, 2008, 2010, 2013b, 2014), which provide a series of categories for different applications and various methodologies.

15.6.3 Discrete Dipole Approximation

Most of the particles in nature are found in agglomerated form. From air pollution to nanotechnology, one can find many examples of colloids, which are one of the most important forms of matter for engineering applications. Agglomerates have much larger area/volume ratios than spherical or other irregular particles. Because of this, their absorption and scattering profiles are significantly different; they cannot be simply modeled as effective spheres or any smooth particles. Therefore, the use of Lorenz–Mie or the T-matrix methods does not yield accurate results. Instead, it is preferable to consider them as agglomerates (or an irregular shaped particle) as composed of many smaller volume elements. Indeed, many of the agglomerates are considered as fractals (Mandelbrot 1975, 1983) including atmospheric soot Sorensen 1997, 2001, Mishchenko 2009, Mishchenko et al. 2013) and nanowires, nanotubes, and nanoparticles (Saltiel et al. 2004, 2005, Kozan and Mengüç 2008a,b). Then the absorption and scattering by such a particle is calculated by considering the cumulative effect of all these volumes. In the limit, these volumes can be considered as electric diploes and their interaction with each other can be accounted for in a rigorous way, as suggested by Purcell and Pennypacker (1973). Each of these volume elements are considered polarizable and still contain a large number of molecules that make up the particle itself.

There are several versions and modifications of the method that are available since the original formulation by Purcell and Pennypacker (1973). They include the coupled dipole approximation (Singham and Bohren 1987, 1988, 1993), discrete dipole approximation (Draine 1998, Draine and Goodman 1993, Draine and Flatau 1994, 2008, 2009), digitized Green's function (Goedecke and O'Brien 1987), the volume integral equation formulation (Iskander et al. 1989), another VIE based algorithm called AGGLOME by Manickavasagam and Mengüç (1997, 1998), and Amsterdam DDA (ADDA) by Yurkin (2007).

The integral equation methods consist of expressing the external field around an agglomerate or irregular-shaped object in terms of an integral equation. The equation contains an internal field term that is unknown, which is represented at each point as a sum of the incident field and the field induced by all other interior points. Once the internal field is obtained, the external field is calculated with the help of the integral equation. In the model, the response of each dipole to the incoming electromagnetic field is represented by its complex polarizability α, which is given by Clasius–Mossotti relation (Bohren and Huffman 1983, Jackson 1989, Draine 1998):

$$\alpha_{CM} = \frac{3}{N}\frac{\bar{\varepsilon}(r_i)-1}{\bar{\varepsilon}(r_i)+2} \tag{15.69}$$

where N is the number of polarizable elements per unit volume. Each of these N elements (or dipoles) is exposed to the incident field as well as to the scattered field originating from all other elements. The details of formulation are not to be given here. The following brief discussion is simply to introduce the reader to the idea of "discretization" of an irregular-shaped particle or an agglomerate. The formulation is followed to find the electric field on each dipole. With this in mind, the total electric field at a point on a dipole located at r is given as (Draine and Flatau 1994, 2008, Draine 1998):

$$E_i(r) = E_o(r) + \sum_{j\neq i=1}^{N} E_{ij}(r); \quad r \in V_i; \quad i = 1,\ldots,N \tag{15.70}$$

If the agglomerate is small compared to the wavelength of incident radiation, the r-dependence of $E_i(r)$ can be replaced by an average E_i. Then

$$E_i = E_{o,i} + \sum_{j\neq i=1}^{N} E_{ij}; \quad i = 1,\ldots,N \tag{15.71}$$

Following this, E_{ij} can be represented as a function of the local electric field on the particle j

$$E_{ij} = C_{ij} \cdot E_j \tag{15.72}$$

where

$$C_{ij} = \begin{cases} [A_{ij}I + B_{ij}(n_{ji} \otimes n_{ji})] \cdot \alpha_j & i \neq j \\ 0 & i = j \end{cases} \tag{15.73}$$

where
n_{ji} is the unit vector pointing from particle j to particle i
$\otimes$ is the tensor product
α_j is the polarizability tensor of particle j

The A_{ij} and B_{ij} functions are defined as

$$A_{ij} = \frac{\exp(ikr_{ij})}{r_{ij}} \left(k^2 - \frac{1}{r_{ij}^2} + \frac{ik}{r_{ij}} \right) \tag{15.74}$$

$$B_{ij} = \frac{\exp(ikr_{ij})}{r_{ij}} \left(\frac{3}{r_{ij}^2} - k^2 - \frac{3ik}{r_{ij}} \right) \tag{15.75}$$

Equation 15.67 can be written in matrix representation and then solved to obtain local electric fields at each dipole. After calculating electric fields at each particle, the total scattered field can be found from

$$E_s = \frac{1}{4\pi} \sum_{i=1}^{N} \left[A_i I + B_i (n \otimes n) \right] \cdot \alpha_i E_i \tag{15.76}$$

The accuracy of a DDA formulation is validated by comparing the predictions against exact solutions. For that, a spherical particle is represented by a large number of small volume elements each corresponding to a dipole, as originally shown by Purcell and Pennypacker (1973). As expected, if the number of dipoles used to express a particle increases, the DDA solution approaches the exact Lorenz–Mie solution. For this, two parameters are important: d, the inter dipole spacing, and the wavelength of incident radiation or the wavenumber, $k = \omega/c$. In applying DDA, Draine and Flatau (2009) suggest the following requirements:

1. $|n|kd \leq 1$ so that the lattice spacing d is small compared to the wavelength of a plane wave in the target material.
2. The value of d must be small enough (N must be large enough) to describe the target shape satisfactorily. As the size of the dipoles decreases, the number of dipoles representing the target increases, hence the dipole representation approaches the real target geometry.

The advantage of the method is that it has few unknowns and can be applied to anisotropic and inhomogeneous scatterers. The disadvantages include low computational accuracy and rising computational time with increasing size parameters. Two different DDA codes are readily available, which can be adapted for radiative transfer calculations.

DDSCAT 7.0 is an open-source Fortran-90 software package applying the discrete dipole approximation to calculate scattering and absorption of electromagnetic waves by targets with arbitrary geometries and complex refractive index (Draine and Flatau 2008). The targets may be isolated entities (e.g., dust particles), but may also be 1D or 2D periodic arrays of "target unit cells," allowing calculation of absorption, scattering, and electric fields around arrays of nanostructures (Draine and Flatau 1994), who presented an extension to periodic structures and near-field radiation calculations (Draine and Flatau 2008). DDSCAT supports calculations for a variety of target geometries. Target materials may be both inhomogeneous and anisotropic. It is straightforward for the user to "import" arbitrary target geometries into the code. DDSCAT automatically calculates total cross sections for absorption and scattering and selected elements of Mueller scattering intensity matrix. The package can be downloaded from: http://ddscat.wikidot.com/start or from Scatterlib maintained by Flatau.

ADDA is a C software package to calculate the scattering and absorption of electromagnetic waves by particles of arbitrary geometry using the discrete dipole approximation (DDA) (Yurkin 2007, Yurkin et al. 2007, Yurkin and Hoekstra 2011). Its main feature is the ability to run on a multiprocessor system or multicore processors (parallelizing a single DDA simulation). ADDA is intended to be a versatile tool, suitable for a wide variety of applications ranging from interstellar dust and atmospheric aerosols to biological particles; its applicability is limited only by available computer resources. The package can be downloaded from: http://www.science.uva.nl/research/scs/Software/adda/. The program Open-DDA was developed by McDonald et al. (2009) for use with high-performance computers.

Another DDA code for particles on a substrate was recently made available. DDA-SI, developed by Loke and Mengüç (2010), is freely available as a public domain MATLAB toolbox. It can be used for both free-space and half-space-substrate light scattering calculations. As discussed thoroughly in a number of publications by Loke et al. (2010, 2011a,b, 2012, 2013, 2014), the interaction between the dipoles that make up a particle and their image counterpart from the substrate are considered for decomposing the resulting spherical wave into planar and cylindrical components by means of Sommerfeld integration (see Loke and Mengüç 2010, 2011a,b and references cited). Indeed, this idea was suggested by Schmehl et al. (1994, 1997) and discussed by Zhang and Hirleman (2002) for an earlier Fortran implementation of DDA with surface interaction called DDSURF. This work is extensive and likely to impact the nanomanufacturing of designer surfaces with structures or imbedded particles with specific applications. The analysis is long and cannot be covered in detail here. Interested readers can refer to a series of articles by Loke and Mengüç (2010), Loke et al. (2011a,b, 2013, 2014). The extension of DDA-SI is reported in Moghaddam, Erturk and Mengüç (2015a,b). In addition, a recent report provides experimental validation of these approaches (Short et al. 2014).

Here, we emphasize again that in all formulations we have presented so far we have not considered radiative emission. All analyses are for propagation of electromagnetic waves. DDA, for example, is to determine the scattering profiles of electromagnetic waves by arbitrary shapes of particles. In Chapter 16, we show how the emission from small objects can be accounted for in Maxwell equations. Indeed, it is possible to couple that analysis with the DDA formulation. Edalatpour and Francour (2014) present such an innovative approach where the interaction of emitting arbitrary shaped particles can be determined. The so-called thermal discrete dipole approximation (T-DDA) was evaluated for a number of cases.

15.6.4 Finite-Element Method

Solution of radiation scattering problems for complex, yet more realistic systems requires a numerical solution of Maxwell equations. For that purpose, several different commercial codes are available that are usually based on FEM. The FEM is preferred over other numerical approaches because of its ability to handle any shape and medium inhomogeneity. The FEM is, however, limited by computational power, particularly for 3D simulations.

A review of FEM fundamentals and solutions for scattering problems was given by Volakis et al. (2004). Since then, several commercial codes were introduced; however, among them one of the best known code is the COMSOL multiphysics program with a dedicated electromagnetics module (COMSOL 2007). The FEM requires a limited volume for analysis, and within this volume subdomains geometrically define the system. The volume is subdivided (meshed) into areas that are small when compared to both the wavelength and the subdomain. The volume and subdomains are defined by their index of refraction. For regions where the subdomain has a frequency-dependent index of refraction, these values are usually called from a user-defined program. Maxwell's equations are then solved in each of the meshed areas, and the near-field solution can be determined.

In general, an FEM program can solve both the harmonic and transient propagation problems and eigenfrequency cavity problems in 2- or 3D. The solution of partial differential equations for the magnetic or electric fields is obtained by transforming the equations into ordinary differential equations (ODE). The resulting ODE can be solved using FDTD schemes. The FDTD formulation gives a numerical solution for near the field, and solves the time-dependent Maxwell's equations. The electric and magnetic field are written in terms of the spatial and time locales of each point, and then the equations can be solved using matrix numerical solvers.

In COMSOL Multiphysics (2007), the governing equations used for the solution of the electric and magnetic field are available from Jin (2002). These equations are solved with the help of one of the two boundary conditions that were used when constructing the models: (1) the scattering boundary condition and (2) the matched boundary condition. The scattering boundary condition is used for a boundary that is transparent to a scattered or incident wave. The matched boundary condition is used if it is totally nonreflecting. The boundary on the edge of the system does not affect the propagation of waves. The matched boundary condition can be written with E or H as the dependent variable. The electric field is directly used in the solver. Application of COMSOL Multiphysics to a nanoscale atomic force microscopy (AFM) tip heating problem was reported by Huda et al. (2011, 2012) and Huda and Hastings (2013).

15.6.5 Finite Difference Time-Domain Method

The FDTD method is probably the most common approach for simulation of scattering problems (Wriedt 2009). This method was first introduced by Yee (1966), where time-dependent Maxwell equations are solved to calculate electromagnetic scattering in both time and space domains. The derivatives of Maxwell equations in space and time are approximated by a finite difference scheme and discretized in both space and time domains. The equations are solved numerically with appropriate boundary conditions and particle properties using a fully explicit scheme.

FDTD methods were reviewed by Schlager and Schneider (1995) and Taflove (2007). This approach is commercially available, as more than 25 companies were listed by Taflove. There are several other open source development projects, which are listed at the corresponding Wikipedia site. It is also used extensively for simulations of metamaterials as discussed by Veselago et al. (2006). This method computes the solution in a finite domain, like the FEM, and so a far zone transformation has to be invoked to calculate fields in far-field. The method is popular because of its conceptual simplicity and ease of implementation but has disadvantages similar to those of the FEM including limitations in accuracy, mathematical complexity, and the need to repeat computations for different angles of incidence. The FDTD approach can be applied to arbitrary geometries with different properties.

Maxwell equations could be discretized using the FDTD method. In order to present the implementation of FDTD, we give an example on how Maxwell equations are written in terms of electrical displacement and magnetic fields $D_x, D_z,$ and H_y, respectively. They are considered for transverse magnetic (TM) wave, in which the only nonzero component of magnetic field is H_y, that is, propagation along the z-axis. They are

$$\frac{\partial D_x}{\partial t} = -\frac{\partial H_y}{\partial z} \tag{15.77}$$

$$\frac{\partial D_z}{\partial t} = \frac{\partial H_y}{\partial x} \tag{15.78}$$

$$D_x(\omega) = \bar{\varepsilon}_0 \bar{\varepsilon}_r(\omega) E_x(\omega) \tag{15.79}$$

$$D_z(\omega) = \bar{\varepsilon}_0 \bar{\varepsilon}_r(\omega) E_z(\omega) \tag{15.80}$$

$$\frac{\partial H_y}{\partial t} = \frac{1}{\mu_0}\left(\frac{\partial E_z}{\partial x} - \frac{\partial E_x}{\partial z}\right) \tag{15.81}$$

The first order finite difference representation of the fields in Equations 15.77, 15.78, and 15.81 are written in discrete time domain as

$$\frac{D^{n+1/2}_{x i+1/2,k} - D^{n-1/2}_{x i+1/2,k}}{\Delta t} = -\frac{H^n_{y i+1/2,k+1/2} - H^n_{y i+1/2,k-1/2}}{\Delta z} \tag{15.82}$$

$$\frac{D^{n+1/2}_{z i,k+1/2} - D^{n-1/2}_{z i,k+1/2}}{\Delta t} = -\frac{H^n_{y i+1/2,k+1/2} - H^n_{y i-1/2,k+1/2}}{\Delta x} \tag{15.83}$$

$$\frac{H^{n+1}_{y i+1/2,k+1/2} - H^n_{y i+1/2,k+1/2}}{\Delta t} = \frac{E^{n+1/2}_{z i+1,k+1/2} - E^{n+1/2}_{z i,k+1/2}}{\mu_0 \Delta x} - \frac{E^{n+1/2}_{x i+1/2,k+1} - E^{n+1/2}_{x i+1/2,k}}{\mu_0 \Delta z} \tag{15.84}$$

If the Yee cell size is kept small, the central differences are said to have second-order accuracy or second-order behavior (Δx^2 terms can be ignored). Furthermore, we can rewrite Equations 15.82 through 15.84 as

$$D^{n+1/2}_{x i+1/2,k}(k) = D^{n-1/2}_{x i+1/2,k}(k) - \frac{\Delta t}{\Delta z}\left[H^n_{y i+1/2,k+1/2} - H^n_{y i+1/2,k-1/2}\right] \tag{15.85}$$

$$D^{n+1/2}_{z i,k+1/2}(k) = D^{n-1/2}_{z i,k+1/2}(k) - \frac{\Delta t}{\Delta x}\left[H^n_{y i+1/2,k+1/2} - H^n_{y i-1/2,k+1/2}\right] \tag{15.86}$$

$$H^{n+1}_{y i+1.2,k+1/2} = H^n_{y i+1.2,k+1/2} + \frac{\Delta t}{\mu_0 \Delta x}\left[E^{n+1/2}_{z i+1,k+1/2} - E^{n+1/2}_{z i,k+1/2}\right] - \frac{\Delta t}{\mu_0 \Delta z}\left[E^{n+1/2}_{x i+1/2,k+1} - E^{n+1/2}_{x i+1/2,k}\right] \tag{15.87}$$

In the xz plane of interest, the 1D wave equation is

$$\left(\frac{\partial}{\partial x} - \frac{1}{c}\frac{\partial}{\partial t}\right) E_z = 0 \tag{15.88}$$

$$\left(\frac{\partial}{\partial z} - \frac{1}{c}\frac{\partial}{\partial t}\right) E_x = 0 \tag{15.89}$$

They can be easily discretized using only the field components on, or just inside the mesh wall, yielding an explicit finite difference equation at time step $n + 1$:

$$E^{n+1}_{Z1,k+1/2} = E^n_{Z2,k+1/2} + \frac{c\Delta t - \Delta x}{c\Delta t + \Delta x}\left(E^{n+1}_{Z2,k+1/2} - E^n_{Z1,k+1/2}\right) \tag{15.90}$$

$$E_{x_{i+1/2,1}}^{n+1} = E_{x_{i+1/2,2}}^{n} + \frac{c\Delta t - \Delta z}{c\Delta t + \Delta z}\left(E_{x_{i+1/2,2}}^{n+1} - E_{x_{i+1/2,1}}^{n}\right) \tag{15.91}$$

where $E_{z_{i,k+1/2}}^{n}$ and $E_{x_{i+1/2,k}}^{n}$, respectively, refer to the i, kth electric component of Yee cell in z and x direction, and c is the speed of light.

The numerical algorithm for Maxwell's curl equations requires that the time increment Δt have a specific bound relative to the space increments Δx, Δy, and Δz. The time increment has to obey the following bound, known as Courant–Friedrichs–Lewy (CFL) stability criterion:

$$\Delta t \le \min\left\{\left(c\sqrt{\frac{1}{(\Delta x)^2} + \frac{1}{(\Delta y)^2} + \frac{1}{(\Delta z)^2}}\right)^{-1}\right\} \tag{15.92}$$

The difficulty in working with FDTD is due mainly to the imposition of the boundary conditions. The choice of boundary conditions is the key to an accurate FDTD simulation. There are various techniques to achieve the aim of simulating a geometry that extends to infinity in all dimensions, thereby eliminating reflections from the actual physical edges. In FDTD these are known as absorbing boundary conditions or ABCs as they absorb or attenuate waves (electric or magnetic) as they approach the edge of the actual problem geometry. Which method you use (known as Dirichlet, periodic, Mur, Mie/Fang superabsorption, perfectly matched layers (PML), convolution PML (CPML), etc.) will very much depend on the nature of the waves, their incidence angle, and their wavelength relative to the geometry size. Choosing the best ABC for an FDTD simulation is crucial as it can minimize the overall problem space, that is, the number of grid cells in the x, y, and z dimensions, and thereby reduce computation time (Didari and Mengüç 2014, 2015a).

FDTD has found applications in several diverse areas of mechanical engineering such as near-field thermal radiation (see Datas et al. 2013, Didari and Mengüç 2014, 2015a,b), microelectronics, with applications in energy-conversion devices, nanothermal manufacturing, as well as in electrical engineering such as antenna imaging design and bioelectromagnetic device development.

MATLAB by Mathworks is a particularly powerful tool for developing FDTD solutions because of its inherent support for matrices and arrays. FDTD algorithms implemented in MATLAB only need a single iteration per time-step to cover the complete spatial space (1D, 2D, or 3D). A fast Fourier transform (FFT) can be applied at any point in time or space to produce a frequency domain solution without an additional frequency "sweep." This is achieved by applying a broadband excitation pulse at some appropriate point within the problem geometry and then applying an FFT to the resultant time-domain response.

Implementing FDTD programmatically through MATLAB (or similar open source solutions such as GSVIT) allows complete independence over problem parameter definition, particularly in near-field thermal radiation research where arbitrary geometries and multiple dispersive materials may be used (Didari and Mengüç 2014, 2015a,b). In addition, the application of FDTD to light scattering aerosols and the relevant literature review was reported by Sun et al. (2013) and Datas et al. (2013).

In the last few years, MATLAB and other FDTD solvers have embraced parallel computation, both through support for multiple CPU cores and by harnessing the power of graphics processing units (GPU) with their hundreds of processing cores. These developments have made finely gridded simulations, over large 2D or 3D geometries, possible by dramatically reducing simulation times (by factors between 10 and 200). As one of the weaknesses of FDTD is numerical accuracy due to the discrete representations of electromagnetic fields, faster execution time allows proportional reduction in the discretization steps in both space and time.

Recent FDTD studies of near field thermal radiation have shown that modeling geometries separated by nanogaps in orders of a few tens of nanometers is also possible using the FDTD method, which is a promising factor in this field since analytical methods may not be easily available for some geometries due to geometry asperities; however, a tradeoff needs to be found between the accuracy and the simulation time requirements when working in such small scales (Didari and Mengüç 2014, 2015a). Note that, for complex particles, these relations can be used for experimental determination of the phase function coefficients. Agarwal and Mengüç (1991) carried out an extensive numerical and experimental research to determine these parameters from experiments for polystyrene latex particles. A similar experimental study was conducted for coal particles by Mengüç et al. (1994).

15.7 APPROXIMATE ANISOTROPIC SCATTERING PHASE FUNCTIONS

Usually the phase function obtained from any of these approaches is expressed in terms of series expansion in Legendre polynomials to be used for the solution of the RTE. For particles much larger than the wavelength of radiation ($\xi > 5$ or so), the number of terms required in the series expansion becomes large, making the solution of the RTE quite difficult. For that reason, the phase function can be further simplified by replacing it with a delta function to represent the highly forward scattering component and an isotropic component (Viskanta and Mengüç 1987, 1989). Because the exact phase functions obtained for the actual particles with arbitrary shapes, sizes, and size distributions are complicated and difficult to use in radiative transfer analyses, various approximate phase functions have been proposed. Note that they are independent of the solution methods discussed earlier, but can enter into the solution of RTE as approximations. These simpler forms retain the anisotropic characteristics of exact phase functions. The cosine of the scattering angle $\mu_0 = \cos\theta_0$ can be expressed as

$$\mu_0 = \mu\mu' + (1-\mu^2)^{1/2}(1-\mu'^2)^{1/2}\cos(\phi-\phi') \tag{15.93}$$

where $\mu = \cos\theta$, (θ, ϕ) is the angle between the direction of the incident intensity and a coordinate, and $\mu' = \cos\theta'$ (see Figure 15.3).

Lorenz–Mie scattering phase for single scattering into the direction θ_0 relative to the direction of the incident intensity is expressed by an expansion in Legendre polynomials of the first kind, $P_m(\mu_0)$, similar to Equation 15.49:

$$\Phi(\mu_0) = 1 + \sum_{m=1}^{\infty} a_m P_m(\mu_0) \tag{15.94}$$

where (Özişik 1973)

$$\left.\begin{aligned} &P_1(\mu_0) = \mu_0, \quad P_2(\mu_0) = \frac{1}{2}(3\mu_0^2 - 1), \\ &P_3(\mu_0) = \frac{1}{2}(5\mu_0^3 - 3\mu_0), \ldots \end{aligned}\right\} \tag{15.95}$$

The a_m are constants to be determined by curve fitting to Lorenz–Mie (or to any other prediction method) results (see Agarwal and Mengüç 1991 for the theoretical and experimental approaches).

15.7.1 Forward Scattering Phase Function

For a highly forward-scattering medium, the phase function can be approximated (Houf and Incropera 1980, Crosbie and Davidson 1985, Mengüç and Viskanta, 1985, Viskanta and Mengüç 1987, 1989, Agarwal and Mengüç 1991) by

$$\Phi_\delta(\theta,\phi,\theta',\phi') = 4\pi\delta(\cos\theta - \cos\theta')\delta(\phi-\phi') = 4\pi\delta(\mu-\mu')\delta(\phi-\phi') \tag{15.96}$$

where δ is the Dirac delta function, so that $\Phi_\delta(\theta, \phi, \theta', \phi')$ equals infinity when the argument is zero (when $\mu = \mu'$ and $\phi = \phi'$ for forward scattering) and Φ_δ equals zero otherwise.

15.7.1.1 Linear-Anisotropic Phase Function

A general phase function can be simplified using an isotropic scattering first term and a modification term. This is called linearly anisotropic scattering (Özişik 1973, Mengüç and Iyer 1988):

$$\Phi_{la}(\theta,\theta') = 1 + g\mu_0 \tag{15.97}$$

where g is a dimensionless asymmetry factor that can vary between $\pm\infty$. If $g = 0$ it corresponds to the isotropic phase function. This phase function is rotationally symmetric around the direction of travel of the incident intensity and contains the first two terms of the general phase function in terms of Legendre polynomials, Equation 15.82. Integration over all solid angles shows that this phase function is normalized for any value of g.

15.7.1.2 Delta-Eddington Phase Function

The delta-Eddington approximation uses a two-term Legendre polynomial expansion (Equation 15.49) of the actual phase function plus a Dirac delta term to account for forward scattering (Joseph et al. 1976, Viskanta and Mengüç 1987, 1989). For the cosine of the angle of scatter between the incident and scattered directions μ_0, the phase function becomes

$$\Phi_{\delta E}(\mu_0) = 2f\delta(1-\mu_0) + (1-f)(1+3g'\mu_0) \tag{15.98}$$

where

$f = (1/2)\int_{-1}^{1} \Phi(\mu_0)P_2(\mu_0)d\mu_0$

$P_2(\mu_0)$ is the second term of Legendre polynomial Equation 15.95

$g' = (g-f)/(1-f)$ where $g = (1/2)\int_{-1}^{1} \mu_0\Phi(\mu_0)d\mu_0$

When f and g are calculated by this procedure $\Phi_{\delta E}(\mu_0)$ can have negative values for certain ranges of μ_0. This has no physical meaning, and in such a case f and g can be chosen by methods discussed by Crosbie and Davidson (1985). For isotropic scattering, f and g are zero, and hence g' is zero.

15.7.1.3 Henyey–Greenstein Phase Function

Henyey and Greenstein (1940) proposed the phase function

$$\Phi_{HG}(\mu_0) = \frac{(1-g^2)}{(1+g^2-2g\mu_0)^{3/2}} \tag{15.99}$$

where g is a dimensionless asymmetry factor that can vary from 0 (isotropic scattering) to 1 (for forward scattering). If g is negative, a backscattering phase function is produced. This approximate phase function provided good agreement with complete Lorenz–Mie scattering calculations when used in the discrete ordinates and P_N methods (Mengüç and Viskanta, 1983, Viskanta and Mengüç 1987, 1989).

The effect of phase function on radiative flux was investigated by Stockham and Love (1968), who studied radiation at plane and cylindrical boundaries with given reflectivities adjacent to volumes of absorbing, emitting, and scattering media. Experimentally determined values of the scattering phase functions for glass beads and aluminum, carbon, iron, and silica particles were used for comparison of their effect in various energy exchange cases. In some cases, there was

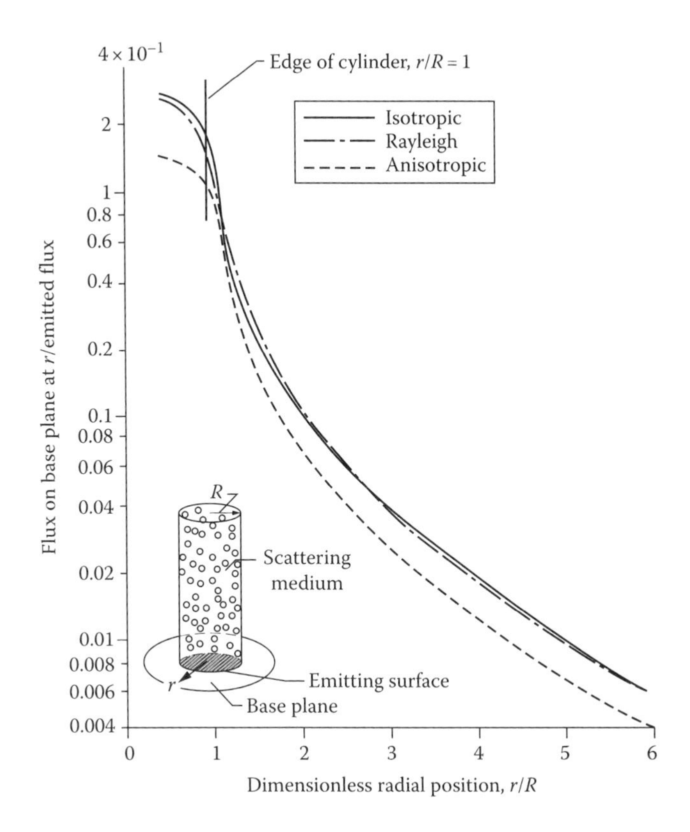

FIGURE 15.17 Effect of scattering phase function on energy scattered back to base plane by cylinder of scattering medium. Optical diameter of cylinder, 2; height to diameter ratio 5. (From Love, T.J. et al., *Radiative Transfer in Absorbing, Emitting and Scattering Media*, Oklahoma University, Norman, OK, December 1967.)

not much difference in the energy transfer using these experimental phase functions as compared with using either Rayleigh or isotropic phase functions. Other examples did show significant anisotropic effects. Agarwal and Mengüç (1991) reported both numerical and experimental procedures to determine g along with other phase function parameters.

In Figure 15.17, results are shown for the energy flux resulting from emitted energy from a black disk that is scattered back to the base plane by a cylinder of scattering medium adjacent to the disk (Love et al. 1967). The results using various scattering phase functions in the medium are in reasonably good agreement. Some of the other cases resulted in larger effects. In parallel-plane geometries the various phase functions gave energy transfer results that had less variation than in the cylindrical geometry. The phase function is important for beam transmission or other situations in which strong sources transmit directionally into a scattering atmosphere. It may also have a larger effect near the boundary of the medium, where there is less equalization by reflections in all directions than in the interior of a medium.

Azad and Modest (1981b) analyzed the effect of anisotropic scattering for a medium within a long cylinder and provided the predictions for a medium at uniform temperature. As might be expected, particles with strong forward scattering tended to enhance the heat flow from the central regions toward the wall, which was at a temperature one-fifth that of the medium. Strong backscattering by the particles decreased the heat flux to the wall. For an optical radius of 2 and a scattering albedo of 0.5, the anisotropy influenced the wall heat flux by about 10%. Larger effects were found when the albedo was increased to 0.9 and 0.95. A similar analysis was carried out by Tong and Swathi (1987) for a medium between two concentric spheres at uniform temperature. Backscattering was found to increase the heat flux at the inner wall and decrease the heat flux at the outer wall. An example using both the delta-Eddington and Henyey–Greenstein phase functions was discussed by Marakis et al. (2000) for the forward scattering behavior of coal combustion particles.

15.8 DEPENDENT ABSORPTION AND SCATTERING

If particles, fibers, and other bodies are in close proximity, scattering can differ from predictions using properties for independent scattering of isolated individual particles. This *dependent scattering* arises from two effects. One is that the internal radiation field in a translucent particle is affected by scattering from surrounding particles; this is the *near-field* effect. A second effect is that scattered radiation from one particle can constructively or destructively interfere with that from another particle; this is the *far-field* effect. These effects become more important as the volume fraction occupied by particles increases (porosity decreases). These interactions have received attention in recent years; the computations are difficult and usually require simplifying assumptions that make the prediction accuracy somewhat suspect.

For spherical particles, scattering is found to be independent if the criterion is met that

$$C + 0.1D > \frac{\lambda_m}{2} \quad \text{or} \quad \xi_c + 0.1\xi > \frac{\pi}{2} \tag{15.100}$$

where

C is the clearance distance between particles

D is the particle diameter, the clearance parameter is $\xi_c = \pi C/\lambda_m$, and the size parameter is defined as usual $\xi = \pi D/\lambda_m$ (Kaviany and Singh 1993)

This result was derived for porosities (≈0.26) typical of rhombohedral packing. If the criterion of Equation 15.100 is not met, dependent scattering should be considered. Dependent scattering in packed beds remains an important effect for bed porosities as high as 0.935, and the effect is most pronounced for opaque particles. Both the porosity requirement and the relation between C and λ_m in Equation 15.100 must be satisfied before the assumption of independent scattering can be used with confidence. An earlier criterion given by Brewster and Tien (1982) for independent scattering is

$$\xi > \pi \frac{C}{\lambda_m} \frac{(1-\varepsilon_p)^{1/2}}{0.95-(1-\varepsilon_p)^{1/2}} \quad (\varepsilon_p = \text{porosity}) \tag{15.101}$$

The dependent scattering region (results deviate more than 5% from independent scattering) is demarcated when $C/\lambda_m = 0.5$ is inserted into Equation 15.101 (Tien 1988); this is based on far-field effects alone. Figure 15.18 shows the regions of dependent and independent scattering by this criterion (Tien and Drolen 1987).

Closely spaced parallel cylinders are of interest in applications such a fibrous insulation and the determination of nanomaterial properties. In the production of structures by winding closely spaced fibers in a resin, it is possible to initiate resin polymerization by infrared radiation from an external source. The fibers are very closely spaced, and the effects on the scattering phase function must be considered of both dependent scattering and the dependence of scattering on incidence angle. Nearly identical approaches including near-field effects are used to determine the radiative properties of arrays of parallel cylinders subject to normal irradiation and dependent scattering (Lee 1988, Chern et al. 1995, Lee and Cunnington 1998). They show the ranges of cylinder volume fraction and size parameter where dependent scattering becomes important for various values of the fiber refractive index (see an example in Figure 15.19). The analysis was further extended into coated and

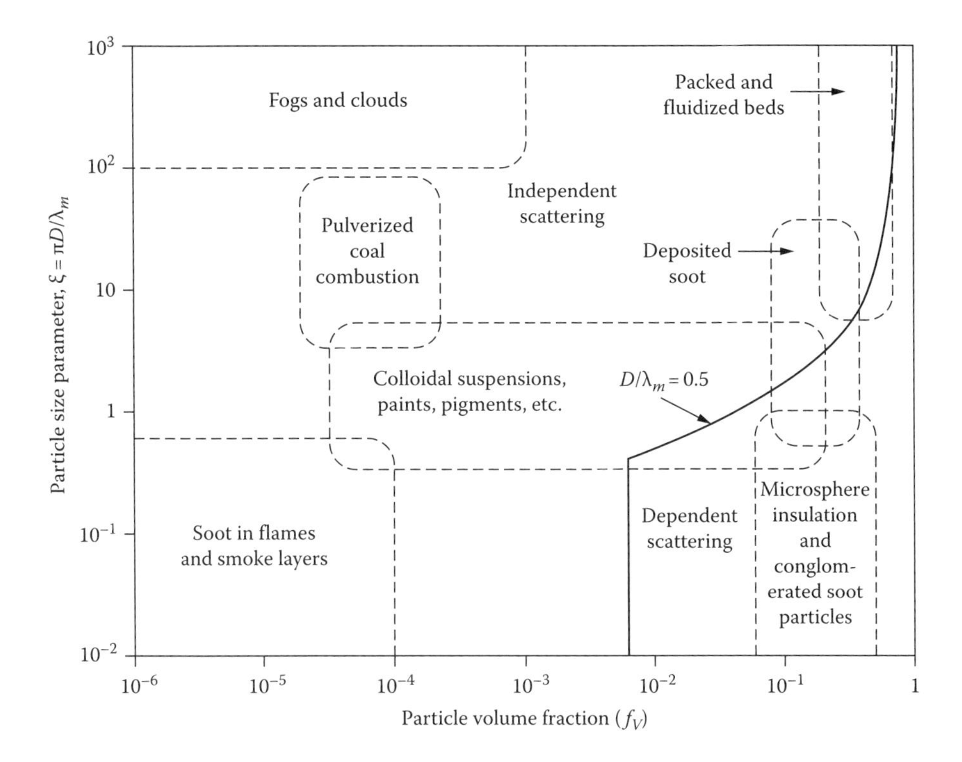

FIGURE 15.18 Regime map of independent and dependent scattering regimes as a function of particle size parameter and volume fraction. (From Tien, C.L. and Drolen, B.L., *Annu. Rev. Num. Fluid Mech. Heat Trans.*, 1, 1–32, Hemisphere, Washington, DC, 1987.)

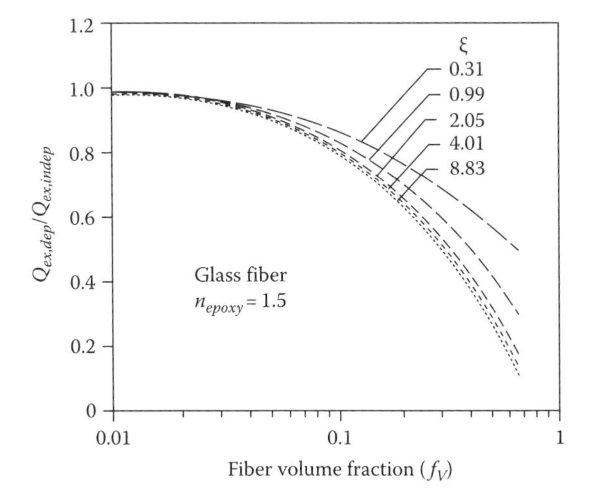

FIGURE 15.19 Variation of dependent extinction efficiency with fiber size parameter $\xi = \pi D/\lambda_m$ and volume fraction, for normal incidence on S-glass fibers dispersed in epoxy resin. (From Chern, B.C. et al., Dependent radiative transfer regime for unidirectional fiber composites exposed to normal incident radiation, *Proceedings of Fourth ASME/JSME Joint Symposium*, Maui, HI, 1995.)

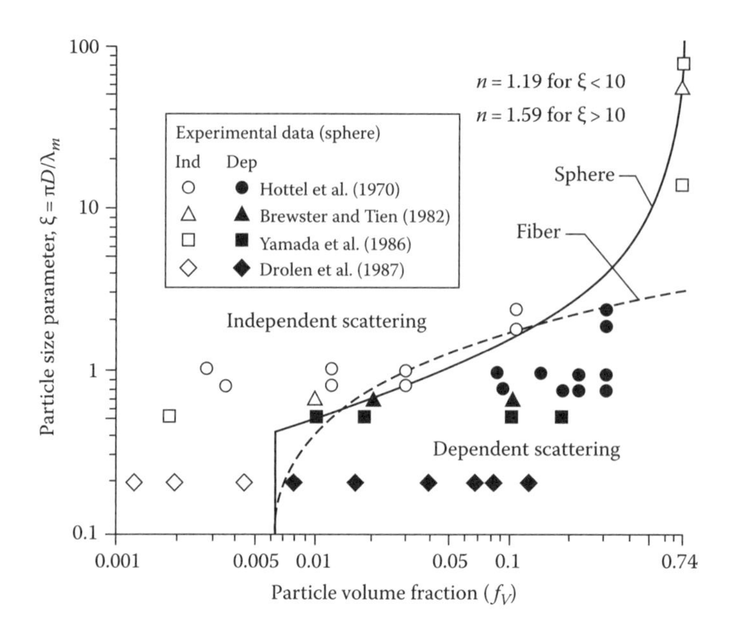

FIGURE 15.20 Regimes of dependent and independent scattering for nonabsorbing spheres and parallel cylinders (fibers); cylinders have normally incident radiation (λ_m is in the medium surrounding the spherical particles or cylindrical fibers). (From Lee, S.-C., *JTHT*, 8, 641, 1994.)

uncoated fibers subject to obliquely incident radiation (Lee and Grzesik 1995). For fiber layers, the interaction of wavelength, refractive index, fiber diameter, and fiber spacing is so complex that no simple criterion is available for deciding whether dependent scattering will be important, and maps must be referred to, similar to Figure 15.18. However, for $\xi = \pi D/\lambda_m$, (based on cylinder diameter, λ_m, being the wavelength in the medium surrounding the fibers) >2, dependent scattering is relatively independent of the fiber refractive index and the size parameter, and dependent scattering is present for porosity <0.925 (Chern et al. 1995).

A regime map of the approximate regions where dependent scattering occurs is shown in Figure 15.20. The map is based on experimental observations for nonabsorbing spheres, on criteria based on far-field effects, and on computations for normal incidence on cylinders, and is only an approximate guide.

An effect of dependent scattering is that the scattering cross section for particles in a dispersion can decrease as the volume fraction of particles increases. Thus, increasing the number of particles does not result in a proportionate increase in the extinction by scattering. It has been observed that an excessive amount of pigment in paint can decrease the hiding ability of the paint. On the other hand, agglomerates of soot particles absorb and scatter light different than those for independent particles. This affects the contribution of soot agglomerates to radiative balance in combustion chambers. To understand this effect, Ivezic and Mengüç (1996, 1997) used DDA and showed that an agglomerate of N particles must be modeled as dependent scatterers if the effective size parameter of N particles ($\xi_e = N\pi d/\lambda$) is between 0.2 and 2. They explored the effects of agglomeration using DDA. They conclude that independent scattering depends on the single particle size parameter and the distance between two particles, such that $c > 2/x_s = 2.4/x_e$

Another effect of increasing particle density is *dependent absorption* (Kumar and Tien 1990). For highly absorbing particles, the particle absorption efficiency was found to increase as the spacing between the particles decreased, while the scattering efficiency declined. The total extinction by combined absorption and scattering increased. About 5% increase in absorption efficiency was found for a particle volume fraction of 0.06. Dependent effects tend to increase the absorption over that predicted by using independent absorption, while dependent scattering tends to decrease scattering from that predicted using independent scattering. Enhanced absorption resulting from dependent scattering is analyzed by Ma et al. (1990).

White and Kumar (1990) solved Maxwell equations for two cases: (1) normal incidence on a layer of evenly spaced coplanar fibers and (2) randomly oriented fibers with a prescribed distribution of fiber spacing. For a single cylinder at normal incidence with $\xi \le 1$, the phase function in a plane normal to the cylinder is

$$\Phi(\theta_0) = \frac{\left|\bar{n}^2 + 1\right|^2 + 4\cos^2\theta_0}{\left|\bar{n}^2 + 1\right|^2 + 2} \tag{15.102}$$

which for $\bar{n} \approx 1$ reduces to

$$\Phi(\theta_0) = \frac{2}{3}(1 + \cos^2\theta_0) \tag{15.103}$$

For evenly spaced fibers of diameter D with spacing a between fiber edges, the scattering efficiency Q_N relative to that for a single fiber Q_1 is found for a small size parameter $\xi = \pi D/\lambda_m$ to be

$$\frac{Q_{\lambda,N}}{Q_{\lambda,1}} = 1 + \frac{1}{N}\sum_{k=1}^{N}\sum_{\substack{j=1\\ j\neq k}}^{N}\left\{J_0[2xA(k-j)] + \frac{1}{3}J_2[2xA(k-j)]\right\} \tag{15.104}$$

where $A = 1 + (a/D)$. Comparison of the analysis with experimental measurements for radiant transmission in a region of silica fibers showed excellent agreement. For the case of high-porosity fiber regions where dependent scattering effects are unimportant, the interaction of conduction and radiation and the effect of fiber orientation are found to be important in radiative transfer (Lee and Cunnington 2000).

Another consideration in analyzing scattering effects is *single* versus *multiple scattering*. In multiple scattering, energy scattered from one particle can hit other particles and be scattered additional times. To account for this, the incoming scattering term is included in the equation of radiative transfer. Although this complicates the analysis, it must be included in many engineering problems. Both multiple and dependent scattering become important for laser diagnostics. In this case a single beam path needs to be explored; of course in radiative transfer many such beams are considered.

Aslan et al. (2006a) conducted an experimental study to monitor both bubble size and gas hold-up in a bubble laden medium. In their experiments, they also explored the regime between the independent and dependent scattering regimes. These data are plotted in Figure 15.21, which also shows the traditional criterion given by Brewster and Tien (1982) where the particle clearance $c/\lambda = 0.3$ (corresponds to $\tau = 2.5$ and shown as a dashed line in Figure 15.21). Their experiments showed that the cut-off line between the single and multiple scattering regimes may vary depending on optical properties of particles/bubbles and the liquid. The demarcation boundary between the single versus multiple scattering regimes was calculated using $\tau = 0.3$ (shown as a solid line in Figure 15.21). This assumption was based on the experimental results presented for latex particles (Agarwal and Mengüç 1991).

The effect of dense bubbles on radiative transfer predictions was discussed by Pilon and Viskanta in two separate accounts (2002, 2003). They also published a regime map to decide which radiative transfer approximations can be used at certain bubble sizes. Similarly, Swamy et al. (2009) worked with slow decaying foams. These studies are to be expanded to practical systems for characterization

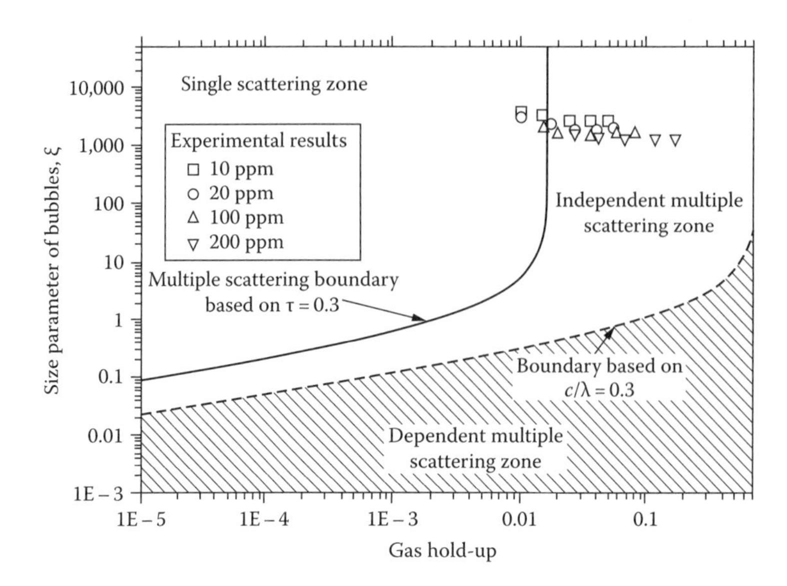

FIGURE 15.21 Single and multiple (independent and dependent) scattering regimes for bubbles in the liquid as a function of size parameter and gas hold-up. Experimental results are shown for the present system. Dependent/independent scattering regime demarcation is based on Brewster and Tien (1982), and the multiple scattering demarcation line based on $\tau = 0.3$ is from Agarwal and Mengüç (1991). (From Aslan, M.M. et al., *JQSRT*, 101, 527, 2006.)

of medium properties from measurement signals. This work would need to be coupled with the inverse analysis. Such a work for dense media involving bubbles was carried out by Gay et al. (2010), where they used a polarized imaging system. They considered the entire Mueller matrix elements made comparisons with the experimental measurements. A similar concept was recently reported by Carmagnola et al. (2014).

HOMEWORK

15.1 The spectral intensity of radiation ($\lambda \approx 0.589$ μm) along a path is to be attenuated by scattering. A proposed scheme is to use very small spherical particles of gold having a characteristic diameter of 200 Å (optical data for gold are in Table 3.2). (Absorption by the particles is being neglected.) The particles are to be suspended in a nonscattering, nonabsorbing medium. Assuming Rayleigh scattering is applicable, what is the particle scattering cross section, C_λ? For the intensity to be 10% attenuated by scattering in a path length of 2 m, approximately what number density of particles would be required? What are the volume fractions of the particles and the mass of the particles per cubic centimeter of the scattering medium?

Answer: 1.35×10^{-5} μm²; 3.90×10^{9} cm⁻³; 1.64×10^{-8}; 36.1×10^{-8} g.

15.2 The radiation intensity in Homework Problem 15.1 is changed to blue light ($\lambda \approx 0.42$ μm) while the scattering particle size and number density are kept the same. Assuming the same optical constants apply at this wavelength, what is the percent attenuation for this beam for a 2 m path length?

Answer: 33.4%.

15.3 Consider Rayleigh scattering for very small copper particles at wavelengths of 0.589 and 10 μm shown in Table 3.2. How do the scattering cross sections C_λ differ at these two λ (note Figure 15.14)? Repeat the calculations for gold particles.

Answer: Ratio of $C_\lambda = 1.67 \times 10^{5}$.

15.4 Verify that the phase function in Equation 15.67 satisfies the normalization specified by Equation 1.71.

15.5 The complex index of refraction for a particular sample of carbon is given by $\bar{n} = 2.2 - 1.2i$ at the wavelength of 2.88 μm. The carbon is ground into fine particles (assumed spherical) that are 0.1 μm in diameter. Compute the efficiency factors for absorption and scattering for these particles. What is the ratio of the absorption cross section to the scattering cross section? The surrounding medium is air.

Answer: $Q_{\lambda,a} = 0.121$; $Q_{\lambda,s} = 2.23 \times 10^{-4}$; $C_{\lambda,a}/C_{\lambda,s} = 544$.

15.6 In Homework Problem 15.5, the scattering for very small carbon particles is found to be small compared with absorption. A dispersion of these particles in air is formed with a particle concentration of $N = 10^{10}$ particles/cm³. The dispersion is at low temperature.

1. Radiation at $\lambda = 2.88$ μm enters the dispersion with an intensity of 0.4×10^4 W/m²·μm·sr and travels through a path length of 0.15 m. What is the intensity at the end of this path length?
2. The dispersion is now heated to 800 K. What is the intensity at the end of the path length for this condition?
3. The dispersion is now heated to a different uniform temperature, and the intensity at the end of the path length is measured as 0.4×10^4 W/m²·μm·sr. What is the temperature of the dispersion?

Answer: 959 W/(m²·μm·sr); 1848 W/(m²·μm·sr); 995 K.

15.7 Emission of particulate matter (PM) from the internal combustion engines of transportation vehicles is an important issue for mega cities. A PM concentration exceeding 50 μg/m³ is unhealthy for the general population. Consider a case where the 1000 m thick air layer above the Earth is filled with 100 nm diameter soot spheres at a concentration of 100 μg/m³. The apparent density of the soot particles can be taken as 2.26 g/cm³. The complex index of refraction of the soot particles can be assumed to be the same as the $2.2 - 1.2i$ value given for carbon particles in Homework Problem 15.5, where the scattering by these particles are found to be small compared to the absorption for a wavelength of 2.88 μm. Consider only the absorption by the soot particles, determine the transmittance and absorptance of the 1000 m thick air layer at the peak wavelength (λ_{max}) of solar radiation and Earth radiation, respectively. The temperatures of the sun and Earth are 5780 and 283 K, respectively.

Answers: $T_{\lambda,solar} = 0.398$; $T_{\lambda,IR} = 0.956$; $\alpha_{\lambda,solar} = 0.602$; $\alpha_{\lambda,IR} = 0.044$.

16 Near-Field Thermal Radiation

16.1 INTRODUCTION

Throughout the following discussion it will be assumed that the linear dimensions of all parts of space considered, as well as the radii of curvature of all surfaces under consideration, are large compared with the wavelengths of the rays considered.

Max Planck
The Theory of Heat Radiation (1906)

In his seminal book *The Theory of Heat Radiation*, Planck presents a clear argument that the dimensions of the surfaces and geometries in the classical approach are always large compared to the primary wavelength of radiation considered in the analysis. This means that the classical analysis of thermal radiation is valid for the far field, at distances several times longer than Wien's wavelength of emission, λ_w. As we discussed in Chapter 1, Equation 1.26 suggests that the peak wavelength of blackbody emission is obtained from Wien's expression, to $\lambda_w T = (hc/k_B)T$, that is, $\lambda_w = 2897.8/T$, where T is in Kelvin (K) and λ_w is in micrometers. However, with the advances in nanotechnology and nanoscale engineering, we can consider much smaller size objects than λ_w and much smaller distances between them for applications ranging from sensing to energy harvesting. When the gap size is below a few hundreds of nanometers, the so-called near-field radiative transfer can be enhanced by several orders of magnitude over that predicted by standard radiation transfer analysis due to tunneling of evanescent waves.

These waves are due to total internal reflection (TIR) and fluctuations of electrons or lattices near the surface. These fluctuations are spectral and functions of material properties and can be very pronounced for polar dielectrics, such as SiC and SiO_2. It is imperative to understand the transfer of energy for such small distances and sizes to design and operate reliable system and processes or tune them for specific applications. These tuning can be achieved by either using multilayer coatings or different nano- and microstructures on the otherwise smooth surfaces. The near-field radiative transfer concepts can be used in tandem with many other physical applications, as discussed by Cahill et al. (2014).

This chapter provides an overview of near-field radiation transfer (NFRT) either between nanosize and larger objects or large objects separated by nanosize gaps. We use primarily the electromagnetic wave analysis in our discussion and do not get into quantum mechanics at all. There is no question that with diminishing size of the emitting objects, the importance of quantum mechanical analysis increases, but it is beyond our scope here. In addition, we briefly discuss metamaterials and their importance in near-field radiative transfer. These are "designer" materials, and their electromagnetic properties and corresponding resonances, which are critical for near-field radiative transfer exchange, can be tuned to an extent.

The effects of electron motion or lattice vibrations in a medium can be sensed if a second medium is at a sufficiently close distance to it. Then, fluctuations in an electron field (plasmons) or vibrating lattices (phonons) can be tunneled to the second medium. The role of the fluctuating near field was completely lost in the classical theory of thermal radiation until the question of "radiant" heat transfer, which was already posed in the 1960s, received the attention of researchers such as Rytov (1959). Rytov pointed out how the effects of a fluctuating field were ignored in the previously published work on heat flux calculations between closely spaced bodies. He established a correlation between the classical theory of thermal radiation and the near-fluctuating fields, which led him to developing the fluctuation–dissipation theorem (FDT). Fluctuational electrodynamics (FE) stems from the FDT, where the origin of thermal radiation is tied to the random movement of charges inside the medium at temperatures above 0 K. All physical objects having finite temperatures other than absolute zero emit a fluctuating electromagnetic field arising from an internal field. This allows the possibility of spontaneous transitions between vibration–rotation levels of molecules. Here, we focus only thermal emission and refrain from considering the other types of electromagnetic emissions.

Let us assume two nonmagnetic, isotropic, homogenous, parallel plates separated from each other by a narrow gap, changing from a few nanometers to a few tens of micrometers. Assume one plate is at a higher temperature than the other one. We know that the transition of thermal radiation takes place in both ways, but net radiative flux will be from the hot to the cold plate. Figure 16.1a shows the geometry considered here. Now consider a beam of radiative energy emanating from deep inside the hot medium. It would travel to the surface, and if the angle at which it is incident on the surface is smaller than the critical angle, it would be transferred out, as depicted. Here, the critical angle refers to the one obtained from Snell's law at interfaces (see Section 14.4.1). On the other hand, if the angle of incidence is greater than the critical angle, then there would be no transferred energy, yet evanescent waves would be present due to the TIR. Evanescent waves are standing waves, naturally oscillating. They do not carry any energy at all. However, they tunnel the energy from the first to the second medium if it is at sufficiently close distance. They do decay rapidly and will therefore not be able to carry energy to the other medium if it is at the far field.

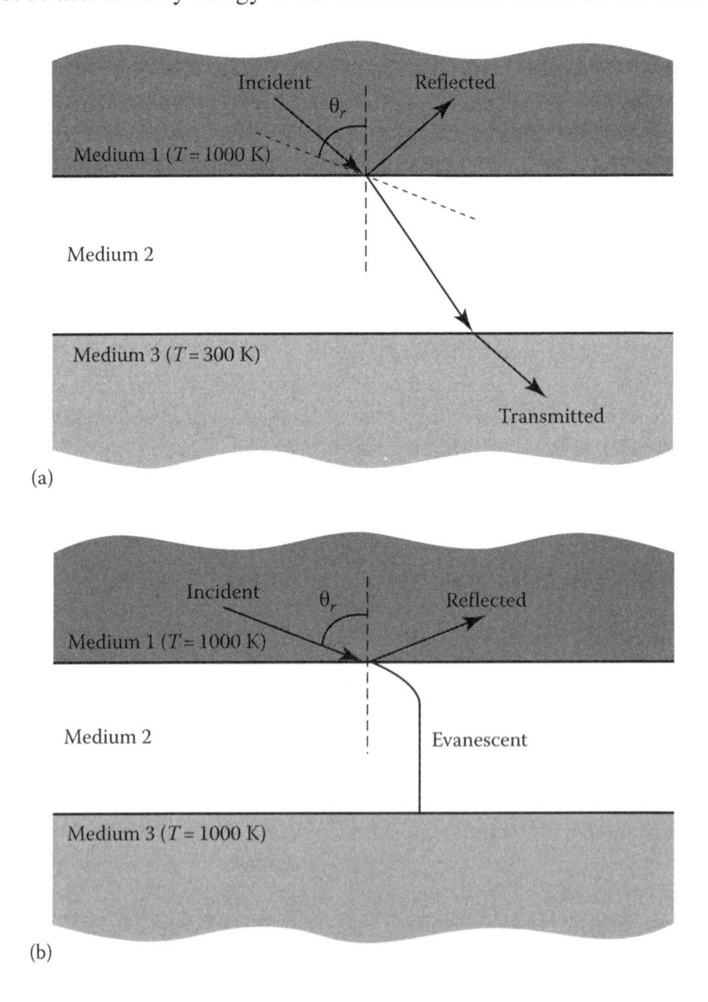

FIGURE 16.1 (a) Far-field thermal radiation transfer for $\theta_i < \theta_c$. (b) Far-field thermal radiation transfer for $\theta_i > \theta_c$.

The transport of energy by electromagnetic radiation generally is categorized into two types: far-field propagating electromagnetic modes and near-field evanescent modes that decay over a few hundred nanometers to micrometer length scales. Again, these near-field modes could originate from the presence of energy carried to the surface or surface phonon polaritons (SPhPs) (vibration of electrons in the matter) or surface plasmons (lattice vibrations). They are the standing waves, resulting from TIR of electromagnetic waves or due to the random oscillations of the so-called phonons and/or electrons. In classical radiative heat transfer, far-field modes are the primary carrier in energy transport and it is in this regime where Planck's blackbody limit is formulated. However, when two bodies are closely spaced, it is possible for surface wave modes to tunnel across the surface to the second medium, as depicted in Figure 16.2. As we discuss during this chapter, it can be shown that the resulting enhancement in radiative transfer can break Planck's blackbody limit (Chen 2005, Zhang 2007, Francoeur and Mengüç 2008, Budaev and Bogy 2011, Otey et al. 2014).

In classical radiative transfer, we are concerned with radiative exchange between objects far from each other. The "closeness" or "proximity" between structures for radiative transfer is a relative term as it should be quantified in relation to the dominant (peak) wavelength of thermal emission, as discussed earlier. Radiant energy exchanges between bodies separated by distances much longer than the dominant emitted wavelength can be described by the radiative transfer equation (RTE) and Planck's blackbody distribution. In that case, transport is assumed to be incoherent (see Section 16.2.5) as thermal radiation is conceptualized as a particle (photon, even though one needs to be careful about this description, Mishchenko 2006). This regime is considered as the "far-field regime." The approximation that radiation is incoherent in the far-field regime is acceptable since the coherence length of a blackbody is of the same order of magnitude as the dominant wavelength of thermal emission (Chen 2005). Here, "coherence" refers to the phases of the EM waves. A laser is a coherent light source where there is phase match between the waves emitted in the laser cavity. Indeed, there is coherence of phases of waves emitted by any point source right after the emission. However, because of random oscillations of the thermal sources, this coherence lasts a few nanoseconds or a few tens of nanometers (Chen 2005). For that reason, we can treat a blackbody source radiation as incoherent at far field. A more detailed discussion of coherence is given in Section 16.2.5.

As the typical size between structures exchanging thermal radiation decreases to a size comparable or below the dominant emitted wavelength, the classical theory of radiative transfer ceases to be valid, and the wave nature needs to be considered in the calculations. Note that in classical radiative transfer discussed in earlier chapters, we have not considered the phase and polarization associated with the electromagnetic waves, except for the cases where we have considered radiative properties in detail. That simplification has allowed us to reduce the problem to the propagation of the so-called photons, and ray-tracing approaches could be used. The wavelength effects are considered through the spectral behavior of the emission (due to Planck's blackbody distribution) and the

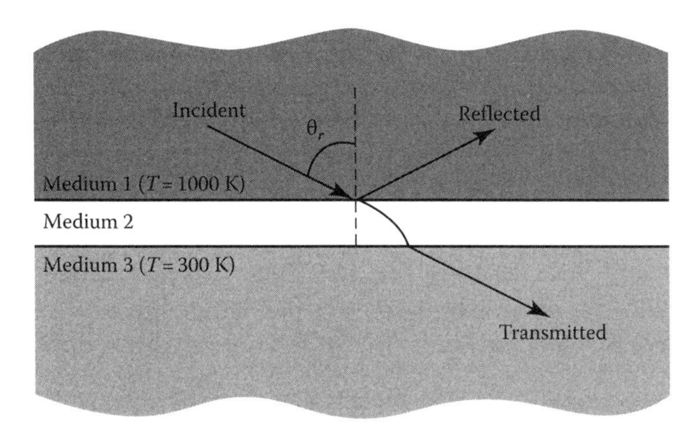

FIGURE 16.2 Configuration considered for near-field thermal radiation transfer through evanescent waves.

spectral radiative properties of the matter. However, if the wave nature is considered in wave propagation, then the phase and the polarization of the waves reveal the rich physics behind the radiative exchange phenomenon. At short distances, both the phase and the polarization variations must be accounted for. Thus, in the "near-field" (NF) regime, radiative energy transfer is correctly described only by the complete Maxwell equations, which describe more general behavior than the photon formulation discussed as a ray-tracing formulation used in classical radiative transfer theory (Chapters 4 through 7). As the typical wavelengths involved in thermal radiation are a few micrometers, near-field effects become dominant when bodies are separated by a few hundreds of nanometers. For this reason, near-field thermal radiation is also referred in the literature as "nanoscale thermal radiation." Near-field effects are observed at nanoscales for thermal radiation only. For other EM-wave spectra, we can still find near-field effects at different length scales. For example for microwaves, the NF is observed within millimeter to centimeter distances (Geffrin et al. 2012).

In the near-field regime, reflection, refraction, diffraction, and transmission are all considered by solution of the electromagnetic wave equations, discussed in Chapter 14. In addition, the RTE can be derived from the electromagnetic theory (Mishchenko et al. 2006, Mishchenko 2014), as briefly discussed in Section 14.6. Yet, the emission of radiation at such scales can deviate significantly from Planck's law and requires separate treatment.

As mentioned earlier, the electromagnetic description of thermal radiation emission from the electrodynamic point of view was pioneered by Rytov (Rytov 1959, Rytov et al. 1989). In Rytov's description, thermal radiation emission is conceptualized as the field generated via chaotic motion of charges within a material, behaving like small radiant dipoles with random amplitudes (Chen 2005, Greffet et al. 2007, Zhang 2007). According to his theory, oscillating dipoles emit waves that carry the radiative energy away from the surface of an emitting body (i.e., propagating waves). They also result in evanescent waves that are confined to very near the surface of the body (Whale 1997, Jackson 2005), which are not important in the far-field calculations. These evanescent waves exist and oscillate along the interface between two materials, while exponentially decaying over a distance of about a wavelength normal to that interface.

The concept of an evanescent wave field is usually discussed in the literature in the context of TIR, where an external light beam propagating through a medium with a large refractive index is reflected at the interface of a medium of lower refractive index (Hecht 2002, Novotny and Hecht 2006). If the angle of incidence of the beam at the interface between the high- and low-refractive-index media is larger than the critical angle as calculated from Snell's law, the light is totally reflected back into the high-refractive-index medium. Still, an evanescent wave forms and propagates at the interface delimiting the high- and low-refractive-index media, while being confined normal to that interface with the field exponentially decaying in the low-refractive-index medium. These evanescent waves decay within a wavelength or so, or within a few hundred nanometers depending on the material properties, wavelength, and the geometry. If a surface is further than this distance, then their effect is unimportant.

From the electromagnetic description of thermal radiation, evanescent waves are generated by chaotic motions of charges and are present at the surface of any material that has a finite temperature ($T > 0$ K). Even though evanescent waves do not propagate to the far-field, energy transfer through these modes can occur if a second body is brought within the evanescent wave field of the emitting material. It is observed from the mathematical treatment of the problem that even though there is no normal component of the Poynting vector at the first interface, there is a nonzero component at the second one, indicating that net energy exchange can occur (Zhang 2007, Basu and Zhang 2009). This mode of radiant energy transfer is usually referred to as radiation tunneling and causes radiative heat transfer in the near field to exceed the values predicted by Planck's blackbody distribution.

A thermal radiation source is an example of an incoherent radiative source. The temporal, or spectral, coherence of a radiative source manifests itself through emission within a narrow spectral band, while emission in a narrow angular band is a manifestation of spatial coherence

(Carminati and Greffet 1999, Henkel et al. 2000, Joulain 2007, Zhang 2007). In the far-field regime, thermal radiation emission can be treated as a broadband phenomenon with quasi-isotropic angular distribution. The opposite of this example is a laser source that has a high degree of both spatial and temporal coherence as the radiation is emitted around one wavelength with narrow angular distribution. In the near field, thermal sources can also exhibit high spatial and temporal coherences due to the presence of *surface waves*, also referred to as *surface modes* or *surface polaritons*.

Surface waves are hybrid modes that arise from coupling of an electromagnetic field and a mechanical oscillation of energy carriers within a material. Here, the hybrid mode refers to the combination of electromagnetic waves with the energy transfer from oscillating electrons or lattice vibrations. The hybrid mode of the collective motion of free electrons and electromagnetic radiation is called a *surface plasmon polariton* (SPP), which arises in metals and doped semiconductors. Similarly, the hybrid mode of lattice vibrations (transverse optical phonons) and an electromagnetic field is an *SPhP*, which is supported by polar crystals (Kittel 2005a). Similar to evanescent waves, surface polaritons propagate along an interface between two materials, but with an evanescent field decaying in both media (Raether 1988, Maier 2007). The surface polaritons greatly modify the coherence properties in the near field of a thermal source (Henkel et al. 2000). Indeed, radiative heat transfer between closely spaced bodies supporting surface waves not only exceeds Planck's distribution but can also become quasi-monochromatic due to the high degree of spectral coherence of these waves (Carminati and Greffet 1999). It is also possible to achieve highly directional thermal sources in the far field by exciting surface waves having a high degree of spatial coherence via, for example, a grating (Maruyama et al. 2001, Greffet et al. 2002, Arnold et al. 2012). A high degree of spatial coherence means that laser-like emission from a thermal source is observed at distances far from the emitting body.

With the recent advances in nanotechnology and nanopatterning procedures, near-field radiation heat transfer is no longer a pure conceptual phenomenon. Near-field thermal radiation problems are becoming increasingly important in the thermal management of microelectromechanical systems (MEMS) and nanoelectromechanical systems (NEMS) devices, in nanoscale-gap thermophotovoltaic power generation (Whale 1997, DiMatteo et al. 2001, Whale and Cravalho 2002, Laroche et al. 2006, Park et al. 2008, Francoeur 2010), in tuning far- and near-field thermal radiation emission (Greffet et al. 2002, Kollyukh et al. 2003, Ben-Abdallah 2004, Luo et al. 2004, Narayanaswamy and Chen 2004, 2005a, Celanovic et al. 2005, Fu et al. 2005, Lee et al. 2005, Lee and Zhang 2006, 2007, Drevillon 2007, Drevillon and Ben-Abdallah 2007a, Francoeur et al. 2008, 2010a,b, 2011a,b,c, 2013, Ben-Abdallah et al. 2009, Francoeur 2010, Biehs et al. 2011, 2013, Zheng and Xuan 2011a,b, Ilic et al. 2013, Lim et al. 2013, Tschikin et al. 2013), for thermal rectification (Wang and Zhang 2013, Yang et al. 2013, Iizuka and Fan 2015), in near-field thermal microscopy (De Wilde et al. 2006) and spectroscopy (Jones and Raschke 2012, Babuty et al. 2013, O'Callahan et al. 2014), and in advanced nanofabrication techniques (Hawes et al. 2007, 2008, Cahill et al. 2014, Francoeur et al. 2015), to name only a few. Although the rich physics behind near-field thermal radiation at the interfaces of quantum mechanics, electrodynamics, and statistical thermodynamics is quite complex, our objective in this chapter is simply to provide the basic tools to summarize near-field thermal radiation calculations for engineering applications. The extension of the NFRT to steady-state systems was discussed by Budaev and Bogy (2011). In addition, its imitations were explored by the same researchers (Budaev and Bogy 2014a,b). In that sense, the focus of this chapter is on the calculation of the near-field radiative heat flux. The intriguing physical phenomena underlying this still emerging field are outlined from an intuitive rather than mathematical point of view, and the relevant references are pointed out for readers who are interested to learn more about a specific subject.

In the next section, we outline the electromagnetic treatment of thermal radiation and briefly introduce the concepts of density of electromagnetic states (DOS) and coherence necessary to understand the physics of near-field radiative heat transfer. Then, evanescent waves and surface polaritons are overviewed. Subsequently, near-field radiative heat transfer between closely spaced bodies is explained in detail through flux calculations in a one-dimensional (1D) layered geometry. Finally, we discuss experimental works related to near-field thermal radiation.

16.2 ELECTROMAGNETIC TREATMENT OF THERMAL RADIATION AND BASIC CONCEPTS

Near-field thermal radiation differs from its far-field counterpart due to the presence of evanescent and surface waves and the fact that incoherent transport can no longer be assumed. Here, we provide the mathematical tools to tackle near-field thermal radiation problems and explain some basic concepts that will help to understand the underlying physics.

16.2.1 Near-Field Thermal Radiation versus Far-Field Thermal Radiation

Radiative heat transfer can be categorized in terms of four fundamental processes: emission, absorption, scattering, and propagation of radiative energy. Generally speaking, emission refers to how the radiative energy emanates from an object, where the internal energy of the body is converted to electromagnetic waves. Absorption is when these waves interact with a body and the energy is converted back into the internal energy of the object. Scattering is defined as the redirection of radiant energy and involves the phenomena of refraction, reflection, transmission, and diffraction, as discussed in Chapter 15.

For far-field calculations, we assume that the typical size of objects exchanging radiant energy is much larger than the dominant wavelength emitted. In that case, the wave nature of radiation can be neglected and the RTE is employed. This means that diffraction and interference effects are negligible, and thermal radiation emission is described by Planck's blackbody distribution. For most engineering calculations, absorption of radiation by an object is determined using an absorption coefficient (see Equation 14.46), which is simply a proportionality constant. It is multiplied by the intensity incident on an object to determine how much energy is absorbed by it. Similarly, scattering is described using an *ad hoc* proportionality constant, the scattering coefficient, to determine the loss of energy from a beam incident on an object via scattering mechanism. The scattering phase function, on the other hand, is a probability distribution function, depicting how the scattered radiation field is distributed in the angular domain, both as a function of azimuthal and zenith angles.

As discussed earlier, the demarcation of the far- and near-field regimes can be achieved by using a critical length scale, which is traditionally determined from Wien's law corresponding to the wavelength of the peak emission according to Planck's blackbody distribution: $\lambda_{max}T = 2897.8\,\mu\text{m}\cdot\text{K}$. This length scale is quite useful as it is easy to remember: however, it is still an approximation, and more precise values have been proposed for some specific cases (Francoeur and Mengüç 2008a).

Since radiative transport is not incoherent in the near field, we need to use the electromagnetic description of thermal radiation based on the macroscopic Maxwell equations (Rytov et al. 1989). Maxwell equations describe the interrelationship between the fields, the sources, and the material properties. Radiation absorption by the medium is included in this set of equations via the imaginary part of the dielectric function or, equivalently, via the imaginary part of the complex refractive index. Scattering of electromagnetic waves is also calculated directly via Maxwell equations by assuming, for example, that the total field is the superposition of incident and scattered fields. However, thermal radiation emission is not included in Maxwell equations. The FE theory of Rytov, discussed in the next section, provides this bridge between the classical Maxwell equations and thermal radiation emission.

16.2.2 Electromagnetic Description of Near-Field Thermal Radiation

Inclusion of a thermal radiation emission term in Maxwell equations is not straightforward. In its most general form, thermal radiation emission is described starting from a quantum mechanical point of view, where the transition of an elementary energy carrier (electrons, molecules, phonons, etc.) from a higher energy level to a lower energy level emits a photon (Carey et al. 2008). To be able to link this phenomenon with electromagnetic waves, we need to consider thermal radiation emission from an electrodynamics point of view (Rytov et al. 1989, Whale 1997, Mulet et al. 2002). Using the electrodynamics viewpoint, far- and near-field radiation (i.e., propagating and evanescent waves) is emitted via the out-of-phase oscillations of charges of opposite signs. A couple of charges of opposite signs is called a dipole (Jackson 1999, Hecht 2002). At any finite temperature ($T > 0$ K), thermal agitation causes a chaotic motion of charged particles inside the body, which induces oscillating dipoles. The random fluctuations of the charges generate in turn a fluctuating electromagnetic field, called the thermal radiation field, as it originates from random thermal motion (Rytov et al. 1989, Mulet et al. 2002). On a macroscopic level, the field fluctuations are due to thermal fluctuations of the volume densities of the charges and current. In other words, the electromagnetic field generated thermally is not the sum of the fields of the individual charges, but is a field produced by sources that are also macroscopic (volume densities of charge and current).

However, FE is built on this simplified macroscopic description. Since it is based on fluctuations around an equilibrium temperature T, the theory is thus applicable to media of any form that are in local thermodynamic equilibrium, where an equilibrium temperature can be defined at any given location inside the body at any instant. The FE approach is also said to be applicable to nonequilibrium conditions, in cases where the transport phenomena required to maintain steady-state conditions are negligible when compared to the energy emitted by the body (Rytov et al. 1989).

In Figure 16.3, we present a flowchart on different steps that need to be taken in order to combine Maxwell equations with fluctuation–dissipation equations, through which the local DOS and near-field thermal radiation can be obtained. We highlight each of these concepts.

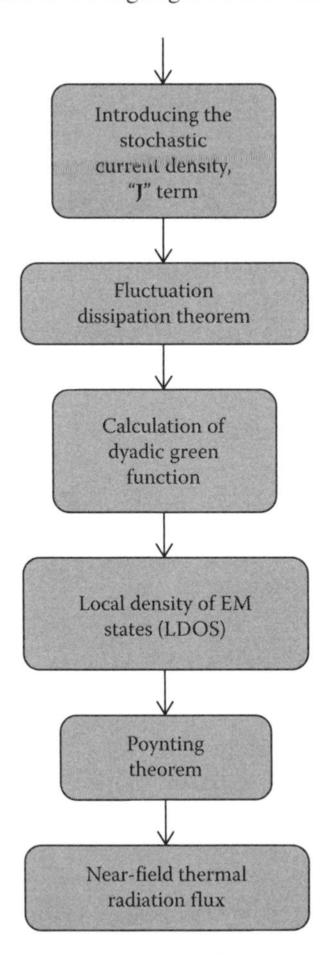

FIGURE 16.3 Near-field thermal radiation flowchart.

Mathematically, FE can be illustrated by first considering the general form of Maxwell equations in the time domain, as given by Equations 14.1 through 14.4. In this chapter, we are dealing with frequency-dependent quantities, and it is convenient to convert Maxwell equations into the frequency domain. The monochromatic components of a signal are related to the temporal signal via a Fourier transform. By assuming that the time-harmonic fields have the form $\exp(-i\omega t)$, Maxwell equations in the frequency domain become

$$\text{Faraday's law: } \nabla \times \mathbf{E}(\mathbf{r},\omega) = i\omega \mathbf{B}(\mathbf{r},\omega) = i\omega\mu_0 \mathbf{H}(\mathbf{r},\omega) \tag{16.1}$$

$$\text{Ampère's law: } \nabla \times \mathbf{H}(\mathbf{r},\omega) = -i\omega \mathbf{D}(\mathbf{r},\omega) + \mathbf{J}(\mathbf{r},\omega) = -i\omega\gamma \mathbf{E}(\mathbf{r},\omega) + \frac{\mathbf{E}(\mathbf{r},\omega)}{r_e}$$

$$= -i\omega\left(\gamma + \frac{i}{\omega r_e}\right)\mathbf{E}(\mathbf{r},\omega) = -i\omega\gamma_0\bar{\varepsilon}\,\mathbf{E}(\mathbf{r},\omega) \tag{16.2}$$

$$\text{Gauss's law: } \nabla \cdot \mathbf{D}(\mathbf{r},\omega) = \nabla \cdot [\gamma \mathbf{E}(\mathbf{r},\omega)] = \rho_e \tag{16.3}$$

$$\text{Gauss's law: } \nabla \cdot \mathbf{B}(\mathbf{r},\omega) = \nabla \cdot [\mu_0 \mathbf{H}(\mathbf{r},\omega)] = 0 \tag{16.4}$$

where we assume that the media are nonmagnetic (i.e., $\mu = \mu_0$). In Ampère's law, the dielectric constant $\bar{\varepsilon}$, or dielectric function, is defined as the ratio of the complex electric permittivity $(\gamma + i/\omega r_e)$ and the electrical permittivity of the vacuum γ_0.

In Chapter 14, the time-harmonic fields were expressed as $\exp(i\omega t)$. In consequence, complex quantities such as the refractive index and dielectric function have been expressed as $a = a' - ia''$, where $a = a'$ and a'' refer, respectively, to the real and imaginary components of the variable a. In this chapter, we assume time-harmonic fields with $\exp(-i\omega t)$, to be consistent with most of the literature published in the field. Then, the complex quantities are given as $a = a' + ia''$. Both formulations are equally valid, as long as we are consistent throughout the calculations.

In addition, the continuity equation given by Equation 14.5 in the frequency domain is

$$\nabla \cdot \mathbf{J}(\mathbf{r},\omega) = i\omega\rho_e \tag{16.5}$$

From here on, we assume that there is no free charge density (i.e., $\rho_e = 0$). From the qualitative description of radiation emission given earlier, thermal fluctuations of a body around an equilibrium temperature T imply random fluctuations of current, which constitutes the source term of thermal radiation. In Ampère's law (Equation 16.2), the current density J is combined (using Ohm's law) with the electrical resistivity, leading to a complex electric permittivity $(\gamma + i/\omega r_e)$. For an insulator, the current density $J \to 0$, and then $(\gamma + i/\omega r_e) \to \gamma$. Therefore, to account for the random thermal fluctuations of current in Maxwell equations that are present regardless of the nature of the materials, an extraneous current density term is added in Ampère's law (Rytov et al. 1989):

$$\nabla \times \mathbf{H}(\mathbf{r},\omega) = -i\omega\gamma_0\varepsilon \mathbf{E}(\mathbf{r},\omega) + \mathbf{J}^r(\mathbf{r},\omega) \tag{16.6}$$

The current density $\mathbf{J}^r$ plays the role of a random external source causing thermal fluctuations of the field. The mean value of this random current density, $\langle \mathbf{J}^r \rangle$, is zero implying that the mean radiated field is also zero. On the other hand, quantities such as Poynting vector (i.e., radiative flux) and energy density are functions of the spatial correlation function of the fluctuating currents, which is a nonzero quantity. By using Ampère's law as given by Equation 16.6, instead of Equation 16.2,

Maxwell equations become stochastic in nature due to the fact that $\mathbf{J}^r$ is a random fluctuating variable and are sometimes referred in the literature as the "stochastic Maxwell equations." These equations set the basis of the FE.

To calculate quantities that are useful for heat transfer engineers, we need to solve the stochastic Maxwell equations. Although different approaches can be used, the most common technique adopted in near-field thermal radiation calculations is the method of dyadic Green's function (DGF). Using the method of potentials (Peterson et al. 1997, Francoeur and Mengüç 2008a), the electric and magnetic fields can be expressed as (Joulain et al. 2005, Narayanaswamy and Chen 2005b, Francoeur and Mengüç 2008a)

$$\mathbf{E}(\mathbf{r},\omega) = i\omega\mu_0 \int_V dV' \overline{\overline{\mathbf{G}}}^E(\mathbf{r},\mathbf{r}',\omega)\cdot \mathbf{J}^r(\mathbf{r}',\omega) \tag{16.7}$$

$$\mathbf{H}(\mathbf{r},\omega) = \int_V dV' \overline{\overline{\mathbf{G}}}^H(\mathbf{r},\mathbf{r}',\omega)\cdot \mathbf{J}^r(\mathbf{r}',\omega) \tag{16.8}$$

where
$\overline{\overline{\mathbf{G}}}^E(\mathbf{r},\mathbf{r}',\omega)$ and $\overline{\overline{\mathbf{G}}}^H(\mathbf{r},\mathbf{r}',\omega)$ are the electric and magnetic DGFs and r
$\mathbf{r}'$ denote field and source point, respectively (Tsang et al. 2000)

The DGF is a 3 × 3 matrix and is alternatively referred to as a Green's tensor. The electric DGF can be written as (Novotny and Hecht 2006)

$$\overline{\overline{\mathbf{G}}}^E(\mathbf{r},\mathbf{r}',\omega) = \left[\overline{\overline{\mathbf{I}}} + \frac{1}{k^2}\nabla\nabla\right] G_0(\mathbf{r},\mathbf{r}',\omega) \tag{16.9}$$

where G_0 is the scalar Green's function, which can be determined by solving the scalar Helmholtz equation for a single point source located at $\mathbf{r} = \mathbf{r}'$ The dyadic $\overline{\overline{\mathbf{I}}}$ is called an idem factor, which is a 3 × 3 identity matrix.

Physically, the DGF can be seen as a spatial transfer function relating the field observed at location $\mathbf{r}$ with a frequency ω due to a vector source located at $\mathbf{r}'$. The magnetic DGF is calculated from the electric DGF by $\overline{\overline{\mathbf{G}}}^H(\mathbf{r},\mathbf{r}',\omega) = \nabla\times\overline{\overline{\mathbf{G}}}^E(\mathbf{r},\mathbf{r}',\omega)$. The physical interpretation of Equations 16.7 and b is quite straightforward, as they imply that the fields observed, or calculated, at location $\mathbf{r}$ are proportional to the sum of currents $\mathbf{J}^r$ distributed at different $\mathbf{r}'$ locations within an emitting body of volume V. Also, as mentioned before, the averaged radiated fields, $\langle\mathbf{E}\rangle$ and $\langle\mathbf{H}\rangle$, are zero since $\langle\mathbf{J}^r\rangle = 0$.

To be able to calculate the radiative heat flux, we need to determine the link between the local temperature of an emitting body and the stochastic current density $\mathbf{J}^r$. This link is provided by the FDT, which makes the bridge between the ensemble average of the spatial correlation function of $\mathbf{J}^r$ and the temperature T (Rytov et al. 1989). In other words, the FDT establishes the relationship between the electromagnetic description of thermal radiation and the usual theory of heat transfer (Greffet et al. 2007).

Derivation of the FDT is not provided here, as its details can be found elsewhere (Landau and Lifshitz 1960, Rytov et al. 1989, Novotny and Hecht 2006). We shall, however, list the assumptions underlying the derivation of the FDT to better understand the concept. For our purposes, the derivation of the FDT is restricted by the following assumptions: (1) the bodies are assumed to be in local thermodynamic equilibrium, at an equilibrium temperature T, which is subject to small fluctuations; (2) all the media considered are isotropic; (3) the media are nonmagnetic and are defined by a frequency-dependent dielectric function $\overline{\varepsilon}(\omega)$; and (4) the dielectric function is local in space (i.e., the polarization at a given point in a medium is directly proportional to the electric field at that point, and does not directly depend on the fields from other points), and consequently, the fluctuations are uncorrelated between neighboring volume elements. While the last three assumptions can be relaxed, the condition of local thermodynamic equilibrium where a temperature can be defined must be satisfied when applying the FDT.

The FDT is not limited to electrodynamics and was applied to thermal radiation by Rytov et al. (1989). Following the assumptions stated earlier, the FDT can be written for time-averaged thermal fluctuations as

$$\left\langle J_\alpha^r(\mathbf{r}',\omega)J_\beta^{r*}(\mathbf{r}'',\omega')\right\rangle = \frac{\omega\gamma_0\overline{\varepsilon}''(\omega)}{\pi}\Theta(\omega,T)\delta(\mathbf{r}'-\mathbf{r}'')\delta(\omega-\omega')\delta_{\alpha\beta} \quad (16.10)$$

where the subscripts α and β refer to orthogonal components indicating the state of polarization of the source. The term $\Theta(\omega,T)$ is the mean (time-averaged) energy of a Planck oscillator in thermal equilibrium at frequency ω and temperature T:

$$\Theta(\omega,T) = \frac{\hbar\omega}{\exp(\hbar\omega/k_bT)-1} \quad (16.11)$$

In Equation 16.10, the Dirac function $\delta(\mathbf{r}'-\mathbf{r}'')$ mathematically translates the assumption of locality for the dielectric constant (i.e., the fluctuations at two different points are correlated in the limit $\mathbf{r}'' \to \mathbf{r}'$). The Dirac function $\delta(\omega-\omega')$ represents the fact that the fluctuating currents are stationary (i.e., the spectral components with different frequencies are totally uncorrelated), and $\delta_{\alpha\beta}$ accounts for the assumption of isotropic media.

Going back to the discussion of Section 16.2.1, we now have all the necessary tools to solve near-field radiative heat transfer problems.

16.2.3 Near-Field Radiative Heat Flux

In heat transfer analysis, one is mostly interested by the radiative heat flux given by the time-averaged Poynting vector, which is the quantity measurable by a detector (Chen 2005):

$$\langle \mathbf{S}(\mathbf{r},\omega)\rangle = 4\times\frac{1}{2}\mathrm{Re}\{\langle \mathbf{E}(\mathbf{r},\omega)\times\mathbf{H}^*(\mathbf{r},\omega)\rangle\} \quad (16.12)$$

This expression of the time-averaged Poynting vector is four times larger than its customary definition, since only the positive frequencies are considered in Fourier decomposition of the time-dependent fields into frequency-dependent quantities (Zhang 2007). Note that Poynting vector is, by definition, the cross-product of the electric and magnetic field vectors and it oscillates with the same frequency as these waves. Since we are interested in measurable quantities, we consider its time-averaged value, which is equivalent to radiative intensity. Evaluation of the time-averaged Poynting vector requires computation of the terms $\langle E_m(\mathbf{r},\omega)H_n^*(\mathbf{r},\omega)\rangle$ expressed as a function of the stochastic current (Equations 16.7 and 16.8):

$$\left\langle E_m(\mathbf{r},\omega)H_n^*(\mathbf{r},\omega)\right\rangle = i\omega\mu_0\int_V dV'\int_V dV'' G_{m\alpha}^E(\mathbf{r},\mathbf{r}',\omega)G_{n\beta}^{H*}(\mathbf{r},\mathbf{r}',\omega)\left\langle J_\alpha^r(\mathbf{r}',\omega)J_\beta^{r*}(\mathbf{r}'',\omega)\right\rangle \quad (16.13)$$

where the subscripts m and n are orthogonal components representing the states of polarization of the fields ($m \neq n$). For example, if we calculate $\langle E_xH_y^*\rangle$, then $m = x$ and $n = y$. To apply the FDT to Equation 16.13, we use the ergodic hypothesis, where it is assumed that averaging over time can be replaced by an ensemble average (Mandel and Wolf 1995, Mishchenko et al. 2006). The underlying idea is that for certain systems, the time average of their properties is equal to the average over the entire space. Then, application of the FDT to Equation 16.13 leads to the following general expression $\langle E_m(\mathbf{r},\omega)H_n^*(\mathbf{r},\omega)\rangle$:

$$\left\langle E_m(\mathbf{r},\omega)H_n^*(\mathbf{r},\omega)\right\rangle = \frac{k_0^2\Theta(\omega,T)}{\pi} i\varepsilon''(\omega)\int_V dV' G_{m\alpha}^E(\mathbf{r},\mathbf{r}',\omega)G_{n\alpha}^{H*}(\mathbf{r},\mathbf{r}',\omega) \tag{16.14}$$

The subscripts m and n represent the state of polarization of the fields observed at $\mathbf{r}$, while α represents the state of polarization of the source at $\mathbf{r}'$. For example, the set of indices $m\alpha$ implies that a summation is performed over all orthogonal components (*e.g.*, $G_{mx}^E G_{nx}^{H*} + G_{my}^E G_{ny}^{H*} + G_{mz}^E G_{nz}^{H*}$).

At this point, the radiative heat flux can be calculated given that the DGFs for the system under study are known. In Section 16.4, we discuss the solution for the radiative heat flux in a 1D geometry.

16.2.4 Density of Electromagnetic States

The concept of DOS is important to understand the underlying physics of near-field thermal radiation. Using the principles of energy quantization and statistical thermodynamics (Chen 2005, Zhang 2007), the density of energy of a system at a given frequency ω is obtained as the product of the DOS and the mean energy of a state at frequency ω and temperature T (Joulain et al. 2003). The mean energy of a state, or mean energy of a Planck oscillator, at ω and T is given by Equation 16.11. Therefore, the DOS can be seen as the number of states, or modes, per unit frequency and per unit volume (Zhang 2007). In vacuum, the DOS is given by (Joulain et al. 2003, Chen 2005)

$$N(\omega) = \frac{\omega^2}{\pi^2 c_0^3} \tag{16.15}$$

The product of the DOS given by Equation 16.15 and the mean energy of a Planck oscillator leads to

$$u(\omega) = \frac{\hbar\omega^3}{\pi^2 c_0^3[\exp(\hbar\omega/k_b T)-1]} \tag{16.16}$$

which is the energy density of a blackbody. Starting from Equation 16.16, we can derive Planck's blackbody intensity and the blackbody emissive power (Chen 2005).

In the near field of a thermal source (i.e., a body at a finite temperature T), we expect an increase of the DOS due the presence of evanescent and surface waves. As these waves depend strongly on the distance from the thermal source, we usually calculate the local DOS (LDOS) at a given location $\mathbf{r}$ in space (Joulain et al. 2003). Assuming that the source at temperature T is emitting in free space, the energy density in vacuum is calculated as the sum of the electric and magnetic energies (Joulain et al. 2003, Zhang 2007):

$$\langle u_\omega(\mathbf{r},\omega,T)\rangle = 4\times\frac{1}{4}\left[\gamma_0\left\langle|\mathbf{E}(\mathbf{r},\omega)|^2\right\rangle + \mu_0\left\langle|\mathbf{H}(\mathbf{r},\omega)|^2\right\rangle\right] \tag{16.17}$$

where again an extraneous factor of four is included for the same reason as in Poynting vector, Equation 16.12. An explicit expression for the energy density, similar to Equation 16.14 for Poynting vector, can be derived by substituting the electric and magnetic field expressions and by applying the FDT; this derivation is left as an exercise. The energy density given by Equation 16.17 is relative to the vacuum energy density, as the vacuum fluctuations are neglected in the mean energy of a Planck oscillator since they do not affect radiative heat flux calculations (Zhang 2007). The LDOS $N(r,\omega)$ in the near field is calculated by dividing Equation 16.17 by $\Theta(\omega,T)$.

16.2.5 Spatial and Temporal Coherence of Thermal Radiation

The coherence properties of an emitted thermal field are directly related to the wave nature of radiation. We briefly discussed the concepts earlier, but a more detailed exposition is in order. Spatial and temporal coherence of an electromagnetic field can be quantified by calculating the correlation function $\langle \mathbf{E}(\mathbf{r}_1,t_1)\, \mathbf{E}(\mathbf{r}_2,t_2)\rangle$, where **r** denotes a spatial location and t represents the time. The fields are perfectly correlated if the field $\mathbf{E}(\mathbf{r}_2,t_2)$ follows the same evolution as $\mathbf{E}(\mathbf{r}_1,t_1)$ (Drevillon 2007). Temporal coherence, or spectral coherence, is the measure of the correlation of the fields at times t_1 and t_2. A temporally coherent radiative source emits in a narrow spectral band for any given direction. The spatial coherence is a measure of the correlation of the fields at location $\mathbf{r}_1$ and $\mathbf{r}_2$ (Carminati and Greffet 1999, Zhang 2007). A radiative source that is spatially coherent emits radiation in a narrow angular band.

As discussed in the previous chapters, thermal radiation is a broadband phenomenon and emission is generally quasi-isotropic. These two observations show that thermal radiation is indeed an incoherent process. In fact, the coherence length of blackbody radiation is about $\lambda/2$, such that coherence properties of thermal radiation are not observed in the far field (Carminati and Greffet 1999). However, by structuring surfaces at nanoscales, it is possible to transmit the near-field coherence of thermal radiation to the far field. In that way, it is possible to achieve quasi-monochromatic and directional thermal radiative sources. This is especially true when the emitting materials support surface waves that show a high degree of temporal and spatial coherence in the near field.

16.3 EVANESCENT AND SURFACE WAVES

In this section, we provide an overview of evanescent and surface waves. Without going through all the mathematical details, we describe the physics of these waves and outline their impact on near-field thermal radiation.

16.3.1 Evanescent Waves and Total Internal Reflection

First, we discuss an evanescent wave through the concept of TIR. Let us assume the interface 1–2 between two lossless dielectric half-spaces with index of refraction $n_1 > n_2$, as shown in Figure 16.1.

A wave propagating in medium 1 at an angle θ_i is incident at the interface 1–2. From the discussion in Chapter 3, we know that the wave is partially reflected in medium 1 at an angle $\theta_r = \theta_i$, while a fraction of the wave is transmitted to medium 2 at an angle θ_t. The angle θ_t is predicted via Snell's law, repeated here for convenience:

$$n_1 \sin\theta_i = n_2 \sin\theta_t \tag{16.18}$$

Since the index of refraction of medium 1 is greater than the index of refraction of medium 2, there is an angle of incidence for which no wave is transmitted to medium 2, such that all the energy is reflected back in medium 1. This phenomenon is called TIR. We can easily predict the critical incident angle for TIR via Snell's law. Indeed, the angle θ_t of the wave transmitted to medium 2 cannot exceed 90°. Imposing this condition in Equation 16.18, we find that the critical angle for TIR is given by $\theta_c = \sin^{-1}(n_2/n_1)$. For example, if $n_1 = 1.5$ and $n_2 = 1$, the critical angle for TIR is 41.8°. However, when TIR occurs, an evanescent wave is generated at the interface 1–2 with an exponentially decaying field in medium 2 (Hecht 2002, Novotny and Hecht 2006).

The presence of the evanescent wave at the interface 1–2 can be better understood by working with wave vectors instead of angles. Let us assume that wave incident in medium 1 is described with an electric field $E_1\exp[i(k_1 \cdot r - \omega t)]$ that can be either TE or TM polarized. Without loss of generality, we can assume that waves are propagating in the xz plane only. The magnitude of the wave vector $k_1 = |\mathbf{k}_1|$ is given by n_1k_0, where k_0 is the magnitude of the wave vector in vacuum = (ω/c_0). (Note that this k is not the same as the imaginary part of the complex index of refraction; indeed, k can be seen as the spatial frequency of the wave in rad/m.) The z-component of the wave vector in medium *1* can thus be written as

$$k_{z1} = \sqrt{n_1^2k_0^2 - k_x^2} \tag{16.19}$$

It is also possible to express the x-component of the wave vector in terms of the angle of incidence:

$$k_x = n_1k_0 \sin\theta_i \tag{16.20}$$

If the angle of incidence is equal to the critical angle for TIR, the x-component of the wave vector is

$$k_x = n_2k_0 \tag{16.21}$$

Equation 16.21 provides the smallest value of k_x for TIR. Also, note that the x-component of the wave vector is conserved from medium 1 to medium 2 due to the assumption that the media are infinite along that direction. Similarly, we can write the z-component of the wave vector in medium 2 as follows:

$$k_{z2} = \sqrt{n_2^2k_0^2 - k_x^2} \tag{16.22}$$

Inspection of Equations 16.21 and 16.22 shows that when $\theta_i > \theta_c$, then $k_x > n_2k_0$, and k_{z2} becomes a pure imaginary number. Assuming that the field in medium 2 has the form $E_2\exp[i(k_2 \cdot r - \omega t)]$, substitution of a pure imaginary k_{z2} in this last expression leads to $E_2\exp[i(k_xx - \omega t)]\exp(-k_{z2}z)$. This equation shows clearly that the wave is propagating along the x-direction, while being evanescently confined at the interface 1–2 in the z-direction. While a field is present at the interface 1–2 in medium 2, it can be shown that the time-averaged Poynting vector of the evanescent wave field is zero, such that there is no net energy flow in medium 2 (Chen 2005). The penetration depth of an evanescent wave, δ, is defined as the distance at which the electric field amplitude has decayed by e^{-1} of its value; therefore, $\delta \approx |k_z|^{-1}$.

The presence of evanescent waves at the surface of a body at temperature T can be conceptualized via an analogy with TIR (Pendry 1999). Referring to Figure 16.1, where we assume from now that medium 2 is a vacuum, we can imagine that electromagnetic propagating radiation is emitted throughout the volume of medium 1. Waves with $k_x < k_0$ are transmitted through the interface 1–2, while waves with $k_x > k_0$ experience TIR. The largest k_x value for an evanescent wave is n_1k_0, since the maximum angle of incidence θ_i is 90° (see Equation 16.20).

If a medium 3 with index of refraction $n_3 = n_1$ is located at a distance d much larger than the dominant wavelength emitted from medium 1, only the waves with $k_x < k_0$ participate in radiative transfer. In that case, we are in the far-field regime and the mathematical tools developed in the previous chapters can be employed. If the separation distance d is less than the dominant wavelength emitted, the evanescent wave field of medium 1 excites the charges within medium 3 and dissipates its energy through Joule heating (Mulet et al. 2002); this mode of energy transfer is referred to as radiation tunneling, or frustrated TIR, as Poynting vector from the evanescent wave is no longer zero (Zhang 2007).

This extraneous energy transfer in the near field leads to radiative flux exceeding the values predicted by Planck's distribution. For the case of lossless dielectric materials with refractive indices n_1, the maximum radiative heat transfer occurs at the limit $d \to 0$, and its achievable value is n_1^2 times the values predicted between blackbodies; this limit comes from the fact that the blackbody intensity in a lossless material is proportional to n_1^2 (see Chapter 1, and Narayanaswamy and Chen 2005b, Carey et al. 2008).

This simple picture for describing near-field radiative heat transfer is, however, inadequate when dealing with materials supporting surface waves, such as metals, doped semiconductors, and polar crystals. As discussed previously in this chapter, thermal radiation emission should be seen from the electrodynamics point of view where oscillating dipoles generate propagating and evanescent waves (Mulet et al. 2002). Indeed, when materials can support surface waves, or surface polaritons, the largest contributing wave vector k_x can greatly exceed the limit $n_1 k_0$ established via analogy with TIR (Pendry 1999, Mulet et al. 2002). Surface waves are discussed in Section 16.3.2.

16.3.2 Surface Waves

Surface waves, or surface polaritons, are the hybrid modes of a mechanical oscillation and an electromagnetic field. In a metal or a doped semiconductor, the out-of-phase longitudinal oscillations of free electrons (i.e., plasma oscillations), relative to the positive ion cores, creates dipoles generating an electromagnetic field (Raether 1988, Mulet 2003, Kittel 2005a, Maier 2007). The near-field component of the spectrum emitted is called a SPP. Similarly, the out-of-phase oscillations of transverse optical phonons in polar crystals, such as silicon carbide (SiC), generate an electromagnetic field, and its near-field component is called a SPhP (Mulet et al. 2002, Mulet 2003, Kittel 2005a). As mentioned in Section 16.1, surface waves propagate along the interface between two media with evanescent fields decaying in both materials (Raether 1988, Maier 2007).

To illustrate the physics of surface waves, we consider the plane interface depicted in Figure 16.2, where both media 1 and 2 are infinite along the x- and y-directions. At $z < 0$, the frequency-dependent dielectric function of the medium 1 is given by $\bar{\varepsilon}_1(\omega)$ while medium 2 at $z > 0$ is assumed to be a vacuum with $\bar{\varepsilon}_2(\omega) = 1$. The assumptions stated in Section 16.2.2 apply to medium 1. Without loss of generality, it is also assumed that surface waves are propagating along the x-direction only.

The impact of surface waves on near-field radiative heat transfer can be understood by analyzing the dispersion relation, which is the relationship between the periodicity of the wave in time (i.e., angular frequency ω) and its periodicity in space (i.e., wave vector k_x here). Such a dispersion relation can be determined by solving Maxwell equations at the interface 1–2 separately for the TE- and TM-polarized waves (Joulain et al. 2005, Maier 2007). An alternative approach to determine surface wave dispersion relation is to find the poles of Fresnel reflection coefficients at the interface 1–2 (Joulain et al. 2005, Francoeur 2010, Petersen et al. 2013). The poles of Fresnel reflection coefficients correspond to the conditions for which these reflection coefficients tend to go to infinity. We adopt this simple method here; more details can be found in Joulain et al. (2005).

Fresnel reflection coefficients in transvere electric (TE) and transverse magnetic (TM) polarizations in terms of wave vectors are given by (Yeh 2005)

$$r_{12}^{TE} = \frac{k_{z1} - k_{z2}}{k_{z1} + k_{z2}} \tag{16.23}$$

$$r_{12}^{TM} = \frac{k_{z1} - \bar{\varepsilon}_1(\omega) k_{z2}}{k_{z1} + \bar{\varepsilon}_1(\omega) k_{z2}} \tag{16.24}$$

In TE polarization, Fresnel reflection coefficient diverges if $k_{z1} + k_{z2} = 0$. We are interested in the dispersion relation of surface waves, with an exponentially decaying field in both media 1 and 2 along the z-direction. Therefore, we deal with surface waves if and only if both k_{z1} and k_{z2} are pure imaginary numbers. Following the convention used in this chapter, the imaginary part of the z-component of the wave vector is always positive, such that no surface wave exists in

TE polarization. It can, however, be shown that surface polaritons can exist in TE polarization if the materials are magnetic (Joulain et al. 2005).

In TM polarization, Fresnel reflection coefficient tends to infinity if $k_{z1} + \bar{\varepsilon}_1(\omega)k_{z2} = 0$. Because the real part of the dielectric function of medium 1 can take negative values, surface waves can therefore exist in TM polarization. Using $k_{zj} = \sqrt{\bar{\varepsilon}k_0^2 - k_x^2}$, the aforementioned condition in terms of k_x is

$$k_x = k_0\sqrt{\frac{\bar{\varepsilon}_1(\omega)}{\bar{\varepsilon}_1(\omega)+1}} \tag{16.25}$$

Equation 16.25 is the dispersion relation at the interface 1–2 and poses two conditions for the existence of surface polaritons. First, since surface waves are propagating along the interface 1–2, the wave vector k_x must be a real number. Moreover, since surface polaritons are evanescent waves along the z-direction, k_x must be greater than the wave vector in vacuum k_0. By combining these two conditions, surface waves exist when the term within the square root in Equation 16.25 is greater than unity, which can happen when $\bar{\varepsilon}_1(\omega) < (-1)$ (Joulain et al. 2005).

To illustrate the dispersion relation of surface waves, we assume that medium 1 is SiC, which is a polar crystal supporting SPhPs in the infrared region. The dielectric function of SiC can be modeled as a damped harmonic oscillator:

$$\bar{\varepsilon}_1(\omega) = \bar{\varepsilon}_\infty\left[\frac{\omega^2 - \omega_{LO}^2 + i\Gamma\omega}{\omega^2 - \omega_{TO}^2 + i\Gamma\omega}\right] \tag{16.26}$$

where

$\bar{\varepsilon}_\infty(\omega)$ is the high-frequency dielectric constant
Γ is the damping factor
ω_{LO} and ω_{TO} are the frequencies of longitudinal and transverse optical phonons, respectively

As an example, these properties for SiC can be listed as $\bar{\varepsilon}_\infty = 6.7$, $\omega_{LO} = 1.825 \times 10^{14}$ rad/s, $\omega_{TO} = 1.494 \times 10^{14}$ rad/s and $\Gamma = 8.966 \times 10^{11}\ \mathrm{s}^{-1}$ (Palik 1998). To plot the SPhP dispersion relation, we neglect the losses in the dielectric function of SiC (i.e., $\Gamma = 0$). The dispersion relation at the interface 1–2 for SiC is shown in Figure 16.4a, where the light line in vacuum $k_x = k_0$, and the frequencies of transverse and longitudinal optical phonons in SiC are identified; the real part of the dielectric function of SiC is plotted in Figure 16.4b.

To better understand the dispersion relation shown in Figure 16.4a and b, it is necessary to identify the zones where the waves are either propagating or evanescent in vacuum. The z-component of the wave vector in vacuum is given by $k_{z2} = \sqrt{k_0^2 - k_x^2}$. When $k_x \le k_0$, then the z-component of the wave vector is a pure real number, and therefore the wave is propagating. As a consequence, the part that is on the left-hand side of the light line in vacuum in Figure 16.4a corresponds to propagating waves. On the other hand, when $k_x > k_0$, k_{z2} becomes a pure imaginary number, and therefore the wave is evanescent, such that the part of the dispersion relation on the right-hand side of the light line in vacuum in Figure 16.4a corresponds to evanescent waves.

To show this more clearly, we plot a part of the dispersion relation that does not correspond to a SPhP (left of light line in vacuum). This curve is obtained by solving Equation 16.25, but does not satisfy the requirement that the term under the square root be greater than unity. This can be seen by examining the real part of the dielectric function of SiC shown in Figure 16.4b; indeed, for frequencies greater than ω_{LO}, the real part of the dielectric function of SiC is greater than −1. The same observation is true for a part of the lower branch of the dispersion relation for frequencies less than ω_{TO}, where the real part of the dielectric function of SiC is greater than −1.

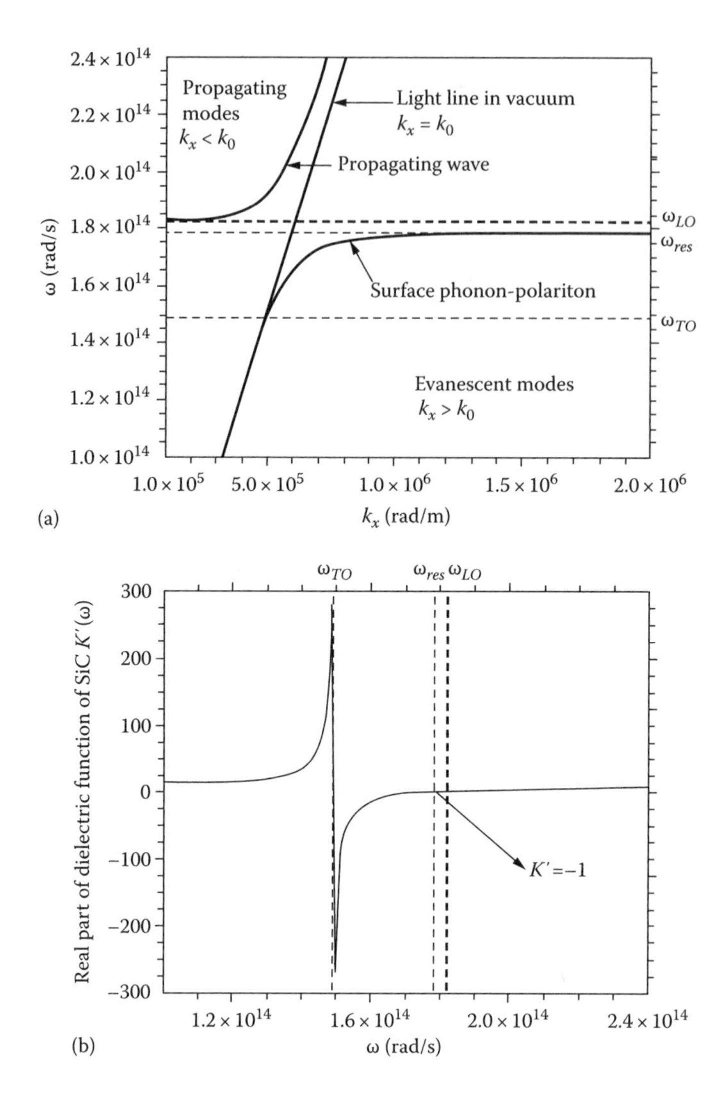

FIGURE 16.4 (a) SPhP dispersion relation at SiC-vacuum interface. (b) Real part of the dielectric function of SiC calculated via Equation 16.26.

On the other hand, the part of the lower branch included in the region between ω_{LO} and ω_{TO} is a SPhP, where $k_x > k_0$. By inspecting Figure 16.4b, it is seen that the real part of the dielectric function of SiC in the spectral region between ω_{LO} and ω_{TO} is less than −1. When the real part of the dielectric function of SiC equals −1, the SPhP dispersion relation reaches an asymptote referred hereafter as ω_{res}, which is defined as the resonant frequency of surface polariton at a single interface. Indeed, when the real part of the dielectric function of SiC is exactly −1, the denominator of Equation 16.25 becomes zero, and then $k_x \rightarrow \infty$. Using the condition, $\bar{\varepsilon}'(\omega) = -1$, we find that the resonant frequency is $\omega_{res} = \sqrt{(\bar{\varepsilon}_\infty \omega_{LO}^2 + \omega_{TO}^2)/(1 + \bar{\varepsilon}_\infty)}$; for the SiC-vacuum interface, $\omega_{res} = 1.786 \times 10^{14}$ rad/s, corresponding to a wavelength of 10.55 μm.

We defined earlier in Section 16.2.4 the DOS, or LDOS, as the number of electromagnetic modes per unit frequency and per unit volume; as a consequence, the LDOS is directly proportional to $|dk_x/d\omega|$. At resonance, $|dk_x/d\omega| \rightarrow \infty$ such that the LDOS and the energy density become very large. We therefore expect radiative heat exchange between materials supporting surface waves to be much greater than the values predicted by Planck's distribution. Also, the fact that an important enhancement takes place around a given frequency ω_{res} implies temporal, or spectral, coherence of the near field (Henkel et al. 2000, Zhang 2007). The spatial coherence of the near field of a thermal source is also greatly modified due to the presence of surface waves (Carminati and Greffet 1999, Henkel et al. 2000). The high degree of spatial coherence very close to an emitting material supporting surface waves can be physically understood by the fact that the mechanical oscillations within the material (plasma oscillations or lattice vibrations) transmit their spatial coherency to the emitted electromagnetic field (Carminati and Greffet 1999). Excitation in the far field of surface waves lead to thermal emission in a narrow spectral band and narrow angular lobe.

Excitation of surface waves is usually discussed in the framework of near-field optics applications. In these applications, surface waves are excited via an external radiation beam that experience TIR. Different techniques exist in order to excite surface polaritons via an external radiation beam, such as Krestchmann and Otto configurations (Raether 1988, Novotny and Hecht 2006, Maier 2007). In thermal radiation, the situation is different as surface waves are excited via the random fluctuation of charges within the emitting material. For typical temperatures involved in thermal radiation applications, SPhPs in SiC can easily be excited as $\Theta(\omega, T)$ reaches its peak value around 10 μm at 300 K, as predicted from Wien's law.

If we consider that medium 1 in Figure 16.2 is a bulk region of gold, then the SPP resonant frequency is around 9.69×10^{15} rad/s, which corresponds to a wavelength of about 0.194 μm. These calculations can be done easily using a Drude model for the dielectric function of gold. For typical thermal radiation temperatures between 300 and 2000 K, $\Theta(\omega,T)$ is very small at this frequency such that the energy density at resonance is also small. Therefore, SPhPs with resonance in the infrared spectrum are usually more interesting from a thermal radiation point of view than SPPs (Mulet et al. 2002). However, materials such as doped silicon support SPPs in the infrared and can thus behave like polar crystals supporting SPhPs (Fu and Zhang 2006, Basu et al. 2010).

16.4 NEAR-FIELD RADIATIVE HEAT FLUX CALCULATIONS

So far, we have presented the electromagnetic description of thermal radiation by discussing FE and the FDT. We have presented some fundamental concepts and have provided an overview of evanescent and surface waves. Using this background, it is now possible to discuss the typical problems encountered by a heat transfer engineer, which almost always require calculation of radiative heat flux.

Despite the fact that near-field thermal radiation seems to be a relatively new subject, the problem of near-field radiative heat flux calculations was addressed in the late 1960s (Cravalho et al. 1967, Boehm and Tien 1970). Their work provided near-field thermal radiation calculation results between two bulk materials separated by a vacuum gap. They used the analogy with TIR discussed in Section 16.3.1 to define the source of thermal radiation, which did not account for all evanescent modes. Polder and Van Hove (1971) reported the first correct radiative heat flux calculations between two bulk materials using FE and the FDT. Mulet et al. (2002) have shown that quasi-monochromatic radiative heat transfer can be achieved between two bulk materials when the materials support SPhPs, as discussed in Section 16.3.2. Near-field radiative heat transfer between two bulk materials has been investigated in many publications (Loomis and Maris 1994, Pendry 1999, Volokitin and Persson 2001, 2004, Joulain et al. 2005, Fu and Zhang 2006, Chapuis et al. 2008b, Francoeur and Mengüç 2008a, Basu and Zhang 2009, Francoeur et al. 2009). Also, numerical predictions of radiative flux in the near field have been investigated for a film emitter (Biehs 2007, Biehs et al. 2007, Francoeur et al. 2008b, 2009, 2010b, Fu and Tan 2009, Francoeur 2010), between a dipole and a surface (Mulet et al. 2001), between two dipoles (Volokitin and Persson 2001, Domingues et al. 2005, Chapuis et al. 2008a), in a cylindrical cavity (Hammonds 2006), between two large spheres (Narayanaswamy and Chen 2008), between a sphere and a surface (Otey and Fan 2011, Kruger et al. 2013), and between a dipole and a structured surface

(Biehs et al. 2008). The solution of near-field thermal radiation problems beyond these relatively simple geometries can be achieved using numerical approaches such as the finite-difference time domain (FDTD), boundary element, and the thermal discrete dipole methods (Wen 2010, Rodriguez et al. 2011, 2012, McCauley et al. 2012, Liu and Shen 2013, Didari and Mengüç 2014, Edalatpour and Francoeur 2014).

We shall first discuss near-field thermal radiation in 1D layered media. The general relations for a medium of N layers are provided, and then we move on the specific case of near-field thermal radiation between two bulk materials separated by a vacuum gap.

16.4.1 Near-Field Radiative Heat Flux in 1D Layered Media

The geometry considered for the calculation of near-field radiative heat flux in 1D layered medium is shown in Figure 16.5, where both Cartesian (x, y, z) and polar (ρ, ϕ, z) coordinate systems are depicted.

We consider $N-2$ layers with finite thicknesses, which are sandwiched between two half-spaces denoted medium 1 ($z < z_1$) and medium N ($z > z_{N-1}$). This layered structure is infinite along the x- and y-directions, and therefore only variations along the z-axis are considered. It is assumed that a given layer l is in local thermodynamic equilibrium at a prescribed temperature T_l, homogeneous, isotropic, nonmagnetic, and described by a frequency-dependent dielectric function local in space denoted by $\bar{\varepsilon}_l(\omega)$. The monochromatic radiative heat flux at location z_c along the z-direction in layer l due to an emitting layer s is found by considering the z-component of the Poynting vector given by Equation 16.12:

$$q_{\omega,sl}(z_c) = \frac{2k_0^2\Theta(\omega,T_s)}{\pi}\mathrm{Re}\left\{iK_s''(\omega)\int_V dV'\left[G_{slx\alpha}^E(\mathbf{r},\mathbf{r}',\omega)G_{sly\alpha}^{H*}(\mathbf{r},\mathbf{r}',\omega) - G_{sly\alpha}^E(\mathbf{r},\mathbf{r}',\omega)G_{slx\alpha}^{H*}(\mathbf{r},\mathbf{r}',\omega)\right]\right\} \tag{16.27}$$

where

the subscript α involves a summation over the three orthogonal components (state of polarization of the source)

T_s, V, and $K_s''(\omega)$ are, respectively, the temperature, volume, and imaginary part of the dielectric function of the emitting layer

The first two subscripts of the DGF terms show explicitly that they are calculated between layers s and l. The only unknowns in Equation 16.27 are the components of the electric and magnetic DGFs.

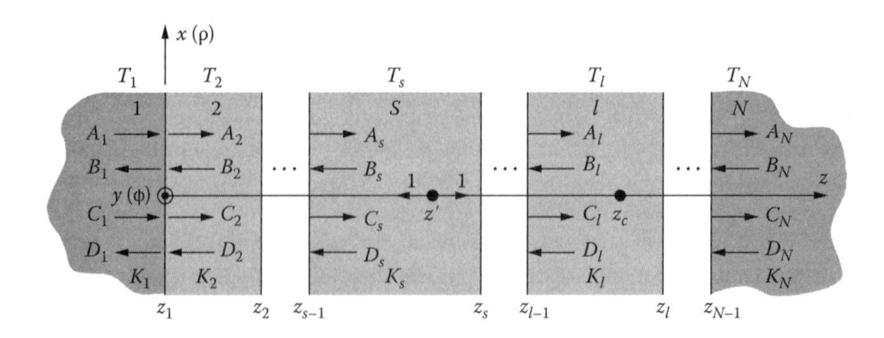

FIGURE 16.5 Schematic representation of the 1D layered medium, where $N-2$ layers of finite thicknesses are sandwiched between two half-spaces. The field patterns in each layer are also identified.

In a 1D geometry, it is convenient to use a plane wave representation of the DGF as there is no need to introduce any unnecessary complexity. For this purpose, we apply a 2D spatial Fourier transform on the DGF (Tsang et al. 2000, Francoeur et al. 2009):

$$\overline{\overline{\mathbf{G}}}_{sl}(\mathbf{r},\mathbf{r}',\omega)=\int_{-\infty}^{\infty}\frac{d\mathbf{k}_{\rho}}{(2\pi)^{2}}\overline{\overline{\mathbf{g}}}_{sl}(\mathbf{k}_{\rho},z_{c},z',\omega)e^{i\mathbf{k}_{\rho}\cdot(\mathbf{R}-\mathbf{R}')} \tag{16.28}$$

where

$\mathbf{k}_{\rho}=k_{x}\hat{\mathbf{x}}+k_{y}\hat{\mathbf{y}}$
$d\mathbf{k}_{\rho}=dk_{x}dk_{y}$
$R=x\hat{\mathbf{x}}+y\hat{\mathbf{y}}$
z' is the location of the source
$\overline{\overline{\mathbf{g}}}_{sl}$ is Weyl component of the DGF

As shown in Equation 16.27, the calculation of the radiative heat flux involves computation of terms $\int_{V}dV'G^{E}_{sli\alpha}G^{H*}_{slj\alpha}$, where i and j refer to x and y ($i\neq j$). Substitution of Weyl development of the DGF in the aforementioned term leads to

$$\int_{V}dV'G^{E}_{sli\alpha}(\mathbf{r},\mathbf{r}',\omega)G^{H*}_{slj\alpha}(\mathbf{r},\mathbf{r}',\omega)=\int_{-\infty}^{\infty}\frac{d\mathbf{k}_{\rho}}{(2\pi)^{2}}\int_{z}dz'g^{E}_{sli\alpha}(\mathbf{k}_{\rho},z_{c},z',\omega)g^{H*}_{slj\alpha}(\mathbf{k}_{\rho},z_{c},z',\omega) \tag{16.29}$$

A polar coordinate system (ρ, ϕ, z) is employed hereafter due to the azimuthal symmetry of the problem. The following transformation is performed:

$$\int_{-\infty}^{\infty}d\mathbf{k}_{\rho}=\int_{-\infty}^{\infty}\int_{-\infty}^{\infty}dk_{x}dk_{y}=\int_{k_{\rho}}^{\infty}\int_{\phi=0}^{2\pi}k_{\rho}dk_{\rho}d\phi=2\pi\int_{0}^{\infty}k_{\rho}dk_{\rho} \tag{16.30}$$

where k_{ρ} stands for any wave vector parallel to the surfaces of the layers. This integral is evaluated over the entire x and y domains, extending to infinity. Substitution of Equations 16.29 and 16.30 into Equation 16.27 in polar coordinates gives the monochromatic radiative heat flux at z_{c} along the z-direction in terms of Weyl components of the DGF:

$$q_{\omega,sl}(z_{c})=\frac{k_{0}^{2}\Theta(\omega,T_{s})}{\pi^{2}}\mathrm{Re}\left\{iK''_{s}(\omega)\int_{0}^{\infty}k_{\rho}dk_{\rho}\int_{z}dz'\left[\begin{array}{l}g^{E}_{sl\rho\alpha}(k_{\rho},z_{c},z',\omega)g^{H*}_{sl\phi\alpha}(k_{\rho},z_{c},z',\omega)\\-g^{E}_{sl\phi\alpha}(k_{\rho},z_{c},z',\omega)g^{H*}_{sl\rho\alpha}(k_{\rho},z_{c},z',\omega)\end{array}\right]\right\} \tag{16.31}$$

We analyze first the general case of a layer s of finite thickness $z_{s}-z_{s-1}$ emitting thermal radiation, as shown in Figure 16.5. Let us consider that waves of unit amplitudes are emitted both in the forward (z-positive) and backward (z-negative) directions from the source point z'. The emitting layer s consists of multiple source points z' distributed along z in the volume $z_{s}-z_{s-1}$. The radiative heat flux calculated at z_{c} in layer l is therefore proportional to the integration of these source points over the volume of layer s. The field in each layer, resulting from multiple reflections within the structure, is divided into four patterns. The coefficients A and B denote the amplitudes of forward and backward traveling waves, respectively, arising from a source emitting in the forward direction. Similarly, the coefficients C and D represent, respectively, the amplitudes of forward and backward traveling waves generated by a source emitting in the backward direction. For the layers j = 2 to $N-1$, these coefficients are calculated on the left boundary of the film at $z=z^{+}_{j-1}$. For the two half-spaces 1 and N, the coefficients are calculated, respectively, at $z=z^{-}_{1}$ and $z=z^{+}_{N-1}$.

The formalism introduced by Sipe (1987) can be used to express Weyl components of the DGF, where TE- and TM-polarized unit vectors are, respectively, defined as

$$\hat{\mathbf{s}} = -\hat{\varphi} \tag{16.32}$$

$$\hat{\mathbf{p}}_i^{\pm} = \frac{1}{k_i}(k_\rho \hat{\mathbf{z}} \mp k_{zi}\hat{\rho}) \tag{16.33}$$

Sipe's unit vectors are used to write Weyl components of the DGF in terms of dyads $\hat{a}\hat{b}$, where $\hat{b}$ represents the polarization of the wave at the source and $\hat{a}$ the polarization of the wave at the point z_c where the fields are calculated. Using the coefficients A, B, C, and D, the electric Weyl representation of the DGF can be written as

$$\overline{\overline{g}}_{sl}^{E}(k_\rho, z_c, z', \omega)$$
$$= \frac{i}{2k_{zs}}\begin{bmatrix} \left(A_l^{TE}\hat{\mathbf{s}}\hat{\mathbf{s}} + A_l^{TM}\hat{\mathbf{p}}_l^{+}\hat{\mathbf{p}}_s^{+}\right)e^{i[k_{zl}(z_c - z_l) - k_{zs}z']} + \left(B_l^{TE}\hat{\mathbf{s}}\hat{\mathbf{s}} + B_l^{TM}\hat{\mathbf{p}}_l^{-}\hat{\mathbf{p}}_s^{+}\right)e^{i[-k_{zl}(z_c - z_l) - k_{zs}z']} \\ +\left(C_l^{TE}\hat{\mathbf{s}}\hat{\mathbf{s}} + C_l^{TM}\hat{\mathbf{p}}_l^{+}\hat{\mathbf{p}}_s^{-}\right)e^{i[k_{zl}(z_c - z_l) + k_{zs}z']} + \left(D_l^{TE}\hat{\mathbf{s}}\hat{\mathbf{s}} + D_l^{TM}\hat{\mathbf{p}}_l^{-}\hat{\mathbf{p}}_s^{-}\right)e^{i[-k_{zl}(z_c - z_l) + k_{zs}z']} \end{bmatrix} \tag{16.34}$$

In Equation 16.34, the term $A_l^{TM}\hat{\mathbf{p}}_l^{+}\hat{\mathbf{p}}_s^{+}$ means that a forward traveling wave with polarization $\hat{\mathbf{p}}_l^{+}$ and amplitude A_l^{TM} is calculated in layer l due to a forward traveling wave emitted in layer s with polarization $\hat{\mathbf{p}}_s^{+}$; the physical interpretation of the other terms is similar. The magnetic Weyl representation of the DGF can be found using $\overline{\overline{g}}^{H} = \nabla \times \overline{\overline{g}}^{E}$.

Equation 16.34 is valid if the point z_c is located in a layer other than the one where the source is located (i.e., if $l \neq s$). Indeed, the coefficients A, B, C, and D can be seen as the amplitudes of the waves after multiple reflections within the 1D structure. If the point z_c is located in layer s (i.e., $l = s$), the DGFs have also to account for the primary wave propagation in unbounded medium or, in other words, the part of the wave reaching the point z_c without being scattered by the boundaries (Tsang et al. 2000, Narayanaswamy and Chen 2005, Francoeur et al. 2009). Therefore, if $l = s$, the electric DGF becomes the superposition of the response of the layered medium, given by Equation 16.34, and becomes the primary wave propagating in unbounded medium. For more clarity, we omit the primary wave term in the expressions of the DGFs, keeping in mind that it should be accounted for when $l = s$.

In the expression of the monochromatic radiative heat flux (Equation 16.31), Weyl components of the DGF are written as a function of ρ, ϕ, z. We therefore need to convert the dyads $\hat{\mathbf{a}}\hat{\mathbf{b}}$ contained in the electric and magnetic DGFs in terms of these coordinates. For example, using Equation 16.32, we find that the dyad $\hat{\mathbf{s}}\hat{\mathbf{s}}$ is given by $\hat{\varphi}\hat{\varphi}$. By regrouping the terms of as a function of $\hat{\rho}\hat{\rho}$, $\hat{\rho}\hat{\varphi}$,..., $\hat{\mathbf{z}}\hat{\mathbf{z}}$, Weyl representation of the electric and magnetic DGF in tensor form are

$$\overline{\overline{g}}_{sl}^{E}(k_\rho, z_c, z', \omega) = \begin{bmatrix} g_{sl\rho\rho}^{E} & 0 & g_{sl\rho z}^{E} \\ 0 & g_{sl\phi\phi}^{E} & 0 \\ g_{slz\rho}^{E} & 0 & g_{slzz}^{E} \end{bmatrix} \tag{16.35}$$

$$\overline{\overline{g}}_{sl}^{H}(k_\rho, z_c, z', \omega) = \begin{bmatrix} 0 & g_{sl\rho\phi}^{H} & 0 \\ g_{sl\phi\rho}^{H} & 0 & g_{sl\phi z}^{H} \\ 0 & g_{slz\phi}^{H} & 0 \end{bmatrix} \tag{16.36}$$

The explicit expressions for the different components of the aforementioned tensors have been given by Francoeur et al. (2009). Using Equations 16.35 and 16.36, the monochromatic radiative heat flux at location $z = z_c$ in layer l along the z-direction, due to a source layer s of volume $z_s - z_{s-1}$, is

$$q_{\omega,sl}(z_c) = \frac{k_0^2\Theta(\omega,T_s)}{\pi^2}\mathrm{Re}\left\{ iK_s''(\omega)\int_0^\infty k_\rho dk_\rho \int_{z_{s-1}}^{z_s} dz' \begin{bmatrix} g_{sl\rho\rho}^E(k_\rho,z_c,z',\omega)g_{sl\phi\rho}^{H*}(k_\rho,z_c,z',\omega) \\ +g_{sl\rho z}^E(k_\rho,z_c,z',\omega)g_{sl\phi z}^{H*}(k_\rho,z_c,z',\omega) \\ -g_{sl\phi\phi}^E(k_\rho,z_c,z',\omega)g_{sl\rho\phi}^{H*}(k_\rho,z_c,z',\omega) \end{bmatrix} \right\} \tag{16.37}$$

The only unknowns in the expression for the radiative heat flux are the field amplitude coefficients A, B, C, and D contained in Weyl representation of the DGF. Therefore, at this point, solution of the near-field radiative heat flux is reduced to an electromagnetic scattering problem in layered media. An efficient way to determine the field amplitude coefficients is to use a transfer matrix approach (Aulender and Hava 1996, Yeh 2005, Zhang 2007). A complete procedure calculate the field amplitude coefficients for both cases of an emitting bulk region and a film can be found in Francoeur et al. (2009).

16.4.2 Near-Field Radiative Heat Transfer between Two Bulk Materials Separated by a Vacuum Gap

The different physical phenomena involved in radiation heat transfer in the near field is discussed in this section with the solution of a simple problem involving two bulk regions of SiC, the material that has been discussed in Section 16.3.2. They are separated by a vacuum gap of thickness d, as depicted in Figure 16.6.

Medium 1 is maintained at 300 K, while medium 3 is a heat sink (0 K). The objective is to calculate the radiative heat flux emitted by medium 1 and absorbed by medium 3 for different vacuum gap thicknesses d. Mathematically, the radiative heat flux absorbed by medium 3 is calculated just after the boundary 2–3 (i.e., $z = z_2^+$), since all energy crossing the interface 2–3 is eventually absorbed by medium 3.

An explicit expression for the near-field radiative heat flux between two bulk materials can be found starting from Equation 16.37, with $s = 1$, $l = 3$, and $z_c = z_2^+$. We therefore need to compute the coefficients A_3, B_3, C_3, and D_3 in both polarization states and substitute them into Weyl representation of the DGF. For the case of the emitting bulk, the coefficients C and D do not need to be calculated, as there is no wave emitted by medium 1 in the backward direction that can propagate in the layered medium; therefore, the coefficients C_3 and D_3 can be set as zero in the DGF. Also, since

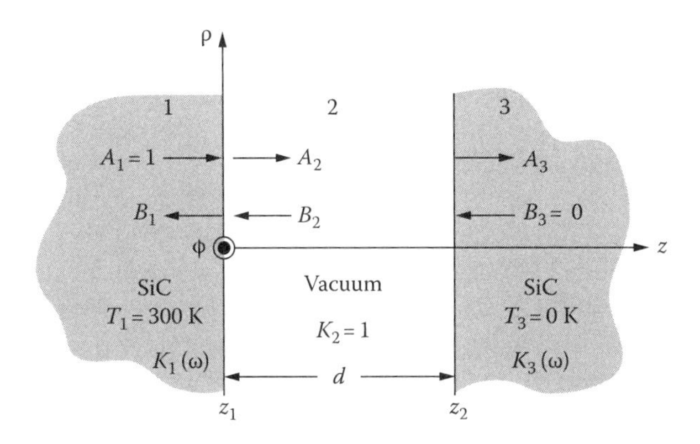

FIGURE 16.6 Schematic representation of two bulk regions of SiC separated by a vacuum gap of thickness d.

medium 3 is unbounded in the positive z-direction, the coefficient B_3 is zero. Assuming that a wave of unit amplitude is emitted from medium 1 in the forward direction, the field in medium 3 is simply given by the transmission coefficient of layer 2 (Yeh 2005, Francoeur et al. 2009):

$$A_3^{TE,TM} = \frac{t_{12}^{TE,TM} t_{23}^{TE,TM} e^{ik_{z2}d}}{1 + r_{12}^{TE,TM} r_{23}^{TE,TM} e^{2ik_{z2}d}} \tag{16.38}$$

where r_{ij} and t_{ij} are, respectively, Fresnel reflection and transmission coefficients. For the particular case of an emitting half-space, the z'-dependence of Weyl components of the DGF can be written as (Francoeur et al. 2009)

$$g_{13i\alpha}^{E}(k_\rho, z_2^+, z', \omega) g_{13j\alpha}^{H*}(k_\rho, z_2^+, z', \omega) = g_{13i\alpha}^{E}(k_\rho, z_2^+, \omega) g_{13j\alpha}^{H*}(k_\rho, z_2^+, \omega) e^{2k_{z1}''z'} \tag{16.39}$$

The integration over z' in Equation 16.39 can be performed easily with $z_{s-1} = -\infty$ and $z_s = 0$. Substitution of Weyl components of the DGF in the spatially integrated monochromatic radiative flux leads to (Francoeur et al. 2009)

$$q_{\omega,13}(z_2^+) = \frac{\Theta(\omega,T_1)}{4\pi^2} \int_0^\infty \frac{k_\rho dk_\rho}{|k_{z1}|^2} \left[\frac{\mathrm{Re}\left(\varepsilon_1 k_{z1}^*\right) \mathrm{Re}\left(\varepsilon_3 k_{z3}^*\right)}{|\bar{n}_1|^2 |\bar{n}_3|^2} |A_3^{TM}|^2 + k_{z1}' k_{z3}' |A_3^{TE}|^2 \right] \tag{16.40}$$

where $\bar{n}_j$ represents the complex refractive index of medium j. Substitution of Equation 16.38 into Equation 16.40 gives the following expressions for the radiative heat flux (Mulet et al. 2002, Francoeur et al. 2009):

$$q_{\omega,13}^{prop}\left(z_2^+\right) = \frac{\Theta(\omega,T_1)}{4\pi^2} \int_0^{k_0} k_\rho dk_\rho \left[\frac{\left(1-\left|r_{21}^{TE}\right|^2\right)\left(1-\left|r_{23}^{TE}\right|^2\right)}{\left|1 - r_{21}^{TE} r_{23}^{TE} e^{2ik_{z2}'d}\right|^2} + \frac{\left(1-\left|r_{21}^{TM}\right|^2\right)\left(1-\left|r_{23}^{TM}\right|^2\right)}{\left|1 - r_{21}^{TM} r_{23}^{TM} e^{2ik_{z2}'d}\right|^2} \right] \tag{16.41}$$

$$q_{\omega,13}^{evan}\left(z_2^+\right) = \frac{\Theta(\omega,T_1)}{\pi^2} \int_{k_0}^{\infty} k_\rho dk_\rho e^{-2k_{z2}''d} \left[\frac{\mathrm{Im}\left(r_{21}^{TE}\right)\mathrm{Im}\left(r_{23}^{TE}\right)}{\left|1 - r_{21}^{TE} r_{23}^{TE} e^{-2k_{z2}''d}\right|^2} + \frac{\mathrm{Im}\left(r_{21}^{TM}\right)\mathrm{Im}\left(r_{23}^{TM}\right)}{\left|1 - r_{21}^{TM} r_{23}^{TM} e^{-2k_{z2}''d}\right|^2} \right] \tag{16.42}$$

Details for the derivation of Equations 16.41 and 16.42 are provided by Francoeur et al. (2009).

The radiative flux is split into two distinct equations. By writing the near- and far-field components of the radiative flux separately, we are able to quantify the impact of near-field effects on the total energy transport more explicitly. Equation 16.41 stands for propagating waves, which are accounted for in the classical theory of thermal radiation, while Equation 16.42 stands for evanescent waves contributing to radiative heat transfer only in the near field. The splitting into propagating and evanescent modes is done by breaking down the integration over the parallel wave vector k_ρ into two parts, as discussed in Section 16.3.2, delimiting the transition of the z-component of the wave vector in vacuum from a pure real number (propagating wave) to a pure imaginary number (evanescent wave). Therefore, the integration over k_ρ for propagating waves is done from 0 up to k_0, while the integration is performed from k_0 to infinity for evanescent waves. From a numerical point of view, however, a cutoff value of k_ρ is needed in order to obtain the evanescent component of the radiative heat flux. The physics behind this cutoff k_ρ value should be explained more. Remember that the decay length of an evanescent wave in medium j can be approximated as $\delta_j \approx |k_{zj}|^{-1}$. Physically, for the problem treated here, where we assume the temperature of medium 3 is 0 K, only evanescent waves emitted by medium 1 with penetration depths in vacuum δ_2 equal or greater than the vacuum gap d can be tunneled in medium 3 and thus contribute to radiant energy transfer. If the medium 3 would be at a finite temperature other than 0 K, we would need to consider the mutual interactions. For evanescent waves with k_ρ much larger than k_0, the z-component of the wave vector in medium j, given by $k_{zj} = \sqrt{\varepsilon_j k_0^2 - k_\rho^2}$, can be approximated by $k_{zj} \approx ik_\rho$. This is quite acceptable, even for moderate values of k_ρ. For example, in vacuum, if $k_\rho = 2k_0$, then $k_{z2} = i1.732k_0$; similarly, if $k_\rho = 5k_0$, then $k_{z2} = i4.90k_0$. By substituting this approximation within the definition of penetration depth of evanescent waves in vacuum, we find that $k_\rho \approx \delta_2^{-1}$. Using the limiting condition $\delta_2 = d$, we find that the largest contributing wave vector to the evanescent radiative heat flux is approximately given by $k_\rho \propto d^{-1}$. This relation shows that as the gap decreases between the two bulk regions, the limiting k_ρ increases, and therefore more energy is transferred via radiation tunneling.

The denominator in Equations 16.41 and 16.42 accounts for multiple wave reflection and interference within the vacuum gap. The terms $(1-|r_{2j}^{TE,TM}|^2)$ in the numerator of Equation 16.41 represents the spectral absorptance of medium j and also play the role of spectral emittance of medium j. The interpretation of $\mathrm{Im}(r_{2j}^{TE,TM})$ in the numerator of Equation 16.42 is similar: this term can be considered as a spectral emittance and absorptance of medium j for evanescent waves (Mulet et al. 2002). The evanescent nature of these modes is also explicitly shown in Equation 16.42 via the exponentially decaying term $e^{-2k''_{z2}d}$.

In the far-field limit, the radiative heat flux between the two bulk materials is independent of the gap thickness d. Indeed, as $d \to \infty$, $e^{-2k''_{z2}d} \to 0$ such that $q_{\omega,13}^{even}(z_2^+) \to 0$. Also, in the far-field limit, radiation heat transfer becomes incoherent, such that the denominator of Equation 16.41 can be written as (Fu and Zhang 2006)

$$\frac{1}{\left|1 - r_{21}^{TE,TM} r_{23}^{TE,TM} e^{2ik'_{z2}d}\right|^2} \to \frac{1}{1 - \rho_{r,21}^{TE,TM} \rho_{r,23}^{TE,TM}} \tag{16.43}$$

where $\rho_{r,ij}^{TE,TM}$ is the reflectivity of the interface i–j, calculated from Fresnel reflection coefficient as $\rho_{r,ij}^{TE,TM} = \left|r_{ij}^{TE,TM}\right|^2$. We can also recast the integration over k_ρ as an integration over the polar angle θ_i using $k_\rho = k_0 \sin\theta_i$. By using the emissivity of medium j ($\varepsilon_j^{TE,TM} = 1 - \rho_{r,ij}^{TE,TM}$ for a bulk material) and Planck's blackbody intensity $[I_{b,\omega}(T) = \Theta(\omega,T)\omega^2/4\pi^3 c_0^2]$, the radiative heat flux in the far-field limit is

$$q_{\omega,13}^{prop}\left(z_2^+\right) = 2\pi I_{b,\omega}\left(T_1\right) \int_{\theta_i=0}^{\pi/2} \frac{\varepsilon_1 \varepsilon_3}{1 - \rho_{r,21}\rho_{r,23}} \cos\theta_i \sin\theta_i d\theta_i \tag{16.44}$$

where it is assumed that TE- and TM-polarized waves have equal contributions to the radiative heat flux.

Spectral distributions of radiative heat flux between two SiC bulk regions are shown in Figure 16.7a for vacuum gaps d of 10 nm, 100 nm, and 1 μm. These results are compared with the flux calculated in the far-field regime (Equation 16.44) and the values obtained for blackbodies. In Figure 16.7b, the TE and TM evanescent contributions to the radiative heat flux are shown for a gap thickness of 10 nm and compared with the radiative flux between blackbodies.

Figure 16.7a shows that the radiative heat flux in the near field exceeds the values predicted for two blackbodies by several orders of magnitude. For small gap thicknesses, the radiative heat flux is quasi-monochromatic due to excitation of SPhPs. That is, at resonance frequencies, we see significant enhancement in the flux. Note that this does not mean that the broadband radiation emission is suppressed. Since there is a significant increase in emission in a very narrow frequency range, the spectral nature of emission shows this quasi-monochromatic behavior. Resonance of the flux arises at 1.786×10^{14} rad/s, where the DOS is very large, as discussed in Section 16.3.2. For a frequency smaller than the resonant frequency of SPhPs, there is also a small peak of radiative heat flux due to tunneling of evanescent waves. Indeed, the radiative heat flux due to regular evanescent waves

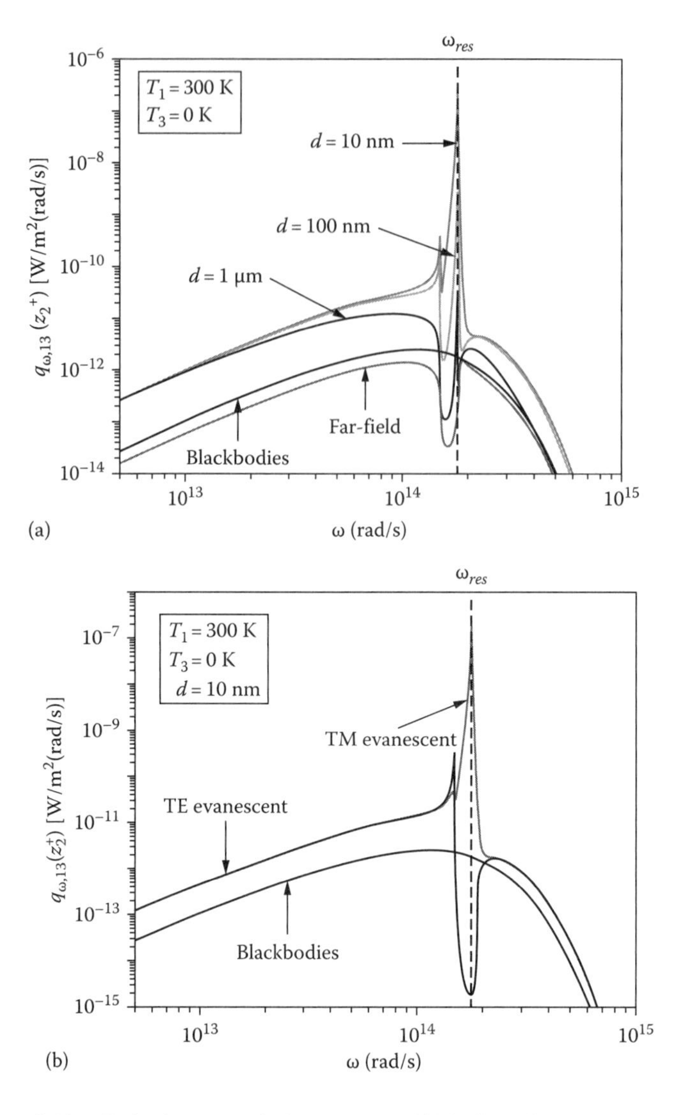

FIGURE 16.7 Near-field radiative heat transfer between two SiC bulk regions separated by a vacuum gap of thickness d: (a) $d = 10$ nm, 100 nm, and 1 μm; results are compared with the fluxes in the far-field regime and between blackbodies. (b) Evanescent TE and TM contributions for $d = 10$ nm; results are compared with the flux between blackbodies.

is directly proportional to the real part of the dielectric function of the emitting material, and the smaller peak corresponds to a frequency where $\bar{\varepsilon}_1'$ takes a very large value (see discussion of Section 16.3.1 and Figure 16.4b). Figure 16.7b shows clearly that the TM evanescent component of the radiative heat flux dominates the energy exchanges, as SPhPs can only be excited in TM polarization for nonmagnetic materials. This is due to the fact that TM polarization results from oscillating electric field on the surface, which yields the evanescent waves (Zhang 2007).

Before closing this section, it is interesting to consider the dependence of the near-field radiative heat flux versus the gap thickness d. As mentioned before, for $k_\rho \approx k_0$, we can approximate $k_{zj} \approx ik_\rho$. Since the radiant energy exchange is dominated by surface waves, we can keep only the TM evanescent contribution of the radiative flux. Using $k_{zj} \approx ik_\rho$, the Fresnel reflection coefficients in TM polarization can be approximated as

$$r_{2j}^{TM} \approx \frac{\bar{\varepsilon}_j - 1}{\bar{\varepsilon}_j + 1} \tag{16.45}$$

By substituting Equation 16.45 in the TM evanescent component of the radiative flux, and by assuming that $d \to 0$, it can be shown after some algebra that (Fu and Zhang 2006, Zhang 2007)

$$q_{\omega,13}\left(z_2^+\right) \approx \frac{\Theta(\omega, T_1)}{\pi^2 d^2} \frac{\mathrm{Im}(\bar{\varepsilon}_1)\mathrm{Im}(\bar{\varepsilon}_3)}{|(\bar{\varepsilon}_1 + 1)(\bar{\varepsilon}_3 + 1)|^2} \tag{16.46}$$

This equation suggests that for distances d smaller than the dominant wavelength emitted, the radiative heat flux increases proportionally to d^{-2}. Using similar arguments, it can be shown that the energy density, and therefore the LDOS, increases proportionally to d^{-3} within close proximity of an emitting material.

16.5 COMPUTATIONAL STUDIES OF NEAR-FIELD THERMAL RADIATION

During the last decade, various computational techniques were suggested for near-field heat transfer studies. A computational technique that can take into account all the near-field modes of any given problem is of crucial importance with increasing interest in nanoscale energy-harvesting devices and nanoscale manufacturing and sensing. In Chapter 15, different solution methods for the EM wave equations were discussed for possible consideration to absorption and scattering problems. Among them, the FDTD method can also be used for near-field thermal radiation problems. Since the FDTD algorithm provides close-to-exact (or full-wave) solution of the EM field in a given region of space, it can always be applied to the near-field analysis of that region. Further, FDTD is a time domain technique that uses the discretized partial differential Maxwell equations to calculate the electric and magnetic field values at any given grid space. However, there are some critical points that necessitate fundamentally different considerations in setting up the solution algorithms. For example, one important difference stems from the change in length scale of the problem. That is, in a conventional problem, one has to choose the numerical dispersion associated with FDTD such that the smallest grid size would be chosen to be smaller than one-tenth of the smallest wavelength of interest ($\lambda_{min}/10$). Considering $\lambda = \lambda_0/n$, where λ_0 is the wavelength of vacuum and n is the refractive index equivalent to the real part of the $\sqrt{\varepsilon_r(\omega)}$, the smallest wavelength that may be used in near-field thermal radiation problems could be a few micrometer, which may be larger than the physical geometry of interest itself. For this reason, a creative approach is needed to find the right trade-off between the computational time, accuracy, and the grid size. When working with plasmonic materials, the permittivity of the material has an effect not only inside the material but also on the grid points adjacent to the material itself. Hence, handling the grid sizes as well as the boundaries in a multilayer structure is of vital importance for the robustness of the solution methodology.

As mentioned earlier, FDTD is a time domain technique. However, when we are working with materials whose properties are frequency dependent, we need to model the frequency-dependent dielectric function of these materials. This is quite a challenging task, because to simulate the permittivity model, we have to devise a creative approximation as there is no known time domain function for it. To circumvent this difficulty, an equivalent recursive expression based on inverse Fourier transform is used to obtain the time domain model of the permittivity model.

Computational techniques for the solution of near-field thermal radiation problems need careful attention to the use of correct boundary conditions (BCs). Setting the BCs in an efficient way is essential in the success of FDTD methods, as the computational domain is finite, yet the wave propagation can continue beyond the bounds of the domain. To attenuate the reflection of energy back into the computational grid, we need to consider some form of absorbing BCs (ABCs), which has the effect of simulating infinite or open-boundary geometry. The right BC hence allows more streamlined simulations to be carried out when working with subwavelength structures. Among BCs that are well suited for such kind of problems are perfectly matched layer (PML) and convolutional PML (CPML) (Didari and Mengüç 2014).

The optimum computational scenario can be achieved if we can couple the near- and far-field FDTD solutions within a single software package. This could be facilitated by the fact that the two scenarios share common description files. Generally, the near-field scenario would require additional input information related to the material and boundary configurations. At the same time, the output could be provided both in the near- and far-field zones per user's request. This means that, in addition to the current separate configurations, one should be able to output near-field data for far-field excitation or far-field data for near-field excitation. Such features would greatly enhance the flexibility and applicability of the FDTD technique to near-field problems (see Didari and Mengüç 2014, 2015b for details).

16.6 EXPERIMENTAL STUDIES OF NEAR-FIELD THERMAL RADIATION

Understanding the fundamental and potential applications of NFRT is important for several emerging fields. Among them, three stand out as immediate engineering applications: (1) energy conversion devices, (2) nanoscale patterning and manufacturing, and (3) near-field imaging. Successful use of NFRT concepts in these areas is likely to provide breakthroughs, yet requires extensive and repeatable experimental evidence and validation of near-field radiative exchange between surfaces and/or finite-sized objects. First and foremost, laboratory-scale measurements need to be conducted carefully to determine the effects of materials, surface conditions, impurities, gap thicknesses, and other basic manufacturing details that may alter the steady and consistent operation of the new processes or devices.

Numerical and theoretical studies on near-field radiative transfer are not extensive and limited to a few cases as discussed earlier. The experimental research on NFRT is far from being extensive, as there are relatively few experimental accounts available in the literature. In this section, we briefly outline recent experimental work that has been performed to validate near-field radiative heat flux predictions. In addition, we discuss the experiments related to atomic-force-microscopy-based patterning applications and near-field imaging. Note that this field is still an emerging one, and new experimental approaches will be needed for different applications.

16.6.1 Overview

The earlier studies on NFRT experiments were conducted in cryogenic environments. The reason for the choice of cryogenic conditions becomes obvious when the characteristic length of radiation is considered. As discussed earlier, Wien's law gives a rough estimate of the important length scale involving near-field exchange. Near-field radiation is observed only if the gap between two objects is comparable to this value, or less. If the temperature of the medium is at about room temperature of 300 K, the dominant wavelength of the radiative exchange peaks around 10 μm. With decreasing temperature, the length scale where the near field is observed increases. On the other hand, at a higher temperature such as 3000 K, the peak wavelength is only about 1 μm. Any two objects that stand at distances smaller than these peak values are affected by the NFRT; the closer they are, the more the effect would be. In the 1970s, before the era of nanopatterning facilities, only micrometer-sized gaps could effectively be built and studied. This necessitated the studies to be conducted at cryogenic temperatures as low as 200 K to facilitate the experiments

The late Professor C. L. Tien's group at the University of California–Berkeley worked on near-field radiative exchange problems at low temperatures, although they did not explicitly consider the effects of surface waves. Domoto et al. (1970) built an apparatus to measure the flux between copper plates in the liquid-helium temperature range. The available laboratory tools provided the ability to vary the gap between 1 mm down to 50 μm. Based on their experiments, they reported fluxes much less than those predicted by the blackbody radiation exchange (close to 3%) and attributed this to the material properties. Nevertheless, the near-field effects were shown to enhance the radiative transfer rate more than 2.5 fold as compared to the far-field values. At about the same time, Hargreaves at Philips Research Laboratories conducted similar experiments (Hargreaves 1969, 1973). In these studies, chromium plates were used, and gap distances from 1 to about 6 μm were explored under vacuum conditions (10^{-5} Torr). He reported results at temperatures down to 145 K. The results showed that the flux was about 40%–50% of the blackbody radiation, and there was a twofold increase in radiative flux at room temperature and at gap distances of about 1 μm. Even though these studies observed enhancement of the flux at near field, it was not beyond blackbody energy exchange values. The reason for the lack of demonstration of the near-field effects was coming from the choice of materials, as at that time the impact of plasmonic and phonon polaritons were not well understood.

Gap distances below 1 μm were first attempted by Xu et al. (1994), who were able to go down to 12 nm gap using a profiler based on a scanning thermal microscope with an indium probe. Their results did not show any strong heat transfer enhancement, which was attributed to the lack of sensitivity of their measurement approach. A scanning tunneling microscope (STM) was also used by Kittel et al. (2005b), who measured the near-field enhancement between a tip and a plate, which was either gold or gallium nitride, separated by a gap smaller than 100 nm, all the way down to 1 nm. Their results, which were obtained under ultrahigh vacuum conditions, supported the predicted theoretical trends for gaps larger than 10 nm. However, below this gap distance, there was disagreement between the results and the theory, which was attributed to existence of a material-dependent small length scale below the macroscopic definitions of dielectric properties. In addition, their tip geometry was complicated, making interpretations of the results in an unambiguous way more difficult. Indeed, recent similar, but less controlled, experiments have also shown increased energy transfer to nanosize gold particles (20–100 nm diameter) that are in a close proximity (less than 20 nm) of an atomic force microscope (AFM) probe (Hawes et al. 2007, 2008). In their case, the near-field exchange was not due to thermal excitement, but due to the evanescent waves formed by a totally reflected laser beam at 532 nm interacting with nanosize gold particles. If an AFM probe, used in tipping or scanning mode, hovered close to these particles, the particles were heated, and even fused or melted. No quantitative near-field radiation flux measurements were provided, but the analysis suggested enhancement of radiative exchange between the particles and the tip.

Experimental studies between two parallel plates were reported by Hu et al. (2008). They considered glass plates in their setup and clearly showed the enhanced near-field radiation heat transfer due to the SPhPs. The radiative flux measured was more than 35% higher than that can be achieved by blackbody radiation and that corresponded to smaller gap distances of about 1.6 μm. The same group also reported experiments between a plate and an AFM tip with attached silica particles with diameters of about 50 μm (Shen et al. 2009). The tip of a triangular SiN/Au cantilever acted as sensitive thermal probe, and the experiments were conducted at low pressures to remove the effect of air conduction; therefore, the heat exchange was due to radiation transfer only. The near-field effects were observed for gap distances smaller than 10 μm. These results suggested more than 20% increase of blackbody radiation exchange between the glass substrate and a silica microsphere, which was attributed to near-field effects.

Near-field effects of radiation are also used to develop advanced near-infrared imaging systems. Such a concept was reported by De Wilde et al. (2006) who used thermal evanescent emission from structured surfaces as the energy source. Again, an oscillating tungsten STM tip was employed that acted as an antenna and scattered the evanescent fields emitted by the surface to the far field for measurement. An approach based on STM described by Kittel et al. (2005b) was also extended to nanoscale imaging. Their unique near-field scanning thermal microscopy concept can be used on detection of evanescent thermal electromagnetic fields emitted by objects in close proximity or on the surface (Kittel et al. 2008). In principle, this apparatus was built around an STM and also allows the heat transfer measurements between the STM probe and a plate of gold or gallium nitride

(GaN). Working under ultrahigh vacuum, the group was able to reach a resolution of 1 nm, which is quite useful to study thermal effects due to near-field interactions. The details of thermal scanning tunneling microscopy are provided by Majumdar for sub-500 nm imaging applications (Majumdar 1999). Several other applications are summarized by Cahill et al. (2014).

Near-field interactions between evanescent waves and nanosize particles were used to develop a characterization tool for real-time in situ measurements of size, shape, and structure of nanoparticles on surfaces (Aslan et al. 2005, Venkata 2006, Francoeur et al. 2007, Venkata et al. 2007, Francoeur 2010). The system is based on elliptically polarized scattering system to characterize the particles from their scattering signals measured at far field. To identify the shape and the structure of these particles, detailed inversion algorithms would be needed as reported by Francoeur et al. (2007), Francoeur (2010), and Charnigo et al. (2007).

It is important to note that the experiments at near field are extremely difficult and demand precision. For that reason, they need to be coupled with the numerical studies to make sure that the key details are not missed. For example, Heltzel et al. (2005, 2007, 2008) used a numerical solution of Maxwell equations in an FDTD code to predict the near-field enhancement of incident laser radiation on silica spheres on a silicon substrate and gold spheres on a silica substrate. Very large enhancement of the incident intensity was predicted to occur beneath the spheres, and eventually predicted melting patterns on the silicon surface and the predicted ablation features on the silica substrate were validated by the experiment. The experiments of Hawes et al. (2007, 2008) were also numerically modeled by Huda et al. (2011) to determine best operating conditions.

16.6.2 Experimental Determination of NFRT Coefficient

Experimental evidence of near-field radiative transfer (NFRT) and discussion of a NFRT coefficient has scarcely been addressed; one of the best works is that by Rousseau et al. (2009). The group has built an apparatus to measure NFRT between a spherical particle and a plate and made measurements between the gaps of 30 nm and 2.5 μm. They used 22 and 42 mm diameter spheres made of sodalime glass and a plate that was borosilicate glass. Since both materials contain more than 70% silica, they assumed their properties are those of amorphous silica, which were obtained from Palik (1998). The near-field radiative heat transfer between these dielectric materials was measured efficiently as both materials support SPhPs. The plate was heated to have temperature difference about 10–20 K. The flux was measured using a silicon-nitride AFM cantilever, with thickness of about 525 nm; it was coated with a chromium layer of about 15 nm and then a gold layer of about 60 nm. The bending of the cantilever was measured with an interferometric technique, which allowed them to monitor a constant distance between the sphere and the plate. The end result of these experiments was a so-called thermal conductance, a proportionality constant. If the so-called Derjaguin approximation is used, the flux between the sphere and the plate is locally approximated as between two parallel plates separated by a distance d, as discussed in Section 16.4. The local near-field radiative heat transfer coefficient h_r is calculated for different gaps and is afterward integrated over the area in order to calculate a theoretical value of the so-called near field radiation coefficient or thermal conductance:

$$G_{theo}(d,T) = \int_0^R h_r(\bar{d}(r),T)2\pi r dr \tag{16.47}$$

where $\bar{d}(r) = d + R - \sqrt{R^2 - r^2}$ is the local distance between the plane and the sphere surface.

This term, $G(d,T)$, is a function of both the gap distance and temperature difference and is similar to thermal conductivity. However, since the measurements are carried out in a vacuum environment (10^{-6} mbar), it is a direct measure of radiative flux. Therefore, this term can be called, more appropriately, the near-field radiation coefficient.

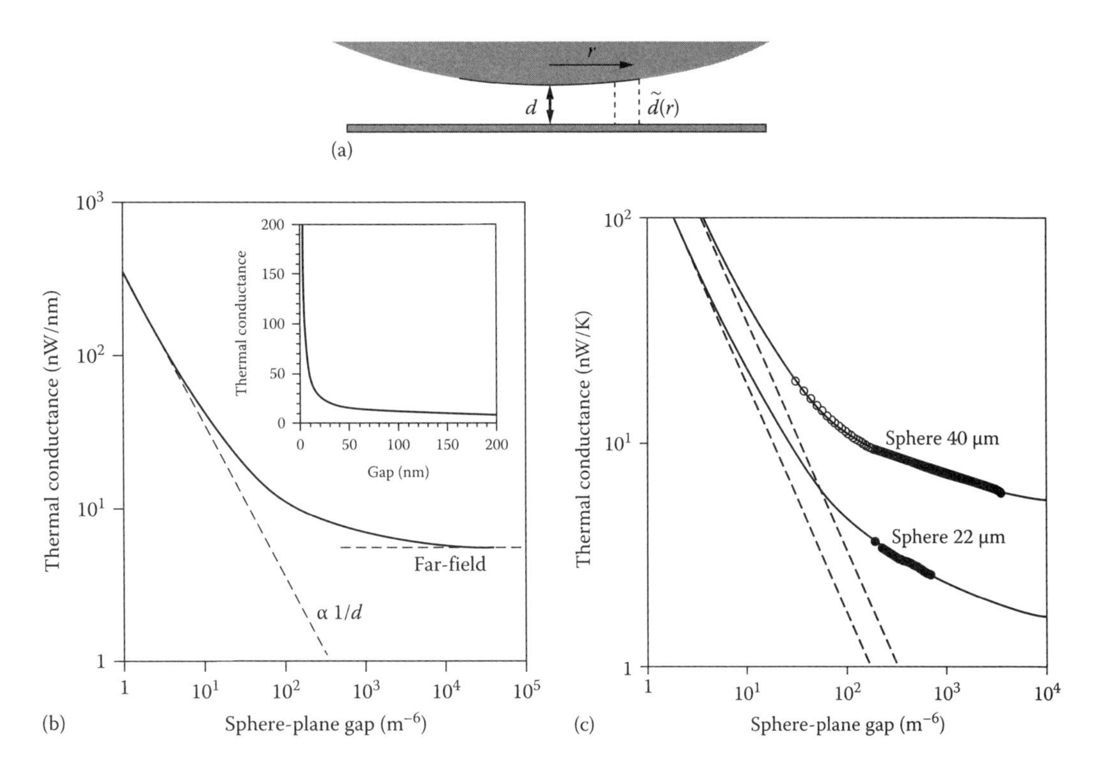

FIGURE 16.8 Near-field radiative transfer coefficient (thermal conductance, as referred to by Rousseau et al. 2009) between a sphere and a plate separated by a vacuum gap of thickness d. (a) Geometry between the sphere and the plate and (b) theoretical results are compared with the fluxes in the far-field regime and between blackbodies for sphere radius of R = 20 nm. (c) Experimental NFRT results compared with the theoretical data for R = 22 and 40 m^{-6}. The $1/d$ regime shown in (c) is called Derjaguin approximation, which is observed below 10 nm gap distances. (Courtesy of J.-J. Greffet.)

At far field (ff), that is, at distances longer than 10 μm, this value is given as $G_{ff} = 2\pi R^2 4\sigma\varepsilon(T)T^3$, where R is the sphere diameter. For the emissivity, the value of $\varepsilon(T)$ = 0.354 is used based on the emissivity of silica (Palik 1998). As the sphere and the plate become closer, the near-field radiation coefficient increases due to the near-field effects

$$G_{exp}(d) = G_{ff} + \frac{H}{\Delta T}\delta(d) = 2\pi R^2 4\sigma\varepsilon(T)T^3 + \frac{H}{\Delta T}\delta(d) \tag{16.48}$$

where H is a proportionality constant due to bending of cantilever at about $\delta(d)$, both of which are experimentally determined quantities. Figures 16.8a and 16.8b show the theoretical results and the experimental data obtained for two different sphere diameters.

Comparing the results obtained from Derjaguin approximation with the $1/d$ dependence of the near-field radiation coefficient in Figure 16.8b, it can be seen that the $1/d$ law is applicable to gaps smaller than about 10 nm. For sphere diameters of 22 and 40 μm, Derjaguin approximation is in excellent agreement with the experimental measurements.

16.6.3 Near-Field Effects on Radiative Properties and Metamaterials

Near-field effects are important not only the way they affect the heat transfer rates between closely spaced bodies but also the way they can be manipulated to change the medium properties. As discussed in Section 16.3.2, surface waves exhibit a high degree of spatial and spectral coherence in the near field of a thermal source. By nanostructuring a surface supporting such surface waves, it is possible to transmit this coherence into the far field. This allows directional emittance by the surfaces (Maruyama et al. 2001, Greffet et al. 2002, Arnold et al. 2012), which has potential impact to directional emission for thermophotovoltaic studies or spectroscopic sensors. A potential applications of these concepts for controlling the light emission was briefly discussed by Greffet (2011).

The Greffet group has shown how the angular profile of emissivity can be altered using a diffraction grating (Greffet et al. 2002). They have fabricated gratings on a SiC substrate, which supports SPhPs as discussed in Section 16.3.2 (see Figure 16.9). The diffraction by the grating of the SPhPs, propagating at the surface of the SiC substrate, leads to emissivity in narrow solid angles, as depicted in Figure 16.10. The emissivity thus obtained is about 20 times larger compared to a flat source. This extraordinary behavior is due to the coherence properties of the SPhPs that are observed only in the near field if the substrate is flat. The recent extension of this work was reported for cross-slit SiC gratings (Arnold et al. 2012). They could achieve 90% p-polarized narrow angle emission as a function the periodicity of the grating.

The observations made by Greffet et al. (2002) are quite important as they show that radiative properties can be significantly modified by nanostructuring the surface of a material. For the specific case of SiC, they have shown that a reflectivity of 94% can be reduced to almost zero in the infrared region due to the presence of the grating that couples surface waves into propagating waves. Other structures and 1D layered media can be used to tune far-field properties of materials

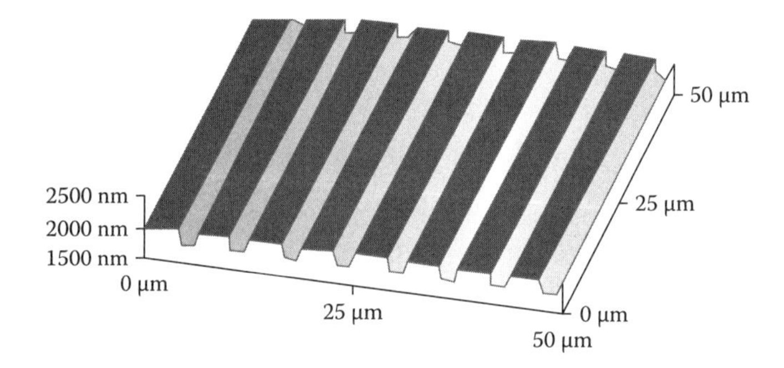

FIGURE 16.9 Schematic image of grating obtained by atomic force microscopy. Period d = 0.55λ (λ = 11.36 μm) chosen so that a surface wave propagating along the interface is coupled to a propagating wave in the range of frequencies of interest. Depth: h = λ/40 was optimized so that the peak emissivity is ε = 1 at λ = 11.36 μm. Fabricated on SiC by standard optical lithography and reactive ion etching. (Courtesy of J.-J. Greffet.)

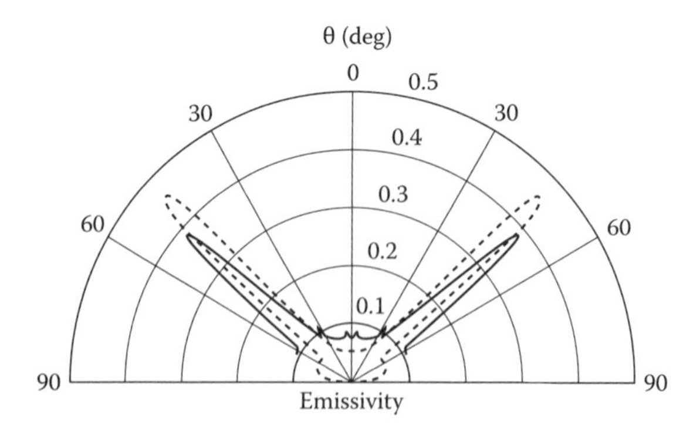

FIGURE 16.10 Emissivity of a SiC grating in p-polarization. Curve at 45° for λ = 11.86 μm. Emissivity deduced from specular reflectivity measurements using Kirchhoff's law. Data taken at ambient temperature using FTIR spectrometer as source and detector mounted on rotating arm. Dashed line with peak at 45° is based on theoretical calculations; the solid line is from the experimental data. (Courtesy of Professor J.-J. Greffet.)

(see, e.g., Greffet et al. 2002, Narayanaswamy and Chen 2004, 2005, Fu and Zhang 2005, Lee et al. 2005, Lee and Zhang 2006, 2007, Drevillon and Ben-Abdallah 2007, Huang et al. 2011, Zhang and Wang 2011, Hajimirza and Howell 2014c, Wang and Zhang 2014). In addition, the nanostructures on surfaces have an impact on far-infrared properties at far field (Ghua et al. 2012, Repheali et al. 2013, Didari and Mengüç 2015b), which allow the effective design and application of different materials to radiation cooling concepts.

The advances in nanotechnology have allowed researchers to design the materials for the desired functions of their devices. This approach allows engineers to optimize not only a process or device but also the materials used for this purpose. These new type of materials are called "metamaterials," and they may have negative electric permittivity or magnetic permeability or both, in case of double-negative metamaterials (DNGs). Of course, with these variations in electromagnetic properties, one can expect the tunability of the NFRT itself.

Joulain et al. (2010) were the first group to tackle this problem where they considered two identical metamaterials. The investigation of near-field radiative transfer in 1D media with any magnetic permeability and electric permittivity property was carried out by Zheng and Xuan (2011a,b). Both groups considered split-ring resonators (SRRs) and wires in their analyses. Note that an SRR is a metamaterial built-in laboratory. It allows strong magnetic coupling of the material to an applied electromagnetic field, such as exposure to light. In this case, negative permeability is produced with a periodic array of SRRs. Zheng and Xuan also considered the exchange between SRR and doped silica and aluminum. Metamaterials are new for engineering applications, and therefore they demand fundamental studies, including the determination of the penetration depth of spectral waves into their bulk. Basu and Francouer (2011) have reported a detailed study for the penetration depth of surface polariton-based NF radiation in SRR-wire metamaterials. One interesting observation they had was that magnetic and electric field penetration depths could be different. The same group extended this study to explore NF radiation exchange between two metamaterials made of potassium bromide (KBr) hosting SiC spheres (Francoeur et al. 2011). All of these studies suggested that TE-mode surface polaritons can also contribute to NF radiation exchange significantly, although all other natural materials contribute only at the TM-mode. In all these studies, effective medium theory is used, which shows some weakness in the case of studies involving nanostructures (Tschikin et al. 2013, Didari and Mengüç 2015b).

Biehs et al. (2011, 2013) discussed how the metamaterials can be considered as the near-field analog of the blackbody. The reason for this analogy is the broadband emission from metamaterials. Again, the same group has questioned if the properties of such broadband source can be obtained using the effective medium theory (Tschikin et al. 2013). On the other hand, not all applications demand broadband emission at all. Near-field radiative transfer is needed in selective fashion in some specific applications such as for energy harvesting using near-field radiative transfer at nano-thermophotovoltaic systems (Francoeur et al. 2011a,b,c). Spencer et al. (2013) presented an analysis to discuss the potential use of metamaterials for such application. They consider 3D metastructures and used the EMT theory and obtained a closed form analytical expression to obtain LDOS due to thermal emission from a metamaterials bulk. They also raised questions about the use of EMT in their analyses; yet, they stated that this approach was sufficiently accurate for understanding of general trends of metamaterials systems in use of near-field radiative transfer applications.

The research on metamaterials have been carried out by different groups as well. The effects of structures much smaller than the wavelength on the surface has been discussed by different researchers. These structures allow the manipulations of light–matter interactions as shown by Smith and Pendry (2004), Soukalis and Wegner (2011), Liu and Zhang (2011), Guha et al. (2012), Repheali et al. (2013), Liu et al. (2014), Wang et al. (2014), and Didari and Mengüç (2015a). It is interesting that the changes in the structural properties allow variations in the spectral nature of emission from even common, cheaper materials at far infrared. These material modifications are likely to impact major bigger scale energy conservation efforts, such as radiative cooling of buildings or electronic equipment.

16.7 CONCLUDING REMARKS

In this chapter, we have outlined the basic tools needed to understand and model radiative heat transfer between bodies separated by subwavelength distances. In this so-called near-field regime, the presence of evanescent and surface waves greatly modifies the radiant energy exchange as compared to the far-field regime that accounts only for the propagating waves. As opposed to the far-field regime where incoherent transport is assumed, the wave nature of thermal radiation needs to be accounted for in the near-field regime to include the effects radiation tunneling and coherence.

Mathematically, near-field thermal radiation problems can be solved using Maxwell equations combined with FE, where the source of thermal radiation emission is modeled as a stochastic current density. The link between the electromagnetic description of thermal radiation and heat transfer theory is provided by the FDT, which is restricted to media in local thermodynamic equilibrium. Using this set of equations, a general expression for the time-averaged Poynting vector has been derived, which is equivalent to the radiative heat flux. Evanescent and surface waves play a key role for the near-field effects. In this chapter, we discussed the materials and metamaterials supporting spatially and spectrally coherent surface polaritons in the infrared region, which greatly modify the near field of a thermal source. We have also outlined discussed analytical and numerical computational techniques for the radiative heat flux in 1D layered media. A limited number of examples, including for two SiC bulk materials, supporting SPhPs and separated by a vacuum gap have been presented, showing clearly that radiant energy exchanges can exceed by a few orders of magnitude the values predicted by Planck's blackbody distribution. We also pointed out in the discussion that quasi-monochromatic radiative heat transfer can be achieved in the near field when surface polaritons are excited.

We have provided a brief overview of recent near-field radiative heat transfer measurements and experimental evidence on the impact of near-field exchange on total heat transfer. Although radiative heat transfer in the near field was addressed long ago, multiple physical and engineering aspects still remain open research areas. Modeling of near-field thermal radiation problems under FE is based on the assumption that the media are in local thermodynamic equilibrium where a temperature can be defined. Therefore, the electromagnetic description of thermal radiation given in this chapter, based on FE and the FDT, is a theory applicable to macroscopic media that are separated by subwavelength distances. For nanosized objects where it becomes questionable to define a local temperature, the use of FE is also questionable. For such small bodies, it might become necessary to analyze the microscopic behavior of charge fluctuations along with nonequilibrium thermodynamic behavior (see the recent discussion of Pérez-Madrid et al. (2009) about near-field radiative heat transfer between two nanoparticles beyond the FDT). That problem is also addressed by Edalatpour and Francoeur (2014) who introduced a computational methodology (T-DDA), starting from the DDSCAT of Draine and Flatau (1994). They used fluctuational dissipation theory to model temperature of nanoparticles and were able to calculate the radiative exchange between the particles (see also Chapter 15).

Engineering applications often require modeling of radiation heat transfer in complex geometries. We have pointed out in this chapter that near-field thermal radiation problems have only been solved for simple geometries for which analytical expressions of the dyadic Green's function can be derived. Usually, effective medium theories are employed to tackle such problems, which may not reveal the rich physics to be observed due to nanostructures. There is a need to develop methodologies allowing solution of near-field thermal radiation in realistic structures. Such methods will allow us to tackle the physics of more sophisticated problems and also allow the application of near-field radiative heat transfer to diverse engineering problems. The exception to these closed form solutions are only recently suggested and summarized (Wen 2010, Didari and Mengüç 2015a, Otey et al. 2014, Wang et al. 2014). A number of studies on the ever-emerging field of near-field radiative transfer is reported on two recent workshops on micro- and nanoscale radiative transfer, which were held in Sandai, Japan, in 2012 and in Shanghai, China, in 2014. The papers from the first workshop was published in a special issue of JQSRT (Zhang et al. 2014), and the second yield another special issue in early 2015.

Research on near-field thermal radiation has so far mainly focused on understanding the different physical aspects of this very challenging problem. From the engineering point of view, near-field thermal radiation finds applications in energy conversion technologies, thermal imaging, and advanced nanomanufacturing processes. Moreover, understanding radiant energy exchanges at the nanoscale allows the engineering design of materials that selectively absorb and emit radiation. Near-field radiative heat transfer could also be potentially used as a novel technique for localized cooling of micro- and nanosized structures. Of course, these theoretical developments cannot be applied to advanced and innovative engineering solutions without extensive experimental validation. This requires further and detailed experiments, which should allow the precise determination of how near-field radiative transfer can be coupled with the material properties, geometry, and other modes of energy transfer.

Experimental studies on near-field exchange concepts are not extensively available in the literature, even though the body of the literature has been growing during the last decade. There is a limited number of experimental configurations considered so far. Surface–surface, sphere–sphere, and sphere–plate radiative transfer have been reported. Without additional and detailed fundamental experiments, the field cannot grow with solid physical support. It is expected that in the near future, more experiments will be conducted for different geometries, materials, and applications. Such studies will help the near-field radiative transfer concepts to play more significant role for the advanced energy-harvesting devices, nanomanufacturing, sensing, and imaging. The application of NFHT with other modes of heat transfer is also expected to grow, as discussed by Wong et al. (2014), who found that NFRT had to be included under specific physical conditions.

We anticipate that with the growing interest and advances in nanotechnologies, near-field radiation heat transfer is likely to be increasingly important and find numerous practical uses. Its many uses were recently summarized by Cahill et al. (2014).

HOMEWORK

16.1 Discuss the difference between surface waves and evanescent waves.

16.2 Consider a 20 nm gold film. You want to excite an SPP via a laser beam. What wavelength are you planning to use (you can give a range)? How would you excite the SPP (explain your setup)? Could you excite the SPP by illuminating the gold film with the propagating wave emanating from the laser? Why?

16.3 Starting with Equation 16.17, derive an expression for the energy density in terms of the mean energy of a Planck oscillator in thermal equilibrium and Weyl components of the dyadic Green's function similar to the expression for Poynting vector given by Equation 16.31.

Answer: $\langle u_\omega(\mathrm{r},T)\rangle = \dfrac{\omega K''(\omega)}{2\pi^2 c_0^2}\Theta(\omega,T)\displaystyle\int_0^\infty k_\rho dk_\rho \int_z dz' \left[k_0^2\left|g_{m\alpha}^E(\mathbf{r},\mathbf{r}',\omega)\right|^2 + \left|g_{m\alpha}^H(\mathbf{r},\mathbf{r}',\omega)\right|^2\right].$

16.4 Consider an interface delimiting two lossless dielectric media. A radiation beam of wavelength 633 nm in vacuum is incident at the interface 1–2 with an angle of incidence θ_i. Medium 2 is vacuum, while medium 1 is quartz with a refractive index of about 1.72 at 633 nm. What is the critical angle for TIR? For an angle of incidence θ_i of 50°, estimate the penetration depth of the evanescent wave in vacuum.

Answer: $\theta_{CR} = 35.5°$; $\delta_2 = 117$ nm.

16.5 Consider an interface delimiting a metal (medium 1) and a vacuum (medium 2). Using a lossless Drude model, $K_1(\omega) = 1 - \omega_p^2/\omega^2$, to describe the dielectric function of the metal, show that the resonant frequency of SPP is $\omega_p/\sqrt{2}$. For gold with a plasma frequency ω_p of 13.71×10^{15} rad/s, plot the dispersion relation [ω (rad/s) vs. k_x (rad/m)]. Identify in the figure the zones where the waves are propagating and evanescent in vacuum.

16.6 Starting with Equation 16.34 along with Equations 16.32 and 16.33, derive the components of Weyl representation of the electric DGF in terms of ρ, ϕ, and z.

16.7 Consider a SiC grating similar to that reported by Greffet et al. (2002) (see Figure 16.9). Comment on the structure of the grating and the emission wavelengths. List your suggestions for the use of such structures.

16.8 Do surface waves exist in TE polarization? Explain your answer starting from Maxwell equations. (Hint: Draw the TE- and TM-polarized waves on an interface and consider their physical importance.)

16.9 Assume that two SiC thin films are separated by a vacuum gap d where $T_1 = 400$ K and $T_2 = 300$ K. Calculate the near-field heat flux for different values of d and comment on your results.

16.10 Plot the dispersion relation at SiO_2–vacuum interface in the infrared region. Calculate the resonant frequency and identify transverse and longitudinal optical frequencies.
Answer: $\omega_{LO} = 1.825 \times 10^{14}$ rad/s, $\omega_{TO} = 1.494 \times 10^{14}$ rad/s

$$\omega_{res} = \sqrt{\frac{\left(\bar{\varepsilon}_\infty \omega_{LO}^2 + \omega_{TO}^2\right)}{(1+\bar{\varepsilon}_\infty)}} = 1.786 \times 10^{14} \text{ rad/s}$$

where $\bar{\varepsilon}_\infty = 6.7$.

17 Radiative Effects in Translucent Solids, Windows, and Coatings

17.1 INTRODUCTION

In the previous chapters, most of the discussion of radiative transfer was for materials that have a simple refractive index $n \approx 1$ (except in Chapter 16). This approximation is valid when the absorbing, emitting, and scattering medium is a gas, since almost all gases have a refractive index very close to 1 (Lide 2008) (Table 17.1). However, many important effects result from the refractive indices $n > 1$ possessed by many common materials, which is our focus in this chapter. The most obvious application is for predicting reflection and refraction profiles at an interface as formulated in Chapters 3 and 14. These phenomena are very important in predicting radiative transfer in translucent coatings, thin films, and multiple windows. Also, local blackbody emission within a medium increases by a factor of n^2, as discussed in Section 1.5.8.

The reflection, transmission, and absorption of solar and environmental radiation through single- and multilayered windows are very important for determining heating and cooling loads of buildings, designing flat-plate solar collectors, and many different optical devices. Multiple reflections from window surfaces can be appreciable in reducing transmission, especially if more than one window layer is present, as in a two-cover-plate solar collector. Transmission losses through multiple windows can be significant and relations for their calculation are required. Transient temperature distributions are important during manufacturing operations for glass and other translucent materials where there is internal absorption and emission of radiation in the heated materials, combined with heat conduction.

If one or more reflecting translucent layers are directly attached to an opaque surface or to other reflecting translucent layers, a coating system can be formed that has desirable radiative properties. A coating can be either thick or thin compared with the wavelength of the incident radiation. Thin films produce wave interference effects between incident and reflected waves in the film, thus influencing their reflection and transmission characteristics. Thin-film coating systems with high or low reflectivity can also be fabricated. A composite window can be formed that is selectively transmitting, or a coated surface that is selectively absorbing can be produced. Examples are the transparent heat mirror and the solar selective surfaces discussed in Section 3.6. Films that are thick relative to the radiation wavelength can also be used to modify surface radiative characteristics.

The formulations in this chapter provide the necessary analytical tools for treating radiation transfer in enclosures with windows and for determining the radiative behavior of coatings. Since windows and coatings are often thin, temperature variations within them are not considered in part of the development presented here. There are applications, however, in which the temperature distribution must be obtained within a translucent medium, including the effects of interface reflections. This requires solving the transfer equations within the medium, as in Chapter 10, with the addition of reflection boundary conditions at the interfaces. Some important applications are heat treating of glass plates, temperature distributions in glass-melting tanks, high-temperature solar components, laser heating of windows and lenses, heating of spacecraft and aircraft windows, radiation-scattering heat shields, and some thermal barrier coatings. The theory for these applications is in Sections 17.5 and 17.6.

TABLE 17.1
Refractive Indices of Some Common Substances

Material	Refractive Index n
Air (at $\lambda = 0.589$ μm)	1.00028
Liquids (at $\lambda = 0.589$ μm)	
Chlorine	1.3834
Ethyl alcohol	1.36 (25°C)
Oxygen	1.2243
Water	1.334–1.318 (0°C–100°C)
Solids (at λ given)	
Glass:	
Crown	1.50–1.53 (0.36–1 μm)
Light flint	1.56–1.62 (0.36–1 μm)
Heavy flint	1.71–1.81 (0.36–1 μm)
Ice	1.31 (0.589 μm)
Quartz	1.45–1.47 (0.36–1 μm)
Rock salt	1.61–1.532 (0.30–1 μm)

Source: Lide, D.R. (ed.), *Handbook of Chemistry and Physics*, 88th edn., CRC Press, Boca Raton, FL, 2008.

17.2 TRANSMISSION, ABSORPTION, AND REFLECTION OF WINDOWS

Enclosures can have windows that are partially transparent to radiation. A window can be of a single material, or it can have one or more transmitting coatings on it. The transparency of a window is a function of the glass, plastic, and coating thicknesses and can be strongly wavelength dependent, as illustrated by the transmittance of glass in Figures 9.20 and 9.21. Some applications are glass or plastic cover plates for flat-plate solar collectors and thick- and thin-film coatings to provide modified reflection and transmission properties for camera lenses and solar cells. This section is concerned with the transmission of incident radiation as modulated by surface reflections and absorption within the window. Scattering is not considered in these applications.

As discussed in Section 3.2.1 and illustrated in Figure 17.1, radiation incident on a surface is reflected and refracted. For a layer of thickness D, the refracted portion travels a distance $D/\cos\chi$ and is then partially reflected from the inside of the second surface. To analyze the radiative behavior of a layer such as in Figure 17.1, the reflectivities are needed for radiation striking the outside and the inside of an interface. For a smooth interface, reflectivity relations are given by Equations 3.4 and 3.5, since windows are often dielectrics with a small extinction (absorption) index, k. Since these relations depend only on the square of the terms containing the angles $(\theta - \chi)$, the θ and χ can be interchanged, and hence along the incident and refracted paths, the reflectivity is the same for radiation incident on an interface either from the outside or from within the material. For a constant absorption coefficient κ within the material, the transmittance along the path length within the window is $t = e^{-\kappa D/\cos\chi}$, where κ is a function of wavelength as shown by the transmission spectrum of glass in Figures 9.21 and 9.22 (for simplicity in notation, the functional dependence on λ is omitted here as well as in most of the formulations that follow).

Fresnel reflection relations in Equations 3.4 and 3.5 show that the reflectivity at a smooth interface not only is a function of incidence angle but is also different for each of the two components of polarization (see Hecht 2002). In what follows, the path of radiation will be traced through partially transparent windows where the interface reflections are specular (i.e., mirrorlike). For precise results, the resulting formulas are applied at each incidence angle and for each component

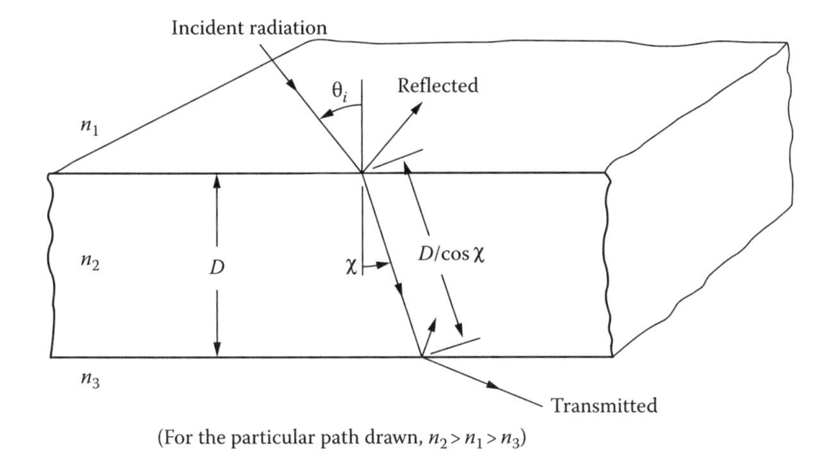

FIGURE 17.1 Reflection and transmission of incident radiation by a partially transmitting layer.

of polarization. If the incident radiation is nonpolarized, half the energy is in each component of polarization. For diffuse incident radiation, the fraction of energy incident in each θ direction within the increment $d\theta$ is $2\sin\theta\cos\theta d\theta$. Then, when integrating to find the total reflectivity, the results for each direction are weighted in the integration according to the amount of incident energy in each $d\theta$ at θ and in each component of polarization.

17.2.1 Single Partially Transmitting Layer with Thickness $D \gg \lambda$ (No Wave Interference Effects)

When a window or transmitting layer is very thin, so that its thickness is comparable to the radiation wavelength, there can be interference between incident and reflected waves. This is discussed later. Here, consider a window pane where D is at least several wavelengths thick so that the interference effects need not be considered. We consider the ray-tracing method and then the net-radiation method to predict transmission through windows in the next sections.

17.2.1.1 Ray-Tracing Method

Referring to Figure 17.2, consider a unit intensity incident on the upper boundary, and apply ray-tracing methods. At contact with the first interface, a fraction ρ is reflected so that the fraction $(1-\rho)$ of the incident energy enters into the material. Of this, the fraction $(1-\rho)\tau$ is transmitted (hence, $(1-\rho)(1-\tau)$ is absorbed along the path) to interface 3 where $\rho(1-\rho)\tau$ is reflected, and consequently

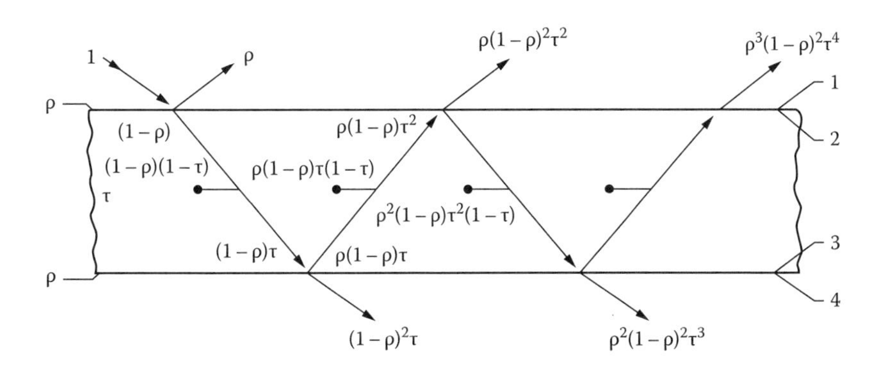

FIGURE 17.2 Multiple internal reflections for radiation incident on a window.

$(1-\rho)^2\tau$ passes out of the window through the lower boundary. As the process continues, the fraction of incident energy reflected by the window is the sum of the terms leaving surface 1:

$$R=\rho\left[1+(1-\rho)^2\tau^2\left(1+\rho^2\tau^2+\rho^4\tau^4+\cdots\right)\right]=\rho\left[1+\frac{(1-\rho)^2\tau^2}{1-\rho^2\tau^2}\right] \tag{17.1}$$

The fraction transmitted is the sum of terms leaving surface 4:

$$T=\tau(1-\rho)^2\left[1+\rho^2\tau^2+\rho^4\tau^4+\cdots\right]=\frac{\tau(1-\rho)^2}{1-\rho^2\tau^2}=\tau\frac{1-\rho}{1+\rho}\frac{1-\rho^2}{1-\rho^2\tau^2} \tag{17.2}$$

The last factor on the right, $\frac{1-\rho^2}{1-\rho^2\tau^2}$, is often close to 1. Therefore, $T\approx\tau(1-\rho)/(1+\rho)$. The fraction of energy absorbed is

$$A=(1-\rho)(1-\tau)\left[1+\rho\tau+\rho^2\tau^2+\rho^3\tau^3+\cdots\right]=\frac{(1-\rho)(1-\tau)}{1-\rho\tau} \tag{17.3}$$

In the limit when absorption in the window can be neglected, $\tau=1$, $R=2\rho/(1+\rho)$, $T=(1-\rho)/(1+\rho)$ and $A=0$. In each of these relations, ρ and τ are functions of the incident (or refracted) angle.

17.2.1.2 Net-Radiation Method

From the enclosure theory presented in Chapter 5, it is evident that the net-radiation method is a powerful analytical tool that can, in many situations, be much less difficult to apply than the ray-tracing method. The net-radiation method (Siegel 1973a) can be used to derive the radiation characteristics of a partially absorbing window. Referring to Figure 17.3, the outgoing flux at each interface can be written in terms of the incoming fluxes to yield the following equations for the conditions of a unit incoming flux from a single direction at surface 1 and a zero incoming flux at surface 4:

$$J_1=\rho G_1+(1-\rho)G_2=\rho+(1-\rho)G_2 \tag{17.4}$$

$$J_2=(1-\rho)G_1+\rho G_2=(1-\rho)+\rho G_2 \tag{17.5}$$

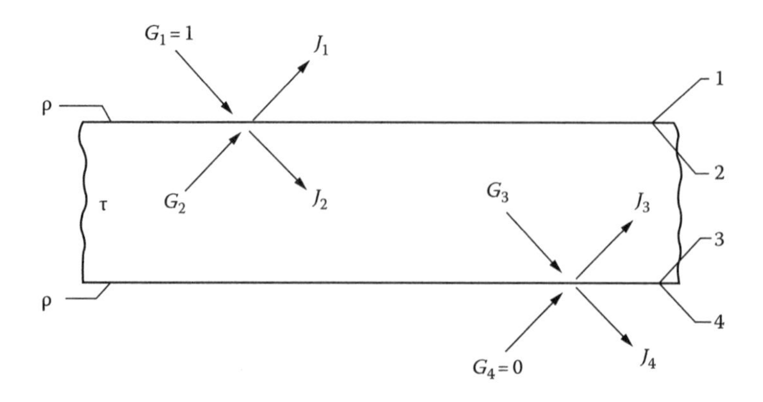

FIGURE 17.3 Net-radiation method applied to partially transmitting layer.

$$J_3 = \rho G_3 + (1-\rho)G_4 = \rho G_3 \tag{17.6}$$

$$J_4 = (1-\rho)G_3 + \rho G_4 = (1-\rho)G_3 \tag{17.7}$$

The transmittance of the layer is used to relate the internal incident radiative flux G and outgoing radiative flux (radiosity) J to give $G_2 = J_3\tau$ and $G_3 = J_2\tau$. These are used to eliminate the G from Equations 17.4 and 17.5, and the resulting equations are solved for the J to yield

$$J_1 = \rho\left[1 + \frac{(1-\rho)^2\tau^2}{1-\rho^2\tau^2}\right]; \quad J_2 = \frac{1-\rho}{1-\rho^2\tau^2}$$

$$J_3 = \frac{\rho\tau(1-\rho)}{1-\rho^2\tau^2}; \quad J_4 = \frac{\tau(1-\rho)^2}{1-\rho^2\tau^2}$$

For $G_1 = 1$, the fractions reflected and transmitted by the plate are J_1 and J_4 so that

$$R = J_1 = \rho\left[1 + \frac{(1-\rho)^2\tau^2}{1-\rho^2\tau^2}\right] = \rho(1+\tau T) \tag{17.8}$$

$$T = J_4 = \frac{\tau(1-\rho)^2}{1-\rho^2\tau^2} = \tau\left(\frac{1-\rho}{1+\rho}\right)\left(\frac{1-\rho^2}{1-\rho^2\tau^2}\right) \tag{17.9}$$

The fraction absorbed is

$$A = (J_2 + J_3)(1-\tau) = \frac{(1-\rho)(1-\tau)}{1-\rho\tau} \tag{17.10}$$

These results agree, as they should, with those obtained by the ray-tracing method. If the ρ at the upper and lower surfaces are not equal, the results for R and T are left as an exercise in Homework Problem 17.6. These relations were used by Nicolau and Maluf (2001) to determine the properties of commercial tinted glass.

Example 17.1

What is the fraction of externally incident unpolarized radiation that is transmitted through a glass window in air? The window is $D = 0.75$ cm thick, radiation is incident at $\theta = 50°$, $n_{glass} = 1.53$, and $\kappa_{glass} = 0.1\ \text{cm}^{-1}$.

To find the path length through the glass, evaluate $\chi = \sin^{-1}(\sin\theta/n) = \sin^{-1}(\sin 50°/1.53) = 30°$. The path length is $S = 0.75/\cos\chi = 0.866$ cm. The transmittance is $\tau = \exp(-\kappa S) = \exp(-0.1 \times 0.866) = 0.917$. The surface reflectivities for the two components of polarization are

$$\rho_\parallel = \frac{\tan^2(\theta-\chi)}{\tan^2(\theta+\chi)} = 0.00412; \quad \rho_\perp = \frac{\sin^2(\theta-\chi)}{\sin^2(\theta+\chi)} = 0.1206$$

Then, the overall transmittance for each component is

$$T_\parallel = \tau\frac{1-\rho_\parallel}{1+\rho_\parallel}\frac{1-\rho_\parallel^2}{1-\rho_\parallel^2\tau^2} = 0.917\frac{0.9959}{1.0041}\frac{1-0.00002}{1-0.00001} = 0.9095$$

$$T_\perp = \tau \frac{1-\rho_\perp}{1+\rho_\perp} \frac{1-\rho_\perp^2}{1-\rho_\perp^2 \tau^2} = 0.917 \frac{0.8794}{1.1206} \frac{1-0.0145}{1-0.0122} = 0.7179$$

For unpolarized incident radiation, one-half the energy is in each component. Hence, $T = (T_\parallel + T_\perp)/2 = 0.814$.

17.2.2 Multiple Parallel Windows

Consider now more complicated multilayer formulations. We analyze a system of m and n plates by the net-radiation method. Different layers with different properties separated by air or a vacuum are analyzed such as for double or triple glazed windows. Using the notation in Figure 17.4, the outgoing radiation terms J are written in terms of the incoming radiation fluxes G as

$$J_{m1} = R_m + G_{m2} T_m \tag{17.11}$$

$$J_{m2} = G_{m2} R_m + T_m \tag{17.12}$$

$$J_{n1} = G_{n1} R_n \tag{17.13}$$

$$J_{n2} = G_{n1} T_n \tag{17.14}$$

The G are further related to the J by using the relations $G_{m2} = J_{n1}$ and $G_{n1} = J_{m2}$. The G are then eliminated, and solving for the J yields

$$T_{m+n} = J_{n2} = \frac{T_m T_n}{1 - R_m R_n} \tag{17.15}$$

$$R_{m+n} = J_{m1} = R_m + \frac{R_n T_m^2}{1 - R_m R_n} \tag{17.16}$$

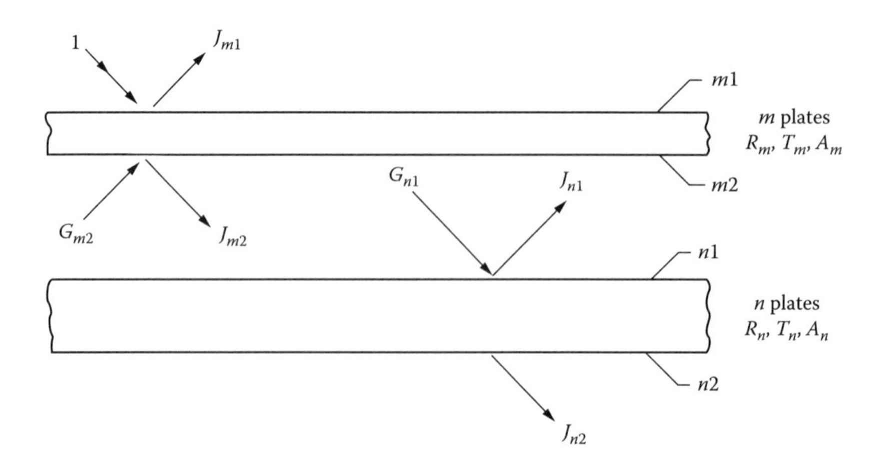

FIGURE 17.4 Net-radiation method for the system of multiple parallel transmitting plates.

The fractions of energy absorbed in the group of m plates and in the group of n plates within the system of $m + n$ plates are

$$A_m^{(m+n)} = A_m + G_{m2}A_m = A_m\left(1 + \frac{T_m R_n}{1 - R_m R_n}\right) \tag{17.17}$$

$$A_n^{(m+n)} = G_{n1}A_n = \frac{T_m A_n}{1 - R_m R_n} \tag{17.18}$$

The $A_m^{(m+n)}$ is the absorption in the system of m plates when it is a part of a total system of $m + n$ plates. The A_m is the absorption in a system of m plates that is by itself and is not a part of a larger system.

Note that T_{m+n} shows symmetry; that is, the m and n subscripts can be exchanged and the expression remains the same; hence, $T_{m+n} = T_{n+m}$. The T_{m+n} is the transmission for the system of $m + n$ plates for radiation incident first on the m plates, while T_{n+m} is for incidence first on the n plates. From Equation 17.16, however, $R_{m+n} \neq R_{n+m}$; the system reflectance depends on whether the radiation is incident first on the group of m plates or on the group of n plates.

If there is a stack of N plates that can all be different, as in Figure 17.5, then the fraction of incident energy absorbed by the top plate (a system of 1 plate) is, from Equation 17.17

$$A_1^{(N)} = A_1\left(1 + \frac{T_1 R_n}{1 - R_1 R_n}\right) \tag{17.19}$$

where n is the system composed of plates: 2, 3, … N. The fraction absorbed in plate 2 is then

$$A_{in\ plate\ 2}^{(N)} = A_m^{(N)} - A_1^{(N)} = A_m\left(1 + \frac{T_m R_n}{1 - R_m R_n}\right) - A_1^{(N)} \tag{17.20}$$

where

m is the system composed of plates: 1 and 2

n is the system composed of plates: 3, 4, … N

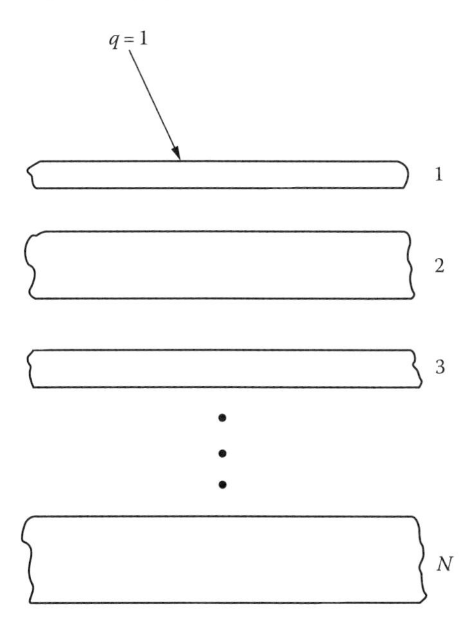

FIGURE 17.5 A stack of N partially transmitting parallel plates.

This process can be continued to determine the amount of radiant energy absorbed in each plate. The absorption in each plate is analyzed by Edwards (1977) for a system of partially transmitting parallel plates with a parallel opaque absorber plate.

17.2.3 Transmission through Multiple Parallel Glass Plates

Transmission through multiple glass plates is of interest in the design of flat-plate solar collectors. The glass surface reflectivity, as given by Equations 3.4 and 3.5, depends on the angle of incidence and the component of polarization. Since reflections from within the glass are assumed to be specular, the same angles of reflection and refraction are maintained throughout the multiple reflection process. Equation 17.15 can be used to calculate the overall transmittance through one and three parallel glass plates with an index of refraction $n = 1.5$. Neglecting absorption within the glass, results from Shurcliff (1974) are in Figure 17.6 for the two components of polarization and as a function of the radiation incidence angle θ. As $\theta \rightarrow 90°$, the transmission goes to zero; this is because dielectrics have perfect reflectivity at grazing incidence (Figure 3.2). For incidence at Brewster's angle, the overall transmittance for the parallel component becomes 1, as there is zero reflection at this angle and losses by absorption are being neglected.

Incident solar radiation is unpolarized and hence has equal energy in the parallel and perpendicular components. The transmittance is then the average of the two transmittance values computed by individually using $\rho_{\parallel}$ and $\rho_{\perp}$. This is shown in Figure 17.7 for the limiting case of nonabsorbing plates and for absorbing plates having a product of absorption coefficient

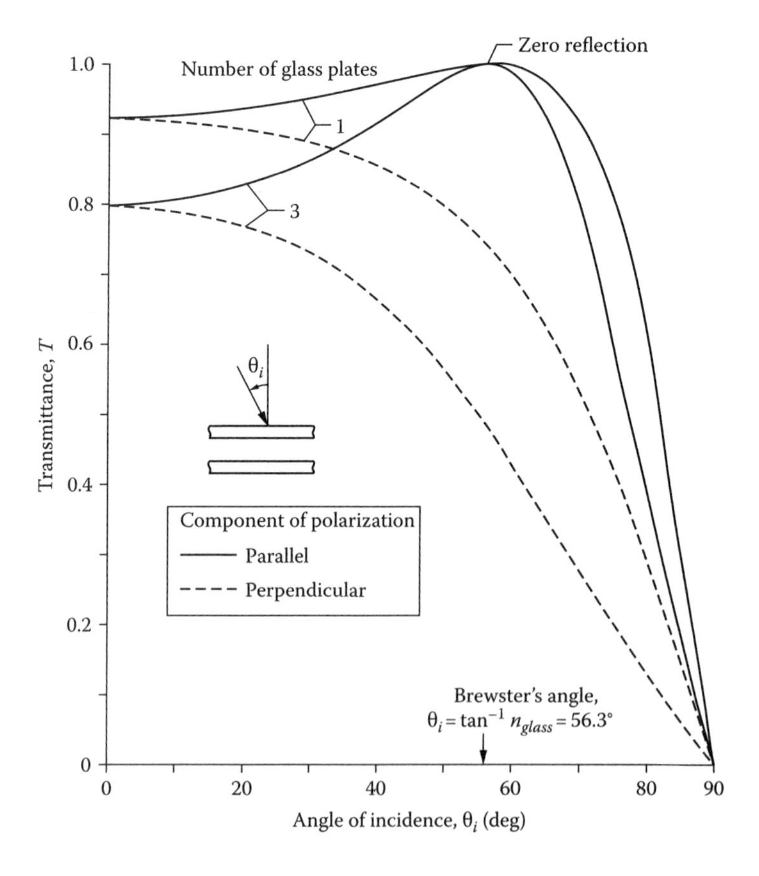

FIGURE 17.6 Overall transmittance of radiation in two components of polarization for nonabsorbing parallel glass plates; $n_{glass} = 1.5$. (From Shurcliff, W.A., *Solar Energ.*, 16, 149, 1974.)

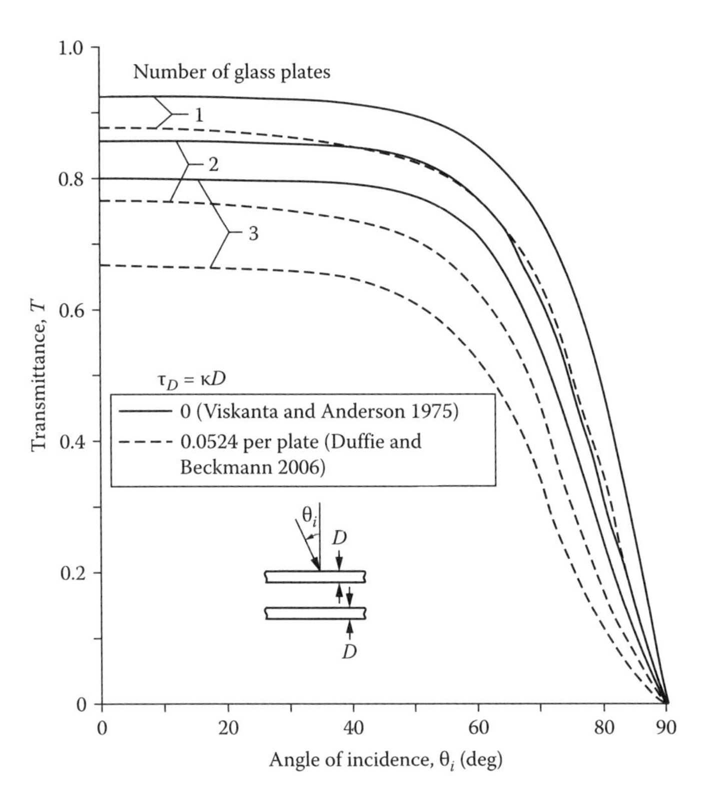

FIGURE 17.7 Effect of incidence angle and absorption on overall transmittance of multiple parallel glass plates; $n_{glass} = 1.5$.

and thickness of 0.0524 per plate. For angles near normal incidence, the effect of absorption reduces transmission by about 5% for each plate.

Reflection and transmission as a function of wavelength are important for the proper choice of window glasses for buildings. Both architectural constraints (effects of window spectral properties as they affect both glass color and the changes in observed color when looking through the windows) and thermal engineering constraints from the point of view of energy efficiency dictate the need for such information.

17.2.4 Interaction of Transmitting Plates with Absorbing Plate

A flat-plate solar collector usually consists of one or more parallel transmitting windows covering an opaque absorber plate, as in Figure 17.8. It is desired to obtain the fraction A_c of incident energy that is absorbed by the opaque collector plate. At the collector plate, the absorbed flux is

$$A_c = G_c - J_c \tag{17.21}$$

$$J_c = (1 - \alpha_c) G_c \tag{17.22}$$

Across the space between the transmitting plates and the opaque collector plate

$$J_c = G_n \tag{17.23}$$

$$J_n = G_c \tag{17.24}$$

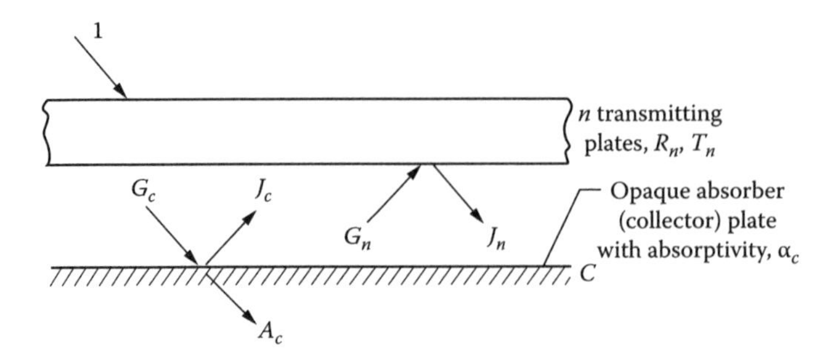

FIGURE 17.8 Interaction of transmitting windows and an absorbing collector plate.

For the system of n transmitting plates,

$$J_n = T_n + G_n R_n \tag{17.25}$$

The system of Equations 17.21 through 17.25 is solved to yield the fraction of incident energy absorbed by the opaque collector plate:

$$A_c = \frac{\alpha_c T_n}{1-(1-\alpha_c)R_n} \tag{17.26}$$

So far, the fundamentals have been developed for analyzing the reflection, transmission, and absorption behavior of window systems for incident radiation, and some numerical results have been given. Many additional aspects are treated in the literature (Siegel 1973a,b, Shurcliff 1974, Viskanta and Anderson 1975, Wijeysundera 1975, Edwards 1977, Viskanta et al. 1978, Mitts and Smith 1987, Duffie and Beckman 2006). In Mitts and Smith, the model includes both beam and diffuse incident energy and the conversion of beam to diffuse energy that occurs in some of the semitransparent layers.

With regard to radiative transfer between parallel surfaces, there are additional effects when reflecting layers have a very close spacing. These were examined by Cravalho et al. (1967) for the geometry in Figure 17.9 of two semi-infinite dielectric media having refractive indices n_1 and

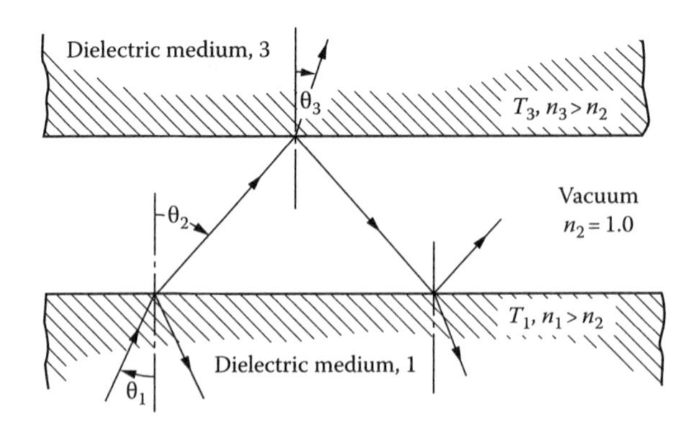

FIGURE 17.9 Reflection and transmission of electromagnetic wave in gap between two dielectrics; $T_3 < T_1$.

n_3, separated by a vacuum gap. In the usual analysis for radiative transfer between two spectrally dependent surfaces such as surface 1 and surface 3, the heat flux transferred across the gap is given by Equation 6.1.2 as $q_1 = \int_0^\infty \{[E_{\lambda b,1}(T_1) - E_{\lambda b,3}(T_3)]/[1/\varepsilon_{\lambda,1}(T_1) + 1/\varepsilon_{\lambda,3}(T_3) - 1]\} d\lambda$, and the spacing between the surfaces does not appear. When the spacing is very small, however, two effects enter that depend on spacing. One effect is wave interference, where a wave reflecting back and forth between two dielectrics may undergo cancellation or reinforcement as described in Section 17.4.2 for thin film layers.

For ordinary behavior at an interface as in Figure 17.9, some of the radiation in medium 1 traveling toward region 2 will undergo total internal reflection at the interface when $n_1 > n_2$, as discussed in Section 17.5.2. For ordinary radiative behavior, this occurs when the incidence angle θ_1 is equal to or larger than the angle for total reflection, $\theta_1 \geq \sin^{-1}(n_2/n_1)$. When region 2 in Figure 17.9 is sufficiently thin, however, electromagnetic theory predicts that, even for an intensity incident at θ_1 greater than $\sin^{-1}(n_2/n_1)$, total internal reflection does not occur. Rather, part of the incident intensity propagates across the thin region 2 and enters medium 3. This effect is the essence of radiation tunneling as discussed in Chapter 16. This tunneling stems from the surface due to electron or lattice vibrations. They are called either surface plasmon polaritons (SPP, for electrons) or surface phonon polaritons (SPhP, for phonons). The effect has been verified experimentally (Xu et al. 1994) and has been used in producing photon scanning tunneling microscopes and near-field optical thermometers that can have subwavelength resolution (Goodson and Ashegi 1997). Additional details about recent developments are given in Chapter 16.

Malcolm (2009) studied the augmented near-field enhancement of radiative transfer between parallel plates as affected by plate spacing, and the results are in Figure 17.10. It is seen that radiative transfer is predicted to be greatly enhanced by near-field effects at very small spacing. More detailed information on near-field radiation transfer including the effects of surface plasmons and surface phonon polaritons is in Chapter 16.

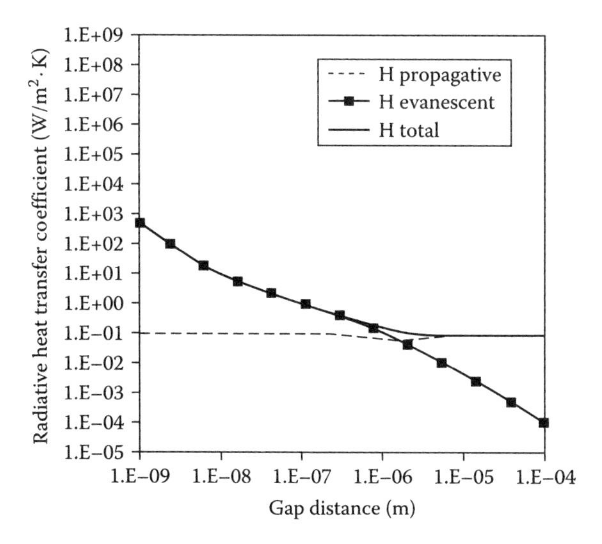

FIGURE 17.10 Effect of plate spacing near-field (evanescent) radiation on radiative transfer between parallel plates as embodied in the radiative heat transfer coefficient: plate 1 aluminum and plate 2 amorphous silicon dioxide. (From Malcolm, N.P., Simulation of a plasmonic nanowire waveguide, MS thesis, Department of Mechanical Engineering, The University of Texas at Austin, Austin, TX, May 2009.)

17.3 ENCLOSURE ANALYSIS WITH PARTIALLY TRANSPARENT WINDOWS

The enclosures considered in Chapters 5 through 8 have opaque walls or one or more openings such as at the end of a cylindrical cavity. A more general enclosure may contain partially transmitting windows, as in Figure 17.11. Only a simplified case for such an enclosure is considered here; more information is in Siegel (1973a). For this simplified case, the window properties are assumed independent of wavelength, and the radiation transmitted through and reflected from the window is assumed diffuse. A window such as a smooth glass plate reflects specularly, so the reflected portion of the energy incident from each direction will not leave the surface in a diffuse manner. However, within an enclosure, there are usually many multiple reflections, and the directionality of each reflection loses its importance in contributing to the energy fluxes at the boundaries. Hence, the assumption of diffuse reflection is often satisfactory even though the enclosure has multiple surfaces including smooth windows. When a window is hot enough to radiate appreciably, the analysis is restricted to a window that is thin enough so that it is essentially at a uniform temperature throughout. A symmetric window is considered, so the radiative properties are the same on both sides; for example, if the window is coated, the same coating is on both sides. A two-band semigray analysis is given by Siegel (1973a) to account for the large difference in spectral transmission properties on either side of the infrared cutoff wavelength, as shown for glass in Figure 9.22.

For the energy flux quantities in Figure 17.11, an overall energy balance at window k yields

$$q_k = J_k - G_k + q_{l,k} - q_{e,k} \tag{17.27}$$

The q_k is the energy flux supplied to the window by a means other than radiation from *both* the inside and outside of the window, such as by convective heating and/or by heating wires within the semitransparent window. For convective *cooling* of the window, the contribution to q_k is negative. The radiative flux leaving the inside surface of the window consists of emitted energy, reflected incoming energy, and transmitted externally incident flux:

$$J_k = E_{w,k}\sigma T_k^4 + R_{w,k}G_k + T_{w,k}q_{e,k} \tag{17.28}$$

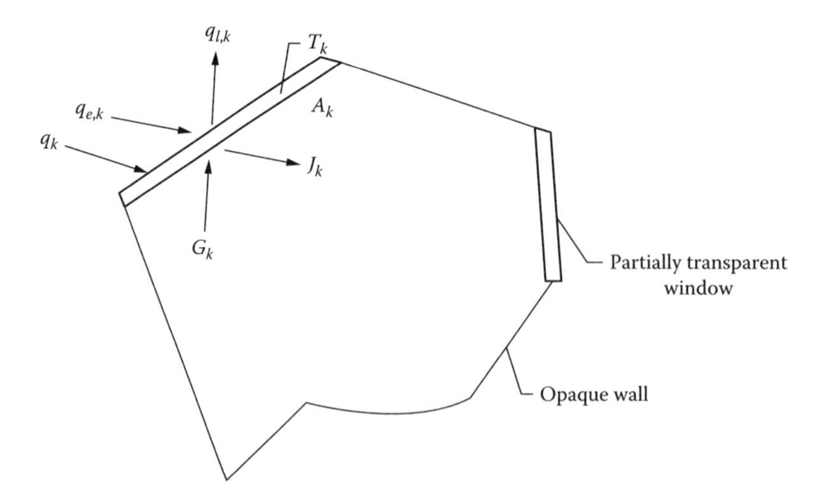

FIGURE 17.11 Enclosure with partially transparent windows.

where E_w, R_w, and T_w are the overall emittance, reflectance, and transmittance of the window (T without a w subscript is the window temperature). Similarly, the radiative flux leaving the outside surface of the window is

$$q_{l,k} = \tilde{E}_{w,k}\sigma T_k^4 + R_{w,k}q_{e,k} + T_{w,k}q_{i,k} \tag{17.29}$$

For a gray window, $\tilde{E}_{w,k} = A_{w,k} = 1 - T_{w,k} - R_{w,k}$. The use of an emittance implies that the window temperature can be considered uniform throughout its thickness. The $q_{l,k}$ is eliminated from Equations 17.27 and 17.29 to give

$$J_k = q_k - \tilde{E}_{w,k}\sigma T_k^4 + (1 - T_{w,k})G_k + (1 - R_{w,k})q_{e,k} \tag{17.30}$$

where q_k and $q_{e,k}$ are prescribed. The incoming flux is obtained from the usual enclosure relation in terms of outgoing fluxes and configuration factors:

$$G_k = \sum_{j=1}^{N} J_j F_{k-j} \tag{17.31}$$

Equations 17.28, 17.30, and 17.31 provide three equations relating J, G, and T for each partially transparent surface.

The G_k can be eliminated to reduce the three relations to two equations. The first equation is obtained by solving Equation 17.28 for G_k and inserting it into Equation 17.30. The second results from substituting Equation 17.31 into Equation 17.30. After rearrangement, the two relations are

$$q_k - \tilde{E}_{w,k}\sigma T_k^4 + (1 - R_{w,k})q_{e,k} = \frac{\tilde{E}_{w,k}}{R_{w,k}}\left[\left(1 - T_{w,k}\right)\sigma T_k^4 - J_k\right] + \left(1 - T_{w,k}\right)\frac{T_{w,k}}{R_{w,k}}q_{e,k} \tag{17.32}$$

$$q_k - \tilde{E}_{w,k}\sigma T_k^4 + (1 - R_{w,k})q_{e,k} = J_k - (1 - T_{w,k})\sum_{j=1}^{N} J_j F_{k-j} \tag{17.33}$$

The left sides are the external heat input to the inside surface of the window. This is q_k minus the emission leaving through the outside surface and augmented by the amount of external radiation that passes into the window. For an opaque wall, q_k was defined to include all the energy quantities other than radiation at the surface inside the enclosure. Hence, for an opaque wall, the left sides of Equations 17.32 and 17.33 become q_k, and, with the window transmittance $T_w = 0$ on the right side, the equations become the same as Equations 5.16 and 5.17. By eliminating J from Equations 17.32 and 17.33, a result that directly reduces to Equation 5.21 is obtained:

$$\sum_{j=1}^{N} \frac{1}{\tilde{E}_{w,j}}\left(\delta_{kj} - R_{w,j}F_{k-j}\right)\left[q_j - \tilde{E}_{w,j}\sigma T_j^4 + \left(1 - R_{w,j}\right)q_{e,j}\right]$$
$$= \sum_{j=1}^{N}\left(\sigma T_j^4 + \frac{T_{w,j}}{\tilde{E}_{w,j}}q_{e,j}\right)\left[\delta_{kj} - \left(1 - T_{w,j}\right)F_{k-j}\right] \tag{17.34}$$

where

T_j is the temperature

$T_{w,j}$ is the window transmittance

17.4 EFFECTS OF COATINGS OR THIN FILMS ON SURFACES

The radiative behavior of a surface can be modified by depositing on it one or more very thin layers of other materials. The coatings can be dielectric or metallic and can be thick or thin relative to the radiation wavelength. It is possible to obtain high or low absorption by the coated surface, and its characteristics will depend on the radiation wavelength. Thus, it is possible to use thin-film coatings to tailor surfaces to have a desired wavelength-selective behavior.

17.4.1 Coating without Wave Interference Effects

The geometry in Figure 17.12 consists of a coating of thickness D on a thick substrate. The coating has transmittance τ and reflectivities ρ_1 and ρ_2 at the first and second interfaces. The film is thick enough for it not to be necessary to consider interference between waves reflected from the two interfaces. It will be determined to what extent the film alters the reflection characteristics from that of the substrate alone. By the net-radiation method, as derived for multiple layers, the fraction of incident radiation that is reflected is

$$R = \frac{\rho_1 + \rho_2(1-2\rho_1)\tau^2}{1-\rho_1\rho_2\tau^2} \tag{17.35}$$

This is used to calculate the behavior of a few different types of coatings.

17.4.1.1 Nonabsorbing Dielectric Coating on Nonabsorbing Dielectric Substrate

Consider the relatively simple case of a dielectric film on a dielectric substrate. There is incident radiation in the normal direction through a surrounding dielectric medium with a refractive index n_s. The film and substrate have refractive indices n_1 and n_2. Although the film is thick relative to the radiation wavelength, it is still physically quite thin, and since it is also a dielectric, the effect of absorption within it is assumed very small. Then $\tau \cong 1$, and Equation 17.35 reduces to

$$R = \frac{\rho_1 + \rho_2(1-2\rho_1)}{1-\rho_1\rho_2} \tag{17.36}$$

For normal incidence, the interface reflectivities are given by ($k_1 = k_2 = k_s \approx 0$):

$$\rho_1 = \left(\frac{n_1 - n_s}{n_1 + n_s}\right)^2 \tag{17.37}$$

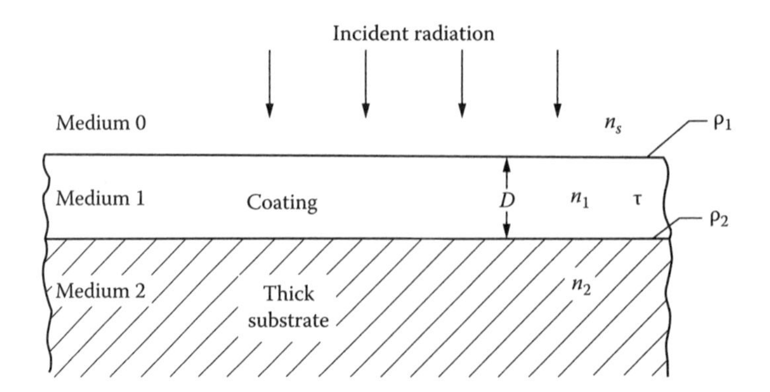

FIGURE 17.12 Coating of thickness D on a thick substrate.

$$\rho_2 = \left(\frac{n_2 - n_1}{n_2 + n_1}\right)^2 \tag{17.38}$$

Substituting Equations 17.37 and 17.38 into Equation 17.36 gives, after simplification,

$$R = 1 - \frac{4n_s n_1 n_2}{(n_1^2 + n_s n_2)(n_s + n_2)} \tag{17.39}$$

Dielectric coatings can be used to provide reduced reflection at the surface and hence maximize the radiation passing into the substrate. The proper n_1 to minimize reflection is obtained by letting $dR/dn_1 = 0$. This yields

$$n_1 = \sqrt{n_s n_2} \tag{17.40}$$

Using this n_1 in Equation 17.39 gives the R for minimum reflection:

$$R = 1 - \frac{2\sqrt{n_s n_2}}{n_s + n_2} \tag{17.41}$$

If there is no coating, the reflectivity of the substrate by itself for normal incidence is $\rho_{sub} = (n_2 - n_s)^2/(n_2 + n_s)^2$. The ratio of the minimum R to ρ_{sub} can be simplified to

$$\frac{R}{\rho_{sub}} = \frac{n_s + n_2}{n_s + 1\sqrt{n_s n_2} + n_2} \tag{17.42}$$

For an optimum antireflection coating on a glass substrate in air, $n_s \approx 1$ and $n_2 \approx 1.53$, which yields $R/\rho_{sub} = 0.506$. Thus for these materials, the dielectric coating can, at best, reduce surface reflection to about half the uncoated value. As will be shown, much better results can be obtained by using thin films. The square root of $n_2 = 1.53$ is 1.24, which from Equation 17.40 is the optimum n_1, and it is difficult to find a suitable coating material with a refractive index this low. Some commonly used materials are magnesium fluoride, $n = 1.38$, or cryolite (sodium aluminum fluoride), $n = 1.36$. Also used are lithium fluoride, $n = 1.36$, and aluminum fluoride, $n = 1.39$.

17.4.1.2 Absorbing Coating on Metal Substrate

The same geometry is considered, Figure 17.12, but now the coating is attenuating ($\tau < 1$) and is on a metallic substrate. An example is a coated metal absorber plate for a flat-plate solar collector. The external radiation is normally incident in air, $n_s \approx 1$. The complex index of refraction for the coating is $n_1 - ik_1$, and for the substrate $n_2 - ik_2$. The film transmittance is $\tau = \exp(-\kappa_1 D)$, where $\kappa_1 = 4\pi k_1/\lambda$ (this equality is useful for obtaining absorption coefficients, κ, from handbook values of the optical absorption index k, the imaginary part of the complex index of refraction). The κ, k, and n are wavelength dependent, so the following can be regarded as a spectral calculation. For normal incidence (Equation 3.20),

$$\rho_1 = \frac{(n_1 - 1)^2 + k_1^2}{(n_1 + 1)^2 + k_1^2} \tag{17.43}$$

$$\rho_2 = \frac{(n_2 - n_1)^2 + (k_2 - k_1)^2}{(n_2 + n_1)^2 + (k_2 + k_1)^2} \tag{17.44}$$

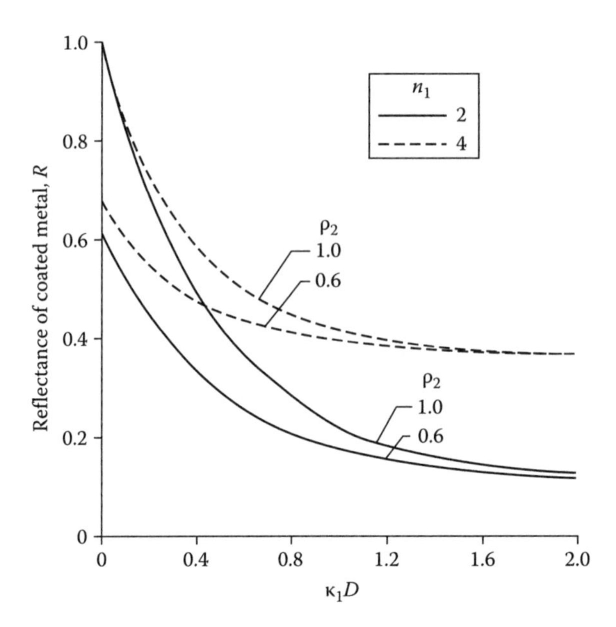

FIGURE 17.13 Reflectivity of thick attenuating film on metal substrate; no wave interference effects; $\kappa_1 = 10^3\ \text{cm}^{-1}$; $\lambda = 0.7\ \mu\text{m}$.

Equation 17.35 applies, and this becomes

$$R = \frac{\rho_1 + \rho_1(1-2\rho_1)\exp(-2\kappa_1 D)}{1-\rho_1\rho_2\exp(-2\kappa_1 D)} \tag{17.45}$$

As an example, if $\kappa_1 = 10^3\ \text{cm}^{-1}$ and $\lambda = 0.7\ \mu\text{m}$, the results in Figure 17.13 are obtained for various coating thicknesses, for two substrate reflectivities and two refractive indices of the coating. For high reflectance R of the metal, it is evident that the substrate reflectivity ρ_2 should be high and the absorptivity of the coating κ_1 should be low; the coating refractive index n_1 is not important. To obtain low reflectance, $\kappa_1 D > 1$ and $n_1 < 2$ are desired; the substrate reflectivity ρ_2 is not important when $\kappa_1 D$ is greater than about 1.

17.4.2 Thin Film with Wave Interference Effects

17.4.2.1 Nonabsorbing Dielectric Thin Film on Nonabsorbing Dielectric Substrate

When a film coating is very thin, of order of the wavelength λ of the incident radiation, interference effects occur between waves reflected from the first and second surfaces of the film. As given in Section 14.5, the amplitude reflection coefficients for the two components of polarization of the incident radiation are

$$\frac{E_{\parallel,r}}{E_{\parallel,i}} = r_\parallel = \frac{\tan(\theta-\chi)}{\tan(\theta+\chi)} \tag{17.46}$$

$$\frac{E_{\perp,r}}{E_{\perp,i}} = r_\perp = -\frac{\sin(\theta-\chi)}{\sin(\theta+\chi)} \tag{17.47}$$

When $n_2 > n_1$ for a wave incident from medium 1, then $\theta > \chi$ and $r_\perp$ is negative; that is, there is a phase change of π upon reflection. The $\tan(\theta - \chi)$ is positive, but $\tan(\theta + \chi)$ becomes negative for $\theta + \chi > \pi/2$, and $r_\parallel$ then yields a phase change of π. In a similar fashion, for transmitted radiation

$$\frac{E_{\parallel,t}}{E_{\parallel,i}} = t_\parallel = \frac{2\sin\chi\cos\theta}{\sin(\theta+\chi)\cos(\theta-\chi)} \tag{17.48}$$

$$\frac{E_{\perp,t}}{E_{\perp,i}} = t_\perp = \frac{2\sin\chi\cos\theta}{\sin(\theta+\chi)} \tag{17.49}$$

In going from medium 2 to medium 1, the χ and θ values are interchanged in these relations (the χ and θ remain the angles in 2 and 1, respectively), and the reflection coefficients are equal to $-r_\parallel$ and $-r_\perp$. In this instance, the transmission coefficients are called t' and are equal to

$$t'_\parallel = \frac{2\sin\theta\cos\chi}{\sin(\chi+\theta)\cos(\chi-\theta)} \tag{17.50}$$

$$t'_\perp = \frac{2\sin\theta\cos\chi}{\sin(\chi+\theta)} \tag{17.51}$$

For the simplified case of normal incidence and for radiation going from medium 1 to medium 2, these expressions reduce to

$$\frac{E_r}{E_i} = r_\parallel = r_\perp = r = \frac{n_1 - n_2}{n_1 + n_2} \tag{17.52}$$

$$\frac{E_t}{E_i} = t_\parallel = t_\perp = t = \frac{2n_1}{n_1 + n_2} \tag{17.53}$$

Formally, $r_\parallel$ has a negative sign as $E_{\parallel,r}$ and $E_{\parallel,i}$ point in opposite directions. This sign is not significant in the present discussion. For normally incident radiation going from medium 2 into medium 1

$$r = \frac{n_2 - n_1}{n_2 + n_1} \tag{17.54}$$

$$t' = \frac{2n_2}{n_2 + n_1} \tag{17.55}$$

For simplicity, the following discussion is limited to *normal incidence* on the thin film. Figure 17.14 shows the radiation reflected from the first and second interfaces (*note:* for clarity in showing each path, the paths are drawn at an angle). The beams a and b can interfere with each other. For normal incidence, beam b reflected from the second interface travels $2D$ farther than beam a, which is reflected from the first interface. Hence reflected beam b originated at time $-2D/c_1$ earlier than reflected beam a, where c_1 is the propagation speed in the film. If beam a originated at time 0, then beam b originated at time $-2D/c_1$. If the two waves originated from the same vibrating source, the phase of b relative to a is $e^{i\omega\tau} = e^{-i\omega 2D/c_1}$. The circular frequency can be written as $\omega = 2\pi c_0/\lambda_0$, where λ_0 is the wavelength in vacuum. Then, $e^{i\omega\tau} = e^{-i4\pi n_1 D/\lambda_0}$ where $n_1 = c_0/c_1$ is the film refractive index.

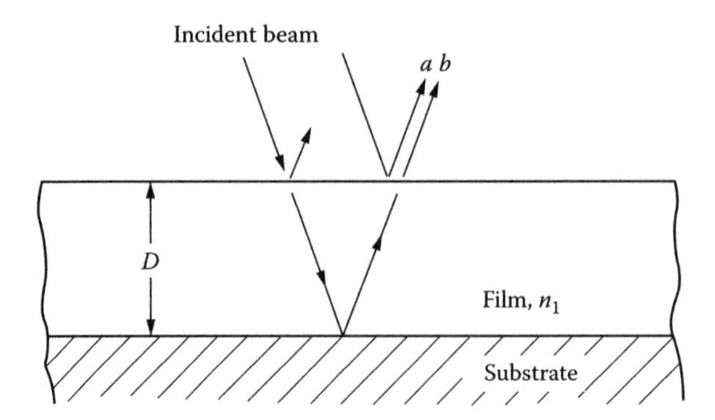

FIGURE 17.14 Reflection from the first and second interfaces of a thin film. This is for *normal* incidence; paths are drawn at an angle for clarity.

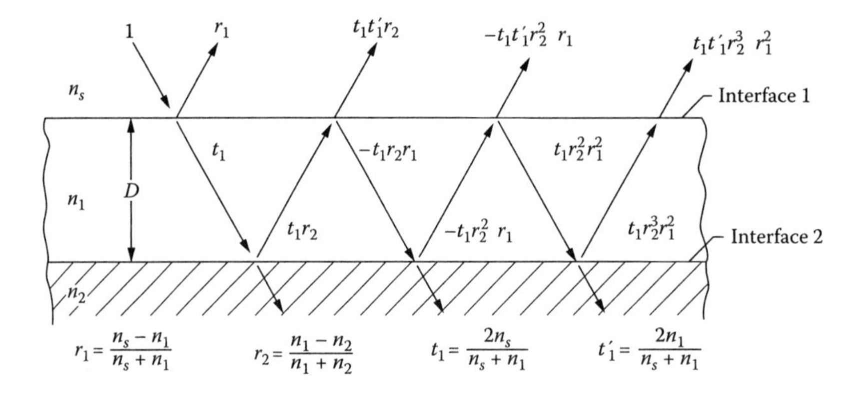

FIGURE 17.15 Multiple reflections within a thin nonattenuating film for normal incidence; paths are drawn at an angle for clarity.

Consider a thin nonabsorbing film of refractive index n_1 on a substrate with index n_2. For a normally incident wave of unit amplitude, the reflected radiation is shown in Figure 17.15. Taking into account the phase relationships and defining $\gamma_1 \equiv 4\pi n_1 D/\lambda_0$, the reflected amplitude is

$$R_M = r_1 + t_1t_1'r_2e^{-i\gamma_1} - t_1t_1'r_1r_2^2e^{-2i\gamma_1} + t_1t_1'r_1^2r_2^3e^{-3i\gamma_1} - \cdots = r_1 + \frac{t_1t_1'r_2e^{-i\gamma_1}}{1 + r_1r_2e^{-i\gamma_1}} \tag{17.56}$$

Note that $t_1t_1' = 1 - r_1^2$, so this can be reduced to

$$R_M = \frac{r_1 + r_2e^{-i\gamma_1}}{1 + r_1r_2e^{-i\gamma_1}} \tag{17.57}$$

An important application for a thin coating is to obtain low reflection from a surface to reduce reflection losses during transmission through a series of lenses in optical equipment. To have zero reflected amplitude, $R_M = 0$, requires $r_1 = -r_2e^{-i\gamma_1}$. This can be obtained if $r_1 = r_2$ and $e^{-i\gamma_1} = -1$. Since $e^{-i\pi} = -1$, this yields $D = \lambda_0/4n_1$. The quantity λ_0/n_1 is the wavelength of the radiation within the film. Hence the film thickness for zero reflection at normal incidence is one-quarter of the wavelength within the film. The required condition is $(n_s - n_1)/(n_s + n_1) = (n_1 - n_2)/(n_1 + n_2)$, which reduces to $n_1 = \sqrt{n_sn_2}$. Thus, for normal incidence onto a quarter-wave film from a dielectric medium with index of refraction n_s, the index of refraction of the film for zero reflection should be $n_1 = \sqrt{n_s n_2}$, the geometric mean of the n values on either side of the film. The thin film provides better performance than the thick film, as it is possible to achieve zero reflectivity. However, this result is only for normal incidence at one wavelength. To obtain more than one condition of zero reflectivity, it is necessary to use multilayer films. The optimization of the design of low-reflectivity multilayer coatings is discussed by Thornton and Tran (1978). For a system of two nonabsorbing quarter-wave films, zero reflection is obtained for $n_1^2 n_3 = n_2^2 n_s$, where n_1 and n_2 are for the coatings (the coating with n_2 is next to the substrate) and n_3 is for the substrate.

The previous expressions have been for the reflected amplitude from a thin film; now the reflected energy for normal incidence is considered. From Section 14.5, the reflected energy depends on $|\mathbf{E}|^2$. Since $R_m = E_r/E_i$, the reflectivity for energy is $R = |R_M|^2 = R_M R_M^*$ where R_M^* is the complex conjugate of R_M. From Equation 17.57, the reflectivity of the film is

$$R = \frac{r_1 + r_2 e^{-i\gamma_1}}{1 + r_1 r_2 e^{-i\gamma_1}} \frac{r_1 + r_2 e^{i\gamma_1}}{1 + r_1 r_2 e^{i\gamma_1}}$$

After multiplication and simplification, this becomes

$$R = \frac{r_1^2 + r_2^2 + 2r_1 r_2 \cos\gamma_1}{1 + r_1^2 r_2^2 + 2r_1 r_2 \cos\gamma_1} \tag{17.58}$$

Inserting $r_1 = (n_s - n_1)/(n_s + n_1)$ and $r_2 = (n_1 - n_2)/(n_1 + n_2)$, using the identity $\cos\gamma_1 = 1 - 2\sin^2(\gamma_1/2)$, and simplifying gives

$$R = \frac{n_1^2(n_s - n_2)^2 - (n_s^2 - n_1^2)(n_1^2 - n_2^2)\sin^2(2\pi n_1 D/\lambda_0)}{n_1^2(n_s + n_2)^2 - (n_s^2 - n_1^2)(n_1^2 - n_2^2)\sin^2(2\pi n_1 D/\lambda_0)} \tag{17.59}$$

For a quarter-wave film, $D = \lambda_0/4n_1$ and this reduces to

$$R = \left(\frac{n_s n_2 - n_1^2}{n_s n_2 + n_1^2}\right)^2 \tag{17.60}$$

The reflectivity becomes zero when $n_1 = \sqrt{n_s n_2}$. If n_1 is high, R is increased, and this behavior can be used to obtain *dielectric mirrors*. For various film materials of refractive index n_1 on glass ($n_2 = 1.5$), the reflectivity becomes, for incidence in air ($n_s \approx 1$),

Film	n_1	n_2	R
None	1	1.5	0.04
ZnS	2.3	1.5	0.31
Ge	4.0	1.5	0.69
Te	5.0	1.5	0.79

Multilayer films can be used to obtain reflectivities very close to 1. An application is for reflection of high-intensity laser beams where the fractional absorption of energy must be kept very small to avoid damage from heating the mirror.

Thin films of superconducting material are predicted to have absorptivity approaching zero out to very large wavelengths (Zeller 1990). This high reflectivity gives some promise of producing nearly ideal (zero heat loss) radiation shields, particularly for preventing energy loss from cryogenic storage systems in outer space. This technology will be of increasing usefulness as the critical temperature T_c of superconductors reaches higher values with new materials. A thin film on a rough surface was investigated by Tang et al. (1999a,b). Paretta et al. (1999) provide measurements of bidirectional spectral reflectivity for various surface treatments to the surface of photovoltaic cells.

17.4.2.2 Absorbing Thin Film on a Metal Substrate

In this instance, the complex refractive indices $n_1 - ik_1$ and $n_2 - ik_2$ replace the n_1 and n_2 of the previous section. For *normal incidence*, the $R = |R_M|^2$ becomes, by use of Equation 17.57

$$R = \left| \frac{r_1 + r_2 e^{-i\gamma_1}}{1 + r_1 r_2 e^{-i\gamma_1}} \right|^2 \tag{17.61}$$

where r_1 and r_2 now contain $\bar{n}_1$ and $\bar{n}_2$ and γ_1 contains $\bar{n}_1$. The behavior of Equation 17.61 is studied by Snail (1985) to obtain a selective surface for absorption of solar energy. The R values of some coatings on an aluminum substrate are in Figure 17.16.

The two complex indices of refraction of the coating in Figure 17.16 illustrate a trade-off between a low reflectivity for solar energy (high solar absorption) and the width of the transition region between the visible and infrared regions. Coating 1, which is thicker and has a lower n, has a significantly higher solar absorptance than coating 2, but it also has a wider transition region and would reemit more in the infrared region, thereby increasing energy loss by a solar collector. The lower n for coating 1 provides lower reflection losses at the front surface of the coating, but its higher value of κD produces a wider transition region. A method for improving the performance of the film–substrate combination for use as a solar collection surface is to have the refractive index n vary within the film, from the substrate value at the interface with the substrate to the value for the incident medium (air) at the front surface. This is explored by Snail (1985), and references are given for analytical solutions for the use of a graded index of refraction. A similar strategy can be used to produce fiber-optic cables with low losses by using radially varying refractive index.

This development has shown how the net-radiation method and ray tracing can be used to analyze the radiative behavior of partially transparent coatings. Extensive information is available on coatings and thin films. In addition to the references already discussed, further information is in Hsieh and Coldewey (1974), Musset and Thelen (1970), Forsberg and Domoto (1972), Taylor and Viskanta (1975), Heavens (1965, 1978), Roux et al. (1974), Palik et al. (1978), and Palik (1998).

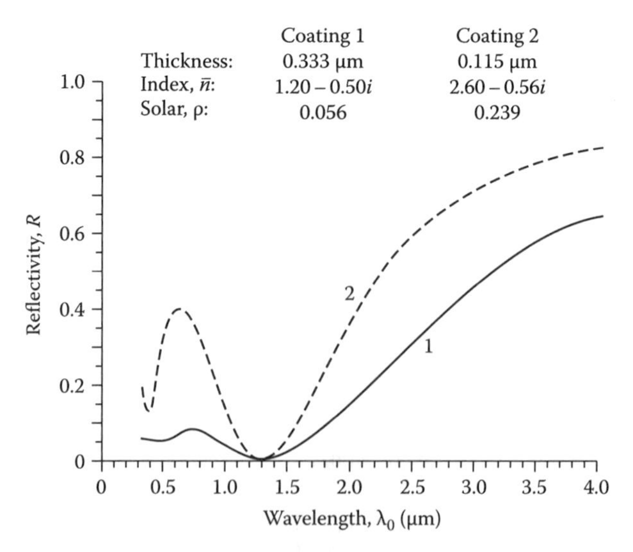

FIGURE 17.16 Spectral reflectivity for two homogeneous films on an aluminum ($\bar{n} = 1.50 - 10i$) substrate for normal incidence. (From Snail, K.A., *Solar Energ. Mater.*, 12, 411, 1985.)

Palik has reported extensive spectral data for n and k for metals, semiconductors, and dielectrics. By use of the relation $\kappa = 4\pi k/\lambda$, the spectral absorption coefficient can be found from the k values in the complex refractive index.

17.4.3 Films with Partial Coherence

When an incident wave enters a thin film, it is refracted and propagates into the film. The incident wave is joined at the first interface by the wave that has been reflected from the substrate and rereflected from the front surface interface. This rereflected wave has a time lag relative to the refracted incident wave. The two portions of the forward-propagating wave are thus partially coherent. In the limits of a thick film, the two portions become completely incoherent, and the theory presented in Section 17.4.1 (geometric optics) applies. If the film is thin, with thickness $\approx \lambda$ or less, the theory in Section 17.4.2 for a coherent wave applies. However, there is a significant portion of the wavelength range for a given film thickness where neither approach provides accurate results. This is the region of *partial coherence*. This region is analyzed by Chen and Tien (1992), and relations are presented for determining the film transmittance and absorptance. The region of partial coherence for nearly monochromatic incident radiation with frequency range $\Delta\nu$ is bounded by $1.13 \le f \le 2.59$, where $f = 4\pi n D\Delta\nu/c_0$ and c_0 is the speed of light in a vacuum. Above this range for f, geometric optics can be used; below this value of f, coherent optics needs to be considered, meaning that Maxwell and Fresnel equations should be used. Information is in Phelan et al. (1992) on the effect of thickness on radiative performance of a superconducting thin film on a substrate, and Fu et al. (2006) provide an analytical formulation. For thin films or coatings on substrates exposed to radiation with very narrow spectral spans, such as a laser beam, the partially coherent theory becomes important in predicting the reflectance and absorptance of the films. To examine other wave and tunneling effects in thin films, particularly when the gap between the plates is smaller than the wavelength of the incident radiation, it is necessary to include near-field effects (Chapter 16).

17.5 REFRACTIVE INDEX EFFECTS ON RADIATION IN A PARTICIPATING MEDIUM

Each of the partially transmitting layers in Sections 17.2 through 17.4 is assumed to have a uniform temperature throughout their thicknesses. It is also assumed that their temperatures are low enough that internal radiation emission from the layers is not significant. The behavior of layers at elevated temperatures and with internal temperature variations can be analyzed by using the radiative transfer equations given in Chapters 10 through 13, but modifying them for $n \neq 1$, along with appropriate conditions at the boundaries that include refractive index effects. These effects are now discussed for use in detailed heat transfer analyses.

17.5.1 Effect of Refractive Index on Intensity Crossing an Interface

Consider radiation with intensity $I_{\lambda,1}$ in an ideal dielectric medium of refractive index n_1. Let the radiation in solid angle $d\Omega_1$ pass into an ideal dielectric medium of refractive index n_2 as in Figure 17.17. As a result of the differing refractive indices, the rays change direction as they pass into medium 2. The radiation in the solid angle $d\Omega_1$ at incidence angle θ_1 passes into solid angle $d\Omega_2$ at angle of refraction θ_2. After allowing for reflection, the radiative energy is conserved in crossing the interface. From the definition of intensity, this energy conservation is given by

$$I_{\lambda,1}[1-\rho_\lambda(\theta_1)]\cos\theta_1 dA d\Omega_1 d\lambda_1 = I_{\lambda,2}\cos\theta_2 dA d\Omega_2 d\lambda_2 \tag{17.62}$$

where

$\rho_\lambda(\theta_1)$ is the directional–hemispherical reflectivity of the interface
dA is an area element in the plane of the interface

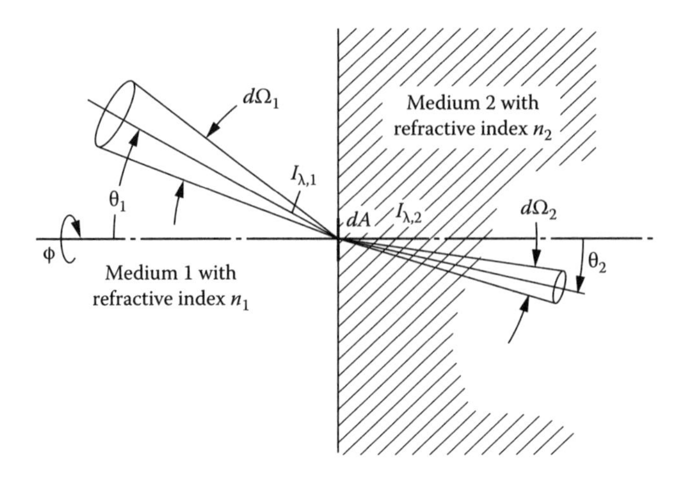

FIGURE 17.17 Radiation with intensity $I_{\lambda,1}$ crossing interface between two ideal dielectric media having unequal refractive indices.

The λ_1 and λ_2 are related by $\lambda_2 = (n_1/n_2)\lambda_1$. Note that $I_{\lambda,2}$ does not include internal radiation reflected from the inside of the interface. Using the relation for solid angle, $d\Omega = \sin\theta d\theta d\phi$, Equation 17.61 can be modified as (noting that the increment of circumferential angle $d\phi$ is not changed in crossing the interface)

$$I_{\lambda,1}[1-\rho_\lambda(\theta_1)]\sin\theta_1\cos\theta_1 d\theta_1 d\lambda_1 = I_{\lambda,2}\sin\theta_2\cos\theta_2 d\theta_2 d\lambda_2 \tag{17.63}$$

From Equation 3.3, Snell's law relates the indices of refraction to the angles of incidence and refraction by $n_1/n_2 = \sin\theta_2/\sin\theta_1$ and by differentiation $n_1\cos\theta_1 d\theta_1 = n_2\cos\theta_2 d\theta_2$. Substituting into Equation 17.63 gives

$$\frac{I_{\lambda,1}[1-\rho_\lambda(\theta_1)]d\lambda_1}{n_1^2} = \frac{I_{\lambda,2}d\lambda_2}{n_2^2} \tag{17.64}$$

For spectral calculations with variable n, it is better to work with *frequency* than with wavelength; this avoids introducing relations between λ and n. Frequency does not change with n, so Equation 17.64 becomes

$$\frac{I_{\nu,1}[1-\rho_\nu(\theta_1)]}{n_1^2} = \frac{I_{\nu,2}}{n_2^2} \tag{17.65}$$

17.5.2 Effect of Angle for Total Reflection

Consider a volume element dV inside a medium with refractive index n_2 as in Figure 17.18. Suppose that diffuse radiation of intensity I_1 is incident upon the boundary of this region from a medium having refractive index n_1, where $n_1 < n_2$. Radiation incident at grazing angles to the interface ($\theta_1 \approx 90°$) is refracted into medium 2 at a maximum value of θ_2 given by

$$\sin\theta_{2,\max} = \frac{n_1}{n_2}\sin 90° = \frac{n_1}{n_2} \tag{17.66}$$

Hence, the volume element in medium 2 receives direct radiation from medium 1 only at angular directions within the range $0 \le \theta_2 \le \theta_{2,\max}$, where $\theta_{2,\max} = \sin^{-1}(n_1/n_2)$.

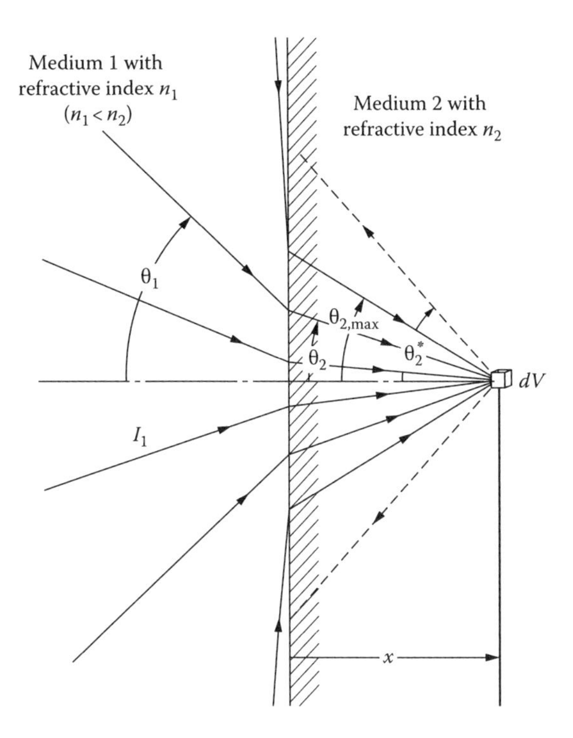

FIGURE 17.18 Effect of refraction on radiation transport in media with nonunity refractive index.

Now, consider emission from dV. The portion of this emission that can exit from region 2 is along paths found by reversing the arrows on the solid lines in Figure 17.18. However, there is also radiation emitted from dV that travels along paths such as shown by the dashed lines in Figure 17.18 that are incident on the interface at angles θ_2^*, where $\sin\theta_2^* > n_1/n_2$. From Snell's law (Equation 3.3), this means that such a ray would enter medium 1 at an angle given by $\sin\theta_1 = (n_2/n_1)\sin\theta_2^* > (n_2/n_1)(n_1/n_2) = 1$. Since sin θ_1 *cannot* be greater than 1, any ray incident on the interface from medium 2 at an angle greater than $\theta_{2,\max} = \sin^{-1}(n_1/n_2)$ cannot enter medium 1 and has total internal reflection at the interface. The $\theta_{2,\max}$ defined by Equation 17.66 is the *angle for total internal reflection.*

From Section 1.5.8, the blackbody spectral intensity emitted locally *inside* a medium with $n \neq 1$ and with n a function of frequency is

$$I_{\nu b,m} d\nu = \frac{2n_\nu^2 C_1 \nu^3}{c_o^4 (e^{C_2\nu/c_o T} - 1)} d\nu \tag{17.67}$$

where C_1 and C_2 are the coefficients defined in Chapter 1. If spectral variations of n_o are important, integration over frequency should be performed during the solution to obtain total energy quantities, provided that data for n_ν as a function of ν are available. If the refractive index is constant with frequency, integrating Equation 17.67 over all ν yields the local total emitted blackbody intensity inside a medium:

$$I_{b,m} = n^2 I_b \tag{17.68}$$

where I_b in this relation is for $n = 1$. Consequently, for an absorbing–emitting gray medium (n not a function of frequency) with absorption coefficient κ, the total energy dQ_e emitted by a volume element dV at temperature T has an n^2 factor:

$$dQ_e = 4n^2 \kappa\sigma T^4 dV \tag{17.69}$$

From Equation 17.68, it might appear that because $n > 1$, the intensity radiated from a dielectric medium into air could be larger than blackbody radiation $I_b = \sigma T^4/\pi$. This is not the case, as some of the energy emitted within the medium is reflected back into the emitting body at the inside surface of the medium–air interface. Consider a thick ideal dielectric medium ($k = 0$) at uniform temperature and with refractive index n. The maximum intensity received at an element dA on the interface from all directions within the medium is $n^2 I_b$. Only the energy received within a cone having a vertex angle θ_{max} relative to the normal of dA can penetrate through the interface; for incidence angles larger than θ_{max}, the energy is totally reflected into the medium. Hence, the maximum amount of energy received at dA that can leave the medium is $\int_{\theta=0}^{\theta_{max}} 2\pi n^2 I_b dA \cos\theta \sin\theta d\theta = 2\pi n^2 I_b dA\left(\sin^2\theta_{max}/2\right)$. From Equation 17.66, with $n_1 = 1$ and $n_2 = n$ in this case, $\sin\theta_{max} = 1/n$, so the total hemispherical emissive power leaving the interface is $2\pi n^2 I_b/2n^2 = \pi I_b$. Dividing by π gives I_b, as the maximum diffuse intensity that can leave the interface, which is the expected blackbody radiation intensity. For a real interface, there is partial internal reflection for the angles where $0 \le \theta \le \theta_{max}$, and the intensity leaving the outside of the interface is less than I_b.

17.5.3 Effects of Boundary Conditions for Radiation Analysis in a Plane Layer

In Section 11.2, the plane layer of semitransparent medium with $n = 1$ was analyzed. Several results presented in Chapter 11 were for a layer bounded by diffuse or black walls, so the intensities at the boundaries of the translucent layer with $n = 1$ were $1/\pi$ times the diffuse fluxes leaving the walls. Here, we consider a plane layer within a surrounding medium with a different refractive index, such as a glass plate in air or in water, or a layer of ceramic in a high-temperature gaseous environment. Radiation is incident from the surrounding medium, and some of it crosses the boundaries into the plane layer. Expressions are obtained for the intensities inside the plane layer at the boundaries. The analysis of Chapter 11 can then be applied using these internal boundary conditions.

17.5.3.1 Layer with Nondiffuse or Specular Surfaces

The layer in Figure 17.19 has smooth surfaces that are not diffuse reflectors. As discussed in connection with Equation 17.65, it is preferable to use frequency as the spectral variable when radiation crosses an interface between media having different refractive index n. The reason for this was discussed in Chapter 1, where we noted that the frequency is independent of the medium properties; therefore, the same value of frequency is valid in all media. On the other hand, the wavelength within a medium changes inversely with n, as $\lambda = \lambda_0/n$. Both the plane layer and the surrounding medium are assumed to be ideal dielectrics with regard to interface behavior. A subscript s (for "surroundings") is used to designate conditions *outside* the layer. The angles θ and ϕ give the direction within the medium. The θ_i and ϕ_i fix the incident directions at the boundaries. The $q_{\nu,s}\,d\nu$ are the spectral fluxes incident on the layer from within the surrounding medium; these fluxes are assumed uniform over the layer boundary.

The intensity $I_\nu^+(0,\theta,\phi)$ leaving boundary 1 *inside* the medium in Figure 17.19 is composed of a transmitted portion from $I_{\nu,s}(0,\theta_i,\phi_i)$ and the reflected portion of $I_\nu^-(0,\theta_i,\phi_i)$. The bidirectional spectral reflectivity Equation 2.33 relates the reflected and incident intensities by

$$\rho_\nu(\theta,\phi,\theta_i,\phi_i) = \frac{I_{\nu,r}(\theta,\phi,\theta_i,\phi_i)}{I_{\nu,i}(\theta_i,\phi_i)\cos\theta_i d\Omega} \tag{17.70}$$

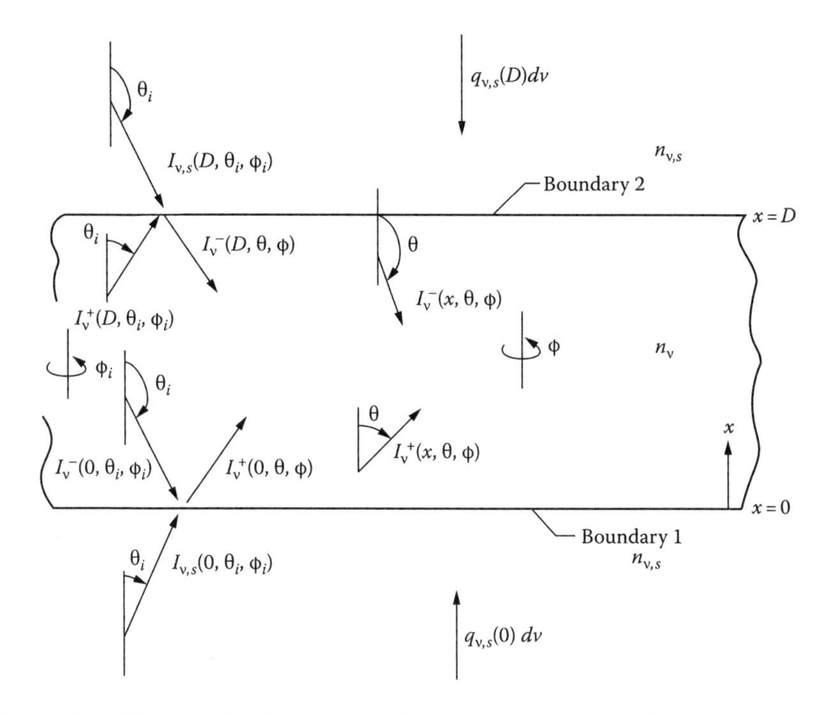

FIGURE 17.19 Intensities in a plane layer surrounded by a medium with a different refractive index.

Similarly, a bidirectional transmissivity of the interface is defined (Equation 2.61) as

$$\tau_\nu(\theta,\phi,\theta_i,\phi_i) = \frac{I_{\nu,\tau}(\theta,\phi,\theta_i,\phi_i)}{I_{\nu,i}(\theta_i,\phi_i)\cos\theta_i d\Omega_i} \tag{17.71}$$

where $I_{\nu,\tau}$ is the intensity in the direction θ, ϕ obtained by transmission. The τ_ν contains a factor of $(n_\nu/n_{\nu,s})^2$ to account for the effect in Section 17.5.1. By integrating over all incident solid angles at boundary 1, the intensity inside the layer leaving the boundary is

$$I_\nu^+(0,\theta,\phi) = \int_{\phi_i=0}^{2\pi}\int_{\theta_i=0}^{\pi/2} \tau_{\nu,1}(\theta,\phi,\theta_i,\phi_i) I_{\nu,s}(0,\theta_i,\phi_i)\cos\theta_i \sin\theta_i d\theta_i d\phi_i$$
$$+ \int_{\phi_i=0}^{2\pi}\int_{\theta_i=\pi/2}^{\pi} \rho_{\nu,1}(\theta,\phi,\theta_i,\phi_i) I_\nu^-(0,\theta_i,\phi_i)\cos\theta_i \sin\theta_i d\theta_i d\phi_i \tag{17.72}$$

Similarly, inside the medium at boundary 2

$$I_\nu^-(D,\theta,\phi) = \int_{\phi_i=0}^{2\pi}\int_{\theta_i=\pi/2}^{\pi} \tau_{\nu,2}(\theta,\phi,\theta_i,\phi_i) I_{\nu,s}(D,\theta_i,\phi_i)\cos\theta_i \sin\theta_i d\theta_i d\phi_i$$
$$+ \int_{\phi_i=0}^{2\pi}\int_{\theta_i=0}^{\pi/2} \rho_{\nu,2}(\theta,\phi,\theta_i,\phi_i) I_\nu^+(D,\theta_i,\phi_i)\cos\theta_i \sin\theta_i d\theta_i d\phi_i \tag{17.73}$$

As a special case, let the media be ideal dielectrics and the layer have optically smooth surfaces with incident energy fluxes $q_{\nu,s}(0)d\nu$ at $x = 0$ and $q_{\nu,s}(D)\,d\nu$ at $x = D$ that are diffuse. As discussed in Chapter 3, reflection at an interface depends on the component of polarization. The analysis should consider the portion of radiation in each polarization component and then add the two energies to obtain the total quantity. For simplicity, this effect is neglected at present, and average values of the surface properties are used. The two components of polarization can be included by using Equations 3.7 and 3.8, as illustrated in Example 17.2.

The internal reflections from the optically smooth interfaces are specular; hence $\theta = \pi - \theta_i$ and $\phi = \phi_i + \pi$, and, from Equation 3.9, at $x = D$

$$\frac{I_\nu^-(D,\theta,\phi)}{I_\nu^+(D,\theta_i,\phi_i)} = \frac{1}{2}\frac{\sin^2(\theta_i+\chi)}{\sin^2(\theta_i+\chi)}\left[1+\frac{\cos^2(\theta_i+\chi)}{\cos^2(\theta_i-\chi)}\right] \tag{17.74}$$

and similarly for $I_\nu^+(0,\theta,\phi)/I_\nu^-(0,\theta_i,\phi_i)$ where $\pi - \theta_i$ is used on the right in place of θ_i. The χ is determined from Snell's law, so that, for the internal reflections, $\sin\chi/\sin(\pi - \theta_i) = \sin\chi/\sin\theta_i = n_\nu/n_{\nu,s}$ at $x = 0$ (boundary 1) and $\sin\chi/\sin\theta_i = n_\nu/n_{\nu,s}$ at $x = D$ (boundary 2), where the subscript s designates the surroundings outside the layer. For some conditions and directions, there is total internal reflection, so the intensity reflected from the interface has the same magnitude as the incident intensity. This occurs when the index of refraction inside the medium is greater than that outside, $n_\nu > n_{\nu,s}$. There is total reflection when $\theta_i > \theta_{max}$, where $\theta_{max} = \sin^{-1}(n_{\nu,s}/n_\nu)$.

For the transmitted intensity, the factor $(n_\nu/n_{\nu,s})^2$ is included to account for the effect discussed in Section 17.5.1, and incidence is from the surrounding medium onto the layer. Then, using Equation 3.9 for either boundary 1 or 2 gives

$$\frac{I_\nu^+(0,\theta,\phi)}{I_{\nu,s}(0,\theta,\phi_i)} = \frac{I_\nu^-(D,\theta,\phi)}{I_{\nu,s}(D,\theta_i,\phi_i)} = \left\{1-\frac{\sin^2(\theta_i-\theta)}{2\sin^2(\theta_i+\theta)}\left[1+\frac{\cos^2(\theta_i+\theta)}{\cos^2(\theta_i-\theta)}\right]\right\}\left(\frac{n_\nu}{n_{\nu,s}}\right)^2 \tag{17.75}$$

where θ and θ_i are related by $\sin\theta/\sin\theta_i = n_{\nu,s}/n_\nu$.

For diffuse incident fluxes in Figure 17.19, the externally incident intensities are given by

$$I_{\nu,s}(0,\theta_i,\phi_i) = \frac{q_{\nu,s}(0)}{\pi} \tag{17.76}$$

$$I_{\nu,s}(D,\theta_i,\phi_i) = \frac{q_{\nu,s}(D)}{\pi} \tag{17.77}$$

Example 17.2

A volume element dV is located inside glass at $x = 3$ cm from an optically smooth planar interface in air, as in Figure 17.20. A diffuse–gray radiation flux $q_i = 40$ W/cm^2 in the air is incident on the glass surface. The absorption coefficient of the glass is assumed constant at $\kappa = 0.08$ cm^{-1} and the refractive index of the glass is $n = 1.52$. Scattering is neglected. Determine the energy absorption rate per unit volume in dV.

Since the incident energy is diffuse and unpolarized, the incident intensity in each component of polarization is $G/2\pi$. The fraction of the intensity transmitted through the interface depends on the angle of incidence and is $1 - \rho(\theta)$, where $\rho(\theta)$ is given for each component of polarization by Equations 3.7 and 3.8. From Equation 17.64, the intensity inside the medium for each component of polarization requires multiplying the incident intensity by n^2 in addition to a reflectivity factor itself. The intensity in the medium in direction χ then becomes $(G/2\pi)[1-\rho(\theta(\chi))]n^2$, where χ is the

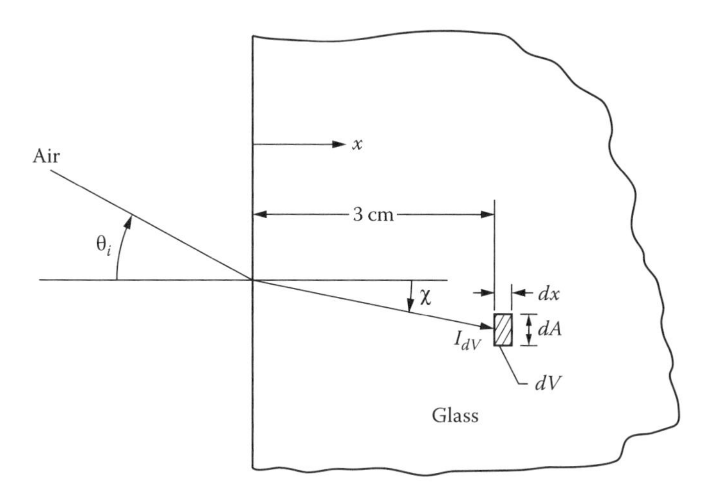

FIGURE 17.20 Schematic for Example 17.2.

angle of refraction. The path length traveled from the interface to the volume element is $x/\cos\chi$. Using Bouguer's law, the fraction reaching dV is $e^{-\kappa x/\cos\chi}$. The fraction $\kappa dx/\cos\chi$ of this incident energy at dV, $I_{dV}(\chi)dA\cos\chi$, is then absorbed in the volume element $dA\,dx$. The energy absorption for all directions of the arriving energy is found by integrating over $0 \le \chi \le \chi_{max}$, where χ_{max} is given by Snell's law as $\sin^{-1}(l/n)$. The integration over all incident solid angles introduces the factor for solid angle, $2\pi\sin\chi\,d\chi$. The energy absorbed per unit volume in dV is obtained by doing the integration for each component of polarization and summing the results. This yields

$$\frac{dQ}{dV} = \kappa G n^2 \int_0^{\chi_{max}} \left[2-\rho_{\parallel}(\chi)-\rho_{\perp}(\chi)\right] e^{-\kappa x/\cos\chi} \sin\chi\, d\chi$$

From Equations 3.7 and 3.8, the reflectivities are given by

$$\rho_{\parallel}(\chi) = \left[\frac{n^2\cos\theta_i - (n^2-\sin^2\theta_i)^{1/2}}{n^2\cos\theta_i + (n^2-\sin^2\theta_i)^{1/2}}\right]^2$$

$$\rho_{\perp}(\chi) = \left[\frac{(n^2-\sin^2\theta_i)^{1/2} - \cos\theta_i}{(n^2-\sin^2\theta_i)^{1/2} + \cos\theta_i}\right]^2$$

where the $\theta_i = \theta_i(\chi)$ is given by Snell's law as $\theta_i(\chi) = \sin^{-1}(n\sin\chi)$. Using the specified values, the integration is carried out numerically to give $dQ/dV = 3.38$ W/cm^3.

17.5.3.2 Diffuse Surfaces

For high-temperature applications such as in combustion chambers for advanced aircraft engines, it may be necessary to use ceramic parts or ceramic coatings to protect metal components. Some ceramics are partially transparent for thermal radiation, and they may be somewhat crystalline. They scatter strongly and their surface may not be smooth enough to use the specular (mirrorlike) reflection assumption. Instead, it is assumed that they reflect randomly in all directions, and for such a surface, the transmitted and reflected external or internal radiation would be diffuse, as in the case of frosted glass. We assume radiation from the interior of the medium that reaches a boundary is diffuse. Note that there is a difference between specular (mirrorlike) boundaries and diffuse boundaries. When the boundary is specular, part of the radiation reaching to the surface from high index of refraction side would be trapped inside the medium due to total internal reflection. However, for the diffuse surface, part of the radiation leaving the rough interface going into a material of lower refractive index can be from within the angular range where there would be total internal reflection for an optically smooth interface. This affects the path lengths being followed by reflected radiation within the layer. For ceramics, n can be large enough, such as 1.5–2.5, for there to be significant effects of internal reflections.

At a diffuse boundary, the outgoing flux inside a layer of medium equals the transmission of externally incident flux and the reflection of internal incoming flux as illustrated in Figure 17.21 (superscripts o and i designate outside and inside an interface):

$$J_\nu(0) = (1-\rho^o_{\nu,dif})q_{\nu r1} + \rho^i_{\nu,dif}G_\nu(0) \tag{17.78}$$

$$J_\nu(D) = (1-\rho^o_{\nu,dif})q_{\nu r2} + \rho^i_{\nu,dif}G_\nu(D) \tag{17.79}$$

The radiative flux at the interior of each surface is related to the outgoing and incoming fluxes within the layer by

$$q_{\nu r}(0) = J_\nu(0) - G_\nu(0) \tag{17.80}$$

$$q_{\nu r}(D) = -J_\nu(D) + G_\nu(D) \tag{17.81}$$

The diffuse reflectivity at the outside of each surface, $\rho^o_{dif}(n)$, is estimated from a hemispherically averaged Fresnel relation as $\rho^o_{dif}(n) = 1-\varepsilon(n)$, where $\varepsilon(n)$ is given by Equation 3.13. This assumes that the medium behaves like a perfect dielectric where the effect on reflectivity can be neglected of the absorption index k in the complex refractive index. This is a good assumption unless the absorption index is large, as shown by Cox (1965) and Hering and Smith (1968). For reflections inside the layer, each roughness facet is assumed to reflect in a specular manner, but the random orientation of the roughness elements results in a diffuse reflection. For each reflection from a roughness element, some of the energy has total internal reflection, and the rest is reflected as predicted by Fresnel relations for the smooth surface of each element. This results in the reflectivity at the inside of the interface being given in terms of the reflectivity at the outside of the interface by $\rho^i_{dif} = 1-(1/n^2)\times(1-\rho^o_{dif})$, as derived by Richmond (1963).

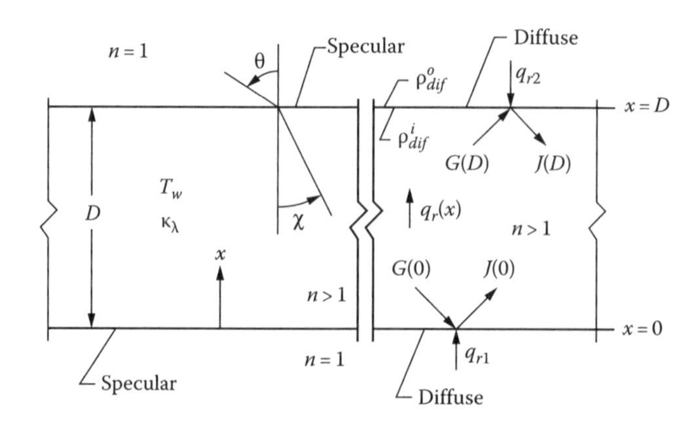

FIGURE 17.21 Plane layer of thickness D with $n > 1$ and with specular or diffuse surfaces.

17.5.4 Emission from a Translucent Layer ($n > 1$) at Uniform Temperature with Specular or Diffuse Boundaries

To illustrate the application and effect of specular or diffuse boundary conditions, the emission is analyzed from a translucent nonscattering plane layer at uniform temperature with $n > 1$, into surroundings with $n = 1$. For a layer at *uniform* temperature, a layer emittance can be defined as the emission from the layer relative to that from a blackbody at the same temperature. For blackbody radiation, emission is diffuse, but from an emitting translucent layer with smooth specular boundaries, the radiation intensity is not uniform over all directions. From Kirchhoff's laws, the emittance into a given direction equals the absorptance for energy from that direction. For diffuse incident radiation, the emittance including all directions equals the absorptance including all directions, which provides the proper comparison with blackbody emission, which is diffuse.

First, consider the absorptance of a nonscattering layer with refractive index $n > 1$ and with specular boundaries. The layer has diffuse incident energy that is partially absorbed as shown in Figure 17.21. For specularly reflecting surfaces, radiation in each incident direction θ is refracted and internally reflected as given by Fresnel relations. The fraction of energy absorbed is given by Equation 17.3 as $\alpha(\theta) = [1-\rho^o(\theta)][1-\tau(\theta)]/[1-\rho^o(\theta)\tau(\theta)]$, where the internal transmittance is $\tau(\theta) = e^{-\kappa D/\cos\chi(\theta)}$ and the angle of refraction, from Equation 3.3, is $\chi(\theta) = \sin^{-1}[(\sin\theta)/n]$. The reflectivity $\rho^o(\theta)$ is a function of the polarization component, so $\alpha(\theta)$ has different values for the perpendicular and parallel components. For incident diffuse unpolarized radiation, the absorptance of a layer with specular surfaces is obtained by integrating the absorbed energy for all incident solid angles. This gives the following relation that for absorptance, which, by virtue of Kirchhoff's law, equals the emittance of the uniform temperature layer:

$$\alpha_s(n,\kappa D) = \varepsilon_s(n,\kappa D) = \int_0^{\pi/2} [\alpha_\perp(\theta) + \alpha_\parallel(\theta)]\cos\theta\sin\theta d\theta \tag{17.82}$$

The interface reflectivities $\rho^o(\theta)$ needed to obtain $\alpha_\perp(\theta)$ and $\alpha_\parallel(\theta)$ are given by Equations 3.7 and 3.8. With all quantities in Equation 17.82 in terms of θ, the integration is performed numerically for various n and optical thicknesses κD (this expression for α_s is also given by Gardon (1956)).

The boundaries are now assumed *diffuse*. The diffuse reflectivity at a boundary is obtained from the hemispherically integrated Fresnel relation in Equation 3.13 by using $\rho^o_{dif}(n) = 1 - \varepsilon(n)$. The layer emittance ε_{dif}, which is a comparison with blackbody emission that is diffuse, is found by ε_{dif} being equal to the layer absorptance that is derived for diffuse incident radiation. Consider the interaction of the layer with a unit incident diffuse flux, $q_{rl} = 1$, on the boundary at $x = 0$ in Figure 17.21. The outgoing flux from the interior side of each diffuse boundary is, from Equations 17.78 and 17.79

$$J(0) = 1 - \rho^o_{dif} + \rho^i_{dif}G(0); \quad J(D) = \rho^i_{dif}G(D) \tag{17.83}$$

where ρ^i_{dif} is at the interior side of a diffuse boundary. At the interior side, the diffuse reflectivity is obtained by locally accounting for ordinary reflection from each interface element for incidence angles less than those for total internal reflection and totally reflected energy for larger incidence angles (see Siegel and Spuckler 1994b). As a result of transmission within the layer, the incoming and outgoing fluxes at the interior sides of the two boundaries are related by

$$G(0) = 2J(D)E_3(\kappa D); \quad G(D) = 2J(0)E_3(\kappa D) \tag{17.84}$$

as obtained by using the terms without emission or scattering in Equations 11.36 and 11.37. Equations 17.83 and 17.84 are solved for $J(0)$, $J(D)$, $G(0)$, and $G(D)$. The fraction of incident energy that is absorbed is $\alpha_{dif} = 1-(1-\rho^i_{dif})[G(0)+G(D)]$, which yields for the absorptance or emittance of the layer (Siegel and Spuckler 1994b):

$$\alpha_{dif}(n,\kappa D) = \varepsilon_{dif}(n,\kappa D) = (1-\rho^o_{dif})\frac{1-2E_3(\kappa D)}{1-[1-(1/n^2)(1-\rho^o_{dif})]2E_3(\kappa D)} \quad (17.85)$$

The ε_s and ε_{dif} for the isothermal layer from Equations 17.82 and 17.85 are spectral quantities since the absorption coefficient, κ_ν, depends on frequency (the subscript ν has been omitted to simplify the notation).

For specular or diffuse boundaries, the layer emittance was evaluated from Equations 17.82 and 17.85, and comparisons (from Siegel and Spuckler 1994b) are in Figure 17.22 for refractive indices n = 1–4 and optical thicknesses $\kappa_\nu D$ = 0.2–4. For $\kappa_\nu D > 4$, almost all incident energy that is not reflected from the first surface is absorbed in the layer; the layer spectral absorptance and emittance then approach the asymptotic value, $1-\rho^o_\nu$. The type of surface reflection is very significant when $\kappa_\nu D$ is less than about 3. For the limit n = 1, there are no surface reflections, so the results are the same for the solid and dashed curves. As n increases, the layer ε_υ are increasingly different, so for n = 4 with $\kappa_\nu D < 3$, the $\varepsilon_{\nu,s}$ and $\varepsilon_{\nu,dif}$ are substantially different. For specular boundaries, $\varepsilon_{\nu,s}$ decreases as the angle of refraction becomes smaller with larger n, thus providing shorter path lengths within the layer for the emitted energy that leaves the layer. For diffuse boundaries, the increase in $\varepsilon_{\nu,dif}$ with n is caused by internal radiation along paths at angles larger than the critical angle for total reflection being diffused at a boundary into directions at less than the critical angle; part of this radiation can then leave the layer. Emission is further increased by this as it occurs multiple times as radiation undergoes successive internal reflections.

The results in Equations 17.82 and 17.85 are for a translucent layer that has a uniform temperature, and the temperature is high enough for the layer to emit significant energy. For general heating conditions, the temperature distribution would usually be nonuniform, so that a more detailed

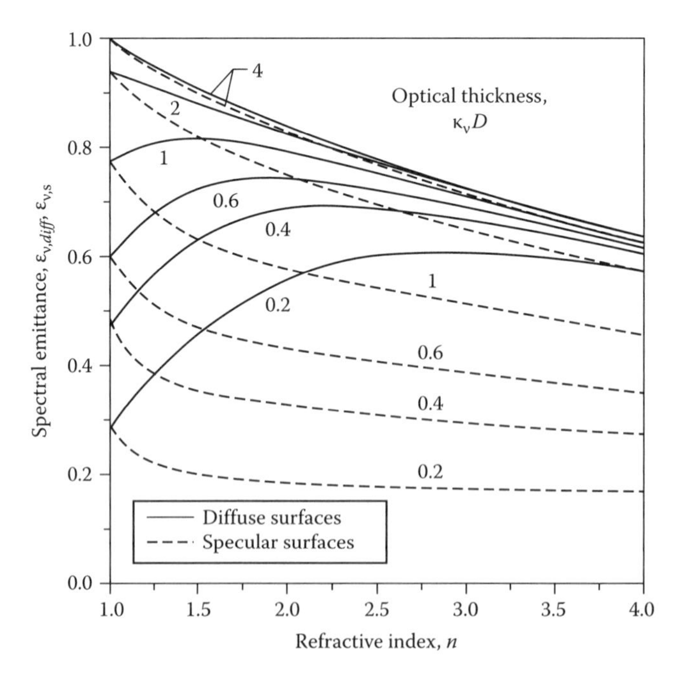

FIGURE 17.22 The effect of diffuse or specular surfaces on spectral emittance of a translucent plane layer at uniform temperature as a function of its optical thickness and refractive index. (From Siegel, R. and Spuckler, C.M., *JHT*, 116, 787, 1994b.)

analysis within the layer is required. Using boundary conditions that account for the refractive index of the layer being larger than one, the intensities within the layer at the boundaries can be related to the external heating conditions. Then the radiative transfer and energy equations are solved as described in Chapter 13 that treated a layer with $n = 1$. Klar et al. (2005) used the SP_N method to model transient radiation cooling in glass.

The equations used are presented in the next section where, in addition, the interface conditions are considered between two layers for analyzing composites of multiple translucent layers.

17.6 MULTIPLE PARTICIPATING LAYERS WITH HEAT CONDUCTION

The previous sections in this chapter have focused on the amount of energy transmitted through single- or multilayered windows and the effect on transmission of multiple reflections between interfaces. If temperature distributions are needed within multiple layers while including heat conduction, emission of energy, absorption, scattering, and internal heat generation, it is necessary to use both the energy and radiative transfer equations as in Chapter 13 with the n^2 index of refraction factor included in the internal blackbody emission relations.

Research papers and some review articles dealing with computing temperatures in hot translucent layers and other geometric shapes are listed in a short literature survey at the end of Section 17.6.3. A few relevant citations are briefly considered here. The effects of the thermal conditions at the boundaries of a translucent layer were investigated by Schwander et al. (1990) for a gray layer and for a layer of molten glass with four spectral bands. The gray layer was between opaque walls at differing temperatures, and there was either contact with both walls, one wall, or neither wall. The effect of diffuse or specular reflections was considered at the layer surfaces, and transients were investigated for a layer initially at a uniform temperature equal to the temperature of one of the walls. For an optical thickness of 1, only a small effect of the type of surface reflections was found for the gray layer for the processes studied, and the effect became very small for an optical thickness of 5. The same behavior was found for the layer of molten glass. This helps to define when it is necessary to be concerned with the directional details of the surface reflections and the transmission at the boundaries. The directional behavior is difficult to define for moderately rough surfaces, and it is helpful if directional details can be omitted for some conditions. Another analysis for heating a glass window is in Su and Sutton (1995), where one boundary was heated for 5 s by convection in high-speed flow. The maximum temperature that was located at this boundary is of interest for predicting when the glass will start to soften and can become optically distorted. Sixteen spectral bands were used in the calculations. In these types of analyses, a difficulty is that some of the spectral bands may be optically thick while others are optically thin. For the optically thick bands, it may be helpful to use the diffusion approximation. An analytical method is in Siegel and Spuckler (1994a) for dealing with both thick and thin bands in a layer where the temperature distribution is to be determined.

For cooling a glass plate, comparisons of predicted temperatures with experiment are in Field and Viskanta (1993). The plate was heated to about 800 K and then allowed to cool in an environment with both convective and radiative boundary conditions. Transient temperatures were measured by thermocouples fused into the glass. Good agreement of calculations and experimental results was obtained using five spectral bands. Internal radiation was found to be very important for the temperature distributions during cooling. Yao and Chung (1999) obtained transient solutions by numerical methods. The boundary conditions for a detailed thermal analysis must include continuity of temperature and conduction heat flux across each interface (see, e.g., discussions by Amlin and Korpela (1979) and Chan et al. (1983)). The detailed directional reflection and transmission effects at an interface can become complicated; Rokhsaz and Dougherty (1989) and Reguigui and Dougherty (1992) give formulations including Fresnel reflection at a boundary. Simplifications are often made, such as assuming that the interfaces act as diffuse surfaces. Some examples using this approximation are in Ho and Özişik (1987a,b), Özişik and Shouman (1981), Timoshenko and Trenev (1986), Tsai and Nixon (1986), and Tarshis et al. (1969).

17.6.1 Formulation for Multiple Participating Plane Layers

To illustrate the analysis for a multilayer configuration, which can also be applied for a single layer, consider the transient thermal behavior of a three-layer system bounded by two opaque diffuse–gray walls as in Figure 17.23. All of the interfaces are assumed diffuse, and each layer is gray with isotropic scattering and constant properties. The transient energy Equation 17.86 is written for each layer by using Equations 10.1 and 10.52. The n^2 appears in the emission term within the translucent medium. For the first layer (layer I)

$$\rho c \frac{\partial T}{\partial t} = k \frac{\partial^2 T}{\partial x^2} + 4\frac{\beta}{\omega}[\pi \hat{I}(\tau) - n^2 \sigma T^4(\tau)] \tag{17.86}$$

where τ is the optical coordinate measured from the left boundary of the layer and all properties are for layer I. The integral equation for $\hat{I}(\tau)$ is given by Equation 11.44 with an n^2 factor as

$$\hat{I}(\tau) = (1-\omega)n^2 \frac{\sigma T^4(\tau)}{\pi} + \frac{\omega}{2}\left[\frac{J_1}{\pi}E_2(\tau) + \frac{J_2}{\pi}E_2(\tau_D - \tau) + \int_0^{\tau_D} \hat{I}(\tau^*)E_1(|\tau^* - \tau|)d\tau^*\right] \tag{17.87}$$

Since Equation 17.86 has a second derivative in x, two spatial boundary conditions are needed. At each interface, there is continuity of both temperature (contact resistance is assumed zero) and conduction heat flux so that at interface 2–3 between layers I and II (see Song and Viskanta [1990] for additional discussion):

$$T_2 = T_3 \quad \text{and} \quad k_I \left.\frac{\partial T}{\partial x}\right|_2 = k_{II} \left.\frac{\partial T}{\partial x}\right|_3 \tag{17.88}$$

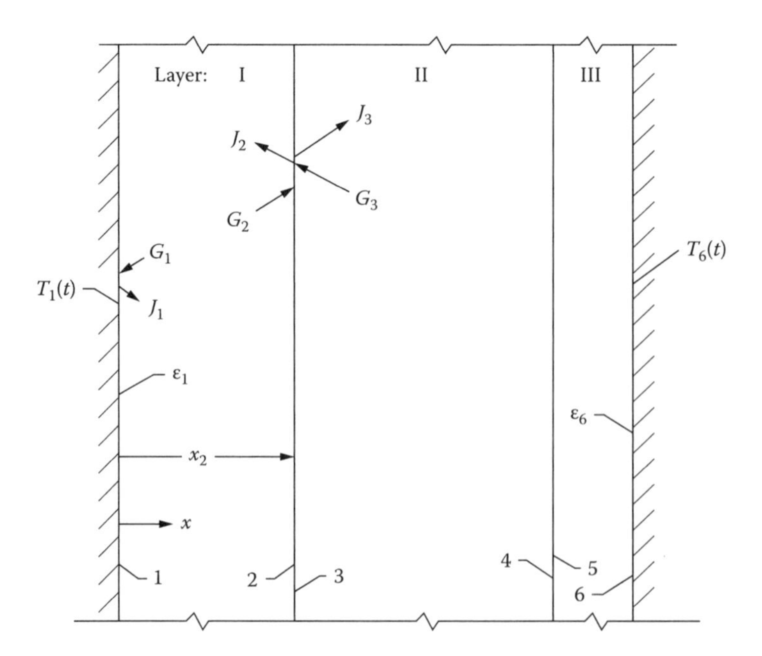

FIGURE 17.23 Multiple translucent plane layers between parallel opaque walls.

The wall temperature $T_1(\tau)$ is a specified function of time. Similar equations and boundary conditions are written for the other layers and interfaces. The first derivative in time in Equation 17.85 requires that the initial temperature distribution be specified.

Expressions for the J values are needed at the opaque walls and at the transmitting internal interfaces. It is assumed that at each internal interface, there is no absorption of energy, only reflection and transmission. Then, at wall 1 and interfaces 2 and 3

$$J_1 = \varepsilon_1 n_1^2 \sigma T_1^4 + (1-\varepsilon_1)G_1 \tag{17.89}$$

$$J_2 = G_3(1-\rho_3) + G_2\rho_2 \tag{17.90}$$

$$J_3 = G_2(1-\rho_2) + G_3\rho_3 \tag{17.91}$$

and similarly at interfaces 4 and 5 and wall 6. The wall emissivity is for emission into a medium; this may differ from that into air because the refractive index of the medium is different from 1 (this is illustrated by Figure 3.2, where the emissivity depends on the ratio of refractive indices of the emitting material and the surrounding medium). Note that for a black boundary, the total emission into a surrounding medium with index of refraction n is $n^2\sigma T_{wall}^4$ (Section 1.5.8).

Equations 17.89 through 17.91 have introduced the incident fluxes G, each of which can be found from the flux leaving the other interface of the layer and from the source function within the layer (emitted and scattered energy). From the relations used in Equations 11.36 and 11.37, there is obtained for layer I

$$G_1 = 2J_2E_3(\tau_D) + 2\pi \int_{\tau=0}^{\tau_D} \hat{I}(\tau)E_2(\tau)d\tau \tag{17.92}$$

$$G_2 = 2J_1E_3(\tau_D) + 2\pi \int_{\tau=0}^{\tau_D} \hat{I}(\tau)E_2(\tau_D-\tau)d\tau \tag{17.93}$$

where τ_D is the optical thickness $(\kappa_I + \sigma_{sI})x_2$ of layer I. The G is written in a similar fashion for the other layers, using the optical distances in the individual layers. By using Equations 17.89 through 17.93, the J can be eliminated from Equation 17.87. The resulting energy and source function Equations 17.86 and 17.87 are solved simultaneously by numerical procedures as described in Chapter 13, Ho and Özişik (1987a,b), Tsai and Nixon (1986), and Spuckler and Siegel (1992).

17.6.2 Translucent Layer on a Metal Wall

A way of shielding metal surfaces from a hot gas and/or a hot external environment is to coat the metal with a thin ceramic layer that can withstand high temperatures. These *thermal barrier coatings* may be used in combustion chambers, on turbine blades, or in diesel engine cylinders. A radiative analysis is now considered for a translucent layer on a metal wall. Zirconia is a common coating material found to have favorable heat conduction and thermal expansion properties. Zirconia is partially transparent to thermal radiation, and radiation effects are increased as temperatures are raised to improve engine efficiency. Radiation transfer can be treated using the two-flux method in Section 12.3.1. This method allows the large scattering of zirconia to be included without solving a separate source function Equation 11.43. The two-flux method usually provides good results for plane layers. For conditions in an engine, convection and conduction are very important, and the relatively smaller radiative effects can be calculated approximately without producing a large

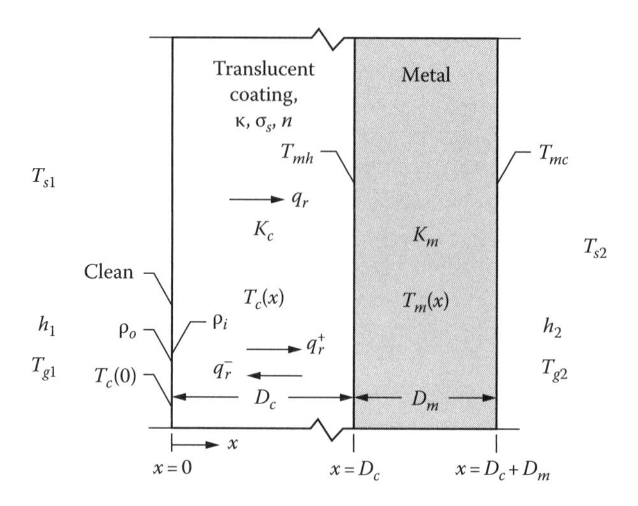

FIGURE 17.24 Geometry and nomenclature for a translucent thermal barrier coating on a metal wall.

error in the total heat transfer results. However, an accurate determination of the extent of radiation effects on the metal surface temperature can be quite important.

A composite is considered of a translucent thermal barrier coating on a metal wall with a clean coating surface at $x = 0$ (Figure 17.24); in practice, the coating surface may have soot on it for some applications. The coating is exposed to a black environment at T_{s1} and to convection by a hot gas at T_{g1}. The heat flux within the coating is the sum of conduction and radiation, as obtained by integrating the energy Equation 13.1 or 13.3:

$$q_{tot}(x) = -k_c \frac{dT_c(x)}{dx} + \int_{\nu=0}^{\infty} q_{\nu r}(x)d\nu \tag{17.94}$$

The radiative flux $q_{\nu r}(x)d\nu$ is given by Milne–Eddington two-flux Equation 12.108, $q_{\nu r}(x)d\nu = -(4\pi/3\beta_\nu)(d\bar{I}_\nu/dx)d\nu$. After substituting into Equation 17.94, the result is integrated from $x = 0$ to x to yield the coating temperature distribution as

$$T_c(x) = T_c(0) - \frac{1}{k_c}\left[q_{tot}x - \frac{4}{3}\int_{\nu=0}^{\infty} \frac{\bar{I}_\nu(0) - \bar{I}_\nu(x)}{\beta_\nu} d\nu\right] \tag{17.95}$$

The two-flux equation for $\bar{I}_\nu$ is a second-order differential Equation 12.50, written here for a material with refractive index $n > 1$ (Spuckler and Siegel 1996):

$$\frac{d^2\bar{I}_\nu(x)}{dx^2} - 3\beta_\nu^2(1-\omega_\nu)\bar{I}_\nu(x) = -\frac{3}{\pi}\beta_\nu^2(1-\omega_\nu)n^2 E_{\nu b}(x) \tag{17.96}$$

The local blackbody spectral emission, $n^2E_{\nu b}[T(x)]$, depends on the local coating temperature where $E_{\nu b}(T)d\nu$ is given by Equation 1.16. Two boundary conditions are required for solving Equation 17.95, since a differential representation of the radiative transfer is being used rather than an integral form that would already contain the boundary conditions. For a clean external surface, the boundary condition for $\bar{I}_\nu$ at $x = 0$ is derived by starting with Equation 17.78. The incident radiative spectral flux is $q_{\nu r i}d\nu = E_{\nu b}(T_{s1})d\nu$, and Milne–Eddington Equations 12.44 and 12.47 are used that $q_{\nu o}(0) = \pi\bar{I}_\nu(0) + q_{\nu r}(0)/2$, $q_{\nu i}(0) = \pi\bar{I}_\nu(0) - q_{\nu r}(0)/2$, and $q_{\nu r}(0) = -(4\pi/3\beta_\nu)(d\bar{I}_\nu/dx|_0)$. After substituting into Equation 17.78, the result is rearranged into the boundary condition:

$$-\frac{2}{3\beta_\nu}\frac{d\bar{I}_\nu}{dx}\bigg|_{x=0} + \frac{1-\rho_i}{1+\rho_i}\bar{I}_\nu(0) = \frac{1}{\pi}\frac{1-\rho_o}{1+\rho_i}E_{\nu b}(T_{s1}) \tag{17.97}$$

A somewhat similar derivation gives the boundary condition for $\bar{I}_\nu$ at $x = D_c$, but here the opaque surface of the metal that has an emissivity ε_m enters the relation. The radiative flux in the negative direction is the emission from the metal and the reflected incoming radiation; this gives, for a gray wall, $q^-_{\nu r}(D_c) = \varepsilon_m n^2 E_{\nu b}(D_c) + (1-\varepsilon_m)q^+_{\nu r}(D_c)$. The two-flux relations in terms of $\bar{I}$ are used to replace the q^- and q^+, and the result is the boundary condition for $\bar{I}$ at the interface of the translucent layer and the metal wall (Siegel 1997b):

$$\frac{2}{3\beta_\nu}\frac{d\bar{I}_\nu}{dx}\bigg|_{x=D_c} + \frac{\varepsilon_m}{2-\varepsilon_m}\bar{I}_\nu(D_c) = \frac{1}{\pi}\frac{\varepsilon_m}{2-\varepsilon_m}n^2 E_{\nu b}(D_c) \tag{17.98}$$

At this interface, continuity of temperature gives $T_c(D_c) = T_m(D_c) \equiv T_{mh}$ (the metal temperature on the hot side). Within the metal wall, energy is transferred only by conduction, so $q_{tot} = k_m(T_{mh} - T_{mc})/D_m$. At the cooled side of the metal wall, there is convection to the cooling gas and radiation to the surroundings. If large surroundings at T_{s2} are assumed, $q_{tot} = h_2(T_{mc} - T_{g2}) + \varepsilon_m\sigma(T^4_{mc} - T^4_{s2})$.

For zirconia, the translucent spectral region extends from small λ to $\lambda \approx 5$ μm; for larger λ, the extinction coefficient becomes quite large. This behavior is appropriate to demonstrate a two-band calculation; the band for large wavelengths (small frequencies) is considered opaque. To obtain the temperature distribution in the thermal barrier coating, Equation 17.95 is integrated over the translucent band (large frequencies). The $\bar{I}(x)$ was found by solving Equation 17.96 subject to the boundary conditions of Equations 17.97 and 17.98. Siegel (1997a) derived a Green's function solution for this equation and boundary conditions, and it is applied here using the properties and blackbody energy in the translucent spectral band; this yields $\bar{I}_L(x)$, where the subscript L designates the band with large frequencies. An iterative solution can be developed, where the opaque heat conduction solution is used as a first guess for $T_c(x)$ (Siegel 1997a). Then the $\bar{I}_L(x)$ can be evaluated since it depends on the temperature distribution. Relations between the temperatures and the total heat flow are developed. For example, at a clean exposed zirconia surface, q_{tot} consists of radiation in the translucent band, radiation at the surface in the opaque band, and convection at the surface. Using the two-flux relation following Equation 17.94 for the radiation in the translucent band gives

$$\frac{d\bar{I}_L}{dx}\bigg|_{x=0} = \frac{3}{4\pi}\beta_{cL}[h_1[1-T_c(0)+(1-\rho^o)(T^4_{s1}[1-F(T_{s1})]-T_c(0)^4\{1-F[T_c(0)]\})-q_{tot}] \tag{17.99}$$

These relations can be used to develop an iterative procedure where successive iterative values are adjusted to conform to energy conservation; this provides rapid convergence. For the details of the method, see Siegel (1997a).

Illustrative temperatures are in Figure 17.25 for a zirconia coating on a high-alloy steel combustion liner. The solid lines are for an uncoated metal wall; the lower line is for oxidized metal ($\varepsilon_m = 0.6$) on both sides. In both instances, the temperatures are above the metal melting point, so a thermal barrier protective coating must be used. A limiting calculation is for a zirconia coating assumed to be opaque (long dashed lines). For a 1 mm thick coating, the metal temperature is substantially reduced. If the coating can be kept clean, there may be a further benefit if the coating is semitransparent, although this effect depends on properties such as the index of refraction; an increase in n will shift the positions of the results. The high scattering of zirconia reflects much of the incident radiation so that the temperatures in the zirconia are decreased (lower short dashed line) and the hot side of the metal is reduced about 20 K.

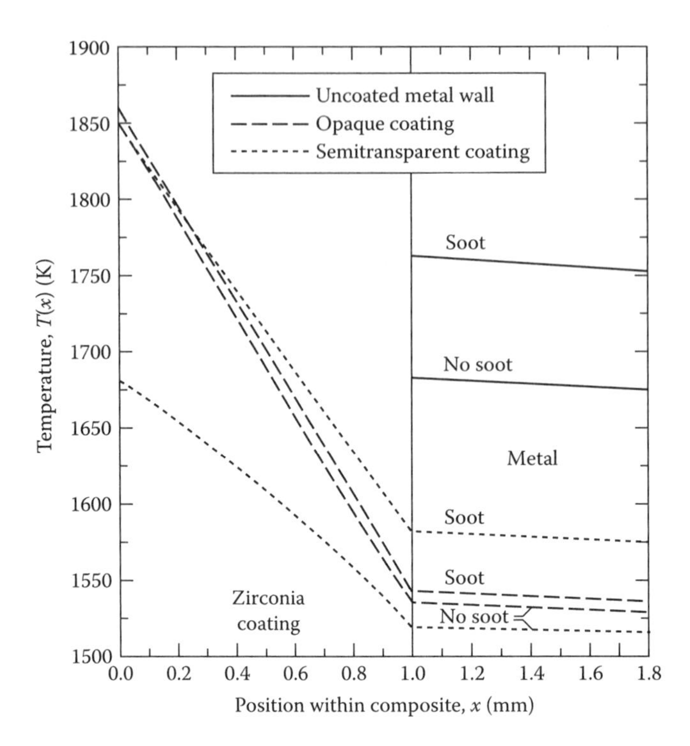

FIGURE 17.25 Combustor liner wall temperature distributions for oxidized metal without a coating, metal with an opaque thermal barrier coating, and metal with a semitransparent thermal barrier coating; results with and without soot on the exposed surface; on the cooled side of the metal, radiation is to large surroundings. Parameters: h_1 =250, h_2 = 110 W/(m$^2\cdot$K); k_c = 0.8, k_m = 33 W/(m$\cdot$K); $D_c = 10^{-3}$; $D_m = 0.794 \times 10^{-3}$ m; n = 1.58; κ = 30 m^{-1} and $\sigma_s = 10^4$ m^{-1} for $\lambda < 5$ μm; $T_{s1} = T_{gl}$ = 2000 K; $T_{s2} = T_{g2}$ = 800 K. (From Siegel, R., *JHT*, 11, 533, 1997b.)

17.6.3 Composite of Two Translucent Layers

The radiative analysis is now described for a composite layer of two gray translucent absorbing and scattering materials with thicknesses D_1 and D_2, as in Figure 17.26, which have unequal refractive indices larger than 1 (Spuckler and Siegel 1994). To provide general boundary conditions, each side of the composite is heated by radiation and convection. The diffuse radiation incident from the surroundings is q_{r1}^o and q_{r2}^o on the two outer boundaries $x_1 = 0$ and $x_2 = D_2$. External convective heat transfer is provided by gas flows at T_{g1} and T_{g2} with heat transfer coefficients h_1 and h_2. For convenience, assume $q_{r1}^o > q_{r2}^o$ and $T_{g1} > T_{g2}$ in the present analysis. Energy is transferred internally by conduction, emission, absorption, and isotropic scattering. The layers have absorption and scattering coefficients κ_1, σ_{s1} and κ_2, σ_{s2}. The external surfaces of the composite and its internal interface are diffuse, which is intended to model composite ceramics that have not been polished and are bonded together. Internal reflections are included at the boundaries and at the internal interface. Inside each layer there are outgoing and incoming diffuse fluxes, J and G, at each interface.

17.6.3.1 Temperature Distribution Relations from Energy Equation

Within each layer, energy is transferred by conduction and radiation according to the energy Equation 13.3

$$k_j \frac{d^2T_j}{dx_j^2} - \frac{dq_{rj}}{dx_j} = 0 \quad (j = 1,2) \tag{17.100}$$

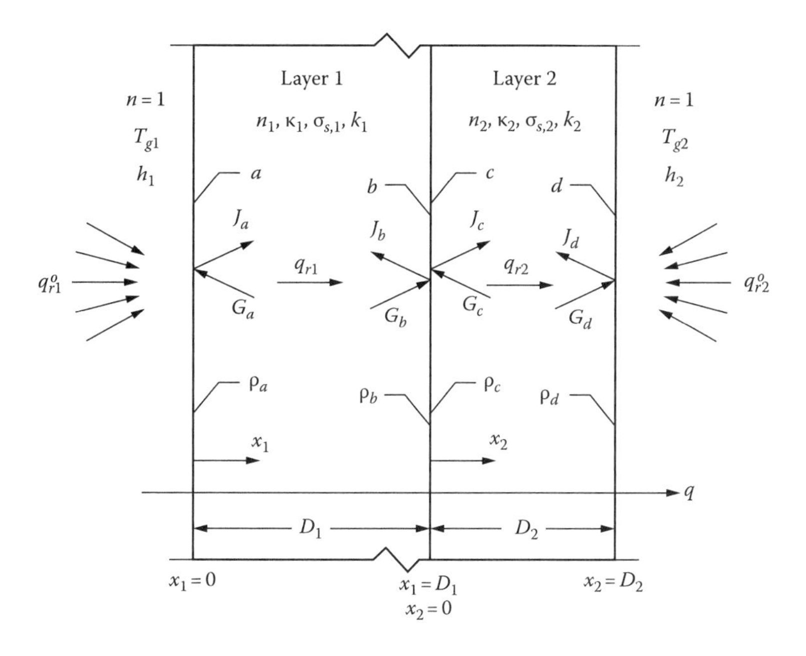

FIGURE 17.26 Geometry, coordinate system, and nomenclature for radiative fluxes for absorbing and scattering composite consisting of two gray translucent layers with differing properties.

Since the energy flux by combined radiation and conduction through the composite is a constant, Equation 17.100 can be integrated with respect to χ and then equated to its values at $x_1 = 0$ and $x_2 = D_2$ to evaluate the constant of integration; this gives

$$k_1 \left.\frac{dT_1}{dx_1}\right|_{x1} - q_{r1}(x_1) = k_2 \left.\frac{dT_2}{dx_2}\right|_{x2} - q_{r2}(x_2) = k_1 \left.\frac{dT_1}{dx_1}\right|_{0} - q_{r1}(x_1 = 0) = k_2 \left.\frac{dT_2}{dx_2}\right|_{D_2} - q_{r2}(x_2 = D_2) \quad (17.101)$$

After partial reflection at each outer boundary, the externally incident radiation q_{rj}^o passes into the composite and interacts internally. There is no absorption at the exact plane of an outer boundary since an interface does not have any volume. Hence, the conduction derivative terms at the boundaries in Equation 17.101 are equal to only the external convection, and Equation 17.101 can be written as

$$\begin{aligned} h_1[T_{g1} - T_1(0)] + q_{r1}(x_1 = 0) &= -k_1 \left.\frac{dT_1}{dx_1}\right|_{x_1} + q_{r1}(x_1) \\ &= -k_2 \left.\frac{dT_2}{dx_2}\right|_{x_2} + q_{r2}(x_2) \\ &= h_2[T_2(D_2) - T_{g2}] + q_{r2}(x_2 = D_2) \end{aligned} \quad (17.102)$$

Equation 17.102 is integrated to give the following relations for the layer temperature distributions:

$$T_1(x_1) = T_1(x_1 = 0) - \frac{h_1}{k_1}[T_{g1} - T_1(x_1 = 0)]x_1 - \frac{x_1}{k_1} q_{r1}(x_1 = 0) + \frac{1}{k_1}\int_0^{x_1} q_{r1}\left(x_1^*\right) dx^* \quad (17.103)$$

$$T_2(x_2) = T_2(x_2 = 0) - \frac{h_2}{k_2}[T_2(x_2 = D_2) - T_{g2}]x_2 - \frac{x_2}{k_2} q_{r2}(x_2 = D_2) + \frac{1}{k_2}\int_0^{x_2} q_{r2}\left(x_2^*\right) dx_2^* \quad (17.104)$$

Equations 17.103 and 17.104 are evaluated, respectively, at the boundaries $x_1 = D_1$ and $x_2 = D_2$, and the overall temperature difference for the composite $T_1(x_1 = 0)$–$T_2(x_2 = D_2)$ is found by addition noting that, at the internal interface, $T_1(x_1 = D_1) = T_2(x_2 = 0)$. Then, by using the equality of the first and last sets of terms in Equation 17.102, the $T_2(x_2 = D_2)$ or $T_1(x_1 = 0)$ is eliminated to yield the two surface temperatures as

$$T_1(x_1 = 0) = T_{g1} + \left[1 + \frac{h_1}{h_2} + h_1\left(\frac{D_1}{k_1} + \frac{D_2}{k_2}\right)\right]^{-1} \left\{T_{g2} - T_{g1} + \left[\frac{1}{h_2} + \frac{D_1}{k_1} + \frac{D_2}{k_2}\right] q_{r1}(x_1 = 0) - \frac{1}{h_2} q_{r2}(x_2 = D_2) - \frac{1}{k_1}\int_0^{D_1} q_{r1}(x_1)dx_1 - \frac{1}{k_2}\int_0^{D_2} q_{r2}(x_2)dx_2\right\} \quad (17.105)$$

$$T_2(x_2 = D_2) = T_{g2} - \left[1 + \frac{h_2}{h_1} + h_2\left(\frac{D_1}{k_1} + \frac{D_2}{k_2}\right)\right]^{-1} \left[T_{g2} - T_{g1} + \left(\frac{1}{h_1} + \frac{D_1}{k_1} + \frac{D_2}{k_2}\right) q_{r2}(x_2 = D_2) - \frac{1}{h_1} q_{r1}(x_1 = 0) - \frac{1}{k_1}\int_0^{D_1} q_{r1}(x_1)dx_1 - \frac{1}{k_2}\int_0^{D_2} q_{r2}(x_2)dx_2\right] \quad (17.106)$$

17.6.3.2 Relations for Radiative Flux

The temperature relations of Equations 17.103 through 17.106 contain the radiative fluxes. In scattering layers, they depend on the unknown radiative source function $\hat{I}_j\left(x_j\right)$ and are given in each layer by Equation 11.63:

$$q_{rj}(x_j) = 2J(x_j = 0)E_3(\beta_j x_j) - 2J(x_j = D_j)E_3[\beta_j(D_j - x_j)] + 2\pi\beta_j\left\{\int_0^{x_j} \hat{I}_j(x_j^*)E_2[\beta_j(x_j - x_j^*)]dx_j^* - \int_{x_j}^{D_j} \hat{I}_j(x_j^*)E_2[\beta_j(x_j^* - x_j)]dx_j^*\right\} \quad (17.107)$$

Equation 17.107 contains the diffuse fluxes $J(x_j = 0) \equiv J_a$ and $J(x_j - D_j) = J_d$ leaving the *internal* surface of each boundary (Figure 17.26). These fluxes are expressed in terms of the fluxes incident from outside of each layer to provide coupling with the external radiation and with the energy crossing the internal interface. By using the relations given previously between diffuse reflectivities on the two sides of an interface (see information following Equation 17.81), the outgoing flux is written at each diffuse interface in terms of transmitted and reflected energy fluxes (see Figure 17.26):

$$J_a = q_{r1}^o n_1^2(1 - \rho_a) + G_a\rho_a \quad (17.108)$$

$$J_b = G_c\left(\frac{n_1}{n_2}\right)^2(1 - \rho_b) + G_b\rho_b \quad (17.109)$$

$$J_d = q_{r2}^o n_2^2(1-\rho_d) + G_d\rho_d \tag{17.110}$$

Because there are four internal sides of interfaces (a, b, c, d), another independent relation is needed. This is obtained from continuity of radiative flux across the internal interface, which gives

$$G_b - J_b = J_c - G_c \tag{17.111}$$

For j = 1, 2, by using $q_{rj}(x_j = 0) = J - G$ at interfaces a and c and $q_{rj}(x_j = D_j) = G - J$ at interfaces b and d, Equation 17.107 is used to obtain the internal incoming fluxes at the four internal boundaries as

$$G_a = 2J_bE_3(\beta_1D_1) + 2\pi\beta_1 \int_{x_1=0}^{D_1} \hat{I}_1(x_1)E_2(\beta_1x_1)dx_1 \tag{17.112}$$

$$G_b = 2J_aE_3(\beta_1D_1) + 2\pi\beta_1 \int_{x_1=0}^{D_1} \hat{I}_1(x_1)E_2[\beta_1(D_1-x_1)]dx_1 \tag{17.113}$$

$$G_c \quad 2J_dE_3(\beta_2D_2) + 2\pi\beta_2 \int_{x_2=0}^{D_2} \hat{I}_2(x_2)E_2(\beta_2x_2)dx_2 \tag{17.114}$$

$$G_d = 2J_cE_3(\beta_2D_2) + 2\pi\beta_2 \int_{x_2=0}^{D_2} \hat{I}_2(x_2)E_2[\beta_2(D_2-x_2)]dx_2 \tag{17.115}$$

The G are eliminated between Equations 17.108 through 17.115, and the resulting equations are solved simultaneously for J at each internal boundary to yield

$$J_b = \frac{(A_2C_3 + A_3C_1 + C_2)[1-(A_4^2/\rho_d)] + A_2A_4[(A_1C_1/\rho_a) + (A_4C_3/\rho_d) + C_4]}{(1-A_1A_3)[1-(A_4^2/\rho_d)] + A_2A_4[1-(A_1^2/\rho_a)]} \tag{17.116}$$

$$J_a = C_1 + A_1J_b \tag{17.117}$$

$$J_c = \frac{-(A_2C_3 + A_3C_1 + C_2)[1-(A_1^2/\rho_a)] + (1-A_1A_3)[(A_1C_1/\rho_a) + (A_4C_3/\rho_d) + C_4]}{(1-A_1A_3)[1-(A_4^2/\rho_d)] + A_2A_4[1-(A_1^2/\rho_a)]} \tag{17.118}$$

$$J_d = C_3 + A_4J_c \tag{17.119}$$

where

$$A_1 = 2\rho_aE_3(\beta_1D_1) \tag{17.120}$$

$$A_2 = 2(1-\rho_b)\left(\frac{n_1}{n_2}\right)^2 E_3(\beta_2 D_2) \quad (17.121)$$

$$A_3 = 2\rho_b E_3(\beta_1 D_1) \quad (17.122)$$

$$A_4 = 2\rho_d E_3(\beta_2 D_2) \quad (17.123)$$

$$C_1 = (1-\rho_a)n_1^2 q_{r1}^o + 2\rho_a \pi \beta_1 \int_{x_1=0}^{D_1} \hat{I}_1(x_1) E_2(\beta_1 x_1)dx_1 \quad (17.124)$$

$$C_2 = 2\pi\left\{(1-\rho_b)(n_1/n_2)^2 \beta_2 \int_{x_2=0}^{D_2} \hat{I}_2(x_2)E_2(\beta_2 x_2)dx_2 + \rho_b \beta_1 \int_{x_1=0}^{D_1} \hat{I}_1(x_1)E_2[\beta_1(D_1-x_1)]dx_1\right\} \quad (17.125)$$

$$C_3 = (1-\rho_b)n_2^2 q_{r2}^o + 2\rho_d \pi \beta_2 \int_{x_2=0}^{D_2} \hat{I}_2(x_2)E_2[\beta_2(D_2-x_2)]dx_2 \quad (17.126)$$

$$C_4 = 2\pi\left\{\beta_1 \int_{x_1=0}^{D_1} \hat{I}_1(x_1)E_2[\beta_1(D_1-x_1)]dx_1 + \beta_2 \int_{x_2=0}^{D_2} \hat{I}_2(x_2)E_2(\beta_2 x_2)dx_2\right\} \quad (17.127)$$

17.6.3.3 Equation for the Source Function

The source function $\hat{I}_j(x_j)$ required for the radiative flux is obtained in each layer from Equation 11.44 with the addition of an n_j^2 factor:

$$\hat{I}_j(x_j) = (1-\omega)n_j^2 \frac{\sigma T_j^4(x_j)}{\pi} + \frac{\omega_j}{2}\left\{\frac{J(x_j=0)}{\pi}E_2(\beta_j x_j) + \frac{J(x_j=D_j)}{\pi}E_2[\beta_j(D_j-x_j)] + \beta_j \int_{x_j^*=0}^{D_j} \hat{I}_j(x_j^*)E_1(\beta_j |x_j - x_j^*|)dx_j^*\right\} \quad (j=1,2) \quad (17.128)$$

17.6.3.4 Solution Procedure and Typical Results

For diffuse interfaces, the relations in Section 17.5.3 are used for the interface reflectivities. An iterative solution method is used by first assuming temperature and source function distributions in both layers. The $A_1 \ldots A_4$ and $C_1 \ldots C_4$ are evaluated from Equations 17.120 through 17.127 and are used to calculate the J from Equations 17.116 through 17.119. New $\hat{I}_1(x_1)$ and $\hat{I}_2(x_2)$ are then evaluated by iterating Equation 17.128 using the assumed $T_1(x_1)$ and $T_2(x_2)$, and the flux distributions $q_{r1}(x_1)$ and $q_{r2}(x_2)$ are evaluated from Equation 17.107. New boundary temperatures $T_1(x_1 = 0)$ and $T_2(x_2 = D_2)$ are calculated from Equations 17.105 and 17.106 and new temperature distributions $T_1(x_1)$ and $T_2(x_2)$ from Equations 17.103 and 17.104. The new $T_1(x_1)$ and $T_2(x_2)$ and $T_1(x_1)$ and $T_2(x_2)$ are used to start a new iteration; the process is continued until they converge. Underrelaxation factors are usually required, as discussed in Section 7.8.3.

Representative results are displayed in Figure 17.27, where the refractive index for the first layer is $n_1 = 1.5$ and for the second layer is either $n_2 = 1.5$ or $n_2 = 3$. The $\vartheta = T/T_{g1}, N_j = k_j/\sigma T_{g1}^3 D_j$,

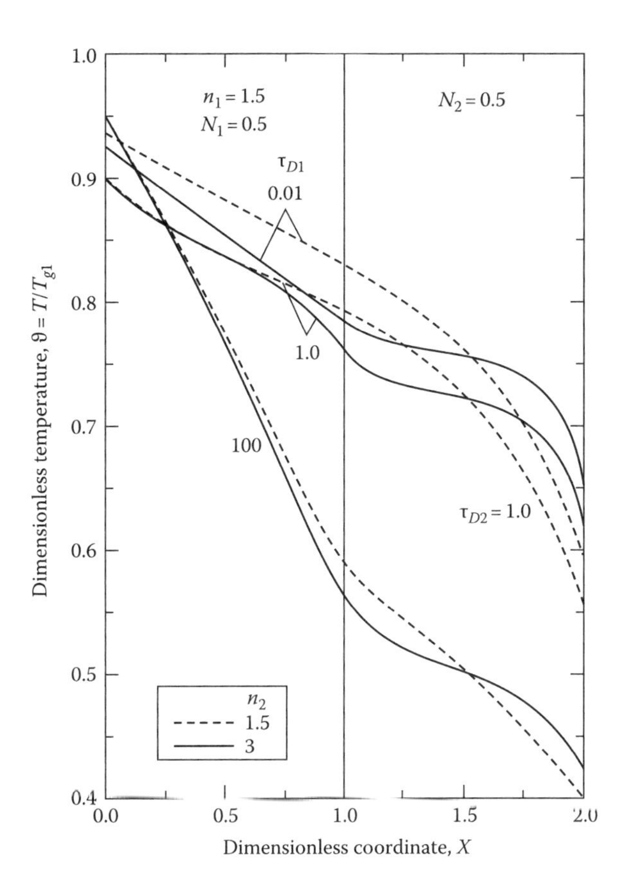

FIGURE 17.27 Effect on temperature distributions of different optical thicknesses and refractive indices in the two translucent layers of a composite; $\vartheta_{st} = \vartheta_{g1} = 1$, $\vartheta_{s2} = \vartheta_{g2} = 0.25$, $H_1 = H_2 = 1$, $\tau_{D1} = 0.01, 1$, and 100, $\tau_{D2} = 1$. (From Spuckler, C.M. and Siegel, R., *JTHT*, 8(2), 193, 1994.)

and $H_j = h_j/\sigma T_{g1}^3$. For $n_2 = n_1 = 1.5$, there are no reflections at the internal interface. Because there is no scattering, the optical thicknesses are the physical thicknesses multiplied by the layer absorption coefficients. The second layer has $\tau_{D2} = 1.0$, while the first layer has $\tau_{D1} = 0.01$, 1, and 100. A large τ_{D1} produces a large temperature decrease in the first layer, giving reduced temperatures in the second layer. For $n_2 = 3$, the temperature profiles become more uniform in the central portion of the second layer as a result of increased internal reflections. Additional steady and transient analysis and results for a two-layer composite with absorption and scattering are available in Tan et al. (2000b).

For radiation in translucent solids, a review of the literature and a summary of the theory are in Viskanta and Anderson (1975). The review by Siegel (1998) discusses transient effects that have been investigated in translucent media since the 1950s, and it also summarizes some of the analytical relations used for radiative transfer during transients. Some noteworthy contributions to the analysis of translucent media are those by Gardon (1956, 1958, 1961, 1995b), who was a pioneer in the analysis of radiative transfer in hot glass where the refractive index and surface reflections are important along with internal emission. Perpendicular and parallel polarization contributions of the radiation, as discussed for the reflectivity relations in Chapter 3, are in the original analysis in Gardon (1956). In Gardon (1958, 1995b), a comprehensive analysis of the heat treatment of glass is given. Heat treatment is important for manufacturing safety glass in automobile windows, and its analysis includes the effects of conduction within the glass and convection at the surface coupled with internal emission and radiative transport. Gardon (1961) reviews radiant heat transfer as studied by researchers in the glass industry with a digest of much of the literature on the subject up to 1961. A later review of radiation effects occurring in the glass industry in early days was given by Condon (1968), and a recent overview is by Frank and Klar (2008).

With regard to the thermal conductivity of a translucent solid, it has been pointed out (e.g., Gardon 1961, Araki 1990) that for heat transfer by combined radiation and conduction in translucent media, care must be taken in an analysis to use values for the thermal conductivity k that include only the molecular conductivity. At elevated temperatures, measurements of heat flux to determine k in translucent materials must be corrected to eliminate the radiative transfer that produces an apparent increase in k that increases strongly as the temperature is raised. A discussion of glass thermal conductivities and their determination from dynamic temperature measurements is in Mann et al. (1992).

In Siegel and Spuckler (1994b), Caren and Liu (1971), Baba and Kanayama (1975), Armaly et al. (1973), Anderson (1975), and Isard (1980), the emittance is given from layers of translucent materials with $n > 1$ that are semi-infinite in thickness (Armaly et al. 1973, Isard 1980), of finite thickness (Siegel and Spuckler 1994b), bounded on one side by a substrate (Caren and Liu 1971, Baba and Kanayama 1975, Anderson 1975), and have internal opaque radiation barriers Siegel (1999b). The radiation and conduction heat transfer through a high-temperature semitransparent layer of slag is analyzed by Viskanta and Kim (1980). Refractive index effects were analyzed by Spuckler and Siegel (1992) for plane layers exposed to external radiation and convection. Internal reflections tended to make the temperature distributions more uniform in the central portions of the layer as compared with a layer with $n = 1$, as demonstrated in Figure 17.27. This is also shown by Siegel and Spuckler (1992) for an absorbing and scattering radiating layer; in the limit where radiation is dominant so that conduction and convection are very small, the energy flows and temperatures for $n > 1$ are shown to be directly related to those for $n = 1$. This was extended by Siegel and Spuckler (1996) to a multilayer lamination of translucent materials, which provides a means for examining the effect of a variable refractive index within a plane layer. Inhomogeneous films may provide useful absorption properties for solar energy collection (Fan 1978, Heavens 1978). In Heping et al. (1991), Andre and Degiovanni (1995), Tan et al. (2000a), and Sakami et al. (2002), the response to an external radiation pulse was analyzed. Anisotropic scattering effects are examined by Wu and Wu (2000) and Liu and Dougherty (1999). An interesting example of ray tracing in a nonplanar geometry with $n > 1$ is that of radiation absorption by spheres (water droplets) Harpole (1980). The emittance is analyzed by Wu and Wang (1990) of a sphere of scattering medium with Fresnel conditions at its boundary. Transient radiation effects in two-dimensional (2D) rectangular regions are reviewed by Siegel (1998), and an analysis of cooling of a 2D axisymmetric glass disk is in Lee and Viskanta (1998). Transient heat transfer in a highly backscattering heat shield is analyzed by Cornelison and Howe (1992) for use during reentry into a planetary atmosphere. Ceramic coatings may be used for thermal protection on the walls in a channel in which there is flow of a high-temperature gas. Some ceramics are translucent, and if the walls are at different temperatures, there is radiative exchange between the coatings; analyses of temperature distributions in the coatings that include these conditions are in Siegel (1999a,b) and Wang et al. (2000).

Some situations such as antireflection coatings with graded refractive index require treatment of the variation in refractive index within a medium (Siegel and Spuckler 1993a,b, Ben Abdallah and LeDez 2000a). Lemmonier and LeDez (2005) discussed a varying refractive index in a slab using the discrete ordinates method. Huang et al. (2006) studied a layer with sinusoidally varying n. In Ben-Abdallah et al. (2001), a cylindrical region is analyzed that has a spatially variable index of refraction. Liu (2006) presents a solution for a 2D solid with graded refractive index, and Ferwerda (1999) presented the effect of a varying refractive index on scattering within a medium. Zhao et al. (2015) use Monte Carlo to study radiation in planar gradient-index media and include the effects of polarization and of Mie or Rayleigh scattering in the medium. Coupled radiation and conduction in media with varying refractive index is treated in Ben-Abdallah and DeVez (2000a,b), Xia et al. (2002), and Huang et al. (2004, 2006).

17.7 LIGHT PIPES AND FIBER OPTICS

Perfect internal reflection at an interface within medium 2 with $n_2 > n_1$ is predicted to occur (Equation 17.66) whenever the incident angle $\theta > \sin^{-1}\left(\frac{n_2}{n_1}\right)$. The radiation within the higher refractive index material will undergo 100% reflection at the interface. This means that once radiation enters a perfectly transmitting material (a perfect dielectric) such as an optical fiber in which no radiation absorption occurs, the radiation will propagate without loss along the fiber. The perfect wall reflectivity allows no losses by refraction through the fiber surface. This phenomenon is observed in a planar geometry by swimmers (and fish), who can only see through the water surface above them within a cone of angles described by

$$\theta < \sin^{-1}\left(\frac{n_{air}}{n_{water}}\right) = \sin^{-1}\left(\frac{1}{1.33}\right) = \sin^{-1}(0.75) = 48.6°$$

At greater angles, the water surface appears to be a mirror; at lesser angles, although all directions in the hemisphere are viewed from within medium 2, the refraction through the surface produces considerable distortion of the view.

For a light pipe or fiber optic, radiation enters the flat end of a long circular cylinder. If the refractive index ratio n_2/n_1 exceeds $\sqrt{2}$, all radiation entering the light-pipe end will encounter the internal cylindrical surface at greater than the critical angle (Qu et al. 2007b). An example of the pattern of radiation transmitted through a light pipe is shown in Figure 17.28. Because of the increasing reflectivity (decreasing transmissivity) of dielectrics with incident angle (Equations 3.7 and 3.8), the radiation actually entering the light-pipe end is chiefly from near-normal angles.

Barker and Jones (2003a,b) and Frankman et al. (2006) examined the temperature distribution that may arise internally in the light pipe because of absorption by inclusions or the inherent absorption by the light-pipe material. The first two papers looked at the particular case when the light-pipe tip exposed to radiation was coated with a highly absorbing material so that it acts as a blackbody absorber, emitting into the light pipe in proportion to the absorbed signal.

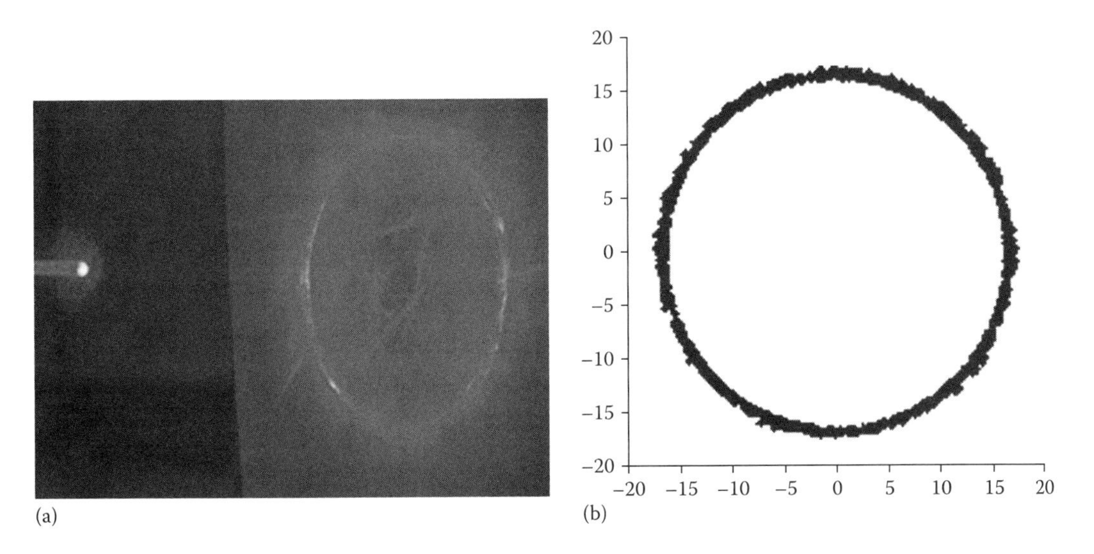

FIGURE 17.28 (a) Experimental and (b) predicted patterns of laser energy incident at exit of a fused quartz light pipe (n = 1.45843) for laser energy incident on the inlet end at 20° to the light-pipe axis. Light-pipe diameter 4 mm and length 16 cm. (From Qu, Y. et al., *IEEE Trans. Semiconduc. Manuf.*, 20(1), 26, 2007b.)

Meyer (2002), Kreider et al. (2003), Qu et al. (2007a,b), and Ertürk et al. (2008a,b) have analyzed the errors that arise because of various signal loss mechanisms that can occur in a light pipe, such as the presence of very minor surface imperfections, internal inclusions in the light-pipe material that can act as scattering centers, the effects of blocking and shadowing of reflected energy from the surface being measured by the light pipe itself, changing the temperature of the measured object by radiative losses to the colder probe, and any environmental radiation that may enter through the light-pipe walls (rather than the tip) and enter the internal signal path by encountering surface imperfections or scattering centers within the light pipe. Arnaoutakis et al. (2013) examine the effect of adding claddings or coatings and of tapered entries to light pipes for solar energy applications.

The signal entering the detector at the end of a light pipe radiation thermometer (LPRT) or optical fiber consists of emitted plus reflected energy from the object being viewed. If the environment is cold relative to the viewed object so that reflected energy can be neglected and the detector is only sensitive in a small wavelength range around a particular wavelength λ, then the detected emission is proportional to, from Equation 2.7

$$E_\lambda(T_{act}) = \varepsilon_\lambda E_\lambda(T_{act}) = \varepsilon_\lambda \frac{2\pi C_1}{\lambda^5\left[\exp\left(\dfrac{C_2}{\lambda T_{act}}\right)-1\right]} \tag{17.129}$$

However, rather than the actual temperature T_{act}, the detector reads an apparent temperature T_{app} that would appear to originate from a blackbody:

$$E_\lambda(T_{app}) = \frac{2\pi C_1}{\lambda^5\left[\exp\left(\dfrac{C_2}{\lambda T_{app}}\right)-1\right]} \tag{17.130}$$

Equating the two emitted energy rates relates the actual and apparent temperatures as

$$\varepsilon_\lambda \frac{2\pi C_1}{\lambda^5\left[\exp\left(\dfrac{C_2}{\lambda T_{act}}\right)-1\right]} = \frac{2\pi C_1}{\lambda^5\left[\exp\left(\dfrac{C_2}{\lambda T_{app}}\right)-1\right]} \tag{17.131}$$

For most engineering conditions, the factor of (−1) in the denominator can be neglected relative to the exponential term (resulting in Wien's spectral distribution [Equation 1.22]), and the actual temperature in terms of the apparent absolute temperature then becomes

$$T_{act} = \frac{T_{app}}{\left[1+\dfrac{\lambda T_{app}}{C_2}\ln\varepsilon_\lambda\right]} \tag{17.132}$$

As $\varepsilon_\lambda \rightarrow 1$, the apparent and actual temperatures approach one another. Equation 17.132 can be used for many spectrally based temperature measurements, remembering the proviso that the environmental temperature must be low in order to get an accurate temperature measurement from an optical pyrometer or light-pipe radiation thermometer.

17.8 FINAL REMARKS

In this chapter, we have discussed several models and formulations for radiative transfer through translucent materials, windows, and coatings. Because of space constraints, many other radiative transfer applications and the corresponding details are omitted. A natural extension would be to paints, for example, which are important for energy efficiency applications in buildings. Similarly, there are many other dispersed media, from foams to food preparation requirements (cheese making or baking), where radiation transfer plays a crucial role. The concepts discussed in this book constitute a good starting point for analysis of such systems, but the details need a careful study. A discussion of radiative transfer in porous, dispersed, and foamy materials is in the online appendix (Appendix E) to this book at http://www.thermalradiation.net.

HOMEWORK

17.1 A horizontal glass plate 0.225 cm thick is covered by a plane layer of water 0.625 cm thick. Radiation is incident from within air onto the upper surface of the water at an incidence angle of 60°. What is the path length of this radiation through the glass? (Let $n_{H2O} = 1.33$, and $n_{glass} = 1.57$.)
Answer: 0.270 cm.

17.2 Do Example 10.10 modified to include surface reflection and refraction effects. (Let $n_{glass} = 1.53$.)
Answer: 750 W/(m²·sr).

17.3 A radiation flux q_s is incident from the normal direction on a series of two glass sheets, each 0.35 cm thick, in air. What is the fraction T that is transmitted? ($n_{glass} = 1.53$, and $\kappa_{glass} = 9.5$ m^{-1}.) What is the fraction transmitted for a single plate 0.7 cm thick in air?

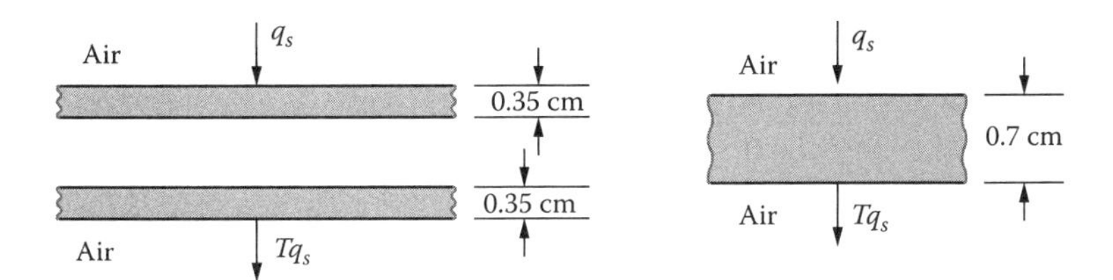

Answer: 2 plates, $T = 0.790$; 1 plate, $T = 0.857$.

17.4 Prove that for an absorbing–emitting layer with $L > \lambda$, $T + A + R = 1$.

17.5 Derive an analytical expression that shows whether or not the total absorption in a system of n plates and m plates, $A(m + n)$, is independent of whether radiation is first incident on the n or the m plates.

17.6 As a result of surface treatment, a partially transparent plate has a different reflectivity at each surface. For radiation incident on surface 1 in a single direction, obtain an expression for the overall reflectance and transmittance in terms of ρ_1, ρ_2, and τ.

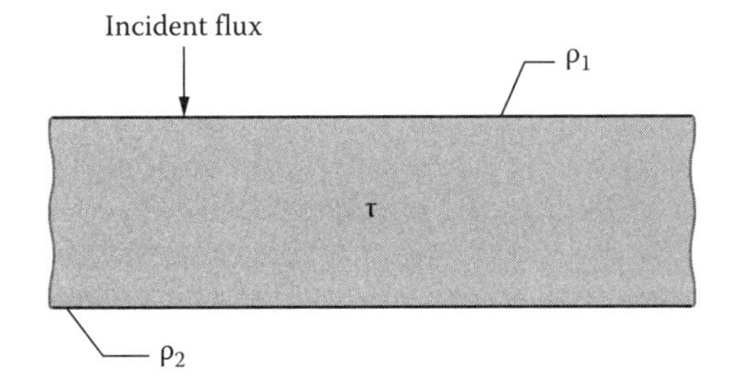

Answer: $R = \dfrac{\rho_1 + \rho_2(1-2\rho_1)\tau^2}{1-\rho_1\rho_2\tau^2}$; $T = \dfrac{(1-\rho_1)(1-\rho_2)\tau}{1-\rho_1\rho_2\tau^2}$

17.7 A double-panel glass window consists of two 5 mm parallel glass plates separated by a 10 mm air gap. The outside glass plate is made of crown glass with a refractive index of 1.52. The inside glass plate is made of flint glass with a refractive index of 1.62. Absorption in the glass plates can be ignored. Calculate the transmittance for normal incidence on the outer plate.
Answer: 0.828

17.8 For the two-plate system in Homework Problem 17.7, determine the maximum transmittance for parallel-polarized radiation and the incidence angle that gives rise to that maximum transmittance. Also, find the transmittance through the plate assembly for unpolarized incident radiation at this angle. Compare with the transmittance found for normal incidence in Homework Problem 17.7.
Answer: $T_{top+bottom} = 0.725$.

17.9 Two parallel, partially transparent plates have different τ values and a different ρ at each surface. Obtain an expression for the overall transmittance of the two-plate system for radiation incident from a single direction.

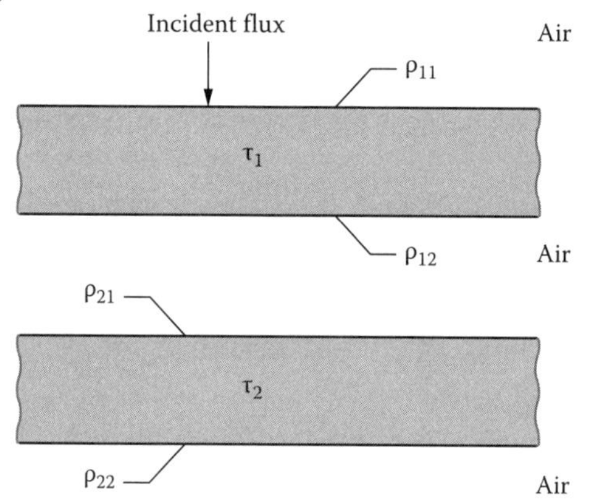

Answer:

$$T = \frac{T_1 T_2}{1 - R_{12}R_{21}}, \quad \text{where } T_n = \frac{(1-\rho_{n1})(1-\rho_{n2})\tau_n}{1-\rho_{n1}\rho_{n2}\tau_n^2}$$

$$\text{and } R_{nm} = \frac{\rho_{nm} + \rho_{nn}(1-2\rho_{nm})\tau_n^2}{1-\rho_{nm}\rho_{nn}\tau_n^2}$$

17.10 In a solar still, a thin layer of condensed water is flowing down a glass plate. The plate has a transmittance τ, and the water layer is assumed nonabsorbing. Obtain an expression for the overall transmittance T_L of the system for radiation incident on the glass in a single direction.

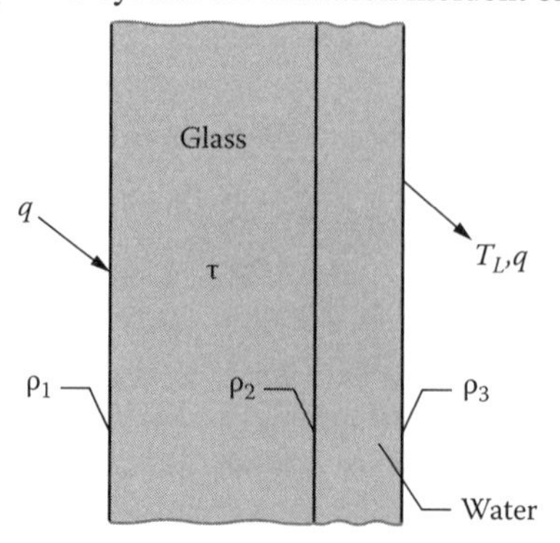

Answer: $T_L = \dfrac{(1-\rho_1)(1-\rho_2)(1-\rho_3)\tau}{(1-\rho_2\rho_3)(1-\rho_1\rho_2\tau^2)-\rho_1\rho_3(1-\rho_2)^2\tau^2}$

17.11 The two glass plates in Homework Problem 17.3 are followed by a collector surface with absorptivity $\alpha_c = 0.95$. What is the fraction A_c absorbed by the collector surface for normally incident radiation?

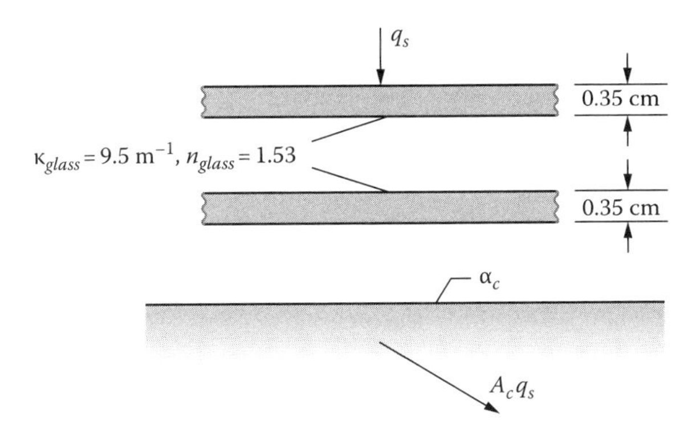

Answer: $A_c = 0.756$.

17.12 Derive Equation 17.35 for a thick dielectric film on a dielectric substrate.

Answer: $R = \dfrac{\rho_1 + \rho_2(1-\rho_1)\tau^2}{1-\rho_1\rho_2\tau^2}$

17.13 Extend Homework Problem 17.12 to a system of two differing dielectric layers coated onto a dielectric substrate.

Answer: $R = \rho_1 + \dfrac{(1-\rho_1)^2\tau_1^2\left[\rho_2 + (1-2\rho_2)\rho_3\tau_2^2\right]}{(1-\rho_1\rho_2\tau_1^2)(1-\rho_2\rho_3\tau_2^2)-(1-\rho_2)^2\rho_1\rho_3\tau_1^2\tau_2^2}$

17.14 Two opaque gray walls have a transparent (nonabsorbing) plate between them. The transparent plate has surface reflectivities ρ. Derive a relation for the heat transfer from wall $w1$ to wall $w2$. Neglect heat conduction in the transparent plate, and neglect the fact that ρ depends on the angle of incidence.

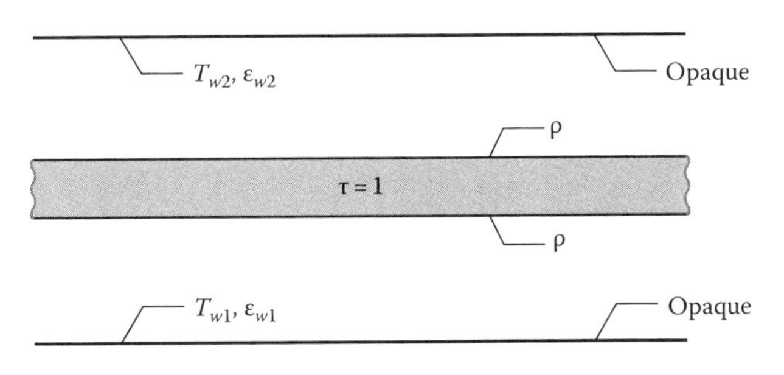

Answer: $q_{w1\to w2} = \dfrac{\sigma\left(T_{w1}^4 - T_{w2}^4\right)}{\dfrac{1}{\varepsilon_{w1}} + \dfrac{1}{\varepsilon_{w2}} - 1 + \dfrac{2\rho}{1-\rho}}$

17.15 Water is flowing between two identical glass plates adjacent to an opaque wall. Derive an expression for the fraction of radiation incident from a single direction that is absorbed by the water in terms of the ρ and τ values of the interfaces and layers. (Include only radiative energy transfer.)

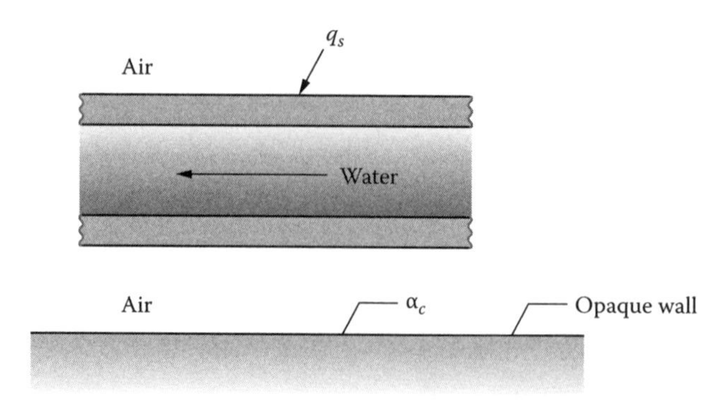

17.16 Do Homework Problem 12.5 with the index of refraction for glass equal to 1.53, and for the liquid equal to 1.35.
Answer: 503 W/m^2.

17.17 The sun shines on a flat-plate solar collector that has a single 0.5 cm thick glass cover (use the data in Figure 9.22 and assume spectral properties are constant below $\lambda = 1$ μm) over a silicon oxide-coated aluminum absorber plate (use data from Figure 3.46 and assume properties are constant below $\lambda = 0.4$ μm). Determine the effective solar absorptivity of the collector for normal incidence (i.e., the fraction of incident solar energy that is absorbed by the collector plate.)
Answer: 0.54.

17.18 A diffuse radiation flux q_{air} is incident on a dielectric medium having an absorption coefficient κ and a simple index of refraction n. The air–medium interface is slightly rough so that it acts like a diffuse surface. It has a directional–hemispherical reflectivity of $\rho(\theta)$ that is assumed to be independent of θ. Derive a relation to determine the amount of incident energy directly absorbed per unit volume, dQ/dV, in the medium as a function of depth x from the surface. For $n = 1.52$, obtain an estimate for the surface reflectivity for diffuse incident radiation by using Figure 3.3. Using this value and the other conditions as in Example 17.2, compute dQ/dV. How does the result compare with that in Example 17.2 and what is the reason for the difference?

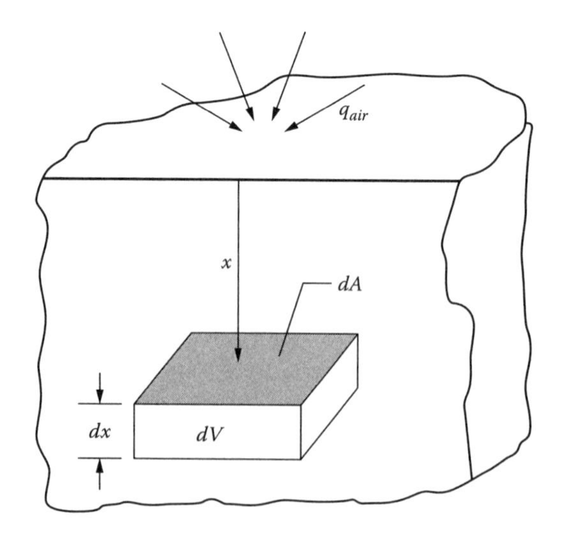

Answer: 3.08 W/cm^3.

17.19 A semitransparent layer of fused silica has a complex index of refraction of $n - ik = 1.42 - i \times 1.45 \times 10^{-4}$ at the wavelength $\lambda = 5.20$ μm. Unpolarized diffuse radiation in air at this wavelength is incident on the layer. What is the value of the overall transmittance of the layer for this diffuse incident radiation?
Answer: $T = 0.148$.

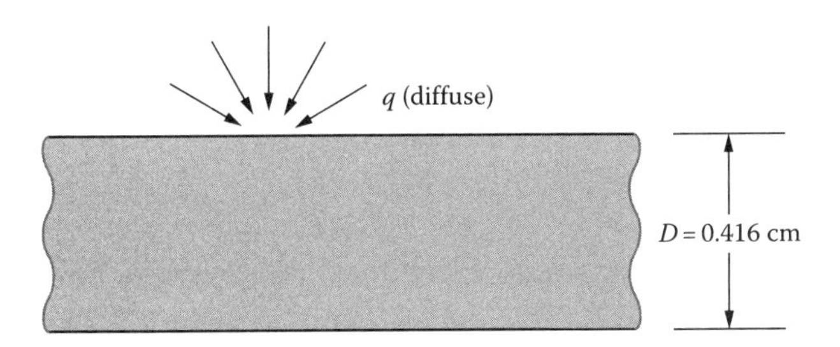

17.20 A series of three parallel glass plates is being used to absorb incident infrared radiation. The incident radiation is at wavelength $\lambda = 4$ μm. The complex index of refraction of the glass at this wavelength is $n - ik = 1.40 - 5.8 \times 10^{-5}i$. The plates have optically smooth surfaces. The radiation is incident from air in a direction normal to the plates. What fraction of the incident radiation is absorbed by the center plate?
Answer: 0.237.

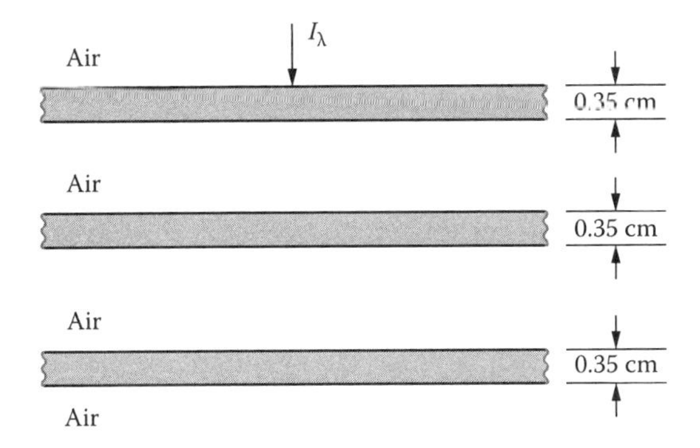

17.21 A plane layer of semitransparent emitting and absorbing hot material is at uniform temperature. The surfaces of the layer are optically smooth, and the layer material is nonscattering. The simple index of refraction of the layer is n and the absorption coefficient is κ (the layer is assumed gray). Obtain an expression for the emittance of the layer into vacuum.

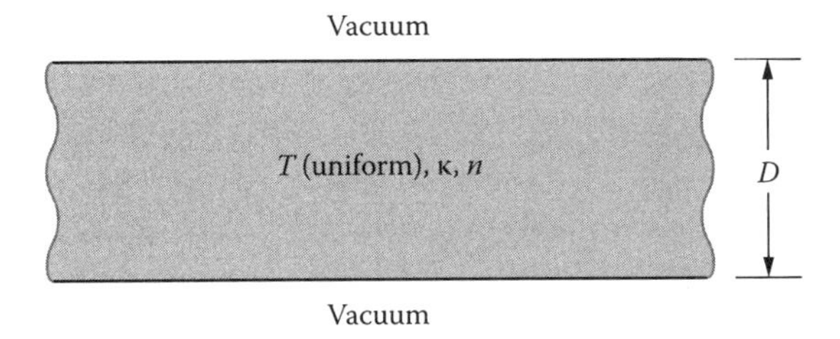

17.22 Two very thick dielectric regions with small absorption coefficients are in perfect contact at an optically smooth interface A_i. Prior to being placed in contact, region 1 was heated to a uniform temperature T_1, while region 2 was kept very cold at $T_2 \ll T_1$. The indices of refraction

are such that $n_1 > n_2 > 1$. Derive an expression for the heat flux emitted across A_1 from region 1 into region 2. Evaluate the result for $n_1 = 2.35$, $n_2 = 1.54$, $T_1 = 627$ K, and $T_2 \approx 0$.
Answer: 18,800 W/m^2.

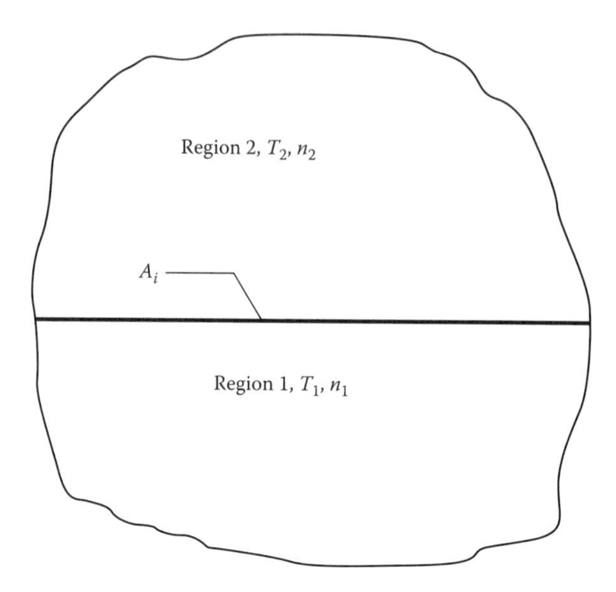

17.23 Two very thick dielectric regions with small absorption coefficients are separated by a plane layer of a third dielectric material that is of thickness $D = 2$ cm. The two interfaces are optically smooth. The plane layer between the two thick regions is perfectly transparent. This plane layer is at low temperature, having been suddenly inserted between the two dielectric regions. For the temperatures shown, calculate the net heat flux being transferred by radiation from medium 1 into medium 2. Include the effects of interface reflections. The effect of heat conduction is neglected. The indices of refraction are given in the figure.
Answer: 13.0 kW/m^2.

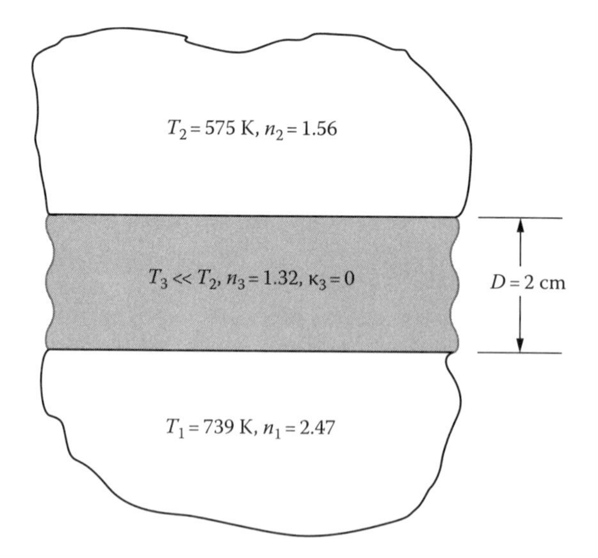

17.24 A layer of transmitting glass is over a parallel opaque cold substrate with gray absorptivity α. The glass has a surface reflectivity of ρ, and the glass layer has transmittance of τ. Derive an expression for the reflectance R of the glass/substrate assembly.

Answer: $R=\rho+\dfrac{(1-\rho)^2\tau\left[1-\rho-\alpha+2\rho\alpha\right]}{1-\rho+\rho\alpha-\rho\tau^2\left[1-\rho-\alpha+2\rho\alpha\right]}$

17.25 A layer of transmitting glass (silica) is separated from a silicon wafer by a layer of polymer. A laser provides monochromatic energy incident on the assembly of layers. The silicon wafer is opaque at the laser wavelength and has surface absorptivity $\alpha = 0.694$. The silica has a refractive index that gives a surface reflectivity for the laser energy of $\rho_{SiO_2} = 0.04$ and has transmittance for laser energy of $\tau_{SiO_2} = 0.95$ Similarly, the polymer layer has a surface reflectivity for the laser energy of $\rho_{poly} = 0.0452$ and has transmittance for laser energy of $\tau_{poly} = 0.90$. Derive an expression for the reflectance R for the laser energy of the glass/substrate assembly, and also determine the fraction of the incident laser energy that is deposited in the silica and polymer layers.
Answer: 0.302.

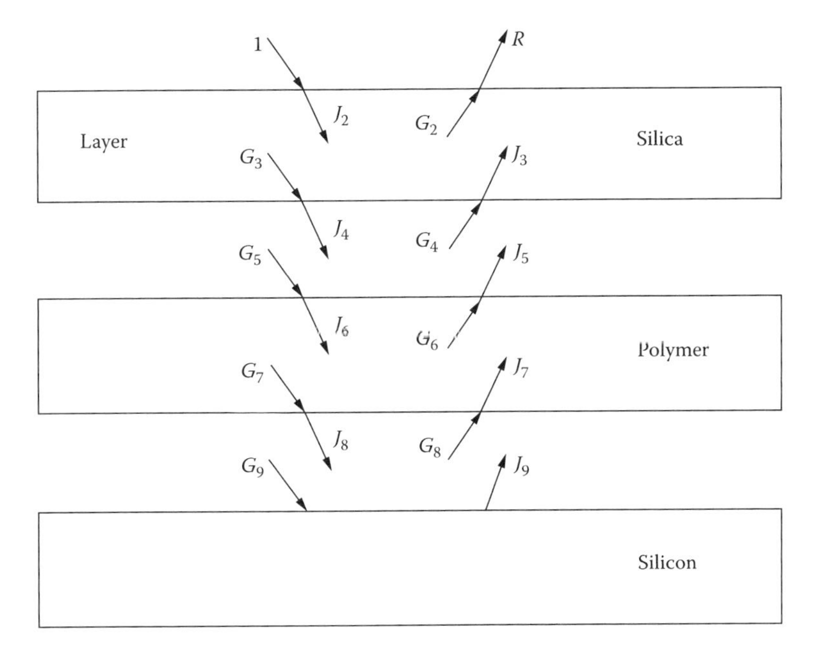

Appendix A: Conversion Factors, Radiation Constants, and Blackbody Functions

Table A.1 lists some fundamental constants of radiation physics. Tables of conversion factors between the International System of Units (SI) and other systems of units are in Tables A.2 and A.3. In Table A.4, values of various radiation constants are given in both SI and U.S. conventional system (USCS) engineering units. All values in Tables A.1, A.4, and A.5 are based on values of the physical constants given by Mohr et al. (2012).

The emission properties for a blackbody are presented in Chapter 1. Fifty years ago, when the study of radiative transfer was stimulated by applications in outer space, the lack of convenient computer capability sometimes made it inconvenient to evaluate the blackbody emission function and the fraction of blackbody energy in a wavelength band. Extensive tables of radiation functions had been calculated and published such as in Lowan (1941) and Pivovonsky and Nagel (1961), and an extended table was included in previous editions of this text. Tables are no longer as useful because the functions can now be readily evaluated numerically for engineering radiation calculations. A shortened Table A.5 is included for quick reference and for help in checking calculations. It lists blackbody emission properties as functions of λT in both SI and USCS units. The $F_{0\to\lambda T}$ were calculated using the series obtained by integrating the blackbody radiation function using integration by parts as in Chapter 1. This series was used to evaluate the blackbody integrals in Lowan and is also given by Chang and Rhee (1984). The relation is repeated here for convenience:

$$F_{0\to\lambda T} = \frac{15}{\pi^4}\sum_{m=1}^{\infty}\left[\frac{e^{-m\zeta}}{m}\left(\zeta^3 + \frac{3\zeta^2}{m} + \frac{6\zeta}{m^2} + \frac{6}{m^3}\right)\right]$$

where $\zeta = C_2/\lambda T$. This series is quite useful for computer solutions to provide calculation of the $F_{0\to\lambda T}$ function.

TABLE A.1
Values of the Fundamental Physical Constants

Definition	Symbol and Value
Bohr electron radius	$a_0 = (4\pi/\mu_0 c_0^2)(\hbar^2/m_e e^2)$ $= 0.52917721092 \times 10^{-10}$ m
Speed of light in vacuum	$c_0 = 2.99792458 \times 10^8$ m/s
Electronic charge	$e = 1.6021764565 \times 10^{-19}$ C
Planck's constant	$h = 6.62606957 \times 10^{-34}$ J·s $\hbar = h/2\pi = 1.054571726 \times 10^{-34}$ J·s
Boltzmann's constant	$k = 1.3806488 \times 10^{-23}$ J/K
Electron rest mass	$m_c = 9.10938291 \times 10^{-31}$ kg
Avogadro's number	$N_a = 6.02214129 \times 10^{26}$ particles/kg·mol
Classical electron radius	$r_0 = (m_0/4p)\,(e^2/m_e)$ $= 2.8179403267 \times 10^{-15}$ m
Thomson cross section	$\sigma_T = 8\pi r_0^2/3 = 0.6652458734 \times 10^{-28}$ m^2
Permeability of vacuum	$\mu_0 = 4\pi \times 10^{-7}$ N·s^2/C^2
Electron volt	1 eV $= 1.602176565 \times 10^{-19}$ J
Ionization potential of hydrogen atom (Rydberg's constant)	$R = (c_0^2\mu_0/4\pi)(e^2/2a_0)$ $= [c_0^4\mu_0^2/(4\pi)^2](e^4 m_e/2\hbar^2)$ $= 13.60569253$ eV

Source: Mohr, P.J. et al., *Rev. Mod. Phys.*, 84(4), 1527, 2012.

TABLE A.2
Conversion Factors for Lengths

	Mile (mi)	Kilometer (km)	Meter (m)	Foot (ft)	Inch (in.)
1 mile =	1	1.609	1609	5280	6.336×10^{4}
1 kilometer =	0.6214	1	10^{3}	3.281×10^{3}	3.937×10^{4}
1 meter =	6.214×10^{-4}	10^{-3}	1	3.281	39.37
1 foot =	1.894×10^{-4}	3.048×10^{-4}	0.3048	1	12
1 inch =	1.578×10^{-5}	2.540×10^{-5}	2.540×10^{-2}	8.333×10^{-2}	1
1 centimeter =	6.214×10^{-6}	10^{-5}	10^{-2}	3.281×10^{-2}	0.3937
1 millimeter =	6.214×10^{-7}	10^{-6}	10^{-3}	3.281×10^{-3}	0.03937
1 micrometer =	6.214×10^{-10}	10^{-9}	10^{-6}	3.281×10^{-6}	3.937×10^{-5}
1 nanometer =	6.214×10^{-13}	10^{-12}	10^{-9}	3.281×10^{-9}	3.937×10^{-8}
1 angstrom =	6.214×10^{-14}	10^{-13}	10^{-10}	3.281×10^{-10}	3.937×10^{-9}
	Centimeter (cm)	**Millimeter (mm)**	**Micrometer (μm)**	**Nanometer (nm)**	**Angstrom (Å)**
1 mile =	1.609×10^{-5}	1.609×10^{6}	1.609×10^{9}	1.609×10^{12}	1.609×10^{13}
1 kilometer =	10^{5}	10^{6}	10^{9}	10^{12}	10^{13}
1 meter =	10^{2}	10^{3}	10^{6}	10^{9}	10^{10}
1 foot =	30.48	3.048×10^{2}	3.048×10^{5}	3.048×10^{8}	3.048×10^{9}
1 inch =	2.540	25.40	2.540×10^{4}	2.540×10^{7}	2.540×10^{8}
1 centimeter =	1	10	10^{4}	10^{7}	10^{8}
1 millimeter =	10^{-1}	1	10^{3}	10^{6}	10^{7}
1 micrometer =	10^{-4}	10^{-3}	1	10^{3}	10^{4}
1 nanometer =	10^{-7}	10^{-6}	10^{-3}	1	10
1 angstrom =	10^{-8}	10^{-7}	10^{-4}	10^{-1}	1

Source: Mechtly, E.A., The International System of Units, Physical Constants and Conversion Factors, NASA SP-7012, 2nd revision, 1973.

TABLE A.3
Useful Conversion Factors

Area

1 ft^2 = 0.0929030 m^2
1 $in.^2$ = 6.4516 × 10^{-4} m^2
1 m^2 = 10.7639 ft^2

Volume

1 ft^3 = 0.028317 m^3
1 m^3 = 35.315 ft^3

Energy

(1 kJ = 1 kW·s)
1 kJ = 0.94782 Btu[a] = 0.23885 kcal[a]
1 Btu = 1.0551 kJ = 0.25200 kcal
1 kcal = 4.1868 kJ = 3.9683 Btu
1 kW·h = 3.60 × 10^6 J
1 electron volt = 1 eV
= 1.602176565 × 10^{-19} J

Energy Rate

1 W = 3.4121 Btu[a]/h = 0.85985 kcal[a]/h
1 Btu/h = 0.29307 W = 0.25200 kcal/h
1 kcal/h = 1.1630 W = 3.9683 Btu/h

Thermal Conductivity

1 W/(m·K) = 0.57779 Btu/(h·ft·°R)
= 0.85985 kcal/(h·m·K)
1 Btu/(h·ft·°R) = 1.7307 W/(m·K)
= 1.4882 kcal/(h·m·K)
1 kcal/(h·m·K) = 1.1630 W/(m·K)
= 0.67197 Btu/(h·ft·°R)

Energy Per Unit Mass

1 kJ/kg = 0.42992 Btu/lb = 0.23885 kcal/kg
1 Btu/lb = 2.3260 kJ/kg = 0.55556 kcal/kg
1 kcal/kg = 4.1868 kJ/kg = 1.8000 Btu/lb

Pressure

1 Pa = 1 N/m^2 = 1 × 10^{-5} bar
= 9.8692 × 10^{-6} atm
1 kPa = 0.14504 psia
1 bar = 10^5 Pa = 0.9862 atm
= 14.504 psia
1 lbf/in^2 (psia) = 6894.8 Pa
= 6.8948 × 10^{-2} bar = 0.068046 atm
1 atm = 101.325 kPa = 14.696 psia
= 1.0133 bar

Mass

1 lbm = 0.453592 kg
1 kg = 2.20462 lbm

Density

1 lbm/ft^3 = 16.0185 kg/m^3
1 kg/m^3 = 0.062428 lbm/ft^3

Energy Rate Per Unit Area

1 W/m^2 = 0.31700 Btu[a]/h·ft^2)
= 0.85985 kcal[a]/(h·m^2)
1 Btu/(h·ft^2) = 3.1546 W/m^2
= 2.7125 kcal/(h·m^2)
1 kcal/(h·m^2) = 1.1630 W/m^2
= 0.36867 Btu/(h·ft^2)

Heat Transfer Coefficient

1 W/(m^2·K) = 0.17611 Btu[a]/(h·ft^2·°R)
= 0.85985 kcal[a] (h·m^2·K)
1 Btu/(h·ft^2·°R) = 5.6783 W/(m^2·K)
= 4.8824 kcal/(h·m^2·K)
1 kcal/(h·m^2·K) = 1.1630 W/(m^2·K)
= 0.20482 Btu/(h·ft^2·°R)

Specific Heat

1 kJ/(kg·K) = 0.23885 Btu/(lb·°R)
= 0.23885 kcal/(kg·K)
1 Btu/(lb·°R) = 4.1868 kJ/(kg·K)
= 1.0000 kcal/(kg·K)
1 kcal/(kg·K) = 4.1868 kJ/(kg·K)
= 1.0000 Btu/(lb·°R)

Temperature

$K = (5/9)°R = (5/9)°F + 459.67 = °C + 273.15$
$°R = (9/5)K = (9/5)°C + 273.15 = °F + 459.67$
$°F = (9/5)°C + 32$
$°C = (5/9)°F - 32$

Source: Mechtly, E.A., The International System of Units, Physical Constants and Conversion Factors, NASA SP-7012, 2nd revision, 1973.

[a] International Steam Table (for all Btu and kcal).

TABLE A.4
Radiation Constants

Symbol	Definition	Value
C_1	Constant in Planck's spectral energy (or intensity) distribution	0.18878×10^8 Btu·μm^4/(h·ft^2·sr)
		0.59552200×10^8 W·μm^4/(m^2·sr)
		$0.59552200 \times 10^{-16}$ W·m^2/sr
C_2	Constant in Planck's spectral energy (or intensity) distribution	25,897.986 μm·°R
		14,387.770 μm·K
		0.014387770 m·K
C_3	Constant in Wien's displacement law	5,215.9833 μm·°R
		2,897.7721 μm·K
		0.0028977721 m·K
C_4	Constant in equation for maximum blackbody intensity	6.8761×10^{-14} Btu/(h·ft^2·μm·°R^5·sr)
		4.09570×10^{-12} W/(m^2·μm·K^5·sr)
σ	Stefan–Boltzmann constant	0.17123×10^{-8}Btu/(h·ft·°R^4)
		5.670373×10^{-8} W/(m^2·K^4)
q_{solar}	Solar constant	433 Btu/(h·ft^2)
		1,366 W/m^2
T_{solar}	Effective surface radiating temperature of the sun	5,780 K, 10,400°R

Source: Mohr, P.J. et al., *Rev. Mod. Phys.*, 84(4), 1527, 2012.

TABLE A.5
Blackbody Functions

Wavelength–Temperature Product, λT		Blackbody Hemispherical–Spectral Emissive Power Divided by the Fifth Power of Temperature, $E_{\lambda b}/T^5$		Blackbody Fraction
μm·K	μm·°R	W/(m²·μm·K⁵)	Btu/(h·ft²·μm·°R⁵)	$F_{0\to\lambda T}$
600	1,080	185.40E–18	311.04E–20	9.29E–8
800	1,440	176.59E–16	296.26E–18	1.64E–5
1,000	1,800	211.13E–15	354.20E–17	3.21E–4
1,200	2,160	933.41E–15	156.59E–16	0.00213
1,400	2,520	239.45E–14	401.72E–16	0.00779
1,600	2,880	443.82E–14	744.56E–16	0.01972
1,800	3,240	669.05E–14	112.24E–15	0.03934
2,000	3,600	879.01E–14	147.47E–15	0.06673
2,200	3,960	105.04E–13	176.22E–15	0.10089
2,400	4,320	117.37E–13	196.90E–15	0.14026
2,500	4,500	121.72E–13	204.19E–15	0.16136
2,600	4,680	124.93E–13	209.58E–15	0.18312
2,700	4,860	127.09E–13	213.21E–15	0.20536
2,800	5,040	128.30E–13	215.24E–15	0.22789
2,897.7685[a]	5,216	128.67E–13	215.86E–15	0.25006
2,900	5,220	128.67E–13	215.86E–15	0.25056
3,000	5,400	128.30E–13	215.24E–15	0.27323
3,100	5,580	127.30E–13	213.56E–15	0.29578
3,200	5,760	125.76E–13	210.98E–15	0.31810
3,300	5,940	123.77E–13	207.63E–15	0.34011
3,400	6,120	121.41E–13	203.67E–15	0.36173
3,500	6,300	118.75E–13	199.21E–15	0.38291
3,600	6,480	115.86E–13	194.36E–15	0.40360
3,800	6,840	109.59E–13	183.85E–15	0.44337
4,000	7,200	102.97E–13	172.75E–15	0.48087
4,200	7,560	962.61E–14	161.49E–15	0.51600
4,400	7,920	896.45E–14	150.39E–15	0.54878
4,600	8,280	832.48E–14	139.66E–15	0.57926
4,800	8,640	771.49E–14	129.43E–15	0.60754
5,000	9,000	713.96E–14	119.78E–15	0.63373
5,200	9,360	660.11E–14	110.74E–15	0.65795
5,400	9,720	609.99E–14	102.33E–15	0.68034
5,600	10,080	563.56E–14	945.44E–16	0.70102
5,800	10,440	520.67E–14	873.49E–16	0.72013
6,000	10,800	481.16E–14	807.21E–16	0.73779
6,400	11,520	411.45E–14	690.26E–16	0.76920
6,800	12,240	352.71E–14	591.71E–16	0.79610
7,200	12,960	303.27E–14	508.78E–16	0.81918
7,600	13,680	261.65E–14	438.96E–16	0.83907
8,000	14,400	226.55E–14	380.07E–16	0.85625
8,400	15,120	196.87E–14	330.28E–16	0.87116
8,800	15,840	171.71E–14	288.06E–16	0.88413
9,200	16,560	150.30E–14	252.14E–16	0.89547

(*Continued*)

TABLE A.5 (*Continued*)
Blackbody Functions

Wavelength–Temperature Product, λT		Blackbody Hemispherical–Spectral Emissive Power Divided by the Fifth Power of Temperature, $E_{\lambda b}/T^5$		Blackbody Fraction
μm·K	μm·°R	W/(m²·μm·K⁵)	Btu/(h·ft²·μm·°R⁵)	$F_{0\rightarrow\lambda T}$
9,600	17,280	132.02E–14	221.48E–16	0.90541
10,000	18,000	116.37E–14	195.22E–16	0.91416
11,000	19,800	860.92E–15	144.43E–16	0.93185
12,000	21,600	649.08E–15	108.89E–16	0.94505
13,000	23,400	497.78E–15	835.10E–17	0.95509
14,000	25,200	387.67E–15	650.37E–17	0.96285
15,000	27,000	306.13E–15	513.58E–17	0.96893
16,000	28,800	244.80E–15	410.68E–17	0.97377
18,000	32,400	161.78E–15	271.41E–17	0.98081
20,000	36,000	111.03E–15	186.26E–17	0.98555
25,000	45,000	492.47E–16	826.18E–18	0.99217
30,000	54,000	250.2 IE–16	419.76E–18	0.99529
35,000	63,000	140.12E–16	235.07E–18	0.99695
40,000	72,000	844.11E–17	141.61E–18	0.99792
45,000	81,000	538.22E–17	902.93E–19	0.99852
50,000	90,000	359.11E–17	602.45E–19	0.99890
60,000	108,000	117.57E–17	297.90E–19	0.99935
70,000	126,000	975.65E–18	163.68E–19	0.99959

[a] $\lambda_{max}T$.

Appendix B: Radiative Properties

Tables B.1 and B.2 of normal total emissivities and normal total absorptivities for incident solar radiation, respectively, are provided here for convenience in working problems and to give the reader an indication of the magnitudes to be expected. As discussed in Chapter 3, many factors such as roughness and oxidation can strongly affect the radiative properties. No attempt is made here to describe in detail the condition of the material sample; hence, the values given here are only reasonable approximations in some instances. For detailed information on radiative properties including sample descriptions and results from many sources, the reader is referred to the collections in Gubareff et al. (1960), Wood et al. (1964), Touloukian and Ho (1970, 1972a,b), and Palik (1998). The list of properties given by Touloukian and Ho in three volumes is very extensive. Some additional information is in Svet (1965). Henninger (1984) has information on spacecraft materials and absorption for the solar spectrum. Additional properties are available on the web, with varying degrees of credibility. As will be seen from the references cited, for the same material, there can be considerable differences in the property values measured by different investigators.

TABLE B.1
Normal Total Emissivity

Metal	Surface Temperature,[a] °F (K)	ε_n
Aluminum:		
Highly polished plate	400–1100 (480–870)	0.038–0.06
Bright foil	70 (295)	0.04
Polished plate	212 (373)	0.095
Heavily oxidized	200–1000 (370–810)	0.20–0.33
Antimony, polished	100–500 (310–530)	0.28–0.31
Bismuth, bright	176 (350)	0.34
Brass:		
Highly polished	500–700 (530–640)	0.028–0.031
Polished	200 (370)	0.09
Dull	120–660 (320–620)	0.22
Oxidized	400–1000 (480–810)	0.60
Cadmium	77 (298)	0.02
Chromium, polished	100–2000 (310–1370)	0.08–0.40
Copper:		
Highly polished	100 (310)	0.02
Polished	100–500 (310–530)	0.04–0.05
Scraped, shiny	100 (310)	0.07
Slightly polished	100 (310)	0.15
Black oxidized	100 (310)	0.78
Dow metal	0–600 (255–590)	0.15
Gold:		
Highly polished	200–1100 (370–870)	0.018–0.035
Polished	266 (400)	0.018
Haynes alloy X, oxidized	600–2000 (590–1370)	0.85–0.88
Iron:		
Highly polished, electrolytic	100–500(310–530)	0.05–0.07
Polished	800–900 (700–760)	0.14–0.38
Freshly rubbed with emery	100 (310)	0.24
Wrought iron, polished	100–500 (310–530)	0.28
Cast iron, freshly turned	100 (310)	0.44
Iron plate, pickled, then rusted red	68 (293)	0.61
Cast iron, oxidized at 1100°F	400–1100(480–870)	0.64–0.78
Cast iron, rough, strongly oxidized	100–500 (310–530)	0.95
Lead:		
Polished	100–500 (310–530)	0.06–0.08
Rough unoxidized	100 (310)	0.43
Oxidized at 1100°F	100 (310)	0.63
Magnesium, polished	100–500 (310–530)	0.07–0.13
Mercury, unoxidized	40–200 (280–370)	0.09–0.12
Molybdenum:		
Polished	100–500 (310–530)	0.05–0.08
Polished	1000–2500 (810–1640)	0.10–0.18
Polished	5000 (3030)	0.29
Monel:		
Polished	100 (310)	0.17
Oxidized at 1100°F	1000 (810)	0.45

(Continued)

TABLE B.1 (*Continued*) Normal Total Emissivity

Metal	Surface Temperature,[a] °F (K)	ε_n
Nickel:		
Electrolytic	100–500 (310–530)	0.04–0.06
Technically pure, polished	440–710 (500–650)	0.07–0.087
Electroplated on iron, not polished	68 (293)	0.11
Plate oxidized at 1100°F	390–1110 (470–870)	0.37–0.48
Nickel oxide	1200–2300 (920–1530)	0.59–0.86
Platinum:		
Electrolytic	500–1000 (530–810)	0.06–0.10
Polished plate	440–1160 (500–900)	0.054–0.104
Silver, polished	100–1000 (310–810)	0.01–0.03
Stainless steel:		
Type 304 foil (1 mil)	80 (300)	0.05
Inconel X foil (1 mil)	80 (300)	0.10
Inconel X, polished	−300 to 900 (90–760)	0.19–0.20
Inconel B, polished	−300 to 900 (90–760)	0.19–0.22
Type 301, polished	75 (297)	0.16
Type 310, smooth	1500 (1090)	0.39
Type 316, polished	400–1900 (480–1310)	0.24–0.31
Steel:		
Polished sheet	−300 to 0 (90–273)	0.07–0.08
Polished sheet	0–300 (273–420)	0.08–0.14
Mild steel, polished	500–1200 (530–920)	0.27–0.31
Sheet with skin due to rolling	70 (295)	0.66
Sheet with rough oxide layer	70 (295)	0.81
Tantalum	2500–5000 (1640–3030)	0.2–0.3
Foil	80 (300)	0.05
Tin:		
Polished sheet	93 (310)	0.05
Bright tinned iron	76 (298)	0.043–0.064
Tungsten:		
Polished	80 (300)	0.03
Clean	100–1000 (310–810)	0.03–0.08
Filament	80 (300)	0.032
Filament	6000 (3590)	0.39
Zinc:		
Polished	100–1000 (310–810)	0.02–0.05
Galvanized sheet, fairly bright	100 (310)	0.23
Gray oxidized	70 (295)	0.23–0.28
Dielectric	**Surface Temperature,[a] °F (K)**	ε_n
Alumina on inconel	1000–2000 (810–1370)	0.65–0.45
Asbestos:		
Cloth	199 (365)	0.90
Paper	100 (310)	0.93
Board	100 (310)	0.96
Asphalt pavement	100 (310)	0.93

(*Continued*)

TABLE B.1 (*Continued*)
Normal Total Emissivity

Dielectric	Surface Temperature,[a] °F (K)	ε_n
Brick:		
White refractory	2000 (1370)	0.29
Fireclay	1800 (1260)	0.75
Rough red	100 (310)	0.93
Carbon, lampsoot	100 (310)	0.95
Ceramic, glazed earthenware	70 (295)	0.90
Clay, fired	158 (340)	0.91
Concrete, rough	100 (310)	0.94
Corundum, emery rough	200 (370)	0.86
Cotton cloth	68 (293)	0.77
Granite	70 (295)	0.45
Ice:		
Smooth	32 (273)	0.966
Rough crystals	32 (273)	0.985
Magnesium oxide, refractory	300–900 (420–760)	0.69–0.55
Marble, white	100 (310)	0.95
Mica	100 (310)	0.75
Paint:		
Oil, all colors	212 (373)	0.92–0.96
Red lead	200 (370)	0.93
Lacquer, flat black	100–200 (310–370)	0.96–0.98
Silicone, flat black	−149 to 212 (173–273)	0.90
Velostat black plastic	27 (300)	0.85
Paper:		
Roofing	100 (310)	0.91
White	100 (310)	0.95
Plaster	100 (310)	0.91
Porcelain, glazed	70 (295)	0.92
Rokide A on molybdenum	600–1500 (590–1090)	0.79–0.60
Rubber, hard	68 (293)	0.92
Sand	68 (293)	0.76
Sandstone	100–500 (310–530)	0.83–0.90
Silicon carbide	300–1200 (420–920)	0.83–0.96
Silk cloth	68 (293)	0.78
Slate	100 (310)	0.67–0.80
Snow	20 (270)	0.82
Soil:		
Black loam	68 (293)	0.66
Plowed field	68 (293)	0.38
Soot, candle	200–500 (370–530)	0.95
Water, deep	32–212 (273–373)	0.96
Wood:		
Sawdust	100 (310)	0.75
Oak, planed	70 (295)	0.90
Beech	158 (340)	0.94

[a] When temperatures and emissivities both have ranges, linear interpolation can be used over these values.

TABLE B.2
Normal Total Absorptivity for Incident Solar Radiation (Receiving Material at 300 K)

Metal	ε_n
Aluminum:	
Highly polished	0.10
Polished	0.20
Chromium, electroplated	0.40
Copper:	
Highly polished	0.18
Clean	0.25
Tarnished	0.64
Oxidized	0.70
Galvanized iron	0.38
Gold, bright foil	0.29
Iron:	
Ground with fine grit	0.36
Blued	0.55
Sandblasted	0.75
Magnesium, polished	0.19
Nickel:	
Highly polished	0.15
Polished	0.36
Electrolytic	0.40
Platinum, bright	0.31
Silver:	
Highly polished	0.07
Polished	0.13
Commercial sheet	0.30
Stainless steel #301, polished	0.37
Tungsten, highly polished	0.37
Dielectric	ε_n
Aluminum oxide (Al_2O_3)	0.06–0.23
Asphalt pavement, dust-free	0.93
Brick, red	0.75
Clay	0.39
Concrete roofing tile:	
Uncolored	0.73
Brown	0.91
Black	0.91
Earth, plowed field	0.75
Felt, black	0.82
Graphite	0.88
Grass	0.75–0.80
Gravel	0.29
Leaves, green	0.71–0.79
Magnesium oxide (MgO)	0.15
Marble, white	0.46

(*Continued*)

TABLE B.2 (*Continued*)
Normal Total Absorptivity for Incident Solar Radiation (Receiving Material at 300 K)

Dielectric	ε_n
Paint:	
Aluminum	0.55
Oil, zinc white	0.30
Oil, light green	0.50
Oil, light gray	0.75
Oil, black on galvanized iron	0.90
Silicone, flat black	0.95–0.97
Velostat black plastic	0.96
Paper, white	0.28
Slate, blue gray	0.88
Snow, clean	0.2–0.35
Soot, coal	0.95
Titanium dioxide (TiO_2)	0.12
Zinc oxide	0.15
Zinc sulfide (ZnS)	0.21
Coating	ε_n
Black coatings:	
Anodize black	0.88
Carbon black paint NS-7	0.96
Ebonol C black	0.97
Martin black velvet paint	0.91
3M black velvet paint	0.97
Tedlar black plastic	0.94
Velostat black plastic	0.96
White coatings:	
Barium sulfate with polyvinyl alcohol	0.06
Catalac white paint	0.23
Dow Corning white paint DC-007	0.19
Magnesium oxide white paint	0.09
Potassium fluorotitanate white paint	0.15
Tedlar white plastic	0.39
Titanium oxide white paint with methyl silicone	0.20
Zinc oxide with sodium silicate	0.15
Conversion coatings (values can vary significantly with coating thickness):	
Alzac A-2	0.16
Black chrome	0.96
Black copper	0.98
Black iridite	0.62
Black nickel	0.91
Vapor-deposited coatings on glass substrates:	
Aluminum	0.08
Chromium	0.56
Gold	0.19
Nickel	0.38
Silver	0.04

(*Continued*)

TABLE B.2 (*Continued*)
Normal Total Absorptivity for Incident Solar Radiation (Receiving Material at 300 K)

Coating	ε_n
Titanium	0.52
Tungsten	0.60
Plastic films with metal backing:	
Mylar film, 3 mil aluminum backing	0.17
Tedlar film, 1 mil gold backing	0.26
Teflon film, 2 mil aluminum backing	0.08
Teflon film, 1 mil gold backing	0.22
Teflon film, 2 mil silver backing	0.08

[a] Receiving material at 300 K.

Appendix C: Catalog of Selected Configuration Factors*

1

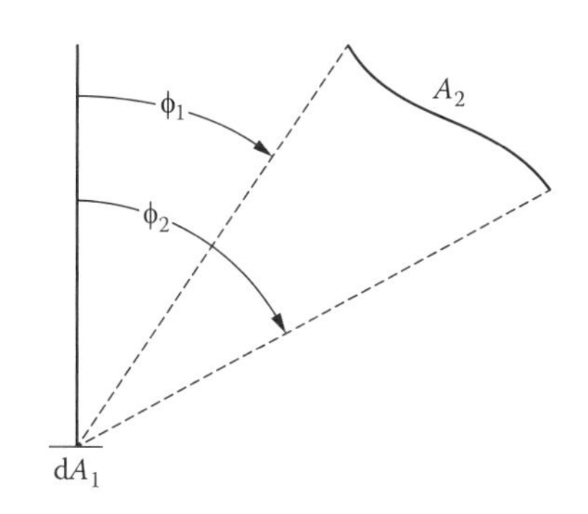

Area dA_1 of differential width and any length, to infinitely long strip dA_2 of differential width and with parallel generating line to dA_1

$$dF_{d1-d2} = \frac{\cos\phi}{2} d\phi = \frac{1}{2} d(\sin\phi)$$

2

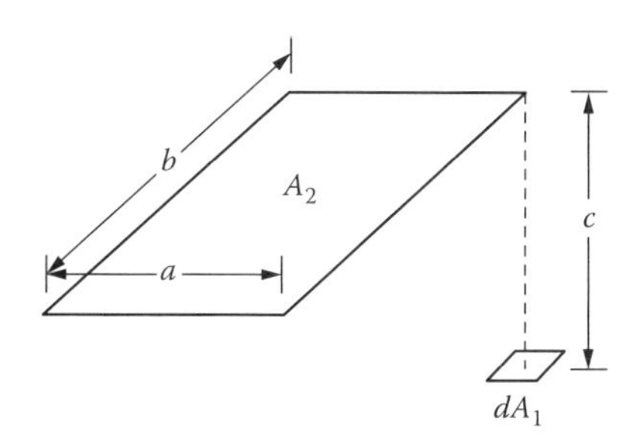

Plane element dA_1 to plane parallel rectangle; normal to element passes through the corner of a rectangle

$$X = \frac{a}{c}; \quad Y = \frac{b}{c}$$

$$F_{d1-d2} = \frac{1}{2\pi}\left(\frac{X}{\sqrt{1+X^2}}\tan^{-1}\frac{Y}{\sqrt{1+X^2}} + \frac{Y}{\sqrt{1+Y^2}}\tan^{-1}\frac{X}{\sqrt{1+Y^2}}\right)$$

3

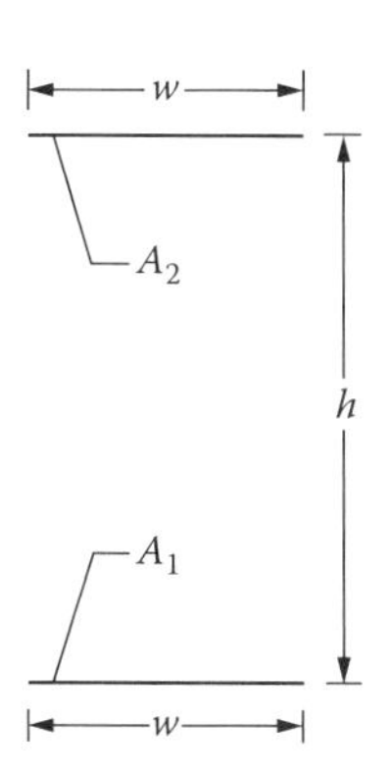

Two infinitely long, directly opposed parallel plates of the same finite width

$$H = \frac{h}{w}$$

$$F_{1-2} = F_{2-1} = \sqrt{1+H^2} - H$$

* A catalog of over 300 factors is available at http:www.thermalradiation.net/IndexCat.html.

4

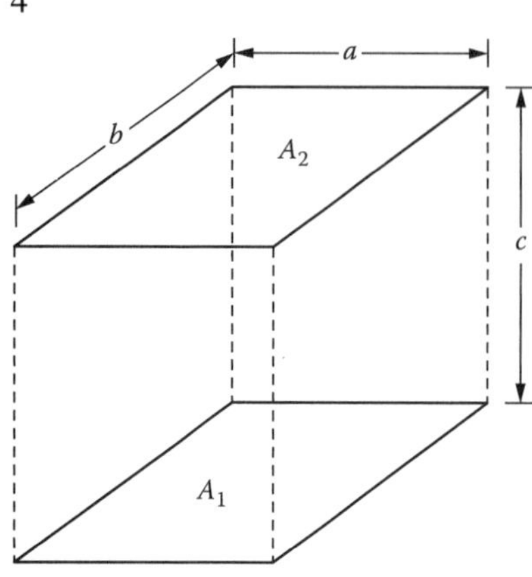

Identical, parallel, directly opposed rectangles

$$X = \frac{a}{c}; \quad Y = \frac{b}{c}$$

$$F_{1-2} = \frac{2}{\pi XY}\left\{ \begin{aligned} &\ln\left[\frac{(1+X^2)(1+Y^2)}{1+X^2+Y^2}\right]^{1/2} \\ &+ X\sqrt{1+Y^2}\tan^{-1}\frac{X}{\sqrt{1+Y^2}} \\ &+ Y\sqrt{1+X^2}\tan^{-1}\frac{Y}{\sqrt{1+X^2}} \\ &- X\tan^{-1}X - Y\tan^{-1}Y \end{aligned} \right\}$$

5

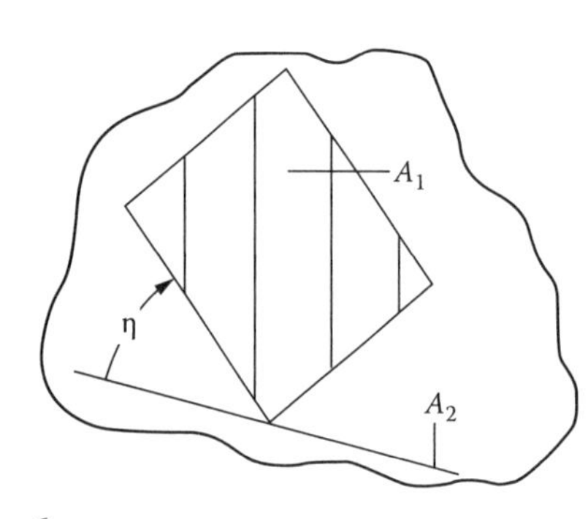

Finite rectangle A_1 of any size, tilted at angle η relative to an infinite plane A_2

$$F_{1-2} = \frac{1}{2}(1 - \cos\eta)$$

6

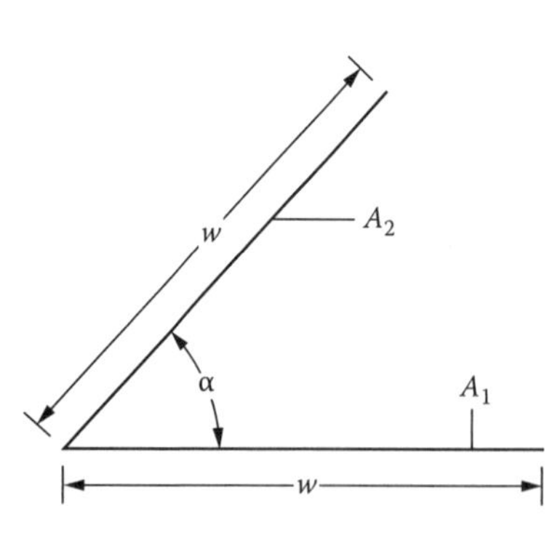

Two infinitely long plates of equal finite width w, having one common edge and having an included angle α to each other

$$F_{1-2} = F_{2-1} = 1 - \sin\frac{\alpha}{2}$$

7

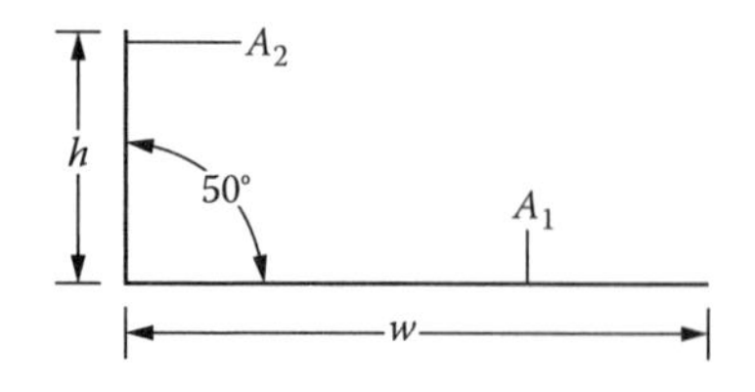

Two infinitely long plates of unequal widths h and w, having one common edge and having an angle of 90° to each other.

$$H = \frac{h}{w}$$

$$F_{1-2} = \frac{1}{2}(1 + H - \sqrt{1+H^2})$$

8

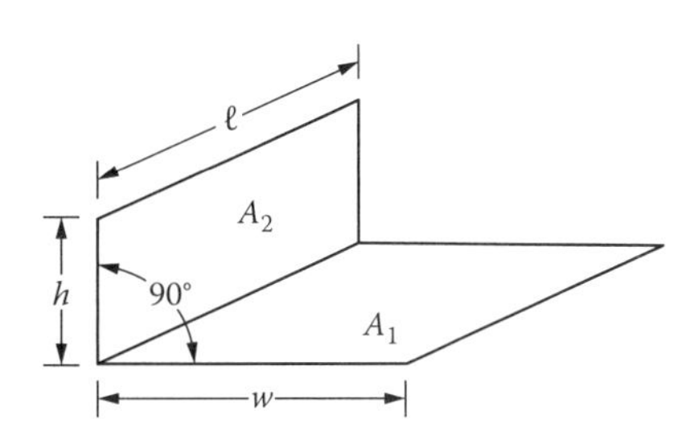

Two finite rectangles of the same length, having one common edge and having an angle of 90° to each other

$$H = \frac{h}{l}; \quad W = \frac{w}{l}$$

$$F_{1-2} = \frac{1}{\pi W}\left(\begin{aligned} & W\tan^{-1}\frac{1}{W} + H\tan^{-1}\frac{1}{H} \\ & -\sqrt{H^2+W^2}\tan^{-1}\frac{1}{\sqrt{H^2+W^2}} \\ & +\frac{1}{4}\ln\left\{ \begin{aligned} & \frac{(1+W^2)(1+H^2)}{1+W^2+H^2} \\ & \times\left[\frac{W^2(1+W^2+H^2)}{(1+W^2)(W^2+H^2)}\right]^{W^2} \\ & \times\left[\frac{H^2(1+H^2+W^2)}{(1+H^2)(H^2+W^2)}\right]^{H^2} \end{aligned} \right\} \end{aligned} \right)$$

9

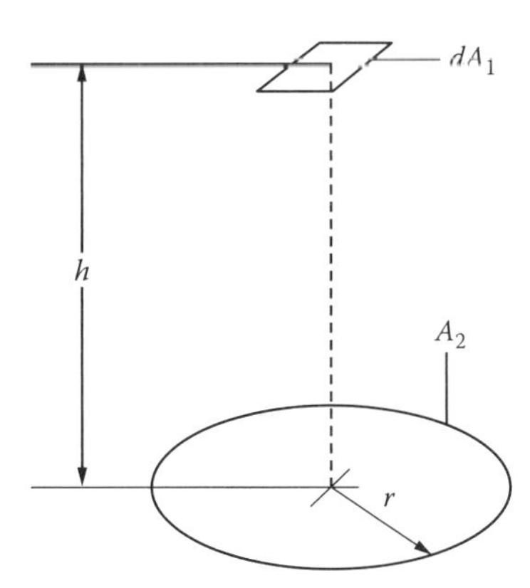

Plane element dA_1 to circular disk in plane parallel to element; normal to element passes through the center of a disk

$$F_{d1-2} = \frac{r^2}{h^2+r^2}$$

10

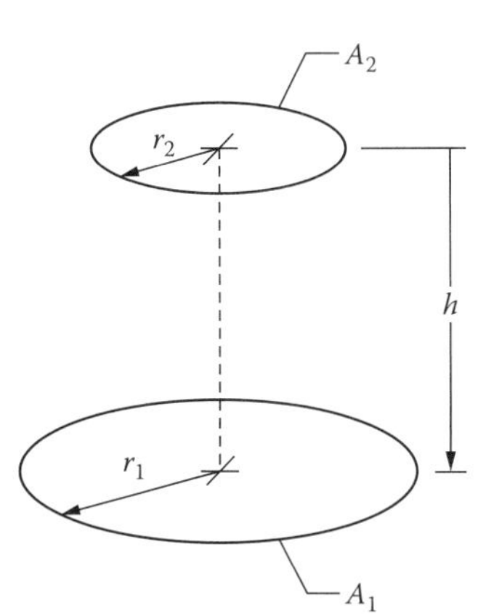

Parallel circular disks with centers along the same normal

$$R_1 = \frac{r_1}{h}; \quad R_2 = \frac{r_2}{h}$$

$$X = 1 + \frac{1+R_2^2}{R_1^2}$$

$$F_{1-2} = \frac{1}{2}\left[X - \sqrt{X^2 - 4\left(\frac{R_2}{R_1}\right)^2}\right]$$

11

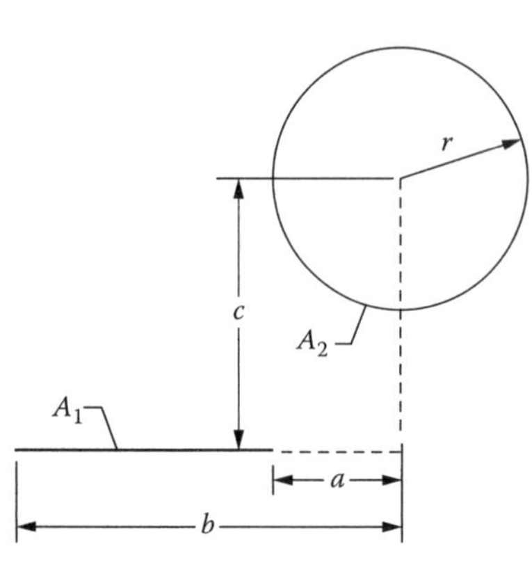

Infinitely long plane of finite width to parallel infinitely long cylinder

$$F_{1-2} = \frac{r}{b-a}\left(\tan^{-1}\frac{b}{c} - \tan^{-1}\frac{a}{c}\right)$$

12

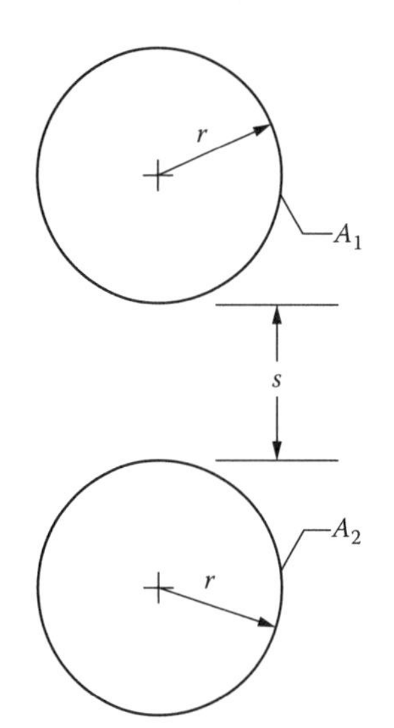

Infinitely long parallel cylinders of the same diameter

$$X = 1 + \frac{s}{2r}$$

$$F_{1-2} = F_{2-1} = \frac{1}{\pi}\left(\sqrt{X^2 - 1} + \sin^{-1}\frac{1}{X} - X\right)$$

13

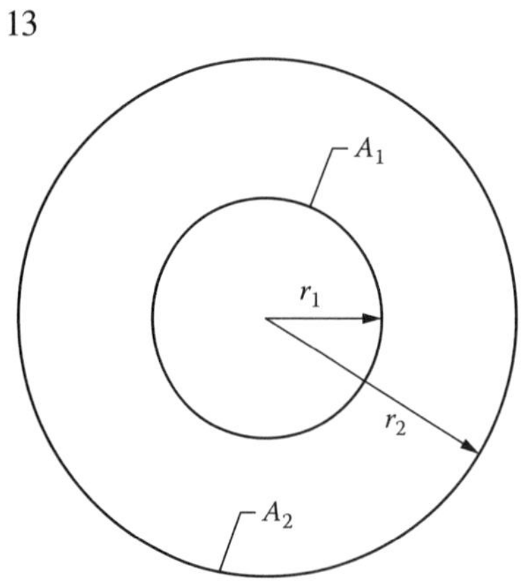

Concentric cylinders of infinite length

$$F_{1-2} = 1$$

$$F_{2-1} = \frac{r_1}{r_2}$$

$$F_{2-2} = 1 - \frac{r_1}{r_2}$$

14

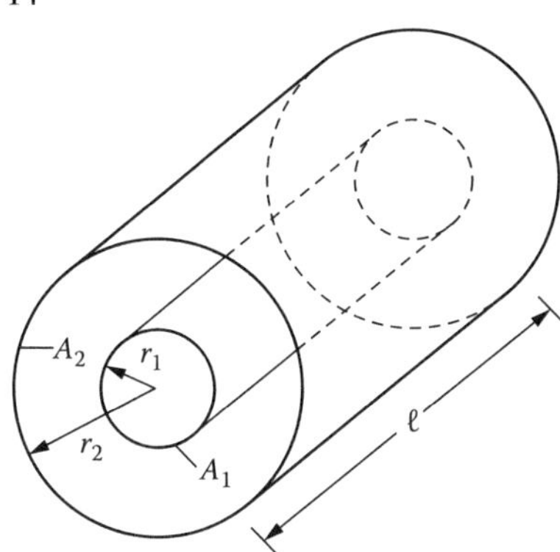

Two concentric cylinders of the same finite length

$$R = \frac{r_2}{r_1};\quad L = \frac{l}{r_1}$$

$$A = L^2 + R^2 - 1$$

$$B = L^2 - R^2 + 1$$

$$F_{2-1} = \frac{1}{R} - \frac{1}{\pi R}\left\{\cos^{-1}\frac{B}{A} - \frac{1}{2L}\times\left[\begin{matrix}\sqrt{(A+2)^2-(2R)^2}\cos^{-1}\dfrac{B}{RA} \\ +B\sin^{-1}\dfrac{1}{R} - \dfrac{\pi A}{2}\end{matrix}\right]\right\}$$

$$F_{2-2} = 1 - \frac{1}{R} + \frac{2}{\pi R}\tan^{-1}\frac{2\sqrt{R^2-1}}{L} - \frac{L}{2\pi R}$$

$$\times\left[\begin{matrix}\dfrac{\sqrt{4R^2+L^2}}{L}\sin^{-1}\dfrac{4(R^2-1)+(L^2/R^2)(R^2-2)}{L^2+4(R^2-1)} \\ -\sin^{-1}\dfrac{R^2-2}{R^2} + \dfrac{\pi}{2}\left(\dfrac{\sqrt{4R^2+L^2}}{L} - 1\right)\end{matrix}\right]$$

where for any argument ξ

$$-\frac{\pi}{2} \le \sin^{-1}\xi \le \frac{\pi}{2}$$

$$0 \le \cos^{-1}\xi \le \pi$$

15

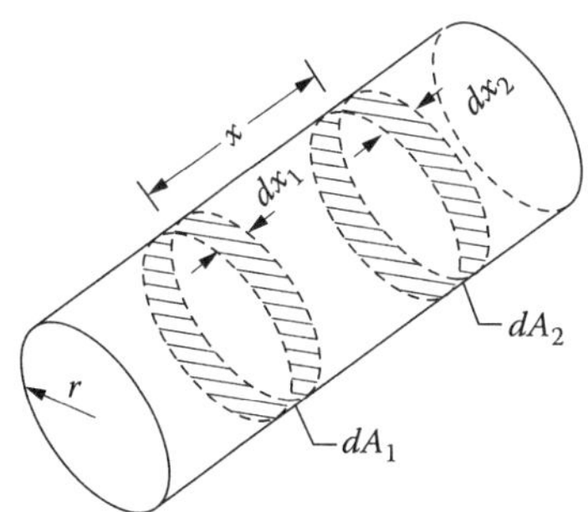

Two ring elements on the interior of a right circular cylinder

$$X = \frac{x}{2r}$$

$$dF_{d1-d2} = \left[1 - \frac{2X^3+3X}{2(X^2+1)^{3/2}}\right]dX_2$$

16

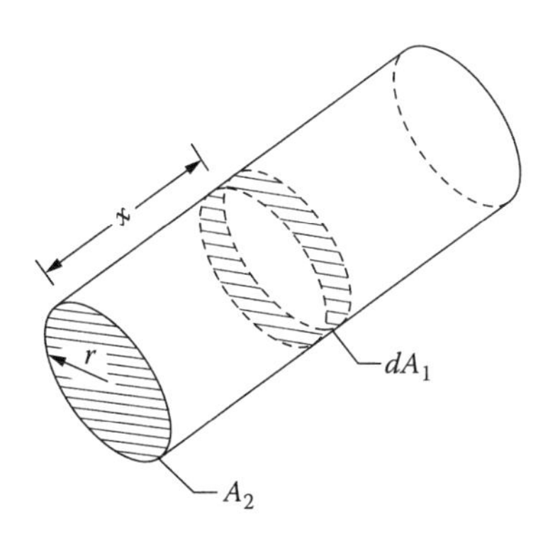

Ring element dA_1 on interior of right circular cylinder to circular disk A_2 at the end of a cylinder

$$X = \frac{x}{2r}$$

$$F_{d1-2} = \frac{X^2+(1/2)}{\sqrt{X^2+1}} - X$$

17

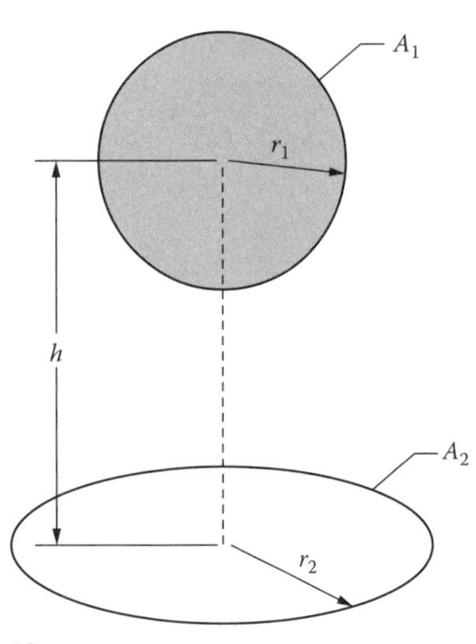

Sphere of radius r_1 to disk of radius r_2; normal to center of disk passes through the center of a sphere

$$R_2 = \frac{r_2}{h}$$

$$F_{1-2} = \frac{1}{2}\left(1 - \frac{1}{\sqrt{1+R_2^2}}\right)$$

18

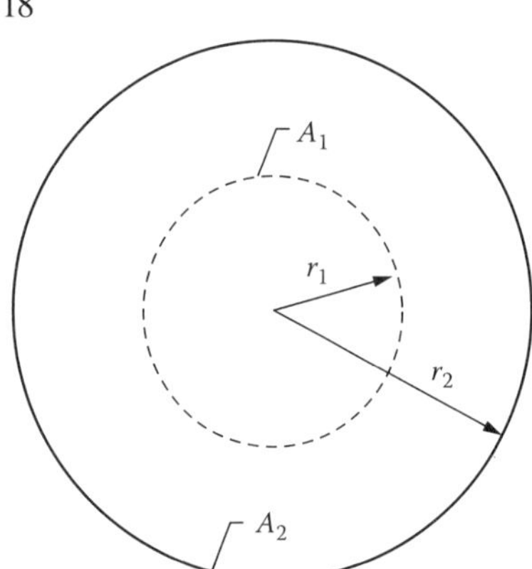

Concentric spheres

$$F_{1-2} = 1$$

$$F_{2-1} = \left(\frac{r_1}{r_2}\right)^2$$

$$F_{2-2} = 1 - \left(\frac{r_1}{r_2}\right)^2$$

19

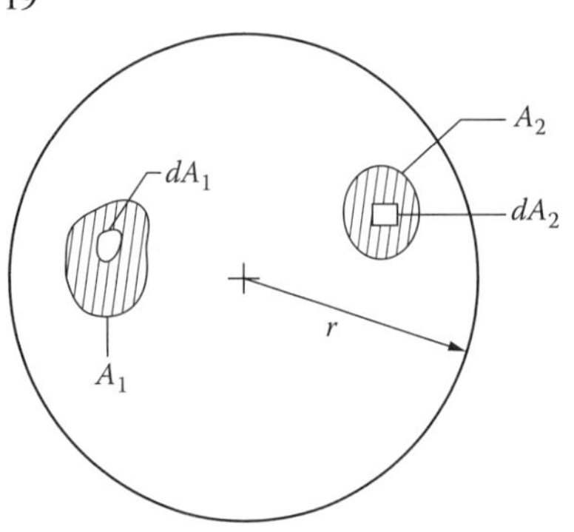

Differential or finite areas on the inside of a spherical cavity

$$dF_{d1-d2} = dF_{1-d2} = \frac{dA_2}{4\pi r^2}$$

$$F_{d1-2} = F_{1-2} = \frac{A_2}{4\pi r^2}$$

Appendix D: Exponential Integral Relations and Two-Dimensional Radiation Functions

D.1 EXPONENTIAL INTEGRAL RELATIONS

A summary of some useful exponential integral relations is presented here. Additional relations are in Chandrasekhar (1960), Kourganoff (1963), and Abramowitz and Stegun (1965).

For positive real arguments, the nth exponential integral is defined as

$$E_n(x) \equiv \int_0^1 \mu^{n-2} \exp\left(\frac{-x}{\mu}\right) d\mu \tag{D.1}$$

and only positive integral values of n will be considered here. An alternative form is

$$E_n(x) = \int_1^\infty \frac{1}{t^n} \exp(-xt) dt \tag{D.2}$$

By differentiating Equation D.1 under the integral sign, the recurrence relation is obtained:

$$\begin{aligned} \frac{d}{dx} E_n(x) &= -E_{n-1}(x) \quad x \geq 2 \\ \frac{d}{dx} E_1(x) &= -\frac{1}{x} \exp(-x) \end{aligned} \tag{D.3}$$

Another recurrence relation obtained by integration is

$$nE_{n+1}(x) = \exp(-x) - xE_n(x) = \exp(-x) + x\frac{d}{dx}E_{n+1}(x) \quad n \geq 1 \tag{D.4}$$

Also, integration results in

$$\int E_n(x) dx = -E_{n+1}(x) \tag{D.5}$$

By use of Equation D.4, all exponential integrals can be reduced to the first exponential integral given by

$$E_1(x) = \int_0^1 \mu^{-1} \exp\left(\frac{-x}{\mu}\right) d\mu \tag{D.6}$$

Alternative forms of $E_1(x)$ are

$$E_1(x) = \int_1^\infty t^{-1} \exp(-xt)\, dt = \int_x^\infty t^{-1} \exp(-t)\, dt \tag{D.7}$$

For $x = 0$, the exponential integrals are equal to

$$E_n(0) = \frac{1}{n-1} \quad n \geq 2$$

$$E_1(0) = +\infty \tag{D.8}$$

For large values of x there is the asymptotic expansion

$$E_n(x) = \frac{\exp(-x)}{x}\left[1 - \frac{n}{x} + \frac{n(n+1)}{x^2} - \frac{n(n+1)(n+2)}{x^3} + \cdots\right] \tag{D.9}$$

Therefore, as $x \to \infty$, $E_n(x) \to \exp(-x)/x \to 0$.

Series expansions are of the form

$$E_1(x) = -\gamma - \ln x + x - \frac{x^2}{2 \times 2!} + \frac{x^3}{3 \times 3!} - \cdots$$

$$= -\gamma - \ln x - \sum_{n=1}^{\infty} (-1)^n \frac{x^n}{x \times n!}$$

$$E_2(x) = 1 + (\gamma - 1 + \ln x)x - \frac{x^2}{1 \times 2!} + \frac{x^3}{2 \times 3!} - \cdots$$

$$E_3(x) = \frac{1}{2} - x + \frac{1}{2}\left(-\gamma + \frac{3}{2} - \ln x\right)x^2 + \frac{x^3}{1 \times 3!} - \cdots \tag{D.10}$$

where $\gamma = 0.577216$ is Euler's constant. The general series expansion given by Abramowitz and Stegun (1965) is

$$E_n(x) = \frac{(-x)^{n-1}}{(n-1)!}[-\ln x + \psi(n)] - \sum_{\substack{m=0 \\ (m \neq n-1)}}^{\infty} \frac{(-x)^m}{(m-n+1)m!} \tag{D.11}$$

where $\Psi(1) = -\gamma$ and

$$\psi(n) = -\gamma + \sum_{m=1}^{n-1} \frac{1}{m} \quad n \geq 2$$

Some approximations are

$$E_3(x) \approx \frac{1}{2}\exp(-1.8x)$$

Using Equation D.3 gives

$$E_2(x) \approx 0.9\exp(-1.8x)$$

In Cess and Tiwari (1972), the following approximations are used:

$$E_2(x) \approx \frac{3}{4}\exp\left(-\frac{3}{2}x\right)$$

$$E_3(x) \approx \frac{1}{2}\exp\left(-\frac{3}{2}x\right)$$

Tabulations of $E_n(x)$ are in Kourganoff (1963) and Abramowitz and Stegun (1965). An abridged listing is given in Table D.1 for convenience. Breig and Crosbie (1974) present forms of a generalized exponential integral function that are convenient for numerical computation. The generalized forms include, as a special case, the $E_n(x)$ discussed here.

D.2 TWO-DIMENSIONAL RADIATION FUNCTIONS

For two-dimensional (2D) problems, the exponential integral functions are replaced by another set of integral functions. These are the S_n functions and can be defined in a number of equivalent ways. An expression that is easily evaluated by numerical integration is

$$S_n(x) = \frac{2}{\pi}\int_0^{\pi/2} \exp\left(-\frac{x}{\cos\theta}\right)\cos^{n-1}\theta d\theta \tag{D.12}$$

An alternative form is

$$S_n(x) = \frac{2}{\pi}\int_1^{\infty} \frac{\exp(-xt)}{t^n(t^2-1)^{1/2}} dt \quad x \geq 0, n = 0,1,2,\ldots \tag{D.13}$$

Another form that has been used is

$$S_n(a\tau) = \frac{\tau^n}{\pi}\int_{-\infty}^{\infty} \frac{\exp\left[-a(x^2+\tau^2)^{1/2}\right]}{(x^2+\tau^2)(n+1)/2} dx, \quad a \geq 0, \tau \geq 0, n = 0,1,2,\ldots \tag{D.14}$$

TABLE D.1
Values of Exponential Integrals $E_n(x)$

x	$E_1(X)$	$E_2(x)$	$E_3(x)$	$E_4(x)$	$E_5(x)$
0	∞	1.00000	0.50000	0.33333	0.25000
0.01	4.03793	0.94967	0.49028	0.32838	0.24669
0.02	3.35471	0.91311	0.48097	0.32353	0.24343
0.03	2.95912	0.88167	0.47200	0.31876	0.24022
0.04	2.68126	0.85354	0.46332	0.31409	0.23706
0.05	2.46790	0.82784	0.45492	0.30949	0.23394
0.06	2.29531	0.80405	0.44676	0.30499	0.23087
0.07	2.15084	0.78184	0.43883	0.30056	0.22784
0.08	2.02694	0.76096	0.43112	0.29621	0.22486
0.09	1.91874	0.74124	0.42361	0.29194	0.22191
0.10	1.82292	0.72255	0.41629	0.28774	0.21902
0.15	1.46446	0.64104	0.38228	0.26779	0.20514
0.20	1.22265	0.57420	0.35195	0.24945	0.19221
0.25	1.04428	0.51773	0.32468	0.23254	0.18017
0.30	0.90568	0.46912	0.30004	0.21694	0.16893
0.35	0.79422	0.42671	0.27767	0.20250	0.15845
0.40	0.70238	0.38937	0.25729	0.18914	0.14867
0.50	0.55977	0.32664	0.22160	0.16524	0.13098
0.60	0.45438	0.27618	0.19155	0.14463	0.11551
0.70	0.37377	0.23495	0.16606	0.12678	0.10196
0.80	0.31060	0.20085	0.14432	0.11129	0.09007
0.90	0.26018	0.17240	0.12570	0.09781	0.07963
1.00	0.21938	0.14850	0.10969	0.08606	0.07045
1.20	0.15841	0.11110	0.08393	0.06682	0.05525
1.40	0.11622	0.08389	0.06458	0.05206	0.04343
1.60	0.08631	0.06380	0.04991	0.04068	0.03420
1.80	0.06471	0.04882	0.03872	0.03187	0.02698
2.00	0.04890	0.03753	0.03013	0.02502	0.02132
2.25	0.03476	0.02718	0.02212	0.01855	0.01592
2.50	0.02491	0.01980	0.01630	0.01378	0.01191
2.75	0.01798	0.01449	0.01205	0.01027	0.00892
3.00	0.01305	0.01064	0.00893	0.00767	0.00670
3.25	0.00952	0.00785	0.00664	0.00573	0.00504
3.50	0.00697	0.00580	0.00495	0.00430	0.00379

Source: Kourganoff, V., *Basic Methods in Transfer Problems*, Dover, New York, 1963.

From the definition of $S_n(x)$, the values at $x = 0$ can be found from ($n > 0$):

$$S_n(0) = \frac{1}{\pi^{1/2}} \frac{\Gamma(n/2)}{\Gamma[(n+1)/2]} \tag{D.15}$$

This yields $S_1(0) = 1$, $S_2(0) = 2/\pi$, $S_3(0) = 1/2$, and $S_4(0) = 4/(3\pi)$. The $S_0(x)$ is related to the modified Bessel function by

$$S_0(x) = \frac{2}{\pi} K_0(x) \tag{D.16}$$

TABLE D.2
Values of the Two-Dimensional Radiation Function $S_n(x)$

x	$S_0(x)$	$S_1(x)$	$S_2(x)$	$S_3(x)$	$S_4(x)$
0	∞	1.00000	0.63662	0.50000	0.42441
0.002	4.03015	0.99067	0.63463	0.49873	0.42341
0.005	3.44684	0.97958	0.63167	0.49683	0.42192
0.02	2.56460	0.93598	0.61732	0.48746	0.41454
0.04	2.12411	0.88960	0.59908	0.47530	0.40491
0.07	1.76969	0.83169	0.57329	0.45772	0.39092
0.10	1.54512	0.78217	0.54910	0.44089	0.37744
0.15	1.29236	0.71168	0.51180	0.41438	0.35607
0.20	1.11581	0.65169	0.47776	0.38965	0.33597
0.25	0.98135	0.59940	0.44651	0.36656	0.31708
0.30	0.87374	0.55312	0.41772	0.34496	0.29929
0.35	0.78477	0.51172	0.39111	0.32475	0.28256
0.40	0.70953	0.47441	0.36648	0.30582	0.26680
0.45	0.64484	0.44059	0.34362	0.28807	0.25195
0.5	0.58850	0.40979	0.32237	0.27143	0.23797
0.6	0.49499	0.35580	0.28417	0.24115	0.21237
0.7	0.42050	0.31016	0.25093	0.21443	0.18962
0.8	0.35991	0.27124	0.22191	0.19082	0.16939
0.9	0.30986	0.23783	0.19650	0.16993	0.15137
1.0	0.26803	0.20899	0.17419	0.15142	0.13532
1.2	0.20277	0.16227	0.13728	0.12042	0.10826
1.5	0.13611	0.11218	0.09661	0.08571	0.07764
2.0	0.07251	0.06183	0.05442	0.04900	0.04484
3.0	0.02212	0.01964	0.01778	0.01633	0.01516
4.0	0.00710	0.00646	0.00595	0.00554	0.00520
5.0	0.00235	0.00217	0.00202	0.00190	0.00180

The derivative of $S_n(x)$ is given by

$$\frac{dS_n(x)}{dx} = -S_{n-1}(x), \quad n \geq 1 \tag{D.17}$$

Hence, the integral of S_n is

$$\int S_n(x)dx = -S_{n+1}(x), \quad n \geq 0 \tag{D.18}$$

Values of the first five S_n functions were computed numerically from Equation D.12 and are in Table D.2. Additional information is in Yuen and Wong (1983).

Appendix E: References

Note: The following abbreviations are used in the references:

IJHMT: International Journal of Heat and Mass Transfer
JHT: ASME Journal of Heat Transfer
JTHT: AIAA Journal of Thermophysics and Heat Transfer
JOSA: Journal of the Optical Society of America
JQSRT: Journal of Quantitative Spectroscopy and Radiative Transfer

Abarbanel, S. S.: Time dependent temperature distribution in radiating solids, *J. Math. Phys.*, 39(4), 246–257, 1960.

Abbasi, M. H. and Evans, J. W.: Monte Carlo simulation of radiant transport through an adiabatic packed bed or porous solid, *AIChE J.*, 28(5), 853–854, 1982.

Abdelrahman, M., Fumeaux, P., and Suter, P.: Study of solid-gas suspensions used for direct absorption of concentrated solar radiation, *Sol. Energy*, 22(1), 45–48, 1979.

Abdulkadir, A. and Birkebak, R. C.: Spectral directional emittance of rough metal surfaces: Comparison between semi-random and pyramidal surface approximations, AIAA Paper 78–848, *Second AIAA/ASME Thermophysics and Heat Transfer Conference*, Palo Alto, CA, 1978.

Abhyankar, K. D. and Fymat, A. L.: Relations between the elements of the phase matrix for scattering, *J. Math. Phys.*, 10, 1935–1938, 1969.

Abramowitz, M. and Stegun, I. A.: *Handbook of Mathematical Functions*, Dover, New York, 1965.

Abu-Romia, M. M. and Tien, C. L.: Appropriate mean absorption coefficients for infrared radiation in gases, *JHT*, 89(4), 321–327, 1967.

Adhya, S.: Thermal re-radiation modelling for the precise prediction and determination of spacecraft orbits, PhD thesis, University of London, London, U.K., 2005.

Adrianov, V. N. and Polyak, G. I.: Differential methods for studying radiant heat transfer, *IJHMT*, 6(5), 355–362, 1963.

Adrianov, V. N. and Shorin, S. N.: Radiative transfer in the flow of a radiating medium, Trans. TT-1, Purdue University, West Lafayette, IN, February 1961.

Aernouts, B., Van Beers, R., Watte, R., and Lammertyn, J.: Dependent scattering in Intralipid phantoms in the 600–1850 nm range, *Opt. Express*, 22(5), 6086–6098, 2014.

Agarwal, B. M. and Mengüç, M. P.: Single and multiple scattering of collimated radiation in an axisymmetric system, *IJHMT*, 34(3), 633–647, 1991.

Al Abed, A. and Sacadura, J.-F.: A Monte Carlo-finite difference method for coupled radiation-conduction heat transfer in semitransparent media, *JHT*, 105(4), 931–933, 1983.

Al-Bahadili, H. and Wood, J.: Calculation of radiative heat transfer through a grey gas on parallel computer architecture, in A. E. Fincham and B. Ford (eds.), *Parallel Computation*, pp. 135–157, Clarendon Press, Oxford, U.K., 1993.

Alciatore, D., Lipp, S., and Janna, W. S.: Closed-form solution of the general three dimensional radiation configuration factor problem with microcomputer solution, *Proceedings of the 26th National Heat Transfer Conference*, Philadelphia, PA, August 1989.

Alfano, G.: Apparent thermal emittance of cylindrical enclosures with and without diaphragms, *IJHMT*, 15(12), 2671–2674, 1972.

Alfano, G. and Sarno, A.: Normal and hemispherical thermal emittances of cylindrical cavities, *JHT*, 97(3), 387–390, 1975.

Alifanov, O. M.: *Inverse Heat Transfer Problems*, Springer, Berlin, Germany, 1994.

Alifanov, O. M., Artyukhin, E. A., and Rumyantsev, S. V.: *Extreme Methods for Solving Ill-Posed Problems with Applications to Inverse Heat Transfer Problems*, Begell House, New York, 1995.

Altaç, Z.: Integrals involving Bickley and Bessel functions in radiative transfer, and generalized exponential integral functions, *JHT*, 118(3), 789–792, 1996.

Altes, D., Breitbach, G., and Szimmat, J.: Application of the finite element method to the calculation of heat transfer by radiation, *Proceedings of the International Finite Element Method Congress*, Baden-Baden, Germany, November 1986.

Ambarzumian, V. A.: Diffusion of light by planetary atmospheres, *Astron. Zh.*, 19, 30–41, 1942.

Ambirajan, A. and Look, D. C., Jr.: A backward Monte Carlo estimator for the multiple scattering of a narrow light beam, *JQSRT*, 56(3), 317–336, 1996.

Amiri, H., Mansouri, S. H., Safavinejad, A., and Coelho, P. J.: The optimal number and location of discrete radiant heaters in enclosures with the participating media using the micro genetic algorithm, *Num. Heat Trans. Part A Appl.*, 60(5), 461–483, 2011.

Amlin, D. W. and Korpela, S. A.: Influence of thermal radiation on the temperature distribution in a semi-transparent solid, *JHT*, 101(1), 76–80, 1979.

Anderson, E. E.: Estimating the effective emissivity of nonisothermal diatherminous coatings, *JHT*, 97, 480–482, 1975.

Anderson, E. E., Viskanta, R., and Stevenson, W. H.: Heat transfer through semitransparent solids, *JHT*, 95(2), 179–186, 1973.

Anderson, K. M. and Hadvig, S.: Geometric mean beam lengths for a space between two coaxial cylinders, *JHT*, 111(3), 811–813, 1989.

Andre, S. and Degiovanni, A.: A theoretical study of the transient coupled conduction and radiation heat transfer in glass: Phonic diffusivity measurements by the flash technique, *IJHMT*, 38(18), 3401–3412, 1995.

André, F., Hou, L., Roger, M., and Vaillon, R.: The multispectral gas radiation modeling: A new theoretical framework based on a multidimensional approach to *k*-distribution methods, *JQSRT*, 147, 178–195, 2014a.

André, F., Solovjov, V. P., Hou, L., Vaillon, R., and Lemonnier, D.: The generalized *k*-moment method for the modeling of cumulative *k*-distributions of H_2O at high temperature, *JQSRT*, 143, 92–99, 2014b.

André, F. and Vaillon, R.: The spectral-line moment-based (SLMB) modeling of the wide-band and global blackbody weighted transmission function and cumulative distribution function of the absorption coefficient in uniform gaseous media, *JQSRT*, 109(14), 2401–2416, 2008.

André, F. and Vaillon, R.: A database for the SLMB modeling of the full spectrum radiative properties of CO_2, *JQSRT*, 111(2), 325–330, 2010.

André, F. and Vaillon, R.: Generalization of the *k*-moment method using the maximum entropy principle: Application to the NBKM and full spectrum SLMB gas radiation models, *JQSRT*, 113(12), 1508–1520, 2012.

Ang, J. A., Pagni, P. J., Mataga, T. G., Margie, J. M., and Lyons, V. L.: Temperature and velocity profiles in sooting free convection diffusion flames, *AIAA J.*, 26(3), 323–329, 1988.

Araki, N.: Effect of radiation heat transfer on thermal diffusivity measurements, *Int. J. Thermophys.*, 11(2), 329–337, 1990.

Argento, C. and Bouvard, D.: A ray tracing method for evaluating the radiative heat transfer in porous media, *IJHMT*, 39(15), 3175–3180, 1996.

Arimilli, R. V. and Ketkar, S. P.: Radiation heat transfer in enclosures with discrete heat sources, *Int. Commun. Heat Mass Transfer*, 15, 31–40, 1988.

Armaly, B. F., Lam, T. T., and Crosbie, A. L.: Emittance of semi-infinite absorbing and isotropically scattering medium with refractive index greater than unity, *AIAA J.*, 11(11), 1498–1502, 1973.

Arnaoutakis, G. E., Marques-Hueso, J., Mallick, T. K., and Richards, B. S.: Coupling of sunlight into optical fibres and spectral dependence for solar energy applications, *Solar Energy*, 93, 235–243, 2013.

Arnold, C., Marquier, F., Garin, M., Pardo, F., Collin, S., Bardou, N., Pelouard, J.-L., and Greffet, J.-J., Coherent thermal emission by two-dimensional silicon carbide gratings, *Phys. Rev. B*, 86, 035316, 2012.

Arpaci, V. S. and Bayazitoglu, Y.: Thermal stability of radiating fluids: Asymmetric slot problem, *Phys. Fluids*, 16(5), 589–593, 1973.

Arpaci, V. S. and Gözüm, D.: Thermal stability of radiating fluids: The Benard problem, *Phys. Fluids*, 16(5), 581–588, 1973.

Artvin, Z.: Fabrication of nanostructured samples for the investigation of near field radiation transfer, MS thesis, Middle East Technical University, Ankara, Turkey, 2012.

Ascher, U., Christiansen, J., and Russell, R. D.: Collocation software for boundary value ODE's, in B. Childs et al. (eds.), *Codes for Boundary Value Problems in Ordinary Differential Equations*, pp. 164–185, Springer-Verlag, Berlin, Germany, 1979.

Aschkinass, E.: Heat radiation of metals, *Ann. Phys.*, 17(5), 960–976, 1905.

Aslan, M., Yamada, J., Mengüç, M. P., and Thomasson, J. A.: Characterization of individual cotton particles via light scattering experiments, *JTHT*, 17(4), 442–449, 2003.

Aslan, M. M., Mengüç, M. P., Manickavasagam, S., and Saltiel, C.: Size and shape prediction of colloidal metal oxide MgBaFeO particles from light scattering measurements, *J. Nanopart. Res.*, 8(6), 981–994, 2006.

Aslan, M. M., Mengüç, M. P., and Videen, G.: Characterization of metallic nano-particles via surface wave scattering: B. Physical concept and numerical experiments, *JQSRT*, 93, 207–217, 2005.

Aulender, M. and Hava, S.: Scattering-matrix propagation algorithm in full-vectorial optics of multilayer grating structures, *Opt. Lett.*, 21(21), 1765–1767, 1996.

Avery, L. W., House, L. L., and Skumanich, A.: Radiative transport in finite homogeneous cylinders by the Monte Carlo technique, *JQSRT*, 9, 519–531, 1969.

Azad, F. H. and Modest, M. F.: Combined radiation and convection in absorbing, emitting and anisotropically scattering gas-particulate tube flow, *IJHMT*, 24(10), 1681–1698, 1981a.

Azad, F. H. and Modest, M. F.: Evaluation of the radiative heat flux in absorbing, emitting and linear-anisotropically scattering cylindrical media, *JHT*, 103(2), 350–356, 1981b.

Aziz, A. and Benzies, J. Y.: Application of perturbation techniques to heat transfer problems with variable thermal properties, *IJHMT*, 19, 271–276, 1976.

Baba, H. and Kanayama, A.: Directional monospectral emittance of dielectric coating on a flat metal substrate, Paper 75–664, *AIAA 10th Thermophysics Conference,* Denver, May, 1975.

Babikian, D. S., Edwards, D. K., Karam, S. E., Wood, C. P., and Samuelsen, G. S.: Experimental mass absorption coefficients of soot in spray combustor flames, *JTHT*, 4(1), 8–15, 1990.

Babuty, A., Joulain, K., Chapuis, P.-O., Greffet, J.-J., and De Wilde, Y.: Blackbody spectrum revisited in the near field, *Phys. Rev. Lett.*, 110, 146103, 2013.

Badari Narayana, K. and Kumari, S. U.: Two-dimensional heat transfer analysis of radiating plates, *IJHMT*, 31(9), 1767–1774, 1988.

Baek, S. W., Byun, D. Y., and Kang, S. J.: The combined Monte-Carlo and finite-volume method for radiation in a two-dimensional irregular geometry, *IJHMT*, 43, 2337–2344, 2000.

Baker, C. T. H.: *The Numerical Treatment of Integral Equations*, Chapters 4 and 5, Clarendon Press, Oxford, U.K., 1977.

Balaji, C. and Venkateshan, S. P.: Combined surface radiation and free convection in cavities, *JTHT*, 8(2), 373–376, 1994a.

Balaji, C. and Venkateshan, S. P.: Correlations for free convection and surface radiation in a square cavity, *Int. J. Heat Fluid Flow*, 15(3), 249–251, 1994b.

Balaji, C. and Venkateshan, S. P.: Natural convection in L corners with surface radiation and conduction, *JHT*, 118(1), 222–225, 1996.

Balakrishnan, A. and Edwards, D. K.: Molecular gas radiation in the thermal entrance region of a duct, *JHT*, 101(3), 489–495, 1979.

Bannerot, R. B. and Wierum, F.: Approximate configuration factors for a gray nonisothermal gas-filled conical enclosure, in R.G. Hering (ed.), *Thermophysics and Spacecraft Thermal Control,* vol. 35 *of Progress in Astronautics and Aeronautics*, MIT Press, Cambridge, MA, pp. 41–63, 1974.

Bansal, A. and Modest, M.: Multiscale part-spectrum *k*-distribution database foratomic radiation in hypersonic nonequilibrium flows, *JHT*, 133, 122701–122707, December 2011.

Bansal, A., Modest, M., and Levin, D.: Correlated *k*-method for atomic radiation in hypersonic nonequilibrium flows, Paper 2009–1027, *47th AIAA Aerospace Sciences Meeting*, Orlando, January 2009a.

Bansal, A., Modest, M., and Levin, D.: Narrow-band *k*-distribution database for atomic radiation in hypersonic nonequilibrium flows, ASME Paper No. HT2009–88120, *ASME Summer Heat Transfer Conference*, San Francisco, CA, 2009b.

Bansal, A., Modest, M., and Levin, D.: *k*-Distributions for gas mixtures in hypersonic nonequilibrium flows, *48th AIAA Aerospace Sciences Meeting*, Orlando, FL, January 4–7, 2010a.

Bansal, A., Modest, M., and Levin, D.: Multigroup correlated *k*-distribution method for nonequilibrium atomic radiation, *JTHT*, 24, 638–646, 2010b.

Bansal, A., Modest, M., and Levin, D.: Multi-scale *k*-distribution model for gas mixtures in hypersonic nonequilibrium flows, *JQSRT*, 112, 1213–1221, 2011.

Baranoski, G. V. G., Rokne, J. G., and Xu, G.: Applying the exponential Chebyshev inequality to the nondeterministic computation of form factors, *JQSRT*, 69(4), 447–467, May 2001.

Barber, P. W. and Hill, S. S.: *Light Scattering by Particles: Computational Methods*, World Scientific, Singapore, 1990.

Barker, D. G. and Jones, M. R.: Inversion of spectral emission measurements to reconstruct the temperature profile along a blackbody optical fiber, *Inverse Probs. Eng.*, 11(6), 495–513, December 2003a.

Barker, D. G. and Jones, M. R.: Temperature measurements using a high-temperature blackbody optical fiber thermometer, *JHT*, 125, 471–477, June 2003b.

Barnes, J. W. and Chambers, J. B.: Solving the job shop scheduling problem using tabu search, *IIE Trans.*, 27, 257–263, 1995.

Barr, E. S.: Historical survey of the early development of the infrared spectral region, *Am. J. Phys.*, 28(1), 42–54, 1960.

Basu, S., Chen, Y. B., and Zhang, Z. M.: Microscale radiation in thermophotovoltaic devices—A review, *Int. J. Energy Res.*, 31(6–7), 689–716, 2007.

Basu, S. and Francoeur, M.: Penetration depth in near field radiative heat transfer between metamaterials, *Appl. Phys. Lett.*, 99, 143107, 2011.

Basu, S., Lee, B. J., and Zhang, Z. M.: Near-field radiation calculated with an improved dielectric function model for doped silicon, *JHT*, 132, 023302, 2010.

Basu, S. and Zhang, Z. M.: Maximum energy transfer in near-field thermal radiation at nanometer distances, *J. Appl. Phys.*, 105, 093535, 2009.

Bates, D. R. (ed.): *Atomic and Molecular Processes*, Academic Press, New York, 1962.

Bauman, P. T., Stogner, R., Carey, G. F., Schulz, K. W., Updadhyay, R., and Maurente, A.: Loose-coupling algorithm for simulating hypersonic flows with radiation and ablation, *J. Spacecraft Rockets*, 48(1), 72–80, 2011. doi: 10.2514/1.50588.

Bayat, N., Mehraban, S., and Hosseini Sarvari, S. M., Inverse boundary design of a radiant furnace with diffuse-spectral design, *Int. Commun. Heat Mass Transfer*, 37, 103–110, 2010.

Bayazitoglu, Y. and Higenyi, J.: The higher-order differential equations of radiative transfer: P_3 approximation, *AIAA J.*, 17(4), 424–431, 1979.

Bayazitoglu, Y., Kharadmand, S., and Tullius, T. K.: An overview of nanoparticle assisted laser therapy, *IJHMT*, 67, 469–486, 2013.

Bayvel, L. P. and Jones, A. R.: *Electromagnetic Scattering and Its Applications*, Applied Science, London, U.K., pp. 5–6, 1981.

Beck, J. V., Blackwell, B., and St. Clair, Jr., C. R.: *Inverse Heat Conduction: Ill-Posed Problems*, Wiley-Interscience, New York, 1995.

Beckman, F. S.: The solution of linear equations by the conjugate gradient method, in A. Ralston and H. S. Wilf (eds.), *Mathematical Methods for Digital Computers*, pp. 62–72, John Wiley & Sons, New York, 1960.

Beckmann, P. and Spizzichino, A.: *The Scattering of Electromagnetic Waves from Rough Surfaces*, Macmillan, New York, 1963.

Bedford, R. E.: Calculation of effective emissivities of cavity sources of thermal radiation, in D. P. DeWitt and G. D. Nutter (eds.), *Theory and Practice of Radiation Thermometry*, Chapter 12, John Wiley & Sons, Inc., New York, 1988.

Ben, X., Yi, H. L., and Tan, H. P.: Polarized radiative transfer in an arbitrary multilayer semitransparent medium, *Appl. Optics* 53(7), 1427–1441, 2014.

Ben-Abdallah, P.: Thermal antenna behavior for thin-film structures, *JOSA*, 21(7), 1368–1371, 2004.

Ben-Abdallah, P., Fumeron, S., Le Dez, V., and Charette, A.: Integral form of the radiative transfer equation inside refractive cylindrical media, *JTHT*, 25(2), 184–189, 2001.

Ben-Abdallah, P., Joulain, K., Drevillon, J., and Domingues, G.: Tailoring the local density of states of nonradiative field at the surface of nanolayered material, *Appl. Phys. Lett.*, 94, 153117, 2009.

Ben-Abdallah, P. and Le Dez, V.: Temperature field inside an absorbing-emitting semitransparent slab at radiative equilibrium with variable spatial refractive index, *JQSRT*, 65(4), 595–608, 2000a.

Ben-Abdallah, P. and Le Dez, V.: Radiative flux field inside an absorbing-emitting semitransparent slab with variable spatial refractive index at radiative conductive coupling, *JQSRT*, 67(2), 125–137, 2000b.

Ben-Abdallah, P. and Ni, B.: Single-defect Bragg stacks for high power narrow-band thermal emission, *J. Appl. Phys.*, 97, 104910, 2005.

Berg, M. J., Sorenson, C. M., and Chakabarti, A.: A new explanation of the extinction paradox, *RAD VI, Proceedings of the VI Radiation Conference*, Antalya, Turkey, June 2010.

Berlad, A. L., Tangirala, V., Ross, H, and Facca, L.: Radiative structures of lycopodium-air flames in low gravity, *J. Propul. Power*, 7(1), 5–8, 1991.

Berthon, C., Dubois, J., and Turpault, R.: Numerical approximation of the M1-model, *Mathematical Models and Numerical Methods for Radiative Transfer Conference*, Nantes, France, July–August 2007.

Bertsekas, D. P.: *Nonlinear Programming*, 2nd edn., Athena Scientific, Belmont, MA, 1999.

Bevans, J. T. and Edwards, D. K.: Radiation exchange in an enclosure with directional wall properties, *JHT*, 87(3), 388–396, 1965.

Bhanti, D., Manickavasagam, S., and Mengüç, M. P.: Identification of non-homogeneous spherical particles from their scattering matrix elements, *JQSRT*, 56(4), 591–608, 1996.

Bhattacharjee, S., King, M., Cobb, W., Altenkirch, R. A., and Wakai, K.: Approximate two-color emission pyrometry, *JHT*, 122(1), 15–20, 2000.

Bi, L., Yang, P., Liu, C., Yi, B., Baum, B. A., Diedenhoven, B. V., and Iwabuchi, H.: Assessment of the accuracy of the conventional ray-tracing technique: Implications in remote sensing and radiative transfer involving ice clouds, *JQSRT*, 146, 158–174, 2014.

Bialecki, R. A.: Solving heat radiation problems using the boundary element method, in C. A. Brebbia and J. J. Connor (eds.), *Topics in Engineering*, p. 15, Computational Mechanics Publications, Boston, MA, 1992.

Biehs, S.-A.: Thermal heat radiation, near-field energy density and near-field radiative heat transfer of coated materials, *Eur. Phys. J. B*, 58, 423–431, 2007.

Biehs, S.-A., Huth, O., and Rüting, F.: Near-field radiative heat transfer for structured surfaces, *Phys. Rev. B*, 78, 085414, 2008.

Biehs, S.-A., Reddig, D., and Holthaus, M.: Thermal radiation and near-field energy density of thin metallic films, *Eur. Phys. J. B*, 55, 237–251, 2007.

Biehs, S.-A., Tschikin, M., and Ben-Abdallah, P.: Hyperbolic metamaterials as an analog of a blackbody in the near-field, *Phys. Rev. B*, 84, 195459, 2011.

Biehs, S.-A., Tschikin, M., Messina, R., and Ben-Abdallah, P.: Super-Planckian near-field thermal emission with phonon-polaritonic hyperbolic metamaterials, *Appl. Phys. Lett.*, 102(13), 131106, 2013.

Billings, R. L., Barnes, J. W., and Howell, J. R.: Markov analysis of radiative transfer in enclosures with bidirectional reflections, *Num. Heat Transfer A*, 19(3), 101–114, 1991a.

Billings, R. L., Barnes, J. W., Howell, J. R., and Slotboom, O. E.: Markov analysis of radiative transfer in specular enclosures, *JHT*, 113(2), 429–436, 1991b.

Birkebak, R. C.: Spectral emittance of Apollo-12 lunar fines, *JHT*, 94(3), 323–324, 1972.

Birkebak, R. C.: Thermophysical properties of lunar materials: Part 1. Thermal radiation properties of lunar materials from the Apollo missions, in J. P. Hartnett and T. F. Irvine, Jr. (eds.), *Advances in Heat Transfer*, p. 10, Academic Press, New York, 1974.

Birkebak, R. C. and Abdulkadir, A.: Random rough surface model for spectral directional emittance of rough metal surfaces, *IJHMT*, 19(9), 1039–1043, 1976.

Birkebak, R. C. and Eckert, E. R. G.: Effects of roughness of metal surfaces on angular distribution of monochromatic reflected radiation, *JHT*, 87(1), 85–94, 1965.

Black, W. Z.: Optimization of the directional emission from V-groove and rectangular cavities, *JHT*, 95(1), 31–36, 1973.

Blechschmidt, D.: Monte Carlo study of light transmission through a cylindrical tube, *J. Vac. Sci. Technol.*, 11(3), 570–574, 1974.

Bobco, R. P.: Radiation heat transfer in semigray enclosures with specularly and diffusely reflecting surfaces, *JHT*, 86(1), 123–130, 1964.

Bobco, R. P.: Directional emissivities from a two-dimensional, absorbing-scattering medium, the semi-infinite slab, *JHT*, 89(4), 313–320, 1967.

Bobco, R. P., Allen, G. E., and Othmer, P. W.: Local radiation equilibrium temperatures in semigray enclosures, *J. Spacecraft Rockets*, 4(8), 1076–1082, 1967.

Bobco, R. P. and Drolen, B. L.: Engineering model of surface specularity: Spacecraft design implications, *JTHT*, 3(3), 289–296, 1989.

Bobco, R. P. and Starkovs, R. P.: Rectangular thermal doublers of uniform thickness, *AIAA J.*, 23(12), 1970–1977, 1985.

Boehm, R. F. and Tien, C. L.: Small spacing analysis of radiative transfer between parallel metallic surfaces, *JHT*, 92C(3), 405–411, 1970.

Bohren, C. W. and Clothiaux, E. E.: *Fundamentals of Atmospheric Radiation*, John Wiley & Sons, New York, 2006.

Bohren, C. and Huffman, D.: *Absorption and Scattering of Light by Small Particles*, Wiley-Interscience, New York, 1983.

Boles, M. A. and Özişik, M. N.: Simultaneous ablation and radiation in an absorbing, emitting and isotropically scattering medium, *JQSRT*, 12, 838–847, 1972.

Boltasseva, A. and Shalaev, V.: Fabrication of optical negative-index metamaterials: Recent advances and outlook, *Metamaterials*, 2, 1–17, 2008.

Boltzmann, L.: Ableitung des Stefan'schen Gesetzes, betreffend die Abhängigkeit der Wärmestrahlung von der Temperatur aus der electromagnetischen Lichttheorie, *Ann. Phys.*, *Ser.* 2, 22, 291–294, 1884.

Bone, W. A. and Townsend, D. T. A.: *Flame and Combustion in Gases*, Longmans, London, U.K., 1927.

Bonnell, D. W., Treverton, J. A., Valerga, A. J., and Margrave, J. L.: The emissivities of liquid metals at their fusion temperatures, *Fifth Symposium on Temperature: Its Measurement and Control in Science and Industry*, June 1971, vol. 4, pp. 483–487, American Institute of Physics, College Park, MD, 1972.

Bonnis-Sassi, M., Leduc, G., Thellier, F., Lacarrière, B., and Monchoux, F.: Regulation of a human thermal environment by inverse method, *Appl. Thermal Eng.*, 26(17–18), 2176–2183, December 2006.

Boothroyd, S. A. and Jones, A. R.: A comparison of radiative characteristics for fly ash and coal, *IJHMT*, 29(11), 1649–1654, 1986.

Borjini, M. N., Mbow, C., and Daguenet, M.: Etude numerique de l'influence du transfert radiatif sur la convection naturelle laminaire, bidimensionnelle, permanente, dans un espace annulaire d'axe horizontal, délimité par deux cylindres circulaires isothermes, *Int. J. Thermal Sci.*, 37, 475–487, 1998.

Born, M., Wolf, E., and Bhatia, A. B.: *Principles of Optics*, 7th expanded edn., Cambridge University Press, Cambridge, U.K., 1999.

Bourdon, A. and Bultel, A.: Numerical simulation of stagnation line nonequilibrium airflows for reentry applications, *JTHT*, 22(2), 168–177, 2008.

Brandenberg, W. M.: The reflectivity of solids at grazing angles, in J. C. Richmond (ed.), *Measurement of Thermal Radiation Properties of Solids*, NASA SP-31, Washington, DC, pp. 75–82, 1963.

Brandenberg, W. M. and Clausen, O. W.: The directional spectral emittance of surfaces between 200°C and 600°C, *Symposium on Thermal Radiation of Solids*, NASA SP-55, Washington, DC, pp. 313–320, 1965.

Brandis, A. N., Johnston, C. O., Cruden, B. A., Prabhu, D. K., and Bose, D.: Uncertainty analysis of NEQAIR and HARA predictions of air radiation measurements obtained in the EAST facility, AIAA Paper 2011–3478, *42nd AIAA Thermophysics Conference*, Honolulu, HI, June 27–30, 2011.

Brandis, A. N., Johnston, C. O., Cruden, B. A., Prabhu, D. K., Wray, A. A., Liu, Y., Schwenke, D. W., and Bose, D.: Validation of CO 4th positive radiation for mars entry, *JQSRT*, 121, 91–104, 2013.

Brannon, R. R., Jr. and Goldstein, R. J.: Emittance of oxide layers on a metal substrate, *JHT*, 92(2), 257–263, 1970.

Branstetter, J. R.: *Radiant Heat Transfer between Nongray Parallel Plates of Tungsten*, NASA TN D-1088, Washington, DC, 1961.

Brantley, P. S. and Larsen, E.: The simplified P3 approximation, *Nucl. Sci. Eng.*, 134, 1–21, 2000.

Braun, M. M. and Pilon, L.: Effective optical properties of nano-absorbing nano-porous media, *Thin Solid Films*, 496(2), 505–514, 2006.

Breene, R. G.: *Theories of Spectral Line Shapes*, John Wiley & Sons, New York, 1981.

Breig, W. F. and Crosbie, A. L.: Numerical computation of a generalized exponential integral function, *Math. Comput.*, 28(126), 575–579, 1974.

Brewster, M. Q.: *Thermal Radiative Transfer and Properties*, John Wiley & Sons, New York, pp. 502–510, 1992.

Brewster, M. Q. and Tien, C. L.: Radiative transfer in packed fluidized beds: Dependent versus independent scattering, *JHT*, 104(4), 573–579, 1982.

Brockmann, H.: The use of generalized configuration factors for calculating the radiant interchange between nondiffuse surfaces, *IJHMT*, 40(18), 4473–4486, 1997.

Brunner, T. A. and Holloway, J. P.: One-dimensional Riemann Solvers and the maximum entropy closure, *JQSRT*, 69, 543–566, 2001.

Buckalew, C. and Fussell, D.: Illumination networks: Fast realistic rendering with general reflectance functions, *Comput. Graph.* (*ACM SIGGRAPH'89 Proc.*), 23(3), 89–98, 1989.

Buckius, R. O. and Tien, C. L.: Infrared flame radiation, *IJHMT*, 20(2), 93–106, 1977.

Buckley, H.: On the radiation from the inside of a circular cylinder: Part I, *Philos. Mag.*, 4, 753–762, 1927.

Buckley, H.: On the radiation from the inside of a circular cylinder: Part II, *Philos. Mag.*, 6, 447–457, 1928.

Budaev, B. V. and Bogy, D. B.: Extension of Planck's law of thermal radiation to systems with a steady heat flux, *Ann. Phys.*, 523(10), 791–804, 2011.

Budaev, B. V. and Bogy, D. B.: Computation of radiative heat transport across a nanoscale vacuum gap, *Appl. Phys. Lett.*, 104(6), 061109-1–061109-4, 2014a.

Budaev, B. V. and Bogy, D. B.: On thermal radiation across nanoscale gaps, *Proc. R. Soc. A*, submitted for publication, 2014b.

Burger, M. and Osher, S. J.: A survey on level set methods for inverse problems and optimal design, *Eur. J. Appl. Math.*, 16, 263–301, 2005.

Burns, P. J. and Pryor, D. V.: Vector and parallel Monte Carlo radiative heat transfer simulation, *Numer. Heat Transfer B*, 16(1), 97–124, 1989.

Burns, P. J. and Pryor, D. V.: Surface radiative transport at large scale via Monte Carlo, in C.-L. Tien (ed.), *Annual Review of Heat Transfer*, IX, Chapter 7, pp. 79–158, Begell House, New York, 1998.

Burns, S. P., Howell, J. R., and Klein, D. E.: Application of the finite element method to the solution of combined natural convection-radiation in a horizontal cylindrical annulus, in R. W. Lewis and P. Durbetaki (eds.), *Numerical Methods in Thermal Problems*, vol. IX, pp. 327–338, Pineridge Press, Swansea, U.K., 1995a.

Burns, S. P., Howell, J. R., and Klein, D. E.: Empirical evaluation of an important approximation for combined-mode heat transfer in a participating medium using the finite-element method, *Numer. Heat Transfer B*, 27, 309–322, 1995b.

Burns, S. P., Howell, J. R., and Klein, D. E.: Finite element solution for radiative heat transfer with nongray, nonhomogeneous radiative properties, *ASME HTD*-315, 3–10, 1995c.

Bushinskii, A. V.: Determination of the geometric-optics coefficients of thermal radiation by the Monte Carlo method, *Inzhenerno-Fizicheskii Zh.*, 30(1), 160–166, 1976.

Byun, D., Lee, C., and Baek, S. W.: Radiative heat transfer in discretely heated irregular geometry with an absorbing, emitting, and anisotopically scattering medium using combined Monte-Carlo and finite volume method, *IJHMT*, 47, 4195–4203, 2004.

Byun, K. H. and Smith, T. F.: View factors for rectangular enclosures using the direct discrete-ordinates method, *JTHT*, 11(4), 593–595, 1997.

Cahill, D. G., Braun, P. V., Chen, G. et al.: Nanoscale thermal transport II. 2003–2012, *Appl. Phys. Rev.*, 1, 011305, 2014.

Cahill, D. G., Ford, W. K., Goodson, K. E., Mahan, G. D., Majumdar, A., Maris, H. J., Merlin, R., and Phillpot, S. R.: Nanoscale thermal transport, *J. Appl. Phys.*, 93, 793–818, 2003.

Cai, G.-B., Zhang, X.-Y., and Zhu, D.-Q.: Modeling the radiation of participating media with coupled finite volume method, *JTHT*, 21(1), 239–243, 2007.

Cai, J., Lei, S., Dasgupta, A., Modest, M. F., and Haworth, D. C.: Radiative heat transfer in high-pressure laminar hydrogen-air diffusion flames using spherical harmonics and *k*-distributions, in J. Cai, S. Lei, and M. F. Modest (eds.), Paper No. IMECE2012-88507, *ASME 2012 International Mechanical Engineering Congress and Exposition*, vol. 7, pp. 1771–1777, Houston, TX, November 9–15, 2012.

Cai, J., Marquez, R., and Modest, M. F.: Comparisons of radiative heat transfer calculations in a jet diffusion flame using spherical harmonics and k-distributions, *JHT*, 136(11), 112702-1--112702-9, 2014.

Cai, W. and Shalaev, V.: *Optical Metamaterials: Fundamentals and Applications*, Springer, New York, 2010.

Caldas, M. and Semião, V.: Modelling of scattering and absorption coefficients for a polydispersion, *IJHMT*, 42(24), 4535–4548, 1999.

Caliot, C., Flamant, G., El Hafi, M., and le Maoult, Y.: Assessment of the single-mixture gas assumption for the correlated *k*-distribution fictitious gas method in H_2O–CO_2–CO mixture at high temperature, *JHT*, 130, 10501-1–10501-6, October 2008.

Campbell, A. T., III and Fussell, D. S.: Adaptive mesh generation for global diffuse illumination, *Comput. Graph.*, 24(4), 155–164, August 1990.

Campo, A.: Variational techniques applied to radiative-convective fins with steady and unsteady conditions, *Wärme Stoffüber Trag.*, 9, 139–144, 1976.

Canuto, C., Quarteroni, A., Hussaini, M. Y., and Zang, T. A., *Spectral Methods: Fundamentals in Single Domains,* Springer, Berlin, Germany, 2006.

Caren, R. P. and Liu, C. K.: Effect of inhomogeneous thin films on the emittance of a metal substrate, *JHT*, 93(4), 466–468, 1971.

Carey, V. P., Chen, G., Grigoropoulos, C., Kaviany, M., and Majumdar, A.: A review of heat transfer physics, *Nanosc. Microsc. Therm.*, 12, 1–60, 2008.

Carlson, B. G.: *Transport Theory: Discrete Ordinates Quadrature over the Unit Sphere*, Los Alamos Scientific Laboratory Report LA-4554, Los Alamos, NM, 1970.

Carlson, B. G.: On a more precise definition of discrete ordinates methods, *Proceedings of the Second Conference Transport Theory*, Los Alamos, NM, 1971.

Carlson, B. G. and Lathrop, K. D.: Transport theory—The method of discrete ordinates, in H. Greenspan, C. N. Kelber, and D. Okrent (eds.), *Computing Methods in Reactor Physics*, Chapter 3, Gordon & Breach, New York, 1968.

Carlson, B. G. and Lee, C. E.: *Mechanical Quadrature and the Transport Equation*, Los Alamos Scientific Laboratory Report LAMS-2573, Los Alamos, NM, 1961.

Carminati, R. and Greffet, J.-J.: Near-field effects in spatial coherence of thermal sources, *Phys. Rev. Lett.*, 82(8), 1660–1663, 1999.

Carminati, R., Greffet, J.-J., and Sentenac, A.: A model for the radiative properties of rough surfaces, *Proceedings of the 11th International Heat Transfer Conference*, Kyong Ju, Korea, vol. 7, pp. 427–432, Taylor & Francis, New York, 1998.

Carnagnola, F., Sanz, J. M., and Saiz, J. M.: Development of a Mueller matrix imaging system for detecting objects embedded in a turbid media, *JQSRT*, 146, 199–206, 2014.

Carpenter, J. R., Briggs, D. G., and Sernas, V.: Combined radiation and developing laminar free convection between vertical flat plates with asymmetric heating, *JHT*, 98(1), 95–100, 1976.

Carslaw, H. S. and Jaeger, J. C.: *Conduction of Heat in Solids*, 2nd edn., Clarendon Press, Oxford, U.K., 1959.
Cartigny, J. D.: Mean beam length for a scattering medium, *Proceedings of the Eighth International Heat Transfer Conference*, vol. 2, pp. 769–772, Hemisphere, Washington, DC, 1986.
Carvalho, M. da G., Farias, T., and Fontes, P.: Multidimensional modeling of radiative heat transfer in scattering media, *JHT*, 115(2), 486–489, 1993.
Case, K. M.: Transfer problems and the reciprocity principle, *Rev. Mod. Phys.*, 29, 651–663, 1957.
Cashwell, E. D. and Everett, C. J.: *A Practical Manual on the Monte Carlo Method for Random Walk Problems*, Pergamon, New York, 1959.
Cassell, J. S. and Williams, M. M. R.: An integral equation for radiation transport in an infinite cylinder with Fresnel reflection, *JQSRT*, 105, 12–31, 2007.
Cassol, F., Brittes, R., França, F. H. R., and Ezekoye, O. A.: Application of the weighted-sum-of-gray-gases model for media composed of arbitrary concentrations of H_2O, CO_2 and soot, *IJHMT*, 79, 796–806, 2014.
Cassol, F., Schneider, P. S., França, F. H. R., Silva Neto, A. J.: Multi-objective optimization as a new approach to illumination design of interior spaces, *Building Environ.*, 46, 331–338, 2011.
Castro, R. O. and Trelles, J. P.: Spatial and angular finite element method for radiative transfer in participating media, *JQSRT*, 157, 81–105, 2015.
Celanovic, I., Perreault, D., and Kassakian, J.: Resonant-cavity enhanced thermal emission, *Phys. Rev. B*, 72, 075127, 2005.
Centeno, F. R., daSilva, C. V., and França, F. H. R.: The influence of gas radiation on the thermal behavior of a 2D axisymmetric turbulent non-premixed methane–air flame, *Energy Convers. Manage.*, 79, 405–414, 2014.
Cess, R. D.: The effect of radiation upon forced-convection heat transfer, *Appl. Sci. Res.*, 10(A), 430–438, 1961.
Cess, R. D.: The interaction of thermal radiation with conduction and convection heat transfer, in T. F. Irvine, Jr. and J. P. Hartnett (eds.), *Advances in Heat Transfer*, vol. 1, pp. 1–50, Academic Press, New York, 1964.
Cess, R. D.: The interaction of thermal radiation in boundary layer heat transfer, *Proceedings of the Third International Heat Transfer Conference AIChE*, Chicago, vol. 5, pp. 154–163, 1966a.
Cess, R. D.: The interaction of thermal radiation with free convection heat transfer, *IJHMT*, 9(11), 1269–1277, 1966b.
Cess, R. D. and Lian, M. S.: A simple parameterization for the water vapor emissivity, *JHT*, 98(4), 676–678, 1976.
Chai, J. C., Lee, H. S., and Patankar, S. V.: Ray effect and false scattering in the discrete ordinates method, *Numer. Heat Transfer B Fundam.*, 24(4), 373–389, 1993.
Chai, J. C., Lee, H. S., and Patankar, S. V.: Improved treatment of scattering using the discrete ordinates method, *JHT*, 116(1), 260–263, 1994.
Chai, J. C., Parthasarathy, G., Lee, H. S., and Patankar, S. V.: Finite volume radiative heat transfer procedure for irregular geometries, *JTHT*, 9(3), 410–415, 1995.
Chai, J. C. and Patankar, S. V.: Finite volume method for radiation heat transfer, in W. Minkowycz, and E. Sparrow (eds.), *Advances in Numerical Heat Transfer*, p. 2, Taylor & Francis, London, U.K., 2000.
Chambers, R. L. and Somers, E. V.: Radiation fin efficiency for one-dimensional heat flow in a circular fin, *JHT*, 81(4), 327–329, 1959.
Chambré, P. L.: Nonlinear heat transfer problem, *J. Appl. Phys.*, 30(11), 1683–1688, 1959.
Chan, S. and Ge, X. S.: An improvement on the prediction of optical constants and radiative properties by introducing an expression for the damping frequency in drude model, *Int. J. Thermophys.*, 21(1), 269–280, 2000.
Chan, S. H., Cho, D. H., and Kocamustafaogullari, G.: Melting and solidification with internal radiative transfer—A generalized phase change model, *IJHMT*, 26(4), 621–633, 1983.
Chandrasekhar, S.: *Radiative Transfer*, Dover, New York, 1960.
Chang, L. C., Yang, K. T., and Lloyd, J. R.: Radiation-natural convection interactions in two-dimensional complex enclosures, *JHT*, 105(1), 89–95, 1983.
Chang, S. L. and Rhee, K. T.: Blackbody radiation functions, *Int. Commun. Heat Mass Transfer*, 11, 451–455, 1984.
Chang, Y.-P., and Smith, R. S. Jr.: Steady and transient heat transfer by radiation and conduction in a medium bounded by two coaxial cylindrical surfaces, *IJHMT*, 13(1), 69–80, 1970.
Chapuis, P.-O., Laroche, M., Volz, S., and Greffet, J.-J.: Radiative heat transfer between metallic nanoparticles, *Appl. Phys. Lett.*, 92, 201906, 2008a.
Chapuis, P.-O., Volz, S., Henkel, C., Joulain, K., and Greffet, J.-J.: Effects of spatial dispersion in near-field radiative heat transfer between two parallel metallic surfaces, *Phys. Rev. B*, 77, 035431, 2008b.
Charalampopoulos, T. T.: Morphology and dynamics of agglomerated particulates in combustion systems using light scattering techniques, *Prog. Energy Combust. Sci.*, 18, 13–45, 1992.

Charalampopoulos, T. T. and Chang, H.: Agglomerate parameters and fractal dimension of soot using light scattering: Effects on surface growth, *Combust. Flame*, 87, 89–99, 1991.

Charalampopoulos, T. T. and Felske, J. D.: Refractive indices of soot particles deduced from in-situ laser light scattering measurements, *Combust. Flame*, 68, 283–294, 1987.

Charette, A., Larouche, A., and Kocaefe, Y. S.: Application of the imaginary planes method to three-dimensional systems, *IJHMT*, 33(12), 2671–2681, 1990.

Charle, M.: *Les Manuscripts de Léonard de Vinci, Manuscripts C, E, et K de la Bibliothèque de l'Institute Publiés en Facsimilés Phototypiques*, Ravisson-Mollien, Paris, 1888. (Referenced in Knowles Middleton, W. E.: Note on the invention of photometry, *Am. J. Phys.*, 31(3), 177–181, 1963.)

Charnigo, R., Francoeur, M., Kenkel, P., Mengüç, M. P., Hall, B., and Srinivasan, C.: Estimating quantitative features of nanoparticles using multiple derivatives of scattering profiles, *JQSRT*, 112(8), 1369–1382, May 2011.

Charnigo, R., Francoeur, M., Kenkel, P., Mengüç, M. P., Hall, B., and Srinivasan, C.: Credible intervals for nanoparticle characteristics, *JQSRT*, 113(2), 182–193, 2012.

Charnigo, R., Francoeur, M., Mengüç, M. P., Brock, A., Leichter, M., and Srinivasan, C.: Derivatives of scattering profiles: Tools for nanoparticle characterization, *JOSA A*, 24(9), 2578–2589, 2007.

Chato, J. C. and Khodadadi, J. M.: Optimization of cooled shields in insulations, *JHT*, 106(4), 871–875, 1984.

Chawla, T. C. and Chan, S. H.: Combined radiation convection in thermally developing Poiseuille flow with scattering, *JHT*, 102(2), 297–302, 1980.

Chen, G.: Heat transfer in micro- and nanoscale photonic devices, in C.-L. Tien (ed.), *Annual Review of Heat Transfer*, VII, Chapter 1, pp. 1–57, Begell House, New York, 1995.

Chen, G.: *Nanoscale Energy Transport and Conversion*, Oxford University Press, New York, 2005.

Chen, G. and Tien, C. L.: Partial coherence theory of thin film radiative properties, *JHT*, 114(3), 636–643, 1992.

Chen, J. and Ge., X.-S.: An improvement on the prediction of optical constants and radiative properties by introducing an expression for the damping frequency in drude model, *Int. J Thermophys.*, 21(1), 269–280, 2000.

Chen, J. C. and Churchill, S. W.: Radiant heat transfer in packed beds, *AIChE J.*, 9, 35–41, 1963.

Cheng, P.: Exact solutions and differential approximation for multi-dimensional radiative transfer in cartesian coordinate configurations, *Prog. Astronaut. Aeronaut.*, 31, 269, 1972.

Cherkaoui, M., Dufresne, J.-L., Fournier, R., Grandpiex, J.-Y., and Lahellec, A.: Monte Carlo simulation of radiation in gases with a narrow-band model and a net-exchange formulation, *JHT*, 118(2), 401–407, 1996.

Chern, B. C., Moon, T. J., and Howell, J. R.: Thermal analysis of in-situ curing for thermoset, hoop-wound structures using infrared heating: Part I—Numerical predictions using independent scattering, *JHT*, 117(3), 674–680, 1995a.

Chern, B. C., Moon, T. J., and Howell, J. R.: Thermal analysis of in-situ curing for thermoset, hoop-wound structures using infrared heating: Part II—Dependent scattering effects, *JHT*, 117, 681–686, 1995b.

Chern, B. C., Moon, T. J., and Howell, J. R.: Dependent radiative transfer regime for unidirectional fiber composites exposed to normal incident radiation, *Proceedings of the Fourth ASME/JSME Joint Symposium*, Maui, HI, 1995c.

Chern, B.-C., Moon, T. J., and Howell, J. R.: On-line processing of unidirectional fiber composites using radiative heating: I. Model, *J. Compos. Mater.*, 36(16), 1905–1934, August 2002a.

Chern, B.-C., Moon, T. J., and Howell, J. R.: On-line processing of unidirectional fiber composites using radiative heating: II. Radiative properties, experimental validation and process parameter selection, *J. Compos. Mater.*, 36(16), 1935–1665, August 2002b.

Chiang, W. C. and Russell, R. A.: Simulated annealing metaheuristics for the vehicle routing problem with time windows, *Ann. Operat. Res.*, 63, 3–27, 1996.

Chiba, Z. and Greif, R.: Heat transfer to steam flowing turbulently in a pipe, *IJHMT*, 16, 1645–1648, 1973.

Chin, J. H., Panczak, T. D., and Fried, L.: Spacecraft thermal modeling, *Int. J. Numer. Methods Eng.*, 35, 641–653, 1992.

Chin, J. S. and Lefebvre, A. H.: Influence of fuel composition on flame radiation in gas turbine combustors, *J. Propul.*, 6(4), 497–503, 1990.

Chiou, J. S.: Combined radiation-convection heat transfer in a pipe, *JTHT*, 7(1), 178–180, 1993.

Choi, B. C. and Churchill, S. W.: A technique for obtaining approximate solutions for a class of integral equations arising in radiative heat transfer, *Int. J. Heat Fluid Flow*, 6(1), 42–48, 1985.

Chopade, R. P., Mishra, S. C., Mahanta, P., and Marayama, S.: Estimation of power of heaters in a radiant furnace for uniform thermal conditions, *IJHMT*, 55, 4340–4351, 2012.

Chou, Y. S. and Tien, C. L.: A modified moment method for radiative transfer in non-planar systems, *JQSRT*, 8, 919–933, 1968.

Christiansen, C.: Absolute determination of the heat emission and absorption capacity, *Ann. Phys. Wied.*, 19, 267–283, 1883.

Chu, H., Liu, F., and Consalvi, J.-L.: Relationship between the spectral line based weighted-sum-of-gray-gases model and the full spectrum *k*-distribution model, *JQSRT*, 143, 111–120, 2014.

Chu, H.-S. and Weng, L.-C.: Transient combined conduction and radiation in anisotropically scattering spherical media, *JTHT*, 6(2), 553–556, 1992.

Chui, E. H., Raithby, G. D., and Hughes, P. M. J.: Prediction of radiation transfer in cylindrical enclosures with the finite volume method, *JTHT*, 6(4), 605–611, 1992.

Chung, T. J.: Integral and integrodifferential systems, in W. J. Minkowitz, E. M. Sparrow, G. E. Schneider, and R. H. Pletcher (eds.), *Handbook of Numerical Heat Transfer*, Chapter 14, Wiley, New York, 1988.

Chung, T. J. and Kim, J. Y.: Two-dimensional combined-mode heat transfer by conduction, convection and radiation in emitting, absorbing and scattering media—Solution by finite elements, *JHT*, 106(2), 448–452, 1984.

Chung, B. T. F. and Zhang, B. X.: Optimization of radiating fin array including mutual irradiations between radiator elements, *JHT*, 113(4), 814–822, 1991.

Clarke, F. J. J. and Parry, D. J.: Helmholtz reciprocity: Its validity and application to reflectometry, *Light. Res. Technol.*, 17(1), 1–11, 1985.

Clarksean, R. and Solbrig, C.: Minimization of the effect of errors in approximate radiation view factors, *Nucl. Eng. Des.*, 1994, 431, 1994.

Clements, A. G., Porter, R., Pranzitelli, A., and Pourkashanian, M.: Evaluation of FSK models for radiative heat transfer under oxyfuel conditions, *JQSRT*, 151, 67–75, 2015.

Coakley Jr, J. A. and Yang, P.: *Atmospheric Radiation: A Primer with Illustrative Solutions*, John Wiley & Sons, New York, 2014.

Cobble, M. H.: Theoretical concentrations for solar furnaces, *Sol. Energy*, 5(2), 61–72, 1961.

Coelho, P. J.: The role of ray effects and false scattering on the accuracy of the standard and modified discrete ordinates methods, in M. P. Mengüç and N. Selçuk (eds.), *Radiation III: Third International Symposium on Radiative Transfer*, Begell House, New York, 2001.

Coelho, P. J.: A comparison of spatial discretization schemes for differential solution of the radiative transfer equation, *Proceedings of the InterNational Symposium on Radiation Transfer V*, Bodrum, Turkey, July 2007.

Coelho, P. J.: Turbulence–radiation interaction: From theory to application in numerical simulations, *JHT*, 134(3), 031001, 2012a.

Coelho, P. J.: A theoretical analysis of the influence of turbulence on radiative emission in turbulent diffusion flames of methane, *J. Phys. Conf. Ser.*, 369, Paper 012010, 2012b.

Coelho, P. J.: Advances in the discrete ordinates and finite volume methods for the solution of radiative heat transfer problems in participating media, *JQSRT*, 145, 241–246, 2014.

Coelho, P. J. and Carvalho, M. G.: Modeling of soot formation and oxidation in turbulent diffusion flames, *JTHT*, 9(4), 644–652, 1995.

Coelho, P. J. and Carvalho, M. G.: A conservative formulation of the discrete transfer method, *JHT*, 119(1), 118–128, 1997.

Coelho, P. J., Perez, P., and El Hafi, M.: Benchmark numerical solutions for radiative heat transfer in two-dimensional axisymmetric enclosures with non-gray sooting media, *Numer. Heat Transfer B Fundam.*, 43(5), 425–444, 2003.

Cohen, M. F., Chen, S. E., and Wallace, J. R.: A progressive refinement approach to fast radiosity image generation, *Comput. Graph.*, 22(4), 315–324, 1988.

Cohen, M. F. and Greenberg, D. P.: The hemi-cube: A radiosity solution for complex environments, in K. I. Joy, C. W. Grant, N. L. Max, and L. Hatfield (eds.), *Computer Graphics: Image Synthesis*, pp. 254–263, Computer Society Press, IEEE, New York, 1985.

Cohn, D. W., Tang, K., and Buckius, R. O.: Comparison of theory and experiments for reflection from microcontoured surfaces, *IJHMT*, 40(13), 3223–3235, 1997.

COMSOL: *COMSOL Multiphysics User Guide*, COMSOL, Burlington, MA, 2007.

Condon, E. U.: Radiative transport in hot glass, *JQSRT*, 8(1), 369–385, 1968.

Coquard, R. and Baillis, D.: Radiative characteristics of opaque spherical particles beds: A new method of prediction, *JTHT*, 18(2), 178–186, June 2004.

Coquard, R., Rochais, D., and Baillis, D.: Experimental investigations of the coupled conductive and radiative heat transfer in metallic/ceramic foams, *IJHMT*, 52, 4907–4918, 2009.

Coquard, R., Rochais, D., and Baillis, D.: Conductive and radiative heat transfer in ceramic and metal foams at fire temperatures, *Fire Technol.,* 48(3), 699–732, 2012.

Corana, A., Marchesi, M., Martini, C., and Ridella, S.: Minimizing multimodal functions of continuous variables with the "simulated annealing" algorithm, *ACM Trans. Math. Softw.*, 13, 262–280, 1987.

Corbitt, S. J., Francoeur, M., and Raeymaekers, B.: Implementation of optical dielectric materials: A review, *JQSRT* (in press), 2015.

Corlett, R. C.: Direct Monte Carlo calculation of radiative heat transfer in vacuum, *JHT*, 88(4), 376–382, 1966.

Cornelison, C. J. and Howe, J. T.: Analytic solution of the transient behavior of radiation-backscattering heat shields, *JTHT*, 6(4), 612–617, 1992.

Cort, G. E., Graham, H. L., and Johnson, N. L.: Comparison of methods for solving non-linear finite element equations in heat transfer, ASME Paper 82-HT-40, *Proc. 21st National Heat Transfer Conf.*, Seattle, August 1982.

Costello, F. A. and Schrenk, G. L.: Numerical solution to the heat-transfer equations with combined conduction and radiation, *J. Comput. Phys.*, 1, 541–543, 1966.

Cox, R. L.: Fundamentals of thermal radiation in ceramic materials, in S. Katzoff (ed.), *Thermal Radiation in Solids*, NASA SP-55, pp. 83–101, 1965.

Cox, R. L.: Radiant emission from cavities in scattering and absorbing media, SMU Research Report 68–2, Southern Methodist University Institute of Technology, Dallas, TX, October 1968.

Craighead, H. G., Bartynski, R., Buhrman, R. A., Wojcik, L., and Sievers, A. J.: Metal/insulator composite selective absorbers, *Solar Energy Mater.*, 1(1–2), 105–124, 1979.

Cravalho, E. G., Tien, C. L., and Caren, R. P.: Effect of small spacings on radiative transfer between two dielectrics, *JHT*, 89(4), 351–358, 1967.

Crepeau, J.: A brief history of the T^4 law, Paper HT2009-88060, *Proceedings of the HT 2009; 2009 Summer Heat Transfer Conference*, San Francisco, CA, July 2009.

Crofcheck, C. L., Payne, F. A., and Mengüç, M. P.: Characterization of milk properties using a radiative transfer model, *Appl. Opt.*, 41(10), 2028–2037, 2002.

Crofcheck, C. L., Wade, J., Aslan, M. M., and Mengüç, M. P.: Effect of fat and casein particles in milk on the scattering of elliptically-polarized light, *Trans. ASAE*, 48(3), 1147–1155, 2005.

Crosbie, A. L. and Davidson, G. W.: Dirac-delta function approximations to the scattering phase function, *JQSRT*, 33(4), 391–409, 1985.

Crosbie, A. L. and Pattabongse, M.: Transient conductive and radiative transfer in a planar layer with Arrhenius heat generation, *JQSRT*, 37, 319–329, 1987.

Crosbie, A. L. and Sawheny, T. R.: Application of Ambarzumian's method to radiant interchange in a rectangular cavity, *JHT*, 96, 191–196, 1974.

Crosbie, A. L. and Sawheny, T. R.: Radiant interchange in a nonisothermal rectangular cavity, *AIAA J.*, 13(4), 425–431, 1975.

Crosbie, A. L. and Schrenker, R. G.: Exact expressions for radiative transfer in a three-dimensional rectangular geometry, *JQSRT*, 28, 507–526, 1982.

Crosbie, A. L. and Schrenker, R. G.: Radiative transfer in a two-dimensional rectangular medium exposed to diffuse radiation, *JQSRT*, 31(4), 339–372, 1984.

Cruden, B. A., Prabhu, D., and Martinez, R.: Absolute radiation measurement in Venus and Mars entry conditions, *J. Spacecraft Rockets*, 49(6), 1069–1079, November–December 2012.

Cumber, P. S.: Improvements to the discrete transfer method of calculating radiative heat transfer, *IJHMT*, 38(12), 2251–2258, 1995.

Cumber, P. S. and Onokpe, O.: Turbulent radiation interaction in jet flames: Sensitivity to the PDF, *IJHMT*, 57, 250–264, 2013.

Cunnington, G. R. and Lee, S. C.: Radiative properties of fibrous insulations: Theory versus experiment, *JTHT*, 10(3), 460–466, 1996.

d'Aguillon, F., S. J.: Opticorum Libri Sex, Antwerp, 1613. (Referenced in Knowles Middleton, W. E.: Note on the invention of photometry, *Am. J. Phys.*, 31(3), 177–181, 1963.)

Dalzell, W. H. and Sarofim, A. F.: Optical constants of soot and their application to heat-flux calculations, *JHT*, 91(1), 100–104, 1969.

Darvishvand, L., Kowsary, F., and Hadi Jafar, P.: Optimization of 3-D radiant enclosures with the objective of uniform thermal conditions on 3-D design bodies, *Heat Transfer Engineering*, Accepted, 2015.

Das, R., Mishra, S. C., and Uppaluri, R.: Multiparameter estimation in a transient conduction-radiation problem using the lattice Boltzmann method and the finite-volume method coupled with the genetic algorithms, *Numer. Heat Trans. Part A-Appl.*, 53(12), 1321–1338, 2008.

Datas, A., Hirashima, D., and Hanamura, K.: FTDT simulation of near-field radiative heat transfer between thin films supporting surface phonon polaritons: Lessons learned, *J. Thermal Sci. Technol.*, 8(1), 91–105, 2013.

Daun, K. J., Ertürk, H., and Howell, J. R.: Inverse design methods for high-temperature systems, *Arabian J. Sci. Technol.*, 27(2C), 3–48, 2003a.

Daun, K. J., Ertürk, H., Howell, J. R., Gamba, M., and Hosseini Sarvari, M.: The use of inverse methods for the design and control of radiant sources, *JSME Int. J., Ser. B*, 46(4), 470–478, 2003b.

Daun, K. J., França, F., Larsen, M., Leduc, G., and Howell, J. R.: Comparison of methods for inverse design of radiant enclosures, *JHT*, 128(3), 269–282, March 2006.

Daun, K. J. and Hollands, K. G. T.: Infinitesimal-area radiative analysis using parametric surface representation, through NURBS, *JHT*, 123(2), 249–256, 2001.

Daun, K. J. and Howell, J. R.: Optimization of transient heater settings to provide spatially uniform transient heating in manufacturing processes involving radiant heating, *Numer. Heat Transfer A*, 46(7), 651–668, October 2004.

Daun, K. J. and Howell, J. R.: Inverse design methods for radiative transfer systems, *JQSRT*, 93(1–3), 43–60, June 2005.

Daun, K. J., Howell, J. R., and Morton, D. P.: Design of radiant enclosures using inverse and non-linear programming techniques, *Inverse Probl. Eng.*, 11, 541–560, 2003c.

Daun, K. J., Howell, J. R., and Morton, D. P.: Geometric optimization of radiative enclosures through nonlinear programming, *Numer. Heat Transfer B*, 43, 203–219, March 2003d.

Daun, K. J., Howell, J. R., and Morton, D. P.: Geometric optimization of radiant enclosures containing specular surfaces, *JHT*, 125, 845–851, 2003e.

Daun, K. J., Howell, J. R., and Morton, D. P.: Smoothing Monte Carlo exchange factors through constrained maximum likelihood estimation, *JHT*, 127(10), 1124–1128, October 2005.

Davies, H.: The reflection of electromagnetic waves from a rough surface, *Proc. Inst. Elec. Eng. Lond.*, 101, 209–214, 1954.

Davisson, C. and Weeks, J. R., Jr.: The relation between the total thermal emissive power of a metal and its electrical resistivity, *JOSA*, 8(5), 581–605, 1924.

de Bastos, R.: Computation of radiation configuration factors by contour integration, MS thesis, Oklahoma State University, Stillwater, OK, 1961.

De Champlain, A., Kretschmer, D., and Tsogo, J.: Prediction of soot emissions in gas-turbine combustors, *J. Propul. Power*, 13(1), 117–122, 1997.

de Groh III, H. C. and Kassemi, M.: Effect of radiation on convection in a top-heated enclosure, *JTHT*, 7(4), 561–568, 1993.

de Lataillade, Defresne, J. L., El Hafi, M., Eymet, V., and Fournier, R.: A net exchange Monte Carlo approach to radiation in optically thick systems, *JQSRT*, 75(4), 563–584, 2002.

De Vos, J. C.: A new determination of the emissivity of tungsten ribbon, *Physica*, 20, 690–714, 1954.

De Wilde, Y., Formanek, F., Carminati, R., Gralak, B., Lemoine, P. A., Joulain, K., Mulet. J. P., Chen, Y., and Greffet, J.-J.: Thermal radiation scanning tunneling microscopy, *Nature*, 444(7120), 740–743, 2006.

Dehghan, A. A. and Behnia, M.: Combined natural convection-conduction and radiation heat transfer in a discretely heated open cavity, *JHT*, 118(1), 56–64, 1996.

Deissler, R. G.: Diffusion approximation for thermal radiation in gases with jump boundary condition, *JHT*, 86(2), 240–246, 1964.

Deiveegan, M., Balaji, C., and Venkateshan, S. P.: Comparison of various methods for simultaneous retrieval of surface emissivities and gas properties in gray participating media, *JHT*, 128, 829–837, August 2006.

Delatorre, J., Rolland, J. Y., Roger, M., El Hafi, M., Fournier, R., Blanco, S., and Bezian, J. J.: Monte Carlo estimations of domain-deformation sensitivities for optimal geometric design, *Proceedings of the Eurotherm83-Computational Thermal Radiation in Participating Media III*, Lisbon, Portugal, April 2009.

Delatorre, J., et al.: Monte Carlo advances and concentrated solar applications, *Solar Energy*, 103, 653–681, 2014.

DeMarco, A. G. and Lockwood, F. C.: A new flux model for the calculation of three-dimensional radiation heat transfer, *Riv. Combust.*, 5, 184, 1975.

Dembele, S. and Wen, J. X.: Investigation of a spectral formulation for radiative heat transfer in one-dimensional fires and combustion systems, *IJHMT*, 43(21), 4019–4030, 2000.

Demont, P., Huetz-Aubert, M., and Tran N'Guyen, H.: Experimental and theoretical studies of the influence of surface conditions on radiative properties of opaque materials, *Int. J. Thermophys.*, 3(4), 335–364, 1982.

Denison, M. K. and Webb, B. W.: A spectral line-based weighted-sum-of-gray-gases model for arbitrary RTE solvers, *JHT*, 115(4), 1004–1012, 1993a.

Denison, M. K. and Webb, B. W.: An absorption-line blackbody distribution function for efficient calculation of gas radiative transfer, *JQSRT*, 50, 499–510, 1993b.

Denison, M. K. and Webb, B. W.: Development and application of an absorption-line blackbody distribution for CO_2, *IJHMT*, 38(10), 1813–1821, 1995a.

Denison, M. K. and Webb, B. W.: The spectral-line weighted-sum-of-gray-gases model for H_2O/CO_2 mixtures, *JHT*, 117, 788–798, 1995b.

Denison, M. K. and Webb, B. W.: The absorption-line blackbody distribution function at elevated pressure, in M. P. Mengüç (ed.), *Radiative Transfer—I, Proceedings of the First International Symposium on Radiation Transfer*, pp. 228–238, Begell House, New York, 1996a.

Denison, M. K. and Webb, B. W.: The spectral line weighted-sum-of-gray-gases model—A review, in M. P. Mengüç (ed.), *Radiative Transfer—I, Proceedings of the First International Symposium on Radiative Transfer*, pp. 193–208, Begell House, New York, 1996b.

Dennar, E. A. and Sibulkin, M.: An evaluation of the differential approximation for spherically symmetric radiative transfer, *JHT*, 91(1), 73–76, 1969.

Denning, P. J.: The science of computing: The evolution of parallel processing, *Am. Sci.*, 73(5), 414–416, 1985.

DeSilva, A. A. and Jones, B. W.: Bidirectional spectral reflectance and directional-hemispherical spectral reflectance of six materials used as absorbers of solar energy, *Solar Energy Mater.*, 15, 391–401, 1987.

deSoto, S.: The radiation from an axisymmetric, real gas system with a nonisothermal temperature distribution, *Chem. Eng. Progr. Symp. Ser.*, 61(59), 138–154, 1965.

deSoto, S. and Edwards, D. K.: Radiative emission and absorption in nonisothermal nongray gases in tubes, in A. F. Charwat (ed.), *Proceedings of the Heat Transfer Fluid Mechanical Institute*, Stanford University, pp. 358–372, 1965.

Desrayaud, G. and Lauriat, G.: Radiative influence on the stability of fluids enclosed in vertical cavities, *IJHMT*, 31(5), 1035–1048, 1988.

DeVahl Davis, G.: Finite difference method for natural and mixed convection in enclosures, *Proceedings of the Eighth International Heat Transfer Conference*, vol. 1, pp. 101–109, San Francisco, CA, 1986.

Diaz, L. and Viskanta, R.: Experiments and analysis on the melting of a semitransparent material by radiation, *Warme-Stoffübertrag.*, 20, 311–321, 1986.

Didari, A. and Mengüç, M. P.: Analysis of near-field radiation transfer within nano-gaps using FDTD method, *JQSRT*, 146, 214–226, 2014.

Didari, A. and Mengüç, M. P.: Near-field thermal emission between corrugated surfaces separated by nano-gaps, *JQSRT*, 158, 43–51, 2015.

DiMatteo, R. S., Greiff, P., Finberg, S. L., Young-Waithe, K. A., Choy, H. K. H., Masaki, M. M. and Fonstad, C. G.: Enhanced photogeneration of carriers in a semiconductor via coupling across a nonisothermal nanoscale vacuum gap, *Appl. Phys. Lett.*, 79(12), 1894–1896, 2001.

Dimenna, R. A. and Buckius, R. O.: Quantifying specular approximations for angular scattering from perfectly conducting random rough surface, *JTHT*, 8, 393–399, 1994a.

Dimenna, R. A. and Buckius, R. O.: Electromagnetic theory predictions of the directional scattering from triangular surfaces, *JHT*, 116(3), 639–645, 1994b.

Dixon, M. J., Schoegl, I., Hull, C. B., and Ellzey, J. L.: Experimental and numerical conversion of liquid heptane to syngas through combustion in porous media, *Combust. Flame*, 154(1–2), 217–231, July 2008.

Docherty, P.: Prediction of gas emissivity for a wide range of process conditions, Paper R5, *Proceedings of the Seventh International Heat Transfer Conference*, vol. 2, pp. 481–485, Munich, Germany, 1982.

Doicu, A. and Wriedt, T.: T-matrix method for electromagnetic scattering from scatterers with complex structure, *JQSRT*, 70, 663–673, 2001.

Doicu, A., Wriedt, T., and Eremin, Y. A.: *Light Scattering by Systems of Particles*, Springer, Heidelberg, Germany, 2006.

Dombrovsky, L. A. and Baillis, D.: *Thermal Radiation in Disperse Systems: An Engineering Approach*, Begell House, Redding, PA, 2010.

Domingues, G., Volz, S., Joulain, K., and Greffet, J.-J.: Heat transfer between two nanoparticles through near field interaction, *Phys. Rev. Lett.*, 94(8), 085901, 2005.

Domoto, G. A.: Frequency integration for radiative transfer problems involving homogeneous non-gray gases: The inverse transmission function, *JQSRT*, 14, 935–942, 1974.

Domoto, G. A., Bohem, R. F., and Tien, C. L.: Experimental investigation of radiative transfer between metallic surfaces at cryogenic temperatures, *JHT*, 92, 412–417, 1970.

Dorigon, L. J., Duciak, G., Brittes, R., Cassol, F., Galarca, M. M., and França, F. H. R.: WSGG correlations based on HITEMP2010 for computation of thermal radiation in non-isothermal, non-homogeneous H_2O/CO_2 mixtures, *IJHMT*, 64, 863–873, 2013.

Dorn, O. and Lesselier, D.: Level set methods for inverse scattering, *Inverse Probl.*, 22, R67–R131, 2006.

Draine, B. T.: The discrete dipole approximation an its application to interstellar graphite grains, *Astrophy. J.*, 333, 848–872, 1998.

Draine, B. T. and Flatau, P. J.: The discrete-dipole approximation for scattering calculations, *JOSA A*, 11, 1491–1499, 1994.

Draine, B. T. and Flatau, P. J.: The discrete dipole approximation for periodic targets I: Theory and tests, *JOSA*, 25, 2693–2703, 2008.

Draine, B. T. and Goodman, J.: Beyond Clausius-Mossotti–wave-propagation on a polarizable point lattice and the discrete dipole approximation, *Astrophys. J.*, 405(2), 685–697, 1993.

Draper, J. W.: On the production of light by heat, *Phil. Mag., Ser. 3*, 30, 345–360, 1847.

Drevillon, J.: Design ab-initio de matériaux micro et nanostructurés pour l'emission thermique cohérente en champ proche et en champ lointain, PhD thesis, Université de Nantes, Nantes, France, 2007 (in French).

Drevillon, J. and Ben-Abdallah, P.: Ab initio design of coherent thermal sources, *J. Appl. Phys.*, 102, 114305, 2007a.

Drevillon, J. and Ben-Abdallah, P.: Inverse design of quasi monochromatic light sources in the visible range, *Proceedings of the Fifth International Symposium on Radiation Transfer*, Bodrum, Turkey, June 2007b.

Drolen, B. L.: Bidirectional reflectance and specularity of twelve spacecraft thermal control materials, *JTHT*, 6(4), 672–679, 1992.

Drolen, B. L., Kumar, S., and Tien, C. L.: Experiments on dependent scattering of radiation, *Proc. AIAA 22nd Thermophysics Conf.*, Honolulu, AIAA Paper 87–1485, June 1987.

Dua, S. S. and Cheng, P.: Multi-dimensional radiative transfer in non-isothermal cylindrical media with non-isothermal bounding walls, *IJHMT*, 18, 254–259, 1975.

Dubovik, O., Labonnote, L., Litvinov, P., Parol, F., Mishchenko, M. I.: Electromagnetic and light scattering by nonspherical particles XIV, *JQSRT*, 146, 1–3, 2014.

Duffie, J. A. and Beckman, W. A.: *Solar Energy Thermal Processes*, 3rd edn., Wiley, New York, 2006.

Dunkle, R. V.: Thermal radiation characteristics of surfaces, in J. A. Clark (ed.), *Theory and Fundamental Research in Heat Transfer*, pp. 1–31, Pergamon Press, New York, 1963.

Dunkle, R. V.: Geometric mean beam lengths for radiant heat-transfer calculations, *JHT*, 86(1), 75–80, 1964.

Dunkle, R. V. and Bevans, J. T.: Radiant interchange within an enclosure. Part 3: A method for solving multinode networks and a comparison of the band energy and gray radiation approximations, *JHT*, 82(1), 14–19, 1960.

Dunn, S. T., Richmond, J. C., and Parmer, J. F.: Survey of infrared measurement techniques and computational methods in radiant heat transfer, *J. Spacecraft Rockets*, 3(7), 961–975, 1966.

Dupoirieux, F., Tessé, L., Avila, S., and Taine, J.: An optimized reciprocity Monte Carlo method for the calculation of radiative transfer in media of various optical thicknesses, *IJHMT*, 49(7–8), 1310–1319, April 2006.

Eckert, E.: Das strahlungsverhaltnis von flachen mit einbuchtungen und von zylindrischen bohrungen, *Arch. Waermewirtsch.*, 16(5), 135–138, 1935.

Eckert, E. R. G. and Drake, R. M. Jr.: *Heat and Mass Transfer*, 2nd edn., McGraw–Hill, New York, 1959.

Eckert, E. R. G. and Sparrow, E. M.: Radiative heat exchange between surfaces with specular reflection, *IJHMT*, 3(1), 42–54, 1961.

Edalatpour, S. and Francoeur, M.: The thermal discrete dipole approximation (T-DDA) for near-field radiative heat transfer simulations in three-dimensional arbitrary geometries, *JQSRT*, 133, 364–373, 2014.

Eddington, A. S.: *The Internal Constitution of the Stars*, Cambridge University Press, Cambridge, U.K., 1926.

Edwards, D. K.: Radiative transfer characteristics of materials, *JHT*, 91, 1–15, 1969.

Edwards, D. K.: Molecular gas band radiation, in T. F. Irvine, Jr. and J. P. Hartnett (eds.), *Advances in Heat Transfer*, vol. 12, pp. 115–193, Academic Press, New York, 1976.

Edwards, D. K.: Solar absorption by each element in an absorber-coverglass array, *Solar Energy*, 19, 401–402, 1977.

Edwards, D. K.: Numerical methods in radiation heat transfer, in T. M. Shih (ed.), *Numerical Properties and Methodologies in Heat Transfer*, Hemisphere, Washington, DC, 1983.

Edwards, D. K. and Babikian, D. S.: Radiation from a nongray scattering, emitting, and absorbing solid rocket motor plume, *JTHT*, 4(4), 446–453, 1990.

Edwards, D. K. and Balakrishnan, A.: Slab band absorptance for molecular gas radiation, *JQSRT*, 12, 1379–1387, 1972a.

Edwards, D. K. and Balakrishnan, A.: Volume interchange factors for nonhomogeneous gases, *JHT*, 94(2), 181–188, 1972b.
Edwards, D. K. and Bayard de Volo, N.: Useful approximations for the spectral and total emissivity of smooth bare metals, in S. Gratch (ed.), *Advances in Thermophysical Properties at Extreme Temperatures and Pressures*, pp. 174–188, ASME, New York, 1965.
Edwards, D. K. and Bertak, I. V.: Imperfect reflections in thermal radiation heat transfer, in J. W. Lucas (ed.), *Heat Transfer and Spacecraft Thermal Control*, pp. 143–165, AIAA Progress in Astronautics and Aeronautics Series 24, MIT Press, Cambridge, MA, 1971.
Edwards, D. K. and Catton, I.: Radiation characteristics of rough and oxidized metals, in S. Gratch (ed.), *Advances in Thermophysical Properties at Extreme Temperatures and Pressures*, pp. 189–199, ASME, New York, 1964.
Edwards, D. K. and Matavosian, R.: Scaling rules for total absorptivity and emissivity of gases, *JHT*, 106, 684–689, 1984.
Edwards, D. K. and Menard, W. A.: Comparison of models for correlation of total band absorption, *Appl. Opt.*, 3(5), 621–625, 1964.
Edwards, D. K., Sakurai, Y., and Babikian, D. S.: A two-particle model for rocket plume radiation, *JTHT*, 1(1), 13–20, 1987.
Edwards, D. K. and Tobin, R. D.: Effect of polarization on radiant heat transfer through long passages, *JHT*, 89(2), 132–138, 1967.
Egan, W. and Hilgeman, T.: Spectral reflectance of particulate materials: A Monte Carlo model including asperity scattering, *Appl. Opt.*, 17(2), 245–252, 1978.
Einstein, T. H.: *Radiant Heat Transfer to Absorbing Gases Enclosed between Parallel Flat Plates with Flow and Conduction*, NASA TR R-154, Washington, DC, 1963a.
Einstein, T. H.: *Radiant Heat Transfer to Absorbing Gases Enclosed in a Circular Pipe with Conduction, Gas Flow, and Internal Heat Generation*, NASA TR R-156, Washington, DC, 1963b.
El-Wakil, S. A. and Abulwafa, E. M.: Variational-iterative method for conductive-radiative heat transfer in spherical inhomogeneous medium, *JTHT*, 14(4), 612–615, 2000.
Emery, A. F.: VIEW—A radiation view factor program with interactive graphics for geometry definition (Version 5.5.3), 1986, available through Open Channel Foundation, http://www.openchannelfoundation.org/projects/VIEW/ (accessed April 27, 2015).
Emery, A. F., Johansson, O., Lobo, M., and Abrous, A.: A comparative study of methods for computing the diffuse radiation viewfactors for complex structures, *JHT*, 113(2), 413–422, 1991.
Enoch, I. E., Ozil, E., and Birkebak, R. C.: Polynomial approximation solution of heat transfer by conduction and radiation in a one-dimensional absorbing, emitting, and scattering medium, *Numer. Heat Transfer*, 5, 353–358, 1982.
Erickson, W. D., Williams, G. C., and Hottel, H. C.: Light scattering measurements on soot in a benzene-air flame, *Combust. Flame*, 8(2), 127–132, 1964.
Ertürk, H., Ezekoye, O. A., and Howell, J. R.: Inverse design of a three-dimensional furnace with moving design environment, *Proceedings of the ASME IMECE*, New York, 2001.
Ertürk, H., Ezekoye, O. A., and Howell, J. R.: The application of an inverse formulation in the design of boundary conditions for transient radiating enclosures, *JHT*, 124(6), 1095–1102, 2002a.
Ertürk, H., Ezekoye, O. A., and Howell, J. R.: Comparison of three regularized solution techniques in a three-dimensional inverse radiation problem, *JQSRT*, 73, 307–316, 2002b.
Ertürk, H., Ezekoye, O. A., and Howell, J. R.: The use of inverse formulation in design and control of transient thermal systems, *Proceedings of the 12th International Heat Transfer Conference*, pp. 729–734, Grenoble, France, 2002c.
Ertürk, H., Ezekoye, O. A., and Howell, J. R.: Boundary condition design to heat a moving object at uniform transient temperature using inverse formulation, *J. Manuf. Sci. Eng.*, 126, 619–626, August 2004. doi: 10.1115/1.1763179.
Ertürk, H., Ezekoye, O. A., and Howell, J. R.: Reverse Monte Carlo modeling of signal transport in light-pipe radiation thermometers, Paper HT2008-56180, *ASME Summer Heat Transfer Conference*, Jacksonville, FL, August 2008a.
Ertürk, H., Gamba, M., Ezekoye, O. A., and Howell, J. R.: Validation of inverse boundary condition design in a thermometry test bed, *JQSRT*, 109(2), 317–326, January 2008b.
Eryou, N. D. and Glicksman, L. R.: An experimental and analytical study of radiative and conductive heat transfer in molten glass, *JHT*, 94(2), 224–230, 1972.
Eslinger, R. and Chung, B.: Periodic heat transfer in radiating and convecting fins or fin arrays, *AIAA J.*, 17(10), 1134–1140, 1979.

Esposito, L. W. and House, L. L.: Radiative transfer calculated from a Markov chain formalism, *Astrophys. J.*, 219, 1058–1067, 1978.

Evans, T. M., Urbatsch, T. J., Lichtenstein, H., and Morel, J. E.: A residual Monte Carlo method for discrete thermal radiative diffusion, *J. Comput. Phys.*, 189, 539–556, 2003.

Eymet, V., Poitou, D., Galtier, M., El Hafi, M., Terrée, G., and Fournier, R.: Null-collision meshless Monte-Carlo: Application to the validation of fast radiative transfer solvers embedded in combustion simulators, *JQSRT*, 129, 145–157, 2013.

Ezekoye, O. A. and Zhang, Z.: Convective and radiative coupling in a burner-supported diffusion flame, *JTHT*, 11(2), 239–245, 1997.

Fan, D.: Evaluation of maximum entropy moment closure for solution to radiative heat transfer equation, MS thesis, Department of Aerospace Science, University of Toronto, Toronto, Ontario, Canada, 2012.

Fan, J. C. C.: Selective-black absorbers using sputtered cermet films, *Thin Solid Films*, 54, 139–148, 1978.

Fan, J. C. C. and Bachner, F. J.: Transparent heat mirrors for solar-energy applications, *Appl. Opt.*, 15(4), 1012–1017, 1976.

Farias, T. L., Carvalho, M. G., and Köylü, Ü. Ö.: Radiative heat transfer in soot-containing combustion systems with aggregation, *IJHMT*, 41(17), 2581–2587, 1998.

Farias, T. L., Carvalho, M. G., Köylü, Ü. Ö., and Faeth, G. M.: Computational evaluation of approximate Rayleigh-Debye-Gans/Fractal-aggregate theory for the absorption and scattering properties of soot, *JHT*, 117(1), 152–159, 1995a.

Farias, T. L., Köylü, Ü. Ö., and Carvalho, M. G.: Effects of polydispersity of aggregates and primary particles on radiative properties of simulated soot, *JQSRT*, 55, 357, 1995b.

Farmer, J. T. and Howell, J. R.: Hybrid Monte Carlo/diffusion method for enhanced solution of radiative transfer in optically thick non-gray media, in Y. Bayazitoglu et al. (eds.), *Radiative Transfer: Current Research*, vol. 276, pp. 203–212, *AIAA/ASME Heat Transfer Conf.*, Colorado Springs, 1994a.

Farmer, J. T. and Howell, J. R.: Monte Carlo algorithms for predicting radiative heat transport in optically thick participating media, *Proceedings of the 10th International Heat Transfer Conference*, vol. 2, pp. 37–42, Brighton, U.K., August 1994b.

Farmer, J. T. and Howell, J. R.: Monte Carlo prediction of radiative heat transfer in inhomogeneous, anisotropic, non-gray media, *JTHT*, 8(1), 133–139, 1994c.

Farmer, J. T. and Howell, J. R.: Monte Carlo strategies for radiative transfer in participating media, in J. P. Hartnett and T. Irvine (eds.), *Advances in Heat Transfer*, vol. 31, pp. 1–97, Acadamic Press, San Diego, CA, 1998.

Farrell, R.: Determination of configuration factors of irregular shape, *JHT*, 98(2), 311–313, 1976.

Fedorov, A. G., Lee, K. H., and Viskanta, R. V.: Inverse optimal design of the radiant heating in materials processing and manufacturing, *J. Mater. Eng. Perform.*, 7, 719–726, 1999.

Feingold, A.: Radiation-interchange configuration factors between various selected plane surfaces, *Proc. R. Soc. Lond., Ser. A*, 292(1428), 51–60, 1966.

Felske, J. D. and Charalampopoulos, T. T.: Gray gas weighting coefficients for arbitrary gas-soot mixtures, *IJHMT*, 25(12), 1849–1855, 1982.

Felske, J. D., Charalampopoulos, T. T., and Hura, H. S.: Determination of the refractive indices of soot particles from the reflectivities of compressed soot pellets, *Combust. Sci. Technol.*, 37, 263–284, 1984.

Felske, J. D., Chu, Z. Z., and Ku, J. C.: Mie scattering subroutines (DBMIE and MIEVO): A comparison of computational times, *Appl. Opt.*, 22(15), 2240–2241, 1983.

Felske, J. D. and Tien, C. L.: Calculation of the emissivity of luminous flames, *Combust. Sci. Technol.*, 7(1), 25–31, 1973.

Felske, J. D. and Tien, C. L.: The use of the Milne-Eddington absorption coefficient for radiative heat transfer in combustion systems, *JHT*, 99(3), 458–465, 1977.

Fernandes, R. and Francis, J.: Combined conductive and radiative heat transfer in an absorbing, emitting, and scattering cylindrical medium, *JHT*, 104(4), 594–601, 1982.

Fernandes, R., Francis, J., and Reddy, J. N.: A finite element approach to combined conductive and radiative heat transfer in a planar medium, in A. L. Crosbie (ed.), *Heat Transfer and Thermal Control*, pp. 93–109, AIAA, New York, 1981.

Fernandez, M. and Howell, J. R.: Radiative drying of model porous materials, *Drying Technol.*, 15(10), 2377–2399, 1997.

Ferwerda, H. A.: Radiative transfer equation for scattering media with a spatially varying refractive index, *J. Opt. A Pure Appl. Opt.*, 1(3), L1–L2, 1999.

Feynman, R.: *The Feynman Lectures on Physics*, vol. 1, Eq. 45.17, California Institute of Technology, Pasadena, CA, 1963.

Field, R. E. and Viskanta, R.: Measurement and prediction of dynamic temperatures in unsymmetrically cooled glass windows, *JTHT*, 7(4), 616–623, 1993.

Fiveland, W. A.: Discrete ordinates solutions of the radiative transport equation for rectangular enclosures, *JHT*, 106(4), 699–706, 1984.

Fiveland, W. A.: Discrete ordinate methods for radiative heat transfer in isotropically and anisotropically scattering media, *JHT*, 109(3), 809–812, 1987.

Fiveland, W. A.: Three-dimensional radiative heat transfer solutions by the discrete-ordinates method, *JTHT*, 2(4), 309–316, 1988.

Fiveland, W. A., Cornelius, D. K., and Oberjohn, W. J.: COMO: A numerical model for predicting furnace performance in axisymmetric geometries, ASME Paper 84-HT-103, *Proc. 1984 National Heat Transfer Conf.*, Niagara Falls, August 1984.

Fiveland, W. A. and Jamaluddin, A. S.: Three-dimensional spectral radiative heat transfer solutions by the discrete-ordinates method, *JTHT*, 5(3), 335–339, 1991.

Fiveland, W. A. and Jessee, J. P.: Finite element formulation of the discrete-ordinates method for multidimensional geometries, *JTHT*, 8(3), 426–433, 1994.

Fiveland, W. A. and Jessee, J. P.: Comparison of discrete ordinates formulations for radiative heat transfer in multidimensional geometries, *JTHT*, 9(1), 47–54, 1995.

Fiveland, W. A. and Jessee, J. P.: Acceleration schemes for the discrete ordinates method, *JTHT*, 10(3), 445–451, 1996.

Flatau, P. J.: SCATTERLIB (http://atol.ucsd.edu/scatlib/scatterlib.htm, accessed May 26, 2015), 2009.

Foote, P. D.: The emissivity of metals and oxides, III. The total emissivity of platinum and the relation between total emissivity and resistivity, *NBS Bull.*, 11(4), 607–612, 1915.

Ford, J. N., Tang, K., and Buckius, R. O.: Fourier transform infrared system measurement of the bidirectional reflectivity of diffuse and grooved surfaces, *JHT*, 117(4), 955–962, 1995.

Forsberg, C. H. and Domoto, G. A.: Thermal-radiation properties of thin metallic films on dielectrics, *JHT*, 94(4), 467–472, 1972.

Fox, M: *Optical Properties of Solids*, Oxford University Press, Oxford, U.K., 2001.

França, F., Ezekoye, O. A., and Howell, J. R., Inverse determination of heat source distribution in radiative systems with participating media, *Proceedings of the ASME National Heat Transfer Conference*, Albuquerque, NM, July 1999.

França, F. H. R., Ezekoye, O. A., and Howell, J. R.: Inverse boundary design combining radiation and convection heat transfer, *JHT*, 123, 884–891, 2001.

França, F. H. R. and Howell, J. R.: Transient inverse design of radiative enclosures for thermal processing of materials, *Inverse Probl. Sci. Eng.*, 14(4), 423–435, June 2006.

França, F. H. R., Howell, J. R., Ezekoye, O. A., and Morales, J. C.: Inverse design of thermal systems, in J. P. Hartnett and T. F. Irvine (eds.), *Advances in Heat Transfer*, vol. 36, 1, pp. 1–110, Academic Press, Waltham, MA, 2002.

França, F., Oguma, M., and Howell, J. R., Inverse radiative heat transfer with nongray, nonisothermal participating media, ASME HTD-vol. 361–365, in R. A. Nelson, T. Chopin, and S. T. Thynell (eds.), *1998 IMECE*, pp. 145–151, Anaheim, CA, November 1998.

Francoeur, M: Near-field radiative transfer: Thermophotovoltaic power generation, and optical characterization, PhD dissertation, University of Kentucky, Lexington, KY, 2010.

Francoeur, M., Basu, S., and Petersen, S. J.: Electric and magnetic surface polariton mediated near-field radiative heat transfer metamaterials made of silicon carbide particles, *Opt. Express*, 19, 18774–18788, 2011.

Francoeur, M. and Mengüç, M. P.: Role of fluctuational electrodynamics in near-field radiative heat transfer, *JQSRT*, 109, 280–293, 2008.

Francoeur, M., Mengüç, M. P., and Vaillon, R.: Near-field radiative heat transfer enhancement via surface phonon polaritons coupling in thin films, *Appl. Phys. Lett.*, 93, 043109, 2008.

Francoeur, M., Mengüç, M. P., and Vaillon, R.: Solution of near-field thermal radiation in one-dimensional layered media using dyadic Green's functions and the scattering matrix method, *JQSRT*, 110, 2002–2018, 2009.

Francoeur, M., Mengüç, M. P., and Vaillon, R.: Local density of electromagnetic states within a nanometric gap formed between two thin films supporting surface phonon-polaritons, *J. Appl. Phys.*, 107, 034313, 2010a.

Francoeur, M., Mengüç, M. P., and Vaillon, R.: Spectral tuning of near-field radiative heat flux between two thin silicon carbide films, *J. Phys. D: Appl. Phys.*, 43, 075501, 2010b.

Francoeur, M., Mengüç, M. P., and Vaillon, R.: Local density of electromagnetic states within a nanometric gap formed between two thin films supporting surface phonon-polaritons, *J. Phys. D: Appl. Phys.*, 43(7), 075501, February 24, 2010c.

Francoeur, M., Vaillon, R., and Mengüç, M. P., Impacts of thermal effects on the performances of nanoscale-gap thermophotovoltaic power generators, *IEEE Trans. Energy Convers.*, 26(2), 686–698, June 2011a.

Francouer, M. F., Vaillon, R., and Mengüç, M. P., Control of near-field radiative heat transfer via surface phonon-polariton coupling in thin films, *J. Appl. Phys. A*, 103(3), 547–550, June 2011b.

Francouer, M. F., Vaillon, R., and Mengüç, M. P.: Coexistence of multiple regimes for near-field thermal radiation between two layers supporting surface phonon polaritons in the infrared, *Phys. Rev. B*, 84(7), 075436, August 2011c.

Francoeur, M., Venkata, P. G., and Mengüç, M. P.: Sensitivity analysis for characterization of gold nanoparticles and agglomerates via surface plasmon scattering patterns, *JQSRT*, 106, 44–55, 2007.

Frank, M.: Approximate models for radiative transfer, *Bull. Inst. Math. Acad. Sin. (New Ser.)*, 2(2), 409–432, 2007.

Frank, M. and Klar, A.: Radiative heat transfer and applications for glass production processes, in A. Farina, A. Klar, R. M. M. Mattheijj, A. Mikelić, N. Siedow, and A. Fasano (eds.), *Mathematical Models in the Manufacturing of Glass, Proceedings of the C.I.M.E. Summer School*, pp. 57–134, Springer, Terme, Italy, 2008.

Frankel, J. I. and Silvestri, J. J.: Green's function solution to radiative heat transfer between longistudinal gray fins, *JTHT*, 5(1), 120–122, 1991.

Frankel, J. L. and Wang, T. P.: Radiative exchange between gray fins using a coupled integral equation formulation, *JTHT*, 2(4), 296–302, 1988.

Frankman, D. J., Webb, B. W., and Jones, M. R.: Investigation of lightpipe volumetric radiation effects in RTP thermometry, *JHT*, 128, 132–141, February 2006.

Fu, C. J. and Tan, W. C.: Near-field radiative heat transfer between two plane surfaces with one having a dielectric coating, *JQSRT*, 110, 1027–1036, 2009.

Fu, C. J. and Zhang, Z. M.: Nanoscale radiation heat transfer for silicon at different doping levels, *IJHMT*, 49, 1703–1718, 2006.

Fu, C. J., Zhang, Z. M., and Tanner, D. B.: Planar heterogeneous structures for coherent emission of radiation, *Opt. Lett.*, 30(14), 1873–1875, 2005.

Fu, K. and Hsu, P.-F.: Modeling the radiative properties of microscale randomly rough surfaces, *JHT*, 129(1), 71–79, January 2006.

Fu, K. and Hsu, P.-F.: Radiative property of gold surfaces with one-dimensional microscale Gaussian random roughness, *Int. J. Thermophys.*, 28(2), 598–615, April 2007.

Fu, K. and Hsu, P.-F.: New regime map of the geometric optics approximation for scattering from random rough surfaces, *JQSRT*, 109(2), 180–188, January 2008.

Fu, K., Hsu, P.-F., and Zhang, Z. M.: Unified analytical formulations of thin-film radiative properties including partial coherence, *Appl. Opt.*, 54(4), 653–661, February 2006.

Fu, X., Viskanta, R., and Gore, J. P., A model for the volumetric radiation characteristics of cellular ceramics, *Int. Commun. Heat Mass Transfer*, 24, 1069–1082, 1997.

Furukawa, M.: Practical method for calculating radiation incident upon a panel in orbit, *JTHT*, 6(1), 173–177, 1992.

Fusegi, T. and Farouk, B.: Laminar and turbulent natural convection-radiation interactions in a square enclosure filled with a nongray gas, *Numer. Heat Transfer*, 15(3), 303–322, 1989.

Fusegi, T. and Farouk, B.: A computational and experimental study of natural convection and surface/gas radiation interactions in a square cavity, *JHT*, 112(3), 802–804, 1990.

Galarça, M. M., Mossi, A., and França, F. H. R.: A modification of the cumulative wavenumber method to compute the radiative heat flux in non-uniform media, *JQSRT*, 112(3), 384–393, 2011.

Galatry, L.: Simultaneous effect of Doppler and foreign gas broadening on spectral lines, *Phys. Rev.*, 122, 1218, 1961.

Galtier, M., Blanco, S., Caliot, C. et al.: Integral formulation of null-collision Monte Carlo algorithms, *JQSRT*, 125, 57–68, 2013a.

Galtier, M., El Hafi, M., Eymet, V., Fournier, R., and Terrée, G.: Null collision Monte Carlo algorithms: A meshless technique to deal with radiative problems in heterogeneous media, *Proceedings of the Seventh International Symposium on Radiative Transfer, RAD-13*, Kuşadasi, Turkey, June 2–8, 2013b.

Gamba, M., Pavy, T., and Howell, J. R.: Modeling of a radiative RTP-type furnace through an inverse design: Mathematical model and experimental results, *Proceedings of the 2002 ASME International Mechanical Engineering Conference and Exhibition*, New Orleans, LA, November 17–22, 2002.

Gamba, M., Pavy, T., and Howell, J. R.: Inverse methods for design and control of thermal systems: Validation in a 2-D visible light enclosure, *Proceedings of the 2003 ASME International Mechanical Engineering Conference and Exhibition*, Washington, DC, November 15–21, 2003.

Ganesan, R., Narayanawamy, R., and Perumal, K.: Mixed convection and radiation heat transfer in a horizontal duct with variable wall temperature, *Heat Transfer Eng.*, 36, 335–345, 2015.

Garbuny, M.: *Optical Physics*, 2nd printing, Academic Press, New York, 1967.

Gardon, R.: The emissivity of transparent materials, *J. Am. Ceram. Soc.*, 39(8), 278–287, 1956.

Gardon, R.: Calculation of temperature distributions in glass plates undergoing heat-treatment, *J. Am. Ceram. Soc.*, 41(6), 200–209, 1958.

Gardon, R.: A review of radiant heat transfer in glass, *J. Am. Ceram. Soc.*, 44(7), 305–312, 1961.

Ge, W., Marquez, R., Modest, M. F., and Roy, S. P.: Implementation of high-order spherical harmonics methods for radiative heat transfer on OPENFOAM, *JHT* 137(5), 052701-1--052701-9, 2015.

Ge, W., Modest, M. F., and Marquez, R.: Two-dimensional axisymmetric formulation of high-order spherical harmonics methods for radiative heat transfer, *JQSRT*, 156, 58–66, 2015.

Gebhart, B.: Surface temperature calculations in radiant surroundings of arbitrary complexity—For gray, diffuse radiation, *IJHMT*, 3(4), 341–346, 1961.

Gebhart, B.: *Heat Transfer*, 2nd edn., McGraw-Hill, New York, pp. 150–163, 1971.

Geffrin, J. M., García-Cámara, B., Gómez-Medina, R., Albella, P., Froufe-Pérez, L. S., Eyraud, C., Litman, A., Vaillon, R., González, F., Nieto-Vesperinas, M., Sáenz, J. J., Moreno, F., Magnetic and electric coherence in forward- and back-scattered electromagnetic waves by a single dielectric subwavelength sphere, *Nature Communications* 3, Article Number: 1171, November 2012.

Gelbard, E. M.: *Simplified Spherical Harmonics Equations and Their Use in Shielding Problems*, WAPD-T-1182 (Rev. 1), Bettis Atomic Power Laboratory, West Mifflin, PA, February 1961.

Gentle, A. R., Aguilar, J. L. C., and Smith, G. B.: Optimized cool roofs: Integrating albedo and thermal emittance with R-value, *Solar Energy Mater. Solar Cells*, 95, 3207–3215, 2011.

Gentle, A. R. and Smith, G. B.: Radiative heat pumping from the earth using surface phonon resonant nanoparticles, *Nano Lett.*, 10, 373–379, 2010.

Genz, A. C. Numerical multiple integration on parallel computers, *Comput. Phys. Commun.*, 26, 349–352, 1982.

Gerencser, D. S. and Razani, A.: Optimization of radiative-convective arrays of pin fins including mutual irradiation between fins, *IJHMT*, 38(5), 899–907, 1995.

Ghannam, B.: Fast modeling of radiation and conduction heat transfer and application example, Doctoral thesis, l'École Nationale Supérieure des Mines Paris, Paris, France, 2012.

Ghannam, B., Nemer, M., El Khoury, K., and Yuen, W.: Experimental validation of the multiple absorption coefficient zonal method (MACZM) in a dynamic modeling of a steel reheating furnace, *Numer. Heat Transfer A*, 58, 1–19, 2010.

Ghannam, B., Nemer, M., El Khoury, K., and Yuen, W.: An efficient CPU-GPU implementation of the multiple absorption coefficient zonal method (MACZM), *Numer. Heat Transfer B*, 62(6), 439–461, 2012.

Gibson, M. M. and Monahan, J. A.: A simple model of radiation heat transfer from a cloud of burning particles in a confined gas stream, *IJHMT*, 14, 141–147, 1971.

Giestas, M., Pina, H., and Joyce, A.: The influence of radiation absorption on solar pond stability, *IJHMT*, 39(18), 3873–3885, 1996.

Gille, J. and Goody, R.: Convection in a radiating gas, *J. Fluid Mech.*, 20(1), 47–79, 1964.

Gilpin, R. R., Roberton, R. B., and Singh, B.: Radiative heating in ice, *JHT*, 99, 227–232, 1977.

Glass, M. W.: *CHAPARRAL: A Library for Solving Large Enclosure Radiation Heat Transfer Problems*, Sandia National Laboratory Report SAND95-2049, Albuquerque, NM, August 1995.

Glass, N. E., Maradudin, A. A., and Celli, V.: Surface plasmons on a large-amplitude doubly periodically corrugated surface, *Phys. Rev. B*, 26(10), 5357–5365, 1982.

Glicksman, L., Schuetz, M., and Sinofsky, M.: Radiation heat transfer in foam insulation, *IJHMT*, 30(1), 187–197, 1987.

Glover, F.: Future paths for integer programming and links to artificial intelligence, *Comput. Operat. Res.*, 5, 533–549, 1986.

Glover, F.: Tabu search—Part I, *ORSA J. Comput.*, 1(3), Summer, 191–206, 1989.

Göbel, G., Lippek, A., Wreidt, T., and Bauckhage, K., Monte Carlo simulation of light scattering by inhomogeneous spheres, in M. P. Mengüç (ed.), *Radiative Transfer II; Proceedings of the Second International Symposium on Radiation Transfer*, pp. 367–376, Begell House, New York, 1998.

Goedecke, G. H. and O'Brien, S. G.: Scattering by irregular inhomogeneous particles via the digitized green's function algorithm, *Appl. Opt.*, 27, 2431–2438, 1987.

Goffe, W. L., Ferrier, G. D., and Rogers, J.: Global optimization of statistical functions with simulated annealing, *J. Econom.*, 60(1–2), 65–100, 1994.

Golchert, B., Chang, S. L., and Zhou, C. Q.: Modeling of radiation heat transfer in glass melts, Paper IMECE2002-33891, *Proceedings of the ASME IMECE 2002*, New Orleans, LA, November 2002.

Golden, S. A.: The Doppler analog of the Elsasser band model I, *JQSRT*, 7(3), 483–494, 1967.
Golden, S. A.: The Doppler analog of the Elsasser band model, II, *JQSRT*, 8, 877–897, 1968.
Golden, S. A.: The Voigt analog of an Elsasser band, *JQSRT*, 9(8), 1067–1081, 1969.
Goldstein, M. E. and Howell, J. R.: *Boundary Conditions for the Diffusion Solution of Coupled Conduction-Radiation Problems*, NASA TN D-4618, Washington, DC, 1968.
Gonome, H., Baneshi, M., Okajima, J., Komiya, A., and Maruyama, S.: Controlling the radiative properties of cool black-color coatings pigmented with CuO submicron particles, *JQSRT*, Special Issue on *Micro- and Nano-scale Radiative Transfer*, 132, 90–98, 2014.
González, M., Garcıa-Fernández, C, and Velarde, P.: 2D numerical comparison between Sn and M1 radiation transport methods, *Ann. Nuclear Energy*, 36(7), 886–895, 2009.
González, M., Velarde, P., and Garcıa-Fernández, C.: First comparison of two radiative transfer methods: M_1 and Sn techniques, in N. V. Pogorelov, E. Audit, and G. P. Zank (eds.), *Numerical Modeling of Space Plasma Flows: Astronum 2007*, Paris, ASP Conference Series, vol. 385, 2008.
Goodman, S: Radiant-heat transfer between nongray parallel plates, *J. Res. Natl. Bur. Stand.*, 58(1), 37–40, 1957.
Goodson, K. E. and Ashegi, M.: Near-field optical thermometry, *Microscale Thermophys. Eng.*, 1(3), 225–235, 1997.
Goody, R., West, R., Chen, L., and Crisp, D.: The correlated-*k* method for radiation calculations in nonhomogeneous atmospheres, *JQSRT*, 42, 539–550, 1989.
Goody, R. M. and Yung, Y. L.: *Atmospheric Radiation*, 2nd edn., Oxford University Press, New York, 1989.
Gordon, I. E., Dothe, H., and Rothman, L. S.: The resurrection of the HITEMP database and its application to the study of stellar and planetary atmospheres, *Geophys Res. Abst.*, 9, 01799, 2007.
Gordon, I. E., Rotger, M., and Tennyson, J., (eds): HITRAN 2012 Special Issue, *JQSRT*, 130, 1–3, 2013.
Gorthala, R., Harris, K. T., Roux, J. A., and McCarty, T. A.: Transient conductive, radiative heat transfer coupled with moisture transport in attic insulations, *JTHT*, 8(1), 125–132, 1994.
Gosman, A. D., Pun, W. M., Runchal, A. K., Spalding, D. B., and Wolfshtein, M.: *Heat and Mass Transfer in Recirculating Flows*, Academic Press, London, U.K., 1969.
Gouesbet, G. and Lock, J. A.: List of problems for future research in generalized Lorenz–Mie theories and related topics, review and prospectus, *Appl. Opt.*, 52(5), 897–916, 2013.
Goutiere, V., Charette, A., and Kiss, L.: Non-gray gas modeling in complex enclosures: Application of the SNB-CK method, in M. P. Mengüc and N. Selçuk (eds.), *Radiation III: Third International Symposium on Radiative Transfer*, Begell House, New York, 2001.
Goutiere, V., Charette, A., and Liu, F.: Modelling radiative transfer in real gases: An assessment of existing methods in 2D enclosures, Paper NHTC 99–53, *Proceedings of the 1999 ASME National Heat Transfer Conference*, Albuquerque, NM, August 1999.
Govindan, R., Manickavasagam, S., and Mengüç, M. P.: On measuring the Mueller matrix elements of soot agglomorates, *Radiation I: Proceedings of the First International Symposium on Radiative Transfer*, Kusadasi, Turkey, August 1995. (Published by Begell House, New York, 1996.)
Grant, I. P.: On the representation of frequency dependence in non-grey radiative transfer, *JQSRT*, 5(1), 227–243, 1965.
Gray, W. A. and Muller, R.: *Engineering Calculations in Radiative Heat Transfer*, Pergamon Press, New York, 1974.
Greffet, J. J.: Controlled incandescence, *Nature*, 478, 191–192, 2011.
Greffet, J.-J., Carminati, R., Joulain, K., Mulet, J.-P., Mainguy, S., and Chen, Y.: Coherent emission of light by thermal sources, *Nature*, 416, 61–64, March 7, 2002.
Greffet, J.-J., Chapuis, P. O., Carminati, R., Laroche, M., Marquier, F., Volz, S., and Henkel, C.: Thermal radiation revisited in the near field, *Proceedings of the Fifth International Symposium on Radiative Transfer*, Bodrum, Turkey, 2007.
Greffet, J.-J. and Henkel, C.: Coherent thermal radiation, *Contemp. Phys.*, 48(4), 183–194, July–August 2007.
Greffet, J.-J. and Nieto-Vesperinas, M.: Field theory for generalized bidirectional reflectivity: Derivation of Helmholtz's reciprocity and Kirchhoff's law, *JOSA*, 15(10), 2735–2744, 1998.
Greif, R.: Laminar convection with radiation: Experimental and theoretical results, *IJHMT*, 21(4), 477–480, 1978.
Greif, R. and Clapper, G. P.: Radiant heat transfer between concentric cylinders, *Appl. Sci. Res. A*, 15, 469–474, 1966.
Greif, R. and Willis, D. R.: Heat transfer between parallel plates including radiation and rarefaction effects, *IJHMT*, 10, 1041–1048, 1967.

Grewal, N. S. and Kestoras, M.: Total normal emittance of dolomitic limestone, *IJHMT*, 31(1), 207–209, 1988.

Griem, H. R.: *Principles of Plasma Spectroscopy*, Cambridge University Press, Cambridge, U.K., 2005.

Grigoropoulos, C. P., Dutcher, W. E. Jr., and Barclay, K. E.: Radiative phenomena in CW laser annealing, *JHT*, 113(3), 657–662, 1991.

Gubareff, G. G., Janssen, J. E., and Torborg, R. H.: *Thermal Radiation Properties Survey*, 2nd edn., Honeywell Research Center, Minneapolis-Honeywell Regulator Co., Minneapolis, MN, 1960.

Guglielmini, G., Nannei, E., and Tanda, G.: Natural convection and radiation heat transfer from staggered vertical fins, *IJHMT*, 30(9), 1941–1948, 1987.

Guha, B., Otey, C., Poitras, C. B., Fan, S., and Lipson, M.: Near-field radiative cooling of nanostructures, *Nano Lett.*, 12, 4546–4550, 2012.

Guo, Z. and Kumar, S.: Three-dimensional discrete ordinates method in transient radiative transfer, *JTHT*, 16(3), 289–296, 2002.

Guo, Z., Kumar, S., and San, K.-C.: Multidimensional Monte Carlo simulation of short-pulse laser transport in scattering media, *JTHT*, 14(4), 504–511, 2000.

Guo, Z. and Maruyama, S.: Scaling anisotropic scattering in radiative transfer in three-dimensional nonhomogeneous media, *IJHMT*, 26(7), 997–1007, 1999.

Guo, Z. and Maruyama, S.: Radiative heat transfer in inhomogeneous, nongray, and anisotropically scattering media, *IJHMT*, 43, 2325–2336, 2000.

Gupta, R. P., Wall, T. F., and Truelove, J. S.: Radiative scatter by fly ash in pulverized-coal-fired furnaces: Application of the Monte Carlo method to anisotropie scatter, *IJHMT*, 26(11), 1649–1660, 1983.

Güven, O. and Bayazitoglu, Y.: The radiative transfer solution of a rectangular enclosure using angular domain discrete wavelets, *IJHMT*, 46, 687–694, 2003.

Habib, I. S.: Solidification of a semitransparent cylindrical medium by conduction and radiation, *JHT*, 95(1), 37–41, 1973.

Habib, Z. G. and Vervisch, P.: On the refractive index of soot at flame temperature, *Combust. Sci. Technol.*, 59, 261–274, 1988.

Hadley, M. A., Morris, G. J., Richards, G., and Upadhyay, P. C.: Frequency response of bodies with combined convective and radiative heat transfer, *IJHMT*, 42(23), 4287–4297, 1999.

Hagen, E. and Rubens, H.: Metallic reflection, *Ann. Phys.*, 1(2), 352–375, 1900. (See also E. Hagen and H. Rubens: Emissivity and electrical conductivity of alloys, *Deutsch. Phys. Ges. Verhandl.*, 6(4), 128–136, 1904.)

Hajimirza, S., El Hitti, G., Heltzel, A., and Howell, J.: Using inverse analysis to find optimum nanoscale radiative surface patterns to enhance solar cell performance, *Int. J. Thermal Sci.*, 12, 93–102, 2011. doi: 10.1016/j.ijthermalsci.2011.12.011.

Hajimirza, S., El Hitti, G., Heltzel, A., and Howell, J.: Specification of micro-nanoscale radiative patterns using inverse analysis for increasing solar panel efficiency, *JHT*, 134, 102702-1–102702-8, October 2012.

Hajimirza, S. and Howell, J.: Inverse optimization of plasmonic and antireflective grating in thin film PV cells, *J. Phys. Conf. Ser.*, 369, 2012. doi: 10.1088/1742-6596/369/1/012015.

Hajimirza, S. and Howell, J.: Design and analysis of spectrally selective patterned thin film cells, *Int. J. Thermophysics,* 34(10), 1930–1952, October 2013a. doi: 10.1007/s10765-013-1495-y.

Hajimirza, S. and Howell, J.: Statistical analysis of surface nanopatterned thin film solar cells obtained by inverse optimization, *JHT*, Special Issue on *Micro/Nanoscale Heat and Mass Transfer*, 135(9), 091501-1–091501-9, July 2013b. doi: 10.1115/1.4024464.

Hajimirza, S. and Howell, J.: Computational and experimental study of a multi-layer absorptivity enhanced thin film silicon solar cell, *JQSRT*, 143, 56–62, 2014a.

Hajimirza, S. and Howell, J.: Flexible nano-texture structures for thin film PV cells using wavelet functions, *ASME 2012 International Mechanical Engineering Congress and Exposition,* Denver, November 2012, and *IEEE Trans. Nanotechnol.*, 2014b.

Hajimirza, S. and Howell, J. R.: Computational and experimental study of a multi-layer absorptivity enhanced thin film silicon solar cell, *JQSRT*, 143, 56–62, 2014c.

Haji-Sheikh, A.: Monte Carlo methods, in W. J. Minkowycz, E. M. Sparrow, G. E. Schneider, and R. H. Pletcher (eds.), *Handbook of Numerical Heat Transfer*, Chapter 16, pp. 672–723, Wiley Interscience, New York, 1988.

Haji-Sheikh, A. and Howell, J. R.: Monte Carlo methods, in W. J. Minkowycz, E. M. Sparrow, and J. Y. Murthy (eds.), *Handbook of Numerical Heat Transfer*, 2nd edn., Chapter 8, pp. 249–296, Wiley Interscience, New York, 2006.

Haji-Sheikh, A. and Sparrow, E. M.: Probability distributions and error estimates for Monte Carlo solutions of radiation problems, in T. F. Irvine, W. E. Ibele, J. P. Hartnett, and R. J. Goldstein (eds.), *Progress in Heat and Mass Transfer*, vol. 2, pp. 1–12, Pergamon Press, Oxford, U.K., 1969.

Hale, G. M. and Querry, M. R.: Optical constants of water in the 200 nm to 200 μm wavelength region, *Appl. Opt.*, 12(3), 555–563, 1973.

Hale, M. J. and Bohn, M. S.: Measurement of the radiative transport properties of reticulated alumina foams, *Proceedings of the ASME/ASES Joint Solar Energy Conference*, Washington, DC, pp. 507–515, 1993.

Hall, D. F. and Fote, A. A.: Thermal control coatings performance at near geosynchronous altitude, *JTHT*, 6(4), 665–671, 1992.

Halton, J. H.: A retrospective and prospective survey of the Monte Carlo method, *SIAM Rev.*, 12(1), 1–63, 1970.

Hammersley, J. M. and Handscomb, D. C.: *Monte Carlo Methods*, Wiley, New York, 1964.

Hammonds Jr., J. S.: Thermal transport via surface phonon polaritons across a two-dimensional pore, *Appl. Phys. Lett.*, 88, 041912, 2006.

Hanrahan, P., Salzman, D., and Aupperle, L.: A rapid hierarchical radiosity algorithm, *Comput. Graph.*, 25(4), 197–206, 1991.

Hansen, P. C.: *Rank-Deficient and Discrete Ill-Posed Problems: Numerical Aspects of Linear Inversion*, SIAM, Philadelphia, PA, 1998.

Hargreaves, C. M.: Anomalous radiative transfer between closely spaced bodies, *Phys. Lett. A*, 30, 491, 1969.

Hargreaves, C. M.: Radiative transfer between closely spaced bodies, *Philips Res. Rep.*, 5(Suppl.), 1–80, 1973.

Harpole, G. M.: Radiative absorption by evaporating droplets, *IJHMT*, 23, 17–26, 1980.

Harris, J. A.: Solution of the conduction/radiation problem with linear-anisotropic scattering in an annular medium by the spherical harmonics method, *JHT*, 111(1), 194–197, 1989.

Harrison, J. K.: Non-diffuse infrared emission from the lunar surface, *IJHMT*, 12, 689–697, 1969.

Harrison, W. H., Richmond, J. C., Plyler, E. K., Stair, R., and Skramstad, H. K.: *Standardization of Thermal Emittance Measurements*, WADC Report 59-510 Part II, Wright Air Development Center, Wright-Patterson Air Force Base, OH, February 1961.

Hartung, L. C. and Hassan, H. A.: Radiation transport around axisymmetric blunt body vehicles using a modified differential approximation, *JTHT*, 7(2), 220–227, 1993.

Hasegawa, S., Echigo, R., and Fukuda, K.: Analytical and experimental studies on simultaneous radiative and free convective heat transfer along a vertical plate, *Proc. Jpn. Soc. Mech. Eng.*, 38(315), 2873–2882, 1972; 39(317), 250–257, 1973.

Hassab, M. A. and Özişik, M. N.: Effects of radiation and convective boundary conditions on the stability of fluid in an inclined slender slot, *IJHMT*, 22, 1095–1105, 1979.

Hassanzadeh, P. and Raithby, G. D.: The efficient iterative solution of the P_1 equation, *JHT*, 131(1), 014504-1–014504-3, January 2009.

Hassanzadeh, P., Raithby, G. D., and Chui, E. H.: Efficient calculation of radiation heat transfer in participating media, *JTHT*, 22(2), 129–139, April–June 2008.

Hattori, N., Takasu, H., and Iguchi, T.: Emissivity of liquid sodium, *Heat Transfer Jpn. Res.*, 13(1), 30–40, 1984.

Havstad, M. A., McLean, W. II, and Self, S. A.: Measurements of the thermal radiative properties of liquid uranium, *JHT*, 115(4), 1013–1020, 1993.

Havstad, M. A. and Qiu, T.: Thermal radiative properties of liquid metal alloys from measurement data of Hall coefficients and DC resistivities, *Exp. Heat Transfer*, 9, 1–10, 1996.

Hawes, E. A., Hastings, J. T., Crofcheck, C., and Mengüç, M. P.: Spectrally selective heating of nanosized particles by surface plasmon resonance, *JQSRT*, 104, 199–207, 2007.

Hawes, E. A., Hastings, J. T., Crofcheck, C., and Mengüç, M. P.: Spatially selective melting and evaporation of nanosized gold particles, *Opt. Lett.*, 33, 1383–1385, 2008.

Hawking, S. W.: Black hole explosions? *Nature*, 248 (5443), 30–31, 1974.

Hawksley, P. G. W.: The methods of particle size measurements, Part 2, optical methods and light scattering, *Brit. Coal Utiliz. Res. Assoc. Monthly Bull.*, 16(4–5), 134–209, 1952.

He, X. D., Torrance, K. E., Sillion, F. X., and Greenberg, D. P.: A comprehensive physical model for light reflection, *Comput. Graph.*, 25(4), 175–186, 1991.

Heaslet, M. A. and Lomax, H.: *Numerical Predictions of Radiative Interchange between Conducting Fins with Mutual Irradiations*, NASA TR R-116, Washington, DC, 1961.

Heaslet, M. A. and Warming, R. F.: Radiative transport and wall temperature slip in an absorbing planar medium, *IJHMT*, 8(7), 979–994, 1965.

Heaslet, M. A. and Warming, R. F.: Theoretical predictions of radiative transfer in a homogeneous cylindrical medium, *JQSRT*, 6(6), 751–774, 1966.

Heavens, O. S.: *Optical Properties of Thin Solid Films*, Dover, New York, 1965.

Heavens, O. S.: Optical properties of thin films—Where to? *Thin Solid Films*, 50, 157–161, 1978.

Hecht, E.: *Optics*, 4th edn., Pearson Education, Inc., Upper Saddle River, NJ, 2002.

Heinemann, U., Caps, R., and Fricke, J.: Radiation-conduction interaction: An investigation on silica aerogels, *IJHMT*, 39(10), 2115–2130, 1996.

Heinisch, R. P., Sparrow, E. M., and Shamsundar, N.: Radiant emission from baffled conical cavities, *JOSA*, 63(2), 152–158, 1973.

Helton, J. C.: *Conceptual and Computational Basis for the Quantification of Margins and Uncertainty*, Sandia National Laboratories Report SAND2009-3055, Albuquerque, NM, June 2009.

Heltzel, A., Battula, A., Howell, J. R., and Chen, S.: Nanostructuring borosilicate glass with near-field enhanced energy using a femtosecond laser pulse, *J. Heat Transfer*, Special Issue on *Nanoscale Heat Transfer*, 129, 53–59, January 2007.

Heltzel, A., Chen, S., and Howell, J. R.: Surface plasmon-based nanopatterning assisted by gold nanospheres, *Nanotechnology*, 19(2), January 2008.

Heltzel, A., Theppakuttai, S., Howell, J. R., and Chen, S.: Analytical and experimental investigation of laser-microsphere interaction for nanoscale surface modification, *J. Heat Transfer*, 127(11), 1231–1235, 2005.

Hendricks, T. J. and Howell, J. R.: Inverse radiative analysis to determine spectral radiative properties using the discrete ordinates method, *Proceedings of the 10th International Heat Transfer Conference*, Brighton, U.K., 1994.

Hendricks, T. J. and Howell, J. R.: Absorption/scattering coefficients and scattering phase functions in reticulated porous ceramics, *JHT*, 118, 79–87, 1996.

Henkel, C., Joulain, K., Carminati, R., and Greffet, J.-J.: Spatial coherence of thermal near fields, *Opt. Commun.*, 186, 57–67, 2000.

Henninger, J. H.: *Solar Absorptance and Thermal Emittance of Some Common Spacecraft Thermal-Control Coatings*, NASA Reference Publication 1121, Washington, DC, 1984.

Henson, J. C. and Malalasekera, W. M. G.. Benchmark comparisons with discrete transfer method solutions for radiative heat transfer in three-dimensional, nongray, scattering media, in M. P. Mengüç (ed.), *Radiative Transfer II: Proceedings of the Second International Symposium on Radiative Transfer*, Begell House, New York, 1998.

Henson, J. C., Malasekekera, W. M. G., and Dent, J. C.: Comparison of the discrete transfer and Monte Carlo methods for radiative heat transfer in three-dimensional, nonhomogeneous participating media, *Numer. Heat Transfer A*, 32(1), 19–36, July 1997.

Henyey, L. G. and Greenstein, J. L.: Diffuse radiation in the galaxy, *Astrophys. J.*, 88, 70–83, 1940.

Heping, T., Maestre, B., and Lallemand, M.: Transient and steady-state combined heat transfer in semi-transparent materials subjected to a pulse or a step irradiation, *JHT*, 113(1), 166–173, 1991.

Hering, R. G.: Radiative heat exchange between specularly reflecting surfaces with direction-dependent properties, *AIChE J.*, 5, 200–206, 1966a.

Hering, R. G.: Theoretical study of radiant heat exchange for non-gray non-diffuse surfaces in a space environment, Report No ME-TN-036-1, NASA CR-81653, University of Illinois, Champaign, IL, September 1966b.

Hering, R. G. and Smith, T. F.: Surface radiation properties from electromagnetic theory, *IJHMT*, 11(10), 1567–1571, 1968.

Hering, R. G. and Smith, T. F.: Surface roughness effects on radiant transfer between surfaces, *IJHMT*, 13(4), 725–739, 1970.

Hering, R. G. and Smith, T. F.: Surface roughness effects on radiant energy interchange, *JHT*, 93(1), 88–96, 1971.

Herschel, W.: Investigation of the powers of the prismatic colours to heat and illuminate objects, *Trans. R. Soc. Lond.*, 90(2), 255–283, 1800.

Herzberg, G.: *Molecular Spectra and Molecular Structure*, 2nd edn., 3 vols., Krieger, Malabar, FL, 1992.

Hesheth, P. J., Gebhart, B., and Zemel, J. N.: Measurements of the spectral and directional emission from microgrooved silicon surfaces, *JHT*, 110(3), 680–686, 1988.

Hibbard, R. R.: Equilibrium temperatures of ideal spectrally selective surfaces, *Sol. Energy*, 5(4), 129–132, 1961.

Higenyi, J.: Higher order differential approximation of radiative energy transfer in a cylindrical gray medium, PhD dissertation, Rice University, Houston, TX, 1979.

Hildebrand, F. B.: *Methods of Applied Mathematics*, 2nd edn., Dover, New York, 1992.

Hirano, M.: Enhancement of radiative heat transfer in the laminar channel flow of non-gray gases, *IJHMT*, 31(2), 367–374, 1988.

Ho, C.-H. and Özişik, M. N.: Combined conduction and radiation in a two-layer planar medium with flux boundary condition, *Numer. Heat Transfer*, 11, 321–340, 1987a.
Ho, C.-H. and Özişik, M. N.: Simultaneous conduction and radiation in a two-layer planar medium, *JTHT*, 1(2), 154–161, 1987b.
Ho, C.-H. and Özişik, M. N.: Combined conduction and radiation in a two-dimensional rectangular enclosure, *Numer. Heat Transfer*, 13, 229–239, 1988.
Hogan, R. E. and Gartling, D. K.: Solution strategies for coupled conduction/radiation problems, *Commun. Numer. Methods Eng.*, 24(6), 523–542, June 2008.
Hollands, K. G. T.: On the superposition rule for configuration factors, *JHT*, 117(1), 241–245, 1995.
Horvath, H. (ed.): Light scattering: Mie and more–commemorating 100 years of Mie's 1908 publication, *JQSRT*, 110(11), 783–786, 2009a.
Horvath, H.: Gustav Mie and the scattering and absorption of light by particles: Historic development and basics, *JQSRT*, 110(11), 787–799, July 2009b.
Hosseini Sarvari, S. M.: Optimum placement of heaters in a radiant furnace using the genetic algorithm, Paper RAD-01, *Proceedings of the 13th International Heat Transfer Conference*, Sydney, New South Wales, Australia, August 2006.
Hosseini Sarvari, S. M.: Inverse reconstruction of path-length *k*-distribution in a plane-parallel radiative medium, *Numer. Heat Transfer A*, 68(3), 336–354, 2015.
Hosseini Sarvari, S. M., Howell, J. R., and Mansouri, S. M. H.: Inverse boundary design radiation problem in absorbing-emitting media with irregular geometry, *Numer. Heat Transfer A*, 43(5), 565–584, 2003a.
Hosseini Sarvari, S. M., Howell, J. R., and Mansouri, S. M. H.: Inverse boundary design conduction-radiation problem in irregular two-dimensional domains, *Numer. Heat Transfer A Appl.*, 44, 1–16, 2003b.
Hosseini Sarvari, S. M., Howell, J. R., and Mansouri, S. M. H.: Inverse design of three-dimensional enclosures with transparent and absorbing-emitting media using an optimization technique, *Int. Commun. Heat Mass Transfer*, 30, 149–162, 2003c.
Hottel, H. C.: Radiant heat transmission, Chapter 4, in W. H. McAdams (ed.), *Heat Transmission*, 3rd edn., McGraw-Hill, New York, 1954.
Hottel, H. C. and Cohen, E. S.: Radiant heat exchange in a gas-filled enclosure: Allowance for nonuniformity of gas temperature, *AIChE J.*, 4(1), 3–14, 1958.
Hottel, H. C. and Sarofim, A. F.: The effect of gas flow patterns on radiative transfer in cylindrical furnaces, *IJHMT*, 8(8), 1153–1169, 1965.
Hottel, H. C. and Sarofim, A. F.: *Radiative Transfer*, McGraw–Hill, New York, 1967.
Hottel, H. C. Sarofim, A. F., Vasalos, I. A. and Dalzell, W. H.: Multiple scatter: Comparison of theory with experiment, *JHT*, 92(2), 285–291, 1970.
Hottel, H. C., Sarofim, A. F., Wankat, P. C., Noble, J. J., Silcox, G. D., and Knaebel, K. S.: Heat and mass transfer, in D. W. Green and R. H. Perry (eds.), *Perry's Chemical Engineer's Handbook*, 8th edn., Chapter 5, pp. 5-16–5-43, McGraw-Hill, New York, 2008.
Houchens, A. F. and Hering, R. G.: Bidirectional reflectance of rough metal surfaces, *Progr. Astronaut. Aeronaut.: Thermophys. Spacecraft Planet. Bodies*, 20, 65–89, 1967.
Houf, W. G. and Incropera, F. P.: An assessment of techniques for predicting radiation heat transfer in aqueous media, *JQSRT*, 23, 101–115, 1980.
Howarth, C. R., Foster, P. J., and Thring, M. W.: The effect of temperature on the extinction of radiation by soot particles, *Proceedings of the Third International Heat Transfer Conference AIChE*, vol. 5, pp. 122–128, New York, 1966.
Howe, J. T. and Sheaffer, Y. S.: Spectral radiative transfer approximations for multicomponent gas mixtures, *JQSRT*, 7(4), 695–701, 1967.
Howell, J. R.: Determination of combined conduction and radiation of heat through absorbing media by the exchange factor approximation, *Chem. Eng. Progr. Symp. Ser.*, 61(59), 162–171, 1965.
Howell, J. R.: Monte Carlo treatment of data uncertainties in thermal analysis, *J. Spacecraft Rockets*, 10(6), 411–114, 1973.
Howell, J. R.: Modern computational methods in radiative heat transfer, in K. T. Yang and W. Nakayama (eds.), *Computers and Computing in Heat Transfer Science and Engineering*, pp. 153–171, CRC Press, Boca Raton, FL, December 1992.
Howell, J. R.: The Monte Carlo method in radiative heat transfer, *JHT*, 120(3), 547–560, 1998.
Howell, J. R. (ed.), Editorial on nomenclature and symbols in heat transfer, *JHT*, 121(4), 770–773, November 1999.

Howell, J. R.: Non-equilibrium radiative transfer models: *k*-Distribution, von Karman Institute Lecture Series STO-AVT-218-VKI, Radiation and gas-surface interaction phenomena in high-speed re-entry, University of Illinois at Urbana Champagne, Champaign, IL, April 2014.

Howell, J. R. and Bannerot, R. B.: Trapezoidal grooves as moderately concentrating solar energy collectors, in A. E. Smith (ed.), *Radiative Transfer and Thermal Control*, vol. 49, pp. 277–289, AIAA Progress in Aeronautics and Astronautics Series, New York, 1976.

Howell, J. R. and Durkee, R. E.: Radiative transfer between surfaces in a cavity with collimated incident radiation: A comparison of analysis and experiment, *JHT*, 93(2), 129–135, 1971.

Howell, J. R. and Goldstein, M. E.: Effective slip coefficients for coupled conduction-radiation problems, *JHT*, 91(1), 165–166, 1969.

Howell, J. R., Hall, M. J., and Ellzey, J. L.: Combustion of hydrocarbon fuels within porous inert media, *Prog. Energy Comb. Sci.*, 22, 121–145, 1996.

Howell, J. R. and Mengüç, M. P.: The JQSRT web-based configuration factor catalog: A listing of relations for common geometries, *JQSRT*, 112(5), 910–912, March 2011.

Howell, J. R. and Perlmutter, M.: *Directional Behavior of Emitted and Reflected Radiant Energy from a Specular, Gray, Asymmetric Groove*, NASA TN D-1874, Washington, DC, 1963.

Howell, J. R. and Perlmutter, M.: Monte Carlo solution of radiant heat transfer in a nongrey nonisothermal gas with temperature dependent properties, *AIChE J.*, 10(4), 562–567, 1964a.

Howell, J. R. and Perlmutter, M.: Monte Carlo solution of thermal transfer through radiant media between gray walls, *JHT*, 86(1), 116–122, 1964b.

Howell, J. R. and Renkel, H. E.: *Analysis of the Effects of a Seeded Propellant Layer on Thermal Radiation in the Nozzle of a Gaseous-Core Nuclear Propulsion System*, NASA TN D-3119, Washington, DC, 1965.

Hsieh, C. K. and Coldewey, R. Q.: Study of thermal radiative properties of antireflection glass for flat-plate solar collector covers, *Solar Energy*, 16, 63–72, 1974.

Hsieh, C. K. and Su, K. C.: Thermal radiative properties of glass from 0.32 to 206 μm, *Solar Energy*, 22(1), 37–43, 1979.

Hsu, P.-F. and Farmer, J. T.: Benchmark solutions of radiative heat transfer within nonhomogeneous participating media using the Monte Carlo and YIX method, *JHT*, 119(1), 185–188, 1997.

Hsu, P.-F. and Howell, J. R.: Measurements of thermal conductivity and optical properties of porous partially stabilized zirconia, *Exp. Heat Transfer*, 5, 293–313, 1992.

Hsu, P.-F. and Ku, J. C.: Radiative heat transfer in finite cylindrical enclosures with nonhomogeneous participating media, *JTHT*, 8(3), 434–440, 1994.

Hsu, P.-F. and Tan, Z.: Recent benchmarkings of radiative heat transfer within nonhomogeneous participating media and the improved YIX method, in M. P. Mengüç (ed.), *Radiative Transfer I: Proceedings of the First International Symposium Radiative Transfer*, Begell House, New York, 1996.

Hsu, P.-F. and Tan, Z.: Radiative and combined-mode heat transfer within l-shaped nonhomogeneous and nongray participating media, *Numer. Heat Transfer A*, 31, 819–835, 1997.

Hsu, P.-F., Tan, Z.-M., Wu, S.-H., and Wu, C.-Y.: Radiative heat transfer in the finite cylindrical homogeneous and nonhomogeneous scattering media exposed to collimated radiation, *Numer. Heat Transfer*, 31(8), 819–836, 1999.

Hu, L., Narayanaswamy, A., Chen, X. Y., and Chen, G.: Near-field thermal radiation between two closely spaced glass plates exceeding Planck's blackbody radiation law, *Appl. Phys. Lett.*, 92, 133106, 2008.

Huang, J. M. and Lin, J. D.: Combined radiative and forced convective heat transfer in thermally developing laminar flow through a circular pipe, *Chem. Eng. Commun.*, 101, 147–164, 1991a.

Huang, J. M. and Lin, J. D.: Radiation and convection in circular pipe with uniform wall heat flux, *JTHT*, 5(4), 502–507, 1991b.

Huang, Y., Xia, X. L., and Tan, H. P.: Coupled radiation and conduction in a graded index layer with specular surfaces, *JTHT*, 18(2), 281–285, 2004.

Huang, Y., Xia, X. L., and Tan, H. P.: Radiative intensity solution and thermal emission analysis of a semitransparent medium with a sinusoidal refractive index, *JQSRT*, 74, 217–233, 2006.

Huang, Z. F., Hsu, P.-F., Wang, A.-H., Chen, Y.-B., Liu, L.-H., and Zhou, H.-C.: Wavelength selective infrared absorptance of heavily doped silicon complex gratings with geometric modifications, *JOSA B*, 28, 929–936, 2011.

Huda, G. M., Donev, E. U., Mengüç, M. P., and Hastings, J. H.: Effects of a silicon probe on gold nanoparticles on glass under evanescent illumination, *Opt. Express*, 19(13), 12679–12687, June 2011.

Huda, G. M. and Hastings, J. T.: Absorption modulation of plasmon resonant nanoparticles in the presence of an AFM tip, *IEEE J. Sel. Top. Quant. Electron.*, 19(3), article 4602306, May/June 2013.

Huda, G. M., Mengüç, M. P., and Hastings, J. T.: Absorption suppression of silver nanoparticles in the presence of an AFM tip: A harmonic oscillator model, *Proceedings of the Fifth International Workshop on Theoretical and Computational Nano-Photonics AIP Conference*, Bad Honnef, Germany, vol. 1475, pp. 134–136, 2012.

Hunter, B. and Guo, Z.: Numerical smearing, ray effect, and angular false scattering in radiation transfer computation, *IJHMT*, 81, 63–74, 2015.

Hurst, C.: The emission constants of metals in the near infra-red, *Proc. R. Soc. Lond., Ser. A*, 142(847), 466–490, 1933.

Hutchinson, J. E. and Richards, R. F.: Effect of nongray gas radiation on thermal stability in carbon dioxide, *JTHT*, 13(1), 25–32, 1999.

Iizuka, H. and Fan, S.: Consideration of thermal rectification using metamaterial models, *JQSRT*, 148, 156–164, November 2014.

Im, K. H. and Ahluwalia, R. K.: Radiation properties of coal combustion products, *IJHMT*, 36(2), 293–302, 1993.

Immel, D. S., Cohen, M. F., and Greenberg, D. P.: A radiosity method for nondiffuse environments, *Comput. Graph., SIGGRAPH'86 Proc.*, 20(4), 133–142, 1986.

Irvine, W. M. and Pollack, J. B.: Infrared optical properties of water and ice spheres, *Icarus,* 8, 324–360, 1968.

Isard, J. O.: Surface reflectivity of strongly absorbing media and calculation of the infrared emissivity of glasses, *Infrared Phys.*, 20, 249–256, 1980.

Iskander, M. F., Chen, H. Y., and Penner, J. E.: Optical scattering and absorption by branched chains of aerosols, *Appl. Opt.*, 28, 3083–3091, 1989.

Ivezic, Z. and Mengüç, M. P.: An investigation of dependent/independent scattering regimes for soot particles using discrete dipole approximation, *IJHMT*, 39(7), 811–822, 1996.

Ivezic, Z., Mengüç, M. P., and Knauer, T. G.: A procedure to determine the onset of soot agglomeration from multiwavelength experiments, *JQSRT*, 57(6), 859–865, 1997.

Iyer, I. K. and Mengüç, M. P.: Quadruple spherical harmonics approximation for radiative transfer in two-dimensional, rectangular enclosures, *JTHT*, 3, 266–273, 1989.

Jackson, J. D.: *Classical Electrodynamics*, 3rd edn., John Wiley & Sons, Inc., New York, 1998.

Jaeger, J. C.: Conduction of heat in a solid with a power law of heat transfer at its surface, *Cambridge Philos. Soc. Proc.*, 46(4), 634–641, 1950.

Jakob, M.: *Heat Transfer*, vol. II, Wiley, New York, 1957.

Jaluria, Y. and Torrance, K. E.: *Computational Heat Transfer*, Hemisphere, Washington, DC, 1986.

Jamaluddin, A. S. and Fiveland, W. A.: Radiative transfer in multidimensional enclosures with specularly reflecting walls, *AIAA/ASME Thermophysics and Heat Transfer Conference*, ASME HTD-vol. 137, pp. 95–100, Seattle, WA, June 1990.

Jamaluddin, A. S. and Smith, P. J.: Discrete-ordinates solution of radiative transfer equation in nonaxisymmetric cylindrical enclosures, *JTHT*, 6(2), 242–245, 1992.

Jarell, J. J.: An adaptive angular discretization method for neutral-particle transport in three-dimensional geometries, PhD dissertation, Texas A&M University, College Station, TX, 2010.

Jarro, C. A., Donev, U., Mengüç, M. P., and Hastings, J. H.: Silver patterning using an atomic force microscope tip and laser-induced chemical deposition from liquids, *J. Vacuum Sci. Technol. B*, 30(6), 06FD02, November 2012. doi: 10.1116/14764093.

Jarvinen, P. O.: Heat mirrored solar energy receivers, Paper No. 77–728, *AIAA 12th Thermophysics Conference*, Albuquerque, NM, 1977.

Jeans, Sir. J.: On the partition of energy between matter and the ether, *Phil. Mag.*, 10, 91–97, 1905.

Jellison, G. E., Jr. and Modine, F. A.: Optical functions of silicon at high temperatures, *J. Appl. Phys.*, 76(6), 3758–3761, September 15, 1994.

Jeng, D. R., Lee, E. J., and DeWitt, K. J.: A study of two limiting cases in convective and radiative heat transfer with nongray gases, *IJHMT*, 19(6), 589–596, 1976.

Jessee, J. P. and Fiveland, W. A.: Bounded, High-resolution differencing schemes applied for the discrete ordinates method, *JTHT*, 11(4), 540–548, 1997.

Jiang, Y. Y.: A vector form exchange-area-based method for computation of anisotropic radiative transfer, *JHT*, 131(1), 012701-1–012701-7, January 2009.

Jin, J.: *The Finite Element Method in Electrodynamics*, 2nd edn., John Wiley & Sons, New York, 2002.

Johnston, C. O. and Gnoffo, P. A.: The influence of ablation on radiative heating for earth entry, Paper AIAA 2008-4107, *40th Thermophysics Conference*, Seattle, WA, June 2008.

Johnston, C. O., Hollis, B. R., and Sutton., K.: Spectrum modeling for air shock-layer radiation at lunar-return conditions, *J. Spacecraft Rockets*, 45(5), 865–878, September–October 2008a.

Johnston, C. O., Hollis, B. R., and Sutton, K.: Non-Boltzmann modeling for air shock layers at lunar return conditions, *J. Spacecraft Rockets*, 45(5), 879–890, September–October 2008b.

Jones, A. C. and Raschke, M. B.: Thermal infrared near-field spectroscopy, *Nano Lett.*, 12, 1475–1481, 2012.

Jones, P. D., Dorai-Raj, D. E., and McLeod, D. G.: Spectral-directional emittance of oxidized copper, *JTHT*, 10(2), 343–349, 1996.

Joseph, D., Perez, P., El Hafi, M., and Cuenot, B.: Discrete ordinates and Monte Carlo methods for radiative transfer simulation applied to computational fluid mechanics combustion modeling, *JHT*, 131(5), 052701-1–052701-9, May 2009.

Joseph, J. H., Wiscombe, W. J., and Weinman, J. A.: The delta-Eddington approximation for radiative flux transfer, *J. Atmos. Sci.*, 33(12), 2452–2459, 1976.

Joulain, K.: Radiative transfer on short length scales, *Topics in Applied Physics: Microscale and Nanoscale Heat Transfer*, vol. 107, pp. 107–131, Springer, Berlin, Germany, 2007.

Joulain, K., Carminati, R., Mulet, J.-P., and Greffet, J.-J.: Definition and measurement of the local density of electromagnetic states close to an interface, *Phys. Rev. B*, 68, 245405, 2003.

Joulain, K., Drevillon, J., and Ben-Abdallah, P.: Non-contact heat transfer between two metamaterials, *Phys. Rev. B*, 81, 165119, 2010.

Joulain, K., Mulet, J.-P., Marquier, F., Carminati, R., and Greffet, J.-J.: Surface electromagnetic waves thermally excited: Radiative heat transfer, coherence properties and casimir forces revisited in the near field, *Surf. Sci. Rep.*, 57, 59–112, 2005.

Ju, Y., Masuya, G., Liu, F., Hattori, Y., and Riechelmann, D.: Asymptotic analysis of radiation extinction of stretched premixed flames, *IJHMT*, 43(2), 231–239, 2000.

Kadanoff, L. P.: Radiative transport within an ablating body, *JHT*, 83(2), 215–225, 1961.

Kahn, H.: Applications of Monte Carlo, Report No. RM-1237-AEC (AEC No. AECU-3259), Rand Corporation, Santa Monica, CA, April 1956.

Kahnert, F. M.: Numerical methods in electromagnetic scattering theory, *JQSRT*, 79, 775–824, 2003.

Kahnert, M.: Modeling radiometric properties of inhomogeneous mineral dust particles: Applicability and limitations of effective medium theories, *JQSRT*, 152, 16–27, 2015.

Kajiya, J. T.: Anisotropie reflection models, *Computer Graphics, SIGGRAPH'85 Proceedings*, San Francisco, vol. 19(4), pp. 15–21, 1985.

Kamada, O.: Theoretical concentration and attainable temperature in solar furnaces, *Solar Energy*, 9(1), 39–47, 1965.

Kaminski, D. A.: Radiative transfer from a gray, absorbing-emitting, isothermal medium in a conical enclosure, *J. Solar Energy Eng.*, 111(4), 324–329, 1989.

Kaminski, D. A., Fu, X. D., and Jensen, M. K.: Numerical and experimental analysis of combined convective and radiative heat transfer in laminar flow over a circular cylinder, *IJHMT*, 38(17), 3161–3169, 1995.

Kamiuto, K.: Combined conduction and nongray radiation heat transfer in carbon dioxide, *JTHT*, 10(4), 701–704, 1996.

Kamkari, B. and Darvishvand, L.: Design and optimization of 3-D radiant furnaces using genetic algorithms and Monte Carlo method, *Int. J. Adv. Design Manuf. Tech.*, 2014.

Kanayama, K.: Apparent directional emittance of v-groove and circular-groove rough surfaces, *Heat Transfer Jpn. Res.*, 1(1), 11–22, 1972.

Kang, S. H. and Song, T.-H.: Finite element formulation of the first- and second-order discrete ordinates equations for radiative heat transfer calculation in three-dimensional participating media, *JQSRT*, 109, 2094–2107, 2008.

Kangwanpongpan, T., França, F. H. R., da Silva, R. C., Schneider, P. S., and Krautz, H. J.: New correlations for the weighted-sum-of-gray-gases model in oxy-fuel conditions based on HITEMP 2010 Database, *IJHMT*, 55(25/26), 7419–7433, 2012.

Kassemi, M. and Duval, W. M. B.: Interaction of surface radiation with convection in crystal growth by vapor transport, *JTHT*, 4(4), 454–461, 1990.

Kassemi, M. and Naraghi, M. H. N.: Analysis of radiation-natural convection interaction in 1-g and low-g environments using the discrete exchange factor method, *IJHMT*, 36(17), 4141–4149, 1993.

Katika, K. M. and Pilon, L.: Modified method of characteristics for transient radiative transfer, *JQSRT*, 98(2), 220–237, 2006.

Keller, H. H. and Holdredge, E. S.: Radiation heat transfer for annular fins of trapezoid profile, *JHT*, 92(6), 113–116, 1970.

Kellett, B. S.: Transmission of radiation through glass in tank furnaces, *J. Soc. Glass Technol.*, 36, 115–123, 1952.

Kendall, M. G. and Smith, B. B.: Randomness and random sampling numbers, *R. Stat. Soc. J.*, 1, 147–166, 1938.

Kerker, M.: *The Scattering of Light and Other Electromagnetic Radiation*, Academic Press, New York, 1961.

Kerker, M. (ed.): *Proceedings of the Interdisciplinary Conference on Electromagnetic Scattering*, Potsdam, NY, August 1962, Pergamon Press, New York, 1963.

Keshock, E. G. and Siegel, R.: Combined radiation and convection in an asymmetrically heated parallel plate flow channel, *JHT*, 86(3), 341–350, 1964.

Kesten, A. S.: Radiant heat flux distribution in a cylindrically symmetric nonisothermal gas with temperature-dependent absorption coefficient, *JQSRT*, 8(1), 419–434, 1968.

Khalil, E. E.: *Modelling of Furnaces and Combustors*, Energy and Engineering Science Series, Gupta, A.K. and Lilley, D.G. (eds.), Abacus Press, Tunbridge Wells, U.K., 1982.

Kholopov, G. K.: Radiation of diffuse isothermal cavities, *Inzh. Fiz. Zh.*, 25(6), 1112–1120, 1973.

Kidd, R., Ardini, J., and Anton, A.: Evolution of the modern photon, *Am. J. Phys.*, 57(1), 27–35, January 1989.

Kiefer, J., and Wolfowitz, J.: Stochastic estimation of the maximum of a regression function, *Ann. Math. Stat.*, 23, 462–466, 1952.

Kim, D. M. and Viskanta, R.: Interaction of convection and radiation heat transfer in high pressure steam, *IJHMT*, 27(6), 939–941, 1984.

Kim, K. W. and Baek, S. W.: Inverse surface radiation analysis in an axisymmetric cylinder enclosure using a hybrid genetic algorithm, *Numer. Heat Transfer A*, 46, 367–381, 2004.

Kim, L.-K. and Kim, W.-S.: A hybrid spatial differencing scheme for discrete ordinates method in 2D rectangular enclosures, *IJHMT*, 44(3), 575–586, 2001.

Kim, S. H. and Huh, K. Y.: A new angular discretization scheme of the finite volume method for 3-D radiative heat transfer in absorbing, emitting and anisotropically scattering media, *IJHMT*, 43(7), 1233–1242, 2000.

Kim, S. S. and Baek, S. W.: Radiation affected compressible turbulent flow over a backward facing step, *IJHMT*, 39(16), 3325–3332, 1996a.

Kim, T. Y. and Baek, S. W.: Analysis of combined conductive and radiative heat transfer in a two-dimensional rectangular enclosure using the discrete ordinates method, *IJHMT*, 34(9), 2265–2273, 1991.

Kim, T. Y. and Baek, S. W.: Thermal development of radiatively active pipe flow with nonaxisymmetric circumferential convective heat loss, *IJHMT*, 39(14), 2969–2976, 1996b.

Kim, T.-K. and Lee, H. S.: Radiative transfer in two-dimensional anisotropie scattering media with collimated incidence, *JQSRT*, 42(3), 225–238, 1989.

Kim, T.-K. and Lee, H. S.: Two-dimensional anisotropie scattering radiation in a thermally developing poiseuille flow, *JTHT*, 4(3), 292–298, 1990.

Kimes, D. S., Newcomb, W. W., Nelson, R. F., and Schutt, J. B.: Directional reflectance distributions of a haldwood and pine forest canopy, *IEEE Trans. Geosci. Remote Sens.*, GE-24, 281–293, 1986.

Kirkpatrick, S., Gelatt, Jr., C. C., and Vecchi, M. P.: Optimization by simulated annealing, *Science*, 220(4598), 71–680, 1983.

Kittel, A., Müller-Hirsch, W., Parisi, J., Biehs, S.-A., Reddig, D., and Holthaus, M.: Near-field heat transfer in scanning thermal microscope, *Phys. Rev. Lett.*, 95, 224301, 2005.

Kittel, A., Wischnath, U. F., Welker, J., Huth, O., Rüting, F., and Biehs, S.-A.: Near-field thermal imaging of nanostructured surfaces, *Appl. Phys. Lett.*, 93, 193109, 2008.

Kittel, C.: *Introduction to Solid State Physics*, John Wiley & Sons, Hoboken, NJ, 2005.

Klar, A., Lang, J., and Seaid, M.: Adaptive solutions of the SP_N-approximations to radiative heat transfer in glass, *Int. J. Thermal Sci.*, 44, 1013–1023, 2005.

Kobiyama, M.: Reduction of computing time and improvement of convergence stability of the Monte Carlo Method applied to radiative heat transfer with variable properties, *JHT*, 111(1), 135–140, 1989.

Koch, R. and Becker, R.: Evaluation of the quadrature schemes for the discrete ordinates method, *JQSRT*, 84, 423–435, 2004.

Koch, R., Krebs, W., Wittig, S., and Viskanta, R.: Discrete ordinates quadrature schemes for multidimensional radiative transfer, *JQSRT*, 53(4), 353–372, 1995.

Koenigsdorff, R., Miller, F., and Ziegler, R.: Calculation of scattering fractions for use in radiative flux models, *IJHMT*, 34(10), 2673–2676, 1991.

Koh, J. C. Y.: Radiation of spherically shaped gases, *IJHMT*, 8(2), 373–374, 1965.

Kollyukh, O. G., Liptuga, A. I., Morozhenko, V., and Pipa, V. I.: Thermal radiation of plane-parallel semitransparent layers, *Opt. Commun.*, 225(4–6), 349–352, 2003.

Kondratyev, Ya. K.: *Radiation in the Atmosphere*, Academic Press, New York, 1969.

Koo, H.-M., Cheong, K.-B., and Song, T.-H.: Schemes and applications of first and second-order discrete ordinates interpolation methods to irregular two-dimensional geometries, *JHT*, 119(4), 730–737, 1997.

Koptelov, R., Malikov, G., Lisienko, V., and Viskanta, R.: On a priori method for choosing surface view factor evaluation, Paper IMECE2012-85808, *ASME 2012 International Mechanical Engineering Conference and Exposition*, Houston, TX, November 2012.

Koptelov, R., Malikov, G., Lisienko, V., and Viskanta, R.: Accelerated method for surface view factor evaluation based on error estimation, *JHT*, 136(7), 074502-1–074502-4, April 2014.

Korolev, Y. M. and Yagola, A. G.: On inverse problems in partially ordered spaces with a priori information, *J. Inverse Ill-Posed Probl.*, 20, 567–573, 2012. doi: 10.1515/jip-2012-0022.

Korolev, Y. M., Yagola, A. G., Johnson, J., and Brinkerhoff, D.: Methods of error estimation in inverse problems on compact sets in Banach lattices—Theory and applications in ice sheet modeling, *Inverse Problems, Design and Optimization Symposium,* Albi, France, June 26–28, 2013.

Kounalakis, M. E., Gore, J. P., and Faeth, G. M.: Turbulence/radiation interactions in nonpremixed hydrogen/air flames, *22nd Symposium (International) on Combustion*, pp. 1281–1290, The Combustion Institute, Pittsburg, PA, 1988.

Kourganoff, V.: *Basic Methods in Transfer Problems*, Dover, New York, 1963.

Kowsary, F.: A computational efficient method for Monte Carlo simulation of diffuse radiant emission or reflection, *IJHMT*, 42, 193–195, 1999.

Köylü, Ü. Ö. and Faeth, G. M.: Radiative properties of flame-generated soot, *JHT*, 115(2), 409–417, 1993.

Köylü, Ü. Ö. and Faeth, G. M.: Optical properties of soot in buoyant laminar diffusion flames, *JHT*, 116(4), 971–979, 1994.

Köylü, Ü. Ö. and Faeth, G. M.: Spectral extinction coefficients of soot aggregates from turbulent diffusion flames, *JHT*, 118(2), 415–421, 1996.

Kozan, M. and Mengüç, M. P.: Exploration of fractal nature of WO_3 nanowire aggregates, *JQSRT*, 109, 380–393, 2008.

Kozan, M., Thangala, J., Bogale, R., Mengüç, M. P., and Sunkara, M. K.: In-situ characterization of dispersion stability of WO_3 nanoparticles and nanowires, *J. Nanoparticle Res.*, 10(4), 599–612, 2008.

Kramer, S., Gritzki, R., Perschk, A., Rösler, M., and Felsmann, C.: Numerical simulation of radiative heat transfer in indoor environments on programmable graphics hardware, *Int. J. Thermal Sci.*, In press, corrected proof, 2015.

Kramida, A., Ralchenko, Y., Reader, J. and NISTASDTeam: NIST Atomic Spectra Database (version 5.0), [Online]. Available: http://physics.nist.gov/asd (accessed April 27, 2015), National Institute of Standards and Technology, Gaithersburg, MD, 2012.

Kraus, A. D., Aziz, A., and Welty, J.: *Extended Surface Heat Transfer*, Wiley, New York, 2001.

Kreider, K. G., Chen, D. H., DeWitt, D. P., Kimes, W. A., and Tsai, B. K.: Lightpipe proximity effects on si wafer temperature in rapid thermal processing tools, *11th Annual Conference Advanced Thermal Processing of Semiconductors-RTP 2003*, Charleston, SC, pp. 125–129, 2003.

Krishnan, K. S. and Sundaram, R.: The distribution of temperature along electrically heated tubes and coils, I. Theoretical, *Proc. R. Soc. Lond., Ser. A*, 257(1290), 302–315, 1960.

Krishnaprakas, C. K.: Optimum design of radiating rectangular plate fin array extending from a plane wall, *JHT*, 118(2), 490–493, 1996. (See also discussion by C. Balaji in *JHT*, 119(2), 393–394, 1997.)

Krishnaprakas, C. K.: Optimum design of radiating longitudinal fin array extending from a cylindrical surface, *JHT*, 119(4), 857–860, 1997.

Krishnaprakas, C. K.: Combined conduction and radiation heat transfer in a cylindrical medium, *JTHT*, 12(4), 605–608, 1998a.

Krishnaprakas, C. K.: View-factor evaluation by quadrature over triangles, *JTHT*, 12(1), 118–120, 1998b.

Krishnaprakas, C. K., Narayana, K. B., and Dutta, P.: Combined convective and radiative heat transfer in turbulent tube flow, *JTHT*, 13(3), 390–394, 1999a.

Krishnaprakas, C. K., Narayana, K. B., and Dutta, P.: Interaction of radiation with natural convection, *JTHT*, 13(3), 387–390, 1999b.

Krook, M.: On the solution of equations of transfer, I, *Astrophys. J.*, 122(3), 488–497, 1955.

Krüger, M., Emig, T., and Kardar, M.: Nonequilibrium electromagnetic fluctuations: Heat transfer and interactions, *Phys. Rev. Lett.*, 106, 210404, 2011.

Ku, J. C. and Felske, J. D.: The range of validity of the Rayleigh limit for computing mie scattering and extinction efficiencies, *JQSRT*, 31(6), 569–574, 1984.

Ku, J. C. and Shim, K.-H.: Optical diagnostics and radiative properties of simulated soot agglomerates, *JHT*, 113(4), 953–958, 1991.

Kudo, K., Li, B., and Kuroda, A.: Analysis on radiative energy transfer through fibrous layer considering fibrous orientation, *ASME HTD*, 315, 37–43, 1995.

Kudo, K., Taniguchi, H., Kim, Y.-M., and Yang, W.-J.: Transmittance of radiative energy through three-dimensional packed spheres, in J. Lloyd and Y. Kurosoki (eds.), *Proceedings of the 1991 ASME/JSME Thermal Engineering Joint Conference*, Reno, pp. 35–42, 1991a.

Kudo, K., Taniguchi, H., Kuroda, A., Oath, M., and Dakota, H.: Improvement of analytical method on radiative heat transfer in nongray media by Monte Carlo method, *Heat Transfer Jpn. Res.*, 22(6), 559–572, 1993.

Kudo, K., Taniguchi, H., Kuroda, A., Sasaki, T., and Yamamoto, T.: Development of general purpose computer code for two/three dimensional radiation heat transfer analysis, in R. L. Lewis, J. H. Chin, and G. M. Homsy (eds.), *Proceedings of the Seventh International Conference on Numerical Methods in Thermal Problems*, vol. VII, pt. 1, pp. 698–708, Pineridge Press, Swansea, NY, 1991b.

Kumar, S. and Felske, J. D.: Radiative transport in a planar medium exposed to azimuthally unsymmetric incident radiation, *JQSRT*, 35(3), 187–212, 1986.

Kumar, S., Majumdar, A., and Tien, C. L.: The differential-discrete-ordinate method for solutions of the equation of radiative transfer, *JHT*, 112(2), 424–429, 1990.

Kumar, S. and Mitra, K.: Transient radiative transfer, in M. P. Mengüç (ed.), *Radiative Transfer I, Proceedings of the First International Symposium Radiative Transfer*, pp. 488–504, Begell House, New York, 1995.

Kumar, S. and Tien, C. L.: Dependent absorption and extinction of radiation by small particles, *JHT*, 112(1), 178–185, 1990.

Kumar, S. and White, S. M., Dependent scattering properties of woven fibrous insulations for normal incidence, *JHT*, 117(1), 160–166, February 1996.

Kunitomo, T.: Luminous flame emission under pressure up to 20 Atm, in N. H. Afgan and J. M. Beer (eds.), *Heat Transfer in Flames*, pp. 271–281, Scripta, Washington, DC, 1974.

Kunitomo, T.: Present status of research on radiative properties of materials, *Int. J. Thermophys.*, 5(1), 73–90, 1984.

Kunitomo, T. and Sato, T.: Experimental and theoretical study on the infrared emission of soot particles in luminous flame, *Fourth International Heat Transfer Conference*, Paris-Versailles, France, September 1970.

Kuo, C. H. and Kulkarni, A. K.: Analysis of heat flux measurements by circular gages in a mixed convection/radiation environment, *JHT*, 113(4), 1037–1040, 1991.

Kuo, D.-C., Morales, J. C., and Ball, K. S.: Combined natural convection and volumetric radiation in a horizontal annulus: Spectral and finite volume predictions, *JHT*, 121(3), 610–615, 1999.

Kuriyama, M., Katayama, K., Takuma, Y., and Hasegawa, Y.: The effect of radiation heat transfer in the measurement of thermal conductivity for the semitransparent medium, *Bull. JSME*, 19(134), 973–979, 1976.

Kushner, H. J. and Clark, D. S.: *Stochastic Approximation Methods for Constrained and Unconstrained Systems*, Springer, New York, 1978.

Kyle, T. G.: Absorption of radiation by uniformly spaced doppler lines, *Astrophys. J.*, 148(3), 845–848, 1967.

Lacis, A. A. and Hansen, J. E.: A parameterization for the absorption of solar radiation in the Earth's atmosphere, *J. Atm. Sci.*, 31, 118–133, 1973.

Lage, J. L., Lim, J. S., and Bejan, A.: Natural convection with radiation in a cavity with open top end, *JHT*, 114(2), 479–486, 1992.

Lalich, S., Enguehard, F., and Baillis, D.: Experimental determination and modeling of the radiative properties of silica nanoporous matrices, *JHT*, 131, 082701-1–082701-12, August 2009.

Lamb, W. E., Jr.: AntiPhoton, *Appl. Phys. B*, 60, 68–84, 1995.

Lamet, J., Babou, Y., Rivière, P., Perrin, M. Y., and Soufiani, A.: Radiative transfer in gases under thermal and chemical nonequilibrium conditions: Application to earth atmospheric reentry, *JQSRT*, 109(2), 235–244, 2008.

Lamet, J., Rivière, P., Perrin, M., and Soufiani, A.: Narrow-band model for nonequilibrium air plasma radiation, *JQSRT*, 111(1), 87–104, 2010.

Lan, C.-H., Ezekoye, O. A., and Howell, J. R.: Transitions and bifurcations to chaos in combined radiation and natural convection in a two-dimensional participating medium, in L. C. Witte (ed.), *Proceedings of the 1999 ASME IMECE*, HTD-364–1, pp. 53–59, Nashville, TN, November, 1999.

Lan, C.-H., Ezekoye, O. A., and Howell, J. R.: Computation of radiative combined-mode heat transfer in a two-dimensional rectangular participating medium using the spectral method, in S. C. Yao and A. Jones (eds.), Paper 2000–1214 (on CD), *Proceedings of the 2000 National Heat Transfer Conference*, Pittsburgh, PA, 2000.

Lan, C.-H., Ezekoye, O. A., Howell, J. R., and Ball, K. S.: Stability analysis for three dimensional Rayleigh-Bénard convection with radiatively participating medium using spectral methods, *IJHMT*, 46, 1371–1383, 2003.

Landau, L. D. and Lifshitz, E. M.: *Electrodynamics of Continuous Media*, Addison-Wesley, Reading, MA, 1960.

Landram, C. S., Greif, R., and Habib, I.. S.: Heat transfer in turbulent pipe flow with optically thin radiation, *JHT*, 91(3), 330–336, 1969.
Langley, S. P.: Experimental determination of wave-lengths in the invisible prismatic spectrum, *Mem. Natl. Acad. Sci.*, 2, 147–162, 1883.
Lanzo, C. D. and Ragsdale, R. G.: Heat transfer to a seeded flowing gas from an arc enclosed by a quartz tube, in W. H. Giedt and S. Levy (eds.), *Proceedings of the 1964 Heat Transfer Fluid Mechanics Institute*, pp. 226–244, Berkeley, CA, 1964.
Laroche, M., Carminati, R., and Greffet, J.-J.: Near-field thermophotovoltaic energy conversion, *J. Appl. Phys.*, 100, 063704, 2006.
Larsen, E. W., Morel, J. E., and McGee, J. M.: The simplified P_N equations as an asymptotic limit of the transport equation, *Trans. Am. Nuc. Soc.*, 66, 231–232, 1993.
Larsen, E. W., Thömmes, G., and Klar, A.: New frequency-averaged approximations to the equations of radiative transfer, *SIAM J. Appl. Math.*, 64(2), 565–582, 2003.
Larsen, E. W., Thömmes, G., Klar, A., Seaïd, M., and Götz, T.: Simplified P_N approximations to the equations of radiative transfer and applications, *J. Comp. Phys.*, 183, 652–675, 2002.
Larsen, M. E.: Use of contact resistance algorithm to implement jump boundary conditions for the radiation diffusion approximation, Paper HT2005-72561, *Proceedings 2005 ASME Summer Heat Transfer Conference*, San Francisco, CA, July 2005.
Larsen, M. E. and Howell, J. R.: The exchange factor method: An alternative zonal formulation of radiating enclosure analysis, *JHT*, 107(4), 936–942, 1985.
Larsen, M. E. and Howell, J. R.: Least squares smoothing of direct exchange areas in zonal analysis, *JHT*, 18(1), 239–242, 1986.
Larsen, M. E. and Porter, J.: Theory and experimental validation of SPLASH, Sandia Report SAND2005-2947, Sandia National Laboratories, Albuquerque, NM, June 2005.
Lataillade, A., Blanco, S., Clergent, Y., Dufresne, J. L., El Hafi, M., and Fournier, R.: Monte Carlo methods and sesitivity estimation, *JQSRT*, 75, 529–538, 2002.
Lathrop, K. D.: Use of discrete ordinate methods for solution of photon transport problems, *Nucl. Sci. Eng.*, 24, 381–388, 1966.
Lathrop, K. D.: Ray effects in discrete ordinates equations, *Nucl. Sci. Eng.*, 32, 357–369, 1968.
Lathrop, K. D.: Remedies for ray effects, *Nucl. Sci. Eng.*, 45, 255–268, 1971.
Lathrop, K. D.: The early days of the Sn method, Stanford Linear Accelerator Center Report SLAC-PUB 5829, Stanford University, Stanford, CA, December 1992.
Lathrop, K. D. and Carlson, B. G.: Discrete ordinates angular quadrature of the neutron transport equation, Los Alamos Scientific Laboratory Report LA 3186, Los Alamos, NM, 1965.
Laux, C. O.: Radiation and nonequilibrium collisional-radiative models, in D. Fletcher, J.-M. Charbonnier, G. S. R. Sarma, and T. Magin (eds.), *Physico-Chemical Modeling of High Enthalpy and Plasma Flows*, von Karman Institute Lecture Series 2002-07, Rhode-Saint-Genèse, Belgium, 2002. [SPECAIR User Manual, ver. 3.0, 2012 (latest version on line)].
Laux, C. O.: Spectroscopic challenges in the modeling and diagnostics of high temperature air plasma radiation for aerospace applications, *Proceedings of the Fifth International Conference on Atomic and Molecular Data and their Applications* (*ICAMDATA*), vol. 901, pp. 191–203, Meudon, France, 2006.
Lawson, D. A.: An improved method for smoothing approximate exchange areas, *IJHMT*, 38(16), 3109–3110, 1995.
Le Dez, V., Lemonnier, D., and Sadat, H.: Restitution of the temperature field inside a cylinder of semitransparent dense medium from directional intensity data, *JHT*, 131(11), 112701-1–112701-14, 2009.
Le Dez, V. and Sadat, H.: Radiative transfer in a semi-transparent medium enclosed in a cylindrical annulus, *JQSRT*, 113(1), 96–116, 2012a.
Le Dez, V. and Sadat, H.: Corrigendum to radiative transfer in a semi-transparent medium enclosed in a cylindrical annulus, *JQSRT*, 113(1), 96–116, 2012; *JQSRT*, 113, 816–817, 2012b.
Le Dez, V. and Sadat, H.: Radiative transfer in a semitransparent medium enclosed in a spherical annulus, *Int. J. Thermal Sci*, 2015.
Lebedev, V. A.: Equations relating integral radiation configuration factors in cylindrical emitting systems, *Sov. J. Appl. Phys.*, 2(6), 11–17, 1988.
Leckner, B.: Radiation from flames and gases in a cold wall combustion chamber, *IJHMT*, 13(1), 185–197, 1970.
Leckner, B.: Spectral and total emissivity of water vapor and carbon dioxide, *Combust. Flame,* 19, 33–48, 1972.
Leduc, G., Monchoux, F., and Thellier, F.: Inverse radiative design in human thermal environment, *IJHMT*, 47(14–16), 3291–3300, 2004a.

Leduc, G., Monchoux, F., and Thellier, F.: Optimal design of complex human environment by inverse radiative method, *International Symposium on Radiative Transfer IV*, Istanbul, Turkey, June 2004b.
Lee, B. J., Fu, C. J., and Zhang, Z. M.: Coherent thermal emission from one-dimensional photonic crystals, *Appl. Phys. Lett.*, 87, 071904, 2005.
Lee, B. J. and Zhang, Z. M.: Design and fabrication of planar multilayer structures with coherent thermal emission characteristics. *J. Appl. Phys.*, 100, 063529, 2006.
Lee, B. J. and Zhang, Z. M.: Coherent thermal emission from modified periodic multilayer structures, *JHT*, 129(1), 17–27, 2007.
Lee, C. E.: Discrete SN approximation to transport theory, Los Alamos Scientific Laboratory Report LA-2595, Los Alamos, NM, March 1962.
Lee, H. K. and Viskanta, R.: Transient conductive-radiative cooling of an optical quality glass disk, *IJHMT*, 41(14), 2083–2096, 1998.
Lee, H. S., Menart, J. A., and Fakheri, A.: Multilayer radiation solution for boundary-layer flow of gray gases, *JTHT*, 4(2), 180–185, 1990.
Lee, K.-B. and Howell, J. R.: Effect of radiation on the laminar convective heat transfer through a layer of highly porous medium, in T. Tong and M. Modest (eds.), *Radiation in Energy Systems*, ASME HTD-55, pp. 51–59, ASME, New York, 1986.
Lee, K. Y., Zhong, Z. Y., and Tien, C. L.: Blockage of thermal radiation by the soot layer in combustion of condensed fuels, *Proceedings of the 12th International Symposium on Combustion*, pp. 1629–1636, The Combustion Institute, Pittsburg, PA, 1984.
Lee, S.-C.: Dependent scattering by parallel fibers: Effects of multiple scattering and wave interference, *JTHT*, 6(4), 589–595, 1992.
Lee, S.-C.: Dependent vs independent scattering in fibrous composites containing parallel fibers, *JTHT*, 8, 641–646, 1994.
Lee, S.-C. and Cunnington, G.: Conduction and radiation heat transfer in high-porosity fiber thermal radiation, *JTHT*, 14(2), 121–136, 2000.
Lee, S.-C. and Cunnington, G. R.: Theoretical models for radiative transfer in fibrous media, in C.-L. Tien (ed.), *Annual Reviews of Heat Transfer*, vol. IX, Chapter 3, pp. 159–218, Begell House, New York, 1998.
Lee, S.-C. and Grzesik, J. A.: Scattering characteristics of fibrous media containing closely spaced parallel fibers, *JTHT*, 9, 403–409, 1995.
Lee, S.-C. and Tien, C. L.: Optical constants of soot in hydrocarbon flames, *Proceedings of the 18th International Symposium on Combustion*, pp. 1159–1166, The Combustion Institute, Pittsburg, PA, 1981.
Lee, S. H. K. and Jaluria, Y.: Effects of variable properties and viscous dissipation during optical fiber drawing, *JHT*, 118, 350–358, 1996.
Lemmonier, D. and LeDez, V.: Discrete ordinate solution of radiative transfer across a slab with variable refractive index, *JQSRT*, 72(2–5), 195–204, 2005.
Levermore, C. D.: Relating Eddington factors to flux limiters, *JQSRT*, 31(2), 149–160, 1984.
Lewis, E. E. and Miller, W. F., Jr.: *Computational Methods of Neutron Transport*, American Nuclear Society, Inc., La Grange Park, IL, 1993.
Lewis, H. R.: Einstein's derivation of Planck's radiation law, *Am. J. Phys.*, 41(1), 38–14, 1973.
Li, B., Kudo, K., and Kuroda, A.: Study on radiative heat transfer through fibrous layer, *Proceedings of the Third KSME-JSME Thermal Engineering Conference*, vol. III, pp. 279–284, KyongJu, Korea, October 1996.
Li, B.-W., Sun, Y. S., and Yu, Y.: Iterative and direct Chebyshev collocation spectral methods for one-dimensional radiative heat transfer, *IJHMT*, 51, 5887–5894, 2008.
Li, B.-W., Sun, Y.-S., and Zhang, D.-W.: Chebyshev collocation spectral methods for coupled radiation and conduction in a concentric spherical participating medium, *JHT*, 131, 062701-1–062701-9, June 2009.
Li, B.-W., Yao, Q., Cao, X.-Y., and Cen, K.-F.: A new discrete ordinates quadrature scheme for three-dimensional radiative heat transfer, *JHT*, 120(3), 514–518, 1998.
Li, B. X., Yu, X. J., and Liu, L. H.: Backward Monte Carlo simulation for apparent directional emissivity of non-isothermal semitransparent slab, *JQSRT*, 91(2), 173–179, March 1, 2005.
Li, H. Y.: Inverse radiation problem in two-dimensional rectangular media, *JTHT*, 11(4), 556–561, 1993.
Li, H. Y.: Estimation of thermal properties in combined conduction and radiation, *IJHMT*, 42, 565–572, 1999.
Li, H. Y. and Yang, C. Y.: A genetic algorithm for inverse radiation problems, *IJHMT*, 40, 1545–1549, 1997.
Licciulli, A., Diso, D., Torsello, G., and Tundo, S.: The challenge of high-performance selective emitters for thermophotovoltaic applications, *Semicond. Sci. Technol.*, 18, 174–183, 2003.

Lick, W.: Energy transfer by radiation and conduction, in A. Roshko, B. Sturtevant, and D. R. Bartz (eds.), *Proceedings of the 1963 Heat Transfer Fluid Mechanics Institute*, Stanford University, pp. 14–26, 1963.

Lide, D. R. (ed.): *Handbook of Chemistry and Physics*, 88th edn., CRC Press, Boca Raton, FL, 2008.

Liebert, C. H.: *Spectral Emittance of Aluminum Oxide and Zinc Oxide on Opaque Substrates*, NASA TN D-3115, Washington, DC, 1965.

Liebert, C. H. and Hibbard, R. R.: *Spectral Emittance of Soot*, NASA TN D-5647, Washington, DC, 1970.

Liebert, C. H. and Thomas, R. D.: Spectral emissivity of highly doped silicon, in G. B. Heller (ed.) *Thermophysics of Spacecraft and Planetary Bodies*, 20, *AIAA Progress in Astronautics and Aeronautics*, pp. 17–40. Academic Press, New York, 1967. (Also, NASA TN D-4303, April 1968.)

Lienhard, V. J. H.: Thermal radiation in Rayleigh-Benard instability, *JHT*, 112(1), 100–109, 1990.

Lilienfeld, P.: Gustav Mie: The person, *Appl. Opt.*, 30(33), 4696–4698, 1991.

Lim, M., Lee, S. S., Lee, B. J.: Near field thermal radiation between grapheme-covered doped silicon plates, *Opt. Express*, 21(19), 022173, 2013.

Lin, E. I., Stultz, J. W., and Reeve, R. T.: Effective emittance for cassini multilayer insulation blankets and heat loss near seams, *JTHT*, 10(2), 357–363, 1996.

Lin, J.-D.: Exact expressions for radiative transfer in an arbitrary geometry exposed to radiation, *JQSRT*, 37(6), 591–601, 1987.

Lin, J.-D.: Radiative transfer within an arbitrary isotropically scattering medium enclosed by diffuse surfaces, *JTHT*, 2(1), 68–74, 1988.

Lin, S. H. and Sparrow, E. M.: Radiant interchange among curved specularly reflecting surfaces—Application to cylindrical and conical cavities, *JHT*, 87(2), 299–307, 1965.

Linhua, L., Heping, T., and Qizheng, Y.: Inverse radiation problem of temperature field in three-dimensional rectangular furnaces, *Int. Commun. Heat Mass Transfer*, 26(2), 239–248, 1998.

Liou, K.-N.: *An Introduction to Atmospheric Radiation*, 2nd edn., Academic Press, Amsterdam, the Netherlands, 2002.

Lipps, F. W.: Geometric configuration factors for polygonal zones using Nusselt's unit sphere, *Solar Energy*, 30(5), 413–419, 1983.

Liu, B. and Shen, S.: Broadband near-field radiative thermal emitter/absorber based on hyperbolic metamaterials: Direct numerical simulation by the Wiener chaos expansion method, *Phys. Rev. B*, 87, 115403, 2013.

Liu, B., Shi, J., Liew, K., and Shen, S: Near-field radiative heat transfer for Si based metamaterials, *Opt. Commun.*, 314, 57–65, 2014.

Liu, C. C. and Dougherty, R. L.: Anisotropically scattering media having a reflective upper boundary, *JTHT*, 13(2), 177–184, 1999.

Liu, F., Garbett, E. S., and Swithenbank, J.: Effects of anisotropic scattering on radiative heat transfer using the P_1-approximation, *IJHMT*, 35(10), 2491–2499, 1992a.

Liu, F., Smallwood, G. J., and Gülder, Ö. L.: Application of the statistical narrow-band correlated-*k*: Method to low-resolution spectral intensity and radiative heat transfer calculations—Effects of the quadrature scheme, *IJHMT*, 43(17), 3119–3135, 2000a.

Liu, F., Smallwood, G. J., and Gülder, Ö. L.: Band lumping strategy for radiation heat transfer calculations using a narrowband model, *JTHT*, 14(2), 278–281, 2000b.

Liu, F., Smallwood, G. J., and Gülder, Ö. L.: Application of the statistical narrow-band correlated-*k* method to non-grey gas radiation in CO_2–H_2O mixtures; approximate treatments of overlapping bands, *JQSRT*, 68, 401–417, 2001.

Liu, F., Swithenbank, J., and Garbett, E. S.: The boundary condition of the P_N -approximation used to solve the radiative transfer equation, *IJHMT*, 35(8), 2043–2052, 1992b.

Liu, H.-P. and Howell, J. R.: Scale modeling of radiation in enclosures with absorbing/emitting and isotropically scattering media, *JHT*, 109(2), 470–477, 1987.

Liu, J. and Tiwari, S. N.: Investigation of radiative transfer in nongray gases using a narrow band model and Monte Carlo simulation, *JHT*, 116(1), 160–166, 1994.

Liu, J. and Tiwari, S. N.: Radiative heat transfer effects in chemically reacting nozzle flows, *JTHT*, 10(3), 436–444, 1996.

Liu, L. H.: Benchmark numerical solutions for radiative heat transfer in two-dimensional medium with graded index distribution, *JQSRT*, 102, 293–303, 2006.

Liu, L. H., Li, B. X., Tan, H. P., and Yu, Q. Z.: Emissive power of semitransparent spherical particle with non-uniform temperature, *IJHMT*, 45, 4907–4910, 2002.

Liu, L. H., Tan, H. P. and Yu, Q. Z.: Inverse radiation problem of sources and emissivities in one-dimensional semitransparent media, *IJHMT*, 44(1), 63–72, 2001.

Lock, J. A. and Gouesbet, G.: Generalized Lorenz-Mie theory and applications, *JQSRT*, 110(11), 800–807, 2009.

Lockwood, F. C. and Shah, N. G.: A new radiation solution method for incorporation in general combustion prediction procedures, *Proceedings of the 18th International Symposium on Combustion*, pp. 1405–1414, The Combustion Institute, Pittsburg, PA, 1981.

Loehrke, R. I., Dolaghan, J. S., and Burns, P. J.: Smoothing Monte Carlo exchange factors, *JHT*, 117(2), 524–526, 1995.

Loke, V., Huda, G. M., Donev, E. U., Schmidt, V., Hastings, J. T., Mengüç, M. P., and Wriedt, T.: Comparison between discrete dipole approximation and other modeling methods for the plasmonic response of gold nanospheres, *Appl. Phys. B: Lasers Opt.*, 115(2), 237–246, 2014.

Loke, V. and Mengüç, M. P.: Surface waves and atomic force microscope probe-particle near-field coupling: Discrete dipole approximation with surface interaction, *JOSA A*, 27(10), 2293–2303, 2010.

Loke, V. L. Y., Donev, E. U., Huda, G. M., Hastings, J. T., Mengüç, M. P., Wriedt, T.: Discrete dipole approximation of gold nanospheres on substrates: Considerations and comparison with other discretization methods, *AAPP/Phys. Math. Nat. Sci.*, 89(S1), 060-1, 2011a.

Loke, V. L. Y., Nieminen, T. A., and Mengüç, M. P.: Discrete dipole approximation with surface interaction: Computational toolbox for MATLAB, *JQSRT*, 112(11), 1711–1725, July 2011b.

Loke, V. L. Y., Nieminen, T. A., Heckenberg, N. R., and Rubinsztein-Dunlop, H.: T-matrix calculation via discrete-dipole approximation, point-matching and exploiting symmetry, *JQSRT*, 110, 1460–1471, 2009.

Loke, V. L. Y., Nieminen, T. A., Parkin, S. J., Heckenberg, N. R., and Rubinsztein-Dunlop, H.: FDFD/T-matrix hybrid method, *JQSRT*, 106, 274–284, 2007.

Long, R. L.: A review of recent air force research on selective solar absorbers, *J. Eng. Power*, 87(3), 277–280, 1965.

Longoni, G.: Advanced quadrature sets, acceleration and preconditioning techniques for the discrete ordinates method in parallel computing environments, PhD dissertation, University of Florida, Gainesville, FL, 2004.

Longtin, J. P. and Tien, C.-L.: Efficient laser heating of transparent liquids using multiphoton absorption, *IJHMT*, 40, 951–959, 1997.

Loomis, J. J. and Maris, H. J.: Theory of heat transfer by evanescent electromagnetic states, *Phys. Rev. B*, 50, 18517, 1994.

Lorenz, L.: Lysbevaegelsen i og uden for en af plane Lysbolger belyst Kugle, Det Kongelige Danske Videnskabernes Selskabs Skrifter, 6. Raekke, 6. Bind, vol. 1, pp. 1–62, 1890 (see Classical Papers at http://www.t-matrix.de/).

Lorenz, L: Sur la lumière réfléchie et réfractée par une sphère (surface) transparente, in Oeuvres scientifiques de L. Lorenz." revues et annotées par H. Valentiner. Tome Premier, Libraire Lehmann & Stage, Copenhague, Denmark, pp. 403–529, 1898 (see Classical Papers at http://www.t-matrix.de/, accessed April 26, 2015.).

Love, T. J., Stockham, L. W., Lee, F. C., Munter, W. A., and Tsai, Y. W.,: *Radiative Transfer in Absorbing, Emitting and Scattering Media*, ARL-67-0210 (DDC no. AD 6664227), Oklahoma University, Norman, OK, December 1967.

Lowan, A. N.: *Planck's Radiation Functions and Electronic Functions*, Federal Works Agency Works Projects Administration for the City of New York, under the sponsorship of the U.S. National Bureau of Standards Computation Laboratory, New York, 1941.

Loyalka, S. K.: Radiative heat transfer between parallel plates and concentric cylinders, *IJHMT*, 12, 1513–1517, 1969.

Lu, X. and Hsu, P.-F.: Reverse Monte Carlo simulations of light pulse propagation in nonhomogeneous media, *JQSRT*, 93, 349–367, 2005.

Ludwig, C. B., Malkmus, W., Reardon, J. E., and Thomson, J. A. L.: *Handbook of Infrared Radiation from Combustion Gases*, NASA SP-3080, Washington, DC, 1973.

Luo, C., Narayanaswamy, A., Chen, G., and Joannopoulos, J. D.: Thermal radiation from photonic crystals: A direct calculation, *Phys. Rev. Lett.*, 93, 213905, 2004.

Ma, A. K.: Generalized zoning method in one-dimensional participating media, *JHT*, 117(2), 520–523, 1995.

Ma, C. Y., Zhao, J. M., Liu, L. H., and Zhang, L.: GPU-accelerated inverse identification of radiative properties of particle suspensions in liquid by Monte Carlo, *Proc. 5th Int. Symp. Computational Thermal Radiation in Participating Media,* Albi, France, April 1–3, 2015.

Ma, J., Li, B.-W., and Howell, J. R.: Thermal radiation heat transfer in one- and two-dimensional enclosures using the spectral collocation method with full spectrum k-distribution model, *IJHMT*, 71, 35–41, 2014.

Ma, Y., Varadan, V. K., and Varadan, V. V.: Enhanced absorption due to dependent scattering, *JHT*, 112(2), 402–407, 1990.

Mackay, D. B.: *Design of Space Powerplants*, Prentice-Hall, Englewood Cliffs, NJ, 1963.

Mackowski, D. W., Altenkirch, R. A., and Mengüç, M. P.: A comparison of electromagnetic wave and radiative transfer equation analyses of a coal particle surrounded by a soot cloud, *Combust. Flame*, 76, 415–420, 1989a.

Mackowski, D. W., Altenkirch, R. A., and Mengüç, M. P.: Internal absorption cross sections in a stratified sphere, *Appl. Opt.*, 29(10), 1551–1559, 1990.

Mackowski, D. W., Altenkirch, R. A., Mengüç, M. P., and Saito, K.: Radiative properties of chain-agglomerated soot formed in hydrocarbon diffusion flames, *Proceedings of the 22nd International Symposium on Combustion*, pp. 1263–1269, The Combustion Institute, Pittsburgh, PA, 1989b.

Mackowski, D. W. and Mishchenko, M. I.: Calculation of the T matrix and the scattering matrix for ensembles of spheres, *JOSA*, 13, 2266–2278, 1996.

Magnussen, B. F. and Hjertager, B. H.: On mathematical modeling of turbulent combustion with special emphasis on soot formation and combustion, *Proceedings of the 16th International Symposium on Combustion*, pp. 719–729, The Combustion Institute, Pittsburgh, PA, 1977.

Mahan, J. R., Kingsolver, J. B., and Mears, D. T.: Analysis of diffuse-specular axisymmetric surfaces with application to parabolic reflectors, *JHT*, 101(4), 689–694, 1979.

Mahapatra, S. K.: Mixed convection inside a differentially heated enclosure and its interaction with radiation—An exhaustive study, *Heat Transfer Eng.*, 35(1), 74–83, 2014.

Maier, S. A.: *Plasmonics: Fundamentals and Applications*, Springer, New York, 2007.

Majumdar, A.: Scanning thermal microscopy, *Ann. Rev. Mater. Sci.*, 29, 505–585, 1999.

Makino, T.: Radiation thermal spectroscopy for heat transfer science and for engineering surface diagnosis, in J. Taine (ed.), *Heat Transfer 2002, Proceedings of the 2002 International Heat Transfer Conference*, vol. 1, pp. 55–56, Elsevier, Paris, France, 2002

Makino, T. and Kaga, K.: Scattering of radiation at a rough surface modeled by a three-dimensional superimposition technique, *Proceedings of the Third ASME/JSME Thermal Engineering Joint Conference*, vol. 4, pp. 27–33, Reno, NV, 1991.

Makino, T., Kaga, K., and Murata, H.: Numerical experiment on transient behavior in reflection characteristics of a real surface of a metal, *Proceedings of the 28th National Symposium Heat Transfer*, vol. 2, pp. 568–570, Fukuoka, Japan, 1991.

Makino, T., Sotokawa, O., and Iwata, Y.: Transient behaviors in thermal radiation characteristics of heat-resisting metals and alloys in oxidation processes, *Int. J. Thermophys.*, 9(6), 1121–1130, 1988.

Makino, T. and Wakabayashi, H.: Thermal radiation spectroscopy diagnosis for temperature and microstructure of surfaces, *JSME Int. J., Ser. B*, 46(4), 500–509, 2003.

Malalasekera, W. M. G. and James, E. H.: Calculation of radiative heat transfer in three-dimensional complex geometries, in Y. Bayazitoglu, D. Kaminski, and P. D. Jones (eds.), *Proceedings of the 1995 National Heat Transfer Conference*, ASME HTD-vol. 315, pp. 53–61, Portland, OR, 1995.

Malcolm, N. P.: Simulation of a plasmonic nanowire waveguide, MS thesis, Department of Mechanical Engineering, The University of Texas at Austin, Austin, TX, May 2009.

Malkmus, W.: Random lorentz band model with exponential-tailed s^{-1} line intensity distribution function, *JOSA*, 57(3), 323–329, 1967.

Maltby, J. D. and Burns, P. J.: Performance, accuracy, and convergence in a 3-D Monte Carlo radiative heat transfer simulation, *Numer. Heat Transfer B*, 19(3), 191–209, 1991.

Mandel, L. and Wolf, E.: *Optical Coherence and Quantum Optics*, Cambridge University Press, Cambridge, U.K., 1995.

Mandelbrot, B. B.: *Les Objets Fractals: Forme, Hasard et Dimension*, Flammarion, Paris, France, 1975.

Mandelbrot, B. B.: *The Fractal Geometry of Nature*, W. H. Freeman, San Francisco, CA, 1983.

Mandell, D. A. and Miller, F. A.: Comparison of exact and mean beam length results for a radiating hydrogen plasma, *JQSRT*, 13, 49–56, 1973.

Manickavasagam, S., Govindan, R., and Mengüç, M. P.: Estimation of the morphology of soot agglomerates by measuring their scattering matrix elements, *ASME HTD*, 352, 29–32, 1997.

Manickavasagam, S. and Mengüç, M. P.: Scattering matrix elements of fractal-like soot agglomerates, *Appl. Opt.*, 36(6), 1337–1351, 1997. (Correction/Addition: *AO*, 36(27), 7008, 1997.)

Manickavasagam, S. and Mengüç, M. P.: Scattering matrix elements of coated infinite-length cylinders, *Appl. Opt.*, 37(12), 2473–2482, 1998.

Mann, D., Field, R. E., and Viskanta, R.: Determination of specific heat and true thermal conductivity of glass from dynamic temperature data, *Wärme- und Stoffübertrag.*, 27, 225–231, 1992.

Mann, T., Heltzel, A., and Howell, J. R.: Metamaterial window glass for adaptable energy efficiency, Paper HT2013-17511, *Proceedings of the ASME 2013 Summer Heat Transfer Conference*, Minneapolis, MN, July 14–19, 2013.

Mannan, K. D. and Cheema, L. S.: Compound-wedge cylindrical stationary concentrator, *Solar Energy,* 19(6), 751–754, 1977.

Manohar, S. S., Kulkarni, A. K., and Thynell, S. T.: In-depth absorption of externally incident radiation in nongray media, *JHT*, 117(1), 146–151, 1995.

Marakis, J. G., Papapavlou, C., and Kakaras, E.: A parametric study of radiative heat transfer in pulverised coal furnaces, *IJHMT*, 43(16), 2961–2971, 2000.

Marin, O. and Buckius, R.: Wideband correlated-*k* method applied to absorbing, emitting and scattering media, *JTHT*, 10(2), 364–371, 1996. (See also Errata to this reference, *JTHT*, 11(4), 598, 1997.)

Marin, O. and Buckius, R.: A simplified wide band model of the cumulative distribution function for water vapor, *IJHMT*, 41, 2877–2892, 1998a.

Marin, O. and Buckius, R.: A simplified wide band model of the cumulative distribution function for carbon dioxide, *IJHMT*, 41, 3881–3897, 1998b.

Mark, J. C: The spherical harmonic method—Parts I and II, National Research Council of Canada, Ottawa, Ontario, Canada, Atomic Energy Reports MT 92-1944 and MT 97-1945, 1945.

Markham, J. R., Best, P. E., Solomon, P. R., and Yu, Z. Z.: Measurement of radiative properties of ash and slag by ft-ir emission and reflection spectroscopy, *JHT*, 114(2), 458–464, 1992.

Marschall, J. and Milos, F. S.: The calculation of anisotropic extinction coefficients for radiation diffusion in solid fibrous ceramic insulations, *IJHMT*, 40, 627–634, 1997.

Marshak, R. E.: Note on the spherical harmonics methods as applied to the Milne problem for a sphere, *Phys. Rev.*, 71, 443–446, 1947.

Martin, J. K. and Hwang, C. C.: Combined radiant and convective heat transfer to laminar stream flow between gray parallel plates with uniform heat flux, *JQSRT*, 15, 1071–1081, 1975.

Martynenko, O. G., German, M. L., Nekrasov, V. P., and Nogotov, E. F.: The radiation transfer in emitting, absorbing and scattering media of complex geometric form, *IJHMT*, 41(17), 2697–2704, 1998.

Maruyama, S.: Uniform isotropic emission from an aperture of a collector, *Proceedings of the Fourth ASME/JSME Thermal Engineering Joint Conference*, vol. 4, pp. 47–53, Reno, NV, 1991.

Maruyama, S.: Uniform isotropie emission from an involute reflector, *JHT*, 115(2), 492–495, 1993.

Maruyama, S. and Aihara, T.: Radiative heat transfer of arbitrary 3-D participating media and surfaces with non-participating media by a generalized numerical method REM^2, in M. P. Mengüç (ed.), *Radiative Transfer I—Proceedings of the First International Symposium Radiation Transfer*, pp. 153–167, Begell House, New York, 1996.

Maruyama, S., Kashiwa, T., Yugami, H., and Esashi, M.: Thermal radiation from two-dimensionally confined modes in microcavities, *Appl. Phys. Lett.*, 79, 1393, 2001.

Maslowski, S. I., Simovski, C. R., and Trtyakov, S. A.: Equivalent circle model of radiative transfer, *Phys. Rev. B.*, 87, 1555124-1–1555124-15, 2013.

Maslowski, W., Kinney, J. C., Higgens, M., and Roberts, A.: The future of Arctic sea ice, *Ann. Rev. Earth Planet. Sci.*, 40, 625–654, 2012.

Masuda, H.: Radiant heat transfer on circular-finned cylinders, *Rep. Inst. High Speed Mech. Tohoku Univ.*, 27(255), 67–89, 1973. (See also *Trans. Jpn. Soc. Mech. Eng.*, 38, 3229–3234, 1972.)

Masuda, H. and Higano, M.: Measurement of total hemispherical emissivities of metal wires by using transient calorimetric technique, *JHT*, 110(1), 166–172, 1988.

Mathes, R., Blumenberg, J., and Keller, K.: Radiative heat transfer in insulations with random fibre orientation, *IJHMT*, 33(4), 767–770, 1990.

Mathur, S. R. and Murthy, J. Y.: Coupled ordinates method for multigrid acceleration of radiation calculations, *JTHT*, 13(4), 467–473, 1999.

Mathur, S. R. and Murthy, J. Y.: Unstructured finite volume methods for multi-mode heat transfer, in W. J. Mincowycz and E. M. Sparrow (eds.), *Advances in Numerical Heat Transfer*, vol. 2, pp. 37–70, Taylor & Francis, New York, 2000.

Matsushima, H. and Viskanta, R.: Effects of internal radiative transfer on natural convection and heat transfer in a vertical crystal growth configuration, *IJHMT*, 33(9), 1957–1968, 1990.

Matthews, L. K., Viskanta, R., and Incropera, F. P.: Combined conduction and radiation heat transfer in porous materials heated by intense solar radiation, *J. Solar Energy Eng.*, 107, 29–34, 1985.

Mattis, D. C. and Bardeen, J.: Theory of the anomalous skin effect in normal and superconducting metals, *Phys. Rev.*, 111, 412–417, 1958.

Maurente, A., França, F., Miki, K., and Howell, J.: Application of joint cumulative *k*-distributions to FSK radiation heat transfer in multicomponent high temperature non-LTE plasmas, *JQSRT*, 113(12), 1521–1535, August 2012.

Maurente, A., Vielmo, H. A., and França, F. H. R: A Monte Carlo implementation to solve radiation heat transfer in non-uniform media with spectrally dependent properties, *JQSRT*, 108(2), 295–307, 2007.

Maurente, A., Vielmo, H. A., and França, F. H. R.: Comparison of the standard weighted-sum-of-gray-gases with the absorption-line blackbody distribution function for the computation of radiative heat transfer in H_2O/CO_2 mixtures, *JQSRT*, 109(10), 1758–1770, 2008.

Mavroulakis, A. and Trombe, A.: A new semianalytical algorithm for calculating diffuse plane view factors, *JHT*, 120(1), 279–282, 1998.

Maxwell, J. C.: A dynamical theory of the electromagnetic field, in W. D. Niven (ed.), *The Scientific Papers of James Clerk Maxwell*, vol. 1, Cambridge University Press, Cambridge, U.K., 1890.

Mazumder, S.: A new numerical procedure for coupling radiation in participating medi with other modes of heat transfer, *JHT*, 127(9), 1037–1045, September 2005.

Mazumder, S.: Methods to accelerate ray tracing in the Monte Carlo method for surface-to-surface radiation transport, *JHT*, 128(9), 945–952, September 2006.

Mazumder, S. and Modest, M. F.: A PDF approach to modelling turbulence–radiation interactions in nonluminous flames, *IJHMT*, 42, 971–991, 1998.

Mbiock, A. and Weber, R.: *Radiation in Enclosures: Elliptic Boundary Value Problem*, Springer-Verlag, Berlin, Germany, 2000.

McCauley, A. P., Reid, M. T. H., Krüger, M., and Johnson, S. G.: Modeling near-field radiative heat transfer from sharp objects using a general three-dimensional numerical scattering technique, *Phys. Rev. B*, 85, 165104, 2012.

McConnell, D. G.: Radiant energy transport within cryogenic condensates, *Institute of Environmental Science and Annual Meeting Equipment Exposition*, San Diego, CA, April 11–13, 1966.

McDonald, J., Golden, A., and Jennings, G.: OpenDDA: A novel high-performance computational framework for the discrete dipole approximation, *Int. J. High Perf. Comp. Appl.* 23(1), 42–46, 2009.

McHugh, J., Burns, P. J., Hittle, D., and Miller, B.: Daylighting design via Monte Carlo, in S. T. Thynell et al. (eds.), *Developments in Radiative Heat Transfer*, ASME HTD-203, pp. 129–136, ASME, New York, 1992.

Mechtly, E. A.: *The International System of Units, Physical Constants and Conversion Factors*, 2nd revision, NASA SP-7012, Washington, DC, 1973.

Men, A. A.: Radiative-conductive heat transfer in a medium with a cylindrical geometry. I, *Inz. Fiz. Zh.*, 24(6), 984–991, 1973.

Menart, J. A. and Lee, H. S.: Nongray gas analysis for reflecting walls utilizing a flux technique, *JHT*, 115(3), 645–652, 1993.

Menart, J. A., Lee, H. S., and Kim, T.-K.: Discrete ordinates solutions of nongray radiative transfer with diffusely reflection walls, *JHT*, 115(1), 184–193, 1993.

Mengüç, M. P.: Modeling of radiative heat transfer in multidimensional enclosures using spherical harmonics approximation, PhD dissertation, Department of Mechanical Engineering, Purdue University, Lafayette, IN, 1985.

Mengüç, M. P.: Engineering with and for light absorption and scattering, Invited Review paper at ELS XIII, Taormina, Italy, *Atti. Accad. Pelorit. Pericol.* CI. Sci. Fis. Mat. Nat., 89(1), C1V89S1P007, 2011.

Mengüç, M. P., Cummings, W. G. III, and Viskanta, R.: Radiative Transfer in a Gas Turbine Combustor, *J. Propulsion and Power*, 2, 241–247, 1986a.

Mengüç, M. P, and Dutta, P.: Scattering tomography and application to sooting diffusion flames, *JHT*, 116(1), 144–151, 1994.

Mengüç, M. P. and Iyer, R. K.: Modeling of radiative transfer using multiple spherical harmonics approximations, *JQSRT*, 39(6), 445–461, 1988.

Mengüç, M. P. and Manickavasagam, S.: Inverse radiation problem in axisymmetric cylindrical scattering media, *JTHT*, 7(3), 479–486, 1993.

Mengüç, M. P. and Manickavasagam, S.: Characterization of size and structure of agglomerates and inhomogeneous particles via polarized light, *Int. J. Eng. Sci.*, Special issue in memory of S. Chandrasekhar, 36, 1569–1593, 1998.

Mengüç, M. P., Manickavasagam, S., and D'sa, D. A.: Determination of radiative properties of pulverized coal particles from experiments, *FUEL*, 73(4), 613–625, 1994.

Mengüç, M. P. and Viskanta, R.: Comparison of radiative transfer approximations for a highly forward scattering planar medium, *JQSRT*, 29(5), 381–394, 1983.

Mengüç, M. P. and Viskanta, R.: Radiative transfer in three-dimensional rectangular enclosures containing inhomogeneous, anisotropically scattering media, *JQSRT*, 33(6), 533–549, 1985.

Mengüç, M. P. and Viskanta, R.: Radiative transfer in axisymmetric, finite cylindrical enclosures, *JHT*, 108(2), 271–276, 1986a.

Mengüç, M. P. and Viskanta, R.: Radiation transfer in a cylindrical vessel containing high temperature corium aerosols, *Nuc. Sci. Engng.*, 92, 570–583, 1986b.

Mengüç, M. P., and Viskanta, R.: Effect of fly-ash particles on spectral and total radiation blockage, *Comb. Sci. and Technology,* 60, 97–115, 1988.

Mengüç, M. P., Viskanta, R., and Ferguson, C. R.: Multidimensional modeling of radiative heat transfer in diesel engines, *SAE Trans.*, SAE Technical Paper 850503, Detroit, MI, 1985.

Mengüç, M. P., Yener, Y., and Ozisik, M. N.: Interaction of radiation and convection in thermally developing laminar flow in a parallel-plate channel, ASME Paper 83-HT-035, *ASME National Heat Transfer Conference*, Seattle, WA, August 1983.

Merriam, R. L. and Viskanta, R.: Radiative characteristics of cryodeposits for room temperature black body radiation, *Cryogenic Engineering Conference*, Case-Western Reserve University, Cleveland, OH, August 1968.

Metropolis, N., Rosenbluth, A., Rosenbluth, M., Teller, A., and Teller, E.: Equation of state calculations by fast computing machines, *J. Chem. Phys.*, 21, 1087–1092, 1953.

Metropolis, N. and Ulam, S.: The Monte Carlo method, *J. Am. Stat. Assoc.*, 44(247), 335–341, 1949.

Meyer, C. W.: Effects of extraneous radiation on the performance of lightpipe radiation thermometers, in B. Fellmuth, J. Seidel, and G. Scholz (eds.), *Proceedings of the TEMPMEKO-2001*, pp. 937–942, VDE Verlag GMBH, Berlin, Germany, 2002.

Mie, G.: Beiträge zur Optik trüber Medien, speziell kolloidaler Metallösungen, *Ann. Phys.*, 330, 377–445, 1908a.

Mie, G.: Optics of turbid media, *Ann. Phys.*, 25(3), 377–445, 1908b.

Mie, G.: *Contributions on the Optics of Turbid Media, Particularly Colloidal Metal Solutions—Translation.* Sandia Laboratories, Albuquerque, NM, 1978, SAND78-6018, National Translation Center, Chicago, IL, Translation 79-21946.

Mihail, R. and Maria, G.: The emitter-receptor geometrical configuration influence on radiative heat transfer, *IJHMT*, 26(12), 1783–1789, 1983.

Miliauskas, G.: Regularities of unsteady radiative-conductive heat transfer in evaporating semitransparent liquid droplets, *IJHMT*, 44(4), 785–798, 2001.

Milne, F. A.: Thermodynamics of the stars, in G. Eberhard, A. Kohlschütter, H. Ludendorff, E. A. Milne, A. Pannekoek, S. Rosseland, and W. Westphal (eds.), *Handbuch der Astrophysik*, 3, pp. 65–255, Springer-Verlag, OHG, Berlin, Germany, 1930.

Minerbo, G. N.: Maximum entropy Eddington factors, *JQSRT*, 20, 541–545, 1978.

Minkowycz, W. J., Sparrow, E. M., and Murthy, J. Y. (eds.): *Handbook of Numerical Heat Transfer*, 2nd edn., Wiley, New York, 2006.

Miranda, P. C: A Calculation and experimental verification of the infrared transmission coefficient of straight cylindrical metal tubes, *JHT*, 118(2), 495–497, 1996.

Mishchenko, M. I.: Light scattering by randomly oriented axially symmetric particles, *JOSA A*, 8, 871–872, 1991.

Mishchenko, M. I.: Vector radiative transfer equation for arbitrarily shaped and arbitrarily oriented particles: A microphysical derivation from statistical electromagnetics, *Appl. Opt.*, 41, 7114–7134, 2002a.

Mishchenko, M. I.: Microphysical approach to polarized radiative transfer: Extension to the case of an external observation point, *Appl. Opt.*, 42, 4963–4967, 2002b.

Mishchenko, M. I.: Electromagnetic scattering by nonspherical particles: A tutorial review, *JQSRT*, 110, 808–832, 2009.

Mishchenko, M. I.: Directional radiometry and radiative transfer: The convoluted path from centuries-old phenomenology to physical optics, *JQSRT*, 146, 4–33, 2014.

Mishchenko, M. I., Hovenier, J. W., and Travis, L. D.: *Light Scattering by Nonspherical Particles*, Academic Press, New York, 2000.

Mishchenko, M. I., Kahnert, M., Mackowski, D. W., and Wriedt, T.: Peter Waterman and his scientific legacy, *JQSRT*, 123, 1, 2013a.

Mishchenko, M. I. and Travis, L. D.: Gustav Mie and the evolving discipline of electromagnetic scattering by particles, *Bull. Am. Meteorol. Soc.*, 89, 1853–1861, 2008.

Mishchenko, M. I., Travis, L. D., and Mackowski, D. W.: T-matrix computations of light scattering by nonspherical particles: A review, *JQSRT*, 55, 535–575, 1996.
Mishchenko, M. I., Travis, L. D., and Lacis, A. A.: *Scattering, Absorption, and Emission of Light by Small Particles*, Cambridge University Press, Cambridge, U.K., 2002.
Mishchenko, M. I., Travis, L. D., and Lacis, A. A.: *Multiple Scattering of Light by Particles: Radiative Transfer and Coherent Back Scattering*, Cambridge Press, New York, 2006.
Mishchenko, M. I., Videen, G., Babenko, V. A. et al.: T-matrix theory of electromagnetic scattering by particles and its applications: A comprehensive reference data base, *JQSRT*, 88, 357–406, 2004.
Mishchenko, M. I., Videen, G., Babenko, V. A. et al.: Comprehensive T-matrix database: A 2004–2006 update, *JQSRT*, 106, 304–324, 2007.
Mishchenko, M. I., Videen, G., Babenko, V. A. et al.: Comprehensive T-matrix database: A 2006–2007 update, *JQSRT*, 109, 1447–1460, 2008.
Mishchenko, M. I., Videen, G., Khlebtsov, N. G., and Wriedt, T.: Comprehensive T-matrix reference database: A 2012–2013 update, *JQSRT*, 123, 145–152, 2013b.
Mishchenko, M. I., Zakharova, N., Videen, G., Khlebtsov, N., and Wriedt, T.: Comprehensive T-matrix database: A 2007–2009 update, *JQSRT*, 111, 650–658, 2010.
Mishchenko, M. I., Zakharova, N. T., Khlebtsov, N. G., Wriedt, T., and Videen, G.: Comprehensive thematic T-matrix reference database: A 2013–2014 update, *JQSRT*, 146, 349–354, 2014.
Mishra, C. S., Hari Krishna, Ch., and Kim, M. Y.: Analysis of conduction and radiation heat transfer in a 2-D cylindrical medium using the modified discrete ordinate method and the lattice Boltzmann method, *Numer. Heat Transfer A Appl.*, 60(3), 254–287, 2011.
Mishra, C. S. and Lankadasu, A.: Transient conduction-radiation heat transfer in participating media using the lattice Boltzmann method and the discrete transfer method, *Numer. Heat Transfer A Appl.*, 47(9), 935–954, 2005.
Mishra, S. C., Poonia, H., Vernekar, R. R., and Das, A. K.: Lattice Boltzmann method applied to radiative transport analysis in a planar participating medium, *Heat Transfer Eng.*, 35(14–15), 1267–1278, 2014.
Mital, R., Gore, J. P., and Viskanta, R.: Measurements of radiative properties of cellular ceramics at high temperatures, *JTHT*, 10, 33–38, 1996.
Mitalas, G. P. and Stephenson, D. G.: FORTRAN IV programs to calculate radiant interchange factors, National Research Council of Canada, Division of Building Research Report DBR-25, Ottawa, Ontario, Canada, 1966.
Mitra, K., Lai, M.-S., and Kumar, S.: Transient radiation transport in participating media within a rectangular enclosure, *JTHT*, 11(3), 409–414, 1997.
Mitts, S. J. and Smith, T. F.: Solar energy transfer through semitransparent plate systems, *JTHT*, 1(4), 307–312, 1987.
Mochida, A., Kudo, K., Mizutani, Y., Hattori, M., and Nakamura, Y.: Transient heat transfer analysis in vacuum furnace by radiant tube burners, *Proceedings of the RAN 95: International Symposium on Advanced Energy Conversion Systems*, Society of Chemical Engineers, Nagoya, Japan, December 1995.
Modest, M. F.: Three-dimensional radiative exchange factors for nongray, nondiffuse surfaces, *Numer. Heat Transfer B*, 1, 403–416, 1978.
Modest, M. F.: A simple differential approximation for radiative transfer in non-gray gases, *JHT*, 101(4), 735–736, 1979.
Modest, M. F.: Modified differential approximation for radiative transfer in general three-dimensional media, *JTHT*, 3, 283–288, 1989.
Modest, M. F.: The improved differential approximation for radiative transfer in multidimensional media, *JHT*, 112(3), 819–821, 1990.
Modest, M. F.: The weighted-sum-of-gray-gases model for arbitrary solution methods in radiative transfer, *JHT*, 113(3), 650–656, 1991.
Modest, M. F.: The Monte Carlo method applied to gases with spectral line structure, *Numer. Heat Transfer B*, 22, 273–284, 1992.
Modest, M. F.: Backward Monte Carlo simulations in radiative heat problems with arbitrary radiation sources, including small collimated beams, point sources, etc., in media of arbitrary optical thickness, *JHT*, 125(1), 57–62, February 2003.
Modest, M. F.: The treatment of nongray properties in radiative heat transfer: From past to present, *JHT*, 135(6), 061801-1–061801-12, 2013. doi: 10.1115/1.40235.
Modest, M. F. and Singh, V.: Engineering correlations for full-spectrum *k*-distributions of H_2O from the HITEMP spectroscopic database, *JQSRT*, 93, 263–271, 2005.
Modest, M. F. and Zhang, H.: The full-spectrum correlated *k*-distribution for thermal radiation from molecular gas-particulate mixtures, *JHT*, 124(1), 30–38, February 2002.

Moghaddam, S. N., Ertürk, H., and Mengüç, M. P.: Light Scattering from defected nano-sized objects on a surface with DDA-SI, *Electomagnetic Wave and Light Scattering Conference*, ELS-XV, Leipzig, Germany, June 2015a.

Moghaddam, S. N., Ertürk, H., and Mengüç, M. P.: Heating of noble metal nanostructures on a dielectric surface due to plasmonic resonances and the effect of a probe, *Proc. 1st Thermal and Fluid Engineering Summer Conf.*, TFESC, ASTFE, New York City, August 2015b.

Mohamad, A. A.: Local analytical discrete ordinate method for the solution of the radiative transfer equation, *IJHMT*, 39(9), 1859–1864, 1996.

Mohr, P. J., Taylor, B. N., and Newell, D. B.: CODATA recommended values of the fundamental physical constants: 2010, *Rev. Mod. Phys.*, 84(4), 1527, 2012.

Moon, P.: *The Scientific Basis of Illuminating Engineering*, Revised edn., Dover, New York, 1961.

Moore, S. W.: Solar absorber selective paint research, *Solar Energy Mater.*, 12, 435–447, 1985a.

Moore, S. W.: Progress on solar absorber selective paint research, *Solar Energy Mater.*, 12, 449–460, 1985b.

Morales, J. C. and Campo, A.: Radiative effects on natural convection of gases confined in horizontal isothermal annuli, in S. T. Thynell et al. (eds.), *Developments in Radiative Transfer*, pp. 2312–2318, ASME, New York, 1992.

Morel, J. E., McGhee, J. M., and Larsen, E. W.: A 3-D Time-dependent unstructured tetrahedral-mesh SPN method, *Nucl. Sci. Eng.*, 123(5), 467–474, 1979.

Morizumi, S. J.: Comparison of analytical model with approximate models for total band absorption and its derivative, *JQSRT*, 22(5), 467–474, 1979.

Morozov, V. A.: *Methods for Solving Incorrectly Posed Problems*, Springer-Verlag, New York, 1984.

Moscowitz, C. M., Stretz, L. A., and Bautista, R. G.: The spectral emissivities of lanthanum, cerium, and praseodymium, *High Temp. Sci.*, 4(5), 372–378, 1972.

Mossi, A. C., Vielmo, H. A., França, F. H. R., and Howell, J. R., Inverse design involving combined radiative and turbulent convective heat transfer, *IJHMT*, 51(11–12), 3217–3222, March 2008.

Mott, N. F. and Zener, C.: The optical properties of metals, *Cambridge Philos. Soc. Proc.*, 30(2), 249–270, 1934.

Moura, L. M., Baillis, D., and Sacadura, J. F.: Identification of thermal radiation properties of dispersed media: Comparison of different strategies, in J. S. Lee (ed.), *Heat Transfer 1998: Proceedings of the 11th International Heat Transfer Conference*, vol. 7, pp. 489–414, Taylor & Francis, Levittown, NY, 1998.

Moutsoglou, A., Rhee, J. H., and Won, J. K.: Natural convection-radiation cooling of a vented channel, *IJHMT*, 35(11), 2855–2863, 1992.

Moutsoglou, A. and Wong, Y. H.: Convection-radiation interaction in buoyancy-induced channel flow, *JTHT*, 3(2), 175–181, 1989.

Mugnai, A. and Wiscombe, J. W.: Scattering from nonspherical Chebyshev particles I: Cross sections, single-scattering albedo, asymmetry factor, and backscattered fraction, *Appl. Opt.*, 25, 1235–1244, 1986.

Mulet, J. P., Joulain, K., Carminati, R., and Greffet, J.-J.: Nanoscale radiative heat transfer between a small particle and a plane surface, *Appl. Phys. Lett.*, 78(19), 2931–2933, 2001.

Mulet, J.-P., Joulain, K., Carminati, R., and Greffet, J.-J.: Enhanced radiative heat transfer at nanometric distances, *Microscale Thermophys. Eng.*, 6, 209–222, 2002.

Mulet, J.-P.: Modélisation du Rayonnement Thermique par une Approche Electromagnétique. Rôle des Ondes de Surfaces dans le Transfert d'Energie aux Courtes Echelles et dans les Forces de Casimir, PhD thesis (in French), Université Paris-Sud 11, Paris, France, 2003.

Müller-Hirsch, W., Kraft, A., Hirsch, M. T., Parisi, J., and Kittel, A.: Heat transfer in ultrahigh vacuum scanning thermal microscopy, *J. Vac. Sci. Technol. A.*, 17(4), 1205–1210, 1999.

Munch, B.: Directional distribution in the reflection of heat radiation and its effect in heat transfer, PhD thesis, Swiss Technical College of Zurich, Zürich, Switzerland, 1955.

Murthy, J. Y. and Mathur, S. R.: Finite volume method for radiative heat transfer using unstructured meshes, *JTHT*, 12(3), 313–321, 1998.

Murty, C. V. S. and Murty, B. S. N.: Significance of exchange area adjustment in zone modelling, *IJHMT*, 34(2), 499–503, 1991.

Musset, A. and Thelen, A.: Multilayer antireflection coatings, in E. O. DuBois (ed.), *Progress in Optics*, vol. 8, pp. 203–237, American Elsevier, New York, 1970.

Nakouzi, S.: *Modélisation du Procédé de Cuisson de Composites Infusés par Chauffage Infra Rouge*, PhD Thesis, École des Mines Albi, Albi, France, 2012.

Naraghi, M. H. N. and Chung, B. T. F.: A stochastic approach for radiative exchange in enclosures with non-participating medium, *JHT*, 106(4), 690–698, 1984.

Naraghi, M. H. N. and Chung, B. T. F.: A unified matrix formulation for the zone method: A stochastic approach, *IJHMT*, 28(1), 245–251, 1985.

Naraghi, M. H. N. and Chung, B. T. F.: A stochastic approach for radiative exchange in enclosures with directional-bidirectional properties, *JHT*, 108(2), 264–270, 1986.
Naraghi, M. H. N. and Kassemi, M.: Analysis of radiative transfer in rectangular enclosures using a discrete exchange factor method, *JHT*, 111(4), 1117–1119, 1989.
Narayanaswamy, A. and Chen, G.: Surface modes for near field thermophotovoltaics, *Appl. Phys. Lett.*, 82(20), 3544–3546, 2003.
Narayanaswamy, A. and Chen, G.: Thermal emission control with one-dimensional metallodielectric photonic crystals, *Phys. Rev. B*, 70, 125101, 2004.
Narayanaswamy, A. and Chen, G.: Thermal radiation in 1D photonic crystals, *J. Quant. Spectrosc. Radiat. Transfer*, 93, 175–183, 2005a.
Narayanaswamy, A. and Chen, G.: Direct computation of thermal emission from nanostructures, *Ann. Rev. Heat Transfer*, 14, 169–195, 2005b.
Narayanaswamy, A. and Chen, G.: Thermal near-field radiative transfer between two spheres, *Phys. Rev. B*, 77, 075125, 2008.
Narayanaswamy, A., Shen, S., Hu, L., Chen, X., and Chen, G., Breakdown of the Planck blackbody radiation law at nanoscale gaps, *Appl Phys*, 96, 357–362, 2009.
Nelson, D. A.: A study of band absorption equations for infrared radiative transfer in gases. I, transmission and absorption functions for planar media, *JQSRT*, 14, 69–80, 1974.
Nelson, D. A.: On the uncoupled superposition approximation for combined conduction-radiation through infrared radiating gases, *IJHMT*, 18(5), 711–713, 1975.
Nelson, D. A.: Band radiation of isothermal gases within diffuse-walled enclosures, *IJHMT*, 27(10), 1759–1769, 1984.
Nelson, H. F.: Radiative transfer through carbon ablation layers, *JQSRT*, 13, pp. 427–445, 1973.
Ness, A. J.: Solution of equations of a thermal network on a digital computer, *Solar Energy*, 3(2), p. 37, 1959.
Neuer, G. and Jaroma-Weiland, G.: Spectral and total emissivity of high-temperature materials, *Int. J. Thermophys.*, 19(3), 917–929, 1998.
Nice, M. L.: Application of finite element method to heat transfer in a participating medium, in T. M. Shih (ed.), *Numerical Properties and Methodologies in Heat Transfer*, pp. 497–514, Hemisphere, Washington, DC, 1983.
Nichols, L. D.: *Surface-Temperature Distribution on Thin-Walled Bodies Subjected to Solar Radiation in Interplanetary Space*, NASA TN D-584, Washington, DC, 1961.
Nicodemus, F. E., Richmond, J. C., Hsia, J. J., Ginsberg, I. W., and Limperis, T.: *Geometrical Considerations and Nomenclature for Reflectance*, NBS Monograph 160, National Bureau of Standards, United States Department of Commerce, Washington, DC, 1977.
Nicolau, V. de P. and Maluf, F. P.: Determination of radiative properties of commercial glass, *Proceedings of the 18th Conference on Passive and Low Energy Architecture*, Florianopolis, Brazil, November 2001.
Nicolau, V. de P., Raynaud, M., and Sacadura, J. F.: Spectral radiative properties identification of fiber insulating materials, *IJHMT*, 37(Suppl. 1), 311–324, 1994.
Nieminen, T. A., Loke, V. L. Y., Stilgoe, A. B., Knöner, G., Branczyk, A. M., Heckenberg, N. R., and Rubinsztein-Dunlop, H.: Optical tweezers computational toolbox, *J. Opt. A*, 9, S196–S203, 2007.
Nishimura, T., Shiraishi, M., Nagasawa, F., and Kawamura, Y.: Natural convection heat transfer in enclosures with multiple vertical partitions, *IJHMT*, 31(8), 1679–1686, 1988.
Noble, J. J.: The zone method: Explicit matrix relations for total exchange areas, *IJHMT*, 18(2), 261–269, 1975.
Novotny, J. L. and Hecht, B.: *Principles of Nano-Optics*, Cambridge University Press, Cambridge, U.K., 2006.
Novotny, J. L. and Kelleher, M. D.: Free-convection stagnation flow of an absorbing-emitting gas, *IJHMT*, 10(9), 1171–1178, 1967.
NREL/ASTM G173-03 Tables of Reference Solar Spectral Irradiance. National Renewable Energy Center Renewable Resource Data Center, On line at 2012. http://www.nrel.gov/rredc/ (accessed May 1, 2015).
Nunes, E. M., Modi, V., and Naraghi, M. H. N.: Radiative transfer in arbitrarily-shaped axisymmetric enclosures with anisotropic scattering media, *IJHMT*, 43(18), 3275–3285, 2000.
Nusselt, W.: Graphische bestimmung des winkelverhaltnisses bei der wärmestrahlung, *VDI Z.*, 72, 673, 1928.
O'Callahan, B. T., Lewis, W. E., Jones, A. C., and Raschke, M. B.: Spectral frustration and spatial coherence in thermal near-field spectroscopy, *Phys. Rev. B*, 89, 245446, 2014.
O'Neill, P., Ignatiev, A., and Doland, C.: The dependence of optical properties on the structural composition of solar absorbers: Gold black, *Solar Energy*, 21(6), 465–468, 1978.
Oden, T., Moser, R., and Ghattas, O.: Computer predictions with quantified uncertainty, Part I, *SIAM News, News J. Soc. Ind. Appl. Math.*, 43(9), 3 pp., November 2010a.

Oden, T., Moser, R., and Ghattas, O.: Computer predictions with quantified uncertainty, Part II, *SIAM News, News J. Soc. Ind. Appl. Math.*, 43(10), 4 pp., December 2010b.

Ody Sacadura, J. F.: Influence de la rugosité sur le rayonnement thermique émis par les surfaces opaques: Essai de modèle (Influence of surface roughness on the radiative heat emitted by opaque surfaces: A test model), *IJHMT*, 15(8), 1451–1465, 1972.

Oguma, M. and Howell, J. R.: Solution of two-dimensional blackbody inverse radiation by an inverse Monte Carlo method, *Proceedings of the Fourth ASME/JSME Joint Symposium*, Maui, HI, March 1995.

Ogut, E., Mengüç, M. P., and Sendur. K.: Integrating magnetic heads with plasmonic nanostructures in multilayer configurations, *IEEE Trans. Magnet.*, 49(7), 3687–3690, July 2013.

Ohlsen, P. E. and Etamad, G. A.: Spectral and total radiation data of various aircraft materials, Report NA57-330, North American Aviation, Los Angeles, CA, July 23, 1957.

Okamoto, Y.: Temperature distribution and efficiency of a single sheet of radiative and convective fin accompanied by internal heat source, *Bull. JSME*, 7(28), 751–758, 1964.

Okamoto, Y.: Temperature distribution and efficiency of a plate and annular fin with constant thickness, *Bull. JSME*, 9(33), 143–150, 1966a.

Okamoto, Y.: Thermal performance of radiative and convective plate-fins with mutual irradiation, *Bull. JSME*, 9(33), 150–165, 1966b.

Oks, E.: *Stark Broadening of Hydrogen Hydrogenlike Spectral Lines in Plasmas: The Physical Insight*, 1st edn., Alpha Science International, Oxford, U.K., 2006.

Olson, G. L., Auer, L. H., and Hall, M. L.: Diffusion, *P*1, and other approximate forms of radiation transport, *JQSRT*, 64, 619–634, 2000.

Omori, T., Nagata, T., Taniguchi, H., and Kudo, K.: Three-dimensional heat transfer analysis of a steel heating furnace, in R. L. Lewis, J. H. Chin, and G. M. Homsy (eds.), *Proceedings of the Seventh International Conference on Numerical Methods in Thermal Problems*, vol. VII, pt. 2, pp. 1346–1356, Pineridge Press, Swansea, U.K., 1991.

Oppenheim, A. K.: Radiation analysis by the network method, *Trans. ASME*, 78(4), 725–735, 1956.

Orlande, H. R. B., Wellele, O., Ruperti, N. Jr., Colaço, M. J., and Delmas, A.: Identification of the thermophysical properties of some transparent materials, Paper RAD-20, *Proceedings of the 13th International Heat Transfer Conference*, Sydney, New South Wales, Australia, August 2006.

Orlova, N. S.: Photometric relief of the lunar surface, *Astron. Z.*, 33(1), 93–100, 1956.

Otey, C. and Fan, S.: Numerically exact calculations of electromagnetic heat transfer between a dielectric sphere and a plate, *Phys. Rev. B*, 84, 245431, 2011.

Otey, C. R., Zhu, L., Sandhu, S., and Fan, S.: Fluctuational electrodynamics calculations of near-field heat ttransfer in non-planar geometries: A brief overview, *JQSRT*, Special Issue on *Micro- and Nano-scale Radiative Transfer*, 132, 3–11, 2014.

Ottens, R. S., Quetschke, V., Wise, S., Alemi, A. A., Lundock, R., Mueller, G., Reitze, D. H., Tanner, D. B., and Whiting, B. F.: Near-field radiative heat transfer between macroscopic planar surfaces, *Phys. Rev. Lett.*, PRL107, 014301, 2011.

Ouyang, X. and Varghese, P. L.: Reliable and efficient program for fitting Galatry and Voigt profiles to spectral data on multiple lines, *Appl. Opt.*, 28(8), 1538–1545, 1989.

Özişik, M. N.: *Radiative Transfer, and Interactions with Conduction and Convection*, John Wiley & Sons, New York, 1973.

Özişik, M. N. and Orlande, H. R. B.: *Inverse Heat Transfer: Fundamentals and Applications*, Taylor & Francis, New York, 2000.

Özişik, M. N. and Shouman, S. M.: Radiative transfer in an isotropically scattering two-region slab with reflecting boundaries, *JQSRT*, 26, 1–9, 1981.

Pal, G. and Modest, M. F.: A narrow-band based multiscale multi-group full-spectrum *k*-distribution method for radiative transfer in nonhomogeneous gas-soot mixtures, *ASME-JHT*, 132(2), article 023307, 2010.

Pal, G., Modest, M. F., and Wang, L.: Hybrid full-spectrum correlated *k*-distribution method for radiative transfer in nonhomogeneous gas mixtures, *JHT*, 130(8), 082701-1–082701-8, August 2008.

Palik, E. D. (ed.): *Handbook of Optical Constants of Solids*, vols. I–IV, Elsevier, New York, 1998.

Palik, E. D., Ginsburg, N., Rosenstock, H. B., and Holm, R. T.: Transmittance and reflectance of a thin absorbing film on a thick substrate, *Appl. Opt.*, 17(21), 3345–3347, 1978.

Pandey, D. K.: Combined conduction and radiation heat transfer in concentric cylindrical media, *JTHT*, 3(1), 75–82, 1989.

Pantokratoras, A.: Natural convection along a vertical iisothermal plate with linear and non-linear Rosseland thermal radiation, *Int. J. Thermal Sci.*, 84(1), 151–157, 2014.

Park, K., Basu, S., King, W. P., and Zhang, Z. M.: Performance analysis of near-field thermophotovoltaic devices considering absorption distribution, *JQSRT*, 109, 305–316, 2008.
Park, S. H. and Kim, S. S.: Thermophoretic deposition of absorbing, emitting and isotropically scattering particles in laminar tube flow with high particle mass loading, *IJHMT*, 36(14), 3477–3485, 1993.
Park, W.-H. and Kim, T.-K.: Development of the WSGGM using a gray gas regrouping technique for the radiative solutions within a 3-D enclosure filled with nonuniform gas mixtures, *JSME Int. J., Ser. B*, 48(2), 310–315, 2005.
Parker, W. J. and Abbott, G. L.: Theoretical and experimental studies of the total emittance of metals, *Symposium on Thermal Radiation Solids*, pp. 11–28, NASA SP-55, San Francisco, CA, 1964.
Parretta, A., Sarno, A., Tortora, P., Yakubu, H., Maddalena, P., Zhao, J., and Wang, A.: Angle-dependent reflectance measurements on photovoltaic and solar cells, *Optics Comms*., 172, 139–151, 1999.
Patankar, S. V.: *Numerical Heat Transfer and Fluid Flow*, Hemisphere, Washington, DC, 1980.
Patankar, S. V. and Spalding, D. B.: A calculation procedure for heat, mass, and momentum transfer in three-dimensional parabolic flows, *IJHMT*, 15, 1787–1806, 1972.
Patch, R. W.: *Approximation for Radiant Energy Transport in Nongray, Nonscattering Gases*, NASA TN D-4001, Washington, DC, 1967a.
Patch, R. W.: Effective absorption coefficients for radiant energy transport in nongrey, nonscattering gases, *JQSRT*, 7(4), 611–637, 1967b.
Pearce, B. L. and Emery, A. F.: Heat transfer by thermal radiation and laminar forced convection to an absorbing fluid in the entry region of a pipe, *JHT*, 92(2), 221–230, 1970.
Pearson, J. T., Webb, B. W., Solovjov, V. P., and Ma, J.: Updated correlation of the absorption line blackbody distribution function for H_2O based on the HITEMP2010 database, *JQSRT*, 128, 10–17, 2013.
Pearson, J. T., Webb, B. W., Solovjov, V. P., and Ma, J.: Efficient representation of the absorption line blackbody distribution function for H_2O, CO_2, and CO at variable temperature, mole fraction, and total pressure, *JQSRT*, 138, 82–96, 2014a.
Pearson, J. T., Webb, B. W., Solovjov, V. P., and Ma, J.: Effect of total pressure on the absorption line blackbody distribution function and radiative transfer in H_2O, CO_2, and CO, *JQSRT*, 143, 100–110, August 2014b.
Pelevin, V. N. and Rostovtseva, V. V.: Modelling of bio-optical parameters of open ocean waters, *Oceanologia*, 143(4), 469–477 2001.
Pendry, J. B.: Radiative exchange of heat between nanostructures, *J. Phys. Condens. Matter*, 11, 6621–6633, 1999.
Penndorf, R. B.: Scattering and extinction coefficients for small absorbing and nonabsorbing aerosols, *JOSA*, 52(8), 896–904, 1962.
Penner, S. S.: *Quantitative Molecular Spectroscopy and Gas Emissivities*, Addison-Wesley, Reading, MA, 1959.
Penner, S. S. and Olfe, D. B.: *Radiation and Reentry*, Academic Press, New York, 1968.
Pérez-Madrid, A., Lapas, L. C., and Rubi, J. M.: Heat exchange between two interacting nanoparticles beyond the fluctuation-dissipation regime, *Phys. Rev. Lett.*, 103, 048301, 2009.
Perlmutter, M. and Howell, J. R.: A strongly directional emitting and absorbing surface, *JHT*, 85(3), 282–283, 1963.
Perlmutter, M. and Howell, J. R.: Radiant transfer through a gray gas between concentric cylinders using Monte Carlo, *JHT*, 86(2), 169–179, 1964.
Perlmutter, M. and Siegel, R.: Heat transfer by combined forced convection and thermal radiation in a heated tube, *JHT*, 84(4), 301–311, 1962.
Perlmutter, M. and Siegel, R.: Effect of specularly reflecting gray surface on thermal radiation through a tube and from its heated wall, *JHT*, 85(1), 55–62, 1963.
Pessoa-Filho, J. B. and Thynell, S. T.: Approximate solution to radiative transfer in two-dimensional cylindrical media, *JTHT*, 10(3), 452–459, 1996.
Peterson, A. F., Ray, S. L., and Mittra, R.: *Computational Methods for Electromagnetics*, Wiley-IEEE Press, New York, 1997.
Petersen, S. J., Basu, S., and Francoeur, M.: Near-field thermal emission from metamaterials, *Photon. Nanostruct. Fundam. Appl.*, 11, 167–181, 2013.
Petrov, D., Synelnyk, E., Shkuratov, Y., and Videen, G.: The T-matrix technique for calculations of scattering properties of ensembles of randomly oriented particles with different size, *JQSRT*, 102, 85–110, 2006.
Petrov, V. A.: Combined radiation and conduction heat transfer in high temperature fiber thermal insulation, *IJHMT*, 40(9), 2241–2247, 1997.
Petty, G. W.: *A First Course in Atmospheric Radiation*, 2nd edn., Sundog Publishing, Madison, WI, 2006.

Phelan, P. E., Chen, G., and Tien, C. L.: Thickness-dependent radiative properties of Y-Ba-Cu-O thin films, *JHT*, 114(1), 227–233, 1992.

Phelan, P. E., Flik, M. I., and Tien, C. L.: Radiative properties of superconducting Y-Ba-Cu-O thin films, *JHT*, 113(2), 487–493, 1991.

Phillips, D. L.: A Treatment for the numerical solution of certain integral equations of the first kind, *J. Assoc. Comput. Mach.*, 9, 84–97, 1962.

Pilon, L. and Viskanta, R.: Apparent radiation characteristics of semitransparent media containing gas bubbles, *Proceedings of the 12th International Heat Transfer Conference*, Grenoble, France, 2002.

Pilon, L. and Viskanta, R.: Radiation characteristics of glass containing gas bubbles, *J. Am. Ceramic Soc.*, 86(8), 1313–1320, 2003.

Pincus, R. and Evans, F. E.: Computational cost and accuracy in calculating three-dimensional radiative transfer: Results for new implementations of Monte Carlo and SHDOM, *J. Atmos. Sci.*, 66(10), 3131–3146, October 2009.

Pinkley, L. W., Sethna, P. P., and Williams, D.: Optical constants of water in the infrared: Influence of temperature, *JOSA*, 67(4), 494–499, 1977.

Pivovonsky, M. and Nagel, M. R.: *Tables of Blackbody Radiation Functions*, Macmillan, New York, 1961.

Plamondon, J. A. and Horton, T. E.: On the determination of the view function to the images of a surface in a nonplanar specular reflector, *IJHMT*, 10(5), 665–679, 1967.

Plamondon, J. A. and Landram, C. S.: Radiant heat transfer from nongray surfaces with external radiation. Thermophysics and temperature control of spacecraft and entry vehicles, *Prog. Astronaut. Aeronaut.*, 18, 173–197, 1966.

Planck, M.: Distribution of energy in the spectrum, *Ann. Phys.*, 4(3), 553–563, 1901.

Plass, G. N.: Mie scattering and absorption cross sections for absorbing particles, *Appl. Opt.*, 5(2), 279–285, 1966.

Polder, D. and Van Hove, M.: Theory of radiative heat transfer between closely spaced bodies, *Phys. Rev. B.*, 4(10), 3303–3314, 1971.

Polgar, L. G. and Howell, J. R.: *Directional Thermal-Radiative Properties of Conical Cavities*, NASA TN D-2904, Washington, DC, 1965.

Polgar, L. G. and Howell, J. R.: Directional radiative characteristics of conical cavities and their relation to lunar phenomena, in G. B. Heller (ed.), *Thermophysics and Temperature Control of Spacecraft and Entry Vehicles*, pp. 311–323, Academic, New York, 1966.

Poljak, G.: Analysis of heat interchange by radiation between diffuse surfaces, *Tech. Phys. USSR* 1(5–6), 555–590, 1935.

Pomraning, G. C.: Flux limiters and Eddington factors, *JQSRT*, 27(5), 517–530, 1982.

Porter, J., Larsen, M., Barnes, J., and Howell, J.: Optimization of discrete heater arrays in radiant furnaces, *JHT*, 128(10), 1031–1040, October 2006.

Porter, J. M., Larsen, M. E., and Howell, J. R.: Discrete optimization of radiant heaters with simulated annealing, *Proceedings of the ASME Summer Heat Transfer Conference* (on CD), San Francisco, CA, July 2005.

Porteus, J. O.: Relation between the height distribution of a rough surface and the reflectance at normal incidence, *JOSA*, 53(12), 1394–1402, 1963.

Postlethwait, M. A., Sikka, K. K., Modest, M. F., and Hellmann, J. R.: High-temperature, normal spectral emittance of silicon carbide based materials, *JTHT*, 8(3), 412–418, 1994.

Pourshaghaghy, A., Pooladvand, K., Kowary, F., and Karimi-Zand, K.: An inverse radiation boundary design problem for an enclosure filled with an emitting, absorbing and scattering media, *Int. Commun. Heat Mass Transfer*, 33, 381–390, 2006.

Prahl, S.: Mie scattering calculator, http://omlc.ogi.edu/calc/mie_calc.html (accessed April 16, 2015). Oregon Medical Laser Center, Portland, OR, 2009.

Preface to the HITRAN 2012 special issue, I. E. Gordon, M. Rotger, and J. Tennyson (eds.) *JQSRT*, 130, 1–3, 2013.

Price, D. J.: The emissivity of hot metals in the infra-red, *Proc. Phys. Soc. Lond., Ser. A*, 59(1), 118–131, 1947.

Purcell, E. M. and Pennypacker, C. R.: Scattering and absorption of light by nonspherical dielectric grains, *Astrophys. J.*, 186, 705–714, 1973.

Qiao, H., Ren, Y., and Zhang, B.: Approximate solution of a class of radiative heat transfer problems, *JHT*, 122(3), 606–612, 2000.

Qiu, T. Q. and Tien, C. L.: Short-pulse laser heating on metals, *IJHMT*, 35(3), 719–726, 1992.

Qu, Y., Howell, J. R., and Ezekoye, O. A.: Monte Carlo modeling of a light-pipe radiation thermometer, *IEEE Trans. Semicond. Manuf.*, 20(1), 39–50, 2007a.

Qu, Y., Puttitwong, E., Howell, J. R., and Ezekoye, O. A.: Errors associated with light-pipe radiation thermometer temperature measurements, *IEEE Trans. Semicond. Manuf.*, 20(1), 26–38, 2007b.

Rabl, A.: Radiation transfer through specular passages—A simple approximation, *IJHMT*, 20(4), 323–330, 1977.

Rabl, A. and Nielsen, C. E.: Solar ponds for space heating, *Solar Energy*, 17, 1–12, 1975.

Radzevicius, S. J. and Daniels, J. J.: Ground penetrating radar polarization and scattering from cylinders, *J. App. Geophys.*, 45(2), 111–125, September 2000.

Raether, H.: *Surface Plasmons on Smooth and Rough Surfaces and on Gratings*, Springer-Verlag, Berlin, Germany, 1988.

Raithby, G. D. and Chui, E. H.: A finite-volume method for predicting a radiant heat transfer in enclosures with participating media, *JHT*, 112(2), 415–423, 1990.

Ramesh, N. and Venkateshan, S. P.: Effect of surface radiation on natural convection in a square enclosure, *JTHT*, 13(3), 299–301, 1999.

Randrianalisoa, J. and Baillis, D.: Radiative properties of densely packed spheres in semitransparent media: A new geometric optics approach, *JQSRT*, 111(10), 1372–1388, 2010.

Randrianalisoa, J. H., Dombrovsky, L. A., Lipiński, W., and Timchenko, V.: Effects of short-pulsed laser radiation on transient heating of superficial human tissues, *IJHMT,* 78, 488–497, November 2014.

Rao, V. R. and Sastri, V. M. K.: Efficient evaluation of diffuse view factors for radiation, *IJHMT*, 39(6), 1281–1286, 1996.

Ratzel, A. and Howell, J. R.: Two-dimensional energy transfer in radiatively participating media with conduction by the P-N approximation, *Proceedings of the 1982 International Heat Transfer Conference*, vol. 2, pp. 535–540, Munich, Germnay, September 1982.

Ratzel, A. C.: P-N differential approximation for solution of one- and two-dimensional radiation and conduction energy transfer in gray participating media, PhD dissertation, Department of Mechanical Engineering, University of Texas, Austin, TX, 1981.

Rayleigh, Lord: On scattering of light by small particles, *Philos. Mag.*, 41, 447–454, 1871.

Rayleigh, Lord: On the incidence of electric and aerial waves upon small obstacles in the form of ellipsoids or elliptic cylinders and on the passage of electric waves through a circular aperture in a conducting screen, *Phil. Mag.*, 44, 28–52, 1897.

Rayleigh, Lord: The law of complete radiation, *Phil. Mag*., 49, 539–540, 1900.

Razzaque, M. M., Howell, J. R., and Klein, D. E.: Finite element solution of heat transfer for a gas flow through a tube, *AIAA J.*, 20(7), 1015–1019, 1982.

Razzaque, M. M., Howell, J. R., and Klein, D. E.: Coupled radiative and conductive heat transfer in a two-dimensional rectangular enclosure with gray participating media using finite elements, *JHT*, 106(3), 613–619, 1984.

Razzaque, M. M., Klein, D. E., and Howell, J. R.: Finite element solution of radiative heat transfer in a two-dimensional rectangular enclosure with gray participating media, *JHT*, 105(4), 933–934, 1983.

Reed, B., Biaglow, J., and Schneider, S.: *Advanced Materials for Radiation-Cooled Rockets*, vol. II, pp. 115–118, NASA Propulsion Engineering Research Center, Annual Report 199, NASA Lewis Research Center, N94-28052, Washington, DC, September 1993.

Reguigui, N. M. and Dougherty, R. L.: Two-dimensional radiative transfer in a cylindrical layered medium with reflective interfaces, *JTHT*, 6(2), 232–241, 1992.

Rephaeli, E., Raman, A., and Fan, S.: Ultrabroadband photonic structures to achieve high-performance daytime radiative cooling, *Nanoletters*, 13, 1457–1461, March 5, 2013.

Rhyming, I. L.: Radiative transfer between two concentric spheres separated by an absorbing-emitting gas, *IJHMT*, 9, 315–324, 1966.

Richmond, J. C.: Relation of emittance to other optical properties, *J. Res. Natl. Bureau Stand.*, 67C(3), 217–226, 1963.

Ripoll, J.-F. and Wray, A. A.: *Macroscopic Models of Radiative Transfer as Applied to Computation of the Radiation Field in the Solar Atmosphere*, Center for Turbulence Research, Annual Research Briefs, Stanford University, Stanford, CA, 2003.

Rivière, P., Langlois, S., Soufiani, A., and Taine, J.: An approximate data base of H_2O infrared lines for high temperature applications at low resolution: Statistical narrow band parameters, *JQSRT*, 53, 221–234, 1995.

Rivière, P. and Soufiani, A.: Updated band model parameters for H_2O, CO_2, CH_4 and CO radiation at high temperature, *IJHMT*, 55(13–14), 3349–3358, 2012.

Rizk, N. K. and Mongia, H. C.: Three-dimensional analysis of gas turbine combustors, *J. Propul. Power*, 7(3), 445–451, 1991.

Robin, L., Delmas, A., Lanternier, T., Oelhoffen, F., and Ducamp, V.: Experimental and theoretical determination of spectral heat flux emitted by a ceramic under axial temperature gradient, Paper RAD-10, *Proceedings of the 13th International Heat Transfer Conference*, Sydney, New South Wales, Australia, August 2006.

Rodigheiro, C. and de Socio, L. M.: Some aspects of natural convection in a corner, *JHT*, 105(1), 212–214, 1983.

Rodriguez, A. W., Ilic, O., Bermel, P., Celanovic, I., Joannopoulos, J. D., Soljačić, M., and Johnson, S. G.: Frequency-selective near-field radiative heat transfer between photonic crystal slabs: A computational approach for arbitrary geometries and materials, *Phys. Rev. Lett.*, 107, 114302, 2011.

Rodriguez, A. W., Reid, M. T. H., and Johnson, S. G.: Fluctuating-surface-current formulation of radiative heat transfer: Theory and applications, *Phys. Rev. B*, 88, 054305, 2013.

Roger, M., Caliot, C., Crouseilles, N., and Coelho, P. J.: A hybrid transport-diffusion model for radiative transfer in absorbing and scattering media, *J. Comp. Phys.*, 275, 346–362, 2014.

Rokhsaz, F. and Dougherty, R. L.: Radiative transfer within a finite plane-parallel medium exhibiting Fresnel reflection at a boundary, in R. K. Shah, (ed.), *Heat Transfer Phenomena in Radiation, Combustion and Fires*, New York, ASME HTD-vol. 106, pp. 1–8, 1989.

Rolling, R. E. and Tien, C. L.: Radiant heat transfer for nongray metallic surfaces at low temperatures, Paper No. 67–335, *Proc. AIAA Thermophsics Specialist Conf.*, New Orleans, LA, April 1967.

Rose, M. F., Adair, P., and Schroeder, K.: Selective emitters for thermophotovoltaic power systems for use in aerospace applications, *J. Propul. Power*, 12(1), 83–88, 1996.

Rosseland, S.: *Theoretical Astrophysics: Atomic Theory and the Analysis of Stellar Atmospheres and Envelopes*, Clarendon Press, Oxford, U.K., 1936.

Rother, T.: *Electromagnetic Wave Scattering on Nonspherical Particles, Basic Methodology and Simulations*, Springer Series in Optical Sciences, Heidelberg, Germany, 2009.

Rothman, L. S. Editorial, *HITRAN Newslett.*, 5(4), 1–4, 1996.

Rothman, L. S., Gordon, I. E., Barber, R. J., Dothe, H., Gamache, R. R., Goldman, A., Perevalov, V., Tashkun, S. A., and Tennyson, J.: HITEMP, the high-temperature molecular spectroscopic database, *JQSRT*, 111(15), 2139–2150, 2010.

Rothman, L. S., Gordon, I. E., Barbe, A. et al., The HITRAN 2012 molecular spectroscopic database, *JQSRT* 130, 4–50, 2013. (For most recent version (2012) see http://www.cfa.harvard.edu/hitran//. Accessed April 27, 2015).

Rothman, L. S., Jacquemart, D., Barbe, A. et al.: The HITRAN 2004 molecular spectroscopic database, *JQSRT*, 96, 139–204, 2005.

Rousseau, E., Siria, A., Jourdan, G., Volz, S., Comin, F., Chevrier, J., and Greffet, J.-J.: Radiative heat transfer at the nanoscale, *Nat. Photon.*, 3(9), 514–517, 2009. doi: 10.1038/NPHOTON.2009.144.

Roux, J. A., Smith, A. M., and Shahrokhi, F.: Effect of boundary conditions on the radiative reflectance of dielectric coatings, in R. G. Hering (ed.), *Thermophysics and Spacecraft Thermal Control, Progress in Astronautics and Aeronautics,* vol. 35, pp. 131–144, MIT Press, Cambridge, MA, 1974.

Rukolaine, S. A.: Shape gradient of the least squares objective functional in optimal shape design radiative transfer problems, radiative transfer V, *Proceedings of the Fifth International Conference on Radiative Transfer*, Bodrum, Turkey, June 2007.

Rukolaine, S. A.: Shape optimization of radiant enclosures with specular-diffuse surfaces by means of a random search and gradient minimization, *JQSRT*, 151, 174–191, January 2015.

Rukolaine, S. A. and Yuferev, V. S.: Discrete ordinatates quadrature schemes based on the angular interpolation of radiation intensity, *JQSRT*, 69, 257–275, 2001.

Rulko, R. P. and Larsen, E. W.: Variational derivation and numerical analysis of P2 theory in planar geometry, *Nuclear Sci. Eng.*, 114, 271–285, 1993.

Ruperti, N. J., Jr., Raynaud, M., and Sacadura, J. F.: A method for the solution of the coupled inverse heat conduction-radiation problem, *JHT*, 118(1), 10–17, 1996.

Rushmeier, H. E., Baum, D. R., and Hall, D. E.: Accelerating the hemi-cube algorithm for calculating radiation form factors, *JHT*, 113(4), 1044–1047, 1991.

Russell, L. D., and Chapman, A. J.: Analytical solution of the "known-heat-load" space radiator problem, *J. Spacecraft Rockets*, 4(3), 311–315, 1967.

Rytov, S. M.: *Theory of Electric Fluctuations and Thermal Radiation*, Air Force Cambridge Research Center, Bedford, MA, 1959.

Rytov, S. M., Kravtsov, Y. A., and Tatarskii, V. I.: *Principles of Statistical Radiophysics 3: Elements of Random Fields*, Springer, Berlin, Germany, 1989.

Saari, J. M. and Shorthill, R. W.: Review of lunar infrared observations, in S. F. Singer (ed.), *Physics of the Moon*, AAS Science and Technology Series, 13, 57–99, 1967.

Saatdjian, E.: A cell model that estimates radiative heat transfer in a particle-laden flow, *JHT*, 109, pp. 256–258, 1987.

Sacadura. J.-F.: Thermal radiative properties of complex media: Theoretical predictions versus experimental identification, *Heat Transfer Engineering*, 32(9), 754, 2011.

Sacadura, J.-F. and Baillis, D.: Experimental characterization of thermal radiation properties of dispersed media, *Int. J. Therm. Sci*, 41, 699–707, 2002.

Sadooghi, P.: Transient thermal effects of radiant energy in semitransparent materials, *JTHT*, 20(3), 607–611, 2006.

Sadykov, B. S.: Temperature dependence of the radiating power of metals, *High Temp.*, 3(3), 352–356, 1965.

Safavinejad, A., Mansouri, S. H., and Hosseini Sarvari, M.: Inverse boundary design of 2-D radiant enclosures with absorbing-emitting media using micro-genetic algorithm, *IASME Trans.*, 2(8), 1558–1567, 2005.

Safavinejad, A., Mansouri, S. H., Sakurai, A., and Maruyama, S.: Optimal number and location of heater in 2-d radiant enclosures composed of specular and diffuse surfaces using micro-genetic algorithm, *Appl. Thermal Eng.*, 29(5–6), 1075–1085, 2009.

Sakami, M., Mitra, K. P., and Hsu, F.: Analysis of light pulse transport through two-dimensional scattering and absorbing media, *JQSRT*, 73, pp. 169–179, 2002.

Sakurai, A., Mishra, S. C., and Maruyama, S.: Radiation element method coupled with the lattice Boltzmann method applied to the analysis of transient conduction and radiation heat transfer problem with heat generation in a participating medium, *Numer. Heat Transfer A Appl.*, 57(5), 346–368, 2010.

Saltiel, C., Chen, Q., Manickavasagam, S., Schandler, L. S., Siegel, R. W., and Mengüç, M. P.: Identification of dispersion behavior of surface-treated nano-scale powders, *J. Nanoparticle Res.*, 6, 35–46, 2004.

Saltiel, C., Manickavasagam, S., Mengüç, M. P., and Andrews, R.: Light scattering and dispersion behavior of multi-walled carbon nanotubes, *J. Opt. Soc. Am.-A*, 22(8), 1546–1554, 2005.

Saltiel, C. J. and Kolibal, J.: Adaptive grid generation for the calculation of radiative configuration factors, *JTHT*, 7(1), 175–178, 1993.

Sambegoro, P. L., Near-field radiation in nanoscale gaps, Thesis, Mechanical Engineering, MIT, Cambridge, MA, 2011.

Sampson, D. H.: Choice of an appropriate mean absorption coefficient for use in the general grey gas equations, *JQSRT*, 5(1), 211–225, 1965.

Sànchez, A. and Smith, T. F.: Surface radiation exchange for two-dimensional rectangular enclosures using the discrete-ordinates method, *JHT*, 114(2), 465–472, 1992.

Santoro, R. J., Semerjian, H. G., and Dobbins, R. A.: Soot particle measurements in diffusion flames, *Combust. Flame*, 51, 203–218, 1983.

Santoro, R. J., Yeh, T. T., Horvath, J. J., and Semerjian, H. G.: The transport and growth of soot particles in laminar diffusion flames, *Combust. Sci. Technol.*, 53, 89–115, 1987.

Sarofim, A. F. and Hottel, H. C.: Radiative exchange among non-Lambert surfaces, *JHT*, 88(1), 37–44, 1966.

Sarofim, A. F. and Hottel, H. C.: Radiative transfer in combustion chambers: Influence of alternative fuels, *Sixth International Heat Transfer Conference*, vol. 6, pp. 199–217, Toronto, Onatrio, Canada, August 1978.

Sasse, C., Koenigsdorff, R., and Frank, S.: Evaluation of an improved hybrid six-flux/zone model for radiative transfer in rectangular enclosures, *IJHMT*, 38(18), 3423–3431, 1995.

Sato, T., Kunitomo, T., Yoshi, S., and Hashimoto, T.: On the monochromatic distribution of the radiation from the luminous flame, *Bull. Jpn. Soc. Mech. Eng.*, 12, 1135–1143, 1969.

Saunders, O. A.: Notes on some radiation heat transfer formulae, *Proc. Phys. Soc. Lond.*, 41, 569–575, 1929.

Schimmel, W. P., Novotny, J. L., and Olsofka, F. A.: Interferometric study of radiation-conduction interaction, *Fourth International Heat Transfer Conference*, Paris, France, September1970.

Schlichting, H. (J. Kestin, trans.): *Boundary Layer Theory*, 4th edn., p. 116, McGraw-Hill, New York, 1960.

Schmehl, R.: The coupled-dipole method for light scattering from particles on plane surfaces, MS thesis, Arizona State University, Tempe, AZ, 1994.

Schmehl, R., Nebeker, B. M., and Hirleman, E. D.: Discrete-dipole approximation for scattering by features on surfaces by means of a two-dimensional fast Fourier transform technique, *JOSA A*, 14(11), 3026–3036, 1997.

Schmid-Burgk, J.: Radiant heat flow through cylindrically symmetric media, *JQSRT*, 14, 979–987, 1974.

Schmidt, E. and Eckert, E. R. G.: Über die richtungsverteilung der wärmestrahlung von oberflächen, *Forsch. Geb. Ingenieurwes.*, 6(4), 175–183, 1935.

Schnurr, N. M., Shapiro, A. B., and Townsend, M. A.: Optimization of radiating fin arrays with respect to weight, *JHT*, 98(4), 643–648, 1976.

Schoegl, I. and Ellzey, J.: Superadiabatic combustion in conducting tubes and heat exchangers of finite length, *Combust. Flame*, 151(1–2), 142–159, October 2007.

Scholand, E. and Schenkel, P.: A solution of mean beam lengths of radiating gases in rectangular parallelepiped enclosures, *Proceedings of the Eighth International Heat Transfer Conference*, vol. 2, pp. 763–768, Hemisphere, Washington, DC, 1986.

Schornhorst, J. R. and Viskanta, R.: An experimental examination of the validity of the commonly used methods of radiant heat transfer analysis, *JHT*, 90(4), 429–436, 1968.

Schreider, Yu. A. (ed.): *Method of Statistical Testing—Monte Carlo Method*, American Elsevier, New York, 1964.

Schuerman, D. W. (ed.): *Light Scattering by Irregularly Shaped Particles*, Plenum Press, State University of New York at Albany, Albany, NY, 1979.

Schultz, W. and Bejan, A.: Exergy conservation in parallel thermal insulation systems, *IJHMT*, 26(3), 335–340, 1983.

Schuster, A.: Radiation through a foggy atmosphere, *Astrophys. J.*, 21, 1–22, 1905.

Schwander, D., Flamant, G., and Olalde, G.: Effects of boundary properties on transient temperature distributions in condensed semitransparent media, *IJHMT*, 33(8), 1685–1695, 1990.

Schwarzschild, K.: Equilibrium of the sun's atmosphere, *Ges. Wiss. Gottingen Nachr., Math-Phys. Klasse*, 1, 41–53, 1906.

Scutaru, D., Rosenmann, L., Taine, J., Wattson, R. B., and Rothman, L. S.: Measurements and calculations of CO_2 absorption at high temperature in the 4.3 and 2.7 μm regions, *JQSRT*, 50(2), 179–191, 1993.

Seban, R. A.: *Thermal Radiation Properties of Materials*, pt. III, WADD-TR-60-370, University of California, Berkeley, CA, August 1963.

Seki, N., Sugawara, M., and Fukusako, S.: Back-melting of a horizontal cloudy ice layer with radiative heating, *JHT*, 101(1), 90–95, 1979.

Selamet, A.: Visible and infra-red sensitivity of Rayleigh limit and Penndorf extension to complex refractive index of soot, *IJHMT*, 35(12), 3479–3484, 1992.

Selamet, A. and Arpaci, V. S.: Rayleigh limit—Penndorf extension, *IJHMT*, 32(110), 1809–1820, 1989.

Selçuk, N.: Evaluation of flux models for radiative transfer in cylindrical furnaces, *IJHMT*, 32(3), 620–624, 1989.

Selçuk, N. and Kayakol, N.: Evaluation of discrete ordinates method for radiative transfer in rectangular furnaces, *IJHMT*, 40(2), 213–222, 1997.

Selçuk, N. and Kayakol, N.: An improvement in the prediction of source terms in the discrete transfer method, in J. S. Lee (ed.), *Heat Transfer 1998, Proceedings of the 11th International Heat Transfer Conference*, pp. 331–336, vol. 7, KyongJu, Korea, Taylor & Francis, New York, 1998.

Selçuk, N. and Siddall, R. G.: Two-flux spherical harmonic modelling of two-dimensional radiative transfer in furnaces, *IJHMT*, 19, 313–321, 1976.

Sendur, K., Koşar, A., and Mengüç, M. P., Localized radiative energy transfer from a plasmonic bow-tie nanoantenna to a magnetic thin-film stack, *J. Appl. Phys. A*, 103(3), 703–707, June 2011.

Sentenac, A. and Greffet, J.-J.: Design of surface microrelief with selective radiative properties, *IJHMT*, 37(4), 553–558, 1994.

Shafey, H. M., Abd El-Ghany, A. M., and Nassib, A. M.: An analysis of the combined conductive-radiative heat transfer between a surface and a gas fluidized bed at high temperature, *IJHMT*, 36(9), 2281–2292, 1993.

Shafey, H. M., Tsuboi, Y., Fuijita, M., Makino, T., and Kunitomo, T.: Experimental study on spectral reflective properties of a painted layer, *AIAA J.*, 20(12), 1747–1753, 1982.

Shaffer, L. H.: Wavelength-dependent (selective) processes for the utilization of solar energy, *J. Solar Energy Sci. Eng.*, 2(3–4), 21–26, 1958.

Shamsundar, N., Sparrow, E. M., and Heinisch, R. P.: Monte Carlo radiation solutions—Effect of energy partitioning and number of rays, *IJHMT*, 16(3), 690–694, 1973.

Shapiro, A. B.: FACET—A computer view factor computer code for axisymmetric, 2d planar, and 3d geometries with shadowing, UCID-19887, University of California, Lawrence Livermore National Laboratory, Livermore, CA, August 1983 (LLNL Methods Development Group, http://www.oecd-nea.org/tools/abstract/detail/nesc9578/, accessed April 27, 2015).

Shapiro, A. B.: Computer implementation, accuracy, and timing of radiation view factor algorithms, *JHT*, 107(3), 730–732, 1985.

Shaughnessy, B. M. and Newborough, M.: A new method for tracking radiative paths in Monte Carlo simulations, *JHT*, 120, 792–795, 1998.

Shen, J. and Tang, T. S., *Spectral and High-Order Methods with Applications*, Science Press, Beijing, People's Republic of China, 2006.

Shen, S., Mavrokefalos, A., Sambegoro, P., and Chen, G., Nanoscale thermal radiation between two gold surfaces, *Appl. Phys. Lett.*, 100(233114), 1–4, 2012.
Shen, S., Narayanaswamy, A., and Chen, G.: Surface phonon polaritons mediated energy transfer between nanoscale gaps, *Nano Lett.*, 9(8), 2909–2913, 2009.
Shih, T. M., Hayes, L. J., Minkowycz, W. J., Yang, K. T., and Aung, W.: Parallel computations in heat transfer, *Numer. Heat Transfer*, 9, 639–662, 1986a.
Shih, T. M., Hsu, I. C., and Cunnington, G. R.: Combined conduction and radiation with phase change in teflon slabs, in T. W. Tong and M. F. Modest (eds.), *Radiation in Energy Systems*, pp. 25–31, ASME, New York, 1986b.
Shimoji, S.: Local temperatures in semigray nondiffuse cones and V-grooves, *AIAA J.*, 15(3), 289–290, 1977.
Shlager, K. L. and Schneider, J. B.: A selective survey of the finite-difference time domain literature, *IEEE Antennas Propag. Mag.*, 37, 39–57, 1995.
Shokair, I. R. and Pomraning, G. C.: Boundary conditions for differential approximations, *JQSRT*, 25(4), 325–337, 1981.
Shokouhi, A. A., Payan, S., Shokouhi, A., and Hosseini Sarvari, S. M.: Inverse boundary design problem of turbulent forced convection between parallel plates with surface radiation exchange, *Heat Transfer Eng.*, 36, 488–497, 2015.
Short, M. R., Geffrin, J. M., Vaillon, R., Tortel, H., Lacroix, B., and Francoeur, M.: Evanescent wave scattering by particles on a surface: Validation of the discrete dipole approximation with surface interaction against microwave analog experiments, *JQSRT*, 146, 452–458, 2014.
Shouman, A. R.: An exact solution for the temperature distribution and radiant heat transfer along a constant cross sectional area fin with finite equivalent surrounding sink temperature, *Proceedings of the Ninth Midwestern Mechanics Conference*, pp. 175–186, Madison, WI, August 16–18, 1965.
Shouman, A. R.: *Nonlinear Heat Transfer and Temperature Distribution through Fins and Electric Filaments of Arbitrary Geometry with Temperature-Dependent Properties and Heat Generation*, NASA TN D-4257, Washington, DC, 1968.
Shurcliff, W. A.: Transmittance and reflection loss of multi-plate planar window of a solar-radiation collector: Formulas and tabulations of results for the case of $n = 1.5$, *Solar Energy,* 16, 149–154, 1974.
Shvarev, K. M., Baum, B. A., and Gel'd, P. V.: Optical properties of liquid silicon, *Sov. Phys. Solid State*, 16(11), 2111–2112, 1975.
Shvartsburg, A. M.: Error estimation in differential approximation to equation of transfer, *Zh. Prikl. Mekh. Tekh. Fiz.*(5), 9–13, 1976.
Siedow, N., Lochegnies, D., Béchet, F., Moreau, P., Wakatsuki, H., and Inoue, N.: Axisymmetric modeling of the thermal cooling, including radiation, of a circular glass disk, *IJHMT*, 2015, in press.
Siegel, R.: *Net Radiation Method for Enclosure Systems Involving Partially Transparent Walls*, NASA TN D-7384, Washington, DC, August 1973a.
Siegel, R.: Net radiation method for transmission through partially transparent plates, *Solar Energy*, 15, 273–276, 1973b.
Siegel, R.: *Radiative Behavior of a Gas Layer Seeded with Soot*, NASA TN D-8278, Washington, DC, July 1976.
Siegel, R.: Separation of variables solution for non-linear radiative cooling, *IJHMT*, 30(5), 959–965, 1987a.
Siegel, R.: Transient radiative cooling of a droplet-filled layer, *JHT*, 109(1), 159–164, 1987b.
Siegel, R.: Transient radiative cooling of a layer filled with solidifying drops, *JHT*, 109(4), 977–982, 1987c.
Siegel, R.: Transient radiative cooling of an absorbing and scattering cylinder—A separable solution, *JTHT*, 2(2), 110–117, 1988.
Siegel, R.: Solidification by radiative cooling of a cylindrical region filled with drops, *JTHT*, 3(3), 340–344, 1989a.
Siegel, R.: Some aspects of transient cooling of a radiating rectangular medium, *IJHMT*, 32(10), 1955–1966, 1989b.
Siegel, R.: Transient radiative cooling of an absorbing and scattering cylinder, *JHT*, 111(1), 199–203, 1989c.
Siegel, R.: Emittance bounds for transient radiative cooling of a scattering rectangular region, *JTHT*, 4(1), 106–114, 1990.
Siegel, R.: Analytical solution for boundary heat fluxes from a radiating rectangular medium, *JHT*, 113(1), 258–261, 1991a.
Siegel, R.: Transient cooling of a square region of radiating medium, *JTHT*, 5(4), 495–501, 1991b.
Siegel, R.: Boundary fluxes for spectral radiation from a uniform temperature rectangular medium, *JTHT*, 6(3), 543–546, 1992a.
Siegel, R.: Finite difference solution for transient cooling of a radiating-conducting semitransparent layer, *JTHT*, 6(1), 77–83, 1992b.

Siegel, R.: Relations for local radiative heat transfer between rectangular boundaries of an absorbing-emitting medium, *JHT*, 115(1), 272–276, 1993.

Siegel, R.: Internal radiation effects in zirconia thermal barrier coatings, *JTHT*, 10(4), 707–709, 1996.

Siegel, R.: Green's function to determine temperature distribution in a semitransparent thermal barrier coating, *JTHT*, 11(2), 315–318, 1997a.

Siegel, R.: Temperature distribution in a composite of opaque and semitransparent spectral layers, *JTHT*, 11(4), 533–539, 1997b.

Siegel, R.: Transient thermal effects of radiant energy in translucent materials, *JHT*, 120(1), 4–23, 1998.

Siegel, R.: Radiative exchange in a parallel-plate enclosure with translucent protective coatings on its walls, *IJHMT*, 42(1), 73–84, 1999a.

Siegel, R.: Transient thermal analysis for heating a translucent wall with opaque radiation barriers, *JTHT*, 13(3), 277–284, 1999b.

Siegel, R.: Transient thermal analysis of parallel translucent layers by using Green's functions, *JTHT*, 13(1), 10–17, 1999c.

Siegel, R. and Keshock, E. G.: *Wall Temperatures in a Tube with Forced Convection, Internal Radiation Exchange, and Axial Wall Heat Conduction*, NASA TN D-2116, Washington, DC, 1964.

Siegel, R. and Molls, F. B.: Finite difference solution for transient radiative cooling of a conducting semitransparent square region, *IJHMT*, 35(10), 2579–2592, 1992.

Siegel, R. and Perlmutter, M.: Convective and radiant heat transfer for flow of a transparent gas in a tube with a gray wall, *IJHMT*, 5, 639–660, 1962.

Siegel, R. and Spuckler, C. M.: Effect of index of refraction on radiation characteristics in a heated absorbing, emitting, and scattering layer, *JHT*, 114(3), 781–784, 1992.

Siegel, R. and Spuckler, C. M.: Refractive index effects on radiation in an absorbing, emitting, and scattering laminated layer, *JHT*, 115(1), 194–200, 1993a.

Siegel, R. and Spuckler, C. M.: Variable refractive index effects on radiation in semitransparent scattering multilayered regions, *JTHT*, 7(4), 624–630, October–December 1993b.

Siegel, R. and Spuckler, C. M.: Approximate solution methods for spectral radiative transfer in high refractive index layers, *IJHMT*, 37(Suppl. 1), 403–413, 1994a.

Siegel, R. and Spuckler, C. M.: Effects of refractive index and diffuse or specular boundaries on a radiating isothermal layer, *JHT*, 116, 787–790, 1994b.

Sievers, A. J.: Thermal radiation from metal surfaces, *JOSA*, 68(11), 1505–1516, 1978.

Siewert, C. E.: The F_N method for solving radiative transfer problems in plane geometry, *Astrophys. Space Sci.*, 58, 131–137, 1978.

Sika, J.: Evaluation of direct-exchange areas for a cylindrical enclosure, *JHT*, 113(4), 1040–1044, 1991.

Sillion, F. X., Arvo, J. R., Westin, S. H., and Greenberg, D. P.: A global illumination solution for general reflectance distributions, *Comput. Graph.*, 25(4), 187–196, 1991.

Simovski, C., Maslovski, S., Tretyakov, S., Nefedov, I., Kosulnikov, S., and Belov, P: Super-Planckian thermal emission from a hyperlens, *Photon. Nanostruct. Fundam. Appl.*, June 4, 2014.

Sin, E. H., Ong, C. K., and Tan, H. S.: Temperature dependence of interband optical absorption of silicon at 1152, 1064, 750, and 694 nm, *Phys. Status Solidi*, 85, 199–204, 1984.

Singh, B. P. and Kaviany, M.: Independent theory versus direct simulation of radiative heat transfer in packed beds, *IJHMT*, 34, 2869–2881, 1991.

Singham, S. B. and Bohren, C. F.: Light-scattering by an arbitrary particle—A physical reformulation of the coupled-dipole method, *Opt. Lett.*, 12(1), 10–12, 1987.

Singham, S. B. and Bohren, C. F.: Light-scattering by an arbitrary particle—The scattering-order formulation of the coupled-dipole method, *JOSA A*, 5(11), 1867–1872, 1988.

Singham, S. B. and Bohren, C. F.: Scattering of unpolarized and polarized light by particle aggregates of different size and fractal dimension, *Langmuir*, 9, 1431–1435, 1993.

Sipe, J. E.: New Green-function formalism for surface optics, *JOSA B*, 4(4), 481–489, 1987.

Sivathanu, Y. R. and Gore, J. P.: A discrete probability function method for the equation of radiative transfer, *JQSRT*, 49(3), 269–280, 1993.

Sivathanu, Y. R. and Gore, J. P.: A discrete probability function method for radiation in enclosures and comparison with the Monte Carlo method, in Y. Bayazitoglu et al. (eds.), *Radiative Heat Transfer: Current Research*, ASME HTD-vol. 276, pp. 213–218, ASME, New York, 1994.

Sivathanu, Y. R. and Gore, J. P.: Radiative heat transfer inside a cylindrical enclosure with nonparticipating media using a deterministic statistical method, in P. D. Jones and U. Grigull (ed.), *Proceedings of the ASME Heat Transfer Division*, ASME HTD-vol. 332, pp. 145–152, ASME, New York, 1996.

Sivathanu, Y. R. and Gore, J. P.: Effects of surface properties on radiative transfer in a cylindrical tube with a nonparticipating medium, *JHT*, 119(3), 495–501, 1997.
Sivathanu, Y. R., Gore, J. P., Janssen, J. M., and Senser, D. W.: A study of in situ specific absorption coefficients of soot particles in laminar flat flames, *JHT*, 115(3), 653–658, 1993.
Skocypec, R. D. and Buckius, R. O.: Total hemispherical emittances for CO_2 or H_2O including particulate scattering, *IJHMT*, 27(1), 1–13, 1984.
Skocypec, R. D., Walters, D. V., and Buckius, R. O.: Spectral emission measurements from planar mixtures of gas and particulates, *JHT*, 109, 151–158, 1987.
Slutz, S. A., Gauntt, R. O., Harms, G. A., Latham, T., Roman, W., and Rodgers, R. J.: Thermal radiation in gas core nuclear reactors for space propulsion, *J. Propul. Power*, 10(3), 419–424, 1994.
Smith, G.: Greening: From nanostructures to buildings, cities, and farms, 2010, http://spie.org/x43138.xml, accessed 26 April, 2015.
Smith, G. B.: Amplified radiative cooling via optimised combinations of aperture geometry and spectral emittance profiles of surfaces and the atmosphere, *Solar Energy Mater. Solar Cells*, 93, 1696–1701, 2009.
Smith, G. B., Gentle, A. R., and Edmonds, I.: Urban growth, albedo and global warming, *48th AuSES Annual Conference*, Canberra, Australian Capital Territory, Australia, December 1–3, 2010.
Smith, R. C. and Baker, K. S.: Optical properties of the clearest natural waters (200–800 nm), *Appl. Optics*, 20(2), 171–184, January 1981.
Smith, T. F., Al-Turki, A. M., Byun, K.-H., and Kim, T. K.: Radiative and conductive transfer for a gas/soot mixture between diffuse parallel plates, *JTHT*, 1(1), 50–55, 1987.
Smith, T. F. and Hering, R. G.: Comparison of bidirectional measurements and model for rough metallic surfaces, *Proceedings of the Fifth Symposium on Thermophysical Properties*, ASME, Boston, MA, 1970.
Snail, K. A.: Analytical solutions for the reflectivity of homogeneous and graded selective absorbers, *Solar Energy Mater.*, 12, 411–424, 1985
Snyder, W. C., Wan, Z., and Li, X.: Thermodynamic constraints on reflectance reciprocity and Kirchhoff's Law, *Appl. Opt.*, 37(16), 3464–3470, 1998.
Sohal, M. and Howell, J. R.: Determination of plate temperature in case of combined conduction, convection, and radiation heat exchange, *IJHMT*, 16, 2055–2066, 1973.
Sohal, M. and Howell, J. R.: Thermal modeling of a plate with coupled heat transfer modes, in R. G. Hering (ed.), *Thermophysics and Spacecraft Thermal Control*, vol. 35 of *Progress in Astronautics and Aeronautics*, MIT Press, Cambridge, MA, 1974.
Sohn, I., Bansal, A., and Levin, D. A.: Advanced radiation calculations of hypersonic reentry flows using efficient databasing schemes, *JTHT*, 24(3), 623–637, July/September 2010.
Sokman, C. N. and Razzaque, M. M.: Finite element analysis of conduction-radiation heat transfer in an absorbing-emitting and scattering medium contained in an enclosure with heat flux boundary conditions, in Y. Jaluria, V. P. Carey, W. A. Fiveland, and W. Yuen (eds.), *Radiation, Phase Change Heat Transfer, and Thermal Systems*, ASME HTD-vol. 81, pp. 17–23, ASME, New York, 1987.
Solovjov, V. P., Lemonnier, D., and Webb, B. W.: Extension of the exact SLW model to non-isothermal gaseous media, *JQSRT*, 143, 83–91, 2014.
Solovjov, V. P. and Webb, B. W.: SLW modeling of radiative transfer in multicomponent gas mixtures, *JQSRT*, 65, 655–672, 2000.
Solovjov, V. P. and Webb, B. W.: A local-spectrum correlated model for radiative transfer in non-uniform gas media, *JQSRT*, 73, 361–373, 2002.
Solovjov, V. P. and Webb, B. W.: The cumulative wavenumber method for modeling radiative transfer in gas mixtures with soot, *JQSRT*, 93, 273–287, 2005.
Solovjov, V. P. and Webb, B. W.: Multilayer modeling of radiative transfer by SLW and CW methods in nonisothermal gaseous medium, *JQSRT*, 109, 245–257, 2008.
Song, B. and Viskanta, R.: Deicing of solids using radiant heating, *JTHT*, 4(3), 311–317, 1990.
Song, M., Ball, K. S., and Bergman, T. L.: A model for radiative cooling of a semi-transparent molten glass jet, *JHT*, 120(4), 931–938, 1998.
Song, R., Ye, X., and Chen, X.: Reconstruction of scatterers with four different boundary conditions by T-matrix method, *Inverse Probl. Sci. Eng.*, 23(4), 601–616, June 2014. doi: 10.1080/17415977.2014.923418.
Song, T. H. and Viskanta, R.: Interaction of radiation with turbulence - application to a combustion system, *JTHT*, 1(1), 56–62, 1987. doi: 10.2514/3.7.
Sorensen, C. M.: Scattering and absorption of light by particles and aggregates, in K. S. Birdi (ed.), *Handbook of Surface and Colloid Chemistry*, pp. 533–558, CRC Press, Boca Raton, FL, 1997.
Sorensen, C. M.: Light scattering by fractal aggregates: A review, *Aerosol Sci. Technol.*, 35(2), 648–687, 2001.

Soufiani, A. and Taine, J.: Application of statistical narrow-band model to coupled radiation and convection at high temperature, *IJHMT*, 30(3), 437–447, 1987.

Soufiani, A. and Taine, J.: Experimental and theoretical studies of combined radiative and convective transfer in CO_2 and H_2O laminar flows, *IJHMT*, 32(3), 477–486, 1989.

Soufiani, A. and Taine, J.: High temperature gas radiative property parameters of statistical narrow band model for H_2O, CO_2 and CO and correlated-*k* model for H_2O and CO_2, *IJHMT*, 40(4), 987–991, 1997.

Sowell, E. F. and O'Brien, P. F.: Efficient computation of radiant-interchange configuration factors within an enclosure, *JHT*, 94(3), 326–328, 1972.

Sparrow, E. M.: Application of variational methods to radiation heat-transfer calculations, *JHT*, 82(4), 375–380, 1960.

Sparrow, E. M.: A new and simpler formulation for radiative angle factors, *JHT*, 85(2), 81–88, 1963a.

Sparrow, E. M.: On the calculation of radiant interchange between surfaces, in W. Ibele (ed.), *Modern Developments in Heat Transfer*, pp. 181–212, Academic Press, New York, 1963b.

Sparrow, E. M. and Albers, L. U.: Apparent emissivity and heat transfer in a long cylindrical hole, *JHT*, 82(3), 253–255, 1960.

Sparrow, E. M. and Cess, R. D.: *Radiation Heat Transfer*, augmented edition, Hemisphere, Washington, DC, 1978.

Sparrow, E. M. and Eckert, E. R. G.: Radiant interaction between fin and base surfaces, *JHT*, 84(1), 12–18, 1962.

Sparrow, E. M. and Heinisch, R. P.: The normal emittance of circular cylindrical cavities, *Appl. Opt.*, 9(11), 2569–2572, 1970.

Sparrow, E. M. and Jonsson, V. K.: *Absorption and Emission Characteristics of Diffuse Spherical Enclosures*, NASA TN D-1289, Washington, DC, 1962.

Sparrow, E. M. and Lin, S. H.: Radiation heat transfer at a surface having both specular and diffuse reflectance components, *IJHMT*, 8(5), 769–779, 1965.

Sparrow, E. M., Miller, G. B., and Jonsson, V. K.: Radiating effectiveness of annular-finned space radiators, including mutual irradiation between radiator elements, *J. Aerosp. Sci.*, 29(11), 1291–1299, 1962.

Sparrow, E. M., Szel, J. V., Gregg, J. L., and Manos, P.: Analysis, results, and interpretation for radiation between some simply arranged gray surfaces, *JHT*, 83(2), 207–214, 1961.

Spuckler, C. M. and Siegel, R.: Refractive index effects on radiative behavior of a heated absorbing-emitting layer, *JTHT*, 6(4), 596–604, 1992.

Spuckler, C. M. and Siegel, R.: Refractive index and scattering effects on radiation in a semitransparent laminated layer, *JTHT*, 8(2), 193–201, 1994.

Spuckler, C. M. and Siegel, R.: Two-flux and diffusion methods for radiative transfer in composite layers, *JHT*, 118(1), 218–222, 1996.

Sri Jayaram, K., Balaji, C., and Venkateshan, S. P.: Interaction of surface radiation and free convection in an enclosure with a vertical partition, *JHT*, 119(3), 641–645, 1997.

Stasiek, J. and Collins, M. W.: Radiant and convective heat transfer for flow of a radiation gas in a heated/cooled tube with a gray wall, *IJHMT*, 36(14), 3633–3645, 1993.

Stasiek, J. A.: Application of the transfer configuration factors in radiation heat transfer, *IJHMT*, 41(19), 2893–2907, 1998.

Stefan, J.: Über die beziehung zwischen der wärmestrahlung und der temperatur, *Sitzber. Akad. Wiss. Wien*, 79(2), 391–428, 1879.

Stevens, N. J.: Method for estimating atomic oxygen surface erosion in space environments, *J. Spacecraft Rockets*, 27(1), 93–95, 1990.

Stewart, J. C.: Non-grey radiative transfer, *JQSRT*, 4(5), 723–729, 1964.

Stockham, L. W. and Love, T. J.: Radiative heat transfer from a cylindrical cloud of particles, *AIAA J.*, 6(10), 1935–1940, 1968.

Stockman, N. O. and Kramer, J. L.: *Effect of Variable Thermal Properties on One-Dimensional Heat Transfer in Radiating Fins*, NASA TN D-1878, Washington, DC, 1963.

Stokes, G. G.: On the composition and resolution of streams of polarized light from different sources, *Trans. Camb. Phil. Soc.*, 9, 399–424, 1852. (Reprinted in *Mathematical and Physical Papers*, vol. 3, pp. 233–258, Cambridge University Press, London, U.K., 1901.)

Stone, J. M.: *Radiation and Optics*, McGraw–Hill, New York, 1963.

Stone, P. H. and Gaustad, J. E.: The application of a moment method to the solution of non-gray radiative-transfer problems, *Astrophys. J.*, 134(2), 456–468, 1961.

Su, M.-H. and Sutton, W. H.: Transient conductive and radiative heat transfer in a silica window, *JTHT*, 9(2), 370–373, 1995.

Subramaniam, S. and Mengüç, M. P.: Solution of the inverse radiation problem for inhomogeneous and anisotropically scattering media using a Monte Carlo technique, *IJHMT*, 34(1), 253–266, 1991.

Sun, B., Zheng, D., Klimpke, B., and Yildir, B.: Modified boundary element method for radiative heat transfer analysis in emitting, absorbing and scattering media, *Eng. Anal. Bound. Elem.*, 21, 93–104, 1998.

Sun, B. K., Zhang, X., and Grigoropoulis, C. P.: Spectral optical functions of silicon in the range of 1.13–4.96 eV at elevated temperatures, *IJHMT*, 40(7), 1591–1600, 1997.

Sun, W., Videen, G., Fu, Q., and Hu, Y.: Scattered-field FDTD and PSTD algorithms with CPML absorbing boundary conditions for light scattering by aerosols, *JQSRT*, 131, 166–174, 2013.

Sun, Y. S., Ma, J., and Li, B. W., Chebyshev collocation spectral method for three-dimensional transient coupled radiative-conductive heat transfer, *JHT*, 134, 92701, 2012.

Sunden, B.: Transient conduction in a cylindrical shell with a time-varying incident surface heat flux and convective and radiative surface cooling, *IJHMT*, 32(3), 575–584, 1989.

Suram, S., Bryden, K. M., and Ashlock, D. A.: Quantitative trait loci-based solution of an inverse radiative heat transfer problem, *IEEE Transactions on Evolutionary Computations, CEC 2004, Congress on Evolutionary Computation*, Portland, vol. 1, pp. 427–432, June 2004.

Svet, D. I.: *Thermal Radiation; Metals, Semiconductors, Ceramics, Partly Transparent Bodies, and Films*, Consultants Bureau, Plenum Publishing, New York, 1965.

Swamy, J. N., Crofcheck, C., and Mengüç, M. P.: A Monte Carlo ray tracing study of the polarized light propagation in liquid foams, *JQSRT*, 101, 527–539, 2007.

Swamy, J. N., Crofcheck, C., and Mengüç, M. P.: Time dependent scattering properties of slow decaying foams, *Colloids Surf. A Physicochem. Eng. Aspects*, 338(1–3), 80–86, April 15, 2009.

Tabanfar, S. and Modest, M. F.: Radiative heat transfer in a cylindrical mixture of non-gray particulates and molecular gases, *JQSRT*, 30(6), 555–570, 1983.

Tabanfar, S. and Modest, M. F.: Combined radiation and convection in absorbing, emitting, nongray gas-particulate tube flow, *JHT*, 109(2), 478–484, 1987.

Taflove, A.: *Advances in FDTD Techniques and Applications in Photonics, Photonics North 2007*, Ottawa, Ontario, Canada, June 4, 2007.

Taitel, Y. and Hartnett, J. P.: Application of Rosseland approximation and solution based on series expansion of the emission power to radiation problems, *AIAA J.*, 6(1), 80–89, 1968.

Talukdar, P., Steven, M., Isindorff, F. V., and Trivis, D.: Finite volume method in 3-D curvilinear coordinates with multiblocking procedure for radiative transport problems, *IJHMT*, 48, 4657–4666, 2005.

Tan, H., Ruan, L., and Tong, T. W.: Temperature response in absorbing, isotropie scattering medium caused by laser pulse, *IJHMT*, 43(2), 311–320, 2000a.

Tan, H.-P., Wang, P.-Y., and Xia, X.-L.: Transient coupled radiation and conduction in an absorbing and scattering composite layer, *JTHT*, 14(1), 77–87, 2000b.

Tan, J. Y., Liu, L. H., and Li, B. X.: Least-squares collocation meshless approach for transient radiative transfer, *JTHT*, 20(4), 912–918, October–December, 2006.

Tan, Z.: Combined radiative and conductive heat transfer in two-dimensional emitting, absorbing, and anisotropically scattering square media, *Int. Commun. Heat Mass Transfer*, 16, 391–401, 1989a.

Tan, Z.: Radiative heat transfer in multidimensional emitting, absorbing and anisotropic scattering media—Mathematical formulation and numerical method, *JHT*, 111(1), 141–147, 1989b.

Tan, Z. and Howell, J. R.: Combined radiation and natural convection in a participating medium between horizontal concentric cylinders, in R. K. Shah (ed.), *Heat Transfer Phenomena in Radiation, Convection, and Fires*, ASME HTD-vol. 106, pp. 87–94, ASME, New York, 1989a.

Tan, Z. and Howell, J. R.: *Radiation Heat Transfer in a Partially Divided Square Enclosure with a Participating Medium*, R. K. Shah (ed.), ASME HTD-vol. 106, pp. 199–203, ASME, New York, 1989b.

Tan, Z. and Howell, J. R.: New numerical method for radiation heat transfer in nonhomogeneous participating media, *JTHT*, 4(4), 419–424, 1990a.

Tan, Z. and Howell, J. R.: *Two-Dimensional Radiative Heat Transfer in an Absorbing, Emitting, and Linearly Anisotropically Scattering Medium Exposed to a Collimated Source*, ASME HTD-vol. 137, pp. 101–106, ASME, New York, 1990b.

Tan, Z. and Howell, J. R.: Combined radiation and natural convection in a two-dimensional participating square medium, *IJHMT*, 34(3), 785–793, 1991.

Tan, Z., Przekwas, A. J., Wang, D., Srinivasan, K., and Sun, R.: Numerical simulation of coupled radiation and convection for complex geometries, Paper AIAA-98-2677, *Seventh AIAA/ASME Joint Thermophysics and Heat Transfer Confeerence*, Albuquerque, NM, June 1998.

Tan, Z.-M., and Hsu, P.-F.: Transient radiative transfer in three-dimensional homogeneous and non-homogeneous participating media, *JQSRT*, 73, 181–194, 2002.
Tan, Z.-M., Hsu, P.-F., Wu, S.-H., and Wu, C.-Y.: Modified YIX method and pseudoadaptive angular quadrature for ray effects mitigation, *JTHT*, 14(3), 289–296, 2000.
Tancrez, M. and Taine, J.: Identification of absorption and scattering coefficients and phase function of a porous medium by a Monte Carlo technique, *IJHMT*, 47, 373–383, 2004.
Tang, K. C. and Brewster, M. Q.: *k*-Distribution analysis of gas radiation with nongray, emitting, absorbing, and anisotropie scattering, *JHT*, 116(4), 980–985, 1994.
Tang, K. C. and Brewster, M. Q.: Analysis of molecular gas radiation: Real gas property effects, *JTHT*, 13(4), 460–466, 1999.
Tang, K. and Buckius, R. O.: The geometric optics approximation for reflection from two-dimensional random rough surfaces, *IJHMT*, 41(13), 2037–2047, 1998.
Tang, K., Dimenna, R. A., and Buckius, R. O.: Regions of validity of the geometric optics approximation for angular scattering from very rough surfaces, *IJHMT*, 40, 49–59, 1977.
Tang, K., Kawka, P. A., and Buckius, R. O.: Geometric optics applied to rough surfaces coated with an absorbing thin film, *JTHT*, 13(2), 169–176, 1999a.
Tang, K., Yang, Y., and Buckius, R. O.: Theory and experiments on scattering from rough interfaces, in C.-L. Tien (ed.), *Annual Review of Heat Transfer*, vol. X, Chapter 3, pp. 101–140, Begell House, New York, 1999b.
Tarshis, L. A., O'Hara, S., and Viskanta, R.: Heat transfer by simultaneous conduction and radiation for two absorbing media in intimate contact, *IJHMT*, 12(3), 333–347, 1969.
Taussig, R. T. and Mattick, A. T.: Droplet radiator systems for spacecraft thermal control, *J. Spacecraft Rockets*, 23(1), 10–17, 1986.
Taussky, O. and Todd, J.: Generating and testing of pseudo-random numbers, in H. A. Meyer (ed.), *Symposium on Monte Carlo Methods*, pp. 15–28, Wiley, New York, 1956.
Taylor, R. P. and Luck, R.: Comparison of reciprocity and closure enforcement methods for radiation view factors, *JTHT*, 9(4), 660–666, 1995.
Taylor, R. P., Luck, R., Hodge, B. K., and Steele, W. G.: Uncertainty analysis of diffuse-gray radiation enclosure problems, *JTHT*, 9(1), 63–69, 1995.
Taylor, R. P. and Viskanta, R.: Spectral and directional radiation characteristics of thin-film coated isothermal semitransparent plates, *Wärme und Stoffübertrag.*, 8, 219–227, 1975.
Tencer, J.: Error analysis for radiation transport, PhD dissertation, Department of Mechanical Engineering, The University of Texas at Austin, Austin, TX, December 2013.
Tencer, J. and Howell, J. R.: A parametric study of the accuracy of several radiative transfer solution methods for a set of 2-D benchmark problems, *Proceedings of the ASME 2013 Summer Heat Transfer Conference*, Minneapolis, MN, July 14–19, 2013a.
Tencer, J. and Howell, J. R.: On multilayer modeling of radiative transfer for use with the multisource *k*-distribution method for in homogeneous media, *JHT*, 136, 062703–062703-7, 2014a.
Tencer, J. T.: The impact of reference frame orientation on discrete ordinates solutions in the presence of ray effects and a related mitigation technique, Paper IMECE2014-40445, *Proceedings of the 2014 IMECE,* Montreal, Quebec, Canada, November 2014.
Tencer, J. T. and Howell, J. R.: A multi-source full spectrum *k*-distribution method for 1-D inhomogeneous media, *JQSRT*, 129, 308–315, November 2013b.
Tencer, J. T. and Howell, J. R.: Quantification of model-form uncertainty in the correlated-*k* distribution method for radiation heat transfer, *JQSRT*, 143, 73–82, 2014a.
Tessé, L., Dupoirieux, F., Zamuner, B., and Taine, J.: Radiative transfer in real gases using reciprocal and forward Monte Carlo methods and correlated-*k* approach, *IJHMT*, 45, 2797–2814, 2002.
Thekaekara, M. P.: *Survey of the Literature on the Solar Constant and the Spectral Distribution of Solar Radiant Flux*, NASA SP-74, Washington, DC, 1965.
Thibault, W. and Naylor, B.: Set operations on polyhedra using binary space partitioning trees, *Comput. Graph.*, 21(3), 315–324, 1987.
Thomas, D. L.: Problems in applying the line reversal method of temperature measurement to flames, *Combust. Flame*, 12(6), 541–549, 1968.
Thomas, M. and Rigdon, W. S.: A simplified formulation for radiative transfer, *AIAA J.*, 2(11), 2052–2054, 1964.
Thornton, B. S. and Tran, Q. M.: Optimum design of wideband selective absorbers with provisions for specified included layers, *Solar Energy*, 20(5), 371–378, 1978.

Thurgood, C., Pollard, A., and Rubini, P.: Development of TN quadrature sets and heart solution method for calculating radiative heat transfer, *International Symposium on Steel Reheat Furnace Technology*, Hamilton, Ontario, Canada, 1990.
Thynell, S. T.: Interaction of conduction and radiation in anisotropically scattering, spherical media, *JTHT*, 4(3), 299–304, 1990a.
Thynell, S. T.: Radiation due to CO_2 or H_2O and particulates in cylindrical media, *JTHT*, 4(4), 436–445, 1990b.
Thynell, S. T.: Effect of linear-anisotropic scattering on spectral emission from cylindrical plumes, *JTHT*, 6(2), 224–231, 1992.
Thynell, S. T. and Özişik, M. N.: Radiation transfer in an isotropically scattering solid sphere with space dependent albedo $\omega(r)$, *JHT*, 107, 732–734, 1985.
Thynell, S. T. and Özişik, M. N.: Radiation transfer in isotropically scattering, rectangular enclosures, *JTHT*, 1(1), 69–76, 1987.
Tian, W. and Chiu, W. K. S.: Calculation of direct exchange areas for nonuniform zones using a reduced integration scheme, *JHT*, 125, 839–844, October 2003.
Tian, W. and Chiu, W. K. S.: Radiative absorption in an infinitely long hollow cylinder with Fresnel surfaces, *JQSRT*, 98, 249–263, 2006.
Tien, C. L.: Thermal radiation properties of gases, in T. F. Irvine, Jr. and J. P. Hartnett (eds.), *Advances in Heat Transfer*, vol. 5, pp. 253–324, Academic Press, New York, 1968.
Tien, C. L. and Cunnington, G. R.: Cryogenic insulation heat transfer, in T. F. Irvine, Jr. and J. P. Hartnett (eds.), *Advances in Heat Transfer*, vol. 9, pp. 349–417, Academic Press, New York, 1973.
Tien, C. L. and Drolen, B. L.: Thermal radiation in particulate media with dependent and independent scattering, in *Annual Review of Numerical Fluid Mechanics and Heat Transfer*, vol. 1, pp. 1–32, Hemisphere, Washington, DC, 1987.
Tikhonov, A. N.: Solution of incorrectly formulated problems and the regularization method, *Soviet Math. Dokl.*, 4, 1035–1038, 1963. [Engl. trans. *Dokl. Akad. Nauk. SSSR*, 151, 501–504, 1963.]
Timoshenko, V. P. and Trenev, M. G.: A Method for evaluating heat transfer in multilayered semitransparent materials, *Heat Transfer Sov. Res.*, 18(5), 44–57, 1986.
Tiwari, S. N.: Band models and correlations for infrared radiation, in R. G. Hering (ed.), *Radiative Transfer and Thermal Control*, vol. 49 of Progress in Astronautics and Aeronautics Series, pp. 155–182, AIAA, New York, 1976.
Tomašević, D. I. and Larsen. E. W.: The simplified P2 approximation, *Nucl. Sci. Eng.*, 122, 309–325, 1996.
Tong, T. W. and Skocypec, R. D.: Summary on comparison of radiative heat transfer solutions for a specified problem, ASME HTD-203, pp. 253–264, ASME, New York, 1992.
Tong, T. W. and Swathi, P. S.: Radiative heat transfer in emitting-absorbing-scattering spherical media, *JTHT*, 1(2), 162–170, 1987.
Tong, T. W. and Tien, C. L.: Radiative heat transfer in fibrous insulations—Part I: Analytical study, *JHT*, 105(1), 70–75, 1983.
Tong, T. W., Yang, Q. S., and Tien, C. L.: Radiative heat transfer in fibrous insulations—Part II: Experimental study, *JHT*, 105, 76–81, 1983.
Toor, J. S.: Radiant heat transfer analysis among surfaces having direction dependent properties by the Monte Carlo method, MS thesis, Purdue University, Lafayette, IN, 1967.
Toor, J. S. and Viskanta, R.: A numerical experiment of radiant heat exchange by the Monte Carlo method, *IJHMT*, 11(5), 883–897, 1968a.
Toor, J. S. and Viskanta, R.: Effect of direction dependent properties on radiant interchange, *J. Spacecraft Rockets*, 5(6), 742–743, 1968b.
Toor, J. S. and Viskanta, R.: A critical examination of the validity of simplified models for radiant heat transfer analysis, *IJHMT*, 15, 1553–1567, 1972.
Torrance, K. E. and Sparrow, E. M.: Off-specular peaks in the directional distribution of reflected thermal radiation, *JHT*, 88(2), 223–230, 1966.
Torrance, K. E. and Sparrow, E. M.: Theory for off-specular reflection from roughened surfaces, *JOSA*, 57(9), 1105–1114, 1967.
Toscano, W. M. and Cravalho, E. G.: Thermal radiative properties of the noble metals at cryogenic temperatures, *JHT*, 98(3), 438–445, 1976.
Touloukian, Y. S. and Ho, C. Y. (eds.): Thermophysical properties of matter, TRPC data services, in Y. S. Touloukian and D. P. DeWitt (eds.), *Thermal Radiative Properties: Metallic Elements and Alloys*, vol. 7, 1970; Y. S. Touloukian, and D. P. DeWitt (eds.), *Thermal Radiative Properties: Nonmetallic Solids*, vol. 8, 1972a; Y. S. Touloukian, and D. P. DeWitt (eds.), Y. S. Touloukian, D. P. DeWitt, and R. S. Hernicz (eds.), *Thermal Radiative Properties: Coatings*, vol. 9, 1972b, Plenum Press, New York.

Townsend, A. A., The effects of radiative transfer on turbulent flow of a stratified fluid, *J. Fluid Mech.* 4(4), 361–375, August 1958.

Traugott, S. C. and Wang, K. C.: On differential methods for radiant heat transfer, *IJHMT*, 7(2), 269–273, 1964.

Trelles, J. P., Heberlein, J. V. R., and Pfender, E.: Non-equilibrium modeling of arc plasma torches, *J. Phys. D Appl. Phys.*, 40, 5937–5952, 2007

Tremante, A. and Malpica, F.: Contribution of thermal radiation to the temperature profile of ceramic composite materials, *J. Eng. Gas Turbines Power*, 116(3), 583–586, 1994.

Trivic, D. N.: Modeling of 3-D non-gray gases radiation by coupling the finite volume method with weighted sum of gray gases model, *IJHMT*, 47, 1367–1382, 2004.

Trivic, D. N., O'Brien, T. J., and Amon, C. H.: Modeling the radiation of anisotropically scattering media by coupling Mie theory with the finite volume method, *IJHMT*, 47, 5765–5780, 2004.

Truelove, J. S.: Discrete-ordinate solutions of the radiation transport equation, *JHT*, 109(4), 1048–1051, 1987.

Tsai, C.-F. and Nixon, G.: Transient temperature distribution of a multilayer composite wall with effects of internal thermal radiation and conduction, *Numer. Heat Transfer*, 10, 95–101, 1986.

Tsai, J. R. and Özişik, M. N.: Transient, combined conduction and radiation in an absorbing, emitting, and anisotropically scattering solid sphere, *JQSRT*, 38(4), 243–251, 1987.

Tsai, D. S. and Strieder, W.: Radiation across a spherical cavity having both specular and diffuse reflectance components, *Chem. Eng. Sci.*, 40(1), 170–173, 1985.

Tsai, D. S. and Strieder, W.: Radiation across and down a cylindrical pore having both specular and diffuse reflectance components, *Ind. Eng. Chem. Fundam.*, 25, 244–249, 1986.

Tsang, L., Kong, J. A., and Ding, K. H.: *Scattering of Electromagnetic Waves*, Wiley, New York, 2000.

Tsao, B.-H., Ramalingam, M. L., Moraga, N. O., and Jacobson, D. L.: Normal spectral emittance of W-Re-HfC and W-Re-ThO_2 alloys, *JTHT*, 6(4), 680–684, 1992.

Tschikin, M., Biehs, S.-A, Messina, R., and Ben-Abdallah, P., On the limits of the effective description of hyperbolic metamaterials in the presence of surface waves, *J. Opt.*, 15, 105101-1–105101-7, 2013.

Tseng, C.-J. and Howell, J. R.: Combustion of liquid fuels in a porous ceramic burner, *Combut. Sci. Technol.*, 112, 141–161, 1996.

Tucker, R. J.: Direct exchange areas for calculating radiation transfer in rectangular furnaces, *JHT*, 108, 707–710, 1986.

Tuntomo, A. and Tien, C. L.: Transient heat transfer in a conducting particle with internal radiant absorption, *JHT*, 114(2), 304–309, 1992.

Turyshev, S. G., Toth, V. T., Kinsella, G., Lee, S.-C, Lok, S. M., and Ellis, J.: Support for the thermal origin of the pioneer anomaly, *Phys. Rev. Lett.*, 108, 241101, 2012.

Usiskin, C. M. and Siegel, R.: Thermal radiation from a cylindrical enclosure with specified wall heat flux, *JHT*, 82(4), 369–374, 1960.

Utreja, L. R. and Chung, T. J.: Combined convection-conduction-radiation boundary layer flows using optimal control penalty finite elements, *JHT*, 111(2), 433–437, 1989.

Vaillon, V., Wong, B. T., and Mengüç, M. P.: Polarized radiative transfer in a particle laden transparent medium via Monte Carlo method, *JQSRT* 84, 383–394, 2004.

Van de Hülst, H. C.: *Light Scattering by Small Particles*, Wiley, New York, 1957; Dover Publications, New York, 1981.

van Doormall, J. P. and Raithby, G. D.: Enhancements of the simple method for predicting incompressible fluid flows, *Numer. Heat Transfer*, 7, 147–163, 1983.

Van Laarhove, P. J. M., Aarts, E. H. L., and Lenstra, J. K.: Job shop scheduling by simulated annealing, *Oper. Res.*, 40(1), 113–125, 1992.

van Leersum, J.: A method for determining a consistent set of radiation view factors from a set generated by a nonexact method, *Int. J. Heat Fluid Flow*, 10(1), 83–85, 1989.

Varanasi, P.: A critical appraisal of the current spectroscopic databases used in atmospheric and other radiative transfer applications, in M. P. Mengüç and N. Selçuk (eds.), *Radiation III: Third International Symposium on Radiative Transfer*, Begell House, New York, 2001.

Varghese, P. L. and Hanson, R. K.: Measured collisional narrowing effect on spectral line shapes at high resolution, *Appl. Opt.*, 23, 2376, 1984.

Venkata, G.: Characterization of nano-size particles near metallic surfaces via surface plasmon scattering, MSME thesis, University of Kentucky, Lexington, KY, 2006.

Venkata, P. G., Aslan, M. M., Mengüç, M. P. and Videen, G.: Surface plasmon scattering patterns of gold nanoparticles and 2D agglomerates, *ASME J. Heat Transfer*, 129, 60–70, 2007.

Vercammen, H. A. J. and Froment, G. F.: An improved zone method using Monte Carlo techniques for the simulation of radiation in industrial furnaces, *IJHMT*, 23(3), 329–336, 1980.

Veselago, V., Braginsky, L., Shklover, V., and Hafner, C.: Negative refractive index materials, *J. Comput. Theoret. Nanosci.*, 3, 1–30, 2006.

Villeneuve, P. V., Chapman, D. D., and Mahan, J. R.: Use of the Monte-Carlo ray-trace method as a design tool for jet engine visibility suppression, in Y. Bayazitoglu et al. (eds.), *Radiative Heat Transfer: Current Research*, ASME HTD-vol. 276, pp. 59–71, ASME, New York, 1994.

Vincenti, W. G. and Kruger Jr., C. H.: *Introduction to Physical Gas Dynamics*, Corrected edn., Krieger Publishing Company, Malabar, FL, 1986.

Viskanta, R.: Interaction of heat transfer by conduction, convection, and radiation in a radiating fluid, *JHT*, 85(4), 318–328, 1963.

Viskanta, R.: Heat transfer by conduction and radiation in absorbing and scattering materials, *JHT*, 87(1), 143–150, 1965.

Viskanta, R.: Radiation transfer and interaction of convection with radiation heat transfer, in T. F. Irvine, Jr. and J. P. Hartnett (eds.), *Advances in Heat Transfer*, vol. 3, pp. 175–251, Academic Press, New York, 1966.

Viskanta, R.: Radiation heat transfer: Interaction with conduction and convection and approximate methods in radiation, in U. Grigull (ed.), *Heat Transfer 1982, Proceedings of the Seventh International Heat Transfer Conference*, vol. 1, pp. 103–121, Hemisphere, Washington, DC, 1982.

Viskanta, R.: *Radiative Transfer in Combustion Systems: Fundamentals and Applications,* Begell House, New York, 2005.

Viskanta, R. and Anderson, E. E.: Heat transfer in semi-transparent solids, in J. P. Hartnett and T. F. Irvine, Jr. (eds.), *Advances in Heat Transfer*, vol. 11, pp. 317–441, Academic Press, New York, 1975.

Viskanta, R. and Bathla, P. S.: Unsteady energy transfer in a layer of gray gas by thermal radiation, *Z. Angew. Math. Phys.*, 18(3), 353–367, 1967.

Viskanta, R. and Grosh, R. J.: Effect of surface emissivity on heat transfer by simultaneous conduction and radiation, *IJHMT*, 5, 729–734, 1962a.

Viskanta, R. and Grosh, R. J.: Heat transfer by simultaneous conduction and radiation in an absorbing medium, *JHT*, 84(1), 63–72, 1962b.

Viskanta, R. and Grosh, R. J.: Boundary layer in thermal radiation absorbing and emitting media, *IJHMT*, 5, 795–806, 1962c.

Viskanta, R. and Kim, D. M.: Heat transfer through irradiated, semi-transparent layers at high temperature, *JHT*, 102(1), 182–184, 1980.

Viskanta, R. and Lall, P. S.: Transient cooling of a spherical mass of high-temperature gas by thermal radiation, *J. Appl. Mech.*, 32(4), 740–746, 1965.

Viskanta, R. and Lall, P. S.: Transient heating and cooling of a spherical mass of gray gas by thermal radiation, in M. A. Saad and J. A. Miller (eds.), *Proceedings of the Heat Transfer and Fluid Mechanics Institute*, University of Santa Clara, pp. 181–197, 1966.

Viskanta, R. and Merriam, R. L.: Shielding of surfaces in couette flow against radiation by transpiration of an absorbing-emitting gas, *IJHMT*, 10(5), 641–653, 1967.

Viskanta, R. and Merriam, R. L.: Heat transfer by combined conduction and radiation between concentric spheres separated by radiating medium, *JHT*, 90(2), 248–256, 1968.

Viskanta, R. and Mengüç, M. P.: Radiation heat transfer in combustion systems, *Prog. Energy Combust. Sci.*, 13, 97–160, 1987.

Viskanta, R., Schornhorst, J. R., and Toor, J. S.: *Analysis and Experiment of Radiant Heat Exchange between Simply Arranged Surfaces*, AFFDL-TR-67-94, DDC No. AD-655335, Purdue University, Lafayette, IN, June 1967.

Viskanta, R., Siebers, D. L., and Taylor, R. P.: Radiation characteristics of multiple-plate glass systems, *IJHMT*, 21, 815–818, 1978.

Vogel, C. R.: *Computational Methods for Inverse Problems*, SIAM, Philadelphia, PA, 2002.

Volakis, J. L., Chatterjee, A., and Kempel, L. C.: Review of the finite-element method for 3-dimensional electromagnetic scattering, *JOSA A Opt. Image Sci. Vision*, 11(4m), 1422–1433, 2004.

Volokitin, A. I. and Persson, B. N. J.: Radiative heat transfer between nanostructures, *Phys. Rev. B*, 63(20), 205404, 2001.

Volokitin, A. I. and Persson, B. N. J.: Resonant photon tunneling enhancement of the radiative heat transfer, *Phys. Rev. B*, 69(20), 045417, 2004.

Vortmeyer, D.: Radiation in packed solids, in J. T. Rogers (ed.), *Heat Transfer 1978: Proceedings of the Sixth Interntional Heat Transfer Conference*, vol. 6, pp. 525–539, Hemisphere, Washington, DC, 1978.

Voshchinnikov, N. V.: Electromagnetic scattering by homogeneous and coated spheroids: Calculations using the separation of variables method, *JQSRT*, 55, 627–636, 1996.

Voshchinnikov, N. V. and Farafonov, V. G.: Optical properties of spheroidal particles, *Astrophys. Space Sci.*, 204, 19–86, 1993.

Vujičić, M. R., Lavery, N. P., and Brown, S. G. R.: Numerical sensitivity and view factor calculation using the Monte Carlo method, *Proc. IMechE* C: *J. Mech. Eng. Sci.*, 220, 697–702, 2006a.

Vujičić, M. R., Lavery, N. P., and Brown, S. G. R.: View factor calculation using the Monte Carlo method and numerical sensitivity, *Commun. Numer. Methods Eng.*, 22(3), 197–203, 2006b.

Wahiduzzaman, S. and Morel, T.: *Effect of Translucence of Engineering Ceramics on Heat Transfer in Diesel Engines*, Oak Ridge National Laboratory, Report ORNL/Sub/88–22042/2, Oak Ridge, TN, April 1992.

Walker, T.: The use of primatives in the calculation of radiative view factors, PhD dissertation, School of Chemical and Biomolecular Engineering. University of Sydney, Sydney, New South Wales, Australia, December 2013.

Walker, T., Xue, S.-C., and Barton, G. W: Numerical determination of radiative view factors using ray tracing, *JHT*, 132(7), 072702-1–072702-6, 2010.

Walker, T., Xue, S.-C., and Barton, G. W: A robust Monte Carlo-Based ray-tracing approach for the calculation of view factors in arbitrary three- dimensional geometries, *Comput. Thermal Sci.*, 4(5), 425–442, 2012.

Walters, D. V. and Buckius, R. O.: Normal spectral emission from nonhomogeneous mixtures of CO_2 gas and $A1_2O_3$ particulate, *JHT*, 113(1), 174–184, 1991.

Walters, D. V. and Buckius, R. O.: Rigorous development for radiation heat transfer in nonhomogeneous absorbing, emitting and scattering media, *IJHMT*, 35(12), 3323–3333, 1992.

Walters, D. V. and Buckius, R. O.: Monte Carlo methods for radiative heat transfer in scattering media, in C.-L. Tien (ed.), *Annual Review of Heat Transfer*, vol. 5, Chapter 3, pp. 131–176, CRC Press, Boca Raton, FL, 1994.

Walthall, C. L., Norman, J. M., Welles, J. M., Campbell, G., and Blad, B. L.: Simple equation to approximate the bidirectional reflectance from vegetative canopies and bare soil surfaces, *Appl. Opt.*, 24, 383–387, 1985.

Wang, C.H., Ai, Q., Yi, H.L., and Tan, H.P.: Transient radiative transfer in a graded index medium with specularly reflecting surfaces, *Num. Heat Transfer, Pt. A-Applications*, 67(11), 1232–1252, June 3, 2015.

Wang, K. Y. and Tien, C. L.: Radiative transfer through opacified fibers and powders, *JQSRT*, 30(3), 213–223, 1983.

Wang, L., Haider, A., and Zhang, Z.: Effect of magnetic polaritons on the radiative properties of inclined plate arrays, *JQSRT*, Special Issue on *Micro- and Nano-Scale Radiative Transfer*, 132, 52–60, 2014.

Wang, L. and Modest, M. F.: Narrow-band based multiscale full-spectrum *k*-distribution method for radiative transfer in inhomogeneous gas mixtures, *JHT*, 127, 740–748, July 2005a.

Wang, L. and Modest, M. F.: High-accuracy, compact database of narrow-band *k*-distributions for water vapor and carbon dioxide, *JQSRT*, 93, 245–261, 2005b.

Wang, L. and Modest, M. F.: Treatment of wall emission in the narrow-band based multiscale full-spectrum *k*-distribution method, *JHT*, 129, 743–748, June 2007.

Wang, L., Yang, J., Modest, M. F., and Haworth, D. C.: Application of the full-spectrum *k*-distribution method to photon Monte Carlo solvers, *JQSRT*, 104, 297–304, March 2007.

Wang, L. P. and Zhang, Z. M.: Thermal rectification enabled by near-field radiative heat transfer between intrinsic silicon and dissimilar material, *Nanoscale Microscale Thermophys. Eng.*, 17, 337–348, 2013.

Wang, L. S. and Tien, C. L.: A study of various limits in radiation heat-transfer problems, *IJHMT*, 10, 1327–1338, 1967.

Wang, P.-W., Tan, H.-P., and Liu, L.-H.: Coupled radiation and conduction in a scattering composite layer with coatings, *JTHT*, 14(4), 512–522, 2000.

Warming, R. F. and Beam, R. M.: On the construction and application of implicit factored schemes for conservation laws, in H. B. Keller (ed.), *Computational Fluid Dynamics, SIAM-AMS Proceedings*, New York, vol. 11, pp. 85–129, 1978.

Wassel, A. T. and Edwards, D. K.: Mean beam lengths for spheres and cylinders, *JHT*, 98, 308–309, 1976a.

Wassel, A. T. and Edwards, D. K.: Molecular gas radiation in a laminar or turbulent pipe flow, *JHT*, 98(1), 101–107, 1976b.

Waterman, P. C.: Matrix formulation for electromagnetic scattering, *Proc. IEEE*, 53, 805, 1965.

Waterman, P. C.: Symmetry, unitarity, and geometry in electromagnetic scattering, *Phys. Rev. D*, 3, 825–839, 1971.

Watt, A. and Watt, M.: *Advanced Animation and Rendering Techniques; Theory and Practice*, Section 11.2.3, Addison Wesley, New York, 1992.

Webb, B. W.: Interaction of radiation and free convection on a heated vertical plate: Experiment and analysis, *JTHT*, 4(1), 117–121, 1990.

Webb, B. W. and Viskanta, R.: Analysis of radiation-induced natural convection in rectangular enclosures, *JTHT*, 1(2), 146–153, 1987.

Webb, K., Near-field radiative transfer experiments, MS thesis, Bogazici University, Istanbul, Turkey, 2012.
Webb, K. D., Artvin, Z., Khosroshahi, F. K., Ertürk, H., Okutucu, T., and Mengüç, M. P., Near-field radiative heat transfer measurements between parallel plates, *Proceedings of the Seventh International Symposium on Radiative Transfer (RAD-13)*, Kusadasi, Turkey, RAD-13-P-19, June 2–8, 2013.
Webb, K. D., Mengüç, M. P., Ertürk, H., and Başım, B.: Technique For measurement of near-field radiation heat transfer between parallel planes with nano-scale spacing, *Proceedings of the Sixth International Symposium on Radiative Transfer (RAD-10)*, Antalya, Turkey, June 13–19, 2010.
Wen, S.-B.: Direct numerical simulation of near field thermal radiation based on Wiener chaos expansion of thermal fluctuating current, *JHT*, 132, 072704, 2010.
Wendlandt, B. C. H.: Temperature in an irradiated thermally conducting medium, *J. Phys. D*, 6, 657–660, 1973.
Whale, M. D.: A fluctuational electrodynamic analysis of microscale radiative transfer and the design of microscale thermophotovoltaic devices, PhD thesis, MIT, Cambridge, MA, 1997.
Whale, M. D. and Cravalho, E. G.: Modeling and performance of microscale thermophotovoltaic energy conversion devices, *IEEE Trans. Energy Convers.*, 17(1), 130–142, 2002.
White, S. M. and Kumar, S.: Interference effects on scattering by parallel fibers at normal incidence, *JTHT*, 4(3), 305–310, 1990.
Whiting, E. E., Park, C., Liu, Y., Arnold, J., and Paterson, J.: *NEQAIR96, Nonequilibrium and Equilibrium Radiative Transfer and Spectra Program: User's Manual*, NASA RP-1389, Moffett Field, CA, December 1996.
Wien, W.: Temperatur und Entropie der Strahlung, *Ann. Phys., Ser. 2*, 52, 132–165, 1894.
Wien, W.: Über die Energievertheilung im Emissionsspectrum eines schwarzen Körpers, *Ann. Phys., Ser. 3*, 58, 662–669, 1896.
Wijewardane, S. and Goswami, D. Y.: A review on surface control of thermal radiation by paints and coatings for new energy applications, *Renew. Sust. Energy Rev.*, 16, 1863–1873, 2012.
Wijeysundera, N. E.: A net radiation method for the transmittance and absorptivity of a series of parallel regions, *Solar Energy*, 17, 75–77, 1975.
Wilbers, A. T. M., Beulens, J. J., and Schram, D. C.: Radiative energy loss in a two-temperature argon plasma, *JQSRT*, 46(5), 385–392, 1991.
Wilkins, J. E, Jr.: Minimum-mass thin fins and constant temperature gradients, *J. Soc. Ind. Appl. Math.*, 10(1), 62–73, 1962.
Williams, D. A., Lappin, T. A., and Duffie, J. A: Selective radiation properties of particulate coatings, *J. Eng. Power*, 85(3), 213–220, 1963.
Williams, J. J. and Dudley, D. P.: The radiative contribution to heat transfer in metalized propellant exhausts, *Fifth Symposium on Thermophysical Properties*, ASME, Boston, MA, September 30–October 2, 1970.
Wing, G. W.: *A Primer on Integral Equations of the First Kind*, SIAM, Philadelphia, PA, 1991.
Winston, R.: Principles of solar concentrators of a novel design, *Solar Energy*, 16(2), 89–95, 1974.
Wirgin, A. and Maradudin, A. A.: Resonant enhancement of the electric field in the grooves of bare metallic gratings exposed to s-polarized light, *Phys. Rev. B*, 31(8), 5573–5576, 1985.
Wolff, L. B. and Kurlander, D. J.: Ray tracing with polarization parameters, *IEEE Comput. Graph. Appl.*, 10(6), 44–55, 1990.
Wong, B. and Mengüç, M. P.: Depolarization of radiation by foams, *JQSRT*, 73(2–5), 273–284, 2002.
Wong, B. T., Francoeur, M., Bong, V. N.-S., and Mengüç, M. P.: Coupling of near-field thermal radiative heating and phonon Monte Carlo simulation: Assessment of temperature gradient in doped silicon thin film, *JQSRT*, 143, 46–55, 2014.
Wong, B. T., Francoeur, M., and Mengüç, M. P.: A Monte Carlo simulation for phonon transport within silicon structures at nanoscales with heat generation, *IJHMT*, 54(9–10), 1825–1838, April 2011.
Wong, B. T. and Mengüç, M. P.: Comparison of Monte Carlo techniques to predict the propagation of a collimated beam in participating media, *Num. Heat Transfer B*, 42, 119–140, 2002.
Wong, B. T. and Mengüç, M. P.: Monte Carlo methods in radiative transfer and electron beam processing, *JQSRT*, 84, 437–450, 2004.
Wong, B. T. and Mengüç, M. P.: *Thermal Transport for Applications in Micro- and Nanomachining*, Springer, Berlin, Germany, 2008.
Wong, B. T. and Mengüç, M. P.: A unified Monte Carlo treatment of the transport of electromagnetic energy, electrons, and phonons in absorbing and scattering media, *JQSRT*, 111(3), 399–419, 2010.
Wood, W. D., Deem, H. W., and Lucks, C. F.: *Thermal Radiative Properties*, Plenum Press, New York, 1964.
Wriedt, T.: Mie theory 1908, on the mobile phone 2008, *JQSRT*, 109, 1543–1548, 2008.
Wriedt, T.: Light scattering theories and computer codes, *JQSRT*, 110, 833–843, 2009.
Wriedt, T.: Electromagnetic scattering programs, http://www.t-matrix.de/, accessed April 27, 2015, 2010a.
Wriedt, T.: Light scattering information portal, http://www.scattport.org/, accessed April 27, 2015, 2010b.

Wriedt, T. and Hellmers, J.: New scattering information portal for the light scattering community, *JQSRT*, 109(8), 1536–1542, 2008.

Wu, C.-Y.: Exact integral formulation for radiative transfer in an inhomogeneous scattering medium, *JTHT*, 4(4), 425–431, 1990.

Wu, C. Y. and Fu, M. N.: Radiative transfer in a coating on a rectangular corner: Allowance for shadowing, *IJHMT*, 33(12), 2735–2741, 1990.

Wu, C.-Y. and Liou, B.-T.: Discrete-ordinate solutions for radiative transfer in a cylindrical enclosure with fresnel boundaries, *IJHMT*, 40(10), 2467–2475, 1997.

Wu, C.-Y. and Ou, N.-R.: Transient two-dimensional radiative and conductive heat transfer in a scattering medium, *IJHMT*, 37(17), 2675–2686, 1994.

Wu, C.-Y. and Wang, C.-J.: Emittance of a finite spherical scattering medium with fresnel boundary, *JTHT*, 4(2), 250–252, 1990.

Wu, C.-Y. and Wu, S.-C.: Radiative transfer in a two-layer medium on a cylinder, *IJHMT*, 36(5), 1147–1158, 1993.

Wu, C.-Y. and Wu, S.-H.: Integral equation formulation for transient radiative transfer in an anisotropically scattering medium, *IJHMT*, 43(11), 2009–2020, 2000.

Wu, S.-H., Wu, C.-Y., and Hsu, P.-F.: Solutions of radiative heat transfer in nonhomogeneous participating media using the quadrature method, *ASME HTD*, 332(1), 101–108, 1996.

Wu, S. T., Ferguson, R. E., and Altgilbers, L. L.: Application of finite element techniques to the interaction of conduction and radiation in participating media, in A. L. Crosbie (ed.), *Heat Transfer and Thermal Control*, pp. 61–92, AIAA, New York, 1981.

Wu, W. J. and Mulholland, G. P.: Two dimensional inverse radiation heat transfer analysis using Monte Carlo techniques, in S. T. Thynell et al. (eds.), *Developments in Radiative Heat Transfer*, ASME HTD-203, pp. 181–190, ASME, New York, 1989.

Xia, X. L., Tan, H. P., and Huang, Y.: Simultaneous radiation and conduction heat transfer in a graded index semitransparent slab with gray boundaries, *IJHMT*, 45(13), 2673–2688, 2002.

Xia, Y. and Strieder, W.: Complementary upper and lower truncated sum, multiple scattering bounds on the effective emissivity, *IJHMT*, 37, 443–450, 1994a.

Xia, Y. and Strieder, W.: Variational calculation of the effective emissivity for a random bed, *IJHMT*, 37, 451–460, 1994b.

Xu, R.: *Particle Characterization: Light Scattering Methods*, Kluwer Academic Press, Dordrecht, the Netherlands, 2001.

Xu, J.-B., Läuger, K., Möller, R., Dransfeld, K., and Wilson, I. H.: Heat transfer between two metallic surfaces at small distances, *J. Appl. Phys.*, 76(11), 7209–7216, 1994.

Yagi, S. and Inoue, H.: Radiation from soot particles in luminous flames, *Proceedings of the Eighth International Symposium on Combustion*, Baltimore, MD, pp. 288–293, 1962.

Yamada, J. and Kurosaki, Y.: Radiative characteristics of fibers with a large size parameter, *IJHMT*, 43(6), 981–991, 2000.

Yamada, Y.: Combined radiation and free convection heat transfer in a vertical channel with arbitrary wall emissivities, *IJHMT*, 31(2), 429–440, 1988.

Yamada, Y., Cartigny, J. D., and Tien, C. L.: Radiative transfer with dependent scattering by particles: Part 2—Experimental investigation, *JHT*, 108, 614–618, 1986.

Yan, W.-M. and Li, H.-Y.: Radiation effects on mixed convection heat transfer in a vertical square duct, *IJHMT*, 44(7), 1401–1410, 2001.

Yan, W.-M., Li, H.-Y., and Lin, D.: Mixed convection heat transfer in a radially rotating square duct with radiation effects, *IJHMT*, 42(1), 35–47, 1999.

Yang, K. T.: Numerical modelling of natural convection-radiation interactions in enclosures, in *Heat Transfer 1986, Proceedings of the Eighth International Heat Transfer Conference*, vol. 1, pp. 131–140, Hemisphere, Washington, DC, 1986.

Yang, P. and Liou, K. N.: An "exact" geometric-optics approach for computing the optical properties of large absorbing particles, *JQSRT*, 110, 1162–1177, 2009.

Yang, W.-J., Taniguchi, H., and Kudo, K.: Radiative heat transfer by the Monte Carlo method, in J. P. Hartnett and T. F. Irvine (eds.), *Advances in Heat Transfer*, vol. 27, pp. 1–215, Academic Press, San Diego, CA, 1995.

Yang, W.-M. and Leu, M.-C.: Instability of radiation-induced flow in an inclined slot, *IJHMT*, 36(12), 3089–3098, 1993.

Yang, Y., Basu, S., and Wang, L.: Radiation-based near-field thermal rectification with phase transition materials, *Appl. Phys. Lett.*, 103, 163101, 2013.

Yang, Y. and Buckius, R. O.: Surface length scale contributions to the directional and hemispherical emissivity and reflectivity, *JTHT*, 9(4), 653–659, 1995.

Yang, Y. S., Howell, J. R., and Klein, D. E.: Radiative heat transfer through randomly packed bed of spheres by the Monte Carlo method, *JHT*, 105(2), 325–332, 1983.

Yao, C. and Chung, B. T. F.: Transient heat transfer in a scattering-radiating-conducting layer, *JTHT*, 13(1), 18–24, 1999.

Yarbrough, D. W. and Lee, C.-L.: Monte Carlo calculation of radiation view factors, in F. R. Payne et al. (eds.), *Integral Methods in Science and Engineering*, vol. 85, pp. 563–574, Hemisphere, Washington, DC, 1985.

Yee, K. S.: Numerical solution of initial value problems involving Maxwell's equations in isotropic media, *IEEE Trans. Antennas Propag.*, 14, 302–307, 1966.

Yeh, P.: *Optical Waves in Layered Media*, John Wiley & Sons, Hoboken, NJ, 2005.

Yi, H. L., Ben, X., and Tan, H. P.: Transient radiative transfer in a scattering slab considering polarization, *Optics Express*, 21(22), 26693–26713, 2013.

Yi, H. L., Wang, C. H., and Tan, H. P.: Transient radiative transfer in a complex refracting medium by a modified Monte Carlo simulation, *IJHMT*, 79, 437–449, 2014.

Yih, K. A.: Effect of radiation on natural convection about a truncated cone, *IJHMT*, 42(23), 4299–4305, 2000.

Yoon, K.-B., Park, S.-J., and Kim, T.-K.: Study on inverse estimation of radiative reflection properties in mid-wavelength infrared region by using the repulsive particle swarm optimization algorithm, *Appl. Opt.*, 52(22), 5533–5538, 2013.

Yoshida, H., Yun, J. H., Echigo, R., and Tomimura, T.: Transient characteristics of combined conduction, convection and radiation heat transfer in porous media, *IJHMT*, 33(5), 847–857, 1990.

Yu, E. and Joshi, Y. K.: Heat transfer in discretely heated side-vented compact enclosures by combined conduction, natural convection, and radiation, *JHT*, 121(4), 1002–1010, 1999.

Yu, Y. J., Baek, S. W., and Park, J. H.: An extension of the weighted sum of gray gases non-gray gas radiation model to a two phase mixture of non-gray gas with particles, *IJHMT*, 43(10), 1699–1713, 2000.

Yu, Z., Sergeant, N. P., Skauli, T., Zhang, G., Wang, H., and Fan, S.: Enhancing far-field thermal emission with thermal extraction, *Nat. Commun.*, 4, 1730, April 2014.

Yücel, A. and Williams, M. L.: Heat transfer by combined conduction and radiation in axisymmetric enclosures, *JTHT*, 1(4), 301–306, 1987.

Yuen, W. W.: The multiple absorption coefficient zonal method (MACZM), an efficient computational approach for the analysis of radiative heat transfer in a multidimensional inhomogeneous nongray media, *Numer. Heat Transfer B*, 49, 89–103, 2006.

Yuen, W. W. and Ho, C. F.: Analysis of two-dimensional radiative heat transfer in a gray medium with internal heat generation, *IJHMT*, 28(1), 17–23, 1985.

Yuen, W. W. and Takara, E. E.: Analysis of combined conductive-radiative heat transfer in a two-dimensional rectangular enclosure with a gray medium, *JHT*, 110(2), 468–474, 1988.

Yuen, W. W. and Takara, E. E.: Development of a generalized zonal method for the analysis of radiative heat transfer in absorbing and anisotropically scattering media, in K. Vafai and J. L. S. Chen (eds.), *Numerical Heat Transfer*, ASME HTD-vol. 130, pp. 123–132, ASME, New York, 1990a.

Yuen, W. W. and Takara, E. E.: Superposition technique for radiative equilibrium in rectangular enclosures with complex boundary conditions, *IJHMT*, 33(5), 901–915, 1990b.

Yuen, W. W. and Tien, C. L.: A simple calculation scheme for the luminous-flame emissivity, *Proceedings of the 16th International Symposium on Combustion*, pp. 1481–1487, The Combustion Institute, Pittsburgh, PA, 1977.

Yuen, W. W. and Wong, L. W.: Heat transfer by conduction and radiation in a one-dimensional absorbing, emitting and anisotropically scattering medium, *JHT*, 102(2), 303–307, 1980.

Yuen, W. W. and Wong, L. W.: Numerical computation of an important integral function in two-dimensional radiative transfer, *JQSRT*, 29(2), 145–149, 1983.

Yuen, W. W. and Wong, L. W.: Analysis of radiative equilibrium in a rectangular enclosure with gray medium, *JHT*, 106(2), 433–440, 1984.

Yüksel, A.: Design, characterization and optimization of high-efficiency thermophotovoltaic (TPV) device using near-field thermal energy conversion, MS thesis, Department of Mechanical Engineering, The University of Texas at Austin, Austin, TX, December 2013.

Yurkin, M. A.: Discrete dipole simulations of light scattering by blood cells, PhD thesis, University of Amsterdam, Amsterdam, the Netherlands, 2007.

Yurkin, M. A., and Hoekstra, A. G.: The discrete-dipole-approximation code ADDA: Capabilities and known limitations, *JQSRT*, 112(13), 2234–2247, 2011.

Yurkin, M. A., Maltsev, V. P., and Hoekstra, A. G.: The discrete dipole approximation for simulation of light scattering by particles much larger than the wavelength, *JQSRT*, 106, 546–557, 2007.

Zakhidov, R. A.: Mirror system synthesis for radiant energy concentration—An inverse problem, *Solar Energy*, 42(6), 509–513, 1989.

Zaworski, J. R., Welty, J. R., and Drost, M. K.: Measurement and use of bidirectional reflectance, *IJHMT*, 39(6), 1149–1156, 1996a.

Zaworski, J., Welty, J. R., Palmer, B. J., and Drost, M. K.: Comparison of experiment with Monte Carlo simulations on a reflective gap using a detailed surface properties model, *JHT*, 118(2), 388–393, 1996b.

Zeeb, C. N. and Burns, P. J.: Performance enhancements of Monte Carlo particle tracing algorithms for large, arbitrary geometries, Paper NHTC99-31, *Proceedings of the 1999 ASME National Heat Transfer Conference*, San Francisco, CA, August 1999.

Zel'dovich, Ya. B. and Raizer, Yu. P.: *Physics of Shock Waves and High-Temperature Hydrodynamic Phenomena*, vol. 1, pt. II, Academic Press, New York, 1966.

Zeller, A. F.: High T_c superconductors as thermal radiation shields, *Cryogenics*, 30, 545–546, 1990.

Zerefos, C. S., Tetsis, P., Kazantzidis, A., Amiridis, V., Zerefos, S.C., Luterbacher, J., Eleftheratos, K., Gerasopoulos, E., Kazadzis, S., and Papayannis, A.: Further evidence of important environmental information content in red-to-green ratios as depicted in paintings by great masters, *Atmos. Chem. Phys.*, 14, 2987–3015, 2014.

Zhang, H. and Hirleman. E. D.: Prediction of light scattering from particles on a filmed surface using discrete-dipole approximation, *Proceedings of the SPIE Conference,* Santa Clara, CA, vol. 4692, *Design, Process Integration, and Characterization for Microelectronics*, pp. 38–45, 2002.

Zhang, H. and Modest, M. M.: Multi-group full-spectrum *k*-distribution database for water vapor mixtures in radiative transfer calculations, *IJHMT*, 46, 3593–3603, 2003a.

Zhang, H. and Modest, M. M.: Full-spectrum *k*-distribution correlations for carbon dioxide mixtures, *JTHT*, 17(2), 259–263, April–June 2003b.

Zhang, H. and Modest, M. M.: Scalable multi-group full-spectrum correlated-*k* distributions for radiative transfer calculations, *JHT*, 125, 454–461, June 2003c.

Zhang, J.-M. and Sutton, W. H.: Predictions of radiative transfer in two-dimensional nonhomogeneous participating cylindrical media, *JTHT*, 10(1), 47–53, 1996.

Zhang, Q. C., Sinko, T. M., Dey, C. J., Collins, R. E., and Turner, G. M.: The measurement and calculation of radiative heat transfer between uncoated and doped tin oxide coated glass surfaces, *IJHMT*, 40(1), 61–71, 1997.

Zhang, Y., Ma, Y., Yi, H.-L., and Tan, H.-P.: Natural element method for solving radiative transfer with or without conduction in three-dimensional complex geometries, *JQSRT*, 129, 118–130, November 2013a.

Zhang, Y., Ragusa, J. C., and Morel, J. E.: Iterative performance of various formulations of the SP_N equations, *J. Comput. Phys.*, 252, 558–572, 2013b.

Zhang, Z., Maruyama, S., Sakurai, A., and Mengüç, M. P., Special issue on micro- and nano-scale radiative transfer, *JQSRT*, 132, 1–2, 2014.

Zhang, Z. M.: *Nano/Microscale Heat Transfer*, McGraw-Hill, New York, 2007.

Zhang, Z. M. and Wang, L. P.: Measurements and modeling of the spectral and directional radiative properties of micro/nano-structured materials, *Int. J. Thermophys.*, 32, 1–34, 2011.

Zhao, J. M. and Liu, L. H.: Spectral element approach for coupled radiative and conductive heat transfer in semitransparent medium, *JHT*, 129, 1417–1424, October 2007.

Zhao, J. M., Tan, J. Y., and Liu, L. H.: Monte Carlo method for polarized radiative transfer in gradient-index media, *JQSRT*, 152, 114–126, 2015.

Zhao, Z., Poulikakos, D., and Ren, Z.: Combined natural convection and radiation from heated cylinders inside a container, *JTHT*, 6(4), 713–720, 1992.

Zheng, Z. H. and Xuan, Y. M.: Theory of near-field radiative heat transfer for stratified magnetic media, *IJHMT*, 54, 1101–1110, 2011a.

Zheng, Z. H. and Xuan, Y. M.: Enhancement or suppression of the near-field radiative heat transfer between two materials, *Nanoscale Microscale Thermophys. Eng.*, 15, 237–251, 2011b.

Zhou, X., Li, S., and Stamnes, K.: Geometrical-optics code for computing the optical properties of large dielectric spheres, *Appl. Opt.*, 42(21), 4295–4306, 2003.

Zhou, Y. H. and Zhang, Z. M.: Radiative properties of semitransparent silicon wafers with rough surface, *JHT*, 125, 462–470, June 2003.

Zhu, L., Raman, A., and Fan, S.: Color-preserving daytime radiative cooling, *Appl. Phys. Lett.*, 103, 223902, 2013.

Zhu, L., Raman, A., Wang, K. X., Anoma, M. A., and Fan, S.: Radiative cooling of solar cells, *Optica*, 1(1), 32–38, July 2014.

Zigrang, D. J.: Statistical treatment of data uncertainties in heat transfer, *Proceedings 1975 National Heat Transfer Conference*, Portland, OR, AIAA Paper 75–710, May 1975.

Index

D

E

F

G

N